The Crust

Citations
Please use the following example for citations:
Rudnick R.L. and Gao S. (2003) Composition of the continental crust, pp. 1–64. In *The Crust* (ed. R.L. Rudnick) Vol. 3 *Treatise on Geochemistry* (eds. H.D. Holland and K.K. Turekian), Elsevier–Pergamon, Oxford.

Cover photo: Tarawera volcano, New Zealand: View of crater rim looking north, toward White Island. Note fumeroles in plains. The volcano last erupted on June 10th, 1886, killing 153 people. Though within a rhyolite complex, the eruption was basaltic in composition, New Zealand is an example of crust formation in a convergent margin setting. (Photograph provided by Roberta L. Rudnick)

The Crust

Edited by

R. L. Rudnick
University of Maryland, MD, USA

TREATISE ON GEOCHEMISTRY
Volume 3

Executive Editors

H. D. Holland
Harvard University, Cambridge, MA, USA

and

K. K. Turekian
Yale University, New Haven, CT, USA

ELSEVIER

2005

AMSTERDAM – BOSTON – HEIDELBERG – LONDON – NEW YORK – OXFORD
PARIS – SAN DIEGO – SAN FRANCISCO – SINGAPORE – SYDNEY – TOKYO

ELSEVIER B.V.	ELSEVIER Inc.	**ELSEVIER Ltd**	ELSEVIER Ltd
Radarweg 29	525 B Street, Suite 1900	**The Boulevard, Langford Lane**	84 Theobalds Road
P.O. Box 211, 1000 AE Amsterdam	San Diego, CA 92101-4495	**Kidlington, Oxford OX5 1GB**	London WC1X 8RR
The Netherlands	USA	**UK**	UK

First edition 2005

Library of Congress Cataloging in Publication Data
A catalog record is available from the Library of Congress.

British Library Cataloguing in Publication Data
A catalogue record is available from the British Library.

ISBN: 0-08-044847-X (Paperback)

Printed and bound by CPI Group (UK) Ltd, Croydon, CR0 4YY
Transferred to Digital Print 2012

DEDICATED
TO

VICTOR GOLDSCHMIDT
(1888–1947)

Photograph provided by Gerhard Woerner, Geowissenschaftliches Zentrum Göttingen

Contents

Executive Editors' Foreword

H. D. Holland

Harvard University, Cambridge, MA, USA

and

K. K. Turekian

Yale University, New Haven, CT, USA

Geochemistry has deep roots. Its beginnings can be traced back to antiquity, but many of the discoveries that are basic to the science were made between 1800 and 1910. The periodic table of elements was assembled, radioactivity was discovered, and the thermodynamics of heterogeneous systems was developed. The solar spectrum was used to determine the composition of the Sun. This information, together with chemical analyses of meteorites, provided an entry to a larger view of the universe.

During the first half of the twentieth century, a large number of scientists used a variety of methods to determine the major-element composition of the Earth's crust, and the geochemistries of many of the minor elements were defined by V. M. Goldschmidt and his associates using the then new technique of emission spectrography. V. I. Vernadsky founded biogeochemistry. The crystal structures of most minerals were determined by X-ray diffraction techniques. Isotope geochemistry was born, and age determinations based on radiometric techniques began to define the absolute geologic timescale. The intense scientific efforts during World War II yielded new analytical tools and a group of people who trained a new generation of geochemists at a number of universities. But the field grew slowly. In the 1950s, a few journals were able to report all of the important developments in trace-element geochemistry, isotopic geochronometry, the exploration of paleoclimatology and biogeochemistry with light stable isotopes, and studies of phase equilibria. At the meetings of the American Geophysical Union, geochemical sessions were few, none were concurrent, and they all ranged across the entire field.

Since then the developments in instrumentation and the increases in computing power have been spectacular. The education of geochemists has been broadened beyond the old, rather narrowly defined areas. Atmospheric and marine geochemistry have become integrated into solid Earth geochemistry; cosmochemistry and biogeochemistry have contributed greatly to our understanding of the history of our planet. The study of Earth has evolved into "Earth System Science," whose progress since the 1940s has been truly dramatic.

Major ocean expeditions have shown how and how fast the oceans mix; they have demonstrated the connections between the biologic pump, marine biology, physical oceanography, and marine sedimentation. The discovery of hydrothermal vents has shown how oceanography is related to economic geology. It has revealed formerly unknown oceanic biotas, and has clarified the factors that today control, and in the past have controlled the composition of seawater.

Seafloor spreading, continental drift and plate tectonics have permeated geochemistry. We finally understand the fate of sediments and oceanic crust in subduction zones, their burial and their exhumation. New experimental techniques at temperatures and pressures of the deep Earth interior have clarified the three-dimensional structure of the mantle and the generation of magmas.

Moon rocks, the treasure trove of photographs of the planets and their moons, and the successful search for planets in other solar systems have all revolutionized our understanding of Earth and the universe in which we are embedded.

Geochemistry has also been propelled into the arena of local, regional, and global anthropogenic problems. The discovery of the ozone hole came as a great, unpleasant surprise, an object lesson for optimists and a source of major new insights into the photochemistry and dynamics of the atmosphere. The rise of the CO_2 content of the atmosphere due to the burning of fossil fuels and deforestation has been and will continue to be at the center of the global change controversy, and will yield new insights into the coupling of atmospheric chemistry to the biosphere, the crust, and the oceans.

The rush of scientific progress in geochemistry since World War II has been matched by organizational innovations. The first issue of *Geochimica et Cosmochimica Acta* appeared in June 1950. The Geochemical Society was founded in 1955 and adopted *Geochimica et Cosmochimica Acta* as its official publication in 1957. The International Association of Geochemistry and Cosmochemistry was founded in 1966, and its journal, *Applied Geochemistry*, began publication in 1986. *Chemical Geology* became the journal of the European Association for Geochemistry.

The Goldschmidt Conferences were inaugurated in 1991 and have become large international meetings. Geochemistry has become a major force in the Geological Society of America and in the American Geophysical Union. Needless to say, medals and other awards now recognize outstanding achievements in geochemistry in a number of scientific societies.

During the phenomenal growth of the science since the end of World War II an admirable number of books on various aspects of geochemistry were published. Of these only three attempted to cover the whole field. The excellent *Geochemistry* by K. Rankama and Th.G. Sahama was published in 1950. V. M. Goldschmidt's book with the same title was started by the author in the 1940s. Sadly, his health suffered during the German occupation of his native Norway, and he died in England before the book was completed. Alex Muir and several of Goldschmidt's friends wrote the missing chapters of this classic volume, which was finally published in 1954.

Between 1969 and 1978 K. H. Wedepohl together with a board of editors (C. W. Correns, D. M. Shaw, K. K. Turekian and J. Zeman) and a large number of individual authors assembled the *Handbook of Geochemistry*. This and the other two major works on geochemistry begin with integrating chapters followed by chapters devoted to the geochemistry of one or a small group of elements. All three are now out of date, because major innovations in instrumentation and the expansion of the number of practitioners in the field have produced valuable sets of high-quality data, which have led to many new insights into fundamental geochemical problems.

At the Goldschmidt Conference at Harvard in 1999, Elsevier proposed to the Executive Editors that it was time to prepare a new, reasonably comprehensive, integrated summary of geochemistry. We decided to approach our task somewhat differently from our predecessors. We divided geochemistry into nine parts. As shown below, each part was assigned a volume, and a distinguished editor was chosen for each volume. A tenth volume was reserved for a comprehensive index:

(i) *Meteorites, Comets, and Planets*: Andrew M. Davis

(ii) *Geochemistry of the Mantle and Core*: Richard Carlson

(iii) *The Earth's Crust*: Roberta L. Rudnick

(iv) *Atmospheric Geochemistry*: Ralph F. Keeling

(v) *Freshwater Geochemistry, Weathering, and Soils*: James I. Drever

(vi) *The Oceans and Marine Geochemistry*: Harry Elderfield

(vii) *Sediments, Diagenesis, and Sedimentary Rocks*: Fred T. Mackenzie

(viii) *Biogeochemistry*: William H. Schlesinger

(ix) *Environmental Geochemistry*: Barbara Sherwood Lollar

(x) *Indexes*

The editor of each volume was asked to assemble a group of authors to write a series of chapters that together summarize the part of the field covered by the volume. The volume editors and chapter authors joined the team enthusiastically. Altogether there are 155 chapters and 9 introductory essays in the Treatise. Naming the work proved to be somewhat problematic. It is clearly not meant to be an encyclopedia. The titles *Comprehensive Geochemistry* and *Handbook of Geochemistry* were finally abandoned in favor of *Treatise on Geochemistry*.

The major features of the Treatise were shaped at a meeting in Edinburgh during a conference on Earth System Processes sponsored by the Geological Society of America and the Geological Society of London in June 2001. The fact that the Treatise is being published in 2003 is due to a great deal of hard work on the part of the editors, the authors, Mabel Peterson (the Managing Editor), Angela Greenwell (the former Head of Major Reference Works), Diana Calvert (Developmental Editor, Major Reference Works),

Bob Donaldson (Developmental Manager), Jerome Michalczyk and Rob Webb (Production Editors), and Friso Veenstra (Senior Publishing Editor). We extend our warm thanks to all of them. May their efforts be rewarded by a distinguished journey for the Treatise.

Finally, we would like to express our thanks to J. Laurence Kulp, our advisor as graduate students at Columbia University. He introduced us to the excitement of doing science and convinced us that all of the sciences are really subdivisions of geochemistry.

Contributors to Volume 3

J. J. Ague
Yale University, New Haven, CT, USA

P. A. Candela
University of Maryland, College Park, MD, USA

T. Elliott
University of Bristol, UK

G. L. Farmer
University of Colorado, Boulder, CO, USA

S. Gao
China University of Geosciences, Wuhan, People's Republic of China and
Northwest University, Xi'an, People's Republic of China

A. R. Greene
Western Washington University, Bellingham, WA, USA

K. Hanghøj
Woods Hole Oceanographic Institution, MA, USA

C. J. Hawkesworth
University of Bristol, UK

K. V. Hodges
Massachusetts Institute of Technology, Cambridge, MA, USA

B. M. Jahn
National Taiwan University, Taipei, Republic of China

C. Jaupart
Institut de Physique du Globe de Paris, France

P. B. Kelemen
Woods Hole Oceanographic Institution, MA, USA

A. I. S. Kemp
University of Bristol, UK

A. C. Kerr
Cardiff University, Wales, UK

E. M. Klein
Duke University, Durham, NC, USA

M. J. Kohn
University of South Carolina, Columbia, SC, USA

J. G. Liou
Stanford University, CA, USA

J.-C. Mareschal
GEOTOP Université du Québec à Montréal, Canada

C. Oppenheimer
University of Cambridge, UK

P. J. Patchett
University of Arizona, Tucson, AZ, USA

S. Poli
Universita di Milano, Italy

M. R. Reid
University of California at Los Angeles, CA, USA and
Northern Arizona University, Flagstaff, AZ, USA

R. L. Rudnick
University of Maryland, College Park, MD, USA

D. Rumble
Geophysical Laboratory, Washington, DC, USA

S. D. Samson
Syracuse University, NY, USA

M. W. Schmidt
ETH Zürich, Switzerland

M. Spiegelman
Columbia University, New York, NY, USA

H. Staudigel
University of California at San Diego, La Jolla, CA, USA

Volume Editor's Introduction

R. L. Rudnick
University of Maryland, College Park, MD, USA

The rocky layer at the surface of our planet, the crust, is important in many ways. Perhaps most significantly, the continental crust, in concert with the atmosphere and hydrosphere, provides the nurturing habitat in which our species evolved. Furthermore, it may be no coincidence that Earth is the only planet in our solar system that has both liquid water and a topographically bimodal crust consisting of low-lying, higher-density basaltic oceanic crust and high-standing, lower-density andesitic continental crust. The presence of water may be intimately linked to the development of plate tectonics and thus the processes that produced the continental crust (Campbell and Taylor, 1985; Richards *et al.*, 2001).

Although the continental crust is insignificant in terms of mass (it constitutes only 0.6% of the silicate earth), it forms an important reservoir for many of the scarce (or trace) elements on our planet (Chapter 3.01), including the heat producing elements (Chapter 3.02). It also provides us with a rich geologic history; the oldest crustal rocks dated formed within 500 Ma of Earth accretion (Chapter 3.10). Some of the earliest geochemical studies focused on defining the composition of the Earth's continental crust (Clarke, 1889; Clarke and Washington, 1924), including the important works of "the father of geochemistry," V. M. Goldschmidt (1933, 1958), to whom this volume is dedicated. It is interesting that our understanding of the major-element composition of the continental crust has not changed significantly during the hundred-plus years that it has been studied; only the estimates of the minor- and trace-element concentrations of the continental crust have varied appreciably with time and in different models. Some of the scarce elements are concentrated by fluid and magmatic processes to form ore deposits in both oceanic and continental environments (Chapter 3.12). Human societies rely on the discovery and mining of ores to produce the raw materials needed to build the edifices of civilization. Hence, understanding the geochemistry of crustal processes is required for the development and maintenance of our society.

Unlike the continental crust, the oceanic crust records only the last 200 Ma of Earth history, but the production and recycling of oceanic crust is of fundamental importance in the Earth's geochemical cycles and in determining the nature of the underlying mantle (Volume 2). Mid-ocean ridge basaltic magmatism has been an ongoing process during most of geological history and is the primary way in which the Earth loses its heat (Chapter 3.13). Uranium-series isotope studies have been important in unraveling mid-ocean ridge igneous processes, and provide insights into the timing and processes of oceanic crust formation (Chapter 3.14), which is better understood than the formation of the continents. Recycling of oceanic crust at subduction zones gives rise to arc magmas that ultimately contribute to the formation of continental crust (Chapter 3.18). Areas of thickened oceanic crust occur in oceanic plateaus, which are probably manifestations of plume volcanism (Chapter 3.16). The production of these plateaus is also an important mode of heat loss from the Earth's interior and may contribute to the development of continental crust.

The Earth's crust is also important because it is the site of chemical exchange between the hydrosphere, atmosphere, and the Earth's interior. This is observed most spectacularly at the hydrothermal vents along mid-oceanic ridges (Chapter 3.15) and at convergent plate margins where arc volcanoes pump out gaseous emissions at a prodigious rate (Chapter 3.04). Exchange processes at both of these locations have profound effects on the chemistry and evolution of the oceans (Volume 6) and atmosphere (Volume 4). However, even more significant mass fluxes occur in regions where their direct observation is more difficult. Significant mass redistribution within the continental crust occurs during both magmatic differentiation (Chapter 3.11) and fluid flow (Chapter 3.06) processes. These processes may be accompanied by mass transfer to the mantle and atmosphere, respectively. Arguably the greatest mass flux between geochemical reservoirs occurs as oceanic lithosphere subducts into the mantle along with its retinue of hydrothermally altered basaltic crust and overlying sediments (Chapters 3.17 and 2.11).

Although recyclying of continental crust into the Earth's mantle is traditionally considered to occur by subduction of terrigenous sediments (e.g., Plank and Langmiur, 1998), preservation of ultra-high pressure mineral assemblages, including minerals such as coesite and diamond, attest to the metamorphism of continental rocks at mantle depths and thus bear witness to continental subduction (Chapter 3.09). The amount of continental material that is recycled into the mantle is uncertain. However, a number of lines of evidence suggest that continental recycling occurs, especially recycling of mafic to ultramafic materials from the lower crust into the mantle, which is needed to explain the evolved composition of the bulk continental crust (see Chapters 3.01 and 3.18; Plank, 2003).

The geologic record indicates that continent formation occurred episodically, starting with the earliest crustal vestiges preserved 4 Ga ago, to the present. Geochronology therefore plays a leading role in revealing Earth history (Chapter 3.10). Application of high precision chronometers to magmatic minerals defines the time scales involved in intracrustal igneous differentiation (Chapter 3.05). Other important geochemical processes occur within mountain belts, producing changes in both the physical and chemical properties of pre-existing rocks (Chapters 3.06 and 3.07). Through the study of trace element zoning in metamorphic minerals (Chapter 3.07) and thermochronology (Chapter 3.08) the physical conditions and timescales of metamorphism can be quantified.

Despite intensive study, the processes by which continents form and are modified are still only partially understood. Igneous additions to the continents occur at both convergent margin (Chapters 3.18 and 3.11) and intraplate settings (Chapter 3.03) and may have varied over Earth history (Chapter 3.11). Intraplate magmatism provides additional insights into the composition and evolution of the continents through the study of deep-seated xenoliths that occur in some intraplate magmas (Chapter 2.05). No matter what the tectonic setting of magmatic addition, magmatic differentiation, probably accompanied by lower crust recycling, is the main process driving the crust toward its andesititc bulk composition.

The chapters in this volume explore the composition and evolution of the oceanic and continental crust as well as those of arcs and oceanic platforms. Geochemical studies of these regions and the processes that occur within them have given us a better understanding of the processes that formed the Earth as we see it today. Each chapter was written by an expert or experts in the field and provides an overview of the topic. Some chapters emphasize a personal perspective, others are more generic. All are informative, but it should be apparent that our understanding of the Earth's crust is far from complete. There is still much unexplored territory, and the many unresolved problems provide fertile grounds for research by future generations of geochemists.

ACKNOWLEDGMENTS

It has been a pleasure to work with all of the authors of the chapters in this volume. They have approached this monumental task with dedication and good humor that have survived (for the most part) to the very end. An equally important job was performed by the reviewers who typically went above and beyond the call of duty in providing insightful critiques on sometimes very short notice. I thank the executive editors of the Treatise for the initiation of this work and their dedication to achieving the highest quality of the final product. Finally, Mabel Peterson at Yale and the editors and managers at Elsevier are thanked for their efficiency and patience.

REFERENCES

Campbell I. H. and Taylor S. R. (1985) No water, no granites—no oceans, no continents. *Geophys. Res. Lett.* **10**, 1061–1064.

Clarke F. W. (1889) The relative abundance of the chemical elements. *Phil. Soc. Washington Bull.* **XI**, 131–142.

Clarke F. W. and Washington H. S. (1924) The composition of the Earth's crust. USGS Professional Paper 127, USGS, Washington, DC, 117pp.

Goldschmidt V. M. (1933) Grundlagen der quantitativen Geochemie. *Fortschr. Mineral. Kristallog. Petrogr.* **17**, 112.

Goldschmidt V. M. (1958) *Geochemistry*. Oxford University Press, London, 730pp.

Plank T. (2003) Constraints from Th/La on sediment recycling at subduction zones and the evolution of the continents. *J. Petrol.* (in review).

Plank T. and Langmiur C. H. (1998) The chemical composition of subducting sediment and its consequences for the crust and mantle. *Chem. Geol.* **145**, 325–394.

Richards M. A., Yang W. S., Baumgardner J. R., and Bunge H. P. (2001) Role of a low-viscosity zone in stabilizing plate tectonics: implications for comparative terrestrial planetology. *Geochem. Geophys. Geosys.* **2** (article no. 2000GC000115).

3.01
Composition of the Continental Crust

R. L. Rudnick

University of Maryland, College Park, MD, USA

and

S. Gao

China University of Geosciences, Wuhan, People's Republic of China and Northwest University, Xi'an, People's Republic of China

3.01.1 INTRODUCTION

The Earth is an unusual planet in our solar system in having a bimodal topography that reflects the two distinct types of crust found on our planet. The low-lying oceanic crust is thin (~7 km on average), composed of relatively dense rock types such as basalt and is young (≤200 Ma old) (see Chapter 3.13). In contrast, the high-standing continental crust is thick (~40 km on average), is composed of highly diverse lithologies (virtually every rock type known on Earth) that yield an average intermediate or "andesitic" bulk composition (Taylor and McLennan (1985) and references therein), and contains the oldest rocks and minerals yet observed on Earth (currently the 4.0 Ga Acasta gneisses (Bowring and Williams, 1999) and 4.4 Ga detrital zircons from the Yilgarn Block, Western Australia (Wilde *et al.*, 2001)), respectively. Thus, the continents preserve a rich geological history of our planet's evolution and understanding their origin is critical for understanding the origin and differentiation of the Earth.

The origin of the continents has received wide attention within the geological community, with hundreds of papers and several books devoted to the topic (the reader is referred to the following general references for further reading: Taylor and McLennan (1985), Windley (1995), and Condie (1997). Knowledge of the age and composition of the continental crust is essential for understanding its origin. Patchett and Samson (Chapter 3.10) review the present-day age distribution of the continental crust and Kemp and Hawkesworth (Chapter 3.11) review secular evolution of crust composition. Moreover, to understand fully the origin and evolution of continents requires an understanding of not only the crust, but also the mantle lithosphere that formed more-or-less contemporaneously with the crust and translates with it as the continents move across the Earth's surface. The latter topic is reviewed in Chapter 2.05.

This chapter reviews the present-day composition of the continental crust, the methods employed to derive these estimates, and the implications of the continental crust composition for the formation of the continents, Earth differentiation, and its geochemical inventories.

3.01.1.1 What is the Continental Crust?

In a review of the composition of the continental crust, it is useful to begin by defining the region under consideration and to provide some generalities regarding its structure. The continental crust, as considered here, extends vertically from the Earth's surface to the Mohorovicic discontinuity, a jump in compressional wave speeds from ~7 km s^{-1} to ~8 km s^{-1} that is interpreted to mark the crust–mantle boundary. In some regions the Moho is transitional rather than discontinuous and there may be some debate as to where the crust–mantle boundary lies (cf. Griffin and O'Reilly, 1987; McDonough *et al.*, 1991). The lateral extent of the continents is marked by the break in slope on the continental shelf. Using this definition, ~31% of continental area is submerged beneath the oceans (Figure 1; Cogley, 1984), and is thus less accessible to geological sampling. For this reason, most estimates of continental crust composition derive from exposed regions of the continents. In some cases the limited geophysical data for submerged continental shelves reveal no systematic difference in bulk properties between the shelves and exposed continents; the shelves simply appear to be thinned regions of the crust. In other cases, such as volcanic rifted margins, the submerged continent is characterized by high-velocity layers interpreted to represent massive basaltic intrusions associated with continental breakup (Holbrook and Kelemen, 1993). Depending on the extent of the latter type of continental margin (which is yet to be quantified), crust compositional estimates derived from exposed regions may not be wholly representative of the total continental mass.

The structure of the continental crust is defined seismically to consist of upper-, middle-, and

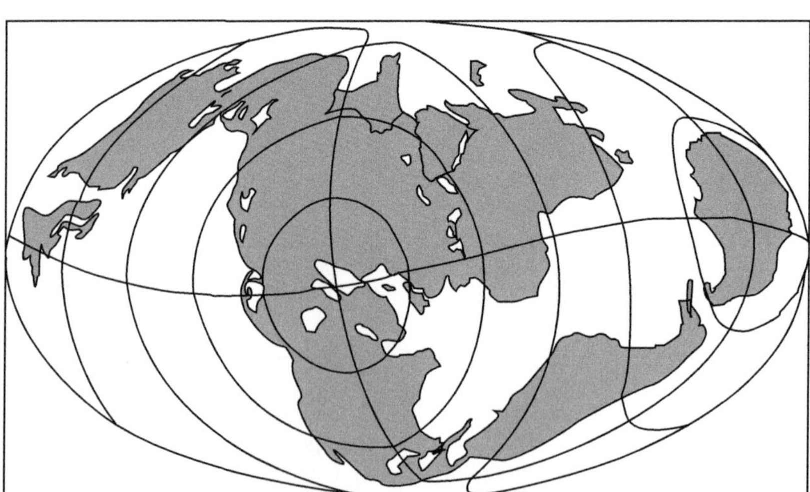

Figure 1 Map of continental regions of the Earth, including submerged continents (Cogley (1984); reproduced by permission of American Geophysical Union from *Rev. Geophys. Space Phys.*, **1984**, *22*, 101–122).

lower crustal layers (Christensen and Mooney, 1995; Holbrook *et al.*, 1992; Rudnick and Fountain, 1995). The upper crust is readily accessible to sampling and robust estimates of its composition are available for most elements (Section 3.01.2). These show the upper crust to have a granodioritic bulk composition, to be rich in incompatible elements, and generally depleted in compatible elements. The deeper reaches of the crust are more difficult to study. In general, three probes of the deep crust are employed to unravel its composition: (i) studies of high-grade metamorphic rocks (amphibolite or granulite facies) exposed in surface outcrops (Bohlen and Mezger, 1989) and, in some cases, in uplifted cross-sections of the crust reaching to depths of 20 km or more (Fountain *et al.*, 1990a; Hart *et al.*, 1990; Ketcham, 1996; Miller and Christensen, 1994); (ii) studies of granulite-facies xenoliths (foreign rock fragments) that are carried from great depths to the Earth's surface by fast-rising magmas (see Rudnick (1992) and references therein); and (iii) remote sensing of lower crustal lithologies through seismic investigations (Christensen and Mooney, 1995; Holbrook *et al.*, 1992; Rudnick and Fountain, 1995; Smithson, 1978) and surface heat-flow studies (see Chapter 3.02). Collectively, the observations from these probes show that the crust becomes more mafic with depth (Section 3.01.3). In addition, the concentration of heat-producing elements drops off rapidly from the surface downwards. This is due, in part, to an increase in metamorphic grade but is also due to increasing proportions of mafic lithologies (see Chapter 3.02). Thus, the crust is vertically stratified in terms of its chemical composition.

In addition to this stratification, the above studies also show that the crust is heterogeneous from place to place, with few systematics available for making generalizations about crustal structure and composition for different tectonic settings. For example, the crust of Archean cratons in some regions is relatively thin and has low seismic velocities, suggesting an evolved composition (e.g., Yilgarn craton (Drummond, 1988); Kaapvaal craton (Durrheim and Green, 1992; Niu and James, 2002); and North China craton (Gao *et al.*, 1998a,b)). However, in other cratons, the crust is thick (40–50 km) and the deep crust is characterized by high velocities, which imply mafic-bulk compositions (Wyoming craton (Gorman *et al.*, 2002) and Baltic shield (Luosto and Korhonen, 1986; Luosto *et al.*, 1990)). The reasons for these heterogeneities are not fully understood and we return to this topic in Section 3.01.3. Similar heterogeneities are observed for Proterozoic and Paleozoic regions (see Rudnick and Fountain (1995) and references therein). Determining an average composition of such a heterogeneous mass is difficult and, at first glance, may seem like a futile endeavor. Yet it is just such averages that allow insights into the relative contribution of the crust to the whole Earth-chemical budget and the origin of the continents. Thus, deriving average compositions is critical to studies of the continents and the whole Earth.

3.01.1.2 The Importance of Determining Crust Composition

Although the continental crust constitutes only ~0.6% by mass of the silicate Earth, it contains a very large proportion of incompatible elements (20–70%, depending on element and model considered; Rudnick and Fountain (1995)), which include the heat-producing elements and members of a number of radiogenic-isotope systems (Rb–Sr, U–Pb, Sm–Nd, Lu–Hf). Thus the continental crust factors prominently in any mass-balance calculation for the Earth as a whole and in estimates of the thermal structure of the Earth (Sclater *et al.*, 1980).

In addition, knowledge of the bulk composition of the crust and determining whether this composition has changed through time is important for: (i) understanding the processes by which the crust is generated and modified and (ii) determining whether there is any secular evolution in crust generation and modification processes (see Chapter 3.11). The latter has important implications for the evolution of our planet as a whole.

In this chapter we review the composition of the upper, middle, and lower continental crust (Sections 3.01.2 and 3.01.3). We then examine the bulk crust composition and the implications of this composition for crust generation and modification processes (Sections 3.01.4 and 3.01.5). Finally, we compare the Earth's crust with those of the other terrestrial planets in our solar system (Section 3.01.6) and speculate about what unique processes on Earth have given rise to this unusual crustal distribution.

3.01.2 THE UPPER CONTINENTAL CRUST

The upper continental crust, being the most accessible part of our planet, has long been the target of geochemical investigations (Clarke, 1889). There are two basic methods employed to determine the composition of the upper crust: (i) establishing weighted averages of the compositions of rocks exposed at the surface and (ii) determining averages of the composition of insoluble elements in fine-grained clastic sedimentary rocks or glacial deposits and using these to infer upper-crust composition.

The first method was utilized by F. W. Clarke and colleagues over a century ago (Clarke, 1889; Clarke and Washington, 1924) and entails

Table 1 Major element composition[a] (in weight percent oxide) of the upper continental crust. Columns 1–9 represent averages of surface exposures and glacial clays. Columns 10–11 are derivative compositions from these data. Column 12 shows our recommended values.

Element	*1* *Clarke (1889)*	*2* *Clarke and Washington (1924)*	*3* *Goldschmidt (1933)*	*4* *Shaw et al. (1967)*	*5* *Fahrig and Eade (1968)*	*6* *Ronov and Yaroshevskiy (1976)*	*7* *Condie (1993)*	*8* *Gao et al. (1998a)*	*9* *Borodin (1998)*	*10* *Taylor and McLennan (1985)*	*11* *Wedepohl (1995)*	*12* *This Study*[b]
SiO_2	60.2	60.30	62.22	66.8	66.2	64.8	67.0	67.97	67.12	65.89	66.8	66.62
TiO_2	0.57	1.07	0.83	0.54	0.54	0.55	0.56	0.67	0.60	0.50	0.54	0.64
Al_2O_3	15.27	15.65	16.63	15.05	16.10	15.84	15.14	14.17	15.53	15.17	15.05	15.40
FeO_T[c]	7.26	6.70	6.99	4.09	4.40	5.78	4.76	5.33	4.94	4.49	4.09	5.04
MnO	0.10	0.12	0.12	0.07	0.08	0.10		0.10	0.00	0.07	0.07	0.10
MgO	4.59	3.56	3.47	2.30	2.20	3.01	2.45	2.62	2.10	2.20	2.30	2.48
CaO	5.45	5.18	3.23	4.24	3.40	3.91	3.64	3.44	3.51	4.19	4.24	3.59
Na_2O	3.29	3.92	2.15	3.56	3.90	2.81	3.55	2.86	3.21	3.89	3.56	3.27
K_2O	2.99	3.19	4.13	3.19	2.91	3.01	2.76	2.68	3.01	3.39	3.19	2.80
P_2O_5	0.23	0.31	0.23	0.15	0.16	0.16	0.12	0.16	0.00	0.20	0.15	0.15
Mg#	53.0	48.7	46.9	50.1	47.4	48.1	47.9	46.7	43.2	46.6	50.1	46.7

Mg# = molar $100 \times Mg/(Mg + Fe_{tot})$.
[a] Major elements recast to 100% anhydrous. [b] See Table 3 for derviation of this estimate. [c] Total Fe as FeO.

large-scale sampling and weighted averaging of the wide variety of rocks that crop out at the Earth's surface. All major-element (and a number of soluble trace elements) determinations of upper-crust composition rely upon this method.

The latter method is based on the concept that the process of sedimentation averages wide areas of exposed crust. This method was originally employed by Goldschmidt (1933) and his Norwegian colleagues in their analyses of glacial sediments to derive average composition of the crystalline rocks of the Baltic shield and has subsequently been applied by a number of investigators, including the widely cited work by Taylor and McLennan (1985) to derive upper-crust composition for insoluble trace elements. In the following sections we review the upper-crust composition determined from each of these methods, then provide an updated estimate of the composition of the upper crust.

3.01.2.1 Surface Averages

In every model for the composition of the upper-continental crust, major-element data are derived from averages of the composition of surface exposures (Table 1). Several surface-exposure studies have also provided estimates of the average composition of a number of trace elements (Table 2). For soluble elements that are fractionated during the weathering process (e.g., sodium, calcium, strontium, barium, etc.), this is the only way in which a reliable estimate of their abundances can be obtained.

The earliest of such studies was the pioneering work of Clarke (1889), who, averaging hundreds of analyses of exposed rocks, determined an average composition for the crust that is markedly similar to present-day averages of the bulk crust (cf. Tables 1 and 9). Although Clarke's intention was to derive the average crust composition, his samples are limited to the upper crust; there was little knowledge of the structure of the Earth when these studies were undertaken; oceanic crust was not distinguished as different from continental and the crust was assumed to be only 16 km thick. Clarke's values are, therefore, most appropriately compared to upper crustal estimates. Later, Clarke, joined by H. S. Washington, used a larger data set to determine an average composition of the upper-crust that is only slightly different from his original 1889 average (Clarke and Washington, 1924; Table 1). Compared to more recent estimates of upper-crust composition, these earliest estimates are less evolved (lower silicon, higher iron, magnesium, and calcium), but contain similar amount of the alkali elements, potassium and sodium.

The next major undertakings in determining upper-crust composition from large-scale surface

Table 2 Estimates of the trace-element composition of the upper continental crust. Columns 1–4 represent averages of surface exposures. Columns 5–8 are estimates derived from sedimentary and loess data. Column 9 is a previous estimate, where bracketed data are values derived from surface exposure studies. Column 10 is our recommended value (see Table 3).

Element	*Units*	*1* *Shaw et al. (1967, 1976)*	*2* *Eade and Fahrig (1973)*	*3* *Condie (1993)*	*4* *Gao et al. (1998a)*	*5* *Sims et al. (1990)*	*6* *Plank and Langmuir (1998)*	*7* *Peucker-Eherenbrink and Jahn (2001)*	*8* *Taylor and McLennan (1985, 1995)*	*9* *Wedepohl (1995)*[a]	*10* *This study*[b]
Li	$\mu g\ g^{-1}$	22			20				20	[22]	21
Be	"	1.3			1.95				3	3.1	2.1
B	"	9.2			28				15	17	17
N	"									83	83
F	"	500			561					611	557
S	"	600			309					953	621
Cl	"	100			142					640	370
Sc	"	7	12	13.4	15				13.6[c]	[7]	14.0
V	"	53	59	86	98				107[c]	[53]	97
Cr	"	35	76	112	80				85[c]	[35]	92
Co	"	12		18	17				17[c]	[12]	17.3
Ni	"	19	19	60	38				44[c]	[19]	47
Cu	"	14	26		32				25	[14]	28
Zn	"	52	60		70				71	[52]	67
Ga	"	14			18				17	[14]	17.5
Ge	"				1.34				1.6	1.4	1.4
As	"				4.4	5.1			1.5	2	4.8
Se	"				0.15				0.05	0.083	0.09
Br	"									1.6	1.6
Rb	"	110	85	83	82				112	110	84
Sr	"	316	380	289	266				350	[316]	320
Y	"	21	21	24	17.4				22	[21]	21
Zr	"	237	190	160	188				190	[237]	193
Nb	"	26		9.8	12		13.7		12[c]	[26]	12
Mo	"				0.78	1.2			1.5	1.4	1.1
Ru	$ng\ g^{-1}$							0.34			0.34
Pd	"				1.46			0.52	0.5		0.52
Ag	"				55				50	55	53
Cd	$\mu g\ g^{-1}$	0.075			0.079				0.098	0.102	0.09
In	"								0.05	0.061	0.056
Sn	"				1.73				5.5	2.5	2.1
Sb	"				0.3	0.45			0.2	0.31	0.4

(continued)

Table 2 (continued).

Element	Units	1 *Shaw et al. (1967, 1976)*	2 *Eade and Fahrig (1973)*	3 *Condie (1993)*	4 *Gao et al. (1998a)*	5 *Sims et al. (1990)*	6 *Plank and Langmuir (1998)*	7 *Peucker-Eherenbrink and Jahn (2001)*	8 *Taylor and McLennan (1985, 1995)*	9 *Wedepohl (1995)*[a]	10 *This study*[b]
I	"									1.4	1.4
Cs	"				3.55		7.3		4.6[c]	5.8	4.9
Ba	"	1070	730	633	678				550	668	624
La	"	32.3	71	28.4	34.8				30	[32.3]	31
Ce	"	65.6		57.5	66.4				64	[65.7]	63
Pr	"								7.1	6.3	7.1
Nd	"	25.9		25.6	30.4				26		27
Sm	"	4.61		4.59	5.09				4.5	4.7	4.7
Eu	"	0.937		1.05	1.21				0.88	0.95	1.0
Gd	"			4.21					3.8	2.8	4.0
Tb	"	0.481		0.66	0.82				0.64	[0.5]	0.7
Dy	"	2.9							3.5	[2.9]	3.9
Ho	"	0.62							0.8	[0.62]	0.83
Er	"								2.3		2.3
Tm	"								0.33		0.30
Yb	"	1.47		1.91	2.26				2.2	[1.5]	2.0
Lu	"	0.233		0.32	0.35				0.32	[0.27]	0.31
Hf	"	5.8		4.3	5.12				5.8	[5.8]	5.3
Ta	"	5.7		0.79	0.74		0.96		1.0[c]	1.5	0.9
W	"				0.91	3.3			2	1.4	1.9
Re	ng g^{-1}							0.198	0.4		0.198
Os	"							0.031	0.05		0.031
Ir	"	0.02						0.022	[0.02]		0.022
Pt	"							0.51			0.5
Au	"	1.81			1.24				[1.8]		1.5
Hg	μg g^{-1}	0.096			0.0123					0.056	0.05
Tl	"	0.524			1.55				0.75	0.75	0.9
Pb	"	17	18	17	18				17[c]	17	17
Bi	"	0.035			0.23				0.13	0.123	0.16
Th	"	10.3	10.8	8.6	8.95				10.7	[10.3]	10.5
U	"	2.45	1.5	2.2	1.55				2.8	[2.5]	2.7

[a] Wedepohl's upper crust is largely derived from the Canadian Shield composites of Shaw *et al.* (1967, 1976). Values taken directly from Shaw *et al.* are shown in brackets. [b] See Table 3 for derviation of this estimate.
[c] Updated in McLennan (2001b).

sampling campaigns did not appear until twenty years later in studies centered on the Canadian, Baltic, and Ukranian Shields. It is these studies that form the foundation on which many of the more recent estimates of upper-crust composition are constructed (e.g., Taylor and McLennan, 1985; Wedepohl, 1995).

Shaw *et al.* (1967, 1976, 1986) and Eade and Fahrig (1971, 1973) independently derived estimates for the average composition of the Canadian Precambrian shield. Both studies created composites from representative samples taken over large areas that were weighted to reflect their surface outcrop area. The estimates of Shaw *et al.* are based on a significantly smaller number of samples than that of Eade and Fahrig's (i.e., ~430 versus ~14,000) and cover different regions of the shield, but the results are remarkably similar (Figure 2). All major elements agree to within ~10% except for CaO, which is ~20% higher, and MnO, which is 15% lower in the estimates of Shaw *et al.* estimates.

Shaw *et al.* (1967, 1976, 1986) also measured a number of trace elements in their shield composites and these are compared to the smaller number of trace elements determined by Eade and Fahrig in Table 2 and Figure 3. As might be expected, considering the generally greater variability in trace-element concentrations and the greater analytical challenge, larger discrepancies exist between the two averages. For example, scandium, chromium, copper, lanthanum, and

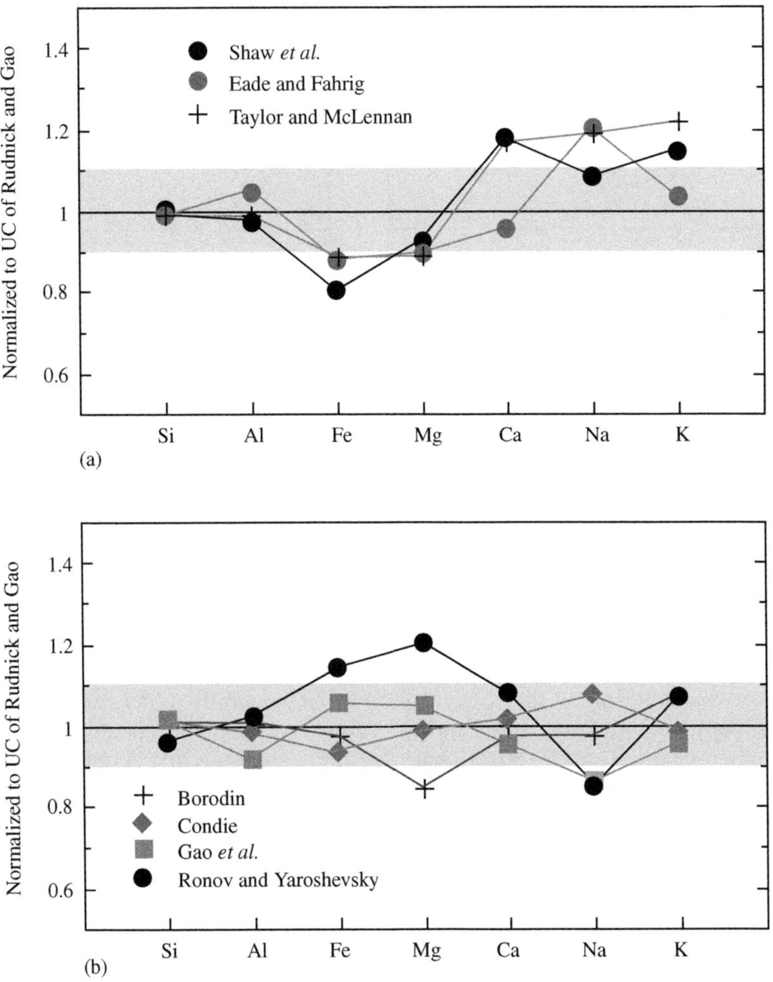

Figure 2 Comparison of different models for the major-element composition of the upper continental crust. All values normalized to the new composition provided in Table 3. Gray shaded field represents ±10% variation from this value. (a) Compositions derived from Canadian Shield samples (Shaw *et al.*, 1967, 1976, 1986; Fahrig and Eade, 1968; Eade and Fahrig, 1971, 1973) and the Taylor and McLennan model (1985, 1995, as modified by McLennan, 2001b). (b) Compositions derived from surface sampling of the former Soviet Union (Ronov and Yaroshevsky, 1967, 1976; Borodin, 1998) and China (Gao *et al.*, 1998a) and a global compilation of upper crustal rock types weighted in proportion to their areal distribution (Condie, 1993). The Canadian shield averages appear to be more evolved (having lower Mg, Fe, and higher Na and K) than other estimates.

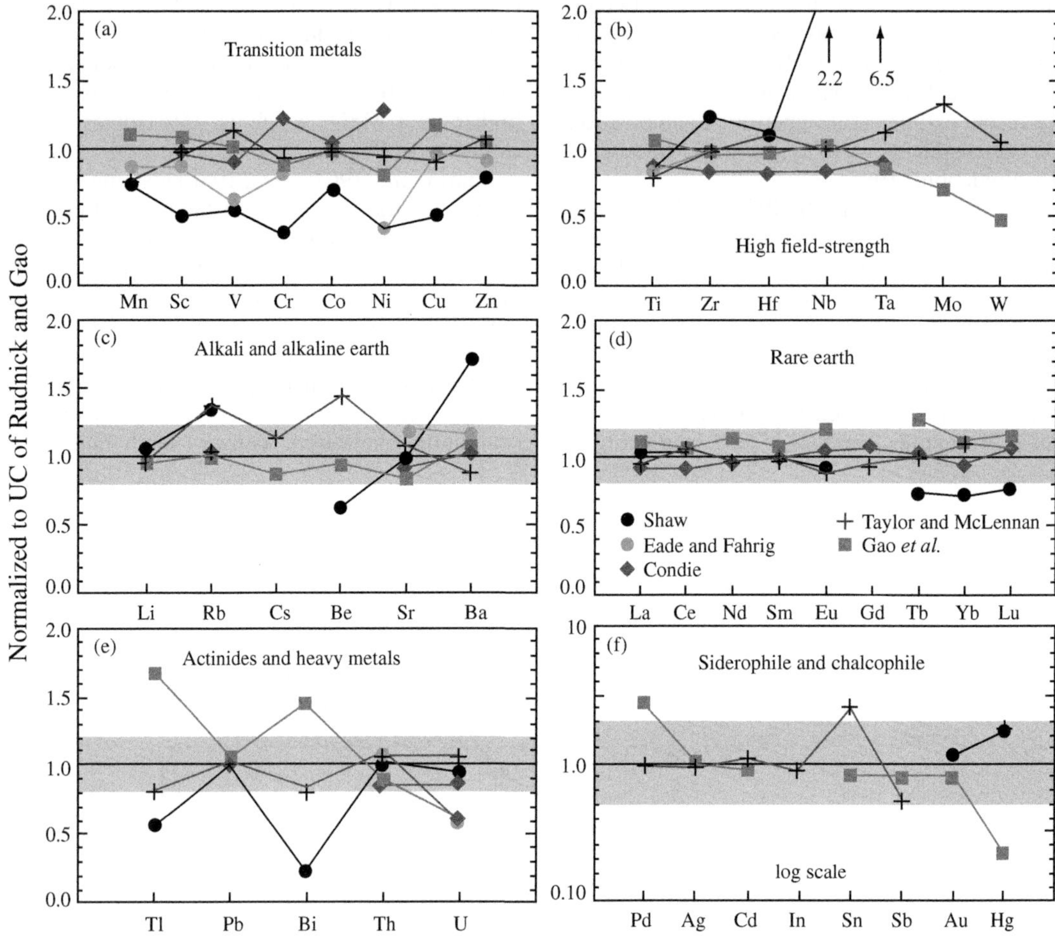

Figure 3 Comparison of different models for the trace-element composition of the upper-continental crust. All values normalized to the new composition provided in Table 3. Gray shaded field represents ±20% variation from this value for all panels except (f), in which gray field represents a factor of two variation. Trace elements are divided into the following groups: (a) transition metals, (b) high-field strength elements, (c) alkali, alkaline-earth elements, (d) REEs, (e) actinides and heavy metals, and (f) highly siderophile and chalcophile elements (note log scale). Data from Tables 1 and 2; lanthanum estimate from Eade and Fahrig (1973) is omitted from panel D.

uranium values vary by ~50% or more. In some cases this may reflect compromised data quality (e.g., lanthanum was determined by optical-emission spectroscopy in the Eade and Fahrig study) and in other cases it may reflect real differences in the composition between the two averages. However, for a number of trace elements (e.g., vanadium, nickel, zinc, rubidium, strontium, zirconium, and thorium), the averages agree within 30%.

In a similar study, Ronov and Yaroshevsky (1967, 1976) determined the average major-element composition of the upper crust based on extensive sampling of rocks from the Baltic and Ukranian shields and the basement of the Russian platform (Table 1). While the SiO_2, Al_2O_3, and K_2O values fall within 5% of those of the average Canadian Shield, as determined by Eade and Fahrig (1971, 1973), FeO_T, MgO, and CaO are ~10–30% higher, and Na_2O is ~30% lower than the Canadian average, suggesting a slightly more mafic composition.

The generally good correspondence between these independent estimates of the composition of shield upper crust lends confidence in the methodologies employed. However, questions can be raised about how representative the shields are of the global upper continental crust. For example, Condie (1993) suggests that shield averages may be biased because (i) shields are significantly eroded and thus may not be representative of the 5–20 km of uppermost crust that has been removed from them and (ii) they include only Precambrian upper crust and largely ignore any Phanerozoic contribution to upper crust. Condie (1993) derived an upper-crust composition based on over 3,000 analyses of upper crustal rock types weighted according to their distributions on geologic maps and stratigraphic sections, mainly covering regions of North America, Europe, and

Australia. He utilized two methods in calculating an average upper-crust composition: (i) using the map distributions, irrespective of level of erosion and (ii) for areas that have been significantly eroded, restoring the eroded upper crust, assuming it has a ratio of supracrustal rocks to plutonic rocks similar to that seen in uneroded upper crustal regions. The latter approach was particularly important for his study, as one of his primary objectives was to evaluate whether there has been any secular change in upper crust composition. However, in this review, we are interested in the present-day composition of the upper crust (eroded or not), so it seems most appropriate to consider his "map model" for comparisons with other models (for a discussion of the secular evolution of the continents, see Chapter 3.11).

Condie's "map model" is compared with other estimates of the upper crust in Tables 1 and 2 and Figures 2 and 3. For major elements, his upper crust composition is within 10% of the Canadian Shield values of Eade and Fahrig. It is also within 10% of some of the major elements estimated by Shaw, but has generally higher magnesium and iron, and lower calcium and potassium compared to Shaw's estimate (Figure 2). Many trace elements in Condie's upper-crust composition are similar (i.e., within 20%) to those of Shaw's Canadian Shield composites (Figure 3), including the light rare-earth elements (LREEs), strontium, yttrium, thorium, and uranium. However, several trace elements in Condie's average vary by $\geqslant$50% from those of Shaw *et al.* (1967, 1976, 1986) as can be seen in the figure. These include transition metals (scandium, vanadium, chromium, and nickel), which are considerably higher in Condie's upper crust, and niobium, barium and tantalum, which are significantly lower in Condie's upper crust compared to Shaw's. These differences may reflect regional variations in upper crust composition (i.e., the Canadian Shield is not representative of the worldwide upper crust) or inaccuracies in either of the estimates due to data quality or insufficiency. As will be discussed below, it is likely that Condie's values for transition metals, niobium, tantalum, and barium are the more robust estimates of the average upper crust composition.

A recent paper by Borodin (1998) provides an average composition of the upper crust that includes much Soviet shield and granite data not included in most other worldwide averages. For this reason, it makes an interesting comparison with other data sets. Like other upper crustal estimates, major elements in the Borodin average upper crust (Table 1 and Figure 2) fall within 10% of the Eade and Fahrig average for the Canadian Shield, except for TiO_2 and FeO, which are ~13% higher, and Na_2O, which is ~20% lower than the Canadian average. Borodin's limited trace element averages (for chromium, nickel, rubidium, strontium, zirconium, niobium, barium, lanthanum, thorium, and uranium—not given in table or figures) fall within 50% of Shaw's Canadian Shield values except for niobium, which, like other upper crustal estimates, is about a factor of 2 lower than the Canadian average.

The more recent and comprehensive study of upper-crust composition derived from surface exposures was carried out by Gao *et al.* (1998a). Nine hundred and five composite samples were produced from over 11,000 individual rock samples covering an area of 9.5×10^5 km^2 in eastern China, which includes samples from Precambrian cratons as well as Phanerozoic fold belts. The samples comprised both crystalline basement rocks and sedimentary cover, the thickness of which was determined from seismic and aeromagnetic data. Averages were derived by combining compositions of individual map units weighted according to their thicknesses (in the cases of sedimentary cover) and areal exposure, for shields. The upper crust is estimated to be ~15 km thick based on seismic studies (Gao *et al.*, 1998a) and the crystalline rocks exposed at the surface area assumed to maintain their relative abundance through this depth interval. Average upper crust was calculated both as a grand average and on a carbonate-free basis; carbonates comprise a significant rock type (7–22%) in many of the areas sampled (e.g., Yangtze craton). The grand average (including carbonate) has a significantly different bulk composition than other estimates of the upper crust (Gao *et al.*, 1998a; Table 2). Most of the latter are derived from crystalline shields and so a difference is expected. However, Condie's map model incorporates sedimentary cover as well as crystalline basement. The differences between Condie's map model and Gao *et al.* grand-total upper crust suggest that the carbonate cover in eastern China is thicker than most other areas. For this reason, we use Gao *et al.* (1998a) carbonate-free compositions in further discussions, but with the caveat that carbonates may be an overlooked upper crustal component in many upper crustal estimates.

The Gao *et al.* (1998a) major- and trace-element results are presented in Tables 1 and 2 and plotted in Figures 2 and 3, respectively. Unlike the model of Condie (1993), several of the major elements fall beyond 10% of Eade and Fahrig's Canadian Shield data (Figures 2 and 3). These include TiO_2, FeO, MnO, and MgO, which are higher, and Na_2O, which is lower in the eastern China upper crust compared to the Canadian Shield. Gao *et al.* (1998a) attribute these differences to erosional differences between the two areas. Whereas the Canadian Shield composites comprised mainly metamorphic rocks of the amphibolite facies, the eastern China composites contain large proportions of unmetamorphosed supracrustal units that are considered to have, on average, higher proportions of mafic volcanics. In this respect, the Gao *et al.* model composition compares favorably to Condie's map model and the Russian estimates for all major elements. However, the Na_2O content of the eastern China upper crust is one of lowest of all (~20% lower than Condie's average and 10% lower than Borodin's values, but similar to Ronov and Yaroshevsky's average) (Figure 2).

The trace-element composition estimated by Gao *et al.* (1998a) for the Chinese upper crust is very similar to that of Condie (1993). Like the latter model, many lithophile trace elements in the Gao *et al.* model are within 50% of the Canadian Shield averages of Shaw *et al.* (e.g., LREEs, yttrium, rubidium, strontium, zirconium, hafnium, thorium, and uranium), and the Chinese average has significantly higher transition metals and lower niobium, barium, and tantalum than the Canadian Shield average. In addition, Gao *et al.* (1998a) provide values for some of the less well-constrained element concentrations. Of these, averages for lithium, beryllium, zinc, gallium, cadmium, and gold fall within 40% of the Shaw *et al.* averages, but boron, thallium, and bismuth are significantly higher, and mercury is significantly lower in Gao's average than in Shaw's. There is too little information for these elements in general to fully evaluate the significance of these differences.

Several generalizations can be made from the above studies of surface composites.

(i) Major element data are very consistent from study to study, with most major-element averages falling within 10% of Eade and Fahrig's Canadian Shield average. When differences do occur, they appear to reflect a lower percentage of mafic lithologies in the Canadian averages: all other estimates (including the Russian shield data) have higher FeO and TiO_2 than the Canadian averages and most also have higher CaO and MgO (Figures 2 and 3). The Eade and Fahrig average also has higher Na_2O than all other estimates (including Shaw's estimate for the Canadian Shield).

(ii) Trace elements show more variation than major elements from study to study, but some lithophile trace elements are relatively constant: rare earth elements (REEs), yttrium, lithium, rubidium, caesium, strontium, zirconium, hafnium, lead, thorium, and uranium do not vary beyond 50% between studies. Transition metals (scandium, cobalt, nickel, chromium, and vanadium) are consistently lower in the Canadian Shield estimates than in other studies, which may also be attributed to a lower percentage of mafic lithologies in the Canadian Shield (a conclusion supported by studies of sediment composition, as discussed in the next section). Barium is ~40% higher in the Shaw *et al.* average than in all other averages, including that of Eade and Fahrig, suggesting that this value is too high. Finally, niobium and tantalum are both about a factor of 2 higher in the Shaw *et al.* average than in any other average, suggesting that the former is not representative of the upper continental crust, a conclusion reached independently by Plank and Langmuir (1998) and Gallet *et al.* (1998) based on the composition of marine sediments and loess (see next section).

3.01.2.2 Sedimentary Rocks and Glacial Deposit Averages

While the large-scale sampling campaigns outlined above are the primary means by which the major-element composition of the upper continental crust has been determined, many estimates of the trace-element composition of the upper crust rely on the natural wide-scale sampling processes of sedimentation and glaciation. These methods are used primarily for elements that are insoluble during weathering and are, therefore, transported quantitatively from the site of weathering/glacial erosion to deposition. This methodology has been especially useful for determining the REE composition of the upper crust (see Taylor and McLennan (1985) and references therein). The averages derived from each of these natural large-scale samples are discussed in turn. When the upper crustal concentration of elements is discussed, the element name is printed in italic text so that the reader can quickly scan the text to the element of interest.

3.01.2.2.1 Sedimentary rocks

Processes that produce sedimentary rocks include weathering, erosion, transportation, deposition, and diagenesis. Elemental fractionation during weathering is discussed in detail by Taylor and McLennan (1985) (see also Chapters 5.01 and 7.01) and the interested reader is referred to these works for more extensive information. Briefly, elements with high solubilities in natural waters (Figure 4) have greater potential for being fractionated during sedimentary processing; thus, their concentration in fine-grained sedimentary rocks may not be representative of their source region. These elements include the alkali and alkaline-earth elements as well as boron, rhenium, molybdenum, gold, and uranium.

In contrast, a number of elements have very low solubilities in waters. Their concentrations in sedimentary rocks may, therefore, provide robust

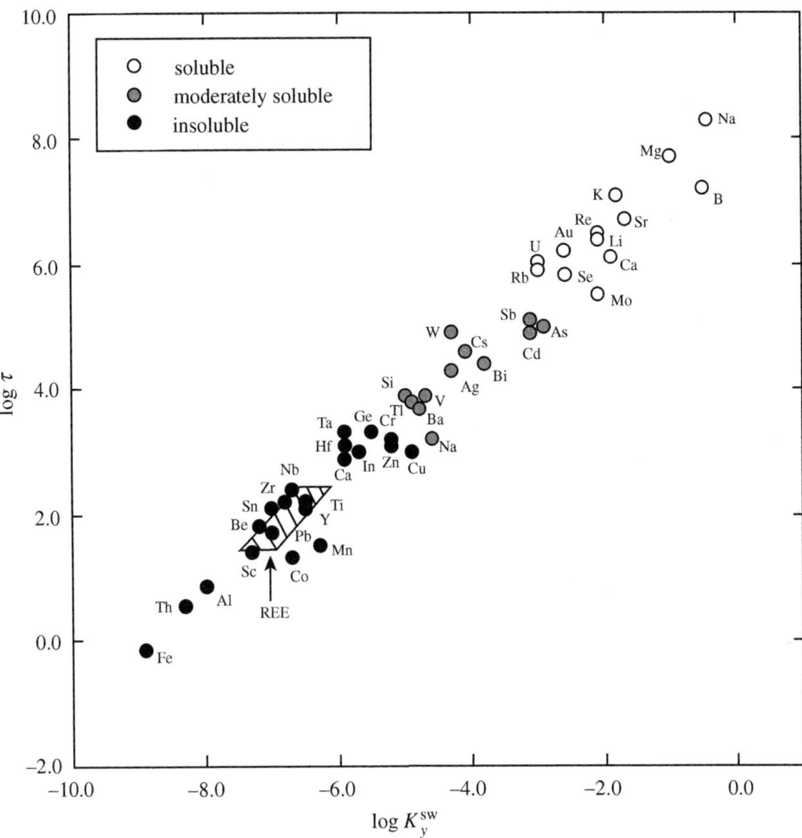

Figure 4 Plot of residence time (expressed as log τ) against seawater upper crust partition coefficient (expressed as K_y^{sw}) (source Taylor and McLennan, 1985).

estimates of the average composition of their source regions (i.e., average upper-continental crust). Taylor and McLennan (1985) identified that REEs, yttrium, scandium, thorium, and possibly cobalt as being suitably insoluble and thus providing useful information on upper crust composition.

The *REE* patterns for post-Archean shales show striking similarity worldwide (Figure 5): they are light REE enriched, with a negative europium anomaly and relatively flat heavy REEs. This remarkable consistency has led to the suggestion that the REE patterns of shales reflect that of the average upper-continental crust (Taylor and McLennan (1985) and references therein). Thus, Taylor and McLennan's (1985) upper crustal REE pattern is parallel to average shale, but lower in absolute abundances due to the presence of sediments with lower REE abundances such as sandstones, carbonates, and evaporites. Using a mass balance based on the proportions of different types of sedimentary rocks, they derive an upper crustal REE content that is 80% that of post-Archean average shale.

Comparison of various upper crustal REE patterns is provided in Figure 5. All estimates, whether from shales, marine sediments, or surface sampling, agree to within 20% for the LREEs and ~50% for the heavy rare-earth elements (HREEs). The estimate of Shaw *et al.* (1976) has the lowest HREEs and if these data are excluded, the HREEs agree to within 15% between the models of Condie (1993), Gao *et al.* (1998a), and Taylor and McLennan (1985). Thus, the REE content of the upper continental crust is established to within 10–25%, similar to the uncertainties associated with its major-element composition.

Once the REE concentration of the upper crust has been established, values for other insoluble elements can be determined from their ratios with an REE. Using the constant ratios of La/Th and La/Sc observed in shales, McLennan *et al.* (1980) and Taylor and McLennan (1985) estimated the upper crustal *thorium* and *scandium* contents at 11 ppm and 10.7 ppm, respectively. The scandium value increased slightly (to 13.7 ppm) and the thorium value remained unchanged when a more comprehensive sediment data set was employed by McLennan (2001b). The sediment-derived scandium and thorium averages agree to within 20% of the surface-sample averages (Table 2 and Figure 3).

Other insoluble elements include the high-field strength elements (HFSEs—titanium, zirconium,

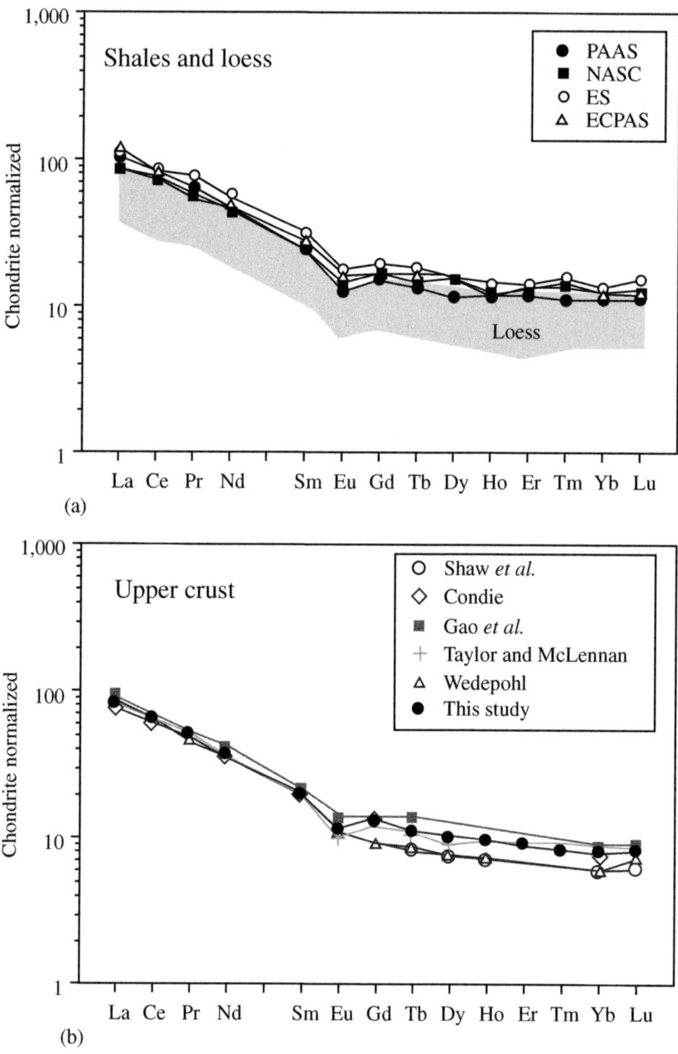

Figure 5 Comparison of REE patterns between (a) average post-Archean shales and loess and (b) various estimates of the upper continental crust composition. PAAS = post-Archean Australian Shale (Taylor and McLennan, 1985); NASC = North American shale composite (Haskin *et al.*, 1966); ES = European shale composite (Haskin and Haskin, 1966); ECPAS = Eastern China post-Archean shale (Gao *et al.*, 1998a). The loess range includes samples from China, Spitsbergen, Argentina, and France (Gallet *et al.*, 1998; Jahn *et al.*, 2001). Chondrite values are from Taylor and McLennan (1985).

hafnium, niobium, tantalum, molybdenum, tungsten), beryllium, aluminum, gallium, germanium, indium, tin, lead, and a number of transition metals (chromium, cobalt, nickel, copper, and zinc). Taylor and McLennan (1985) noted that some of these insoluble elements (e.g., HFSEs) may be fractionated during sedimentary processing if they reside primarily in heavy minerals. More recent evaluations have suggested that this effect is probably not significant for niobium and tantalum, and fractionations of zirconium and hafnium due to heavy mineral sorting are only really apparent in loess (Barth *et al.*, 2000; McLennan, 2001b; Plank and Langmuir, 1998). Plank and Langmuir (1998) noted that the *niobium*, *tantalum*, and *titanium* concentrations derived for the upper crust using marine sediments are considerably different from those of the Taylor and McLennan's upper continental crust composition. As oceanic processes are unlikely to fractionate these elements, Plank and Langmuir (1998) suggested that marine sediments provide a reliable estimate of the average composition of the upper continental crust. Using correlations between Al_2O_3 and niobium, they derived a niobium concentration for the upper crust of 13.7 ppm, and tantalum of 0.96 ppm (assuming Nb/Ta = 14); these values are about a factor of 2 lower than Taylor and McLennan's (1985) upper crustal estimates. Taylor and McLennan (1985) adopted their niobium value from Shaw *et al.* (1976), and their tantalum value was derived by assuming a Nb/Ta ratio of ~12 for the upper crust. The Plank and Langmuir niobium and tantalum values are similar to those derived from the surface-sampling studies of Condie (1993), Gao *et al.* (1998a), and Borodin (1998) and from more recent evaluations of these elements in shales, loess and other terrigenous sedimentary rocks (Barth *et al.*, 2000; Gallet *et al.*, 1998; McLennan, 2001b). All of these estimates range between 10 ppm and 14 ppm niobium and 0.74 ppm to 1.0 ppm tantalum, an overall variation of ~30%. Thus, niobium and tantalum concentrations now appear to be nearly as well constrained as the REE in the upper continental crust.

Plank and Langmuir (1998) also suggested, from their analyses of marine sediments, increasing the upper crustal TiO_2 values by ~40% (from 0.5 wt. % to 0.76 wt. %). Thus the TiO_2 content of the upper-continental crust probably lies between 0.55 wt.% and 0.76 wt.%, a difference of ~30%.

Of the remaining insoluble elements, recent evaluation of *zirconium* and *hafnium* concentrations derived from terrigenous sediment (McLennan, 2001b) show no significant differences with Taylor and McLennan's estimates, whose upper crustal zirconium value derives from the *Handbook of Geochemistry* (Wedepohl, 1969–1978), with hafnium determined from an assumed Zr/Hf ratio of 33. These values lie within ~20% of the surface-exposure averages (Table 2, Figure 3).

For the insoluble transition metals *chromium*, *cobalt*, and *nickel*, McLennan's (2001b) recent evaluation suggests approximate factor of 2 increases in average upper crustal values over those of Taylor and McLennan (1985). Taylor and McLennan's (1985) values were taken from a variety of sources (see Table 1 of Taylor and McLennan, 1981) and are similar to the Canadian Shield averages, which appear to represent a more felsic upper-crust composition, as discussed above. Even after eliminating these lower values, 30–40% variation exists for chromium, cobalt, and nickel between different estimates (Table 2 and Figure 3), and the upper crustal concentrations of these elements remains poorly constrained relative to REE.

McLennan (2001b) evaluated the upper crustal *lead* concentration from sediment averages and suggested a slight (~15%) downward revision (17 ppm) from the value of Taylor and McLennan (1985), whose value derives from a study by Heinrichs *et al.* (1980). McLennan's value is identical to that of surface averages (Table 2) and collectively these should be considered as a robust estimate for the lead content of the upper crust. For the remaining insoluble elements—*beryllium*, *copper*, *zinc*, *gallium*, *germanium*, *indium*, and *tin*—no newer data are available for terrigenous sediment averages. Estimates for some elements (e.g., zinc, gallium, germanium, and indium) vary by only ~20–30% between different studies, but others (beryllium, copper, and tin) vary by a factor of 2 or more (Table 2 and Figure 3).

It may also be possible to derive average upper crustal abundances of elements that have intermediate solubilities (e.g., vanadium, arsenic, silver, cadmium, antimony, caesium, barium, tungsten, and bismuth) using their concentrations in fine-grained sedimentary rocks, if they show significant correlations with lanthanum. Using this method McLennan (2001b) derived estimates of the upper crustal composition for *barium* (550 ppm) and *vanadium* (107 ppm). McLennan's barium value does not differ from that of Taylor and McLennan (1985), which derives from the *Handbook of Geochemistry* (Wedepohl, 1969–1978). This value is ~10% to a factor of 2 lower than the shield estimates; 630–700 ppm seems to be the most common estimate for barium from surface exposures. McLennan's vanadium estimate is ~50% higher than that of Taylor and McLennan (1985), which was derived from a 50 : 50 mixture of basalt : tonalite (Taylor and McLennan, 1981) and is similar to the Canadian Shield averages (Table 2). The revised vanadium value is similar to the surface-exposure averages from eastern China (Gao *et al.*, 1998a) and Condie's (1993) global average.

Several studies have used data for sedimentary rocks to derive the concentration of *caesium* in the upper crust. McDonough *et al.* (1992) found that a variety of sediments and sedimentary rocks (including loess) have an Rb/Cs ratio of 19 (± 11, 1σ), which is lower than the value of 30 in Taylor and McLennan's (1985) upper crust. Using this ratio and assuming a rubidium content of 110 ppm (from Shaw *et al.*, 1986; Shaw *et al.*, 1976; Taylor and McLennan, 1981), led them to an upper crustal caesium concentration of ~6 ppm. Data for marine sediments compiled by Plank and Langmuir (1998) also support a lower Rb/Cs ratio of the upper crust. Using the observed Rb/Cs ratio of 15 and a rubidium concentration of 112 ppm, they derived an upper crustal caesium concentration of 7.8 ppm. Although caesium data show only a poor correlation with lanthanum, the apparent La/Cs ratio of sediments led McLennan (2001b) to a revised caesium estimate of 4.6 ppm, which yields an Rb/Cs ratio of 24. Very few data exist for caesium from shield composites. Gao *et al.* determined a value of 3.6 ppm caesium, which is very similar to the estimate of Taylor and McLennan (1985). However, the Gao *et al.* rubidium estimate (83 ppm) is lower than Taylor and McLennan's (112 ppm), leading to an Rb/Cs ratio of 23 in the upper crust of eastern China. Caesium concentrations in all estimates vary by up to 70% and there thus appears to be substantial uncertainty in the upper crust's caesium concentration. Further evidence for the caesium content of the upper continental crust is derived from loess (see next section).

The upper crustal abundances of *arsenic*, *antimony*, and *tungsten* were determined by Sims *et al.* (1990), based on measurements of these elements in loess and shales. They find As/Ce to be rather constant at 0.08, leading to an arsenic content of 5.1 ± 1 ppm. In a similar fashion they estimate the upper crustal antimony content to be 0.45 ± 0.08 ppm and tungsten to be 3.3 ± 1.1 ppm. The antimony and arsenic values are factors of 2 and 3 higher, respectively, than the values given by Taylor and McLennan (1985), and the tungsten contents are a factor of 2 lower than Taylor and McLennan's (1985), which were adopted from the *Handbook of Geochemistry* (Wedepohl, 1969–1978). For all three elements, the Sims *et al.* estimates lie within uncertainty of the values given by Gao *et al.* (1998a) for the upper crust of eastern China, and these new estimates can thus be considered as representative of the upper crust to within ~30% uncertainty.

For the remaining moderately soluble elements *silver*, *cadmium*, and *bismuth*, there are no data for sedimentary composites. Taylor and McLennan (1985) adopted values from Heinrichs *et al.* (1980) for cadmium and bismuth and from the *Handbook of Geochemistry* (Wedepohl, 1969–1978) for silver. The only other data come from the study of Gao *et al.* (1998a). So essentially there are only two studies that address the concentrations of these elements in the upper crust: Gao *et al.* (1998a) and Wedepohl (1995) (which incorporates data from the *Handbook of Geochemistry* and Heinrichs *et al.* (1980)). For silver and cadmium, the two estimates converge: silver is identical and cadmium varies by 25% between Gao *et al.* and Wedepohl *et al.* estimates. In contrast, bismuth shows a factor of 2 of variation, with the Gao *et al.* estimates being higher.

3.01.2.2.2 Glacial deposits and loess

The concept of analyzing glacial deposits in order to determine average upper crustal composition originated with Goldschmidt (1933, 1958). The main attraction of this approach is that glaciers mechanically erode the rock types that they traverse, giving rise to finely comminuted sediments that represent averages of the bedrock lithologies. Because the timescale between erosion and sedimentation is short, glacial sediments experience little chemical weathering associated with their transport and deposition. In support of this methodology for determining upper crust composition, Goldschmidt noted that the major-element composition of composite glacial loams from Norway (analysed by Hougen *et al.*, 1925, as cited in Goldschmidt, 1933, 1958), which sample ~2×10^5 km^2 of Norwegian upper crust, compares favorably with the average igneous-rock composition determined by Clarke and Washington (1924) (Table 1). It would take another fifty years before geochemists returned to this method of determining upper crustal composition.

More recent studies using glacial deposits to derive average upper-crust composition have focused on the chemical composition of loess—fine-grained eolian sediment derived from glacial outwash plains (Taylor *et al.*, 1983; Gallet *et al.*, 1998; Peucker-Ehrenbrink and Jahn, 2001; Hattori *et al.*, 2003). This can be accomplished in two ways: either using the average composition of loess as representative of the upper continental crust or, if an element correlates with an insoluble element such as lanthanum whose upper concentration is well established, using the average X/La ratio of loess (where "X" is the element of interest), and assuming an upper crustal lanthanum value to determine the concentration of "X" (e.g., McLennan, 2001b). In this and subsequent discussion of loess, we derive upper crustal concentrations for particular elements using this method and assuming an upper crustal lanthanum value of 31 ppm, and compare these to previous estimates for these elements. The quoted uncertainty reflects 1σ on that ratio.

Loess is rich in SiO_2 (most carbonate-free loess has 73 wt.% to 80 wt.% SiO_2 (Taylor *et al.*, 1983; Gallet *et al.*, 1998), which probably reflects both the preferential eolian transport of quartz into loess and sedimentary recycling processes. This enrichment causes other elemental concentrations to be diluted. In addition, some other elements may be similarly fractionated during eolian processing. For example, loess shows anomalously high concentrations of zirconium and hafnium (Taylor *et al.*, 1983; Barth *et al.*, 2000), which, like the SiO_2 excess, have been attributed to size sorting through eolian concentration of zircon (Taylor *et al.*, 1983). Thus, loess Zr/La and Hf/La are enriched relative to the upper continental crust and cannot be used to derive upper crustal zirconium and hafnium concentrations. In addition, a recent study of rhenium and osmium in loess suggests that osmium contents are enhanced in loess compared to its source regions (Hattori *et al.*, 2003). This is explained by Hattori *et al.* (2003) as being due to preferential sampling of the fine sediment fraction by the wind, which may be enriched in mafic minerals that are soft and hence more easily ground to finer, transportable particle sizes. Mafic-mineral enhancement could give rise to similar fractionations between lanthanum and elements that are found primarily in mafic minerals (e.g., nickel, vanadium, scandium, chromium, cobalt, manganese, etc.). In such cases neither averages nor La/X ratios can be used to determine a reliable estimate of upper crustal composition. However, it is not apparent that eolian processing has significantly fractionated incompatible elements from lanthanum (e.g., barium, strontium, potassium, rubidium, niobium, thorium, etc.) that are not hosted primarily in mafic minerals. Indeed, the close correspondence of the thorium content of the upper crust derived from loess La–Th correlations (10.5 ± 1 ppm; Figure 6) to that deduced from shales (10.7 ppm, Taylor and McLennan, 1985) suggests that upper crustal concentrations of these elements derived from loess La–X correlations are not significantly affected by eolian processing.

Taylor *et al.* (1983), and later Gallet *et al.* (1998), determined the trace-element composition of a variety of loess samples from around the world and found that their REE patterns are remarkably constant and similar to that of average shales (see previous section and Figure 5). Likewise, niobium, tantalum, and thorium show strong positive correlations with the REE (Figure 6; Barth *et al.*, 2000; Gallet *et al.*, 1998). Thus, it appears that loess provides a robust estimate of average upper crustal composition for insoluble, incompatible trace elements.

Because loess is glacially derived, weathering effects are significantly reduced compared to shales (Taylor *et al.*, 1983), raising the possibility that loess may provide robust upper crustal estimates for the more soluble trace elements. However, examination of the major-element compositions of loess shows that all bear the signature of chemical weathering (Gallet *et al.*, 1998). Gallet *et al.* attributed this to derivation of

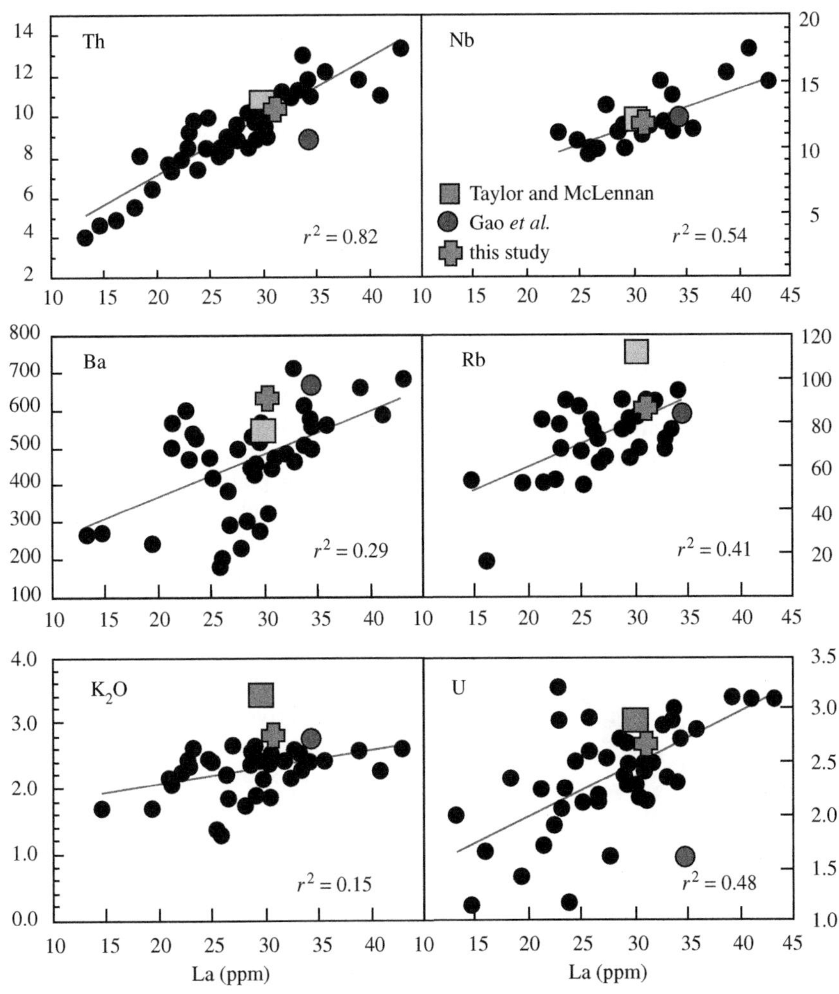

Figure 6 Lanthanum versus moderately to highly soluble incompatible trace elements in loess. Although loess is derived in part from weathered source regions, the positive correlations suggest that weathering has not completely obliterated the original, upper crustal mixing trends. Lines represent linear fit to data. Various models for the average upper crustal composition are superimposed (Taylor and McLennan, 1995, as modified by McLennan, 2001b; Gao *et al.*, 1998a and this study—Table 3) (sources Taylor *et al.*, 1983; Gallet *et al.*, 1998; Barth *et al.*, 2000; Jahn *et al.*, 2001; Peucker-Ehernbrink and Jahn, 2001).

loess particles from rocks that had previously experienced sedimentary differentiation. Likewise, Peucker-Ehernbrink and Jahn (2001) noted a positive correlation between $^{87}Rb/^{86}Sr$ and $^{87}Sr/^{86}Sr$ in loess, indicating that the weathering-induced fractionation is an ancient feature, and therefore inherited from the glacially eroded bedrocks. Even so, the degree of weathering in loess, as measured by the "chemical index of alteration" (CIA = molar $Al_2O_3/(Al_2O_3 + CaO + Na_2O + K_2O)$ Nesbit and Young (1984)), is small relative to that seen in shales (Gallet *et al.*, 1998), and it is likely that loess would provide a better average upper crustal estimate for moderately soluble trace elements (e.g., arsenic, silver, cadmium, antimony, caesium, barium, tungsten, and bismuth) than shales. Unfortunately, few measurements of these elements in loess are available (Barth *et al.*, 2000; Gallet *et al.*, 1998; Jahn *et al.*, 2001; Taylor *et al.*, 1983). Barium data show a scattered, positive correlation with lanthanum, yielding an upper crustal average of 510 ± 139 ppm (Figure 6). This value of barium concentration is similar to the one adopted by Taylor and McLennan (1985) and is within the uncertainty of all the other estimates save those of the Canadian Shield, which are significantly higher (Table 2). Caesium also shows a positive, scattered correlation with lanthanum, yielding an uncertain upper crustal caesium content of 4.8 ± 1.6 ppm, which is similar to that recently suggested by McLennan (2001b). However, caesium shows a better correlation with rubidium (Figure 7), defining an Rb/Cs ratio of ~17 in loess. Thus, if the upper crustal rubidium concentration can be determined, better constraints on the caesium content can be derived.

The highly soluble elements (lithium, potassium, rubidium, strontium, and uranium) show

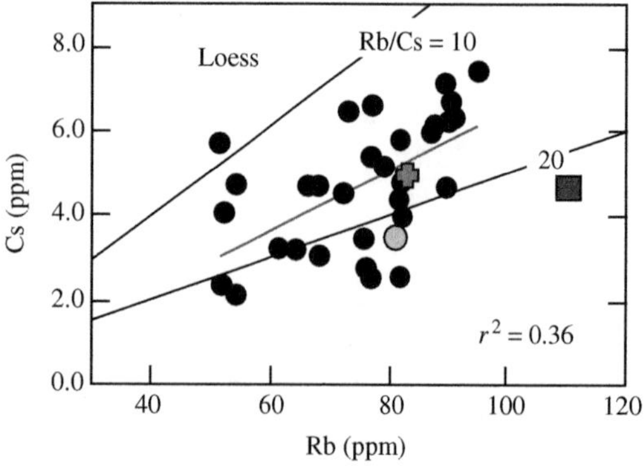

Figure 7 Rubidium versus caesium concentrations in loess samples. Short line is linear regression of data, thin, labeled lines represent constant Rb/Cs ratios. Symbols for crustal models and data sources as in Figure 6.

variable degrees of correlation with lanthanum in loess. Strontium shows no correlation with lanthanum, which is likely due to variable amounts of carbonate in the loess samples (Taylor *et al.*, 1983). Teng *et al.* (2003) recently reported lithium contents and isotopic compositions of shales and loess. Lithium contents of loess show no correlation with lanthanum, but fall within a limited range of compositions (17–41 ppm), yielding an average of 29 ± 10 ppm ($n = 14$). A similar value is derived using the correlation observed between lithium and niobium in shales. Thus, Teng *et al.* (2002) estimated the upper crustal lithium content at 31 ± 10 ppm, which is within error of previous estimates (Shaw *et al.*, 1976; Taylor and McLennan, 1985; Gao *et al.*, 1998a).

Potassium and rubidium show scattered, positive correlations with lanthanum (the Rb–La correlation is better, and the K–La is worse than the Ba–La correlation) (Figure 6). These correlations yield an upper crustal rubidium concentration of 84 ± 17 ppm. This rubidium value is identical to those derived from surface sampling by Eade and Fahrig (1973) , Condie (1993), and Gao *et al.* (1998a), but is lower than the widely used value of Shaw *et al.* (1976) at 110 ppm. The latter was adopted by both Taylor and McLennan (1985) and Wedepohl (1995) for their upper crustal estimates. The weak K–La correlation yields an upper crustal K_2O value of 2.4 ± 0.5 wt.%. This is within error of the surface-exposure averages of Fahrig and Eade (1968), Condie (1993), and Gao *et al.* (1998a), but is lower than the Shaw *et al.* surface averages of the Canadian shield (Shaw *et al.*, 1967), values for the Russian platform and the value adopted by Taylor and McLennan (1985) based on K/U and Th/U ratios. The loess-derived K/Rb ratio is 238, which is similar to the "well established" upper crustal K/Rb ratio of 250 (Taylor and McLennan, 1985). Because both potassium and rubidium are highly soluble elements, and loess shows evidence for some weathering, the potassium and rubidium contents derived from loess are best viewed as minimum values for the upper crust.

Uranium shows a reasonable correlation with lanthanum (Figure 6), which yields an upper crustal uranium content of 2.7 ± 0.6 ppm. This value is within error of the averages derived from surface exposures, except for the value of Gao *et al.* (1998a) and Eade and Fahrig (1973), which are distinctly lower. The loess-derived K/U ratio of 7,400 is lower than that assumed for the upper crust of 10,000 (Taylor and McLennan, 1985), and may reflect some potassium loss due to weathering, as discussed above.

Peucker-Ehrenbrink and Jahn (2001) analyzed loess in order to determine the concentrations of the platinum-group element (PGE) in the upper continental crust. To do this they examined PGE-major-element trends and used previously determined major-element compositions of the upper continental crust to infer the PGE concentrations (Table 2). Of the elements analyzed (ruthenium, palladium, osmium, iridium, and platinum), they found positive correlations for ruthenium, palladium, osmium, and iridium with major and trace elements for which upper crustal values had previously been established, leading to suggested upper crustal abundances of 340 ppt, 520 ppt, 31 ppt, and 22 ppt, respectively. They found no correlation between platinum contents and other elements, and so they simply used the average loess platinum content (510 ppt) as representative of the upper continental crust. We have estimated uncertainty on these values (shown in Table 2) by using the 95% confidence limit on the correlations published by Peucker-Ehrenbrink and Jahn (2001) and, for platinum, the standard deviation of the mean (Table 3). Recently, Hattori *et al.* (2003) suggested that preferential sampling of mafic minerals in loess may lead to enhancement of PGE and thus, loess-derived estimates may represent maximum concentrations for the upper crust. Based on samples of glacially derived desert sands and glacial moraines, Hattori *et al.* (2003) estimated an upper crustal osmium abundance of ~10 ppt.

Prior to these studies, few estimates were available for the PGE content of the upper continental crust. Peucker-Ehrenbrink and Jahn's loess-derived palladium value is similar to the value published by Taylor and McLennan (1985), which derives from the *Handbook of Geochemistry* (S. R. Taylor, personal communication), but is a factor of 3 smaller than that determined by Gao *et al.* (1998a) for the upper crust of eastern China. Peucker-Ehrenbrink and Jahn's (2001) loess-derived osmium abundance is ~65% lower than the estimate of Esser and Turekian (1993), which Peucker-Ehrenbrink and Jahn attribute to the hydrogenous uptake of osmium by the riverine sediments used in that study. Furthermore, the desert-sand and glacial moraine-derived osmium value of Hattori *et al.* (2003) is a factor of 3 lower than the estimate of Peucker-Ehrenbrink and Jahn (2001). Peucker-Ehrenbrink and Jahn's (2001) loess-derived iridium content is the same as that published for the Canadian Shield by Shaw *et al.* (1976). Thus, the upper crustal concentration of some PGE may be reasonably well constrained (e.g., palladium and iridium), while considerable uncertainty remains for others (e.g., platinum and osmium).

Rhenium is a highly soluble element that is easily leached during weathering, so the rhenium abundances of loess cannot be used directly to infer its upper crustal abundance. Following Esser and Turekian (1993), Peucker-Ehrenbrink and Jahn (2001) used the average $^{187}Os/^{188}Os$ ratio, osmium concentration, and average neodymium-model age of the crust to calculate the rhenium content of the upper continental crust. Their value (198 ppt) is about half that reported in Taylor and McLennan (1985) and calculated by Esser and Turekian (1993), who used the higher osmium abundance in their calculation. Using a similar methodology and osmium-isotopic composition, and the lower osmium abundance determined for glacially derived desert sands, Hattori *et al.* (2003) determined an upper crustal $^{187}Re/^{188}Os$ ratio of 35, which (assuming an average neodymium model age of 2.2 Ga for the crust) corresponds to a rhenium content of 74 ppt, about a third of the concentration determined by Peucker-Ehrenbrink and Jahn from loess data. Sun *et al.* (2003) used the rhenium contents of undegassed arc lavas to estimate the rhenium content of the bulk continental crust, assuming that the crust grows primarily by arc accretion. Their value of 2.0 ± 0.1 ppb is over an order of magnitude higher than that estimated by Peucker-Ehrenbrink and Jahn (2001) and Hattori *et al.* (2003) and is ~5 times higher than the Esser and Turekian (1993) and Taylor and McLennan (1985) values. Because rhenium is a moderately incompatible element, the rhenium concentration of the upper crust should be comparable to or higher than the bulk crust value (similar to ytterbium). However, this extreme rhenium concentration would require an order of magnitude higher osmium concentration in the crust or an extremely radiogenic crust composition, neither of which are consistent with any current estimates. Sun *et al.* (2003) suggest that rhenium may be lost from the continents by either rhenium degassing during arc volcanism or continental rhenium deposition into anoxoic sediments that are recycled into the mantle. Thus the value of 2 ppb rhenium is a maximum value for the upper continental crust and our knowledge of the rhenium content of the upper crust remains uncertain.

3.01.2.3 An Average Upper-crustal Composition

In Table 3 we present our best estimate for the chemical composition of the upper continental crust. The footnote provides detailed information on how the value for each element was derived. In general, major-element values represent averages of the different surface-exposure studies, and errors represent one standard deviation of the mean. Because two independent studies are available for the Canadian Shield, and because it appears the Canadian Shield has lower abundances of mafic lithologies and higher abundances of sodium-rich tonalitic–trondhjemitic granitic gneisses compared to other areas (see Section 3.01.2.1),

Table 3 Recommended composition of the upper continental crust. Major elements in weight percent.

Element	*Units*	*Upper crust*	*1 Sigma*	*%*	*Source*[a]	*Element*	*Units*	*Upper crust*	*1 Sigma*	*%*	*Source*[a]
SiO_2	wt.%	66.6	1.18	2	1	Ag	ng g^{-1}	53	3	5	4
TiO_2	"	0.64	0.08	13	2	Cd	μg g^{-1}	0.09	0.01	15	4
Al_2O_3	"	15.4	0.75	5	1	In	"	0.056	0.008	14	4
FeO_T	"	5.04	0.53	10	1	Sn	"	2.1	0.5	26	14
MnO	"	0.10	0.01	13	1	Sb	"	0.4	0.1	28	12
MgO	"	2.48	0.35	14	1	I	"	1.4		50	5
CaO	"	3.59	0.20	6	1	Cs	"	4.9	1.5	31	15
Na_2O	"	3.27	0.48	15	1	Ba	"	628	83	13	16
K_2O	"	2.80	0.23	8	3	La	"	31	3	9	4
P_2O_5	"	0.15	0.02	15	1	Ce	"	63	4	6	4
Li	μg g^{-1}	24	5	21	11	Pr	"	7.1			4
Be	"	2.1	0.9	41	4	Nd	"	27	2	8	4
B	"	17	8	50	4	Sm	"	4.7	0.3	6	4
N	"	83			5	Eu	"	1.0	0.1	14	4
F	"	557	56	10	4	Gd	"	4.0	0.3	7	4
S	"	62	33	53	4	Tb	"	0.7	0.1	21	4
Cl	"	370	382	103	4	Dy	"	3.9			17
Sc	"	14.0	0.9	6	6	Ho	"	0.83			17
V	"	97	11	11	6	Er	"	2.3			4
Cr	"	92	17	19	6	Tm	"	0.30			17
Co	"	17.3	0.6	3	6	Yb	"	1.96	0.4	18	4
Ni	"	47	11	24	6	Lu	"	0.31	0.05	17	4
Cu	"	28	4	14	7	Hf	"	5.3	0.7	14	4
Zn	"	67	6	9	7	Ta	"	0.9	0.1	13	11
Ga	"	17.5	0.7	4	8	W	"	1.9	1	54	18
Ge	"	1.4	0.1	9	4	Re	ng g^{-1}	0.198			13
As	"	4.8	0.5	10	9	Os	"	0.031	0.009	29	13
Se	"	0.09	0.05	54	4	Ir	"	0.022	0.007	32	13
Br	"	1.6			5	Pt	"	0.5	0.5	95	13
Rb	"	84	17	20	10	Au	"	1.5	0.4	26	4
Sr	"	320	46	14	4	Hg	μg g^{-1}	0.05	0.04	76	4
Y	"	21	2	11	4	Tl	"	0.9	0.5	57	4
Zr	"	193	28	14	4	Pb	"	17	0.5	3	4
Nb	"	12	1	12	11	Bi	"	0.16	0.06	38	19
Mo	"	1.1	0.3	28	12	Th	"	10.5	1.0	10	20
Ru	ng g^{-1}	0.34	0.02	6	13	U	"	2.7	0.6	21	20
Pd	"	0.52	0.02	3	13		"				

[a] Sources: (1) Average of all surface exposure data from Table 1, excluding Shaw *et al.* (1967), which is replicated by Fahrig and Eade (1968). (2) As (1) above, but including sediment-derived data from Plank and Langmuir (1998) and McLennan (2001b). (3) As (1) above, but also including K_2O value derived from loess (see text). (4) Average of all values in Table 2, excluding Wedepohl (1995) value or Taylor and McLennan (1985) value for Au, if it is derivitive from Shaw *et al.* (1976) and Taylor and McLennan (1985). (5) Wedepohl (1995). (6) Average of all surface composite data in Table 2, excluding Shaw *et al.* (1976), and including additional data from sediments (McLennan, 2001b). (7) Average of all surface composite data in Table 2, excluding Shaw *et al.* (1976), and including Taylor and McLennan (1985) values. (8) Average of all surface composite data in Table 2, excluding Shaw *et al.* (1976) due to their fractionated Ga/Al ratio. (9) Average of sedimentary data from Table 2 (Sims *et al.*, 1990) and Gao *et al.* (1998a) surface averages. (10) Dervied from La/Rb correlation in loess (see text). Value is identical to surface exposure data except for the Shaw *et al.* (1976) values. Data from Handbook of Geochemistry are about a factor of 2 lower than the latter and are not included in the average. (11) Average of all surface exposure data in Table 2 (minus Shaw *et al.* (1976), values) plus data from sediments and loess (Plank and Langmuir, 1998; Barth *et al.*, 2000; McLennan, 2001b; Teng *et al.*, 2003). (12) Average of all data in Table 2, excluding Taylor and McLennan (1985), which derive from same source as Wedephol's. (13) From Peucker-Ehrenbrink and Jahn (2001); see text for origin of error estimates. (14) Average of all data in Table 2, excluding Taylor and McLennan (1985), which is a factor of two higher than all other estimates. (15) Derived from Rb/Cs = 17 and upper crustal Rb value (see text). (16) Average of all data in Table 2, excluding the Shaw *et al.* (1976) and including additional data from loess (see text). (17) Value interpolated from REE pattern. (18) Average of all values in Table 2, plus correlation from Newsom *et al.* (1996), assuming W/Th = 0.2 (19) Average of all values in Table 2, excluding the Shaw *et al.* (1976) value, which is a factor of 5 lower than the others. (20) From loess correlations with La (see text). Both values are within error of the average of all surface exposure data and other sedimentary data (Taylor and McLennan, 1985; McLennan, 2001b).

we include only values from one of these studies—the Fahrig and Eade (1968) study, which encompasses a greater number of samples compared to that of Shaw *et al.* (1967). We also incorporate TiO_2 values derived from recent sedimentary studies (McLennan, 2001b; Plank and Langmuir, 1998) and the K_2O value from loess (Section 3.01.2.2) into the upper crustal averages (note that including the latter value in the average does not significantly change it). The standard deviation for most major-element averages is 10% or less. Only the ferromagnesian elements (iron, manganese, magnesium, and titanium), Na_2O, and P_2O_5 vary by up to 15%.

The trace-element abundances shown in Table 3 derive from different methods depending on their solubility. For most insoluble elements (see Section 3.01.2.2 and Figure 4 for definitions of solubility), we average the surface composites in addition to sediment or loess-derived estimates to derive the upper crustal composition. The uncertainty reported represents 1σ from the mean of all estimates. For moderately and highly soluble elements, we use the data derived from loess, if the elements show correlations with lanthanum ($r^2 > 0.4$), to infer their concentrations. In this case, the error represents the SD of the X/La ratio (where X is the element of interest). For elements that show no or only a poor correlation with lanthanum in loess and sediments (e.g., K_2O, Li, Ba, and Sr), we use the average of surface composites and sedimentary data (if some correlations exist with lanthanum) to derive an average. In most cases, the loess or sediment-derived values are within error of the surface-composite averages and these are noted in the footnote. Caesium is a special case. The loess caesium data show a poor correlation with lanthanum, but good correlation with rubidium. We thus use the observed Rb/Cs ratio of 17 (which is similar to the previous determination of this ratio in sedimentary rocks (McDonough *et al.*, 1992)) and the upper crustal rubidium concentration of 84 ppm to derive the caesium concentration of 4.9 ppm in the upper crust. The error on this estimate derives from the standard deviation of the Rb/Cs ratio. For some elements, only single estimates are available (e.g., bromine, nitrogen, iodine), and these are adopted as reported. The uncertainty of these estimates is likely to be very high, but there is no way to estimate uncertainty quantitatively with such few data. Remarkably, the SD on a large number of trace elements is below 20%, and the concentrations of a few (flourine, scandium, vanadium, cobalt, zinc, gallium, germanium, arsenic, yttrium, niobium, LREEs, tantalum, lead, and thorium) would appear to be known within ~10% (Table 3). However, in a number of these cases (e.g., flourine, cobalt, gallium, germanium, and arsenic), the small uncertainties undoubtedly reflect the fact that there have been few independent estimates made of the upper crust composition for these elements. It is likely that the true uncertainty for these elements is considerably greater than expressed in Table 3.

The upper crustal composition in Table 3 has many similarities to the widely used estimate of Taylor and McLennan (1985, with recent revision by McLennan (2001b)), but also some notable differences (Figure 8). Most of the elements that vary by more than 20% from the estimate of Taylor and McLennan are elements for which new data are recently available and few data exist overall (i.e., beryllium, arsenic, selenium, molybdenum, tin, antimony, rhenium, osmium, iridium, thallium, and bismuth). However, a number of estimates exist for K_2O, P_2O_5, and rubidium contents of the upper crust and our estimates are significantly lower (by 20–40%) than Taylor and McLennan's upper crust. The difference in P_2O_5 may simply be due to rounding errors. Taylor and McLennan (1985) report P_2O_5 of 0.2 wt.% versus 0.15 wt.% in our and other estimates of the upper crust—Tables 2 and 3). Taylor and McLennan (1985) derived their upper crustal K_2O indirectly from thorium abundances by assuming Th/U = 3.8 and K/U = 10,000. The resulting K_2O value is the highest of any of the estimates

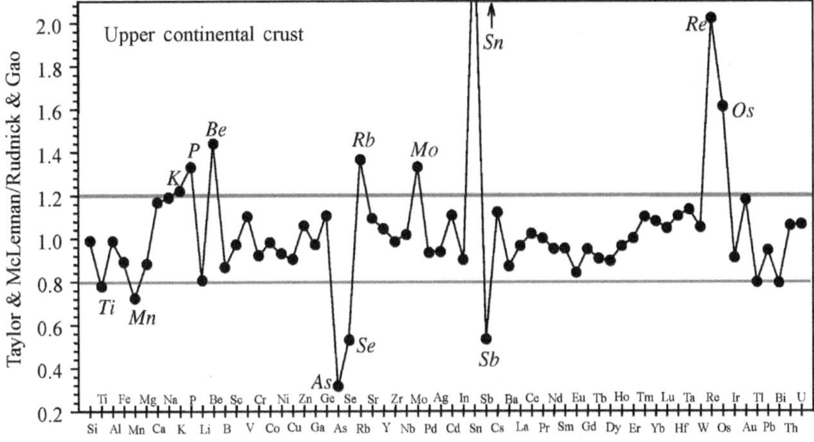

Figure 8 Plot of upper crustal compositional estimate of Taylor and McLennan (1995) (updated with values from McLennan, 2001b), divided by recommended values from this study. Horizontal lines mark 20% variation. Most elements fall within the ±20% bounds; elements falling beyond these bounds are labeled. Of the elements that differ by over 20%, potassium and rubidium are probably the most significant, since these elements are commonly analyzed to high precision in crustal rocks.

(Table 2 and Figure 2). Likewise, Taylor and McLennan's rubidium value was determined from their K_2O content, assuming a K/Rb ratio of the upper crust of 250. Their rubidium concentration matches the Canadian Shield value of Shaw *et al.* (1976), but is higher than all other surface-exposure studies, including Fahrig and Eade's Canadian Shield estimate (Table 2 and Figure 3). In contrast, the remaining surface-exposure studies match the rubidium value we derived from the loess Rb–La correlation (Figure 6). We conclude that the upper crust may have lower potassium and rubidium contents than estimated by Taylor and McLennan (1985). This finding has implications for total crustal heat production (see Section 3.01.4 and Chapter 3.02).

3.01.3 THE DEEP CRUST

The deep continental crust is far less accessible than the upper crust and consequently, estimates of its composition carry a greater uncertainty. Compared to the upper crust, the earliest estimates of the composition of the deep crust are relatively recent (i.e., 1950s and later) and derive from both seismological and geological studies.

On the basis of observed isostatic equilibrium of the continents and a felsic upper crust composition, Poldervaart (1955) suggested a two-layer crust with granodioritic upper crust underlain by a basaltic lower crust. The topic of deep crustal composition doesn't seem to have been considered again until ~20 years later, when a series of works in the 1970 s and 1980 s made significant headway into the nature of the deep continental crust. On the basis of surface heat-flow, geochemical studies of high-grade metamorphic rocks and seismological data, Heier (1973) proposed that the deep crust is composed of granulite-facies rocks that are depleted in heat-producing elements. A similar conclusion was reached by Holland and Lambert (1972) based on their studies of the Lewisian complex of Scotland. Smithson (1978) used seismic reflections and velocities to derive both structure and composition of the deep crust. He divided the crust into three, heterogeneous regions: (i) an upper crust composed of supracrustal metamorphic rocks intruded by granites, (ii) a migmatitic middle crust and (iii) a lower crust composed of a heterogenous mixture of igneous and metamorphic rocks ranging in composition from granite to gabbro, with an average intermediate (dioritic) composition. This three-layer model of the crust survives today in most seismologically based studies. Weaver and Tarney (1980, 1981, 1984) derived a felsic and intermediate composition for the Archean middle and lower crust, respectively, based on studies of amphibolite to granulite-facies rocks exposed in the Lewisian complex, Scotland. R. W. Kay and S. M. Kay (1981) were one of the first to stress the importance of xenolith studies to unravelling deep-crustal composition. They highlighted the heterogeneous nature of the deep crust and suggested its composition should vary depending on tectonic setting, cautioning against the use of singular cross sections or deep-crustal exposures to derive global models. Taylor and McLennan (1985) considered the lower crust to be the portion of the crust from 10 km depth to the Moho. Their "lower crust" thus includes both middle and lower crust, as used here (see Section 3.01.3.1). Taylor and McLennan's (1985) lower-crust composition was derived by subtracting the upper crust from their total-crust composition (see Section 3.01.4). The Taylor and McLennan (1985) lower crust is thus not based on observed lower crustal rock compositions, but rather on models of upper- and total-crust compositions and assumptions about the origin of surface heat flow.

More recent attempts to define deep crust composition have relied upon linking geophysical data (principally seismic velocities) to deep crustal lithologies and their associated compositions to derive the bulk composition of the deep crust as a function of tectonic setting (Christensen and Mooney, 1995; Rudnick and Fountain, 1995; Wedepohl, 1995; Gao *et al.*, 1998a,b). Despite the attendant large uncertainties in deriving composition from velocity (Rudnick and Fountain, 1995; Brittan and Warner, 1996, 1997; Behn and Kelemen, 2003) and the lack of thorough geochemical sampling of the deep crust in many regions, these efforts nevertheless provide the best direct estimates of present-day deep crustal composition.

In this section we examine the composition of the deep crust by first defining its structure and lithology and the methods employed to determine deep crust composition. We then examine observations on middle and lower crustal samples, average seismic velocities and the resulting models of deep crust composition.

3.01.3.1 Definitions

Following recent compilations of the seismic-velocity structure of the continental crust, we divide the deep crust into middle and lower crust (Holbrook *et al.*, 1992; Christensen and Mooney, 1995; Rudnick and Fountain, 1995). Holbrook *et al.* (1992) defined the middle crust as: (i) the middle-third, where the velocity structure suggests a natural division of the crust into thirds; (ii) the region beneath the upper crust and above a Conrad discontinuity, if there is a layer beneath the Conrad; and (iii) the region immediately beneath the Conrad if there are two distinct velocity layers beneath a Conrad discontinuity.

The lower crust is thus the layer beneath the middle crust and above the Moho.

For a ~40 km thick average global continental crust (Christensen and Mooney, 1995; Rudnick and Fountain, 1995), the middle crust is 11 km thick and ranges in depth from 12 km, at the top, to 23 km at the bottom (Gao *et al.* (1998b) based on the compilations of data for crustal structure in various tectonic settings by Rudnick and Fountain (1995)). The average lower crust thus begins at 23 km depth and is 17 km thick. However, the depth and thickness of both middle and lower crust vary from setting to setting. In fore-arcs, active rifts, and rifted margins, the crust is generally thinner: middle crust extends from 8 km to 17 km depth and lower crust from 17 km to 27 km depth. In Mesozoic–Cenozoic orogenic belts the crust is thicker and middle crust extends from 16 km to 27 km depth and the lower crust from 27 km to 51 km depth (Rudnick and Fountain, 1995).

3.01.3.2 Metamorphism and Lithologies

Studies of exposed crustal cross-sections and xenoliths indicate that the middle crust is dominated by rocks metamorphosed at amphibolite facies to lower granulite facies, while the lower crust consists mainly of granulite facies rocks (Fountain *et al.*, 1990a; Fountain and Salisbury, 1981; Mengel *et al.*, 1991; Weber *et al.*, 2002). However, exceptions to these generalities do occur. For thin crust in rifted areas, greenschist-facies and amphibolite-facies rocks may predominate in the middle and lower crust, respectively. In overthickened Mesozic and Cenozoic orogenic belts (e.g., Alps, Andes, Tibet, and Himalyas), and paleo-orogenic belts that now have normal crustal thicknesses (e.g., Appalachains, Adirondacks, Variscan belt), granulite-facies and eclogite-facies rocks may be important constituents of the middle and lower crust (Leech, 2001; LePichon *et al.*, 1997; Lombardo and Rolfo, 2000). In contrast, amphibolite-facies lithologies may be present in the deep crust of continental arcs (Aoki, 1971; Miller and Christensen, 1994; Weber *et al.*, 2002), where hydrous fluids are fluxed from the subducting slab and the water contents of underplating magmas are high.

Lithologically, both middle and lower crust are highly heterogeneous, as seen in surface exposures of high-grade metamorphic rocks, crustal cross-sections, and deep-crustal xenolith suites. However, there is a general tendency for the middle crust to have a higher proportion of evolved rock compositions (as observed in cross-sections and granulite-facies terranes) while the lower crust has a higher proportion of mafic rock types (as observed in xenolith suites (Bohlen and Mezger, 1989)). Metasedimentary lithologies are often present, albeit in small proportions. The exact proportions of felsic to mafic lithologies in the deep crust varies from place to place and can only be established through the study of crustal cross-sections or inferred from seismic velocity profiles of the crust (Christensen and Mooney, 1995; Rudnick and Fountain, 1995; Wedepohl, 1995; Gao *et al.*, 1998b).

3.01.3.3 Methodology

There are three approaches to derive the composition of the deep crust (see Rudnick and Fountain (1995) for a review).

(i) *By studying samples derived from the deep crust.* These occur as surface outcrops of high-grade metamorphic terranes (e.g., Bohlen and Mezger, 1989; Harley, 1989), tectonically uplifted crustal cross-sections (e.g., Fountain and Salisbury, 1981; Percival *et al.*, 1992), and as deep-crustal xenoliths carried in volcanic pipes (Rudnick, 1992; Downes, 1993).

(ii) *By correlating seismic velocities with rock lithologies* (Christensen and Mooney, 1995; Rudnick and Fountain, 1995; Wedepohl, 1995, Gao *et al.*, 1998a,b).

(iii) From surface heat-flow measurements (see Chapter 3.02).

As pointed out by Jaupart and Mareschal (see Chapter 3.02), surface heat flow is the only geophysical parameter that is a direct function of crustal composition. In general, however, heat flow provides only very broad constraints on deep-crust composition due to the ambiguity involved in distinguishing the amount of surface heat flow arising from crustal radioactivity versus the Moho heat flux (see Chapter 3.02); Rudnick *et al.*, 1998). Most models of the deep-crust composition fall within these broad constraints. The exception is the global model of Wedepohl, 1995, which produces more heat than the average surface heat flow in the continents, thereby allowing no mantle heat flux into the base of the crust (Rudnick *et al.*, 1998). In addition, the regional model of Gao *et al.* (1998a) for eastern China produces too much heat to be globally representative of the continental crust composition (see discussion in Rudnick *et al.* (1998) and Jaupart and Mareschal (Chapter 3.02)). However, the Gao *et al.* composition may be representative of the continental crust of eastern China, where the crust is relatively thin (30–35 km) and the heat flow is high (>60 mW m^{-2}). In the remaining discussion of deep-crust composition, we rely most heavily on methods (i)–(ii), above, but return to the question of heat flow when considering the bulk crust composition in Section 3.01.4.

In addition to mineralogy, which is in turn a function of bulk composition and metamorphic grade, factors affecting the seismic velocities of the continental crust include temperature, pressure, and the presence or absence of volatiles, fractures, and mineralogical anisotropy. It is generally assumed that cracks and fractures are closed under the ambient confining pressures of the middle to lower crust (0.4–1.2 GPa). In addition, although evidence for volatile transport is present in many rocks derived from the deep crust (see Chapters 3.06 and 3.09), the low density of these fluids allows for their escape to the upper crust shortly after their formation. Hence, most studies assume the deep crust does not, in general, contain an ambient, free volatile phase (Yardley, 1986).

Some minerals are particularly anisotropic with respect to seismic-wave speeds (e.g., olivine, sillimanite, mica (Christensen, 1982)), which can lead to pronounced seismic anisotropy in rocks if these minerals are crystallographically aligned through deformational processes (Meltzer and Christensen, 2001). This, in turn, could lead to over- or underestimation of representative seismic velocities of the deep crust if deformed rocks with such anisotropic minerals occur there. Olivine is not commonly stable in the deep crust, but other strongly anisotropic minerals are (e.g., mica, which is predominantly stable in the middle crust, and sillimanite, which is found in metapelitic rocks in the middle-to-lower crust). Some of the largest seismic anisotropies have been recorded in mica schists and gneisses, which can have average anisotropies over 10% (Christensen and Mooney, 1995; Meltzer and Christensen, 2001). Amphibole is also anisotropic and the average anisotropies for amphibolite are also ~10% (Christensen and Mooney, 1995; Kern *et al.*, 1996). In general, anisotropy is expected to be highest in metapelitic rocks and amphibolites, which contain the highest proportions of anisotropic minerals. These lithologies appear to be subordinate in middle-crustal sections and outcrops (described in the next section) compared to felsic gneisses, which typically have low anisotropies (<5%). In contrast, studies of xenoliths show metapelite to be a common lithology in the lower crust, albeit proportionally minor, and amphibolite may be important in some regions (Section 3.01.3.5.1). Thus, seismic anisotropy could be especially important in regions having large amounts of metasedimentary rocks (e.g., accretionary wedges) and amphibolite (arc crust?) in the deep crust, but is less likely to be important in crust dominated by felsic metaigneous rocks or mafic granulites.

Changes in *P*-wave velocity of a rock as a function of temperature and pressure are generally assumed to be on the order of -4×10^{-4} km s^{-1} °C^{-1} and 2×10^{-4} km s^{-1} MPa^{-1} (see Rudnick and Fountain (1995 and references therein). Because most laboratory measurements of ultrasonic velocities are carried out at confining pressures of 0.6–1.0 GPa, no pressure correction needs to be made in order to compare field and laboratory-based velocity measurements. However, temperature influence on seismic-wave speeds can be significant, especially when comparing laboratory data collected at room temperature to field-based measurements in areas of high heat flow (e.g., rifts, arcs, extentional settings). The decrease in compressional wave velocities in the deep crust under these high geotherms can be as much as 0.3 km s^{-1} (see Rudnick and Fountain, 1995, figure 1). For these reasons, Rudnick and Fountain (1995) used regional surface heat flow and assumed a conductive geothermal gradient, to correct the field-based velocities to room-temperature conditions. In this way, direct comparisons can be made between velocity profiles and ultrasonic velocities of lower-crustal rock types measured in the laboratory. Another benefit of this correction is that deep-crustal velocities from areas with grossly different geotherms can be considered directly in light of possible lithologic variations. In subsequent sections we quote deep-crustal velocities corrected to room-temperature conditions as "temperature-corrected velocities."

3.01.3.4 The Middle Crust

3.01.3.4.1 Samples

The best evidence for the compositional makeup of the middle crust comes from studies of high-grade metamorphic terranes and crustal cross-sections. There are far fewer studies of amphibolite-facies xenoliths derived from mid-crustal depths (Grapes, 1986; Leeman *et al.*, 1985; Mattie *et al.*, 1997; Mengel *et al.*, 1991; Weber *et al.*, 2002) compared to their granulite-facies counterparts. This may be due to the fact that it can be difficult to distinguish such xenoliths from the exposed or near-surface amphibolite-facies country rocks through which the xenolith-bearing volcanic rocks erupted. For this reason, xenolith studies have not been employed to any large extent in understanding the composition of the middle crust, and most information about the middle crust comes from studies of high-grade terranes, crustal cross-sections, and seismic profiles.

Interpreting the origin of granulite-facies terranes and hence their significance towards determining deep-crustal composition depends on unraveling their pressure–temperature–time history (see Chapters 3.07 and 3.08). Those showing evidence for a "clockwise" P–T path (i.e., heating during decompression) are often interpreted as having been only transiently in the lower crust; they represent upper crustal assemblages that passed through high $P-T$ conditions on their way back to the surface during continent-scale collisional orogeny. In contrast, granulite terrains showing evidence for isobaric cooling can have extended lower-crustal histories, and thus may shed light on deep-crustal composition (see discussion in Rudnick and Fountain (1995)). Bohlen and Mezger (1989) pointed out that isobarically cooled granulite-facies terranes show evidence of equilibration at relatively low pressures (i.e., $\leq$0.6–0.8 GPa), corresponding to mid-crustal depths ($\leq$25 km). Although a number of high-pressure and even ultra-high-pressure metamorphic belts (Chapter 3.09) have been recognized since their study, it remains true that the majority of isobarically cooled granulite-facies terranes show only moderate equilibration depths and, therefore, may provide evidence regarding the composition of the middle crust.

Although lithologically diverse, the average composition of rocks analyzed from granulite terrains is evolved (Rudnick and Presper, 1990), with median compositions corresponding to granodiorite/dacite (64–66 wt.% SiO_2, 4.1–5.2 wt.% $Na_2O + K_2O$, based on classification of Le Bas and Streckeisen (1991)). Rudnick and Fountain (1995) suggested that isobarically cooled granulite terrains have a higher proportion of mafic lithologies than granulites having clockwise $P-T$ paths. However, the median composition of rocks analyzed from isobarically cooled terranes is indistinguishable (62 wt.% SiO_2, 4.6 wt.% $Na_2O + K_2O$) from the median composition of the entire granulite-terrane population given in Rudnick and Presper (1990). Collectively, these data point to a chemically evolved mid-crustal composition.

Observations from crustal cross-sections also point to an evolved mid-crust composition (Table 4). Most of these cross-sections have been exposed by compressional uplift due to thrust faulting (e.g., Kapuskasing, Ivrea, Kohistan, and Musgrave). Other proposed origins for the uplift include wide, oblique transitions (Pikwitonei), impactogenesis (Vredefort), and transpression (Sierra Nevada) (Percival *et al.*, 1992). In nearly all these sections, sampling depth ranges from upper to middle crust; only a few (e.g., Vredefort, Ivrea, Kohistan) appear to penetrate into the lower crust. In the following paragraphs we review the insights into middle (and lower) crust lithologies gained from the studies of these crustal cross-sections.

The Vredefort dome represents a unique, upturned section through ~36 km of crust of the Kaapvaal craton, possibly exposing a paleo-Moho at its base (Hart *et al.*, 1981, 1990; Tredoux *et al.*, 1999; Moser *et al.*, 2001). The origin of this structure is debated, but one likely scenario is that it was produced by crustal rebound following meteorite impact. The shallowest section of basement (corresponding to original depths of 10–18 km depth) is composed of amphibolite-facies rocks consisting of granitic gneiss (the outer granite gneiss). The underlying granulite-facies rocks (original depths of 18–36 km) are composed of charnockites and leucogranofels with ~10% mafic and ultramafic granulites (the Inlandsee Leucogranofels terrain). The mid-crust, as defined here, is thus composed of amphibolite-facies felsic gneisses in fault contact with underlying charnockites and mixed felsic granulites and mafic/ultramafic granulites (Hart *et al.*, 1990). The lower crust, which is only partially exposed, consists of mixed felsic and mafic/ultramafic granulites, with the proportion of mafic rocks increasing with depth. The mantle beneath the proposed paleo-Moho, as revealed by borehole drilling, is dominated by 3.3–3.5 Ga serpentinized amphibole-bearing harzburgite (Tredoux *et al.*, 1999).

The Kapuskasing Structural Zone represents an exposed middle-to-lower crustal section through a greenstone belt of the Archean Canadian Shield, where the middle crust is represented by the amphibolite-facies Wawa gneiss dome and lower granulite-facies litihologies along the Kapuskasing uplift. Altogether, ~25 km of crust are exposed out of a total crustal thickness of 43 km (Fountain *et al.*, 1990b; Percival and Card, 1983). The Wawa gneiss dome is dominated by tonalite–granodiorite gneisses and their igneous equivalents (87%), but also contains small amounts of paragneiss (5%) and mafic gneiss and intrusives (8%) (Burke and Fountain, 1990; Fountain *et al.*, 1990b; Shaw *et al.*, 1994). The slightly deeper-level Kapuskasing Structural Zone has a greater proportion of paragneisses and mafic lithologies. It contains 35% mafic or anorthositic gneisses, 25% dioritic gneisses, 20% paragneiss, and only 20% tonalite gneisses.

Like the high-grade rocks of the Kapusksasing Structural Zone, those in the Pikwitonei crustal cross-section represent high-grade equivalents of granite–gneiss–greenstone successions (Fountain and Salisbury, 1981; Percival *et al.*, 1992). Approximately 25 km of upper-to-middle crust is exposed in this section out of a total-crustal thickness of 37 km (Fountain *et al.*, 1990b). Both amphibolite- and granulite-facies rocks are dominated by tonalitic gneiss with minor mafic gneiss, and metasedimentary rocks.

The Wutai–Jining Zone is suggested to be an exposed cross-section through the Archean North China craton (Kern *et al.*, 1996). Rocks from this exposure equilibrated at depths of up to ~30 km, thus sampling middle and uppermost lower crust, but leaving the lowermost 10 km of crust unexposed (Kern *et al.*, 1996). Like the previously described cross-sections, felsic gneisses dominate

Table 4 Chemical and petrological composition of crustal cross-sections.

	Reference[a]	*Age*	*Setting, (uplift origin)*[b]	*Current crustal thickness* (km)	*Maximum depth* (km)	*Middle crust lithologies*	*Lower crust lithologies*
Archean							
Vredefort Dome	1–3	2.6–3.6 Ga	Kaapvaal craton, (3)	36	36 (w/paleo Moho)	Amphibolite-facies granitic gneiss	Granulite-facies charnockites, leucogranofels, mafic, and ultramafic granulites
Kapuskasing Uplift	4–5	2.5–2.7 Ga	Superior craton, (1)	43	25	*Amphibolite-facies* *87% felsic* 8% mafic-intermediate 5% metasediment	*Lower granulite-facies* *35% mafic/anorthositic* 25% diorite 20% metasediments 20% felsic
Pikwitonei granulite domain	6–9	2.5–3.1 Ga	Superior craton, (2)	37	25	*Amphibolite-lower granulite facies.* Dominately tonalite gneiss, minor mafic gneisses, quartzites, anorthosites.	*Granulite-facies.* Predominantly silicic to intermediate gneiss, with minor paragneiss, mafic-ultramafic bodies and anorthosites
Wutai-Jining zone	10	2.5–2.8 Ga	North China Craton, (2)	40	30	*Amphibolite-lower granuilte facies.* *89% tonalitic-trondhjemitic-granodioritic-granitic gneiss* 8% amphibolite and mafic granulite 3% metapelite	*Granulite-facies.* *54% tonalitic-trondhjemitic-granodioritic-granitic gneiss* 32% mafic granulite 6% metapelite 8% metasandstone
Proterozoic							
Musgrave ranges	6, 8, 11	1.1–2.0 Ga	Central Australia, (1)	40	Unknown	Quartzofeldspathic gneiss, amphibolite, metapelite, marble, calc-silicate gneiss	Silicic to intermediate gneiss, mafic granulite, layered mafic-ultramafic intrusions
S. Norway	12–13	1.5–2.0 Ga	Baltic Shield, (2)	35	Unknown	Quartzofeldspathic gneiss, amphibolite, metasediments	Felsic granulite, mafic granulite, metasediments
Phanerozoic							
Ivrea-Verbano zone	14–17	Permian	Alps, (1)	35	30	*Amphibolite-facies.* felsic gneiss, amphibolite, metapelite (kinzigite), marble	*Granulite facies.* mafic intrusives and ultramafic cumulates, resistic metapelite (stronalite), diorite
Sierra Nevada, California	8, 18–20	Cretaceous	Continental arc, (4)	27–43	30	Mafic to felsic gneiss, amphibolite, diorite–tonalite	Granofels, mafic granulite, graphite-bearing metasediments
Kohistan, Pakistan	8, 21	Late Jurassic-Eocene	Oceanic arc, (1)	Unknown	45	Diroite, metadiorite, gabbronorite	Amphibolite, metagabbro, gabbronorite, garnet gabbro, garnet hornblendite, websterite
Talkeetna, Alaska	22–23	Jurassic	Oceanic arc, (1)	25–35	13	Gabbro, tonalite, diorite	Garnet gabbro, amphibole gabbro, dunite, wehrlite, pyroxenite

[a] References: 1. Hart *et al.* 1990, 2. Tredoux *et al.*, 1999, 3. Moser *et al.*, 2001, 4. Fountain *et al.*, 1990a, 5. Shaw *et al.*, 1994, 6. Fountain and Salisbury, 1981, 7. Fountain *et al.*, 1987, 8. Percival *et al.*, 1992, 9. Fountain and Salisbury 1995, 10. Kern *et al.*, 1996, 11. Clitheroe *et al.*, 2000, 12. Pinet and Jaupart, 1987, 13. Alirezaei and Cameron, 2002, 14. Mehnert, 1975, 15. Fountain, 1976, 16. Voshage *et al.*, 1990, 17. Mayer *et al.*, 2000, 18. Ross, 1985, 19. Saleeby, 1990, 20. Ducea, 2001, 21. Miller and Christensen, 1994, 22. Pearcy *et al.*, 1990, 23. see Chapter 3.18. [b] The different mechanisms responsible for uplift of these crustal cross sections include (1) compressional uplifts along thrust faults, (2) wide, oblique transitions, which are also compressional in origin, but over wide transitions, with no one thrust fault obviously responsible for their uplift, (3) meteorite impact, and (4) transpressional uplifts, which are vertical uplifts along a transcurrent faults (Percival *et al.*, 1992).

the middle crust; tonalitic–trondhjemitic–granodioritc, and granitic gneisses comprise 89% of the dominant amphibolite to granulite-facies Henshan-Fuping terrains, the remaining lithologies are amphibolite-mafic granulite (8%) and metapelite (3%). Tonalitic–trondhjemitic–granodioritc and granitic gneiss (54%) are less significant but still dominant in the lower-crustal Jining terrain.

The Musgrave Range (Fountain and Salisbury, 1981; Percival *et al.*, 1992) and the Bamble Sector of southern Norway (Pinet and Jaupart, 1987; Alirezaei and Cameron, 2002)) represent two crustal sections through Proterozoic crust of central Australia and the Baltic Shield, respectively. In both sections, the middle crust is dominated by quartzofeldspathic gneiss. The lower crust consists of silicic to intermediate gneiss, felsic granulite, and mafic granulite with layered mafic and ultramafic intrusions being important lithological components in the Musgrave Range and metasediments being important in the lower crust of southern Norway.

The Ivrea–Verbano Zone in the southern Alps of Italy was the first to be proposed as an exposed deep-crustal section by Berckhemer (1969) and has subsequently been the focus of extensive geological, geochemical, and geophysical studies (e.g., Mehnert, 1975; Fountain, 1976; Dostal and Capedri, 1979; Voshage *et al.*, 1990; Quick *et al.*, 1995). The Paleozoic rocks of the Ivrea zone are unusual when compared with Precambrian granulite outcrops because they contain a large proportion of mafic lithologies and, as such, closely resemble granulite xenoliths in composition (Rudnick, 1990b). Amphibolite-facies rocks of the middle crust consist of felsic gneiss, amphibolite, metapelite (kinzigite), and marble, whereas the lower crustal section comprises mafic granulite and diorite, which formed by intrusion and subsequent fractionation of basaltic melts that partially melted the surrounding metasediments (now resistic stronalite) (Mehnert, 1975; Dostal and Capedri, 1979; Fountain *et al.*, 1976; Voshage *et al.*, 1990). Detailed mapping by Quick *et al.* (1995) demonstrated that mantle peridotites in the southern Ivrea Zone are lenses that were tectonically interfingered with metasedimentary rocks prior to intrusion of the gabbroic complex and the present exposures reside an unknown distance above the pre-Alpine contiguous mantle. Thus reference to the section as a complete crust–mantle transition could be misleading. Altogether, the exposed rocks represent ~30 km of crust with ~5 km lowermost crust remaining unexposed (Fountain *et al.*, 1990a). The similarity in isotope composition and age between the Ivrea zone cumulates and Hercynian granites in the upper crust led Voshage *et al.* (1990) to speculate that these granites were derived from lower-crustal magma chambers similar to those in the Ivrea Zone, suggesting that basaltic underplating may be important in the formation and modification of the lower continental crust (Rudnick, 1990a).

Three sections through Mesozoic arcs show contrasting bulk compositions, depending on their settings (continental versus oceanic). In the southern Sierra Nevada, a tilted section exposes the deeper reaches of the Sierra Nevada batholith, which is part of a continental arc formed during the Mesozoic. This section is dominated by arc-related granitoids to depths of ~30 km, which have a tonalitic bulk chemistry (Ducea and Saleeby, 1996; Ducea, 2001). At the deepest structural levels, the mafic Tehachapi Complex comprises mafic and felsic gneiss, amphibolite, diorite, tonalite, granulite, and rare metasediments (Percival *et al.*, 1992; Ross, 1985). In contrast, two sections through accreted intraoceanic arcs have considerably more mafic middle-crust compositions. In the Jurassic Talkeetna section of southeastern Alaska, the middle crust comprises gabbro and tonalite (4.5 km), which is underlain by variably deformed garnet gabbro and gabbro with cumulate dunite, wehrlite, and pyroxenite (2.2 km) in the lower crust (Pearcy *et al.*, 1990; see Chapter 3.18). The upper, middle, and lower crustal units are estimated to have an average SiO_2 of 57%, 52%, and 44–45%, respectively. The Late Jurassic–Eocene Kohistan arc of Pakistan represents a 45 km thick reconstructed crustal column through a deformed, intruded intraoceanic arc sequence exposed in the Himalayan collision zone (Miller and Christensen, 1994). The depth interval from 10 km to 18 km is dominated by diorite and metadiorite. Rocks below this level, from ~18 km to the Moho, are dominated by metamorphosed mafic to ultramafic rocks from a series of layered mafic intrusions.

In summary, exposed amphibolite- to granulite-facies terranes and middle crustal cross-sections contain a wide variety of lithologies, including metasedimentary rocks, but they are dominated by igneous and metamorphic rocks of the diorite–tonalite–trondhjemite–granodiorite (DTTG), and granite suites. This is true not only for Precambrian shields but also for Phanerozoic crust and continental arcs, as documented in the crustal cross-sections described above. However, intraoceanic arcs may contain substantially greater proportions of mafic rocks in the middle and lower crust, as illustrated by the Kohistan and Talkeetna arc sections (Pearcy *et al.*, 1990; Miller and Christensen, 1994; see Chapter 3.18).

3.01.3.4.2 Seismological evidence

The samples described above provide evidence of the lithologies likely to be present in the middle crust. By definition, however, these samples no longer reside in the middle crust and additional information is required in order to determine the composition of the present-day middle crust. For this, we turn to seisomological data for continental crust from a variety of tectonic settings.

Except for active rifts and some intra-oceanic island arcs, which exhibit the highest middle-crust *P*-wave velocities (6.7 ± 0.3 km s^{-1} (Rudnick and Fountain, 1995) and 6.8 ± 0.2 km s^{-1} (data from Holbrook *et al.*, 1992) corrected to room temperature), other continental tectonic units have room-temperature middle-crustal *P*-wave velocities between 6.4 km s^{-1} and 6.6 km s^{-1} (Rudnick and Fountain, 1995). This range overlaps the average velocity of *in situ* middle crust, which was determined by Christensen and Mooney (1995) to be from 6.3 km s^{-1} to 6.6 ± 0.3 km s^{-1}, with an average of 6.5 ± 0.2 km s^{-1} over the depth range of 15–25 km. When corrected for temperature (an increase of 0.1–0.2 km s^{-1}, depending on the regional geotherm), these average middle-crustal velocities are similar to the room-temperature velocities considered by Rudnick and Fountain (1995). Thus, the middle crust has average, room-temperature-corrected velocity between 6.4 km s^{-1} and 6.7 km s^{-1}.

Amphibolite-facies felsic gneisses have room temperature *P*-wave velocities of 6.4 ± 0.1 km s^{-1} (Rudnick and Fountain, 1995). This compares well with the room-temperature velocity of average biotite (tonalite) gneiss at 6.32 ± 0.17 km s^{-1} (20 km depth; Christensen and Mooney, 1995). Granitic gneiss has a slightly lower velocity (6.25 ± 0.11 km s^{-1}; Christensen and Mooney (1995)), but is within uncertainty of the tonalite. A mixture of such gneisses with 0–30% amphibolite or mafic gneiss of the same metamorphic grade ($Vp = 7.0$ km s^{-1}; Rudnick and Fountain, 1995, Christensen and Mooney, 1995) yields *P*-wave velocities in the range observed for most middle crust. The above seismic data are thus consistent with the observations from granulite terranes and crustal cross-sections, and suggest that the middle crust is dominated by felsic gneisses.

3.01.3.4.3 Middle-crust composition

Compared to other regions of the crust (upper, lower, and bulk), few estimates have been made of the composition of the middle crust (Table 5, and Figures 9 and 10). Moreover, these estimates provide data for a far more limited number of elements, and large differences exist between different estimates. The estimates of Weaver and Tarney (1984), Shaw *et al.* (1994) and Gao *et al.* (1998a) are based on surface sampling of amphibolite-facies rocks in the Lewisian Complex, the Canadian Shield, and Eastern China, respectively. Rudnick and Fountain (1995) modeled the middle crust as 45% intermediate amphibolite-facies gneisses, 45% mixed amphibolite and felsic amphibolite-facies gneisses, and 10% metapelite. This mixture is very similar to that of Christensen and Mooney (1995), who proposed a middle crust of 50% tonalitic gneiss, 35% amphibolite, and 15% granitic gneiss. Unfortunately, compositional data are not available for Christensen and Mooney's samples and so the chemical composition of their middle crust cannot be calculated.

The estimates of Rudnick and Fountain (1995) and Gao *et al.* (1998a) show a broad similarity, although the latter is more evolved, having higher SiO_2, K_2O, barium, lithium, zirconium, and LREEs and La_N/Yb_N and lower total FeO, scandium, vanadium, chromium, and cobalt with a significant negative europium anomaly (Figures 9 and 10). These differences are expected, based on the slightly higher compressional velocity of Rudnick and Fountain's global middle crust compared to that of Eastern China (6.6 km s^{-1} versus 6.4 km s^{-1}; Gao *et al.*, 1998b). The consistency is surprising considering that the two estimates are based on different sample bases and different approaches, one global and the other regional.

The middle-crustal compositions of Weaver and Tarney (1984) and Shaw *et al.* (1994) deviate from the above estimates by being markedly higher in SiO_2 and lower in TiO_2, FeO, MgO, and CaO. Moreover, these middle-crust compositions are more felsic (based on the above elements) than all estimates of the upper-continental crust composition given in Table 1. Thus, it is unlikely that the Weaver and Tarney (1984) and Shaw *et al.* (1994) compositions are representative of the global average middle crust, as both heat flow and seismic observations require that the crust becomes more mafic with depth. It should be noted, however, that heat production for Shaw's middle-crust composition is indistinguishable from those of Rudnick and Fountain (1995) and Gao *et al.* (1998a) at ~1.0 μW m^{-3}, due largely to the very high K/Th and K/U of Shaw *et al.* estimate. The middle crust of Weaver and Tarney (1984) has significantly higher heat production, at 1.4 μW m^{-3}.

Generally speaking, it would be best to derive the middle-crust composition from observed seismic-wave speeds and chemical analyses of amphibolite-facies rocks. However, few such data sets exist. Only two studies attempt to define the global average seismic-wave speeds for the middle crust (Christensen and Mooney, 1995; Rudnick and Fountain, 1995) and neither provides chemical data for amphibolite facies samples. Rudnick and Fountain used compiled chemical

Table 5 Compositional estimates of the middle continental crust. Major elements in weight percent. Trace element concentration units the same as in Table 2.

	1 *Weaver and Tarney (1984)*	*2* *Shaw et al. (1994)*	*3* *Rudnick and Fountain (1995)*	*4* *Gao et al.* (1998a)	*5* *This study*[a]	*1 Sigma*[a]	%
SiO_2	68.1	69.4	62.4	64.6	63.5	2	2
TiO_2	0.31	0.33	0.72	0.67	0.69	0.04	6
Al_2O_3	16.33	16.21	15.96	14.08	15.0	1	9
FeO_T^b	3.27	2.72	6.59	5.45	6.02	0.8	13
MnO	0.04	0.03	0.10	0.11	0.10	0.00	2
MgO	1.43	1.27	3.50	3.67	3.59	0.1	3
CaO	3.27	2.96	5.25	5.24	5.25	0.01	0
Na_2O	5.00	3.55	3.30	3.48	3.39	0.1	4
K_2O	2.14	3.36	2.07	2.52	2.30	0.3	14
P_2O_5	0.14	0.15	0.10	0.19	0.15	0.06	43
Mg#	43.8	45.5	48.6	54.5	51.5		
Li		20.5	7	16	12	6	55
Be				2.29	2.29		
B		3.2		17	17		
N							
F				524	524		
S				20	20		
Cl				182	182		
Sc		5.4	22	15	19	5	27
V		46	118	95	107	16	15
Cr	32	43	83	69	76	10	13
Co		30	25	18	22	5	23
Ni	20	18	33	34	33.5	0.7	2
Cu		8	20	32	26	8	33
Zn		50	70	69	69.5	0.7	1
Ga			17	18	17.5	0.7	4
Ge				1.13	1.13		
As				3.1	3.1		
Se				0.064	0.064		
Br							
Rb	74	92	62	67	65	4	5
Sr	580	465[c]	281	283	282	1	1
Y	9	16	22	17.0	20	4	18
Zr	193	129	125	173	149	34	23
Nb	6	8.7	8	11	10	2	22
Mo		0.3		0.60	0.60		
Ru							
Pd				0.76	0.76		
Ag				48	48		
Cd				0.061	0.061		
In							
Sn				1.30	1.30		
Sb				0.28	0.28		
I							
Cs		0.98	2.4	1.96	2.2	0.3	14
Ba	713	1376	402	661	532	183	34
La	36	22.9	17	30.8	24	10	41
Ce	69	42.1	45	60.3	53	11	21
Pr			5.8		5.8		
Nd	30	18.3	24	26.2	25	2	6
Sm	4.4	2.8	4.4	4.74	4.6	0.2	5
Eu	1.09	0.78	1.5	1.20	1.4	0.2	16
Gd		2.11	4.0		4.0		
Tb	0.41	0.28	0.58	0.76	0.7	0.1	19
Dy		1.54	3.8		3.8		
Ho			0.82		0.82		

(continued)

Table 5 (continued).

	1 *Weaver and Tarney (1984)*	*2* *Shaw et al. (1994)*	*3* *Rudnick and Fountain (1995)*	*4* *Gao et al.* (1998a)	*5* *This study*[a]	*1 Sigma*[a]	%
Er			2.3		2.3		
Tm	0.14				0.32		
Yb	0.76	0.63	2.3	2.17	2.2	0.09	4
Lu	0.1	0.12	0.41	0.32	0.4	0.06	17
Hf	3.8	3.3	4.0	4.79	4.4	0.6	13
Ta		1.8	0.6	0.55	0.6	0.04	6
W				0.60	0.60		
Re							
Os							
Ir							
Pt				0.85	0.85		
Au				0.66	0.66		
Hg				0.0079	0.0079		
Tl				0.27	0.27		
Pb	22	9.0	15.3	15	15.2	0.2	1
Bi				0.17	0.17		
Th	8.4	6.4	6.1	6.84	6.5	0.5	8
U	2.2	0.9	1.6	1.02	1.3	0.4	31

Units for trace elements are the same as in Table 2. Major elements recast to 100% anhydrous.
[a] Averages and standard deviations of middle crustal composition by Rudnick and Fountain (1995) and Gao *et al.* (1998a), or from either of these two studies if data from the other one are unavailable. [b] Total Fe as FeO. [c] Recalculated from original data given by Shaw *et al.* (1994; Table 4), due to a typographical error in the published table. Mg# = molar 100 × Mg/(Mg + Fe_{tot}).

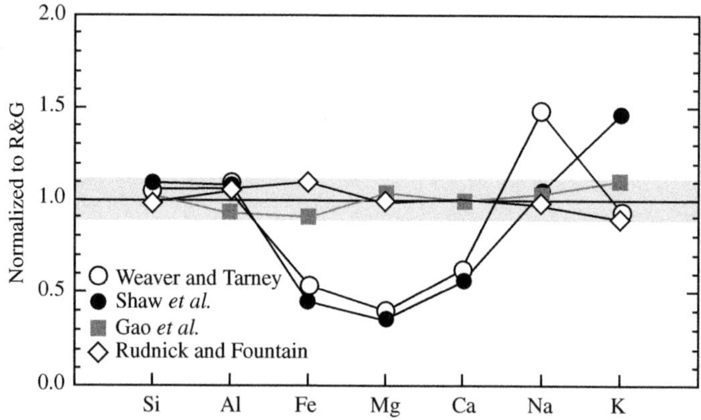

Figure 9 Comparison of the major-element composition of the middle continental crust as determined by sampling of surface exposures (Shaw *et al.*, 1994; Weaver and Tarney, 1984) and inferred from middle-crustal seismic velocities combined with surface and xenolith samples (Rudnick and Fountain, 1995; Gao *et al.*, 1998a). All values normalized to the new composition provided in Table 5 ("R&G"), which is an average between the values of Gao *et al.* (1998a) and Rudnick and Fountain (1995). Gray shaded field represents ±10% variation from this value.

data for granulite-facies rocks and inferred the concentrations of fluid-mobile elements (e.g., rubidium, uranium) of their amphibolite facies counterparts, while Christensen and Mooney (1995) did not publish their chemical data for the amphibolite-facies rocks they studied. For this reason, we have chosen to estimate the middle-crust composition by averaging the estimates of Rudnick and Fountain (1995) and Gao *et al.* (1998a) (Table 5), where corresponding data are available. Although the latter study is regional in nature, its similarity to the global model of Rudnick and Fountain (1995) suggests that it is not anomalous from a global perspective (unlike the lower crust of Eastern China as described in Section 3.01.3.5) and it provides additional estimates for little-measured trace elements.

This middle crust has an intermediate composition with lower SiO_2 and K_2O concentrations and higher FeO, MgO, and CaO concentrations than average upper crust (Table 1), consistent with the geophysical evidence (cited above) of

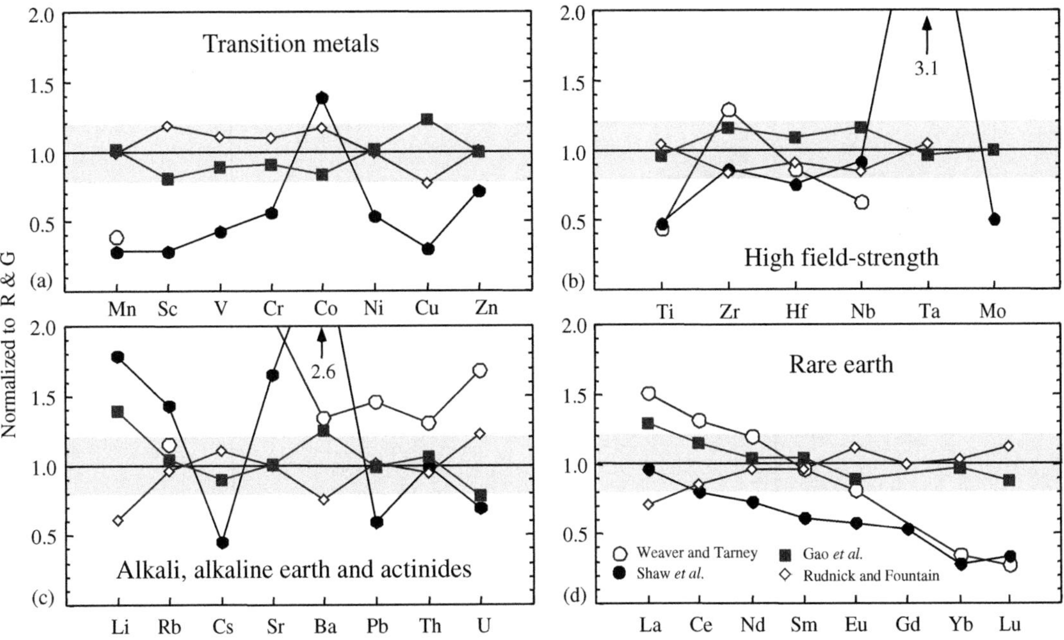

Figure 10 Comparison of the trace-element composition of the middle continental crust as determined by sampling of surface exposures (Shaw *et al.*, 1994; Weaver and Tarney, 1984) and inferred from middle-crustal seismic velocities combined with surface and xenolith samples (Rudnick and Fountain, 1995; Gao *et al.*, 1998a). All values normalized to the new composition provided in Table 5 ("R&G"), which is an average of the values of Gao *et al.* (1998a) and Rudnick and Fountain (1995). Gray shaded field represents ±20% variation from this value. (a) transition metals, (b) high-field strength elements, (c) alkali, alkaline earth and actinides, and (d) REEs.

a chemically stratified crust. Differences in trace-element concentrations between these two estimates are generally less than 30%, with the exceptions of P_2O_5, lithium, copper, barium, lanthanum, and uranium (Figure 10). The concentrations of these elements are considered to be less constrained. The middle crust is LREE enriched and exhibits the characteristic depletion of niobium relative to lanthanum and enrichment of lead relative to cerium seen in all other parts of the crust (Figure 11).

In summary, our knowledge of middle-crustal composition is limited by the small number of studies that have focused on the middle crust and the ambiguity in deriving chemical compositions from seismic velocities. Thus, the average composition given in Table 5 is poorly constrained for a large number of elements. Seismological and heat-flow data suggest an increase in seismic-wave speeds and a decrease in heat production with depth in the crust. Studies of crustal cross-sections show the middle crust to be dominated by felsic gneisses of tonalitic bulk composition. The average middle-crust composition given in Table 5 is consistent with these broad constraints and furthermore suggests that the middle crust contains significant concentrations of incompatible trace elements. However, the uncertainty on the middle-crust composition, particularly the trace elements, remains large.

3.01.3.5 The Lower Crust

3.01.3.5.1 Samples

Like the middle crust, the lower crust also contains a wide variety of lithologies, as revealed by granulite xenoliths, exposed high-pressure granulite terranes and crustal cross-sections. Metaigneous lithologies range from granite to gabbro, with a predominance of the latter in most lower crustal xenolith suites. Exceptions include xenolith suites from Argentina (Lucassen *et al.*, 1999) and central Spain (Villaseca *et al.*, 1999), where the xenoliths are dominated by intermediate to felsic granulites and the Massif Central (Leyreloup *et al.*, 1977; Downes *et al.*, 1990) and Hannuoba, China (Liu *et al.*, 2001), where intermediate to felsic granulites comprise nearly half the population. Metapelites occur commonly in both terranes and xenoliths, but only rarely do other metasedimentary lithologies occur in xenolith suites; unique xenolith localities have been documented with meta-arenites (Upton *et al.*, 1998) and quartzites (Hanchar *et al.*, 1994), but so far marbles occur only in terranes. The reason for their absence in lower crustal xenolith suites is uncertain—they may be absent in the lower crust sampled by volcanoes, they may not survive transport in the hot magma, or they may simply have been overlooked by xenolith investigators.

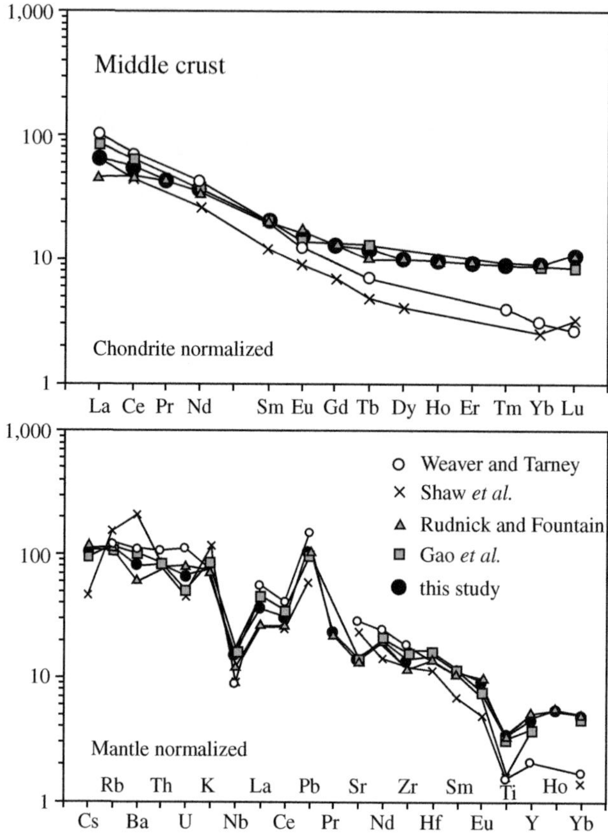

Figure 11 REE (upper) and multi-element plot (lower) of the compositions of the middle crust given in Table 5. Chondrite values from Taylor and McLennan (1985) and primitive mantle values from McDonough and Sun (1995).

These issues related to the representativeness of xenolith sampling are the reason why robust estimates of lower-crustal composition must rely on a grand averaging technique, such as using seismic velocities to infer composition.

Information on the lower crust derived from crustal cross-sections has been given in Section 3.01.3.4.1 and only the main points are summarized here. All crustal cross-sections show an increase in mafic lithologies with depth and most of those in which possible crust–mantle boundaries are exposed reveal a lower crust that is dominated by mafic compositions. For example, in the Ivrea Zone, Italy, the lower crust is dominated by mafic granulite formed from basaltic underplating of country rock metapelite (Voshage *et al.*, 1990). The same is true for the Kohistan sequence, Pakistan (Miller and Christensen, 1994), although here metapelites are lacking. Although the crust–mantle boundary is not exposed in the Wutai-Jining terrain, the granulite-facies crust exposed in this cross-section has a more mafic composition than the rocks of the middle-crust section. Even in the Vredefort and Sierra Nevada cross-sections, which are dominated by granitic rocks throughout most of the crustal sections (Ducea, 2001; Hart *et al.*, 1990), the deepest reaches of exposed crust are characterized by more mafic lithologies (Ross, 1985; Hart *et al.*, 1990; Table 4).

There have been a number of studies of granulite-facies xenoliths since the reviews of Rudnick (1992) and Downes (1993) and a current tabulation of xenolith studies is provided in Table 6, which provides a summary of most lower crustal xenolith studies published through 2002. Perhaps most significant are the studies of lower-crustal xenoliths from Archean cratons, which had been largely lacking prior to 1992 (Kempton *et al.*, 1995, 2001; Davis, 1997; Markwick and Downes, 2000; Schmitz and Bowring, 2000, 2003a,b; Downes *et al.*, 2002). These studies reveal a great diversity in lower-crustal lithologies beneath Archean cratons, which appear to correlate with seismic structure of the crust.

Lower-crustal xenoliths from the Archean part of the Baltic (or Fennoscandian) Shield, like their post-Archean counterparts, are dominated by mafic lithologies (Kempton *et al.*, 1995, 2001; Markwick and Downes, 2000; Hölttä *et al.*, 2000). Most equilibrated at depths of 22–50 km and contain hydrous phases (amphibole ± biotite). Partial melting and restite development is evident in some migmatitic xenoliths, but cumulates are

Table 6 Geochemical and mineral chemical studies of lower crustal xenoliths.

Locality	*Host*	*Xenolith types*	*Types of analyses*	*Pipe age*	*Crust age*	*Age*	*References*
North America							
Nunivak Island, Alaska	AB	MG	ME, Min.	<5 Ma	Phanerozoic		Francis (1976)
Central Slave Province, Canada	K	MGG, FG, MG	U–Pb	50–70 Ma	Archean	2.5 Ga, 1.3 Ga (meta)	Davis (1997)
Kirkland Lake, Ontario, Superior craton	K	AN, MG	U–Pb	160 Ma	Archean	2.6–2.8 Ga, 2.4–2.5 Ga (meta)	Moser and Heaman (1997)
Ayer's cliff, Quebec	K	MG, PG	Min.	~100 Ma	Proterozoic		Trzcienski and Marchildon (1989)
Popes Harbour[a], Nova Scotia	K	MG, PG, FG	ME, TE, Min.	<400 Ma	Phanerozoic		Owen (1988), Eberz (1991)
Snake River Plains[a], Idaho	Evol. B	PG	Sr, Nd, Pb	<2 Ma	Archean	~2.8 Ga	Leeman *et al.* (1985, 1992)
Bearpaw Mts., Montana	K	MGG, MG, FG	ME, TE, O, U–Pb	45 Ma	Archean	~2.6 Ga	Collerson *et al.* (1989), Kempton and Harmon (1992), Moecher *et al.* (1994), Rudnick *et al.* (1999)
Simcoe Volcanic Field, Washington Cascades	AB	MG	FI	<1 Ma	Phanerozoic		Ertan and Leeman (1999)
Riley County, Kansas	K	MGG, EC	Min.	<230 Ma	Proterozoic		Meyer and Brookins (1976)
Central Sierra Nevada, California	AB	MGG, EC, FG, MP	ME, TE, Min., Sr, Nd, O, U–Pb	8–11 Ma	Proterozoic	180 Ma	Dodge *et al.* (1986, 1988), Domenick *et al.* (1983), Ducea and Saleeby (1996, 1998)
Colorado/Wyoming	K	MGG, MG	ME, Min.		Proterozoic		Bradley and McCallum (1984)
Mojave Desert, California	AB	MP, FG, IG, QZ, MGG	ME, TE, Min., Sr, Nd, Pb, U–Pb		Archean	~1,7 Ga	Hanchar *et al.* (1994)
Navajo Volcanic Field, Colorado Plateau	K	MGG, FG, AM, EC, MP	ME, TE, Min., Sr, Nd	25–30 Ma	Proterozoic	~1.8 Ga	Ehrenberg and Griffin (1979), Broadhurst (1986), Wendlandt *et al.* (1993, 1996), Mattie *et al.* (1997), Condie *et al.* (1999)
Camp Creek, Arizona	L	MGG, AM, EC	ME, TE, Min., Sr, Nd, Pb	23–27 Ma	Proterozoic	1.2–1.9 Ga	Esperanca *et al.* (1988)

(continued)

Table 6 (continued).

Locality	*Host*	*Xenolith types*	*Types of analyses*	*Pipe age*	*Crust age*	*Age*	*References*
Chino Valley, Arizona	L	MGG, AM	ME, TE, Min.	25 Ma	Proterozoic		Arculus and Smith (1979), Schulze and Helmstadt (1979), Arculus *et al.* (1988)
San Francisco Volcanic Field	AB	MG, IG	ME, TE, Min., Sr, Nd	<20 Ma	Proterozoic	~1.9 Ga	Chen and Arculus (1995)
Geronimo Volcanic Field, New Mexico	AB	MG, IG	ME, TE, Min., Sr, Nd, Pb, O	<3 Ma	Proterozoic	1.1–1.4 Ga	Kempton *et al.* (1990), Kempton and Harmon (1992)
Kilbourne Hole, New Mexico	AB	MG, AN, PG, FG	Min., ME, Sr, Nd, Pb, O, U–Pb	<1 Ma	Proterozoic	1.5 Ga	Padovani and Carter (1977), Davis and Grew (1977), James *et al.* (1980), Padovani *et al.* (1982), Reid *et al.* (1989), Leeman *et al.* (1992), Scherer *et al.* (1997).
Elephant Butte, New Mexico	AB	MG	Min., ME	<3 Ma	Proterozoic		Baldridge (1979)
Engle Basin, New Mexico	AB	MG	Min.	<3 Ma	Proterozoic		Warren *et al.* (1979)
West Texas	AB	MGG, IG, FG	ME, Sr, Nd, Pb, U–Pb	<40 Ma	Proterozoic	1.1 Ga	Cameron and Ward (1998)
Northern Mexico	AB	MG, PG, FG	ME, TE, Min., Sr, Nd, C, U–Pb	<25 Ma	Proterozoic	From 1 to 1400 Ma	Nimz *et al.* (1986), Ruiz *et al.* (1988a,b), Roberts and Ruiz (1989), Hayob *et al.* (1989), Rudnick and Cameron (1991), Cameron *et al.* (1992), Moecher *et al.* (1994), Smith *et al.* (1996), Scherer *et al.* (1997)
Central Mexico	AB	MG	ME, Min, Nd		Phanerozoic	1.5 Ga	Urrutia-Fucagauchi and Uribe-Cifuentes (1999)
San Luis Potosi, Central Mexico	AB	MG, MGG, IG	ME, TE, Min., Sr, Nd	<1 Ma	Proterozoic	~1.2 Ga	Schaaf *et al.* (1994)
South America							
Mercaderes, SW Columbia	AB	MG, MGG, IG, HB,	ME, TE, Min., Sr, Nd, Pb	<10 Ma	Phanerozoic		Weber *et al.* (2002)
Salta Rift, NW Argentina	AB	FG, MG	Min., ME, TE, Sr, Nd, Pb	Mesozoic	Proterozoic	~1.8 Ga	Lucassen *et al.* (1999)
Calbuco Volcano, Chile	AND	MG	ME, TE, Min., Sr, Nd	<1,000 yr	Paleozoic		Hickey-Vargas *et al.* (1995)
Pali Aike, Southern Chile	AB	MG	ME, Min., FI	<3 Ma	Phanerozoic		Selverstone and Stern (1983)

Europe							
Scotland, Northern Uplands	AB	MG, AN, IG, MP, MGG, HB	ME, TE, Min., Nd, Sr, U–Pb	~300 Ma	Archean/ Proterozoic	360 Ma, 1.8 Ga	van Breeman and Hawkesworth (1980), Upton *et al.* (1983), Halliday *et al.* (1984), Hunder *et al.* (1984), Upton *et al.* (1998, 2001)
Eastern Finland	K	MG, MGG, AM	ME, TE, Min., Nd, U–Pb	525 Ma	Archean	1.7–2.6 Ga	Hölttä *et al.* (2000)
Arkhangelsk Kimberlite, Baltic shield, Russia	K	MGG	ME, TE, Min., Sr, Nd	360 Ma	Archean	1.7–1.9 Ga	Markwick and Downes (2000)
Elovy island, Baltic shield, Russia	K	MGG, EC, FG, AM	ME, TE, Min., Sr, Nd, Pb, U–Pb	360–380 Ma	Archean	~1.8 Ga, 2.4–2.5 Ga	Kempton *et al.* (1995, 2001), Downes *et al.* (2002)
Belarus, Russia	K	MGG, EC, HB	ME, TE, Min., Sr, Nd	370 Ma			Markwick *et al.* (2001)
Pannonian basin, W. Hungary	AB	MG, MGG	ME, TE, Min., Sr, Nd, O	2–5 Ma	Phanerozoic		Embey-Isztin *et al.* (1990), Kempton *et al.* (1997), Embey-Isztin *et al.* (2003), Dobosi *et al.* (2003)
Kampernich, E. Eifel, Germany	AB	MGG, AM	ME, TE, Min., Sr, Nd, Hf, Pb, O	<1 Ma	Phanerozoic	1.5 Ga or ~450 Ma?	Okrusch *et al.* (1979), Stosch and Lugmair (1984), Rudnick and Goldstein (1990), Loock *et al.* (1990), Kempton and Harmon (1992), Sachs and Hansteen (2000)
Wehr Volcano[a], E. Eifel, Germany	AB	AM	ME, TE, Min.	<1 Ma	Phanerozoic		Worner *et al.* (1982), Grapes (1986)
N. Hessian Depression, Germany	AB	MGG, MG, PG, FG	ME, TE, Min., O	<50 Ma	Phanerozoic		Mengel and Wedepohl (1983), Mengel (1990)
Massif Central, France	AB	MGG, MG, PG, FG	ME, TE, Nd, Sr, Pb, O, C	<5 Ma	Phanerozoic	~350 Ma	Leyreloup *et al.* (1977), Dostal *et al.* (1980), Vidal and Postaire (1985), Downes and Leyreloup (1986), Kempton and Harmon (1992), Downes *et al.* (1991), Moecher *et al.* (1994)
Tallante, Spain	AB	PG	Min.	<20 Ma	Phanerozoic		Vielzeuf (1983)
Central Spain	K	IG, FG, MP	ME, TE, Min.	Early Mesozoic	Proterozoic		Villaseca *et al.* (1999)
Sardinia, Italy	AB	MG	Min.	~3 Ma	Phanerozoic		Rutter (1987)

(continued)

Table 6 (continued).

Locality	Host	Xenolith types	Types of analyses	Pipe age	Crust age	Age	References
Africa							
Hoggar, Algeria	AB	MG, AN, PG	ME, TE	<20 Ma	Proterozoic		Leyreloup *et al.* (1982)
Man Shield, Sierra Leone	K	MGG, AN, EC	ME, Min., U–Pb	90–120 Ma	Archean		Toft *et al.* (1989), Barth *et al.* (2002)
Lashaine, Tanzania	AB	MGG, EC, AN	ME, TE, Min., Sr, Nd, Pb, C	<20 Ma	Proterozoic		Dawson (1977), Jones *et al.* (1983), Cohen *et al.* (1984), Moecher *et al.* (1994)
Fort Portal, Uganda	AB	MGG, MG	Min., ME, TE	<3Ma	Proterozoic		Thomas and Nixon (1987)
Free State Kimberlites, Kaapvaal Craton, South Africa	K	PG	Min., U–Pb	90–140 Ma	Archean	~2.7 Ga (meta)	Dawson and Smith (1987), Dawson *et al.* (1997), Schmitz and Bowring (2003a,b)
Newlands Kimberlite, Kaapvaal Craton, South Africa	K	PG	U–Pb	114 Ma	Archean	~2.7 Ga	Schmitz and Bowring (2003a,b)
Lesotho, South Africa	K	MGG, MG, FG, MP, EC	ME, TE, Min., Sr, Nd, Pb, U–Pb	90–140 Ma	Proterozoic	1.4 Ga, 1.1–1.0 Ga (meta)	Davis (1977), Rogers and Hawkesworth (1982), Griffin *et al.*, 1979), Rogers (1977), van Calsteren *et al.* (1986), Huang *et al.* (1995), Schmitz and Bowring (2003a)
Orapa, Zimbabwe Craton, Botswana	K	MP	U–Pb	93 Ma	Proterozoic	2.0 Ga (meta), 1.24 Ga (meta)	Schmitz and Bowring (2003a)
Central Cape Province, Eastern Namaqualand, South Africa	K	MGG, AM	ME, TE, Min., Sr, Nd, Pb, U–Pb	90–140 Ma	Proterozoic	~1.1 Ga (meta)	van Calsteren *et al.* (1986), Pearson *et al.* (1995), Schmitz and Bowring (2000), Schmitz and Bowring (2003a)
Middle East							
Mt. Carmel, Israel	AB	MGG	Min.	~100 Ma	Phanerozoic		Esperanca and Garfunkel (1986), Mittlefehldt (1986)
Birket Ram, Israel	AB	MGG, Am	TE	10,000 yr	Phanerozoic		Mittlefehldt (1984)
Jordan	AB	MG	Min.	<2 Ma	Proterozoic		Nasir (1992), Nasir (1995)
Shamah volcanic fields, Syria	AB	MG					Nasir and Safarjalani (2000)

Asia							
Udachnaya, Siberia, Russia	K	MGG	Min.		Archean		Shatsky *et al.* (1990, 1983)
Tariat Depression, Central Mongolia	AB	MG, MGG, IG, AM	ME, TE, Min., Sr, Nd, Pb, O	<5 Ma	Proterozoic		Kempton and Harmon (1992), Kopylova *et al.* (1995), Stosch *et al.* (1995)
Hannuoba, North China Craton	AB	FG, IG, MG, MGG, AN, MP	ME, TE, Sr, Nd, Pb, U–Pb	14–27 Ma	Archean	~2.5, 1.9, 0.4, 0.22 Ga	Gao *et al.* (2000), Liu *et al.* (2001), Chen *et al.* (2001), Zhou *et al.* (2002)
Xinyang, North China craton	AB	MGG	ME, TE, Min.	Mesozoic	Archean		Zheng *et al.* (2003)
Penghu Islands, SE China	AB	MG					Lee *et al.* (1993)
Southeastern China	AB	MG, MGG, IG, FG	ME, TE, Min., Sr, Nd	<20Ma	Archean/ Proterozoic		Yu *et al.* (2003)
Ichinomegata, Japan	AND	AM	ME, TE, Min., Sr	10,000 yr	Phanerozoic		Kuno (1967), Aoki (1971), Zashu *et al.* (1980), Tanaka and Aoki (1981)
Deccan Traps, India	K	MG	ME, Min.				Dessai *et al.* (1999), Dessai and Vasseli (1999)
Tibetan plateau			Min.				Hacker *et al.* (2000)
Australia–New Zealand–Antarctica							
McBride Province, N. Queensland	AB	MGG, MG, PG, FG	ME, TE, Min., Sr, Nd, Pb, O, C, U–Pb	<3 Ma	Proterozoic	300 Ma and ~1.6 Ga	Kay and Kay (1983), Rudnick and Taylor (1987), Rudnick and Williams (1987), Stolz and Davies (1989), Stolz (1987), Rudnick (1990), Rudnick and Goldstein (1990), Kempton and Harmon (1992), Moecher *et al.* (1994)
Chudleigh Province, N. Queensland	AB	MGG, MG	ME, TE, Min., Sr, Nd, Pb, O	<1 Ma	Phanerozoic	<100 Ma	Kay and Kay (1983), Rudnick *et al.* (1986), Rudnick and Taylor (1991), O'Reilly *et al.* (1988), Rudnick and Goldstein (1990), Kempton and Harmon (1992)
Central Queensland	AB	MG, MGG	ME, TE, Min., Sr, Nd	≤50 Ma	Phanerozoic		Griffin *et al.* (1987), O'Reilly *et al.* (1988)
Gloucester, NSW	AB	MGG, MG	ME, TE, Sr, Nd	≤50 Ma	Phanerozoic		Griffin *et al.* (1986), O'Reilly *et al.* (1988)
Sydney Basin	AB	MG	ME, TE, Sr, Nd	≤50 Ma	Phanerozoic		Griffin *et al.* (1986), O'Reilly *et al.* (1988)
Boomi Creek, NSW	AB	MG	ME, TE, Min.	≤50 Ma	Phanerozoic		Wilkinson (1975), Wilkinson and Taylor (1980)

(continued)

Table 6 (continued).

Locality	*Host*	*Xenolith types*	*Types of analyses*	*Pipe age*	*Crust age*	*Age*	*References*
Delegate, NSW	AB	MG, MGG, FG, EC	ME, TE, Min., Sr, Nd, U–Pb	~140 Ma	Phanerozoic	400 Ma	Lovering and White, (1964, 1969), Griffin and O'Reilly (1986), O'Reilly *et al.* (1988), Arculus *et al.* (1988), Chen *et al.* (1998)
Jugiong, NSW	K	MG, MGG	Min.	<17 Ma	Phanerozoic		Arculus *et al.* (1988)
White Cliffs, NSW	K	MGG	Min.	~260 Ma	Proterozoic		Arculus *et al.* (1988)
Anakies, Victoria	AB	MG, MGG	ME, TE, Min., Sr, Nd	<2 Ma	Phanerozoic		Sutherland and Hollis (1982), Wass and Hollis (1983), O'Reilly *et al.* (1988)
El Alamein, South Australia	K	MGG, EC	ME, TE, Min.	~170 Ma	Proterozoic		Edwards *et al.* (1979), Arculus *et al.* (1988)
Calcutteroo, South Australia	K	MGG, FG, EC	ME, TE, Min., Sr, Nd, U–Pb	~170 Ma	Proterozoic	1.6–1.5 Ga, 780 Ma, 620 Ma, 330 Ma	McCulloch *et al.* (1982), Arculus *et al.* (1988), Chen *et al.* (1994)
Banks Penninsula, New Zealand	AB	MG	ME, TE, Min				Sewell *et al.* (1993)
Mt. Erebus Volcanic Field,	AB	MG, MGG	ME, TE, Min., Sr, O	<5 Ma	Phanerozoic/ Proterozoic		Kyle *et al.* (1987), Kalamarides *et al.* (1987), Berg *et al.* (1989)
Indian Ocean							
Kergulen Archipelago	AB	MG, SG	ME, Min.		Phanerozoic/ Proterozoic		McBirney and Aoki (1973), Gregoire *et al.* (1998, 1994)

Only papers in which data are reported are listed here.

Abbreviations

Host types: AB = alkali basaltic association; AND = andesite; K = kimberlitic association (including lamproites, minettes, kimberlites), Evol. B. = evolved basalt; L = latite.

Xenolith types: AM = amphibolite; AN = anorthosite; EC = elcogite; FG = felsic granuilte; HB = hornblendite; IG = intermediate granulite; MG = mafic granulite; MGG = mafic garnet granulite; MP = metapelite, PG = paragneiss, QZ = quartzite, SG = saphirine granulites.

Types of analyses: FI = fluid inclusions, ME = major element analyses; Min = mineral analyses; TE = trace element analyses; Sr = Sr isotope analyses; Nd = Nd isotope analyses; Pb = Pb isotope analyses, O = oxygen isotope analyses, C = carbon isotope analyses, U–Pb = U–Pb geochronology on accessory phases (zircon, rutile, titanite, etc.).

[a] Xenoliths from these localities are probably derived from mid-crustal levels based on either: equilibration pressures, lack of mantle derived xenoliths in the same hosts and/or chemically evolved character of the host.

absent (Kempton *et al.*, 1995, 2001; Hölttä *et al.*, 2000). A curious feature of these samples is the common occurrence of potassic phases (e.g., potassium feldspar, hornblende, biotite) in otherwise mafic granulites. These mafic xenoliths have been interpreted to represent gabbroic intrusions that underplated the Baltic Shield during the Paleoproterozoic flood-basalt event (2.4–2.5 Ga) and later experienced potassium-metasomatism coincident with partial melting at ~1.8 Ga, a major period of granitic magmatism in this region (Kempton *et al.*, 2001; Downes *et al.*, 2002). The dominately mafic compositions of these xenoliths is consistent with the thick layer of high-velocity ($\geq$7 km s^{-1}) material imaged beneath the Archean crust of the Baltic Shield (Luosto *et al.*, 1989, 1990). The xenolith studies suggest that this layer formed during Paleoproterozoic basaltic underplating and is not part of the original Archean architecture of this Shield.

In contrast to the Baltic Shield, mafic granulites appear to be absent in lower-crustal xenolith suites from the Archean Kaapvaal craton, which are dominated by metapelite and unique ultra-high-temperature granulites of uncertain petrogenesis (Dawson *et al.*, 1997; Dawson and Smith, 1987; Schmitz and Bowring, 2003a,b). These xenoliths derive from depths of $>$30 km and show evidence for multiple thermal metamorphic overprints starting with ultrahigh temperature metamorphism at ~2.7 Ga, which is associated with Ventersdorp magmatism (Schmitz and Bowring, 2003a,b). The absence of mafic granulites is consistent with the relatively low *P*-wave velocities in the lower crust of the Kaapvaal craton (Durrheim and Green, 1992; Nguuri *et al.*, 2001; Niu and James, 2002), but it is not clear whether the lack of a mafic lower crust reflects the original crustal structure of this Archean craton (Nguuri *et al.*, 2001) or reflects loss of a mafic complement some time after crust formation in the Archean (Niu and James, 2002).

Lower crustal xenoliths from the Hannuoba basalts, situated in the central zone of the North China Craton, show a diversity of compositions ranging from felsic to mafic metaigneous granulites and metapelites (Gao *et al.*, 2000; Chen *et al.*, 2001; Liu *et al.*, 2001; Zhou *et al.*, 2002); approximately half the xenoliths have evolved compositions (Liu *et al.*, 2001). All granulite xenoliths equilibrated under high temperatures (700–1,000 °C), corresponding to depths of 25–40 km (Chen *et al.*, 2001), but mafic granulites yield higher temperatures than metapelitic xenoliths, suggesting their derivation from deeper crustal levels (Liu *et al.*, 2001). Liu *et al.* used regional seismic refraction data and the lithologies observed in the Hannuoba xenoliths to infer the lower-crust composition in this part of the North China craton. They describe a layered lower crust in which the upper portion (from 24 km to ~38 km, $V_p \sim 6.5$ km s^{-1}), consists largely of felsic granulites and metasediments, and is underlain by a "lowermost" crust (38–42 km, $V_p \sim 7.0$ km s^{-1}) composed of intermediate granulites, mafic granulites, pyroxenite, and peridotite. Thus, the bulk lower crust in this region is intermediate in composition, consistent with the relatively large proportion of evolved granulites at Hannuoba. Zircon geochronology shows that mafic granulites and some intermediate granulites were formed by basaltic underplating in the Cretaceous. This mafic magmatism intruded pre-existing Precambrian crust consisting of metapelites that had experienced high-grade metamorphism at 1.9 Ga (Liu *et al.* (2001) and references therein).

Fragmentary xenolithic evidence for the composition of the lower crust is available for three other Archean cratons. Two mafic garnet granulites from the Udachnaya kimberlite in the Siberian craton yield Archean lead–lead and Proterozoic samarium–neodymium mineral isochrons (Shatsky, Rudnick and Jagoutz, unpublished data). It is likely that that lead–lead isochrons are frozen isochrons yielding anomalously old ages due to ancient uranium loss; the best estimate of the true age of these mafic granulites is Proterozoic. Moser and Heaman (1997) report Archean uranium–lead ages for zircons derived from mafic lower-crustal xenoliths from the Superior Province, Canada. They suggest these samples represent the mafic lower crust presently imaged seismically beneath the Abitibi greenstone belt, but which is not exposed in the Kapuskasing uplift. These granulites experienced an episode of high-grade metamorphism at 2.4 Ga, which Moser and Heaman (1997) attribute to underplating of basaltic magmas associated with the opening of the Matachewan Ocean. Davis (1997) reports mafic to felsic granulite xenoliths from the Slave craton, Canada, that have Archean to Proterozoic uranium–lead zircon ages. The mafic granulites appear to derive from basaltic magmas that underplated the felsic-Archean crust during the intrusion of the 1.3 Ga McKenzie dike swarm.

The above case studies illustrate the utility of lower crustal xenolith studies in defining the age, lithology, and composition of the lower crust beneath Archean cratons. When viewed collectively, an interesting generality emerges: when mafic granulites occur within the lower crust of Archean cratons they are generally inferred to have formed from basaltic underplating related to post-Archean magmatic events (In addition to the studies mentioned above is the case of the thick, high-velocity lower crust beneath the Archean Wyoming Province and Medicine Hat Block, western North America, which is also inferred to have formed by Proterozoic underplating based on uranium–lead zircon ages from lower crustal xenoliths (Gorman *et al.*, 2002)). Only granulites from the Superior Province appear to represent Archean mafic lower crust (Moser and Heaman, 1997). This generality is based on still only a handful of studies of xenoliths from Archean cratons and more such studies are clearly needed. However, if this generality proves robust, it implies that the processes responsible for generation of crust in most Archean cratons did not leave behind a mafic lower crust, the latter of which is commonly observed in post-Archean regions (Rudnick (1992) and references therein). It may be that this mafic lower crust was never produced, or that it formed but was removed from the crust, perhaps via density foundering (R. W. Kay and S. M. Kay, 1991; Gao *et al.*, 1998b; Jull and Kelemen, 2001). In either case, the apparent contrast in lower-crustal composition between Archean and post-Archean regions, originally pointed out by Durrheim and Mooney (1994), suggests different processes may have been operative in the formation of Archean crust (see Chapter 3.11). We rcturn to the issue of what crust composition tells us about crustal generation processes in Section 3.01.5.

In summary, despite the uncertainties regarding the representativeness of any given lower-crustal xenolith suite (Rudnick, 1992), the above studies show that an accurate picture of the deep crust can be derived from such studies, especially when xenolith studies are combined with seismological observations of lower-crust velocities, to which we now turn.

3.01.3.5.2 Seismological evidence

The *P*-wave velocity of the lower crust varies from region to region, but average, temperature-corrected velocities for lower crust from a variety of different tectonic settings are high (6.9–7.2 km s^{-1}; Rudnick and Fountain, 1995; Christensen and Mooney, 1995). Such velocities are consistent with the dominance of mafic lithologies (mafic granulite and/or amphibolite) in these lower-crustal sections. High-grade metapelite, in which much of the quartz and feldspars have been removed by partial melting, is also characterized by high seismic velocities and thus may also be present (Rudnick and Fountain, 1995). Although seismically indistinct, some limit on the amount of metapelite in these high-velocity layers can be made on the basis of heat-flow and xenolith studies; these suggest that metapelite is probably a minor constituent of the lower crust (i.e., <10%; Rudnick and Fountain, 1995). In addition, average *P*-wave velocities for mafic granulite or amphibolite are higher than those observed in many lower-crustal sections (corrected to room-temperature velocities). Average room-temperature *P*-wave velocities for a variety of mafic lower crustal rock types are generally equal to or higher than 7 km s^{-1}: 7.0 ± 0.2 km s^{-1} for amphibolite, 7.0 to 7.2 ± 0.2 km s^{-1} for garnet-free mafic granulites, and 7.2 to 7.3 ± 0.2 km s^{-1} for garnet-bearing mafic granulites at 600 MPa (Rudnick and Fountain, 1995; Christensen and Mooney, 1995). Lower-crustal sections having temperature-corrected *P*-wave velocities of 6.9–7.0 km s^{-1} (e.g., Paleozoic orogens and Mesozoic/Cenozoic extensional and contractional terranes), are thus likely to have lower-velocity rock types present (up to 30% intermediate to felsic granulites), in addition to mafic granulites or amphibolites (Rudnick and Fountain, 1995).

Although the average lower-crustal seismic sections discussed above show high velocities, some sections are characterized by much lower velocities, indicating a significantly more evolved lower-crust composition. For example, the crust of a number of Archean cratons is relatively thin (~35 km) with low seismic velocities in the lower crust (6.5–6.7 km s^{-1}), suggesting an evolved composition (e.g., Yilgarn craton (Drummond, 1988), Kaapvaal craton (Durrheim and Green, 1992; Niu and James, 2002), and North China craton (Gao *et al.*, 1998a,b)). As discussed above, it is not clear whether these thin and relatively evolved regions of Archean crust represent the original crustal architecture, formed by processes distinct from those responsible for thicker and more mafic crustal regions (e.g., Nguuri *et al.*, 2001), or reflect loss of a mafic layer from the base of the original crust (Gao *et al.*, 1998b; Niu and James, 2002). In addition, some Cenozoic–Mesozoic extensional and contractional regions, Paleozoic orogens, and active rifts show relatively slow lower-crustal velocities of 6.7–6.8 km s^{-1} and may contain >40% felsic and intermediate granulites (Rudnick and Fountain, 1995). Two extreme examples are the southern Sierra Nevada and Central Andean backarc. In both cases, the entire crustal columns are characterized by *P*-wave velocities of less than 6.4 km s^{-1} (Beck and Zandt, 2002; Wernicke *et al.*, 1996). A relatively high-velocity ($Vp = 6.4$–6.8 km s^{-1}) layer of <5 km in thickness occurs only at the base of the Central Andean backarc at ~60 km depth.

In summary, the seismic velocity of the lower crust is variable from region to region, but is generally high, suggesting a dominance of mafic lithologies. However, most seismic sections require the presence of evolved compositions in addition to mafic lithologies in the lower crust (up to 30% for average velocity of 6.9 km s^{-1}) and a few regions (e.g., continental arcs and some Archean cratons) are characterized by slow lower crust, indicating a highly evolved average composition. This diversity of lithologies is consistent with that seen in both crustal cross-sections and lower-crustal xenolith suites and also provides a mechanism (lithological layering) to explain the common occurrence of seismic reflections observed in many seismic reflection profiles (Mooney and Meissner, 1992).

3.01.3.5.3 Lower-crust composition

Table 7 lists previous estimates of the composition of the lower crust. These estimates include averages of exposed granulites (columns 1 and 2; Weaver and Tarney, 1984; Shaw *et al.*, 1994), averages of individual lower-crustal xenolith suites (columns 3–6, Condie and Selverstone, 1999; Liu *et al.*, 2001; Rudnick and Taylor, 1987; Villaseca *et al.*, 1999), the median composition of lower-crustal xenoliths (column 7; updated from Rudnick and Presper (1990), with data from papers cited in Table 6 (the complete geochemical database for lower crustal xenoliths is available on the GERM web site http://earthref.org/cgi-bin/erda.cgi?n = 1,2,3,8 and also on the Treatise web site), averages derived from linking seismic velocity data for the lower crust with the compositions of lower-crustal rock types (columns 8–10; Rudnick and Fountain, 1995; Wedepohl, 1995; Gao *et al.*, 1998a), and Taylor and McLennan's model lower crust (column 11). It is readily apparent from this table and Figures 12 and 13 that, compared to estimates of the upper-crust composition (Table 1), there is much greater variability in estimates of the lower-crust composition. For example, TiO_2, MgO, FeO_T, and Na_2O all vary by over a factor of 2, CaO varies by almost a factor of 7, and K_2O varies by over an order of magnitude between the different estimates (Figures 12 and 13). Trace elements show correspondingly large variations (Figure 13). In contrast, modern estimates of major elements in the upper crust generally fall within 20% of each other (Table 1 and Figure 2—gray shading) and

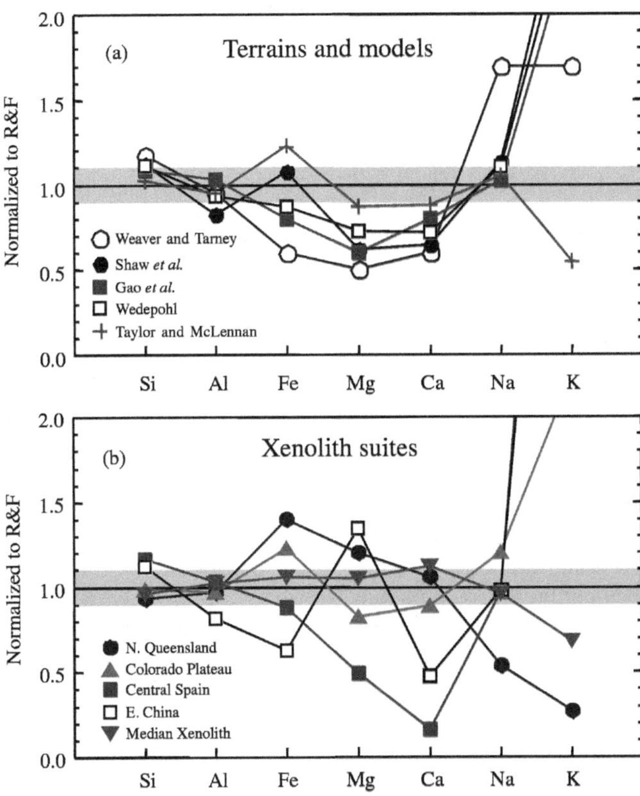

Figure 12 Comparison of different major-element estimates of the composition of the lower continental crust. All data normalized to the lower-crust composition of Rudnick and Fountain (1995), which is adopted here. Gray shaded field represents ±10% variation from the model of Rudnick and Fountain (1995). (a) Models based on granulite terrains (Scourian granulites: Weaver and Tarney, 1984; Kapuskasing Structure Zone: Shaw *et al.*, 1986), seismological models (Eastern China: Gao *et al.*, 1998a,b; western Europe: Wedepohl, 1995) and Taylor and McLennan (1985, 1995; modified by McLennan, 2001b) model lower crust. (b) Models based on weighted averages of lower crustal xenoliths. These include: Northern Queensland, Australia (Rudnick and Taylor, 1987); Colorado Plateau, USA (Condie and Selverstone, 1999); Central Spain (Villaseca *et al.*, 1999); eastern China (Liu *et al.*, 2001) and the median global lower crustal xenolith composition, updated from Rudnick and Presper (1990). Note that K data for eastern China and Central Spain are co-incident on this plot, making it hard to distinguish the separate lines.

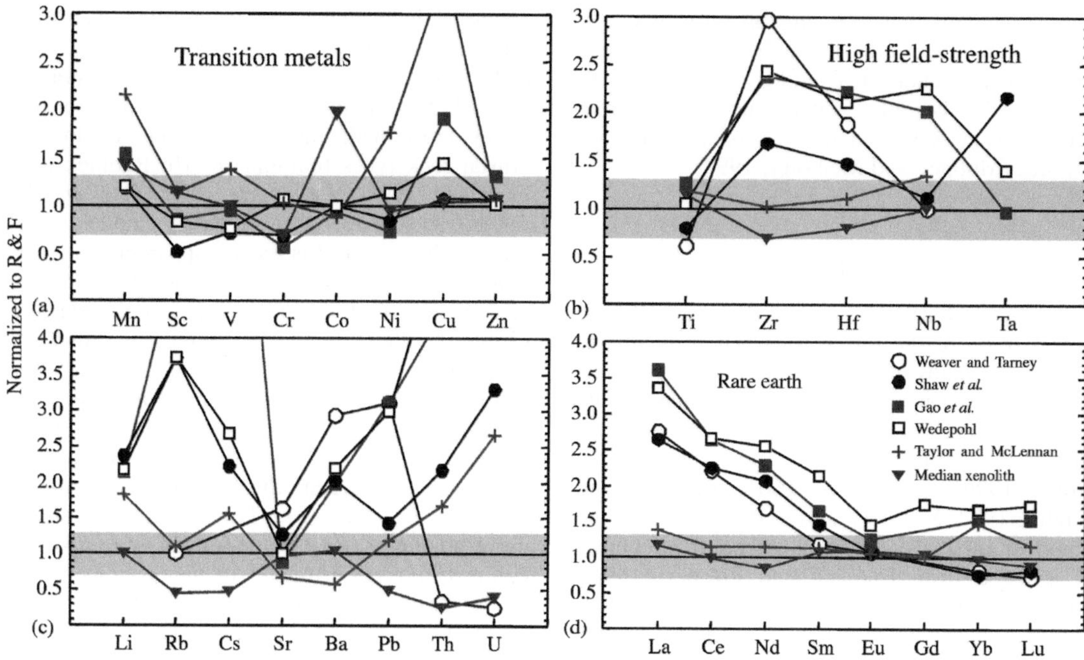

Figure 13 Comparison of different models of the trace-element composition of the lower continental crust. All values normalized to the lower-crust composition of Rudnick and Fountain (1995), which is adopted here as the "best estimate" of the global lower crust. Gray-shaded field represents ±30% variation from this value. Trace elements are divided into the following groups: (a) transition metals, (b) high-field strength elements, (c) alkali, alkaline earth, and actinides, and (d) REEs.

most trace elements fall with 50%. We now explore the possible reasons for these variations, with an eye towards determining a "best estimate" of the global lower-crust composition.

The two lower-crustal estimates derived from averages of surface granulites are generally more evolved than other estimates (Table 7, and Figures 12 and 13). Weaver and Tarney's (1984) lower-crustal estimate derives from the average of Archean Scourian granulites in the Lewisian complex, Scotland. It is one of the most evolved compositions given in the table and is characterized by a steeply fractionated REE pattern, which is characteristic of Archean granitoids of the tonalite–trondhjemite–granodiorite assemblage (see Chapter 3.11) and severe depletions in the large-ion lithophile elements, in addition to thorium and uranium (Rudnick *et al.*, 1985). The estimate of Shaw *et al.* (1994) derives from a weighted average of the granulite-facies rocks of the Kapuskasing Structure zone, Canadian Shield. The average is intermediate in overall composition. As discussed in Section 3.01.3.4.1, the Kapuskasing cross-section provides samples down to depths of ~25 km, leaving the lower 20 km of lower crust unexposed. Seismic-velocity data show this unexposed deepest crust to be mafic in bulk composition, consistent with the limited data for lower-crustal xenoliths from the Superior province (Moser and Heaman, 1997). Thus, the lower-crustal estimates of Weaver and Tarney (1984) and Shaw *et al.* (1994) may be representative of evolved lower crust in Archean cratons lacking a high-velocity lower crust, but are unlikely to be representative of the global continental lower crust (Archean cratons constitute only ~7% of the total area of the continental crust (Goodwin, 1991)).

The lower-crustal estimates derived from particular xenolith suites (columns 3–6, Table 7) were selected to illustrate the great compositional heterogeneity in the deep crust. The average weighted composition of lower-crustal xenoliths from central Spain (Villaseca *et al.*, 1999) is one of the most felsic compositions in Table 7 (with ~63 wt.% SiO_2, Figure 12(b)). It has higher K_2O content than nearly every estimate of the upper crust composition (Table 1), and has such a high heat production (0.8 $\mu W\ m^{-3}$), that a 40 km thickness of crust with average upper- and middle-crustal compositions given in Tables 3 and 5, respectively, would generate a surface heat flow of 41 $mW\ m^{-2}$. This is equivalent to 100% of surface heat flow through Archean crust, 85% of surface heat flow through Proterozoic crust, and 71% of the surface heat flow through Paleozoic crust (see Chapter 3.02). Assuming the heat flux through the Moho is ~17 $mW\ m^{-2}$ (see Chapter 3.02) this lower-crust composition could thus be representative of the lower crust in Phanerozoic regions with high surface heat flow, but clearly cannot be representative of the global

Table 7 Compositional estimates of the lower continental crust. Major elements in weight percent.

	1 *Weaver and Tarney (1984)*	2 *Shaw et al. (1994)*	3 *Rudnick and Taylor (1987)*	4 *Condie and Selverstone (1999)*	5 *Villaseca et al. (1999)*	6 *Liu et al. (2001)*	7 *Updated from Rudnick and Presper (1990)*	8 *Rudnick and Fountain (1995)*	9 *Wedepohl (1995)*	10 *Gao et al. (1998a)*	11 *Taylor and McLennan (1985, 1995)*
SiO_2	62.9	58.3	49.6	52.6	62.7	59.6	52.0	53.4	59.0	59.8	54.3
TiO_2	0.5	0.65	1.33	0.95	1.04	0.60	1.13	0.82	0.85	1.04	0.97[b]
Al_2O_3	16.0	17.4	16.4	16.4	17.4	13.9	17.0	16.9	15.8	14.0	16.1
FeO_T[a]	5.4	7.09	12.0	10.5	7.52	5.44	9.08	8.57	7.47	9.30	10.6
MnO	0.08	0.12	0.22	0.16	0.10	0.08	0.15	0.10	0.12	0.16	0.22
MgO	3.5	4.36	8.72	6.04	3.53	9.79	7.21	7.24	5.32	4.46	6.28
CaO	5.8	7.68	10.1	8.50	1.58	4.64	10.28	9.59	6.92	6.20	8.48
Na_2O	4.5	2.70	1.43	3.19	2.58	2.60	2.61	2.65	2.91	3.00	2.79
K_2O	1.0	1.47	0.17	1.37	3.41	3.30	0.54	0.61	1.61	1.75	0.64[b]
P_2O_5	0.19	0.24		0.21	0.16	0.13	0.13	0.10		0.21	
Mg#	53.4	52.3	56.5	50.5	45.6	76.2	58.6	60.1	55.9	46.1	51.4
Li		14				3.3	5	6	13	13	11
Be									1.7	1.1	1.0
B		3.2							5	7.6	8.3
N									34		
F									429	703	
S									408	231	
Cl									278	216	
Sc		16	33	28	17	20	29	31	25	26	35[b]
V		140	217		139	100	189	196	149	185	271[b]
Cr	88	168	276	133	178	490	145	215	228	123	219[b]
Co		38	31	20	22	31	41	38	38	36	33[b]
Ni	58	75	141	73	65	347	80	88	99	64	156[b]
Cu		28	29		40		32	26	37	50	90
Zn		83			83	89	85	78	79	102	83
Ga						15	17	13	17	19	18
Ge									1.4	1.24	1.6
As									1.3	1.6	0.8
Se									0.17	0.17	0.05
Br									0.28		
Rb	11	41	12	37	90	51	7	11	41	56	12[b]
Sr	569	447	196	518	286	712	354	348	352	308	230
Y	7	16	28		40	8	20	16	27	18	19
Zr	202	114	127	86	206	180	68	68	165	162	70
Nb	5	5.6	13	7.75	15	6.4	5.6	5.0	11	10	6.7[b]
Mo			0.8				0.8		0.6	0.54	0.8
Ru											
Pd										2.78	1
Ag									80	51	90

(continued)

Table 7 (continued).

	1 Weaver and Tamey (1984)	*2* Shaw et al. (1994)	*3* Rudnick and Taylor (1987)	*4* Condie and Selverstone (1999)	*5* Villaseca et al. (1999)	*6* Liu et al. (2001)	*7* Updated from Rudnick and Presper (1990)	*8* Rudnick and Fountain (1995)	*9* Wedepohl (1995)	*10* Gao et al. (1998a)	*11* Taylor and McLennan (1985, 1995)
Cd									0.101	0.097	0.098
In									0.052		0.050
Sn							1.3		2.1	1.34	1.5
Sb									0.30	0.09	0.2
I									0.14		
Cs		0.67	0.07			0.15	0.19	0.3	0.8	2.6	0.47[b]
Ba	757	523	212	564	994	1434	305	259	568	509	150
La	22	21	12	22	38	18	9.5	8	27	29	11
Ce	44	45	28	46	73	36	21	20	53	53	23
Pr			3.6				[2.1]		7.4		2.8
Nd	19	23	16	24	30	14	13.3	11	28	25	13
Sm	3.3	4.1	4.1	5.17	6.6	2.59	3.40	2.8	6.0	4.65	3.17
Eu	1.18	1.18	1.36	1.30	1.8	0.97	1.20	1.1	1.6	1.39	1.17
Gd			4.31	4.67	6.8		3.6	3.1	5.4		3.13
Tb	0.43	0.28	0.79	0.72		0.33	0.50	0.48	0.81	0.86	0.59
Dy			5.05		6.7		3.9	3.1	4.7		3.6
Ho			1.12				0.6	0.68	0.99		0.77
Er			3.25				2.0	1.9			2.2
Tm	0.19										0.32
Yb	1.2	1.13	3.19	2.09	4.0	0.79	1.70	1.5	2.5	2.29	2.2
Lu	0.18	0.2		0.37	0.65	0.12	0.30	0.25	0.43	0.38	0.29
Hf	3.6	2.8	3.3	1.9		4.6	1.9	1.9	4.0	4.2	2.1
Ta		1.3		0.5	2.1	0.3	0.5	0.6	0.8	0.6	0.7[b]
W			0.5				0.5		0.6	0.51	0.6[b]
Re											0.4
Os											0.05
Ir											0.13
Pt										2.87	
Au										1.58	3.4
Hg									0.021	0.0063	
Tl									0.26	0.38	0.23
Pb	13	6	3.3	9.8		12.9	4.1	4	12.5	13	5.0[b]
Bi									0.037	0.38	0.038
Th	0.42	2.6	0.54	1.64	5.74	0.49	0.50	1.2	6.6	5.23	2.0[b]
U	0.05	0.66	0.21	1.38	0.47	0.18	0.18	0.2	0.93	0.86	0.53[b]

[a] Total Fe as FeO. [b] Value from McLennan (2001b).

1. Weighted average of Scourian granulites, Scotland, from Weaver and Tarney (1984). 2. Weighted average of Kapuskasing Structural Zone granulites, from Shaw *et al.* (1994). 3. Average lower crustal xenoliths from the McBride Province, Queensland, Australia from Rudnick and Taylor (1987). 4. Average lower crustal xenoliths from the four corners region, Colorado Plateau, USA from Condie and Selverstone (1999). 5. Weighted mean composition calculated from lithologic proportions of lower crustal xenoliths from Central Spain from Villaseca *et al.*, (1999). 6. Weighted average of lower crustal xenoliths from Hannuoba according to seismic crustal model of North China Craton from Liu *et al.* (2001). 7. Median worldwide lower crustal xenoliths from Rudnick and Presper (1990), updated with data from more recent publications. Complete database available at http://earthref.org/cgi-bin/erda.cgi?*n* = 1, 2, 3, 8, and on Treatise website. 8. Average lower crust derived from global average seismic velocities and granulites from Rudnick and Fountain (1995). 9. Average lower crust in western Europe derived from seismic data and granulite xenolith compositions from Wedepohl (1995). 10. Average lower crust derived from seismic velocities and granulite data from the North China craton from Gao *et al.* (1998a). 11. Average lower crust from Taylor and McLennan (1985, 1995), updated by McLennan and Taylor (1996) and McLennan (2001b). Mg# = molar 100 × Mg/(Mg + Fe_{tot}).

average lower crust. Likewise, the "evolved" xenolith-derived lower crustal estimate of Liu *et al.* (2001) for the central zone of the North China craton, also has a high-K_2O content that exceeds that in most modern estimates of the upper continental crust (Table 1 and Figure 12(b)). Total crustal heat production calculated as described above using the Liu *et al.* composition for the lower crust yields a value of 0.95 $\mu W\ m^{-3}$, which corresponds to a surface heat flow of 34 $mW\ m^{-2}$. This composition is thus unlikely to be representative of the lower crust in Archean cratons, where average surface heat flow is $41 \pm 1\ mW\ m^{-2}$ (Nyblade and Pollack, 1993). However, surface heat flow through this part of the North China craton is unusually high (50 $mW\ m^{-2}$; Hu *et al.*, 2000), thus permitting a more radiogenic lower crust in this region. Other peculiarities of this bulk composition include an extreme Mg# of 76 (Mg# = 100 * molar Mg/(Mg + Fe)) and extreme nickel (347 ppm) and chromium (490 ppm) contents. These values are especially unusual given the rather felsic bulk composition of this estimate and reflect the very high Mg# mafic granulites present in this suite and the inclusion of up to 25% peridotite within the lower-crustal mixture modeled by Liu *et al.* (2001). This estimate also has the highest strontium and barium contents of all estimates (712 ppm and 1,434 ppm, respectively) and is the most HREE depleted (Table 7). The two remaining average lower-crustal xenolith suites (Rudnick and Taylor, 1987 and Condie and Selverstone, 1999) both have mafic compositions that more closely approximate the global average-xenolith composition (column 7 in Table 7).

It has been shown repeatedly from numerous xenolith studies that the majority of lower-crustal xenoliths are mafic in composition (Rudnick and Presper (1990), Rudnick (1992), and Downes (1993) and references therein). Thus, the "best estimate" of the lower crust made on the basis of xenolith studies is found in column 7 of Table 7, which gives the median composition of all analyzed lower-crustal xenoliths. Yet it remains unclear to what degree xenolith compositions reflect average lower crust. Uncertainties include the degree to which volcanic pipes sample a representative cross-section of the deep crust and whether certain xenoliths (e.g., felsic xenoliths and meta-carbonates) suffer preferential disaggregation or dissolution in the host magmas. In addition, large compositional variations are apparent from place to place (Figure 12(b)) and xenolith data are only available for limited regions of the continents.

For these reasons, the best estimates of lower-crustal composition rely on combining seismic velocities of the lower crust with compositions of "typical" lower-crustal lithologies to derive the bulk composition (Christensen and Mooney, 1995; Rudnick and Fountain, 1995; Wedepohl, 1995; Gao *et al.*, 1998a). Wedepohl (1995) used seismic data from the European Geotraverse and Gao *et al.* (1998a) used data from the North China craton to estimate lower-continental crust composition. The resulting compositions derived from these studies (Table 7, and Figures 12 and 13) reflect the thin and more evolved crust in these regions relative to global averages, and produce too much heat to be representative of global lower crust (Rudnick *et al.*, 1998; see Chapter 3.02).

The studies of Rudnick and Fountain (1995) and Christensen and Mooney (1995) are probably best representative of the global deep continental crust, as these authors used overlapping, but not identical, global seismic data sets and independent geochemical data sets to derive the bulk-crust composition. Christensen and Mooney (1995) do not provide the compositional data they used for their lower-crustal assemblages and they do not report a lower-crust composition; thus, one cannot derive an independent estimate of the bulk lower crust from their work. However, they do model the global lower crust (25–40 km depth) as containing ~7% tonalite gneiss and 93% mafic lithologies (including amphibolite, mafic granulite, and mafic-garnet granulite). Such a mafic-bulk composition is consistent with the results of Rudnick and Fountain (1995) (Table 7). Thus, we adopt the lower-crust composition of Rudnick and Fountain (1995) as the best available model of global lower-crust composition, with the proviso that our understanding will evolve as more extensive and detailed information becomes available about the seismic velocity structure of the lower crust. This composition has higher iron, magnesium, and calcium, and considerably lower potassium than most other estimates of the lower continental crust (Figure 12), reflecting the overall high *P*-wave velocities in the lower crust on a worldwide basis and the use of lower crustal xenoliths to derive the average mafic granulite composition. Reliance on xenolith data also accounts for the lower concentrations of highly incompatible trace elements in this composition (e.g., LREEs, rubidium, caesium, barium, thorium, and uranium) compared to most other estimates of the lower crust (Figure 13). For trace elements not considered by Rudnick and Fountain (1995), we adopt the averages of values given in Gao *et al.* (1998a), Wedepohl (1995), or studies focused specifically on the lower-crustal composition of particular trace elements (e.g., antimony, arsenic, molybdenum (Sims *et al.*, 1990); boron (Leeman *et al.*, 1992); tungsten (Newsom *et al.*, 1996); rhenium; and osmium (Saal *et al.*, 1998)). The resulting lower-crustal composition is given in Table 8.

Table 8 Recommended composition of the lower continental crust. Major elements in weight percent. Trace element concentration units the same as in Table 2.

Element	*Lower crust*	*Source*[a]	*Element*	*Lower crust*	*Source*[a]
SiO_2	53.4	1	Ag	65	2
TiO_2	0.82	1	Cd	0.1	2
Al_2O_3	16.9	1	In	0.05	4
FeOT	8.57	1	Sn	1.7	2
MnO	0.10	1	Sb	0.1	5
MgO	7.24	1	I	0.1	4
CaO	9.59	1	Cs	0.3	1
Na_2O	2.65	1	Ba	259	1
K_2O	0.61	1	La	8	1
P_2O_5	0.10	1	Ce	20	1
Li	13	2	Pr	2.4	7
Be	1.4	2	Nd	11	1
B	2	3	Sm	2.8	1
N	34	4	Eu	1.1	1
F	570	2	Gd	3.1	1
S	345	2	Tb	0.48	1
Cl	250	2	Dy	3.1	1
Sc	31	1	Ho	0.68	1
V	196	1	Er	1.9	1
Cr	215	1	Tm	0.24	7
Co	38	1	Yb	1.5	1
Ni	88	1	Lu	0.25	1
Cu	26	1	Hf	1.9	1
Zn	78	1	Ta	0.6	1
Ga	13	1	W	0.6	8
Ge	1.3	2	Re	0.18	9
As	0.2	5	Os	0.05	9
Se	0.2	2	Ir	0.05	10
Br	0.3	4	Pt	2.7	6
Rb	11	1	Au	1.6	11
Sr	348	1	Hg	0.014	2
Y	16	1	Tl	0.32	2
Zr	68	1	Pb	4	1
Nb	5	1	Bi	0.2	2
Mo	0.6	5	Th	1.2	1
Ru	0.75	6	U	0.2	1
Pd	2.8	4			

[a] Sources: 1. Rudnick and Fountain (1995). 2. Average of values given in Wedepohl (1995) and Gao *et al.* (1998a). 3. Leeman *et al.* (1992). 4. Wedepohl (1995). 5. Calculated assuming As/Ce = 0.01, Sb/Ce = 0.005 and Mo/Ce = 0.03 (Sims *et al.*, 1990). 6. Assuming Ru/Ir ratio and Pt/Pd ratios equal to that of upper continental cust. 7. Value interpolated from REE pattern. 8. Average of all values in Table 7, plus correlation from Newsom *et al.* (1996), using W/Th = 0.5. 9. Saal *et al.* (1998). 10. Taylor and McLennan (1985). 11. Gao *et al.* (1998a)

It is interesting to note that the lower-crust composition of Rudnick and Fountain (1995) is quite similar to that of Taylor and McLennan (1985, 1995). However, this similarity is deceptive as Taylor and McLennan's "lower crust" actually represents the crust below the "upper crust" or between ~10 km depth and the Moho. Thus, Taylor and McLennan's lower crust is equivalent to the combined middle and lower crust given here, and is considerably less evolved than the crustal models adopted here (see Section 3.01.4 for a description of how Taylor and McLennan's crust composition was derived).

In summary, our knowledge of lower-crustal composition, like the middle crust, is limited by the ambiguity in deriving chemical compositions from seismic velocities, the lack of high-quality data for a number of trace elements and by the still fragmentary knowledge of the seismic structure of the continental crust. Although the various lower-crustal compositional models in Table 7 show large variations, the true uncertainty in the global model is likely to fall within the seismologically constrained estimates and thus the uncertainty is on the order of ≤30% for most major elements. Uncertainties in trace-element abundances are generally higher (Figure 13). Whereas concentrations of the transition metals between the different estimates generally fall within ~60%, uncertainties in the highly incompatible trace elements (e.g., caesium and thorium; Figure 13) and highly siderophile elements (PGE) can be as large as an order of magnitude. Despite these rather large uncertainties, there are some conclusions that can be drawn from this analysis. The lower crust has a mafic composition and is strongly depleted in potassium and other highly incompatible elements relative to higher levels of the crust. The lower crust is LREE enriched and probably has a positive europium anomaly (Figure 14). Like the upper and middle crust, it is also characterized by enrichment in lead relative to cerium and praseodymium and depletion in niobium relative to lanthanum. It is also likely to be enriched in strontium relative to neodymium (Figure 14).

3.01.4 BULK CRUST COMPOSITION

The earliest estimates of the continental crust composition were derived from analyses and observed proportions of upper crustal rock types (Clarke, 1889; Clarke and Washington, 1924; Ronov and Yaroshevsky, 1967). These estimates do not take into account the changes in both lithological proportions and metamorphic grade that are now recognized to occur with depth in the crust (see Section 3.01.3) and are thus more appropriately regarded as estimates of upper-crust composition. (It is interesting to note, however, the remarkably good correspondence of these earliest estimates with those of today (cf. Tables 1 and 9)) Taylor (1964) used a different approach to estimate bulk crust composition. Following Goldschmidt (1933), he assumed that the nearly constant REE pattern of sedimentary rocks reflected the REE pattern of the crust as a whole, and recreated that pattern by mixing "average" felsic- and mafic-igneous rocks in approximately equal proportions. His composition (Table 9) is also remarkably similar to more modern estimates

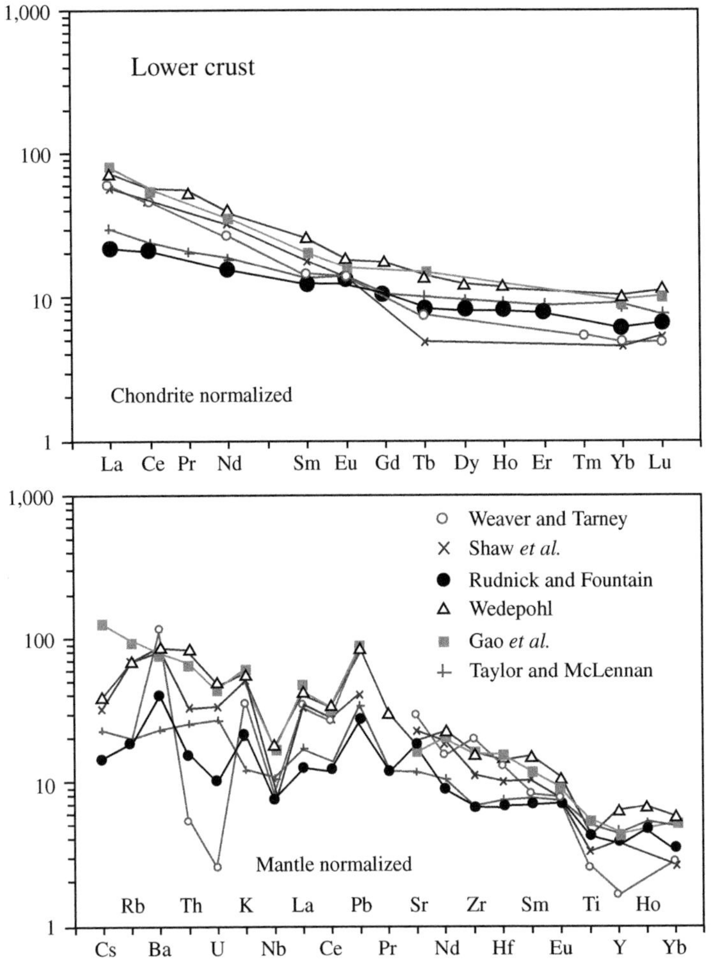

Figure 14 REE (upper) and multi-element plot (lower) of the compositions of the lower crust given in Table 7. Compositions derived from individual xenolith suites not shown. Chondrite values from Taylor and McLennan (1985) and primitive mantle values from McDonough and Sun (1995).

for a large number of elements, but like the earliest estimates, is more appropriately considered an upper crustal estimate since sediments derive strictly from upper crustal sources. Following the plate-tectonic revolution, Taylor (1967) modified his crust-composition model. Recognizing that the present site of continental growth is at convergent-plate margins, he developed the "island arc" or "andesite" model for crustal growth and hence crust composition. In this model (Taylor, 1967, 1977), the crust is assumed to have a composition equal to average convergent-margin andesite. Taylor and McLennan (1985) discussed the difficulties with this approach. Moreover, it is now recognized that basalts dominate present intra-oceanic arcs (see Chapter 3.18, and references therein).

Crust-composition estimates made since the 1970s derive from a variety of approaches. Smithson (1978) was the first to use seismic velocities to determine the lithological makeup of the deep crust. His crust composition is similar to other estimates, save for the very high alkali element contents (4 wt.% Na_2O and 2.7 wt.% K_2O), which presumably reflects the choice of granitic rocks used in his calculations. Holland and Lambert (1972), Weaver and Tarney (1984), and Shaw *et al.* (1986) recognized the importance of granulite-facies rocks in the deep crust and based their crustal models on the composition of rocks from high-grade terranes exposed at the Earth's surface and previous estimates of upper crustal composition.

Taylor and McLennan (1985, 1995), like Taylor's previous estimates (Taylor, 1967, 1977), derived their crust composition using an approach based on assumptions about its formation processes. They assumed that 75% of the crust grew during the Archean from bimodal volcanism and the remaining 25% originated from post-Archean accretion of island arcs having an average andesite composition (from Taylor, 1977). To constrain the proportions of mafic- to felsic-Archean volcanics, they used heat-flow data

Table 9 Compositional estimates of the bulk continental crust. Major elements in weight percent. Trace element concentration units the same as in Table 2.

	1 Taylor (1964)	*2* Ronov and Yaroshevsky (1967)	*3* Holland and Lambert (1972)	*4* Smithson (1978)	*5* Weaver and Tarney (1984)	*6* Shaw et al. (1986)	*7* Christensen and Mooney (1995)	*8* Rudnick and Fountain (1995)	*9* Wedepohl (1995)	*10* Gao et al. (1998a)	*11* Taylor and McLennan (1985, 1995)	*12* This study[a]
SiO_2	60.4	62.2	62.8	63.7	63.9	64.5	62.4	60.1	62.8	64.2	57.1	60.6
TiO_2	1.0	0.8	0.7	0.7	0.6	0.7	0.9	0.7	0.7	0.8	0.9	0.72
Al_2O_3	15.6	15.7	15.7	16.0	16.3	15.1	14.9	16.1	15.4	14.1	15.9	15.9
FeO_T[b]	7.3	6.3	5.5	5.3	5.0	5.7	6.9	6.7	5.7	6.8	9.1	6.71
MnO	0.12	0.10	0.10	0.10	0.08	0.09	0.10	0.11	0.10	0.12	0.18	0.10
MgO	3.9	3.1	3.2	2.8	2.8	3.2	3.1	4.5	3.8	3.5	5.3	4.66
CaO	5.8	5.7	6.0	4.7	4.8	4.8	5.8	6.5	5.6	4.9	7.4	6.41
Na_2O	3.2	3.1	3.4	4.0	4.2	3.4	3.6	3.3	3.3	3.1	3.1	3.07
K_2O	2.5	2.9	2.3	2.7	2.1	2.4	2.1	1.9	2.7	2.3	1.3	1.81
P_2O_5	0.24		0.20		0.19	0.14	0.20	0.20		0.18		0.13
Mg#	48.7	47.0	50.9	49.0	50.5	50.1	44.8	54.3	54.3	48.3	50.9	55.3
Li	20							11	18	17	13	17
Be	2.8								2.4	1.7	1.5	1.9
B	10					9.3			11	18	10	11
N	20								60			56
F	625								525	602		553
S	260								697	283		404
Cl	130								472	179		244
Sc	22					13		22	16	19	30	21.9
V	135					96		131	98	128	230	138
Cr	100				56	90		119	126	92	185	135
Co	25					26		25	24	24	29	26.6
Ni	75				35	54		51	56	46	105	59
Cu	55					26		24	25	38	75	27
Zn	70					71		73	65	81	80	72
Ga	15							16	15	18	18	16
Ge	1.5								1.4	1.25	1.6	1.3
As	1.8								1.7	3.1	1.0	2.5
Se	0.05								0.12	0.13	0.05	0.13
Br	2.5								1.0			0.88
Rb	90				61	76		58	78	69	37[c]	49
Sr	375				503	317		325	333	285	260	320
Y	33				14	26		20	24	17.5	20	19
Zr	165				210	203		123	203	175	100	132
Nb	20				13	20		8[d]	19	11	8[d]	8
Mo	1.5								1.1	0.65	1.0	0.8

Ru					0.1			0.57
Pd					0.4	1.74	1	1.5
Ag	70				70	52	80	56
Cd	0.20				0.10	0.08	0.10	0.08
In	0.1				0.05		0.05	0.05
Sn	2.0				2.3	1.5	2.5	1.7
Sb	0.2				0.3	0.2	0.2	0.2
I	0.5				0.8			0.7
Cs	3.0			2.6	3.4	2.8	1.5[c]	2
Ba	425	707	764	390	584	614	250	456
La	30	28		18	30	31.6	16	20
Ce	60	57		42	60	60.0	33	43
Pr	8.2				6.7		3.9	4.9
Nd	28	23		20	27	27.4	16	20
Sm	6	4.1		3.9	5.3	4.84	3.5	3.9
Eu	1.2	1.09		1.2	1.3	1.27	1.1	1.1
Gd	5.4				4.0		3.3	3.7
Tb	0.9	0.53		0.56	0.65	0.82	0.60	0.6
Dy	3				3.8		3.7	3.6
Ho	1.2				0.80		0.78	0.77
Er	2.8				2.1		2.2	2.1
Tm	0.48	0.24			0.30		0.32	0.28
Yb	3.0	1.5		2.0	2.0	2.2	2.2	*1.9*
Lu	0.50	0.23		0.33	0.35	0.35	0.30	0.30
Hf	3	4.7	5	3.7	4.9	4.71	3.0	3.7
Ta	2		4	0.7[d]	1.1	0.6	0.8[c]	0.7
W	1.5				1.0	0.7	1.0	1
Re					0.4		0.4	0.19
Os					0.05		0.05	0.041
Ir					0.05		0.10	0.037
Pt					0.4	1.81		0.5
Au	40				2.5	1.21	3.0	1.3
Hg	0.08				0.040	0.009		0.03
Tl	0.45				0.52	0.39	0.36	0.50
Pb	12.5	15	20	12.6	14.8	15	8.0	11
Bi	0.17				0.085	0.27	0.06	0.18
Th	9.6	5.7	9	5.6	8.5	7.1	4.2	5.6
U	2.7	1.3	1.8	1.4	1.7	1.2	1.1	1.3

Major elements recast to 100% anhydrous.
[a] See Table 10 for derviation of this estimate. [b] Total Fe as FeO. Mg# = molar $100 \times Mg/(Mg + Fe_{tot})$. [c] Updated by McLennan (2001b). [d] Updated by Barth *et al.* (2000).

from Archean cratons, which is relatively low and uniform at ~40 mW m^{-2} (see Chapter 3.02). Assuming that half the surface heat flow derives from the mantle yielded a mafic to felsic proportion of 2 : 1. This dominantly mafic Archean-crustal component is reflected in the major- and trace-element composition of their crust. Their crust has low SiO_2 and K_2O, high MgO and CaO, and very high FeO content. It also has the highest transition-metal concentrations and lowest incompatible element concentrations of all the models presented in Table 9.

Refinements of the Taylor and McLennan (1985) model are provided by McLennan and Taylor (1996) and McLennan (2001b). The latter is a modification of several trace-element abundances in the upper crust and as such, should not affect their compositional model for the bulk crust, which does not rely on their upper crustal composition. Nevertheless, McLennan (2001b) does provide modified bulk-crust estimates for niobium, rubidium, caesium, and tantalum (and these are dealt with in the footnotes of Table 9). McLennan and Taylor (1996) revisited the heat-flow constraints on the proportions of mafic and felsic rocks in the Archean crust and revised the proportion of Archean-aged crust to propose a more evolved bulk crust composition. This revised composition is derived from a mixture of 60% Archean crust (which is a 50 : 50 mixture of mafic and felsic end-member lithologies), and 40% average-andesite crust of Taylor (1977). McLennan and Taylor (1996) focused on potassium, thorium, and uranium, and did not provide amended values for other elements, although other incompatible elements will be higher (e.g., rubidium, barium, LREEs) and compatible elements lower in a crust composition so revised.

More recently, a number of studies have estimated the bulk-crust composition by deriving lithological proportions for the deep crust from seismic velocities (as discussed in Section 3.01.3) with upper crustal contributions based on data for surface rocks or previous estimates of the upper crust (Christensen and Mooney, 1995; Rudnick and Fountain, 1995; Wedepohl, 1995; Gao *et al.*, 1998a) (Table 9). Like previous estimates of the crust, all of these show intermediate bulk compositions with very similar major-element contents. The greatest differences in major elements between these recent seismologically based estimates are for MgO, CaO, and K_2O, which show ~30% variation, with the Rudnick and Fountain (1995) estimate having the highest MgO and CaO, and lowest K_2O (Table 9, Figure 15). Most trace elements from these estimates fall within 30% total variation as well (Figure 16); the exceptions are trace elements for which very limited data exist (i.e., sulfur, chlorine, arsenic, tin, mercury, bismuth, and the PGEs).

Table 10 Recommended composition of the bulk continental crust.

Element	*Units*		*Element*	*Units*	
SiO_2	wt.%	60.6	Ag	ng g^{-1}	56
TiO_2	"	0.7	Cd	μg g^{-1}	0.08
Al_2O_3	"	15.9	In	"	0.052
FeOT	"	6.7	Sn	"	1.7
MnO	"	0.10	Sb	"	0.2
MgO	"	4.7	I[a]	"	0.7
CaO	"	6.4	Cs	"	2
Na_2O	"	3.1	Ba	"	456
K_2O	"	1.8	La	"	20
P_2O_5	"	0.1	Ce	"	43
Li	μg g^{-1}	16	Pr	"	4.9
Be	"	1.9	Nd	"	20
B	"	11	Sm	"	3.9
N[a]	"	56	Eu	"	1.1
F	"	553	Gd	"	3.7
S	"	404	Tb	"	0.6
Cl	"	244	Dy	"	3.6
Sc	"	21.9	Ho	"	0.77
V	"	138	Er	"	2.1
Cr	"	135	Tm	"	0.28
Co	"	26.6	Yb	"	1.9
Ni	"	59	Lu	"	0.30
Cu	"	27	Hf	"	3.7
Zn	"	72	Ta	"	0.7
Ga	"	16	W	"	1
Ge	"	1.3	Re[a]	ng g^{-1}	0.188
As	"	2.5	Os[a]	"	0.041
Se	"	0.13	Ir[a]	"	0.037
Br[a]	"	0.88	Pt	"	1.5
Rb	"	49	Au	"	1.3
Sr	"	320	Hg	μg g^{-1}	0.03
Y	"	19	Tl	"	0.50
Zr	"	132	Pb	"	11
Nb	"	8	Bi	"	0.18
Mo	"	0.8	Th	"	5.6
Ru[a]	ng g^{-1}	0.6	U	"	1.3
Pd	"	1.5			

The total-crust composition is calculated according to the upper, middle and lower-crust compositions obtained in this study and corresponding weighing factors of 0.317, 0.296 and 0.388. The weighing factors are based on the layer thickness of the global continental crust, recalculated from crustal structure and areal proportion of various tectonic units given by Rudnick and Fountain (1995).
[a] Middle crust is not considered due to lack of data.

Niobium and tantalum also show >30% total variation, but this is due to the very high niobium and tantalum contents of Wedepohl's estimate, which reflects his reliance on the old Canadian Shield data, for which niobium content is anomalously high (see discussion in Section 3.01.2.1). These elements were also compromised in the Rudnick and Fountain (1995) crust composition, which relied (indirectly) on the Canadian Shield data for the upper crust by adopting the Taylor and McLennan upper-crust composition (see discussions in Plank and Langmuir (1998) and Barth *et al.* (2000)). The values for niobium and tantalum in

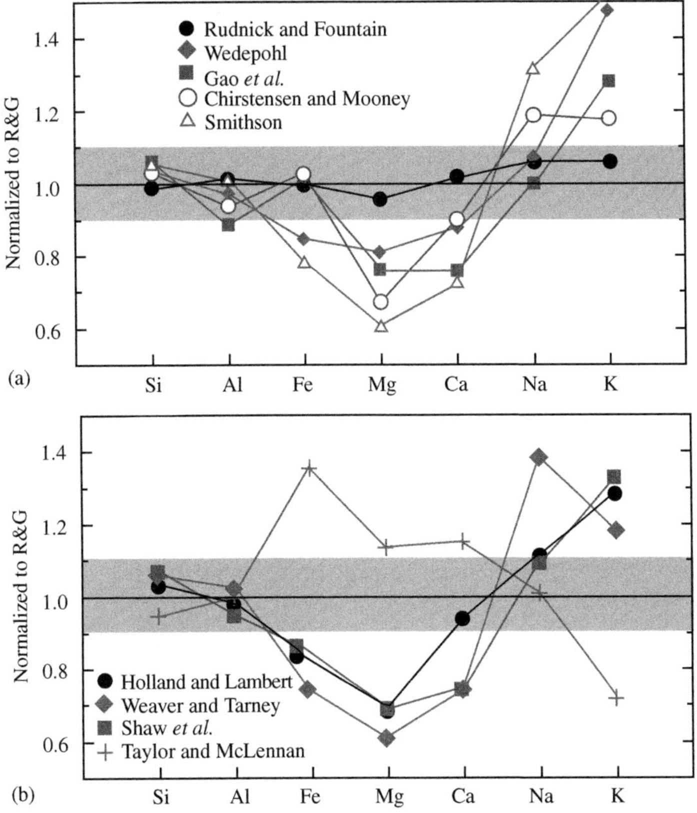

Figure 15 Comparison of different estimates of the major-element composition of the bulk continental crust. All data normalized to the new composition given here (Table 10, "R&G"); gray shading depicts 10% variation from this composition. (a) Models based on seismological data (Rudnick and Fountain, 1995; Wedepohl, 1995; Gao *et al.*, 1998a; Christensen and Mooney, 1995 and Smithson, 1978). (b) Models based on both surface exposures (Holland and Lambert, 1972; Weaver and Tarney, 1984; Shaw *et al.*, 1986) and Taylor and McLennan's (1985, 1995) model-generated crust composition.

the Rudnick and Fountain (1995) model given in Table 9 have been updated by Barth *et al.* (2000). These values are similar to those of Gao *et al.* (1998a) and thus the concentrations of these elements are known to within 30% in the bulk crust.

Of all the estimates in Table 9, that of Taylor and McLennan (1985, 1995) stands out as being the most mafic overall (Figures 15 and 16). This mafic composition stems from their model for Archean crust, which constitutes 75% of their crust and is composed of a 2 : 1 mixture of mafic- to felsic-igneous rocks. This relatively mafic crust composition was necessitated by their inferred low heat production in Archean crust and the inferred large proportion of the Archean-aged crust. However, such a high proportion of mafic rocks in the Archean crust is at odds with seismic data (summarized in Section 3.01.3), which show that the crust of most Archean cratons is dominated by low velocities, implying the presence of felsic (not mafic) compositions, even in the lower crust. In addition, some of the assumptions used by Taylor and McLennan (1985) regarding heat flow are not very robust, as recognized by McLennan and Taylor (1996) (see also discussion in Rudnick *et al.*, 1998). First, the 20 mWm^{-2} of mantle heat flow they assumed for Archean cratons is probably too high (see Chapter 3.02), thus allowing for more heat production in the crust. Second, granulite-facies felsic rocks are often depleted in heat-producing elements (the Scourian granulites are an extreme example), thereby allowing a greater proportion of felsic rocks in the crust. Third, it is unlikely that 75% of the present continents were formed in the Archean (see Chapter 3.10) and, importantly, the observed low surface heat flow that is the rationale for Taylor and McLennan's (1985, 1995) dominantly mafic crust composition is restricted to Archean cratons, which constitute only ~7% of the present continental crust (Goodwin, 1991). Indeed, heat production of the Taylor and McLennan (1985, 1995) crustal model falls outside the range estimated for average continental crust by Jaupart and Mareschal (see Chapter 3.02)

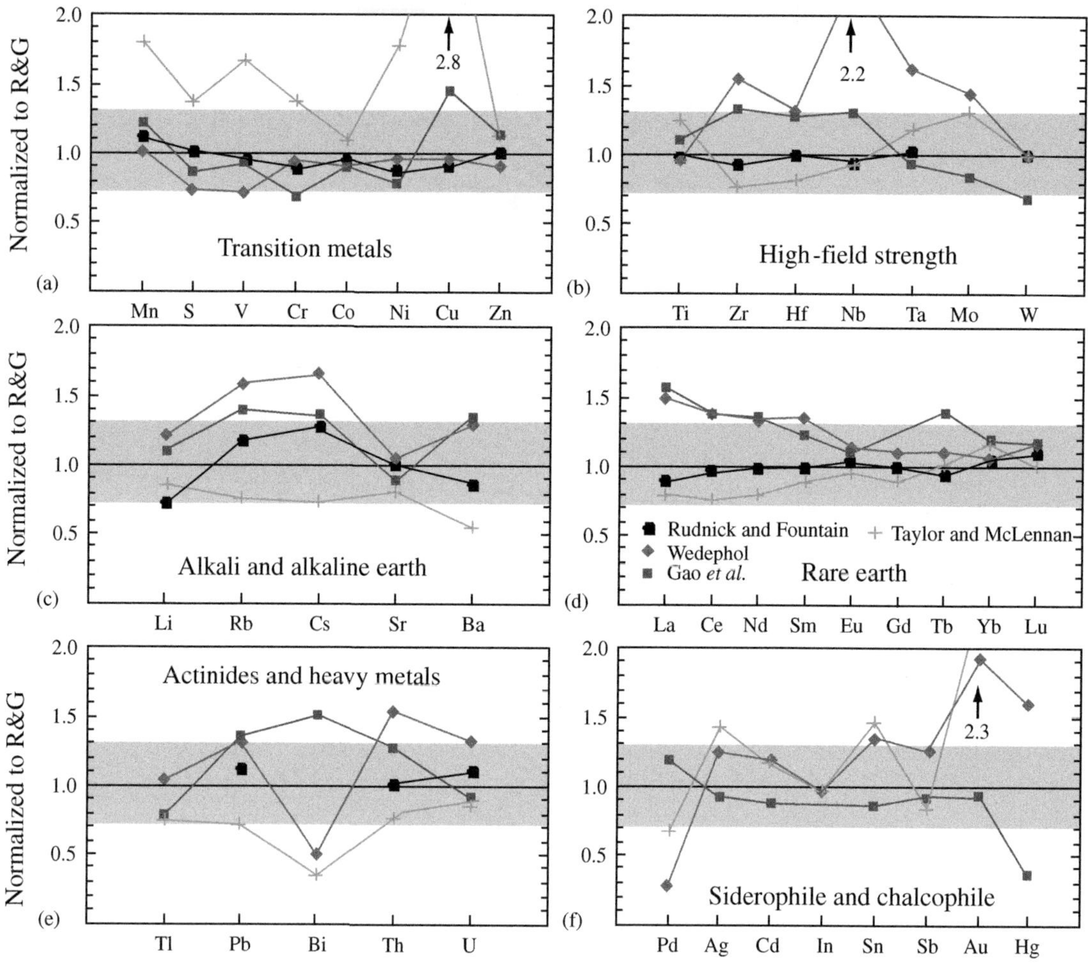

Figure 16 Comparison of the trace-element composition of bulk continental crust from seismological and model-based approaches. All data normalized to the new composition given here (Table 10, "R&G"). Gray shading depicts ±30% variation from Rudnick and Gao composition (this work). (a) Transition metals, (b) high-field strength elements, (c) alkali and alkaline earth metals, and (d) REEs, (e) actinides and heavy metals, and (f) siderophile and chalcophile elements.

(i.e., 0.58 μW m^{-2} versus 0.79–0.99 μW m^{-2}). The modifications made to this model by McLennan and Taylor (1996) help to reconcile their model with the above observations, but the proportion of mafic rocks in Archean-aged crust still appears to be high (based on observed seismic velocities) and the total heat production (at 0.70 μW m^{-2}) may still be somewhat low.

3.01.4.1 A New Estimate of Crust Composition

In column 12 of Table 9 we present a new estimate of the bulk crust composition. This composition derives from our estimates of upper, middle, and lower crust given in Tables 3, 5, and 8, mixed in the proportions derived from the global compilation of Rudnick and Fountain (1995): 31.7% upper, 29.6% middle, and 38.8% lower crust. Our new crustal estimate thus relies heavily on the previously derived lower crust of Rudnick and Fountain (1995), the middle crust of Rudnick and Fountain and Gao *et al.* (1998a) and the new estimate of upper-crust composition provided here (Table 3). The latter is very similar to the upper crust of Taylor and McLennan (1985) (Figure 8), which was used by Rudnick and Fountain (1995) in calculating their bulk crust composition. The main differences lie in the concentrations of K_2O, rubidium, niobium, and tantalum, which are lower in the new estimate of the upper continental crust provided here. Accordingly, this new estimate has many similarities with that of Rudnick and Fountain (1995), but contains lower potassium, rubidium, niobium, and tantalum, and considers a wider range of trace elements than given in that model. The heat production of this new estimate is 0.89 μW m^{-2}, which falls in the middle of the range estimated for average-crustal heat production by Jaupart and Mareschal (see Chapter 3.02).

Figures 15–17 show how this composition compares to other estimates of crust composition. Figure 15 shows that our new composition has generally higher MgO, CaO, and FeO, and lower Na_2O and K_2O than most other seismically based models. The differences between our model and that of Wedepohl (1995) and Gao *et al.* (1998a) likely reflect the regional character of these latter models (western Europe, eastern China), where the crust is thinner and more evolved than the global averages (Chirstensen and Mooney, 1995, and Rudnick and Fountain, 1995). The lower MgO and higher alkali elements in Christensen and Mooney's model compared to ours must stem from the differences in the chemical databases used to construct these two models, as the lithological proportions of the deep crust are very similar (Section 3.01.3). Rudnick and Fountain (1995) (and hence our current composition) used the compositions of lower-crustal xenoliths to constrain the mafic end-member of the deep crust. These xenoliths have high Mg# and low alkalis (Table 7), and thus may be chemically distinct from mafic rocks exposed on the Earth's surface (the chemical data used by Christensen and Mooney, 1995).

The variations between the different seismological-based crust compositions can be considered representative of the uncertainties that exist in our understanding of the bulk crust composition. Some elemental concentrations (e.g., silicon, aluminum, sodium) are known to within 20% uncertainty. The remaining major-element and many trace-element (transition metals, high-field strength elements, most REE) concentrations are known to within 30% uncertainty. Still some trace element concentrations are yet poorly constrained in the crust, including many of the highly siderophile elements (Figure 16).

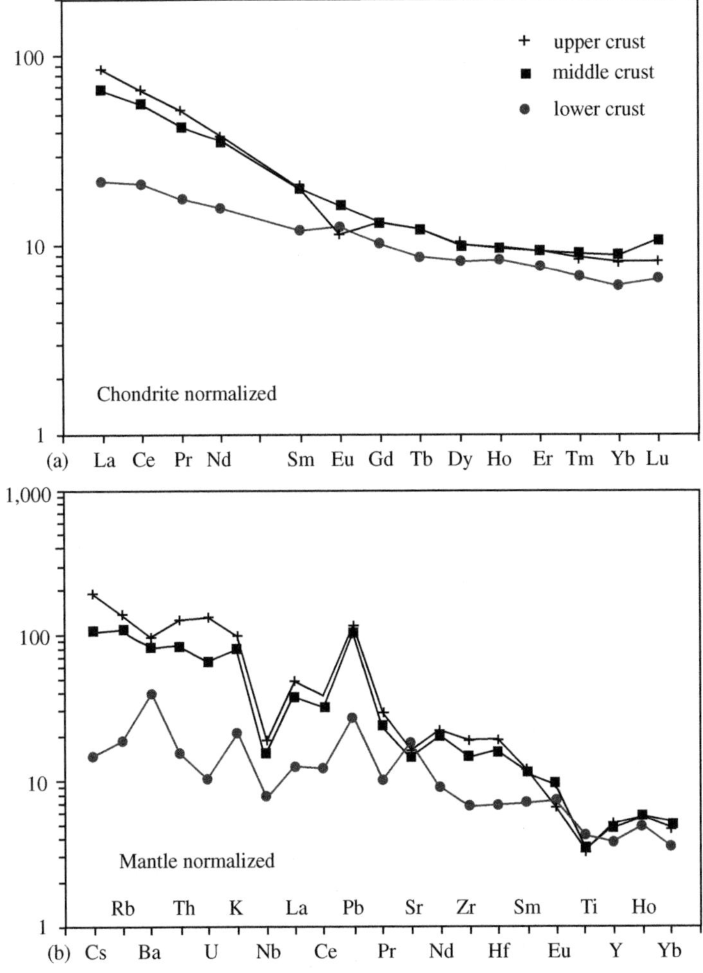

Figure 17 Comparison of (a) rare-earth and (b) additional trace-element compositions of the upper, middle, and lower crust recommended here. Chondrite values from Taylor and McLennan (1985), mantle-normalizing values from McDonough and Sun (1995).

3.01.4.2 Intracrustal Differentiation

Table 11 provides the composition of the upper, middle, lower, and bulk crust for comparison purposes and Figure 17 compares their respective REE and extended trace-element patterns. The upper crust has a large negative europium anomaly (Eu/Eu*, Table 11) that is largely complemented by the positive europium anomaly of the lower crust; the middle crust has essentially no europium anomaly. Similar complementary anomalies exist for strontium. These features, in addition to the greater LREE enrichment of the upper crust relative to the lower crust, suggests that the upper crust is largely the product of intracrustal magmatic differentiation in the presence of plagioclase (see Taylor and McLennan, 1985; see Chapter 3.11). That is, the upper crust is dominated by granite that differentiated from the lower crust through partial melting, crystal fractionation and mixing processes. The middle crust has an overall trace-element pattern that is very similar to the upper crust, indicating that it too is dominated by the products of intracrustal differentiation. All segments of the crust are characterized by an overall enrichment of the most incompatible elements, as well as high La/Nb and low Ce/Pb ratios. These are characteristics of convergent margin magmas (see Chapter 3.18, and references therein) and thus have implications for the processes responsible for generation of the continental crust as discussed in the next section.

3.01.5 IMPLICATIONS OF THE CRUST COMPOSITION

Despite the uncertainties in estimating crust composition discussed in the previous section, there are a number of similarities that all crust-compositional models share and these may be important for understanding the origin of the crust. The crust is characterized by an overall *intermediate* igneous-rock composition, with relatively high Mg#. It is enriched in incompatible elements (Figure 18), and contains up to 50% of the silicate Earth's budget of these elements (Rudnick and Fountain, 1995). It is also well established that the crust is depleted in niobium relative to lanthanum, and has a subchondritic Nb/Ta ratio. These features are not consistent with formation of the crust by single-stage melting of peridotitic mantle, as discussed in Rudnick (1995), Kelemen (1995) and Rudnick *et al.* (2000) (see also Chapter 3.18).

If one assumes that the crust grows ultimately by igneous processes (i.e., magmatic transport of mass from the mantle into the crust), then the disparity between crust composition and the composition of primary mantle melts requires the operation of additional processes to produce the present crust composition. As reviewed in Rudnick (1995) and Kelemen (1995), these additional processes could include (but are not limited to):

(i) *Recycling of mafic/ultramafic lower crust and upper mantle via density foundering (often referred to as delamination within the geochemical literature).* In this process, lithologically stratified continental crust is thickened during an orogenic event, causing the mafic lower crust to transform to eclogite, which has a higher density than the underlying mantle peridotite. Provided the right temperatures and viscosities exist (i.e., hot and goey), the base of the lithosphere will sink into the underlying asthenosphere. Numerical simulations of this process show that it is very likely to occur at the time of arc–continent collision (and in fact, may be impossible to avoid)(Jull and Kelemen, 2001).

(ii) *Production of crust from a mixture of silicic melts derived from subducted oceanic crust, and basaltic melts from peridotite.* This process is likely to have been more prevalent in a hotter, Archean Earth (although see Chapter 3.18 for an alternative view) and would have involved extensive silicic melt–peridotite reaction as the slab melts traverse the mantle wedge (Kelemen, 1995). The abundance of Archean-aged granitoids of the so-called "TTG" suite (trondhjemite, tonalite, granodiorite) are often cited as the surface manifestations of these processes (Drummond and Defant, 1990; Martin, 1994; see Chapter 3.11).

(iii) *Weathering of the crust, with preferential recycling of Mg ± Ca into the mantle via hydrothermally altered mid-ocean ridge basalt (Albarede, 1998; Anderson, 1982).* This hypothesis states that during continental weathering, soluble cations such as Ca^{2+}, Mg^{2+}, and Na^{+} are carried to the oceans while silicon and aluminum remain behind in the continental regolith. Whereas other elements (e.g., sodium) may be returned to the continents via arc magmatism, magnesium may be preferentially sequestered into altered seafloor basalts and returned to the mantle via subduction, producing a net change in the crust composition over time. However, one potential problem with this hypothesis is that examination of altered ocean-floor rocks suggests that magnesium may not be significantly sequestered there (see Chapter 3.15).

A fourth possibility, that ultramafic cumulates representing the chemical complement to the andesitic crust are present in the uppermost mantle, is not supported by studies of peridotite xenoliths, which show a predominance of restitic peridotite over cumulates (e.g., Wilshire *et al.*, 1988). If such cumulates were originally there,

Table 11 Comparison of the upper, middle, lower and total continental crust compositions recommended here.

Element	*Upper crust*	*Middle crust*	*Lower crust*	*Total crust*
SiO_2	66.6	63.5	53.4	60.6
TiO_2	0.64	0.69	0.82	0.72
Al_2O_3	15.4	15.0	16.9	15.9
FeO_T	5.04	6.02	8.57	6.71
MnO	0.10	0.10	0.10	0.10
MgO	2.48	3.59	7.24	4.66
CaO	3.59	5.25	9.59	6.41
Na_2O	3.27	3.39	2.65	3.07
K_2O	2.80	2.30	0.61	1.81
P_2O_5	0.15	0.15	0.10	0.13
Total	100.05	100.00	100.00	100.12
Mg#	46.7	51.5	60.1	55.3
Li	24	12	13	16
Be	2.1	2.3	1.4	1.9
B	17	17	2	11
N	83		34	56
F	557	524	570	553
S	621	249	345	404
Cl	294	182	250	244
Sc	14.0	19	31	21.9
V	97	107	196	138
Cr	92	76	215	135
Co	17.3	22	38	26.6
Ni	47	33.5	88	59
Cu	28	26	26	27
Zn	67	69.5	78	72
Ga	17.5	17.5	13	16
Ge	1.4	1.1	1.3	1.3
As	4.8	3.1	0.2	2.5
Se	0.09	0.064	0.2	0.13
Br	1.6		0.3	0.88
Rb	82	65	11	49
Sr	320	282	348	320
Y	21	20	16	19
Zr	193	149	68	132
Nb	12	10	5	8
Mo	1.1	0.60	0.6	0.8
Ru	0.34		0.75	0.57
Pd	0.52	0.76	2.8	1.5
Ag	53	48	65	56
Cd	0.09	0.061	0.10	0.08
In	0.056		0.05	0.052
Sn	2.1	1.30	1.7	1.7
Sb	0.4	0.28	0.10	0.2
I	1.4		0.14	0.71
Cs	4.9	2.2	0.3	2
Ba	628	532	259	456
La	31	24	8	20
Ce	63	53	20	43
Pr	7.1	5.8	2.4	4.9
Nd	27	25	11	20
Sm	4.7	4.6	2.8	3.9
Eu	1.0	1.4	1.1	1.1
Gd	4.0	4.0	3.1	3.7
Tb	0.7	0.7	0.48	0.6
Dy	3.9	3.8	3.1	3.6
Ho	0.83	0.82	0.68	0.77
Er	2.3	2.3	1.9	2.1

(continued)

Table 11 (continued).

Element	*Upper crust*	*Middle crust*	*Lower crust*	*Total crust*
Tm	0.30	0.32	0.24	0.28
Yb	2.0	2.2	1.5	1.9
Lu	0.31	0.4	0.25	0.30
Hf	5.3	4.4	1.9	3.7
Ta	0.9	0.6	0.6	0.7
W	1.9	0.60	0.60	1
Re	0.198		0.18	0.188
Os	0.031		0.05	0.041
Ir	0.022		0.05	0.037
Pt	0.5	0.85	2.7	1.5
Au	1.5	0.66	1.6	1.3
Hg	0.05	0.0079	0.014	0.03
Tl	0.9	0.27	0.32	0.5
Pb	17	15.2	4	11
Bi	0.16	0.17	0.2	0.18
Th	10.5	6.5	1.2	5.6
U	2.7	1.3	0.2	1.3
Eu/Eu*	0.72	0.96	1.14	0.93
Heat production ($\mu W\ m^{-3}$)	1.65	1.00	0.19	0.89
Nb/Ta	13.4	16.5	8.3	12.4
Zr/Hf	36.7	33.9	35.8	35.5
Th/U	3.8	4.9	6.0	4.3
K/U	9475	15607	27245	12367
La/Yb	15.4	10.7	5.3	10.6
Rb/Cs	20	30	37	24
K/Rb	283	296	462	304
La/Ta	36	42	13	29

they must have been subsequently removed via a process such as density foundering.

All of the above processes require return of mafic to ultramafic lithologies to the convecting mantle. These lithologies are the chemical complement of the present-day andesitic crust. Thus crustal recycling, in various forms, must have been important throughout Earth history.

Another implication of the distinctive trace element composition of the continental crust is that the primary setting of crust generation is most likely to be that of a convergent margin. The characteristic depletion of niobium relative to lanthanum seen in the crust (Figure 18) is a ubiquitous feature of convergent margin magmas (see review of Kelemen *et al.* (Chapter 3.18)) and is virtually absent in intraplate magmas. Simple mixing calculations indicate that the degree of niobium depletion seen in the crust suggests that at least 80% of the crust was generated in a convergent margin (Barth *et al.*, 2000; Plank and Langmuir, 1998).

3.01.6 EARTH'S CRUST IN A PLANETARY PERSPECTIVE

The other terrestrial planets show a variety of crustal types, but none that are similar to that of the Earth. Mercury has an ancient, heavily cratered crust with a high albedo (see review of Taylor and Scott (Chapter 1.18)). Its brightness plus the detection of sodium, and more recently the refractory element calcium, in the Mercurian atmosphere (Bida *et al.*, 2000) has led to the speculation that Mercury's crust may be anorthositic, like the lunar highlands (see Taylor, 1992 and references therein). The MESSENGER mission (http://messenger.jhuapl.edu/), currently planned to rendezvous with Mercury in 2007, should considerably illuminate the nature of the crust on Mercury.

In contrast to Mercury's ancient crust, high-resolution radar mapping of Venus' cloaked surface has revealed an active planet, both tectonically and volcanically (see review of Fegley (Chapter 1.19) and references therein). Crater densities are relatively constant, suggesting a relatively young surface (~300–500 Ma, Phillips *et al.*, 1992; Schaber *et al.*, 1992; Strom *et al.*, 1994). It has been suggested that this statistically random crater distribution may reflect episodes of mantle overturn followed by periods of quiescence (Schaber *et al.*, 1992; Strom *et al.*, 1994). Most Venusian volcanoes appear to erupt basaltic magmas, but a few are pancake-shaped, which may signify the eruption of a highly viscous lava such as rhyolite (e.g., Ivanov and Head, 1999). The unimodal topography of Venus is

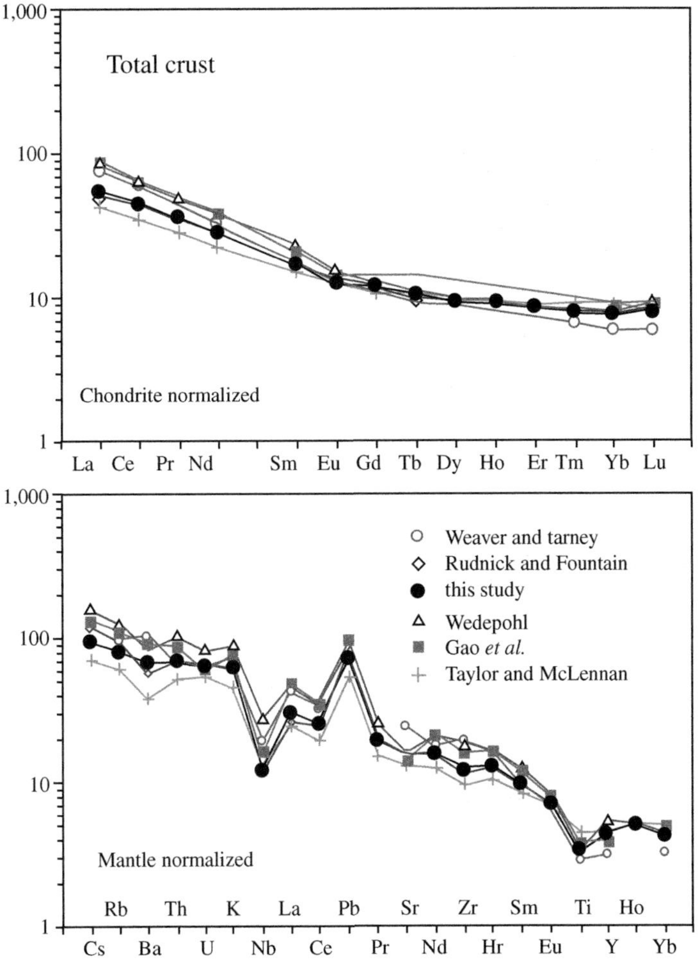

Figure 18 REE (upper) and multi-element plot (lower) of the compositions of the continental crust given in Table 9. Chondrite values from Taylor and McLennan (1985), mantle-normalizing values from McDonough and Sun (1995).

distinct from that of the Earth and there appear to be no equivalents to Earth's oceanic and continental dichotomy. It is possible that the high elevations on Venus were produced tectonically by compression of basaltic rocks made rigid by the virtual absence of water (Mackwell *et al.*, 1998).

Of the terrestrial planets, only Mars has the bimodal topographic distribution seen on the Earth (Smith *et al.*, 1999). In addition, evolved igneous rocks, similar to the andesites found in the continents on Earth, have also been observed on the Martian surface, although their significance and relative abundance is a matter of contention (see review by McSween (Chapter 1.22)). However, the bimodal topography of Mars appears to be an ancient feature (Frey *et al.*, 2002), unlike the Earth's, which is a product of active plate tectonics. It remains to be seen whether the rocks that compose the high-standing southern highlands of Mars bear any resemblance to those of Earth's continental crust (McLennan, 2001a; Wanke *et al.*, 2001).

3.01.7 SUMMARY

The crust is the Earth's major repository of incompatible elements and thus factors prominently into geochemical mass-balance calculations for the whole Earth. For this reason, and to understand the processes by which it formed, determining the composition of the continental crust has been a popular pursuit of geochemists from the time the first rocks were analyzed.

It has been known for over a century that the continental crust has an average composition approximating to andesite (when cast as an igneous rock type) (Clarke, 1889, Clarke and Washington, 1924). The myriad studies on continental crust composition carried out in the intervening years have refined our picture of the crust's composition, particularly for trace elements.

Based on seismic investigations the crust can be divided into three regions: upper, middle, and lower continental crust. The upper crust is the most accessible region of the solid earth and its composition is estimated from both weighted averages of surface samples and studies of shales and loess. The latter is a particularly powerful means of estimating the average upper crustal concentrations of insoluble to moderately soluble trace elements. Most estimates of the major-element composition of the upper continental crust fall within 20% standard deviation of the mean and thus the composition of this important reservoir appears to be reasonably well known. The concentrations of some trace elements also appear to be known to within 20% (most of the transition metals, rubidium, strontium, yttrium, zirconium, niobium, barium, REE, hafnium, tantalum, lead, thorium, and uranium), whereas others are less-precisely known. In particular, very few estimates have been made of the upper crust's halogen, sulfur, germanium, arsenic, selenium, indium, and platinum-group element concentration.

Lacking the access and widescale natural sampling by sediments afforded the upper crust, the composition of the deep crust must be inferred from more indirect means. Both heat flow and seismic velocities have been employed towards this end. Heat flow provides bounds on the potassium, thorium, and uranium content of the crust, and seismic-wave speeds can be interpreted, with some caveats, in terms of rock types, whose compositions are derived from averages of appropriate deep-crustal lithologies. The middle crust is perhaps the least well characterized of the three crustal regions. This is due to the lack of systematic geochemical studies of amphibolite-facies crustal lithologies. In contrast, the lower crust has been the target of a number of geochemical investigations, yet there is wide variation in different estimates of lower-crust composition. This reflects, in part, the highly heterogeneous character of this part of the Earth. However, some generalities can be made. Heat production must decrease and seismic velocities are observed to increase with depth in the crust. Thus the lower crust is, on an average, mafic in composition and depleted in heat-producing elements. Curiously, the lower crust of many Archean cratons, where heat flow is lowest, has relatively slow *P*-wave velocities. Such low velocities imply the dominance of evolved rock types and thus these rocks must be highly depleted in potassium, thorium, and uranium compared to their upper crustal counterparts.

The andesitic continental crust composition is difficult to explain if the crust is generated by single-stage melting of peridotitic mantle, and additional processes must therefore be involved in its generation. All of these processes entail return of mafic or ultramafic crustal material (which is complementary to the present continental crust) to the convecting mantle. Thus crustal recycling, in various forms, must have been important throughout Earth history and is undoubtedly related to the plate-tectonic cycle on our planet. Crustal recycling, along with the presence of abundant water to facilitate melting (Campbell and Taylor, 1985), may be the major factor responsible for our planet's unique crustal dichotomy.

ACKNOWLEDGMENTS

We thank Sandy Romeo and Yongshen Liu for their able assistance in the preparation of this manuscript and updating of the lower-crustal xenolith database. Reviews by Kent Condie and Scott McLennan and comments from Herbert Palme and Rich Walker improved the presentation. This work was supported by NSF grant EAR 99031591 to R.L.R., a National Nature Science Foundation of China grant (40133020) and a Chinese Ministry of Science and Technology grant (G1999043202) to S.G.

REFERENCES

Albarede F. (1998) The growth of continental crust. In *Continents and their Mantle Roots*, Tectonophysics (eds. A. Vauchez and R. O. Meissner). Elsevier, Amsterdam, vol. 296 (1–2), pp. 1–14.

Alirezaei S. and Cameron E. M. (2002) Mass balance during gabbro-amphibolite transition, Bamble sector, Norway: implications for petrogenesis and tectonic setting of the gabbros. *Lithos* **60**(1–2), 21–45.

Anderson A. T. (1982) Parental basalts in subduction zones: implications for continental evolution. *J. Geophys. Res.* **87**, 7047–7060.

Aoki K.-I. (1971) Petrology of mafic inclusion from Itinome-gata, Japan. *Contrib. Mineral. Petrol.* **30**, 314–331.

Arculus R. J. and Smith D. (1979) Eclogite, pyroxenite, and amphibolite inclusions in the Sullivan Buttes latite, Chino Valley, Yavapai County, Arizona. In *The Mantle Sample: Inclusions in Kimberlites and Other Volcanics* (eds. F. R. Boyd and H. O. A. Meyer). American Geophysical Union, Washington, DC, pp. 309–317.

Arculus R. J., Ferguson J., Chappell B. W., Smith D., McCulloch M. T., Jackson I., Hensel H. D., Taylor S. R., Knutson J., and Gust D. A. (1988) Trace element and isotopic characteristics of eclogites and other xenoliths derived from the lower continental crust of southeastern Australia and southwestern Colorado Plateau USA. In *Eclogites and Eclogite-Facies Rocks* (ed. D. C. Smith). Elsevier, Amsterdam, pp. 335–386.

Baldridge W. S. (1979) Mafic and ultramafic inclusion suites from the Rio Grande Rift (New Mexico) and their bearing on the composition and thermal state of the lithosphere. *J. Volcanol. Geotherm. Res.* **6**, 319–351.

Barth M., McDonough W. F., and Rudnick R. L. (2000) Tracking the budget of Nb and Ta in the continental crust. *Chem. Geol.* **165**, 197–213.

Barth M., Rudnick R. L., Carlson R. W., Horn I., and McDonough W. F. (2002) Re–Os and U–Pb geochronological constraints on the eclogite-tonalite connection in the Archean Man Shield, West Africa. *Precamb. Res.* **118**(3–4), 267–283.

Beck S. L. and Zandt G. (2002) The nature of orogenic crust in the central Andes. *J. Geophys. Res. Solid Earth* **107**(B10) (article no. 2230).

Behn M. D. and Kelemen P. B. (2003) The relationship between seismic P-wave velocity and the composition of anhydrous igneous and meta-igneous rocks. *Geochem. Geophys. Geosys.* (**4**)1041, doi: 10.1029/2002GC000393.

Berckhemer H. (1969) Direct evidence for the composition of the lower crust and Moho. *Tectonophysics* **8**, 97–105.

Berg J. H., Moscati R. J., and Herz D. L. (1989) A petrologic geotherm from a continental rift in Antarctica. *Earth Planet. Sci. Lett.* **93**, 98–108.

Bida T. A., Killen R. M., and Morgan T. H. (2000) Discovery of calcium in Mercury's atmosphere. *Nature* **404**, 159–161.

Bohlen S. R. and Mezger K. (1989) Origin of granulite terranes and the formation of the lowermost continental crust. *Science* **244**, 326–329.

Borodin L. S. (1998) Estimated chemical composition and petrochemical evolution of the upper continental crust. *Geochem. Int.* **37**(8), 723–734.

Bowring S. A. and Williams I. S. (1999) Priscoan (4.00–4.03 Ga) orthogneisses from northwestern Canada. *Contrib. Mineral. Petrol.* **134**(1), 3–16.

Bradley S. D. and McCallum M. E. (1984) Granulite facies and related xenoliths from Colorado-Wyoming kimberlite. In *Kimberlites: II. The Mantle and Crust–Mantle Relationships* (ed. J. Kornprobst). Elsevier, Amsterdam, vol. 11B, pp. 205–218.

Broadhurst J. R. (1986) Mineral reactions in xenoliths from the Colorado Plateau: implications for lower crustal conditions and fluid composition. In *The Nature of the Lower Continental Crust*, Geol. Soc. Spec. Publ. (eds. J. B. Dawson, D. A. Carswell, J. Hall, and K. H. Wedepohl). London, vol. 24, pp. 331–349.

Brittan J. and Warner M. (1996) Seismic velocity, heterogeneity, and the composition of the lower crust. *Tectonophysics* **264**, 249–259.

Brittan J. and Warner M. (1997) Wide-angle seismic velocities in heterogeneous crust. *Geophys. J. Int.* **129**, 269–280.

Burke M. M. and Fountain D. M. (1990) Seismic properties of rocks from an exposure of extended continental crust—new laboratory measurements from the Ivrea zone. *Tectonophysics* **182**, 119–146.

Cameron K. L. and Ward R. L. (1998) Xenoliths of Grenvillian granulite basement constrain models for the origin of voluminous Tertiary rhyolites, Davis Mountains, west Texas. *Geology* **26**(12), 1087–1090.

Cameron K. L., Robinson J. V., Niemeyer S., Nimz G. J., Kuentz D. C., Harmon R. S., Bohlen S. R., and Collerson K. D. (1992) Contrasting styles of pre-Cenozoic and mid-Tertiary crustal evolution in northern Mexico: evidence from deep crustal xenoliths from La Olivina. *J. Geophys. Res.* **97**, 17353–17376.

Campbell I. H. and Taylor S. R. (1985) No water, no granites—no oceans, no continents. *Geophys. Res. Lett.* **10**, 1061–1064.

Chen S., O'Reilly S. Y., Zhou X., Griffin W. L., Zhang G., Sun M., Feng J., and Zhang M. (2001) Thermal and petrological structure of the lithosphere beneath Hannuoba, Sino-Korean craton, China: evidence from xenoliths. *Lithos* **56**, 267–301.

Chen W. and Arculus R. J. (1995) Geochemical and isotopic characteristics of lower crustal xenoliths, San Francisco volcanic field, Arizona, USA. *Lithos* **36**(3–4), 203–225.

Chen Y. D., O'Reilly S. Y., Kinny P. D., and Griffin W. L. (1994) Dating lower crust and upper-mantle events—an ion microprobe study of xenoliths from kimberlitic pipes, South-Australia. *Lithos* **32**(1–2), 77–94.

Chen Y. D., O'Reilly S. Y., Griffin W. L., and Krogh T. E. (1998) Combined U–Pb dating and Sm–Nd studies on lower crustal and mantle xenoliths from the delegate basaltic pipes, southeastern Australia. *Contrib. Mineral. Petrol.* **130**(2), 154–161.

Christensen N. I. (1982) Seismic velocities. In *Handbook of Physical Properties of Rocks* (ed. R. S. Carmichael). CRC Press, Boca Raton, FL, vol. II, pp. 1–228.

Christensen N. I. and Mooney W. D. (1995) Seismic velocity structure and composition of the continental crust: a global view. *J. Geophys. Res.* **100**(B7), 9761–9788.

Clarke F. W. (1889) The relative abundance of the chemical elements. *Phil. Soc. Washington Bull.* **XI**, 131–142.

Clarke F. W. and Washington H. S. (1924) The composition of the Earth's crust. *USGS Professional Paper* **127**, 117pp.

Clitheroe G., Gudmundsson O., and Kennett B. L. N. (2000) The crustal thickness of Australia. *J. Geophys. Res. Solid Earth* **105**(B6), 13697–13713.

Cogley J. G. (1984) Continental margins and the extent and number of the continents. *Rev. Geophys. Space Phys.* **22**, 101–122.

Cohen R. S., O'Nions R. K., and Dawson J. B. (1984) Isotope geochemistry of xenoliths from East Africa: implications for development of mantle reservoirs and their interaction. *Earth Planet. Sci. Lett.* **68**, 209–220.

Collerson K. D., Hearn B. C., MacDonald R. A., Upton B. F., and Park J. G. (1988) Granulite xenoliths from the Bearpaw mountains, Montana: constraints on the character and evolution of lower continental crust. *Terra Cognita* **8**, 270.

Condie K. C. (1993) Chemical composition and evolution of the upper continental crust: contrasting results form surface samples and shales. *Chem. Geol.* **104**, 1–37.

Condie K. C. (1997) *Plate Tectonics and Crustal Evolution.* Butterworth-Heinemann, Oxford, UK.

Condie K. C. and Selverstone J. (1999) The crust of the Colorado plateau: new views of an old arc. *J. Geol.* **107**(4), 387–397.

Condie K. C., Latysh N., Van Schmus W. R., Kozuch M., and Selverstone J. (1999) Geochemistry, Nd and Sr isotopes, and U/Pb zircon ages of Navajo Volcanic Field, Four Corners area, southwestern United States. *Chem. Geol.* **156**(1–4), 95–133.

Davis G. L. (1977) The ages and uranium contents of zircons from kimberlites and associated rocks. *Carnegie Inst. Wash. Yearbook* **76**, 631–635.

Davis G. L. and Grew E. S. (1977) Age of zircon from a crustal xenolith, Kilbourne Hole, New Mexico. *Carnegie Inst. Wash. Yearbook* **77**, 897–898.

Davis W. J. (1997) U–Pb zircon and rutile ages from granulite xenoliths in the Slave province: evidence for mafic magmatism in the lower crust coincident with Proterozoic dike swarms. *Geology* **25**(4), 343–346.

Dawson J. B. (1977) Sub-cratonic crust and upper-mantle models based on xenolith suites in kimberlite and nephelinitic diatremes. *J. Geol. Soc.* **134**, 173–184.

Dawson J. B. and Smith J. V. (1987) Reduced sapphirine granulite xenoliths from the Lace Kimberlite, South Africa: implications for the deep structure of the Kaapvaal Craton. *Contrib. Mineral. Petrol.* **95**, 376–383.

Dawson J. B., Harley S. L., Rudnick R. L., and Ireland T. R. (1997) Equilibration and reaction in Archaean quartz-sapphirine granulite xenoliths from the Lace Kimberlite pipe, South Africa. *J. Metamorph. Geol.* **15**(2), 253–266.

Dessai A. G. and Vaselli O. (1999) Petrology and geochemistry of xenoliths in lamprophyres from the Deccan Traps: implications for the nature of the deep crust boundary in western India. *Min. Mag.* **63**(5), 703–722.

Dessai A. G., Knight K., and Vaselli O. (1999) Thermal structure of the lithosphere beneath the Deccan Trap along the western Indian continental margin: evidence from xenolith data. *J. Geol. Soc. India* **54**(6), 585–598.

Dobosi G., Kempton P. D., Downes H., Embey-Isztin A., Thirlwall M., and Greenwood P. (2003) Lower crustal granulite xenoliths from the Pannonian Basin, Hungary: Part 2. Sr–Nd–Pb–Hf and O isotope evidence for formation of continental lower crust by tectonic emplacement of oceanic crust. *Contrib. Mineral. Petrol.* **144**, 671–683.

Dodge F. C. W., Lockwood J. P., and Calk L. C. (1988) Fragments of the mantle and crust from beneath the Sierra Nevada batholith: xenoliths in a volcanic pipe near Big Creek, California. *Geol. Soc. Am. Bull.* **100**, 938–947.

Dodge F. W., Calk L. C., and Kistler R. W. (1986) Lower crustal xenoliths, Chinese Peak lava flow, central Sierra Nevada. *J. Petrol.* **27**, 1277–1304.

Domenick M. A., Kistler R. W., Dodge F. W., and Tatsumoto M. (1983) Nd and Sr isotopic study of crustal and mantle inclusions from the Sierra Nevada and implications for batholith petrogenesis. *Geol. Soc. Am. Bull.* **94**, 713–719.

Dostal J. and Capedri S. (1979) Rare earth elements in high-grade metamorphic rocks from the western Alps. *Lithos* **12**, 41–49.

Dostal J., Dupuy C., and Leyreloup A. (1980) Geochemistry and petrology of meta-igneous granulitic xenoliths in Neogene volcanic rocks of the Massif Central, France—implications for the lower crust. *Earth Planet. Sci. Lett.* **50**, 31–40.

Downes H. (1993) The nature of the lower continental crust of Europe petrological and geochemical evidence from xenoliths. *Phys. Earth Planet. Inter.* **79**(1–2), 195–218.

Downes H., and Leyreloup A. F. (1986) Granulitic xenoliths from the French massif central—petrology, Sr and Nd isotope systematics and model age estimates. In *The Nature of the Lower Continental Crust* (eds. B. Dawson, D. A. Carswell, J. Hall, and K. H. Wedepohl). Geological Society of London, London, pp. 319–330.

Downes H., Dupuy C., and Leyreloup A. F. (1990) Crustal evolution of the Hercynian belt of western Europe: evidence from lower-crustal granulitic xenoliths (French Massif-Central): *Chem. Geol.* **83**(3–4), 209–231.

Downes H., Kempton P. D., Briot D., Harmon R. S., and Leyreloup A. F. (1991) Pb and O isotope systematics in granulite facies xenoliths, French Massif-Central—implications for crustal processes. *Earth Planet. Sci. Lett.* **102** (3–4), 342–357.

Downes H., Peltonen P., Manttari I., and Sharkov E. V. (2002) Proterozoic zircon ages from lower crustal granulite xenoliths, Kola Peninsula, Russia: evidence for crustal growth and reworking. *J. Geol. Soc.* **159**, 485–488.

Drummond B. J. (1988) A review of crust upper mantle structure in the Precambrian areas of Australia and implications for Precambrian crustal evolution. *Precamb. Res.* **40**(1), 101–116.

Drummond M. S. and Defant M. J. (1990) A model for trondhjemite-tonalite-dacite genesis and crustal growth via slab melting: archean to modern comparisons. *J. Geophys. Res.* **95**(B13), 21503–21521.

Ducea M. (2001) The California arc: thick granitic batholiths, eclogitic residues, lithospheric-scale thrusting, and magmatic flare-ups. *GSA Today* **11**(11), 4–10.

Ducea M. N. and Saleeby J. B. (1996) Buoyancy sources for a large, unrooted mountain range, the Sierra Nevada, California: evidence from xenolith thermobarometry. *J. Geophys. Res. Solid Earth* **101**(B4), 8229–8244.

Ducea M. N. and Saleeby J. B. (1998) The age and origin of a thick mafic-ultramafic keel from beneath the Sierra Nevada batholith. *Contrib. Mineral. Petrol.* **133**(1–2), 169–185.

Durrheim R. J. and Green R. W. E. (1992) A seismic refraction investigation of the Archaean Kaapvaal craton, South Africa, using mine tremors as the energy source. *Geophys. J. Int.* **108**, 812–832.

Durrheim R. J. and Mooney W. D. (1994) The evolution of the Precambrian lithosphere: seismological and geochemical constraints. *J. Geophys. Res.* **99**, 15359–15374.

Eade K. E. and Fahrig W. F. (1971) Chemical Evolutionary Trends of Continental Plates—preliminary Study of the Canadian Shield. *Geol. Sur. Can. Bull.* **179**, 51pp.

Eade K. E. and Fahrig W. F. (1973) Regional, Lithological, and Temporal Variation in the Abundances of some Trace Elements in the Canadian Shield. Geol. Sur. Canada Paper 72–46, Ottawa, Ontario.

Eberz G. W., Clarke D. B., Chatterjee A. K., and Giles P. S. (1991) Chemical and isotopic composition of the lower crust beneath the Meguma Lithotectonic Zone, Nova Scotia: evidence from granulite facies xenoliths. *Contrib. Mineral. Petrol.* **109**, 69–88.

Edwards A. C., Lovering J. F., and Ferguson J. (1979) High pressure basic inclusions from the Kayrunnera kimberlitic breccia pipe in New South Wales, Australia. *Contrib. Mineral. Petrol.* **69**, 185–192.

Ehrenberg S. N. and Griffin W. L. (1979) Garnet granulite and associated xenoliths in minette and serpentinite diatremes of the Colorado Plateau. *Geology* **7**, 483–487.

Embey-Isztin A., Scharbert H. G., Deitrich H., and Poultidis H. (1990) Mafic granulite and clinopyroxenite xenoliths from the Transdanubian volcanic region (Hungary): implication for the deep structure of the Pannonian Basin. *Min. Mag.* **54**, 463–483.

Embey-Isztin A., Downes H., Kempton P. D., Dobosi G., and Thirlwall M. (2003) Lower crustal granulite xenoliths from the Pannonian Basin, Hungary: Part 1. Mineral chemistry, thermobarometry and petrology. *Contrib. Mineral. Petrol.* **144**, 652–670.

Ertan I. E. and Leeman W. P. (1999) Fluid inclusions in mantle and lower crustal xenoliths from the Simcoe volcanic field, Washington. *Chem. Geol.* **154**(1–4), 83–95.

Esperança S. and Garfunkel Z. (1986) Ultramafic xenoliths from the Mt. Carmel area (Karem Maharal Volcano), Israel. *Lithos* **19**, 43–49.

Esperança S., Carlson R. W., and Shirey S. B. (1988) Lower crustal evolution under central Arizona: Sr, Nd, and Pb isotopic and geochemical evidence from the mafic xenoliths of Camp Creek. *Earth Planet. Sci. Lett.* **90**, 26–40.

Esser B. K. and Turekian K. K. (1993) The osmium isotopic composition of the continental crust. *Geochim. Cosmochim. Acta* **57**, 3093–3104.

Fahrig W. F., and Eade K. E. (1968) The chemical evolution of the Canadian Shield. *Geochim. Cosmochim. Acta* **5**, 1247–1252.

Fountain D. M. (1976) The Ivrea-Verbano and Strona-Ceneri zones, northern Italy: a cross-section of the continental crust—new evidence from seismic velocities of rock samples. *Tectonophysics* **33**, 145–165.

Fountain D. M. and Salisbury M. H. (1981) Exposed cross-sections through the continental crust: implications for crustal structure, petrology, and evolution. *Earth Planet. Sci. Lett.* **56**, 263–277.

Fountain D. M., and Salisbury M. H. (1995) Seismic properties of rock samples from the Pikwitonei granulite belt—God's lake domain crustal cross section, Manitoba. *Can. J. Earth Sci.* **33**(5), 757–768.

Fountain D. M., Percival J., and Salisbury M. H. (1990a) Exposed cross sections of the continental crust-synopsis. In *Exposed Cross-sections of the Continental Crust* (eds. M. H. Salisbury and D. M. Fountain). Kluwer, Amsterdam, pp. 653–662.

Fountain D. M., Salisbury M. H., and Percival J. (1990b) Seismic structure of the continental crust based on rock velocity measurements from the Kapuskasing uplift. *J. Geophys. Res.* **95**, 1167–1186.

Fountain D. M., Salisbury M. H., and Furlong K. P. (1987) Heat production and thermal conductivity of rocks from the Pikwitonei-Sachigo continental cross section, central Manitoba: implications for the thermal structure of Archean crust. *Can. J. Earth Sci.* **24**, 1583–1594.

Francis D. M. (1976) Corona-bearing pyroxene granulite xenoliths and the lower crust beneath Nunivak Island, Alaska. *Can. Mineral.* **14**, 291–298.

Frey H. V., Roark J. H., Shockey K. M., Frey E. L., and Sakimoto S. E. H. (2002) Ancient lowlands on Mars. *Geophys. Res. Lett.* **29**(10), 22-1–22-4.

Gallet S., Jahn B.-M., van Vliet Lanoë B., Dia A., and Rossello E. (1998) Loess geochemistry and its implications for particle origin and composition of the upper continental crust. *Earth Planet. Sci. Lett.* **156**, 157–172.

Gao S., Luo T.-C., Zhang B.-R., Zhang H.-F., Han Y.-W., Hu Y.-K., and Zhao Z.-D. (1998a) Chemical composition of the continental crust as revealed by studies in east China. *Geochim. Cosmochim. Acta* **62**, 1959–1975.

Gao S., Zhang B.-R., Jin Z.-M., Kern H., Luo T.-C., and Zhao Z.-D. (1998b) How mafic is the lower continental crust? *Earth Planet. Sci. Lett.* **106**, 101–117.

Gao S., Kern H., Liu Y. S., Jin S. Y., Popp T., Jin Z. M., Feng J. L., Sun M., and Zhao Z. B. (2000) Measured and calculated seismic velocities and densities for granulites from xenolith occurrences and adjacent exposed lower crustal sections: a comparative study from the North China craton. *J. Geophys. Res. Solid Earth* **105**(B8), 18965–18976.

Goldschmidt V. M. (1933) Grundlagen der quantitativen Geochemie. *Fortschr. Mienral. Kirst. Petrogr.* **17**, 112.

Goldschmidt V. M. (1958) *Geochemistry*. Oxford University Press, Oxford.

Goodwin A. M. (1991) *Precambrian Geology*. Academic Press, London.

Gorman A. R., Clowes R. M., Ellis R. M., Henstock T. J., Spence G. D., Keller G. R., Levander A., Snelson C. M., Burianyk M. J. A., Kanasewich E. R., Asudeh I., Hajnal Z., and Miller K. C. (2002) Deep probe: imaging the roots of western North America. *Can. J. Earth Sci.* **39**(3), 375–398.

Grapes R. H. (1986) Melting and thermal reconstitution of pelitic xenoliths, Wehr volcano, East Eifel, West Germany. *J. Petrol.* **27**, 343–396.

Gregoire M., Mattielli N., Nicollet C., Cottin J. Y., Leyrit H., Weis D., Shimizu N., and Giret A. (1994) Oceanic mafic granulite xenoliths from the Kerguelen archipelago. *Nature* **367**, 360–363.

Gregoire M., Cottin J. Y., Giret A., Mattielli N., and Weis D. (1998) The meta-igneous granulite xenoliths from Kerguelen archipelago: evidence of a continent nucleation in an oceanic setting. *Contrib. Mineral. Petrol.* **133**(3), 259–283.

Griffin W. L. and O'Reilly S. Y. (1986) The lower crust in eastern Australia: xenolith evidence. In *The Nature of the Lower Continental Crust*, Geol. Soc. London Spec. Publ. (eds. B. Dawson, D. A. Carswell, J. Hall, and K. H. Wedepohl). London, vol. 25, pp. 363–374.

Griffin W. L. and O'Reilly S. Y. (1987) Is the continental Moho the crust–mantle boundary? *Geology* **15**, 241–244.

Griffin W. L., Carswell D. A., and Nixon P. H. (1979) Lower-crustal granulites and eclogites from Lesotho, southern Africa. In *The Mantle Sample: Inclusions in Kimberlites* (eds. F. R. Boyd and H. O. A. Meyer). American Geophysical Union, Washington, DC, pp. 59–86.

Griffin W. L., Sutherland F. L., and Hollis J. D. (1987) Geothermal profile and crust–mantle transition beneath east-central Queensland: volcanology, xenolith petrology and seismic data. *J. Volcanol. Geotherm. Res.* **31**, 177–203.

Griffin W. L., Jaques A. L., Sie S. H., Ryan C. G., Cousens D. R., and Suter G. F. (1988) Conditions of diamond growth: a proton microprobe study of inclusions in West Australian diamonds. *Contrib. Mineral. Petrol.* **99**, 143–158.

Hacker B. R., Gnos E., Ratschbacher L., Grove M., McWilliams M., Sobolev S. V., Wan J., and Wu Z. H. (2000) Hot and dry deep crustal xenoliths from Tibet. *Science* **287**(5462), 2463–2466.

Halliday A. N., Aftalion M., Upton B. G. J., Aspen P., and Jocelyn J. (1984) U–Pb isotopic ages from a granulite-facies xenolith from Partan Craig in the Midland Valley of Scotland. *Trans. Roy. Soc. Edinburgh: Earth Sci.* **75**, 71–74.

Hanchar J. M., Miller C. F., Wooden J. L., Bennett V. C., and Staude J.-M. G. (1994) Evidence from xenoliths for a dynamic lower crust, eastern Mojave desert, California. *J. Petrol.* **35**, 1377–1415.

Harley S. L. (1989) The origin of granulites: a metamorphic perspective. *Geol. Mag.* **126**, 215–247.

Hart R. J., Nicolaysen L. O., and Gale N. H. (1981) Radioelement concentrations in the deep profile through Precambrian basement of the Vredefort structure. *J. Geophys. Res.* **86**, 10639–10652.

Hart R. J., Andreoli M. A. G., Tredoux M., and Dewit M. J. (1990) Geochemistry across an exposed section of Archean crust at Vredefort, South Africa with implications for midcrustal discontinuities. *Chem. Geol.* **82**(1–2), 21–50.

Haskin M. A. and Haskin L. A. (1966) Rare earths in European shales: a redetermination. *Science* **154**, 507–509.

Haskin L. A., Wildeman T. R., Frey F. A., Collins K. A., Keedy C. R., and Haskin M. A. (1966) Rare earths in sediments. *J. Geophys. Res. B: Solid Earth* **71**(24), 6091–6105.

Hattori Y., Suzuki K., Honda M., and Shimizu H. (2003) Re–Os isotope systematics of the Taklimakan Desert sands, moraines and river sediments around the Taklimakan Desert, and of Tibetan soils. *Geochim. Cosmochim. Acta* **67**, 1195–1206.

Hayob J. L., Essene E. J., Ruiz J., Ortega-Gutiérrez F., and Aranda-Gómez J. J. (1989) Young high-temperature granulites form the base of the crust in central Mexico. *Nature* **342**, 265–268.

Heier K. S. (1973) Geochemistry of granulite facies rocks and problems of their origin. *Phil. Trans. Roy. Soc. London* **A273**, 429–442.

Heinrichs H., Schulz-Dobrick B., and Wedepohl K. H. (1980) Terrestrial geochemistry of Cd, Bi, Tl, Pb, Zn, and Rb. *Geochim. Cosmochim. Acta* **44**, 1519–1533.

Hickey-Vargas R., Abdollahi M. J., Parada M. A., Lopezescobar L., and Frey F. A. (1995) Crustal xenoliths from Calbuco volcano, Andean southern volcanic zone—implications for crustal composition and magma–crust interaction. *Contrib. Mineral. Petrol.* **119**(4), 331–344.

Holbrook W. S. and Kelemen P. B. (1993) Large igneous province on the US Atlantic margin and implications for magmatism during continental breakup. *Nature* **364**, 433–436.

Holbrook W. S., Mooney W. D., and Christensen N. I. (1992) The seismic velocity structure of the deep continental crust. In *Continental Lower Crust* (eds. D. M. Fountain, R. Arculus, and R. W. Kay). Elsevier, Amsterdam, pp. 1–44.

Holland J. G. and Lambert R. S. J. (1972) Major element chemical composition of shields and the continental crust. *Geochim. Cosmochim. Acta* **36**, 673–683.

Hölttä P., Huhma H., Manttari I., Peltonen P., and Juhanoja J. (2000) Petrology and geochemistry of mafic granulite xenoliths from the Lahtojoki kimberlite pipe, eastern Finland. *Lithos* **51**(1–2), 109–133.

Hu S., He L., and Wang J. (2000) Heat flow in the continental area of China: a new data set. *Earth Planet. Sci. Lett.* **179**, 407–419.

Huang Y. M., van Calsteren P., and Hawkesworth C. J. (1995) The evolution of the lithosphere in southern Africa—a perspective on the basic granulite xenoliths from kimberlites in South-Africa. *Geochim. Cosmochim. Acta* **59**(23), 4905–4920.

Hunter R. H., Upton B. G. J., and Aspen P. (1984) Meta-igneous granulite and ultramafic xenoliths from basalts of the Midland Valley of Scotland: petrology and mineralogy of the lower crust and upper mantle. *Trans. Roy. Soc. Edinburgh* **75**, 75–84.

Ivanov M. A. and Head J. W. (1999) Stratigraphic and geographic distribution of steep-sided domes on Venus: preliminary results from regional geological mapping and implications for their origin. *J. Geophys. Res. Planet.* **104**(E8), 18907–18924.

Jahn B. M., Gallet S., and Han J. M. (2001) Geochemistry of the Xining, Xifeng, and Jixian sections, Loess Plateau of China: eolian dust provenance and paleosol evolution during the last 140 ka. *Chem. Geol.* **178**(1–4), 71–94.

James D. E., Padovani E. R., and Hart S. R. (1980) Preliminary results on the oxygen isotopic composition of the lower crust, Kilbourne Hole Maar, New Mexico. *Geophys. Res. Lett.* **7**, 321–324.

Jones A. P., Smith J. V., Dawson J. B., and Hansen E. C. (1983) Metamorphism, partial melting, and K-metasomatism of garnet-scapolite-kyanite granulite xenoliths from Lashaine, Tanzania. *J. Geol.* **91**, 143–166.

Jull M. and Kelemen P. B. (2001) On the conditions for lower crustal convective instability. *J. Geophys. Res. B: Solid Earth* **106**(4), 6423–6446.

Kalamarides R. I., Berg J. H., and Hank R. A. (1987) Lateral isotopic discontinuity in the lower crust: an example from Antarctica. *Science* **237**, 1192–1195.

Kay R. W. and Kay S. M. (1981) The nature of the lower continental crust: inferences from geophysics, surface geology, and crustal xenoliths. *Rev. Geophys. Space Phys.* **19**, 271–297.

Kay R. W. and Kay S. M. (1991) Creation and destruction of lower continental crust. *Geol. Rundsch.* **80**, 259–278.

Kay S. M. and Kay R. W. (1983) Thermal history of the deep crust inferred from granulite xenoliths, Queensland, Australia. *Am. J. Sci.* **283**, 486–513.

Kelemen P. B. (1995) Genesis of high Mg# andesites and the continental crust. *Contrib. Mineral. Petrol.* **120**, 1–19.

Kempton P. D. and Harmon R. S. (1992) Oxygen-isotope evidence for large-scale hybridization of the lower crust during magmatic underplating. *Geochim. Cosmochim. Acta* **55**, 971–986.

Kempton P. D., Harmon R. S., Hawkesworth C. J., and Moorbath S. (1990) Petrology and geochemistry of lower crustal granulites from the Geronimo volcanic field, southeastern Arizona. *Geochim. Cosmochim. Acta* **54**, 3401–3426.

Kempton P. D., Downes H., Sharkov E. V., Vetrin V. R., Ionov D. A., Carswell D. A., and Beard A. (1995) Petrology and geochemistry of xenoliths from the northern Baltic shield: evidence for partial melting and metasomatism in the lower crust beneath an Archaean terrane. *Lithos* **36** (3–4), 157–184.

Kempton P. D., Downes H., and Embey-Isztin A. (1997) Mafic granulite xenoliths in Neogene alkali basalts from the western Pannonian Basin: insights into the lower crust of a collapsed orogen. *J. Petrol.* **38**(7), 941–970.

Kempton P. D., Downes H., Neymark L. A., Wartho J. A., Zartman R. E., and Sharkov E. V. (2001) Garnet granulite xenoliths from the northern Baltic shield the underplated lower crust of a palaeoproterozoic large igneous province. *J. Petrol.* **42**(4), 731–763.

Kern H., Gao S., and Liu Q.-S. (1996) Seismic properties and densities of middle and lower crustal rocks exposed along the North China geoscience transect. *Earth Planet. Sci. Lett.* **139**, 439–455.

Ketcham R. A. (1996) Distribution of heat-producing elements in the upper and middle crust of southern and west central Arizona: evidence from core complexes. *J. Geophys. Res.* (B) **101**, 13611–13632.

Kopylova M. G., O'Reilly S. Y., and Genshaft Y. S. (1995) Thermal state of the lithosphere beneath Central Mongolia: evidence from deep-seated xenoliths from the Shavaryn-Saram volcanic centre in the Tariat depression, Hangai, Mongolia. *Lithos* **36**, 243–255.

Kuno H. (1967) Mafic and ultramafic nodules from Itinome-gata, Japan. In *Ultramafic and Related Rocks* (ed. P. J. Wiley). Wiley, New York, pp. 337–342.

Kyle P. R., Wright A., and Kirsch I. (1987) Ultramafic xenoliths in the late Cenozoic McMurdo volcanic group, western Ross Sea embayment, Antarctica. In *Mantle Xenoliths* (ed. P. H. Nixon). Wiley, New York, pp. 287–294.

Le Bas M. J. and Streckeisen A. L. (1991) The IUGS systematics of igneous rocks. *J. Geol. Soc. London* **148**, 825–833.

Lee C.-Y., Chung S. L., Chen C.-H., and Hsieh Y. L. (1993) Mafic granulite xenoliths from Penghu Islands: evidence for basic lower crust in SE China continental margin. *J. Geol. Soc. China* **36**(4), 351–379.

Leech M. L. (2001) Arrested orogenic development: eclogitization, delamination, and tectonic collapse. *Earth Planet. Sci. Lett.* **185**(1–2), 149–159.

Leeman W. P., Menzies M. A., Matty D. J., and Embree G. F. (1985) Strontium, neodymium, and lead isotopic compositions of deep crustal xenoliths from the Snake River Plain: evidence for Archean basement. *Earth Planet. Sci. Lett.* **75**, 354–368.

Leeman W. P., Sisson V. B., and Reid M. R. (1992) Boron geochemistry of the lower crust-evidence from granulite terranes and deep crustal xenoliths. *Geochim. Cosmochim. Acta* **56**(2), 775–788.

LePichon X., Henry P., and Goffe B. (1997) Uplift of Tibet: from eclogites to granulites-implications for the Andean Plateau and the Variscan belt. *Tectonophysics* **273**(1–2), 57–76.

Leyreloup A., Dupuy C., and Andriambololona R. (1977) Catazonal xenoliths in French Neogene volcanic rocks: constitution of the lower crust: 2. Chemical composition and consequences of the evolution of the French Massif Central Precambrian crust. *Contrib. Mineral. Petrol.* **62**, 283–300.

Leyreloup A., Bodinier J. L., Dupuy C., and Dostal J. (1982) Petrology and geochemistry of granulite xenoliths from Central Hoggar (Algeria)—implications for the lower crust. *Contrib. Mineral. Petrol.* **79**, 68–75.

Liu Y. S., Gao S., Jin S. Y., Hu S. H., Sun M., Zhao Z. B., and Feng J. L. (2001) Geochemistry of lower crustal xenoliths from Neogene Hannuoba basalt, North China craton: implications for petrogenesis and lower crustal composition. *Geochim. Cosmochim. Acta* **65**(15), 2589–2604.

Lombardo B. and Rolfo F. (2000) Two contrasting eclogite types in the Himalayas: implications for the Himalayan orogeny. *J. Geodynam.* **30**(1–2), 37–60.

Loock G., Seck H. A., and Stosch H.-G. (1990) Granulite facies lower crustal xenoliths from the Eifel, West Germany: petrological and geochemical aspects. *Contrib. Mineral. Petrol.* **105**, 25–41.

Lovering J. F. and White A. J. R. (1964) The significance of primary scapolite in granulitic inclusions from deep-seated pipes. *J. Petrol.* **5**, 195–218.

Lovering J. F. and White A. J. R. (1969) Granulitic and eclogitic inclusions from basic pipes at Delegate, Australia. *Contrib. Mineral. Petrol.* **21**, 9–52.

Lucassen F., Lewerenz S., Franz G., Viramonte J., and Mezger K. (1999) Metamorphism, isotopic ages and composition of lower crustal granulite xenoliths from the Cretaceous Salta Rift, Argentina. *Contrib. Mineral. Petrol.* **134**(4), 325–341.

Luosto U. and Korhonen H. (1986) Crustal structure of the baltic shield based on off-fennolora refraction data. *Tectonophysics* **128**, 183–208.

Luosto U., Flüh E. R., Lund C.-E., and Group W. (1989) The crustal structure along the POLAR profile from seismic refraction investigations. *Tectonophysics* **162**, 51–85.

Luosto U., Tiira T., Korhonen H., Azbel I., Burmin V., Buyanov A., Kosminskaya I., Ionkis V., and Sharov N. (1990) Crust and upper mantle structure along the DSS Baltic profile in SE Finland. *Geophys. J. Int.* **101**, 89–110.

Mackwell S. J., Zimmerman M. E., and Kohlstedt D. L. (1998) High-temperature deformation of dry diabase with application to tectonics on Venus. *J. Geophys. Res. B: Solid Earth* **103**, 975–984.

Markwick A. J. W. and Downes H. (2000) Lower crustal granulite xenoliths from the Arkhangelsk kimberlite pipes: petrological, geochemical and geophysical results. *Lithos* **51**(1–2), 135–151.

Markwick A. J. W., Downes H., and Veretennikov N. (2001) The lower crust of SE Belarus: petrological, geophysical, and geochemical constraints from xenoliths. *Tectonophysics* **339**(1–2), 215–237.

Martin H. (1994) The Archean grey gneisses and the genesis of continental crust. In *Archean Crustal Evolution* (ed. K. C. Condie). Elsevier, Amsterdam, pp. 205–259.

Mattie P. D., Condie K. C., Selverstone J., and Kyle P. R. (1997) Origin of the continental crust in the Colorado Plateau: geochemical evidence from mafic xenoliths from the Navajo volcanic field, southwestern USA. *Geochim. Cosmochim. Acta* **61**(10), 2007–2021.

Mayer A., Mezger K., and Sinigoi S. (2000) New Sm–Nd ages for the Ivrea-Verbano Zone, Sesia and Sessera valleys (Northern-Italy). *J. Geodynamics* **30**(1–2), 147–166.

McBirney A. R. and Aoki K.-I. (1973) Factors governing the stability of plagioclase at high pressures as shown by spinel-gabbro xenoliths from the Kerguelen archipelago. *Am. Mineral.* **58**, 271–276.

McCulloch M. T., Arculus R. J., Chappell B. W., and Ferguson J. (1982) Isotopic and geochemical studies of nodules in kimberlite have implications for the lower continental crust. *Nature* **300**, 166–169.

McDonough W. F. and Sun S.-S. (1995) Composition of the Earth. *Chem. Geol.* **120**, 223–253.

McDonough W. F., Rudnick R. L., and McCulloch M. T. (1991) The chemical and isotopic composition of the lower eastern Australian lithosphere: a review. In *The Nature of the Eastern Australian Lithosphere*, Geol. Soc. Austral. Spec. Publ. (ed. B. Drummond). Sydney, vol. 17, pp. 163–188.

McDonough W. F., Sun S.-S., Ringwood A. E., Jagoutz E., and Hofmann A. W. (1992) Potassium, rubidium, and cesium in the Earth and Moon and the evolution of the mantle of the Earth. *Geochim. Cosmochim. Acta* **56**, 1001–1012.

McLennan S. M. (2001a) Crustal heat production and the thermal evolution of Mars. *Geophys. Res. Lett.* **28**(21), 4019–4022.

McLennan S. M. (2001b) Relationships between the trace element composition of sedimentary rocks and upper continental crust. *Geochem. Geophys. Geosys.* **2** (article no. 2000GC000109).

McLennan S. M. and Taylor S. R. (1996) Heat flow and the chemical composition of continental crust. *J. Geol.* **104** 396–377.

McLennan S. M., Nance W. B., and Taylor S. R. (1980) Rare earth element-thorium correlations in sedimentary rocks, and the composition of the continental crust. *Geochim. Cosmochim. Acta* **44**, 1833–1839.

Mehnert K. R. (1975) The Ivrea zone: a model of the deep crust. *Neus Jahrb. Mineral. Abh.* **125**, 156–199.

Meltzer A. and Christensen N. (2001) Nanga Parbat crustal anisotrophy: implication for interpretation of crustal velocity structure and shear-wave splitting. *Geophys. Res. Lett.* **28**(10), 2129–2132.

Mengel K. (1990) Crustal xenoliths from Tertiary volcanics of the northern Hessian depression: petrological and chemical evolution. *Contrib. Mineral. Petrol.* **104**, 8–26.

Mengel K. and Wedepohl K. H. (1983) Crustal xenoliths in Tertiary volcanics from the northern Hessian depression. In *Plateau Uplift* (eds. K. Fuchs, *et al.*). Springer, Berlin, pp. 332–335.

Mengel K., Sachs P. M., Stosch H. G., Worner G., and Loock G. (1991) Crustal xenoliths from Cenozoic volcanic fields of West Germany implications for structure and composition of the continental crust. *Tectonophysics* **195**(2–4), 271.

Meyer H. O. A. and Brookins D. G. (1976) Sapphirine, sillimanite, and garnet in granulite xenoliths from Stockdale kimberlite, Kansas. *Am. Mineral.* **61**, 1194–1202.

Miller J. D. and Christensen N. I. (1994) Seismic signature and geochemistry of an island arc: a multidisciplinary study of the Kohistan accreted terrane, northern Pakistan. *J. Geophys. Res.* (B) **99**, 11623–11642.

Mittlefehldt D. W. (1984) Genesis of clinopyroxene-amphibole xenoliths from Birket Ram: trace element and petrologic constraints. *Contrib. Mineral. Petrol.* **88**, 280–287.

Mittlefehldt D. W. (1986) Petrology of high pressure clinopyroxenite series xenoliths, Mount Carmel, Israel. *Contrib. Mineral. Petrol.* **94**, 245–252.

Moecher D. P., Valley J. W., and Essene E. J. (1994) Extraction and carbon isotope analysis of CO_2 from scapolite in deep crustal granulites and xenoliths. *Geochim. Cosmochim. Acta* **58**(2), 959–967.

Mooney W. D. and Meissner R. (1992) Multi-genetic origin of crustal reflectivity: a review of seismic reflection profiling of the continental lower crust and Moho. In *Continental Lower Crust* (eds. D. M. Fountain, R. Arculus, and R. W. Kay). Elsevier, pp. 45–80.

Moser D. E. and Heaman L. M. (1997) Proterozoic zircon growth in Archean lower crust xenoliths, southern Superior craton: a consequence of Matachewan ocean opening. *Contrib. Mineral. Petrol.* **128**, 164–175.

Moser D. E., Flowers R. M., and Hart R. J. (2001) Birth of the Kaapvaal tectosphere 3.08 billion years ago. *Science* **291**(5503), 465–468.

Nasir S. (1992) The lithosphere beneath the northwestern part of the Arabian plate (Jordan)—evidence from xenoliths and geophysics. *Tectonophysics* **201**(3–4), 357–370.

Nasir S. (1995) Mafic lower crustal xenoliths from the northwestern part of the Arabian plate. *Euro. J. Mineral.* **7**(1), 217–230.

Nasir S. and Safarjalani A. (2000) Lithospheric petrology beneath the northern part of the Arabian plate in Syria: evidence from xenoliths in alkali basalts. *J. African Earth Sci.* **30**(1), 149–168.

Nesbitt H. W. and Young G. M. (1984) Prediction of some weathering trends of plutonic and volcanic rocks based on thermodynamic and kinetic considerations. *Geochim. Cosmochim. Acta* **48**, 1523–1534.

Newsom H. E., Sims K. W. W., Noll P. D., Jr., Jaeger W. L., Maehr S. A., and Beserra T. B. (1996) The depletion of tungsten in the bulk silicate Earth: constraints on core formation. *Geochim. Cosmochim. Acta* **60**, 1155–1169.

Nguuri T. K., Gore J., James D. E., Webb S. J., Wright C., Zengeni T. G., Gwavava O., and Snoke J. A. (2001) Crustal structure beneath southern Africa and its implications for the formation and evolution of the Kaapvaal and Zimbabwe cratons. *Geophys. Res. Lett.* **28**(13), 2501–2504.

Nimz G. J., Cameron K. L., Cameron M., and Morris S. L. (1986) Petrology of the lower crust and upper mantle beneath southeastern Chihuahua, Mexico. *Geofísica Int.* **25**, 85–116.

Niu F. L. and James D. E. (2002) Fine structure of the lowermost crust beneath the Kaapvaal craton and its implications for crustal formation and evolution. *Earth Planet. Sci. Lett.* **200**(1–2), 121–130.

Nyblade A. A. and Pollack H. N. (1993) A global analysis of heat flow from Precambrian terrains: implications for the thermal structure of Archean and Proterozoic lithosphere. *J. Geophys. Res.* **98**, 12207–12218.

Okrusch M., Schröder B., and Schnütgen A. (1979) Granulite-facies metabasite ejecta in the Laacher Sea area, Eifel, West Germany. *Lithos* **12**, 251–270.

O'Reilly S. Y., Griffin W. L., and Stabel A. (1988) Evolution of Phanerozoic eastern Australian lithosphere: isotopic evidence for magmatic and tectonic underplating. In *Oceanic and Continental Lithosphere: Similarities and Differences*, J. Petrol. Spec. Vol. (eds. M. A. Menzies and K. G. Cox). Oxford University Press, Oxford, pp. 89–108.

Owen J. V., Greenough J. D., Hy C., and Ruffman A. (1988) Xenoliths in a mafic dyke at Popes Harbour, Nova Scotia: implications for the basement to the Meguma Group. *Can. J. Earth Sci.* **25**, 1464–1471.

Padovani E. R. and Carter J. L. (1977) Aspects of the deep crustal evolution beneath south central New Mexico. In *The Earth's Crust* (ed. J. G. Heacock). American Geophysical Union, Washington, DC, pp. 19–55.

Padovani E. R. and Hart S. R. (1981) Geochemical constraints on the evolution of the lower crust beneath the Rio Grande rift. In *Conference on the Processes of Planetary Rifting*. Lunar and Planetary Science Institute, pp. 149–152.

Padovani E. R., Hall J., and Simmons G. (1982) Constraints on crustal hydration below the Colorado Plateau from V_p measurements on crustal xenoliths. *Tectonophysics* **84**, 313–328.

Pearcy L. G., DeBari S. M., and Sleep N. H. (1990) Mass balance calculations for two sections of island arc crust and implications for the formation of continents. *Earth Planet. Sci. Lett.* **96**, 427–442.

Pearson N. J., O'Reilly S. Y., and Griffin W. L. (1995) The crust–mantle boundary beneath cratons and craton margins: a transect across the south-west margin of the Kaapvaal craton. *Lithos* **36**(3–4), 257–287.

Percival J. A. and Card K. D. (1983) Archean crust as revealed in the Kapuskasing uplift, Superior province, Canada. *Geology* **11**, 323–326.

Percival J. A., Fountain D. M., and Salisbury M. H. (1992) Exposed cross sections as windows on the lower crust. In *Continental Lower Crust* (eds. D. M. Fountain, R. Arculus, and R. W. Kay). Elsevier, Amsterdam, pp. 317–362.

Peucker-Ehrenbrink B. and Jahn B.-M. (2001) Rhenium-osmium isotope systematics and platinum group element concentations: loess and the upper continental crust. *Geochem. Geophys. Geosys.* **2**, 2001GC000172.

Phillips R. J., Raubertas R. F., Arvidson R. E., Sarkar I. C., Herrick R. R., Izenberg N., and Grimm R. E. (1992) Impact craters and Venus resurfacing history. *J. Geophys. Res. Planet.* **97**(E10), 15923–15948.

Pinet C. and Jaupart C. (1987) The vertical distribution of radiogenic heat production in the Precambrian crust of Norway and Sweden: geothermal implications. *Geophys. Res. Lett.* **14**, 260–263.

Plank T. and Langmuir C. H. (1998) The chemical composition of subducting sediment and its consequences for the crust and mantle. *Chem. Geol.* **145**, 325–394.

Poldervaart A. (1955) The chemistry of the Earth's crust. *Geol. Soc. Am. Spec. Pap.* **62**, 119–144.

Quick J. E., Sinigoi S., and Mayer A. (1995) Emplacement of mantle peridotite in the lower continental crust, Ivrea-Verbano zone, northwest Italy. *Geology* **23**(8), 739–742.

Reid M. R., Hart S. R., Padovani E. R., and Wandless G. A. (1989) Contribution of metapelitic sediments to the composition, heat production, and seismic velocity of the lower crust of southern New Mexico. *Earth Planet. Sci. Lett.* **95**, 367–381.

Roberts S. and Ruiz J. (1989) Geochemical zonation and evolution of the lower crust in Mexico. *J. Geophys. Res.* **94**, 7961–7974.

Rogers N. W. (1977) Granulite xenoliths from Lesotho kimberlites and the lower continental crust. *Nature* **270**, 681–684.

Rogers N. W. and Hawkesworth C. J. (1982) Proterozoic age and cumulate origin for granulite xenoliths, Lesotho. *Nature* **299**, 409–413.

Ronov A. B. and Yaroshevsky A. A. (1967) Chemical structure of the Earth's crust. *Geokhimiya* **11**, 1285–1309.

Ronov A. B. and Yaroshevsky A. A. (1976) A new model for the chemical structure of the Earth's crust. *Geokhimiya* **12**, 1761–1795.

Ross D. C. (1985) Mafic gneissic complex (batholithic root?) in the southernmost Sierra Nevada, California. *Geology* **13**, 288–291.

Rudnick R. L. (1990a) Continental crust: growing from below. *Nature* **347**, 711–712.

Rudnick R. L. (1990b) Nd and Sr isotopic compositions of lower crustal xenoliths from North Queensland, Australia: implications for Nd model ages and crustal growth processes. *Chem. Geol.* **83**, 195–208.

Rudnick R. L. (1992) Xenoliths—samples of the lower continental crust. In *Continental Lower Crust* (eds. D. M. Fountain, R. Arculus, and R. W. Kay). Elsevier, Amsterdam, pp. 269–316.

Rudnick R. L. (1995) Making continental crust. *Nature* **378**, 571–578.

Rudnick R. L. and Cameron K. L. (1991) Age diversity of the deep crust in northern Mexico. *Geology* **19**, 1197–1200.

Rudnick R. L. and Goldstein S. L. (1990) The Pb isotopic compositions of lower crustal xenoliths and the evolution of lower crustal Pb. *Earth Planet. Sci. Lett.* **98**, 192–207.

Rudnick R. L. and Fountain D. M. (1995) Nature and composition of the continental crust: a lower crustal perspective. *Rev. Geophys.* **33**(3), 267–309.

Rudnick R. L. and Presper T. (1990) Geochemistry of intermediate to high-pressure granulites. In *Granulites and Crustal Evolution* (eds. D. Vielzeuf and P. Vidal). Kluwer, Amsterdam, pp. 523–550.

Rudnick R. L. and Taylor S. R. (1987) The composition and petrogenesis of the lower crust: a xenolith study. *J. Geophys. Res.* **92**(B13), 13981–14005.

Rudnick R. L. and Taylor S. R. (1991) Petrology and geochemistry of lower crustal xenoliths from northern Queensland and inferences on lower crustal composition. In *The Eastern Australian Lithosphere*, Geol. Soc. Austral. Spec. Publ. (ed. B. Drummond), 189–208.

Rudnick R. L. and Williams I. S. (1987) Dating the lower crust by ion microprobe. *Earth Planet. Sci. Lett.* **85**, 145–161.

Rudnick R. L., McLennan S. M., and Taylor S. R. (1985) Large ion lithophile elements in rocks from high-pressure granulite facies terrains. *Geochim. Cosmochim. Acta* **49**, 1645–1655.

Rudnick R. L., McDonough W. F., McCulloch M. T., and Taylor S. R. (1986) Lower crustal xenoliths from Queensland, Australia: evidence for deep crustal assimilation and fractionation of continental basalts. *Geochim. Cosmochim. Acta* **50**, 1099–1115.

Rudnick R. L., McDonough W. F., and O'Connell R. J. (1998) Thermal structure, thickness and composition of continental lithosphere. *Chem. Geol.* **145**, 399–415.

Rudnick R. L., Ireland T. R., Gehrels G., Irving A. J., Chesley J. T., and Hanchar J. M. (1999) Dating mantle metasomatism: U–Pb geochronology of zircons in cratonic mantle xenoliths from Montana and Tanzania. In *Proceedings of the VIIth International Kimberlite Conference* (eds. J. J. Gurney, J. L. Gurney, M. D. Pascoe, and S. R. Richardson). Red Roof Design, Cape Town, pp. 728–735.

Rudnick R. L., Barth M., Horn I., and McDonough W. F. (2000) Rutile-bearing refractory eclogites: missing link between continents and depleted mantle. *Science* **287**, 278–281.

Ruiz J., Patchett P. J., and Arculus R. J. (1988a) Nd–Sr isotope composition of lower crustal xenoliths—evidence for the origin of mid-Tertiary felsic volcanics in Mexico. *Contrib. Mineral. Petrol.* **99**, 36–43.

Ruiz J., Patchett P. J., and Ortega-Gutierrez F. (1988b) Proterozoic and Phanerozoic basement terranes of Mexico from Nd isotopic studies. *Geol. Soc. Am. Bull.* **100**, 274–281.

Rutter M. J. (1987) The nature of the lithosphere beneath the Sardinian continental block: mantle and deep crustal inclusions in mafic alkaline lavas. *Lithos* **20**, 225–234.

Saal A. E., Rudnick R. L., Ravizza G. E., and Hart S. R. (1998) Re–Os isotope evidence for the composition, formation and age of the lower continental crust. *Nature* **393**, 58–61.

Sachs P. M. and Hansteen T. H. (2000) Pleistocene underplating and metasomatism of the lower continental crust: a xenolith study. *J. Petrol.* **41**(3), 331–356.

Saleeby J. B. (1990) Progress in tectonic and petrogenetic studies in an exposed cross-section of young (c. 100 Ma) continental crust southern Sierra Nevada, California. In *Exposed Cross-sections of the Continental Crust* (eds. M. H. Salisbury, and D. M. Fountain). Kluwer Academic, Norwell, MA, pp. 137–159.

Schaaf P., Heinrich W., and Besch T. (1994) Composition and Sm–Nd isotopic data of the lower crust beneath San-Luis-Potosi, Central Mexico—evidence from a granulite-facies xenolith suite. *Chem. Geol.* **118**(1–4), 63–84.

Schaber G. G., Strom R. G., Moore H. J., Soderblom L. A., Kirk R. L., Chadwick D. J., Dawson D. D., Gaddis L. R., Boyce J. M., and Russell J. (1992) Geology and distribution of impact craters on Venus-what are they telling us. *J. Geophys. Res. Planet.* **97**(E8), 13257–13301.

Scherer E. K., Cameron K. L., Johnson C. M., Beard B. L., Barovich K. M., and Collerson K. D. (1997) Lu–Hf geochronology applied to dating Cenozoic events affecting lower crustal xenoliths from Kilbourne Hole, New Mexico. *Chem. Geol.* **142**, 63–78.

Schmitz M. D. and Bowring S. A. (2000) The significance of U–Pb zircon dates in lower crustal xenoliths from the southwestern margin of the Kaapvaal craton, southern Africa. *Chem. Geol.* **172**, 59–76.

Schmitz M. D. and Bowring S. A. (2003a) Constraints on the thermal evolution of continental lithosphere from U–Pb accessory mineral thermochronometry of lower crustal xenoliths, southern Africa. *Contrib. Mineral. Petrol.* **144**, 592–618.

Schmitz M. D. and Bowring S. A. (2003b) Ultrahigh-temperature metamorphism in the lower crust during Neoarchean Ventersdorp rifting and magmatism, Kaapvaal craton, southern Africa. *Geol. Soc. Am. Bull.* **115**, 533–548.

Schulze D. J. and Helmstaedt H. (1979) Garnet pyroxenite and eclogite xenoliths from the Sullivan Buttes latite, Chino valley, Arizona. In *The Mantle Sample: Inclusions in Kimberlites and Other Volcanics* (eds. F. R. Boyd and H. O. A. Meyer). American Geophysics Union, Washington, DC, pp. 318–329.

Sclater J. G., Jaupart C. J., and Galson D. (1980) The heat flow through oceanic and continental crust and the heat loss of the earth. *Rev. Geophys. Space Phys.* **18**, 269–311.

Selverstone J. and Stern C. R. (1983) Petrochemistry and recrystallization history of granulite xenoliths from the Pali-Aike volcanic field, Chile. *Am. Mineral.* **68**, 1102–1111.

Sewell R. J., Hobden B. J., and Weaver S. D. (1993) Mafic and ultramafic mantle and deep-crustal xenoliths from Banks Peninsula, South-Island, New-Zealand. *NZ J. Geol. Geophys.* **36**(2), 223–231.

Shatsky V., Rudnick R. L., and Jagoutz E. (1990) Mafic granulites from Udachnaya pipe, Yakutia: samples of Archean lower crust? *Deep Seated Magmatism and Evolution of Lithosphere of the Siberian Platform*, 23–24.

Shatsky V. S., Sobolev N. V., and Pavlyuchenko V. S. (1983) Fassaïte-garnet-anorthite xenolith from the Udachnaya kimberlite pipe, Yakutia. *Dokl. Akad. Nauk. SSSR* **272**(1), 188–192.

Shaw D. M., Reilly G. A., Muysson J. R., Pattenden G. E., and Campbell F. E. (1967) An estimate of the chemical composition of the Canadian Precambrian shield. *Can. J. Earth Sci.* **4**, 829–853.

Shaw D. M., Dostal J., and Keays R. R. (1976) Additional estimates of continental surface Precambrian shield composition in Canada. *Geochim. Cosmochim. Acta* **40**, 73–83.

Shaw D. M., Cramer J. J., Higgins M. D., and Truscott M. G. (1986) Composition of the Canadian Precambrian shield and the continental crust of the Earth. In *The Nature of the Lower Continental Crust* (eds. J. B. Dawson, D. A. Carswell, J. Hall, and K. H. Wedepohl). Geol. Soc. London, London, vol. 24, pp. 257–282.

Shaw D. M., Dickin A. P., Li H., McNutt R. H., Schwarcz H. P., and Truscott M. G. (1994) Crustal geochemistry in the Wawa-Foleyet region, Ontario. *Can. J. Earth Sci.* **31**(7), 1104–1121.

Sims K. W. W., Newsom H. E., and Gladney E. S. (1990) Chemical fractionation during formation of the Earth's core and continental crust: clues from As, Sb, W, and M., and In *Origin of the Earth* (eds. H. E. Newsom, J. H. Jones, and J. H. Newson). Oxford University Press, Oxford, pp. 291–317.

Smith D. E., Zuber M. T., Solomon S. C., Phillips R. J., Head J. W., Garvin J. B., Banerdt W. B., Muhleman D. O., Pettengill G. H., Neumann G. A., Lemoine F. G., Abshire J. B., Aharonson O., Brown C. D., Hauck S. A., Ivanov A. B., McGovern P. J., Zwally H. J., and Duxbury T. C. (1999) The global topography of Mars and implications for surface evolution. *Science* **284**(5419), 1495–1503.

Smith R. D., Cameron K. L., McDowell F. W., Niemeyer S., and Sampson D. E. (1996) Generation of voluminous silicic magmas and formation of mid-Cenozoic crust beneath north-central Mexico: evidence from ignimbrites, associated lavas, deep crustal granulites, and mantle pyroxenites. *Contrib. Mineral. Petrol.* **123**, 375–389.

Smithson S. B. (1978) Modeling continental crust-structural and chemical constraints. *Geophys. Res. Lett.* **5**(9), 749–752.

Stolz A. J. (1987) Fluid activity in the lower crust and upper mantle: mineralogical evidence bearing on the origin of amphibole and scapolite in ultramafic and mafic granulite xenoliths. *Min. Mag.* **51**, 719–732.

Stolz A. J. and Davies G. R. (1989) Metasomatized lower crustal and upper mantle xenoliths from north Queensland: chemical and isotopic evidence bearing on the composition and source of the fluid phase. *Geochim. Cosmochim. Acta* **53**, 649–660.

Stosch H.-G. and Lugmair G. W. (1984) Evolution of the lower continental crust: granulite facies xenoliths from the Eifel, West Germany. *Nature* **311**, 368–370.

Stosch H.-G., Ionov D. A., Puchtel I. S., Galer S. J. G., and Sharpouri A. (1995) Lower crustal xenoliths from Mongolia and their bearing on the nature of the deep crust beneath cental Asia. *Lithos* **36**, 227–242.

Strom R. G., Schaber G. G., and Dawson D. D. (1994) The global resurfacing of Venus. *J. Geophys. Res. Planet.* **99**(E5), 10899–10926.

Sun W., Bennett V. C., Eggins S. M., Kamenetsky V. S., and Arculus R. J. (2003) Evidence for enhanced mantle to crust rhenium transfer from undegassed arc magmas. *Nature* **422**, 294–297.

Sutherland F. L. and Hollis J. D. (1982) Mantle–lower crust petrology from inclusions in basaltic rocks in eastern Australia—an outline. *J. Volcanol. Geotherm. Res.* **14**, 1–29.

Tanaka T. and Aoki K.-I. (1981) Petrogenetic implications of REE and Ba data on mafic and ultramafic inclusions from Itinome-gata, Japan. *J. Geol.* **89**, 369–390.

Taylor S. R. (1964) Abundance of chemical elements in the continental crust—a new table. *Geochim. Cosmochim. Acta* **28**, 1273–1285.

Taylor S. R. (1967) The origin and growth of continents. *Tectonophysics* **4**, 17–34.

Taylor S. R. (1977) Island arc models and the composition of the continental crust. In *Island Arcs, Deep Sea Trenches and Back-Arc Basins* (ed. M. Talwani). American Geophysical Union, Washington, DC, pp. 325–336.

Taylor S. R. (1992) *Solar System Evolution*. Cambridge University Press, Cambridge.

Taylor S. R. and McLennan S. M. (1981) The composition and evolution of the continental crust: rare Earth element evidence from sedimentary rocks. *Phil. Trans. Roy. Soc. London* **A301**, 381–399.

Taylor S. R. and McLennan S. M. (1985) *The Continental Crust: Its Composition and Evolution*. Blackwell, Oxford.

Taylor S. R., and McLennan S. M. (1995) The geochemical evolution of the continental crust. *Rev. Geophys.* **33**, 241–265.

Taylor S. R., McLennan S. M., and McCulloch M. T. (1983) Geochemistry of loess, continental crustal composition and crustal model ages. *Geochim. Cosmochim. Acta* **47**, 1897–1905.

Teng F., McDonough W. F., Rudnick R. L., Dalpé C., Tomascak P. B., Chappell B. W., and Gao S. (2003) Lithium isotopic composition and concentration of the upper continental crust. *Geochim. Cosmochim. Acta* (submitted).

Thomas C. W. and Nixon P. H. (1987) Lower crustal granulite xenoliths in carbonatite volcanoes of the western rift of East Africa. *Min. Mag.* **51**, 621–633.

Toft P. B., Hills D. V., and Haggerty S. E. (1989) Crustal evolution and the granulite to eclogite transition in xenoliths from kimberlites in the West African craton. *Tectonophysics* **161**, 213–231.

Tredoux M., Hart R. J., Carlson R. W., and Shirey S. B. (1999) Ultramafic rocks at the center of the Vredefort structure:

further evidence for the crust on edge model. *Geology* **27**(10), 923–926.

Trzcienski W. E. and Marchildon N. (1989) Kyanite-garnet-bearing Cambrian rocks and Grenville granulites from the Ayers Cliff, Quebec, Canada, Lamprophyre Dike Suite—deep crustal fragments from the northern appalachians. *Geology* **17**(7), 637–640.

Upton B. G. J., Aspen P., and Chapman N. A. (1983) The upper mantle and deep crust beneath the British Isles: evidence from inclusion suites in volcanic rocks. *J. Geol. Soc. London* **140**, 105–122.

Upton B. G. J., Aspen P., Rex D. C., Melcher F., and Kinny P. (1998) Lower crustal and possible shallow mantle samples from beneath the Hebrides: evidence from a xenolithic dyke at Gribun, western Mull. *J. Geol. Soc.* **155**, 813–828.

Upton B. G. J., Aspen P., and Hinton R. W. (2001) Pyroxenite and granulite xenoliths from beneath the Scottish northern Highlands terrane: evidence for lower-crust/upper-mantle relationships. *Contrib. Mineral. Petrol.* **142**(2), 178–197.

Urrutia-Fucugauchi J. and Uribe-Cifuentes R. M. (1999) Lower-crustal xenoliths from the Valle de Santiago maar field, Michoacan-Guanajuato volcanic field, central Mexico. *Int. Geol. Rev.* **41**(12), 1067–1081.

van Breeman O. and Hawkesworth C. J. (1980) Sm–Nd isotopic study of garnets and their metamorphic host rocks. *Trans. Roy. Soc. Edinburgh* **71**, 97–102.

van Calsteren P. W. C., Harris N. B. W., Hawkesworth C. J., Menzies M. A., and Rogers N. W. (1986) Xenoliths from southern Africa: a perspective on the lower crust. In *The Nature of the Lower Continental Crust*, Geol. Soc. London Spec. Publ. (eds. J. B. Dawson, D. A. Carswell, J. Hall, and K. H. Wedepohl). London, vol. 25, pp. 351–362.

Vidal P. and Postaire B. (1985) Étude par la méthode Pb–Pb de roches de haut grade métamorphique impliquées dans la chaîne Hercynienne. *Chem. Geol.* **49**, 429–449.

Vielzeuf D. (1983) The spinel and quartz associations in high grade xenoliths from Tallante (S. E. Spain) and their potential use in geothermometry and barometry. *Contrib. Mineral. Petrol.* **82**, 301–311.

Villaseca C., Downes H., Pin C., and Barbero L. (1999) Nature and composition of the lower continental crust in central Spain and the granulite-granite linkage: inferences from granulitic xenoliths. *J. Petrol.* **40**(10), 1465–1496.

Voshage H., Hofmann A. W., Mazzucchelli M., Rivalenti G., Sinigoi S., Raczek I., and Demarchi G. (1990) Isotopic evidence from the Ivrea zone for hybrid lower crust formed by magmatic underplating. *Nature* **347**, 731–736.

Wanke H., Bruckner J., Dreibus G., Rieder R., and Ryabchikov I. (2001) Chemical composition of rocks and soils at the pathfinder site. *Space Sci. Rev.* **96**(1–4), 317–330.

Warren R. G., Kudo A. M., and Keil K. (1979) Geochemistry of lithic and single-crystal inclusions in basalts and a characterization of the upper mantle-lower crust in the Engle Basin, Rio Grande Rift, New Mexico. In *Rio Grande Rift: Tectonics and Magmatism* (ed. R. E. Riecker). American Geophysical Union, Washington, DC, pp. 393–415.

Wass S. Y. and Hollis J. D. (1983) Crustal growth in south-eastern Australia—evidence from lower crustal eclogitic and granulitic xenoliths. *J. Metamorph. Geol.* **1**, 25–45.

Weaver B. L. and Tarney J. (1980) Continental crust composition and nature of the lower crust: constraints from mantle Nd–Sr isotope correlation. *Nature* **286**, 342–346.

Weaver B. L. and Tarney J. (1981) Lewisian gneiss geochemistry and Archaean crustal development models. *Earth Planet. Sci. Lett.* **55**, 171–180.

Weaver B. L. and Tarney J. (1984) Empirical approach to estimating the composition of the continental crust. *Nature* **310**, 575–577.

Weber M. B. I., Tarney J., Kempton P. D., and Kent R. W. (2002) Crustal make-up of the northern Andes: evidence based on deep crustal xenolith suites, Mercaderes, SW Colombia. *Tectonophysics* **345**(1–4), 49–82.

Wedepohl H. (1995) The composition of the continental crust. *Geochim. Cosmochim. Acta* **59**, 1217–1239.

Wedepohl K. H. (1969–1978) *Handbook of Geochemistry*. Springer, Berlin.

Wendlandt E., DePaolo D. J., and Baldridge W. S. (1993) Nd and Sr isotope chronostratigraphy of Colorado Plateau lithosphere: implications for magmatic and tectonic underplating of the continental crust. *Earth Planet. Sci. Lett.* **116**, 23–43.

Wendlandt E., DePaolo D. J., and Baldridge W. S. (1996) Thermal history of Colorado Plateau lithosphere from Sm–Nd mineral geochronology of xenoliths. *Geol. Soc. Am. Bull.* **108**(7), 757–767.

Wernicke B., Clayton R., Ducea M., Jones C. H., Park S., Ruppert S., Saleeby J., Snow J. K., Squires L., Fliedner M., Jiracek G., Keller R., Klemperer S., Luetgert J., Malin P., Miller K., Mooney W., Oliver H., and Phinney R. (1996) Origin of high mountains in the continents: the southern Sierra Nevada. *Science* **271**, 190–193.

Wilde S. A., Valley J. W., Peck W. H., and Graham C. M. (2001) Evidence from detrital zircons for the existence of continental crust and oceans on the Earth 4.4 Gyr ago. *Nature* **409**(6817), 175–178.

Wilkinson J. F. G. (1975) An Al-spinel ultramafic-mafic inclusion suite and high pressure megacrysts in an analcimite and their bearing on basaltic magma fractionation at elevated pressures. *Contrib. Mineral. Petrol.* **53**, 71–104.

Wilkinson J. F. G. and Taylor S. R. (1980) Trace element fractionation trends of thoeiitic magma at moderate pressure: evidence from an Al-spinel ultramafic-mafic inclusion suite. *Contrib. Mineral. Petrol.* **75**, 225–233.

Wilshire H. W., Meyer C. E., Nakata J. K., Calk L. C., Shervais J. W., Nielson J. E., and Schwarzman E. C. (1988) *Mafic and Ultramafic Xenoliths from Volcanic Rocks of the Western United States*. Prof. Paper, USGS, Washington, DC, West Sussex, UK.

Windley B. F. (1995) *The Evolving Continents*. Wiley.

Wörner G., Schmincke H.-U., and Schreyer W. (1982) Crustal xenoliths from the Quaternary Wehr volcano (East Eifel). *Neus. Jahrb. Mineral. Abh.* **144**(1), 29–55.

Yardley B. W. D. (1986) Is there water in the deep continental crust? *Nature* **323**, 111.

Yu. J. H., O'Reilly S. Y., Griffin W. L., Xu X. S., Zhang M., and Zhou X. M. (2003) The thermal state and composition of the lithospheric mantle beneath the Leizhou Peninsula South China. *J. Volcanol. Geotherm. Res.* **122**(3–4), 165–189.

Zashu S., Kaneoka I., and Aoki K.-I. (1980) Sr isotope study of mafic and ultramafic inclusions from Itinome-gata, Japan. *Geochem. J.* **14**, 123–128.

Zheng J. P., Sun M., Lu F. X., and Pearson N. (2003) Mesozoic lower crustal xenoliths and their significance in lithospheric evolution beneath the Sino-Korean craton. *Tectonophysics* **361**(1–2), 37–60.

Zhou X. H., Sun M., Zhang G. H., and Chen S. H. (2002) Continental crust and lithospheric mantle interaction beneath North China: isotopic evidence from granulite xenoliths in Hannuoba, Sino-Korean craton. *Lithos* **62**(3–4), 111–124.

3.02
Constraints on Crustal Heat Production from Heat Flow Data

C. Jaupart

Institut de Physique du Globe de Paris, France

and

J.-C. Mareschal

GEOTOP Université du Québec à Montréal, Canada

NOMENCLATURE

A	Radiogenic heat production per unit volume (μW m^{-3})
D	Slope of linear relation between heat flow density and heat production (km)
h_C	Crustal thickness (km)
h_L	Lithosphere thickness (km)
H	Radiogenic heat production per unit mass (W kg^{-1})
k	Thermal conductivity (W m^{-1} K^{-1})
k_x, k_y, k_z	Wave numbers in (x, y, z) directions (km^{-1})
[K]	Potassium concentration (%)
pW	Picowatts (10^{-12} W)
P_A	Energy spectrum of heat production distribution ((mW m^{-2})^2 m^2)
P_q	Energy spectrum of heat flow distribution ((μW m^{-3})^2 m^2)
Q	Heat flow density at Earth's surface (mW m^{-2})
Q_B	Heat flux at the base of the continental lithosphere (mW m^{-2})
Q_C	Amount of heat produced in Earth's crust (mW m^{-2})
Q_L	Amount of heat produced in mantle part of continental lithosphere (mW m^{-2})
Q_T	Transient component of heat flow due to thermal relaxation (mW m^{-2})
T_B	Temperature at base of lithosphere (°C)
T_M	Temperature at base of crust (°C)
[Th]	Thorium concentration (ppm)
TW	Terrawatts (10^{12} W)
[U]	Uranium concentration (ppm)
β	Slope of energy spectrum
Δ	Thickness of crustal layer (km)
λ	Wavelength (km)

3.02.1 INTRODUCTION

The continental crust is an important repository of highly incompatible elements such as uranium and thorium. Exactly how much it contains is a key issue for the thermal regime of continents and for understanding how the Earth's mantle has evolved through geological time due to crust extraction. Recent estimates of the average uranium, thorium, and potassium concentrations in the continental crust vary by almost a factor of 2 (Wedepohl, 1995; Rudnick and Fountain, 1995; Taylor and McLennan, 1995; see also Chapter 3.01). These estimates are based on different assumptions regarding crustal structure and rely on different types of crustal samples, ranging from xenoliths to shales. They require an extrapolation in scale from tiny specimens to the whole crust of a geological province. Uranium and thorium tend to be located in accessory minerals and on grain boundaries, which are not related simply to bulk chemical composition. Thus, their concentrations vary on the scale of a petrological thin section, a hand sample, an outcrop, and a whole massif. In a geological province, abundant rocks such as gneisses and metasedimentary rocks are usually under-studied because of their complex origin and metamorphic history. A final difficulty is to evaluate the composition of intermediate and lower crustal levels, which are as heterogeneous as the shallow ones (e.g., Fountain and Salisbury, 1981; Clowes *et al.*, 1992).

Independent estimates of the amount of uranium and thorium in the continental crust can be obtained from heat flow data. The energy produced by the decay of these radioactive elements accounts for a large fraction of the heat flow at the surface of continents (Birch, 1954; Wasserburg *et al.*, 1964; Clark and Ringwood, 1964; Sclater *et al.*, 1980; Taylor and McLennan, 1995). This may be the only case where geophysical data bear directly on geochemical budgets. Since the mid-1970s, there has been much progress in our understanding of continental heat flow. The relationship between variations in heat flow and crustal heat production has been investigated systematically (England *et al.*, 1980; Jaupart, 1983a; Vasseur and Singh, 1986; Ketcham, 1996; Jaupart and Mareschal, 1999). Heat flow determinations on continents have been multiplied by almost a factor of 10 between the compilations by Jessop *et al.* (1976) and Pollack *et al.* (1993). Since the last compilation, a large number of high-quality data have been obtained for the poorly studied Precambrian Shield areas of Canada and India (Mareschal *et al.*, 2000a,b; Roy and Rao, 2000; Rolandone *et al.*, 2002; Lewis *et al.*, 2003).

3.02.2 ESTIMATES OF BULK CRUSTAL HEAT PRODUCTION

3.02.2.1 Heat Production Rate due to Uranium, Thorium, and Potassium

For ease of comparison with heat flow data, we shall not discuss in detail the concentration of uranium, thorium, and potassium in crustal rocks and shall refer instead to the heat production rate. Table 1 lists the key characteristics of the main radiogenic isotopes ^{238}U, ^{235}U, ^{232}Th, and ^{40}K as well as their heat production rates (from Rybach, 1988). Using present-day isotopic ratios, the bulk heat production of a rock sample is calculated (in W kg^{-1}) by summing the contributions of each element as follows:

$$H = 10^{-11}\,(9.52[\mathrm{U}] + 2.56[\mathrm{Th}] + 3.48[\mathrm{K}]) \quad (1)$$

Table 1 Heat production constants.

Isotope/element	*Natural abundance* (%)	*Half-life* (yr)	*Energy per atom* ($\times 10^{-12}$ J)	*Heat production per unit mass of isotope/element* (W kg^{-1})
^{238}U	99.27	4.46×10^9	7.41	9.17×10^{-5}
^{235}U	0.72	7.04×10^8	7.24	5.75×10^{-4}
U				9.52×10^{-5}
^{232}Th	100	1.40×10^{10}	6.24	2.56×10^{-5}
Th				2.56×10^{-5}
$^{40}K^a$	0.0117	1.26×10^9	0.114	2.97×10^{-5}
K				3.48×10^{-9}

[a] β and γ decay only.

Table 2 Some estimates of bulk continental crust heat production.

$\langle A \rangle$ (μW m^{-3})	$\langle Q_C \rangle^a$ (mW m^2)	*References*
0.74–0.86	30–34	Allègre *et al.* (1988) and O'Nions *et al.* (1979)
0.83	33	Furukawa and Shinjoe (1997)
0.92	38	Weaver and Tarney (1984)
0.58	24	Taylor and McLennan (1985)
1.31	54	Shaw *et al.* (1986)
1.25	51	Wedepohl (1995)
0.93	37	Rudnick and Fountain (1995)
0.70	29	McLennan and Taylor (1996)
0.55–0.68	21–26	Gupta *et al.* (1991)
0.94	39	Nicolaysen *et al.* (1981) and Jones (1988)
0.84–1.15	34–47	Gao *et al.* (1998)
0.70	28	Jaupart *et al.* (1998)

[a] Crustal component of heat flow for a 41 km thick crust.

where [U] and [Th] are the uranium and thorium concentrations in ppm and [K] the potassium concentration in percent. The heat production per unit volume, A, is given by $A = \rho H$, where ρ is the sample density. For an average crustal density of 2,700 kg m^{-3}, one has:

$$A = 0.257[\mathrm{U}] + 0.069[\mathrm{Th}] + 0.094[\mathrm{K}] \quad (2)$$

where A is given (in μW m^{-3}). For typical crustal rocks, potassium accounts for less than 20% of the total, and hence heat production may serve as a proxy for uranium and thorium. Using average [Th]/[U] = 4 and [K]/[U] = 1.2 ratios, we obtain an estimate of the average [U] or [Th] values from the heat production value:

$$[\mathrm{U}] \approx 1.5A, \qquad [\mathrm{Th}] \approx 6A \quad (3)$$

Table 2 lists several estimates of the average heat production in the continental crust. It illustrates that a wide range of heat production models are compatible with the different data sets.

We review how these various estimates have been obtained and emphasize the key assumptions and address some pitfalls. To avoid repeating what is covered in other chapters (see Chapter 3.01), we shall not present a comprehensive summary but shall restrict our discussion to the main sources of uncertainty in standard estimates of crustal heat production. We shall then discuss the advantage of using heat flow data to determine crustal heat production.

3.02.2.2 Geochemical and Petrological Estimates of Crustal Heat Production

Bulk estimates have relied on the hypothesis that continental crust is extracted from the mantle by processes operating in well-defined environments and leading to specific magma types.

3.02.2.2.1 Crust/mantle chemical budgets

Bulk continental crust deduced from global crust/mantle chemical budgets provides a useful reference. Various studies lead to a small range of 0.74–0.86 μW m^{-3} for the average rate of crustal heat production (O'Nions *et al.*, 1979; Allègre *et al.*, 1983, 1988; Galer *et al.*, 1989). These estimates of crustal heat production are obtained by redistributing the heat producing elements in the bulk silicate earth (see Chapter 2.01) between the continental crust and various reservoirs in the mantle. They require assumptions regarding the structure of the convecting mantle, the composition and homogeneity of the different reservoirs, as well as the composition of the bulk silicate earth.

3.02.2.2.2 Genetic crustal models

Another approach is to follow a genetic model for continental crust formation, for example in island arcs, oceanic plateaus or Archean granite-greenstone terrains. For example, the full vertical crustal sequence of a mature island-arc has been reconstructed in the Hidaka belt in Hokkaido, Japan over a total thickness of 30 km. This arc contains high-grade metamorphic rocks representative of lower crustal material and its average rate of crustal heat generation is 0.83 $\mu W\ m^{-3}$ (Furukawa and Uyeda, 1989; Furukawa and Shinjoe, 1997).

This approach assumes that the continental crust is a closed system. Following extraction from the mantle and internal differentiation, continental crust is continuously modified by intrusions of mafic melts, thrusting and deformation in collision belts as well as basin formation in extension zones. From a geochemical standpoint, the genesis of average felsic continental crustal material requires several events, including magmatic episodes to produce chemically evolved granitic melts and the elimination of depleted residue (Rudnick, 1995). The latter process probably involves some form of delamination or gravitational instability at the base of the crust (Kay and Mahlburg Kay, 1991; Jull and Kelemen, 2001). Thus, it is not clear that a single magma type, such as andesite, is representative of proto-continental crust.

3.02.2.3 Heat Production Measurements

Various attempts have been made to use the vast quantity of chemical and geophysical data available for estimating crustal heat production.

3.02.2.3.1 Global compilations

Bulk crustal composition has been estimated from a set of representative rock types and their proportions in crustal columns derived from geophysical profiles (Haack, 1983; Condie, 1993; Wedepohl, 1995; Borodin, 1999). One difficulty with this method is that there is no straightforward relationship between uranium and thorium concentrations and bulk physical properties such as seismic velocity and density (Fountain, 1985; Kukkonen and Peltoniemi, 1998). This is because uranium and thorium are incompatible trace elements that tend to be located in accessory minerals and at grain boundaries, i.e., sites that depend weakly on major element composition. As a consequence, one may not use a crustal cross-section derived from geophysical data in a geological province to calculate crustal heat production without a knowledge of radioelement concentrations in the local rocks. Furthermore, it is now clear that there is no such thing as a "typical" crustal column that is valid worldwide (Mooney *et al.*, 1998).

3.02.2.3.2 Regional studies

Large-scale heat production data sets have been collected over wide regions in several provinces (Table 3). Such surface sampling leads to a bias towards upper crustal rocks. Some of the data sets were compiled for petrological studies and have gone unnoticed by many geoscientists, including the majority of the heat flow community. For this reason, we show in Figure 1 some histograms of heat production in Precambrian provinces, which emphasize the very wide range of these heat production distributions. In the New Quebec region of the Archean Superior Province, the average heat production was obtained by systematic sampling on a regular grid, which leads to truly unbiased average concentrations (Eade and Fahrig, 1971). In other cases, the various lithologies present at the surface have been carefully mapped and their proportions determined. Average concentrations and heat production have been calculated using a weighted average of values for all of the rock types (Ashwal *et al.*, 1987; Fountain *et al.*, 1987; Shaw *et al.*, 1994; Gao *et al.*, 1998). Table 3 lists estimates for three different regions of the Archean Superior Province, Canada, as well as for several other provinces of the same age. From these data, it is clear that the local heat production average varies over large distances within a single geological province and is not a function of age.

The estimates of Table 3 are heavily biased towards the upper crust and are systematically higher than estimates for the whole crust (Table 1). For a bulk crustal estimate, samples need to be incorporated in a generic crustal cross-section. Rocks may be ranked according to metamorphic grade and restored in their original crustal position as a function of pressure and structural level. Data on deep crustal material have recently been added in the Canadian Shield (Ashwal *et al.*, 1987; Fountain *et al.*, 1987;

Table 3 Average surface heat production for different geological provinces obtained by systematic regional sampling over large areas.

	$\langle A\rangle^{a}$ (μW m^{-3})	$\sigma_A{}^{b}$	N^{c}	*References*
Yilgarn (Archean, Western Australia)	3.3	3.3	540	Heier and Lambert (1978)
Superior (Archean, Canada)				
New Québec region	1.22	d	3,085	Eade and Fahrig (1971)
Wawa Gneiss Terrane	1.01	e	56	Shaw *et al.* (1994)
Sachigo subprovince	1.04	e	20	Fountain *et al.* (1987)
"Churchill" (Archean to Proterozoic, Canada)	2.0	d	1,510	Eade and Fahrig (1971)
Gawler (Proterozoic, Central Australia)	3.6	3.4	90	Heier and Lambert (1978)
Baltic Shield (Proterozoic, Finland + Russia)	1.2	1.2	284	f
Central East China (Archean to Neogene, China)	1.22[g]	d	11,451	Gao *et al.* (1998)

[a] Mean heat production. [b] Standard deviation on the heat production distribution. [c] Number of sites. [d] Analyses were made on mixed powders, implying that the standard deviation of the analyses underestimates the true spread of values for individual rock samples. [e] Average calculated by weighting according to the abundances of the different rock types. [f] Baltic Shield data compiled from Hanski (1992), Eilu (1994), Salonsaari (1995), and Lahtinen (1996). [g] Heat production value recalculated for a bulk density of 2,700 kg m^{-3}.

Shaw *et al.*, 1994), the Baltic Shield (Jõeleht and Kukkonen, 1998), and the Indian Shield (Roy and Rao, 2000). Nevertheless, these efforts remain limited to a few exceptional areas (Rudnick and Fountain, 1995). One outstanding problem is that the lowermost crust cannot be sampled directly, except perhaps in the Ivrea Zone, Italy, or in the Kohistan arc, Pakistan (Miller and Christensen, 1994). In seismic Shield models, the lowermost crust, identified by seismic *P*-wave velocities in the range 6.8–7.2 km s^{-1}, makes up 16 km out of 45 km of the crustal column (Durrheim and Mooney, 1991; Christensen and Mooney, 1995). This large fraction of the crust remains elusive.

3.02.2.3.3 Lower crustal xenoliths

Lower crustal rocks are extremely rare at the Earth's surface and regional studies have been supplemented by analyses of deep crustal xenoliths. One crucial issue is that they may not represent a random sample of all lithologies present at depth (Rudnick, 1992). Kimberlite pipes and basaltic dykes are found only in specific areas of a geological province. Another bias comes from the very geological process responsible for xenolith sampling, the ascent of basaltic magmas through the crust. The ascent is buoyancy driven, implying that basaltic magmas are not able to rise through rocks of lower density, i.e., of felsic composition. Thus, one must expect poor sampling efficiency for evolved rocks with enriched compositions, and hence a bias towards mafic compositions. Xenolith model compositions in Rudnick and Fountain (1995) are indeed systematically more mafic and depleted in radioelements than estimates from terrain exposures.

3.02.2.3.4 Summary

The approaches outlined above involve assumptions regarding the process of scaling upwards a small number of analyses on a suite of different rock types. Seismic data show that the lowermost crust is extremely heterogeneous (Durrheim and Mooney, 1991). Sampling deep crustal levels is a serious challenge, because outcrops of fossil crust–mantle transition zones are very rare. They are limited in a real extent and in the number of represented lithologies. Sampling by basaltic magmas may not be random and may be biased towards mafic rock types.

3.02.2.5 Sampling Continental Crust

Our knowledge of crustal composition obviously depends on the number and spatial coverage of the available measurements. Uranium and thorium concentrations can now be determined routinely using modern techniques, and the number of data published each year is very large. Yet there are several systematic biases in the data, and these must be borne in mind.

3.02.2.5.1 How well is the crust sampled?

Data are seldom collected for gneisses and metasedimentary rocks, presumably because these rocks have complex origins and phase assemblages. However, these rock types account for a

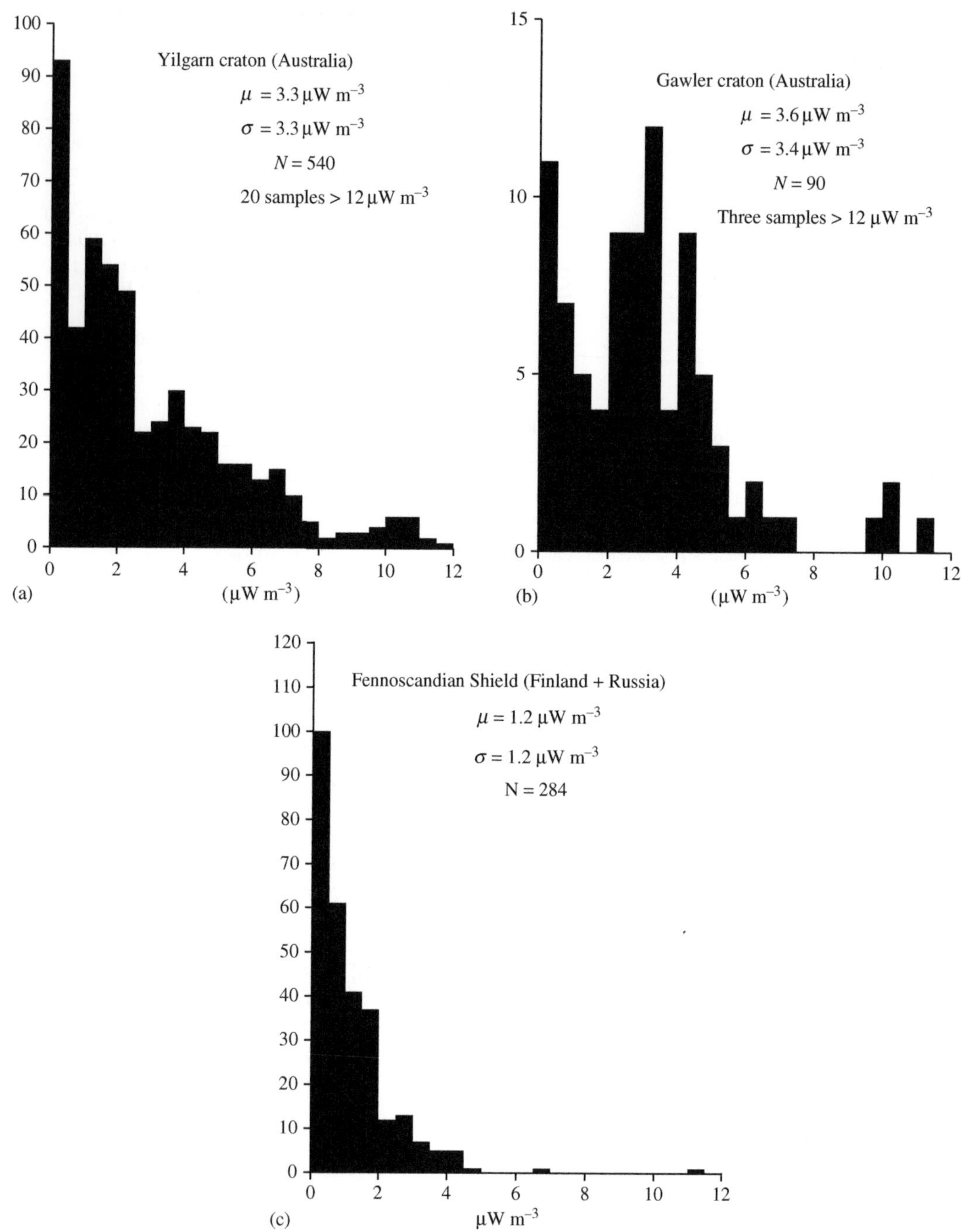

Figure 1 Heat production data for several well-sampled provinces: (a) the Archean Yilgarn craton of Western Australia, (b) the Proterozoic Gawler craton of Central Australia, and (c) the Proterozoic Fennoscandian Shield (sources Australian data—Heier and Lambert (1978); Fennoscandian data—Hanski (1992), Eilu (1994), Salonsaari (1995), and Lahtinen (1996)).

large fraction of the crust. For example, gneisses account for more than 60% of the exposed rock types in the upper crust of the Archean Superior Province in the Canadian Shield (Fountain *et al.*, 1987; Ashwal *et al.*, 1987). In the Appalachians, gneissic basement and metamorphic rocks account for more than 50% of the outcrops, yet the majority of the available analyses are of granitic plutons (Jaupart *et al.*, 1982; Chamberlain and Sonder, 1990). By nature, these rocks are extremely heterogeneous and require extensive sampling for reliable averages. Figure 2 shows data for gneisses from Australia. Their heat production covers a wide range and has high mean values.

A problem of a different type occurs on a large scale. Some geological provinces contain

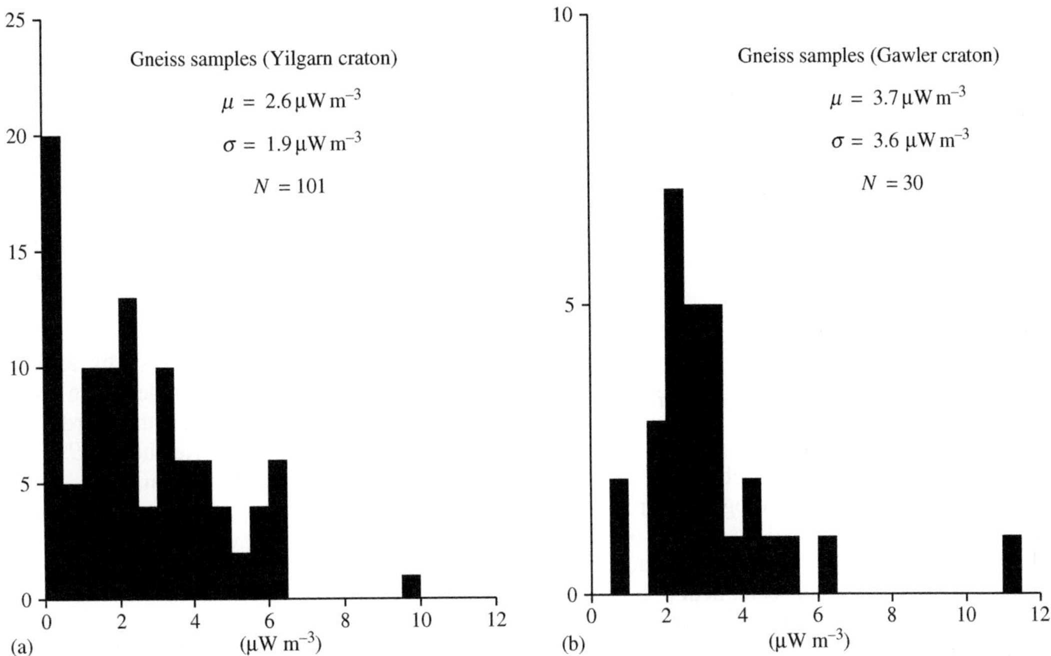

Figure 2 Heat production in gneisses of the Australian Shields: (a) Archean gneisses of the Yilgarn craton and (b) Proterozoic gneisses of the Gawler craton (source Heier and Lambert, (1978)).

extremely enriched granitic rocks which sometimes form distinctive belts, as in Norway and Sweden (Killeen and Heier, 1975a), in southern England (Lee *et al.*, 1987) and in New Hampshire, USA (Jaupart *et al.*, 1982). For example, the Precambrian Fla, Iddefjord, and Bohus granites represent a large uranium and thorium enriched belt within the Fennoscandian Shield. Within this belt heat production rates are higher than 6 μW m^{-3} (Killeen and Heier, 1975a). Such anomalous areas are likely to be overlooked in global crustal studies. Even if they were not, their radioelement content is likely to be underestimated as enriched rock units can also be anomalously thick. For example, the Cornish batholith in southern England is enriched to a depth of more than 15 km (Lee *et al.*, 1987)! In such extreme cases, the thickness of the units can be determined directly from local heat flow data without detailed geophysical surveys.

3.02.2.5.2 How representative are surface samples?

Evolved granites may contain uranium and thorium in very high concentrations and hence may account for a large fraction of their total content in the crust. One difficulty is that they are open to water circulation in the near-surface environment, which leads to uranium depletion (Oversby, 1976; Jaupart *et al.*, 1981; A. Forster and H.-J. Forster, 2000). These effects are well documented and have been demonstrated even in samples which look fresh and unaltered (Zielinski *et al.*, 1978). This problem has two complementary aspects. First, shallow and leached samples are not representative of rocks from the same massif at greater depth. Second, shallow levels cannot be neglected because late-stage uranium remobilization may be effective over thicknesses >1 km (Smellie and Stuckless, 1985; Wollenberg and Flexser, 1985; A. Forster and H.-J. Forster, 2000). A chemist usually selects unaltered samples, which is valid for studies of magmatic evolution. For a bulk chemical budget of the crust, this is inappropriate and leads to an overestimate of the concentration of uranium in the upper crust.

Heat flow data circumvent this problem, because they record the total amount of heat generated in a massif, including deep levels which have not been affected by late-stage surficial processes. They may even be used to estimate the vertical extent of leached rocks (Jaupart *et al.*, 1981).

3.02.2.6 Scaling of Heat Sources

Heat production varies on all sorts of scales. When dealing with such large volumes, one must pay attention to the averaging procedure.

3.02.2.6.1 Variations within a single pluton

Within a single pluton, radioelement concentrations may be quite variable in both vertical and horizontal directions (Rogers *et al.*, 1965; Killeen and Heier, 1975b; Landstrom *et al.*, 1980). These variations may be due to many different causes, such as facies changes, fundamental heterogeneity of the source material, fluid migration, and late-stage alteration. For example, in the Bohus granite, Sweden, concentrations of the relatively immobile thorium vary by a factor of 5 over horizontal distances as small as a few tens of meters and as large as a few kilometers (Landstrom *et al.*, 1980).

3.02.2.6.2 Variations within a geological province

Radioelement concentrations obviously change according to lithology and a geological map allows a rough idea of the spatial variation of heat production. However, this is not sufficient for a precise estimate. On a local scale, heat production within Cordilleran core complexes varies significantly over distances of a few kilometers (Ketcham, 1996). Variogram analysis of the extensive data set available there demonstrates that the correlation distance is <10 km. In other words, proper sampling should be carried out with a grid-spacing <10 km to achieve a statistically robust average. On a larger scale, heat production of a given rock type may vary significantly within a single province, which does not allow extrapolation from one unit to another. Examples may be drawn from New Hampshire, USA which forms part of the Appalachians province. There, heat production in metasediments and in the gneissic basement varies by a factor 3 (Chamberlain and Sonder, 1990). Similarly, there are large differences in thorium abundance amongst plutons of the same magma series, e.g., in the White Mountain series (Rogers and Adams, 1969).

Even Precambrian Shield regions exhibit remarkable variability in heat production. For example, the exposed part of the Proterozoic Trans-Hudson Orogen in Saskatchewan and Manitoba (≈ 500 km $\times$ 500 km) is a mosaic of different belts of different origins and compositions with different heat production values (Rolandone *et al.*, 2002). Another example is provided by the Abitibi province of eastern Canada, which stands out as a low-heat-production region within the Archean Superior Province (Mareschal *et al.*, 2000a).

3.02.2.6.3 Vertical variations

On a vertical scale of ≈ 10 km, measurements on samples from the deep boreholes at Kola, Russia (Arshavskaya *et al.*, 1987), and the KTB, Germany (Clauser *et al.*, 1997) indicate no systematic variation of heat production with depth. At KTB, heat production between 8 km and 9 km depth is the same as between 1 km and 2 km, and higher than above 1 km (Figure 3). At Kola, heat generation in the Archean rocks between 8 km and 12 km is higher (1.47 μW m^{-3}) than in the shallower Proterozoic section (0.4 μW m^{-3}) (Kremenentsky *et al.*, 1989). Over a crustal thickness scale, heat production is lower in mid-crustal assemblages than in the upper crust (Fountain *et al.*, 1987). This vertical variation is not monotonous by any means and cannot be described by a simple function valid everywhere, as shown by exposed crustal sections such as the Vredefort in South Africa (Nicolaysen *et al.*, 1981), the Cordilleran core complexes of Arizona (Ketcham, 1996), or the Pikwitonei–Sachigo (Fountain *et al.*, 1987) and Kapuskasing–Wawa areas of Canada (Ashwal *et al.*, 1987; Shaw *et al.*, 1994).

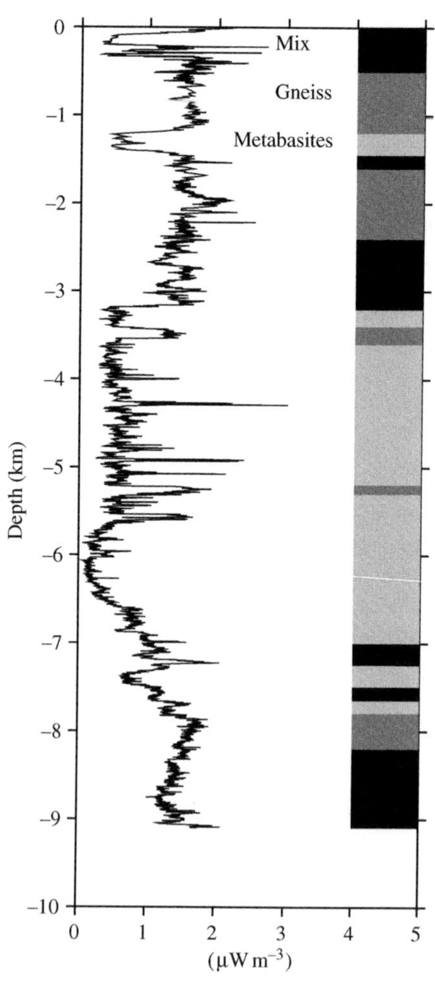

Figure 3 Vertical distribution of heat production in the KTB deep borehole, Germany, and schematic lithological log of the borehole (interlayered gneiss and metabasites (black), gneiss (dark gray) metabasites (light gray)) (source KTB web site).

3.02.3 HEAT FLOW AND CRUSTAL HEAT PRODUCTION

The amount of heat that escapes through Earth's surface is due to radiogenic heat production and secular cooling. In general, continental heat flow is a superposition of four components:

$$Q = Q_C + Q_L + Q_B + Q_T \tag{4}$$

where Q_B is the heat input at the base of the lithosphere due to mantle convection, Q_T is a long-term transient due to cooling after a major tectonic or magmatic perturbation, Q_L is the radiogenic heat production in the mantle part of the lithosphere, and Q_C is the radiogenic heat production of the crust. We are only interested in Q_C, and this requires evaluating the other three components.

The contribution of radiogenic heat production in the mantle part of the lithosphere is poorly constrained (Rudnick *et al.*, 1998; Jaupart and Mareschal, 1999; Russell *et al.*, 2001). The transient component depends on thermal relaxation following a major perturbation such as rifting, extension, crustal thickening, and magmatic underplating. The timescales involved depend on the type of perturbation. Thermal effects of crustal perturbations such as magmatic intrusion or underplating decay in less than 100 Myr. Induced heat flow anomalies at the Earth surface are characterized by relatively small length scales and are associated with detectable tectonic or magmatic structures. Thus, they can be easily identified. The low heat flow (40 mW m^{-2}) in the ca. 65 Ma Deccan traps region (Roy and Rao, 2000) is the most striking observation supporting that even for extreme events affecting the crust, the thermal transient is short. Lithospheric perturbations decay over longer timescales that may be as large as 500 Myr (Nyblade and Pollack, 1993; Hamdani *et al.*, 1991; Kaminski and Jaupart, 2000). Thermal relaxation of such deep thermal anomalies has two important characteristics, as discussed by Jaupart *et al.* (1998). First, lateral variations are efficiently smoothed out by heat conduction and hence do not lead to spatial variations of surface heat flow over distances $<$500 km. Second, this long thermal transient induces changes of lithosphere thickness, implying that the mantle heat flow can be determined from constraints on the local lithosphere thickness.

For geological provinces older than ~100 Ma, one may lump together the three components Q_L, Q_B, and Q_T into a single parameter which will be called the mantle heat flow Q_M. For all practical purposes, this parameter can be assumed constant over a geological province.

3.02.3.1 Heat Flow and Surface Heat Production

Early work on continental heat flow focused on a linear relationship between the local values of heat flow and heat production (Birch *et al.*, 1968):

$$Q = Q_r + AD \tag{5}$$

where Q and A stand for the observed heat flow and the local heat production of rocks of the crystalline basement, respectively. The slope D, which has dimension of length and is usually ≈10 km, is related to the thickness of a surficial heat-producing layer. Q_r is called the reduced heat flow. Within the relatively large errors in both heat flow and heat production values, this relationship seemed to be valid in several provinces. Among the many heat source distributions that fit this relationship, the exponentially decreasing one was favored because it is independent of the erosion level (Lachenbruch, 1970). This implies that the total heat production in the crust is AD and that $Q_M = Q_r$.

The simple mathematical models based on this linear relationship imply that the vertical distribution of heat production can be described by a single universal function within a province. This conclusion, as well as the linear relationship itself, have been challenged by many studies. As discussed above, measurements of heat production in deep boreholes have not shown that the concentration of heat sources systematically decreases with depth, as would be required by the heat flow heat production relationship. Studies of exposed crustal sections suggest a general trend of decreasing heat production with depth, but this trend is not due to a ubiquitous mechanism that affects all crustal rocks in the same manner everywhere. We know that it is not due to fluid redistribution accompanying metamorphic reactions (Bingen *et al.*, 1996; Bea and Montero, 1999). The uranium and thorium content of lower crustal rocks depends weakly on pressure and temperature conditions and depends mostly on prior geological history. There is no reason to expect that tectonic and magmatic processes distribute heat production in a systematic fashion on a scale of a few kilometers. Thus, there is no simple stratigraphic control on crustal heat production and no universal function describing the vertical distribution of heat production. In the Abitibi and in the FlinFlon belts of the Superior Province and Trans-Hudson Orogen (both in Canada), the upper crust consists of depleted mafic volcanics and intrusives that overlay more radiogenic basement (Pinet *et al.*, 1991; Guillou *et al.*, 1994; Mareschal *et al.*, 2003). This is also the case in the Kola peninsula, where the Proterozoic supracrustal rocks have lower heat production (0.4 μW m^{-3}) than the Archean basement (1.47 μW m^{-3}) (Kremenentsky *et al.*, 1989). In such areas, heat production actually increases with depth in the upper crust.

On a large scale, the crust is stably stratified, with density and hence mafic assemblages increasing with depth. This gravity control causes the general trend of decreasing heat production with depth. However, it is not valid on a local scale and there is a wide range of rock types at all depths. Evolved rocks enriched in radioelements can be found just above the Moho (Fountain and Salisbury, 1981). The crust is heterogeneous on all scales in both the horizontal and vertical directions.

With the large data sets now available, it is clear that the correlation between local values of surface heat flow and heat production only holds over exposed plutons enriched in radioactive elements (Birch *et al.*, 1968). Over other rock types, it is weak at best, as demonstrated by data from large Precambrian provinces of India (Roy and Rao, 2000), Canada (Mareschal *et al.*, 1999), and South Africa (Jones, 1987, 1988). As discussed below, theory shows that, for the rather small wavelengths involved, surface heat flow is only sensitive to shallow heat production contrasts (Jaupart, 1983a; Vasseur and Singh, 1986). In fact, the linear relationship is an artifact, because horizontal heat transport smoothes out differences in deep heat production rates. Values of D are not related to other physical dimensions in a geological province, such as the pluton thickness, and hence have no geological or petrological meaning (Jaupart, 1983b).

Despite these well-known problems with the linear heat flow–heat production relationship, many authors still rely on it in global studies of heat flow and lithosphere structure (e.g., Artemieva and Mooney, 2001). In addition, the very few provinces where the linear relation is roughly valid have been used to establish a "global" correlation between surface heat flow and heat production, including areas such as the Precambrian of North America, where the linear relation does not work at all (Mareschal *et al.*, 1999). One of the reasons for the persistence of this questionable approach is the lack of a database for worldwide heat production.

3.02.3.2 Sampling with Heat Flow

Heat transport by conduction is effective in both the vertical and horizontal directions. Surface heat flow values are not sensitive to lateral heat production variations at depth. For the sake of discussion, we consider a series of crustal layers with thickness $\Delta = 10$ km at depths (in km) $z = 0, 10, 20, 30$, and 40. The distribution of heat production in each layer can be considered as the sum of components with different wavelengths λ, denoted $A(\lambda)$. The amount of heat generated by each component is $H(\lambda) = \Delta A(\lambda)$. For a layer at depth z, the component of the surface heat flow $Q(\lambda)$ is given by

$$Q(\lambda) = \frac{H(\lambda)}{2\pi}\frac{\lambda}{\Delta} \times \exp\left(\frac{-2\pi z}{\lambda}\right) \times \left(1 - \exp\left(\frac{-2\pi\Delta}{\lambda}\right)\right) \quad (6)$$

Figure 4 shows the amplitude of the contribution of each component to surface heat flow as a function of wavelength and depth. For the surficial layer ($z = 0$ km), all wavelengths shorter than 10 km are eliminated in the surface heat flow field (i.e., $Q/H < 0.1$). For a deep layer ($z = 30$ km), the effect is even more dramatic, and there is no surface effect for wavelengths shorter than 90 km. This shows that heat diffusion smoothes out lateral heat production variations, and that surface heat flow provides an estimate of the horizontally averaged heat production. The deeper the crustal layer, the wider the averaging window. These results illustrate the point made above that surface heat flow variations are due to shallow heat production contrasts.

In Equation (6), it is assumed that the heat production remains coherent over a layer of thickness Δ, which is unlikely, as shown by the vertical heat production profiles (Figure 3). The amplitude of each component of the surface heat flow will be even smaller when heat production variations are not coherent in a vertical column. General results allowing for vertical changes in heat production are given in Appendix A.

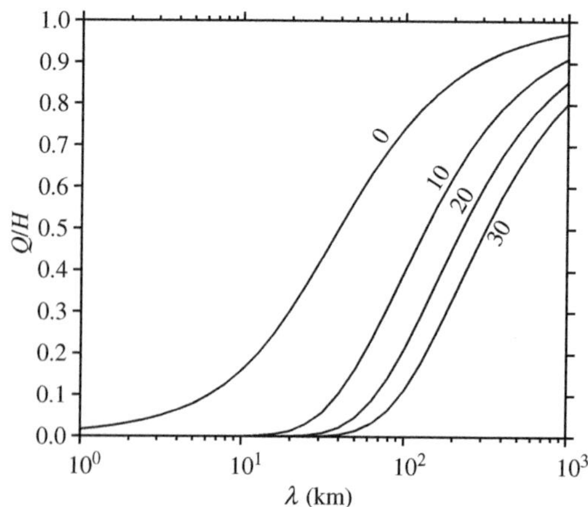

Figure 4 Amplitude of surface heat flow variations due to heat production variations in a 10 km thick layer as a function of wavelength. Curves are shown for four layer depths (in km), 0, 10, 20, and 30.

3.02.3.3 Heat Flow Data

With these principles in mind, we turn to heat flow data. Table 4 summarizes data for several well-studied provinces throughout the world in three different age groups, Archean, Proterozoic, and Paleozoic. Data are also available for younger regions, but are demonstrably affected by transient effects due to magmatism and tectonic deformation (Sclater *et al.*, 1980; Pollack *et al.*, 1993). Table 4 also contains values from the global compilations by Nyblade and Pollack (1993) for the Precambrian and Pollack *et al.* (1993) for younger provinces. For Precambrian provinces, Nyblade and Pollack (1993) rejected data from regions affected by young tectonic and magmatic events. In other words, they were not interested in basement ages but in the ages of the last recorded tectonothermal events and they excluded possible post-Precambrian thermal anomalies which are superimposed on the steady-state lithospheric thermal structure. This explains why their values for the same age groups are lower than those reported by Pollack *et al.* (1993) and included in Stein (1995). Because we are interested in steady-state conditions, we shall use the Nyblade and Pollack (1993) values for the Precambrian. For the Paleozoic, there is no such problem, because basement ages and perturbation ages are usually identical and well known.

One must first assess the quality of the sampling in presently available heat flow data. For the Precambrian of North America and Australia, the data coverage is very good (see Table 4). In North America, the number of heat flow data in the Shield has more than doubled

Table 4 Mean heat flow and heat production in major provinces.

	$\langle Q\rangle^a$ (mW m^{-2})	σ_Q[b]	N_Q[c]	$\langle A\rangle^a$ (μW m^{-3})	σ_A[b]	N_A[c]	*References*
Archean							
Dharwar (India)	36 ± 2.1	6	8				Roy and Rao (2000)
Witwatersrand (S. Africa)	51 ± 0.7	6	81				Ballard *et al.* (1987) and Jones (1992)
Lesotho (S. Africa)	61 ± 5.3	16	9				Jones (1988)
Western Shield (Australia)	49 ± 2.1	13	37	2.6	2.1	12	Cull (1991)
Superior (N. America)	42 ± 1.0	10	57	0.95	1.0	44	Mareschal *et al.* (2000a)
Slave (N. America)	54 ± 1.0	10	2	2.3	1.0	20	Lewis *et al.* (2003)
Total Archean[e]	41 ± 0.8	11	188				Nyblade and Pollack (1993)
Proterozoic							
Aravalli (India)	68 ± 4.9	13	7				Roy and Rao (2000)
Namaqua (S. Africa)	61 ± 2.5	11	20	2.3		10	Jones (1987)
Central Shield (Australia)	78 ± 2.2	19	75	3.6	1.9	38	Cull (1991)
Baltic Shield	31 ± 1.2	9	68				Cermak (1993)
Ukrainian Shield	32 ± 0.7	7	106				Cermak (1993)
Trans-Hudson (N. America)	42 ± 2.0	11	49	0.73	0.50	47	Rolandone *et al.* (2002)
Wopmay (N. America)	90 ± 1.0	11	12	4.8	1.0	20	Lewis *et al.* (2003)
Grenville (N. America)	41 ± 2.0	11	30	0.80	d	17	Mareschal *et al.* (2000a)
Undisturbed North China Massif (China)	50 ± 1.1	11	99				Hu *et al.* (2000)
Upper Yangtze Block (China)	53 ± 0.8	9	119				Hu *et al.* (2000)
Tarim Block (China)	44 ± 2.1	10	22				Hu *et al.* (2000)
Qaidam Basin (China)	52 ± 3.4	13	16				Hu *et al.* (2000)
Total Proterozoic[e]	48 ± 0.8	16	675				Nyblade and Pollack (1993)
Paleozoic							
Appalachians (N. America)	57 ± 1.5	13	79	2.6	1.9	50	Jaupart and Mareschal (1999)
Mainland United Kingdom	54[d]	12	188				Lee *et al.* (1987)
Dnieper aulacogen (Russia)	45 ± 0.6	8	165				Cermak (1993)
Pripyan depression (Russia)	66 ± 2.2	17	59				Cermak (1993)
Platforms (Russia)	68 ± 0.9	16	323				Cermak (1993)
Altay-Ergula Belt (China)	60 ± 2.2	9	18				Hu *et al.* (2000)
Junggar-Higgan Belt (China)	47 ± 2.1	10	22				Hu *et al.* (2000)
Total Paleozoic[e]	58.3 ± 0.5	22	2,213				Pollack *et al.* (1993)

[a] Mean ± one standard error. [b] Standard deviation on the distribution. [c] Number of sites. [d] Area-weighted average value. [e] Total in the compilation by Nyblade and Pollack (1993) excluding the more recent measurements included here.

since the global compilation by Pollack *et al.* (1993), but the average heat flow values for the different Provinces of the Shield have not changed. Significant future change in these average values is thus very unlikely. Heat flow measurement sites may not be evenly distributed. However, in well-sampled areas, the straight average of individual data points cannot be distinguished from an area-weighted spatial average. For example, this was demonstrated in the Grenville province (Mareschal *et al.*, 2000a). Another way of demonstrating this is to consider the Appalachian province in eastern North America. Two different data sets have been obtained in the American and Canadian parts of this province that are quite consistent with each other (Table 5). This shows that the heat flow coverage is adequate.

The data illustrate two important points (Table 4). The first is that, within a single age group, e.g., the Proterozoic, there are significant variations of mean heat flow. However, where surface heat production data are available, it is clear that the differences are due to differences of surface heat production. For example, in Canada, the Wopmay Orogen has a higher mean heat flow than the Trans-Hudson Orogen (90 mW m^{-2} versus 42 mW m^{-2}), but it also has much higher mean heat production (4.8 μW m^{-3} versus 0.7 μW m^{-3}) (Tables 3 and 4). The difference in heat flow between these two provinces can be accounted for by the difference in heat production in the upper (12 km) crust. The second point is that there is a general trend of decreasing heat flow with age. However, this trend is only valid for the global data set and does not hold within several continents. In North America, both heat flow and production are lower in the Proterozoic Trans-Hudson Orogen than in the Archean Superior Province. In the Baltic Shield, the average heat production in the truly Proterozoic crust (0.4 μW m^{-3}) is much lower than in the Archean basement (1.47 μW m^{-3}) (Kremenentsky *et al.*, 1989) and the average heat flow is the lowest value reported for a single province worldwide (31 mW m^{-2}). Both in the Baltic Shield and in the Trans-Hudson Orogen, most of the crustal heat production comes from the Archean basement.

In general, high heat flow can be accounted for solely by high radioactivity in the upper crust. This is obviously true for the Paleozoic provinces compared to some of their older counterparts. For example, the Appalachians consist of highly radiogenic upper crust, with an average heat production of 2.6 μW m^{-3}. This also is observed in Australia, where the Proterozoic central province has both higher mean heat flow and heat production than the Archean western province (78 mW m^{-2} versus 49 mW m^{-2} and 3.6 μW m^{-3} versus 2.6 μW m^{-3}, respectively, Tables 3 and 4). The global trend of decreasing mean heat flow with age can be attributed to a decrease of crustal radioactivity (Morgan, 1985) and there is no need to call for changes in deep mantle input. In the next section, we shall return to this issue using different arguments and independent constraints on lithospheric structure.

Surface heat production data are not available for all provinces, and it is difficult to study their relationship with heat flow on a global basis. Data are available for North America (Table 6). Sub-provinces with particular geological and petrological characteristics are also individually considered. The Abitibi province differs from the rest of the Superior Province because it has large volumes of greenstones (altered basic flows and intrusives). The Thompson belt in the Trans-Hudson Orogen, is singular, because it consists of reworked Archean sediments and metamorphic rocks at the edge of the Superior craton. In the rest of this Orogen, the upper crust consists of juvenile rocks of true Proterozoic age. We can define six provinces with contrasting heat flow and heat production characteristics. Figure 5 emphasizes that heat flow and surface heat production are statistically correlated but that the correlation is far from perfect. This reinforces the point made earlier that heat flow variations reflect variations in crustal heat production.

Table 5 Heat flow statistics for the Appalachians with the number of heat flow sites, mean heat flow and standard deviation, number of heat production values, mean heat production and standard deviation.

Province	N_Q	$\langle Q \rangle \pm \sigma_Q$ (mW m^{-2})	N_A	$\langle A \rangle \pm \sigma_A$ (μW m^{-3})	*References*[a]
Canadian Appalachians	43	56 ± 12	14	2.6 ± 2	(1), (2), (3)
US Appalachians	36	58 ± 13	36	2.5 ± 1.9	(4), (5)
Total Appalachians	79	57 ± 13	50	2.6 ± 1.9	

[a] (1) Pinet *et al.* (1991) and references therein; (2) Guillou-Frottier *et al.* (1995) and references therein; (3) Drury *et al.* (1987); (4) Birch *et al.* (1968); (5) Jaupart *et al.* (1982).

Table 6 Mean heat flow and heat production in different belts and provinces of North America.

	$\langle Q\rangle^a$ (mW m^{-2})	σ_Q[b]	N_Q[c]	$\langle A\rangle^a$ (μW m^{-3})	σ_A[b]	N_A[d]
Superior (>2.5 Ga) (excl. Abitibi)	45 ± 2.4	12	26	1.4 ± 0.26	1.2	21
Abitibi (>2.5 Ga)	37 ± 1	7	26	0.41 ± 0.07	0.33	21
Trans-Hudson (1.8 Ga) (juvenile crust only)	37 ± 1.4	7	38	0.6 ± 0.08	0.48	36
Thompson Belt (reworked Archean in the Trans-Hudson)	53 ± 1.6	5	10	1.12 ± 0.10	0.32	11
Grenville (1.1 Ga)	41 ± 2	11	30[d]	0.80[e]		
Appalachians (<0.5 Ga)	57 ± 1.5	13	79	2.6 ± 0.27	1.9	50

Sources: Jaupart and Mareschal (1999), Mareschal *et al.* (2000a), and Rolandone *et al.* (2002).
[a] Mean ± one standard error. [b] Standard deviation on the distribution. [c] Number of sites. [d] Number of heat production values, where each value is based on many samples. [e] Area-weighted average.

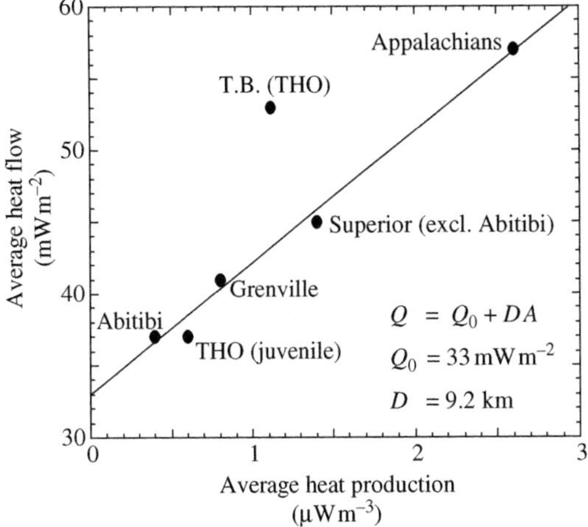

Figure 5 Relationship between mean surface heat flow and heat production in the major provinces of North America (data from Table 6).

3.02.4 CONTRIBUTION OF THE CONTINENTAL CRUST TO THE HEAT BUDGET

Table 2 shows the bulk crustal heat production estimated by different authors. These estimates were obtained by weighting the results of sampling according to the methods described above and in Chapter 3.01. We also give the average crustal component of the heat flow for a 41 km thick crust. Table 2 emphasizes that the range of estimates is wide and that some chemical/petrological estimates are clearly incompatible with the heat flow data. Comparison with heat flow is meaningful only for stable continents where only crustal heat production and mantle heat flow are important. This is the case for provinces older than 200 Ma, where the area weighted world average heat flow is 51 mW m^{-2} (e.g., Stein, 1995). The global estimate of Wedepohl (1995) requires that the mantle heat flow at the base of the continental crust is 0. Local estimates for several geological provinces differ from one another (Nicolaysen *et al.*, 1981; Shaw *et al.*, 1986; Gupta *et al.*, 1991; Jaupart *et al.*, 1998; Gao *et al.*, 1998). The value of Shaw *et al.* (1986) for the Canadian Shield is not consistent with the heat flow data. Thus, one may not apply one estimate for one province to another province. A further difficulty is that the crustal thickness varies significantly. For example, the crust of the North China craton is thinner than average (≈35 km; Gao *et al.*, 1998).

Alternatively, one can attempt to estimate the bulk crustal heat production directly from the heat flow data and local studies of crustal structure. This requires estimates of the mantle heat flow Q_M, which can be obtained in several ways.

3.02.4.1 Precambrian Provinces

In several parts of the Canadian Shield, heat flow values as low as 22 mW m^{-2} have been measured (Jaupart and Mareschal, 1999; Mareschal *et al.*, 2000b). Similar values have also been reported for the Norwegian Shield (Swanberg *et al.*, 1974), and Western Australia (Cull, 1991). These are areas where the crustal contribution is smallest and hence provide an upper bound to the mantle heat flow. One may refine this estimate further by subtracting a lower bound for crustal heat production. As shown above, surface heat flow records a large-scale average of heat production, and hence one should consider a representative crustal assemblage, and not a single rock type, such as gabbro. Heat production estimates are not lower than 0.1 μW m^{-3} for any crustal material (Pinet and Jaupart, 1987; Rudnick and Fountain, 1995; Jõeleht and Kukkonen, 1998). Over an average thickness of ≈40 km, the contribution of the crust must be at least 4 mW m^{-2}. Hence the mantle heat flow must be <18 mW m^{-2}. A different argument was used by Rolandone *et al.* (2002) to obtain a lower bound on mantle heat flow. In high heat flow areas of the Canadian Shield, crustal rocks are at high temperatures today and were still hotter in the past when radiogenic heat production was higher. The condition of thermal stability, such that melting conditions are not attained in the crust in the absence of tectonic events and magmatic intrusions, provides a lower bound of 11 mW m^{-2} on the mantle heat flow. Combining the two independent arguments leads to a range of 11–18 mW m^{-2} for the mantle heat flow beneath the Canadian Shield. Arguments different from these have led to the same range of values in other Precambrian areas (Jones, 1988; Guillou-Frottier *et al.*, 1995; Gupta *et al.*, 1991).

The above estimates were derived using local geophysical and heat production data in several provinces and rely on a knowledge of crustal structure. It is useful to evaluate alternative ways to estimate the mantle heat flow. Independent determinations of the mantle heat flow may be obtained by considering the lithospheric thickness determined by seismic and xenolith studies. To obtain an estimate of the amount of heat conducted in the mantle part of the lithosphere, one must estimate the temperature difference across the mantle lithosphere. In Shield areas, temperatures at the Moho range between 300 °C and 500 °C and that at the base of the lithosphere is 1,400 ± 50 °C (Jaupart *et al.*, 1998; Rolandone *et al.*, 2002). The latest tomographic models indicate that shields do not extend deeper than 250 km, with an uncertainty of ~50 km (S. Grand, personal communication, 2002; Gung *et al.*, 2003). This yields a temperature gradient >4–5 K km^{-1} in the mantle. In shield areas, "low T" deep mantle xenoliths, which are representative of the unperturbed lithosphere, never record temperatures higher than 1,300 °C above 175 km depth, including the Kaapvaal craton, South Africa (Russell and Kopylova, 1999). This implies a temperature gradient <8 K km^{-1}. For a mantle thermal conductivity of 3 W m^{-1} K^{-1}, seismic tomography data require the mantle heat flow to be >12 mW m^{-2} and xenolith data require it to be <24 mW m^{-2}.

These arguments can be refined. In Appendix B, we calculate the mantle heat flow as a function of lithosphere thickness and show how it depends on the surface heat flow. Pressure and temperature estimates from mantle xenoliths may be combined to determine a best-fit geotherm consistent with heat transport by conduction. Mantle heat flow estimates obtained in this manner are in the following ranges: 7–15 mW m^{-2} beneath the Fennoscandian Shield (Kukkonen and Peltonen, 1999), 17–25 mw m^{-2} for the Kalahari craton, South Africa (Rudnick and Nyblade, 1999), and 12–24 mW m^{-2} for the Slave craton, Canada (Russell and Kopylova, 1999; Russell *et al.*, 2001). "Best-fit" estimates are 18 mW m^{-2} and 15 mW m^{-2} for the Kalahari and Slave cratons, respectively.

Allowing for the uncertainties and requiring consistency with low-heat-flow measurements, we shall consider the range of 11–18 mW m^{-2}. For this range, the differences of average heat flow values between geological provinces cannot be accounted for by changes of mantle heat flow and hence must be attributed to changes in crustal heat production. The same conclusion holds for the variation of average heat flow as a function of age. In other words, the crustal component of heat flow varies among provinces within a single age group and is a function of age. This conclusion is consistent with the data of Tables 2 and 3.

Over the whole Precambrian, a large volume of continental crust is sampled and local heterogeneities are smoothed out. Furthermore, the average Precambrian crust accounts for more than 65% of the total crustal volume (Goodwin, 1997). From Table 4, the average heat flow through Precambrian continents is 46 mW m^{-2}, with a negligible error of 2 mW m^{-2} due to the uncertainty in the proportions of Archean and Proterozoic crust (see the discussion in Rudnick and Fountain, 1995). Removing the mantle heat flow leads to a range of 28–35 mW m^{-2} for crustal heat generation. For an average Precambrian crustal thickness of 41 km (Mooney *et al.*, 1998), this translates into an estimate of 0.77 ± 0.08 μW m^{-3} for the average heat production rate of Precambrian crust. For the Canadian Shield, the average heat flow for all Precambrian provinces is 41 mW m^{-2} and we have estimated the mantle heat flow to range between 11–15 mW m^{-2} (Guillou-Frottier *et al.*, 1995; Jaupart *et al.*, 1998). The average crustal contribution is 29 ± 2 mW m^{-2} which gives a crustal heat production of 0.70 ± 0.05 μW m^{-3} for the Precambrian crust of North America.

3.02.4.2 Paleozoic Provinces

Younger provinces are difficult to analyze, because they are smaller and tend to be located at craton margins. In such areas, it is difficult to determine the local lithosphere thickness. It is likely that, in some areas, the lithosphere is thinner than in Precambrian provinces (Goes and van der Lee, 2002). There is no evidence for significant thermal transients in Paleozoic provinces. This is certainly true for the Appalachians where, as shown by Jaupart and Mareschal (1999), the sharp transition at the boundary with the Grenville province indicates that heat flow differences are due to changes of crustal composition, and not to some deep-seated component. There can be no doubt that there is a large change of mean surface heat production between the two provinces (Table 4). Figure 5 emphasizes that the Appalachians follow the same global heat flow trend as the Precambrian provinces of North America. According to recent high-resolution tomographic studies (Goes and van der Lee, 2002), the lithospheric thickness beneath Paleozoic regions falls within a small range 150–200 km. Using the equations of Appendix B, we obtain a range of 15–21 mW m^{-2} for the mantle heat flow. This leads to a mean crustal heat flow component between 37 mW m^{-2} and 43 mW m^{-2}. Paleozoic areas have quite variable crustal thickness (Mooney *et al.*, 1998). We take an average value of 39 km, with an error which is at most a few kilometers (~10%). Thus, the average heat production for Paleozoic crust is 1.03 ± 0.08 μW m^{-3}.

The area weighted heat flow for all provinces older than 200 Ma is 51 mW m^{-2} (Stein, 1995). Estimates of the mantle heat flow by various authors vary between 11 mW m^{-2} and 18 mW m^{-2} (Jaupart and Mareschal, 1999). After removing the mantle heat flow, the average contribution of the crust to the surface heat flow in stable continental regions is between 33 mW m^{-2} and 40 mW m^{-2}. This implies that the bulk crustal heat production is 0.9 ± 0.1 μW m^{-3}.

3.02.4.3 Mesozoic and Cenozoic Provinces—Continental Margins

For provinces younger than the Paleozoic, the compilation of Pollack *et al.* (1993) indicates mean heat flow values that are larger than that for the Paleozoic (64 mW m^{-2} for the Mesozoic and up to 97 mW m^{-2} for the Cenozoic, compared to the mean Paleozoic value of 58 mW m^{-2}). Such values are clearly affected by recent magmatic and tectonic events and hence cannot be used for steady-state calculations. We may consider with little chance of error that the mean Paleozoic value reflects average crustal heat production up to the present day. For the Mesozoic, the difference is small and within the uncertainty range.

Continental margins account for an important fraction of the continental surface. They are characterized by gradual crustal thinning towards oceanic basins, which implies a lateral variation of the crustal heat flow component. For a bulk estimate of heat production, one can extend the heat flow statistics to the margins, but this requires evaluating the thinning characteristics, i.e., whether it is homogeneous over the whole vertical extent of the crust. Heat flow data are consistent with homogeneous thinning (e.g., Louden *et al.*, 1991). We may consider that, on a global scale, the crust of continental margins decreases from an average thickness of 41 km to ~10 km, the thickness of oceanic crust (see Chapter 3.13). If we further assume that the age distribution of continental margins is identical to that of the rest of the continental surface, they do not affect the statistics.

3.02.4.4 Summary

Table 7 lists the values of heat production obtained for three age groups—Archean, Proterozoic, and Phanerozoic. As explained above, the values for the Phanerozoic are assumed equal to those for the Paleozoic. There is a clear decrease with age. It is worth emphasizing that this trend only holds for the global data set and that, within a single age group, there are large variations in the average heat production among geological provinces from different areas and continents. Most of this scatter can be attributed to the distinct tectonic and magmatic history of each province, as illustrated by the Abitibi subprovince in the Superior Province of Canada. The global value for an age group lumps together different types of crustal structures and hence should not be used for a particular geological province.

Heat production for the bulk continental crust in Table 7 is calculated using the crustal age distribution of model 2 in Rudnick and Fountain (1995) and is in the upper end of the range of values from global crust/mantle budgets (Table 2).

The total contribution of radioactive decay in the continental crust of the Earth's energy budget can be estimated from these numbers. The total volume of the Earth's crust is calculated using the total continental area of 210 × 10^6 km^2 (Sclater *et al.*, 1980; Cogley, 1984). In this total, continental margins account for 30% (Cogley, 1984). Using an average crustal thickness of 40 km for provinces above sea level (Mooney *et al.*, 1998) and 25 km for margins (i.e., the mean value between 10 km and 40 km), the total volume of continental crust is 7.3 × 10^{18} m^3. For a range of average heat production of 0.79–0.95 μW m^{-3}, the continental crust contributes 5.8–6.9 TW to the total energy budget of the Earth (44 TW; Pollack *et al.*, 1993). Active provinces and continental margins now represent about 30% of the total volume of the crust; 50%

Table 7 Estimates of bulk continental crust heat production from heat flow data.

Age group	*A*[a] (μW m^{-3})	Q_C[b] (mW m^{-2})	*% area*[c]
Archean	0.56–0.73	23–30	9
Proterozoic	0.73–0.90	30–37	56
Phanerozoic	0.95–1.10	37–43	35
Total continents	0.79–0.95	32–38	

Source: Model 2 in Rudnick and Fountain (1995).
[a] Range of heat production in μW m^{-3}. [b] Range of the crustal heat flow component in mW m^{-2}. [c] Fraction of total continental surface.

error on their heat production would translate to a 15% error in the global budget.

The total energy from radioactivity in the Earth cannot be determined, but the bulk silicate earth composition provides a useful reference model (McDonough and Sun, 1995). In this model, the average heat production of primitive mantle is 5 pW kg^{-1}, which contributes 20 TW to the energy budget of the Earth. Thus, the crust contains 29–34% of the heat producing elements of the bulk silicate earth. The remaining heat producing elements in the mantle account for an average heat production rate of ≈3.5 pW kg^{-1}. This is larger than the estimates for the upper (depleted) mantle (≈1.5 pW kg^{-1}), which suggests the existence of a nondepleted reservoir in the mantle.

3.02.5 CONCLUSIONS

We have reviewed the main standard methods for estimating the bulk crustal heat production. Geochemical models based on crust–mantle budgets yield a global crustal heat production of 0.8 μW m^{-3}. Petrological models based on island arc compositions yield 0.83 μW m^{-3}. Estimates based on rocks available at the Earth's surface, either on the outcrop or carried by volcanic flows, strongly depend on sampling and on the assumptions underlying the extrapolation from small-scale surface samples to the composition of the entire continental crust. Published values vary by a factor >2 between 0.55 μW m^{-3} and 1.31 μW m^{-3}. This shows the limits and illustrates the lack of robustness of these estimates.

Heat flow data are not very sensitive to small-scale horizontal variations of heat production and integrate the heat production in the entire crustal column, and hence alleviate many of the sampling problems encountered when using only geochemical data. They are sensitive to heat production in the whole crustal column, including surficial rocks, which may have been affected by late-stage leaching and alteration processes, as well as deeper material which has remained untouched. Further, they are sensitive to the deepest crustal levels, which can seldom be sampled.

In stable continents, the crustal contribution is obtained by removing the mantle heat flow from the surface heat flow. The mantle heat flow is constrained by the heat flow and heat production data and by the required consistency of thermal data with lithospheric thickness from seismology and xenolith studies. Results from heat flow studies yield an average heat production of 0.77 ± 0.08 μW m^{-3} for the Precambrian crust and 1.03 ± 0.08 μW m^{-3} for the Phanerozoic. The global value for each age group lumps together different types of crustal structures. Within an age group, there are large variations of the average heat production among geological provinces, because they experienced different tectonic and magmatic histories.

For the whole continental crust, the average heat production is in the range 0.79–0.95 μW m^{-3}. This range lies in the upper part of the bracket of "global" chemical models and is not consistent with several heat-producing element-rich estimates which have been proposed for the bulk crust. For Th/U = 4 and K/U = 1.2×10^4, average uranium and thorium concentrations are 1.30 ± 0.11 ppm and 5.19 ± 0.45 ppm, respectively.

The amount of crustal heat production is a key parameter for the distribution of temperatures in continental crust. Further constraints may be obtained by studying thermal aspects of models for the generation of "average crustal material," which involve thermally activated processes in one form or another, such as delamination and gravitational instability (Jull and Kelemen, 2001).

APPENDIX A: POWER SPECTRA

The Fourier transform of the surface heat flow $q(k_x, k_y)$ is related to $A(k_x, k_y, z)$ the Fourier transform of the heat source distribution at each depth z by the relationship (e.g., Mareschal, 1985):

$$q(k_x, k_y) = \int_0^{\infty} A(k_x, k_y, z) \exp(-Kz)\, \mathrm{d}z \qquad (7)$$

where wavenumber $K = \sqrt{k_x^2 + k_y^2}$ is inversely proportional to wavelength λ ($K = 2\pi/\lambda$). In terms of the three-dimensional Fourier transform of the heat source distribution, this expression becomes

$$q(k_x, k_y) = \int_{-\infty}^{\infty} A(k_x, k_y, k_z) \frac{\mathrm{d}k_z}{K + \mathrm{i}k_z} \qquad (8)$$

where i is the pure imaginary number. The energy spectrum for the surface heat flow, $P_q(k_x, k_y)$, is obtained from the three-dimensional spectrum of the heat source distribution $P_A(k_x, k_y, k_z)$ as

$$P_q(k_x, k_y) = \int_0^{\infty} \frac{P_A(k_x, k_y, k_z)}{k_x^2 + k_y^2 + k_z^2} \mathrm{d}k_z \qquad (9)$$

An isotropic distribution of heat sources with a power-law power spectrum is such that

$$P_A(k_x, k_y, k_z) = C\left(k_x^2 + k_y^2 + k_z^2\right)^{-\beta_3/2} \qquad (10)$$

where C is a constant and β_3 the three-dimensional power spectral slope. From this, the power

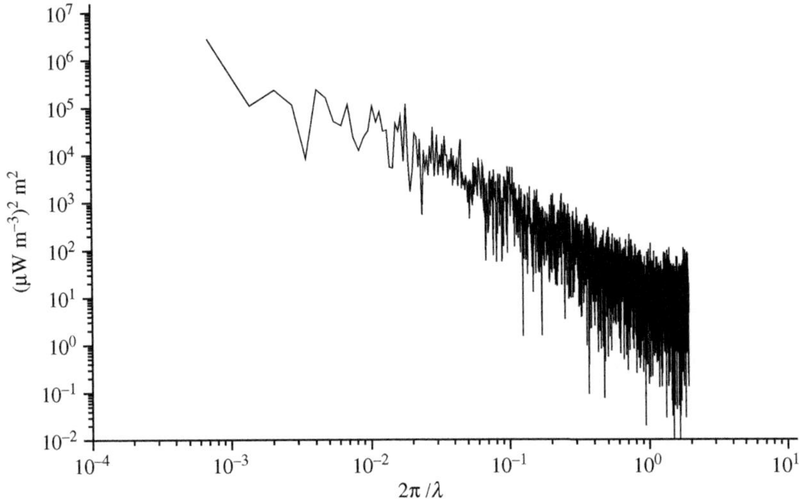

Figure 6 Power spectrum of heat production variations with depth in the KTB borehole.

spectrum of the surface heat flow is

$$P_q(k_x, k_y) = C\int_0^\infty \frac{dk_z}{\left(k_x^2 + k_y^2 + k_z^2\right)^{(1+\beta_3/2)}} = \frac{C}{K^{(1+\beta_3)}}\int_0^\infty \frac{du}{(1+u^2)^{(1+\beta_3/2)}} \quad (11)$$

For the 9,000 m deep KTB scientific hole, heat production measurements were made with an average sampling interval of 3 m. The power spectrum of the vertical (one-dimensional) heat production profile appears to follow a power law $\propto K^{-\beta_1}$ with $\beta_1 = 1.73$ (Figure 6). Using this value for the one dimensional spectral slope, we obtain the exponent β_3 of the three-dimensional power spectrum using the equation $\beta_3 = \beta_1 + 2$. (e.g., Maus and Dimri, 1994). We then find that the power spectrum of the surface heat flow $P_q(K) \propto K^{-4.73}$. This is a very "smooth" function, which exhibits extremely small fluctuations at the shorter scale (<1 km), and has very long range (>10 km) correlations.

APPENDIX B: MANTLE HEAT FLOW, MOHO TEMPERATURE AND LITHOSPHERE THICKNESS

There has been some confusion in the past regarding the causes of changes in lithosphere thickness. They have usually been attributed to changes of mantle heat flow, reflecting differences in the vigor of mantle convection. This idea is valid for active regions such as extension zones, but may not be applicable to the base of the cratonic lithosphere. Here, we emphasize that changes of crustal heat production are as important as changes of deep convective input to surface heat flow. We also discuss how to use constraints on lithosphere thickness to estimate the mantle heat flow.

At steady state, the temperature at the Moho depends on the surface heat flow Q_0 and on the vertical distribution of heat sources. For the sake of simplicity, we shall assume a uniform distribution of heat sources. The Moho temperature T_m may then be written as

$$T_m = \left(\frac{Q_C}{2} + Q_M\right)\frac{h_C}{k_C} \quad (12)$$

where Q_C and Q_M are the crustal and mantle heat flow ($Q_0 = Q_C + Q_M$), and h_C and k_C are the crustal thickness and mean thermal conductivity, respectively.

The base of the lithosphere may be considered to lie along a well-mixed mantle isentropic profile (but, see Jaupart *et al.*, 1998). Heat is transported by conduction in the lithosphere, and hence

$$T_B = T_m + \frac{Q_M}{k_M}(h_L - h_C) \quad (13)$$

where k_M is the mean mantle conductivity, T_B the basal lithospheric temperature, and h_L the lithosphere thickness. We can see that, for fixed lithosphere thickness, an increase of crustal radioactivity leads to an increase of Moho temperature and to a decrease of mantle heat flow. For the sake of simplicity, we assume $k_M = k_C = k$. Combining the equations above, we obtain the mantle heat flow as follows:

$$Q_M = \frac{2kT_B - Q_0 h_C}{2h_L - h_C} \quad (14)$$

This shows that, for constant lithospheric thickness and basal temperature, an increase in surface heat flow can come only from crustal heat production, and leads to a decrease in mantle heat flow. If the lithospheric thickness varies, Equation (14) shows that an increase of surface heat flow and a decrease of lithospheric thickness can compensate one another with little change in mantle heat flow.

ACKNOWLEDGMENTS

The authors thank Jessica Tastet who with as much patience as good humor compiled some of the heat production data. This research was supported by NSERC (Canada), the FQRNT, previously FCAR (Québec), and the INSU (CNRS, France).

REFERENCES

Allègre C. J., Hart S. R., and Minster J.-F. (1983) Chemical structure and evolution of the mantle and continents determined by inversion of Nd and Sr isotopic data: II. Numerical experiments and discussion. *Earth Planet. Sci. Lett.* **66**, 191–213.

Allègre C. J., Lewin E., and Dupré B. (1988) A coherent crust-mantle model for the uranium–thorium–lead isotopic system. *Chem. Geol.* **70**, 211–234.

Arshavskaya N. I., Galdin N. E., Karus E. W., Kuznetsov O. L., Lubimova E. A., Milanovski S. Y., Nartikoev V. D., Semaskko S. A., and Smirnova E. V. (1987) Geothermic investigations. In *The Superdeep Well of the Kola Peninsula* (ed. Y. A. Kozlovsky). Springer, New York pp. 387–393.

Artemieva I. M. and Mooney W. D. (2001) Thermal thickness and evolution of Precambrian lithosphere: a global study. *J. Geophys. Res.* **106**, 16387–16414.

Ashwal L. D., Morgan P., Kelley S. A., and Percival J. (1987) Heat production in an Archean crustal profile and implications for heat flow and mobilization of heat producing elements. *Earth Planet. Sci. Lett.* **85**, 439–450.

Ballard S., Pollack H. N., and Skinner N. J. (1987) Terrestrial heat flow in Botswana and Namibia. *J. Geophys. Res.* **92**, 6291–6300.

Bea F. and Montero P. (1999) Behavior of accessory phases and redistribution of Zr, REE, Y, Th, and U during metamorphism and partial melting of metapelites in the lower crust: an example from the Kinzigite Fromation of Ivrea-Verbano, NW Italy. *Geochim. Cosmochim. Acta* **63**, 1133–1153.

Bingen B., Demaiffe D., and Hertogen J. (1996) Redistribution of rare earth elements, thorium, and uranium over accessory minerals in the course of amphibolite to granulite facies metamorphism: the role of apatite and monazite in orthogneisses from southwestern Norway. *Geochim. Cosmochim. Acta* **60**, 1341–1354.

Birch F. (1954) Heat from radioactivity. In *Nuclear Geology* (ed. H. Faul). Wiley, pp. 148–174.

Birch F., Roy R. F., and Decker E. R. (1968) Heat flow and thermal history in New England and New York. In *Studies of Appalachian Geology* (ed. E. An-Zen). Wiley-Interscience, pp. 437–451.

Borodin L. S. (1999) Estimated chemical composition and petrochemical evolution of the upper continental crust. *Geochem. Int.* **37**, 723–734.

Cermak V. (1993) Lithospheric thermal regimes in Europe. *Phys. Earth Planet. Int.* **79**, 179–193.

Chamberlain C. P. and Sonder L. J. (1990) Heat-producing elements and the thermal and baric patterns of metamorphic belts. *Science* **250**, 763–769.

Christensen N. I. and Mooney W. D. (1995) Seismic velocity structure and composition of the continental crust: a global view. *J. Geophys. Res.* **100**, 9761–9788.

Clark S. P., Jr. and Ringwood A. E. (1964) Density distribution and constitution of the mantle. *Rev. Geophys.* **2**, 35.

Clauser C., Gieses P., Huenges E., Kohl T., Lehmann H., Rybach L., Safanda J., Wilhelm H., Windlow K., and Zoth G. (1997) The thermal regime of the crystalline continental crust: implications from the KTB. *J. Geophys. Res.* **102**, 18417–18441.

Clowes R. M., Cook F. A., Green A. G., Keen C. E., Ludden J. N., Percival J. A., Quinlan G. M., and West G. F. (1992) LITHOPROBE—new perspectives on crustal evolution. *Can. J. Earth Sci.* **29**, 1831–1864.

Cogley J. G. (1984) Continental margins and the extent and number of continents. *Rev. Geophys. Space Phys.* **22**, 101–122.

Condie K. C. (1993) *Plate Tectonics and Crustal Evolution*. Pergamon, London.

Cull J. P. (1991) Heat flow and regional geophysics in Australia. In *Terrestrial Heat Flow and the Lithosphere Structure* (eds. V. Cermak and L. Rybach). Springer, New York, pp. 486–500.

Drury M. J., Jessop A. M., and Lewis T. J. (1987) The thermal nature of the Canadian Appalachians. *Tectonophysics* **113**, 1–14.

Durrheim R. V. and Mooney W. D. (1991) Archean and Proterozoic crustal evolution: evidence from crustal seismology. *Geology* **19**, 606–609.

Eade K. E. and Fahrig W. F. (1971) Geochemical evolutionary trends of continental plates, a preliminary study of the Canadian Shield. Geol. *Surv. Canada Bull.* **179**, 1–59.

Eilu P. (1994) Hydrothermal alteration in volcano-sedimentary rocks in the central Lapland greenstone belt, Finland, *Geol. Surv. Finland Bull.* **374**, 145pp.

England P. C., Oxburg E. R., and Richardson S. W. (1980) Heat refraction in and around granites in north-east England. *Geophys. J. Roy. Astron. Soc.* **62**, 439–455.

Forster A. and Forster H.-J. (2000) Crustal composition and mantle heat flow: implications from surface heat flow and radiogenic heat production in the Variscan Erzgebirge (Germany). *J. Geophys. Res.* **105**, 27917–27938.

Fountain D. M. (1985) Is there a relationship between seismic velocity and heat production for crustal rocks? *Earth Planet. Sci. Lett.* **79**, 145–150.

Fountain D. M. and Salisbury M. H. (1981) Exposed cross-sections of the continental crust: implications for crustal structure petrology, and evolution. *Earth Planet. Sci. Lett.* **56**, 263–277.

Fountain D. M., Salisbury M. H., and Furlong K. P. (1987) Heat production and thermal conductivity of rocks from the Pikwitonei–Sachigo continental cross section, central Manitoba: implications for the thermal structure of Archean crust. *Can. J. Earth Sci.* **24**, 1583–1594.

Furukawa Y. and Shinjoe H. (1997) Distribution of radiogenic heat generation in the arc's crust of the Hokkaido island, Japan. *Geophys. Res. Lett.* **24**, 1279–1282.

Furukawa Y. and Uyeda S. (1989) Thermal state under the Tohoku arc with consideration of crustal heat generation. *Tectonophysics* **164**, 175–187.

Galer S. J. G., Goldstein S. L., and O'Nions R. K. (1989) Limits on chemical and convective isolation in the Earth's interior. *Chem. Geol.* **75**, 257–290.

Gao S., Luo T.-C., Zhang B.-R., Zhang H.-F., Han Y.-W., Hu Y.-K., and Zhao Z.-D. (1998) Chemical composition of the continental crust as revealed by studies in East China. *Geochim. Cosmochim. Acta* **62**, 1959–1975.

Goes S. and van der Lee S. (2002) Thermal structure of the North American uppermost mantle inferred from seismic tomography. *J. Geophys. Res.* **107**, ETG, 2-1–2-13.

Goodwin A. (1997) *Precambrian Geology: The Dynamic Evolution of the Continental Crust*. Academic Press, New York.

Guillou L., Mareschal J.-C., Jaupart C., Gariépy C., Bienfait G., and Lapointe R. (1994) Heat flow gravity and structure of the Abitibi belt, Superior Province, Canada: implications for mantle heat flow. *Earth Planet. Sci. Lett.* **122**, 103–123.

Guillou-Frottier L., Mareschal J.-C., Jaupart C., Gariépy C., Lapointe R., and Bienfait G. (1995) Heat flow variations in the Grenville Province, Canada. *Earth Planet. Sci. Lett.* **136**, 447–460.

Gupta M. L., Sundar A., and Sharma S. R. (1991) Heat flow and heat generation in the Archean Dharwar cratons and implications for the southern Indian Shield geotherm and lithospheric thickness. *Tectonophysics* **194**, 107–122.

Gung Y., Panning M., and Romanowicz B. (2003) Global anisotropy and the thickness of the continents. *Nature* **422**, 707–711.

Haack U. (1983) On the content and vertical distribution of K, Th, and U in the continental crust. *Earth Planet. Sci. Lett.* **62**, 360–366.

Hamdani Y., Mareschal J.-C., and Arkani-Hamed J. (1991) Phase change and thermal subsidence in intracontinental sedimentary basins. *Geophys. J. Int.* **106**, 657–665.

Hanski E. J. (1992) Petrology of the Pechenga ferropicrites and cogenetic Ni-bearing gabbro-wherlite intrusions, Kola peninsula, Russia. *Geol. Surv. Finland Bull.* **367**, 192pp.

Heier K. S. and Lambert I. B. (1978) *A Compilation of Potassium, Uranium and Thorium Abundances and Heat Production of Australian Rocks*. Technical Report, Research School of Earth Science, Australian National University, Canberra.

Hu S., He L., and Wang J. (2000) Heat flow in the continental area of China: a new data set. *Earth Planet. Sci. Lett.* **179**, 407–419.

Jaupart C. (1983a) Horizontal heat transfer due to radioactivity contrasts: causes and consequences of the linear heat flow-heat production relationship. *Geophys. J. Roy. Astron. Soc.* **75**, 411–431.

Jaupart C. (1983b) The effects of alteration and the interpretation of heat flow and radioactivity data—a reply to R.U.M. Rao. *Earth Planet. Sci. Lett.* **62**, 430–438.

Jaupart C. and Mareschal J.-C. (1999) The thermal structure of continental roots. *Lithos* **48**, 93–114.

Jaupart C., Sclater J. G., and Simmons G. (1981) Heat flow studies: constraints on the distribution of uranium, thorium and potassium in the continental crust. *Earth Planet. Sci. Lett.* **52**, 328–344.

Jaupart C., Mann J. R., and Simmons G. (1982) A detailed study of the distribution of heat flow and radioactivity in New Hampshire (USA). *Earth Planet. Sci. Lett.* **59**, 267–287.

Jaupart C., Mareschal J.-C., Guillou-Frottier L., and Davaille A. (1998) Heat flow and thickness of the lithosphere in the Canadian Shield. *J. Geophys. Res.* **103**, 15269–15286.

Jessop A. M., Hobart M., and Sclater J. G. (1976) The World heat flow data compilation—1975. *Geothermal Ser. No. 5*. Dept. Energy, Mines and Resources, Ontario, Canada.

Jõeleht A. and Kukkonen I. T. (1998) Thermal properties of granulite facies rocks in the Precambrian basement of Finland and Estonia. *Tectonophysics* **291**, 195–203.

Jones M. Q. W. (1987) Heat flow and heat production in the Namaqua mobile belt, South Africa. *J. Geophys. Res.* **92**, 6273–6289.

Jones M. Q. W. (1988) Heat flow in the Witwatersrand Basin and environs and its significance for the South African shield geotherm and lithosphere thickness. *J. Geophys. Res.* **93**, 3243–3260.

Jones M. Q. W. (1992) Heat flow anomaly in Lesotho: implications for the southern boundary of the Kaapvaal craton. *Geophys. Res. Lett.* **19**, 2031–2034.

Jull M. and Kelemen P. B. (2001) On the conditions for lower crustal convective instability. *J. Geophys. Res.* **106**, 6423–6446.

Kaminski E. and Jaupart C. (2000) Lithosphere structure beneath the Phanerozoic intracratonic basins of North America. *Earth Planet. Sci. Lett.* **178**, 139–149.

Kay R. W. and Mahlburg Kay S. (1991) Creation and destruction of lower continental crust. *Geol. Rundsch.* **80**, 259–278.

Ketcham R. A. (1996) Distribution of heat-producing elements in the upper and middle crust of southern and west central Arizona: evidence from the core complexes. *J. Geophys. Res.* **101**, 13611–13632.

Killeen P. G. and Heier K. S. (1975a) A uranium and thorium enriched province of the Fennoscandian Shield in southern Norway. *Geochim. Cosmochim. Acta* **39**, 1515–1524.

Killeen P. G. and Heier K. S. (1975b) Trend surface analysis of Th, U and K, and heat production in three related granitic plutons, Farsund area, south Norway. *Chem. Geol.* **15**, 163–176.

Kremenentsky A. A., Milanovsky S. Y., and Ovchinnikov L. N. (1989) A heat generation model for the continental crust based on deep drilling in the Baltic Shield. *Tectonophysics* **159**, 231–246.

Kukkonen I. T. and Peltonen P. (1999) Xenolith-controlled geotherm for the central Fennoscandian Shield: implications for lithosphere–asthenosphere relations. *Tectonophysics* **304**, 301–315.

Kukkonen I. T. and Peltoniemi S. (1998) Relationships between thermal and other petrophysical properties of rocks in Finland. *Phys. Chem. Earth* **23**, 341–349.

Lachenbruch A. H. (1970) Crustal temperature and heat production: implications of the linear heat flow heat production relationship. *J. Geophys. Res.* **73**, 3292–3300.

Lahtinen R. (1996) Geochemistry of supracrustal and Plutonic rocks. *Geol. Surv. Finland, Bull.* **389**, 113pp.

Landstrom O., Larson S. A., Lind G., and Malmqvist D. (1980) Geothermal investigations in the Bohus granite area in southwestern Sweden. *Tectonophysics* **64**, 131–162.

Lee M. K., Brown G. C., Webb P. C., Wheildon J., and Rollin K. E. (1987) Heat flow, heat production and thermo-tectonic setting in mainland UK. *J. Geol. Soc. London* **144**, 35–42.

Lewis T. J., Hyndman R. D., and Fluck P. (2003) Heat flow, heat generation and crustal temperatures in the northern Canadian cordillera: thermal control on tectonics. *J. Geophys. Res.* 108B6, doi# 10.1029/2002JB002090.

Louden K. E., Sibuet J.-C., and Foucher J.-P. (1991) Variations of heat flow across the Goban Spur and Galicia Bank continental margins. *J. Geophys. Res.* **96**, 16131–16150.

Mareschal J.-C. (1985) Inversion of potential field data in Fourier transform domain. *Geophysics* **50**, 685–691.

Mareschal J.-C., Jaupart C., Cheng L.-Z., Rolandone F., Gariepy C., Bienfait G., Guillou-Frottier L., and Lapointe R. (1999) Heat flow in the Trans Hudson Orogen of the Canadian Shield: implications for Proterozoic continental growth. *J. Geophys. Res.* **104**, 29007–29024.

Mareschal J.-C., Jaupart C., Gariépy C., Cheng L.-Z., Guillou-Frottier L., Bienfait G., and Lapointe R. (2000a) Heat flow and deep thermal structure near the edge of the Canadian Shield. *Can. J. Earth Sci.* **37**, 399–414.

Mareschal J.-C., Poirier A., Rolandone F., Bienfait G., Gariepy C., Lapointe R., and Jaupart C. (2000b) Low mantle heat flow at the edge of the North American continent, Voisey Bay, Labrador. *Geophys. Res. Lett.* **27**, 823–826.

Mareschal J.-C., Jaupart C., Rolandone F., Gariépy C., Fowler M., Bienfait G., Carbonne C., and Lapointe R. (2003) Heat flow, thermal regime, and rheology of the lithosphere in the Trans-Hudson Orogen. *Can. J. Earth Sci.* (submitted for publication).

Maus S. and Dimri V. (1994) Scaling properties of potential fields due to scaling sources. *Geophys. Res. Lett.* **21**, 891–894.

McDonough W. F. and Sun S. S. (1995) The composition of the Earth. *Chem. Geology* **120**, 223–253.

McLennan S. M. and Taylor S. R. (1996) Heat flow and the chemical composition of continental crust. *J. Geol.* **104**, 377–396.

Miller D. J. and Christensen N. I. (1994) Seismic signature and geochemistry of an island arc: a multidisciplinary study of the Kohistan accreted terrane, northern Pakistan. *J. Geophys. Res.* **99**, 11623–11642.

Mooney W. D., Laske G., and Masters G. T. (1998) CRUST 5.1: a global crustal model at 5° × 5°. *J. Geophys. Res.* **103**, 727–747.

Morgan P. (1985) Crustal radiogenic heat production and the selective survival of ancient continental crust. *J. Geophys. Res.* **90**, C561–C570.

Nicolaysen L. O., Hart R. J., and Gale N. H. (1981) The Vredefort radioelement profile extended to supracrustal strata at Carletonville, with implications for continental heat flow. *J. Geophys. Res.* **86**, 10653–10661.

Nyblade A. A. and Pollack H. N. (1993) A global analysis of heat flow from Precambrian terrains: implications for the thermal structure of Archean and Proterozoic lithosphere. *J. Geophys. Res.* **98**, 12207–12218.

O'Nions R. K., Evensen N. M., and Hamilton P. J. (1979) Geochemical modeling of mantle differentiation and crustal growth. *J. Geophys. Res.* **84**, 6091–6101.

Oversby V. M. (1976) Isotopic ages and geochemistry of Archean and igneous rocks from the Pilbara, Western Australia. *Geochim. Cosmochim. Acta* **40**, 817–829.

Pinet C. and Jaupart C. (1987) The vertical distribution of radiogenic heat production in the Precambrian crust of Norway and Sweden: geothermal implications. *Geophys. Res. Lett.* **14**, 260–263.

Pinet C., Jaupart C., Mareschal J.-C., Gariépy C., Bienfait G., and Lapointe R. (1991) Heat flow and structure of the lithosphere in the eastern Canadian Shield. *J. Geophys. Res.* **96**, 19941–19963.

Pollack H. N., Hurter S. J., and Johnson J. R. (1993) Heat flow from the Earth's Interior: analysis of the global data set. *Rev. Geophys.* **31**, 267–280.

Rogers J. J. W. and Adams J. A. S. (1969) Thorium. In *Handbook of Geochemistry* (ed. K. H. Wedephol). Springer, pp. 90B–90O.

Rogers J. J. W., Adams J. A. S., and Gatlin B. (1965) Distribution of thorium, uranium and potassium in three cores from the Conway granite, New Hampshire. *Am. J. Sci.* **263**, 817–822.

Rolandone F., Jaupart C., Mareschal J.-C., Gariépy C., Bienfait G., Carbonne C., and Lapointe R. (2002) Surface heat flow, crustal temperatures and mantle heat flow in the Proterozoic Trans-Hudson Orogen, Canadian Shield. *J. Geophys. Res.* **107**, 2341, doi:10.1029/2001JB000698.

Roy S. and Rao R. U. M. (2000) Heat flow in the Indian shield. *J. Geophys. Res.* **105**, 25587–25604.

Rudnick R. L. (1992) Xenoliths—samples of the lower continental crust. In *Continental Lower Crust* (ed. D. M. Fountain, R. Arculees, and R. W. Kay). Elsevier, Amsterdam, pp. 269–316.

Rudnick R. L. (1995) Making continental crust. *Nature* **378**, 571–578.

Rudnick R. L. and Fountain D. M. (1995) Nature and composition of the continental crust: a lower crustal perspective. *Rev. Geophys.* **33**, 267–309.

Rudnick R. L. and Nyblade A. A. (1999) The thickness and heat production of Archean lithosphere: constraints from xenolith thermobarometry and surface heat flow. In *Mantle Petrology: Field Observations and High Pressure Experimentation: A Tribute to Francis R. (Joe) Boyd* (eds. Y. Fei, C. M. Bertka, and B. O. Mysen). The Geochemical Society, pp. 3–12.

Rudnick R. L., McDonough W. F., and O'Connell R. J. (1998) Thermal structure, thickness and composition of continental lithosphere. *Chem. Geol.* **145**, 395–411.

Russell J. K. and Kopylova M. G. (1999) A steady-state conductive geotherm for the north central Slave, Canada: inversion of petrological data from the Jericho Kimberlite pipe. *J. Geophys. Res.* **104**, 7089–7101.

Russell J. K., Dipple G. M., and Kopylova M. G. (2001) Heat production and heat flow in the mantle lithosphere, Slave craton, Canada. *Phys. Earth Planet. Int.* **123**, 27–44.

Rybach L. (1988) Determination of heat production rate. In *Handbook of Terrestrial Heat-flow Density Determination* (eds. R. Haenel, L. Rybach, and L. Stegena). Kluwer, pp. 125–142.

Salonsaari P. T. (1995) Hybridization in the subvolcanic Jaala-Iitti complex. *Bull. Geol. Soc. Finland* **67**, 1–104.

Sclater J. G., Jaupart C., and Galson D. (1980) The heat flow through oceanic and continental crust and the heat loss from the Earth. *Rev. Geophys.* **18**, 269–311.

Shaw D. M., Cramer J. J., Higgins M. D., and Truscott M. G. (1986) Composition of the Canadian Precambrian Shield and the continental crust of the Earth. In *Nature of the Lower Continental Crust* (ed. J. B. Dawson *et al.*). Geological Society of London, pp. 257–282.

Shaw D. M., Dickin A. P., Li H., McNutt R. H., Schwarcz H. P., and Truscott M. G. (1994) Crustal geochemistry in the Wawa-Foleyet region, Ontario. *Can. J. Earth Sci.* **31**, 1104–1121.

Smellie J. A. T. and Stuckless J. S. (1985) Element mobility studies of two drill-cores from the Gtemar granite (Krakemala test site), southern Sweden. *Chem. Geol.* **51**, 55–78.

Stein C. A. (1995) Heat flow of the Earth. In *Global Earth Physics. A Handbook of Physical constants. AGU Reference Shelf 1* (ed. T. J. Ahrens). American Geophysical Union, pp. 144–158.

Swanberg C. A., Chessman M. D., Simmons G., Smithson S. B., Gronlie G., and Heier K. S. (1974) Heat flow—heat generation studies in Norway. *Tectonophysics* **23**, 31–48.

Taylor S. R. and McLennan S. M. (1985) *The Continental Crust: Its Composition and Evolution*. Blackwell.

Taylor S. R. and McLennan S. M. (1995) The geochemical evolution of the continental crust. *Rev. Geophys.* **33**, 241–265.

Vasseur G. and Singh R. N. (1986) Effects of random horizontal variations in radiogenic heat source distribution on its relationship with heat flow. *J. Geophys. Res.* **91**, 10397–10404.

Wasserburg G. J., MacDonald G. J. F., Hoyle F., and Flower W. A. (1964) Relative contributions of uranium, thorium and potassium to heat production in the Earth. *Science* **143**, 465–467.

Weaver B. L. and Tarney J. (1984) Empirical approach to estimating the composition of the continental crust. *Nature* **310**, 575–577.

Wedepohl K. H. (1995) The composition of the continental crust. *Geochim. Cosmochim. Acta* **59**, 1217–1239.

Wollenberg H. A. and Flexser S. (1985) The distribution of uranium and thorium in the Stripa quartz monzonite, Sweden. *Uranium* **2**, 155–167.

Zielinski R. A., Peterman Z. E., Stuckless J. S., Rosholt J. N., and Nkomo I. T. (1978) The chemical and isotopic record of water-rock interaction in the Sherman granite. *Contrib. Mineral. Petrol.* **78**, 209–219.

3.03
Continental Basaltic Rocks

G. L. Farmer

University of Colorado, Boulder, CO, USA

3.03.1 INTRODUCTION

During the past few decades, geochemical studies of continental basaltic rocks and their petrologic kin have become mainstays of studies of the continental lithosphere. These igneous rocks have taken on such an important role largely because the chemical and isotopic composition of continental basaltic rocks and their mantle (see Chapter 2.05) and crustal xenoliths (see Chapter 3.01) provide the best proxy record available to earth scientists for the chemical and physical evolution of the deep continental lithosphere and underlying mantle, areas that are otherwise resistant to direct study. Keeping this in mind, the primary goal of this chapter is to illustrate how geochemical data can be used both to assess the origin of these rocks and to study the evolution of the continental lithosphere.

A complete overview of continental basaltic rocks will not be attempted here, because continental "basalts" come in too wide a range of compositions, and because of the sheer volume of geochemical data available for such rocks worldwide. The scope of the chapter is limited to a discussion of a select group of ultramafic to mafic composition "intraplate" continental igneous rocks consisting primarily of kimberlites, potassic and sodic alkali basalts, and continental flood basalts. Igneous rocks forming at active continental margins, such as convergent or transform plate margins, are important examples of continental magmatism but are not directly discussed here (convergent margin magmas are discussed in Chapters 2.11, 3.11, and 3.18). The geochemistry of intraplate igneous rocks of the ocean basins are covered in Chapters 2.04 and 3.16. Although basaltic magmatism has occurred throughout the Earth's history, the majority of the examples presented here are from Mesozoic and Cenozoic volcanic fields due to the more complete preservation of younger continental mafic igneous rocks.

While considerable effort has been expended in studying the chemical differentiation of mafic magmas, the present discussion concentrates on the least differentiated basaltic rocks in a given location. Such rocks generally provide the best estimate of the compositions of "primary" magmas generated beneath a given volcanic field, and primary magmas provide the most direct insights into the nature of the magma source regions.

3.03.2 GENERAL PRINCIPLES

Intraplate continental ultramafic to mafic magmatism ranges in volume from <5 km^3 of erupted material for kimberlite pipes and alkali basalt volcanic centers to >2,500 km^3 for individual continental flood basalt flows associated with "large igneous provinces" (Sigurdsson, 2000). Regardless of the volume involved, petrologists are faced with similar questions when attempting to define the origin of any continental igneous rock: where, and why, did the melting occur that ultimately led to the formation of that rock? This information is not simple to extract from the composition of a given igneous rock, because its composition is influenced both by the composition of the magmatic source region, which, in the case of the continental lithosphere, can be highly heterogeneous, and through the processes by which the magma was generated and then modified during transport to shallower depths in the lithosphere. Nevertheless, because mafic continental igneous rocks could have been derived from or have extensively interacted with continental lithosphere, they contain information regarding the chemical and isotopic composition and the physical state of the deep lithosphere.

Although defining the origin and evolution of mafic continental magmatism is a complex task, there are certain basic tenets regarding the possible source regions of such magmas and the factors influencing magma chemical and isotopic compositions that can be used to help constrain the origin of a given igneous rock. These tenets are outlined in the following sections.

3.03.2.1 Potential Sources of Continental Mafic Magmatism

It is generally conceded that ultramafic to mafic magmas are the products of partial melting of ultramafic rocks in the Earth's upper mantle (Kushiro, 2001; see Chapter 2.08). The composition of primary magmas generated by partial melting of ultramafic mantle rocks are, in turn, controlled by the mineralogy and chemical composition of the ultramafic rocks, as well as by the pressure, temperature, and extent of partial melting (Grove, 2000). A "primary" magma, for our purposes, is a magma produced at depth in the Earth, the composition of which is controlled by the mineralogic and chemical makeup of the source and the conditions of melting, and has not been modified by differentiation during ascent to shallower depths (Winter, 2001). Basalts representing primary magmas derived from the melting of the upper mantle are simple to recognize, if not commonly found, because magmas that were in equilibrium with magnesium-rich olivine (mg# = Mg/(Mg + Fe^{2+}) = 0.90) in the mantle will also have high mg# (~0.73; Grove (2000)).

The ability of the mantle to produce basaltic magmas, generally referred to as its "fertility," is a fundamental parameter controlling where primary continental basaltic rocks can be generated. For a typical lherzolitic mantle, consisting of olivine, orthopyroxene, clinopyroxene, and an aluminous phase (either plagioclase, spinel, or garnet, depending on depth), extraction of a basaltic magma results in a decrease in whole-rock aluminum, silicon, iron, sodium, calcium, titanium, and chromium contents, and a relative increase in magnesium concentrations (and an increase in the whole-rock mg#; Frey and Green (1974) and Griffin *et al.* (1999)). Mineralogically, basaltic magma production from lherzolite generally consumes clinopyroxene, resulting in a mantle residuum of either harzburgite (olivine + orthopyroxene) or even dunite (i.e., olivine) when the percentage of partial melting is high (Winter, 2001). As a result, continental basalts must either be produced from portions of the mantle that have remained fertile for basalt generation throughout Earth history, or that have been subsequently refertilized with basaltic components through metasomatism by infiltrating fluids and/or magmas (Kelemen *et al.*, 1998; Griffin *et al.*, 1999).

Where are the potential fertile mantle sources for continental basalt generation? In simplest terms, two possibilities exist: the continental lithospheric mantle (CLM) and the underlying upper mantle. The latter is sometimes referred to as the "asthenosphere" by geochemists (Figure 1). Unfortunately, the definitions of lithosphere and asthenosphere used for geochemical purposes have been the source of some confusion (Anderson, 1995). Adopted here is the definition of continental "lithosphere" (White, 1988), in which the lithosphere is considered to be the "outer shell of the Earth where there is a conductive temperature gradient, overlying the well-mixed adiabatic interior." In this definition the adiabatic "interior" includes the convecting, sublithospheric, upper mantle, embedded in which is the shallow, weak, and presumably partially molten asthenosphere.

The above definition of the lithosphere equates the lithosphere with a thermal boundary layer

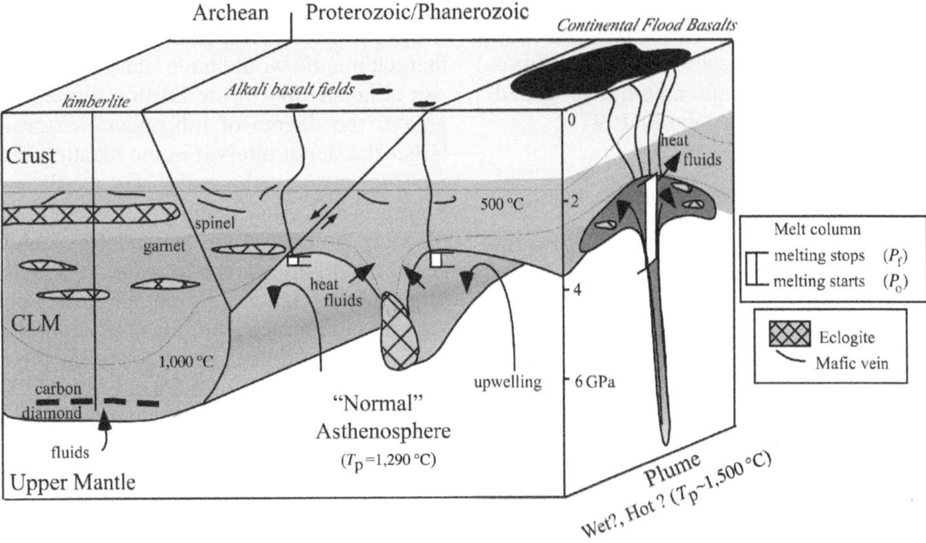

Figure 1 Cartoon depicting continental crust and mantle lithosphere (CLM), underlying sublithospheric mantle, and potential sources of ultramafic to mafic composition continental magmatism.

(TBL), a layer in which heat is transferred by conduction and not by advection (Anderson, 1995). As a result, the base of the lithosphere by this definition essentially represents the depth at which the conductive geotherm of the lithosphere intersects the adiabat describing temperature as a function of depth in the underlying convecting mantle (Rudnick *et al.*, 1998). Note that the lithosphere, as originally defined on the basis of its high viscosity and high strength (Barrell, 1914), represents at best only the upper half of the TBL (Anderson, 1995). Therefore, the base of the lithosphere, as defined here, is not necessarily strong and permanent. In fact, some workers have argued that the base of the lithosphere, if hydrated, becomes weak and subject to flow (Anderson, 1995).

The sublithospheric mantle is involved in the production of voluminous basaltic magmatism at mid-ocean ridges (see Chapter 3.13), and so is obviously fertile for basalt production. The fertility of continental mantle lithosphere, in contrast, is more variable. Continental mantle lithosphere, unlike its oceanic counterpart, may not have been isolated from the convecting portions of the mantle solely through progressive conductive cooling through time. It has long been argued that the mantle portions of the Archean continental lithosphere may be compositionally distinct from the mantle lithosphere associated with younger crustal segments or oceanic lithosphere due to the extraction of large melt fractions from the upper mantle during Archean crust formation (Jordan, 1978; Jordan, 1988). Komatiites, which are primarily Archean, highly magnesian (>18 wt.% MgO; Le Bas, 2000) extrusive igneous rocks, likely represent the products of such large degrees of mantle melting (either in anhydrous, hot plume or in wet, Archean subduction zones; Arndt *et al.* (1998) and Parman *et al.* (2001)). Extraction of komatiitic magma increases the buoyancy of the mantle due to the preferential removal of iron (relative to magnesium) into the magma. The resulting decrease in the density of the residual mantle could account both for the preferential preservation of Archean mantle and the great thickness (>200 km; Figure 1) proposed for many Archean cratons (Jordan, 1988). However, thick, unmetasomatized Archean mantle lithosphere is also likely to be refractory, and therefore infertile when it comes to the production of basaltic magmas. In contrast, post-Archean CLM could be sufficiently fertile (Jordan, 1988) to spawn additional basaltic magmatism. It should also be noted that variations in CLM compositions affect lithospheric strength (Lee *et al.*, 2001), and as a result fertile mantle lithosphere may be more susceptible to thinning, either through stretching or the development of lithospheric mantle "drips" (Figure 1).

Although ultramafic lithologies in the CLM and underlying mantle are the principal sources of continental basaltic magmas, both lithospheric and sublithospheric mantles are heterogeneous and can contain mafic lithologies that might become involved in magma production. For example, pyroxenites, websterites, eclogites, etc., exist as veins and/or discrete layers in portions of the continental mantle (Wilshire *et al.*, 1991; see also Chapters 2.04 and 2.05). Although these lithologies are generally thought to produce intermediate to silicic, and not mafic, magmas when partially melted (Leeman and Harry, 1993) (see Pertermann and Hirschmann (2003) for an alternative view), magmatic contributions by mafic lithologies can influence both the composition and the volume of mafic magmas produced in the upper mantle (Cordery *et al.*, 1997).

3.03.2.2 Trigger Mechanisms for Mantle Melting

Mantle melting simply requires that the mantle temperature, at a given pressure, exceeds its solidus temperature. This can be induced in the mantle either by heating, decompression, or a change in chemical composition, particularly the addition of volatiles, which greatly reduces the mantle solidus temperature (see also Chapter 2.07; Figure 2). Each of these processes can be relevant in the formation of continental magmatism. For example, consider melting in the sublithospheric mantle. Decompression melting of this mantle can be induced by lithospheric thinning associated with continental extension, particularly during the formation of continental rifts (Figure 1). Melting occurs when deeper mantle flows upwards to compensate for the decreased lithospheric thickness. In a seminal paper, McKenzie and Bickle (1988) numerically modeled this process and

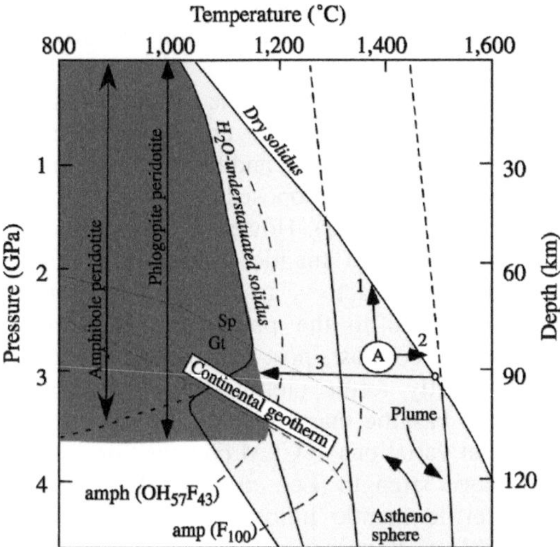

Figure 2 Pressure versus temperature plot for the upper mantle, showing both dry and H_2O undersatured solidii for lherzolite. Melting of dry mantle originally at point "A" can be induced either by adiabatic upwelling (path "1"), conductive heating (path "2"), or addition of volatiles, which lowers the mantle solidus temperature (path "3"). The diagram also illustrates that amphibole and phlogopite are unlikely to be present in either normal asthenosphere, or hot mantle plumes because these phases break down at higher temperatures. Both phases are apparently stable only in colder lithospheric mantle. Fine dashed lines show stability limits of amphiboles with various fluorine contents. Shaded area is stability range of phlogopite. Sp = spinel, Gt = garnet (after Class and Goldstein, 1997; le Roex *et al.*, 2001).

demonstrated that the higher the sublithospheric mantle-potential temperature (T_p, the temperature that the mantle would have should it be allowed to rise adiabatically to the Earth's surface), and the greater the degree of lithospheric extension, the larger the depth interval in the mantle over which melting occurs (the polybaric "melt column," Figure 1) and, consequently, the larger the volume of melt produced. In general, the thicker the continental lithosphere and the colder the sublithospheric mantle, the smaller the volume of magma produced by sublithospheric mantle melting. For example, decompression melting of "normal" potential temperature (1,290 °C) sublithospheric mantle induced by lithospheric extension may occur through only a short pressure interval because the melt column is effectively capped by the base of the CLM (Figure 1). As a result, only small melt volumes are generated, as observed in alkali basalt volcanic fields (Figure 1) and melting may be restricted to the garnet peridotite stability field. Conversely, the hotter the sublithospheric mantle, the greater the depth of initiation of melting, and the greater the volume of magma produced in upwelling mantle, even in the absence of thinning lithosphere. Hence, the likely importance of high potential temperature mantle plumes in the generation of large-volume continental flood basalts (Figure 1). Decompression melting in the sublithospheric mantle can also be induced within convective instabilities in the upper mantle resulting from original variations in lithospheric thickness (King and Anderson, 1998) or lithospheric "delamination" (Elkins and Hager, 2000).

Decompression melting in the continental mantle lithosphere, in contrast, is difficult to induce, even in fertile mantle. Harry and Leeman (1995), for example, demonstrated that lithospheric extension does not typically bring either dry, or water saturated peridotite into supersolidus conditions (Figure 3). Instead, melting of lithospheric mantle seems to require the addition of fluids and/or heat from the underlying mantle (Harry and Leeman, 1995). Turner *et al.* (1996b) demonstrated that conductive heating of hydrated lithospheric mantle by an upwelling mantle plume can produce significant volumes of melt depending, in part, on the original thickness of hydrated mantle.

Clearly, then, the physical and chemical characteristics of the continental lithosphere play an important role in determining the sources and volumes of continental mafic magmatism. Therefore, if the sources of a given continental igneous rock can be determined, then inferences can be made regarding the dynamic processes in the subcontinental mantle during magma generation. The difficulty is to determine unambiguously the source of a given continental basalt. For this purpose,

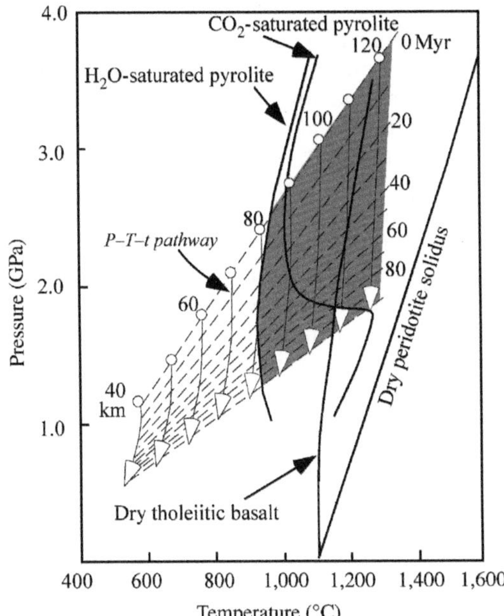

Figure 3 Pressure versus temperature plot showing calculated temperatures of lithospheric mantle as a function of time during an episode of lithospheric extension. The shaded area is mantle with pressures and temperatures above the H_2O-saturated solidus. In the example shown, lithospheric mantle that originally ($t = 0$) spanned the depth interval from 40 km to 125 km undergoes extension and thinning at a rate of 5 mm yr^{-1}. The diagram illustrates that dry peridotite mantle does not cross its solidus during ascent associated with lithospheric extension. As a result, extension cannot generally induce melting in dry, lithospheric peridotite. Similarly, hydrated, shallow (<80 km) peridotite does not cross the lower-temperature wet solidus during extension. Melting of lithospheric peridotite instead requires hydration of initially dry mantle and/or conductive heating from the underlying mantle, for example, during plume impact (Turner *et al.*, 1996b). However, melting of mafic lithologies within CLM can occur if the mantle crosses the basalt solidus during extension (reproduced by permission of American Geophysical Union from *J. Geophys. Res.*, **1995**, *100*, 10255–10269).

the geochemist must rely on indirect information regarding magma source regions provided by a basalt's major, trace element and isotopic composition.

3.03.2.3 Factors Influencing Magma Major Element Compositions

Petrologic and major element studies of Cenozoic continental basaltic rocks have defined two primary magma "series" in intraplate basaltic igneous rocks: alkalic, and tholeiitic (the latter comprising, along with "calc-alkalic," the subalkalic series of Iddings (1892)). Alkalic continental basalts typically have lower silicon and higher sodium and potassium contents than tholeiitic basalts, while the latter show a characteristic pattern of initial iron enrichment during differentiation (Winter, 2001). A primary goal of modern studies in igneous petrology has been to determine what factors lead to the production of these two magma series, in particular to assess the relative importance of mantle source composition and the physical parameters of melting in controlling melt composition. One line of thought is that the two magma series represent different degrees and depths of melting of dry peridotite. Experimental studies have shown that small degrees of melting (<5% by weight) of peridotite at pressures greater than 3 GPa produces primary alkali magmas of basaltic composition, while larger degrees of melting of the same source at shallower depths can generate tholeiitic series basaltic magmas (Hirose and Kushiro, 1993; see Chapter 2.08). Other workers, however, argue that compositional variations in the mantle source are required to produce the various magma series. In particular, the generation of highly potassic alkaline magmas has been attributed to melting within ultramafic mantle containing metasomatically introduced, potassium-rich, hydrous mineral phases such as phlogopite or potassic amphibole (K-richterite, for example) (Foley, 1992). The metasomatic addition of alkali elements and volatile constituents, principally CO_2 or H_2O, to the lithospheric mantle via fluids and/or melts from sublithospheric mantle is key to the generation of other alkaline continental igneous rocks, including kimberlites, carbonatites, and lamproites (Wyllie and Huang, 1976; see Chapter 2.07).

Understanding the range of major element compositions produced by polybaric decompression melting of the sublithospheric mantle has also been a focus of recent research efforts, in part because theoretical and experimental studies in the 1990s have provided a detailed picture of the compositions of primary magmas produced through melting of spinel and garnet peridotites (cf. Baker and Stolper, 1994). Most notably Langmuir *et al.* (1992) used information on the partitioning of magnesium and iron between olivine and magma (Roeder and Emslie, 1970) and sodium (an incompatible element) partitioning between mantle clinopyroxene and melt to model the composition of magmas as a function of the depth of melt initiation and the extent of melting. The results of these calculations illustrate that iron contents of primary magmas derived from mantle peridotite increase with increasing depth of melt initiation, and that the sodium content decreases regularly with the extent of mantle melting (Figure 4). The sodium and iron contents of a given suite of basaltic rocks can therefore provide constraints on the physical

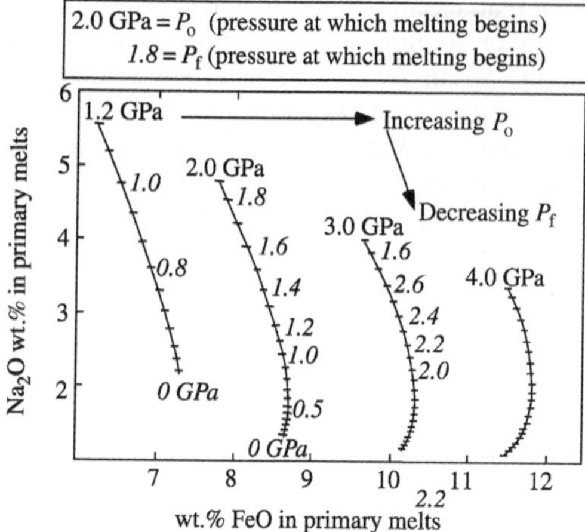

Figure 4 Plot of Na_2O versus FeO (wt.%) for products of polybaric melting of peridotitic mantle using the model of Langmuir *et al.* (1992). Each line shows a polybaric melting "path" during adiabatic decompression from the depth (pressure) where melting commenced, P_o, to a final depth of melting, P_f. The compositions of fractional melts produced during decompression melting of the mantle are accumulated and the resulting bulk compositions are plotted at intervals of 0.1 GPa (tick marks on each line; numbers in italics are values for P_f). The results of these calculations illustrate that dry peridotite of uniform composition will show increasing Fe contents as P_o increases (essentially as potential temperature of mantle increases) while Na contents decrease with increasing extent of melting (i.e. as P_f decreases) (reproduced by permission of American Geophysical Union from *J. Geophys. Res.*, **2002**, *107*).

conditions of decompression mantle melting. Several workers have attempted to use these systematics to investigate the depth of origin and the extent of mantle melting of certain continental basaltic rocks, including alkali basalts (Wang *et al.*, 2002) and continental flood basalts (Basu *et al.*, 1998).

3.03.2.4 Factors Influencing Magma Trace-element Abundances

The trace-element abundances in basaltic rocks can also be used to gain insights into the composition of the basalt source regions and the physical processes involved in melting and magmatic differentiation (see Rollinson (1993)), for a comprehensive review). Trace elements, which are defined as elements that are present at concentrations <0.1 wt.%, are usually not sufficiently abundant to form their own mineral phases in the mantle but substitute to varying degrees in existing mineral structures, depending largely on their ionic size and charge (see Chapter 2.09). During partial melting the trace elements partition between the residual mineral phases and coexisting liquid. Incompatible elements partition dominantly into the liquid phase, and so have low-mineral/melt partition coefficients, K_d, which are functions of the physical conditions during melting and of the melt composition (K_d = mineral/melt partition coefficient = (concentration of element, i, in mineral)/(concentration of i in coexisting melt). Compatible elements, in contrast, have high K_d values and partition into the solid phase.

A simple application of these systematics in studies of basalt petrogenesis involves constructing ratios of incompatible to potentially compatible (or, at least, less incompatible) elements during mantle melting. Such ratios, when corrected for modifications that may have occurred during magmatic differentiation, can provide information regarding the mineral phases that might have remained in the mantle source at the time the basaltic magmas were generated. This information can, in turn, provide information regarding the composition and depth of the mantle source. One popular ratio used for this purpose is $(La/Yb)_N$ (where "N" refers to fact this is the ratio of lanthanum and ytterbium abundance normalized to the abundance of these elements in chondritic meteorites). Lanthanum, a light rare earth element (LREE), is highly incompatible in minerals of spinel and garnet peridotite, while ytterbium, a heavy rare earth element (HREE), is incompatible in spinel but not garnet (K_d for ytterbium in garnet is >3; Thirlwall *et al.* (1994)). As a result, a high $(La/Yb)_N$ (>~5; Thirlwall *et al.* (1994)), or low Lu/Hf (Beard and Johnson, 1997), in primary basaltic magmas can be attributed to melting in the presence of residual garnet and therefore requires that at least some portion of the melting must have occurred at depths below the spinel to garnet transition in mantle peridotites (Figures 1 and 2). Fractionation of thorium from uranium in basaltic magmas, as manifested in the uranium-series disequilibria observed in some young continental basalts (Asmerom, 1999), may also be controlled by the presence of residual garnet, due to the greater compatibility of uranium in this mineral (Beattie, 1993), although recent experimental studies have demonstrated that clinopyroxene can also fractionate uranium from thorium, even during melting of spinel peridotite (Landwehr *et al.*, 2001).

Other elements are also uniquely partitioned into specific mantle mineral phases. For example, potassium, rubidium, and barium partition preferentially into phlogopite (Schmidt *et al.*, 1999). Relative depletions in the abundances of these elements on primitive mantle normalized trace-element plots (see later sections) can therefore serve as evidence of residual phlogopite in the source region of basaltic rocks. Amphibole can also sequester potassium and barium, as well as strontium and the middle REE (gadolinium through erbium), and both phases can incorporate titanium, niobium, and tantalum (i.e., high-field-strength elements, HFSE; Dalpé and Baker, 2000). Recognition of the trace-element signatures of residual phlogopite and amphibole in basalts is significant, because both mineral phases are unstable at the high temperatures of the convecting upper mantle and therefore imply basalt derivation from, or interaction with, lithospheric mantle (Figure 2; Class and Goldstein (1997)).

It is also possible to use the ratios of highly incompatible elements in basaltic rocks to determine these ratios in their mantle source regions, given that elements having similar K_Ds produce incompatible element ratios that are independent of the degree of partial melting of the mantle source or the amount of subsequent magmatic differentiation (Hanson, 1989; Hofmann, 1997). The ratio of niobium to uranium has proved useful in fingerprinting ocean-island basalts (OIBs) and for distinguishing these basalts from those produced in subduction environments (island-arc basalts (IAB)). The latter (together with continental crust) have uniquely low Nb/U, as well as low Rb/Cs, and Ce/Pb ratios, relative to both OIB and mid-ocean ridge basalts (MORBs) (Figure 5; McDonough *et al.* (1992) and Hofmann (1997)). The origin of the relative depletions in niobium, tantalum, and titanium, and the enrichments in lead and the large-ion lithophile elements' (LILEs; rubidium, barium, caesium, etc.) abundances in IABs remains a matter of debate, but most likely is due to a combination of the high mobility of lead and LILE in fluids produced during the dehydration of subducting oceanic lithosphere and the sequestering of some HFSE in an Fe–Ti mineral stable in the down-going slab (Ryerson and Watson, 1987; Pearce and Peate, 1995).

The incompatible element ratios of oceanic basalts provide a convenient template against which to compare the trace-element characteristics of continental basaltic rocks. Intraplate continental basalts can also be distinguished on the basis of their Nb/U and Ce/Pb ratios, with "low-titanium" potassic alkali basalts (loosely defined as containing <2 wt.% TiO_2, Figure 6, and Ti/Y <310; Peate (1997)) having significantly lower ratios than kimberlites and "high-titanium" alkali basalts (Figure 5). This observation implies that intraplate continental basalts may be derived from mantle source regions that are chemically similar to the sources of oceanic basalts and IABs, although enthusiasm for such a straightforward conclusion must be tempered by the fact that contamination of primary mantle-derived magmas by continental crust can lower both ratios (see Section 3.03.2.6).

More elaborate modeling of the relative trace-element abundances of basaltic rocks has long been undertaken, using both forward (Shaw, 1970) and inverse techniques (McKenzie and O'Nions, 1991). These studies have demonstrated that the relative trace-element abundances of primary melts of the mantle are influenced not only by the pertinent mineral/melt partition coefficients, but also by the initial trace-element abundances of the rock, and the style and degree of melting (i.e., batch versus fractional melting, congruent versus incongruent melting (Zou and Reid, 2001)). As a result, while quantitative models continue to provide important insights into the origin of continental igneous rocks (e.g., Tainton and McKenzie, 1994), their interpretation remains fraught with uncertainties due to the number of unconstrained variables involved in the calculations.

3.03.2.5 Factors Influencing Magma Isotopic Compositions

Isotopic data are an essential part of all studies of the source of continental basaltic rocks, and are

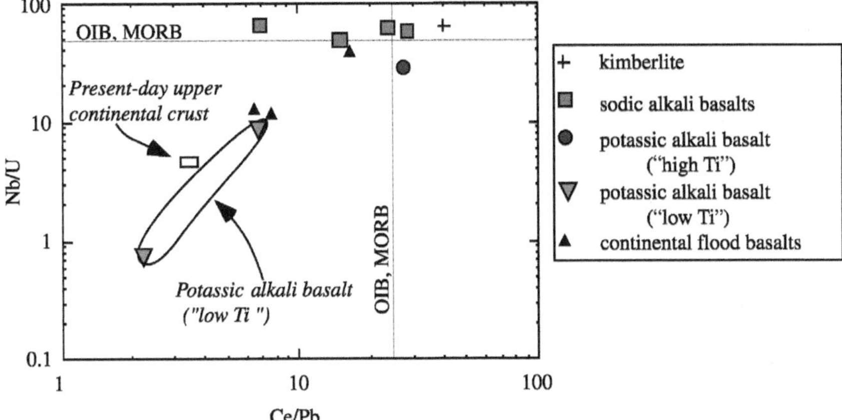

Figure 5 Average Ce/Pb and Nb/U from selected kimberlites, continental alkali basalts, and continental flood basalts. Present-day continental crust values from Condie (1993), and MORB and OIB values from Hofmann (1997) (data from Tables 1–3).

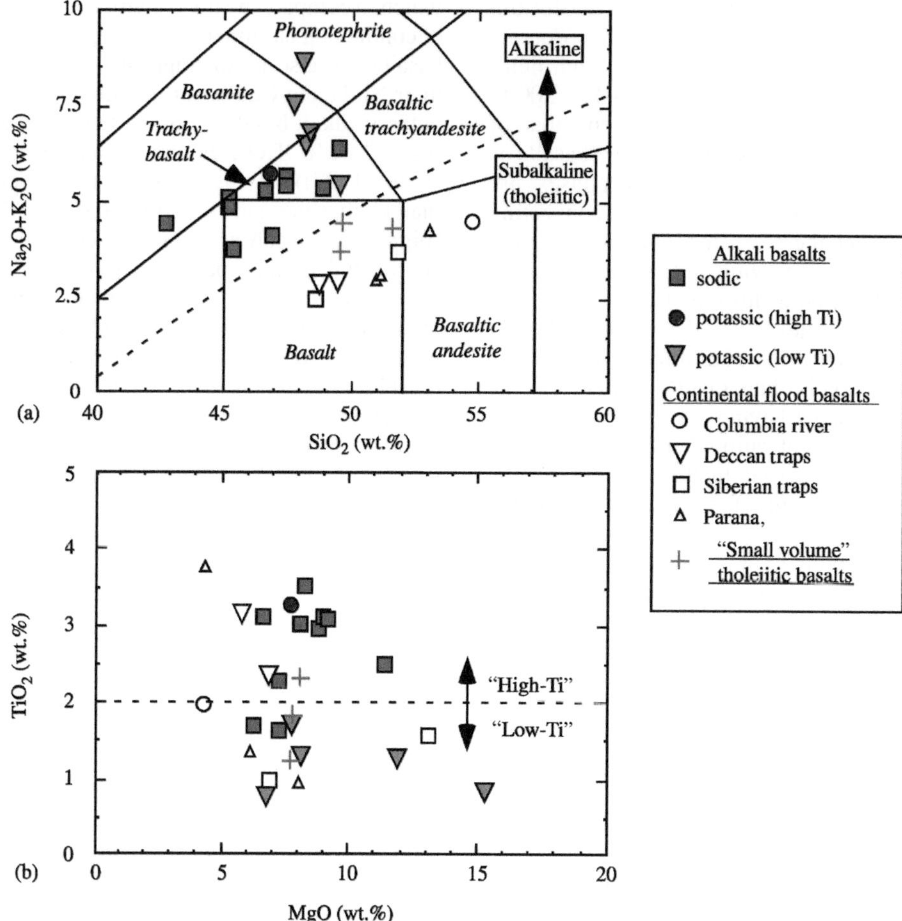

Figure 6 Major element compositions of alkali basalts and continental flood basalts: (a) total alkali elements versus wt.% SiO_2. Dashed line separating alkaline from subalkaline (tholeiitic) basalts from Irvine and Baragar (1971) and (b) wt.% TiO_2 versus wt.% MgO. Data from Tables 2 and 3.

particularly valuable in attempts to distinguish between sublithospheric and lithosphere sources of basaltic magmas and to assess whether the magmas have interacted with the crust. For example, Cenozoic continental basalts derived from sublithospheric sources, which have not assimilated lithosphere, are expected to have strontium, neodymium, hafnium, osmium, lead, and/or oxygen-isotopic compositions identical to those of oceanic basalts. For the latter, there is a menu of potential mantle sources, including the high ε_{Nd} $(= ((^{143}Nd/^{144}Nd)_{sample(t)}/(^{143}Nd/^{144}Nd)_{chondrite(t)}) - 1) \times 10{,}000)$, low $^{87}Sr/^{86}Sr$ upper mantle sources of MORB and a variety of deeper, isotopically distinct mantle sources that have been defined for OIBs (Figure 7(b); Zindler and Hart, 1986; see Chapter 2.03). In terms of their noble gas isotopic compositions, continental basalts derived from the sublithospheric mantle are expected to have $^3He/^4He$ ranging from relatively low and constant values $(R_a = (^3He/^4He)_{sample}/(^3He/^4He)_{atmosphere} \sim 9)$ for the upper mantle MORB source to higher and more variable values (R_a from 4 to as high as 43) for the deeper mantle sources of OIBs (Moreira and Kurz, 2001; see Chapter 2.06).

What are the possible isotopic compositions of basaltic magmas derived, at least in part, from the CLM, and can they be distinguished from magmas derived from sublithospheric sources? There are now sufficient isotopic data from continental lithosphere-derived mantle xenoliths to demonstrate that the CLM typically has lower ε_{Nd} (<0), higher $^{87}Sr/^{86}Sr$ ratios (>0.7045), and higher $^{207}Pb/^{204}Pb$ ratios at a given $^{206}Pb/^{204}Pb$, compared to sublithospheric mantle (Pearson, 1999; Chapter 2.05) particularly ancient (Precambrian), LREE, and rubidium-enriched lithospheric mantle, or CLM that has been metasomatized by fluids and/or melts derived from subducted oceanic lithosphere and overlying sediments (Nelson, 1992). Basaltic magmas derived from this type of lithosphere inherit these isotopic characteristics and can be distinguished on this basis from many basaltic rocks derived from the sublithospheric mantle.

Osmium-isotopic data also hold promise for distinguishing lithospheric- and

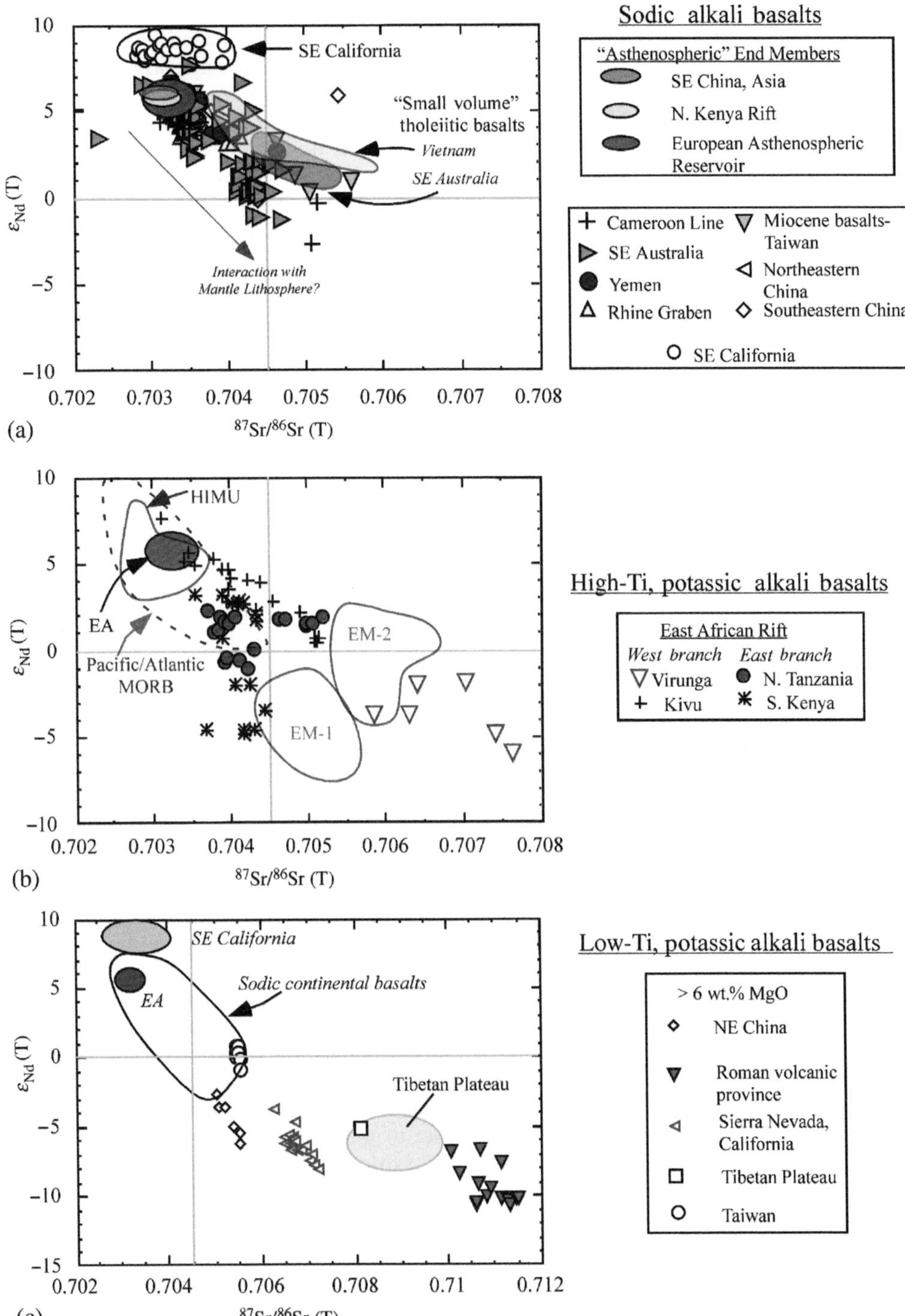

Figure 7 Initial ε_{Nd} versus $^{87}Sr/^{86}Sr$ for continental alkali basalts. (a) Sodic alkali basalts. (b) "High-Ti" potassic alkali basalts. Present-day MORB, EM1, EM2, HIMU mantle sources from Hofmann (1997). (c) "Low-Ti" potassic alkali basalts. EA = European asthenospheric reservoir (data from Table 2).

sublithospheric-derived basaltic rocks. Archean CLM has been shown to have a nonradiogenic osmium-isotopic composition (present-day $^{187}Os/^{188}Os = 0.105-0.129$; Shirey and Walker (1998)) relative to sublithospheric mantle ($^{187}Os/^{188}Os = 0.123-0.152$), because rhenium is more incompatible than osmium during mantle melting, resulting in a very low Re/Os ratio in Archean mantle, which is probably a product of large degrees of melt extraction (Walker *et al.*, 1989). However, because the Archean mantle is generally refractory, few continental basalts are expected to have exceptionally low $^{187}Os/^{188}Os$. Younger CLM does not necessarily have osmium-isotopic characteristics distinct from the sublithospheric mantle (Lee *et al.*, 2000; Schaefer *et al.*, 2000). Osmium-isotopic data have proved useful, however, in identifying the contribution to continental mafic magmas from olivine-poor, mafic lithologies, because these lithologies have a high Re/Os and, if sufficiently old, can contribute osmium with high $^{187}Os/^{188}Os$ (>0.2) to continental basaltic magmas (Carlson and Nowell, 2001).

Several other isotopic techniques have been brought to bear on the issue of lithospheric versus sublithospheric sources for basaltic rocks, but have not yet been widely applied. Studies of helium isotopes in olivine phenocrysts have demonstrated that CLM-derived magmas characteristically have more radiogenic helium ($R_a \sim 6$) than most sublithosphere-derived basalts (with the exception of HIMU basalts), presumably due to a high (Th + U)/^{3}He ratio in the continental mantle (Reid and Graham, 1996; Dodson *et al.*, 1998). In Quaternary continental basaltic rocks, uranium-series data have also been obtained (Reid and Ramos, 1996). Some workers suggest that large ^{230}Th excesses (10–40%) in some sublithosphere-derived basaltic rocks reflect fractionation of uranium from thorium by garnet and hence a relatively deep derivation for these magmas (Asmerom, 1999).

Light stable-isotope data are relatively rare from intraplate basaltic rocks, in part because the whole-rock stable-isotope compositions of these rocks are susceptible to modification by groundwater during low-temperature, post-emplacement alteration (Taylor, 1968). Laser fluorination oxygen-isotope analyses of unaltered basalt phenocrysts do provide a method for determining magmatic oxygen-isotopic compositions in spite of alteration (Eiler *et al.*, 1997), but the few studies of continental basalts that exist suggest that there is no discernible difference between the oxygen-isotopic composition of lithospheric and sublithosphere mantle (both having $\delta^{18}O \sim 5.3‰$) and magmas derived therefrom (Dobosi *et al.*, 1998).

3.03.2.6 Role of Crustal Contamination

Determining the chemical and isotopic characteristics of primary basaltic magmas is complicated due to the fact that mantle-derived mafic magmas can interact with the continental crust during ascent. Unfortunately, there is no simple recipe for unequivocally recognizing the effects of crustal contamination on the chemical and isotopic compositions of continental basaltic rocks, given the wide range of crustal lithologies that can be assimilated. Interaction between mafic magmas and upper continental crust, for example, often results in the presence of partially assimilated and disaggregated crustal xenoliths in basaltic rocks. Such rocks are relatively easy to avoid during sampling (Farmer *et al.*, 2002). However, assimilation of lower crust and/or melt derived therefrom can be more difficult to recognize unless sufficient isotopic contrast exists between the primary basaltic magma and the assimilating crust, so that isotopic variations correlate with some index of magma differentiation (e.g., wt.% MgO). In order to avoid the geochemical effects of such crustal interaction, only the most primitive basaltic rocks in a given location are used to infer the isotopic composition of the original mafic magmas. However, even these magmas may not retain their original isotopic compositions, depending on the chemical and isotopic compositions of the crustal assimilant and the nature of the assimilation process (Reiners *et al.*, 1995; Bohrson and Spera, 2001). In addition, assimilation of pre-existing mafic crust by mafic magmas may produce only minor variations in the chemical and isotopic composition of magma, variations that are only discernible with detailed geochemical studies of multiple rock samples (Glazner and Farmer, 1992).

The problem of crustal contamination is particularly acute for low mg# continental flood basalts and smaller volume continental tholeiitic basalts, both of which have low trace-element concentrations (see Sections 3.03.3.2.3 and 3.03.3.3). The issue is less critical for many smaller volume continental rocks, such as kimberlites and alkali basalts, which have much higher abundances of many trace elements. As a result of their high strontium and neodymium content, for example, the isotopic compositions of these elements in kimberlites and alkali basalts are relatively insensitive to modification during crustal contamination. Conversely, the osmium and lead concentration of basaltic magmas are so low that these isotope systems are particularly vulnerable to modification by interaction with crustal rocks (McBride *et al.*, 2001; Chesley *et al.*, 2002); hence these systems provide relatively sensitive indicators of crustal assimilation.

3.03.3 CONTINENTAL EXTRUSIVE IGNEOUS ROCKS

The above discussion illustrates that the interplay between the thickness and composition of the lithosphere and the composition and potential temperature of the underlying upper mantle control the composition, source, and volumes of Phanerozoic intraplate continental magmatism. Exactly how these factors are related to the production of continental magmas is best illustrated by considering the formation of specific continental igneous rocks types.

3.03.3.1 Kimberlites

Kimberlites are small-volume, volatile-rich (CO_2 and/or H_2O), potassic, ultramafic igneous rocks that occur worldwide as dikes, sills, and diatremes within Precambrian cratons. These rocks have been studied extensively, in part, because kimberlites represent the primary source of naturally occurring diamonds, but also because they are unusually enriched in LREE, LILE (barium, strontium), and HFSE (niobium, tantalum) and have an ultramafic composition that demands a mantle origin for their parental magmas (Table 1; Mitchell (1995)). Investigations of the origin of their chemical characteristics have been hampered by the fact that kimberlites are clearly hybrid rocks, containing both primary components (i.e., magmatic phenocrysts and groundmass) and xenocrysts (e.g., diamonds) derived from the conduit through which the kimberlitic magma passed during eruption. For example, all kimberlites have unusually high nickel contents (>1,000 ppm; Table 1). These high concentrations do not reflect the values in primary kimberlitic magmas but are due to the incorporation of olivine xenocrysts derived from the disaggregation of peridotite wall rocks (Mitchell, 1995). Another complication arises from the general occurrence of pervasive hydrothermal and/or deuteric alteration in these rocks (Mitchell, 1995).

Despite the difficulty of defining the chemical characteristic of kimberlites, and even in naming the rocks (Woolley *et al.*, 1996), some aspects of the compositions and origins of kimberlites have been broadly accepted. For example, it is now generally accepted that there are two classes of kimberlites, kimberlites (*sensu stricto*) and orangeites (also referred to as "group 1" and "group 2" kimberlites, respectively). The former are CO_2 rich and are characterized by a macrocryst assemblage of olivine, magnesium-ilmenite, pyrope, diopside, phlogopite, enstatite, and chromite, set in a fine-grained matrix of olivine and a variety of other minerals, including perovskite and apatite (Mitchell, 1995). Orangeites, in contrast, are H_2O-rich rocks that contain less olivine but significantly more phlogopite than kimberlites, both as macrocrysts and microphenocrysts (hence the original name for these rocks of "micaceous" kimberlites; Dawson (1967)). The distinction between these two kimberlite types was originally recognized amongst the abundant Cretaceous kimberlitic rocks that perforate the Archean Kaapvaal craton in southern Africa, on the basis of the higher initial $^{87}Sr/^{86}Sr$ (0.707–0.711) and low ε_{Nd} values (-6 to -13) of orangeites compared to kimberlites (Figure 8; Smith (1983)). However, while Proterozoic to Tertiary kimberlites worldwide also have relatively high initial ε_{Nd} values (>0), micaceous kimberlites outside South Africa do not always have low initial ε_{Nd} values (cf. the Late Proterozoic Aries kimberlites in western Australia; Edwards *et al.* (1992)). Both kimberlite types have high MgO contents (22–29 wt.% MgO), and low aluminum and silicon (2.3–3.8 wt.% Al_2O_3 and 33–37 wt.% SiO_2), but orangeites have considerably higher K/Na ratios than kimberlites (17–22 versus 3.3–6.2; Table 1). Finally, both kimberlites and orangeites are strongly enriched in LIL, LREE, and HSFE relative to OIB, and, unlike typical OIB, have prominent negative potassium anomalies on primitive mantle normalized trace-element plots (Figure 9).

The immediate source of kimberlitic rocks is either CO_2 (for kimberlites) or H_2O (for orangeites) metasomatized peridotite at some depth greater than that of the graphite/diamond transition (~140 km; Figure 1), given the occurrence of xenocrystic diamonds in these rocks (Mitchell, 1995). The high $(La/Yb)_N$ ratios (100–300) of all kimberlitic rocks (Mitchell, 1995), combined with their low aluminum contents, further suggest that kimberlitic rocks represent the products of small degrees of melting (<1%) of an LREE-enriched, garnet-bearing ultramafic rock (Tainton and McKenzie, 1994). A deep source is also consistent with recent experimental data, which suggest that kimberlites equilibrated with carbonated, garnet harzburgite at ~180 km depth (Girnis *et al.*, 1995). Several important questions remain, however, including: Do kimberlites and orangeites represent the products of melting in the lithospheric or the sublithospheric mantle? Where did the mantle-metasomatizing fluids originate? What mechanisms are involved in triggering kimberlitic magmatism?

Many authors have suggested that the high ε_{Nd} values of kimberlites (*sensu stricto*) require a source in the convecting, sublithospheric mantle (Smith, 1983; Edwards *et al.*, 1992; Taylor *et al.*, 1994; Mahotkin *et al.*, 2000). The occurrence of majoritic garnet in kimberlites and in entrained diamonds further supports models of an ultra-deep origin of the magmas parental to kimberlites, possibly even in the transition zone (Sautter *et al.*, 1991; Ringwood *et al.*, 1992). In contrast, the low ε_{Nd} of orangeites requires a source in long-lived (>1 Ga), LREE-enriched mantle, presumably representing lithospheric mantle. Such a conclusion is consistent with major and trace-element modeling of orangeites by Tainton and McKenzie (1994), who suggested that the aluminum and HREE abundance of these rocks can only be reconciled if they were derived from a mantle that had undergone extensive partial melting, as expected for mantle beneath Archean cratons,

Table 1 Average major element, trace element and isotopic data for selected orangeites and kimberlites.[a]

	Kimberlite[b] *Sierra Leone*	*Group 1A kimberlite*[c] *global avg.*	*Orangeities*[d] *South Africa*	*Micaceous kimberlites*[c] *global avg.*
Age (Ma)	115		165–110	
$n =$	22	35	114	32
SiO_2	32.5	32.6	32.5	37.0
TiO_2	1.84	1.70	0.88	1.40
Al_2O_3	2.68	2.70	2.27	3.80
MnO	0.17	0.20	0.14	0.20
MgO	27.8	29.1	22.1	26.7
Fe_2O_3t	10.9	10.2	7.18	8.9
CaO	7.49	8.30	5.16	6.90
Na_2O	0.26	0.30	0.11	0.20
K_2O	1.63	1.00	2.31	3.30
P_2O_3	0.59	0.70	0.74	1.00
LOI			9.38	
CO_2	4.53	5.30		3.60
H_2O	7.63			
K_2O/Na_2O	6.2	3.3	22	17
Ni	1,177		1,224	
Cr	1,305		1,624	
Zn	66.5		65.3	
Cu			26.8	
Co			80.0	
V	76		104	
Y	10.0			
Sc	14.0		18.8	
Rb	75.0			
Sr	601		1,105	
Cs				
Ba	1,641		3,133	
Zr	179		273	
Hf	5.71		7.00	
Ga	8.40			
Nb	239		115	
Ta	15.72		8.50	
U	3.91		5.00	
Th	26.3		27.5	
Pb	8.29			
La	179		175	
Ce	331		337	
Pr			45.5	
Nd	110		123	
Sm	13.7		14.3	
Eu	3.12		3.44	
Gd			11.4	
Tb	0.71		1.01	
Dy			3.59	
Ho	0.50		0.57	
Er			1.36	
Tm			0.18	
Yb	0.66		1.23	
Lu	0.08		0.14	
$(La/Yb)_N$	175.6		92.2	
$n_i =$ [e]	2			
$^{87}Sr/^{86}Sr(T)$	0.7040–0.7043		0.7071–0.7109	
$\varepsilon_{Nd}(T)$	−104 to +2.7		−6.2 to −13.4	
$^{206}Pb/^{204}Pb$			17.21–18.24	
$^{207}Pb/^{204}Pb$			15.51–15.58	
$^{208}Pb/^{204}Pb$			37.45–38.23	

[a] Major elements reported as weight percent oxides (hydrous), trace elements in ppm. [b] Averge of dikes and pipe analyses from Koidu kimberlites, Sierra Leone. Data from table 1, Taylor *et al.* (1994). [c] Global average from table 4, Taylor *et al.* (1994). [d] Data from Swartuggens, Finisch, Bellsbank and Sover kimberlite localities, as complied in tables 3.1, 3.5, 3.7, and 3.10 in Mitchell (1995). [e] n_i is the number of samples used for radiogenic isotope averages.

that was subsequently metasomatized by fluids enriched in LILE, LREE, and HFSE.

It is important to note, however, that the negative potassium anomalies characteristic of kimberlites and orangeites require that both have equilibrated with lithospheric mantle containing a residual potassic phase, i.e., either phlogopite or potassium-rich amphibole. As a result, the difference in the isotopic composition of kimberlites and orangeites may simply indicate that kimberlites were derived from lithospheric mantle that was metasomatized at or near the time of kimberlite formation, while a much longer time interval occurred between metasomatism and kimberlitic magmatism in the case of orangeites (Mitchell, 1995).

Exactly what triggers these metasomatic events, or the production of the kimberlite magmatism, remains enigmatic. Several workers have suggested that similarities in the age and spatial patterns of kimberlite magmatism and various hot-spot tracks support the idea that kimberlite magmatism is ultimately related to decompression melting of mantle plumes beneath continents (le Roex, 1986; Heaman and Kjarsgaard, 2000). Others have questioned whether any relationship exists between hot spots and kimberlite magmatism (Mitchell, 1995), particularly because episodes of kimberlite magmatism can occur in the same area, separated in time by hundreds of millions of years (Lester *et al.*, 2001). However, regardless of the trigger mechanism involved in kimberlite magmatism, kimberlites provide evidence that mantle lithosphere, particularly long-lived, deep-rooted, Archean mantle lithosphere, has been periodically metasomatized by $CO_2 \pm H_2O$, LREE, LILE, and HFSE bearing fluids derived from the underlying convecting mantle. The derivation of these fluids from the deep mantle is at least consistent with the similarity in the Nb/U and Ce/Pb ratios of kimberlites and OIB (Figure 5).

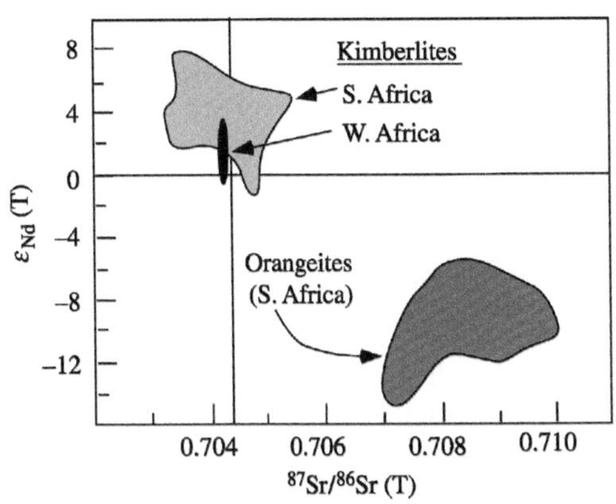

Figure 8 Initial ε_{Nd} versus $^{87}Sr/^{86}Sr$ for African kimberlites and orangeites (group 2 kimberlites) (after Mitchell, 1995).

3.03.3.2 Alkali Basalts

Cenozoic alkali basalt provinces occur in every continent (Figure 10), often, but not exclusively, associated with active lithospheric extension in localized continental rifts (e.g., Baikal, East African and Rio Grande rifts) or over broader continental regions (western USA, eastern China, southeast Asia, southeastern Australia). Continental rifts have long been the focus of studies of continental basaltic rocks, due, in part, to the large volume of basaltic magmatism that can occur both on and off the rift axis, and because rift magmatism provides a window into the processes involved in rifting and, potentially, in continental breakup (Baldridge *et al.*, 1995). In general, alkaline magmatism occurs early along the axis of rifts, where it can be supplanted later by larger volumes of tholeiitic magmatism (cf. Rio Grande rift; Baldridge *et al.* (1995)), and along the rift flanks (Thompson and Gibson, 1994).

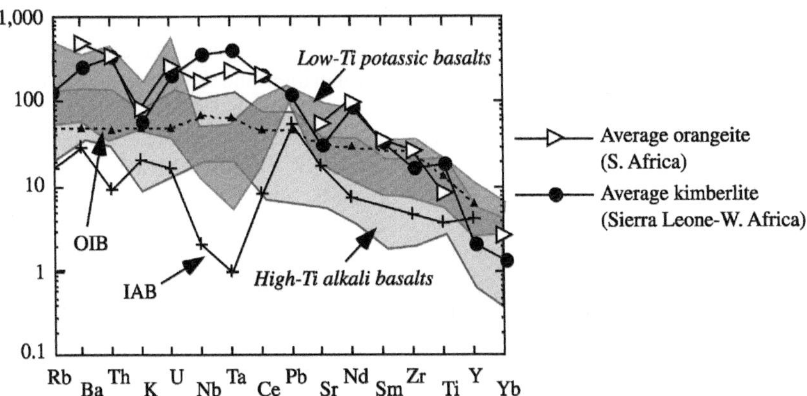

Figure 9 Primitive mantle normalized trace-element abundances for kimberlites and orangeites. Normalization values from Sun and McDonough (1989). Average OIB and IAB data from Sun and McDonough (1989) and McCulloch and Gamble (1991), respectively. High- and low-Ti basalt fields from Figure 11 (data from Table 1).

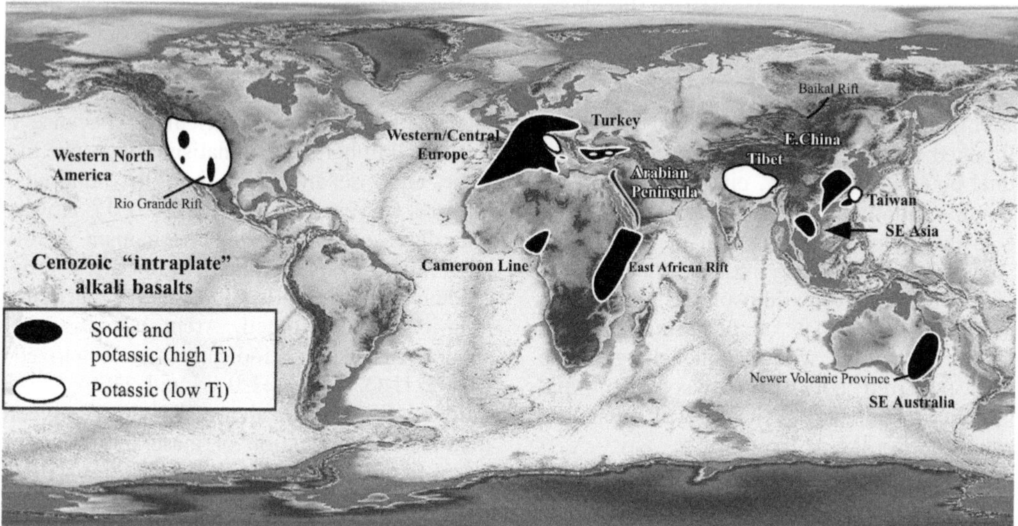

Figure 10 Location of sodic and potassic "intraplate" alkali basalt volcanic fields worldwide. Shown are volcanic fields mentioned in text and for which data are tabulated in Table 2.

Total magmatic volumes vary substantially from rift to rift. The East African Rift is generating the largest volume of magma among present-day rifts ($\sim 2.2 \times 10^5$ km^3 for the Kenya Rift alone; Mohr (1992)).

In broader areas of continental extension, alkali basalts comprise multiple volcanic fields that collectively can cover up to 1,000 km^2 or more, with up to a hundred or more separate volcanoes (Connor and Conway, 2000). The volcanic vents in these regions are often monogenetic (i.e., formed during a single eruptive event), and individual volcanic fields generally produce less than 10 km^3 of volcanic rock during a few thousand years to 5 Myr (Connor and Conway, 2000). Eruptive rates are low, as a result, amounting to <1 km^3 kyr^{-1} (as opposed to the present-day flux of basaltic magma at mid-ocean ridges of ~ 18 km^3 yr^{-1}; Sigurdsson (2000)). Despite their relatively small volume, continental alkali basalt fields have been studied intensively, in large part because this type of volcanism can be triggered by lithospheric deformation, and hence provides insights into the composition and physical behavior of the deep continental lithosphere, and because they carry mantle xenoliths.

Geochemical studies of alkali basalts have been aided over the past 10–15 yr by an explosion in the literature of ICP-MS-based trace-element and radiogenic-isotope data for continental alkali basalts. Representative chemical analyses of the two main alkali basalt groups, sodic and potassic ($K_2O/Na_2O < 1$ and ≥ 1, respectively), are listed in Tables 2(a) and (b), respectively, derived from the average composition of lavas at each locality having <51 wt.% SiO_2 and >5 wt.% MgO. No attempt was made to account for crystal accumulation or fractionation, and as a result, these averages should not be misconstrued as estimates of "primary" magma compositions. Nevertheless, they do serve to demonstrate the basic chemical distinction between the two alkali basalt varieties.

3.03.3.2.1 *Sodic alkali basalts*

Cenozoic sodic alkali basalts represent the dominant basalt type in many continental alkali basalt provinces, including northern Europe, eastern Australia, and eastern China (Zhang and O'Reilly, 1997; Zou *et al.*, 2000; Wilson and Patterson, 2001). Sodic alkali basalts are similar compositionally worldwide (Table 2(a)) and correspond to basalts and trachybasalts with relatively high TiO_2 (~ 1.5–4 wt.%; Figure 6) and high absolute sodium contents (~ 2–4 wt.%). The sodic basalts have primitive-mantle normalized trace-element patterns similar to those of kimberlites and OIBs; they are characterized by relatively high LREE contents, low LILE/HFSE ratios, and, in many cases, a negative potassium anomaly (Figure 11(a)). Although the sodic basalts show a range of radiogenic isotopic compositions, they typically have high initial ε_{Nd} (>0), low $^{87}Sr/^{86}Sr$ ratios (<0.705), and lead-isotopic compositions that overlap, or parallel (at high $^{207}Pb/^{204}Pb$ for a given $^{206}Pb/^{204}Pb$), the northern hemisphere reference line (Figure 12).

Both the trace-element and the isotopic composition of sodic basalts are typically cited as evidence that their parental magmas were produced by relatively small degrees of melting

Table 2(a) Average major element, trace element and isotope data for Cenozoic continental sodic alkali basalts and associated small volume tholeiitic basalts.[a]

	Sodic alkali basalts										*Tholeiitic basalts*		
	SE Australian[b] *Dubbo V.F.*	*Arabian Peninsula*[c] *Yemen*	*Africa-N. Tanzania*[d] *East African Rift*	*West Africa*[e] *Cameroon Line*		*Turkey*[f] *Central Anatolia*		*Europe*[g] *Rhine graben*	*Taiwan*[h]	*NE China*[i]	*Vietnam*[j]	*SE Australia*[k] *Newer V.P.*	*Taos Plateau*[l] *Rio Grande Rift*
				Low Sr	*High Sr*	*Early Miocene*	*Late Miocene*						
Age (Ma)	13	<1	<1–8	5–23	5–23	18–20	9.5	18–22	20–23	–15	163–>6.3	<.010–45	3.5–5
$n =$	8	44	23	8	6	3	5	16	3	4	9	7	12
SiO_2	45.3	47.4	46.6	45.1	45.1	48.8	49.4	42.7	47.4	46.8	49.6	51.6	49.5
TiO_2	2.50	2.30	3.02	3.12	3.53	1.64	1.70	3.07	3.12	2.96	2.33	1.81	1.22
Al_2O_3	13.4	16.6	12.4	14.5	14.5	16.6	17.4	13.4	15.6	12.9	14.0	14.3	16.3
MnO	0.19	0.19	0.20	0.17	0.19	0.15	0.16	0.19	0.18	0.96	0.16	0.16	0.18
MgO	11.3	7.2	8.0	8.9	8.2	7.2	6.2	9.1	6.6	8.7	8.1	7.8	7.7
Fe_2O_3t	12.9	11.6	13.7	12.4	12.6	9.6	9.0	12.5	2.5	13.2	12.0	11.3	12.1
CaO	9.98	8.74	10.2	10.0	10.1	10.0	9.09	11.6	8.15	9.68	8.88	8.48	9.06
Na_2O	2.79	4.27	3.61	3.62	3.41	3.86	3.94	3.30	3.82	2.78	2.74	3.38	3.15
K_2O	1.00	1.15	1.69	1.49	1.46	1.51	2.50	1.12	1.87	1.35	1.72	0.93	0.56
P_2O_3	0.64	0.62	0.56	0.76	0.97	0.72	0.61	0.76	0.83	0.56	0.57	0.34	0.18
K_2O/Na_2O	0.36	0.27	0.47	0.41	0.43	0.39	0.64	0.34	0.19	0.48	0.63	0.27	0.18
Ni	252	90	195	153	130	153	86	138	99	191	174	153	156
Cr	377	153	334	343	232	193	99	248	108	233	224	266	225
Zn	98.5	81.7	541			74.3	71.0		108		104	112	
Cu	61.4	54.1	109			40.7	40.4			51.4	58.8	61.9	
Co									41.0				51.4
V	211	196	316			133	101	297	205	234		161	157
Y	25.8	34.5	28.3	31.4	35.2	26.7	23.4	29.4	32.3	24.6	24.4	27.1	21.5
Sc	22.0	29.5	19.2			16.0	14.2		17.0	20.6	20.1	19.3	25.3
Rb	23.1	22.1	46.7	38.9	38.0	35.7	42.2	67.7	37.7	24.0	36.3	23.4	54
Sr	765	644	886	959	1470	927	730	926	881	915	740	436	329
Cs												1	
Ba	648	382	670	571	630	805	526	791	730	433	514	278	259
Zr	216	244	283	247	195	183	183	287	295	246	194	147	96
Hf	4.60								6.80	5.48	540	3.37	2.49
Ga	18.3	17.6										20.1	
Nb	62.4	43	90.7	91	74	30	45	82	69.0	69.0	40	24	7
Ta	3.77								4	4.19	2.64		0.45

(continued)

Table 2(a) (continued).

	Sodic alkali basalts										Tholeiitic basalts		
	SE Australian[b] Dubbo V.F.	Arabian Peninsula[c] Yemen	Africa-N. Tanzania[d] East African Rift	West Africa[e] Cameroon Line		Turkey[f] Central Anatolia		Europe[g] Rhine graben	Taiwan[h]	NE China[i]	Vietnam[j]	SE Australia[k] Newer V.P.	Taos Plateau[l] Rio Grande Rift
				Low Sr	High Sr	Early Miocene	Late Miocene						
U		0.75	1.53						1.13	1.41	0.92	1.00	
Th	7.23	3.9	9.91			11.7	9.2		5.20	5.33	4.3	2.9	0.7
Pb		2.45	5.90							5.59		3.4	
La	39.1	33.2	65.6	55.4	48.2	43.1	25.0	62.0	50.3	39.9	32.7	22.1	9.1
Ce	77.9	69.6	143	107	97	85.8	49.1	124	101.5	83	67	40	21
Pr						8.69	5.23			9.58		5.3	
Nd	37.6	34.3	60.0	49.4	50.9	32.3	20.6	56.4	52.3	37.5	31.0	21.5	
Sm	7.56	7.0		9.5	10.7	5.95	4.52	11.7	9.8	7.87	6.8	5.2	3.3
Eu	2.31	2.27		3.01	3.51	1.84	1.54	3.59	3.1	2.49	2.14	1.73	1.16
Gd		6.75		7.75	8.03	5.56	4.69	8.43		6.92		6.08	
Tb	0.957								1.17	0.99	1.03	0.85	0.67
Dy		5.84		5.52	5.65	4.79	4.28	6.01		4.95	4.72	5.26	
Ho	0.986					0.84	0.74			0.85		0.97	
Er		3.25		2.35	2.43	2.28	2.10	2.64		2.08		2.50	
Tm										0.27			
Yb	1.71	2.86		1.85	2.07	2.28	2.06	1.83	2.2	1.46	1.68	1.86	2.07
Lu	0.23	0.43		0.28	0.29	0.39	0.35	0.18	0.32	0.20	0.22	0.27	0.32
$(La/Yb)_N$	14.8	7.5		19.3	15.1	12.2	7.9	21.9	14.8	17.7	12.6	7.7	2.8
$n_i =$[m]	7	25	19	8	6	3	1	26	3	4	8	7	3
$^{87}Sr/^{86}Sr(T)$	0.7042	0.7035	0.7041	0.7033	0.7035	0.7047	0.7034	0.7036	0.7036	0.7038	0.7046	0.7054	0.7043
ε_{Nd} (T)	0.04	5.14	0.65	4.06	4.17	3.0	6.9	3.67	5.44	4.72	3.35	2.09[n]	
$^{206}Pb/^{204}Pb$		18.563	19.255						18.746	18.139	18.533	18.726[n]	17.958
$^{207}Pb/^{204}Pb$		15.584	15.587						15.568	15.507	15.594	15.620	15.515
$^{208}Pb/^{204}Pb$		38.624	39.645						38.859	38.233	38.588	38.840	37.482

[a] Major elements reported as weight percent oxides (anhydrous), trace elements in ppm. Average analyses for sodic alkali basalts ($K_2O/Na_2O < 1$) constructed from lavas at each locality having <51 wt.% SiO_2 and >5 wt.% MgO. [b] Average basanite and alkali olivine basalt, Dubbo volcanic province, eastern Australia. Data from Zhang and O'Reilly (1997). [c] Average Quaternary alkali basalt, western Yemen. Data from Baker *et al.* (1997). [d] Combined average of "older extrusive" and alkali basalts from Loolmalasin, Ketumbeine and Monduli volcanoes in East African rift, northern Tanzania. Data from Paslick *et al.* (1995). [e] Average, "high" and "low" Sr basanites and alkali basalts from Cameroon line calculated for rocks with >6 wt.% MgO. Data from Marzoli *et al.* (2000). [f] Average alkali basalt, Galatia volcanic province, central Anatolia. Data from Wilson *et al.* (1997). [g] Data from Jung and Hoernes (2000). [h] Miocene basalts, northern Taiwan. Data from Chung *et al.* (1995). [i] Data from Chung (1999). [j] Average olivine tholeiite from Buon Ma Thuol and Plieku volcanic areas. Data from Hoang and Flower (1998). [k] Average tholeiitic plains basalt from Newer Volcanic Province, southeastern Australia. Data from Price *et al.* (1997). [l] Average composition of middle member of Servilleta Basalt. Data from Dungan *et al.* (1986). [m] n_i is number of samples used for radiogenic isotope averages. [n] Pb and Nd isotope compositions are average of four olivine tholeiites reported by McBride *et al.* (2001).

Table 2(b) Average major element, trace element and isotopic data for Cenozoic continental potassic alkali basalts.[a]

	Africa[b] *Virunga V.F.* *High Ti*	*China*[c] *Tibetan Plateau* *Low Ti*	*Taiwan*[d] *Mt. Tsaoling* *Low Ti*	*Italy*[e] *Roman V.F.* *Low Ti*	*Western US*[f] *Sierra Nevada* *Low Ti*	*Aegean Sea*[g] *Dodecanese V.F.* *Low Ti*
Age (Ma)	<1	<1–13	0–2.8	0.4–0.6	3.5	8.6–12
$n =$	16	6	10	7	27	13
SiO_2	46.8	47.7	48.3	48.1	48.2	49.5
TiO_2	3.26	1.70	0.83	0.78	1.27	1.30
Al_2O_3	13.3	13.2	12.2	15.6	12.6	15.9
MnO	0.18	0.15	0.13	0.14	0.15	0.14
MgO	7.7	7.7	15.3	6.8	11.9	8.1
Fe_2O_3t	12.2	10.7	7.6	8.3	8.9	8.8
CaO	10.3	10.2	7.28	11.1	9.50	10.15
Na_2O	2.37	3.54	1.84	1.16	2.93	2.73
K_2O	3.36	3.97	4.96	7.47	3.52	2.73
P_2O_3	0.56	1.04	1.58	0.55	1.13	0.63
K_2O/Na_2O	1.42	1.12	2.70	6.45	1.20	1.00
Ni	64	102	481	74	271	103
Cr	273	296	1,082	134	665	268
Zn	91.6		49.2	38.1	108	80.4
Cu	35.2		32.9	47.9		
Co	42.2	197	45.2	32.8	40.4	29.9
V	312		154	181		213
Y	27.9	17.0	13.9	35.7	20.6	27.2
Sc	30.4	19.5	27.4	12.5	23.7	27.3
Rb	120	164	1,099	607	69.7	160
Sr	993	3,216	676	1,469	2,223	1,014
Cs	1		116			38
Ba	1,322	2,564	937	1,483	2,927	1,276
Zr	323	585	114	313	341	218
Hf	7.38	10.0	3.20	3.03	7.15	6.66
Ga		17.0	12.5	7.6		
Nb	105		16	14	14	20
Ta	7.36	2.48	0.88	0.30	0.62	1.24
U	3.76		19.1		1.57	6.09
Th	15.3	58.7	20.8	29.6	5.8	19.6
Pb	6.90	30.0	22.9		18.2	
La	93.1	28.2	25.0	95.4	57.0	51.1
Ce	181	465	50	219	122	97
Pr		37	6.03		14.7	
Nd	72.2	273	24.7	95.3	58.9	45.5
Sm	11.9	28.8	4.7	18.7	9.7	9.0
Eu	3.05	7.02	1.14	3.47	2.44	2.39
Gd		12	3.53		6.97	7.09
Tb	1.23	2.2	0.50	1.53	0.82	0.86
Dy		6.30	2.59		3.87	4.57
Ho			0.52		0.68	
Er		2.7	1.47		1.84	2.27
Tm			0.22		0.25	
Yb	2.58	2.20	1.33	2.37	1.44	1.92
Lu	0.38	0.33	0.20	0.40	0.19	0.35
$(La/Yb)_N$	23.4	82.8	12.2	26.1	25.6	17.3
$n_i =$[h]	7	1	10		27	10
$^{87}Sr/^{86}Sr(T)$	0.7069	0.7081	0.7055	0.71007[i]	0.7068	0.7062
$\varepsilon_{Nd}(T)$	−3.93	−5.23	0.25	−8.30	−6.40	−1.62
$^{206}Pb/^{204}Pb$	19.423	18.680		18.76	19.566	
$^{207}Pb/^{204}Pb$	15.758	15.656			15.720	
$^{208}Pb/^{204}Pb$	40.746	39.015			39.197	

[a] Potassic basalts defined herein as having $K_2O/Na_2O > 1$. Average analyses constructed from lavas at each locality having <51 wt.% SiO_2 and >5 wt.% MgO. [b] Average "low silica" basaltic rocks from Muhavura and Gahinga, eastern Virunga Province, Rwanda. Data from Rogers *et al.* (1998). [c] Average calculated for volcanic rocks with >5.5 wt.% MgO. Data from Turner *et al.* (1996a). [d] Average Tsaolingshan rocks. Data from Chung *et al.* (2001). [e] Average calculated for volcanic rocks with >6 wt.% MgO in Sabatini Volcanic District in central Italy. Data from Conticelli *et al.* (1997). [f] Average Pliocene basaltic rock from central Sierra Nevada, California. Data from Farmer *et al.* (2002). [g] Average mafic igneous rock from Bodrum volcanic complex. Data from Robert *et al.* (1992). [h] n_i is number of samples used for radiogenic isotope averages. [i] Isotopic data are those reported for Roman Volcanic Province in Peccerillo (1999).

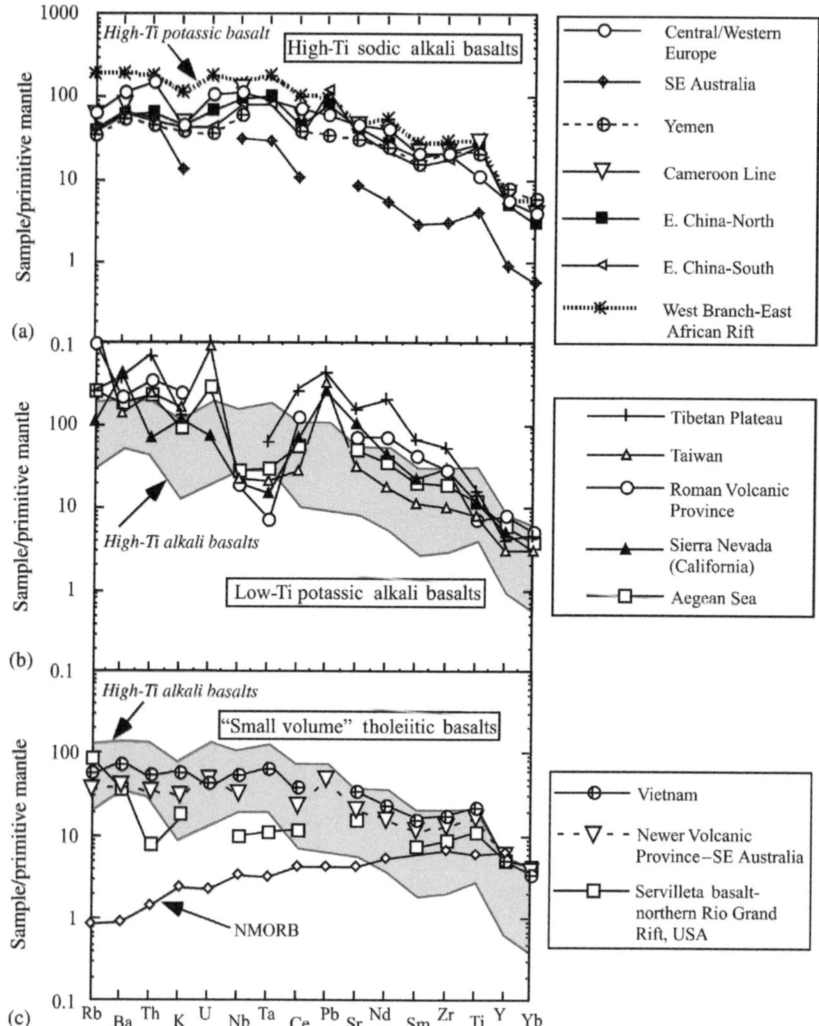

Figure 11 Average primitive mantle normalized trace-element abundances for alkali basalts. Normalization values from Sun and McDonough (1989): (a) high-titanium alkali basalts, including sodic basalts worldwide and potassic basalts from the western branch of the East African Rift; (b) low-titanium potassic alkali basalts; and (c) "small volume" tholeiitic basalts. N-MORB data from Sun and McDonough (1989) (data from sources given in Table 2).

(<5 wt.%) of either active (plume) or passively upwelling "asthenospheric" mantle (Fitton and Dunlop, 1985). In western and central Europe most workers favor models in which Tertiary and Quaternary magmatism is produced by active mantle upwelling (Wedepohl and Baumann, 1999; Wilson and Patterson, 2001). Wilson and Patterson (2001) suggested that domal uplift, lithospheric extension, and Cenozoic volcanism throughout western and central Europe is the result of the impact of five or more relatively narrow mantle diapirs ("hot fingers") derived from a common reservoir at the base of the upper mantle, which is considered by these authors to be akin isotopically and chemically to HIMU OIBs (Figures 7(a) and 12). Other models have been proposed concerning the role of lithospheric thinning and plume upwelling in producing the temporal and spatial patterns in European alkalic magmatism during the Cenozoic, including channeling of upwelling plume material along thinspots in the European lithosphere (Oyarzun *et al.*, 1997), but most workers agree that mantle lithosphere was not the primary source of the alkalic magmatism. However, interaction between plume-derived melts and low ε_{Nd}, higher $^{87}Sr/^{86}Sr$ mantle lithosphere could have contributed to some of the isotopic heterogeneities observed in the European sodic basalts (Figure 7(a); Wilson and Patterson (2001)).

Melting within actively upwelling mantle plumes has been suggested as the ultimate origin of high ε_{Nd} sodic alkali basalts in many other regions around the world, including the northern portions of the East African Rift (Class *et al.*, 1994), western Africa (Cameroon Line; (Marzoli *et al.*, 2000), and southeast Australia (McDonough *et al.*, 1985; Zhang and O'Reilly, 1997).

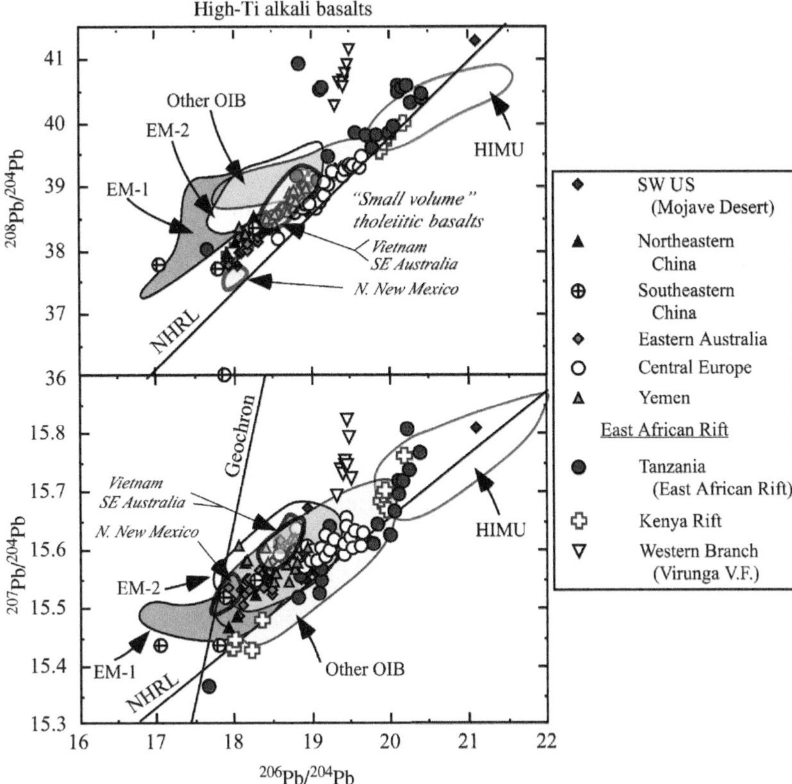

Figure 12 Pb-isotopic compositions for high-Ti sodic alkali basalts worldwide. Data from sources listed in Table 2 and from Zhang *et al.* (2001). NHRL = Northern Hemisphere Reference Line (source Hart, 1984).

In southeast Australia, this conclusion has been supported by recent osmium-isotopic studies of the Quaternary Newer Volcanic Province, which reveal that sodic alkali basalts there have osmium-isotopic compositions indistinguishable from OIB, again supporting a sublithospheric origin for their parental magmas (McBride *et al.*, 2001). Helium-isotope data are available primarily from sodic basalts in the northern half of Africa (Cameroon Line, Sudan). These basalts have R/R_a values (~5.1–7.5) consistent with a derivation from an HIMU-type plume source (Barfod *et al.*, 1999; Franz *et al.*, 1999).

While few workers would dispute that the high ε_{Nd} values of sodic alkali basalts require that their parental magmas were dominantly comprised of components recently derived from the sublithospheric mantle, there has been debate whether the sodic basalts represent direct products of sublithospheric mantle melting. An alternative is that these basalts are derived from melting of lithospheric mantle (particularly "young" high ε_{Nd} mantle lithosphere) that had been metasomatized just prior to basalt formation by incompatible element enriched fluids/ melts derived from small degrees of melting of upwelling asthenosphere. Small melt fractions are likely to freeze within the continental lithosphere (McKenzie, 1989), producing a veined mantle prone to subsequent melting. Such a lithospheric mantle source has been proposed for Quaternary sodic basalts on the Arabian Peninsula (Figure 10), one of the largest alkali basalt fields in the world (Baker *et al.*, 1997). Modeling of very incompatible versus moderately incompatible element ratios (e.g., Ce/Y, Nb/Zr) suggests that the magmas parental to these basalts were derived primarily from ~5% melting of relatively shallow, LREE enriched, amphibole-bearing, spinel peridotites within the mantle lithosphere (Baker *et al.*, 1997). Given the high ε_{Nd} values for the basalts (average $\varepsilon_{Nd} \sim 5.0$; Table 2 and Figure 7(a)), the metasomatism of the lithospheric mantle must have been relatively recent. Baker *et al.* (1997) suggest that the metasomatism occurred during percolation of small melt fractions through the lithosphere. These melts were derived from the high potential temperature and potentially wet Afar plume, which is responsible for the nearby Oligocene flood basalt volcanism. It is interesting to note that the negative potassium anomaly found in the Arabian basalts, which can be taken as evidence of an amphibole-bearing lithospheric mantle source for their primary magmas, is actually common to many continental sodic basalts (Figure 11(a)), and may indicate that magmatic processes occurring in the mantle lithosphere represent an important step in the generation of all these rocks.

The addition to the lithospheric mantle of small melt fractions derived from the shallow mantle source of MORB (DMM; Zindler and Hart, 1986) is also possible (Ellam and Cox, 1991). A particularly strong case can be made in this regard for sodic basalts in the Cima volcanic field in the southwestern USA. These Late Cenozoic basalts have among the highest ε_{Nd} values of any sodic basalts worldwide (up to 9.5), values that overlap those of Pacific MORBs (Figure 7(a); Farmer *et al.* (1995)). Farmer *et al.* (1995) argued that the high ε_{Nd} values of the Cima volcanic field basalts, coupled with their high LREE contents, require that the mantle source of these basalts underwent a recent enrichment in LREE from melts or fluids derived from high ε_{Nd} MORB source mantle. One possibility is that lithospheric mantle beneath this region was metasomatized by small melt fractions derived from the DMM source. These melts may have been generated during mantle upwelling caused by the Late Cenozoic opening of a slab window through subducting oceanic lithosphere, and do not require the presence of a mantle plume. Such a scenario is not uncommon, and has been suggested for the Antarctic Peninsula and British Columbia (Hole, 1988; Hole *et al.*, 1991). Thus, mantle plumes are not always required for the production of small volume, sodic alkali basalts.

3.03.3.2.2 Potassic alkali basalts

Cenozoic potassic alkali basalts are less common than their sodic counterparts. They are the dominant basalt type in relatively few regions, such as the Tibetan Plateau (Turner *et al.*, 1996a; Miller *et al.*, 1999), Italy (the famous Roman volcanic province; Conticelli *et al.* (1997) and Peccerillo (1999)), and portions of eastern China (Zhang *et al.*, 1995), Taiwan (Chung *et al.*, 2001), Turkey (Robert *et al.*, 1992), and western North America (Kempton *et al.*, 1991) (Figure 10). In each of these areas, the most mafic potassic basalts are characterized by high potassium contents (~3–7.5 wt.% K_2O) and low titanium contents (0.8–1.7 wt% TiO_2; Table 2 and Figure 6). The high magnesium contents (>10 wt.% MgO; Table 2(b)) of many of these basalts clearly require their derivation from an ultramafic mantle source (Turner *et al.*, 1996a). The most distinctive aspect of these rocks, however, is that in addition to their high LILE and LREE abundances, the potassic basalts have pronounced negative niobium and tantalum anomalies, and positive lead anomalies (Figure 11(b)). In addition, the magmas parental to these rocks were highly oxidizing (~3–7 log f_{O_2} units greater than typical "asthenospheric" mantle) and contained high volatile contents (H_2O and F greater than 2 wt.%; Feldstein and Lange (1999)). All these chemical characteristics are similar to those of modern IABs (McCulloch and Gamble, 1991). As a result, while none of the low-titanium potassic basalts listed in Table 2(b) formed during active subduction, the mantle sources of these rocks appear to have been metasomatized by hydrous, high-potassium, low-HFSE, fluids derived from the dehydration of subducted oceanic lithosphere. Such an origin is consistent with the fact that many potassic basalts, including the basalts in the Tibetan Plateau, Taiwan, Italy, and the western USA, erupted soon after the cessation of subduction beneath the underlying continental lithosphere.

The primary exception to a "post-subduction" origin for mantle sources of potassic basalts are those occurring in the southern portions of both the eastern and western branches of the East African Rift (Furman, 1995; Paslick *et al.*, 1995; Rogers *et al.*, 1998; Furman and Graham, 1999), an area that has not witnessed subduction for at least 500 Ma (Petter, 1991). However, unlike potassic basalts elsewhere, these basalts have high titanium contents (>2 wt.% TiO_2; Table 2) and lack negative HFSE anomalies (Figure 11(a)). Instead, they have trace-element patterns similar to those of kimberlites, OIBs, and sodic continental alkali basalts (Paslick *et al.*, 1995).

From an isotopic perspective, both low- and high-titanium potassic basalts are distinct from sodic basalts in having generally lower ε_{Nd} values (~0 to −8), higher $^{87}Sr/^{86}Sr$ ratios (~0.704–0.708; Figures 7(b) and 13) and more variable lead-isotopic compositions (Figure 13). The low ε_{Nd} values of the potassic basalts are generally taken as evidence that the basalts were derived from ancient (i.e., Precambrian age), LREE enriched, lithospheric mantle sources (Rogers *et al.*, 1998; Miller *et al.*, 1999). It is important to note, however, that because of the mobility of lead and strontium in slab fluids, and lead, strontium, and neodymium in melts produced from subducted oceanic lithosphere and overlying sediments (Pearce and Peate, 1995), mantle metasomatized above subducted lithosphere could have low ε_{Nd}, high $^{87}Sr/^{86}Sr$, and variable lead-isotopic composition as the result of addition of such components from the subducted lithosphere, itself (Nelson, 1992). As a result, the isotopic composition of the low-titanium potassic basalts do not necessarily require melting in ancient lithospheric mantle but could represent melting of young lithospheric mantle (Farmer *et al.*, 2002). The African high-titanium potassic basalts, in contrast, represent melting of ancient lithospheric mantle originally metasomatized by HFSE-rich fluids that were ultimately derived from a deeper portion of the sublithospheric mantle. This scenario is

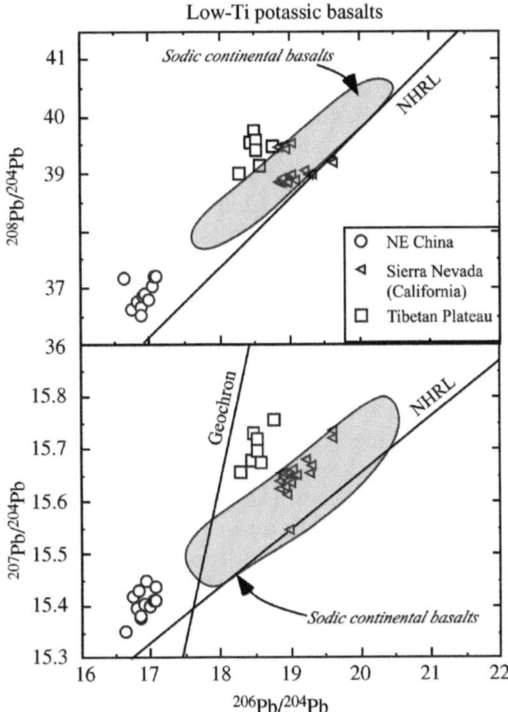

Figure 13 Pb-isotopic compositions for low-Ti potassic alkali basalts worldwide (data from sources listed in Table 2).

similar to that suggested for the metasomatized sources of orangeites.

Given the relatively low melting temperature of hydrated mantle of any age (Figure 3) (Harry and Leeman, 1995), it is not surprising that potassic continental basaltic magmatism derived from such mantle can follow closely the cessation of subduction. But what triggers the magmatism? In the Tibetan Plateau, several authors have suggested that Cenozoic potassic magmatism is derived from relatively shallow lithospheric mantle (Turner *et al.*, 1996a; Chung *et al.*, 1998). This mantle was apparently heated and melted as a result of the removal of the originally underlying lithospheric mantle via convective erosion by the asthenosphere, or "delamination," shortly after collision of India and Asia. Similarly, in the Sierra Nevada of the western USA, the abrupt onset and short duration of Cenozoic potassic magmatism has been taken as evidence that shallow (~40 km deep), hydrous lithospheric mantle melted in response to abrupt delamination of the underlying lithosphere (Farmer *et al.*, 2002). Therefore, potassic magmatism may provide information regarding the timing and extent of major deformational events in the deep-continental lithosphere, events that led not only to the potassic magmatism but also to major episodes of continental mountain building (i.e., the Tibetan Plateau and the Sierra Nevada).

3.03.3.2.3 Small volume tholeiitic basalts

Cenozoic alkalic basalt volcanism worldwide is often accompanied by the eruption of tholeiitic (subalkaline) basalts. In some continental regions, such as eastern Australia, tholeiitic basalts can comprise up to 50% or more of the total erupted material (Johnson *et al.*, 1989). In SE Asia, where plateaus dominantly comprise tholeiitic basalts covering an area of $\sim 2.3 \times 10^4$ km^2 (Figure 10), a total of ~8,000 km^3 of basalt erupted during the Late Cenozoic, at rates estimated at ~3 km^3 kyr^{-1} (Hoang and Flower, 1998). Although significant, even these magmatic volumes and eruptive rates are several orders of magnitude smaller than those of the typical continental flood basalts (see following section).

These "small volume" tholeiitic basalts are compositionally heterogeneous in any given volcanic area. They range in composition from olivine to quartz tholeiites, with phenocrysts assemblages of olivine ± clinopyroxene ± plagioclase ± Fe–Ti oxides (Dungan *et al.*, 1986; Price *et al.*, 1997; Wedepohl, 2000). Olivine tholeiites are generally considered the most primitive of the tholeiitic basalts, but their relatively high SiO_2 contents (~50 wt.%; Table 2(a) and Figure 6) and low mg numbers (<~0.66) indicate that even these rocks crystallized from magmas that had undergone significant differentiation prior to eruption, most likely through the removal of early formed olivine crystals (Dungan *et al.*, 1986). Such protracted magmatic differentiation not only obscures the original chemical composition of the parent magmas, but also suggests that these magmas may have resided sufficiently long in the continental crust to experience substantial crustal contamination. A slow ascent through the crust is consistent with the general lack of entrained mantle-derived xenoliths in the tholeiitic rocks. The high rubidium, barium, and potassium contents, relative to thorium (Figure 11(c)), of the Servilleta basalts at the northern end of the Rio Grande Rift in the southern USA (Figure 10), along with their low $^{206}Pb/^{204}Pb$ (~18) and $^{208}Pb/^{204}Pb$ (~37.5) ratios (Figure 12), have all been taken as evidence that these rocks interacted significantly with the lower continental crust in this region (Dungan *et al.*, 1986). The trend towards higher $^{87}Sr/^{86}Sr$, at a given ε_{Nd} value, for "small volume" Cenozoic tholeiitic rocks in both the Newer Volcanic Province in SE Australia and in Vietnam (Figure 7(a)) can also be interpreted as evidence of interaction with low ε_{Nd}, high $^{87}Sr/^{86}Sr$, continental crust. In addition, the osmium-isotopic compositions of tholeiitic basalts in the Newer Volcanic Province are considerably more radiogenic and variable than the compositions of contemporaneously erupted alkali basalts (γ_{Os} $((^{187}Os/^{188}Os)_{sample(t)}/(^{187}Os/^{188}Os)_{chondrite(t)})-1)*100$ =42–251 versus 5.6–7.6), and so provide evidence that the olivine tholeiites assimilated continental crust in amounts up to 2.5% of original magma mass (McBride *et al.*, 2001).

The fact that "small volume" tholeiites may have generally interacted with continental crust considerably complicates the issue of determining the sources of their parental magmas on the basis of chemical and isotopic data. As a result, while many workers accept the conclusion that "small volume" tholeiitic basalts represent larger degrees (~10%) of partial melting of peridotite than alkali basalts and last equilibrated at shallow levels (<~45 km) in the mantle (Hoang and Flower, 1998; DePaolo and Daley, 2000; Wedepohl, 2000), there is no consensus regarding the relative roles of CLM and sublithospheric mantle in their generation. In the northern half of the Rio Grande rift, for example, some authors have suggested that the relatively high $^{87}Sr/^{86}Sr$ (~0.705) and low ε_{Nd} (~0 to +5; Baldridge *et al.* (1991)) of tholeiitic basalts require a source in the lithospheric mantle (Perry *et al.*, 1987). Others have argued for melting in upwelling asthenosphere, potentially initiated in the garnet stability field (Beard and Johnson, 1993), and subsequent interaction with either continental crust, or lithospheric mantle (Thompson and Gibson, 1994). Interaction between tholeiitic magmas and mafic veins within the lithospheric mantle has also been proposed to account for the radiogenic osmium-isotopic compositions of Late Cenozoic high-alumina tholeiites in the northwest USA (Hart *et al.*, 1997). There is a similar diversity of opinion regarding the sources of tholeiitic basalts in the Newer Volcanic Province, both lithospheric (Price *et al.*, 1997) and sublithospheric (McDonough *et al.*, 1985) sources having been proposed.

Distinguishing between lithospheric and sublithospheric mantle as a source for the "small volume" tholeiites ultimately involves assessing whether lithospheric mantle beneath a given area is sufficiently fertile to spawn basaltic magmas at all, let alone basalts of the appropriate chemical and isotopic compositions and, if not, whether the CLM has been eroded and/or thinned to allow asthenospheric mantle of a given potential temperature to rise to the shallow depths required to produce the observed compositions and volumes of tholeiitic magmatism. The conflicting interpretations of the origin of "small volume" tholeiitic basalts worldwide makes addressing these issues a difficult task. A similar difficulty exists in determining the sources of large-volume tholeiitic basalts found on continents, as described in the following section.

3.03.3.3 Continental Flood Basalts

Large igneous provinces (LIPs) are the most massive short-lived igneous events on Earth (Coffin and Eldholm, 1994). These events produce large volumes of mafic lavas and intrusive rocks, and are responsible for the formation of many oceanic plateaus (see Chapter 3.16), volcanic passive continental margins, and continental flood basalts. Post-250 Ma continental flood basalts, because of their better preservation and because of their potential link to Phanerozoic mass extinctions (Renne *et al.*, 1995), are the most thoroughly studied of continental LIPs (Hooper, 2000). The archetypal Mesozoic and younger continental flood basalts (CFBs) include those of the Paraná (Brazil)–Etendeka (NW Namibia) Province, the Deccan Traps, the Siberian Traps, the Karoo (Lesotho, S. Africa)–Ferrar (Antarctica)–Tasmanian Province, and the Columbia River basalts (NW USA; Figure 14). Estimates of eruptive volume of flood basalt provinces are huge. They range from a low of $\sim 2\times 10^5$ km^3 for the Columbia River basalts to more than 2×10^6 km^3 for the Siberian Traps (Hooper, 2000; Reichow *et al.*, 2002). Even more remarkable is the relatively short time span apparently involved in the eruption of the bulk of the CFB in a given province which, according to some estimates, could be as short as ~1–3 Ma (Renne and Basu, 1991; Renne *et al.*, 1992; Courtillot *et al.*, 1999). This implies eruption rates as high as 10^3 km^3 kyr^{-1}.

Compositionally, CFBs are all tholeiitic basalts with a simple phenocryst assemblage consisting of plagioclase, clinopyroxene, and Fe–Ti oxides (Hooper, 2000). The high Fe/Mg ratios for these rocks (wt.% $Fe_2O_{3T}/MgO \sim 1-3$) and their relatively high silicon contents (wt.% $SiO_2 = 49-57$; Table 3 and Figure 6) are generally interpreted as evidence that CFBs are not primary mantle melts but represent magmas that have undergone significant magmatic differentiation, presumably through crystal–liquid separation, prior to eruption. However, Lange (2002) has demonstrated, at least in the case of the Columbia River basalts, that iron-rich CFB magmas are too dense to ascend through the continental crust unless the pre-eruptive volatile contents of these magmas are much higher than generally assumed ($CO_2 + H_2O$ of at least 4 wt.%). If so, CFB must be an even more significant source of atmospheric volatiles, including greenhouse gases, than previously recognized, and phenocrysts present in CFB may have been produced during eruption and degassing of their parental magmas, in which case the role of crystal fractionation in producing the apparently evolved compositions of CFB is called into question (Lange, 2002).

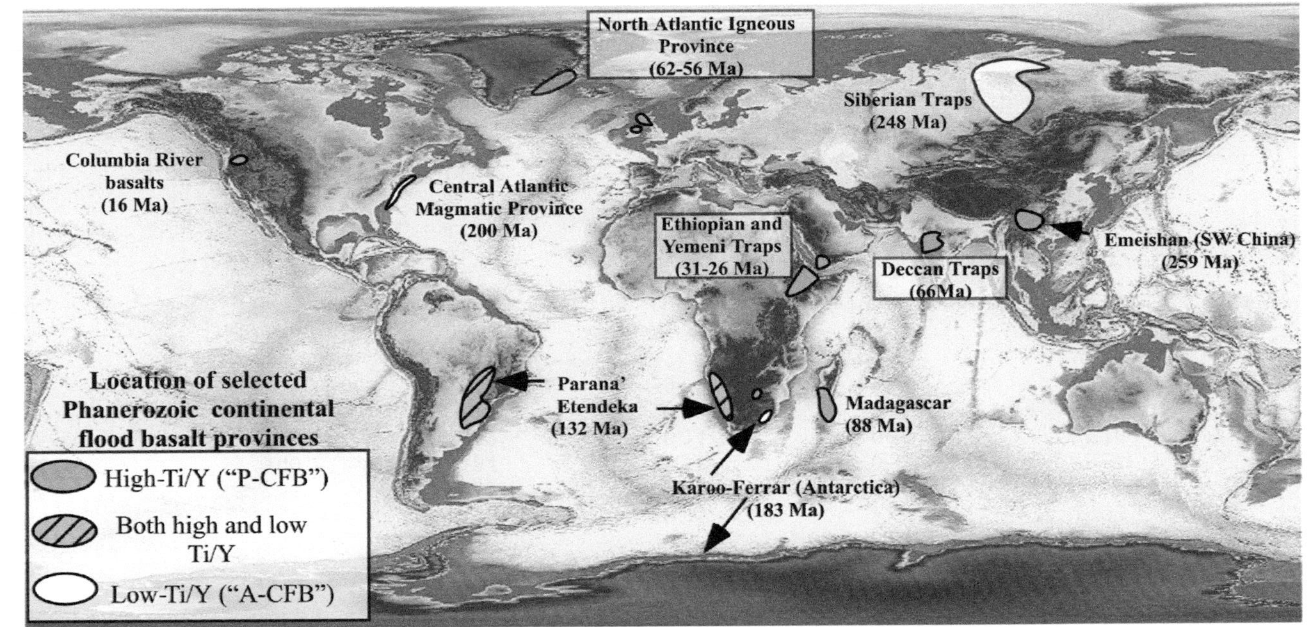

Figure 14 Locations of selected major Phanerozoic continental flood basalts worldwide. References as in Table 3, with exception of Madgascar (Storey *et al.*, 1997), Emeishan (Chung and Jahn, 1995; Zhou *et al.*, 2002), North Atlantic Igneous Province (Saunders *et al.*, 1997), Ethiopian and Yemeni traps (Menzies *et al.*, 1997; Pik *et al.*, 1999), and Central Atlantic magmatic province (CAMP) (Hames *et al.*, 2000). The separation of CFB provinces into high Ti/Y ("P" or "plume" type) and low Ti/Y ("A" or "arc" type) from Puffer (2001). However, most provinces include examples of both "types" of basaltic rocks (cf. the Paraná-Etendeka provinces). A more complete compilation of terrestrial large igneous provinces can be found in Ernst and Buchan (2001).

Table 3 Major element, trace element, and isotopic compositions of continental flood basalts.[a]

	Columbia River Basalts[b] *NW US* *High Ti*[f]	*Deccan Traps*[c] *Mahabaleshwar* *High Ti*	*Deccan Traps*[c] *Kolhapur* *Low Ti*	*Siberian Traps*[d] *Nadezhdinsky* *High Ti*	*Siberian Traps*[d] *Gudchikhinsky* *Low Ti*	*Paraná* *Esmeralda* *High Ti*	*Paraná*[e] *Gramado* *Low Ti*	*Paraná*[e] *Urubici* *High Ti*
Age (Ma)	16	60	60	250	250	127–137	127–137	127–137
$n =$	36	6	18	9	7	1	1	1
SiO_2	54.7	49.5	48.7	51.8	48.5	51.1	50.9	53.0
TiO_2	1.97	2.34	3.16	0.97	1.58	1.37	0.95	3.76
Al_2O_3	14.0	14.1	13.4	15.1	11.6	13.8	14.9	12.9
MnO	0.25	0.18	0.23	0.17	0.17	0.21	0.17	0.19
MgO	4.20	6.9	5.7	6.8	13.0	6.1	8.0	4.3
Fe_2O_3t	12.1	13.4	15.1	10.8	13.8	13.4	10.3	12.7
CaO	7.99	10.45	10.28	10.45	8.69	10.73	11.61	8.30
Na_2O	3.06	2.51	2.51	2.37	2.12	2.55	2.44	2.57
K_2O	1.44	0.40	0.34	1.35	0.38	0.54	0.51	1.70
P_2O_3	0.35	0.23	0.39	0.14	0.15	0.16	0.16	0.58
K_2O/Na_2O	0.47	0.16	0.14	0.57	0.18	0.21	0.21	0.66
Ni	28	134	69	59	589	58	99	58
Cr	39		151	137	595	94	307	75
Zn	11.3	109	124	84.3	112	89.0	72.0	107
Cu	39.6	142	266	60.8	105	169	99.0	267
Co				41.3	64.7	50.0	44.0	36.0
V	332		405			323	221	343
Y	35.6	30.0	42.0	27.0	24.0	29.0	23.0	39.0
Sc	34.4			32.2	24.1	42.0	40.0	29.0
Rb	35.6	7.0	13.0	40.2	14.7	19.0	10.0	30.0
Sr	323	283	234	259	307	163	216	764
Cs	0.93			0.77	1.40			
Ba	546	175	123	412	110	163	243	600
Zr	165	138	198	129	102	100	92	307
Hf	4.24	3.70	5.40	2.94	2.46	2.71	2.15	7.97
Ga	20.9					21.0	17.0	26.0
Nb	13.7	15	17	11	14	6	9	27
Ta	0.92	0.80	1.30	0.48	0.47	0.37	0.50	1.92
U	1.08			0.99	0.36	0.71		1.34

Th	3.71	2.1	2.3	2.92	1.07	1.75	2.04	4.25
Pb	6.94			4.72	1.26		3.39	
La	22.1	16.5	17.5	16.7	7.94	8.35	10.6	42.5
Ce	44.6	37.6	43.8	36.0	20.4	20.8	22.9	90.4
Pr	5.79							
Nd	25.2	23.2	29.2	17.5	13.5	14.5	12.8	54.3
Sm	6.45	5.7	7.7	4.01	3.63	3.97	3.18	11.6
Eu	2.01	1.74	2.47	1.10	1.24	1.41	1.11	3.53
Gd	6.56			4.09	3.91			
Tb	1.12			0.65	0.62	0.90	0.63	1.54
Dy	6.90							
Ho	1.40			0.93	0.74			
Er	3.99							
Tm	0.55			0.39	0.27			
Yb	14.43	2.47	3.68	2.39	1.55	2.92	2.09	2.99
Lu	0.54	0.39	0.58	0.35	0.21	0.47	0.35	0.44
$(La/Yb)_N$	0.99	4.32	3.08	4.51	3.31	1.85	3.28	9.20
Ti/Y	332	468	451	216	395	283	248	578
$n_i =$ [g]	31	6	18	5	3	1	1	1
$^{87}Sr/^{86}Sr(T)$	0.7063	0.7066	0.7045	0.7078	0.7063	0.7058	0.7073	0.7060
ε_{Nd} (T)	1.64	−0.18	4.54	−8.10	3.20	0.70	−3.80	−3.40
$^{206}Pb/^{204}Pb$ [h]	18.911	18.370	17.820	18.296	18.726	18.661($n = 13$)	18.696($n = 11$)	
$^{207}Pb/^{204}Pb$	15.614	15.520	15.421	15.570	15.570	15.639	15.649	
$^{208}Pb/^{204}Pb$	38.788	38.880	28.244	38.341	38.341	38.706	38.824	

[a] Major element compositions reported as weight percent oxides (anhydrous), trace element abundances in ppm. [b] Grande Ronde basalt average. Data from Hooper and Hawkesworth (1993). [c] Average compositions of Mahabeleshwar and Kothapur basalts from Lightfoot *et al.* (1990). [d] Average compositions of Nadezhdinsky and Gudehikhinsky basalts from Wooden *et al.* (1993). [e] Representative analyses from Esmeralda (sample # DSMO6), Gramado (#DUP30) and Urbici (# DUP35) from Peate (1997). [f] High and low Ti basalts defined following Peate (1997), the former having wt.% $TiO_2 > \sim 2$ and Ti/Y > 310. [g] n_i number of samples used for radiogenic isotope averages. [h] Pb isotopic compositions are measured values.

Despite their similar major element compositions, CFBs show a wide range of trace-element and isotopic compositions. They can be divided into two groups based on their trace- and minor-element abundances: the "high-titanium" (>2 wt.% TiO_2) and "low-titanuium," or, alternatively, high Ti/Y (>310) and the low Ti/Y groups (see Peate (1997) and references therein, and Hornig (1993)). The "high-titanium" CFBs, such as the Deccan Traps, have low LIL/HFSE ratios similar to, albeit with lower overall LIL and LREE abundances, those of ocean island basalts, kimberlitic rocks, and sodic alkali basalts. "Low-titanium" (or low Ti/Y) CFBs, such as some CFB flows within the Siberian Traps and the Paraná provinces, have prominent depletions in HSFE relative to LIL, and high lead contents similar to the relative trace-element abundances of IABs and potassic alkali basalts (Figure 15). In general, the "high-titanium" CFB also tend to have higher initial ε_{Nd} and lower $^{87}Sr/^{86}Sr$ than their "low-titanium" counterparts (0 to +4.5 versus −3 to −8, and 0.7045–0.7066 versus 0.706–0.708, respectively; Table 3; Figure 16).

The previous discussion of kimberlites and alkali basalts indicates that it would be reasonable to suggest that "high-titanium" CFBs represent melts of upwelling asthenosphere (or young lithospheric mantle recently metasomatized by fluids/melts derived thereof), while "low-titaniumf" CFBs are the products of melting of Precambrian or younger lithospheric mantle that had been metasomatized by fluids/melts derived from subducted oceanic lithosphere and/or sediments (Puffer, 2001). However, it is important to note that any model for the formation of CFB must also take into account the large volume of magma generated during the formation of CFBs and the fact that CFBs are not primary magmas and may have differentiated within, and interacted extensively with, the continental crust. As a result, suggesting that high-titanium and low-titanium CFBs are derived sublithospheric and lithospheric sources, respectively, is not as reasonable an assertion as it was for the much smaller volume kimberlites and alkali basalts.

The large volume of magma represented by CFBs, and large igneous provinces in general, favor models of CFB generation within mantle plumes with potential temperatures up to 300 °C higher than typical upper mantle (Morgan, 1971). Upwelling of such mantle plumes, particularly when accompanied by lithospheric thinning, leads to thick mantle melting columns that extend to shallow mantle depths (Figure 1). This provides the conditions required to produce the large

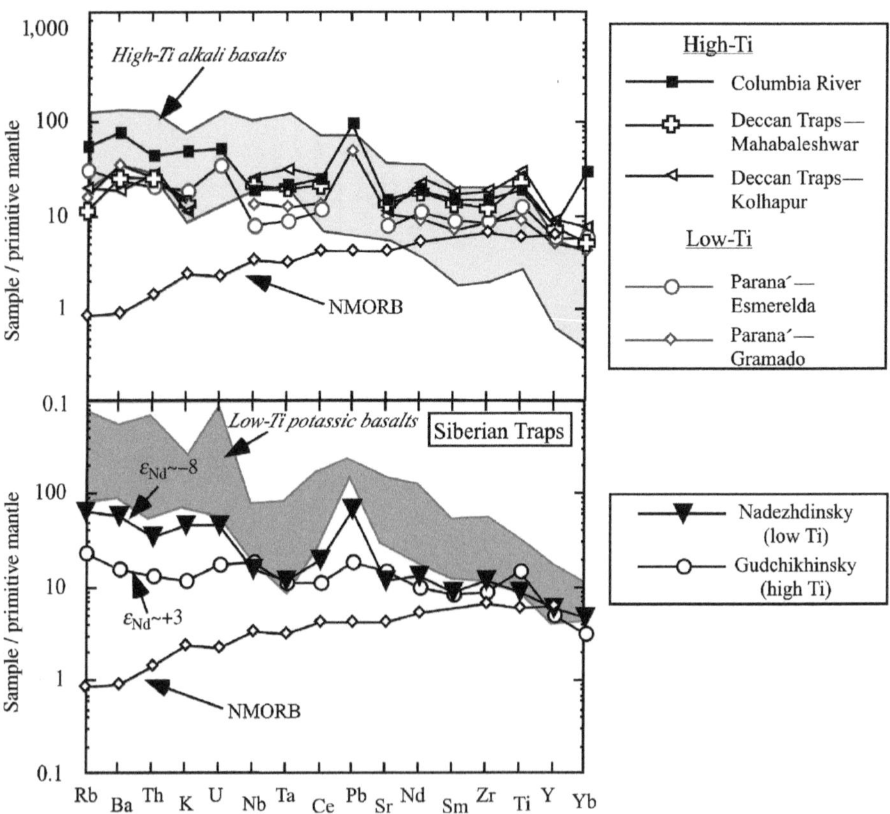

Figure 15 Primitive mantle normalized trace-element abundance for continental flood basalts (data from Table 3). NMORB data from Sun and McDonough (1989).

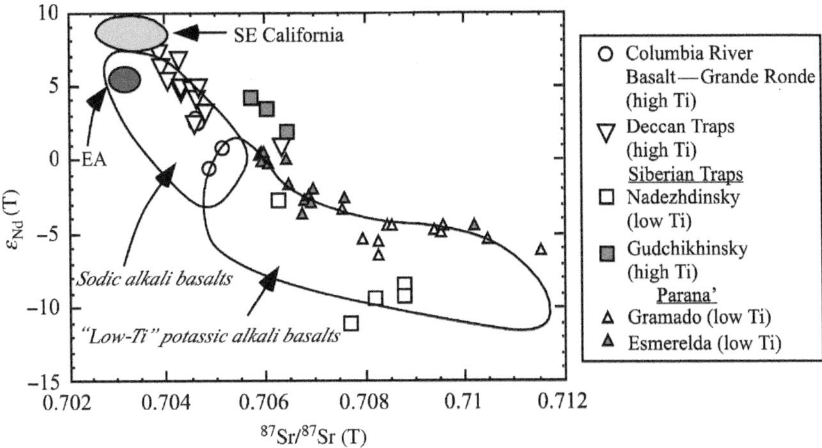

Figure 16 Initial ε_{Nd} versus $^{87}Sr/^{86}Sr$ for continental flood basalts. EA = European asthenospheric reservoir (Wilson and Patterson, 2001). Data from Table 3.

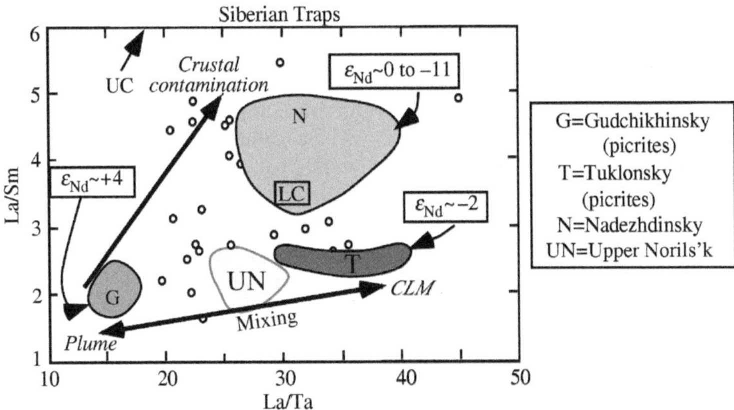

Figure 17 La/Sm versus La/Ta for the Siberian Traps. UC = upper crust, LC = lower crust (reproduced by permission of American Geophysical Union from *Large Igneous Provinces*, **1997**, *100*, 335–355).

volumes of tholeiitic magmas represented by CFBs (McKenzie and Bickle, 1988). From a geochemical standpoint, melting of an upwelling mantle plume can account for the chemical and isotopic composition of "high-titanium" CFBs, most notably the ~64 Ma Deccan Traps (Figures 15 and 17; Puffer (2001)). Many "high-titanium" Deccan Trap flows have trace-element and isotopic compositions consistent with derivation by decompression melting of the Reunion plume (cf. Mahoney, 1988). The high $^3He/^4He$ ratios of mafic alkaline rocks associated with the Deccan flood basalts ($R_a \sim 13.9$) are similar to those of oceanic volcanic rocks derived from the Reunion plume. This observation also supports the hypothesis that the impingement of this plume on the base of the Indian lithosphere was responsible for the CFB magmatism (Basu *et al.*, 1993).

A plume source, with or without an upper mantle (MORB source) contribution, is also likely for some of the Siberian CFBs (Sharma, 1997), particularly the "high-titanium," older Gudchiniksy basalts (Table 3) and associated highly magnesian (up to 22 wt.% MgO) picrites, which have high ε_{Nd} values (~+4) and low LIL/HFSE ratios (Figures 15 and 17; Wooden *et al.* (1993)). Picrites are rare, highly magnesian lavas (>12 wt.% MgO) that occur early in the eruptive sequence of several CFBs (Gibson, 2002) and are interpreted as the products of mantle melting at pressures above ~3 GPa (Herzberg and O'Hara, 1998). In the Siberian Traps, picrites also have high initial γ_{Os} (3.4–6.5; Horan *et al.*, 1995) and olivine phenocrysts within early alkaline mafic and ultramafic volcanic rocks elsewhere in the Siberian Traps have high $^3He/^4He$ (up to ~13 times the atmospheric values) (Basu *et al.*, 1995). Both observations are consistent with the derivation of these rocks from melting in a mantle plume that is compositionally similar to those spawning some OIBs. Even the major- and trace-element compositions of the low-titanium Putorana basalts, which represent 90% of the Siberian Trap volcanism, can be modeled as requiring the derivation of these rocks from primary picritic magmas generated by 12–18% melting of an upwelling mantle plume at shallow (spinel stability field) depths in the sublithospheric mantle beneath a thinned lithosphere (Basu *et al.*, 1998).

Picritic rocks are also of interest in CFB provinces because of their compositional heterogeneities. Some of the early picrites have very high iron contents, inconsistent with their derivation by melting of peridotite, even at great depth (Gibson *et al.*, 2000). The high iron contents are instead interpreted as evidence of compositional heterogeneities in the plume head rising beneath continents. The high iron content is taken as evidence of melting of mafic lithologies, or of peridotite metasomatized by melts derived therefrom (Cordery *et al.*, 1997; Korenaga and Kelemen, 2000; Gibson, 2002).

Regardless of the degree of chemical heterogeneity, derivation of low-titanium CFBs by direct melting of an upwelling plume is difficult to reconcile with their high-LIL/HFSE ratios, particularly those CFBs with ε_{Nd} values less than zero (Figure 16), because these chemical and isotopic characteristics are so distinct from those of plume-derived OIBs. A lively debate has developed in the literature about whether the isotopic and trace-element compositions of these CFBs reflect derivation, at least in part, from the lithospheric mantle, metasomatized by subduction related fluids/melts, or whether it reflects extensive crustal contamination of plume-derived magmas (Ellam and Cox, 1991; Hergt *et al.*, 1991; Arndt and Christensen, 1992; Lassiter and DePaolo, 1997). Low-titanium, low-ε_{Nd} CFBs from the Norils'k area of the Siberian Traps, for example, have been interpreted as the products of extensive interaction between plume-derived magmas and continental crust during magmatic differentiation (Wooden *et al.*, 1993). Others, however, have noted that the Norils'k area "low-titanium" (and high-La/Ta) CFBs and picrites do not have high La/Sm ratios, as would be expected from interaction with high La/Sm upper crust (Figure 17; Lassiter and DePaolo, 1997). The low-titanium basalts are instead attributed to mixing between high-titanium, high-ε_{Nd} plume-derived magmas, and a low-ε_{Nd}, high-La/Ta component derived presumably from metasomatized portions of the lithospheric mantle. A similar model was proposed earlier to account for the chemical and isotopic compositions of the Karoo CFBs in southern Africa (Ellam and Cox, 1991; Ellam *et al.*, 1992). Still other workers have suggested that the low-titanium Norils'k area CFBs were derived exclusively from CLM contaminated by subducted sediments, although melting of the lithosphere may have been triggered by conductive heating from the Siberian plume when it impinged upon the base of the lithosphere (Lightfoot *et al.*, 1993).

The controversy regarding the origin of low-ε_{Nd}, low-titanium CFBs has also been intense for the ~130 Ma Paraná CFBs. The Paraná CFB province covers an area of at least 1.2×10^6 km^2, and consists of a central area of voluminous tholeiitic flood basalts surrounded by a smaller volume of alkalic intrusions and lavas (Peate, 1997). Amongst the flood basalts, low-titanium, low-ε_{Nd} basalts (the Gramado and Esmeralda basalts; Table 3) cover the southern half of the province, while high-titanium, higher-ε_{Nd} CFBs (e.g., Urubici basalts; Table 3) are present in the northern half (Peate, 1997). Some authors have suggested that the two magma types were generated in a common upwelling plume source. These workers consider the high-titanium basalts to have been derived from deeper (within garnet stability field) and smaller degrees of partial melting than low-titanium basalts. The contrast in ε_{Nd} values between the two basalt types is interpreted as the result of differing amounts of crustal contamination. Others have suggested that while evidence of minor crustal contamination exists for all the Paraná CFB (such as increasing $^{87}Sr/^{86}Sr$ with increasing Th/Ta ratios), the chemical and isotopic contrasts between the high- and low-titanium CFB are retained even in the least-contaminated rocks, suggesting that distinctly different parental magmas are required for the different CFBs (Peate and Hawkesworth, 1996). However, neither basalt type has chemical or isotopic compositions similar to basalts derived from the Tristan plume, which is probably responsible in some fashion for both CFB formation in the Paraná and Etendeka and for the opening of the southern Atlantic (Peate, 1997). If they are not plume derived, all of the Paraná CFBs may be products of melting within the lithospheric mantle. If so, the compositional differences between the high- and low-titanium CFBs may represent regular north–south differences between the chemical and isotopic composition of the lithospheric mantle (Peate, 1997). Deriving such large volumes of magma from the lithospheric mantle is unlikely, if the mantle is initially cold and dry (Arndt and Christensen, 1992), but is theoretically possible if the mantle is initially wet and melts when conductively heated from below by an upwelling plume (Gallagher and Hawkesworth, 1992; Turner *et al.*, 1996b). However, mafic magmatism occurred again in southern Brazil during the Late Cretaceous, ~50 Myr after the emplacement of the Paraná CFB, possibly as a result of the impact of the Trinidade plume at the base of the mantle lithosphere (Gibson *et al.*, 1999). Gibson *et al.* (1999) argue that the alkaline magmatism produced at this time was derived from CLM that was metasomatically enriched in the Proterozoic. The preservation of such low-melting T-components in the lithospheric mantle, at least until the Late Cretaceous, suggests that the CLM beneath this region was not pervasively melted during the formation of the earlier Paraná CFB, as Turner *et al.* (1996b) had suggested. Clearly, the source of the Paraná basalts is still an open question.

The above discussion shows that the role of crustal contamination in the chemical and isotopic compositions of CFB remains an important issue. Some workers have suggested that this type of contamination can be most readily recognized through osmium-isotopic data, given the large osmium-isotopic contrast between mantle-derived magmas and the generally high $^{187}Os/^{188}Os$ of the continental crust (Chesley and Ruiz, 1998). Another promising approach towards resolving the contamination issue is through the use of phenocryst oxygen-isotope data. A recent study of olivine, clinopyroxene, and plagioclase separates from Oligocene flood basalts in Yemen showed that these minerals have $\delta^{18}O$ values that deviate considerably from the values expected for either a lithospheric or sublithospheric mantle-derived magma undergoing fractional crystallization alone (Baker *et al.*, 2000). Olivine $\delta^{18}O$ values, for example, range from 5.1‰, to as high as 6.2‰. This increase correlates with a small but significant increase in whole-rock $^{87}Sr/^{86}Sr$, decrease in ε_{Nd}, and increased variability in whole-rock lead-isotopic compositions. These variations are clearly due to the assimilation of the high $^{87}Sr/^{86}Sr$ and $\delta^{18}O$, low ε_{Nd} Proterozoic basement of the Arabian Peninsula. Furthermore, because oxygen represents ~50% of a typical mantle or crustal rock (Taylor, 1968), the increase in magmatic $\delta^{18}O$ recorded in the phenocrysts requires a substantial amount of assimilation (up to ~25 wt.%, depending on the crustal $\delta^{18}O$). Addition of such large amounts of crust can shift the trace-element compositions of the basaltic magma dramatically, and could even produce a shift from the low-LILE/HFSE ratio typical of OIB and high-titanium CFB, to the high-LILE/HFSE ratio of low-titanium CFB and potassic alkali basalt (due to the high-LILE/HFSE ratios typical of intermediate to felsic composition crust; Baker *et al.* (2000)). While this is a provocative study, it remains to be demonstrated that phenocryst oxygen-isotope studies can provide similar insights into the importance of crustal assimilation in the generation of CFB worldwide. Granulite facies lower continental crust, for example, can have low, mantle-like $\delta^{18}O$ values (Valley, 1986). Lower crustal assimilation, therefore, need not always produce a significant increase in magmatic $\delta^{18}O$.

3.03.3.4 Case Example—Western USA

Although our understanding of the how and why of continental basaltic rock formation is incomplete, their generation clearly provides an opportunity to study the dynamic behavior of both lithospheric and sublithospheric mantle beneath the continents. A good example of the extremes to which this approach can take is the study of Late Cenozoic basaltic magmatism in the Basin and Range and adjacent regions of western North America. Late Cenozoic volcanism within the Basin and Range commenced at ~30 Ma after the subduction of oceanic lithosphere beneath the region had ceased and at the onset of extensional tectonism within the continental lithosphere (Stewart, 1998). There is general agreement that extensional tectonism was involved in some fashion in inducing the melting of sublithospheric and/or lithospheric mantle that led to the Late Cenozoic basaltic magmatism (Fitton *et al.*, 1991). As a result, spatial and temporal patterns in the chemical and isotopic compositions of the Late Cenozoic basalts should provide insights into changes in the sources of magmas within the Basin and Range and vicinity, information that, in turn, could be related to the mechanisms involved in lithospheric thinning. Because the volume of data available for these rocks is so large, it has become possible to attempt a detailed reconstruction of the Late Cenozoic history of the deep continental lithosphere in this region.

The first step in this reconstruction involves a detailed characterization of the chemical and isotopic variations of the basaltic rocks themselves. Both sodic and potassic basalts erupted in the Basin and Range in the Late Cenozoic. Sodic basalts with $\varepsilon_{Nd} > 0$ are found primarily within the Basin and Range proper. Generally they have low-LIL/HFSE ratios, as well as low $^{87}Sr/^{86}Sr$, and lead-isotopic compositions that overlap those of northern hemisphere oceanic basalts (Kempton *et al.*, 1991). Sodic and potassic basalts with $\varepsilon_{Nd} < 0$, in contrast, are generally preserved within the western Great Basin (including the previously discussed Sierra Nevada basaltic rocks) and along the margins of the present-day Basin and Range (Kempton *et al.*, 1991). These basalts have high-LIL/HFSE ratios as well as generally high $^{87}Sr/^{86}Sr$, and lead-isotopic compositions characterized by high $^{207}Pb/^{204}Pb$ ratios at a given $^{206}Pb/^{204}Pb$ value relative to northern hemisphere oceanic basalts (Farmer *et al.*, 1989; Kempton *et al.*, 1991).

Because of the generally high LREE contents of all the western USA alkali basalts, their neodymium-isotopic compositions are likely to represent those of their mantle source regions. The high-ε_{Nd} (low-LIL/HFSE) and low-ε_{Nd} (high-LILE/HFSE) alkali basalts are typically, but not universally, interpreted as being derived from sublithospheric and slab-metasomatized lithospheric sources, respectively. There remains, however, the usual difficulty of defining whether mantle melting was provoked by active upwelling of a high-temperature mantle plume, or induced in passively upwelling upper mantle and/or in overlying lithospheric mantle as a consequence of lithospheric extension. An important observation in this regard is the fact that the hafnium- and neodymium-isotopic compositions of alkali basalts plot on an array distinct from that for OIB (Beard and Johnson, 1997). Beard and Johnson (1997) argue cogently that high-ε_{Nd} sodic basalts have combined neodymium- and hafnium-isotopic compositions that are consistent with their derivation not from a mantle plume but from a MORB upper mantle source that had undergone an ancient melt extraction in the spinel peridotite stability field (Beard and Johnson, 1997). Other workers have argued against a plume source based on the relatively low volume of Late Cenozoic magmatism produced in much of the Basin and Range (Bradshaw *et al.*, 1993).

As with the high-ε_{Nd} basalts, the low-ε_{Nd} basalts also plot off the OIB Hf–Nd array, which, along with their relatively radiogenic helium ($R_a \sim 6$; Reid and Graham (1996) and Dodson *et al.* (1998)), and lack of uranium-series disequilibria in the youngest basalts (Asmerom, 1999) argues for a lithospheric mantle source potentially metasomatized by subducted sediment just prior to stabilization of the Precambrian continental mantle lithosphere in this region (Fitton *et al.*, 1988). The sole dissenting voice is Wang *et al.* (2002), who suggest that high-iron content of low-ε_{Nd} basalts in southern Nevada requires a deep source for their parental magmas (Figure 4), below the geophysically defined base of the mantle lithosphere in this region. Melting at such depths requires a mantle potential temperature of at least 1,500 °C. These authors, therefore, suggest that the major-element composition of the basalts provides evidence for a mantle plume under the region.

Having more or less defined the sources of the Basin and Range basalts, the next step in extracting geodynamic information from the basalts is to determine what space–time patterns exist in the eruption of sublithosphere versus lithosphere derived basaltic rocks. In several areas within the Basin and Range, and particularly in the Rio Grande Rift, low-ε_{Nd} alkali basalts are followed by higher-ε_{Nd} basalts, including tholeiites (Perry *et al.*, 1987). A similar transition from low to high ε_{Nd} is found in alkali basalts of other regions experiencing large amounts of lithospheric extension, such as the Lake Mead extensional corridor in southern Nevada (DePaolo and Daley, 2000). In the Rio Grande Rift, this transition is interpreted as a direct response to lithospheric extension. Initial extension provoked melting in the lithosphere. This was followed by decompression melting of upwelling mantle directly beneath the rift axis (Perry *et al.*, 1987). The latter is manifested along the axis of the central Rio Grande Rift by the eruption of sodic alkali basalts and tholeiitic basalts with ε_{Nd} values as high as +5 (Figure 18; Gibson *et al.*, 1993). Lithospheric mantle is preserved, however, along the rift flanks, where it undergoes conductive heating from the upwelling mantle and spawns low volume, low-ε_{Nd} volcanism (Thompson and Gibson, 1994). This spatial and temporal pattern in the source of mafic magmatism along the rift is consistent with an overall "pure-shear" model for lithospheric extension, in which maximum crustal and mantle lithosphere extension are spatially coincident (Baldridge *et al.*, 1991).

It has also been suggested that extension within the Basin and Range may be accommodated in some regions by "simple shear," in which case the regions of maximum crustal and mantle lithosphere thinning are significantly displaced from one another (Wernicke *et al.*, 1988). In this case, the spatial–temporal pattern of basalt chemical and isotopic composition might be quite different from the pure-shear case, with basalt forming along the axis of maximum crustal thinning and undergoing a shift from lithosphere to sublithospheric source at a later time than those found in "off-axis" positions (Farmer *et al.*, 1989). Such a

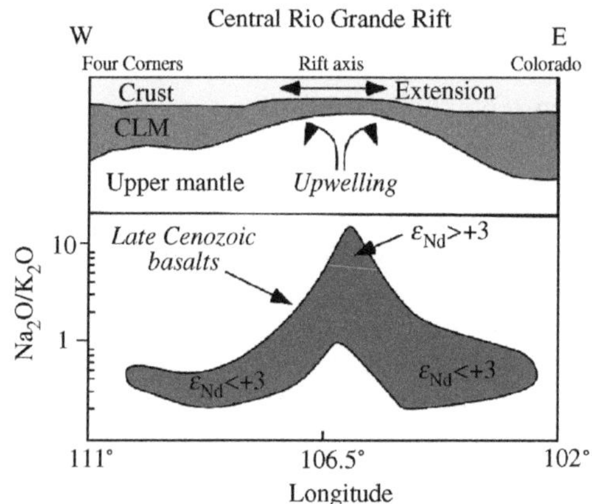

Figure 18 Cartoon depicting Late Cenozoic pure shear extension of central Rio Grande Rift, northern New Mexico, and range of Na_2O/K_2O for Late Cenozoic basaltic rocks (ultrapotassic, alkali, and tholeiitic composition) along a transect perpendicular to the rift axis. Thinning of mantle lithosphere beneath the rift axis triggers melting of upwelling asthenosphere, producing sodic composition, high ε_{Nd} basaltic magmas, while along the rift flanks potassic magmas are generated via conductive heating within preserved, K-metasomatized portions of the pre-extension CLM (after Gibson *et al.*, 1993).

displacement could account for the contemporaneous eruption of low-ε_{Nd} basalt in southern Nevada and high-ε_{Nd} basalt in adjacent portions of Owens Valley to the west (Farmer *et al.*, 1989; DePaolo and Daley, 2000). Regardless of the exact interpretation involved, however, it is clear that regular spatial–temporal patterns do exist in the chemical and isotopic composition of Late Cenozoic basalts in the western USA, and integration of these observations into geodynamic models can provide important insights into the physical evolution of the deep continental lithosphere.

3.03.4 INTRUSIVE EQUIVALENTS OF CONTINENTAL BASALTIC ROCKS

Because of the relatively high density of basaltic magmas (~2,600 kg m^{-3} at 10^{-4} GPa; Spera, 2000), it is likely that the bulk of mafic magmas produced beneath continents, particularly if anhydrous, will not reach the surface, but will instead crystallize at depth in the continental lithosphere. As a result, intrusive mafic magmatism may represent an important mechanism of continental growth (Johnson, 1991). One obvious manifestation of the intrusion and crystallization of mafic magmas in the continental crust is the presence of massive dike swarms, such as the Proterozoic McKenzie dike swarm in North America, which could represent the eroded underpinnings of pre-Mesozoic LIPs (Ernst and Buchan, 1997). Layered mafic intrusions also occur worldwide and range in age from Precambrian to Tertiary. These intrusions are significant both for establishing the crystallization history of mafic magmas and as an economic source of platinum-group metals (Cawthorn, 1999). Many of these intrusions may also be related to large igneous provinces. The oft-studied Tertiary Skaergård intrusion in southeastern Greenland (McBirney, 2002) probably represents an intrusive equivalent of the basalts comprising portions of the North Atlantic igneous province, while the Precambrian Muskox intrusion in Canada could be related to the large magmatic event that produced the McKenzie dike swarm (Ernst and Buchan, 2001). Furthermore, in southern Africa, seismic evidence suggests that the magmatism that resulted in the Archean Bushveld layered mafic intrusion also produced a wholesale modification of the crustal and mantle lithosphere, including the introduction of iron into CLM during magmatism. This demonstrates the significance of mafic magmatism in modifying the volume and composition of the lithosphere (James *et al.*, 2001).

It has also been recognized that much of the lower continental crust worldwide is mafic in composition (Rudnick, 1992). This leads to the question of whether the continents have been extensively underplated by mafic magmas, or whether mafic lower crust represents an original characteristic of the continents. Investigations of this issue have largely involved studies of mafic lower crustal xenoliths entrained in younger basalts or ultramafic magmas (Rudnick, 1992) and have included studies of both Phanerozoic and Precambrian lower crust.

Perhaps the best evidence for Phanerozoic underplating of the continental crust comes from mafic, granulite facies xenoliths entrained in Plio-Pleistocene basalts in north Queensland, Australia (Rudnick *et al.*, 1986). Here, xenoliths show a negative correlation between incompatible element content and mg#. Furthermore, the xenoliths' ε_{Nd} decrease, and $^{87}Sr/^{86}Sr$ increase, with decreasing mg#. Both the trace-element and isotopic data are consistent with the notion that the xenoliths represent cumulates from Cenozoic basaltic magmas that intruded the lower crust and underwent concomitant fractional crystallization and crustal assimilation. This example demonstrates that underplating of mafic magmas can occur, but because xenoliths represent a nonrepresentative and incomplete sampling of the lower crust, it is not clear whether this underplating represents a significant contribution to the overall mass of the continental crust in this region.

Xenoliths of Precambrian mafic lower crust have been studied in various locations worldwide, such as the western USA, northern Europe, and southern Africa (Selverstone *et al.*, 1999; Carlson *et al.*, 2000; Kempton *et al.*, 2001). Extracting unambiguous age and compositional information regarding the lower crust from these rocks is difficult, because they were extracted from continental crust that has potentially undergone multiple episodes of high-pressure ± high-temperature metamorphism, partial melting, and metasomatism. As a result, it is not easy to decide whether these rocks represent mafic lithologies that were present in the lower crust at the time of crustal stabilization, or whether they represent subsequently underplated material. Kempton *et al.* (2001) argue that plagioclase-bearing garnet granulites entrained in Devonian lamprophyres from the Baltic Shield represent solidified tholeiitic melts. These authors suggest that the xenoliths represent portions of 2.4 Gyr–2.5 Gyr old lower crust emplaced during the development of a large, plume-related Late Archean igneous province. Other authors have suggested that mantle plumes, perhaps related to episodes of mantle overturn, can produce significant crustal growth (Stein and Hoffman, 1994). The chemical composition of lower crustal xenoliths, however, is not always consistent with an origin via underplating by plume related magmas. Many xenolith suites, including those from the Colorado Plateau and the southern margin of the Archean Wyoming craton in the western USA, have low-nickel contents and high-La/Nb ratios, as expected for arc-related magmas, or magmas that had interacted with continental crust (Selverstone *et al.*, 1999). These rocks apparently represent mafic lower crust that formed at or near the time of crustal stabilization by subduction related magmas. There is the possibility, then, that mafic lower crust of two general flavors exist, one produced by subduction-related processes at or near the time of crust formation, and a second with low-La/Nb ratios and higher-nickel contents produced during plume-related (?) episodes of mafic magma underplating (Condie, 1999). It remains for future studies to assess the validity of this suggestion, and to determine the relative proportion of mafic lower crust formed by subduction- and plume-related processes.

3.03.5 CONCLUDING REMARKS

Considerable progress has been made during the past few decades in identifying the processes involved in the production of continental basaltic magmatism, but some fundamental questions remain. For example, if the continental mantle lithosphere is a source of continental magmatism, how critical are mafic lithologies in CLM to the production of such magmas and in controlling their chemical composition? Do reactions between ultramafic lithologies and partial melts of mafic lithologies produce the immediate sources of some continental basalts (Carlson and Nowell, 2001)? Is metasomatism of the CLM by fluids/melt derived from the sublithospheric mantle a necessary immediate precursor to the production not only of kimberlites but also of both sodic and potassic continental alkali basalts? Does pervasively metasomatized mantle exist at the base, or some other depth, within the CLM, and if so, what are its physical properties? Under what conditions can hydrated CLM be preserved, and under what circumstances does such mantle get involved in continental magmatism?

These questions do not just involve a few details in an overall known picture of continental basaltic magma formation. They represent fundamental issues regarding the generation of these magmas. The basic issue remains whether the geochemistry of continental basaltic rocks can distinguish rocks derived primarily from melting of the lithospheric mantle or from those derived primarily from the sublithospheric mantle. A complication discovered by recent studies is the increasing recognition that the lithospheric mantle is dynamic, can change thickness through time, and contains components of varying chemical and isotopic composition that were introduced at different times during the history of a given segment of CLM. If a better understanding of the sources of continental basaltic rocks can be achieved, these rocks will provide additional insights into the evolution of the deep continental lithosphere.

ACKNOWLEDGMENTS

This chapter benefited greatly from comments by Dawnika Blatter and from the patient editorial handling by Roberta Rudnick.

REFERENCES

Anderson D. L. (1995) Lithosphere, asthenosphere, and perisphere. *Rev. Geophys.* **33**, 125–149.

Arndt N. T. and Christensen U. (1992) The role of lithospheric mantle in continental flood volcanism: thermal and geochemical constraints. *J. Geophys. Res.* **97**, 10967–10981.

Arndt N. T., Czamanske G. K., Wooden J. L., and Fedorenko V. A. (1993) Mantle and crustal contributions to continental flood volcanism. *Tectonophysics* **223**, 39–52.

Arndt N. T., Ginibre C., Chauvel C., Albarede F., Cheadle M., Herzberg C., Jenner G., and Lahaye Y. (1998) Were komatiites wet? *Geology* **26**, 739–742.

Asmerom Y. (1999) Th–U fractionation and mantle structure. *Earth Planet. Sci. Lett.* **166**, 163–175.

Baker J. A., Menzies M. A., Thirlwall M. F., and MacPherson C. G. (1997) Petrogenesis of Quaternary intraplate volcanism, Sana'a, Yemen: implications for plume–lithosphere interaction and polybaric melt hybridization. *J. Petrol.* **38**, 1359–1390.

Baker J. A., MacPherson C. G., Menzies M. A., Thirlwall M. F., Al-Kadasi M., and Mattey D. P. (2000) Resolving crustal and mantle contributions to continental flood volcanism, Yemen; constraints from mineral oxygen isotope data. *J. Petrol.* **41**, 1805–1820.

Baker M. B. and Stolper E. M. (1994) Determining the composition of high-pressure mantle melts using diamond aggregates. *Geochem. Cosmochim. Acta* **58**, 2811–2827.

Baldridge W. S., Perry F. V., Vaniman D. T., Nealey L. D., Leavy B. D., Laughlin A. W., Kyle P., Bartov Y., Steinitz G., and Gladney E. S. (1991) Middle to late Cenozoic magmatism of the southeastern Colorado Plateau and central Rio Grande rift (New Mexico and Arizona, USA): a model for continental rifting. *Tectonophysics* **197**, 327–354.

Baldridge W. S., Keller G. R., Haak V., Wendlandt E., Jiracek G. R., and Olsen K. H. (1995) The Rio Grande rift. In *Continental Rifts: Evolution, Structure, Tectonics*, Developments in Geotectonics (ed. K. H. Olsen). Elsevier, Amsterdam, vol. 25, pp. 233–275.

Barfod D. N., Ballentine C. J., Halliday A. N., and Fitton J. G. (1999) Noble gases in the Cameroon line and the He, Ne, and Ar isotopic compositions of high μ (HIMU) mantle. *J. Geophys. Res.* **104**, 29509–29527.

Barrell J. (1914) The strength of the Earth's crust. *J. Geol.* **22**, 425–433.

Basu A. R., Renne P. R., Dasgupta D. K., Teichmann F., and Poreda R. J. (1993) Early and late alkali igneous pulses and a high-(super 3) He plume origin for the Deccan flood basalts. *Science* **261**, 902–906.

Basu A. R., Poreda R. J., Renne P. R., Teichmann F., Vasiliev Y. R., Sobolev N. V., and Turrin B. D. (1995) High-3He plume origin and temporal-spatial evolution of the Siberian flood basalts. *Science* **269**, 822–825.

Basu A. R., Hannigan R. E., and Jacobsen S. B. (1998) Melting of the Siberian mantle plume. *Geophys. Res. Let.* **25**, 2209–2212.

Beard B. L. and Johnson C. M. (1993) Hf isotope composition of late Cenozoic basaltic rocks from northwestern Colorado,

USA: new constraints on mantle enrichment processes. *Earth Planet. Sci. Lett.* **119**, 495–509.

Beard B. L. and Johnson C. M. (1997) Hafnium isotope evidence for the origin of Cenozoic basaltic lavas from the southwestern United States. *J. Geophys. Res.* **102**, 20149–20178.

Beattie P. (1993) Uranium–thorium disequilibria and partitioning on melting of garnet peridotite. *Nature* **363**, 63–65.

Bohrson W. A. and Spera F. J. (2001) Energy-constrained open-system magmatic processes: I. General model and energy-constrained assimilation and fractional crystallization (EC-AFC) formulation. *J. Petrol.* **42**, 1019–1041.

Bradshaw T. K., Hawkesworth C. J., and Gallagher K. (1993) Basaltic volcanism in the southern Basin and Range: no role for a mantle plume. *Earth Planet. Sci. Lett.* **116**, 45–62.

Carlson R. W. and Nowell G. M. (2001) Olivine-poor sources for mantle-derived magmas: Os and Hf isotopic evidence from potassic magmas of the Colorado Plateau. *Geochem. Geophys. Geosyst.* **2**, 200GC000128.

Carlson R. W., Boyd F. R., Shirey S. B., Janney P. E., Grove T. L., Bowring S. A., Schmitz M. D., Dann J. C., Bell D. R., Gurney J. J., Richardson S. H., Tredoux M., and Menzies A. H. (2000) Continental growth, preservation, and modification in Southern Africa. *GSA Today* **10**, 1–7.

Cawthorn R. G. (1999) Platinum-group element mineralization in the Bushveld Complex: a critical reassessment of geochemical models. *S. Afr. J. Geol.* **102**, 268–281.

Chesley J., Ruiz J., Righter K., Ferrari L., and Gomez-Tuena A. (2002) Source contamination versus assimilation: an example from the Trans-Mexican Volcanic Arc. *Earth Planet. Sci. Lett.* **195**, 211–221.

Chesley J. T. and Ruiz J. (1998) Crust–mantle interaction in large igneous provinces: implications from the Re–Os isotope systematics of the Columbia River flood basalts. *Earth Planet. Sci. Lett.* **154**, 1–11.

Chung S.-L. (1999) Trace element and isotope characteristics of Cenozoic basalts around the Tanlu fault with implications for the eastern plate boundary between north and south China. *J. Geol.* **107**, 301–312.

Chung S.-L. and Jahn B.-M. (1995) Plume–lithosphere interaction in generation of the Emeishan flood basalts at the Permian–Triassic boundary. *Geology* **23**, 889–892.

Chung S.-L., Jahn B.-M., Chen S.-J., Lee T., and Chen C.-H. (1995) Miocene basalts in northwestern Taiwan: evidence for EM-type mantle sources in the continental lithosphere. *Geochem. Cosmochim. Acta* **59**, 549–555.

Chung S.-L., Lo C.-H., Lee T.-Y., Zhang Y., Xie Y., Xinhua L., Wang K.-L., and Wan P.-L. (1998) Diachronous uplift of the Tibetan plateau starting 40 Myr ago. *Nature* **394**, 769–773.

Chung S.-L., Wang K.-L., Crawford A. J., Kamenetsky V. S., Chen C.-H., Lan C.-Y., and Chen C.-H. (2001) High-Mg potassic rocks from Taiwan: implications for the genesis of orogenic potassic lavas. *Lithos* **59**, 153–170.

Class C. and Goldstein S. L. (1997) Plume–lithosphere interactions in the ocean basins: constraints from the source mineralogy. *Earth Planet. Sci. Lett.* **150**, 245–260.

Class C., Altherr F. V., Eberz C., and McCulloch M. T. (1994) Geochemistry of Pliocene to Quaternary alkali basalts from the Huri Hills, northern Kenya. *Chem. Geol.* **113**, 1–22.

Coffin M. F. and Eldholm O. (1994) Large igneous provinces: crustal structure, dimension, and external consequences. *Rev. Geophys.* **32**, 1–36.

Condie K. C. (1993) Chemical composition and evolution of the upper continental crust; contrasting results from surface samples and shales. *Chem. Geol.* **104**, 1–37.

Condie K. C. (1999) Mafic crustal xenoliths and the origin of the lower continental crust. *Lithos* **46**, 95–101.

Connor C. B. and Conway F. M. (2000) Basaltic volcanic fields. In *Encylopedia of Volcanoes* (ed. H. Sigurdsson). Academic Press, San Diego, pp. 331–343.

Conticelli S., Francalianci L., Manetti P., Cioni R., and Sbrana A. (1997) Petrology and geochemistry of the ultrapotassic rocks from the Sabatini volcanic district, central Italy: the role of evolutionary process in the genesis of variably enriched alkaline magmas. *J. Volcanol. Geotherm. Res.* **7–5**, 107–136.

Cordery M. J., Davies G. F., and Campbell I. H. (1997) Genesis of flood basalts from eclogite-bearing mantle plumes. *J. Geophys. Res.* **102**, 20179–20197.

Courtillot V., Besse J., Vandamme D., Montigny R., Jaeger J. J., and Cappeta H. (1999) Deccan flood basalts at the Cretacous/Tertiary boundary? *Earth Planet. Sci. Lett.* **80**, 361–374.

Dalpé C. and Baker D. R. (2000) Experimental investigation of large-ion-lithophile-element-, high-field-strength-element- and rare-earth-element-partitioning between calcic amphibole and basaltic melt: the effects of pressure and oxygen fugacity. *Contrib. Mineral. Petrol.* **140**, 233–250.

Dawson J. B. (1967) A review of the geology of kimberlites. In *Ultramafic and Related Rocks* (ed. P. J. Wyllie). Wiley, New York, pp. 241–251.

DePaolo D. J. and Daley E. E. (2000) Neodymium isotopes in basalts of the southwest basin and range and lithospheric thinning during continental extension. *Chem. Geol.* **169**, 157–185.

Dobosi G., Downes H., Mattey D., and Embey-Isztin A. (1998) Oxygen isotope ratios of phenocrysts from alkali basalts of the Pannonian basin: evidence for an O-isotopically homgeneous upper mantle beneath a subduction-influenced area. *Lithos* **42**, 213–223.

Dodson A., DePaolo D. J., and Kennedy B. M. (1998) Helium isotopes in lithospheric mantle: evidence from Tertiary basalts of the western USA. *Geochem. Cosmochim. Acta* **62**, 3775–3787.

Dungan M. A., Lindstrom M. M., McMillan N. J., Moorbath S., Hoefs J., and Haskin L. A. (1986) Open system magmatic evolution of the Taos Plateau volcanic field, northern New Mexico: 1. The petrology and geochemistry of the Servilleta Basalt. *J. Geophys. Res.* **91**, 5999–6028.

Edwards D., Rock N. M. S., Taylor W. R., Griffin W. L., and Ramsay R. R. (1992) Mineralogy and petrology of the Aries diamondiferous kimberlite pipe, central Kimberley Block, western Australia. *J. Petrol.* **33**, 1157–1191.

Eiler J. M., Farley K. A., Valley J. W., Hauri E. H., Harmon C., Hart S. R., and Stolper E. M. (1997) Oxygen isotope variations in ocean island basalt phenocrysts. *Geochem. Cosmochim. Acta* **61**, 2281–2293.

Elkins L. T. and Hager B. H. (2000) Melt intrusion as a trigger for lithospheric foundering and the eruption of the Siberian flood basalts. *Geophys. Res. Lett.* **27**, 3937–3940.

Ellam R. M. and Cox K. G. (1991) An interpretation of Karoo picrite basalts in terms of interaction between asthenospheric magmas and the mantle lithosphere. *Earth Planet. Sci. Lett.* **105**, 330–342.

Ellam R. M., Carlson R. W., and Shirey S. B. (1992) Evidence from Re–Os isotopes for plume–lithosphere mixing in Karoo flood basalt genesis. *Nature* **359**, 718–721.

Ernst R. E. and Buchan K. L. (1997) Giant radiating dyke swarms: their use in identifying pre-Mesozoic large igneous provinces and mantle plumes. In *Large Igneous Provinces*, Geophysical Monograph 100 (eds. J. J. Mahoney and M. F. Coffin). American Geophysical Union, Washington, DC, pp. 297–333.

Ernst R. E. and Buchan K. L. (2001) Large mafic magmatic events through time and links to mantle-plume heads. In *Mantle Plumes: Their Identification through Time*, Geol. Soc. Amer. Spec. Paper 252 (eds. R. E. Ernst and K. L. Buchan). Geological Society of America, Boulder, CO, pp. 483–575.

Farmer G. L., Perry F. V., Semken S., Crowe B., Curtis D., and DePaolo D. J. (1989) Isotopic evidence on the structure and origin of subcontinental lithospheric mantle in southern Nevada. *J. Geophys. Res.* **94**, 7885–7898.

Farmer G. L., Glazner A. F., Wilshire H. G., Wooden J. L., Pickthorn W. J., and Katz M. (1995) Origin of late Cenozoic basalts at the Cima volcanic field, Mojave Desert, California. *J. Geophys. Res.* **100**, 8399–8415.

Farmer G. L., Glazner A. F., and Manley C. R. (2002) Did lithospheric delamination trigger late Cenozoic potassic volcanism in the southern Sierra Nevada, California? *Geol. Soc. Am. Bull.* **114**, 754–768.

Feldstein S. N. and Lange R. A. (1999) Pliocene potassic magmas from the Kings River region, Sierra Nevada, California: evidence for melting of a subduction-modified mantle. *J. Petrol.* **40**, 1301–1320.

Fitton J. G. and Dunlop H. M. (1985) The Cameroon Line, West Africa, and is bearing on the origin of oceanic and continental alkali basalt. *Earth Planet. Sci. Lett.* **72**, 23–38.

Fitton J. G., James D., Kempton P. D., Ormerod D. S., and Leeman W. P. (1988) The role of lithospheric mantle in the generation of late Cenozoic basic magmas in the western United States. *J. Petrol.* (Spec. Lithosphere Issue), 331–349.

Fitton J. G., Dodie J., and Leeman W. P. (1991) Basic magmatism associated with late Cenozoic extension in the Western United States: compositional variations in space and time. *J. Geophys. Res.* **96**, 13693–13711.

Fodor R. V. (1987) Low- and high-TiO_2 flood basalts of southern Brazil: origin from picritic parentage and a common mantle source. *Earth Planet. Sci. Lett.* **84**, 423–430.

Foley S. (1992) Vein-plus-wall-rock melting mechanisms in the lithosphere and the origin of potassic alkaline magmas. *Lithos* **28**, 435–453.

Franz G., Steiner G., Volker F., Pudlo D., and Hammerschmidt K. (1999) Plume-related alkaline magmatism in central Africa--the Meidob Hills (W. Sudan). *Chem. Geol.* **157**, 27–47.

Frey F. A. and Green D. H. (1974) Mineralogy, geochemistry and origin of lherzolite inclusions in Victorian basanites. *Geochem. Cosmochim. Acta* **38**, 1023–1059.

Furman T. (1995) Melting of metasomatized subcontinental lithosphere: undersaturated mafic lavas from Rungwe, Tanzania. *Contrib. Mineral. Petrol.* **122**, 97–115.

Furman T. and Graham D. (1999) Erosion of lithospheric mantle beneath the East African Rift system: geochemical evidence from the Kivu volcanic province. *Lithos* **48**, 237–262.

Gallagher K. and Hawkesworth C. (1992) Dehydration melting and the generation of continental flood basalts. *Nature* **258**, 57–59.

Gibson S. A. (2002) Major element heterogeneity in Archean to Recent mantle plume starting-heads. *Earth Planet. Sci. Lett.* **195**, 59–74.

Gibson S. A., Thompson R. N., Leat P. T., Morrison M. A., Hendry G. L., Dickin A. P., and Mitchell J. G. (1993) Ultrapotassic magmas along the flanks of the Oligo-Miocene Rio Grande rift, USA—monitors of the zone of lithospheric mantle extension and thinning beneath a continental rift. *J. Petrol.* **34**, 187–228.

Gibson S. A., Thompson R. N., Leonardos O. H., Dickin A. P., and Mitchell J. G. (1999) The limited extent of plume–lithosphere interactions during continental flood-basalt genesis: geochemical evidence from Cretaceous magmatism in southern Brazil. *Contrib. Mineral. Petrol.* **137**, 147–169.

Gibson S. A., Thompson R. N., and Dickin A. P. (2000) Ferropicrites: geochemical evidence for Fe-rich streaks in upwelling mantle plumes. *Earth Planet. Sci. Lett.* **173**, 355–374.

Girnis A. V., Brey G. P., and Ryabchikov I. D. (1995) Origin of Group 1A kimberlites: fluid-saturated melting experiments at 45–55 kbar. *Earth Planet. Sci. Lett.* **134**, 283–296.

Glazner A. F. and Farmer G. L. (1992) Production of isotopic variability in continental basalts by cryptic crustal contamination. *Science* **255**, 72–74.

Griffin W. L., O'Reilly S. Y., and Ryan C. G. (1999) The composition and origin of sub-continental lithospheric mantle. In *Mantle Petrology: Field Observations and High Pressure Experimentation*, Special Publication no. 6 (eds. Y. Fei, C. M. Bertka, and B. O. Mysen). The Geochemical Society, Houston, pp. 13–45.

Grove T. L. (2000) Origin of magmas. In *Encyclopedia of Volcanoes* (ed. H. Sigurdsson). Academic Press, San Diego, pp. 133–147.

Hames W. E., Renne P. R., and Ruppel C. (2000) New evidence for geologically instantaneous emplacement of earliest Jurassic Central Atlantic magmatic province basalts on the North American margin. *Geology* **28**, 859–862.

Hanson G. N. (1989) An approach to trace element modeling using a simple igneous system as an example. In *Geochemistry and Mineralogy of Rare Earth Elements*, Reviews in Mineralogy 21 (eds. B. R. Lipin and G. A. McKay). Mineralogical Society of America, Washington, DC, pp. 79–89.

Harry D. L. and Leeman W. P. (1995) Partial melting of melt metasomatized subcontinental mantle and the magma source potential of the lower lithosphere. *J. Geophys. Res.* **100**, 10255–10269.

Hart S. R. (1984) A large-scale isotope anomaly in the Southern Hemisphere mantle. *Nature* **309**, 753–757.

Hart W. K., Carlson R. W., and Shirey S. B. (1997) Radiogenic Os in primitive basalts from the northwestern USA: implications for petrogenesis. *Earth Planet. Sci. Lett.* **150**, 103–116.

Heaman L. M. and Kjarsgaard B. A. (2000) Timing of eastern North American kimberlite magmatism: continental extension of the Great Meteor hotspot track? *Earth Planet. Sci. Lett.* **178**, 253–268.

Hergt J. M., Peate D. W., and Hawkesworth C. J. (1991) The petrogenesis of Mesozoic Gondwana low-Ti flood basalts. *Earth Planet. Sci. Lett.* **105**, 134–148.

Herzberg C. and O'Hara M. J. (1998) Phase equilibrium constraints on the origin of basalts, picrites, and komatiites. *Earth Sci. Rev.* **44**, 39–79.

Hirose K. and Kushiro I. (1993) Partial melting of dry peridotites at high pressures: determinations of compositions of melts segregated from peridotite using aggregates of diamond. *Earth Planet. Sci. Lett.* **114**, 477–489.

Hoang N. and Flower M. (1998) Petrogenesis of Cenozoic basalts from Vietnam: implications for origins of a "diffuse igneous province". *J. Petrol.* **39**, 369–395.

Hofmann A. W. (1997) Mantle geochemistry: the message from oceanic magmatism. *Nature* **385**, 219–229.

Hole M. J. (1988) Post-subduction alkaline volcanism along the Antarctic Peninsula. *J. Geol. Soc. London* **145**, 985–998.

Hole M. J., Rogers G., Saunders A. D., and Storey M. (1991) Relation between alkalic volcanism and slab-window formation. *Geology* **19**, 657–660.

Hooper P. R. (2000) Flood basalt provinces. In *Encyclopedia of Volcanoes* (ed. H. Sigurdsson). Academic Press, San Diego, pp. 345–359.

Hooper P. R. and Hawkesworth C. J. (1993) Isotopic and geochemical constraints on the origin and evolution of the Columbia River Basalt. *J. Petrol.* **34**, 1203–1246.

Horan M. F., Walker R. J., Fedorenko V. A., and Czamanske G. K. (1995) Osmium and neodymium isotopic constraints on the temporal and spatial evolution of Siberian flood basalt sources. *Geochem. Cosmochim. Acta* **59**, 5159–5168.

Hornig I. (1993) High-Ti and low-Ti tholeiites in the Jurassic Ferrar Group, Antarctica. *Geol. Jahr. Reihe E: Geophysik* **47**, 335–369.

Iddings J. P. (1892) The origin of igneous rocks. *Bull. Phil. Soc. Wash.* **12**, 89–213.

Irvine T. N. and Baragar W. R. A. (1971) A guide to the chemical classification of the common volcanic rocks. *Can. J. Earth Sci.* **8**, 523–548.

James D. E., Fouch M. J., VanDecar J. C., and van der Lee S. (2001) Tectospheric structure beneath southern Africa. *Geophys. Res. Lett.* **28**, 2485–2488.

Johnson C. M. (1991) Large-scale crust formation and lithosphere modification beneath Middle to Late Cenozoic calderas and volcanic fields, western North America. *J. Geophys. Res.* **96**, 13485–13507.

Johnson R. W., Knutson J., and Taylor S. R. (1989) *Intraplate Volcanism in Eastern Australia and New Zealand*. Cambridge University Press, Cambridge.

Jordan T. H. (1978) Composition and development of continental tectosphere. *Nature* **274**, 544–548.

Jordan T. H. (1988) Structure and formation of the continental tectosphere. *J. Petrol.* (Spec. Lithosphere Issue), 11–37.

Jung S. and Hoernes S. (2000) The major- and trace-element and isotope (Sr, Nd, O) geochemistry of Cenozoic alkaline rift-type volcanic rocks from the Rhon area (central Germany): petrology, mantle source characteristics and implications for asthenosphere–lithosphere interactions. *J. Volcan. Geotherm. Res.* **99**, 27–53.

Kelemen P. B., Hart S. R., and Bernstein S. (1998) Silica enrichment in the continental upper mantle via melt/rock reaction. *Earth Planet. Sci. Lett.* **164**, 387–406.

Kempton P. D., Fitton J. G., Hawkesworth C. J., and Ormerod D. S. (1991) Isotopic and trace element constraints on the composition and evolution of the lithosphere beneath the southwestern United States. *J. Geophys. Res.* **96**, 13713–13735.

Kempton P. D., Downes H., Neymark L. A., Wartho J. A., Zartman R. E., and Sharkov E. V. (2001) Garnet granulite xenoliths from the northern Baltic Shield--the underplated lower crust of a Palaeoproteroic large igneous province. *J. Petrol.* **42**, 731–763.

King S. D. and Anderson D. L. (1998) Edge-driven convection. *Earth Planet. Sci. Lett.* **160**, 289–296.

Korenaga J. and Kelemen P. B. (2000) Major element heterogeneity in the mantle source of the North Atlantic igneous province. *Earth Planet. Sci. Lett.* **184**, 251–268.

Kushiro I. (2001) Partial melting experiments on peridotite and origin of mid-ocean ridge basalt. *Ann. Rev. Earth Planet. Sci.* **29**, 71–107.

Landwehr D., Blundy J., Chamorro-Perez E., Hill E., and Wood B. (2001) U-series disequilibria generated by partial melting of spinel lherzolite. *Earth Planet. Sci. Lett.* **188**, 329–348.

Lange R. A. (2002) Constraints on the preeruptive volatile concentrations in the Columbia River basalts. *Geology* **30**, 179–182.

Langmuir C. H., Klein E. M., and Plank T. (1992) Petrological systematics of mid-ocean ridge basalts: constraints on melt generation beneath ocean ridges. In *Mantle Flow and Melt Generation at Mid-ocean Ridges*, Geophysical Monograph 71 (eds. J. P. Morgan, D. K. Blackman, and J. M. Sinton). American Geophysical Union, Washington, DC, pp. 183–280.

Lassiter J. C. and DePaolo D. J. (1997) Plume/lithosphere interaction in the generation of continental and oceanic flood basalts: chemical and isotopic constraints. In *Large Igneous Provinces*, Geophysical Monograph 100 (eds. J. J. Mahoney and M. F. Coffin). American Geophysical Union, Washington, DC, pp. 335–355.

Le Bas M. J. (2000) IUGS reclassification of the high-Mg and picritic volcanic rocks. *J. Petrol.* **41**, 1467–1470.

le Roex A. P. (1986) Geochemical correlation between southern African kimberlites and South Atlantic hotspots. *Nature* **324**, 243–245.

le Roex A. P., Spath A., and Zartman R. E. (2001) Lithospheric thickness beneath the southern Kenya Rift: implications from basalt geochemistry. *Contrib. Mineral. Petrol.* **142**, 89–106.

Lee C. T., Yin Q., Rudnick R., Chesley J. T., and Jacobsen S. B. (2000) Osmium isotopic evidence for Mesozoic removal of lithospheric mantle beneath the Sierra Nevada, California. *Science* **289**, 1912–1916.

Lee C. T., Yin Q., Rudnick R., and Jacobsen S. B. (2001) Preservation of ancient and fertile lithospheric mantle beneath the southwestern United States. *Nature* **411**, 69–72.

Leeman W. P. and Harry D. L. (1993) A binary source model for extension-related magmatism in the Great Basin, western North America. *Science* **262**, 1550–1554.

Lester A. P., Larson E. E., Farmer G. L., Stern C. R., and Funk J. A. (2001) Neoproterozoic kimberlite emplacement in the Front Range, Colorado. *Rocky Mount. Geol.* **36**, 1–12.

Lightfoot P. C., Hawkesworth C. J., Devey C. W., Rogers N. W., and Van Calsteren P. W. C. (1990) Source and differentiation of Deccan Trap lavas: implications of geochemical and mineral chemical variations. *J. Petrol.* **31**, 1165–1200.

Lightfoot P. C., Hawkesworth C. J., Hergt J., Naldrett A. J., Gorbachev N. S., Fedorenko V. A., and Doherty W. (1993) Remobilization of continental lithosphere by a mantle plume: major-trace-element, and Sr-, Nd- and Pb-isotopic evidence from picritic and tholeiitic lavas of the Norils'k District, Siberian Trap, Russia. *Contrib. Mineral. Petrol.* **114**, 171–188.

Mahoney J. J. (1988) Deccan Traps. In *Continental Flood Basalts* (ed. J. D. Macdougall). Kluwer Academic, Dordrecht, 341pp.

Mahotkin I. L., Gibson S. A., Thompson R. N., Zhuravlev D. Z., and Zherdev P. U. (2000) Late Devonian diamondiferous kimberlite and alkaline picrite (proto-kimberlite) magmatism in the Arkhangelsk region, NW Russia. *J. Petrol.* **41**, 201–227.

Marzoli A., Piccirillo E. M., Renne P. R., Bellieni G., Iacumin M., Nyobe J. B., and Tongwa A. T. (2000) The Cameroon Volcanic Line revisted: petrogenesis of continental basaltic magmas from lithospheric and asthenospheric mantle sources. *J. Petrol.* **41**, 87–109.

McBirney A. R. (2002) The Skaergaard layered series: Part VI. Excluded trace elements. *J. Petrol.* **43**, 535–556.

McBride J. S., Lambert D. D., Nicholls I. A., and Price R. C. (2001) Osmium isotopic evidence for crust–mantle interaction in the genesis of continental intraplate basalts from the Newer Volcanic Province southeastern Australia. *J. Petrol.* **42**, 1197–1218.

McCulloch M. T. and Gamble J. A. (1991) Geochemical and geodynamical constraints on subduction zone magmatism. *Earth Planet. Sci. Lett.* **102**, 358–374.

McDonough W. F., McCulloch M. T., and Sun S. S. (1985) Isotopic and geochemical systematics in Tertiary–Recent basalts from southeastern Australia and implications for the evolution of the subcontinental mantle. *Geochem. Cosmochim. Acta* **49**, 2051–2067.

McDonough W. F., Sun S. S., Ringwood A. E., Jagoutz E., and Hofmann A. W. (1992) Potassium, rubidium, and cesium in the Earth and Moon and the evolution of the mantle of the Earth. *Geochem. Cosmochim. Acta* **56**, 1001–1012.

McKenzie D. (1989) Some remarks on the movement of small melt fractions in the mantle. *Earth Planet. Sci. Lett.* **95**, 53–72.

McKenzie D. and Bickle M. J. (1988) The volume and composition of melt generated by extension of the lithosphere. *J. Petrol.* **29**, 625–679.

McKenzie D. and O'Nions R. K. (1991) Partial melt distributions from inverstion of rare earth element concentrations. *J. Petrol.* **32**, 1021–1091.

Menzies M. A., Baker J. A., Chazot G., and Al'Kadasi M. (1997) Evolution of the Red Sea volcanic margin, western Yemen. In *Large Igneous Provinces*, Geophysical Monograph 100 (eds. J. J. Mahoney and M. F. Coffin). American Geophysical Union, Washington, DC, pp. 29–44.

Miller C., Schuster R., Klotzli U., Frank W., and Purtscheller F. (1999) Post-collisional potassic and ultrapotassic magmatism in SW Tibet: geochemical and Sr–Nd–Pb–O isotopic constraints for mantle source characteristics and petrogenesis. *J. Petrol.* **40**, 1399–1424.

Mitchell R. H. (1995) *Kimberlites, Orangeites, and Related Rocks*. Plenum, New York.

Mohr P. (1992) Nature of the crust beneath magmatically active continental rifts. *Tectonophysics* **213**, 269–284.

Moreira M. and Kurz M. D. (2001) Subducted oceanic lithosphere and the origin of the "high μ" basalt helium signature. *Earth Planet. Sci. Lett.* **189**, 49–57.

Morgan W. J. (1971) Convection plumes in the lower mantle. *Nature* **230**, 42–43.

Nelson D. R. (1992) Isotopic characteristics of potassic rocks: evidence for the involvement of subducted sediments in magma genesis. *Lithos* **28**, 403–420.

Oyarzun R., Doblas M., Lopez-Ruiz J., and Cebria J. M. (1997) Opening of the central Atlantic and asymmetric mantle upwelling phenomena: implications for long-lived magmatism in western North Africa and Europe. *Geology* **25**, 727–730.

Parman S. W., Grove T. L., and Dann J. C. (2001) The production of Barbeton komatiites in an Archean subduction zone. *Geophys. Res. Lett.* **28**, 2513–2516.

Paslick C., Halliday A., James D., and Dawson J. B. (1995) Enrichment of the continental lithosphere by OIB melts: isotopic evidence from the volcanic province of northern Tanzania. *Earth Planet. Sci. Lett.* **130**, 109–126.

Pearce J. A. and Peate D. W. (1995) Tectonic implications of the composition of volcanic arc magmas. *Ann. Rev. Earth Planet. Sci.* **23**, 252–285.

Pearson D. G. (1999) Evolution of cratonic lithospheric mantle: an isotopic perspective. In *Mantle Petrology: Field Observations and High-pressure Experimentation* (eds. Y. Fei, C. M. Bertka, and B. O. Mysen). The Geochemical Society, Houston, vol. 6, pp. 57–78.

Peate D. W. (1997) The Paraná-Etendeka Province. In *Large Igneous Provinces*, Geophysical Monograph 100 (eds. J. J. Mahoney and M. F. Coffin). American Geophysical Union, Washington, DC, pp. 217–245.

Peate D. W. and Hawkesworth C. J. (1996) Lithospheric to asthenospheric transition in low-Ti flood basalts from southern Paraná, Brazil. *Chem. Geol.* **127**, 1–24.

Peccerillo A. (1999) Multiple mantle metasomatism in central-southern Italy: geochemical effects, timing and geodynamic implications. *Geology* **27**, 315–318.

Perry F. V., Baldridge W. S., and DePaolo D. J. (1987) Role of asthenosphere and lithosphere in the genesis of late Cenozoic basaltic rocks from the Rio Grande rift and adjacent regions of the southwestern United States. *J. Geophys. Res.* **92**, 9193–9213.

Pertermann M. and Hirschmann M. M. (2003) Partial melting experiments on a MORB-like pyroxenite between 2 and 3 GPa: constraints on the presence of pyroxenite in basalt source regions from solidus location and melting rate. *J. Geophys. Res.* 10.1029/2000JB000118.

Petter S. W. (1991) *Regional Geology of Africa*. Springer, Berlin, 722p.

Pik R., Deniel C., Coulon C., Yirgu G., and Marty B. (1999) Isotopic and trace element signatures of Ethiopian flood basalts: evidence for plume–lithosphere interactions. *Geochem. Cosmochim. Acta* **63**, 2263–2279.

Price R. C., Gray C. M., and Frey F. A. (1997) Strontium isotopic and trace element heterogeneity in the plains basalts of the Newer Volcanic Province, Victoria, Australia. *Geochem. Cosmochim. Acta* **61**, 171–192.

Puffer J. H. (2001) Contrasting high field strength element contents of continental flood basalts from plume versus reactivated-arc sources. *Geology* **29**, 675–678.

Reichow M. K., Saunders A. D., White R. S., Pringle M. S., Al'Mukhamedov A. I., Medvedev A. I., and Kirda N. P. (2002) 40Ar/39Ar dates from the West Siberian Basin: Siberian flood basalt province doubled. *Science* **296**, 1846–1849.

Reid M. R. and Graham D. W. (1996) Resolving lithospheric and sub-lithospheric contributions to helium isotope variations in basalts from the southwestern US. *Earth Planet. Sci. Lett.* **144**, 213–222.

Reid M. R. and Ramos R. (1996) Chemical dynamics of enriched mantle in the southwestern United States: thorium isotope evidence. *Earth Planet. Sci. Lett.* **138**, 239–254.

Reiners P. W., Nelson B. K., and Ghiorso M. S. (1995) Assimilation of felsic crust by basaltic magma: thermal limits and extents of crustal contamination of mantle-derived magmas. *Geology* **23**, 563–566.

Renne P. R. and Basu A. R. (1991) Rapid eruption of the Siberian Traps flood basalts at the Permo-Triassic boundary. *Science* **253**, 176–179.

Renne P. R., Ernesto M., Pacca I. G., Coe R. S., Glen J. M., Prevot M., and Perring M. (1992) The age of the Paraná flood volcanism, rifting of Gondwanaland, and the Jurassic–Cretaceous boundary. *Science* **246**, 975–979.

Renne P. R., Zichao Z., Richards M. A., Black M. T., and Basu A. R. (1995) Synchrony and causal relations between Permian–Triassic boundary crises and Siberian flood volcanism. *Science* **269**, 1413–1416.

Ringwood A. E., Kesson S. E., Hibberson W., and Ware N. (1992) Origin of kimberlites and related magmas. *Earth Planet. Sci. Lett.* **113**, 521–538.

Robert U., Foden J., and Varne R. (1992) The Dodecanese Province, SE Aegean: a model for tectonic control on potassic magmatism. *Lithos* **28**, 241–260.

Roeder P. L. and Emslie R. F. (1970) Olivine–liquid equilibrium. *Contrib. Mineral. Petrol.* **29**, 275–289.

Rogers N. W., James D., Kelley S. P., and de Mulder M. (1998) The generation of potassic lavas from the eastern Virunga Province, Rwanda. *J. Petrol.* **39**, 1223–1247.

Rollinson H. (1993) *Using Geochemical Data: Evaluation, Presentation, Interpretation*. Longman, Singapore.

Rudnick R., McDonough W. F., and O'Connell R. J. (1998) Thermal structure, thickness and composition of continental lithosphere. *Chem. Geol.* **145**, 395–411.

Rudnick R. L. (1992) Xenoliths-samples of the lower continental crust. In *Continental Lower Crust*, Developments in Geotectonics 23 (eds. D. M. Fountain, R. J. Arculus, and R. W. Kay). Elsevier, Amsterdam, pp. 363–390.

Rudnick R. L., McDonough W. F., McCulloch M. T., and Taylor S. R. (1986) Lower crustal xenoliths from Queensland, Australia: evidence for deep crustal assimilation and fractionation of continental basalts. *Geochem. Cosmochim. Acta* **50**, 1099–1115.

Ryerson F. J. and Watson E. B. (1987) Rutile saturation in magmas: implications for Ti–Nb–Ta depletion in island-arc basalts. *Earth Planet. Sci. Lett.* **86**, 225–239.

Saunders A. D., Fitton J. G., Kerr A. C., Norry M. J., and Kent R. W. (1997) The North Atlantic igneous province. In *Large Igneous Provinces*, Geophysical Monograph 100 (eds. J. J. Mahoney and M. F. Coffin). American Geophysical Union, Washington, DC, pp. 45–94.

Sautter V., Haggerty S. E., and Field S. (1991) Ultra-deep (>300 km) ultramafic xenoliths: new petrologic evidence from the transition zone. *Science* **252**, 827–830.

Schaefer B. F., Turner S. P., Rogers N. W., Hawkesworth C. J., Williams H. M., Pearson D. G., and Nowell G. M. (2000) Re–Os isotope characteristics of postorogenic lavas: implications for the nature of young lithospheric mantle and its contribution to basaltic magmas. *Geology* **28**, 563–566.

Schmidt K. H., Botazzi P., Vannucci R., and Mengel K. (1999) Trace element partitioning between phlogopite, clinopyroxene and leucite lamproite melt. *Earth Planet. Sci. Lett.* **168**, 287–299.

Selverstone J., Pun A., and Condie K. C. (1999) Xenolithic evidence for Proterozoic crustal evolution beneath the Colorado Plateau. *Geol. Soc. Am. Bull.* **111**, 590–606.

Sharma M. (1997) Siberian Traps. In *Large Igneous Provinces*, Geophysical Monograph 100 (eds. J. J. Mahoney and M. F. Coffin). American Geophysical Union, Washington, DC, pp. 273–295.

Shaw D. M. (1970) Trace element fractionation during anatexis. *Geochem. Cosmochim. Acta* **34**, 237–243.

Shirey S. B. and Walker R. J. (1998) The Re–Os isotope system in cosmochemistry and high-temperature geochemistry. *Ann. Rev. Earth Planet. Sci.* **26**, 423–500.

Sigurdsson H. (2000) Volcanic episodes and rates of volcanism. In *Encyclopedia of Volcanoes* (ed. H. Sigurdsson). Academic Press, San Diego, pp. 271–282.

Smith C. B. (1983) Pb, Sr and Nd isotopic evidence for sources of southern African Cretaceous kimberlites. *Nature* **304**, 51–54.

Spera F. J. (2000) Physical properties of magmas. In *Encyclopedia of Volcanoes* (ed. H. Sigurdsson). Academic Press, San Diego, pp. 171–190.

Stein M. and Hoffman A. W. (1994) Mantle plumes and episodic crustal growth. *Nature* **372**, 63–68.

Stewart J. H. (1998) Regional characteristics, tilt domains, and extensional history of the later Cenozoic Basin and Range Province, western North America. In *Accommodation Zones and Transfer Zones: The Regional Segmentation of the Basin and Range Province*, Geol. Soc. Amer. Spec. Paper 323 (eds. J. E. Faulds and J. H. Stewart). Geological Society of America, Boulder, CO, pp. 47–74.

Storey B. C., Mahoney J. J., and Saunders A. D. (1997) Cretaceous basalts in Madagascar and the transition between plume and continental lithosphere mantle sources. In *Large Igneous Provinces*, Geophysical Monograph 100 (eds. J. J. Mahoney and M. F. Coffin). American Geophysical Union, Washington, DC, pp. 95–122.

Sun S.-S. and McDonough W. F. (1989) Chemical and isotopic systematics of oceanic basalts: implications for mantle composition and processes. In *Magmatism in the Ocean Basins* (eds. A. D. Saunders and M. J. Norry). Geological Society, Oxford, pp. 313–345.

Tainton K. M. and McKenzie D. (1994) The generation of kimberlites, lamproites, and their source rocks. *J. Petrol.* **35**, 787–817.

Taylor H. P., Jr. (1968) The oxygen isotope geochemistry of igneous rocks. *Contrib. Mineral. Petrol.* **19**, 1–71.

Taylor W. R., Tompkins L. A., and Haggerty S. E. (1994) Comparative geochemistry of West African kimberlites: evidence for a micaceous kimberlite endmember of sublithospheric origin. *Geochem. Cosmochim. Acta* **58**, 4017–4037.

Thirlwall M. F., Upton B. G. J., and Jenkins C. (1994) Interaction between continental lithosphere and the Iceland Plume-Sr–Nd–Pb isotope geochemistry of Tertiary basalts, NE Greenland. *J. Petrol.* **35**, 839–879.

Thompson R. N. and Gibson S. A. (1994) Magmatic expression of lithospheric thinning across continental rifts. *Tectonophysics* **233**, 41–68.

Turner S., Arnaud N., Liu J., Rogers N., Hawkesworth C., Harris N., Kelley S., Van Calsteren P., and Deng W. (1996a) Post-collision, shoshonitic volcanism on the Tibetan Plateau: implications for convective thinning of the lithosphere and the source of ocean island basalts. *J. Petrol.* **37**, 45–71.

Turner S., Hawkesworth C. J., Gallagher K., Stewart K., Peate D. W., and Mantovani M. (1996b) Mantle plumes, flood basalts, and thermal models for melt generation beneath continents: assessment of a conductive heating model and application to the Paraná. *J. Geophys. Res.* **101**, 11503–11518.

Valley J. W. (1986) Stable isotope geochemistry of metamorphic rocks. In *Stable Isotopes in High Temperature Geological Processes*, Reviews in Mineralogy 16 (ed. P. H. Ribbe). Mineralogical Society of America, Chelsea, MI, pp. 445–490.

Walker R. J., Carlson R. W., Shirey S. B., and Boyd F. R. (1989) Os, Sr, Nd, and Pb isotope systematics of southern African peridotite xenoliths: implications for the chemical evolution of subcontinental mantle. *Geochem. Cosmochim. Acta* **53**, 1583–1595.

Wang K., Plank T., Walker J. D., and Smith E. I. (2002) A mantle melting profile across the Basin and Range SW USA. *J. Geophys. Res.* **107**, 10.1029/2001JB000209.

Wedepohl K. H. (2000) The composition and formation of Miocene tholeiites in the Central European Cenozoic plume volcanism (CECV). *Contrib. Mineral. Petrol.* **140**, 180–189.

Wedepohl K. H. and Baumann A. (1999) Central European Cenozoic plume volcanism with OIB characteristics and indications of a lower mantle source. *Contrib. Mineral. Petrol.* **136**, 225–239.

Wernicke B. P., Axen G. T., and Snow J. S. (1988) Basin and range extensional tectonics at the latitude of Las Vegas, Nevada. *Geol. Soc. Am. Bull.* **100**, 1738–1757.

White R. S. (1988) The Earth's crust and lithosphere. *J. Petrol.* (Spec. Lithosphere Issue), 1–10.

Wilshire H. W., McGuire A. V., Noller J. S., and Turrin B. D. (1991) Petrology of lower crustal and upper mantle xenoliths from the Cima Volcanic Field, California. *J. Petrol.* **32**, 169–200.

Wilson M. and Patterson R. (2001) Intraplate magmatism related to short-wavelength convective instabilities in the upper mantle: evidence from the Tertiary–Quaternary volcanic province of western and central Europe. In *Mantle Plumes: Their Identification through Time*, Geol Soc. Amer. Spec. Paper 352 (eds. R. E. Ernst and K. L. Buchan). Geological Society of America, Boulder, CO, pp. 37–58.

Wilson M., Tankut A., and Gulec N. (1997) Tertiary volcanism of the Galatia province, north-west Central Anatolia, Turkey. *Lithos* **42**, 105–122.

Winter J. D. (2001) *An Introduction to Igneous and Metamorphic Petrology*. Prentice Hall, Upper Saddle River, NJ.

Wooden J. L., Czamanske G. K., Fedorenko V. A., Arndt N. T., Chauvel C., Bouse R. M., King B. W., Knight R. J., and Siems D. F. (1993) Isotopic and trace-element constraints on mantle and crustal contributions to Siberian continental flood basatls, Noril'sk area, Siberia. *Geochem. Cosmochim. Acta* **57**, 3677–3704.

Woolley A. R., Bergman S. C., Edgar A. D., Le Bas M. J., Mitchell R. G., Rock N. M. S., and Scott-Smith B. H. (1996) Classification of lamprophyres, lamproites, kimberlites, and the kalsilitic, melitic, and leucitic rocks. *Can. Mineral.* **34**, 175–186.

Wyllie P. J. and Huang W. L. (1976) Carbonation and melting reactions in the system $CaO–MgO–SiO_2–CO_2$ at mantle pressures with geophysical and petrologic applications. *Contrib. Mineral. Petrol.* **54**, 79–107.

Zhang M. and O'Reilly S. Y. (1997) Multiple sources for basaltic rocks from Dubbo, eastern Australia: geochemical evidence for plume–lithosphere interaction. *Chem. Geol.* **136**, 33–54.

Zhang M., Suddaby P., Thompson R. N., Thirlwall M. F., and Menzies M. A. (1995) Potassic volcanic rocks in NE China: geochemical constraints on mantle source and magma genesis. *J. Petrol.* **36**, 1275–1303.

Zhang M., Stephenson P. J., O'Reilly S. Y., McCulloch M. T., and Norman M. (2001) Petrogenesis and geodynamic implications of Late Cenozoic basalts in North Queensland, Australia: trace element and Sr–Nd–Pb isotope evidence. *J. Petrol.* **42**, 685–719.

Zhou M. F., Malpas J., Song S. H., Robinson P. T., Sun M., Kennedy A. K., Lesher C. M., and Keays R. R. (2002) A temporal link between the Emeishan large igneous province (SW China) and the end-Guadalupian mass extinction. *Earth Planet. Sci. Lett.* **196**, 113–122.

Zindler A. and Hart S. (1986) Chemical geodynamics. *Ann. Rev. Earth Planet. Sci.* **14**, 493–571.

Zou H. and Reid M. R. (2001) Quantitative modeling of trace element fractionation during incongruent dynamic melting. *Geochem. Cosmochim. Acta* **65**, 153–162.

Zou H., Zindler A., Xu X., and Qi Q. (2000) Major, trace element, and Nd, Sr and Pb isotope studies of Cenozoic basalts in SE China: mantle sources, regional variations, and tectonic significance. *Chem. Geol.* **171**, 33–47.

3.04
Volcanic Degassing

C. Oppenheimer
University of Cambridge, UK

NOMENCLATURE

r_{eff}	effective aerosol radius
D	diffusion coefficient of given volatile in melt
I	solar irradiance at given level in atmosphere
I_o	solar irradiance at top of the atmosphere
K_R	constant in vesicularity equation (2)
N	constant in solubility equation (1)
P	pressure
Pe	Peclet number
R	bubble radius
s	constant in solubility equation (1)
V_g/V_l	ratio between the volumes of gas and melt at given pressure
X_{H_2O}	water solubility in melt
$X^r_{H_2O}$	residual melt water content at given pressure
$X^0_{H_2O}$	initial melt water content at saturation pressure
η	melt dynamic viscosity
τ	aerosol optical depth
τ_d	timescale of volatile diffusion
τ_η	timescale of viscous relaxation
ΔP	oversaturation pressure
Θ	solar zenith angle

Nature only reveals her secrets if we ask the right questions and listen, and listening in geochemistry means sampling, analyzing, plotting. (Werner Giggenbach, 1992b)

3.04.1 INTRODUCTION

Humans have long marveled at the odorous and colorful manifestations of volcanic emissions, and, in some cases, have harnessed them for their economic value (Figure 1). Moreover, the degassing of magma that is responsible for them is one of the key processes influencing the timing and nature of volcanic eruptions, and the emissions of these volatiles to the atmosphere can have profound effects on the atmospheric and terrestrial environment, and climate, at timescales ranging from a few years to >1 Myr, and spatial scales from local to global (Oppenheimer *et al.*, 2003a). Even more fundamental are the relationships between the history of planetary outgassing, differentiation of the Earth's interior, chemistry of the atmosphere and hydrosphere, and the origin and evolution of life (e.g., Kelley *et al.*, 2002).

This chapter focuses on the origins, composition and flux, and the environmental impacts of volcanic volatile emissions. This introductory section sets the scene by considering the general context and significance of volcanic degassing. Several chapters in this volume interface with this one on volcanic degassing, and in particular the reader is referred to the chapters on hydrothermal systems (Chapter 3.16) and ore formation (Chapter 3.13).

3.04.1.1 Earth Outgassing, Atmospheric Evolution and Global Climate

Volcanic emissions have occurred throughout Earth history, and have provided the inventory of volatile elements that take part in the major geochemical cycles involving the lithosphere, hydrosphere, atmosphere, and biosphere (Sections 3.04.2 and 3.04.5; Holland, 1984; Arthur, 2000; see Chapter 4.11). The mantle is an important reservoir for volatiles, and its concentration of carbon, sulfur, hydrogen, oxygen, and halogens has changed through Earth history as a result of differentiation (see Chapter 2.08). Anhydrous minerals such as olivine, pyroxene, and garnet can hold structurally bound OH^-, while molecular water is present in amphibole, phologopite, and apatite. Carbon is present in carbonate minerals or in elemental form (e.g., diamond and graphite), and sulfur in sulfide minerals. Volatiles probably also exist in the mantle in intergranular films. Since the volatile species are incompatible, they partition into the melt phase during partial melting of the mantle. In this way, magmagenesis plays a key role in transferring volatiles between the mantle and the crust. Magma evolution then partitions volatiles between the crust and the atmosphere/hydrosphere via degassing and eruption (with important feedbacks on magma differentiation), and plate recycling ensures a return flux of a proportion of the volatiles back to the mantle.

Major changes have occurred in atmospheric composition and in greenhouse gas forcing over Earth history, in part coupled to the evolution of life, and interacting with changes in solar flux and planetary albedo to control global climate. Over timescales exceeding 1 Myr, the carbon cycle operates as a climate thermostat on the Earth. The Archean atmosphere was anoxic, even after the onset of oxygenic photosynthesis ~2.7 Gyr ago. This has been attributed to an excess of reductants (e.g., CH_4 and H_2) able to scavenge out O_2 accumulating in the atmosphere (Holland, 2002; Catling and Kasting, 2003). Bacterial consumption of hydrogen also contributed methane, such that the Archean atmosphere may have contained as much as hundreds or thousands of ppm of CH_4

Figure 1 "The burning valley called Vulcan's Cave near Naples" or Solfatara (Campi Flegrei), from Bankes's New System of Geography (~1800).

(Catling *et al.*, 2001). The greenhouse forcing associated with these high concentrations of atmospheric methane may account for the absence of prolonged freezing of the Earth that should otherwise have arisen from the reduced solar luminosity (30% lower than today) early in the Earth's history (the "early faint Sun paradox"; Sagan and Chyba (1997)).

Hydrogen escape from the Earth's atmosphere has been proposed as a mechanism for inexorable oxidation of the mantle, tilting the redox balance of volcanic emissions today in favor of oxidized gases (Kasting *et al.*, 1993; Catling and Kasting, 2003). If correct, this shifting redox balance could have led to the rise of $O_2 \sim 2.3$ Gyr ago (Catling *et al.*, 2001). However, the mechanisms for oxygenation resulting from methane-induced hydrogen escape remain controversial (e.g., Towe, 2002). Furthermore, geochemical evidence argues strongly against any substantial increase in the oxidation state of the mantle over the past 3.5 Gyr—the Archean mantle was already oxidized (Delano, 2001; Canil, 2002).

Major changes in atmospheric composition have, in turn, been held responsible for so-called "snowball Earth" events, characterized by low-latitude glaciation and effective shutdown of the hydrological cycle, ~2.3 Gyr (coinciding with the rise of oxygen), and 750 Myr and 600 Myr ago (e.g., Kirschvink, 1992; Hoffman *et al.*, 1998; Kirschvink *et al.*, 2000; Hoffman and Schrag, 2002). Loss of CH_4 accompanying the rise of oxygen could explain the onset of the glacial conditions that characterize these events, with the "snowball" being ultimately thawed by the action of the carbonate–silicate inorganic cycle: under the glacial conditions, weathering rates would fall and CO_2 from volcanism would accumulate in the atmosphere, providing a long-term climatic recovery via greenhouse feedbacks. Between the Paleoproterozoic and Neoproterozoic "snowball" events, climate was warm and very stable, possibly due to methane emissions from biogenic sources (e.g., Catling *et al.*, 2001; Catling and Kasting, 2003).

On short timescales, years to decades, possibly centuries, individual volcanic eruptions, or conceivably a burst of eruptions from several volcanoes, are capable of perturbing global climate by the release of sulfur gases into the upper atmosphere. These oxidize to form a veil of minute sulfuric acid particles that can girdle the globe at heights of 20 km or more, scattering back into space some of the sunlight that would ordinarily penetrate the tropopause, and heat the lower atmosphere and Earth's surface. The complex spatial and temporal patterns of radiative forcing that result are reflected in regional climate change, and globally averaged surface temperature decreases. The 1991 eruption of Mt. Pinatubo in the Philippines has, as of early 2000s, provided the clearest evidence of these interactions, and an outstanding opportunity for climatologists to test both radiative and dynamical aspects of general circulation models

(Sections 3.04.6.1 and 3.04.6.2). Such work has contributed significantly to the development of our understanding of the importance of aerosols in climate change, and has helped to unravel the natural variability from anthropogenic forcing.

3.04.1.2 Magma Evolution and Dynamics, and Volcanic Eruptions

The partition of magmatic volatiles from the melt to the gas phase, and their subsequent separation—collectively referred to as degassing—exert fundamental controls on magma overpressure, viscosity and density, and thereby on the chemical evolution, storage, and transport (notably ascent rate) of magmas, and the style, magnitude, and duration of eruptions (Section 3.04.3; Huppert and Woods, 2002). Viscosity is a critical factor and is strongly dependent on temperature, melt composition (especially silica content (Figure 2(a)) and dissolved volatile content (Figure 2(b)), which control polymerization of the melt), applied stress, bubble fraction, and crystal content. It is also weakly dependent on pressure, with a general trend of decreasing viscosity at higher pressure. Kinetic factors, including cooling rate and diffusion, are also important.

The expansion of bubbles of gas accompanying decompression of magmas plays a key role in over-pressuring of magmatic systems, and hence in eruption triggering. It also provides the energy capable of propelling pyroclasts several kilometers into the atmosphere. For example, from the ideal gas law, 5 wt.% of dissolved water in 1 m^3 of melt would occupy ~700 m^3 in the vapor phase at atmospheric pressure. Such a volatile rich magma may experience explosive fragmentation on eruption if the exsolving gas cannot escape fast enough. Alternatively, if the gas is able to segregate with relative ease from the melt prior to, or during, eruption, as is generally the case with lower viscosity magmas, then lava lakes or flows are more likely to form, accompanied perhaps by strombolian or hawaiian style activity. Degassing can result in orders of magnitude changes in magma viscosity (Figure 2(b)), because: (i) the presence of bubbles has a strong effect on rheology (that is difficult to model because of steep viscosity gradients in the melt close to bubble walls, and changing bubble size distributions); (ii) it may induce crystallization; and (iii) dissolved water helps to polymerize silicate melts (Lange, 1994). Degassing, therefore, strongly influences rheological behavior, and can initiate feedbacks between magma dynamics and eruptive style that have been documented at several volcanoes (e.g., Voight *et al.*, 1999).

Vesiculation also has a profound effect on the permeability of magmas, and can reach the point where bubbles are sufficiently interconnected to permit gas loss from deeper to shallower levels in a conduit, or through the conduit walls. In this way, even highly viscous silicic and intermediate magmas can degas non-explosively, as seen in the case of lava dome eruptions such as that of Soufrière Hills Volcano (Montserrat). Understanding the rheological consequences of bubble growth, crystallization, magma ascent, and gas loss is clearly critical to understanding volcanic behavior.

3.04.1.3 Volcanic Hazards and Volcano Monitoring

Volcanic gases have been described as "telegrams from the Earth's interior" (Matsuo, 1975)—if the messages can be intercepted and interpreted, they can be used to aid in forecasting and prediction of volcanic activity. To this end, many investigators have developed and applied methods to measure the gas mixtures and emission rates from volcanoes, and to understand what controls these parameters (Section 3.04.4). While volcano surveillance efforts still largely concentrate on seismological and geodetic approaches, gas geochemistry is widely recognized as an important and highly desirable component of multidisciplinary monitoring, and increasingly sophisticated remote sensing techniques are becoming available to measure volcanic volatile emissions. These include a range of spectroscopic instruments that can be deployed from the ground, aircraft, and satellites. Improved forecasting of volcano behavior offers immediate benefits in mitigating the risks of eruptions to society.

In addition to offering a means for assessing eruption hazards, volcanic gas and aerosol emissions can pose a direct hazard (Sections 3.04.6.3–3.04.6.5). At local and regional scales, several gas species and aerosols (including sulfate and fine ash) can affect the health of humans and animals. In 1984 and 1986, catastrophic releases of CO_2 from two volcanic crater lakes in Cameroon claimed several thousand lives in nearby villages (Le Guern and Tazieff, 1989). Acid gas species and their aerosol products can be transported over ranges of hundreds or thousands of kilometers, and can have various impacts on respiratory and cardiovascular health. Fluorine deposited to the ground on ash has been responsible for large losses of grazing animals during a number of eruptions (e.g., Cronin *et al.*, 2003). Emissions of radioactive species, principally radon, have also given rise to speculation over potential health risks. Several components of volcanic emissions, including heavy metals as well as the acid species, can damage vegetation, though in some cases, leachable

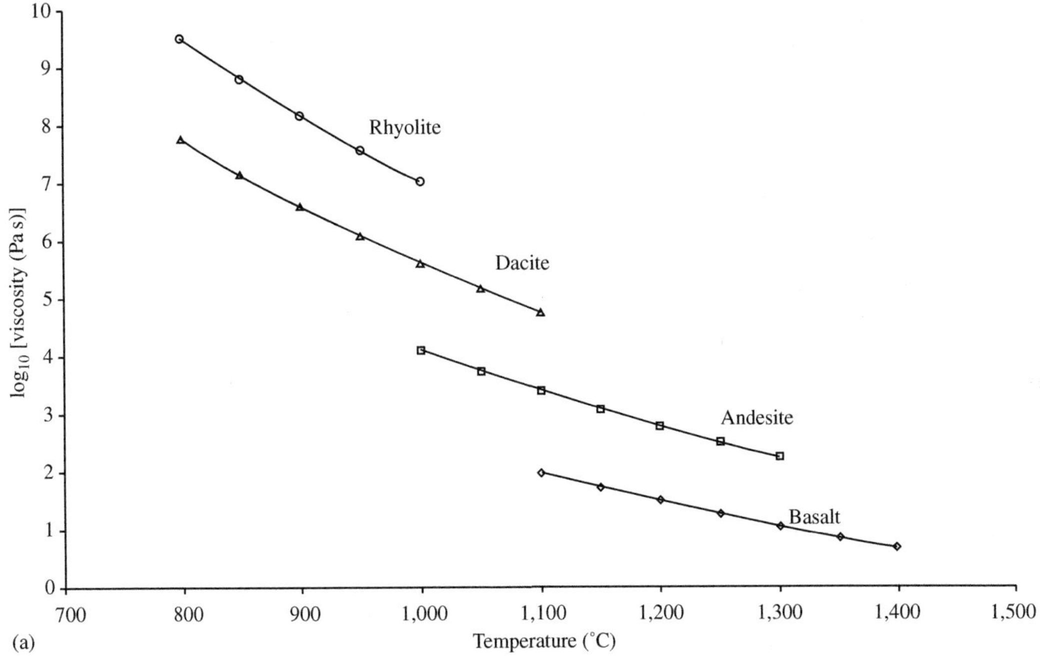

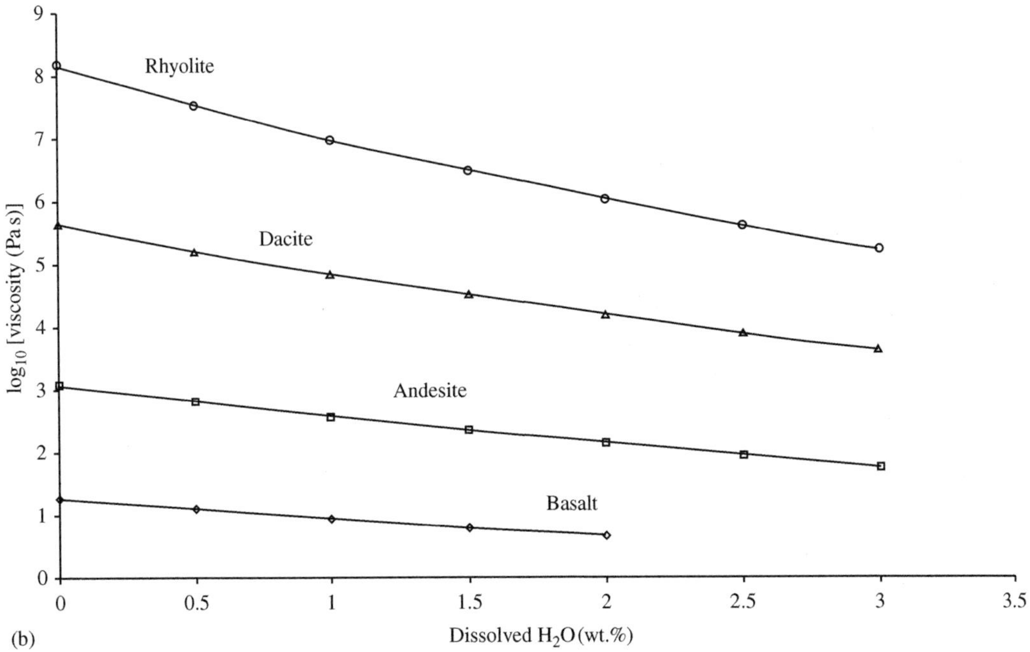

Figure 2 Viscosities at 1 bar pressure for natural melts as a function of: (a) temperature and (b) dissolved water content based on model of Shaw (1972) (see Spera (2000) for a review).

components of ashfalls can provide beneficial nutrients to terrestrial and aquatic ecosystems.

3.04.2 ORIGIN, SPECIATION, AND ABUNDANCE OF VOLATILES

"Next to nothing is known about the sources of the volatile components of magmas or how they are distributed and transported between the mantle and shallow levels of the crust." This is how Williams and McBirney (1979) open the chapter on volcanic gases in their influential textbook *Volcanology*. Though there is much still to learn, thanks to recent investigations spanning experimental phase petrology, analysis of glass inclusions, isotope geochemistry, thermodynamical and fluid dynamical modeling, and satellite and ground-based remote sensing, this pessimistic view can be replaced today by a much more encouraging outlook of our understanding of the origins, emissions, and impacts of volcanic gases (see Carroll and Holloway, 1994). The benefits of this are significant, not least because of the fundamental relationships between volcanic degassing, crystallization, magma dynamics, eruption style, mineral deposition, and atmospheric and climate change. This section reviews the sources and distribution of volatiles in magmas. The means by which they escape to the atmosphere are then explored in Section 3.04.3.

3.04.2.1 Sources and Abundances of Volatiles in Magmas

The origin of magmatic volatiles lies in the Earth's interior but, as a result of global tectonics, various species can be recycled through the lithosphere–hydrosphere–atmosphere system, and sourced more recently from the mantle, subducted slab, crust or hydrosphere. Interactions between magma bodies and hydrothermal fluids (with meteoric and/or seawater contributions) can also contribute to the complement of volatiles in magmas. The relative contributions from these different sources for a given volatile species are difficult to ascertain, for an individual volcano, for an arc, or for the whole Earth, but general trends exist between magmatic volatile content and tectonic environment (Table 1). These reflect magmatic differentiation in the crust, and concentration of incompatible volatiles in the melt, as well as the contributions of slab, depleted vs. undepleted mantle, and crustal assimilation (e.g., decarbonation and dehydration of crustal rocks).

The principal volatile species in magma is water, usually followed in abundance by carbon dioxide, hydrogen sulfide, sulfur dioxide, hydrogen chloride, and hydrogen (Section 3.04.4.2). Many other trace species have been identified in high-temperature volcanic gas discharges, including heavy metals. Analyses of trace elements and isotopic composition of volcanic rocks and fluids have provided some of the clearest indications of the source of volatiles in magmas. In particular, the emissions of arc volcanoes are thought to derive the larger part of their volatile budgets from the subducted slab (see Chapter 2.12), either through dehydration reactions occurring in the altered oceanic crust (see Chapter 3.18), release of pore water contained in sediments, or dehydration of pelagic clay minerals.

The abundance of slab-derived components is partly revealed by straightforward comparisons of N_2/He and CO_2/He ratios. For nonarc basalts, these are typically in the range 0–200 and $(2-6) \times 10^4$, respectively, whereas andesites are characterized by ranges of 1000–10,000 and $(6.5-9) \times 10^4$, respectively (Giggenbach, 1992a,b). Of the magmatic component in andesitic fluids, more than 95% is estimated to be sourced from the slab. Evidence for this is provided by oxygen-isotopic composition. Giggenbach (1992a) has promoted the idea of a common "andesitic water" feeding arc volcanoes because of characteristic $\delta^{18}O$ (around +10‰) values. The consistency in isotopic composition strongly suggests that this fluid is ultimately derived from seawater, and Giggenbach identified water released from pelagic sediments as the likely dominant source, though others believe that altered oceanic crust provides more water. According to his view, "arc magmatism is effectively driven by the inability of the mantle to accommodate the subducted water. Andesitic magmas represent simply the most suitable vehicle for the 'unwelcome' water to travel through the crust back to the surface, andesitic volcanoes 'vent holes' for excess subducted volatiles" (Giggenbach, 1992a). In other words, he suggested that most of the water dissolved in andesitic magmas derived from recycled seawater, along with high proportions of carbon, nitrogen, and chlorine (see Giggenbach, 1992a, figure 4). It is worth qualifying, though, that most andesites are thought to be the products of differentiation in the crust of basalts generated by partial melting of the mantle wedge.

Further work to characterize volatile sources has been carried out by Fischer *et al.* (1998). They analyzed carbon-, helium-, and nitrogen-isotopic ratios in emissions from Kudriavy, a basaltic andesite volcano in the Kurile arc (Table 2). Combined with measurements of sulfur dioxide flux from the volcano (see Section 3.04.4.3.2 for details on how this can be done), they estimated that 84% of the 2,700 mol yr^{-1} flux of helium was mantle derived (the remainder being radiogenic, and derived from the crust), and only 12% of the CO_2, and almost none of the N_2, was sourced from the mantle. For CO_2, they estimated that most (67%) was derived from inorganic carbonate sediments and hydrothermal veins in altered oceanic crust in the slab. Subducted organic nitrogen can account for all the observed nitrogen flux. More recently, Fischer *et al.* (2002) have examined the fate of nitrogen subducted along different segments of the Central American volcanic arc. They demonstrated efficient transfer of shallow marine sedimentary nitrogen carried in subduction zones back to the atmosphere via arc volcanism, thereby limiting recycling of nitrogen to the deep mantle.

3.04.2.2 Solubility and Speciation of Volatiles

Fundamental controls on the solubility of volatiles in melt include pressure and temperature, and the presence of nonvolatile phases (in the

Table 1 Summary of typical pre-eruptive volatile abundances of magmas in different tectonic settings. Note that volatiles can be contained in solution, in crystallized phases, and in a separate vapor phase. The values reported in the table are generally for the melt phase (dissolved) though true bulk volatile abundances for a magma would sum all these potential reservoirs. In particular, some magmas show abundant evidence for a substantial fluid phase co-existing with the melt prior to eruption, which is often not represented in melt inclusion-based estimates of pre-eruptive volatiles.[a]

MORB (e.g., Mid Atlantic Ridge, East Pacific Rise)	
H_2O	<0.4–0.5 wt.%, typically 0.1–0.2 wt.%; enriched mid-ocean-ridge basalt (E-MORB) up to 1.5 wt.%
CO_2	50–400 ppm; typically saturated at eruption (gas phase almost pure CO_2) leading to vesiculation
S	800–1,500 ppm; immiscible Fe–S–O liquids indicate saturation at eruption
Cl	20–50 ppm in most primitive MORB, occasionally much higher due to assimilation of hydrothermally altered rocks
F	100–600 ppm
Ocean island basalts (OIB) (e.g., Kīlauea, Galapagos, Réunion)	
H_2O	0.2–1 wt.%; e.g., Hawai'i 0.4–0.9 wt.% (Dixon *et al.*, 1991)
CO_2	2,000–6,500 ppm for Hawai'i
S	up to 3,000 ppm; 200–1,900 ppm for Hawai'i
Cl	comparable to MORB; Kīlauea estimate around 90 ppm
F	comparable to MORB; Kīlauea estimate around 35 ppm
Arc basalt (island arc basalts and continental margin basalts, e.g., Cerro Negro, Marianas)	
H_2O	up to 4–6 wt.% (e.g., see Roggensack *et al.* (1997), on Cerro Negro), largely sourced from subducted slab; crustal assimilation another potential source, especially for arcs built on continental crust
Back arc basin basalt (BABB) (e.g., Lau Basin)	
H_2O	1–3 wt.%; generally speaking, intermediate between MORB and island arc basalts
Andesites (e.g., Soufrière Hills Volcano)	
Note that it is particularly difficult to quantify pre-eruptive volatile contents of andesites because most are erupted subaerially (i.e., at atmospheric pressure) after significant degassing has taken place, and contain abundant phenocrysts (e.g., >30 wt.%) such that liquid compositions are more silicic (often rhyolitic) than bulk rock. Also, good host minerals for melt inclusions (e.g., olivine and quartz) are rare, and mineral disequilibria hamper experimental work.	
H_2O	>3 wt.%
CO_2	10–1,200 ppm
S	<1,000 ppm; typically 200–400 ppm
Cl	can be high, e.g., 1,500 ppm not unusual; (5,000 ppm or more in phonolites)
F	<500 ppm
Dacites and rhyolites (e.g., Mount St. Helens 1980, Pinatubo, 1991, Bishop Tuff)	
H_2O	typically 3–7 wt.%; e.g., 4.6 wt.% dissolved for Mount St. Helens 1980, 6–7 wt.% for Pinatubo 1991; there is strong evidence for vertical gradients in both dissolved and exsolved H_2O and CO_2 in pre-eruptive magmas (e.g., Wallace *et al.*, 1995; Wallace, 2001)
CO_2	often below detection limits
S	typically <200 ppm (75 ppm for Pinatubo, 1991) but melt often saturated with sulfide (pyrrhotite) or sulfate (anhydrite) crystalline phases
Cl	600–2,700 ppm in metaluminous dacites and rhyolites; 6,700 ppm in trachytes; 9,000 ppm in peralkaline rhyolites (pantellerites); 1,100 ppm for Pinatubo 1991
F	200–1,500 ppm in metaluminous dacites and rhyolites; up to 1.5 wt.% in peralkaline rhyolites (pantellerites)

[a] Assimilated from the detailed reviews of Johnson *et al.* (1993) and Wallace and Anderson (2000). See also Signorelli and Carroll (2000, 2002) for Cl solubility data.

case of sulfur, these could be sulfates or sulfides; in the case of chlorine, metal chlorides). As a first approximation, the solubility of water in silicate melt, X_{H_2O}, is roughly proportional to the square root of pressure, P (Burnham, 1979):

$$X_{H_2O} = nP^s \tag{1}$$

where n and s for water have values of around 0.34 and 0.54, respectively.

Analytical work on natural and synthetic melts has helped to establish an understanding of the speciation of volatiles in both the melt and vapor phase (e.g., Stolper, 1982, 1989; Silver *et al.*, 1990; Ihinger *et al.*, 1999). Dissolved water exists in the melt in the form of OH^- groups or as H_2O molecules that are structurally bound to the aluminosilicate network of the melt (McMilan, 1994). Likewise, CO_2 dissolves as molecular CO_2 and CO_3^{2-} (Fine and Stolper, 1985; Blank and Brooker, 1994). Speciation is a function of structure of the silicate melt and oxidation state of the magma (availability of cations). In experiments, OH^- reacts strongly with the silicate melt, lowering viscosity. Molecular water is less reactive and does not disrupt the polymerization of a melt as much. Water speciation thereby has an important effect on melt viscosity. The interactions of multiple phases (e.g., CO_2 and H_2O) are not well understood but are important (Papale, 1999, Papale and Polacci, 1999; Moretti *et al.*, 2003). Newman and Lowenstern (2002) have developed a simple-to-use code to model H_2O–CO_2-melt equilibrium for rhyolite and basalt systems (Figure 3).

Sulfur solubility is complicated due to multiple valancies including S_2, H_2S, SO_2, SO_3 (in the gas phase), and nonvolatile solid phases (e.g., pyrrhotite and anhydrite) or liquid phases. Redox equilibria, therefore, play a major role in determining speciation of sulfur. Predominant phases, though, are thought to be H_2S and SO_2 in the gas phase, and S^{2-} and SO_4^{2-} in the melt (Carroll and Webster, 1994). Métrich *et al.* (2002) have shown that sulfite (SO_3^{2-}) is also an important species in arc basalt melts.

Fluorine and chlorine speciation and solubility in magmas are complex and interdependent to a degree, though some general processes are recognized, including an inverse relationship between chlorine solubility and pressure in water-saturated systems, direct proportionality of phosphorus and chlorine solubility in water-poor, brine saturated systems, and a typically strong dependence of chlorine solubility on melt structure and composition (Carroll and Webster, 1994; Signorelli and Carroll, 2000, 2002; Webster *et al.*, 1999, 2001). Further experimental studies on speciation are required in order to develop predictive models of halogen solubility as a function of melt composition and physical parameters. These could prove highly fruitful in the interpretation of observed halogen content through time in volcanic gas emissions, with applications in volcano monitoring and hazard assessment (Villemant *et al.*, 2003).

Table 2 Percentage (molar) contribution to estimated fluxes of volatiles to the atmosphere from the high temperature (up to 920 °C) fumaroles of Kudriavy volcano, Kuril islands.

	Mantle	*Slab inorganic*	*Slab organic*	*Crust*	*Flux* (mol yr^{-1})
CO_2	12%	67%	21%	0	4.32×10^8
^3He	100%	0	0	0	2.5×10^{-2}
^4He	84%	0	0	16%	2.7×10^3
N_2	2%	0	98%	0	5.1×10^6
SO_2					3.47×10^8
HCl					1.06×10^8

Source: Fischer *et al.* (1998).

3.04.3 DEGASSING

Volcanic degassing begins with the expulsion or *exsolution* of volatiles from the melt and formation of gas bubbles. This is the first

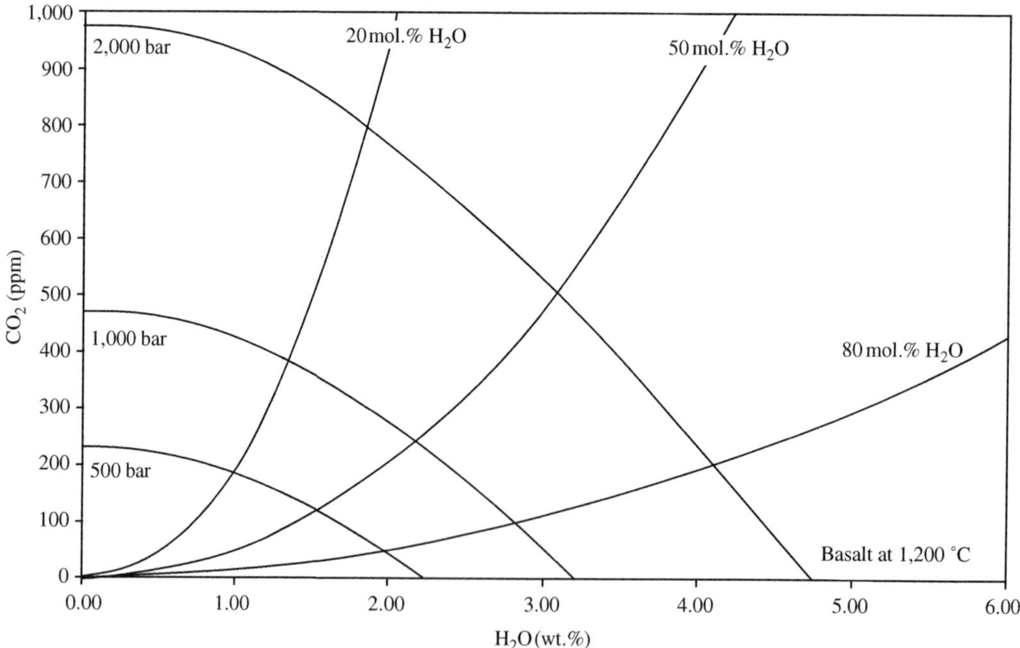

Figure 3 Isobars representing locus of values for dissolved H_2O and CO_2 in basaltic melt in equilibrium with H_2O–CO_2 vapor at 1,200 °C and selected pressures. Similarly, isopleths represent locus of basaltic melt compositions in equilibrium with given vapor compositions (20 mol.%, 50 mol.%, and 80 mol.% H_2O) at 1,200 °C (source Newman and Lowenstern, 2002). See also: http://wrgis.wr.usgs.gov/docs/geologic/jlwnstrn/other/software_jbl.html.

stage in *vesiculation*; the subsequent stages of bubble evolution are described in Sections 3.04.3.3–3.04.3.6. Vesiculated magma has very different physical properties to bubble-free melt, and exsolution can provide the trigger for magma intrusion and eruption. The term "degassing" refers to vesiculation, followed by separation of the gas phase from the melt. This section examines, in essence, how volatiles segregate from magma and ultimately reach the atmosphere. This field is of tremendous importance in linking eruptive behavior to magma dynamics (e.g., Eichelberger *et al.*, 1986; Eichelberger, 1995; Jaupart and Allègre, 1991; Sparks, 1997), and throws up many puzzles, such as the "excess sulfur" conundrum (Section 3.04.3.7).

3.04.3.1 Saturation

Because the pressure, temperature, and composition (redox state) of the magma change through time, volatiles in solution may reach and exceed their saturation points and enter the vapor phase. In the broadest terms, H_2O and CO_2 exsolution are effectively controlled by the phase diagram, and trace elements by partition coefficients. Because of its low solubility (roughly two orders of magnitude less than H_2O), CO_2 plays an important role in vapor saturation at crustal pressures (Anderson, 1975; Newman and Lowenstern, 2002). Formation of a CO_2- and H_2O-rich vapor phase in a magma body can result in strong partitioning of other species, including sulfur and halogens, into it. For example, Keppler (1999) showed from experimentally determined fluid–melt partition coefficients for sulfur that development and accumulation of a hydrous vapor phase could efficiently extract most of the dissolved sulfur in the melt. He went on to suggest that this could explain the high sulfur yield of the 1991 eruption of Mt. Pinatubo (though his experimental work was on a calcium- and iron-free haplogranite melt rather than actual Pinatubo dacite; Scaillet *et al.* (1998) and (Scaillet and Pichavant, 2003).

Pressure exerts a first-order control on the solubility of volatile species (Equation (1)) because of the great increase in molar volume going from the melt to vapor phase. Therefore, decompression, which can accompany magma ascent, or failure of confining rock, is a key process leading to saturation of volatiles and exsolution. Decompression also promotes undercooling, driving crystallization, and exsolution.

Magmatic systems have sufficiently long lifetimes that magma mixing occurs between more and less evolved batches of melt. Assimilation of the country rocks that confine magma reservoirs may also provide a source of volatiles. Juxtaposition, mingling, or mixing of different magmas can induce degassing by supplying fresh volatiles, e.g., from a mafic intrusion into an intermediate or silicic magma reservoir. When a dense, mafic magma is injected into an intermediate or silicic reservoir, the induced cooling and crystallization of the mafic melt causes volatile exsolution, which lowers its bulk density (due to the presence of bubbles), potentially leading to overturn of the mafic and silicic magmas. Alternatively, the bubbles may rise and form a foam at the interface between the two magmas. What actually happens will depend on the viscosity of the mafic magma with respect to the ascent rate of the bubbles forming within it (Hammer and Rutherford, 2002).

Magma mixing may also promote degassing by affecting the stability of volatile species in both the melt and solid phase. For example, Kress (1997) invoked mixing of comparatively reduced, sulfide-bearing basalt with oxidized, anhydrite-bearing dacitic magma as responsible for sulfur degassing prior to the 1991 Pinatubo eruption. Sulfur solubility reaches a minimum at the precise oxidation state that he calculated to result from mixing these two end-members, promoting exsolution. In addition, he showed that anhydrite and pyrrhotite stability would decrease, also liberating sulfur into the vapor phase. This could have induced magma ascent, decompression, and further degassing, leading to eruption. His argument provides a mechanism for exsolving the sulfur-rich vapor phase that Westrich and Gerlach (1992) and Wallace and Gerlach (1994), among others, suggested had formed prior to the eruption.

Crystallization (of anhydrous or water-poor phases) due to decompression or substantial cooling increases the fraction of volatiles dissolved in melts. This can also lead to oversaturation and exsolution, a process termed "second boiling." In the case of closed-system, isobaric crystallization of a volatile-saturated silicic magma, because CO_2 is less soluble than H_2O, it will preferentially exsolve. Residual melts formed during progressive crystallization, therefore, contain decreasing amounts of dissolved CO_2 and increasing amounts of H_2O. Degassing and crystallization are, therefore, very closely linked—the one can induce the other—and both have profound effects on the viscosity of magma. As the crystal fraction increases to ~40 vol.%, bulk viscosity also increases dramatically. In the case of erupting intermediate and silicic magmas, extensive microlite crystallization within the eruption conduit (e.g., Cashman, 1992; Hammer *et al.*, 1999; Blundy and Cashman, 2001) is thought to play an important role in magma dynamics by dramatically increasing bulk viscosity (Lejeune and Richet, 1995; Sparks, 1997), thereby strongly influencing rheology and transport.

3.04.3.2 Supersaturation and Nucleation

When the melt becomes saturated or oversaturated in a given volatile phase in solution, the subsequent fate of the phase in question depends on the kinetics of bubble nucleation, and the growth of bubbles. Because there is an energy cost in creating a bubble–melt interface, a degree of supersaturation in the melt is typically required to overcome it. The energy gained by moving the volatile into the new gas bubble is proportional to the volume of the bubble nucleus. This means that there is a critical size of nucleus for which the energy terms balance. Above this size, bubble nucleation is spontaneous. Since water is the principal volatile species, considerable work has gone into understanding its diffusion in silicate melts (see Watson, 1994).

Much of the theoretical work on bubble formation has considered homogeneous nucleation but, in real magmas, the presence of crystalline phases can be important in providing nucleation sites for bubbles, with both the sizes and compositions of those phases playing a role in the exsolution (Navon and Lyakhovsky, 1998). Because the surface energy associated with a gas–crystal interface can be lower than a gas–melt interface, the degree of oversaturation required for heterogeneous nucleation can be much lower than for homogeneous nucleation. Pre-existing bubbles and other inhomogeneities in the fluid may also play a role in promoting heterogeneous nucleation. For arc magmas, which appear to be mostly fluid saturated prior to eruption, the importance of heterogeneous versus homogeneous nucleation for eruption regimes is probably limited.

3.04.3.3 Bubble Growth and Magma Ascent

The growth of bubbles is controlled by the rates at which volatiles in the melt can diffuse towards the bubbles, and the opposing viscous forces. Near a bubble, volatiles are depleted such that melt viscosity increases dramatically, and diffusivities drop, making it harder for volatiles to diffuse through and grow the bubble. These opposing factors are described by the nondimensional Peclet number (*Pe*), which is the ratio of the characteristic timescales of volatile diffusion ($\tau_d = r^2/D$, where r is the bubble radius and D the diffusion coefficient of the volatile in the melt) and of viscous relaxation ($\tau_\eta = \eta/\Delta P$ where η is the melt dynamic viscosity and ΔP the oversaturation pressure, i.e., $Pe = \tau_d/\tau_\eta$; Dingwell (1998) and Navon and Lyakhovsky (1998)). For $Pe > 1$ (i.e., mafic melts) bubble growth is controlled by volatile diffusion, and for $Pe \leq 1$ (i.e., viscous intermediate and silicic melts), viscous resistance dominates, resulting in bubble overpressure. The higher viscosity of the immediate shell of magma surrounding bubbles can impart a significant control on τ_η and should be taken into account when modeling conduit flow dynamics. The coupling between viscosity and diffusivity in melts has also been explored in detail by Lensky *et al.* (2001, 2002) and Blower *et al.* (2001). When magma is rising, growth models for bubbles also have to consider the falling pressure, adding a decompression timescale to bubble nucleation (Lensky *et al.*, 2003).

3.04.3.4 Bubble Coalescence

When bubbles have formed in magma, their subsequent evolution and role depend on their buoyancy, and the magma transport and rheology. If bubbles nucleate through decompression and diffusion, their size may be insufficient for them to move relative to the magma. If the magma ascends, the bubbles grow; as the vesicularity of the magma increases, bubbles can interact and coalesce. At high gas volume fractions, foams may develop.

Foams are typically unstable and reduce gravitational and surface energy by collapse. The failure of bubble walls is promoted by thinning as capillary and gravitational forces drain melt from the films between adjacent bubbles. Capillary forces dominate over gravity when the vesicularity exceeds 74% (Mader, 1998). Once bubbles have coalesced the new, larger bubbles tend to relax and gain a spherical shape. A fundamental control on gas separation, once a bubble forms, is the melt viscosity. Other factors being equal, hot mafic magmas permit more efficient separation of gas from melt than cooler intermediate and silicic magmas. However, the fate of bubbles will also depend on several other factors, including the motion of the magma itself (e.g., ascent or descent in a conduit, or convection in the chamber), the degree of coalescence between bubbles, the flow regime (and extent of loss of gas through conduit walls).

Bubbly magma has very different rheological and physical properties from those of dense, nonvesiculated or slightly vesiculated magma. Thus, as degassing affects magma rheology and overpressure, these, in turn, influence degassing, resulting in strong feedbacks that ultimately impact eruption style (Sparks, 2003).

3.04.3.5 Gas Separation

A very strong control on degassing is the degree to which gas–melt separation proceeds to the point that the gas leaves its host magma. Two end-member states can be envisaged—closed-system degassing in which the exsolved vapor does not leave the melt, and open-system degassing in which the separated gas phase produced at each stage in the magma body is expelled into country rock (possibly through a hydrothermal system) or directly, via the magmatic plumbing system, to the atmosphere (in the case of a lava–air interface such as characterizes a number of persistently active volcanoes like Stromboli and Mount Etna, where bubbles can rise much faster than the melt—Section 3.04.4.1.2). These states exert a strong influence on partitioning of volatiles into the vapor phase, and result in differing isotopic shifts (e.g., between ^{16}O and ^{18}O) between melt and vapor. Magma ascent rate will also determine the extent to which degassing is incomplete (high ascent rates) or is in equilibrium (low ascent rates).

Two equations roughly describe the evolution of the gas phase in ascending magmas for a closed system. Water solubility and exsolution are described by (1), and vesiculation by the ideal gas law:

$$V_g/V_l = K_R\,(X^0_{H_2O} - X^r_{H_2O})/P \qquad (2)$$

where the term on the left of the equation is the volumetric ratio of gas to melt (vesicularity) at pressure P, $X^0_{H_2O}$ is the initial melt water content at saturation pressure, $X^r_{H_2O}$ the residual melt water content at pressure P, and K_R is a constant. In the case of open-system degassing, the evolution of the water content is still controlled by water solubility (1), but the vesicles are no longer maintained by the internal pressure of bubbles, and they collapse. In this case, the bulk evolution of the system is closer to a distillation process than an equilibrium process—each small quantity of fluid that is produced is expelled from the melt. Models for open-system degassing, therefore, incorporate the Rayleigh distillation law, but complications arise because of degassing-induced crystallization, particularly in intermediate and silicic systems, and where there is a competition between extraction of the vapor phase and mineral phases during vapor exsolution (Villemant and Boudon, 1998, 1999; Villemant *et al.*, 2003). Extensive microlite formation (of anhydrous minerals like plagioclase) acts to increase the volatile fraction in the residual melt, promoting further exsolution (Sparks, 1997). When crystallization is advanced, it impedes the expansion of bubbles, enhancing gas pressures during decompression. In intermediate and silicic systems, the highest overpressures are generally reached in the uppermost part of the conduit (Melnik and Sparks, 1999).

Theoretical models and experiments (e.g., Eichelberger *et al.*, 1986; Klug and Cashman, 1996) have indicated that magmas become permeable when the gas volume fraction reaches between 30% and 60%. In reality, deformation of bubbles may be necessary to achieve permeability at low vesicularities. In this case, elongation of bubbles would be flow parallel, and hence conduit parallel, promoting vertical permeability but not the horizontal permeability needed for degassing to take place through conduit walls. However, Jaupart (1998) suggests that fracturing in overpressured magma in the conduit may be a more likely mechanism for exporting gas into the country rock, or possibly nonlaminar flow in the conduit, and sufficient turbulence to bring significant quantities of melt into contact with the conduit walls on ascent. Studies of a fossil eruption conduit at Mule Creek, New Mexico, revealed a fractured conduit wall, and low vesicularity of the lava close to the conduit walls, indicating efficient gas loss (Jaupart, 1998). Massol and Jaupart (1999) found from models of conduit flow that significant horizontal pressure gradients can develop due to larger bubble overpressures developing in faster rising magma at the center of a conduit, exporting gas to the conduit walls given sufficient interconnection of bubbles. Permeable magma filling a conduit can permit gas contained in a deeper reservoir to flow to the surface, as has been suggested to account for high SO_2 fluxes from Soufrière Hills Volcano, Montserrat, during a hiatus in its ongoing eruption (Edmonds *et al.*, 2001; Oppenheimer *et al.*, 2002a). Crystallization can strongly promote high degrees of permeability in conduit magma because the bubbles that form are effectively squeezed along and between crystal boundaries, helping to establish an interconnected framework that promotes gas loss (Melnik and Sparks, 1999).

One of the best-understood low-viscosity magmatic plumbing systems is that of Kilauea. Gerlach and Graeber (1985) have shown how observed gas emissions (analyses for Kilauea in Table 3) indicate the differentiation of volatiles between the summit magma chamber and the sites of flank eruptions fed by intrusions from the main reservoir. Volatile-rich, mantle-derived magma arrives in the shallow summit chamber, and is saturated in CO_2. The exsolved CO_2-rich, H_2O-poor fluid degasses through summit lava lakes (when present) or through the hydrothermal system (type I gas). The magma, now equilibrated in the summit chamber, passes laterally via dikes that may feed subaerial or submarine eruption sites along the rift systems. These emit a distinct type II gas, with reversed H_2O and CO_2 contents compared with type I gas, reflecting the second degassing stage of the now CO_2-depleted magma. Degassing-driven fractionation of sulfur and halogens has been recently documented at Mount Etna, and has been related to open-system degassing and geometries and branching of

Table 3 Representative compositional analyses obtained by direct sampling of hot gas vents at different volcanoes.

Volcano	*Mt. St. Helens, USA, 1980*	*Mt. St. Augustine, USA, 1979*	*Momotombo, Nicaragua, 1980*	*Kiluaea, USA, 1918*	*Kiluaea, USA, 1983*	*Erta 'Ale, Ethiopia, 1974*	*Oldoinyo Lengai, Tanzania, 1999*
Magma type and tectonic association	Dacite (continental margin subduction)	Andesite (island arc)	Basalt (continental margin)	OIB summit (Type I gas)	OIB rift (Type II gas)	Transitional MORB (incipient plate boundary)	Carbonatite (continental intraplate)
Temperature (°C)	802	648	820	1,170	1,010	1,130	600
H_2O (mol.%)	91.58	97.23	97.11	37.09	79.8	77.24	75.6
H_2 (mol.%)	0.8542	0.381	0.7	0.49	0.9025	1.39	
CO_2 (mol.%)	6.924	1.90	1.44	48.90	3.15	11.26	24.4
CO (mol.%)	0.06	0.0035	0.0096	1.51	0.0592	0.44	0.0787
SO_2 (mol.%)	0.2089	0.006	0.50	11.84	14.9	8.34	0.0197
H_2S (mol.%)	0.3553	0.057	0.23	0.04	0.622	0.68	
S_2 (mol.%)	0.0039		0.0003	0.02	0.309	0.21	
HCl (mol.%)		0.365	2.89	0.08	0.1	0.42	
HF (mol.%)		0.056	0.259		0.19		
OCS (mol.%)	0.0008				0.0013		

[a] Data all for thermodynamically calculated equilibrium compositions from Symonds *et al.* (1994) except for Oldoinyo Lengai (from Oppenheimer *et al.*, 2002b)

the magmatic conduits (Aiuppa *et al.*, 2002; Burton *et al.*, 2003).

3.04.3.6 Fragmentation

When the product of viscosity and strain rate exceeds some critical value, and the strength of the magma is exceeded, fragmentation of magma occurs, driving an explosive eruption. In the case, at least, of rhyolitic systems, this occurs at the "glass transition" (Dingwell, 1998). Fragmentation can occur as the bubbly flow in a volcanic conduit accelerates and disintegrates, or due to sudden decompression of already vesiculated conduit magma, for instance resulting from a lava dome collapse and propagation of a rarefaction wave down the conduit (Cashman *et al.*, 2000; Melnik and Sparks, 2002a). Variations in, and pressurization feedback between, magma ascent rate, crystallization, and open- versus closed-system degassing may account for the rapid transitions between explosive and effusive eruption style that seem to characterize a number of intermediate and silicic eruptions (e.g., Melnik and Sparks, 1999, 2002b; Slezin, 1995, 2003; Barmin *et al.*, 2002; Sparks, 2003). Although volcanic systems are unlikely ever to be strictly deterministic, such nonlinearities have important implications for predicting volcanic behavior during periods when certain modes and timescales of behavior are dominant. Also, explosive eruptions are obviously capable of releasing very large quantities of volatiles almost instantaneously (Sections 3.04.4.1.2 and 3.04.6).

3.04.3.7 Excess Degassing

Numerous studies in the 1980s and 1990s identified the conundrum of "excess sulfur" or more generally "excess degassing" in which measured gas yields could not be provided by syneruptive degassing of observed quantities of erupted melt (Francis *et al.*, 1993). However, one way that excess degassing has been identified is really just an artifact of the measurement technique. For example, eruptions like Pinatubo 1991 and El Chichón 1985 produced far more sulfur than could be accounted for by petrological estimates of the volatile yields of the eruption (Section 3.04.4.3.4). However, in these cases, the excess sulfur was observed because the petrological method was inappropriate. Estimates of pre-eruptive volatile contents based on analyses of melt inclusions were invalid, because large amounts of volatiles had already exsolved prior to entrapment of the inclusions. Several investigations have yielded strong evidence for the existence of pre-eruptive vapor phases, e.g., for Pinatubo 1991 (e.g., Wallace and Gerlach, 1994).

Another way of resolving the excess degassing issue is to realize that when the sum of the partial pressures of dissolved volatiles equals the local confining pressure, then a separate multicomponent gas phase will exist in equilibrium with the melt. This gas can decouple from melt at many levels in the crust and potentially leak to the surface. A large body of unerupted melt can then be the source of volatiles released into the atmosphere. The degassing only seems excessive, because the mass of unerupted magma that supplied the volatiles is not being taken into account. Whether the unerupted melt exsolves gas and delivers it to the erupting magma syneruptively, or whether it has supplied vapor to the shallow parts of the pre-eruptive magma reservoir over periods of decades, centuries, or more, is uncertain, but several arguments favor the latter explanation (Wallace, 2001).

Extreme examples of excess degassing can be seen in the emissions from volcanoes such as Mt. Etna and Stromboli (Francis *et al.*, 1993; Allard *et al.*, 1994). Allard (1997) proposed that during the period 1975–1995, the sulfur observed to be degassing from Mt. Etna derived from 3.5 km^3 to 5.9 km^3 of magma, but only 10–20% of this was actually erupted. The remainder probably accretes as part of the plutonic complex within the sedimentary basement beneath the volcano. Allard *et al.* (1994) estimated an even more extreme ratio of eruptive-to-intrusive magma for Stromboli. They estimated that the observed SO_2 flux from 1980 to 1993 implied degassing of 0.01–0.02 $km^3\ yr^{-1}$ of magma, exceeding by a factor of 100–200 the volume of material actually erupted. It should be borne in mind, however, that the extent to which degassing appears excessive will depend on the time period to which observations pertain. Volcanoes can substantially catch up on erupted mass with major events outside the observation period.

Several interesting models have been proposed to account for excess degassing, some of which show how shallow degassing drives the convection in the volcanic conduit, permitting the emptying of large fractions of chamber volatiles without major eruption. Kazahaya *et al.* (1994) and Stevenson and Blake (1998) developed similar models based on poiseuille flow, and applied them to explain the behavior of quite diverse volcanoes, including Izu-Oshima (Japan) and Stromboli (Italy) (both basaltic), Sakurajima (Japan) (andesitic), and Mount St. Helens (USA) (dacitic). Stevenson and Blake (1998) suggested that conduit convection involves concentric flow of upwelling and downwelling magma. They showed that where convective overturn controls the gas supply, the gas flux is a function of the density difference between gas-rich and degassed magma, the conduit radius, and the magma viscosity (which is itself a function of gas mass fraction). Thus, efficient degassing induces a negative feedback by increasing viscosity of the downwelling magma, and thereby slowing the supply of gas-rich magma from the reservoir. Evidence for significant mixing of magma degassed at shallow depths with "fresh" volatile-rich magma is provided for Kilauea from analyses of volatile concentrations in samples dredged from the submarine Puna ridge at depths of as much as 5.5 km. Dixon *et al.* (1991) found ranges of CO_2, H_2O, and S contents in the tholeiitic glasses that could only be explained by deep recycling of magma that had already lost its gas at shallow depths, possibly even subaerially in a lava lake.

3.04.4 EMISSIONS

Volcanic emissions of gases and aerosol to the atmosphere take many different forms—from geothermal/hydrothermal manifestations, to massive syn-eruptive releases, such as that of Mount Pinatubo in 1991. This section illustrates this spectrum, highlighting the importance of understanding the fluxes and composition of gases, and their time–space distribution, for assessing and predicting the atmospheric, environmental, and climatic impacts of volcanic degassing. It describes also the techniques available to analyze volatile emissions, and to interpret them in the context of volcano surveillance efforts.

3.04.4.1 Styles of Surface Emissions

The manifestations of volatile release from volcanoes vary tremendously—from the diffuse leaks of CO_2 on both active and dormant volcanoes to the highly concentrated, sporadic injections of water and acid gases into the upper atmosphere by major explosive eruptions (Figure 4).

3.04.4.1.1 Noneruptive emissions

Noneruptive volcanic emissions are most apparent when they are exhaled from discrete vents, either in gas, liquid, or gas/liquid state. However, significant fluxes of gases can be discharged over wide areas in a much more distributed fashion. Broadly speaking, these are the manifestations of magmatic–hydrothermal systems, whose constituents are derived, in varying proportions, from magma, the crust, and groundwaters of assorted provenance.

In the case of *hot springs*, the steam and gas flux is subordinate to the liquid water flux. *Geysers* are

(a) (c) (b) (d) (e)

Figure 4 Illustrations of emission styles from volcanoes: (a) Old Faithful geyser in Yellowstone National Park, USA; (b) fumaroles at Kawah Ijen (Indonesia)—the pipes visible between the steam are used to condense sulfur, which is collected by miners working in the crater; (c) diffuse degassing of CO_2 produced this tree-kill zone at Mammoth Mountain, USA; (d) Kawah Ijen's crater lake—such lakes are typically enriched in acid species and have elevated temperatures; (e) open-vent degassing from Masaya (Nicaragua)—this volcano emits high fluxes of sulfur and halogens directly to the atmosphere from the surface of a magma-filled conduit or "open vent" opening on to the floor of the crater; (f) when low viscosity lava fills a crater, a lava lake forms, such as this long-lived example at Erta 'Ale (Ethiopia; Oppenheimer and Yirgu, 2002)—these can efficiently degas large volumes of subsurface magma by convective circulation; (g) slugs of gas bursting through an open vent generate strombolian eruptions, as demonstrated here at Stromboli volcano (Italy); (h) the lava dome of Soufrière Hills Volcano (Montserrat), which has sustained a considerable gas flux since its emergence in 1995 (Edmonds *et al.*, 2001, 2003a,b,c); (i) dome collapses (as seen here at Soufrière Hills Volcano) can degas lava as it fragments, fuelling eruption plumes—this is at the low-scale end of a spectrum of explosive degassing behavior that spans up to releases of well in excess of 10^7 kg s^{-1} of volatiles.

(f) (h) (g) (i)

Figure 4 (continued).

spectacular examples of hot springs (Figure 4(a)). When gaseous emissions predominate, the term *fumarole* is usually applied. Emission temperatures of fumaroles, therefore, typically exceed the local boiling point. Long-lived fumarole fields are sometimes termed *solfataras* or *soufrieres* (Figure 4(b)), their longevity giving rise to substantial alteration of host rock and deposition of sublimates. Solfatara in the Campi Flegrei north of Napoli is the classic example (Figure 1). Fumarole emissions are very often composed of magmatic gases and hydrothermal gases (which result from complex interactions between magmatic fluids, meteoric water, seawater, and rock).

CO_2-rich emissions at temperatures below the boiling point of water are sometimes referred to as *mofettes* when they are localized. More diffuse emissions of CO_2 (and radon) can also occur over wide areas, reflecting the exsolution of these gases from magma bodies at depth (Figure 4(c)). Such emissions can present a hazard close to the ground and in depressions, especially in calm atmospheric conditions, as CO_2 concentrations can exceed several percent (Section 3.04.6.5.3).

Numerous volcanoes discharge acid gases that are condensed in crater lakes (Figure 4(d)). The lake water derives, therefore, both from volcanic and meteoric inputs. Such lakes provide valuable opportunities to evaluate the heat and mass budgets of the host volcanoes, but also represent particular hazards since even small eruptions can displace large quantities of water over crater rims, triggering catastrophic lahars (Varekamp and Rowe, 2000). The black smokers associated with active oceanic ridges are another well-known manifestation of subaqueous volatile discharge.

3.04.4.1.2 Eruptive emissions

When magma reaches the surface, it can release gases directly into the atmosphere. Magma-filled conduits (often referred to as *open vents*; Figure 4(e)), lava lakes (Figure 4(f)), and lava domes (Figure 4(g)) can all discharge large amounts of gas. If they do this nonexplosively, the efficient segregation of gas typically reflects low-viscosity magmas in which coalescence of bubbles permits them to rise faster than the magma, or the development of significant permeability by vesiculation in more viscous intermediate or silicic magmas. Gas fluxes from open vents can be very high (Section 3.04.5). Various circumstances can lead to explosive fragmentation of magma, witnessed in a wide spectrum of eruptive behavior (Figures 4(h) and (i)) including discrete strombolian explosions, lava fountains, and sustained eruption columns typified by the plinian eruptions of Mt. St. Helens (1980) or Mt. Pinatubo (1991). In the intermediate and silicic cases, as bubbly magma ascends in the conduit, progressive vesiculation and inhibited gas loss can culminate in fragmentation of the accelerating mixture into a gas–particle mixture. The eruption of this mixture increases the pressure drop between the vent and chamber, typically raising the eruption intensity (mass discharge rate of magma at the vent) further. In this way, significant fractions of magma reservoirs may be erupted in a matter of hours.

3.04.4.2 Chemical Composition of Volcanic Gases

The major and trace element composition of volcanic gases varies widely (Table 3). In general terms, the composition of volcanic gas represents the complex sources, histories, and processes of magma generation, mixing and ascent (e.g., variable contributions from the mantle wedge or slab in the case of arc magmas; time-dependent vapor–melt separation as a function of evolving magma composition and physical conditions), and the interactions of the gas phase after it separates from the host magma, for example, with rocks and fluids in the crust (Giggenbach, 1996). Degassing is a continuous process, and the distribution of volatiles between vapor and melt phases varies strongly as a function of depth and time. Magma bodies and their plumbing systems have ample time to evolve and can have significant vertical extents. The latter point implies that deep degassing, shallow degassing, and anything in between, can contribute to the mixture of magmatic fluids observed at the surface. Superimposed on these magmatic complexities are the effects of shallow re-equilibration of fluids due to cooling and dilution by groundwater (e.g., meteoric water, seawater, or hydrothermal fluids). Isotopic ratios of various fluid species can be used to model magmatic and hydrothermal processes. In general terms, isotopes of carbon, hydrogen, and nitrogen are good tracers of magmatic processes, while oxygen isotopes are useful for tracing the source (see Valley and Cole, 2001, for reviews).

3.04.4.3 Measurement of Volatiles

The conventional way to measure volcanic emissions is by direct sampling, either by *in situ* collection of samples from fumarole vents and active lava bodies using "Giggenbach bottles," filter packs and condensing systems, or within atmospheric plumes (sometimes from aircraft) using various kinds of sampling apparatus and onboard analyzers. A range of spectroscopic, gravimetric, isotopic, and chromatographic techniques is available to determine chemical concentrations in real time or subsequently in the laboratory (Symonds *et al.*, 1994). While such direct sampling can deliver very detailed and precise analyses, it is difficult to sustain routine surveillance in this way, and to compete with seismological and geodetic monitoring techniques in terms of sampling rate. However, an expanding array of remote sensing methods is available to provide the data streams needed to characterize the composition and fluxes of volatiles from volcanoes. In particular, both field-based and satellite spectroscopic methods are increasingly in use to measure gas composition remotely.

This section briefly reviews the principal measurement techniques (Figure 5), including direct sampling, ground-based, or airborne ultraviolet spectroscopy (correlation spectrometer and successors), ground-based infrared spectroscopy (Fourier transform spectroscopy and other infrared spectroscopic analysers), and spaceborne methods, including the important role of the total ozone mapping spectrometer (TOMS), and

(a) (b)

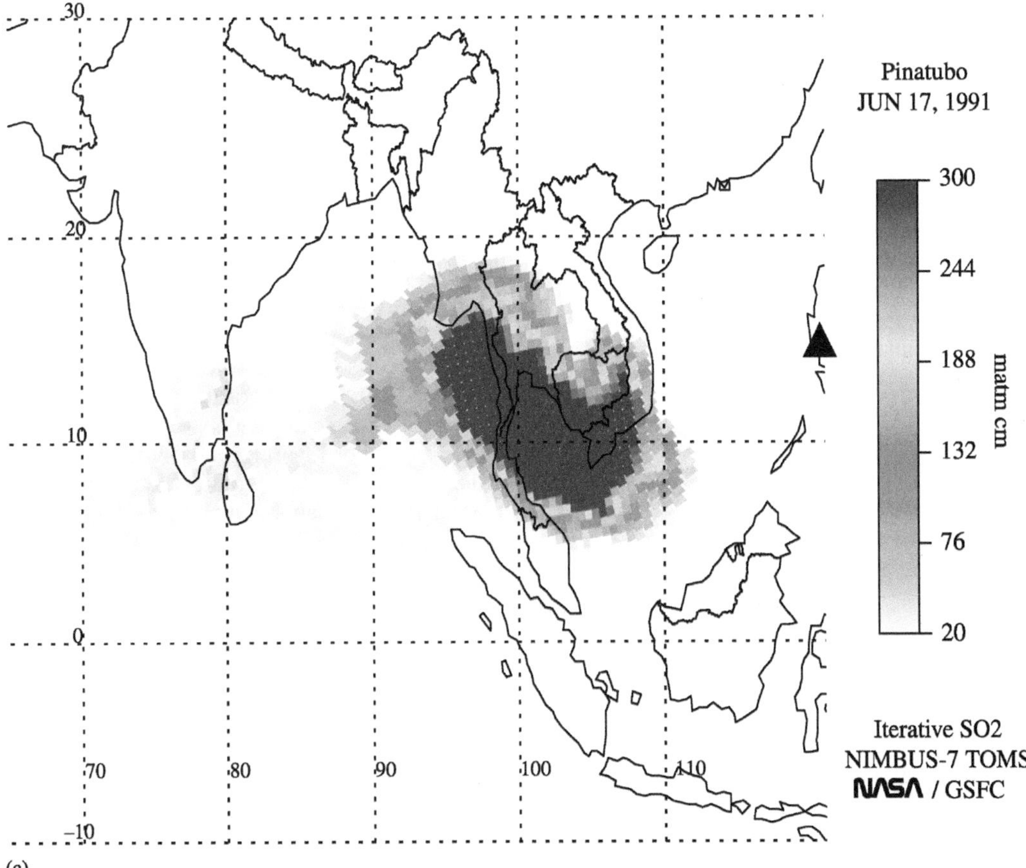

(c)

Figure 5 Techniques for measuring volcanic volatiles: (a) direct sampling of fumarole vent at Ol Doinyo Lengai Volcano (Tanzania) using Giggenbach bottle; (b) ground-based remote sensing techniques—shown here in the flanks of Mount Etna are the ultraviolet sensing COSPEC instrument (for SO_2 measurements) and a Fourier transform infrared spectrometer (for SO_2, HCl, and HF measurements); (c) satellite-based TOMS observations of the Mt. Pinatubo stratospheric SO_2 emission (courtesy of Simon Carn, TOMS Volcanic Emissions Group); (d) glassy melt inclusion in quartz and its associated shrinkage bubble, from the ignimbrite unit of the 1912 eruption at the Valley of Ten Thousand Smokes, Alaska—such inclusions can provide valuable information on volatile contents (micrograph and caption courtesy of Jake Lowenstern (Volcano Hazards Team, US Geological Survey); and (e) ice core chemical stratigraphy has been used to estimate sulfur yields from major eruptions (for review see Zielinski, 2000)—this section of the North GRIP core from Greenland shows the 10.14 kyr BP (ice stratigraphic age) Icelandic Saksunarvatn ash layer (thin horizon in the center of the core shown) (courtesy of Sune Olander Rasmussen, Trine Ebbensgaard and Sigfus Johnsen, Glaciology Group, University of Copenhagen; Dahl-Jensen *et al.*, 2002).

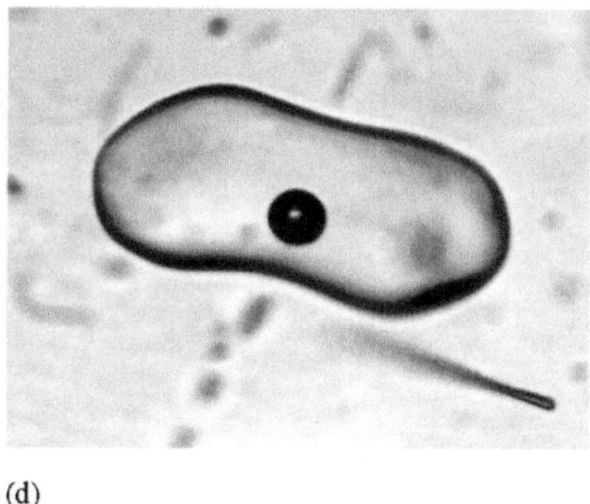

(d)

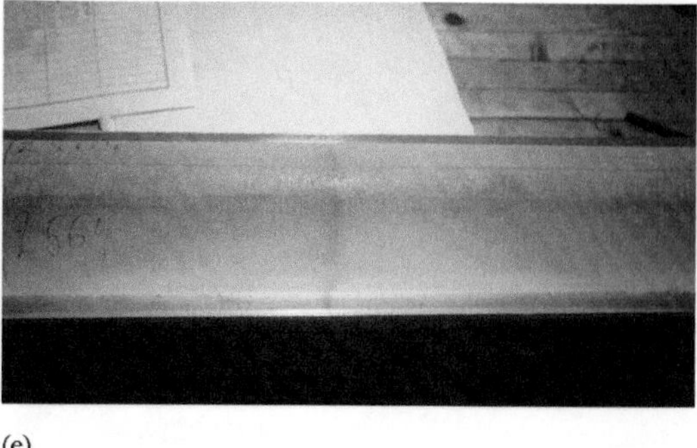

(e)

Figure 5 (continued).

similar instruments, for measuring the SO_2 emissions from major eruptions. It also covers petrological approaches to characterizing volatile abundances and emissions.

3.04.4.3.1 In situ *sampling and analysis*

Conventional analyses of volcanic gases have been made by collection of samples directly from fumarole vents using evacuated bottles and caustic solutions, and subsequent laboratory analysis. The classic "Giggenbach" bottle consists of an evacuated glass vessel partially filled with NaOH (Figure 5(a); Giggenbach, 1975; Giggenbach and Matsuo, 1991). On the volcano, the gas stream is allowed to bubble through the solution via tubing inserted into the volcanic vent. Acid species condense according to reactions such as

$$CO_2 + 2OH^-_{(aq)} = CO^{2-}_{3(aq)} + H_2O \qquad (3)$$

$$4SO_2 + 7OH^-_{(aq)} = 3SO^{2-}_{4(aq)} + HS^-_{(aq)} + 3H_2O \qquad (4)$$

and are analyzed by ion chromatography. The remaining species collect in the headspace and are usually analyzed by gas chromatography. Base-treated filters can also be used to trap acid species, and can be deployed around crater rims and in the vicinity of gas sources. Such filter-based methods have been extended to characterization of aerosol size distribution and chemistry (e.g., Vié le Sage, 1983; Allen *et al.*, 2000, 2002). The volatiles scavenged out of eruption clouds by ash particles, which then sediment to the ground, can also be studied analytically by leaching samples with distilled water (e.g., Edmonds *et al.*, 2003a).

Although such approaches offer very high sensitivity, measurements can be difficult and often dangerous to obtain, and there can be problems of postcollection reactions. Also, the inevitable delays in obtaining results limit their value in volcanic crises. Remote sensing methods can overcome these difficulties, and, importantly, several are able to constrain gas fluxes, which are hard to derive by point sampling.

3.04.4.3.2 Portable remote sensing systems

Since the early 1970s, ground-based, optical remote-sensing techniques have been increasingly used for volcanic gas and aerosol monitoring (Figure 5(b)). In particular, the correlation spectrometer (COSPEC), which operates in the ultraviolet region of the spectrum, using scattered skylight as a source, has been used routinely by volcano observatories worldwide to measure SO_2 fluxes (Stoiber *et al.*, 1983). COSPEC has seen active service in numerous volcanic crises, crucially helping to ascertain whether or not new magma pathways are opening up to shallow levels beneath a volcano. More recently, a much smaller ultraviolet spectrometer (the size of a pack of cards) has been applied to measurements of SO_2 flux (Galle *et al.*, 2003; McGonigle and Oppenheimer, 2003; McGonigle *et al.*, 2002, 2003). The device is set to revolutionize ground-based sensing of volcanic gas emissions because of its extreme portability, ease of operation, and suitability for automated scanning measurements at high temporal resolution (Edmonds *et al.*, 2003b; McGonigle *et al.*, 2002, 2003).

The reasons for focusing measurements on SO_2 are that: (i) the background levels of this gas are very low in the atmosphere (typical volume mixing ratios are $\ll$1 ppb) while mixing ratios in volcanic plumes can easily exceed 1 ppm, and (ii) several strong absorption bands for SO_2 are found in the ultraviolet and infrared regions of the spectrum. In other words, the spectroscopic observation of this species is straightforward, despite its comparatively low abundance in volcanic emissions. In contrast, H_2O or CO_2, which are the principal components of volcanic gases, are difficult to measure accurately because of the high, and in the case of H_2O, rapidly varying, atmospheric background concentrations.

However, it remains highly desirable, for the purposes of volcano surveillance, to be able to measure other components of the gas emission, and to be able to follow these through time. To this end, broad-band infrared measurements have been carried out using Fourier transform spectrometers. These are capable of simultaneously sensing several gas molecules of interest, including HCl, HF, CO_2, CO, OCS, SiF_4, and H_2O, as well as SO_2 (Oppenheimer *et al.*, 1998a; Burton *et al.*, 2000; Horrocks *et al.*, 2001). Measurements can even be made from moving vehicles in order to build up cross-sections of plume burdens of SO_2 and halogens that can then be integrated and multiplied by plume speed to yield estimates of the fluxes of these species (Duffell *et al.*, 2001).

Aerosol concentrations and size distributions can be investigated remotely using sun-photometry. Characterization of volcanic aerosol is important in studies of plume chemistry, atmospheric radiation, and the environmental and health impacts of particle emissions. Watson and Oppenheimer (2000, 2001) used a portable sun-photometer to observe tropospheric aerosol emitted by Mt. Etna. They found distinct aerosol optical signatures for the several plumes emitted from Etna's different summit craters, and apparent coagulation of particles as the plume aged. More recently, Porter *et al.* (2002) have obtained sun-photometer and pulsed lidar data for the plume from Pu'u O'o vent on Kilauea, Hawaii, from a moving vehicle in order to build profiles of sulfate concentration.

3.04.4.3.3 Satellite remote sensing

The larger releases of volcanic volatiles to the atmosphere defy synoptic measurements from the ground. Major advances in our understanding of explosive volcanism and its impact on the atmosphere and climate have been achieved thanks to satellite observations. Again, sulfur dioxide is the most readily measured species, and a number of spaceborne sensors operating in the ultraviolet (electronic lines), infrared (roto-vibrational lines), and microwave (rotational lines) have been utilized for measurements of this gas. TOMS operating in the ultraviolet region has provided the most comprehensive database on the release of SO_2 to the upper atmosphere (Section 3.04.5.1; Figure 5(c); Krueger *et al.*, 1995). It has detected many of the larger silicic and intermediate composition explosive eruptions that have taken place since 1979, and some mafic eruptions, prominently including those from Nyiragongo and Nyamuragira in the Great Lakes region of central Africa (Carn *et al.*, 2003). TOMS provided early estimates of the initial sulfur yield to the stratosphere of the 1991 Mt. Pinatubo eruption (~20 Tg of SO_2), but other instruments, including NASA's Microwave Limb Sounder (MLS), were able to track the SO_2 clouds as they were depleted over the following weeks (Read *et al.*, 1993).

Recently, Prata *et al.* (2002) have pointed out the suitability of an infrared system—the high-resolution infrared radiation sounder (HIRS)—carried by NOAA satellites, for measuring volcanic SO_2 above altitudes of ~5 km. This could prove valuable in filling in gaps in the existing TOMS record of volcanic emissions due to instrument problems (Carn *et al.*, 2003). Hitherto, satellites have only really been capable of measuring the larger releases of SO_2 during eruptions. However, two newer infrared instruments with higher spatial resolution—the moderate resolution imaging spectrometer (MODIS) and advanced spaceborne thermal emission and reflection radiometer (ASTER)—both carried on-board NASA's Terra satellite (launched in late 1999)—show considerable promise for monitoring smaller yields of SO_2 to the troposphere. Researchers at Michigan Technological University have used imagery from both sensors to estimate sulfur dioxide fluxes from volcanoes in Guatemala, suggesting that these and future similar sensors could be used for routine volcano monitoring purposes (W. I. Rose, personal communication).

3.04.4.3.4 Petrological methods

There are various "petrological" approaches to determining volatile concentrations in magma (melt, vapor phase, and crystals) or volatile yields to the atmosphere. Most of the data on pre-eruptive volatile abundances reported in Table 1, for example, have been obtained by analyses of quenched glasses. These include the chilled glassy rinds of lavas or pyroclasts, and melt inclusions trapped inside phenocrysts (Figure 5(d)). In some cases, differencing melt inclusion and matrix glass concentrations for specific volatile species permits estimates of degassing yield, if it can be assumed that the former represents pre-eruptive concentrations, and the latter degassed melt. Other approaches to characterizing volatile abundance in magmas include the application of experimentally determined phase equilibria (for both natural and synthetic melts), and thermodynamic calculations based on mineral compositions (see Johnson *et al.* (1993) and Wallace and Anderson (2000) for overviews of these methods).

Petrological estimates of eruption yields of sulfur and, to a lesser extent, halogens have received much attention, partly because they offer a means to assess the volatile yield for historic and ancient eruptions. The search for suitable host phenocrysts requires careful microscopic examination of thin sections and even more careful analysis of the inclusions themselves, which can measure only a few microns across. While the method appears to work well for some systems (e.g., Thordarson *et al.*, 2003), it is not universally applicable. For example, the crystals may not always be leak-proof containers for their high-pressure melt samples, and inclusions can interact with their host minerals. Most problematic, however, is the possibility of volatile exsolution into a fluid phase *prior* to crystal growth and entrapment of the melt, leading to the inclusion recording a lower volatile content than that which the melt started with. This very likely explains why the petrological technique singularly fails to explain the sulfur release during Mount Pinatubo's eruption in 1991 (Section 3.04.3.7). The sulfur contents in glass and inclusions are more or less the same, which would suggest a sulfur-free eruption plume. Instead, the eruption is known to have released some 20 Tg of sulfur dioxide.

Wallace *et al.* (1995) and Wallace (2001) have explored this issue in detail, and Scaillet *et al.* (1998, 2003) and Scaillet and Pichavant (2003) have advanced petrological modeling approaches to circumvent the problem by estimating the volatile contents contained in the vapor phase prior to eruption. Another problem with petrological estimates of volatile yields is that they scale linearly with the eruption magnitude, which is often only poorly constrained from the rock record. This is especially true in cases where tephra dispersal is very widespread, perhaps largely at sea, and where substantial burial, erosion or redeposition limit efforts to identify original thicknesses of sediment in the field.

3.04.4.3.5 Ice cores

The cryosphere provides an important repository of volcanic volatiles from past eruptions in the form of sulfate layers deposited within a few years of eruption (Hammer *et al.*, 1980; Zielinski *et al.*, 1994; Zielinski, 2000). Ice core stratigraphy is used to date the eruption year, and in some cases, tephra particles can be fingerprinted chemically to the products of known eruptions. The most productive cores for identifying volcanic markers have come from Greenland and Antarctica, including the Greenland Ice Sheet Project 2 (GISP2) and Greenland Icecore Project (GRIP and NorthGRIP) efforts (Figure 5(e)).

Zielinski *et al.* (1996a) have published a 110 kyr record of volcanism as recorded by the GISP2 core, providing one of the most intriguing records of palaeovolcanism during the Late Quaternary (Figure 6). By determining the flux of sulfate that formed the layer (in kg m^{-2}), it is possible to estimate the total atmospheric mass of sulfur by calibrating against fallout from sources of known magnitude (Hammer *et al.*, 1980; Zielinski, 1995). The most widely used calibration is based on β-activity measurements of the layers of radioactive fallout in the ice cores that resulted from nuclear weapons tests in the atmosphere conducted in the 1950s and 1960s. Regressions can be obtained between the flux of fallout, say for the Greenland ice core, and the known yield of the explosions, and then applied to the volcanic markers. This, of course, assumes that the volcanic sulfate was transported by comparable atmospheric dynamics, from similar latitude, and so on, which is difficult to gauge for ancient eruptions (usually, the responsible volcano is unidentified). Very low deposition rates, and post-depositional effects, such as densification, diffusion, and aeolian deflation or redeposition, and other potential sources of sulfur (including marine sources), also pose difficulties in interpreting sulfate layers in the ice cores. Nevertheless, multiple estimates of sulfur yields of eruptions such as Tambora 1815, whose sulfate marker is found in both Arctic and Antarctic cores, are reasonably consistent with each other, and with estimates obtained by other methods (e.g., based on astronomical observations of atmospheric optical thickness), lending some confidence in the approach.

3.04.4.3.6 Application of geochemical surveillance to volcano monitoring

Time-series chemical and isotopic measurements, and flux observations can be interpreted with respect to volcano behavior, and the inter-relationships between degassing, eruptive character, and other geophysical and geodetic parameters. Gas geochemistry, therefore, plays an important role in volcanic hazard assessment. Temporal changes in gas chemistry and flux, in particular, are widely regarded as potential indicators of future volcanic activity, and many volcano observatories worldwide carry out some kind of geochemical surveillance as part of their overall monitoring efforts. The basic tasks are to identify volatile sources, magmatic-hydrothermal system interactions, the dynamics of degassing, and changes in these through time. Unfortunately, interpretation of the observations is far from straightforward because of the multiple intensive parameters that control magmatic volatile content (mantle melting, slab contributions, wall–rock assimilation, etc.), exsolution and gas separation of different volatile species from magma, and the subsequent chemical and physical interactions of the exsolved fluids, for example, with crustal rocks and hydrothermal fluids, as they ascend to

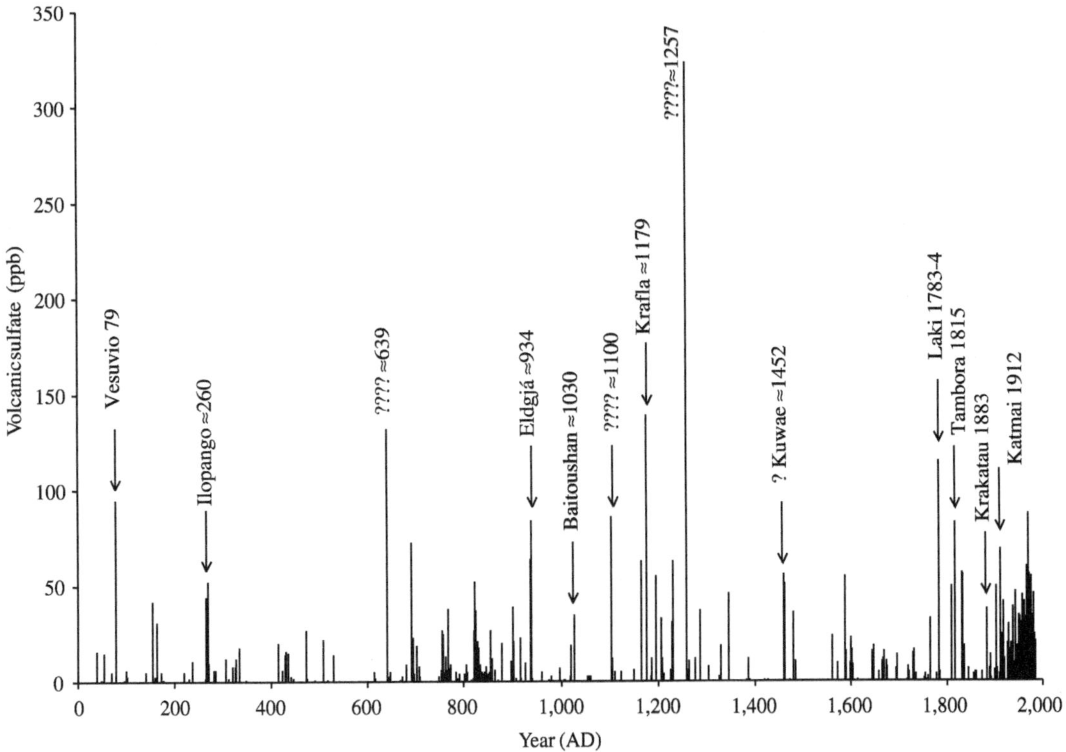

Figure 6 GISP2 volcanic sulfate markers for the past 2,000 yr based on statistical analysis (Zielinski *et al.*, 1994). Several large anomalies have not been traced to the responsible volcanoes, including the prominent AD 640 and 1259 peaks. Data provided by the National Snow and Ice Data Center, University of Colorado at Boulder, and the WDC-A for Paleoclimatology, NGDC, Boulder, CO, USA.

the surface (Figure 7). Some general observations on the interpretation of the principal volcanic gas species are made in Table 4.

In the broadest terms, Matuso (1960) recognized that the composition of volcanic gas reflects the balance of contributions from magmatic degassing and from the crust. An oft-quoted sequence of volatile release with decreasing pressure, based largely on experimental observations of solubility in the melt, is

$$C \Rightarrow S \Rightarrow Cl \Rightarrow H_2O \Rightarrow F$$

However, this greatly oversimplifies the exsolution behavior of real magmas (which, of course, vary enormously in chemical composition, viscosity, etc.) and their degassing histories, and is of limited use in trying to interpret measurements of, for example, increasing Cl/S ratio or decreasing SO_2 flux of fumarolic emissions. Also, in general terms, the relative proportions of magmatic carbon, sulfur, and halogen compounds are dictated by the pressure- and redox-controlled partition coefficients between vapor and melt, the mixing or mingling of different magma sources, and the dynamics of degassing. For example, important redox equilibria affecting volcanic gas composition include

$$SO_2 + 3H_2 = H_2S + 2H_2O \quad (5)$$

$$CO_2 + H_2 = CO + H_2O \quad (6)$$

$$CO_2 + 4H_2 = CH_4 + 2H_2O \quad (7)$$

Reaction (5) is most likely to represent the major gas redox buffer because of the comparable abundances of the two sulfur-bearing gas species. The other effective geochemical buffer is the $FeO-Fe_2O_3$ "rock buffer," which can affect the redox equilibria above in magma, and as gas and wall–rock interact.

Although thermodynamical codes have been developed to restore observed analyses of gas composition to equilibrium compositions (e.g., Gerlach, 1993; Symonds *et al.*, 1994), unravelling these complex and highly nonlinear processes from a patchy record of surface observations, and identifying precise magmatic and hydrothermal processes, remain real challenges. Indeed, a casual perusal of the literature on gas geochemical monitoring will reveal conflicting interpretations of ostensibly similar observations—for instance, decreasing SO_2 fluxes could be ascribed to: (i) depletion of volatiles in a magma body, or (ii) a decrease in the permeability of the plumbing

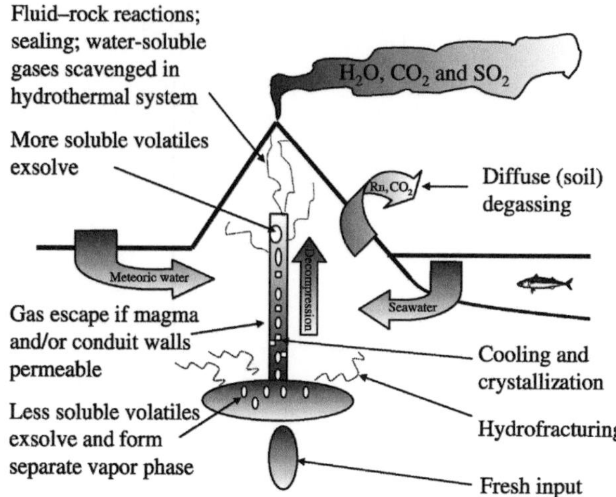

Figure 7 Potential physical and chemical processes occurring in a magmatic–hydrothermal system, including the influence of magma dynamics in the chamber-conduit plumbing system, and interactions between magmatic fluids and the crust. These can strongly modulate the speciation and flux of various magmatic components emitted into the atmosphere, complicating the interpretation of geochemical measurements of surface emissions.

Table 4 General observations on the information content of principal volcanic gas species.

H_2O	Several origins in volcanic discharges—principally magmatic, hydrothermal, meteoric, seawater (volcanic islands)—hence difficult to interpret water contents without measurements of isotopic composition.
CO_2	Predominantly magmatic origin, and comparatively inert in hydrothermal systems and the atmosphere. Fluxes can be measured by some spectrometric techniques.
SO_2	Common in high temperature volcanic gases and a good indicator of magmatic degassing. Fluxes readily measured by remote sensing methods.
H_2S	Typical of lower temperature volcanic vents and indicative of hydrothermal contributions.
HCl	Typically magmatic but potentially sourced by hydrolysis/volatilisation of chloride compounds or dissociation of brines. Readily scavenged by hydrothermal systems so changes in HCl emission can reflect both magmatic and hydrothermal processes.
He	Total He content and $^3He/^4He$ ratios are good tracers of mantle contribution and mixing with atmospheric He. Unaffected by secondary processes.
Ar	A good tracer of atmospheric and air-saturated meteoric water mixing in volcanic gas samples.
H_2	Hydrogen content increases with increasing temperature and provides the basis for gas geothermometers.
N_2	Either of atmospheric origin or from subducted slab (in which case, N_2/Ar usually >83, the value for air).
CH_4	Nonmagmatic component generally formed in hydrothermal environment and therefore a good indicator of hydrothermal system. Also favored magmatic species according to control of calcite-anhydrite buffer ($CaCO_3 + H_2S + CH_4 + 5H_2O = CaSO_4 + 2CO_2 + 8H_2$) or the H_2S/SO_2 and rock buffers: $2Fe_3O_4 + 3SiO_2 + CO_2 + 2H_2S = 3Fe_2SiO_4 + CH_4 + 2SO_2$.
CO	Controlled by temperature and redox potential and a useful indicator of these parameters.

Source: Giggenbach *et al.* (2001).

system. Process (i) might be taken to indicate diminished eruption likelihood, while (ii), perhaps induced by sealing of bubble networks by hydrothermal precipitation, which would increase overpressure (e.g., Edmonds *et al.*, 2003c), could increase the chance of an eruption. Thus, the same observation can be interpreted in different ways with contradictory hazard implications.

Nevertheless, high SO_2 fluxes remain a reliable indicator of the presence of magma during new episodes of unrest at volcanoes (e.g., seismicity, changes in fumarole emissions), and help to discriminate between magmatic, and tectonic or hydrothermal causes of unrest. In particular, COSPEC measurements are generally recognized as having contributed significantly to hazard assessment prior to the 1991 eruption of Mt. Pinatubo (Daag *et al.*, 1996; Hoff, 1992). Immediately prior to the eruption the measured SO_2 flux increased by an order of magnitude over two weeks, mirroring seismic unrest. These observations were interpreted as evidence of shallow intrusion of magma, increasing the estimates of the probability of an impending eruption. More recently, COSPEC observations have supported conclusions of magmatic unrest prior to eruptions of Soufrière Hills Volcano (Montserrat), Popocatèpetl (Mexico), and Tungurahua (Ecuador) (Dan Miller, USGS, personal communication, 2002). For ongoing eruptions, combining observations of gas emissions with melt inclusion constraints on melt volatile contents can offer valuable insights into the sources of gases and mechanisms of degassing. For example, Edmonds *et al.* (2001) used both kinds of information to conclude that, at Soufrière Hills Volcano, the HCl content of the summit plume has been largely supported by shallow, crystallization-induced degassing of the andesite approaching and forming the lava dome, while SO_2 fluxes have been sustained, even through noneruptive periods, by degassing of a deep, possibly mafic, source.

Decreasing SO_2 fluxes accompanying diminishing posteruptive activity have also been observed at many volcanoes, including Mt. St. Helens from 1980 to 1988 (McGee, 1992), where a decline in CO_2 flux and increase in H_2O flux were also observed following the 1980 eruption. This decrease in the CO_2 and SO_2 emissions suggested that the magma reservoir was not being replenished, consistent with the decreased eruption rates (see also Stevenson, 1993; Stevenson and Blake, 1998). The increase in water vapor emission was interpreted as the result of groundwater permeating the conduit system (Gerlach and Casadevall, 1986).

Because of its deep exsolution, and comparative chemical inertness in the crust and hydrothermal systems, and atmosphere, CO_2 flux measurements are seen as a particularly valuable indicator of magmatic degassing and unrest. Infrared analyzers have been used to determine CO_2 fluxes from the ground and in atmospheric plumes (e.g., Gerlach *et al.*, 1997; McGee and Gerlach, 1998) but the available methods are time consuming (ground surveys) and expensive (where aircraft deployments are required), with the possible exception of open-path Fourier transform infrared spectroscopy combined with a method for determining SO_2 fluxes (Burton *et al.*, 2000).

Villemant *et al.* (2003) have examined in detail the implications of closed-system versus open-system degassing (Section 3.04.3.5) on the compositional evolution of the gas phase, focusing on halogens, since their behavior is not particularly affected by the oxidation state of the magma (e.g., Symonds and Reed, 1993), and their concentrations are relatively easy to measure in lava and tephra samples. Open-system degassing is more efficient at extracting halogens from the melt than closed-system degassing. HCl is less bound to the silicate chains in melts than HF, and is therefore more efficiently extracted by a water-vapor phase. HCl/HF ratios in an open system are therefore predicted to fall with increased degassing of water. In other words, gas–rock interaction, cooling, and decompression should not affect halogen ratios strongly. However, kinetic effects can also be significant in halogen partitioning during shallow degassing (e.g., Signorelli and Carroll, 2001), and scrubbing in hydrothermal systems (Oppenheimer, 1996) can dissolve acid species from the gas phase, stripping fumarole gases of their halogen content (e.g., Symonds *et al.*, 2001). Duffell *et al.* (2003) observed an increase in SO_2/HCl ratios prior to a minor phreatic eruption of Masaya volcano, Nicaragua, which they suggested could have reflected hydrothermal scavenging of HCl. Increased magma–water interaction may have played a role in triggering the eruption, indicating the potential predictive value of routine monitoring of sulfur and halogen ratios. Second boiling was thought to explain a *decrease* in S/Cl ratios in gases emitted from the Showa–Shinzan lava dome, several years after its emplacement, as the slow crystallization of the thick lava body delayed the eventual exsolution of chlorine from the melt (Symonds *et al.*, 1996).

Isotopic signatures can be of great value in identifying magmatic components of volcanic fluids. In particular, the ratio of $^3He/^4He$ is useful in distinguishing the contribution from primordial (mantle) ^3He from that of radiogenic (crustal) ^4He (e.g., Sorey *et al.*, 1993, 1998; Tedesco, 1995). Nitrogen-isotopic composition is a useful tracer of sedimentary input into arc volcanic gases (Section 3.04.2.1; Fischer *et al.*, 2002). $\delta^{18}O$ and δD compositions of water arc important indicators of mixing of magmatic, hydrothermal, marine, and meteoric sources (Section 3.04.2.1). Stable isotopes of carbon and sulfur can also be used to identify contributions of magmatic volatiles, assuming that isotopic compositions of surface reservoirs are constrained. For example, $\delta^{13}C_{CO_2}$ and $\delta^{34}S_{SO_2}$ measurements of gas samples collected from the lava lake of Erta 'Ale volcano (Ethiopia) indicated the mantle origin of emitted carbon and sulfur ($\delta^{13}C \approx -4‰$ and $\delta^{34}S \approx 0‰$), whereas hot fumaroles on arc volcanoes display more variable $\delta^{13}C$ $-12‰$ to $+2.5‰$) and $\delta^{34}S$ (0‰ to $+10‰$) indicative of contamination by slab and crustal sources (summarized by Delmelle and Stix (2000)). Very little work has been carried out on chlorine-isotopic distribution in volcanic gases, and indeed, the mantle $\delta^{37}Cl$ value is arguably still not well constrained, but improved analytical techniques could open up considerable potential to study chlorine distribution and cycling. Eggenkamp (1994) reported variations in $\delta^{37}Cl$ between about $-2.5‰$ for sublimates and $+10‰$ for gases from Lewotolo volcano (Indonesia), which were thought to indicate surficial and deep processes, respectively, demonstrating significant chlorine-isotopic fractionation in volcanic systems.

3.04.5 FLUXES OF VOLCANIC VOLATILES TO THE ATMOSPHERE

Inventories of the spatial and temporal distribution of volcanic emissions to the atmosphere are important for studies of tropospheric and stratospheric chemistry, and the Earth's radiation budget. This section summarizes estimates of global fluxes and distribution of several key species. As pointed out in the IPCC Report (2001), referring in this case to the volcanic sulfur budget, "…estimates are highly uncertain because only very few of the potential sources have ever been measured and the variability between sources and between different stages of activity of the sources is considerable." Crucial parameters for climate studies include the quantity and height of entrainment of sulfur and other species into the atmosphere (large explosive eruptions, e.g., Pinatubo are uniquely capable of more or less instantaneous injections of sulfur into the upper atmosphere), and zonal and seasonal controls on atmospheric circulation (Section 3.04.6).

The more soluble volatile components of volcanic plumes (e.g., halogens) can be removed from eruption clouds rapidly by adsorption on to tephra, or deposition in hydrometeors. The estimates of atmospheric burdens of volcanic volatiles in eruption clouds, therefore, depend on the time after eruption that the measurements were made, and can, in general, be lower than the total volatile release. Surprisingly, little work has been undertaken to quantify the total volatile budgets of eruptions (see De Hoog *et al.*, 2001, for an exception).

This section focuses on subaerial emissions. Submarine fluxes of volatiles are also important in understanding global geochemical cycles but are rather poorly constrained. They will only be referred to briefly.

3.04.5.1 Sulfur

There is considerable interest in volcanic emissions of sulfur compounds because of the role of atmospheric sulfur chemistry in atmospheric radiation and climate, the hydrological cycle, acid precipitation, and air quality (see Chapter 8.14). Early theories on the climatic effects of eruptions considered that ash particles were responsible for raising the planetary albedo, but it is now clear that even fine tephra sediment rapidly from the atmosphere, and that the main protagonist in volcanic forcing of climate is the sulfate aerosol formed by oxidation of sulfur gases released to the upper atmosphere.

Most estimates of the volcanic source strength of sulfur to the atmosphere are based on compilations of COSPEC and related observations of lesser emissions from individual volcanoes (many exhibiting long-term degassing; Tables 5 and 6), and TOMS measurements of the larger, near instantaneous, and mostly explosive releases of SO_2 to the upper troposphere and stratosphere (Table 7). These data sources are patchy and go back to the 1970s only, and the statistical distribution of emissions is not well constrained, though some volcanoes appear responsible for substantial fractions of the total volcanic source strength. These include Mt. Etna, which is exceptional not just within Europe but globally as one of the most prodigious sources of volcanic gases to the troposphere (Tables 5 and 6).

Table 5 The top ten SO_2 emitters among "continuously erupting" volcanoes (Andres and Kasgnoc, 1998). Note that several of these volcanoes have very rarely had their SO_2 emissions measured (including those in Melanesia—four volcanoes in the top ten; and Láscar), and SO_2 fluxes from an individual volcano can and do vary on timescales of months or years (e.g., Mount Etna's SO_2 emission fell to around 0.3–0.4 Tg for the 12 months after July 2001).

Volcano	*SO_2 flux* ($Tg\ yr^{-1}$)
Etna, Italy	1.5
Bagana, PNG	1.2
Láscar, Chile	0.88
Ruíz, Colombia	0.69
Sakurajima, Japan	0.69
Manam. PNG	0.34
Yasur, Vanuatu	0.33
Kilauea East Rift, Hawaii	0.29
Masaya, Nicaragua	0.29
Stromboli, Italy	0.27

Table 6 Measured annual emissions from two continuously degassing volcanoes: Mt. Etna (Italy) and Merapi (Indonesia).

Species (units)	*Mt. Etna*	*Merapi*
H_2O ($Tg\ yr^{-1}$)	55	7.7
CO_2 ($Tg\ yr^{-1}$)	13	0.88
SO_2 ($Tg\ yr^{-1}$)	1.5	0.15
H_2S ($Gg\ yr^{-1}$)	100	73
HCl ($Gg\ yr^{-1}$)	110	33
HF ($Gg\ yr^{-1}$)	5.5	1.1
Pb ($Mg\ yr^{-1}$)	130	2.6
Hg ($Mg\ yr^{-1}$)	27	0.6
Cd ($Mg\ yr^{-1}$)	10	0.07
Ag ($Mg\ yr^{-1}$)	3.3	0.32
Au ($kg\ yr^{-1}$)	700	0.6

Courtesy of Patrick Allard.

Table 7 Top 10 recent stratospheric releases of SO_2, based on TOMS data. Typical errors on TOMS estimates are 30%. See Bluth *et al.* (1993, 1997) and Krueger *et al.* (1995) for details of methods.

Volcano	*Eruption date* (s)	*TOMS SO_2* (Tg)
Pinatubo	June 12–15, 1991	20.2[a]
El Chichón	March 28–April 4, 1982	8.1[a]
Sierra Negra	November 13, 1979	4.5
Cerro Hudson	August 8–15, 1991	4.0[a]
Nyamuragira	December 25, 1981	3.3
Mauna Loa	March 25, 1984	2.0
Galunggung	April 5–September 19, 1982	1.73[a]
Alaid	April 27, 1981	1.1
Nyamuragira	October 17, 1998	1.1
Wolf	August 28, 1982	1.1

Courtesy of Simon Carn (TOMS group, NASA).
[a] Cumulative totals of several eruptive episodes. The largest single release was that of Pinatubo on June 15, 1991 (20 Tg of SO_2).

Table 8 Estimated volcanic emissions of sulfur species to the atmosphere between the early 1970s and 1997.

Emission	*SO_2* (Mg d^{-1})	*SO_2* (Tg yr^{-1})	*SO_2* (Tg yr^{-1})
Sustained SO_2	26,200	9.6	8
Sporadic SO_2	200	0.07	0.04
TOMS SO_2	10,100	3.7	1.9
Total SO_2	36,500	13.4	6.7
Other S species[a]			3.7
Total volcanic S			10.4

Source: Andres and Kasgnoc (1998).
[a] Includes OCS, H_2S, H_2SO_4, etc., partitioned between sustained, sporadic and TOMS in the same proportions indicated in the first three rows.

Its average SO_2 emission rate (~4,000–5,000 Mg d^{-1}) is similar to the total industrial sulfur flux from France (Allard *et al.*, 1991) and must result in substantial elevations in tropospheric sulfate in southern Italy (Graf *et al.*, 1998). It has been suggested that these emissions have caused pollution events in mainland Italy, and even that they have been responsible for deterioration of Roman monuments (Camuffo and Enzi, 1995). Etna also pumps an astonishing estimated 700 kg of gold into the atmosphere every year (Table 6).

There is a reasonable consensus regarding the magnitude of annual volcanic source strengths of sulfur, though difficulties arise in time-averaging the sporadic but large magnitude releases to the stratosphere from explosive eruptions, and in extrapolating field data for a comparatively small number of observed tropospheric volcanic plumes to the global volcano population. The most widely used global data set is that compiled for the global emissions inventory activity (GEIA) by Andres and Kasgnoc (1998). This arrives at a global annual flux of sulfur from all sources that exceeds 10.4 Tg (Table 8). More recently, Halmer *et al.* (2002) have estimated the global volcanic SO_2 emission to the atmosphere as 15–21 Tg yr^{-1} for the period 1972–2000. Their figures for all sulfur species add considerable uncertainty to the total volcanic sulfur flux (9–46 Tg of sulfur) mainly because of a very large uncertainty in the H_2S emission (1.4–35 Tg of sulfur). For comparison, the IPCC (2001) estimates of annual emissions of other sources of sulfur include anthropogenic (76 Tg), biomass burning (2.2 Tg), and dimethyl sulfide (DMS, 25 Tg, mainly from the oceans). Most of the volcanic source strength is from the continuous degassing of many volcanoes worldwide (Table 8; Sections 3.04.4.1.1–3.04.4.1.2). Using the TOMS data set, Pyle *et al.* (1996) estimated the medium-term (~10^2 yr) annual flux of volcanic sulfur to the stratosphere to be ~1 Tg (range of 0.3–3 Tg). This does not take into account the larger releases of sulfur that are indicated by historic and prehistoric eruptions such as that of Toba 74 kyr BP, and Tambora 1815 AD (Table 9).

The sulfur gases released by volcanoes are either deposited at the Earth's surface or are oxidized to form sulfate aerosol. An important point that has emerged from recent work is the disproportionate contribution of volcanic sulfur emissions to the global atmospheric sulfate budget compared with other sources of sulfur including anthropogenic and oceanic emissions. Episodic, large magnitude eruptions are the principal perturbation to stratospheric aerosol levels (e.g., the 30 Mt of sulfate injected by the 1991 Pinatubo eruption). In the troposphere, the picture is less clear, but modeling suggests that up to 40% of the global tropospheric sulfate burden may be volcanogenic (Graf *et al.*, 1997), though Stevenson *et al.* (2003a) and Chin and Jacob (1996) obtained lower figures (14% and 18%, respectively), in part due to use of a lower volcanic sulfur source strength. In any case, these figures all exceed the fraction of the sulfur source to the atmosphere that is volcanogenic (around 10%) because of the generally higher altitudes of entrainment of volcanic sulfur compared with biogenic (DMS) or anthropogenic sources, and hence the longer residence time of volcanic SO_2 compared with other sources. This is largely because of lower deposition rates, and results in more conversion of SO_2 to sulfate, and a longer residence time of the higher altitude aerosol. Sulfate aerosol plays a significant role in the Earth's radiation budget, because it may both backscatter incoming shortwave solar radiation and absorb outgoing longwave radiation, the competition of these processes

Table 9 Estimates of sulfur yield from selected major historic and prehistoric eruptions (magnitude $>10^{13}$ kg).

Eruption year and volcano	*Magnitude* (kg)[a]	*Sulfur yield* (Tg of S)[b]	*Northern hemisphere summer cooling* (K)[c]
≈74 kyr BP	7×10^{15}	35–3,300	>1
≈AD 181 Taupo	7.7×10^{13}	?6.5	0.4
≈AD 1028 Baitoushan	5.8×10^{13}	>2	0.5
≈1257 Unknown	10^{14}–10^{15}?	>100	?
≈1452 Kuwae	$>8\times10^{13}$	?40	0.5
1600 Huaynaputina	2.1×10^{13}	23–55[d]	0.8
1815 Tambora	1.4×10^{14}	28	0.5
1883 Krakatau	3.0×10^{13}	15	0.3
1902 Santa Maria	2.2×10^{13}	11[e]	Not detected
1912 Katmai	2.5×10^{13}	10	0.4
1991 Pinatubo	1.3–1.8×10^{13}	10.1	<0.5

[a] Total eruption magnitude for multiple phases of eruption and combining plinian and phoenix cloud ashfall and associated pyroclastic flow deposits where applicable, data mainly from Carey and Sigurdsson (1989), Chesner and Rose (1991), Monzier *et al.* (1994), Holasek *et al.* (1996), Pyle (2000), Horn and Schmincke (2000), Adams *et al.* (2001), and Oppenheimer (2003b). [b] Stratospheric sulfur yield from Table 6, and from Zielinski (1995), de Silva and Zielinski (1998). [c] Estimated northern hemisphere summertime temperature anomaly derived from tree-ring chronologies reported by Briffa *et al.* (1998) for eruptions before Mt. St. Helens (note that other records do indicate a≈0.2 K northern hemisphere summer cooling in 1903). [d] Costa *et al.* (2003). [e] Since this estimate is based on ice core sulfate deposition, it may reflect the cumulative aerosol fallout of other notable 1902 eruptions, i.e., Mont Pelée (Martinique) and Soufrière (St. Vincent) as well as the Santa Maria event.

depending strongly on particle size. In addition, sulfate aerosol can have a secondary, and possibly more profound, radiative effect by promoting cloud condensation or modification of the microphysical properties and longevity of existing clouds (Graf *et al.*, 1997). Changes in this "background" emission in time and space could represent an important forcing that has yet to be characterized.

3.04.5.2 Carbon and Water

As mentioned in Section 3.04.4.3.2, SO_2 is the most readily measured volcanic volatile in the atmosphere. Thus, most flux estimates of other components have been based on measuring their ratios to SO_2 (often obtained by *in situ* sampling methods), and multiplying them by the SO_2 flux measured using COSPEC. Other approaches include scaling estimates of ^{3}He flux against measured C/^{3}He ratios, and extrapolation of the few available direct observations of CO_2 flux. Arthur (2000) provides a review of current estimates of global subaerial volcanic CO_2 flux, which range from 15 Tg yr^{-1} to 130 Tg yr^{-1}. This is swamped by the anthropogenic source of carbon to the atmosphere and is not considered globally significant on short timescales, though CO_2 emissions are important as a local hazard (Section 3.04.6.5). At Kilauea, a "hot spot" volcano, the exhaled CO_2 is derived directly from the mantle (Gerlach and Taylor, 1990), but carbon-isotope studies show that at arc volcanoes most of the CO_2 is recycled from subducted organic and carbonate sediments (Section 3.04.2.1; Fischer *et al.*, 1998). Estimates of the CO_2 outgassing to the oceans by mid-ocean ridge basalt (MORB) volcanism are mostly in the range 90–350 Tg yr^{-1}.

While a few estimates of water budgets have been attempted for individual volcanoes and eruptions (e.g., 3 Tg yr^{-1} for White Island, New Zealand (Rose *et al.*, 1986); 13 Tg yr^{-1} for Masaya, Nicragua (Burton *et al.*, 2000)), meaningful estimates of the global volcanic flux of water to the atmosphere are unavailable as of early 2000s. Interpretation of water emission rates from volcanoes is complicated by the likelihood, in many cases, that a substantial fraction of the emitted water has been derived from groundwaters (e.g., Taran *et al.*, 1995).

3.04.5.3 Halogens

In contrast to carbon, volcanic emissions of halogens are thought to be substantial compared with other sources. However, the fluxes remain very poorly constrained. They are important as they play a critical role in atmospheric chemistry—in boundary layer ozone depletion, tropospheric hydrocarbon oxidation, the oxidizing capacity of the troposphere, and stratospheric ozone depletion (Section 3.04.6.1.2). Realistic estimates of the anthropogenic and natural emissions of halogen species are essential in order to model and assess such processes accurately. Very little work exists to improve upon the emission inventory for volcanic sources elaborated by Cadle (1975, 1980). The more recent work indicates greater than one to two orders of magnitude of uncertainty in halogen fluxes (Table 10).

Measurements of gas samples collected at Augustine Volcano, Alaska, indicate typical orders of magnitude concentrations of different halogen species in volcanic emissions (Table 11). With peak SO_2 emission rates of nearly 280 kg s^{-1} during minor eruptive episodes at

Table 10 Estimated annual mean global emissions of HCl, HF, and HBr from volcanoes.

Volcanic source	*HCl* (Tg)	*HF* (Tg)	*HBr* (Gg)
Cadle (1980)	7.8	0.4	78
Symonds *et al.* (1988)	0.4–11	0.06–6	
Halmer *et al.* (2002)[a]	1.2–170	0.7–8.6	2.6–43.2

[a] For period 1972–2000.

Table 11 Volcanic halogens at Mt. Augustine, all figures as mole fractions.

HCl	6.0×10^{-2}
HF	5.4×10^{-4}
SiF_4	8.0×10^{-5}
NaCl	1.4×10^{-5}
KCl	8.7×10^{-6}
HBr	6.8×10^{-6}
$SiOF_2$	2.1×10^{-6}
$FeCl_2$	1.6×10^{-6}

Source: Andres and Rose (1995).

Augustine, even the HBr emission corresponds to ~0.5 kg s^{-1}. These emissions are significant because they are, at many volcanoes, released directly into the free troposphere, where they can initiate and catalyze a large range of processes. Again, Mt. Etna is a prodigious source of halogens, continuously emitting ~2 kg s^{-1} of HF, and >8 kg s^{-1} of HCl (Francis *et al.*, 1998).

Bureau *et al.* (2000) considered the input of bromine to the atmosphere from explosive eruptions, based on experimental data for synthetic (albite) melts (Table 12). Their results suggested even stronger partitioning of both bromine and iodine into the fluid phase compared with chlorine. Assuming a Cl/Br mass ratio of ~300 based on measurements of a range of volcanic rocks, and scaling against estimated chlorine emissions for various explosive eruptions, they suggested that volcanoes could be a significant source for stratospheric bromine. They pointed out, however, that the behavior of bromine in eruption columns is not known at present, so it is not clear what fraction of bromine degassed would actually cross the tropopause and remain in the stratosphere.

3.04.5.4 Trace Metals

Volcanoes are an important source of trace metals to the atmosphere—for some species (e.g., arsenic, cadmium, copper, lead, and selenium) they may be the principal natural source (e.g., Nriagu, 1998). Measurements of fluxes are complicated because of rapid condensation of the vapor phases carrying the

Table 12 Estimated chlorine and bromine yields of eruptions.

Eruption	*Cl yield* (Tg)	*Br yield* (Gg)
Toba 74 kyr BP	400–4,000	1,460–33,700
Tambora 1815	216	790–1,820
Krakatau 1883	3.75	14–31
Mt. St. Helens 1980	0.67	2.4–5.6
El Chichón 1982	0.04	0.15–0.4
Mt. Pinatubo 1991	>3	>11–25

Source: Bureau *et al.* (2000).

Table 13 Comparison of volcanic, total natural and anthropogenic fluxes to the atmosphere of selected species.

Element	*Volcanic* (Gg yr^{-1})	*Total natural* (Gg yr^{-1})	*Anthropogenic* (Gg yr^{-1})
Al	13,280[a]	48,900[b]	7,200[b]
Co	0.96[c]	6.1[c]	4.4[b]
Cu	1.0[d], 4.7[c], 15[e], 22[f]	28[c]	35[c]
Zn	4.8[c], 7.2[d], 8.5[f]	45[c]	132[c]
Pb	0.9[d], 1.7[c], 2.5[e], 4.1[f]	12[c]	332[c]
As	1.9[c]	12[c]	19[c]
Se	0.3[f], 0.5[c]	9.3[c]	6.3[c]
Mo	0.2[c]	3.0[c]	3.3[c]
Cd	0.4[c]	1.3[c]	7.6[c]

Source: Mather *et al.* (2003a).
[a] Symonds *et al.* (1988). [b] Lantzy and Mackenzie (1979).
[c] Nriagu (1989). [d] Hinkley *et al.* (1999). [e] Lambert *et al.* 1988.
[f] Le Cloarec and Marty (1991).

trace metals, but estimates have been attempted, based on scaling of metal/sulfur ratios by known sulfur fluxes or by using the flux of a radioactive volatile metal (^{210}Po) as the normalizing factor (e.g., Lambert *et al.*, 1988). These methods have provided metal flux estimates for individual volcanoes (e.g., Erebus (Zreda-Gostynska *et al.*, 1997), Stromboli (Allard *et al.*, 2000), Etna (Gauthier and Le Cloarec, 1998), Vulcano (Cheynet *et al.*, 2000)), and estimates of global source strengths (e.g., Nriagu, 1989; Hinkley *et al.*, 1999; Table 13).

3.04.6 IMPACTS

The atmospheric, climatic, environmental, and health effects of volcanic volatile emissions depend on several factors but importantly on fluxes of sulfur and halogens. As discussed in Section 3.04.5.1, intermittent explosive eruptions can pump >10^{10} kg of sulfur into the stratosphere, against a background of continuous fumarolic and "open-vent" emission into the troposphere. The episodic, large explosive eruptions are the principal perturbation to stratospheric aerosol levels (e.g., 30 Mt of sulfate due to the 1991 eruption of

Pinatubo), and can result in global climate forcing and stratospheric ozone depletion (Sections 3.04.6.1 and 3.04.6.2). Large lava eruptions, such as that of Laki in Iceland in 1783–1784, which released ≈120 Tg of SO_2, 7.0 Tg of HCl, and 15 Tg of HF into the upper troposphere–lower stratosphere region, have resulted in major pollution episodes responsible for regional-scale extreme weather, damage to farming and agriculture, and elevated human morbidity and mortality (Thordarson *et al.*, 1996; Thordarson and Self, 2002). Individual continuously degassing volcanoes such as Mt. Etna and Masaya volcano (Nicaragua) can also represent major polluters. The adverse environmental and health impacts observed downwind of many degassing volcanoes are widely recognized if poorly investigated (e.g., Baxter *et al.*, 1982; Delmelle *et al.*, 2001, 2002; Delmelle, 2003). Tropospheric volcanic emissions and their impacts are discussed in Sections 3.04.6.3–3.04.6.5.

3.04.6.1 Stratospheric Chemistry and Radiative Impacts of Volcanic Plumes

On April 2, 1991, steam explosions were observed on a little known volcano called Mount Pinatubo on the island of Luzon, the Philippines. Within 11 weeks, on the afternoon of June 15, and following a crescendo in activity, the volcano erupted $(1.3-1.8)\times10^{13}$ kg (8.4–10.4 km^3 bulk volume) of pumice. It devastated a 400 km^2 area and mantled much of Southeast Asia with ash. This was the second largest magnitude eruption of the twentieth century after that of Katmai in Alaska in 1912. The tropopause was at ~17 km altitude at the time of the eruption and was punctured both by the central eruption column fed directly by the vent and by co-ignimbrite or "phoenix" clouds that lofted above immense pyroclastic currents moving down the west and south flanks of the volcano. The eruption intensity peaked between around 13:40 h and 16:40 h local time on June 15, based on infrasonic records from Japan, with vent exit velocities of ~280 m s^{-1}, and discharge rates estimated at 1.6×10^9 kg s^{-1}. A wide variety of observations, especially from satellite instruments, have quantified the eruption's emissions and their impacts on the atmosphere and climate. The eruption serves as a benchmark in our understanding of the impacts of major eruptions on the Earth system, and is therefore discussed in detail below.

3.04.6.1.1 Formation of stratospheric sulfate aerosol veil

The stratospheric umbrella cloud formed during the Pinatubo eruption attained a vertical thickness of 10–15 km, extending from the tropopause up to ~35 km above sea level. Weather satellites tracked the cloud for two days, after which other spaceborne instruments, including TOMS, were able to continue monitoring dispersal of the plume. It took 22 days for the cloud to circumnavigate the globe. Its estimated initial SO_2 yield, ~17–20 Tg, remains the largest measured (Bluth *et al.*, 1992; Read *et al.*, 1993). The amount of SO_2 decreased daily through oxidation by ·OH radicals (Coffey, 1996):

$$SO_2 + \cdot OH \rightarrow HOSO_2 \qquad (8)$$

Various chemical pathways then promoted the formation of sulfuric acid (H_2SO_4):

$$HOSO_2 + O_2 \rightarrow SO_3 + HO_2 \qquad (9)$$

$$SO_3 + H_2O \rightarrow H_2SO_4 \qquad (10)$$

This scheme conserves HO_x and results in a predicted e-folding lifetime of SO_2 in the stratosphere of 38 days (i.e., the time taken for the abundance of SO_2 to drop by 1/e of its starting amount), only slightly longer than the observed time of 33–35 days (Figure 8; Read *et al.*, 1993). The aerosol consisted of around 25% water and 75% sulfuric acid by weight.

The aerosol cloud was tracked and measured by several instruments, including the Stratospheric Aerosol and Gas Experiment (SAGE) satellite (McCormick *et al.*, 1995). Independent estimates of the total mass of aerosol generated (~30 Tg) are slightly higher than the amount expected from the initial SO_2 load (Baran *et al.*, 1993; Baran and Foot, 1994). The cloud was so thick at its peak that no sunlight was transmitted and there are missing data from the record. Gradually, the aerosol sedimented back to the surface, and, by the end of 1993, only ~5 Tg of aerosol remained airborne. The observed e-folding time was ~12 months, comparable to that measured after the previous major stratospheric aerosol perturbation due to the 1982 eruption of El Chichón in Mexico.

3.04.6.1.2 Impacts on ozone chemistry

Following the Pinatubo eruption, global stratospheric ozone levels began to show a strong downturn (McCormick *et al.*, 1995). Ozone levels decreased 6–8% in the tropics in the first months after the eruption. These figures for the total vertical column of ozone hide local depletions of up to 20% at altitudes of 24–25 km. By mid-1992, total column ozone was lower than at any time in the preceding 12 yr, reaching a low point in April 1993 when the global deficit was ~6% compared with the average. Losses were greatest in the northern hemisphere. For example, total ozone above the USA dropped 10% below average with

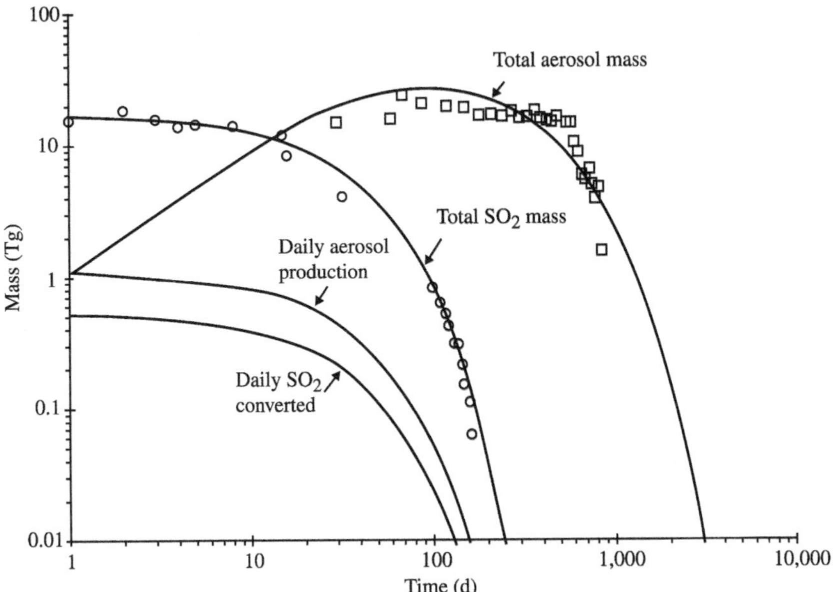

Figure 8 Oxidation of SO_2 to SO_4 in the Pinatubo stratospheric cloud. The "total SO_2 mass" curve indicates the loss of SO_2 by oxidation to sulfate aerosol according to an initial loading of 17 Tg of SO_2 (after Read *et al.* (1993), and 3 Tg lower than the TOMS-only estimate of Bluth *et al.* (1992)) and a 33 d e-folding time. The circles indicate satellite measurements of stratospheric SO_2 burden from Read *et al.* (1993). The "total aerosol mass" curve is obtained by modeling the aerosol mass generated by sulfate oxidation and an e-folding time for aerosol loss of 1 yr. The squares show estimates of the stratospheric aerosol mass from Baran *et al.* (1993). Daily rates of SO_2 conversion and aerosol production are shown by the two other curves (labeled).

the strongest depletion observed between 13 km and 33 km altitude. These decreases have been attributed to complex chemical reactions exploiting the presence of the aerosol surface. These result in a shift of stratospheric chlorine from stable compounds (HCl and $ClONO_2$) into more reactive ones (e.g., HOCl) that can destroy ozone (Coffey, 1996). First, the aerosol surface promotes reactions that remove gaseous NO_x compounds. For example, N_2O_5 is removed as follows:

$$N_2O_5 + H_2O \rightarrow 2HNO_3 \qquad (11)$$

$$N_2O_5 + HCl \rightarrow ClONO + HNO_3 \qquad (12)$$

In the months following the eruption, increases in HNO_3 were detected (e.g., Koike *et al.*, 1994) providing compelling evidence for the transfer of nitrogen from comparatively reactive (NO_x) to more inert species (HNO_3). Depletion of N_2O_5, in turn, reduces NO_2 levels. Airborne measurements obtained above the USA and Mexico after the eruptions show strong depletions in NO_2 levels (Coffey and Mankin, 1993). Remarkably, by January 1992, seven months after the eruption, NO_2 concentrations had halved. This is important because NO_2 ordinarily captures reactive chlorine monoxide, forming the inert species $ClONO_2$:

$$ClO + NO_2 \rightarrow ClONO_2 \qquad (13)$$

In other words, if stratospheric NO_x levels are lowered, chlorine monoxide levels increase, increasing rates of ozone destruction.

Although enhanced stratospheric HCl levels were reported after the El Chichón eruption (with up to 40% increases over background levels observed within the cloud), it is generally considered that most volcanic HCl is "scrubbed" out of eruption columns as they ascend through the troposphere, due to its high solubility. Pinatubo emitted an estimated 3 Tg of chlorine, but measurements indicate that it was rapidly removed, perhaps assisted by the estimated 250–300 Tg of steam erupted, and the comparable amount of tropospheric moisture entrained into the eruption column. From analyses of the sulfate marker in the Greenland ice core for the 1815 Tambora eruption, Delmas *et al.* (1992) found no evidence for changes in atmospheric chlorine associated with this much larger event, suggesting that the massive amount of chlorine thought to have been released by the eruption (Table 12) was rapidly and efficiently scavenged by the troposphere as the eruption clouds ascended.

Although simple eruption-column physical and chemical models have indicated that halogens are readily scavenged in eruption clouds by hydrometeors (Tabazadeh and Turco, 1993), more recent, and more sophisticated, models have suggested that limited scavenging by hydrometeors in a dry troposphere could permit substantial quantities of HCl and HBr to reach the stratosphere (Textor *et al.*, 2003). The observation of chlorine-rich fallout layers in the GISP2 Greenland ice core linked to another great eruption, that of Mt. Mazama (USA) in ≈5677 BC (Zdanowicz *et al.*, 1999), lends weight to the idea that there are some atmospheric environmental conditions that permit transport of substantial quantities of halogens across the tropopause.

3.04.6.1.3 Optical and radiative effects

The widespread dispersal of Pinatubo aerosol led to many spectacular optical effects in the atmosphere, including vividly colored sunsets and sunrises, crepuscular rays and a hazy, whitish appearance to the Sun. These phenomena occurred as the aerosol veil absorbed and scattered sunlight. An objective measure of the fraction of solar energy removed as it travels down through an aerosol layer is the optical depth, τ, defined by the relationship between the initial irradiance, I_0, at the top of the layer, and the radiation received at the bottom of the layer, I:

$$I = I_0 \exp(-\tau/\cos\theta) \qquad (14)$$

where θ is the solar zenith angle (measured between the vertical and the Sun). τ depends on the concentration and thickness (path length) of the aerosol as well as its absorbing or scattering properties. In the months following the Pinatubo eruption, optical depths in the stratosphere were the highest ever recorded by modern techniques, peaking at ~0.5 in the visible region in August–September 1991.

The appearance of these optical effects of Pinatubo's aerosol veil necessarily implies that it was interrupting the transmission of visible solar radiation to the Earth's surface. In detail, the radiative effects of stratospheric aerosol veils are highly complex, because they can consist of variable proportions of minute glassy ash fragments and sulfuric acid droplets of different compositions, sizes, and shapes (and hence optical properties). The various components also accumulate and sediment out at different rates according to their masses and aerodynamic properties, so that any effects on the Earth's radiation budget can be expected to change through time. As a result, the effective radius of the aerosol veil should decrease in the months after an eruption. Measurements following the El Chichón and Pinatubo eruptions agree with this picture (McCormick *et al.*, 1995). The particles scatter incoming solar ultraviolet and visible radiation, directing some back into space but also sideways and forwards. The aerosol can also absorb radiation—short wavelength from the Sun or long wavelength from the Earth—and warm up. The net effect on the Earth's radiation budget is not straightforward to determine.

Observations by the Earth Radiation Budget Experiment (ERBE) satellite sensor revealed significant increases in the reflectivity of the Earth (Minnis *et al.*, 1993). By August 1991, the backscattering of solar radiation by the aerosol had increased the global albedo to ~0.25, some 5 SDs above the 5 yr mean of 0.236. Corresponding reductions of the direct solar beam were ~25–30%. ERBE was also able to show that the albedo increase was not uniform across the Earth but was most pronounced in normally low-albedo, cloud-free regions, including the Australian deserts and the Sahara, and in typically high-albedo regions associated with convective cloud systems in the tropics such as the Congo Basin and around New Guinea. The latter observation is initially puzzling, because over regions that have a naturally high albedo, the percentage increase due to volcanic aerosol in the stratosphere is small. It appears that transport of Pinatubo aerosol across the tropopause seeded clouds, or modified the optical properties of existing upper tropospheric cirrus clouds. Satellite observations support this interpretation since they indicate a correlation between Pinatubo aerosol and increased cirrus clouds, persisting for more than 3 yr after the eruptions. The enhancement was especially noticeable in mid-latitudes ~6 months after Pinatubo's eruption, consistent with the likely time lag before accumulation and initial sedimentation into the troposphere of the sulfate aerosol. The mid-latitudes are dominant sites for transfer between stratosphere and troposphere across tropospheric folds associated with the jet stream and cyclonic systems.

The albedo changes corresponded to radiative forcing, as observed by ERBE (Minnis *et al.*, 1993). In July 1991, the short-wave flux increased dramatically over the tropics. The corresponding net flux decreased by 8 W m^{-2} in August 1991, twice the magnitude of any other monthly anomaly. A similar but weaker trend was observed between 40° S and 40° N, with the net forcing for August 1991 amounting to −4.3 W m^{-2}, nearly 3 SDs from the 5 yr mean. Unfortunately, ERBE was not operating pole-ward of 40° latitude but even if there were no aerosol forcing at higher latitudes, then the globally averaged volcanic forcing still amounts to −2.7 W m^{-2}. This should represent the minimum global forcing, because enhanced stratospheric aerosols were observed at higher latitudes by mid-August 1991. The net radiation flux anomalies seen by ERBE remain the largest that have been observed by satellites. These results represented the first unambiguous, direct observations of eruption-induced radiative forcing (Stenchikov *et al.*, 1998).

Stratospheric aerosols can also have a "greenhouse" effect on the Earth if they are effective in absorbing upwelling long-wave radiation. For this warming effect to outbalance the cooling due to increased albedo, the aerosol's effective radius, r_{eff} (its mean radius weighted by surface area), should exceed 2 μm. Prior to Pinatubo, the r_{eff} of stratospheric aerosol was ~0.2 μm. Pinatubo's aerosol pushed the stratospheric r_{eff} up to more than 0.5 μm compatible with the observed negative forcing.

3.04.6.2 Climatic Impacts of Major Volcanic Eruptions

The impacts of volcanism on climate are complex in both space and time (Robock, 2000). They can also depend on complex feedback mechanisms; for example, the lower global mean surface temperatures following the 1991 Pinatubo eruption resulted in less evaporation, and so less atmospheric heating due to water vapor (Soden *et al.*, 2002). Best understood now are the short-term climatic perturbations following high sulfur-yield explosive eruptions like Pinatubo. Scaling up to the largest known explosive eruptions (e.g., the Younger Toba Tuff, 74 kyr BP) poses considerable difficulties, not least because of uncertainties in the microphysics and chemistry of very large releases of sulfur to the atmosphere. No large mafic eruption has occurred in modern times, but there is ample documentary evidence to implicate the Laki 1783–1784 eruption in strong climatic effects, at least at regional scale. On timescales up to 10^6 yr, large igneous provinces (see Chapter 3.17) may be responsible for climate change, though even less is understood about the volatile yields or dynamics of such eruptions. At timescales $>10^6-10^7$ yr, variations in the rate of oceanic crust growth are associated with major global climate change via their control on eustatic sea level and atmospheric concentrations of CO_2, and hence the carbonate–silicate cycle (Section 3.04.1.1). This section focuses on the volcanic forcing of climate on the shorter timescales ($1-10^3$ yr).

3.04.6.2.1 Intermediate to silicic eruptions

Pinatubo's negative radiative forcing exceeded for two years the positive forcings due to anthropogenic greenhouse gases (carbon dioxide, methane, CFCs and N_2O; McCormick *et al.* (1995)). The net drop in global mean tropospheric temperature in 1992 was ~0.2 K compared with the baseline from 1958 to 1991. This may not sound much, but it is a globally averaged figure that hides much larger regional and temporal variations, with pockets of abnormally strong surface heating as well as cooling. For example, the Siberian winter was 5 K warmer, and the North Atlantic was 5 K cooler than average. Furthermore, the actual global troposphere temperature decrease due to Pinatubo is closer to 0.4 K if a correction is made for the tropospheric warming associated with the 1992 El Niño-Southern Oscillation event.

The satellite-borne Microwave Sounding Unit (MSU) records lower stratospheric temperatures. Global mean increases of up to 1.4 K were apparent following both the Pinatubo and El Chichón eruptions due to the local heating of the volcanic aerosol (Parker *et al.*, 1996). Interestingly, for the Pinatubo case, the temperature anomaly decreased as aerosol sedimented back into the troposphere, and, by early 1993, below average, lower stratospheric temperatures were observed. This could be due to cooling coincident with the destruction of stratospheric ozone (Section 3.04.6.2.2).

Kelly *et al.* (1996) examined surface temperature anomalies following the Pinatubo eruption from worldwide observatory records of surface air temperature and nighttime marine air temperature obtained from ships and buoys. The observed global cooling was initially rapid but punctuated by a warming trend, predominantly over land, between January and March 1992. Cooling resumed, and by 1 yr after the eruption amounted to ~0.5 K. In 1992, the USA experienced its third coldest and wettest summers in 77 years. The Mississippi flooded its banks spectacularly while drought desiccated the Sahel. There were further relative warmings in early 1993 and 1995, and the mid-year cooling reduced each year. Globally averaged sea-surface temperatures indicated a slightly lower maximum cooling ~0.4 K, which can be accounted for by the high thermal inertia of the oceans. These trends are more or less mirrored by mean lower tropospheric temperatures determined by the MSU and have been fitted quite successfully by climate models (e.g., Hansen *et al.*, 1997). Indeed, the Pinatubo case has provided an important test of the capabilities of climate models for the simulation of atmospheric radiation and dynamics.

The winter warmth after the Pinatubo eruption was concentrated over Scandinavia and Siberia and central North America. These temperature anomalies were associated with marked departures in sea-level pressure patterns in the first northern winter. There was a pole-ward shift and strengthening of North Atlantic westerlies at ~60° N, associated with corresponding shifts in the positions and strengths of the Iceland Low and Azores High. These effects have been modeled as a result of changes to the atmospheric circulation around the Arctic (the Arctic Oscillation; Thompson and Wallace (1998)) arising from the differential heating effects of the volcanic aerosol in the upper atmosphere (Graf *et al.*, 1993; Kirchner *et al.*, 1999; Stenchikov *et al.*, 2002). Strong stratospheric heating in the tropics due to the presence of volcanic aerosol establishes a steeper meridional thermal gradient at the tropopause, enhancing geostrophic winds, and the polar vortex. At the same time, surface cooling in the subtropics leads to weaker planetary waves, and surface warming of mid-high latitudes, reducing tropospheric temperature gradients. This also helps to strengthen the polar vortex, and the combined effects amplify the Arctic Oscillation, leading to winter warming. The stronger northern hemisphere response, compared with the southern hemisphere, reflects the greater landmass area.

One apparent impact of the volcanic forcing induced by the Pinatubo eruption was a reduction in the melt area of the Greenland Ice Sheet observed by passive microwave satellite imagery (Abdalati and Steffen, 1997). Melting and freezing are of great significance in ice sheet thermal dynamics, because they represent strong positive feedback mechanisms. When snow melts its albedo drops from about 0.9 to 0.7. Thus, it absorbs even more incident radiation, warms up more, and melts more. The reverse situation occurs as wet snow freezes. The combination of Greenland's high albedo and size (its ice sheet covers $1.75 \times 10^6\,km^2$) accounts for a strong cooling contribution to Arctic climate. Superimposed on a (poorly understood) 3 yr cycle in melt extent was a longer-term increasing trend in melt area of over 4% per year up to 1991. However, then, in 1992, there was a drop of almost $1.5 \times 10^5\,km^2$ in the melt extent. This pattern is apparent also in coastal temperature records for Greenland, which indicate around a 2 K cooling between 1991 and 1992. The calculated decrease of melting extent in 1992 amounts to a decrease in absorption of solar radiation by the ice sheet of $\sim 10^{13}$ W. If Pinatubo were responsible for this cooling of Greenland, then it has important implications for the amplification of volcanic aerosol forcing by ice-albedo feedbacks, and the potential enhancement of regional and global cooling following large eruptions. Meanwhile, cooling of the Red Sea is thought to have enhanced mixing in the water column, bringing nutrients to the surface and stimulating algal blooms (Genin *et al.*, 1995). These, in turn, resulted in coral mortality.

The largest known historic eruption, that of Tambora in 1815, yielded an estimated 60 Tg of sulfur to the stratosphere (based mostly on ice core sulfate deposition records; reviewed by Oppenheimer (2003a)). This is ~6 times the release of Pinatubo. Considerable efforts have been expended to quantify the climatic effects of this eruption, and have spanned investigations of surface temperature measurements and proxy climate indicators such as tree rings (Harington, 1992). Anomalously cold weather hit the northeastern USA, maritime provinces of Canada, and Europe in the boreal summer of 1816, which came to be known as the "year without a summer" in these regions. Widespread crop failures occurred in Europe and North America in 1816, and the eruption has been implicated in accelerated emigration from New England, widespread outbreaks of epidemic typhus, and what Post (1977) termed "the last great subsistence crisis in the western world."

Briffa and Jones (1992) reconstructed 1816 weather across Europe from contemporary meteorological observations and tree-ring data. They showed that the summer temperatures across much of Europe were 1–2 K cooler than the average for the period 1810–1819, and up to 3 K cooler than the mean for 1951–1970. Rainfall was also high across much of Europe in the summer of 1816. More recent dendrochronological studies have confirmed the distribution of these cool summers on both sides of the North Atlantic (Briffa *et al.*, 1998). These reconstructions of northern hemisphere summer temperatures indicate 1816 as one of the very coldest of the past six centuries, second only to 1601 (the year after the eruption of Huaynaputina in Peru). The estimated mean northern hemisphere (land and marine) surface temperature anomalies in the summers of 1816, 1817, and 1818 are −0.51 K, −0.44 K, and −0.29 K, respectively. Despite Tambora's substantially greater sulfur yield to the atmosphere, it does not appear to have had a correspondingly larger effect on global surface temperatures, highlighting the nonlinear scaling between volatile yield and climatic impact.

Assessing the impacts of the largest known Quaternary "super-eruption," that of the Younger Toba Tuff (YTT), Sumatra, at 74 kyr BP, is highly challenging given uncertainties in the key parameters for the eruption (intensity, height, and magnitude), and amounts of gaseous sulfur species released (Oppenheimer, 2002). At the high end of impact claims, Rampino and Self (1992) have argued that Toba caused a "volcanic winter"—a global mean surface temperature drop of 3–5 K (Table 12). However, there are considerable challenges in modeling the climatic impacts of such large eruptions, in particular arising from uncertainties in cloud microphysics. Pinto *et al.* (1989) argued that the chemical and physical processes occurring in very large volcanic clouds may act in a "self-limiting" way. They applied simplified aerosol photochemical and microphysical models to show that, for a 100 Tg injection of SO_2, condensation and coagulation produce larger aerosol particles, which are less effective at scattering incoming sunlight, and also sediment more rapidly. Combined, these effects lessen the expected magnitude and duration of climate forcing expected from super-eruptions, certainly compared with linear extrapolations of observed climate response following eruptions like Pinatubo 1991.

Zielinski *et al.* (1996b) reported finding the Toba sulfate marker in the GISP2 core. Their high temporal resolution measurements showed that while the Toba anomaly coincides with a 1 kyr cool period between interstadials 19 and 20, it is separated by the 2 kyr long, and dramatic warming event of interstadial 19, from the prolonged (9 kyr) major glacial which began ~67.5 kyr BP. This undermines earlier suggestions that the YTT eruption played a role in initiating the last glaciation, though leaves open the possibility that it is implicated in the ~1 kyr cold period prior to interstadial 19.

Efforts to model the climatic impacts of super-eruptions were reported at a Chapman Conference on Santorini in 2002. Jones and Stott (2002) showed that simulations of a 2,000 Tg injection of SO_2 in the tropical stratosphere (100 × Pinatubo) resulted in a 5–7 yr forcing with a peak of -60 W m^{-2} (global annual mean at the tropopause), corresponding to a peak global mean monthly temperature anomaly near the surface of -10 K, remaining about -2 K for 5–7 yr, and remaining around -0.5 K for up to 50 yr due to the thermal inertia of the oceans. Fifteen months after the eruption, the maximum cooling in parts of Africa and North America exceeded 20 K in the model, resulting in a large increase in land and sea ice, and reduced evaporation and precipitation. Interestingly, the snow and ice cover were neither extensive enough nor persistent enough in the model to cause an ice-albedo effect and an ice age transition.

3.04.6.2.2 *Mafic eruptions*

At first consideration, it might seem that mafic eruptions, typically more effusive in character, would not have much impact on hemispheric- to global-scale climate because of their lower-altitude sulfur emission. There are some compensating factors, however. First, mafic magma tends to contain high sulfur contents, and high-intensity lava eruptions can propel fire fountains to heights in excess of 1 km altitude, and can persist for days, weeks, months, and even years, compared with the several hour bursts of an eruption like Pinatubo's. Also, there are some rare instances of basaltic plinian eruptions, e.g., Masaya, >20 and ≈6.5 kyr BP (Williams, 1983), Mt. Etna, 122 BC (Coltelli *et al.*, 1998), and Tarawera, 1886 AD (Walker *et al.*, 1984), that may have combined high eruption intensities and plume heights with high sulfur yields.

Interest in the potential global impacts of effusive volcanism has been fuelled by the coincidence of some of the greatest mass extinctions that punctuate the fossil record with the massive outpourings of lava during flood basalt episodes (Rampino and Stothers, 1988). Estimated sulfur yields of such provinces are certainly very high (e.g., Thordarson and Self, 1996), and Ar–Ar dates have now demonstrated that the immense volumes of lava (of order 10^6 km^3) are erupted in comparatively short time periods (~1 Myr, e.g., the 30 Myr old Ethiopian Plateau basalts; Hofmann *et al.* (1997)), suggesting sustained high annual fluxes of sulfur and other volatiles to the atmosphere.

On a vastly smaller scale, the Laki fissure eruption on Iceland in 1783–1784 (which emitted ~15 km^3 of magma) yielded an estimated 122 Tg of SO_2, distributed between the troposphere and stratosphere (Thordarson and Self, 2002). Fire fountains reached an estimated 1,400 m in height and are believed to have fed convective columns to up to 10–13 km in height (the upper troposphere–lower stratosphere region). Recent chemistry-transport models for a Laki-like eruption have indicated that a surprisingly high fraction (60–70%) of the SO_2 in such a release may be deposited at the surface before it is oxidized to sulfate aerosol in the atmosphere (Stevenson *et al.*, 2003b). This suggests that assumptions of complete conversion of SO_2 to aerosol might result in overestimation of the radiative effects of the volcanic clouds. However, such results have important implications for SO_2 air quality over distances of up to thousands of kilometers from the source.

Various meteorological and proxy records do indicate regional climate anomalies in 1783–1784 (e.g., Briffa *et al.*, 1998), but the patterns are not yet well understood or modeled. More generally, it is not clear how readily volcanic plumes generated by fissure and flood basalt eruptions can reach, and entrain sulfur into, the stratosphere, and considerable work is still required to understand the circumstances by which effusive activity can compete in the climate stakes with explosive eruptions.

3.04.6.3 Tropospheric Chemistry of Volcanic Plumes

Thanks largely to the work following Pinatubo, there is now a good understanding of the stratospheric chemistry of volcanic eruption clouds, at least for emissions on this scale (Section 3.04.6.1). In contrast, the tropospheric chemistry of volcanic plumes is rather poorly known. This partly stems from a wide spectrum of emissions and emission styles (e.g., continuous degassing, minor eruptions), rapid transport of some components to the Earth's surface, and the greater concentration and variability of H_2O in the troposphere. Reactive sulfur, chlorine, and fluorine compounds may be present in both the gas and particle phase in volcanic plumes, and are typically co-emitted with many other volatile species, including water vapor, and with silicate ash particles (Figure 9; see Chapter 4.04 and 4.09).

Once in the troposphere, physical and chemical processes convert gaseous SO_2 to sulfate, while gaseous HF and HCl establish equilibria with aqueous phase H^+, F^-, and Cl^-. SO_2 reacts with ·OH radicals during daylight, giving an SO_2 lifetime of around a week in the troposphere (substantially shorter than in the stratosphere). In volcanic plumes, aqueous phase oxidation of SO_2 by H_2O_2 is likely to be limited by low H_2O_2 availability. Ozone is another potential oxidant, but its effectiveness decreases rapidly as the aerosol acidifies due to dissociation of the acid gases. Observations indicate a wide variation in the lifetime of volcanic SO_2 in the troposphere (e.g., Oppenheimer *et al.*, 1998b).

Increasing interest in the impacts of tropospheric volcanic emissions on terrestrial and aquatic ecology and on human and animal health is driving more research into the tropospheric chemistry and transport of volcanic plumes. This will be of particular importance in understanding the long-range pollution impacts of volcanic clouds (that might be expected from future eruptions like that of Laki 1783–1784).

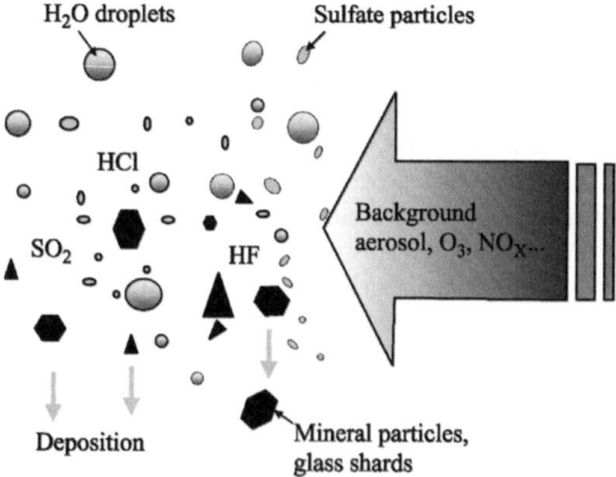

Figure 9 Volcanic clouds are typically composed of gases and particles and diluted by the background atmosphere. Various chemical and physical processes and transformations acting during plume transport further modify plume composition. The chemical and physical form of plume components, their spatial and temporal distribution, and their deposition are therefore strongly controlled by atmospheric chemistry and transport of the plume.

3.04.6.4 Impacts of Volcanic Volatiles on Vegetation and Soils

Volcanic volatiles emitted into the atmosphere are ultimately wet or dry deposited at the Earth's surface. As discussed in Section 3.04.6.3, the chemical and physical form in which they are deposited, and the spatial and temporal distribution of deposition are strongly controlled by atmospheric chemistry and transport of the volcanic plume. Various components of volcanic emissions (including acid species and heavy metals) can be taken up by plants, and can have both harmful and beneficial effects. The detrimental effects are generally either mediated through acidification of soils (by dry or wet deposition) or by direct fumigation of foliage (e.g., respiration of acid gases through stomata; see Chapter 9.11). Rarely, the diffuse emissions of CO_2 described in Section 3.04.4.1.1 can damage plant communities. The gas crisis at Mammoth Mountain, California resulted in an extensive tree kill zone (Figure 4(c); Farrar *et al.*, 1995).

Chemical burning of leaves and flowers of vegetation downwind of the crater is a common observation during degassing crises of Masaya volcano (Figure 4(e)), with substantial economic impact from the loss of coffee crops (Delmelle *et al.*, 2001; Delmelle, 2003). SO_2 can cause direct damage to plants once it is taken up by the foliage (Smith, 1990). The response, however, can be very variable, depending on dosage, atmospheric conditions, and leaf type (e.g., Linzon *et al.*, 1979; Winner and Mooney, 1980; Smith, 1990). In general, chronic exposures to SO_2 concentrations of a few tens or hundreds ppb are sufficient to affect plant ecosystems, decrease agricultural productivity, and cause visible foliar chlorosis and necrosis (Winner and Mooney, 1980).

However, other gas species (e.g., HF and HCl) can be important, as well as the extent of soil acidification due to wet and dry deposition. The impacts of sulfate deposition on soils have been investigated widely in the context of anthropogenic pollution, indicating that the SO_4^{2-} retention capacity of soils varies widely and can be expected to dictate the ecosystem disturbance of volcanogenic sulfur deposition. Anion sorption in soils is directly related to the soil mineralogy and hence to the soil parent material. Delmelle *et al.* (2001) estimated that the amount of SO_2 and HCl dry-deposited within 44 km of Masaya volcano generates an equivalent H^+ flux ranging from $<1\ mg\ m^{-2}\ d^{-1}$ to $30\ mg\ m^{-2}\ d^{-1}$. Sustained acid loading at these rates can severely impact soil chemistry, reflected downwind of Masaya in low pH and depressed base-saturation contents of soils (Parnell, 1986).

Several studies have indicated benefits of fresh ashfall in supplying nutrients. For example, analyses of leachates of fresh ashfall from the modest 2000 eruption of Hekla in Iceland suggest that ash fallout can actually fertilize the oceans by supplying macronutrients and "bioactive" trace metals (Frogner *et al.*, 2001). Cronin *et al.* (1998) also documented beneficial additions of soluble sulfur and selenium to agricultural soils during the 1995–1996 eruption of Ruapehu, New Zealand.

In the case of explosive eruptions such as Pinatubo, 1991, there is some evidence that despite the increase in planetary albedo and resulting global surface cooling, photosynthesis could be encouraged in some regions by reduction of the direct solar irradiance at the surface (which, when very intense, results in some varieties turning or closing their leaves to avoid damage) and increase of the diffuse flux (which can penetrate more of the canopy than direct light). Following Pinatubo, measurements at the Mauna Loa Observatory, Hawaii, indicate a decrease in the direct flux from about 520 W m^{-2} to 400 W m^{-2}, but an increase in the diffuse flux from 40 W m^{-2} to 140 W m^{-2}. Increased photosynthesis helps to explain a slowing of the rate of CO_2 increase in the atmosphere in the 2 yr following the eruption (Gu *et al.*, 2002).

3.04.6.5 Impacts of Volcanic Pollution on Animal and Human Health

Several volcanic volatile species are harmful on contact with the skin, if taken into the lungs, or ingested (Williams-Jones and Rymer, 2000). In particular, sulfur species (in both gaseous and aerosol form) can affect respiratory and cardiovascular health in humans, fluorine can contaminate pasture, leading to poisoning of grazing animals, and catastrophic releases of volcanic CO_2 have resulted in several disasters in recent times.

3.04.6.5.1 Gaseous and particulate sulfur and air quality

Sulfur dioxide has well-known effects on humans—sometimes called "tear gas," its utility in dispersing rioters is regularly witnessed on television news bulletins. Downwind of Masaya Volcano, SO_2 levels exceed background levels at the surface across an area of 1,250 km^2, and an estimated 50,000 people are exposed to concentrations of SO_2 exceeding WHO air quality standards (Baxter *et al.*, 1982; Baxter, 2000; Delmelle *et al.*, 2001, 2002). Symptoms of respiratory illness are commonly reported anecdotally, though formal epidemiological studies have yet to be undertaken. Measurements in Mexico City have indicated the direct influence of Popocatépetl Volcano's emissions on urban air quality: Raga *et al.* (1999) found SO_2 quadrupled, and sulfate concentrations doubled in the city when affected by the volcanic plume. This hints at the potential impacts of future eruptions like Laki, 1783–1784, with long-range transport of SO_2 and aerosol reaching European cities, where air quality standards are already surpassed at certain times of the year. Air quality in Hawaii is reportedly affected by "vog" (volcanic fog) associated with SO_2 and sulfate aerosol from Kilauea's plume, and "laze" (lava haze), composed of HCl-rich droplets formed when active lavas enter the sea (Sutton and Elias, 1993; Mannino *et al.*, 1996).

More recent attention has focused on the abundance of very fine (submicron) and very acidic (pH ~ 1) aerosol emitted from volcanoes (Allen *et al.*, 2002; Mather *et al.*, 2003b). The few reports available of the human health effects of fine sulfuric acid aerosol are confined to industrial incidents, e.g., the case of a community exposed to emissions from a titanium dioxide plant in Japan in the 1960s, in which some 600 individuals living within 5 km of the plant reported asthmatic symptoms (Kitagawa, 1984).

3.04.6.5.2 Fluorine

Several recent eruptions have been remarkable for the quantities of fluorine distributed on tephra. HF appears to be readily adsorbed on to ash surfaces, and is efficiently scavenged out of the gas-particle plume as it erupts and dilutes in the atmosphere. The sedimentation of tephra thereby carries significant quantities of volatiles to the Earth's surface. Grazing animals not only consume contaminated foliage, they can also directly ingest large quantities of tephra (an average cow on an average pasture ingests over a kg of soil per day). Much of the fluorine is contained in bioavailable compounds such as NaF and CaF_2 and can lead rapidly to skeletal deformity, bone lesions, and deformation of teeth. When bones become saturated in fluorine, soft tissues become flooded with fluorine, leading to death. Many animals were lost during the eruptions of Hekla (1970; Óskarsson, 1980), Lonquimay (Chile, 1989–1990; Araya *et al.*, 1990, 1993), and Ruapehu (New Zealand, 1995–1996; Cronin *et al.*, 2003). During the 1783–1784 eruption of Laki, 50% of the livestock on Iceland perished, many probably as a result of fluorine poisoning (Thorarinsson, 1979). It is possible that chronic fluorosis may even have affected parts of the human population through contamination of drinking water.

3.04.6.5.3 Carbon dioxide

Due to its low solubility in magma, CO_2 is able to exsolve from magma even at high pressures. Diffuse carbon dioxide emissions can be dangerous, particularly when there is limited air circulation (e.g., in basements and excavations). Deaths from CO_2 asphyxia are not uncommon—two people died on the flanks of Cosiguina Volcano (Nicaragua) in 1999 while descending a water well in order to carry out repairs. More catastrophic releases have occurred in recent times, twice in Cameroun (at Lakes Monoun and Nyos in 1984 and 1986, respectively; Le Guern and Tazieff (1989) and Sigurdsson *et al.* (1987)), and on the Dieng Plateau (Java; Le Guern *et al.*, 1982). The Nyos disaster claimed ~1,800 lives, and is generally thought to have resulted from sudden overturn of CO_2-rich waters near the base of the lake, possibly triggered by a landslide into the lake. As the CO_2-rich water ascended and decompressed, it released CO_2 into the gas phase. A controlled degassing project has been initiated at Nyos to pump the deep waters to the surface continuously, allowing them to lose their CO_2, and preventing the buildup that could lead to a future catastrophic overturn.

3.04.7 CONCLUSIONS AND FUTURE DIRECTIONS

Volcanic volatile emissions play vital roles in the major geochemical cycles of the Earth system, and have done so throughout Earth history. Their short-term (months/years) impacts on the atmosphere, climate, and environment are strongly controlled by fluxes and emission altitudes of sulfur gases (principally SO_2 or H_2S depending on oxidation state of the magma, and hydrothermal influences), which form sulfate aerosol. Halogen emissions may also be important. Episodic explosive eruptions are the principal perturbation to stratospheric aerosol. In the troposphere, the picture is less clear but recent analyses suggest that up to 40% of the global tropospheric sulfate burden may be volcanogenic. Sulfate aerosol influences the Earth's radiation budget by scattering and absorption of short-wave and long-wave radiation, and by seeding or modifying clouds. When they are brought to the boundary layer and Earth's surface, volcanic sulfur and halogens can result in profound environmental and health impacts.

The global impacts of only one large eruption (magnitude $>10^{13}$ kg) have been studied with any instrumental detail—the 1991 eruption of Mt. Pinatubo. More than a decade after the eruption, new findings are still being published on the climatic, environmental, and ecological consequences of the volatile emissions from this eruption. Despite the tremendous insights afforded by this event, it represents a very small sample of the range of volcanic eruption styles, geographic locations, atmospheric conditions, etc., that could combine to produce significant perturbations to atmospheric radiation and dynamics. This begs the question: how ready is the scientific community to record the next major climate-forcing eruption, and what measures can be taken now to ensure optimal observations of such an event? More generally, substantial further work is required to constrain the temporal and spatial distribution of gas and particle emissions (including sulfur, halogens, and trace metal species) to the atmosphere from all erupting and dormant volcanoes.

In recent years, immense progress has also been made in understanding the physics of volcanic plumes (e.g., Sparks *et al.*, 1997). This has been extremely influential for many reasons, not least since it provides a link between observable phenomena (e.g., height of an eruption column) and critical processes (e.g., the switch between lava effusions and explosive eruption), and hence a quantitative basis for volcanic hazard assessment and mitigation. The time is ripe for comparable investigations of the *chemistry* of volcanic plumes, which would substantially improve our understanding of the environmental, atmospheric, climatic, and health impacts of volcanism. Apart from studies of Pinatubo's stratospheric cloud, such work barely exists. In addition to improved observational data on the spatial and temporal distributions of volcanic volatiles to the atmosphere, further studies are required to characterize the physical and chemical interactions of gases and particles in the atmosphere. This will be essential for the realistic application of numerical models describing the transport and chemical evolution of plumes, and will contribute to a better understanding of volcanogenic pollution and improvements in options for risk mitigation.

Surveillance of gas composition and flux are essential for interpretation of volcanic activity, since the nature of degassing exerts a strong control on eruption style, and is closely associated with volcano seismicity and ground deformation. New optical remote sensing techniques are emerging for the monitoring of volcanic emissions such as the miniaturized ultraviolet spectrometers described by Galle *et al.* (2003). Proliferation of such technologies will also help in efforts to improve the global database of volcanic volatile source distribution. Unfortunately, the modeling frameworks for interpretation of geochemical data remain poorly developed, limiting the application of such data in hazard assessment. Advances in this area will benefit from development and validation of comprehensive physico-chemical models for volcanic degassing based on the integration of results from experiments on the controls on distribution of volatiles in synthetic and natural melts, analysis of dissolved volatiles preserved in melt inclusions, and observed volcanic gas geochemistry. Ultimately, such models can be applied to integrated geophysical, geodetic, and geochemical monitoring data to support eruption forecasting.

Research on volcanic degassing is a rich and diverse field affording many opportunities for interdisciplinary and multidisciplinary efforts at many scales. Broadly, the investigator of volcanic volatiles may follow the path from which the emissions came, to probe the chemistry, dynamics, and evolution of the source regions, storage zones, and volcanic plumbing systems; or consider the interactions of the emissions with the atmosphere, environment, and climate.

ACKNOWLEDGMENTS

The author gratefully acknowledge recent research support from the European Commission 5th Framework programs "MULTIMO" and "DORSIVA," the UK Natural Environment Research Council, grants GR9/4655 and GR29250, the Italian Gruppo Nazionale per la Vulcanologia grant "Development of an integrated spectroscopic system for remote and continuous monitoring of volcanic gas," and NASA grant NAG5-10640. The author is very grateful to Bruno Scaillet, Oded Navon, and David Catling for comments on the original manuscript, and Simon Carn for the TOMS information (Table 7 and Figure 5(c)), Jake Lowenstern for Figure 5(d), and Sune Olander Rasmussen, Trine Ebbensgaard, and Sigfus Johnsen for Figure 5(e). The author thanks all the emissions enthusiasts he has worked with in recent years for their generous collaboration and friendship, including Marie Edmonds, Lisa Horrocks, Andrew McGonigle, David Pyle, Mike Burton, Tamsin Mather, Matt Watson, Hayley Duffell, Bo Galle, Claire Witham, Bill Rose, Frank Tittel, Dirk Richter, Andrew Allen, Peter Baxter, Steve Sparks, Jenni Barclay, Pierre Delmelle, Patrick Allard, Bruno Scaillet, Benoît Villemant, Pete Mouginis-Mark, Keith Horton, John Porter, and Roy Harrison. Lastly, I remember Peter Francis, who set me on sulfurous journeys a decade ago.

REFERENCES

Abdalati W. and Steffen K. (1997) The apparent effects of the Pinatubo eruption on the Greenland ice sheet melt extent. *Geophys. Res. Lett.* **24**, 1795–1797.

Adams N. K., de Silva S. L., Self S., Salas G., Schubring S., Permenter J. L., and Arbesman K. (2001) The physical volcanology of the 1600 eruption of Huaynaputina, southern Peru. *Bull. Volcanol.* **62**, 493–518.

Aiuppa A., Federico C., Paonita A., Pecoraino G., and Valenza M. (2002) S, Cl, and F degassing as an indicator of volcanic dynamics: the 2001 eruption of Mount Etna. *Geophys. Res. Lett.* **29**, DOI 10.1029/2002GL015032.

Allard P. (1997) Endogenous magma degassing and storage at Mount Etna. *Geophys. Res. Lett.* **24**, 2219–2222.

Allard P., Carbonelle J., Metrich N., and Zettwoog P. (1991) Eruptive and diffuse emissions of carbon dioxide from Etna volcano. *Nature* **351**, 38–391.

Allard P., Carbonnelle J., Métrich N., Loyer H., and Zettwoog P. (1994) Sulphur output and magma degassing budget of Stromboli volcano. *Nature* **368**, 326–330.

Allard P., Aiuppa A., Loyer H., Carrot F., Gaudry A., Pinte G., Michel A., and Dongarrà G. (2000) Acid gas and metal emission rates during long-lived basalt degassing at Stromboli volcano. *Geophys. Res. Lett.* **27**, 1207–1210.

Allen A. G., Baxter P. J., and Ottley C. J. (2000) Gas and particle emissions from Soufrière Hills volcano, Montserrat WI: characterization and health hazard assessment. *Bull. Volcanol.* **62**(1), 6–17.

Allen A. G., Oppenheimer C., Ferm M., Baxter P. J., Horrocks L., Galle B., McGonigle A. J. S., and Duffell H. J. (2002) Primary sulphate aerosol and associated emissions from Masaya volcano, Nicaragua. *J. Geophys. Res.* (4), Doi 10.1029/2002JD0 (4 December 2002).

Anderson A. T. (1975) Some basaltic and andesitic gases. *Rev. Geophys. Space Phys.* **13**, 37–55.

Andres R. J. and Kasgnoc A. D. (1998) A time-averaged inventory of subaerial volcanic sulfur emissions. *J. Geophys. Res.* **103**, 25251–25261.

Andres R. J. and Rose W. I. (1995) Remote sensing spectroscopy of volcanic plumes and clouds. In *Monitoring Active Volcanoes: Strategies, Procedures and Techniques* (eds. B. McGuire, C. Kilburn, and J. Murray). UCL Press, London, pp. 301–314.

Araya O., Wittwer F., Villa A., and Ducom C. (1990) Bovine fluorosis following volcanic activity in the southern Andes. *Vet. Rec.* **126**, 641–642.

Araya O., Wittwer F., and Villa A. (1993) Evolution of fluoride concentrations in cattle and grass following a volcanic eruption. *Vet. Hum. Toxicol.* **35**, 437–440.

Arthur M. A. (2000) Volcanic contributions to the carbon and sulfur geochemical cycles and global change. In *Encyclopedia of Volcanoes* (eds. H. Sigurdsson, B. F. Houghton, S. R. McNutt, H. Rymer, and J. Stix). Academic Press, San Diego, pp. 1046–1056.

Baran A. and Foot J. S. (1994) New application of the operational sound HIRS in determining a climatology of acid aerosol from the Pinatubo eruption. *J. Geophys. Res.* **99**, 25673–25679.

Baran A. J., Foot J. S., and Dibben P. C. (1993) Satellite detection of volcanic sulfuric acid aerosol. *Geophys. Res. Lett.* **20**, 1799–1801.

Barmin A., Melnik O., and Sparks R. S. J. (2002) Periodic behaviour in lava dome eruptions. *Earth Planet. Sci. Lett.* **199**, 173–184.

Baxter P. J. (2000) Impacts of eruptions on human health. In *Encyclopedia of Volcanoes* (eds. H. Sigurdsson, B. F. Houghton, S. R. McNutt, H. Rymer, and J. Stix). Academic Press, San Diego, pp. 1035–1043.

Baxter P. J., Stoiber R. E., and Williams S. N. (1982) Volcanic gases and health: Masaya volcano, Nicaragua. *Lancet* **2**, 150–151.

Blank J. G. and Brooker R. A. (1994) Experimental studies of carbon dioxide in silicate melts: solubility, speciation, and stable carbon isotope behavior. *Rev. Mineral.* **30**, 157–186.

Blower J. D., Mader H. M., and Wilson S. D. R. (2001) Coupling of viscous and diffusive controls on bubble growth during explosive volcanic eruptions. *Earth Planet. Sci. Lett.* **193**, 47–56.

Blundy J. and Cashman K. V. (2001) Ascent-driven crystallization of dacite magmas of Mount St Helens, 1980–1986. *Contrib. Mineral. Petrol.* **140**, 631–650.

Bluth G. J. S., Doiron S. D., Schnetzler C. C., Krueger A. J., and Walter L. S. (1992) Global tracking of the SO_2 clouds from the June, 1991 Mount Pinatubo eruptions. *Geophys. Res. Lett.* **19**, 151–154.

Bluth G. J. S., Schnetzler C. C., Krueger A. J., and Walter L. S. (1993) The contribution of explosive volcanism to global atmospheric sulphur dioxide concentrations. *Nature* **366**, 327–329.

Bluth G. J. S., Rose W. I., Sprod I. E., and Krueger A. J. (1997) Stratospheric loading of sulfur from explosive volcanic eruptions. *J. Geol.* **105**, 671–684.

Briffa K. R. and Jones P. D. (1992) The climate of Europe during the 1810s with special reference to 1816. In *The Year without a Summer? World Climate in 1816* (ed. C. R. Harington). Canadian Museum of Nature, Ottawa, pp. 372–391.

Briffa K. R., Jones P. D., Schweingruber F. H., and Osborn T. J. (1998) Influence of volcanic eruptions on northern hemisphere summer temperature over the past 600 years. *Nature* **393**, 450–455.

Bureau H., Keppler H., and Métrich N. (2000) Volcanic degassing of bromine and iodine: experimental fluid/melt partitioning data and applications to stratospheric chemisty. *Earth Planet. Sci. Lett.* **183**, 51–60.

Burnham C. W. (1979) The importance of volatile constituents. In *The Evolution of Igneous Rocks: Fiftieth Anniversary Perspectives.* (ed. H. S. Yoder, Jr.). Princeton University Press, Princeton, NJ.

Burton M., Allard P., Murè F., and Oppenheimer C. (2003) FTIR remote sensing of fractional magma degassing at Mt. Etna, Sicily. In *Volcanic Degassing*, Geological Society of London Special Publication 213 (eds. C. Oppenheimer, D. M. Pyle, and J. Barclay). Geological Society of London, pp. 281–293.

Burton M. R., Oppenheimer C., Horrocks L. A., and Francis P. W. (2000) Remote sensing of CO_2 and H_2O emission rates from Masaya volcano, Nicaragua. *Geology* **28**, 915–918.

Cadle R. D. (1975) Volcanic emissions of halides and sulfur compounds to the troposphere and stratosphere. *J. Geophys. Res.* **80**, 1650–1652.

Cadle R. D. (1980) A comparison of volcanic with other fluxes of atmospheric trace gas constituents. *Rev. Geophys. Space Phys.* **18**, 746–752.

Camuffo D. and Enzi S. (1995) Impact of clouds of volcanic aerosols in Italy during the last seven centuries. *Nat. Hazards* **11**(2), 135–161.

Canil D. (2002) Vanadium in peridotites, mantle redox and tectonic environments: Archean to present. *Earth Planet. Sci. Lett.* **195**, 75–90.

Carey S. and Sigurdsson H. (1989) The intensity of plinian eruptions. *Bull. Volcanol.* **51**, 28–40.

Carn S. A., Krueger A. J., Bluth G. J. S., Schaefer S. J., Krotkov N. A., Watson I. M., and Datta S. (2003) Volcanic eruption detection by the Total Ozone Mapping Spectrometer (TOMS) instruments: a 22-year record of sulfur dioxide and ash emissions. In *Volcanic Degassing*. Geological Society of London Special Publication 213 (eds. C. Oppenheimer, D. M. Pyle, and J. Barclay). Geological Society of London, 177–202.

Carroll M. R. and Holloway J. R. (eds). (1994) *Volatiles in Magmas.* Reviews in Mineralogy, 30. Mineralogical Society of America, Washington, DC, 517pp.

Carroll M. R. and Webster J. D. (1994) Solubilities of sulfur, noble gases, nitrogen, chlorine, and fluorine in magmas. *Rev. Mineral.* **30**, 187–230.

Cashman K. V. (1992) Groundmass crystallization of Mount St. Helens dacite, 1980–1986—a tool for interpreting shallow magmatic processes. *Contrib. Mineral. Petrol.* **109**, 431–449.

Cashman K. V., Sturtevant B., Papale P., and Navon O. (2000) Magmatic fragmentation. In *Encyclopedia of Volcanoes* (eds. H. Sigurdsson, B. F. Houghton, S. R. McNutt, H. Rymer, and J. Stix). Academic Press, San Diego, CA, pp. 421–430.

Catling D. and Kasting J. F. (2003) Planetary atmospheres and life. In *The Emerging Science of Astrobiology* (eds. J. Baross and W. Sullivan). Cambridge University Press, (in press).

Catling D. C., Zahnle K. J., and McKay C. P. (2001) Biogenic methane, hydrogen escape, and the irreversible oxidation of the early earth. *Science* **393**, 839–843.

Chesner C. A. and Rose W. I. (1991) Stratigraphy of the Toba tuffs and the evolution of the Toba caldera complex, Sumatra, Indonesia. *Bull. Volcanol.* **53**, 343–356.

Cheynet B., Dall'Aglio M., Garavelli A., Grasso M. F., and Vurro F. (2000) Trace elements from fumaroles at Vulcano Island (Italy): rates of transport and a thermochemical model. *J. Volcanol. Geotherm. Res.*, **95**, 273–283.

Chin M. and Jacob D. J. (1996) Anthropogenic and natural contributions to tropospheric sulfate: a global model analysis. *J. Geophys. Res.* **101**, 18691–18699.

Coffey M. T. (1996) Observations of the impact of volcanic activity on stratospheric chemistry. *J. Geophys. Res.* **101**, 6767–6780.

Coffey M. T. and Mankin W. G. (1993) Observations of the loss of stratospheric NO_2 following volcanic eruptions. *Geophys. Res. Lett.* **29**, 2873–2876.

Coltelli M., Del Carlo P., and Vezzoli L. (1998) Discovery of a plinian basaltic eruption of Roman age at Etna volcano, Italy. *Geology* **26**, 1095–1098.

Costa F., Scaillet B., Gourgaud A. (2003). Massive atmospheric sulfur loading of the AD 1600 Huaynaputina eruption and implications for petrologicl sulfur estimates. *Geophys. Res. Lett.* **30**(2), 10.102912002GL016402.

Cronin S. J., Hedley M. J., Neall V. E., and Smith G. (1998) Agronomic impact of tephra fallout from 1995 and 1996 Ruapehu volcano eruptions, New Zealand. *Environ. Geol.* **34**, 21–30.

Cronin S. J., Neall V. E., Lecointre J. A., Hedley M. J., and Loganathan P. (2003) Environmental hazards of fluoride in volcanic ash: a case study from Ruapehu volcano, New Zealand. *J. Volcanol. Geotherm. Res.* **121**, 271–191.

Daag A. S., Tubianosa B. S., Newhall C. G., Tuñgol N. M., Javier D., Dolan M. T., Delos Reyes P. J., Arboleda R. A., Martinez M. L., and Regalado T. M (1996) Monitoring sulfur dioxide emission at Mount Pinatubo. In *Fire and Mud: Eruptions and Lahars of Mount Pinatubo Philippines* (eds. C. G. Newhall and R. S. Punongbayan). Philippine Institute of Volcanology and Seismology, Quezon City/ University of Washington Press, Seattle, pp. 409–434.

Dahl-Jensen D., Gundestrup N. S, Miller H., Watanable O., Johnson S. J., Steffensen J. P., Clausen H. B., Svensson A., and Larsen L. B. (2002) The NorthGRIP deep drilling program. *Ann. Glaciol.* **35**, 1–4.

De Hoog J. C. M., Koetsier G. W., Bronto S., Sriwana T., and van Bergen M. J. (2001) Sulphur and chlorine degassing from primitive arc magmas: temporal changes during the 1982–1983 eruptions of Galunggung (West Java, Indonesia). *J. Volcanol. Geotherm. Res.* **108**, 55–83.

Delano J. W. (2001) Redox history of the earth's interior since ~3900 Ma: implications for prebiotic molecules. *Origins Life Evol. Biosphere* **31**, 311–341.

Delmas R. J., Kirchner S., Palais J. M., and Petit J. R. (1992) 1,000 years of explosive volcanism recorded at the South Pole. *Tellus* **44B**, 335–350.

Delmelle P. (2003) Environmental impacts of tropospheric volcanic gas plumes. In *Volcanic Degassing*, Geological Society of London Special Publication 213. (eds. C. Oppenheimer, D. M. Pyle, and J. Barclay). Geological Society of London, pp. 381–399.

Delmelle P. and Stix J. (2000) Volcanic gases. In *Encyclopedia of Volcanoes* (eds. H. Sigurdsson, B. F. Houghton, S. R. McNutt, H. Rymer, and J. Stix). Academic Press, San Diego, CA, pp. 803–815.

Delmelle P., Stix J., Bourque C. P. A., Baxter P. J., Garcia-Alvarez J., and Barquero J. (2001) Dry deposition and heavy acid loading in the vicinity of Masaya volcano, a major sulfur and chlorine source in Nicaragua. *Environ. Sci. Technol.* **35**, 1289–1293.

Delmelle P., Stix J., Baxter P. J., Garcia-Alvarez J., and Barquero J. (2002) Atmospheric dispersion, environmental effects and potential health hazard associated with the low-altitude gas plume of Masaya volcano, Nicaragua. *Bull. Volcanol.* **64**, 423–434.

Dingwell D. (1998) Recent experimental progress in the physical description of silicic magma relevant to explosive volcanism. In *The Physics of Explosive Volcanic Eruptions*. Geology Society of London Special Publication 145 (eds. J. S. Gilbert and R. S. J. Sparks). Geological Society of London, pp. 9–26.

de Silva S. L. and Zielinski G. A. (1998) Global influence of the AD 1600 eruption of Huaynaputina Peru. *Nature* **393**, 455–458.

Dixon J. E., Clague D. A., and Stolper E. M. (1991) Degassing history of water, sulfur, and carbon in submarine lavas from Kilauea volcano, Hawaii. *J. Geol.* **99**, 371–394.

Duffell H., Oppenheimer C., and Burton M. (2001) Volcanic gas emission rates measured by solar occultation spectroscopy. *Geophys. Res. Lett.* **28**, 3131–3134.

Duffell H. J., Oppenheimer C., Pyle D., Galle B., McGonigle A. J. S., and Burton M. R. (2003) Geochemical precursors to a minor explosive eruption at Masaya volcano, Nicaragua. *J. Volcanol. Geotherm. Res* **126**, 327–339.

Edmonds M., Pyle D., and Oppenheimer C. (2001) A model for degassing at the Soufrière Hills Volcano, Montserrat, West Indies, based on geochemical data. *Earth Planet. Sci. Lett.* **186**, 159–173.

Edmonds M., Oppenheimer C., Pyle D. M., and Herd R. A. (2003a) Rainwater and ash leachate analysis as a proxy for plume chemistry at Soufrière Hills Volcano, Montserrat. In *Volcanic Degassing*, Geological Society of London Special Publication 213. (eds. C. Oppenheimer, D. M. Pyle, and J. Barclay). Geological Society of London, pp. 203–218.

Edmonds M., Herd R. A., Galle B., and Oppenheimer C. (2003b) Automated, high time-resolution measurements of SO_2 flux at Soufrière Hills Volcano, Montserrat. *Bull. Volcanol.* D.o.i. 10.1007/s00445-003-0286-x.

Edmonds M., Oppenheimer C., Pyle D. M., Herd R. A., and Thompson G. (2003c) SO_2 emissions from Soufrière Hills Volcano and their relationship to conduit permeability, hydrothermal interaction and degassing regime. *J. Volcanol. Geotherm. Res.* **124**, 23–43.

Eggenkamp H.G.M. (1994). The geochemistry of chlorine isotopes. PhD Thesis, University of Utrecht, The Netherlands, 150pp (unpublished).

Eichelberger J. C. (1995) Silicic volcanism: ascent of viscous magmas from crustal reservoirs. *Ann. Rev. Earth Plant. Sci. Lett.* **23**, 41–63.

Eichelberger J. C., Carrigan C. R., Westrich H. R., and Price R. H. (1986) Non-explosive silicic volcanism. *Nature* **323**, 598–602.

Farrar C. D., Sorey M. L., Evans W. C., Howle J. F., Kerr B. D., Kennedy B. M., King C.-Y., and Southon J. R. (1995) Forest-killing diffuse CO_2 emissions at Mammoth Mountain as a sign of magmatic unrest. *Nature* **376**, 675–678.

Fine G. J. and Stolper E. M. (1985) The speciation of carbon dioxide in sodium aluminosilicate glasses. *Contrib. Mineral. Petrol.* **91**, 105–121.

Fischer T. P., Giggenbach W. F., Sano Y., and Williams S. N. (1998) Fluxes and sources of volatiles discharged from Kudryavy, a subduction zone volcano, Kurile Islands. *Earth Planet. Sci. Lett.* **160**, 81–96.

Fischer T. P., Hilton D. R., Zimmer M. M., Shaw A. M., Sharp Z. D., and Walker J. A. (2002) Subduction and recycling of nitrogen along the Central American margin. *Science* **297**, 1154–1157.

Francis P., Burton M., and Oppenheimer C. (1998) Remote measurements of volcanic gas compositions by solar FTIR spectroscopy. *Nature* **396**, 567–570.

Francis P. W., Oppenheimer C., and Stevenson D. (1993) Endogenous growth of persistently active volcanoes. *Nature* **366**, 554–557.

Frogner P., Gislason S. R., and Oskarsson N. (2001) Fertilizing potential of volcanic ash in ocean surface water. *Geology* **29**, 487–490.

Galle B., Oppenheimer C., Geyer A., McGonigle A., Edmonds M., and Horrocks L. A. (2003) A miniaturised ultraviolet spectrometer for remote sensing of SO_2 fluxes: a new tool for volcano surveillance. *J. Volcanol. Geotherm. Res.* **119**, 241–254.

Gauthier P. J. and Le Cloarec M.-F. (1998) Variability of alkali and heavy metal fluxes released by Mt. Etna volcano, Sicily, between 1991 and 1995. *J. Volcanol. Geotherm. Res.* **81**, 311–326.

Genin A., Lazar B., and Brenner S. (1995) Vertical mixing and coral death in the Red Sea following the eruption of Mt. Pinatubo. *Nature* **377**, 507–510.

Gerlach T. M. (1993) Thermodynamic evaluation and restoration of volcanic gas analyses; an example based on modern collection and analytical methods. *Geochem. J.* **27**, 305–322.

Gerlach T. M. and Casadevall T. J. (1986) Fumarole emissions at Mount St Helens volcano, June 1980 to October 1981: degassing of a magma-hydrothermal system. *J. Volcanol. Geotherm. Res.* **28**, 141–160.

Gerlach T. M. and Graeber E. J. (1985) Volatile budget of Kilauea volcano. *Nature* **313**, 273–277.

Gerlach T. M. and Taylor B. E. (1990) Carbon isotope constraints on degassing of carbon dioxide from Kilauea Volcano. *Geochim. Cosmochim.* Acta **54**, 2051–2058.

Gerlach T. M., Delgado H., McGee K. A., Doukas M. P., Venegas J. J., and Cardenas L. (1997) Application of the LI-COR CO_2 analyzer to volcanic plumes: a case study, volcan Popocatépetl, Mexico, June 7 and 10, 1995. *J. Geophys. Res.* **102**(B4), 8005–8019.

Giggenbach W. F. (1975) A simple method for the collection and analysis of volcanic gas samples. *Bull. Volcanol.* **39**, 132–145.

Giggenbach W. F. (1992a) Isotopic shifts in waters from geothermal and volcanic systems along convergent plate boundaries and their origin. *Earth Planet. Sci. Lett.* **113**, 495–510.

Giggenbach W. F. (1992b) Magma degassing and mineral deposition in hydrothermal systems along convergent plate boundaries. *Econ. Geol.* **87**, 1927–1944.

Giggenbach W. F. (1996) Chemical composition of volcanic gases. In *Monitoring and Mitigation of Volcano Hazards* (eds. R. Scarpa and R. I. Tilling). Springer, Berlin, pp. 221–256.

Giggenbach W. F. and Matsuo S. (1991) Evaluation of results from second and third IAVCEI field workshops on volcanic gases, Mt. Usu, Japan and White Island, New Zealand. *Appl. Geochem.* **6**, 125–141.

Giggenbach W. F., Tedesco D., Sulistiyo Y., Caprai A., Cioni R., Favara R., Fischer T. P., Hirabayashi J.-I., Korzhinsky M., Martini M., Menyailov I., and Shinohara H. (2001) Evaluation of results from the fourth and fifth IAVCEI field workshops on volcanic gases, Vulcano island, Italy and Java, Indonesia. *J. Volcanol. Geotherm. Res.* **108**, 157–172.

Graf H. F., Kirchner I., Robock A., and Schult I. (1993) Pinatubo eruption winter climate effects: model versus observations. *Clim. Dyn.* **9**, 81–93.

Graf H.-F., Feichter J., and Langmann B. (1997) Volcanic sulfur emissions: estimates of source strength and its contribution to the global sulfate distribution. *J. Geophys. Res.* **102**, 10727–10738.

Graf H.-F., Langmann B., and Feichter J. (1998) The contribution of earth degassing to the atmospheric sulfur budget. *Chem. Geol.* **147**, 131–145.

Gu L., Baldocchi D., Verma S. B., Black T. A., Vesala T., Falge E. M., and Dowty P. R. (2002) Advantages of diffuse radiation for terrestrial ecosystem productivity. *J. Geophys. Res.* **107**, 1–23.

Halmer M. M., Schmincke H.-U., and Graf H.-F. (2002) The annual volcanic gas input into the atmosphere, in particular into the stratosphere: a global data set for the past 100 years. *J. Volcanol. Geotherm. Res.*, **115**, 511–528.

Hammer C. U., Clausen H. B., and Dansgaard W. (1980) Greenland ice sheet evidence of post-glacial volcanism and its climatic impact. *Nature* **288**, 230–235.

Hammer J. E. and Rutherford M. J. (2002) An experimental study of the kinetics of decompression-induced crystallization in silicic melt. *J. Geophys. Res.* **107**(131), 10.1029/2001JB000281.

Hammer J. E., Cashman K. V., Hoblitt R. P., and Newman S. (1999) Degassing and microlite crystallization during pre-climactic events of the 1991 eruption of Mt. Pinatubo, Phillipines. *Bull. Volcanol.* **60**, 355–380.

Hansen J., Sato M., Ruedy R., Lacis A., Asamoah K., Beckford K., Borenstein S., Brown E., Cairns B., Carlson B., Curran B., de Castro S., Druyan L., Etwarrow P., Ferede T., Fox M., Gaffen D., Glascoe J., Gordon H., Hollandsworth S., Jiang X., Johnson C., Lawrence N., Lean J., Lerner J., Lo K., Logan J., Luckett A., McCormick M. P., McPeters R., Miller R., Minnis P., Ramberran I., Russell G., Russell P., Stone P., Tegen I., Thomas S., Thomason L., Thompson A., Wilder J., Willson R., and Zawodny J. (1997) Forcings and chaos in interannual to decadal climate change. *J. Geophys. Res.* **102**, 25679–25720.

Harington C. R. (ed.) (1992) *The Year without a Summer? World Climate in (1816)*. Canadian Museum of Nature, Ottawa, 576 pp.

Hinkley T. K., Lamothe P. J., Wilson S. A., Finnegan D. L., and Gerlach T. M. (1999) Metal emissions from Kilauea, and a suggested revision of the estimated worldwide metal output by quiescent degassing of volcanoes. *Earth Planet. Sci. Lett.* **170**, 315–325.

Hoff R. M. (1992) Differential SO_2 column measurements of the Mt. Pinatubo volcanic plume. *Geophys. Res. Lett.* **19**, 175–178.

Hoffman P. F. and Schrag D. P. (2002) The snowball earth hypothesis: testing the limits of global change. *Terra Nova* **14**, 129–155.

Hoffman P. F., Kaufman A. J., Halverson G. P., and Schrag D. P. (1998) A neoproterozoic snowball earth. *Science* **281**, 1342–1346.

Hofmann C., Courtillot V., Feraud G., Rochette P., Yirgu G., Ketefo E., and Pik R. (1997) Timing of the Ethiopian flood basalt event and implications for plume birth and global change. *Nature* **389**, 838–841.

Holasek R. E., Self S., and Woods A. W. (1996) Satellite observations and interpretation of the 1991 Mount Pinatubo eruption plumes. *J. Geophys. Res.* **101**, 27635–27655.

Holland H. D. (1984) *The Chemical Evolution of the Atmosphere and Oceans*. Princeton University Press, Princeton, 598 pp.

Holland H. D. (2002) Volcanic gases, black smokers, and the great oxidation event. *Geochim. Cosmochim. Acta* **66**, 3811–3826.

Horn S. and Schmincke H.-U. (2000) Volatile emission during the eruption of Baitoushan volcano (China/North Korea) ca. 969 AD. *Bull. Volcanol.* **61**, 537–555.

Horrocks L. A., Oppenheimer C., Burton M. R., Duffell H. R., Davies N. M., Martin N. A., and Bell W. (2001) Open-path Fourier transform infrared spectroscopy of SO_2: an empirical error budget analysis, with implications for volcano monitoring. *J. Geophys. Res.* **106**, 27647–27659.

Huppert H. E. and Woods A. W. (2002) The role of volatiles in magma chamber dynamics. *Nature* **420**, 493–495.

Ihinger P. D., Zhang Y., and Stolper E. M. (1999) The speciation of dissolved water in rhyolitic melt. *Geochim. Cosmochim. Acta* **63**, 3567–3578.

Intergovernmental Panel on Climate Change Working Group (IPCC Report) (2001) *Climate Change 2001: The Scientific Basis*, Contribution of Working Group I to the Third Assessment Report of the Intergovernmental Panel on Climate Change (J. T. Houghton, Y. Ding, D. J. Griggs, M. Noguer, P. J. van der Linden, D. Xiaosu).

Jaupart C. (1998) Gas loss through conduit walls during eruption. In *The Physics of Explosive Volcanic Eruptions*. Geological Society of London Special Publication 145 (eds. J. S. Gilbert and R. S. J. Sparks). Geological Society of London, pp. 73–90.

Jaupart C. and Allegre C. J. (1991) Gas content, eruption rate and instabilities of eruption regime in silicic volcanoes. *Earth Planet. Sci. Lett.* **102**, 413–429.

Johnson M. C., Anderson A. T., Jr., and Rutherford M. J. (1993) Pre-eruptive volatile contents of magmas. In *Volatiles in Magmas*, Reviews in Mineral 30 (eds. M. R. Carroll and J. R. Holloway). Mineralogical Society of America, Washington, DC, pp. 281–330.

Jones G.S., and Stott P.A., (2002). Simulation of climate response to a super eruption. In *American Geophysical Union Chapman Conference on Volcanism and the Earth's Atmosphere, Santorini 17–21 June, (2002)*. Abstract volume, 45p.

Kasting J. F., Eggler D. H., and Raeburn S. P. (1993) Mantle redox evolution and the oxidation state of the Archean atmosphere. *J. Geol.* **101**, 245–257.

Kazahaya K., Shinohara H., and Saito G. (1994) Excessive degassing of Izu-Oshima volcano: magma convection in a conduit. *Bull. Volcanol.* **56**, 207–216.

Kelley D. S., Baross J. A., and Delaney J. R. (2002) Volcanoes, fluids, and life at mid-ocean ridge spreading centers. *Ann. Rev. Earth Planet. Sci.* **30**, 385–491.

Kelly P. M., Jones P. D., and Pengqun J. L. A. (1996) The spatial response of the climate system to explosive volcanic eruptions. *Int. J. Climatol.* **16**, 537–550.

Keppler H. (1999) Experimental evidence for the source of excess sulfur in explosive volcanic eruptions. *Science* **284**, 1652–1654.

Kirchner I., Stenchikov G., Graf H.-F., Robock A., and Antuna J. (1999) Climate model simulation of winter warming and summer cooling following the 1991 Mount Pinatubo volcanic eruption. *J. Geophys. Res.* **104**, 19039–19055.

Kirschvink J. L. (1992) Late proterozoic low-latitude glaciation: the snowball earth. In *The Proterozoic Biosphere* (eds. J. W. Schopf and C. Klein). Cambridge University Press, New York, pp. 51–52.

Kirschvink J. L., Gaidos E. J., Bertani L. E., Beukes N. J., Gutzmer J., Maepa L. N., and Steinberger R. E. (2000) Paleoproterozoic snowball earth: extreme climatic and geochemical global change and its biological consequences. *Proc. Natl Acad. Sci.* **97**, 1400–1405.

Kitagawa T. (1984) Cause analysis of the Yokkaichi asthma episode in Japan. *J. Air Pollut. Control Assoc.* **34**, 743–746.

Klug C. and Cashman K. V. (1996) Permeability development in vesiculating magmas: implications for fragmentation. *Bull. Volcanol.* **58**, 87–100.

Koike M., Jones N. B., Matthews W. A., Johnston P. V., McKenzie R. L., Kinnison D., and Rodriguez J. (1994) Impact of pinatubo aerosols on the partitioning between NO_2 and HNO_3. *Geophys. Res. Lett.* **21**, 597–600.

Kress V. C. (1997) Magma mixing as a source for Pinatubo sulphur. *Nature* **389**, 591–593.

Krueger A. J., Walter L. S., Bhartia P. K., Schnetzler C. C., Krotkov N. A., Sprod I., and Bluth G. J. S. (1995) Volcanic sulfur dioxide measurements from the Total Ozone Mapping Spectrometer (TOMS) instruments. *J. Geophys. Res.* **100**, 14057–14076.

Lambert G., Le Cloarec M.-F., and Pennisi M. (1988) Volcanic output of SO_2 and trace metals: a new approach. *Geochim. Cosmochim. Acta* **52**, 39–42.

Lange R. A. (1994) The effect of H_2O, CO_2, and F on the density and viscosity of silicate melts. *Rev. Mineral.* **30**, 331–369.

Lantzy R. J. and Mackenzie F. T. (1979) Atmospheric trace metals: global cycles and assessment of man's impact. *Geochim. Cosmochim. Acta* **43**, 511–525.

Le Cloarec M.-F. and Marty B. (1991) Volatile fluxes from volcanoes. *Terra Nova* **3**, 17–27.

Le Guern, F., Tazieff, H. (1989) Lake Nyos. *J. Volcanol. Geotherm. Res.* **39** (special issue).

Le Guern F., Tazieff H., and Faivre-Perret R. (1982) An example of health hazard: people killed by gas during a phreatic eruption, Dieng Plateau (Java, Indonesia), February 20th (1979). *Bull. Volcanol.* **45**, 153–156.

Lejeune A.-M. and Richet P. (1995) Rheology of crystal bearing silicate melts: an experimental study at high viscosities. *J. Geophys. Res.* **100**, 4215–4229.

Lensky N., Lyakhovsky V., and Navon O. (2001) Radial variations of melt viscosity around growing bubbles and gas overpressure in vesiculating magmas. *Earth Planet. Sci. Lett.* **186**, 1–6.

Lensky N., Lyakhovsky V., and Navon O. (2002) Expansion dynamics of volatile-saturated fluid and bulk viscosity of bubbly magmas. *J. Fluid Mech.* **460**, 39–56.

Lensky N., Navon O., and Lyakhovsky V. (2003) Bubble growth during decompression of magma: experimental and theoretical investigation. *J. Volcanol. Geotherm. Res.* (in review).

Linzon S. N., Temple P. J., and Pearson R. G. (1979) Sulfur concentrations in plant foliage and related effects. *J Air Pollut. Control Assoc.* **29**, 520–525.

Mader H. M. (1998) Conduit flow and fragmentation. In *The Physics of Explosive Volcanic Eruptions*. Geological Society of London Special Publication 145 (eds. J. S. Gilbert and R. S. J. Sparks), Geological Society of London, pp. 51–71.

Mannino D. M., Ruben S., Holschuh F. C., Holschuh T. C., Wilson M. D., and Holschuh T. (1996) Emergency department visits and hospitalizations for respiratory disease on the island of Hawaii, 1981 to 1991. *Hawaii Med. J.* **55**(3), 48–53.

Massol H. and Jaupart C. (1999) The generation of gas overpressure in volcanic eruptions. *Earth Planet. Sci. Lett.* **166**, 57–70.

Mather T. A., Pyle D. M., and Oppenheimer C. (2003a) Volcanic aerosol in the troposphere. In *Volcanism and the Earth's Atmosphere* (eds. A. Robock and C. Oppenheimer). American Geophysical Union Monograph (in press).

Mather T. A., Allen A. G., Oppenheimer C., Pyle D. M., and McGonigle A. J. S. (2003b) Size-resolved particle compositions of the tropospheric plume of Masaya volcano, Nicaragua. *J. Atmos. Chem.* (in press).

Matuso S. (1960) On the origin of volcanic gases. *J. Earth Sci. Nagoya Univ.* **8**, 222–245.

Matsuo S. (1975) Chemistry of volcanic gases. *Kazan (Bull. Volcanol. Soc. Japan)* **20**, 319–329.

McCormick M. P., Thomason L. W., and Trepte C. R. (1995) Atmospheric effects of the Mt. Pinatubo eruption. *Nature* **373**, 399–404.

McGee K. A. (1992) The structure, dynamics and chemical composition of non-eruptive plumes from Mt. St. Helens, 1980–88. *J. Volcanol. Geotherm. Res.* **51**, 269–282.

McGee K. A. and Gerlach T. M. (1998) Annual cycle of magmatic CO_2 in a tree-kill soil at Mammoth mountain, California: implications for soil acidification. *Geology* **26**, 463–466.

McGonigle A. J. S. and Oppenheimer C. (2003) Optical sensing of volcanic gas and aerosol. In *Volcanic Degassing*, Geological Society of London Special Publication, 213 (eds. C. Oppenheimer, D. M. Pyle, and J. Barclay). Geological Society of London pp. 149–168.

McGonigle A. J. S., Oppenheimer C., Galle B., Mather T., and Pyle D. (2002) Walking traverse and scanning DOAS measurements of volcanic gas emission rates. *Geophys. Res. Lett.* Doi 10.1029/2002GL015827 (October 26, 2002).

McGonigle A. J. S., Oppenheimer C., Hayes A. R., Galle B., Edmonds M., Caltabiano T., Salerno G., Burton M., and Mather T. A. (2003) Sulphur dioxide fluxes from Mount Etna, Vulcano, and Stromboli measured with an automated scanning ultraviolet spectrometer. *J. Geophys. Res.* **180**, D.o.i. 10.1029/2002JB002261 (30 September 2003).

McMilan P. F. (1994) Water solubility and speciation models. *Rev. Mineral.* **30**, 131–156.

Melnik O. and Sparks R. S. J. (1999) Nonlinear dynamics of lava extrusion. *Nature* **402**, 37–41.

Melnik O. and Sparks R. S. J. (2002a) Modelling of conduit flow dynamics during explosive activity at Soufrière Hills Volcano, Montserrat. In *The Eruption of Soufrière Hills Volcano, Montserrat, from 1995 to 1999*. Geological Society of London Memoir 21 (eds. T. H. Druitt and P. Kokelaar). Geological Society of London, pp. 307–317.

Melnik O. and Sparks R. S. J. (2002b) Dynamics of magma ascent and lava extrusion at Soufrière Hills Volcano, Montserrat. In *The Eruption of Soufrière Hills Volcano, Montserrat, from 1995 to 1999*, Geological Society of London Memoir 21 (eds. T. H. Druitt and P. Kokelaar). Geological Society of London, pp. 153–171.

Métrich N., Bonnin-Mosbah M., Susini J., Menez B., and Galoisy L. (2002) Presence of sulfite (S^{IV}) in arc magmas: implications for volcanic sulfur emissions. *Geophys. Res. Lett.* **29** DOI 10.1029/2001GL014607.

Minnis P., Harrison E. F., Stowe L. L., Gibson G. G., Ddenn F. M., Doelling D. R., and Smith W. L., Jr. (1993) Radiative climate forcing by the Mount Pinatubo eruption. *Science* **259**, 1411–1415.

Monzier M., Robin C., and Eissen J.-P. (1994) Kuwae (1425 AD); the forgotten caldera. *J. Volcanol. Geotherm. Res.* **59**, 207–218.

Moretti R., Papale P., and Ottonello G. (2003) A model for the saturation of C–O–H–S fluids in silicate melts. In *Volcanic Degassing*, Geological Society of London Special Publication 213 (eds. C. Oppenheimer, D. M. Pyle, and J. Barclay). Geological Society of London, pp. 81–101.

Navon and Lyakhovsky (1998) Vesiculation processes in silicic magmas. In *The Physics of Explosive Volcanic Eruptions*, Geological Society of London Special Publication 145 (eds. J. S. Gilbert and R. S. J. Sparks). Geological Society of London, pp. 27–50.

Newman S. and Lowenstern J. B. (2002) VolatileCalc: a silicate melt-H_2O–CO_2 solution model written in visual basic for excel. *Comput. Geosci.* **28**, 597–604.

Nriagu J. O. (1989) A global assessment of natural sources of atmospheric trace metals. *Nature* **338**, 47–49.

Oppenheimer C. (1996) On the role of hydrothermal systems in the transfer of volcanic sulfur to the atmosphere. *Geophys. Res. Lett.* **23**, 2057–2060.

Oppenheimer C. (2002) Limited global change due to largest known Quaternary eruption, Toba ≈74 kyr BP? *Quat. Sci. Rev.* **21**, 1593–1609.

Oppenheimer C. (2003a) Climatic, environmental, and human consequences of the largest known historic eruption: Tambora volcano (Indonesia) 1815. *Prog. Phys. Geogr.* **27**, 230–259.

Oppenheimer C. (2003b) Ice core and palaeoclimatic evidence for the great volcanic eruption of (1257). *Int. J. Climatol.*, **23**, 417–426.

Oppenheimer C. and Yirgu G. (2002) Imaging of an active lava lake: Erta 'Ale volcano. *Ethiopia. Int. J. Remote Sensing* **23**, 4777–4782.

Oppenheimer C., Francis P., Burton M., Maciejewski A., and Boardman L. (1998a) Remote measurement of volcanic gases by Fourier transform infrared spectroscopy. *Appl. Phys.* **B67**, 505–515.

Oppenheimer C., Francis P., and Stix J. (1998b) Depletion rates of sulfur dioxide in tropospheric volcanic plumes. *Geophys. Res. Lett.* **25**, 2671–2674.

Oppenheimer C., Edmonds M., Francis P., and Burton M. R. (2002a) Variation in HCl/SO_2 gas ratios observed by Fourier transform spectroscopy at Soufrière Hills Volcano, Montserrat. In *The Eruption of Soufrière Hills Volcano, Montserrat, from 1995 to 1999*, Geological Society of

London Memoir 21 (eds. T. H. Druitt and P. Kokelaar). Geological Society of London, pp. 621–639.

Oppenheimer C., Burton M. R., Durieux J., and Pyle D. M. (2002b) Open-path Fourier transform spectroscopy of gas emissions from a carbonatite volcano: Oldoinyo Lengai, Tanzania. *Opt. Lasers Eng.* **37**, 203–214.

Oppenheimer C., Pyle D. M., and Barclay J. (eds.) (2003) *Volcanic Degassing*. Geological Society of London Special Publication 213, 420pp.

Óskarsson N. (1980) The interaction between volcanic gases and tephra, fluorine adhering to tephra of the 1970 Hekla eruption. *J. Volcanol. Geotherm. Res.* **8**, 251–266.

Papale P. (1999) Modelling of the solubility of a two-component $H_2O + CO_2$ fluid in silicate liquid. *Am. Mineral.* **84**, 477–492.

Papale P. and Polacci M. (1999) Role of carbon dioxide in the dynamics of magma ascent in explosive eruptions. *Bull. Volcanol.* **60**, 583–594.

Parker D. E., Wilson H., Jones P. D., Christy J. R., and Folland C. K. (1996) The impact of Mount Pinatubo on world-wide temperatures. *Int. J. Climatol.* **16**, 487–197.

Parnell R. A. (1986) Processes of soil acidification in tropical Durandepts, Nicaragua. *Soil Sci.* **42**, 43–55.

Pinto J. P., Turco R. P., and Toon O. B. (1989) Self-limiting physical and chemical effects in volcanic eruption clouds. *J. Geophys. Res.* **94**, 11165–11174.

Porter J. N., Horton K., Mouginis-Mark P., Lienert B., Lau E., Sutton A. J., Elias T., and Oppenheimer C. (2002) Sun photometer and lidar measurements of the plume from the Hawaii Kilauea volcano Pu'u O'o vent: estimates of aerosol flux rates and SO_2 lifetime. *Geophys. Res. Lett.* Doi 10.1029/2002GL014744 (23 August 2002).

Post J. D. (1977) *The Last Great Subsistence Crisis in the Western World*. The John Hopkins University Press, Baltimore, 240 pp.

Prata, A. J., Self, S., Rose, W. I., O'Brien, D. M., (2002). Global, long-term sulphur dioxide measurements from TOVS data: a new tool for studying explosive volcanism and climate. In *American Geophysical Union Chapman Conference on Volcanism and the Earth's Atmosphere, Santorini June 17–21, 2002*. Abstract volume, 33.

Pyle D. M. (2000) Sizes of volcanic eruptions. In *Encyclopedia of Volcanoes* (eds. H. Sigurdsson, B. F. Houghton, S. R. McNutt, H. Rymer, and J. Stix). Academic Press, San Diego, CA, pp. 263–269.

Pyle D. M., Beattie P. D., and Bluth G. J. S. (1996) Sulphur emissions to the stratosphere from explosive volcanic eruptions. *Bull. Volcanol.* **57**, 663–671.

Raga G. B., Kok G. L., and Baumgardner D. (1999) Evidence for volcanic influence on Mexico City aerosols. *Geophys. Res. Lett.* **26**, 1149–1152.

Rampino M. R. and Self S. (1992) Volcanic winter and accelerated glaciation following the Toba super-eruption. *Nature* **359**, 50–52.

Rampino M. R. and Stothers R. B. (1988) Flood basalt volcanism during the past 250 million years. *Science* **241**, 663–668.

Read W. G., Froidevaux L., and Waters J. W. (1993) Microwave limb sounder measurements of stratospheric SO_2 from the Mt. Pinatubo eruption. *Geophys. Res. Lett.* **20**, 1299–1302.

Robock A. (2000) Volcanic eruptions and climate. *Rev. Geophys.* **38**(2), 191–219.

Roggensack K. R., Hervig R. L., McKnight S. B., and Williams S. N. (1997) Explosive basaltic volcanism from Cerro Negro volcano: influence of volatiles on eruption style. *Science* **277**, 1639–1642.

Rose W. I., Chuan R. L., Giggenbach W. F., Kyle P. R., and Symonds R. B. (1986) Rates of sulfur dioxide and particle emissions from White Island volcano, New Zealand, and an estimate of the total flux of major gaseous species. *Bull. Volcanol.* **48**, 181–188.

Sagan C. and Chyba C. (1997) The early faint Sun paradox: organic shielding of ultraviolet-labile greenhouse gases. *Science* **276**, 1217–1221.

Scaillet B. and Pichavant M. (2003) Experimental constraints on volatile abundances in arc magmas and their implications for degassing processes. In *Volcanic Degassing*. Geological Society of London Special Publication 213 (eds. C. Oppenheimer, D. M. Pyle, and J. Barclay). Geological Society of London, pp. 23–52.

Scaillet B., Clemente B., Evans B. W., and Pichavant M. (1998) Redox control of sulfur degassing in silicic magmas. *J. Geophys. Res.* **103**, 23937–23949.

Scaillet B., Luhr J., and Carroll M. (2003) Petrological and volcanological constraints on volcanic sulfur emissions to the atmosphere. In *Volcanism and the Earth's Atmosphere* (eds. A. Robock and C. Oppenheimer). American Geophysical Union Monograph (in press).

Shaw H. R. (1972) Viscosities of magmatic silicate liquids: an empirical method of prediction. *Am. J. Sci.* **272**, 870–893.

Signorelli S. and Carroll M. R. (2000) Solubility and fluid-melt partitioning of Cl in hydrous phonolitic melts. *Geochim. Cosmochim. Acta* **64**, 2851–2862.

Signorelli S. and Carroll M. R. (2001) Experimental constraints on the origin of chlorine emissions at the Soufriere Hills Volcano. *Montserrat. Bull. Volcanol.* **62**, 431–440.

Signorelli S. and Carroll M. R. (2002) Experimental study of Cl solubility in hydrous alkaline melts: constraints on the theoretical maximum amount of Cl in trachytic and phonolitic melts. *Contrib. Mineral. Petrol.* **143**, 209–218.

Sigurdsson H., Devine J. D., Tchoua F. M., Presser T. S., Pringle M. K. W., and Evans W. C. (1987) Origin of the lethal has burst from Lake Monoun, Cameroon. *J. Volcanol. Geotherm. Res.* **31**, 1–16.

Silver L. A., Ihinger P. D., and Stolper E. M. (1990) The influence of bulk composition on the speciation of water in silicate glasses. *Contrib. Mineral. Petrol.* **104**, 142–162.

Slezin Y. B. (1995) Principal regimes of volcanic eruptions. *Volcanol. Seismol.* **17**, 193–205.

Slezin Y. B. (2003) The mechanism of volcanic eruption (steady state approach). *J. Volcanol. Geotherm. Res.* **122**, 7–50.

Smith W. H. (1990) *Air Pollution and Forests: Interaction between Air Contaminants and Forest Ecosystems*. Springer, New York, 618pp.

Soden B. J., Wetherald R. T., Stenchikov G. L., and Robock A. (2002) Global cooling following the eruption of Mt. Pinatubo: a test of climate feedback by water vapor. *Science* **296**, 727–730.

Sorey M. L., Kennedy B. M., Evans W. C., Farrar C. D., and Suemnicht G. A. (1993) Helium isotope and gas discharge variations associated with crustal unrest in Long Valley caldera, California. *J. Geophys. Res.* **98**, 15871–15889.

Sorey M. L., Evans W. C., Kennedy B. M., Farrar C. D., Hainsworth L. J., and Hausback B. (1998) Carbon dioxide and helium emissions from a reservoir of magmatic gas beneath Mammoth Mountain, California. *J. Geophys. Res.* **103**, 15303–15323.

Sparks R. S. J. (1997) Causes and consequences of pressurisation in lava dome eruptions. *Earth Planet. Sci. Lett.* **150**, 177–189.

Sparks R. S. J. (2003) Dynamics of magma degassing. In *Volcanic Degassing*, Geological Society of London Special Publication 213 (eds. C. Oppenheimer, D. M. Pyle, and J. Barclay). Geological Society of London, pp 5–22.

Sparks R. S. J., Bursik M. I., Carey S. N., Gilbert J. S., Glaze L., Sigurdsson H., and Woods A. W. (1997) *Volcanic Plumes*. Wiley, Chichester, England, 557pp.

Spera F. J. (2000) Physical properties of magma. In *Encyclopedia of Volcanoes* (eds. H. Sigurdsson, B. F. Houghton, S. R. McNutt, H. Rymer, and J. Stix). Academic Press, San Diego, CA, pp. 171–190.

Stenchikov G. L., Kirchner I., Robock A., Graf H.-F., Antuna J. C., Grainger R. G., Lambert A., and Thomason L. (1998)

Radiative forcing from the 1991 Mount Pinatubo volcanic eruption. *J. Geophys. Res.* **103**, 13837–13857.

Stenchikov G., Robock A., Ramaswamy V., Schwarzkopf M. D., Hamilton K., and Ramachandran S. (2002) Arctic Oscillation response to the 1991 Mount Pinatubo eruption: effects of volcanic aerosols and ozone depletion. *J. Geophys. Res.* Doi 10.1029/2002JD002090 (28 December 2002).

Stevenson D. S. (1993) Physical models of fumarolic flow. *J. Volcanol. Geotherm. Res.* **57**, 139–156.

Stevenson D. S. and Blake S. (1998) Modelling the dynamics and thermodynamics of volcanic degassing. *Bull. Volcanol.* **60**, 307–317.

Stevenson D. S., Johnson C. E., Collins W. J., and Derwent R. G. (2003a) The tropospheric sulphur cycle and the role of volcanic SO_2. In *Volcanic Degassing*, Geological Society of London Special Publication 213 (eds. C. Oppenheimer, D. M. Pyle, and J. Barclay). Geological Society of London, pp. 295–305.

Stevenson D. S., Johnson C. E., Highwood E. J., Ganei V., Collins W. J., and Derwent R. G. (2003b) Atmospheric impact of the 1783–1784 Laki eruption: Part 1. Chemistry modelling, *Atmos. Chem. Phys.* **3**, 487–507.

Stoiber R. E., Malinconico L. L., Jr., and Williams S. N. (1983) Use of the correlation spectrometer at volcanoes. In *Forecasting Volcanic Events* (eds. H. Tazieff and J.-C. Sabroux). Elsevier, Amsterdam, pp. 425–444.

Stolper E. (1989) Temperature dependence of the speciation of water in rhyolitic melts and glasses. *Am. Mineral.* **74**, 1247–1257.

Stolper E. M. (1982) The speciation of water in silicate melts. *Geochim. Cosmochim. Acta* **46**, 2609–2620.

Sutton A. J. and Elias T. (1993) Volcanic gases create air pollution in the Island of Hawaii. *Earthquakes Volcanoes* **24**, 178–196.

Symonds R. B. and Reed M. H. (1993) Calculation of multicomponent chemical equilibria in gas–solid–liquid systems: calculation methods, thermochemical data and applications to studies of high-temperature volcanic gases with examples from Mount St. Helens. *Am. J. Sci.* **293**, 758–864.

Symonds R. B., Rose W. I., and Reed M. H. (1988) Contribution of Cl- and F-bearing gases to the atmosphere by volcanoes. *Nature* **334**, 415–418.

Symonds R. B., Rose W. I., Bluth G. J. S., and Gerlach T. M. (1994) Volcanic gas studies: methods, results and applications. In *Volatiles in Magmas*, Revienes in Mineralogy, 30 (eds. M. R. Carroll and J. R. Hollaway) Mineralogical Society of America, Washington, DC, pp. 1–66.

Symonds R. B., Mizutani Y., and Briggs P. H. (1996) Long-term geochemical surveillance of fumaroles at Showa-Shinzan dome, Usu volcano, Japan. *J. Volcanol. Geotherm. Res.* **73**, 177–211.

Symonds R. B., Gerlach T. M., and Reed M. H. (2001) Magmatic gas scrubbing: implications for volcano monitoring. *J. Volcanol. Geotherm. Res.* **108**, 303–341.

Tabazadeh A. and Turco R. P. (1993) Stratospheric chlorine injection by volcanic eruptions; HCl scavenging and implications for ozone. *Science* **260**, 1082–1086.

Taran Y. A., Hedenquist J. W., Korzhinsky M. A., Tkachenko S. I., and Shmulovich K. I. (1995) Geochemistry of magmatic gases from Kudryavy volcano, Iturup, Kuril Islands. *Geochim. Cosmochim. Acta* **59**, 1749–1761.

Tedesco D. (1995) Monitoring fluids and gases at active volcanoes. In *Monitoring Active Volcanoes* (eds. B. McGuire and J. Murray). University College London Press, London, pp. 315–345.

Textor C., Sachs P. M., Graf H.-F., and Hansteen T. (2003) The scavenging of sulphur and halogen gases in a plinian volcanic plume similar to the Laacher see eruption 12,900 yr BP. In *Volcanic Degassing*, Geological Society of London Special Publication 213 (eds. C. Oppenheimer, D. M. Pyle, and J. Barclay). Geological Society of London, pp. 307–328.

Thompson D. W. J. and Wallace J. M. (1998) The Arctic Oscillation signature in the wintertime geopotential height and temperature fields. *Geophys. Res. Lett.* **25**, 1297–1300.

Thorarinsson S. (1979) On the damage caused by volcanic eruptions with special reference to tephra and gases. In *Volcanic Activity and Human Ecology* (eds. P. D. Sheets and D. K. Grayson). Academic Press, New York, pp. 125–159.

Thordarson T. and Self S. (1996) Sulfur, chlorine, and fluorine degassing and atmospheric loading by the Roza eruption, Columbia River Basalt group, Washington, USA. *J. Volcanol. Geotherm. Res.* **74**, 49–73.

Thordarson T. and Self S. (2002) Atmospheric and environmental effects of the 1783–84 Laki eruption: a-review and reassessment. *J. Geophys. Res.* **108** (DI), 4011.

Thordarson T., Self S., Oskarsson N., and Hulsebosch T. (1996) Sulfur, chlorine, and flourine degassing and atmospheric loading by the 1783–1784 AD Laki (Skaftar fires) eruption in Iceland. *Bull. Volcanol.* **58**, 205–225.

Thordarson T., Self S., Miller D. J., Larsen G., and Vilmundardóttir E. G. (2003) Sulphur release from flood lava eruptions in the Veidivötn, Grímsvötn, and Katla volcanic systems, Iceland. In *Volcanic Degassing*, Geological Society of London Special Publication 213, (eds. C. Oppenheimer, D. M. Pyle, and J. Barclay). Geological Society of London, pp. 103–121.

Towe K. M. (2002) The problematic rise of Archean oxygen. *Science* **295**, 1419.

Valley J. W. and Cole D. (eds.) (2001) Stable isotope geochemistry, Reviews in Mineralogy, 43, Mineralogical Society of America, Washington, DC, 662pp.

Varekamp J. C. and Rowe G. L., Jr. (eds.) (2000) Crater Lakes In *Journal of Volcanology and Geothermal Research*, Special Issue 97, Elsevier, 508pp.

Vié le Sage R. (1983) Chemistry of the volcanic aerosol. In *Forecasting Volcanic Events* (eds. H. Tazieff and J.-C. Sabroux). Elsevier, Amsterdam, pp. 445–474.

Villemant B. and Boudon G. (1998) Transition between dome-building and plinian eruptive styles: H_2O and Cl degassing behaviour. *Nature* **392**, 65–69.

Villemant B. and Boudon G. (1999) H_2O and halogen (F, Cl, Br) behaviour during shallow magma degassing processes. *Earth Planet. Sci. Lett.* **168**, 271–286.

Villemant B., Boudon G., Nougrigat S., Poteaux S., and Michel A. (2003) H_2O and halogens in volcanic clasts: tracers of degassing processes during plinian and dome-forming eruptions. In *Volcanic Degassing*, Geological Society of London Special Publication 213 (eds. C. Oppenheimer, D. M. Pyle, and J. Barclay). Geological Society of London, pp. 63–79.

Voight B. V., Sparks R. S. J., Miller A. D., Stewart R. C., Hoblitt R. P., Clarke R. P., Ewart J., Aspinall W., Baptie B., Druit T. H., Herd R., Jackson P., Lockhart A. B., Loughlin S. C., Luckett R., Lynch L., McMahon J., Norton G. E., Robertson R., Watson I. M., and Young S. R. (1999) Magma flow instability and cyclic activity at Soufriere Hills volcano, Montserrat BWI. *Science* **283**, 1138–1142.

Walker G. P. L., Self S., and Wilson L. (1984) Tarawera 1886, New Zealand, a basaltic plinian eruption. *J. Volcanol. Geotherm. Res.* **21**, 61–78.

Wallace P. and Anderson A. T. (2000) Volatiles in magmas. In *Encyclopedia of Volcanoes* (eds. H. Sigurdsson, B. F. Houghton, S. R. McNutt, H. Rymer, and J. Stix). Academic Press, San Diego, CA, pp. 149–170.

Wallace P. J. (2001) Volcanic SO_2 emissions and the abundance and distribution of exsolved gas in magma bodies. *J. Volcanol. Geotherm. Res.* **108**, 85–106.

Wallace P. J. and Gerlach T. M. (1994) Magmatic vapor source for the sulfur dioxide released during volcanic eruptions: evidence from Mount Pinatubo. *Science* **265**, 497–499.

Wallace P. J., Anderson A. T., Jr., and Davis A. M. (1995) Quantification of pre-eruptive exsolved gas contents in silicic magmas. *Nature* **377**, 612–616.

Watson B. E. (1994) Diffusion in volatile-bearing magmas. *Rev. Mineral.* **30**, 371–411.

Watson I. M. and Oppenheimer C. (2000) Particle size distributions of Mt. Etna's aerosol plume constrained by sun-photometry. *J. Geophys. Res. Atmos.* **105**, 9823–9830.

Watson I. M. and Oppenheimer C. (2001) Particle-size distributions of ash-rich volcanic plumes determined by sun photometry. *Atmos. Environ.* **35**, 3561–3572.

Webster J. D., Kinzler R. J., and Mathez E. A. (1999) Chloride and water solubility in basalt and andesite melts and implications for magma degassing. *Geochim. Cosmochim. Acta* **63**, 729–738.

Webster J. D., Raia F., De Vivo B., and Rolandi G. (2001) The behavior of chlorine and sulfur during differentiation of the Mt. Somma-Vesuvius magmatic system. *Mineral. Petrol.* **73**, 177–200.

Westrich H. R. and Gerlach T. M. (1992) Magmatic gas source for the stratospheric SO_2 cloud from the June 15, 1991 eruption of Mount Pinatubo. *Geology* **20**, 867–870.

Williams S. N. (1983) Plinian airfall deposits of basaltic composition. *Geology* **11**, 211–214.

Williams H. and McBirney A. R. (1979) *Volcanology*. Freeman, Cooper and Company, San Francisco, CA 397pp.

Williams-Jones G. and Rymer H. (2000) Hazards of volcanic gases. In *Encyclopedia of Volcanoes* (eds. H. Sigurdsson, B. F. Houghton, S. R. McNutt, H. Rymer, and J. Stix). Academic Press, San Diego, CA, pp. 997–1013.

Winner W. E. and Mooney H. A. (1980) Responses of Hawaiian plants to volcanic sulfur dioxide: stomatal behavior and foliar injury. *Science* **210**, 789–791.

Zdanowicz C. M., Zielinski G. A., and Germani M. S. (1999) Mt. Mazama eruption: calendrical age verified and atmospheric impact assessed. *Geology* **27**, 621–624.

Zielinski G. A. (1995) Stratospheric loading and optical depth estimates of explosive volcanism over the last 2100 years derived from the Greenland ice sheet project 2 ice core. *J. Geophys. Res.* **100**, 20937–20955.

Zielinski G. A. (2000) Use of paleo-records in determining variability within the volcanism-climate system. *Quat. Sci. Rev.* **19**, 417–438.

Zielinski G. A., Mayewski O. A., Meeker L. D., Whitlow S., Twickler M. S., Morrison M., Meese D. A., Gow A. J., and Alley R. B. (1994) Record of volcanism since 7000BC from the GISP2 Greenland ice core and implications for the volcano-climate system. *Science* **264**, 948–952.

Zielinski G. A., Mayewski P. A., Meeker L. D., Whitlow S., and Twickler M. S. (1996a) A 110,000-yr record of explosive volcanism from the GISP2 (Greenland) ice core. *Quat. Res.* **45**, 109–118.

Zielinski G. A., Mayewski P. A., Meeker L. D., Whitlow S., Twickler M. S., and Taylor K. (1996b) Potential atmospheric impact of the Toba mega-eruption ~71,000 years ago. *Geophys. Res. Lett.* **23**, 837–840.

Zreda-Gostynska G., Kyle P. R., Finnegan D. L., and Prestbo K. M. (1997) Volcanic gas emissions from Mount Erebus and their impact on the Antarctic environment. *J Geophys. Res.* **102**, 15039–15055.

3.05
Timescales of Magma Transfer and Storage in the Crust

M. R. Reid

University of California at Los Angeles, CA, USA and Northern Arizona University, Flagstaff, AZ, USA

3.05.1 INTRODUCTION

The influx of magma from the mantle into the crust is the principal mechanism by which Earth's crust grows. Understanding the assembly of igneous rocks, whether those ejected at Earth's surface or those solidified within the crust, is therefore critical to our ability to delineate crustal evolution. Processes of magmatic differentiation, including crystal–liquid separation, crustal melting, and other thermal and mass exchanges between ascending magmas and pre-existing crust, are intimately linked to the development of compositional heterogeneities and their distribution within the crust. Understanding the kinematics of these differentiation processes, therefore, provides insight into the dynamics of crustal growth. From a volcanological perspective, the crust provides the staging area from which magmas rise to the surface. Understanding the dynamic balance between magma supply and evolution will lead to better understanding of the factors that determine whether magmas will erupt or not and, when they do erupt, whether they will erupt explosively.

Many steps may limit the supply of magmas to their sites of crustal emplacement. These steps include melt production and segregation, magma ascent and emplacement into regions of storage, and episodes of outright magma storage. In this chapter, I address the topic of magma storage—and specifically the question of how long magmas may be stored in the crust before erupting or solidifying at depth—by reviewing the manifold ways in which magma storage times have been quantified. This study complements and expands on recent reviews by Hawkesworth *et al.* (2000) and Condomines *et al.* (2003). Magma systems representative of all settings are included here, but the focus is on those associated with the continental crust; Elliott and Spiegelman (see Chapter 3.14) review how U-series investigations specifically illuminate igneous processes in the oceanic crust.

3.05.1.1 Physical Aspects of Magma Transport and Storage

Percolative flow of magma is important for melt segregation at both the sites of melt generation and during differentiation (Brandeis and Jaupart, 1986; Kelemen *et al.*, 1997; Sisson and Bacon, 1999; Tait and Jaupart, 1992), but ascent of magma through the crust is probably dominated by flow that is channelized in some fashion. Magmas may ascend along faults or propagating fractures, as diapirs, or by heterogeneous brittle–ductile processes (e.g., Brown and Solar, 1998; Korenaga and Kelemen, 1998; Miller and Paterson, 1999; Petford, 1996); dike ascent, in particular, is both thermally efficient and rapid (Clemens, 1998). Ascent rates of 0.001 m s^{-1} to >0.1 m s^{-1} have been obtained for various magmas, based on theoretical and empirical evidence (e.g., Brandon *et al.*, 1996; Devine *et al.*, 1998; Klugel, 1998; Roggensack, 2001a; Rutherford and Hill, 1993) and could imply that only a matter of months are required for magmas to traverse the continental crust unless ascent stalls and magmas accumulate.

Episodes of magma storage may be attendant, although not universal features of the transfer of magma through the crust, and may govern the average rate of magma ascent (Bons *et al.*, 2001; Hansteen *et al.*, 1998; Klugel, 1998; Singer *et al.*, 1998). Magma reservoirs can develop when ascending melts become neutrally buoyant or encounter a permeability or rheological barrier, or when instabilities within or around the ascending melts are amplified (Brown, 2001; Glazner and Ussler, 1988; Putirka, 1997). Crustal accommodation of melt reservoirs may occur by ductile flow or by lifting or lateral migration of wall rock, depending on the tectonic setting, depth of emplacement, and rheology of the magma (Brown and Solar, 1998; Cruden and McCaffrey, 2001; Glazner and Miller, 1997; Wiebe and Collins, 1998; Yin and Kelty, 2000). Given possible magma ascent rates, timescales of magma reservoir growth could also be geologically fast (≪1 Myr), even for large plutons (Clemens and Mawer, 1992; Cruden and McCaffrey, 2001; Petford, 1996), unless growth is limited by magma supply or by the mechanics of pluton emplacement (e.g., Clemens, 1998; Johnson *et al.*, 2001).

The composite characteristic of many plutons show that larger magma reservoirs can be constructed piecemeal from numerous magma pulses (e.g., McNulty *et al.*, 2000; Robinson and Miller, 1999; Wiebe *et al.*, 2001). Evidence of inputs from compositionally and/or isotopically distinct injections may be recorded at a variety of scales—from submillimeter to hundreds of meters (e.g., Bacon and Druitt, 1986; Davidson and Tepley, 1997; Nakada *et al.*, 1994; Reid *et al.*, 1983; Wiebe, 1996). The rate of accumulation of single magma inputs may, therefore, be a subordinate component of the overall timescale of magma emplacement. As long as a system persists at supersolidus conditions, the duration of magma recharge may also be insignificant relative to that of magma storage. Accordingly, even though the duration of magma storage in the crust is the primary focus of this chapter, it is not always possible to differentiate this duration from that of magma emplacement or even ascent. For this reason, in more ambiguous cases, magma ascent, emplacement, and storage will collectively be referred to as magma transfer.

Evidence that magmas presently "reside" in the crust is provided by geophysical imaging of probable magma in a variety of tectonic settings, including under fast- and slow-spreading mid-ocean ridges and ridge-adjacent volcanoes (e.g., Dunn *et al.*, 2000; Sinha *et al.*, 1998; West *et al.*, 2001), ocean islands (Dawson *et al.*, 1999), volcanic arcs (Chmielowski *et al.*, 1999; Masturyono *et al.*, 2001; Zandt *et al.*, 2003), intracontinental rifts and calderas (Miller and Smith, 1999; Sheetz and Schlue, 1992; Steck and Prothero, Jr., 1994), and even within continental collision zones (Kind *et al.*, 1996; Nelson *et al.*, 1996). It is, however, notable in the context of this chapter that magma reservoirs have not been detected beneath many otherwise active volcanoes. Multiple levels of magma intrusion beneath individual volcanoes have been inferred from geophysical and geochemical observations (e.g., Blundy and Cashman, 2001; Condomines *et al.*, 1995; Crawford *et al.*, 1999; Dunn *et al.*, 2000; Haslinger *et al.*, 2001; Ryan, 1988). Intrusive igneous rocks also provide unequivocal evidence for magma storage at crustal levels at times in the past.

Magma storage must also play an important, if not universal, petrologic role in the evolution of magmas as they transit the crust. Even though the mantle is the fundamental source of magmatism and magmas derived largely by crustal melting may contain mantle-derived components (e.g., Coleman *et al.*, 1995; Hildreth and Moorbath, 1988; Patino Douce, 1999; Sisson *et al.*, 1996; Wiebe, 1996), most magmas are not in equilibrium with the mantle. Some locus for differentiation and/or hybridization of magma is thus implied. Evidence that magmas can differentiate in regions of magma ponding is provided by appreciable volumes of crystal cumulates in numerous intrusive complexes (e.g., Robinson and Miller, 1999; Sisson *et al.*, 1996; Wiebe, 1996) and in lower crustal xenolith suites (see Rudnick and Presper, 1990, and references therein). The duration of magma storage can, therefore, help to constrain timescales of differentiation, and vice versa.

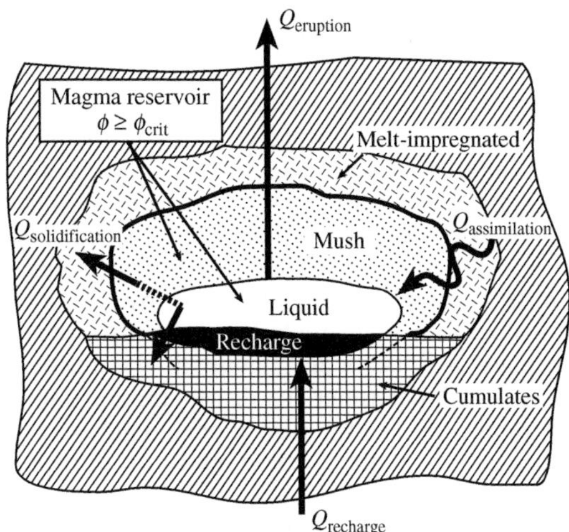

Figure 1 Cartoon of simplified magma reservoir representing the relationship between different rheological regimes associated with a magma system and some of the fluxes (Q) associated with the magma reservoir. The magma reservoir consists of liquid and mush regimes and is separated from the melt-impregnated regime by a critical porosity, ϕ_{crit}, beyond which the magma is quite rigid and unlikely to be eruptable. Much of the liquid regime could be relatively crystal-free if crystallization occurs dominantly near the transition to the mush regime. The cumulate pile illustrated at the base of the reservoir may consist of phases that settled through the liquid, especially those present when the magma was initially emplaced, and/or mush-zone crystals from which melt has been partially expelled. The melt-impregnated regions may also consist of cumulates if melts migrated from them, especially while in the mush regime. Input fluxes are those of melts supplied from depth (recharge) and from assimilation; in this case, the recharge liquid is shown as unmixed with resident liquid, as might occur in a stratified magma reservoir, but mixing between these magmas or between different liquid domains within the magma reservoir is possible. Outputs include eruption and subsurface crystallization, either in the solidified margins of the magma reservoir or in associated dike and sills (not shown); with respect to geochemical estimates of magma reservoir volumes, the melt-impregnated regime and some or all of the mush may also effectively be part of the output (after Koyaguchi and Kaneko, 2000).

3.05.1.2 What Constitutes a "Magma Reservoir" and Its Storage Time?

Where ponding and cooling of magma occurs, the associated *magma reservoir* can be considered to consist of two portions, liquid and mush (e.g., Marsh, 1989; Sinton and Detrick, 1992; Vigneresse *et al.*, 1996). The *liquid* regime (Figure 1) behaves as a liquid in a mechanical sense and may contain up to 25–40% crystals (e.g., Philpotts *et al.*, 1998; Saar *et al.*, 2001). The *mush* regime consists of a dense suspension containing up to 50–75% crystals that can resist shear deformation and therefore eruptive withdrawal (e.g., Lejeune and Richet, 1995; Marsh, 1981; Rutter and Neumann, 1995; Ryerson *et al.*, 1988). Accordingly, "magma reservoir" refers to a region of magma supply and storage. This reference is distinct from that of "magma chamber" which, because it connotes an enclosed space or cavity, might more aptly describe only the liquid regime of the reservoir. As defined here, the magma reservoir does not include the melt-impregnated regime (<25–50% melt) that is closer to the magma's solidus than the mush state. In this near-solidus regime, the interconnectivity of crystals is great and chemical diffusion is significantly slower than thermal migration. This regime may, however, be imaged geophysically and may cause magmatic anomalies to be detected in space and time well beyond where and when they cease to be magma reservoirs as defined here (e.g., Sharp *et al.*, 1980).

As delimited here, the boundary of the magma reservoir broadly corresponds to the conditions at which differential movement of crystals and melt, and therefore appreciable chemical modification of the magma, is likely to cease (e.g., Trial and Spera, 1990; Vigneresse *et al.*, 1996). Just as the thermal anomaly associated with magmatism is a transient crustal feature, so are the locations of the boundaries between a cooling magma reservoir's liquid and mush regimes, and between the reservoir and largely solid material. Besides the effect of temperature in determining the relative proportion of crystals and melt, boundaries between the various rheological regimes also depend on the viscosity and volatile content of the melt, and on the extent of syn-magmatic deformation (e.g., Barboza and Bergantz, 2000; Renner *et al.*, 2000; Rutter and Neumann, 1995; Saar *et al.*, 2001; Vigneresse *et al.*, 1996). Supersolidus temperatures can be maintained by repeated intrusions of new melt which, depending on their volumes and locations relative to the resident magma, will determine the distribution of different rheological and compositional regimes. For example, at relatively slow rates of intrusion relative to cooling, magmatic systems may consist mostly of mush and relatively more evolved melt (Shaw, 1985; Sinton and Detrick, 1992; Smith, 1979). At higher rates of melt intrusion, there may be greater penetration of magma recharge into a system, establishment of a liquid reservoir, and appreciable magma mixing; generation of evolved melt may, at the highest rates, be suppressed (e.g., Rhodes and Hart, 1995; Rubin *et al.*, 2001). Repeated magma recharge may sustain regions of silicic liquid and/or mush in the crust (Koyaguchi and Kaneko, 2000, and references therein). Magma recharge can also cause the boundaries between the rheological regimes within reservoirs to migrate if they scour the virtual "walls" created by the liquid–mush transition (Bergantz, 2000) or invade earlier solidified or near-solidified intrusions.

For the purposes of this chapter, no rigorous definition of "magma storage time" is implied, because it is not always possible to uniquely distinguish what part of a magma system is being characterized in a particular study or the effects of various rate-controlling processes that govern the accumulation and solidification of magma in the crust. Rather, the approach taken here is to examine the various bounds that can be placed on how long volumes of magma are present within the crust. Magma residence times so described may, for example, pertain to the duration of magma emplacement (e.g., inflation of a magma reservoir that is not supply limited) or, where multiple pulses of magma are involved, magma accumulation, or to the duration of cooling from a liquid to a semi-solid state. Timescales may also be delimited by rates of magma transfer, as defined above. Most of the estimates for magmatic timescales summarized here and elsewhere are based on volcanoes and their eruptive products, largely because these provide the best petrological, geochemical, and geophysical constraints. Storage timescales do not strictly correspond to the magma reservoir as defined here: where based on the eruptive products, estimates generally pertain to the more mobile portions of the magma reservoir, i.e., the liquid and, to variable degrees, mush regimes of the magma reservoir; where based on geophysical imaging of subvolcanic systems, estimates may include regions larger than the magma reservoir. Timescales have also been inferred from plutonic rocks (e.g., Paterson and Tobisch, 1992; Petford *et al.*, 2000), but these are generally less certain, insofar as pluton records are clouded by greater opportunity for subsolidus re-equilibration, the heretofore lack of sensitive chronometers, and uncertainties about the relative proportion of crystals and melt at various stages of pluton evolution. Recent work promises future insights in this area (e.g., Coulson *et al.*, 2002; Friedrich *et al.*, 1999; McNulty *et al.*, 1996; Miller *et al.*, 1999; Miller and Paterson, 2001; Wallace and Bergantz, 2002).

In this review, I am primarily concerned with describing constraints on the interval of time that a particular pulse of magma is transferred and/or stored in the crust. In the broadest sense, the crustal residence times of magmas could also be construed to be the interval of time that magma associated with a particular magmatic center occupies the crust. Approximately 0.5–1 Myr life spans characterize many individual volcanic and plutonic systems (e.g., Coulson *et al.*, 2002; Hildreth and Lanphere, 1994; Miller and Paterson, 2001; Quane *et al.*, 2000; Singer *et al.*, 1997), and loci of magmatic activity may last for several million years (e.g., Christiansen, 2001; Mankinen *et al.*, 1986; Miller *et al.*, 1999; Stewart *et al.*, 2002; Zou *et al.*, 2002). It is within these million-year time frames, then, that the specific magma batches that are the subject of this review transit the crust.

3.05.2 GEOPHYSICAL AND TIME-SERIES ESTIMATES FOR RESIDENCE TIMES AND VOLUMES OF MAGMAS

One simple conceptualization of a magma reservoir is a storage container to which mass is supplied at a particular rate, Q_{in}, and from which mass is extracted at a particular rate, Q_{out} (Figure 1). In the limit, some magma reservoirs may be thought of as undergoing essentially continuous efflux of magma due to eruption (exogenous output) and plutonic rock formation (endogenous output), and influx due to magma replenishment. In such cases, they may attain a quasi-steady-state condition where magma input and output rates are equal, i.e., $Q_{out} = Q_{in}$. For these, the magma renewal or residence time (*sensu stricto*), τ, or average duration of magma storage in the reservoir prior to eruption, can be strictly defined as mass $\times\, Q^{-1}$ (or, in volumetric terms, $V \times Q^{-1}$). Fewer than 30% and likely <10% of active subaerial volcanoes may approximate these conditions (cf. Pyle, 1992;

Wadge, 1980, 1982), but these include familiar examples such as Kilauea (Hawaii), Etna prior to 1970 (Italy), and Piton de la Fournaise (Reunion Island). For other volcanoes, a mean magma storage rather than residence time may be estimated by assuming that a particular magma volume is filled or emptied at an average magma flux rate.

Some measure of volcanic effluxes is provided by time-integrated eruptive rates and by rates of volcanic edifice construction. It should be borne in mind that eruption rates may be relatively invariant or variant, depending on the observational timescale (Dvorak and Dzurisin, 1993; Harris *et al.*, 2000; Hildreth and Lanphere, 1994; Quane *et al.*, 2000; Singer *et al.*, 1997). Additionally, for most volcanoes, eruption is not the sole output of the magma reservoir: there is also subsurface solidification of magma. Thus, the relationship between volumes and residence times is only an approximate one when magmatic flux rates are based on eruption rates alone. In favorable cases, magmatic flux rates may be estimated from influxes instead, based on gas and thermal fluxes (e.g., Francis *et al.*, 1993) or, for mid-ocean ridges, the rate of crust formation. These, too, may lead to inaccurate estimates of the volumes or storage times of resident magmas if the influx is distributed in multiple rather than single reservoirs.

Volumes of magma reservoirs can be delimited from seismic tomographic and geodetic inversions as well as from the distribution of gravity and electromagnetic anomalies. At eruption rates characteristic of each volcano, mean ages for magma reservoirs beneath central vent volcanoes have been estimated to be tens of years to several thousands of years or more (Figure 3). For spreading centers, timescales of less than a few thousands of years have also been inferred (Barth *et al.*, 1994; Hooft *et al.*, 1997). The accuracy of these estimates depends in part on the scale of the imaged volcanic system and the distribution of liquid within it, and on the connectivity between that liquid and erupted magma. Because $Q_{eruption} \leq Q_{input}$ in most cases, residence times inferred from volumes and eruption rates are likely to be maxima. For example, if rates of endogenous output exceed those of exogeneous output by as much as an order of magnitude or more (cf. Crisp, 1984; Francis *et al.*, 1993; Smith, 1979), then residence times could be commensurately lower.

For quasi-steady-state magma reservoirs, secular chemical fluctuations in the eruptive products provide a means of estimating the storage time of magma directly, from which the volume of resident magma can also be delimited (Figure 2; Cortini and Scandone, 1982 Albarede, 1993; Pyle, 1992). Progressive mixing between compositionally distinct input and resident magmas within a magma reservoir (or "chemical reactor"; cf. Albarede, 1993) will be expressed in extrusive lavas by chemical variations, whose magnitudes are inversely proportional to magma residence time and volume of resident magma. The input and resident magmas may differ in their incompatible trace element and/or isotopic ratios as a result of secular chemical variations in the input magma, as illustrated in Figure 2, or, in the case of U-series disequilibria (described further in Sections 3.05.3 and 3.05.4 below), as a result of radioactive decay in the resident magma. Provided that the compositional ranges of magma inputs can be delimited and that the reservoir is chemically homogeneous (i.e., the mixing time is significantly less than residence time), an average duration of pre-eruptive magma storage can be obtained. The latter caveat means that magmas must be capable of mixing so that magma residence times—and volumes derived from them—reflect only the low-viscosity portion of the magma reservoir and not liquid contained in the crystal mush. Accordingly, time-series geochemical estimates for the residence times of the more fluid portions of magma are generally tens of years (Figure 3(a)). Corresponding volumes are less than a few cubic kilometers (except for the deep reservoir beneath Etna) and uniformly less than volumes obtained by geophysical approaches where both have been applied to the same volcano. Notably, even in cases where Q_{in} and Q_{out} deviate appreciably from one another, resulting uncertainties in residence times for a given magma reservoir may still be subordinate to other sources of error in the timescale (cf. Albarede, 1993).

Even though quantitative estimates of reservoir volumes may not be obtained from the chemical variations of magma systems that deviate significantly from steady state, chemical fluctuations (or the lack thereof) may still provide temporal markers that delimit the duration of magma storage. Chemical changes can often be correlated with changes in rates of effusion or volcano inflation, as observed at Etna (Condomines *et al.*, 1995), Kilauea (Pietruszka and Garcia, 1999), and Unzen (Chen *et al.*, 1993), suggesting that the chemical changes are due to introduction of chemically distinct magmas into the reservoir rather than, for example, turnover or sequential tapping of compositionally zoned magma reservoirs. Progressive changes in the isotopic compositions of effusives from Etna and Stromboli suggest that a common magma reservoir may be sequentially tapped over minimum periods of tens of years beneath stratovolcanoes (Condomines *et al.*, 1995; Francalanci *et al.*, 1999). Petrologic, chemical, and/or isotopic affinities between lavas erupted intermittently over periods of tens of thousands (e.g., Soufriere Hills andesite; Murphy *et al.*, 2000) or hundreds of thousands (e.g., Glass Mountain rhyolites; Halliday *et al.*, 1989) of years could indicate repeated sampling of a much more

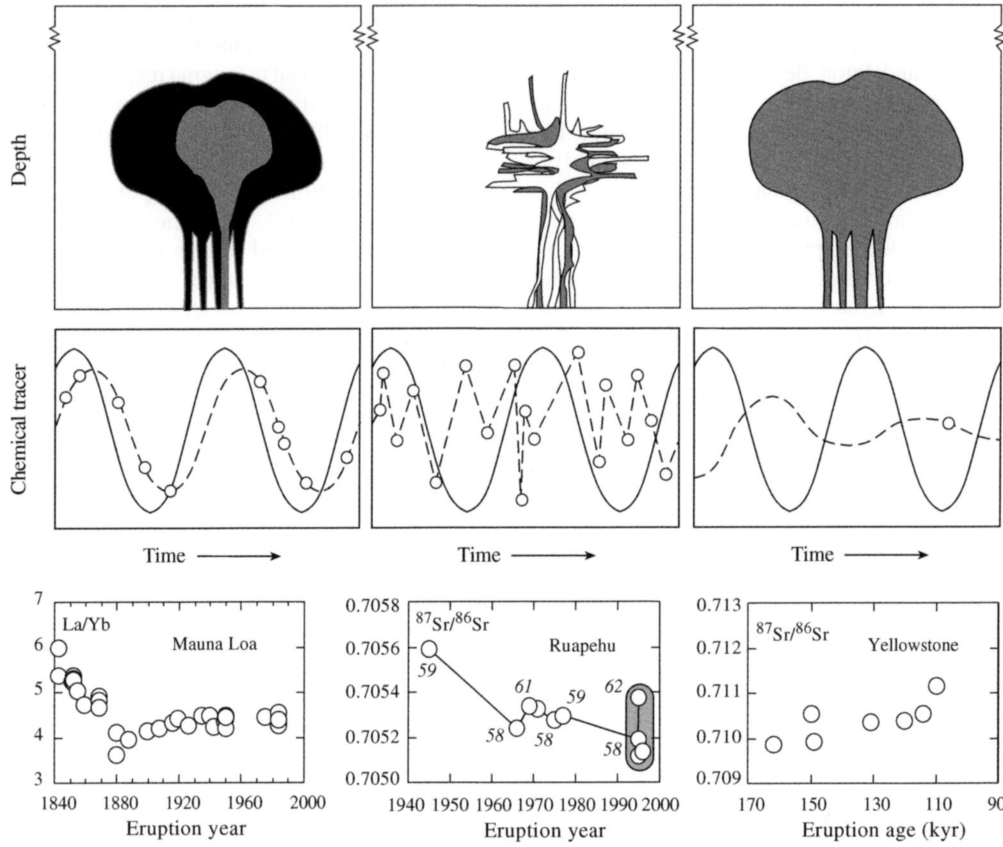

Figure 2 Various views on aspects of magma reservoirs. **Top panels**: cartoons of magma reservoirs with different geometries. *Left*: mush-dominated system, the most general magma reservoir configuration (cf. Figure 1), inferred to exist beneath mid-ocean ridges, ocean islands, and, increasingly, voluminous silicic centers. *Center:* plexus of dikes and sills, inferred to characterize many central vent volcanic systems, especially arcs. *Right:* more conventional view of a magma reservoir as a largely liquid-filled domain, often taken to characterize voluminous silicic magma systems. The magma reservoir configurations shown here are largely end-member ones and aspects of all three, e.g., liquid-dominated dikes and sills within a mush-dominated reservoir, may characterize magmatic systems in time and space. **Middle panels**: illustrations of how chemical tracers (i.e., chemical and/or isotopic ratios) in erupted lavas (circles) might relate to secular evolution in the magma reservoir (dashed lines) and temporally varying chemical signatures of input magmas (solid lines), whether crust- or mantle-derived. *Left:* steady-state magma system. Secular chemical variations in the output magmas are derivative of variations in input magmas but are offset in time and reduced in magnitude or, in the case of U-series decay, balance the effects of magma input and radioactive decay during magma residence. *Center:* coexisting reservoirs of dikes and sills in which magmas are intermittently stored results in output magma variations that do not correlate in any straightforward way with the secular chemical variations of input magmas. *Right:* accumulating magma reservoir in which input magma variations are increasingly damped by continuous mixing. If this system were, instead, at steady state with respect to magmatic fluxes, a situation more like that of the left panel would a pertain, but the oscillations would be even more offset in time and the magnitude of chemical heterogeneity would be even more damped from those of the input magmas. Note that for more silicic magmas, secular damping of chemical variations could reflect isolation of stored magma from input magmas due to stratification. **Bottom panels**: Examples of time-series chemical variations in different magmatic systems. *Left:* basalts from Mauna Loa, Hawaii (Rhodes and Hart, 1995). *Center:* andesites from Ruapehu volcano, New Zealand (Hobden *et al.*, 1999); italicized numbers are silica contents (in weight percent). *Right:* rhyolites from the Central Plateau Member of the Plateau Rhyolite, Yellowstone volcanic system (Vazquez and Reid, 2002). Note the longer time interval represented by the Central Plateau Member rhyolites compared to the first two. Also, the largely liquid-filled reservoir illustrated in the top panel *might* better characterize the climatic eruptions at Yellowstone (for which the significance of time-series analyses is equivocal) and not the reservoir responsible for the Central Plateau Member rhyolites (for which the increasing isotopic compositions is largely attributable to assimilation and radioactive decay in a long-standing magma reservoir) (Vazquez and Reid, 2002). Figure based, in part, on Pietruszka and Garcia (1999).

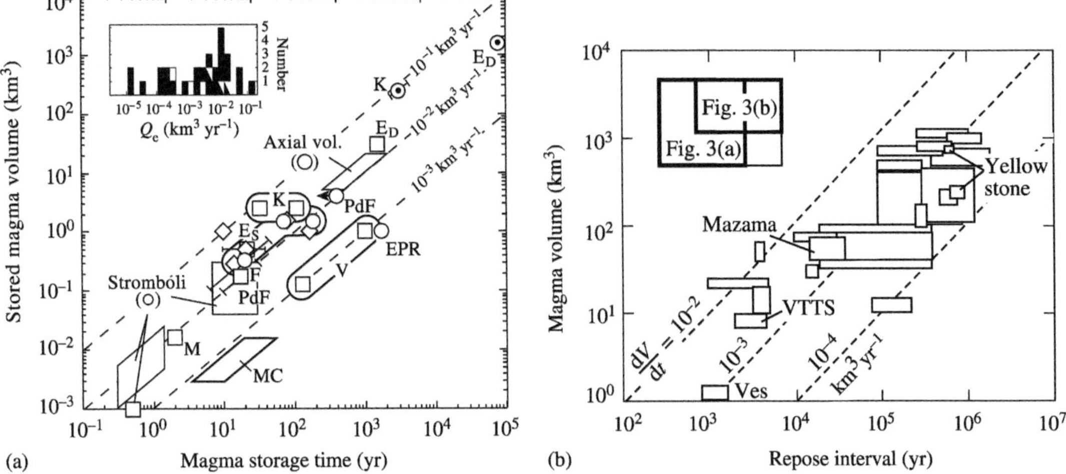

Figure 3 Illustration of inferred magma residence times and volumes of volcanic rocks. (a) Calculated residence time versus estimated volume for a variety of volcanoes. Inferences are based mainly on coupling magma efflux rates to geophysical imaging of magma volumes or time-series analyses of volcanic products. See text for more details. Fields encircle estimates for individual volcanoes representing different eruptive intervals studied by the same technique or similar estimates obtained using different techniques. Abbreviations are M: Merapi; MC: Mount Cameroon; F: Fuego; PdF: Piton de la Fournaise; E: Etna (subscripts D and S for deep and shallow magma reservoirs, respectively); K: Kilauea; V: Vesuvius; EPR: East Pacific Rise. *Circles*: Geophysical estimates of magma volumes; storage times calculated from volcanic efflux rates except for Axial volcano and highest Kilauea estimate for which estimates of total magma fluxes were used (sources Barth *et al.*, 1994; Denlinger, 1997; Pyle, 1992; Sharp *et al.*, 1980; West *et al.*, 2001). *Squares*: geochemical estimates of magma volumes; magma storage times are estimated from volcanic efflux. The two lowest estimates for storage times are constrained by the effects of degassing and therefore reflect the duration of magma storage at relatively shallow depths and/or of ascent (sources Albarede, 1993; Cortini and Scandone, 1982; Fitton *et al.*, 1983; Francalanci *et al.*, 1999; Gauthier and Condomines, 1999; Pietruszka and Garcia, 1999; Pyle, 1992). *Diamonds*: other, mainly geochemical estimates of magma storage times; stored magma volumes estimated from volcanic efflux. No labels or errors shown for these, generally qualitative, estimates (sources Chen *et al.*, 1993; Fitton *et al.*, 1983; Hobden *et al.*, 1999). Contours are magma flux rates that are used to calculate residence times in the case of geophysical estimates or those from which volumes are calculated in the case of geochemical estimates. *Inset*: histogram of volcanic efflux rates for basaltic magmas from Crisp (1984). (b) Plot of volume of explosively erupted and compositionally zoned magmas relative to repose time, i.e., the time elapsed between two successive compositionally heterogeneous pyroclastic eruptions. Boxes show range of uncertainty in volume and repose interval. Abbreviations are Ves: Vesuvius and VTTS: Valley of Ten Thousand Smokes (after Trial and Spera, 1990).

persistent magma reservoir. Protracted storage and repeated eruptions may be facilitated by magma recharge with little mixing but significant thermal input.

In many cases, lavas may exhibit nearly as much geochemical variability over eruptive intervals of only tens to hundreds of years as the range exhibited by lavas erupted over the significant fraction of the volcano's history (e.g., Gamble *et al.*, 1999; Hobden *et al.*, 1999; Quane *et al.*, 2000; Rhodes and Hart, 1995). For arc stratovolcanoes especially, successive eruptions can be compositionally distinct and bear no obvious chemical affinity to immediately foregoing eruptions. This could reflect derivation from magma reservoirs that are strongly influenced by assimilation (Gamble *et al.*, 1999) besides being subject to magmatic recharge (Figure 1; Murphy *et al.*, 2000; Nakamura, 1995a; Pallister *et al.*, 1992). By analogy to steady-state magma reservoirs, the strong responsiveness of these systems to the chemical effects of recharge, assimilation, and/or fractional crystallization implies relatively small magma reservoirs and therefore relatively short residence times. An alternative interpretation is that erupted magmas are stored in separate, noncommunicating crustal magma chambers, as might apply if the magma reservoir system is a plexus of dikes and sills (e.g., Brophy and Dreher, 2000; Gamble *et al.*, 1999; Myers *et al.*, 2002; Figure 2). In either case, magma input apparently never appreciably exceeds output, so that magma reservoirs are essentially ephemeral features of volcanic plumbing systems. Persistent magma reservoirs are frequently concluded to be small (<1 km^3) and, storage timescales, where inferred, are generally on the order of no more than a few thousand years and perhaps only years (Dungan *et al.*, 2001; Gamble *et al.*, 1999; Hobden *et al.*, 1999).

When time-integrated Q_{in} exceeds Q_{out}, magma will accumulate and reservoirs will inflate. That voluminous quantities of magmas can accumulate is evident from caldera-forming eruptions in particular, where tens to thousands of cubic kilometers of magma are evacuated rapidly and the caldera footprint correlates with, and likely mirrors, the areal extent of the magma reservoir (Smith, 1979; Spera and Crisp, 1981). Accumulation times of thousands to hundreds of thousands of years would pertain to these systems if one assumes average rates of magma influx like that of mafic magmatism (0.01 $km^3\ yr^{-1}$) (Crisp, 1984) those are required to sustain a high-level, mostly liquid reservoir (Shaw, 1985). Given that production of evolved magma by differentiation and/or crustal melting is relatively inefficient, attendant mass losses due to crystallization would mean still larger fluxes and therefore accumulation times (or input rates). Inferred accumulation times would also be minima if, as is often deduced, only a portion of the stored magma is evacuated in any particular eruption.

A broad correlation between the volumes of caldera-forming eruptions and the interval of repose between them implies an average production rate 0.001 $km^3\ yr^{-1}$ for differentiated magmas that erupt explosively (Shaw, 1985) or about an order magnitude less than typical of mafic magmas (Crisp, 1984; Smith, 1979). This relationship is also observed when only those explosive eruptions that are compositionally zoned are compared to repose intervals (Figure 3(b)) which suggests that magmas not only accumulate but differentiate internally during the repose interval between eruptions. Storage times might, therefore, be approximately as long as the repose interval, and therefore range from thousands of years (e.g., Valley of Ten Thousand Smokes; Hildreth and Fierstein, 1983) to approximately a million years (e.g., Lava Creek Tuff, Yellowstone; Lanphere *et al.*, 2002), with mean storage times being half that. However, if silicic magmas are, for example, the product of multistage intrusion and subsequent remelting (e.g., Huppert and Sparks, 1988; Koyaguchi and Kaneko, 1999), the actual storage time for silicic magmas may be a fraction of the nominal accumulation times.

3.05.3 GENERAL CONSTRAINTS ON THE DURATION OF MAGMA TRANSFER FROM U-SERIES DISEQUILIBRIA

In most magmas some of the longer-lived members of the two U-decay series are not in statistical equilibrium, and the magnitudes of these disequilibria provide valuable chronometers of magmatic timescales. To produce these disequilibria, elements involved in the decay chain must be chemically fractionated even though, with notable exceptions, the affected elements are comparably incompatible during crystallization of melts. Melting and melt extraction is a potentially effective means of fractionating the U-series nuclides, and the largest degrees of disequilibria are found in relatively more mafic rocks (Figure 4); disequilibria in certain silicic magmas may also reflect fractionation during crustal anatexis. Consequently, most U-series clocks are started when melt is extracted from its source and expire thousands or several hundreds of thousands of years later. Upper limits for magma transfer (magma ascent and storage) times can, therefore, be obtained by delimiting the maximum time elapsed since melt extraction (e.g., Allègre and Condomines, 1976; Oversby and Gast, 1968).

Radioactive disequilibria in magmatic rocks is generally observed between ^{238}U and its daughter ^{230}Th ($t_{1/2} \sim 75$ kyr), between ^{235}U and its daughter ^{231}Pa ($t_{1/2} \sim 33$ kyr), and between ^{230}Th and its daughter ^{226}Ra ($t_{1/2} \sim 1.6$ kyr) (half-lives are those of the daughter nuclides). Because the half-lives of the daughter nuclides dictate the rate of recovery of each pair to statistical equilibrium, secular disequilibria involving the foregoing pairs will generally vanish within 8 kyr ($^{230}Th-^{226}Ra$) to 400 kyr ($^{238}U-^{230}Th$) of any fractionation. Considering the general magnitude of disequilibria observed in mafic lavas from a variety of tectonic settings, despite the nominally similar incompatibility of the elements involved (Figure 4), a zeroth-order conclusion is that magma transfer (ascent, emplacement, and storage) times are apparently less than the time for ^{226}Ra to recover to equilibrium with ^{230}Th, or <8 kyr. In some instances ^{230}Th and ^{238}U are in, or nearly in, secular equilibrium, but the presence of rather pronounced disequilibria between $^{226}Ra-^{230}Th$ and/or $^{231}Pa-^{235}U$ (e.g., Asmerom *et al.*, 2000; Pickett and Murrell, 1997) shows that this is likely due to a lack of U–Th fractionation during melting rather than to the passage of several hundreds of thousands of years. The much shorter-lived radionuclide pair, $^{228}Ra-^{232}Th$ ($t_{1/2} = 5.8$ yr) is, in contrast, almost invariably in secular equilibrium. Because the other, longer-lived isotopes of these elements (^{226}Ra and ^{230}Th) are not in equilibrium, it is evident that the $^{228}Ra-^{232}Th$ equilibrium must have been similarly disrupted initially, from which it can be concluded that magma transfer times must be long enough for the $^{228}Ra-^{232}Th$ pair to recover to secular equilibrium, i.e., a minimum of tens of years. Disequilibrium between ^{226}Ra and its longest-lived radioactive daughter, ^{210}Pb ($t_{1/2} = 22$ yr), is also sometimes observed. In these cases, either magma transfer times are relatively short (e.g., Sigmarsson, 1996) or, because lead is relatively volatile, degassing plays a significant, potentially quantifiable (Figure 3; Gauthier and Condomines, 1999; Gauthier *et al.*, 2000), role in fractionating radium and lead at shallow magma depths.

If the magnitude of radioactive decay at the time of eruption is to be used to bracket the timescale of magmatic transfer and storage, the absolute age of eruption must be known for prehistoric eruptions, eruption ages depend on radiometric dating, the certainty of which can be problematic, as illustrated by well-known (e.g., Fish Canyon Tuff: Schmitz and Bowring, 2001 and Lathrop Wells: Heizler *et al.*, 1999) and lesser-known (e.g., Rockland Tuff: Lanphere *et al.*, 1999 and the Torihama dacite: Sano *et al.*, 2002) examples of discordant eruption ages for the same rocks. For the purpose of this review, reported eruption ages and analytical errors are used, wherever possible, to establish the uncertainty on any particular magma storage time, but it should be borne in mind that the accuracy of eruption ages is a compounding source of uncertainty.

3.05.4 TIMESCALES OF MAGMA DIFFERENTIATION

Magma ascent and storage may be accompanied by opportunities for cooling and crystal–liquid separation, i.e., differentiation. The time elapsed during this ascent and storage will also serve to bring melting-induced disequilibria towards secular equilibrium. Durations of mass transfer might, therefore, be expected to correlate with indices of differentiation if the duration of differentiation represents a significant fraction of those transfer times (e.g., Allègre and Condomines, 1976; Gill *et al.*, 1993).

3.05.4.1 Timescales of Differentiation from U-series Disequilibrium

Differences in the relative extents of U-series disequilibria in consanguineous suites of lavas (Figure 4) provide a means of broadly constraining the differential magma storage times for more evolved magmas. For the suites of arc and mid-ocean ridge lavas illustrated in Figures 4(a) and (b), the magnitudes of $^{226}Ra-^{230}Th$ disequilibria broadly decrease with increases in differentiation indices, e.g., increasing SiO_2 or decreasing Mg#. If the magnitude of disequilibria in the mafic lavas is representative of that at the onset of differentiation and variations result from decay alone, then these differences in $^{226}Ra/^{230}Th$ delimit relative differences in overall transfer times as a function of composition. For example, taking the initial $^{226}Ra/^{230}Th$ ratio of the island arc lavas to be ~5, like that of the more radium-enriched mafic lavas ($SiO_2 \sim 50\%$), apparent transfer times would be >3,000 yr longer for melts containing 55% SiO_2 than for their mafic precursors (Figure 4(a)). Some fraction of these time differences would, therefore, appear to be associated with the duration of magma storage and not just the duration of upward migration of melts.

Figure 4(c) shows specific differentiation timescales derived from compositional end-members of heterogeneous single eruptions (e.g., Vigier *et al.*, 1999) and/or geographically proximal eruptions (e.g., Claude-Ivanaj *et al.*, 1998), the affinities between which have been inferred on the basis of isotopic and incompatible element similarities. Differentiation timescales based on the magnitude of disequilibria between $^{238}U-^{230}Th$ as well as $^{230}Th-^{226}Ra$ are illustrated. Given the frequency with which $^{230}Th^{226}Ra$ disequilibria are observed in young lavas, ages of >8 kyr may seem surprising: changes in $^{230}Th/^{238}U$ arising from decay would not be expected to be detectable in such a relatively short time. Nonetheless, the lack of appreciable ^{230}Th excesses over ^{238}U in trachytes and phonolites could indicate tens to hundreds of thousands of years of differentiation if they evolved from ^{230}Th-enriched alkalic basalts like those with which they are associated (Figure 4; Bourdon *et al.*, 1994; Hawkesworth *et al.*, 2000; Widom *et al.*, 1992).

3.05.4.2 Effects Other than Time on U-decay Series Differentiation Ages

Although the variation in the magnitude of U-series disequilibria with magma composition offers considerable promise for determining differential timescales of magma transfer, and therefore of differentiation, its rigor is tempered by several competing factors. These are discussed in the following sections.

3.05.4.2.1 Effect of variation in initial ratios on differentiation ages

The magnitudes of initial disequilibria exhibited by parental magmas may vary considerably and this presents the most severe—and least constrained—source of uncertainty in assigning a specific transfer time to a specific composition. Partial melting dynamics, metasomatic effects (especially in the case of arcs), the time elapsed since extraction from the source, and/or preferential interaction with or melting of minerals open to radium diffusion are factors that must be considered (e.g., Claude-Ivanaj *et al.*, 1998, 2001; Feineman *et al.*, 2002; Lundstrom *et al.*, 2000; Pietruszka *et al.*, 2001; Saal *et al.*, 2002; Sigmarsson *et al.*, 1998). For example, an alternative model for the variable ^{226}Ra excesses of the MORB lavas shown in Figure 4(b) is mixing of melts whose variations in $^{226}Ra/^{230}Th$ and Mg# reflect equilibration at different mantle depths

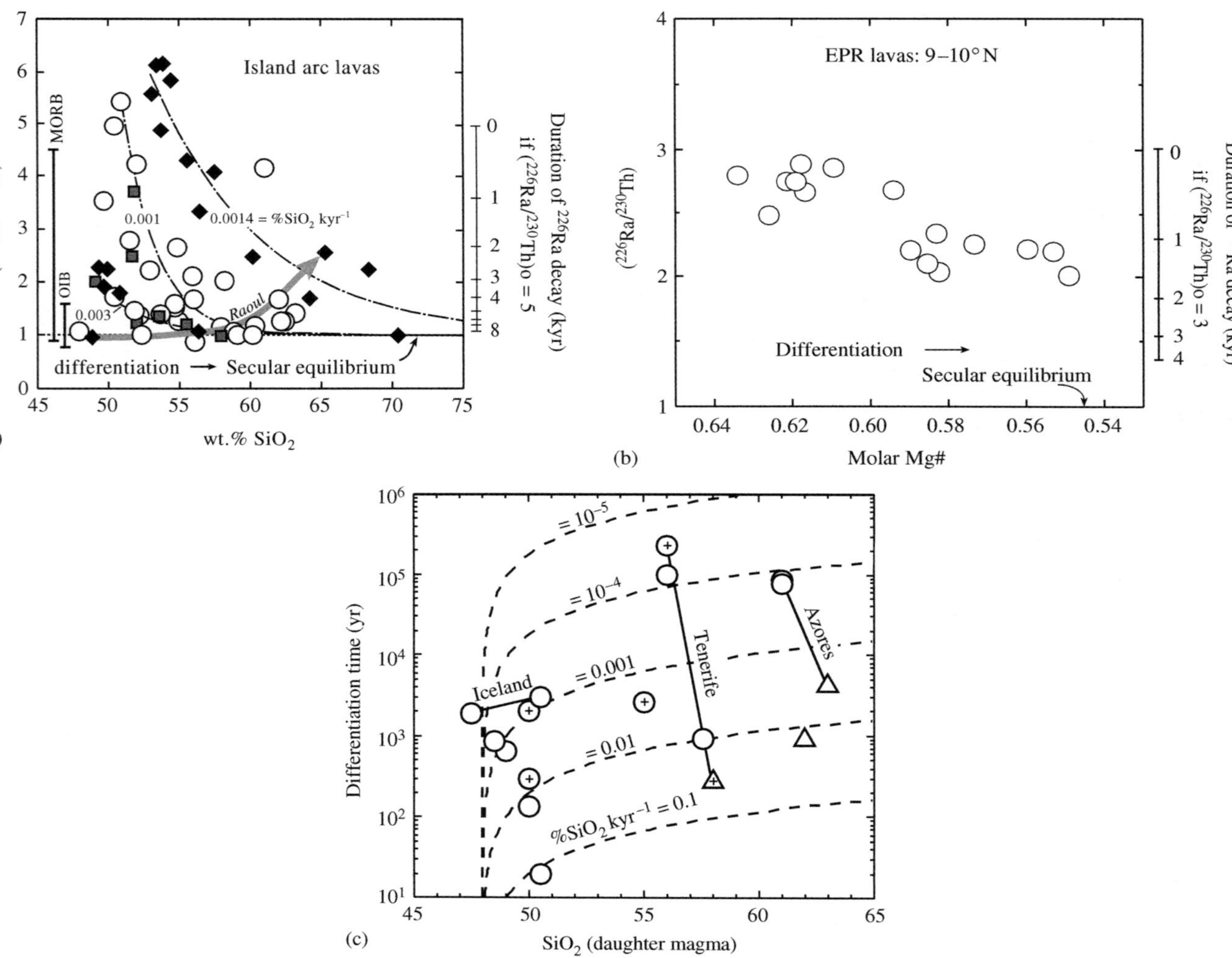

Island arc lavas
MORB
OIB
0.001
0.003
0.0014 = %SiO$_2$ kyr^{-1}
Raoul
differentiation → Secular equilibrium
(^{226}Ra/^{230}Th)
wt.% SiO$_2$
Duration of ^{226}Ra decay (kyr)
if (^{226}Ra/^{230}Th)o = 5
(a)
EPR lavas: 9–10° N
Differentiation →
Secular equilibrium
(^{226}Ra/^{230}Th)
Molar Mg#
Duration of ^{226}Ra decay (kyr)
if (^{226}Ra/^{230}Th)o = 3
(b)
= 10^{-5}
= 10^{-4}
= 0.001
= 0.01
%SiO$_2$ kyr^{-1} = 0.1
Iceland
Tenerife
Azores
Differentiation time (yr)
SiO$_2$ (daughter magma)
(c)

(Sims *et al.*, 2002). An alternative model for the protracted residence time illustrated for the Azores trachytes in Figure 4(c) is much more rapid differentiation from parents with similar $^{230}Th/^{238}U$ ratios, like those of recently reported Azores basalts (Claude-Ivanaj *et al.*, 2001). Initial parent–daughter ratios can sometimes be better delimited by reference to other interelement ratios (e.g., Sigmarsson *et al.*, 2002; Turner *et al.*, 2000, 1998), but these remain model dependent even in favorable cases.

3.05.4.2.2 *Effect of fractional crystallization on differentiation ages*

Preferential incorporation of the parent or daughter nuclide into a fractionating assemblage may serve to modulate the magnitude of U-series disequilibria, but this is generally a subordinate source of uncertainty in differentiation ages. For Th–Ra disequilibria, the most important mineral phases to consider are the feldspars, although biotite may also be important in certain circumstances. Radium is concentrated preferentially over thorium in feldspars, and ages for differentiation based solely on radium excesses will be overestimated if feldspars are fractionated, or underestimated if feldspars are accumulated. Plagioclase is likely to influence the magnitude of radium excesses slightly (e.g., Blundy and Wood, 1994; Cooper *et al.*, 2003, 2001; Sigmarsson, 1996; Vigier *et al.*, 1999; Zellmer *et al.*, 2000); fractionation of an assemblage containing K-feldspar could, however, account for ratios of $^{226}Ra–^{230}Th$ activities less than unity in some evolved rocks (e.g., Zellmer *et al.*, 2000). For $^{238}U–^{230}Th$ disequilibria, the behavior of accessory phases, such as zircon, allanite, monazite, and, to a lesser degree, apatite and magnetite, are mainly responsible for fractionation of uranium and thorium. A corollary of this is that evolved, and especially silicic magmas, might be expected to be susceptible to changes in the magnitude of $^{238}U–^{230}Th$ disequilibria because of crystal fractionation.

3.05.4.2.3 *Effect of crustal melting and open-system processes (contamination, recharge, and crustal melting) on differentiation ages*

Crustal melting and the effects of open-system magmatic behavior on U-series disequilibria

Figure 4 Magma storage times based on timescales of magma differentiation. (a) Plot showing variation in the activity ratio of ^{226}Ra to ^{230}Th in island arc lavas at the time of eruption, plotted as a function of SiO_2. Island arc lavas are chosen because mafic, possibly parental, magmas can have very high $^{226}Ra/^{230}Th$ ratio and because the effects of crustal contamination may be less acute than in continental arc lavas. The time required for $^{226}Ra/^{230}Th$ to decay from an initial ratio of 5 is illustrated on the right-hand side of the diagram; if the initial ratio is lower, longer decay times are required for a given $^{226}Ra/^{230}Th$. Curved trends illustrate constant differentiation rates, assuming that lavas more enriched in SiO_2 reflect greater cooling and intervals of storage in crustal magma reservoirs; trends are approximate fits to data for a subset of samples from Tonga-Kermadec (filled diamonds: 0.0014 wt.% SiO_2 kyr^{-1}, Marianas (not distinguished: 0.001 wt.% SiO_2 kyr^{-1}), and Sunda (filled squares: 0.003 wt.% SiO_2 kyr^{-1}), respectively. Note that the scatter in the Tonga-Kermadec and Sunda, data are typical for individual arcs. Arrow connects three samples from Raoul island, Kermadec arc, erupted over a time interval of <1 kyr; the positive trend shows that the compositional dependence of U-series disequilibrium within specific volcanoes is not always as predicted from progressively longer differentiation times for more evolved lavas. Rest of data (open circles) are from Turner *et al.* (2001) and references therein, and represent Lesser Antilles, Vanuatu, the Philippines, Aleutians, and Kamchatka arcs. Also shown are ranges of $^{226}Ra/^{230}Th$ in MORB (Sims *et al.* (2002) and references therein) and OIB lavas (Bourdon *et al.*, 1998; Claude-Ivanaj *et al.*, 1998, 2001; Sigmarsson *et al.*, 1998; Sims *et al.*, 1999; Thomas *et al.*, 1999). In this and other estimates of timescales based on U-series data presented here, isotopic abundances determined by mass spectrometry rather than by disintegration counting are emphasized, because mass spectrometric methods are typically more precise and, because they require smaller sample sizes, are more likely to represent greater sample purity. (b) Plot similar to that of "(a)" except that data for MORBs from 9° N to 10° N on the East Pacific Rise (Sims *et al.*, 1999) are shown and Mg# rather than SiO_2 is the measure of increasing differentiation. Note that the correlation between Ra/Th and Mg# has also been interpreted to reflect mixing between melts extracted from different mantle depths instead of radioactive decay alone (see text). (c) Plot showing differentiation ages determined from differences in U-series disequilibria in the case studies involving co-magmatic or otherwise presumably related lavas from a variety of tectonic settings. Ages are plotted against the SiO_2 content of the derivative magma and represent differentiation from basaltic parents (circles) and from evolved (>55 wt.% SiO_2) parents (triangles). Tielines connect two different stages in the evolution of individual suites of lavas. SiO_2 contents approximated in some cases. Differentiation times are maximum ages except for those marked with crosses, which are mean ages. Ages >10 kyr are from magnitudes of $^{238}U–^{230}Th$ disequilibria; rest are mostly from magnitudes of $^{226}Ra–^{230}Th$ but also include $^{238}U–^{230}Th$ and $^{210}Pb–^{226}Ra$ disequilibria. Curves show trends of differentiation at various rates from a parental magma with 48 wt.% SiO_2 (sources Black *et al.*, 1998a; Bourdon *et al.*, 1994; Claude-Ivanaj *et al.*, 1998; Hawkesworth *et al.*, 2000; Sigmarsson, 1996; Sims *et al.*, 2002; Turner *et al.*, 2000; Vigier *et al.*, 1999; Widom *et al.*, 1992; Zellmer *et al.*, 2000).

may also compound uncertainties in differentiation timescales. If silicic magmas are produced by crustal melting rather than differentiation, the magnitude of associated U-series disequilibria would lead to erroneous differentiation times when attributed to evolution from mafic parental magmas. Contamination by such crustal melts could either increase or decrease the degree of U-series disequilibrium with increasing indices of differentiation and have little to do with the duration of differentiation. However, wholesale assimilation of crust is likely to reduce the magnitude of disequilibrium and therefore lead to an overestimation of differentiation times. Some studies have suggested contamination of magmas by radium-rich components, including volatiles, fluids, and/or salts (e.g., Black *et al.*, 1998b; Condomines *et al.*, 1995; Hemond *et al.*, 1994), in which case the degree of disequilibrium would underestimate the actual timescale of differentiation.

Infusion of the reservoir by magma more freshly derived from the source will also tend to reset the radiometric clock, resulting in ages for magma storage, and hence transfer, which, strictly speaking, are characteristic neither of the resident nor of the recharge magma. Because radioactive decay results in exponential changes in the ratios between radionuclides, apparent ages will tend to be biased towards the more youthful end-member of the mixture, i.e., the recharge (Hughes and Hawkesworth, 1999). Actual magma storage times could be considerably longer than the approach to secular equilibrium would appear to indicate, especially compared to magma that evolves by differentiation alone. In the limit, magma reservoirs may approach steady-state conditions and exhibit perpetually invariant $^{226}Ra/^{230}Th$ or $^{230}Th/^{238}U$, as proposed for magmas parental to pre-1970 AD Etna lavas (Condomines *et al.*, 1995); the volumes and residence times of steady-state magmas are summarized earlier in this chapter (Figure 3).

3.05.4.3 Timescales of Differentiation from Other Absolute Ages

In particularly favorable cases, inferences about magma residence times can be derived from other isotopic systematics, most notably $^{87}Rb-^{87}Sr$ (Davies *et al.*, 1994; Halliday *et al.*, 1989). Extremely evolved magmas can have sufficiently high Rb/Sr for Sr isotope signatures to evolve detectably on timescales commensurate with magmatic processes, enabling the time since differentiation to high Rb/Sr to be delimited. Correlations between diverse Rb/Sr and $^{87}Sr/^{86}Sr$ in some suites of silicic lavas may be dominantly those of *in situ* decay rather than open-system processes, in which case they delimit the differentiation events responsible for apparently co-magmatic melts. Interpreted in this way, the spectrum of eruption ages would, therefore, imply that a compositionally heterogeneous magma reservoir was tapped repeatedly. As constrained by such Rb–Sr isochrons, differentiation and, therefore, magma storage could have preceded eruption by several thousand years in African comendites (Heumann and Davies, 2002) to several hundreds of thousands of years in some high silica rhyolites (Heumann *et al.* (2002) and references therein; see also Section 3.05.5.3).

3.05.5 TIMESCALES OF CRYSTALLIZATION

3.05.5.1 U-series Constraints on Crystallization Ages of Mineral Populations

A fundamental source of information about the chemical evolution of magmas is provided by the chemical and isotopic compositions of minerals they contain. Mineral assemblages that are essentially those expected for equilibrium crystallization could have precipitated from their host magma or, at the very least, from magmas of composition like those in which they are found. Consequently, ages obtained for these crystals should complement those of differentiation by providing insight into the duration of magmatic storage since derivation of a particular melt composition. In detail, however, the crystal record is often more complex and can reveal compositional heterogeneities within and between crystals that provide evidence for varying physiochemical conditions of crystallization as well as for open-system processes such as recharge, mixing, assimilation, and incorporation of mush crystals precipitated at an earlier stage in a volcano's evolution (Figure 5).

From the earliest studies onward (e.g., Cerrai *et al.*, 1965; Condomines, 1997; Fukuoka, 1974; Kigoshi, 1967; Peate *et al.*, 1996; Taddeucci *et al.*, 1967), crystal ages based on U-series disequilibria in mineral separates have been used to constrain the eruption ages of prehistoric volcanic rocks. The utility of these ages as measures of the timing of eruption depends on the extent to which crystallization precedes eruption (cf. Allègre, 1968; Taddeucci *et al.*, 1967). For those lavas where eruption ages are independently known, U-series dating of minerals from volcanic eruptions can delimit how long before eruption crystallization occurred. U-series ages for mineral separates are summarized in Figure 6, where they are presented as ages before eruption and plotted with respect to the compositions of their host lavas. Crystal ages based on U-series disequilibria in single crystals have also been obtained and these are described in the next section. Many of the crystal ages in Figure 6 are based on isochrons

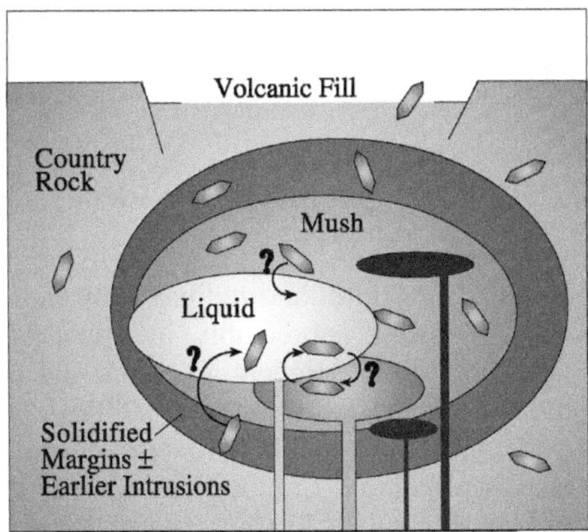

Figure 5 Cartoon of a magma reservoir illustrating the possible sources of crystals contained in erupted magmas. Besides crystals suspended in the liquid portions of a magma reservoir or entrained by magmas that recharge that magma reservoir, crystals in erupted magmas may also be derived from the mush regime, from plutonic and volcanic rocks created during earlier episodes of magmatic activity, and/or from the country rocks in which the magma reservoir is located. Arrows illustrate possible sources of cogenetic crystals that might be difficult to distinguish from co-magmatic crystals. Transfer of crystals between different domains is likely to be equally important in plutonic rocks (e.g., Wallace and Bergantz, 2002).

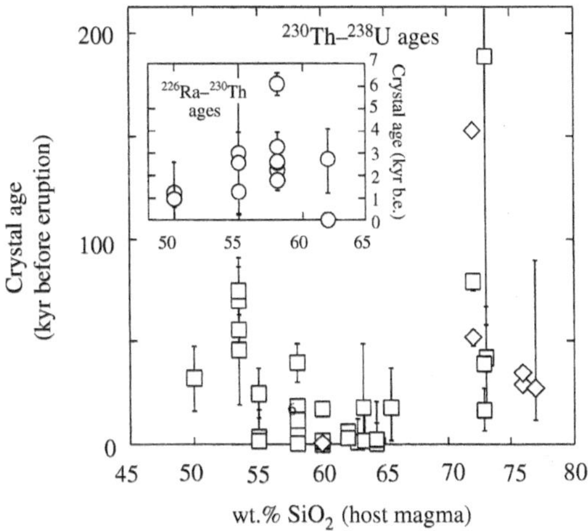

Figure 6 U-series disequilibria ages obtained for aliquots of crystals separated from lavas, presented in terms of age before eruption. Most crystal ages are defined by isochron arrays, but some are model ages for individual phases derived by assuming initial nuclide abundances based on those of the host rocks or associated glasses or groundmasses. Eruption ages are delimited either by historic records or by radiometric dating techniques other than U-series disequilibrium. See text for details. Main figure: U–Th crystal ages plotted as a function of SiO_2; ages that are largely constrained by zircon are indicated by diamonds; inset: Ra–Th crystal ages plotted as a function of SiO_2 (note very different age range). Age data for Kilauea and Mt. St. Helens (shaded circles: Cooper *et al.*, 2001, 2003; Cooper and Reid, 2003) allow for differential Ra–Ba fractionation (see text for details) (sources Black *et al.*, 1998a,b; Bourdon *et al.*, 1994; Charlier and Zellmer, 2000; Condomines, 1997; Cooper and Reid, 2003; Cooper *et al.*, 2001; Heath *et al.*, 1998; Heumann and Davies, 2002; Heumann *et al.*, 2002; Pyle, 1992; Reagan *et al.*, 1992; Schaefer *et al.*, 1993; Volpe and Cashman, 1992; Volpe and Hammond, 1991; Zellmer *et al.*, 2000).

defined by aliquots of various individual mineral phases, glasses, matrices, and/or whole rocks. Some model or "two-point isochron" ages for separates of individual mineral phases are also presented in Figure 6, where the initial isotopic characteristics of the minerals are delimited with reference to the isotopic compositions of host lavas or matrices. For Ra–Th isochron ages, it is essential to note that, because there is no long-lived isotope of radium, ages are obtained by using the relative concentrations of barium in the various separates as a chemically similar proxy for the distribution of initial ^{226}Ra.

Given that the record of crystallization might be expected to be commensurate with or even shorter than the duration of differentiation, it is perhaps not surprising that Ra–Th disequilibrium in minerals is the rule rather than the exception, although such data are still sparse. Mineral ages based on Ra–Th disequilibria range up to ~6 kyr before eruption but are generally ~1–3 kyr before eruption (inset, Figure 6). In apparent contradiction to the relatively youthful ages given by Ra–Th disequilibrium in the mineral separates are ages obtained or inferred for many of the same mineral separates based on ^{238}U–^{230}Th disequilibria that are 10 kyr to >50 kyr before eruption (Figure 6). In somewhat more than half of the few instances where both Ra–Th and U–Th data are available for the same mineral aliquots, discordant U–Th and Th–Ra crystal ages have been found (e.g., Black *et al.*, 1998b; Volpe and Cashman, 1992; Volpe and Hammond, 1991).

Evidence that the distribution of barium between mineral phases and/or melt is not perfectly analogous to that of radium has only recently been included in the derivation of Th–Ra mineral ages (Cooper *et al.*, 2001, 2003; Cooper and Reid, 2003). Compared to ages derived by assuming no Ba–Ra fractionation, the model ages for a given mineral could be either younger or older, depending on whether the mineral initially has a radium excess or deficit; this effect becomes increasingly subordinate to other sources of uncertainty as apparent ages approach the time required for ^{226}Ra–^{230}Th equilibrium (i.e., ~8 kyr). Considering the limited constraints on Ba–Ra fractionation in minerals, the difficulty of correcting for initial radium when mineral separates are impure (Cooper *et al.*, 2001), and the possible complication of rapid crystallization (Cooper and Reid, 2003), a re-evaluation of the Ra–Th mineral ages would be complex and is beyond the scope of this review. Uncertainties in the exact Th–Ra ages of minerals notwithstanding, the discordant $^{238}U-^{230}Th$ and $^{226}Ra-^{230}Th$ ages obtained for some lavas could be the product of mixing between different age mineral populations or, in some cases, failure of minerals to close with respect to diffusion of the U-series nuclides. As such, discordant ages offer potential insight into how mineral ages should be interpreted.

3.05.5.1.1 Effect of mixing of different aged populations on U-series crystallization ages

Two aspects of crystal mixtures warrant special recognition; in describing these here, the weighting effect of concentration differences between the mixing end-members is neglected. First, once a crystal is close to secular equilibrium (~400 kyr for U–Th and ~8 kyr for Ra–Th), its leverage on the isotopic composition of a mixture is no different than that of a crystal that is millions or billions of years older. A tangible benefit of this is that recycling of a few basement crystals will not skew U-series ages as much as might be expected if the effect of such crystals on mixtures was linearly related to age. A corollary is that for U-series ages appreciably older than those of eruption to be the product of crystal recycling, a significant fraction of "old" crystals must be involved, whatever their source (cf. Cooper *et al.*, 2001). Second, because disequilibrium between ^{230}Th and ^{226}Ra diminishes much more rapidly than that between ^{238}U and ^{230}Th, different apparent age populations for the two decay systems would be the logical result of mixing between relatively young and "old" crystal populations. Possible scenarios for mixed age populations are, in the case of closed systems, young crystal growth superimposed on growth that is at least several thousand years old or, in the case of open systems, mixing between "young" recharge magmas and their crystals, and crystals from "old" resident liquids, mushes, and/or wall rocks (Figure 5; cf. Cooper and Reid, 2003). In light of the fact that many host magmas (as represented by whole rock and groundmass analyses) are themselves characterized by $^{226}Ra-^{230}Th$ disequilibria (cf. Figure 4), open-system interpretations appear to pertain. Th–Ra and U–Th ages would, therefore, provide a minimum *magma* residence time only in the case where mixing involves a young, radium-enriched liquid and the contribution of crystals foreign to the magma reservoir is negligible.

3.05.5.1.2 Effect of diffusion on U-series crystallization ages

Of uranium, thorium, and radium, only radium is likely to diffuse appreciably in minerals at magmatic conditions. Recent experimental and theoretical estimates suggest that $^{226}Ra-^{230}Th$ ages for plagioclase and pyroxene could be significantly affected by radium diffusion if they are held at $T > 1{,}100\,°C$, even when they are phenocrysts ($r > 0.15$ mm), or at $T > 900\,°C$ when the crystals are microphenocrysts ($r > 0.015$ mm; cf. Cherniak, 1996; Van Orman *et al.*, 1998, 2001). $^{226}Ra-^{230}Th$ apparent ages may be either younger or older than crystallization ages as a result of diffusion, depending on whether crystals continuously reside in the same magma or are introduced into magmas with higher or lower radium contents (Cooper and Reid, 2003). Some constraints on the magnitude of diffusive re-equilibration might be derived from the extent to which other chemical heterogeneities within the crystals have been preserved despite potential for diffusion (cf. Cooper and Reid, 2003). Because most crystal age data are for phenocrysts, only the Ra–Th crystal ages for the mafic and, therefore, relatively high-temperature magmas shown in Figure 6 could reasonably have been affected by diffusion and should thus be regarded with particular caution.

3.05.5.2 U-series Constraints on Crystallization Ages of Individual Minerals

Ages of individual crystals and subdomains within crystals can be suitably precise to assess magma storage times when concentrations of uranium and/or its radiogenic daughter nuclides in mineral phases are high. This is often the case for zircon, but dating of individual crystals of less-common allanite, monazite, chevkinite, and titanite is also possible. Most studies have used ion microprobe analyses (e.g., Brown and Fletcher, 1999; Dalrymple *et al.*, 1999; Reid *et al.*, 1997), but *in situ* analyses by laser ablation inductively coupled mass spectrometry, especially with a multicollector instrument (e.g., Hirata and Nesbitt, 1995; Horn *et al.*, 2000; Willigers *et al.*, 2002), and analyses of chemically processed single crystals should also be possible in favorable cases (e.g., Schmitz and Bowring, 2001). For zircon in particular, U–Th and U–Pb ages are not readily equilibrated at the temperatures of many of their host magmas (Cherniak and Watson, 2001; Reid *et al.*, 1997). Therefore, in instances where zircons crystallize sufficiently long before eruption, dating of zircon by U–Pb or U–Th (for young eruptions) methods can yield ages resolvable from those of eruption. The results are, for the most part, specifically applicable to more evolved magmas but provide insights into processes that affect the crystal record generally.

In most cases, U–Pb and U–Th ages for accessory phases are relatively robust to uncertainties associated with corrections for inherited lead or thorium. Moreover, the ability to determine the ages of individual crystals or parts of crystals means that the distribution of crystal ages is revealed, thus providing a potent dating technique. The principal uncertainty associated with how to interpret these age distributions generally revolves around the affinity of the accessory phases to their host magmas. Basement-aged zircons in rhyolites studied to date are remarkably rare, which suggests that crystal–melt separation can be extremely efficient at some stage in silicic magmagenesis. Consequently, zircon separates that yield pre-eruption ages (cf. Figure 6) may not simply reflect admixtures of ancient zircons with eruption- or near-eruption-age zircons. More often than not, when resorption surfaces are identified within crystals, the cores of these crystals yield ages that fall within the magmatic interval characteristic of a particular volcano (e.g., Brown and Fletcher, 1999; Vazquez and Reid, 2002). The age-zoning of such crystals could be evidence of strong thermochemical fluctuations in the magma reservoir or could represent recycling of crystals from plutonic roots of a volcano (cf. Bachmann *et al.*, 2002) and/or from earlier volcanic products (cf. Bindeman *et al.*, 2001). Recycling of minerals from prior episodes of activity within magmatic centers has also been identified on the basis of inherited argon (e.g., Gansecki *et al.*, 1998; Gardner *et al.*, 2002; Spell *et al.*, 2001) and combined oxygen and U–Pb (Bindeman *et al.*, 2001) isotope characteristics. Crystallization of zircons that, at least on petrographic grounds, are cognate to the system appears to peak at least tens of thousands of years before eruption in many silicic systems (Figure 7). Because the scatter in the individual zircon ages is commonly more than expected solely from analytical uncertainties, the onset of crystallization may be even earlier than the mean age of the crystal population. Compositionally distinct but otherwise co-magmatic magmas that have similarly old zircon populations may erupt intermittently over at least several tens of thousands of years (e.g., Central Plateau Member, Yellowstone; Western Moat, Long Valley; Figure 7), showing that the zircon memory of early crystallization is not completely erased by subsequent differentiation (Vazquez and Reid, 2002) and that related magma reservoirs may be tapped repeatedly (Reid *et al.*, 1997; Vazquez and Reid, 2002).

3.05.5.3 Timescales of Crystallization from Other Absolute Age Constraints

Besides providing insights into timescales of differentiation, the high Rb/Sr of some highly evolved melts enables minerals and their hosts to rapidly become isotopically distinct from one another and thereby to permit model mineral ages to be determined (e.g., Christensen and DePaolo, 1993; Davies *et al.*, 1994). Minimum ages of a few thousand years to several hundreds of thousands of years before eruption have been obtained for plagioclase and sanidine, as well as for other phenocrysts (Figure 7). Strontium diffusion in feldspar is sufficiently slow (Cherniak, 1996; Giletti and Casserly, 1994) that the isotopic compositions of these minerals are only modestly affected by diffusive re-equilibration with their host (Christensen and DePaolo, 1993). Notably, the oldest model ages for these crystals are like those obtained for differentiation based on strontium isotope variations (Figure 7). Biotite and sometimes feldspar can yield both distinctly older and younger (including negative) model ages (e.g., Davies and Halliday, 1998; Heumann and Davies, 2002). Besides possible diffusional re-equilibration of biotite or real differences in the timing of crystallization (Davies *et al.*, 1994), the possibility remains that some or all of the crystals are not cognate to the liquids in which they are found. More generally, even where the crystallization sequence inferred from strontium model ages agrees with that observed petrographically, it should be borne in mind that the absolute values of these ages still critically depend on the affinity between minerals and glass, because the isotopic compositions of the latter largely govern these ages.

3.05.5.4 Timescales of Crystallization Based on Kinetic Phenomena

Kinetic phenomena can also be used to delimit the timescales of magmatic processes and, unlike radiometric ages, do not require absolute constraints on the timing of eruption. Two important kinetic controls on crystal properties are crystal growth and diffusional relaxation of compositional heterogeneities in minerals. Rates of crystal settling are not described here but have also been used to delimit crystal storage times (e.g., Anderson *et al.*, 2000; Resmini and Marsh, 1995).

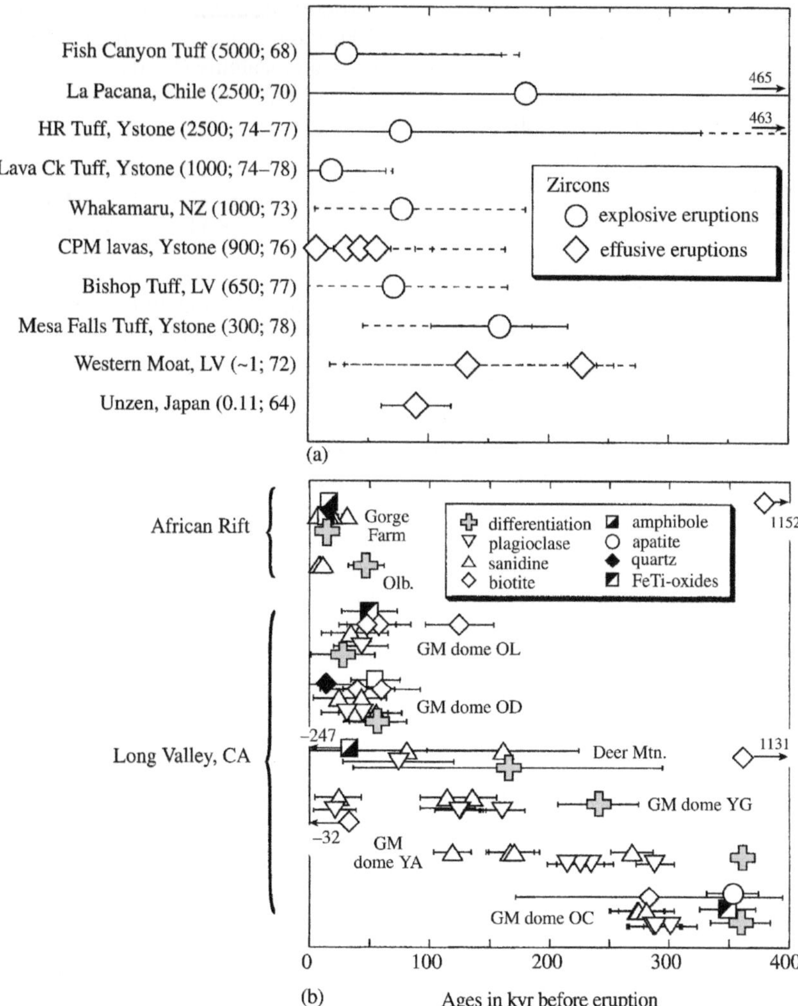

Figure 7 Crystal ages for silicic magmas obtained from U–Pb, U–Th, and Rb–Sr isotope systems, in thousands of years before eruption. (a) Mean zircon ages determined by either *in situ* or single crystal U–Pb or U–Th isotope analyses (versus the zircon aliquots shown in Figure 6). Solid error bars give uncertainty in mean pre-eruption crystal ages. Dashed error bars show maximum and minimum pre-eruption crystal ages where model ages for individual spots are determined; ages of resorbed cores are not included. The limiting ages shown here are based on measured maximum and minimum values, even though they may be within error of other, less extreme ages. Age distributions obtained for the Central Plateau Member lavas from Yellowstone are remarkably similar to one another, as are those for the two western moat samples from Long Valley; the spread in residence times within each suite are therefore the result of diverse eruption ages. Results for rhyolites erupted immediately after caldera-forming eruptions at Yellowstone are not shown because of independent evidence for crystal recycling (Bindeman *et al.*, 2001). Eruptive volumes (in km^3) and representative SiO_2 contents (in weight percent) have been given in parentheses. Abbreviations are Ystone: Yellowstone; HR: Huckleberry Ridge; CPM: Central Plateau Member; and LV: Long Valley (sources Bachmann *et al.*, 2002; Bindeman *et al.*, 2001; Brown and Fletcher, 1999; Reid and Coath, 2000; Reid *et al.*, 1997; Sano *et al.*, 2002; Schmitt *et al.*, 2003; Schmitz and Bowring, 2001; Vazquez and Reid, 2002). (b) Plot of Sr model ages for crystals and Sr isochron ages for differentiation, in thousands of years before eruption. Model crystal ages are calculated from the Sr isotope compositions of minerals and host glasses; model ages for both mineral aggregates and individual crystals are shown. Differentiation ages are calculated from isochrons defined by glasses and whole rocks samples from lavas inferred to be cogenetic. Uncertainties in eruption ages (where reported) have been included in the uncertainty in differentiation/crystal ages. Note that ages for three biotites and one amphibole plot outside this diagram (ages indicated adjacent to arrows). Not shown are ages for rims of crystals, ages for which uncertainties are larger than the absolute age itself, and, because a significant proportion of them are negative, most crystal ages for the Bishop Tuff (cf. Christensen and DePaolo (1993) and Davies and Halliday (1998)) (sources Davies and Halliday, 1998; Davies *et al.*, 1994; Heumann and Davies, 2002; Heumann *et al.*, 2002).

3.05.5.4.1 Crystal size distributions

The apparent times required for crystals to attain their sizes can be bracketed simply by assuming some value for, and temperal dependence of (e.g., linear) crystal growth rates (e.g., Resmini and Marsh, 1995; Roggensack, 2001b). To test whether the distributions of crystal sizes reflect those of growth rather than selective accumulation and/or fractionation, the size distribution can be compared to the crystal population densities (Cashman, 1990; Marsh, 1988, 1998). Provided that estimates for crystal growth rates are available, such crystal size distributions (CSDs) can be used to determine when crystals began to form before eruption, in the case of a closed system, or a mean residence time for the crystals, in the case of an open, steady-state system (Mangan, 1990; Resmini and Marsh, 1995).

Most CSD studies have focused on plagioclase and olivine, because they are relatively common and because growth rates for these minerals have been delimited (Armienti *et al.*, 1994; Cashman, 1993; Jambon *et al.*, 1992). Studies of CSD for alkali feldspar, quartz, zircon, and opaque oxides are rarer but have also been reported (Figure 8(a)). For plagioclase and olivine phenocrysts and microphenocrysts, both linear and nonlinear correlations between population density and crystal sizes have been obtained. Deviations from linear CSDs may reflect mixing between different age populations (Higgins, 1996b), nucleation and crystallization under changing magmatic conditions (Armienti *et al.*, 1994; Knesel *et al.*, 1999), and/or dissolution of smaller crystals and coarsening of larger ones (Ostwald ripening) (Bindeman and Valley, 2002; Marsh, 1998). Plagioclase microlite size distributions have been attributed to largely syn-eruptive crystallization, as illustrated by the rhyolitic groundmasses of Merapi (Hammer *et al.*, 2000), Mt. St. Helens (Cashman, 1992; Geschwind and Rutherford, 1995), and Pinatubo (Hammer *et al.*, 1999) lavas.

Magma storage times based on the durations of crystal growth, as estimated for a variety of lavas, are shown in Figure 8(a). Measured CSD slope values are proportional to the duration and rate of crystal growth and vary by more than two orders of magnitude. Most estimates of crystal growth

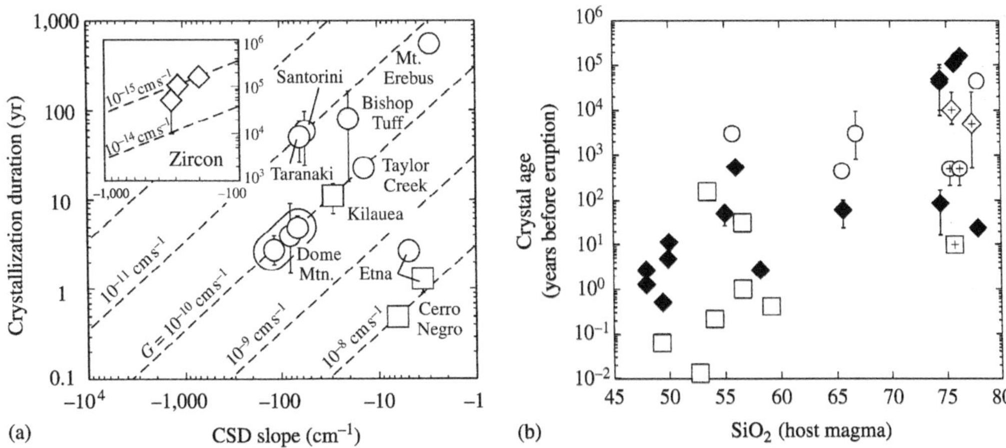

Figure 8 Illustration of magma storage times based on kinetic controls on crystal properties. (a) Mean duration of crystal growth obtained from crystal size distributions. Circles: quartz and feldspars; squares: olivine; open diamonds: zircon. Diagram is contoured for growth rates (G) from which the displayed crystallization times were calculated. Note that residence times obtained from crystal growth are proportional to growth rates and observed plagioclase growth rates range from 5×10^{-12} cm s^{-1}or lower to 7×10^{-9} cm s^{-1}, excluding relatively fast rates only observed for microlites (up to 5×10^{-8} cm s^{-1}), but a range of 10^{-11}–10^{-10} cm s^{-1} is preferred (Cashman, 1993); observed olivine growth rates range from 2×10^{-9} cm s^{-1} to 6×10^{-5} cm s^{-1} (Armienti *et al.*, 1994; Jambon *et al.*, 1992). Note that at least some of the ages could reflect the duration of magma ascent rather than storage (e.g., Roggensack, 2001a) (sources Armienti *et al.*, 1994; Bindeman and Valley, 2001, 2002; Cashman, 1990; Higgins, 1996a,b; Knesel *et al.*, 1999; Mangan, 1990; Reagan *et al.*, 1992; Resmini and Marsh, 1995; Roggensack, 2001a). (b) Mineral ages obtained for individual crystals from diffusional relaxation (open symbols) in addition to those for crystal populations from their size distributions (closed diamonds), plotted as a function of SiO_2. A few of the diffusional relaxation ages reflect averages obtained on a small (generally $n \leq 5$) population of grains. Crosses signify minerals that are either xenocrystic or restitic phases. Minerals dated by diffusion relaxation are quartz and feldspars (open circles), ferromagnesian phases (open squares), of which all are olivine except for the amphibole xenocrysts (square with cross), and zircons (open diamonds). In all likelihood, xenocrysts yield minimum ages for magma storage; see text for possible interpretations of phenocryst ages (sources, in addition to those above: Bindeman and Valley, 2001; Danyushevsky *et al.*, 2002; Davidson *et al.*, 2001; Gardner *et al.*, 2002; Ginibre *et al.*, 2002b; Knesel and Davidson, 1999; Nakamura, 1995b; Zellmer *et al.*, 1999).

rates also vary by two to three orders of magnitude, and uncertainties in actual crystallization times could therefore be commensurate to this. True megacrysts (>~5 mm) are rarely observed in volcanic rocks, from which it follows that crystal sizes would necessarily give residence times of less than a few thousand years for silicate phases, if continuous growth is assumed. Crystallization intervals imaged by CSD might therefore be finite, perhaps because crystallization occurs largely in solidification fronts at the margins of the mush regimes (e.g., Marsh, 1996) or because crystals are continuously lost from the magma because of crystal settling. Estimates for crystal growth rates inferred from radiometric ages, in studies such as those described above (Christensen and DePaolo, 1993; Davies *et al.*, 1994), are in some cases several orders of magnitude slower than those from which CSD ages are usually calculated. Potentially related to this is uncertainty about whether crystals grow continuously. For plagioclase in particular, phenocrysts often show evidence for resorption, and resorption may be more frequent than usually recognized (Ginibre *et al.*, 2002a). If, therefore, crystal growth rates vary more than currently inferred, the few-thousand-year limit for apparent crystal growth would be relaxed.

3.05.5.4.2 *Diffusional relaxation*

Chemical and/or isotopic heterogeneities may arise within or between crystals if they fail to continuously equilibrate with a liquid that is changing in composition or if crystals are transferred from one liquid to another. Chemical heterogeneities may also arise if kinetic, rather than equilibrium processes, affect the chemical composition of the crystal. With time, these heterogeneities will be eliminated by diffusive re-equilibration. The rate dependence of diffusion means that if both initial and final boundary conditions can be delimited in some fashion, the distribution and diffusivity of an element within a crystal can constrain the time elapsed since entrapment of the chemical heterogeneities. The sensitivity of diffusion relaxation calculations to the duration of crystal storage depends, therefore, on the host crystal and its temperature, the diffusing species, and the spatial resolution of analyses compared to chemical or isotopic gradients. For compositionally complex minerals, final boundary conditions for trace components (e.g., strontium in compositionally zoned feldspar; Zellmer *et al.*, 1999) may be local chemical equilibrium which, along with diffusion coefficients, must be estimated from experimental and/or theoretical studies (e.g., Blundy and Wood, 1994; Cherniak, 1996).

In contrast to the ages of crystal populations represented by CSD, ages for individual crystals can be obtained from diffusional relaxation. Estimates for the duration of diffusional relaxation in individual or small numbers of crystals and the implied duration of magma storage are shown in Figure 8(b), where they are compared to the silica content of their host. Estimates for the crystal residence times of both phenocrysts and xenocrysts are illustrated. Ages based on CSD are also shown in the same diagram. Solid-state diffusion coefficients for chemical relaxation generally depend on host magma temperatures. Extreme initial boundary conditions are typically assumed (e.g., an initial step-function in chemical zoning) and so, to the extent that the crystals studied are representative, crystal ages are likely to be maxima. However, if crystals are isolated in cooler portions of the mush for some fraction of their evolution, ages may be underestimated. Additionally, crystals grown in equilibrium with the liquid cannot be easily distinguished from those that were initially zoned but equilibrated diffusionally; neglecting this possibility will also lead to an underestimation of crystal residence times. Gravitational settling, especially for dense crystals in low-viscosity magmas, could limit the residence times of large crystals (cf. Danyushevsky *et al.*, 2002), and might, in part, account for the observation that ages obtained from diffusional gradients in olivine from mafic magmas tend to be younger than those from plagioclase and olivine CSDs (Figure 8(b)). These uncertainties notwithstanding, crystal ages constrained by the two kinetic methods vary by more than six orders of magnitude across the compositional spectrum. Taking these ages at face value, crystals from intermediate, and especially silicic, magmas are older than those from mafic magmas. Either crystals are preferentially retained in more silicic magmas or, if crystals grew largely after differentiation, then storage, in addition to differention times, are longer for evolved magmas (cf. Section 3.05.4).

3.05.6 DISCUSSION AND SUMMARY

The lack of magma storage estimates based on complementary approaches for more than a few volcanoes makes it difficult to make a comparison between the plumbing systems associated with different magmatic centers, or, for that matter, between different parts of the same system. Besides differences associated with the part of the system imaged by each approach, there may be differences associated with the location of magma storage as well as related factors such as the thermal state of the crust and the rate of melt supply that may affect how efficiently a magma chamber is sustained. Not all magma is necessarily stored at shallow levels or even in the crust (e.g., Brophy *et al.*, 1999; Klugel *et al.*, 2000; Kuritani, 1999; Roggensack, 2001a). One consequence of this is that differentiation may occur at a variety of depths and storage times may vary from one reservoir to another (e.g., Claude-Ivanaj *et al.*, 1998; Condomines *et al.*, 1995; Hawkesworth *et al.*, 2001). All else being equal, magma residence times would be expected to be shorter when there is a greater thermal contrast between the magma and its surroundings, because cooling and crystallization would occur at a faster rate. For this reason, larger volumes of magma might be stored in the lower crust. In addition, larger volumes of magma might also be stored in areas that have simply experienced a protracted prior history of magmatism.

The foregoing vagaries of the volcanic record notwithstanding, some general observations about apparent magma storage times will illustrate the latitude in how magma residence times might be interpreted. Even though differentiation times for mafic magmas are highly sensitive to uncertainties in the magnitude of initial U-series disequilibria of parental magmas and relatively few absolute crystal ages have been obtained, it appears that magma residence in mafic systems is on the order of a few thousand years at most and generally years to hundreds of years (Figures 2, 4, and 8). These relatively short periods of basaltic residence could reflect the absence or small volumes of shallow magma reservoirs (e.g., Brophy *et al.*, 1999; Kuritani, 1999; Roggensack, 2001a) or, where an upper magma reservoir is present, high turnover rates in these reservoirs. For example, estimates for Etna seem to be ~10–30 yr for the shallow magma reservoir, but possibly 1,500 yr for the deeper magma reservoir (Albarede, 1993; Armienti *et al.*, 1994; Condomines *et al.*, 1995); estimates for Karthala, Comores, are ~100 yr for the shallow reservoir and several hundred years for the deeper one.

The fidelity of the various geochemical and geophysical approaches to understanding different aspects of a magmatic system is well demonstrated by the range of storage estimates obtained for magma associated with Kilauea. Basaltic magma from what is likely to represent the most active conduit system beneath the summit of Kilauea has been estimated to have had a residence time of 30–40 yr during the late twentieth century based on steady-state modeling (Pietruszka and Garcia, 1999). Only 10 yr of storage is inferred for the 1959 Kilauea Iki eruption based on the distribution of crystal sizes (Mangan, 1990), but this magma may have bypassed the main shallow magma reservoir; a minimum estimate of at least 550 yr was obtained by ^{230}Th–^{226}Ra dating of crystals in a more evolved, rift-zone lava (Cooper *et al.*, 2001). Geophysical estimates based on volumes of magma reservoirs range from 100 yr to 120 yr based on geodetic data (Decker, 1987) to several thousands of years based on hydrostatic magma pressure (Denlinger, 1997). The largest estimate would represent the entire interconnected melt volume of the magma reservoir system beneath Kilauea, whereas the others likely represent the main liquid portion of the reservoir that contributes to lava output. Still larger, unconnected domains of melt could be present under the volcano, like those domains imaged geophysically elsewhere under ocean islands (e.g., West *et al.*, 2001). Perhaps, like a significant fraction of the interconnected volume, these disconnected bodies would contribute to the endogenous output associated with Kilauea.

Differentiation and storage estimates for intermediate (basaltic andesite to dacitic) magmas associated with stratovolcanoes are more problematic because of the evident effects of mineral fractionation, crystal recycling, magma recharge, and assimilation. In addition, phenocrysts may differentially grow or be resorbed, in the case of ascent of relatively hydrous arc magmas (Cashman and Blundy, 2000). Crystal size distributions suggest both simple and mixed crystallization histories but, as described above, radiometric ages obtained for crystals can predate eruption considerably more than permitted by apparent crystal growth rates (Figure 6 versus Figure 8). This apparent contradiction, as well as the radium enrichment of many arc magmas and their crystals (Figures 4 and 6), could require either frequent mixing and mingling between recharge melts and reservoir magmas and/or entrainment of crystals from stagnant, mush-dominated magma regimes (Figure 5). Furthermore, crystal and compositional heterogeneities such as these in arc lavas, especially phenocryst-rich ones, suggest that many arc volcano reservoirs are small, and storage times for their liquid portions are short (no more than a few thousands of years; Figure 4). Even so, to reconcile this timescale with the secular heterogeneity of the volcanic outputs on timescales of years (Figure 2) may alternatively and/or additionally require multiple magma domains to co-evolve in limited geographical areas, as might occur in a subvolcanic plexus of dikes and sills. Such scenarios may have broader applicability than simply to stratovolcanoes but may be more easily detectable in these, more compositionally diverse, settings.

The development of compositional heterogeneities within silicic alkaline magma systems may occur in no more than a few thousand years, at least based on ^{238}U–^{230}Th analyses (Figure 4; Black *et al.*, 1998a; Bourdon *et al.*, 1994;

Hawkesworth *et al.*, 2000; Widom *et al.*, 1992). Timescales for the generation of trachytes from their mafic precursors are, in contrast, inferred to be considerably longer, i.e., tens to hundreds of kyr. In the absence of strong constraints it remains uncertain, however, how to partition the evolution from mafic to evolved compositions between differentiation, storage in a magma reservoir, and remelting of the underplated mafic magmas (e.g., Bohrson and Reid, 1998).

A significant fraction of the magmatic rocks that compose continental crust are also silicic in composition and these dominantly subalkaline magmas deserve special discussion. Rb–Sr mineral ages obtained for small volume silicic eruptives agree with those obtained from U–Th or U–Pb mineral dating (cf. Figures 6 and 7; Heumann and Davies, 2002; Heumann *et al.*, 2002; Reid *et al.*, 1997). This concordancy is particularly notable, because U–Th and U–Pb ages of accessory phases are dictated by the isotopic compositions of the minerals, whereas Rb–Sr ages are dictated by the isotopic composition of the liquid (i.e., glass). To the extent that comparisons have been possible, there is concordance between ages of differentiation and those of the oldest crystals and therefore, like their alkaline counterparts, differentiation may occur rather rapidly. A straightforward interpretation of these observations is, therefore, that storage times for silicic magmas can be relatively long, i.e., several tens to a few hundreds of thousands of years (Figures 6 and 7); these time frames are, however, short lived with respect to repose intervals between the more voluminous silicic eruptions and therefore the duration of silicic activity at any particular volcanic system (cf. Figure 3). These observations do not, however, preclude the possibility that once differentiated, silicic magmas solidify and are stored in the crust, only to be later completely remobilized by reheating (Mahood, 1990). For the case of long-lived silicic magma reservoirs, the same reservoir may be tapped repeatedly as also inferred for steady-state or quasi-steady-state mafic magma centers. Unlike these centers, however, effusive silicic eruptions appear to sometimes tap essentially closed-system reservoirs and therefore, rather than the balance between supply and eruption, the associated timescales of magma storage are dictated simply by when storage is interrupted by eruption.

In summary, each of the various approaches by which magma systems can be imaged provides insights into the duration and dynamics of magma storage that are unique, because they are necessarily biased in some fashion. It should be reiterated that because these estimates are largely based on volcanic rocks and/or their effusion rates, an even more fundamental bias to the majority of estimates for the durations of magma storage presented here may pertain. As such, the record may be skewed by the fact that magmas are required to be able to erupt which, particularly for geochemical approaches, means that the most fluid portions of a magma reservoir will be preferentially imaged. Nonetheless, these estimates are central to volcanology and petrology and can have consequences for the dynamics and longevity of plutonic systems as well.

Notwithstanding the variation in ages obtained in work to date—whether related to analytical technique, magma composition, or tectonic setting—it appears that, at least for the majority of magmas, storage timescales are no more than a few thousands of years and many are on the order of several years. Individual magma batches would, therefore, be very transient features of the $\sim$0.5 Myr to several million year life spans of many individual volcanic and plutonic systems. The liquid (and mush) portions of reservoirs responsible for these magma batches may be equally ephemeral features or may be more long-standing, because they are repeatedly renewed/replenished. Exceptions to this generalization about the absolute longevity of magma reservoirs are reflected in the differentiation and crystallization times of silicic magmas. It remains to be determined just how each line of investigative inquiry relates to specific aspects of a magma reservoir and its evolution, and how to reconcile the various age estimates obtained for individual volcanoes. The increasing availability and application of spatially resolved analytical techniques will be invaluable in this regard, especially when coupled with insights gained from experimentally determined phase equilibria and from dynamical modeling of the time-dependent physical properties of magmatic systems.

ACKNOWLEDGMENTS

The content of this chapter was significantly improved as a result of the efforts of several reviewers and I am sincerely grateful to them. My thanks to Francis Albarede, George Bergantz, Wendy Bohrson, and Allen Glazner for thoughtful comments on specific sections of this chapter, to Aaron Pietruszka, Roberta Rudnick, and Justin Simon for their substantive reviews of the entire chapter, and especially to Kari Cooper and Jorge Vazquez who also reviewed all of the chapter—twice. My thanks also to participants in the 2002 igneous petrology seminar at UCLA, and especially to Bill Moore, for continuing my education in various aspects of magmatic systems. This work was supported by National Science Foundation grants EAR-9980646 and EAR-0003601.

REFERENCES

Albarede F. (1993) Residence time analysis of geochemical fluctuations in volcanic series. *Geochim. Cosmochim. Acta* **57**, 615–621.

Allègre C. J. (1968) ^{230}Th dating of volcanic rocks: a comment. *Earth Planet. Sci. Lett.* **5**, 209–210.

Allègre C. J. and Condomines M. (1976) Fine chronology of volcanic processes using ^{238}U–^{230}Th systematics. *Earth Planet. Sci. Lett.* **28**, 395–406.

Anderson A. T., Davis A. M., and Lu F. Q. (2000) Evolution of Bishop Tuff rhyolitic magma based on melt and magnetite inclusions and zoned phenocrysts. *J. Petrol.* **41**, 449–473.

Armienti P., Pareschi M. T., Innocent I. F., and Pompilio M. (1994) Effects of magma storage and ascent on the kinetics of crystal-growth-the case of the 1991–93 Mt. Etna eruption. *Contrib. Mineral. Petrol.* **115**, 402–414.

Asmerom Y., Cheng H., Thomas R., Hirschmann M., and Edwards R. L. (2000) Melting of the Earth's lithospheric mantle inferred from protactinum–thorium–uranium isotopic data. *Nature* **406**, 293–296.

Bachmann O., Dungan M. A., and Lipman P. W. (2002) The fish canyon magma body, San Juan volcanic field, Colorado: rejuvenation and eruption of an upper-crustal batholith. *J. Petrol.* **43**, 1469–1503.

Bacon C. R. and Druitt T. H. (1986) Magmatic inclusions in silicic and intermediate volcanic-rocks. *J. Geophys. Res.: Solid Earth* **91**, 6091–6112.

Barboza S. A. and Bergantz G. W. (2000) Metamorphism and anatexis in the mafic complex contact aureole, Ivrea zone, northern Italy. *J. Petrol.* **41**, 1307–1327.

Barth G. A., Kleinrock M. C., and Helz R. T. (1994) The magma body at Kilauea-Iki lava lake-potential insights into mid-ocean ridge magma chambers. *J. Geophys. Res.: Solid Earth* **99**, 7199–7217.

Bergantz G. W. (2000) On the dynamics of magma mixing by reintrusion: implications for pluton assembly processes. *J. Struct. Geol.* **22**, 1297–1309.

Bindeman I. N. and Valley J. W. (2001) Low δ^{18}O rhyolites from Yellowstone: magmatic evolution based on analyses of zircons and individual phenocrysts. *J. Petrol.* **42**, 1491–1517.

Bindeman I. N. and Valley J. W. (2002) Oxygen isotope study of the long valley magma system, California: isotope thermometry and convection in large silicic magma bodies. *Contrib. Mineral. Petrol.* **144**, 185–205.

Bindeman I. N., Valley J. W., Wooden J. L., and Persing H. M. (2001) Post-caldera volcanism: *in situ* measurement of U–Pb age and oxygen isotope ratio in Pleistocene zircons from Yellowstone caldera. *Earth Planet. Sci. Lett.* **189**, 197–206.

Black S., Macdonald R., Barreiro B. A., Dunkley P. N., and Smith M. (1998a) Open system alkaline magmatism in northern Kenya: evidence from U-series disequilibria and radiogenic isotopes. *Contrib. Mineral. Petrol.* **131**, 364–378.

Black S., Macdonald R., DeVivo B., Kilburn C. R. J., and Rolandi G. (1998b) U-series disequilibria in young (AD 1944) Vesuvius rocks: preliminary implications for magma residence times and volatile addition. *J. Volcanol. Geotherm. Res.* **82**, 97–111.

Blundy J. and Cashman K. (2001) Ascent-driven crystallisation of dacite magmas at Mount St Helens, 1980–1986. *Contrib. Mineral. Petrol.* **140**, 631–650.

Blundy J. and Wood B. (1994) Prediction of crystal-melt partition coefficients from elastic moduli. *Nature* **372**, 452–454.

Bohrson W. A. and Reid M. R. (1998) Genesis of evolved ocean island magmas by deep- and shallow-level basement recycling, Socorro Island, Mexico: constraints from Th and other isotope signatures. *J. Petrol.* **39**, 995–1008.

Bons P. D., Dougherty-Page J., and Elburg M. A. (2001) Stepwise accumulation and ascent of magmas. *J. Metamorph. Geol.* **19**, 625–631.

Bourdon B., Zindler A., and Worner G. (1994) Evolution of the Laacher-see magma chamber-evidence from SIMS and TIMS measurements of U–Th disequilibria in minerals and glasses. *Earth Planet. Sci. Lett.* **126**, 75–90.

Bourdon B., Joron J. L., Claude-Ivanaj C., and Allegre C. J. (1998) U–Th–Pa–Ra systematics for the Grande Comore volcanics: melting processes in an upwelling plume. *Earth Planet. Sci. Lett.* **164**, 119–133.

Brandeis G. and Jaupart C. (1986) On the interaction between convection and crystallization in cooling magma chambers. *Earth Planet. Sci. Lett.* **77**, 345–361.

Brandon A., Creaser R., and Chacko T. (1996) Constraints on rates of granitic magma transport from epidote dissolution kinetics. *Science* **271**, 1845–1848.

Brophy J. G. and Dreher S. T. (2000) The origin of composition gaps at South Sister volcano, central Oregon: implications for fractional crystallization processes beneath active calc-alkaline volcanoes. *J. Volcanol. Geotherm. Res.* **102**, 287–307.

Brophy J. G., Whittington C. S., and Park Y. R. (1999) Sector-zoned augite megacrysts in Aleutian high alumina basalts: implications for the conditions of basalt crystallization and the generation of calc-alkaline series magmas. *Contrib. Mineral. Petrol.* **135**, 277–290.

Brown M. (2001) Crustal melting and granite magmatism: key issues. *Phys. Chem. Earth Part A: Solid Earth Geodesy* **26**, 201–212.

Brown M. and Solar G. S. (1998) Granite ascent and emplacement during contractional deformation in convergent orogens. *J. Struct. Geol.* **20**, 1365–1393.

Brown S. J. A. and Fletcher I. R. (1999) SHRIMP U–Pb dating of the preeruption growth history of zircons from the 340 ka Whakamaru Ignimbrite, New Zealand: evidence for >250 ky magma residence times. *Geology* **27**, 1035–1038.

Cashman K. V. (1990) Textural constraints on the kinetics of crystallization of igneous rocks. In *Understanding Magmatic Processes* (eds. J. Nicholls and J. K. Russell). MSA, Washington, DC, vol. 24, pp. 259–314.

Cashman K. V. (1992) Groundmass crystallization of Mount St Helens dacite, 1980–1986 a tool for interpreting shallow magmatic processes. *Contrib. Mineral. Petrol.* **109**, 431–449.

Cashman K. V. (1993) Relationship between plagioclase crystallization and cooling rate in basaltic melts. *Contrib. Mineral. Petrol.* **113**, 126–142.

Cashman K. V. and Blundy J. (2000) Degassing and crystallization of ascending andesite and dacite. *Phil. Trans. Roy. Soc. London Ser. a: Math. Phys. Eng. Sci.* **358**, 1487–1513.

Cerrai E., Dugnani Lonati F., Gazzarrini F., and Tongiorgi T. (1965) Il metodo ionio-uranio per la determinazione dell'eta dei minerali vulcanici recenti. *Rend. Soc. Min. Ital.* **21**, 47–62.

Charlier B. and Zellmer G. (2000) Some remarks on U–Th mineral ages from igneous rocks with prolonged crystallisation histories. *Earth Planet. Sci. Lett.* **183**, 457–469.

Chen C. H., DePaolo D. J., Nakada S., and Shieh Y. N. (1993) Relationship between eruption volume and neodymium isotopic composition at Unzen volcano. *Nature* **362**, 831–834.

Cherniak D. J. (1996) Strontium diffusion in sanidine and albite, and general comments on strontium diffusion in alkali feldspars. *Geochim. Cosmochim. Acta* **60**(24), 5037–5043.

Cherniak D. J. and Watson E. B. (2001) Pb diffusion in zircon. *Chem. Geol.* **172**, 5–24.

Chmielowski J., Zandt G., and Haberland C. (1999) The central Andean Altiplano-puna magma body. *Geophys. Res. Lett.* **26**, 783–786.

Christensen J. N. and DePaolo D. J. (1993) Timescales of large volume silicic magma systems-Sr isotopic systematics of phenocrysts and glass from the Bishop Tuff, Long Valley, California. *Contrib. Mineral. Petrol.* **113**, 100–114.

Christiansen R. L. (2001) The Quaternary and Pliocene Yellowstone Plateau volcanic field of Wyoming, Idaho, and Montana. US Geological Survey Professional Paper, vol. 729-G, 145p.

Claude-Ivanaj C., Bourdon B., and Allegre C. J. (1998) Ra–Th–Sr isotope systematics in Grande Comore Island: a case study of plume-lithosphere interaction. *Earth Planet. Sci. Lett.* **164**, 99–117.

Claude-Ivanaj C., Joron J. L., and Allegre C. J. (2001) ^{238}U–^{230}Th–^{226}Ra fractionation in historical lavas from the Azores: long-lived source heterogeneity vs. metasomatism fingerprints. *Chem. Geol.* **176**, 295–310.

Clemens J. D. (1998) Observations on the origins and ascent mechanisms of granitic magmas. *J. Geol. Soc.* **155**, 843–851.

Clemens J. D. and Mawer C. K. (1992) Granitic magma transport by fracture propagation. *Tectonophysics* **204**, 339–360.

Coleman D. S., Glazner A. F., Miller J. S., Bradford K. J., Frost T. P., Joye J. L., and Bachl C. A. (1995) Exposure of a late cretaceous layered mafic–felsic magma system in the Central Sierra-Nevada batholith, California. *Contrib. Mineral. Petrol.* **120**, 129–136.

Condomines M. (1997) Dating recent volcanic rocks through ^{230}Th–^{238}U disequilibrium in accessory minerals: example of the Puy de Dome (French Massif Central). *Geology* **25**, 375–378.

Condomines M., Tanguy J. C., and Michaud V. (1995) Magma dynamics at Mt. Etna-constraints from U–Th–Ra–Pb radioactive disequilibria and Sr isotopes in historical lavas. *Earth Planet. Sci. Lett.* **132**, 25–41.

Condomines M., Gauthier P.-J., and Sigmarsson O. (2003) Timescales of magma chamber processes and dating of young volcanic rocks. In *Uranium-series Geochemistry*, Reviews in Mineralogy and Geochemistry (eds. B. Bourdon, G. M. Henderson, C. C. Lundstrom, and S. P. Turner). Mineralogical Society of America, vol. 52, pp. 124–171.

Cooper K. M. and Reid M. R. (2003) Re-examination of crystal in recent ages. Mt. St. Helens lavas. *Earth. Planet. Sci. Lett.* **213,** 149–167.

Cooper K. M., Reid M. R., Murrell M. T., and Clague D. A. (2001) Crystal and magma residence at Kilauea Volcano, Hawaii: ^{230}Th–^{226}Ra dating of the 1955 east rift eruption. *Earth Planet. Sci. Lett.* **184**, 703–718.

Cooper K. M., Goldstein S. J., Sims K. W. W., and Murrell M. T. (2003) Uranium-series chronology of Gorda ridge volcanism: new evidence from the 1996 eruption. *Earth Planet. Sci. Lett.* **206**, 459–475.

Cortini M. and Scandone R. (1982) The feeding system of Vesuvius between 1754 and 1944. *J. Volcanol. Geotherm. Res.* **12**, 393–400.

Coulson I. M., Villeneuve M. E., Dipple G. M., Duncan R. A., Russell J. K., and Mortensen J. K. (2002) Timescales of assembly and thermal history of a composite felsic pluton: constraints from the Emerald Lake area, northern Canadian Cordillera, Yukon. *J. Volcanol. Geotherm. Res.* **114**, 331–356.

Crawford W. C., Webb S. C., and Hildebrand J. A. (1999) Constraints on melt in the lower crust and Moho at the East Pacific Rise, 9°48' N, using seafloor compliance measurements. *J. Geophys. Res.: Solid Earth* **104**, 2923–2939.

Crisp J. A. (1984) Rates of magma emplacement and volcanic output. *J. Volcanol. Geotherm. Res.* **20**, 177–211.

Cruden A. R. and McCaffrey K. J. W. (2001) Growth of plutons by floor subsidence: implications for rates of emplacement, intrusion spacing and melt-extraction mechanisms. *Phys. Chem. Earth Part A: Solid Earth Geodesy* **26**, 303–315.

Dalrymple G. B., Grove M., Lovera O. M., Harrison T. M., Hulen J. B., and Lanphere M. A. (1999) Age and thermal history of the Geysers plutonic complex (felsite unit), geysers geothermal field, California: a Ar-40/Ar-39 and U–Pb study. *Earth Planet. Sci. Lett.* **173**, 285–298.

Danyushevsky L. V., Sokolov S., and Falloon T. J. (2002) Melt inclusions in olivine phenocrysts: using diffusive re-equilibration to determine the cooling history of a crystal, with implications for the origin of olivine-phyric volcanic rocks. *J. Petrol.* **43**, 1651–1671.

Davidson J., Tepley F., Palacz Z., and Meffan-Main S. (2001) Magma recharge, contamination, and residence times revealed by *in situ* laser ablation isotopic analysis of feldspar in volcanic rocks. *Earth Planet. Sci. Lett.* **184**, 427–442.

Davidson J. P. and Tepley F. J. (1997) Recharge in volcanic systems: evidence from isotope profiles of phenocrysts. *Science* **275**, 826–829.

Davies G. R. and Halliday A. N. (1998) Development of the Long Valley rhyolitic magma system: strontium and neodymium isotope evidence from glasses and individual phenocrysts. *Geochim. Cosmochim. Acta* **62**, 3561–3574.

Davies G. R., Halliday A. N., Mahood G. A., and Hall C. M. (1994) Isotopic constraints on the production-rates, crystallization histories and residence times of pre-caldera silicic magmas, Long Valley, California. *Earth Planet. Sci. Lett.* **125**, 17–37.

Dawson P. B., Chouet B. A., Okubo P. G., Villasenor A., and Benz H. M. (1999) Three-dimensional velocity structure of the Kilauea caldera. *Hawaii. Geophys. Res. Lett.* **26**, 2805–2808.

Decker R. W. (1987) Dynamics of Hawaiian volcanoes: an overview. In *Volcanism in Hawaii* (eds. R. W. Decker, T. L. Wright, and P. H. Stauffer). US Geological Survey Professional Paper, vol. 1350, pp. 997–1018.

Denlinger R. P. (1997) A dynamic balance between magma supply and eruption rate at Kilauea Volcano, Hawaii. *J. Geophys. Res.: Solid Earth* **102**, 18091–18100.

Devine J. D., Rutherford M. J., and Gardner J. E. (1998) Petrologic determination of ascent rates for the 1995–1997 Soufriere hills volcano andesitic magma. *Geophys. Res. Lett.* **25**, 3673–3676.

Dungan M. A., Wulff A., and Thompson R. (2001) Eruptive stratigraphy of the Tatara-San Pedro complex, 36°S, southern volcanic zone, Chilean Andes: reconstruction method and implications for magma evolution at long-lived arc volcanic centers. *J. Petrol.*, 555–626.

Dunn R. A., Toomey D. R., and Solomon S. C. (2000) Three-dimensional seismic structure and physical properties of the crust and shallow mantle beneath the East Pacific Rise at 9°30′N. *J. Geophys. Res.: Solid Earth* **105**, 23537–23555.

Dvorak J. J. and Dzurisin D. (1993) Variations in magma supply rate at Kilauea Volcano, Hawaii. *J. Geophys. Res.* **98**, 22255–22268.

Feineman M. D., DePaolo D. J., and Ryerson F. J. (2002) Steady-state ^{226}Ra/^{230}Th disequilibrium in hydrous mantle minerals. *Geochim. Cosmochim. Acta* **66**, A228.

Fitton J. G., Kilburn C. R. J., Thirlwall M. F., and Hughes D. J. (1983) 1982 eruption of Mount Cameroon, West Africa. *Nature* **306**, 327–332.

Francalanci L., Tommasini S., Conticelli S., and Davies G. R. (1999) Sr isotope evidence for short magma residence time for the 20th century activity at Stromboli volcano, Italy. *Earth Planet. Sci. Lett.* **167**, 61–69.

Francis P., Oppenheimer C., and Stevenson D. (1993) Endogenous growth of persistently active volcanos. *Nature* **366**, 554–557.

Friedrich A. M., Bowring S. A., Martin M. W., and Hodges K. V. (1999) Short-lived continental magmatic arc at Connemara, western Irish Caledonides: implications for the age of the Grampian orogeny. *Geology* **27**, 27–30.

Fukuoka T. (1974) Ionium dating of acidic volcanic rocks. *Geochem. J.* **8**, 109–116.

Gamble J. A., Wood C. P., Price R. C., Smith I. E. M., Stewart R. B., and Waight T. (1999) A fifty year perspective of magmatic evolution on Ruapehu Volcano, New Zealand: verification of open-system behaviour in an arc volcano. *Earth Planet. Sci. Lett.* **170**, 301–314.

Gansecki C. A., Mahood G. A., and McWilliams M. (1998) New ages for the climactic eruptions at Yellowstone: single-crystal $^{40}Ar/^{39}Ar$ dating identifies contamination. *Geology* **26**, 343–346.

Gardner J. E., Layer P. W., and Rutherford M. J. (2002) Phenocrysts versus xenocrysts in the youngest Toba Tuff: implications for the petrogenesis of 2800 km^3 of magma. *Geology* **30**, 347–350.

Gauthier P. J. and Condomines M. (1999) $^{210}Pb-^{226}Ra$ radioactive disequilibria in recent lavas and radon degassing: inferences on the magma chamber dynamics at Stromboli and Merapi volcanoes. *Earth Planet. Sci. Lett.* **172**, 111–126.

Gauthier P. J., Le Cloarec M. F., and Condomines M. (2000) Degassing processes at Stromboli volcano inferred from short-lived disequilibria ($^{210}Pb-^{210}Bi-^{210}Po$) in volcanic gases. *J. Volcanol. Geotherm. Res.* **102**, 1–19.

Geschwind C. H. and Rutherford M. J. (1995) Crystallization of microlites during magma ascent-the fluid-mechanics of 1980–1986 eruptions at Mount St. Helens. *Bull. Volcanol.* **57**, 356–370.

Giletti B. J. and Casserly J. E. D. (1994) Strontium diffusion kinetics in plagioclase feldspars. *Geochim. Cosmochim. Acta* **58**, 3785–3793.

Gill J. B., Morris J. D., and Johnson R. W. (1993) Timescale for producing the geochemical signature of island-arc magmas U–Th–Po and Be–B systematics in recent Papua-New Guinea lavas. *Geochim. Cosmochim. Acta* **57**, 4269–4283.

Ginibre C., Kronz A., and Worner G. (2002a) High-resolution quantitative imaging of plagioclase composition using accumulated backscattered electron images: new constraints on oscillatory zoning. *Contrib. Mineral. Petrol.* **142**, 436–448.

Ginibre C., Worner G., and Kronz A. (2002b) Minor- and trace-element zoning in plagioclase: implications for magma chamber processes at Parinacota volcano, northern Chile. *Contrib. Mineral. Petrol.* **143**, 300–315.

Glazner A. F. and Miller D. M. (1997) Late-stage sinking of plutons. *Geology* **25**, 1099–1102.

Glazner A. F. and Ussler W. (1988) Trapping of magma at midcrustal density discontinuities. *Geophys. Res. Lett.* **15**, 673–675.

Halliday A. N., Mahood G. A., Holden P., Metz J. M., Dempster T. J., and Davidson J. P. (1989) Evidence for long residence times of rhyolitic magma in the Long Valley magmatic system the isotopic record in precaldera lavas of Glass Mountain. *Earth Planet. Sci. Lett.* **94**, 274–290.

Hammer J. E., Cashman K. V., Hoblitt R. P., and Newman S. (1999) Degassing and microlite crystallization during pre-climactic events of the 1991 eruption of Mt. Pinatubo, Philippines. *Bull. Volcanol.* **60**, 355–380.

Hammer J. E., Cashman K. V., and Voight B. (2000) Magmatic processes revealed by textural and compositional trends in Merapi dome lavas. *J. Volcanol. Geotherm. Res.* **100**, 165–192.

Hansteen T. H., Klugel A., and Schmincke H. U. (1998) Multi-stage magma ascent beneath the Canary Islands: evidence from fluid inclusions. *Contrib. Mineral. Petrol.* **132**, 48–64.

Harris A. J. L., Murray J. B., Aries S. E., Davies M. A., Flynn L. P., Wooster M. J., Wright R., and Rothery D. A. (2000) Effusion rate trends at Etna and Krafla and their implications for eruptive mechanisms. *J. Volcanol. Geotherm. Res.* **102**, 237–270.

Haslinger F., Thurber C., Mandernach M., and Okubo P. (2001) Tomographic image of P-velocity structure beneath Kilauea's east rift zone and South Flank: seismic evidence for a deep magma body. *Geophys. Res. Lett.* **28**, 375–378.

Hawkesworth C. J., Blake S., Evans P., Hughes R., MacDonald R., Thomas L. E., and Turner S. P. (2000) Time scales of crystal fractionation in magma chambers—integrating physical, isotopic and geochemical perspectives. *J. Petrol.* **41**, 991–1006.

Hawkesworth C. J., Blake S., Thomas L., and Turner S. P. (2001) Volcanic power outputs and U-series age constraints: a missing link between crystallization processes and magma differentiation. *EOS* **82**(261), 264–265.

Heath E., Turner S. P., Macdonald R., Hawkesworth C. J., and van Calsteren P. (1998) Long magma residence times at an island are volcano (SoufriereSt. Vincent) in the Lesser Antilles: evidence from $^{238}U-^{230}Th$ isochron dating. *Earth Planet. Sci. Lett.* **160**, 49–63.

Heizler M. T., Perry F. V., Crowe B. M., Peters L., and Appelt R. (1999) The age of Lathrop Wells volcanic center: An $^{40}Ar/^{39}Ar$ dating investigation. *J. Geophys. Res.: Solid Earth* **104**, 767–804.

Hemond C., Hofmann A. W., Heusser G., Condomines M., Raczek I., and Rhodes J. M. (1994) U–Th–Ra systematics in Kilauea and Mauna-Loa basalts, Hawaii. *Chem. Geol.* **116**, 163–180.

Heumann A. and Davies G. R. (2002) U–Th disequilibrium and Rb–Sr age constraints on the magmatic evolution of peralkaline rhyolites from Kenya. *J. Petrol.* **43**, 557–577.

Heumann A., Davies G. R., and Elliott T. (2002) Crystallization history of rhyolites at Long Valley, California, inferred from combined U-series and Rb–Sr isotope systematics. *Geochim. Cosmochim. Acta* **66**, 1821–1837.

Higgins M. D. (1996a) Crystal size distributions and other quantitative textural measurements in lavas and tuff from Egmont volcano (Mt Taranaki), New Zealand. *Bull Volcanol.* **58**, 194–204.

Higgins M. D. (1996b) Magma dynamics beneath Kameni volcano, Thera, Greece, as revealed by crystal size and shape measurements. *J. Volcanol. Geotherm. Res.* **70**, 37–48.

Hildreth W. and Fierstein J. (1983) The compositionally zoned eruption of 1912 in the Valley of 10 thousand smokes, Katmai National Park, Alaska. *J. Volcanol. Geotherm. Res.* **18**, 1–56.

Hildreth W. and Lanphere M. A. (1994) Potassium–argon geochronology of a basalt–andesite–dacite arc system the Mount Adams volcanic field, cascade range of southern Washington. *Geol. Soc. Am. Bull.* **106**, 1413–1429.

Hildreth W. and Moorbath S. (1988) Crustal contributions to arc magmatism in the Andes of central Chile. *Contrib. Mineral. Petrol.* **98**, 455–489.

Hirata T. and Nesbitt R. W. (1995) U–Pb isotope geochronology of zircon-evaluation of the laser probe-inductively coupled plasma-mass spectrometry technique. *Geochim. Cosmochim. Acta* **59**, 2491–2500.

Hobden B. J., Houghton B. F., Davidson J. P., and Weaver S. D. (1999) Small and short-lived magma batches at composite volcanoes: time windows at Tongariro volcano, New Zealand. *J. Geol. Soc.* **156**, 865–868.

Hooft E. E. E., Detrick R. S., and Kent G. M. (1997) Seismic structure and indicators of magma budget along the southern East Pacific Rise. *J. Geophys. Res.: Solid Earth* **102**, 27319–27340.

Horn I., Rudnick R. L., and McDonough W. F. (2000) Precise elemental and isotope ratio determination by simultaneous solution nebulization and laser ablation ICP–MS: application to U–Pb geochronology. *Chem. Geol.* **164**, 281–301.

Hughes R. D. and Hawkesworth C. J. (1999) The effects of magma replenishment processes on $^{238}U-^{230}Th$ disequilibrium. *Geochim. Cosmochim. Acta* **63**, 4101–4110.

Huppert H. E. and Sparks R. S. J. (1988) The generation of granitic magmas by intrusion of basalt into continental crust. *J. Petrol.* **29**, 599–642.

Jambon A., Lussiez P., Clocchiatti R., Weisz J., and Hernandez J. (1992) Olivine growth rates in a tholeiitic basalt-an experimental study of melt inclusions in plagioclase. *Chem. Geol.* **96**, 277–287.

Johnson S. E., Albertz M., and Paterson S. R. (2001) Growth rates of dike-fed plutons: are they compatible with observations in the middle and upper crust? *Geology* **29**, 727–730.

Kelemen P. B., Hirth G., Shimizu N., Spiegelman M., and Dick H. J. B. (1997) A review of melt migration processes in the adiabatically upwelling mantle beneath oceanic spreading ridges. *Phil. Trans. Roy. Soc. London Ser. A: Math. Phys. Eng. Sci.* **255**, 283–318.

Kigoshi K. (1967) Ionium dating of igneous rocks. *Science* **156**, 932–934.

Kind R., Ni J., Zhao W. J., Wu J. X., Yuan X. H., Zhao L. S., Sandvol E., Reese C., Nabelek J., and Hearn T. (1996) Evidence from earthquake data for a partially molten crustal layer in southern Tibet. *Science* **274**, 1692–1694.

Klugel A. (1998) Reactions between mantle xenoliths and host magma beneath La Palma (Canary Islands): constraints on magma ascent rates and crustal reservoirs. *Contrib. Mineral. Petrol.* **131**, 237–257.

Klugel A., Hoernle K. A., Schmincke H. U., and White J. D. L. (2000) The chemically zoned 1949 eruption on La Palma (Canary Islands): petrologic evolution and magma supply dynamics of a rift zone eruption. *J. Geophys. Res.: Solid Earth* **105**, 5997–6016.

Knesel K. M. and Davidson J. P. (1999) Sr isotope systematics during melt generation by intrusion of basalt into continental crust. *Contrib. Mineral. Petrol.* **136**, 285–295.

Knesel K. M., Davidson J. P., and Duffield W. A. (1999) Evolution of silicic magma through assimilation and subsequent recharge: evidence from Sr isotopes in sanidine phenocrysts, Taylor Creek Rhyolite, NM. *J. Petrol.* **40**, 773–786.

Korenaga J. and Kelemen P. B. (1998) Melt migration through the oceanic lower crust: a constraint from melt percolation modelling with finite solid diffusion. *Earth Planet. Sci. Lett.* **156**, 1–11.

Koyaguchi T. and Kaneko K. (1999) A two-stage thermal evolution model of magmas in continental crust. *J. Petrol.* **40**, 241–254.

Koyaguchi T. and Kaneko K. (2000) Thermal evolution of silicic magma chambers after basalt replenishments. *Trans. Roy. Soc. Edinburgh: Earth Sci.* **91**, 47–60.

Kuritani T. (1999) Phenocryst crystallization during ascent of alkali basalt magma at Rishiri volcano, northern Japan. *J. Volcanol. Geotherm. Res.* **88**, 77–97.

Lanphere M. A., Champion D. E., Clynne M. A., and Muffler L. J. P. (1999) Revised age of the Rockland tephra, northern California: implications for climate and stratigraphic reconstructions in the western United States. *Geology* **27**, 135–138.

Lanphere M. A., Champion D. E., Christiansen R. L., Izett G. A., and Obradovich J. D. (2002) Revised ages for tuffs of the Yellowstone Plateau volcanic field: assignment of the Huckleberry ridge tuff to a new geomagnetic polarity event. *Geol. Soc. Am. Bull.* **114**(5), 559–568.

Lejeune A.-M. and Richet P. (1995) Rheology of crystal-bearing silicate melts: an experimental study of high viscosities. *J. Geophys. Res.: Solid Earth* **100**, 4215–4229.

Lundstrom C. C., Gill J., and Williams Q. (2000) A geochemically consistent hypothesis for MORB generation. *Chem. Geol.* **162**, 105–126.

Mahood G. A. (1990) Evidence for long residence times of rhyolitic magma in the Long Valley magma in the Long Valley magmatic system-the isotopic record in the pre-caldera lavas of Glass Mountain-reply. *Earth Planet. Sci. Lett.* **99**, 395–399.

Mangan M. T. (1990) Crystal size distribution systematics and the determination of magma storage times—the 1959 eruption of Kilauea Volcano, Hawaii. *J. Volcanol. Geotherm. Res.* **44**, 295–302.

Mankinen E. A., Gromme C. S., Dalrymple G. B., Lanphere M. A., and Bailey R. A. (1986) Paleomagnetism and K–Ar ages of volcanic rocks from Long Valley caldera, California. *J. Geophys. Res.* **91**, 633–652.

Marsh B. D. (1981) On the crystallinity, probability of occurrence, and rheology of lava and magma. *Contrib. Mineral. Petrol.* **78**, 85–98.

Marsh B. D. (1988) Crystal size distribution (CSD) in rocks and the kinetics and dynamics of crystallization: 1. Theory. *Contrib. Mineral. Petrol.* **99**, 277–291.

Marsh B. D. (1989) Magma chambers. *Ann. Rev. Earth Planet. Sci.* **17**, 437–474.

Marsh B. D. (1996) Solidification fronts and magmatic evolution. *Min. Mag.* **60**, 5–40.

Marsh B. D. (1998) On the interpretation of crystal size distributions in magmatic systems. *J. Petrol.* **39**, 553–599.

Masturyono, McCaffrey R., Wark D. A., Roecker S. W., Fauzi, Ibrahim G., and Sukhyar (2001) Distribution of magma beneath the Toba caldera complex, north Sumatra, Indonesia, constrained by three-dimensional P wave velocities, seismicity, and gravity data. *Geochem. Geophys. Geosys.* **2**, U1–U24.

McNulty B. A., Tong W. X., and Tobisch O. T. (1996) Assembly of a dike-fed magma chamber: the Jackass lakes pluton, central Sierra Nevada, California. *Geol. Soc. Am. Bull.* **108**, 926–940.

McNulty B. A., Tobisch O. T., Cruden A. R., and Gilder S. (2000) Multistage emplacement of the Mount Givens pluton, central Sierra Nevada batholith, California. *Geol. Soc. Am. Bull.* **112**, 119–135.

Miller B. V., Samson S. D., and D'Lemos R. S. (1999) Time span of plutonism, fabric development, and cooling in a Neoproterozoic magmatic arc segment: U–Pb age constraints from syn-tectonic plutons, Sark, Channel Islands, UK. *Tectonophysics* **312**, 79–95.

Miller D. S. and Smith R. B. (1999) P and S velocity structure of the Yellowstone volcanic field from local earthquake and controlled-source tomography. *J. Geophys. Res.: Solid Earth* **104**, 15105–15121.

Miller R. B. and Paterson S. R. (1999) In defense of magmatic diapirs. *J. Struct. Geol.* **21**, 1161–1173.

Miller R. B. and Paterson S. R. (2001) Construction of mid-crustal sheeted plutons: examples from the north Cascades, Washington. *Geol. Soc. Am. Bull.* **113**, 1423–1442.

Murphy M. D., Sparks R. S. J., Barclay J., Carroll M. R., and Brewer T. S. (2000) Remobilization of andesite magma by intrusion of mafic magma at the Soufriere Hills Volcano, Montserrat, West Indies. *J. Petrol.* **41**, 21–42.

Myers J. D., Marsh B. D., Frost C. D., and Linton J. A. (2002) Petrologic constraints on the spatial distribution of crustal magma chambers, Atka Volcanic Center, central Aleutian arc. *Contrib. Mineral. Petrol.* **143**, 567–586.

Nakada S., Bacon C. R., and Gartner A. E. (1994) Origin of phenocrysts and compositional diversity in pre-Mazama rhyodacite lavas, Crater Lake, Oregon. *J. Petrol.* **35**, 127–162.

Nakamura M. (1995a) Continuous mixing of crystal mush and replenished magma in the ongoing Unzen eruption. *Geology* **23**, 807–810.

Nakamura M. (1995b) Residence time and crystallization history of nickeliferous olivine phenocrysts from the northern Yatsugatake volcanoes, central Japan: application of a growth and diffusion model in the system Mg–Fe–Ni. *J. Volcanol. Geotherm. Res.* **66**, 81–100.

Nelson K. D., Zhao W. J., Brown L. D., Kuo J., Che J. K., Liu X. W., Klemperer S. L., Makovsky Y., Meissner R., Mechie J., Kind R., Wenzel F., Ni J., Nabelek J., Chen L. S., Tan H. D., Wei W. B., Jones A. G., Booker J., Unsworth M., Kidd W. S. F., Hauck M., Alsdorf D., Ross A., Cogan M., Wu C. D., Sandvol E., and Edwards M. (1996) Partially molten middle crust beneath southern Tibet: synthesis of project INDEPTH results. *Science* **274**, 1684–1688.

Oversby V. M. and Gast P. W. (1968) Lead isotope compositions and uranium decay series disequilibrium in recent volcanic rocks. *Earth Planet. Sci. Lett.* **5**, 199–206.

Pallister J. S., Hoblitt R. P., and Reyes A. G. (1992) A basalt trigger for the 1991 eruptions of Pinatubo Volcano. *Nature* **356**, 426–428.

Paterson S. R. and Tobisch O. T. (1992) Rates of processes in magmatic arcs-implications for the timing and nature of pluton emplacement and wall rock deformation. *J. Struct. Geol.* **14**, 291–300.

Patino Douce A. E. (1999) What do experiments tell us about the relative contributions of crust and mantle to the origin of granitic magmas? In *Understanding Granites: Integrating New and Classical Techniques*. Special Publications (eds. A. Castro, C. Fernandez, and J. L. Vigneresse). Geological Society, London, vol. 168, pp. 55–75.

Peate D., Chen J., Wasserburg G., Papanastassiou D., and Geissman J. (1996) U-238–Th-230 dating of a geomagnetic excursion in quaternary basalts of the Albuquerque volcanoes field, New Mexico (USA). *Geophys. Res. Lett.* **23**, 2271–2274.

Petford N. (1996) Dykes or diapirs? *Trans. Roy. Soc. Edinburgh: Earth Sci.* **87**, 105–114.

Petford N., Cruden A. R., McCaffrey K. J. W., and Vigneresse J. L. (2000) Granite magma formation, transport and emplacement in the Earth's crust. *Nature* **408**, 669–673.

Philpotts A. R., Shi J., and Brustman C. (1998) Role of plagioclase crystal chains in the differentiation of partly crystallized basaltic magma. *Nature* **395**, 343–346.

Pickett D. A. and Murrell M. T. (1997) Observations of ^{231}Pa/^{235}U disequilibrium in volcanic rocks. *Earth Planet. Sci. Lett.* **148**, 259–271.

Pietruszka A. J. and Garcia M. O. (1999) The size and shape of Kilauea Volcano's summit magma storage reservoir: a geochemical probe. *Earth Planet. Sci. Lett.* **167**, 311–320.

Pietruszka A. J., Rubin K. H., and Garcia M. O. (2001) ^{226}Ra–^{230}Th–^{238}U disequilibria of historical Kilauea lavas (1790–1982) and the dynamics of mantle melting within the Hawaiian plume. *Earth Planet. Sci. Lett.* **186**, 15–31.

Putirka K. (1997) Magma transport at Hawaii: inferences based on igneous thermobarometry. *Geology* **25**, 69–72.

Pyle D. M. (1992) The volume and residence time of magma storage beneath active volcanos determined by decay-series disequilibria methods. *Earth Planet. Sci. Lett.* **112**, 61–73.

Quane S. L., Garcia M. O., Guillou H., and Hulsebosch T. P. (2000) Magmatic history of the east rift zone of Kilauea Volcano, Hawaii based on drill core from SOH: 1. *J. Volcanol. Geotherm. Res.* **102**, 319–338.

Reagan M. K., Volpe A. M., and K V. C. (1992) ^{238}U-series and ^{232}Th-series chronology of phonolite fractionation at Mounta Erebus, Antarctica. *Geochim. Cosmochim. Acta* **56**, 1401–1407.

Reid J. B., Evans O. C., and Fates D. G. (1983) Magma mixing in granitic rocks of the central Sierra Nevada, California. *Earth Planet. Sci. Lett.* **66**, 243–261.

Reid M. R. and Coath C. D. (2000) *In situ* U–Pb ages of zircons from the Bishop Tuff: no evidence for long crystal residence times. *Geology* **28**, 443–446.

Reid M. R., Coath C. D., Harrison T. M., and McKeegan K. D. (1997) Prolonged residence times for the youngest rhyolites associated with Long Valley caldera: ^{230}Th^{238}U ion microprobe dating of young zircons. *Earth Planet. Sci. Lett.* **150**, 27–39.

Renner J., Evans B., and Hirth G. (2000) On the rheologically critical melt fraction. *Earth Planet. Sci. Lett.* **181**, 585–594.

Resmini R. G. and Marsh B. D. (1995) Steady-state volcanism, paleoeffusion rates, and magma system volume inferred from plagioclase crystal size distributions in mafic lavas Dome Mountain, Nevada. *J. Volcanol. Geotherm. Res.* **68**, 273–296.

Rhodes J. M. and Hart S. R. (1995) Episodic trace element and isotopic variations in historical Mauna Loa lavas: implications for magma and plume dynamics. *Am. Geophys. Union Monogr.* **92**, 263–288.

Robinson D. M. and Miller C. F. (1999) Record of magma chamber processes preserved in accessory mineral assemblages, Aztec Wash pluton, Nevada. *Am. Mineral.* **84**, 1346–1353.

Roggensack K. (2001a) Sizing up crystals and their melt inclusions: a new approach to crystallization studies. *Earth Planet. Sci. Lett.* **187**, 221–237.

Roggensack K. (2001b) Unraveling the 1974 eruption of Fuego volcano (Guatemala) with small crystals and their young melt inclusions. *Geology* **29**, 911–914.

Rubin K. H., Smith M. C., Bergmanis E. C., Perfit M. R., Sinton J. M., and Batiza R. (2001) Geochemical heterogeneity within mid-ocean ridge lava flows: insights into eruption, emplacement and global variations in magma generation. *Earth Planet. Sci. Lett.* **188**, 349–367.

Rudnick R. T. and Presper T. (1990) Geochemistry of intermediate- to high-pressure granulites. In *Granulites and Crustal Evolution* (eds. D. Vielzeuf and P. Vidal). Kluwer Academic, Dordrecht, vol. 311, pp. 523–550.

Rutherford M. J. and Hill P. M. (1993) Magma ascent rates from amphibole breakdown an experimental study applied to the 1980–1986 Mount St. Helens eruptions. *J. Geophys. Res.: Solid Earth* **98**, 19667–19685.

Rutter E. H. and Neumann D. H. K. (1995) Experimental deformation of partially molten Westerly granite under fluid-absent conditions, with implications for the extraction of granitic magmas. *J. Geophys. Res.: Solid Earth Planets* **100**, 15697–15715.

Ryan M. P. (1988) The mechanics and 3-dimensional internal structure of active magmatic systems—Kilauea Volcano, Hawaii. *J. Geophys. Res.: Solid Earth Planets* **93**, 4213–4248.

Ryerson F. J., Weed H. C., and Piwinskii A. J. (1988) Rheology of subliquidus magmas: I. Picritic compositions. *J. Geophys. Res.: Solid Earth* **93**, 3421–3436.

Saal A. E., Van-Orman J. A., Hauri E. H., Langmuir C. H., and Perfit M. R. (2002) An alternative hypothesis for the origin of the high ^{226}Ra excess in mid-ocean ridge basalts. *Geochim. Cosmochim. Acta* **66**, A659.

Saar M. O., Manga M., Cashman K. V., and Fremouw S. (2001) Numerical models of the onset of yield strength in crystal-melt suspensions. *Earth Planet. Sci. Lett.* **187**, 367–379.

Sano Y. J., Tsutsumi Y., Terada K., and Kaneoka I. (2002) Ion microprobe U–Pb dating of quaternary zircon: implication for magma cooling and residence time. *J. Volcanol. Geotherm. Res.* **117**, 285–296.

Schaefer S. J., Sturchio N. C., Murrell M. T., and Williams S. N. (1993) Internal ^{238}U-series systematics of pumice from the November 13, 1985, eruption of Nevado del Ruiz, Colombia. *Geochim. Cosmochim. Acta* **57**, 1215–1219.

Schmitt A. K., Lindsay J. M., de Silva S., and Trumbull R. B. (2003) U–Pb zircon chronostratigraphy of Early-Pliocene ignimbrites from La Pacana, north Chile: implications for the formation of stratified magma chambers. *J. Volcanol. Geotherm. Res.* **120**, 43–53.

Schmitz M. D. and Bowring S. A. (2001) U–Pb zircon and titanite systematics of the fish canyon tuff: an assessment of high-precision U–Pb geochronology and its application to young volcanic rocks. *Geochim. Cosmochim. Acta* **65**, 2571–2587.

Sharp A. D. L., Davis P. M., and Gray F. (1980) A low velocity zone beneath Mount Etna and magma storage. *Nature* **287**, 587–591.

Shaw H. R. (1985) Links between magma-tectonic rate balances, plutonism, and volcanism. *J. Geophys. Res.: Solid Earth Planets.* **90**, 11275–11288.

Sheetz K. E. and Schlue J. W. (1992) Inferences for the Socorro magma body from teleseismic receiver functions. *Geophys. Res. Lett.* **19**, 1867–1870.

Sigmarsson O. (1996) Short magma chamber residence time at an Icelandic volcano inferred from U-series disequilibria. *Nature* **382**, 440–442.

Sigmarsson O., Carn S., and Carracedo J. C. (1998) Systematics of U-series nuclides in primitive lavas from the 1730-36 eruption on Lanzarote, Canary Islands, and implications for the role of garnet pyroxenites during oceanic basalt formations. *Earth Planet. Sci. Lett.* **162**, 137–151.

Sigmarsson O., Chmeleff J., Morris J., and Lopez-Escobar L. (2002) Origin of ^{226}Ra–^{230}Th disequilibria in arc lavas from southern Chile and implications for magma transfer time. *Earth Planet. Sci. Lett.* **196**, 189–196.

Sims K. W. W., DePaolo D. J., Murrell M. T., Baldridge W. S., Goldstein S., Clague D., and Jull M. (1999) Porosity of the melting zone and variations in the solid mantle upwelling rate beneath Hawaii: inferences from $^{238}U-^{230}Th-^{226}Ra$ and $^{235}U-^{231}$ Pa disequilibria. *Geochim. Cosmochim. Acta* **63**, 4119–4138.

Sims K. W. W., Goldstein S. J., Blichert-Toft J., Perfit M. R., Kelemen P., Fornari D. J., Michael P., Murrell M. T., Hart S. R., DePaolo D. J., Layne G., Ball L., Jull M., and Bender J. (2002) Chemical and isotopic constraints on the generation and transport of magma beneath the East Pacific Rise. *Geochim. Cosmochim. Acta* **66**, 3481–3504.

Singer B. S., Thompson R. A., Dungan M. A., Feeley T. C., Nelson S. T., Pickens J. C., Brown L. L., Wulff A. W., Davidson J. P., and Metzger J. (1997) Volcanism and erosion during the past 930 ky at the Tatara San Pedro complex, Chilean Andes. *Geol. Soc. Am. Bull.* **109**, 127–142.

Singer B. S., Wijbrans J. R., Nelson S. T., Pringle M. S., Feeley T. C., and Dungan M. A. (1998) Inherited argon in a Pleistocene andesite lava: $^{40}Ar/^{39}Ar$ incremental-heating and laser-fusion analyses of plagioclase. *Geology* **26**, 427–430.

Sinha M. C., Constable S. C., Peirce C., White A., Heinson G., MacGregor L. M., and Navin D. A. (1998) Magmatic processes at slow spreading ridges: implications of the RAMESSES experiment at 57°45' N on the mid-Atlantic Ridge. *Geophys. J. Int.* **135**, 731–745.

Sinton J. M. and Detrick R. S. (1992) Mid-ocean ridge magma chambers. *J. Geophys. Res. Solid Earth* **97**, 197–216.

Sisson T. W. and Bacon C. R. (1999) Gas-driven filter pressing in magmas. *Geology* **27**, 613–616.

Sisson T. W., Grove T. L., and Coleman D. S. (1996) Hornblende gabbro sill complex at onion valley, California, and a mixing origin for the Sierra Nevada batholith. *Contrib. Mineral. Petrol.* **126**, 81–108.

Smith R. L. (1979) Ash-flow magmatism. In *Ash-flow Tuffs*, Geological Society of America Special Paper (eds. C. E. Chapin and W. E. Elston). Geological Society of America, Boulder, CO, pp. 5–27.

Spell T. L., Smith E. I., Sanford A., and Zanetti K. A. (2001) Systematics of xenocrystic contamination: preservation of discrete feldspar populations at McCullough Pass caldera revealed by $^{40}Ar/^{39}Ar$ dating. *Earth Planet. Sci. Lett.* **190**, 153–165.

Spera F. J. and Crisp J. A. (1981) Eruption volume, periodicity, and caldera area-relationships and inferences on development of compositional zonation in silicic magma chambers. *J. Volcanol. Geotherm. Res.* **11**, 169–187.

Steck L. K. and Prothero W. A., Jr., (1994) Crustal structure beneath long valley caldera from modelling of teleseismic P wave polarizations and Ps converted waves. *J. Geophys. Res. Solid Earth* **99**, 6881–6898.

Stewart M. A., Klein E. M., and Karson J. A. (2002) Geochemistry of dikes and lavas from the north wall of the Hess deep rift: insights into the four-dimensional character of crustal construction at fast spreading mid-ocean ridges. *J. Geophys. Res. Solid Earth* **107**, article no. 223.

Taddeucci A., Broecker W. S., and Thurber D. L. (1967) ^{230}Th dating of volcanic rocks. *Earth Planet. Sci. Lett.* **3**, 338–342.

Tait S. and Jaupart C. (1992) Compositional convection in a reactive crystalline mush and melt differentiation. *J. Geophys. Res. Solid Earth* **97**, 6735–6756.

Thomas L., Hawkesworth C., Van Calsteren P., Turner S., and Rogers N. (1999) Melt generation beneath ocean islands: a U–Th–Ra isotope study from Lanzarote in the Canary Islands. *Geochim. Cosmochim. Acta* **63**, 4081–4099.

Trial A. F. and Spera F. J. (1990) Mechanisms for the generation of compositional heterogeneities in magma chambers. *Geol. Soc. Am. Bull.* **102**, 353–367.

Turner S., McDermott F., Hawkesworth C., and Kepezhinskas P. (1998) A U-series study of lavas from Kamchatka and the Aleutians: constraints on source composition and melting processes. *Contrib. Mineral. Petrol.* **133**, 217–234.

Turner S., Bourdon B., Hawkesworth C., and Evans P. (2000) $^{226}Ra-^{230}Th$ evidence for multiple dehydration events, rapid melt ascent and the timescales of differentiation beneath the Tonga-Kermadec island arc. *Earth Planet. Sci. Lett.* **179**, 581–593.

Turner S., Evans P., and Hawkesworth C. (2001) Ultrafast source-to-surface movement of melt at island arcs from $^{226}Ra-^{230}Th$ systematics. *Science* **292**, 1363–1366.

Van Orman J. A., Grove T. L., and Shimizu N. (1998) Uranium and thorium diffusion in diopside. *Earth Planet. Sci. Lett.* **160**, 505–519.

Van Orman J. A., Grove T. L., and Shimizu N. (2001) Rare earth element diffusion in diopside: influence of temperature, pressure, and ionic radius, and an elastic model for diffusion in silicates. *Contrib. Mineral. Petrol.* **141**, 687–703.

Vazquez J. A. and Reid M. R. (2002) Timescales of magma storage and differentiation of voluminous high-silica rhyolites at Yellowstone caldera, Wyoming. *Contrib. Mineral. Petrol.* **144**, 274–285.

Vigier N., Bourdon B., Joron J. L., and Allegre C. J. (1999) U-decay series and trace element systematics in the 1978 eruption of Ardoukoba, Asal rift: timescale of magma crystallization. *Earth Planet. Sci. Lett.* **174**, 81–97.

Vigneresse J. L., Barbey P., and Cuney M. (1996) Rheological transitions during partial melting and crystallization with application to felsic magma segregation and transfer. *J. Petrol.* **37**, 1579–1600.

Volpe A. M. and Hammond P. E. (1991) $^{238}U-^{230}Th-^{226}Ra$ disequilibria in young Mount St. Helens rocks—time constraint for magma formation and crystallization. *Earth Planet. Sci. Lett.* **107**, 475–486.

Volpe A. M. and Cashman K. V. (1992) $^{238}U-^{230}Th-^{226}Ra$ disequilibrium in young Mt. Shasta andesites and dacites. *J. Volcanol. Geotherm. Res.* **53**, 227–238.

Wadge G. (1980) Output rate of magma from active central volcanos. *Nature* **288**, 253–255.

Wadge G. (1982) Steady-state volcanism—evidence from eruption histories of polygenetic volcanos. *J. Geophys. Res.* **87**, 4035–4049.

Wallace G. S. and Bergantz G. W. (2002) Wavelet-based correlation (WBC) of zoned crystal populations and magma mixing. *Earth Planet. Sci. Lett.* **202**, 133–145.

West M., Menke W., Tolstoy M., Webb S., and Sohn R. (2001) Magma storage beneath axial volcano on the Juan de Fuca mid-ocean ridge. *Nature* **413**, 833–836.

Widom E., Schmincke H. U., and Gill J. B. (1992) Processes and timescales in the evolution of a chemically zoned trachyte-Fogo-A, Sao Miguel, Azores. *Contrib. Mineral. Petrol.* **111**, 311–328.

Wiebe R. A. (1996) Mafic-silicic layered intrusions: the role of basaltic injections on magmatic processes and the evolution of silicic magma chambers. *Trans. Roy. Soc. Edinburgh: Earth Sci.* **87**, 233–242.

Wiebe R. A. and Collins W. J. (1998) Depositional features and stratigraphic sections in granitic plutons: implications for the emplacement and crystallization of granitic magma. *J. Struct. Geol.* **20**, 1273–1289.

Wiebe R. A., Frey H., and Hawkins D. P. (2001) Basaltic pillow mounds in the Vinalhaven intrusion, Maine. *J. Volcanol. Geotherm. Res.* **107**, 171–184.

Willigers B. J. A., Baker J. A., Krogstad E. J., and Peate D. W. (2002) Precise and accurate *in situ* Pb–Pb dating of apatite, monazite, and sphene by laser ablation multiple-collector ICP–MS. *Geochim. Cosmochim. Acta* **66**, 1051–1066.

Yin A. and Kelty T. K. (2000) An elastic wedge model for the development of coeval normal and thrust faulting in the Mauna Loa-Kilauea rift system in Hawaii. *J. Geophys. Res. Solid Earth* **105**, 25909–25925.

Zandt G., Leidig M., Chmielowski J., Baumont D., and Yuan X. H. (2003) Seismic detection and characterization of the Altiplano-Puna magma body, central Andes. *Pure Appl. Geophys.* **160**, 789–807.

Zellmer G., Turner S., and Hawkesworth C. (2000) Timescales of destructive plate margin magmatism: new insights from Santorini, Aegean volcanic arc. *Earth Planet. Sci. Lett.* **174**, 265–281.

Zellmer G. F., Blake S., Vance D., Hawkesworth C., and Turner S. (1999) Plagioclase residence times at two island arc volcanoes (Kameni Islands, Santorini, and Soufriere, St. Vincent) determined by Sr diffusion systematics. *Contrib. Mineral. Petrol.* **136**, 345–357.

Zou H. B., Zindler A., and Niu Y. L. (2002) Constraints on melt movement beneath the East Pacific Rise from Th-230–U-238 disequilibrium. *Science* **295**, 107–110.

3.06
Fluid Flow in the Deep Crust

J. J. Ague

Yale University, New Haven, CT, USA

NOMENCLATURE

General Symbols and those for Fluid and Heat Flow

$2b$	distance between fracture walls (m)
B	thermal Peclet number
$C_{P,\mathrm{f}}$	heat capacity of fluid (J kg^{-1} K^{-1})
f_{m}	mass fluid released per unit mass solid (kg (fluid) kg^{-1} (solid))
f_{mv}	mass fluid released per unit volume rock (kg (fluid) m^{-3} (rock))
g	acceleration of gravity (m s^{-2})
k	permeability (constant) (m^2)
$\tilde{k}$	permeability tensor (m^2)
k_{f}	permeability due to fractures (m^2)
$K_{T,\mathrm{r}}$	thermal conductivity of rock (J m^{-1} s^{-1} K^{-1})
L_{c}	length of crustal column (m)
n_{fr}	frequency of fractures (fractures m^{-1})
P	pressure
$\vec{q}_{\mathrm{D}}$	Darcy flux vector (m^3 m^{-2} s^{-1})
q_{fr}	fluid flux through fractured rock (m^3 m^{-2} s^{-1})
q_{TI}	time-integrated fluid flux (m^3 m^{-2})
t	time (s)
Δt	total time of fluid–rock interaction (s)
T	temperature
x'	coordinate which is parallel to, and increases in, direction of flow

x, y, z	subscripts denoting x, y, z axes
Z	vertical reference coordinate
μ	dynamic viscosity of fluid (Pa s)
$\vec{v}$	pore velocity vector (m s^{-1})
ρ_f	fluid density (kg m^{-3})
ρ_r	rock density (kg m^{-3})
ρ_s	solid density (kg m^{-3})
ϕ	porosity (m^3 (fluid) m^{-3} (rock))
ω	constant for porosity–permeability law (m^2)

Symbols for Chemical Mass Transfer and Reaction Rates

a_i	activity of species i
A_θ	reactive surface area of θ (m^2 m^{-3})
$\bar{A}_{l,m}$	surface area of rate-limiting mineral l in reaction m (m^2 m^{-3})
C_i	concentration of i in fluid (mol m^{-3})
C_i^{Solid}	concentration of i in solid (mol m^{-3})
$D_{i,f}$	diffusion coefficient for i in fluid (m^2 s^{-1})
$\tilde{D}_{HD,i}$	hydrodynamic dispersion tensor (m^2 s^{-1})
$\tilde{D}_{MD}$	mechanical dispersion tensor (m^2 s^{-1})
$D_{HD,i}$	hydrodynamic dispersion coefficient (m^2 s^{-1})
E_a	activation energy (J mol^{-1})
$\vec{F}_i$	flux vector for species i (mol m^{-2} s^{-1})
ΔG	Gibbs free energy change for reaction (J mol^{-1})
i	subscript denoting fluid species i
K_v	equilibrium fluid/solid partition coefficient by volume $(\text{mol m}^{-3})^{\text{Fluid}}/(\text{mol m}^{-3})^{\text{Solid}}$
K_θ	equilibrium constant (dissolution reaction for phase θ)
L	characteristic length scale (m)
L_{GF}	distance of geochemical front propagation (m)
m	subscript denoting reaction m
$\dot{n}$	nucleation rate
n_i	moles i produced/consumed per unit volume rock (mol m^{-3})
N_m	reaction order for reaction m
p, M, N	constants for rate expressions
P_e	Peclet number
Q_θ	ion activity product
r_θ	dissolution/precipitation rate for θ (mol m^{-3} s^{-1})
R	gas constant (J mol^{-1} K^{-1})
$R_{i,m}$	consumption/production rate of i (mol m^{-3} s^{-1})
s	sign
T^0	kinetic reference temperature (K)
T_{eq}	equilibrium temperature (K)
$\bar{V}_f$	molar volume of fluid (m^3 mol^{-1})
$\bar{V}_{Qtz}$	molar volume of quartz (m^3 mol^{-1})
X_i	mole fraction of i in fluid
X_{eq}	equilibrium mole fraction in fluid
α_L	longitudinal dispersivity (m)
α_T	transverse dispersivity (m)
θ	subscript denoting solid phase θ
κ	rate constant (s^{-1})
κ_m	intrinsic rate constant for reaction m (mol m^{-2} s^{-1} (J $mol^{-1})^{-1}$)
κ_θ	intrinsic dissolution/precipitation rate constant for phase θ (mol m^{-2} s^{-1})
$\upsilon_{i,m}$	stoichiometric coefficient for i in reaction m
τ	tortuosity (constant)
$\tilde{\tau}$	tortuosity tensor

3.06.1 INTRODUCTION

The heating and burial of rock masses during mountain building drives chemical reactions that liberate volatile fluid species (Figure 1). These volatiles, including H_2O, CO_2, and CH_4, are much less dense and viscous than the surrounding rock and will, therefore, have a strong tendency to migrate along grain boundaries or fractures through the Earth's crust. Fluids released in the deep crust interact geochemically with their surroundings (Rye *et al.*, 1976) as they ascend to shallow levels where they invade hydrothermal and groundwater systems and, ultimately, interact with the hydrosphere and atmosphere. This flux of fluid from active mountain belts to the surface is a major contributor to planetary volatile cycling and is estimated to be currently in excess of $\sim 10^{17}$ kg Myr^{-1} (based on Kerrick and Caldeira, 1998; Wallmann, 2001a,b).

The deep crust is composed largely of metamorphic rock (cf. Rudnick and Fountain, 1995; Wedepohl, 1995; see Chapter 3.01). Fluids and magmas are the primary agents of chemical mass transport through the deep crust; fluid flow dominates at temperatures $<\sim 600$ °C and can be important at much higher temperatures as well—even in the granulite facies. As a consequence, an understanding of the fundamental controls exerted by metamorphic fluids on mass and heat transfer, mineral reactions, and rock rheology is critical for determining the geochemical and petrological evolution of the crust. Moreover, metamorphic fluids impact directly many problems of societal relevance, including ore deposit formation (see Chapter 3.12), global release of greenhouse gases, seismic hazards, and arc magma genesis (see Chapter 3.18) and the associated volcanic hazards. This chapter first examines basic fluid flow, mass transfer, and reaction concepts. This discussion is followed by a review of selected natural examples of fluid transport during active metamorphism. The focus is on deeper levels of the crust ($>\sim 15$ km depth),

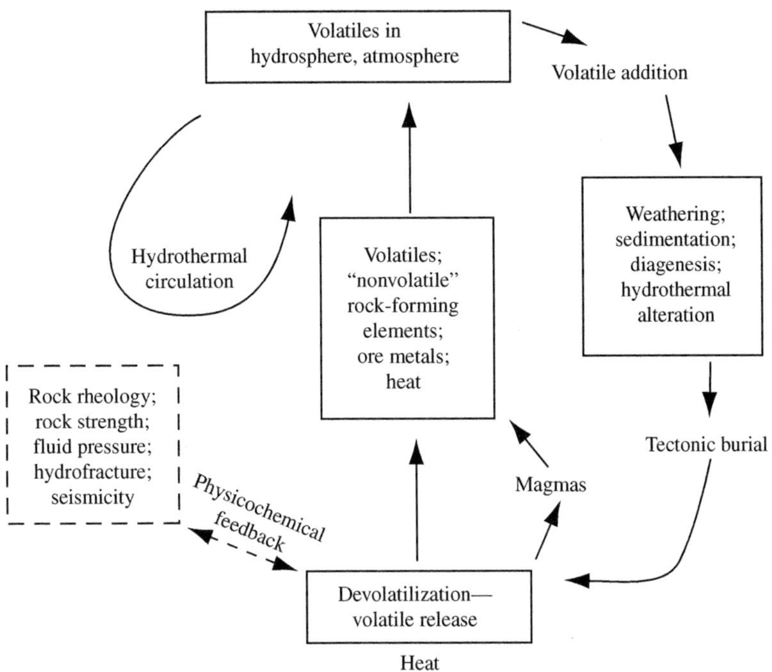

Figure 1 Diagram of crustal fluid cycling.

although many of the concepts discussed are general and also apply to shallower levels.

3.06.2 EVIDENCE FOR DEEP CRUSTAL FLUIDS

The body of evidence for deep crustal fluids has grown substantially in the last several decades, and includes the following. (i) High-density fluid inclusions that were trapped in metamorphic minerals that grew under deep crustal conditions. (ii) High-pressure metamorphic vein minerals characterized by habits diagnostic of growth into fluid-filled fractures with macroscopic apertures. (iii) Formation of veins and fractures under conditions where fluid pressure in cracks exceeded the minimum principal stress and the tensile strength of the rock. (iv) General consistency between the mineral assemblages observed in exhumed deep crustal settings, and the phase assemblages developed during laboratory experiments done under high fluid pressures. (v) Alteration of the chemical and isotopic compositions of rocks due to mass transfer driven by infiltrating fluids. (vi) Petrologic and isotopic evidence indicating volumetric fluid/rock ratios greater than one that exceed rock porosity and, thus, require, infiltration of fluids. (vii) Large time-integrated fluid fluxes based on petrologic and isotopic evidence from rocks, as well as numerical models of orogenesis. Note that (v)–(vii) require both the presence of fluids and considerable mass transport via a fluid phase.

A number of these topics are addressed in more detail below and, for additional perspectives, the reader is referred to Zimmermann and Poty (1970), Fyfe *et al.* (1978), Etheridge *et al.* (1983, 1984), Peacock (1983, 1990), Rumble (1989), Ferry (1994a), Ague (1995), Person and Baumgartner (1995), Young (1995), Hanson (1997), and Ferry and Gerdes (1998). Fluids have been demonstrated to be an integral part of prograde and retrograde metamorphism, but it is important to point out that they need not be present continuously throughout an orogenic episode (Thompson, 1983).

3.06.3 POROUS MEDIA AND FRACTURE FLOW

Fluid flow through rocks is commonly referred to as being "pervasive" or "channelized," although some overlap exists in the definitions of these terms. Fluid migration around individual mineral grains through an interconnected porosity is known as "pervasive" or "porous media" flow (Figure 2). "Channelized" or "focused" flow implies preferential fluid motion in one or more high-permeability conduits (Figure 3). These include highly permeable layers, lithologic contacts, fracture sets, or individual fractures. Note that flow in a permeable layer could still be "pervasive" at the grain scale within the layer, whereas flow in a fracture is much more strongly localized to the open space between the crack walls. Some metamorphic systems involve both channelized and pervasive flow components (e.g., Rumble *et al.*, 1991; Oliver, 1996, Thompson, 1997).

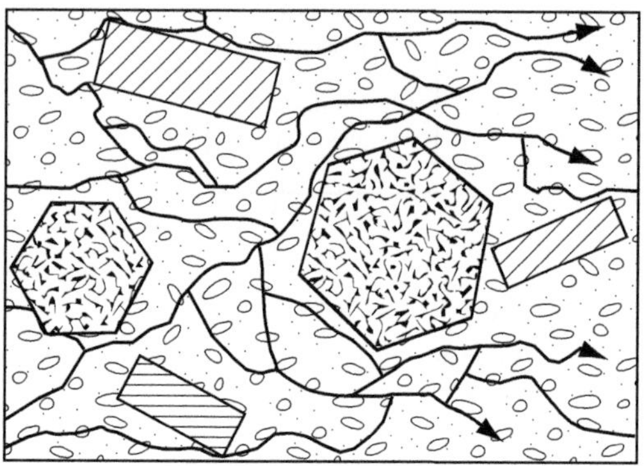

Figure 2 Schematic representation of metamorphic porphyroblasts and matrix, illustrating the concept of pervasive, grain-scale flow. Flow paths around grains denoted by black arrows.

3.06.3.1 Pervasive Flow and Darcy's Law

Pervasive fluid flow through a porous, permeable medium is described by Darcy's law, written here for three dimensions (cf. Bear, 1972):

$$\vec{q}_{\mathrm{D}} = \begin{pmatrix} q_x \\ q_y \\ q_z \end{pmatrix} = -\frac{\tilde{k}}{\mu}\cdot\left(\begin{pmatrix} \dfrac{\partial P}{\partial x} \\ \dfrac{\partial P}{\partial y} \\ \dfrac{\partial P}{\partial z} \end{pmatrix} + \rho_{\mathrm{f}} g \begin{pmatrix} \dfrac{\partial Z}{\partial x} \\ \dfrac{\partial Z}{\partial y} \\ \dfrac{\partial Z}{\partial z} \end{pmatrix}\right) = -\frac{\tilde{k}}{\mu}\cdot(\nabla P + \rho_{\mathrm{f}} g \nabla Z) \tag{1}$$

where x, y, and z are Cartesian spatial coordinates; $\vec{q}_{\mathrm{D}}$ is the fluid flux vector or Darcy flux (q_x, q_y, and

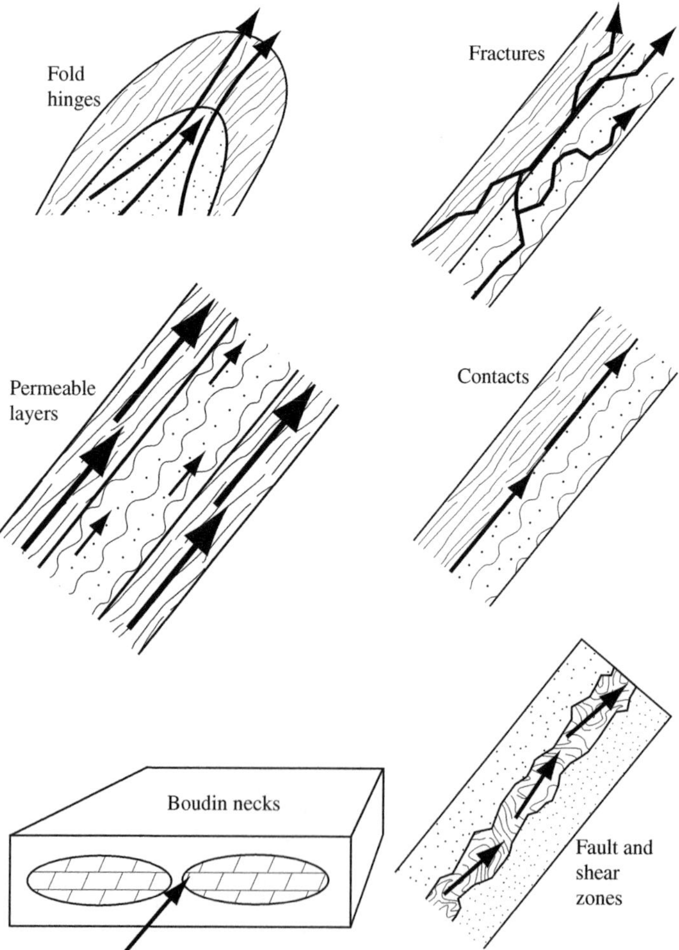

Figure 3 Examples of channelized flow.

q_z are the components of the flux in the x, y, and z directions, respectively); $\tilde{k}$ is the intrinsic permeability tensor; μ is the dynamic viscosity of the fluid; P is fluid pressure; ρ_f is fluid density; g is the acceleration of gravity expressed as a constant ($9.81\ m\ s^{-2}$); and Z is a vertical reference coordinate axis that increases upward. Darcy's law is valid only when the flow is laminar, not turbulent (cf. Bear, 1972, pp. 125–129). Note that fluid flux is a vector, having both magnitude and direction. Clearly, the flux is increased by increasing permeability, increasing fluid pressure gradients, and/or decreasing fluid viscosity. These key geologic variables are examined in the following paragraphs.

3.06.3.2 Fluid Flux, Fluid Velocity, and Porosity

$\vec{q}_D$ is sometimes referred to as the "Darcy velocity," but it is really a flux expressed in terms of volume of fluid passing over unit surface area per unit time. The true average pore fluid velocity is found by dividing the flux by the amount of interconnected porosity, ϕ, through which the fluid flows. Porosity for a fully saturated porous medium is the fluid volume in pores per unit volume rock (volume rock = volume solids + volume pore space; ϕ is expressed as a fraction and is assumed here to represent interconnected pores). Thus,

$$\frac{\vec{q}_D}{\phi} = \vec{v} = \begin{pmatrix} v_x \\ v_y \\ v_z \end{pmatrix} \tag{2}$$

where $\vec{v}$ is the pore fluid velocity vector. The theoretical estimates of syn-metamorphic porosity of Connolly (1997) are $\sim 10^{-3}$–10^{-4}, and Hiraga *et al.* (2001) found similar values based on direct observation of relic pores preserved in schist. The theoretical analysis of isotopic profiles across lithologic contacts in nearly pure calcite marbles, however, suggests values as small as 10^{-4}–10^{-6} (Bickle and Baker, 1990), although these estimates are subject to considerable uncertainties (Lewis *et al.*, 1998). Because grain-scale porosities are likely to be small during metamorphism, the magnitude of the pore velocity will be much larger than that of the Darcy flux. For example, if $q_x = 10^{-3}\ m^3_{(fluid)}$ per $m^2_{(rock)}$ per year and $\phi = 10^{-3}$, the pore velocity, v_x, is $1\ m\ yr^{-1}$. If the total porosity includes some "dead-end" pores which are not interconnected and do not transmit fluid, then ϕ in Equation (2) must be reduced by multiplying it by the fraction of interconnected pore space.

Porosity evolves as a result of deformation and fluid–rock reactions. Four general pathways of porosity evolution are commonly recognized. First, deformation can collapse porosity and drive fluids out, or produce cracking at the grain scale or larger to create porosity. The low porosity values estimated by Bickle and Baker (1990) may reflect the relative ease with which calcite can deform plastically (cf. Rutter, 1995) and choke off porosity. Second, increases in fluid pressure will tend to expand pore spaces and increase porosity, whereas decreases in pressure will do the opposite (cf. Walder and Nur, 1984). Third, the mineral products of prograde reactions are typically denser and occupy less volume than the reactants, so increases in porosity may accompany fluid infiltration and devolatilization if fluid pressure is sufficient to keep the pore space from collapsing (cf. Rumble and Spear, 1983; Ague *et al.*, 1998; Balashov and Yardley, 1998; Zhang *et al.*, 2000). The coupled metamorphic–rheological models of Connolly (1997) suggest that devolatilization reactions generate pulses of fluid that travel upward in the form of porosity waves, leaving trails of interconnected pore space in their wake. Finally, infiltrating metasomatic fluids will destroy porosity if they precipitate new minerals in the pore spaces, or create porosity if they dissolve existing minerals (Balashov and Yardley, 1998; Bolton *et al.*, 1999). In addition to the above four processes, Nakamura and Watson (2001) have recently shown experimentally that interfacial energy-driven infiltration of water or NaCl aqueous solution into quartzite can create high porosity zones that propagate through rock much like traveling waves. Nakamura and Watson (2001) suggest that this mechanism may contribute significantly to fluid fluxes in high-grade metamorphism.

3.06.3.3 Fluid Pressure Gradients

Fluid motion occurs in a direction of decreasing pressure and is due ultimately to the net pressure gradients described by the $(\nabla P + \rho_f g \nabla Z)$ term in Darcy's law. Note that the $\rho_f g \nabla Z$ term is necessitated by gravity. Pressure increases downward in a column of motionless fluid according to the hydrostatic gradient ($= -\rho_f g$), yet there is no flow. To drive flow upward, the total pressure gradient must exceed the hydrostatic gradient. For convenience, the z-axis of the coordinate system is commonly oriented vertically so that it coincides exactly with the vertical reference Z-axis. Then, $\partial Z/\partial x$ and $\partial Z/\partial y$ vanish, and $\partial Z/\partial z = 1$. For example, the net pressure gradient driving the vertical component of flow would be given by the difference between the total pressure gradient, $\partial P/\partial z$, and $-\rho_f g$, thus yielding: $\partial P/\partial z - (-\rho_f g)$ or, equivalently, $\partial P/\partial z + \rho_f g$.

Brittle deformation involves fracturing on the scale of individual mineral grains or larger, whereas ductile (plastic) deformation occurs without fracturing (cf. Passchier and Trouw, 1996). In the shallow crust, rocks are in the brittle regime, have substantial strength, and can support open pore networks over km scale distances. Fluid pressure gradients are close to the hydrostatic gradient, and free convection cells may develop if permeability and thermal gradients are large enough, as is often the case around cooling intrusions (cf. Norton and Knight, 1977; Norton and Dutrow, 2001). Furthermore, groundwater can circulate down into sedimentary basins to depths of several km by gravity-driven (or topography-driven) flow involving fluid input into high-elevation parts of foreland basins, subhorizontal flow for tens or even hundreds of kilometers, and discharge into lower-elevation areas (Garven and Freeze, 1984a,b).

The transition between hydrostatic and deeper regimes is thought to occur near 10 km (cf. Manning and Ingebritsen, 1999), but is not precisely constrained and may be considerably deeper. At deeper levels, where plastic deformation becomes more important, rocks are considerably weaker and tend to collapse around fluid filled pores, producing larger pressure gradients that may approach lithostatic gradients ($\mathrm{d}P/\mathrm{d}Z = -\rho_r g$; ρ_r is rock density). Thus, for vertical, upward flow under the lithostatic gradient, the pressure gradient term in Darcy's law is the difference between the lithostatic and hydrostatic gradients ($-\rho_r g + \rho_f g \approx -1.7 \times 10^4$ to -2.0×10^4 Pa m$^{-1} \approx -0.17$ to -0.2 bar m^{-1}). In general, deep crustal fluid pressure regimes that drive flow are thought to be closer to lithostatic than hydrostatic (cf. Hanson, 1997), but much uncertainty remains. For example, the subhorizontal flow constrained by nearly flat-lying lithologic layering inferred on petrologic grounds for regional metamorphism in northern New England by Ferry (1992, 1994b) could have been driven by very small gradients—even smaller than the hydrostat ($<-10^4$ Pa m^{-1}).

3.06.3.4 Permeability

The intrinsic permeability is a property of the porous medium only, and is a quantitative measure of how readily a fluid can flow through the medium. Permeability varies over a remarkable 16 orders of magnitude in the Earth's crust, from values as high as 10^{-7} m^2 in gravels to 10^{-23} m^2 in some shales and "crystalline" igneous and metamorphic rocks (cf. Freeze and Cherry, 1979; Brace, 1980; Connolly, 1997; Manning and Ingebritsen, 1999). In the general case, $\tilde{k}$ is a second-rank tensor because permeability varies with direction. Metamorphic foliations defined by inequidimensional minerals, particularly sheet silicates, are a primary source of anisotropy (cf. Zhang *et al.*, 2001). The measurements of Huenges *et al.* (1997) reveal mean permeability parallel to foliation as much as ~10 times greater than perpendicular to it, consistent with field-based studies that suggest fluid fluxes are greatest parallel to layering and foliation (cf. Rye *et al.*, 1976; Rumble and Spear, 1983; Ferry, 1987, 1994b; Ganor *et al.*, 1989; Baker, 1990; Oliver *et al.*, 1990; Cartwright *et al.*, 1995). Oriented fracture sets are another source of anisotropy. If, however, the medium is isotropic and can transmit fluid equally well in all directions, then $\tilde{k}$ reduces to a constant (k). Permeability also varies from one layer to the next and even within individual layers, regardless of the degree of anisotropy, producing permeability contrasts that can exceed two orders of magnitude (Baumgartner and Ferry, 1991; Oliver, 1996; Baumgartner *et al.*, 1997). Comparisons of inferred metamorphic fluid fluxes suggest that metapelitic rocks are often more permeable than metapsammites or very pure calcite marbles (e.g., Rye *et al.*, 1976; Chamberlain and Conrad, 1991; Skelton *et al.*, 1995; Oliver *et al.*, 1998).

Following the approach of Baumgartner and Ferry (1991), Manning and Ingebritsen (1999) estimated a mean k of $10^{-18.5\pm1}$ m^2 for rocks deeper than ~12 km and combined this result with permeability data for shallower geothermal systems to yield the depth–permeability relation: $\log k \approx -3.2 \log$ (depth in km) -14. The deep crustal k estimates are for rocks that underwent substantial fluid flow. Other, probably lower-k rock types that had little fluid–rock interaction are not represented. Nonetheless, the estimates strongly suggest that considerable permeability is possible even at the base of the continental crust.

In general, permeability increases as the amount of interconnected pore space increases, resulting in strong coupling between porosity and permeability. Thus, because porosity is time-dependent, permeability is as well. A number of porosity–permeability relationships, such as the Kozeny–Carman equation, have been proposed; these commonly include a strong (often cubic) dependence of permeability on porosity (cf. Bear, 1972, p. 166; Walder and Nur, 1984; Bickle and Baker, 1990; David *et al.*, 1994; Bolton *et al.*, 1999; Wong and Zhu, 1999, and references therein). For example, Connolly (1997) described deep crustal permeability using: $k = \omega\phi^3$, where the constant $\omega = 10^{-13}$ m^2. While calculated porosity–permeability relationships are still subject to major uncertainties of order of magnitude scale or larger, it is clear that porosity–permeability feedback can control spatial patterns of flow. For example, increases in porosity due to infiltration and devolatilization reaction can increase permeability, causing more flow to focus into the reacting area (cf. Balashov and Yardley, 1998), whereas precipitation of minerals that occlude the porosity can decrease permeability and divert flow away.

3.06.3.5 Dynamic Viscosity

The dynamic viscosity is the viscosity of a moving fluid and depends on T, P, and fluid composition. Values for pure H_2O, CO_2, and, by extension, H_2O-CO_2 mixtures are similar and vary relatively little compared to properties like porosity and permeability; a representative value for the middle and lower crust is $\sim 1.5 \times 10^{-4}$ Pa·s (cf. Walther and Orville, 1982). However, it should be pointed out that the effects of solute species, as well as viscosities at very high pressures (1–2 GPa), remain to be fully explored.

3.06.3.6 Crack Flow

The deformational behavior of rocks, whether brittle or ductile, depends mainly on temperature, fluid pressure, rock pressure, mineralogy, grain size, and strain rate. Temperature is one of the main controls on deformation behavior. For example, for slow strain rates, common minerals like quartz are brittle at $T < \sim 300$ °C, but ductile deformation involving dislocation glide, and creep becomes increasingly important at higher T. The transition from dominantly brittle to dominantly ductile behavior is thought to occur at depths corresponding to temperatures of ~ 300 °C—usually around 10–13 km for typical crustal geotherms (e.g., Sibson, 1983; Scholz, 1990; Yeats *et al.*, 1997). These depths are consistent with measured and inferred pressure gradients near hydrostatic in much of the shallow crust.

Nonetheless, brittle behavior is not restricted to shallow levels. If the fluid pressure exceeds the sum of the tensile strength of the rock and the least principal stress, then hydrofracturing, transient fluid release, and associated drops in fluid pressure will occur (e.g., Hubbert and Willis, 1957; Etheridge, 1983; Yardley, 1986). The maximum tensile strength of most rocks is only ~ 0.01 GPa (~ 100 bar), so even modest fluid overpressures will cause hydrofracturing. Elevated pore fluid pressures generated by metamorphic devolatilization reactions (cf. Ague *et al.*, 1998), as well as by deformation and collapse of pore space (cf. Sibson *et al.*, 1975; Walder and Nur, 1984; Sibson, 1992; Wong *et al.*, 1997), can lead to rock weakening and hydrofracture (Yardley, 1986). With time, the porosity and permeability created by a hydrofracturing event are reduced as cracks are sealed and pores collapse. If permeability reaches low enough levels ($< \sim 10^{-20}$ m^2), fluid pressure can once again build up and ultimately produce another hydrofracturing event. The episodes of fracturing and healing preserved in veins attest to this cyclic behavior (Ramsay, 1980; Rye and Bradbury, 1988; Fisher and Brantley, 1992; Kirschner *et al.*, 1993). Oxygen isotope disequilibrium suggests transient timescales of fluid–rock interaction as short as 10^3 yr to 5×10^4 yr in and around some veins (cf. Palin, 1992; Young and Rumble, 1993; van Haren *et al.*, 1996).

Rocks that are ductile at low strain rates can undergo brittle deformation at larger strain rates. For example, earthquakes release massive amounts of energy in seconds or minutes during fault slippage, and are capable of producing regionally extensive brittle deformation. Data from several recent, damaging earthquakes, including the Northridge and Loma Prieta events in California and the Kobe event in Japan, demonstrate that rupture and brittle deformation occur well below 10–13 km (cf. Davis and Namson, 1994; Lees and Lindley, 1994; Zhao *et al.*, 1996). For the Northridge event, the main shock occurred at 17–18 km, and some aftershocks extended to ~ 25 km. In fact, from April 1980 to February 1994, nearly 1,100 seismic events were recorded in the 20–35 km depth range in the Los Angeles, California area alone (Ague, 1995). If rapid devolatilization and hydrofracturing occur within seismically active areas, the rock failure may trigger earthquakes that recur on human timescales (e.g., Ague *et al.*, 1998). The fluid-filled earthquake hypocenters that have been inferred on the basis of seismic evidence for both the Loma Prieta (Lees and Lindley, 1994) and Kobe (Zhao *et al.*, 1996) events strongly suggest links between the presence of fluids and seismicity.

The above evidence establishes that fracturing and seismic behavior can extend well into the zone of mid to lower crustal metamorphism at rock pressures of ~ 0.5–1 GPa. Veins preserve a valuable record of this brittle deformation; they are fractures in which mineral mass has been deposited. The most common vein-forming minerals are quartz, calcite, and the feldspars, but a huge variety of other minerals are also observed. Fractures tend to focus flow, because they are zones of elevated permeability. Fracture flow is commonly approximated using the well-known expression from fluid mechanics for laminar flow between two parallel plates (e.g., White, 1979). For a set of parallel fractures, the flux is approximated by (e.g., Norton and Knapp, 1977):

$$q_{fr} = -\frac{(2b)^3 n_{fr}}{12\mu}\left(\frac{dP}{dx'} + \rho_f g \frac{dZ}{dx'}\right) \quad (3)$$

where q_{fr} is the fluid flux through the fractures in the rock mass, the coordinate x' is parallel to, and increases in, the direction of flow, $2b$ is the distance between the fracture walls (or crack aperture), and n_{fr} is the frequency of the fractures. Equation (3) has been shown experimentally to be applicable to real fractures, even those with rough walls and many points of contact (asperities), if $2b$ is taken as the average crack aperture. By comparison with Equation (1), it is clear that the $(2b)^3 n_{fr}/12$ grouping is directly analogous to the permeability in Darcy's law. Consequently, the "fracture permeability," k_{fr}, of a rock mass can be estimated if the average number and aperture of fractures are known (cf. Norton and Knapp, 1977; Ague, 1995):

$$k_{fr} = \frac{(2b)^3 n_{fr}}{12} \quad (4)$$

Metamorphic fracture apertures range from the micrometer scale (Ramsay, 1980; Etheridge *et al.*, 1984) to mm or cm scales (Ague, 1995).

Even small amounts of fracturing can affect the permeability markedly. If an unfractured rock with low permeability, say 10^{-23} m^2, is deformed to produce, on average, just one 10^{-5} m aperture fracture per meter of rock ($n_{fr} = 10^{-5}$ m^{-1}), then Equation (4) gives $k_{fr} \sim 8 \times 10^{-17}$ m^2—over seven orders of magnitude greater than 10^{-23} m^2. The permeability systematics of "crystalline" rocks suggest a scale dependence; laboratory measurements done on unfractured rock cores generally yield the smallest values, whereas regional field tests indicate the largest (Brace, 1984). At least some of this discrepancy is probably due to natural fractures that increase considerably the permeability of the field test sites (Manning and Ingebritsen, 1999).

3.06.4 OVERVIEW OF FLUID CHEMISTRY

Metamorphic fluids are chemically complex and are able to transport "molecular" species like H_2O, CO_2, CH_4, and H_2S, as well as solutes including $(H_4SiO_4)^0$, Na^+, $NaCl^0$, and many others (Figure 4). This section provides a brief review of some common fluid constituents.

Deep crustal H_2O is released mostly by prograde devolatilization of sheet silicates and amphiboles and, in some cases, by degassing magmas, whereas CO_2 is released mostly by devolatilization of carbonate minerals. Because H_2O and CO_2 are such fundamental constituents of crustal fluids, considerable attention has been focused on their physicochemical properties. Thermodynamic treatment is complicated, because both pure H_2O and pure CO_2 deviate strongly from ideal gas behavior, and because mixing of H_2O and CO_2 is also nonideal (e.g., Kerrick and Jacobs, 1981; Ferry and Baumgartner, 1987; Shi and Saxena, 1992; Aranovich and Newton, 1999; Blencoe *et al.*, 1999; Schmidt and Bodnar, 2000). The expressions for fugacity and activity coefficients given by Kerrick and Jacobs (1981) for pure H_2O, pure CO_2, and H_2O–CO_2 mixtures have been used widely, and are accurate for $P \leq \sim 1$ GPa. Similar results for pure H_2O and pure CO_2 are obtained using Haar *et al.* (1984) and Mäder and Berman (1991), respectively. Expressions valid for pure species to pressures well in excess of 1 GPa include Holland and Powell (1991, 1998), Shi and Saxena (1992), and Sterner and Pitzer (1994). Aranovich and Newton (1999) provide activity–composition relations for H_2O–CO_2 mixtures valid for temperatures and pressures in the range ~600–1,000 °C and 0.6–1.4 GPa, respectively.

H_2O and CO_2 are extremely important, but deep crustal fluids contain many other constituents. For example, at low enough oxygen fugacities in the presence of reactive graphite, CH_4 can be significant (e.g., French, 1966; Skippen and Marshall, 1991; Figure 4(a)). Furthermore, progressive heating promotes desulfidation of pyrite to produce pyrrhotite and liberate S (e.g., Carpenter, 1974; Mohr and Newton, 1983). Thermodynamic treatments of geologically important species, including CH_4, CO, H_2, S_2, H_2S, and COS, can be found in, for example, Jacobs and Kerrick (1981) and Shi and Saxena (1992).

Rock-forming and ore-forming metals are key components of aqueous, chloride-bearing metamorphic fluids. Salinity varies from near zero to as much as several molal in typical metamorphic environments (e.g., Rich, 1979; Hollister and Crawford, 1981; Crawford and Hollister, 1986; Roedder, 1984; Yardley, 1997; Smith and Yardley, 1999), and can reach extreme levels in granulites (cf. Crawford and Hollister, 1986; Markl *et al.*, 1998). Standard state thermodynamic properties for many aqueous species of interest may be calculated to ~1,000 °C and ~0.5 GPa using the internally consistent methods and data sets of Pokrovskii and Helgeson (1995), Shock *et al.* (1997), and Sverjensky *et al.* (1997). These data have also been extrapolated to somewhat higher pressures with reasonable results (e.g., Dipple and Ferry, 1992a). Holland and Powell (1998) advanced an alternative method and data set for calculating standard-state properties relevant for the deep crust. The extended Debye–Hückel equation is commonly used to estimate the activities of charged species, although it cannot be applied to highly concentrated brines (Sverjensky, 1987). Activity coefficients for neutral species are assumed to be unity or, in some cases, are modeled using the Setchénow equation (cf. Sverjensky, 1987; Xie and Walther, 1993). The activity of H_2O remains close to unity if the salt content is low (Sverjensky, 1987), but decreases markedly in concentrated brines (Aranovich and Newton, 1996, 1997). An important area of future research is the experimental and theoretical investigation of activity–composition relations for aqueous species in fluids that contain considerable quantities of CO_2, CH_4, and H_2S.

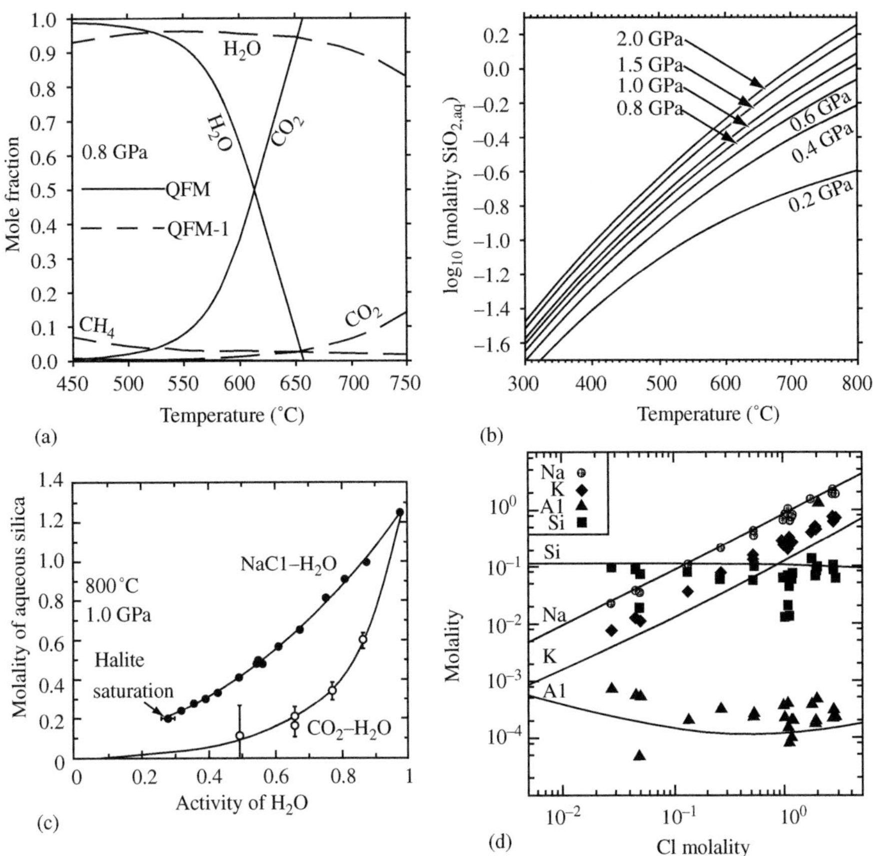

Figure 4 Examples of metamorphic fluid compositions. (a) Species in C–O–H fluids at 0.8 GPa and oxygen fugacities equivalent to the quartz–fayalite–magnetite (QFM) buffer, and 1 $\log_{10}$ unit below QFM (QFM-1). Computed following Ague *et al.* (2001). Graphite is unstable above ~660 °C for QFM. (b) Quartz solubility computed using the expression of Manning (1994). (c) Effect of NaCl and CO_2 on quartz solubility (reproduced by permission of Elsevier from *Geochim. Cosmochim. Acta*, **2000a**, *64*, pp. 2993–3005). Experimental data for H_2O–NaCl fluids (filled symbols) and H_2O–CO_2 fluids (open symbols) shown. (d) Fluid composition coexisting with quartz, microcline, albite, and andalusite as a function of total chlorine molality at 0.2 GPa and 600 °C (reproduced by permission of Elsevier from *Geochim. Cosmochim. Acta*, **2001**, *65*, pp. 4493–4507, figure 6(d)). Symbols: experimental data; lines: theoretical calculations. Note that total molalities of potassium and sodium increase with increasing chlorine content, unlike silicon and aluminum which do not form significant chloride complexes.

Experimental, fluid inclusion, and field-based evidence indicates the following four generalized groupings of elemental abundances, from largest to smallest, for fluids in typical quartz-bearing rocks: (i) chlorine, sodium, potassium, and silicon; (ii) calcium, iron, and magnesium; (iii) aluminum; (iv) high field strength elements (like zirconium and titanium), and rare earth elements (REEs). The concentrations of silicon, potassium, and sodium are relatively large in crustal fluids that coexist with quartz, feldspars, and/or micas. Aqueous silica concentrations in equilibrium with quartz increase markedly with T and P (cf. Kennedy, 1950; Weill and Fyfe, 1964; Anderson and Burnham, 1965). Manning (1994) performed experiments at high P and T and combined the results with those of previous studies to obtain an expression for silica molality at temperatures ~900 °C and at least 2 GPa (Figure 4(b)). At deep crustal conditions, quartz solubility decreases as the NaCl content of the fluid increases, and drops very sharply with the addition of CO_2; thus, if immiscible brine–CO_2 fluids exist in the deep crust, silica will partition strongly into the brine (Figure 4(c); Walther and Orville, 1983; Newton and Manning, 2000a,b). The effective solubility of albite also increases with P and T, and decreases with increasing NaCl, although it should be noted that its dissolution is incongruent (Shmulovich *et al.*, 2001). The behavior of aqueous silica, which is present mainly as the neutral $(H_4SiO_4)^0$ complex, differs from that of sodium and potassium, which are present mostly as the charged species Na^+ and K^+ and the neutral chloride complexes $NaCl^0$ and KCl^0. Consequently, the total concentrations of sodium and potassium generally increase as the total concentration of chlorine increases in typical mica-bearing quartzofeldspathic rocks

(Figure 4(d)). Calcium, magnesium, and iron are also present, mostly as charged species or chlorine complexes, and can reach relatively high concentrations in some fluids, particularly those coexisting with metaultramafic and calcium-silicate-bearing metacarbonate rocks (cf. Vidale, 1969; Ferry and Dipple, 1991; Dipple and Ferry, 1992a). Study of the incongruent dissolution of diopside shows that its effective solubility increases with increasing NaCl content owing mostly to complexing of calcium and magnesium with chlorine in solution (Shmulovich *et al.*, 2001). The retrograde solubility of calcite at low P and T is well known but, surprisingly, much remains to be learned about calcite behavior in the deep crust. The results of Caciagli and Manning (1999) indicate that calcite solubility in H_2O at 500 °C and 1 GPa is considerable—approximately 0.03 molal.

The concentrations of aluminum, titanium, zirconium, and REEs are traditionally regarded as small in aqueous fluids. However, recent work on the solubility of Al_2SiO_5 in the presence of quartz indicates total aluminum molalities at 0.8–1.0 GPa of $\sim10^{-3}$ m at 500 °C, increasing to $\sim10^{-2}$ m at 800 °C (Manning, 2001). These values are still relatively small compared to those of constituents like aqueous silica, but are apparently sufficient for calc-silicate formation (Ague, 2000b, 2003) and for formation of aluminosilicates and other aluminum-rich phases in some syn-metamorphic veins (Kerrick, 1990; Widmer and Thompson, 2001). The aluminum is thought to be present largely as uncharged $HAlO_2^0$ or, possibly, as polymerized Si–Al species (Pokrovskii and Helgeson, 1995; Manning, 2001). The experimentally determined solubilities of rutile and zircon in quartz-saturated H_2O–CO_2 fluids are small. The transport of titanium and zirconium, probably mostly as $Ti(OH)_4^0$ and $Zr(OH)_4^0$ complexes, is therefore very limited in many deep crustal environments (Ayers and Watson, 1991). A number of field-based studies also indicate negligible zircon solubility (e.g., Carson *et al.*, 2002; Steyrer and Sturm, 2002). Zircon dissolves incongruently to form baddeleyite in quartz-undersaturated H_2O at high P and T, but the amount of zirconium in solution remains small (Ayers and Watson, 1991). Despite the low measured solubilities of rutile and zircon, field evidence suggests that significant titanium and zirconium may be present in some subduction zone fluids (cf. Philippot and Selverstone, 1991; Bröcker and Enders, 2001; Heaman *et al.*, 2002). It appears that REEs can be mobilized under appropriate circumstances, but relatively little is known about the processes and fluid chemistry involved (cf. Grauch, 1989; Ague, 2003). The REE hosts monazite and apatite have relatively low solubilities in H_2O, although their solubilities increase with increasing HCl in the fluid (Ayers and Watson, 1991). F^- or other agents might complex titanium, zirconium, and REEs, thus enhancing their transport in crustal fluids; radiation damage in zircon might increase its solubility, and titanite and ilmenite might be more soluble than rutile, but much experimental work needs to be done to assess these possibilities (cf. Gieré, 1990, 1993; Ayers and Watson, 1991; Gieré and Williams, 1992; Pan and Fleet, 1996).

Recent laboratory and field-based studies indicate that fluids and silicate melts can be completely miscible under appropriate conditions (see also Chapter 3.17). For example, Bureau and Keppler (1999) observed experimentally complete miscibility between silicate melt and hydrous fluids for a variety of melt compositions, and concluded that complete miscibility is possible in all but the shallowest parts of the upper mantle. Their results imply that amphibole breakdown in subduction zones should produce the mobile, hydrous fluids necessary for arc magma genesis, whereas breakdown of lawsonite or phengite deeper in subduction zones should produce much less mobile, silicate-rich fluids. While the results of Bureau and Keppler (1999) apply to relatively high-pressure settings, Thomas *et al.* (2000) showed that melt-fluid miscibility is possible even in low pressure (~0.1 GPa) pegmatite environments if the system is rich in fluorine, boron, and phosphorus. The recognition of significant miscibility opens up a host of new research avenues for melt-hydrous fluid systems, including their composition, phase relations, physical and transport properties, and impact on the chemical evolution of the crust and mantle.

3.06.5 CHEMICAL TRANSPORT AND REACTION

3.06.5.1 Mass Fluxes

The processes of advection, diffusion, and mechanical dispersion transport chemical species in fluids. For a porous medium, the flux, $\vec{F}_i$, of species i in the x, y, and z coordinate directions (mol $m_{(rock)}^{-2}$ s^{-1}) can be written as

$$\vec{F}_i = \vec{v}\phi C_i - D_{i,f}\tilde{\tau}\phi\nabla C_i - \tilde{D}_{MD}\phi\nabla C_i \quad (5)$$

where C_i is concentration of i in mol $m_{(fluid)}^{-3}$, $D_{i,f}$ is the diffusion coefficient for i through a free fluid, $\tilde{\tau}$ is the tortuosity tensor, and $\tilde{D}_{MD}$ is the mechanical dispersion tensor. The first, second, and third terms on the right-hand side describe fluxes due to fluid flow (advection), diffusion (Fick's first law), and mechanical dispersion, respectively. Diffusion through the solids is assumed negligible relative to the other transport processes. Diffusion and mechanical dispersion are known collectively as "hydrodynamic dispersion" and are discussed below.

Values of $D_{i,\mathrm{f}}$ for aqueous species under metamorphic conditions are typically of the order of $10^{-8}\ \mathrm{m^2\ s^{-1}}$ (cf. Oelkers and Helgeson, 1988). Taking $\sqrt{D_{i,\mathrm{f}}t}$ as a characteristic length scale for diffusion in a free fluid yields ~0.6 m for one year of diffusion, to ~600 m for one million years! In rocks, however, diffusive fluxes are limited by porosity and tortuosity (Equation (5)). The pathways for diffusion between grains through an interconnected porosity are not straight, but complex and *tortuous*. Consequently, diffusion through rock is "slower" than that through a free fluid; the tortuosity tensor is introduced to describe this behavior. Tortuosity is expressed as a tensor because it can vary with direction. Tortuosity systematics for metamorphic rocks are largely unknown, but it seems likely that pathways will be less tortuous parallel to penetrative fabrics than perpendicular to them (Ague, 2000a). Ague (1997a) found that diffusion adjacent to a cross-cutting quartz vein occurred more readily in metapelitic layers than metapsammitic ones; the difference may reflect less tortuosity parallel to micaceous fabrics in the metapelites relative to the metapsammites. In theory, tortuosity can vary between 1 (perfectly "straight" pathways) and near 0. Values measured in porous media range from about 0.3 to 0.6 (cf. Bear, 1972, p. 111), consistent with measurements of diffusion through pores in granodiorite (Fisher and Elliott, 1973), but much remains to be learned regarding tortuosity during active metamorphism.

Length scales of coupled diffusion and reaction can be considerable. Diffusion of mass across contacts between metapelitic and metacarbonate rocks can drive reactions that produce calc–silicate reaction zones on the centimeter to meter scale (e.g., Vidale and Hewitt, 1973; Ague, 2003). Bickle *et al.* (1997) estimated a ~5–10 m length scale for the diffusional component of oxygen isotope exchange across a lithologic contact. Characteristic length scales for reactive strontium diffusion across the lithologic contacts studied by Bickle *et al.* (1997) and Baxter and DePaolo (2000) are ~2 m and ~0.7 m, respectively. Numerical models suggest ~1–10 m length scales may be relevant for exchange of H_2O and CO_2 between adjacent layers, particularly if the product $\phi\tau > \sim 5\times10^{-5}$ (Ague and Rye, 1999; Ague, 2000a, 2002).

Diffusion transports mass down concentration gradients (from regions of high concentration to low concentration) according to Equation (5), although it is actually driven by gradients in chemical potential (cf. Shewmon, 1969). The concentration gradient-based approach turns out to be accurate for most tracers. It is generally used for species at higher concentrations as well, but complications, including the dependence of diffusion coefficients on fluid composition and diffusion up concentration gradients, can arise as the diffusing species interact subject to their chemical potential gradients and mass and charge balance constraints. While little is known of more complex diffusional behavior during metamorphism, the papers of Graf *et al.* (1983) and Liang *et al.* (1996) illustrate some of the problems that need to be solved in future studies.

Mechanical dispersion of transported species occurs during fluid flow, increases as flow velocity increases, and arises because: (i) fluid in adjacent porous pathways will be moving at slightly different velocities and, (ii) the tortuous nature of the flow paths causes mixing. Because of its dissipative nature, mechanical dispersion is generally modeled using a mathematical form identical to that for diffusion, although the processes underlying diffusion and mechanical dispersion are very different. Mechanical dispersion is a function of both the intrinsic properties of the porous medium and the fluid velocity, and is described using a directional framework even for homogeneous, isotropic media because dispersion in the direction of flow (longitudinal) tends to be greater than that perpendicular to flow (transverse). For example, assume that flow occurs with velocity v_x parallel to the x-direction. Then,

$$\tilde{D}_{\mathrm{MD}} = \begin{pmatrix} \alpha_{\mathrm{L}}|v_x| & 0 & 0 \\ 0 & \alpha_{\mathrm{T}}|v_x| & 0 \\ 0 & 0 & \alpha_{\mathrm{T}}|v_x| \end{pmatrix} \tag{6}$$

can be used, where α_{L} and α_{T} are the coefficients of longitudinal and transverse dispersivity, respectively (cf. Bear, 1972). The $\alpha_{\mathrm{L}}|v_x|$ term describes mechanical dispersion parallel to x, whereas the $\alpha_{\mathrm{T}}|v_x|$ terms describe it parallel to y and z. Longitudinal dispersivity can vary over several orders of magnitude in natural geologic materials (cf. Garven and Freeze, 1984b). It is another variable that is not well known for metamorphism but, by analogy with low-permeability rocks like shales, values between near 0 m and ~10 m appear reasonable (Ague, 2000a). The transverse dispersivity is expected to be as much as two orders of magnitude smaller (Garven and Freeze, 1984b). Note that if the medium is anisotropic, then significant complexities are introduced into the tensor representation of mechanical dispersion (Bear, 1972).

3.06.5.2 Reaction Rates

The fluid infiltration histories of fossil metamorphic flow systems are recorded when the fluid reacts with the rock. Obviously, if there was no reaction during flow, then fluxes could be large but the rock would not preserve evidence of the flow. The rates of chemical reactions link the timescales of mineralogical and fluid composition changes to those of mass and heat transfer (see also Chapter 3.07). The rate of precipitation or dissolution is dependent on the rate that ions can be attached to or removed from the mineral surface, and the rate at which ions can be transported to and from the surroundings to the mineral surface. Complete description of all these processes in multicomponent systems is generally not possible with current rate data, so the usual approach is to cast the rates in terms of the slowest or rate-limiting step (e.g., Berner, 1980). When the rate of attachment or removal of ions at precipitating or dissolving areas of the mineral is slow relative to transport rates through the solution surrounding the grain, the rate is said to be *surface controlled.* The opposite case of fast attachment or removal relative to transport is called *transport controlled* or sometimes *diffusion controlled.* Consider a quartz crystal bathed in a hydrothermal fluid. For surface-controlled rates, the concentration of aqueous silica species at the surface of the mineral is identical to that in the surrounding bulk solution. Note that these concentrations need not be equilibrium values. However, for transport control by diffusion, the concentrations of aqueous silica species at the surface of the quartz are at equilibrium values, but concentration gradients exist between the solution composition at the surface and the composition in the surroundings. Either surface or transport control can be dominant, depending upon the nature of the reaction-transport system (e.g., Lüttge and Metz, 1991; Sanchez-Navas, 1999), but a number of studies suggest dominance of surface-control in many common fluid–rock settings (e.g., Steefel and Lasaga, 1994).

A wide array of experimental studies indicates that rates of mineral dissolution and precipitation in aqueous solution are controlled primarily by reactive surface area, rate constants, activities of catalyzing/inhibiting agents, and departures of Gibbs free energy away from equilibrium. The relationships between these factors can be described by the following generalized rate law, which is consistent with transition state theory (cf. Steefel and Lasaga, 1994, and references therein):

$$\mathrm{Rate} = A_\theta \kappa_\theta f(a_i) f(\Delta G) \tag{7}$$

where A_θ is the reactive surface area of mineral θ in units of mineral area per unit volume of rock; κ_θ is an intrinsic rate constant in units of moles mineral per unit area per time; $f(a_i)$ is a function of the activities of aqueous species i in solution which act to either catalyze or inhibit reaction, and $f(\Delta G)$ is some function of the free energy of the system. A more specific form of Equation (7) that is generally applicable to surface-controlled reactions among common minerals is

$$r_\theta = sA_\theta \kappa_\theta \left(\prod_i a_i^p \right) \left| \left(\frac{Q_\theta}{K_\theta} \right)^M - 1 \right|^N \tag{8}$$

where r_θ is the dissolution or precipitation rate for θ in moles per unit volume rock per unit time; p is an experimentally determined constant; M and N are two positive numbers also generally determined by experiment; Q_θ is the ion activity product; K_θ is the equilibrium constant for the reaction (Q_θ and K_θ are equal at equilibrium); and s is, by convention, negative if the solution is undersaturated with respect to θ, and positive if supersaturated (Steefel and Lasaga, 1994). If M and N are both one, then the rate law is said to be *linear*, otherwise it is *nonlinear*. The intrinsic dissolution or precipitation rate constant κ_θ is strongly T dependent, and is usually expressed as an Arrhenius-type equation (cf. Oxtoby *et al.*, 1999):

$$\kappa_\theta = \kappa_\theta^0 \exp\left(\frac{-E_a}{R} \left[\frac{1}{T} - \frac{1}{T^0} \right] \right) \tag{9}$$

in which E_a is the activation energy and T^0 is a reference temperature (often 298.15 K). With this kinetic behavior, reaction rates will increase by a factor of ~10 per kilometer for a geothermal gradient of 30 °C km^{-1} (holding all other rate terms constant). Consequently, departures from local equilibrium are expected to diminish as T increases. E_a is not tightly constrained but appears to be limited to the range ~40–90 kJ mol^{-1} for common silicate and carbonate minerals, averaging around ~60 kJ mol^{-1}. The dissolution or precipitation reaction corresponding to the rate expression given by (9) is usually written for 1 mol of mineral, e.g., for K-feldspar we could write: $KAlSi_3O_8 + 4H^+ = K^+ + Al^{3+} + 3SiO_{2,aq} + 2H_2O$. Dissolution proceeds from left-to-right and precipitation from right-to-left; at equilibrium both rates are equal. The rates of reaction among the species in solution are commonly assumed to be instantaneous (cf. Helgeson, 1979; Sverjensky, 1987).

The overall rate of a fluid–rock reaction can also be modeled, rather than computing the dissolution and precipitation of each solid separately. For example, one could write an overall reaction between solids and fluids such as: Muscovite + Quartz = Sillimanite + K-feldspar + H_2O. The model for overall reactions in metamorphic rocks advanced by Lasaga and Rye (1993) includes all the basic parts of Equation (7) except the catalysis/inhibitor term (Ague, 1998):

$$R_{i,m} = \left(\frac{1}{\phi}\right) s\kappa_m \upsilon_{i,m} \bar{A}_{l,m} |\Delta G_m|^{N_m} \quad (10)$$

where $R_{i,m}$ is the rate in moles of i per unit volume fluid per unit time for reaction m; κ_m is the intrinsic reaction rate constant; $\upsilon_{i,m}$ is the stoichiometric coefficient for i; $\bar{A}_{l,m}$ is the surface area of the rate-limiting mineral l per volume rock in reaction m; $|\Delta G_m|$ is the absolute value of the Gibbs free energy change of reaction m at the T and P of interest; N_m is the reaction order; and s is, by convention, $+1$ if ΔG_m is negative and -1 otherwise (cf. Ague, 1998). If $N_m = 1$, the rate law is *linear*; otherwise it is *nonlinear*. Lasaga and Rye (1993) provide intrinsic rate constant expressions (Equation (9)) for both linear and nonlinear descriptions of the rate of muscovite dehydration based on the experiments of Schramke *et al.* (1987). These expressions are thought to be representative of a wide range of metamorphic devolatilization reactions, although much experimental and theoretical work needs to be done to map out the rate systematics of metamorphism. Note that Equation (10) gives the production/consumption rate for moles of a fluid species i, whereas Equation (8) is written for moles of solid. The net production/consumption rate for i is obtained by summing over the rate expressions for all reactions m.

In general, the rate-limiting surface area, $\bar{A}_{l,m}$, in the overall rate equation, (10), is determined by the mineral with the slowest surface reaction kinetics and the lowest surface area in contact with fluid. $\bar{A}_{l,m}$ is thus a function of critical rock physical properties including the rock porosity structure and the size, shape, and abundance of mineral grains. Much progress has been made by estimating mineral surface areas using simple geometric shapes like spheres or cubes which grow or shrink during reaction (cf. Steefel and van Cappellen, 1990; Bolton *et al.*, 1999). However, the amount of reactive surface area remains as one of the major uncertainties in modeling reaction rates.

A simpler but useful expression has been widely used in geochemistry to model reaction rates:

$$R_i = \kappa\left(C_i^{\mathrm{eq}} - C_i\right)^N \quad (11)$$

The reaction rate is proportional to the difference in concentration between the fluid in contact with the mineral assemblage at a particular place in the system (C_i), and the fluid composition that would be in equilibrium with the mineral assemblage (C_i^{eq}). Thus, the net rate is zero when fluid and rock are at chemical equilibrium ($C_i = C_i^{\mathrm{eq}}$). The rate "constant" κ actually combines several key rate variables and, for linear rates ($N = 1$), can be cast in terms of the product of the intrinsic reaction rate constant, a reactive surface area term, and the derivative of the ΔG of reaction with respect to concentration (Lasaga and Rye, 1993).

In real rocks, dissolution, precipitation, and devolatilization reactions proceed only if there is some departure from local fluid–rock equilibrium, a condition known as "overstepping." Models of fluid–rock reaction that use the local equilibrium approximation assume that reaction rates are so fast as to be essentially instantaneous. However, for some reactions, such as those with small intrinsic rate constants and small reactive surface areas, the rate of reaction near equilibrium is slow. Consequently, the T, P, and/or fluid composition must depart significantly from equilibrium before ΔG (the "driving force" for reaction) is large enough to produce significant rates. The magnitude of the overstepping may be particularly large if the reaction rate law is nonlinear (cf. Lasaga and Rye, 1993).

Overall rate expressions like Equation (10) assume that reactants and products are present and reacting, with the degree of overstepping being controlled by the rate at which the transformation of reactants into products occurs. Equilibrium is also overstepped if the product solids fail to nucleate at all. Classical theory holds that when a new phase nucleates, extra energy is necessary to form the grain boundary between the new phase and the phases from which it is growing (e.g., Shewmon, 1969). At equilibrium, this extra energy is unavailable, and no growth of reaction products occurs. Overstepping of the equilibrium condition, however, provides the energy necessary to nucleate and grow reaction products and to decrease the overall free energy of the system. For the cases of T and fluid composition (X) overstepping, the nucleation rate $\dot{n}$ is proportional to (Ridley and Thompson, 1986):

$$\dot{n} \propto \exp(T - T_{\mathrm{eq}})^2 \quad (12)$$

and

$$\dot{n} \propto \exp(X - X_{\mathrm{eq}})^2 \quad (13)$$

where T_{eq} and X_{eq} are equilibrium temperature and fluid composition, respectively. Thus, once the products do nucleate in an overstepped reaction, it is likely that they will do so rapidly, given the exponential and power terms in these expressions. It appears that some phases, particularly garnet solid solutions, continue to nucleate well after the exponential stage, albeit at considerably reduced rates (cf. Carlson, 1989).

The rates of metamorphic reactions and the magnitude of departures from local chemical equilibrium are important and controversial issues in the earth sciences today. Fluid fluxes, P–T–time evolution, and reaction histories estimated assuming local equilibrium models would clearly be in error if the actual processes operated far from equilibrium. Some examples of chemical disequilibrium, such as sluggish phase transformations among the Al_2SiO_5 polymorphs and selective retrograde reaction along pathways of fluid infiltration, have been well documented (cf. Kerrick, 1990; Giorgetti *et al.*, 2000). Nonetheless, concrete examples of chemical disequilibrium for prograde metamorphic devolatilization reactions and fluid flow are still relatively rare. It is common to observe variable grain sizes for a given mineral in a metamorphic rock. Such textures are inconsistent with equilibrium (Thompson, 1987), but it is unclear if they indicate large energetic departures from equilibrium. Laboratory evidence strongly suggests that the degree of prograde T overstepping due to nucleation problems may be of the order 10–100 K and 1–10 kJ, with the smallest values for devolatilization reactions and the largest for reactions with small entropy changes such as solid–solid reactions (Ridley and Thompson, 1986). P oversteps may be 0.1 GPa or more (Ridley and Thompson, 1986; Ernst and Banno, 1991). When multiple product phases must nucleate, oversteps are likely to be large and may exceed ~0.7 GPa (Rubie, 1998). If devolatilization reactions are overstepped significantly in T, then the subsequent rapid reaction that occurs upon nucleation and mineral growth may generate large fluid pressures sufficient to drive hydrofracture and fluid flow (cf. Walther, 1996; Ague *et al.*, 1998) over short timescales of $10-10^3$ yr (Ague *et al.*, 1998). Calculations suggest that slow rates may cause significant reaction overstepping in metacarbonate rocks, consistent with observed oxygen and carbon isotopic disequilibrium in some contact aureoles (Lasaga and Rye, 1993). A particularly insidious problem here is that sequences of mineral assemblages produced in the field relatively far from equilibrium can mimic local equilibrium sequences. Baxter and DePaolo (2000, 2002a,b) measured mineral chemistry and strontium isotope systematics for garnet and whole-rock across a lithologic contact near the Simplon Pass, Switzerland, and concluded that rates of reaction during cooling from ~610 °C to ~500 °C were extremely small, amounting to $\sim 10^{-7}$ g solid reacted per gram of rock per year. The slow reaction rates may reflect, in part, the cooling regime of retrograde metamorphism when fluids are not abundant, but Baxter and DePaolo (2000) also argue that prograde rates could not have been fast either. One provocative implication is that the chemical systematics of minerals that participate in such slow reactions may be unable to track changes in fluid chemistry, P, and T (Baxter and DePaolo, 2002b). In summary, mounting field, laboratory, and theoretical evidence indicates that chemical kinetics may be an important control on metamorphic processes, so it is prudent to examine the assumption of equilibrium before using it.

3.06.5.3 Advection–Dispersion–Reaction Equation

A fundamental task of fluid–rock studies is to determine the infiltration and reaction histories of rocks at any given point in a flow system. Imagine that the flow region of interest comprises infinitesimally small, cube-shaped building blocks or "control volumes" that are interconnected. We seek to quantify the changes in the masses of fluid species within each control volume due to net advection and hydrodynamic dispersion of fluid into or out of the volume, as well as the consumption or production of species within the volume due to internal chemical reactions. The required partial differential equation describing mass conservation for a fully saturated porous medium has been derived by many workers (cf. DeGroot and Mazur, 1969; Bear, 1972; Fletcher and Hofmann, 1973; Garven and Freeze, 1984a; Guenther and Lee, 1988). It is based on the flux equation, (5), and includes a term for chemical reaction:

$$\frac{\partial(C_i\phi)}{\partial t} = \underset{\text{(a)}}{-\nabla\cdot(\vec{v}\phi C_i)} + \underset{\text{(b)}}{\nabla\cdot\left(\tilde{D}_{\mathrm{HD},i}\phi\nabla C_i\right)} + \underset{\text{(c)}}{\phi\sum_m R_{i,m}} \tag{14}$$

where $R_{i,m}$ is the production rate (positive) or consumption rate (negative) for species i in reaction m (e.g., Equation (10)), and diffusion and mechanical dispersion have been combined into a single hydrodynamic dispersion tensor ($\tilde{D}_{\mathrm{HD},i} = D_{i,\mathrm{f}}\tilde{\tau} + \tilde{D}_{\mathrm{MD}}$). Terms (a), (b), and (c) describe advection, hydrodynamic dispersion, and reaction, respectively, and the left-hand side gives the total change in the moles of species i in the fluid per unit volume rock per unit time. Equation (14) is a partial differential equation known as the advection–dispersion–reaction equation and, as written, it has an infinite number of solutions. It can be solved for individual cases by specifying *initial conditions* and *boundary conditions* that describe the flow system. The rock medium through which the fluid flows is assumed to be stationary; additional terms are required if the rock moves as well.

3.06.6 GEOCHEMICAL FRONTS

One common reaction–transport scenario arises when fluid that is out of equilibrium with a rock mass of interest infiltrates across a boundary and drives reaction. The boundary could, for example, be a lithologic contact between two chemically and/or isotopically distinct kinds of rock, or the contact between the rock and a fracture through which fluid flows. Equation (14) is valid for general transport–reaction problems, but even a one-dimensional version with constant pore velocity, hydrodynamic dispersion, porosity, and rate constant exhibits significant complexity and illustrates fundamental principles:

$$\frac{\partial C_i}{\partial t} = -v_x \frac{\partial C_i}{\partial x} + D_{\mathrm{HD},i} \frac{\partial^2 C_i}{\partial x^2} + \kappa (C_i^{\mathrm{eq}} - C_i) \quad (15)$$

The concentration of i in the solid, C_i^{Solid}, is given by

$$\frac{\partial C_i^{\mathrm{Solid}}}{\partial t} = -\kappa \frac{\phi}{1-\phi}(C_i^{\mathrm{eq}} - C_i) \quad (16)$$

For nonreactive transport with $\kappa = 0$, a constant input concentration of $C_{i,x=0}$ at the $x = 0$ boundary (boundary condition), and an initial concentration of $C_i = 0$ throughout the flow domain (initial condition), the analytical solution to Equation (15) is well known (Fried and Combarnous, 1971):

$$\frac{C_{i,x}}{C_{i,x=0}} = \frac{1}{2}\left(\mathrm{erfc}\left[\frac{x - v_x t}{2\sqrt{D_{\mathrm{HD},i} t}}\right] + \exp\left[\frac{v_x t}{D_{\mathrm{HD},i}}\right]\mathrm{erfc}\left[\frac{x + v_x t}{2\sqrt{D_{\mathrm{HD},i} t}}\right]\right) \quad (17)$$

where erfc is the complimentary error function. Equation (17) gives the concentration of i at any point and time in the flow region if there is no reaction, and is a useful approximation for transport of i from one layer to another across a model "lithologic contact" at $x = 0$. Several important concepts can be illustrated by first assuming that transport occurs only by flow, and that hydrodynamic dispersion is negligible. As flow proceeds, the input fluid displaces more and more of the initial fluid in the direction of flow; the boundary between the input and displaced fluids is known as the "infiltration front" or "hydrodynamic front" and is marked by a sharp change in concentration referred to as a "concentration front" or "solute front" (Figure 5(a)). For constant v_x, the distance of front travel over a time interval Δt is: $v_x \Delta t$. The time-integrated fluid flux (q_{TI}) is the total amount of fluid flow that passes across an area of interest during a given time interval; for this problem it is simply: $v_x \phi \Delta t$ ($\mathrm{m}^3_{(\mathrm{fluid})}$ per $\mathrm{m}^2_{(\mathrm{rock})}$). Pioneering studies, including Baumgartner and Ferry (1991), Ferry and Dipple (1991), and Bickle (1992), demonstrated that the time-integrated fluid flux is invaluable for quantifying fluid–rock interactions

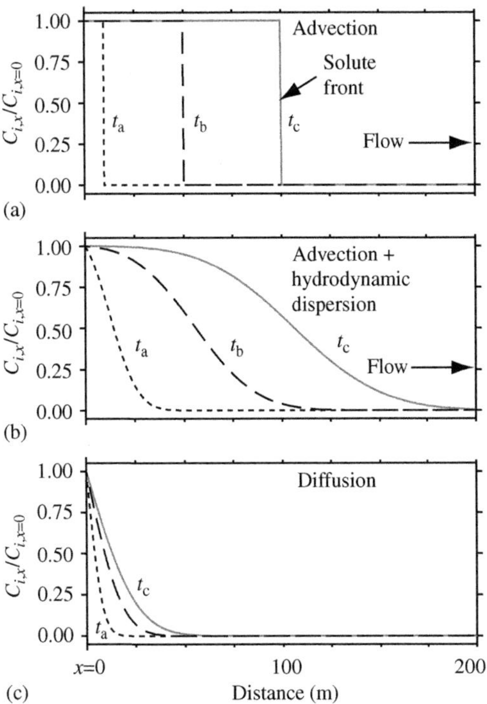

Figure 5 Fluid composition as a function of distance for three different model times ($t_a = 100$ yr; $t_b = 500$ yr; $t_c = 1{,}000$ yr) computed using Equation (17). No chemical reaction. $v_x = 0.1\ \mathrm{m\ yr^{-1}}$; $D_{i,\mathrm{f}} = 10^{-8}\ \mathrm{m^2\ s^{-1}}$; $\tau = 0.6$; $\alpha_{\mathrm{L}} = 5$ m. (a) Propagation of solute front by advection. (b) Combined advection and hydrodynamic dispersion. Note front broadening with increasing time and distance. (c) Transport by diffusion.

(much more will be said about q_{TI} below). Hydrodynamic dispersion acts to "smooth" sharp concentration fronts, and the degree of front broadening increases the farther the front travels (Figure 5(b)). Finally, the characteristic concentration profiles for pure diffusion are shown in Figure 5(c).

The relative importance of advection relative to hydrodynamic dispersion is often assessed using the dimensionless Peclet number: $P_e = v_x L / D_{\mathrm{HD},i}$, in which L is the length scale of interest. Hydrodynamic dispersion tends to dominate for small L, whereas advection dominates for large L, say on the scale of a mountain belt. The dissipative effects of hydrodynamic dispersion probably operate over length scales of $<\sim 100$ m (Bickle and McKenzie, 1987; Bickle, 1992).

The concepts of steady state and equilibrium are important to clarify for reactive systems. If, for example, the fluid composition is at steady state in a reacting flow system, the fluid composition is not changing with time. The system is not at equilibrium, however, since equilibrium would require that no transport and no reaction were occurring. Instead, at steady state the transport and reaction processes can be thought of as being in balance, so that the fluid composition remains constant. Transient changes in porosity, fluid velocity, and other geologic factors make it unlikely that natural systems ever reach perfect steady states, although the steady-state approximation has proven useful for constraining average time-integrated fluid fluxes and general processes of fluid–rock interaction.

Importantly, the concentration of i in the solid is not at steady state even if the fluid composition is. Consider a schematic example. A rock comprising corundum and a small amount of kyanite in equilibrium with an aqueous pore fluid reacts with a fluid input at $x = 0$ that is in equilibrium with just kyanite. The input fluid has a larger activity of aqueous silica, so as it enters the flow region, silica is consumed as corundum breaks down by reactions such as: $SiO_{2,(aq)}$ + Corundum = Kyanite. Reaction is removing i (aqueous silica) from the fluid, so the concentration of i in the solid continually increases (e.g., Figure 6). The fluid composition, however, is at steady state because the amount of i consumed at any given point is replenished by the amount transported to that point by advection and hydrodynamic dispersion. These relations illustrate that one or more processes in a system may be in steady state, but others need not be. The boundary between the reacted and unreacted regions in the rock is known as a "reaction front" or a "geochemical front." Eventually, all the corundum at the point of fluid input would be converted to kyanite, and the reaction and solute fronts would then begin to propagate out in the direction of flow (e.g., Figure 7). As a consequence, fossil geochemical fronts can provide valuable clues regarding the direction of fluid flow (cf. Rye *et al.*, 1976; Bickle, 1992; Skelton *et al.*, 1995). Note that when both fronts are moving, neither fluid nor rock compositions are at steady state. For advection-dominated systems with small porosity ($<\sim 0.01$), the time-integrated fluid flux is approximated by (Ague, 1998)

$$q_{\mathrm{TI}} \approx L_{\mathrm{GF}} \frac{n_i}{C_i^{\mathrm{eq}} - C_{i,x=0}^{\mathrm{Input}}} \tag{18}$$

where L_{GF} is the distance of geochemical front propagation in the rock. For our example, $C_{i,x=0}^{\mathrm{Input}}$ is the concentration of i (silica) in the input fluid, whereas n_i is the number of moles of i produced (+) or consumed (−) per unit volume rock (mol m^{-3}). Of course, this highly reduced example neglects many factors, including the speciation of the multicomponent fluid and reaction mechanisms. In most cases, these complexities are not

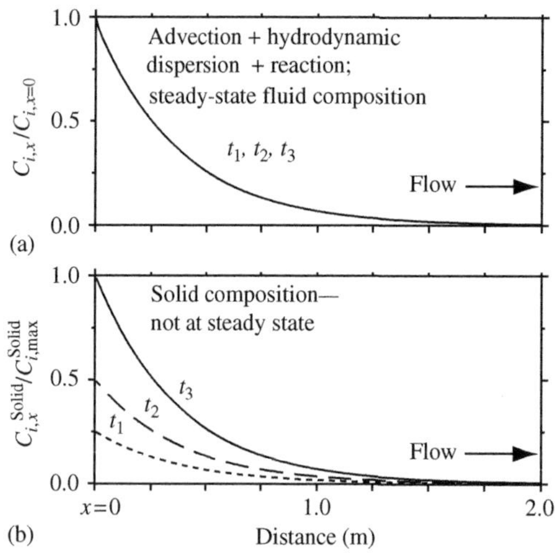

Figure 6 Steady and unsteady states for advection + hydrodynamic dispersion + reaction for three different model times ($t_1 < t_2 < t_3$; these do not correspond to times in Figure 5). $v_x = 0.1$ m yr^{-1}; $D_{i,\mathrm{f}} = 10^{-8}$ m^2 s^{-1}; $\tau = 0.6$; $\alpha_{\mathrm{L}} = 5$ m. Rate constant $\kappa = 5$ yr^{-1} used for illustration purposes and does not necessarily correspond to any specific reaction. $C_{i,\mathrm{max}}^{\mathrm{Solid}}$ is maximum concentration of i possible for solid. For convenience, concentration of i in fluid in equilibrium with initial rock (C_i^{eq}) and initial concentration of i in solid assumed negligible relative to $C_{i,x=0}$ and $C_{i,\mathrm{max}}^{\mathrm{Solid}}$, respectively. Equations (15) and (16) solved using numerical methods described in Ague (1998). (a) Steady-state fluid composition. Note fluid composition does not change with time. (b) Solid compositions corresponding to times t_1, t_2, and t_3. Concentration of i in solid increases with time due to its removal from fluid; thus, solid composition is not at steady state.

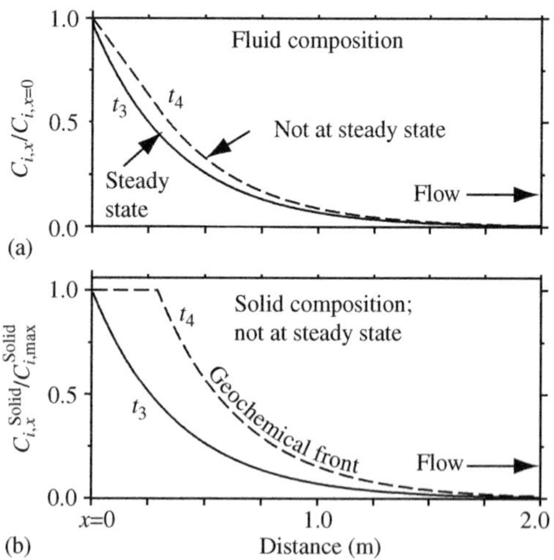

Figure 7 Front propagation for two different model times ($t_3 < t_4$). Time t_3 corresponds to time t_3 in Figure 6. See Figure 6 for calculation details. At t_3, fluid composition (a) is at steady state, but solid composition (b) is not (cf. Figure 6). At t_4, both solute and geochemical fronts are propagating, and neither fluid nor rock compositions are in steady state.

tractable with analytical solutions and require numerical modeling (cf. Steefel and Lasaga, 1994; Ague, 1998, 2000a; Ague and Rye, 1999).

The shape of the solute front depends strongly on the reaction rate (Figure 8). If rates are fast, then i is consumed rapidly and the solute front dies away close to the boundary where reactive fluid is input. In contrast, fronts broaden considerably as the rate decreases. Such front broadening can occur even if hydrodynamic dispersion is absent, and can be significant at >100 m length scales (Bickle, 1992; Lasaga and Rye, 1993). Dimensionless Damköhler numbers are often used to assess the relative roles of transport and reaction. For the case of advective transport the Damköhler number is $\kappa L/v_x$, and for transport by diffusion it is $\kappa L^2/D_{i,f}$ (Boucher and Alves, 1959). In either case, solute and geochemical fronts sharpen as the Damköhler number increases. If the rate of reaction is very fast (nearly instantaneous) relative to the rate of transport, then fluid and rock are essentially in chemical equilibrium at any given point along the flow path (the local equilibrium condition).

A somewhat different form of the reaction rate term in Equations (11) and (15) has been widely used for modeling transport of tracers, particularly isotopic tracers (cf. Ogata, 1964; Lassey and Blattner, 1988; Blattner and Lassey, 1989; Bickle, 1992; DePaolo and Getty, 1996; Bickle *et al.*, 1997):

$$\text{Rate} = \kappa \frac{1-\phi}{\phi}\left(C_i^{\text{Solid}} - \frac{C_i}{K_v}\right) \qquad (19)$$

where K_v is the equilibrium fluid/solid partition coefficient by volume for species i, and κ has been interpreted as the rate constant for fluid–rock exchange (Bickle, 1992) or for dissolution and precipitation of mineral material (DePaolo and Getty, 1996). Here, as the bulk composition of the solid changes, so too does the composition of the fluid in equilibrium with the solid. For advective transport, tracer exchange according to Equation (19), and constant porosity and fluid velocity, the

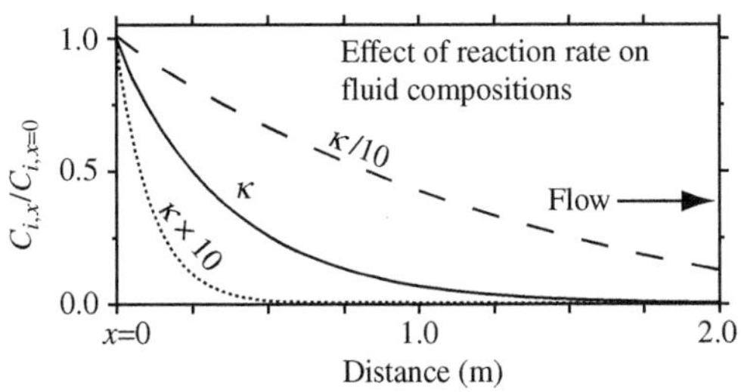

Figure 8 Effect of reaction rate. Solid line computed for reaction rate constant $\kappa = 5\ \text{yr}^{-1}$, dotted line for $\kappa = 50\ \text{yr}^{-1}$, and dashed line for $\kappa = 0.5\ \text{yr}^{-1}$. Note that front sharpens as reaction rate increases. See Figure 6 for calculation details.

time-integrated fluid flux (q_{TI}) is approximated by L_{GF}/K_v (Bickle, 1992). The q_{TI} expression for transport by advection with coupled local fluid–rock equilibrium exchange ± hydrodynamic dispersion has exactly the same form (cf. Bickle, 1992; Dipple and Ferry, 1992a,b).

For example, if an oxygen isotopic front has propagated 1,000 m and the K_v for oxygen is 0.6, the q_{TI} estimate is $\sim$1,700 $\text{m}^3_{(\text{fluid})}\ \text{m}^{-2}_{(\text{rock})}$. Fronts for tracers with smaller K_v (e.g., Sr) would propagate smaller distances for the same flux, and vice versa. Interpretations become more complicated if K_v changes with T or the individual mineral grains become isotopically zoned during reaction (cf. Bowman *et al.*, 1994; Lewis *et al.*, 1998; Graham *et al.*, 1998). Radioactive decay is not considered in Equation (19) and must be handled with an additional term (cf. DePaolo and Getty, 1996).

3.06.7 FLOW AND REACTION ALONG GRADIENTS IN TEMPERATURE AND PRESSURE

In the examples thus far, fluid that is out of equilibrium with a rock mass of interest infiltrates across some type of lithologic boundary, drives reaction, and forms geochemical fronts. Reaction will also occur if fluid flows along gradients in T and P because fluid compositions coexisting with minerals change as T and P change. A classic example is the precipitation of quartz in fractures to form veins (Fyfe *et al.*, 1978; Walther and Orville, 1982; Yardley, 1986; Ferry and Dipple, 1991). Precipitation occurs because the concentration of aqueous silica coexisting with quartz must decrease as T and P decrease along a flow path (Figure 4(b)). Imagine silica-saturated fluid ascending and cooling though a cubic control volume 1 m on a side. Fluid enters the bottom face of the cube, and exits through the top. The concentration of aqueous silica at the inlet is higher than that at the outlet, since quartz is precipitating from the fluid within the control volume. Assume that this drop in concentration is 5×10^{-2} mol m^{-3} for a typical geothermal gradient (more will be said about this below). The corresponding concentration gradient (5×10^{-2} mol m^{-3} m^{-1}) is small, so fluxes due to hydrodynamic dispersion can be neglected. The amount of quartz needed to fill up the control volume is equivalent to the inverse of the molar volume of quartz ($\sim 4.4 \times 10^4$ mol m^{-3}). The time-integrated fluid flux can then be estimated by reinterpreting Equation (18) slightly:

$$q_{\text{TI}} \approx (1\ \text{m}) \frac{4.4 \times 10^4\ \text{mol m}^{-3}}{5 \times 10^{-2}\ \text{mol m}^{-3}} \approx 9 \times 10^5 \text{m}^3\ \text{m}^{-2} \qquad (20)$$

Ferry and Dipple (1991) derived a more formal one-dimensional, local equilibrium, steady-state expression based on Equation (14) that neglects hydrodynamic dispersion and explicitly accounts for changes in concentration along the flow path due to T and P:

$$q_{\mathrm{TI}} = \frac{-(1/\bar{V}_{\mathrm{Qtz}})}{\left(\dfrac{\partial C_{\mathrm{SiO_2,aq}}}{\partial T}\dfrac{\partial T}{\partial x'} + \dfrac{\partial C_{\mathrm{SiO_2,aq}}}{\partial P}\dfrac{\partial P}{\partial x'}\right)} \quad (21)$$

where $\bar{V}_{\mathrm{Qtz}}$ is the molar volume of quartz and the coordinate direction x' increases in, and is parallel to, the direction of flow. The $\partial C_{\mathrm{SiO_2,aq}}/\partial T$ and $\partial C_{\mathrm{SiO_2,aq}}/\partial P$ terms can be calculated using the known solubility of quartz (Figure 4(b)), whereas estimation of $\partial T/\partial x'$ and $\partial P/\partial x'$ requires some knowledge of thermal and baric gradients in the direction of flow. Since quartz solubility varies more strongly with T than with P along typical geotherms (Figure 4(b)), the $\partial C_{\mathrm{SiO_2,aq}}/\partial T$ term generally dominates $\partial C_{\mathrm{SiO_2,aq}}/\partial P$. The calculations do not account for kinetic effects. Nonetheless, in quartz–saturated rock sequences at metamorphic temperatures, concentration gradients in kinetically limited systems will still tend to approach local equilibrium gradients, even if the absolute concentration values depart from equilibrium (cf. Ague, 1998). Equation (21) is inadequate, however, for shallow hydrothermal systems with high flow rates at relatively low T (Bolton *et al.*, 1999).

Enormous fluxes of the order of $10^6\ \mathrm{m^3\ m^{-2}}$ imply that a column of fluid ~1,000 km long flowed across each square meter of vein cross-section! A similar flux would be required to dissolve large amounts of quartz out of a rock, but here the fluid would have to flow in a direction of increasing T ("up-T" flow; Feehan and Brandon, 1999). The fluxes are large because changes in aqueous silica concentration along typical crustal geotherms are small. However, concentration gradients could be much steeper if quartz solubility drops due to decreases in water activity (Figure 4(c); cf. Walther and Orville, 1983; Newton and Manning, 2000a). For example, local decreases in water activity due to increased fluid CO_2 content near marbles could dramatically lower quartz solubility and, thus, produce quartz veins with a much lower flux than predicted by Equation (21). In addition, quartz veins can form by diffusion-dominated processes that require little or no fluid flow (see below).

The advective treatment for quartz veins can be extended to other types of metasomatic reactions. For example, alkali metasomatism is possible if a large amount of fluid flow occurs along gradients in T and P (cf. Orville, 1962; Dipple and Ferry, 1992a; Ague, 1994b, 1997a). Total concentrations of potassium increase and sodium decrease with increasing T for chlorine-bearing fluids in typical quartzofeldspathic and micaceous rocks (Figure 9; P effects are considerably smaller). Consequently, up-T fluid flow will tend to destroy micas and/or K-feldspar so as to remove potassium from the rock and, at the same time, produce a sodium-bearing phase like plagioclase and add sodium ("sodium metasomatism"). Down-T flow will do the opposite, favoring the growth of potassium-rich phases and destroying sodium-rich ones ("potassium metasomatism").

The expression describing the steady-state advection and reaction is similar to Equation (21) (Dipple and Ferry, 1992a):

$$q_{\mathrm{TI}} = \frac{n_{\mathrm{Na}}}{\left(\dfrac{\partial C_{\mathrm{Na,aq}}}{\partial T}\dfrac{\partial T}{\partial x'} + \dfrac{\partial C_{\mathrm{Na,aq}}}{\partial P}\dfrac{\partial P}{\partial x'}\right)} \quad (22)$$

in which n_{Na} is the total moles of sodium produced (+) or consumed (−) per unit volume of rock. An analogous expression can be written for potassium. As shown in Figure (9), the concentrations of sodium and potassium increase and $\partial C_{\mathrm{Na,aq}}/\partial T$ changes as the total amount of chlorine in the fluid increases. Thus, chlorine molality must be known to evaluate the denominator of the expression. Furthermore, an unaltered starting rock or "protolith" composition is needed so that sodium gains or losses due to alteration can be quantified to provide an estimate for the numerator. These metasomatic gains and losses are evaluated using mass balance methods (cf. Gresens, 1967; Grant, 1986; Ague, 1994a; Ague and van Haren, 1996). For typical total chlorine molalities of ~1 m and reasonable estimates for T and P gradients in the

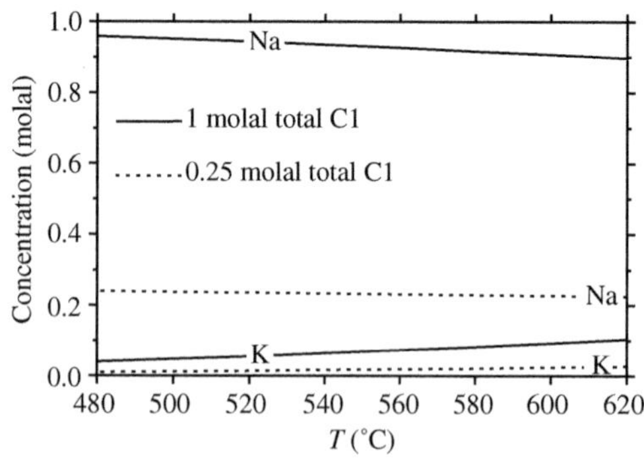

Figure 9 Total concentrations of sodium and potassium species in an aqueous solution coexisting with quartz, albite, muscovite, and kyanite for Cl concentrations of 1 molal (solid lines) and 0.25 molal (dotted lines). Computed following Ague (1997a), along a geothermal gradient of 20 °C km^{-1}. Note that total concentration of sodium decreases with increasing temperature, whereas the concentration of potassium increases. These trends become more pronounced as the total chlorine concentration in the fluid increases.

direction of flow, the time-integrated fluxes needed to cause alkali metasomatism are $\sim 10^4\ m^3\ m^{-2}$ (Ferry and Dipple, 1991; Dipple and Ferry, 1992a; Ague, 1997a). These fluxes are substantially smaller than those required to make quartz veins by advective flow (Equation (20)). Of course, metasomatism involving other elements that are transported effectively in chlorine-bearing fluids, including calcium, magnesium, and iron, can also be treated using Equation (22). Metal leaching due to reaction with H^+ or HCl^0 ("hydrogen metasomatism") is generally possible only if fluxes are well in excess of $10^4\ m^3\ m^{-2}$, due to the small concentrations of hydrogen species in typical fluids (e.g., HCl^0 + Albite = 0.5 Kyanite + 2.5 Quartz + $NaCl^0$ + 0.5 H_2O; cf. Yardley, 1986; Ague, 1994b).

Reactions involving H_2O–CO_2 fluids in metacarbonate rocks ("mixed volatile reactions") are invaluable for assessing fluid fluxes and mass transfer processes. Common prograde reactions release CO_2, so fluids in closed or nearly closed systems should get richer in CO_2 during heating (cf. Greenwood, 1975). In many metasedimentary sequences, however, reactions proceeded at relatively low X_{CO_2}, implying that H_2O was also being input into the rocks (e.g., New England, USA; cf. Hewitt, 1973; Baumgartner and Ferry, 1991; Ferry, 1992, 1994a,b; Ague, 2002). Input of external H_2O-bearing fluids, such as those derived from dehydrating schists or degassing magmas, can drive many prograde reactions (Figure 10(a); cf. Hewitt, 1973; Ague and Rye, 1999). This type of infiltration has been treated quantitatively for advection-dominated systems (cf. Dipple and Ferry, 1992b; Ferry, 1996; Ague and Rye, 1999; Evans and Bickle, 1999) as well as for systems in which hydrodynamic dispersion is important (Ague and Rye, 1999; Ague, 2000a, 2002).

Prograde reaction and CO_2 release can also occur, however, if fluids flow along gradients in T and P, as pointed out by Baumgartner and Ferry (1991). For a simple H_2O–CO_2 fluid, the analogue of Equations (21) and (22) for mixed volatile reactions can be written as (Baumgartner and Ferry, 1991; Ferry, 1992, 1994b; Léger and Ferry, 1993)

$$q_{TI} = \frac{\bar{V}_f\left(n_i - X_i\left(n_{CO_2} + n_{H_2O}\right)\right)}{\left(\dfrac{\partial X_i}{\partial T}\dfrac{\partial T}{\partial x'} + \dfrac{\partial X_i}{\partial P}\dfrac{\partial P}{\partial x'}\right)} \tag{23}$$

where $\bar{V}_f$ is the molar volume of the fluid, i is either CO_2 or H_2O, and n_i is the total moles of i produced (+) or consumed (−) per unit volume rock. The expression can be evaluated given estimates for P, T, fluid composition, total volatile production/consumption, and gradients in T and P along the flow path. Volatile production/consumption is typically quantified using reaction progress methods (cf. Brimhall, 1979; Ferry, 1983). For many common reactions proceeding under water-rich conditions, Equation (23) requires that prograde reaction and CO_2 release be driven by up-T fluid flow (Baumgartner and Ferry, 1991). For this case, fluid should get progressively more CO_2-rich along regional up-T flow paths; down-T fluid flow, in contrast, would tend to drive retrogression and remove CO_2 from the fluid (Figure 10(b)). The possibility of widespread up-T flow is discussed in a later section.

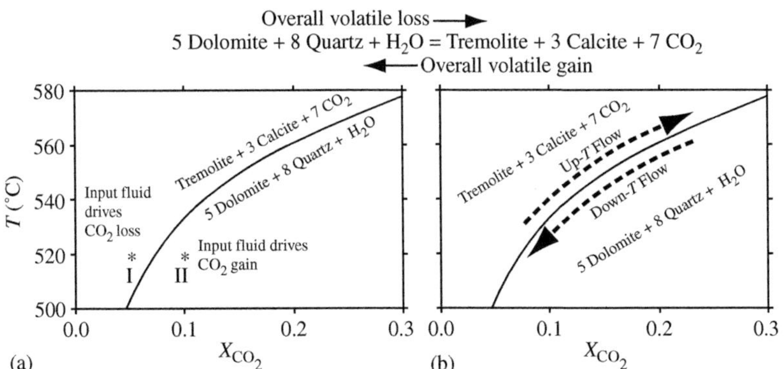

Figure 10 Two ways to drive CO_2 loss or gain for a common mineral assemblage in metacarbonate rocks. Computed for 0.7 GPa following Ague (2000a). (a) Fluid composition I is more water-rich than the equilibrium fluid (solid line) at 520 °C, and is in the stability field of the reaction products tremolite and calcite. Input of this fluid into a metacarbonate layer would thus drive prograde reaction and CO_2 loss. Input of fluid II would drive retrograde CO_2 gain. Fluid could be input by, for example, direct flow through a metacarbonate layer or diffusion/mechanical dispersion across lithologic contacts or vein margins. (b) Fluid flowing in a direction of increasing temperature ("up-T" flow) at or near local equilibrium with the assemblage dolomite + quartz + tremolite + calcite must get progressively richer in CO_2; it does so by driving prograde reaction and CO_2 release from the rock. Down-T flow would produce retrograde CO_2 gain.

3.06.8 EXAMPLES OF MASS AND HEAT TRANSFER

Bickle and McKenzie (1987) identified three broad crustal mass and heat transport regimes. In the first, fluid flow is limited, and chemical transport by diffusion and heat transport by conduction prevail. In the second, fluid fluxes are large enough such that advection of mass by fluid flow dominates diffusive transport. However, because conduction is relatively efficient, heat conduction through the rock still dominates heat advection by the fluid. Many deep crustal systems are inferred to have formed in this regime (e.g., Manning and Ingebritsen, 1999). In the third regime, fluxes are very large, and both chemical and heat transport are predominantly by fluid flow. In addition to these three categories, hybrid modes of transport are possible. For example, regional scale fluid flow along fractures can be coupled to local scale diffusion to and from the wallrock adjacent to the fractures. The mass and heat transfer literature is vast and cannot be reviewed fully here; thus, the following sections explore some selected field examples that illustrate a representative spectrum of processes.

3.06.8.1 Regional Devolatilization and Directions of Fluid Motion

The processes of metamorphic volatile release and the directions of fluid motion are basic questions in deep crustal petrology (Figure 11). The "single pass" flow model holds that fluids generated during devolatilization move upward toward the surface (cf. Walther and Orville, 1982). The total flux of fluid is constrained by the thickness of the metamorphic pile undergoing devolatilization, and is $\sim 10^3\ m^3\ m^{-2}$ in continental collision settings (Hanson, 1997). This overall average can be exceeded if fluid is channelized into regional structures like fold hinges (cf. D. M. Rye and R. O. Rye, 1974; Skelton *et al.*, 1995) or quartz vein networks (cf. Ague, 1994b). However, since fluid is diverted into the conduits from the surrounding area, fluxes outside the conduits will tend to be lower than they would be in the absence of channelization (cf. Breeding *et al.*, 2003). If the fluid is replenished by processes such as subduction, then regional average time-integrated fluxes can be considerably larger than $\sim 10^3\ m^3\ m^{-2}$ (cf. Nur and Walder, 1990; Peacock, 1990; Breeding and Ague, 2002).

Recirculation of fluid by convection ("multi-pass" flow) is another possible regime (cf. Etheridge *et al.*, 1983). While convection is widely recognized in shallow hydrothermal systems, the conventional wisdom is that permeability is too small in the deep crust to allow the downward penetration of fluid necessary for convection (cf. Walther and Orville, 1982; England and Thompson, 1984; Hanson, 1997; Manning and Ingebritsen, 1999). Nonetheless, study of active mountain belts in New Zealand and Pakistan has shown that shallow fluids can penetrate to at least mid-crustal levels near the brittle–ductile transition (cf. Koons and Craw, 1991; Templeton *et al.*, 1998; Poage *et al.*, 2000). For example, fluid

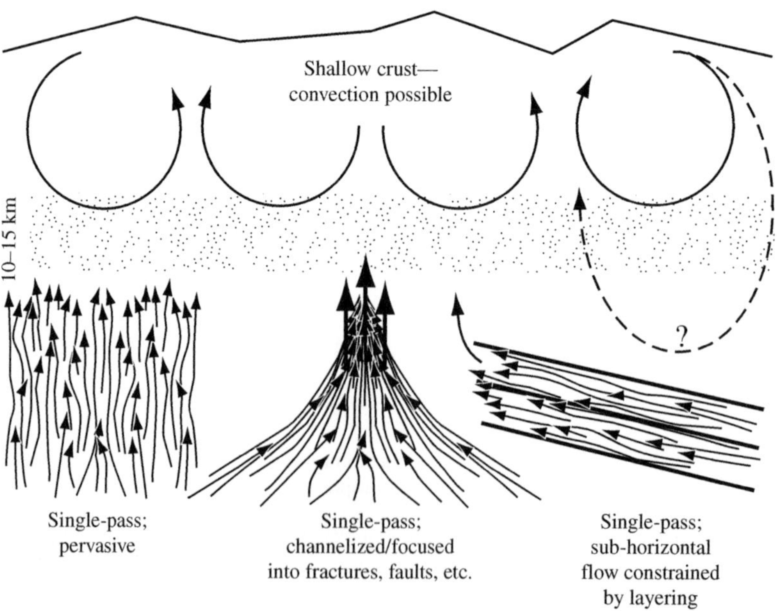

Figure 11 Schematic cross-section through the crust illustrating some possible modes of regional fluid flow. Contains some elements modified after figure 9 in Etheridge *et al.* (1983).

inclusion and stable isotopic systematics in Nanga Parbat, Pakistan, led Poage *et al.* (2000) to conclude that meteoric fluids can penetrate downward to depths of ~10 km, where they then mix with CO_2-rich fluids of metamorphic origin in the rapidly uplifting core of the mountain belt. Overall fluid fluxes, however, are inferred to be small, except in fault zones. Wickham and Taylor (1990) concluded that massive penetration of marine fluid to depths of at least 10–12 km resulted in widespread homogenization of stable isotope ratios in Hercynian basement of the Pyrenees. Convection is also one possible explanation for mid-crustal thermal anomalies preserved in New Hampshire (Chamberlain and Rumble, 1989; see below).

Metasedimentary rocks in New England, USA, preserve a valuable record of fluid–rock interaction that bears on the problem of flow direction. If pervasive advection was the only significant transport mode for fluid species, then prograde CO_2 loss from calcite-bearing metasediments indicates that fluid flow was largely in a direction of increasing T (up-T; Figure 10(b)) over 10–100 km length scales in the Acadian orogen (Baumgartner and Ferry, 1991; Ferry, 1992, 1994b; Léger and Ferry, 1993). X_{CO_2} increased with metamorphic grade, and the flow is inferred to have been layer-parallel (Ferry, 1992, 1994b). Estimated time-integrated fluid fluxes increase from ~10^2 m^3 m^{-2} in the greenschist facies to ~7×10^3 m^3 m^{-2} in the amphibolite facies. Fluxes were generally insufficient to cause major element metasomatism, although local alkali loss occurred in the amphibolite facies in the vicinity of plutons (Léger and Ferry, 1993). Ferry (1992) concluded that the flow comprised a giant metamorphic hydrothermal system in which fluids moved up-T sub-horizontally inward from the flanks of the orogen toward a central high-grade core intruded by plutons, and then ascended and cooled sub-vertically forming a regional network of quartz veins. Oxygen isotope systematics from a number of orogens around the world are also consistent with pervasive, regional up-T flow (cf. Dipple and Ferry, 1992b; Cartwright *et al.*, 1995; Oliver *et al.*, 1998).

These provocative results have generated considerable productive debate. For example, based on numerical models, Hanson (1997) argued that fluid flow under prograde metamorphic conditions will be dominantly upward, and that sub-horizontal flow is unlikely. However, these models do not take into account anisotropy due to layering and foliations. Fluids could have been constrained to follow flat-lying flow paths if layering was sub-horizontal, given that permeability can be over an order of magnitude greater parallel to layering than perpendicular to it (Figure 11; cf. Ingebritsen and Manning, 1999). In addition, transport–reaction processes other than layer-parallel, up-T advection can potentially drive at least some of the prograde volatile loss. As discussed above, infiltration of fluid with lower X_{CO_2} than the equilibrium value for a given rock can drive reaction (Figure 10(a)). For example, H_2O derived from dehydrating metapelites that infiltrates across lithologic contacts or vein selvages by hydrodynamic dispersion (Hewitt, 1973; Ague and Rye, 1999; Ague, 2000a) or by advection (Ague and Rye, 1999; Evans and Bickle, 1999) can potentially drive substantial prograde CO_2 loss from metacarbonate rocks without the requirement of up-T flow. Advective flow of dehydration fluids directly along layers will also drive CO_2 loss (cf. Legér and Ferry, 1993; Ague and Rye, 1999). Degassing magmas are another significant source of H_2O that can, at least in part, account for water-rich diopside zone conditions (Palin, 1992; Léger and Ferry, 1993; Ague and Rye, 1999; Ague, 2002). These models of external fluid infiltration can drive prograde CO_2 loss with time-integrated fluid fluxes that are considerably smaller than models based solely on up-T advection (Ague and Rye, 1999; Evans and Bickle, 1999). Recently, Wing and Ferry (2002) used a potentially groundbreaking new method, akin to those developed for tracing ocean circulation patterns (cf. Lee and Veronis, 1989), to estimate regional time-integrated fluid fluxes and flow directions based on a three-dimensional inversion of reaction progress, δ^{18}O, and δ^{13}C data. Their results suggest that the metamorphic fluid flow was highly complex locally, and included upward, downward, up-T, and down-T flow regimes. On the regional scale, however, flow was dominantly upward and parallel to regional lithologic layering.

3.06.8.2 Mass Transfer in Veins and Shear Zones

Veins (mineralized fractures) are present in regionally metamorphosed rocks worldwide and are unambiguous indicators of mass transfer. The nature and scale of mass transfer, however, vary strongly, depending on the vein-forming process. Vein minerals may be precipitated from large-scale fluid flow through regional fracture systems, local diffusion of mass to (and from) fractures through an essentially stagnant pore fluid, and local-scale fluid flow from wallrocks to fractures (cf. Walther and Orville, 1982; Yardley, 1986; Ague, 1994b, 1997b; Widmer and Thompson, 2001). The latter mechanism is probably too inefficient to form common types of metamorphic veins (cf. Yardley, 1975; Fisher and Brantley, 1992). The fluid fluxes required to precipitate vein minerals by regional flow are large, but local formation by diffusional mass transfer may occur with little or no flow. Consequently, vein formation mechanisms must be assessed before any conclusions about fluid fluxes can be drawn.

Fracturing tends to increase rock permeability. Thus, fluids can be focused effectively into cracks, the resultant fluid fluxes can be large, and element transport can be extensive. For example, Barrovian-style metamorphism of the Wepawaug Schist (Connecticut, USA), was accompanied by significant fracture-controlled mass transfer (Tracy *et al.*, 1983; Palin, 1992; Ague, 1994b, 2002; Ague and Rye, 1999). Vein abundance increases from a few vol.% in the chlorite zone to 20–30% in the amphibolite facies (Ague, 1994b). Veins cutting metapelites are surrounded by cm to dm thick, highly aluminous, silica-depleted alteration zones or "selvages," particularly in the upper greenschist facies and amphibolite facies (Ague, 1994b). Based on three independent types of mass balance analysis, Ague (1994b) concluded that ~70% of the volume of the average amphibolite facies vein could be accounted for by local, diffusion-dominated movement of silica from the selvages to the veins, but ~30% must have been derived from external fluids. This 30% still requires a huge flux; Equation (21), appropriate for regional escape of fluids through cracks down regional T and P gradients in these quartz-saturated rocks, yields a time-integrated fluid flux of $\sim 2.8 \times 10^5\ m^3\ m^{-2}$ for the average amphibolite facies vein. Given that ~25 vol.% of the amphibolite facies rock mass consists of veins, the average regional flux across the amphibolite facies is $\sim 6 \times 10^4\ m^3\ m^{-2}$ (Ague, 1994b). This large flux strongly suggests that fluids derived from prograde devolatilization deeper in the metasedimentary sequence were focused into the schist through a regional fracture network. Ferry (1992) inferred similarly large fluxes ($\sim 10^5\ m^3\ m^{-2}$) for fracture-controlled discharge of fluids in the axial portions of regional antiforms in eastern Vermont.

van Haren *et al.* (1996) found that the oxygen in garnet rims was as much as ~2‰ heavier than garnet cores in an alteration selvage adjacent to an amphibolite facies vein cutting metapelite in the Wepawaug Schist. They concluded that the garnets record progressive increases in the $\delta^{18}O$ of the selvage resulting from isotopic exchange between the wallrock and large fluxes of regional devolatilization fluids that were ascending through fractures (down-T flow will tend to partition the heavy oxygen into the rock; cf. Dipple and Ferry, 1992b). Garnets outside the selvages contain significantly less zonation, indicating smaller degrees of fluid–rock reaction away from veins.

Local and regional-scale vein-forming processes (Figure 12) operated in the Wepawaug Schist (Ague, 1994b, 1997b). Failure caused by, e.g., hydrofracture or tectonic stress, opened up fractures that were zones of elevated permeability and which channeled fluid flow. Some precipitation of silica occurred as the fluids flowed down P and T gradients in the fractures. This flow was accompanied and/or followed by fracture sealing due to local diffusional transport of silica from wallrocks to fractures. Crack-seal textures (cf. Ramsay, 1980) and other evidence of repeated cycles of fracturing and healing indicate that the proposed crack–flow–seal model sequence was repeated many times to form large quartz veins (cf. Fisher and Brantley, 1992; Ague, 1994b; van Haren *et al.*, 1996).

Element transport by regional, fracture-controlled flow can influence mineral assemblage development. Alkali metasomatism is one of the most common types recognized adjacent to veins, consistent with the substantial concentrations of sodium and potassium that can coexist with mica and feldspar-bearing assemblages (Figure 9). For example, Yardley (1986) documents retrograde alkali metasomatism and the resultant formation of aluminous mineral assemblages including tourmaline, staurolite, garnet, and andalusite in Knockaunbaun, Ireland. Ague (1994b) concluded that alkali loss from the vein alteration selvages in metapelites of the Wepawaug Schist described above played a key role in stabilizing the regional index minerals staurolite and kyanite (Figure 13). A variety of other elements were mobile as well, including phosphorus (essential for forming geochronologically important minerals like monazite), which was lost from the selvages. Oliver *et al.* (1998) documented mid-crustal amphibolite facies alkali and calcium loss and the growth of aluminous mineral assemblages in vein selvages related to regional fluid flow in the vicinity of the Kanmantoo ore deposit, Australia. Ague (1997a) found metasomatic increases in Na/K, Ca/K, and Sr/Rb corresponding to plagioclase growth and muscovite destruction adjacent to mid-crustal, Barrovian garnet zone veins near Stonehaven, Scotland. This open system mass transfer, coupled to a metasomatic increase in Mg/Fe, stabilized the index mineral garnet in a number of lithologies that originally had bulk compositions unsuitable for the growth of this key index mineral. The metasomatism can be explained by fluid flow along fractures in a direction of increasing T, coupled to diffusional exchange of sodium and potassium between the vein fluids and the wallrock (Ague, 1997a). In this scenario, the amount of quartz dissolved by up-T fluid flow ($q_{TI} \sim 3 \times 10^4\ m^3\ m^{-2}$) was much smaller than the amount derived by diffusion of silica from local vein selvages to the fractures. However, sodium-metasomatism may have also occurred as fluids from sodium-rich, spillitized oceanic crust ascended into and reacted with the overlying

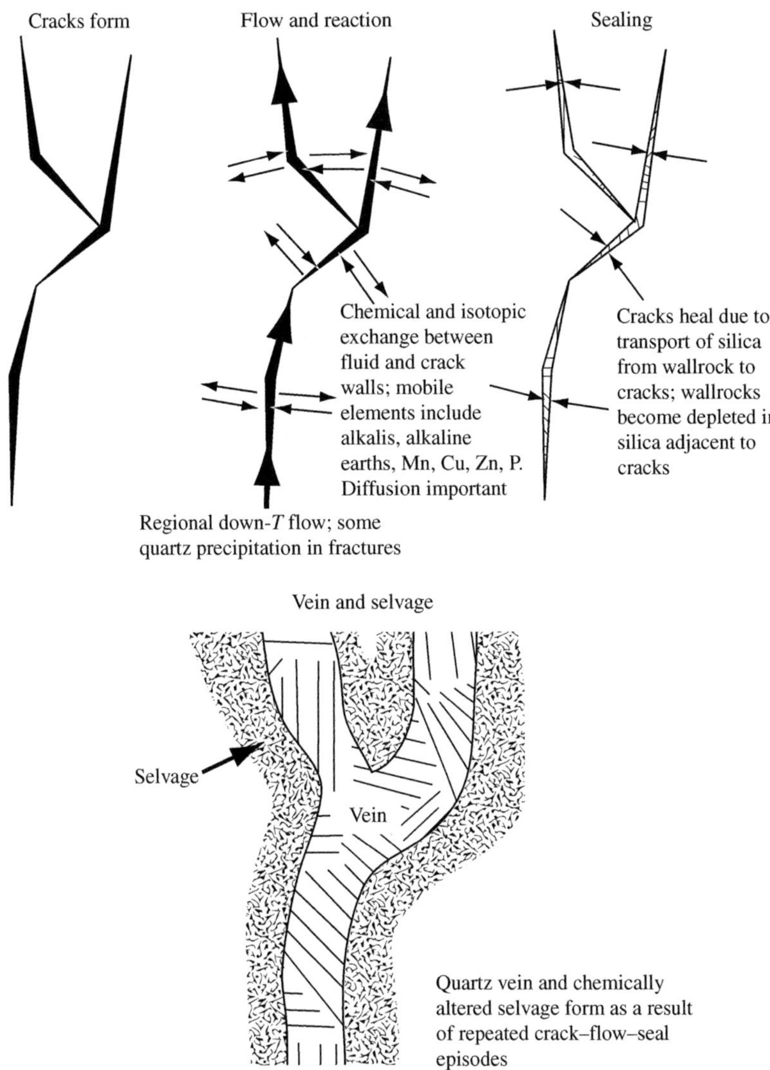

Figure 12 Illustration of crack–flow–seal model of vein formation derived for Wepawaug Schist, Connecticut, USA. Repeated crack–flow–seal episodes build up large quartz veins, and produce chemically and isotopically altered zones (or "selvages") strongly depleted in silica in wallrock adjacent to cracks (after Ague, 1997b).

metasediments (e.g., Breeding and Ague, 2002). An important point to emphasize is that although selvages may only be centimeter to meter scale in thickness, they can line flow conduits at the kilometer scale. Thus, selvages are critical loci of interaction between regionally migrating fluids and their enclosing wallrocks, facilitating ion exchange and modifying the composition of fluids substantially along crustal flow paths.

Major element mass transfer can be coupled to isotopic shifts that allow the fluid source to be traced. Tracy *et al.* (1983) found extreme metasomatism, including the loss of nearly all potassium and sodium, for metacarbonate rock adjacent to a quartz vein in the Wepawaug Schist. The infiltrating fluid had lower X_{CO_2} than fluid in equilibrium with the wallrock and thus drove extensive decarbonation of the selvage (e.g., Figure 10(a)). The $\delta^{18}O$ of the vein quartz (and selvage calcite) is smaller than in metacarbonate rock beyond the selvage margins (Tracy *et al.*, 1983), consistent with infiltration of external fluid derived from or equilibrated with syn-metamorphic intrusions or underlying mafic metavolcanic rocks (Palin, 1992; van Haren *et al.*, 1996). This fluid was distinct from the regional devolatilization fluid that produced the aluminous selvages in metapelites described above. Other cases show that fluid fluxes can be large enough to cause isotopic shifts, but not elemental metasomatism. For example, Chamberlain and Conrad (1991) documented that the isotopic signature of metapelites was

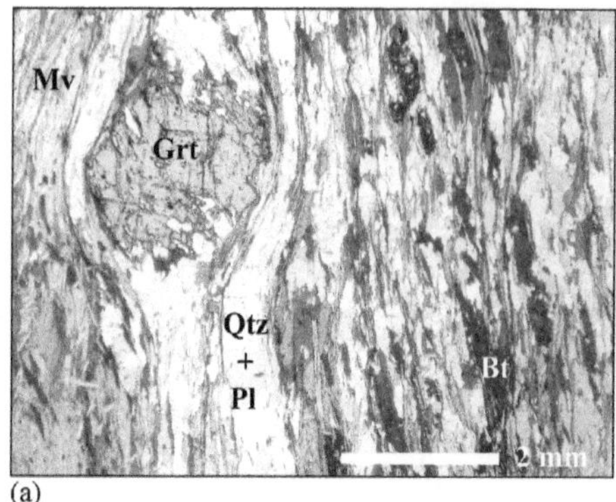

(a)

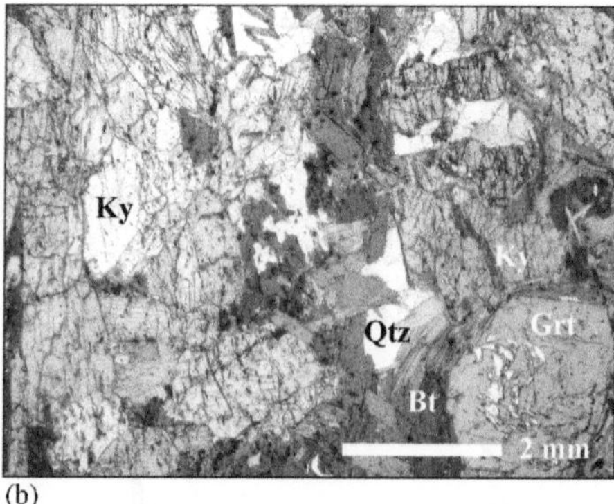

(b)

Figure 13 Photomicrographs illustrating vein selvage–wallrock relations for metapelite layer in Wepawaug Schist (Barrovian Kyanite zone; JAW-114 locale (Ague, 1994a)). Plane-polarized light. (a) Little-altered schist far-removed (dm scale) from ~5 cm wide vein of pure quartz. Schist composed mostly of quartz (Qtz), biotite (Bt), muscovite (Mv), Plagioclase (Pl), and garnet (Grt). (b) Highly altered, alkali- and silica-depleted, aluminous selvage at contact with vein. Derived from rock analogous to that shown in part (a). Dominated by kyanite (Ky), Bt, and Grt. Note that schistose fabric in (a) almost entirely obliterated in matrix due to metasomatic alteration, although relic fabric preserved as quartz inclusions in garnet (lower right).

carried into quartzites along fractures (New Hampshire, USA), although the total fluxes were small. Finally, in some cases, isotope ratios for a given mineral in veins and enclosing wallrocks are essentially identical, suggesting limited fluid–rock interaction and local derivation of vein mass (cf. Yardley and Bottrell, 1992).

A variety of mechanisms have been proposed to explain local diffusion-dominated transport of vein-forming constituents like silica from wallrocks into fractures; some of these are reviewed here. For example, if matrix grains are small and strained, whereas grains that grow in the veins are larger and less strained, then material will diffuse down chemical potential gradients toward the veins through an interconnected pore fluid to reduce strain energy and surface free energy (Yardley, 1975). In this way, the small, strained matrix grains dissolve and the larger, less strained vein crystals grow. Alternatively, if fracturing results in a fluid pressure drop across the fracture walls (e.g., Etheridge *et al.*, 1984), then the solubility of vein-forming minerals like quartz will be higher in the wallrock (where solid and possibly fluid pressure are higher) than in the fracture, and diffusion of material toward the fracture will occur (a form of "pressure solution"; cf. Elias and Hajash, 1992; Fisher and Brantley, 1992). Fluid inclusion evidence suggests that the initial fluid pressure drop upon fracturing can be as great as ~0.1 GPa in accretionary prism settings (Vrolijk, 1987), although such large drops are likely to be short-lived (cf. Fisher and Brantley, 1992). Recently, Widmer and Thompson (2001) proposed that disequilibrium overstepping of reactions in wallrocks could set up large chemical potential differences between the wallrocks and precipitation sites in veins. They argue that aluminum can be mobilized locally without significant fluid flow to form segregation veins in high-P (~2.5 GPa) environments.

Shear zones can be loci of focused fluid flow and major element metasomatism and associated volume loss (e.g., O'Hara, 1988; Selverstone *et al.*, 1991; Dipple and Ferry, 1992a; Ring, 1999). In their survey of ductile shear zones, Dipple and Ferry (1992a) concluded that time-integrated fluid fluxes of $\sim 2 \times 10^4$ m^3 m^{-2} attended down-T flow or up-T flow, leading to considerable alkali mass transfer and other major element metasomatism.

3.06.8.3 Accretionary Prisms and Subduction Zones

Dehydration of downgoing oceanic lithosphere can release tremendous volumes of fluid to accretionary prisms and, at deeper levels, the mantle wedge (see Chapter 3.17). Breeding and Ague (2002) estimated regional time-integrated fluid fluxes of $\sim 10^4$–10^5 m^3 m^{-2} for fracture-controlled flow into the Otago Schist accretionary prism, New Zealand, based on a mass balance analysis of silica addition to regional quartz vein sets. These values are consistent with theoretically estimated dehydration fluxes (cf. Peacock, 1990) for the ~100 Ma timescale of subduction relevant for the Otago Schist. The fluid fluxes were large enough to produce gold mineralization in Otago and in accretionary environments elsewhere (cf. Craw and Norris, 1991; Jia and Kerrich, 2000), and provide a mechanism for the long-term bulk silica enrichment of the continents (Breeding and Ague, 2002).

The quantities and directions of fluid flow within subducted crust under high-P low-T (HP–LT) metamorphic conditions remain controversial (see also Chapter 3.09). Field studies of stable isotope systematics (e.g., Bebout and Barton, 1989, 1993) and trace element mobility (e.g., Sorensen and Grossman, 1989) provide strong evidence for regional, kilometer-scale fluid migration during Mesozoic subduction in the Cordillera (California, USA). Field and theoretical studies also suggest that large-scale fluid flow may transport significant mass of "nonvolatile" elements including silicon, aluminum, and alkali and alkaline earth metals, producing metasomatic alteration features such as regional vein systems and silicification of the mantle wedge (e.g., Bebout and Barton, 1989, 1993; Manning, 1997). In contrast, mounting evidence from the Cycladic Archipelago (Greece) and the Alps indicates that subduction during Alpine orogenesis may have involved limited fluid flow and fluid–rock interaction (e.g., Philippot and Selverstone, 1991; Bröcker *et al.*, 1993; Getty and Selverstone, 1994; Barnicoat and Cartwright, 1995; Ganor *et al.*, 1996; Putlitz *et al.*, 2000). For example, little evidence has been found for fluid–rock interactions during subduction on the Cycladic island of Naxos; the classic metamorphic sequence there formed during a later Barrovian-style overprint of original HP–LT assemblages (cf. Rye *et al.*, 1976). Putlitz *et al.* (2000) found no evidence favoring the large-scale release or flow of fluids during HP–LT metamorphism of subducted oceanic crust based on oxygen and hydrogen isotope studies of metabasalts and metagabbros in the Cyclades. One possibility is extreme channelization of fluids into high permeability structures such as fractures, but regionally extensive vein systems or other types of conduits, if present, are rarely recognized in the field. However, Breeding *et al.* (2003) concluded that lithologic contacts can act as high-flux conduits. Strong metasomatic interactions between metamorphosed ultramafic, volcanic, and sedimentary rocks occurred in mélange zones in both the California and Alpine/Cycladic subduction complexes, but fluid fluxes, whether small or large, have yet to be quantified (Dixon and Ridley, 1987; Putlitz *et al.*, 2000; Bröcker and Enders, 2001; Bebout and Barton, 2002; Catlos and Sorensen, 2003). Although the unusual isotopic and chemical compositions of the famous ultrahigh-pressure Dora Maira whiteschists (western Alps) may be the result of extreme metasomatism of an orthogneiss protolith during subduction, interpretations remain controversial and the role of fluids is uncertain (Sharp *et al.*, 1993).

It is critical to clarify the fluid flow picture during HP–LT metamorphism since subduction of sediment and hydrothermally altered oceanic crust and mantle is the primary means by which reactive volatiles including H_2O and CO_2 are returned to the deep Earth, ultimately giving rise to arc magma genesis at ~100–150 km depth (c.f. Stolper and Newman, 1994; Hawkesworth *et al.*, 1997; Johnson and Plank, 1999; Kent *et al.*, 2002).

3.06.8.4 Heat Transport by Fluids

Rocks can conduct heat fairly readily, so fluid fluxes must be large, channelized, and/or transient for advective heat transport to be important (cf. Bickle and McKenzie, 1987; Brady, 1988; Hoisch, 1991). For example, a series of ten granulite facies thermal anomalies or "hot spots" measuring 10–30 km^2 are spread out in a belt ~150 km long in part of the Acadian orogen, New Hampshire, USA (Chamberlain and Rumble, 1988). Chamberlian and Rumble (1988, 1989) proposed that the hot spot near the town of Bristol is an area where large volumes of ascending hot fluid were focused through a network of quartz veins, thereby perturbing regional thermal and oxygen isotope systematics. The large fluxes could have been achieved by focusing of fluids generated by metamorphic devolatilization or magmatic degassing into the comparatively small area of the hot spot (Brady, 1988; Chamberlain and Rumble, 1989), or by recycling fluids in a convective flow system (Chamberlain and Rumble, 1989). The timescale of flow must have been less than ~10^6 yr, otherwise the surroundings would have heated up, destroying the steep thermal gradients observed in the field (Brady, 1988; Chamberlain and Rumble, 1989). Another alternative is that heat was transported up through the hot spots mostly by magmas, rather than hot fluids, although no direct evidence for such magmas has been found.

Ferry (1992) and Ague (1994b) used the dimensionless thermal Peclet number (B) of Brady (1988) to assess whether or not the large fluxes needed to make regional quartz vein sets elsewhere in the Acadian orogen of New England may have also transported heat:

$$B = \frac{(q_{\mathrm{TI}}/\Delta t)L\rho_{\mathrm{f}}C_{P,\mathrm{f}}}{K_{T,\mathrm{r}}} \tag{24}$$

where L is the length scale, $\rho_{\mathrm{f}}C_{P,\mathrm{f}}$ is the product of the density and heat capacity of the fluid, $K_{T,\mathrm{r}}$ is the thermal conductivity of the rock, and Δt is the total time of fluid flow. B estimates the relative importance of heat transfer by advection (numerator) and conduction (denominator). For example, a B of ~2.7 is obtained by using the q_{TI} for the higher grade parts of the Wepawaug Schist ($= 6\times10^4$ m^3 m^{-2}), a regional $L = 10$ km, $\Delta t = 10^7$yr, $K_{T,\mathrm{r}} = 2.5$ W m^{-1} K^{-1}, and $\rho_{\mathrm{f}}C_{P,\mathrm{f}} = 3.5\times10^6$ J m^{-3} K^{-1} (Ague, 1994b). $B > 1$ suggests a significant role for heat transport by fluid flow, and that at least some of the heat required for amphibolite facies metamorphism was supplied by ascending fluids channelized in the regional quartz vein network. Ferry (1992) came to a similar conclusion for rocks in northern New England.

In the Barrovian type locality, Scotland, Baxter *et al.* (2002) determined precise Sm/Nd ages for garnet growth which, when combined with the age data of Oliver *et al.* (2000), indicate that the difference in peak T attainment between the garnet and sillimanite zones was only 2.8 ± 3.7 Ma (2σ; statistically indistinguishable from zero). This short time span (~464–468 Ma) is inconsistent with the larger intervals predicted by conduction-only thermal relaxation of variably overthickened crust (e.g., Thompson and England, 1984), and strongly suggests the involvement of an additional, advective component of heat transfer. Baxter *et al.* (2002) concluded that considerable heat was supplied by syn-metamorphic magmas, although hydrothermal fluids, perhaps exsolved from the crystallizing intrusions, may have also played a role. It is worthwhile to note that Barrow (1893) himself first proposed that intrusions provided, at least in part, the heat required for metamorphism.

3.06.8.5 Fluids in the Granulite Facies

Phase relations and fluid inclusion evidence indicate greatly reduced water activity (a_{H_2O}) in the granulite facies (cf. Lamb and Valley, 1984; Crawford and Hollister, 1986; B. R. Frost and C. D. Frost, 1987; Newton, 1995; Aranovich and Newton, 1996, 1997, 1998, and numerous references cited therein). Concentrated aqueous solutions of strong electrolytes, including (Na,K)Cl, would have low a_{H_2O}, as would CO_2-rich fluids. If both chlorine and CO_2 contents are high, then it is likely that immiscible CO_2-rich and chlorine-rich fluids would coexist at high P and T (Duan *et al.*, 1995; Schmidt and Bodnar, 2000), consistent with fluid inclusion evidence for multi-phase granulite facies fluids (Touret, 1985; Crawford and Hollister, 1986). The flow of dense CO_2-rich fluid would be limited by the inability of CO_2 to wet grain boundaries effectively, unlike chlorine-rich brines (Watson and Brenan, 1987). Consequently, brines are probably more effective deep crustal transport agents, and have the potential to cause metasomatic effects, including alkali metasomatism and regional rubidium depletion, owing to their high chlorine contents and alkali exchange capacities (cf. Harlov *et al.*, 1997; Smit and Van Reenen, 1997). Nonetheless, direct expulsion of CO_2 from crystallizing intrusions could lower a_{H_2O} dramatically, promote granulite facies metamorphism and, in some cases, result in graphite precipitation (Farquhar and Chacko, 1991). A variety of mechanisms for generating low a_{H_2O} fluids have been proposed, including infiltration of connate brines or fluids equilibrated with meta-evaporites, loss of H_2O to anatectic melts leaving behind residual fluids enriched in salts and CO_2 (Fyfe, 1973; Philippot, 1993), release of brines and CO_2 from deep crustal intrusions (Hansen *et al.*, 1995), and loss of H_2O to retrograde rehydration reactions (Markl *et al.*, 1998).

3.06.9 CONCLUDING REMARKS

The concerted efforts of a diverse spectrum of Earth scientists have made it possible to estimate the amounts of fluid that flow through the crust during metamorphism, a goal that was unattainable just 10–20 yr ago. Figure 14 presents a selection of time-integrated flux estimates from the literature. Each of these has large uncertainties, perhaps an order of magnitude or more, yet a distinct pattern emerges. Regional devolatilization fluxes for zones of continental collision in which pervasive flow is dominant are less than $\sim 10^4\ m^3\ m^{-2}$, averaging around $500\ m^3\ m^{-2}$ (Figure 14). It is probably not coincidental that this flux magnitude corresponds to that expected for regional devolatilization of a metamorphic pile. Consider a crustal control volume having a uniform cross-sectional area of $1\ m^2$, length L_c, and constant fluid-filled porosity. The mass of fluid produced by devolatilization per unit initial volume of rock (f_{mv}) is

$$f_{mv} = f_m \rho_s (1 - \phi) \qquad (25)$$

in which ρ_s is the density of the solid (not including porosity) and f_m is the mass of fluid released per unit initial mass solid. If all the fluid produced exits by unidirectional flow out of the top of the column, and if small volume changes due to devolatilization reactions are ignored, then the time-integrated fluid flux across the top surface of the column is estimated by

$$q_{TI} = \frac{f_{mv} L_c}{\rho_f} \qquad (26)$$

where ρ_f is the density of the fluid exiting the column. For example, say a metasedimentary pile (or degassing magma) loses 3 wt.% water ($f_m \sim 0.03$; Ague, 1994b), and the densities of fluid and solid are $950\ kg\ m^{-3}$ and $2{,}800\ kg\ m^{-3}$, respectively. If we take a representative crustal-scale column 10 km long having $\varphi = 0.01$, then q_{TI} is $\sim 875\ m^3\ m^{-2}$. The flux estimate is larger for smaller porosity and vice versa, but is relatively insensitive to reasonable variations in porosity. This rough estimate for q_{TI} is clearly consistent with the range shown in Figure 14, and indicates that devolatilization of crustal columns of order 10^1 km long can easily provide the large

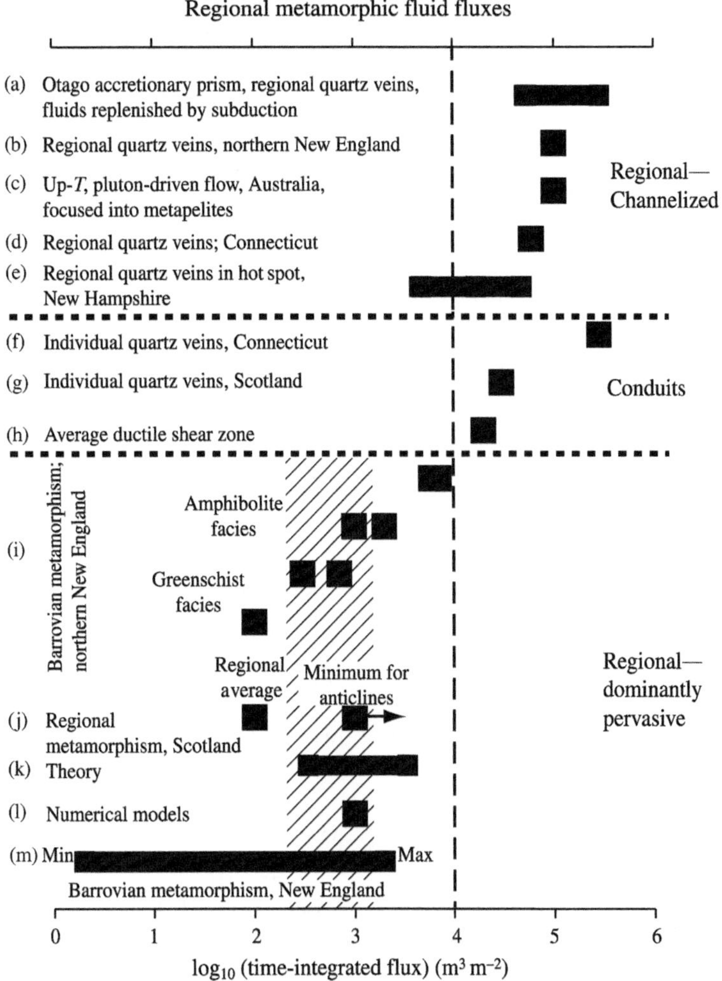

Figure 14 Selection of time-integrated fluid fluxes from the literature. The average regional, pervasive flow-dominated flux of $10^{2.7\pm0.5}$ $m^3_{(fluid)}$ $m^{-2}_{(rock)}$ (2σ) denoted with diagonal ruled bar (computed using geometric mean; cf. Ague, 1994a) (sources (a) Breeding and Ague (2002); (b) Ferry (1992); (c) Ague (1994b); (d) Oliver *et al.* (1998); (e) Chamberlain and Rumble (1989). Range for (e) computed using average flux of 1.5×10 m^3 m^{-2} s^{-1} for 10^5 yr and 10^6 yr. (f) Ague (1994b); (g) Ague (1997a); (h) Dipple and Ferry (1992a); (i) Ferry (1992), and Léger and Ferry (1993); (j) Skelton *et al.* (1995); (k) Walther and Orville (1982), and Walther (1990). Range for (k) computed using total timescales of fluid flow of 10^6 yr and 10^7 yr. (l) Hanson (1997); (m) Evans and Bickle (1999)).

amounts of fluid necessary for many regionally pervasive flow settings.

Structural features that focus flow including veins and ductile shear zones may transfer significantly greater fluxes of fluid, ranging from $\sim10^4$ m^3 m^{-2} to in excess of 10^5 m^3 m^{-2}. These fluxes are large enough to cause significant mass transfer of rock-forming elements. If the focused flow conduits deprive other areas of the rock mass of fluid, then regionally averaged fluxes could, of course, be less than $\sim10^4$ m^3 m^{-2}.

The maximum large-scale fluxes are associated with areas of long-term regional flow channelization into fracture sets, or with flow patterns set up around cooling intrusions. Fracture-controlled, regional outflow during Barrovian metamorphism, metamorphic "hot-spot" genesis, and devolatilization of subducted slabs beneath accretionary prisms can involve large-scale fluxes of $\sim10^4$ m^3 m^{-2} to $>10^5$ m^3 m^{-2}. Fluid flow toward cooling intrusions may also produce fluxes in this range, although examples to date are mostly from mid-crustal levels ($<\sim15$ km depth). Large-scale focusing of flow is one way to produce regional time-integrated fluxes well in excess of 10^4 m^3 m^{-2}. For example, areal focusing of devolatilization fluids derived from underlying crust by a factor of $\sim10-20$ could have produced the large fluxes recorded by the regional vein systems of the Wepawaug Schist (Ague, 1994b). Elevated permeability caused by the fracturing associated with vein formation served to focus the flow. Subduction can also generate large time-integrated fluid fluxes. Here, the supply of fluid is continually replenished as new oceanic crust and sediment is fed into the subduction zone and undergoes devolatilization during burial. As a consequence, accretionary prisms above long-lived subduction zones can be subjected to extreme fluxes (Breeding and Ague, 2002). In addition, large fluxes could potentially be achieved by convective recycling of fluids, or by the long-term circulation of fluids derived from near surface reservoirs down into the deep crust. These latter two possibilities seem unlikely for the deep crust, given current understanding, but future research may prove otherwise.

Progress as of early 2000s has been substantial, but many questions remain. For example, what are the directions and fluxes of fluid in deeply subducted crust and ultrahigh-pressure metamorphic rocks? (see also Chapter 3.09). What fraction of this deep fluid enters the mantle, and how does it influence the chemistry and isotopic systematics of mantle-derived melts? What is the role of melt–hydrous fluid miscibility in subduction zones? (see also Chapter 3.17). Time-integrated fluxes can be estimated, but over what timescales are the fluids evolved? Rapid CO_2 release, for instance, may perturb the climate system toward higher global average temperatures (Kerrick and Caldeira, 1998), whereas long-term sequestering of CO_2 may lead to cooling (Selverstone and Gutzler, 1993). A related question concerns the rates of metamorphic reactions. Are reactions "fast" enough to track accurately evolving P, T, t, and fluid composition during dynamic metamorphism? How deep do surficial waters penetrate into metamorphic belts? How do large fluid fluxes modify the chemical and isotopic composition of the crust?

ACKNOWLEDGMENTS

I would like to thank D. M. Rye, M. T. Brandon, B. J. Skinner, K. K. Turekian, J. Park, E. F. Baxter, C. M. Breeding, C. J. Carson, E. W. Bolton, A. Luttge, J. L. M. van Haren, and R. L. Masters for stimulating discussions and collaborations over the course of the past decade at Yale. B. A. Wing and R. L. Rudnick provided thoughtful and constructive reviews. Support from Department of Energy grant DE-FG02-01ER15216 and National Science Foundation grant EAR-0105927 is gratefully acknowledged.

REFERENCES

Ague J. J. (1994a) Mass transfer during Barrovian metamorphism of pelites, south-central Connecticut: I. Evidence for changes in composition and volume. *Am. J. Sci.* **294**, 1061–1134.

Ague J. J. (1994b) Mass transfer during Barrovian metamorphism of pelites, south-central Connecticut: II. Channelized fluid flow and the growth of staurolite and kyanite. *Am. J. Sci.* **294**, 1061–1134.

Ague J. J. (1995) Deep crustal growth of quartz, kyanite, and garnet into large aperture, fluid-filled fractures, north-eastern Connecticut, USA. *J. Metamorph. Geol.* **13**, 299–314.

Ague J. J. (1997a) Crustal mass transfer and index mineral growth in Barrow's garnet zone, northeast Scotland. *Geology* **25**, 73–76.

Ague J. J. (1997b) Compositional variations in metamorphosed sediments of the Littleton Formation, New Hampshire (Discussion). *Am. J. Sci.* **297**, 440–449.

Ague J. J. (1998) Simple models of coupled fluid infiltration and redox reactions in the crust. *Contrib. Mineral. Petrol.* **132**, 180–197.

Ague J. J. (2000a) Release of CO_2 from carbonate rocks during regional metamorphism of lithologically heterogeneous crust. *Geology* **28**, 1123–1126.

Ague J. J. (2000b) Aluminum mobility during metamorphism of carbonate rocks. *Geol. Soc. Am. Abstr. Progr.* **32**, A-296.

Ague J. J. (2002) Gradients in fluid composition across metacarbonate layers of the Wepawaug Schist, Connecticut, USA. *Contrib. Mineral. Petrol.* **143**, 38–55.

Ague J. J. (2003) Fluid infiltration and transport of major, minor, and trace elements during regional metamorphism of carbonate rocks, Wepawaug Schist, Connecticut, USA. *Am. J. Sci.* **303**(9).

Ague J. J. and Rye D. M. (1999) Simple models of CO_2 release from metacarbonates with implications for interpretation of directions and magnitudes of fluid flow in the deep crust. *J. Petrol.* **40**, 1443–1462.

Ague J. J. and van Haren J. L. M. (1996) Assessing metasomatic mass and volume changes using the bootstrap, with application to deep crustal hydrothermal alteration of marble. *Econ. Geol.* **91**, 1169–1182.

Ague J. J., Park J. J., and Rye D. M. (1998) Regional metamorphic dehydration and seismic hazard. *Geophys. Res. Lett.* **25**, 4221–4224.

Ague J. J., Baxter E. F., and Eckert J. O., Jr. (2001) High f_{O_2} during sillimanite zone metamorphism of part of the Barrovian type locality, Glen Clova, Scotland. *J. Petrol.* **42**, 1301–1320.

Anderson G. M. and Burnham C. W. (1965) The solubility of quartz in supercritical water. *Am. J. Sci.* **263**, 494–511.

Aranovich L. Y. and Newton R. C. (1996) H_2O activity in concentrated NaCl solutions at high pressures and temperatures measured by the brucite-periclase equilibrium. *Contrib. Mineral. Petrol.* **125**, 200–212.

Aranovich L. Y. and Newton R. C. (1997) H_2O activity in concentrated KCl and KCl–NaCl solutions at high temperatures and pressures measured by the brucite–periclase equilibrium. *Contrib. Mineral. Petrol.* **127**, 261–271.

Aranovich L. Y. and Newton R. C. (1998) Reversed determination of the reaction: Phlogopite + quartz = enstatite + potassium feltspar + H_2O in the ranges 750–875 °C and 2–12 kbar at low H_2O activity with concentrated KCl solutions. *Am. Mineral.* **127**, 261–271.

Aranovich L. Y. and Newton R. C. (1999) Experimental determination of CO_2–H_2O activity-composition relations at 600–1,000 °C and 6–14 kbar by reversed decarbonation and dehydration reactions. *Am. Mineral.* **84**, 1319–1332.

Ayers J. C. and Watson E. B. (1991) Solubility of apatite, monazite, zircon, and rutile in supercritical aqueous fluids with implications for subduction zone geochemistry. *Phil. Trans. Roy. Soc. London A* **335**, 365–375.

Baker A. J. (1990) Stable isotopic evidence for fluid–rock interactions in the Ivrea zone, Italy. *J. Petrol.* **31**, 243–260.

Balashov V. N. and Yardley B. W. D. (1998) Modeling metamorphic fluid flow with reaction–compaction–permeability feedbacks. *Am. J. Sci.* **298**, 441–470.

Barnicoat A. C. and Cartwright I. (1995) Focused fluid flow during subduction: oxygen isotope data from high-pressure ophiolites of the western Alps. *Earth Planet. Sci. Lett.* **132**, 53–61.

Barrow G. (1893) On an intrusion of muscovite–biotite gneiss in the south-eastern Highlands of Scotland, and its accompanying metamorphism. *Quart. J. Geol. Soc. London* **49**, 330–354.

Baumgartner L. P. and Ferry J. M. (1991) A model for coupled fluid-flow and mixed-volatile mineral reactions with applications to regional metamorphism. *Contrib. Mineral. Petrol.* **106**, 273–285.

Baumgartner L. P., Gerdes M. L., Person M. A., and Roselle G. T. (1997) Porosity and permeability of carbonate rocks during contact metamorphism. In *Fluid Flow and Transport in Rocks: Mechanisms and Effects* (eds. B. Jamtveit and B. W. D. Yardley). Chapman and Hall, London, pp. 83–98.

Baxter E. F. and DePaolo D. J. (2000) Field measurement of slow metamorphic reaction rates at temperatures of 500 ° to 600 °C. *Science* **288**, 1411–1414.

Baxter E. F. and DePaolo D. J. (2002a) Field measurement of high temperature bulk reaction rates: I. Theory and technique. *Am. J. Sci.* **302**, 442–464.

Baxter E. F. and DePaolo D. J. (2002b) Field measurement of high temperature bulk reaction rates: II. Interpretation of results from a field site near Simplon Pass, Switzerland. *Am. J. Sci.* **302**, 465–516.

Baxter E. F., Ague J. J., and DePaolo D. J. (2002) Prograde temperature–time evolution in the Barrovian type-locality constrained by Sm/Nd garnet ages from Glen Clova, Scotland. *J. Geol. Soc. London* **159**, 71–82.

Bear J. (1972) *Dynamics of Fluids in Porous Media*. Dover, New York.

Bebout G. E. and Barton M. D. (1989) Fluid flow and metasomatism in a subduction zone hydrothermal system: Catalina Schist terrane, California. *Geology* **17**, 976–980.

Bebout G. E. and Barton M. D. (1993) Metasomatism during subduction: products and possible path in the Catalina Schist, California. *Chem. Geol.* **108**, 61–91.

Bebout G. E. and Barton M. D. (2002) Tectonic and metasomatic mixing in a high-*T*, subduction-zone mélange—insights into the geochemical evolution of the slab–mantle interface. *Chem. Geol.* **187**, 79–106.

Berner R. A. (1980) *Early Diagenesis: A Theoretical Approach*. Princeton University Press, Princeton.

Bickle M. J. (1992) Transport mechanisms by fluid flow in metamorphic rocks: oxygen and strontium decoupling in the Trois Seigneurs Massif—a consequence of kinetic dispersion? *Am. J. Sci.* **292**, 289–316.

Bickle M. J. and Baker J. (1990) Advective-diffusive transport of isotopic fronts: an example from Naxos, Greece. *Earth Planet. Sci. Lett.* **97**, 78–93.

Bickle M. J. and McKenzie D. (1987) The transport of heat and matter by fluids during metamorphism. *Contrib. Mineral. Petrol.* **95**, 384–392.

Bickle M. J., Chapman H. J., Ferry J. M., Rumble D., III, and Fallick A. E. (1997) Fluid flow and diffusion in the Waterville Limestone, south-central Maine: constraints from strontium, oxygen, and carbon isotope profiles. *J. Petrol.* **38**, 1489–1512.

Blattner P. and Lassey K. R. (1989) Stable isotope exchange fronts, Damköhler numbers and fluid to rock ratios. *Chem. Geol.* **78**, 381–392.

Blencoe J. G., Seitz J. C., and Anovitz L. M. (1999) The CO_2–H_2O system: II. Calculated thermodynamic mixing properties for 400 °C, 0–400 MPa. *Geochim. Cosmochim. Acta* **63**, 2393–2408.

Bolton E. W., Lasaga A. C., and Rye D. M. (1999) Long-term flow/chemistry feedback in a porous medium with heterogeneous permeability: kinetic control of dissolution and precipitation. *Am. J. Sci.* **299**, 1–68.

Boucher D. F. and Alves G. E. (1959) Dimensionless numbers for fluid mechanics, heat transfer, mass transfer, and chemical reaction. *Chem. Eng. Prog.* **55**, 55–64.

Bowman J. R., Willett S. D., and Cook S. J. (1994) Oxygen isotopic transport and exchange during fluid flow: one-dimensional models and applications. *Am. J. Sci.* **294**, 1–55.

Brace W. F. (1980) Permeability of crystalline and argillaceous rocks. *Int. J. Rock Mech. Min. Sci. Geomech. Abstr.* **17**, 241–251.

Brace W. F. (1984) Permeability of crystalline rocks: new *in situ* measurements. *J. Geophys. Res.* **89**, 4327–4330.

Brady J. B. (1988) The role of volatiles in the thermal history of metamorphic terranes. *J. Petrol.* **29**, 1187–1213.

Breeding C. M. and Ague J. J. (2002) Slab-derived fluids and quartz-vein formation in an accretionary prism, Otago Schist, New Zealand. *Geology* **30**, 499–502.

Breeding C. M., Ague J. J., Bröcker M., and Bolton E. W. (2003) Blueschist preservation in a retrograded HP-LT terrane, Tinos, Greece: implications for fluid flow paths in subduction zones. *Geochem. Geophys. Geosys.* **4**, 9002, doi: 10.1029/2002GC000380, 11p.

Brimhall G. H. (1979) Lithologic determination of mass transfer mechanisms of multiple-stage porphyry copper mineralization at Butte, Montana: vein formation by hypogene leaching and enrichment of potassium silicate protore. *Econ. Geol.* **74**, 556–589.

Bröcker M. and Enders M. (2001) Unusual bulk-rock compositions in eclogite-facies rocks from Syros and Tinos (Cyclades, Greece): implications for U–Pb zircon geochronology. *Chem. Geol.* **175**, 581–603.

Bröcker M., Kreuzer H., Matthews A., and Okrusch M. (1993) $^{40}Ar/^{39}Ar$ and oxygen isotope studies of polymetamorphism from Tinos Island, Cycladic blueschist belt, Greece. *J. Metamorph. Geol.* **11**, 223–240.

Bureau H. and Keppler H. (1999) Complete miscibility between silicate melts and hydrous fluids in the upper mantle: experimental evidence and geochemical implications. *Earth Planet. Sci. Lett.* **165**, 187–196.

Caciagli N. and Manning C. E. (1999) Calcite solubility in water at high temperature and pressure: implications for the geochemistry of dense aqueous fluids. *Trans., AGU* **80**, F1157.

Carlson W. D. (1989) The significance of intergranular diffusion to the mechanisms and kinetics of porphyroblast crystallization. *Contrib. Mineral. Petrol.* **103**, 1–24.

Carson C. J., Ague J. J., Grove M., Coath C. D., and Harrison T. M. (2002) U–Pb isotopic behaviour of zircon during upper-amphibolite facies fluid infiltration in the Napier Complex, east Antarctica. *Earth Planet. Sci. Lett.* **199**, 287–310.

Carpenter R. A. (1974) Pyrrhotite isograd in southeastern Tennessee and southwestern North Carolina. *Geol. Soc. Am. Bull.* **85**, 451–456.

Cartwright I., Vry J., and Sandiford M. (1995) Changes in stable isotope ratios of metapelites and marbles during regional metamorphism, Mount Lofty Ranges, South Australia: implications for crustal scale fluid flow. *Contrib. Mineral. Petrol.* **120**, 292–310.

Catlos E. J. and Sorensen S. S. (2003) Phengite-based chronology of K- and Ba-rich fluid flow in two paelosubduction zones. *Science* **299**, 92–95.

Chamberlain C. P. and Conrad M. E. (1991) The relative permeabilities of quartzites and schists during active metamorphism at midcrustal levels. *Geophys. Res. Lett.* **18**, 959–962.

Chamberlain C. P. and Rumble D. (1988) Thermal anomalies in a regional metamorphic terrane: an isotopic study of the role of fluids. *J. Petrol.* **29**, 1215–1232.

Chamberlain C. P. and Rumble D. (1989) The influence of fluids on the thermal history of a metamorphic terrain: New Hampshire, USA. In *Evolution of Metamorphic Belts* Geological Society Special Publication 43, (eds. J. S. Daly, R. A. Cliff, and B. W. D. Yardley). Blackwell, Oxford, pp. 203–213.

Connolly J. A. D. (1997) Devolatilization-generated fluid pressure and deformation-propagated fluid flow during prograde regional metamorphism. *J. Geophys. Res.* **102**, 18149–18173.

Craw D. and Norris R. J. (1991) Metamorphogenic Au–W veins and regional tectonics: mineralisation throughout the uplift history of the Haast Schist, New Zealand. *NZ J. Geol. Geophys.* **34**, 373–383.

Crawford M. L. and Hollister L. S. (1986) Metamorphic fluids, the evidence from fluid inclusions. In *Fluid–Rock Interactions during Metamorphism* (eds. J. V. Walther and B. J. Wood). Springer, New York, pp. 1–35.

David C., Wong T., Zhu W., and Zhang J. (1994) Laboratory measurement of compaction-induced permeability change in porous rock: implications for the generation and maintenance of pore pressure excess in the crust. *Pure Appl. Geophys.* **143**, 425–456.

Davis T. L. and Namson J. S. (1994) A balanced cross-section of the 1994 Northridge earthquake, southern California. *Nature* **372**, 167–169.

DeGroot S. R. and Mazur P. (1969) *Non-equilibrium Thermodynamics*. North-Holland, Amsterdam.

DePaolo D. J. and Getty S. R. (1996) Models of isotope exchange in reactive fluid–rock systems: implications for geochronology in metamorphic rocks. *Geochim. Cosmochim. Acta* **60**, 3933–3947.

Dipple G. M. and Ferry J. M. (1992a) Metasomatism and fluid flow in ductile fault zones. *Contrib. Mineral. Petrol.* **112**, 149–164.

Dipple G. M. and Ferry J. M. (1992b) Fluid flow and stable isotopic alteration in rocks at elevated temperatures with applications to metamorphism. *Geochim. Cosmochim. Acta* **56**, 3539–3550.

Dixon J. E. and Ridley J. R. (1987) Syros. In *Chemical Transport in Metasomatic Processes* (ed. H. C. Helgeson). Reidel, Dordrecht, pp. 489–501.

Duan Z., Moller N., and Weare J. H. (1995) Equation of state for the $NaCl–H_2O–CO_2$ system: prediction of phase equilibria and volumetric properties. *Geochim. Cosmochim. Acta* **59**, 2869–2882.

Elias B. P. and Hajash A., Jr. (1992) Changes in quartz solubility and porosity due to effective stress: an experimental investigation of pressure solution. *Geology* **20**, 451–454.

England P. C. and Thompson A. B. (1984) Pressure–temperature–time paths of regional metamorphism: I. Heat transfer during the evolution of regions of thickened continental crust. *J. Petrol.* **25**, 894–928.

Ernst W. G. and Banno S. (1991) Neoblastic jadeitic pyroxene in Franciscan metagreywackes from Pacheco Pass, central Diablo Range, California, and implications for the inferred metamorphic *P–T* trajectory. *NZ J. Geol. Geophys.* **34**, 285–292.

Etheridge M. A. (1983) Differential stress magnitudes during regional deformation and metamorphism: upper bound imposed by tensile fracturing. *Geology* **11**, 213–234.

Etheridge M. A., Wall V. J., and Vernon R. H. (1983) The role of the fluid phase during regional metamorphism and deformation. *J. Metamorph. Geol.* **1**, 205–226.

Etheridge M. A., Wall V. J., Cox S. F., and Vernon R. H. (1984) High fluid pressure during regional metamorphism and deformation. *J. Geophys. Res.* **89**, 4344–4358.

Evans K. A. and Bickle M. J. (1999) Determination of time-integrated metamorphic fluid fluxes from the reaction progress of multivariant assemblages. *Contrib. Mineral. Petrol.* **134**, 277–293.

Farquhar J. and Chacko T. (1991) Isotopic evidence for involvement of CO_2-bearing magmas in granulite formation. *Nature* **354**, 60–63.

Feehan J. G. and Brandon M. T. (1999) Contribution of ductile flow to exhumation of low-temperature, high-pressure metamorphic rocks: San Juan-Cascade nappes, NW, Washington State. *J. Geophys. Res.* **104**, 10883–10902.

Ferry J. M. (1983) On the control of temperature, fluid composition, and reaction progress during metamorphism. *Am. J. Sci.* **283-A**, 201–232.

Ferry J. M. (1987) Metamorphic hydrology at 13-km depth and 400–500 °C. *Am. Mineral.* **72**, 39–58.

Ferry J. M. (1992) Regional metamorphism of the Waits River Formation, eastern Vermont: delineation of a new type of giant metamorphic hydrothermal system. *J. Petrol.* **33**, 45–94.

Ferry J. M. (1994a) A historical review of metamorphic fluid flow. *J. Geophys. Res.* **99**, 15487–15498.

Ferry J. M. (1994b) Overview of the petrologic record of fluid flow during regional metamorphism in northern New England. *Am. J. Sci.* **294**, 905–988.

Ferry J. M. (1996) Prograde and retrograde fluid flow during contact metamorphism of siliceous carbonate rocks from the Ballachulish aureole, Scotland. *Contrib. Mineral. Petrol.* **124**, 235–254.

Ferry J. M. and Baumgartner L. P. (1987) Thermodynamic models of molecular fluids at the elevated pressures and temperatures of crustal metamorphism. In *Thermodynamic Modeling of Geological Materials: Minerals, Fluids, and Melts*, Reviews in Mineralogy (eds. I. S. E. Carmichael and H. P. Eugster). Mineralogical Society of America, Washington, DC, vol. 17, pp. 323–365.

Ferry J. M. and Dipple G. M. (1991) Fluid flow, mineral reactions, and metasomatism. *Geology* **19**, 211–214.

Ferry J. M. and Gerdes M. L. (1998) Chemically reactive fluid flow during metamorphism. *Ann. Rev. Earth Planet. Sci.* **26**, 255–287.

Fisher D. M. and Brantley S. L. (1992) Models of quartz overgrowth and vein formation: deformation and episodic fluid flow in an ancient subduction zone. *J. Geophys. Res.* **97**, 20043–20061.

Fisher G. W. and Elliott D. (1973) Criteria for quasi-steady diffusion and local equilibrium in metamorphism. In *Geochemical Transport and Kinetics*, Carnegie Institution of Washington Publication 634 (eds. A. W. Hoffman, B. J. Giletti, H. S. Yoder, Jr. and R. A. Yund). Washington, DC, pp. 231–341.

Fletcher R. C. and Hofmann A. W. (1973) Simple models of diffusion and combined diffusion-infiltration metasomatism. In *Geochemical Transport and Kinetics*, Carnegie Institution of Washington Publication 634 (eds. A. W. Hoffman, B. J. Giletti, H. S. Yoder, Jr., and R. A. Yund). Washington, DC, pp. 243–259.

Freeze R. A. and Cherry J. A. (1979) *Groundwater*. Prentice-Hall, Englewood Cliffs, NJ.

French B. M. (1966) Some geological implications of equilibrium between graphite and a C–O–H gas phase at high temperatures and pressures. *Rev. Geophys.* **4**, 223–253.

Fried J. J. and Combarnous M. A. (1971) Dispersion in porous media. In *Advances in Hydroscience* (ed. V. T. Chow). Academic Press, New York, vol. 7, pp. 170–282.

Frost B. R. and Frost C. D. (1987) CO_2, melts and granulite metamorphism. *Nature* **327**, 503–506.

Fyfe W. S. (1973) The granulite facies, partial melting and the Archean crust. *Phil. Trans. Roy. Soc. London Ser. A* **273**, 457–461.

Fyfe W. S., Price N. J., and Thompson A. B. (1978) *Fluids in the Earth's Crust*. Elsevier, Amsterdam.

Ganor J. A., Matthews A., and Paldor N. (1989) Constraints on effective diffusivity during oxygen isotope exchange at a marble-schist contact, Sifnos (Cyclades), Greece. *Earth Planet. Sci. Lett.* **94**, 208–216.

Ganor J., Matthews A., Schliestedt M., and Garfunkel Z. (1996) Oxygen isotope heterogeneities of metamorphic rocks: an original tectonostratigraphic signature, or an imprint of exotic fluids? A case study of Sifnos and Tinos Islands (Greece). *Euro. J. Mineral.* **8**, 719–731.

Garven G. and Freeze R. A. (1984a) Theoretical analysis of the role of groundwater flow in the genesis of stratabound ore deposits: 1. Mathematical and numerical model. *Am. J. Sci.* **284**, 1085–1124.

Garven G. and Freeze R. A. (1984b) Theoretical analysis of the role of groundwater flow in the genesis of stratabound ore deposits: 2. Quantitative results. *Am. J. Sci.* **284**, 1125–1174.

Getty S. and Selverstone J. (1994) Stable isotope and trace element evidence for restricted fluid migration in 2 GPa eclogites. *J. Metamorph. Geol.* **12**, 747–760.

Gieré R. (1990) Hydrothermal mobility of Ti, Zr, and REE: examples from the Bergell and Adamello contact aureoles (Italy). *Terra Nova* **2**, 60–67.

Gieré R. (1993) Transport and deposition of REE in H_2S-rich fluids—evidence from accessory mineral assemblages. *Chem. Geol.* **110**, 251–268.

Gieré R. and Williams C. T. (1992) REE-bearing minerals in a Ti-rich vein from the Adamello contact aureole (Italy). *Contrib. Mineral. Petrol.* **112**, 83–100.

Giorgetti G., Tropper P., Essene E. J., and Peacor D. R. (2000) Characterization of non-equilibrium and equilibrium occurrences of paragonite/muscovite intergrowths in an eclogite from the Sesia-Lanzo Zone (Western Alps, Italy). *Contrib. Mineral. Petrol.* **138**, 326–336.

Graf D. L., Anderson D. E., and Woodhouse J. E. (1983) Ionic diffusion in naturally-occurring aqueous solutions: transition-state models that use either empirical expressions or statistically-derived relationships to predict mutual diffusion coefficients in the concentrated-solution regions of 8 binary systems. *Geochim. Cosmochim. Acta* **47**, 1985–1998.

Graham C. M., Valley J. W., Eiler J. M., and Wada H. (1998) Timescales and mechanisms of fluid infiltration in a marble: an ion microprobe study. *Contrib. Mineral. Petrol.* **132**, 371–389.

Grant J. A. (1986) The isocon diagram—a simple solution to Gresens' equation for metasomatic alteration. *Econ. Geol.* **81**, 1976–1982.

Grauch R. L. (1989) Rare earth elements in metamorphic rocks. In *Geochemistry and Mineralogy of Rare Earth Elements*, Reviews in Mineralogy 21 (eds. B. R. Lipin and G. A. Mckay). Mineralogical Society of America, pp. 147–167.

Greenwood H. J. (1975) Buffering of pore fluids by metamorphic reactions. *Am. J. Sci.* **275**, 573–593.

Gresens R. L. (1967) Composition-volume relations of metasomatism. *Chem. Geol.* **2**, 47–65.

Guenther R. B. and Lee J. W. (1988) *Partial Differential Equations of Mathematical Physics and Integral Equations*. Prentice Hall, Englewood Cliffs, NJ.

Haar C., Gallagher J. S., and Kell G. S. (1984) *NBS/NRC Steam Tables*. Hemisphere, Washington, DC.

Hansen E. C., Newton R. C., Janardhan A. S., and Lindenberg S. (1995) Differentiation of Late Archean crust in the eastern Dharwat Craton, South India. *J. Geol.* **103**, 629–651.

Hanson R. B. (1997) Hydrodynamics of regional metamorphism due to continental collision. *Econ. Geol.* **92**, 880–891.

Harlov D. E., Newton R. C., Hansen E. C., and Janardhan A. S. (1997) Oxide and sulfide minerals in highly oxidized, Rb-depleted, Archaean granulites of the Shevaroy Hills massif, South India: oxidation states and the role of metamorphic fluids. *J. Metamorph. Geol.* **15**, 701–717.

Hauzenberger C. A., Baumgartner L. P., and Pak T. M. (2001) Experimental study on the solubility of the "model"-pelite mineral assemblage albite + K-feldspar + andalusite + quartz in supercritical chloride-rich aqueous solutions at 0.2 GPa and 600 °C. *Geochim. Cosmochim. Acta* **65**, 4493–4507.

Hawkesworth C., Turner S., Peate D., McDermott F., and van Calsteren P. (1997) Elemental U and Th variations in island arc rocks: Implications for U-series isotopes. *Chem. Geol.* **139**, 207–221.

Heaman L. M., Creaser R. A., and Cookenboo H. O. (2002) Extreme enrichment of high field strength elements in Jericho eclogite xenoliths: a cryptic record of Paleoproterozoic subduction, partial melting, and metasomatism beneath the Slave craton, Canada. *Geology* **30**, 507–510.

Hewitt D. A. (1973) The metamorphism of micaceous limestones from south-central Connecticut. *Am. J. Sci.* **273-A**, 444–469.

Helgeson H. C. (1979) Mass transfer among minerals and hydrothermal solutions. In *Geochemistry of Hydrothermal Ore Deposits* (ed. H. L. Barnes). Wiley, New York, pp. 568–610.

Hiraga T., Nishikawa O., Nagase T., and Akizuki M. (2001) Morphology of intergranular pores and wetting angles in pelitic schists studied by transmission electron microscopy. *Contrib. Mineral. Petrol.* **141**, 613–622.

Hoisch T. D. (1991) The thermal effects of pervasive and channelized fluid flow in the deep crust. *J. Geol.* **99**, 69–80.

Holland T. J. B. and Powell R. (1991) A compensated Redlich-Kwong equation for volumes and fugacities of CO_2 and H_2O in the range 1 bar to 50 kbar and 100–1,600 °C. *Contrib. Mineral. Petrol.* **109**, 265–273.

Holland T. J. B. and Powell R. (1998) An internally consistent data set for phases of petrological interest. *J. Metamorph. Geol.* **16**, 309–343.

Hollister L. S. and Crawford M. L. (1981) Fluid inclusions: applications to petrology. *Mineral. Assoc. Canada Short Course Handbook* **6**, 153–58.

Hubbert M. K. and Willis D. G. (1957) Mechanics of hydraulic fracturing. *Trans. Am. Inst. Min., Metall. Pet. Eng.* **210**, 153–168.

Huenges E., Erzinger J., Kuck J., Engeser B., and Kessels W. (1997) The permeable curst: geohydraulic properties down to 9,101 m depth. *J. Geophys. Res.* **102**, 18255–18265.

Ingebritsen S. E. and Manning C. E. (1999) Geological implications of a permeability-depth curve for the continental curst. *Geology* **27**, 1107–1110.

Jacobs G. K. and Kerrick D. M. (1981) Methane: an equation of state with application to the ternary system $H_2O–CO_2–CH_4$. *Geochim. Cosmochim. Acta* **45**, 607–614.

Jia Y. and Kerrich R. (2000) Giant quartz vein systems in accretionary orogenic belts: the evidence for a metamorphic fluid origin from $\delta^{15}N$ and $\delta^{13}C$ studies. *Earth Planet. Sci. Lett.* **184**, 211–224.

Johnson M. C. and Plank T. (1999) Dehydration and melting experiments constrain the fate of subducted sediments. *Geochemistry, Geophysics, Geosystems* **1**, 1999GC000014, 29p.

Kennedy G. C. (1950) A portion of the system silica–water. *Econ. Geol.* **45**, 629–653.

Kent A. J. R., Peate D. W., Newman S., Stolper E. M., and Pearce J. A. (2002) Chlorine in submarine glasses from the Lau Basin: seawater contamination and constraints on the composition of slab-derived fluids. *Earth Planet. Sci. Lett.* **202**, 361–377.

Kerrick D. M. (1990) *The Al_2SiO_5 Polymorphs*. Reviews in Mineralogy, **22** Mineralogical Society of America.

Kerrick D. M. and Caldeira K. (1998) Metamorphic CO_2 degassing from orogenic belts. *Chem. Geol.* **145**, 213–232.

Kerrick D. M. and Jacobs G. K. (1981) A modified Redlich-Kwong equation for H_2O, CO_2, and $H_2O–CO_2$ mixtures at elevated temperatures and pressures. *Am. J. Sci.* **281**, 735–767.

Kirschner D. L., Sharp Z. D., and Teyssier C. (1993) Vein growth mechanisms and fluid sources revealed by oxygen isotope laser microprobe. *Geology* **21**, 85–88.

Koons P. O. and Craw D. (1991) Evolution of fluid driving forces and composition within collisional orogens. *Geophys. Res. Lett.* **18**, 935–938.

Lamb W. and Valley J. W. (1984) Metamorphism of reduced granulites in a low-CO_2 vapour-free environment. *Nature* **312**, 56–58.

Lasaga A. C. and Rye D. M. (1993) Fluid flow and chemical reaction kinetics in metamorphic systems. *Am. J. Sci.* **293**, 361–404.

Lassey K. R. and Blattner P. (1988) Kinetically controlled oxygen isotope exchange between fluid and rock in one-dimensional advective flow. *Geochim. Cosmochim. Acta* **52**, 2169–2175.

Léger A. and Ferry J. M. (1993) Fluid infiltration and regional metamorphism of the Waits River Formation, northeast Vermont, USA. *J. Metamorph. Geol.* **11**, 3–29.

Lee J. H. and Veronis G. (1989) Determining velocities and mixing coefficients from tracers. *J. Phys. Oceanogr.* **19**, 487–500.

Lees J. L. and Lindley G. T. (1994) Three dimensional attenuation tomography at Loma Prieta: inversion of t^* for Q. *J. Geophys. Res.* **99**, 6843–6863.

Lewis S., Holness M., and Graham C. (1998) Ion microprobe study of marble from Naxos Greece: grain-scale fluid pathways and stable isotope equilibration during metamorphism. *Geology* **26**, 935–938.

Liang Y., Richter F. M., and Watson E. B. (1996) Diffusion in silicate melts: II. Multicomponent diffusion in $CaO-Al_2O_3-SiO_2$ at 1,500 C and 1 GPa. *Geochim. Cosmochim. Acta* **60**, 5021–5035.

Lüttge A. and Metz P. (1991) Mechanism and kinetics of the reaction 1 dolomite + 2 quartz = 1 diopside + $2CO_2$ investigated by powder experiments. *Can. Mineral.* **29**, 803–821.

Mäder U. K. and Berman R. G. (1991) An equation of state for carbon dioxide to high pressure and temperature. *Am. Mineral.* **76**, 1547–1559.

Manning C. E. (1994) The solubility of quartz in the lower crust and upper mantle. *Geochim. Cosmochim. Acta* **58**, 4831–4839.

Manning C. E. (1997) Coupled reaction and flow in subduction zones: silica metasomatism in the mantle wedge. In *Fluid Flow and Transport in Rocks: Mechanisms and Effects* (eds. B. Jamtveit and W. D. Yardley). Chapman and Hall, London, pp. 139–148.

Manning C. E. (2001) Experimental studies of fluid–rock interaction at high pressure: the role of polymerization and depolymerization of solutes. *11th Annual V. M. Goldschimidt Conference*, Roanoke, Virginia, USA.

Manning C. E. and Ingebritsen S. E. (1999) Permeability of the continental crust: implications of geothermal data and metamorphic systems. *Rev. Geophys.* **37**, 127–150.

Markl G., Ferry J., and Bucher K. (1998) Formation of saline brines and salt in the lower crust by hydration reactions in partially retrogressed granulites from the Lofoten Islands, Norway. *Am. J. Sci.* **298**, 705–757.

Mohr D. W. and Newton R. C. (1983) Kyanite–staurolite metamorphism in sulfidic schists of the Anakeesta Formation, Great Smoky Mountains, North Carolina. *Am. J. Sci.* **283**, 97–134.

Nakamura M. and Watson E. B. (2001) Experimental study of aqueous fluid infiltration into quartzite: implications for the kinetics of fluid redistribution and grain growth driven by interfacial energy reduction. *Geofluids* **1**, 73–89.

Newton R. C. (1995) Simple-system mineral reactions and high-grade metamorphic fluids. *Euro. J. Mineral.* **7**, 861–881.

Newton R. C. and Manning C. E. (2000a) Quartz solubility in H_2O–NaCl and H_2O-CO_2 solutions at deep crust–upper mantle pressures and temperatures: 2–15 kbar and 500–900 °C. *Geochim. Cosmochim. Acta* **64**, 2993–3005.

Newton R. C. and Manning C. E. (2000b) Metasomatic phase relations in the system $CaO-MgO-SiO_2-H_2O$–NaCl at high temperature and pressures. *Int. Geol. Rev.* **42**, 152–162.

Norton D. and Dutrow B. L. (2001) Complex behavior of magma-hydrothermal processes: role of supercritical fluid. *Geochim. Cosmochim. Acta* **65**, 4009–4017.

Norton D. and Knapp R. (1977) Transport phenomena in hydrothermal systems: the nature of porosity. *Am. J. Sci.* **277**, 913–936.

Norton D. and Knight J. (1977) Transport phenomena in hydrothermal systems: cooling plutons. *Am. J. Sci.* **277**, 937–981.

Nur A. and Walder J. (1990) Time-dependent hydraulics of the Earth's crust. In *The Role of Fluids in Crustal Processes* (Geophysical Study Committee, Commission on Geosciences, Environment, and Resources, National Research Council). National Academy Press, Washington, DC, pp. 113–127.

Oelkers E. H. and Helgeson H. C. (1988) Calculation of the thermodynamic and transport properties of aqueous species at high pressures and temperatures: aqueous tracer diffusion coefficients of ions to 1,000 °C and 5 kb. *Geochim. Cosmochim. Acta.* **52**, 63–85.

Ogata A. (1964) Mathematics of dispersion with linear absorption isotherm. *US Geol. Surv. Prof. Pap.* **441-H**.

O'Hara K. D. (1988) Fluid flow and volume loss during mylonitization: an origin for phyllonite in an overthrust setting. *Tectonophysics* **156**, 21–34.

Oliver G. J. H., Chen F., Buchwaldt R., and Hegner E. (2000) Fast tectonometamorphism and exhumation in the type area of the Barrovian and Buchan zones. *Geology* **28**, 459–462.

Oliver N. H. S. (1996) Review and classification of structural controls on fluid flow during regional metamorphism. *J. Metamorph. Geol.* **14**, 477–492.

Oliver N. H. S., Valenta R. K., and Wall V. J. (1990) The effect of heterogeneous stress and strain on metamorphic fluid flow, Mary Kathleen, Australia, and a model for large scale fluid circulation. *J. Metamorph. Geol.* **8**, 311–332.

Oliver N. H. S., Dipple G. M., Cartwright I., and Schiller J. (1998) Fluid flow and metasomatism in the genesis of the amphibolite-facies, pelite-hosted Kanmantoo copper deposit, South Australia. *Am. J. Sci.* **298**, 181–218.

Orville P. M. (1962) Alkali metasomatism and the feldspars. *Norsk Geologisk Tidsskrift* **42**, 283–316.

Oxtoby D. W., Gillis H. P., and Nachtrieb N. H. (1999) *Principles of Modern Chemistry* 4th edn. Saunders College, Fort Worth.

Palin J. M. (1992) Stable isotope studies of regional metamorphism in the Wepawaug Schist, Connecticut. PhD thesis, Yale University.

Pan Y. and Fleet M. E. (1996) Rare earth element mobility during prograde granulite facies metamorphism: significance of fluorine. *Mineral. Petrol.* **123**, 251–262.

Passchier C. W. and Trouw R. A. J. (1996) *Microtectonics.* Springer, Berlin.

Peacock S. M. (1983) Numerical constraints on rates of metamorphism, fluid production, and fluid flux during regional metamorphism. *Geol. Soc. Am. Bull.* **101**, 476–485.

Peacock S. M. (1990) Fluid processes in subduction zones. *Science* **248**, 329–337.

Person M. and Baumgartner L. (1995) New evidence for long-distance fluid migration within the Earth's crust. *Rev. Geophys. Part 2, Suppl. S* **33**, 1083–1091.

Philippot P. (1993) Fluid–melt–rock interaction in mafic eclogites and coesite-bearing metasediments: constraints on volatile recycling during subduction. *Chem. Geol.* **108**, 93–112.

Philippot P. and Selverstone J. (1991) Trace-element-rich brines in eclogitic veins: implications for fluid composition and transport during subduction. *Contrib. Mineral. Petrol.* **106**, 417–430.

Putlitz B., Matthews A., and Valley J. W. (2000) Oxygen and hydrogen isotope study of high-pressure metagabbros and metabasalts (Cyclades, Greece): implications for the subduction of oceanic crust. *Contrib. Mineral. Petrol.* **138**, 114–126.

Poage M. C., Chamberlain C. P., and Craw D. (2000) Massif-wide metamorphism and fluid evolution at Nanga Parbat, northern Pakistan. *Am. J. Sci.* **300**, 463–482.

Pokrovskii V. A. and Helgeson H. C. (1995) Thermodynamic properties of aqueous species and the solubilities of minerals at high pressures and temperatures: the system $Al_2O_3-H_2O$–NaCl. *Am. J. Sci.* **295**, 1255–1342.

Ramsay J. (1980) The crack-seal mechanism of rock deformation. *Nature* **284**, 135–139.

Rich R. A. (1979) Fluid inclusion evidence for Silurian evaporites in southeastern Vermont. *Geol. Soc. Am. Bull.* **90**, 1628–1643.

Ridley J. and Thompson A. B. (1986) The role of mineral kinetics in the development of metamorphic microtextures. In *Fluid–Rock Interactions during Metamorphism* (eds. J. V. Walther and B. J. Wood). Springer, New York, pp. 154–193.

Ring U. (1999) Volume loss, fluid flow, and coaxial versus noncoaxial deformation in retrograde, amphibolite facies shear zones, northern Malawi, east-central Africa. *Geol. Soc. Am. Bull.* **111**, 123–142.

Roedder E. (1984) Fluid inclusions. *MSA Rev.* **12**, 644.

Rubie D. C. (1998) Disequilibrium during metamorphism: the role of nucleation kinetics. In *What Drives Metamorphism and Metamorphic Reactions?* Special Publications, 138 (eds. P. J. Treloar and P. J. O'Brien). Geological Society, London, pp. 199–214.

Rudnick R. L. and Fountain D. M. (1995) Nature and composition of the continental crust: a lower crustal perspective. *Rev. Geophys.* **33**, 267–309.

Rumble D., III (1989) Evidences for fluid flow during regional metamorphism. *Euro. J. Mineral.* **1**, 731–737.

Rumble D., III and Spear F. S. (1983) Oxygen isotope equilibration and permeability enhancement during regional metamorphism. *J. Geol. Soc. London* **140**, 619–628.

Rumble D., III, Oliver N. H. S., Ferry J. M., and Hoering T. C. (1991) Carbon and oxygen isotope geochemistry of chlorite-zone rocks of the Waterville Limestone, Maine, USA. *Am. Mineral.* **76**, 857–866.

Rutter E. H. (1995) Experimental study of the influence of stress, temperature, and strain on the dynamic recrystallization of Carrara marble. *J. Geophys. Res.* **100**, 24651–24663.

Rye D. M. and Bradbury H. J. (1988) Fluid flow in the crust: an example from a Pyrenean thrust ramp. *Am. J. Sci.* **288**, 197–235.

Rye D. M. and Rye R. O. (1974) Homestake gold mine, South Dakota: I. Stable isotope studies. *Econ. Geol.* **69**, 293–317.

Rye R. O., Schuiling R. D., Rye D. M., and Jansen J. B. H. (1976) Carbon, hydrogen and oxygen isotope studies of the regional metamorphic complex at Naxos, Greece. *Geochim. Cosmochim. Acta* **40**, 1031–1049.

Sanchez-Navas A. (1999) Sequential kinetics of a muscovite-out reaction: a natural example. *Am. Mineral.* **84**, 1270–1286.

Schmidt C. and Bodnar R. J. (2000) Synthetic fluid inclusions: XVI. PVTX properties in the system H_2O–NaCl–CO_2 at elevated temperatures, pressures, and salinities. *Geochim. Cosmochim. Acta* **64**, 3853–3869.

Scholz C. H. (1990) *The Mechanics of Earthquakes and Faulting*. Cambridge University Press, Cambridge.

Schramke J. A., Kerrick D. M., and Lasaga A. C. (1987) The reaction muscovite + quartz = andalusite + K-feldspar + water: Part I. Growth kinetics and mechanism. *Am. J. Sci.* **287**, 517–559.

Selverstone J. and Gutzler D. S. (1993) Post-125 Ma carbon storage associated with continent–continent collision. *Geology* **21**, 885–888.

Selverstone J., Morteani G., and Staude J. M. (1991) Fluid channeling during ductile shearing: transformation of granodiorite into aluminous schist in the Tauern Window, eastern Alps. *J. Metamorph. Geol.* **9**, 419–431.

Sharp Z. D., Essene E. J., and Hunziker J. C. (1993) Stable isotope geochemistry and phase equilibria of coesite-bearing whiteschists, Dora Maira Massif, western Alps. *Contrib. Mineral. Petrol.* **114**, 1–12.

Shewmon P. G. (1969) *Transformations in Metals*. McGraw-Hill, New York.

Shi P. and Saxena S. K. (1992) Thermodynamic modeling of the C–H–O–S fluid system. *Am. Mineral.* **77**, 1038–1049.

Shmulovich K., Graham C., and Yardley B. W. D. (2001) Quartz, albite, and diopside solubilities in H_2O–CO_2 fluids at 0.5–0.9 GPa. *Contrib. Mineral. Petrol.* **141**, 95–108.

Shock E. L., Sassani D. C., Willis M., and Sverjensky D. A. (1997) Inorganic species in geologic fluids: correlations among standard molal thermodynamic properties of aqueous ions and hydroxide complexes. *Geochim. Cosmochim. Acta* **61**, 907–950.

Sibson R. H. (1983) Continental fault structure and the shallow earthquake source. *J. Geol. Soc. London* **140**, 741–767.

Sibson R. H. (1992) Implications of fault valve behavior for rupture nucleation and recurrence. *Tecnophysics* **211**, 283–293.

Sibson R. H., McMoore J., and Rankin R. H. (1975) Seismic pumping—a hydrothermal fluid transport mechanism. *J. Geol. Soc. London* **131**, 653–659.

Skelton A. D. L., Graham C. M., and Bickle M. J. (1995) Lithological and structural controls on regional 3-D fluid flow patterns during greenschist facies metamorphism of the Dalradian of the SW Scottish Highlands. *J. Petrol.* **36**, 563–586.

Skippen G. B. and Marshall D. D. (1991) The metamorphism of granulites and devolatilization of the lithosphere. *Can. Mineral.* **29**, 693–705.

Smit C. A. and Van Reenen D. D. (1997) Deep crustal shear zones, high-grade tectonites, and associated metasomatic alteration in the Limpopo Belt, South Africa: Implications for deep crustal processes. *J. Geol.* **105**, 37–58.

Smith M. P. and Yardley B. W. D. (1999) Fluid evolution during metamorphism of the Otago Schist, New Zealand: (I) Evidence from fluid inclusions. *J. Metamorph. Geol.* **17**, 173–186.

Sorensen S. S. and Grossman J. N. (1989) Enrichment of trace elements in garnet-amphibolites from a paleo-subduction zone: Catalina Schist, Southern California. *Geochim. Cosmochim. Acta* **53**, 3155–3177.

Steefel C. I. and Lasaga A. C. (1994) A coupled model for transport of multiple chemical species and kinetic precipitation/dissolution reactions with application to reactive flow in single phase hydrothermal systems. *Am. J. Sci.* **249**, 529–592.

Steefel C. I. and van Cappellen P. (1990) A new kinetic approach to modeling water–rock interaction: the role of nucleation, precursors, and Ostwald ripening. *Geochim. Cosmochim. Acta* **54**, 2657–2677.

Sterner S. M. and Pitzer K. S. (1994) An equation of state for carbon dioxide valid from zero to extreme pressures. *Contrib. Mineral. Petrol.* **117**, 362–374.

Steyrer H. P. and Sturm R. (2002) Stability of zircon in a low-grade ultramylonite and its utility for chemical mass balancing: The shear zone at Miéville, Switzerland. *Chem. Geol.* **187**, 1–19.

Stolper E. and Newman S. (1994) The role of water in the petrogenesis of Mariana trough magmas. *Earth Planet. Sci. Lett.* **121**, 293–325.

Sverjensky D. A. (1987) Calculation of the thermodynamic properties of aqueous species and the solubilities of minerals in supercritical electrolyte solutions. In *Thermodynamic Modeling of Geological Materials: Minerals, Fluids, and Melts*, Reviews in Mineralogy (eds. I. S. E. Carmichael and H. P. Eugster). Mineralogical Society of America, Washington, DC, vol. 17, pp. 177–209.

Sverjensky D. A., Shock E. L., and Helgeson H. C. (1997) Prediction of the thermodynamic properties of aqueous metal complexes to 1,000 °C and 5 kb. *Geochim. Cosmochim. Acta* **61**, 1359–1412.

Templeton A. S., Chamberlain C. P., Koons P. O., and Craw D. (1998) Stable isotopic evidence for mixing between metamorphic fluids and surface-derived waters during recent uplift of the southern Alps, New Zealand. *Earth Planet. Sci. Lett.* **154**, 73–92.

Thomas R., Webster J. D., and Heinrich W. (2000) Melt inclusions in pegmatite quartz: complete miscibility between silicate melts and hydrous fluids at low pressure. *Contrib. Mineral. Petrol.* **139**, 394–401.

Thompson A. B. (1983) Fluid absent metamorphism. *J. Geol. Soc. London* **140**, 533–547.
Thompson A. B. (1997) Flow and focusing of metamorphic fluids. In *Fluid Flow and Transport in Rocks: Mechanisms and Effects* (eds. B. Jamtveit and B. W. D. Yardley). Chapman and Hall, London, pp. 297–314.
Thompson A. B. and England P. C. (1984) Pressure–temperature–time paths of regional metamorphism: II. Their inference and interpretation using mineral assemblages in metamorphic rocks. *J. Petrol.* **25**, 929–955.
Thompson J. B., Jr. (1987) A simple thermodynamic model for grain interfaces: some insights on nucleation, rock textures, and metamorphic differentiation. In *Chemical Transport in Metasomatic Processes* (ed. H. C. Helgeson). Reidel, Dordrecht, pp. 169–188.
Touret J. L. R. (1985) Fluid regime in Southern Norway: the record of fluid inclusions. In *The Deep Proterozoic Crust in the North Atlantic Provinces* (eds. A. C. Tobi and J. R. L. Touret). Reidel, Dordrecht, pp. 517–549.
Tracy R. J., Rye D. M., Hewitt D. A., and Schiffries C. M. (1983) Petrologic and stable-isotopic studies of fluid–rock interactions, south-central Connecticut: I. The role of infiltration in producing reaction assemblages in impure marbles. *Am. J. Sci.* **283A**, 589–616.
van Haren J. L. M., Ague J. J., and Rye D. M. (1996) Oxygen isotope record of fluid infiltration and mass transfer during regional metamorphism of pelitic schist, Connecticut, USA. *Geochim. Cosmochim. Acta* **60**, 3487–3504.
Vidale R. (1969) Metasomatism in a chemical gradient and the formation of calc-silicate bands. *Am. J. Sci.* **267**, 857–874.
Vidale R. and Hewitt D. A. (1973) "Mobile" components in the formation of calc-silicate bands. *Am. Mineral.* **58**, 991–997.
Vrolijk P. (1987) Tectonically driven fluid flow in the Kodiak accretionary complex, Alaska. *Geology* **15**, 466–469.
Walder J. and Nur A. (1984) Porosity reduction and crustal pore pressure development. *J. Geophys. Res.* **89**, 11539–11548.
Wallmann K. (2001a) The geological water cycle and Cenozoic evolution of marine $\delta^{18}O$ values. *Geochim. Cosmochim. Acta* **65**, 2469–2485.
Wallmann K. (2001b) Controls on the Cretaceous and Cenozoic evolution of seawater composition, atmospheric CO_2 and climate. *Geochim. Cosmochim. Acta* **65**, 3005–3025.
Walther J. V. (1990) Fluid dynamics during progressive regional metamorphism. In *The Role of Fluids in Crustal Processes* (Geophysics Study Committee, Commission on Geosciences, Environment, and Resources, National Research Council). National Academy Press, Washington, DC, pp. 64–71.
Walther J. V. (1996) Fluid production and isograd reactions at contacts of carbonate-rich and carbonate-poor layers during progressive metamorphism. *J. Metamorph. Geol.* **14**, 351–360.
Walther J. V. and Orville P. M. (1982) Volatile production and transport in regional metamorphism. *Contrib. Mineral. Petrol.* **79**, 252–257.
Walther J. V. and Orville P. M. (1983) The extraction-quench technique for determination of the thermodynamic properties of solute complexes. Application to quartz solubility in fluid mixtures. *Am. Mineral.* **68**, 731–741.
Watson E. B. and Brenan J. M. (1987) Fluids in the lithosphere: 1. Experimentally determined wetting characteristics of CO_2–H_2O fluids and their implications for fluid transport, host-rock physical properties, and fluid inclusion formation. *Earth Planet. Sci. Lett.* **85**, 497–515.
Wedepohl K. H. (1995) The composition of the continental crust. *Geochim. Cosmochim. Acta* **59**, 1217–1232.
Weill D. F. and Fyfe W. S. (1964) The solubility of quartz in H_2O in the range 1,000–4,000 bars and 400–550 °C. *Geochim. Cosmochim. Acta* **28**, 1243–1255.
White F. M. (1979) *Fluid Mechanics*. McGraw-Hill, New York.
Wickham S. M. and Taylor H. P., Jr. (1990) Hydrothermal systems associated with regional metamorphism and crustal anatexis: examples from the Pyrenees, France. In *The Role of Fluids in Crustal Processes* (Geophysical Study Committee, Commission on Geosciences, Environment and Resources, National Research Council). National Academy Press, Washington, DC, pp. 96–112.
Widmer T. and Thompson A. B. (2001) Local origin of high pressure vein material in eclogite facies rocks of the Zermatt-Saas zone, Switzerland. *Am. J. Sci.* **301**, 627–656.
Wing B. A. and Ferry J. M. (2002) Three-dimensional geometry of metamorphic fluid flow during Barrovian regional metamorphism from an inversion of combined petrologic and stable isotopic data. *Geology* **30**, 639–643.
Wong T. and Zhu W. (1999) Brittle faulting and permeability evolution: hydromechanical measurement, microstructural observation, and network modeling. In *Faults and Subsurface Fluid Flow in the Shallow Crust*, Geophysical Monograph 113 (eds. W. C. Haneberg, P. S. Mozley, J. C. Moore, and L. B. Goodwin). American Geophysical Union, Washington, DC, pp. 83–99.
Wong T., Ko S., and Olgaard D. L. (1997) Generation and maintenance of pore pressure excess in a dehydrating system 2. Theoretical analysis. *J. Geophys. Res.* **102**, 841–852.
Xie Z. and Walther J. V. (1993) Quartz solubilities in NaCl solutions with and without wollastonite at elevated temperatures and pressures. *Geochim. Cosmochim. Acta* **57**, 1947–1955.
Yardley B. W. D. (1975) On some quartz-plagioclase veins in the Connemara Schists, Ireland. *Geol. Mag.* **112**, 183–190.
Yardley B. W. D. (1986) Fluid migration and veining in the Connemara Schists, Ireland. In *Fluid–Rock Interactions during Metamorphism* (eds. J. V. Walther and B. J. Wood). Springer, New York, pp. 109–131.
Yardley B. W. D. (1997) The evolution of fluids through the metamorphic cycle. In *Fluid Flow and Transport in Rocks* (eds. B. Jamtveit and B. W. D. Yardley). Chapman and Hall, London, pp. 139–147.
Yardley B. W. D. and Bottrell S. H. (1992) Silica mobility and fluid movement during metamorphism of the Connemara schists, Ireland. *J. Metamorph. Geol.* **10**, 453–464.
Yeats R. S., Sieh K., and Allen C. R. (1997) *The Geology of Earthquakes*, Oxford, New York.
Young E. D. (1995) Fluid flow in metamorphic environments. *Rev. Geophys. Part 1, Suppl. S* **33**, 41–52.
Young E. D. and Rumble D., III (1993) The origin of correlated variations in *in-situ* $^{18}O/^{16}O$ and elemental concentrations in metamorphic garnet from southeastern Vermont, USA. *Geochim. Cosmochim. Acta* **57**, 2585–2597.
Zhang S. Q., FitzGerald J. D., and Cox S. F. (2000) Reaction-enhanced permeability during decarbonation of calcite + quartz → wollastonite + carbon dioxide. *Geology* **28**, 911–914.
Zhang C. F., Tullis T. J., and Scruggs V. J. (2001) Implications of permeability and its anisotropy in a mica gouge for pore pressures in fault zones. *Tectonophysics* **335**, 37–50.
Zhao D. H., Kanamori H., Negishi H., and Weins D. (1996) Tomography of the source area of the 1995 Kobe earthquake: evidence for fluids at the hypocenter? *Science* **274**, 1891–1894.
Zimmermann J. L. and Poty B. (1970) Etude par spectrométrie de masse de la composition des fluids dans les cavités alpines du massif du Mont Blanc. *Bulletin suisse de Minéralogie et Pétrographie* **50**, 99–108.

3.07
Geochemical Zoning in Metamorphic Minerals

M. J. Kohn
University of South Carolina, Columbia, SC, USA

3.07.1 INTRODUCTION

Rock's encode the sum of Earth processes that affected them during their "lifetimes," and the purpose of most geological studies is to invert that information to refine our understanding of those processes. A metamorphic rock records not just a peak $P-T$ condition, a single cooling rate, or a simple texture, but rather has undergone an evolving history of changes in P and T, mineral abundances, rim compositions, and textures, acting over its metamorphic lifespan, in response to heat flow, stress and strain, and inter- and intragranular movement of material. The greatest advances in understanding metamorphic rocks have been achieved through a recognition that metamorphism is a continuum, and by collection of data and development of models that directly address these continuum processes. Of the many approaches for investigating and interpreting metamorphic rocks, one of the most important is the characterization and quantitative modeling of geochemical zoning in metamorphic minerals. Geochemical zoning is particularly useful, because it is a quasicontinuous record of these metamorphic processes.

There have been several previous reviews of chemical zoning in metamorphic minerals (e.g., Tracy, 1982; Loomis, 1983; Chakraborty and Ganguly, 1991; Spear, 1993). This review differs from them in three ways.

(i) A catalogue was not compiled of all the minerals that exhibit zonation, and of the elements that are zoned. Although some minerals may be more obviously zoned than others, all minerals must be zoned in some element or isotope at some scale, and it is simply a matter of time before that zonation is described.

(ii) Emphasis is placed on theoretical models of zoning, and these are illustrated with one or two of the best natural examples. Some other examples of zonation are also discussed which are particularly relevant to other endeavors, or hold special promise for future research.

(iii) A large body of information, mostly collected since the early 1990s, is presented for trace element, stable isotope, and radiogenic isotope zoning.

Garnet (Grt) is overwhelmingly favored for geochemical zoning studies, in part because it is commonly zoned in major and trace elements as well as in stable and radiogenic isotopes, but also because its chemistry, geochemical partitioning behavior, and physical shape are readily modeled theoretically. Other minerals are also zoned, but most are not as amenable to study or modeling. For these reasons, garnet is a major focus of this review. Emphasis is placed on theory and principles rather than on documentation, so extrapolation to other minerals and other chemical systems should be possible. The information is organized into four sections: major elements, stable isotopes, trace elements, and radiogenic isotopes. Most theory has been developed for the distribution of the major elements, but studies of the other types of zoning are becoming increasingly common, and deserve separate subsections in this review.

This chapter does not cover analytical techniques, and assumes that the reader is at least moderately familiar with the electron microprobe, ion microprobe, and laser ablation–inductively coupled plasma mass spectrometry (LA–ICPMS), as well as with the distinction between X-ray mapping and spot analysis by electron microprobe. Mineral abbreviations are after Kretz (1983).

3.07.2 MAJOR ELEMENTS

3.07.2.1 Rayleigh Distillation

Rayleigh distillation describes the partitioning of a chemical species (element, molecular species, isotope, etc.) between two reservoirs during a distillation process, i.e., the preferential transfer of material from one reservoir to another (Rayleigh, 1896). If a geochemical species is especially compatible within a mineral, then as the mineral grows, the concentration of that species in the matrix and toward the rim of the mineral will decrease during metamorphism. The rock matrix and the growing crystals are the reservoirs from which and into which the species is being transferred. The rock matrix is compositionally homogeneous and if the mineral undergoes perfect fractional crystallization, so that it preserves a perfect compositional record of its growth, its compositions will conform to Equation (1) (Hoefs, 1997):

$$C_i^{\mathrm{M}} = \alpha_i C_{i,\mathrm{o}}^{\mathrm{R}} f^{\alpha_i - 1} \tag{1}$$

where C_i^{M} is the concentration of the species i in the mineral, α_i is the fractionation factor (partition coefficient) for species i in the mineral relative to the rock, $C_{i,\mathrm{o}}^{\mathrm{R}}$ is the initial concentration in the rock matrix, and f is the fraction of the species i remaining in the rock. For $\alpha_i > 1$ (i.e., the species partitions into the mineral), the instantaneous concentration of the species in the mineral will exponentially decrease, asymptotically approaching 0. A linear traverse across such a grain should be characteristically "bell shaped." For $\alpha_i < 1$ (i.e., the species partitions into the matrix), the instantaneous concentration will increase, because mineral growth depletes the matrix in other components, increasing the concentration of the less-preferred species.

Hollister (1966, 1969) first proposed that manganese zoning in garnets could be understood in terms of Rayleigh distillation. He recast Equation (1) in terms of weight percent MnO in the garnet ($C_{\mathrm{Mn}}^{\mathrm{Grt}}$) and in the manganese-bearing minerals of the rock ($C_{\mathrm{Mn,o}}^{\mathrm{R}}$), and the weight fraction of garnet (W_{Grt}) relative to the other manganese-bearing minerals in the rock (W_{o}):

$$C_{\mathrm{Mn}}^{\mathrm{Grt}} = \alpha_{\mathrm{Mn}} C_{\mathrm{Mn,o}}^{\mathrm{R}} \left(1 - \frac{W_{\mathrm{Grt}}}{W_{\mathrm{o}}}\right)^{\alpha_{\mathrm{Mn}} - 1} \tag{2}$$

He then showed that the manganese zoning profiles in garnets from Kwoiek, British Columbia could be fit with Rayleigh distillation models for α_{Mn} values in the range 15–40 (Figure 1). Although manganese in garnet is the type example of Rayleigh distillation during metamorphism, other minerals that strongly partition an element can also show qualitatively similar chemical depletions from cores to rims (Figure 2).

Hollister's work was critical to later interpretations, because Rayleigh distillation implies perfect preservation of compositions, i.e., that compositions were not modified during or after mineral growth by processes such as diffusion, dissolution–reprecipitation, and recrystallization. Ever since Hollister's classic paper, "bell-shaped" Mn-profiles in garnets have been cited as *prima facie* evidence for the preservation of growth zonation. Although bell-shaped profiles are expected for Rayleigh distillation, they are not proof positive that a mineral has faithfully retained its original composition, because compositions may change during or after growth, and still exhibit a general trend of decreasing manganese from core to rim. Furthermore, in Rayleigh distillation *sensu stricto* the value of α is constant. Partitioning of elements depends on temperature (T), and because changes in T will drive mineral growth, T and α are likely to change as the mineral grows. In fact, the best fit to the Kwoiek profiles has a continuously decreasing value of α_{Mn} from core to rim, which Hollister (1969) ascribed to an increase in T as the garnets grew. Compositional patterns in minerals (e.g., Figures 1 and 2) are usually the sum of several contributing processes; their formation is therefore better termed "fractional crystallization" than "Rayleigh distillation."

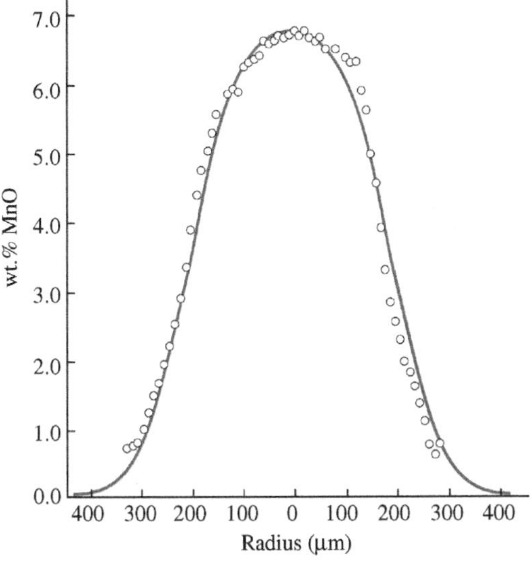

Figure 1 Plot of wt.% MnO versus radial distance in garnets from Kwoiek, British Columbia (Hollister, 1966) (reproduced by permission of American Association for the Advancement of Science from *Science* **1966**, *154*, 1647–1651). MnO can be fitted with a Rayleigh distillation model, with $\alpha_{Mn} \sim 23$ (O, analyzed points) (source Spear, 1993).

3.07.2.2 Equilibrium Partitioning and Growth Zoning Models

The matrix of a metamorphic rock is not well stirred and homogeneous. It is better described as an assembly of chemically distinct minerals that simultaneously change mode and composition during metamorphism. Furthermore, if growth is driven by a change in pressure (P) or T, element partitioning changes as the mineral grows. Modeling these effects requires assessing the changes in abundance and composition of each mineral in a rock, as well as any P or T changes attending mineral growth. This was first accomplished in a fully quantitative fashion by Spear (1988), using a differential thermodynamic technique that he dubbed the Gibbs' method (Spear *et al.*, 1982). In this approach, changes in mineral composition and mode are determined from a starting set of compositions, modes, and P–T conditions, as a function of the thermodynamic properties of the minerals (entropy, volume, and mixing parameters), compositions, and changes in P and T. Numerous publications have refined these models for interpreting geochemical zoning patterns (e.g., Spear and Selverstone, 1983; Spear, 1988; Spear *et al.*, 1990a,b, 1995, 1999; Wang and Spear, 1991; Spear and Florence, 1992; Kohn, 1993a; Young, 1993; Menard and Spear, 1996; Kohn *et al.*, 1997; Spear and Markussen, 1997; Pyle and Spear, 2003). An "integrated" form of this approach is also possible (Spear, 1993) and was implemented by Powell *et al.* (1998) as an extension of their work on internally consistent

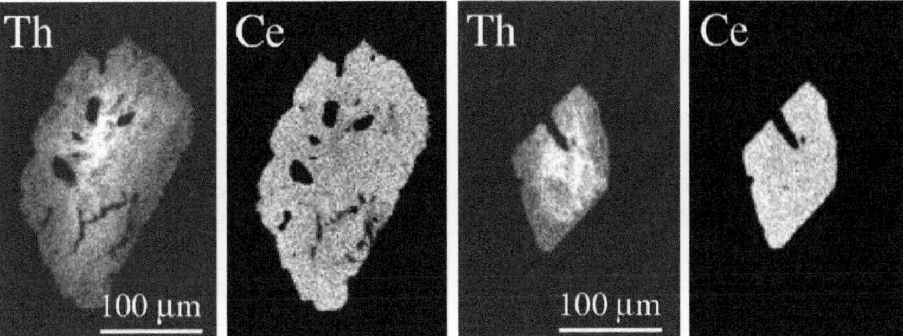

Figure 2 X-ray maps of monazite from the Great Smoky Mountains, North Carolina, showing cores with high Th and low Ce, and rims with low Th and high Ce, as expected from Rayleigh distillation of Th during monazite growth.

thermodynamic data for metamorphic minerals (Holland and Powell, 1985, 1990, 1998; Powell and Holland, 1985, 1988).

In thermodynamically based models, it is critical to account for the *P*- and *T*-dependence of the partitioning of elements among minerals, as well as the mass balance of reactions among minerals. A set of equations corresponds to each of these requirements. The principles have been detailed elsewhere (Spear, 1993), and so the equations are simply presented here.

The equations that describe element partitioning are the differentials of the thermodynamic expression of equilibrium for each independent reaction:

$$dG = 0 = \left(R \ln K_{eq} - \Delta\bar{S} + RT\frac{\partial \ln K_{eq}}{\partial T}\right)dT + \left(\Delta\bar{V} + RT\frac{\partial \ln K_{eq}}{\partial P}\right)dP + RT\sum_j \frac{\partial \ln K_{eq}}{\partial X_j}dX_j \tag{3}$$

where $\Delta\bar{S}$ and $\Delta\bar{V}$ and are the changes in molar entropy and volume for each reaction, R is the gas constant, dT, dP, and dX_j are the changes in temperature, pressure, and mole fraction of the phase component j, and K_{eq} is the equilibrium constant:

$$K_{eq} = \prod_j (a_j)^{\nu_j} \tag{4}$$

where a_j and ν_j are the activity and the stoichiometric coefficient of the phase component j involved in the reaction. Each mineral is also subject to a stoichiometric constraint, because the sum of mole fractions must equal 1, which in differential form is

$$\sum_j dX_j = 0 \tag{5}$$

The variance of the combined set of equations of the types shown in Equations (3) and (5) is equivalent to the thermodynamic degrees of freedom ("the number of chemical components" plus 2 (for *P* and *T*) minus "the number of minerals"). This set of equations involves only intensive variables, and is important for inferring *P–T* paths from chemical zoning.

It is often convenient to assume a closed chemical system, which imposes one mass balance constraint for each chemical component. In differential form, each equation is

$$dm_i = 0 = \sum_k M_k \sum_j n^k_{i,j} dX_{j,k} + \sum_k \left(\sum_j n^k_{i,j} X_{j,k}\right) dM_k \tag{6}$$

where m_i and $n^k_{i,j}$ are the number of moles of each chemical component i (e.g., SiO_2, Al_2O_3, etc.) in the rock and in each phase component (j) of each mineral (k), respectively, M_k is the number of moles of the mineral k in the rock (i.e., proportional to its mode), $X_{j,k}$ is the mole fraction of the phase component j in the mineral k, and the summation is over all minerals in the rock. This equation simply reflects the transfer of material as accomplished via changes to mineral compositions (first summation in $dX_{j,k}$) and modes (second summation in dM_k). Combining equations of the type shown in Equation (6) with the equilibrium and stoichiometric constraints yields a set of equations with a linear algebraic and thermodynamic variance of only 2, a result known as Duhem's theorem (Spear, 1988). Duhem's theorem implies that, after specifying starting conditions (*P*, *T*, *X*, and *M*), one can, for example, specify a ΔP and ΔT and predict changes in mineral composition and mode with thermodynamic and mass balance rigor. This permits contouring of *P–T* space with mineral compositions and modes, to see graphically how compositions and abundances change with changing *P–T* conditions (Figure 3; e.g., Spear *et al.*, 1982, 1990b; Spear and Selverstone, 1983; Kohn, 1993a; Spear, 1993; Spear and Markussen, 1997; Pyle and Spear, 2003).

Strictly speaking, a zoning profile that is developed during growth depends on the *P–T* path of the rock because fractional crystallization is path dependent. Therefore, contour diagrams are completely accurate only for equilibrium crystallization models, in which minerals are compositionally homogeneous. Although such diagrams may not predict chemical zoning precisely, nonetheless they reveal compositional and modal trends that help interpret metamorphic mineral growth and composition change. To a first order, Rayleigh distillation would appear to be a dominant process controlling X_{Sps} and X_{Prp} (where Sps and Prp represent spessartine ($Mn_3Al_2Si_3O_4$) and pyrope ($Mg_3Al_2Si_3O_{12}$), respectively), because neither show maxima or minima over the contoured region (Figure 3). However, X_{Alm} (Alm = almandine ($Fe_3Al_2Si_3O_{12}$)) shows obvious non-Rayleigh behavior, even though iron is strongly preferred by garnet. In fact, the trends in X_{Alm}, X_{Sps}, and X_{Prp} are described quite well in terms of iron, magnesium, and manganese phase equilibria. The end-member reactions that form garnet from chlorite + quartz occur at low, intermediate, and high temperatures for the manganese-, iron-, and magnesium endmembers, respectively (Wang and Spear, 1991). Thus, with increasing *T*, phase equilibria demand that manganese continually decreases, magnesium continually increases, and iron shows a maximum in the vicinity of the iron-endmember reaction.

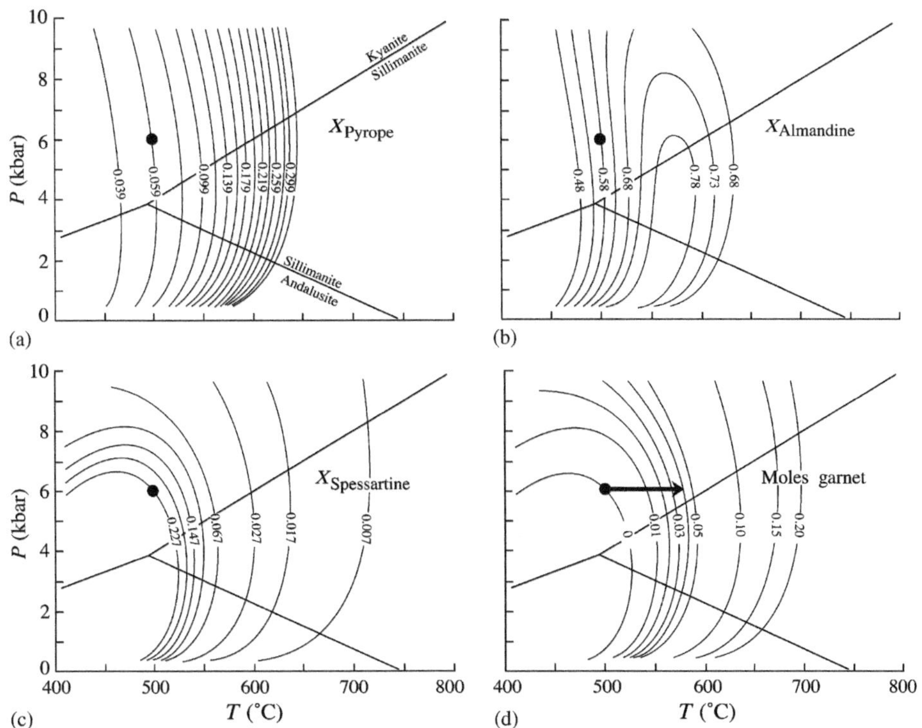

Figure 3 Contours of garnet composition for the assemblage garnet + biotite + chlorite + plagioclase + quartz + muscovite + H_2O as modeled in the system $MnO-Na_2O-CaO-K_2O-FeO-MgO-Al_2O_3-SiO_2-H_2O$. Black dot shows reference $P-T$ conditions. Horizontal arrow corresponds to isobaric $P-T$ path used for modeling garnet compositions in Figure 4. Note systematic increase in X_{Prp} and decrease in X_{Sps} with increasing T, but maximum exhibited by X_{Alm} at ~550 °C (source Spear, 1993).

These diagrams illustrate that Rayleigh distillation is not particularly useful for explaining Fe–Mg–Mn systematics, even though there are major differences in the partitioning of these elements between garnet and the matrix minerals. Instead, reaction locations in the Fe–Mg–Mn garnet system better explain the overall trends. More realistic models of intracrystalline zoning combine thermodynamic equilibrium and mass balance with fractional crystallization (Spear, 1988, 1993). An example involving an isobaric temperature increase (Figure 4) illustrates how fractional crystallization profiles deviate from equilibrium models. Most importantly, manganese decreases faster than predicted by equilibrium models, causing iron and magnesium to increase faster. The actual zoning patterns reflect both phase equilibria and chemical kinetics.

Although no theoretical model can capture the idiosyncracies of every rock, there are now hundreds of examples of garnets whose chemical zonation follows the basic patterns depicted in Figure 4 of decreasing manganese, increasing magnesium, and decreasing Fe/(Fe + Mg) from core to rim. This implies that thermodynamic equilibrium and fractional crystallization are plausible drivers of chemical zonation, and that both thermodynamics and mass balance must be

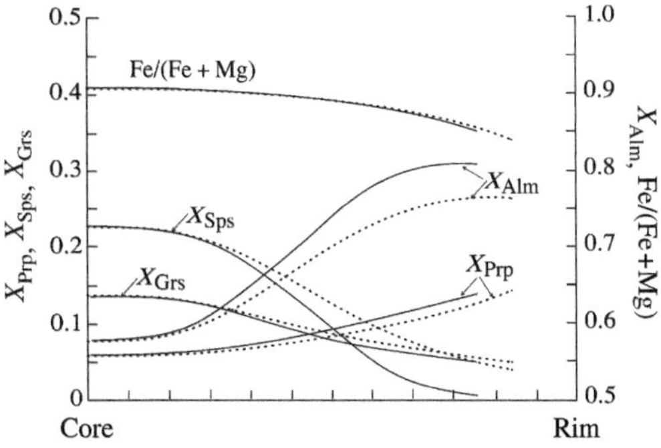

Figure 4 Plot of garnet composition versus radius for garnet growth between 500 °C and 577 °C at 6 kbar. Assemblage and chemical system are the same as in Figure 3. Dotted lines show trace of equilibrium compositions, which result from phase equilibrium effects. Solid lines show fractional crystallization model. Similarity of curves for all components (not just Mn) indicates that phase equilibria strongly control garnet chemical compositional changes, but significant differences between curves, especially for Mn, indicate that fractional crystallization is also important (source Spear, 1993).

accounted for to explain compositional zonation patterns quantitatively. The occurrence of profiles that have a Rayleigh-like appearance in one or more elements (Figure 1) is evidence that fractional crystallization has occurred, but not that compositions are perfectly preserved, or that simple Rayleigh distillation models are accurate.

3.07.2.2.1 *P–T paths*

The retrieval of metamorphic *P–T* paths is a major application of quantitative modeling of chemical zonation in metamorphic garnets. If the chemical zonation encoded in a garnet is a reflection of the changes in *P* and/or *T* that drove garnet growth, then in principle one can invert the zonation profile to infer those *P* and *T* changes. That is, if ΔP and ΔT cause $\Delta X_{j,k}$, one should be able to use the $\Delta X_{j,k}$ that are preserved as chemical zoning to infer the original ΔP and ΔT. This approach is computationally simple if mass balance constraints are imposed. Duhem's theorem requires a variance of only 2, so zonation in only two components of the garnet (two $\Delta X_{j,k}$) are necessary to retrieve ΔT and ΔP. However, it is difficult to verify that the mass of a chemical component has remained constant. For example, the whole-rock Na_2O and CaO contents depend on bulk plagioclase content and composition. If plagioclase cores are nonreactive, the effective bulk composition that the garnet "sees" as it grows may be substantially depleted in Na_2O and/or CaO compared to the whole rock. Plagioclase nonreactivity affects calcium mass balance, which influences X_{Grs} trends in garnet (Spear *et al.*, 1990b). This problem may be evident in plagioclase compositional zoning, but difficult to model quantitatively. Consequently, most workers use only the equations that involve intensive variables (dP, dT, and $dX_{j,k}$). The linear algebraic variance of the set of equations is then equivalent to the thermodynamic variance, which ordinarily means that changes in several mole fractions must be measured to resolve dP and dT. In the case of typical garnet-grade metapelites, the thermodynamic variance is 4, so four independent dXs must be measured. Only three of the major mole fractions in garnet are independent, so changes in the mole fraction of another mineral component would need to be measured to model a *P–T* path. The calculated path can be very sensitive to the choice of the fourth dX (Kohn, 1993b). For example, dX_{Ann} (Ann = annite ($KFe_3AlSi_3O_{10}(OH)_2$)) and dX_{Phl} (Phl = phlogopite ($KMg_3AlSi_3O_{10}(OH)_2$)) in biotite are poor choices numerically because they do not resolve changes in pressure, whereas dX_{An} (An = anorthite ($CaAl_2Si_2O_8$) in plagioclase is numerically preferable.

The first detailed *P–T* path determined by inversion of a zoning profile was from the Tauern Window, Austria (Figure 5; Selverstone *et al.*, 1984). The rock has an unusual, low-variance assemblage, so in fact only two independent $dX_{j,k}$ were needed to retrieve a path. Two logical pairings are X_{Alm}–X_{Grs} and X_{Sps}–X_{Grs}, because X_{Sps} and X_{Alm} are both sensitive to temperature (e.g., Figure 3), whereas X_{Grs} is moderately sensitive to pressure. The different path calculations show very similar *P–T* evolution—exhumation by ~3 kbar with heating of ~20 °C (Figure 5(b); open symbols)—suggesting that the model assumptions are valid, i.e., that chemical equilibrium was attained among minerals, and

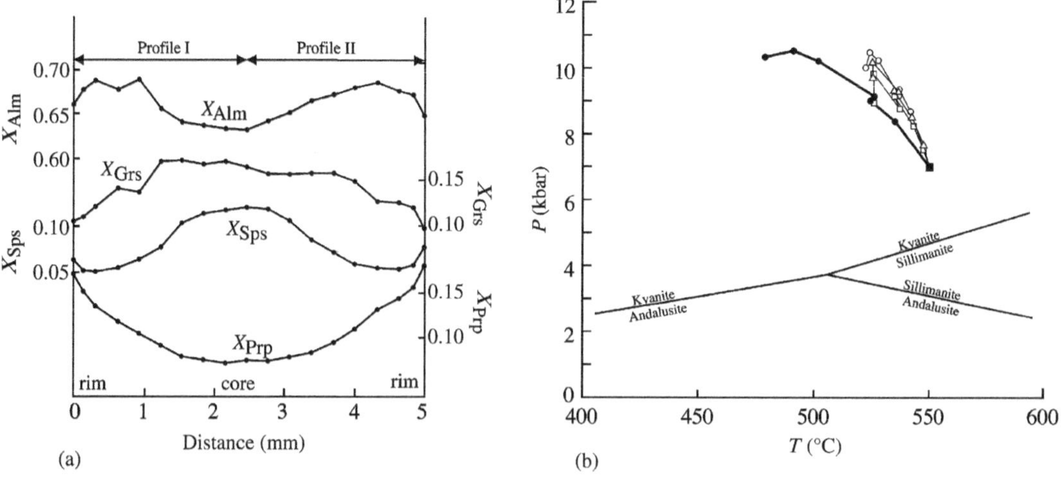

Figure 5 (a) Compositional zoning profile collected across a single garnet from sample FH-1M, Tauern Window, Austria. (b) *P–T* path calculations for sample FH-1M, from the Tauern Window, Austria. The mineral assemblage in this rock has an unusually low variance (2) which permits a *P–T* path to be inferred solely from garnet compositional changes. Paths labeled I and II (△ profile I, X_{Alm} and X_{Grs}; □ profile I, X_{Sps} and X_{Grs}; ○ profile II, X_{Alm} and X_{Grs}) correspond to different halves of the zoning profile. Recalculation of the original path by using improved thermodynamic data corroborates ΔP, but suggests a larger ΔT (● profile II, X_{Alm} and X_{Grs}) (after Selverstone *et al.*, 1984; Spear, 1993). Aluminosilicate diagram after Holdaway (1971) for reference.

garnet faithfully retained a record of changing composition. Subsequent updates to mixing models and thermodynamic data (Berman, 1988; Holland and Powell, 1998; Pattison *et al.*, 2002), especially for water, suggest that the change in ΔT was ~75 °C, although differences between calculations using the thermodynamic databases and models are now imperceptible. Most importantly, none of these modifications affect the original conclusion of Selverstone *et al.* that the garnet grew during exhumation with heating (Figure 5(b)).

The recovery of the Tauern Window $P-T$ path (Selverstone *et al.*, 1984) was revolutionary because the geodynamics community had predicted that exhumation with heating would be the result of overthickening, relaxation of isotherms, and exhumation in collisional orogens (England and Richardson, 1977). England and Thompson (1984) published a now-classic thermal modeling study the same year as Selverstone *et al.* (1984), in which they investigated numerically the $P-T$ implications of overthickening during orogenesis. The convergent series of publications in petrology and geodynamics (Spear and Selverstone, 1983; Spear *et al.*, 1984; Selverstone *et al.*, 1984; England and Thompson, 1984; Thompson and England, 1984) have linked these two fields inseparably, so that today most metamorphic petrologists justify $P-T$ path research on the basis of its ability to inform geodynamic and tectonic processes.

3.07.2.3 Diffusion

Diffusion is a thermally activated process, whereby a chemical or isotopic species moves down a chemical potential gradient (usually from high to low concentration), at a rate dependent on the diffusion coefficient, D. This applies to the progressive decrease of growth zoning in a mineral as a rock heats, and to diffusive fluxes into or out of the surface of a mineral as processes in the matrix alter the rim composition relative to the mineral interior. The diffusion coefficient is formally defined as the proportionality constant between flux rate (J) and concentration gradient (∇C):

$$J = -D\nabla C \tag{7}$$

and is an exponential function of temperature:

$$D = \mathrm{D_o e}^{-E/RT} \tag{8}$$

where D_o is a pre-exponential constant, E is the activation energy, R is the gas constant, and T is absolute temperature. For cations, E generally depends weakly on pressure and f_{O_2} (for minerals that contain iron and other redox sensitive elements). Commonly, the P-dependence is expressed as an activation volume contribution to E, and results are adjusted to a particular f_{O_2} buffer, for example, the graphite–O_2 buffer (Loomis, 1978a,b; Loomis *et al.*, 1985; Chakraborty and Ganguly, 1991, 1992; Chakraborty and Rubie, 1996; Ganguly *et al.*, 1998a). These corrections are especially important for reconciling experimental data that are collected at different P and f_{O_2}.

The diffusion coefficient can also depend on composition, and this gives rise to three different types of diffusion coefficients (e.g., see Chakraborty and Ganguly (1991)):

(i) tracer diffusion describes movement of an element at infinite dilution (e.g., iron in "pure" forsterite);

(ii) self-diffusion describes movement of an element in the absence of a chemical potential gradient (e.g., iron in pure fayalite), as determined via measurement of isotope diffusion rates; and

(iii) interdiffusion describes movement of one element in exchange for another (e.g., Fe–Mg interdiffusion in a forsterite–fayalite solid solution).

Tracer and self-diffusion do not cause appreciable chemical compositional changes. Consequently, geologic applications to major and minor element zoning ubiquitously involve interdiffusion. However, there is generally not a single interdiffusion coefficient in a multicomponent mineral. Rather, for an n-component mineral, the concentration gradient (∇C) and chemical flux rate (J) are $(n-1)$ vectors related via an $(n-1)\times(n-1)$ diffusion coefficient matrix (Lasaga, 1979). Each term in ∇C and J refers to a chemically independent component, and the dimensionality is reduced because the flux rate of one component can be expressed in terms of rates of the other $(n-1)$ components. Lasaga (1979) described how to derive the diffusion coefficient matrix from self-diffusion coefficients, if the thermodynamic mixing properties of the different components are known. Assuming ideal solid solutions, each component in the matrix (D_{ij}) is

$$D_{ij} = D_i^*\delta_{ij} - \left[\frac{D_i^* C_i z_i z_j}{\sum_k D_k^* C_k z_k^2}\right]\left[D_j^* - D_n^*\right] \tag{9}$$

where z_i is the charge on the cation i, $\delta_{ij} = 1$ if $i = j$ and $\delta_{ij} = 0$ if $i \neq j$, and D^* is the self-diffusion coefficient. More complex equations apply for nonideal solid solutions. Some geologic problems can be reduced essentially to binary exchange, for example, for Fe–Mg interdiffusion in silicates. For a binary nonideal solution, the expression for the interdiffusion coefficient is

$$D_{ij} = \left[\frac{D_i^* D_j^*}{X_i D_i^* + X_j D_j^*}\right]\left[1 + \left(\frac{\partial \ln \gamma_i}{\partial \ln X_i}\right)_{P,T}\right] \tag{10}$$

where i and j are the interdiffusing species, X_i and X_j are the mole fractions of components i and j, respectively, and γ_i is the activity coefficient of component i. Thus, most generally, cation diffusion calculations must account for the T-, P-, f_{O_2}-, and X-dependence of the diffusion coefficient, including the mixing properties of the mineral. Few, if any, applications actually account explicitly for all these effects; instead, simplifications are made, for example, assuming that solutions are ideal, that f_{O_2} is buffered, and/or that D_i^* and D_j^* are approximately equal (which eliminates cross-terms in Equation (9)). Chakraborty and Ganguly (1991, 1992) suggested using an effective binary diffusion coefficient with a single D_0 and E that describes interdiffusion of two species over the composition range in the sample studied. In part, these simplifications are useful because even the best diffusion experiments, when extrapolated to metamorphic conditions, yield uncertainties in D of $\pm 1-2$ orders of magnitude. Errors arising from other assumptions are relatively small. Furthermore, many diffusion calculations are limited not by model simplifications, but by uncertainties in the assignment of initial concentrations or changes in rim compositions.

3.07.2.3.1 Diffusional flattening of manganese growth profiles in garnet

If an increase in temperature drives garnet growth, fractional crystallization implies that the garnet will develop a core–rim manganese decrease. However, increasing temperature also increases the diffusion rates, and this affects the overall manganese pattern and composition of the rim region. There is competition between fractional crystallization and the core–rim diffusive flux to decrease and increase manganese at the rim. At a sufficiently high temperature, garnet compositions are homogenized. The degree of homogenization can be plotted geographically and, in principle, related to the temperature–time history. Several studies have documented progressively flatter manganese profiles in garnets at progressively higher metamorphic grades (e.g., Anderson and Olimpio, 1977; Woodsworth, 1977; Yardley, 1977; Dempster, 1985; Carlson and Schwarze, 1997).

Investigation of polymetamorphic rocks of the Llano Uplift (Carlson and Schwarze, 1997; Carlson, 2002) illustrates both diffusional flattening of high-manganese cores and near-rim diffusion profiles imposed via thermal overprinting and mineral resorption. For garnet crystals with radii of ~0.5–1 mm, the degree of manganese flattening depends on peak metamorphic temperatures and grain size (Figure 6). Samples that were collected directly adjacent to plutons have completely flattened core profiles, and the smaller garnets in a single rock exhibit flatter manganese profiles than the larger garnets (Figure 6). Both results are consistent with growth of garnets with manganese-enriched cores that were subsequently flattened as temperature increased, either during the first metamorphic event for samples far from plutons, or during a contact metamorphic overprint for samples at pluton margins. The increase in manganese at the rims of the garnets is the result of postpeak metamorphic garnet resorption, and the repartioning of manganese among matrix minerals and the garnet rim. Rim zoning impacts many applications, including "geospeedometry" (the use

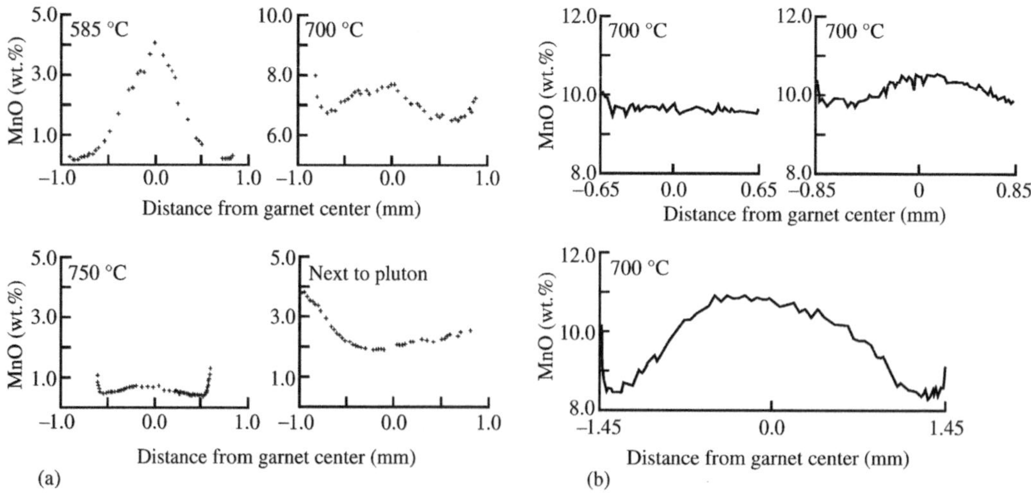

Figure 6 (a) Plots of garnet MnO content (wt.%) versus distance across central sections of crystals from different localities in the Llano Uplift polymetamorphic block, Texas. Garnets show increased core–rim flattening with increasing temperature (either peak conditions for prograde event or proximity to plutons during static overprinting event). (b) Garnet MnO content versus distance across central sections of crystals of different sizes from a single locality. Garnets show increased core–rim flattening with decreasing grain size. All data are consistent with diffusive modification of earlier-formed growth profiles (sources Carlson and Schwarze, 1997; Carlson, 2002).

of diffusion profiles for estimating cooling rates), and thermobarometry (the determination of metamorphic $P-T$ conditions).

3.07.2.3.2 Geospeedometry

Lasaga (1983) outlined a technique that he dubbed "geospeedometry" for inverting chemical diffusion profiles at the rims of minerals to infer cooling rates. The basic diffusion equation is

$$\frac{\partial c}{\partial t} = D(t)\frac{\partial^2 c}{\partial x^2} \tag{11}$$

where c is the concentration of the species of interest (e.g., magnesium in garnet), D is the diffusion coefficient, t is time, and x is distance. The difficulty in solving this equation is that D is a function of T (Equation (8)), which changes with time. However, Equation (11) can be made analytically tractable through a series of transformations, if linear cooling is assumed (Lasaga, 1983; Lindström *et al.*, 1991). First, Lasaga (1983) introduced a compressed time variable (t'):

$$t' = \frac{1}{\gamma}(1 - e^{-\gamma t}) \tag{12}$$

where t is time, and γ is given by

$$\gamma = \frac{E{\cdot}s}{RT_{\text{init}}^2} \tag{13}$$

E is the activation energy, s is the cooling rate ("speed" of cooling), R is the gas constant, and T_{init} is the initial temperature. He then nondimensionalized time (t') and distance (x_r) as

$$t_r' = \frac{D_{\text{init}}}{a^2}t' \tag{14}$$

$$x_r' = \frac{x}{a} \tag{15}$$

where a is the characteristic distance of the diffusing grain (e.g., the grain radius). These substitutions transform Equation (11) to

$$\frac{\partial c}{\partial t_r'} = \frac{\partial^2 c}{\partial x_r'^2} \tag{16}$$

In principle, one must solve Equation (16) for all minerals in the rock that are diffusionally limited from equilibrating. However, many applications involve Fe–Mg exchange between garnet and biotite, and because biotite is almost always homogeneous with respect to Fe/Mg, one need only solve for the diffusion profile in the garnet.

Equation (16) can be solved for γ and hence s subject to the constraint of equilibrium partitioning on the rim of the garnet due to exchange with another mineral:

$$K_D(t) = K_{D,\text{init}}\exp\left(-\left(\frac{\Delta H{\cdot}s}{RT_{\text{init}}^2}t\right)\right) \tag{17}$$

where K_D is the distribution coefficient for the exchange (e.g., for Fe/Mg between garnet and biotite), and ΔH is the enthalpy of reaction. Lasaga's original approach has been improved both theoretically and computationally (Lindström *et al.*, 1991; Lasaga and Jiang, 1995; Ganguly and Tirone, 1999; Jaoul and Sautter, 1999; Ganguly *et al.*, 2000).

Two fundamental cautions are, however, warranted. First, the diffusion profile really results from an integrated diffusional history (e.g., Ganguly *et al.*, 2000):

$$\gamma = \frac{D(T_o)}{\int_0^t D(\tau)\mathrm{d}\tau} \tag{18}$$

which corresponds to an infinite number of cooling paths. Linear cooling (i.e., Equation (13)) is only one solution of Equation (18). Thus, one must interpret the results within the context of the most likely thermal history, for example, linear cooling versus steadily increasing or decreasing rates. Second, Lasaga's transformation of variables essentially converts the problem into a description of the position-dependent closure temperature $[T_c(x)]$. If the matrix remains compositionally homogeneous (e.g., very small and very large amounts of garnet and biotite, respectively), then a general analytical solution exists (Dodson, 1986; McDougall and Harrison, 1988):

$$T_c(x) = \frac{E/R}{\ln[\varepsilon RT_c^2 D_o/a^2/E{\cdot}s] + 4S_2(x)} \tag{19}$$

where ε is the exponential of Euler's constant, $4S_2(x)$ is a position-dependent term, x is the fractional distance from the mineral center, and the other terms are as described above. The position-dependent temperatures (and hence compositions) recorded along the zoning profile depend on the logarithm of cooling rate. The insensitivity of the composition profile to s leads to large uncertainties in retrieved s. This problem is further compounded by the large uncertainties in D because of the large extrapolation of experimental data to metamorphic conditions. For many problems, the uncertainty in the calculated cooling rate may amount to several orders of magnitude (Lindström *et al.*, 1991).

Ganguly *et al.* (2000) have provided an example of geospeedometric calculations for the rocks of the Sikkim Himalaya (NE India). They emphasized that composition profiles should be measured perpendicular to the grain edge, whereas edges may "dip" with respect to the plane of the section, thus requiring correction of

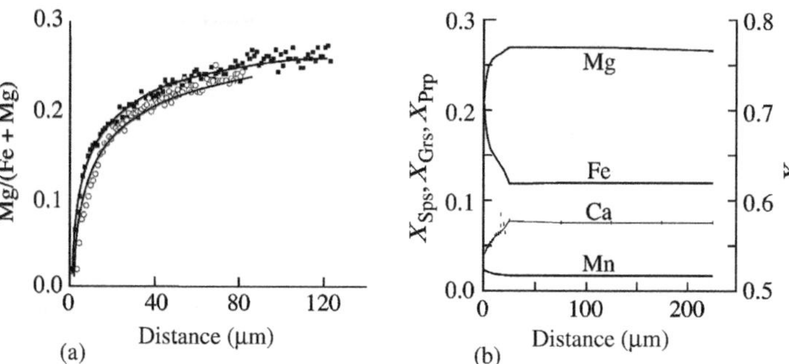

Figure 7 Composition versus distance profiles for high-T garnets. (a) Mg/(Mg + Fe) ratio in garnet versus distance from the rim in contact with biotite, showing smoothly decreasing values towards rim. Filled and open squares are for profiles measured in two different garnets (G1 and G2), and the difference is interpreted to result because edges of garnets are not necessarily perpendicular to the plane of the section. The two profiles can be reconciled if profile from G2 is rotated 46° relative to profile from G1. Profiles can be fit with a diffusion model with cooling rates of ~20 °C Myr^{-1}, assuming diffusion coefficients from Ganguly *et al.* (1998a). Note that different diffusion coefficients and their uncertainties can yield widely different cooling rates (source Ganguly *et al.*, 2000). (b) All garnet components are zoned in outer rim of garnets, suggesting diffusion is not the only process affecting rim garnet compositions.

the diffusion profile for the crystal boundary orientation (Figure 7). However, for correctly oriented grains it is relatively straightforward to solve for $\gamma \sim 2.1 \times 10^{-14}$. One could then use Equation (13) to infer $s \sim 20$ °C Myr^{-1}, but a uniform cooling rate may not be applicable over the entire temperature range during which Fe–Mg exchange is evident. Instead, Ganguly *et al.* (2000) used the product $a^2 \cdot \gamma / D(T_o)$ as a constraint in a one-dimensional thermal model linking cooling with exhumation. From this, they inferred accelerated cooling from an initial rate of ~15 °C Myr^{-1} at 800 °C, to >100 °C Myr^{-1} by ~450 °C. The retrieved cooling rate depends directly on the diffusion coefficient, and because experimental uncertainties in D for divalent cations in garnet are well over an order of magnitude at $T \leq 800$ °C, this directly translates into an error in the retrieved cooling rate. That is, if D is not known to better than a factor of 10, neither is s (Lindström *et al.*, 1991).

3.07.2.4 Combination of Retrograde Diffusional Exchange and Reaction

Interpretation of near-rim composition profiles depends critically on the nature of retrograde reactions (Robinson, 1991; Spear and Florence, 1992; Figure 8). These belong to two classes. First, exchange reactions (ERs) involve the exchange of two elements between two minerals, for example, Fe–Mg exchange between garnet and biotite (almandine + phlogopite = annite + pyrope). Retrograde exchange reactions do not significantly change mineral modes, and cause divergence of mineral compositions. Second, net-transfer reactions (NTRs) involve production and consumption of minerals, i.e., a net transfer of material among mineral reservoirs. Unlike ERs, NTRs cause mineral compositions to shift in the same direction. Thus, a decrease in magnesium and increase in iron towards the rim of a garnet (e.g., Figure 7) could result from retrograde Fe–Mg exchange between garnet and biotite (an ReER), or retrograde reaction between several minerals, such that they all shift towards iron-richer compositions (an ReNTR; Figure 8). The physical difference between them is that an ReER does not change the position of mineral rims perceptibly, whereas an ReNTR causes mineral resorption or growth. Therefore, it is critical in any geospeedometry study either to verify petrologically that ReNTRs have not occurred, or to characterize the amount by which the boundaries of minerals have moved.

One way of identifying ReNTRs in garnets is to examine their manganese profiles. Because garnet partitions manganese so strongly, growth of garnet causes manganese to decrease towards the rim, whereas resorption will cause it to increase (Figure 8). Therefore, higher manganese at rims is *prima facie* evidence for the dissolution of garnet via ReNTRs, whereas a flat or decreasing manganese profile at the rim indicates only ReERs. This is evident in diffusion profiles in garnet around biotite inclusions, where iron and magnesium are zoned, but manganese is not (e.g., Spear and Parrish, 1996).

Carlson (2002) ascribed composition profiles in garnet rims from the Llano Uplift, Texas, to both ReERs and ReNTRs. These rocks are unusual, because spectacular retrograde coronas permit identification of the original rim locations (Figure 9). Composition profiles show increases in iron and decreases in magnesium towards garnet rims, as expected for diffusional re-equilibration of the garnet rim during cooling (Figure 10).

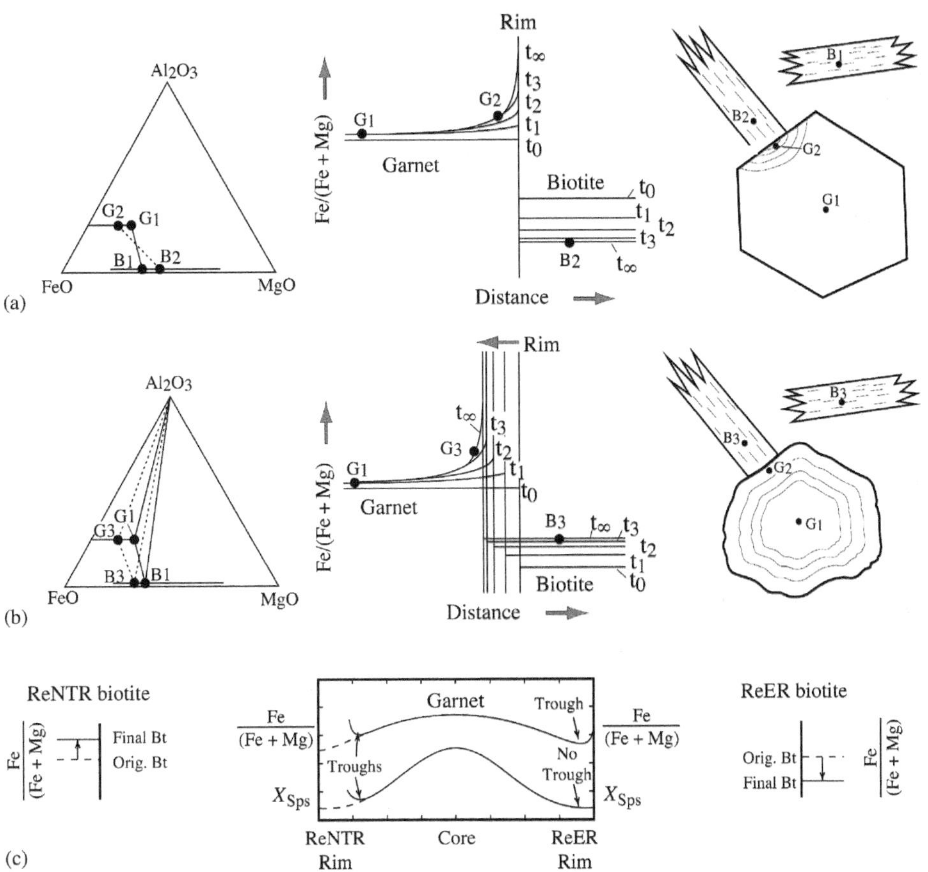

Figure 8 Diagrams illustrating the change in Fe/(Fe + Mg) for garnet and biotite during retrograde reactions (Spear, 1993; Kohn and Spear, 2000). (a, b) High-grade, diffusionally homogenized garnets. G1 and B1 are peak compositions; G2 and B2 are retrograde compositions. (a) Effects of retrograde exchange reactions (ReERs). On an Al_2O_3–FeO–MgO diagram, garnet and biotite compositions diverge with decreasing temperature. The compositions corresponding most closely to peak conditions are the garnet with highest Mg/Fe and biotite with highest Fe/Mg. Because of slow D_{Fe-Mg} in garnet, but fast D_{Fe-Mg} in biotite, biotite remains homogeneous while a diffusion profile is established at the garnet rim. Pairing of garnet core (G1) with matrix biotite (B2) yields a temperature below the peak. (b) Effects of retrograde net transfer reactions (ReNTRs). On an Al_2O_3–FeO–MgO diagram, garnet rim and biotite compositions move towards higher Fe/(Fe + Mg) during retrograde garnet dissolution. The compositions corresponding most closely to peak conditions are the garnet with highest Mg/Fe and biotite with lowest Fe/Mg. However, the original biotite may not be present, and pairing of matrix biotite (B3) with garnet core (G1) yields too high a temperature. (c) Diagrams for lower-grade conditions illustrating the change to Fe/(Fe + Mg) and Mn in garnet, and Fe/(Fe + Mg) in biotite during retrograde reactions. Dashed line shows original profile at peak conditions. General decrease in X_{Sps} towards rim is result of prograde growth. ReERs (right side) cause an increase in garnet Fe/(Fe + Mg) at rim, which forms a compositional trough near the grain edge (right side). No change in Mn occurs. Pairing of garnet trough composition with matrix biotite yields a temperature below the peak because garnet and biotite have shifted to higher and lower Fe/(Fe + Mg), respectively. ReNTRs (left side) cause an increase in both Fe/(Fe + Mg) and Mn in garnet, which forms compositional troughs near the grain edge in both profiles (left side). Pairing of garnet trough composition with matrix biotite may yield a temperature that is too high because both garnet and biotite shift to higher Fe/(Fe + Mg) (Kohn and Spear, 2000) (reproduced by permission of GSA from *Geology*, **2000**, *28*, 1127–1130).

However, the coronal textures and the increase in manganese towards the garnet rims (Figure 10) both unequivocally indicate operation of one or more garnet-consuming ReNTRs, so that the rim of the garnet has not remained stationary. If this fact was not recognized, application of geospeedometry to the Fe–Mg profiles would yield an apparent cooling rate that is too fast. An estimate of the amount of garnet resorption from the size of the coronas permits a rigorous accounting of mass fluxes (particularly manganese), and fitting of the composition profiles. By accounting quantitatively for both ReERs and ReNTRs, Carlson (2002) refined the relative diffusion rates for calcium, iron, magnesium, and manganese, which in turn permitted reconciliation of the observed profiles with the independently constrained cooling rate.

Most garnets lack coronas, and resorption may only be evident either texturally from embayed margins or geochemically from near-rim manganese increases. Without direct markers of original rim locations, the extent of reaction is difficult to characterize quantitatively. Therefore, Carlson's work cautions against application of geospeedometry except in special cases where the garnet rim has not moved, or the amount of garnet resorption is independently determinable. In the case of the garnets from Sikkim analyzed by Ganguly *et al.* (2000), the small increases in manganese towards garnet rims (Figure 7(b)) imply that garnet has dissolved somewhere in the rock (albeit not necessarily at those particular rims). If dissolution has shortened the near-rim profiles, then calculated cooling rates assuming no resorption will be too fast.

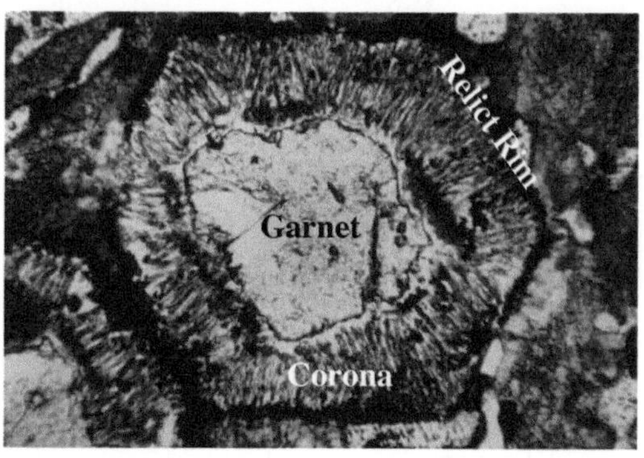

Figure 9 Photomicrograph of spectacularly well-developed corona around relict garnet from Llano Uplift, Texas. The coronas permit identification of the original rim position of the garnet (assuming isovolumetric replacement), which in turn allows quantification of the amount of garnet that was resorbed during cooling (source Carlson and Schwarze, 1997).

3.07.2.5 Thermobarometric Implications

Thermobarometry, or the calculation of $P-T$ conditions from mineral compositions, is not based on mineral zoning, nor does it directly inform processes that cause geochemical zoning. Nonetheless, $P-T$ calculations in metamorphic rocks usually require selecting compositions from minerals that are compositionally heterogeneous. How one selects compositions from a zoned mineral can strongly influence retrieved P and T, and that issue is worth discussion.

$P-T$ conditions are ordinarily calculated based on the T-dependence of a cation partitioning equilibrium (e.g., Fe–Mg exchange between garnet and biotite), and the P-dependence of a net-transfer equilibrium (e.g., anorthite = grossular + aluminosilicate + quartz). Most barometers have moderate slopes, so the calculated P depends moderately on T. The basic problem

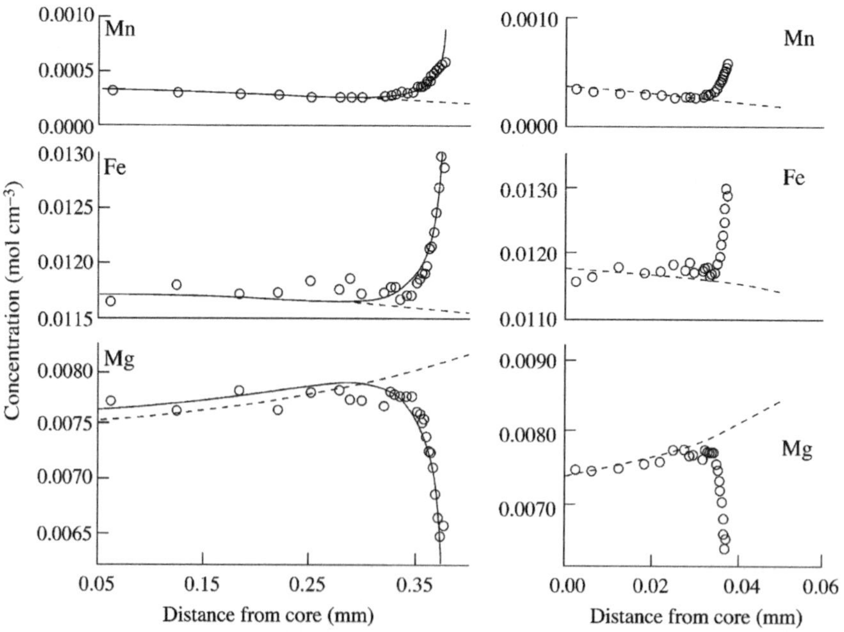

Figure 10 Zoning profiles collected at the outer edges of garnets from rocks with much less pronounced coronas show diffusional patterns over the outer ~50 μm. However, the coronal texture and increase in Mn towards rim unequivocally indicate operation of an ReNTR, which must have consumed part of the diffusion profile as it was growing. Thus, the apparent extent of diffusion is much lower than if the rim had been static. These profiles can be fitted to infer relative rates of cation diffusion (source Carlson, 2002).

lies in choosing the best compositions from which one calculates T (Robinson, 1991; Spear, 1991, 1993, chapter 17; Spear and Florence, 1992; Kohn and Spear, 2000, figure 8). A higher Mg/Fe ratio in garnet or Fe/Mg ratio in biotite will result in a higher computed T. If only ReERs have occurred, then compositions have diverged and all calculated temperatures will be below the peak. Therefore, the highest T will be the best estimate of peak conditions, and, of all compositions expressed in a rock, will correspond to the highest Mg/Fe in garnet and highest Fe/Mg in biotite. In contrast, if ReNTRs have occurred, then compositions have shifted simultaneously. Consumption of (iron-rich) garnet causes biotite and the garnet rim to shift to more iron-rich compositions. If the iron-enriched matrix biotite is then paired with relict, magnesium-rich garnet, temperatures exceeding peak conditions can be calculated (Spear, 1991; Spear and Parrish, 1996; Kohn and Spear, 2000). Commonly, matrix biotite is rather homogeneous compositionally, so the main difficulties are selecting garnet compositions and assessing how much biotite compositions could have changed. In high-grade garnets, the highest Mg/Fe (lowest Fe/(Fe + Mg)) ordinarily occurs in the core (Figures 7 and 8). In lower-grade garnets, the lowest Fe/(Fe + Mg) occurs in an Fe/(Fe + Mg) trough, whose location results from competition between prograde growth of garnet, which causes Fe/(Fe + Mg) to decrease, and ReERs or ReNTRs, which commonly cause Fe/(Fe + Mg) to increase (Figure 8(c)). Similar troughs are observed in manganese for the same reasons (Figure 8(c)).

Within this context, Kohn and Spear (2000) developed a semiquantitative "ReNTR insurance policy." Because the increases in manganese towards garnet rims predominantly result from garnet resorption, the excess manganese in the rim region relative to the lowest manganese reflects the amount of garnet that was resorbed. The manganese-based estimate of this resorbed volume permits resurrection of the original composition of the matrix biotite prior to the ReNTR. This revised biotite composition can then be paired with the highest Mg/Fe composition garnet to better estimate peak $P-T$ conditions prior to garnet resorption. This approach is analogous to Carlson's inference of material fluxes and resulting ReNTR correction of mineral compositions (Carlson, 2002), except that his calculations were texturally based, whereas Kohn and Spear's were based on X-ray maps of manganese distributions. Application of ReNTR insurance to most rocks in the author's research collection indicates relatively small corrections: 10–20 °C compared to uncorrected compositions. However, in some cases corrections can be as large as 200 °C (Kohn and Spear, 2000). Virtually all garnets exhibit retrograde manganese increases towards their rims, so ReNTR insurance should be adopted for most rocks, as it permits identification and remedial petrologic action, if necessary. Although both Carlson (2002) and Kohn and Spear (2000) developed their approaches for amphibolite-facies ReNTRs, the implications can be generalized to all metamorphic facies in which diffusion and NTRs can affect mineral compositions during cooling (Spear and Florence, 1992).

3.07.2.6 Kinetically Limited Transport within the Rock Matrix

Compositional zoning in a mineral depends on rates of crystal growth, "nutrient" supply, and "waste" removal. In the previous discussions, it was implicitly assumed that matrix compositions are homogeneous and that equilibrium among grain edges occurs on scales that are large compared to individual crystals. Although this may be true, limited transport of some elements through the matrix can profoundly impact chemical zoning. If diffusion of nutrients limits the rate of crystal growth (diffusion control), then there should be a simple relationship for the spacing between compositional contours (c) versus radius (C), as normalized to the largest grain in the rock (Kretz, 1974). A different rate-limiting mechanism, such as adsorption and addition of chemical species to the growing crystal surface (interface control), yields a different normalized radius-rate (c^* versus C^*) relationship. This concept has been developed extensively by Carlson and co-workers (Carlson, 1989, 1991; Carlson *et al.*, 1995; Denison and Carlson, 1997; Denison *et al.*, 1997; Chernoff and Carlson, 1997, 1999), who have shown that c^* versus C^* diagrams commonly conform to diffusion control. It is important to recognize that a c^* versus C^* plot assumes different diffusivities for different elements (Figure 11). Specifically some element must be homogeneous over the sampled region of the rock, and so can be used to monitor the growth rate of different crystals as a function of crystal size, i.e., serve as a geochemical proxy for time (Kretz, 1974; Carlson, 1989), whereas transport of another element diffusionally limits growth (Carlson *et al.*, 1995). For example, slow aluminum diffusion could cause diffusion limited growth, whereas the fast diffusion of manganese provides a time marker. The success of diffusion control in explaining many radius-rate data apparently supports this paradigm.

Other zoning studies further suggest radically different intergranular diffusivities for different elements, and a strong dependence of the

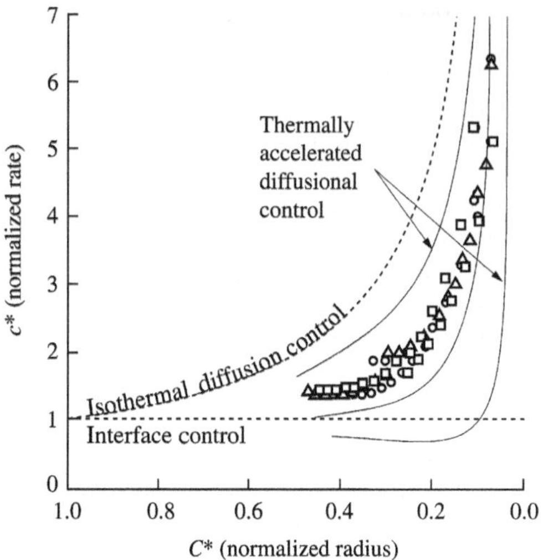

Figure 11 Plot of normalized growth rate (c^*), as determined from chemical gradients, versus normalized radius (C^*). Normalization is relative to composition gradients and radius of the largest crystal in a rock. Interface control implies that c^* will be constant for all radii, whereas diffusion control implies an increasing c^* with decreasing C^*. Thermally accelerated diffusion control (Carlson, 1989) shifts the diffusion-control curve to the right and permits c^* values less than 1. If garnets grow over a larger range of T, then they should plot further to the right of isothermal diffusion control. Data from a pair of garnets are better described by thermally accelerated diffusion control than by either interface control or isothermal diffusion control. (○, FeO: □, MgO; △, MnO) (sources Carlson, 1989; Denison and Carlson, 1997).

behavior of particular elements on temperature. For example, X-ray maps collected by Yang and Rivers (2001) show that some elements in garnet behave systematically, for example, manganese, whereas others have patterns that mimic the original texture of the rock, for example, chromium. If intergranular diffusion of manganese is fast, then the rim of growing garnets always has the same manganese content, and grains are zoned systematically. But if intergranular diffusion of chromium is extremely slow, then local mineralogical differences in chromium content due to textural variability in mica and/or oxide abundance directly control the local supply of chromium to different edges of growing garnets, and grains become zoned heterogeneously. At low grade, even manganese zoning can be quite heterogeneous (Daniel and Spear, 1998; Spear and Daniel, 1998, 2001). This presumably reflects extremely slow intergranular rates of diffusion, which causes locally heterogeneous supply of manganese.

Rates of intergranular chemical transport are extremely important because they govern the length scales over which equilibrium occurs. For example, the generally systematic behavior of manganese zoning in most garnets (albeit not garnet cores studied by Daniel and Spear (1998)) implies that equilibrium can be assumed over several grain diameters, and that equilibrium thermodynamics can be applied to the rock as a whole. However, the heterogeneous zonation of chromium implies that its distribution is dominated by local effects, and hence is much less easily modeled. These differences in element mobility become increasingly important for many trace elements which are believed to have transport behavior more like chromium and less like manganese, and for accessory minerals, whose occurrence is controlled by the distribution of trace elements.

3.07.2.7 Dissolution–Reprecipitation

Theoretical models suggest that some minerals are driven thermodynamically to change composition, yet are prevented from re-equilibrating via diffusion by extremely slow cation diffusivities. One common example is plagioclase. Anorthite and albite end-members are linked via a coupled substitution ($Ca^{IV}AlNa_{-1}Si_{-1}$), and so diffusive re-equilibration of X_{An} requires exchange of tetrahedral aluminum and silicon, a process which is extremely slow (Grove *et al.*, 1984). Therefore, the ubiquitous zonation in X_{An} must be caused by a mechanism other than diffusion. The texture of many plagioclase grains suggests that they change composition via dissolution of old grains and reprecipitation of new grains or as overgrowths (Figures 12 and 13).

Plagioclase compositional change is intimately tied to garnet growth and consumption as required by calcium mass balance in the rock (Figure 14). Garnet and plagioclase are commonly the only calcium-bearing silicates in a rock. Growth or consumption of calcium-bearing garnet, therefore, consumes or produces the anorthite component of plagioclase. Plagioclase cannot adjust composition diffusionally, so thermodynamics drives grain dissolution and reprecipitation. Such direct modal and compositional links among minerals must be accounted for when interpreting chemical zoning patterns. Even if the overall mode and composition of reactive plagioclase changes systematically, local dissolution or precipitation can happen continuously or sporadically, leading to different textures in different grains (Figures 12 and 13). Compositional zoning in a single grain can be perplexing or even misleading unless X-ray maps or backscattered electron images of many grains are collected and the overall sense of zoning compared with theoretical expectations. If a mineral contains an essential constituent that is shared by other minerals in the rock,

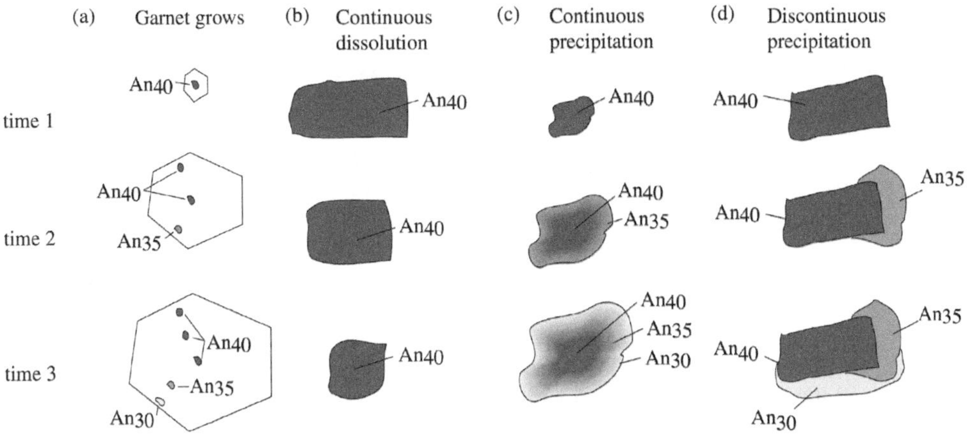

Figure 12 Sketch of plagioclase zoning patterns and compositional trends. Plagioclase cannot change composition significantly via diffusion, and so must dissolve and reprecipitate to maintain equilibrium with other minerals in a rock. (a) Growth of garnet drives plagioclase compositional changes. Local textural and chemical variations can cause disparate zoning patterns in different plagioclase grains, which can be reflected in compositional differences among plagioclase grains at the same radial distance from the garnet core. (b) Plagioclase dissolves continuously. (c) Plagioclase precipitates continuously. (d) Plagioclase precipitates discontinuously (source Spear, 1993).

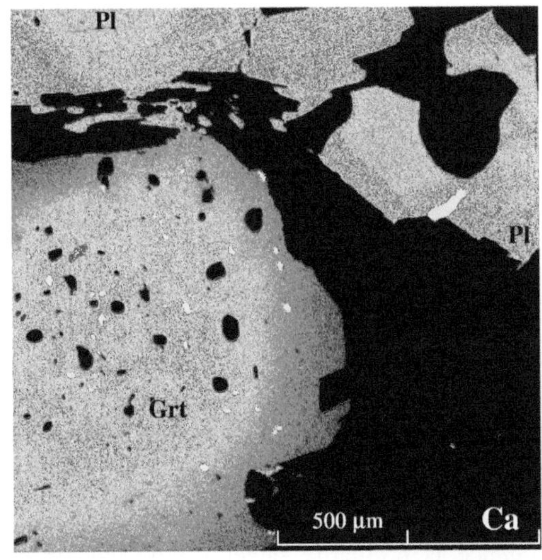

Figure 13 X-ray map of Ca in garnet gneiss from the Greater Himalayan Sequence, central Nepal, showing compositional systematics in plagioclase (upper grains). A garnet grain (lower left, with inclusions) is also present. Upper left grain of plagioclase is quasicontinuously zoned towards higher X_{An} (see Figure 12(c)). Upper right grain has continuous zoning in one region and a compositionally distinct high X_{An} overgrowth on one side (see Figure 12(d)). Retrograde consumption of garnet fluxed Ca to matrix, stabilizing plagioclase with higher X_{An}, and causing both continuous and discontinuous growth of these plagioclase grains (Kohn *et al.*, 2001) (reproduced by permission of GSA from *Geology*, **2001**, *29*, 571–574).

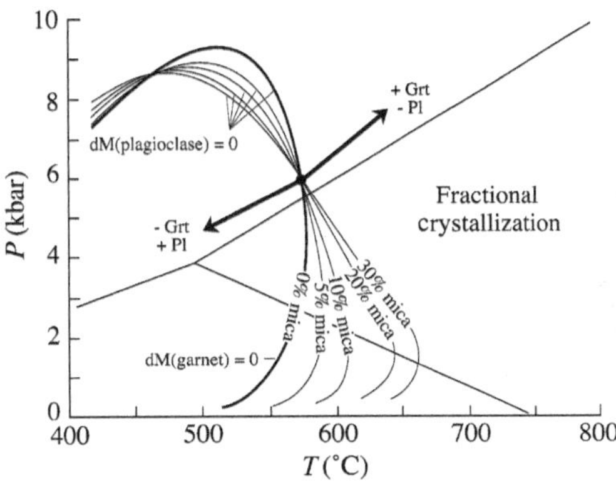

Figure 14 Modal isopleths of plagioclase in the assemblage garnet + biotite + chlorite + plagioclase + quartz + white mica (muscovite) + H_2O as modeled in the system $MnO–Na_2O–CaO–K_2O–FeO–MgO–Al_2O_3–SiO_2–H_2O$, assuming fractional crystallization of garnet. Different curves for plagioclase correspond to different abundance of white mica, which affects plagioclase abundances because of the paragonite component's influence on Na mass balance. Black dot shows reference $P–T$ conditions. Black arrows show that garnet growth generally causes a decrease in plagioclase abundance, whereas garnet consumption causes an increase in plagioclase abundance. Changes in X_{An} broadly follow plagioclase modal abundances, so garnet growth generally causes a decrease in X_{An}, whereas garnet consumption generally causes an increase in X_{An} (source Spear *et al.*, 1990b).

and if that constituent has an extremely slow intracrystalline diffusivity, then chemical or modal changes in the other minerals will drive dissolution–reprecipitation reactions.

3.07.3 STABLE ISOTOPES

There are relatively few studies of stable isotope zoning in metamorphic minerals, because the required measurements are more difficult than for major elements. This section will focus on oxygen, because there are more studies of oxygen isotope zoning, and because oxygen transport is more readily modeled. Growth zoning, diffusion zoning, and dissolution–reprecipitation are best documented for this element.

Isotopic compositions are usually expressed in per mil notation, relative to a standard, such as Standard Mean Ocean Water (SMOW):

$$\delta^{18}\mathrm{O}(i) = \left(\frac{R_i - R_{\mathrm{SMOW}}}{R_{\mathrm{SMOW}}}\right) \times 1,000 \quad (20)$$

where R_i and R_{SMOW} are the ratios of ^{18}O to ^{16}O in the material i and SMOW, respectively. Partitioning between two materials, i and j, is usually expressed in terms of a fractionation factor, α_{i-j}, where

$$1,000 \ln \alpha_{i-j} \sim \delta^{18}\mathrm{O}(i) - \delta^{18}\mathrm{O}(j) = \Delta_{ij} \quad (21)$$

Partitioning is almost solely dependent on absolute temperature:

$$\Delta_{ij} = \frac{A_{ij}}{T^2} + \frac{B_{ij}}{T} + C_{ij} \quad (22)$$

where A, B, and C are constants. For the partitioning of oxygen among many silicates and oxides, B and C are nearly zero and can be ignored, but for fluids all terms must be used (e.g., see summary by Chacko *et al.* (2001)). Pressure dependence is strongly subsidiary to the effect of temperature, and is similar in most minerals (Polyakov and Kharlashina, 1994), so that mineral–mineral Δs are nearly P-independent. Equation (22) indicates that the isotopic compositions of minerals tend to converge with increasing temperature.

3.07.3.1 Growth Zoning

Many orthosilicates have slow oxygen diffusion rates and can readily develop growth zoning, whereas many sheet and tectosilicates have fast diffusion rates and control the oxygen isotope mass balance of a rock. A rigorous theoretical approach for predicting changes in the isotopic composition of minerals was derived independently (Kohn, 1993a; Young, 1993), by incorporating a set of oxygen isotope partitioning equilibria and mass balance equations into the differential thermodynamics method. These equations maintain partitioning equilibrium:

$$\mathrm{d}\Delta_{ij} = \left(\frac{-2A_{ij}}{T^3} + \frac{-B_{ij}}{T^2}\right)\mathrm{d}T \quad (23)$$

(ignoring pressure dependence) and oxygen isotope mass balance:

$$\mathrm{d}\delta^{18}\mathrm{O}_{\mathrm{sys}}N_{\mathrm{sys}} = \sum_{k=1}^{n} M_k N_k \mathrm{d}\delta^{18}\mathrm{O}_k + \sum_{k=1}^{n} N_k \delta^{18}\mathrm{O}_k \mathrm{d}M_k = 0 \quad (24)$$

where $\delta^{18}\mathrm{O}_{\mathrm{sys}}$ is the whole-rock oxygen isotopic composition, N_{sys} is the total number of moles of oxygen in the system, M_k and N_k are the number of moles and the number of moles of oxygen in the kth mineral, and $\delta^{18}\mathrm{O}_k$ is the δ^{18}O of the kth mineral. For a rock with n minerals, there are n new variables ($\delta^{18}\mathrm{O}_k$), and n new equations—$(n-1)$ in the form of Equation (23), and one in the form of Equation (24). Thus, the total linear algebraic variance of the combined system of thermodynamic mass balance and isotope partitioning equations remains 2, and changes in only two variables (e.g., dP and dT) need to be specified to model changes in $\delta^{18}\mathrm{O}_k$.

When mineral modes change slowly due to continuous reaction over a large temperature range, the temperature dependence of isotopic partitioning controls compositions (Figure 15). For example, garnet in a model metapelite that contains biotite and chlorite simply increases in δ^{18}O with increasing temperature because it has a lower δ^{18}O value than the whole rock, and mineral compositions converge with increasing temperature. However, at nearly discontinuous reactions, such as

$$\text{garnet} + \text{chlorite} + \text{muscovite} = \text{biotite} + \text{staurolite} + \text{quartz} + \mathrm{H_2O} \quad (25)$$

mineral modes change so rapidly that isotopic mass balance (Equation (24)) controls compositions, as indicated by the coincidence of compositional and modal slopes for the staurolite-in reaction. However, such reactions also rapidly exhaust a reactant, so their isotopic influence is ordinarily rather small. Models of growth zoning can be constructed by incorporating fractional crystallization. Generally, in a closed system, the strongest prograde isotope zoning will result from continuous reactions that occur over a large temperature range. Even so, predicted growth zoning is subdued, and resolution of the small changes in isotopic composition indicated by the models requires an analytically and spatially precise technique. Such profiles could not be routinely measured until the advent of miniaturized laser fluorination systems (Sharp, 1990, 1992).

Oxygen isotope zoning in garnet was documented by the early 1990s (Chamberlain and Conrad, 1991, 1993; Young and Rumble, 1993; Jamtveit and Hervig, 1994), but Kohn *et al.* (1993) described the first isotopic zoning profiles that clearly conformed with independent predictions of growth models. The observed monotonic ~0.5‰ increase

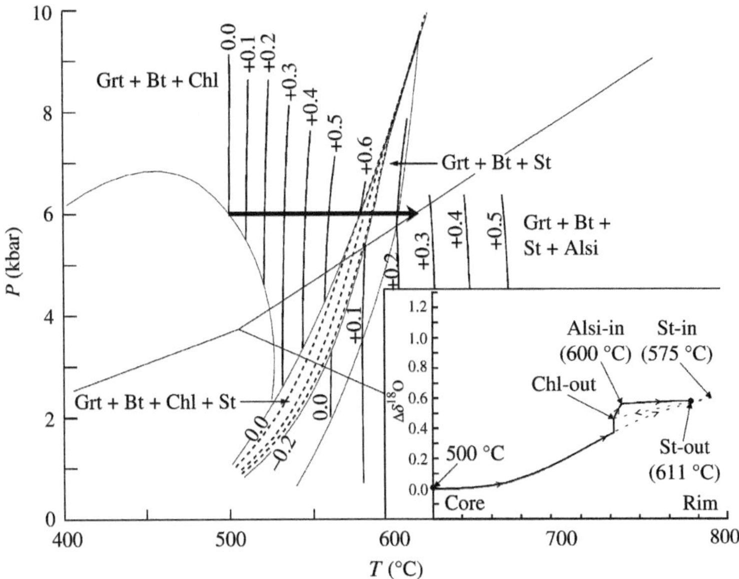

Figure 15 Isopleths of garnet $\delta^{18}O$ in a model metapelite, showing nearly *P*-invariant orientations in garnet + biotite + chlorite and garnet + biotite + staurolite assemblages, and slopes nearly parallel to reaction boundaries in a garnet + biotite + chlorite + staurolite assemblage. Absolute compositions are arbitrary, and are zeroed at a particular isopleth to illustrate compositional differences within each assemblage field. Inset shows isotopic zoning predicted for a garnet that grows isobarically from 500 °C to 611 °C at 6 kbar, showing development of zoning and how mineral reactions affect zoning profile (source Kohn, 1993a).

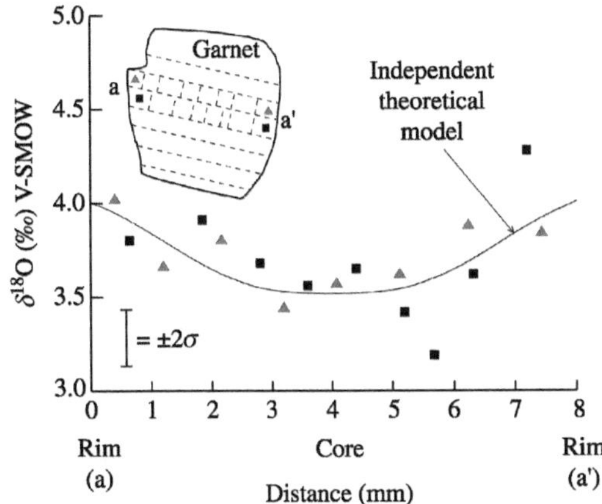

Figure 16 Oxygen isotope profiles across a garnet from Tierra del Fuego, Chile, showing general ~0.5‰ increase in $\delta^{18}O$ from core to rim. This zoning is consistent with independent calculations of oxygen isotope growth zoning in a closed chemical and isotopic system. Squares and triangles correspond to individual composition measurements (source Kohn *et al.*, 1993).

in $\delta^{18}O$ from core to rim (Figure 16) is consistent with prograde growth over the 75 °C temperature interval that had been inferred from major element zoning. Staurolite from the same sample had no resolvable zoning in its interior, which again was consistent with closed-system models. The internal consistency between the theoretical models and observed zonation in this study implies that isotopic partitioning was maintained during prograde metamorphism, and that the rock was closed to infiltration by an isotopically disequilibrium fluid. Conversely, one of the main objectives of measuring oxygen isotope zoning profiles is to identify open system processes by seeking out core–rim isotopic trends that cannot be explained by closed-system models (Chamberlain and Conrad, 1991, 1993; Young and Rumble, 1993; Jamtveit and Hervig, 1994; Kohn and Valley, 1994; Van Haren *et al.*, 1996; Crowe *et al.*, 2001; Skelton *et al.*, 2002).

3.07.3.2 Diffusion

Different minerals have radically different oxygen diffusivities (see review by Cole and Chakraborty (2001)), so there is potentially a wealth of petrologic information encoded in the stable isotopic products of diffusion. The effects of diffusive exchange on bulk mineral compositions are widely studied. This is, in large part, an outgrowth of the seminal work of Giletti (1986). Giletti (1986) used the closure temperature concept of Dodson (1973) together with the temperature dependence of isotope partitioning and oxygen isotope mass balance to describe the changes to bulk mineral composition that occur in a rock during cooling. To a first order, his model is accurate for slowly cooled rocks. However, the analytical expression of Dodson (1973) requires a uniform matrix reservoir, whereas the matrix adjusts its partitioning character as minerals successively close to isotope exchange. Therefore, Eiler *et al.* (1992, 1993) modified Giletti's model to account for modes, grain sizes, and diffusive profiles. The Eiler *et al.* models are more rigorous, explain more observations, especially from mineralogically uncommon rocks, and predict isotopic zoning profiles in each mineral. Jenkin *et al.* (1994) independently modeled and explored mode and grain size effects using a similar approach.

An important feature of oxygen diffusion rates is their very strong dependence on water fugacity, f_{H_2O} (for a detailed discussion, see Cole and Chakraborty (2001, pp. 126–155)), although there are several alternative explanations for this behavior (e.g., Elphick and Graham, 1988; Zhang *et al.*, 1991; Graham and Elphick, 1990; McConnell 1995; Doremus, 1998, 1999). Regardless, the f_{H_2O} history strongly influences diffusion rates and isotopic compositions, and should be taken into account explicitly in diffusion models (Kohn, 1999). Conversely, if grain size, grain geometry, and cooling rate are independently determined, differences in isotopic composition or zoning patterns can potentially reveal variations in f_{H_2O} that are difficult to ascertain from mineralogy alone (e.g., Edwards and Valley, 1998).

There are few published oxygen isotope diffusion profiles in natural samples, mainly due to analytical difficulties. The length scale of a natural diffusion profile is commonly of the order of several μm to ~100 μm, and the magnitude of the isotopic change is small (<1‰) except in the outer few microns. Thus, excellent spatial and/or analytical precision is required to demonstrate that an isotopic profile exists, let alone that it is the product of diffusion. Ion microprobe analysis has excellent spatial resolution (ca. 10 μm), but is relatively imprecise analytically (±1‰), whereas microsampling in combination with laser fluorination is analytically precise (±0.1‰), but has much poorer spatial resolution (~250 μm). Both techniques underwent rapid analytical improvements at the end of the twentieth century (e.g., Young *et al.*, 1998a,b; Valley *et al.*, 1998) and may soon permit relatively routine measurement of natural diffusion profiles. Oxygen zoning trends reportedly due to diffusional exchange have been measured in calcite (Wada, 1988; Arita and Wada, 1990; Graham *et al.*, 1998; Wada *et al.*, 1998), magnetite (Valley and Graham, 1991; Eiler *et al.*, 1995), and garnet (Burton *et al.*, 1995).

Granulite facies calcite from the Hida metamorphic belt, Japan, exhibits one of the best-documented isotope profiles that is at least in part due to diffusional exchange (Wada, 1988; Arita and Wada, 1990; Graham *et al.*, 1998). The marbles are modally dominated by extremely coarse-grained calcite. Graphite is the only other mineral present. No oxygen diffusion profile is to be expected, because there is no other mineral with which the marble could exchange. The discovery of near-rim $\delta^{18}O$ decreases (Wada, 1988), therefore, implies that the rock was infiltrated by an isotopically distinct fluid. Profiles for un-recrystallized calcite margins are well matched by error-function curves (Figure 17), as expected for diffusion-controlled isotopic equilibration (Graham *et al.*, 1998).

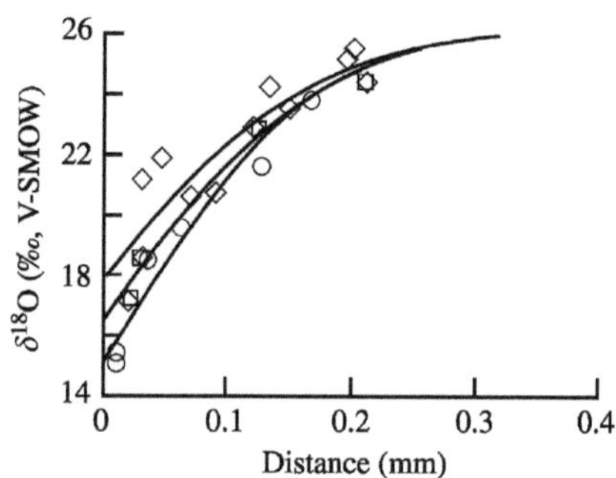

Figure 17 Oxygen isotope composition versus distance for unaltered calcite grains from the Hida metamorphic belt, Japan. Each set of symbols reflects a separate zoning profile measurement. Data represented by circles have been corrected for an estimated 30° dip of the grain boundary. These profiles are well described by diffusional exchange (fitted curves) (source Graham *et al.*, 1998).

3.07.3.3 Dissolution–Reprecipitation

Hot fluids readily dissolve and transport material, both in hydrothermal systems (see Chapter 3.15) and in deeper metamorphic environments (see Chapter 3.06). If such fluids are isotopically distinct, then stable isotopes can help define fluid–rock reactions. Isotopic processes at deeper levels are likely to be rather subtle, because temperatures are high, timescales are long, and different oxygen reservoirs may not be very distinct isotopically. Therefore, isotopic studies of contact metamorphism provide some of the best examples of the mechanisms by which fluids and rocks interact and exchange oxygen.

The Tertiary Skye igneous complex has been well studied isotopically. It contains multiple, shallow-level intrusions, which established successive hydrothermal systems involving low $\delta^{18}O$ meteoric water (Forester and Taylor, 1977; Elsenheimer and Valley, 1993; Valley and Graham, 1996). Ion microprobe analyses of quartz from an older, highly altered pluton (Valley and Graham, 1996) elucidate exchange mechanisms (Figure 18). Cathodoluminscence images reveal healed microcracks, and the lowest $\delta^{18}O$ values are either in or adjacent to these healed cracks,

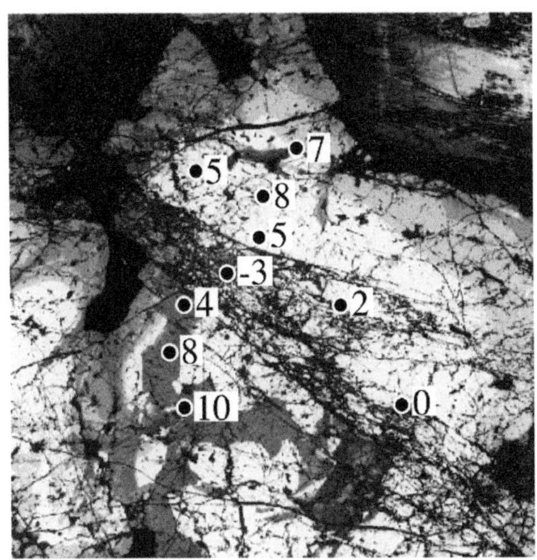

Figure 18 Digitized cathodoluminescence image of Skye quartz grain analyzed for $\delta^{18}O$ via ion microprobe. Dots are analytical spots; numbers are $\delta^{18}O$ in per mil (V-SMOW). Analytical error is approximately $\pm 1‰$. Backscattered electron images and oxygen X-ray maps show that the cracks are now completely healed, and so the cathodoluminscence contrast reflects different generations of quartz. Isotopic compositions are extremely low in the region dominated by healed microfractures, indicating the importance of newly precipitated quartz, possibly coupled with diffusive exchange of surrounding quartz with low $\delta^{18}O$ contact metamorphic fluids (source Valley and Graham, 1996).

indicating that igneous quartz grains were fractured and infiltrated by a fluid with very low $\delta^{18}O$ value. Quartz adjacent to these fractures probably exchanged oxygen diffusively. The fractures were then sealed with very low $\delta^{18}O$ quartz precipitated from the fluid. These data imply that in metamorphic settings with high strain rates and low temperatures, dissolution–reprecipitation is a likely mechanism for producing isotopic heterogeneities.

3.07.4 TRACE ELEMENTS

Trace element zoning is probably ubiquitous in metamorphic minerals. Pioneering analysis by Hickmott and co-workers established trace element zoning as a sensitive monitor of P–T histories (Hickmott *et al.*, 1987; Hickmott and Spear, 1992), changes in major mineral assemblages (Hickmott and Spear, 1992), abrupt breakdown of trace element-rich minerals (Hickmott and Shimizu, 1990), and infiltration of fluids (Hickmott *et al.*, 1992; Hervig and Peacock, 1989). However, the need for an ion microprobe probably deterred most petrologists from making such measurements. Renewed interest has been spurred by the recognition that some trace elements can be investigated with the electron microprobe (Spear and Kohn, 1996) or LA–ICPMS (e.g., Bea *et al.*, 1996), and by an increased petrologic and geochronologic focus on accessory minerals whose stability depends on the availability of trace elements.

Unlike major elements, which can homogenize at high T via simple interdiffusion and exchange with matrix minerals, trace elements often have different charges than the major elements for which they substitute. These differences are expected to impede diffusion. For example, to maintain charge balance, the ^{IV}P–^{IV}Si, ^{VIII}Na–^{VIII}Mg, ^{VIII}Y–^{VIII}Mg, and ^{VI}Ti–^{VI}Al exchanges all require coupled substitution of other cations on other sites, such as ^{IV}Al–^{IV}Si. This may permit the preservation of trace element zonation in minerals that lack obvious major element zonation.

3.07.4.1 Growth Zoning

Pyle and Spear (2003) have presented the only thermodynamically based petrogenetic models for the distribution of trace elements, specifically for yttrium in garnet coexisting with monazite, xenotime, apatite, and the typical silicates of a calcium-poor metapelite. Their differential thermodynamic models expand the theoretical treatment described above for major elements, to include yttrium and phosphates. Of particular interest is the zonation of yttrium predicted in garnet in a prograde metamorphic sequence, and the demise of xenotime as an accessory mineral. Garnet is a common target of yttrium zoning studies because it readily accommodates yttrium as "YAG" ($Y_3Al_5O_{12}$), where substitution of Y^{3+} for divalent cations in the cubic site is charge balanced by substitution of Al^{3+} for Si^{4+} in the tetrahedral site. Dramatic yttrium zoning in one garnet (Figure 19) is correlated with the rim-ward disappearance of xenotime inclusions, and could be modeled in reference to changing accessory mineral abundances (Pyle and Spear, 1999, 2003). Xenotime buffers yttrium in garnet, but garnet growth also consumes xenotime. If the mode of xenotime is sufficiently low, it will eventually disappear, and further growth of garnet then rapidly depletes the remaining matrix minerals in yttrium. These effects yield high yttrium in garnet cores, while xenotime is present, and a sharp dropoff in yttrium as matrix xenotime disappears.

A full theoretical prediction for a specific bulk composition (Figure 20; Pyle and Spear, 2003) has intriguing implications for the control exerted by major minerals on accessory mineral stability (e.g., Ferry, 2000), and for the repartitioning of trace elements among major and accessory minerals at different P and T. For example, the increase in monazite abundance at the expense of apatite with decreasing P accords with

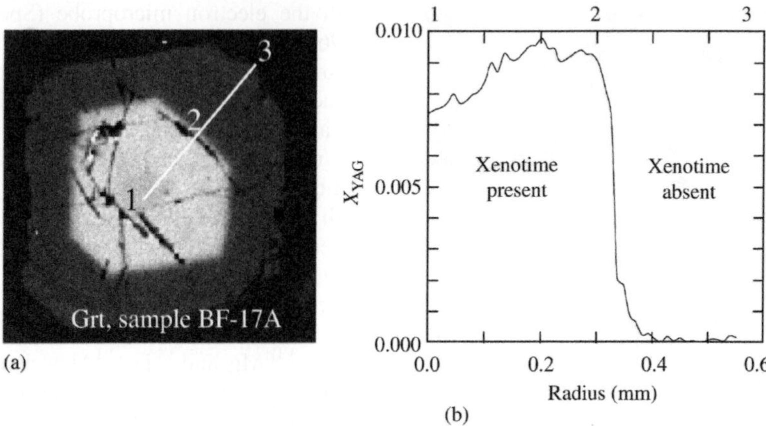

Figure 19 (a) X-ray map and (b) compositional profile of yttrium across a garnet-zone garnet, showing slight increase in yttrium outward from the core, followed by a dramatic drop approximately halfway towards the rim. Yttrium then remains low to the rim (source Pyle and Spear, 2003).

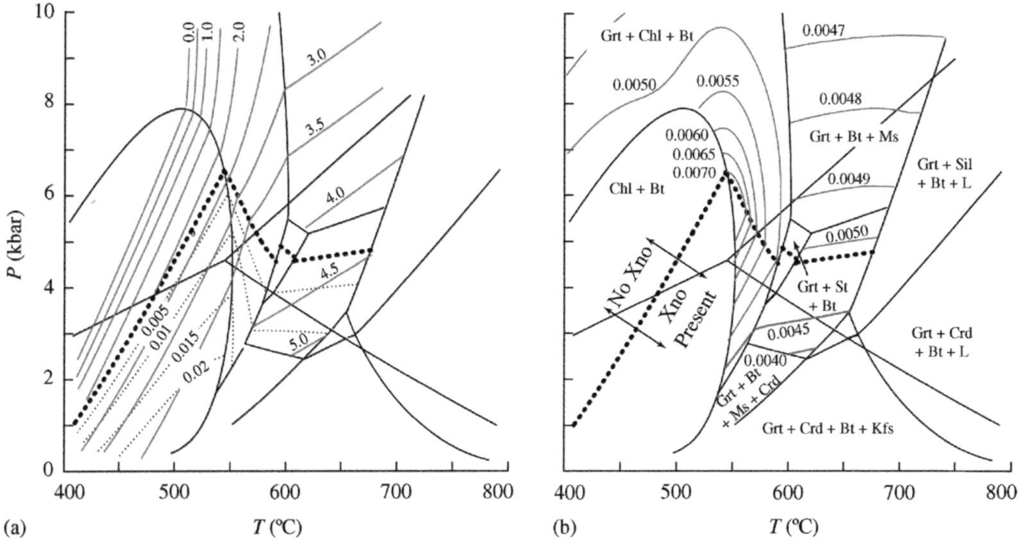

Figure 20 Pseudosection of: (a) moles of monazite and xenotime and (b) X_{YAG} in garnet, in pelitic assemblages. $M_{Monazite}$ contours are thin lines, $M_{Xenotime}$ contours are dotted lines. All models include plagioclase, quartz, and apatite, as well as muscovite, K-feldspar, or melt (depending on P–T conditions). Xenotime is only stable at relatively low P, and monazite abundance decreases at higher P relative to apatite. X_{YAG} contours are strongly dependent on major mineral assemblages (sources Spear *et al.*, 2002; Pyle and Spear, 2003).

observations in ultra-high pressure metamorphic (UHP) terranes that monazite exsolves from apatite during exhumation (Liou *et al.*, 1998). However, other studies suggest that trace element zoning in many garnets is strongly affected by other processes, and that equilibrium models, while useful as a benchmark, may not be as widely applicable for trace elements as for major elements or stable isotopes.

3.07.4.2 Diffusion

No clear examples of diffusion profiles in garnet have been identified. This could reflect either extremely slow diffusivities (e.g., Cherniak, 1998), or simply a lack of natural samples amenable to investigation combined with difficulty of measuring trace element profiles compared to those of major elements. However, Hervig and Peacock (1989) presented an excellent case for lithium diffusion in porphyroclastic quartz from a mylonite (Figure 21). Ion microprobe analysis reveals a "bell-shaped" distribution, with relatively high concentrations in the center and near-zero concentrations on the rim. This profile probably reflects the initial growth of lithium-enriched quartz, followed by diffusive loss of the lithium to an infiltrating, lithium-poor fluid during deformation. Preservation of the profile is surprising because

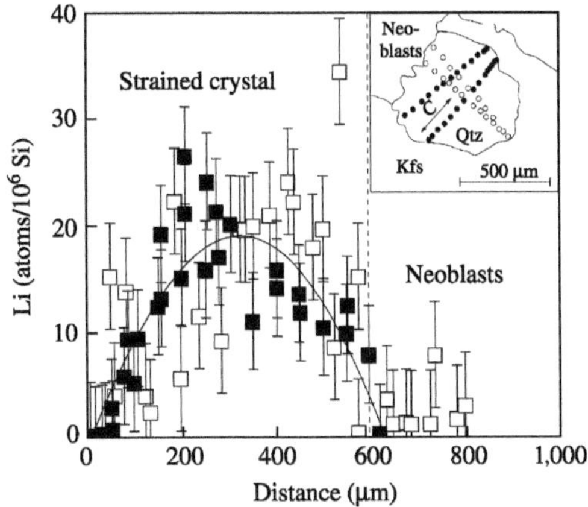

Figure 21 Plot of Li concentration versus distance across a large quartz porphyroclast from the Santa Catalina mylonite zone, California. Solid symbols are parallel to the *c*-axis; open symbols are perpendicular to it. The systematic decrease in Li content towards rim is interpreted to reflect diffusive loss to an intergranular fluid that was depleted in Li. Inset shows sketch of quartz grain in thin section and analyses; "C" and arrows indicate crystallographic orientation of *c*-axis (source Hervig and Peacock, 1989).

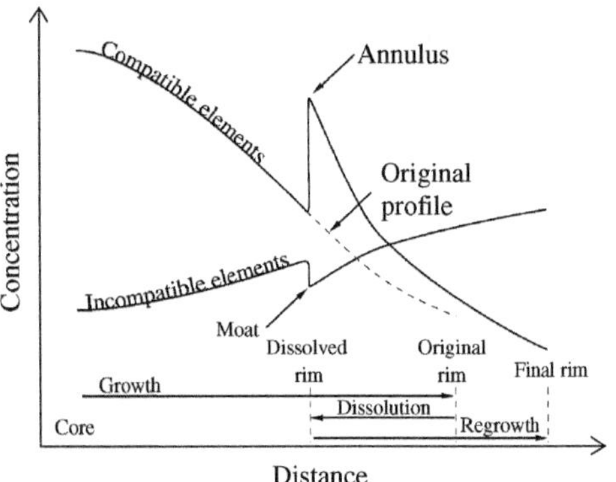

Figure 22 Schematic plot of concentration versus distance for compatible versus incompatible elements in a mineral that is resorbed and regrown. Solid line shows final profile. Dashed line shows original profile for compatible element, prior to dissolution. If trace element zoning is radially distributed, this process would lead to an annulus and moat in compatible and incompatible elements, respectively (source Yang and Rivers, 2002).

experimentally determined lithium diffusivities in quartz are extremely fast (Verhoogen, 1952; White, 1970). Either laboratory determinations of lithium diffusivities are nearly 10 orders of magnitude too fast, because of a different diffusion mechanism, or fluid is required to enhance diffusion, and was present only intermittently during mylonitization. If lithium diffusion rates were better known, the duration of fluid infiltration during mylonitization could potentially be recovered.

3.07.4.3 Dissolution–Reprecipitation

Trace elements can provide a sensitive monitor of dissolution and reprecipitation or regrowth. As summarized by Yang and Rivers (2002; Figure 22), if an element is compatible in a mineral, then resorption will increase the concentration of the element at the rim. Subsequent regrowth will cause a depletion, leading to an annulus in that element (see also Hickmott *et al.* (1987), Spear *et al.* (1990a), Kohn *et al.* (1997), and Pyle and Spear (1999)). If instead an element is incompatible, resorption will cause its decrease at the mineral rim. Subsequent regrowth will produce an increase, leading to a moat. The development and preservation of an annulus or moat depends on the partition coefficient (the degree of compatibility), the amount of material dissolved and then regrown, and trace element diffusivities. Of particular interest is that if diffusivities are extremely slow, the step on the inner side of the annuli and moats should be extremely sharp, whereas if diffusion is non-negligible, back-diffusion towards the mineral core will round the profile.

A staurolite-grade garnet from New Hampshire exhibits a yttrium annulus in garnet (Pyle and Spear, 1999; Figure 23). Most garnets in staurolite-bearing rocks are predicted to have originally grown in an assemblage containing garnet + chlorite + biotite (Spear *et al.*, 1990b). At the staurolite-in isograd, garnet and chlorite are consumed via the staurolite-in reaction (Equation (25)). After chlorite is completely consumed, garnet can regrow, via continuous staurolite and biotite breakdown. This provides a thermodynamic driving force to produce the observed yttrium annulus. Interestingly, the annulus is not completely sharp at the inner boundary, but is rounded on a scale of 10 μm to ~100 μm. This likely implies that yttrium diffusivity cannot be several orders of magnitude slower than that of divalent cations, and that REE diffusion rates determined in natural-composition garnets (Coghlan, 1990; Ganguly *et al.*, 1998b; Van Orman *et al.*, 2002) are more likely to be applicable to rocks than the rates determined in $Y_3Al_5O_{12}$ by Cherniak (1998).

3.07.4.4 Kinetically Limited Transport within the Rock Matrix

Trace element annuli and spikes in garnets have been found over a range of metamorphic grades (Lanzirotti, 1995; Spear and Kohn, 1996; Pyle and Spear, 1999; Chernoff and Carlson, 1999; Pyle *et al.*, 2001; Yang and Rivers, 2002). In some

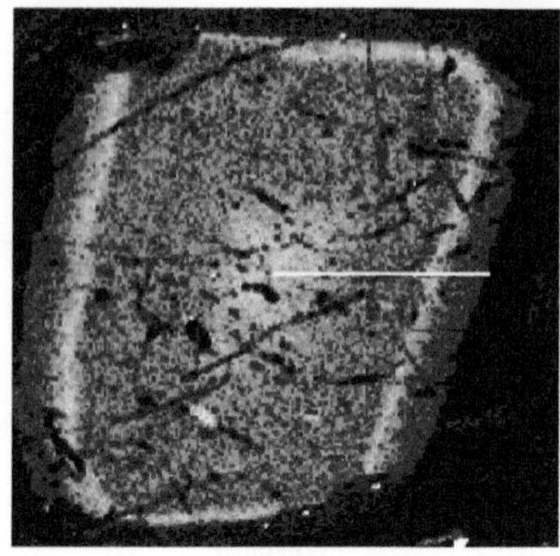

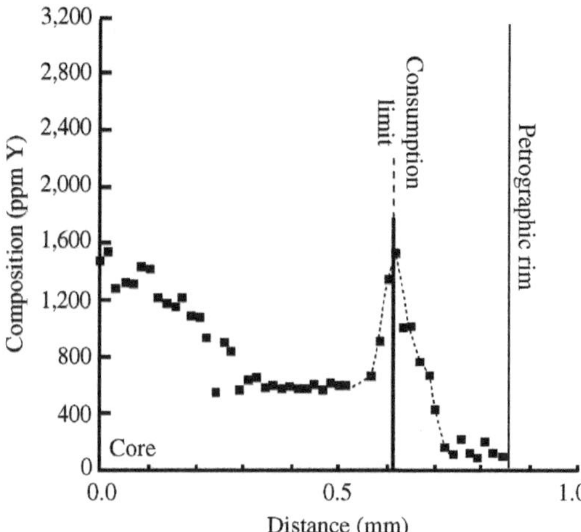

Figure 23 X-ray map of yttrium zoning in a staurolite-grade garnet, illustrating a yttrium annulus, likely caused by prograde dissolution of garnet at the staurolite-in isograd, followed by regrowth. White bar is location of composition traverse. Composition versus distance plot illustrates trend expected conceptually from Figure 22. Note that the annulus does not show a simple step on its inner side, likely indicating some diffusive re-equilibration of Y after rim growth (source Pyle and Spear, 1999).

instances they are ascribed to transport limitations. For example, garnets from Picuris quartzites have ytterbium, yttrium, phosphorous, titanium, and scandium annuli or moats that are spatially coincident with calcium spikes (Chernoff and Carlson, 1999; Figure 24). Kinetically limited intergranular transport of these elements is expected to produce chemically and mineralogically distinct zones concentric about each growing porphyroblast (Chernoff and Carlson, 1997, 1999). As the zones grow outward through time, they begin to impinge on each other, possibly causing abrupt changes in local mineralogy. If a trace phosphate or other mineral within one of these zones suddenly reacts out, the local concentration of many elements—including phosphorous, calcium, and REEs—can change suddenly. Petrologic interpretations may be further refined by using the shape of the profiles (Figure 22), which elements are enriched, and whether element enrichments versus depletions are systematic with respect to garnet compatibility (Yang and Rivers, 2002). For example, a sudden enrichment in LREEs could reflect the breakdown of apatite, allanite, or monazite, enrichment in MREEs could reflect breakdown of epidote, and enrichment in HREEs could reflect the breakdown of xenotime or zircon (Yang and Rivers, 2002). Identification of the reactions which occurred in a particular rock requires careful characterization of textures, zoning patterns, and mineral inclusion suites.

3.07.5 RADIOGENIC ISOTOPES (AGE VARIABILITY)

Radiogenic isotopes are, of course, commonly measured in metamorphic minerals to obtain mineral isochron ages (see Chapter 3.08). However, sample size requirements restrict most isotopic measurements to bulk mineral separates; core versus rim isotopic variability is rarely investigated. Nonetheless, some inferences regarding isotopic trends are obtainable for measured concentration variations in parent isotopes, or from direct measurements of two or more isotopic compositions from large crystals that exhibit systematic zoning of other elements.

3.07.5.1 Growth Zoning

Because most radiogenic isotopes and their parent isotopes are trace elements and/or have large ionic radii, charge balance and/or physical arguments imply that they should have slow diffusivities. Some of the most extensively studied systems include Rb–Sr, Sm–Nd, U–Pb, and Lu–Hf in garnet, and U–Th–Pb in monazite, and experimental studies indeed indicate slow diffusion rates, and hence a high potential to retain original growth compositions (Coghlan, 1990; Smith and Giletti, 1997; Ganguly *et al.*, 1998b; Cherniak *et al.*, 2000; Van Orman *et al.*, 2002). Although growth zoning of parent isotopes is sometimes obvious and easily resolvable (e.g., thorium in monazite; Figure 2), direct measurements of growth profiles in daughter isotopes have not been made with even remotely similar spatial resolution.

Strontium isotopes in garnet provide an interesting means of investigating the timescale over which a mineral grows. Garnet contains very little rubidium, and strontium diffusion rates are extremely slow (Coghlan, 1990; Burton *et al.*, 1995). If isotopic partitioning takes place at equilibrium, garnet will record the matrix $^{87}Sr/^{86}Sr$ during

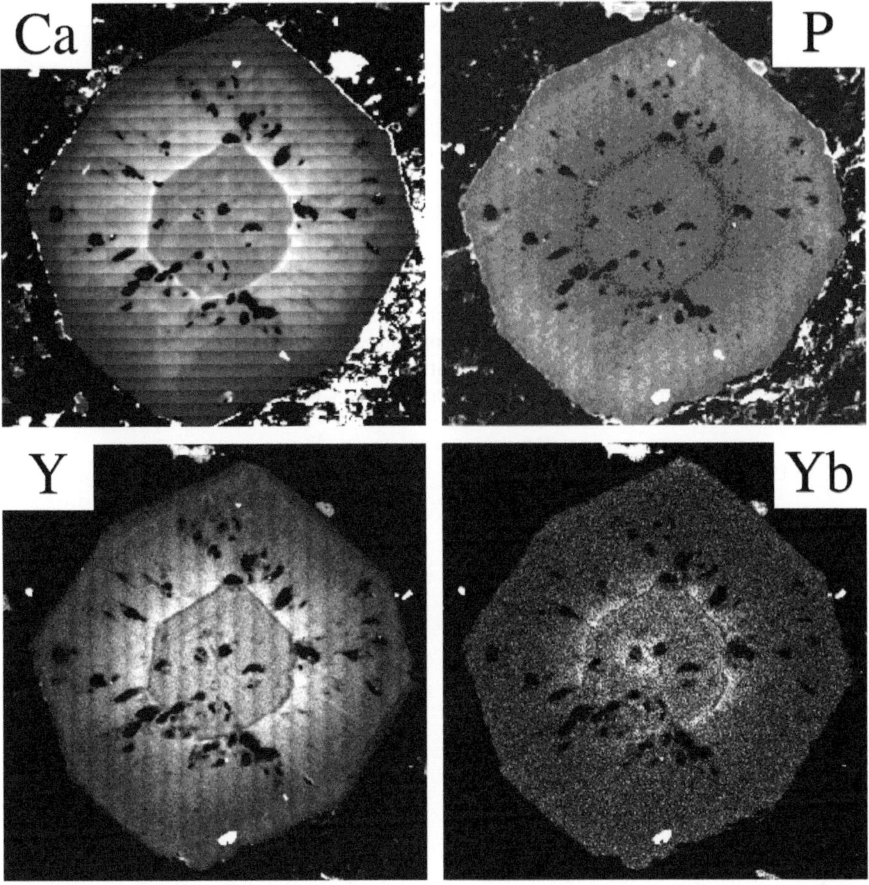

Figure 24 X-ray maps of trace elements and Ca from garnet-bearing quartzite, showing spatially coincident, pronounced spikes. The location of these spikes is systematically different in different-sized garnets, relative to Fe–Mg–Mn systematics (Chernoff and Carlson, 1997, 1999). Diffusionally produced haloes around growing crystals can cause sudden stabilization or destabilization of minerals (Johnson and Carlson, 1990). The coincidence of spikes in many trace elements as well as Ca is interpreted to reflect modal changes in a mineral like apatite or allanite. A common, diffusional mechanism for intergranular diffusion of trace elements and Ca may also be required, possibly by complexing with a common, slow-diffusing "carrier" element (Chernoff and Carlson, 1999) (reproduced by permission of GSA from *Geology*, **1999**, *27*, 555–558).

growth. A high $^{87}Rb/^{86}Sr$ ratio in the matrix assures that its $^{87}Sr/^{86}Sr$ ratio increases with time. Progressive growth of a garnet should, therefore, record a progressively higher $^{87}Sr/^{86}Sr$ ratio towards the rim. The magnitude of the $^{87}Sr/^{86}Sr$ change reflects the $^{87}Rb/^{86}Sr$ ratio of the matrix, and the duration of garnet growth (Δt). Even if garnet contains a small amount of rubidium, or contains small inclusions of rubidium-bearing minerals, its isotopic composition will change slowly, and the intersection of the garnet and matrix growth curves will define the time of garnet growth. Christensen *et al.* (1989, 1994) were able to exploit $^{87}Sr/^{86}Sr$ differences among different segments of individual garnet grains to determine the Δt for crystal growth (Figure 25). Their results demonstrate that regional metamorphic garnets grow over a period of 5–30 Myr, in accordance with thermal model predictions for the timescale of temperature increases of $\sim$50 °C in overthrust terranes (e.g., England and Thompson, 1984).

3.07.5.2 Diffusion

The principles of diffusional zoning in radiogenic isotopes follow many of those for major elements and stable isotopes. Just as for stable isotopes, measurement of diffusion profiles in radiogenic isotopes is much more difficult than growth zoning, and instead bulk analysis of different coexisting minerals has been used to assess the magnitude of diffusional re-equilibration (Giletti and Casserly, 1994; Jenkin *et al.*, 1995, 2001; Jenkin, 1997). Similarly, core versus rim isotopic compositions of garnet coronas adjacent to orthopyroxene and plagioclase have been used

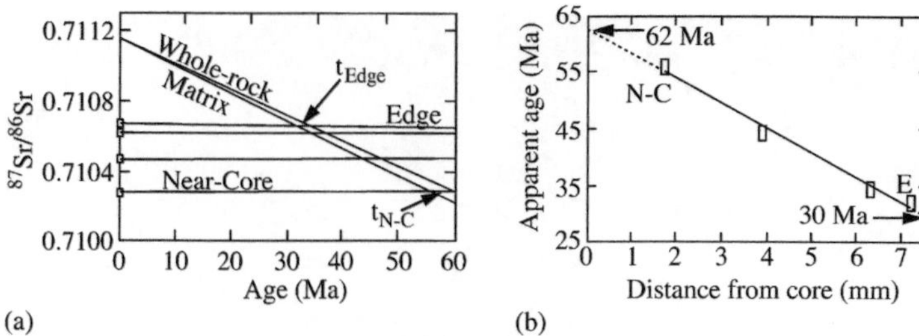

Figure 25 (a) Sr isotope composition of segments from a large garnet and its rock matrix, illustrating very slow change in isotopic composition of garnet relative to matrix. Intersection of garnet and matrix Sr-evolution curves yields ages. (b) Apparent age versus distance from core for same garnet, showing a ~20 Myr time difference. This implies a ~0.23 mm Myr^{-1} radial growth rate, and nucleation and rim ages of ~62 Ma and ~30 Ma, respectively. Rectangles reflect measurement errors (source Christensen *et al.*, 1994).

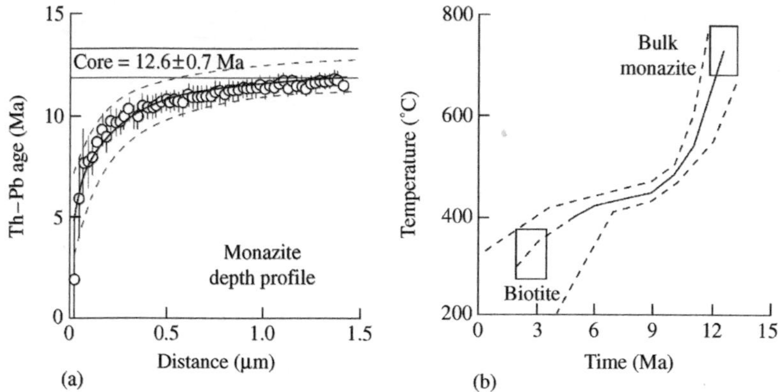

Figure 26 (a) Age variation versus depth for ion microprobe depth profiles collected from the faces of Himalayan monazite grains, showing apparent diffusive loss of ^{208}Pb towards rim, and consequent age increase towards core. Solid curve is best fit to the data. Dashed lines show implied compositional trends for bounds on cooling history. Vertical bars are analytical errors. (b) Inferred temperature–time history, based on composition profile in (a). Solid line is best fit to data, dashed lines are bounds. White boxes are temperature–time points based on peak metamorphic conditions, and $^{40}Ar/^{39}Ar$ cooling age from biotite. Size of boxes reflects age and temperature uncertainties (Grove and Harrison, 1999) (reproduced by permission of GSA from *Geology*, **1999**, *27*, 487–490).

to infer closure temperatures for the Sm–Nd and Rb–Sr systems relative to U–Pb (Burton *et al.*, 1995).

In contrast to these relatively crude studies, Grove and Harrison (1999) showed that diffusional zoning of ^{208}Pb in monazite could be measured with high spatial and analytical resolution via ion microprobe. Because monazite has such a high concentration of thorium, sufficient ^{208}Pb is produced to permit relatively straightforward measurements (Harrison *et al.*, 1995). Grove and Harrison (1999) analyzed the cores of crystals using sectioned, polished grains, and the outer few microns of other crystals, using depth profiling. In depth profiling, grains with good crystal faces are pressed into a soft mounting medium. The primary ion beam is used to sputter a crater into the face of a crystal, and isotopic compositions are collected continuously. These compositions are periodically averaged, and since the sputter rate is known (from the depth of the crater and total sputter time), the depth corresponding to each of these compositions can be determined. Depth profiling into the faces of crystals from a rock from the Greater Himalayan Sequence, Nepal, yields a monotonically increasing apparent age (Figure 26(a)), which can be inverted to obtain a temperature–time history (Figure 26(b)). The retrieved cooling path is generally compatible with other chronologic data, and reveals details of the cooling history that are otherwise difficult to resolve.

In principle, diffusion zoning profiles for radiogenic isotopes are inherently much better than major element profiles at resolving cooling histories. Each point on a composition profile fundamentally reflects *temperature* for major elements via cation exchange thermometry with matrix minerals versus *time* for radiogenic isotopes via the radioactive decay equation. The distance of that point from the rim defines *time* for major elements versus *temperature* for radiogenic elements (specifically $T_c(x)$, Equation (19)). Retrieval of t from zoning profiles depends *directly* on temperature, cooling rate, and diffusion parameters, so has large uncertainty. $T_c(x)$ depends on their *logarithm*, so has much smaller uncertainty. Thus, cooling rates inferred from major elements are much more uncertain than those inferred from radiogenic isotopes.

3.07.5.3 Dissolution–Reprecipitation

Accessory minerals commonly contain high concentrations of radioactive elements, and are a common target of radiogenic isotope measurements. Specific elements include uranium (zircon, apatite, titanite, monazite, xenotime, allanite) and thorium (monazite and allanite). Each accessory mineral is stabilized in a rock via a single element or suite of related elements, specifically phosphorous (apatite), REE (allanite, monazite, xenotime), zirconium (zircon), and titanium (titanite). Trace elements also occur in the major minerals (particularly phosphorous, zirconium, and titanium), so accessory minerals participate directly in major mineral reactions (Pyle and Spear, 1999, 2000, 2003; Ferry, 2000; Pyle *et al.*, 2001; Spear and Pyle, 2002; Wing *et al.*, 2003; Kohn and Malloy, in press). However, the slow diffusion of many elements in accessory minerals as verified experimentally (e.g., Cherniak, 1995, 2000; Cherniak *et al.*, 1997) implies that dissolution–reprecipitation must be a common mechanism for changing compositions, and that different zones within single grains should have different elemental and isotopic compositions.

Monazite ubiquitously exhibits this type of behavior. Backscattered electron images and yttrium, thorium, and uranium X-ray maps nearly always reveal complex zonation (e.g., Parrish, 1990; DeWolf *et al.*, 1993; Zhu *et al.*, 1997; Zhu and O'Nions, 1999; Williams *et al.*, 1999; Pyle *et al.*, 2001; Townsend *et al.*, 2001; Williams and Jercinovic, 2002; see Figures 27 and 28), and several studies have demonstrated significant age differences between these chemically distinct domains (e.g., DeWolf *et al.*, 1993; Zhu *et al.*, 1997; Zhu and O'Nions, 1999; Williams *et al.*, 1999; Townsend *et al.*, 2001; Figures 27 and 28). Extreme compositional and age heterogeneity implies that the analysis of a bulk mineral separate or even of a single grain is not very useful

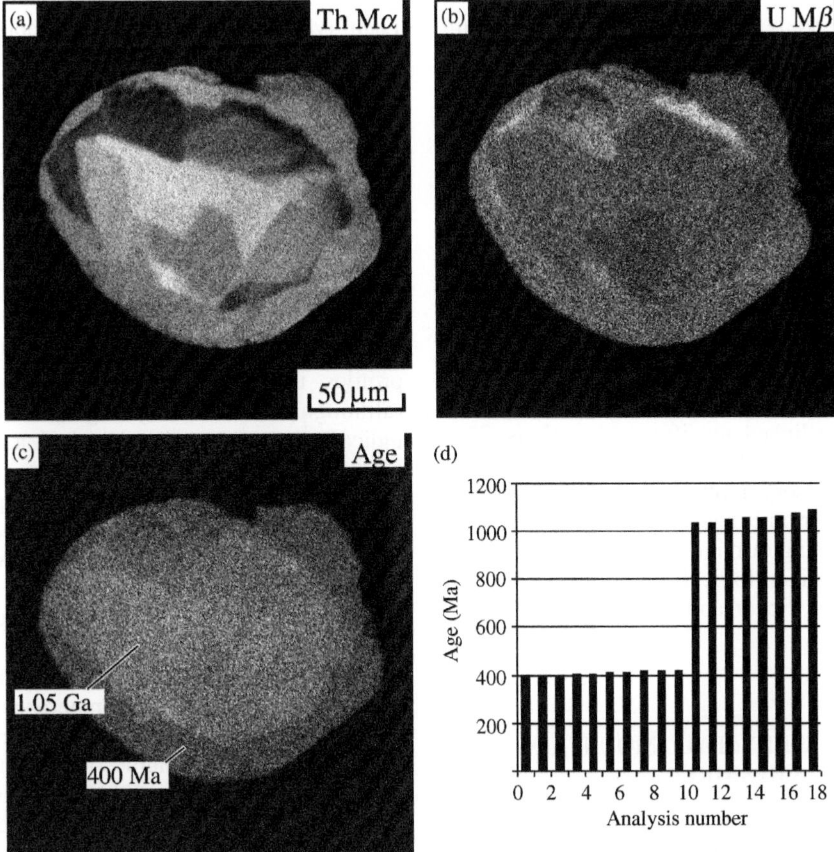

Figure 27 (a, b) Th and U X-ray maps; (c) age map; and (d) age histogram for monazite grain from the western gneiss region, Norway, as measured via the electron microprobe, showing correlation of ages (and implied isotopic composition) with some chemically delineated zones within grains (Williams *et al.*, 1999) (reproduced by permission of GSA from *Geology*, **1999**, *27*, 1023–1026).

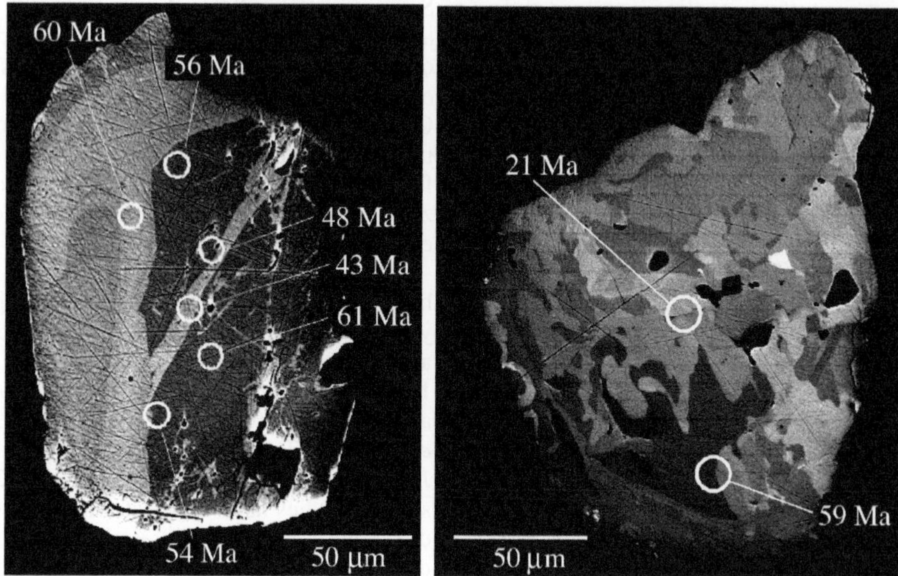

Figure 28 Backscattered electron image and age measurements for complexly zoned monazite crystals from contact metamorphosed Ireteba granite, Nevada. Crystallization age of the pluton is ~65 Ma. Subsequent infiltration of fluids perhaps as young as ~16 Ma apparently dissolved and reprecipitated new monazite in an extremely complex fashion. Ages determined via ion microprobe, and have typical uncertainties of a few percent (source Townsend *et al.*, 2001).

geochronologically, because data from zones of different origin and age are then averaged. By analogy, geochronologists have long since recognized the domainal behavior of zircon and its ages, and we cannot now imagine analyzing bulk zircon separates, or even single whole grains without first characterizing their internal chemistry via backscattered electron imaging or cathodoluminescence. Conversely, because there is a direct link between major mineral reactions and accessory mineral abundance and chemistry (Pyle and Spear, 1999, 2000, 2003; Ferry, 2000; Wing and Ferry, in press; Kohn and Malloy, in press), different zones within an accessory mineral grain can potentially be linked to different reactions. The most fruitful research will be to link chemical zones in accessory minerals with metamorphic reactions, determine ages for those zones, and deduce the timing of mineral reactions and hence the overall mineralogical evolution of metamorphic rocks (Spear and Pyle, 2002; Harrison *et al.*, 2002).

3.07.6 CASE STUDY: FALL MOUNTAIN, NEW HAMPSHIRE

Garnets in rocks from the Fall Mountain nappe, southwestern New Hampshire, have been analyzed for intracrystalline zoning in major elements, trace elements, and stable isotopes (Spear *et al.*, 1990a; Spear and Kohn, 1996; Kohn *et al.*, 1997; Pyle and Spear, 1999, 2000). Each type of zoning reveals important details about thermal evolution, tectonism (loading), and mineral reactions. Major and trace elements show evidence for at least four and possibly five generations of garnet growth, whereas oxygen isotope profiles monitor fluid infiltration and/or open system behavior (Figures 29–31). All generations of garnets, the important reactions that produce or consume garnet, and oxygen isotope trends are summarized in Figure 32.

Garnet cores (Grt_1 + Grt_2) are believed to have grown at low pressure (andalusite field), but flat profiles in manganese, iron, and magnesium (although not calcium; Figures 29 and 30) indicate that these rocks reached a sufficiently high temperature to homogenize manganese, iron, and magnesium by diffusion (Spear *et al.*, 1990a, 1995). The existence of Grt_1 is suggested by patchy zoning in calcium in garnet cores (Figure 29), by phase equilibrium arguments (Spear *et al.*, 1995), and by trace-element compositions in some relict cores (Pyle and Spear, 1999). Grt_3 grew during partial melting, which is indicated both macroscopically in migmatitic textures, and by the occurrence of trace element discontinuities towards the rims of some garnets (Figure 29). Most importantly, the abrupt increase in chromium (Figure 29) is diagnostic of a mica-breakdown reaction, specifically the muscovite + plagioclase + quartz dehydration–melting reaction. Although this reaction does not produce or consume garnet, it increases the chromium content of the matrix

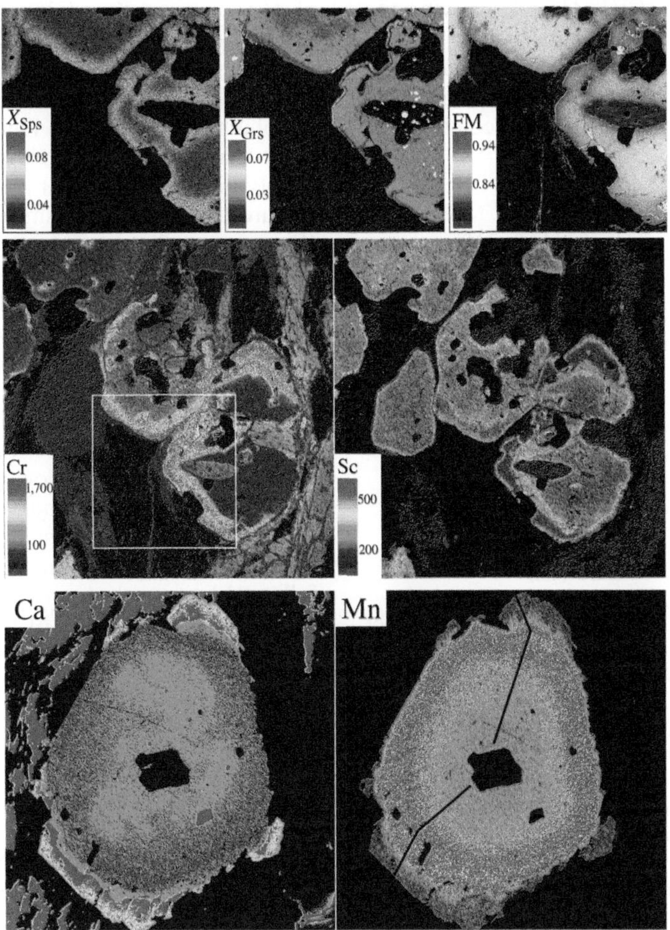

Figure 29 X-ray maps of Mn, Ca, Fe/(Fe + Mg) ("FM"), Cr, and Sc in garnets from Fall Mountain, New Hampshire. White box indicates area of major element maps. Different generations of growth can be linked to prograde reaction history, to better understand mineralogical changes attending metamorphism and anatexis, and overall P–T evolution. Black line shows location of composition traverse plotted in Figure 30 (source Kohn *et al.*, 1997).

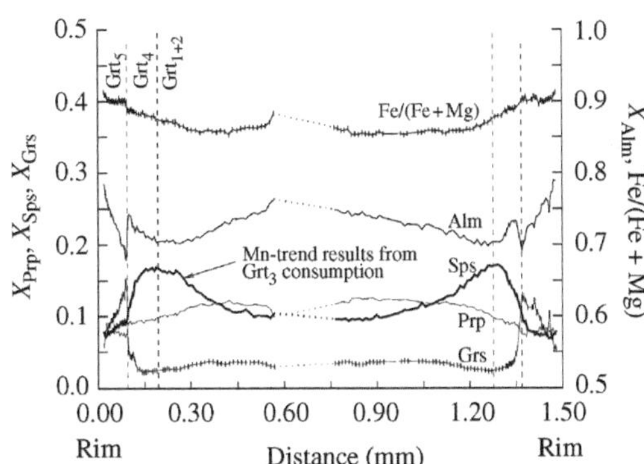

Figure 30 Electron microprobe compositional traverse across garnet in Figure 29, showing relatively flat core, pronounced Mn "humps," and high-Ca rims. Vertical lines separate different garnet generations, as inferred compositionally. Third-generation garnet is not present, but is evident from Cr-zoning in other garnets (Figure 29) (source Kohn *et al.*, 1997).

phases, and further heating produces garnet with increasing chromium contents (Figure 29) via continuous biotite dehydration–melting (Spear and Kohn, 1996; Kohn *et al.*, 1997; Spear *et al.*, 1999). The occurrence of these melting reactions after earlier, low-P metamorphism, rather than dehydration reactions that produce K-feldspar, implies an increase in pressure. Because garnet cores are unzoned in $\delta^{18}O$ (Figure 31), this loading and garnet core growth apparently occurred nearly isothermally.

On the retrograde path, cooling and melt crystallization produced biotite at the expense of garnet. This ReNTR consumed most Grt_3, causing an increase in manganese concentration at the rims of garnets, which diffused back towards garnet cores. At the muscovite dehydration–melting reaction, muscovite reformed (the melt fully crystallized), and in the resulting solid-state assemblage garnet started to grow again with cooling, to produce Grt_4. Garnet growth was accompanied by a decrease in manganese content towards the rim, and differential thermodynamic modeling of this zonation indicates isobaric cooling (Spear *et al.*, 1990a). The combination of Grt_3 consumption followed by Grt_4 growth resulted in the pronounced manganese "hump" (Spear *et al.*, 1990a; Kohn *et al.*, 1997), just as annuli in trace elements can result from dissolution followed by regrowth (Figure 22). The decrease in $\delta^{18}O$ towards garnet rims is simply the result of garnet growth during cooling (Figure 31). High-calcium garnet overgrowths (Grt_5) are believed to reflect late-stage fluid infiltration, as emplacement of the nappe caused dewatering of structurally lower metapelites. The $\delta^{18}O$ composition of quartz, muscovite, and biotite in different rocks provides additional evidence for low-T infiltration.

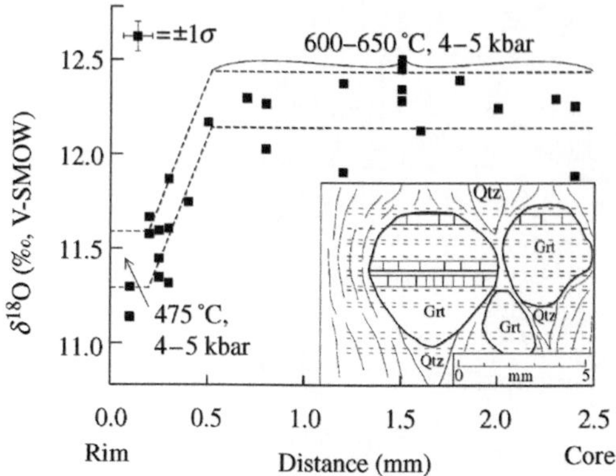

Figure 31 Oxygen isotope profile across garnet porphyroblasts, showing homogeneous core, and pronounced decrease in $\delta^{18}O$ towards rims. The homogeneous core implies growth at relatively constant temperature, whereas the decreasing $\delta^{18}O$ is interpreted to result from growth during cooling. Solid boxes show individual measurements. Dashed line is theoretical prediction assuming an absence of infiltration by an isotopically distinct fluid. Inset shows sketch of garnets analyzed. Horizontal lines are saw cuts used to subsample garnets. Short vertical lines show size of each subsample (source Kohn *et al.*, 1997).

What is interesting about this work is that no one zoning study could possibly have resolved all the details of the mineralogical and $P-T$ evolution. Specifically, (a) major elements indicated general $P-T$ conditions, high temperatures (compositional homogenization), retrograde resorption followed by growth during isobaric cooling, and late-stage overgrowths; (b) trace elements indicated the importance of dehydration–melting reactions; and (c) stable isotopes indicated an isothermal pressure increase, cooling during Grt_4 growth, and a general absence of open system behavior throughout most of the rocks' history. Clearly, a combination of zoning studies provides a much better picture of the $P-T$, mineralogical, and tectonic evolution than any one zoning study. Two particularly interesting results of this work are given as follows.

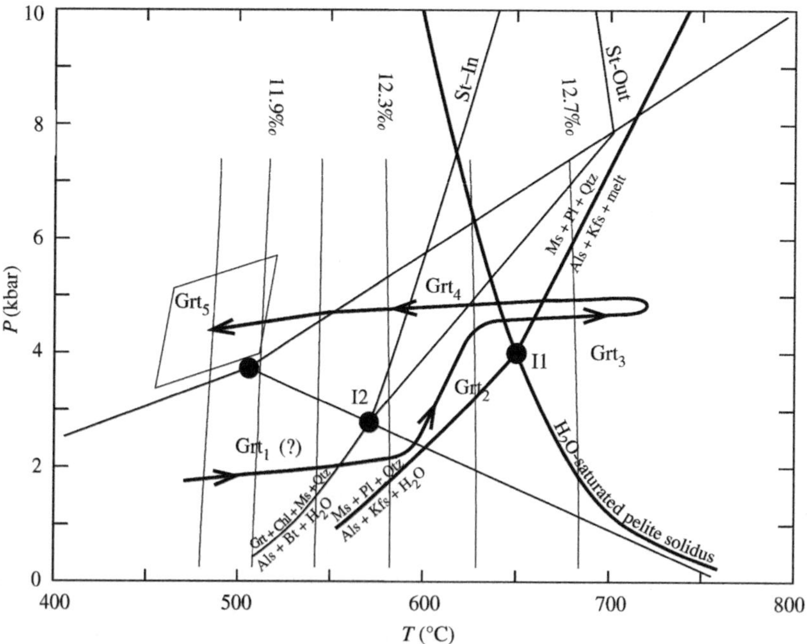

Figure 32 $P-T$ diagram showing important reactions, isopleths of oxygen isotope compositions (thin vertical lines), and inferred $P-T$ path, as deduced from the major element, trace element, and oxygen isotope zoning in garnet porphyroblasts. The parallelogram at the end of the $P-T$ path is the thermobarometrically derived estimate of final $P-T$ conditions (source Kohn *et al.*, 1997).

(i) Fluids were "recycled" during high-grade metamorphism from hydrous minerals to dispersed melt pockets during heating, and then back to hydrous minerals during cooling (Kohn *et al.*, 1997).

(ii) Accessory mineral abundance and the trace element compositions of major silicates are very strongly influenced by melting reactions. This is perhaps not surprising given the high phosphorus solubility in most felsic melts.

3.07.7 DISCUSSION AND CONCLUSIONS

It is perhaps self-evident from this review that the most powerful application of geochemical zoning involves combining theoretical models (both forward and inverse) with detailed geochemical measurements. Theory readily accommodates only end-member processes, however, and recent research has confirmed that minerals do not exhibit end-member equilibrium or disequilibrium behavior with respect to element partitioning and geochemical zoning. It has long been recognized that minerals do not grow without overstepping a reaction (e.g., Lasaga, 1986, 1989). The degree to which a reaction is overstepped and compositions are equilibrated within a rock volume clearly depends on many variables, and reaction overstepping can be quite small for some elements in regional metamorphic rocks, or quite profound, especially in contact metamorphic rocks (Lasaga, 1986, 1989). Thus, some elements may equilibrate on the scale of a thin section (e.g., iron, magnesium, manganese in high-grade rocks), whereas others may not (e.g., REEs). Deciphering these effects from the geochemical record first requires understanding the basic phase equilibria of a rock, and how elements partition and repartition in an equilibrium sense as mineral abundances and species change. Equilibrium processes do explain some chemical behavior quite well, and inversion of composition profiles can yield information on $P-T$ evolution, cooling rates, etc. Conversely, trace elements show local concentration heterogeneities, indicating poor equilibration, at least over length scales of millimeters. In isotope studies, equilibrium partitioning is difficult to verify, and is commonly simply assumed. Any behavior outside equilibrium models is then assigned to open system effects.

The fastest growing areas, in the early 2000s, are studies of geochemical zoning that deal with trace and radiogenic elements. Interest in this field is driven in part by a propensity of minerals to retain growth zoning in trace elements, but also because of geochronologic interests. One highlight of this work is the realization that all minerals, not just accessories, control trace element behavior. Accessory mineral compositions (and therefore preserved ages) are undoubtedly, yet complexly, related to major mineral reactions. The advent of geochronologic microanalysis of inclusions and zones in grains has inspired a new generation (at least) of petrologic research. However, assigning chronologic significance to an accessory mineral grain or even a single zone within it requires identifying the mineral reaction(s) responsible for its growth. In metamorphic rocks whose petrologic and reaction histories are unknown, geochronologic interpretation is perilous or, in some cases, pointless.

Several areas deserve additional study, especially in kinetics and in the phase equilibria of accessory minerals. Some minerals and elements probably equilibrate rapidly because of fast inter- and intragranular diffusion rates (e.g., oxygen). For other elements (e.g., strontium, neodymium, etc.) there is, in the early 2000s, very little information available. If different elements do not equilibrate on the scale of a thin section, then different regions on this scale could have different isotopic compositions. There are as yet no systematic data for the effects of water fugacity on cation diffusion rates. There could be major differences between diffusive behavior during heating when water is being liberated and f_{H_2O} is high, and during cooling, when water is being consumed and f_{H_2O} rapidly decreases. The links among major and trace mineral reactions, modal abundances, and compositions are also paramount for understanding metamorphic geochemistry. There are several examples of chemical zoning, such as annuli, that are extremely common in trace elements, but rare in major elements. Some annuli can be explained by equilibrium and mass balance processes, and some cannot. A significant effort is required to explore the systematics of accessory mineral chemistry and its relation to major mineral reactions and abundances.

New technologies always yield new insights. The increasing availability of laser ablation ICPMS and ion microprobes, and more general application of electron microprobes for mapping and analyzing minor and trace elements, implies that petrologists have a substantially larger suite of elements and isotopes at their disposal than ever before. Zoning patterns in trace elements and isotopes will undoubtedly reveal insights into growth processes, diffusion, and kinetics. Nonetheless, interpretation of the new data will require a full understanding of the many processes, equilibrium or otherwise, that lead to geochemical zoning in metamorphic minerals.

ACKNOWLEDGMENTS

The author thanks Frank Spear for discussing petrologic issues these many years and for sending original figures, and Bill Carlson for sending original photographs and figures. The author also thanks R. Rudnick and B. Wing for reviews. This work was funded by NSF grant EAR 0073803.

REFERENCES

Anderson D. E. and Olimpio J. C. (1977) Progressive homogenization of metamorphic garnets, South Morar, Scotland: evidence for volume diffusion. *Can. Mineral.* **15**, 205–216.

Arita Y. and Wada H. (1990) Stable isotopic evidence for migration of metamorphic fluids along grain boundaries of marbles. *Geochem. J.* **24**, 173–186.

Bea F., Montero P., Stroh A., and Baasner J. (1996) Microanalysis of minerals by an Excimer UV–LA–ICP–MS system. *Chem. Geol.* **133**, 145–156.

Berman R. G. (1988) Internally-consistent thermodynamic data for minerals in the system Na_2O–K_2O–CaO–MgO–FeO–Fe_2O_3–Al_2O_3–SiO_2–TiO_2–H_2O–CO_2. *J. Petrol.* **29**, 445–522.

Burton K. W., Kohn M. J., Cohen A. S., and O'Nions R. K. (1995) The relative diffusion of Pb, Nd, Sr, and O in garnet. *Earth Planet. Sci. Lett.* **133**, 199–211.

Carlson W. D. (1989) The significance of intergranular diffusion to the mechanism and kinetics of porphyroblast crystallization. *Contrib. Mineral. Petrol.* **103**, 1–24.

Carlson W. D. (1991) Competitive diffusion-controlled growth of porphyroblasts. *Min. Mag.* **55**, 317–330.

Carlson W. D. (2002) Scales of disequilibrium and rates of equilibration during metamorphism. *Am. Mineral.* **87**, 185–204.

Carlson W. and Schwarze E. (1997) Petrological significance of prograde homogenization of growth zoning in garnet: an example from the Llano Uplift. *J. Metamorph. Geol.* **15**, 631–644.

Carlson W. D., Denison C., Ketcham R. A., and Boyle A. P. (1995) Controls on the nucleation and growth of porphyroblasts: kinetics from natural textures and numerical models. *Geol. J.* **30**, 207–225.

Chacko T., Cole D. R., and Horita J. (2001) Equilibrium oxygen, hydrogen and carbon isotope fractionation factors applicable to geologic systems. *Rev. Mineral.* **43**, 1–81.

Chakraborty S. and Ganguly G. (1991) Compositional zoning and cation diffusion in garnets. In *Diffusion, Atomic Ordering, and Mass Transport: Selected Topics in Geochemistry* (ed. J. Ganguly). Springer, New York.

Chakraborty S. and Ganguly J. (1992) Cation diffusion in aluminosilicate garnets: experimental determination in spessartine–almandine diffusion couples, evaluation of effective binary, diffusion coefficients, and applications. *Contrib. Mineral. Petrol.* **111**, 74–86.

Chakraborty S. and Rubie D. C. (1996) Mg tracer diffusion in aluminosilicate garnets at 750–850 degrees C, 1 atm. and 1300 degrees C, 8.5 GPa. *Contrib. Mineral. Petrol.* **122**, 406–414.

Chamberlain C. P. and Conrad M. E. (1991) Oxygen isotope zoning in garnet. *Science* **254**, 403–406.

Chamberlain C. P. and Conrad M. E. (1993) Oxygen isotope zoning in garnet: a record of volatile transport. *Geochim. Cosmochim. Acta* **57**, 2613–2629.

Cherniak D. J. (1995) Sr and Nd diffusion in titanite. *Chem. Geol.* **125**, 219–232.

Cherniak D. J. (1998) Rare earth element and gallium diffusion in yttrium aluminum garnet. *Phys. Chem. Mineral.* **26**, 156–163.

Cherniak D. J. (2000) Rare earth element diffion in apatite. *Geochim. Cosmochim. Acta* **64**, 3871–3885.

Cherniak D. J., Hanchar J. M., and Watson E. B. (1997) Rare-earth diffusion in zircon. *Chem. Geol.* **134**, 289–301.

Cherniak D. J., Watson E. B., Harrison T. M., and Grove M. (2000) Pb diffusion in monazite: a progress report on a combined RBS/SIMS study. *EOS, Trans., AGU* **81**, S25.

Chernoff C. B. and Carlson W. D. (1997) Disequilibrium for Ca during growth of pelitic garnet. *J. Metamorph. Geol.* **15**, 421–438.

Chernoff C. B. and Carlson W. D. (1999) Trace element zoning as a record of chemical disequilibrium during garnet growth. *Geology* **27**, 555–558.

Christensen J. N., Rosenfeld J. L., and DePaolo D. J. (1989) Rates of tectonometamorphic processes from rubidium and strontium isotopes in garnet. *Science* **244**, 1465–1469.

Christensen J. N., Selverstone J., Rosenfeld J. L., and DePaolo D. J. (1994) Correlation of Rb–Sr geochronology of garnet growth histories from different structural levels within the Tauern Window, Eastern Alps. *Contrib. Mineral. Petrol.* **118**, 1–12.

Coghlan R. A. N. (1990) Studies in diffusional transport: grain boundary transport of oxygen in feldspars, diffusion of oxygen, strontium, and the REE's in garnet, and thermal histories of granitic intrusions in south-central Maine using oxygen isotopes. PhD Thesis, Brown University.

Cole D. R. and Chakraborty S. (2001) Rates and mechanisms of isotopic exchange. *Rev. Mineral.* **43**, 83–223.

Crowe D. E., Riciputi L. R., Bezenek S., and Ignatiev A. (2001) Oxygen isotope and trace element zoning in hydrothermal garnets: windows into large-scale fluid-flow behavior. *Geology* **29**, 479–482.

Daniel C. G. and Spear F. S. (1998) Three-dimensional patterns of garnet nucleation and growth. *Geology* **26**, 503–506.

Dempster T. J. (1985) Garnet zoning and metamorphism of the Barrovian type area, Scotland. *Contrib. Mineral. Petrol.* **89**, 30–38.

Denison C. and Carlson W. D. (1997) Three-dimensional quantitative textural analysis of metamorphic rocks using high-resolution computed X-ray tomography: Part II. Application to natural samples. *J. Metamorph. Geol.* **15**, 45–57.

Denison C., Carlson W. D., and Ketcham R. A. (1997) Three-dimensional quantitative textural analysis of metamorphic rocks using high-resolution computed X-ray tomography: Part I. Methods and techniques. *J. Metamorph. Geol.* **15**, 29–44.

DeWolf C. P., Belshaw N. S., and O'Nions R. K. (1993) A metamorphic history from micron-scale $^{207}Pb/^{206}Pb$ chronometry of Archean monazite. *Earth Planet. Sci. Lett.* **120**, 207–220.

Dodson M. H. (1973) Closure temperature in cooling geochronological and petrological systems. *Contrib. Mineral. Petrol.* **40**, 259–274.

Dodson M. H. (1986) Closure profiles in cooling systems. *Mater. Sci. Forum* **7**, 145–154.

Doremus R. H. (1998) Diffusion of water and oxygen in quartz: reaction–diffusion model. *Earth Planet. Sci. Lett.* **163**, 43–51.

Doremus R. H. (1999) Diffusion of water in crystalline and glassy oxides: diffusion–reaction model. *J. Material. Res.* **14**, 3754–3758.

Edwards K. J. and Valley J. W. (1998) Oxygen isotope diffusion and zoning in diopside: the importance of water fugacity during cooling. *Geochim. Cosmochim. Acta* **62**, 2265–2277.

Eiler J. M., Baumgartner L. P., and Valley J. W. (1992) Intercrystalline stable isotope diffusion: a fast grain boundary model. *Contrib. Mineral. Petrol.* **112**, 543–557.

Eiler J. M., Valley J. W., and Baumgartner L. P. (1993) A new look at stable isotope thermometry. *Geochim. Cosmochim. Acta* **57**, 2571–2583.

Eiler J. M., Valley J. W., Graham C. M., and Baumgartner L. P. (1995) The oxygen isotope anatomy of a slowly cooled metamorphic rock. *Am. Mineral.* **80**, 757–764.

Elphick S. C. and Graham C. M. (1988) The effect of hydrogen on oxygen diffusion in quartz: evidence for fast proton transients? *Nature* **335**, 243–245.

Elsenheimer D. and Valley J. W. (1993) Sub-millimeter scale zonation of $\delta^{18}O$ in quartz and feldspar, Isle of Skye, Scotland. *Geochim. Cosmochim. Acta* **57**, 3669–3676.

England P. C. and Richardson S. W. (1977) The influence of erosion upon the mineral facies of rocks from different metamorphic environments. *J. Geol. Soc. London* **134**, 201–213.

England P. C. and Thompson A. B. (1984) Pressure–temperature–time paths of regional metamorphism: Part I. Heat transfer during the evolution of regions of thickened continental crust. *J. Petrol.* **25**, 894–928.

Ferry J. M. (2000) Patterns of mineral occurrence in metamorphic rocks. *Am. Mineral.* **85**, 1573–1588.

Forester R. W. and Taylor H. P., Jr. (1977) $^{18}O/^{16}O$, D/H, and $^{13}C/^{12}C$ studies of the Tertiary igneous complex of Skye, Scotland. *Am. J. Sci.* **277**, 136–177.

Ganguly J. and Tirone M. (1999) Diffusion closure temperature and age of a mineral with arbitrary extent of diffusion: theoretical formulation and applications. *Earth Planet. Sci. Lett.* **170**, 131–140.

Ganguly J., Cheng W., and Chakraborty S. (1998a) Cation diffusion in aluminosilicate garnets: experimental determination in pyrope–almandine diffusion couples. *Contrib. Mineral. Petrol.* **131**, 171–180.

Ganguly J., Tirone M., and Hervig R. L. (1998b) Diffusion kinetics of samarium and neodymium in garnet, and a method for determining cooling rates of rocks. *Science* **281**, 805–807.

Ganguly J., Dasgupta S., Cheng W., and Neogi S. (2000) Exhumation history of a section of the Sikkim Himalayas, India: records in the metamorphic mineral equilibria and compositional zoning of garnet. *Earth Planet. Sci. Lett.* **183**, 471–486.

Giletti B. J. (1986) Diffusion effects on oxygen isotope temperatures of slowly cooled igneous and metamorphic rocks. *Earth Planet. Sci. Lett.* **77**, 218–228.

Giletti B. J. and Casserly J. E. D. (1994) Strontium diffusion kinetics in plagioclase feldspars. *Geochim. Cosmochim. Acta* **58**, 3785–3793.

Graham C. M. and Elphick S. C. (1990) Some experimental constraints on the role of hydrogen in oxygen and hydrogen diffusion and Al–Si interdiffusion in silicates. In *Diffusion, Atomic Ordering, and Mass Transport, Selected Topics in Geochemistry*, Advances in Physical Geochemistry (ed. J. Ganguly). Springer, New York, vol. 8, pp. 248–285.

Graham C. M., Valley J. W., Eiler J. M., and Wada H. (1998) Timescales and mechanisms of fluid infiltration in a marble: an ion microprobe study. *Contrib. Mineral. Petrol.* **132**, 371–389.

Grove M. and Harrison T. M. (1999) Monazite Th–Pb age depth profiling. *Geology* **27**, 487–490.

Grove T. L., Baker M. B., and Kinzler R. J. (1984) Coupled CaAl–NaSi diffusion in plagioclase feldspar: experiments and applications to cooling rate speedometry. *Geochim. Cosmochim. Acta* **48**, 2113–2121.

Harrison T. M., McKeegan K. D., and LeFort P. (1995) Detection of inherited monazite in the Manaslu leucogranite by 208Pb/232Th ion microprobe dating: crystallization age and tectonic significance. *Earth Planet. Sci. Lett.* **133**, 271–282.

Harrison T. M., Catlos E. J., and Montel J.-M. (2002) U–Th–Pb dating of phosphate minerals. *Rev. Mineral.* **48**, 523–558.

Hervig R. L. and Peacock S. M. (1989) Implications of trace element zoning in deformed quartz from the Santa Catalina mylonite zone. *J. Geol.* **97**, 343–350.

Hickmott D. D. and Shimizu N. (1990) Trace element zoning in garnet from the Kwoiek area, British Columbia: disequilibrium partitioning during garnet growth? *Contrib. Mineral. Petrol.* **104**, 619–630.

Hickmott D. D. and Spear F. S. (1992) Major- and trace-element zoning in garnets from calcareous pelites in the NW Shelburne Falls Quadrangle, Massachusetts: garnet growth histories in retrograded rocks. *J. Petrol.* **33**, 965–1005.

Hickmott D. D., Shimizu N., Spear F. S., and Selverstone J. (1987) Trace element zoning in a metamorphic garnet. *Geology* **15**, 573–576.

Hickmott D. D., Sorensen S. S., and Rogers P. S. Z. (1992) Metasomatism in a subduction complex: constraints from microanalysis of trace elements in minerals from garnet amphibolite from the Catalina Schist. *Geology* **20**, 347–350.

Hoefs J. (1997) *Stable Isotope Geochemistry*. Springer, Berlin.

Holdaway M. J. (1971) Stability of andalusite and the aluminium silicate phase diagram. *Am. J. Sci.* **271**, 97–131.

Holland T. J. B. and Powell R. (1985) An internally consistent thermodynamic dataset with uncertainties and correlations: 2. Data and results. *J. Metamorph. Geol.* **3**, 343–370.

Holland T. J. B. and Powell R. (1990) An enlarged and updated internally consistent thermodynamic dataset with uncertainties and correlations: the system K_2O–Na_2O–CaO–MgO–MnO–FeO–Fe_2O_3–Al_2O_3–TiO_2–SiO_2–C–H_2–O_2. *J. Metamorph. Geol.* **8**, 89–124.

Holland T. J. B. and Powell R. (1998) An internally consistent thermodynamic data set for phases of petrological interest. *J. Metamorph. Geol.* **16**, 309–343.

Hollister L. S. (1966) Garnet zoning: an interpretation based on the Rayleigh fractionation model. *Science* **154**, 1647–1651.

Hollister L. S. (1969) Contact metamorhpism in the Kwoiek Area of British Columbia: an end member of the metamorphic process. *Geol. Soc. Am. Bull.* **80**, 2465–2494.

Jamtveit B. and Hervig R. L. (1994) Constraints on transport and kinetics in hydrothermal systems from zoned garnet crystals. *Science* **263**, 505–508.

Jaoul O. and Sautter V. (1999) A new approach to geospeedometry based on the "compensation law". *Phys. Earth Planet. Inter.* **110**, 95–114.

Jenkin G. R. T. (1997) Do cooling paths derived from mica Rb–Sr data reflect true cooling paths? *Geology* **25**, 907–910.

Jenkin G. R. T., Farrow C. M., Fallick A. E., and Higgins D. (1994) Oxygen isotope exchange and closure temperatures in cooling rocks. *J. Metamorph. Geol.* **12**, 221–235.

Jenkin G. R. T., Rogers G., Fallick A. E., and Farrow C. M. (1995) Rb–Sr closure temperatures in bi-mineralic rocks: a mode effect and test for different diffusion models. *Chem. Geol.* **122**, 227–240.

Jenkin G. R. T., Ellam R. M., Rogers G., and Stuart F. M. (2001) An investigation of closure temperature of the biotite Rb–Sr system: the importance of cation exchange. *Geochim. Cosmochim. Acta* **65**, 1141–1160.

Johnson C. D., and Carlson W. D. (1990) The origin of olivine–plagioclase coronas in metagabbros from the Adirondack Mountains, NY. *J. Metamorph. Geol.* **8**, 697–717.

Kohn M. J. (1993a) Modeling of prograde mineral $\delta^{18}O$ changes in metamorphic systems. *Contrib. Mineral. Petrol.* **113**, 24–39.

Kohn M. J. (1993b) Uncertainties in differential thermodynamic (Gibbs Method) P–T paths. *Contrib. Mineral. Petrol.* **113**, 249–261.

Kohn M. J. (1999) Why most "dry" rocks should cool "wet". *Am. Mineral.* **84**, 570–580.

Kohn M. J. and Malloy M. A. (in press) Formation of monazite via prograde metamorphic reactions among common silicates: implications for age determinations. *Geochim. Cosmochim. Acta*.

Kohn M. J. and Spear F. S. (2000) Retrograde net transfer reaction insurance for pressure–temperature estimates. *Geology* **28**, 1127–1130.

Kohn M. J. and Valley J. W. (1994) Oxygen isotope constraints on metamorphic fluid flow, Townshend Dam, Vermont, USA. *Geochim. Cosmochim. Acta* **58**, 5551–5566.

Kohn M. J., Valley J. W., Elsenheimer D., and Spicuzza M. J. (1993) Oxygen isotope zoning in garnet and staurolite:

evidence for closed system mineral growth during regional metamorphism. *Am. Mineral.* **78**, 988–1001.

Kohn M. J., Spear F. S., and Valley J. W. (1997) Dehydration melting and fluid recycling during metamorphism: Rangeley Formation, New Hampshire, USA. *J. Petrol.* **38**, 1255–1277.

Kohn M. J., Catlos E. J., Ryerson F. J., and Harrison T. M. (2001) Pressure–temperature–time path discontinuity in the Main Central thrust zone, central Nepal. *Geology* **29**, 571–574.

Kretz R. (1974) Some models for the rate of crystallization of garnet in metamorphic rocks. *Lithos* **7**, 123–131.

Kretz R. (1983) Symbols for rock-forming minerals. *Am. Mineral.* **68**, 277–279.

Lanzirotti A. (1995) Yttrium zoning in metamorphic garnets. *Geochim. Cosmochim. Acta* **59**, 4105–4110.

Lasaga A. C. (1979) Multicomponent exchange and diffusion in silicates. *Geochim. Cosmochim. Acta* **43**, 455–469.

Lasaga A. C. (1983) Geospeedometry: an extension of geothermometry. In *Kinetics and Equilibrium in Mineral Reactions* (ed. S. K. Saxena). Springer, pp. 81–114.

Lasaga A. C. (1986) Metamorphic reaction rates laws and development of isograds. *Min. Mag.* **50**, 359–373.

Lasaga A. C. (1989) Fluid flow and chemical reaction kinetics in metamorphic systems: a new simple model. *Earth Planet. Sci. Lett.* **94**, 417–424.

Lasaga A. C. and Jiang J. (1995) Thermal history of rocks: P–T–t paths for geospeedometry, petrologic data, and inverse theory techniques. *Am. J. Sci.* **295**, 697–741.

Lindström R., Viitanen M., Juhanoja J., and Holtta P. (1991) Geospeedometry of metamorphic rocks: examples in the Rantasalmi–Sulkava and Kiuruvesi areas, eastern Finland; biotite–garnet diffusion couples. *J. Metamorph. Geol.* **9**, 181–190.

Liou J. G., Zhang R. Y., Ernst W. G., Rumble D., III, and Maruyama S. (1998) High-pressure minerals from deeply subducted metamorphic rocks. *Rev. Mineral.* **37**, 33–96.

Loomis T. P. (1978a) Multicomponent diffusion in garnet: I. Formulation of isothermal models. *Am. J. Sci.* **278**, 1099–1118.

Loomis T. P. (1978b) Multicomponent diffusion in garnet: II. Comparison of models with natural data. *Am. J. Sci.* **278**, 1119–1137.

Loomis T. P. (1983) Compositional zoning of crystals: a record of growth and reaction history. In *Kinetics and Equilibrium in Mineral Reactions*, Advances in Physical Geochemistry (ed. S. K. Saxena). Springer, New York, vol. 3, pp. 1–60, .

Loomis T. P., Ganguly J., and Elphick S. C. (1985) Experimental determinations of cation diffusivities in aluminosilicate garnets: II. Multicomponent simulation and tracer diffusion coefficients. *Contrib. Mineral. Petrol.* **90**, 45–51.

McConnell J. D. C. (1995) The role of water in oxygen isotope exchange in quartz. *Earth Planet. Sci. Lett.* **136**, 97–107.

McDougall I. and Harrison T. M. (1988) Geochronology and thermochronology by the $^{40}Ar/^{39}Ar$ method. *Oxford Monographs Geol. Geophys.* **9**, 212pp.

Menard T. and Spear F. S. (1996) Interpretation of plagioclase zonation in calcic pelitic schist, south Strafford, Vermont, and the effects on thermobarometry. *Can. Mineral.* **34**, 133–146.

Parrish R. R. (1990) U–Pb dating of monazite and its application to geological problems. *Can. J. Earth Sci.* **27**, 1431–1450.

Pattison D. R. M., Spear F. S., Debuhr C. L., Cheney J. T., and Guidotti C. V. (2002) Thermodynamic modeling of the reaction muscovite + cordierite → Al_2SiO_5 + biotite + garnet + H_2O: constraints from natural assemblages and implications for the metapelitic petrogenetic grid. *J. Metamorph. Geol.* **20**, 99–118.

Polyakov V. B. and Kharlashina N. N. (1994) Effect of pressure on equilibrium isotopic fractionation. *Geochim. Cosmochim. Acta* **58**, 3750–4739.

Powell R. and Holland T. J. B. (1985) An internally consistent thermodynamic dataset with uncertainties and correlations: 1. Methods and a worked example. *J. Metamorph. Geol.* **3**, 327–342.

Powell R. and Holland T. J. B. (1988) An internally consistent thermodynamic dataset with uncertainties and correlations: 3. Applications to geobarometry, worked examples and a computer program. *J. Metamorph. Geol.* **6**, 173–204.

Powell R., Holland T., and Worley B. (1998) Calculating phase diagrams involving solid solutions via non-linear equations, with examples using THERMOCALC. *J. Metamorph. Geol.* **16**, 577–588.

Pyle J. M. and Spear F. S. (1999) Yttrium zoning in garnet: coupling of major and accessory phases during metamorphic reactions. *Geol. Mat. Res.* **1**, http://gmr.minsocam.org/ContentsLink.html.

Pyle J. M. and Spear F. S. (2000) An empirical garnet (YAG) xenotime thermometer. *Contrib. Mineral. Petrol.* **138**, 51–58.

Pyle J. M. and Spear F. S. (2003) Formation, consumption, and compositional zonation of pelitic monazite and xenotime: metamorphic whole-rock reaction controls. *J. Metamorph. Geol.* (in review).

Pyle J. M., Spear F. S., Rudnick R. L., and McDonough W. F. (2001) Monazite–xenotime–garnet equilibrium in metapelites and a new monazite–garnet thermometer. *J. Petrol.* **42**, 2083–2107.

Robinson P. (1991) The eye of the petrographer, the mind of the petrologist. *Am. Mineral.* **76**, 1781–1810.

Rayleigh J. W. S. (1896) Theoretical considerations respecting the separation of gases by diffusion and similar processes. *Philos. Mag.* **42**, 493.

Selverstone J., Spear F. S., Franz G., and Morteani G. (1984) High pressure metamorphism in the SW Tauern window, Austria: P–T paths from hornblende–kyanite–staurolite schists. *J. Petrol.* **25**, 501–531.

Sharp Z. D. (1990) A laser-based microanalytical method for the *in situ* determination of oxygen isotope ratios of silicates and oxides. *Geochim. Cosmochim. Acta* **54**, 1353–1357.

Sharp Z. D. (1992) *In situ* laser microprobe techniques for stable isotope analysis. *Chem. Geol.* **101**, 3–19.

Skelton A., Annersten H., and Valley J. (2002) $\delta^{18}O$ and yttrium zoning in garnet: time markers for fluid flow? *J. Metamorph. Geol.* **20**, 457–466.

Smith H. A. and Giletti B. J. (1997) Lead diffusion in monazite. *Geochim. Cosmochim. Acta* **61**, 1047–1055.

Spear F. S. (1988) The Gibbs method and Duhem's theorem: the quantitative relationships among P, T, chemical potential, phase composition and reaction progress in igneous and metamorphic systems. *Contrib. Mineral. Petrol.* **99**, 249–256.

Spear F. S. (1991) On the interpretation of peak metamorphic temperatures in light of garnet diffusion during cooling. *J. Metamorph. Geol.* **9**, 379–388.

Spear F. S. (1993) *Metamorphic Phase Equilibria and Pressure–Temperature–Time Paths.* Mineralogical Society of America, Washington, DC.

Spear F. S. and Daniel C. G. (1998) Three-dimensional imaging of garnet porphyroblast sizes and chemical zoning: nucleation and growth history in the garnet zone. *Geol. Mat. Res.* **1** http://gmr.minsocam.org/ContentsLink.html.

Spear F. S. and Daniel C. G. (2001) Diffusion control of garnet growth, Harpswell Neck, Maine, USA. *J. Metamorph. Geol.* **19**, 179–195.

Spear F. S. and Florence F. P. (1992) Thermobarometry in granulites: pitfalls and new approaches. *J. Precamb. Res.* **55**, 209–241.

Spear F. S. and Kohn M. J. (1996) Trace element zoning in garnet as a monitor of crustal melting. *Geology* **24**, 1099–1102.

Spear F. S. and Markussen J. C. (1997) Mineral zoning, P–T–X–M phase relations, and metamorphic evolution of some Adirondack granulites, New York. *J. Petrol.* **38**, 757–783.

Spear F. S. and Parrish R. R. (1996) Petrology and cooling rates of the Valhalla Complex, British Columbia, Canada. *J. Petrol.* **37**, 733–765.

Spear F. S. and Pyle J. M. (2002) Apatite, monazite and xenotime in metamorphic rocks. *Rev. Mineral.* **48**, 293–335.

Spear F. S. and Selverstone J. (1983) Quantitative P–T paths from zoned minerals: theory and tectonic applications. *Contrib. Mineral. Petrol.* **83**, 348–357.

Spear F. S., Ferry J. M., and Rumble D. (1982) Analytical formulation of phase equilibria: the Gibbs method. *Rev. Mineral.* **10**, 105–152.

Spear F. S., Selverstone J., Hickmott D., Crowley P., and Hodges K. V. (1984) P–T paths from garnet zoning: a new technique for deciphering tectonic processes in crystalline terranes. *Geology* **12**, 87–90.

Spear F. S., Hickmott D. D., and Selverstone J. (1990a) Metamorphic consequences of thrust emplacement, Fall Mountain, New Hampshire. *Geol. Soc. Am. Bull.* **102**, 1344–1360.

Spear F. S., Kohn M. J., Florence F., and Menard T. (1990b) A model for garnet and plagioclase growth in pelitic schists: implications for thermobarometry and P–T path determinations. *J. Metamorph. Geol.* **8**, 683–696.

Spear F. S., Kohn M. J., and Paetzold S. (1995) Petrology of the regional sillimanite zone, west-central New Hampshire USA, with implications for the development of inverted isograds. *Am. Mineral.* **80**, 361–376.

Spear F. S., Kohn M. J., and Cheney J. T. (1999) P–T paths from anatectic pelites. *Contrib. Mineral. Petrol.* **134**, 17–32.

Spear F. S., Chemey J. T., Pyle J. M., and Storm L. C. (2002) *Monazite: Another Rosetta Stone? Goldschmidt Conference Abstracts*. (http://www.goldschmidt-conference.com/2002/gold2002/).

Thompson A. B. and England P. C. (1984) Pressure–temperature–time paths of regional metamorphism: II. Their influence and interpretation using mineral assemblages in metamorphic rocks. *J. Petrol.* **25**, 929–955.

Townsend K. J., Miller C. F., D'Andrea J. L., Ayers J. C., Harrison T. M., and Coath C. D. (2001) Low temperature replacement of monazite in the Ireteba Granite, southern Nevada: geochronological implications. *Chem. Geol.* **172**, 95–112.

Tracy R. J. (1982) Compositional zoning and inclusions in metamorphic minerals. *Rev. Mineral.* **10**, 355–397.

Valley J. W. and Graham C. M. (1991) Ion microprobe analysis of oxygen isotope ratios in granulite facies magnetites: diffusive exchange as a guide to cooling history. *Contrib. Mineral. Petrol.* **109**, 38–52.

Valley J. W. and Graham C. M. (1996) Ion microprobe analysis of oxygen isotope ratios in quartz from Skye granite: healed micro-cracks, fluid flow, and hydrothermal exchange. *Contrib. Mineral. Petrol.* **124**, 225–234.

Valley J. W., Graham C. M., Harte B., Eiler J. M., and Kinny P. D. (1998) Ion microprobe analysis of oxygen, carbon, and hydrogen isotope ratios. *Rev. Econ. Geol.* **7**, 73–98.

Van Haren J. L. M., Ague J. J., and Rye D. M. (1996) Oxygen isotope record of fluid infiltration and mass transfer during regional metamorphism of pelitic schist, Connecticut, USA. *Geochim. Cosmochim. Acta* **60**, 3487–3504.

Van Orman J. A., Grove T. L., Shimizu N., and Layne G. D. (2002) Rare earth element diffusion in a natural pyrope single crystal at 2.8 GPa. *Contrib. Mineral. Petrol.* **142**, 416–424.

Verhoogen J. (1952) Ionic diffusion and electrical conductivity in quartz. *Am. Mineral.* **37**, 637–655.

Wada H. (1988) Microscale isotopic zoning in calcite and graphite crystals in marble. *Nature* **331**, 61–63.

Wada H., Ando T., and Suzuki M. (1998) The role of the grain boundary at chemical and isotopic fronts in marble during contact metamorphism. *Contrib. Mineral. Petrol.* **132**, 309–320.

Wang P. and Spear F. S. (1991) A field and theoretical analysis of garnet + chlorite + chloritoid + biotite assemblages from the tri-state (MA, CT, NY) area, USA. *Contrib. Mineral. Petrol.* **106**, 217–235.

White S. (1970) Ionic diffusion in quartz. *Nature* **225**, 375–376.

Williams M. L. and Jercinovic M. J. (2002) Microprobe monazite geochronology: putting absolute time into microstructural analysis. *J. Struct. Geol.* **24**, 1013–1028.

Williams M. L., Jercinovic M. J., and Terry M. P. (1999) Age mapping and dating of monazite on the electron microprobe: deconvoluting multistage tectonic histories. *Geology* **27**, 1023–1026.

Wing B. N., Ferry J. M., and Harrison T. M. (2003) Prograde destruction and formation of monazite and allanite during contact and regional metamorphism of pelites: petrology and geochronology. *Contrib. Mineral. Petrol.* (in press).

Woodsworth G. J. (1977) Homogenization of zoned garnets from pelitic schists. *Can. Mineral.* **15**, 230–242.

Yang P. and Rivers T. (2001) Chromium and manganese zoning in pelitic garnet and kyanite: spiral, overprint, and oscillatory (?) zoning patterns and the role of growth rate. *J. Metamorph. Geol.* **19**, 455–474.

Yang P. and Rivers T. (2002) The origin of Mn and Y annuli in garnet and the thermal dependence of P in garnet and Y in apatite in calc-pelite and pelite, Gagnon terrane, western Labrador. *Geol. Mat. Res.* **4** http://gmr.minsocam.org/contents.html.

Yardley B. W. D. (1977) An empirical study of diffusion in garnet. *Am. Mineral.* **62**, 793–800.

Young E. D. (1993) On the $^{18}O/^{16}O$ record of reaction progress in open and closed metamorphic systems. *Earth Planet. Sci. Lett.* **117**, 147–167.

Young E. D. and Rumble D. (1993) The origin of correlated variations in *in-situ* $^{18}O/^{16}O$ and elemental concentrations in metamorphic garnet from southeastern Vermont, USA. *Geochim. Cosmochim. Acta* **57**, 2585–2597.

Young E. D., Fogel M. L., Rumble D., III, and Hoering T. C. (1998a) Isotope-ratio-monitoring of O_2 for microanalysis of $^{18}O/^{16}O$ and $^{17}O/^{16}O$ in geological materials. *Geochim. Cosmochim. Acta* **62**, 3087–3094.

Young E. D., Coutts D. W., and Kapitan D. (1998b) UV laser ablation and irm-GCMS microanalysis of $^{18}O/^{16}O$ and $^{17}O/^{16}O$ with application to a calcium–aluminium-rich inclusion from the Allende Meteorite. *Geochim. Cosmochim. Acta* **62**, 3161–3168.

Zhang Y., Stolper E. M., and Wasserburg G. J. (1991) Diffusion of a multi-species component and its role in oxygen and water transport in silicates. *Earth Planet. Sci. Lett.* **103**, 228–240.

Zhu X. K. and O'Nions R. K. (1999) Zonation of monazite in metamorphic rocks and its implications for high temperature thermochronology: a case study from the Lewisian terrain. *Earth Planet. Sci. Lett.* **171**, 209–220.

Zhu X. K., O'Nions R. K., Belshaw N. S., and Gibb A. J. (1997) Significance of *in situ* SIMS chronometry of zoned monazite from the Lewisian granulites, Northwest Scotland. *Chem. Geol.* **135**, 35–53.

3.08
Geochronology and Thermochronology in Orogenic Systems

K. V. Hodges

Massachusetts Institute of Technology, Cambridge, MA, USA

NOMENCLATURE

a	effective diffusion dimension
C	Euler constant (~0.57721547)
C_0	initial concentration of a diffusant
C_ξ	concentration of a diffusant at distance ξ along a gradient
C_{Pb}	concentration of Pb
C_{Th}	concentration of Th
C_U	concentration of U
D_i	diffusivity at infinite temperature
D_t	diffusivity at time t
D_T	diffusivity at temperature T
E	activation energy
erf^{-1}	inverse error function
G_{av}	average value of the closure function for a specific grain geometry

G_x	closure function, evaluated for radial position x in a grain
k	proportionality constant
n_d	number of daughter isotopes
n_{d0}	number of non-radiogenic daughter isotopes
n_{d*}	number of radiogenic daughter isotopes
n_{FT}	number of fission tracks
n_p	number of parent isotopes
n_{p0}	initial number of parent isotopes
P	pressure
R	universal gas constant (8.3144 J mol^{-1} K^{-1})
t	time
t_{cb}	bulk cooling age
t_{cx}	cooling age for position x in a grain
t_{init}	time of initiation of an unroofing event
t_p	present time
t_0	time of crystallization
$t_{1/2}$	half-life of parent isotope
T	temperature
T_{cb}	bulk closure temperature
T_{cx}	closure temperature as a function of position within a grain
T_t	temperature at time t
T_0	crystallization temperature
x	distance, expressed as a fraction of the radius of a crystal
X_{phl}	mole fraction of phlogopite component in biotite
z	depth
γ	geothermal gradient
ε	elevation
λ_p	decay constant of parent isotope
ξ	distance along a diffusion gradient
σ	sample standard deviation
τ	duration of a diffusion experiment
Φ	constant of proportionality between synthetic ^{39}Ar and naturally occurring ^{40}K, related to the efficiency of the neutron bombardment reaction $^{39}K \rightarrow {}^{39}Ar$ and the essentially constant ratio $^{39}K/^{40}K$

3.08.1 INTRODUCTION

An important part of modern tectonic research is determining the rates of deformational, thermal, and erosional processes that define the evolution of orogenic systems. Such endeavors fall into three categories: studies of the crystallization ages of rocks and minerals (geochronology); studies of the thermal history of rocks (thermochronology); and studies of the exposure ages of geomorphic surfaces (cosmogenic nuclide dating). This chapter focuses on the first two types of study; additional material regarding geochronology may be found in Chapter 3.10. Here a brief treatment of the basic concepts of radioactive decay and isotopic dating is followed by a discussion of how limited open-system behavior of radiogenic isotopes in minerals, as well as the partial annealing of radiation damage in mineral structures, may be used to explore the thermal histories of orogens. Subsequent sections describe specific applications of these techniques to a variety of problems encountered in orogenic systems, ranging from the frequency of magmatism to the timescales of landscape evolution. The final section suggests potential avenues for future research, both to improve the quality of geochronologic and thermochronologic data and to evaluate their capacity to constrain orogenic processes. Readers interested in learning more about such topics may consult isotope geology textbooks such as those by Faure (1986) or Dickin (1995).

3.08.2 BASIC CONCEPTS OF GEOCHRONOLOGY

The process of radioactive decay is usually described in terms of the spontaneous transformation of one isotope (the "parent") into another (the "daughter"). It occurs at a rate that is proportional to the number of parent isotopes (n_p) present in a rock or mineral at any given time:

$$\frac{-dn_p}{dt} = \lambda_p n_p \quad (1)$$

where λ_p is the decay constant for the parent and t represents time. This equation may be integrated to yield a second equation that describes the number of parent isotopes at a given time as a function of the number of parent isotopes present at the time of crystallization (n_{p0}):

$$n_p = n_{p0} e^{-\lambda_p t} \quad (2)$$

After a period of time indicated by t, the decay process will have produced a number of radiogenic daughter isotopes (n_{d^*}) given by

$$n_{d^*} = n_p(e^{\lambda_p t} - 1) \quad (3)$$

Most samples contain one (or more) components of the daughter isotope that were not produced by *in situ* decay of the parent. Instead, these "non-radiogenic" isotopes were trapped during crystallization or during postcrystallization alteration. The total number of the daughter isotopes in a sample at time t (or n_d) is thus the sum of the radiogenic (n_{d^*}) and non-radiogenic atoms (n_{d0}):

$$n_d = n_{d0} + n_p(e^{\lambda_p t} - 1) \quad (4)$$

When solved for t, this relationship yields the *fundamental equation of geochronology*:

$$t = \left(\frac{1}{\lambda_p}\right)\ln\left[\left(\frac{n_d - n_{d0}}{n_p}\right) + 1\right] \quad (5)$$

which may be used to calculate the elapsed time since crystallization of a rock or mineral when: (i) n_p and n_d are measured, present-day abundances; (ii) n_{d0} can be assumed reasonably; and (iii) the sample has remained a closed system with respect to gain or loss of parent and daughter isotopes since the time of crystallization. Table 1 shows the decay systems that are commonly exploited in geochronology. In general, the age ranges over which these chronometers are used depends on the precision and accuracy with which measurements for a particular system can be made, as well as the decay constant; parent isotopes with larger decay constants produce more daughter isotopes over a given time interval.

3.08.2.1 Effects of Branched, Sequential, and Multiparent Decay

The preceding equations hold for decay schemes that involve the simple transformation of one isotope to another. Three types of decay schemes useful for geochronology are special cases that yield somewhat more complicated age equations. *Branched decay* leads to the production of two different daughter isotopes from a single radioactive isotope. A familiar example is the spontaneous conversion of ^{40}K to both ^{40}Ar (through electron capture) and ^{40}Ca (through β decay). In such cases, the age equations must be modified by factors that correct for the fact that not all decays result in the production of the daughter isotope of interest. For the $^{40}K \rightarrow {}^{40}Ar$ decay mechanism, Equation (4) becomes

$$^{40}Ar = {}^{40}Ar_0 + {}^{40}K\left(\frac{\lambda_{40K(ec)}}{\lambda_{40K}}\right)(e^{\lambda_{40K}t} - 1) \quad (6)$$

where ^{40}Ar is the total number of ^{40}Ar isotopes in the sample, $^{40}Ar_0$ is the number of non-radiogenic ^{40}Ar isotopes, ^{40}K is the number of ^{40}K isotopes, $\lambda_{40K(ec)}$ is the decay constant for the $^{40}K \rightarrow {}^{40}Ar$ branch (generally assumed to be $5.81 \times 10^{-11}\ yr^{-1}$; Steiger and Jäger, 1977), and λ_{40K} is the total decay constant for ^{40}K ($5.543 \times 10^{-10}\ yr^{-1}$).

Some radioactive isotopes undergo *sequential decay* through a series of intermediate, radioactive daughter isotopes before finally yielding a stable daughter product. (U–Th)/Pb geochronology results from three such decay series: $^{238}U \rightarrow {}^{206}Pb$, $^{235}U \rightarrow {}^{207}Pb$, and $^{232}Th \rightarrow {}^{208}Pb$. Fortunately, the intermediate daughter isotopes are short-lived compared to the typical timescales studied through (U–Th)/Pb geochronology (10^6–10^9 yr), such that a condition of "secular equilibrium" is reached: the rate of decay of the parent and intermediate daughter isotopes is essentially the same. In such cases, the decay series can be treated mathematically like a simple one-step decay mechanism. For the $^{238}U \rightarrow {}^{206}Pb$ mechanism, Equation (4) is thus

$$^{206}Pb = {}^{206}Pb_0 + {}^{238}U(e^{\lambda_{238U}t} - 1) \quad (7)$$

where ^{206}Pb is the total number of ^{206}Pb isotopes in the sample, $^{206}Pb_0$ is the number of non-radiogenic ^{206}Pb isotopes, ^{238}U is the number of ^{238}U isotopes, and λ_{238U} is the decay constant for ^{238}U ($1.55125 \times 10^{-10}\ yr^{-1}$).

Finally, some daughter isotopes of geochronologic interest result from *multiparent decay*. For example, virtually all ^4He is produced as particles released from the series decay of ^{232}Th, ^{235}U, and ^{238}U. Taking into consideration the number of α particles released as part of each decay series, Equation (4) for ^4He production takes the form

$$\begin{aligned}{}^4He = {}^4He_0 &+ 8^{238}U(e^{\lambda_{238U}t} - 1)\\ &+ 7^{235}U(e^{\lambda_{235U}t} - 1)\\ &+ 6^{232}Th(e^{\lambda_{232Th}t} - 1)\end{aligned} \quad (8)$$

where ^4He is the total number of ^4He isotopes in the sample, $^4He_0$ is the number of non-radiogenic

Table 1 Decay schemes frequently used for geochronology and thermochronology.

Reaction	*Minerals*[a]	λ_p (yr^{-1})	$t_{1/2}$ (Gyr)[b]	*Source*
$^{147}Sm \rightarrow {}^{143}Nd$	Grt, Cpx	6.54×10^{-12}	106.0	Lugmair and Marti (1978)
$^{87}Rb \rightarrow {}^{87}Sr$	Ms, Bt, Phl, Kfs	1.42×10^{-11}	48.8	Steiger and Jäger (1977)
$^{176}Lu \rightarrow {}^{176}Hf$	Grt	1.865×10^{-11}	37.2	Scherer *et al.* (2001)
$^{232}Th \rightarrow {}^{208}Pb^c$	Mz	4.9475×10^{-11}	14.0	Steiger and Jäger (1977)
$^{40}K \rightarrow {}^{40}Ar$	Hbl, Ms, Bt, Phl, Kfs	5.8×10^{-11}	11.9	Steiger and Jäger (1977); Min *et al.* (2000)
$^{238}U \rightarrow {}^{206}Pb^c$	Zcn, Mz, Xn, Ttn, Rt, Ap	1.55125×10^{-10}	4.5	Steiger and Jäger (1977)
^{238}U fission	Zcn, Ap	$\sim 7 \times 10^{-10}$	~1.0	Naeser *et al.* (1989)
$^{235}U \rightarrow {}^{207}Pb^c$	Zcn, Mz, Xn, Ttn, Rt, Ap	9.8485×10^{-10}	0.7	Steiger and Jäger (1977)

[a] Abbreviations: Ap—Apatite; Bt—Biotite; Cpx—Clinopyroxene; Grt—garnet; Hbl—Hornblende; Kfs—K-feldspar; Ms—Muscovite; Mz—Monazite; Phl—phlogopite; Rt—rutile; Ttn—titanite; Xn—xenotime; Zcn—zircon. The half-life of a parent isotope, is equal to $0.693/\lambda_p$.
[b] $t_{1/2}$, the half-life of a parent isotope, is equal to $0.693/\lambda_p$. [c] These decay schemes also produce ^4He, forming the basis for (U–Th)/He geochronology (Equation (7)). Commonly analyzed minerals include Ap, Ttn, and Zcn.

^{4}He isotopes, ^{238}U is the number of ^{238}U isotopes, ^{235}U is the number of ^{235}U isotopes, and ^{232}Th is the number of ^{232}Th isotopes. The decay constant for ^{238}U (λ_{238U}) was given above; for ^{232}Th and ^{235}U (λ_{232Th} and λ_{235U}), they are $4.9475 \times 10^{-11}\,\text{yr}^{-1}$ and $9.8485 \times 10^{-10}\,\text{yr}^{-1}$, respectively.

3.08.2.2 Fission-track Geochronology

One geochronologic technique is based on the secondary effects of radioactive decay rather than the accumulation of daughter isotopes. Although most ^{238}U decay in a mineral involves α-particle emission, a very small proportion occurs through spontaneous fission. Each fission reaction releases ~200 MeV of kinetic energy that propels the resulting fragments through the mineral, disrupting its crystal structure. The number of these disrupted zones, or *fission tracks*, in a sample (n_{FT}) depends on the amount of ^{238}U present and age:

$$n_{FT} = \left(\frac{\lambda_{238U(FT)}}{\lambda_{238U}}\right)^{238}U(e^{\lambda_{238U}t} - 1) \quad (9)$$

where $\lambda_{238U(FT)}$ is the decay constant for ^{238}U fission ($\sim 7 \times 10^{-17}\,\text{yr}^{-1}$). Equation (9) forms the basis for calculating dates for geologic samples by determining uranium content and counting the number of accumulated tracks.

3.08.2.3 Chemical Pb Dating

Another special technique takes advantage of the fact that high-uranium and high-thorium accessory minerals typically contain very small amounts of non-radiogenic lead. If it is assumed that all lead in a sample is radiogenic, and that the abundances of ^{238}U, ^{235}U, and ^{232}Th are essentially constant in natural uranium and thorium, then the total concentration of lead (C_{Pb}) is related to the concentrations of uranium (C_U) and thorium (C_{Th}):

$$\begin{aligned} C_{Pb} \approx\ & 0.897C_{Th}(e^{\lambda_{232Th}t} - 1) \\ & + 0.006C_U(e^{\lambda_{235U}t} - 1) \\ & + 0.859C_U(e^{\lambda_{238U}t} - 1) \end{aligned} \quad (10)$$

where concentrations are in parts per million (ppm) (Montel *et al.*, 1996). From Equation (10), an approximate *chemical Pb date* can be calculated knowing only the concentrations of uranium, thorium, and lead.

3.08.3 ANALYTICAL METHODS

Except for the fission-track and chemical lead methods, all geochronologic techniques involve the use of quadrupole or magnetic-sector mass spectrometers to separate atoms by mass and determine their absolute or relative abundances. Isotopic ratios can be measured more precisely than isotopic abundances on these instruments, so it is common practice to recast many of the above equations in terms of isotopic ratios by dividing both sides through by the number of some stable, non-radiogenic isotope of the daughter element. For example, Equation (7) can be divided by the number of ^{204}Pb isotopes (^{204}Pb) to yield:

$$\frac{^{206}Pb}{^{204}Pb} = \left(\frac{^{206}Pb}{^{204}Pb}\right)_0 + \frac{^{238}U}{^{204}Pb}\left(e^{\lambda_{238U}t} - 1\right) \quad (11)$$

where $(^{206}Pb/^{204}Pb)_0$ is the *initial ratio* of non-radiogenic ^{206}Pb to ^{204}Pb. In most cases, however, the absolute abundance of at least one isotope still must be determined for an age calculation, and this is accomplished by spiking the sample with a solution of known, artificial isotopic composition in preparation for isotope dilution analysis (Dickin, 1995; Albarède, 1995).

For the majority of geochronologically important decay schemes, measurements of parent and daughter isotopic ratios and abundances can be made on a single aliquot of sample with the same, solid-source or plasma-source mass spectrometer. The principal complication with such measurements is the need to separate different elements prior to analysis in order to prevent mass interferences, and this is accomplished through standard ion-exchange chemistry (Harland, 1994). For solid-source mass spectrometry, the purified sample is then loaded onto a metallic filament. Under vacuum, this filament is heated by passing an electrical current through it, inducing ionization of the sample in preparation for mass analysis. For plasma-source mass spectrometry, the sample is introduced using an inert carrier gas. Isotope dilution and thermal ionization mass spectrometry (ID-TIMS) is the most widely used method of isotope geochronology when parent and daughter isotopes are in the solid state at Earth surface conditions.

For (U–Th)/He and K/Ar, where the daughter products are noble gases, the analytical process is more involved. Most (U–Th)/He experiments begin by driving off helium (and other gases) from the sample, either by vacuum heating in a furnace or by laser irradiation. After a purification process to remove reactive gases from the mixture, a gas-source mass spectrometer is used to measure the isotopic composition of the helium. Subsequently, the sample is removed from the vacuum system, dissolved, spiked, and analyzed for uranium and thorium with an inductively coupled, plasma-source mass spectrometer (ICP-MS). This approach to (U–Th)/He dating avoids the potential problems with heterogeneity that might arise if the sample was split into separate aliquots for He and U–Th analysis. Such problems are frequently encountered in conventional K–Ar geochronology, where argon isotopes are measured by gas-source, isotope-dilution mass spectrometry and potassium is typically measured on a separate aliquot by flame photometry. As a consequence, most K–Ar studies now employ the $^{40}Ar/^{39}Ar$ method (Merrihue and Turner, 1966; McDougall and Harrison, 1988), which permits the determination of dates by measuring only argon isotopic ratios. If a sample is bombarded with high-energy neutrons in a nuclear reactor prior to analysis, naturally occurring ^{39}K isotopes can be induced to convert to artificial ^{39}Ar with an efficiency that depends on the neutron flux and the duration of irradiation. The ratio $^{39}K/^{40}K$ is effectively constant in geologic materials because the half-life of ^{40}K is so long. Consequently, the amount of ^{39}Ar produced by neutron bombardment is also proportional to the amount of ^{40}K in the sample prior to irradiation. If Φ is the constant of this proportionality, the fundamental equation for $^{40}Ar/^{39}Ar$ geochronology can be written as

$$t = \left(\frac{1}{\lambda_{40K}}\right) \ln\left[\frac{(^{40}Ar/^{36}Ar) - (^{40}Ar/^{36}Ar)_0}{\Phi(^{39}Ar/^{36}Ar)(\lambda_{40K(ec)}/\lambda_{40K})} + 1\right] \quad (12)$$

(The stable isotope ^{36}Ar appears in Equation (12) in order to permit the calculation of a date from isotopic ratios rather than isotopic concentrations.) In practice, samples of known $^{40}Ar/^{39}Ar$ age are co-irradiated with the sample to monitor ^{39}Ar production efficiency and allow the determination of Φ using a version of Equation (12).

Fission-track geochronology requires the counting of tracks as well as some method of determining ^{238}U concentration. Because fission tracks are too small to be seen effectively with petrographic microscopes, standard analytical protocols involve the chemical etching of a polished face of a sample, which enlarges tracks intersecting the surface and makes them easier to resolve. Rather than measure ^{238}U with a mass spectrometer, fission-track geochronologists generally employ neutron irradiation. By bombarding the sample with low-energy neutrons, some fraction of ^{235}U can be induced to undergo a fission reaction that produces new tracks, such that the ^{235}U abundance—and thus the ^{238}U abundance, if a constant natural ratio is assumed—can be calculated by counting the new tracks. One of the most popular approaches involves the use of an external detector, in which a uranium-free material is attached to the sample and the two are irradiated together (e.g., Hurford and Carter, 1991; Gallagher *et al.*, 1998). Fission of ^{235}U produces fragments that cross the interface and leave tracks in the external detector. With appropriate corrections for the efficiency of this process, dates can be calculated solely by counting spontaneous ^{238}U fission tracks in the sample prior to irradiation and induced ^{235}U fission tracks in the external detector after irradiation.

3.08.3.1 Microanalytical Techniques

A major step forward in geochronology since the 1980s has been the development of various microanalytical techniques. Such tools allow us to explore isotopic variations within individual crystals, to distinguish grains of different origins in polygenetic samples, to date mineral inclusions and place absolute age constraints on pressure–temperature histories, and to determine the apparent ages of minerals in petrographic context.

Analytical protocols for ID-TIMS have improved to the point that accurate and precise dates may be obtained for samples as small as a few nanograms, such that even this "conventional" method of geochronology is limited principally by the minimum sample size that can be manipulated for sample dissolution, spiking, and ion-exchange chemistry. Various mechanical microsampling tools, ranging from obsidian knives to microdrills, have been employed to extract datable fragments from single crystals on scales of tens of microns (e.g., Christensen *et al.*, 1989; Hawkins and Bowring, 1997; Müller *et al.*, 2001).

Other microanalytical techniques employ high-energy beams of photons, ions, or electrons to excavate material from the sample or induce the production of characteristic X-rays from the sample surface. *Laser microprobes* of various wavelength are commonly used to melt or ablate samples at spatial resolutions ranging from a few tenths of a micron to (more typically) several hundreds of microns (e.g., York *et al.*, 1981; Maluski and Schaeffer, 1982; Sutter and Hartung, 1984; Feng *et al.*, 1993; Fryer *et al.*, 1993; Halliday *et al.*, 1998; Hodges, 1998a; Horn *et al.*, 2000; Poitrasson *et al.*, 2000; Bruguier *et al.*, 2001; Li *et al.*, 2001; Kosler *et al.*, 2001, 2002; Machado and Simonetti, 2001; Willigers *et al.*, 2002). *Ion microprobes* are based on the principle that ion milling stimulates the emission of secondary ions from a sample's surface. These ions can be filtered and analyzed through a process referred to as secondary-ion mass spectrometry (SIMS). With the advent of ion microprobes having especially high sensitivity and mass resolution, SIMS U/Pb and Th/Pb geochronology of accessory phases has become increasingly popular (e.g., Compston *et al.*, 1986; DeWolf *et al.*, 1993; Harrison *et al.*, 1995b; Williams *et al.*, 1996; Zhu *et al.*, 1997b; Williams, 1998; Stern and Berman, 2001). The spatial resolution afforded by ion microprobes is generally somewhat better than that of laser microprobes, but neither can approach the micron-scale resolution of *electron microprobes*. These instruments enable the nondestructive analysis of uranium, thorium, and lead concentrations for chemical lead dating (Suzuki and Adachi, 1991; Montel *et al.*, 1996; Cocherie *et al.*, 1998; Crowley and Ghent, 1999; Williams *et al.*, 1999; Geisler and Schleicher, 2000; Terry *et al.*, 2000; French *et al.*, 2002; Kempe, 2003).

It is a fundamental principle of analytical geochemistry that higher concentrations of an element or specific isotope can be measured with greater precision than lower concentrations. As a consequence, most microanalytical methods sacrifice high analytical precision for high spatial resolution. Even with the trend toward single-crystal and crystal-fragment studies, ID-TIMS remains the most precise geochronologic method, in part because somewhat larger sample sizes are used and in part because the technique is the oldest and, consequently, the most refined. For example, single crystals of accessory minerals such as monazite often yield ID-TIMS U/Pb dates with 2σ uncertainties of less than 0.5%, even for materials of Late Tertiary age (e.g., Viskupic and Hodges, 2001). SIMS and laser-ablation ICP-MS U/Pb and Th/Pb dates for comparably aged samples typically have 2σ uncertainties that are at least an order of magnitude larger (e.g., Harrison *et al.*, 1996; Kosler *et al.*, 2001). The current generation of electron microprobes has such low abundance sensitivities for uranium, thorium, and lead that the only Tertiary mineral dated by the chemical lead method thus far is the uranium oxide uraninite (Hurtado *et al.*, in preparation), with 2σ uncertainties of $\sim 10\%$. Most applications of this technique have involved the dating of Paleozoic or Precambrian monazites, with typical uncertainties of a few percent.

3.08.4 THE INTERPRETATION OF DATES AS CRYSTALLIZATION AGES

The quantity t as calculated from the above equations—which will be referred to here as a *date*—may or may not have geologic significance. The interpretation of a date as a *crystallization age* is justified if:

- decay constants do not change with time;
- the non-radiogenic component of the daughter isotope is known or can be determined empirically; and
- the sample has been closed with respect to gain or loss of parent or daughter isotopes since crystallization.

If sequential or multiparent decay is involved in the production of the final daughter isotope, the second criterion must be amended to include all parents and all intermediate daughter isotopes. In some applications, isotopic data for more than one sample are used in an effort to characterize the non-radiogenic daughter component. Examples of this include the analysis of presumably co-genetic, whole-rock samples from a single igneous intrusion, and the analysis of different minerals with different parent/daughter ratios from a single rock sample (e.g., Schreiner, 1958; Fairbairn *et al.*, 1961; Lanphere *et al.*, 1964; Weatherill *et al.*, 1968). Such practices require an additional assumption:

- All samples used for the determination of t must have been in isotopic equilibrium with one another at the time of crystallization.

When the date of interest refers to the crystallization of rocks or minerals from a melt and all of the above criteria are met, the interpreted age is best referred to as an *igneous age*. When minerals in a single rock are the products of solid-state recrystallization during metamorphism and all the above criteria are met, then the calculated date may be referred to as a *metamorphic age*.

3.08.5 OPEN-SYSTEM BEHAVIOR: THE ROLE OF DIFFUSION

Of all the criteria necessary for the interpretation of dates as crystallization ages, the assumption of closed-system behavior is the most problematic. Many daughter isotopes and some important parent isotopes may be mobile during metamorphism, hydrothermal alteration, and deformation, and a surprising number of high-precision isotopic dates for samples from orogenic settings are demonstrably younger than the crystallization ages of the samples. Among the mechanisms potentially responsible for such open-system behavior are:

- metamorphic "net-transfer" reactions that lead to the crystallization of new minerals or the elimination of others as a consequence of changes in temperature and pressure;
- dynamic recrystallization of minerals as a consequence of deformation; and
- metamorphic "exchange" reactions that lead to changes in the compositions of minerals but not their modal proportions, as a consequence of changes in pressure and temperature (Spear, 1993).

The first two mechanisms leave direct textural evidence for open-system behavior. Samples displaying such evidence yield complex geochronologic data that are difficult to interpret without a detailed understanding of the metamorphic and structural context (e.g., Hames and Cheney, 1997; Arnaud and Eide, 2000; Hoskin and Black, 2000; Di Vincenzo *et al.*, 2001;

Rubatto *et al.*, 2001; Stern and Berman, 2001; Townsend *et al.*, 2001). However, variable mineral dates also characterize many samples that show no obvious textural evidence of disturbance. In most cases, these dates differ in a predictable way (Hart, 1964; Armstrong *et al.*, 1966; Wagner *et al.*, 1977). For example, a typical amphibolite-facies pelitic schist will yield a (U–Th)/Pb monazite date older than its $^{40}Ar/^{39}Ar$ muscovite date, which will be, in turn, older than its $^{40}Ar/^{39}Ar$ biotite date. Given the lack of textural evidence for other causes, the most likely mechanism to produce such inconsistencies is isotopic exchange between minerals and their surroundings.

Isotopic exchange can be thought of as a transport process governed by concentration gradients. Consider the simplest case of a mineral that crystallizes with some amount of parent isotopes and that is surrounded by an intergranular medium, such as a fluid, containing none of the parent species. As daughter isotopes are produced in a mineral by radioactive decay, a concentration gradient develops between the mineral and its surroundings and there is a natural tendency for the daughter isotopes in the mineral to move out of the crystal. (This process can be thought of as an exchange of daughter isotope for a crystallographic vacancy in the mineral.) Because the mobility of virtually all isotopes would be greater in an intergranular fluid than in a mineral, any daughter that leaves the mineral is rapidly transported away from the interface and the fluid may be thought of as an infinite sink for the daughter isotope. Under such circumstances, the rate of isotopic exchange is largely dependent on the rate of atomic migration or *diffusion* of the daughter isotope in the mineral. A wealth of experimental evidence (Section 3.08.5.2) demonstrates that the rate of diffusion (or *diffusivity*) of isotopes in minerals differs in a way that is consistent with the notion that the sequence of older-to-younger dates obtained for many samples reflects a sequence of greater-to-lesser diffusivities for the daughter isotopes used to determine the dates.

Thermochronology is predicated on the notion that minerals that have experienced loss of radiogenic daughter isotopes after crystallization may still provide useful information about the thermal history of a sample because the diffusive-loss mechanism depends strongly on temperature. Establishing a quantitative basis for thermochronology requires developing effective models for the combined effects of three time-dependent processes: the production of daughter isotopes through radioactive decay of parent isotopes (as discussed in Section 3.08.2), the loss of the daughter isotopes through diffusion, and the cooling of the system. Sections 3.08.5.1 and 3.08.5.2 focus on the second process, addressing the nature of diffusion in minerals and presenting diffusion data for mineral-isotopic systems frequently used for thermochronology.

3.08.5.1 Modes of Diffusion

If mineral crystals were perfect, the only intracrystalline atomic migration we would have to be concerned with would be the three-dimensional or *volume diffusion* of an isotope. Unfortunately, natural samples contain internal imperfections, such as subgrain boundaries or line defects formed by dislocation arrays, that can serve as fast-diffusion pathways. This suggests that a reasonable conceptual model for isotope migration in natural crystals is one that involves interactive volume diffusion at one rate in intact domains and "short-circuit" diffusion at a different rate in the pathways that separate them (e.g., Lee and Aldama, 1992; Lee, 1995). An alternative to this "multipath" model, referred to by Lovera *et al.* (1989) as the "multidiffusion domain" model, regards a crystal as a collection of discrete, non-interactive domains of different size with different volume diffusion characteristics. Recently, Lovera *et al.* (2002) suggested a third, "heterogeneous diffusion" model that features spatially variable diffusivity but no discrete domains. Of the three, the multipath model probably provides the most realistic *physical* model of diffusion in many minerals (Parsons *et al.*, 1999), but the multi-diffusion domain model appears to provide an adequate *mathematical* model to explain the diffusive behavior of natural crystals under experimental conditions (Lovera *et al.*, 1997, 2002). This is significant because the multidomain diffusion model has provided a remarkably successful mathematical protocol for extracting cooling histories from $^{40}Ar/^{39}Ar$ K-feldspar data, as will be discussed further in Section 3.08.7.2.

Despite the fact that open-system behavior in mineral-isotopic systems is governed by a combination of volume and short-circuit diffusional processes, most thermochronologists make the simplifying assumption that volume diffusion alone controls the open- to closed-system transition that is so important to thermochronologic theory. There are a variety of reasons to believe that this assumption is reasonable. First, the structure of short-circuit pathways is such that they should be characterized by much faster diffusion than the intact crystal structure that surrounds them; if so, then the rate of daughter isotope loss should be limited by the rate of diffusion out of intact domains and into short-circuit pathways. Second, the volumetric proportion of short-circuit pathways to intact domains is small in all but the most strongly deformed natural crystals, implying that their contribution to bulk diffusive loss is minor. For example, maps of the distribution of argon isotopes in natural crystals display topologies consistent with volume diffusion at the grain scale with relatively minor modification by fast-diffusion pathways (e.g., Hames and Hodges, 1993; Hodges and Bowring, 1995; Reddy *et al.*, 1996; Pickles *et al.*, 1997; Wartho *et al.*, 1999). Finally, the fact that different thermochronometers yield predictable sequences of apparent ages implies that diffusivity, while temperature-dependent, is largely an intrinsic property of a mineral. We should not expect this to be the case if short-circuit diffusion exerts a controlling influence on mineral ages because the development of fast-diffusion pathways is related to the specific thermal and deformational history of a crystal. For the remainder of the chapter, we will assume that volume diffusion is the principal mode of daughter isotope loss, although practitioners of thermochronology are well-advised to evaluate each data set for evidence of significant loss by other mechanisms, both diffusive and non-diffusive; if important, these processes will have rendered the data set inappropriate for thermochronology.

3.08.5.2 Experimental Constraints on Daughter-isotope Diffusion for Useful Minerals

The volume diffusivity of isotopes in minerals depends on temperature through the Arrhenius relationship:

$$D_T = D_{\rm i}{\rm e}^{-E/RT} \qquad (13)$$

where D_T is diffusivity at the temperature of interest (T), $D_{\rm i}$ is the diffusivity at infinite temperature, E is the activation energy of the process, and R is the gas constant. Since the mid-1970s, there has been a concerted effort to characterize volume diffusion of daughter elements in minerals commonly used for geochronology and thermochronology by quantifying $D_{\rm i}$ and E.

The methods employed to do this typically fall into two categories: "bulk" experiments, in which fractional losses of an element or specific isotope are measured; and experiments in which induced diffusion profiles are measured. Both bulk diffusion and diffusion-profile experiments can be done in a vacuum or hydrothermally. Bulk hydrothermal experiments, frequently used to recover the diffusivity of gaseous species such as argon (e.g., Giletti, 1974; Grove and Harrison, 1996), involve the heating of samples at atmospheric or higher pressures for sufficient time to cause the loss of a fraction of the isotope of interest from the bulk sample. This method is required if the sample is hydrous and must be kept stable during the experiment. The fraction of the initial isotopic concentration lost by heating depends on the effective diffusion dimension (a), the diffusion geometry, and D_T (Fechtig and Kalbitzer, 1966; Crank, 1975). A series of experiments is conducted at different temperatures in order to define an array of points in $\ln(D_T/a^2)$ versus $1/T$ space. If loss occurred through volume diffusion over a single diffusion dimension, this array should be linear and diffusion parameters can be recovered by fitting a function derived by linearizing Equation (13) to the array. In many experimental studies, a is equivalent to the physical grain half-size (e.g., Grove and Harrison, 1996; Reiners and Farley, 1999; Wartho *et al.*, 1999; Farley, 2000). When this is the case, the diffusion parameter $D_{\rm i}$ can be determined directly from the $\ln(D_T/a^2)$-intercept of the fitted line. A few experiments seem to suggest the existence of an effective diffusion dimension much smaller than the physical grain half-size (Harrison, 1981; Harrison *et al.*, 1985). In those cases, a is generally approximated by conducting experiments on samples crushed to a variety of grain sizes.

For minerals that are thought to be stable at elevated temperatures in a vacuum, an alternative approach to determining diffusivities is to perform bulk *in vacuo* experiments (e.g., Foland, 1994; Wolf *et al.*, 1996). This approach has proved to be especially valuable for studying argon diffusion in K-feldspar, the only mineral commonly used for $^{40}Ar/^{39}Ar$ thermochronology that is stable during vacuum heating. Bulk *in vacuo* diffusion experiments on individual feldspar samples often reveal subparallel linear arrays in $\ln(D_T/a^2)$ versus $1/T$ space, which are consistent with the existence of multiple diffusion domains with different dimensions (Gillespie *et al.*, 1982; Zeitler, 1987; Lovera *et al.*, 1989). The fact that a single stepwise-heating experiment on K-feldspar provides sufficient information to characterize argon diffusivity in these domains, as well as their closure dates, means that $^{40}Ar/^{39}Ar$ data are capable of providing remarkably detailed information about the cooling history of rocks *if* the diffusion characteristics of K-feldspar are the same in the laboratory and in nature during cooling (Lovera *et al.*, 2002).

Many recent diffusion experiments have relied on various microanalytical techniques to quantify diffusion gradients produced by heating under hydrothermal conditions or in a vacuum (e.g., Cherniak, 1993; Cherniak and Watson, 2000). Assuming simple, one-dimensional diffusion, these gradients are related to D_T through the inverse error function (${\rm erf}^{-1}$):

$${\rm erf}^{-1}\left(\frac{C_0 - C_\xi}{C_0}\right) = \left(\sqrt{4D_T\tau}\right)\xi \qquad (14)$$

where C_0 is the initial concentration of the diffusant, C_ξ is the concentration at distance ξ

along the gradient, and τ is the duration of the experiment (Crank, 1975). A series of such experiments at different temperatures yields different values of D_T that can be used to recover D_i and E through linear regression of $\ln D_i$ versus $1/T$. Measurements of induced diffusion gradients are particularly useful for two reasons. The first is that heating over timescales and temperatures typically used in diffusion experiments produces little bulk loss of slow-diffusing species such as lead; as a consequence, measurements of bulk fractional losses of such species can have relatively large uncertainties, while careful measurements of diffusion profiles can yield higher-precision data. The second is that diffusion profiles can be measured orthogonal to carefully prepared surfaces or clean crystal faces, minimizing the effects of fast-diffusion pathways and increasing the probability that only volume diffusivity is responsible for loss of the element or isotope of interest.

Table 2 lists diffusion coefficients for many mineral-isotopic systems currently used for thermochronometry. Diffusivities of some important isotopes depend on major-element composition; e.g., ^{40}Ar diffusion in phlogopite is ~88% faster than in iron-rich biotite (Giletti, 1974; Grove and Harrison, 1996). In general, however, such variations are small in comparison with mineral-to-mineral differences in the diffusivity of a specific isotope. These differences for the basis for quantitative thermochronology.

3.08.6 CLOSURE TEMPERATURE THEORY

As defined in Section 3.08.4, igneous or metamorphic ages may be obtained for samples that retain all radiogenic daughter isotopes produced subsequent to crystallization. In contrast, the dates measured for samples that experienced some degree of daughter loss substantially underestimate the igneous or metamorphic age. Until the 1960s, these dates were regarded as geologically meaningless, but the eventual realization that daughter isotope loss is a predominantly diffusive process led to the proposition that they may be powerful probes of the cooling history of igneous and metamorphic terrains.

We can explore the concept of *cooling ages* and the temperatures they represent by considering an idealized, spherical mineral grain of radius a, which crystallizes instantaneously from a melt at time t_0 and at temperature T_0. Let us begin by assuming that our hypothetical grain grows with a spatially uniform initial concentration of parent isotopes (n_{p0}), but no non-radiogenic daughter isotopes (n_{d0}). With no atomic migration of parent or daughter within the grain and no loss of either species to the surroundings, the daughter–parent ratio (n_d/n_p) increases with time according to Equation (3), and it will be the same at all positions within the grain at any given time. Figure 1(a) schematically illustrates n_d/n_p as a function of radial position for $t = t_0$ and three subsequent times: t_1, t_2, and t_3. Measurements of n_d/n_p at any of these times would yield the actual crystallization age of the mineral.

Now suppose that T_0, the temperature of crystallization, is sufficiently high to permit diffusive loss of daughter into the intergranular medium surrounding the grain as rapidly as it was produced by radioactive decay. (Since concentration gradients are necessary to drive diffusive transfer, let us surround our hypothetical grain with a fluid that has no initial concentration of daughter isotopes, but a diffusivity of the daughter that is much greater than the corresponding diffusivity in the mineral at any given temperature. This condition ensures that the concentration of daughter in the fluid remains effectively zero through time.) In the absence of cooling from T_0, the grain will never accumulate daughter isotopes, such that n_d/n_p would remain at zero throughout the crystal at t_0, t_1, t_2, and t_3. However, let us allow the grain to cool at some reasonable rate. An assumption commonly made in thermochronologic studies is that reciprocal temperature increases linearly with time during cooling:

$$\frac{1}{T_t} = \frac{1}{T_0} + kt \tag{15}$$

where T_t is the temperature of the grain at time t, and k is some constant of proportionality. Remembering that volume diffusion depends on temperature through the Arrhenius relationship (Equation (13)), we can see that Equation (15) implies a time dependence for diffusion in a cooling mineral:

$$\ln D_t = \ln D_i - \left(\frac{E}{R}\right)\left(\frac{1}{T_0} + kt\right) \tag{16}$$

where D_t is diffusivity at time t. Given sufficient time, the diffusivity of the daughter isotope in our hypothetical sample should eventually drop to the point that none is lost to the fluid and the system becomes closed. Just prior to that, the sample will experience a transitional period of partially open-system behavior, during which n_d/n_p will increase, but at a rate less than that dictated by Equation (3).

Although diffusion is a "random-walk" process, the net transfer of matter by diffusion is driven by concentration gradients. In our model sample, the steepest gradients in daughter-isotope concentration will occur near the grain margin, and we should expect that diffusive loss of that isotope over time will be greatest at the grain

Table 2 Diffusion data and bulk closure temperatures for important mineral-isotopic systems.

Mineral-isotopic system	D_i ($cm^2\ s^{-1}$)	E ($kcal\ mol^{-1}$)	*Geometry*	a (μm)[a]	T_{cb} (°C)[b]	*References*
Monazite (U–Th)/Pb	9.4×10^3	141	Spherical(?)	50	987	Cherniak *et al.* (2002)
Zircon (U–Th)/Pb	7.8×10^2	130	Spherical	50	942	Cherniak and Watson (2000)
Garnet Sm–Nd						
Almandine ($Alm_{75}Py_{22}$)	4.7×10^{-5}	62	Spherical	500	676	Ganguly *et al.* (1998)
Rutile (U–Th)/Pb	1.6×10^{-6}	58	Spherical	250	671	Cherniak (2000)
Titanite (U–Th)/Pb	1.1×10^0	79	Spherical	500	659	Cherniak (1993)
Hornblende $^{40}Ar/^{39}Ar$						
Magnesio-hornblende	2.4×10^{-2}	64	Spherical	500	557	Harrison (1981)
K-feldspar Rb–Sr	6.0×10^{-3}	68	Spherical	10	487	Cherniak and Watson (1992)
Apatite (U–Th)/Pb	2.0×10^{-4}	55	Spherical	50	446	Cherniak *et al.* (1991)
Phlogopite $^{40}Ar/^{39}Ar$	7.5×10^{-1}	58	Cylindrical	500	433	Giletti (1974)
Muscovite $^{40}Ar/^{39}Ar$	3.9×10^{-4}	43	Cylindrical	500	366	Robbins (1972); Hames and Bowring (1994)
Biotite $^{40}Ar/^{39}Ar$						
Fe–Mg biotite ($X_{phl} = 0.29$)[d]	4.0×10^{-1}	51	Cylindrical	500	359	Grove and Harrison (1996)
Fe–Mg biotite ($X_{phl} = 0.46$)[d]	7.5×10^{-2}	47	Cylindrical	500	335	Harrison *et al.* (1985)
Muscovite Rb–Sr	1.0×10^{-9}	25	Cylindrical	500	316	Chen *et al.* (1996)
Biotite Rb–Sr	2.0×10^{-9}	25	Cylindrical	500	299	Jenkin *et al.* (1995)
K-feldspar $^{40}Ar/^{39}Ar$						
Low sanidine	3.7×10^{-2}	47	Spherical	10	237	Wartho *et al.* (1999)
Orthoclase	9.8×10^{-3}	44	Spherical	10	218	Foland (1994)
Zircon fission track[c]					227	Brandon *et al.* (1998)
Titanite (U–Th)/He	5.9×10^1	45	Spherical	250	206	Reiners and Farley (1999)
Apatite fission track[c]					110	Laslett *et al.* (1987); Brandon *et al.* (1998)
Apatite (U–Th)/He	3.16×10^1	33	Spherical	50	63	Farley (2000)

[a] Nominal grain half-sizes or subgrain diffusion dimensions. Other values would raise or lower T_{cb} as described by Equation (19). [b] T_{cb} calculated assuming a cooling rate of 5 °C Myr^{-1}. [c] Closure temperatures estimated using both experimental and empirical data. [d] X_{phl}-mole fraction of phlogopite component in Fe–Mg biotite.

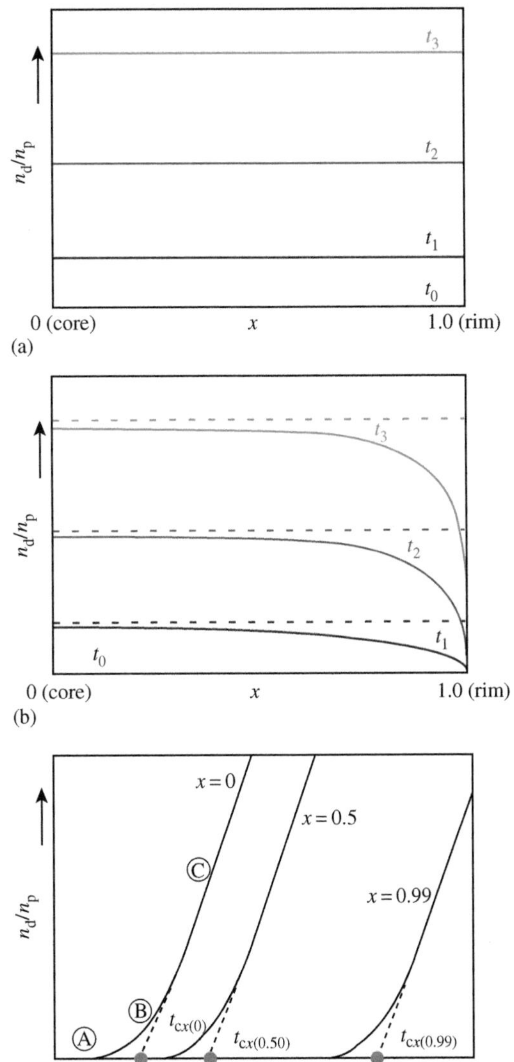

Figure 1 The evolution of daughter/parent (n_d/n_p) ratios with time in a cooling mineral-isotopic system. (a) In a model system that is closed with respect to diffusive loss of daughter isotopes, the ratio n_d/n_p is uniform throughout the crystal at any time. Curves are shown for four successive times (t_0, t_1, t_2, and t_3). In this frame, as in (b), the horizontal axis represents a faction of the radius of an idealized spherical grain (x). (b) With diffusive loss of daughter isotopes during cooling, n_d/n_p at any time is less than the value predicted by Equation (3) by an amount that differs depending on radial position. (c) Changes in the ratio n_d/n_p with time are shown for three different radial positions in the crystal: the core ($x=0$), the rim ($x=0.99$), and an intermediate position ($x=0.5$). The $x=0$ curve is labeled to indicate the three stages of n_d/n_p evolution during cooling: complete open-system behavior (A), partial open-system behavior (B), and complete closed-system behavior (C). Extrapolation of the closed-system behavior segments of three curves to the time axis designates the closure ages for the three positions in the crystal ($t_{cx(0)}$, $t_{cx(0.5)}$, $t_{cx(0.99)}$).

margin and progressively less toward the center. Thus, the concentration of the daughter in the sample at any given time depends on the competition over time between a spatially independent process (daughter production by radioactive decay) and a spatially dependent process (daughter loss by diffusion). This is illustrated schematically in Figure 1(b) for t_0, t_1, t_2, and t_3; note that, for all times subsequent to t_0 and all positions within the crystal, the value of n_d/n_p is always less than Equation (3) predicts.

Another consequence of the spatial dependence of diffusive loss is that the open- to closed-system transition during cooling will take place at different radial positions in the crystal at different times. Figure 1(c) illustrates the evolution of n_d/n_p with time at three different radial positions (x) within the grain: the core ($x=0$), a near-rim position ($x=0.99$), and half-way between the core and rim ($x=0.5$). (Note the convention of expressing radial distance as a fraction of the radius.) The evolution curve for each of these positions can be thought of in terms of three segments corresponding to: (i) purely open-system behavior, during which $n_d/n_p=0$ (curve section A in Figure 1(c)); (ii) purely closed-system behavior, during which n_d/n_p grows at a rate of $e^{\lambda_p t}-1$ (curve section C in Figure 1(c)); and (iii) transitional behavior characterized by the partial retention of daughter isotopes (curve section B in Figure 1(c)). If we measure the n_d/n_p ratio at any one of these positions at the present time (t_p) and calculate a date, we are, in effect, extrapolating backwards along the closed-system evolution curve in Figure 1(c) to its t-intercept. Note that this date—the *cooling age*—does not correspond to the crystallization age, the initiation age of partial daughter retention, or the initiation age of closed-system behavior. The cooling age (t_{cx}) instead corresponds to the time at which a specific position within the grain cooled through a specific temperature known as its *closure temperature* (also sometimes called the *blocking temperature* after a similar concept in the field of rock magnetism).

3.08.6.1 Quantitative Estimates of Closure Temperatures

Dodson (1986) presented a mathematical formula for calculating closure temperatures for various positions within a cooling crystal:

$$T_{cx} = \frac{E}{R\left[\ln(RD_i T_{cx}^2/Ea^2(\mathrm{d}T/\mathrm{d}t)) + G_x\right]} \quad (17)$$

where T_{cx} is the closure temperature at fractional position x, and $\mathrm{d}T/\mathrm{d}t$ is the cooling rate. (Note that this equation must be solved iteratively because T_{cx} appears on both sides.) G_x, which Dodson referred to as the "closure function," depends on diffusion geometry; for the case of our hypothetical spherical grain:

$$G_x = C + 4\sum_{n=1}^{\infty}\frac{(-1)^{n+1}\sin(n\pi x)\ln(n\pi)}{n\pi x} \quad (18)$$

where C is the Euler constant (~ 0.57721547) and the trigonometric argument is in radians. Analogous expressions for plane sheet and cylindrical geometries, as well as useful approximations of G_x as a function of x, may be found in Dodson (1986).

Equation (17) provides a theoretical basis for reconstructing a significant proportion of the time–temperature history of a single crystal during cooling. The thermochronologist first measures n_d/n_p at various positions within a crystal using some microanalytical technique, then uses those data to calculate cooling ages, and finally determines the closure temperatures corresponding to those ages. An arbitrary cooling rate is assumed to calculate T_{cx} for two or more positions, these temperatures and their corresponding dates are then used to make a better estimate of the cooling rate (assuming that $dT/dt \approx \Delta T/\Delta t$, where the deltas represent differences), and the process is repeated until convergence.

Although microanalytical dating techniques are developing at a rapid pace, most thermochronologic data still are obtained for entire single crystals or aggregates of crystals; these will be referred to here as *bulk cooling ages* (t_{cb}). Fortunately, we can recover the *bulk closure temperature* (T_{cb}) of a crystal (Dodson, 1973) from the weighted mean of T_{cx}:

$$T_{cb} = \frac{E}{R\ln(e^{G_{av}}RD_iT_{cb}^2/a^2E(dT/dt))} \quad (19)$$

G_{av}, the means value of G_x, depends only on the assumed geometry of diffusion: 4.00660 for radial diffusion in a sphere, 3.29506 for radial diffusion in a cylinder, and 2.15821 for diffusion across a plane sheet (Dodson, 1986). Equation (19) may be used in conjunction with t_{cb}'s for multiple mineral-isotopic systems, or for multiple grain sizes of the same mineral-isotopic system, in order to determine cooling histories for rocks.

Equations (17) and (19) are strictly valid only if it is assumed that: (i) postcrystallization loss of daughter isotopes occurs exclusively through volume diffusion; (ii) the medium surrounding the mineral serves as an infinite sink for daughter isotopes lost from the grain; (iii) temperature changes linearly with $1/t$, as described by Equation (15), and (iv) that no part of the grain—not even the core—retains all of the daughter isotope produced by radioactive decay of the parent at that position since crystallization. The first of these assumptions is a basic tenet of thermochronology, but the others invite closer scrutiny.

The second assumption is likely to be true in most instances, but exceptions can occur when an intergranular fluid contains high concentrations of the daughter isotope. In some cases, these concentrations are so high that the daughter isotope may diffuse into the mineral as an "excess" component. Excess ^{40}Ar, resulting in anomalously old and geologically meaningless ^{40}Ar/^{39}Ar and K/Ar dates, has been documented through many studies, as reviewed by Kelley (2002). Although slight excess ^{40}Ar contamination may be successfully corrected for using various methods (Roddick *et al.*, 1980; Harrison *et al.*, 1994), samples that have been heavily contaminated are unsuitable for thermochronometry. For some mineral-isotopic systems (e.g., Rb/Sr muscovite, biotite, and K-feldspar), the concentrations of non-radiogenic daughter isotopes are sufficiently unpredictable that many geochronologists follow the classical approach of Nicolaysen (1961) and Fairbairn *et al.* (1961) by combining isotopic data for more than one mineral, or a mineral and the whole rock from which it was separated, to determine both a date and an initial ratio. This *internal isochron* approach requires the assumption that the concentration of daughter isotopes in different minerals are interdependent, a notion that is inconsistent with the second core assumption of Dodson's closure temperature theory. Models more sophisticated than Equation (19) have been proposed to estimate the bulk closure temperatures corresponding to dates obtained with internal isochrons (Giletti, 1991; Jenkin *et al.*, 1995; Chen *et al.*, 1996).

Violations of the third assumption implicit in Equations (17) and (19) are relatively insignificant. Through a series of mathematical experiments, Lovera *et al.* (1989) showed that Dodson's formalism yields correct results for natural samples with virtually any monotonic cooling history. Finally, the fourth assumption may not hold for mineral-isotopic systems characterized by extremely low daughter diffusivities at geologically reasonable temperatures (e.g., Sm/Nd or Lu/Hf in garnet, and (U–Th)/Pb for zircon, monazite, and xenotime). Ganguly and Tirone (1999) have derived alternative forms of Equations (17) and (19) that are more appropriate for such systems, but they require *a priori* knowledge of the crystallization temperature T_0. The Ganguly and Tirone equations yield closure temperature estimates that converge on the results of calculations made with Dodson's equations when T_0 is high (>750 °C) and cooling rates are relatively slow (<5 °C my^{-1}). When T_0 is lower than ~650 °C, low-diffusivity mineral-isotopic systems typically remain closed throughout their cooling history. As a consequence, the Ganguly and Tirone formulations are most useful for thermochronologic studies of igneous and granulite facies metamorphic rocks that cooled quickly from maximum temperatures in the 650–750 °C range.

3.08.6.2 The Influence of Input Parameters on Closure Temperature Calculations

The specific formulation of G_x used for a mineral-isotopic system depends on crystal morphology, the geometry of subgrain features that might define diffusion domains, and whether or not laboratory experiments on the system yield direct evidence for diffusive anisotropy. For example, experimental studies (Giletti, 1974; Harrison *et al.*, 1985; Grove and Harrison, 1996) and empirical observations (Hames and Bowring, 1994; Hodges *et al.*, 1994; Pickles *et al.*, 1997) suggest that daughter isotope loss from micas is anisotropic, with the most rapid diffusion occurring parallel to the (0 0 1) crystallographic plane. For these minerals, daughter loss is best modeled with a cylindrical diffusion geometry. For most other minerals used in thermochronology, it is common practice to assume a spherical geometry.

The choice of an appropriate diffusion dimension (a) for Equations (17) and (19) naturally depends on diffusion geometry as well. If a spherical geometry is assumed, then half of the crystal's physical grain size is a reasonable estimate of a. For micas with cylindrical geometries, half the crystal dimension measured parallel to the (0 0 1) cleavage plane is reasonable. Although the value of a is often assumed to be related to be related to physical grain size, a few careful experiments have documented that diffusive loss in some samples is independent of grain size until the samples are crushed to a dimension much smaller than the natural grain size (e.g., Harrison, 1981; Harrison *et al.*, 1985). A reasonable interpretation of these data is that the experimental starting materials contained intact domains surrounded by fast-diffusion pathways, such that progressive crushing eventually reduced each sample to the dimension of the intact domains. This suggests that using an "effective" diffusion dimension for a may be more appropriate than using the half-grain size *if* that dimension can be known *a priori*. Harrison (1981) and Harrison *et al.* (1985) have suggested that the effective diffusion dimension for radiogenic ^{40}Ar loss in hornblende and biotite may be intrinsic properties of the minerals themselves, such that a single value of a may be used for all samples of a particular mineral. However, this seems unlikely given that subgrain structures in minerals are demonstrably a function of their thermal and deformational histories. Other researchers have found evidence that a does correspond to the half-grain size for most mineral-isotopic systems (Goodwin and Renne, 1991; Onstott *et al.*, 1991; Hess *et al.*, 1993; Hames and Bowring, 1994; Hodges *et al.*, 1994; Hodges and Bowring, 1995, Farley, 2000; Reiners and Farley, 1999, 2001), but this relationship breaks down in samples subjected to intensive ductile or brittle deformation (e.g., Arnaud and Eide, 2000; Kramar *et al.*, 2001; Mulch *et al.*, 2002). In general, it seems prudent to assume that a is related to the physical grain size when applying Equations (17) and (19) unless samples show textural evidence for the extensive development of subgrain boundaries that may act as fast diffusion pathways, or—in the case of K-feldspar—show direct evidence of the existence of multiple diffusion domains during incremental heating experiments.

Figure 2 illustrates the sensitivity of T_{cb} to variations in parameters a and dT/dt using volume diffusion parameters for radiogenic ^4He in apatite (Farley, 2000). Effective diffusion dimension has a strong influence on bulk closure temperature, implying that it is possible to recover a significant proportion of the time–temperature history of a single structural level in an orogen using only one thermochronometer if a range of grain sizes of the mineral is available for study (e.g., Hess *et al.*, 1993; Reiners and Farley, 2001). Closure

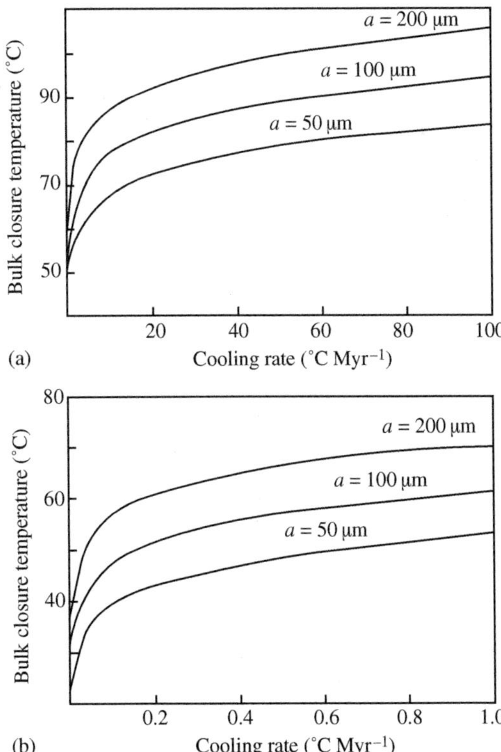

Figure 2 Bulk closure temperature for the (U–Th)/He thermochronometer as a function of cooling rate for effective diffusion dimensions of 50 μm, 100 μm, and 200 μm, based on calculations made with Equation (19) using diffusion coefficients as in Table 2. (a) Variation in T_{cb} for cooling rates from 0 °C Myr^{-1} to 100 °C Myr^{-1}. (b) For cooling rates from 0 °C to 1°C. Note the increased dependence of T_{cb} on dT/dt for very slow cooling rates.

temperature is less dependent on dT/dt. Typical postcrystallization cooling rates in orogenic settings range from about 1–100 °C Myr^{-1}, and solutions of Equation (19) over this entire range would yield T_{cb} estimates for any specific thermochronometer that vary by less than 50%. However, Equation (19) does imply a strong dependence of T_c on dT/dt for metamorphic terrains that cool at <1 °C my^{-1} (Figure 2(b)). Although documented in only a few settings thus far (e.g., the Proterozoic orogen of the southwestern United States: Hodges *et al.*, 1994; Hodges and Bowring, 1995), such slow cooling rates may have been more common than generally believed in Archean–Proterozoic terrains characterized by long-term tectonic stability.

3.08.6.3 Qualitative Estimates of Closure Temperatures

Unfortunately, high-quality diffusion data are not available for all useful mineral-isotopic systems. Approximate closure temperatures for these thermochronometers can be estimated empirically by applying them to natural samples for which cooling ages have been determined using better characterized systems (e.g., Mezger *et al.*, 1992; Scherer *et al.*, 2000; Williger *et al.*, 2001). Some T_{cb} values approximated in this way are shown in Table 2. An alternative approach is based on the notion that volume diffusion in a mineral is related to its *ionic porosity*. Ionic porosity was defined by Dowty (1980) as the percentage of the unit-cell volume not occupied by ions. Fortier and Giletti (1989) showed that experimental data for the diffusivity of argon and oxygen in silicate minerals defined positive correlations with ionic porosity. Dahl (1996a,b) used such relationships to predict variations in $^{40}Ar/^{39}Ar$ T_{cb} with chemical composition for micas and amphiboles. Ionic porosity also provides a useful tool for predicting the relative closure temperatures for the same daughter isotopes in different minerals. For example, Dahl (1997) found a log–linear relationship between ionic porosity and experimentally determined lead diffusivity in various minerals. Furthermore, he suggested a similar control of ionic porosity on another kinetic process of great importance to thermochronology: fission-track annealing.

3.08.6.4 Fission-track Closure Temperatures and the Partial Annealing Zone

If a mineral crystallizes at relatively high temperatures, any damage to the crystal structure produced by ^{238}U fission rapidly heals. This annealing occurs by atomic reorganization of the crystal structure and, thus, is a diffusive process. By analogy with the diffusive loss of daughter isotopes from a cooling mineral (Figure 1(c)), we can also define closure temperatures for fission-track thermochronometers based on the concept that the evolution of fission tracks in a mineral with time has three stages: (i) a period of high temperatures when fission tracks anneal as rapidly as they develop; (ii) a period of low temperatures when tracks accumulate and do not anneal perceptibly; and (iii) a period of intermediate temperatures when tracks accumulate but also partially anneal (Haack, 1977; Dodson, 1979; Brandon *et al.*, 1998). Approximate bulk closure temperatures for the commonly used fission-track zircon and apatite thermochronometers, made with the arbitrary assumption that monotonic cooling results in 50% annealing of the tracks in a sample, are provided in Table 1. However, such estimates have proven less valuable in fission-track studies than estimates of the entire temperature range of 0–100% annealing. This *partial annealing zone* has been estimated from experimental data (e.g., Brandon *et al.*, 1998) and from empirical studies of samples from deep bore holes with known thermal structures (e.g., Gleadow and Duddy, 1981) to be ~60–120 °C for apatite and ~200–300 °C for zircon (Gallagher *et al.*, 1998; Brandon *et al.*, 1998). Examples of how fission-track data are used to identify the temporal evolution of this temperature zone in orgoenic settings are discussed in Section 3.08.7.5.

3.08.7 APPLICATIONS

The advent of modern methods of geochronology and thermochronology has revolutionized the way earth scientists study the evolution of orogenic systems. This section explores some of the uses of these techniques in tectonics research, and provides specific examples of especially successful applications.

3.08.7.1 Determining Timescales of Granitic Magmatism

Crustal anatexis and the migration of granitic melts strongly influence the thermal and deformational histories of orogenic systems (DeYoreo *et al.*, 1989; Brown and Solar, 1998: Sandiford *et al.*, 2002). As a consequence, determining the crystallization ages of granitic rocks has become an important part of many research projects in continental tectonics. The principal chronometers used for such studies are (U–Th)/Pb zircon, monazite, and xenotime. All three minerals are common accessory phases in granitic rocks, and all contain large concentrations of uranium and thorium, such that easily measurable concentrations of radiogenic lead are produced after only a few million years of decay. Most importantly, experimental studies of lead diffusivity in zircon and monazite (Cherniak and Watson, 2000; Cherniak *et al.*, 2002) and predictions of lead diffusivity in xenotime from its ionic porosity (Dahl, 1997) imply that nominal closure temperatures for these chronometers greatly exceed the solidus temperatures for granitic rocks commonly encountered in orogenic settings (Thompson, 1982; Vielzeuf and Holloway, 1988; Patiño Douce and Johnston, 1991; Holland and Powell, 2001).

Unfortunately, the interpretation of (U–Th)/Pb geochronologic data for granitic rocks is not always straightforward. Many granitic rocks are derived through the melting of protoliths that contain pre-existing accessory minerals. These refractory phases survive the melting process to be preserved as *inherited* grains in the final melt products. Moreover, the lead diffusivity of minerals like zircon, monazite, and xenotime is such that low-temperature melting may produce little or no resetting of the (U–Th)/Pb chronometer. A typical orogenic granite contains numerous inherited grains in addition to new *magmatic* grains that crystallized from the melt. In many cases, overgrowths of neoblastic magmatic rims on inherited cores are common. Conventional ID-TIMS (U–Th)/Pb analysis of these granites using multigrain separates of zircon or orthophosphates yields highly discordant $^{208}Pb/^{232}Th$, $^{207}Pb/^{235}U$, and $^{206}Pb/^{238}U$ dates that, at best, provide low-precision estimates of the granite crystallization age. Single-crystal analyses are more effective, but the results are still sometimes difficult to interpret if inherited grains with neoblastic overgrowths cannot be avoided. The most successful strategies involve: (i) careful backscattered electron, cathodoluminesence, and/or X-ray compositional mapping of grains to understand their internal complexity; (ii) hand picking of crystals with a minimum of structural and chemical complexity; and (iii) the application of microanalytical dating techniques, either the mechanical separation and ID-TIMS dating of "clean" crystal fragments, or ion, electron, or laser microprobe dating of specific regions in individual crystals. The choice of microanalytical method depends on the level of analytical precision desired. As a general rule, ID-TIMS is the favored method unless the accessory phases are so complex at such a small scale that mechanical separation of clean fragments is impossible. Examples of applications of single-crystal and microanalytical (U–Th)/Pb techniques to the dating of granitic rocks in orogenic settings may be found in Parrish and Tirrrul (1989), Parrish (1990), Harrison *et al.* (1995b), Hawkins and Bowring (1997), Coleman (1998), Hodges *et al.* (1998), Anczkiewicz *et al.* (2001), Viskupic and Hodges (2001), and Aleinikoff *et al.* (2002).

3.08.7.2 Constraining the Cooling Histories of Igneous Rocks

Intrusive rocks of felsic and intermediate composition contain many thermochronologically useful minerals, and their study provides a remarkably detailed accounting of the postmagmatic thermal evolution of continental arc terrains. High-temperature (>400 °C) cooling histories can be constrained with (U–Th)/Pb thermochronometers such as titanite and apatite (Corfu and Stone, 1998; Frost *et al.*, 2000; Chamberlain and Bowring, 2001). Although both minerals are notorious for containing high concentrations of non-radiogenic lead, the analysis of comagmatic feldspars permits adequate corrections for this component in many cases (e.g., Housh and Bowring, 1991). For lower temperatures, the thermal history is typically constrained through a combination of $^{40}Ar/^{39}Ar$, (U–Th)/He, and fission-track techniques. Particularly useful in this regard are the $^{40}Ar/^{39}Ar$ K-feldspar and apatite fission-track thermochronometers. Despite controversies regarding the physical nature of and developmental mechanism of intragranular diffusion domains in feldspars (Villa, 1994; Parsons *et al.*, 1999), numerous thermochronologic studies of granitic rocks have employed multidomain diffusion modeling of argon data for feldspars to reconstruct detailed temperature–time (*Tt*) paths over tens to hundreds of degrees (e.g., Richter *et al.*, 1991; Krol *et al.*, 1996; Quidelleur *et al.*, 1997; Kirby *et al.*, 2002). The results are, for the most part, both geologically reasonable and consistent with the results of other, less-controversial methods. Unfortunately, feldspars that show evidence for alteration or extensive contamination with excess ^{40}Ar cannot be used to extract reliable thermal histories. In a recent compilation of the results of nearly 200 K-feldspar experiments from a variety of settings, Lovera *et al.* (1989) concluded that only about half were appropriate for thermal history modeling.

Apatite fission-track data also provide substantial segments of *Tt* paths through track length analysis. As fission-tracks anneal, they become progressively shorter, such that the frequency distribution of track lengths in a sample should be a sensitive indicator of cooling history (e.g., Gleadow *et al.*, 1986; Carlson, 1990). Several sophisticated algorithms permit the inversion of track length distributions to recover cooling histories (Corrigan, 1991; Gallagher, 1995; Willett, 1997; Ketcham *et al.*, 2000).

3.08.7.3 Calibrating Metamorphic Histories

Element partitioning and fluid inclusion thermobarometry, in concert with thermodynamic modeling of porphyroblast zoning, are powerful tools for reconstructing the pressure–temperature (*PT*) paths followed by metamorphic rocks during orogenesis (see Chapter 3.07). Placing these paths in temporal context—thus permitting the reconstruction of *PTt* paths—has become a fundamental goal of geochronology and thermochronology in tectonic settings.

The metamorphic samples best suited for thermochronology are those displaying textural and chemical evidence for a single progressive metamorphic event with no appreciable retrograde re-equilibration among the constituent minerals. In such samples, porphyroblasts like garnet typically display major- and trace-element zoning patterns consistent with prograde growth (Hollister, 1966; Tracy *et al.*, 1976; Spear, 1993). Well-conceived thermochronologic studies of these rocks begin with careful petrographic observation, preferably augmented with X-ray compositional mapping of thin sections, to evaluate the paragenetic sequence of minerals suitable for thermobarometry or thermodynamic modeling, as well as those suitable for thermochronology. If peak metamorphic temperatures can be estimated through thermobarometry, this result and the paragenetic sequence can serve as a guide for developing an effective thermochronologic strategy.

For example, suppose mineral rim compositions in a pelitic schist suggest a peak metamorphic temperature of ~500 °C, and textures suggest that the final equilibrium assemblage included the rock-forming minerals garnet + biotite + muscovite and the accessory minerals monazite and apatite. The most reliable way to estimate the age of peak metamorphism is by dating one of these minerals using a decay scheme with a nominal T_{cb} safely above the estimated peak temperature. Both (U–Th)/Pb monazite and Sm/Nd garnet would be appropriate choices, but correct interpretation of the dates obtained using either technique would require care. In the case of (U–Th)/Pb monazite dating, a special concern is that the sample may contain a component of detrital monazite manifested as cores with metamorphic overgrowths; since peak metamorphic temperatures were too low to induce substantial diffusive loss of radiogenic lead in detrital monazite, a conventional (U–Th)/Pb ID-TIMS date of a single polygenetic crystal would be an overestimate of the metamorphic age. Fortunately, the internal structure of accessory minerals can be revealed through backscatter electron, cathodoluminesence, and X-ray compositional mapping. In metamorphic samples, accessory minerals always should be characterized using one or more of these tools prior to dating as a guide to proper interpretation of the results. Under ideal circumstances, lower-precision microanalytical techniques—such as electron microprobe, ion microprobe, or ICP-MS laser microprobe (U–Th)/Pb dating—could be used to establish the approximate ages of different monazite components, and the results might permit us to target specific unzoned metamorphic grains (or grain fragments) for higher-precision ID-TIMS dating. For samples with multicomponent accessory mineral populations, especially those displaying complex, small-scale intergrowths, it may be impossible to extract sufficient material for ID-TIMS analysis and some precision must be sacrificed to ensure having readily interpretable data.

Sm–Nd dating of the garnet in our hypothetical sample also might be problematic. One concern is the likelihood of trace-element growth zoning, which would preclude dating single garnet crystals by Sm/Nd ID-TIMS unless we make the assumption that the growth rate of the garnet was so rapid that the elapsed time permitted no appreciable accumulation of radiogenic Nd. An even greater problem is the probability that the garnet will contain accessory mineral inclusions that would compromise the results; even tiny amounts of monazite would contain very large concentrations of the light rare earths and would dominate the Sm–Nd isotopic characteristics of the "garnet" (DeWolf *et al.*, 1996; Vance *et al.*, 1998; Prince *et al.*, 2000; Thoni, 2002). As a consequence, the isolation of inclusion-free rims would be required for effective Sm/Nd garnet geochronology. In addition, some method must be employed to estimate the non-radiogenic ^{143}Nd component in the sample, which in turn has an impact on our confidence in the T_{cb} estimate in Table 2 as well as the precision of the calculated date.

For our hypothetical sample, it is probable that (U–Th)/Pb ID-TIMS dating of the monazite would provide the most reliable and most precise estimate of the age of peak metamorphism *if* the metamorphic component could be physically separated. If the metamorphic monazite could be identified through some imaging technique but could not be physically separated from detrital monazites, (U–Th)/Pb *in situ* microanalytical dating would be the preferred methodology. However, if our petrographic characterization of the rock left doubts regarding the relative age of monazite growth and peak metamorphism, Sm/Nd dating of the garnet rim may be the most reliable way to estimate the age of peak metamorphism. Unfortunately, no *in situ* microanalytical technique with sufficiently high analytical precision has been developed for Sm/Nd dating of garnet, and some method of mechanical separation of the garnet rim would be necessary to enable ID-TIMS dating.

Microanalytical methods may provide age constraints on the *PT* path prior to peak metamorphism. Chapter 3.07 illustrates how thermodynamic calculations based on major- or trace-element zoning in garnet can be used to model differential pressure and temperature over the period of garnet growth. If a modeled garnet contains distributed accessory mineral inclusions useful for microanalytical (U–Th)/Pb dating, the resulting *PT* evolution can be placed in a temporal context (DeWolf *et al.*, 1993; Harrison *et al.*, 1997; Zhu *et al.*, 1997a). However, accessory mineral inclusions are rarely distributed *throughout* their host porphyroblasts, and it would be extremely fortuitous to be able to assign more than one or two absolute ages to *PT* points along a reconstructed path.

An unfortunate reality of *PTt* studies of metamorphic rocks is that the prograde *PT* path can be reconstructed with higher fidelity than the retrograde *PT* path, but the inverse is true for the prograde and retrograde *Tt* paths (Hodges, 1991). Our hypothetical sample provides many opportunities for calibrating its cooling history. In addition to the (U–Th)/Pb dating of monazite discussed above, U/Pb, (U–Th)/He, and fission-track dating of apatite and $^{40}Ar/^{39}Ar$ and Rb/Sr dating of muscovite and biotite permit *Tt* path reconstruction from peak temperatures to ~60–70 °C. If the sample cooled very slowly, microanalytical studies of isotopic zoning or dating different grain sizes might provide even more detail. In contrast, secondary fluid inclusion thermobarometry is the only tool available for retrograde *PT* path reconstruction for a sample such as this displaying no textural evidence of retrograde resorbtion or re-equilibration of principal rock forming minerals.

Samples that do show evidence for retrogression or polymetamorphism provide special challenges and special opportunities for *PTt* path reconstruction. As discussed in Chapter 3.07, sophisticated modeling techniques have been developed to gain remarkably detailed insights into the often complex *PT* evolution of such samples. Most of these involve the topical application of traditional techniques to specific subassemblages of minerals that are thought to have been in equilibrium with one another at a specific time in a sample's history. An analogous approach to geochronology and thermochronology is possible—we can date minerals of different generations. However, as is the case with *PT* path studies, particular attention must be paid to the potential geochemical impact of later prograde or retrograde metamorphic events on minerals crystallized during early events.

As an example, consider the case of a pelitic schist sample (with the same mineral assemblage as our previous hypothetical sample) that experienced early prograde metamorphism (M_1) at ~600 °C and a second episode of metamorphism (M_2) at ~450 °C. Suppose further that this sample contains two generations of garnet—expressed as M_2 rims on M_1 cores—and two generations of all other geochronologically important minerals that occur as discrete M_1 and M_2 crystals. In this case, *in situ* microanalysis of inclusions of M_1 and M_2 monazite in the cores and rims of garnets might help elucidate the prograde histories of both M_1 and M_2. Peak M_1 and M_2 temperatures could be dated reliably using the (U–Th)/Pb monazite geochronometer. The age of peak M_2 metamorphism also might be confirmed through (U–Th)/Pb dating of M_2 apatites, but the significance of such dates for M_1 apatites is less clear. Do they date a temperature on the cooling path subsequent to M_1? Were they reset during M_2 metamorphism? Or did partial diffusive lead loss during M_2 render the results geologically meaningless? Thermochronometers with closure temperatures below 450 °C would be useful for reconstructing the post-M_2 cooling history regardless of whether or not the minerals involved grew during M_1 or M_2. Unfortunately, this sample provides no viable options for establishing an unambiguous record of the *Tt* path of the sample between M_1 and M_2.

Such thought experiments emphasize two important points about the thermochronology of metamorphic rocks. The first is that it is virtually impossible to draw robust conclusions about the thermal history of a sample by thermochronologic methods without concomitant petrologic study. The second is that, since multiple generations of datable minerals are likely to be found in metamorphic rocks, the most successful thermochronologic studies employ microanalytical techniques to ensure that the results are placed in proper petrographic context. Some examples include DeWolf *et al.* (1993), Cocherie *et al.* (1998), Hawkins and Bowring (1999), Schaltegger *et al.* (1999), Vavra and Schaltegger (1999), Simpson *et al.* (2000), Terry *et al.* (2000), Catlos *et al.* (2001), Di Vincenzo *et al.* (2001), Rubatto *et al.* (2001), Rubatto (2002), Catlos *et al.* (2002), and Zeck and Whitehouse (2002).

3.08.7.4 Calibrating Deformational Histories

In orogenic settings where multiple generations of intrusive or extrusive rocks occur, field relationships frequently provide constraints on their age relative to deformational fabrics and structures. By documenting that an igneous unit is post- or pre-kinematic with respect to a particular deformational feature, and by determining the crystallization age of that unit, it is possible to assign a minimum or maximum age to the feature. Under fortuitous circumstances, it can be possible to "bracket" the age of a structure by dating both pre- and post-kinematic igneous rocks. For landscapes exhumed from deeper structural levels, comparative thermochronometry of the hanging walls and footwalls of major post-metamorphic or post-intrusive thrust and normal faults provides a way to estimate the age of such structures as the time of convergence of their *Tt* paths. If *Tt* paths can be reconstructed only for the hanging wall or footwall but not both, estimates of the developmental temperatures of fault-related rocks might be compared to the hanging wall or footwall *Tt* path to constrain the maximum age of faulting. Such approaches to calibrating deformational histories have been used by Pan and Kidd (1992), Hodges and Applegate (1993), Applegate and Hodges (1995), Harrison *et al.* (1997), Hodges *et al.* (1996, 1998), Coleman and Hodges (1998), Hartz *et al.* (2000), Wells *et al.* (2000), Gans *et al.* (2001), and Murphy *et al.* (2002).

Major normal fault systems that root into the middle and lower crust, sometimes referred to as "detachments," have been identified in tectonic regimes as varied as the Basin and Range Province of western North America to the Himalayan orogen of South Asia (Burchfiel *et al.*, 1992; Wernicke, 1992). Large-magnitude displacement on these structures has the effect of rapidly cooling their footwalls (e.g., Ruppel *et al.*, 1988). Because of this, many researchers have used thermochronologic techniques to date the onset of rapid footwall cooling and have interpreted that date as a close approximation of the inception age of slip on the detachment (e.g., Dokka *et al.*, 1986; Davis and Lister, 1988; Harrison *et al.*, 1995a; Miller *et al.*, 1999; Murphy *et al.*, 2002). However, such interpretations require caution, because both theoretical and empirical studies indicate that they may underestimate the true age of fault initiation by up to several million years (House and Hodges, 1994; Ruppel and Hodges, 1994). Some of the most interesting studies of detachments involve the development of large thermochronologic databases from well-exposed footwalls and the use of cooling age patterns to deduce fault kinematics and slip rates (Foster and John, 1999; Ehlers *et al.*, 2001; Stockli *et al.*, 2001; Brady, 2002).

A more direct approach to dating deformational features involves the application of isotopic geochronology to minerals that grew initially or were dynamically recrystallized during deformation at temperatures below their nominal closure temperatures (Reuter and Dallmeyer, 1989; Gromet, 1991; West and Lux, 1993; Coleman and Hodges, 1995; Resor *et al.*, 1996; Dunlap, 1997; Anderson *et al.*, 2001; Dallmeyer *et al.*, 2001; Müller *et al.*, 2001; Reddy *et al.*, 2001). Many deformed rocks display multiple generations of tectonite fabrics, and techniques such as laser microprobe $^{40}Ar/^{39}Ar$ dating of polished thin sections permit the dating of specific minerals in structural context and thus determination of the absolute ages of different fabric-forming events. Although care must be taken to ensure that the dates for older grains are not compromised by deformation-induced loss of radiogenic ^{40}Ar (Reddy and Potts, 1999; Dunlap and Kronenberg, 2001; Kramar *et al.*, 2001; Mulch *et al.*, 2002), some successful studies have been documented (e.g., Hames and Cheney, 1997; Chan *et al.*, 2000; Di Vincenzo *et al.*, 2001) and this approach ultimately may prove to be particularly valuable for the study of the duration of and interval between fabric-forming events during orogenesis.

A relatively recent development has been the ability to date brittle faulting using the $^{40}Ar/^{39}Ar$ method. Quenched frictional melts (pseudotachylytes) have been dated using both step-heating and laser microprobe techniques (Kelley *et al.*, 1994; Magloughlin *et al.*, 2001; Müller *et al.*, 2001; Sherlock and Hetzel, 2001; White and Hodges, 2002). In addition, van der Pluijm *et al.* (2001) and Parry *et al.* (2001) have shown how careful $^{40}Ar/^{39}Ar$ dating of synkinematic illites can be used to determine the age of fault gouge development during slip on near-surface faults.

3.08.7.5 Estimating Unroofing Rates

Although thermochronologic studies can yield remarkably detailed reconstructions of the cooling histories of rocks, more useful information—from the perspective of continental tectonics—would be a better understanding of how these rocks changed their positions with respect to Earth's surface over time. Since temperatures are known to increase with depth (z) in the stable continental crust, it seems logical that the *Tt* path of a sample should reflect the history of its transport toward Earth's surface. When unroofing can be attributed to tectonic denudation, thermochronology can help constrain the kinematic evolution of structurally higher normal fault systems. When it can be attributed to erosion, such data improve our understanding of sedimentary processes during orogenesis.

One approach to estimating unroofing histories has been to relate the cooling rate ($\mathrm{d}T/\mathrm{d}t$) determined for a single sample to its unroofing rate ($\mathrm{d}z/\mathrm{d}t$) through an assumed geothermal gradient (γ):

$$\frac{\mathrm{d}z}{\mathrm{d}t} \approx \left(\frac{1}{\gamma}\right)\frac{\mathrm{d}T}{\mathrm{d}t} \qquad (20)$$

The cooling rate can be determined either by applying at least two thermochronometers and approximating the cooling rate as differences in T_{cb}/t_{cb}, or by determining core–rim variations in t_{cx} and T_{cx} in individual isotopically zoned crystals. A popular alternative is to apply a single thermochronometer to samples collected over a range of modern elevations. For example, suppose samples are collected over a range of elevations ($\varepsilon_1, \varepsilon_2, \varepsilon_3$) and they yield a range of (U–Th)/He closure dates ($t_{cb1}, t_{cb2}, t_{cb3} \ldots$). If T_{cb} is presumed to be the same for all of the apatites, then

$$\frac{dz}{dt} \approx \frac{d\varepsilon}{dt_{cb}} \tag{21}$$

Age–elevation data also may be helpful in determining when a particular denudation episode began. Consider the case of a stable crustal section in which the upper and lower bounding isotherms of the apatite fission-track partial annealing zone (Section 3.08.6.4) occur at depths z_a and z_b. If an episode of unroofing begins at time t_{init} and continues to the present day at a uniform rate, modern surface exposures may—if topographic relief is high enough and the unroofing is rapid enough—include samples from below and within the partial annealing zone (Figure 3). Age–elevation profiles developed from samples collected from such a terrain will be characterized by an inflection point that marks the approximate position of paleodepth z_b; rocks below this elevation yield an age–elevation gradient indicative of the denudation rate, whereas the gradient above is a reflection of the pre-unroofing thermal structure. Profiles interpreted as exhumed portions of fission-track partial annealing zones—or conceptually similar (U–Th)/He "partial retention zones" (Wolf *et al.*, 1998)—have been described by Fitzgerald *et al.* (1995), Fitzgerald and Stump (1997), Reiners *et al.* (2000), Stockli *et al.* (2000), Xu and Kamp (2000), Bullen *et al.* (2001), and Crowley *et al.* (2002).

All methods by which thermochronologic data are used to estimate unroofing rates require a series of assumptions that are unlikely to be valid in all orogenic settings and thus must be made with caution. One is that temperatures in the crust increase monotonically toward deeper structural levels at a constant rate, with isothermal surfaces remaining at least roughly parallel to Earth's surface at all levels. Another is that the thermal structure of the crust is invariant during the unroofing interval. Applications that involve the application of Equation (20) require the assumption of a geothermal gradient. Although the "age–elevation" method avoids this, it requires a different assumption: that the samples collectively moved as a block, without rotation, through the closure isotherm for the applied chronometer.

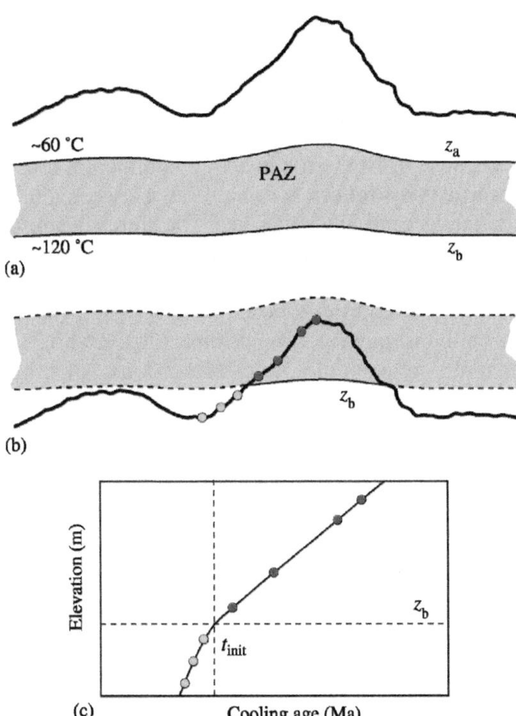

Figure 3 Exhumed apatite fission-track partial annealing zone (PAZ) and its value for unroofing rate studies. (a) Diagrammatic cross-section showing the PAZ (gray) beneath a steady-state topography prior to an unroofing episode. The top and base of the PAZ (z_a and z_b) represent the approximate positions of the 60 °C and 120 °C isotherms, respectively. Only a slight effect of topography on the shape of isotherms is shown for the sake of simplicity. (b) Position of the PAZ after exhumation by erosion. Samples may be collected along a modern age–elevation profile. Those collected above the structural level that once was z_b are shown in red; those below are shown in blue. (c) Elevation versus apatite fission-track cooling ages for samples collected along the transect. An inflection in the slope of the best-fit curve indicates the approximate initiation age of the unroofing event (t_{init}) and the position of the base of the exhumed PAZ.

At the most fundamental level, these assumptions seem inconsistent with our expectations that the thermal structures in active orogens are both complex and transient (e.g., Huerta *et al.*, 1996; Batt and Braun, 1997; Jamieson *et al.*, 2002). Nevertheless, an increasing number of tectonic studies suggest that major structural configurations in orogenic systems may be active over tens of millions of years (e.g., Hodges *et al.*, 2001). This condition naturally leads to the development of a thermodynamic steady state that may persist over relatively long time intervals, a conclusion that should not be surprising given the natural tendency toward self-organization in simpler dynamical systems (Hodges, 1998b). Indeed, steady-state thermal structures are predicted by numerical and analytical models of the thermomechanical evolution of orogenic systems (e.g., Huerta *et al.*, 1996; Batt and Braun, 1997; Jamieson *et al.*, 2002).

The topologies of isotherms in orogens at a thermodynamic steady state are more complex than those of isotherms in stable continental crust because they reflect the interplay between both conductive and advective heat transfer, but these differences become less significant toward shallower crustal levels (i.e., towards the temperature boundary condition imposed by Earth's surface). As a consequence, Equations (20) and (21) provide reasonable approximations of the relations between cooling rate and unroofing rate if steady-state conditions have been achieved and if data from thermochronometers with relatively low T_{cb}'s (roughly $<500\,^{\circ}C$) are used for the calculations. Estimates of denudation rate with higher fidelities may be obtained through more sophisticated inverse methods that account for advective processes explicitly (e.g., Moore and England, 2001).

Unfortunately, it is not possible to know *a posteriori* if, when, and for how long an orogen may have been at steady state. Rather than presume the achievement of steady state for calculating unroofing rates, a better strategy may be to evaluate the consistency of unroofing rates calculated with Equation (20) using samples collected at different structural levels, or with Equation (21) using more than one thermochronometer. Such consistency is a direct indication of the achievement of steady-state conditions and, as noted by Willett and Brandon (2002), is an important monitor of the developmental maturity of an orogenic system.

3.08.7.6 Monitoring the Evolution of Topography

One manifestation of a thermodynamic steady state during orogeny can be an erosional steady state, during which the rate of rock uplift toward Earth's surface is matched by the erosion rate at the surface. Numerical models of landscape evolution in idealized orogens frequently predict the development of erosional steady states (Kooi and Beaumont, 1996; Braun and Sambridge, 1997; Willett *et al.*, 2001). A true erosional steady state implies that the surface topography is static because landscapes evolve as a consequence of lateral variations in erosion rate with time. While this extreme condition is unrealistic, it offers a valuable point of departure for exploring how topography affects near-surface thermal structures and, thus, how the distributions of cooling ages across a varied topography can be used to understand the antiquity of the landscape.

It has been known for many years that isotherms at shallow crustal levels mimic surface topography (Bullard, 1938; Birch, 1950; Turcotte and Schubert, 1982), but it was not until recently that Stüwe *et al.* (1994) and Mancktelow and Grasemann (1997) demonstrated the effect of this phenomenon on thermochronology. In particular, they showed that high-wavelength topographic variations and high exhumation rates—both of which are characteristic of the orogenic settings where thermochronology is most often practiced—can lead to substantial errors in calculations based on equations such as (20) and (21). This effect is most pronounced in the near-surface, such that exhumation rates estimated from apatite fission-track and (U–Th)/He apatite data may be especially unreliable unless an appropriate "terrain correction" can be made (e.g., Stüwe *et al.*, 1994).

Like many obstacles in the earth sciences, this one can be turned to our advantage: if apatite closure ages are so strongly influenced by topography, then it should be possible to use spatial patterns in closure ages to reconstruct the evolution of topography. As an example of how this might be done, consider the steady-state thermal effects of periodic topographic variation in an orogen with constant erosion at $1\ \mathrm{mm\ yr^{-1}}$ (Figure 4). Sample A, starting at a nominal depth of 9 km beneath a topographic high and uplifted through the illustrated thermal structure at $1\ \mathrm{mm\ yr^{-1}}$, would pass through the nominal closure isotherm for (U–Th)/He in apatite after 8 Myr. Sample B, starting at the same structural level but beneath a topographic low, reaches the closure isotherm after 6.6 Myr. The 1.4 Myr difference in (U–Th)/He apatite cooling age between samples A and B—if it could be resolved within the limits of analytical imprecision—would be a direct indication of the magnitude and distribution of topographic relief over the recovered cooling interval. With this technique, House *et al.* (1998) found that the thermal influences of major transverse drainages on the western flank of the Sierra Nevada of California had been established by Late Cretaceous time. Note that the effects of surface topography on isotherms are damped with depth in Figure 4. As a consequence, the difference between (U–Th)/He titanite closure ages for A and B would be far less than that between (U–Th)/He apatite closure ages, and the difference in muscovite Rb/Sr closure ages would be negligible. At higher exhumation rates and with higher relief, topographic effects extend to deeper levels, such that studies integrating (U–Th)/He, fission-track, $^{40}Ar/^{39}Ar$ and Rb/Sr data for a variety of minerals from the several structural levels can recover not just the magnitude of topography, but also how that topography changed in time and space during the orogenic process.

While simplified, steady-state models with periodic topography are valuable for understanding

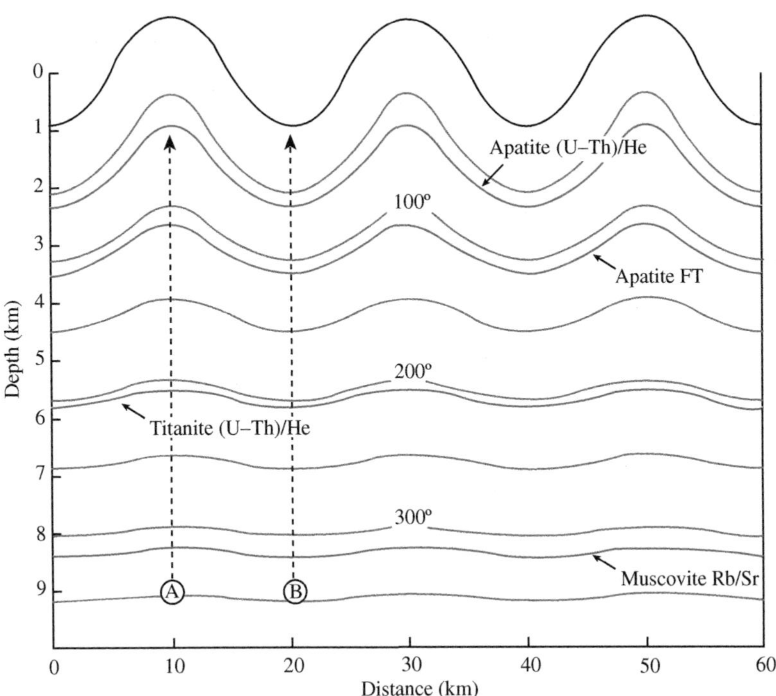

Figure 4 Steady-state temperature structure beneath a periodic topography as calculated using the algorithms of Mancktelow and Grasemann (1997). This simulation assumes a topographic relief of 1.5 km, a topographic wavelength of 20 km, and a uniform denudation rate of 1 mm yr^{-1}. The topmost sinusoidal line indicates the land surface (defined as having a temperature of 0 °C). Shaded contours are steady-state isotherms from 50 °C to 350 °C at 50 °C intervals. The red contours represent nominal bulk closure isotherms for the muscovite Rb/Sr, apatite fission track, and apatite and titanite (U–Th)/He thermochronometers. Positions A and B and their dashed unroofing paths are discussed in the text.

the fundamental physics of the problem, they do not provide a sufficiently sophisticated mathematical foundation for the themochronologic study of landscape evolution. First steps toward the next generation of models include the work of Stüwe and Hintermüller (2000) and Braun (2002b), who considered the impact of an evolving topography, and Braun (2002a), who introduced a spectral analysis approach to data inversion that provides estimates of mean exhumation rates and relief changes without requiring the assumption of a geothermal gradient.

3.08.7.7 Reconstructing Regional Patterns of Deformation and Erosion

With the advent of increasingly sophisticated thermal models, many researchers are developing regionally extensive data sets in an effort to invert patterns of bedrock cooling ages for information about how deformational structures and topography co-evolve during orogenesis. Such studies have been done recently along the San Andreas fault zone in the San Bernadino Mountains of southern California (Spotila *et al.*, 2001), the Olympic Mountains of the northwestern United States (Batt *et al.*, 2001), the Southern Alps of New Zealand (Tippett and Kamp, 1993, 1995; Batt *et al.*, 2000), the Coast Mountains of British Columbia, Canada (Farley *et al.*, 2001), the central Alaska Range (Fitzgerald *et al.*, 1995), the Nanga Parbat syntaxis of Pakistan (Zeitler *et al.*, 2001), and the eastern Tibetan Plateau (Xu and Kamp, 2000; Kirby *et al.*, 2002).

Sedimentary deposits provide a valuable record of the erosional history of an orogen. The crystallization or cooling ages of detrital minerals in a basin can be used to determine provenance (e.g., Hurford *et al.*, 1984; Baldwin *et al.*, 1986; Hurford and Carter, 1991; Kelley and Bluck, 1992; von Eynatten *et al.*, 1996; Gray and Zeitler, 1997; Adams and Kelley, 1998; Kalsbeek *et al.*, 2000; Watt and Thrane, 2001; Kosler *et al.*, 2002). By comparing depositional ages with the cooling ages for detrital minerals, it is possible to ascertain the timescales for erosion and transport in orogenic settings (Cerveny *et al.*, 1988; Copeland and Harrison, 1990; Brandon and Vance, 1992; Corrigan and Crowley, 1992; Garver and Brandon, 1994; Bullen *et al.*, 2001). If the distribution of bedrock cooling ages are known for source regions, the thermochronology of detrital minerals provide a valuable tool for the reconstruction of regional patterns of erosion (Renne *et al.*, 1990; DeCelles *et al.*, 1998; Lonergan and Johnson, 1998; Garver *et al.*, 1999;

Stuart, 2002). A relatively recent development has been the use of thermochronologic data for modern stream sediment samples to limit the bedrock cooling age distribution and relief (Stock and Montgomery, 1996) or average erosion rate (Brewer, 2001; Bullen *et al.*, 2001) for the stream's catchment. Such studies may yield a sense of how relief developed over time within specific catchments, and catchment-to-catchment variations may, in turn, reveal the nature of topographic change over time.

3.08.8 DIRECTIONS FOR FUTURE RESEARCH

Modern tectonics research is heavily dependent on the tools of geochronology and thermochronology and, as a consequence, has become an important driver of analytical innovation. Understanding the pace of thermal, deformational, and erosional processes requires the development of chronometers that are both accurate and precise, and it must be possible to apply these tools selectively to features formed during specific tectonic events. These three needs—accuracy, specificity, and precision—help define the frontiers of geochronology and thermochronology research.

The *accuracy* of isotopic chronometry depends, in part, on an exact knowledge of decay constants. While most of the constants shown in Table 1 are reasonably well constrained, the need to resolve small segments of time when studying orogenic processes and the need to compare the results of different mineral-isotopic systems demand that we minimize the uncertainties in decay constants even more. Relatively little attention has been paid over the past two decades to the need to know these values better, and recent comparative geochronology studies suggest that their refinement must be a priority for future research. Especially problematic in this regard are: (i) the $^{40}K \rightarrow {}^{40}Ca$ decay scheme, for which nuclear physicists and geologists typically assume a different decay constant (Min *et al.*, 2000; Renne, 2000); (ii) the $^{147}Sm \rightarrow {}^{143}Nd$ decay scheme, for which there is a large uncertainty in the accepted decay constant (Lugmair and Marti, 1978); and (iii) the $^{176}Lu \rightarrow {}^{176}Hf$ decay scheme, for which the decay constant is actively debated (Scherer *et al.*, 2000, 2001).

Other limits to the accuracy of chronologic data involve uncertainties in the nature of what is being dated, and the corrections necessary for the application of a specific chronometer. Some techniques (e.g., Sm/Nd and Lu/Hf dating of garnet) can be seriously compromised by inclusions of accessory phases, such that accurate geochronology and thermochronology require ultra-pure mineral separates. In addition, many commonly dated minerals display complex compositional zoning (see Chapter 3.07) and contain defects that influence the intracrystalline transport of daughter isotopes and, thus, closure temperatures. Efforts to develop new microsampling and sample characterization protocols should be encouraged as a way to minimize such problems. A concern specific to $^{40}Ar/^{39}Ar$ thermochronology is the need to have high-purity flux monitors for which ages are known well. Recent studies have cast doubt on the ages of some of the most widely used monitors (Renne *et al.*, 1998; Lanphere and Baadsgaard, 2001; Schmitz and Bowring, 2001). While problems of this kind have no impact on applications that do not require a comparison of $^{40}Ar/^{39}Ar$ results with those made with other isotopic chronometers, they must be resolved through further research before absolute ages based on the $^{40}Ar/^{39}Ar$ technique can be considered truly reliable.

A limitation specific to the (U–Th)/He decay scheme is the need to correct calculated cooling ages for ^{4}He lost due to α-particle emission (Farley *et al.*, 1996). In many cases, these corrections—based on the size and shape of a dated crystal and assumptions about the stopping distances of α-particles in various compounds—amount to several tens of percent of the apparent age, making them the principle contributor to the uncertainty in a (U–Th)/He cooling age. The development of microanalytical methods to eliminate the need for the α-ejection correction—or at least better methods of characterizing crystal morphology and stopping distances in important minerals in order to minimize the uncertainty in the correction—should have high priority for future research.

The accuracy with which we calculate closure temperatures depends on having high-quality diffusion data. For some systems (e.g., Lu/Hf in garnet), reliable diffusion data do not exist. For many others, we have no good understanding of the effects of natural compositional variability or deformational history on diffusivity. The experiments necessary to improve this situation are difficult and time-consuming, but they are of crucial importance.

Specificity refers to our ability to determine crystallization ages for minerals that grew at a specific time during the history of a sample. For a typical metamorphic sample from the internal regions of an orogen, it may be desirable to date minerals that define two or three fabrics, or specific zones within a porphyroblast that has a prolonged growth history, or multiple generations of minerals that grew during different metamorphic episodes. Many of the microanalytical methods reviewed in Section 3.08.3.1 were developed specifically to address such needs. However, most of these techniques provide high-spatial resolution at the expense of analytical *precision*. This is a significant limitation because, as we learn more and more about the evolution of orogenic systems, it becomes clear that deformational and erosional process transitions typically occur over short timescales that can be resolved only with extremely high-precision chronometers. The problem is particularly acute for Cenozoic orogens. For example, much of the tectonic architecture that has characterized the Himalaya for the past 20 Myr appears to have been established over a period of ~5 Myr, yet many published ages for Himalayan samples are reported with uncertainties in excess of 2 Myr at the ~95% confidence level. Clearly, this level of imprecision is unacceptably large for many avenues of Himalayan research. In the future, we can expect to see more attention paid to the development of analytical techniques that represent a practical compromise between the desire for improved spatial resolution and the equally pressing need for improved analytical precision.

Finally, as the availability of high-quality geochronologic and thermochronologic data increases over the next few years, we can expect to see an increase in the number of inverse modeling techniques developed to extract from those data information about the thermal, deformational, and erosional histories of an orogen. Unfortunately, the success of these initiatives will be limited by the idiosyncratic behavior of complex systems such as orogens. For example, many tectonics researchers currently favor a two-sided developmental model of convergent orogens (Willett *et al.*, 1993) because it predicts many of the first-order characteristics of well-studied active orogens (Beaumont *et al.*, 1996, 2001; Willett, 1999; Jamieson *et al.*, 2002). Without a doubt, such forward models provide valuable insights regarding appropriate interpretations of geochronologic and thermochronologic data, but many time–temperature trajectories predicted for specific structural horizons in these models are dependent on a large number of assumptions that may be impossible to justify for real orogens. As orogenic models become increasingly sophisticated, their sensitivity to specific assumptions becomes more difficult to quantify. Ultimately, they may begin to behave like the complex systems they were designed to simulate. A key characteristic of such systems is their capacity to self-organize into behavioral patterns that are sustained by largely unpredictable process interactions. If such "emergent behavior" occurs, it would be impossible to trace an effect to a specific cause—the principal goal of inverse modeling—in either numerical simulations or real orogens. One of the great challenges before us is to attain the scientific sophistication to understand not only how much we can learn about orogenic processes from geochronology and thermochronology, but also the limitations of what we can learn.

ACKNOWLEDGMENTS

Geochronology and thermochronology are vibrant fields, and no short review can do justice to the work being done today by dozens of research groups worldwide. This chapter is intended as a (relatively!) objective overview of current methods and the many ways in which they are used for tectonic studies. I apologize in advance for any errors of omission.

Much of the review of basic concepts in this chapter derives from the notes for graduate-level subjects that I have taught at MIT, and I appreciate very much how the participating students have helped me to refine the presentation of these concepts. My perceptions about geochronology and thermochronology have benefited greatly from collaborations with my wonderful graduate students and postdoctoral advisees; with my MIT colleagues Sam Bowring, Clark Burchfiel, Leigh Royden, and Kelin Whipple; and with many other colleagues from other universities. In particular, Daniele Cherniak and Ken Farley kindly provided preprints of manuscripts in progress. This work, as well as my research programs in general, have been supported by grants from the US National Science Foundation. Its continuing support is gratefully acknowledged.

REFERENCES

Adams C. J. and Kelley S. (1998) Provenance of Permian–Triassic and Ordovician metagraywacke terranes in New Zealand: evidence from $^{40}Ar/^{39}Ar$ dating of detrital micas. *Geol. Soc. Am. Bull.* **110**, 422–432.

Albarède F. (1995) *Introduction to Geochemical Modeling.* University Press, Cambridge, England.

Aleinikoff J. N., Wintsch R. P., Fanning C. M., and Dorais M. J. (2002) U–Pb geochronology of zircon and polygenetic titanite from the Glastonbury Complex, Connecticut, USA: an integrated SEM, EMPA, TIMS, and SHRIMP study. *Chem. Geol.* **188**, 125–147.

Anczkiewicz R., Oberli F., Burg J. P., Villa I. M., Gunther D., and Meier M. (2001) Timing of normal faulting along the Indus suture in Pakistan Himalaya and a case of major Pa-231/U-235 initial disequilibrium in zircon. *Earth Planet. Sci. Lett.* **191**, 101–114.

Anderson S. D., Jamieson R. A., Reynolds P. H., and Dunning G. R. (2001) Devonian extension in northwestern Newfoundland: $^{40}Ar/^{39}Ar$ and U–Pb data from the Ming's Bight area, Baie Verte peninsula. *J. Geol.* **109**, 191–211.

Applegate J. D. R. and Hodges K. V. (1995) Mesozoic and Cenozoic extension recorded by metamorphic rocks in the Funeral Mountains, California. *Geol. Soc. Am. Bull.* **107**, 1063–1076.

Armstrong R. L., Jäger E., and Eberhardt P. (1966) A comparison of K–Ar and Rb–Sr ages on Alpine biotites. *Earth Planet. Sci. Lett.* **1**, 13–19.

Arnaud N. O. and Eide E. A. (2000) Brecciation-related argon redistribution in alkali feldspars: an in naturo crushing study. *Geochim. Cosmochim. Acta* **64**(18), 3201–3215.

Baldwin S. L., Harrison T. M., and Burke K. (1986) Fission track evidence for the source of accreted sandstones, Barbados. *Tectonics* **5**, 457–468.

Batt G. E. and Braun J. (1997) On the thermomechanical evolution of compressional orogens. *Geophys. J. Inter.* **128**, 364–382.

Batt G., Braun J., Kohn B., and McDougall I. (2000) Thermochronological analysis of the dynamics of the southern Alps, New Zealand. *Geol. Soc. Am. Bull.* **112**, 250–266.

Batt G. E., Brandon M. T., Farley K. A., and Roden-Tice M. (2001) Tectonic synthesis of the Olympic Mountains segment of the Cascadia wedge, using two-dimensional thermal and kinematic modeling of thermochronological ages. *J. Geophys. Res. Solid Earth* **106**(B11), 26731–26746.

Beaumont C., Kamp P., Hamilton J., and Fullsack P. (1996) The continental collision zone, South Island, New Zealand: comparison of geodynamical models and observations. *J. Geophys. Res.* **101**(B2), 3333–3359.

Beaumont C., Jamieson R. A., Nguyen M. H., and Lee B. (2001) Himalayan tectonics explained by extrusion of a low-viscosity crustal channel coupled to focused surface denudation. *Nature* **414**, 738–742.

Birch F. (1950) Flow of heat in the front range, Colorado. *Geol. Soc. Am. Bull.* **61**, 567–630.

Brady R. J. (2002) Very high slip rates on continental extensional faults: new evidence from (U–Th)/He thermochronometry of the Buckskin Mountains, Arizona. *Earth Planet. Sci. Lett.* **197**, 95–104.

Brandon M. T. and Vance J. A. (1992) Fission-track ages of detrital zircon grains: implications for the tectonic evolution of the Cenozoic Olympic subduction complex. *Am. J. Sci.* **292**, 565–636.

Brandon M. T., Roden-Tice M. K., and Garver J. I. (1998) Late Cenozoic exhumation of the Cascadia accretionary wedge in the Olympic Mountains, northwest Washington State. *Geol. Soc. Am. Bull.* **110**, 985–1009.

Braun J. (2002a) Estimating exhumation rate and relief evolution by spectral analysis of Age–elevation datasets. *Terra Nova* **14**, 210–214.

Braun J. (2002b) Quantifying the effect of recent relief changes on Age-elevation relationships. *Earth Planet. Sci. Lett.* **200**, 331–343.

Braun J. and Sambridge M. (1997) Modelling landscape evolution on geological time scales: a new method based on irregular spatial discretization. *Basin Res.* **9**, 27–52.

Brewer I. D. (2001) Detrital-mineral thermochronology: investigations of Orogenic denudation in the Himalaya of central Nepal. PhD, The Pennsylvania State University.

Brown M. and Solar G. S. (1998) Shear-zone systems and melts: feedback relations and self-organization in orogenic belts. *J. Struct. Geol.* **20**(2–3), 211–227.

Bruguier O., Telouk P., Cocherie A., Fouillac A. M., and Albarede F. (2001) Evaluation of Pb–Pb and U–Pb laser ablation ICP–MS zircon dating using matrix-matched calibration samples with a frequency quadrupled (266 nm) Nd-YAG laser. *Geostand. Newslett.: J. Geostand. Geoanal.* **25**(2–3), 361–373.

Bullard E. C. (1938) The disturbance of the temperature gradient in the Earth's crust by inequalities of height. *Mon. Not. Roy. Astron. Soc.: Geophys. Suppl.* **4**, 360–362.

Bullen M. E., Burbank D. W., Garver J. I., and Abdrakhmatov K. Y. (2001) Late Cenozoic tectonic evolution of the northwestern Tien Shan: new age estimates for the initiation of mountain building. *Geol. Soc. Am. Bull.* **113**(12), 1544–1559.

Burchfiel B. C., Chen Z., Hodges K. V., Liu Y., Royden L. H., Deng C., and Xu J. (1992) *The South Tibetan Detachment System, Himalayan Orogen: Extension Contemporaneous with and Parallel to Shortening in a Collisional Mountain Belt.* Geological Society of America, Boulder, CO.

Carlson W. D. (1990) Mechanism and kinetics of apatite fission-track annealing. *Am. Mineral.* **75**, 1120–1139.

Catlos E. J., Harrison T. M., Kohn M. J., Grove M., Ryerson F. J., Manning C. E., and Upreti B. N. (2001) Geochronologic and thermobarometric constraints on the evolution of the main central thrust, central Nepalese Himalaya. *J. Geophys. Res.* **106**, 16177–16204.

Catlos E. J., Gilley L. D., and Harrison T. M. (2002) Interpretation of monazite ages obtained via *in situ* analysis. *Chem. Geol.* **188**, 193–215.

Cerveny P. F., Naeser N. D., Zeitler P. K., Naeser C. W., and Johnson N. M. (1988) History of uplift and relief of the Himalaya during the past 18 million years: evidence from sandstones of the Siwalik group. In *New Perspectives in Basin Analysis* (eds. K. L. Kleinspehn and C. Paola). Springer, New York, pp. 43–61.

Chamberlain K. R. and Bowring S. A. (2001) Apatite–feldspar U–Pb thermochronometer: a reliable, mid-range (~450°C), diffusion-controlled system. *Chem. Geol.* **172**(1–2), 173–200.

Chan Y.-C., Crespi J. M., and Hodges K. V. (2000) Dating cleavage formation in slates and phyllites with the $^{40}Ar/^{39}Ar$ laser microprobe; an example from the western New England appalachians, USA. *Terra Nova* **12**, 264–271.

Chen C.-H., DePaolo D. J., and Lan C.-Y. (1996) Rb–Sr microchrons in the Manaslu granite: implications for Himalayan thermochronology. *Earth Planet. Sci. Lett.* **143**, 125–135.

Cherniak D. J. (1993) Lead diffusion in titanite and preliminary results on the effects of radiation damage on Pb transport. *Chem. Geol.* **110**, 177–194.

Cherniak D. J. (2000) Pb diffusion in rutile. *Contributions to Mineralogy and Petrology* **139**(2), 198–207.

Cherniak D. J. and Watson E. B. (1992) A study of strontium diffusion in K-feldspar, Na–K feldspar and anorthite using Rutherford backscattering spectroscopy. *Earth and Planetary Science Letters* **113**, 411–425.

Cherniak D. J. and Watson E. B. (2000) Pb diffusion in zircon. *Chem. Geol.* **172**, 5–24.

Cherniak D. J., Lanford W. A. and Ryerson F. J. (1991) Lead diffusion in apatite and zircon using ion implantation and Rutherford backscattering techniques. *Geochim. Cosmochim. Acta* **55**, 1663–1673.

Cherniak D. J., Watson E. B., Grove M., and Harrison T. M. (2002) Pb diffusion in monazite. *Geol. Soc. Am. Abstr. Prog.* **34**(6).

Christensen J. N., Rosenfeld J. L., and DePaolo D. J. (1989) Rates of techonometamorphic processes from Rubidium and Strontium isotopes in garnet. *Science* **244**, 1465–1469.

Cocherie A., Legendre O., Peucat J. J., and Kouamelan A. N. (1998) Geochronology of polygenetic monazites constrained by *in situ* electron microprobe Th-U-total lead determination: implications for lead behavior in monazite. *Geochim. Cosmochim. Acta* **62**, 2475–2497.

Coleman M. E. (1998) U–Pb constraints on Oligocene–Miocene deformation and anatexis within the central Himalaya, Marsyandi valley, Nepal. *Am. J. Sci.* **298**(7), 553–571.

Coleman M. and Hodges K. (1995) Evidence for Tibetan Plateau uplift before 14 Myr ago from a new minimum age for east–west extension. *Nature* **374**(6517), 49–52.

Coleman M. E. and Hodges K. V. (1998) Contrasting Oligocene and Miocene thermal histories from the hanging wall and footwall of the South Tibetan detachment in the central Himalaya from $^{40}Ar/^{39}Ar$ thermochronology, Marsyandi Valley, central Nepal. *Tectonics* **17**(5), 726–740.

Compston W., Kinny P. D., Williams I. S., and Foster J. J. (1986) The age and Pb loss behavior of zircons from the Isua supracrustal belt as determined by ion microprobe. *Earth Planet. Sci. Lett.* **80**, 71–81.

Copeland P. and Harrison T. M. (1990) Episodic rapid uplift in the Himalaya revealed by $^{40}Ar/^{39}Ar$ analysis of detrital K-feldspar and muscovite, Bengal Fan. *Geology* **18**, 354–357.

Corfu F. and Stone D. (1998) The significance of titaniite and apatite U–Pb ages: constraints for the post-magmatic thermal-hydrothermal evolution of a batholithic complex,

Berens River area, northwestern superior province, Canada. *Geochim. Cosmochim. Acta* **62**, 2979–2995.

Corrigan J. (1991) Inversion of apatite fission track data for thermal history information. *J. Geophys. Res.* **96**, 10347–10360.

Corrigan J. D. and Crowley K. D. (1992) Unroofing of the Himalayas: a view from apatite fission-track analysis of Bengal Fan sediments. *Geophys. Res. Lett.* **19**, 2345–2348.

Crank J. (1975) *The Mathematics of Diffusion*. Oxford University Press, Oxford, England.

Crowley J. L. and Ghent E. D. (1999) An electron microprobe study of the U–Th–Pb systematics of metamorphosed monazite: the role of Pb diffusion versus overgrowth and recrystallization. *Chem. Geol.* **157**, 285–302.

Crowley P. D., Reiners P. W., Reuter J. M., and Kaye G. D. (2002) Laramide exhumation of the Bighorn Mountains, Wyoming: an apatite (U–Th)/He thermochronology study. *Geology* **30**(1), 27–30.

Dahl P. S. (1996a) The crystal-chemical basis for Ar retention in micas: inferences from interlayer partitioning and implications for geochronology. *Contrib. Mineral. Petrol.* **123**, 22–39.

Dahl P. S. (1996b) The effect of composition on retentivity of Ar and O in hornblende and related amphiboles: a field-tested empirical model. *Geochim. Cosmochim. Acta* **60**, 3687–3700.

Dahl P. S. (1997) A crystal-chemical basis for Pb retention and fission-track annealing systematics in U-bearing minerals, with implications for geochronology. *Earth Planet. Sci. Lett.* **150**(3–4), 277–290.

Dallmeyer R. D., Strachan R. A., Rogers G., Watt G. R., and Friend C. R. L. (2001) Dating deformation and cooling in the Caledonian thrust nappes of north Sutherland, Scotland: insights from $^{40}Ar/^{39}Ar$ and Rb–Sr chronology. *J.Geol. Soc. London.* **158**, 501–512.

Davis G. A. and Lister G. S. (1988) Detachment faulting in continental extension: perspectives from the southwestern US Cordillera. In *Processes in Continental Lithospheric Deformation* (eds. S. P. Clark, B. C. Burchfiel, and J. Suppe). Geological Society of America, vol. 218, pp. 133–159.

DeCelles P. G., Gehrels G. E., Quade J., Ojha T. P., Kapp P. A., and Upreti B. N. (1998) Neogene foreland deposits, erosional unroofing, and the kinematic history of the Himalayan fold-thrust belt, western Nepal. *Geol. Soc. Am. Bull.* **110**, 2–21.

DeWolf C. P., Belshaw N., and O'Nions R. K. (1993) A metamorphic history from micron-scale $^{207}Pb/^{206}Pb$ chronometry of Archean monazite. *Earth Planet. Sci. Lett.* **120**, 207–220.

DeWolf C. P., Zeissler C. J., Halliday A. N., Mezger K., and Essene E. J. (1996) The role of inclusions in U–Pb and Sm–Nd garnet geochronology: stepwise dissolution experiments and trace uranium mapping by fission track analysis. *Geochim. Cosmochim. Acta* **60**, 121–134.

DeYoreo J. J., Lux D. R., and Guidotti C. V. (1989) The role of crustal anatexis and magma migration in the in the thermal evolution of regions of thickened crust. In *Evolution of Metamorphic Belts*, Special Publications 43 (eds. J. S. Daly, R. A. Cliff, and B. W. D. Yardley). Geological Society of London, London, England, pp. 187–202.

Di Vincenzo G., Ghiribelli B., Giorgetti G., and Palmeri R. (2001) Evidence of a close link between petrology and isotope records: constraints from SEM, EMP, TEM and *in situ* Ar-40–Ar-39 laser analyses on multiple generations of white micas (Lanterman Range, Antarctica). *Earth Planet. Sci. Lett.* **192**(3), 389–405.

Dickin A. P. (1995) *Radiogenic Isotope Geology*. Cambridge University Press, Cambridge, England.

Dodson M. H. (1973) Closure temperature in cooling geochronological and petrological systems. *Contrib. Mineral. Petrol.* **40**, 259–274.

Dodson M. H. (1979) Theory of cooling ages. In *Lectures in Isotope Geology* (eds. E. Jäger and J. C. Hunziker). Springer, Berlin, pp. 194–202.

Dodson M. H. (1986) Closure profiles in cooling systems. *Mater. Sci. Forum* **7**, 145–154.

Dokka R. K., Mahaffie M. J., and Snoke A. W. (1986) Thermochronologic evidence of major tectonic denudation associated with detachment faulting, northern Ruby Mountains: East Humbolt range, Nevada. *Tectonics* **5**, 995–1006.

Dowty E. (1980) Crystal-chemical factors affecting the mobility of ions in minerals. *Am. Mineral.* **65**, 174–182.

Dunlap W. J. (1997) Neocrystallization or cooling? $^{40}Ar/^{39}Ar$ ages of white micas from low-grade mylonites. *Chem. Geol.* **143**, 181–203.

Dunlap W. J. and Kronenberg A. K. (2001) Argon loss during deformation of micas: constraints from laboratory deformation experiments. *Contrib. Mineral. Petrol.* **141**, 174–185.

Ehlers T. A., Armstrong P. A., and Chapman D. S. (2001) Normal fault thermal regimes and the interpretation of low-temperature thermochronometers. *Phys. Earth Planet. Inter.* **126**, 179–194.

Fairbairn H. W., Hurley P. M., and Pinson W. H. (1961) The relation of discordant Rb–Sr mineral and rock ages in an igneous rock to its time of subsequent Sr87/Sr86 metamorphism. *Geochim. Cosmochim. Acta* **23**, 135–144.

Farley K. A. (2000) Helium diffusion from apatite: general behavior as illustrated by Durango fluorapatite. *J. Geophys. Res.* **105**, 2903–2914.

Farley K. A., Wolf R. A., and Silver L. T. (1996) The effects of long alpha-stopping distances on (U–Th)/He ages. *Geochim. Cosmochim. Acta* **60**(21), 4223–4229.

Farley K. A., Rusmore M. E., and Bogue S. W. (2001) Post-10 Ma uplift and exhumation of the northern Coast Mountains, British Columbia. *Geology* **29**(2), 99–102.

Faure G. (1986) *Principles of Isotope Geology*. Wiley, New York.

Fechtig H. and Kalbitzer S. (1966) The diffusion of argon in potassium-bearing solids. In *Potassium–argon dating* (eds. O. Schaeffer and J. Zähringer). Springer, pp. 68–107.

Feng R., Machado N., and Ludden J. (1993) Lead geochronology of zircon by laserprobe-inductively coupled plasma mass spectrometry (LP-ICPMS). *Geochim. Cosmochim. Acta* **57**, 3479–3486.

Fitzgerald P. G. and Stump E. (1997) Cretaceous and Cenozoic episodic denudation of the Transantarctic Mountains, Antarctica: new constraints from apatite fission track thermochronology in the Scott Glacier region. *J. Geophys. Res. B: Solid Earth Planets* **102**(4), 7747–7765.

Fitzgerald P. G., Sorkhabi R. B., Redfield T. F., and Stump E. (1995) Uplift and denudation of the central Alaska range: a case study in the use of apatite fission track thermochronology to determine absolute uplift parameters. *J. Geophys. Res.* **100**, 20175–20191.

Foland K. A. (1994) Argon diffusion in feldspars. In *Feldspars and their Reactions* (ed. I. Parsons). Kluwer, Amsterdam, pp. 415–447.

Fortier S. M. and Giletti B. J. (1989) An empirical model for predicting diffusion coefficients in silicate minerals. *Science* **245**, 1481–1484.

Foster D. A. and John B. E. (1999) Quantifying tectonic exhumation in an extensional orogen with thermochronology: examples from the southern basin and range province. In *Exhumation Processes: Normal Faulting, Ductile Flow and Erosion*, Special Publications 154 (eds. U. Ring, M. T. Brandon, G. S. Lister, and S. D. Willett). Geological Society of London, London, pp. 343–364.

French J. E., Heaman L. M., and Chacko T. (2002) Feasibility of chemical U–Th–total Pb baddeleyite dating by electron microprobe. *Chem. Geol.* **188**, 85–104.

Frost B. R., Chamberlain K. R., and Schumacher J. C. (2000) Sphene (titanite): phase relations and role as a geochronometer. *Chem. Geol.* **172**, 131–148.

Fryer B. J., Jackson S. E., and Longerich H. (1993) The application of laser ablation microprobe-inductively coupled plasma-mass spectrometry (LAM–ICP–MS) *to in situ* (U)–Pb geochronology. *Chem. Geol.* **109**, 1–8.

Gallagher K. (1995) Evolving temperature histories from apatite fission-track data. *Earth Planet. Sci. Lett.* **136**, 421–435.

Gallagher K., Brown R. L., and Johnson C. (1998) Fission track analysis and its application to geological problems. *Ann. Rev. Earth Planet. Sci.* **26**, 519–572.

Ganguly J. and Tirone M. (1999) Diffusion closure temperature and age of a mineral with arbitrary extent of diffusion: theoretical formulation and applications. *Earth Planet. Sci. Lett.* **170**(1–2), 131–140.

Ganguly J., Tirone M., and Hervig R. L. (1998) Diffusional Kinetics of Samarium and neodymium in garnet, and a method for determining cooling rates of rocks. *Science* **281**, (5378), 805–807.

Gans P., Seedorff E., Fahey P. L., Hasler R. W., Maher D. J., Jeanne R. A., and Shaver S. A. (2001) Rapid Eocene extension in the Robinson district, White Pine County, Nevada: constraints from $^{40}Ar/^{39}Ar$ dating. *Geology* **29**, 475–478.

Garver J. I. and Brandon M. T. (1994) Erosional exhumation of the British Columbia coast ranges as determined from fission-track ages of detrital zircon from the Tofino basin, Olympic Peninsula, Washington. *Geol. Soc. Am. Bull.* **106**, 1398–1412.

Garver J. I., Brandon M. T., Roden-Tice M., and Kamp P. J. J. (1999) Erosional exhumation determined by fission-track ages of detrital apatite and zircon. In *Exhumation Processes: Normal Faulting, Ductile Flow, and Erosion*, Special Publication 154 (eds. U. Ring, M. T. Brandon, G. S. Lister, and S. D. Willett). Geological Society of London, London, pp. 283–304.

Geisler T. and Schleicher H. (2000) Improved U–Th–total Pb dating of zircons by electron microprobe using a simple new background modelling procedure and Ca as a chemical criterion of fluid-induced U–Th–Pb discordance in zircon. *Chem. Geol.* **163**, 269–285.

Giletti B. J. (1974) Studies in diffusion: I. Argon in phlogopite mica. In *Geochemical Transport and Kinetics*, Carnegie Institute of Washington Publication (eds. A. W. Hofmann, B. J. Giletti, H. S. Yoder, and R. A. Yund), vol. 634, pp. 107–115.

Giletti B. J. (1991) Rb and Sr diffusion in alkali feldspars, with implications for cooling histories of rocks. *Geochim. Cosmochim. Acta* **55**, 1331–1343.

Gillespie A. R., Huencke J. C., and Wasserburg G. J. (1982) An assessment of $^{40}Ar/^{39}Ar$ dating of incompletely degassed xenoliths. *J. Geophys. Res.* **87**, 9247–9257.

Gleadow A. J. W. and Duddy I. R. (1981) A natural long-term track annealing experiment for apatite. *Nuclear Tracks* **5**, 169–174.

Gleadow A. J. W., Duddy I. R., Green P. F., and Lovering J. F. (1986) Confined fission track lengths in apatite: a diagnostic tool for thermal history analysis. *Contrib. Mineral. Petrol.* **94**, 405–415.

Goodwin L. B. and Renne P. R. (1991) Effects of progressive mylonitization on Ar retention in biotites from the Santa Rosa mylonite zone, California, and thermochronologic implications. *Contrib. Mineral. Petrol.* **108**, 283–297.

Gray M. B. and Zeitler P. K. (1997) Comparison of clastic wedge provenance in the Appalachian foreland using U/Pb ages of detrital zircons. *Tectonics* **16**, 151–160.

Gromet L. P. (1991) Direct dating of deformational fabrics. In *Applications of Radiogenic Isotope Systems to Problems in Geology* (eds. L. Heaman and J. N. Ludden). Mineralogical Association of Canada, Ottawa, Canada, pp. 167–189.

Grove M. and Harrison T. M. (1996) $^{40}Ar^*$ diffusion in Fe-rich biotite. *Am. Mineral.* **81**, 940–951.

Haack U. (1977) The closing temperature for fission track retention in minerals. *Am. J. Sci.* **277**, 459–464.

Halliday A. N., Christensen J. N., Lee D.-C., Hall C. M., Ballentine C. J., Rehkämper M., Yi W., Luo X., and Barford D. (1998) ICP multiple-collector mass spectrometry and *in situ* high-precision isotopic analysis. In *Applications of Microanalytical Techniques to Understanding Mineralizing Processes*, Rev. Econ. Geol. 7 (eds. M. A. McKibben and W. C. Shanks). Society of Economic Geologists, pp. 37–51.

Hames W. E. and Bowring S. A. (1994) An empirical evaluation of the argon diffusion geometry in muscovite. *Earth Planet. Sci. Lett.* **124**, 161–167.

Hames W. E. and Cheney J. T. (1997) On the loss of ^{40}Ar from muscovite during polymetamorphism. *Geochim. Cosmochim. Acta* **61**(18), 3863–3872.

Hames W. E. and Hodges K. V. (1993) Laser (40)Ar/(39)Ar evaluation of slow cooling and episodic loss of (40)Ar from a sample of polymetamorphic muscovite. *Science* **261**(5129), 1721–1723.

Harland C. E. (1994) *Ion Exchange: Theory and Practice.* Royal Society of Chemistry, London.

Harrison T. M. (1981) Diffusion of ^{40}Ar in hornblende. *Contrib. Mineral. Petrol.* **78**, 324–331.

Harrison T. M., Duncan I., and McDougall I. (1985) Diffusion of ^{40}Ar in biotite: temperature, pressure, and compositional effects. *Geochim. Cosmochim. Acta* **49**, 2461–2468.

Harrison T. M., Heizler M. T., Lovera O. M., Chen W., and Grove M. (1994) A chlorine disinfectant for excess argon released from K-feldspar during step heating. *Earth Planet. Sci. Lett.* **123**, 95–104.

Harrison T. M., Copeland P., Kidd W. S. F., and Lovera O. (1995a) Activation of the Nyainquentanghla Shear Zone: implications for uplift of the southern Tibetan Plateau. *Tectonics* **14**, 658–676.

Harrison T. M., McKeegan K. D., and LeFort P. (1995b) Detection of inherited monazite in the Manaslu leucogranite by $^{208}Pb/^{232}Th$ ion microprobe dating: crystallization age and tectonic implications. *Earth Planet. Sci. Lett.* **133**, 271–282.

Harrison T. M., Ryerson F. J., McKeegan K. D., Le Fort P., and Yin A. (1996) Th–Pb monazite ages of Himalayan metamorphic and leucogranitic rocks: constraints on the timing of inverted metamorphism and slip on the MCT and STD. In *11th Himalaya–Karakoram–Tibet Workshop Abstracts* (eds. A. M. Macfarlane, R. B. Sorkhabi, and J. Quade), pp. 58–59.

Harrison T. M., Ryerson F. J., Le Fort P., Yin A., Lovera O., and Catlos E. J. (1997) A Late Miocene–Pliocene origin for the central Himalayan inverted metamorphism. *Earth Planet. Sci. Lett.* **146**, E1–E7.

Hart S. R. (1964) The petrology and isotopic-mineral age relations of a contract zone in the front range, Colorado. *J. Geol.* **72**, 493–525.

Hartz E. H., Andresen A., Martin M. W., and Hodges K. V. (2000) U–Pb and $^{40}Ar/^{39}Ar$ constraints on the Fjord region detachment zone: a long-lived extensional fault in the East Greenland Caledonides. *J. Geol. Soc. London* **157**, 795–809.

Hawkins D. P. and Bowring S. A. (1997) U–Pb systematics of monazite and xenotime: case studies from the Paleoproterozoic of the Grand Canyon, Arizona. *Contrib. Mineral. Petrol.* **127**, 87–103.

Hawkins D. P. and Bowring S. A. (1999) U–Pb monazite, xenotime, and titanite geochronological constraints on the prograde to post-peak metamorphic thermal history of Paleoproterozoic migmatites from the Grand Canyon, Arizona. *Contrib. Mineral. Petrol.* **134**, 150–169.

Hess J. C., Lippolt H. J., Gurbanov A. G., and Michalski I. (1993) The cooling history of the Late Pliocene Eldzhurtinskiy granite (Caucasus, Russia) and the thermochronological potential of grain-size/age relationships. *Earth Planet. Sci. Lett.* **117**, 393–406.

Hodges K., Bowring S., Davidek K., Hawkins D., and Krol M. (1998) Evidence for rapid displacement on Himalayan normal faults and the importance of tectonic denudation in the evolution of mountain ranges. *Geology* **26**, 483–486.

Hodges K. V. (1991) Pressure–Temperature–time paths. *Ann. Rev. Earth Planet. Sci.* **19**, 207–236.

Hodges K. V. (1998a) $^{40}Ar/^{39}Ar$ geochronology using the laser microprobe. In *Applications of Microanalytical Techniques to Understanding Mineralizing Processes*, Rev. Econ. Geol. (eds. M. A. McKibben and W. C. Shanks). Society of Economic Geologists, Tuscaloosa, AL, pp. 53–72.

Hodges K. V. (1998b) The thermodynamics of Himalayan orogenesis. In *What Drives Metamorphism and Metamorphic Reactions?*. Geological Society of London, Special Publication 138 (eds. P. J. Treloar and P. O'Brien). Geological Society of London, London, pp. 7–22.

Hodges K. V. and Applegate J. D. R. (1993) Age of Tertiary extension, Bitterroot metamorphic core complex, Montana-Idaho. *Geology* **21**, 161–164.

Hodges K. V. and Bowring S. A. (1995) $^{40}Ar/^{39}Ar$ thermochronology of isotopically zoned micas: insights from the southwestern USA Proterozoic orogen. *Geochim. Cosmochim. Acta* **59**(15), 3205–3220.

Hodges K. V., Hames W. E., and Bowring S. A. (1994) $^{40}Ar/^{39}Ar$ age gradients in micas from a high-temperature–low-pressure metamorphic terrain: evidence for very slow cooling and implications for the interpretation of age spectra. *Geology* **22**(1), 55–58.

Hodges K. V., Parrish R. R., and Searle M. P. (1996) Tectonic evolution of the central Annapurna range, Nepalese Himalayas. *Tectonics* **15**, 1264–1291.

Hodges K. V., Hurtado J. M., and Whipple K. X. (2001) Southward extrusion of Tibetan crust and its effect on Himalayan tectonics. *Tectonics* **20**, 799–809.

Holland T. and Powell R. (2001) Calculation of phase relations involving haplogranitic melts using an internally consistent thermodynamic dataset. *J. Petrol.* **42**(4), 673–683.

Hollister L. S. (1966) Garnet zoning: an interpretation based on the Rayleigh fractionation model. *Science* **154**, 1647–1651.

Horn I., Rudnick R. L., and McDonough W. F. (2000) Precise elemental and isotope ratio determination by simultaneous solution nebulization and laser ablation-ICP-MS: application to U–Pb geochronology. *Chem. Geol.* **164**(3–4), 281–301.

Hoskin P. W. O. and Black L. P. (2000) Metamorphic zircon formation by solid-state recrystallization of protolith igneous zircon. *J. Metamorph. Geol.* **18**(4), 423–439.

House M. A. and Hodges K. V. (1994) Limits on the tectonic significance of rapid cooling events in extensional settings: insights from the Bitterroot metamorphic core complex, Idaho-Montana. *Geology* **22**, 1007–1010.

House M. A., Wernicke B. P., and Farley K. A. (1998) Dating topography of the Sierra Nevada, California, using apatite (U–Th)/He ages. *Nature* **396**, 66–69.

Housh T. B. and Bowring S. A. (1991) Lead isotopic heterogeneities within alkali feldspars: implications for the determination of lead isotopic compositions. *Geochim. Cosmochim. Acta* **55**, 2309–2316.

Huerta A. D., Royden L. H., and Hodges K. V. (1996) The interdependence of deformational and thermal processes in mountain belts. *Science* **273**, 637–639.

Hurford A. J. and Carter A. (1991) The role of fission track dating in discrimination of provenance. In *Developments in Sedimentary Provenance Studies*, Geological Society of London Special Publication 57 (eds. A. C. Morton, S. P. Todd, and P. D. W. Haughton). Geological Society of London, London, pp. 67–78.

Hurford A. J., Fitch F. J., and Clarke A. (1984) Resolution of the age structure of detrital zircon populations of two lower Cretaceous sandstones from the Weald of England by fission track dating. *Geol. Mag.* **121**, 285–317.

Hurtado J. M., Chatterjee N., Ramezani J., Hodges K. V., and Bowring S. A. (xxxx) Electron microprobe chemical dating of uraninite as a reconaissance tool for leucogranite geochronology. *Earth Planet. Sci. Lett.* (in preparation).

Jamieson R. A., Beaumont C., Nguyen M. H., and Lee B. (2002) Interaction of metamorphism, deformation, and exhumation in large convergent orogens. *J. Metamorph. Geol.* **20**, 9–24.

Jenkin G. R. T., Rogers G., Fallick A. E., and Farrow C. M. (1995) Rb–Sr closure temperatures in bi-mineralic rocks: a mode effect and test for different diffusion models. *Chem. Geol.* **122**, 227–240.

Kalsbeek F., Thrane K., Nutman A. P., and Jepsen H. F. (2000) Late Mesoproterozoic to early Neoproterozic history of the East Greenland Caledonides: evidence for Grenvillian orogenesis? *J. Geol. Soc.* **157**, 1215–1225.

Kelley S. (2002) Excess argon in K–Ar and Ar–Ar geochronology. *Chem. Geol.* **188**, 1–22.

Kelley S. P. and Bluck B. J. (1992) Laser $^{40}Ar–^{39}Ar$ ages for individual detrital muscovites in the southern Uplands of Scotland UK. *Chem. Geol.* **101**(1–2), 143–156.

Kelley S. P., Reddy S. M., and Maddock R. (1994) Laser-probe $^{40}Ar/^{39}Ar$ investigation of a pseudotachylyte and its host rock from the outer Isles thrust, Scotland. *Geology* **22**(5), 443–446.

Kempe U. (2003) Precise electron microprobe age determination in altered uraninite: consequences on the instrusion age and the metallogenic significance of the Kirchberg granite (Erzgebirge, Germany). Contributions to Mineralogy and Petrology **145**, 107–118.

Ketcham R. A., Donelick R. A., and Donelick M. B. (2000) AFTSolve: a program for multikinetic modeling of apatite fission-track data. *Geol. Mater. Res.* **2**, 1–32.

Kirby E., Reiners P. W., Krol M. A., Whipple K. X., Hodges K. V., Farley K. A., Tang W. Q., and Chen Z. L. (2002) Late Cenozoic evolution of the eastern margin of the Tibetan Plateau: inferences from $^{40}Ar/^{39}Ar$ and (U–Th)/He thermochronology. *Tectonics* **21**, 3–22.

Kooi H. and Beaumont C. (1996) Large-scale geomorphology: classical concepts reconciled and integrated with contemporary ideas via a surface processes model. *J. Geophys. Res.* **101**, 3361–3386.

Kosler J., Tubrett M. N., and Sylvester P. J. (2001) Application of laser ablation ICP–MS to U–Th–Pb dating of monazite. *Geostand. Newslett.: J. Geostand. Geoanal.* **25**(2–3), 375–386.

Kosler J., Fonneland H., Sylvester P., Tubrett M., and Pedersen R. B. (2002) U–Pb dating of detrital zircons for sediment provenance studies: a comparison of laser ablation ICPMS and SIMS techniques. *Chem. Geol.* **182**(2–4), 605–618.

Kramar N., Cosca M. A., and Hunziker J. C. (2001) Heterogeneous $^{40}Ar^*$ distributions in naturally deformed muscovite: *in situ* UV-laser ablation evidence for micro structurally controlled intragrain diffusion. *Earth Planet. Sci. Lett.* **192**(3), 377–388.

Krol M. A., Zeitler P. K., Poupeau G., and Pêcher A. (1996) Temporal variations in the cooling and denudation history of the Hunza plutonic complex, Karakoram Batholith, revealed by $^{40}Ar/^{39}Ar$ thermochronology. *Tectonics* **15**, 403–415.

Lanphere M. A. and Baadsgaard H. (2001) Precise K–Ar, Ar-40/Ar-39, Rb–Sr and U/Pb mineral ages from the 27.5 Ma Fish Canyon Tuff reference standard. *Chem. Geol.* **175**(3–4), 653–671.

Lanphere M. A., Wasserburg G. J., Albee A. L., and Tilton G. R. (1964) Redistribution of strontium and rubidium isotopes during metamorphism, World Beater Complex, Panamint range, California. In *Isotopic and Cosmic Chemistry* (eds. H. Craig, S. L. Miller, and G. J. Wasserburg). North Holland, Amsterdam, pp. 269–320.

Laslett G. M., Green P. F., Duddy I. R., and Gleadow A. J. W. (1987) Thermal annealing of fission tracks in apatite. *Chemical Geology; Isotope Geoscience Section* **65**(1), 1–13.

Lee J. K. W. (1995) Multipath diffusion in geochronology. *Contrib. Mineral. Petrol.* **120**, 60–82.

Lee J. K. W. and Aldama A. A. (1992) Multipath diffusion—a general numerical-model. *Comput. Geosci.* **18**(5), 531–555.

Li X. H., Liang X. R., Sun M., Guan H., and Malpas J. G. (2001) Precise Pb-206/U-238 age determination on zircons by laser ablation microprobe-inductively coupled plasma-mass spectrometry using continuous linear albation. *Chem. Geol.* **175**(3–4), 209–219.

Lonergan L. and Johnson C. (1998) A novel approach for reconstructing the denudation histories of mountain belts: with an example from the Betic Cordillera (S. Spain). *Basin Res.* **10**, 353–364.

Lovera O. M., Richter F. M., and Harrison T. M. (1989) $^{40}Ar/^{39}Ar$ geochronometry for slowly cooled samples having a distribution of diffusion domain size. *J. Geophys. Res.* **94**, 17917–17936.

Lovera O. M., Grove M., Harrison T. M., and Mahon K. I. (1997) Systematic analysis of K-feldspar $^{40}Ar/^{39}Ar$ step heating results: I. Significance of activation energy determinations. *Geochim. Cosmochim. Acta* **61**, 3171–3192.

Lovera O. M., Grove M., and Harrison T. M. (2002) Systematic analysis of K-feldspar Ar-40/Ar-39 step heating results: II. Relevance of laboratory argon diffusion properties to nature. *Geochim. Cosmochim. Acta* **66**(7), 1237–1255.

Lugmair G. W. and Marti K. (1978) Lunar initial $^{143}Nd/^{144}Nd$: differential evolution of the lunar crust and mantle. *Earth Planet. Sci. Lett.* **39**, 349–357.

Machado N. and Simonetti A. (2001) U–Pb dating and Hf isotopic composition of zircon by laser ablation-MC-ICP-MS. In *Laser-Ablation-ICPMS in the Earth Sciences: Principles and Applications* (ed. P. Sylvester). Mineralogical Association of Canada, Ottawa, pp. 121–146.

Magloughlin J. F., Hall C. M., and van der Pluijm B. A. (2001) $^{40}Ar/^{39}Ar$ geochronometry of pseudotachylytes by vacuum encapsulation: North Cascade mountains, Washington, USA. *Geology* **29**, 51–54.

Maluski H. and Schaeffer O. A. (1982) ^{39}Ar–^{40}Ar laser probe dating of terrestrial rocks. *Earth Planet. Sci. Lett.* **59**, 21–27.

Mancktelow N. S. and Grasemann B. (1997) Time-dependent effects of heat advection and topography on cooling histories during erosion. *Tectonophysics* **270**, 167–195.

McDougall I. and Harrison T. M. (1988) *Geochronology and Thermochronology by the $^{40}Ar/^{39}Ar$ Method.* Oxford University Press, New York.

Merrihue C. M. and Turner G. (1966) Potassium–argon dating by activation with fast neutrons. *J. Geophys. Res.* **71**, 2852–2857.

Mezger K., Essene E. J., and Halliday A. N. (1992) Closure temperature of the Sm–Nd system in metamorphic garnets. *Earth Planet. Sci. Lett.* **113**, 397–409.

Miller E. L., Dumitru T. A., Brown R. W., and Gans P. B. (1999) Rapid Miocene slip on the Snake range–Deep Creek range fault system, east-central Nevada. *Geol. Soc. Am. Bull.* **111**, 886–905.

Min K. W., Mundil R., Renne P. R., and Ludwig K. R. (2000) A test for systematic errors in Ar-40/Ar-39 geochronology through comparison with U/Pb analysis of a 1.1-Ga rhyolite. *Geochim. Cosmochim. Acta* **64**(1), 73–98.

Montel J., Foret S., Veschambre M., Nicollet C., and Provost A. (1996) Electron microprobe dating of monazite. *Chem. Geol.* **131**, 37–53.

Moore M. A. and England P. C. (2001) On the inference of denudation rates from cooling ages of minerals. *Earth Planet. Sci. Lett.* **185**, 265–284.

Mulch A., Cosca M. A., and Handy M. R. (2002) *In-situ* UV-laser $^{40}Ar/^{39}Ar$ geochronology of a micaceous mylonite: an example of defect-enhanced argon loss. *Contrib. Mineral. Petrol.* **142**(6), 738–752.

Müller W., Prosser G., Mancktelow N. S., Villa I. M., Kelley S. P., Viola G., and Oberli F. (2001) Geochronological constraints on the evolution of the Peradriatic fault system (Alps). *Int. J. Earth Sci.: Geol. Rundsch.* **90**, 623–653.

Murphy M. A., Yin A., Kapp P., Harrison T. M., Manning C. E., Ryerson F. J., Ding L., and Guo J. H. (2002) structural evolution of the Gurla Mandhata detachment system, southwest Tibet: implications for the eastward extent of the Karakoram fault system. *Geol. Soc. Am. Bull.* **114**(4), 428–447.

Naeser N. D., Naeser C. W., and McCulloh T. H. (1989) The application of fission-track dating to the depositional and thermal history of rocks in the sedimentary basins. In *Thermal history of sedimentary basins: methods and case studies* (ed N. D. Naeser and T. H. McCulloh). Springer-Verlag, pp. 157–180.

Nicolaysen L. O. (1961) Graphic interpretation of discordant age measurements on metamorphic rocks. *Ann. NY Acad. Sci.* **91**, 198–206.

Onstott T. C., Phillips D., and Pringle-Goodell L. (1991) Laser microprobe measurement of chlorine and argon zonation in biotite. *Chem. Geol.* **90**, 145–168.

Pan Y. and Kidd W. S. F. (1992) Nyainqentanglha shear zone: a late Miocene extensional detachment in the southern Tibetan plateau. *Geology* **20**, 775–778.

Parrish R. (1990) U–Pb dating of monazite and its application to geological problems. *Can. J. Earth Sci.* **27**, 1431–1450.

Parrish R. and Tirrul R. (1989) U–Pb age of the Baltoro granite and implications for zircon inheretance. *Geology* **17**, 1076–1079.

Parry W. T., Bunds M. P., Bruhn R. L., Hall C. M., and Murphy J. M. (2001) Mineralogy, $^{40}Ar/^{39}Ar$ dating and apatite fission track dating along the Castle Mountain fault, Alaska. *Tectonophysics* **337**, 149–172.

Parsons I., Brown W. L., and Smith J. V. (1999) $^{40}Ar/^{39}Ar$ thermochronology using alkali feldspars: real thermal history or mathematical mirage of microtexture? *Contrib. Mineral. Petrol.* **136**, 92–110.

Patiño Douce A. E. and Johnston A. D. (1991) Phase equilibria and melt productivity in the pelitic system: implications for the origin of peraluminous granitoids and aluminous granites. *Contrib. Mineral. Petrol.* **107**, 202–218.

Pickles C. S., Kelley S. P., Reddy S. M., and Wheeler J. (1997) Determination of high spatial resolution argon isotope variations in metamorphic biotites. *Geochim. Cosmochim. Acta* **61**(18), 3809–3833.

Poitrasson F., Chenery S., and Shepherd T. J. (2000) Electron microprobe and LA-ICP-MS study of monazite hydrothermal alteration: implications for U–Th–Pb geochronology and nuclear ceramics. *Geochim. Cosmochim. Acta* **64**, 3283–3297.

Prince C. I., Kosler J., Vance D., and Gunther D. (2000) Comparison of laser ablation ICP-MS and isotope dilution REE analyses: implications for Sm–Nd garnet geochronology. *Chem. Geol.* **168**(3–4), 255–274.

Quidelleur X., Grove M., Lovera O. M., Harrison T. M., Yin A., and Ryerson F. J. (1997) Thermal evolution and slip history of the Renbu-Zedong thrust, southeastern Tibet. *J. Geophys. Res.* **102**, 2659–2679.

Reddy S. M. and Potts G. J. (1999) Constraining absolute deformation ages: the relationship between deformation mechanisms and isotope systematics. *J. Struct. Geol.* **21**, 1255–1265.

Reddy S. M., Kelley S. P., and Wheeler J. (1996) A $^{40}Ar/^{39}Ar$ laser probe study of micas from the Sesia zone, Italian Alps: implications for metamorphic and deformational histories. *J. Metamorph. Geol.* **14**, 493–508.

Reddy S. M., PottsG J., and Kelley S. P. (2001) $^{40}Ar/^{39}Ar$ ages in deformed potassioum feldspar: evidence of microstructural control on Ar isotope systematics. *Contrib. Mineral. Petrol.* **141**, 186–200.

Reiners P. W. and Farley K. A. (1999) Helium diffusion and (U–Th)/He thermochronometry of titanite. *Geochim. Cosmochim. Acta* **63**, 3845–3859.

Reiners P. W. and Farley K. A. (2001) Influence of crystal size on apatite (U–Th)/He thermochronology: an example from the Bighorn Mountains, Wyoming. *Earth Planet. Sci. Lett.* **188**(3–4), 413–420.

Reiners P. W., Brady R., Farley K. A., Fryxell J. E., Wernicke B., and Lux D. (2000) Helium and argon thermochronometry of the Gold Butte block, south Virgin Mountains, Nevada. *Earth Planet. Sci. Lett.* **178**(3–4), 315–326.

Renne P. R. (2000) Ar-40/Ar-39 age of plagioclase from Acapulco meteorite and the problem of systematic errors in cosmochronology. *Earth Planet. Sci. Lett.* **175**(1–2), 13–26.

Renne P. R., Becker T. A., and Swapp S. M. (1990) $^{40}Ar/^{39}Ar$ laser-probe dating of detrital micas from the Montgomery creek formation, northern California: clues to provenance, tectonics, and weathering processes. *Geology* **18**(6), 563–566.

Renne P. R., Swisher C. C., Deino A. L., Karner D. B., Owens T., and DePaolo D. J. (1998) Intercalibration of standards, absolute ages and uncertainties in $^{40}Ar/^{39}Ar$ dating. *Chem. Geol.: Isotope Geosci.* **145**, 117–152.

Resor R. G., Chamberlain K. R., Frost C. D., Snoke A. W., and Frost B. R. (1996) Direct dating of deformation: U–Pb age of syndeformational sphene growth in the Proterozoic Laramie peak shear zone. *Geology*, 623–626.

Reuter A. and Dallmeyer R. D. (1989) K–Ar and $^{40}Ar/^{39}Ar$ dating of cleavage formed during very low-grade metamorphism: a review. In *Evolution of Metamorphic Belts*, Special Publication 43 (eds. J. S. Daly, R. A. Cliff, and B. W. D. Yardley). Geological Society of London, London, pp. 161–171.

Richter F. M., Lovera O. M., Harrison T. M., and Copeland P. C. (1991) Tibetan tectonics from $^{40}Ar/^{39}Ar$ analysis of a single K-feldspar sample. *Earth Planet. Sci. Lett.* **105**, 266–278.

Robbins G. A. (1972) Radiogenic argon diffusion in muscorite under hydrothermal conditions. M. S., Brown University.

Roddick J. C., Cliff R. A., and Rex D. C. (1980) The evolution of excess argon in Alpine biotites-A $^{40}Ar–^{39}Ar$ analysis. *Earth Planet. Sci. Lett.* **48**, 185–208.

Rubatto D. (2002) Zircon trace element geochemistry: partitioning with garnet and the link between U–Pb ages and metamorphism. *Chem. Geol.* **184**, 123–138.

Rubatto D., Williams I. S., and Buick I. S. (2001) Zircon and monazite response to prograde metamorphism in the Reynolds Range, central Australia. *Contrib. Mineral. Petrol.* **140**(4), 458–468.

Ruppel C. and Hodges K. V. (1994) Pressure–Temperature–Time paths from two-dimensional thermal models: prograde, retrograde, and inverted metamorphism. *Tectonics* **13**, 17–44.

Ruppel C., Royden L., and Hodges K. V. (1988) Thermal modeling of extensional tectonics: application to pressure–temperature-time histories of metamorphic rocks. *Tectonics* **7**, 947–957.

Sandiford M., McLaren S., and Neumann N. (2002) Long-term thermal consequences of the redistribution of heat-producing elements associated with large-scale granitic complexes. *J. Metamorph. Geol.* **20**, 87–98.

Schaltegger U., Fanning C. M., Günther D., Maurin J. C., Schulmann K., and Gebauer D. (1999) Growth, annealing and recrystallization of zircon and preservation of monazite in high-grade metamorphism: conventional and *in-situ* U–Pb isotope, cathodoluminesence and microchemical evidence. *Contrib. Mineral. Petrol.* **134**, 186–201.

Scherer E., Munker C., and Mezger K. (2001) Calibration of the lutetium–hafnium clock. *Science* **293**(5530), 683–687.

Scherer E. E., Cameron K. L., and Blichert-Toft J. (2000) Lu–Hf garnet geochronology: closure temperature relative to the Sm–Nd system and the effects of trace mineral inclusions. *Geochim. Cosmochim. Acta* **64**(19), 3413–3432.

Schmitz M. D. and Bowring S. A. (2001) U–Pb zircon and titanite systematics of the Fish Canyon Tuff: an assessment of high-precision U–Pb geochronology and its application to young volcanic rocks. *Geochim. Cosmochim. Acta* **65**(15), 2571–2587.

Schreiner G. D. L. (1958) Comparison of the Rb-87/Sr-87 age of the red granite of the Bushveld complex from measurements on the total rock and separated mineral fractions. *Proc. Roy. Soc. London Ser. A* **245**, 112–117.

Sherlock S. C. and Hetzel R. (2001) A laser-probe $^{40}Ar/^{39}Ar$ study of pseudotachylite from the Tambach fault zone, Kenya: direct isotopic dating of brittle faults. *J. Struct. Geol.* **23**, 33–44.

Simpson R. L., Parrish R. R., Searle M. P., and Waters D. J. (2000) Two episodes of monazite crystallization during metamorphism and crustal melting in the Everest region of the Nepalese Himalaya. *Geology* **28**, 403–406.

Spear F. S. (1993) *Metamorphic Phase Equilibria and Pressure–Temperature–Time Paths*. Mineralogical Society of America, Washington, DC.

Spotila J. A., Farley K. A., Yule J. D., and Reiners P. W. (2001) Near-field transpressive deformation along the San Andreas fault zone in southern California, based on exhumation constrained by (U–Th)/He dating. *J. Geophys. Res.-Solid Earth* **106**(B12), 30909–30922.

Steiger R. H. and Jäger E. (1977) Subcommission on geochronology: convention on the use of decay constants in geo- and cosmochronology. *Earth Planet. Sci. Lett.* **36**, 359–362.

Stern R. A. and Berman R. G. (2001) Monazite U–Pb and Th–Pb geochronology by ion microprobe, with an application to *in situ* dating of an Archean metasedimentary rock. *Chem. Geol.* **172**(1–2), 113–130.

Stock J. D. and Montgomery D. R. (1996) Estimating paleorelief from detrital mineral age ranges. *Basin Res.* **8**, 317–327.

Stockli D. F., Farley K. A., and Dumitru T. A. (2000) Calibration of the apatite (U–Th)/He thermochronometer on an exhumed fault block, White Mountains, California. *Geology* **28**(11), 983–986.

Stockli D. F., Linn J. K., Walker J. D., and Dumitru T. (2001) Miocene unroofing of the Canyon range during extension along the Sevier desert detachment, west central Utah. *Tectonics* **20**, 289–307.

Stuart F. M. (2002) The exhumation history of orogenic belts from $^{40}Ar/^{39}Ar$ ages of detrital micas. *Min. Mag.* **66**, 121–135.

Stüwe K. and Hintermüller M. (2000) Topography and isotherms revisited: the influence of laterally migrating drainage divides. *Earth Planet. Sci. Lett.* **184**, 287–303.

Stüwe K., White L., and Brown R. (1994) The influence of eroding topography on steady-state isotherms: application to fission track analysis. *Earth Planet. Sci Lett.* **124**, 63–74.

Sutter J. F. and Hartung J. B. (1984) Laser microprobe $^{40}Ar/^{39}Ar$ dating of mineral grains *in situ*. *Scan. Electr. Micros.* **4**, 1525–1529.

Suzuki K. and Adachi M. (1991) Precambrian provenance and Silurian metamorphism of the Tsubonosawa paragneiss in the South Kitakami terrane, northeast Japan, revealed by the chemical Th–U–total Pb isochron ages of monazite, zircon and xenotime. *Geochem. J.* **25**, 357–376.

Terry M. P., Robinson P., Hamilton M. A., and Jercinovic M. J. (2000) Monazite geochronology of UHP and HP metamorphism, deformation, and exhumation, Nordoyane, western Gneiss region, Norway. *Am. Mineral.* **85**(11–12), 1651–1664.

Thompson A. B. (1982) Dehydration melting of pelitic rocks and the generation of H_2O-undersaturated granitic liquids. *Am. J. Sci.* **282**, 1567–1595.

Thoni M. (2002) Sm–Nd isotope systematics in garnet from different lithologies (eastern Alps): age results, and an evaluation of potential problems for garnet Sm–Nd chronometry. *Chem. Geol.* **185**(3–4), 255–281.

Tippett J. M. and Kamp P. J. J. (1993) Fission track analysis of late Cenozoic vertical kinematics of continental Pacific crust, South Island, New Zealand. *J. Geophys. Res.* **98**, 16119–16148.

Tippett J. M. and Kamp P. J. J. (1995) Geomorphic evolution of the Southern Alps, New Zealand. *Earth Surf. Process. Landforms* **20**, 177–192.

Townsend K. J., Miller C. F., D'Andrea J. L., Ayers J. C., Harrison T. M., and Coath C. D. (2001) Low temperature replacement of monazite in the Ireteba granite, southern Nevada: geochronological implications. *Chem. Geol.* **172**(1–2), 95–112.

Tracy R. J., Robinson P., and Thompson A. B. (1976) Garnet composition and zoning in the determination of temperature and pressure of metamorphism, central Massachusetts. *Am. Mineral.* **61**, 762–775.

Turcotte D. L. and Schubert G. (1982) *Geodynamics: Applications of Continuum Physics to Geological Problems*. Wiley, New York.

van der Pluijm B. A., Hall C. M., Vrolijk P. J., Pevear D. R., and Covey M. C. (2001) The dating of shallow faults in the Earth's crust. *Nature* **412**, 172–175.

Vance D., Meier M., and Oberli F. (1998) The influence of high U–Th inclusions on the U–Th–Pb systematics of almandine-pyrope garnet: results of a combined bulk dissolution, stepwise-leaching, and SEM study. *Geochim. Cosmochim. Acta* **62**(21–22), 3527–3540.

Vavra G. and Schaltegger U. (1999) Post-granulite facies monazite growth and rejuvenation during Permian to Lower Jurassic thermal and fluid events in the Ivrea zone (southern Alps). *Contrib. Mineral. Petrol.* **134**, 405–414.

Vielzeuf D. and Holloway J. R. (1988) Experimental determination of the fluid-absent melting relations in the pelitic system. Consequences for crustal differentiation. *Contrib. Mineral. Petrol.* **98**, 257–276.

Villa I. M. (1994) Multipath Ar transport in K-feldspar deduced from isothermal heating experiments. *Earth Planet. Sci. Lett.* **122**, 393–401.

Viskupic K. and Hodges K. V. (2001) Monazite–xenotime thermochronometry: methodology and an example from the Nepalese Himalaya. *Contrib. Mineral. Petrol.* **141**, 233–247.

von Eynatten H., Gaup R., and Wijbrans J. R. (1996) ^{40}Ar/^{39}Ar laser-probe dating of detrital white micas from Cretaceous sedimentary rocks of eastern Alps: evidence for Variscan high-pressure metamorphism and implications for Alpine Orogeny. *Geology* **24**, 691–694.

Wagner G. A., Reimer G. M., and Jäger E. (1977) Cooling ages derived by apatite fission track, mica Rb–Sr, and K–Ar dating: the uplift and cooling history of the Central Alps. *Mem. Univ. Padova* **30**, 1–27.

Wartho J.-A., Kelley S. P., Brooker R. A., Carroll M. R., Villa I. M., and Lee M. R. (1999) Direct measurement of Ar diffusion profiles in a gem-quality Madagascar K-feldspar using the ultra-violet laser ablation microprobe (UVLAMP). *Earth Planet. Sci. Lett.* **170**, 141–153.

Watt G. R. and Thrane K. (2001) Early Neoproterozoic events in East Greenland. *Precamb. Res.* **110**(1–4), 165–184.

Weatherill G. W., Davis G. L., and Lee-Hu C. (1968) Rb–Sr measurements on whiole-rocks and separated minerals from the Baltimore Gneiss, Maryland. *Geol. Soc. Am. Bull.* **79**, 757–762.

Wells M. L., Snee L. W., and Blythe A. E. (2000) Dating of major normal fault systems using thermochronology: an example from the Raft river detachment, basin and range, western United States. *J. Geophys. Res.* **105**, 16303–16327.

Wernicke B. (1992) Cenozoic extensional tectonics of the US Cordillera. In *The Cordilleran Orogen: Conterminous United States* (eds. B. C. Burchfiel, P. W. Lipman, and M. L. Zoback). Geological Society of America, vol. G-3, Boulder, CO, pp. 553–581.

West D. P. and Lux D. R. (1993) Dating mylonitic deformation by the ^{40}Ar–^{39}Ar method: an example from the Norumbega fault zone, Maine. *Earth Planet. Sci. Lett.* **120**, 221–237.

White A. P. and Hodges K. V. (2002) Multistage extensional evolution of the central East Greenland Caledonides. *Tectonics* **21**, 101029/2001TC001308.

Willett S., Beaumont C., and Fullsack P. (1993) Mechanical model for the tectonics of doubly vergent compressional orogens. *Geology* **21**, 371–374.

Willett S. D. (1997) Inverse modeling of annealing of fission tracks in apatite 1: a controlled random search method. *Am. J. Sci.* **297**, 939–969.

Willett S. D. (1999) Orogeny and orography: the effects of erosion on the structure of mountain belts. *J. Geophys. Res.* **104**, 28957–28981.

Willett S. D. and Brandon M. T. (2002) On steady states in mountain belts. *Geology* **30**, 175–178.

Willett S. D., Slingerland R., and Hovius N. (2001) Uplift, shortening and steady state topography in active mountain belts. *Am. J. Sci.* **301**, 455–485.

Williams I. S. (1998) U–Th–Pb geochronology by ion microprobe. In *Applications of Microanalytical Techniques to Understanding Mineralizing Processes,* Rev. Econ. Geol. 7 (eds. M. A. McKibben and W. C. Shanks). Society of Economic Geologists, Iuscaloosa, AL, pp. 1–35.

Williams I. S., Buick I. S., and Cartwright I. (1996) An extended episode of early Mesoproterozoic metamorphic fluid flow in the Reynolds range, central Australia. *J. Metamorph. Geol.* **14**, 29–47.

Williams M. L., Jercinovic M. J., and Terry M. P. (1999) Age mapping and dating of monazite on the electron microprobe: deconvoluting multistage tectonic histories. *Geology* **27**, 1023–1026.

Willigers B. J. A., Krogstad E. J., and Wijbrans J. R. (2001) Comparison of thermochronometers in a slowly cooled granulite terrain: Nagssugtoqidian orogen, West Greenland. *J. Petrol.* **42**(9), 1729–1749.

Willigers B. J. A., Baker J. A., Krogstad E. J., and Peate D. W. (2002) Precise and accurate *in situ* Pb–Pb dating of apatite, monazite, and sphene by laser ablation multiple-collector ICP-MS. *Geochim. Cosmochim. Acta* **66**(6), 1051–1066.

Wolf R. A., Farley K. A., and Silver L. T. (1996) Helium diffusion and low temperature thermochronometry of apatite. *Geochim. Cosmochim. Acta* **60**, 4231–4240.

Wolf R. A., Farley K. A., and Kass D. M. (1998) Modeling of the temperature sensitivity of the apatite (U–Th)/He thermochronometer. *Chem. Geol.* **148**(1–2), 105–114.

Xu G. Q. and Kamp P. J. J. (2000) Tectonics and denudation adjacent to the Xianshuihe fault, eastern Tibetan plateau: constraints from fission track thermochronology. *J. Geophys. Res.-Solid Earth* **105**(B8), 19231–19251.

York D., Hall C. M., Yanase Y., Hanes J. A., and Kenyon W. J. (1981) ^{40}Ar/^{39}Ar dating of terrestrial minerals with a continuous laser. *Geophys. Res. Lett.* **8**, 1136–1138.

Zeck H. P. and Whitehouse M. J. (2002) Repeated age resetting in zircons from Hercynian-Alpine polymetamorphic schists (Betic-Rif tectonic belt S, Spain)—a U–Th–Pb ion microprobe study. *Chem. Geol.* **182**(2–4), 275–292.

Zeitler P. K. (1987) Argon diffusion in partially outgassed alkali feldspars: insights from ^{40}Ar/^{39}Ar analysis. *Isotope Geosci.* **65**, 167–181.

Zeitler P. K., Koons P. O., Bishop M. P., Chamberlain C. P., Craw D., Edwards M. A., Hamidullah S., Jan M. Q., Khan M. A., Khattak M. U. K., Kidd W. S. F., Mackie R. L., Meltzer A. S., Park S. K., Pêcher A., Poage M. A., Sarker G., Schneider D. A., Seeber L., and Shroder J. F. (2001) Crustal reworking at Nanga Parbat, Pakistan: metamorphic consequences of thermal-mechanical coupling facilitated by erosion. *Tectonics* **20**(5), 712–728.

Zhu X. K., O'Nions R. K., Belshaw N. S., and Gibb A. J. (1997a) Lewisian crustal history from *in situ* SIMS mineral chronometry and related metamorphic textures. *Chem. Geol.* **136**, 205–218.

Zhu X. K., O'Nions R. K., Belshaw N. S., and Gibb A. J. (1997b) Significance of *in situ* SIMS chronometry of zoned monazite from the Lewisian granulites, NW Scotland. *Chem. Geol.* **135**, 35–53.

3.09
Continental Crust Subduction and Ultrahigh Pressure Metamorphism

D. Rumble
Geophysical Laboratory, Washington, DC, USA
J. G. Liou
Stanford University, CA, USA
and
B. M. Jahn
National Taiwan University, Taipei, Republic of China

3.09.1 INTRODUCTION

Evidence of continental collision is as accessible as the nearest copy of the "Times Atlas." A glance at the map shows the northern edge of the Indian subcontinent festooned with mountain chains rising as India crashes northward into Eurasia. The processes of continental crust subduction and ultrahigh pressure (UHP) metamorphism are necessary consequences of collision but evidence for them is to be found in the disciplines of petrology, structural geology, and geochemistry, not physiography. Exhumation of subducted, buoyant continental rock is a physically inevitable process but one that is especially important to geologists as it exposes these remarkable probes of the upper mantle for study. Geochemistry is important for understanding continental crust subduction and UHP metamorphism because it is uniquely capable of providing a geochronological framework for determining the sequence and duration of collision, subduction, and metamorphism (see Chapter 3.08). Geochemistry is equally important in its capacity to identify protolith signatures and to trace them through the complex events along the margins of converging tectonic plates. Petrology, in combination with geochemistry, traces the $P-T$ time path of rocks through subduction and exhumation and provides a complete characterization of the geodynamics of continental collision (see Chapter 3.07).

The continuing discovery of new UHP metamorphic belts, since the seminal publications of Chopin (1984) and Smith (1984), has unexpectedly and repeatedly re-invigorated the entire field. The authors of this review thus face a daunting task in presenting a coherent view of rapidly developing and ever-increasing research results on a topic that is global in scope, diverse in scientific origins, and of great importance to many specialists striving to understand the growth of continental plates. A search of the bibliographic database "GeoRef" on the keyword "UHP" found over 200 references since 1995. A number of book length reviews of the field are available: Smith (1988), Carswell (1990), Coleman and Wang (1995b), Cong (1996), Bebout *et al.* (1996), Hacker and Liou (1998), Ernst and Liou (2000), and Parkinson *et al.* (2002). In addition, there are thematic issues of Journals focused on UHP metamorphism, including Liou *et al.* (1995), Schreyer and Stockhert (1997), Liou *et al.* (1998), Carswell (2000), Liou and Banno (2000), Liou and Carswell (2000), Liou *et al.* (in press-a), and Liou *et al.* (in press-b). Godard (2001) has presented a historical review of the study of eclogites; Chopin's (2003) review is recommended. Studies of UHP metamorphism in relation to the tectonics and geodynamics of continental crust subduction and exhumation are given by Hacker and Peacock (1995), Hacker (1996), O'Brien (2001), Peacock (1996), and Roselle and Engi (2002). The objective of the present chapter is to present a compact introduction to the literature of continental crust subduction and UHP metamorphism and to illustrate recent geochemical research with selected case studies.

Primary evidence for the subduction of continental crust is found in the similarity of lithologic successions of UHP metamorphic terranes with stratigraphic sequences in unmetamorphosed sedimentary rocks. Interlayered quartzite, marble, mica schist, and gneiss resemble sedimentary sequences of sandstone, limestone, shale, and graywacke, respectively, deposited along continental margins. Concordant layers of eclogite in quartzite, marble, schist, and biotite gneiss suggest basaltic sills or lava flows. Geochemistry plays an important role in recognizing Earth surface features in subducted rocks. The low $\delta^{18}O$ ($\delta^{18}O = [\{(^{18}O/^{16}O)_{sample}/(^{18}O/^{16}O)_{VSMOW}\}-1]\times 1{,}000$ where VSMOW stands for Vienna standard mean ocean water) and low δD rocks of Dabie–Su-Lu, China, for example, demonstrate that UHP gneisses, eclogites, and quartzites were altered by meteoric water under a cold climate in a geothermal system at the Earth's surface prior to subduction (Yui *et al.*, 1995; Zheng *et al.*, 1996; Baker *et al.*, 1997; Rumble and Yui, 1998).

UHP metamorphism is defined by occurrences of the index minerals coesite and diamond in regional eclogite-facies metamorphic rocks. Coesite is found typically as inclusions, tens of microns in diameter, in metamorphic porphyroblasts of carbonate, epidote, garnet, kyanite, omphacite, zircon, and zoisite. Diamond occurs as micron-sized inclusions in garnet, diopside, kyanite, and zircon. The lower limit of $P-T$ conditions for UHP metamorphism is $P > 2.7$ GPa and $T > 600$ °C. Recent reports of relict majoritic garnet and other "super" UHP mineralogical indicators suggest exhumation from even greater depths, where pressures exceed those at the graphite–diamond equilibrium boundary.

3.09.2 WORLDWIDE DISTRIBUTION AND AGES OF UHP METAMORPHIC BELTS

Outcrops of UHP metamorphic rocks occur in regional belts located principally in the Eurasian landmass. They are also known to occur in Africa, Greenland, Indonesia, South America, and Antarctica (Figure 1 and Table 1). The UHP belts mark the margins of continental plates that are either colliding presently or did so at some time in the geologic past. Among the common characteristics of UHP occurrences are the following:

- UHP mineralogical records are preserved mainly in pods and slabs of eclogite and garnet peridotite enclosed within quartzofeldspathic gneiss. Coesite and diamond inclusions, however, are recognized in metamorphic porphyroblasts from a growing list of rock types. Especially significant are occurrences of coesite inclusions in zircons extracted from quartzofeldspathic gneisses, a lithology that dominates many UHP belts but otherwise carries no signature of UHP metamorphism (Tabata *et al.*, 1998; Ye *et al.*, 2000b; J.-B. Liu *et al.*, 2001; F. Liu *et al.*, 2001a, 2002; Katayama *et al.*, 2000a).
- Lithologies are largely continental in chemical and isotopic composition (Jahn, 1998).
- Exhumed UHP units are now present in the upper continental crust as thin subhorizontal

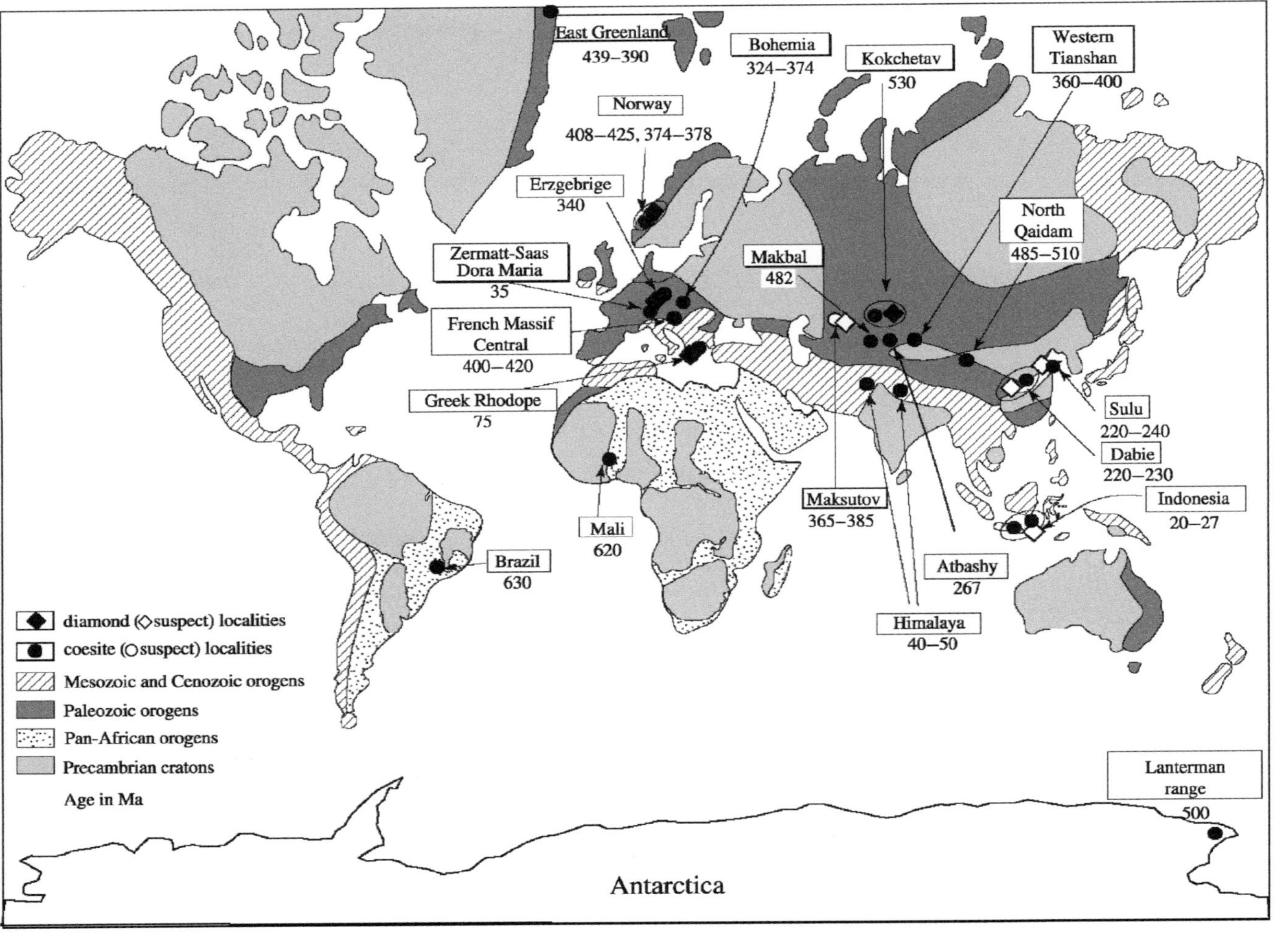

Figure 1 Worldwide distribution of UHP localities with diamond (filled diamond) and coesite (filled circle) occurrences. Age of UHP metamorphism in millions of years.

Table 1 Worldwide occurrences of coesite-eclogite and diamond-eclogite facies rocks.

Locality	*UHP index minerals*	*P–T conditions*	*Age of UHP metamorphism*	*References*
Sulawesi, Indonesia	Coesite in zircon in quartzite	T: 800 ± 100 °C; P: >40 kb	20–27 Ma	Parkinson and Katayama (1999)
Dora Maira, Italian Alps	Coesite in pyrope from quartzite	T: >700 °C; P: >28 kb	35–38 Ma	Chopin (1984); Nowlan *et al.* (2000)
Gittidas, Pakistan Himalaya	Coesite in omphacite from eclogite	T: 690–750 °C; P: 27–29 kb	40–50 Ma	O'Brien *et al.* (2001); Kaneko *et al.* (2001)
Zermatt-Saas, W. Alps	Coesite, aragonite	T: 600–630 °C; P: 27–29 kb	50–70 Ma	Reinecke (1998)
Rhodope, Greece	Diamond, majorite, quartz pseudomorph after coesite	T: 600–900 °C; P: 31–39 kb	75 Ma	Mposkos and Kostopoulos (2001)
Dabieshan-Su-Lu Belt, Eastern China	Coesite in garnet, omphacite, zircon; diamond in garnet	T: 600–900 °C; P: 27–50 kb	220–240 Ma	Liou *et al.* (1996); Li *et al.* (2000); Xu *et al.* (1992, 2003)
Saxonian Erzegebirge, Germany	Micro-diamond included in garnet, kyanite and zircon from quartz-feldspar gneiss; TiO_2 phase with α-PbO_2 structure	T: >900 °C; P: >40 kb	340 Ma	Massonne (1999); Hwang *et al.* (2000)
Maksyutov Complex, Ural Mountains	Quartz pseudomorph after coesite; Graphite pseudomorphs after diamond	T: 694–637 °C; P: 15–17 kb (annealed?)	375–400 Ma	Chesnokov and Popov (1965); Beane *et al.* (1995); Leech and Ernst (1998)
Massif Central, France	Coesite in garnet from kyanite eclogite	T: 700–800 °C; P: >28 kb	400–420 Ma	Lardeaux *et al.* (2001)
E. Greenland Caledonides	Palisaded quartz pseudomorph after coesite	T: 972 °C; P: 36 kb	390–439 Ma	Gilotti and Krogh-Ravna (2002)
Western Gneiss Region, Norway	Coesite in clinopyroxene and garnet from eclogite; diamond; majorite.	T: >800 °C; P: >32 kb	400–410 Ma	Smith (1984); Dobrzhinetskaya *et al.* (1995); Cuthbert *et al.* (2000); van Roermund *et al.* (2002)
N. Qaidam, Western China	Coesite	T: 624–735 °C; P: 28–35 kb	495 Ma	Yang *et al.* (2001)
Lanterman Range, Antarctica	Coesite in garnet from eclogite	T: 850 °C; P: >29 kb	500 Ma	Ghiribelli *et al.* (2002)
Kokchetav Massif, Kazakhstan	Diamond in garnet, clinopyroxene, and zircon from gneiss and metacarbonate; coesite in garnet from whiteschist	T: 800–1000 °C; P: 40–60 kb	530 Ma	Sobolev and Shatsky (1990); Shatsky *et al.* (1999); Katayama *et al.*, (2001)
Brazil	Coesite	?	630 Ma	Parkinson *et al.* (2001)
Pan-African Belt, Mali, W. Africa	Coesite in omphacite from marble	T: 690–750 °C; P: >27 kb	620 Ma	Caby (1994); Jahn *et al.* (2001a)

slabs bounded by normal faults on top and reverse faults on the bottom, and sandwiched in between high pressure (HP) or lower-grade metamorphic units (Xue *et al.*, 1996; Kaneko *et al.*, 2000).
- Coeval island-arc volcanic and plutonic rocks do not occur, whereas post-collisional or late-stage A-type granitic plutons are common in some UHP belts (Chen *et al.*, 2002).

The Dora Maira Massif, one of the two type discovery sites of regional metamorphic coesite (Chopin, 1984), consists of Late Paleozoic and older continental basement rocks, metamorphosed under UHP and HP conditions as a result of Mesozoic–Cenozoic convergence of the European and African plates. The HP + UHP belt is a stack of four tectonic units separated by low-angle faults. These units have been affected by Cenozoic HP metamorphism and later re-equilibration at lower pressures. The UHP rocks consist of dark- and light-colored eclogites, pyrope quartzites with phengite schist, and jadeite-rich rock inclusions, as well as orthogneiss country rocks and undeformed metagranites. Boudins of pyrope quartzite contain UHP relics such as coesite and other UHP minerals (Chopin *et al.*, 1991; Compagnoni *et al.*, 1995). Kyanite-eclogites with relict coesite and coesite pseudomorphs have estimated peak P–T conditions of 3.7 GPa and 790 °C, within the diamond stability field. Eclogites re-equilibrated at lower-P were metamorphosed at 1.5 GPa and 500 °C (Nowlan *et al.*, 2000). Analyses of U–Pb in zircon with a SHRIMP ion microprobe give core ages of 240–275 Ma and rims of 35 Ma (Gebauer *et al.*, 1997).

The western gneiss region (WGR) of Norway, the second type discovery site (Smith, 1984) lies along an Early Paleozoic collisional zone between the Fennoscandian and Greenland plates. The WGR is 300 km long and 150 km wide; it consists of interlayered pelite and migmatite, marble, quartzite, and amphibolite, with tectonic inclusions of gabbro and peridotite. The WGR exhibits mainly amphibolite-facies assemblages, but relics of UHP assemblages do occur. Coesite and coesite pseudomorphs have been found as inclusions in metamorphic porphyroblasts in eclogites, kyanite-eclogites, and adjacent gneissic rocks (Smith, 1984; Wain, 1997; Cuthbert *et al.*, 2000; Wain *et al.*, 2000). Microdiamond grains 20–50 μm in diameter have been described in mineral separates from two gneisses (Dobrzhinetskaya *et al.*, 1995) and in a mantle-derived peridotite lens from Bardane (van Roermund *et al.*, 2002). Carswell *et al.* (in press) provide new petrographic evidence and a review of the latest radiometric age data, and conclude that the UHP metamorphism of eclogites within WGR occurred at 400–410 Ma, significantly younger than the previous, widely accepted age of 425 Ma. A two-stage exhumation process was suggested: an initial exhumation to ~35 km depth by ~395 Ma at a mean rate of ~10 mm yr^{-1}, and subsequent exhumation to 8–10 km by ~375 Ma at a rate of ~1.3 mm yr^{-1}. The recent discoveries of relict majoritic garnets containing exsolution lamellae of pyroxene located in peridotite bodies on the islands of Otrogy and Flemsoy (van Roermund and Drury, 1998, van Roermund *et al.*, 2001) are startling. The presence of relict majoritic garnet implies that the peridotites may have been exhumed from depths greater than 185 km. A UHP terrane complementary to WGR and separated from it by the opening of the Atlantic Ocean is located on the east coast of Greenland (Gilotti and Krogh-Ravna, 2002).

The Kokchetav Massif, northern Kazakhstan, is the type discovery site for the UHP diamond-eclogite regional metamorphic facies (Sobolev and Shatsky, 1990). Diamonds occur as minute inclusions in garnet, diopside, and zircon porphyroblasts from pyroxene–carbonate–garnet rock, and garnet-biotite gneiss and schist. Inclusions of coesite and coesite pseudomorphs are found in garnet from eclogite and diamond-bearing gneiss. The UHP/HP unit extends NW–SE for at least 80 km and is ~17 km wide. The UHP slab is structurally overlain by a weak- to low-grade metamorphic unit and is underlain by a low-P metamorphic andalusite–staurolite facies unit (Kaneko *et al.*, 2000). The UHP slab is composed of felsic gneiss with locally abundant eclogite lenses and minor orthogneiss, metacarbonate, and rare garnet peridotite with a range of Proterozoic protolith ages. Radiometric age dating yields 530 ± 7 Ma for UHP metamorphism for eclogite, biotite gneiss, and diamondiferous metasedimentary rocks (Shatsky *et al.*, 1999; Katayama *et al.*, 2001).

The Dabie–Su-Lu terrane, east-central China, occupies a Triassic collision zone between the Sino-Korean and Yangtze plates. The central UHP coesite- and diamond-bearing eclogite belt ($P =$ 2.7–5 GPa) is flanked to the north by a migmatite zone ($P < 2$ GPa), and to the south by a belt of blueschist, epidote amphibolite, and eclogite facies rocks ($P = 0.5$–1.2 GPa). Prevalent rock types include felsic gneiss, orthogneiss, marble, and quartzite. Protolith ages range from Proterozoic to Ordovician but the majority of the reported ages are Neoproterozoic (Li, 1996; Hacker *et al.*, 2000). Blocks, boudins, and layers of eclogites and garnet peridotities occur as enclaves in gneisses in the UHP unit. The Dabie–Su-Lu UHP rocks show several important features. (i) Widespread occurrences of coesite with hydrous minerals such as talc, zoisite/epidote, and phengite are found in eclogites. (ii) There are several garnet peridotites of mantle origin; others are of crustal origin, formed as crystal cumulates in layered intrusions (Chavagnac and Jahn, 1996; Zhang *et al.*, 1998, 2000). (iii) Abundant exsolution textures were identified in UHP minerals from garnet peridotite and eclogite. Diamond inclusions in garnet eclogite and garnet peridotite have been reported (Xu *et al.*, 1992, 2003). Some garnet peridotites record much higher pressures than associated coesite-bearing eclogites, reaching 5–6 GPa (Yang *et al.*, 1993; Zhang *et al.*, 2000). Various isotopic age dating methods for UHP minerals and rocks from the Dabie–Su-Lu terrane gives 220–240 Ma (Ames *et al.*, 1996; Li *et al.*, 1993; Rowley *et al.*, 1997; Li *et al.*, 2000; Hacker *et al.*, 2000; Ayers *et al.*, 2002; Chavagnac and Jahn, 1996; Chavagnac *et al.*, 2001).

The Erzgebirge crystalline complex (ECC), a new example of rocks in the diamond-eclogite facies of regional metamorphism, is situated at the northern margin of the Bohemian Massif and is in fault contact with a low-grade Paleozoic sequence at its margins. The Variscan ECC consists of abundant ortho- and para-gneisses that include numerous lenses of eclogite. The "gneiss-eclogite unit" at the core of this massif contains some UHP rocks with inclusions of coesite in garnets and omphacites from eclogite (Massonne, 2001) and inclusions of microdiamonds in garnet, kyanite, and zircon from gneisses that contain abundant quartz, plagioclase, phengite, and minor rutile and graphite (Massonne, 1999). Thus far, the diamondiferous gneisses have only been found in a 1-km long strip near the eastern shore of the Saidenbach reservoir. In addition to completely preserved, 1–25 μm diamonds, partially graphitized diamonds and graphite pseudomorphs after diamond occur in the host minerals. The microdiamond inclusions in garnet and zircon are similar to those in the diamondiferous biotite gneiss of the Kokchetav Massif described by Sobolev and Shatsky (1990). Some of these inclusions contain additional fluid-bearing phases including apatite, phengitic mica, and possible fluid inclusions. The morphology of the inclusions and their mineral association suggest that microdiamonds from both the Erzgebirge and the Kokchetav may have crystallized from supercritical fluids under UHP conditions (Stockhert *et al.*, 2001; Hwang *et al.*, 2001a; Dobrzhinetskaya *et al.*, 2001, 2003).

Schmadicke *et al.* (1995) reported Sm–Nd isochrons for garnet–clinopyroxene–whole rock at 360 ± 7 Ma for eclogite and 353 ± 6 Ma for garnet pyroxenites from the Erzgebirge UHP rocks. $^{40}Ar/^{39}Ar$ spectra of phengite from two eclogite samples give plateau ages of 348 ± 2 Ma and 355 ± 2 Ma. Recently, diamond-bearing zircons were dated by SHRIMP, and yielded 336±2 Ma (Massonne, 2001). The similarity in the age of garnet peridotites and eclogite enclosed in gneiss suggests a coeval Variscan UHP *in situ* metamorphism of the ECC around 340–360 Ma.

Other UHP terranes have partially preserved trace index minerals in strong containers such as zircon or garnet (Figure 1 and Table 1; for a review, see Liou *et al.*, 1998). The Maksyutov Complex contains cuboid graphitized diamond in pelitic schist (Leech and Ernst, 1998). Inclusions of coesite and coesite pseudomorphs occur in the Zermatt-Sass area of the Western Alps, Makbal and Atbashy of the Krghyzstan Tien-Shan, Mali of Western Africa, and Sulawesi of Indonesia (Table 1). The latest recognized UHP terranes include: (i) coesite in the French Massif Central (Lardeaux *et al.*, 2001), (ii) coesite and diamond from the Greek Rhodope metamorphic province (Mposkos and Kostopoulos, 2001), (iii) coesite pseudomorphs in the Northeast Greenland Eclogite Province (Gilotti and Krogh-Ravna, 2002), (iv) coesite in granulite-facies overprinted eclogites of SE Brazil (Parkinson *et al.*, 2001), (v) coesite in Himalayan eclogite from the Upper Kaghan Valley, Pakistan (O'Brien *et al.*, 2001; Kaneko *et al.*, 2003) and coesite from the Tso-Morari crystalline complex of India (Sachan *et al.*, 2003), (vi) coesite inclusions in gneissic rocks from the North Qaidam belt, western China (Yang *et al.*, 2001), and (vii) inclusions of quartz pseudomorphs and minor coesite in garnets of mafic ecolgites from the Lanterman Range in Antarctica (Ghiribelli *et al.*, 2002). Among all UHP terranes, the oldest has been dated at ca. 620 Ma for coesite-bearing rocks from the Pan-African belt in northern Mali (Caby, 1994; Jahn *et al.*, 2001a).

3.09.3 EXOTIC VERSUS *IN SITU* ORIGIN OF UHP METAMORPHIC ROCKS

A typical outcrop in a UHP metamorphic belt consists of a lens of eclogite enveloped in quartzofeldspathic gneisses. Petrographic examination of thin sections reveals that the eclogite carries UHP index mineral assemblages, whereas the gneiss is amphibolite facies and may show retrograde features. Outcrop relations like these are seen worldwide and have inspired a spirited debate between those who argue that the eclogites are exotic tectonic blocks whose metamorphism is unrelated to that of the gneisses and those who advocate *in situ* UHP metamorphism. According to proponents of the exotic hypothesis the eclogites were exhumed from great depth and interleaved tectonically with their wall rocks following UHP metamorphism; the quartzofeldspathic gneisses were never under UHP conditions. Proponents of the hypothesis of *in situ* UHP metamorphism hold that both eclogites and gneisses experienced the same metamorphic history, including UHP conditions, but that the gneisses failed to develop UHP indicator minerals owing to an unfavorable bulk chemical composition, or that evidence of UHP was erased from gneisses by retrogradation during exhumation (see Coleman and Wang, 1995a; Harley and Carswell, 1995; Krogh and Carswell, 1995; Schreyer, 1995; Smith, 1995; Cong and Wang, 1996; Ernst and Peacock, 1996; Liou *et al.*, 1996).

The question of the exotic versus the *in situ* origin of UHP metamorphics has tectonic implications. If UHP terranes are exposed at the surface as dismembered mélanges of continental, oceanic, and mantle slivers, disruptive deformation scenarios are evoked. But if UHP rocks are subducted and exhumed as thin but laterally coherent slabs, appeal must be made to quite different mechanisms. There is a logical connection between the exotic versus *in situ* discussion and hypotheses of exhumation. In the absence of plausible, buoyancy-driven exhumation processes, it was appealing to suggest penetrative faulting for the emplacement and interleaving of UHP and non-UHP rock units. However, with the advent of physical models of exhumation that are driven by body forces rather than surface tractions, it has become possible to consider the exhumation of coherent slabs as a reasonable working hypothesis (Chemenda *et al.*, 1995; Ernst *et al.*, 1997).

The debate about exotic versus *in situ* origin has inspired much research. Five different types of evidence supports the hypothesis of *in situ* origin. (i) Field mapping demonstrates that UHP belts contain thin nappes of structurally coherent rock slivers (*Dora Maira*: Compagnoni *et al.*, 1995; Michard *et al.*, 1995; *Dabieshan*: Cong *et al.*, 1995; Xue *et al.*, 1996; *Kokchetav*: Kaneko *et al.*, 2000). (ii) Petrologic and mineralogical studies have found examples of quartz-feldspar gneisses from UHP belts with evidence of UHP metamorphism. A number of systematic searches for UHP indicator minerals preserved as inclusions in zircon have shown that coesite is widespread in UHP gneisses, despite the observation that their matrix mineral assemblages are consistent with amphibolite facies metamorphism. Identification of coesite has been confirmed with micro-Raman spectroscopy. In Dabieshan, Tabata *et al.* (1998) found coesite in zircon from orthogneisses whose matrix mineral assemblages are amphibolite-facies. Similar results are reported from Su-Lu (Ye *et al.*, 2000b) and in southern Dabie near the mapped boundary between coesite-bearing and coesite-free eclogites (J.-B. Liu *et al.*, 2001). Detailed sampling of two drill cores from pre-pilot holes of the Chinese Continental Scientific Drilling Project (CCSDP) near Donghai, in Su-Lu, found coesite inclusions in zircons from granitic gneisses extending to depths of 432 m and 1,000 m, respectively (F. Liu *et al.*, 2001a,b, 2002). Similar findings are reported from Kokchetav (Katayama *et al.*, 2000a). (iii) A re-evaluation of thermobarometry in quartz–feldspar–garnet–phengite–epidote–biotite–titanite orthogneisses from Dabieshan resulted in revised $P-T$ estimates of 3.6 GPa and 690–715 °C, well within the UHP regime (Carswell *et al.*, 2000). (iv) Geochronology has shown that UHP eclogites and their wall-rock gneisses have the same thermal history and similar metamorphic ages (*Dabieshan and Su-Lu*: Ames *et al.*, 1996; Rowley *et al.*, 1997; Hacker *et al.*, 1998; Li *et al.*, 2000; Chavagnac *et al.*, 2001; Ayers *et al.*, 2002; Rumble *et al.*, 2002; *Kokchetav*: Claoue-Long *et al.*, 1991; Shatsky *et al.*, 1999; Katayama *et al.*, 2001). (v) The presence of a pre-subduction, cold-climate geothermal area outcropping contiguously over hundreds of square kilometers in both Dabieshan and Su-Lu is recorded by unusually low $\delta^{18}O$ and δD values in minerals of eclogites, quartzites, para- and ortho-gneisses (Yui *et al.*, 1995, 1997; Zheng *et al.*, 1996, 1998a, 1999; Baker *et al.*, 1997; Fu *et al.*, 1999). That the geothermal system affected both UHP rocks and their wall rocks equally suggests that both units experienced the same geologic history, and supports the hypothesis of *in situ* UHP metamorphism (Rumble, 1998; Rumble and Yui, 1998; Rumble *et al.*, 2002). Analyses of drill cores from a bore hole near Donghai, in Su-Lu, demonstrate that the thickness of the geothermal system was at least 432 m, with epidote separated from amphibolite at that depth having a $\delta^{18}O_{VSMOW}$ of −7.9‰ (Zhang and Rumble, unpublished).

Geochemical investigations also play a role in assessing exotic versus *in situ* hypotheses of the origin of UHP metamorphism. The interleaving of eclogites whose protoliths were ocean-floor basalts with continental metasediments would, for example, suggest tectonic juxtaposition rather then *in situ* UHP metamorphism (Zhai and Cong, 1996). The geochemical evidence is discussed more fully below.

The balance of evidence now favors the hypothesis of *in situ* UHP metamorphism. The shift towards acceptance of an *in situ* origin has been accelerated by the development of physically plausible models of the subduction and exhumation of continental crust. The current focus of debate has, in fact, shifted towards evaluation of evidence of "super" ultrahigh pressure metamorphism. These topics will be discussed below.

3.09.4 MINERALOGIC INDICATORS OF "SUPER" UHP METAMORPHISM

The discovery of coesite in regionally metamorphosed, eclogite-facies rocks by Chopin (1984) and Smith (1984) was initially greeted with some skepticism. How could high-pressure minerals be preserved metastably, it was asked, during what must have been an arduous ascent from the upper mantle? The discovery of microdiamonds in eclogite-facies gneisses by Sobolev and Shatsky (1990) stimulated even greater skepticism. Validation of the discoveries by *in situ* Raman spectroscopic measurements has removed any doubts regarding possible mineral misidentification (Nasdala and Massonne, 2000). Mineralogical indicators of UHP metamorphism have been reviewed by Chopin and Sobolev (1995) and Liou *et al.* (1998). Exsolution features of UHP minerals have been reviewed by Green *et al.* (2000) and Zhang and Liou (2000). Recent discoveries of new mineralogical evidence of "super" UHP metamorphism have sparked a new wave of controversy. Some of the new findings suggest pressures of 6.0 GPa, or more; some samples may have been exhumed from depths greater than 200 km. We list, below, novel mineralogical indicators that have been discovered recently. Relict majoritic garnet has been found in the WGR (van Roermund *et al.*, 1998, 2001), Dabie–Su-Lu, China (Ye *et al.*, 2000a, and Rhodope, Greece (Mposkos and Kostopoulos, 2001) suggesting an origin at $P > 6.0$ GPa and depths >185 km. High-*P* clinoenstatite has been reported in Alpe Arami (Bozhilov *et al.*, 1999) and Dabie–Su-Lu garnet peridotites (Zhang *et al.*, 2002b). Garnet megacrysts with inclusions of exsolved clinopyroxene + garnet ± ilmenite interpreted to be majoritic garnet have been recently suggested from the Granulitgebirge, Germany (Massonne and Bautsch, 2002) and Su-Lu of eastern China (Zhang and Liou, in press). Supersilicic clinopyroxene and silica exsolution is present in eclogite and pelitic gneiss from the Kokchetav massif, Kazakhstan (Katayama *et al.*, 2000b). Ogasawara *et al.* (2002) report coesite exsolution from supersilicic titanite in UHP marble from the Kokchetav Massif, northern Kazakhstan. Nanometer-size inclusions of α-PbO_2-type TiO_2 have been found in garnet from diamond-eclogite facies gneisses in the Erzgebirge, Germany, indicating $P > 4.0$ GPa; $T > 900$ °C (Hwang *et al.*, 2000). Unusual defect microstructures including nanometer-scale twin lamellae in omphacite, kyanite, and spinel and microcleavages in garnet from Su-Lu eclogite are said to indicate rapid exhumation from depth (Hwang *et al.*, 2001b). Given the application of new determinative methods including micro-Raman spectroscopy, electron microscopy, and intense, focused X-ray beams at synchrotron laboratories, it is likely that new discoveries will push the depth of subduction even further.

3.09.5 GEOCHEMISTRY OF UHP METAMORPHISM

Eclogites and peridotites have received a disproportionate share of the attention of researchers despite their volumetric insignificance in UHP metamorphic terranes, because they are the principal rock types that bear evidence of UHP metamorphism. An enhanced level of study of mafic and ultramafic rocks is, moreover, justified because of their importance in the interpretation of the tectonic history of UHP terranes. The evaluation of a hypothesis of continental collision initiated by oceanic crust subduction followed by exhumation triggered by break-off of the oceanic slab, for example, is greatly facilitated by distinguishing eclogites and peridotites originating in continental crust from those of oceanic crust or mantle origin. Radiometric dating provides indispensable tests of tectonic models. Examples of both protolith studies and age dating are given in the following sections.

3.09.5.1 Chemical Compositions of Eclogites and Ultramafic Rocks

Eclogites are mainly basaltic in composition, but they show a wide range of major and trace element abundances. This suggests multiple origins and heterogeneous sources. In some cases, the data reflect the fact that analyses were made on banded rocks produced by metamorphic segregation, thus biasing the results. Eclogites and ultramafic rocks that recrystallized under UHP metamorphic conditions often have a coarse-grained texture with distinct mineral banding of variable scale. It is often difficult to obtain a truly representative bulk chemical composition of the protolith of a banded eclogite. Fine-grained or homogenous textured eclogites also occur, but they seem to be minor in comparison with heterogeneous textured facies.

Eclogites from the Dabie and Su-Lu terranes exhibit a range of SiO_2 contents from 36% to 60%, although the majority still have basaltic or gabbroic compositions with $SiO_2 = 45–52\%$ (Figure 2; Jahn, 1998). Higher silica eclogites ($SiO_2 \geq 53\%$) suggest a more differentiated protolith, while lower silica eclogites ($\leq$45%) imply a cumulate protolith or represent a banded eclogite rich in low-SiO_2 phases (e.g., garnet, epidote, rutile, etc.). Garnetite is a good example of metamorphic differentiation; it cannot be used to discuss the nature of its protolith. Many of the chemical compositions of eclogites cannot be matched by reasonable magmatic protoliths.

Numerous petrological studies have established that the protoliths of eclogites underwent progressive dehydration during prograde metamorphism, and that many were later rehydrated to

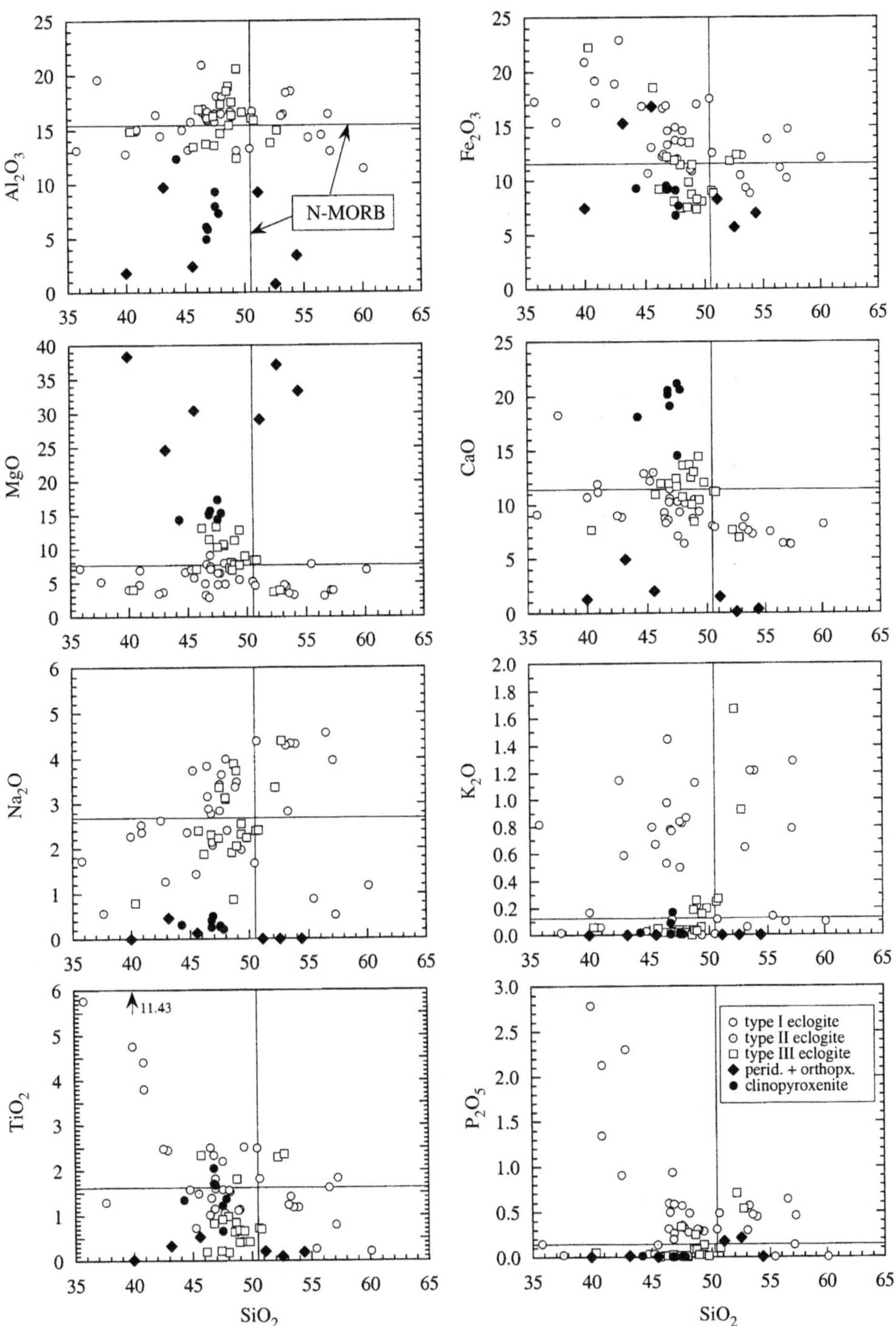

Figure 2 Major element variations of eclogites and associated ultramafic rocks from the Su-Lu and Dabie terranes of east-central China (Jahn, 1998). Eclogites are generally basaltic and quartz-normative, with some showing the cumulate nature of their protoliths. Type I is gneiss-hosted, type II, marble-associated. Type III are members of layered intrusions or associated with ultramafic rocks. Also plotted are data from peridotites (Perid.), orthopyroxenites (Orthopx.), and clinopyroxenites.

some extent during retrograde metamorphism when the rocks were exhumed (Liou *et al.*, 1998). These processes are not isochemical. LIL elements (potassium, rubidium, caesium, strontium, barium), uranium, and to some extent, lanthanum and cerium, can be mobilized in a fluid phase together with volatile species such as H_2O and CO_2. In the subduction of oceanic crust, as in the circum-Pacific environment, dehydration of subducted amphibolite and serpentinite produces hydrous fluids which move upward, enriched in these elements, but leaving behind niobium and tantalum, and metasomatizing the mantle wedge (Tatsumi and Eggins, 1995). Magmas produced by the melting of such metasomatized mantle would show enrichment in LIL elements, uranium, and probably also La–Ce, but depletion in niobium and tantalum. This is the characteristic geochemical signature of arc magmas formed in subduction zones (see Chapter 3.09).

Subduction of continental crust, has a somewhat different chemical consequence than the subduction of oceanic crust, because continental crust is globally less hydrated. This is true particularly of the lower crust, which is composed mainly of anhydrous, granulite facies rocks. The protolith of an eclogite could also be derived from the upper to middle crust. It may occur as a basic dike or enclave, or as a layered intrusion (e.g., Bixiling or Maowu of the Dabieshan; Chavagnac and Jahn, 1996; Zhang *et al.*, 1998) within granitic gneisses, metamorphosed in the greenschist to amphibolite facies. Such mafic rocks are often observed in Precambrian terranes. During continental subduction, the greenschist or amphibolite units would follow a similar route as the oceanic crust, and undergo dehydration reactions. The patterns of elemental loss should be rather similar to that in a subducted oceanic lithosphere.

The present abundances of LIL elements, uranium, and REE patterns in eclogites cannot be easily used to argue for or against chemical mobility. The concentration of these elements shows a great deal of variation in basaltic rocks from diverse tectonic settings, hence any difference in their concentration could be related to original differences in the protoliths, and might not be related to gains or losses during metamorphism. Nevertheless, the clearest evidence of elemental loss during eclogitization is supplied by studies of Rb–Sr isotopic systematics. Analyses of eclogites from western Norway (Griffin and Brueckner, 1985) and from the Dabie–Su-Lu terranes (Ames *et al.*, 1996; Jahn, 1998) indicate that many eclogites have highly radiogenic $^{87}Sr/^{86}Sr$ ratios that are "unsupported" by their Rb/Sr ratios (Figure 3). This is particularly true for eclogites that occur as enclaves within quartzofeldspathic gneisses (Griffin and Brueckner, 1985; Jahn, 1998). The most plausible explanation is that rubidium is preferentially lost relative to strontium during prograde eclogite facies metamorphism. Alternatively, high $^{87}Sr/^{86}Sr$ ratios may result from the assimilation of continental crust by mafic magmas. By analogy, elements of similar geochemical behavior, such as potassium, caesium, and probably barium, could have been lost together with rubidium during the same process.

Retrograde metamorphism is often observed in HP–UHP eclogites during the exhumation of deeply subducted continental crust. Retrograde rehydration of UHP rocks may replace the mobile

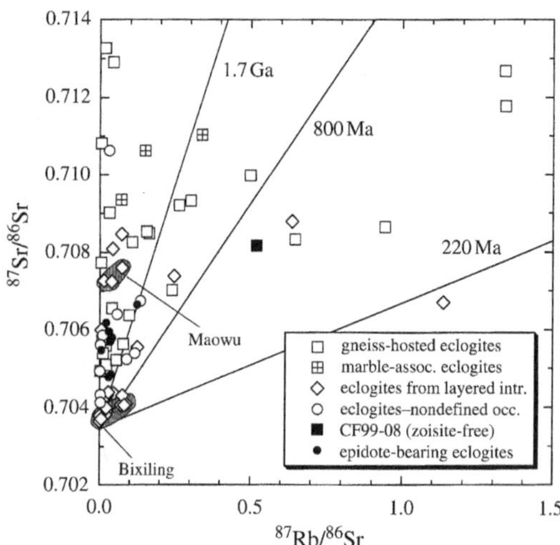

Figure 3 Plot of $^{87}Sr/^{86}Sr$ versus $^{87}Rb/^{86}Sr$ for UHP eclogites from the Su-Lu and Dabie trerranes. Eclogites are classified as gneiss-hosted, marble associated, from layered intrusions, and as "nondefined" occurrences. Three reference isochrons based on the protolith ages of the Weihai eclogites (1.7 Ga; Jahn *et al.*, 1996), granitic gneisses of the Su-Lu and Dabie terranes (~800 Ma) and the UHP metamorphic event (~220 Ma) are shown for assessment of Rb–Sr isotopic systems in different types of eclogites. Data sources: Jahn (1998), Li *et al.* (1999), and Ames *et al.* (1996). Note that the data are highly dispersed and many samples show "unsupported" high Sr isotopic ratios (data left of the 1.7 Ga isochron), implying depletion of Rb and alkali elements during the metamorphic processes.

elements lost during prograde metamorphism. However, the process should not be considered as reversible with respect to prograde metamorphism. In the Dabie orogen, many studies have shown that during the course of continental subduction and exhumation, a hydrous fluid phase was not pervasively present. This is shown by the preservation of the world's lowest $\delta^{18}O$ values (−10‰) in high-temperature rocks (Yui *et al.*, 1995; Zheng *et al.*, 1996; Baker *et al.*, 1997; Rumble and Yui, 1998; Zheng *et al.*,1998; Fu *et al.*, 1999), the presence of the world's highest ε_{Nd}($\varepsilon_{Nd}=[\{(^{143}Nd/^{144}Nd)_{sample}/(^{143}Nd/^{144}Nd)_{CHUR}\}-1]\times10^4$, where CHUR stands for chondritic uniform reservoir) values (+270) in eclogites of the Weihai area (Jahn *et al.*, 1996) and the general absence of syntectonic granites in the orogen.

Eclogites from the European Hercynides and the Alpine chain are known to have an affinity to oceanic basalts (or MORB); whereas those from the Caledonides and the Dabie orogen have an affinity to continental basalts (Figure 4; see Jahn, 1999, and references therein). REE patterns are invariably LREE-enriched in the eclogites of Dabieshan and Su-Lu (Jahn, 1998), but LREE-depleted, MORB-like patterns are abundant in the Hercynian and Alpine chains. Although LREE,

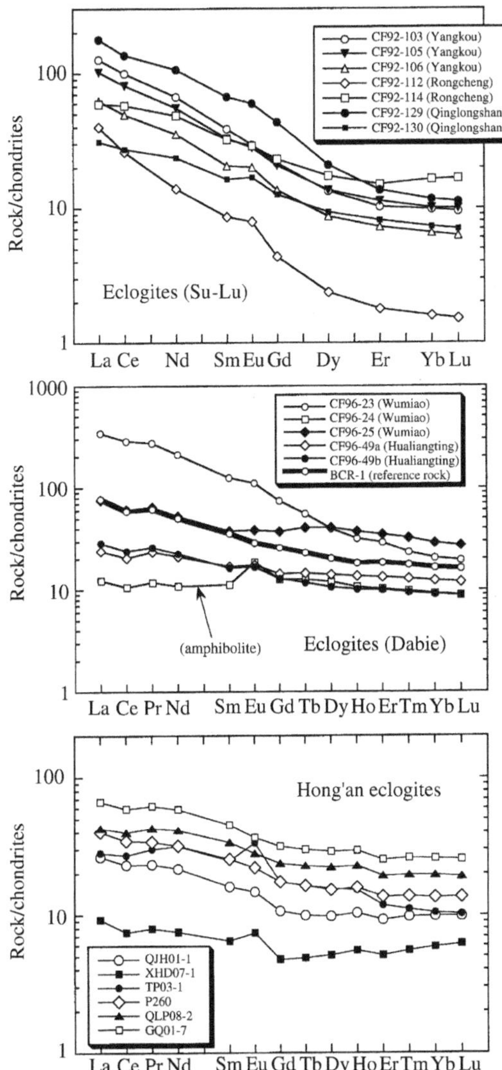

Figure 4 REE distribution patterns of eclogites from the Chinese Su-Lu, Dabie, and Hong'an terranes. Localities from Su-Lu and Dabie are given in parenthesis. All eclogites are gneiss-hosted except two from Rongcheng (Su-Lu). The common feature is the enrichment in light REE, which contrasts with N-MORB but resembles continental mafic rocks (basalt, amphibolite, or basic granulite). Source: Jahn (1998, for Su-Lu and Dabie; unpublished, for Hong'an).

especially lanthanum, might be susceptible to mobilization in the presence of a hydrous fluid, the stable isotope evidence for the absence of such fluid in Dabieshan makes identification of protolith REE patterns credible (Figure 4).

Jahn (1999) compiled samarium and neodymium concentration data for 223 analyses of eclogites and associated ultramafic rocks. Samarium concentrations range from <1 ppm to >15 ppm but most rocks contain ≤ 8 ppm (grand average $= 3.1 \pm 2.5$, 1σ). Neodymium ranges from <1 ppm to 65 ppm, and most eclogites contain ≤ 30 ppm Nd (grand average $= 11.3 \pm 11.4$, 1σ). The Sm/Nd ratios for the majority of eclogites lie between 0.20 (for alkali basalts, continental basalts, or average continental crust) and 0.36 (for depleted N-MORB). Many rocks of low samarium and neodymium concentrations have Sm/Nd ratios higher than that of depleted N-MORBs. This suggests that the protoliths of these eclogites are probably not common basaltic rocks, but represent garnet-rich cumulate portions of metamorphic assemblages or simply pyroxene cumulates.

3.09.5.2 Sm–Nd, Lu–Hf, and Rb–Sr Isochron Ages and Nd–Sr Isotope Tracers

Geochronology plays a crucial role in our understanding of the subduction and exhumation of the continental crust. When dealing with UHP metamorphic rocks, key questions related to the timing of events include: the ages of the protoliths, the ages of peak and retrograde metamorphism, timing of later thermal disturbances, and episodes of fluid–rock interaction. Rates of exhumation can be estimated if precise age information can be obtained and linked to a specific stage of an orogenic process (see Chapter 3.08). The availability of comprehensive age patterns for an entire metamorphic terrane make it possible to reconstruct its tectonic evolution (e.g., *Dabie*: Hacker *et al.*, 1998, 2000; *WGR*: Carswell *et al.*, 2003; *Kokchetav*: Katayama *et al.*, 2001; Shatsky *et al.*, 1999).

Different events can be dated by different chronometers. The most commonly used chronometers for dating UHP events include U–Pb on zircon or monazite, Sm–Nd on garnet, Ar–Ar on white mica or hornblende, and Rb–Sr on biotite or white mica. Lu–Hf has also been used for garnet dating, and it has been argued that this method may be superior to Sm–Nd in some cases (Duchêne *et al.*, 1997; de Sigoyer *et al.*, 2000; Scherer *et al.*, 2000). However, until now only a few results have been published. The U–Pb system in zircon and monazite yields a great amount of chronological information on a variety of thermal events. However, as pointed out cogently by Thöni in his excellent summary of geochronological problems of the Alpine metamorphic rocks (Thöni, 2002), even though the zircon and monazite U–Pb chronometers usually provide exact time information, their use is somewhat hampered by the fact that the mechanisms of zircon and monazite (re)-crystallization are still poorly understood in the context of the P–T evolution of metamorphic rocks (but see Chapter 3.07). In HP and UHP metamorphic terranes, zircon may grow during prograde as well as retrograde recrystallization (Tilton *et al.*, 1991; Gebauer, 1996; Thöni, 2002). Metamorphic recrystallization of zircon and resetting of its

U–Pb systematics could be enhanced by fluids even at moderate temperatures (<600 °C; Rubatto *et al.*, 1999; Liati *et al.*, 2000). Consequently, zircon may yield no clear petrological record of the $P-T$ path, and it may be difficult to link SHRIMP U–Pb ages with a specific stage of metamorphic evolution.

The technique of mineral isochrons is based on the *in situ* isotopic evolution of an equilibrated mineral assemblage. In dating HP–UHP eclogite, only the minerals that are formed in the eclogite facies metamorphism are useful; these include garnet, omphacite, phengitic mica, etc. Presence of relict minerals or retrograde metamorphic products tend to disturb the isochron relationship between different phases and their whole rock (WR). The mineral isochron method not only provides age information but also yields initial isotopic ratios, which are most valuable in tracing protolith evolution (see below). Among the high-grade metamorphic minerals, garnet is known to be the most useful in metamorphic geochronology. Garnet is ubiquitous; it occurs in a wide range of metamorphic and igneous rocks of crustal and mantle origin. In metamorphic petrology, garnet is often used in thermobarometry, and is most useful in the reconstruction of $P-T$ paths. Since the mid-1980s, garnet has become one of the most powerful metamorphic chronometers because of the high rate of the radiogenic growth of its Nd-isotope ratios (Griffin and Brueckner, 1985; Vance and O'Nions, 1990, 1992; Mezger *et al.*, 1992; Li *et al.*, 1993; Chavagnac and Jahn, 1996; Thöni, 2002).

As emphasized by Thöni (2002), the major advantage of garnet dating is the ability to extract both microstructural and thermobarometric information. The combination of textural evidence, $P-T$ estimates and age data allows a much better understanding of the evolution of a metamorphic terrane and hence of orogenic processes. Garnet is most useful in Sm–Nd and Lu–Hf dating, but the interpretation of garnet Sm–Nd ages is not always straightforward, because they are influenced by: isotopic disequilibrium and the presence of LREE-enriched inclusions.

3.09.5.2.1 REE partitioning

Several recent studies of elemental abundances in eclogite minerals have established the value of measuring elemental partitioning between minerals as indicators of mineral equilibrium. These studies mainly used secondary ion mass spectrometry (SIMS) and laser ablation ICP-MS. The data for the Maowu eclogite-pyroxenite body are used to illustrate this point, because the coherent U–Pb and Sm–Nd age data seem to indicate that REE in coexisting clinopyroxene and garnet attained chemical and Nd-isotopic equilibrium (Rowley *et al.*, 1997; Jahn *et al.*, 2002). The general patterns for WR and minerals observed for the Maowu rocks are very similar to those for the Norwegian eclogites (Griffin and Brueckner, 1985), the diamondiferous eclogites from Yakutia, Siberia (Jerde *et al.*, 1993; Taylor *et al.*, 1996) and the eclogites of the Adula Nappe of the central Alps reported by Bocchio *et al.* (2000). They are known to represent chemically equilibrated assemblages.

The values of clinopyroxene/garnet partition coefficients (K_D = concentration of element A in clinopyroxene/concentration of element A in garnet) are shown in Figure 5. The K_D's of the LREE have greater variation than that of HREE, up to three orders of magnitude for lanthanum, but all K_D's decrease regularly from lanthanum (or cerium) to lutetium. Similar patterns have been observed for ecologites from the central Alps (Bocchio *et al.*, 2000). The K_D's are also quite comparable with those obtained in high-pressure (2–3 GPa) and high temperature (1,300–1,470 °C) experiments of Johnson (1994). Consequently, our results strongly indicate that chemical equilibrium was achieved during UHP metamorphism. The only green amphibole shows a REE pattern almost identical to that of the coexisting clinopyroxene, so that the REE K_D values between amphibole and clinopyroxene are close to unity. Thus, the amphibole appears to have been transformed entirely from clinopyroxene during retrograde

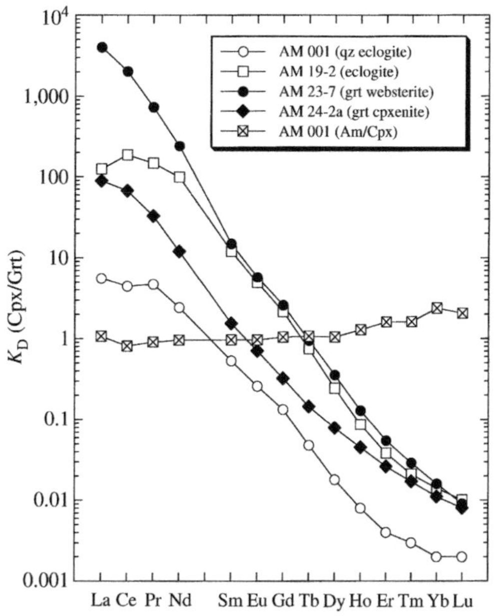

Figure 5 REE partition coefficients (K_D) between clinopyroxene and garnet. The data were obtained on the eclogite–pyroxenite complex of Maowu, Dabieshan, China (data source: Jahn *et al.*, in press). K_D values for LREE vary by three orders of magnitude, whereas those for HREE by only one order.

metamorphism and did not achieve equilibrium with coexisting clinopyroxene (see also Barth *et al.*, 2002).

The REE K_D values for clinopyroxene/garnet show a dependence on the calcium content of the host phases, with K_D's decreasing with decreasing K_D(Ca) (Figure 6). This behavior has also been found in eclogite xenoliths (O'Reilly and Griffin, 1995; Harte and Kirkley, 1997), which have much higher equilibrium temperatures of 1,000–1,200 °C than the Maowu UHP rocks (700–800 °C). The temperature difference appears to have little effect on the K_D values, but this view is not shared by Bocchio *et al.* (2000) on the basis of their analysis of Alpine eclogites. Sassi *et al.* (2000) also observed that the patterns of REE K_D's for all UHP eclogites from Dabieshan are similar, but the absolute values vary a great deal, from three orders of magnitude for lanthanum to about one order for lutetium. Furthermore, a weak correlation between K_D(Zr) and K_D(Ca) for both mantle and Dabieshan eclogites was noticed by Harte and Kirkley (1997) and Sassi *et al.* (2000).

3.09.5.2.2 Incorrect geochronologic ages

Excess Ar in phengitic mica of HP–UHP rocks has been documented frequently and is the cause of aberrant ages (Li *et al.*, 1994, 2000; Giorgis *et al.*, 2000; Jahn *et al.*, 2001a). Incorrect Sm–Nd mineral isochron ages have also been observed, particularly for "low temperature" (≤600 °C) HP–UHP metamorphic rocks. The most notorious examples are from the Alps and the Himalayas (Luais *et al.*, 2001; Thöni, 2002). The following discussion considers examples of "correct" versus "incorrect" Sm–Nd ages. The main cause for this chronometric problem is the lack of isotopic equilibrium between garnet and coexisting minerals.

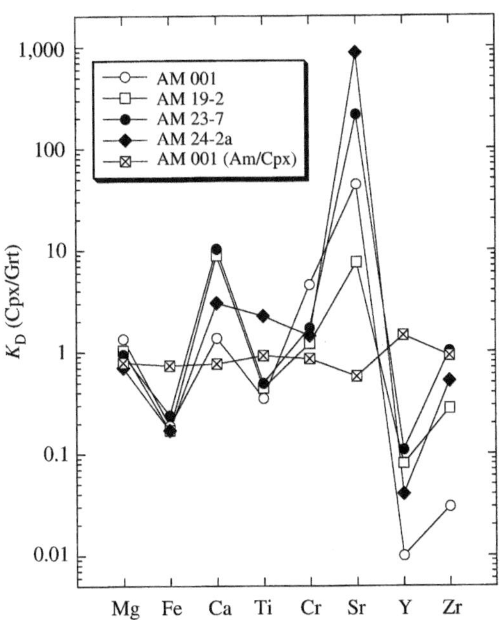

Figure 6 Selected elemental partition coefficients between clinopyroxene and garnet. The data were obtained on the eclogite-pyroxenite complex of Maowu, Dabieshan, China (data source: Jahn *et al.*, in press). Note that the K_D (REE) values are positively correlated with K_D(Ca), and Zr–Y, like HREE, are preferentially partitioned in garnet.

Sm–Nd isotopic equilibrium/disequilibrium depends more on temperature than on pressure. The temperature effect is related to the blocking temperature, T_B (Dodson, 1973), which does not have a unique value and is influenced by several factors including mineral grain size, rate of cooling, the duration of metamorphic reactions at temperature higher than T_B, the presence of a fluid phase and its composition, and penetrative deformation (e.g., Thöni, 2002). The value of T_B ranges from 850 °C to 650 °C for the garnet Sm–Nd isotope chronometer (Humphries and Cliff, 1982; Jagoutz, 1988; Mezger *et al.*, 1992; Burton *et al.*, 1995; Zhou and Hensen, 1995; Günther and Jagoutz, 1997; Thöni, 2002). Garnet out of isotopic equilibrium often has nonequilibrated trace element patterns. In some cases, the Sm/Nd ratios of garnets are even lower than that of WR samples; in other cases, there is little difference between the Sm/Nd ratios of all coexisting minerals, so that they form a cluster in an isochron diagram. In still other cases, garnet and omphacite may have a negative slope isochron relationship. This has been found in eclogites of Tso Morari of the Himalayas and the Sesia zone of the western Alps (Luais *et al.*, 2001), and the Hong'an Block of China (Jahn, unpublished).

Porphyroblastic garnets often contain mineral inclusions. Besides the celebrated coesite, many of them are REE-rich phases, such as monazite, epidote, allanite, titanite, apatite, and zircon. Except zircon, all these phases are highly enriched in LREE and have very low Sm/Nd ratios. Thus, a tiny amount of these inclusions could significantly lower the Sm/Nd ratios of garnets, hence affecting isochrons based on Grt–Cpx–WR and other minerals. Inclusions must have undergone the same metamorphic P–T path as garnet, Cpx, and other major phases. If the metamorphic temperature exceeds the T_B of the inclusions (moderate T_B minerals, such as epidote, apatite, and titanite), the Nd-isotopic compositions of the inclusions can be expected to reach isotopic equilibrium with the host garnet. In this case, a correct isochron age is to be expected, but the spread between parent/daughter ratios would be reduced, and hence the statistical error of the age increased. This has been most frequently observed in the "high-temperature" eclogites of the Dabie and Su-Lu terranes. On the other hand, mineral inclusions of very high T_B (e.g., zircon and monazite) may not be easily reset isotopically, with growing garnet producing a disequilibrium isochron relationship.

In conclusion, the failure to produce correct Sm–Nd isochron ages is due to isotope disequilibrium between garnet and its coexisting minerals. The disequilibrium results from two processes: (i) garnet preserves the isotopic composition of precursor mineral (plagioclase) because the rate of Nd-isotope exchange is more sluggish than the rate of chemical and mineral phase reconstitution; and (ii) garnet contains high-temperature refractory LREE-rich inclusions, which have never re-equilibrated isotopically with the host garnet, though the garnet may have attained equilibrium with coexisting Cpx and other principal phases. The above considerations do not include open system behavior during retrograde metamorphism under the potentially strong influence of hydrothermal activity.

3.09.5.2.3 Coupled Nd- and O-isotopic disequilibrium

One of the most interesting results in isotopic studies of UHP eclogites is the demonstration of a direct correspondence in equilibrium (or disequilibrium) state between the Sm–Nd and O-isotopic systems in eclogite minerals (Zheng *et al.*, 2002). The state of O-isotope equilibrium may provide a critical test for the validity of Sm–Nd mineral isochron ages. Zheng *et al.* (2002) analyzed O- and Sm–Nd isotopic compositions of minerals and WRs in six samples from four eclogite outcrops in southern Su-Lu in east-central China (Figure 7). In four of six samples, Nd–Sm mineral isochrons give ages of 208–226 Ma that agree, within error, with the age of UHP metamorphism as measured by other methods (Ames *et al.*, 1996; Li *et al.*, 1994, 1999). The four samples with consistent Sm–Nd isochron ages also have high temperature $^{18}O/^{16}O$ fractionations between garnet and omphacite that are in agreement with petrological estimates of 700–890 °C (Enami *et al.*, 1993; Zhang *et al.*, 1995a,b; Yang *et al.*, 1998). On the other hand, one of two samples from Yakou and the one sample from Yangkou have Sm–Nd isochrons of 280 Ma and 338 Ma and oxygen isotope fractionations between garnet and omphacite that are not in equilibrium under UHP conditions (Zheng *et al.*, 2002).

Tie-lines between coexisting garnet and omphacite plotted as $\varepsilon_{Nd}(t)$ versus $\delta^{18}O_{SMOW}$ are shown in Figure 7 (redrawn from Zheng *et al.*, 2002). In such a plot, values of $\varepsilon_{Nd}(t)$ calculated for the age of UHP metamorphism are expected to be identical for minerals experiencing isotopic equilibration under UHP conditions. The differences between values of $\delta^{18}O$ for minerals equilibrated

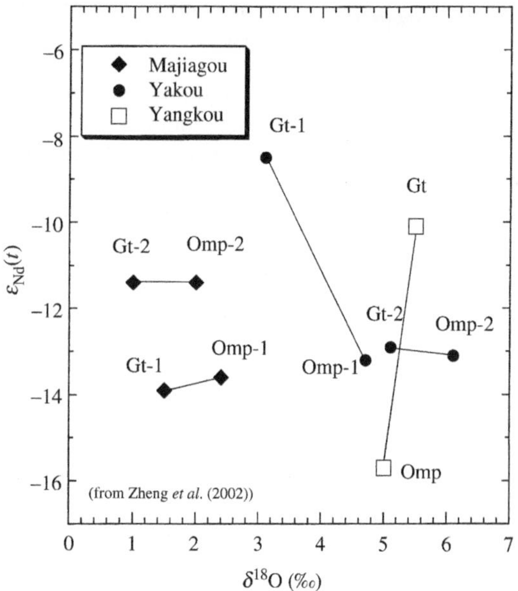

Figure 7 Tie-lines between coexisting garnet (Gt) and omphacite (Omp) in eclogites from the Su-Lu terrane, China, plotted on a $\varepsilon_{Nd}(t)$ versus $\delta^{18}O$ diagram. Samples identified by names of localities where collected. See text for discussion (after Zheng *et al.*, 2002).

under high temperature, UHP conditions, $\Delta^{18}O_{omphacite\text{-}garnet}$, should be in the range 0.8–1.0‰ with $\delta^{18}O_{omphacite} > \delta^{18}O_{garnet}$ (Zheng *et al.*, 2002). These conditions are met by two samples from Majiagou and one sample from Yakou (Figure 7). A fourth sample from Qinglongshan also shows the expected isotopic equilibrium features but is not plotted. In contrast, sample Gt-1/Omp-1 from Yakou gives disparate values of $\varepsilon_{Nd}(t)$ for an age of 211 Ma and its oxygen isotope fractionation records a sub-UHP temperature of 545 °C. The sample from Yangkou is in blatant oxygen isotope disequilibrium with reversed $^{18}O/^{16}O$ partitioning, $\delta^{18}O_{garnet} > \delta^{18}O_{omphacite}$, and unequal values of $\varepsilon_{Nd}(t)$ calculated for a UHP age of 220 Ma. The disequilibrium characteristics of the Yakou sample (Gt-1/Omp-1) are attributed to retrograde metamorphic alteration during exhumation whereas the isotope relationships in the Yangkou sample are caused by incomplete mineral reactions during prograde UHP metamorphism (Zheng *et al.*, 2002).

Analytical work on Sm–Nd and $^{18}O/^{16}O$ relationships in coexisting minerals provides an independent criterion for evaluating geochronologic ages as well as insight into the kinetics of isotope disequilibrium. A survey of published data suggests that rates of oxygen self-diffusion in garnet and omphacite are lower than, or close to, those of neodymium diffusion at the same temperatures (Zheng *et al.*, 2002). Consequently, attainment of high-temperature oxygen isotope exchange equilibrium by garnet-omphacite mineral pairs indicates that Nd-isotopic compositions may have been homogenized under prograde metamorphic conditions. Preservation of high-temperature oxygen isotope fractionation despite possible retrograde alteration suggests the likelihood that Sm–Nd isotope equilibrium may have been preserved during exhumation from upper mantle depths. The analysis of both radiogenic and stable isotopes in the same samples of minerals and rocks holds the promise of providing more definitive criteria for the interpretation of geochronologic ages (Zheng *et al.*, 2002).

3.09.5.3 Comparative Radiogenic Geochemistry of Eclogites from UHP Metamorphic Belts

A comparison of the radiogenic geochemistry of classic UHP belts demonstrates the capabilities and ambiguities of distinguishing oceanic from continental protoliths.

(i) The most celebrated UHP rocks, jadeite-kyanite or garnet quartzites (also termed whiteschists) of Dora Maira, are shown to have negative $\varepsilon_{Nd}(t)$ values from -5 to -8 (Figure 8). This is consistent with their ultimate derivation from middle Proterozoic continental crust as estimated from their T_{DM} ages of 1.3–1.8 Ga (Tilton *et al.*, 1989, 1991). Except for these metasediments, the entire population of eclogites from the Alpine chain possesses positive $\varepsilon_{Nd}(T)$ values (Figure 8), suggesting that all of them had an origin as subducted oceanic crust, or directly from the depleted upper mantle such as the garnet peridotites of the Alpe Arami (Becker, 1993). This conclusion is largely supported by available geochemical and petrological data (Paquette *et al.*, 1989; Thöni and Jagoutz, 1992; Miller and Thöni, 1995; Dobrzhinestskaya *et al.*, 1996; von Quadt *et al.*, 1997).

(ii) In the Hercynian belt, most eclogites possess positive $\varepsilon_{Nd}(t)$ values, suggesting a dominantly oceanic derivation (Figure 8; Bernard-Griffiths and Cornichet, 1985; Bernard-Griffiths *et al.*, 1985; Stosch and Lugmair, 1990; Beard *et al.*, 1992, 1995; Medaris *et al.*, 1995). The few that exhibit negative $\varepsilon_{Nd}(t)$ values probably reflect limited crustal contamination (Figure 8; Stosch and Lugmair, 1990; Kalt *et al.*, 1994). A few eclogites from the Bohemian Massif also show negative $\varepsilon_{Nd}(t)$ values (-3 to -6, Medaris *et al.*, 1995), but these isotopic signatures were interpreted as due to the involvement of oceanic clay in the melting of subducted oceanic lithosphere, and the eclogites were considered to represent high-pressure cumulate of "enriched" basaltic liquid (Medaris *et al.*, 1995). In any case, the oceanic crust origin of the eclogites from the Hercynian belt is clearly demonstrated, and the few negative $\varepsilon_{Nd}(t)$ values do not imply derivation from deeply subducted continental crust.

(iii) Eclogites from the Caledonides have a much more complex history. Two eclogites from NW Scotland (Sanders *et al.*, 1984) and over half of the UHP rocks (including about equal proportions of eclogites and garnet pyroxenites (Griffin and Brueckner, 1980, 1985; Mork and Mearns, 1986; Mearns, 1986; Jamtveit *et al.*, 1991) have negative $\varepsilon_{Nd}(t)$ values (down to -10) (Figure 8). It appears that the Norwegian and Scottish eclogites have both oceanic and "continental" affinities. Although compositionally similar to mantle rocks, some, if not all, were emplaced into the continental crust in Precambrian time(s) and have been an integral part of the continental crust since then. This represents the first stage of evolution. Note that many eclogites have $^{147}Sm/^{144}Nd$ ratios lower than the chondritic values of 0.1967, suggesting that they were emplaced first as LREE-enriched continental basalts, later metamorphosed to amphibolites, and finally became an integral part of the continental crust (Figure 8). This further implies that an ancient crustal segment was subducted to mantle depths and UHP metamorphic assemblages were produced during the Caledonian orogeny. In WGR, eclogites recorded only the Caledonian thermal event of 410–440 Ma (Griffin and Brueckner, 1980, 1985; Mork and Mearns, 1986; Mearns, 1986), whereas garnet peridotites and garnet pyroxenites often recorded a Proterozoic event of $\sim$1.7 Ga (Jamtveit *et al.*, 1991). It is possible that the isotopic systems registering an age of 1.7 Ga in the garnet peridotites at mantle $P-T$ conditions have not been erased or homogenized during the Caledonian event. It is equally possible that these garnets were not isotopically equilibrated, hence that the ages of 1.7–1.0 Ga have no strict geological significance. Nevertheless, the newly formed garnets from basaltic protoliths appear to have faithfully recorded the Caledonian orogeny at 410–400 Ma. Thus, the evolution of the European Caledonides is clearly different from that of the Alpine and Hercynian chains of western Europe.

(iv) The Dabie and Su-Lu terranes have a diagnostic "continental" isotopic signature. Eclogites occurring as enclaves or blocks in granitic gneisses have low $\varepsilon_{Nd}(t)$ values (-6 to -20) at the time of peak metamorphism at $\approx$220–230 Ma (Figure 9). These values are among the lowest ever recorded in eclogites and are significantly different from those of eclogites from the Alpine and Hercynian chains in western Europe. Moreover, most rocks have $^{147}Sm/^{144}Nd$ ratios lower than the chondritic value, reflecting their LREE-enriched geochemical characteristics. The isotopic data and geochemical characteristics indicate that the eclogite protoliths of the Dabie

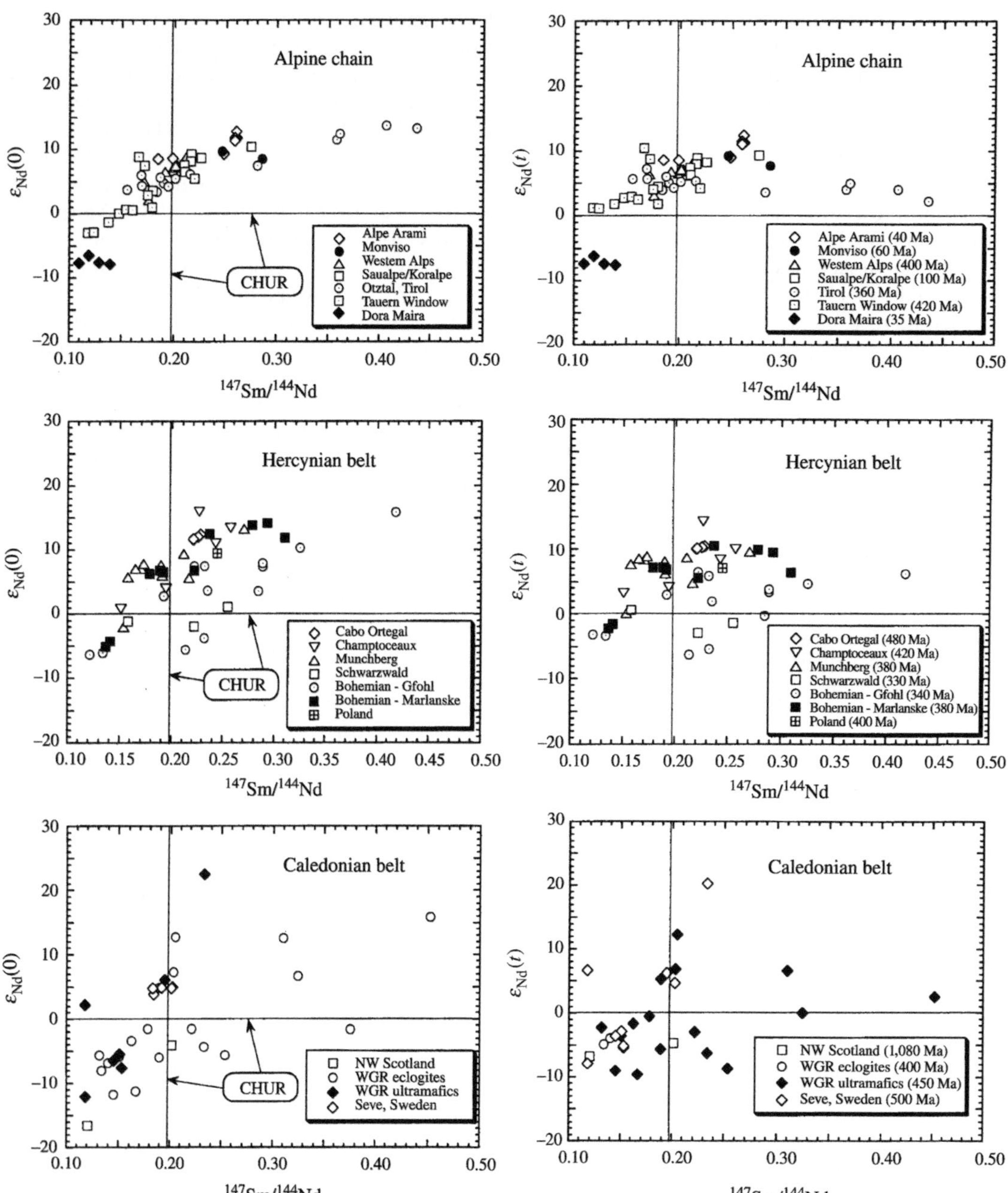

Figure 8 Plots of $\varepsilon_{Nd}(0)$ versus $^{147}Sm/^{144}Nd$ (left diagrams) and $\varepsilon_{Nd}(t)$ versus $^{147}Sm/^{144}Nd$ (right diagrams) for eclogites and garnet peridotities from the Alpine, Hercynian, and Caledonian belts. Data source: see Appendices of Jahn (1999). For the Alpine belt, except the UHP metasediments of Dora Maira, all the rocks show positive $\varepsilon_{Nd}(t)$ values. Most eclogites and garnet peridotites from the Hercynian belt show positive $\varepsilon_{Nd}(t)$ values, suggesting their oceanic affinity. Those having negative values have been interpreted as due to crustal contamination by reinjection of crustal rocks (oceanic clays) into the upper mantle. As for the eclogites and garnet peridotites from the European Caledonide, the majority of the rocks show negative $\varepsilon_{Nd}(t)$ values, in strong contrast to those of the Alpine and Hercynian chains. These rocks appear to have resided in the continental crust for a lengthy period of time before their UHP metamorphism.

Orogen have a "continental" affinity and could not have been produced from a subducted Tethyan oceanic slab. The very low $\varepsilon_{Nd}(t)$ values further require the protoliths of these eclogites to have formed long before the Triassic collision, probably in mid-Proterozoic times. It appears that, in addition to the UHP metasedimentary rocks of Dora Maira, Dabieshan (including Su-Lu) and WGR are two regions where coesite- and diamond-bearing eclogites and ultramafic rocks show clear evidence for the subduction of ancient and cold continental crustal blocks to mantle depths.

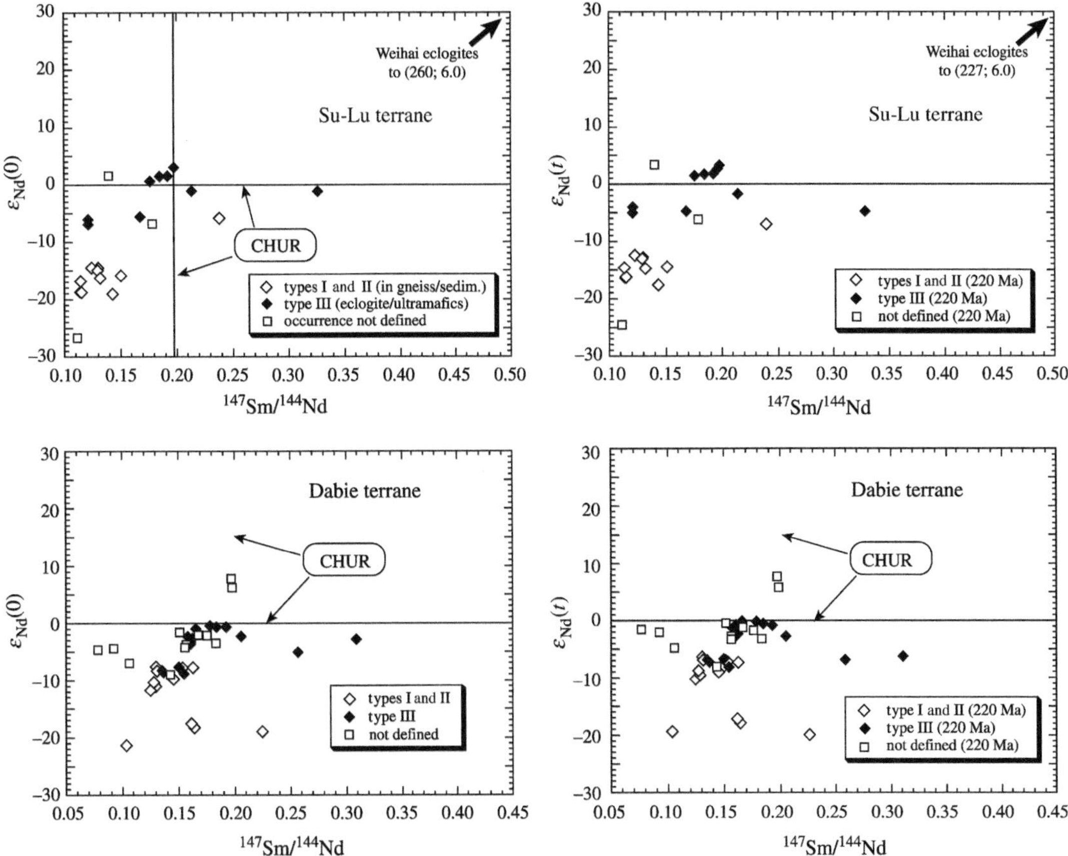

Figure 9 $\varepsilon_{Nd}(0)$ versus $^{147}Sm/^{144}Nd$ (left diagrams) and $\varepsilon_{Nd}(t)$ versus $^{147}Sm/^{144}Nd$ plots (right diagrams) for eclogites from the Su-Lu and Dabie terranes of China. All types I and II eclogites (gneiss and marble-hosted) are characterized by negative $\varepsilon_{Nd}(t)$ values, suggesting their continental affinity. Type III eclogites are associated with ultramafic complexes. The data of the Weihai eclogites are out of the scale, and their ε_{Nd} and $^{147}Sm/^{144}Nd$ ratios are given in parentheses. Data sources: see appendices of Jahn (1999).

The eclogite-ultramafic suites in the Su-Lu and Dabie terranes have different tectonic origins. The first suite, represented by the Bixiling and Maowu complexes, consists of layered intrusions initially emplaced at crustal levels and later subjected to UHP metamorphism as a result of continental subduction (Okay, 1994; Zhang *et al.*, 1998; Chavagnac and Jahn, 1996; Fan *et al.*, 1996; Liou and Zhang, 1998). The second suite consists of mantle rocks exhumed together with crustal UHP metamorphic rocks; they are considered here as "tectonic enclaves of mantle origin" within granitic gneisses. A probable example are the garnet pyroxenites of Rizhao. Even though these pyroxenites represent igneous cumulate rocks from a mantle-derived liquid, the magmatic differentiation probably occurred within the mantle. The isotopic systematics of the Maowu layered intrusion (Figure 9; $\varepsilon_{Nd}(t) \approx -5$ to -6, $I_{sr} \approx 0.707$–0.708; Jahn *et al.*, in press) differ from those of the Bixiling complex. Upper crustal contamination during magma chamber processes is likely to have occurred in the Maowu intrusion (Jahn *et al.*, in press), whereas the Bixiling complex has undergone an assimilation-fractional-crystallization (AFC) process in the lower crust (Chavaganc and Jahn, 1996).

In conclusion, basaltic (or gabbroic) eclogites of continental origin are best represented by the Dabie and Su-Lu terranes in east-central China and by the WGR. Their identification provides strong evidence for subduction of large blocks of continental crust. On the other hand, most eclogites from the Alps and Hercynian belts are of oceanic origin and many have a MORB-like affinity. Their occurrence is not sufficient to indicate that these eclogite-bearing gneiss terranes have also been subducted to mantle depths, unless it can be proven that the oceanic rocks were tectonically emplaced in a continental setting, and that the ensemble was subducted and metamorphosed at great depths.

3.09.6 THE FLUID PHASE OF UHP METAMORPHISM

Fluids in HP and UHP metamorphic rocks are known to exist as intergranular phases, and, must have been presented to facilitate the attainment of mm- to cm-scale cation and stable isotope exchange equilibrium between adjacent mineral grains during prograde metamorphism (Philippot and Rumble, 2000; Scambelluri and Philippot, 2001). Discussion of HP together with UHP fluid–rock interactions is included in this section in order to provide a more complete review of the evidence of fluid behavior during continental collision and subduction. Intergranular fluid phases progressively expelled throughout subduction and metamorphism are not present in sufficient amount, however, to result in the complete obliteration of UHP index minerals and mineral assemblages by retrograde metamorphism during exhumation. Micron-diameter fluid inclusions in matrix and porphyroblastic minerals provide samples of fluids from UHP (De Corte *et al.*, 1998; Fu *et al.*, 2001, 2002; Xiao *et al.*, 2000, 2002) and HP (Nadeau *et al.*, 1993; Giaramita and Sorensen, 1994; Philippot, 1993, 1996; Philippot and Selverstone, 1991; Philippot *et al.*, 1995, 1998; Scambelluri *et al.*, 1997, 1998) metamorphic belts. The composition of the inclusion fluids ranges from concentrated brines to fresh, meteoric water. The oldest generations of fluid inclusions may represent peak metamorphic fluids, and, possibly, record premetamorphic fluids.

3.09.6.1 Fluid Immobility during Prograde Metamorphism

Owing to the diversity of protoliths undergoing subduction, there is ample textural, chemical, and isotopic evidence of fluid interactions during HP and UHP metamorphism. Fluid mobility is limited during prograde metamorphism, as indicated by the failure of rocks to attain textural equilibrium. Preservation of gabbroic textures and primary igneous minerals including plagioclase, orthopyroxene, and biotite at the core of a 3-m coesite-bearing eclogite block in Su-Lu (Zhang and Liou, 1997) and the occurrence of intergranular coesite (Liou and Zhang, 1996; Ye *et al.*, 1996) suggest a lack of fluids to facilitate equilibration. Similarly, Oberhansli *et al.* (2002) document cover–basement relationships in the Dabie UHP terrane, which in part preserve a primary unconformity, pillow structures, and mineral fabrics. Of particular interest is the variable P–T overprint of these units, interpreted as due to incomplete or absent mineral reactions.

Additional evidence of limited fluid mobility during prograde HP and UHP metamorphism is seen in the failure of dissimilar protoliths to achieve stable isotope equilibrium on mm-, cm-, and m-scales despite extreme metamorphic conditions (Sharp *et al.*, 1993; Rumble and Yui, 1998; Baker *et al.*, 1997; Zheng *et al.*, 1998a,b 1999, 2003; Fu *et al.*, 1999; Philippot and Selverstone, 1991; Selverstone *et al.*, 1992; Getty and Selverstone, 1994; Nadeau *et al.*, 1993; Fruh-Green, 1994; Barnicoat and Cartwright, 1997; Miller *et al.*, 2001). The failure to equilibrate is a boon to those seeking to unravel the geological history of HP and UHP rocks, because remnants of premetamorphic stable isotopic signatures diagnostic of protolith origins may persist. Recognition that vestiges of distinctive Earth surface environments and processes have been preserved in HP and UHP metamorphic rocks, despite subduction into the upper mantle, remains a source of astonishment but, more importantly, greatly facilitates the interpretation of the geodynamic cycle of continental collisions (Rumble, 1998). In the European Alps, stable isotopic evidence of ocean floor and ophiolitic hydrothermal alteration has been observed in HP eclogites (Philippot *et al.*, 1998; Barnicoat and Cartwright, 1997; Miller *et al.*, 2001; Scambelluri and Philippot, 2001). A record of Neoproterozoic surface conditions has been found in Chinese UHP rocks including low $\delta^{18}O$ and δD values indicating a cold climate (Yui *et al.*, 1995; Zheng *et al.*, 1996, 1998b; Baker *et al.*, 1997; Rumble and Yui, 1998; Fu *et al.*, 1999). Dabieshan marbles carry the high $\delta^{13}C$ values of carbonates that are typical of the Neoproterozoic (Yui *et al.*, 1997; Baker *et al.*, 1997; Zheng *et al.*, 1998a; Rumble *et al.*, 2000; Shen, 2002). New U–Pb dating and $\delta^{18}O$ analyses of zircons from orthogneisses show that a cold-climate, geothermal system covering hundreds of square kilometers existed along the rifted northern margin of the Yangtze craton during the Neoproterozoic, consistent with snowball Earth conditions (Rumble *et al.*, 2002). Limits on fluid mobility are indicated not only by stable isotope data but also by fluid inclusion studies and argon isotope geochemistry. Low salinity, primary fluid inclusions reported from low $\delta^{18}O$ eclogites at Qinglongshan, China, may be samples of Neoproterozoic meteoric water that survived subduction and exhumation (Fu, 2002). The presence of excess argon in UHP eclogites from Qinglongshan, China, correlates mineralogically with early-formed phengite and epidote and later barroisite (Li *et al.*, 1994; Giorgis *et al.*, 2000). Furthermore, excess argon is more prevalent in low $\delta^{18}O$ eclogites than in less anomalous eclogites, suggesting argon inheritance and retention throughout subduction, metamorphism, and exhumation (D. Giorgis, 2001, personal communication).

The juxtaposition of textural and stable isotope evidence of limited fluid mobility during HP and UHP metamorphism with mineralogical evidence of dehydration and decarbonation, forces one to ask how devolatilization was accomplished. Channelized fluid flow has been found in a number of studies in well-exposed regions such as the WGR (Austrheim, 1987; Jamtveit *et al.*, 1989; Mattey *et al.*, 1994; Van Wick *et al.*, 1996). Extensive pseudotachylite veins may record the initiation by metamorphic devolatilization of earthquake faulting and consequent expulsion of fluids (Austrheim and Boundy, 1994, 1996; cf. Kirby *et al.*, 1996). Localization of subduction related deformation along serpentinized mylonite shear zones records fluid flow while favoring preservation of pre-subduction mantle and low temperature (oceanic) alteration assemblages within undeformed metaperidotite (Fruh-Green *et al.*, 2001). Regional-scale fluid mobility as well as mechanical mixing is seen in mélange units of the HP Franciscan Schists, Catalina Island, California (Bebout, 1991; Bebout and Barton, 2002). Other writers have commented on a discrepancy between mass-balance calculations of voluminous devolatilization mineral reactions and the absence of outcrop evidence of fluid flow (Barnicoat and Cartwright, 1997; Miller *et al.*, 2001). An appeal is made to channelized fluid pathways along bedding surfaces, unconformities, and shear zones to resolve the discrepancy, but in many regions specific fluid flow aquifers remain unmapped. Rock rheology, deformation history, reaction-enhanced, and deformation-enhanced permeability clearly must play an important role in fluid mobility. Equally clearly, there is a need for additional field study of the pathways by which fluids leave the subducting slab during HP and UHP metamorphism (see also Chapter 3.06).

3.09.6.2 Slab–Fluid–Mantle Interactions

The foregoing examples of limited fluid mobility during HP and UHP metamorphism caused by subduction raise questions about interactions between slab fluids and the mantle during continental collision. How deeply into the mantle can mineral-bound water be carried by a subducting slab? Over what depth interval is water transferred from slab to mantle? By what mechanism is the transfer of water into the mantle accomplished? Complete answers to these questions cannot as yet be given.

Field observations show that assemblages containing hydrous and carbonate minerals persist in coesite- and diamond-eclogite facies rocks in Dabieshan (Liou *et al.*, 1995), Su-Lu (Enami and Zang, 1988; Hirajima *et al.*, 1992; Yang *et al.*, 1993; Zhang *et al.*, 1995a, 2002a,b), Erzgebirge (Massonne, 1999; Hwang *et al.*, 2001a,b; Stockhert *et al.*, 2001) and Kokchetav (Sobolev and Shatsky, 1990; Shatsky *et al.*, 1995; Zhang *et al.*, 1997; Ogasawara *et al.*, 2000; Okamoto *et al.*, 2000), equivalent to depths of 130 km, or deeper (Hwang *et al.*, 2000). Recent detailed studies of UHP belts reveal rare occurrences of hydroxyl in clinopyroxene of concentrations up to 3,000 ppm (Katayama and Nakashima, 2003) and OH-topaz (Zhang *et al.*, 2002a) whose $P-T$ stability limits would permit water transport to greater depths. Experimental determination of the stability of hydrous minerals in mineral assemblages of bulk compositions appropriate to oceanic crust as well as metasediments of continental origin show that water may remain stably bound in minerals to depths of 300 km or more (Schmidt and Poli, 1998; Schreyer, 1995).

The evidence of fluid immobility in slabs under UHP conditions is inconsistent with a massive release of fluid and its transfer to the mantle in a single, cataclysmic episode. Fluid release could not have been simultaneous over such a wide range of rock types, because local variations in bulk composition would have controlled the timing of devolatilization reactions in response to changing $P-T$ conditions. The overall devolatilization history is determined by specific subduction $P-T$ trajectories and their intersection with the stability limits of local mineral assemblages. Local fluid compositions and amounts of evolved fluids changed episodically as different mineral assemblages reacted in response to changing $P-T$ conditions. Fluid evolution from the slab and transfer into the mantle was distributed over a significant depth interval, small in amount at any given instant, but with a substantial fluid : rock ratio integrated over the duration of devolatilization (Philippot and Rumble, 2000). Existing outcrop-scale studies inevitably emphasize local effects, whereas estimating the overall devolatilization budget of a subducting slab will require regional-scale investigations.

Investigating the transfer of slab fluids into the mantle is even more difficult than estimating the integrated devolatilization mass balance of slab fluids. It might be thought that the contact relationships of garnet peridotite slices interleaved with metasediments in UHP metamorphic terranes would record the mechanisms of interaction between slab fluids and mantle during continental collision. Study of peridotites in UHP metamorphic belts shows two groups of rocks: one of pre-subduction age that experienced crustal contamination prior to UHP metamorphism and one that was emplaced in UHP metasediments during subduction (Zhang *et al.*, 2000). Some of the latter group (type A-2; Zhang *et al.*, 2000) retain mantle Nd-, Sr-, and O-isotopic compositions and show little evidence of interaction with slab fluids. Other peridotites of type A have low Sm/Nd ratio (0.2), low $\varepsilon_{Nd}(t)$ of -5, and high $^{87}Sr/^{86}Sr$ (0.708) and may have been metasomatized by crustal fluids derived from the descending slab (Zhang *et al.*, 2000). Searches for additional exposures for study of contacts between mantle peridotites and UHP metasediments should be made and geochemical investigations undertaken to strengthen our present understanding. Innovative experiments on the connectivity and composition of fluids in eclogitic and mantle mineral assemblages under UHP and mantle $P-T$ conditions promise to shed light on possible mechanisms of slab–fluid–mantle interaction (Mibe *et al.*, 1998, 1999, 2002a,b; Yoshino *et al.*, 2002). These experiments present data on the interconnectivity of a grain boundary fluid phase, percolation along grain boundaries, and will lead to improved understanding of diffusion through a connected grain boundary fluid phase.

3.09.6.3 Compositions of Fluids

New data from both field and experimental studies suggest that high density, super-critical, silicate-rich fluids may play a role in UHP metamorphism. Such fluids are transitional between hydrothermal solutions and silicate melts. They contain higher solute concentrations than aqueous solutions, but higher water contents than silicate magmas (Scambelluri *et al.*, 2001). The fluids would have an enhanced capacity to transport material within a subducting slab and from a slab into an overlying mantle wedge (see discussions by Philippot and Selverstone, 1991; Philippot, 1993; Philippot and Rumble, 2000; Scambelluri and Philippot, 2001).

The discovery of microdiamonds possibly co-precipitated with silicate and oxide minerals as daughter crystals from C–O–H-silicate fluids focuses new attention on these discussions. Microdiamonds included in garnet and zircon from garnet-biotite gneiss of the Kokchetav Massif, Kazakhstan, contain C–O–H–N fluid inclusions (De Corte *et al.*, 1998) and nanometric inclusions with stoichiometries of $MgCO_3$, TiO_2, $BaSO_4$, ThO_2, and Cr_2O_3 (Dobrzhinetskaya *et al.*, 2001, 2003). Microdiamonds included in garnet porphyroblasts of a quartz–phengite–plagioclase–garnet gneiss from the Erzgebirge, Germany (Massonne, 1999) occur in polyphase assemblages that may include phlogopite, quartz, paragonite, phengite, kyanite, apatite, rutile, and sodic plagioclase (Hwang *et al.*, 2001a,b; Stockhert *et al.*, 2001). A recent finding of microdiamond together with monazite, phlogopite, spinel, and $(Ba,Ca,Mg)CO_3$ as multiphase inclusions in spinel in a megacrystic garnet-websterite pod from the WGR provides additional evidence of diamond growth from infiltrating fluids during deep continental subduction (van Roermund *et al.*, 2002). Experiments show diamond synthesis within the diamond stability field in the presence of a $CO_2–H_2O$ fluid at H_2O-rich compositions (Akaishi *et al.*, 2001). A sandwich experiment using interlayered carbonate and pelite starting materials by Hermann and Green (2001) indicates that diamond could be produced when $CO_2–H_2O–K_2O–CaO$ fluids from carbonates interact with pelites at UHP conditions. Clearly, these novel fluids under extreme $P-T$ conditions present a compelling target for more detailed study.

3.09.7 CONCLUSIONS

Understanding of orogenic belts with UHP metamorphic conditions has advanced greatly in the two decades since their discovery in 1984. Coesite- or diamond-eclogite facies rocks are known from every continent except Australia and North America (Figure 1). Ages of UHP metamorphism range from the Neoproterozoic to the Miocene. Realistic physical mechanisms of continental collision, subduction, and exhumation have been proposed. Some long-standing debates concerning outcrop relationships such as that of the *in situ* versus exotic origin of UHP metamophism appear to be progressing towards a consensus within the research community. Identification of protoliths grows ever more exacting as researchers employ a wide range of geochemical methods. Radiometric ages are being refined by the application of a variety of radiogenic isotope systems. Examples discussed in this chapter emphasize that definitive research results are far more likely to be forthcoming in studies that break new ground in using a multidisciplinary approach, that combines a battery of geochemical and petrological methods.

A number of opportunities for exciting and fruitful research have been identified in this review. The study of mineralogical indicators of "super" UHP metamorphism is likely to continue to be productive as more sophisticated methods of identification and imaging are employed. Application of a variety of geochemical tests to the same rock and mineral samples will yield more definitive protolith signatures and will improve the interpretation of geochronological data. Investigating the role of supercritical fluids in the growth of microdiamond crystals should increase our understanding of material transport under mantle conditions. Finally, a synthesis of geochemical data emerging from studies of the subduction of oceanic crust will lead to a better understanding of the impact of continental collisions on mantle geochemistry.

ACKNOWLEDGMENTS

We are grateful to Roberta Rudnick for her advice, which surely improved this review. Preparation of the review was supported in part by the National Science Foundation, EAR Continental Dynamics program, EAR-0003276 (Rumble) and EAR-0003355 (Liou).

REFERENCES

Akaishi M., Kumar M. D. S., and Yamaoka S. (2001) Crystallization of diamond from graphite or carbonaceous materials in the presence of C–O–H fluids at high pressure and high temperature of 7.7 GPa and 1,300–2,000 deg C. In *Fluid/Slab/Mantle Interactions and Ultrahigh-P Minerals* (eds. Y. Ogasawara, S. Maruyama, and J. G. Liou). Waseda University, Tokyo, Japan, pp. 26–30.

Ames L., Zhou G., and Xiong B. (1996) Geochronology and isotopic character of ultrahigh-pressure metamorphism with implications for collision of the Sino-Korean and Yangtze cratons, central China. *Tectonics* **15**, 472–489.

Austrheim H. (1987) Eclogitization of lower crustal granulites by fluid migration through shear zones. *Earth Planet. Sci. Lett.* **81**, 221–232.

Austrheim H. and Boundy T. M. (1994) Pseudotachylytes generated during seismic faulting and eclogitization of the deep crust. *Science* **265**, 82–83.

Austrheim H. and Boundy T. M. (1996) Garnets recording deep crustal earthquakes. *Earth Planet. Sci. Lett.* **139**, 223–238.

Ayers J. C., Dunkle S., Gao S., and Miller C. F. (2002) Constraints on timing of peak and retrograde metamorphism in the Dabie Shan ultrahigh-pressure metamorphic belt, east-central China, using U–Th–Pb dating of zircon and monazite. *Chem. Geol.* **186**, 315–331.

Baker J., Matthews A., Mattey D., Rowley D. B., and Xue F. (1997) Fluid–rock interaction during ultra-high pressure metamorphism, Dabie Shan, China. *Geochim. Cosmochim. Acta* **61**, 1685–1696.

Barnicoat A. C. and Cartwright I. (1997) Focused fluid flow during subduction: oxygen isotope data from high pressure ophiolites of the western Alps. *Earth Planet. Sci. Lett.* **132**, 53–61.

Barth M. G., Rudnick R. L., Horn I., McDonough W. F., Spicuzza M. J., Valley J. W., and Haggerty S. E. (2002) Geochemistry of xenolithic eclogites from West Africa: Part 2. Origin of high MgO eclogites. *Geochim. Cosmochim. Acta* **66**, 4325–4345.

Beane R. J., Liou J. G., Coleman R. G., and Leech M. L. (1995) Petrology and retrograde *P–T* path for eclogites of the Maksyutov Complex, southern Ural Mountains, Russia. *Island Arc* **4**, 254–266.

Beard B. L., Medaris L. G., Johnson C. M., Brueckner H. K., and Misar Z. (1992) Petrogenesis of Variscan high-temperature Group A eclogites from the Moldanubian zone of the Bohemian Massif, Czechoslovakia. *Contrib. Mineral. Petrol.* **111**, 468–483.

Beard B. L., Medaris L. G., Johnson C. M., Jelinek E., Tonika J., and Riciputi L. R. (1995) Geochronology and geochemistry of eclogites from the Marianske Lazne Complex, Czech Republic: implications for Variscan orogenesis. *Geol. Rundsch.* **84**, 552–567.

Bebout G. E. (1991) Geometry and mechanisms of fluid flow at 15 to 45 kilometer depths in an early Cretaceous accretionary complex. *Geophys. Res. Lett.* **18**, 923–926.

Bebout G. E. and Barton M. D. (2002) Tectonic and metasomatic mixing in a high-T, subduction-zone melange: insights into the geochemical evolution of the slab-mantle interface. *Chem. Geol.* **187**, 79–106.

Bebout G. E., Scholl D. W., Kirby S. H., and Platt J. P. (eds.) (1996) *Subduction—Top to Bottom*. American Geophysical Union, Washington.

Becker H. (1993) Garnet peridotite and eclogite Sm–Nd mineral ages from the Lepontine dome (Swiss Alps): new evidence for Eocene high-pressure metamorphism in the central Alps. *Geology* **21**, 599–602.

Bernard-Griffiths J. and Cornichet J. (1985) Origin of eclogites from South Brittany, France: a Sm–Nd isotopic and REE study. *Chem. Geol.* **52**, 185–201.

Bernard-Griffiths J., Peucat J. J., Cornichet J., Ponce de Leon M. I., and Ibarguchi J. I. G. (1985) U–Pb, Nd isotopic and REE geochemistry in eclogites from the Cabo Ortegal Complex, Galicia, Spain: an example of REE immobility conserving MORB-like patterns during high-grade metamorphism. *Chem. Geol.* **52**, 217–225.

Bocchio R., De Capitan L., Ottolini L., and Cella F. (2000) Trace element distribution in eclogites and their clinopyroxene/garnet pair: a case study from Soazza (Switzerland). *Euro. J. Mineral.* **12**, 147–161.

Bozhilov K. N., Green H. W., and Dobrzhinetskaya L. F. (1999) Clinoenstatite in Alpi Arami peridotite: additional evidence of very high pressure. *Science* **284**, 128–132.

Burton K. W., Kohn M. G., Cohen A. S., and O'Nions R. K. (1995) The relative diffusion of Pb, Nd, Sr and O in garnet. *Earth Planet. Sci. Lett.* **133**, 199–211.

Caby R. (1994) Precambrian coesite from N. Mali: first record and implications for plate tectonics in the trans-Saharan segment of the Pan-African belt. *Euro. J. Mineral.* **6**, 235–244.

Carswell D. A. (ed.) (1990) *Eclogite Facies Rocks*. Chapman and Hall, New York.

Carswell D. A. (ed.) (2000) Ultra-high pressure metamorphic rocks. *Lithos* **52**, 1–278.

Carswell D. A., Wilson R. N., and Zhai M. (2000) Metamorphic evolution, mineral chemistry, and thermobarometry of schists and orthogneisses hosting ultra-high pressure eclogites in the Dabieshan of central China. *Lithos* **52**, 121–155.

Carswell D. A., Brueckner H. K., Cuthbert S. J., Mehta K., and O'Brien P. J. (2003) The timing of stabilization and the exhumation rate for ultra-high pressure rocks in the Western Gneiss region of Norway. *J. Metamorph. Geol.* (in press).

Chavagnac V. and Jahn B. M. (1996) Coesite-bearing eclogites from the Bixiling Complex, Dabie Mtns., China: Sm–Nd ages, geochemical characteristics, and tectonic implications. *Chem. Geol.* **133**, 29–51.

Chavagnac V., Jahn B. M., Villa I., Whitehouse M. J., and Liu D. (2001) Multichronometric evidence for an *in-situ* origin of the ultrahigh-pressure metamorphic terrane of Dabieshan, China. *J. Geol.* **109**, 633–646.

Chemenda A. I., Mattauer M., Malavieille J., and Bokun A. N. (1995) A mechanism for syn-collisional rock exhumation and associated normal faulting; results from physical modelling. *Earth Planet. Sci. Lett.* **132**, 225–232.

Chen B., Jahn B. M., and Wei C. (2002) Petrogenesis of Mesozoic granitoids in the Dabie UHP complex, Central China: trace element and Nd–Sr isotope evidence. *Lithos* **60**, 67–88.

Chesnokov B. V. and Popov V. A. (1965) Increasing volume of quartz grains in eclogites of the South Urals. *Dokl. Akad. Nauk. SSSR* **62**, 909–910.

Chopin C. (1984) Coesite and pure pyrope in high-grade blueschists of the Western Alps: a first record and some consequences. *Contrib. Mineral. Petrol.* **86**, 107–118.

Chopin C. (2003) Ultrahigh-pressure metamorphism: tracing continental crust into the mantle. *Earth Planet Sci. Lett.* **212**, 1–14.

Chopin C. and Sobolev N. V. (1995) Principal mineralogic indicators of UHP in crustal rocks. In *Ultrahigh Pressure Metamorphism* (eds. R. G. Coleman and X. Wang). Cambridge University Press, Cambridge, pp. 96–131.

Chopin C., Henry C., and Michard A. (1991) Geology and petrology of the coesite-bearing terrane, Dora-Maira massif, Western Alps. *Euro. J. Mineral.* **3**, 263–291.

Claoue-Long J. C., Sobolev N. V., Shatsky V. S., and Sobolev A. V. (1991) Zircon response to diamond-pressure metamorphism in the Kokchetav massif. *Geology* **19**, 710–713.

Coleman R. G. and Wang X. (1995a) Overview of the geology and tectonics of UHPM. In *Ultrahigh Pressure Metamorphism* (eds. R. G. Coleman and X. Wang). Cambridge University Press, Cambridge, pp. 1–32.

Coleman R. G. and Wang X. (eds.) (1995b) *Ultrahigh Pressure Metamorphism*. Cambridge Univeristy Press, Cambridge.

Compagnoni R., Hirajima T., and Chopin C. (1995) Ultra-high-pressure metamorphic rocks in the Western Alps. In *Ultrahigh Pressure Metamorphism* (eds. R. G. Coleman and X. Wang). Cambridge University Press, Cambridge, pp. 206–243.

Cong B. (ed.) (1996) *Ultrahigh-pressure Metamorphc Rocks in the Dabieshan-Su-Lu Region of China* Kluwer Academic, Dordrecht.

Cong B. and Wang Q. (1996) A review on researches of UHPM rocks in the Dabieshan-Su-Lu region. In *Ultrahigh-pressure Metamorphic Rocks in the Dabieshan-Su-Lu Region of China* (ed. B. Cong). Science Press; Kluwer Academic, Dordrecht, pp. 1–7.

Cong B., Zhai M., Carswell D. A., Wilson R. N., Wang Q., Zhao Z., and Windley B. F. (1995) Petrogenesis of UHP rocks and their country rocks at Shuanghe in Dabie Shan, central China. *Euro. J. Mineral.* **7**, 119–138.

Cuthbert S. J., Carswell D. A., Krogh-Ravna E. J., and Wain A. (2000) Eclogites and eclogites in the Western Gneiss Region, Norwegian Caledonides. *Lithos* **52**, 165–195.

De Corte K., Cartigny P., Shatsky V. S., Sobolev N. V., and Javoy M. (1998) Evidence of fluid inclusions in metamorphic microdiamonds from the Kokchetav massif, northern Kazakhstan. *Geochim. Cosmochim. Acta* **62**, 3765–3773.

de Sigoyer J., Chavagnac V., Blichert-Toft J., Villa I., Luais B., Guillot S., Cosca M., and Mascle G. (2000) Dating the Indian continental subduction and collisional thickening in the northwest Himalaya: multichronology of the Tso Morari eclogites. *Geology* **28**, 487–490.

Dobrzhinetskaya L. F., Eide E. A., Larsen R. B., Sturt B. A., Tronnes R. G., Smith D. C., Taylor W. R., and Posukhova T. V. (1995) Microdiamond in high-grade metamorphic rocks of the Western Gneiss region, Norway. *Geology* **23**, 597–600.

Dobrzhinetskaya L. F., Green H. W., and Wang S. (1996) Alpe Arami: a peridotite massif from depths more than 300 km. *Science* **271**, 1841–1845.

Dobrzhinetskaya L. F., Green H. W., Mitchell T. E., and Dickerson R. M. (2001) Metamorphic diamonds: mechanism of growth and inclusion of oxides. *Geology* **29**, 263–266.

Dobrzhinetskaya L. F., Green H. W., Bozhilov K. N., Mitchell T. E., and Dickerson R. M. (2003) Crystallization environment of Kazakhstan microdiamond evidence from manometric inclusions and mineral associations. *J. Metamorph. Geol.* **21**, 425–438.

Dodson M. H. (1973) Closure temperature in cooling geochronological and petrological systems. *Contrib. Mineral. Petrol.* **40**, 259–274.

Duchêne S., Blichert-Toft J., Luais B., Télouk P., Lardeaux J. M., and Albarède F. (1997) The Lu–Hf dating of garnets and the ages of the Alpine high-pressure metamorphism. *Nature* **387**, 586–589.

Enami M. and Zang Q. (1988) Magnesian staurolite in garnet corundum rocks and eclogite from the Donghai district, Jiangsu province, E. China. *Am. Mineral.* **73**, 48–56.

Enami M., Zang Q., and Yin Y. (1993) High-pressure eclogites in northern Jiangsu–southern Shangdong province, eastern China. *J. Metamorph. Geol.* **11**, 589–603.

Ernst W. G. and Liou J. G. (eds.) (2000) *Ultrahigh-pressure Metamorphism and Geodynamics in Collision-type Orogenic Belts.* Geological Society of America, Denver.

Ernst W. G. and Peacock S. M. (1996) A thermotectonic model for preservation of ultrahigh-pressure phases in metamorphosed continental crust. In *Subduction Top to Bottom* (eds. G. E. Bebout, D. W. Scholl, S. H. Kirby, and J. P. Platt). American Geophysical Union, Washignton, pp. 171–178.

Ernst W. G., Maruyama S., and Wallis S. R. (1997) Buoyancy-driven, rapid exhumation of ultrahigh-pressure metamorphosed continental crust. *Proc. Natl. Acad. Sci. USA* **94**, 9532–9537.

Fan Q., Liu R., Ma B., Zhao D., and Zhang Q. (1996) Protolith and ultrahigh-pressure metamorphism of Maowu mafic–ultramafic rocks, Dabieshan. *Acta Petrol. Sin.* **12**, 29–47.

Fruh-Green G. L. (1994) Interdependence of deformation, fluid infiltration, and reaction progress recorded in eclogitic metagranitoids (Sesia Zone, western Alps). *J. Metamorph. Geol.* **12**, 327–343.

Fruh-Green G. L., Scambelluri M., and Vallis F. (2001) O–H isotope ratios of high pressure ultramafic rocks: implications for fluid sources and mobility in the subducted hydrous mantle. *Contrib. Mineral. Petrol.* **141**, 145–159.

Fu B. (2002) Fluid regime during high- and ultrahigh-pressure metamorphism in the Dabie–Su-Lu terranes, Eastern China. PhD, Vrije Universiteit, Amsterdam, pp. 1–144.

Fu B., Zheng Y.-F., Wang Z., Xiao Y., Gong B., and Li S. (1999) Oxygen and hydrogen isotope geochemistry of gneisses associated with ultrahigh pressure eclogites at Shuanghe in the Dabie Mountains. *Contrib. Mineral. Petrol.* **134**, 52–66.

Fu B., Touret J. L. R., and Zheng Y.-F. (2001) Fluid inclusions in coesite-bearing eclogites and jadeite quartzite at Shuanghe, Dabie Shan, China. *J. Metamorph. Geol.* **19**, 529–545.

Fu B., Zheng Y.-F., and Touret J. L. R. (2002) Petrological, isotopic and fluid inclusion studies of eclogites from Sujiahe, NW Dabie Shan (China). *Chem. Geol.* **187**, 107–128.

Gebauer, D. (1996) A $P-T-t$ path for an (ultra ?-)high pressure ultramafic/mafic rock association and its felsic country rocks based on SHRIMP dating of magmatic and metamorphic zircon domains. Example: Alpe Arami (Central Swiss Alps). *Earth Processes: Reading the Isotopic Code. Geophysical Monograph* **95**, 307–329.

Gebauer D., Schertl H. P., Brix M., and Schreyer W. (1997) 35 Ma old ultrahigh-pressure metamorphism and evidence for very rapid exhumation in the Dora Maira Massif, western Alps. *Lithos* **41**, 5–24.

Getty S. R. and Selverstone J. (1994) Stable isotope and trace element evidence for restricted fluid migration in 2 GPa eclogites. *J. Metamorph. Geol.* **12**, 747–760.

Ghiribelli B., Frezzotti M.-L., and Palmeri R. (2002) Coesite in eclogites of the Lanterman range (Antarctica): evidence from textural and Raman studies. *Euro. J. Mineral.* **14**, 355–360.

Giaramita M. J. and Sorensen S. S. (1994) Primary fluids in low-temperature eclogites: evidence from two subduction complexes (Dominican Republic and California, USA). *Contrib. Mineral. Petrol.* **117**, 279–292.

Gilotti J. A. and Krogh-Ravna E. J. (2002) First evidence for ultrahigh-pressure metamorphism in the Northeast Greenland Caledonides. *Geology* **30**, 551–554.

Giorgis D., Cosca M., and Li S. (2000) Distribution and significance of extraneous argon in UHP eclogite (Su-Lu terrain, China): insight from *in situ* $^{40}Ar/^{39}Ar$ UV-laser ablation analysis. *Earth Planet. Sci. Lett.* **181**, 605–615.

Godard G. (2001) Eclogites and their geodynamic interpretation: a history. *J. Geodynam.* **32**, 165–203.

Green H. W., Dobrzhinetskaya L. F., and Bozhilov K. N. (2000) Mineralogical and experimental evidence for very deep exhumation from subduction zones. *J. Geodynam.* **30**, 61–76.

Griffin W. L. and Brueckner H. K. (1980) Caledonian Sm–Nd ages and a crustal origin for Norwegian eclogites. *Nature* **285**, 319–321.

Griffin W. L. and Brueckner H. K. (1985) REE, Rb–Sr and Sm–Nd studies of Norwegian eclogites. *Chem. Geol.* **52**, 249–271.

Günther M. and Jagoutz E. (1997) Isotope disequilibria (Sm/Nd, Rb/Sr) between minerals of coarse grained, low temperature garnet peridotites from Kimberley floors, Southern Africa. In *Proc. 5th Int. Kimberlite Conf.* 1, CPRM Spec. Publ. 1A (Brasilia) (eds. H. O. A. Meyer and O. H. Leonardos). Brasilia, pp. 354–365.

Hacker B. R. (1996) Eclogite formation and the rheology, buoyancy, seismicity, and H_2O content of oceanic crust. In *Subduction Top to Bottom* (eds. G. E. Bebout, D. W. Scholl, S. H. Kirby, and J. P. Platt). American Geophysical Union, Washington, pp. 337–346.

Hacker B. R. and Liou J. G. (eds.) (1998) *When Continents Collide: Geodynamics and Geochemistry of Ultrahigh-pressure Rocks*. Kluwer Academic, Dordrecht.

Hacker B. R. and Peacock S. M. (1995) Creation, preservation, and exhumation of UHPM rocks. In *Ultrahigh Pressure Metamorphism* (eds. R. G. Coleman and X. Wang). Cambridge University Press, Cambridge, pp. 159–181.

Hacker B. R., Ratschbacher L., Webb L., Ireland T., Walker D., and Shuwen D. (1998) U/Pb zircon ages constrain the architecture of the ultrahigh-pressure Qinling–Dabie Orogen, China. *Earth Planet. Sci. Lett.* **161**, 215–230.

Hacker B. R., Ratschbacher L., Webb L., McWilliams M. O., Ireland T., Calvert A., Shuwen D., Wenk H.-R., and Chateigner D. (2000) Exhumation of ultrahigh-pressure continental crust in east-central China: Late Triassic–Early Jurassic tectonic unroofing. *J. Geophys. Res.* **105**, 13339–13364.

Harley S. L. and Carswell D. A. (1995) Ultradeep crustal metamorphism: a prospective view. *J. Geophys. Res.* **100**, 8367–8380.

Harte B. and Kirkley M. B. (1997) Partitioning of trace elements between clinopyroxene and garnet: data from mantle eclogites. *Chem. Geol.* **136**, 1–24.

Hermann J. and Green D. H. (2001) Experimental constraints on melt-carbonate interaction at UHP conditions: a clue for metmorphic diamond formation? In *Fluid/Slab/Mantle Interactions and Ultrahigh-P Minerals* (eds. Y. Ogasawara, S. Maruyama, and J. G. Liou). Waseda University, Tokyo, pp. 31–34.

Hirajima T., Zhang R., Li J., and Cong B. (1992) Petrology of the nyboite bearing eclogite in the Donghai area, Jiangsu province, E. China. *Min. Mag.* **56**, 37–46.

Humphries F. J. and Cliff R. A. (1982) Sm–Nd dating and cooling history of Scourian granulites, Sutherland. *Nature* **295**, 515–517.

Hwang S.-L., Shen P., Chu H. T., and Yui T.-F. (2000) Nanometer-size a-PbO_2-type TiO_2 in garnet: a thermobarometer for ultrahigh-pressure metamorphism. *Science* **288**, 321–324.

Hwang S.-L., Shen P., Chu H. T., Yui T.-F., and Lin C.-C. (2001a) Genesis of microdiamonds from melt and associated multiphase inclusions in garnet of ultrahigh-pressure gneiss from Erzegebirge, Germany. *Earth Planet. Sci. Lett.* **188**, 9–15.

Hwang S.-L., Shen P., Yui T.-F., and Chu H.-T. (2001b) Defect microstructures of minerals as a potential indicator of extremely rapid and episodic exhumation of ultrahigh-pressure metamorphic rock: implication to continental collision orogens. *Earth Planet. Sci. Lett.* **192**, 57–63.

Jagoutz E. (1988) Nd and Sr systematics in an eclogite xenolith from Tanzania: evidence for frozen mineral equilibria in the continental lithosphere. *Geochim. Cosmochim. Acta* **52**, 1285–1293.

Jahn B. M. (1998) Geochemical and isotopic characteristics of UHP eclogites and ultramafic rocks of the Dabie orogen: implications for continental subduction and collisional tectonics. In *When Continents Collide: Geodynamics and Geochemistry of Ultrahigh-pressure Rocks* (eds. B. R. Hacker and J. G. Liou). Kluwer Academic, Dordrecht, pp. 203–239.

Jahn B. M. (1999) Sm–Nd isotope tracer study of UHP metamorphic rocks: implications for continental subduction and collisional tectonics. *Int. Geol. Rev.* **41**, 859–885.

Jahn B. M., Cornichet J., Cong B., and Yui T. F. (1996) Ultrahigh ε_{Nd} eclogites from an UHP metamorphic terrane of China. *Chem. Geol.* **127**, 61–79.

Jahn B. M., Caby R., and Monie P. (2001a) The oldest UHP eclogites of the World: age of UHP metamorphism, nature of protoliths and tectonic implications. *Chem. Geol.* **178**, 143–158.

Jahn B. M., Chen B., Li H. Y., and Potel S. (2001b) Continental subduction and mantle metasomatism: consequence on the Cretaceous magmatism and implications for the architecture of the Dabie Orogen. In *UHPM Workshop 2001 at Waseda University* (eds. Y. Ogasawara, S. Maruyama, and J. G. Liou). Waseda University, Tokyo, pp. 147–154.

Jahn B. M., Fan Q. C., Yang J. J., and Henin O. (2002) Petrogenesis of the Maowu pyroxenite–eclogite body from the UHP metamorphic terrane of Dabieshan: chemical and isotopic constraints. *Lithos* (in press).

Jamtveit B., Bucher-Nurminen K., and Austrheim H. (1989) Fluid controlled eclogitization of granulites in deep crustal shear zones, Bergen Arcs, Western Norway. *Contrib. Mineral. Petrol.* **104**, 184–193.

Jamtveit B., Carscoell D. A., and Mearns E.W. (1991) Chronology of the high-pressure metamorphism of Norwegian garnet peridotities/pyroxenes. *J. Metamorph. Geol.* **9**, 125–139.

Jerde E. A., Taylor L. A., Crozaz G., Sobolev N. V., and Sobolev V. N. (1993) Diamondiferous eclogites from Yakutia, Siberia: evidence for a diversity of protoliths. *Contrib. Mineral. Petrol.* **114**, 189–202.

Johnson K. T. M. (1994) Experimental cpx/ and garnet/melt partitioning of REE and other trace elements at high pressures: petrogenetic implications. *Min. Mag.* **58A**, 454–455.

Kalt A., Hanel M., Schleicher H., and Kranum U. (1994) Petrology and geochronology of eclogites from the Variscan Schaeavzwald. *Contrib. Mineral. Petrol.* **115**, 287–302.

Kaneko Y., Katayama I., Yamamoto H., Misawa K., Ishikawa M., Rehman H. U., Kausar A. B., and Shiraishi K. (2003) Timing of Himalayan ultrahigh-pressure metamorphism: sinking rate and subduction angle of the Indian continental crust beneath Asia. *Jour. Metamorphic Geol.* **21**.

Kaneko Y., Maruyama S., Terabayashi M., Yamamoto H., Ishikawa M., Anma R., Parkinson C. D., Ota T., Nakajima Y., Katayama I., Yamamoto J., and Yamauchi K. (2000) Geology of the Kokchetav UHP–HP metamorphic belt, Northern Kazakhstan. *Island Arc* **9**, 264–283.

Katayama I. and Nakashima S. (2003) Hydroxyl in clinopyroxene from the deep subducted crust: evidence for H_2O transport into the mantle. *Am. Mineral.* **88**, 229–234.

Katayama I., Zayachkovsky A. A., and Maruyama S. (2000a) Prograde pressure–temperature records from inclusions in zircons from ultrahigh-pressure-high-pressure rocks of the Kokchetav Massif, northern Kazakhstan. *Island Arc* **9**, 417–427.

Katayama I., Parkinson C. D., Okamoto K., Nakajima Y., and Maruyama S. (2001b) Supersilicic clinopyroxene and silica exsolution in UHPM eclogite and pelitic gneiss from the Kokchetav massif, Kazakhstan. *Am. Mineral.* **85**, 1368–1374.

Katayama I., Maruyama S., Parkinson C. D., Terada K., and Sano Y. (2001) Ion micro-probe U–Pb zircon geochronology of peak and retrograde stages of ultrahigh-pressure metamorphic rocks from the Kokchetav massif, northern Kazakhstan. *Earth Planet. Sci. Lett.* **188**, 185–198.

Kirby S. H., Engdahl E. R., and Denlinger R. (1996) Intermediate-depth intraslab earthquakes and arc volcanism as physical expressions of crustal and uppermost mantle metamorphism in subducting slabs. In *Subduction Top to Bottom* (eds. G. E. Bebout, D. W. Scholl, S. H. Kirby, and J. P. Platt). American Geophysical Union, Washington, pp. 195–214.

Krogh E. J. and Carswell D. A. (1995) HP and UHP eclogites and garnet peridotites in the Scandinavian Caledonides. In *Ultrahigh Pressure Metamorphism* (eds. R. G. Coleman and X. Wang). Cambridge University Press, Cambridge, pp. 244–298.

Lardeaux J. M., Ledru P., Daniel I., and Duchene S. (2001) The Variscan French Massif Central—a new addition to the ultrahigh pressure metamorphic "club": exhumation processes and geodynamic consequences. *Tectonophysics* **332**, 143–167.

Leech M. L. and Ernst W. G. (1998) Graphite pseudomorphs after diamond? A carbon isotope and spectroscopic study of graphite cuboids from the Maksyutov complex, south Ural Mountains, Russia. *Geochim. Cosmochim. Acta* **62**, 2143–2154.

Li S. (1996) Isotopic geochronology. In *Ultrahigh-pressure Metamorphic Rocks in the Dabieshan–Su-Lu Region of China* (ed. B. Cong). Science Press; Kluwer Academic, Dordrecht, pp. 90–105.

Li S., Xiao Y., Liou D., Chen Y., Ge N., Zhang Z., Sun S., Cong B., Zhang R., Hart S. R., and Wang S. (1993) Collision of the N. China and Yangtze Blocks and formation of coesite-bearing eclogite: timing and processes. *Chem. Geol.* **109**, 89–111.

Li S., Wang S., Chen Y., Liu D., Qiu J., Zhou H., and Zhang Z. (1994) Excess Ar in phengite from eclogite: evidence from dating of eclogite minerals by Sm–Nd, Rb–Sr, and $^{40}Ar/^{39}Ar$ methods. *Chem. Geol.* **112**, 343–350.

Li S.-G., Jagoutz E., Lo C. H., Chen Y. Z., Li Q. L., and Xiao Y. L. (1999) Sm–Nd, Rb–Sr, and 40Ar/39Ar isotopic systematics of the ultrahigh pressure metamorphic rocks in the Dabie–Su-Lu belt, central China: a retrospective view. *Int. Geol. Rev.* **41**, 1114–1124.

Li S.-G., Jagoutz E., Chen Y.-Z., and Li Q.-L. (2000) Sm–Nd and Rb–Sr isotopic chronology and cooling history of ultrahigh pressure metamorphic rocks and their country rocks at Shuanghe in the Dabie Mountains, Central China. *Geochim. Cosmochim. Acta* **64**, 1077–1093.

Liati A., Gebauer D., and Fanning M. (2000) U–Pb SHRIMP dating of zircon from the Novate granite (Bergell, Central Alps): evidence for Oligocene–Miocene magmatism, Jurassic/Cretaceous continental rifting and opening of the Valais trough. *Schweiz. Mineral. Petrogr. Mitt.* **80**, 305–316.

Liou J. G. (ed.) (1995) Ultrahigh-pressure metamorphism and tectonics. *Island Arc* **4**, 233–283.

Liou J. G. (ed.) (1998) Geodynamics for high- and ultrahigh-pressure metamorphism. *Island Arc* **7**, pp. 1–5.

Liou J. G. and Banno S. (2000) Petrotectonic characteristics of the Kokchetav Massif, northern Kazakhstan. *Island Arc* **9**, 259–263.

Liou J. G. and Carswell D. A. (eds.) (2000) Garnet Perdotites and ultrahigh-pressure minerals. *J. Metamorph. Geol.* **18**, 121–219.

Liou J. G. and Zhang R. Y. (1996) Occurrences of intergranular coesite in ultrahigh-P rocks from the Su-Lu region, eastern China: implications for lack of fluid during exhumation. *Am. Mineral.* **81**, 1217–1221.

Liou J. G. and Zhang R. Y. (1998) Petrogenesis of an ultrahigh-pressure garnet-bearing ultramafic body from Maowu, Dabie Mountains, east-central China. *Island Arc* **7**, 115–134.

Liou J. G., Zhang R. Y., and Ernst W. G. (1995) Occurrences of hydrous and carbonate phases in ultrahigh-pressure rocks from east-central China: implications for the role of volatiles deep in cold subduction zones. *Island Arc* **4**, 362–375.

Liou J. G., Zhang R. Y., Wang X., Eide E. A., Ernst W. G., and Maruyama S. (1996) Metamorphism and tectonics of high-pressure and ultra-high-pressure belts in the Dabie–Su-Lu region, China. In *The Tectonic Evolution of Asia* (eds. A. Yin and T. M. Harrison). Cambridge University Press, Cambridge, pp. 300–344.

Liou J. G., Zhang R. Y., Ernst W. G., Rumble D., and Maruyama S. (1998) High-pressure minerals from deeply subducted metamorphic rocks. *Rev. Mineral. Geochem.* **37**, 33–96.

Liou J. G., Ernst W. G., and Ogasawara Y. (eds.) Petrochemical and tectonic processes of UHP/HP terranes I. *Int. Geol. Rev.* **44**, 765ff. (in press-a).

Liou J. G., Ernst W. G., and Ogasawara Y. (eds.) Petrochemical and tectonic processes of UHP/HP terranes II. *J. Metamorph Geol.* **21**, 510ff. (in press-b).

Liu F., Xu Z., Katayama I., Yang J. S., Maruyama S., and Liou J. G. (2001a) Mineral inclusions in zircons of para- and orthogneiss from pre-pilot drillhole CCSD-PP1, Chinese Continental Scientific Drilling Project. *Lithos* **59**, 199–215.

Liu F., Xu Z. Q., Yang J., Maruyama S., Liou J. G., Katayama I., and Masago H. (2001b) Mineral inclusions of zircon and UHP metamorphic evidence from paragneiss and orthogneiss of pre-pilot drillhole CCSD-PP2 in North Jiangsu Province, China. *Chinese Sci. Bull.* **46**, 1037–1042.

Liu J.-B., Ye K., Maruyama S., Cong B., and Fan H.-G. (2001) Mineral inclusions in zircon from gneisses in the ultrahigh-pressure zone of the Dabie Mountains, China. *J. Geol.* **109**, 523–535.

Liu F., Xu Z., Liou J. G., Katayama I., Masago H., Maruyama S., and Yang J. (2002) Ultrahigh-P mineral inclusions in Zireous from gneissic core samples of the Chinese Continental Scientific Drilling Site in eastern China. *Eur. J. Mineral.* **14**, 499–512.

Luais B., Duchene S., and de Sigoyer J. (2001) Sm–Nd disequilibrium in high-pressure, low-temperature Himalayan and Alpine rocks. *Tectonophysics* **342**, 1–22.

Massonne H. J. (1999) A new occurence of microdiamonds in quartzofeldspathic rocks of the Saxonian Erzgebirge, Germany, and their metamorphic evolution. In *Proc. 7th Int. Kimberlite Conf. Cape Town* (eds. J. J. Gurney, M. D. Pascoe, and S. H. Richardson). Red Roof Design, Capetown, pp. 533–539.

Massonne H. J. (2001) First find of coesite in ultrahigh-pressure metamorphic region of the Central Erzegebirge, Germany. *Euro. J. Mineral.* **13**, 565–570.

Massonne H. J. and Bautsch H. J. (2002) An unusual garnet pyroxenite from the Granulitgebirge, Germany: Origin in the transition zone (>400 km depth) or in a shallower upper mantle region? *Int. Geol. Rev.* **44**, 779–796.

Mattey D., Jackson D. H., Harris N. B. W., and Kelley S. (1994) Isotopic constraints on fluid infiltration from an eclogite facies shear zone, Holsenoy, Norway. *J. Met. Geol.* **12**, 311–325.

Mearns E. W. (1986) Sm–Nd ages for Norwegian garnet peridotite. *Lithos* **19**, 269–278.

Medaris L. G., Jr, Beard B. L., Johnson C. M., Valley J. W., Spicuzza M. J., Jelinek E., and Misar Z. (1995) Garnet pyroxenite and eclogite in the Bohemian Massif: geochemical evidence for Variscan recycling of subducted lithosphere. *Geol. Rundsch.* **84**, 489–505.

Mezger K., Essene E. J., and Halliday A. N. (1992) Closure temperatures of the Sm–Nd system in metamorphic garnets. *Earth Planet. Sci. Lett.* **113**, 397–409.

Mibe K., Fujii T., and Yasuda A. (1998) Connectivity of aqueous fluid in the Earth's upper mantle. *Geophys. Res. Lett.* **25**, 1233–1236.

Mibe K., Fujii T., and Yasuda A. (1999) Control of the location of the volcanic front in island arcs by aqueous fluid connectivity in the mantle wedge. *Nature* **401**, 262–269.

Mibe K., Yoshino T., Ono S., Yasuda A., and Fujii T. (2002b) Connectivity of aqueous fluid in eclogite and its implications for fluid migration in the Earth's interior. *J. Geophys. Res.* **108**, 2295–2305.

Mibe K., Fujii T., and Yasuda A. (2002a) Composition of aqueous fluid coexisting with mantle minerals at high pressure and its bearing on the differentiation the Earth's mantle. *Geochim. Cosmochim. Acta* **66**, 2273–2285.

Mibe K., Yoshino T., Ono S., Yasuda A., and Fujii T., (2002b) Connectivity of aqueous fluid in eclogite and its implications for fluid migration in the Earth's interior. *Jour. Geophys. Res.* **108**, 2295–2305.

Michard A., Henry C., and Chopin C. (1995) Structures in UHPM rocks: a case study from the Alps. In *Ultrahigh*

Pressure Metamorphism (eds. R. G. Coleman and X. Wang). Cambridge University Press, Cambridge, pp. 132–158.

Miller C. and Thöni M. (1995) Origin of eclogites from the Austroalpine Otztal basement (Tirol, Austria): geochemistry and Sm–Nd vs. Rb–Sr isotope systematics. *Chem. Geol.* **122**, 199–225.

Miller J. A., Cartwright I., Buick I. S., and Barnicoat A. C. (2001) An O-isotope profile through the HP-LT Corsican ophiolite, France, and its implications for fluid flow during subduction. *Chem. Geol.* **178**, 43–69.

Mork M. B. E. and Mearns E. W. (1986) Sm–Nd isotopic systematics of a gabbro-eclogite transition. *Lithos* **19**, 255–267.

Mposkos E. D. and Kostopoulos D. K. (2001) Diamond, former coesite, and supersilicic garnet in metasedimentary rocks from the Greek Rhodope: a new ultrahigh-pressure metamorphic province established. *Earth Planet. Sci. Lett.* **192**, 497–506.

Nadeau S., Philippot P., and Pineau F. (1993) Fluid inclusion and mineral isotopic compositions (H–C–O) in eclogite rocks as tracers of local fluid migration during high pressure metamorphism. *Earth Planet. Sci. Lett.* **114**, 431–448.

Nasdala L. and Massonne H. J. (2000) Microdiamonds from the Saxonian Erzgebirge, Germany: *in situ* micro-Raman characterization. *Euro. J. Mineral.* **12**, 495–498.

Nowlan E. U., Schertl H. P., and Schreyer W. (2000) Garnet-omphacite-phengite thermobarometry of eclogites from the coesite-bearing unit of the southern Dora-Maira Massif, Western Alps. *Lithos* **52**, 197–214.

Oberhansli R., Martinottie G., Schmid R., and Liu X. (2002) Preservation of primary volcanic textures in the ultrahigh-pressure terrain of Dabie Shan. *Geology*, **30**, 699–702.

O'Brien P. J. (2001) Subduction followed by collision: Alpine and Himalayan examples. *Phys. Earth. Planet. Int.* **127**, 277–291.

O'Brien P. J., Zotov N., Law R., Khan M. A., and Jan M. Q. (2001) Coesite in Himalayan eclogite and implications for models of India–Asia collision. *Geology* **29**, 435–438.

Ogasawara Y., Ohta M., Fukasawa K., Katayama I., and Maruyama S. (2000) Diamond-bearing and diamond-free metacarbonate rocks from Kumdy-Kol in the Kokchetav massif, northern Kazakhstan. *Island Arc* **9**, 400–416.

Ogasawara Y., Fukasawa K., and Maruyama S. (2002) Coesite exsolution from supersilicic titanite in UHP marble from the Kokchetav Massif, northern Kazakhstan. *Am. Mineral.* **87**, 454–461.

Okamoto K., Liou J. G., and Ogasawara Y. (2000) Petrology of the diamond-grade eclogite in the Kokchetav Massif, northern Kazakhstan. *Island Arc* **9**, 379–399.

Okay A. I. (1994) Sapphirine and Ti-clinohumite in ultra-high-pressure garnet-pyroxenite and eclogite from Dabie Shan, China. *Contrib. Mineral. Petrol.* **110**, 1–13.

O'Reilly S. Y. and Griffin W. L. (1995) Trace-element partitioning between garnet and clinopyroxene in mantle-derived pyroxenites and eclogites: *P–T–X* controls. *Chem. Geol.* **121**, 105–130.

Paquette J. L., Menot R. P., and Peucat J. J. (1989) REE, Sm–Nd and U–Pb zircon study of eclogites from the Alpine External Massif (western Alps): evidence for crustal contamination. *Earth Planet. Sci. Lett.* **96**, 181–198.

Parkinson C. D. and Katayama I. (1999) Present-day ultrahigh-pressure conditions of coesite inclusions in zircon and garnet: evidence from laser Raman microspectroscopy. *Geology* **27**, 979–982.

Parkinson C. D., Motoki A., Onishi C. T., and Maruyama S. (2001) Ultrahigh-pressure pyrope-kyanite granulites and associated eclogites in Neoproterozoic nappes of Southeast Brazil. In *Fluid/Slab/Mantle Interactions and Ultrahigh-P Minerals* (eds. Y. Ogasawara, S. Maruyama, and J. G. Liou). Waseda University, Tokyo, pp. 87–90.

Parkinson C. D., Katayama I., Liou J. G., and Maruyama S. (eds.) (2002) *The Diamond-bearing Kokchetav Massif, Kazakhstan: Petrochemistry and Tectonic Evolution of an Unique Ultra-high Pressure Metamorphic Terrane* Universal Academy Press, Tokyo.

Peacock S. M. (1996) Thermal and petrologic structure of subduction zones. In *Subduction Top to Bottom* (eds. G. E. Bebout, D. W. Scholl, S. H. Kirby, and J. P. Platt). American Geophysical Union, Washington, pp. 119–134.

Philippot P. (1993) Fluid–melt–rock interaction in mafic eclogites and coesite-bearing metasediments: constraints on volatile recycling during subduction. *Chem. Geol.* **108**, 93–112.

Philippot P. (1996) The composition of high-pressure fluids from the Alps: natural observations vs. experimental constraints. *Earth Sci. Front.* **3**, 39–52.

Philippot P. and Rumble D. (2000) Fluid–rock interactions during high-pressure and ultrahigh-pressure metamorphism. *Int. Geol. Rev.* **42**, 312–327.

Philippot P. and Selverstone J. (1991) Trace-element brines in eclogite veins: implications for fluid composition and transport during subduction. *Contrib. Mineral. Petrol.* **106**, 417–430.

Philippot P., Chevallier P., Chopin C., and Dubessy J. (1995) Fluid composition and evolution in coesite bearing rocks (Dora Maira massif, western Alps): implications for element recycling during subduction. *Contrib. Mineral. Petrol.* **121**, 29–44.

Philippot P., Agrinier P., and Scambelluri M. (1998) Chlorine cycling during subduction of altered oceanic crust. *Earth Planet. Sci. Lett.* **161**, 33–44.

Reinecke T. (1998) Prograde high- to ultrahigh-pressure metamorphism and exhumation of oceanic sediments at Lago di Cignana, Zermatt-Saas Zone, Western Alps. *Lithos* **42**, 147–189.

Roselle G. T. and Engi M. (2001) Ultra high pressure (UHP) terrains: lessons from thermal modeling. *Am. J. Sci.* **302**, 410–441.

Rowley D. B., Xue F., Tucker R. D., Peng Z. X., Baker J., and Davis A. (1997) Ages of ultrahigh pressure metamorphism and protolith orthogneisses from the eastern Dabie Shan: U/Pb zircon geochronology. *Earth Planet. Sci. Lett.* **151**, 191–203.

Rubatto D., Gebauer D., and Compagnoni R. (1999) Dating of eclogite-facies zircons: the age of Alpine metamorphism in the Sesia-Lanzo Zone (Western Alps). *Earth Planet. Sci. Lett.* **167**, 141–158.

Rumble D. (1998) Stable isotope geochemistry of ultrahigh-pressure rocks. In *When Continents Collide: Geodynamics and Geochemistry of Ultrahigh-pressure Rocks* (eds. B. R. Hacker and J. G. Liou). Kluwer Academic, Dordrecht, pp. 241–259.

Rumble D. and Yui T. F. (1998) The Qinglongshan oxygen and hydrogen isotope anomaly near Donghai in Jiangsu province, China. *Geochem. Cosmochim. Acta* **62**, 3307–3321.

Rumble D., Wang Q. C., and Zhang R. Y. (2000) Stable isotope geochemistry of marbles from the coesite UHP terrains of Dabieshan and Su-Lu, China. *Lithos* **52**, 79–95.

Rumble D., Giorgis D., Ireland T., Zhang Z., Xu H., Yui T.-F., Yang J., Xu Z., and Liou J. G. (2002) Low delta ^{18}O zircons, U–Pb dating, and the Age of the Qinglongshan oxygen and hydrogen isotope anomaly near Donghai in Jiangsu Province, China. *Geochim. Cosmochim. Acta* **66**, 2299–2306.

Sachan H. K., Ogasawara Y., Muko A., and Yoshioka N. (2003) Carbonate-bearing UHPM rocks from the Ts0-Morari region, Ladakh, India: petrological implications. *Int. Geol. Rev.* **45**, 49–69.

Sanders I. S., van Calsteren P. W. C., and Hawkesworth C. J. (1984) A Grenville Sm–Nd age for the Glenelg eclogite in northwest Scotland. *Nature* **312**, 439–440.

Sassi R., Harte B., Carswell D. A., and Han Y. (2000) Trace element distribution in central Dabie eclogites. *Contrib. Mineral. Petrol.* **139**, 298–315.

Scambelluri M. and Philippot P. (2001) Deep fluids in subduction zones. *Lithos* **55**, 213–227.

Scambelluri M., Piccardo G., Philippot P., Robbiano A., and Negretti L. (1997) High-salinity brines from recycled seawater in deeply subducted serpentiinite. *Earth Planet. Sci. Lett.* **148**, 485–499.

Scambelluri M., Pennacchioni G., and Philippot P. (1998) Salt rich aqueous fluids formed during eclogitisation of metabasites in the alpine continental crust (Austroalpine Mt. Emilius unit, Italian western Alps). *Lithos* **43**, 151–167.

Scambelluri M., Bottazzi P., Trommsdorff V., Vannucci R., Hermann J., Gomez-Pugnaire M. T., and Vizcaino V. L. S. (2001) Incompatible element-rich fluids released by antigorite breakdown in deeply subducted mantle. *Earth Planet. Sci. Lett.* **192**, 457–470.

Scherer E. E., Cameron K. L., and Blichert-Toft J. (2000) Lu–Hf garnet geochronology: closure temperature relative to the Sm–Nd system and the effects of trace mineral inclusions. *Geochim. Cosmochim. Acta* **64**, 3413–3432.

Schmadicke E., Mezger K., Cosca M. A., and Okrush M. (1995) Variscan Sm–Nd and Ar–Ar ages of eclogite facies rocks from the Erzegebirge, Bohemian Massif. *J. Met. Geol.* **13**, 537–552.

Schmidt M. W. and Poli S. (1998) Experimentally based water budgets for dehydrating slabs and consequences for arc magma generation. *Earth Planet. Sci. Lett.* **163**, 361–379.

Schreyer W. (1995) Ultradeep metamorphic rocks: the retrospective viewpoint. *J. Geophys. Res.* **100**, 8353–8366.

Schreyer W. and Stockhert B. (eds.) (1997) High pressure metamorphism in nature and experiments. *Lithos* **41**, 1–266.

Selverstone J., Franz G., Thomas S., and Getty S. R. (1992) Fluid variability in 2 GPa eclogites as an indicator of fluid behaviour during subduction. *Contrib. Mineral. Petrol.* **112**, 341–357.

Sharp Z. D., Essene E. J., and Hunziger J. C. (1993) Stable isotope geochemistry and phase equilibria of coesite-bearing whiteschists, Dora Maira massif, western Alps. *Contrib. Mineral. Petrol.* **114**, 1–12.

Shatsky V. S., Sobolev N. V., and Vavilov M. A. (1995) Diamond-bearing metamorphic rocks of the Kokchetav massif (N. Kazakhstan). In *Ultrahigh Pressure Metamorphism* (eds. R. G. Coleman and X. Wang). Cambridge University Press, Cambridge, pp. 427–455.

Shatsky V. S., Jagoutz E., Sobolev N. V., Kozmenko O. A., Parkhomenko V. S., and Troesch M. (1999) Geochemistry and age of ultrahigh pressure metamorphic rocks from the Kokchetav massif (Northern Kazakhstan). *Contrib. Mineral. Petrol.* **137**, 185–205.

Shen Y. (2002) C-isotope variations and paleoceanographic changes during the late Neoproterozoic on the Yangtze Platform, China. *Precamb. Res.* **113**, 121–133.

Smith D. C. (1984) Coesite in clinopyroxene in the Caledonides and its implications for geodynamics. *Nature* **310**, 541–544.

Smith D. C. (ed.) (1988) *Eclogites and Eclogite-facies Rocks.* Elsevier, Amsterdam.

Smith D. C. (1995) Microcoesites and microdiamonds in Norway. In *Ultrahigh pressure Metamorphism* (eds. R. G. Coleman and X. Wang). Cambridge University Press, Cambridge, pp. 299–355.

Sobolev N. V. and Shatsky V. S. (1990) Diamond inclusions in garnets from metamorphic rocks: a new environment for diamond formation. *Nature* **343**, 742–746.

Stockhert B., Duyster J., Trepmann C., and Massonne H. J. (2001) Microdiamond daughter crystals precipitated from supercritical COH + silicate fluids included in garnet, Erzegebirge, Germany. *Geology* **29**, 391–394.

Stosch H. G. and Lugmair G. W. (1990) Geochemistry and evolution of MORB-type eclogites from the Münchberg Massif, southern Germany. *Earth Planet. Sci. Lett.* **99**, 230–249.

Tabata H., Yamauchi K., Maruyama S., and Liou J. G. (1998) Tracing the extent of a UHP metamorphic terrane: mineral-inclusion study of zircons in gneisses from the Dabie Shan. In *When Continents Collide: Geodynamics and Geochemistry of Ultrahigh-pressure Rocks* (eds. B. R. Hacker and J. G. Liou). Kluwer Academic, Dordrecht, pp. 261–273.

Tatsumi Y. and Eggins S. (1995) *Subduction Zone Magmatism.* Blackwell, Oxford, 211pp.

Taylor L. A., Snyder G. A., Crozaz G., Sobolev V. N., Yefimova E. S., and Sobolev N. V. (1996) Eclogitic inclusions in diamonds: evidence of complex mantle processes over time. *Earth Planet. Sci. Lett.* **142**, 535–551.

Thöni M. (2002) Sm–Nd isotope systematics in garnet from different lithologies (Eastern Alps): age results, and an evaluation of potential problems for garnet Sm–Nd chronometry. *Chem. Geol.* **185**, 255–281.

Thöni M. and Jagoutz E. (1992) Some new aspects of dating eclogites in orogenic belts: Sm–Nd, Rb–Sr and Pb–Pb isotopic results from the Austroalpine Saualpe and Koralpe type-locality (Carinthia/Styria, southeastern Austria). *Geochim. Cosmochim. Acta* **56**, 347–368.

Tilton G. R., Schreyer W., and Schertl H. P. (1989) Pb–Sr–Nd behavior of deeply subducted crustal rocks from the Dora Maira Massif, western Alps, Italy. *Geochim. Cosmochim. Acta* **53**, 1391–1400.

Tilton G. R., Schreyer W., and Schertl H. P. (1991) Pb–Sr–Nd isotopic behavior of deeply subducted crustal rocks from the Dora Maira Massif, western Alps, Italy: II: What is the age of the ultrahigh-pressure metamorphism? *Contrib. Mineral. Petrol.* **108**, 22–33.

Vance D. and O'Nions R. K. (1990) Isotope chronometry of zoned garnets: growth kinetics and metamorphic histories. *Earth Planet. Sci. Lett.* **97**, 227–240.

Vance D. and O'Nions R. K. (1992) Prograde and retrograde thermal histories from the central Swiss Alps. *Earth Planet. Sci. Lett.* **114**, 113–129.

van Roermund H. L. M. and Drury M. R. (1998) Ultra-high pressure (P>6 GPa) garnet peridotites in western Norway: exhumation of mantle rocks from >185 km deptha. *Terra Nova* **10**, 295–301.

van Roermund H. L. M., Drury M. R., Barnhoorn A., and DeRonde A. (2001) Relict majoritic garnet microstructures from ultra-deep orogenic peridotites in Western Norway. *J. Petrol.* **42**, 117–130.

van Roermund H. L. M., Carswell D. A., Drury M. R., and Heijboer T. C. (2002) Micro-diamonds in a megacrystic garnet-websterite pod from Bardane on the island of Fjortoft, western Norway. *Geology* **30**, 959–962.

Van Wick N., Valley J. W., and Austrheim H. (1996) Oxygen and carbon isotopic constraints on the development of eclogites, Holsnoy, Norway. *Lithos* **38**, 129–145.

Von Quadt A., Günther D., Frischknecht R., Zimmermann R., and Franz G. (1997) The evolution of pre-Variscan eclogites of the Tauern Window (eastern Alps): a Sm/Nd, conventional and laser ICP-MS zircon U–Pb study. *Schweiz. Mineral. Petrogr. Mitt.* **77**, 265–279.

Wain A. (1997) New evidence for coesite in eclogite and gneisses: defining an ultrahigh pressure province in the Western Gneiss Region of Norway. *Geology* **25**, 927–930.

Wain A., Waters D., Jephcoat A., and Olijynk H. (2000) The high-pressure to ultrahigh-pressure eclogite transition in the Western Gneiss Region, Norway. *Euro. J. Mineral.* **12**, 667–687.

Xiao Y., Hoefs J., van den Kerkhof A. M., Fiebig J., and Zheng Y.-F. (2000) Fluid history of UHP metamorphism is Dabie Shan, China: a fluid inclusion and oxygen isotope study on the coesite-bearing eclogite from Bixiling. *Contrib. Mineral. Petrol.* **139**, 1–16.

Xiao Y., Hoefs J., van den Kerkhof A. M., Simon K., Fiebig J., and Zheng Y.-F. (2002) Fluid evolution during HP and UHP metamorphism in Dabie Shan, China: constraints from mineral chemistry, fluid inclusions, and stable isotopes. *J. Petrol.* **43**, 1505–1527.

Xu S., Okay A. I., Ji S., Sengor A. M. C., Su W., Liu Y., and Jiang L. (1992) Diamond from the Dabie Shan metamorphic rocks and its implication for tectonic setting. *Science* **256**, 80–82.

Xu S., Liu Y., Chen G., Compagneni R., Rolfo F., He M., and Liu H. (2003) New findings of micro-diamonds in eclogites from Dabie–Su-Lu region in central–eastern China. *Chinese Sci. Bull.* **48**, 988–994.

Xue F., Rowley D. B., and Baker J. (1996) Refolded syn-ultrahigh-pressure thrust sheets in the south Dabie complex, China: field evidence and tectonic implications. *Geology* **24**, 455–458.

Yang J. J., Godard G., Kienast J. R., Lu Y., and Sun, J. (1993) Ultrahigh-pressure (60 kbar) magnesite-bearing garnet peridotite from Northeastern Jiangsu, China. *J. Geol.* **101**, 541–554.

Yang J. J., Godard G., and Smith D. C. (1998) K-feldspar-bearing coesite pseudomorphs in an eclogite from Lanshantou (eastern China). *Euro. J. Mineral.* **10**, 969–985.

Yang J. S., Xu Z., Song S., Zhang J., Shi R., Li H., and Brunel M. (2001) Discovery of coesite in the North Qaidam Early Paleozoic ultrahigh pressure (UHP) metamorphic belt, NW China. *Comptes Rendus de l'Academie des Sciences, Paris, Sciences de la Terre et des Planets* **333**, 719–724.

Ye K., Hirajima T., Ishiwatari A., Gou J., and Zhai M. (1996) The discovery of intergranular coesite in eclogite from Yangkou, Qingdao. *Chinese Sci. Bull.* **41**, 1407–1408.

Ye K., Cong B., and Ye D. (2000a) The possible subduction of continental material to depths greater than 200 km. *Nature* **407**, 734–736.

Ye K., Yao Y., Katayama I., Cong B., Wang Q., and Maruyama S. (2000b) Areal extent of ultra-high pressure metamorphism in the Su-Lu terrane of east China: evidence from coesite inclusions in zircon from country rock granitic gneiss. *Lithos* **52**, 157–164.

Yoshino T., Mibe K., Yasuda A., and Fujii T. (2002) Wetting properties of anorthite aggregates: implications for fluid connectivity in continental lower crust. *J. Geophys. Res.* **107**, ECV 10-1–ECV 10-8.

Yui T. F., Rumble D., and Lo C. H. (1995) Unusually low $\delta^{18}O$ ultrahigh-pressure metamorphic rocks from the Su-Lu terrain, eastern China. *Geochim. Cosmochim. Acta* **59**, 2859–2864.

Yui T. F., Rumble D., Chen C. H., and Lo C. H. (1997) Stable isotope characteristics of eclogites from the ultra-high-pressure metamorphic terrain, east-central China. *Chem. Geol.* **137**, 135–147.

Zhai M. and Cong B. (1996) Major and trace element geochemistry of eclogites and related rocks. In *Ultrahigh-pressure Metamorphic Rocks in the Dabieshan–Su-Lu Region of China* (ed. B. Cong). Science Press; Kluwer Academic, Dordrecht, pp. 69–89.

Zhang R. Y. and Liou J. G. (1997) Partial transformation of gabbro to coesite-bearing eclogite from Yangkou, the Su-Lu terrane, eastern China. *J. Metamorph. Geol.* **15**, 183–202.

Zhang R. Y. and Liou J. G. (2000) Exsolution lamellae in minerals from ultrahigh-pressure rocks. In *Ultrahigh-pressure Metamorphism and Geodynamics in Collision-type Orogenic Belts* (eds. W. G. Ernst and J. G. Liou). Geological Society of America, Denver, pp. 216–228.

Zhang R. Y. and Liou J. G. Origin and evolution of garnet exsolution in clinopyroxene from the Su-Lu garnet Clinopyroxenite, *China. Am. Mineral.* (in press).

Zhang R. Y., Liou J. G., and Cong B. (1995a) Talc-, magnesite- and Ti-clinohumite-bearing ultrahigh-pressure meta-mafic and ultramafic complex in the Dabie Mountains, China. *J. Petrol.* **36**, 1011–1037.

Zhang R. Y., Hirajima T., Banno S., Cong B., and Liou J. G. (1995b) Petrology of ultrahigh-pressure rocks from the southern Su-Lu region, eastern China. *J. Metamorph. Geol.* **13**, 659–675.

Zhang R. Y., Liou J. G., Ernst W. G., Coleman R. G., Sobolev N. V., and Shatsky V. S. (1997) Metamorphic evolution of diamond-bearing and associated rocks from the Kokchetav Massif, northern Kazakhstan. *J. Met. Geol.* **15**, 479–496.

Zhang R. Y., Rumble D., Liou J. G., and Wang Q. C. (1998) Low $\delta^{18}O$, ultrahigh-P garnet-bearing mafic and ultramafic rocks from Dabie Shan, China. *Chem. Geol.* **150**, 161–170.

Zhang R. Y., Liou J. G., Yang J. S., and Yui T. F. (2000) Petrochemical constrains for dual origin of garnet peridotites from the Dabie–Su-Lu UHP terrane, eastern-central China. *J. Metamorph. Geol.* **18**, 149–166.

Zhang R. Y., Liou J. G., and Shu J. F. (2002a) Hydroxyl-rich topaz in high-pressure and ultrahigh-pressure kyanite quartzites, with retrograde woodhouseite, from the Su-Lu terrane, eastern China. *Am. Mineral.* **87**, 445–453.

Zhang R. Y., Shau Y. H., and Liou J. G. (2002b) Discovery of clinoenstatite in garnet pyroxenite from the Dabie–Su-Lu UHP terrane, east-central, China. *Am. Mineral.* **87**, 867–874.

Zheng Y. F., Fu B., Gong B., and Li S. (1996) Extreme ^{18}O depletion in eclogite from the Su–Lu terrane in east China. *Euro. J. Mineral.* **8**, 317–323.

Zheng Y. F., Fu B., Gong B., and Wang Z.-R. (1998a) Carbon isotope anomaly in marbles associated with eclogites from the Dabie Mountains in China. *J. Geol.* **106**, 97–104.

Zheng Y. F., Fu B., Li Y., Xiao Y., and Li S. (1998b) Oxygen and hydrogen isotope geochemistry of ultrahigh-pressure eclogites from the Dabie Mountains and the Su-Lu terrane. *Earth Planet. Sci. Lett.* **155**, 113–129.

Zheng Y. F., Fu B., Xiao Y., Li Y., and Gong B. (1999) Hydrogen and oxygen isotope evidence for fluid-rock interactions in the stages of pre- and post-UHP metamorphism in the Dabie Mountains. *Lithos* **46**, 677–693.

Zheng Y.-F., Wang Z.-R., Li S.-G., and Zhao Z.-F. (2002) Oxygen isotope equilibrium between eclogite minerals and its constraints on mineral Sm–Nd chronometer. *Geochim. Cosmochim. Acta* **66**, 625–634.

Zheng Y.-F., Fu B., Gong B., and Li L. (2003) Stable isotope geochemistry of ultrahigh pressure metamorphic rocks from the Dabie–Su-Lu orogen in China: implications for geodynamics and fluid regime. *Earth Sci. Rev.* **1276**, 1–57.

Zhou B. and Hensen B. J. (1995) Inherited Sm/Nd isotope components preserved in monazite inclusions within garnets in leucogneiss from East Antarctica and implications for closure temperature studies. *Chem. Geol.* **121**, 317–326.

3.10
Ages and Growth of the Continental Crust from Radiogenic Isotopes

P. J. Patchett
University of Arizona, Tucson, AZ, USA

and

S. D. Samson
Syracuse University, NY, USA

3.10.1 SCOPE OF AVAILABLE METHODS AND DATA

The development and application of radiogenic isotopes to dating of geologic events, and to questions of growth, evolution, and recycling processes in the continental crust are mature areas of scientific inquiry. By this we understand that many of the approaches used to date rocks and constrain the evolution of the continents are well established, even routine, and that the scope of data available on age and evolution of continents is very large. This is not to say that new approaches have not been developed in recent years, or that new approaches and/or insights cannot be developed in the future. However, the science of continental crustal evolution is definitely a domain where many of the problems are well defined, the power of the techniques used to solve them are well known, and the limitations of field and laboratory databases, as well as the preserved geologic record, are understood.

From the very early days of crustal evolution studies, it was innovations and improvements in laboratory techniques that drove the pace of discovery (e.g., Holmes, 1911; Nier, 1939). This remained true through all the increments in capability reviewed in this chapter, up to the present day. Thus, continental crustal evolution is an area of Earth science where a species of very laboratory-oriented investigator, the "radiogenic isotope geologist" or "geochronologist," has made major advances, even breakthroughs, in understanding. This is true in spite of the fact that many of the individuals of the species may have lacked field expertise, or even more than a primitive level of geologic background. Because design and building of instruments like radiation detectors or mass spectrometers requires a knowledge of physics, many of the early practitioners of rock dating were physicists, like Alfred Nier (cited above). Since the 1970s, essentially all mass spectrometers have been constructed by specialized commercial firms, and the level of physics expertise among isotope geologists has been lower. These firms, based mainly (but not exclusively) in and around Manchester, England, and Bremen, Germany, have spearheaded technical innovations in mass spectrometry. Isotope geology researchers are one group of consumers for this technology, along with chemists, nuclear-weapon laboratories and nuclear-power generating facilities. Today, the vast majority of isotope geology researchers are derived from geological backgrounds.

This chapter will briefly review historical aspects of the development of radiogenic isotope geology as applied to continents. Some details, references and cross-references to other chapters in this volume will be provided for most major radiogenic isotopic methods, and for applications of these. However, this chapter will ultimately concentrate on two major approaches that dominate the research field today: (i) crustal tectonic and magmatic ages from U–Pb dating of accessory minerals like zircon; and (ii) crustal differentiation and growth from neodymium isotopic determinations on total rocks.

The sheer amount of data available from continental areas, and the pace of data acquisition today, places any geographically constrained compilation in the impossible category for a chapter like this one. In this chapter, the state of the art in both geochronology and crustal origins will be illustrated by selected examples, not by global compilations or comprehensive discussions of each region that has been studied. Nevertheless, some remarks about the availability of data from different parts of the world need to be made. As with other areas of geology, biology, and botany, the parts of the world that have been longest settled by western civilization have the best data coverage for both ages of continental rocks, and their origins and evolution. Thus, Western Europe, Canada, the USA and Australia have generally somewhat thorough coverage. More limited data have become available from Eastern Europe, Greenland, Central and South America, Africa, Asia, and Antarctica. Some of these data have been produced by groups based in those regions, but much of the data published between 1970 and today have been driven by studies based in, and funded from, western countries. Availability of state-of-the-art results for ages of crust and its evolution are low in Antarctica and Greenland, where climate and ice cover limit work, and in South America, Africa, and parts of Asia, where studies have been sporadic, and are certainly limited in some cases by political instability. This general situation is now changing, however. Two parts of the world where results for ages of orogenic belts and for crustal evolution in general are accumulating more rapidly since about 1990 are China and Russia; this is connected with growth of modern isotope geology facilities in those countries. Understanding of global continent growth and evolution is limited in critical respects by the large regions of the world that are poorly dated. Therefore, gradual improvement of the state of knowledge of less accessible parts of the continents will bring significant benefits, even though the conceptual issues may be understood to a large extent.

3.10.2 DETERMINATION OF AGES OF IGNEOUS EVENTS

3.10.2.1 Early Developments in U–Th–Pb Geochronology

Of all the different geochronological decay schemes that have been employed in crustal studies the U–Pb system is unique in that there are two radioactive parent isotopes (^{238}U and ^{235}U) that decay to two different daughter isotopes (^{206}Pb and ^{207}Pb). Because of this unique dual decay scheme the U–Pb system has long been of considerable interest as a geochronological technique. However, because of the extreme mobility of uranium, most U–Pb dates for whole-rock samples were shown to be unreliable and the focus went to analyzing uranium-bearing accessory minerals that were far less susceptible to uranium loss during weathering. Zircon, because of is extremely low initial lead content, its common occurrence in igneous rocks and its extreme chemical and physical resistance were a particular target for geochronological investigation, with the first analyses performed in the 1950s (Vinogradov *et al.*, 1952; Tilton *et al.*, 1955). Because zircon has such an extremely refractory nature, it is one of the most difficult minerals to dissolve; thus early studies used a borax fusion technique to ensure digestion. However, the very high lead content of the borax (200–1,000 ng g^{-1} according to Krogh, 1973) minimized the utility of the technique. In 1971 it was demonstrated that zircon could be completely dissolved in HF in high pressure Teflon®-lined vessels (Krogh, 1971a,b) with a reduction in lead blank by three orders of magnitude (Krogh, 1973). That analytical breakthrough, combined with the demonstration that the effects of lead loss in zircons could be minimized by mechanically abrading zircons (Krogh, 1982), resulted in the exponential rise of the technique in a wide array of geological studies. U–Pb dating of accessory minerals, particularly zircon and baddeleyite, is now considered *the* most reliable method for determining the crystallization ages of plutonic rocks, even those that have suffered multiple episodes of metamorphism. U–Th–Pb dating of monazite and U–Pb dating of titanite have become extremely important methods for constraining the timing of metamorphism. *In situ* methods of measuring U–Th–Pb isotopes, combined with imaging techniques such as cathodoluminescence, have made it possible to analyze metamorphic overgrowths on accessory minerals dramatically increasing our ability to precisely date metamorphic events. Overall, the U–Th–Pb technique now dominates the field of geochronology as applied to crustal evolution. Since the 1980s several different methodologies for determining U–Th–Pb dates have been developed and these are briefly reviewed below.

3.10.2.2 U–Th–Pb Dating by TIMS—The Isotope Dilution Method

The most common method of determining the U–Pb date of an accessory mineral is the determination of uranium and lead isotopic abundances via isotope dilution and thermal ionization mass spectrometry (ID-TIMS). Most measurements in the 1970s and 1980s required that an aliquot of the dissolved mineral solution be made, with one portion being "spiked" with an enriched ^{235}U and ^{208}Pb tracer solution, and the other aliquot directly measured for lead isotopic abundance. Following the first major production of ^{205}Pb from ^{205}Bi, produced by proton bombardment of enriched ^{206}Pb (Krogh and Davis, 1975), a second, larger production and worldwide distribution of high-purity ^{205}Pb occurred (Parrish and Krogh, 1987). Use of ^{205}Pb eliminated the need for aliquoting dissolved samples, and virtually all modern laboratories have since adopted the use of a ^{205}Pb spike. This is of particular use to U–Th–Pb dating of monazite, as it often contains such a large abundance of $^{208}Pb^*$ (the asterisk denotes radiogenic lead) that a spike other than one enriched in ^{208}Pb is needed.

While the general procedures for mineral dissolution, separation of uranium and lead, and mass spectrometry have not changed substantially since 1973, there has been a constant drive to reduce the laboratory contamination level of lead, combined with improvements in detection of very small Pb^+ ion beams, to allow for an ever-decreasing amount of sample required for high-precision analysis. It is now common for many U–Pb laboratories to report procedural lead blanks of just a few picograms, and some labs have reduced lead blanks to the subpicogram level (e.g., Samson and D'Lemos, 1999; Corfu and Easton, 2000; Ayer *et al.*, 2002; Samson *et al.*, 2003). This reduction in blank level, combined with the development of an emitter solution that produces much stronger and more stable Pb^+ beams than conventional silica gel (Gerstenberger and Haase, 1997), allows for a relatively precise isotopic measurement of as little as 15–20 pg of radiogenic lead. This amount is equivalent to the typical lead content of very young (<10 Ma) or very small (<50 μm) single zircon crystals, or of carefully extracted portions of single crystals.

3.10.2.3 Zircon Evaporation Method

A variation of the method for determining $^{207}Pb-^{206}Pb$ dates of whole single zircons by TIMS analysis that eliminates the need for zircon dissolution and chemical separation of lead was introduced by Kober (1986), with a slight, yet important, modification described a year later (Kober, 1987). In the modified method, usually referred to as the zircon evaporation method, a single zircon crystal is placed into a folded side rhenium filament (the evaporation filament), which is positioned opposite of a blank rhenium ionization filament. The evaporation filament is heated for a short time to evaporate lead onto the target ionization filament. The current to the evaporation filament is then turned off and the ionization filament is heated until Pb^+ ionization begins and lead isotopic ratios are measured in the normal fashion. Current to the ionization filament is then turned off, the evaporation filament heated to a slightly higher temperature than previously, and the process continually repeated until either an adequate number of ratios have been collected or the lead in the zircon is exhausted. The main advantages of the technique are that ultra-clean chemistry is not required, the measured $^{206}Pb/^{204}Pb$ ratio is usually higher than ID-TIMS measurements of single zircons (as minimal lead blank is introduced), and no time is spent waiting for zircons to dissolve or in performing chemical separations. The main disadvantage is that only the age based on $^{207}Pb-^{206}Pb$ can be obtained, with no U–Pb information, and thus the degree of discordance of the analysis cannot be determined. For zircons that experienced only modern-day lead loss, this is not a problem, but for zircons that have suffered non-modern lead loss, the evaporation dates would be inaccurate. Critical to the technique, therefore, is that identical dates are determined from each heating step and that reproducible results are obtained from more than one zircon (Söderlund, 1996). Although some laboratories have embraced the technique, most modern TIMS laboratories involved in U—Pb geochronological studies still use the ID-TIMS method, with the goal of continuing to try to lower lead blanks to allow ever smaller samples to be analyzed with high precision.

3.10.2.4 U–Th–Pb Dating by Ion Microprobe

Although tremendous advances were made in geochronology with the advent of U–Pb dating using chemical separation procedures and TIMS analyses, there was an obvious need for a method of *in situ* isotopic analysis of accessory minerals to more fully exploit the age information preserved in complex crystals (i.e., ones containing metamorphic overgrowths and/or xenocrystic regions, etc.) The first response to this need was the measurement of $^{207}Pb-^{206}Pb$ dates using a secondary ion microprobe mass analyzer (Anderson and Hinthorne, 1973). Following this work, a very high mass resolution secondary ion mass spectrometer (SIMS) was developed in Australia in the late 1970s (Clement *et al.*, 1977). This instrument, coined the SHRIMP (for sensitive high resolution ion microprobe), is particularly suited for determining the age domains within complex crystals, as very small regions of a mineral can be analyzed (e.g., Stern, 1997 reports that minimal "spot sizes" with surface dimensions of 4×6 μm and depths of <1 μm can be obtained). The first age determinations using the SHRIMP were published in 1982 (Compston *et al.*, 1982), and only a year later the instrument had been used to determine the age of the first pre-4.0 Ga terrestrial mineral (Froude *et al.*, 1983).

High-resolution ion microprobes (IMP) are used for *in situ* U–Th–Pb dating by bombarding the surface of a highly polished crystal, usually ground to half its original thickness, with a primary beam of O_2^- ions. These bombarding ions, produced by oxygen gas discharge within a hollow nickel cathode, sputter a small portion of the zircon producing a wide variety of secondary ions. The secondary ions are doubly focused, first through an electrostatic analyzer, which filters them by their values of kinetic energy, then these ions are focused a second time by entering a magnetic sector, which discriminates between the ions based solely on their mass. The magnetic sector is similar to that of thermal ionization mass spectrometers, but has a much larger radius and is capable of operating at a much high mass resolution (>5,000 compared to ~300) to allow discrimination of the uranium and lead isotopes from ions with nominally similar masses produced during the sputtering process (e.g., $^{96}Zr^{94}Zr^{16}O^+$ ions versus $^{206}Pb^+$ ions for zircon analyses). The secondary ion currents generated are very low and thus are measured using a single secondary electron multiplier. $^{207}Pb-^{206}Pb$ dates can be directly determined for each elliptical region analyzed by directly comparing the $^{207}Pb^+$ and $^{206}Pb^+$ ion beam currents, after correcting for common lead components (usually by measuring the $^{204}Pb^+$ peak). Determining $^{206}Pb^*/^{238}U$ and $^{208}Pb^*/^{232}Th$ dates (Pb^* denotes radiogenic lead) cannot be accomplished by a direct comparison of $^{238}U^+$, $^{232}Th^+$, and $^{206}Pb^+$ ion currents. This is because the measured $^{206}Pb^+/^{238}U^+$ and $^{208}Pb^+/^{232}Th^+$ ratios are not the same as the true ratios but vary between 2–5 times greater than the actual ratios (see Stern (1997) for a discussion of this phenomenon). This variation can be corrected because the ion beams display a systematic relationship between $^{206}Pb^+/^{238}U^+$ and UO^+/U^+, and between $^{208}Pb^+/^{232}Th^+$ and ThO^+/Th^+ (Hinthorne *et al.*, 1979). Thus, by including a mineral of known age and of uniform Pb/U (or Th/Pb) in the target mount with the unknowns, a standardization curve can be established based on the measured Pb/U and UO/U (or Pb/Th and ThO/Th) of the standard, so that a normalization factor can be applied to the measured Pb/U,Th ratios of the minerals of unknown ages. The need for well-calibrated external mineral standards to determine ion microprobe U–Th–Pb dates is thus directly analogous to the determination of $^{40}Ar/^{39}Ar$ dates.

The uncertainty in the Pb/U–UO/U calibration curve must be propagated along with all of the other analytical uncertainties in the estimation of realistic errors of the calculated dates, rather than quoting errors based only on counting statistics, for example. For minerals containing low Pb^* contents, either because they have low U–Th contents or because they are geologically young, counting statistics may have the dominant influence in error propagation; for minerals containing high Pb^* content uncertainties in the calibration curve may have a larger influence on the precision of the U–Pb dates. Although a variety of factors may control the ion microprobe precision on any given zircon, typical levels of precision that can be obtained can be estimated. For most zircons the 2σ uncertainty in a *single* $^{238}U/^{206}Pb^*$ date is typically not more precise than 1.5–2%. This level of precision is ~10–20 times larger than that obtainable via ID-TIMS methods (see Figure 1). For zircons younger than 1 Ga $^{207}Pb-^{206}Pb$ dates are typically not more precise than ~5%, and typical values for zircon <500 Ma are more likely to be in the range of 10–20%. However, for zircons ≥1.5 Ga the relative precision of a $^{207}Pb-^{206}Pb$ date is considerably higher, with values of ~0.5% obtainable.

3.10.2.5 U–Th–Pb Dating by ICP-MS

In the early 1980s quadrupole mass spectrometers using argon plasma as the ionization source, i.e., inductively coupled plasma mass spectrometers (ICP-MS), were developed. Although these instruments were designed primarily for measuring the concentrations of trace elements, many studies have employed them for U–Pb dating. $^{206}Pb^*/^{238}U$, $^{207}Pb^*/^{235}U$, and $^{208}Pb^*/^{232}Th$ dates can be determined, in addition to the more simply determined $^{207}Pb-^{206}Pb$ dates,

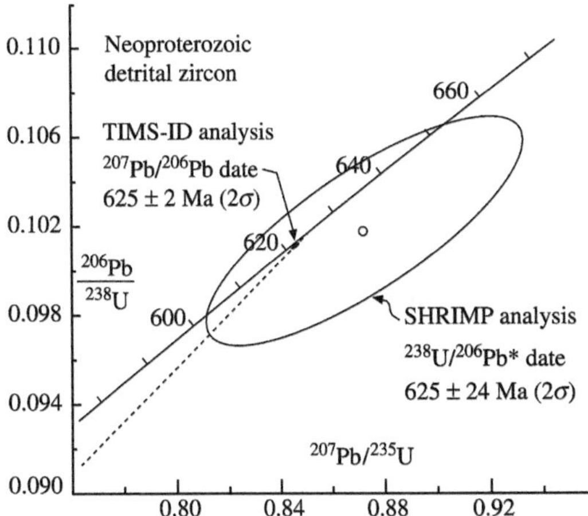

Figure 1 U–Pb Concordia diagram showing the results of an analysis of a detrital zircon crystal by ion microprobe (SHRIMP) followed by analysis of the same crystal using TIMS. Both error ellipses are plotted at 2σ. The best estimate of the age of crystallization of the zircon is identical for both techniques; however, the TIMS analysis is an order of magnitude more precise than that obtained using the ion microprobe (source Samson *et al.*, 2003).

using laser-ablation ICP-MS techniques. The obvious appeal of U–Th–Pb dating by laser ablation ICP-MS is the elimination of the need for ultra-low blank dissolution and U–Th–Pb separation procedures, the speed of the analysis (<10 min per analysis), and the possibility of *in situ* analysis. However, there are several analytical obstacles to obtaining accurate U–Th–Pb dates via laser ablation that must be overcome. One of the most difficult problems is that there is significant elemental (i.e., U/Pb and Th/Pb) fractionation that occurs during laser ablation. That is, measured Pb/U ratios are lower than actual ratios by tens of percent, and this effect is variable with ablation time (see figure 4 in Horn *et al.*, 2000). A second effect, common to all mass spectrometric measurements, is that there is an instrumental mass bias, or discrimination. This bias is several times higher than the bias that occurs during TIMS measurements and thus would be a significant source of error if not corrected. A third potential difficulty is determining the amount of common lead in an analysis, as the argon gas used in ICP-MS contains enough mercury to cause isobaric interference of ^{204}Pb from ^{204}Hg. However, continued improvements are being made as this technique evolves (see below) and it may begin to approach ID-TIMS analysis in the future, at least for accessory minerals with relatively high radiogenic lead contents.

Early attempts to directly date zircon crystals using ICP-MS techniques involved the use of Nd–YAG lasers, operating at a 1,064 nm wavelength, to ablate the zircon crystals (Feng *et al.*, 1993; Fryer *et al.*, 1993). Because of significant variations of U/Pb isotopic ratios, these early laser-ablation studies concentrated on determining $^{207}Pb-^{206}Pb$ dates, which yielded precision between 0.5–6.0% (e.g., Fryer *et al.*, 1993; Jackson *et al.*, 1996). However, elemental fractionation during laser ablation decreases with decreasing wavelength (Geersten *et al.*, 1994), and thus by quadrupling the frequency of Nd–YAG lasers (266 nm wavelength), or using gas-based lasers operating in the deep UV (such as Ar–F excimer lasers which produce light at 193 nm (e.g., Eggins *et al.*, 1998)), more reproducible U/Pb ratios could be measured compared to the earlier analyses using larger wavelengths. To counterbalance the effects of laser-induced elemental fractionation, as well as minimize temporal variations, many workers analyze externally calibrated standard minerals and unknowns under identical operating conditions, and then apply correction factors to the unknowns (e.g., Fernández-Suárez *et al.*, 1999; Knudsen *et al.*, 2001). In this respect, the technique shares strong similarities with ion microprobe U–Th–Pb age determinations. A major difference between the techniques, however, is that the volume of the pit excavated by a laser is much larger than the spot produced by the ion microprobe (Figure 2), and thus laser ablation must be considered a destructive technique.

Precision of $^{206}Pb/^{238}U$ dates using external-standard correction methods with frequency quadrupled lasers is partly dependent on sufficiently high uranium and lead concentrations, but typical values appear to be in the range of several percent. Even higher reproducibility of U/Pb ratios (~2% for $^{206}Pb/^{238}U$ of a Paleozoic zircon standard) from ablated zircons was reported by Horn *et al.* (2000). These workers used an excimer laser (193 nm wavelength), which presumably results in less laser-induced elemental fractionation than studies employing 266 nm wavelength lasers. In addition, Horn *et al.* (2000) combined the laser ablation with solution nebulization of known quantities of Tl and enriched ^{235}U. Comparing the measured $^{205}Tl/^{235}U$ with the known ratio allowed for a correction factor for instrumental mass bias to be used on the measured $^{206}Pb/^{238}U$ ratio. By comparing the measured $^{203}Tl/^{205}Tl$ ratio of the introduced Tl with its natural ratio, a correction factor was applied to the measured lead isotopic ratios to further increase accuracy, an approach first described by Longerich *et al.* (1987). Elemental fractionation produced at the laser-ablation site is still significant, however, and a correction must be applied to obtain accurate data. Horn *et al.* (2000) demonstrated that such fractionation is a function of pit geometry, thus allowing corrections to be made by establishing an empirical correction curve using a standard zircon.

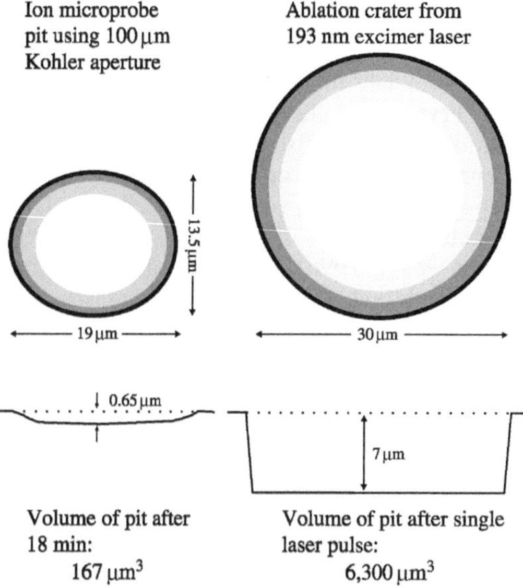

Figure 2 Size of a typical pit produced in an accessory mineral using an ion microprobe during an 18 min analytical run (from Stern, 1997) compared to the size of an ablation crater made from a single pulse of an excimer laser (from Horn *et al.*, 2000). Bottom drawings show generalized cross-sections of the spots made from the two techniques. Note the considerably smaller volume of mineral excavated during the ion microprobe analysis.

In the 1990s, mass spectrometers with a magnetic sector and a full array of Faraday detectors were coupled with a plasma source. These multi collector instruments (MC–ICP–MS) produce the same flat-topped peaks produced by TIMS and thus are capable of higher-precision isotope ratio measurements than quadrupole-type instruments. There is considerable interest in using these new-generation instruments for U–Th–Pb dating by directly analyzing zircon and monazite crystals via laser ablation, largely following the techniques originally developed for quadrupole instruments. However, because MC–ICP–MS is such a new technique, few U–Pb geochronological studies have so far been published discussing the results obtainable using these multicollector instruments (Machado and Simonetti, 2001). Based on the limited data currently available, it appears that typical 2σ uncertainties for U–Pb dates on zircon using multicollector instruments (e.g., 1–2%) are better, but not yet substantially so, than those obtainable using quadrupole instruments. However, use of multicollector ICP–MS for U–Pb dating is in its infancy and thus the technique may hold considerable potential as a geochronological tool.

3.10.2.6 U–Th–Pb Dating of Monazite Using Only Uranium, Thorium, and Lead Concentrations

The mineral monazite, a uranium- and thorium-rich phosphate of a rare-earth element (REE), is a common accessory mineral in a variety of felsic igneous rocks and is a common trace constituent in many metamorphic rocks, particularly metapelites. Because of its high uranium and thorium content, and fairly low common lead content, it has often been dated using ID-TIMS and IMP techniques. In the early 1990s the potential of determining the age of monazite using an electron microprobe (EMP) was investigated (Suzuki and Adachi, 1991; Montel *et al.*, 1994; Suzuki and Adachi, 1994). The so-called "chemical age" of a monazite crystal can be determined solely by measuring its thorium, uranium, and lead contents (no isotopic measurement) if the amount of common lead is negligible and no post-crystallization loss or gain of uranium, thorium and lead occurred (see review in Montel *et al.*, 1996). If these conditions are met, then within ~100–200 million years the amount of accumulated Pb^* in a typical monazite is high enough that it can be measured accurately and an age calculated. Precision of the chemical age is largely governed by lead content (as is true with all *in situ* techniques) and thus is largely age dependent, but 2σ precisions of 10–20 Myr are now obtainable for very lead-rich (>2,000 ppm) crystals, with errors around double those values for monazite with lower lead contents (Williams and Jercinovic, 2002). The main advantage of the EMP technique is the excellent spatial resolution that can be obtained; *in situ* analysis of monazites as small as 5 μm can be obtained from polished thin sections (Montel *et al.*, 1996). This feature is of limited utility in dating relatively large (radius = 30 μm) monazites with simple histories, but is invaluable for studies where *in situ* analysis is critical, such as determining the age of monazites that occur as inclusions in porphyroblasts (e.g., Williams *et al.*, 1999; Montel *et al.*, 2000). Employing such high spatial resolution must be used with considerable caution, however, as accurate ages can only be obtained if no lead diffusion has occurred within the restricted region of the crystal being analyzed.

The current emphases of the technique are to constrain the timing of multiple metamorphic events, to determine directly the timing of deformational events, and to provide links between metamorphism and deformation (see review by Williams and Jercinovic, 2002). The main limitation of chemical dating of monazite by EMP, aside from the necessity of perfect closed-system behavior of the region of the mineral being analyzed, is the detection limit of lead, typically a few hundred ppm for most instruments. This limitation usually precludes the analysis of monazites younger than ~100–200 Ma. The feasibility of determining chemical ages of young monazites containing only a few tens of ppm lead was demonstrated by Cheburkin *et al.* (1997) by using a newly designed X-ray fluorescence microprobe. Improvements on the original instrument design allow for chemical dating of monazite as young as 15 Ma and as small as 50 μm (Engi *et al.*, 2002); however, the monazites could not be measured *in situ*. In a companion study, Scherrer *et al.* (2002) determined chemical ages of monazites that were first optically examined in thin section, thus still preserving full textural context of the analysis, but were then removed from the thin sections by drilling with a diamond microdrill. Although this is a labor-intensive procedure compared to EMP dating, and still cannot be done on very small monazites, an X-ray microprobe age of 55.3 ± 2 Ma was determined for a 54 ± 1 Ma monazite ($^{208}Pb/^{232}Th$ TIMS date), demonstrating the much higher precision of the X-ray microprobe technique compared to the EMP technique.

3.10.3 DETERMINATION OF AGES OF METAMORPHISM

From the very beginning of the pioneering days of geochronology, it was noted that different minerals from a single rock had different apparent ages, suggesting that different minerals retained different proportions of radiogenic daughter nuclides (e.g., Wetherill *et al.*, 1955), thus setting the stage for future thermochronologic studies. By 1959, the use of radioactive decay schemes to estimate the timing of metamorphism had been specifically discussed (Compston and Jeffery, 1959; Tilton *et al.*, 1959). Subsequently, a large number of papers were published involving Rb–Sr and K–Ar dating of different minerals within metamorphic rocks, establishing the beginning of a database of the history of continental crustal deformation. With the introduction of the $^{40}Ar/^{39}Ar$ technique (Merrihue and Turner, 1966) and the seminal discussion of the concept of mineral closure temperature (Dodson, 1973) the field of thermochronology became firmly established and its impact on our understanding of the tectonic evolution of orogenic belts has been profound. Because Chapter 3.08 is devoted to the discussion of crustal metamorphism, only a brief review of current thermochronologic techniques and their applications to continental crustal evolution is given here. Because the U–Th–Pb system is discussed in detail above, a separate section on its application specifically to constraining the timing of metamorphism is beyond the scope of this chapter. Reviews of U–Pb geochronology applied to metamorphic studies can be found in Heaman and Parrish (1991) and Mezger and Krogstad (1997). Recent discussions of the formation of metamorphic zircon domains and interpretation of geochronologic data can be found in Fraser *et al.* (1997) and Bingen *et al.* (2001), and reference therein.

3.10.3.1 $^{40}Ar/^{39}Ar$ Thermochronology

$^{40}Ar/^{39}Ar$ dating is a variation of conventional K–Ar technique in that potassium-bearing samples are irradiated with neutrons to produce ^{39}Ar from ^{39}K, thereby eliminating the need for separate measurements of potassium and argon on two separate aliquots of a sample (see McDougall and Harrison, 1999 for a detailed review). In the 1970s most studies of metamorphosed crustal regions utilized the technique in a similar fashion to previous K–Ar studies, i.e., determining $^{40}Ar/^{39}Ar$ dates of different metamorphic minerals and inferring the cooling history based on estimates of the argon closure temperature in the mineral (e.g., Lanphere and Albee, 1974; Dallmeyer, 1975; Dallmeyer *et al.*, 1975, and many others). In these studies milligram-sized mineral separates were step-heated in a furnace and the argon gas released during each temperature step was isotopically analyzed. Lasers were used to heat smaller samples, first on lunar rocks (e.g., Megrue, 1973), and subsequently on terrestrial samples (e.g., York *et al.*, 1981; Maluski and Schaeffer, 1982; Sutter and Hartung, 1984), although most of these early studies produced only total-fusion $^{40}Ar/^{39}Ar$ dates. Layer *et al.* (1987) demonstrated that detailed age spectra could be determined from single hornblende and biotite crystals using a defocused continuous laser beam.

The advantage of laser microprobe $^{40}Ar/^{39}Ar$ analyses is the ability of *in situ* analysis. Spatial resolution of 50–100 μm can be achieved using lasers in the visible and near-infrared wavelengths, although these are best for minerals that are strong absorbers of such wavelengths such as biotite, phlogopite, and hornblende (see review by Kelley, 1995). With the employment of UV lasers in $^{40}Ar/^{39}Ar$ thermochronology spatial resolution increased considerably ($\sim$10 μm width) as well the ability to analyze most silicate minerals, including white mica and feldspar that are poor absorbers of higher-wavelength energy (Kelley *et al.*, 1994). Important applications of this technique to crustal metamorphic studies include the direct dating of deformation fabrics (e.g., Reddy *et al.*, 1996), dating different portions of single P–T paths (e.g., DiVincenzo *et al.*, 2001) and dating of mineral inclusions in porphyroblasts (Kelley *et al.*, 1997). Further discussions of the modern applications of $^{40}Ar/^{39}Ar$ dating to constrain the timing of metamorphic events are given by Hodges in Chapter 3.08.

3.10.3.2 Rb–Sr Dating

It was recognized early on that the Rb–Sr system was particularly useful for constraining the timing of metamorphic events because of the significant degree of rubidium and strontium diffusion that occurs between minerals during metamorphism. By constructing Rb–Sr mineral isochrons the timing of diffusion (i.e., metamorphism) can be determined, assuming complete isotopic re-equilibration occurred during a discrete metamorphic event and the system remained closed to any further disturbance (e.g., Fairbairn *et al.*, 1961). Important early work in contact metamorphic zones constrained the behavior of Rb–Sr in mineral systems (Hart, 1964; Hanson and Gast, 1967). The Rb–Sr method continues to play a very important role in studies of the deformational and metamorphic history of crustal regions, but the focus has shifted towards determining the ages of minerals within a well-defined textural context to better interpret the significance of the constructed isochrons. Some recent examples are given below.

Because of its common occurrence in metamorphic rocks, garnet separates have been an obvious choice of one of the components to be incorporated in mineral Rb–Sr isochrons in metamorphic studies (see above). Building on that previous work, Christensen *et al.* (1989) sliced large (3 cm) single garnets into separate pieces for Rb–Sr isotopic analysis along with the rock matrix between the garnets. The outer portions of the garnets were sufficiently higher in $^{87}Sr/^{86}Sr$ compared to central rim portions that growth rates of the garnets (1.0–1.7 mm Myr^{-1}) and duration of total garnet growth (9–13 Myr^{-1}) could be determined. Vance and O'Nions (1990) followed a similar procedure measuring both Rb–Sr and Sm–Nd isotopic parameters on single garnet sections obtained by sawing garnet crystals into inner, middle, and outer portions. The $\sim$448 Ma isochrons that were obtained presumably established the timing of prograde growth of the garnets from that region of Newfoundland.

In a different approach to determining the timing of prograde metamorphism, Burton and O'Nions (1991) analyzed the isotopic composition of small (1 mm) garnets from interlayered metasedimentary rocks that experienced a common P–T–t history, but one in which garnets formed at two different P–T conditions, as a function of different H_2O activities. Based on the determined Rb–Sr isochrons the lower P–T garnets formed at 437.3 $\pm$ 11.4 Ma and the higher P–T garnets at 423.5 $\pm$ 4.7 Ma, in excellent agreement with the determined Sm–Nd isochrons (434.1 $\pm$ 1.2 Ma and 424.6 $\pm$ 1.2 Ma, respectively). These ages, when combined with the paleothermometric and barometric data, provided a significant amount of information on the rate of metamorphic processes in this region of Norway. These types of studies in similar regions of crustal thickening should provide an advance in our understanding of the lithospheric thermal response during collisional tectonics.

With the demonstration that extremely small amounts of meteoritic material could be dated via Rb–Sr "microchrons" (Papanastassiou and Wasserburg, 1981) the path was opened for microanalysis of terrestrial samples, allowing much more control over the Rb–Sr dating of metamorphic material than previously possible. By analyzing small quantities of white mica from metamorphosed rocks the Rb–Sr system is capable of providing ages of formation, rather than cooling ages, thus establishing the time of at least one specific event in the metamorphic history of an area (e.g., Cliff, 1994; Chen *et al.*, 1996). These types of studies have been further refined and the more recent microchron methods employ a micro-drill and petrographic microscope to allow very specific areas of a geological thick section (~50 μm) to be sampled with complete textural control. Müller *et al.* (2000a) demonstrated the power of this technique by determining Rb–Sr dates of white mica from mylonites, which developed under greenschist facies metamorphic conditions, from shear zones in the eastern Alps.

In an even more novel approach, the timing of the duration of shearing was established by determining Rb–Sr dates of micromilled samples of crystal fibers that developed in the strain fringe around pyrite grains from a fault zone in the northern Pyrenees (Müller *et al.*, 2000b). These types of studies are just at their beginning stages but appear to be poised to revolutionize our ability to determine the timing, and possibly duration, of different deformational events, a critical step to more fully understanding all aspects of crustal evolutionary processes.

3.10.3.3 Sm–Nd and Lu–Hf Dating

The main target of Sm–Nd and Lu–Hf dating in metamorphic studies continues to be the mineral garnet. Garnet is a major constituent in many metamorphic rocks; it preferentially incorporates heavy rare earth elements, and hence can have very high $^{147}Sm/^{144}Nd$ (e.g., Stosch and Lugmair, 1987) and $^{176}Lu/^{177}Hf$ ratios (e.g., Duchêne *et al.*, 1997), and it has been widely used in thermobarometric studies, thus potentially providing a direct link between time and $P-T$ conditions. In addition, garnet has a relatively high closure temperature for both the Sm–Nd (Mezger *et al.*, 1992; Ganguly *et al.*, 1998) and Lu–Hf systems (Scherer *et al.*, 2000), thus increasing its attraction as a useful mineral in determining the timing of metamorphic events. Early studies were geared towards determining the timing of garnet growth during prograde metamorphism and relied on analyzing very large garnets (Vance and O'Nions, 1990; Mezger *et al.*, 1992; Getty *et al.*, 1993). Caution must be used in interpretation of these types of analyses, however, as it may be inclusions of REE-rich accessory minerals, and not the garnet itself, that dominates the Sm–Nd budget (see De Wolf *et al.*, 1996 for a discussion). These types of analyses can provide information about garnet growth only if the inclusions and host garnet grew simultaneously or if the inclusions were isotopically equilibrated with the host matrix. In a more recent study, cores of up to 50 single garnets were mechanically isolated and then combined for Sm–Nd analysis in an effort to minimize the effects of averaging of different zones and thus provide better constraints on the timing of peak metamorphism (Argles *et al.*, 1999). With the increased availability of computer-controlled microdrilling devices, ones capable of isolating very narrow regions of silicate minerals (e.g., Müller *et al.*, 2000a), it is likely that future studies will focus on selecting even more specific regions within single garnets for thermochronology.

The ability to select specific intracrystalline regions for analysis (e.g., Ducea *et al.*, 2003), combined with increasingly sophisticated leaching techniques to minimize the effect of micro-inclusion contamination (e.g., Amato *et al.*, 1999; Anczkiewicz *et al.*, 2002), should increase the accuracy and precision of garnet thermochronology. Similarly, the increasing number of isotope laboratories with MC–ICP–MS instruments capable of analyzing very small quantities of hafnium will likely cause a dramatic increase in the number of metamorphic studies employing Lu–Hf garnet dating. When such studies are combined with the recent advances in experimental studies of REE diffusion and reexamination of concepts of closure ages (Ganguly *et al.*, 1998; Ganguly and Tirone, 1999, 2001; Albarède, 2003), substantial progress should be made in our understanding of the timing and duration of prograde versus retrograde metamorphic reactions. $P-T-t$ studies will thus become ever more important to crustal evolution studies as a whole as the link between geochronology and metamorphic textural context becomes increasingly strengthened (e.g., Müller, 2003).

3.10.4 DETERMINATION OF AGES OF UPLIFT OR EXHUMATION

Determining the magnitude and timing of crustal uplift or exhumation of orogenic belts are critical to our understanding of the crustal evolution of the regions investigated. Under favorable conditions, the magnitude of the exhumation of part of the crust can be estimated by geobarometry. If geobarometric information is combined with measured "cooling ages" of different minerals with very different closure temperatures, then the average rate of crustal exhumation can be estimated. Of most interest to the majority of uplift/exhumation studies are relatively low-temperature (50–300 °C) thermochronologic techniques. Three such techniques, from highest to lowest closure temperature, are discussed.

3.10.4.1 $^{40}Ar/^{39}Ar$ Dating of Potassium Feldspar

Most common potassium-bearing minerals lose variable amounts of radiogenic argon at geologically modest temperatures, which at first glance would appear to make $^{40}Ar/^{39}Ar$ dating of limited geochronological use. However, argon diffusion appears to be a thermally activated process thus making the $^{40}Ar/^{39}Ar$ technique an excellent and widely used thermochronometer (see Section 3.10.3.1 and Chapter 3.08). One of the most important recent applications of $^{40}Ar/^{39}Ar$ dating to apparent uplift/exhumation studies is potassium feldspar thermochronology. An early assumption was that single closure temperatures (T_c) could be determined for potassium feldspars from slowly cooled plutons by using the $^{40}Ar/^{39}Ar$ data from the lower temperature (<600 °C) steps of a step-heating analysis (e.g., Heizler *et al.*, 1988), consistent with single diffusion domains within the feldspars. Lovera *et al.* (1989) demonstrated inconsistencies with cooling rates determined by the specific closure temperatures and those calculated by examining the release spectra and thus suggested that the feldspar diffusion domains were of variable size, a suggestion also made by Zeitler (1988). In subsequent work, Lovera and Richter (1991) further demonstrated the multidomain behavior of feldspars by demonstrating the same phenomenon even when analyzing single crystals. Of particular importance was the demonstration that modeling the thermal history of feldspar was largely independent of domain geometry, size, and volume fraction. Thus, by measuring the ^{39}Ar release spectra from a large number of heating steps of potassium feldspar from a single rock, and generating Arrhenius plots from those data, an apparently very robustly modeled cooling history can be obtained. A wide number of such cooling studies have now been made (see McDougall and Harrison (1999) for full references). Such studies when used in conjunction with other low-temperature techniques such as apatite fission track (FT) analysis and (U–Th)/He dating (see below) can provide a very substantial percentage of the full cooling path encountered by specific regions of crust.

3.10.4.2 FT Dating of Apatite

In 1962 it was demonstrated that by chemically etching a uranium-bearing mineral the paths, or tracks, traveled by the fragments arising from the spontaneous fission of ^{238}U could easily be viewed with an optical microscope (Price and Walker, 1962). A year later the first FT date of a mineral had been published (Price and Walker, 1963). The track, caused by disruption of the crystal lattice from the oppositely moving fission fragments, is initially ~16–17 μm in length in apatite crystals (after etching), but becomes increasingly shortened, or annealed, with both increasing time and temperature (e.g., Gleadow *et al.*, 1986). The maximum temperatures that can be reached and still retain abundant FTs vary with different minerals, in the same way as does the closure temperature for retention of a specific radiogenic daughter isotope. For apatite, such a closure temperature with regard to FTs is generally quoted at ~100 °C (e.g., 105 ± 10 °C; Parrish, 1983). Assigning an FT closure temperature is much less straightforward than the retention of a daughter isotope, however, as there are no geological temperatures, even 20 °C, at which annealing can be considered negligible (Donelick *et al.*, 1990). Also, chemically different apatites can exhibit significantly different annealing behaviors (e.g., Green *et al.*, 1989).

Because FT formation can be viewed as a continuous, constant process each track has the potential to experience a different segment of the thermal history of the rock containing the apatite crystal. Thus each track could become shortened (annealed) to varying degrees, depending on the thermal history, and if accurate models of FT annealing can be made then the distribution of track lengths can provide considerable insight into that thermal history (e.g., Green *et al.*, 1986, and references therein). A primary objective of annealing experiments is therefore to establish a thermal model that describes the behavior of FT systematics over geological time scales (e.g., Laslett *et al.*, 1987, and many others). Fitting FT data from natural samples to such models may allow significant constraints to be placed on the past uplift/exhumation history of a crustal region, one of the most significant goals of FT thermochronology. Such studies have been applied to the unroofing history of a large number of orogenic belts (e.g., Fitzgerald *et al.*, 1995; Gallagher *et al.*, 1998).

A potential problem with some of the earlier annealing models, upon which most FT thermal studies of sedimentary basins have been based, is that they characterize only a single type of apatite (i.e., the models are monokinetic), which may not always be applicable given the demonstration of different annealing properties of apatites of different composition (Green *et al.*, 1989; Carlson *et al.*, 1999). Significant advances have being made in multikinetic thermal modeling of apatite FT annealing (Ketcham *et al.*, 1999), which has the potential to significantly advance the level of modeling of sedimentary basin thermal evolution as well as refine further crustal uplift and exhumation studies.

3.10.4.3 (U–Th)/He Dating of Apatite

Some of the earliest attempts at dating uranium-bearing minerals were made by measuring the accumulation of helium in crystals from the α-decay of uranium and thorium. Two efforts that immediately predated the development of modern geochronology were Hurley (1954) and Damon (1957). Because of the ease of helium diffusion, however, the dates calculated were shown to be too young in most cases and the technique was soon abandoned in favor of

U–Th–Pb isotopic techniques. A resurgence of interest in the technique began in the late 1980s with a more thorough quantitative understanding of the diffusive behavior of helium in different minerals (Zeitler *et al.*, 1987). It has now been demonstrated that the mineral apatite has a closure temperature for helium of ~70 °C (Wolf *et al.*, 1996) and thus is well suited as a very low-temperature thermochronometer, that can further extend information relating to a variety of uplift and shallow crustal studies (see reviews by Farley, 2002; Ehlers and Farley, 2003). The typical method of determining a (U–Th)/He date is by extraction of helium gas either with a furnace (e.g., Zeitler *et al.*, 1987; Lippolt *et al.*, 1994; Wolf *et al.*, 1996) or by laser (e.g., House *et al.*, 2000) followed by mass spectrometric analysis, usually with small quadrupole-based instruments. After helium extraction, the apatite grains are recovered and dissolved for measurement of uranium and thorium abundances. There does not appear to be any loss of thorium or uranium during the vacuum extraction of the helium. Before the timing of helium closure can be calculated from the collected data, a correction factor must be applied because of the phenomenon of α ejection (see Farley *et al.*, 1996). The α particles produced from uranium and thorium decay can travel 20 μm through an apatite crystal lattice, thus α particles will be ejected (i.e. helium loss) when the parent nucleus occurs near the edge of the crystal. Corrections for this helium loss must be estimated, and are based on the assumptions of idealized crystal geometry and near homogeneous distribution of uranium and thorium. The correction factors to the age, based on the size of the crystal, are typically between 1.2 and 1.5 (Ehlers and Farley, 2003), with uncertainties in the correction factors increasing with decreasing crystal size. For moderate to large apatite crystals (i.e., those occurring in typical plutonic rocks) the reproducibility of helium cooling ages, combining analytical errors with uncertainties in α ejection correction factors, is ~±5% (Ehlers and Farley, 2003).

3.10.5 NEODYMIUM ISOTOPES AND CHEMICAL AGE OF CRUST

3.10.5.1 Sm–Nd Methodology

The Sm–Nd isotopic method depends upon the decay of ^{147}Sm, comprising ~15% of natural samarium, to ^{143}Nd by α-decay. With a reasonably well-known decay constant of $6.54 \times 10^{-12}\ \mathrm{yr}^{-1}$ (Lugmair and Marti, 1978; Begemann *et al.*, 2001), production of ^{143}Nd is slow. The ratio ^{143}Nd/^{144}Nd, which is measured by isotope geologists, changed from ~0.50687 at the birth of the Earth to ~0.51264 today (Jacobsen and Wasserburg, 1984), with the highest values of major rock reservoirs under mid-ocean ridges possessing values ~0.5132. Because the precision of measurement on this ratio is typically $\sim 5 \times 10^{-6}$ from modern mass spectrometers, subtle variations of ^{143}Nd/^{144}Nd are actually quite easily discernible.

Samarium and neodymium are both REEs. They therefore belong to a series of elements that have a very important role in geochemistry, due to progressive chemical fractionations that occur between the lighter and heavier REE. Although samarium and neodymium are adjacent elements in the naturally occurring REE series, fractionation between them is a little larger than for two elements with sequential atomic numbers, because there is a missing element between them, promethium, which has no stable isotope. The utility of the Sm–Nd isotopic method in crustal history is driven by the fact that upper continental crust acquires, due to igneous differentiation, a parent/daughter ratio ^{147}Sm/^{144}Nd that is ~45% lower than that of undifferentiated Earth, and lower still compared to typical depleted upper mantle sources. This fractionation is due to mantle minerals such as garnet, clinopyroxene, and orthopyroxene having lower distribution coefficients for light REE than for heavy REE, so that neodymium is partitioned into magmas slightly more strongly than samarium when mantle sources are melted. Because the ^{147}Sm/^{144}Nd ratio tends to be reasonably constant in average upper crustal rocks like granite, felsic gneiss and shale, and always ~45% lower than upper mantle values, the evolution of ^{143}Nd/^{144}Nd in upper crustal rocks slows down compared to the undifferentiated Earth and to upper mantle reservoirs, always by about the same amount. Thus, the ^{143}Nd/^{144}Nd ratio measured in a crustal rock is usually a good reflection of the average age of mantle separation of the materials in the rock.

The rather constant fractionation of Sm/Nd ratios in upper continental crustal rock reservoirs is the basis for the widely applied neodymium model age that is illustrated in Figure 3. The Sm–Nd systematics of chondritic meteorites serve as a reference for the parent/daughter ratio of the undifferentiated Earth (Jacobsen and Wasserburg, 1984), labeled as CHUR for "chondritic uniform reservoir." The evolution of this undifferentiated Earth is the basis for calculation of CHUR model ages (McCulloch and Wasserburg, 1978), while the neodymium isotopic evolution of the depleted upper part of the mantle is a more valid reference for most crustal materials, resulting in the DM model age (DePaolo, 1981). Neodymium isotopic compositions are usually given by $\varepsilon_{\mathrm{Nd}}$, where the deviations of ^{143}Nd/^{144}Nd above or below

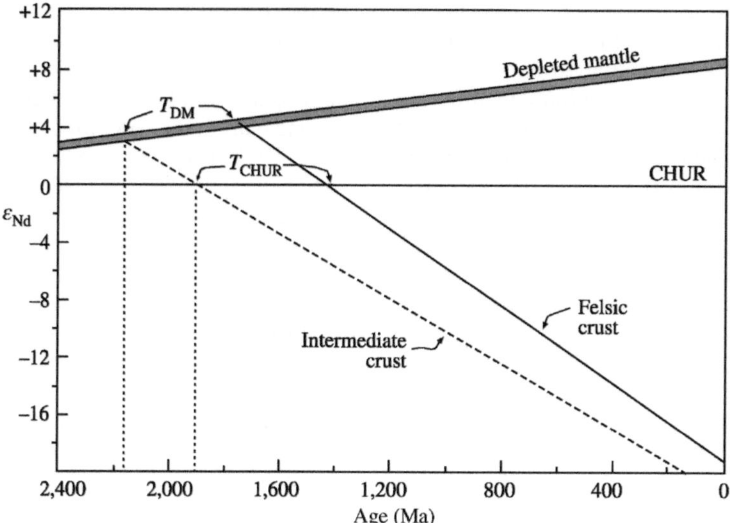

Figure 3 Sm–Nd systematics for crustal model ages and ε_{Nd} values. The parameter $\varepsilon_{Nd} = 10^4.[^{143}Nd/^{144}Nd(sample) - ^{143}Nd/^{144}Nd(CHUR)]/^{143}Nd/^{144}Nd(CHUR)$, where all $^{143}Nd/^{144}Nd$ values are specified at the age of interest (t). CHUR Nd isotopic values can be obtained from the following equation: $^{143}Nd/^{144}Nd(CHUR,t) = 0.512638\text{-}0.1966.(e^{\lambda t}\text{-}1)$, where $\lambda = 6.54 \times 10^{-12}\ a^{-1}$. All crustal materials evolve towards negative ε_{Nd} with time, because the evolution of $^{143}Nd/^{144}Nd$ is slower than the chondritic or bulk silicate Earth reference (CHUR). Nd model ages of crust can be calculated based on the chondritic reference (T_{CHUR}) or on the intersection with the approximate evolution of depleted upper mantle sources (T_{DM}). T_{DM} is generally more meaningful for juvenile crustal rocks produced during plate-tectonic cycles, because it models the average separation age from the type of mantle that is commonly observed as a source of both oceanic crust and island arcs today. Note that the depleted mantle evolution is here shown as a straight line, but is very slightly concave-upwards in the model of DePaolo (1981).

CHUR are given as parts in 10,000 (see Figure 3 and caption). Because of their Sm/Nd ratios below CHUR (see Chapter 3.01), all continental crustal reservoirs evolve towards negative ε_{Nd} with time.

It is important to understand that whether neodymium model ages are an explicit part of the discussion (as in DePaolo, 1981), or are de-emphasized in favor of an interpretation based on ε_{Nd} values (e.g., Patchett *et al.*, 1999), the utility of the Sm–Nd method in crustal evolution is ultimately based on the reproducible Sm/Nd fractionation that occurs between crust and mantle. Note that the Sm/Nd fractionation of more primitive crust of intermediate composition (Figure 3), such as basaltic andesite or andesite, does not evolve to negative ε_{Nd} so rapidly as fully differentiated upper continental felsic crust. In this way, the neodymium isotopic composition of a crustal rock depends on the average mantle differentiation age of the components that went into making it. The origin of some crustal rocks can be very complicated, e.g., a granitoid melted from complex lower crustal sources (Chapter 3.11), or a sedimentary rock derived from erosion of multiple terrains. Thus, a single model age for a granitoid or a sedimentary rock can be quite misleading in terms of real geological events and processes. Nevertheless, it is the differences in age of separation from mantle sources that bestow usefulness on neodymium isotopic variations, and no matter how complex the origin of a crustal rock, it is those age differences that are the ultimate basis for any interpretation. This is clearly seen, e.g., in the similarity of Sm–Nd systematics in many of the world's major rivers today (Goldstein *et al.*, 1984).

Lu–Hf isotope systematics provide an important complement to Sm–Nd in the study of the crust and mantle (e.g., Patchett *et al.*, 1981; Salters and Hart, 1991; Vervoort and Blichert-Toft, 1999). In the crustal context, Lu–Hf is extremely important because of the ~1% hafnium content of zircon, and the consequent ability to isotopically characterize the hafnium within grains that have been U–Pb dated (Patchett *et al.*, 1981; Corfu and Stott, 1996; Vervoort *et al.*, 1996; Amelin *et al.*, 1999). However, the Lu–Hf isotopic system is currently overshadowed by a controversy over the decay constant. For many years, a value for the ^{176}Lu decay constant of $1.94 \times 10^{-11}\ yr^{-1}$, based on the eucrite meteorite isochron of Patchett and Tatsumoto (1980) and Tatsumoto *et al.* (1981) was used. More recent physical determinations reviewed by Begemann *et al.* (2001) have high dispersion, but do not seem to corroborate the 1.94×10^{-11} value. At the present time, there is a discrepancy between values based on U–Pb-dated terrestrial Precambrian REE-rich minerals, such as apatite, which suggest a decay constant of 1.865×10^{-11} (Scherer *et al.*, 2001), and meteorite isochrons, that suggest values of $(1.93–1.98) \times 10^{-11}$ (Bizzarro *et al.*, 2003; Blichert-Toft *et al.*, 2002). For this reason, we mostly do not include Lu–Hf isotopic data in discussion of crustal age and origins. Lu–Hf data are of considerable importance in studies of early Archean rocks (see Section 3.10.6.4), and this uncertainty should be resolved as rapidly as possible.

3.10.5.2 Juvenile Crust Production versus Intracrustal Recycling

It is fundamental to the neodymium isotopic approach that neodymium isotopes are able to distinguish between material added newly to the Earth's continents and material that is merely recycled older crustal rock. This is important because sedimentary rocks derived from erosion of older continent may appear quite similar to those derived from erosion of young island-arc terrain, and unless they show marked S-type characteristics, granitoids are notoriously similar to each other, regardless of their ultimate origin. Thus, in a world where all regions of continents had been characterized for both orogenic ages and neodymium isotopic characteristics, one could draw two global maps. One would show orogenic-belt ages, representing times of consolidation of regions of the continents, while the other would show generally older neodymium-based average ages, that would represent the true differentiation age of the crust. Some orogenic belts might consist of dominantly juvenile crust, with neodymium ages similar to the orogenic age, but other belts might be dominated by materials recycled from older crustal terrains. Because the sedimentary system is a powerful mover of crustal detritus over large geographic scales, and because melting to produce granitoid batholiths often averages large domains of lower continental crust, orogenic belts very often have mixed origins in terms of the crustal age of their components. Regional maps of crustal age based on neodymium model ages, even without the references to the isotopic work, appear quite often in the literature and in presentations (e.g., Karlstrom *et al.*, 1999). The coverage of U–Pb ages and neodymium isotopic data is not yet sufficient to draw robust global maps, and because the current pace of data accumulation is high, we do not attempt to compile global maps in this chapter. Instead, the approaches will be illustrated with examples.

3.10.5.3 Juvenile Crust Production at 1.9–1.7 Ga

The abundant crust that was produced in the 1.9–1.7 Ga interval has been the subject of numerous studies, and the evolution of that work illustrates important elements in study of the continental crust. Following the demonstration of a juvenile origin for the ~1.8 Ga crustal assemblage in Colorado (DePaolo, 1981), there followed a period in which neodymium isotopic data were gathered for numerous terrains in the northern continents (Figure 4). In North America, studies by Nelson and DePaolo (1984, 1985), Bennett and DePaolo (1987),

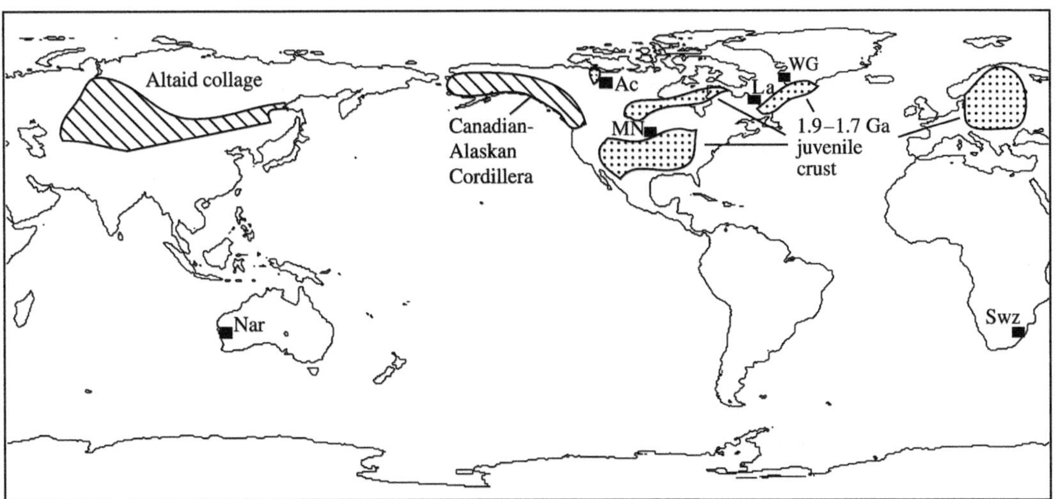

Figure 4 Regions of crust of different ages described in this chapter. Dot pattern represents 1.9–1.7 Ga juvenile crust of North America, Greenland, and Europe. Diagonal lines represent juvenile crust of the Altaid collage of orogenic belts and of the Canadian–Alaskan Cordillera. Black rectangles are locations of small regions of pre-3 Ga crust: WG—west Greenland; La—Labrador; MN—Minnesota; Ac—Acasta Gneiss Complex, northwestern Canada; Swz—Swaziland; Nar—Narryer terrane, western Australia.

Chauvel *et al.* (1987), Barovich *et al.* (1989), and Bowring and Podosek (1989) added rapidly to the database. In Greenland and Europe, work by Patchett and Bridgwater (1984), Skiöld and Cliff (1984), Kalsbeek and Taylor (1985), Patchett and Kouvo (1986), Huhma (1986), Kalsbeek *et al.* (1987), Claesson (1987) and Patchett *et al.* (1987) performed the same function. All of the studies cited above documented a high proportion of juvenile crust production in the 1.9–1.7 Ga orogenic belts of the northern continents, as highlighted in the reviews by Patchett and Arndt (1986) and Condie (1990). Large volumes of new crust were added, at least to the studied regions, during 1.9–1.7 Ga. It appears that the 1.9–1.7 Ga crust of the northern continents is approximately equal in mass to what could be produced by present-day island-arc generation rates over 200 Myr (Reymer and Schubert, 1984; Patchett and Arndt, 1986). Patchett and Chase (2002) modeled this in terms of accumulation of juvenile terrains in restricted regions of the globe due to transform motions associated with large-scale plate tectonics.

Since the late 1980s, two important changes have occurred in the studies of the growth of 1.9–1.7 Ga continental crust. The first is that neodymium isotopic data have become a very common component of any tectonic study of orogenic belts, granitoid complexes or sedimentary sequences, particularly in the Precambrian, meaning that neodymium isotopic data are now routinely gathered as part of characterization of regions not studied before. The second change is that the pace of data acquisition has slowed a little, because regions that were well-characterized in terms of geology and geochronology were studied in the 1980s, while fresh mapping and geochronology are needed to open up new areas for credible neodymium isotopic study. Thus, more recent studies incorporating neodymium isotopes are not super-regional reconnaissance studies of regions for which geology and ages are already known, but are targeted to newer topical problems and areas in Proterozoic geology, such as ophiolites in Quebec and Finland (Hegner and Bevier, 1991; Peltonen *et al.*, 1998), collisional tectonics (Öhlander *et al.*, 1999), or granitoid petrogenesis (Valbracht *et al.*, 1994; Krogstad and Walker, 1996; Rämö *et al.*, 2001).

In regions away from North America, Europe and Australia, the pace at which the volume both of well-dated Precambrian terrain and of neodymium isotopic information increases is often much lower. This leaves large parts of the continents poorly known for neodymium crustal age. The very important conclusion, that the 1.9–1.7 Ga period saw a sequence of relatively rapid and high-volume additions to the continental crust, is not likely to change. However, it is not yet really clear if the 1.9–1.7 Ga crust of the northern continents could represent juvenile terrain gathered from the rest of the globe, as suggested by Patchett and Arndt (1986) and Patchett and Chase (2002). That would be a model where the 1.9–1.7 Ga period could have shown normal crustal growth rates, but where plate tectonics had grouped the juvenile crustal products into restricted regions. If the continental regions that are currently poorly known for their crustal age also show significant juvenile crust of 1.9–1.7 Ga age, then this event must have been larger than presently documented in the northern continents, and would represent a distinct, positive spike in crustal growth. Such a situation might correspond to crustal growth initiated by mantle plume activity, as hypothesized by authors like Boher *et al.* (1992) and Stein and Hofmann (1994).

3.10.5.4 Juvenile Crust Production in the Canadian Cordillera

The Canadian–Alaskan Cordillera (Figures 4 and 5) has become the typical example of a Phanerozoic orogenic belt that contains significant amounts of juvenile crust. It is distinctive because of all the global history of mountain-building events from Cambrian to Cretaceous time, only three orogenic zones seem to display significant juvenile additions. Two of these are the South Island of New Zealand (e.g., Frost and Coombs, 1989), which is small in size, and the very large Altaid orogenic collage of central Asia (e.g., Sengör *et al.*, 1993), for which isotopic data have only recently become available. The Canadian–Alaskan Cordillera has been studied for radiogenic isotopes by several groups against a backdrop of rather well-constrained geology.

In the 1970s, the Canadian Cordillera became a center of tectonic attention because of the demonstration of the distinctness of the belts of rocks from which it was constructed (Monger *et al.*, 1972). This region became the typical example for the concept of tectonostratigraphic terranes and their assembly by accretion to form a mountain belt (Coney *et al.*, 1980). Subsequently, discussions were initiated about the apparent transpressional emplacement of many of the terranes, and northward transform faulting of them after emplacement (Gabrielse, 1985; Umhoefer, 1987). A recent geological review of Cordilleran geology and the development of ideas are found in Monger (1993). In parallel with the geological investigations, geochronology and strontium isotopic tracer investigations were carried out by Armstrong and co-workers (Armstrong, 1988; Armstrong and Ward, 1993). Because initial $^{87}Sr/^{86}Sr$ ratios of

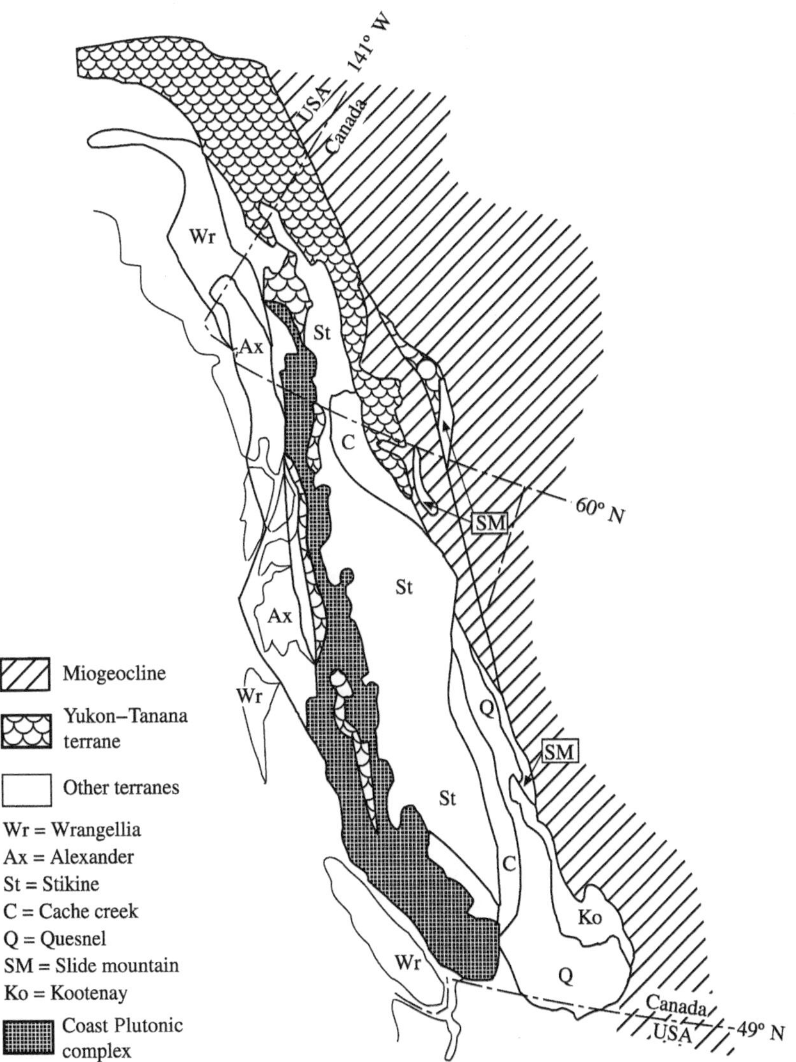

Figure 5 Terranes of the Canadian Cordillera. Older continental components are represented by the Miogeocline, which rests on North American Precambrian basement, and the Yukon–Tanana terrane. Of the other terranes, those near to the Miogeocline, such as Kootenay, have major older crustal components. Juvenile characteristics increase in the direction of the Pacific Ocean (after Patchett *et al.*, 1998; Butler *et al.*, 2001).

plutonic rocks in the outboard parts of the Canadian Cordillera tended to be low, and because geological associations suggested that a high proportion of intra-oceanic rock assemblages had been accreted to North America, there was a general supposition that juvenile crustal elements from the paleo-Pacific basin would be abundant in the Canadian Cordillera (Monger *et al.*, 1972; Coney *et al.*, 1980; Armstrong, 1988). However, many of the terranes, even outboard ones, contain Paleozoic rock assemblages that were viewed as "continental" (Monger *et al.*, 1972).

It was against this background that the present authors, in collaboration with George Gehrels, initiated neodymium isotopic characterization of the Canadian Cordillera, beginning with the Alexander terrane (Figure 5). In this outboard terrane, tightly folded Neoproterozoic and Paleozoic felsic igneous assemblages form the older components, and were some of the rocks referred to by Monger *et al.* (1972) as "continental." At the time when the first research proposal was written, it seemed to the present first author that these assemblages would yield neodymium isotopic data typical of older Precambrian continental crust. The outcome was very different, because all components of the Alexander terrane, including the oldest ones, turned out to have positive initial ε_{Nd} values (Samson *et al.*, 1989). This result became the keynote for much of subsequent isotopic study by our group (Samson *et al.*, 1990, 1991a,b). The isotopic work was consistent with outboard terranes being generally more juvenile, but in addition it was consistent with ancient sedimentary assemblages having been incorporated into the Coast Plutonic Complex (Samson *et al.*, 1991a,b; Boghossian and Gehrels, 2000; Patchett *et al.*, 1998). By no means did these authors have a monopoly of isotopic study in the Canadian–Alaskan Cordillera, and important data for outboard terranes, the Coast Plutonic Complex (Figure 5) and related areas were published by Arth *et al.* (1988), Farmer *et al.* (1993), Mezger *et al.* (2001), Cui and Russell (1995), Friedman *et al.* (1995) and Mahoney *et al.* (1995). One important finding of the last three cited papers is to show that the southern part of the Coast Plutonic Complex does not contain the older crustal elements seen further north around the Alaska Panhandle.

Further inboard lies a zone of the Canadian Cordillera that incorporates terranes showing well-known oceanic features, like Cache Creek and Slide Mountain, arc assemblages like Quesnel, and also large amounts of clastic sedimentary rocks, along with the Yukon–Tanana terrane of more continental affinity in the northern part of the Cordillera (Figure 5). This zone is characterized by juvenile neodymium signatures in most of its volcanic rocks and mafic intrusives (Smith and Lambert, 1995; Smith *et al.*, 1995; Piercey *et al.*, 2002). Neodymium isotopic signatures showing a mixture of juvenile and Precambrian continental North American materials are shown by both clastic sedimentary assemblages (Creaser *et al.*, 1997; Patchett and Gehrels, 1998; Unterschutz *et al.*, 2002; Erdmer *et al.*, 2002), and granitoid plutons (Ghosh, 1995; Ghosh and Lambert, 1995; Piercey *et al.*, 2003).

Still further inboard towards North America lies the Omineca Crystalline Belt (Monger *et al.*, 1982) and the Miogeocline, where pre-Cordilleran Neoproterozoic through Late Jurassic sedimentary rocks rest on North America basement. The Omineca crystalline belt is not separately distinguished on Figure 5, because it is a metamorphic/plutonic overprint rather than a separate terrane, but it occupies the general area of the Quesnel and Kootenay terranes. Although juvenile neodymium is sometimes seen, the Omineca belt generally shows neodymium isotopic signatures, in both metasedimentary and plutonic rocks, that correspond to older North American continent (Burwash *et al.*, 1988; Ghosh and Lambert, 1989; Stevens *et al.*, 1996; Brandon and Lambert, 1993, 1994; Brandon and Smith, 1994; Driver *et al.*, 2000). The miogeoclinal sequence appears to have been derived from Precambrian basement before ~450 Ma, but after this time to have been supplied by distant mountains of early- to mid-Paleozoic age in the Canadian Arctic (Boghossian *et al.*, 1996; Garzione *et al.*, 1997; Patchett *et al.*, 1999).

The general picture of the Canadian–Alaskan Cordillera is of a mountain belt characterized by accretion to older continental crust of juvenile crustal elements from the ocean floor, or from island-arc environments, whose proportion increases towards the Pacific Ocean. Generally, work in the Canadian–Alaskan Cordillera has evolved in the same way as work on Proterozoic terrains described above. From initial reconnaissance studies and large-scale views of how the crust grew, research has evolved into more detailed studies with more precise geological control. Nevertheless, some important general issues about crustal evolution come to fore by thinking about the Cordillera at the largest scale. Samson and Patchett (1991) reviewed the then-existing neodymium isotopic database, and concluded that ~50% by mass of the Canadian segment of the Cordillera was juvenile crustal material. Clearly, such estimates are dependent on assumptions about whether terranes continue to lower crustal depth. Seismic data often reveal that juvenile belts are underlain by older continental material, both in Proterozoic and Phanerozoic contexts (Lucas *et al.*, 1993; Clowes *et al.*, 1999). The 50%-juvenile estimate of Samson and Patchett (1991) should certainly be revised downwards in light of seismic results summarized by Clowes *et al.* (1999), as well as more recent field and isotopic work such as that of Erdmer *et al.* (2002). Qualitatively however, the status of the Canadian Cordillera as the locus of juvenile crustal growth remains in spite of these uncertainties.

Samson and Patchett (1991) also viewed the Canadian Cordillera as an analogue for Proterozoic crustal growth by accretion of juvenile terranes, a comparison that was explicitly examined in terms of major- and trace-element compositions by Condie and Chomiak (1996). The present authors, like many others, have always been impressed by the transform faulting that is presently active along the North American margin, which serves to pile juvenile crust into one region that stretches from southern British Columbia to Alaska. Subsequent synthesis of Cordilleran geology (Johnston, 2001) and discussions about crustal growth in general (Patchett and Chase, 2002) have emphasized the northward along-margin transport displayed in the Canadian Cordillera as a critical element in crustal growth models. The rationale for this is that along-margin transport of slices of juvenile crustal material is able to pile them all up in one segment of the continental margin, perhaps even into a restricted region of the globe, creating the impression of a very intense and localized crustal growth period (Patchett and Chase, 2002), for which extraordinary mechanisms might otherwise be required (e.g., Stein and Hofmann, 1994). This process should have been important in most of geologic time, accounting for numerous apparently very intense crustal growth episodes (Patchett and Chase, 2002).

3.10.5.5 Juvenile Crust Production in the Altaid Collage of Central Asia

Another region of Phanerozoic crustal growth is the Altaid orogenic collage of central Asia, located in the eastern parts of the former Soviet Union, and in western China (Figure 4). The rocks range in age dominantly from Neoproterozoic through Early Carboniferous, with final tectonic movements and post-orogenic igneous activity extending through part of Permian time. Thus, the crustal assemblage is older than the bulk of the material in the Canadian Cordillera. In a simplified view, one might say that the Altaid system represents the bulk of global juvenile crustal accretion of Cambrian through Permian time, while the Canadian Cordillera and New Zealand have that role from Triassic through Cretaceous time.

Early work on the tectonic assembly of Asia, which was published in international journals, took place in the 1980s and early 1990s (e.g., Sengör and Hsü, 1984; Windley *et al.*, 1990; Allen *et al.*, 1992). However, the Altaid system came to the attention of the wider scientific community through the bold syntheses of Sengör and co-workers (Sengör *et al.*, 1993; Sengör and Natalin, 1996a,b). These authors described a collage of belts of arc volcanic rocks, ophiolitic assemblages, greywacke-shale sequences and granitoids, that occupies a very large area in central Asia to the north of the Himalayan system, and lying between the Baltic Shield/Russian Platform to the west, the Siberian Shield to the east, and with the North China Craton providing a partial boundary to the south. Tectonic trends are highly variable, and this seems consistent with the interpretations of Sengör and co-workers, that the collage represents the sweepings of a large ocean, with telescoping of accreted fragments not only by orthogonal collision, but with transpression and transform faulting, essentially along the strike of the developing orogen, over very large geographic scales. This resulted in a collage with variable tectonic trends occupying a large region of roughly triangular shape in the center of the Asian continent.

Isotopic analyses of Altaid-collage rocks are not as abundant as those in the Canadian Cordillera, but have appeared in increasing numbers since about 1996. Kovalenko *et al.* (1999, 2003) presented neodymium isotopic data on juvenile Altaid rocks in Russia and Mongolia, while Wu *et al.* (2000), Jahn *et al.* (2000) and Chen and Jahn (2002) made extensive studies of the abundant granitoid plutons in the Chinese part of the Altaid assemblage. Yarmolyuk *et al.* (1999) and Hu *et al.* (2000) used neodymium isotopic results to study the nature of pre-Altaid basement. Modern U–Pb zircon geochronology, which together with detailed modern documentation of tectonic relations in the field, has also been largely lacking in the Altaid crustal collage, is being actively undertaken (Wilde *et al.*, 2000; Salnikova *et al.*, 2001; Windley *et al.*, 2002). All authors concluded that substantial juvenile material was added to Asia during Altaid events, particularly during Early Ordovician through Devonian time. Post-orogenic granitoids of Late Carboniferous to Early Permian age have sources that may appear juvenile, or to consist of older continental crust, depending on the location with respect to the margins of the Altaid collage (Han *et al.*, 1997; Litvinovsky *et al.*, 2002).

What appears to emerge from the ongoing studies is that the Altaid collage consists of terranes of juvenile material juxtaposed with terranes containing pre-existing Proterozoic crustal rocks over the whole extent of the belt. Sengör *et al.* (1993) painted just such a picture, and also suggested that up to 50% of the entire Altaid collage may represent juvenile crust of Paleozoic age. So far, the neodymium isotopic data appear to support juvenile crust of about this magnitude, but much field data, geochronology and neodymium isotopic data remain to be gathered. In addition, estimates for the volume of juvenile crustal material may need revision when better data are available for whether and how the juvenile terranes project into the lower continental crust, as noted above in the context of the Canadian Cordillera. Estimates of the proportion of juvenile crustal growth represented by the Altaid collage can be expected to be refined in the future.

3.10.6 ISOTOPES AND PRE-3 Ga CONTINENTAL CRUST

3.10.6.1 Existence of Ancient Continental Crust

Isotope geology acquires a dominant significance when continental rocks dating from before 3 Ga are studied. This is because fossils are not available, and metamorphism and tectonism are often so severe that original layering of supracrustal sequences may not be visible, and even magmatic contact relationships may be obscured. Gneisses of 3.8 Ga, 2.8 Ga, and 1.8 Ga may look very similar in the field, and isotopic dating is absolutely required to distinguish them. In addition, pre-3 Ga rocks may carry unique information about environments on the early Earth, or about mantle differentiation and layering, and the search for the oldest continental crust has always been a major area of activity. All the areas of pre-3 Ga crust discussed here are indicated on Figure 4.

Early dates for ancient crust were obtained by the Rb–Sr whole-rock isochron method, or by the U–Pb zircon method when it was in its infancy. Approximately 3.5 Ga U–Pb zircon ages were obtained from gneisses in Minnesota by Goldich *et al.* (1970), and gneisses in western Greenland were dated at 3.6–3.8 Ga by Rb–Sr isochrons (Black *et al.*, 1971; Moorbath *et al.*, 1972). These age determinations were the major scientific news of their day in geology, and began a whole field of endeavor in deciphering the history of ancient crustal rocks. Early foci of research on ancient crust were western Greenland (Moorbath, 1978), Labrador (Hurst *et al.*, 1975; Bridgwater and Collerson, 1976) and Swaziland (Davies and Allsopp, 1976). Early studies of these regions using Sm–Nd were by Hamilton *et al.* (1983) and Carlson *et al.* (1983), and these and other isotopic studies documented the juvenile nature of the these earliest crustal gneiss complexes.

Presently, most occurrences of pre-3 Ga rocks have been accurately dated using modern approaches of the U–Pb zircon method (e.g., Kröner *et al.*, 1989; Bowring *et al.*, 1989b; Horstwood *et al.*, 1999), and neodymium isotope data have been widely obtained (e.g., Collerson *et al.*, 1991). Hafnium isotopic data, combined with neodymium, have been obtained for critical samples (Vervoort *et al.*, 1996; Vervoort and Blichert-Toft, 1999; Blichert-Toft *et al.*, 1999; Amelin *et al.*, 1999). The focus of neodymium and hafnium isotopic studies has moved away from documentation of juvenile character towards an attempt to constrain the degree of heterogeneity in the early Earth's mantle, with a view to detecting "primordial" rock reservoirs and effects of the primary differentiation of the Earth (e.g., Bowring and Housh, 1995; Albarède *et al.*, 2000; see Chapter 2.13). Many arguments have developed concerning the reliability of isotopic values for single samples of polymetamorphic gneisses (e.g., Vervoort *et al.*, 1996; Gruau *et al.*, 1996; Moorbath *et al.*, 1997). The case of the Acasta gneisses, to which the above-cited references apply, is detailed later in this chapter.

3.10.6.2 Crustal Growth Events and Recycling into the Mantle

Early in the evolution of isotopic studies of ancient crust, Moorbath (1975, 1978) developed the concepts of (i) major Precambrian crustal growth events in which juvenile crust was made, and (ii) the essential indestructibility of continental crust once consolidated by orogenic events. Models based on neodymium and strontium isotopes that grew the continents through time were made possible by the arrival of the meteoritically constrained Sm–Nd system (O'Nions *et al.*, 1979; Jacobsen and Wasserburg, 1979; DePaolo, 1979). A very readable account of the chemical relationships between crust and mantle is by Hofmann (1988), while one of the more mathematically complete treatments is by Allègre *et al.* (1983a,b). Although a neodymium-based cumulative age curve for global continental crust cannot yet be drawn, the estimates for North America and Europe (Patchett and Arndt, 1986; Condie, 1990) may approximate the global picture.

Alternatively, the modeling by Armstrong (1968, 1981) and Zartman and Doe (1981), initially strongly based on lead isotopes, advocated the position that during crustal growth events or during plate movement in general, continent could also be destroyed by the subduction of sedimentary rocks resting on the ocean floor. It was realized by the community that all crust and mantle geochemistry and isotopes were consistent with large amounts of crustal recycling, provided that more or less complete mixing occurred during mantle convection. Much of the discussion concerns the present-day Earth. However, the period before 3 Ga is critical to the discussion of crustal recycling because it is generally agreed that large amounts (perhaps 40–50%) of presently surviving crust came into existence during the immediately following period, between 3.0 Ga and 2.6 Ga (e.g., Condie, 1990). Either the pre-3 Ga Earth had much less felsic crust than existed after 2.6 Ga, or large amounts of pre-3 Ga crust were destroyed before or during the time that the 3.0–2.7 Ga crust was produced. If it could be shown that massive amounts of pre-3 Ga crust were destroyed, then essentially all objections to large-scale continent recycling in the mantle would disappear.

Attempts to test the effects of sediment subduction (Patchett *et al.*, 1984), or whether pre-3 Ga crust was available to be recycled into later crust (Stevenson and Patchett, 1990), founder on two possibilities. One is that in the early Earth, aided by a hotter mantle and continuing bombardment from asteroidal objects, it may have been possible to destroy continental fragments wholesale, so that little trace of them remained at the surface. Another is that large amounts of sediment subduction would result in a very well averaged geochemical signature entering the mantle, so that distinctive features of oceanic sediments, such as enrichment or depletion in zircon and its unradiogenic hafnium, could not be used to argue against the process (Plank and Langmuir, 1998). At the same time, budgets for the mass of rock entering subduction zones appear to suggest that sedimentary material has to disappear into the mantle (von Huene and Scholl, 1993), an argument used forcefully by Armstrong (1981). Consequently, a discussion that was once framed in terms of "growth or no growth" (e.g., Stevenson and Patchett, 1990; Armstrong, 1991; McCulloch and Bennett, 1994) may now focus on how much continent recycling occurs, and the magnitude of recycling compared to the growth seen in places like the Canadian Cordillera. The magnitudes of these fluxes are very important for crust–mantle evolution, but are quite difficult to determine accurately, because of the global basis required, and the deep parts of subduction zones being difficult to image precisely (Reymer and Schubert, 1984; von Huene and Scholl, 1993; see Chapter 2.11).

3.10.6.3 Acasta Gneisses, Northwest Territories, Canada

The Slave craton in northwestern Canada is an Archean granite–greenstone terrain that is bounded on the east and west by Paleoproterozoic orogenic belts (see geological reviews of Henderson (1985) and Padgham (1985)). Crustal regions within the western portion of Slave craton contain granitoids older than 2.8 Ga (see Bowring *et al.*, 1990, and references therein). The antiquity (>3.5 Ga) of the westernmost portion of the Slave craton has been known since the mid-1980s (Bowring and Van Schmus, 1984), but in 1989 it was demonstrated, via ion microprobe U–Pb dating, that zircons within a tonalitic gneiss (sample BGXM) from the Acasta River area were as old as 3.96 Ga (Bowring *et al.*, 1989a), making the gneiss the oldest known terrestrial rock. An identical age was determined for a second gneiss (sample SP-405) with a granitic composition (Bowring *et al.*, 1990). Stern and Bleeker (1998) reported $^{207}Pb-^{206}Pb$ dates as old as 4.02 Ga from a gneiss collected from within the same outcrop as that sampled by Bowring *et al.* (1989a). More recent U–Pb geochronological work on the gneisses in the Acasta River area (now called the Acasta Gneiss Complex) has extended the known extent of the oldest known surviving continental crustal region, as zircons from two metatonalites and a metagranodiorite have nearly concordant U–Pb dates of 4,002 ± 4 Ma, 4,012 ± 6 Ma, and 4,031 ± 3 Ma, respectively (Bowring and Williams, 1999).

The neodymium isotopic composition of Acasta gneiss BGXM was investigated by Bowring *et al.* (1989b). These authors reported a value of $\varepsilon_{Nd}(3.96\ Ga) = -1.7$ for a whole-rock analysis and $\varepsilon_{Nd}(3.96\ Ga) = +0.7$ for an amphibolitic-rich layer taken from a hand specimen of BGXM. An indistinguishable value, $\varepsilon_{Nd}(3.96\ Ga) = +0.8$, was reported for the granitic gneiss SP-405 (Bowring *et al.*, 1990). The corresponding T_{CHUR} ages of these three samples are 4.1 Ga, 3.92 Ga, and 3.85 Ga, respectively.

The variable ε_{Nd} values of the Acasta samples were interpreted by Bowring *et al.* (1990) as reflecting original protolith heterogeneity, rather than the effects of metamorphic disturbance. The negative ε_{Nd} value and T_{CHUR} age of 4.1 Ga of the tonalitic gneiss was interpreted as evidence that at least this portion of the Acasta Gneiss Complex must have been derived from, or interacted extensively with, a crustal reservoir considerably older than 3.96 Ga. The same argument cannot be easily made for the granite orthogneiss, however, as it has a T_{CHUR} age that is 100 Myr younger than its crystallization age.

Lead isotopic ratios were determined for HF-leaches of alkali feldspars from the granitic gneiss and of plagioclase from the tonalitic gneiss (Bowring *et al.*, 1990). The ^{206}Pb:^{207}Pb:^{208}Pb:^{204}Pb ratios of the least radiogenic leaches, presumed to be closest to the initial lead isotopic composition of the rocks, were 14.12, 15.08, 34.01 and 14.96, 15.29, 33.53, respectively. The $^{206}Pb/^{204}Pb$ and $^{207}Pb/^{204}Pb$ ratios of the three plagioclase leaches lie on a linear array with a slope equivalent to an age of 3.60 ± 0.28 Ga. Because this age overlaps the 3.62 Ga of the unzoned zircon rims from the tonalitic gneiss, the feldspar lead isotopic composition was interpreted as reflecting homogenization of whole-rock lead isotopes during a 3.6 Ga metamorphic event.

The high $^{207}Pb/^{204}Pb$ ratios of the leached feldspars were taken as evidence that the gneisses were derived from a reservoir characterized by a high U/Pb ratio prior to 3.96 Ga, such as significantly older continental crust that had not lost uranium or thorium (Bowring *et al.*, 1990). Implicit in this argument is that the lead isotopic compositions of the HF-leached feldspars accurately represent the lead isotopic composition of the protoliths to the gneisses, and hence accurately reflect the source materials of the protolith.

Hafnium and U–Pb isotopic ratios were determined for single zircons from two gneisses reported to be Acasta gneisses (Amelin *et al.*, 1999). However, no details of the sampling location or composition of the gneisses were given, except that they were from "granitic" and "amphibolitic" gneisses. Two zircons from the amphibolitic unit have $\varepsilon_{Hf}(t)$ values of −4.7 and −0.6, and the zircon from the granitic unit has a reported $\varepsilon_{Hf}(t)$ value of −1.8. However, the significance of these data to the 4.0 Ga Acasta gneisses are uncertain, as the zircons analyzed by Amelin *et al.* (1999) yielded $^{207}Pb/^{206}Pb$ dates of 3,548 Ma to 3,565 Ma, similar to the 3.6 Ga dates determined for some structureless equant zircons and the unzoned outer portion of some of the 4.0 Ga crystals analyzed by Bowring *et al.* (1989a, 1990), which they interpreted as dating the timing of a second generation of zircon growth. Because of the extreme complexity in the zoning of zircons from the Acasta gneisses, hafnium isotopic compositions of whole zircon crystals might reflect only a weighted average of different isotopic compositions.

All in all, the complexity of the isotopic data and their interpretation in the Acasta Gneiss Complex illustrate the difficulties inherent in study of planetary evolution using polymetamorphic gneisses. This is the ultimate cause for the controversies over relaibility of radiogenic isotopic parameters derived from these rocks (e.g., Vervoort *et al.*, 1996; Gruau *et al.*, 1996; Moorbath *et al.*, 1997).

3.10.6.4 Narryer Terrane, Western Australia

The oldest known portion of the Archean Yilgarn craton is the Narryer terrane, occurring in the northwestern part of the craton. The Narryer terrane contains 3.7 Ga gneisses, 3.7 Ga anorthositic rocks, and 3.6 Ga granites (see Myers, 1995, and references therein). Two important belts of metasedimentary rocks, thought to have been deposited ~3 Ga, occur within the Narryer terrane: the Narryer and Jack Hills belts. The metasedimentary rocks within the Narryer belt include metaconglomerates and quartzites that have preserved cross-bedding despite being metamorphosed at upper amphibolite to granulite facies conditions (Froude *et al.*, 1983). Detrital zircons extracted from one of these quartzites were the first pre–4.0 Ga terrestrial minerals identified (Froude *et al.*, 1983). In the original study, four zircons out of 102 crystals analyzed from one quartzite sample yielded ^{207}Pb-^{206}Pb dates of 4.11–4.18 Ga. The other detrital zircons from that sample and a second quartzite yielded ^{207}Pb–^{206}Pb dates between 3.75–3.3 Ga, with two zircons having nearly concordant 3.1 Ga dates.

The Jack Hills belt, broadly similar to the Narryer belt but metamorphosed to upper greenschist facies, contains metabasalts, chert, and banded iron formation interleaved with clastic metasedimentary rocks (Compston and Pidgeon, 1986). Detrital zircons from a ~3.1 Ga chert pebble conglomerate were analyzed using ion microprobe U–Pb techniques as part of a continuing search for ancient zircons. Compston and Pidgeon (1986) found detrital zircons with ^{207}Pb–^{206}Pb dates ranging between 4.28–4.0 Ga, significantly extending the sampling region of the oldest known minerals. Supporting geochronological evidence for the presence of 4 Ga detrital zircons from these areas was provided by both the zircon evaporation method (Kober *et al.*, 1989) and the ID-TIMS method (Amelin, 1998). In a more recent ion microprobe study of detrital zircons from Jack Hills, Mojzsis *et al.* (2001) confirmed the existence of 4.28 Ga zircons by identifying two grains giving concordant U–Pb dates of 4,279 ± 5 Ma and 4,280 ± 5 Ma. An even older detrital zircon, 4,404 ± 8 Ma, has been discovered from the Jack Hills conglomerate (Wilde *et al.*, 2001), making this crystal only ~150 Myr younger than the estimated time of Earth formation.

In addition to geochronological studies of the Earth's oldest zircons, geochemical and isotopic studies have also been performed (Kinny *et al.*, 1991; Maas *et al.*, 1992; Amelin *et al.*, 1999; Mojzsis *et al.*, 2001; Peck *et al.*, 2001; Wilde *et al.*, 2001). Maas *et al.* (1992) demonstrated that both the older and younger detrital zircon suites from the Jack Hills conglomerate have trace-element compositions and mineral inclusions consistent with nucleation from a felsic magma. Kinny *et al.* (1991) were the first to attempt to determine the hafnium isotopic composition of the zircons, but the large uncertainties in the ion microprobe measurements (±(5–7) $\varepsilon_{\rm Hf}$ units) precluded a detailed discussion of the petrogenesis of the parent magmas. In more recent work, Amelin *et al.* (1999) measured 37 Jack Hills detrital zircons for U–Pb ages and hafnium isotopes. The advantage of measuring U–Pb and hafnium isotopes from the same dissolved zircon crystal is that the degree of discordance can be determined, which in turn is important in assessing the likelihood of the presence of a xenocrystic core and/or metamorphic rims on the zircon. Zircons that are either concordant or not too discordant are the least likely crystals to contain multiple domains. Similarly, zircons that have suffered the least lead loss are the least likely to have experienced open system behavior for lutetium or hafnium (see Samson *et al.*, 2003 for discussion). Of the 10 zircons identified by Amelin *et al.* (1999) that are >3.8 Ga, two have $\varepsilon_{\rm Hf}(t)$ values that are further than ±1 $\varepsilon_{\rm Hf}$ units from CHUR. The bulk of these ancient zircons are thus consistent with derivation from a chondritic-like source, which at this age could include very slightly to nondepleted mantle, or juvenile crust that had recently been extracted from such a mantle. Two crystals, with ^{207}Pb–^{206}Pb dates of 3.82 Ga and 3.97 Ga, have $\varepsilon_{\rm Hf}(t)$ values of −2.2 ± 0.6 and −2.7 ± 0.4, respectively. These values are far enough below the CHUR line to suggest that these zircons crystallized from slightly evolved magmas, assuming that the zircons contain only single-age domains and that the ages and isotopic compositions are accurate. If those assumptions are correct, then we are presented with the intriguing possibility that an isotopically enriched reservoir existed as early as 3.97 Ga, perhaps similar to the one that gave rise to the 4.40 Ga Jack Hills zircon, in contrast to the general conclusions reached by Vervoort and Blichert-Toft (1999). However, all these hafnium isotope arguments are complicated by the present controversy over the decay constant of ^{176}Lu, described in Section 3.10.5.1.

The presence of very early evolved crustal sources is reinforced by oxygen isotope data for the ca. 4 Ga Jack Hills detrital zircons. Mojzsis *et al.* (2001) and Wilde *et al.* (2001) presented, independently, the first ion-microprobe oxygen isotope analyses of the >3.9 Ga detrital zircons. Wilde *et al.* (2001) determined $\delta^{18}O$ values of +5‰ and +7.4‰ (mean of two analyses each) for two different regions within the 4.40 Ga crystal and Mojzsis *et al.* (2001) determined $\delta^{18}O$ values of about +6‰ to +10‰ for 4.3–3.9 Ga zircons. Peck *et al.* (2001) confirmed this general range of values, but individual spot measurements did not exceed $\delta^{18}O$ of +8.6‰. The importance of these values is that because there is a fractionation of ~1.5‰ between zircon and granitic magmas (e.g., Valley *et al.*, 1994), the estimated $\delta^{18}O$ values of the parental magmas to the Jack Hills zircons are in the vicinity of +8.5‰ to +9.5‰ (Peck *et al.*, 2001). Such values are well beyond those of typical mantle (+5.5‰ to +6‰), and are most consistent with granitic magmas having been derived in part from materials that were once exposed to supracrustal conditions (i.e., low-temperature weathering or diagenesis). This suggests the very early presence of some sort of hydrosphere, including possibly a 4.4 Ga ocean (e.g., Valley *et al.*, 2002).

ACKNOWLEDGMENTS

The authors are grateful to Clement Chase and Robert Butler for help with map figures, and to Roberta Rudnick and Mark Schmitz for their thorough reviews of this chapter. Samson acknowledges Syracuse University for granting a sabbatical leave, during which this chapter was written. Patchett was supported by NSF-EAR-0003343, and Samson by NSF-EAR-0106853.

REFERENCES

Albarède F. (2003) The thermal history of leaky chronometers above their closure temperature. *Geophys. Res. Lett.* **30**(1), 1015, doi:10.1029/2002GL016484.

Albarède F., Blichert-Toft J., Vervoort J. D., Gleason J. D., and Rosing M. (2000) Hf–Nd isotope evidence for a transient dynamic regime in the early terrestrial mantle. *Nature* **404**, 488–490.

Allègre C. J., Hart S. R., and Minster J. F. (1983a) Chemical structure and evolution of the mantle and continents determined by inversion of Nd and Sr isotopic data: I. Theoretical methods. *Earth Planet. Sci. Lett.* **66**, 177–190.

Allègre C. J., Hart S. R., and Minster J. F. (1983b) Chemical structure and evolution of the mantle and continents determined by inversion of Nd and Sr isotopic data: II. Numerical experiments and discussion. *Earth Planet. Sci. Lett.* **66**, 191–213.

Allen M. B., Windley B. F., and Chi Z. (1992) Palaeozoic collisional tectonics and magmatism of the Chinese Tien Shan, central Asia. *Tectonophysics* **220**, 89–115.

Amato J. M., Johnson C. M., Baumgartner L. P., and Beard B. L. (1999) Rapid exhumation of the Zermatt-Saas ophiolite deduced from high-precision Sm–Nd and Rb–Sr geochronology. *Earth Planet. Sci. Lett.* **171**, 425–438.

Amelin Y. V. (1998) Geochronology of the Jack Hills detrital zircons by precise U–Pb isotope dilution analysis of crystal fragments. *Chem. Geol.* **146**, 25–38.

Amelin Y. V., Lee D.-C., Halliday A. N., and Pidgeon R. T. (1999) Nature of the Earth's earliest crust from hafnium isotopes in single detrital zircons. *Nature* **399**, 252–255.

Anczkiewicz R., Thirlwall M., and Platt J. P. (2002) Influence of inclusions and leaching techniques on Sm–Nd and Lu–Hf garnet chronology. *Geochim. Cosmochim. Acta (Goldschmidt Conf. Abstr.)* **66**, A19.

Anderson C. A. and Hinthorne J. R. (1973) Thermodynamic approach to the quantitative interpretation of sputtered ion in mass spectra. *Anal. Chem.* **45**, 1421–1438.

Argles T. W., Prince E. I., Foster G. L. and Vauee D. (1999) New garnets for old? Cautionary tales from young mountain belts. *Earth Planet. Sci. lett.* **172**, 301–309.

Armstrong R. L. (1968) A model for the evolution of strontium and lead isotopes in a dynamic Earth. *Rev. Geophys.* **6**, 175–199.

Armstrong R. L. (1981) Radiogenic isotopes: the case for crustal recycling on a near-steady-state no-continental-growth Earth. *Phil. Trans. Roy. Soc. London* **A301**, 443–472.

Armstrong R. L. (1988) Mesozoic and early Cenozoic magmatic evolution of the Canadian Cordillera. *Geol. Soc. Am. Spec. Pap.* **218**, 55–91.

Armstrong R. L. (1991) The persistent myth of crustal growth. *Austral. J. Earth Sci.* **38**, 613–630.

Armstrong R. L. and Ward P. L. (1993) Late Triassic to earliest Eocene magmatism in the North American Cordillera: implications for the western interior basin. In *Evolution of the Western Interior Basin*, Geol. Ass. Can. Spec. Pap. (eds. W. G. E. Caldwell and E. G Kauffman). Geological Association of Canada, St. Johns, Newfoundland, vol. 39, pp. 49–72.

Arth J. G., Barker F., and Stern T. W. (1988) Coast batholith and Taku plutons near Ketchikan, Alaska: petrography, geochronology, geochemistry and isotopic character. *Am. J. Sci.* **288A**, 461–489.

Ayer J., Amelin Y., Corfu F., Kamo S., Ketchum J., Kwok K., and Trowell N. (2002) Evolution of the southern Abitibi Greenstone belt based on U–Pb geochronology: autochthonous volcanic construction followed by plutonism, regional deformation and sedimentation. *Precamb. Res.* **115**, 63–95.

Barovich K. M., Patchett P. J., Peterman Z. E., and Sims P. K. (1989) Nd isotopes and the origin of 1.9–1.7 Ga Penokean continental crust of the Lake Superior Region. *Geol. Soc. Am. Bull.* **101**, 333–338.

Begemann F., Ludwig K. R., Lugmair G. W., Min K. W., Nyquist L. E., Patchett P. J., Renne P. R., Shih C.-Y., Villa I. M., and Walker R. J. (2001) Call for an improved set of decay constants for geochronological use. *Geochim. Cosmochim. Acta* **65**, 111–121.

Bennett V. C. and DePaolo D. J. (1987) Proterozoic crustal history of the western United States as determined by neodymium isotopic mapping. *Geol. Soc. Am. Bull.* **99**, 674–685.

Bingen B., Austrheim H., and Whitehouse M. (2001) Ilmenite as a source of zirconium during high-grade metamorphism? Textural evidence from the Caledonides of western Norway and implications for zircon geochronology. *J. Petrol.* **42**, 355–375.

Bizzarro M., Baker J. A., Haack H., Ulfbeck D., and Rosing M. (2003) Early history of Earth's crust–mantle system inferred from hafnium isotopes in chondrites. *Nature* **421**, 931–933.

Black L. P., Gale N. H., Moorbath S., Pankhurst R. J., and McGregor V. R. (1971) Isotopic dating of very early Precambrian amphibolite facies gneisses from the Godthaab district, west Greenland. *Earth Planet. Sci. Lett.* **12**, 245–259.

Blichert-Toft J., Albarède F., Rosing M., Frei R., and Bridgwater D. (1999) The Nd and Hf isotopic evolution of the mantle through the Archean. Results from the Isua supracrustals, west Greenland, and from the Birimian terranes of west Africa. *Geochim. Cosmochim. Acta* **63**, 3901–3914.

Blichert-Toft J., Boyet M., Télouk P., and Albarède F. (2002) $^{147}Sm-^{143}Nd$ and $^{176}Lu-^{176}Hf$ in eucrites and the differentiation of the HED parent body. *Earth Planet. Sci. Lett.* **204**, 245–259.

Boghossian N. D. and Gehrels G. E. (2000) Nd isotopic signature of metasedimentary pendants in the Coast Plutonic Complex between Prince Rupert and Bella Coola, British Columbia. In *Tectonics of the Coast Mountains,* Southeastern Alaska and Coastal British Columbia (eds. H. H Stowell and W. C. McClelland). Geol. Soc. Am. Spec. Pap. 343, 77–87.

Boghossian N. D., Patchett P. J., Ross G. M., and Gehrels G. E. (1996) Nd isotopes and the source of sediments in the Miogeocline of the Canadian Cordillera. *J. Geol.* **104**, 259–277.

Boher M., Abouchami W., Michard A., Albarède F., and Arndt N. T. (1992) Crustal growth in West Africa at 2.1 Ga. *J. Geophys. Res.* **97**, 345–369.

Bowring S. A. and Housh T. (1995) The Earth's early evolution. *Science* **269**, 1535–1540.

Bowring S. A. and Podosek F. A. (1989) Nd isotopic evidence from Wopmay Orogen for 2.0–2.4 Ga crust in western North America. *Earth Planet. Sci. Lett.* **94**, 217–230.

Bowring S. A. and Van Schmus W. R. (1984) U–Pb zircon constraints on evolution of Wopmay Orogen N. W. T. *Geol. Assoc. Can.: Mineral. Assoc. Can. Prog. Abstr.* **9**, 47.

Bowring S. A. and Williams I. S. (1999) Priscoan (4.00–4.03 Ga) orthogneisses from northwestern Canada. *Contrib. Mineral. Petrol.* **134**, 3–16.

Bowring S. A., Williams I. S., and Compston W. (1989a) 3.96 Ga gneisses from the Slave province, Northwest Territories, Canada. *Geology* **17**, 971–975.

Bowring S. A., King J. E., Housh T. B., Isachsen C. E., and Podosek F. A. (1989b) Neodymium and lead isotope evidence for enriched early Archaean crust in North America. *Nature* **340**, 222–225.

Bowring S. A., Housh T. B., and Isachsen C. (1990) The Acasta gneisses: remnant of Earth's early crust. In *Origin of the Earth* (eds. H. E. Newsom and J. H. Jones). Oxford University Press, Oxford, pp. 319–343.

Brandon A. D. and Lambert R. StJ. (1993) Geochemical characterization of mid-Cretaceous granitoids of the Kootenay Arc in the southern Canadian Cordillera. *Can. J. Earth Sci.* **30**, 1076–1090.

Brandon A. D. and Lambert R. StJ. (1994) Crustal melting in the Cordilleran interior: the mid-Cretaceous White Creek batholith in the southern Canadian Cordillera. *J. Petrol.* **35**, 239–269.

Brandon A. D. and Smith A. D. (1994) Mesozoic granitoid magmatism in southeast British Columbia: implications for the origin of granitoid belts in the North American Cordillera. *J. Geophys. Res.* **99**, 11879–11896.

Bridgwater D. and Collerson K. D. (1976) The major petrological and geochemical characters of the 3,600 m.y. Uivak gneisses from Labrador. *Contrib. Mineral. Petrol.* **54**, 43–59.

Burton K. W. and O'Nions R. K. (1991) High-resolution garnet chronometry and the rates of metamorphic processes. *Earth Planet. Sci. Lett.* **107**, 649–671.

Burwash R. A., Cavell P. A., and Burwash E. J. (1988) Source terrains for Proterozoic sedimentary rocks in southern British Columbia: Nd isotopic and petrographic evidence. *Can. J. Earth Sci.* **25**, 824–832.

Butler R. F., Gehrels G. E., Crawford M. L., and Crawford W. A. (2001) Paleomagnetism of the Quottoon plutonic complex in the Coast Mountains of British Columbia and southeastern Alaska: evidence for tilting during uplift. *Can. J. Earth Sci.* **38**, 1367–1384.

Carlson R. W., Hunter D. R., and Barker F. (1983) Sm–Nd age and isotopic systematics of the bimodal suite, ancient gneiss complex, Swaziland. *Nature* **305**, 701–704.

Carlson W. D., Donelick R. A., and Ketcham R. A. (1999) Variability of apatite fission-track annealing kinetics: I. Experimental results. *Am. Mineral.* **84**, 1213–1223.

Chauvel C., Arndt N. T., Kielinczuk S., and Thom A. (1987) Formation of Canadian 1.9 Ga old continental crust: I. Nd isotopic data. *Can. J. Earth Sci.* **24**, 396–406.

Cheburkin A. K., Frei R., and Shotyk W. (1997) An energy-dispersive miniprobe multi-element analyzer (EMMA) for direct analysis of trace elements and chemical age dating of single mineral grains. *Chem. Geol.* **135**, 75–87.

Chen B. and Jahn B. M. (2002) Geochemical and isotopic studies of the sedimentary and granitic rocks of the Altai Orogen of northwest China and their tectonic implications. *Geol. Mag.* **139**, 1–13.

Chen C.-H., DePaolo D. J., and Lan C.-Y. (1996) Rb–Sr microchrons in the Manaslu granite: implications for Himalayan thermochronology. *Earth Planet. Sci. Lett.* **143**, 125–135.

Christensen J. N., Rosenfeld J. L., and DePaolo D. J. (1989) Rates of tectonometamorphic processes from rubidium and strontium isotopes in garnet. *Science* **244**, 1465–1469.

Claesson S. (1987) Nd isotope data on 1.9–1.2 Ga old basic rocks and meta-sediments from the Bothnian Basin, central Sweden. *Precamb. Res.* **35**, 115–126.

Clement S. W. J., Compston W., and Newstead G. (1977) Design of a large, high resolution ion microprobe. In *Proceedings of the International Secondary Ion Mass Spectrometry Conference* (ed. A. Benninghoven). Spinger, Berlin, 12pp.

Cliff R. A. (1994) Rb–Sr dating of white mica- a new potential in metamorphic geochronology. *US Geol. Surv. Circular* **1107**, 62.

Clowes R., Cook F., Hajnal Z., Hall J., Lewry J. F., Lucas S., and Wardle R. (1999) Canada's Lithoprobe project (collaborative, mutildisciplinary geoscience research leads to new understanding of continental evolution). *Episodes* **22**, 3–20.

Collerson K. D., Campbell L. M., Weaver B. L., and Palacz Z. A. (1991) Evidence for extreme mantle fractionation in early Archean ultramafic rocks from northern Labrador. *Nature* **349**, 209–214.

Compston W. and Jeffery P. M. (1959) Anomalous common strontium in granite. *Nature* **184**, 1792–1793.

Compston W. and Pidgeon R. T. (1986) Jack Hills, evidence of more very old detrital zircons in Western Australia. *Nature* **321**, 766–769.

Compston W., Williams I. S., and Clement S. W. J. (1982) U–Pb ages within single zircons using a sensitive high mass resolution ion microprobe. *30th Ann. Conf. Am. Soc. Mass Spectrom.* 393–593.

Coney P. J., Jones D. L., and Monger J. W. H. (1980) Cordilleran suspect terranes. *Nature* **288**, 329–333.

Condie K. C. (1990) Growth and accretion of continental crust: inferences based on Laurentia. *Chem. Geol.* **83**, 183–194.

Condie K. C. and Chomiak B. (1996) Continental accretion: contrasting Mesozoic and Early Proterozoic tectonic regimes in North America. *Tectonophysics* **265**, 101–126.

Corfu F. and Easton R. M. (2000) U–Pb evidence for polymetamorphic history of Huronian rocks within the Grenville front tectonic zone east of Sudbury, Ontario, Canada. *Chem. Geol.* **172**, 149–171.

Corfu F. and Stott G. M. (1996) Hf isotopic composition and age constraints on the evolution of the Archean central Uchi Subprovince, Ontario, Canada. *Precamb. Res.* **78**, 53–63.

Creaser R. A., Erdmer P., Stevens R. A., and Grant S. L. (1997) Tectonic affinity of Nisutlin and Anvil assemblage strata from the Teslin tectonic zone, northern Canadian Cordillera: constraints from neodymium isotope and geochemical evidence. *Tectonics* **16**, 107–121.

Cui Y. and Russell J. K. (1995) Nd–Sr–Pb isotopic studies of the southern Coast Plutonic Complex, southwestern British Columbia. *Geol. Soc. Am. Bull.* **107**, 127–138.

Dallmeyer R. D. (1975) Incremental $^{40}Ar/^{39}Ar$ ages of biotite and hornblende from retrograded basement gneisses of the southern Blue Ridge: their bearing on the age of Paleozoic metamorphism. *Am. J. Sci.* **275**, 444–460.

Dallmeyer R. D., Sutter J. F., and Baker D. J. (1975) Incremental $^{40}Ar/^{39}Ar$ ages of biotite and hornblende from the northeastern Reading Prong: their bearing on late Proterozoic thermal and tectonic history. *Geol. Soc. Am. Bull.* **86**, 1435–1443.

Damon P. E. (1957) Terrestrial helium. *Geochim. Cosmochim. Acta* **11**, 200–201.

Davies R. D. and Allsopp H. L. (1976) Strontium isotopic evidence relating to the evolution of the lower Precambrian granitic crust in Swaziland. *Geology* **4**, 553–556.

DePaolo D. J. (1979) Implications of correlated Nd and Sr isotopic variations for the chemical evolution of the crust and mantle. *Earth Planet. Sci. Lett.* **43**, 201–211.

DePaolo D. J. (1981) Neodymium isotopes in the Colorado front Range and crust–mantle evolution in the Proterozoic. *Nature* **291**, 193–196.

De Wolf C. P., Zeissler C. J., Halliday A. N., Mezger K., and Essene E. J. (1996) The role of inclusion in U–Pb and Sm–Nd garnet chronology: stepwise dissolution experiments and trace uranium mapping by fission track analysis. *Geochim. Cosmochim. Acta* **60**, 121–134.

DiVincenzo G., Ghiribelli G., and Palmeri R. (2001) Evidence of a close link between petrology and isotope records: constraints from SEM, EMP, TEM and *in situ* Ar-40–Ar-39 laser analyses on multiple generations of white micas (Lanterman range, Antarctica). *Earth Planet. Sci. Lett.* **192**, 389–405.

Dodson M. H. (1973) Closure temperature in cooling geochronological and petrological systems. *Contrib. Mineral. Petrol.* **40**, 259–274.

Donelick R. A., Roden M. K., Mooers J., Carpenter B. S., and Miller D. S. (1990) Etchable length reduction of induced fission tracks in apatite at room temperature (23 °C): crystallographic orientation effects and "initial" mean lengths. *Nucl. Tracks Radiat. Meas.* **17**, 261–266.

Driver L. A., Creaser R. A., Chacko T., and Erdmer P. (2000) Petrogenesis of the Cretaceous Cassiar batholith, Yukon-B. C., Canada: implications for magmatism in the North American Cordilleran Interior. *Geol. Soc. Am. Bull.* **112**, 1119–1133.

Ducea M. N., Ganguly J., Rosenberg E., Patchett P. J., Cheng W., and Isachsen C. (2003) Sm–Nd dating of spatially controlled domains of garnet single crystals: a new method of high temperature thermochronology. *Earth Planet. Sci. Lett.* **213**, 31–42.

Duchêne S., Blichert-Toft J., Luais P., Telouk P., Lardeaux J.-M., and Albarède F. (1997) The Lu–Hf dating of garnets and the ages of the Alpine high-pressure metamorphism. *Nature* **387**, 586–589.

Eggins S. M., Kinsley L. P. J., and Shelley J. M. M. (1998) Deposition and element fractionation processes during atmospheric pressure laser sampling for analysis by ICPMS. *Appl. Surf. Sci.* **127–129**, 278–286.

Ehlers T. A. and Farley K. A. (2003) Apatite (U–Th)/He thermochronometry: methods and applications to problems in tectonic and surface processes. *Earth Planet. Sci. Lett.* **206**, 1–14.

Engi M., Cheburkin A. K., and Köppel V. (2002) Nondestructive chemical dating of young monazite using XRF: 1. Design of a mini-probe, age data for samples from the Central Alps, and comparison to U–Pb (TIMS) data. *Chem. Geol.* **191**, 225–241.

Erdmer P., Moore J. M., Heaman L. M., Thompson R. I., Daughtry K. L., and Creaser R. A. (2002) Extending the ancient margin outboard in the Canadian Cordillera: evidence of Proterozoic crust and Paleocene regional metamorphism in the Nicola horst, southeastern British Columbia. *Can. J. Earth Sci.* **39**, 1605–1623.

Fairbairn H. W., Hurley P. M., and Pinson W. H. (1961) The relation of discordant Rb–Sr mineral and rock ages in an igneous rock to its time of subsequent Sr87/Sr86 metamorphism. *Geochim. Cosmochim. Acta* **23**, 135–144.

Farley K. A. (2002) (U–Th)/He dating: techniques, calibrations, and applications. In *Noble Gases in Geochemistry,* Rev. Mineral. Geochem. (eds. P. D. Porcelli, C. J. Ballentine, and R. Wieler). Mineralogical Society of America, Washington, DC, vol. 47, pp. 819–843.

Farley K. A., Wolf R. A., and Silver L. T. (1996) The effects of long alpha-stopping distances on (U–Th)/He ages. *Geochim. Cosmochim. Acta* **60**, 4223–4229.

Farmer G. L., Ayuso R., and Plafker G. (1993) A Coast Mountains provenance for the Valdez and Orca groups, southern Alaska, based on Nd, Sr, and Pb isotopic evidence. *Earth Planet. Sci. Lett.* **116**, 9–21.

Feng R., Machado N., and Ludden J. (1993) Lead geochronology of zircon by laserprobe-inductively coupled plasma mass spectrometry (LA-ICPMS). *Geochim. Cosmochim. Acta* **57**, 3479–3486.

Fernández-Suárez J., Gutierrez A. G., Jenner G. A., and Tubrett M. N. (1999) Crustal sources in lower Paleozoic rocks from NW Iberia: insights from laser-ablation U–Pb ages of detrital zircons. *J. Geol. Soc. London* **156**, 1065–1068.

Fitzgerald P. G., Sorkhabi R. B., Redfield T. F., and Stump E. (1995) Uplift and denudation of the central Alaska Range: a case study in the use of apatite fission track thermochronology to determine absolute uplift parameters. *J. Geophys. Res.* **100**, 20175–20191.

Fraser G., Ellis D., and Eggins S. (1997) Zirconium abundance in granulite-facies minerals, with implications for zircon geochronology in high-grade rocks. *Geology* **25**, 607–610.

Friedman R. M., Mahoney J. B., and Cui Y. (1995) Magmatic evolution of the southern Coast Belt: constraints from Nd–Sr isotopic systematics and geochronology of the southern Coast Plutonic Complex. *Can. J. Earth Sci.* **32**, 1681–1698.

Frost C. D. and Coombs D. S. (1989) Nd isotope character of New Zealand sediments: implications for terrane concepts and crustal evolution. *Am. J. Sci.* **289**, 744–770.

Froude D. O., Ireland T. R., Kinny P. D., Williams I. S., Compston W., Williams I. R., and Myers J. S. (1983) Ion microprobe identification of 4,100-4,200 Myr-old terrestrial zircons. *Nature* **304**, 616–618.

Fryer B. J., Jackson S. E., and Longerich H. P. (1993) The application of laser ablation microprobe-inductively coupled plasma-mass spectrometry (LAM-ICP-MS) to in situ (U)-Pb geochronology. *Chem. Geol.* **109**, 1–8.

Gleadow A. J. W., Duddy I. R., Green P. F., and Lovering J. F., (1986) Confired track lengths in apatiter—a diagnostic tool for thermal arnlysis. **96**, 1–14.

Gabrielse H. (1985) Major dextral transcurrent displacements along the Northern Rocky Mountain Trench and related lineaments in north-central British Columbia. *Geol. Soc. Am. Bull.* **96**, 1–14.

Gallagher K., Brown R. and Johnson C. (1998) Fission track analysis and its application to geologic problems. *Ann. Rev. Earth Planet. Sci.* **26**, 519–572.

Ganguly J. and Tirone M. (1999) Diffusion closure temperature and age of a mineral with arbitrary extent of diffusion: theoretical formulation and applications. *Earth Planet. Sci. Lett.* **170**, 131–140.

Ganguly J. and Tirone M. (2001) Relationship between cooling rate and cooling age of a mineral: theory and applications to meteorites. *Meteorit. Planet. Sci.* **36**, 167–175.

Ganguly J., Tirone M., and Hervig R. L. (1998) Diffusion kinetics of samarium and neodymium in garnet, and a method of determining cooling rates of rocks. *Science* **281**, 805–807.

Garzione C. N., Patchett P. J., Ross G. M., and Nelson J. (1997) Provenance of sedimentary rocks in the Canadian Cordilleran Miogeocline: a Nd isotopic study. *Can. J. Earth Sci.* **34**, 1603–1618.

Geersten C., Briand A., Chartier F., Lacour J.-L., Mauchient P., and Sjöstrom S. (1994) Comparison between infrared and ultraviolet laser ablation at atmospheric pressure-implications for solid sampling inductively coupled plasma spectrometry. *J. Anal. Atom. Spectrom.* **9**, 17–22.

Gerstenberger H. and Haase G. (1997) A highly effective emitter substance for mass spectrometric Pb isotope ratio determinations. *Chem. Geol.* **136**, 309–312.

Getty S. R., Selverstone J., Wernicke B. P., Jacobsen S. B., Aliberti E., and Lux D. R. (1993) Sm–Nd dating of multiple garnet growth events in an arc-continent collision zone, northwestern US Cordillera. *Contrib. Mineral. Petrol.* **115**, 45–57.

Ghosh D. K. (1995) Nd–Sr isotopic constraints on the interactions of the Intermontane superterrane with the western edge of North America in the southern Canadian Cordillera. *Can. J. Earth Sci.* **32**, 1740–1758.

Ghosh D. K. and Lambert R. St. J. (1989) Nd–Sr isotopic study of Proterozoic to Triassic sediments from southeastern British Columbia. *Earth Planet. Sci. Lett.* **94**, 29–44.

Ghosh D. K. and Lambert R. St. J. (1995) Nd–Sr isotope geochemistry and petrogenesis of Jurassic granitoid intrusives, southeast British Columbia, Canada. In *Jurassic Magmatism and Tectonics of the North American Cordillera*, Geol. Soc. Am. Spec. Pap. (eds. D. M. Miller and C. Busby). The Geological society of America, Boulder, CO, vol. 299, pp. 141–157.

Gleadow A. J. W., Duddy I. R., Green P. F., and Lovering J. F. (1986) Confired track lengths in apatite—a diagnostic tool for thermal analysis. *Contrib. Mineral. Petrol.* **94**, 405–415.

Goldich S. S., Hedge C. E., and Stern T. W. (1970) Age of the Morton and Montevideo gneisses and related rocks, southwestern Minnesota. *Geol. Soc. Am. Bull.* **81**, 3671–3695.

Goldstein S. L., O'Nions R. K., and Hamilton P. J. (1984) A Sm–Nd isotopic study of atmospheric dusts and particulates from major river systems. *Earth Planet. Sci. Lett.* **70**, 221–236.

Green P. F., Duddy I. R., Gleadow A. J. W., Tingate P. R., and Laslett G. M. (1986) Thermal annealing of fission tracks in apatite: 1. A qualitative description. *Chem. Geol. (Isot. Geosci. Sect.)* **59**, 237–253.

Green P. F., Duddy I. R., Laslett G. M., Hegarty K. A., Gleadow A. J. W., and Lovering J. F. (1989) Thermal annealing of fission-tracks in apatite: 4. Quantitative modeling techniques and extension to geological time scales. *Chem. Geol. (Isot. Geosci. Sect.)* **79**, 155–182.

Gruau G., Rosing M., Bridgwater D., and Gill R. C. O. (1996) Resetting of Sm–Nd systematics during metamorphism of >3.7 Ga rocks: implications for isotopic models of early Earth differentiation. *Chem. Geol.* **133**, 225–240.

Hamilton P. J., O'Nions R. K., Bridgwater D., and Nutman A. (1983) Sm–Nd studies of Archaean metasediments and metavolcanics from west Greenland and their implications for the Earth's early history. *Earth Planet. Sci. Lett.* **63**, 263–272.

Han B. F., Wang S. G., Jahn B. M., Hong D. W., Hiroo K., and Sun Y. L. (1997) Depleted mantle source for the Ulungur River A-type granites from North Xinjiang, China: geochemistry and Nd–Sr isotopic evidence, and implications for Phanerozoic crustal growth. *Chem. Geol.* **138**, 135–159.

Hanson G. N. and Gast P. W. (1967) Kinetic studies in contact metamorphic zones. *Geochim. Cosmochim. Acta* **31**, 1119–1153.

Hart S. R. (1964) The petrology and isotopic-mineral age relations of a contact zone in the Front Range, Colorado. *J. Geol.* 72, 493–525.

Heaman L. M. and Parrish R. (1991) U–Pb geochronology of accessory minerals. In *Applications of Radiogenic Isotope Systems to Problems in Geology*, Short Course Handbook 19 (eds. L. Heaman and J. N. Ludden). Mineralogical Association of Canada, Toronto, pp. 59–102.

Hegner E. and Bevier M. L. (1991) Nd and Pb isotopic constraints on the origin of the Purtuniq ophiolite and Early Proterozoic Cape Smith belt, northern Quebec, Canada. *Chem. Geol.* **91**, 357–371.

Heizler M. T., Lux D. R., and Decker E. R. (1988) The age and cooling history of the Chain of Ponds and Big Island Pond plutons and the Spider Lake Granite, west-central Maine and Quebec. *Am. J. Sci.* **288**, 925–952.

Henderson J. B. (1985) Geology of the Yellowknife-Hearne Lake Area, District of Mackenzie: segment across an Archean basin. *Geol. Surv. Can. Mem.* **414**, 135pp.

Hinthorne J. R., Andersen C. A., Conrad R. L., and Lovering J. F. (1979) Single grain $^{207}Pb/^{206}Pb$ and U/Pb age determinations with a 10-μm spatial resolution using the ion microprobe mass analyzer. *Chem. Geol.* **25**, 271–303.

Hofmann A. W. (1988) Chemical differentiation of the Earth: the relationship between mantle, continental crust, and oceanic crust. *Earth Planet. Sci. Lett.* **90**, 297–314.

Holmes A. (1911) The association of lead with uranium in rock-minerals, and its application to the measurement of geological time. *Proc. Roy. Soc. London* **A85**, 248–256.

Horn I., Rudnick R. L., and McDonough W. F. (2000) Precise elemental and isotope ratio determination by simultaneous solution nebulization and laser-ablation ICP-MS: application to *in situ* U/Pb geochronology. *Chem. Geol.* **164**, 281–301.

Horstwood M. S. A., Nesbitt R. W., Noble S. A., and Wilson J. F. (1999) U–Pb zircon evidence fo an extesive early Archean craton in Zimbabwe: a reassessment of the timing of craton formation, stabilization, and growth. *Geology* **27**, 707–710.

House M., Farley K. A., and Stöckli D. (2000) Helium chronometry of apatite and titanite using Nd-YAG laser heating. *Earth Planet. Sci. Lett.* **183**, 365–368.

Hu A. Q., Jahn B. M., Zhang G. X., and Zhang Q. F. (2000) Crustal evolution and Phanerozoic crustal growth in northern Xinjiang: Nd–Sr isotopic evidence: Part I. Isotopic characterisation of basement rocks. *Tectonophysics* **328**, 15–51.

Huhma H. (1986) Sm–Nd, U–Pb, and Pb–Pb isotopic evidence for the origin of the Early Proterozoic Svecokarelian crust in Finland. *Geol. Surv. Finland Bull.* **337**, 5.

Hurley P. M. (1954) The helium age method and the distribution and migration of helium in rocks. In *Nuclear Geology* (ed. H. Faul). Wiley, New York, pp. 301–329.

Hurst R. W., Bridgwater D., Collerson K. D., and Wetherill G. W. (1975) 3600-m.y. Rb–Sr ages from very early Archaean gneisses from Saglek Bay, Labrador. *Earth Planet. Sci. Lett.* **27**, 393–403.

Jacobsen S. B. and Wasserburg G. J. (1979) The mean age of mantle and crustal reservoirs. *J. Geophys. Res.* **84**, 7411–7427.

Jacobsen S. B. and Wasserburg G. J. (1984) Sm–Nd isotopic evolution of chondrites and achondrites: II. *Earth Planet. Sci. Lett.* **67**, 137–150.

Jackson S. E., Longerich H. P., Horn I., and Dunning G. R. (1996) The application of laser-ablation microprobe (LAM)-ICP-MS to *in situ* U–Pb zircon geochronology. *J. Conf. Abstr. (V. M. Goldschmidt Conference)* **1**, 283.

Jahn B. M., Wu F. Y., and Chen B. (2000) Massive granitoid generation in central Asia: Nd isotope evidence and implication for continental growth in the Phanerozoic. *Episodes* **23**, 82–92.

Johnston S. T. (2001) The great Alaskan train wreck: reconciliation of paleomagnetic and geological data in the northern Cordillera. *Earth Planet. Sci. Lett.* **193**, 259–272.

Kalsbeek F. and Taylor P. N. (1985) Isotopic and chemical variation in granites across a Proterozoic continental margin- the Ketilidian mobile belt of South Greenland. *Earth Planet. Sci. Lett.* **73**, 65–80.

Kalsbeek F., Pidgeon R. T., and Taylor P. N. (1987) Nagssugtoqidian mobile belt of West Greenland: a cryptic 1850 Ma suture between two Archean continents—chemical and isotopic evidence. *Earth Planet. Sci. Lett.* **85**, 365–385.

Karlstrom K. E., Williams M. L., McLelland J., Geissman J. W., and Åhäll K.-I. (1999) Refining Rodinia: geologic evidence for the Australia-western USA connection in the Proterozoic. *GSA Today* **9**(10), 1–7.

Kelley S. P. (1995) Ar–Ar dating by laser microprobe. In *Microprobe Techniques in the Earth Sciences: Mineralogical Society Series 6* (eds. P. J. Potts, J. F. W. Bowles, S. J. B. Reed, and M. R. Cave). Kluwer, Dordrecht, pp. 123–143.

Kelley S. P., Arnaud N. O., and Turner S. P. (1994) High spatial resolution $^{40}Ar/^{39}Ar$ investigations using an ultraviolet laser probe extraction technique. *Geochim. Cosmochim. Acta* **58**, 3519–3525.

Kelley S. P., Bartlett J. M., and Harris N. B. W. (1997) Premetamorphic Ar–Ar ages from biotite inclusions in garnet. *Geochim. Cosmochim. Acta* **61**, 3873–3878.

Ketcham R. A., Donelick R. A., and Carlson W. D. (1999) Variability of apatite fission-track annealing kinetics: III. Extrapolation to geological timescales. *Am. Mineral.* **84**, 1235–1255.

Kinny P. D., Compston W., and Williams I. S. (1991) A reconnaissance ion-probe study of hafnium isotopes in zircons. *Geochim. Cosmochim. Acta* **55**, 849–859.

Knudsen T.-L., Griffin W. L., Hartz E. H., Andersen A., and Jackson S. E. (2001) *In-situ* hafnium and lead isotope analyses of detrital zircons from the Devonian sedimentary basin of NE Greenland: a record of repeated crustal reworking. *Contrib. Mineral. Petrol.* **141**, 83–94.

Kober B. (1986) Whole-grain evaporation for $^{207}Pb/^{206}Pb$-age-investigations on single zircons using a double-filament thermal ion source. *Contrib. Mineral. Petrol.* **93**, 482–490.

Kober B. (1987) Single-zircon evaporation combined with Pb^{+} emitter-bedding for $^{207}Pb/^{206}Pb$-age investigations using thermal ion mass spectrometry, and implications to zirconology. *Contrib. Mineral. Petrol.* **96**, 63–71.

Kober B., Pidgeon R. T., and Lippolt H. J. (1989) Single-zircon dating by stepwise Pb-evaporation constrains the Archean history of detrital zircons from the Jack Hills, Western Australia. *Earth Planet. Sci. Lett.* **91**, 286–296.

Kovalenko V. I., Yarmolyuk V. V., Kovach V. P., Budnikov S. V., Zhuravlev D. Z., Kozakov I. K., Kotov A. B., Rytsk E. Y., and Salnikova E. B. (1999) Magmatism as factor of crust evolution in the Central Asian Fold Belt: Sm–Nd isotopic data. *Geotectonics* **33**, 191–208.

Kovalenko V. I., Yarmolyuk V. V., Kovach V. P., Kotov A. B., Kozakov I. K., Salnikova E. B., and Larin A. M. (2003) Isotope provinces, mechanisms of generation and sources of the continental crust in the Central Asian Mobile Belt: geological and isotopic evidence. *J. Asian Earth Sci.*, (in press).

Krogh T. E. (1971a) A low contamination method for the decomposition of zircon and the extraction of U and Pb for isotopic age determinations. *Carnegie Inst. Wash. Yearb.* **70**, 258–266.

Krogh T. E. (1971b) A simplified technique for the dissolution of zircon and the isolation of uranium and lead. *Carnegie Inst. Wash. Yearb.* **69**, 342–344.

Krogh T. E. (1973) A low-contamination method for hydrothermal decomposition of zircon and extraction of U and Pb for isotopic age determinations. *Geochim. Cosmochim. Acta* **37**, 485–494.

Krogh T. E. (1982) Improved accuracy of U–Pb zircon ages by the creation of more concordant systems using an air abra sion technique. *Geochim. Cosmochim. Acta* **46**, 637–649.

Krogh T. E. and Davis G. L. (1975) The production and preparation of 205Pb for use as a tracer for isotope dilution analyses. *Carnegie Inst. Wash. Yearb.* **74**, 416–417.

Krogstad E. and Walker R. J. (1996) Evidence of heterogeneous crustal sources: the Harney Peak granite, South Dakota USA. *Trans. Roy. Soc. Edinburgh: Earth Sci.* **87**, 331–337.

Kröner A., Compston W., and Williams I. S. (1989) Growth of early Archaean crust in the Ancient Gneiss Complex of Swaziland as revealed by single zircon dating. *Tectonophysics* **161**, 271–298.

Lanphere M. A. and Albee A. L. (1974) $^{40}Ar/^{39}Ar$ age measurements in the Worcester Mountians: evidence of Ordovician and Devonian metamorphic events in northern Vermont. *Am. J. Sci.* **274**, 545–555.

Laslett G. M., Green P. F., Duddy I. R., and Gleadow A. J. W. (1987) Thermal annealing of fission tracks in apatite: 2. A quantitative analysis. *Chem. Geol. (Isot. Geosci. Sect.)* **65**, 1–13.

Layer P. W., Hall C. M., and York D. (1987) The derivation of $^{40}Ar/^{39}Ar$ age spectra of single grains of hornblende and biotite by laser step-heating. *Geophys. Res. Lett.* **14**, 757–760.

Lippolt H. J., Leitz M., Wernicke R. S., and Hagedorn B. (1994) (U + Th)/He dating of apatite: experience with samples from different geological environments. *Chem. Geol.* **112**, 179–191.

Litvinovsky B. A., Jahn B. M., Zanvilevich A. N., Saunders A., Poulain S., Kuzmin D. V., Reichow M. K., and Titov A. V. (2002) Petrogenesis of syenite-granite suites from the Bryansky Complex (Transbaikalia, Russia): implications for the origin of A-type granitoid magmas. *Chem. Geol.* **189**, 105–133.

Longerich H. P., Fryer B. J., and Strong D. F. (1987) Determination of lead isotope ratios by inductively coupled plasma-mass spectrometry (ICP-MS). *Spectrochem. Acta* **42B**, 39–48.

Lovera O. M. and Richter F. M. (1991) Diffusion domains determined by ^{39}Ar released during step heating. *J. Geophys. Res.* **96**, 2057–2069.

Lovera O. M., Richter F. M., and Harrison T. M. (1989) The $^{40}Ar/^{39}Ar$ thermochronometry for slowly cooled samples having a distribution of diffusion domain sizes. *J. Geophys. Res.* **94**, 17917–17935.

Lucas S. B., Green A., Hajnal Z., White D., Lewry J. F., Ashton K., Weber W., and Clowes R. (1993) Deep seismic profile across a Proterozoic collision zone: surprises at depth. *Nature* **363**, 339–342.

Lugmair G. W. and Marti K. (1978) Lunar initial 143Nd/144Nd: differential evolution of the lunar crust and mantle. *Earth Planet. Sci. Lett.* **39**, 349–357.

Maas R., Kinny P. D., Williams I. S., Froude D. O., and Compston W. (1992) The Earth's oldest known crust: a geochronological and geochemical study of 3,900–4,200 Ma old detrital zircons from Mt. Narryer and Jack Hills, Western Australia. *Geochim. Cosmochim. Acta* **56**, 1281–1300.

Machado N. and Simonetti A. (2001) U–Pb dating and Hf isotopic composition of zircon by laser-ablation MC-ICP-MS. In *Laser-ablation ICP-MS in the Earth Sciences,* Mineral. Ass. Can. Short Course Ser. (ed. P. Sylvester). The mineralogical Association of Canada, Ottawa, ON, vol. 29, pp. 121–146.

Mahoney J. B., Friedman R. M., and McKinley S. D. (1995) Evolution of a Middle Jurassic volcanic arc: stratigraphic, isotopic and geochemical characteristics of the Harrison Lake Formation, southwestern British Columbia. *Can. J. Earth Sci.* **32**, 1759–1776.

Maluski H. and Schaeffer O. A. (1982) ^{39}Ar–^{40}Ar laser probe dating of terrestrial rocks. *Earth Planet. Sci. Lett.* **59**, 21–27.

McCulloch M. T. and Bennett V. C. (1994) Progressive growth of the Earth's continental crust and depleted mantle: geochemical constraints. *Geochim. Cosmochim. Acta* **58**, 4717–4738.

McCulloch M. T. and Wasserburg G. J. (1978) Sm–Nd and Rb–Sr chronology of continental crust formation. *Science* **200**, 1003–1011.

McDougall I. and Harrison T. M. (1999) *Geochronology and Thermochronology by the $^{40}Ar/^{39}Ar$ Method,* 2nd edn. Oxford University Press, Oxford, 269p.

Megrue G. H. (1973) Spatial distribution of $^{40}Ar/^{39}Ar$ ages in lunar breccia 14301. *J. Geophys. Res.* **78**, 3216–3221.

Merrihue C. and Turner G. (1966) Potassium–argon dating by activation with fast neutrons. *J. Geophys. Res.* **71**, 2852–2856.

Mezger J. E., Creaser R. A., and Erdmer P. (2001) Cretaceous back-arc basin along the Coast Belt of the northern Canadian Cordillera: evidence from geochemical and Nd isotopic signatures of the Kluane metamorphic assemblage, SW Yukon. *Can. J. Earth Sci.* **38**, 91–103.

Mezger K. and Krogstad E. J. (1997) Interpretation of discordant U–Pb zircon ages: an evaluation. *J. Metamorph. Geol.* **15**, 127–140.

Mezger K., Essene E. J., and Halliday A. N. (1992) Closure temperatures of the Sm–Nd system in metamorphic garnets. *Earth Planet. Sci. Lett.* **113**, 397–409.

Mojzsis S. J., Harrison T. M., and Pidgeon R. T. (2001) Oxygen-isotope evidence from ancient zircons for liquid water at the Earth's surface 4,300 Myr ago. *Nature* **409**, 178–181.

Monger J. W. H. (1993) Canadian Cordilleran tectonics: from geosynclines to crustal collage. *Can. J. Earth Sci.* **30**, 209–231.

Monger J. W. H., Souther J. G., and Gabrielse H. (1972) Evolution of the Canadian Cordillera: a plate-tectonic model. *Am. J. Sci.* **272**, 577–602.

Monger J. W. H., Price R. A., and Tempelman-Kluit D. J. (1982) Tectonic accretion and the origin of the two major metamorphic and plutonic welts in the Canadian Cordillera. *Geology* **10**, 70–75.

Montel J.-M., Veschambre M., and Nicollet C. (1994) Datation de la monazite à la microsonde électronique. *Compt. Rend. Acad. Sci. Paris* **318**, 1489–1495.

Montel J.-M., Foret S., Veschambre M., Nicollet C., and Povost A. (1996) Electron microprobe dating of monazite. *Chem. Geol.* **131**, 37–53.

Montel J.-M., Kornprobst J., and Vielzeuf D. (2000) Preservation of old U–Th–Pb ages in shielded monazite: example from the Beni Bousera Hercynian kinzigites (Morocco). *J. Metamorph. Geol.* **18**, 335–342.

Moorbath S. (1975) Evolution of Precambrian crust from strontium isotopic evidence. *Nature* **254**, 395–398.

Moorbath S. (1978) Age and isotope evidence for the evolution of continental crust. *Phil. Trans. Roy. Soc. London* **A288**, 401–413.

Moorbath S., O'Nions R. K., Pankhurst R. J., Gale N. H., and McGregor V. R. (1972) Further rubidium–strontium age determinations on the very early Precambrian rocks of the Godthaab district, west Greenland. *Nature* **240**, 78–82.

Moorbath S., Whitehouse M. J., and Kamber B. S. (1997) Extreme Nd-isotope heterogeneity in the early Archaean-fact or fiction? Case histories from northern Canada and west Greenland. *Chem. Geol.* **135**, 213–231.

Müller W. (2003) Strengthening the link between geochronology, textures and petrology. *Earth Planet. Sci. Lett.* **206**, 237–251.

Müller W., Aerden D., and Halliday A. N. (2000a) Isotopic dating of strain fringe increments: duration and rates of deformation in shear zones. *Science* **288**, 2195–2198.

Müller W., Mancktelow N. S., and Meier M. (2000b) Rb–Sr microchrons of synkinematic mica in mylonites: an example from the DAV fault of the eastern Alps. *Earth Planet. Sci. Lett.* **180**, 385–397.

Myers J. S. (1995) The generation and assembly of an Archean supercontinent: evidence from the Yilgarn craton, Western Australia. In *Early Precambrian Processes,* Geol. Soc. Spec. Publ. 95 (eds. M. P. Coward and A. C. Ries). Geological Society of London, London, pp. 143–154.

Nelson B. K. and DePaolo D. J. (1984) 1,700-Myr greenstone volcanic successions in southwestern North America and isotopic evolution of Proterozoic mantle. *Nature* **311**, 143–146.

Nelson B. K. and DePaolo D. J. (1985) Rapid production of continental crust 1.7 to 1.9 b.y. ago: Nd isotopic evidence from the basement of the North America mid-continent. *Geol. Soc. Am. Bull.* **96**, 746–754.

Nier A. O. (1939) The isotopic constitution of radiogenic lead and the measurement of geological time. *Phys. Rev. Ser. 2,* **55**, 153–163.

Öhlander B., Mellqvist C., and Skiöld T. (1999) Sm–Nd isotope evidence of a collisional event in the Pre-Cambrian of northern Sweden. *Precamb. Res.* **93**, 105–117.

O'Nions R. K., Evensen N. M., and Hamilton P. J. (1979) Geochemical modeling of mantle differentiation and crustal growth. *J. Geophys. Res.* **84**, 6091–6101.

Padgham W. A. (1985) Observations and speculations on supracrustal successions in the Slave Structural Province. In *Evolution of Archean Supracrustal Sequences,* Geol. Ass. Can. Spec. Pap. 28 (eds. L. D. Ayres, P. C. Thurston, K. D. Card, and W. Weber). Geological Association of Canada, St. Johns, Newfoundland, pp. 156–167.

Papanastassiou D. A. and Wasserburg G. J. (1981) Microchrons: the $^{87}Rb-^{87}Sr$ dating of microscopic samples. *Proc. 12th Lunar Planet. Sci. Conf. B,* 1027–1038.

Parrish R. R. (1983) Cenozoic thermal evolution and tectonics ofthe Coast Mountains of British Columbia: 1. Fission track dating, apparent uplift rates, and patterns of uplift. *Tectonics* **2**, 601–631.

Parrish R. R. and Krogh T. E. (1987) Synthesis and purification of 205Pb for U–Pb geoghronology. *Chem. Geol. (Isot. Geosci. Sect.)* **66**, 103–110.

Patchett P. J. and Arndt N. T. (1986) Nd isotopes and tectonics of 1.9–1.7 Ga crustal genesis. *Earth Planet. Sci. Lett.* **78**, 329–338.

Patchett P. J. and Bridgwater D. (1984) Origin of continental crust of 1.9–1.7 Ga age defined by Nd isotopes in the Ketilidian terrain of South Greenland. *Contrib. Mineral. Petrol.* **87**, 311–318.

Patchett P. J. and Chase C. G. (2002) Role of transform continental margins in major crustal growth episodes. *Geology* **30**, 39–42.

Patchett P. J. and Gehrels G. E. (1998) Continental influence on Canadian Cordilleran terrains from Nd isotopic study, and significance for crustal growth processes. *J. Geol.* **106**, 269–280.

Patchett P. J. and Kouvo O. (1986) Origin of continental crust of 1.9–1.7 Ga age: Nd isotopes and U–Pb zircon ages in the Svecokarelian terrain of South Finland. *Contrib. Mineral. Petrol.* **92**, 1–12.

Patchett P. J. and Tatsumoto M. (1980) Lu–Hf total-rock isochron for the eucrite meteorites. *Nature* **288**, 571–574.

Patchett P. J., Kouvo O., Hedge C. E., and Tatsumoto M. (1981) Evolution of continental crust and mantle heterogeneity: evidence from Hf isotopes. *Contrib. Mineral. Petrol.* **78**, 279–297.

Patchett P. J., White W. M., Feldmann H., Kielinczuk S., and Hofmann A. W. (1984) Hafnium/rare-earth element fractionation in the sedimentary system and crustal recycling into the Earth's mantle. *Earth Planet. Sci. Lett.* **69**, 365–378.

Patchett P. J., Gorbatschev R., and Todt W. (1987) Origin of continental crust of 1.9–1.7 Ga age: Nd isotopes in the Svecofennian orogenic terrains of Sweden. *Precamb. Res.* **35**, 145–160.

Patchett P. J., Gehrels G. E., and Isachsen C. E. (1998) Nd isotopic characteristics of metamorphic and plutonic rocks of the Coast Mountains near Prince Rupert, British Columbia. *Can. J. Earth Sci.* **35**, 556–561.

Patchett P. J., Ross G. M., and Gleason J. D. (1999) Continental drainage and mountain sources during the Phanerozoic evolution of North America: evidence from Nd isotopes. *Science* **283**, 671–673.

Peck W. H., Valley J. W., Wilde S. A., and Graham C. M. (2001) Oxygen isotope ratios and rare earth elements in 3.3 to 4.4 Ga zircons: ion microprobe evidence for high $\delta^{18}O$ continental crust and the oceans in the Early Archean. *Geochim. Cosmochim. Acta* **65**, 4215–4229.

Peltonen P., Kontinen A., and Huhma H. (1998) Petrogenesis of the mantle sequence of the Jormua ophiolite (Finland): melt migration in the upper mantle during Palaeoproterozoic continental break-up. *J. Petrol.* **39**, 297–329.

Piercey S. J., Mortensen J. K., Murphy D. C., Paradis S., and Creaser R. A. (2002) Geochemistry and tectonic significance of alkalic mafic magmatism in the Yukon-Tanana terrain, Finlayson Lake region, Yukon. *Can. J. Earth Sci.* **39**, 1729–1744.

Piercey S. J., Mortensen J. K., and Creaser R. A. (2003) Neodymium isotope geochemistry of felsic volcanic and intrusive rocks from the Yukon-Tanana terrain in the Finlayson Lake region, Yukon, Canada. *Can. J. Earth Sci.* **40**, 77–97.

Plank T. and Langmuir C. H. (1998) The chemical composition of subducting sediment and its consequences for the crust and mantle. *Chem. Geol.* **145**, 325–394.

Price P. B. and Walker R. M. (1962) Observation of fossil particle tracks in natural micas. *Nature* **196**, 732–734.

Price P. B. and Walker R. M. (1963) Fossil tracks of charged particles in mica and the age of minerals. *J. Geophys. Res.* **68**, 4847–4863.

Rämö T., Vaasjoki M., Mänttäri I., Elliott B. A., and Nironen M. (2001) Petrogenesis of the post-kinematic magmatism of the central Finland granitoid complex: I. Radiogenic isotope constraints and implications for crustal evolution. *J. Petrol.* **42**, 1971–1993.

Reddy S. M., Kelley S. P., and Wheeler J. (1996) Ar-40/Ar-39 laser probe study of micas from the Sesia Zone, Italian Alps: implications for metamorphic and deformation histories. *J. Metamorph. Geol.* **14**, 493–508.

Reymer A. and Schubert G. (1984) Phanerozoic addition rates to the continental crust and crustal growth. *Tectonics* **3**, 63–77.

Salnikova E. B., Kozakov I. K., Kotov A. B., Kröner A., Todt W., Bibikova E. V., Nutman A., Yakovleva S. Z., and Kovach V. P. (2001) Age of Palaeozoic granites and metamorphism in the Tuvino-Mongolian Massif of the Central Asian Mobile Belt: loss of a Precambrian microcontinent. *Precamb. Res.* **110**, 143–164.

Salters V. J. M. and Hart S. R. (1991) The mantle sources of ocean ridges, islands and arcs: the Hf-isotope connection. *Earth Planet. Sci. Lett.* **104**, 364–380.

Samson S. D. and D'Lemos R. S. (1999) A precise late Neoproterozoic U–Pb zircon age for the syntectonic Perelle quartz diorite, Guernsey, Channel Islands, UK. *J. Geol. Soc. London* **156**, 47–54.

Samson S. D. and Patchett P. J. (1991) The Canadian Cordillera as a modern analogue of Proterozoic crustal growth. *Austral. J. Earth Sci.* **38**, 595–611.

Samson S. D., McClelland W. C., Patchett P. J., Gehrels G. E., and Anderson R. G. (1989) Evidence from Nd isotopes for mantle contributions to Phanerozoic crustal genesis in the Canadian Cordillera. *Nature* **337**, 705–709.

Samson S. D., Patchett P. J., Gehrels G. E., and Anderson R. G. (1990) Nd and Sr isotopic characterization of the Wrangellia terrane and implications for crustal growth of the Canadian Cordillera. *J. Geol.* **98**, 749–762.

Samson S. D., Patchett P. J., McClelland W. C., and Gehrels G. E. (1991a) Nd isotopic characterization of metamorphic rocks in the Coast Mountains, Alaskan and Canadian Cordillera: ancient crust bounded by juvenile terranes. *Tectonics* **10**, 770–780.

Samson S. D., McClelland W. C., Patchett P. J., and Gehrels G. E. (1991b) Nd and Sr isotopic constraints on the petrogenesis of the west side of the northern Coast Mountains batholith, Alaskan and Canadian Cordillera. *Can. J. Earth Sci.* **28**, 939–946.

Samson S. D., D'Lemos R. S., Blichert-Toft J., and Vervoort J. D. (2003) U–Pb geochronology and Hf–Nd isotope compositions of the oldest Neoproterozoic crust within the Cadomian Orogen: new evidence for a unique juvenile terrane. *Earth Planet. Sci. Lett.* **208**, 165–180.

Scherer E. E., Cameron K. L., and Blichert-Toft J. (2000) Lu–Hf garnet geochronology: closure temperature relative to the Sm–Nd system and the effects of trace mineral inclusions. *Geochim. Cosmochim. Acta* **64**, 3413–3432.

Scherer E., Münker C., and Mezger K. (2001) Calibration of the Lutetium–Hafnium clock. *Science* **293**, 683–687.

Scherrer N. C., Engi M., Berger A., Parrish R. R., and Cheburkin A. K. (2002) Nondestructive chemical dating of young monazite using XRF: 2. Context sensitive microanalysis and comparison with Th–Pb laser-ablation mass spectrometric data. *Chem. Geol.* **191**, 243–255.

Sengör A. M. C. and Hsü K. J. (1984) The Cimmerides of eastern Asia: history of the eastern end of paleo-Tethys. *Mem. Soc. Géol. France N. S.* **147**, 139–167.

Sengör A. M. C. and Natalin B. A. (1996a) Turkic-type orogeny and its role in the making of the continental crust. *Ann. Rev. Earth Planet. Sci.* **24**, 263–337.

Sengör A. M. C. and Natalin B. A. (1996b) Paleotectonics of Asia: fragments of a synthesis. In *The Tectonic Evolution of Asia* (eds. A. Yin and M. Harrison). Cambridge University Press, Cambridge, UK, pp. 486–640.

Sengör A. M. C., Natalin B. A., and Burtman V. S. (1993) Evolution of the Altaid tectonic collage and Palaeozoic crustal growth in Eurasia. *Nature* **364**, 299–307.

Skiöld T. and Cliff R. A. (1984) Sm–Nd and U–Pb dating of Early Proterozoic mafic–felsic volcanism in northernmost Sweden. *Precamb. Res.* **26**, 1–13.

Smith A. D. and Lambert R. StJ. (1995) Nd, Sr, and Pb isotopic evidence for cotrasting origins of late Paleozoic volcanic rocks from the Slide Mountain and Cache Creek terranes, south-central British Columbia. *Can. J. Earth Sci.* **32**, 447–459.

Smith A. D., Brandon A. D., and Lambert R. StJ. (1995) Nd–Sr isotope systematics of Nicola Group volcanic rocks, Quesnel terrane. *Can. J. Earth Sci.* **32**, 437–446.

Söderlund U. (1996) Conventional U–Pb dating versus single-grain Pb evaporation dating of complex zircons from a pegmatite in the high-grade gneisses of southwestern Sweden. *Lithos* **38**, 93–105.

Stein M. and Hofmann A. W. (1994) Mantle plumes and episodic crustal growth. *Nature* **372**, 63–68.

Stern R. (1997) The GSC sensitive high resolution ion microprobe (SHRIMP): analytical techniques of zircon U–Th–Pb age determinations and performance evaluation: in radiogenic age and isotopic studies. *Geol. Surv. Can. Current Res.* 1997-F(Report 10), 1–31.

Stern R. and Bleeker W. (1998) Age of the world's oldest rocks refined using Canada's SHRIMP: the Acasta Gneiss Complex, Northwest Territories, Canada. *Geosci. Can.* **25**, 27–31.

Stevens R. A., Erdmer P., Creaser R. A., and Grant S. L. (1996) Mississippian assembly of the Nisutlin assemblage: evidence from primary contact relationships and Mississippian magmatism in the Teslin tectonic zone, part of the Yukon-Tanana terrane of south-central Yukon. *Can. J. Earth Sci.* **33**, 103–116.

Stevenson R. K. and Patchett P. J. (1990) Implications for the evolution of continental crust from Hf isotope systematics of Archean detrital zircons. *Geochim. Cosmochim. Acta* **54**, 1683–1697.

Stosch H.-G. and Lugmair G. W. (1987) Geochronology and geochemistry of eclogites from the Münchberg gneiss massif, FRG. *Terra Cognita* **7**, 163.

Sutter J. F. and Hartung J. B. (1984) Laser microprobe $^{40}Ar/^{39}Ar$ dating of mineral grains *in situ*. *Soc. Electron. Microsc.* **4**, 1525–1529.

Suzuki K. and Adachi M. (1991) Precambrian provenance and Silurian metamorphism of the Tsunosawa paragneiss in the South Kitakami terrane, northeast Japan, revealed by the chemical Th–U–total Pb isochron ages of monazite, zircon and xenotime. *Geochem. J.* **25**, 357–376.

Suzuki K. and Adachi M. (1994) Middle Precambrian detrital monazite and zircon from the Hida gneiss on Oki-Dogo Island, Japan: their origin and implications for the correlation of basement gneiss of southwest Japan and Korea. *Tectonophysics* **235**, 277–292.

Tatsumoto M., Unruh D. M., and Patchett P. J. (1981) U–Pb and Lu–Hf systematics of Antarctic meteorites. In *Proc. 6th Symp. Antarctic Meteorites*, Natl. Inst. Polar Res., Tokyo, pp. 237–249.

Tilton G., Patterson C., Brown H., Inghram M., Hayden R., Hess D., and Larsen E. S., Jr. (1955) Isotopic composition and distribution of lead, uranium and thorium in a Precambrian granite. *Geol. Soc. Am. Bull.* **66**, 1131–1148.

Tilton G. R., Davis G. L., Wetherill G. W., Aldrich L. T., and Jager E. (1959) Mineral ages in the Maryland piedmont. *Carnegie Inst. Wash. Yearb.* **58**, 171–178.

Umhoefer P. J. (1987) Northward translation of "Baja British Columbia" along the Late Cretaceous to Paleocene margin of western North America. *Tectonics* **6**, 377–394.

Unterschutz J. L. E., Creaser R. A., Erdmer P., Thompson R. I., and Daughtry K. L. (2002) North American margin origin of Quesnel terrane strata in the southern Canadian Cordillera: inferences from geochemical and Nd isotopic characteristics of Triassic metasedimentary rocks. *Geol. Soc. Am. Bull.* **114**, 462–475.

Valbracht P. J., Oen I. S., and Beunk F. F. (1994) Sm–Nd isotope systematics of 1.9–1.8-Ga granites from western Bergslagen, Sweden: inferences on a 2.1–2.0 Ga crustal precursor. *Chem. Geol.* **112**, 21–37.

Valley J. W., Chiarenzelli J. R., and McLelland J. M. (1994) Oxygen isotope geochemsitry of zircon. *Earth Planet. Sci. Lett.* **126**, 187–206.

Valley J. W., Peck W. H., King E. M., and Wilde S. A. (2002) A cool early Earth. *Geology* **30**, 351–354.

Vance D. and O'Nions R. K. (1990) Isotopic chronometry of zoned garnets: growth kinetics and metamorphic histories. *Earth Planet. Sci. Lett.* **97**, 227–240.

Vervoort J. D. and Blichert-Toft J. (1999) Evolution of the depleted mantle: Hf isotope evidence from juvenile rocks through time. *Geochim. Cosmochim. Acta* **63**, 533–556.

Vervoort J. D., Patchett P. J., Gehrels G. E., and Nutman A. P. (1996) Constraints on early Earth differentiation from hafnium and neodymium isotopes. *Nature* **379**, 624–627.

Vinogradov A. P., Zadorozhnyi I. K., and Zykor S. I. (1952) Isotopic composition of lead and the age of the earth. *Dokl. Akad. Nauk. SSSR* **85**, 1107–1110.

von Huene R. and Scholl D. W. (1993) The return of sialic material to the mantle indicated by terrigenous material subducted at convergent margins. *Tectonophysics* **219**, 163–175.

Wetherill G. W., Aldrich L. T., and Davis G. L. (1955) A40/K40 ratios of feldspars and micas from the same rock. *Geochim. Cosmochim. Acta* **8**, 171–172.

Wilde S. A., Zhang X., and Wu F. (2000) Extension of a newly identified 500 Ma metamorphic terrain in North East China: further U–Pb SHRIMP dating of the Mashan Complex, Heilongjiang Province, China. *Tectonophysics* **328**, 115–130.

Wilde S. A., Valley J. W., Peck W. H., and Graham C. M. (2001) Evidence from detrital zircons for the existence of continental crust and oceans on the Earth 4.4 Gyr ago. *Nature* **409**, 175–178.

Williams M. L. and Jercinovic M. J. (2002) Microprobe monazite geochronology: putting absolute time into microstructural analysis. *J. Struct. Geol.* **24**, 1013–1028.

Williams M. L., Jercinovic M. J., and Terry M. P. (1999) Age mapping and dating of monazite on the electron microprobe: deconvoluting multistage tectonic histories. *Geology* **27**, 1023–1026.

Windley B. F., Allen M. B., Zhang C., Zhao Z. Y., and Wang G. R. (1990) Paleozoic accretion and Cenozoic redeformation of the Chinese Tien Shan Range, central Asia. *Geology* **18**, 128–131.

Windley B. F., Kröner A., Guo J. H., Qu G., Li Y., and Zhang C. (2002) Neoproterozoic to Palaeozoic geology of the Altai orogen, Chinese Central Asia: new zircon age data and tectonic evolution. *J. Geol.* **110**, 719–739.

Wolf R. A., Farley K. A., and Silver L. T. (1996) Helium diffusion and low-temperature thermochronometry of apatite. *Geochim. Cosmochim. Acta* **60**, 4231–4240.

Wu F. Y., Jahn B. M., Wilde S., and Sun D. Y. (2000) Phanerozoic crustal growth: U–Pb and Sr–Nd isotopic evidence from the granites in northeastern China. *Tectonophysics* **328**, 89–113.

Yarmolyuk V. V., Kovalenko V. I., Kovach V. P., Budnikov S. V., Kozakov I. K., Kotov A. B., and Salnikova E. B. (1999) Nd-isotopic systematics of western Transbaikalian crustal protoliths: implications for Riphean crust formation in central Asia. *Geotectonics* **33**, 271–286.

York D., Hall C. M., Yanase Y., Hanes J. A., and Kenyon W. J. (1981) $^{40}Ar/^{39}Ar$ dating of terrestrial minerals with a continuous laser. *Geophy. Res. Lett.* **8**, 1136–1138.

Zartman R. E. and Doe B. R. (1981) Plumbotectonics—the model. *Tectonophysics* **75**, 135–162.

Zeitler P. K. (1988) Argon diffusion in partially outgassed alkali feldspars: insights from $^{40}Ar/^{39}Ar$ analysis. *Chem. Geol.* **65**, 167–181.

Zeitler P. K., Herczig A. L., McDougall I., and Honda M. (1987) U–Th–He dating of apatite: a potential thermochronometer. *Geochim. Cosmochim. Acta* **51**, 2865–2868.

3.11
Granitic Perspectives on the Generation and Secular Evolution of the Continental Crust

A. I. S. Kemp and C. J. Hawkesworth

University of Bristol, UK

3.11.1 INTRODUCTION

Every geologist is acquainted with the principle of "uniformitarianism," which holds that present-day processes are the key to those that operated in the past. But the extent this applies to the processes driving the growth and differentiation of the Earth's continental crust remains a matter of debate. Unlike its dense oceanic counterpart, which is recycled back into the mantle by subduction within 200 Ma (see Chapter 3.13), the continental crust comprises buoyant quartzofeldspathic materials and is difficult to destroy by subduction. The continental crust is, therefore, the principal record of how conditions on the Earth have changed, and how processes of crust generation have evolved through geological time. It preserves evidence of secular variation in crustal compositions, and thus the way in which the crust has formed throughout Earth's history. Exploring the nature and origin of these variations is the focus of this chapter.

Continental rocks are highly differentiated, and so the crust is enriched in incompatible components compared to the primeval chondritic composition (see Chapter 3.01). Of these, water is perhaps the most relevant, both for the origin and evolution of life, and also for many models of crust generation and differentiation. Similarly, the mass of continental crust is just 0.57% of the silicate Earth, and yet it contains ~35% of the potassium (using the crustal composition estimates in Table 1). Continental rocks comprise the buoyant shell that was once thought to float on a basaltic substratum, inferred from the wide distribution of chemically similar continental flood basalts (von Cotta, 1858). The links with the adjacent oceans were perhaps unclear, "the greatest mountains confront the widest oceans" (Dana, 1873). Yet, it has long been argued that the rock that has the most similar composition to the average continental crust, andesite, may be generated by fractional crystallization of basalt (Daly (1914) and Bowen (1928); but see the contrary arguments of Kelemen (1995) and Chapter 3.18). The average age of the continental crust is old, almost 2 Ga, the processes of crust generation may have changed with time, and the early crust may have been generated and destroyed more rapidly than in more recent times (Armstrong, 1991; Bowring and Housh, 1995).

The present consensus is that the modern Earth's continental crust has a bulk andesitic composition (~61% SiO_2), but it is lithologically and chemically stratified, such that a mafic

Table 1 Compositions of the continental crust and selected trace element ratios referred to throughout this chapter (all values are from Chapter 3.01). Major element oxides in wt.% and trace elements in ppm.

	Upper crust	*Middle crust*	*Lower crust*	*Bulk crust*
SiO_2	66.60	63.50	53.40	60.60
TiO_2	0.64	0.69	0.82	0.72
Al_2O_3	15.40	15.00	16.90	15.90
FeOt	5.04	6.02	8.57	6.70
MnO	0.10	0.10	0.10	0.10
MgO	2.48	3.59	7.24	4.66
CaO	3.59	5.25	9.59	6.40
Na_2O	3.27	3.39	2.65	3.07
K_2O	2.80	2.30	0.61	1.81
P_2O_5	0.15	0.15	0.10	0.13
ASI	1.03	0.85	0.75	0.85
Mg#	46.7	51.5	60.1	55.3
Ba	628	532	259	456
Rb	82	65	11	49
Sr	320	282	348	320
Zr	193	149	68	132
Nb	12	10	5	8
Y	21	20	16	19
Sc	14	19	31	21.9
V	97	107	196	138
Cr	92	76	215	135
Co	17.3	22	38	26.6
Ni	47	33.5	88	59
Cu	28	26	26	27
Zn	67	69.5	78	72
Ga	17.5	17.5	13	16
La	31	24	8	20
Ce	63	53	20	43
Pr	7.1	5.8	2.4	4.9
Nd	27	25	11	20
Sm	4.7	4.6	2.8	3.9
Eu	1.0	1.4	1.1	1.1
Gd	4	4	3.1	3.7
Tb	0.7	0.7	0.5	0.6
Dy	3.9	3.8	3.1	3.6
Ho	0.83	0.82	0.7	0.77
Er	2.3	2.3	1.9	2.1
Tm	0.3	0.32	0.24	0.28
Yb	2	2.2	1.5	1.9
Lu	0.31	0.4	0.25	0.3
Hf	5.3	4.4	1.9	3.7
Ta	0.9	0.6	0.6	0.7
Pb	17	15.2	4	11
Th	10.5	6.5	1.2	5.6
U	2.7	1.3	0.2	1.3
Rb/Sr	0.25	0.23	0.03	0.15
Rb/Ba	0.13	0.12	0.04	0.11
Sr/Nd	11.85	11.28	31.63	16.00
Eu/Eu^*	0.72	0.96	1.14	0.93
Eu/Sr	0.0031	0.0050	0.0032	0.0034
$(Gd/Yb)_N$	1.65	1.50	1.70	1.61

lower crust depleted in granitic components underlies an evolved middle and upper crust (Table 1; see Chapter 3.01 for a review of the compositional structure of the continental crust). But how has this been achieved? In the simplest terms, the continental crust grows magmatically, directly from the underlying mantle, and by the tectonic accretion of thicker portions of oceanic crust, such as island arcs (see Chapter 3.18) or ocean plateaus (see Chapter 3.16). However, these materials are predominantly basaltic, whereas the continental crust is andesitic, and neither most andesites, nor especially the large complement of silicic igneous rocks, can be extracted directly from a peridotitic source. Crust formation must therefore occur in at least two stages, first, melting of the mantle to produce basaltic magma, and second, either fractional crystallization or remelting of the basalt ultimately to produce the more evolved rocks of which the continental crust is dominantly composed. The products of the second stage may subsequently undergo further differentiation by one or more cycles of remelting (termed anatexis or "intracrustal melting"), which is integral in the stabilization of new continental crust, or by weathering and erosion at the Earth's surface (see Kelemen (1995), and Rudnick (1995) and references therein).

Uncertainties revolve around the tectonic settings under which these stages occur, the extent to which they may have been nonlinear or episodic through Earth's history (Figure 1), and especially how the processes associated with each have changed with time. It is reasonably well established that the Earth has cooled considerably from the earliest Archean to the present day (Brown, 1986), and so it follows that the sites, rates and processes of magma generation, and consequently of crust formation, have changed too. Intracrustal melting is also likely to have been strongly influenced by changing thermal and geodynamic conditions in the Earth, governing the type of materials likely to undergo fusion, the compositions of derivative melts and the efficiency with which these segregate from their residues. One possible manifestation is the predominance of tonalitic–trondhjemitic batholiths in the Archean, in contrast to the potassium-rich granitic plutons of Phanerozoic orogenic belts (Martin, 1986; Drummond and Defant, 1990). Whether conventional plate tectonic scenarios operated in the Archean, and when these came into existence remains unclear, and this impacts on models of crust generation and recycling of crustal materials back into the mantle. Equally, isotopic studies have shown that the mantle itself has been progressively modified by the parent–daughter fractionation of certain radiogenic

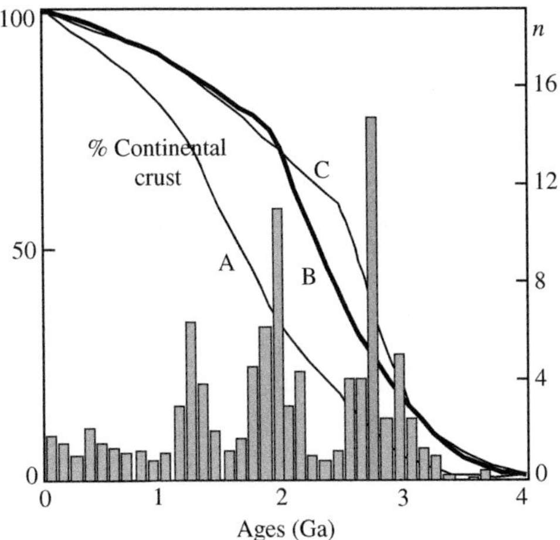

Figure 1 The distribution of U/Pb ages from juvenile crust, with only one age (n) used for each supracrustal succession or pluton to avoid undue weighting from particularly well-studied areas (from Condie, 1998), combined with selected curves representing different models for the changing volumes of stable continental crust with time (A: minimum curve based on Nd isotopes in shales (authors' compilation); B: from Collerson and Kamber (1999), and similar to Kramers and Tolstikhin (1997); and C: Taylor and McLennan (1985)).

isotopes (e.g., Sm–Nd, Rb–Sr, Lu–Hf) that occurs with the removal of crustal materials (e.g., O'Nions *et al.*, 1979; Allegre *et al.*, 1983; see also Chapter 2.13). Hence, the isotope composition of the magmatic flux across the Moho, and therefore of the potential building blocks for the continents, has also changed significantly during planetary evolution.

To investigate these issues, and in keeping with the theme of secular change, the formative stages of the continental crust are addressed sequentially, from generation of the basaltic protolith, through to the subsequent intracrustal and supracrustal recycling processes leading to its differentiation. In this way, the subject material of the chapter will itself become progressively more differentiated, mirroring the secular compositional evolution of the continental crust.

3.11.1.1 Granites and the Continental Crust

3.11.1.1.1 Granite terminology

Given the felsic nature of the exposed continental crust, much of the debate has focused on the origins of relatively evolved rock types, and in particular granite *sensu lato*. This is appropriate, since regional mapping and seismic profiling demonstrates that granitic rocks, their volcanic equivalents, and metamorphic and sedimentary rocks of granitic composition, are easily the most abundant constituents of the upper continental crust (Wedepohl, 1991). The rare earth element (REE) patterns of most post-Archean granitic rocks show the same distinctive features as that of the contemporary upper continental crust, notably enrichment in "light" REE (LREE, lanthanum to samarium), relative depletion in europium and flat "heavy" REE (HREE, holmium to lutetium) trends. Clearly, granitic plutonism may have a fundamental role in shaping the compositional structure of the crust, involving the ascent of buoyant, incompatible element-rich silicic magmas that leave behind residues that either reside in the deep crust or are recycled into the upper mantle. In addition, granitic batholiths formed by intracrustal melting potentially sample large areas of the middle to lower crust and may therefore provide compositional information on this inaccessible part of the Earth.

For all of these reasons, granitic rocks, and their role in crustal evolution, are the emphasis of this chapter. However, the petrogenesis of granitic magmas is controversial, and thus many aspects of this role remain unclear. The literature on granites is complex, and daunting to nonspecialist and specialist alike. This is testament to the lithological and geochemical diversity of granites, which in turn reflects a large variation in their potential mantle and crustal source rocks, conditions of magma formation, evolutionary processes and their emplacement at differing crustal levels in a variety of tectonic environments. It follows that granite classifications should avoid genetic implications, and it is expedient to clarify some granite-related terms at the outset.

For convenience, "granite" or "granitic rock" is here loosely defined to include plutonic rocks with more than 20% modal proportion of quartz relative to the feldspars, which encompasses the alkali feldspar granite through to tonalite lithological fields defined by the IUGS (Le Maitre, 1989). A widely used classification scheme is that of Chappell and White (1974), which subdivides granites into either I-type, sourced from an igneous or infracrustal precursor, and S-type, thought to have been derived by anatexis of sedimentary or supracrustal protoliths that have experienced a weathering cycle. There are also the volumetrically minor "A"-(alkaline, anhydrous, and/or anorogenic) and "M" (mantle-like)-type granites, which are essentially subcategories of the I-types. The I–S scheme is widely used, but it is genetic, and it implies that granitic rocks have a simple crustal source, which can be consistently deduced from mineralogy and geochemistry. In practice, most granitic magmas contain contributions from multiple sources, and the extent to which the compositional features of high level granitic plutons faithfully reflect those of the protolith continues to be vigorously debated. Thus, we favor the use of mineralogical qualifiers (e.g., hornblende granite, cordierite granite) and simple chemical classification schemes, such as the alumina saturation index (ASI) of Shand (1927) that convey no implication for putative protoliths (Table 2). In the latter, granites are either peraluminous, metaluminous, or peralkaline. Peraluminous rocks have an excess of Al_2O_3 over that required to crystallize feldspar and have an alumina saturating phase, metaluminous rocks are just saturated in Al_2O_3, whereas peralkaline granites are deficient in Al_2O_3 relative to the alkalis.

3.11.1.1.2 *Granites and crustal growth*

Since granitic magmas cannot be extracted directly from the upper mantle, there can also be confusion over the language used in the context of models for the generation of new crust. Clearly granitic plutons represent new crust, at least in part, if they contain a component of juvenile mantle-derived material. In practice, this could occur in three ways: (i) fractional crystallization of primary basaltic liquid, (ii) mixing between partial melts of pre-existing crust and mantle-derived magmas, or their differentiates, and (iii) by partial melting of young, mantle-derived mafic protoliths in the crust. Formally, the last case is by remelting within the crust, but it tends to be described in terms of crust generation, particularly when the mafic protolith separated from the mantle during the same tectono-thermal episode that subsequently generated the granite. The implication is that at the time of melting, the mafic protolith had neodymium, strontium, and/or lead isotope compositions indistinguishable from

Table 2 Classification of granitic rocks according to the ASI of Shand (1927), where ASI = molar $[Al_2O_3/(CaO + Na_2O + K_2O)]$.

Category	*Chemical definition*	*Diagnostic mineralogy*
Peraluminous	ASI > 1.0	Aluminous biotite ± muscovite, cordierite, garnet, Al-silicate
Metaluminous	ASI < 1	Subaluminous biotite ± hornblende, clinopyroxene accessory allanite, titanite, epidote
Peralkaline	ASI ≪ 1 Al/(Na + K) < 1	Sodic pyroxene (aegirine) or amphibole (riebeckite–arfvedsonite)

those of its contemporary mantle source, given its short crustal residence time. An example would be the remelting of a lower crustal basaltic sill shortly after its emplacement by the underplating of a successive batch of basaltic magma. However, if the same basaltic sill was to be remelted, say 200 Ma later, then the silicic magma would be derived from older, pre-existing crust, and hence be regarded as crustal remelting. This distinction based on the age of the source is arbitrary, but it is useful to clarify how these terms are used.

3.11.2 BULK CRUSTAL COMPOSITIONS AND THE MANTLE-DERIVED PROTOLITHS

The chemistry of the present-day continental crust is the frame of reference anchoring any discussion of secular evolution of crustal composition, or the processes of its formation. In this chapter we build on the considerable efforts of a number of scientists to develop an integrated geological and geophysical model for the continental crust (see Chapter 3.01 for a detailed treatment of the methodologies employed to infer crustal compositions).

The preferred chemical estimates of the continental crust used throughout this chapter are listed in Table 1. The major element composition of the upper crust is well constrained, since this is the most accessible to sampling, both directly and via erosion and sedimentation, and different studies utilizing diverse databases have yielded remarkably similar results. SiO_2 is ~61%, and Mg number (Mg#, molar Mg/(Mg + Fe)) is ~55 for the bulk continental crust, and so it is more differentiated than any magma in equilibrium with the upper mantle. Trace-element abundances are more variable, as are estimates for the composition and proportion of the middle and lower crust. As we will see below, the latter are critical to any discussion of the mechanisms of crust formation and differentiation.

3.11.2.1 Assessing the Arc–Plume Balance

Despite the elevated silica and potassium values compared to the bulk silicate Earth, the Mg# of the bulk continental crust is relatively high. Thus, the continents cannot be derived by known tholeiite fractionation trends, which involve marked iron enrichment, and the debate has focused on the extent that the continental crust can be generated by calc alkaline differentiation as seen in many orogenic andesite suites, or whether the parental magmas should be high-magnesium andesites (e.g., Kelemen, 1995). The latter may be generated by partial melting of subducted oceanic crust, and subsequent interaction of those melts with the overriding peridotite in the mantle wedge (Kelemen, 1995; see also Chapter 3.18). Such rocks occur, e.g., in the western islands of the Aleutian arc, where the plate convergence rate is extremely low, and the wider context of this setting is difficult to evaluate. The significance of high-magnesium andesites for the evolution of the continental crust is explored by Kelemen *et al.* (2003).

The distinctive trace-element features of the continental crust are it's high incompatible element contents, including uranium, thorium, and potassium, and hence elevated heat production (see Chapter 3.02), the negative mantle normalized anomalies for niobium and tantalum, and high lead contents (Figure 2). Thus, it is characterized by low Nb/La, Ta/La, and Ce/Pb ratios relative to the oceanic crust and the upper mantle. Presumably, the initial stage in the development of continental crust involved the partial melting of primitive spinel and garnet peridotite to form mafic magmas in equilibrium with mantle olivine. Of primary mantle-derived magmas, low Nb/La and Ce/Pb ratios are only a feature of those related to subduction (Hofmann *et al.*, 1986) (or where the mantle source has been overprinted by a subduction-derived component, Section 3.11.4.3.3) and so it is widely assumed that similar processes were responsible for the average composition of the continental crust (e.g., Arculus, 1999; Davidson and Arculus, 2001). However, there are also significant volumes of intraplate magmatism, and a number of authors have sought to assess the balance of intraplate and subduction-related magmatism in the generation of continental crust (e.g., Rudnick, 1995; Plank and Langmuir, 1998; Barth *et al.*, 2000). Intraplate magmas tend to have smoother mantle-normalized trace-element patterns, without the anomalies that characterize subduction-related magmas (Figure 2(b)). Element ratios that reflect the presence or absence of such anomalies therefore provide the basis for assessing the contributions from magmas from the different settings (Figure 3).

Other features of subduction settings are, first, that oceanic island arcs tend to have greater proportions of rocks with high MgO and low silica than their continental counterparts. Second, oceanic arc rocks tend to preserve more of the signature attributed to fluids released from the downgoing slab (e.g., high Ba/Th, U/Th, Sr/Nd), and a smaller subducted sediment contribution. This is most easily seen in the observation that high Ba/Th and U/Th ratios are best developed in the more depleted arc rocks that have relatively low thorium abundances (e.g., McDermott and Hawkesworth, 1991). In contrast, high Th/Ce ratios, which appear to be reasonably diagnostic of a sediment contribution,

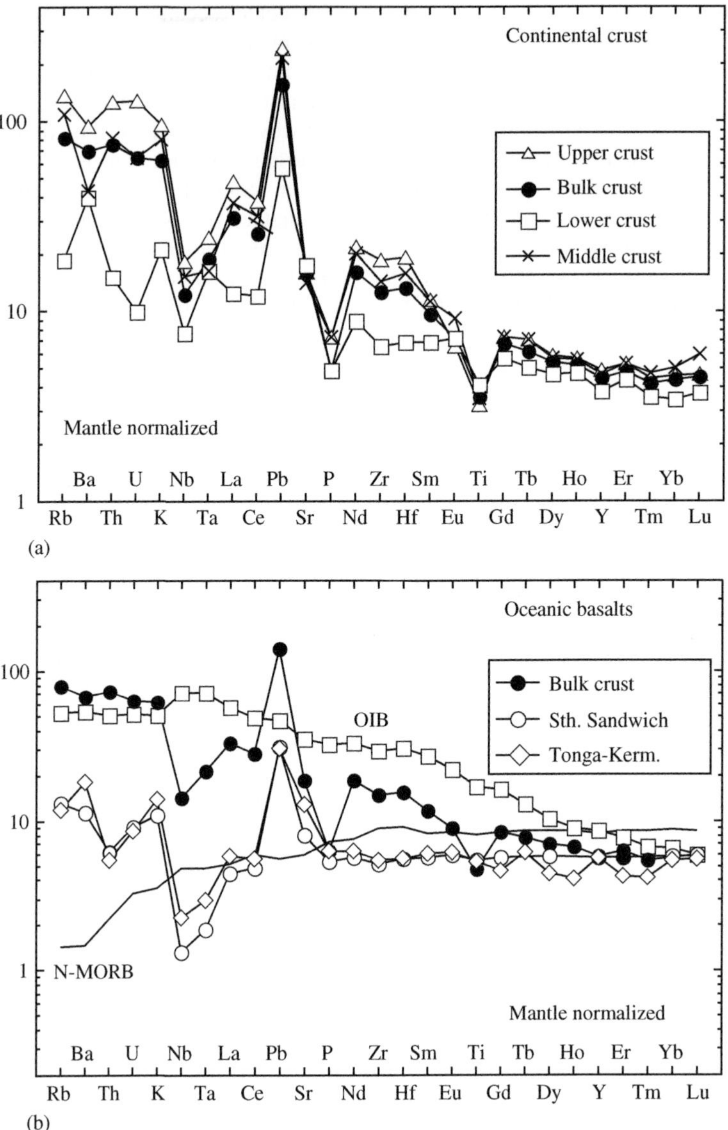

Figure 2 Primitive-mantle normalized minor and trace-element diagrams for: (a) the upper, middle, bulk, and lower continental crust (values from Table 1), and (b) oceanic and island arc basalts and the bulk continental crust (all normalizing values are from McDonough and Sun, 1995). The oceanic basalts (N-MORB, normal mid-ocean ridge basalt and OIB, ocean island basalt) are from Sun and McDonough (1989), whereas the arc basalts are from Turner *et al.* (1997) (Tonga-Kermadec arc) and Pearce *et al.* (1995) (South Sandwich arc).

tend to be better developed in relatively enriched arc rocks with higher incompatible element contents (Hawkesworth *et al.*, 1997). Negative Nb–Ta anomalies are most prominent in the depleted island arc tholeiites, and these are not associated with relatively low titanium values (Figure 2(b); Pearce and Peate (1995) and see discussion of Figure 7(b)), as they are in most other magmatic suites (e.g., see Figure 5). The inference is that negative Nb–Ta anomalies may result from different processes in different magmatic suites (Section 3.11.3.1.1). The more evolved major element compositions in continental arcs are typically attributed to greater degrees of differentiation within the thicker continental crust, often accentuated by crustal contamination (e.g., Davidson *et al.*, 1991). Trace elements may be modified by contamination, but higher abundances are also often linked to larger contributions from subducted sediment, as proposed for lavas from the Aeolian arc, Italy (Ellam *et al.*, 1989; Hawkesworth *et al.*, 1993) or Lesser Antilles (e.g., MacDonald *et al.*, 2000), not least because sediment fluxes at subduction zones are higher near the continents. In some cases this may be accompanied by smaller degrees of melting in the mantle wedge beneath thicker continental lithosphere.

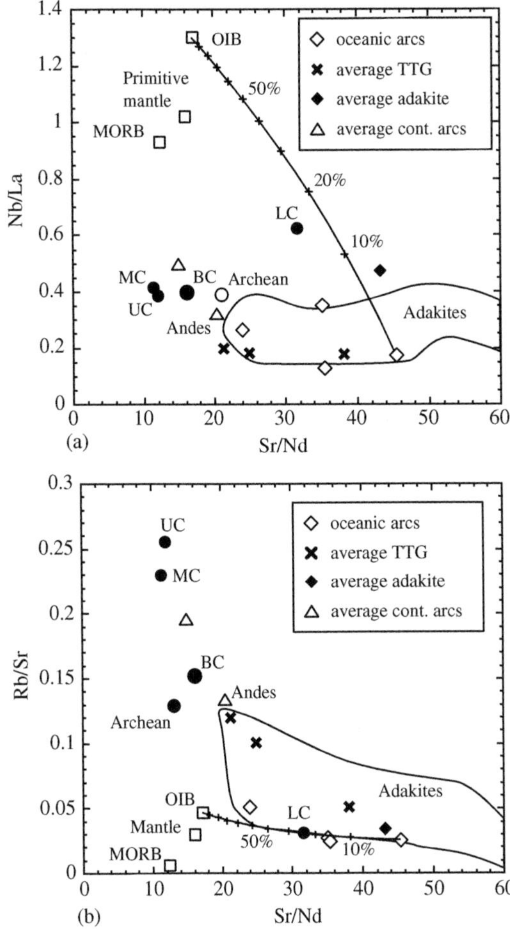

Figure 3 Mixing models for the formation of the bulk continental crust (BC), involving selected mantle-derived protoliths and based on key trace-element ratios (UC, upper crust; MC, middle crust; LC, lower crust). The crustal values for this, and subsequent figures are taken from Chapter 3.01, listed in Table 1 (a) Nb/La versus Sr/Nd and (b) Rb/Sr versus Sr/Nd. The end-members are the average OIB of Sun and McDonough (1989) and a primitive island arc basalt from Vanuatu (from Davidson and Arculus, 2001); other arc basalt data is from Davidson and Arculus (2001) and Turner *et al.* (1997). Crosses on the mixing curves correspond to 10% additions of the OIB component. Also shown for comparison are the primitive mantle and average MORB of Sun and McDonough (1989), the average andesite–dacite–rhyolite suite from the Andes and continental arcs (Drummond *et al.*, 1996) and average Archean TTG compositions (from Martin, 1995 and Smithies, 2000). The adakite field is plotted from the sources quoted by Smithies (2000), and the average adakite composition is taken from Drummond *et al.* (1996). The bulk Archean crust is from Taylor and McLennan (1995).

In the context of this discussion, any assessment of the contributions of subduction and within plate magmatism is sensitive to the choice of the arc end-member. Many authors have selected an oceanic arc as the subduction-related end-member, because these rocks contain less recycled material and hence reflect relatively more new material being added to the crust. Figure 3 illustrates variations in Nb/La, Sr/Nd, and Rb/Sr values for average crustal compositions, the primitive mantle, oceanic basalts (normal mid-ocean ridge basalt (N-MORB); ocean island basalt (OIB)), and island arc lavas (island arc tholeiite (IAT)). Mixing results in near straight arrays on the plot of Nb/La–Sr/Nd, and it is striking that the average crustal compositions do *not* plot on any simple mixing curve between the island arc tholeiite, OIB or primitive mantle (see also Rudnick, 1995). If continental crust primarily consists of some mixture of magmas generated in island arc and intraplate (OIB-like), its *bulk* composition has since been modified by an additional process or processes.

The additional processes are most likely to involve magmatic differentiation and weathering and erosion. However, Nb/La ratios are relatively robust in andesitic magmas, in the sense that they are not readily changed by fractional crystallization, crustal contamination, or weathering. In contrast, Sr/Nd ratios are reduced, and Rb/Sr ratios are increased towards the average crustal values by fractionation of plagioclase in igneous systems or during weathering and sedimentary transport, as strontium is soluble in seawater (we will return to this last point in Section 3.11.5.2). With respect to igneous processes, plagioclase may fractionate during magmatic differentiation, or it may be residual during partial melting, but in either case that fractionation must have taken place *within* the crust, in the stability field of plagioclase. These are longstanding conclusions, with the implication that strontium has been selectively recycled into the mantle, either as a component of magmatic cumulates (and/or residues) or altered oceanic crust (see Chapter 3.15) where strontium is concentrated within carbonate precipitates (Rudnick, 1995).

Alternative solutions to this problem have been proposed, though all tend to founder on the difficulty of generating elevated Rb/Sr ratios directly from the upper mantle. For example, Ellam and Hawkesworth (1988) speculated that most of the continental crust was generated early in Earth's history by processes that were different to those operating above modern active subduction zones, which inherently produce basalts with low average Rb/Sr ratios. In this scenario, the high Rb/Sr component is represented by primitive Archean tonalite–trondhjemite–granodiorite (TTG) associations and andesites, such as those of the 2.9–2.5 Ga Zimbabwe craton, which have Rb/Sr ratios similar to bulk crustal values (Ellam and Hawkesworth, 1988). Unlike arc-related basalts, the TTG were generated from basaltic, rather than peridotitic sources, at depths below the stability field of plagioclase (Section 3.11.3.1.1). Given the ancient average age of the continental crust (~1.8 Ga), there is little doubt that Archean magmatism contributed greatly to the formation of the continents. However, the average TTG composition, estimated from Archean provinces worldwide, has lower Rb/Sr ratios, and high Sr/Nd, than the bulk continental crust (Figure 3(b)). More recent discussions have explored whether the subduction-related component might originate as partial melts of the subducted oceanic crust (i.e., "adakites," Section 3.11.3.1.1), as seen in some recent arc settings like the western Aleutians (Kelemen *et al.*, 2003; see also Chapter 3.18). Yet the Rb/Sr ratios of these lavas only approach those of the average crust in the most differentiated compositions (Kelemen, 2002, personal communication). In fact, the Rb/Sr ratios of global average adakite compositions do not significantly exceed those of arc tholeiites (Figure 3(b)). Thus, this model also requires removal of strontium-rich material from the continents.

Additional evidence that the bulk composition of the continental crust has been modified is that it has a small negative europium anomaly (i.e., europium is depleted relative to neighboring elements samarium and gadolinium in chondrite normalized REE patterns; Table 1) since this is not a normal property of primary mantle-sourced magmas in any tectonic setting (Sun and McDonough, 1989; see Figure 3(b)). As with strontium, and in contrast to the other REE, europium is accommodated by plagioclase, and thus is sensitive to intracrustal differentiation processes. However, as will be demonstrated in Section 3.11.5.2, fractionation of europium from the other REE *cannot* be achieved by dissolution of plagioclase during modern weathering. With the caveat that crustal strontium and europium concentrations have not been underestimated, this suggests that the elevated Rb/Sr and Eu depletion of the bulk crust was accomplished by the removal of what were originally plagioclase-bearing assemblages from the base of the crust. This can only occur in tectonically thickened crustal columns or overthickened arc edifices (>40 km thick), where the intrinsically buoyant plagioclase-rich material transforms with depth into denser garnet-rich granulite or eclogite, which is gravitationally unstable and can founder into mantle peridotite (Jull and Kelemen, 2001). This mechanism has been recently invoked to explain the paucity of early formed primitive (magnesium-rich) mafic to ultramafic cumulates in the Izu-Bonin (Arculus, 1999) and Aleutian (Kelemen *et al.*, 2003) arc systems. In areas of continental collision, the same result may be achieved by convective removal of lower crustal cumulates/residues, if the lithospheric mantle coupled to the lower crust has also been delaminated.

Another point from Figure 3 is that crustal contamination processes will shift mixed OIB–IAT compositions towards the average crust, and it is noticeable that the Andean andesite plots close to the average crustal composition. However, crustal contamination processes do not provide information on the generation of new crust. Rather, the displacement of the crustal compositions away from the OIB–IAT array highlights that the residence time of highly incompatible elements at high levels in the crust is long compared with those of more compatible elements deep in the crust.

Finally, given the envisaged decrease in mantle potential temperature from the Archean to the present day (Campbell and Griffiths, 1992; see also Galer and Mezger, 1998), it seems plausible that the relative supra-Moho fluxes from arc- and intraplate magmas may also have changed through time. Condie (1994, 2000) proposes that the proportion of plume-related oceanic plateau basalts in greenstone successions was greater in the Archean than in younger terranes, where arc-related magmatism became dominant. One might therefore expect to see differences between the bulk composition of the Archean and post-Archean crust with respect to key chemical indices or elemental ratios. However, we note in Figure 3 that the estimated bulk composition of Archean continental crust has slightly lower Nb/La ratios than that of the post-Archean crust. Since high Nb/La ratios are taken to be indicative of an intraplate or plume component (and both are too often taken to be synonymous) in recent volcanic rocks (Hofmann *et al.*, 1986; Condie, 1999), on face value this implies that the Archean crust had about the same, rather than the predicted higher, contribution from plume-derived magmas than that generated subsequently. An alternative, and preferred, interpretation is that the low Nb/La ratios reflect different processes in Archean rocks, and this is discussed further below (Section 3.11.3.1.1).

3.11.2.2 Episodicity in Crust Generation, and Its Implications

It is increasingly recognized (e.g., Boher *et al.*, 1992; McCulloch and Bennett, 1994; Stein and Hofmann, 1994) that the generation of the Earth's continental crust may follow an *episodic*, rather than continuous, pattern through time, and this in turn bears upon unraveling the processes responsible for the extraction of crustal materials from the mantle. For example, a striking feature of age–frequency diagrams for the continental crust, based on U–Pb zircon dates, are the peaks of major episodes of igneous activity, separated by periods of apparent quiescence (Condie, 1998, 2000) (Figure 1). Assuming that this is not simply an artifact of selective preservation of certain crustal segments, the periodicity lies at the heart of current debates on the origins of the Archean and Early Proterozoic continental crust, since it is taken as *prima facie* evidence by some that crust formation was dominated by major thermal pulses associated with the emplacement of mantle plumes. In these scenarios it is considered less likely that a global episodicity in crust forming events would result from the more continuous subduction-related processes (Albarède (1998); but this is discussed further below). In contrast, no such peaks are observed in the last billion years, when it may be reasonably inferred that subduction-related magmatism was dominant. Gurnis and Davies (1986) suggest that this last point might reflect the enhanced erosion of elevated young crust, and thus preferential recycling of this material into the mantle. However, it seems unlikely that the crustal formation ages should be completely erased from the detrital zircon record.

Precise U–Pb zircon analyses have further highlighted that major contemporaneous episodes spread across different continents, such as between 2.74 Ga and 2.66 Ga, and prompted speculation that these are the products of catastrophic turnover events in the Earth's mantle (Stein and Hofmann, 1994; Condie, 1998). Such interpretations are consistent with models of mantle processes in a hotter Earth (Hill, 1993; Davies, 1995). However, Archean rocks are preserved in just 7.5% of the Earth's surface (Goodwin, 1991), and so there is some uncertainty over the extent these age distributions and inferred episodicity are representative of major events in the Archean.

Besides the geochemical constraints, discussed in Section 3.11.2.1, a challenge for models for episodic crustal growth based on mantle plumes is to explain the formation of the evolved silicic rocks that comprise the continental crust. This is because the overwhelming product of mantle plumes is basalt, either as juvenile oceanic crust such as oceanic plateaus (see Chapter 3.16), flood basalt provinces (see Chapter 3.03) or mafic underplates. How then does the episodicity of mantle plume activity translate into a similar periodicity in the generation of evolved continental crust? One possibility is that continental growth occurs by lateral amalgamation of thick, difficult-to-subduct oceanic plateaus against pre-existing continents (Albarède, 1998). Here, the basaltic materials are ultimately reprocessed into felsic continental crust by intracrustal differentiation, involving anatexis and chemical weathering (Albarède, 1998), and typically this has to be in short time periods to ensure the mantle-like initial isotope ratios of many of the felsic rocks. Critically, age peaks in the Archean and the Early Proterozoic in Figure 1 (the "super-events" of Condie (1998)) are observed in both mantle-derived volcanic rocks and in higher silica plutons that are probably crustal melts. Such close temporal links are clearly an important constraint on models of crust generation, and for each pulse of igneous activity there appears to be a consistent delay of 50–80 Myr between the eruption of mantle-derived volcanics and the intrusion of syn-tectonic granites.

Recent examples of felsic magmatism associated with plume activity tend to be temporally associated with large volumes of mafic rocks. The A-type granites of intraplate settings are very different from Archean TTG suites, and their compositions reflect the alkaline to enriched nature of the associated basalts, and the distinctive volatile component (Section 3.11.4.3.2). As we emphasize in Section 3.11.3.1, a number of continental flood basalt provinces are markedly bimodal in their silica distributions, similar to many Archean terranes. However, the felsic rocks of flood basalt provinces tend not to have been generated in equilibrium with residual garnet, which characterizes the Archean TTG (Section 3.11.3.1.1). The key element may therefore be the estimated time lag of 50–80 Myr between the eruption of mantle-derived volcanics and the intrusion of syn-tectonic granites, since it highlights that the mafic and felsic components of Archean crust were generated under different conditions. The tectonic settings may therefore have been different, or at least a particular setting would have evolved in character over the inferred 50–80 Myr.

An alternative to oceanic plateau accretion is that continental growth reflects a transient surge of accelerated subduction, this being a byproduct of enhanced heat flow in the upper mantle attending plume activity. The intrusion of syn-tectonic granites is thought to form through collision or accretion of the juvenile crustal fragments (Condie, 1998). The emplacement of silicic rocks, and thus the growth of continental crust, would, in this model, reflect closure of oceanic basins by subduction. Subduction rates, and thus the volumes of arc magmas, are likely to increase dramatically during periods of high plume flux in the upper mantle, since this will increase the generation rate of juvenile oceanic crust along ridge systems (Condie, 2000). Furthermore, the subducted crust will be younger, which thermally favors slab melting and the production of the felsic TTG that comprise large parts of Archean cratons and the earliest continents. This model has the attraction of coupling mantle plume activity and subduction in the formation of new continental crust.

Another important question is why the first age peak only on Figure 1 occurred at ~2.7 Ga and not before, since ancient detrital zircon ages demonstrate the existence of continental crust dating back to ~4.4 Ga (Wilde *et al.*, 2001). Several authors (e.g., Stein and Hofmann, 1994; Condie, 1998, 2000) propose that the mantle plumes initiating the continent-forming "superevents" are triggered by catastrophic "avalanches" of subducted eclogitic slabs that have accumulated at the 660 km seismic discontinuity. In this scenario, the absence of age peaks prior to ~2.7 Ga might reflect the efficient recycling of subducted oceanic crust into TTG magmas in the hotter early Archean mantle. Continental materials were also likely to be more rapidly returned to the convecting mantle during this period (Armstrong, 1991), and thus have a lower survival rate. Conversely, the 2.7 Ga episode may signal the onset of subduction and the emplacement of large crustal volumes during a short time interval. Clearly, the genesis of the TTG holds the key to understanding the formation of Earth's first continents, and we shall address this problem in Section 3.11.3.

3.11.2.3 Evidence for Secular Changes in the Composition of the Continents

There are clear differences in the common igneous rock types present in Archean and post-Archean regions, and hence in the distribution of major and trace elements, as seen in the preponderance of bimodal silica distributions in Archean cratons compared with those in younger terranes generated along convergent continental margins. Allied with the pattern of episodicity apparent in Figure 1, and the progressive cooling of the Earth, this suggests that we might also expect to resolve some secular changes in crustal compositions. Potentially, the best evidence for this is from detrital continental sediments, since it is generally agreed that they provide good estimates of the average post-Archean upper crust, at least for elements, like scandium, thorium, yttrium, and the REE, that are not soluble in water (Taylor and McLennan (1985, 1995); see also the review by Rudnick and Gao (Chapter 3.01). However, we need to be aware that the extent to which any differences in sedimentary rock geochemistry through time manifest changes in the composition of the bulk continental crust is much more difficult to assess.

Taylor and McLennan (1985) established that compared to post-Archean shales, Archean clastic sedimentary assemblages have systematically lower silicon, potassium, thorium, and scandium contents and higher magnesium, calcium, sodium, chromium, and nickel. This has been largely corroborated by Condie (1993), using cratonic shales, though lower calcium, sodium, and strontium in Archean shales compared to their younger counterparts implies that chemical weathering was more intense in the CO_2-rich atmosphere of the early Earth (Condie, 1993). The differences in compositions between Archean and younger sedimentary rocks documented by Taylor and McLennan (1985) grossly mimic the chemistry of the dominant igneous rocks emplaced during these periods, notably with the preponderance of sodic TTG and magnesium-rich komatiites in the Archean, compared to the more siliceous and potassium-rich granites in younger orogens (Sections 3.11.3 and 3.11.4 respectively). Even allowing for the removal of a large basalt-komatiite component by erosion, Archean shales also have lower titanium, niobium, and tantalum than their post-Archean counterparts (Condie, 1993), which seems to reflect derivation from a provenance dominated by TTG, which are depleted in these elements (see Figure 5(a)). As such, Nb/La ratios are lower in Archean shales (0.36) and the Archean upper crust (0.28), compared to post-Archean shales (0.44–0.50) and contemporary upper crust (0.39–0.46; all values from Condie (1993)), though such differences are less apparent in bulk crustal compositions (Figure 3). The implications of this for crustal growth processes were highlighted in the above section.

However, interpretation of the REE data has provoked more controversy, largely reflecting the choice of Archean sediment from which the estimates derive. Taylor and McLennan (1985, 1995) suggest that, although considerable variability exists, low-grade graywackes of Archean greenstone belts exhibit less LREE enrichment, greater fractionation of the HREE (i.e., extend to higher Gd/Yb ratios) and no negative europium anomaly compared to post-Archean shales, which have marked negative europium anomalies. These features are in turn reflected in contrasting chondrite-normalized REE patterns for the Archean and post-Archean crust (Figure 4; and see Taylor and McLennan (1985), figure 2)). As we shall see in Section 3.11.3.1.1, steep REE patterns lacking significant europium anomalies characterize the Archean TTG. Europium is preferentially partitioned into plagioclase feldspar, and so Taylor and McLennan (1985, 1995), conclude that residual plagioclase was much more of a feature of intracrustal differentiation processes in

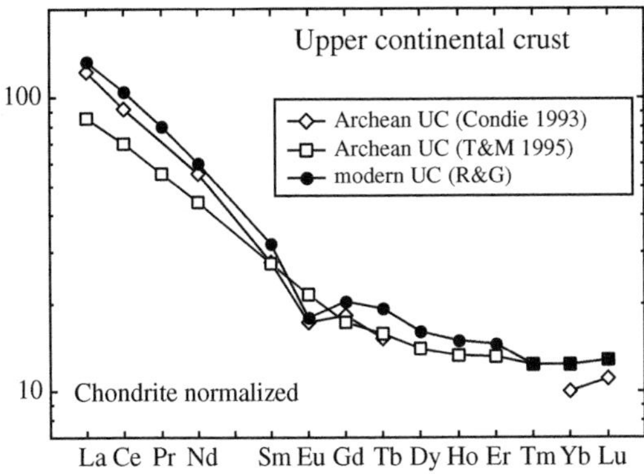

Figure 4 Comparison between the Archean upper crust (UC) and the contemporary upper crust of Rudnick and Gao (see Chapter 3.01) (R&G) with respect to chondrite-normalized REE patterns (normalizing values from Anders and Grevesse, 1989). Archean data are from Condie (1993) and Taylor and McLennan (1995) (T&M 1995).

the post-Archean, compared with the Archean crust. The implication is that intracrustal melting at depths shallow enough to be in the plagioclase stability field was less common in the Archean. Since strontium is also preferentially partitioned into plagioclase relative to rubidium, it may also be inferred that the post-Archean upper crust will have a higher Rb/Sr ratio, and this is evident in Figure 3. The MREE to HREE are fractionated by residual garnet, and overall the Archean and post-Archean upper crusts have similar MREE to HREE patterns. However, the Archean patterns are attributed to variable mixing between flat MREE to HREE greenstone belt volcanic rocks with highly fractionated patterns from the TTG, typically in a ratio of two parts basalt to one part TTG (Taylor and McLennan, 1985). In contrast, the similar patterns in the post-Archean upper crust are taken as evidence of the lack of residual garnet in younger crust differentiation processes, since rocks like the Archean TTG are rare.

Alternatively, Gibbs *et al.* (1986) and Condie (1993) suggest that rather than reflecting the absence of intracrustal differentiation prior to 2.7 Ga, the secular variation in the REE is simply an artifact of comparing Archean graywackes with post-Archean shales that have different tectonic settings. Refuting this, Taylor and McLennan (1995) point out that the negative europium anomaly, and thus the signature of crustal differentiation, persists in modern graywackes of all tectonic settings and is present even in forearc turbidites and pelagic clays remote from the continents. It, therefore, seems inescapable that if mid- to shallow-level intracrustal differentiation were widespread in the Archean, the characteristic europium depletion would be recorded by greenstone belt sedimentary sequences.

Gibbs *et al.* (1986) and Condie (1993) further argue that cratonic (passive margin) shales are more representative of the Archean upper crust than active margin greenstone belt graywackes. Condie (1993) emphasizes that cratonic Archean shales posses negative europium anomalies that overlap statistically with those of post-Archean shales, diminishing the difference between the Archean and post-Archean upper crust with respect to the REE. Nevertheless, even if this approach is justified (see the contrary arguments of Taylor and McLennan (1995)), the Archean upper crust of Condie (1993) still has less relative depletion in europium than the post-Archean upper crust (Figure 4). It seems therefore that these differences are real, and relate to contrasting behavior of major phases (plagioclase, garnet, and possibly amphibole) during magmatism and crustal differentiation in the Archean and post-Archean periods.

The lower crust is much more difficult to sample, and this in turn limits the accuracy of estimates of bulk continental crust in the Archean and younger terranes. The common approaches, summarized by Rudnick and Gao (Chapter 3.01), include studies of xenolith populations in alkali basalts and kimberlites, high-grade metamorphic terranes, and seismic velocity structure. In general, while the evidence for crustal melting in the generation of granites is overwhelming (Section 3.11.4), residual assemblages after partial melting are rarely sampled. Nonetheless, the best estimate of the lower crust is that it is mafic and comprises the residua left after the extraction of the granodioritic upper crust, perhaps subsequently underplated by basaltic magmas (Taylor and McLennan, 1985, 1995). We note that the current lower crustal composition plots near the arc basalt-OIB mixing lines on Figure 3, which is consistent with a mafic component generated in both subduction-related and intraplate settings. There is evidence from xenolith and seismic imaging studies that the Archean continental crust is thinner, and generally lacks the basaltic underplating and development of mafic granulitic lower crust observed in post-Archean regions (see Chapter 3.01). One possible reason for this is that the mafic complement was part of the original crustal architecture but has since been lost to the convecting mantle, as suggested by Nui and James (2002) for the Kaapvaal Craton (southern Africa). Alternatively, mafic lower crust may never have existed in Archean cratons, reflecting a fundamentally different style of crustal generation during this period (Section 3.11.3.1.1).

In summary, sediments and xenolith suites preserve a record of gross secular change in the compositional structure, and perhaps the bulk composition of the continental crust. Given that evolved igneous rocks are the building blocks of the continents, an important step towards understanding the mechanisms driving these changes is to examine silicic igneous activity in the Earth through time, and this is the focus of the following sections. Another aim is to elucidate the tectonic "engine" controlling crustal generation and evolution. It is also necessary to unravel the relative contribution of other differentiation processes, such as weathering and erosion, to the chemistry of the continental crust, and this is treated separately at the end of the chapter.

By way of introduction, average analyses of silicic igneous rocks from various time periods are listed in Table 3, and normalized to the bulk continental crust in Figure 5. Key chemical features are highlighted in bivariate element-element and ratio–ratio plots (Figure 6). Two aspects are immediately evident. First, there are clear differences between Archean tonalites and most Phanerozoic plutonic rocks, particularly with regard to Na/K and the REE (Figure 6). Second, none of the Phanerozoic granite types mimic the current upper crustal composition for the full range of trace elements considered, with the greatest discrepancies shown by the most primitive (oceanic arc) and felsic (leucogranites) rocks. This suggests that other igneous and/or sedimentary processes must be invoked to account for the generation of the uppermost granodioritic crust.

3.11.3 GENERATION OF HIGH SILICA CONTINENTAL ROCKS: I. THE ARCHEAN

3.11.3.1 Archean TTG Associations

The implications from neodymium model ages on upper crustal rocks is that most crust generation occurred relatively early, such that at least 50% of the continental crustal mass was emplaced by the end of the Archean (Taylor and McLennan, 1985; Figure 1, and Chapter 3.10). Although little of this embryonic crust is now exposed at the Earth's surface, bimodal silica distributions in Archean terranes show that the massive igneous activity during this period produced two contrasting suites, high-magnesium basalts (including basaltic komatiites and komatiites) and the relatively felsic TTG. With an average of $\sim$70% SiO_2 (Martin, 1995) the latter are among the oldest high silica igneous rocks preserved and comprise the earliest continental nuclei (e.g., the ca. 4.0 Ga Acasta gneisses of the western Slave Province, Canada, Bowring and Williams, 1999). Although the tectonic setting for Archean magmatism remains unresolved, it is noteworthy that in the recent geologic past bimodal silica suites are a feature of extensional and plume-related settings, as in continental flood basalt-rhyolite provinces like the Parana-Etendeka Province (Garland *et al.*, 1995; Peate, 1997) or the Karoo Igneous Province (Duncan *et al.*, 1984), rather than of convergent plate margins.

3.11.3.1.1 Geochemical constraints on TTG genesis

The TTG are different from most of the evolved plutonic rocks of Proterozoic and Phanerozoic orogenic belts (Figures 5 and 6; Table 3). This is most evident in their typically high Na_2O and Al_2O_3 and low K_2O contents, and steep REE patterns, with strongly elevated strontium and LREE, and depleted HREE, yttrium, and scandium. In contrast to most younger plutons, the most "primitive" TTG also commonly lack negative europium anomalies, and although many TTG in Figure 6 do have small negative europium anomalies, these are generally smaller than that of the bulk continental crust. Strontium and neodymium isotopic ratios of the TTG are primitive and approach mantle values at the time of crystallization (Martin, 1995), though some TTG suites have apparently assimilated older crustal materials (e.g., those of the Superior Province, Canada; Whalen *et al.* (2002)). This striking combination of features is generally attributed to derivation from a hydrated (meta)basaltic source rock at depths below the stability field of plagioclase (i.e., $>$40 km) but in equilibrium with residual garnet, to account for the low HREE contents (Martin, 1986, Martin, 1995; Luais and Hawkesworth, 1994; Rapp, 1997; Wyllie *et al.*, 1997). Isotopic considerations demand a short crustal residence time for the basaltic protolith prior to melting, to preserve the mantle-like $^{87}Sr/^{86}Sr$ and $^{143}Nd/^{144}Nd$ ratios, and thus emplacement of the TTG constitutes growth of new continental crust. However, the tectonic implications are unclear because partial melting may have taken place either in subducted slabs, similar to destructive plate margins at the present day (Drummond and Defant, 1990; Martin, 1995; Condie, 1998), or in underplated basalt beneath thickened crust or oceanic plateau (Kröner, 1985; Smithies, 2000; Whalen *et al.*, 2002). The underplated basalts may have been generated in response to mantle plume activity, or above old subduction zones.

Table 3 Representative analyses of the granite types referred to throughout this chapter and plotted in Figure 5 (see Figure 5 caption for data sources).

Sample	*TTG* (*n* = *355*)	*Archean calc-alkaline*		*Cont. arc. PRB* (*n* = *323*)	*Oceanic arc* (*n* = *8*)	*GRC Leucogran.* (*n* = *8*)	*Himalayan Leucogran.* (*n* = *13*)	*A-type Padthaway* (*n* = *6*)	*A-type LFB*	*LFB Hbl granite* (*n* = *1074*)	*LFB Crd granite* (*n* = *704*)
		Type 1	*Type 2*								
SiO_2	69.79	70.00	71.88	64.63	68.10	74.59	74.16	75.93	72.06	69.50	70.91
TiO_2	0.34	0.40	0.23	0.65	0.48	0.05	0.09	0.17	0.38	0.41	0.44
Al_2O_3	15.56	14.63	14.69	15.94	14.99	14.63	14.74	11.91	12.43	14.21	14.00
Fe_2O_3				1.20	1.77	0.14			1.61	1.01	0.52
FeO				3.19	2.28	0.65			1.55	2.22	2.59
FeOt	2.81	2.72	1.65	6.06	4.32	1.24	0.90	1.93	2.95	3.12	3.06
MnO	0.05	0.05	0.03	0.08	0.10	0.02	0.02	0.07	0.08	0.07	0.06
MgO	1.18	0.84	0.48	2.15	1.29	0.29	0.17	0.12	0.40	1.38	1.24
CaO	3.19	2.28	1.69	5.10	4.36	1.74	0.70	0.40	0.93	3.07	1.88
Na_2O	4.88	3.89	4.45	3.62	3.98	3.03	3.91	3.63	3.94	3.16	2.51
K_2O	1.76	3.58	3.69	1.95	1.38	3.97	4.43	4.91	4.13	3.48	4.09
P_2O_5	0.13	0.17	0.08	0.13	0.09	0.105	0.17	0.02	0.09	0.11	0.15
LOI		0.78	0.81			0.89	0.69	0.29	2.02		
Sum	99.56	98.39	98.79	98.51	98.73	99.98	98.22	97.14	97.51	98.51	98.24
ASI	0.989	1.015	1.021	0.919	0.938	1.176	1.179	0.992	0.983	0.977	1.169
Mg#	42.8	35.5	34.1	38.7	34.8	29.7	25.2	9.7	19.5	44.1	41.9
Ba	690	1300	1210	641	311	1200	223	76	725	519	440
Rb	55	117	125	60	19	125.8	229	132	159	164	245
Sr	454	479	455	375	267	283.1	124	7	95	235	112
Zr	152	218	142	139	112	16.6	35.7	330	460	150	157
Nb	6.4	12	9	6.7	1.3	3.73	9.8	22	24.5	11	13
Y	7.5	21	7	19	22	20.1	10.8	49	79	31	32
Sc	4.7	4.7	1.8	14	14	5		3	15.2	13	11
V	35	28	15	85	63	45		2	8	57	49
Cr	29	26	73	47	2	35		5	2	20	30
Co		13	14		18				4	10	10
Ni	14	12	12	13	2	15		3	1	8	11
Cu		19	13	10	35	11			4	9	9
Zn		59	41	76	50	19			125	48	59
Ga		18	17	18.3	14.8	56		20	20.2	16	18
La	32	71	42	16	6.3	8.86	10.8	147.5	64	31	27
Ce	56	133	71	35	16.3	18.56	23.2	226.93	132	66	61
Nd	21.4	45	29	14	11.3	7.39	10.8	118.41	64	23.7	21.9

(continued)

Table 3 (continued).

Sample	*TTG (n = 355)*	*Archean calc-alkaline*		*Cont. arc. PRB (n = 323)*	*Oceanic arc (n = 8)*	*GRC Leucogran. (n = 8)*	*Himalayan Leucogran. (n = 13)*	*A-type Padthaway (n = 6)*	*A-type LFB*	*LFB Hbl granite (n = 1074)*	*LFB Crd granite (n = 704)*
		Type 1	*Type 2*								
Sm	3.3	7.8	4.4		4.1	2.05	3	13.52	14.9	4.9	5.8
Eu	0.92	1.56	0.79		0.93	1.11	0.63	0.43	2.42		
Gd	2.2	5.1	3.4		4.0	2.34		7.52	12.4		
Tb	0.31	0.73	0.41				0.56				
Dy	1.16	3.2	2.2			3.02		6.48			
Ho					1.2				3.2		
Er	0.59					1.80		3.43			
Tm									1.1		
Yb	0.55	1.55	0.51		2.25	1.67	1.29	3.36	8.2		
Lu	0.12	0.24	0.09		0.50	0.24	0.18		1.27		
Hf	4.5	5.8	4.6		3.3	2.29			11.1		
Ta	0.71	0.93	0.29		0.88	0.45			2.9		
Pb		23	28	10	5	55.8	80	17	22	19	27
Th	6.9	21	22	7.2	0.83	3.42	5	22	19.7	20	19
U	1.6	2.4	3.3	1.5	0.37	2.09		2.9	5	5	5
Rb/Sr	0.12	0.24	0.27	0.16	0.07	0.44	1.84	19.55	1.67	0.70	2.19
Rb/Ba	0.08	0.09	0.10	0.09	0.06	0.10	1.02	1.73	0.22	0.31	0.56
Sr/Nd	21.21	10.64	15.69	26.78	23.73	38.31	11.48	0.06	1.48	9.91	5.11
Eu/Eu^*	1.037	0.751	0.620		0.692	1.537	0.61	0.130	0.541		
Eu/Sr	0.0020	0.0032	0.0017		0.0035	0.0039	0.0051	0.0614	0.0254		
$(Gd/Yb)_N$	3.30	2.72	5.51		1.47	1.15		1.85	1.25		

TTG, Archean tonalite-trondhjemite-granodiorite; PRB, Peninsula Ranges Batholith; GRC, Glenelg River Complex; LFB, Lachlan Fold Belt; Hbl, hornblende granite; Crd, cordierite grantie.

At present there are a number of competing lines of evidence in support of the TTG being formed by remelting at the base of thickened crust, and in the subducted slab. Support for the subduction model is reviewed by Martin (1995), whereas evidence for the role of intracrustal recycling processes in TTG genesis is highlighted by Whalen *et al.* (2002), and references therein. One argument is that the TTG are chemically analogous to modern adakite lavas, which are confined to arc settings and largely attributed to melting of

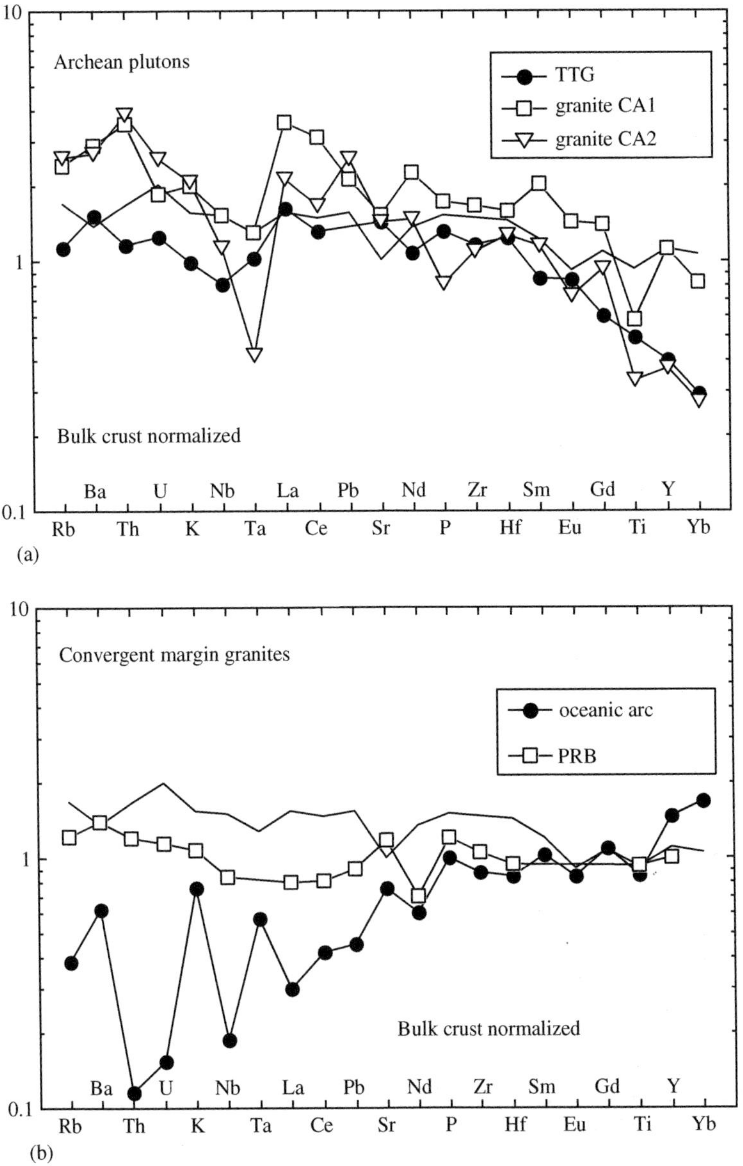

Figure 5 Multi-element diagrams of the different silicic igneous rock types discussed in the text compared to estimates of the upper continental crust (unornamented line), with all compositions are normalized to the bulk crust (values in Table 1). (a) Average of 355 Archean TTG (from Martin, 1995) and average Late Archean "calc-alkaline" (CA) granites (from Sylvester, 1995), (b) convergent margin granites; continental arc, Peninsula Ranges Batholith (PRB) California, from Silver and Chappell (1988), oceanic arc (New Britain) from Whalen (1985), (c) leucogranites; Glenelg River Complex (GRC, southeastern Australia, Kemp, 2001), Himalayan leucogranite from Inger and Harris (1993), (d) A-type granites; average Padthaway Suite granite, southern Australia (from Turner *et al.*, 1992a), Gabo Island metaluminous granite, Lachlan Fold Belt (LFB, southeastern Australia, from Collins *et al.* (1982), peralkaline granite, LFB (Wormald and Price, 1988), (e) average hornblende ($n = 1074$) and cordierite ($n = 704$) granites, LFB (SE Australia), from Chappell and White (1992). Note the higher rubidium, but much lower Sr content of the cordierite granites, consistent with a greater component of weathered (meta)sedimentary material in the source region (Chappell and White, 1992).

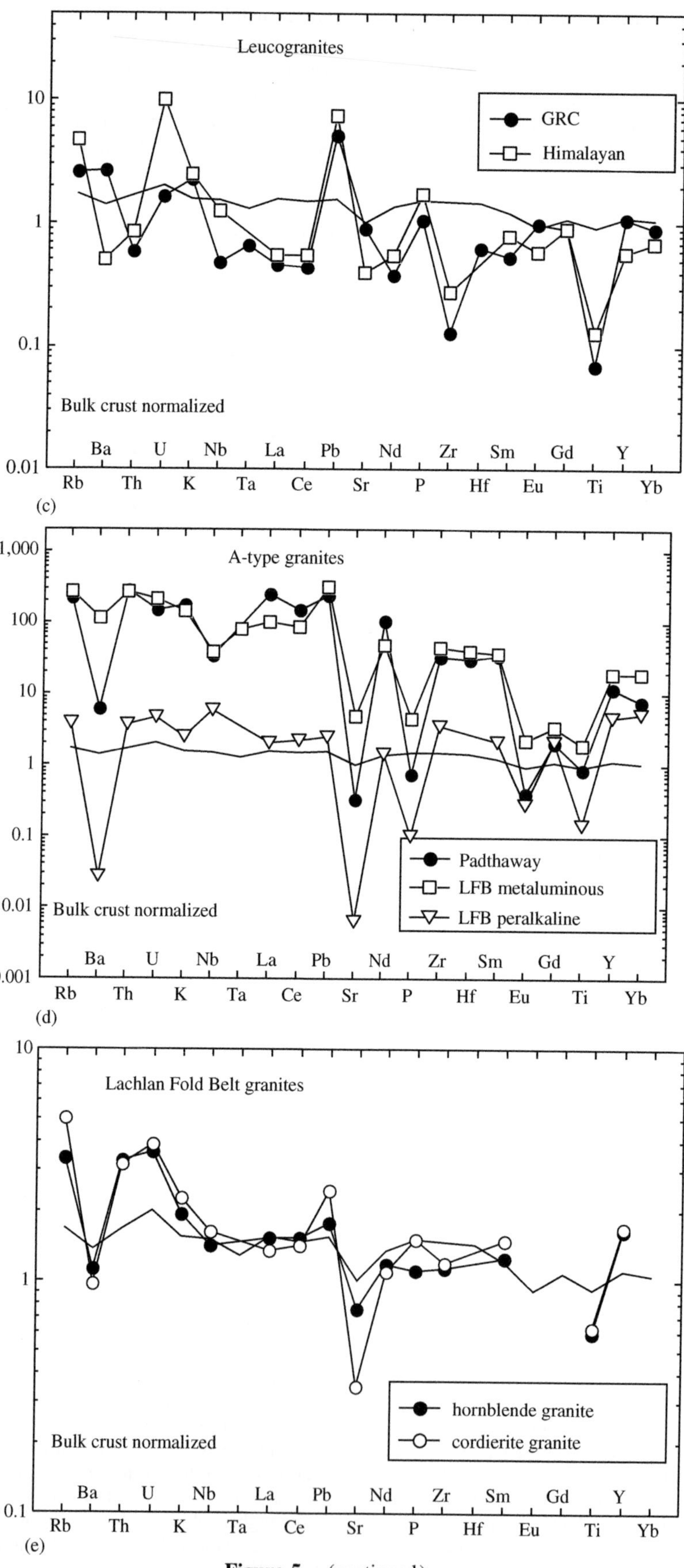

Figure 5 (continued).

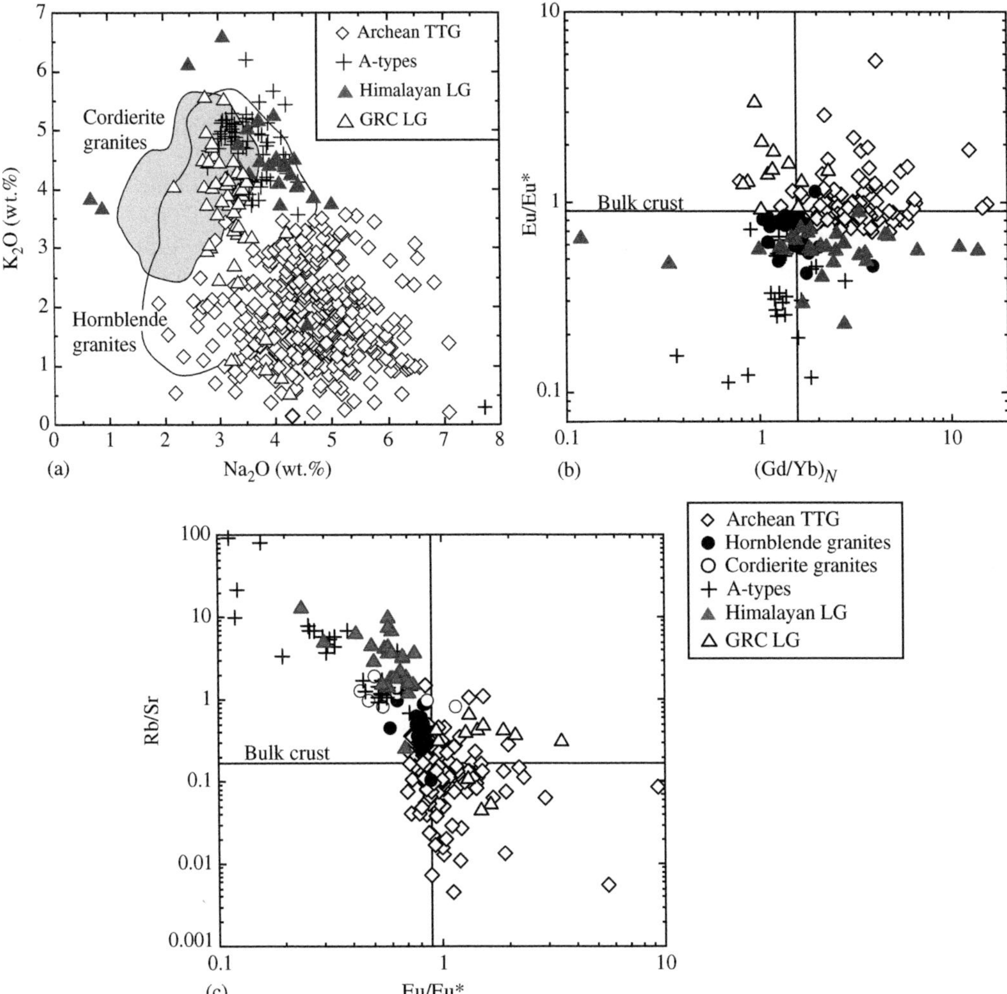

Figure 6 The reference igneous suites shown on plots of (a) K_2O versus Na_2O (in weight percent), (b) Eu/Eu* versus $(Gd/Yb)_N$, and (c) Rb/Sr versus Eu/Eu* (LG, leucogranite; GRC, Glenelg River Complex). The subscript N refers to the abundance of the element in the rock normalized to that of the chondritic composition recommended by Anders and Grevesse (1989). Eu/Eu* is a measure of the normalized abundance of Eu relative to that of Sm and Gd (the "Eu anomaly") and is computed according to $Eu/Eu^* = Eu_N/\sqrt{[(Sm_N)\cdot(Gd_N)]}$. Data sources are: TTG, from the references quoted by Smithies (2000), and including Kamber *et al.* (2002) and Whalen *et al.* (2002); LFB hornblende and cordierite granites are from Anderson (1998), Chappell *et al.* (1991), Chappell and White, 1992; Chen *et al.* (1989, 1990), Elburg (1996b,c), Griffin *et al.* (1978); Hine *et al.* (1978), Kemp (unpublished data), Maas *et al.* (1997), Munksgaard (1988), Soesoo (2000), White and Chappell (1988, 1989), White *et al.* (1977); A-type granites, Collins *et al.* (1982), King *et al.* (1997, 2001), Turner *et al.* (1992a), Whalen *et al.* (1987), Wormald and Price (1988), and some data from the Geoscience Australia *Rockchem* repository; Himalayan leucogranites, Ayres and Harris (1997), Crawford and Windley (1990), Inger and Harris (1993); France-Lanord and Le Fort (1988), Le Fort *et al.* (1987), Searle *et al.* (1997), Vidal *et al.* (1982); Glenelg River Complex leucogranites, Kemp (2001).

the downgoing oceanic lithosphere (see reviews by Defant and Drummond, 1990; Drummond and Defant, 1990; Peacock *et al.*, 1994; Drummond *et al.*, 1996; Martin, 1999). Yet as emphasized by Smithies (2000), sodic plutons that resemble the TTG, at least in terms of major elements, are also a minor component of overthickened continental arcs, such as in the Mesozoic batholiths of the western American Cordillera (Atherton and Petford, 1993; Tepper *et al.*, 1993; Petford and Atherton, 1996), New Zealand (Muir *et al.*, 1998), and Antarctica (Wareham *et al.*, 1997), where they are thought to derive by fusion of a juvenile underplate. A key difference between these two geodynamic models is that melts from subducted oceanic crust migrate through mantle en route to the crust, whereas those from underplated basalts do not. Many TTG, and adakites (e.g., Stern and Kilian, 1996), have higher MgO and Ni contents than experimental melts of basalt (Martin, 1999; Martin and Moyen, 2002). These features have therefore been taken to reflect interaction of TTG magmas with peridotite in the mantle wedge (Drummond *et al.*, 1996; Rapp *et al.*, 1999; Martin and Moyen, 2002), and hence to be supportive of the subduction zone model. Like adakites, TTG extend to lower Nb/Ta and higher Zr/Sm ratios than the primitive mantle and oceanic basalts. Recent experimental work suggests that these characteristics reflect the presence of residual amphibole, and that those amphiboles had low Mg numbers of 40–50 (Foley *et al.*, 2002). This requires melting of low magnesium amphibolite, rather than eclogite or even amphibole-bearing peridotite. Such fractionated basaltic rocks are presumed to be more typical of the upper parts of oceanic crust generated and hydrated at mid-ocean ridges, and later consumed by subduction, rather than the lower crust (Foley *et al.*, 2002), where dry, magnesium-rich cumulates and eclogite are expected to predominate (Saunders *et al.*, 1996).

A striking feature of the TTG is that they have higher incompatible element contents and markedly lower TiO_2, Y, and Yb contents that their inferred basaltic protoliths, and these can be used to constrain the melt generation models. Figure 7(a) summarizes the arguments for the incompatible elements thorium and TiO_2 by illustrating the thorium and TiO_2 contents of the inferred source rocks for different values of partition coefficients, assuming that an average TTG reflects 10% melting. Critically, to reproduce the high thorium contents of primitive TTG (with ~0.4% TiO_2) simple modeling requires the basaltic protolith also to have moderately high thorium contents (>1 ppm) at low TiO_2 for reasonable degrees of melting. At higher degrees of melting, these source thorium contents would also have to be higher, and basalts with 1 ppm thorium contents lie outside the range of contemporary depleted MORB. However, they plot within the upper envelope of the field for Archean basalts and komatiites, and are also similar to the Archean basalt used in the modeling of Foley *et al.* (2002). Many of these high thorium komatiites have Th/Ta ratios

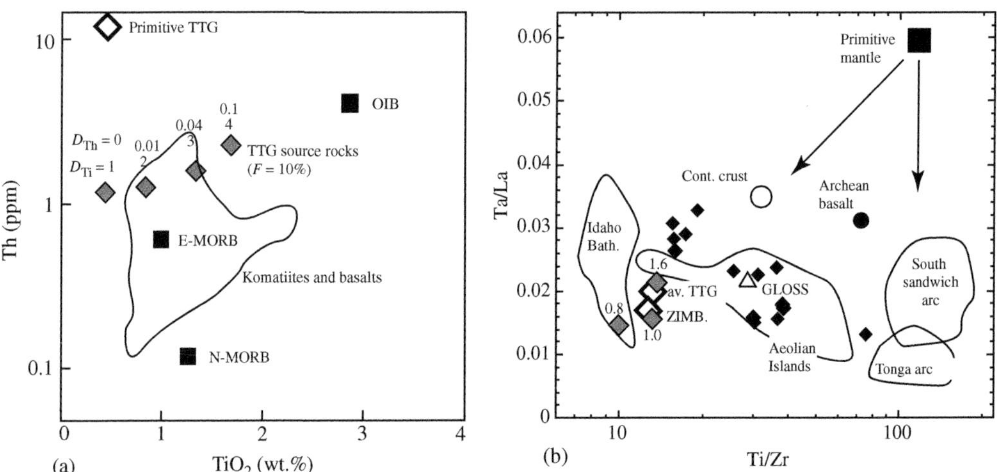

Figure 7 Batch melting models illustrating the possible generation of the Archean TTG from basaltic protoliths (cont. crust is the bulk continental crust from Table 1). (a) Plot of Th versus TiO_2, showing the source composition (gray diamonds) modeled as a function of different bulk partition coefficients required to reproduce the Th and titanium contents of primitive TTG (12 ppm Th, 2,500 ppm Ti) by 10% batch melting. Having higher degrees of melting increases the implied Th content of the source. Various oceanic basalts (from Sun and McDonough, 1989; E-MORB, enriched mid-ocean ridge basalt) and the compositional field of Archean basalts and komatiites are shown for reference. (b) Comparison between oceanic arc basalts (<53% SiO_2), adakites (closed diamonds), TTG, the primitive mantle, and modeled partial melts (gray diamonds) from an Archean basalt (closed circle) at different pressures (0.8 GPa, 1.0 GPa, and 1.6 GPa) in terms of Ta/La versus Ti/Zr (ZIMB = least evolved TTG from Zimbabwe, Luais and Hawkesworth, 1994). The average TTG ($n = 355$) is from Martin (1995). GLOSS represents the average global subducted sediment composition estimated by Plank and Langmuir (1998). Other data sources are as for Figures 2 and 3, with the Aeolian Island field from Ellam *et al.* (1989) and the Idaho Batholith field from Clarke (1990). The partial melts were modeled by 10% batch melting of the average Archean Sula Mountains basalt of Rollinson (1999) (solid circle) using partition coefficients quoted in Foley *et al.* (2002) and references therein. Following Foley *et al.* (2002), the residual assemblages are: 0.8 GPa, 40% amphibole, 45% plagioclase, and 15% orthopyroxene; 1.0 GPa, 32% amphibole, 15% plagioclase, 28% garnet, and 25% clinopyroxene; 1.6 GPa, 17% amphibole, 15% plagioclase, 15% garnet, and 52% clinopyroxene.

similar to the primitive mantle, and on this basis there appears to be little evidence for subduction or crustal contamination processes having contributed to their high thorium contents (Condie, 1994). Such evolved basalts are also less refractory, and so they are likely to yield the greatest amount of melt for a given temperature rise above the solidus in a basaltic pile undergoing anatexis.

A second feature is that the TTG have negative Nb–Ta anomalies (Figure 5(a)), and these are widely regarded as indicative of subduction related magmatism. Figure 7(b) is a plot of Ta/La versus Ti/Zr, and TTG, like most granitic rocks, have both low Ta/La and Ti/Zr reflecting negative anomalies in both tantalum and titanium. A key feature of primitive oceanic arc lavas is that their negative Nb–Ta anomalies are not accompanied by relative depletions in titanium (Figure 2(b)), and thus low Ti/Zr ratios (Figure 7(b)). It follows that their low Ta/La values reflect subduction-related processes that have not fractionated Ti/Zr. In contrast, higher-K island arc suites such as that from the Aeolian Islands (Ellam *et al.*, 1989), which typically have a greater contribution from subducted sediment, have low Ta/La and Ti/Zr values, as do most sediments (see the global subducting sediment (GLOSS) composition from Plank and Langmuir (1998)). Negative Nb–Ta and Ti anomalies are a feature of granitic suites (Figure 5), average crustal compositions, and the Archean TTG (Figure 7(b)). The key question is the extent that such negative Nb–Ta anomalies reflect the same processes as those responsible for the low Ti/Zr values, and whether low Ta/La ratios are formed by different processes in oceanic island arcs and granitic magmas. Clearly, if negative Nb–Ta anomalies are formed by different processes in high silica magmas, they cannot be regarded as a simple test of a subduction-related component in such magmas or in the generation of the bulk continental crust.

The evidence from selected magmatic suites is not straightforward. TiO_2 decreases with increasing SiO_2 in most magmatic suites, but Ta/La fractionates differently. In the intraplate setting of Tenerife in the Canaries, for example, Ta/La increases from 0.09 to 0.13, with little change in Nb/Ta, as SiO_2 increases from 48% to 60% in a basanite–phonolite suite modeled by closed system fractional crystallization (Ablay *et al.*, 1998). In A-type granites Nb/La, and by implication Ta/La, increases in the most differentiated compositions (Figure 5(d)), but in many granitic rocks there is no simple intrasuite relationship between Ta/La and Ti/Zr. Most of the Idaho batholith samples plotted in Figure 7(b) have >60% SiO_2, and the low Ta/La and Ti/Zr values are a feature of the parental crustal melts (Clarke, 1990). Similarly, the TTG have low Ta/La and Ti/Zr and no clear intrasuite correlations between them, although that may, in part, reflect the large-scale sampling of many studies.

An alternative approach is modeling on the basis of experimental data. In principle, low Ta/La and Ti/Zr values might reflect residual, or fractionating, titaniferous phases such as rutile, or amphibole. Rutile is widely invoked as a residual phase in slab melting models (see discussion in Kelemen *et al.* (2003)), but the recent experimental study of Foley *et al.* (2002) suggests that rutile incorporates Ta in preference to Nb and so coexisting melts will have high Nb/Ta. Within many TTG the data are scattered, but decreases in Ta/La are accompanied by increasing Nb/Ta. The TTG considered by Foley *et al.* (2002) had low Nb/Ta, and they argued that in absence of residual rutile, fractionation of titanium from zirconium must be accomplished primarily by amphibole. Figure 7(b) therefore compares the melts produced from an average LREE-enriched Archean tholeiite at different pressures, and thus with different residual assemblages, to natural, undifferentiated TTG, using the partition coefficients for low-Mg amphibole quoted by Foley *et al.* (2002). Although such modeling is sensitive to amphibole chemistry and the composition of the protolith, low-Mg amphibole can also impart the low Ta/La and Ti/Zr of the TTG. This is permissive evidence that the TTG may be formed by low degrees of melting (~10%) of fractionated basaltic rocks leaving a garnet amphibolite residue with little or no rutile, provided that the amphibole has Mg# values less than ~60. An amphibole-bearing residue is also consistent with hornblende being a near-liquidus or fractionating phase in many TTG suites (Martin, 1995).

An intrinsic advantage of the slab-melting model is that with continued subduction the voluminous garnet amphibolite residues (± mafic cumulates) of TTG generation will transform into eclogite and ultimately founder into the deeper parts of the mantle, rather than being retained in the lower crust, as would be the case for derivation from a basaltic underplate. Such efficient "waste disposal" thus provides a ready explanation for the observation (see Chapter 3.01) that the bulk Archean crust appears to be, in general, more felsic than post-Archean crust (Rudnick, 1995). This is because most magmas of post-Archean subduction zones are derived from the mantle wedge rather than the descending oceanic slab (Martin, 1986; see below).

3.11.3.1.2 Secular evolution in TTG composition

If fusion of subducted oceanic crust is the correct origin for many Archean TTG, we might anticipate some secular variation in TTG compositions as the Earth cooled. This is because the attendant decrease in the geothermal gradient along the Benioff plane should lead to an increasing depth of slab melting, and thus TTG generation. The angle of subduction is also likely to increase, since the envisaged decrease in the rate of seafloor spreading from the earliest Archean (Bickle, 1978) leads to an increase in plate dimensions and thus the average age of oceanic crust, and older, colder lithospheric slabs are less buoyant (Christensen, 1997). A direct implication is that the thickness of the overlying mantle wedge that slab-derived magmas must traverse, and potentially interact with, will increase with time, and this is consistent with observed changes in TTG compositions. In a compilation of TTG data, Martin and Moyen (2002) demonstrate that between 4.0 Ga and 2.5 Ga the least differentiated TTG evolve to progressively higher Mg# and Ni contents, consistent with greater assimilation of olivine during passage through the mantle for the younger TTG. In the same period strontium contents also increase appreciably, which is interpreted as evidence for a declining role for residual plagioclase, and thus an increase in the depth of TTG formation from the Early to Late Archean (Martin and Moyen, 2002). It appears that systematic changes in TTG compositions with time in the Archean can be related to likely changes in the conditions of melt generation, if the subducted slab model is adopted. Such secular chemical trends seem less easy to accommodate by models where TTG are formed by remelting basaltic rocks in the lower crust, given that the Mg# and Ni contents of mantle-derived magmas (and thus the potential TTG source rocks) has decreased appreciably with time.

Modern adakites extend to higher strontium and Mg# values than Archean TTG of similar silica contents (Smithies, 2000; Martin and Moyen, 2002). Higher strontium implies lower degrees of partial melting, compatible with a progressively cooling Earth, whereas the elevated Mg# suggests greater interaction with mantle peridotite (Martin and Moyen, 2002). The implication of the latter is that modern adakitic melts are generated even deeper than the TTG, near the threshold between slab melting and slab dehydration. Thermal conditions do not favor slab melting in most modern day subduction zones, except under the exceptional circumstances of anomalously elevated heat flow associated with ridge subduction (Stern and Kilian, 1996), or the consumption of hot, newly created oceanic crust (Martin, 1986; Drummond and Defant, 1990; Peacock *et al.*, 1994) (but see Chapter 3.18 for an alternative view). Hence, the volume of TTG-like rocks emplaced into the continental crust has waned dramatically since the Archean, and in some cases such compositions have been attributed to melt generation at depth within thickened continental arcs (see above). Nevertheless, the continental crust shares the low (subchondritic) Nb/Ta, high Zr/Sm signature relative to the primitive mantle that characterises the TTG (Foley *et al.*, 2002), suggesting that this style of magmatism has made an important imprint on the bulk crustal composition.

3.11.3.1.3 A combined model for TTG formation

On the balance of the evidence discussed above, it seems likely that intraplate plume-related magmatism and subduction-related magmatism were both important aspects of crust generation processes in the Archean (Albarède, 1998). The key component of the TTG represents the second stage of crust generation, as partial melts of basaltic source rocks. It appears that those basalts are typically relatively evolved, in that they had high incompatible element contents (e.g., thorium; Figure 7(a)), and many greenstone belt volcanics have unfractionated Th/Ta suggesting that they were not generated by subduction-related processes, but may have had an intraplate or plume-related origin. It follows that in many cases the negative Nb–Ta anomalies of the TTG were not a feature of their basaltic source rocks, but were instead generated during partial melting of those basalts. Moreover, the low Ta/La ratios of the TTG appear to be linked to low Ti/Zr values, which characterizes most granitic rocks (Figure 7(b)). TTG were generated at depths where residual garnet, and not plagioclase, was stable but it is unclear whether sufficient crustal thickening on a large scale took place in the Archean to permit TTG generation at the base of the crust. Subduction ensures that TTG generation took place at sufficiently high pressures, and interaction of TTG magmas with peridotite in the mantle wedge offers an explanation for their high Ni and Mg#. However, it is curious why there are not more greenstone belt volcanic rocks with clear subduction signals, and the distinctive granite-greenstone belt outcrop patterns are not easy to reconcile with the linear sites of crust generation that characterize recent subduction zones. Models for TTG generation have been developed involving residual titaniferous phases such as rutile, and low Mg# amphiboles, and a distinctive test may be that melts in equilibrium with residual rutile should have high Nb/Ta ratios (Foley *et al.*, 2002).

3.11.3.2 Late Archean Granitic Rocks: The Onset of Major Intracrustal Melting

Towards the end of the Archean, after ~3.0 Ga, greenstone belt volcanism and the TTG were largely supplanted by relatively potassium-rich granitic intrusives (monzogranite and granodiorite with subordinate tonalite), which comprise around 20% of presently exposed Archean cratons (Condie, 1993). As with Phanerozoic granites, "calc-alkaline" (mostly metaluminous to weakly peraluminous), strongly peraluminous and alkaline associations are recognized (Sylvester, 1995). According to Sylvester (1995), these are formed by remelting intermediate igneous and sedimentary compositions at mid- to lower-crustal depths, rather than wholly in the stability field of garnet as inferred for the TTG. The appearance of granitic plutons may therefore reflect stabilization of a critical volume and thickness of buoyant continental crust to permit prolonged weathering processes, formation of sedimentary basins, and ultimately intracrustal melting during collisional orogenesis (see below).

3.11.3.2.1 Generation of the Archean granitic types

Archean granitic rocks are broadly similar to their Phanerozoic counterparts, though some systematic differences can be resolved (Figure 5 and Table 3). In most cases these can be related to small differences in protolith composition and the melting zone lying at deeper crustal levels in the Archean. The Archean *calc-alkaline* association extends to higher Al_2O_3/TiO_2, Na/K, Sr, Sr/Y, La/Yb, and Eu/Eu* than Phanerozoic equivalents and these distinctive features were most likely inherited from a TTG component in the source region. The depleted HREE and yttrium of some Archean calc-alkaline granites requires residual garnet, suggesting formation in the lower crust, and this is much less common in Phanerozoic plutons. Average barium contents of Archean calc-alkaline granites are twice those of Phanerozoic plutons, and this is attributed to greater consumption of biotite during melting under higher thermal gradients in the Archean (Sylvester, 1995). However, high barium contents are not matched by higher rubidium, and it is therefore also possible that the elevated barium contents reflect a paucity of residual feldspar, consistent with a greater melting depth. Interestingly, Phanerozoic calc-alkaline granites have distinctly higher Mg# than those of the Archean. This could reflect the presence of a "juvenile" mantle-derived component in the younger plutons, since, in contrast to the Archean, there is abundant field evidence for the involvement of such material in Phanerozoic granite genesis (see below).

Strongly peraluminous Archean granitic rocks with relatively low CaO, Na_2O, and Sr are most abundant in terranes that contain a major metasedimentary component, such as the Superior Province, Canada (e.g., Percival, 1989). As with their Phanerozoic equivalents, Archean peraluminous rocks are thought to derive largely by anatexis of metasediments (Feng and Kerrich, 1992; Sylvester, 1995). Like the Archean calc-alkaline plutons, the strongly peraluminous Archean plutons have, on average, higher Na/K, Sr, and Eu/Eu*, but lower rubidium, scandium, yttrium, and HREE (Sylvester, 1995). These features directly mirror differences in composition between Archean and Phanerozoic graywackes (Taylor and McLennan, 1985, 1995), and therefore probably reflect secular variation in source composition, perhaps accentuated by a minor amount of residual garnet in the Archean.

Archean *alkaline* plutons are similar in many respects to their Phanerozoic equivalents, with high SiO_2 and rubidium, and prominent negative europium anomalies, though the characteristic high-field-strength element (HFSE) enrichment and CaO–Sr–Ba depletion (see Section 3.11.4.3.1) are less marked. Thorium contents are higher, and Ga/Al ratios are significantly lower in Archean alkaline rocks, the latter being a defining attribute of Paleozoic alkaline plutons (Collins *et al.*, 1982). Given uncertainties with pinpointing the sources of alkaline plutons of Phanerozoic age (Section 3.11.4.3.2) these differences are difficult to evaluate petrogenetically. However, the lower Ga/Al and higher thorium may signify derivation from precursors that were less refractory than those postulated for Phanerozoic alkaline granites, and the higher CaO–Sr–Ba suggest that feldspar fractionation is less important in the evolution of Archean alkaline plutons.

On a more general note, the average Archean *calc-alkaline* and *alkaline* granitic plutons in the compilation of Sylvester (1995) have consistently higher Th/U ratios than Phanerozoic granites (Figure 8), even though this is not a property of the estimated bulk Archean crust (see Condie (1993) and Taylor and McLennan (1995) for compositions of the Archean crust). Ellam *et al.* (1990) also noted that the time-integrated Th/U ratios of Archean granites calculated from their lead isotopes were much higher than those in

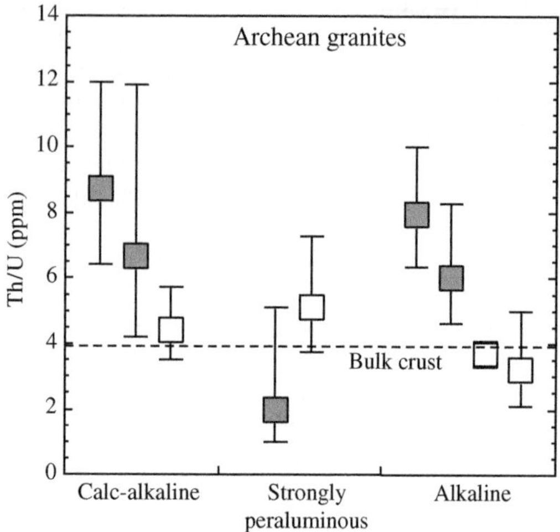

Figure 8 Plot comparing the average and range of measured Th/U ratios in the various Archean granitic types (shaded squares) with those of post-Archean granites of the same chemical group (open squares). All data are from Sylvester (1995), though the range of Th/U ratios indicated for the post-Archean granites incorporates data from the LFB (sources as for Figure 6).

average crustal compositions. However, the *strongly peraluminous* Archean granites appear to have slightly lower average Th/U ratios than their Phanerozoic counterparts. Besides the possibility that this feature is not a product of weathering and/or alteration, there are two potential explanations for these observations. First, it may be that aqueous fluids were not as important in the Archean for the generation of the calc-alkaline and alkaline granite types as in the Phanerozoic. Alternatively, uranium may not have been in the appropriate (oxidized) valence state to be readily mobilized in low f_{O_2} Archean fluids (in contrast to U^{4+}, U^{6+} is highly soluble in aqueous fluids). This second point is consistent with other lines of evidence. For example, to reconcile variations in lead isotope data shown by MORB, numerical simulations by Elliott *et al.* (1999) require that the upper mantle evolves towards markedly lower Th/U ratios after ~2.2 Ga. These authors attribute the decreasing Th/U ratios to a change in the oxidation state, and thus enhanced mobility of uranium in the more oxygen-rich environment at the surface of the post-Archean Earth, such that this element is preferentially recycled into the mantle as a component of altered oceanic crust (see also McCulloch and Bennett, 1994). As a result of the reduced mobility of uranium during continental weathering in the Archean, shales of this period have lower mean Th/U ratios than their younger counterparts (McLennan and Taylor, 1980) and this feature was therefore acquired by strongly peraluminous Archean granitic rocks, which were sourced from such protoliths. The higher Th/U ratios of the calc-alkaline and alkaline Archean granites may partly reflect lack of a sedimentary ingredient, in tandem with a contribution from relatively thorium-enriched mantle-derived magmas and/or crustal fluids during melting.

3.11.3.2.2 *Tectonic implications*

By analogy with "classical" Phanerozoic orogens, the geological context of Archean granites, and the temporal trend of some suites from deformed, calc-alkaline plutons to undeformed alkaline rocks, suggests formation in a collisional to post-collisional tectonic setting (Sylvester, 1995 and references therein). Should this prove correct, it supports the presence of rigid plates in the Late Archean and the operation of conventional plate tectonics by this time. Nevertheless, analogues of the "Cordilleran" type continental arc batholiths, belts of calcic tonalite-dominated suites associated with abundant coeval mafic rocks, are apparently absent in the Archean, and these are one of the hallmarks of convergent plate margins. This point can be construed as either (i) further evidence against subduction-related models for the generation of Archean continental crust, or (ii) reflecting differences in the thermal regime, such that heat input from mantle magmas was less essential in triggering granite genesis in the hotter Archean crust than in young orogens (Section 3.11.4). Typical calc-alkaline igneous suites formed above contemporary subduction zones are thought to have evolved from basaltic parental magmas derived from the mantle wedge, and it has been suggested that this may have been less common in the Late Archean if magmas were largely derived by partial melting of the downgoing slab (Martin, 1986).

The compositional parallels between Archean granitic rocks and their Proterozoic to Phanerozoic counterparts suggest that the style of crustal differentiation was similar during these periods. The implications for crustal evolution are therefore discussed in the next section.

3.11.4 GENERATION OF HIGH SILICA CONTINENTAL ROCKS: II. PROTEROZOIC TO PHANEROZOIC GRANITIC ASSOCIATIONS

The sedimentary record reveals a shift in upper crustal composition near the Archean–Proterozoic boundary (Section 3.11.2.3), suggesting a concomitant change in the degree, and perhaps style of crustal differentiation, and possibly in the composition of new crust. Magmatically, this change is signaled by the emplacement of large volumes of relatively potassic, incompatible-element-rich granitic plutons from about the Late Archean onwards, and the localization of the TTG style of magmatism to specific convergent plate boundary settings (Martin, 1986). Most younger plutons are distinguished from the Archean TTG by less REE fractionation and by having more conspicuous negative europium anomalies across the entire compositional range (Figure 6), translating this character to post-Archean sediments. Notably, Proterozoic and younger granitic suites are typically not bimodal and include less differentiated, more mafic granitic compositions. Although there are provincial differences between some Proterozoic granitic suites and those in the Phanerozoic, no systematic differences in average global sedimentary compositions or near-surface heat flow can be resolved for these time periods (Taylor and McLennan (1985, 1995); see also the compilations of Condie (1993)). Small fluctuations may be masked by sedimentary averaging and recycling processes, but the constancy of sedimentary compositions implies that the processes of crustal generation and differentiation were similar throughout the post-Archean, and that net crustal additions from the Archean onwards were of approximately the same composition.

3.11.4.1 The Importance of Intracrustal Melting, and the Mantle Connection

Since potassic granitic magmatism is usually inferred to be the primary agent of crustal differentiation in Proterozoic and Phanerozoic terranes, a number of issues need to be addressed: (i) the efficiency of elemental fractionation during the generation and evolution of these granitic magmas; (ii) the extent that these can be reconciled with estimates of the gross chemical structure of the continental crust; and (iii) the amount and composition of new crust formed during such granitic episodes. Several lines of evidence indicate that mid- to shallow-level intracrustal melting became significantly more important after the Archean. First, in contrast to the TTG, the near ubiquity of negative europium anomalies and flat HREE patterns indicate that magma generation occurred at relatively shallow, mid-crustal depths with plagioclase, rather than garnet or amphibole, as a residual or fractionating phase. In some cases, the negative europium anomaly may also have been inherited from the source rocks, but this still requires that these have experienced an intracrustal differentiation cycle involving plagioclase, or (in the case of sediments) are derived from such rocks. Second, large discrepancies between crystallization ages and neodymium model ages (e.g., McCulloch and Chappell, 1982) confirm that older (crustal) material was present within the source regions of granitic magmas. Third, there is commonly a component of inherited zircon in post-Archean plutons (e.g., Paterson *et al.*, 1992; Williams *et al.*, 1992), again consistent with a source dominated by recycled crustal materials. Moreover, as phase relationships confirm that intracrustal melting is fundamentally dependent on the availability of solidus-lowering volatiles, post-Archean crustal evolution has been summarized as "no water, no granite; no oceans, no continents" (Campbell and Taylor, 1983). The water required for melting can occur as a free (supercritical) vapor phase ("fluid-present melting"), or be liberated by the peritectic decomposition of hydrous minerals, such as micas and amphiboles, which occurs at higher temperatures and is referred to as fluid-absent or "dehydration melting" (Thompson, 1982). These scenarios have vastly different geochemical consequences for the derivative melts (Section 3.11.4.2). In nature, crustal anatexis is likely to be a progression from fluid-present to fluid-absent conditions as aH_2O decreases during melting, especially where the incipient fluid-rich melts are extracted from the source region.

Although intracrustal melting characterizes the post-Archean Earth, the emplacement of granitic rocks in some oceanic settings remote from continental influences (Section 3.11.4.3) suggests that old crustal rocks are not a prerequisite for silicic magma formation, and that juvenile mantle-derived materials may be important. Moreover, even in continental settings, there is generally some evidence for mantle involvement, either as syn-plutonic basaltic dikes, coeval gabbros, swarms of mafic enclaves, or primitive isotopic signatures. A synergy between intracrustal melting and mantle-derived magmatism in granitic genesis is to be expected, since intrusion of hot, basaltic magma is likely to induce crustal fusion (Huppert and Sparks, 1988), and thus granites may incorporate variable proportions of mantle- and crust-derived components. Thus we envisage three different scenarios for granite generation, involving pure crustal sources, pure mantle sources, and variable combinations of these. Each scenario, and its relevance for post-Archean crustal evolution, is explored in the following sections by reference to specific case studies from well-documented Phanerozoic terranes.

(a)

(b)

Figure 9 Field photographs of the selected granite types (by T. Kemp). (a) Migmatite outcrop from the Cambro-Ordovician Glenelg River Complex (southeastern Australia) showing the deformation-enhanced segregation and accumulation of locally derived partial melts into a dilatant shear zone. Such segregation between partial melt and restite epitomises mid-crustal anatectic terranes (Brown, 1994). In this case, the leucogranite in the shear zone is compositionally equivalent to the kilometer-scale leucogranitic plutons of the area, and supports the viability of a genetic nexus between *in situ* migmatites and granitic bodies. (b) Quenched gabbroic diorite globules hosted by a subvolcanic alkali-feldspar granite on Jersey, Channel Islands (UK). Note the incorporation of alkali feldspar phenocrysts from the granite into the mafic pillow on the bottom left, suggesting that the granitic host was partly solidified prior to injection of the mafic magma. (c) Mingling between a syn-plutonic basaltic dike and the Tuross Head Tonalite, Moruya Batholith, eastern LFB (location indicated on Figure 15). The basaltic dikes and pillows have chilled margins and a tholeiitic to mildly alkaline geochemistry. As with the host tonalite, they have high Sr

3.11.4.2 Peraluminous Leucogranites—Granites Derived from Old Crustal Protoliths

Strongly peraluminous leucogranites (granites or granodiorites with <5 vol.% of mafic minerals; Le Maitre (1989)) with metasedimentary protoliths arguably provide the least ambiguous examples of granites generated exclusively from remelting of pre-existing crustal rocks (Patiño Douce, 1999). They represent the large-scale vertical redistribution of mass within the crust, rather than net growth of the continental crust. These rocks typically occur in continental collisional zones, such as in the Himalayas (Section 3.11.4.2.2), North American Appalachians (e.g., Pressley and Brown, 1999), and the Hercynian fold belts of western Europe (e.g., Strong and Hanmer, 1981; Wickham, 1987). The pluton-forming processes are directly observable in outcrop, typically due to the uplift and exhumation of deeper crustal levels, permitting geochemical and isotopic linkage between protolith and leucogranite. The viability of the process is confirmed by experimental studies, where fusion of the metasedimentary (usually metapelitic) protolith in the laboratory at the relevant P–T-aH_2O conditions consistently reproduces the felsic, strongly peraluminous compositions of the leucogranite (e.g., Patiño Douce and Harris, 1998; Castro *et al.*, 2000). One implication of this is that leucogranites approximate pure anatectic liquid compositions, which is confirmed by field relationships (below).

3.11.4.2.1 Conditions of leucogranite formation

Detailed field and geochemical studies have established that leucogranite emplacement represents the culmination of complexly interacting, multistage melt segregation and extraction processes, commonly controlled by syn-anatectic deformation (see Figure 9(a)), in which fractional crystallization, localized crystal accumulation, and separation of entrained refractory phases have important roles (see Brown, 1994; Solar and Brown, 2001). Isotopic data further indicate that rather than being simply derived from a compositionally homogeneous source, leucogranitic bodies comprise a poorly blended amalgamation of partial melt batches extracted from a range of metasedimentary protoliths (e.g., Deniel *et al.*, 1987; Krogstad and Walker, 1996; Pressley and Brown, 1999; Kemp, 2001). However, there is rarely evidence for the direct involvement of coeval mantle-sourced magmas in leucogranite genesis (an exception is documented by Kemp (2003)). In the absence of heat transfer from mafic liquids (that is, assuming only conductive heat flux from the mantle), modeling suggests that temperatures sufficient for fluid-present partial melting (~650–700 °C) may be achieved in the lower part of the crust by internal radioactive heating and thermal relaxation associated with crustal thickening, even using average continental geotherms (England and Thompson, 1986). However, the same calculations reveal that the higher

---

contents (to 800 ppm) and exhibit primitive isotopic compositions, ranging to $\varepsilon_{Nd} = +8$ (Keay *et al.*, 1997). These basalts are used as the primitive mantle-derived end-member in the mixing models of Keay *et al.* (1997) and Collins (1998). However, the high Na, high Sr signature is only present within granitic suites of the easternmost Bega Batholith. (d) Injection and mingling between dioritic and granitic magmas in the eastern Bega Batholith, LFB. Compared to the basalts in Figure 9(c), the dioritic magmas are compositionally evolved and exhibit certain textural features, notably hornblende-pyroxene-rimmed quartz ocelli and replacement of mafic minerals by biotite, that suggest they had experienced significant hybridization prior to mingling with the granite. (e) Swarms of hornblende-bearing "microgranular" enclaves in a hornblende granite of the eastern Bega Batholith. Note the morphological and textural diversity of the enclaves and the general lack of crenulate (quenched) margins, as is typical of microgranular enclaves in high-level granitic plutons. (f) Close-up view of a microgranular enclave in a hornblende tonalite of the Jindabyne suite, LFB. Enclaves such as this have essentially basaltic compositions, with 49–51% SiO_2, though the enclave population as a whole exhibits the complete spectrum through to intermediate compositions. (g) Array of enclave types in the cordierite-bearing Cowra Granodiorite, LFB. Metasedimentary enclaves range from discrete tabular objects (center left) to more diffuse, partially assimilated bodies (top, near lens cap). Note the irregular microgranular enclave (left) and the darker margin on the spherical microgranular enclave on the center right. The rim is biotite-rich and thought to result from the diffusional influx of K_2O and H_2O from the host magma, promoting biotite crystallization at the expense of pyroxene; relict hypersthene is sometimes preserved in the center of the enclave (Vernon, 1990; Maas *et al.*, 1997). The numerous clots of ferromagnesian minerals imparting the heterogeneous textural aspect are typical. Many of the clots result from the physical disaggregation of metasedimentary and microgranular enclaves, arrested examples of which are conspicuous in outcrops. Other clots, such as those consisting of large cordierite–sillimanite–hercynite ± garnet lumps may be the refractory residue of partial melting. (h) Closer view of a microgranular enclave contained by a lower silica cordierite granite of the Kosciuszko Batholith. The massive structure is typical. Although many enclaves of this type have a superficially "basaltic" appearance, their mineralogy usually mimics that of the host, such that they are rich in biotite and commonly contain cordierite and aluminosilicate.

(c)

(d)

Figure 9 (continued).

temperatures required for fluid-absent, mica dehydration melting reactions (>700 °C) are only attained under anomalous conditions of perturbed (i.e., steeper) geotherms and/or enhanced radiogenic heat flux (England and Thompson, 1986; Thompson and Connelly, 1995). As muscovite (but less so biotite, see Patiño Douce and Harris (1998), figure 3)) dehydration melting equilibria have shallow, positive d*P*/d*T* slopes (Thompson, 1982), much larger melt proportions are generated during adiabatic or near-isothermal decompression, such as occurs during exhumation by normal faulting (England and Thompson, 1984, 1986). This could partly explain the observation of Castro *et al.* (2000) that peraluminous leucogranites of collisional belts tend to be associated with extensional, rather than compressional structures. Muscovite dehydration melting during decompression at mid-crustal depths has been inferred for leucogranite generation in the Iberian massif, Spain (Castro *et al.*, 2000), and the Himalayan orogen (Harris and Massey, 1994; Searle *et al.*, 1997; Section 3.11.4.2.2). However, high precision monazite dating by Harrison *et al.* (1999) reveals that the Manaslu Intrusive Complex of the central Himalaya was emplaced during two pulses of muscovite

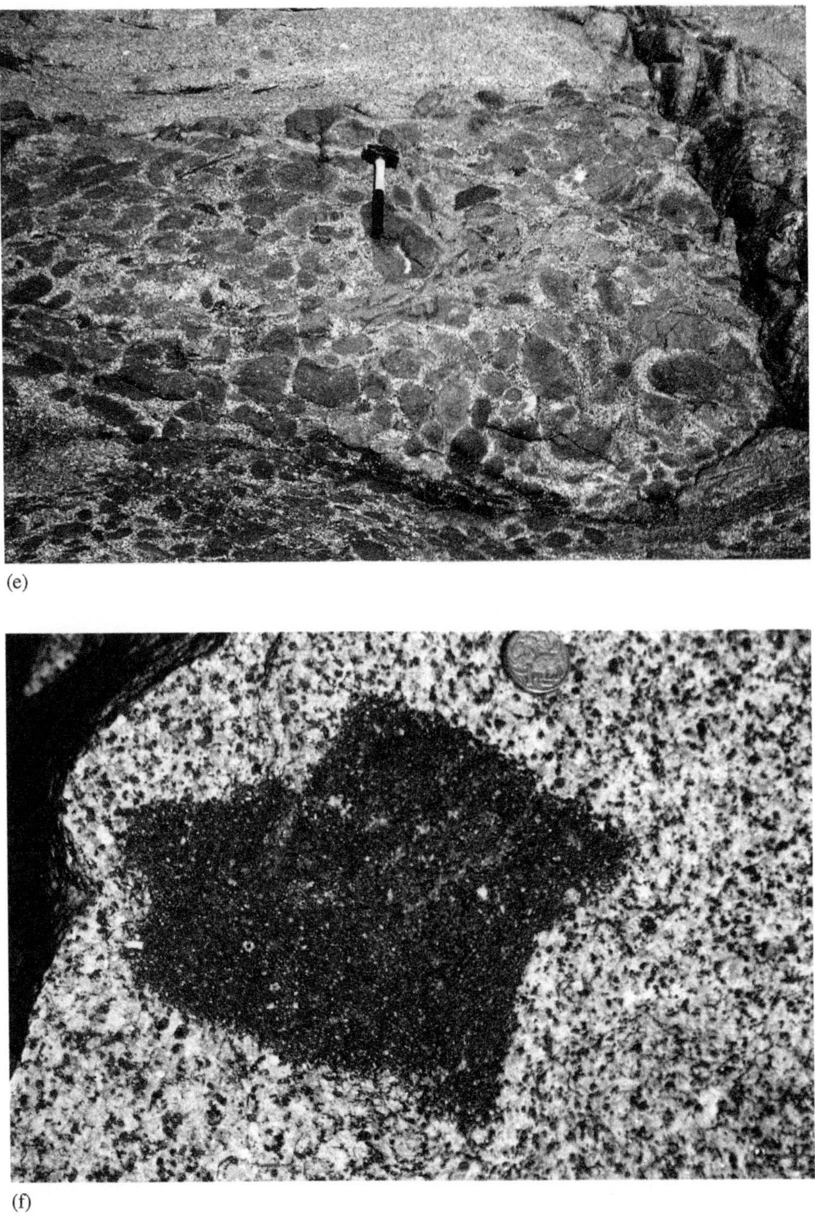

(e)

(f)

Figure 9 (continued).

dehydration melting separated by ~4 Myr, and to achieve this by decompression alone requires unrealistically high denudation rates. On the basis of numerical simulations, Harrison *et al.* (1997, 1999) instead propose that the diachronous leucogranitic magmatism was triggered by shear heating along major Himalayan fault systems. Nabelek and Liu (1999) also conclude that mica dehydration melting leading to Proterozoic leucogranite magmatism in the Black Hills (South Dakota, USA) was promoted by shear heating accompanying tectonic unroofing.

Irrespective of whether crustal anatexis is driven by thermal relaxation attending the collapse of tectonically thickened orogens or frictional heating along active faults, the important point is that leucogranite generation via muscovite-involved melting can be achieved *without* heat advection from mantle-derived magmas. The thermo-tectonic models of England and Thompson (1984, 1986) also predict that P–T paths might intersect higher temperature melting reactions involving breakdown of biotite in metapelites and hornblende in

(g)

(h)

Figure 9 (continued).

amphibolites during crustal thickening where the initial geotherms are abnormally "hot" (see also Thompson, 1999). More mafic (biotite-rich) granitic magmas that contain abundant peritectic reaction products of biotite dehydration, such as cordierite, garnet and/or orthopyroxene, and even metaluminous granitic plutons in the case of amphibolite fusion, could result from this mechanism. However, two additional points stem from this. First, field examples of intracrustal melting in Proterozoic to Phanerozoic metamorphic belts consistently involve metasedimentary or, more rarely, intermediate igneous precursors. This remains true even in the deeper crustal exposures, such as the Ivrea Zone of Northern Italy, where metapelitic rocks experienced granulite-facies anatexis close to the Moho (Voshage *et al.*, 1990). Melting of meta-basic protoliths is rare, exceptions being exhumed subduction complexes where a ready fluid supply is available (e.g., Sørensen, 1988). This is unlike in the Archean, where extensive migmatitic metabasite terranes occur, such as in the Superior Province of the Canadian Shield (e.g., Sawyer, 1991). This presumably reflects the higher solidus temperatures of meta-basic rocks (on average ~822 °C in fluid absent conditions; see Petford and Gallagher (2001)), which were more easily attained in the higher temperature conditions of the Archean.

Second, as noted by several authors (e.g., Clemens and Vielzeuf, 1987), the strongly modified geotherms required for biotite dehydration melting in the middle crust are most easily reconciled by an enhanced heat input from the mantle, either by upwelling of asthenosphere beneath attenuated crust, delamination of mantle lithosphere, or direct incursion of hot, mantle-derived liquids into the crust (underplating or intraplating; see Thompson (1999) for an evaluation of these possibilities). The latter is clearly the case for fusion of metapelites in the Ivrea Zone, which occurred at the roof of a crystallizing basaltic sill (Voshage *et al.*, 1990), and variations of this scenario have been extensively modeled (e.g., Huppert and Sparks, 1988; Bergantz and Dawes, 1994; Petford and Gallagher, 2001). Moreover, in the dynamic environment of multiple pulses of basaltic injection into partially melted crust it seems implausible that the derivative anatectic melts should remain physically and chemically isolated from their heat sources. Indeed, geochemical, isotopic, and experimental evidence, and the content of distinctively igneous-textured microgranular enclaves, attest to a mantle-derived component in many peraluminous granitic plutons generated by biotite dehydration of metasedimentary protoliths, such as the cordierite-rich Hercynian granodiorites of western Europe (Barbarin, 1996; Di Vincenzo *et al.*, 1996; Castro *et al.*, 1999; this aspect is addressed in Section 3.11.4.4). It is for this reason that such plutons are not as suitable as leucogranites for modeling the effects of closed-system intracrustal differentiation.

3.11.4.2.2 Low-strontium leucogranites

The majority of strongly peraluminous leucogranitic plutons documented in the literature contain strontium contents that are less than 200 ppm, and these are hereafter referred to as low-strontium leucogranites. Perhaps the most classic and thoroughly studied example of such rocks are the Miocene leucogranites of the Himalayan orogen (e.g., Le Fort, 1981; Searle and Fryer, 1986; Le Fort *et al.*, 1987; Castelli and Lombardo, 1988; Crawford and Windley, 1990; Scaillet *et al.*, 1990; Inger and Harris, 1993; Searle *et al.*, 1997; Harrison *et al.*, 1999 and references therein). These comprise laccolithic, polyphase plutons associated with numerous dike and sill complexes that crop out intermittently along the 2,000 km strike length of the Himalayan range, and they are emplaced into a partly migmatitic metasedimentary sequence. The size of individual laccolithic plutons varies, but ranges up to 3,000 km^3 for the Manaslu Complex (Guillot and Le Fort, 1995).

Although geochemically evolved (i.e., high silica, potassium, and rubidium contents), the leucogranites are considered to approximate primary minimum melts that are unmodified by fractional crystallization (Inger and Harris, 1993) and disengaged from the source without significant entrainment of refractory phases. Trace-element modeling (Harris and Inger, 1992; Harris *et al.*, 1995), isotopic constraints (Deniel *et al.*, 1987; France-Lanord and Le Fort, 1988; Inger and Harris, 1993), and experimental studies (Scaillet *et al.*, 1995; Patiño Douce and Harris, 1998) suggest that the partial melts that coalesced into the plutons were generated by muscovite dehydration melting of adjacent metapelitic schists at 600–800 MPa and 700–750 °C. Biotite–muscovite leucogranites are thought to result from a larger melt fraction formed at higher temperatures (~700–750 °C) than the tourmaline-bearing leucogranites (<700 °C) (Scaillet *et al.*, 1995; Ayres *et al.*, 1997).

The average biotite and tourmaline leucogranites from the Langtang Valley of the Nepalese Himalaya (Inger and Harris, 1993) are plotted in Figure 10(a), compared to a putative source rock (kyanite schist). Selected trace-element ratios are shown in Figure 11. During partial melting leading to Himalayan leucogranite formation, plagioclase was a significant residual phase, in addition to biotite, peritectic alkali feldspar, and accessory minerals (Harris and Inger, 1992; Harris *et al.*, 1995). The melts are therefore calcium poor and are characterized by depleted strontium and barium, leading to higher Rb/Sr and Rb/Ba, and lower Eu/Eu* compared to their inferred metasedimentary protoliths (Figures 10(a) and 11). The Himalayan leucogranites are also depleted in elements compatible in the other residual phases, such as biotite (titanium, niobium) and accessory zircon (zirconium, hafnium, HREE), monazite (LREE, thorium, uranium), and apatite (phosphorus, yttrium), reflecting the minimal dissolution of these minerals during anatexis. This is a consistent feature of leucogranites worldwide, manifested by their low mafic mineral content. Even lower yttrium and ytterbium contents in some samples reflect the presence of residual garnet (Harris *et al.*, 1995). Most Himalayan leucogranites reached chemical equilibrium with restitic zircon prior to melt extraction, as shown by

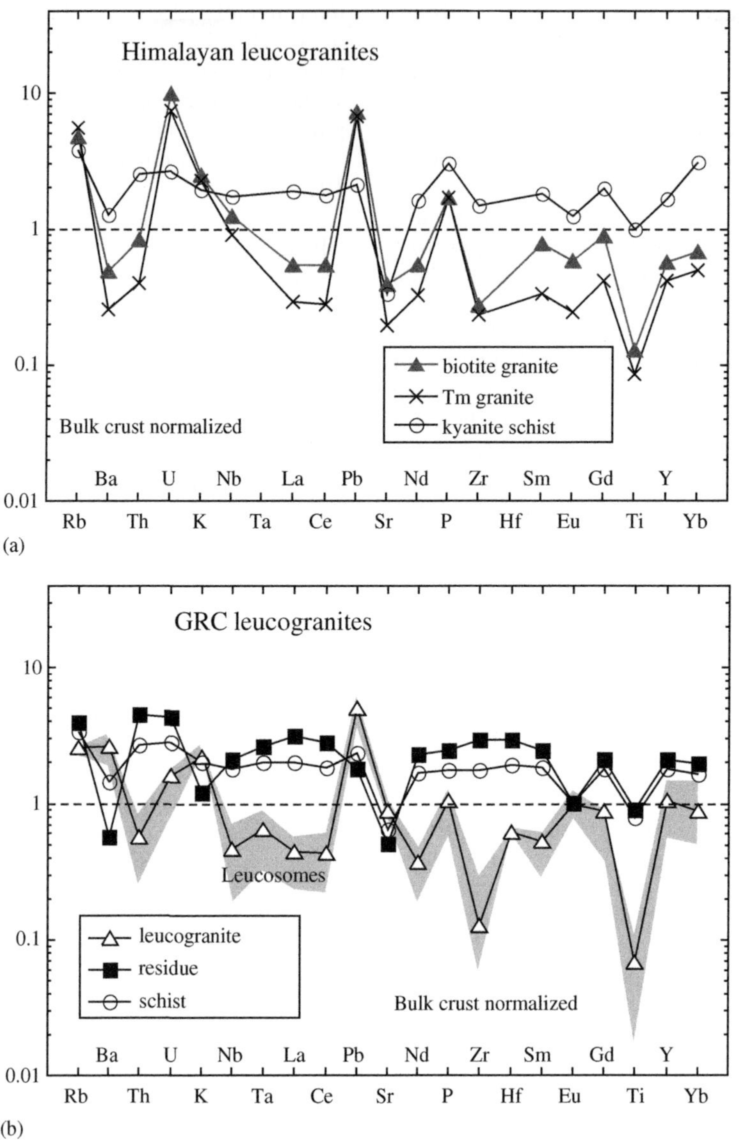

Figure 10 Bulk crust normalized multi-element diagrams comparing the chemical fractionation between source, melt and residue for (a) Himalayan leucogranites (Tm, tourmaline) from the Langtang region, Nepal (data from Inger and Harris, 1993) and that for (b) an average leucogranite from the Glenelg River Complex (GRC) in southeastern Australia; the field of *in situ* migmatite leucosomes is shaded (from Kemp, 2001).

the concordance between their zirconium contents and those predicted by zircon solubility equations (Ayres *et al.*, 1997; Figure 12) However, some leucogranites, including most tourmaline-bearing varieties, contain very low LREE concentrations that imply incomplete equilibration with residual monazite (Figure 12). This probably reflects the marginally lower temperatures of tourmaline granites, and thus more sluggish diffusion rates of the REE, and the short residence time of the melt in the source (Ayres *et al.*, 1997, and see below).

3.11.4.2.3 High-strontium leucogranites

A group of peraluminous leucogranite plutons with low K/Na ratios and unusually elevated strontium contents (250–450 ppm) occur in the Glenelg River Complex of southeastern Australia (Kemp and Gray, 1999; Kemp *et al.*, 2002). These contrast strongly with the predominant low-strontium leucogranite variety discussed above (see Table 3) and thus warrant further examination here.

The Glenelg River Complex comprises part of the Cambro–Ordovician Delamerian Orogen

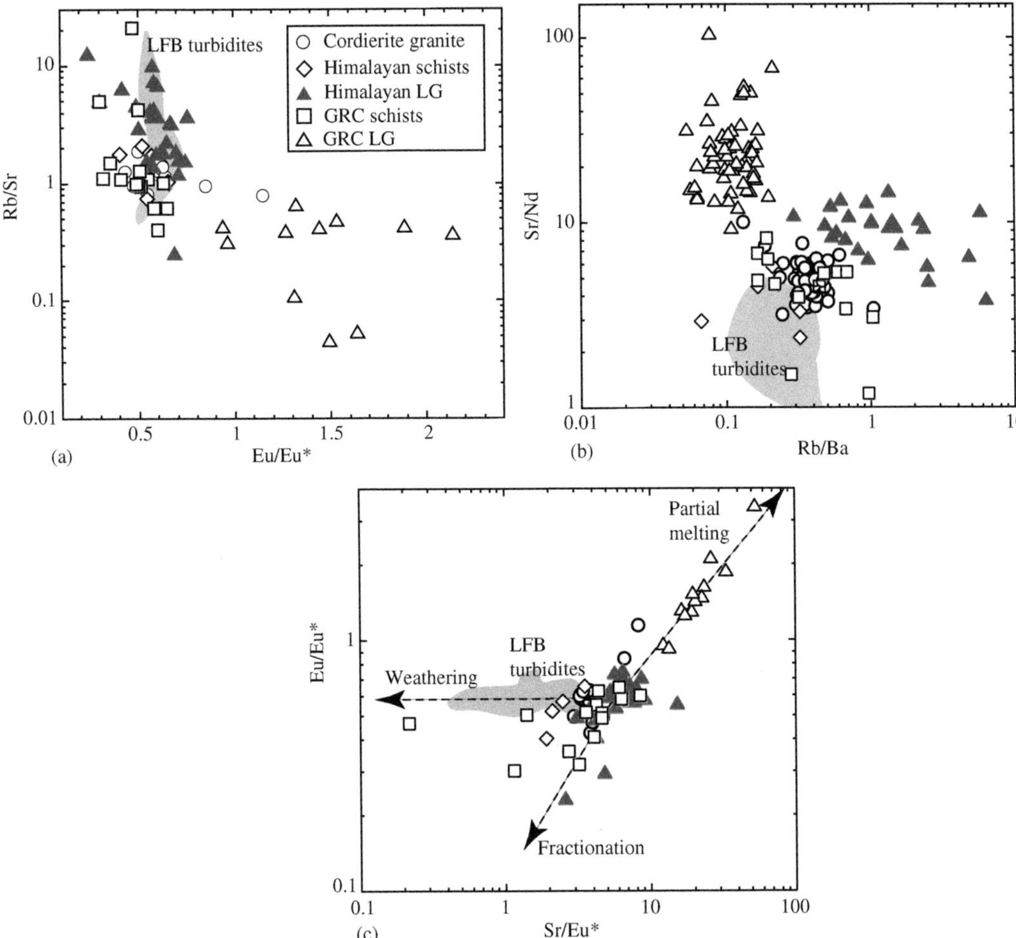

Figure 11 Trace-element ratio plots emphasizing the different chemical effects of fluid-present and dehydration melting during leucogranite generation from metasedimentary protoliths. The LFB cordierite granites (Kemp, 2003, unpublished data), and the field of the Lachlan Ordovician turbidites (B. W. Chappell, unpublished data) are shown for comparison. (a) Rb/Sr versus Eu/Eu*, (b) Sr/Nd versus Rb/Ba, and (c) Eu/Eu* versus Sr/Eu*. The compositional vectors resulting from igneous and sedimentary processes are indicated in (c). Some of the high-grade Glenelg River Complex (GRC) schists plot along the igneous fractionation trend, which cannot reflect weathering but is consistent with extraction of a partial melt component, as seen in the leucogranites of the area. Note the contrasting trends defined by the Himalayan (dehydration melts) and GRC (fluid-present melts) leucogranites, despite the source rocks being similar in composition (see also Figure 10). Also note the limited overlap between the LFB cordierite granites and the Ordovician sedimentary rocks, their potential protoliths.

(equated with the trans-Antarctic Ross Orogen, Stump *et al.* (1986)), which is thought to have formed by arc-continent collision in the Late Cambrian (Kemp, 2003). During the crustal thickening attending collision, a range of quartzo-feldspathic to semi-pelitic metasedimentary rocks (metagraywackes) experienced low temperature anatexis at ~650–680 °C and 500–800 MPa, during which biotite was essentially refractory. The low temperatures, coupled with the persistence of prograde muscovite and paucity of potassium–feldspar–sillimanite assemblages, suggest that partial melting occurred in the presence of excess aqueous fluid, rather than by decomposition of hydrous minerals (Kemp and Gray, 1999). This situation is considered anomalous, given the limited porosity envisaged for metamorphic rocks in the middle crust, and the high solubility of water in silicic melts at these pressures (~10 wt% at 500 MPa (Johannes and Holtz, 1996)) (see discussion in Clemens and Vielzeuf, 1987). Fluid present melting has been advocated in the Trois Seigneurs Massif of the French Pyrenees, though here the derivative leucogranites have low (<100 ppm) strontium contents (Wickham, 1987). A high-strontium (331 ppm) "trondhjemitic" leucogranite intrusion in the Himalayas is attributed to fluid-fluxed melting of a metapelitic source at the onset of anatexis (Patiño Douce and Harris, 1998). In the Glenelg River Complex, melting of the fertile metagraywackes may have been triggered by

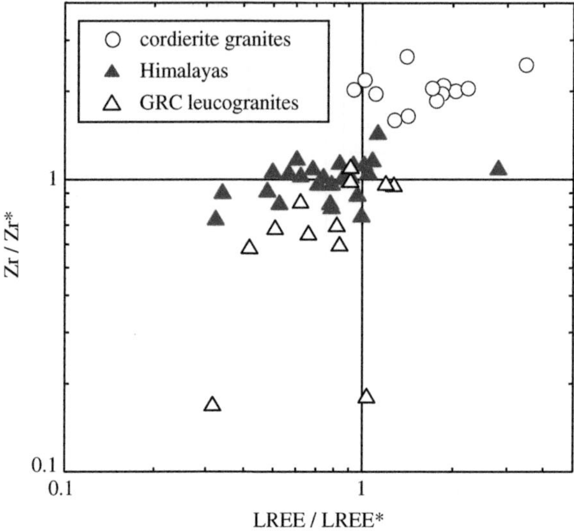

Figure 12 Plot of Zr (ppm measured in the rock)/Zr* (Zr contents in the melt at zircon saturation) versus LREE (La to Gd, excluding Eu, measured)/LREE* (LREE contents in the melt at monazite saturation) to assess and compare the roles of zircon and monazite melting in water-present and dehydration melts (solubility equations of Watson and Harrison (1983) for zircon, and Montel (1993) for monazite). The concentrations of Zr and LREE in the melts are largely buffered by the stability of zircon and monazite during anatexis (Ayres *et al.*, 1997). The "wet" Glenelg River Complex (GRC) melts have very low Zr and LREE concentrations and are therefore undersaturated in zircon and monazite, reflecting the minimal solution of these minerals during low temperature melting (<700 °C), and perhaps some occlusion by residual micas. In contrast, the Zr and LREE contents of the LFB cordierite granites exceed the predicted solubility of zircon and monazite. The cordierite granites are thus oversaturated in these minerals at the inferred crystallization temperature (750 °C, Chappell *et al.*, 2000) and melt water content (5 wt%, from Clemens and Wall, 1981). This is consistent with the physical entrainment of refractory accessory phases by these magmas, which results in unusually high concentrations of those elements that normally have low solubility in granitic melts (Chappell *et al.*, 1987; Chappell, 1996b).

ingress of fluid from an external source, such as dehydration of underthrust lower-grade schists, or cooling of fluid-rich arc magmas at depth.

The muscovite-bearing leucogranitic bodies of the Glenelg River Complex are embedded within the migmatitic metasedimentary sequence, where they have lenticular or sheet-like geometry and may be up to 5 km across (Kemp and Gray, 1999). They have interleaved, or transitional boundaries with the migmatites, and can be traced into veins and patches of *in situ* leucosome via networks of dikes and sills (Figure 9(a)). These field relationships suggest that the leucogranites represent ponded accumulations of partial melts derived from the adjacent metasedimentary sequence. Strong support for this is that the leucogranites are isotopically identical to the quartzofeldspathic migmatites (Kemp and Gray, unpublished data) and geochemically equivalent to the centimeter-scale *in situ* leucosomes hosted by these rocks (Figure 10(b)). Furthermore, due to the efficient, deformation-enhanced segregation and extraction of the constituent partial melts from their biotite-rich residues, it is likely that leucogranites approach liquid compositions.

An average Glenelg River Complex leucogranite is plotted in Figure 10(b), compared to an inferred protolith and complementary residue of partial melting. The distinctively high strontium and barium, but lower Rb/Sr and Rb/Ba, in the leucogranite relative to the protolith (Figure 11) reflect the paucity of residual plagioclase, the stability of which during melting is depressed relative to the micas by excess fluid (Patiño Douce and Harris, 1998). The enhanced consumption of plagioclase during anatexis in the muscovite stability field is therefore responsible for the higher strontium contents of the Glenelg River Complex leucogranites relative to most Himalayan leucogranites (Figures 6(a) and 10(b)). The Glenelg River Complex leucogranites are also depleted in titanium, rubidium, HFSE, and the REE, which are sequestered by residual biotite and accessory minerals. However, unlike many Himalayan leucogranites, both the zirconium and LREE contents of the high-strontium leucogranites are lower than those expected from theoretical zircon and monazite solubility equations (Figure 12). This indicates that in the Glenelg River Complex chemical equilibrium between melt and refractory accessory phases was not attained prior to melt extraction. This could reflect a combination of: (i) the low temperatures of anatexis impeding diffusion of zirconium and the LREE, (ii) occlusion of accessory minerals from the partial melt by enclosure within residual micas, this being actually observed in migmatite melanosomes, and (iii) short melt residence times, such that the rate of melt extraction from the source region exceeded that required for equilibration with dissolving zircon and monazite in the residue. The last of these accords with the field evidence that partial melt segregation and migration are driven by syn-anatectic deformation (see Figure 9(a)) and are thus likely to have been rapid (Harris *et al.*, 2000). Modeling based on monazite dissolution rates (assuming magmatic temperatures of 700 °C) suggests that the tourmaline-bearing Himalayan leucogranites migrated from their sources in less than 10^4 yr (Ayres *et al.*, 1997). Using the higher magmatic temperatures inferred from experimental studies (750 °C), this reduces to $\sim$100 yr (Harris *et al.*, 2000), since under these conditions the leucogranites are also undersaturated in zircon, and zirconium homogenization rates in a melt are three orders of magnitude faster than monazite (Ayres *et al.*, 1997). Although there is large inherent uncertainty in these calculations (mainly in melt water contents and diffusion rates for zirconium and the LREE), similarly short melt residence times are apposite for the Glenelg River Complex leucogranites, given their undersaturation in zircon and monazite. The anomalously low zirconium contents (<10 ppm) in two leucogranites (which are not otherwise chemically unusual) could be due to heterogeneous distribution of zircon grains in the analyzed sample, or that zircon crystals in the protolith were mostly armored by refractory minerals.

Another curious aspect of the high-strontium leucogranites in the Glenelg River Complex is their positive europium anomalies, since most leucogranites, and indeed Phanerozoic granitic rocks in general, have pronounced negative europium anomalies (Figures 11(a) and (c)). Large positive europium anomalies are commonly exhibited by small volume migmatite leucosomes generated under a variety of P–T-aH_2O regimes, where they are variously attributed to near-complete extraction of the feldspathic component of the source during high degrees of melting (Watt and Harley, 1993; Johannes *et al.*, 1995), disequilibrium melting of feldspar (Watt *et al.*, 1996), localized plagioclase accumulation (Sawyer, 1987; Milord *et al.*, 2001), or diffusional equilibration between leucosome and melanosome (residue) during crystallisation (Nabelek, 1999) or protracted cooling (Fourcade *et al.*, 1992). However, positive europium anomalies in kilometer-sized leucogranitic plutons require special circumstances. Diffusional exchange with the residue can be ruled out, since the leucogranitic melts were isolated from melanosomes during crystallisation, and in any case plagioclase was virtually exhausted from the residual assemblage. Textural evidence and the near cotectic compositions are not consistent with the leucogranites being feldspar cumulates (Kemp, 2001), though minor feldspar accumulation cannot be precluded. Instead, the positive europium anomalies are attributed to extensive disequilibrium melting of plagioclase, followed by rapid extraction of the partial melt from the source migmatites. Under these circumstances, europium is preferentially released into the melt and is therefore decoupled from the trivalent REE, bound in the refractory accessory phases that are concentrated in the residue.

A final aspect of this style of leucogranite generation is that the most important radiogenic heat-producing elements (thorium, uranium) are enriched in the residues of melting (i.e., apatite, zircon, and monazite), and thus retained in the deeper crust. This contrasts with the situation modeled by Sandiford *et al.* (2002), in which these elements behave incompatibly during melting and are transported to the upper crust, resulting in long term cooling, and thus stabilization, of the granitic source terrane. However, in the case of leucogranite generation the lower crust remains hot following the evacuation of partial melt, and may even undergo second stage melting, depending on the ambient geotherm and the size of the thermal pulse induced by radioactive decay. The residues from the original leucogranite formation are hydrous (mica-bearing) and thus could yield significant amounts of granitic melt during such an event, via higher temperature ($\sim$850 °C) biotite-dehydration melting (Patiño Douce and Johnston, 1991). Apart from postdating leucogranite emplacement, these melts would be recognized by elevated K/Na ratios, and would have much higher Rb/Sr, but lower Eu/Eu* than the high-strontium leucogranites, in these respects more closely resembling the upper continental crust. This second-stage melting scenario may explain the intrusion of peraluminous magmas in some terranes that postdate the deformational peak.

3.11.4.2.4 Relevance for crustal evolution

The effective separation of the silicic liquid from the metasedimentary source that epitomizes leucogranite generation maximizes the geochemical signature of intracrustal melting (Sawyer, 1998), and yields compositions that are strongly fractionated relative to the bulk continental crust (Figure 5(c)). Operating alone, the extreme chemical fractionation associated with this style of magmatism would tend to drive the upper crust towards a haplogranitic composition, that is quite different from the currently estimated upper crustal composition. In particular, muscovite dehydration melting of pelitic rocks under the conditions of Himalayan leucogranite genesis, while resulting in upwards concentration in lead, rubidium, potassium, and uranium, cannot produce the enrichment in Sr, Ba, REE, Nb–Ta, Zr–Hf that typifies the upper crust relative to the inferred bulk crust composition (Figure 5(c)). This problem is most acute for strontium and barium since these soluble elements will be preferentially removed by seawater and thus their concentrations will be further diluted in the upper continental crust from those observed in pristine igneous rocks (see Section 3.11.6). Fluid-present minimum melts can contribute towards enrichment in Ba and strontium in the upper continental crust, though the low Rb/Sr and positive europium anomaly of these melts contrasts with the characteristic high Rb/Sr and large negative europium anomaly of this reservoir.

We therefore conclude that, although important for enriching the upper crust in the most incompatible elements, intracrustal melting leading to leucogranite generation cannot be the principal agent driving differentiation of the continental crust. This is not unexpected, given that leucogranites are typically volumetrically minor in the context of most orogenic belts, especially those developed at convergent plate margins. In fact, the majority of granitic rocks exposed in the upper crust are considerably more mafic than leucogranites (compare to other Phanerozoic granites in Table 3), suggesting the operation of additional processes and/or derivation from different sources under different anatectic conditions.

3.11.4.3 Granites Formed from Juvenile Mantle-derived Materials

These granites can be generated by extreme differentiation of basaltic magmas, or by remelting mafic mantle-derived materials that have had a short residence time in the crust. Both models effectively result in the formation of new crust and in differentiation, but as the resultant granitic magmas are isotopically primitive, it may be difficult to isolate which of the two processes was responsible for their formation.

The production of broadly granitic magmas from juvenile, mantle-derived materials is most easily demonstrated in oceanic settings remote from continental influences, for example, (i) the development of the volumetrically minor "plagiogranites" of mid-ocean ridge systems, now preserved as part of ophiolite complexes (e.g., Pedersen and Malpas, 1984); (ii) alkaline granites of the Kerguelen plateau in the Indian Ocean (Giret, 1990); (iii) Icelandic dacites and rhyolites (Furman *et al.*, 1992; Jonasson *et al.*, 1992); and (iv) the primitive "M (mantle)-type" quartz diorites of intraoceanic island arcs, such as those of New Britain (Whalen, 1985) or the Aleutians (Perfit *et al.*, 1980).

The last group is potentially of relevance to the continental crust for two reasons. First, they comprise early intrusions in some continental margin arc-related batholiths, such as the Peninsula Ranges Batholith, California (Silver and Chappell, 1988), suggesting that they represent the initial stages of the formation of thick continental crust above subduction zones. Second, they are considered to represent the source rocks for metaluminous granites generated in subsequent thermal events unrelated to subduction (e.g., Chappell and Stephens, 1988). Nevertheless, at the same silica content, oceanic arc granites are most unlike the upper continental crust in having lower rubidium, thorium, uranium, LREE, and markedly depleted niobium (Figure 5(b)). Continental arc granites have higher LILE contents, though niobium and phosphorus remain lower than upper and bulk crustal values. Based on analogy with arc andesites (e.g., those of the Andes, see above), it is likely that some of these features result from crustal contamination, rather than being an intrinsic property of the original magma, and thus their direct relevance for the compositional evolution of the bulk crust is not immediately obvious (see Section 3.11.6 and Figure 21).

3.11.4.3.1 A-type granites

Distinctively silicic, potassium-rich granitic rocks that sometimes have mantle-like isotopic signatures also occur in intracontinental rift zones and postcollisional environments where they usually (but not invariably, see Whalen *et al.*, 1987; King *et al.*, 1997), postdate convergent deformation. They tend to be emplaced at shallow crustal levels, commonly occurring as subvolcanic ring complexes associated with crystal-poor rhyolites (see Bonin, 1986). Such plutons are collectively referred to as "A-type" (Loiselle and Wones, 1979), to reflect their commonly alkaline nature, the content of anhydrous mafic minerals, and their emplacement in intraplate or "anorogenic" settings.

Since the original definition by Loiselle and Wones (1979), a large literature has accumulated on A-types, and it has become increasingly apparent that rocks of this designation are lithologically and compositionally diverse. For example, they may be peralkaline, and contain sodium-rich mafic minerals, such as the aegirine-arfvedsonite granites of Nigeria (Kinnaird *et al.*, 1985), or they can be weakly peraluminous to metaluminous, containing hornblende or ferro-hastingsite, like the Siluro-Devonian A-types of the LFB, eastern Australia (see review by King *et al.*, 1997). Both types may even occur in the same area (e.g., Corsica, Poitrasson *et al.*, 1995). Nevertheless, some common traits are recognized. Many A-type rocks are hypersolvus granites (Tuttle and Bowen, 1958), that contain either a single alkali feldspar, or an unmixed (perthitic) potassic feldspar mantled by albite ("Rapakivi" texture), this being in accord with low-pressure crystallization. A-types typically contain high temperature anhydrous phases, such as pyroxene and fayalite (or their relicts), and late-crystallizing (interstitial) biotite and amphibole, both of which may be fluorine-rich (e.g., Anderson, 1983; Wormald and Price, 1988). Fluorite is a common accessory phase, and zircons in samples studied by ion microprobe mostly lack older, pre-magmatic cores (King *et al.*, 1997). These features suggest formation from high temperature (>900 °C), completely molten magmas with low to moderate water content, an interpretation that is consistent with the shallow emplacement levels and has been confirmed experimentally (Clemens *et al.*, 1986). Many A-type granites contain primary ilmenite and thus have f_{O_2} values below the fayalite–magnetite–quartz (FMQ) buffer (Loiselle and Wones, 1979; Anderson, 1983; Turner *et al.*, 1992a; C. D. Frost and B. R. Frost, 1997), in contrast to the relatively oxidized, magnetite-bearing metaluminous granites that are characteristic of arcs, which typically lie 1–3 log units above FMQ (Ishihara, 1979; Frost and Lindsley, 1991).

Chemically, most A-types are characterized by high K_2O, Fe/(Fe + Mg), and Rb/Sr, and enriched to extreme high field strength element (HFSE, especially zirconium, niobium, and yttrium) and REE^{3+} concentrations, with strongly depleted calcium, barium, strontium, and europium contents (see summary in King *et al.*, 2001). These features implicate either extended fractional crystallization or small degrees of partial melting to generate the incompatible element enrichment (Section 3.11.4.3.2). The large negative europium anomalies also reflect the low f_{O_2} and associated high Eu^{2+}/Eu^{3+} ratio (see discussion of the effects of f_{O_2} on europium partitioning in Section 3.11.5).

A-types are also distinguished by elevated Zn and Ga/Al ratios, and the latter has been taken as diagnostic (Collins *et al.*, 1982; Whalen *et al.*, 1987). However, although the chemical integrity of A-types is marked at the lowest silica contents, towards high silica there is compositional convergence with fractionated metaluminous granitic suites that otherwise lack A-type character (Figure 13(a); see King *et al.*, 1997). Chemical discriminants are most effective when applied to associations of rocks, where the less evolved members are also represented (King *et al.*, 1997). Alternatively, given the difficulties with simple Harker variation diagrams, reasonable separation between A-type and non-A-type granites can be achieved using more complex chemical parameters that emphasize the HFSE enrichment and silica-rich, calcium-poor nature of the former (Figure 13(b)).

Another curious feature of A-types is their uneven distribution throughout the geological record. Rocks of this nature are encountered in Archean cratons (Section 3.11.3.2.1), though they are far subordinate to the TTG and other granitic types. Similarly, A-type magmas are volumetrically minor in Phanerozoic orogenic belts; they comprise only 0.6% of the vast granitic batholiths of the LFB (Chappell *et al.*, 1991), discussed in Section 3.11.4.4.

However, A-type rocks are considerably more abundant in the Proterozoic, e.g., the vast Proterozoic anorogenic granitic provinces of the southwestern US, Adirondack Mountains and eastern Canada (Anderson, 1983), and the Rapakivi-textured granitic sheets of Greenland (Brown *et al.*, 1992). Large volumes of anorogenic, HFSE-enriched granitic (Wyborn *et al.*, 1992) and volcanic (Creaser and White, 1991) units were emplaced across Australian Proterozoic terranes in the 1.8–1.5 Ga interval. Similarly, Rapakivi-textured A-type batholiths intruded the Early Proterozoic crust of southern Finland between 1.65 Ga and 1.55 Ga at the close of the Svecofennian orogeny (Haapala and Ramo, 1992). In all of these areas, silicic A-type rocks are invariably coeval with mafic intrusions, and this is also a feature of some younger A-types (see below), hinting that a mantle input is intrinsic to their genesis.

Interestingly, the largest volumes of anorogenic granites in the Australian Proterozoic and Greenland were emplaced immediately following the 1.9 Ga "super-event" of Condie (1998, 2000; see Figure 2), where massive crustal growth was thought to be triggered by mantle plumes. In keeping with this, some chemical features of the A-type granites, especially the HFSE-enrichment, are reminiscent of plume-related basalts. However, most A-type granites (except the most fractionated varieties), and their coeval mafic rocks, have negative niobium anomalies (i.e., low HFSE/LREE). These are

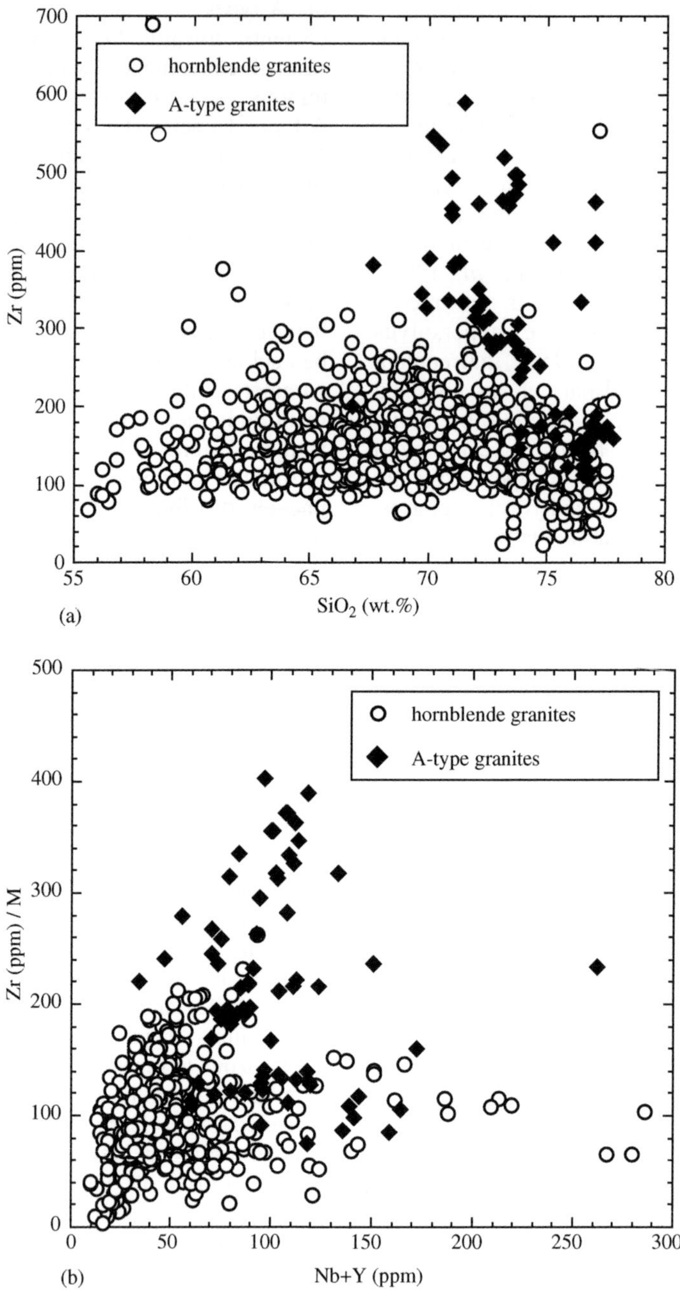

Figure 13 (a) Variation of Zr (ppm) as a function of SiO_2 (weight percent) showing the compositional overlap between the LFB hornblende granites and A-type granites in the most siliceous rocks (data sources as for Figure 6). The steep trend defined by A-type granites is compatible with zircon saturation in felsic rocks of the group (King *et al.*, 2001). Note the comparatively flat trend of the hornblende granites and the downwards inflection at ~74% SiO_2. (b) Plot of Zr (ppm)/M versus Nb + Y, where the contrasting granitic types show minimal overlap. The parameter "M" is the atomic ratio [Na + K + (2·Ca)]/(Si·Al) (from Watson and Harrison, 1983) which tends to be higher in A-type granites than hornblende granites of similar silica content, reflecting the Ca-deficiency of the former. Thus, the Zr/M ratio is directly proportional to the zircon saturation temperature of Watson and Harrison (1983), although this is only rigorously applicable to melts just saturated in zircon, and invalid for peralkaline granites.

not a feature of intraplate basalts, but the low Nb/La ratios of A-type granites are coupled with low Ti/Zr (Figure 5(e)) consistent with fractionation by a titaniferous phase. Some support for a spatial link with mantle plumes is that silicic rocks with A-type affinity are recorded from hotspot settings, including continental (e.g., Yellowstone, Hildreth *et al.* (1991); East African rift,

MacDonald *et al.* (1987); Etendeka province, Namibia, Schmitt *et al.* (2000)) and oceanic, (e.g., Kerguelen, Giret (1990); Ascension Island, Harris (1983)) environments.

3.11.4.3.2 Petrogenetic models for A-type granites

Reflecting their diversity and distinctive compositions, a number of petrogenetic schemes have been proposed for the origins of A-type suites. These essentially fall into two categories, involving crustal and mantle sources, respectively. In either case, fractional crystallization is considered to be the dominant process controlling the geochemical evolution of A-type granite suites, consistent with the inference that magmas were completely liquid at high temperatures, and the extreme degrees of differentiation recorded by the geochemistry.

For models advocating a crustal source, the protolith composition, and the conditions of magma generation are somewhat unusual. These models, underpinned by trace element and experimental evidence, include low degrees of partial melting of a dry, granulitic residue depleted by the prior extraction of granitic melt (Collins *et al.*, 1982; Clemens *et al.*, 1986; Whalen *et al.*, 1987), remelting metaluminous tonalites or granodiorites (Anderson, 1983; Creaser *et al.*, 1991; Anderson and Morrison, 1992; Skjerlie and Johnston, 1993; Patiño Douce, 1997; King *et al.*, 1997, 2001), and fusion of a dehydrated mafic to intermediate "charnockitic" lower crust (Landenberger and Collins, 1996); very low water activities and f_{O_2}, and high temperatures (>900 °C) are intrinsic to all models. The first two of these scenarios have shortcomings (see reviews by Turner *et al.* (1992a) and C. D. Frost and B. R. Frost (1997)) and there is little reason that any should be universally applicable.

The elevated temperatures necessary to generate A-type liquids from refractory or low H_2O crustal sources undoubtedly requires major heat input from mantle-derived magmas (e.g., Landenberger and Collins, 1996), and so many A-types may incorporate a large juvenile component in their lineage. Indeed, physical mingling and hybridization between A-type magmas and coeval basaltic intrusions has been documented (e.g., the Newfoundland Appalachians, Whalen and Currie (1984); Figure 9(b)), but whether mantle-derived magmas have made a material contribution to other "crustally derived" A-types, is difficult to assess and relies heavily on isotopic evidence. Haapala and Ramo (1992) demonstrate that the neodymium and lead isotopic composition of the Finnish Rapakivi granites parallels the evolution of the 1.9 Ga Svecofennian crust, which is consistent with the granites representing anatectic melts of these older rocks. Conversely, the Proterozoic Rapakivi granites of Greenland have low initial $^{87}Sr/^{86}Sr$ and positive epsilon neodymium values, suggesting that juvenile crust and/or mantle magmas were more appropriate sources (Brown *et al.*, 1992). Isotopic evidence seems less conclusive in other cases where A-type granites have ε_{Nd} and initial $^{87}Sr/^{86}Sr$ ratios that straddle crustal and mantle values (Poitrasson *et al.*, 1995) For example, King *et al.* (1997) show that the Siluro–Devonian A-type plutons from the LFB have a large spread in epsilon neodymium values (−3 to +5) at the time of crystallization that overlap the range shown by the least evolved hornblende granites of the area. This has been taken to indicate source heterogeneity and it appears to preclude a significant sedimentary component within these rocks (King *et al.*, 1997), as is true of A-types in general (see the compilation of Turner *et al.*, 1992a). However, some Lachlan A-types have higher ε_{Nd} and lower initial $^{87}Sr/^{86}Sr$ ratios than the isotopic envelope defined by hornblende granites; we shall return to this point in Section 3.11.4.3.3.

In contrast, some A-type granites, including most peralkaline varieties, are ascribed a largely mantle derivation (e.g., Loiselle and Wones, 1979; Bonin, 1986; Bedard, 1990; Turner *et al.*, 1992; Whalen *et al.*, 1996; C. D. Frost and B. R. Frost, 1997; Han *et al.*, 1997; Jahn *et al.*, 2000; Schmitt *et al.*, 2000), and these are the most pertinent in the present context of crustal growth. The basis for many of these models are primitive isotopic compositions (Javoy and Weiss, 1987; Turner *et al.*, 1992a; Whalen *et al.*, 1996; Han *et al.*, 1997; Jahn *et al.*, 2000; see Section 3.11.4.3.3), with trends to more evolved values signifying assimilation or mixing with older crustal materials (Foland and Allen, 1991; Kerr and Fryer, 1993; Poitrasson *et al.*, 1995; Schmitt *et al.*, 2000). Given the distinctive trace-element chemistry of A-type granites, a somewhat "enriched" (intraplate) mafic precursor may be required, and this is variably attributed to small degrees of mantle melting during lithospheric extension (Barbarin, 1999; Jahn *et al.*, 2000), melting lithospheric mantle fertilized by prior subduction (Turner *et al.*, 1992a,b; Whalen *et al.*, 1996), or derivation from a plume source (Schmitt *et al.*, 2000).

3.11.4.3.3 A-type case studies

An example of Phanerozoic crustal growth during A-type magmatism is discussed by

Turner *et al.* (1992a) with reference to the silicic magmas of the bimodal padthaway suite in southern Australia. These magmas were emplaced during postcollisional extension immediately following the convergent deformation of the ca. 500 Ma Delamerian orogeny. The silicic rocks, occurring as shallow level plutons and rhyolitic volcanics, are distinguished mineralogically by Rapakivi textures, interstitial annitic biotite, and sporadic fayalite and Pidgeonite, with accessory fluorite and large euhedral zircons. Pyroxene-olivine assemblages and coexisting Fe–Ti oxides preserve equilibria established at 900–1,000 °C and <100 MPa, and constrain an oxygen fugacity that is 2–3 log units below the FMQ buffer (Turner *et al.*, 1992a). By analogy with the experimental study of Clemens *et al.* (1986), crystallisation proceeded under strongly water-undersaturated conditions, with final H_2O contents below 3% (Turner *et al.*, 1992a). Geochemically, the rocks are highly differentiated and exhibit pronounced A-type affinity, contrasting with the older metaluminous and peraluminous plutons of the same area (Foden *et al.*, 1990; Turner et al., 1992a). In particular, initial isotopic ratios are relatively primitive ($\varepsilon_{Nd} \sim +1$ to -2.7, $^{87}Sr/^{86}Sr < 0.706$) and indistinguishable from associated tholeiitic basalt and gabbro intrusions (Turner *et al.*, 1992a), with which they are intimately intermingled (Turner and Foden, 1996). Accordingly, the silicic A-type magmas are considered to have evolved from these contemporaneous basaltic magmas by protracted (~90%) fractional crystallization, and thus exemplify the addition of new granitic crust. The characteristic iron enrichment trends, low f_{O_2} and high incompatible element concentrations are all explained by this process (Turner *et al.*, 1992a).

Importantly, the enriched incompatible element concentrations of the coeval basaltic rocks, particularly high La/Yb and low Nb/La, coupled with positive epsilon neodymium values, suggests derivation primarily from enriched domains within the subcontinental mantle lithosphere, rather than the asthenosphere (Turner, 1996). The postcollisional magmatism has, therefore, been attributed to delamination or convective thinning of the overthickened lithosphere following tectonic convergence, allowing upwelling of the hot asthenosphere, and facilitating contact melting of the overlying enriched mantle (Turner *et al.*, 1992b; Turner, 1996). A lithospheric mantle derivation has also been invoked for the parental magmas of other postcollisional granitic suites (e.g., the Adamello Massif; Blundy and Sparks, 1992).

A variation of this scenario, proposed for the Proterozoic Rapakivi granites of North America by C. D. Frost and B. R. Frost (1997), is that A-types form by remelting young, underplated tholeiitic basalts, including their evolved, ferrodiorite differentiates, in areas of crustal extension or hot spot activity. This model has the advantage of explaining the bimodal character of A-type suites, though inevitably the hot, dry silicic melts would undergo fractional crystallization upon ascent into the shallow crust, amplifying their distinctive compositional features. Hence, there is conceptual overlap with the model of Turner *et al.* (1992a) and the implications for crustal growth are unchanged. The high-temperature A-type magmas formed in this way are very susceptible to crustal contamination, especially since they have low strontium contents and $^{87}Sr/^{86}Sr$ ratios, which may partly explain the characteristic dispersion of isotopic compositions exhibited by these rocks (Turner *et al.*, 1992a; C. D. Frost and B. R. Frost (1997)).

Although the basaltic progenitors in the C. D. Frost and B. R. Frost (1997) model are depleted tholeiites derived from upwelling asthenosphere, in principle more enriched and isotopically evolved mantle sources could also be involved. In view of this, it is striking that the "aluminous" A-type granites from the LFB fall within the isotopic range of the lithosphere-derived mafic magmas documented by Turner (1996), and thus potentially could be sourced from such magmas rather than pre-existing crust. One sample of King *et al.* (1997) has an ε_{Nd} value of +5 at the time of crystallization, which was clearly inherited from a mantle source, or a juvenile magma derived therefrom. Some syn-plutonic basaltic rocks are associated with A-type granites of the Wangrah suite (King *et al.*, 2001), though unfortunately their isotopic compositions and some key trace-element concentrations (e.g. niobium) are not presented. However, these basalts have enriched zirconium (~250 ppm) and cerium (to 89 ppm) contents and thus could easily yield A-type-like liquids upon high-temperature partial melting, since zirconium and cerium are unlikely to be buffered by residual zircon or allanite in the mafic protolith.

3.11.4.3.4 Relevance for crustal evolution

The average Padthaway suite A-type pluton of Turner *et al.* (1992a), together with metaluminous and peralkaline A-type granite samples from the LFB are plotted in Figure 5(d). The markedly enriched HFSE and REE contents are salient features, and of all granitic types considered, only these magmas can contribute significantly towards the concentration of Nb–Ta and Zr–Hf, and depletion in TiO_2 in the upper crust. Despite the HFSE enrichment, negative Nb–Ta anomalies are present even in isotopically primitive (mantle-like) A-types, and this also mirrors the continental crust (the high Nb/La ratio in the peralkaline rock possibly reflects the late-stage fractionation of REE-rich accessory phases). On the basis of this, and despite their limited areal extent, it is tempting to speculate that A-type magmatism may have made a more important contribution to the HFSE budget and HFSE/LREE ratio of the upper crust than has hitherto been appreciated. However, elevated Zr–Hf and Nb–Ta concentrations are linked to enhanced solubility of these elements in the alkaline melt via high magmatic temperatures, extensive dissolution of accessory phases in the source (King *et al.*, 1997), and the formation of Na–F complexes (Collins *et al.*, 1982), all of which require anomalous circumstances during granitic generation. In the case of a purely mantle heritage, the slightly positive ε_{Nd} values suggest moderate-term enrichment of the source in the LREE. Furthermore, A-type magmas have pronounced depletions in barium, strontium, and phosphorus, and though these are less well developed in the metaluminous A-type in Figure 5(d), such depletions are not typical of the upper continental crust.

In order to assess the broader significance of A-type magmatism for crustal evolution, future studies need to resolve several outstanding issues. These include (i) ascertaining whether partial melting of young mafic protoliths or fractionation from mantle magmas is the most important for the generation of these plutons; (ii) where enriched mantle sources are invoked, constraining the conditions leading to the formation of these; and especially (iii) why A-type magmas are so voluminous in the Proterozoic, as compared to the Archean or Phanerozoic periods.

3.11.4.4 Granites with Juvenile Mantle and Crustal Sources: The Lachlan Case Study

3.11.4.4.1 Mixing or unmixing?

Granitic intrusions of exclusively and unambiguously mantle provenance are rare in Phanerozoic orogenic belts, and in many cases petrological evidence demands derivation of these magmas from mixed sources, which included both mantle-derived materials and older crustal protoliths. However, constraining the nature and proportions of these components has often proved formidable.

Such complexities are now explored in a case study of granitic rocks from the Paleozoic Lachlan Fold Belt (LFB) in southeastern Australia. Although lithologically diverse, these define an apparently simple, overlapping array on the ε_{Nd} versus initial $^{87}Sr/^{86}Sr$ diagram (Figure 14), which has been almost universally used to infer a large scale mixing process between primitive, depleted mantle-like magmas and evolved crustal end-members (e.g., Gray, 1984; Faure, 1990). Yet, such models may not reconcile issues such as why samples with differing mineralogy and major element compositions should have similar trace-element ratios, and how mixing alone can explain large chemical differences between granitic rocks that are isotopically similar. It is also unclear whether the granites represent liquid compositions or contain unmelted refractory material from the source ("restite"), and evolved by the differential "unmixing" of such material White and Chappell, 1977). This has been a particular debate for the Lachlan granites and it has shaped much of the wider debate over granitic generation (e.g., Vernon, 1983; Wall *et al.*, 1987; Chappell *et al.*, 1987; Collins, 1998, 1999 versus Chappell *et al.*, 1999, Chappell *et al.*, 2000).

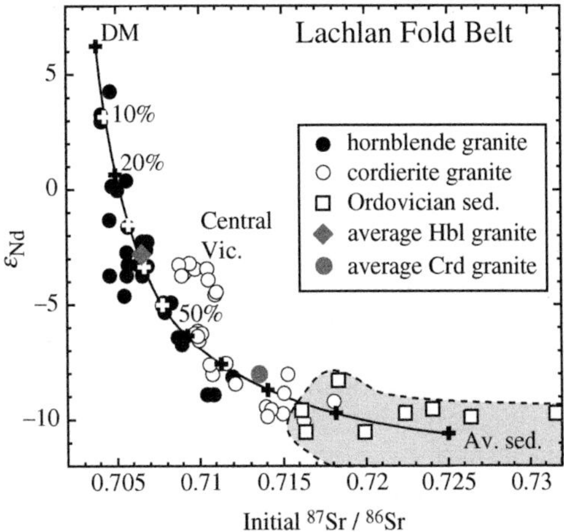

Figure 14 Initial Nd–Sr isotope relations in the LFB granites (data sources are: McCulloch and Chappell, 1982; Chappell *et al.*, 1991; McCulloch and Woodhead, 1993; Elburg, 1995 b,c; Maas *et al.*, 1997; Keay *et al.*, 1997; Anderson, 1998; Waight *et al.*, 2000; Kemp, unpublished data, and unpublished data quoted by Chen *et al.*, 1990). Cordierite granites from the central Victorian portion of the LFB plot away from the main array and appear to comprise a separate petrogenetic system, as is consistent with other geochemical criteria such as anomalously high Ba contents (Anderson, 1998). The hyperbola represents a putative mixing line between the average Ordovician turbidite (av. sed.) and the depleted mantle end-member (DM) determined by McCulloch and Chappell (1982), and encapsulates most hornblende and cordierite granites (crosses represent 10% mixing intervals).

3.11.4.4.2 Two contrasting granite types

Any examination of granites within the LFB is indebted to the pioneering and systematic studies of Chappell and White and co-workers (see Chappell and White, 2001), through which large field, petrographic and geochemical data sets are available and form the basis for petrogenetic models. However, despite intensive research, roughly since the early 1970s, there remain lingering questions and controversial aspects concerning the formation of these granitic rocks. The purpose of this case study is to focus on some of these specific problems in the context of crustal growth and evolution.

The LFB is a 700 km wide segment of a 3,600 km long orogenic system that developed along the eastern Gondwana margin from the Early Ordovician to Devonian (Figure 15). The tectonic setting of the belt remains contentious, although the complex patterns of deformation, magmatism, and metamorphism seem to require multiple, migrating subduction zones (Gray, 1997; Soesoo and Nicholls,1999). The LFB has two main components, a monotonous succession of mature (quartz- and clay-rich) Ordovician turbidites, and an extraordinary volume of granitic rocks. The turbidites apparently accumulated on an oceanic substrate and were subject to episodic deformation, low-grade regional metamorphism and massive igneous intrusion from ~450 Ma to 340 Ma (Gray and Foster, 1997).

Granites and spatially related volcanic units comprise approximately one quarter of the exposed Paleozoic geology of the orogen; gabbros or other relatively mafic rocks are rare. Excluding the volumetrically minor A-types, and following the seminal suggestion of Chappell and White (1974), the granites can be subdivided into two lithological categories, each of which outcrop in approximately equal proportions. The first group consists of metaluminous to weakly peraluminous tonalite, granodiorite and adamellite (ASI < 1.1) that are sometimes mingled with coeval basaltic intrusions (Figures 9(c) and (d)) and contain hornblende-rich, igneous-textured microgranular enclaves (Figures 9(e) and (f)). As these rocks have precipitated hornblende over much of their compositional range, they are here referred to as "hornblende granites." Such plutons are predominant in the easternmost part of the LFB, where they form vast, meridionally trending batholiths (Figure 15).

In contrast, the second group comprises strongly peraluminous granodiorites and adamellites (ASI > 1.0) that mostly contain cordierite, sometimes accompanied by sillimanite, garnet or muscovite, and for convenience are termed cordierite granites. Cordierite granites only crop out some distance inboard of the continental margin (Figure 15), where they occur in composite batholiths with hornblende granites. In the Kosciuszko Batholith they are consistently older on the basis of intrusive relationships (White *et al.*, 1976), though radiometric ages of the two granite types overlap in the Berridale Batholith further east (Williams *et al.*, 1975). The most distinctive field attribute of cordierite granites is that they contain a diverse array of metasedimentary enclaves and mica-rich clots, which become increasingly numerous in the lowest silica samples (Figure 9(g)). These enclaves usually have higher metamorphic grade and more complex structural histories than the host turbidites (Fleming, 1996) and there are migmatitic examples with refractory, melt-depleted compositions (e.g., Chen *et al.*, 1989; Anderson, 1998; Maas *et al.*, 1997). Most metasedimentary enclaves were therefore entrained from a zone of deep crustal anatexis, and may therefore be utilized as structural and compositional "windows" to the unexposed mid-crust (Fleming, 1996; Anderson *et al.*, 1996). Other enclaves include diopside-bearing "calc-silicate" varieties that perhaps represent fragments of metamorphosed marl (Chen *et al.*, 1989). Less abundant are unstructured "micrPogranular" enclaves which are darker and finer-grained than their hosts (Figure 9(h)). The mineralogy of these generally mimics that of the host, though orthopyroxene is common and some enclaves of this type contain actinolitic amphibole (Vernon, 1990; Maas *et al.*, 1997). The origin of the microgranular enclaves continues to be controversial (e.g., Maas *et al.*, 1997 versus White *et al.*, 1999) and in some ways is the key to understanding the formation of the host granite. The complex textures are most easily reconciled by igneous crystallization, implying that the microgranular enclaves are intermingled, variably hybridized globules of a coeval, more mafic magma (Vernon, 1983, 1984, 1990). Nonetheless, field evidence for such interaction involving the cordierite granites has yet to be documented. Furthermore, the micrPogranular enclaves in LFB cordierite granites exhibit an array of lithological and chemical variability, and it seems inevitable that they have multiple origins.

The distinction between the LFB cordierite granites and strongly peraluminous leucogranites is also worth noting. Although both have the same ASI range, cordierite granites extend to considerably more mafic (cordierite- and biotite-rich) compositions (Table 3). At the felsic end of the compositional range the differences are less marked, though cordierite granites retain systematically higher niobium, thorium, REE, titanium,

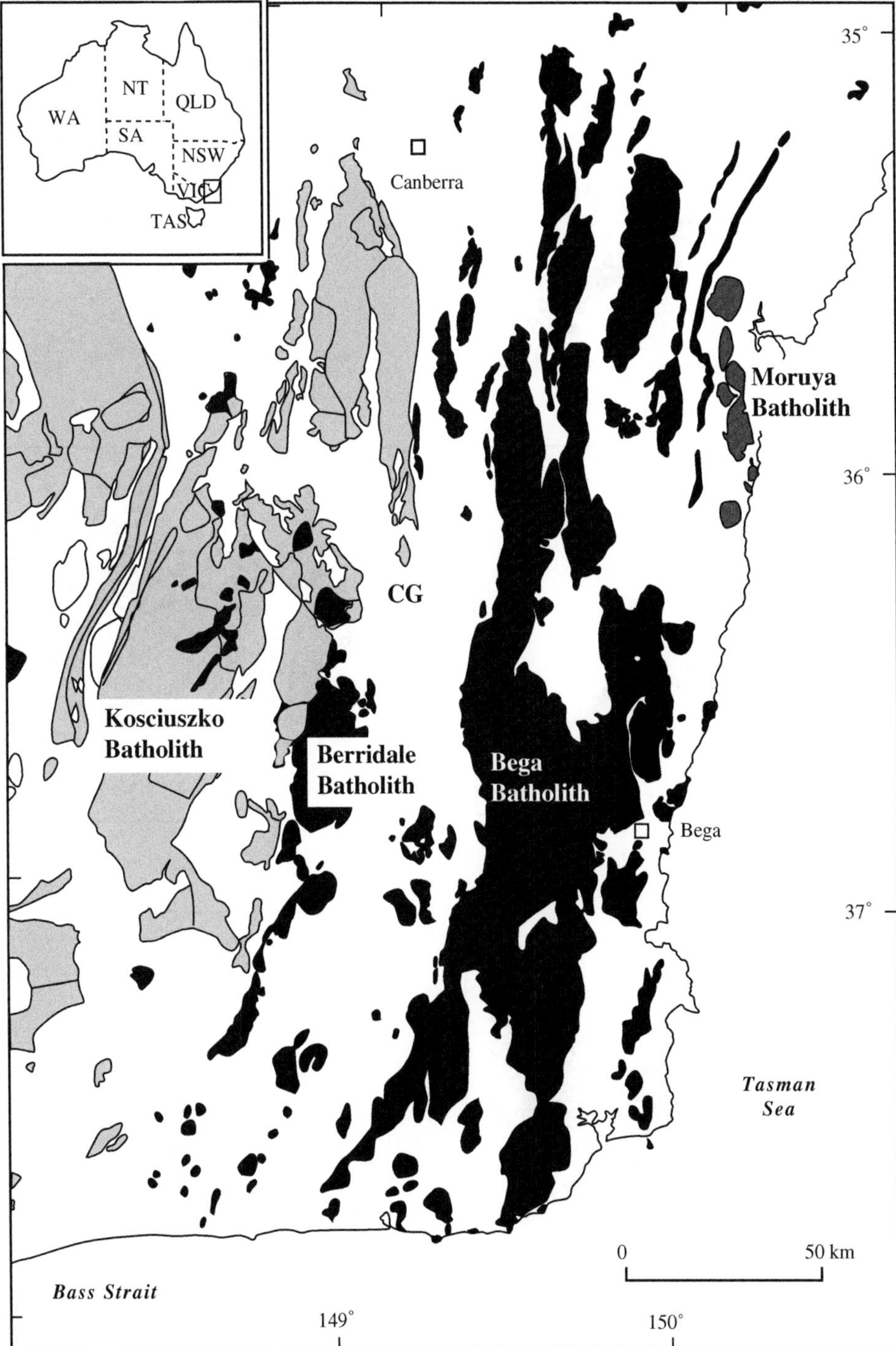

Figure 15 Simplified geological map of southeastern Australia (location in inset), showing the distribution of cordierite (light gray) and hornblende (black and dark gray) granites and equivalent volcanic suites within the eastern part of the LFB, and the location of the specific granitic batholiths and plutons (CG, Cooma Granodiorite) referred to in the text (modified after Chappell, 1996a). The volumetrically minor A-type granites, as well as small gabbro bodies, are grouped with the hornblende granites.

and yttrium abundances than the leucogranites (Figure 5), and show systematic enrichment in zirconium and the LREE compared to the concentrations predicted by zircon and monazite solubility equations (Figure 12). The latter is compatible with the abundance of older, pre-magmatic zircon cores in the LFB cordierite granites, as documented from various plutons by

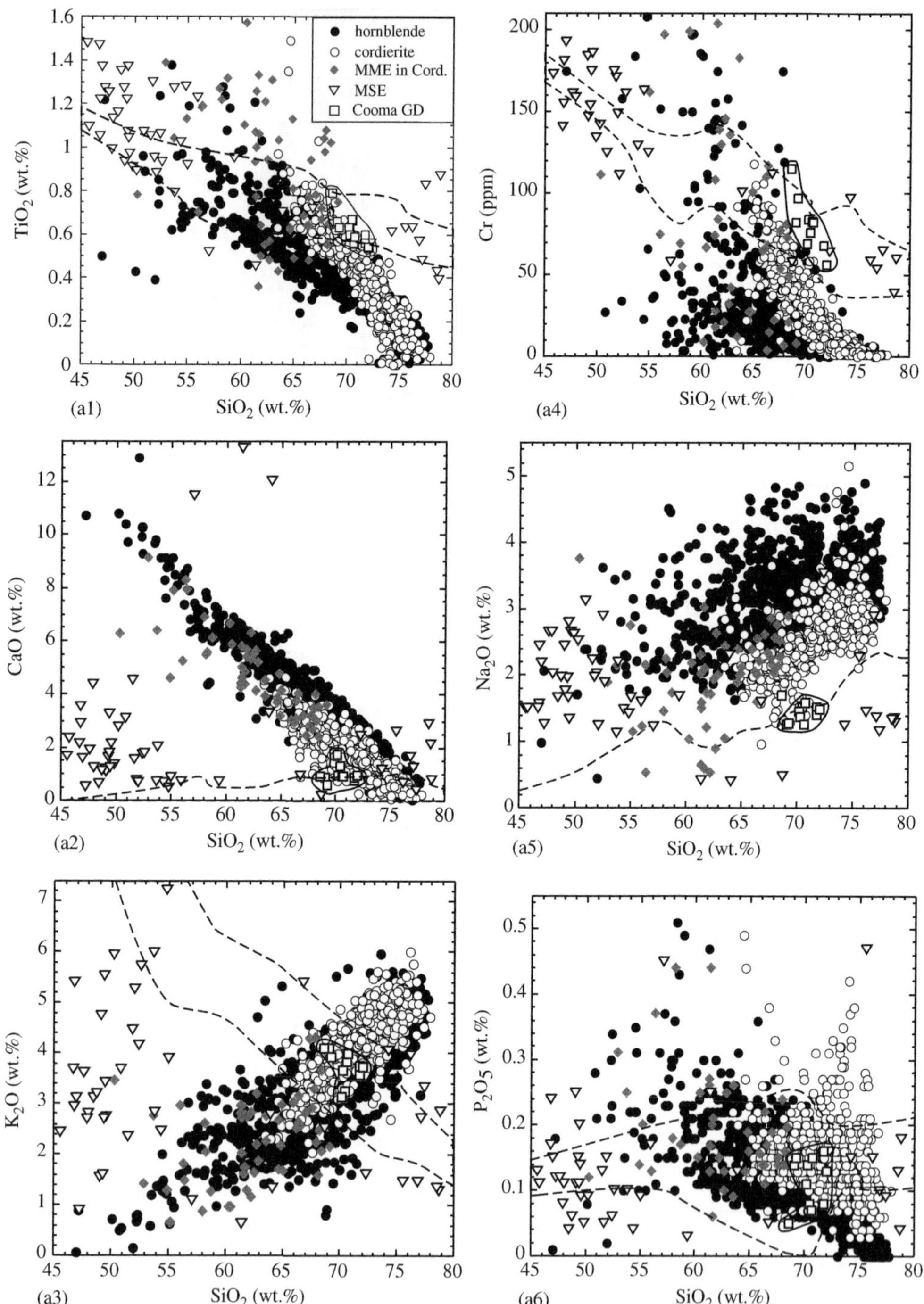

Figure 16 (a) (a1–a14) Chemical variation diagrams for the LFB hornblende and cordierite granites, and microgranular (MME) and metasedimentary enclaves (MSE) contained by cordierite granites. The field of the Ordovician turbidites is enclosed by a thin dashed line, and the Cooma Granodiorite is circled by an unbroken line. Note that the metasedimentary enclaves with very high CaO contents (>10 wt%) are calc-silicates. (b) (b1, b2) Plots of ASI versus SiO_2 (top) and total FeO (bottom). The compositional trends generated from a granitic minimum melt ("MM") by restite entrainment/increasing degrees of melting (thick arrow), contamination by Ordovician turbidites, magma mixing and fractional crystallization (thin dashed arrows) are indicated for the cordierite granites. Scattering of cordierite granites into the Cooma Granodiorite field suggests that partially melted (diatexitic) Ordovician turbidites were assimilated by ascending granitic magmas (see also Collins, 1996). Note the similar trends shown by the cordierite granites on the ASI versus silica and K_2O/Na_2O versus silica plots. That these correlate with initial $^{87}Sr/^{87}Sr$ suggests that the dominant compositional control was source-related for these plutons, specifically reflecting different degrees of weathering of the various sedimentary protoliths (see also discussion of Figure 19). Data sources are; granites, as for Figure 6; Microgranular enclaves, White *et al.* (1977),

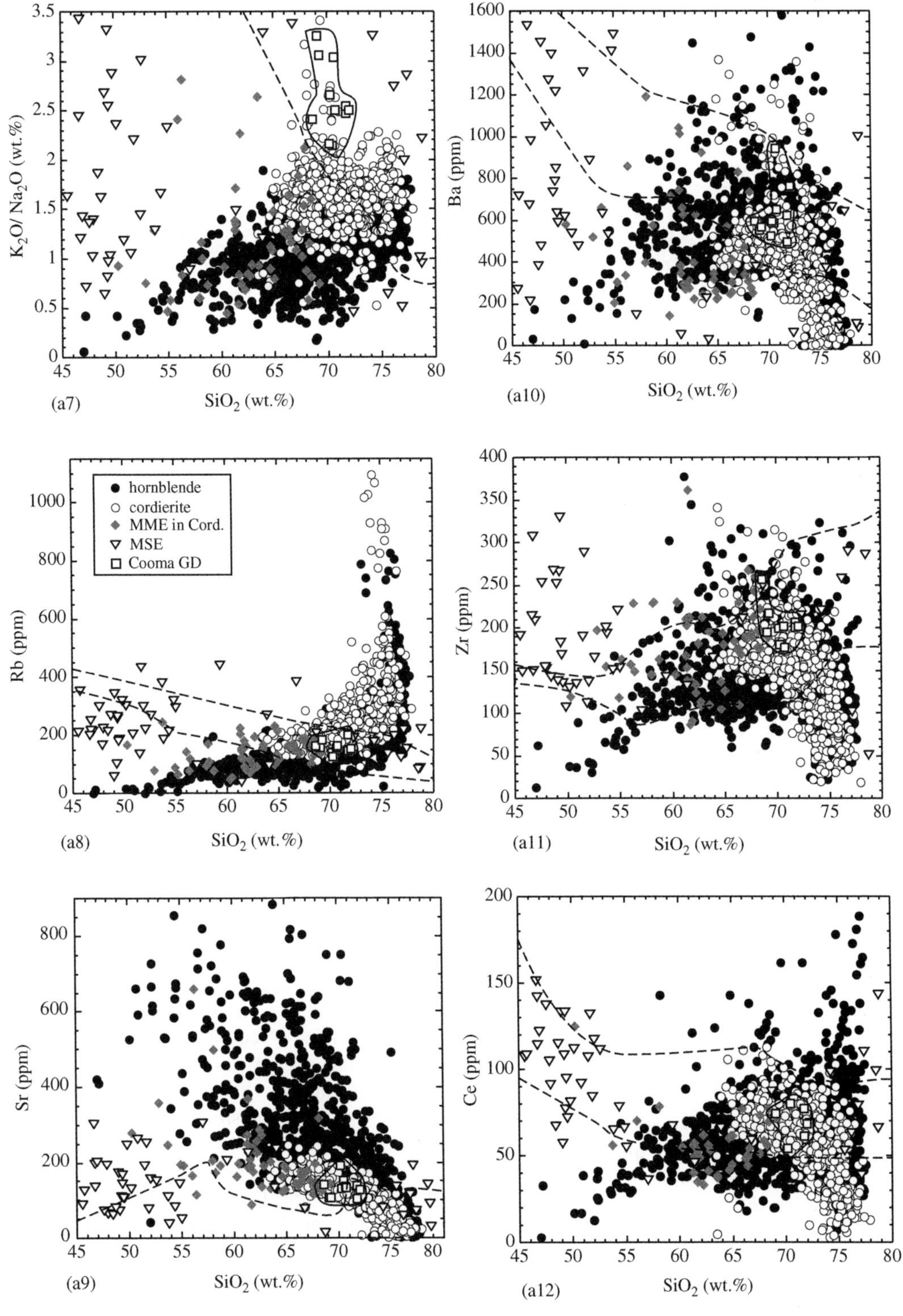

Figure 16 (continued).

Chen *et al.* (1989), Elburg and Nicholls (1995), Elburg (1996b,c), Maas *et al.* (1997), Anderson (1998), Kemp (2003), unpublished); Metasedimentary enclaves; White *et al.* (1977), Chen *et al.* (1989), Maas *et al.* (1997), Anderson (1998), Kemp (2003), unpublished); Ordovician turbidites, Wyborn and Chappell (1983), Munksgaard (1988), Chappell *et al.* (1991), Anderson (1998), B. W. Chappell (unpublished data), Kemp (2003), unpublished data).

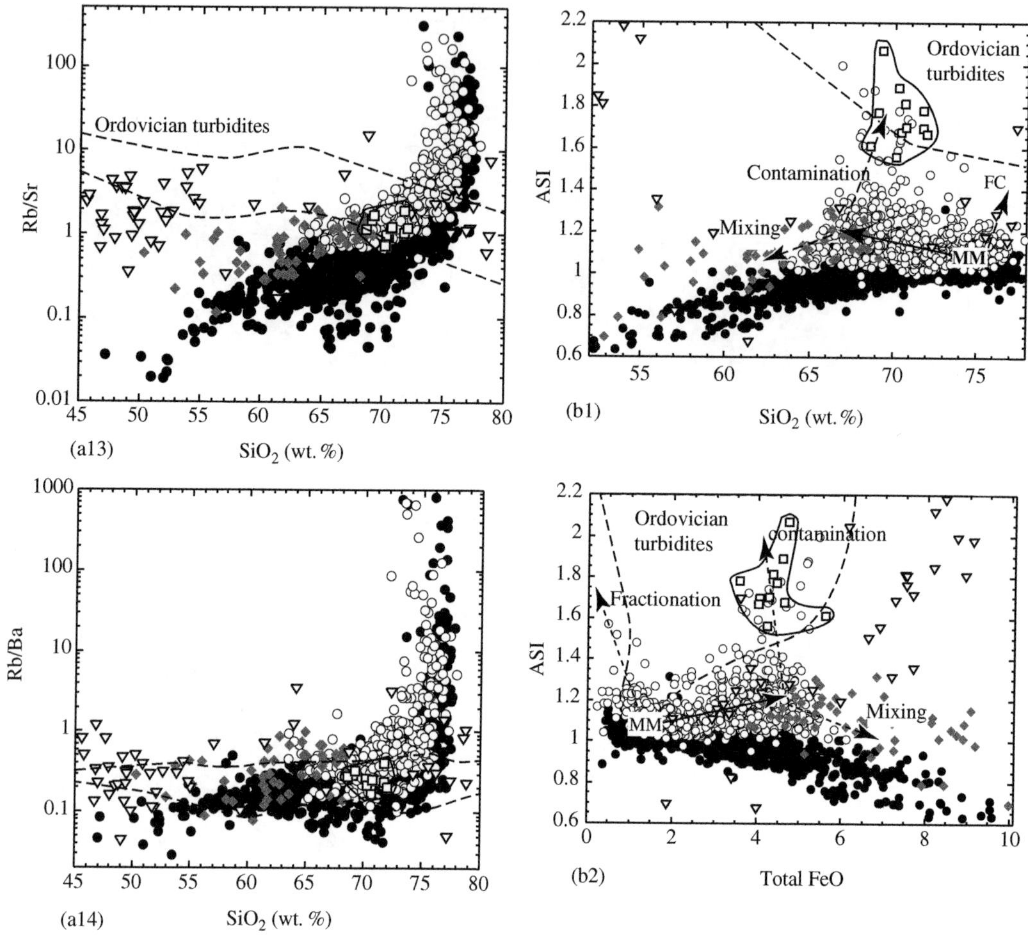

Figure 16 (continued).

Williams (1992), Williams (1995), Anderson *et al.* (1996), Elburg (1996a), and Keay *et al.* (1999).

3.11.4.4.3 Geochemical and isotopic comparisons

In accordance with the mineralogical differences, the LFB cordierite and hornblende granites have different major element compositions, which are shown in an overall sense by Figure 16. The cordierite granites have a relatively restricted silica range (mostly >65%), and towards the lower end they exhibit pronounced clustering on variation diagrams. This contrasts with hornblende granites, which show smooth and continuous compositional variation back to basaltic silica values. More importantly, cordierite granites define tighter chemical arrays that at silica values of ~65% project towards higher TiO_2 FeOt, MgO, Cr, Pb, Rb, K/Na, Rb/Sr, and ASI, and lower Na_2O, CaO, and Sr than most hornblende granites. The ASI is perhaps the best discriminant, as the trends of each granite type diverge and differences are largely maintained over the full compositional range. Differences occur between specific samples on mantle normalized diagrams, where a cordierite granite can have higher contents of most trace elements, and lower strontium than a typical hornblende granites of similar silica (Figure 17). Nevertheless, the patterns and hence the trace-element ratios are strikingly similar, which is perhaps surprising given the ASI difference, and the widespread application of the I- and S-type nomenclature. Chemical contrasts are less apparent at higher silica, where the respective trends defined by the two granite types converge. The exception is P_2O_5, which shows a marked increase at the higher silica contents in cordierite granites, in contrast to a strong decrease for felsic hornblende granites, reflecting differences in the behavior of apatite (Chappell, 1999; see below). It is also important to note that although the geochemical arrays of both groups are approximately linear, at silica contents above ~74% there is an exponential

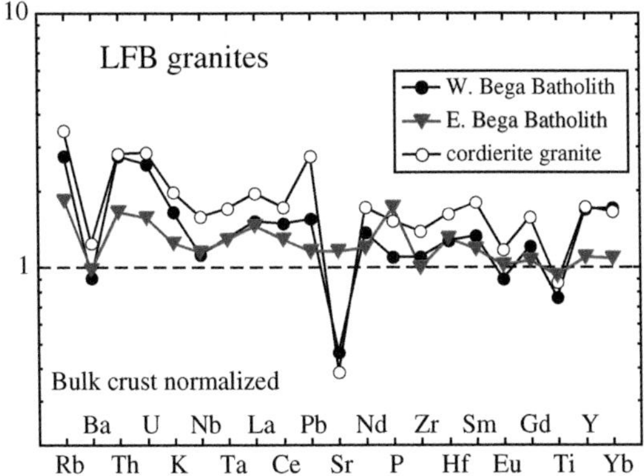

Figure 17 Bulk crust-normalized multi-element diagram comparing a typical cordierite granite of the LFB (the Jillamatong Granodiorite of the Kosciuszko Batholith) and two hornblende granites at ~67% SiO_2. The hornblende granites are from the eastern (Cobargo Granodiorite) and western (Glenbog Granodiorite) sides of the Bega Batholith, respectively (all data from Kemp, 2003, unpublished). The similar shape of the cordierite granite pattern to that of the hornblende granite from the western Bega Batholith is striking, and cannot simply be attributed to compositional convergence at high silica contents. The hornblende granite from the eastern Bega Batholith has less of the signature attributed to a sediment component, notably in higher Sr and Eu, but lower Pb and Rb/Ba.

increase and decrease in elements incompatible (rubidium) and compatible (strontium, barium) in feldspars respectively. Similar inflections are evident for zirconium and the LREE, and coincide with a sharp increase in ASI in the most differentiated cordierite granites.

Cordierite granites extend to more evolved Sr–Nd (and O; O'Neil and Chappell, 1977) isotopic compositions than the hornblende granites. Those evolved compositions overlap with those of the Ordovician turbidites (Figure 14), and yield older, Proterozoic neodymium model ages. It follows that the cordierite granites were derived from source regions that have higher average Rb/Sr ratios than those of the hornblende granites (see Section 3.11.5.1 and Figure 19). Such features are consistent with cordierite granites containing a greater proportion of recycled crust, and this accords with the abundant inherited zircon cores contained by these rocks. Older, pre-magmatic zircons are also recognized in some hornblende granites (Williams *et al.*, 1988, 1992) and their microgranular enclaves (Chen and Williams, 1990). However, cordierite granites extend to more primitive isotopic values than the metasedimentary basement and overlap the range shown by the most evolved hornblende granites. As a result, both granite types define a well-publicized ε_{Nd}–initial $^{87}Sr/^{86}Sr$ hyperbola,

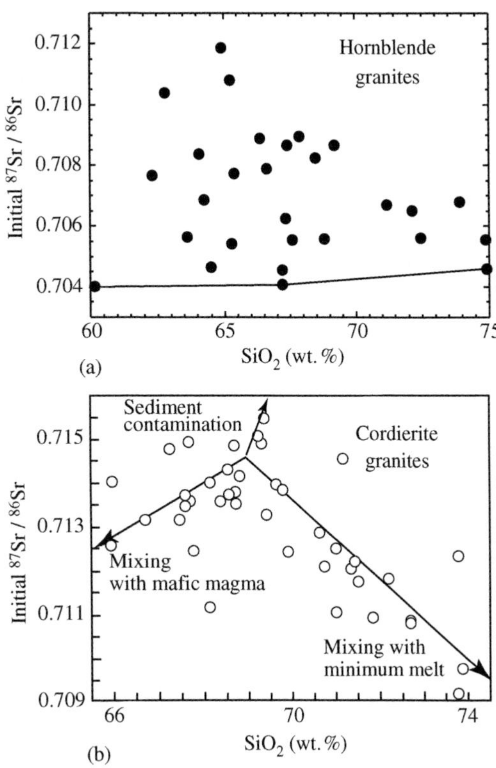

Figure 18 Plot of initial $^{87}Sr/^{86}Sr$ versus silica for (a) LFB hornblende granites (data sources as for Figure 14) and (b) cordierite granites (modified from figure 1 in Chappell *et al.*, 1999), with annotated vectors summarizing the potential petrogenetic processes operating to produce the pattern of isotopic variation. The alternative is that the variation was inherited from a heterogeneous source (see text for discussion). Note the similarity between this figure and the ASI–SiO_2 plot in Figure 16b1. Tie lines in (a) join granitic rocks of the Moruya Suite, whose nearly closed-system isotopic variation could be caused either by fractional crystallization (Collins, 1996, Collins, 1998) or restite unmixing (Griffin *et al.*, 1978), both of which will impart dispersion parallel to the silica axis. There is a vague general trend towards decreasing initial $^{87}Sr/^{86}Sr$ with silica for the hornblende granites, the possible petrogenetic significance of which awaits more detailed study within individual hornblende granite suites.

bracketed between the depleted mantle and Ordovician sediments (Figure 14). However, less well recognized (but see McCulloch and Chappell, 1982) is that there is no simple or systematic relationship between the position of specific rocks on the isotopic array and their whole rock composition. For the hornblende granites, this could, to some extent, reflect the modifying effects of fractional crystallization (Figure 18(a)) Nevertheless, a curious feature is that many of the most mafic hornblende granites are also the most isotopically evolved, such as some plutons from the Bega Batholith (Blue Gum Tonalite, 62.8% SiO_2, ε_{Nd} − 8.9) and

Berridale Batholith (Finister Tonalite, 64% SiO_2, $\varepsilon_{Nd} - 8$) (data sources in Figure 14). The pattern of isotopic variation shown by the cordierite granites is even more complex (Figure 18(b)). Many felsic cordierite granites have lower $^{87}Sr/^{86}Sr$ than the mafic samples, and within an individual cordierite granite suite initial $^{87}Sr/^{86}Sr$ increases with silica to ~0.715 at 68–69% SiO_2 before decreasing steadily to ~0.709 at 74% SiO_2. Such observations impose strict limitations on magma mixing models proposed for the LFB granites.

3.11.4.4.4 Constraints on the origin of the LFB granites

Although the Sr–Nd isotopic array suggests that the LFB cordierite and hornblende granites formed simply by variable degrees of mixing between shared source components (depleted mantle-like magma, and an evolved crustal material), this is precluded by the different geochemical trends defined by the two granitic types. For this reason, it has been persistently argued that the hornblende and cordierite granites were generated separately from contrasting meta-igneous and metasedimentary source rocks, respectively (Chappell and White, 1974; White and Chappell, 1977; Chappell and White, 2001 and references therein). However, it is unclear why rocks with very different ASI, which supposedly reflect the disparate protoliths, should have such similar minor and trace-element patterns, as evident in Figure 17. Thus, we envisage that mantle-derived magmas and older crustal rocks *were* common ingredients in the formation of both the cordierite and hornblende granites, but that these materials were incorporated differently, and in different proportions. As will be discussed below, we also consider that the petrogenesis of each granitic type was coupled during Lachlan orogenesis.

Cordierite granites. The cordierite granites have the greatest crustal affinity, though in some ways their petrogenesis and magmatic evolution is the most problematic. In a general sense, the petrography and geochemistry of these plutons accords with derivation from aluminous metasedimentary protoliths (Chappell and White, 1974). Such a process is evident in several small metamorphic complexes in the central and eastern LFB, where the Ordovician turbidites locally attain migmatite grade and have gradational contacts with small cordierite-rich, diatexite-like plutons. The most thoroughly studied of these, the Cooma Granodiorite, is geochemically and isotopically equivalent to its migmatitic envelope, and demonstrably formed by a closed-system bulk "mobilization" of partially melted quartzofeldspathic metasedimentary rocks (Pidgeon and Compston, 1965; White *et al.*, 1974; Flood and Vernon, 1978; Munksgaard, 1988).

However, the Cooma Granodiorite differs from the "batholithic" cordierite granites in having systematically lower Na_2O, CaO, and Sr, but higher SiO_2 and ASI (Figure 16). It also shows different trends for some elements, particularly for TiO_2, K_2O, and Zr, which mirror the compositional variation of the Ordovician turbidites rather than the array defined by most cordierite granites. The Cooma Granodiorite also lacks the microgranular enclave population present in the lower silica cordierite granites and has the highest strontium and lowest neodymium isotopic ratios (Figure 14). Some other model seems to be required for most of the LFB cordierite granites.

There are two potential resolutions to this problem. One is to suggest that cordierite granites were derived from more feldspathic (i.e., less weathered) sedimentary rocks with higher Na_2O and CaO contents than most of the quartz-rich Ordovician turbidites (Chappell *et al.*, 2000). Additional support for this is that the unfractionated cordierite granites plotted on Figure 11 only show compositional overlap with the least weathered Ordovician sediments that have the lowest Rb/Sr ratios, but highest Sr/Nd and Sr/Eu* (and Eu/Sr, see Figure 20(a)). Such relatively fertile protoliths have yet to be recognized in the LFB and thus would need to be entirely concealed beneath the ubiquitous Ordovician turbidite sequence. The large population of inherited zircons with Ordovician to Late Cambrian ages in the cordierite granites suggest a dominantly Early Paleozoic, rather than Proterozoic age for such a protolith (e.g., Keay *et al.*, 1999). Some evidence for the existence of a more fertile source layer at depth is the generally higher CaO and Na_2O contents of metasedimentary enclaves in the cordierite granites compared to the exposed turbidites (Figure 16) (White *et al.*, 1977; Maas *et al.*, 1997; Chappell *et al.*, 2000). Yet the possibility that some of these are restitic (calcic plagioclase-rich) or modified in some way by the host magma cannot be excluded. The paucity of metasedimentary enclaves with between 60–75% SiO_2 is striking, as this is consistent with the most fertile parts of the sequence having melted to form the granitic magma and thus being unavailable for sampling as enclaves. It is possible that some of the peraluminous "microgranular" enclaves that fall within the 60–75% silica range represent such source material. If the batholithic cordierite granites formed by mobilization of a fertile metasedimentary source in analogous fashion to the Cooma Granodiorite (White *et al.*, 1974), then the protolith must average at least 3% CaO, 3.5% K_2O, and 150 ppm Sr at 66–67% SiO_2, but contain less than 2% Na_2O. Paradoxically, sedimentary rocks of this specific composition do not outcrop on the Australian mainland (Collins, 1998). In fact, in the context of global sedimentary compositions, such a rock seems abnormally calcic and sodium-poor; Phanerozoic graywackes of any tectonic setting with >3% CaO have Na_2O/K_2O ratios exceeding unity, a reflection of their detrital feldspar component (see Taylor and McLennan, 1985 and the compilation of Condie, 1993) However, it is possible that the *bulk* metasedimentary package undergoing melting was of this composition (i.e., represents carbonates mixed with graywackes).

In the absence of a suitable exposed metasedimentary precursor, the second approach is to invoke mixing with an additional magmatic component (e.g., Gray, 1984, 1990; Collins, 1996; Keay *et al.*, 1997). This has the advantage of explaining the pronounced isotopic variation within cordierite granites, since restite separation and fractional crystallization are isotopically closed system processes. Collins (1996) argued that the lower silica cordierite granites formed in the mid-crust by large-scale hybridization between partially melted Ordovician sediments, approximated by Cooma-type magmas (~72% SiO_2), and hotter, ascending tonalite magmas (~60% SiO_2), the proportions being 60–70% Cooma Granodiorite and 30–40% tonalite. However, hornblende tonalites are volumetrically minor in the LFB, and the specific low titanium, low strontium composition required to satisfy the chemical modeling is anomalous amongst hornblende granites. Furthermore, microgranular enclaves in the cordierite granites define trends that converge towards genuinely basaltic compositions, implying the involvement of mantle-derived, rather than tonalitic, magmas (Figure 16). Yet the geochemistry of cordierite granites also seems to prohibit a significant material contribution from basaltic magmas. It is unlikely that the clustered, lowest-silica cordierite granites were generated by simple mixing between basalt and Cooma type diatexite, since the low Na_2O of the granites requires the basalts to have unrealistically high Na_2O contents (~4% at 50% SiO_2). The trends of cordierite granites cannot represent mixing lines between felsic crustal melt and mafic magma, since the cordierite granites become increasingly peraluminous as they become more mafic (Figure 16), which requires a strongly corundum-normative basalt (White and Chappell, 1988). The decrease in initial $^{87}Sr/^{86}Sr$ of cordierite granites as silica increases (Figure 18(b)) is also inconsistent with such bulk mixing. This discussion highlights that although open-system processes are implicated in the genesis and compositional evolution of cordierite granites, these were evidently more complex than binary crust-mantle mixing.

Hornblende granites. Field and chemical evidence, especially the projection of geochemical trends towards coeval gabbros, confers a pivotal role for mafic magmas in the production of the LFB hornblende granite suites. However, many of these plutons have "crust-like" trace-element patterns similar to the cordierite granites (Figure 17), and they contain inherited zircons that have age populations identical to that of the Ordovician turbidites (Williams *et al.*, 1992; Williams, 1995). Thus the hornblende granites also incorporated a sedimentary component (Gray, 1984, 1990; Collins, 1996, 1998; Keay *et al.*, 1997). The inherited zircons are particularly significant, since, assuming that their survival does not reflect sluggish dissolution kinetics, they also suggest that the hornblende granites were not liquid compositions (Chappell *et al.*, 2000), and they had magmatic temperatures lower than that required for the resorption of entrained zircon crystals (i.e., below ~800 °C). In view of this, the hornblende granites cannot have evolved simply by fractional crystallization from basaltic magma, and this is also ruled out by their evolved isotopic compositions relative to mantle values. Instead, as with cordierite granites, it seems that the formation of hornblende granites was linked to crustal melting.

3.11.4.4.5 A petrogenetic model

The spatial and temporal association, as well as the chemical similarities, suggests that generation of the cordierite and hornblende granites in the LFB was a coupled process and involved similar source materials. The preferred model entails the successive emplacement of thick basaltic sills in the lower crust at the onset of Lachlan orogenesis, these being overlain by a thick collage of Ordovician sediments and their oceanic crust substrate. The existence of large amounts of mafic material in the lower parts of the LFB crust is indicated by lower crustal xenolith studies (Chen *et al.*, 1998) and seismic profiling (see Anderson *et al.*, 1998), and confirms that massive basaltic intrusion, and thus crustal growth, accompanied granite genesis. Initially, the sills solidified rapidly, but as the process continued, and the lower crustal region was progressively heated, the larger bodies would have remained partially molten for longer periods, to undergo protracted crystallization. The heat (and perhaps fluid transfer) from these crystallising mafic magmas would have induced fusion of the overlying metasedimentary rocks, and any intercalated screens of such material. The most fertile parts of the sedimentary sequence would yield the greatest proportion of initial partial melt, and segregation of this was possibly responsible for the formation of the felsic cordierite granites. These "minimum melts" would possess relatively low $^{87}Sr/^{86}Sr$, given the feldspar-rich nature of their protoliths. As the lower crustal section continued to heat up, melting would proceed to less fertile lithologies, and this would include fragmentation and assimilation of the more refractory, calcareous horizons. The trends to lower silica of the cordierite granites are thus interpreted in terms of increasing degrees of partial melting and possibly restite contamination, the trajectory of which is a vector towards the *bulk* composition of the turbidite sequence. The increasing restite cargo is strongly suggested by the greater proportion of inherited zircon (Williams, 1995) and metasedimentary enclaves in lower silica cordierite granites. Samples of the clustered compositions were therefore not liquid compositions but probably contained suspended crystals, both refractory and the byproducts of partial melting reactions. In support of this, liquids resembling the cordierite-rich granite compositions have not yet been reproduced experimentally from metasedimentary protoliths under any P–T-aH_2O conditions (Montel and Vielzeof, 1997; Patiño Douce, 1999). The increase in $^{87}Sr/^{86}Sr$ in granites having SiO_2 contents of ~74% to 68% is consistent with greater incorporation of metapelitic rocks, the radiogenic signature of which ($^{87}Sr/^{86}Sr \sim 0.726$ at the time of granitic generation, McCulloch and Woodhead (1993)) is likely to swamp low $^{87}Sr/^{86}Sr$ contributions from the calc-silicates. However, the very high CaO of the latter (11–13%) largely controls the accompanying trend to higher CaO. The numerous metasedimentary enclaves and micaceous clots in the lower silica cordierite granites comprise relatively refractory material and may have been entrained from the metasedimentary source regions (Chappell *et al.*, 1987). Alternatively, they could have been derived from interaction with Cooma-type diatexite during passage through the middle crust, as the scattering of some cordierite granite compositions to high ASI values suggests that this was important locally (see Figure 16).

Periodically, residual mafic liquids from the underlying sills would percolate upwards to mix with the overlying partially melted sediments and crustal magmas. This is considered responsible for driving some cordierite granites towards slightly more mafic, less peraluminous compositions, and for the observed decrease in initial $^{87}Sr/^{86}Sr$ below 68% SiO_2. Most microgranular enclaves were conceivably also dispersed by this process, and their complex chemical variation can be interpreted as reflecting hybridization between variably differentiated basaltic liquids and cordierite granite magma. Given the trends to low Na_2O and high ASI, some enclaves may also have formed by hybridization between basaltic liquids and infertile Ordovician sediments, to be subsequently fragmented and entrained by the ascending granitic magma.

Continuation of the underplating and crystallization process would eventually remelt the earlier, solidified and crustally contaminated basaltic sills, or their evolved, differentiates, and this is considered responsible for the formation of the hornblende granites. The extraction and emplacement of the initial, minimum melts resulted in the felsic hornblende granite plutons that dominate the eastern LFB. More mafic compositions were produced by increasing degrees of melting and also by replenishment of the partially melted sills by more primitive basaltic liquids. The latter process is recorded by the mafic enclaves that are abundant in the lower silica samples. The composition of each derivative hornblende granite suite is therefore a function of the geochemistry of the participating basaltic magma and the amount of crustal material (melt or bulk sediment) assimilated by the crystallizing sills and later remelts. For example, granitic rocks of the Moruya suite in the easternmost LFB have a large compositional range (60–74% SiO_2) but are uniformly isotopically primitive ($\varepsilon_{Nd} \sim +4$) showing that contamination by sedimentary materials was minor (<6%, assuming the isotopic ratios for the basalt and sediment end-members indicated in Figure 14). The isotopic systematics and trace-element patterns of plutons from the eastern edge of the Bega batholith also seem to preclude a substantial sedimentary component (see Figure 17). This has been attributed to the paucity of sedimentary rocks at the depth of granite generation in the eastern LFB, possibly reflecting thinning of the turbidite blanket outboard of the continental margin (Collins, 1998). The distinctively high Na_2O and Sr that characterizes hornblende granite suites of the eastern LFB was clearly imparted by the ancestral mafic magmas, as these features are evident within the coeval basaltic dikes analyses presented by Griffin *et al.* (1978). In this respect, as first pointed out by Chappell and Stephens (1988) and subsequently emphasized by Keay *et al.* (1997) and Collins (1998), the eastern Lachlan granites resemble the sodic plutons of convergent continental margins, which have largely basaltic protoliths and minimum input from evolved supracrustal materials (see discussion in Section 3.11.3.1.1). In contrast, hornblende granites of the Berridale Batholith further west are more potassic and isotopically evolved (ranging to $\varepsilon_{Nd} \sim -8.1$, initial $^{87}Sr/^{86}Sr \sim 0.712$; McCulloch and Chappell, 1982). These rocks are associated with cordierite granites, indicating that a substantial thickness of Ordovician sediments existed at mid-crustal depths, some of which were incorporated by the hornblende granites, or their mafic precursors. This accounts for the population of inherited zircons and similar trace-element patterns to the cordierite granites. Unlike other mixing models, as the mafic and felsic components are genetically related, this scenario has the unique advantage of being able to explain the chemical similarities shown by hornblende granites at opposite extremes of the compositional range. This observation has been previously used as a major objection to the viability of magma mixing in the LFB (e.g., Chappell, 1996a).

The compositional trends of hornblende and cordierite granites are thus explicable in terms of magma mixing and partial melting, respectively, with an ancillary role for restite entrainment in cordierite granites. Fractional crystallization, as advocated for cordierite granites by Collins (1996), is less favored since these rocks were clearly not liquids at the mafic end of their compositional spectrum. Fractionation is also not compatible with the reasonably flat trends for strontium, barium, and rubidium, at least to 74% SiO_2. The role of fractionation in the formation of the hornblende granites in general is difficult to assess, not least because fractionation trends will parallel partial melting trends, and awaits more isotopic data. However, the sharp increases in Rb/Sr and Rb/Ba, and pronounced downward inflections in strontium, CaO, Zr, La, and Ce at greater than 74% SiO_2 (Figure 15(a)) suggest that the final stage in the evolution of both LFB granite types did involve fractional crystallization, controlled predominantly by feldspars and accessory minerals. For cordierite granites, this is consistent with the sharp increase in ASI in the most felsic rocks (i.e., >74% SiO_2), since this is elevated by the removal of feldspars (ASI = 1.0) from peraluminous magmas. The steep P_2O_5 increase is coupled to this increase in ASI, since apatite solubility increases drastically with peraluminosity (Pichavant *et al.*, 1992). In contrast, the decrease in P_2O_5 shown by hornblende granites indicates that apatite saturation was reached for these less peraluminous magmas (Chappell, 1999). An implication of this is that the most evolved compositions in granitic suites should not be used as end-members in mixing models, as any fractionation will produce spurious results, especially on incompatible/compatible element ratio plots.

In summary, the generation and evolution of the contrasting granitic types in the LFB appears to have been a three stage process that was driven by incursion of mantle-derived magmas into the crust. For the hornblende-bearing granites, the stages involved (i) remelting and/or prolonged fractionation of crustally contaminated basaltic sills to produce silicic minimum melts; (ii) increasing degrees of partial melting, combined with mixing with fresh basalt to form more mafic compositions; and (iii) fractional crystallization. For the cordierite granites, these stages are: (i) varying degrees of anatexis of Ordovician sediments; (ii) hybridization between these crustal magmas and basalt to produce the lowest silica compositions; and (iii) fractional crystallization at the highest silica contents. Greater insight into the way that the mantle- and crustally derived source components were combined during generation of the cordierite and hornblende granites awaits more systematic documentation of within-suite isotopic variation. Studies that attempt to unravel the magmatic evolution of individual plutons, for example by tracking changes in hafnium isotopic composition across magmatically zoned zircon grains (e.g., Griffin *et al.*, 2002), should prove to be particularly valuable.

3.11.4.4.6 Relevance for crustal evolution

If the model outlined above is valid, silicic magmatism in the LFB involved net crustal growth, as juvenile mantle-derived liquids, or their differentiates were instrumental in the formation and compositional evolution of hornblende granites, and, to a lesser extent, the cordierite granites. The amount of new crust generated is estimated by determining the overall mantle component present within both granitic types, and this is best done isotopically, since the trace-element ratios are poorly constrained for the potential basaltic end-members.

For the purposes of modeling it is convenient to assign a common depleted mantle-like end-member to all hornblende and cordierite granites (McCulloch and Chappell, 1982). This therefore represents the minimum case for crustal growth, since the mantle-derived end-member could be significantly more evolved than this, especially if generated in a subduction-related setting. Inspection of Figure 14 reveals that the average hornblende granite isotopic composition (initial $^{87}Sr/^{86}Sr = 0.707$, ε_{Nd} -3) falls on a mixing curve between its inferred depleted mantle (basaltic) and crustal end-members, and corresponds to ca. 70% of the mantle component. For cordierite granites, the plotted composition represents the average low silica granite (those with <68% SiO_2), for which an initial $^{87}Sr/^{86}Sr$ of 0.7135 and ε_{Nd} −8 is inferred; such a composition is modeled as ca. 10% depleted mantle component. Assuming that hornblende and cordierite granites occur in equal proportion and constitute 30% of the LFB, and given that the low silica samples thought to incorporate a mantle component comprise ~20% of all cordierite granites, this indicates that ~10% of the presently exposed LFB crust was newly generated during orogenesis, as manifested within granitic intrusions. This estimate is admittedly crude, and serves for illustrative purposes only, but should become more refined as future studies pinpoint the nature of the mantle-derived component and establish the range of isotopic variation shown by the various granitic suites. The percentage of new crust will increase if it can be shown that more enriched basaltic rocks were involved in granite petrogenesis. Although hornblende granites contain a greater mantle component than the cordierite granites, the chemical signature of this ingredient is "swamped" by that of the Ordovician turbidites. This explains the similar trace-element ratios of some hornblende and cordierite granites, despite the inferred differences in petrogenesis.

Differentiation of the newly formed LFB crust was achieved predominantly by intracrustal melting, involving both fusion of the underplated basaltic sills, and mixing between basaltic magmas and relatively evolved magma derived from the pre-existing supracrustal rocks, with the superimposed effects of fractional crystallization. This reinforces the point that crustal growth and intracrustal melting are likely to be coupled during orogenesis, the latter being an inevitable consequence of the heat and (in a subduction setting) volatile transfer from the emplacement of mantle-derived magmas.

However, although similarly related to crustal anatexis, the style of crustal differentiation associated with granitic magmatism in the LFB differs from that of strongly peraluminous leucogranites in three main respects. First, approximately half of the plutons (hornblende granites) are derived from juvenile mantle-derived protoliths, rather than pre-existing sedimentary rocks. Second, in respect of the LFB cordierite granites, the crustal melts also incorporate basalt and some restitic material, partly explaining their more mafic compositions. Hence, they have higher REE contents and have trace-element patterns that more closely resemble the upper crust (Chappell, 1996b). Yet as with the hornblende granites, strontium, and phosphorus contents are lower than the bulk and upper continental crust compositions. Thirdly, heat producing elements are more efficiently transferred to the upper crust via cordierite granite magmas than in leucogranites. This does not reflect greater solubility of refractory zircon and monazite, but rather the physical entrainment of such minerals, probably enclosed within major phases, by the magma (Chappell, 1996b). Sandiford *et al.* (2002) emphasize that scavenging of heat producing elements from the lower parts of the crust will induce long term cooling of these regions, substantially increasing the strength, and also density of the lithosphere. The density increase is predicted to trigger isostatic subsidence, and such subsidence is manifested as thick sedimentary basins in northern Australia that formed following the emplacement of plutons enriched in heat producing elements (Sandiford *et al.*, 2002). The development of Late Devonian sedimentary basins in the LFB immediately after plutonism implies that thermal subsidence related to the redistribution of K–Th–U may also have been important in this area.

3.11.5 CRUSTAL DIFFERENTIATION THROUGH EROSION AND SEDIMENTATION

3.11.5.1 The Link Between Rb/Sr Ratios and Granitic Sources

The discussion this far has highlighted that granite magmatism is a key process in the differentiation and evolution of the continental crust. However, there is no simple match between the composition of common granites and that of the average upper crust, and so some other process(es) must be involved. This section compares the effects of granite formation, erosion and sediment formation on the evolution of the upper crust as expressed through its Rb/Sr ratio. This ratios is apposite because it is the parent/daughter ratio for the Rb–Sr decay scheme, and so it is constrained by measured $^{87}Sr/^{86}Sr$ values, and it is sensitive to the effects of residual plagioclase (which is only really stable within the crust) and weathering and erosion.

Detailed studies of granitic rocks in a number of orogenic belts have identified broad positive arrays between the Rb/Sr ratios of individual granite samples and the time-integrated Rb/Sr ratios of their source regions inferred from their model neodymium ages. The model neodymium ages of the granites are used to estimate the average age when their crustal sources were extracted from the mantle, and then the strontium isotope ratio of the mantle at that time, and the initial strontium isotope ratios of the granites are used to calculate the time-integrated Rb/Sr ratios of the granites' source regions. These are then compared with the measured Rb/Sr ratios of

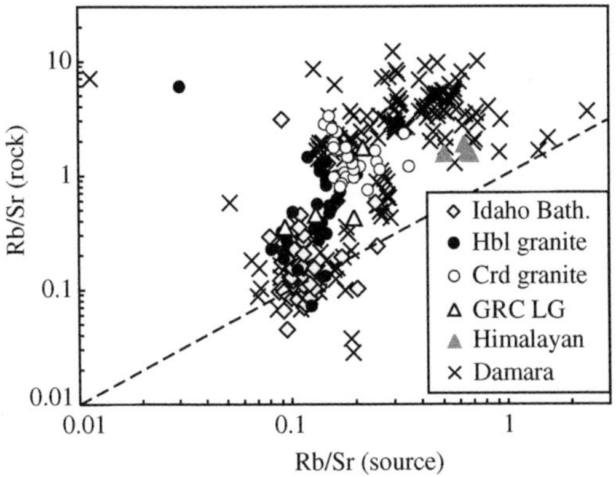

Figure 19 Plot of Rb/Sr in granitic rocks versus that calculated for their putative source regions to illustrate the degree of crust differentiation using Rb/Sr ratios (the 1:1 line is dotted). See text for full explanation. The data are from: Idaho Batholith, Clarke (1990); LFB hornblende (Hbl) and cordierite (Crd) granites, as for Figure 18; Glenelg River Complex leucogranites (GRC LG), Kemp (2001); Himalayan leucogranites, Inger and Harris (1993); Pan-African granitic rocks of the Damara Orogen, Namibia (Hawkesworth and Marlow, 1983; McDermott *et al.*, 1989; McDermott and Hawkesworth, 1990).

individual granite samples in Figure 19. This illustrates that more evolved, and hence high Rb/Sr granites tend to be derived from source regions that are also more evolved, again as reflected in their higher source Rb/Sr ratios. On the basis that the longer rocks remain in the crust, the more likely they are to be involved in further magmatic or sedimentary events (e.g., Davies *et al.* (1985) concluded that the upper crustal rocks of the British Isles were reworked on average every 600 Ma), it might be predicted that the time-integrated source Rb/Sr ratios would increase with their model neodymium ages. However, in practice there is no simple link between the measured or the source Rb/Sr ratios and the age of the source regions for the granites, as reflected in their model neodymium ages. This suggests that the source Rb/Sr ratios primarily reflect the rock types in the granite source regions, rather than the length of time the source rocks, or their precursors, have been in the crust.

3.11.5.2 Resolving the Effects of Igneous and Sedimentary Processes

Weathering and erosion at the Earth's surface preferentially removes those elements that are water soluble (e.g., strontium, calcium, barium, magnesium) from the continental mass, and enriches elements that are adsorbed onto clay minerals (e.g., rubidium, REE^{3+}). One way to isolate the effects of igneous and sedimentary processes is therefore to use trace-element ratios involving europium and strontium. These elements are both sited in plagioclase but they behave differently in igneous and sedimentary environments. During fractional crystallization (or partial melting) within the crust, strontium and europium are both partitioned into plagioclase, and so their concentrations are lower in the coexisting liquids. Europium is also depleted relative to the trivalent REE, leading to the development of progressively larger negative europium anomalies (i.e., Eu/Sm, which is a proxy for europium anomalies, will decrease) as differentiation proceeds. However, the degree to which europium is partitioned into plagioclase depends on the proportion of Eu^{2+} and Eu^{3+}, and therefore on the oxygen fugacity (Drake and Weill, 1975).

In contrast, strontium and europium are decoupled during sedimentary processes. This is because the soluble strontium is released to the oceans during weathering of feldspathic rocks, but europium is retained in the residual weathering profile, as Eu^{2+} is oxidized to Eu^{3+} and thus behaves like the other trivalent, immobile REE. Weathering therefore cannot produce the depletion in europium relative to samarium that fingerprints magmatic differentiation, and this can be used to assess the extent to which strontium depletion in upper crustal rocks is due to magmatic or sedimentary processes.

To explore these effects, Figure 20 shows plots of (a) Rb/Sr versus Eu/Sr and (b) Eu/Sm versus Eu/Sr for various types of granitic and sedimentary rocks, compared to estimates of crustal compositions from Table 1. It is unclear how representative these rocks are of upper crustal processes, but they provide a useful framework for discussion, particularly since the igneous rocks include relatively "primitive" TTG-like plutons (i.e., Idaho Batholith) and highly evolved A-types, and thus encompass the geochemical spectrum of granitic compositions in the crust. On Figure 20(a), the igneous and sedimentary rocks plot in an overlapping positive array, since the Rb/Sr ratio is elevated by weathering (rubidium is retained in clays, and strontium is lost to sea-water) and igneous fractionation (rubidium is incompatible and strontium is compatible in plagioclase). In sedimentary rocks, the increase of Eu/Sr with differentiation reflects the removal of strontium to the oceans, whereas for igneous rocks it indicates that strontium is more compatible than europium in the residual or crystallizing plagioclase feldspar. This is highlighted in Figure 20(b) since europium and samarium are not fractionated during weathering, and so the igneous and sedimentary trends tend to diverge.

Two other points are worth noting. First, apart from the A-type granites, the rocks with elevated Rb/Sr and Eu/Sr ratios in Figure 20 tend to be sediments and sediment-derived granitic rocks.

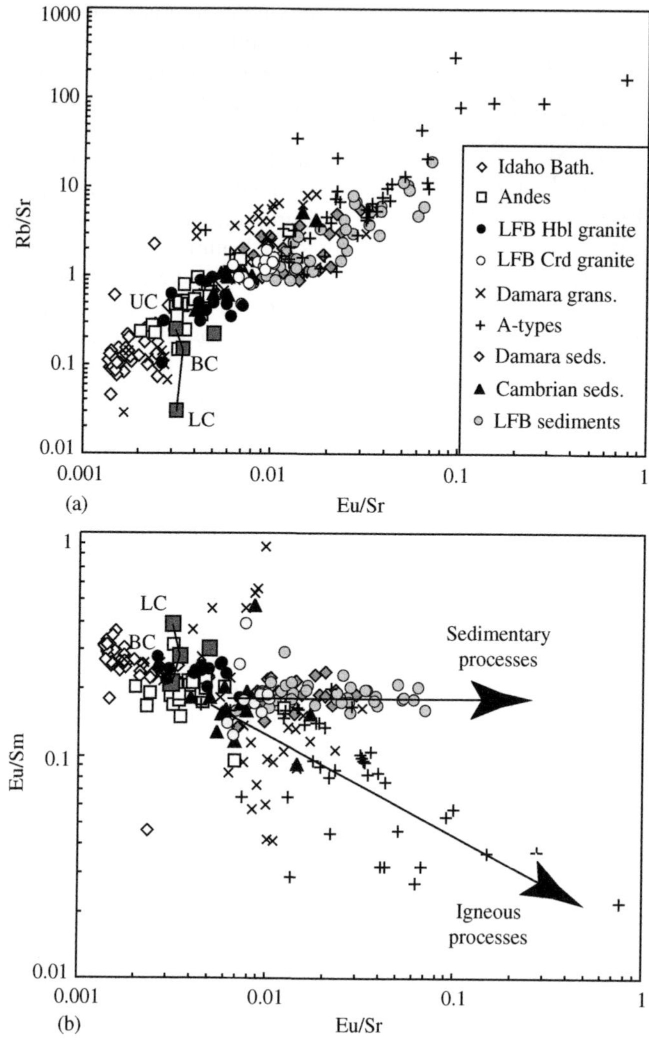

Figure 20 Plots of (a) Rb/Sr versus Eu/Sr and (b) Eu/Sm versus Eu/Sr for the reference igneous and sedimentary suites referred to in this chapter, compared to estimates of the continental crust (UC, upper crust; BC, bulk crust; LC, lower crust, all connected by tie lines). The contrasting trends produced by igneous and sedimentary differentiation processes in (b) are arrowed. Data sources are as for previous figures, with the LFB granites and Cambrian Delamerian Orogen sediments (southeastern Australia) from Kemp (2003), unpublished data), and the granitic and sedimentary rocks of the Damara Orogen from McDermott and Hawkesworth (1990) and McDermott *et al.* (1996, 2000).

This further emphasizes the importance of sedimentary processes in the development of such fractionated Rb/Sr and Eu/Sr ratios. Second, in the context of the LFB, the Ordovician turbidites have progressively lower Eu/Sr ratios than the feldspathic Cambrian sediments from which they were partly derived. Thus, it also appears that the displacement to higher Eu/Sr is linked to the number of weathering and erosion cycles involved in the formation of particular sedimentary rock suites. In the examples in Figure 20, the reduction in strontium content for the most evolved Ordovician sediments is nearly 10-fold, and this clearly results in an associated increase in Rb/Sr, and hence with time in strontium isotope ratios.

3.11.5.3 Relevance for Crustal Differentiation

The fractionated igneous and sedimentary rocks in Figure 20 have Rb/Sr and Eu/Sr ratios that are much higher than those in current estimates for the upper continental crust. The latter can be constrained by the strontium isotope ratios of continental run-off (~0.712), and its model neodymium age (~1.8 Ga). According to this method, a minimum time-integrated upper crustal Rb/Sr ratio of 0.14 is indicated.

Another striking feature of the data in Figure 20 is that the upper, lower and bulk continental crust compositions all have similar Eu/Sr ratios and thus define a distinct, near-vertical trend that is separate from the igneous and sedimentary arrays. There are several potential interpretations for this. It might merely reflect the methods employed to estimate the upper and lower crustal averages. The upper crust represents a mixture between sediments and intermediate to felsic igneous rocks (note how close the upper crustal composition plots to the reference suite array in Figure 20), whereas the lower crust inevitably combines xenolith data from both intraplate and destructive margin settings, not necessarily in representative proportions; the bulk crustal composition is constrained to lie between these extremes. The dispersion in crustal compositions on Figure 20 could therefore be synthetic and petrogenetically meaningless. Alternatively, if the chemical variation between the crustal components results from the differentiation or "unmixing" of a bulk starting composition, the data suggests that neither igneous fractionation nor weathering processes can be wholly responsible for such differentiation, since strongly increasing Eu/Sr is a signature of those processes. In any case, it is clear from Figure 20 that the upper crustal composition has sufficiently low Rb/Sr and Eu/Sr to preclude a significant contribution from the continental sediments, in contrast to the implications drawn from the granite and upper crust trace-element patterns in Figure 5.

One resolution to this conundrum could be related to a shift in the oxidation state of europium (i.e., the proportion of Eu^{2+} to Eu^{3+}) either through time, or in different tectonic settings (Carmichael, 1991). The marked increase in Eu/Sr in igneous suites indicates that D_{Sr} was much larger than D_{Eu} in the fractionating plagioclase. However, D_{Eu} in plagioclase is sensitive to oxygen fugacity (f_{O_2}), and it is very low in oxidizing conditions where europium exists as Eu^{3+} (Drake, 1974). The igneous rocks plotted in Figure 20 can be inferred to have formed under reasonably oxidizing conditions, since many of them originate above subduction zones, where the mantle has been modified by the introduction of hydrous fluids from the subducted slab, or they derive from, or have interacted with, recycled sedimentary rocks in the deep crust.

However, in reducing conditions, D_{Eu} increases until, at $f_{O_2} \sim 10^{-12.5}$ bars, it approaches values similar to that of D_{Sr}. Plagioclase formed under these conditions will therefore not fractionate europium from strontium, and its removal results in vertical arrays on Figure 20, as is shown by the continental crust. This raises the intriguing possibility that the differentiation of the continental crust was primarily achieved under relatively reducing conditions, such as existed in the Archean period where a CO_2-rich atmosphere prevailed, or in intraplate settings. The latter would be marked by magmas with distinctive trace-elements patterns, and in particular no negative Nb–Ta anomalies. Such magmas did contribute to the generation of new crust, they cannot be the dominant component, and so the apparent lack of Eu/Sr fractionation in the crustal compositions may largely reflect processes in the Archean. Independent evidence for a reducing environment at this time includes the presence of banded iron formations, uranium placer deposits, high Th/U ratios in igneous rocks, non-mass dependent sulfur isotope fractionations (Farquhar *et al.*, 2000; see also Chapter 4.04) and the lead isotope composition of the mantle (see Elliott *et al.*, 1999). Under such reducing conditions, intracrustal melting can generate the observed differentiation of the continents, the important point being that the residual plagioclase contained substantially more europium than that of the magmatic reference array in Figure 20. Recycling of small amounts of the residues of melting, approximated by the estimated lower crustal composition, can explain the displacement of the bulk crust from the magmatic array on Figure 20. The fundamental implication from this reasoning is that melting and weathering processes operating in the outer part of the post-Archean Earth have contributed relatively little to the bulk differentiation of the continental crust, consistent with its average age of ~1.8 Ga.

3.11.6 SYNTHESIS AND AVENUES FOR FUTURE STUDY

After decades of effort, we appear to have arrived at robust estimates for the composition of the continental crust that are corroborated by diverse geochemical, isotopic and geophysical approaches. The greatest source of uncertainty is that models for the bulk crust depend on estimates for the lower crust, which are less tightly constrained. However, it is well recognized that although the continental crust has grown throughout Earth's history by material flux from the underlying mantle, the bulk composition is not that of any common primary mantle-derived magma, either basaltic or andesitic. The discrepancy is greatest with elements that behave compatibly during deep (magnesium) and mid- to shallow-level crustal processes (strontium, europium) and thus all current models require preferential recycling of these elements from the continental crust back into the mantle.

Isotopic studies have established that the continental crust is old, in that more than half of it was formed by the end of the Archean. The distinctive igneous components of Archean cratons, coupled with evidence that the crust at this time was overall less mafic, provide clues that crustal growth mechanisms during this period differed significantly from those operating today. The earliest continental nuclei, the Archean TTG, represent high-pressure, second-stage melts of juvenile mantle-derived basalts, analogues of which are rare in post-Archean orogens. However, the tectonic setting in which the TTG, and indeed new crust, were generated in the Archean remains uncertain. This inhibits our understanding of the way in which the crustal protoliths were extracted from mantle sources in the Archean, and how selected components were returned to the mantle. The apparent periodicity in crustal growth, the bimodal nature of Archean magmatism, and the lack of a consistent subduction signal in certain greenstone belt volcanic rocks, are strong evidence for the role of plume-related magmatism. Alternatively, compressive tectonic regimes that may well reflect subduction processes are indicated by the depth of melting of the TTG, and the time interval of 50–80 Ma between the generation of mafic and felsic magmas. Subduction also offers a convenient mechanism for dislocating TTG magmas from their voluminous ultramafic residues, which are recycled into the mantle via foundering slabs of oceanic lithosphere. A satisfactory compromise might therefore involve models whereby the basaltic protoliths to the TTG are formed in a plume-related environment, and are subsequently remelted to form the silicic magmas during subduction of this young mafic crust.

Elemental ratios involving the HFSE can in principle be linked to the behavior of key mineral phases and so place some constraints on the nature of the crust-forming process through time. However, the evidence from HFSE systematics may sometimes be ambiguous. For example, negative Nb–Ta anomalies are widely used to identify subduction related magmatism in ancient orogenic belts, and to distinguish it from what is often inferred to be plume-related intraplate melt generation. However, negative Nb–Ta anomalies may also be generated during intracrustal melting and magma differentiation, and so mask evidence for the tectonic setting of crust generation.

In this chapter, we have emphasized that felsic magmatic rocks, the buoyant building blocks of the continents, show dramatic secular variation, and that this can be related to changing thermal and atmospheric conditions in the Earth. Regardless of the tectonic setting, the thermal conditions necessary for remelting basaltic rocks in the garnet stability field were rarely attained after the Archean. Furthermore, key chemical indices, especially the development of a europium anomaly in the upper crust, indicate that mid to shallow level intracrustal differentiation under high f_{O_2} became more common in the post-Archean Earth. The estimated crustal structure suggests that elements with similar characteristics, such as niobium, tantalum, and titanium, and strontium and europium, were fractionated during formation of the upper crust, such that it is relatively enriched in Nb–Ta and Sr, but depleted in titanium and europium compared to the bulk composition.

Crustal differentiation leading to an evolved upper crust is conventionally attributed to partial melting and granite formation, augmented by erosion and weathering, with the implicit assumption of "closed system" behavior. Furthermore, one might predict that the degree of crustal differentiation should increase progressively with time, given the abundance of broadly granitic rocks emplaced into the upper crust of all time periods.

However, from an examination of the chemical fractionation associated with granitic magmatism and sedimentary processes, the following points emerge.

(i) In the absence of heat transfer from mantle-derived magmas, crustal melting produces silica-rich and strongly peraluminous leucogranites by low temperature, muscovite melting reactions. The extreme chemical fractionation associated with this "haplogranitic" style of magmatism cannot drive crustal differentiation, though it is potentially important in enriching the upper crust in the most incompatible elements (e.g., rubidium and potassium). Regardless of the presence or absence of fluid during melting, the low REE and HFSE of these rocks is buffered by the low solubility of refractory accessory minerals, chiefly zircon and monazite. The heat producing elements uranium and thorium may not therefore be mobilized in this style of granite magmatism.

(ii) From the Late Archean onwards, the formation of relatively mafic granites ($SiO_2 < 70\%$) with elemental ratios more like the continental crust required higher temperature melting reactions and direct heat input from mantle-derived magmas. Chemical and isotopic considerations from specific case studies demand that juvenile mantle-derived magmas have also made a material contribution to granites generated from a range of crustal protoliths; i.e., the Moho has acted as an open interface during post-Archean granitic generation. Granitic rocks should therefore be properly viewed as the products of multistage processes, often involving materials from more than one source and from different geochemical reservoirs. It therefore seems sensible to retain classification schemes for these rocks that are descriptive, rather than generic.

(iii) At the same silica content (~66% SiO_2), the granites of oceanic arcs, continental arcs, and large, probably subduction-related granitic provinces that also include postcollisional magmas (LFB), do not match the estimated upper crustal composition. In arc granites, Nb–Ta and Ti are coupled, and so these magmas cannot enrich the upper crust in Nb–Ta relative to Ti, unless Nb–Ta contents are elevated by an additional process (see point 5 below). Arc granites do however, have high Sr/Nd, and they therefore contribute to maintaining the moderately high strontium contents in the upper crust.

Granites of the LFB have high rubidium, potassium, thorium, HFSE, and REE contents, reflecting variable contributions from mature sediment. Consequently, however, these granites alone do not have sufficient strontium to account for its abundance in the upper crust.

(iv) A-type granites appear mostly to have formed by relatively shallow level melting or fractionation of basalt in the absence of garnet, and thus they represent the other end of the compositional spectrum to the TTG on some chemical diagrams (e.g., Figure 20). These magmas represent the clearest magmatic mechanism for enriching the upper crust in the HFSE relative to titanium, but are associated with extreme strontium, barium, phosphorus, and europium depletion. To understand the role of these granites further in crustal evolution, we need to ascertain why there was a major pulse in A-type generation in the Proterozoic.

(v) Following from points (iii) and (iv) above, the best fit to the average composition of the upper crust is a ~75 : 35 mixture of intermediate to silicic arc magmas (which have relatively high strontium, but low titanium and HFSE) and sedimentary rocks (low strontium and titanium, but high HFSE) (Figure 21). This is consistent with the positioning of the upper crustal composition within the igneous and sedimentary array on Figure 20, and tends to diminish the importance of nonarc magmas for determining the upper crustal composition. The intermediate subduction-related magmas in this model may have evolved from basaltic liquids by high-pressure fractionation and contamination at the base of thickened crust, as suggested for some Andean andesites by Davidson and Arculus (2001).

(vi) Granites with high Rb/Sr ratios are usually attributed to partial melting processes involving mica breakdown and/or later plagioclase fractionation, the latter being most clearly the case for the A-types. However, we contend that the Rb/Sr ratios of granites also strongly reflect the Rb/Sr ratios, and hence the degree of evolution, of their source rocks (Figure 19). This argument is underpinned by isotopic evidence and thus makes no assumption about the melting reactions involved in granite genesis.

(vii) Igneous differentiation in the crust results in decreasing Eu/Sm (proxy for Eu/Eu*) and its effects may therefore be distinguished from those of weathering and erosion, where this ratio is unchanged. Both sedimentary and igneous processes are associated with markedly increasing Eu/Sr, reflecting strontium loss to the oceans in the former, and $D_{Sr} \gg D_{Eu}$ in the latter.

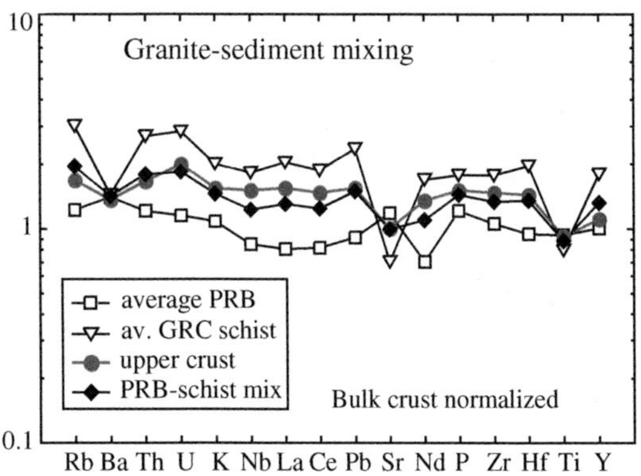

Figure 21 Bulk crust normalized diagram showing simple mixing between the average granitic rock from the subduction-related Peninsula Ranges Batholith, California (PRB, from Silver and Chappell, 1988, with 64.6% SiO_2), and the average of 92 metasedimentary rocks from the Cambro-Ordovician Glenelg River Complex (GRC, from Kemp, unpublished data, with 71.1% SiO_2) in the proportion 75% granite, 35% sediment. The resultant composition has 66.9% SiO_2, comparing favorably with estimates of the upper continental crust (66.6% SiO_2), and closely mimics the upper crust for most trace elements. Phanerozoic arc magmas are most suitable for this type of modeling, owing to their high Sr contents, a reflection of the preferential transport of this element by water-rich fluids from the subducted slab to mantle wedge during magma generation (Pearce and Peate, 1995). Less satisfactory fits are obtained for the TTG, since the resultant mixtures have much lower HREE and Y than the contemporary upper crust.

Strikingly, however, estimates of the upper, middle and lower continental crust do *not* follow such differentiation trends, and in particular there is relatively little fractionation of Eu/Sr with increasing Rb/Sr. If this is not an artifact of how the crustal estimates were derived, it implies that differentiation of the bulk crust took place under different conditions than those prevailing during the formation of the reference magmatic suites in Figure 20. We tentatively suggest that crustal differentiation may have been dominated by conditions in the Archean when oxygen fugacity was less, and thus the relative partitioning of europium and strontium into plagioclase was different than in post-Archean igneous processes.

In conclusion, a number of arguments presented in this chapter have highlighted the pivotal role of Archean processes in shaping the compositional structure of Earth's continental crust. In view of this, and despite the spectacular mode of crustal growth witnessed in modern convergent margin and oceanic island volcanoes, it may be that for crustal studies we need to increasingly focus our attention on the distant past, as the key to understanding the present.

ACKNOWLEDGMENTS

The ideas presented in this manuscript have been focused and strengthened by the advice and thorough review by Roberta Rudnick, and we also appreciate her assistance with editorial matters. Helpful discussions with Simon Turner, Jon Blundy, and Bruce Chappell, who also kindly provided some unpublished LFB data, are acknowledged. Many of the figures were either drafted or resuscitated from earlier versions by Karen Alarcon. TK wishes to thank Richard Arculus for stimulating an interest in the development of the continental crust. This manuscript was prepared while TK held a NERC postdoctoral fellowship, which also supports our ongoing research into the origin of granitic rocks of southeastern Australia.

REFERENCES

Ablay G. J., Carroll M. R., Palmer M. R., Martí J., and Sparks R. S. J. (1998) Basanite-phonolite lineages of the Teide-Pico Viejo volcanic complex, Tenerife, Canary Islands. *J. Petrol.* **39**, 905–936.

Albarède F. (1998) The growth of continental crust. *Tectonophysics* **296**, 1–14.

Allegre C. J., Hart S. R., and Minster J. F. (1983) Chemical structure and evolution of the mantle and continents determined by inversion of Nd and Sr isotopic data: I. theoretical models. *Earth Planet Sci. Lett.* **66** 177-90.

Anders E. and Grevesse N. (1989) Abundances of the elements: meteoric and solar. *Geochim. Cosmochim. Acta* **53**, 197–214.

Anderson J.A.C. (1998) Structural, metamorphic, geochemical and isotopic study of metasedimentary enclaves from the southern Lachlan Fold Belt. PhD Thesis, La Trobe University, Melbourne (unpublished).

Anderson J. A. C., Williams I. S., Price R. C., and Fleming P. D. (1996) U–Pb zircon ages from the Koetong adamellite: implications for granite genesis and the local basement in NE Victoria. *Geol. Soc. Austral. Abstr.* **42**, 1–2.

Anderson J. A. C., Price R. C., and Fleming P. D. (1998) Structural analysis of metasedimentary enclaves: implications for tectonic evolution and granite petrogenesis in the southern Lachlan Fold Belt, Australia. *Geology* **26**, 119–122.

Anderson J. L. (1983) Proterozoic anorogenic granite plutonism of North America. *Geol. Soc. Am. Mem.* **161**, 133–154.

Anderson J. L. and Morrison J. (1992) The role of anorogenic granites in the Proterozoic crustal development of North America. In *Proterozoic Crustal Evolution* (ed. K. Condie). Elsevier, Amsterdam, The Netherlands, pp. 263–299.

Arculus R. J. (1999) Origins of the continental crust. *J. Proc. Roy. Soc. New South Wales* **132**, 83–110.

Armstrong R. L. (1991) The persistent myth of crustal growth. *Austral. J. Earth Sci.* **38**, 613–630.

Atherton M. P. and Petford N. (1993) Generation of sodium-rich magmas from newly underplated basaltic crust. *Nature* **362**, 144–146.

Ayres M., Harris N., and Vance D. (1997) Possible constraints on anatectic melt residence times from accessory mineral dissolution rates: an example from the Himalayan leucogranites. *Min. Mag.* **61**, 29–36.

Barbarin B. (1992) Genesis of the two main types of peraluminous granitoids. *Geology* **24**, 294–298.

Barbarin B. (1996) Genesis of the two main types of peraluminous granitoids. *Geology* **24**, 295–298.

Barbarin B. (1999) A review of the relationships between granitoid types, their origins and their geodynamic environments. *Lithos* **46**, 605–626.

Barth M., McDonough W. F., and Rudnick R. L. (2000) Tracking the budget of Nb and Ta in the continental crust. *Chem. Geol.* **165**, 197–213.

Bedard J. (1990) Enclaves from the A-type granite of the Megantic Complex. White Mountain magma series: clues to granite magma genesis. *J. Geophys. Res.* **95**, 17797–17819.

Bergantz G. W. and Dawes R. (1994) Aspects of magma generation and ascent in the continental lithosphere. In *Magmatic Systems* (ed. M. P. Ryan). Academic Press, San Diego, CA, pp. 291–317.

Bickle M. J. (1978) Heat loss from the Earth: constraint on Archean tectonics from the relationships between geothermal gradients and the rate of plate production. *Earth Planet. Sci. Lett.* **40**, 301–315.

Blundy J. D. and Sparks R. S. J. (1992) Petrogenesis of mafic inclusions in granitoids of the Adamello Massif, Italy. *J. Petrol.* **33**, 1039–1104.

Boher M., Abouchami W., Michard A., Albarede F., and Arndt N. T. (1992) Crustal growth in West Africa at 2.1 Ga. *J. Geophys. Res.* **97**, 345–369.

Bonin B. (1986) *Ring Complex Granites and Anorogenic Magmatism.* North Oxford Academic Publ., Oxford,188pp.

Bowen N. L. (1928) *The Evolution of the Igneous Rocks.* Princeton University Press, Princeton, 32pp.

Bowring S. A. and Housh T. (1995) The Earth's early evolution. *Nature* **269**, 1535–1540.

Bowring S. A. and Williams I. S. (1999) Priscoan (4.00–4.03) orthogneisses from northwestern Canada. *Contrib. Mineral. Petrol.* **134**, 3–16.

Brown G. C. (1986) Processes and problems in the continental lithosphere: geological history and physical implications. In *Geochronology and Geological, Record.* Geol. Soc., Memoir 10 (ed. N. J. Snelling). Geological Society, London, pp. 325.

Brown M. (1994) The generation, segregation, ascent and emplacement of granite magma: the migmatite-to-crustally-derived granite connection in thickened orogens. *Earth Sci. Rev.* **36**, 83–130.

Brown P. E., Dempster T. J., Harrison T. N., and Hutton D. H. W. (1992) The Rapakivi granites of S Greenland-crustal melting in response to extensional tectonics and magmatic underplating. *Trans. Roy. Soc. Edinburgh: Earth Sci.* **83**, 173–178.

Campbell I. H. and Griffiths R. W. (1992) The changing nature of mantle hotspots through time: implications for the chemical evolution of the mantle. *J. Geol.* **92**, 497–523.

Campbell I. H. and Taylor S. R. (1983) No water, no granites-no oceans, no continents. *Geophys. Res. Lett.* **10**, 1061–1064.

Carmichael I. S. E. (1991) The redox states of basic and silicic magmas: a reflection of their source regions? *Contrib. Mineral. Petrol.* **106**, 129–141.

Castelli D. and Lombardo B. (1988) The Gophu La and western Lunana granites: Miocene muscovite leucogranites of the Bhutan Himalaya. *Lithos* **21**, 211–225.

Castro A., Patiño Douce A. E., Corretge L. G., de la Rosa J., El-Biad M., and El Hmidi H. (1999) Origin of peraluminous granites and granodiorites, Iberian massif, Spain: an experimental test of granite petrogenesis. *Contrib. Mineral. Petrol.* **135**, 255–276.

Castro A., Guillermo C., El-Biad M., El-Hmidi H., Fernandez C., and Patiño Douce A. E. (2000) Experimental constraints on Hercynian anatexis in the Iberian Massif, Spain. *J. Petrol.* **41**, 1471–1488.

Chappell B. W. (1996a) Magma mixing and the production of compositional variation within granite suites: evidence from the granites of southeastern Australia. *J. Petrol.* **37**, 449–470.

Chappell B. W. (1996b) Compositional variation within granite suites of the Lachlan Fold Belt: its causes and implications for the physical state of granitic magma. *Trans. Roy. Soc. Edinburgh: Earth Sci.* **88**, 159–170.

Chappell B. W. (1999) Aluminium saturation in I- and S-type granites and the characterization of fractionated haplogranites. *Lithos* **46**, 535–551.

Chappell B. W. and Stephens W. E. (1988) Origin of infracrustal (I-type) granite magmas. *Trans. Roy. Soc. Edinburgh: Earth Sci.* **79**, 71–86.

Chappell B. W. and White A. J. R. (1974) Two contrasting granite types. *Pacific Geol.* **8**, 173–174.

Chappell B. W. and White A. J. R. (1992) I- and S-type granites in the Lachlan Fold Belt. *Trans. Roy. Soc. Edinburgh: Earth Sci.* **83**, 1–26.

Chappell B. W. and White A. J. R. (2001) Two contrasting granite types: 25 years later. *Austral. J. Earth Sci.* **46**, 827–831.

Chappell B. W., White A. J. R., and Wyborn D. (1987) The importance of residual source material (restite) in granite petrogenesis. *J. Petrol.* **28**, 1111–1138.

Chappell B. W., White A. J. R., and Williams I. S. (1991) A Transverse section through granites of the Lachlan Fold Belt: second Hutton symposium excursion guide. Bureau of Mineral Resources, Geology and Geophysics.

Chappell B. W., White A. J. R., Williams I. S., Wyborn D., Hergt J. M., and Woodhead J. D. (1999) Discussion: evaluation of petrogenetic models for Lachlan Fold Belt granitoids: implications for crustal architecture and tectonic models. *Austral. J. Earth Sci.* **46**, 827–831.

Chappell B. W., White A. J. R., Williams I. S., Wyborn D., and Wyborn L. A. I. (2000) Lachlan Fold Belt granites revisited: high- and low-temperature granites and their implications. *Austral. J. Earth Sci.* **47**, 123–138.

Chen Y. and Williams I. S. (1990) Zircon inheritance in mafic inclusions from Bega Batholith granites, southeastern Australia: an ion microprobe study. *J. Geophys. Res.* **95**, 17787–17796.

Chen Y., Price R. C., and White A. J. R. (1989) Inclusions in three S-type granites from southeastern Australia. *J. Petrol.* **30**, 1181–1218.

Chen Y., Price R. C., White A. J. R., and Chappell B. W. (1990) Mafic inclusions from the Glenbog and Blue gum granite suites, southeastern Australia. *J. Geophys. Res.* **95**, 17757–17785.

Chen Y. D., O'Reilly S., Griffin W. L., and Krogh T. E. (1998) Combined U–Pb dating and Sm–Nd studies on lower crustal and mantle xenoliths from the Delegate breccia pipes, southeastern Australia. *Contrib. Mineral. Petrol.* **130**, 154–161.

Christensen U. R. (1997) Influence of chemical buoyancy on the dynamics of slabs in the transition zone. *J. Geophys. Res.* **102**, 22435–22443.

Clarke C.B. (1990) The geochemistry of the Atlanta Lobe of the Idaho Batholith in the western United States Cordillera. PhD Thesis, Open University, 369pp (unpublished).

Clemens J. D. and Vielzeuf D. (1987) Constraints on melting and magma production in the crust. *Earth Planet. Sci. Lett.* **86**, 287–306.

Clemens J. D. and Wall V. J. (1981) Origin and crystallization of some peraluminous (S-type) granitic magmas. *Can. Mineral.* **19**, 111–131.

Clemens J. D., Holloway J. R., and White A. J. R. (1986) Origin of an A-type granite: experimental constraints. *Am. Mineral.* **71**, 317–324.

Collerson K. D. and Kamber B. S. (1999) Evolution of the continents and the atmosphere infrared from Th–U–Nb systematics of depleted mantle. *Science* **283**, 1519–1522.

Collins W. J. (1996) Lachlan Fold Belt granitoids: products of three component mixing. *Trans. Roy. Soc. Edinburgh: Earth Sci.* **88**, 171–181.

Collins W. J. (1998) Evaluation of petrogenetic models for Lachlan Fold Belt granitoids: implications for crustal architecture and tectonic models. *Austral. J. Earth Sci.* **45**, 483–500.

Collins W. J. (1999) Evaluation of petrogenetic models for Lachlan Fold Belt granitoids: implications for crustal architecture and tectonic models—Reply. *Austral. J. Earth Sci.* **46**, 831–836.

Collins W. J., Beams S. D., White A. J. R., and Chappell B. W. (1982) Nature and origin of A-type granites with particular reference to southeastern Australia. *Contrib. Mineral. Petrol.* **80**, 189–200.

Condie K. C. (1993) Chemical composition and evolution of the upper continental crust: contrasting results from surface samples and shales. *Chem. Geol.* **104**, 1–37.

Condie K. C. (1994) Greenstones through time. In *Archean Crustal Evolution* (ed. K. C. Condie). Elsevier, Netherlands, pp. 85–120.

Condie K. C. (1998) Episodic continental growth and supercontinents: a mantle avalanche connection? *Earth Planet. Sci. Lett.* **163**, 97–108.

Condie K. C. (1999) Mafic crustal xenoliths and the origin of the lower continental crust. *Lithos* **46**, 95–101.

Condie K. C. (2000) Episodic continental growth models: afterthoughts and extensions. *Tectonophysics* **322**, 153–162.

Crawford M. B. and Windley B. F. (1990) Leucogranites of the Himalaya/Karakoram: implications for magmatic evolution within collisional belts and the study of collision-related leucogranite petrogenesis. *J. Volcanol. Geotherm. Res.* **44**, 1–19.

Creaser R. A. and White A. J. R. (1991) Yardea Dacite-large volume, high temperature felsic volcanism from the Middle Proterozoic of South Australia. *Geology* **19**, 48–51.

Creaser R. A., Price R., and Wormald R. J. (1991) A-type granites revisited: an assessment of the residual-source model. *Geology* **19**, 163–166.

Daly R. A. (1914) *Igneous Rocks and their Origin*. McGraw-Hill, London.

Dana J. D. (1873) On some results of the earth's contraction from cooling: Part V. Formation of the continental plateaus and oceanic depressions. *Am. J. Sci.* **6** 161–72.

Davidson J. P. and Arculus R. J. (2001) The significance of Phanerozoic arc magmatism in generating continental crust. In *Evolution and Differentiation of the Continental Crust* (eds. M. Brown and T. Rushmer). Cambridge University Press, Cambridge.

Davidson J. P., Harmon R. S., and Worner G. (1991) The source of central Andean magmas: some considerations. In *Andean Magmatism and its Tectonic Setting*. Spec. Pap. 265 (eds. R. S. Harmon and C. W. Rapella). Geological Society of America, Boulder, pp. 233–244.

Davies G. F. (1995) Punctuated tectonic evolution of the Earth. *Earth Planet. Sci. Lett.* **136**, 363–379.

Davies G. R., Gledhill A., and Hawkesworth C. J. (1985) Upper crustal recycling in southern Britain: evidence from Nd and Sr isotopes. *Earth Planet. Sci. Lett.* **75**, 1–12.

Defant M. J. and Drummond M. S. (1990) Derivation of some modern arc magmas by melting of young subducted lithosphere. *Nature* **347**, 662–665.

Deniel C., Vidal P., Fernandez A., Le Fort A., and Peucat J.-J. (1987) Isotopic study of the Manaslu granite (Himalaya Nepal): inferences on the age and source of Himalayan leucogranites. *Contrib. Mineral. Petrol.* **96**, 78–92.

Di Vincenzo G., Andriessen P. A. M., and Ghezzo C. (1996) Evidence of two different components in a Hercynian peraluminous cordierite-bearing granite: the San Basilio intrusion (central Sardinia, Italy). *J. Petrol.* **37**, 1175–1206.

Drake M. J. (1974) The oxidation state of europium as an indicator of oxygen fugacity. *Geochim. Cosmochim. Acta* **39**, 55–64.

Drake M. J. and Weill D. F. (1975) Partition of Sr, Ba, Ca, Y, Eu^{2+}, Eu^{3+}, and other REE between plagioclase feldspar and magmatic liquid: an experimental study. *Geochim. Cosmochim. Acta* **39**, 689–712.

Drummond M. S. and Defant M. J. (1990) A model for trondhjemite-tonalite-dacite genesis and crustal growth via slab melting: archaean to modern comparisons. *J. Geophys. Res.* **95**, 21503–21521.

Drummond M. S., Defant M. J., and Kepezhinskas P. K. (1996) Petrogenesis of slab-derived trondhjemite-tonalite-dacite/adakite magmas. *Trans. Roy. Soc. Edinburgh: Earth Sci.* **87**, 205–215.

Duncan A. R., Erlank A. J., and Marsh J. S. (1984) Regional geochemistry of the Karoo igneous province. *Spec. Publ. Geol. Soc. S. Afr.* **13**, 355–388.

Elburg M. A. (1996a) U–Pb ages and morphologies of zircon in microgranitoid enclaves and peraluminous host granite: evidence for magma mingling. *Contrib. Mineral. Petrol.* **123**, 177–189.

Elburg M. A. (1996b) Genetic significance of multiple enclave types in a peraluminous ignimbrite suite, Lachlan Fold Belt, Australia. *J. Petrol.* **37**, 1385–1408.

Elburg M. A. (1996c) Evidence of isotopic equilibration between microgranitoid enclaves and host granodiorite, Warburton Granodiorite, Lachlan Fold Belt, Australia. *Lithos* **38**, 1–22.

Elburg M. and Nicholls I. A. (1995) The origin of microgranitoid enclaves in the S-type Wilson's Promontory Batholith, Victoria: evidence for magma mingling. *Austral. J. Earth Sci.* **42**, 423–435.

Ellam R. M. and Hawkesworth C. J. (1988) Is average continental crust generated at subduction zones? *Geology* **16**, 314–317.

Ellam R. M., Hawkesworth C. J., Menzies M. A., and Rogers N. W. (1989) The volcanism of southern Italy: the role of subduction and the relationship between potassic and sodic alkaline magmatism. *J. Geophys. Res.* **94**(B4), 4589–4601.

Ellam R. M., Hawkesworth C. J., and McDermott F. (1990) Pb isotope data from Late Proterozoic subduction-related rocks: implications for crust–mantle evolution. *Chem Geol.* **83**, 165–181.

Elliott T., Zindler A., and Bourdon B. (1999) Exploring the kappa conundrum: the role of recycling in the lead isotope evolution of the mantle. *Earth Planet. Sci. Lett.* **169**, 129–145.

England P. C. and Thompson A. B. (1984) Pressure–temperature–time paths of regional metamorphism: I: Heat transfer during the evolution of regions of thickened crust. *J. Petrol.* **25**, 894–928.

England P. C. and Thompson A. B. (1986) Crustal melting in continental collision zones. In *Collision Tectonics*. Geol. Soc. Spec. Publ. No. 19 (eds. M. P. Coward and A. C. Ries) Blackwell Scientific, Oxford, pp. 83–94.

Farquhar J., Bao H., and Thiemens M. (2000) Atmospheric influence of the Earth's earliest sulfur cycle. *Science* **289**, 756–758.

Faure G. (1990) *Principles of Isotope Geology*. Wiley, Singapore.

Feng R. and Kerrich R. (1992) Geochemical evolution of granitoids from the Archean Albitibi southern volcanic zone and Pontiac Subprovince, Quebec, Canada. *Geochim. Cosmochim. Acta* **55**, 3437–3441.

Fleming P. D. (1996) Inherited deformation structures in metasedimentary enclaves in granites as 'windows' into deeper levels of the crust. *Tectonophysics* **267**, 177–185.

Flood R. H. and Vernon R. H. (1978) The Cooma Granodiorite, Australia: an example of *in situ* crustal anatexis? *Geology* **6**, 81–84.

Foden J. D., Turner S. P., and Morrison R. S. (1990) Tectonic implications of Delamerian magmatism in South Australia and western Victoria. *Geol. Soc. Austral. Spec. Publ.* **16**, 465–482.

Foland K. A. and Allen J. C. (1991) Magma sources for Mesozoic anorogenic granites of the White Mountain magma series, New England, USA. *Contrib. Mineral. Petrol.* **109**, 195–211.

Foley S., Tiepolo M., and Vannucci R. (2002) Growth of early continental crust in subduction zones controlled by melting of amphibolite. *Nature* **417**, 837–840.

Fourcade S., Martin H., and De Bremond d'Ars J. (1992) Chemical exchange in migmatites during cooling. *Lithos* **28**, 45–53.

France-Lanord C. and Le Fort P. (1988) Crustal melting and granite genesis during the Himalayan collision orogenesis. *Trans. Roy. Soc. Edinburgh: Earth Sci.* **79**, 183–195.

Frost C. D. and Frost B. R. (1997) Reduced Rapakivi-type granites: the tholeiite connection. *Geology* **25**, 647–650.

Frost B. R. and Lindsley D. H. (1991) Occurrence of iron-titanium oxides in igneous rocks. *Mineral. Soc. Am. Rev. Mineral.* **25**, 433–468.

Furman T., Frey F. A., and Meyer P. S. (1992) Petrogenesis of evolved basalts and rhyolites at Austurhorn, SE Iceland: the role of fractional crystallization. *J. Petrol.* **33**, 1405–1445.

Galer S. J. G. and Mezger K. (1998) Metamorphism, denudation and sea level in the Archean and cooling of the earth. *Precamb. Res.* **92**, 389–412.

Garland F., Hawkesworth C. J. H., and Mantovani M. S. N. (1995) Description and petrogenesis of the Paraná rhyolites, Southern Brazil. *J. Petrol.* **36**, 1193–1227.

Gibbs A. K., Montgomery D. W., O'Day P. A., and Ersler E. A. (1986) The Archean–Proterozoic transition: evidence from the geochemistry of metasedimentary rocks of Guyana and Montana. *Geochim. Cosmochim. Acta* **50**, 2125–2141.

Giret A. (1990) Typology, evolution and origin of the Kerguelen Plutonic series, Indian Ocean: a review. *Geol. J.* **25**, 239–247.

Goodwin A. M. (1991) *Precambrian Geology*. Academic Press, London, 666p.

Gray C. M. (1984) An isotopic mixing model for the origin of granitic rocks in southeastern Australia. *Earth Planet. Sci. Lett.* **70**, 47–60.

Gray C. M. (1990) A strontium isotope traverse across the granitic rocks of southeastern Australia: petrogenetic and tectonic implications. *Austral. J. Earth Sci.* **37**, 331–349.

Gray D. R. (1997) Tectonics of the southeastern Australian Lachlan Fold Belt: structural and thermal aspects. In *Orogeny Through Time*. Geol. Soc. Spec. Publ. No. 121 (eds. J.-P. Burg and M. Ford) pp. 149–177.

Gray D. R. and Foster D. A. (1997) Orogenic concepts-application and definition: Lachlan Fold Belt, eastern Australia. *Am. J. Sci.* **297**, 859–891.

Griffin T. J., White A. J. R., and Chappell B. W. (1978) The Moruya Batholith and geochemical contrasts between the Moruya and Jindabyne suites. *J. Geol. Soc. Austral.* **25**, 235–247.

Griffin W. L., Wang X., Jackson S. E., Pearson N. J., O'Reilly S. Y., Xu X., and Zhou X. (2002) Zircon chemistry and magma mixing, SE China: *in situ* analysis of Hf isotopes, Tonglu and Pingtan igneous complexes. *Lithos* **61**, 237–269.

Guillot S. and Le Fort P. (1995) Geochemical constraints on the bimodal origin of the high Himalayan leucogranites. *Lithos* **35**, 221–234.

Gurnis M. and Davies G. F. (1986) Apparent episodic crustal growth arising from a smoothly evolving mantle. *Geology* **14**, 396–399.

Haapala H. and Ramo O. T. (1992) Tectonic setting and origin of the Proterozoic Rapakivi granites of southeastern Fennoscandia. *Trans. Roy. Soc. Edinburgh: Earth Sci.* **83**, 165–171.

Han B. F., Wang S. G., Jahn B. M., Hong D. W., Kagami H., and Sun Y. (1997) Depleted mantle magma source for the Ulungur River A-type granites from north Xinjiang, China: geochemistry and Sr–Nd isotopic evidence, and implications for Phanerozoic crustal growth. *Chem. Geol.* **138**, 135–159.

Harris C. (1983) The petrology of lavas and associated plutonic inclusions of Ascension Island. *J. Petrol.* **24**, 424.

Harris N. B. W. and Inger S. (1992) Trace element modelling of pelite-derived granites. *Contrib. Mineral. Petrol.* **110**, 46–56.

Harris N. B. W. and Massey J. A. (1994) Decompression and anatexis of Himalayan metapelites. *Tectonics* **13**, 1537–1546.

Harris N., Ayres M., and Massey J. (1995) Geochemistry of granitic melts produced during the incongruent melting of muscovite: implications for the extraction of Himalayan leucogranite magmas. *J. Geophys. Res.* **100**, 15767–15777.

Harris N., Vance D., and Ayres M. (2000) From sediment to granite: timescales of anatexis in the upper crust. *Chem. Geol.* **162**, 155–167.

Harrison T. M., Lovera O. M., and Grove M. (1997) New insights into the origin of two contrasting Himalayan granite belts. *Geology* **25**, 899–902.

Harrison T. M., Grove M., McKeegan K. D., Coath C. D., Lovera O. M., and Le Fort P. (1999) Origin and episodic emplacement of the Manaslu Intrusive Complex, central Himalaya. *J. Petrol.* **40**, 3–19.

Hawkesworth C. J. and Marlow A. (1983) Isotope evolution of the Damara Orogenic Belt. *In Spec. Publ. Geol. Soc. S. Afr.* **11**, 397–407.

Hawkesworth C. J., Gallagher K., Hergt J. M., and McDermott F. (1993) Mantle and slab contributions in arc magmas. *Ann. Revs. Earth Planet. Sci.* **21**, 175–204.

Hawkesworth C. J., Turner S., Peate D., McDermott F., and van Calsteren P. (1997) Elemental U and Th variations in island arc rocks: implications for U-series isotopes. *Chem. Geol.* **139**, 207–222.

Hildreth W., Halliday A. N., and Christiansen S. L. (1991) Isotopic and chemical evidence concerning the genesis and contamination of basaltic and rhyolitic magma beneath the Yellowstone Plateau volcanic field. *J. Petrol.* **32**, 63–138.

Hill R. I. (1993) Mantle plumes and continental tectonics. *Lithos* **30**, 193–206.

Hine R., Williams I. S., Chappell B. W., and White A. J. R. (1978) Contrasts between I- and S-type granitoids of the Kosciuszko Batholith. *J. Geol. Soc. Austral.* **25**, 219–234.

Hofmann A. W., Jochum K. P., Seufert M., and White W. M. (1986) Nb and Pb in oceanic basalts: new constraints on mantle evolution. *Earth Planet. Sci. Lett.* **79**, 33–45.

Huppert H. E. and Sparks R. S. J. (1988) The generation of granitic magmas by intrusion of basalt into continental crust. *J. Petrol.* **29**, 599–624.

Inger S. and Harris N. (1993) Geochemical constraints on leucogranite magmatism in the Langtang valley, Nepal Himalaya. *J. Petrol.* **34**, 345–368.

Ishihara S. (1979) Lateral variation of magnetic susceptibility of the Japanese granitoids. *J. Geol. Soc. Japan* **85**, 509–523.

Javoy M. and Weiss D. (1987) Oxygen isotopic composition of alkaline anorogenic granites as a clue to their origin: the problem of crustal oxygen. *Earth Planet. Sci. Lett.* **84**, 415–422.

Jahn B.-M., Wu F., and Chen B. (2000) Massive granitoid generation in central Asia: Nd isotopic evidence and implication for continental growth in the Phanerozoic. *Episodes* **23**(2), 82–92.

Johannes W. and Holtz F. (1996) *Petrogenesis and Experimental Petrology of Granitic Rocks*. Springer, Berlin.

Johannes W., Holtz F., and Moller P. (1995) REE distribution in some layered migmatites: constraints on their petrogenesis. *Lithos* **35**, 139–152.

Jonasson K., Holm P. M., and Pedersen A. K. (1992) Petrogenesis of silicic rocks from the Kroksfjordur central volcano, NW Iceland. *J. Petrol.* **33**, 1345–1369.

Jull M. and Kelemen P. B. (2001) On the conditions for lower crustal convective instability. *J. Geophys. Res.* **106**, 6423–6446.

Kamber B. S., Ewart A., Collerson K. D., Bruce M. C., and McDonald G. D. (2002) Fluid-mobile trace element constraints on the role of slab melting and implications for Archean crustal growth models. *Contrib. Mineral. Petrol.* **144**, 38–56.

Keay S., Collins W. J., and McCulloch M. T. (1997) A three component Sr–Nd isotopic mixing model for granitoid genesis, Lachlan Fold Belt, eastern Australia. *Geology* **25**, 307–310.

Keay S., Steele D., and Compston W. (1999) Identifying granite sources by SHRIMP U–Pb zircon geochronology: an application to the Lachlan Fold Belt. *Contrib. Mineral. Petrol.* **137**, 323–341.

Kelemen P. B. (1995) Genesis of high Mg# andesites and the continental crust. *Contrib. Mineral. Petrol.* **120**, 1–19.

Kelemen P. B., Yogodzinski G. M., and Scholl D. W. (2003) Along strike variation in lavas of the Aleutian island arc: Implications for the genesis of high Mg# andesite and the continental crust, Chap. 11. In *AGU Monograph*. American Geophysical Union (ed. J. Eiler). American Geophysical Union, (in press).

Kemp A.I.S. (2001) The petrogenesis of granitic rocks: a source-based perspective. PhD Thesis, The Australian National University, Canberra (unpublished).

Kemp A. I. S. (2003) Plutonic boninites in an anatectic setting: tectonic implications for the Delamerian Orogen in southeastern Australia. *Geology* **31**, 371–374.

Kemp A. I. S. and Gray C. M. (1999) Geological context of crustal anatexis and granitic magmatism in the northeastern Glenelg River Complex, western Victoria. *Austral. J. Earth Sci.* **46**, 406–420.

Kemp A. I. S., Gray C. M., Anderson J. A. C., and Ferguson D. J. (2002) The Delamerian Glenelg Tectonic Zone, western Victoria: characterization and synthesis of igneous rocks. *Austral. J. Earth Sci.* **49**, 201–224.

Kerr A. and Fryer B. J. (1993) Nd isotope evidence for crust-mantle interaction in the generation of A-type granitoid suites in Labrador, Canada. *Chem. Geol.* **104**, 39–60.

King P. L., White A. J. R., and Chappell B. W. (1997) Characterization and origin of aluminous A-type granites of the Lachlan Fold Belt, southeastern Australia. *J. Petrol.* **36**, 371–391.

King P. L., Chappell B. W., Allen C. M., and White A. J. R. (2001) Are A-type granites the high-temperature felsic granites? evidence from fractionated granites of the Wangrah Suite. *Austral. J. Earth Sci.* **48**, 501–514.

Kinnaird J. A., Bowden P., and Odling N. W. A. (1985) Mineralogy, geochemistry and mineralization of the Ririwai complex, northern Nigeria. *J. African Earth Sci.* **3**, 185–222.

Kramers J. D. and Tolstikhin I. N. (1997) Two major terrestrial Pb isotope paradoxes, forward transport modeling, core formation and the history of the continental crust. *Chem. Geol.* **139**, 75–110.

Krogstad E. J. and Walker R. J. (1996) Evidence of heterogeneous crustal sources: the Harney Peak granite, South Dakota, USA. *Geochim. Cosmochim. Acta* **57**, 4667–4685.

Kröner A. (1985) Evolution of the Archean continental crust. *Ann. Rev. Earth Planet. Sci.* **13**, 49–74.

Landenberger B. and Collins W. J. (1996) Derivation of A-type granites from a dehydrated charnockitic lower crust: evidence from the Chaelundi Complex, eastern Australia. *J. Petrol.* **37**, 145–170.

Le Fort P. (1981) Manaslu leucogranite: a collision signature of the Himalaya, a model for its genesis and emplacement. *J. Geophys. Res.* **86**, 10545–10568.

Le Fort P., Cuney M., Deniel C., France-Lanord F., Sheppard S. M. F., Upreti B. N., and Vidal P. (1987) Crustal generation of the Himalayan leucogranites. *Tectonophysics* **134**, 39–57.

Le Maitre R. W. (1989) *A Classification of the Igneous Rocks and Glossary of Terms*. Blackwell, Oxford.

Loiselle M. C. and Wones D. R. (1979) Characterization and origin of anorogenic granites. *Geol. Soc. Am. Abstr.* **11**, 468.

Luais B. and Hawkesworth C. J. (1994) The generation of continental crust: an integrated study of crust-forming processes in the Archean of Zimbabwe. *J. Petrol.* **35**, 43–93.

Maas R., Nicholls I. A., and Legg C. (1997) Igneous and metamorphic enclaves in the S-type Deddick Granodiorite, Lachlan Fold Belt, SE Australia: petrographic, geochemical and Nd–Sr isotopic evidence for crustal melting and magma mixing. *J. Petrol.* **38**, 815–841.

Martin H. (1986) Effect of steeper Archean geothermal gradient on geochemistry of subduction-zone magmas. *Geology* **14**, 753–756.

Martin H. (1995) Archean grey gneisses and the genesis of continental crust. In *Archean Crustal Evolution* (ed. K. C. Condie). Elsevier, Netherlands, pp. 205–260.

Martin H. (1999) Adakitic magmas: modern analogues of Archean granitoids. *Lithos* **46**, 411–429.

Martin H. and Moyen J.-F. (2002) Secular changes in tonalite-trondhjemite-granodiorite compositions as markers of the progressive cooling of the Earth. *Geology* **30**, 319–322.

McCulloch M. T. and Bennett V. C. (1994) Progressive growth of the Earth's continental crust and depleted mantle: geochemical constraints. *Geochim. Cosmochim. Acta* **58**, 4717–4738.

McCulloch M. T. and Chappell B. W. (1982) Nd isotopic characteristics of S- and I-type granites. *Earth Planet. Sci. Lett.* **58**, 51–64.

McCulloch M. T. and Woodhead J. D. (1993) Lead isotopic evidence for deep crustal-scale fluid transport during granite petrogenesis. *Geochim. Cosmochim. Acta* **57**, 1837–1856.

McDermott F. and Hawkesworth C. J. (1990) Intracrustal recycling and upper-crustal evolution: a case study from the Pan-African Damara Mobile Belt, central Namibia. *Chem. Geol.* **83**, 263–280.

McDermott F. and Hawkesworth C. J. (1991) Th, Pb, and Sr isotope variations in young island arc volcanics and oceanic sediments. *Earth Planet. Sci. Lett.* **104**, 1–15.

McDermott F., Harris N. B. W., and Hawkesworth C. J. (1989) Crustal reworking in southern Africa: constraints from Sr–Nd isotope studies in Archean to Pan-African terranes. *Tectonophysics* **161**, 257–270.

McDermott F., Harris N. B. W., and Hawkesworth C. J. (1996) Geochemical constraints on crustal anatexis; a case study from the Pan-African Damara granitoids of Namibia. *Contrib. Mineral. Petrol.* **123**, 406–423.

McDermott F., Harris N. B. W., and Hawkesworth C. J. (2000) Geochemical constraints on the petrogenesis of Pan-African A-type granites in the Damara Belt, Namibia. *Commun. Geol. Surv. Namibia* **12**, 139–148.

MacDonald R., Hawkesworth C. J., and Heath E. (2000) The Lesser Antilles volcanic chain: a study in arc magmatism. *Earth Sci. Rev.* **49**, 1–76.

MacDonald R., Davies G. R., Bliss C. M., Leat P. T., Bailey D. K., and Smith R. L. (1987) Geochemistry of high-silica peralkaline rhyolites, Naivasha, Kenya rift valley. *J. Petrol.* **28**, 979–1008.

McDonough W. F. and Sun S.-S. (1995) Composition of the Earth. *Chem. Geol.* **120**, 223–253.

McLennan S. M. and Taylor S. R. (1980) Th and U in sedimentary rocks: crustal evolution and sedimentary recycling. *Nature* **285**, 621–624.

Milord I., Sawyer E. W., and Brown M. (2001) Formation of diatexite migmatite and granite magma during anatexis of semi-pelitic metasedimentary rocks: an example from St Malo, France. *J. Petrol.* **42**, 487–505.

Montel J.-M. (1993) A model for monazite/melt equilibrium and application to the generation of granitic magmas. *Chem. Geol.* **110**, 127–146.

Montel J.-M. and Vielzeuf D. (1997) Partial melting of metagreywackes Part II. Compositions of minerals and melts. *Contrib. Mineral. Petrol.* **128**, 176–196.

Muir R. J., Weaver S. D., Bradshaw J. D., Eby G. N., and Evans J. A. (1998) The Cretaceous separation point batholith, New Zealand: granitoid magmas formed by melting of mafic lithosphere. *J. Geol. Soc. London* **152**, 689–701.

Munksgaard N. C. (1988) Source of the cooma granodiorite, New South Wales—a possible role of fluid–rock interactions. *Austral. J. Earth Sci.* **35**, 363–377.

Nabelek P. I. (1999) Trace element distribution among rock-forming minerals in Black Hills migmatites, South Dakota: a case for solid-state equilibrium. *Am. Mineral.* **84**, 1256–1269.

Nabelek P. I. and Liu M. (1999) Leucogranites in the Black Hills of South Dakota: the consequence of shear heating during continental collision. *Geology* **27**, 523–526.

Nui F. L. and James D.E. (2002) Fine structure of the lowermost crust beneath the Kaapvaal craton and its implications for crustal formation and evolution. *Earth Planet. Sci. Lett.* **200**, 121–130.

O'Neil J. R. and Chappell B. W. (1977) Oxygen and hydrogen isotope relations in the Berridale Batholith. *J. Geol. Soc. London* **133**, 559–571.

O'Nions R. K., Evensen N. M., and Hamilton P. J. (1979) Geochemical modelling of mantle differentiation and crustal growth. *JGR* **84**, 6091–6101.

Paterson B. A., Stephens W. E., Rogers G., Williams I. S., Hinton R. W., and Herd D. A. (1992) The nature of zircon inheritance in two granite plutons. *Trans. Roy. Soc. Edinburgh: Earth Sci.* **83**, 459–471.

Patiño Douce A. E. (1997) Generation of metaluminous A-type granites by low pressure melting of calc-alkaline granitoids. *Geology* **25**, 743–746.

Patiño Douce A. E. (1999) What do experiments tell us about the relative contribution of crust and mantle to the origin of granite magmas? In *Understanding Granites: Integrating New and Classical Techniques*. Spec. Publ., 168 (eds. A. Castro, C. Fernandez, and J. L. Vigneresse). Geological Society, London, pp. 55–76.

Patiño Douce A. E. and Johnston A. D. (1991) Phase equilibria and melt productivity in the pelitic system: implications for the origin of peraluminous granitoids and aluminous granulites. *Contrib. Mineral. Petrol.* **107**, 202–218.

Patiño Douce A. E. and Harris N. (1998) Experimental constraints on Himalayan anatexis. *J. Petrol.* **39**, 689–710.

Peacock S. M., Rushmer T., and Thompson A. B. (1994) Partial melting of subducting oceanic crust. *Earth Planet. Sci. Lett.* **121** 277-244.

Pearce J. A. and Peate D. W. (1995) Tectonic implications of the composition of volcanic arc magmas. *Ann. Rev. Earth Planet. Sci.* **23**, 251–285.

Pearce J. A., Baker P. E., Harvey P. K., and Luff I. W. (1995) Geochemical evidence for subduction fluxes, mantle melting and fractional crystallization beneath the South Sandwich island arc. *J. Petrol.* **36**, 1073–1109.

Peate D. W. (1997) The Parana-Etendeka province. In Large Igneous Provinces: Continental, Oceanic and Planetary Flood Volcanism. (eds. J. Mahoney and M. F. Coffin). *Amer. Geophys. Monogr. Ser. 100* Washington, DC, pp. 217–246.

Pedersen R. B. and Malpas J. (1984) The origin of oceanic plagiogranites from the Karmoy ophiolite, western Norway. *Contrib. Mineral. Petrol.* **88**, 36–52.

Percival J. A. (1989) A regional perspective of the Quetico Metasedimentary Belt, Superior Province, Canada. *Can. J. Earth Sci.* **26**, 677–693.

Perfit M. R., Brueckner H., Lawrence J. R., and Kay R. W. (1980) Trace element and isotopic variations in a zoned pluton and associated rocks, Unalaska Island, Alaska: a model for fractionation in the Aleutian calc-alkaline suite. *Contrib. Mineral. Petrol.* **73**, 69–87.

Petford N. and Atherton M. (1996) Na-rich partial melts from newly underplated basaltic crust: the Cordillera Blanca Batholith. *Peru. J. Petrol.* **37**, 1491–1521.

Petford N. and Gallagher K. (2001) Partial melting of mafic (amphibolitic) lower crust by periodic influx of basaltic magma. *Earth Planet. Sci. Lett.* **193**, 483–499.

Pichavant M., Montel J.-M., and Richard L. R. (1992) Apatite solubility in peraluminous liquids: experimental data and an extension of the Harrison–Watson model. *Geochim. Cosmochim. Acta* **56**, 3855–3861.

Pidgeon R. T. and Compston W. (1965) The age and origin of the Cooma Granite and its associated metamorphic zones, New South Wales. *J. Petrol.* **6**, 193–222.

Plank T. and Langmuir C. H. (1998) The geochemical composition of subducting sediment and its consequences for the crust and mantle. *Chem. Geol.* **145**, 325–394.

Poitrasson F., Duthou J.-L., and Pin C. (1995) The relationship between petrology and Nd isotopes as evidence for contrasting anorogenic granite genesis: example of the Corsican Province (SE France). *J. Petrol.* **36**, 1251–1274.

Pressley R. A. and Brown M. (1999) The Phillips pluton, Maine, USA: evidence of heterogeneous crustal sources and implications for granite ascent and emplacement mechanisms in convergent orogens. *Lithos* **46**, 335–366.

Rapp R. P. (1997) In *Greenstone Belts* (eds. M. J. de Wit and L. D. Ashwal). Clarendon Press, Oxford, pp. 267–279.

Rapp R. P., Shimizu N., Norman M. D., and Applegate G. S. (1999) Reaction between slab-derived melts and peridotite in the mantle wedge: experimental constraints at 3.8 GPa. *Chem. Geol.* **160**, 335–356.

Rollinson H. R. (1999) Petrology and geochemistry of metamorphosed komatiites and basalts from the Sula Mountains Greenstone Belt, Sierra Leone. *Contrib. Mineral. Petrol.* **134**, 86–101.

Rudnick R. L. (1995) Making continental crust. *Nature* **378**, 573–578.

Sandiford M., McLaren S., and Neuman N. (2002) Long term thermal consequences of the redistribution of heat-producing elements associated with large-scale granitic complexes. *J. Metamorph. Geol.* **20**, 87–98.

Saunders A. D., Tarney J., Kerr A. C., and Kent R. W. (1996) The formation and fate of large oceanic igneous provinces. *Lithos* **37**, 81–95.

Sawyer E. W. (1987) The role of partial melting and fractional crystallization in determining discordant migmatite leucosome compositions. *J. Petrol.* **28**, 445–473.

Sawyer E. W. (1991) Disequilibrium melting and the rate of melt-residue separation during migmatization of mafic rocks from the Grenville Front, Quebec. *J. Petrol.* **32**, 701–738.

Sawyer E. W. (1998) Formation and evolution of granite magmas during crustal reworking: the significance of diatexites. *J. Petrol.* **39**, 1147–1167.

Scaillet B., France Lanor C., and Le Fort P. (1990) Badrinath–Gangotri plutons (Garhwal, India): petrological and geochemical evidence for fractionation processes in a high Himalayan leucogranite. *J. Volcanol. Geotherm. Res.* **44**, 163–188.

Scaillet B., Pichavant M., and Roux J. (1995) Experimental crystallization of leucogranite magmas. *J. Petrol.* **36**, 663–705.

Schmitt A. K., Emmermann R., Trumbull R. B., Buhn B., and Henjes-Kunst F. (2000) Petrogenesis and $^{40}Ar/^{39}Ar$ geochronology of the Brandberg Complex. Namibia: evidence for a major mantle contribution in metaluminous and peralkaline granites. *J. Petrol.* **41**, 1207–1239.

Searle M. P. and Fryer B. J. (1986) Garnet, tourmaline and muscovite-bearing leucogranites, gneisses, and migmatites of the higher Himalaya from Zanskar, Kulu, Lahoul, and Kashmir. In *Collision Tectonics*. Geol. Soc. Spec. Publ. No. 19 (eds. M. P. Coward and A. C. Ries). Blackwell Scientific Oxford, pp. 83–94.

Searle M. P., Parrish R. R., Hodges K. V., Hurford A., and Ayres M. W. (1997) Shisha Pangma leucogranite, south Tibetan Himalaya: field relations, geochemistry, age, origin, and emplacement. *J. Geol.* **105**, 295–317.

Skjerlie K. P. and Johnston A. D. (1993) Fluid-absent melting behaviour of an F-rich tonalitic gneiss at mid-crustal pressures: implications for the origin of anorogenic granites. *J. Petrol.* **34**, 785–815.

Shand S. J. (1927) *Eruptive Rocks*. Thomas Murby and Co., London.

Silver L. T. and Chappell B. W. (1988) The Peninsular Ranges Batholith: an insight into the evolution of the Cordilleran batholiths of southwestern North America. *Trans. Roy. Soc. Edinburgh: Earth Sci.* **79**, 105–121.

Smithies R. H. (2000) The Archean tonalite–trondhjemite–granodiorite (TTG) series is not an analogue of Cenozoic adakite. *Earth Planet. Sci. Lett.* **182**, 115–125.

Soesoo A. (2000) Fractional crystallization of mantle-derived melts as a mechanism for some I-type granite petrogenesis: an example from the Lachlan Fold Belt, Australia. *J. Geol. Soc. London* **157**, 135–149.

Soesoo A. and Nicholls I. (1999) Mafic rocks spatially associated with Devonian felsic intrusions of the southern Lachlan Fold Belt: a possible mantle contribution to crustal evolution processes. *Austral. J. Earth Sci.* **46**, 725–734.

Solar G. S. and Brown M. (2001) Petrogenesis of migmatites in Maine, USA: possible source of peraluminous leucogranite in plutons? *J. Petrol.* **42**, 789–823.

Sørensen S. S. (1988) Petrology of amphibolite-facies mafic and ultramafic rocks from the Catalina Schist, southern California: metasomatism and migmatization in a subduction zone metamorphic setting. *J. Metamorph. Geol.* **6**, 405–435.

Stein M. and Hofmann A. W. (1994) Mantle plumes and episodic crustal growth. *Nature* **372**, 63–68.

Stern C. R. and Kilian R. (1996) Role of the subducted slab, mantle wedge and continental crust in the generation of adakites from the Andean Austral Volcanic Zone. *Contrib. Mineral. Petrol.* **123**, 263–281.

Strong B. F. and Hanmer S. K. (1981) The leucogranites of southern Brittany: origin by faulting, frictional heating, fluid flux and fractional melting. *Can. Mineral.* **19**, 163–176.

Stump E., White A.J.R., and Borg S.G. (1986) Reconstruction of Australia and Antarctica: evidence from granites and recent mapping. *Earth Planet. Sci. Lett.* **79**, 348–360.

Sun S.-S. and McDonough W. F. (1989) Chemical and isotopic systematics of oceanic basalts: implications for mantle composition and processes. In *Magmatism in the Ocean Basins*. Geol. Soc. Spec. Publ. 42 (eds. A. D. Saunders and M. J. Norry) Blackwell Scientific, Oxford, pp. 313–345.

Sylvester P. J. (1995) Archean granite plutons. In *Archean Crustal Evolution* (ed. K. C. Condie). Elsevier, Netherlands, pp. 261–314.

Taylor S. R. and McLennan S. M. (1985) *The Continental Crust: its Composition and Evolution*. Blackwell, Malden, Mass.

Taylor S. R. and McLennan S. M. (1995) The geochemical evolution of the continental crust. *Rev. Geophys.* **33**, 241–265.

Tepper J. H., Nelson B. K., Bergantz G. W., and Irving A. J. (1993) Petrology of the Chilliwack batholith, North Cascades, Washington: generation of calc-alkaline granitoids by melting of mafic lower crust with variable water fugacity. *Contrib. Mineral. Petrol.* **113**, 333–351.

Thompson A. B. (1982) Dehydration melting of pelitic rocks and the generation of H_2O-undersaturated granitic liquids. *Am. J. Sci.* **282**, 1567–1595.

Thompson A. B. (1999) Some time–space relationships for crustal melting and granitic intrusion at various depths. In *Understanding Granites: Integrating New and Classical Techniques*. Geol. Soc., Spec. Publ. 168 (eds. A. Castro, C. Fernandez, and J. L. Vigneresse). Geological Society, London, pp. 7–25.

Thompson A. B. and Connelly J. A. D. (1995) Melting of the continental crust: some thermal and petrological constraints on anatexis in continental collision zones and other tectonic settings. *J. Geophys. Res.* **100**, 15565–15579.

Turner S. P. (1996) Petrogenesis of the late-Delamerian gabbroic complex at Black Hill, South Australia: implications for convective thinning of the lithospheric mantle. *Mineral. Petrol.* **56**, 51–89.

Turner S. and Foden J. (1996) Magma mingling in late-Delamerian A-type granites at Mannum, South Australia. *Mineral. Petrol.* **56**, 147–169.

Turner S. P., Foden J. D., and Morrison R. S. (1992a) Derivation of some A-type magmas by fractionation of basaltic magma: an example from the Padthaway ridge, South Australia. *Lithos* **28**, 151–179.

Turner S. P., Sandiford M., and Foden J. (1992b) Some geodynamic and compositional constraints on "postorogenic" magmatism. *Geology* **20**, 931–934.

Turner S. P., Hawkesworth C. J., Rogers N., Bartlett J., Worthington T., and Smith I. (1997) $^{238}U-^{230}Th$ disequilibria, magma petrogenesis, and flux rates beneath the depleted Tonga-Kermadec island arc. *Geochim. Cosmochim. Acta* **61**, 4855–4884.

Tuttle O. F. and Bowen N. L. (1958) Origin of granite in the light of experimental studies in the system $NaAlSi_3O_8-KAlSi_3O_8-SiO_2-H_2O$. *Geol. Soc. Am. Mem.* **74**, 153pp.

Von Cotta B. (1858) Freiberg, Geologische Fragen, 74p.

Vernon R. H. (1983) Restite, xenoliths and microgranitoid enclaves in granites. *J. Proc. Roy. Soc. New South Wales* **116**, 77–103.

Vernon R. H. (1984) Microgranitoid enclaves: globules of hybrid magma quenched in a plutonic environment. *Nature* **304**, 438–439.

Vernon R. H. (1990) Crystallization and hybridism in microgranitoid enclave magmas: microstructural evidence. *J. Geophys. Res.* **95**, 17849–17859.

Vidal P., Cocherie A., and Le Fort P. (1982) Geochemical investigations of the origin of the Manaslu Leucogranite (Himalaya, Nepal). *Geochim. Cosmochim. Acta* **46**, 1061–1072.

Voshage H., Hofmann A. W., Mazzaucchelli M., Rivalenti G., Sinigoi S., Raczek I., and Demarchi G. (1990) Isotopic evidence from the Ivrea Zone for a hybrid lower crust formed by magmatic underplating. *Nature* **347**, 731–736.

Waight T. E., Maas R., and Nicholls I. A. (2000) Fingerprinting feldspar phenocrysts using crystal isotopic composition stratigraphy: implications for crystal transfer and magma mingling in S-type granites. *Contrib. Mineral. Petrol.* **139**, 227–239.

Wall V. J., Clemens J. D., and Clarke D. B. (1987) Models for granitoid evolution and source compositions. *J. Geol.* **95**, 731–749.

Wareham C. D., Millar I. L., and Vaughan A. P. M. (1997) The generation of sodic granite magmas, western Palmer Land, Antarctic Peninsula. *Contrib. Mineral. Petrol.* **128**, 81–96.

Watson E. B. and Harrison M. (1983) Zircon saturation revisited: temperature and composition effects in a variety of crustal magma types. *Earth Planet. Sci. Lett.* **64**, 295–304.

Watt G. R. and Harley S. L. (1993) Accessory phase controls on the geochemistry of crustal melts and restites produced during water-undersaturated partial melting. *Contrib. Mineral. Petrol.* **114**, 550–566.

Watt G. R., Burns I. M., and Graham G. A. (1996) Chemical characteristics of migmatites: accessory phase distribution and evidence for fast melt segregation rates. *Contrib. Mineral. Petrol.* **125**, 100–111.

Wedepohl K. H. (1991) Chemical composition and fractionation of the continental crust. *Geol. Rundsch.* **80**, 207–223.

Whalen J. B. (1985) Geochemistry of an island-arc plutonic suite: the Uasilau-Yau Yau intrusive complex, New Britain PNG. *J. Petrol.* **26**, 603–632.

Whalen J. B. and Currie K. L. (1984) The Topsails Igneous Terrane, Western Newfoundland: evidence for magma mixing. *Contrib. Mineral. Petrol.* **87**, 319–327.

Whalen J. B., Currie K. L., and Chappell B. W. (1987) A-type granites: geochemical characteristics, discrimination and petrogenesis. *Contrib. Mineral. Petrol.* **95**, 407–419.

Whalen J. B., Jenner G. A., Longstaff F. J., Robert F., and Galipey C. (1996) Geochemical and isotopic (O, Nd, Pb, Sr) constraints on A-type granite petrogenesis based on the Topsails igneous suite, Newfoundland Appalachians. *J. Petrol.* **37**, 1463–1489.

Whalen J. B., Percival J. A., McNicoll V. J., and Longstaffe F. J. (2002) A mainly crustal origin for tonalitic granitoid rocks, Superior Province, Canada: implications for Late Archean tectonomagmatic processes. *J Petrol.* **43**, 1551–1570.

White A. J. R. and Chappell B. W. (1977) Ultrametamorphism and granitoid genesis. *Tectonophysics* **43**, 7–22.

White A. J. R. and Chappell B. W. (1988) Some supracrustal (S-type) granites of the Lachlan Fold Belt. *Trans. Roy. Soc. Edinburgh: Earth Sci.* **79**, 169–181.

White A. J. R. and Chappell B.W. (1989) *Geology of the Numbla 1:100 000 Sheet* (8624). Geological Survey of New South Wales, Sydney.

White A. J. R., Chappell B. W., and Cleary J. R. (1974) Geologic setting and emplacement of some Australian Palaeozoic batholiths and implications for intrusive mechanisms. *Pacific Geol.*, **8**, 159–171.

White A. J. R., Williams I. S., and Chappell B. W. (1976) *Geology of the Berridale 1:100 000 Sheet (8625).* Geological Survey of New South Wales, Sydney.

White A. J. R., Williams I. S., and Chappell B. W. (1977) *Geology of the Berridale 1:100,000 sheet.* Geol. Surv. NSW, Sydney, Australia, 138pp.

White A. J. R., Chappell B. W., and Wyborn D. (1999) Application of the restite model to the Deddick Granodiorite and its enclaves—a reinterpretation of the observations and data of Maas *et al.* (1997). *J. Petrol.* **40**, 413–421.

Wickham S. M. (1987) The segregation and emplacement of granitic magmas. *J. Geol. Soc. London* **144**, 281–297.

Wilde S. A., Valley J. W., Peck W. H., and Graham C. M. (2001) Evidence from detrital zircons for the existence of continental crust and oceans on the Earth 4.4 Gyr ago. *Nature* **409**, 175–178.

Williams I. S. (1992) Some observations on the use of zircon U–Pb geochronology in the study of granitic rocks. *Trans. Roy. Soc. Edinburgh: Earth Sci.* **83**, 447–458.

Williams I. S. (1995) Zircon analysis by ion microprobe: the case of the eastern Australian granites. *Proc. L. T. Silver Symp., Pasadena*, 27–31.

Williams I. S., Compston W., Chappell B. W., and Shirahase T. (1975) Rubidium–strontium age determinations on micas from a geologically controlled, composite batholith. *J. Geol. Soc. Australia* **22**, 497–505.

Williams I. S., Chen Y., Chappell B. W., and Compston W. (1988) Dating the sources of the Bega Batholith granites by ion microprobe. *Geol. Soc. Aust. Abstr.* **21**, p424.

Williams I. S., Chappell B. W., Chen Y. D., and Crook K. A. W. (1992) Inherited and detrital zircons—vital clues to the granite protoliths and early igneous history of southeastern Australia. *Trans. Roy. Soc. Edinburgh: Earth Sci.* **83**, 503.

Wormald R. J. and Price R. C. (1988) Peralkaline granites near Temora, southern New South Wales: tectonic and petrological implications. *Austral. J. Earth Sci.* **35**, 209–221.

Wyborn L. A. I. and Chappell B. W. (1983) Chemistry of the Ordovician and Silurian greywackes of the Snowy Mountains, southeastern Australia: an example of chemical evolution of sedim1ents with time. *Chem. Geol.* **39**, 81–92.

Wyborn L. A. I., Wyborn D., Warren R. G., and Drummond B. J. (1992) Proterozoic granite types in Australia: implications for lower crust composition, structure and evolution. *Trans. Roy. Soc. Edinburgh: Earth Sci.* **83**, 201–209.

Wyllie P. J. and Wolf M. B. (1997) In *Greenstone Belts* (eds. M. J. de Wit and L. D. Ashwal). Clarendon Press, Oxford, pp. 256–266.

3.12
Ores in the Earth's Crust

P. A. Candela

University of Maryland, College Park, MD, USA

3.12.1 ORE AND CRUSTAL GEOCHEMISTRY

3.12.1.1 Ores, Mineral Deposits, Geochemical Anomalies, and Crustal Composition

Variations in the major element composition of the crust are manifested in the proportions of rock-forming minerals. Many of the less common elements can be readily accommodated in these phases or is their commonly associated accessories. In some instances, the less common elements may prove to be anomalous in concentration, and a subset of these anomalies may be of economic interest. Much of the geochemical variation found in the crust conforms to this general motif; in other cases, anomalous concentrations of the less common elements are expressed as the appearance of separate minerals or rock units, defining petrologically distinct geologic entities. When the anomalies can be mined for profit, they are referred to as ore deposits, and because of the extensive literature developed on the subject, ores will be a central focus of this discussion.

Oxygen, silicon, aluminum, iron, calcium, sodium, potassium, and magnesium constitute 99% of the Earth's crust. If the next four most abundant elements—titanium, hydrogen, manganese, and phosphorus—are included, then nearly 99.9% of the composition of the crust is explained. That is, most of the industrially important elements, aside from iron and aluminum, make up a very small proportion of the Earth's crust. If we define the scarce elements as those with average abundances less than 1,000 ppm (i.e., elements with abundances less than that of phosphorus), then the 80 or so scarce elements amount to about one-tenth of 1 wt.% (0.1%).

The scarce elements form positive anomalies with varying efficiency. To a first order of approximation, the behavior of the scarce elements is largely controlled by the extent to which they differ in their chemical properties from the top ten or so major rock-forming elements, and the extent to which they enter into solid solution in rock-forming minerals (see Chapter 2.09). Mineralogically, the crust (continental plus oceanic) dominantly comprises quartz, feldspar, olivine, pyroxene, mica, amphibole, and clays (here, fine-grained phyllosilicates) with minor oxides, carbonates, sulfides, and other salts. Ronov and Yaroshevsky (1969) performed a norm calculation on the total crust, and suggested that, by weight, the approximate mineral proportions are: feldspar 60–65%, quartz 10%, ferromagnesian minerals (olivine, clinopyroxene, orthopyroxene, hornblende, and biotite) 25%, and apatite ~0.5% (a few percent clay and carbonate constituted the balance). These constitute the major rock-forming minerals of the crust of the Earth. However, it is worth noting that the dominant phase in most volcanic rocks is glass. Glass can accommodate a wider range of elements than most crystalline solids, and is also rather reactive in hydrothermal environments. The role of volcanic glass as a source of ore elements has not been extensively explored, but may be an important source of ore metals, as well as chlorine and sulfur.

Of the 80 or so scarce elements, many occur in solid solution in major rock-forming minerals (e.g., rubidium, strontium, vanadium, germanium, gallium, scandium, rare earth elements (REEs)), or in solid solution in widely distributed, refractory, or low-solubility accessory phases (e.g., hafnium). Some elements, such as zirconium and phosphorus, form their own minerals. Others, like the base metals (e.g., copper, lead, zinc), semi-metals (arsenic, antimony, bismuth), and precious metals (gold, silver, platinum group elements (PGEs)) do not possess a high affinity for the minerals that make up the bulk of the crust, and are therefore excluded from all but the most limited solid solution in rock-forming minerals. These elements tend to partition into fluid phases (aqueous and brine phases, or silicate and other naturally occurring melts), and ultimately form their own minerals. In most rocks, these minerals may form small, highly dispersed grains. For example, micron-scale grains of chalcopyrite or cupiferous pyrrhotite may host a high proportion of the copper in a rock, gold may occur as submicron grains of native gold, and gold and arsenic may occur in accessory arsenopyrite (cf. Palenik *et al.*, 2002).

If a given volume of the crust is anomalously rich in a given element, but cannot be mined at a profit, we refer to the rock volume as a mineral deposit or a geochemical anomaly. It is tempting to define the term "anomalous" quantitatively as, for example, for a given "number" of standard deviations above the mean. However, most of these definitions are impractical from a geological mining point of view, and a geochemical anomaly is best defined as a significant, positive, geochemical departure from the norm, which may indicate the presence of economic mineralization in a given volume of bedrock. If the elements of interest are concentrated to a sufficient extent to be mined at a profit, then we can call the anomaly an ore. Table 1 lists grades of ores, crustal abundances, Clarke values, and pertinent chemical data for selected ore metals. The Clarke value is the ratio of the concentration of a metal in an ore, relative to its average crustal concentration. These values illustrate the wide range of both crustal concentrations and concentrations in minable ore for these elements. The balance of this paper will examine the nature of the processes that have

Table 1 Data on selected ore metals. Median ore grade for a given element can vary significantly from one deposit type to another. The Clarke value is the ratio of the median grade to the crustal abundance.

Element	*Oxidation state*	*Deposit type*	*Grade*[a] (median)	*Crustal abundance*[b]	*Clarke value*
Cu	1	Porphyry	0.54%	27 ppm	200
Na	1	Halite	40%	2.3%	17
Zn	2	Sedimentary exhalative	5.6%	72 ppm	780
As	3	Sulfide deposits[c]	~0.1%	2.5 ppm	~400
Rb	1	Lepidolite[d]	Up to 3%	49 ppm	~610
Mo	4	Climax	0.19%	0.8 ppm	2,400
W	4 (6)	Skarn	0.66% WO_3	1 ppm	6,600
Pb	2	Sedimentary exhalative	2.8%	11 ppm	2,500
V	3	Layered mafic intrusions	~0.6%	138 ppm	~43
Au	0 (1)	Veins/Homestake	~10 ppm	1.3 ppb	~7,700
Ag	0 (1)	Creed vein	125 ppm	56 ppb	2,200
Ni	2	Komatiite	1.5%	59 ppm	250

[a] All percentages are in weight percent. [b] Continental crust; see Chapter 3.01. [c] Highly variable. [d] In concentrate.

produced the heterogeneous crustal distribution of these and other scarce elements.

Many of the scarce elements are transported in the crust by complexing with sulfur or chlorine in aqueous solution. Furthermore, many metals are precipitated from aqueous solutions or melts as sulfides, including copper, zinc, lead, molybdenum, rhenium, cobalt, PGE, nickel, cadmium, indium, thallium, antimony, bismuth, arsenic, selenium, tellurium, and mercury. Many of these elements rely on chloride as a complexing agent for aqueous transport. Other elements, such as gold, can be complexed in aqueous solutions by sulfur, but may be precipitated as the native element. Silver, which is not commonly transported as a sulfide complex, may precipitate as the native element or as a sulfide. Sulfur occurs in the continental crust at ~700 ppm (Wedepohl, 1995), and is distributed between reduced and oxidized forms. Hydrothermal solutions commonly contain 0.1 m*m* to 0.1 *m* in total reduced sulfur; hence, there is an importance of this element in the transport and deposition of some metals. Chlorine occurs in the crust at concentrations on the order of a few hundred ppm (~300 ppm), and occurs in aqueous solutions over a wide range of concentrations. Chlorine, as well as its heavier congeners, is partitioned into hydrothermal fluids, and ultimately into seawater. Tungsten, uranium, and tin occur in ores almost exclusively as hydrothermally precipitated oxides (Fe, Mn)WO_4, UO_2, and SnO_2, respectively. Most of the ores of these elements are hydrothermal in character. Vanadium, titanium, nickel, and chromium form hydrothermal deposits somewhat less frequently, and are concentrated by crystal settling in mafic intrusions. The noble gases are fugitive, ultimately finding their way into subterranean effluvia or exhalations such as natural gas seeps, volcanic gases, and deep inputs to geothermal systems. Some elements are transitional among groups, usually because of unusual chemical properties. Mercury, for example, concentrates in exhalations, but ultimately can be trapped in organic materials or sulfides as trace cinnabar (HgS). The fluxes of volatile elements in aqueous solutions, as well as the elements transported to sites of deposition by melts, are all affected indirectly by faulting and other elastic manifestations of the upper crust. Therefore, I will devote the next section of this paper to a discussion of fractures in the Earth's crust, and the engines that drive fluid movement through this porous matrix.

3.12.1.2 Physical and Chemical Factors in the Generation of Geochemical Anomalies

Rocks may fail by either fracture or flow. At the low temperatures that obtain at shallow crustal levels, rocks behave elastically and may reach their elastic limit, failing by fracture. With increasing temperature, the resistance of material to flow decreases, and rocks strain by flow before fracturing. With increasing pressure, the elastic strength increases, and so again, rocks will have a tendency to accommodate strain by flow before they reach their elastic limit. For these reasons, rock in the upper-third of the crust behaves differently from deeper rock. In the elastic upper crust, deformation occurs dominantly by frictional sliding on new or pre-existing fractures and by cataclasis, a distributed brittle granulation of rock with progressive decrease in grain size upon deformation. The origin of the stress at a point in the crust can be thought of as a combination of regional (far-field) and localized (near-field) stresses. For example, large-scale regional stresses may be responsible for the shear zones and other fracture networks that allow the upward advection of magmas in active arc zones related to porphyry copper-ore formation. When magma is emplaced at upper levels and continues to evolve physically, it can alter the stress environment through its own thermomechanical evolution (e.g., heating of surrounding rocks, change in volume of the magma due to change in temperature and crystallization, plus volatile exsolution), resulting in a near-field modification of the regional stress state. The diking, fracturing, and concomitant mineralization that forms in and around the crystallizing stock will follow a geometry that is dictated by the time- and space-dependent combination of the near-field and far-field stresses. Generally, these and other upper crustal-fracture networks, comprising new or pre-existing fractures, strongly affect the permeability and porosity of the upper crust, allowing the movement of gases, aqueous solutions, petroleum, and melts.

According to Manning and Ingebritsen (2001), geothermal and metamorphic fluid-flux data show that permeability (k, m^2) decreases with depth (z, km) in the continental crust according to the equation

$$\log k = -14 - 3.2 \log z \qquad (1)$$

above the brittle–ductile transition zone, which crudely corresponds to the upper 10–15 km of the crust. The brittle–ductile transition zone coincides with $\log k \sim -17$, which is an effective upper limit on the permeability at which elevated pore-fluid pressures can be sustained. Thus, Manning and Ingebritsen point out that hydrostatic pressure gradients dominate above the brittle–ductile transition zone. When a fracture forms, fluid-filled void held open by tensile forces or by randomly distributed asperities along the walls of the fracture. The fluids may be at below or above the lithostatic pressure. Pressures above lithostatic will almost always cause fracturing: this is the case around shallow-cooling magmas, such as those that ultimately yield texturally variable, epizonal granites. This fractured, elastic, permeable rock matrix is the mechanical stage upon which the drama of element redistribution plays out in the upper crust.

3.12.1.3 The Generation of Anomalies

The redistribution of chemical elements in the Earth's crust is effected by a wide range of fluids. The defining characteristic of a fluid, of course, is flow. Flow of a medium, whether a tenuous gas or highly viscous magma, is necessary for the redistribution of elements in the Earth on scales of meters or greater. Vein-type and disseminated hydrothermal ores result from processes including, but not limited to, precipitation from hydrothermal solutions brought about by changes in temperature, pressure, wall–rock reaction, or fluid mixing (that occurs dominantly in a fractured rock matrix).

Primarily, magma is transported through the crust by fracture flow. Magma evolution, or differentiation, occurs dominantly by crystal fractionation, and also by a variety of other processes, including assimilation and mingling. One particularly attractive spatial model for differentiation includes the punctuated differentiation model of Marsh (1996) that involves repeated episodes of chambering, fractionation, and removal of liquid (plus some crystals) to higher levels. During these processes, minerals or other phases that are only sparingly soluble in the melt, such as metal oxides, sulfides, PGEs, immiscible oxide and/or sulfide liquids, and saline liquids or vapors may be segregated from the bulk magma by virtue of their significantly different density. Vapor and brine can rise due to their buoyancy. These fluids may be charged with sulfur and chlorine, which can aid in the solubilization of many ore metals, as will be discussed presently. This process appears to be common in high-level intermediate to felsic magmas, and also occurs in mafic systems. In fracture-supported hydrothermal systems, hot fluids rise when surrounded by cooler fluids, or when they are at pressures in excess of the hydrostatic load. Clearly, these and other processes of endogenous element redistribution rely to first order on density differences and upon gradients in the upper crust that are ultimately driven by the irreversible loss of terrestrial heat.

3.12.1.4 The Terrestrial Heat Engine

The fundamental engine of the Earth's endogenous fluid advection system is the irreversible loss of heat from the interior of the planet to the surface, where it is ultimately radiated into space. The idealized Rayleigh–Bernard convection cell that we can use to model convection of Earth systems can be crudely represented as a Carnot Cycle. At the bottom of the convection cell, matter is heated and consequently expands; the fluid buoyantly rises and cools adiabatically upon expansion, because it has been removed physically from the heat source. In this step, the fluid does "$P-V$" (expansion) work, as well as work against gravity. To complete the cycle, the fluid loses heat at a cold (usually upper) boundary, contracts, and consequently sinks. During its downward flow, the fluid is compressed, and also loses gravitational potential energy. The working fluids in "solid–earth" convection systems may be solid mantle, salt (and other sediments), magma, or geothermal fluids; further, upward movement of one fluid (e.g., salt) may be coupled with the downward movement of other "fluids" (e.g., sediment). We can consider the Earth to comprise nested Carnot-convection engines, including the grand convection system of the crust and mantle. The upwelling of magmas and related fluids into the crust can result in the generation of crustally derived magmas. The mantle- and lower crustal-derived magmas are themselves buoyant in the crust, and transport matter to the mid- and upper crust. While in the crust, these magmas can drive the advection and upward transport of magmatic water and locally derived metamorphic water. Further, deep-crustal fluids, as well as more shallowly circulating meteoric fluids are commonly driven to convect, creating large-scale element redistribution systems (see Chapter 3.06).

Convection may transport crustal fluids and their components reversibly, i.e., with no net transfer of fluid or fluid components from one region to another. Alternatively, net transfer of matter may be effected by changes in phase (e.g., solidification of a magma) or by changes in solubility of some fluid-solute components due to changes in temperature, or pressure, or by chemical reaction along the flow path. For minerals that increase in aqueous solubility with increasing temperature (i.e., those with prograde solubility), there is a net transfer of solute materials from high temperature to low temperature. This is the case for quartz (except for some restricted regions of $P-T$ space) and results in the web of quartz veins that are so common in the upper continental crust. However, some minerals, such as anhydrite, decease in solubility with increasing temperature. This phenomenon, referred to commonly as retrograde solubility, results in the precipitation of minerals in the hotter portions of aqueous convective systems (see also Chapter 3.18). For this reason, oxidized sulfur is precipitated at depth in the oceanic crust.

Chemical reaction can fix materials at higher or lower levels. For example, significant concentrations of HCl can occur in magmatic volatiles, and can become reactive upon cooling. Further, magmatic volatiles can also contain SO_2, which upon disproportionation yields H_2S by the reaction

$$4H_2O + 4SO_2 \rightarrow 3H_2SO_4 + H_2S \qquad (2)$$

promoting the fixation of metal sulfides in magmatic–hydrothermal systems.

The potential for focusing fluid flow around a point heat source such as a magma, with or without significant outward advection of magmatic volatiles, provides for strong lateral (i.e., longitudinal and latitudinal) concentration of ore metals, alteration, and associated chemical and isotopic anomalies. Focusing of geothermal fluids appears to be a critical link in the creation of hydrothermal ores. Unfocused hydrothermal fluid flow produces only geothermal systems, not ore deposits. This focusing is important and is shown by the common creation of large positive geochemical anomalies relative to the rather subdued magnitude of the associated negative anomalies (although in the study of ores as geochemical entities, we must always be cognizant of the inherent economic bias toward the study of not only higher grade and tonnage deposits, but toward positive as opposed to negative anomalies). The presence of cupolas and other apophyses in the roof zones of magma chambers creates thermal point sources that can focus fluid flow. Structures such as releasing bends (Henley and Berger, 2000) act to focus upward surges of magma, forming vertically elongated, high aspect ratio cupolas (Sutherland-Brown, 1976). These conditions obtain most efficiently in local environments of dilation in otherwise compressional regimes. In the New Guinea Fold Belt, at least four major magmatic–hydrothermal Cu–Au (Ag) deposits (Grasberg, Porgera, Ok Tedi, and Frieda River) are located along zones of local dilation. Hill *et al.* (2002) proposed that, during orogenesis and crustal thickening, strike-slip motion occurred along fracture zones roughly parallel to the convergence vector. Localized zones of dilation opened where these fracture zones intersected major orthogonal faults and other structural discontinuities, facilitating igneous intrusion and subsequent mineralization. The upward advection and subsequent emplacement of magma in a favorable structural environment led to aqueous fluid advection and focusing, and ultimately element redistribution. Finally, most hydrothermal ore formation can be thought of as a type of metamorphism that results in element redistribution. Whereas the hydrothermal alteration commonly associated with hydrothermal anomalies is metasomatic in nature, metamorphism *sensu stricto* can also be important in ore formation. In the case of greenstone and turbidite hosted mesothermal gold deposits, there is a suggested link between regional metamorphism and the generation of vein-hosted gold mineralization. Metamorphism, together with deformation, can provide a source of fluid, a source of metals, a source of focusing, or all three.

The confluence of thermally induced density variations brought about by melting in the lower crust and upper mantle, together with the structural/tectonic filters operating in the Earth's brittle upper crust, generally provide the near-point sources of heat and matter required for focusing hydrothermal fluid flow. However, whether fed from underlying magma, metamorphic dewatering, or circulating meteoric water, fluid focusing is not sufficient for the formation of a mineralized system. Hobbs and Ord (1997) point out that fluid focusing is ineffective if the fluid-flow vectors are at low angles relative to isotherms and isobars. Yet, magmas and other fluids may be advected nearly adiabatically into the upper crust. Only when high potential gradients are present, as when fluids are brought into close proximity with the surface, will they flow at high angles to isotherms. Further, magmas at lithostatic pressure may be emplaced into rocks wherein the ambient fluids in fractures are near hydrostatic. Under conditions of strong thermal and mechanical gradients, significant changes in solubility of ore and gangue minerals can occur. Of course, strong gradients in chemical potential can also bring about mineral precipitation. Chemical mixing of different fluids or interaction of fluids with rocks with which they are strongly out of equilibrium result in strong gradients of chemical potential (e.g., the formation of skarns at intrusion/limestone contacts). Generally, the combination of thermally induced density changes in a gravity field, together with strong gradients in thermal, mechanical, and chemical potentials, drive hydrothermal element redistribution in the crust. In the next section of this chapter, I will discuss the general issues related to changes in temperature, pressure, and chemical composition affecting the transport and precipitation of ore metals in hydrothermal systems.

3.12.1.5 Ore-mineral Solubility

The overwhelming majority of ore minerals are characterized by very low solubilities in pure water, even at elevated temperatures (Wood and Samson, 1998). Given these low solubilities, it is not clear that pure water could transport sufficient quantities of ore substance to account for the origin of many ores by hydrothermal processes. However, the addition of suitable mineralizers (or complexing agents) can raise the concentrations of many ore metals in hydrothermal fluids to levels required for significant transport. Still, some elements are more commonly represented among hydrothermal deposits than others, even after removing the economic bias. It is certainly tempting to assert that the lack of "substantial" solubility is responsible for the dearth of hydrothermal minerals of a given element; however, this assertion is difficult to support upon detailed analysis. In fact, the opposite is more likely the case: the rarity of halite and sylvite as primary minerals in hydrothermal deposits (their abundance as daughter minerals in fluid inclusions notwithstanding) can be attributed to their high solubility, which not only prevents significant deposition, but also leads to their rapid dissolution if deposited. A similar fate may be suffered by other halides. For example, the large cubic chambers in some fluoride-rich pegmatites may represent casts of NaF crystals now dissolved away by subsurface waters (Roedder, 1984). Moreover, gold has a very low aqueous solubility, and is rare, yet hydrothermal deposits abound. The apparent frequency of hydrothermal mineral-deposit anomalies is a combination of competing factors including crustal abundance of the element in question, the magnitude of the changes in solubility over the common ranges of temperature and pressure and chemical potentials of appropriate reactants, the ability of an element to be geochemically camouflaged, as well as the economic bias that may suppress or amplify the apparent abundance of its deposits.

The extent to which temperature and pressure affect the solubility of ore minerals is strongly influenced by the chemical composition of the aqueous solutions that may transport the dissolved constituent in question (Wood and Samson, 1998). The set of ligands available for ore-metal complexation are strongly influenced by the general temperature regime. At Earth surface temperatures, biogenic ligands, including carboxylic acids, amino acids, porphyrins, and other polydentate molecules including siderophores and related compounds, as well as ammonia and ammonium, form a wide range of complexes that can enhance equilibrium solubilities, as well as inhibit precipitation kinetically in low-temperature aqueous environments. More commonly, especially at the higher temperatures and pressures attained in most hydrothermal environments, the dominant ligands are chloride (Cl^-), bisulfide (HS^-), H_2S, hydroxide (OH^-), sulfate (SO_4^{2-}), and bisulfate (HSO_4^-). However, oxygen is by far the most common anion in the Earth, and elements such as nitrogen, phosphorus and sulfur, as well as carbon and boron, form common anionic complexes with oxygen. These, in turn, can form ionic compounds in solids, and ions or ion pairs in aqueous solutions. Although phosphate, sulfate, fluoride, and carbonate are not uncommon anions, their calcium salts (apatite, anhydrite/gypsum, fluorite, and calcite/aragonite, respectively) are only sparingly soluble, limiting the concentration of these anions in natural waters. Sulfate concentrations in the subsurface are further limited by redox equilibria, fixing some sulfur in sulfide minerals. According to Wood and Samson (1998), hydrothermal solutions range in total reduced sulfur from 0.1 m*m* to 0.1 *m*, and sulfate ranges from below detection to a few tenths of a molal. Carbonate is usually in the millimolal to molal range. In contrast, due to the high solubility of chlorides of the major rock-forming elements, naturally occurring aqueous solutions acquire chloride rather readily. Hence, Cl^- is the dominant anion in many natural waters, followed by reduced and oxidized sulfur and (bi)carbonate. Seawater, with a salinity of 3.5 wt.%, has a chloride concentration of 1.9 wt.% (sulfate and carbonate are 0.65 wt.% and 0.14 wt.%, respectively). Seawater is, of course, a major component of ore fluids in the oceanic realm. In contrast, river water is variable in composition, and is, on average, about two and a half orders of magnitude lower in salinity. During base flow, river water is indicative of the composition of shallow groundwater, another important end-member, both thermally and compositionally, for hydrothermal ore fluids. Fluid inclusions in hydrothermal minerals, especially those from granite-related deposits, can range in salinity up to 70 wt.% salt and beyond. Therefore, hydrothermal fluids range widely in salinity, from less than 0.1 wt.% up to hydrous salt melts possessing salinities of many tens of weight percent salt.

The salinity of a hydrothermal fluid is a strong determinant of the physical and chemical properties of the fluid. A salt (hydrohalite)-saturated $NaCl-H_2O$ solution at 0.1 MPa pressure has a eutectic temperature of −21.1 °C, and a salinity of 23 wt.%; at the same pressure, a saturated solution (~26 wt.%) boils at 108.7 °C. Therefore, even at 0.1 MPa, NaCl can significantly alter the temperature of the phase equilibria of water. Both of these phenomena, freezing point depression and boiling point elevation, are colligative properties, and ultimately are driven by entropic effects that stabilize the mixed aqueous phase relative to the nominally pure solid (ice) and vapor (steam) phase. Two competing effects control the vapor pressure of saturated salt solutions with increasing temperature. Increasing temperature increases the vapor pressure of water; however, a saturated salt solution also increases in salinity with increasing temperature, which tends to decrease the vapor pressure of the solution by dilution of the water component of the liquid. This effect is mostly pronounced for highly soluble salts. In the case of NaCl, the vapor pressure of salt-saturated solutions is always below the critical pressure at any temperature, allowing for an uninterrupted three-phase equilibrium: halite–vapor–liquid (brine at lower temperatures and hydrous salt melt at higher temperatures, see Figure 1). This three-phase curve is continuous from the four-phase peritectic-invariant point: vapor–hydrohalite–halite–liquid near-room temperature, up to its termination at the pure NaCl vapor–halite–liquid triple point, with a pressure maximum of ~36 MPa near 600 °C. The sizable gap in pressure, ranging over 100 MPa at some temperatures, between the critical pressure and the vapor pressure of the halite-saturated solution at any given temperature (see Figure 1), allows for a significant region of vapor–liquid equilibria over P–T–X space. The applicable temperatures span the full hydrothermal range up to 800 °C, and accessible salinities range from seawater up to pure-salt melts. The applicable subcritical pressures essentially define the Earth's epizone. The behavior of KCl is very similar to NaCl. The high solubility of both NaCl and KCl in water, together with the relatively high pressures of the critical curves in the binary salt-water systems, allow for extensive aqueous liquid–vapor equilibrium from Earth-surface conditions up to the temperatures and pressures of the granite solidus, and also leads to the importance of chlorides as complexing agents for ore metals in upper crustal fluids.

Quartz increases in solubility with increasing temperature and density of the fluid. Because of the variations in density that occur around the critical point of water, some reversals occur in the solubility of quartz near those temperatures and pressures. Generally, quartz solubility in water varies from a few tenths of a weight percent near granite solidus conditions in the epizone, to a few ppm near Earth-surface conditions. Wood *et al.* (1987) studied the solubility of the assemblage pyrite, pyrrhotite, magnetite sphalerite, galena, gold, stibnite, bismuthinite, argentite, and molybdenite in aqueous NaCl solutions in the presence of CO_2 from 200 °C to 350 °C. They found that relative solubilities followed the sequence Sb > Fe > Zn > Pb > Ag,Mo > Au,Bi in chloride-free solutions, and Fe > Sb > Zn > Pb > Au > Ag,Mo,Bi in chloride-bearing solutions. Further, Wood *et al.* examined the dependence of ore-metal solubility on the NaCl concentration in the solutions. Their results suggest that iron, gold, and silver are chloride complexed in their chloride-bearing experiments, and that gold and silver are bisulfide complexed in their chloride-free experiments.

Few studies show any significant complexation of iron by sulfide complexes, and $FeCl_2$ appears to be the dominant agent of iron transport at elevated temperatures. These results are in general agreement with later studies summarized by Wood and Samson (1998). The same authors, summarizing recent experimental data from a number of sources, state that silver chloride complexes such as $AgCl^0$, $AgCl_2^-$, and $AgCl_3^{2-}$ are responsible for aqueous silver transport in a wide variety of hydrothermal environments, although bisulfide complexes may be responsible for silver transport in some

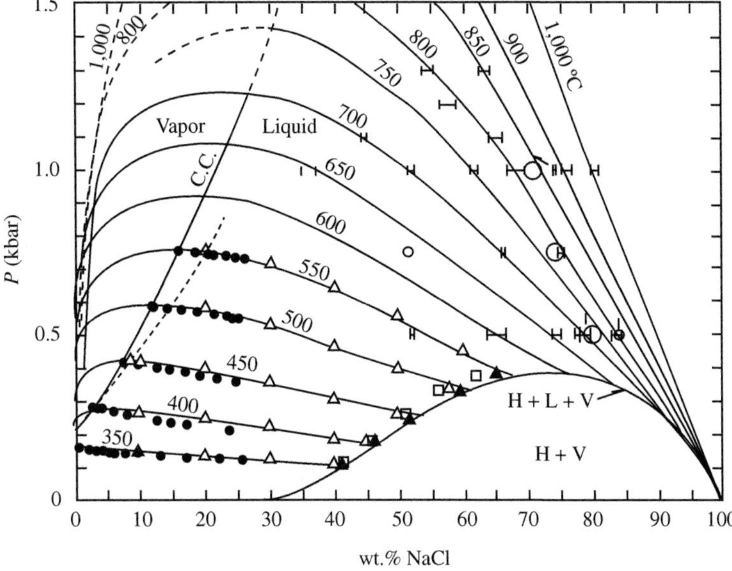

Figure 1 Isotherms of the NaCl–H_2O system illustrating the composition of coexisting vapor (low X_{NaCl}) and brine (high X_{NaCl}) in P–X space. Note that the critical curve (c.c.) is everywhere above the three-phase equilibrium halite–liquid—vapor (source Chou, 1987).

cases (Gammons and Williams-Jones, 1997). Further, Wood and Samson (1998) point out that Au(I)-chloride complexes will be important agents of aqueous gold transport at high chloride activity, high temperature, and moderately high oxygen fugacity, with Au(I)-bisulfide complexes occurring over a wide range of conditions. Generally, we might expect that gold-chloride complexes can be important at near and submagmatic conditions, whereas gold-bisulfide complexes will be important at lower temperatures. Wood and Samson point out that the chloride complexes are probably dominant for copper, as suggested by many workers, and the most likely complexes are probably CuCl and $CuCl_2^-$. Wood and Samson point out that sulfide complexes of copper may play a role, but their stoichiometry and thermodynamics are still a point of contention. They also indicate that tungsten and molybdenum are likely to be transported as tungstate and molybdate species (H_2WO_4 and H_2MoO_4, respectively), although systematic data on molybdenite solubility are still lacking. Zinc appears to be transported in hydrothermal environments by a variety of ligands, and Wood and Samson suggest that chloride, bisulfide, hydroxide, carbonate, and bicarbonate complexes all may play roles in hydrothermal zinc transport in the Earth's crust. Lead appears to be transported as chloride complexes, but carbonate- or sulfide-bearing complexes may be important under some conditions.

Hemley *et al.* (1992) studied the solubility of iron, lead, zinc, and copper sulfides in chloride solutions that were rock-buffered in pH and in oxygen and sulfur fugacity, in the range 300–700 °C; 50–200 MPa. Their results show that iron-, copper-, zinc-, and lead-sulfide mineral solubilities decrease with decreasing temperature and decreasing total chloride (see Figure 2), and with increasing pressure. In nature, the HCl concentration (which is the sum of HCl^0 and some portion of the ionized H^+ and Cl^-) in a hydrothermal solution is controlled by equilibria such as

$$\begin{aligned}&KAl_3Si_3O_{10}(OH)_2^{(\text{white mica})} + 2KCl\\&+ 6SiO_2^{(\text{quartz})} \rightarrow 3KAlSi_3O_8^{(\text{feldspar})}\\&+ 2HCl \qquad (3)\end{aligned}$$

for solutions of a given total KCl concentration, pressure, temperature, and salinity (and activity of the specified mineral components). This equilibrium, and related equilibria relating K-feldspar and aluminum silicate minerals such as andalusite, shifts toward higher total HCl with increasing temperature and decreasing pressure. Examination of the reaction

$$\begin{aligned}&CuFeS_2 + 3HCl + \tfrac{1}{2}H_2 \rightarrow CuCl^0 + FeCl_2^0\\&+ 2H_2S \qquad (4)\end{aligned}$$

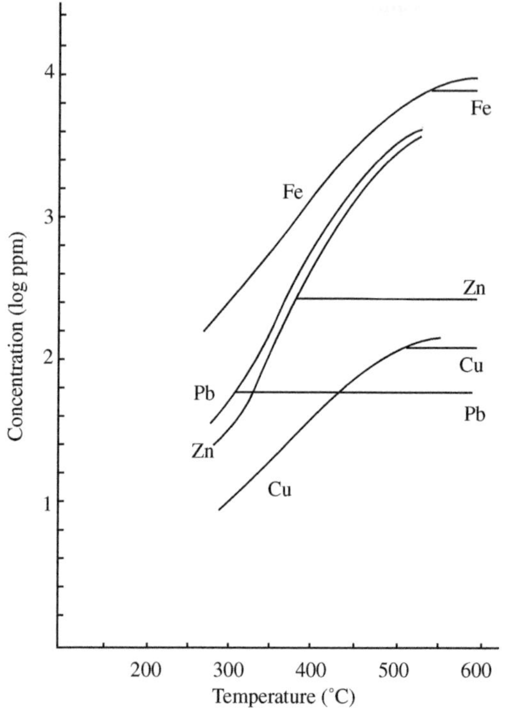

Figure 2 Partly schematic plot of the relationship between solubility of metal sulfides (curves with arrows showing down temperature deposition) and select initial metal concentrations (shown by horizontal lines at temperatures of mineral undersaturation). Drawn for 100 MPa pressure, and for fluids in equilibrium with the quartz–K-feldspar–muscovite buffer with a total chlorinity of 1 *m* (after Hemley and Hunt (1992)).

shows how ore-mineral solubilities increase as a result of increasing HCl activity. Reactions (3) and (4), taken together, account for the increasing solubilities found for sulfide minerals with decreasing pressure in the experiments of Hemley *et al.* (1992). In fact, reactions of type (3) generally account for the generation of acidity in otherwise low-acidity hydrothermal fluids as they are heated along prograde paths. Metasedimentary rocks containing these or analogous assemblages can buffer prograding aqueous solutions to higher values of HCl/KCl, HCl/NaCl, $(HCl)^2/MgCl_2$ relative to their initial values. Similarly, volatiles exsolved from magmas at shallow levels in the Earth's crust possess HCl concentrations that ensure that the fluid will possess significant acidity upon cooling, yielding significant metal-carrying capacity and ensuring significant down-temperature hydrothermal alteration upon cooling (Williams *et al.*, 1997; Frank *et al.*, 1999). In a companion paper, Hemley and Hunt (1992) suggest that fluid flow near cooling igneous bodies is nearly adiabatic. Under these conditions, the effects of adiabatic cooling (wherein the work of expansion upon lowering of pressure consumes internal energy, causing temperature to drop) on solubility are counteracted by the positive effect of pressure on ore-mineral solubility, and mineral precipitation does not occur. When the fluids encounter smaller fractures and colder country rock, they will begin to lose heat, and will deviate from the adiabatic path to what Hemley and Hunt call a geothermal path, along which mineral deposition occurs due to cooling, mixing, and wall–rock reaction.

3.12.1.6 Hydrothermal Redistribution of Gold and Antimony: An Example

Gold occurs in ores primarily as the native metal. In the crust, where its average concentration is ~3 ppb (Taylor and McLennan, 1995), gold occurs primarily as the native element or in limited solution in other minerals, especially in sulfides and possibly oxides. However, even the gold detected in silicate and oxide minerals may be present as metallic inclusions. A number of workers have suggested that gold can be sequestered in sulfide minerals (Cygan and Candela, 1995; Jugo *et al.*, 1999; Simon *et al.*, 2000). In our laboratory, we have conducted many experiments, wherein gold metal has been equilibrated with silicate melts and aqueous fluids at temperatures ~800 °C and at pressures ~100 MPa (Candela *et al.*, 1996; Frank *et al.*, 2002). We have analyzed coexisting phases by a variety of analytical techniques (Frank *et al.*, 2002; Simon *et al.*, 2002). Our experiments show that the solubility of gold in rhyolitic melts is between 0.3 ppm and 1 ppm at the temperatures, pressures, water, sulfur, and oxygen fugacities that occur in upper-crustal magmatic systems (Frank *et al.*, 2002). Connors *et al.* (1993) analyzed 129 glassy silicic-volcanic rocks for gold and found that the majority of them ($n = 113$) contained on the order of 1 ppb. They suggest that granitic systems associated with gold deposits may be elevated in initial gold content, but probably by not more than 4 ppb. Coupling these data with our results on the solubility of gold in a melt at a unit activity of gold, we calculate an activity of gold in those magmas to be on the order of 10^{-3}, or ~1,000 times below that required for gold saturation. If this state of gold saturation is typical for crustal magmas, then the source rocks with which crustal melts are last in equilibrium were strongly undersaturated with gold. From that meager evidence, we can suggest that whereas gold is present in ores primarily in native form, in common rocks it is present in a form that significantly reduces its chemical potential, and that form is like solid solution in minerals at the temperatures and pressures of magma generation. If we take crustal melts as a crude indicator of the chemical potential of gold in crustal rocks in general, then from source region to deposition, the ore-forming process must achieve amplifications of gold activity of three orders of magnitude. The state of saturation of crustal fluids with respect to a given ore mineral varies widely from element to element. For example, some high-temperature oil-field brines, such as those associated with gas plays, are near saturation with fluorite, sphalerite, and either metallic lead or galena (Hartog *et al.*, 2001).

For gold, the amplification of thermodynamic activity may occur by a variety of processes; however, physical concentration of the fluid as well as efficient precipitation are necessary for the formation of an ore deposit. A number of authors have pointed out the difference in the behavior of gold at low (e.g., $<400 \pm 100$ °C), versus high temperatures. In lower-temperature aqueous fluids, gold behaves as a soft (i.e., large and deformable) metal ion, bonding covalently with the soft bisulfide ligand (Wood *et al.*, 1987), whereas at temperatures greater than ~350–400 °C, chloride complexes become important. This effect is probably due to the increasingly ionic nature of bonding within aqueous complexes as the mean interatomic distance grows with temperature, in turn leading to the increasing "hardness" of the gold ion and the preference for its interaction with the harder (i.e., smaller and less deformable) chlorine ion. Hence we find that, above ~400 °C, depending upon the pH, total pressure, total Cl^- concentration, and fugacity of H_2S (see Gammons and Williams-Jones, 1997), $AuCl_2^-$ is the dominant gold ion (Gammons and Williams-Jones, 1997), with $HAuCl_2^0$ dominating in very low dielectric fluids (e.g., in near-magmatic vapors) (Frank *et al.*, 2002). At lower temperatures, $Au(HS)_2^{1-}$ dominates (see Table 2). High-temperature gold-chloride solutions tend to precipitate gold upon cooling, whereas cooling of lower temperature gold-bisulfide solutions does not lead to gold precipitation. Ilchik and Barton (1997) suggest that precipitation of gold from bisulfide solutions results from ore-fluid dilution or wall–rock alteration. Both dilution and wall–rock reaction can also be important in the deposition of gold at high temperatures.

Table 2 Model high-temperature and low-temperature equilibria demonstrating the effects of oxidation and variation of H_2S fugacity, as well as the contrasting stoichiometric effects of changing pH, upon gold solubility.

High T: $AuCl_2^{1-} + 0.5H_2O \rightarrow Au^{0(metal)} + 2Cl^- + H^+ + 0.25O_2$

Low T: $Au(HS)_2^{1-} + 0.5H_2O + H^+ \rightarrow Au^{0(metal)} + 2H_2S + 0.25O_2$

For gold transported at temperatures of less than 400 °C in reduced sulfide-bearing solutions, gold solubility is maximized near the $H_2S-HS^--SO_4^{2-}$ equal predominance point at any given temperature, pressure, and activity of water (see Figure 3). As conditions become more oxidized, the sulfide complexes that stabilize gold in aqueous solution break down, and gold metal precipitates. As pH increases (in the HS^- predominance field), the redox reaction

$$0.5H_2 + Au^+ \rightarrow Au^{0(\mathrm{metal})} + H^+ \quad (5)$$

proceeds to the right, causing gold precipitation. The net equilibrium at gold saturation under these conditions is

$$Au(HS)_2^- + 0.5H_2O \rightarrow Au^{0(\mathrm{metal})} + 2HS^- + H^+ + 0.25O_2 \quad (6)$$

(where the reaction is written involving oxygen and water rather than hydrogen and water so as to show the effect of oxygen fugacity). Note that for a given water activity or fugacity at a given temperature and pressure, the oxygen and hydrogen fugacities vary inversely. If pH decreases into the H_2S-dominant field, the hydrogen–gold redox reaction is coupled with the conversion of HS^- to H_2S, yielding gold precipitation with decreasing pH as illustrated by the low-temperature mass action expression shown in Figure 2. Hence, depending upon whether gold is being transported in the field of H_2S or HS^- predominance under reduced conditions, gold may be precipitated by either pH decrease or pH increase, respectively, as long as the oxidation state of the solution remains in the reduced-sulfur field of predominance. Reduction can also precipitate gold, as is shown by both the gold-bisulfide-bearing equilibria, as long as the oxidation state of the solution remains in the reduced sulfur field of predominance. Only when a sulfide solution is oxidized to sulfate can gold precipitate from an aqueous solution by oxidation. This reaction can be written as

$$Au(HS)_2^- + 0.5H_2O + \tfrac{15}{4}O_2 \rightarrow Au^{0(\mathrm{metal})} + 2SO_4^{2-} + 3H^+ \quad (7)$$

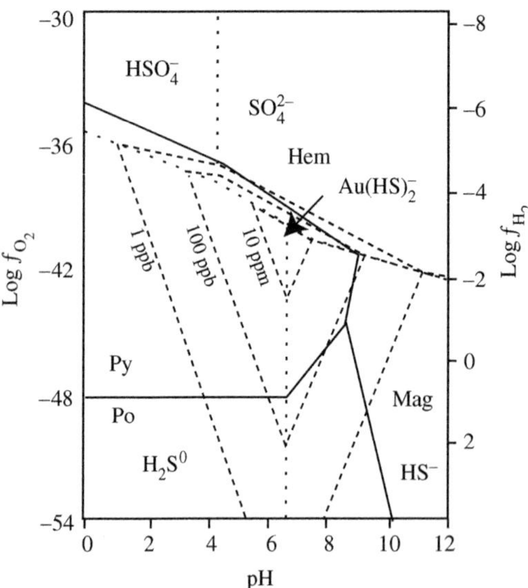

Figure 3 $\log f_{O_2}$–pH diagram for the solubility of gold as $Au(HS)_2^-$, at 200 °C and saturated water vapor pressure. Solid lines delineate mineral-stability fields; dotted lines delineate the fields of dominance for dissolved sulfur species (total dissolved sulfur = 0.01 *m*), and dashed lines show gold solubility contours. Drawn for an activity of water equal to unity (after Wood and Samson, 1998).

Williams-Jones and Normand (1997) suggest that the conditions of maximum gold concentration in hydrothermal fluids correspond closely to the conditions of maximum antimony concentration, where the dominant aqueous antimony species is $HSb_2S_4^-$. Because stibnite (Sb_2S_3) solubility increases rapidly with temperature, and because antimony is a rare element, the ratio of the concentration of antimony in hydrothermal solutions to its solubility limit is usually rather low, until temperature drops to the general range of 150–300 °C. Under the restricted conditions of maximum gold solubility, stibnite precipitation is considered by Williams-Jones and Normand to be controlled by the mass action expression

$$HSb_2S_4^- + H^+ \rightarrow Sb_2S_3 + H_2S \quad (8)$$

suggesting that acidification can cause stibnite precipitation. Williams-Jones and Normand point out that many of the deposits exhibiting a strong antimony–gold association occur in settings which provide for an alkaline environment of transport (talc-carbonate or limestone host rocks), followed by precipitation in an environment capable of aqueous acidification, such as in fractures hosted by phyllosilicate-rich rocks, wherein reactions such as Equation (3) can produce hydrogen ions, promoting both stibnite and gold deposition. Fluids that are more acidic and oxidized can carry antimony as $Sb(OH)_3^0$, precipitating stibnite upon reduction if sufficient sulfur is available, hence the occurrence of many black shale-hosted stibnite deposits. These ore fluids, however, could not transport much gold because of their oxidized nature, unless the transport temperatures were high enough to promote significant chloride complexing of gold. Antimony is characteristic of a class of elements that possess a combination of high solubility at high temperatures, strongly decreasing solubility with decreasing temperature and/or pressure, and low abundance. These characteristics tend to restrict this class of elements, which includes selenium, tellurium, mercury, thallium as well as antimony, and sometimes arsenic and bismuth, to low-temperature deposits and in many cases to igneous-related deposits.

3.12.1.7 Mineral Deposits

The nonlinear interactions in the crust among thermal gradients, mechanical properties of rocks in shallow crustal environments, and irreversible chemical reactions, especially those operating in strong gradients with significant entropy production (dissipative environments), lead to the uneven, power-law, or lognormal distribution of element concentrations in the crust, geochemical anomalies, mineral deposits, and ores (Shen and Zhao, 2001). With regard to ore systems, we refer to the average concentration (by weight) of the metal in an arbitrarily chosen mass of rock as the grade, usually expressed in percent, or ppm (g t^{-1}). The estimated mass of rock that is characterized by that average grade is referred to as the tonnage. Grade and tonnage data were originally used by Lasky (1950) to show the inverse relationship between average grade and the logarithm of the cumulative tonnage of porphyry copper deposits. For comparison, the Earth's continental crust has a volume $\sim 10^{10}$ km^3, with the volume of the upper-one kilometer occupying a little over 10^8 km^3. Using the upper-one kilometer as a standard of comparison (because mining is limited to the upper few kilometers of the Earth's surface), a very large ore deposit with a "tonnage" or mass on the order of 3,000 Mt (e.g., a world class porphyry copper in the top 5%) and with a volume on the order of 1 km^3 would occupy one part in 10^8 of the upper one-kilometer-thick shell of the Earth's land surface (a millionth of a percent). Of course, most ore or mineral deposits are much smaller. If we consider the upper ten kilometers of the Earth's crust rather than just the upper one kilometer, the proportion would drop by an order of magnitude. In terms of the mass proportion of the continental crust, the same giant ore deposit would represent ~0.1 ppb by mass. Clearly, just as mining occupies a vanishingly small proportion of the area of the Earth's surface, ore deposits do not contribute significantly to the bulk inventories of any element. However, whereas ores and related anomalies may not be quantitatively important, they are important as examples of the limits of enrichment that are possible for any given element.

3.12.1.8 Economic Considerations and the Classification of Ore Deposits: The Example of Vanadium–Uranium Deposits

Under the oxidation state conditions that prevail in the crust (a range of a few log units straddling the QFM-NNO region of $\log f_{O_2}-P-T$ space), vanadium occurs dominantly in the trivalent state. The ionic radius of V^{3+} is virtually identical to that of ferric iron, so it is easily accommodated by ferric iron-bearing minerals. There is roughly 400–500 times more iron than vanadium in the Earth's crust. That is, iron-bearing minerals act as "solvents" for trivalent vanadium, and the bulk of the crust's inventory of vanadium resides in ferromagnesian rock-forming minerals. Economic sources of vanadium are commonly considered to be comparatively rare, and include vanadiferous magnetite, by-product vanadium from uranium mining, and recovery from petroleum residues; other sources include vanadium-bearing iron slag, and flyash. The chemical weathering of vanadium-bearing minerals, the release of trivalent vanadium and its subsequent oxidation, lead to the transport and deposition of vanadates in near-surface environments, where pentavalent vanadium minerals precipitate. Much is made in the economic geology literature of the fact that vanadium is a by-product of uranium mining; however, this result is not so much geochemical, as economic. A similar relationship exists between silver and gold in many deposits. As this is a Treatise on Geochemistry, and the subject is ores, it will be instructive to examine how the economic bias in ore deposit information must be considered in a geochemical context. To discuss this issue further, some background in uranium–vanadium deposits is in order.

Sandstone hosted uranium–vanadium deposits, which account for ~20% of the world's uranium, are found in medium to coarse-grained sandstones deposited in continental-fluvial or marginal-marine sedimentary environments. These units acted as aquifers for the mineralizing oxidized aqueous solutions, and are frequently bounded above and below by lower permeability horizons that act as aquacludes. Uranium, vanadium, and other redox sensitive elements precipitate upon encountering reducing conditions that are produced within the sandstone by carbonaceous biogenic material, aqueous or solid sulfide, or other reduced substances. Three main types of sandstone deposits have been recognized: roll-front deposits, which are characterized by arcuate discordant uranium–vanadium mineralized zones; tabular deposits, with irregular, elongate, tabular mineralized zones concordant to the bedding; and structurally controlled deposits that occur in sandstones adjacent to permeable fault zones. These ore bodies are commonly low to medium grade (~0.1% U_3O_8) and their size ranges up to 50 kt of U_3O_8. According to Fitch (1997), deposits of carnotite on the Colorado Plateau were the first major sources of radium from 1912 to 1922 and of vanadium from 1924 to 1945. Uranium was discarded as part of the vanadium mine tailings, and ultimately, these tailings became a source for much of the uranium used for atomic weapons. Mining of uranium ores for military use began in 1947 and continued until 1970. The use of uranium for nuclear power plants began in the early 1950s (Fitch, 1997). As a result of the importance of uranium since 1945, what were once referred to as vanadium deposits became "uranium," or "uranium–vanadium" deposits. In fact, most tabular sandstone uranium deposits in the USA contain more vanadium than uranium, commonly 7–12 times as much; i.e., geochemically, uranium was a by-product of vanadium mining, even though from 1945 up to the time of the Three Mile Island/Chernobyl incidents, these deposits, economically, were uranium mines. The greatly increased demand for vanadium in the late 1990s (spurred in part by the increased demand for steel, worldwide) was demonstrated by the sharp rise in the price of vanadium from $3.30 per pound of V_2O_5 in January 1997 to $6.70 in January 1998; however, the price of U_3O_8 was still $12 per pound at that time; at that point, many tabular sandstone uranium deposits became uranium–vanadium deposits. Therefore, we must be careful how we interpret ore-related information (e.g., vanadium is "only" a by-product of uranium mining) from a geochemical framework.

The next section discusses select deposit types from a variety of crustal settings and shows how density, temperature, pressure, and chemical potential gradients play a role in the genesis of ores by hydrothermal, igneous, or surficial processes.

3.12.2 SOME ORE-METAL ANOMALIES FORMED IN THE OCEANIC REALM

3.12.2.1 Seafloor Hydrothermal Sulfide Deposits

Seafloor hydrothermal deposits include sedimentary exhalative deposits (Sedex), volcanogenic massive sulfide (VMS), and Besshi massive sulfide deposits. Compared to VMS deposits, Sedex deposits are fewer in number but generally an order of magnitude larger (Goodfellow, 2000). Sedex deposits are major sources of zinc and lead, and can be considered as a subgroup of massive sulfide deposits associated with sedimentary rocks. Both the sediment and volcanogenically hosted massive sulfide deposits are characterized by lenses or sheets of ore that comprise at least 60% sulfide, hence the origin of the name. As such, these deposits represent major crustal anomalies of sulfur.

3.12.2.2 Sedimentary Exhalative Deposits

Sedex deposits constitute many of the world's great ore deposits, including the Sullivan mine in Canada, Mount Isa, the Broken Hill protolith, and HYC in Australia, and Rammelsberg in Germany. The ores are characterized by beds and laminations of sulfides that commonly comprise sphalerite, galena, pyrite and pyrrhotite possibly with chalcopyrite, and barite. The lenses or layers of ore range from centimeters to tens of meters in thickness. Multiple layers may occur over a vertical interval that can be on the order of a kilometer, and can extend up to several kilometers along strike. The term Sedex, or "sedimentary exhalative," is a generic name that reflects the current understanding of the genesis of these deposits as having formed by the precipitation of sulfides from hydrothermal fluids vented or "exhaled" on to the seafloor.

According to MacIntyre (1995), the major Sedex metallogenic events are Middle Proterozoic, Early Cambrian, Early Silurian, and Middle to Late Devonian to Mississippian in age. The Middle-Proterozoic and Devonian–Mississippian events are recognized worldwide. Briskey (1986) gives the median tonnage for zinc–lead–gold deposits as 15 Mt, with 10% of deposits in excess of 130 Mt. According to MacIntyre (1995) the median grades worldwide are zinc 5.6%, lead 2.8%, and silver 30 g t^{-1}. The Sullivan deposit, one of the supergiant Sedex deposits, has a total tonnage of 155 Mt with 5.7% zinc, 6.6% lead, and 7 ppm silver. The lead concentration commonly correlates with the silver concentration (Menzie and Mosier, 1986). Lateral Cu → Pb → Zn → Ba zoning sequences extend outward from vent or feeder zones; ammonium anomalies may also be present.

Sedex deposits generally form in restricted sedimentary basins or half grabens bounded by syn-sedimentary growth faults. The exhalative vents generally occur along the faults and the exhaled brine accumulates in adjacent seafloor depressions. As Sangster (2002) points out, most Sedex deposits are vent-distal. He proposes that saline fluids discharged onto the seafloor and flowed away from the vent as "bottom-hugging" fluids. He suggests that their high densities and velocities prevented them from mixing with overlying seawater, and precluded significant cooling and dilution of the ore fluid until they came to rest in a seafloor topographic low. There is a close spatial relationship between Sedex deposits and mafic volcanic rocks. Compared to VMS deposits, however, Sedex deposits are typically lead–zinc rich and copper poor, and formed under lower temperature conditions (i.e., less than 260 °C).

The redox state of the mineralized brines, which in turn is reflected in whether the ore fluid is dominated by sulfate (oxidized) or sulfide (reduced), is important for controlling minor element associations in Sedex deposits. Cooke *et al.* (2000) point out that weakly acidic to weakly alkaline oxidized brines can precipitate siderite but are incapable of carrying significant gold, tin or barium in solution, and that therefore McArthur River-type deposits do not contain anomalous concentrations of these elements. Reduced, acid brines can carry high concentrations of barium or tin; if these reduced brines contained sufficient sulfide, the deposits may contain anomalous or ore-grade gold (Cooke *et al.*, 2000). Across the SO_4^{2-}/H_2S boundary, the solubility of lead and zinc increases rapidly with increasing oxygen fugacity, due to the decreasing activity of reduced sulfur (Large *et al.*, 1998). Temperature decrease and fluid mixing, addition of H_2S, and pH increase can all be effective processes for the deposition of zinc and lead sulfides from reduced brines. In contrast, sulfate reduction and/or addition of H_2S (by way of fluid mixing or interaction with earlier formed sulfides) may be an important process for the deposition of galena and sphalerite from oxidized brines (Cooke *et al.*, 2000).

In some cases, metalliferous hydrothermal fluids may vent into anoxic basins, precipitating a different suite of ore metals, forming, for example, deposits such as the nickel–molybdenum–zinc–PGE-bearing black shales of China (Murowchick *et al.*, 1994). These deposits comprise thin layers or other segregations of sulfide minerals in black-shale sequences. The geochemical anomalies are marked by above-average concentrations of nickel, molybdenum, rhenium, zinc, gold, PGE, organic matter, phosphates (usually as discrete beds), barium, selenium, arsenic, vanadium, uranium, and sulfur (see Table 3).

3.12.2.3 VMS Deposits

VMS deposits are lens-like to sheet-like volumes of sulfide mineral-rich rock bodies associated with volcanic rocks. Generally, the deposits form in deep-sea environments and belong to the larger class of concordant massive sulfide deposits, which includes all massive or semi-massive sulfide deposits formed by the discharge of hydrothermal solutions in near-seafloor environments. They range in age from ~3,500 Ma for deposits of the Pilbara Craton to the modern black smoker deposits of the mid-ocean ridges. The deposits occur at mid-ocean ridge spreading centers, in the fore-arc areas of convergent zones, back-arc basins, and in sub-aqueous continental rift zones. According to Franklin *et al.* (1981), ~80% of the Earth's VMS deposits are in arc-related successions. The major metals in these anomalies include copper, zinc, and lead, as well as gold, silver, and tin. Generally, the Zn : Cu ratio increases upward and outward from the core of the massive sulfide lens. In this chapter, VMS deposits are separated into three classes: the Cyprus-type, Noranda/Kuroko-type, and Besshi-type deposits.

Cyprus-type VMS deposits contain small, stratiform, basalt-hosted, medium-grade ores dominated by copper, and zinc as the prime commodities. The Upper Cretaceous Troodos ophiolite complex in Cyprus is the type locale for this class of VMS deposit. Generally, these deposits take the form of lens-like masses of pyrite, characteristically associated with basaltic pillow lavas and cherts. The related sequence of mafic and ultramafic rocks associated with Cyprus-type VMS deposits are generated at mid-ocean ridge (MOR) or back-arc spreading centers. Cathles (1981) summarizes a number of studies that suggest that these local redistributions of copper, zinc, iron, sulfur, and other metals occur on oceanic crust that is significantly less than 1 Myr old. The geochemical anomalies

Table 3 Summary of tectonic environments and ore components in selected deposit types.

Deposit type	*Tectonic environment*	*Primary ore metals*	*Other metals present*
Sedex	Fault-controlled basins in intracratonic or continental margin settings	Zn, Pb, Ag	Cu, Ba, Sn, Au
VMS-Cyprus	Ophiolite complexes	Cu	S, Fe, Cu, Au, Ag, Zn, Co, Cd, Pb
VMS-Noranda/Kuroko	Local extensional areas of arc settings	Cu, Pb, Zn, Ag, Au	S, Fe, Ba, As, Se, Cd, Sn, Bi, B
Besshi	Extensional environments in the oceanic realm	Cu, Zn, Ag, Co	S, Fe, Pb, Ni, Mn, As, Sb, Sn, Mo, B
Porphyry deposits	Arcs	Cu, Mo, Au	Re, S, Fe, As, Se, Bi, W, B, Sr, Zn, Pb, Co, V, PGE, Sn
Magmatic Ni–S	Mid-continent rifts	Cu, Ni, PGE, Co	S, Fe
Epithermal	Arcs	Cu, Au, Ag	As, Sb, Hg, Pb, Zn, Cu, Ba, F, Mn, Mo, Se
Ni–Mo–Zn–PGE-bearing black shale deposits	Continental platforms	Ni, Mo	Zn, Pt, Pd, Au, P, Ba, Zn, Se, As, S, U, V

comprise sulfur, iron, copper, gold, silver, zinc, cobalt, cadmium, and lead. The main ore zones may be underlain by a pipe-shaped stockwork of copper-rich "stringer-zones" consisting dominantly of quartz-sulfide (and in some cases paragonite-bearing) veins in heavily chloritized basalt. Ore mineralogy includes pyrite + chalcopyrite + magnetite + sphalerite, with minor galena and pyrrhotite. The main ore zones are structurally controlled by normal faults that served to focus fluid discharge onto the seafloor. Contemporary black smoker deposits found at spreading ridges appear to be modern-day analogues of Cyprus-type VMS deposits.

The Noranda and Kuroko types of VMS deposit represent seafloor hydrothermal anomalies of sulfur, iron, copper, zinc, lead (referred to sometimes as copper–zinc ± lead massive sulfide deposits, because these are dominant economic commodities) which form in what are essentially intermediate to felsic-pyroclastic domes in submarine calderas. The Noranda (Kidd Creek) type deposits occur in Archean–Proterozoic greenstone belts and are lead poor. These are to be distinguished from the Cyprus-type VMS deposits that are also lead poor, but which form at MOR spreading centers (copper–zinc massive sulfide deposits), and are associated with mafic rocks. About half of all VMS deposits are associated with intermediate to felsic-volcanic rocks; the balance are associated with mafic rocks, or mixed-volcanic/sedimentary successions. The Kuroko-type deposits, which are associated with intermediate to felsic-volcanic rocks, are stratabound, and are usually stratiform. The host rocks include rhyolite, dacite, and minor basalt. Fine-grained clastic sediments or sedimentary rocks may also be present. Ore mineralogy generally consists of an upper zone of "black ore" which is dominantly pyrite + sphalerite + chalcopyrite ± pyrrhotite ± galena ± barite, and a lower zone of "yellow ore" which is dominantly pyrite + chalcopyrite ± sphalerite ± pyrrhotite ± magnetite. Barite and chert are common gangue minerals. The conical stringer vein or stockwork feeder zone generally contains pyrite + chalcopyrite. Gold and silver, as well as zinc, lead, and barium tend to increase toward the more distal portions of the anomaly. Black chlorite and talc are common alteration minerals. The anomalies usually comprise sulfur, iron, copper, zinc, lead, barium, arsenic, silver, gold, selenium, cadmium, tin, bismuth, and boron.

Besshi deposits comprise thin sheets of massive, well-layered pyrrhotite + chalcopyrite + sphalerite + pyrite with minor galena and cobalt minerals interlayered with terrigenous clastic sedimentary rocks and calc-alkaline basaltic to andesitic volcanic rocks. Ultramafic material can be present. Their grade and tonnage are generally lower than those of other massive sulfides, but cobalt may be present as an additional commodity. Sulfur, iron, copper, zinc, lead, cobalt, nickel and manganese, arsenic, antimony, tin, molybdenum, and boron may be present in the anomaly. As in many other ore deposits, Besshi anomalies can have high-aspect ratio layers that are just meters thick, yet extend laterally on the kilometer scale. The rather eclectic mixture of metals probably reflects the mixed sources that are present near Besshi-type hydrothermal systems. Copper and zinc are typical products of the alteration of mafic volcanic rocks, whereas significant cobalt and nickel may represent an ultramafic component (Candela *et al.*, 1989). However, tin, lead, molybdenum, and boron are more characteristic of continental crust-related sources.

In the formation of a massive sulfide deposit, cold seawater descends along fractures in the far recharge regions of the seafloor hydrothermal system, is heated, and leaches metals and sulfur from the volcanic succession. The fluid can be drawn into the central regions of the convecting system that is focused on the magmatic heat source. The seawater convection cell is driven by the irreversible loss of heat from cooling submarine igneous bodies. Boiling is suppressed during the formation of VMS deposits, allowing the ore fluid to exit onto the seafloor; for this reason, they must form at or above a minimum pressure for any given temperature of the hydrothermal fluid. The minimum pressure to prevent boiling increases as the fluid temperature increases. For the most efficient ore formation in either the Kuroko–Noranda or Cyprus-type VMS deposits, discharge flow is best controlled by a few large fractures, which focus flow through a narrow lateral region. However, recharge favors ore formation if it is diffuse, with long transit times. Here, large through-going fractures would short-circuit the leaching path, reducing the contact between the heating fluid and the rock with which it is strongly out of chemical equilibrium (Cathles, 1981). The fluid that exits onto the seafloor at a submarine hydrothermal vent is saline, reduced, acidic, metalliferous, and sulfidic. During prograde hydrothermal alteration, calcium, sodium, and some iron are leached from the basalt, along with ore metals and sulfur, whereas seawater magnesium is fixed in chlorite and related minerals. Seawater sulfate is reduced by ferrous iron in glass and silicates (basalts are commonly half glass), producing ferric iron and H_2S, with the oxidized iron fixed in magnetite, hematite, or silicates. HCl is also produced by magnesium fixation, as illustrated by the schematic reaction

$$4H_2O + 2Mg^{2+(sw)} + CaAl_2Si_2O_8^{(glass,fs)} \rightarrow Mg_3Si_2O_5(OH)_4^{(chl)} + Al_2Si_{-1}Mg_{-1}^{(chl)} + SiO_2^{(aq)} + Ca^{2+} + 4H^+ \quad (9)$$

and by other reactions. This makes the fluids more aggressive with respect to leaching of metals (manganese, iron, copper, barium, zinc, etc.) at higher temperatures. The mixing of these waters with bottom ocean water near 2 °C causes mineral precipitation, dominantly by cooling upon dilution. Cathles (1981) points out that the salinity of the discharged fluid also determines its fate. Brines are ponded in depressions in the seafloor, as has occurred in the past in the Red Sea.

The hydrothermal fluids may interact with the oceanic crust at any temperature between the temperature of ambient seawater and temperatures on the order of the solidus of basaltic magma (~1,000 °C). However, the exit temperatures on the seafloor in modern hydrothermal vent areas can be on the order of 350–400 °C. Although significant permeability probably does not develop until the rock cools and cracks (by differential thermal contraction) at or near 800 °C (Lister, 1974), other more subtle thermodynamic factors may be involved in controlling the temperature seafloor hydrothermal fluids. Jupp (2000) suggests that this discrepancy between rock and fluid temperatures is due to the variations in the thermodynamic and transport properties of water near the critical point which impose an upper limit of ~400 °C on the temperature of black smokers. Jupp (2000) predicts that temperatures greater than ~400 °C are restricted to a thin boundary layer near the base of the region that supports the aqueous convection cell and which may correspond to the reaction zone observed in ophiolite complexes. He refers to fluids near the critical points as having maximum “flexibility.” Thus, the temperature of venting of the fluids (350–400 °C) appears to be related to the temperatures at which seafloor vent fluids experience maxima in those physical properties that promote flow of the fluid in a convective regime.

3.12.3 CHROMITE AND RELATED MAGMATIC SEGREGATIONS IN THE OCEANIC AND CONTINENTAL REALM

Most of the world’s chromium reserves occur in large, layered mafic/ultramafic intrusions such as the 2.06 Gyr old Bushveld Complex in the Republic of South Africa, where nearly monomineralic layers of chromite appear to result from the mixing of primitive and evolved liquids that lie on or near the chromite/olivine cotectic. However, significant commercial production of chromium also results from ophiolite-hosted podiform chromite or chrome-spinel deposits which occur in alpine-type ultramafic bodies that have been incorporated into the continental crust by obduction. In these deposits, the chromite occurs as podiform bodies found within harzburgite–dunite lithologies. The dunites and harzburgite represent cumulates from the crystallization of MOR basalt, rather than depleted mantle material (otherwise, the banded chromitites are hard to explain). Examples include the deposits of Cyprus, Oman, Newfoundland, and minor showings in many ancient suture zones such as the Appalachians of eastern North America (cf. Wylie and Candela, 1999). The large layered mafic intrusions are gabbroic, and were emplaced into stable cratons, with the economically important bodies all Proterozoic in age.

The intrusions are generally in the form of large upward flaring funnel-shaped bodies, thousands of meters thick, and cover thousands of square kilometers. The uniting factor that may link the chromite deposits of the layered mafic intrusions with the podiform-type deposits is their association with rifting, the former continental (and failed), and the latter oceanic. Ultimately, chromium is concentrated due to the highly refractory (insoluble) nature of chromite in silicate melt systems, which causes chromite to precipitate early from mafic magmas. Irvine (1977) suggested that nearly monomineralic layers of chromite could form by contamination of the magma by felsic material or by mixing of more primitive magma with more evolved residua. Mixing can induce precipitation of chromite only because of the curvature of the chromite/olivine cotectic. Curvature is common for cotectics of sparingly soluble minerals in melts (e.g., chromite, sulfides, and apatite) especially when those minerals differ significantly in composition from the bulk melt composition; curvature of the cotectic crudely reflects changes in solubility of the mineral in the silicate melt with temperature. In general, contamination of less-evolved melt with more evolved material may cause saturation of the melt with respect to chromite (or sulfide, apatite, or other minerals that behave similarly). PGE sulfides may also crystallize as suggested by Kruger *et al.* (2002) for the Bushveld. The PGEs are found associated with minor sulfide-rich bands comprising dominantly pyrrhotite, pyrite, and pentlandite, along with minor but important platinum sulfides. The sulfides are interstitial to the chromite and other minerals of the chromitites. Significant vanadium also occurs in separate magnetite layers within the Bushveld system, representing yet another resource derived from gravitationally segregated precipitates from mafic magmas in rift environments.

Primary nickel sulfide deposits (with associated copper, cobalt, and PGE mineralization) are typically associated with plutonic and volcanic rocks of mafic or ultramafic composition. For example, the Voisey's Bay nickel–copper–cobalt sulfide deposit of Labrador is hosted by the troctolitic Voisey's Bay intrusion, a member of the Proterozoic Nain Plutonic Suite. As of 1999, the total resource was estimated to be 137 Mt of ore containing 1.6% nickel, 0.85% copper, and 0.09% cobalt. Contamination of the mafic magma by a sulfide-bearing paragneiss promoted the development of sulfide liquids, which interacted with successive magma batches resulting in ore-metal enrichment in the sulfide phase (Kerr, 2000). The saturation of a silicate melt with an immiscible sulfide liquid is the fundamental process that forms magmatic sulfide deposits. Droplets of sulfide liquid settle through less dense silicate magma; cobalt, copper, nickel, and PGE partition strongly into the sulfide liquid phase at levels 10–100,000 times those in silicate liquids (Foose *et al.*, 1995). The mineralized zones at Voisey's Bay are associated with fragment-bearing troctolites and olivine gabbros related to the conduit dikes, rather than as basal accumulations at the floor of the chambers (Evans-Lamswood *et al.*, 2002).

In general then, magmatic segregations of oxides, phosphates (nelsonites?), or sulfides can be expected during contamination of the magma, or by mixing of early and late crystallization products. They generally result from the upward advection of magma, contamination of the magma, precipitation of minerals due to a strong decrease in solubility upon compositional evolution of the melt (e.g., chromite and phosphate), concentration of elements by partitioning from large masses of magma into small masses of "getter" phases (e.g., sulfides in the nickel-sulfide deposits), and ultimately, to gravitational density-driven segregation.

3.12.4 SOME ORE-METAL ANOMALIES FORMED IN THE CONTINENTAL REALM: THE EXTENDED FAMILY OF (INTERMEDIATE TO FELSIC) MAGMATIC–HYDROTHERMAL SYSTEMS

Over 65% of all mined copper comes from deposits associated with igneous rocks, i.e., the porphyry, skarn, epithermal, replacement, and massive sulfide deposits, with just over half coming from porphyry type ores (Singer, 1995). The next most important category for contained copper, according to Singer, is sediment-hosted deposits (23%). In contrast, the proportion of zinc from igneous-related systems is half that of copper. Singer calculates that only 19% of mined gold comes from igneous-related systems, and about one-third of that is in epithermal deposits. Sixty percent of all discovered gold is estimated by Singer to come from modern placers and residual deposits, and ancient placers such as the Witwatersrand deposits in South Africa. Molybdenum is even more strongly associated with igneous rocks than is copper, with well over 99% of the world's molybdenum coming from porphyry type deposits (Kirkham and Sinclair, 1996).

Porphyry deposits are the archetype of the magmatic–hydrothermal class of deposits (also known more descriptively as granite-related deposits). They exhibit a strong spatial and temporal relationship between epizonal, porphyritic/variably-textured intermediate to felsic igneous rocks and usually steep-walled, crudely cylindrical sulfide-rich ore bodies. The high-level igneous stocks associated with porphyry and related deposits typically form, as mentioned earlier, the roof zones of shallow plutons where vertically elongated, high-aspect ratio cupolas (Sutherland-Brown, 1976) act to focus upward surges of magma and associated magmatic volatiles. Porphyry copper ore occurs as disseminations or stockwork veins yielding grades of copper on the order of 0.4–1%, with subordinate molybdenum and gold. Porphyry molybdenum deposits have grades of a few tenths of a percent MoS_2, and porphyry gold deposits contain on the order of 1 ppm gold. Characteristic alteration styles are present which represent the combined effects of fluid cooling and the mixing of magmatic and meteoric waters.

The formation of granite-related ore deposits is a by-product of the irreversible transfer of magmatic heat and mass from the Earth's interior to its surface. Magma emplacement at epizonal levels leads to magmatic volatile phase (MVP) saturation if the system is initially volatile-phase undersaturated, or to further exsolution of volatiles if the magma is already volatile saturated (Candela, 1989). The MVP is thought to be a critical agent in ore formation because of the affinity displayed by many ore metals for a chloride-bearing volatile phase relative to rock-forming minerals (Candela and Piccoli, 1995), and because of the high fluidity and the buoyancy of the MVP in the magma. These physical attributes may allow the volatile phase to accumulate near and above suitable apical magmatic structures where it may precipitate a significant proportion of its sparingly soluble load. At epizonal depths, volatile saturation may occur early relative to crystallization progress because of low-prevailing load pressure. MVP saturation occurs when the sum of the vapor pressures of the components of a liquid is equal to or exceeds the load pressure;

boiling can result from a decrease in the load pressure or an increase in the vapor pressure of the magma. A decrease in load pressure for a given vapor pressure of the dissolved constituents (at constant temperature and melt composition) results in first boiling. As many melts are known to possess concentrations of water on the order of 3–6 wt.%, some with CO_2 concentrations of up to a few thousand ppm (Lowenstern, 1994), partial pressures of magmatic gases can sum to a few hundred MPa.

Volatile exsolution can also be brought about isobarically. If a liquid mixture of volatile and relatively nonvolatile components is cooled below its liquidus and begins to crystallize phases that are dominantly anhydrous, the mole fraction of the volatile components in the mixture, and, concomitantly, the vapor pressure of the magma will increase. This second mode of boiling, which is accompanied by cooling, and crystallization of nonvolatile components is termed second boiling.

Some workers have suggested that mafic magma input into shallow felsic magma chambers is an important part of the volatile saturation paradigm for porphyry-type systems (Hattori and Keith, 2002). Indeed, there is ample evidence that mafic and felsic magmas interact in complex manners in subvolcanic environments (Hattori and Keith, 2002). Further, this process may be important in the budgets for chlorine and sulfur (and possibly CO_2) for the ore–magma–volatile system, and significant amounts of sulfur and chlorine can be transferred from subcrustal melts into granite magma chambers, as suggested for chlorine by Piccoli and Candela (1994) on the basis of apatite chemistry in zoned intrusions in arc systems. However, it seems unlikely that the mafic magma is critical for saturation of the system with respect to a water-rich volatile phase.

In summary, we might expect magma, after some differentiation at depth, to experience volatile exsolution and further crystallization upon ascent. During the phase changes occurring at each stage, ore metals and other magmatic constituents will partition among the melt, mineral, and volatile phases (whether in more primitive mafic magmas or more evolved felsic magmas), in accord with the appropriate partitioning and solubility equilibria. Therefore, the timing of volatile saturation relative to crystallization of any of the causative magma batches is critical in our understanding of ore genesis according to the orthomagmatic–hydrothermal model.

The initial water concentration of a melt can only be defined for a given initial melt composition. Note also that the concentration of water in the melt phase is different from the water concentration in the magma. A magma with 2 wt.% water, 50% crystals, and which is volatile-phase undersaturated, contains a melt phase with 4 wt.% water. The competition for ore metals (and other elements) between the magmatic volatile phase (MVP) and the crystallizing magmatic mineral assemblage during crystallization is a first order control on the efficiency with which metals can be removed from magma into an evolving MVP or "ore fluid" (Candela and Holland, 1986). The values of the pertinent crystal/melt and MVP/melt partition coefficients, together with the timing of crystallization relative to saturation with respect to a water-rich MVP, is a major factor in determining whether the MVP or the crystallizing phases take up a higher proportion of a given ore metal. Detailed accounts of ore–metal partitioning in felsic melt–crystal systems have been published (see Candela, 1992; Jugo *et al.*, 1999), and some conclusions are summarized here.

Molybdenum and tungsten can be partitioned into titanium-bearing accessory phases such as ilmenite, magnetite (and other spinels), sphene, and biotite. Both gold and copper are partitioned strongly into some accessory magmatic-sulfide minerals (which are stable at $f_{O_2} < \text{NNO} + 1$; Carroll and Rutherford, 1985) and their protracted crystallization can result in the strong depletion of melts with respect to copper and/or gold. If a significant amount of copper- and gold-bearing sulfide is removed from the magmatic system before saturation of the melt with respect to a volatile phase, the probability of formation of a copper or gold-rich ore or protore will be significantly reduced. In granitic magmas with oxygen fugacities between QFM and NNO, a small amount of pyrrhotite crystallization might remove a significant amount of copper, but not of gold. For some metals such as copper and gold that possess high-MVP/melt partition coefficients (see Candela *et al.*, 1996), the role of f_{O_2} is obviated in cases where hot, primitive melts with high initial water concentrations are emplaced in the epizone. Under these conditions, volatile exsolution may occur before significant fractionation of sulfides, regardless of the oxygen fugacity. However, copper mineralization is usually restricted to relatively oxidized magmas at shallow levels, as even small amounts of pyrrhotite can remove significant amounts of copper from a crystallizing melt (Jugo *et al.*, 1999). Candela and Holland (1986) suggested that magmatic sulfides might, in some cases, react out of the hypersolidus mineral assemblage upon saturation of a magma with a volatile phase, as sulfur is removed from the melt along with other exsolving volatiles. However, Keith *et al.* (1997) argue that metals sequestered in sulfides can be repartitioned into the MVP after volatile-phase saturation, if magmatic sulfides are crystallized at, or brought to, the level of emplacement where volatile exsolution occurs. This process depends on the proportion of ore metals in a magmatic system that are partitioned into these phases during *in situ* crystallization at the level of emplacement versus the proportion of ore metals that are partitioned into crystallizing phases at depths below the level of emplacement and remain at sites distal to MVP exsolution. In the latter case, ore metals are certainly lost from the system.

Metals (excluding REEs) that tend to increase in concentration upon magmatic fractionation include lithium, molybdenum, tungsten, tin, rubidium, beryllium, caesium, bismuth, niobium, tantalum, manganese, uranium, and thorium. These metals therefore dominate ores associated with highly fractionated igneous rocks. However, contrary to conventional wisdom, igneous fractionation is a relatively inefficient means of forming an ore deposit. Consider a highly incompatible element, e.g., rubidium, with an initial melt concentration of 50 ppm and a bulk crystal/melt partition coefficient ~0.2. A highly evolved fractionate resulting from 90% crystallization would be considered "geochemically enriched," with a rubidium concentration of 315 ppm. However, only 63% of the rubidium originally present would remain in the melt phase. That is, fully one-third of the rubidium would have been lost to fractionated crystallized phases. At a value for the bulk partition coefficient for rubidium equal to a more realistic 0.6 (e.g., for a melt crystallizing one-third plagioclase, one-third potassium feldspar, and one-third quartz; Cavazzini, 2001), the rubidium concentration still increases by greater than a factor of 2.5, yet fully three-fourth of the rubidium originally present in the melt would have been lost to the products of crystallization. That is, ore metals are continually dispersed throughout consolidating plutons, and hence the crust, by the action of crystallization. However, some ore metals sequestered and dispersed less readily than others, and the composition of magmatic–hydrothermal ores is dominated by those metals that suffer least in the battle of attrition waged by magmatic crystallization.

In summary, early saturation with a water-rich volatile phase (Candela and Holland, 1986) allows the evolving MVP maximum access to crystal-, or sulfide liquid-compatible ore metals, such as copper or gold. In a relatively dry magma that is crystallizing at a relatively deep level, exsolution of a water-rich MVP will occur *late* in magmatic crystallization progress. Thus, there are two main factors affecting the efficiency of ore-metal removal from magmas: the sequestering of ore metals into crystallizing magmatic phases, and the timing of volatile saturation relative to crystallization (Candela and Holland, 1986), with early volatile exsolution allowing ore metals, especially those that are crystal compatible, to be available for partitioning into the MVP.

3.12.4.1 Epithermal Deposits

Epithermal hydrothermal ore deposits form at shallow levels in volcano-plutonic arcs systems, and fall into two classes: the adularia-sericite type (also called low-sulfidation deposits) and the high-sulfidation state acid-sulfate type. These are sometimes referred to as quartz-adularia and quartz-alunite type deposits, respectively. The acid-sulfate deposits include the Nansatsu district, Japan; Goldfield, Nevada; El Indio, Chile; and Lepanto-Far Southeast, the Philippines. The low-sulfidation deposits include Creede, Colorado; McLauglin, California; Comstock and Round Mountain, Nevada; and Hishikari, Japan. Both types form in near-surface volcanic environments, and may have porphyry-type ore systems at depth. They are commonly found at shallow levels in volcanic environments including volcanic islands, continental magmatic arcs, and extensional regimes (Hedenquist and Lowenstern, 1994). The acid-sulfate deposits comprise shallowly formed veins, stockworks, breccias, or massive replacements associated with vuggy quartz and acid-leached advanced argillic-type alteration. Common metals in the anomalies include gold and copper, with arsenic, silver, zinc, lead, antimony, molybdenum, bismuth, tellurium, tin, boron, and mercury. The ores are commonly rich in pyrite and copper-bearing sulfosalts. The adularia-sericite deposits comprise shallowly formed quartz veins, stockworks, and breccias, with open-space filling textures. Common metals present in the anomalies include gold–silver–arsenic–antimony–mercury–lead–zinc–copper. Barium, fluorine, manganese, molybdenum, and selenium may also be present. Some quartz-adularia deposits have gold tellurides, and the vanadium-bearing mica, roscoelite. Heald *et al.* (1987) point out that the base metal contents of the acid-sulfate-type deposit are relatively high and relatively copper rich, whereas there is great variability in the base-metal contents of the adularia-sericite-type deposits (which tend to be relatively copper poor). Adularia-sericite-type gold–silver deposits are also considerably more abundant than acid-sulfate-type deposits. Heald *et al.* further suggest that the two types of deposits form under similar conditions of depth (1–2 km) and temperature (100–300°) but in different parts of paleogeothermal systems. Acid-sulfate deposits form in the root zones of volcanic structures from acid waters that contain residual magmatic volatiles, whereas the adularia-sericite variants are deposited from neutral to weakly acidic, alkali chloride waters produced by the mixing of surficial waters with deeper, heated saline waters in a lateral flow regime, above and probably offset from the magmatic hearth. Boiling and mixing are two of the more commonly proposed mechanisms for ore deposition in the quartz-adularia deposits. The longer path length for the flow of magmatic fluids promotes both dilution of the fluid by meteoric waters, and equilibration with wall rock, both of which promote a reduction in the sulfur concentration of the ore fluid and a lowering of the oxidation state of the system. In the acid-sulfate systems, cooling magmatic vapors are titrated directly into the overlying hydrological system. Deposition occurs in acid-sulfide systems due to either cooling or mixing (dilution).

3.12.5 (GEOCHEMICALLY SIGNIFICANT) PLACERS/RESIDUAL ORES AND CRUSTAL ANOMALIES (ALUMINUM, NICKEL, GOLD)

Lateritic weathering, promoted by warm-humid environments and low rates of erosion, can enhance some geochemical anomalies to the point where they may be mined at a profit. For example, lateritic weathering of dunites and peridotites or their serpentinized equivalents can produce ores of nickel with 1–2% nickel. Nickel is commonly hosted by either iron oxides or silicates (garnierite). The deposits may be crudely vertically zoned with pisolitic iron-oxide and nickel-bearing zones above richer saprolitic silicate ore.

Aluminum, the third most abundant element in the Earth's crust, is mined as bauxite, a mixure of gibbsite, boehmite, and diaspore. These ores occur dominantly in Cenozoic deposits formed by lateritic weathering of aluminous parent rock, which has broken down to leave a high proportion of aluminum-bearing minerals. These deposits are residual in nature, with the balance of the rock constituents removed by selective leaching. These deposits form most commonly in tropical and subtropical areas.

Another surficial deposit type includes the manganese-rich crusts and nodules that occur on basalt near mid-ocean ridges, and near the sediment–water interface. Whereas the ultimate source of manganese and iron (as well as the associated nickel, copper, and cobalt) is the hydrothermal alteration of MOR basalt, the fields of manganese-rich encrustations and nodules that cover large areas of sediment-starved ocean floor are the result of surficial authigenic upward remobilization of metals and the fixing of those metals at the sediment–water interface.

The final case of surficial ore genesis dealt with here involves the transport of physically disaggregated residuum from physically and chemically weathered materials by wind and water, resulting in placers, mineral deposits formed at the Earth's surface by mechanical concentration of dense minerals during clastic sedimentation. These processes can act as part of the normal river, lacustrine, beach, aeolian, or glacial processes that today form modern sediments, or have acted in the past to produce ancient sediments, sedimentary rocks, or their metamorphic equivalents. For example, moving water can sort particles by density, shape, etc. Common placer minerals include gold and PGE, cassiterite, garnet, corundum, diamond, wolframite, zircon, as well as magnetite, ilmenite, rutile, and other minerals. The minerals may be concentrated by hydraulic processes in marine (beach ridges or offshore bars), glacial (moraine and outwash), or in alluvial fan, fluvial or lacustrine environments. The super giant Witwatersrand detrital gold-uraninite ore fields formed during the Archean, when uraninite could be transported without suffering oxidation before deposition. These deposits are essentially paleofluvial placers, with ore (characterized by gold concentrations on the order of 10 ppm) occurring in extremely thin sheets with breadth to thickness ratios approaching 10^5 (Gilbert and Park, 1986).

3.12.6 EPILOGUE

A confluence of factors determines the general distribution of the scarce metals in the Earth's crust. First, there is the similarity of the element to a major rock-forming mineral. Possibly no element exemplifies this better than rubidium, which does not form any known mineral of its own, is commonly camouflaged in potassium minerals, and consequently is dispersed in the feldspar minerals that characterize the Earth's crust. Even the richest rubidium ores do not contain rubidium minerals, *sensu stricto*; ore-grade rubidium occurs in lepidolite and pollucite (potassium–lithium and caesium minerals, respectively), which occur as minor by-products in the Tanco rare-metal pegmatite deposit of Bernic Lake, Manitoba, Canada (Cerny *et al.*, 1996). The driving force for camouflage, of course, is entropy of mixing on crystalline sites, which, at equilibrium, lowers the overall free energy of the system as long as bonding and size considerations do not preclude it. To a lesser extent, the same fate is suffered by an element such as vanadium, which is camouflaged by ferric iron, and is therefore commonly dispersed though the ferromagnesian constituents of the crust. However, vanadium does form its own minerals, and in fact does form its own deposits, even though it is commonly considered a by-product of uranium mining. This, however, may be seen as something of an economic bias, as the post-World War II industrial boom coincided with the dawn of the "nuclear age," and the importance of uranium.

The group Ib elements—copper, silver, and gold—stand in contradistinction to elements such as rubidium and vanadium. These metals are rather electronegative and soft, and prefer two-coordinate directional bonding. These characteristics are not conducive to the incorporation of these elements into rock-forming oxide matrices, and they therefore form their own minerals, occur in the native state, or reside in trace sulfides in the crust. They also partition readily into fluid phases, promoting their redistribution in the crust, and the formation of ores.

REFERENCES

Briskey J. A. (1986) Descriptive model of sedimentary exhalative Zn–Pb. In *Mineral Deposit Models* (eds. D. P. Cox and D. A. Singer). US Geol. Surv. Bull. 379pp.

Candela P. A. (1989) Felsic magmas, volatiles, and metallogenesis. *Ore Deposit. Assoc. Magmas* **4**, 223–233.

Candela P. A. (1992) Controls on ore metal ratios in granite-related ore systems: an experimental and computational approach. In *The Second Hutton Symposium on the Origin of Granites and Related Rocks,* Proceedings (eds. P. E. Brown and B. W. Chappell). Geological Society of America (GSA), pp. 317–326.

Candela P. A. and Holland H. D. (1986) A mass transfer model for copper and molybdenum in magmatic hydrothermal systems: the origin of porphyry-type ore deposits. *Econ. Geol. Bull. Soc. Econ. Geol.* **81**(1), 1–19.

Candela P. and Piccoli P. (1995) Model ore-metal partitioning from melts into vapor and vapor/brine mixtures. In *Granites, Fluids, and Ore Deposits* (ed. J. F. H. Thompson). Mineral. Assoc., Canada, vol. 23, pp. 101–128.

Candela P. A., Wylie A. G., and Burke T. M. (1989) Genesis of the ultramafic rock associated Fe, Cu, Co, Zn, and Ni deposits of the Sykesville district, Maryland Piedmont. *Econ. Geol.* **84**, 663–675.

Candela P. A., Piccoli P. M., and Williams T. J. (1996) Preliminary study of gold partitioning in a sulfur-free, high oxygen fugacity melt/volatile phase system. In *Geological Society of America*, 28th Annual Meeting, 287 (ed. Anonymous). Geological Society of America (GSA), 402pp.

Carroll M. R. and Rutherford M. J. (1985) Sulfide and sulfate saturation in hydrous silicate melts. In *Proceedings of the 15th Lunar and Planetary Science Conference,* Part 2 (eds. G. Ryder and G. Schubert). American Geophysical Union. pp. C601–612.

Cathles L. M. (1981) Fluid flow and genesis of hydrothermal ore deposits. *Econ. Geol. Bull. Soc. Econ. Geol.* **75**, 424–457.

Cavazzini G. (2001) Modelling Sr isotopic evolution in mineral phases growing from magmatic liquids with changing (super 87) Sr/(super 86) Sr. *Geochem. J.* **35**(6), 421–438.

Cerny P., Ercit T. S., and Vanstone P. T. (1996) Petrology and mineralization of the Tanco rare-element pegmatite, southeastern Manitoba. *Geol. Soc. Can./Mineral. Soc. Can. Field Trip Guide*, A4, Winnipeg, Manitoba.

Chou I.-M. (1987) Phase relations in the system NaCl–KCl–H_2O: III. Solubilities of halite in vapor-saturated liquids above 445°C and redetermination of phase equilibrium properties in the system NaCl–H_2O to 1000°C. *Geochim. Cosmochim. Acta* **51**, 1965–1975.

Connors K. A., Noble D. C., Bussey S. D., and Weiss S. I. (1993) Initial gold contents of silicic volcanic rocks: bearing on the behavior of gold in magmatic systems. *Geology* **21**, 937–940.

Cooke D. R., Bull S. W., Large R. R., and McGoldrick P. J. (2000) The importance of oxidized brines for the formation of Australian Proterozoic stratiform sediment-hosted Pb–Zn (sedex) deposits. *Econ. Geol. Bull. Soc. Econ. Geol.* **95**(1), 1–17.

Cygan G. and Candela P. (1995) Preliminary study of Au partitioning among Pyrrhotite, Pyrite, Magnetite, and Chalcopyrite in Au-saturated chloride solutions at 600 to 700 °C, 140 MPa (1400 bar). In *Magmas, Fluids, and Ore Deposits* (ed. J. F. H. Thompson). Mineralogical Association of Canada, pp. 129–138.

Evans-Lamswood D. M., Butt D. P., Jackson R. S., Lee D. V., Muggridge M. G., Wheeler R. I., and Wilton D. H. D. (2002) Physical controls associated with the distribution of sulfides in the Voisey's Bay Ni–Cu–Co deposit. *Labrador. Econ. Geol.* **95**(June/July), 749–770.

Fitch W. I. (1997) Uranium, its Impact on the National and Global Energy Mix. USGS, Reston, 24p.

Foose M. P., Zientek M. L., and Klein D. P. (1995) Magmatic sulfide deposits (MODELS 1, 2b, 5a, 5b, 6a, 6b, and 7a; Page, 1986a-g). http://geology.cr.usgs.gov/pub/open-file-reports/ofr-95-0831/CHAP4.PDF

Frank M. R., Candela P. A., and Piccoli P. M. (1999) K-feldspar-muscovite-andalusite-quartz-brine phase equilibria: an experimental study at 25 to 60 MPa and 400 to 550 °C. *Geochim. Cosmochim. Acta* **62**, 3717–3727.

Frank M. R., Candela P. A., Piccoli P. M., and Glascock M. D. (2002) Gold solubility and partitioning as a function of HCl in the brine-silicate melt-metallic gold system at 800 °C and 100 MPa. *Geochim. Cosmochim. Acta* **66**, 3719–3732.

Franklin J. M., Sangster D. M., and Lydon J. W. (1981) Volcanic associated massive sulphide deposits. *Econ. Geol.* 485–627 (75th Anniversary Volume).

Gammons C. H. and Williams-Jones A. E. (1997) Chemical mobility of gold in the porphyry-epithermal environment. *Econ. Geol. Bull. Soc. Econ. Geol.* **92**(1), 45–59.

Gilbert J. M. and Park C. F., Jr. (1986) The Geology of Ore Deposits. Freeman and Co, New York, 750p.

Goodfellow W. D. (2000) Sedimentary basinal fluid compositions, anoxic oceans and the origin of SEDEX Zn–Pb deposits. *Can. Soc Explor. Geophys.* http://www.cseg.ca/conferences/2000/329.PDF accessed September 2002.

Hartog F., Jonkers G., Schmidt A., and Schulling R. (2001) Lead deposites in Dutch natural gas systems. In *Third International Symposium on Oilfield Scale*, Aberdeen, Scotland.

Hattori K. H. and Keith J. D. (2002) Contribution of mafic melt to porphyry copper mineralization: evidence from Mount Pinatubo, Philippines, and Bingham Canyon, Utah, USA. *Mineral. Dep.* **36**, 799–806.

Heald P., Foley N. K., and Hayba D. O. (1987) Comparative anatomy of volcanic-hosted epithermal deposits: acid-sulfate and adularia-sericite types (USGS). *Econ. Geol.* **82**, 1–26.

Hedenquist J. W. and Lowenstern J. B. (1994) The role of magmas in the formation of hydrothermal ore deposits. *Nature (London)* **370**(6490), 519–527.

Hemley J. J. and Hunt J. P. (1992) Hydrothermal ore-forming processes in the light of studies in rock-buffered systems: II. Some general geologic applications. *Econ. Geol.* **87**, 23–43.

Hemley J. J., Cygan G. L., Fein J. B., Robinson G. R., and d'Angelo W. M. (1992) Hydrothermal ore-forming processes in the light of studies in rock-buffered systems: I. Iron–copper–zinc–lead sulfide solubility relations. *Econ. Geol.* **87**, 1–22.

Henley R. W. and Berger B. R. (2000) Self-ordering and complexity in epizonal mineral deposits. *Ann. Rev. Earth Planet. Sci.* **28**, 669–719.

Hill K. C., Kendrick R. D., Crowhurst P. V., and Gow P. (2002) Copper-gold mineralisation in New Guinea: tectonics, lineaments, thermochronology and structure. *Austral. J. Earth Sci.* **49**, 737–752.

Hobbs B. E. and Ord A. (1997) Plumbing systems responsible for the formation of giant ore deposits. In *Contributions to the 2nd International Conference on Fluid Evolution, Migration and Interaction in Sedimentary Basins and Orogenic Belts, Belfast* (eds. J. P. Hendry, P. F. Carey, J. Parnell, A. H. Ruffell, and R. H. Worden). Geofluids II '97, pp. 100–102.

Ilchik R. P. and Barton M. D. (1997) An amagmatic origin of carlin-type gold deposits. *Econ. Geol. Bull. Soc. Econ. Geol.* **92**(3), 269–288.

Irvine T. N. (1977) Origin of chromitite layers in the Muskox intrusion and other layered intrusions: a new interpretation. *Geology* **5**, 573–577.

Jugo P. J., Candela P. A., and Piccoli P. M. (1999) Magmatic sulfides and Au:Cu ratios in porphyry deposits: an experimental study of copper and gold partitioning at 850 degrees C, 100 MPa in a haplogranitic melt-pyrrhotite-intermediate solid solution-gold metal assemblage, at gas saturation. In *Granites: Crustal Evolution and Associated Mineralization* (eds. A. N. Sial, W. E. Stephens, and V. P. Ferreira). Elsevier, pp. 573–589.

Jupp T. E. (2000) Fluid flow processes at mid-ocean bridge hydrothermal systems. Thesis, University of Cambridge, Cambridge, UK. 229p.

Keith J. D., Whitney J. A., Hattori K., Ballantyne G. H., Christiansen E. H., Barr D. L., Cannan T. M., and Hook C. J. (1997) The role of magnetic sulfides and mafic alkaline magmas in the Bingham and Tintic mining districts, Utah. In *High Level Silicic Magmatism and Related Hydrothermal Systems* (eds. R. Seltmann, B. Lehmann, J. B. Lowenstern, and P. A. Candela). IAVCEI '97 selected papers. Clarendon Press, pp. 1679–1690.

Kerr A. (2000) Mineral Commodities of Newfoundland 3812 and Labrador (2000). *Geol. Surv. Newfoundland and Labrador.* 12p.

Kirkham R. V. and Sinclair W. D. (1996) Porphyry copper, gold, molybdenum, tungsten, tin and silver. *Geol. Can. Min. Deposit Types.*

Kruger F. J., Kinnaird J. A., Nex P. A. M., and Cawthorn R. G. (2002) Chromite is the Key to PGE. In *9th International Platinum Symposium.* Duke University. http://www.duke.edu/~boudreau/9thPtSymposium/Kruger_Abstract2.pdf. accessed October, 2002.

Large R. R., Bull S. W., Cooke D. R., and McGoldrick P. J. (1998) A genetic model for the H.Y.C. Deposit, Australia: based on regional sedimentology, geochemistry, and sulfide-sediment relationships. *Metall. McArthur River-Mount Isa-Cloncurry Min. Prov.* **93**(8), 1345–1368.

Lasky S. G. (1950) How tonnage and grade relations help predict ore reserves. *Eng. Mining J.* **151**(4), 81–85.

Lister C. R. B. (1974) On the penetration of water into hot rock. *Geophys. J. Roy. Soc.* **39**, 465–509.

Lowenstern J. B. (1994) Chlorine, fluid immiscibility, and degassing in peralkaline magmas from Pantelleria, Italy. *Am. Min.* **79**(3–4), 353–369.

MacIntyre D. (1995) Sedimentary exhalative Zn–Pb–Ag. In *Selected British Columbia Mineral Deposit Profiles* (eds. D. V. Lefebure and G. E. Ray). British Columbia Ministry of Energy, Mines and Petroleum Resources, pp. 37–39.

Manning C. E. and Ingebritsen S. E. (2001) Role of the brittle-ductile transition in large-scale fluid flow in active continental crust. In *Earth System Processes.* Geological Society of America, Edinburgh, Scotland.

Marsh B. D. (1996) Solidification fronts and magmatic evolution. *Min. Mag.* **60**, 5–40.

Menzie W. D. and Mosier D. L. (1986) Grade and tonnage model of sedimentary exhalative Zn–Pb. *Min. deposit models* 1693.

Murowchick J. B., Coveney R. M., Jr., Grauch R. I., Eldridge C. S., and Shelton K. L. (1994) Cyclic variations of sulfur isotopes in Cambrian stratabound Ni–Mo–(PGE-Au) ores of southern China. *Geochim. Cosmochim. Acta* **58**(7), 1813–1823.

Palenik C. S., Utsunomiy A., Kesler S. E., and Ewing R. C. (2002) Gold Nanoparticles in arsenian pyrite from a Carlin-type deposit observed by HRTEM. *GSA Annual Meeting*, abstracts w/ programs, 2002. Paper No. 82-82.

Piccoli P. M. and Candela P. A. (1994) Apatite in felsic rocks: a model for the estimation of initial halogen contents in the Bishop Tuff (Long Valley) and Tuolumne Intrusive Suite (Sierra Nevada Batholith) Magmas. *Am. J. Sci.* **294**, 92–135.

Roedder E. (1984) Fluid inclusions. *Min. Soc. Am.* 644.

Ronov A. B. and Yaroshevsky A. A. (1969) Chemical composition of the Earth's crust. In *Earth's Crust and Upper Mantle.* (ed. J. P. Hart). American Geophysical Union Monograph.

Sangster D. F. (2002) The role of dense brines in the formation of vent-distal sedimentary-exhalative (SEDEX) lead–zinc deposits: field and laboratory evidence. *Mineral. Dep.* (on line).

Shen W. and Zhao P. (2001) Dynamic model of mineralization enrichment with application in mineral resource prediction. Annual Conference of the International Association for Mathematical Geology.

Simon A. C., Candela P. A., Piccoli P. M., Pettke T., and Heinrich C. A. (2002) Gold solubility in magnetite. *Geological Society of America Annual Meeting*, Denver.

Simon G., Kesler S. E., Essene E. J., and Chryssoulis S. L. (2000) Gold in porphyry copper deposits: experimental determination of the distribution of gold in the Cu–Fe–S system at 400–700°C. *Econ. Geol.* 95.

Singer D. A. (1995) World class base and precious metal deposits: a quantitative analysis. *Econ. Geol. Bull. Soc. Econ. Geol.* **90**(1), 88–104.

Sutherland-Brown A. (ed.) (1976) Porphyry Deposits of the Canadian Cordillera. *Canadian Institute of Mining and Metallurgy*, Special Volume 15, 510pp.

Taylor S. R. and McLennan S. M. (1995) the geochemical evolution of the continental crust. *Rev. Geophys.* **33**, 241–265.

Wedepohl K. H. (1995) The composition of the continental crust. *Geochim. Cosmochim. Acta* **59**, 1217–1239.

Williams T. J., Candela P. A., and Piccoli P. M. (1997) Hydrogen-alkali exchange between silicate melts and two-phase aqueous mixtures: an experimental investigation. *Contrib. Mineral. Petrol.* **1282/3**, 114–126.

Williams-Jones A. E. and Normand C. (1997) Controls of mineral parageneses in the system Fe–Sb–S–O. *Econ. Geol. Bull. Soc. Econ. Geol.* **92**(3), 308–324.

Wood S. A. and Samson I. M. (1998) Solubility of ore minerals and complexation of ore metals in hydrothermal solutions. In *Techniques in Hydrothermal Ore Deposits Geology* (eds. J. Richards and P. Larson). Reviews in Economic Geology. pp. 33–80.

Wood S. A., Crerar D. A., and Borcsik M. P. (1987) Solubility of the assemblage pyrite-pyrrhotite-magnetite-sphalerite-galena-gold-stibnite-bismuthinite-argentite-molybdenite in H_2O–NaCl–CO_2 solutions from 200 to 350 °C. *Econ. Geol. Bull. Soc. Econ. Geol.* **82**, 1864–1887.

Wylie A. G. and Candela P. A. (1999) Chromite. In *The Geology of Pennsylvania* (ed. C. H. Shultz). Pennsylvania Geological Survey, pp. 588–595.

3.13
Geochemistry of the Igneous Oceanic Crust

E. M. Klein

Duke University, Durham, NC, USA

NOMENCLATURE

C_0	the initial concentration of an element in the system
C_L	concentration of an element in the melt or liquid
C_s	concentration of an element in a mineral (solid)
C_S	concentration of the element in the total solid
D	bulk distribution coefficient
F	fraction of the system that is melt
K_d	distribution coefficient for an element in a mineral
K_{di}^{j}	distribution coefficient for element i in mineral j
Mg#	100 × molecular [Mg/(Fe + Mg)]
$Na_{8.0}$, $Fe_{8.0}$, etc.	Na_2O, FeO, etc fractionation corrected to 8 wt.% MgO (as discussed in text)
P_o	pressure of intersection of the solidus
P_f	final pressure of melting
X_j	fraction of the mineral j

3.13.1 INTRODUCTION

Approximately 60% of the Earth's surface consists of oceanic crust (Cogley, 1984). New ocean crust is created at divergent plate boundaries called ocean ridges or spreading centers (Figure 1). Once created, the oceanic crust is transported off-axis to each side of the spreading center, accumulating sediment as it ages, and is ultimately consumed at subduction zones and returned in a modified form to the mantle (see Chapters 7.01, 2.11, 3.15, and 3.17).

The oceanic crust plays a key role in the ongoing processes that modify the compositions of major earth reservoirs. As the product of mantle melting, the generation of new oceanic crust continuously changes the composition of the upper mantle from which it forms (e.g., Chapter 2.08). In addition, the crust is the primary interface of exchange between fluids of the Earth's surface and the solid earth below. Hydrothermal circulation of seawater through the ocean crust, for example, is a major factor controlling the chemistry of seawater (see Chapters 3.15 and 6.07). Subduction of hydrothermally altered oceanic crust is believed to initiate arc volcanism, and the particular composition of the subducting crust affects the compositions of the arc magmas (see Chapter 3.18). The deep subduction of altered ocean crust is also the primary means of recycling material back to the mantle where, convectively mixed with ambient mantle, it may form both the source region of some hotspots and dispersed chemical heterogeneities (see Chapters 3.17, 3.18, 2.03, and 2.11). Thus, an understanding of the oceanic crust is central to our elucidation of whole-earth geochemical processes.

This chapter reviews the architecture of the oceanic crust, and the geochemical processes by which it is created, including the nature and origin of its major element, trace element, and radiogenic isotopic composition. More detailed perspectives on these and related topics such as Melting and Melt Percolation Models (see Chapter 3.14), Mantle Heterogeneity (see Chapter 2.03), and the Subduction of Altered Oceanic Crust (see Chapters 3.15, 3.17, 3.18, and 2.11) are presented elsewhere in these volumes.

3.13.2 ARCHITECTURE OF THE OCEANIC CRUST

Our knowledge of the architecture of the oceanic crust derives from four main sources: studies of portions of the oceanic crust that have been obducted onto land (ophiolite complexes); drilling of the ocean crust; exposures of the deeper crust at fracture zones and rare tectonic windows; and geophysical studies of the seismic properties of the ocean crust. These studies have revealed that the general structure of the igneous ocean crust, from top to bottom, consists of a carapace of basaltic lava flows and pillows

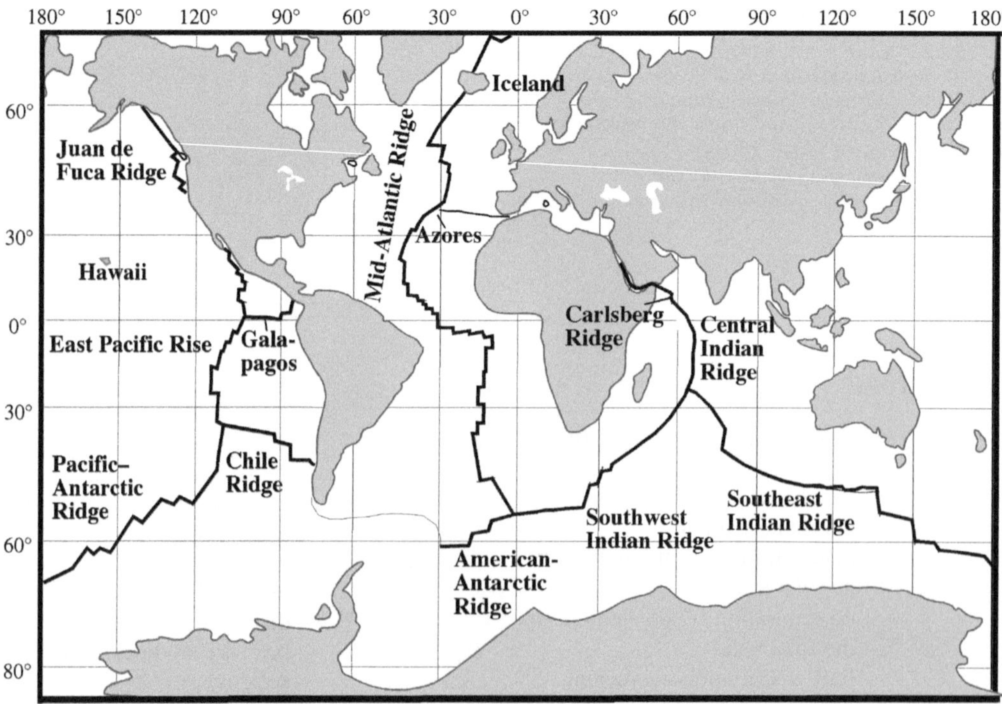

Figure 1 Schematic map of the global ocean ridge system.

(~1–2 km thick), underlain by a sheeted dike complex (~1–2 km thick), a gabbroic unit of diverse textures and lithologies (~3–5 km thick), and finally ultramafic rocks of the lower crust and mantle, including harzburgite, lherzolite, and dunite (Figure 2; e.g., Penrose Conference Participants, 1972; C. Z. Mutter and J. C. Mutter, 1993; Karson, 2002).

A simple view of the creation of each of these units draws upon the results of geophysical studies of melt reservoirs beneath fast-spreading ridges (Figure 3; Detrick, 1987; Sinton and Detrick, 1992) in which a shallow (~1–2 km depth), thin (50–100 m) relatively narrow (0.5–1.5 km) melt lens overlies a wider (>8 km) mush zone (melt + crystals) and partially solid transition zone that extends to the base of the crust (average crustal thickness is ~5–7 km). Melt from the mantle is focussed toward the ridge (e.g., Spiegelman and McKenzie, 1987) and replenishes both the mush zone and the melt lens, where cooling and crystallization take place. Periodically, extensional forces associated with seafloor spreading create a pathway favorable for dike injection above the melt lens and, if the dike pierces the surface, magma erupts on the ocean floor as lava. Continued cooling of the melt lens and the mush zone below leads to the crystallization of the plutonic rocks of the gabbroic sequence. In addition, crystallization of mafic minerals such as olivine and pyroxene at the base of the mush zone may create a basal crustal ultramafic unit. Below the crust lie mantle rocks that often show mineralogic and chemical evidence that they are the residues of a previous melting event that provided the melt that formed the ocean crust above (Dick *et al.*, 1984; Michael and Bonatti, 1985; Johnson and Dick, 1992).

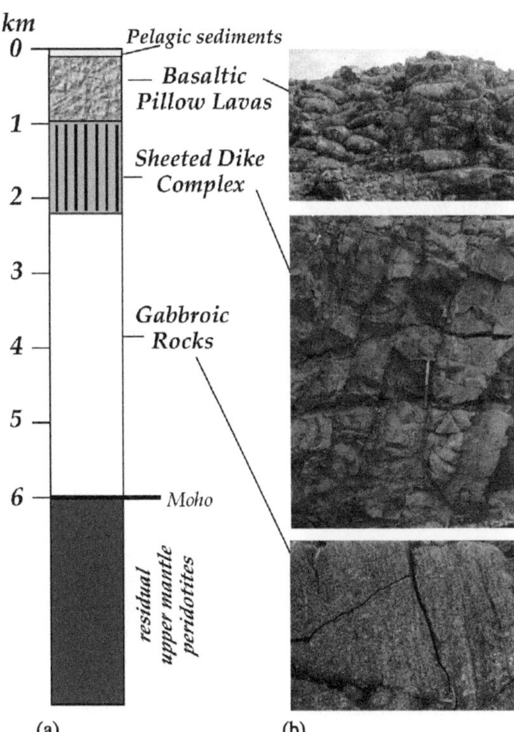

Figure 2 (a) Generalized internal structure and interpretation of the oceanic crust derived from studies of ophiolite complexes and interpretations of marine seismic and geologic data. (b) Outcrop photographs of crustal rocks from ophiolites; top: pillow lavas, Macquarie Island; middle: sheeted dike complex, Oman; bottom: gabbroic rocks, Bay of Islands (Karson, 2002) (reproduced by permission of Annual Reviews from *Annual Reviews of Earth and Planetary Sciences*, **2002**, *30*, 347–384).

3.13.3 MANTLE MELTING: SIMPLE PASSIVE MODEL

In recent years there has been a great deal of experimental and theoretical work on the processes by which the mantle melts to form basaltic magma erupted at mid-ocean ridges (see Chapter 3.14 and Langmuir *et al.*, 1992 for expanded reviews). In the simple passive model for mantle melting (Figure 4; e.g., Oxburgh, 1965, 1980; Lachenbruch, 1976; McKenzie and Bickle, 1988; Plank and Langmuir, 1992), viscous drag associated with seafloor spreading creates a "void" beneath the ridge axis that draws mantle up from depth (in reality, the process is viewed as two continuous conveyor belts, so no actual void space exists). If mantle melting is adiabatic, the mantle largely retains its higher temperature from depth but decompresses as it rises. In pressure–temperature space (Figure 5), this means that the mantle will intersect the solidus at some depth, depending upon its initial subsolidus temperature, and begin to melt; the slope of the melting path differs from that of the solid adiabat, because there is a temperature decrease associated with the heat of fusion. Continued corner flow causes the mantle to rise further, melting more as it ascends; thus, the amount of melting that a parcel of mantle will experience is governed by the difference in pressure between the depth of intersection of the solidus (P_o) and the depth at which it turns the corner and no longer decompresses (P_f, depth of final melting). As melting proceeds, the melt separates from the solid at melt percentages less than 1% (e.g., McKenzie, 1984; Ribe, 1985; Daines and Richter, 1988; Johnson *et al.*, 1990; Johnson and Dick, 1992; Faul, 1997), and is focused toward the ridge by processes that remain a subject of debate (Spiegelman and McKenzie, 1987; Phipps Morgan, 1987; Sparks and Parmentier, 1991).

Much work has been devoted to examining diverse aspects of the assumptions in this model,

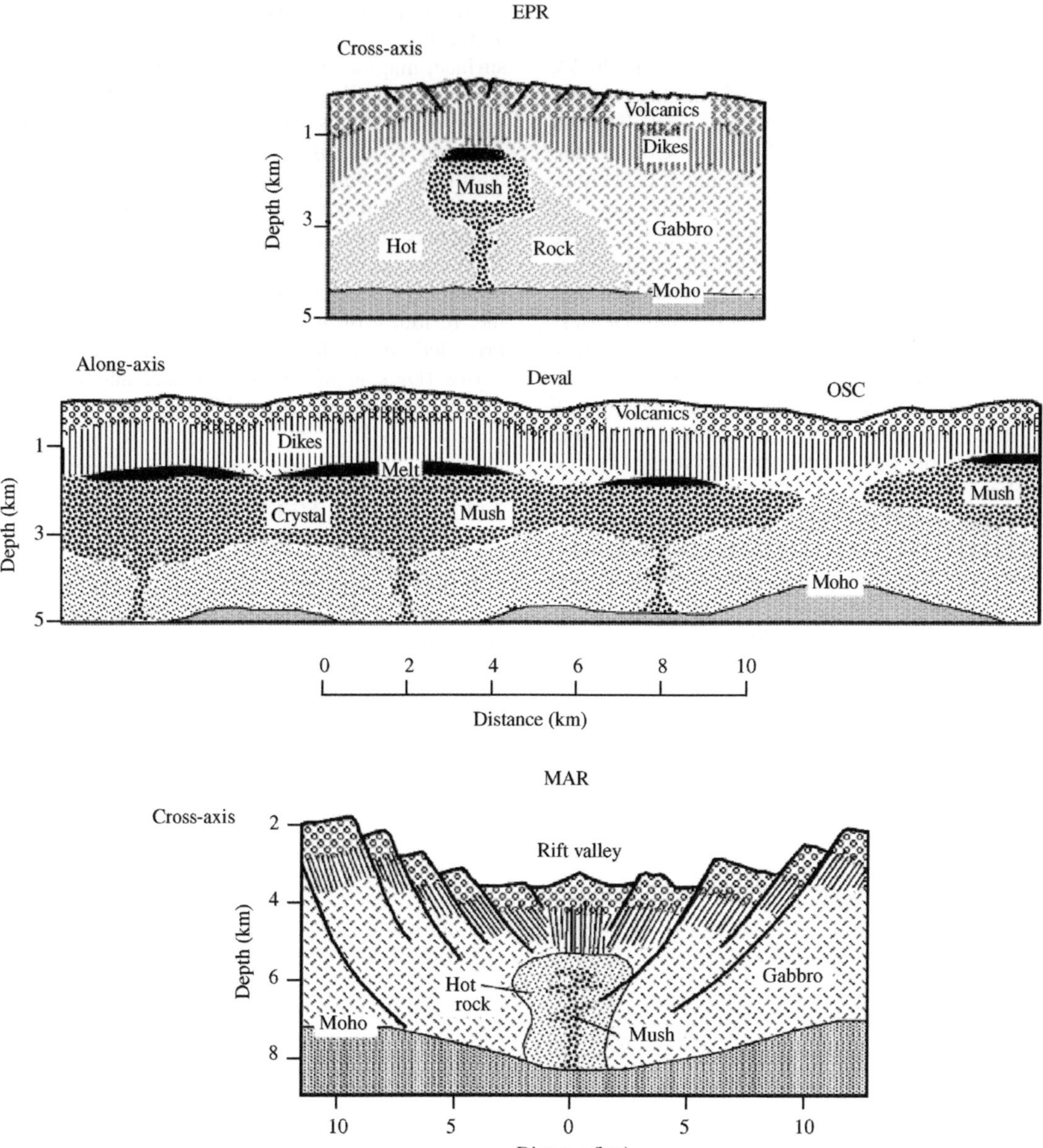

Figure 3 Upper panels: Interpretive model of a magma chamber along a fast-spreading ridge like the EPR based on recent geophysical and petrological constraints. The essential elements of this model are a narrow, sill-like body surrounded by a transition zone to the solidified, but still hot, surrounding rock. The solidus, which defines the limit of magma, can occur anywhere from the boundary of the mush zone to the edges of the axial low-velocity zone (LVZ). Because the solidus may not be isothermal and significant lithologic variations can occur in the lower layered gabbros, isolated pockets of magma with low melt percentages can occur throughout the LVZ. Eruptions will mainly tap the molten, low-viscosity melt lens. The relative volumes of melt must vary along the ridge axis, particularly near ridge axis discontinuities. Lower panel: Intrepretive model of a magma chamber beneath a slow spreading (low magma supply) ridge like the MAR, based on recent geophysical and petrological constraints. Such ridges are unlikely to be underlain by an eruptable magma lens in any steady-state sense. A dike-like mush zone is envisioned beneath the rift valley forming small sill-like intrusive bodies which progressively crystallize to form oceanic crust. Eruptions will be closely coupled in time to injection events of new magma from the mantle. Faults bordering the rift valley may root in the brittle–ductile transition within the partially molten magma chamber (Sinton and Detrick, 1992) (reproduced by permission of American Geophysical Union from *J. Geophys. Res.*, **1992**, *97*, 197–216).

and their implications for the compositions of the melts produced. But, for the purposes of this review, we can use this simple passive model to explore some of the basic aspects of mantle melting and how variables associated with melting (as well as those involved in crystallization and source composition) will affect the compositions of the melts produced.

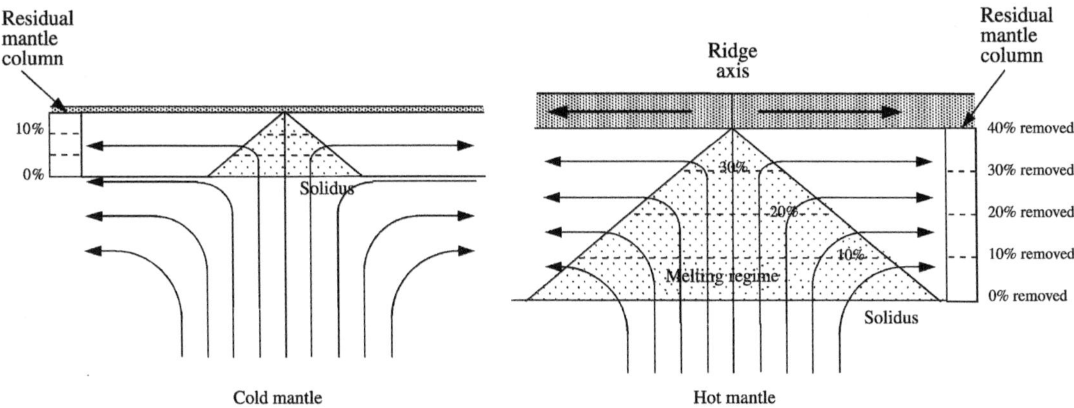

Figure 4 Idealized steady-state melting regimes produced by spreading of the plates. Solid curves with arrows are mantle flow paths through the melting regime. Dashed lines are contours of the extent of melting in the melting regime or the extent of melt removed in the residual mantle column (RMC). The two melting regimes are for two mantle temperatures. Hotter mantle intersects the solidus deeper (see Figure 5), leading to greater extents of melting, a taller RMC, and thicker crust (Langmuir *et al.*, 1992) (reproduced by permission of American Geological Union from *J. Geophys. Res.*, **1992**, *71*, 183–280).

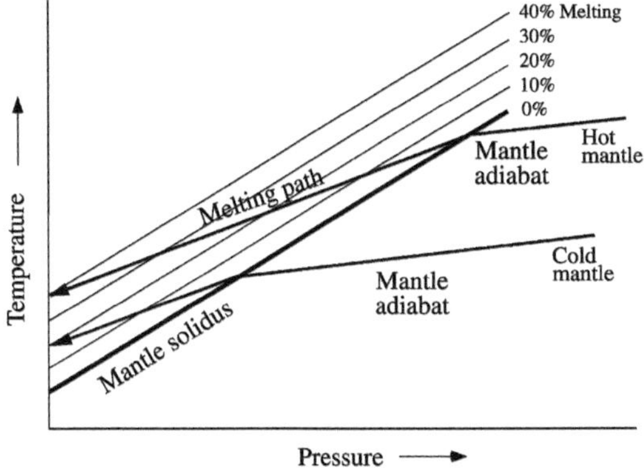

Figure 5 Schematic pressure (P) versus temperature (T) diagram showing melting paths for adiabatically ascending mantle. Melting contours are evenly spaced for illustration only. The two paths are for two different mantle temperatures. The further a mantle parcel ascends, the greater the extent of melting (Langmuir *et al.*, 1992) (reproduced by permission of American Geophysical Union from *J. Geophys. Res.*, **1992**, *71*, 183–280).

3.13.4 WORLDWIDE GEOCHEMICAL VARIATIONS AMONG OCEAN RIDGE BASALTS

The vast majority of lavas erupted along the mid-ocean ridge system are basalts. This uniformity of rock type, however, belies a great diversity in composition resulting from the disparate physical and chemical conditions by which, and from which, the magmas are produced. A number of compilations of worldwide mid-ocean ridge basalt (MORB) compositions now exist, the most complete of which is RidgePetDB (petrologic database of the ocean floor; Lehnert *et al.*, 2000) available at http://petdb.ldeo.columbia.edu/petdb. Other oceanic basalt databases are available through the Smithsonian Institution basalt glass data file (www.hrw.com/science/si-science/physical/geology/mineralsci/research/glass/); and the GEOROC (geochemistry of rocks from the oceans and continents) database at http://georoc.mpch-mainz.gwdg.de/Start.asp. Select examples of various MORB compositions are presented in Tables 1 and 2.

A wide range of processes produces the diversity of magma compositions erupted along mid-ocean ridges. Principal among these are variations in source composition, in the style of mantle melting, and in the crystallization of the magma en route to the surface. These processes produce identifiable chemical signatures in terms of major element, trace element, and radiogenic isotope compositions, which are discussed in the following paragraphs.

A common approach to studying the chemical variations in MORB begins by identifying the effects of the last magmatic processes to occur, namely, *crystallization*. Chemical systematics that cannot be accounted for by crystallization are then examined from the perspective of melting. Finally, those aspects of chemical variability that cannot be attributed to either crystallization or melting are then explored in terms of variations in source composition. To be sure, there is ample evidence for feedback among the three processes: source composition itself likely affects melting systematics, and melting variations can affect the style of crystallization (and additional processes, such as magma chamber contamination by seawater, may occur; e.g., Michael and Cornell, 1998). But it is only by initially examining each of the three main processes in isolation that the

Table 1 Five basaltic glasses from the East Pacific Rise near the Clipperton Transform that, to first order, form a coherent liquid line of descent for fractional crystallization from the highest MgO magma (#1) to lowest (#5).

	1	*2*	*3*	*4*	*5*
	More primitive → more evolved				
SiO_2	50.46	50.65	49.8	50.43	50.93
TiO_2	1.28	1.53	1.64	2.07	2.55
Al_2O_3	15.55	15.15	14.28	14.65	13.51
FeO_T	8.93	9.88	10.45	11.10	13.13
MnO	0.19		0.2	0.20	0.25
MgO	8.37	7.85	7.21	6.87	5.62
CaO	12.21	11.95	11.68	11.40	10.24
Na_2O	2.47	2.54	2.59	2.71	3.2
K_2O	0.11	0.08	0.13	0.19	0.2
P_2O_5	0.11	0.18	0.1	0.25	0.22
Sum	*99.7*	*99.8*	*98.1*	*99.9*	*99.9*
La		3.38	3.9	5.46	5.86
Ce		10.6	12.09	16.8	18.2
Nd			11.1	14.3	16
Sm		3.36	3.77	4.64	5.41
Yb		3.18	3.59	4.14	4.75
Ba	7.43		12.1	20.4	20.7
Nb			3.06	4.89	5.13
Ni		80	72	71	41.4
Sr		120	122	134	133
Y			38.9	39.7	43.9
Zr		100	111	135	155
Mg#	*62.6*	*58.6*	*55.2*	*52.5*	*43.3*
$^{87}Sr/^{86}Sr$			0.702505		0.702566
$^{143}Nd/^{144}Nd$			0.513139		0.513155
$^{206}Pb/^{204}Pb$			18.312		
$^{207}Pb/^{204}Pb$			15.484		
$^{208}Pb/^{204}Pb$			37.722		

Source: samples (left to right) WASRAI2-050-007; WASRAI2-057-006; ODP0142-0864A-001M-003/0-3; MELPHNX-2-068-001; MELPHNX-2-GC083 from RidgePetDB (Lehnert *et al.*, 2000). Major element oxides are in wt.%; trace elements are in ppm. F_eO_T is total iron.

interplay among them (and the influence of other factors), can then be overlaid. Thus, we begin by examining the last magmatic process, crystallization, first; it is also the process that is probably best understood and quantified.

3.13.4.1 Crystallization

3.13.4.1.1 Major elements

It has long been known that suites of MORB samples recovered from a relatively limited spatial area often display regular variations in major element oxides as a function of MgO content (Figure 6) or Mg# (100 × molecular [Mg/(Fe + Mg)]). In addition, the Mg#s of most MORBs (mean Mg# $= 55 \pm 7.6(1\sigma)$; $n \sim 1.3 \times 10^4$) are not in equilibrium with presumed mantle olivine compositions (Mg# ~90; Roeder and Emslie, 1970; Langmuir and Hanson, 1981). These findings, combined with the results of experimental studies on the crystallization of basaltic melts (Bender *et al.*, 1978; Walker *et al.*, 1979; Stolper, 1980; Grove and Bryan, 1983), have led to the recognition that basaltic melts commonly undergo significant and varying extents of crystallization in sub-axial magma reservoirs prior to eruption. Magmas at varying stages of crystallization are periodically tapped from these reservoirs, and erupted at the surface, revealing the evolution of the melt as it cools and crystallizes.

The magnitudes of the changes in each oxide as a function of decreasing MgO result from the crystallization from the melt of stable mineral phases (which, for typical basaltic melts at <50–100 MPa pressure, generally includes olivine ± minor spinel, followed by plagioclase and finally clinopyroxene). Thus, in Figure 6, the compositions and proportions of the minerals that crystallize from each melt drive the residual melt to a different (lower MgO) composition. Because these changes represent the changing composition of the melt as it cools, crystallizes, and evolves, the variations are called the "liquid line of descent" for a particular parental (high MgO) magma composition; a calculated liquid line of descent is shown in Figure 6, which closely approximates the trend of the data.

A number of models now exist that calculate the liquid line of descent for crystallization of a given parental magma composition under specified conditions of pressure, oxygen fugacity, and other variables (e.g., Nielsen, 1985, 1990; Weaver and Langmuir, 1990; Longhi, 1991; various versions of the MELTS program available on the Web: www.geology.washington.edu/~ghiorso/MeltsWWW/Melts.html, Ghiorso and Sack, 1995; Ghiorso *et al.*, 2002). These model calculations suggest that, often, much of the major element variability among MORB from a given region can be accounted for by low (<50 MPa) to intermediate (e.g., 200–300 MPa) pressure fractional crystallization (evidence for high-pressure crystallization, at 600–800 MPa, has also been described (e.g., Elthon and Scarfe, 1984; Grove *et al.*, 1990) but is believed to be less common). During fractional crystallization, the crystallizing minerals separate from the cooling magma as they are formed; this differs from batch or equilibrium crystallization in which minerals and melt remain in contact and continue to equilibrate as cooling and crystallization proceed. Table 1 presents data for magmas that, to first order, appear to result from progressive fractional crystallization from the most primitive (highest MgO) magma in the suite.

3.13.4.1.2 Trace elements

For more than three decades, geochemists have studied the trace element variations in MORB

Table 2 Representative analyses of MORBs. Kolbeinsey Ridge, northern EPR, and mid-Cayman Rise encompass much of the global range in extents and pressures of melting. Also shown are examples of an Indian Ocean MORB and composite analyses of a normal (N-MORB), a transitional (T-type) MORB, and an enriched (E-type) MORB from the MAR.

	Global range: N-MORB ((high F, high P) left to (low F, low P) right)			*Indian Oc. MORB*	*MAR N-MORB*	*MAR T-MORB*	*MAR E-MORB*
	Kolbeinsey ridge	*N. EPR N-MORB*	*Mid-Cayman Rise*				
SiO_2	50.55	50.16	50.98	50.85	50.01	50.88	51.28
TiO_2	0.82	1.47	1.72	1.10	1.11	1.71	1.83
Al_2O_3	14.83	15.79	16.11	16.72	16.31	16.07	15.23
FeO	10.49	9.51	8.745	7.76	9.73	9.73	9.60
MnO	0.18	0.16	0.17	0.15	0.14		0.16
MgO	8.53	7.58	7.23	8.85	8.67	7.39	7.43
CaO	12.99	12.19	10.34	11.11	11.75	11.17	10.59
Na_2O	1.50	2.76	3.56	3.17	2.52	2.89	3.08
K_2O	0.05	0.13	0.24	0.15	0.05	0.11	0.53
P_2O_5	0.06	0.13	0.22	0.13	0.08	0.16	0.26
Sum	*100.0*	*99.9*	*99.3*	*100.0*	*100.4*	*100.1*	*100.0*
La	1.38	3.34	5.77	4.46	1.88	5.34	11.5
Ce	4.10	10.40	17.10	12.21	5.99	13	26
Pr	0.74	1.91					
Nd	4.32	9.62		8.47	6.07	10.7	17.1
Sm	1.69	3.14	4.35	2.72	2.22	3.39	4.38
Eu	0.68	1.18	1.53	1.07	0.9	1.42	1.54
Gd	2.44	3.97		3.49	3.5	4.32	5.26
Tb	0.48	0.72	0.93				
Dy	3.39	4.85		4.04	4.46	5.57	5.24
Ho	0.76	1.03					
Er	2.30	2.72		2.56	2.57	3.24	2.83
Tm	0.34	0.38					
Yb	2.30	2.63	3.42	2.25	2.72	3.28	2.68
Lu	0.340	0.400	0.469	0.354			
Ba	12.0	12.2	22.0	19.3	6.11	31	123
Co	54.0	50.1	37.8				
Cr	161	253	250	446	251		
Cs		0.027		0.014	0.006	0.04	0.08
Cu	137.0	82.3		57.0	68		
Hf	1.44	2.14	3.70		2.9	2.12	2.06

(continued)

Table 2 (continued).

	Global range: N-MORB			Indian Oc. MORB	MAR N-MORB	MAR T-MORB	MAR E-MORB
	Kolbeinsey ridge	N. EPR N-MORB ((high F, high P) left to (low F, low P) right)	Mid-Cayman Rise				
Nb	3.08	2.99	5.02		1.07	5.15	11.2
Ni	81	120	123	137	119	115	91
Pb	0.155	0.359			0.19		0.95
Rb	1.07	1.45		1.08	0.38	2.34	8.85
Sc	48.0	42.3	33.6	30.3	44	43	36
Sr	68	142	188	191	94	129	181
Ta		0.203	0.309		0.13	0.38	0.77
Th		0.141	0.18		0.09	0.45	1.07
U		0.061			0.03	0.11	0.28
V		265	220	200	281	299	288
Y	20.6	27.2	36.0	24.4	25	32	29
Zn	80.0	80.2					
Zr	39	89	153	92	57	100	134
$^{87}Sr/^{86}Sr$	0.7029	0.7026	0.7025	0.7035	0.7025	0.70268	0.70392
$^{143}Nd/^{144}Nd$	0.51319	0.51316	0.51315	0.51299	0.51306	0.51296	0.51271
$^{206}Pb/^{204}Pb$	17.953	18.286	18.21	17.764	18.215	18.71	18.032
$^{207}Pb/^{204}Pb$	15.424	15.464	15.48	15.483	15.535	15.567	15.507
$^{208}Pb/^{204}Pb$	37.608	37.636	37.66	37.803	38.047	38.478	38.237
Latitude	67.98° N	11° 22.8′ N	18.43° N	49.91° S	49.76° S	25.4° N	47.97° S
Longitude	18.31° W	103° 39.6′ N	81.72° W	115.38° E	8° W	45.3° W	10.08° W
Depth (m)	740	3063	4795	3087	3874	3200	2895

Data sources are as follows: Kolbeinsey Ridge: RidgePetDB (Lehnert *et al.*, 2000) for sample POS0158-404-00; major elements and most trace elements on whole rock powders; Pb, Sr, Rb, and isotope ratios on glasses. N. EPR: glass reported by Niu *et al.* (1999); mid-Cayman Rise: glass compositions reported in RidgePetDB for sample KNO0054-027-005, augmented with Ba, V, and Y data on a similar sample reported by Thompson *et al.* (1980) and the sole isotopic analysis of a mid-Cayman Rise basalt from RidgePetDB; Indian Ocean MORB glass reported by Klein *et al.* (1991); N-Type, T-type, and E-Type MAR MORB glass are composite analyses reported in the RidgePetDB database (major and most trace elements for MAR N-type, T-type, and E-type MORB are respectively from samples: EW19309-012-00, VEM0025-001-022 and EW19309-004-002; note that latitude, longitude, and depth for the composite analyses refer to the locations of the samples for which the major element data are reported). Major element data for the Kolbeinsey Ridge, Indian Ocean, and MAR T-type samples have been recalculated to express Fe_2O_3 as FeO. Major element oxides are in wt.%; trace elements are in ppm.

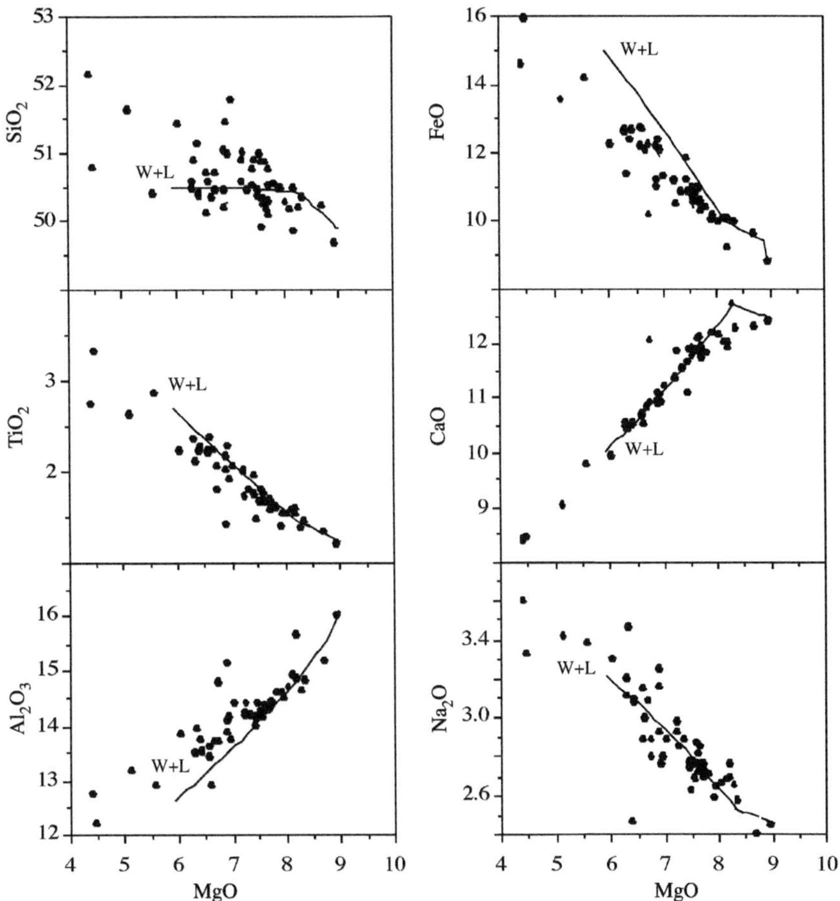

Figure 6 MgO variation diagrams for microprobe data for basalt glasses from near the Clipperton Fracture Zone (from RidgePetDB; Lehnert *et al.*, 2000). Also shown is a calculated liquid line of descent for fractional crystallization at <50 MPa using the program of Weaver and Langmuir (1990) (labeled W + L) for a primitive composition at ~9 wt.% MgO. Oxides in wt.% (Langmuir *et al.*, 1992) (reproduced by permission of American Geophysical Union from *J. Geophys. Res.*, **1992**, *71*, 183–280).

to understand the magmatic processes that occur in their genesis. Pioneering work by Gast (1968), Shaw (1970) and Schilling (1966, 1971) laid much of the theoretical and computational foundation for trace element modeling that continues to be used and expanded upon today. The basic elements of this approach can be summarized as follows. Because much of the chemical variability among the MORB results from the fractionation of melt and minerals, it is useful to approach the partitioning of a given trace element between solid and liquid (melt) in terms of its distribution coefficient, K_d:

$$K_d = C_s/C_L \quad (1)$$

where C_s is the concentration of an element in a particular mineral (solid) and C_L is its concentration in the liquid (melt). Distribution coefficients for various elements between minerals and melts of different compositions are generally determined experimentally and are available for most of the common minerals involved in mantle melting and basalt crystallization (see, e.g., http://earthref.org/GERM/; and Chapter 2.09).

For a multi-mineral system, the bulk distribution coefficient, D, is the combination of the K_d for each mineral weighted by the fraction of each mineral in the system, and is defined as follows:

$$D = [(\text{fraction mineral 1}) \times (K_d \text{ for mineral 1}) + (\text{fraction mineral 2}) \times (K_d \text{ for mineral 2}) + \text{etc.}] = \sum X_j K_{di}^{j} \quad (2)$$

$$D = C_S/C_L \quad (3)$$

where X_j is the fraction of the mineral j; K_{di}^{j} is the distribution coefficient for element i in mineral j; and C_S is the concentration of the element in the total solid.

From the formulation above, it is clear that for a multi-mineral solid/melt system, an element that has a $D > 1$ is preferentially concentrated in the solid; such elements are described as "compatible" because they are such in the mineral assemblage present. An element that has a $D < 1$ is preferentially concentrated in the melt, and is described as an "incompatible" element. (And an element with a $D = 1$ will exist in equal concentration in the solid and melt.)

During any closed-system process, mass balance must be conserved:

$$C_0 = (FC_L) + [(1 - F)C_S] \tag{4}$$

where C_0 is the initial concentration of an element in the system and F is the fraction of the system that is melt. By substituting (3) into (4), we obtain

$$C_L/C_0 = 1/[F + D(1 - F)] \tag{5}$$

This equation, when applied to melting, is called the "equilibrium melting" or "batch melting" equation and, when applied to crystallization, it is called the "equilibrium crystallization" equation. As emphasized by Langmuir *et al.* (1992) this equation, and others like it, is enormously powerful and can be applied to model both major elements and trace elements. This approach has received its most widespread use, however, in modeling trace element variations.

A common application of this equation to trace element modeling is to examine the variations in trace element abundances and ratios for elements with different bulk distribution coefficients (Figure 7). In this plot, F is the fraction of melt: for equilibrium crystallization, F proceeds from 1 (all melt) to 0 (all solid); for equilibrium (or batch) melting, F proceeds from 0 (all solid) to 1 (all melt). During crystallization, a highly compatible element (e.g., $D = 10$) is incorporated in the crystallizing mineral(s) and depleted in the melt, such that the final liquid present (F approaching 0) will have a concentration of $1/D$, less than its starting concentration. In contrast, a highly incompatible element (e.g., $D = 0.01$) during equilibrium crystallization gradually increases in concentration, and at large extents of crystallization (e.g., $F < 0.2$) becomes highly concentrated, again approaching $1/D$, a concentration far greater than the original starting concentration. Compare these results for equilibrium crystallization to those produced by fractional crystallization (Figure 8(a)). Note, for example, that during fractional crystallization, the concentration in the melt of a compatible element is rapidly depleted as the crystallizing assemblage removes this element from the system. In addition, after significant fractional crystallization, moderately incompatible elements (e.g., $D = 0.1$) are enriched in the melt in excess of the amount that can be achieved through equilibrium crystallization. Thus, by examining the concentrations of compatible and incompatible elements as a function of indicators of crystallization (e.g., MgO or Mg#) the effects of crystallization can be identified as well as the type of crystallization that has occurred.

Another common application of this approach is to examine the ratio of two different incompatible elements with different Ds. This involves solving Equation (5) for the ratio of two different

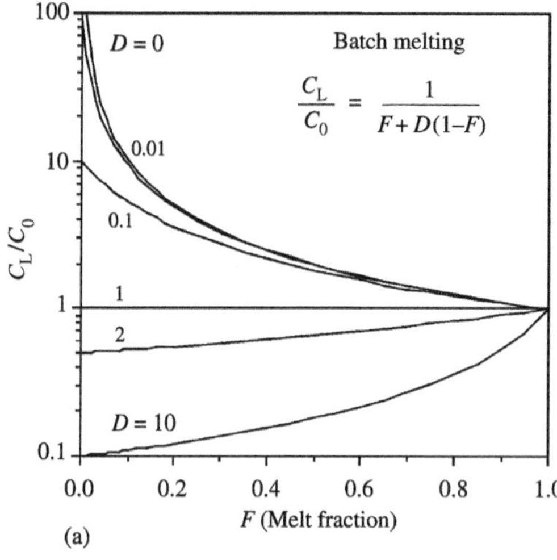

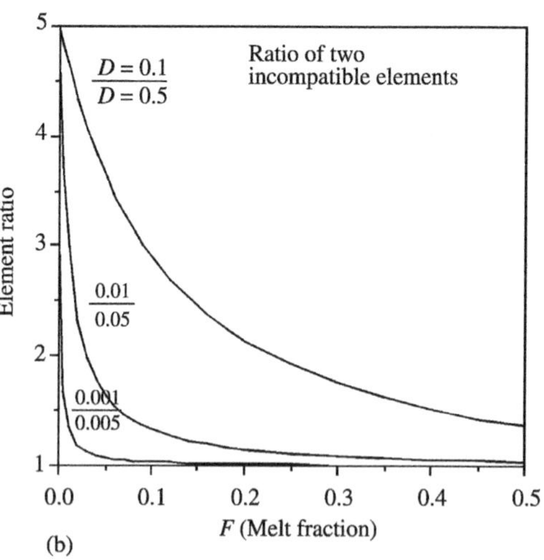

Figure 7 Illustration of the effects of equilibrium (batch) crystallization or melting on trace element abundances. (a) Variation in liquid concentration (C_L) (normalized to unit source concentration $C_0 = 1$) as a function of melt fraction (F) for six "elements" with different bulk distribution coefficients (D). (b) Change in the ratios of incompatible elements with different Ds as a function of F. Each curve is for a different pair of "elements" that have the Ds indicated. Note that when $D < 0.1$, incompatible element ratios can be changed only at very low extents of melting (or high extents of crystallization) (Langmuir *et al.*, 1992) (reproduced by permission of American Geophysical Union from *J. Geophys. Res.*, **1992**, *71*, 183–280).

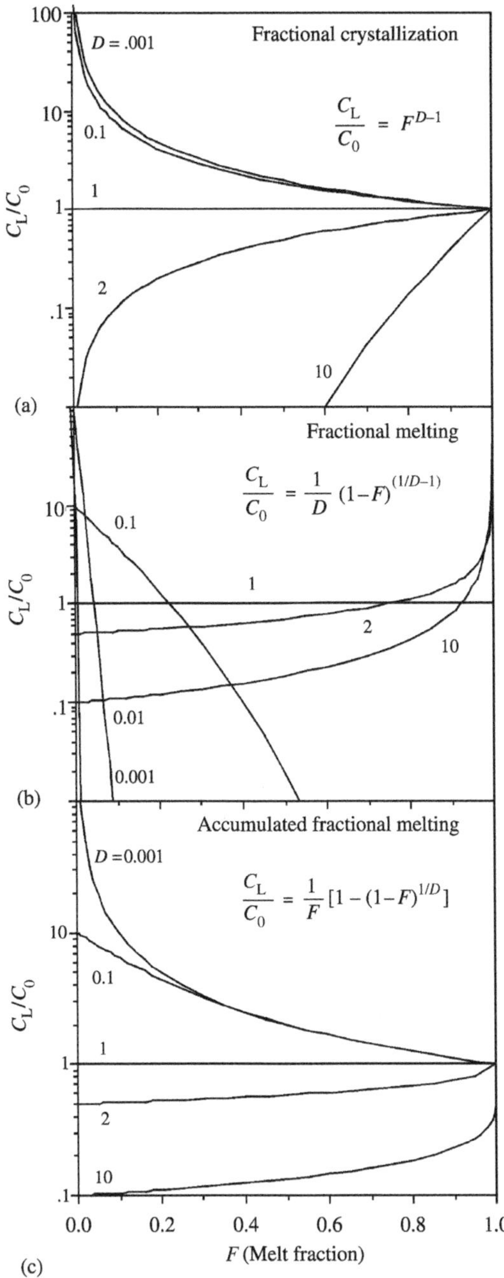

Figure 8 Illustration of the effects of fractional crystallization and melting on trace element abundances. Notation as in Figure 7. (a) Fractional crystallization for five different values of D. (b) Fractional melting. (c) Accumulated fractional melting. Note the very similar results produced by accumulated fractional melting to the equilibrium melting results of Figure 7 (Langmuir *et al.*, 1992) (reproduced by permission of American Geophysical Union from *J. Geophys. Res.*, **1992**, *71*, 183–280).

elements ($C_{L1}/C_{L2} = C_{01}/C_{02}[(F + D_2(1 - F)/(F + D_1(1 - F)]$. The results of this calculation, shown in Figure 7(b), reveal that for equilibrium crystallization, the ratio in the melt of two elements that have similar incompatibility changes little until F is less than ~0.5; at very small remaining melt fractions, the ratios increase abruptly, particularly for ratios of more moderately incompatible elements. Because the ratios of incompatible elements change little during the first 50% of crystallization (most MORB experience less than 50% crystallization), a relative constancy of such ratios with decreasing MgO content is expected. Thus, large variations in this ratio accompanied by major element indices suggesting only modest extents of crystallization (e.g., moderate to high MgO or Mg#), must be attributed to other processes, such as variations in extent of melting or source composition.

A number of more complex crystallization processes have been envisioned. One that has attracted particular attention, in large part because it is inherently linked to a realistic physical model for crystallization, is called *in situ* crystallization (Langmuir, 1989; Sparks *et al.*, 1984; Sparks, 1989). This process recognizes that in a crystallizing magma body, there is likely to be a temperature gradient from hottest in the center of the magma body to coolest along its margins; thus, crystallization is most likely to take place within a boundary layer between magma and country rock. As crystallization proceeds within this boundary layer, more evolved magma may be expelled and mix with the main body of magma, where it is periodically tapped and erupts lavas. Because the magma expelled from the boundary layer may be highly evolved, it may produce unusual enrichments in incompatible trace elements and ratios that exceed those produced by fractional crystallization alone.

3.13.4.1.3 Correcting for crystallization

The discussion above emphasizes that lavas recovered from a limited length of ridge often exhibit fairly regular variations in both major and trace elements as a function of decreasing MgO (or Mg#), and these are commonly attributed to the effects of low (<50 MPa) or intermediate (300 MPa) pressure fractional crystallization. These chemical variations resulting from crystallization obscure differences in parental magma compositions that exist from region to region. Thus, methods have been developed to minimize the effects of fractionation on each oxide or trace element.

One method projects the composition of each magma along a presumed slope of the liquid line of descent to a constant value of MgO, in this case 8 wt.% (Klein and Langmuir, 1987; see also updated algorithms in Castillo *et al.*, 2000). The calculated Na_2O value at 8 wt.% MgO is therefore referred to as $Na_{8.0}$. In theory, if the liquid line of descent is known for any major or trace element, a "fractionation-corrected" value of the element can be calculated (producing calculated values of, e.g., $Fe_{8.0}$, $Al_{8.0}$, $K_{8.0}$, $Ce_{8.0}$, etc.). The few elements whose liquid lines of descent change in the sign of the slope as the fractionating phases change, such as CaO (due to the appearance of calcium-bearing phases on the liquidus), are more difficult to model in this way and therefore produce less reliable fractionation-corrected values. Nevertheless, most major and trace elements display fairly regular liquid lines of descent, and their fractionation-corrected values reveal relative differences in parental magma compositions that result from differences in melting systematics or source composition (see, e.g., differences in average $Na_{8.0}$ for samples from the regions shown in Figure 9).

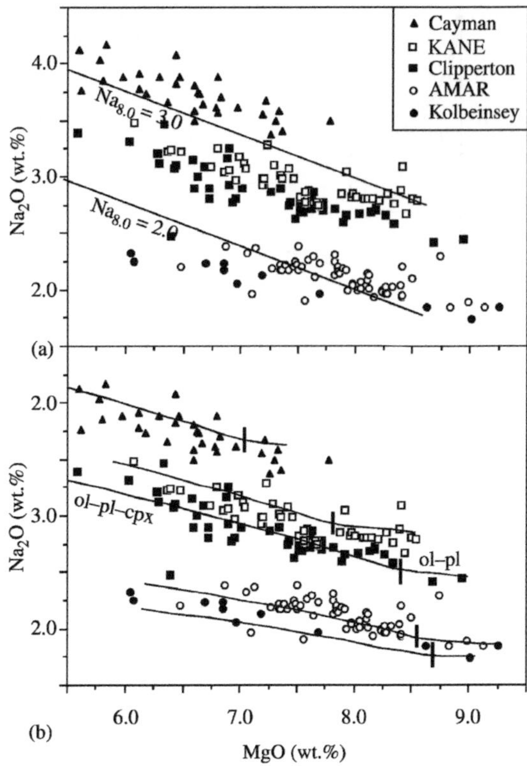

Figure 9 (a) MgO versus Na_2O in MORBs from five regions. Samples are from the Mid-Cayman Rise, from near the Kane fracture zone at 23.22–23.67° N on the MAR, from the AMAR region at 36.42–6.79° N on the MAR, from near the Clipperton fracture zone on the EPR, and from the Kolbeinsey Ridge north of Iceland (data from the Smithsonian catalogue of basalt glasses (Melson *et al.*, 1977), except for Clipperton data (RidgePetDB; Lehnert *et al.*, 2000), and Kolbeinsey data (Schilling *et al.*, 1983)). Lines show slope used by Klein and Langmuir (1987) to calculate $Na_{8.0}$ (Na_2O at 8% MgO). (b) Same data with calculated LLDs for a primitive composition from each of the five suites using the program of Weaver and Langmuir (1990) for fractional crystallization at 50 MPa. Vertical bar indicates the point at which clinopyroxene (cpx) joins olivine and plagioclase (ol–pl) as a fractionating phase, at which point the slope of the LLD steepens (after Langmuir *et al.*, 1992). Original algorithms for calculation of $Na_{8.0}$ and $Fe_{8.0}$ for samples with 5.0–8.5 wt.% MgO are as follows (Klein and Langmuir, 1987): $Na_{8.0} = Na_2 0.373 \times (MgO) - 2.98$; $Fe_{8.0} = FeO + 1.664 \times (MgO) - 13.313$. More recently presented algorithms (that permit calculation for samples with MgO > 5.0 wt.% and that take into account the changing slope of the liquid line of descent are as follows (from Castillo *et al.*, 2000): $Na_{8.0} = 0.6074 - 3.523 \times [(Na_2O + 0.00529 \times MgO^2) - 0.9495)/(MgO - 0.05297 \times (MgO^2) - 8.133)]$; $Fe_{8.0} = 1.825 - 1.529 \times (FeO - 0.03261 \times (MgO^2) + 0.2619)/ \quad (MgO - 0.04467 \times (MgO^2) - 6.67)]$ (Langmuir *et al.*, 1992) (reproduced by permission of American Geophysical Union from *J. Geophys. Res.*, **1992**, *71*, 183–280).

3.13.4.2 Melting

3.13.4.2.1 Major elements

One of the startling findings revealed by examining regional averages of fractionation-corrected values of major element compositions is that the major elements do not vary independently of one another but rather, to first order, correlate in predictable ways. In general, regions with low mean $Na_{8.0}$ are also characterized by high mean $Fe_{8.0}$ (as well as low $Si_{8.0}$ and $Al_{8.0}$, and higher $Ca_{8.0}$), while other regions exhibit the opposite characteristics, as well as a continuum of compositions in between (Klein and Langmuir, 1987; Langmuir *et al.*, 1992). Furthermore, these major element variations have been shown to correlate with physical characteristics of the ridge axis from which they were recovered. Regional averages of $Na_{8.0}$ and $Fe_{8.0}$, for example, show a positive and an inverse correlation, respectively, with the average ridge depth from which the lavas were recovered (Figure 10). In addition, $Na_{8.0}$ was also shown to correlate inversely with seismically and geologically determined estimates of the thickness of the oceanic crust in each region (Figure 11). Thus, some of the most fundamental physical and chemical parameters studied at ocean ridges suggest a common origin in their variability.

The chemical systematics can be understood as the interplay of two main factors affecting the style of mantle melting: the extent of melting and the pressure of melting. Elements such as sodium are moderately incompatible during melting of the mantle minerals ($D \sim 0.02$–0.03), and therefore will be concentrated in the melt at small extents of melting. Iron is well known to vary strongly in the melt as a function of the pressure of melting (e.g., Langmuir and Hanson, 1980). Thus, the inverse correlation between mean $Na_{8.0}$ and mean $Fe_{8.0}$ would suggest that there is a positive correlation between the mean extent of melting

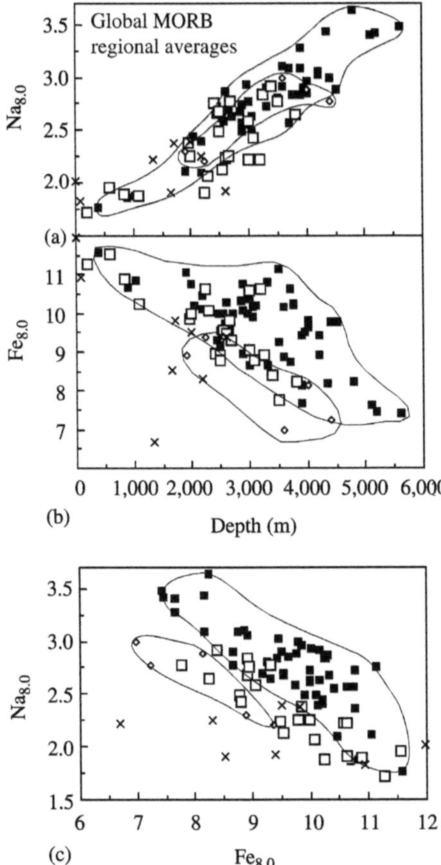

Figure 10 Global correlations between regional averages of smoothed axial depth and the $Na_{8.0}$ and $Fe_{8.0}$ of MORBs. This figure is the same as figure 2 in Klein and Langmuir (1987), except all analyses have been recalculated to sum to 100% (all iron as FeO) and have been corrected for interlaboratory differences. Corrections were applied to all analyses determined at the Smithsonian Institution (Melson *et al.*, 1977) or that were reported to be consistent with the Smithsonian data (i.e., Schilling *et al.*, 1985), based on the factors reported in Klein and Langmuir (1989; their figure 5), so that all data are consistent with those determined at Lamont and by wet chemistry at URI. Data sources as in Klein and Langmuir (1987), with updates from the Smithsonian Catalogue of Basalt Glasses and Ridge-PetDB (Lehnert *et al.*, 2000). Different groups of MORB are distinguished: solid boxes are MORB from "normal" ridge segments; diamonds are from back-arc basins; open squares are from ridges influenced by the Galapagos, Azores, Jan Mayen, Tristan, Iceland, and Bouvet hotspots; Xs are from ridge segments immediately adjacent to these hotspots. Note that Iceland and adjacent ridges are also coincident with the normal MORB array (Langmuir *et al.*, 1992) (reproduced by permission of American Geophysical Union from *J. Geophys. Res.*, **1992**, *71*, 183–280).

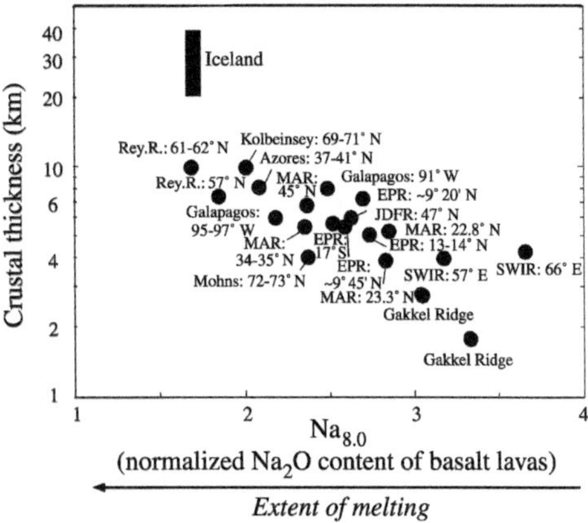

Figure 11 Regional averages of seismically determined crustal thickness versus $Na_{8.0}$ (the Na_2O content of basalts normalized to 8 wt.% MgO; Klein and Langmuir, 1987). Sources for seismic determinations of crustal thickness are from Klein and Langmuir (1987), augmented and/or superceded by the following: Smallwood and White (1998); Navin *et al.* (1998); Darbyshire *et al.* (2000); Detrick *et al.* (2002); Muller *et al.* (1999); Hooft *et al.* (2000); Fowler and Keen (1979); Canales *et al.* (1998); McClain and Lewis (1982); Kodaira *et al.* (1997); Klingelhöfer *et al.* (2000); Jokat *et al.* (2003); Michael *et al.* (2003). Chemical analyses are primarily on basalt glasses from the Ridge Petrologic Database (Lehnert *et al.*, 2000) (reproduced by permission of Nature Publishing Group from *Nature*, **2003**, *423*, 932–933).

and the mean pressure of melting. Melts from bathymetrically deep ridges tend to be produced by smaller extents of melting (high $Na_{8.0}$) and low pressure (low $Fe_{8.0}$), while melting beneath shallower ridges leads to larger extents of melting (low $Na_{8.0}$) at higher pressures of melting (high $Fe_{8.0}$).

It is important to recognize that the parameter "the extent of melting" has real physical implications for the volume of melt produced. Assuming mantle source regions of approximately equal volumes, and that most of the melt produced segregates to form the crust, a region that experiences a small extent of melting would be expected to have thinner oceanic crust than a region that experiences a larger extent of melting. This, in turn, would lead to the observed correlation between a chemical parameter indicative of the extent of melting (e.g., $Na_{8.0}$) and a physical manifestation of the amount of melt, crustal thickness (Figure 11). Furthermore, if the crust is isostatically compensated, thinner crust would lead to greater ridge depth below sea level, and therefore the correlation between chemistry and axial depth (Figure 10).

The "global correlations" among regional averages of major elements and between these elements and physical parameters of the ridge are likely due to lateral variations in the sub-solidus temperature of the mantle (Klein and Langmuir, 1987; McKenzie and Bickle, 1988). Regions underlain by hotter sub-solidus mantle intersect the solidus deeper and melt more upon ascent than regions underlain by cooler sub-solidus mantle temperatures (Figures 3 and 4), leading to the associated correlations with axial depth and crustal thickness. While a portion of the chemical variability among MORB may result from chemical and/or mineralogical heterogeneity of the mantle (e.g., Shen and Forsyth, 1995), the weight of current evidence supports the idea that the first-order trends of the global correlations of regional topography, basalt chemistry, and crustal thickness result from mantle temperature variations.

The discussion of melting above has focused on "mean" values (e.g., mean pressures of melting). It is likely, however, that melting in a given region occurs over a range of pressures, from the solidus to the lithosphere, and that small melt fractions segregate from their source, pool as they rise, and accumulate in shallow crustal magma reservoirs (McKenzie, 1984, 1985; Klein and Langmuir, 1987; McKenzie and Bickle, 1988; Plank and Langmuir, 1992). Thus, melting must be viewed as polybaric; the accumulated melt is not in equilibrium with mantle at any one pressure, but rather represents a mixture of melts derived from various pressures. Accordingly, accumulated melts that exhibit a higher mean pressure of melting than those from another region acquire their higher-pressure signatures by beginning melting at greater depths, hence a deeper solidus. The depth of the solidus also governs the extent of melting; a deeper solidus means that mantle will decompress and melt over a larger range of pressures as it rises (Figures 3 and 4). Accumulated melts from this deeper melting "column" will exhibit chemical characteristics indicative of larger mean extents of melting (e.g., low $Na_{8.0}$) and higher mean pressures of melting (higher $Fe_{8.0}$), as is observed in the global correlation of regional averages (Figure 10).

It is also important to note that the eruption on the surface of melts generated at high pressures (e.g., high $Fe_{8.0}$) suggests that these melts rise from great depths without experiencing significant re-equilibration en route to the surface. One explanation for the apparent absence of re-equilibration may be that the melts do not rise by diffuse porous flow but rather rapidly segregate into semi-isolated channels. Indeed, Kelemen *et al.* (1995) have interpreted dunite channels in the mantle section of the Oman ophiolite as the residual conduits through which chemically isolated melt flow occurred.

A number of studies have used the major element (and trace element) systematics of MORB to constrain the mean extent and mean pressure of melting, as well as the depth and temperature of intersection of the mantle solidus, both for a given region and for the global ocean ridge system as a whole (e.g., Klein and Langmuir, 1987; McKenzie and Bickle, 1988; Langmuir *et al.*, 1992; Kinzler and Grove, 1992; Asimow *et al.*, 2001; Asimow and Langmuir, 2003). A thorough evaluation of these parameters requires numerous assumptions and information that are subject to uncertainty. These include, for example, assumptions about the physical form of the melting regime including variables such as active versus passive upwelling and variations in the final depth of melting (e.g., Scott and Stevenson, 1989; Plank and Langmuir, 1992); the processes of melt extraction and mixing (e.g., batch versus fractional melting, percentages of melt retention, incomplete focusing of melt); a melt generation function (variously calculated as ~2.5–20% melt per GPa pressure release as a function of mineralogy and melt composition; e.g., Ahern and Turcotte, 1979; Cawthorn, 1975; McKenzie, 1984; Hirschmann *et al.*, 1999; Asimow *et al.*, 1997), and the effects of source mineralogy and composition including volatile species (e.g., Langmuir *et al.*, 1992; Asimow *et al.*, 2001; Asimow and Langmuir, 2003). With these caveats in mind, studies noted above have estimated the global range (from cold to hot regions) in the pressure of intersection of the solidus as ~1.5–3.5 GPa, in the temperature of intersection of the solidus as ~1,300–1,550 °C, in the mean extent of melting as ~8–22%, and in the mean pressure of melting as ~0.5–1.6 GPa.

3.13.4.2.2 Trace elements

Moderately incompatible and compatible trace elements also support the conclusions regarding extents and pressures of melting noted above. Regional averages of fractionation-corrected values of cerium (a moderately incompatible element), for example, correlate positively with $Na_{8.0}$ (Figure 12(a)) consistent with regional variations in the extent of melting. In contrast, scandium (an element that is compatible in clinopyroxene in the mantle; Figure 12(b)) shows an inverse correlation with $Na_{8.0}$, reflecting the fact that at small extents of melting (high $Na_{8.0}$) residual clinopyroxene in mantle lherzolite retains much of the scandium; as melting proceeds and clinopyroxene melts out of the residue, scandium increases. (Note that little correlation is found between $Na_{8.0}$ and $Ba_{8.0}$ (Figure 12(c)), reflecting the variations in source composition of the highly incompatible element barium). Trace element studies of abyssal peridotites, the residues of mantle melting, suggest further that melting must be near-fractional ("incremental" with porosities <1%), in order to produce the observed fractionation between elements with similar distribution coefficients such as Ti/Zr or Lu/Hf (Johnson *et al.*, 1990; Johnson and Dick, 1992).

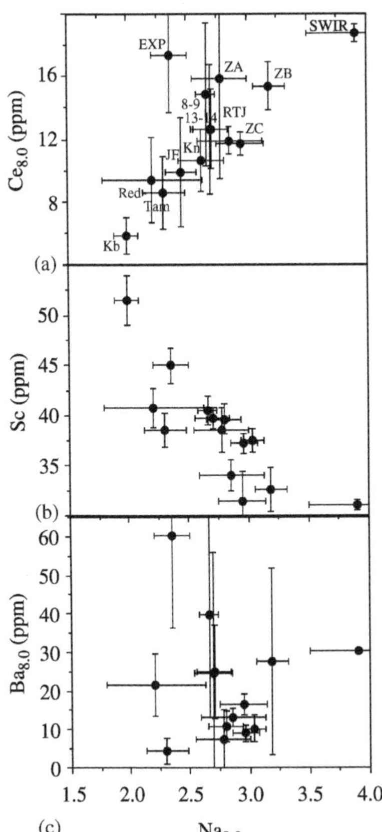

Figure 12 $Na_{8.0}$ versus (a) scandium (in ppm), (b) $Ce_{8.0}$, and (c) $Ba_{8.0}$ for regionally averaged data. Each point is the average of samples from an individual region, and the error bars are one standard deviation of the values. $Ce_{8.0}$ and $Ba_{8.0}$ (considered as highly incompatible elements) are calculated using $\log(Ba_{8.0}$ or $Ce_{8.0}) = \log(Ba$ or $Ce) - 0.11 \times (8 - MgO)$. Regions are: Kolbeinsey Ridge (Kb), 17° N in the Red Sea (Red), near Tamayo transform of the EPR (Tam), Juan Fernandez microplate (JF), near Kane of the MAR (Kn), 13–14° N on EPR (13–14), 8–9° N on EPR (8–9), south Explorer Ridge (EXP), Zones A, B, and C of the Australian Antarctic Discordance (ZA, ZB, ZC), near the Rodriguez or Indian Ocean Triple Junction (RTJ), and around 30° S on the Southwest Indian Ridge (SWIR) (Klein and Langmuir, 1987) (reproduced by permission of American Geophysical Union from *J. Geophys. Res.*, **1987**, *92*, 8089–8115).

3.13.4.2.3 *The shape of the melting regime and the generation of diverse melt compositions*

A paradigm has emerged that combines a view of the physical process of mantle upwelling and melting with the pooling of melts as they are produced. In a simple cornerflow model (Ahern and Turcotte, 1979; Phipps Morgan, 1987; McKenzie and Bickle, 1988; Plank and Langmuir, 1992), melts are produced throughout a wide triangular-shaped melting regime wherever mantle rises (Figure 4). A diversity of melt compositions are produced, resulting from the different pressures, extents and styles of melting, and the changing source composition at each point within the melting regime. Because the residual porosity is believed to be low (<1%), these melts rapidly segregate from the solid mantle and pool and mix into channels as they are focused toward the ridge axis, where they then accumulate and further mix in crustal magma reservoirs. Thus, a diversity of melt compositions are believed to be produced in the mantle, but the melt erupted at the ridge is the pooled product of these diverse melts.

Evidence of the diversity of melt compositions that are generated in the melting regime comes from two main sources. The first is the study of the geochemistry of near-axis seamounts, which are believed to tap the less-pooled products of melting beneath the adjacent ridge (e.g., Allan *et al.*, 1987; Zindler *et al.*, 1984; Graham *et al.*, 1988; Batiza and Vanko, 1984; Batiza *et al.*, 1996; Niu and Batiza, 1997; Karsten *et al.*, 1990; Hekinian *et al.*, 1989; Reynolds *et al.*, 1992). These studies show that lavas erupted along near-axis seamounts record a wider range of magma compositions than are erupted along the adjacent ridge, including both more depleted and more enriched compositions (see, e.g., Figure 13).

The second line of evidence comes from studies of melt inclusions. Melt inclusions are small pockets of melt that are trapped within phenocrysts as the minerals crystallized. Thus, to the extent that the phenocrysts crystallized from a less evolved or less mixed melt, the melt inclusions record the nature of more primitive melt compositions. Studies of melt inclusions have revealed that although they are broadly similar in major element composition to their host lavas, they tend to extend to more primitive (higher MgO or Mg#) compositions, suggesting that indeed they record an earlier point in the evolutionary history of the magma (e.g., Sobolev and Shimizu, 1993; Sobolev, 1996; Sinton *et al.*, 1993; Nielsen *et al.*, 1995; Shimizu, 1998; Sours-Page *et al.*, 1999, 2002). Notably, however, there is often a decoupling of trace element compositions from major element compositions, such that melt inclusions as a whole span a greater range in incompatible element abundances and ratios (Figures 14 and 15). Thus, melt inclusions attest to the greater diversity of melt compositions that exist in the mantle prior to pooling and eruption of magma at the ridge.

3.13.4.3 Mantle Heterogeneity

The discussion above presupposes that mantle heterogeneity does not play such a major role that it obscures or dominates the major element signatures of different extents and pressures of

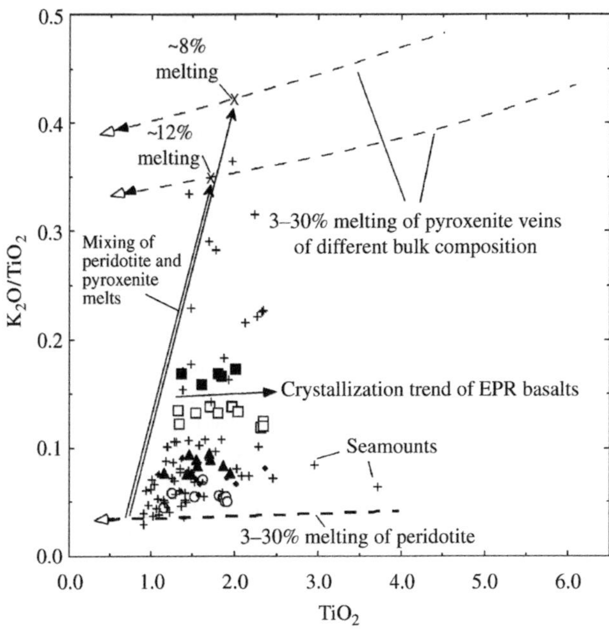

Figure 13 TiO_2 (wt.%) versus K_2O/TiO_2. Large symbols are for samples from the EPR between 11°45′ N and 15° N. Note that the highest MgO lavas in each sample/symbol group have lowest TiO_2; within each sample group, TiO_2 increases with decreasing MgO (increasing extents of crystallization); seamount lavas are shown by crosses. Also shown are model curves for 3–30% batch partial melting of peridotite (lower dashed line) and of two hypothetical enriched pyroxenite compositions (upper dashed curves). Two mixing lines (solid lines) are shown extending from a peridotite melt to the two pyroxenite melting curves; these mixing lines are parallel to a line formed by the most primitive (low TiO_2) lavas within each group, supporting the view that the most primitive lavas from each group may form by variable degrees of mixing between peridotite and pyroxenite melts. However, uncertainty in the composition of the enriched pyroxenite bulk composition (and mineralogy) make it difficult to constrain the extent of melting by which the pyroxenite melt was generated (i.e., for one of the pyroxenite melts, the mixing line interects it at 8% melting, while for the other it intersects it at 12% melting). For melting calculations, the following parameters were used: *D* for K and Ti = 0.0002 and 0.008, respectively; C_o for K_2O and TiO_2 for peridotite are 0.004 wt.% and 0.15 wt.%, respectively; for pyroxenite 1 (upper curve) 0.067 wt.% and 0.17 wt.%, respectively; and for pyroxenite 2 (lower curve) are 0.08 wt.% and 0.225 wt.%, respectively (Castillo *et al.*, 2000) (reproduced by permission of American Geophysical Union from *Geochem. Geophys. Geosys.*, **2000**, *1*, 1999GC000024).

melting that exist from region to region resulting from mantle temperature variations. Nevertheless, evidence for chemical and/or mineralogical heterogeneity of the mantle manifests itself in mantle samples recovered from ophiolites, abyssal and alpine peridotites, and mantle xenoliths (see Chapters 2.04 and 2.05), as well as in the compositions of basaltic melts (see Chapter 2.03; Natland, 1989; Salters and Dick, 2002). Indeed, major element systematics from ridge segments in the vicinity of some hotspots (e.g., the Azores, Galapagos), are anomalous with respect to the global correlations (Klein and Langmuir, 1987; Figure 10), in part due to melting in the presence of increased water in the mantle (Asimow and Langmuir, 2003). In addition, along ridge segments far from hotspots highly incompatible trace element variations are often decoupled from major element and more moderately incompatible or compatible trace element variations (Figure 12(c)); and these highly incompatible trace element enrichments commonly correlate with enrichments in radiogenic isotope compositions (Figure 16). Lastly, the compositions of individual samples from a given region ("local" variations), as opposed to the regional averages, often show chemical systematics that differ from the global trends. Each of these provides evidence of a heterogeneous source, three of which are discussed below (pyroxenite melting, assimilation of altered oceanic crust, and local variations in ocean ridge basalt composition).

3.13.4.3.1 Pyroxenite melting

The commonly observed correlation of isotopic and trace element indices of enrichment attests to the fact that the sub-oceanic mantle is compositionally heterogeneous, even in areas distant from hotspots (see, e.g., Figure 16). One widely discussed theory suggests that the enriched-to-depleted range in MORB (and near-axis seamount) compositions results from melting of enriched

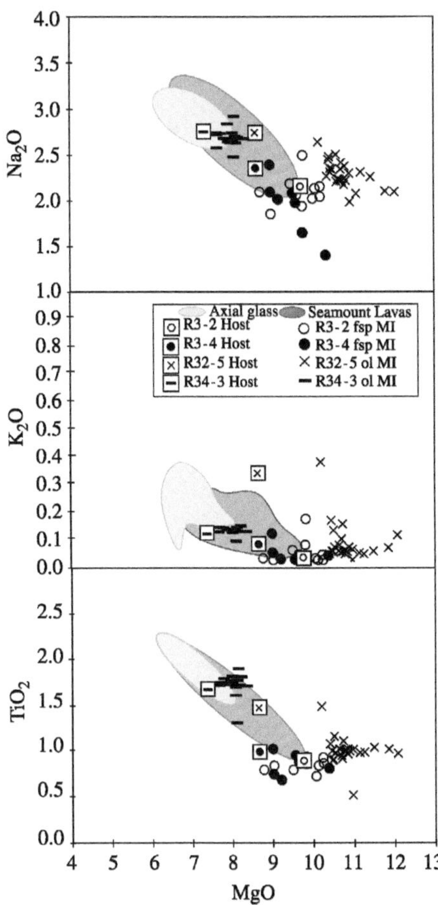

Figure 14 Major and minor elements versus MgO (wt.%) for representative glass inclusions in phenocrysts from axial and seamount lavas from the EPR. Each symbol represents inclusions in either plagioclase (fsp) or olivine (ol) from a particular lava. Shaded fields represent the compositional extent of fresh glass from the lavas of segments 10° 30′ N (light gray) and 11° 20′ N (medium gray); data from Batiza *et al.*, 1996, excluding the E-MORB (K/Ti > 0.15) and FeTi basalts ($FeO^* > 12$ wt.% and $TiO_2 > 2$ wt.%). White boxes with symbols enclosed represent the host lava compositions. Note that the melt inclusions, with the exception of R32-5, are generally similar in composition to their host lavas (Sours-Page *et al.*, 2002) (reproduced by permission of American Geophysical Union from *Chem. Geol.*, **2002**, *183*, 237–261).

pyroxenite veins (Hanson, 1977) and depleted peridotite matrix followed by mixing of these diverse melts to varying degrees (e.g., Prinzhofer *et al.*, 1989; Castillo *et al.*, 2000; Salters and Dick, 2002). It has also been suggested, for example, that these enriched veins may melt preferentially during initial melting, that they may be chemically and/or mineralogically heterogeneous, and that they may vary in size and spatial distribution (e.g., Hanson, 1977; Sleep, 1984; Zindler *et al.*, 1984; Fornari *et al.*, 1988; White, 1993; Lundstrom *et al.*, 1999; Salters and Dick, 2002; Hirschmann and Stolper, 1996). Chemical differences between depleted (N-MORB), transitional (T-MORB) and enriched (E-MORB) ocean ridge basalts (Sun *et al.*, 1979; Sun and McDonough, 1989) are often displayed on chondrite-normalized rare earth element diagrams (Figure 17(a)) or the so-called trace-element "spidergrams" (Figure 17(b)), in which the incompatibility of the trace elements increases from right to left (lower *D*s to left).

Ideally, it would be useful to distinguish the geochemical effects of source heterogeneity from those that result from variations in the extent of melting. This is a long-standing problem in studying the geochemistry of regions that exhibit differing source compositions. Although, in theory, it is possible to constrain plausible extents of melting for peridotite and vein materials, in practice, uncertainties in the composition of the enriched material can overwhelm the constraints on extents of melting. An example of this problem, using the approach of Prinzhofer *et al.* (1989), is shown in Figure 13 with data from the northern East Pacific Rise (EPR; after Castillo *et al.*, 2000). Also shown are the melting trajectories of two enriched pyroxenite veins of different compositions and melts produced from the depleted peridotite matrix. It is clear from this figure that the range in primitive (low TiO_2) EPR magma compositions can be modeled by mixing of depleted and enriched melts. However, because the source compositions are not well known, it is difficult to constrain with confidence the extent of melting of each lithology (although ranges in extents of melting can be plausibly estimated).

3.13.4.3.2 *Assimilation of altered crust*

Various types of mantle heterogeneities have been identified that differ from one another in their major element, trace element, and isotope systematics (see Chapter 2.03). A number of these distinct types of heterogeneities, particularly in isotopic composition, are most strongly manifest by hotspot volcanism (e.g., Hart and Zindler, 1989). In addition, however, it is clear that there are dispersed heterogeneities in the upper mantle that are sampled by melting at mid-ocean ridges far removed from hotspot influences.

A particularly interesting class of this type of heterogeneity is one that appears to display the chemical signatures of recycled oceanic crust (Mahoney *et al.*, 1989; Klein and Karsten, 1995; Rehkämper and Hofmann, 1997; Niu and Batiza, 1997; Sturm *et al.*, 1999; Eiler *et al.*, 2000). One location where this signature has been found is the southern Chile Ridge, an unusual setting in which an actively spreading ridge is being subducted beneath the Chile trench. Klein and Karsten (1995) showed that two of the three ridge segments nearest to the trench display, for

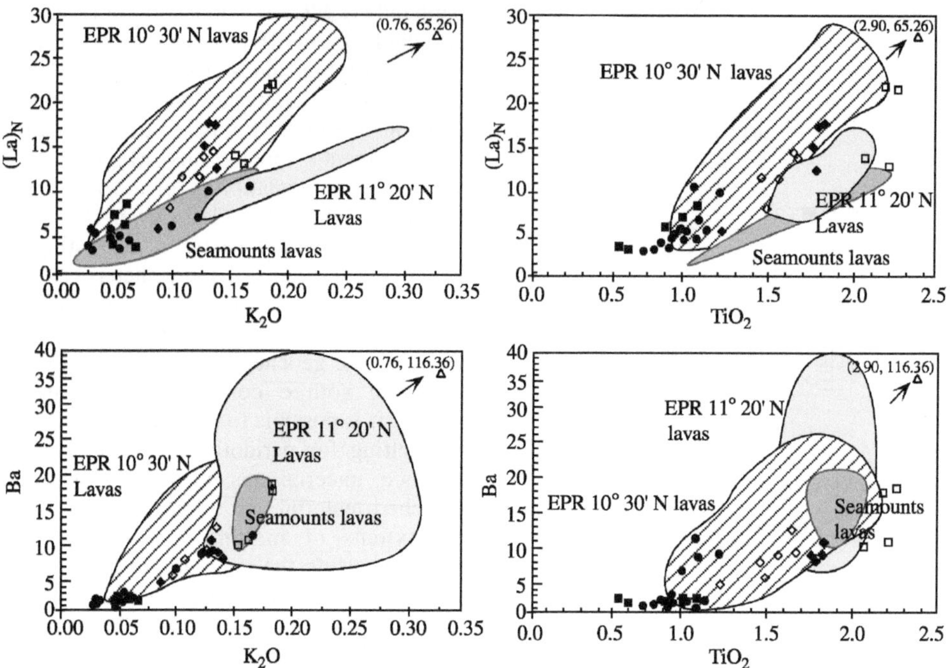

Figure 15 La and Ba (in ppm) versus K_2O and TiO_2 for axial and seamount melt inclusions from the EPR. The three fields represent the three different lava suites: 10° 30′ N (striped), 11° 20′ N (light gray) and seamounts (dark gray). Melt inclusions from seamount lavas are shown as filled symbols; melt inclusions from axial lavas are shown as open symbols. As a group, the melt inclusions show more diversity than the axial lava suites. The most depleted compositions are represented by seamounts melt inclusions (Sours-Page *et al.*, 2002) (reproduced by permission of Elsevier from *Chem. Geol.*, **2002**, *183*, 237–261).

example, low Ce/Pb values more commonly associated with the continental crust (Figure 18; Hofmann *et al.*, 1986). These and other unusual trace element and isotopic systematics suggest the possibility that beneath this ridge segment, the ambient upper mantle has been contaminated by recycled altered oceanic crust and sediment. Similarly, Niu and Batiza (1997) argued that the recycled trace element signatures of some eastern Pacific seamounts result from mixing of subducted oceanic crust into the depleted upper mantle with subsequent upwelling, melting, and dispersal of the enriched material as mobile metasomatic fluids. As is often the case, however, the most extreme manifestation of this recycled type of heterogeneity occurs at some hotspots, where, at least in the case of the Society Islands, it may also be possible to distinguish the signatures of recycled igneous crust from those of recycled marine sediment (e.g., Hémond *et al.*, 1994; Devey *et al.*, 1990; Chauvel *et al.*, 1992).

3.13.4.3.3 Local trends in basalt composition

The discussion above regarding variations in extents and pressures of melting from region to region, and the conclusion that these systematics result, to first order, from regional variations in mantle temperature is based on an examination of the average chemical compositions of lavas erupted over significant lengths of ridge (e.g. ~100 km, depending on sampling density). It has also been recognized, however, that the compositions of individual samples from limited lengths of ridge form trends (the so-called "local trends") that, for slow-spreading ridges, are often orthogonal to the global trends, while for fast-spreading ridges are often parallel to the global array (Klein and Langmuir, 1989; Niu and Batiza, 1993). There has been much discussion about the nature and origin of these different trends. Current evidence supports the idea that the differences in trends formed by individual samples from slow- and fast-spreading ridges is due not to the difference in spreading rate *per se*, but rather to the composition of the mantle between densely sampled slow- and fast-spreading ridges.

The difference in the local trends between slow- and fast-spreading ridges for the major element parameters $Na_{8.0}$ and $Fe_{8.0}$ is displayed in Figures 18 and 19. Individual samples recovered from fast-spreading Pacific ridges (Figure 20) tend to form an inverse trend within the global array. Examination of the isotopic and trace element systematics of these Pacific sample suites shows that variations in $Na_{8.0}$ among the samples also correlates with indices of mantle heterogeneity, such as K_2O/TiO_2 (Figure 16). Thus, it appears that, while the average composition of these

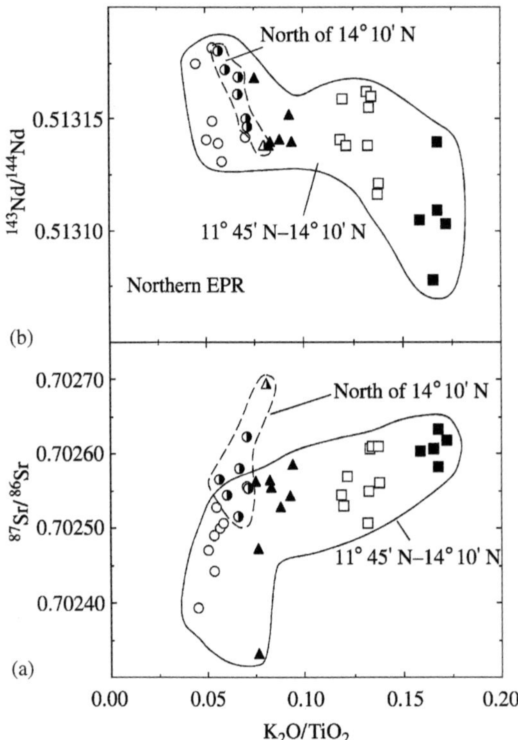

Figure 16 K_2O/TiO_2 versus (a) $^{87}Sr/^{86}Sr$, (b) $^{143}Nd/^{144}Nd$ for samples from the EPR between 11°45′ N and 15° N. EPR samples recovered north of 14° 10′ N are enclosed in a separate field (dashed field) from those recovered south of 14° 10′ N (solid field) (Castillo *et al.*, 2000) (reproduced by permission of American Geophysical Union from *Geochem. Geophys. Geosys.*, **2000**, *1*, 1999GC000024).

samples records the mean extents and pressures of melting, the detailed variations among individual samples from Pacific ridges reflects variations in source composition. This is in keeping with the well-accepted idea that the upper mantle beneath the EPR is heterogeneous on a variety of scales (e.g., Zindler *et al.*, 1984; Langmuir *et al.*, 1986; Hekinian *et al.*, 1989; Prinzhofer *et al.*, 1989; Reynolds *et al.*, 1992; Niu and Batiza, 1997; Castillo *et al.*, 2000).

In contrast to fast-spreading ridges, individual samples from slow-spreading ridges tend to form trends that are orthogonal to the global array (Figure 19). This finding has been interpreted in a number of ways (e.g., Klein and Langmuir, 1989; Langmuir *et al.*, 1992; Kinzler and Grove, 1992; Niu and Batiza, 1993). One general class of models (Klein and Langmuir, 1989) holds that the positive correlation of $Na_{8.0}$ and $Fe_{8.0}$ reflects the imperfect mixing of samples generated throughout the melting regime, such that melts produced by small extents of melting just above the solidus (high $Na_{8.0}$) are also produced at high pressures of melting (high $Fe_{8.0}$), while melts produced from larger extents of melting (or from depleted mantle higher in the melting regime) are also produced at lower pressures of melting. In this sense, the local trend for slow-spreading ridges may reflect with greater fidelity the compositions of individual (less-pooled) melt parcels produced throughout the melting regime.

3.13.4.4 Spatial Variations in Lava Compositions

3.13.4.4.1 Along-axis chemical variations

With increased extent of sampling of the ocean ridge system, as well as more detailed sampling of individual ridge segments, spatial variations in axial lava compositions have been observed on a variety of scales. On the largest scale—that of an ocean basin—it has been shown that lavas erupted along much of the Indian Ocean ridge system display differences in major element, trace element, and isotopic composition compared to the majority of MORB from the Atlantic and Pacific oceans (Hamelin *et al.*, 1986; Dupre and Allegre, 1983; see Table 2). Indian Ocean MORBs, for example, tend to have distinctive isotopic signatures with low $^{206}Pb/^{204}Pb$ and high $^{207}Pb/^{204}Pb$ and $^{87}Sr/^{86}Sr$ (Hamelin *et al.*, 1986; Dupre and Allegre, 1983) as well as different trace element ratios, such as $Ti/Zr < 90$ (Mahoney *et al.*, 1989). There is some question about whether the distinctive Indian Ocean ridge isotopic signature may be related to the Dupal anomaly (Hart, 1984), a globe-encircling belt of anomalous ocean island isotope compositions centered at about 30° S. Large-scale differences in isotopic composition have also been noted within ocean basins, such as the slightly higher $^{87}Sr/^{86}Sr$ and $^{143}Nd/^{144}Nd$ and lower $^{206}Pb/^{204}Pb$ isotopic compositions of northern Pacific ridges compared to southern Pacific ridges (Vlastelic *et al.*, 1999).

One of the earliest recognized large-scale sources of variation in lava composition along individual lengths of ridge is the influence of near-ridge hotspots. In pioneering studies of chemical variations along the northern mid-Atlantic Ridge (MAR), Schilling and co-workers showed that with increasing proximity to the Iceland and Azores plumes, over distances of 500–1,000 km, MORBs display an increasingly pronounced trace element and isotopic signatures characteristic of each of the hotspots (Figure 21; Sun *et al.*, 1975; Schilling *et al.*, 1983; Schilling, 1986). The influence of other near-ridge hotspots on ocean ridge basalt compositions, and what such influences suggest about mantle and melt flow and mixing, continues to be studied at diverse locations of hotspots/ridge interactions throughout the ocean basins (e.g., Schilling *et al.*, 1982; Hanan *et al.*, 1986; Mahoney *et al.*, 1989;

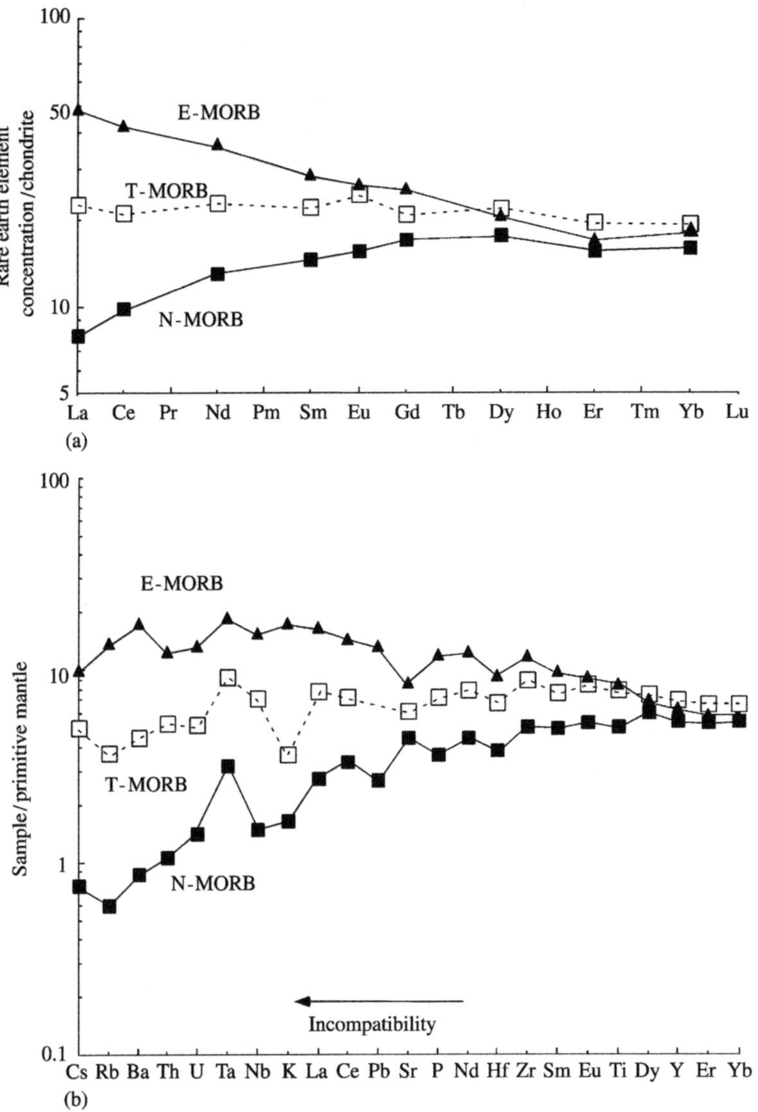

Figure 17 (a) Chondrite-normalized rare earth element patterns and (b) primitive-mantle normalized trace element patterns for N-MORB, T-MORB, and E-MORB from the MAR (see caption to Table 2 for sources; normalizing values from Sun and McDonough, 1989).

Schilling *et al.*, 1992; Kingsley and Schilling, 1998; Douglass *et al.*, 1999; Detrick *et al.*, 2002).

Since the mid-1980s, a great deal of effort has focused on exploring smaller scales of chemical variability. Typically, these studies explore the relationship between the boundaries of discrete domains of similar chemical composition, called magmatic segments, and the various scales of tectonic segmentation of the ridge (e.g., Langmuir *et al.*, 1986; Macdonald and Fox, 1988; Sinton *et al.*, 1991; Batiza, 1996). Defined primarily to describe features observed along fast-spreading ridges, "first-order" tectonic segments are bounded by long-lived transform faults and large-offset overlapping spreading centers and often display significant along-axis depth variations of hundreds to thousands of meters with central humps flanked by deeps at ridge offsets ("overlapping spreading centers" (OSCs) are $<1-10$ km offsets in the continuity of the ridge axis in which the offset limbs of the axis overlap one another). "Second-order" tectonic segments are nested within first-order segments, and are generally bounded by OSCs that typically have axial depth anomalies of a few hundred meters. Smaller scales of tectonic segmentation, "third-" or "fourth-order" segments, truncate the ridge axis into smaller, shorter-lived segments bounded by small-offsets (Macdonald *et al.*, 1988).

Increased density of sampling of the ocean system has made it possible to explore the extent

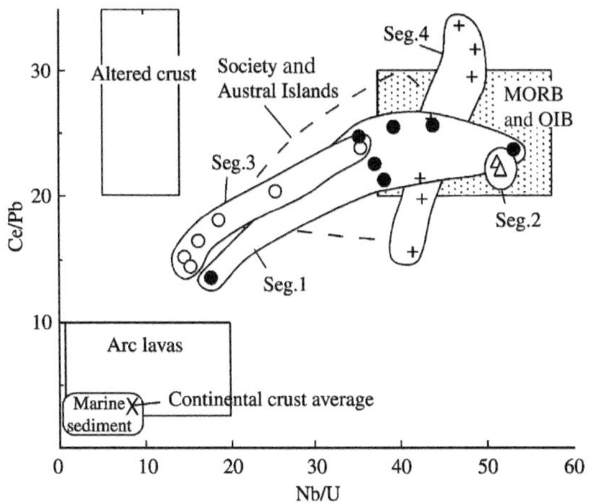

Figure 18 Nb/U versus Ce/Pb for Chile Ridge glasses: segment 1 (filled circles); segment 2 (open triangles); segment 3 (open circles); segment 4 (crosses). Also shown are fields for MORB and ocean islands, marine sediments, arc lavas, altered oceanic crust, average upper continental crust (X labeled CC) and the Society and Austral Islands (which exhibit trace element and isotopic evidence of sediment and altered crust recycling) (Klein and Karsten, 1995) (reproduced by permission of Nature Publishing Group from *Nature*, **1995**, *374*, 52–57).

to which this hierarchy of tectonic segmentation also corresponds to chemically coherent "magmatic segmentation" of the ridge. Using the terminology of Sinton *et al.* (1991), "primary" magmatic segmentation is generally defined as lengths of ridge that erupt magmas of similar isotopic and trace element compositions, suggesting a similar mantle source composition. "Secondary" magmatic segments are defined on the basis of similarities in parental magma composition resulting from both similarities in source composition and melting processes. Finally, "tertiary" and higher-order magmatic segments may encompass lengths of ridge that display similar shallow level magma chamber and/or crustal processes (e.g., Sinton *et al.*, 1991; Batiza, 1996).

In some areas, there is clear evidence of a correspondence between the hierarchical scales of tectonic and magmatic segmentation. In an examination of lavas erupted along the southern EPR, for example, Sinton *et al.* (1991) showed that isotopically coherent domains (primary magmatic segments) several hundred kilometers in length are bounded by large-offset OSCs (first-order tectonic segment boundaries). Nested within these primary magmatic segments are numerous secondary magmatic segments that share a common parental magma composition (Figure 22) and are bounded by smaller offset OSCs (second-order tectonic segments). In a detailed study of variations in lava composition along the northern EPR, Langmuir *et al.* (1986)

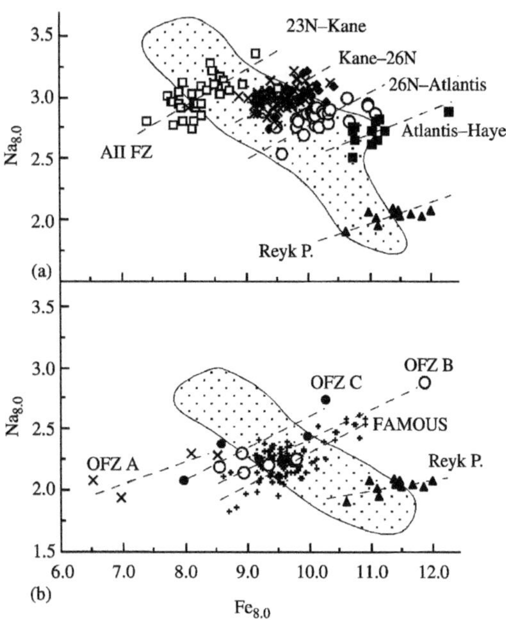

Figure 19 $Fe_{8.0}$ versus $Na_{8.0}$ for individual samples from various densely sampled ridge segments. (a) Samples from "normal" ridge segments. Shaded field is for regional averages of data from "normal" ocean ridges from Figure 10. Solid triangles for Reykjanes Peninsula; Xs for 23° N to the Kane Fracture Zone; filled diamonds from the MAR 23–26° N; open circles from 26° N to Atlantis FZ; filled squares from Atlantis FZ to Hayes FZ; and open squares for Atlantis II FZ. Note that Atlantic data are offset to the high side of the normal field, while the Indian data are on the low side. (b) Samples from ridge segments influenced by hotspots, compared to shaded field for regional averages of hotspot-influenced ridges from Figure 10. Solid triangles for Reykjanes Peninsula as above; + for FAMOUS and AMAR regions of the MAR. Samples from near the Oceanographer Fracture Zone (OFZ) are in three groups based on Ce/Yb as follows Ce/Yb = 9.7–10.7 for A, =7.4–9.1 for B, =4.2–6.5 for C. All data have been renormalized to 100%, and corrected for interlaboratory biases. The $Na_{8.0}$ and $Fe_{8.0}$ algorithms for samples south of Kane and the Reykjanes Peninsula have been adjusted as in Klein and Langmuir (1989; their figure 5) (reproduced by permission of American Geophysical Union from *Mantle Flow and Melt Generation at Mid-ocean Ridges*, **1992**, *71*, 183–280).

showed further that small "deviations from axial linearity" (the so-called "devals," representing third- or fourth-order tectonic segments) often corresponds with the changes in magma composition in terms of fractionation and/or enrichment (Figure 23). It should be noted, however, that even along fast-spreading ridges, where the higher-order tectonic features are most easily identified, not all large-offset transforms correspond to changes in isotopic or source composition (e.g., the Clipperton transform), and there may be variations in composition that grade continuously

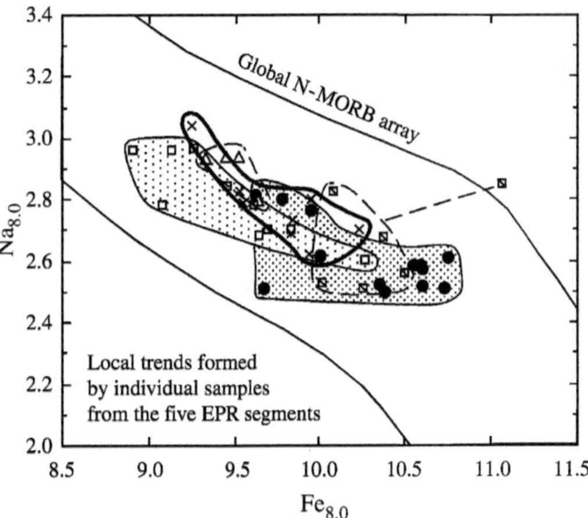

Figure 20 $Fe_{8.0}$ versus $Na_{8.0}$ for individual samples within each of the five transform-bounded or overlapping spreading center- (OSC-) bounded segments studied along the EPR between 11°45′ N and 15°N (Castillo *et al.*, 2000; filled circles). Also shown is the global trend for regional averages of N-MORB (from Langmuir *et al.*, 1992). $Na_{8.0}$ and $Fe_{8.0}$ for EPR samples calculated as described by Castillo *et al.* (2000) and caption to Figure 9 (Castillo *et al.*, 2000) (reproduced by permission of American Geophysical Union from *Geochem. Geophys. Geosys.*, **2000**, *1*, 1999GC000024).

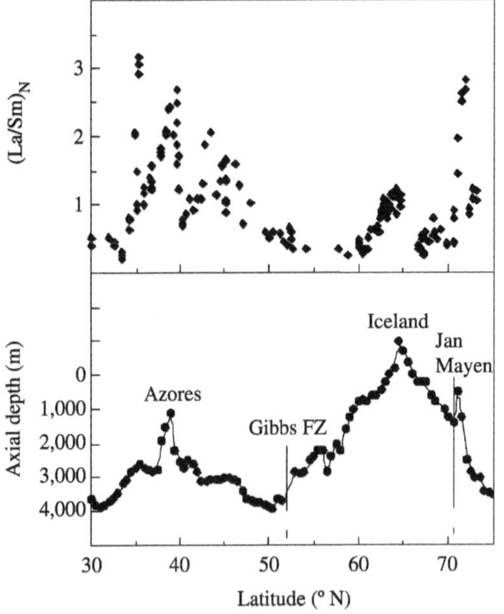

Figure 21 Latitude (° N) versus smoothed axial depth and $(La/Sm)_N$ for the northern MAR. Latitudinal depth variations, taken at 5° intervals, are from LeDouaran and Francheteau (1981) and Vogt (1986); La/Sm data are from Schilling *et al.* (1983) (reproduced by permission of American Geophysical Union from *Mantle Flow and Melt Generation at Mid-ocean Ridges*, **1992**, *71*, 183–280).

across several hundred kilometers of ridge, irrespective of the presence of OSCs or devals (e.g., 12–14° N EPR; Castillo *et al.*, 2000).

3.13.4.4.2 Temporal variations on lava composition

Lavas recovered from the morphologic axis of the ocean ridge system are often referred to as "zero-age" or near "zero-age" MORBs to emphasize their recent eruption. The compositions of these lavas are then interpreted as resulting from the current conditions of melt generation and evolution beneath the ridge. From this perspective, sampling of lavas off-axis, along a traverse that extends from the ridge onto older ocean crust (along a "flowline") reveals information on the changes in lava composition erupted through time and therefore the changing conditions of melting and melt evolution. Since the early 1990s, there has been an increased interest in exploring these temporal variations in lava composition, although such studies are few compared to on-axis sampling.

On a reconnaissance sampling scale, several studies have examined basalt compositions recovered by drilling at widely spaced sites on ocean crust as old as 150 Ma, primarily in the Atlantic and Indian Oceans (Keen *et al.*, 1990; Klein, 1992; Humler *et al.*, 1999). These studies have shown that samples recovered from old ocean crust often differ substantially in composition from zero-age basalts along the same flowline. Humler *et al.* (1999) showed further that the compositions of drill samples >80 Ma suggest substantially hotter sub-axial mantle producing greater extents of melting during the Mesozoic. On the flanks of the ridge axis, where lavas are still exposed above accumulating sediment, samples recovered by dredge or submersible along flowlines have revealed more detailed information on temporal variations (e.g., Reynolds *et al.*, 1992; Perfit *et al.*, 1994; Batiza *et al.*, 1996). Batiza *et al.* (1996), for example, examined closely spaced (1–2 km) samples collected along three flowline traverses between 9° 30′ N and 11° 20′ N on the EPR, and extending >40 km (~800 ka) on each side of the axis. The compositions of the flowline samples were shown to be largely symmetrical about the axis and to display systematic variations in the extent of crystallization through time, suggesting regular variations in recharge and evolution of the sub-ridge AMC. With even more closely spaced sampling in the vicinity of two devals near 12° N on the EPR, Reynolds *et al.* (1992) showed that the along-axis length scale of chemically coherent, deval-bounded segments waxes or wanes with time.

Two factors, however, complicate the straightforward interpretation that flowline sampling

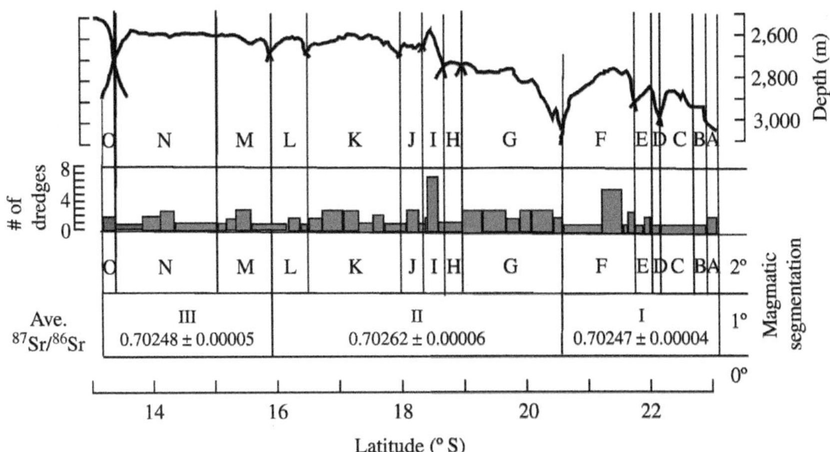

Figure 22 Along-axis magmatic segmentation and sample distribution along the EPR 13–23° S. Primary magmatic segmentation is defined on the basis of isotopic composition; secondary magmatic segmentation is defined on the basis of similar parental magma compositions. Note that most secondary magmatic segment boundaries occur at the more pronounced bathymetric depressions in the axial depth profile (Sinton *et al.*, 1991) (reproduced by permission of American Geophysical Union from *J. Geophys. Res.*, **1991**, *96*, 6133–6155).

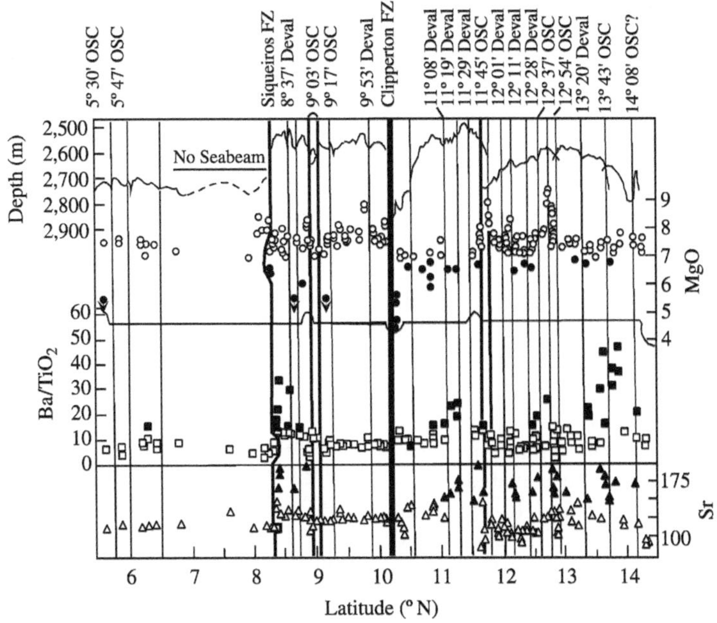

Figure 23 Axial depth and various chemical parameters for zero-age basalts versus latitude for the northern EPR 5.5–14.5° N. Thick vertical lines delineate proposed spreading cell margins at the major OSCs and the two transform faults. Thinner vertical lines occur at small-offset OSCs and "deviations from axial linearity" (devals) along the EPR. Strontium in ppm, MgO in wt.%. Many of the deval-bounded segments form chemically distinct ridge segments (Langmuir *et al.*, 1986) (reproduced by permission of Nature Publishing Group from *Nature*, **1986**, *322*, 422–429).

represents sampling of progressive temporal variations. The first factor derives from uranium-series dating studies of basalts collected off-axis along fast- and intermediate-spreading rate ridges (e.g., Goldstein *et al.*, 1992, 1994; Perfit *et al.*, 1994; see Chapter 3.14). These studies revealed that samples near the axis are generally younger than would be expected based on their distance from the apparent neo-volcanic zone, suggesting that magma may erupt throughout a relatively wide zone extending up to 4 km off-axis. (In contrast, along the slow-spreading mid-Atlantic ridge at 23° N, lavas from the apparently youngest volcanic edifices were dated as tens of thousands of years old (Sturm *et al.*, 1999).) A second factor that complicates flowline studies is that detailed geologic investigations of the spatial extent of lava flows on fast- or intermediate-spreading ridges show that individual eruptions may extend several kilometers both along- and off-axis (e.g., see Perfit and Chadwick, 1998; Sinton *et al.*, 2002). Thus, a certain amount of overprinting of older lavas occurs during new eruptions. Indeed, investigations of the chemical variability of vertically successive lava flows recovered down-hole at ODP drill site 504B reveals a diversity of lava compositions, including variations in radiogenic isotopic composition, suggesting different source compositions (Pedersen and Furnes, 2001).

3.13.5 THE LOWER OCEANIC CRUST

3.13.5.1 Geochemical Systematics of Dikes

Beneath the volcanic unit of the oceanic crust is a unit of sheeted diabase dikes believed to represent the solidified conduits of magmas derived from the melt lens that erupt to form the volcanic unit. It would therefore be expected that the dikes would have compositions similar to those of the lavas. While this is generally true, detailed comparisons of dikes and lavas in both ophiolites and the oceanic crust suggest additional complex relationships.

There are relatively few studies of dikes in the *in situ* ocean crust, primarily because there are relatively few settings in which significant outcrops of the deeper portions of the ocean crust are exposed. One rare "tectonic window" into the lower ocean crust exists at the Hess Deep Rift, which exposes 1 Myr old crust generated at the EPR (Lonsdale, 1988; Francheteau *et al.*, 1990, 1992; Karson *et al.*, 1992, 2002; Gillis, 1995). Stewart *et al.* (2002, 2003) examined the compositions of dikes and lavas at the Hess Deep Rift. While all dikes and lavas are of basaltic composition, the majority of crystalline lavas have lower FeO_T and MgO, and elevated Al_2O_3 and CaO concentrations compared to the dikes (Figure 24). These chemical differences and associated modal variations were attributed to the accumulation of plagioclase in the magmas that ultimately erupt as lavas on the seafloor. The accumulation of plagioclase combined with the fractionation of mafic phases lowers the magma density by more than 0.04 g cm^{-3} relative to most dike magmas. Thus, it appears that lower-density magmas are preferentially erupted because of their increased buoyancy, resulting in the predominance of this magma type as lavas. Conversely, the majority of dikes solidified from a higher-density magma type that is rarely represented by lavas. (This finding also underscores the importance of analyzing basaltic glass compositions in studies of basalt composition, as opposed to magma (whole rock) compositions, which represent a mixture of melt and phenocryts.) These relationships suggest that most dikes never reach the surface and erupt lava. Examination of

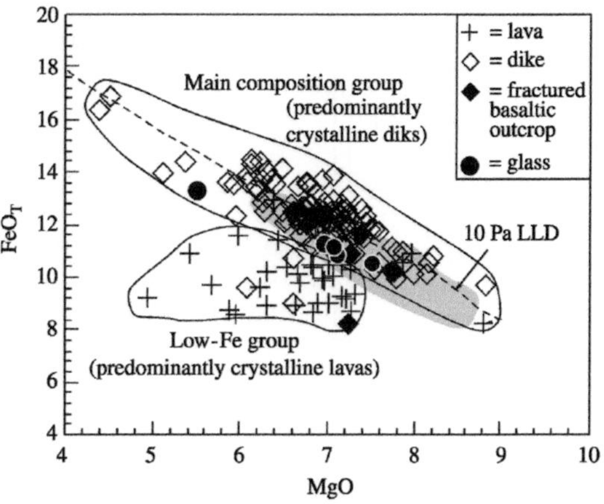

Figure 24 FeO_T versus MgO (wt.%). crosses = lavas; open diamonds = dikes; filled diamonds = samples from fractured basaltic outcrops not distinguishable as lavas or dikes; filled circles = basaltic glass. The fractional crystallization liquid line of descent (10 Pa LLD, shown as dashed line) was calculated using MELTS v2.0.4 with starting composition of a high MgO sample, and f_{O_2} buffered at QFM. Fields enclose the main composition group defined by all but three crystalline dike samples; and the low-Fe group (composed predominantly of crystalline lava samples) offset to lower MgO and FeO_T and defined by samples with MgO < 7.4 wt.% and $FeO_T < 22 - 1.6$MgO. Sample data set is available online at the American Geophysical Union electronic supplement repository. Shaded field shows composition of glasses from the EPR region at ~2.7° N reported by Lonsdale *et al.*, 1992 (reproduced by permission of American Geophysical Union from *J. Geophys. Res.*, **2003**, *108*).

phenocryst contents and modal abundances in MORBs worldwide indicates that magma density variations and crustal heterogeneity may influence magma eruptibility across a wide range of spreading rates. Specifically, lavas from slow-spreading ridges have a wider range of modal phenocryst proportions compared to lavas from intermediate- or fast-spreading ridges. Presumably, lavas from slow-spreading ridges will have a greater range of densities (due to their variable phenocryst contents) and their eruption on the seafloor reflects the heterogeneous upper crust of slow-spreading ridges, where even lower crust and mantle rocks are exposed, compared to that of fast-spreading ridges.

A second finding in the study of dikes exposed at Hess Deep concerns the spatial variation in their compositions (Stewart *et al.*, 2002). Dikes were sampled over an area encompassing 25 km of an east–west flowline, representing ~3.7×10^5 yr of crustal accretion at the EPR. Indices of fractionation (MgO), and incompatible element ratios (La/Sm, Nb/Ti) show no systematic trends along flowline. Rather, over short (<4 m) and long (~25 km) distances, significant variations are observed in major and trace element concentrations and ratios. Modeling of these variations attests to the juxtaposition of dikes of distinct parental magma compositions. These findings, combined with studies of segmentation of the sub-axial magma chamber (AMC) and lateral magma transport in dikes along rift-dominated systems suggest a model in which melts from a heterogeneous mantle feed distinct portions of a segmented axial magma reservoir. Dikes emanating from these distinct reservoirs transport magma along axis, resulting in interleaved dikes and host lavas with different evolutionary histories (Figure 25).

3.13.5.2 Geochemical and Textural Systematics of Gabbros

It is generally agreed that the thick gabbroic unit at the base of the oceanic crust forms by the relatively slow crystallization of basaltic magma at depth. Decades ago, it was envisioned that crystallization along the walls and floor of a large, predominately molten, steady-state magma chamber at depth beneath the ridge would produce the gabbroic layer and, in particular, account for the various types of mineralogic and chemical layering observed in many ophiolite complexes (e.g., Pallister and Hopson, 1981). With improved seismic and tomographic imaging of the region beneath the crust (e.g., Detrick *et al.*, 1987; Toomey *et al.*, 1990; Caress *et al.*, 1992), it has become clear that no large, predominately molten magma body exists beneath the ridge, but rather, at least at fast-spreading ridges, a thin (~50 m), narrow (0.5–1.5 km) melt lens at 1–2 km depth beneath the ocean floor caps a wider region of low seismic velocities, suggesting the presence of melt and significant amounts of crystals (Figure 3). This region, commonly called the "mush zone," may extend to 10 km in width and the fraction of melt within it is believed to decrease both with depth and laterally, ultimately grading laterally into seismic velocities consistent with solidified gabbro (e.g., Sinton and Detrick, 1992).

As discussed by Coogan *et al.* (2002), there are currently two classes of models for the formation of the gabbroic sequence that are, to first order, consistent with available constraints from marine seismic investigations, and structural and petrologic studies of gabbros in ophiolites and oceanic drill cores. In the first class of models (Figure 25(a)), crystallization occurs

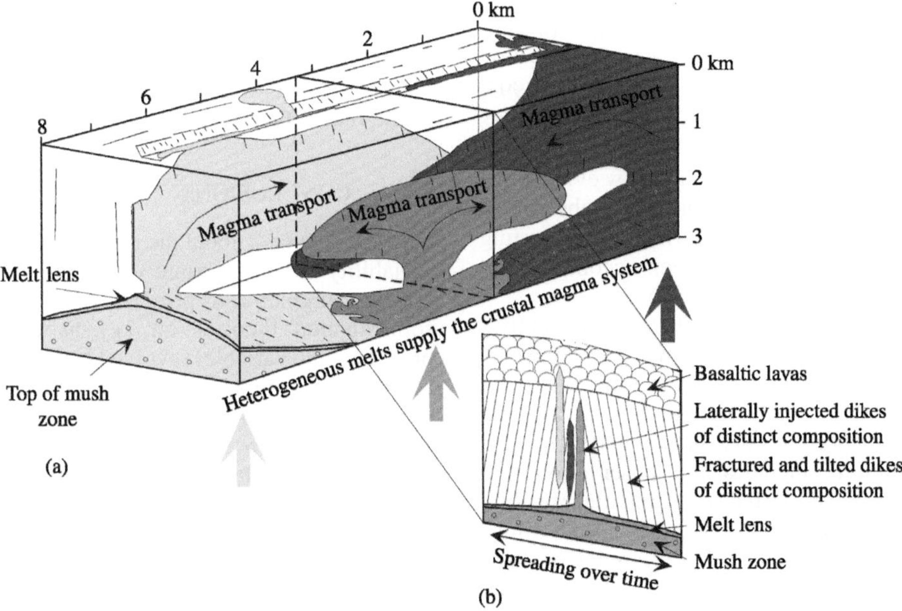

Figure 25 Model for construction of the upper crust at fast-spreading ridges (a), leading to the cross-sectional (flowline) variations in dike compositions (b) exposed on the north wall of the Hess Deep Rift. The melt lens is fed by distinct mantle melts along axis. Incomplete mixing along the length of the AMC for the equatorial EPR produces compositionally distinct portions of the AMC (distinct shading). Dikes are injected in narrow regions near the axis (1–2 km wide), but incomplete mixing in the AMC and lateral (along-axis) magma transport (shown as vectors in the dikes) results in closely spaced dikes of distinct compositions and the absence of a simple trend in composition along a flowline section. Dikes will tend to pierce the surface and erupt lavas over their initial site of injection, but many dikes will not intersect the surface and erupt lavas (see text for discussion). Note that in panels (a) and (b), the dike thickness (commonly ~1 m) has been exaggerated for visual clarity and the top of the melt lens has been placed at 2 km depth, but ranges from ~1–2 km (Stewart *et al.*, 2002) (reproduced by permission of American Geophysical Union from *J. Geophys. Res.*, **2002**, *107*, 2238).

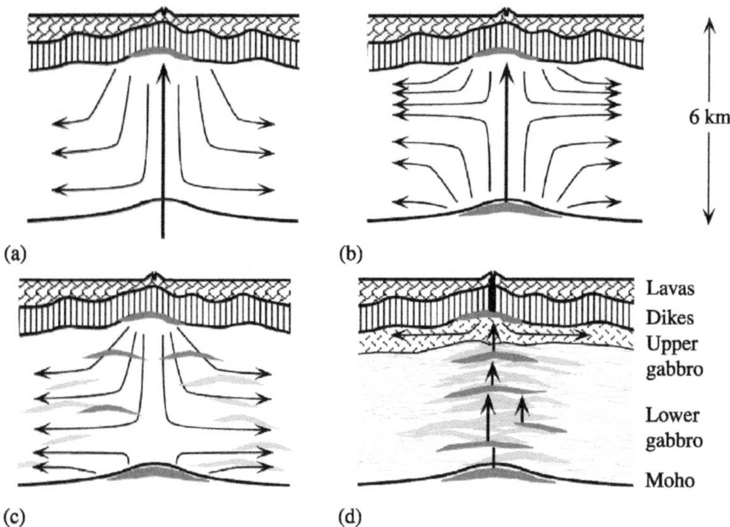

Figure 26 Schematic drawing for recently proposed crustal models. (a) Ductile flow model in its simplest form (e.g., Phipps Morgan and Chen, 1993; Quick and Denlinger, 1993; Nicolas *et al.*, 1988). Ductile flow down and outward from a mid-crustal magma chamber constructs the lower crust. (b) Ductile flow model incorporating a basalt sill (Schouten and Denham, 1995). (c) Hybrid ductile flow model with on- and off-axis sill intrusions (Boudier *et al.*, 1996). (d) Sheeted sill model with melt transport by hydrofracture and *in situ* emplacement of the lower crust by on-axis sill intrusions (Kelemen *et al.*, 1997; Bedard *et al.*, 1988) (reproduced by permission of Elsevier from *Planet. Sci. Lett.*, **1998**, *156*, 1–11).

predominantly within the melt lens (and possibly a second melt lens at the base of the crust; Figure 26(b)). The resulting gabbroic material subsides and/or is transported laterally off-axis (e.g., Sleep, 1974; Dewey and Kidd, 1977; Nicolas *et al.*, 1988; Quick and Denlinger, 1993; Phipps Morgan and Chen, 1993). Evidence for this comes primarily from textural studies of gabbroic rocks in the Oman ophiolite that show evidence of solid-state ductile deformation in the absence of significant crystal–plastic deformation, suggesting that large strains may have been accommodated by deformation of a crystal mush.

In the second class of models (Figure 26(d)), gabbro formation occurs primarily within the mush zone in dispersed sills (Korenaga and Kelemen, 1997, 1998; Kelemen *et al.*, 1997). Evidence supporting this latter model for the Oman ophiolite includes the observation that discrete gabbro sills in the crust–mantle transition zone have textures and chemical compositions similar to those of the modally layered gabbros of the thick, lower gabbro section. In addition, minerals of the lower gabbros are in chemical equilibrium with melt compositions of the upper gabbros, dikes and lavas, suggesting that the latter three formed in large part from melts expelled from the crystallizing lower gabbro sills (Kelemen *et al.*, 1997). Structural and other studies of Oman gabbros support the possibility that both of these processes may be occurring simultaneously (Kelemen and Koregan, 1997), a conclusion also reached by Coogan *et al.* (2002) in a geochemical and textural study of gabbros exposed at Hess Deep. These two models, however, were primarily developed for fast-spreading ridges, and as emphasized by Natland and Dick (2001) may not be applicable to gabbros formed at slow-spreading ridges, such as the petrologically complex gabbros recovered in ODP Hole 735B in the Indian Ocean.

3.13.6 CONCLUSIONS

Much is known about the origin of the chemical variability of the oceanic crust, but much also remains enigmatic. At present, we have relatively sophisticated theories on key aspects of the processes and variables involved in the generation, evolution and emplacement of magma to form the crust. But the linkages between these processes remain speculative. One example of this concerns the enigmatic relationship between the diverse melt compositions produced in the mantle and the various scales of tectonic and magmatic segmentation observed along the ridge axis. Are some of the scales of tectonic and magmatic segmentation a reflection of deep mantle upwelling processes while others merely reflect shallow magmatic plumbing systems? Furthermore, while we have an extensive catalog of the diverse types of heterogeneities in the mantle, no unifying paradigm has emerged to fully explain their relationships to one another, how they developed, and their spatial scales. Lastly, combined geochemical, geophysical, and geological studies are needed to elucidate the processes of crustal accretion that produce the architecture of the ocean crust.

ACKNOWLEDGMENTS

I am grateful to Roberta Rudnick for her insightful editing and unflagging encouragement, and to Terry Plank for her careful review of this manuscript. This work was supported by NSF Grant #OCE- 0099154.

REFERENCES

Ahern J. L. and Turcotte D. L. (1979) Magma migration beneath an ocean ridge. *Earth Planet. Sci. Lett.* **45**, 115–122.

Allan J. F., Batiza R., and Lonsdale P. (1987) Petrology and chemistry of lavas from seamounts flanking the East Pacific Rise, 21° N: implications concerning the mantle source composition for both seamount and adjacent EPR lavas. *Am. Geophys. Union Monogr.* **43**, 255–281.

Asimow P. D. and Langmuir C. H. (2003) The importance of water to oceanic mantle melting regimes. *Nature* **421**, 815–820.

Asimow P. D., Hirschman M. M., and Stolper E. M. (1997) An analysis of variations in isentropic melt productivity. *Phil. Trans. Roy. Soc. London Ser. A* **355**, 255–281.

Asimow P. D., Hirschmann M. M., and Stolper E. M. (2001) Calculation of peridotite partial melting from thermodynamic models of minerals and melts: IV. Adiabatic decompression and the composition and mean properties of mid-ocean ridge basalts. *J. Petrol.* **42**, 963–998.

Batiza R. (1996) Magmatic segmentation of mid-ocean ridges: a review. In *Tectonic, Magmatic, Hydrothermal and Biological Segmentation of Mid-ocean Ridges*, Geol. Soc. Spec. Publ. 118 (eds. C. J. MacLeod, P. Taylor, and C. L. Walker). Geological Society of America, Bath, UK, pp. 103–130.

Batiza R. and Vanko D. (1984) Petrology of young Pacific seamounts. *J. Geophys. Res.* **89**, 11235–11260.

Batiza R., Niu Y., and Zayac W. C. (1990) Chemistry of seamounts near the East Pacific Rise: implications for the geometry of sub-axial mantle flow. *Geology* **18**, 1122–1125.

Batiza R., Niu Y., Karsten J. L., Boger W., Potts E., Norby L., and Butler R. (1996) Steady and non-steady state magma chambers below the East Pacific Rise. *Geophys. Res. Lett.* **23**, 221–224.

Bedard J. H., Sparks R. S. J., Renner R., Cheadle M. J., and Hallworth M. A. (1988) Peridotite sills and metasomatic gabbros in the Eastern Layered Series of the Rhum Complex. *J. Geol. Soc. London* **145**, 207–224.

Bender J. F., Hodges F. N., and Bence A. E. (1978) Petrogenesis of basalts from the project FAMOUS area: experimental study from 0 to 15 kbars. *Earth Planet. Sci. Lett.* **41**, 277–302.

Boudier F., Nicolas A., and Benoit I. (1996) Magma chambers in the Oman Ophiolite: fed from the top and the bottom. *Earth Planet. Sci. Lett.* **144**, 239–250.

Canales J. P., Detrick R. S., Bazin S, Harding A. J., and Orcutt J. A. (1998) Off-axis crustal thickness across and along the East Pacific Rise within the MELT area. *Science* **280**, 1218–1221.

Caress D. W., Burnett M. S., and Orcutt J. A. (1992) Tomographic image of the axial low-velocity zone at 12 degrees 50′ N on the East Pacific Rise. *J. Geophys. Res.* **97**, 9243–9263.

Castillo P. R., Klein E., Bender J., Langmuir C., Shirey S., Batiza R., and White W. (2000) Petrology and Sr, Nd, and Pb isotope geochemistry of mid-ocean ridge basalt glasses from the 11°45′ N to 15°00″ N segment of the East Pacific Rise. *Geochem. Geophys. Geosys.* **1**, 1999GC000024.

Cawthorn R. G. (1975) Degrees of melting in mantle diapirs and the origin of ultrabasic liquids. *Earth Planet. Sci. Lett.* **27**, 113–120.

Chauvel C., Hofmann A., and Vidal P. (1992) HIMU-EM: the French-Polynesian connection. *Earth Planet. Sci. Lett.* **110**, 99–119.

Cogley J. G. (1984) Continental margins and the extent and number of the continents. *Rev. Geophys. Space Phys.* **22**, 101–122.

Coogan L. A., Gillis K. M., MacLeod C. J., Thompson G. M., and Hekinian R. (2002) Petrology and geochemistry of the lower ocean crust formed at the East Pacific Rise and exposed at Hess Deep: a synthesis and new results. *Geochem. Geophys. Geosyst.* **3**, 2001GC000230.

Daines M. J. and Richter F. M. (1988) An experimental method for directly determining the interconnectivity of melt in a partially molten system. *Geophys. Res. Lett.* **15**, 1459–1462.

Darbyshire F. A., White R. S., and Priestley K. F. (2000) Structure of the crust and uppermost mantle of Iceland from a combined seismic and gravity study. *Earth Planet. Sci. Lett.* **181**, 409–428.

Detrick R. S., Buhl P., Vera E., Mutter J., Orcutt J., Madsen J., and Brocher T. (1987) Multi-channel seismic imaging of a crustal magma chamber along the East Pacific Rise. *Nature* **326**, 35–41.

Detrick R. S., Sinton J. M., Ito G., Canales J. P., Behn M., Blacic T., Cushman B., Dixon J. E., Graham D. W., and Mahoney J. J. (2002) Correlated geophysical, geochemical, and volcanological manifestations of plume–ridge interaction along the Galapagos spreading center. *Geochem. Geophys. Geosys.* **3**, doi:10.1029/2002GC000350.

Devey C. W., Albarede F., Cheminée J.-L., Michard A., Mühe R., and Stoffers P. (1990) Active submarine volcanism on the Society hotspot swell (west Pacific): a geochemical study. *J. Geophys. Res.* **95**, 5049–5066.

Dewey J. F. and Kidd S. F. (1977) Geometry of plate accretion. *Geol. Soc. Am. Bull.* **88**, 960–968.

Dick H. J. B., Fisher R. L., and Bryan W. B. (1984) Mineralogic variability of the uppermost mantle along mid-ocean ridges. *Earth Planet. Sci. Lett.* **69**, 88–106.

Douglass J., Schilling J.-G., and Fontignie D. (1999) Plume-ridge interactions of the Discovery and Shona mantle plumes with the southern Mid-Atlantic Ridge (40°–55°S). *J. Geophys. Res.* **104**, 2941–2962.

Dupre B. and Allegre C. J. (1983) Pb–Sr isotope variation in Indian Ocean basalts and mixing phenomena. *Nature* **303**, 142–146.

Eiler J. M., Schiano P., Kitchen N., and Stolper E. M. (2000) Oxygen-isotope evidence for recycled crust in the sources of mid-ocean-ridge basalts. *Nature* **403**, 530–534.

Elthon D. and Scarfe C. M. (1984) High-pressure phase equilibria of a high-magnesia basalt and the genesis of primary oceanic basalts. *Am. Mineral.* **69**, 1–15.

Faul U. H. (1997) Permeability of partially molten upper mantle rocks from experiments and percolation theory. *J. Geophys. Res.* **102**, 10299–10311.

Fornari D. J., Perfit M. R., Allan J. F., Batiza R., Haymon R., Barone A., Ryan W. B. F., Smith J., Srinken J., and Puckman M. (1988) Geochemical and structural studies of the Lamont seamounts: seamounts as indicators of mantle processess. *Earth Planet. Sci. Lett.* **89**, 63–83.

Fowler C. M. R. and Keen C. E. (1979) Oceanic crustal structure: Mid-Atlantic Ridge at 45 degrees N. *Geophys. J. Roy. Astronom. Soc.* **56**, 219–226.

Francheteau J., Armijo R., Cheminee J. L., Hekinian R., Lonsdale P., and Blum N. (1990) 1 Ma East Pacific Rise oceanic crust and uppermost mantle exposed by rifting in Hess Deep (Equatorial Pacific Ocean). *Earth Planet. Sci. Lett.* **101**, 281–295.

Francheteau J., Armijo R., Cheminee J. L., Hekinian R., Lonsdale P., and Blum N. (1992) Dyke complex of the East Pacific Rise exposed in the walls of Hess Deep and the structure of the upper oceanic crust. *Earth Planet. Sci. Lett.* **111**, 109–121.

Gast P. W. (1968) Trace element fractionation and the origin of tholeiitic and alkaline magma types. *Geochim. Cosmochim. Acta* **32**, 1057–1068.

Ghiorso M. S. and Sack R. O. (1995) Chemical mass transfer in magmatic processes: IV. A revised and internally consistent thermodynamic model for the interpolation and extrapolation of liquid-solid equilibria in magmatic systems at elevated temperatures and pressures. *Contrib. Mineral. Petrol.* **119**, 197–212.

Ghiorso M. S., Hirschmann M. M., Reiners P. W., and Kress V. C., III (2002) The pMELTS: a revision of MELTS for improved calculation of phase relations and major element partitioning related to partial melting of the mantle to 3 GPa. *Geochem. Geophys. Geosyst.* **3**, 1030 doi:10.1029/2001GC000217.

Gillis K. M. (1995) Controls on hydrothermal alteration in a section of fast-spreading oceanic crust. *Earth Planet. Sci. Lett.* **134**, 473–489.

Goldstein S. J., Murrell M. T., Janecky D. R., Delaney J. R., and Clague D. A. (1992) Geochronology and petrogenesis of MORB from the Juan de Fuca and Gorda ridges by $^{238}U-^{230}Th$ disequilibrium. *Earth Planet. Sci. Lett.* **109**, 255–272.

Goldstein S. J., Perfit M. R., Batiza R., Fornari D. J., and Murrell M. T. (1994) Off-axis volcanism at the East Pacific Rise detected by uranium-series dating of basalts. *Nature* **367**, 157–159.

Graham D., Zindler A., Kurz M., Jenkins W., Batiza R., and Staudigel H. (1988) He, Pb, Sr, and Nd isotope constraints on magma genesis and mantle heterogeneity beneath young Pacific seamounts. *Contrib. Mineral. Petrol.* **99**, 446–463.

Grove T. L. and Bryan W. B. (1983) Fractionation of pyroxene-phyric MORB at low pressure: an experimental study. *Contrib. Mineral. Petrol.* **84**, 293–309.

Grove T. L., Kinzler R. J., and Bryan W. B. (1990) Natural and experimental phase relations of lavas from Serocki volcano. *Proc. ODP, Sci. Results* **106/109**, 9–17.

Hamelin B., Dupre B., and Allegre C. J. (1986) Pb–Sr–Nd isotopic data of Indian Ocean ridges: new evidence of large-scale mapping of mantle heterogeneities. *Earth Planet. Sci. Lett.* **76**, 288–298.

Hanan B. B., Kingsley R. H., and Schilling J.-G. (1986) Pb isotope evidence in the South Atlantic for migrating ridge-hotspot interactions. *Nature* **322**, 137–144.

Hanson G. N. (1977) Geochemical evolution of the suboceanic mantle. *J. Geol. Soc. London* **134**, 235–260.

Hart S. R. (1984) A large-scale anomaly in the Southern Hemisphere mantle. *Nature* **309**, 753–757.

Hart S. and Zindler A. (1989) Constraints on the nature and development of chemical heterogeneities in the mantle. In *Mantle Convection: Plate Tectonics and Global Dynamics* (ed. W. D. Peltier). Gordon and Breach, New York, pp. 261–387.

Hekinian R., Thompson G., and Bideau D. (1989) Axial and off-axial heterogeneity of basaltic rocks from the East Pacific Rise at 12°35′N–12°51′N and 11°26′N–11°30′N. *J. Geophys. Res.* **94**, 17437–17463.

Hémond C., Devey C. W., and Chauvel C. (1994) Source compositions and melting processes in the Society and Austral plumes (South Pacific Ocean): element and isotope (Sr, Nd, Pb, Th) geochemistry. *Chem. Geol.* **115**, 7–45.

Hirschmann M. M. and Stolper E. M. (1996) A possible role for garnet pyroxenite in the origin of the "garnet signature" in MORB. *Contrib. Mineral. Petrol.* **124**, 185–208.

Hirschmann M. M., Ghiorso M. S., and Stolper E. M. (1999) Calculation of peridotite partial melting from thermodynamic models of minerals and metals: III. Controls on isobaric melt production and the effect of water on melt production. *J. Petrol.* **40**, 831–851.

Hofmann A. W., Jochun K., Seufert M., and White W. M. (1986) Nb and Pb in basalts: new constraints on mantle evolution. *Earth Planet. Sci. Lett.* **79**, 33–45.

Hooft E. E. E., Detrick R. S., Toomey D. R., Collins J. A., and Lin J. (2000) Crustal thickness and structure alone three contrasting spreading segments of the Mid-Atlantic Ridge, 33.5 degrees–3.5 degrees N. *J. Geophys. Res.* **105**, 8205–8226.

Humler E., Langmuir C., and Daux V. (1999) Depth versus age: new perspectives from the chemical compositions of ancient crust. *Earth Planet. Sci. Lett.* **173**, 7–23.

Johnson K. T. M. and Dick H. J. B. (1992) Open system melting and temporal and spatial variation of peridotite and basalt at the Atlantis II fracture zone. *J. Geophys. Res.* **97**, 9219–9241.

Johnson K. T. M., Dick H. J. B., and Shimizu N. (1990) Melting in the oceanic upper mantle: an ion microprobe study of diopsides in abyssal peridotites. *J. Geophys. Res.* **95**, 2661–2678.

Jokat W., Ritzmann O., Scmidt-Aursch M. C., Drachev S., Gauger S., and Snow J. (2003) Geohysical evidence for reduced melt production on the Arctic ultraslow Gakkel mid-ocean ridge. *Nature* **423**, 962–965.

Karson J. A. (2002) Geologic structure of the uppermost oceanic crust created at fast- to intermediate-rate spreading centers. In *Annual Reviews Earth Planet of Sciences*, Annual Reviews (eds. R. Jeanloz, A. L. Albee, and K. C. Burke). Annual Reviews, Palo Alto, CA, vol. 30, pp. 347–384.

Karson J. A., Hurst S. D., and Lonsdale P. (1992) Tectonic rotations of dikes in fast-spread oceanic crust exposed near Hess Deep. *Geology* **20**, 685–688.

Karson J. A., Klein E. M., Hurst S. D., Lee C. E., Rivizzigno P. A., Curewitz D., Morris A. K., and the Hess Deep '99 Scientific Party (2002) Structure of uppermost fast-spread oceanic crust exposed at the Hess deep Rift: implications for subaxial processes at the East Pacific Rise. *Geochem. Geophys. Geosyst.* **3**, 2001GC000155.

Karsten J. L., Delaney J. R., Rhodes J. M., and Liias R. A. (1990) Spatial and temporal evolution of magmatic systems beneath the Endeavour Segment, Juan de Fuca Ridge: Tectonic and petrologic constraints. *J. Geophys. Res.* **95**, 19235–19256.

Keen M. J., Klein E. M., and Melson W. G. (1990) Ocean ridge basalt compositions correlated with palaeobathymetry. *Nature* **345**, 423–426.

Kelemen P. B., Shimizu N., and Salters V. J. M. (1995) Extraction of mid-ocean-ridge basalt from the upwelling mantle by focused flow of melt in dunite channels. *Nature* **375**, 747–753.

Kelemen P. B., Koga K., and Shimizu N. (1997) Geochemistry of gabbro sills in the crust–mantle transition zone of the Oman Ophiolite: implications for the origin of the oceanic lower crust. *Earth Planet. Sci. Lett.* **146**, 475–488.

Kingsley R. H. and Schilling J.-G. (1998) Plume-ridge interaction in the Easter-Salas y Gomez seamount chain-Easter microplate system: Pb isotope evidence. *J. Geophys. Res.* **103**, 24159–24177.

Kinzler R. J. and Grove T. L. (1992) Primary magmas of mid-ocean ridge basalts: 2. Applications. *J. Geophys. Res.* **97**, 6907–6926.

Klein E. M. (1992) Ocean ridge basalt phenocryst compositions reveal variations in extent of melting V. M. Goldschmidt Conference, The Geochemical Society, A-60p.

Klein E. M. (2003) Earth science: spread thin in the Arctic. *Nature* **423**, 932–933.

Klein E. M. and Karsten J. L. (1995) Ocean ridge basalts with convergent margin geochemical affinities from the Chile Ridge. *Nature* **374**, 52–57.

Klein E. M. and Langmuir C. H. (1987) Global correlations of ocean ridge basalt chemistry with axial depth and crustal thickness. *J. Geophys. Res.* **92**, 8089–8115.

Klein E. M. and Langmuir C. H. (1989) Local versus global variations in ocean ridge basalt composition: a reply. *J. Geophys. Res.* **94**, 4241–4252.

Klein E. M., Langmuir C. H., and Staudigel H. (1991) Geochemistry of basalts from the Southeast Indian Ridge, 115°N–138°E. *J. Geophys. Res.* **96**, 2089–2108.

Klingelhöfer F. Géli L., Matias L., Steinsland N., and Mohr J. (2000) Crustal structure of a super-slow spreading centre: a seismic refraction study of Mohns Ridge, 72° N. *Geophys. J. Int.* **141**, 509–526.

Kodaira S., Mjelde R., Gunnarsson K., Shirobara H., and Shimamura H. (1997) Crustal structure of the Kolbeinsey Ridge, North Atlantic, obtained by use of ocean bottom seismographs. *J. Geophys. Res.* **102**, 3131–3151.

Korenaga J. and Kelemen P. B. (1997) Origin of gabbro sills in the Moho transition zone of the Oman Ophiolite: implications for magma transport in the oceanic lower crust. *J. Geophys. Res.* **102**, 27729–27749.

Korenaga J. and Kelemen P. B. (1998) Melt migration through the oceanic lower crust: a constraint from melt percolation modeling with finite solid diffusion. *Earth Planet. Sci. Lett.* **156**, 1–11.

Lachenbruch A. H. (1976) Dyanmics of a passive spreading center. *J. Geophys. Res.* **81**, 1883–1902.

Langmuir C. H. (1989) Geochemical consequences of *in situ* crystallization. *Nature* **340**, 199–205.

Langmuir C. H. and Hanson G. N. (1980) An evaluation of major element heterogeneity in the mantle sources of basalts. *Phil. Trans. Roy. Soc. London* **297**, 383–407.

Langmuir C. H. and Hanson G. N. (1981) Calculating mineral-melt equilibria with stoichiometry, mass balance, and single-component distribution coefficients. In *Thermodynamics of Minerals and Melts, Advances in Physical Geochemistry* (eds. R. C. Newton, A. Navrotsky, and B. J. Wood). Springer, New York, vol. 1, pp. 247–271.

Langmuir C. H., Bender J. F., and Batiza R. (1986) Petrologic and tectonic segmentation of the East Pacific Rise between 5°30′–14°30′ N. *Nature* **322**, 422–429.

Langmuir C. H., Klein E. M., and Plank T. (1992) Petrological systematics of mid-ocean ridge basalts: constraints on melt generation beneath ocean ridges. In *Mantle Flow and Melt Generation at Mid-ocean Ridges*, Am. Geophys. Union, Geophys. Monogr. Ser. (eds. J. P. Morgan, D. K. Blackman, and J. M. Sinton). American Geophysical Union, Washington, DC, vol. 71, pp. 183–280.

LeDouaran S. and Francheteau J. (1981) Axial depth anomalies from 10 to 50 degrees north along the Mid-Atlantic Ridge: correlation with other mantle properties. *Earth Planet. Sci. Lett.* **54**, 29–47.

Lehnert K., Su Y., Langmuir C. H., Sarbas B., and Nohl U. (2000) A global geochemical database structure for rocks. *Geochem. Geophys. Geosyst.* **1**, 10.1029/1999GC000026.

Longhi J. (1991) Comparative liquidus equilibria of hypersthene-normative basalts at low pressure. *Am. Mineral.* **76**, 785–800.

Lonsdale P. (1988) Structural pattern of the Galapagos Microplate and evolution of the Galapagos triple junctions. *J. Geophys. Res.* **93**, 13551–13574.

Lonsdale P., Blum N., and Puchelt H. (1992) The RRR triple junction at the southern end of the Pacific-Cocos East Pacific Rise. *Earth Planet. Sci. Lett.* **109**, 73–85.

Lundstrom C. C., Sampson D. E., Perfit M. R., Gill J., and Williams Q. (1999) Insights into mid-ocean ridge basalt petrogenesis: U-series disequilibria from the Siqueiros Transform, Lamont Seamounts, and East Pacific Rise. *J. Geophys. Res.* **104**, 13035–13048.

Macdonald K. C. and Fox P. J. (1988) The axial summit graben and cross-sectional shape of the East Pacific Rise as indicators of axial magma chambers and recent volcanic eruptions. *Earth Planet. Sci. Lett.* **88**, 119–131.

Macdonald K. C., Fox P., Perram L., Eisen M., Haymon R., Miller S., Carbotte S., Cormier M., and Shor A. (1988) A new view of the mid-ocean ridge from the behavior of ridge axis discontinuities. *Nature* **335**, 217–225.

Mahoney J. J., Natland J. H., White W. M., Poreda R., Bloomer S. H., Fisher R. L., and Baxter A. N. (1989) Isotopic and geochemical provinces of the Western Indian ocean spreading centers. *J. Geophys. Res.* **94**, 4033–4052.

McClain K. J. and Lewis B. T. R. (1982) Geophysical evidence for the absence of a crustal magma chamber under the northern Juan de Fuca Ridge: a contrast with ROSE results. *J. Geophys. Res.* **87**, 8477–8489.

McKenzie D. (1984) The generation and compaction of partially molten rock. *J. Petrol.* **25**, 713–765.

McKenzie D. (1985) The extraction of magma from the crust and mantle. *Earth Planet. Sci. Lett.* **74**, 81–91.

McKenzie D. and Bickle M. J. (1988) The volume and composition of melt generated by extension of the lithosphere. *J. Petrol.* **29**, 625–679.

Melson W. G., Byerly G. R., Nelson J. A., O'Hearn T., Wright T. L., and Vallier T. (1977) A catalogue of the major element chemistry of abyssal volcanic glasses. *Smithsonian Contrib. Earth Sci.* **19**, 31–60.

Michael P. J. and Bonatti E. (1985) Peridotite composition from the North Atlantic: regional and tectonic variations and implications for partial melting. *Earth Planet. Sci. Lett.* **73**, 91–104.

Michael P. J. and Cornell W. C. (1998) Influence of spreading rate and magma supply on crystallization and assimilation beneath mid-ocean ridges: evidence from chlorine and major element chemistry of mid-ocean ridge basalts. *J. Geophys. Res.* **103**, 18325–18356.

Michael P. J., Langmuir C. H., Dick H. J. B., Snow J. E., Goldstein S. L., Graham D. W., Lehnert K., Kurras G., Muhe R., and Edmonds H. N. (2003) Magmatic and amagmatic seafloor spreading at the ultraslow-spreading Gakkel ridge, Arctic Ocean. *Nature* **423**, 956–961.

Muller M. R., Minshull T. A., Timothy A., and White R. S. (1999) Segmentation and melt supply at the Southwest Indian Ridge. *Geology* (Boulder) **27**, 867–870.

Mutter C. Z. and Mutter J. C. (1993) Variations in thickness of Layer 3 dominated oceanic crustal structure. *Earth Planet. Sci. Lett.* **117**, 295–317.

Natland J. H. (1989) Partial melting of a lithologically heterogeneous mantle: inferences from crystallization histories of magnesian abyssal tholeiites from the Siqueiros fracture zone. In *Magmatism in the Ocean Basins*, Geol. Soc. Spec. Publ. 42 (eds. A. D. Saunders and M. J. Norry). Geological Society of London, London, pp. 41–70.

Natland H. and Dick H. J. B. (2001) Formation of the lower ocean crust and the crystallization of gabbroic cumulates at a very slowly spreading ridge. *J. Volcanol. and Geotherm. Res.* **110**, 191–233.

Navin D. A., Peirce C., and Sinha M. C. (1998) The RAMESSES experiment II. Evidence for accumulated melt beneath a slow spreading ridge from wide-angle refraction and multichannel reflection seismic profiles. *Geophys. J. Int.* **135**, 746–772.

Nicolas A., Reuber I., and Benn K. (1988) A new magma chamber model based on structural studies in the Oman Ophiolite. In *The Ophiolites of Oman*, Tectonophysics, **151** (eds. F. Boudier and A. Nicolas). Elsevier, Amsterdam, pp. 87–105.

Nielsen R. L. (1985) EQUIL: a program for the modeling of low pressure differentiation processes in natural mafic magma bodies. *Comput. Geosci.* **11**, 531–546.

Nielsen R. L. (1990) The theory and application of a model of open magma system processes. In *Modern Methods of Petrology,* Reviews in Mineralogy (eds. J. Nicholls and J. K. Russel). Mineralogical Society of America, Washington, DC, vol. 24, pp. 65–106.

Nielsen R. L., Crum J., Bourgeois R., Hascall K., Forsythe L. M., Fisk M. R., and Christie D. M. (1995) Melt inclusions in high—an plagioclase from the Gorda Ridge: an example of the local diversity of MORB parent magmas. *Contrib. Mineral. Petrol.* **122**, 34–50.

Niu Y. and Batiza R. (1993) Chemical variation trends at fast and slow spreading mid-ocean ridges. *J. Geophys. Res.* **98**, 7887–7902.

Niu Y. and Batiza R. (1997) Trace element evidence from seamounts for recycled oceanic crust in the eastern Pacific mantle. *Earth Planet. Sci. Lett.* **148**, 471–483.

Niu Y., Collerson K., Batiza R., Wendt I., and Regelous M. (1999) Origin of enriched-type mid-ocean ridge basalt at ridges far from mantle plumes, the East Pacific Rise at 11°20′ N. *J. Geophys. Res.* **104**, 7067–7087.

Oxburgh E. R. (1965) Volcanism and mantle convection. *Phil. Trans. Roy. Soc. London* **258**, 142–144.

Oxburgh E. R. (1980) Heat flow and magma genesis. In *Physics of Magmatic Processes* (ed. R. B. Hargraves). Princeton University Press, Princeton, NJ, pp. 161–199.

Pallister J. S. and Hopson C. A. (1981) Samail ophiolite plutonic suite: field relations, phase variation, cryptic variation and layering, and a model of a spreading ridge magma chamber. *J. Geophys. Res.* **86**, 2593–2644.

Pedersen R. B. and Furnes H. (2001) Nd- and Pb-isotopic variations through the upper oceanic crust in DSDP/ ODP Hole 504B, Costa Rica Rift. *Earth Planet. Sci. Lett.* **189**, 221–235.

Penrose Conference Participants (1972) Penrose field conference on ophiolites. *Geotimes* **17**, 24–25.

Perfit M. R. and Chadwick W. W., Jr. (1998) Magmatism at mid-ocean ridges: constraints from volcanological and geochemical investigations. In *Faulting and Magmatism at Mid-Ocean Ridges*, Geophys Monogr. Ser. (eds. W. R. Buck *et al.*). American Geophysical Union, Washington, DC, vol. 106, pp. 59–115.

Perfit M. R., Fornari D. J., Ridley W. I., Smith M. C., Bender J. F., Langmuir C. H., and Haymon R. M. (1994) Small-scale spatial and temporal variations in mid-ocean ridge crest magmatic processes. *Geology* **22**, 375–379.

Phipps Morgan J. (1987) Melt migration beneath mid-ocean spreading centers. *Geophys. Res. Lett.* **14**, 1238–1241.

Phipps Morgan J. and Chen Y. J. (1993) The genesis of oceanic crust: magma injection, hydrothermal circulation, and crustal flow. *J. Geophy. Res.* **98**, 6283–6297.

Plank T. and Langmuir C. H. (1992) Effects of the melting regime on the composition of oceanic crust. *J. Geophys. Res.* **97**, 19749–19770.

Prinzhofer A., Lewin E., and Allégre C. J. (1989) Stochastic melting of the marble cake mantle: evidence from local study of the East Pacific Rise at 12°50′ N. *Earth Planet. Sci. Lett.* **92**, 189–206.

Quick J. E. and Denlinger R. P. (1993) Ductile deformation and the origin of layered gabbro in ophiolites. *J. Geophys. Res.* **98**, 14015–14027.

Rehkämper M. and Hofmann A. W. (1997) Recycled ocean crust and sediment in Indian Ocean MORB. *Earth Planet. Sci. Lett.* **147**, 93–106.

Reynolds J. R., Langmuir C. H., Bender J. F., Kastens K. A., and Ryan W. B. F. (1992) Spatial and temporal variability in the geochemistry of basalt from the East Pacific Rise. *Nature* **359**, 493–499.

Ribe N. M. (1985) The deformation and compaction of partial molten zones. *Geophys. J. Roy. Astron. Soc.* **83**, 487–501.

Roeder P. L. and Emslie R. F. (1970) Olivine-liquid equilibrium. *Contrib. Mineral. Petrol.* **29**, 275–289.

Salters J. M. and Dick H. J. B. (2002) Mineralogy of the mid-ocean ridge basalt source from neodymium isotopic composition of abyssal peridotites. *Nature* **418**, 68–72.

Schilling J.-G. (1966) Rare earth fractionation in Hawaii volcanic rocks. PhD Dissertation, Massachusetts Institute of Technology.

Schilling J.-G. (1971) Sea-floor evolution: rare-earth evidence. *Phil. Trans. Roy. Soc. London* **A268**, 663–706.

Schilling J.-G. (1985) Upper mantle heterogeneities and dynamics. *Nature* **314**, 62–67.

Schilling J.-G. (1986) Geochemical and isotopic variation along the Mid-Atlantic Ridge axis from 79°N to 0°N. In *The Western North Atlantic Region*, The Geology of North America (eds. P. R. Vogt and B. E. Tucholke). Geological Society of America, Boulder, CO, pp. 137–156.

Schilling J.-G., Kingsley R. H., and Devine J. D. (1982) Galapagos hot spot-spreading center system: 1. Spatial petrological and geochemical variations (83°W–101°W). *J. Geophys. Res.* **87**, 5593–5610.

Schilling J.-G., Zajac M., Evans R., Johnston T., White W., Devine J. D., and Kingsley R. (1983) Petrological and geochemical variations along the Mid-Atlantic Ridge from 29°N to 73°N. *Am. J. Sci.* **283**, 510–586.

Schilling J.-G., Kingsley R. H., Hanan B. B., and McCully B. L. (1992) Nd–Sr–Pb isotopic variations along the Gulf of Aden: evidence for Afar mantle plume-continental lithosphere interaction. *J. Geophys. Res.* **97**, 10927–10966.

Schouten H. and Denham C. (1995) Virtual ocean crust. *EOS, Trans., AGU* **76**, S48.

Scott D. R. and Stevenson D. J. (1989) A self-consistent model of melting, magma migration, and buoyancy-driven circulation beneath mid-ocean ridges. *J. Geophys. Res.* **94**, 2973–2988.

Shaw D. M. (1970) Trace element fractionation during anatexis. *Geochim. Cosmochim. Acta* **34**, 237–243.

Shen Y. and Forsyth D. W. (1995) Geochemical constraints on initial and final depth of melting beneath mid-ocean ridges. *J. Geophys. Res.* **100**, 2211–2237.

Shimizu N. (1998) The geochemistry of olivine-hosted melt inclusions in a FAMOUS basalt ALV519-4-1. *Phys. Earth Planet. Inter.* **107**, 183–201.

Sinton C. W., Christie D. M., Coombs V. L., Nielsen R. L., and Fisk M. R. (1993) Near primary melt inclusions in anorthite phenocrysts from the Galapagos Platform. *Earth Planet. Sci. Lett.* **119**, 527–537.

Sinton J. M. and Detrick R. S. (1992) Mid-ocean ridge magma chambers. *J. Geophys. Res.* **97**, 197–216.

Sinton J. M., Smaglik S. M., and Mahoney J. J. (1991) Magmatic processes at superfast spreading mid-ocean ridges: glass compositional variations along the East Pacific Rise 13–23 S. *J. Geophys. Res.* **96**, 6133–6155.

Sinton J., Bergmanis E., Rubin K., Batiza R., Gregg T. K. P., Grönvold K., Macdonald K. C., and White S. M. (2002) Volcanic eruptions on mid-ocean ridges: new evidence from the superfast spreading East Pacific Rise, 17°–19°S. *J. Geophys. Res.* **107**, 10.1029/2000JB000090.

Sleep N. H. (1974) Segregation of magma from a mostly crystalline mush. *Geol. Soc. Am. Bull.* **85**, 1225–1232.

Sleep N. H. (1984) Tapping of magmas from ubiquitous mantle heterogeneities: an alternative to mantle plumes? *J. Geophy. Res.* **89**, 10029–10041.

Smallwood J. R. and White R. S. (1998) Crustal accretion at the Reykjanes Ridge, 61 degrees–62 degrees N. *J. Geophys. Res.* **103**, 5185–5201.

Sobolev A. V. (1996) Melt inclusions in minerals as a source of principle petrologic information. *Petrology* **4**, 209–220.

Sobolev A. V. and Shimizu N. (1993) Ultra-depleted primary melt included in an olivine from the Mid-Atlantic Ridge *Nature* **363**, 151–154.

Sours-Page R., Johnson K. T. M., Nielsen R. L., and Karsten J. L. (1999) Local and regional variation of MORB parent magmas: evidence from melt inclusions from the Endeavour Segment of the Juan de Fuca Ridge. *Contrib. Mineral. Petrol.* **134**, 342–363.

Sours-Page R., Nielsen R. L., and Batiza R. (2002) Melt inclusions as indicators of parental magma diversity on the northern East Pacific Rise. *Chem. Geol.* **183**, 237–261.

Sparks D. W. and Parmentier E. M. (1991) Melt extraction from the mantle beneath spreading centers. *Earth Planet Sci. Lett.* **105**, 368–377.

Sparks R. S. J. (1989) Magma chambers: *in situ* differentiation in magma. *Nature* **340**, 187.

Sparks R. S. J., Huppert H. E., and Turner J. S. (1984) Fluid dynamics of involved magma chambers. *Phil. Trans. Roy. Soc. London* **A310**, 511–534.

Spiegelman M. and McKenzie D. (1987) Simple 2-D models for melt extraction at mid-ocean ridges and island arcs. *Earth Planet. Sci. Lett.* **83**, 137–152.

Stewart M. A., Klein E. M., and Karson J. A. (2002) The geochemistry of dikes and lavas from the north wall of the Hess Deep Rift: insights into the four-dimensional character of crustal construction at fast-spreading mid-ocean ridges. *J. Geophys. Res.* **107**, 2238 doi:10.1029/2001JB000545.

Stewart M. A., Klein E. M., Karson J. A., and Brophy J. G. (2003) Geochemical relationships between dikes and lavas at the Hess Deep Rift: implications for magma eruptibility. *J. Geophys. Res.* **108**, 10.1029/2001JB001622.

Stolper E. (1980) A phase diagram for mid-ocean ridge basalts: preliminary results and implications for petrogenesis. *Contrib. Mineral. Petrol.* **74**, 13–27.

Sturm M., Klein E., Graham D., and Karsten J. (1999) Age constraints on crustal recycling to the mantle beneath the southern Chile Ridge: He, Sr, Nd, and Pb isotope systematics. *J. Geophys. Res.* **104**, 5097–5114.

Sun S.-S. and McDonough W. F. (1989) Chemical and isotopic systematics of oceanic basalts: implications for mantle composition and processes. In *Magmatism in the Ocean Basins*, Geol. Soc. Spec. Publ. 42 (eds. A. D. Saunders and M. J. Norry). Geological Society of London, London, pp. 313–345.

Sun S.-S., Tatsumoto M., and Schilling J.-G. (1975) Mantle plume mixing along the Reykjanes ridge axis: lead isotope evidence. *Science* **190**, 143–147.

Sun S.-S., Nesbitt R. W., and Sharaskin A. Y. (1979) Geochemical characteristics of mid-ocean ridge basalts. *Earth Planet. Sci. Lett.* **44**, 118–119.

Thompson G., Bryan W. B., and Melson W. G. (1980) Geological and geophysical investigation of the mid-Cayman rise spreading center: geochemical variations and petrogenesis of basalt glasses. *J. Geol.* **88**, 41–55.

Toomey D. R., Purdy G. M., Solomon S., and Wilcox W. (1990) The three dimensional seismic velocity structure of the East Pacific Rise near latitude 9°30′N. *Nature* **347**, 639–644.

Vlastelic I., Aslanian D., Dosso L., Bougault H., Olivet J.-L., and Geli L. (1999) Large-scale chemical and thermal division of the Pacific mantle. *Nature* **399**, 345–350.

Vogt P. R. (1986) Portrait of a plate boundary: the Mid-Atlantic Ridge axis from the equator to Siberia. In *The Western North Atlantic Region*, The Geology of North America, vol. M., plate 8A (eds. P. R. Vogt and B. E. Tucholke). Geological Society of America, Boulder, CO.

Walker D., Shibata T., and DeLong S. E. (1979) Abyssal tholeiites from the Oceanographer Fraction Zone: II. Phase equilibria and mixing. *Contrib. Mineral. Petrol.* **70**, 111–125.

Weaver J. S. and Langmuir C. H. (1990) Calculation of phase equilibrium in mineral-melt systems. *Comput. Geosci.* **16**, 1–19.

White W. M. (1993) $^{238}U/^{204}Pb$ in MORB and open system evolution of the depleted mantle. *Earth Planet. Sci. Lett.* **115**, 211–226.

Zindler A., Staudigel H., and Batiza R. (1984) Isotope and trace element geochemistry of young Pacific seamounts: implications from the scale of upper mantle heterogeneity. *Earth Planet. Sci. Lett.* **70**, 175–195.

3.14
Melt Migration in Oceanic Crustal Production: A U-series Perspective

T. Elliott

University of Bristol, UK

and

M. Spiegelman

Columbia University, New York, NY, USA

3.14.1 INTRODUCTION

The crust is ultimately the product of mantle melting. To understand its formation requires knowledge of how melts move from depth. Generation of oceanic crust as a consequence of seafloor spreading is clearly related to melt production below. Although melting processes beneath ridges are perhaps the best constrained of any tectonic setting (see Chapter 3.13; Langmuir *et al.*, 1992), neither melt distributions nor rates of melt migration are well resolved. It is necessary to infer dynamic processes hidden within the mantle by using proxy observations. Seismology provides important but static information on the velocity structure of the mantle, which is sensitive to the gross amount and distribution of partial melts (e.g., Forsyth *et al.*, 1998). The geochemistry of erupted melts and their residues provides a different perspective, since their compositions are sensitive to the pathways they follow (Hellebrand *et al.*, 2002; Johnson *et al.*, 1990; Kelemen *et al.*, 1995a; Klein and Langmuir, 1987; Salters and Hart, 1989; Spiegelman, 1996). Furthermore, short-lived radioactive nuclides can provide crucial information on the *rates* of these processes.

The aim of this chapter is to review the observations of uranium series nuclide studies on mid-ocean ridge basalts (MORBs) and discuss their implications for melt transport processes. We note that a recent review by Lundstrom (2003) has a similar remit and makes a useful companion contribution. We initially recap the fundamental behavior of the naturally occurring actinide decay chains, and outline the analytical challenges of their measurement. We then evaluate the gross signatures of disequilibrium observed in MORBs and consider the co-variations of disequilibria with other geochemical and geophysical parameters. This is followed by a detailed discussion of models of melt transport that have been developed to account for these observations, allowing a synthesis of the key constraints that uranium-series (henceforth U-series) measurements provide on the process of melt migration beneath mid-oceanic ridges.

3.14.2 U-SERIES PRELIMINARIES

Ivanovich and Harmon (1992) and Bourdon *et al.* (2003) provide comprehensive reviews of the fundamental behavior of U-series systems. Here we only briefly reiterate these properties, focusing on those relevant to melt migration from the mantle.

3.14.2.1 Naturally Occurring Actinide Decay Chains

The major uranium nuclides ^{238}U and ^{235}U decay to the stable nuclides ^{206}Pb and ^{207}Pb, respectively, through a long chain of intermediate daughter products (Figure 1). The half-lives of ^{238}U and ^{235}U are ~4.5 Gyr and 0.7 Gyr, respectively. All intermediate nuclides have half-lives much shorter than the uranium nuclides at the head of the chain, but still span a huge range of timescales from ~245 kyr to 164 μs. The longer-lived intermediates have half-lives appropriate

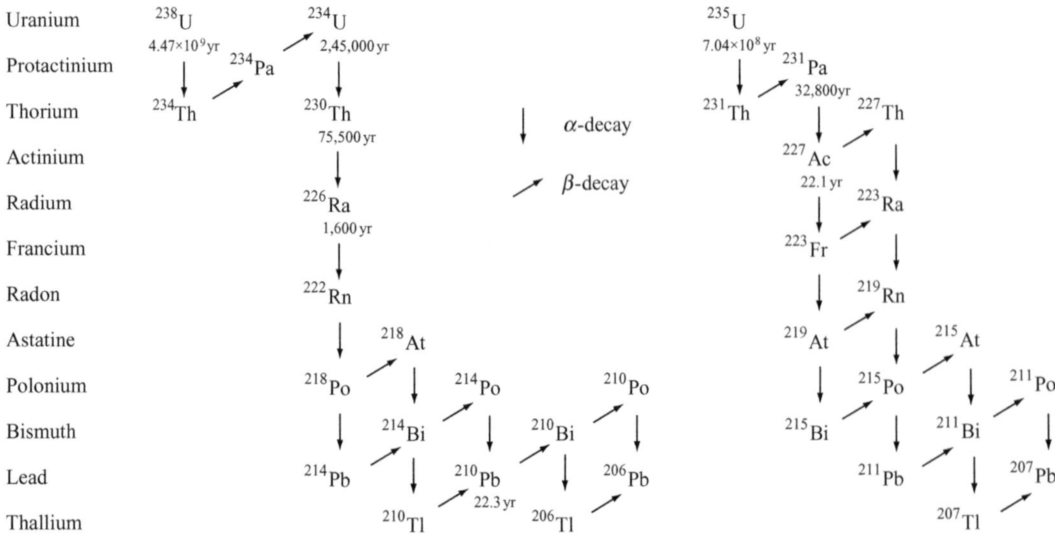

Figure 1 ^{238}U and ^{235}U decay chains. Longer-lived nuclei discussed in the text are shown with their half-lives (in years) listed beneath.

for studying melt migration. The principal nuclides of interest are ^{230}Th ($t_{1/2}$ ~75 kyr) and ^{226}Ra ($t_{1/2}$ ~1.6 kyr) in the ^{238}U chain and ^{231}Pa ($t_{1/2}$ ~33 kyr) in the ^{235}U chain (Figure 1). The longest lived daughter nuclide is actually ^{234}U ($t_{1/2}$ ~ 245 kyr), but this is neither expected (Fleischer *et al.*, 1975) nor observed (see Section 3.14.3.3) to fractionate from ^{238}U in high-temperature igneous systems. Thus, ^{234}U measurements do not provide temporal information, but do provide a useful screen for low temperature alteration (see Section 3.14.3.4).

A third naturally occurring actinide decay chain is headed by ^{232}Th and decays to ^{208}Pb. The intermediate nuclides in this chain have not been used to elucidate melting processes beneath ocean ridges, and here we only concern ourselves with the behavior of the ^{232}Th parent. ^{232}Th is the overwhelmingly dominant Th nuclide, and is important as a reference nuclide for the short-lived ^{230}Th nuclide in the ^{238}U chain. Although the decay of ^{238}U, ^{235}U, and ^{232}Th support chains of short-lived daughter nuclides, the effect of this decay in changing the abundances of the parent nuclides themselves is nevertheless insignificant on the 100 kyr timescales considered here. Thus, ^{238}U, ^{235}U, and ^{232}Th can be paradoxically considered to behave as "stable" reference nuclides on the timescales considered here.

3.14.2.2 Closed-system Models

To understand the behavior of U-series nuclides, we start by discussing closed systems. These models form the basis for more complicated open-system models discussed later. Here we discuss the behavior of generic radioactive decay chains that apply to all the naturally occurring actinide chains.

3.14.2.2.1 Bulk system models

The equations and solutions for closed-system radioactive decay chains have been known since Bateman (1910). To understand the behavior of these systems, however, it is useful to express them as a linear system of ordinary differential equations and use some basic results from linear algebra to discuss the general solutions. This treatment helps to elucidate the ideas of "secular equilibrium" and relaxation to equilibrium.

Consider a general decay chain of nuclides such that element 1 decays to element 2 decays to element 3, etc. (e.g., $^{238}U \rightarrow {}^{230}Th \rightarrow {}^{226}Ra$). In discussing U-series nuclides, it is convenient to represent their abundances in terms of *activities*. The activity of a nuclide is its *rate* of decay, which depends only on the number of atoms of the nuclide and the decay constant λ. Thus, for a chain of nuclides, we can write the (specific) activity of the ith element in the chain ($i = 1, 2, 3, \ldots$) as $a_i = \lambda_i c_i$, where c is atomic concentration by weight. The evolution of activities in a decay chain of length n is then described by

$$\begin{aligned}
\frac{da_1}{dt} &= -\lambda_1 a_1 \\
\frac{da_2}{dt} &= \lambda_2 a_1 - \lambda_2 a_2 \\
\frac{da_2}{dt} &= \lambda_3 a_2 - \lambda_3 a_3 \\
&\vdots \\
\frac{da_n}{dt} &= \lambda_n a_{n-1} - \lambda_n a_n
\end{aligned} \tag{1}$$

For every nuclide (after the parent), the change in activity depends on the difference between decay of the daughter and production by the parent. Using linear algebra, Equation (1) can be written as a vector equation of activities

$$\frac{d}{dt}\begin{bmatrix} a_1 \\ a_2 \\ a_3 \\ \vdots \\ a_n \end{bmatrix} = \begin{bmatrix} -\lambda_1 & & & & \\ \lambda_2 & -\lambda_2 & & & \\ & \lambda_3 & -\lambda_3 & & \\ & & \ddots & \ddots & \\ & & & \lambda_n & -\lambda_n \end{bmatrix} \begin{bmatrix} a_1 \\ a_2 \\ a_3 \\ \vdots \\ a_n \end{bmatrix} \tag{2}$$

or more compactly using matrix-vector notation as

$$\frac{d\boldsymbol{a}}{dt} = A_n \boldsymbol{a} \tag{3}$$

where A_n is the matrix of decay constants in Equation (2). If all the decay constants are distinct then Equation (3) can be solved analytically using standard approaches of linear algebra (e.g., Strang, 1998) as

$$\boldsymbol{a}(t) = S e^{\Lambda t} S^{-1} \boldsymbol{a}_0 \tag{4}$$

where S is the matrix of eigenvectors of A_n, Λ is a diagonal matrix of eigenvalues of A_n, which happen to be the diagonal elements of A_n (i.e., $-\lambda_1, -\lambda_2, \ldots, -\lambda_n$) and $\boldsymbol{a}_0$ is the initial activities at time $t=0$. For this problem, both S and Λ can be found analytically, although for more general problems they can also be found numerically with packages such as MATLAB. (A caution though. For numerical stability, it is important to solve the equations for the *activities* rather than the concentrations because the matrix that arises from the concentration equation is ill-conditioned if the decay constants are wildly different.) Equation (4) can be written more physically as

$$a(t)=c_1 s_1 e^{-\lambda_1 t}+c_2 s_2 e^{-\lambda_2 t}+\cdots + c_n s_n e^{-\lambda_n t} \quad (5)$$

where s_i is the ith eigenvector and $(c_1, c_2,\dots, c_n)= c=S^{-1}a_0$ is the decomposition of the initial condition into its component eigenvectors. Physically, Equation (5) states that the system can be considered composed of n independent vectors that each decay like one of the nuclides. These eigenvectors are the *supported chains*, for example, in the series $^{238}U\rightarrow{}^{230}Th\rightarrow{}^{226}Ra$, any initial set of activities can be decomposed into three chains, one supported by ^{238}U that decays slowly, one supported by ^{230}Th that decays with a half-life of ~75 kyr and one supported by ^{226}Ra that decays rapidly. After a sufficiently long time, the shorter-lived eigenvectors decay away leaving just the most slowly decaying one. This is the state of *secular equilibrium* in which all intermediate nuclides are supported by the parent nuclide at the head of the chain because this is the slowest-decaying species. In addition, if the decay constants are very different in magnitude (e.g., $\lambda_1 \ll \lambda_2 \ll \lambda_3$) then the activities in any given supported chain will be nearly identical such that their activity *ratios* (a_i/a_j) are ~1. The statement that all the ratios of all activities are 1 in secular equilibrium is a convenient short-hand notation for the longer-lived U-series nuclides.

Any disturbance from secular equilibrium decays back toward secular equilibrium through the decay of the shorter-lived supported chains. If the chains have significantly different decay rates, they are reasonably well decoupled and can be used to date processes comparable to the lifetimes of each chain (e.g., Condomines *et al.*, 1988; Rubin *et al.*, 1994; Thompson *et al.*, 2003). As a useful rule of thumb, the time taken for a parent–daughter pair to return to approximate secular equilibrium, is about five half-lives of the shorter lived nuclide. In five half-lives, ~97% of initial disequilibrium has decayed. Whether any detectable disequilibrium actually remains depends on the precision of measurements and degree of initial disequilibrium.

3.14.2.2.2 Multiphase models

The closed-system model is also used to justify the assumption that the solid mantle source is in secular equilibrium before melting begins. This will be the case if the mantle has acted as a closed system for more than about five half-lives of the longest-lived daughter product perturbed in the most recent event. As mentioned above, ^{234}U is not fractionated from ^{238}U at mantle temperatures and so ^{230}Th ($t_{1/2} \sim 75$ kyr) constrains the time for return of the mantle to secular equilibrium (in a closed system) to be only some 400 kyr.

This bulk state of secular equilibrium applies to the total amount of the U-series nuclides, but does not necessarily say where the different elements reside within the system. If the bulk system has a single phase (such as a melt or a monomineralic rock) then that phase will be in secular equilibrium. If the material has multiple phases with different partitioning properties, however, the individual phases can maintain radioactive disequilibria even when the total system is in secular equilibrium. There are two basic sets of models that exploit this fact, the first assumes complete chemical equilibrium between all phases and the second assumes transient diffusion controlled solid exchange.

(i) *Equilibrium models*. In the equilibrium multiphase models, the closed system is assumed to be in both chemical and secular equilibrium at all times (e.g., Allègre and Condomines, 1982; Condomines *et al.*, 1988; Condomines and Sigmarsson, 2000). For example, if we have a rock with N solid phases plus melt where each phases occupies X_i mass fraction and has a melt/mineral equilibrium partition coefficient D_i, then the total activity ratio of $(^{230}Th/^{238}U) = 1$ implies that

$$\left(\frac{^{230}Th}{^{238}U}\right)_{total} = 1 = \frac{\left(\sum_{i=1}^{N} X_i D_i^{Th} + X_f\right)(^{230}Th)_f}{\left(\sum_{i=1}^{N} X_i D_i^{U} + X_f\right)(^{238}U)_f} \quad (6)$$

where $(^{230}Th)_f$ and $(^{238}U)_f$ are the activities of ^{230}Th and ^{238}U *in the melt* and X_f is the mass fraction of melt. Rearranging shows that the activity ratio in the melt is

$$\left(\frac{^{230}Th}{^{238}U}\right)_f = \frac{(1-X_f)D_U + X_f}{(1-X_f)D_{Th} + X_f} \quad (7)$$

where D_U and D_{Th} are the *bulk* melt/solid partition coefficients for U and Th, respectively. Thus, the melt can support activity ratios different from 1 if the melt fraction X_f is smaller than the partition coefficients. In the limit of infinitesimal melt fractions, the largest excess is approximately D_U/D_{Th}. While there is still some debate about the bulk partition coefficients of U-series nuclides in mantle systems (see Section 3.14.4.1), they are all uniformly small ($\lesssim 0.005$). Thus, pure chemical fractionation due to equilibrium batch melting can produce melts with disequilibrium activity ratios but only for very small degrees of melting (i.e., $X_f < D_U$) (e.g., Condomines *et al.*, 1988; Condomines and Sigmarsson, 2000).

These small degree melts should also be extremely enriched in highly incompatible elements and would resemble alkali basalts rather than MORBs (unless the source was extremely depleted). This mechanism cannot be ruled out for some ocean island basalts (OIBs) (Sims *et al.*, 1999; Elliott, 1997). However, it will not work for the significant degrees of melting ($F \gtrsim 10\%$) inferred for MORBs (Klein and Langmuir, 1987; McKenzie and Bickle, 1988; see Chapter 3.13). Moreover, to preserve the excesses produced by this mechanism requires preferential extraction of the enriched phase (e.g., the melt) and extraction on a timescale that is short compared to shortest-lived nuclide of interest.

Nevertheless, this basic mechanism of using specific phases to concentrate nuclides chemically, followed by preferential extraction of those phases, is a process that is common to many of the models proposed for U-series excesses including the "dynamic melting models" (McKenzie, 1985; Williams and Gill, 1989) discussed in Section 3.14.4.3.1 as well as more recently proposed solid state, diffusion controlled models (e.g., Van Orman *et al.*, 2002a; Saal *et al.*, 2002b; Feineman *et al.*, 2002)

(ii) *Diffusion limited models.* Chemical partitioning into specific phases is easily extended to multiphase solids that contain no liquid. For example, Van Orman *et al.* (2002a) and Saal *et al.* (2002a,b) suggest that partitioning of radium into plagioclase in the crust with subsequent re-melting of the plagioclase could be a possible source of radium excesses in MORBs. A similar model is proposed for arcs where the principal phase is assumed to be phlogopite (Feineman *et al.*, 2002). These models allow radioactive disequilibrium in individual solid phases to be preserved in steady state. This arises due to the competing effects of chemical and secular equilibrium between phases in certain mineral assemblages. Three key conditions are required: First the ratios of melt/solid partition coefficients for a parent and daughter nuclide pair are greatly different in two coexisting phases. Second, the solid diffusivity of at least one of the parent or daughter nuclides must be sufficiently large so that the most diffusive element is mobile on a timescale comparable to the decay time of the shortest-lived nuclide. Third, there must be sufficient time for diffusion and ingrowth to take place. These models are reasonably new and need to be tested against their implications for other elements.

3.14.2.3 Measurement and Nomenclature

Activities of U-series nuclides in MORBs have been determined by both nuclear particle counting and mass-spectrometric techniques. In this section, we highlight briefly the key differences in the analytical methods and discuss the associated nomenclature to avoid any confusion such diversity might cause.

U-series nuclide activities can be measured directly by detection of their emitted nuclear particles, e.g., alpha particle counting by solid-state detectors (Ivanovich and Harmon, 1992). In contrast, measurements by mass-spectrometry do not require waiting for Nature to take its course. Atoms of the sample are ionized and accelerated so that charged particles of the nuclides themselves can be measured by Faraday cups or electron multipliers (see Goldstein and Stirling, 2003). Mass-spectrometry is hence a more rapid technique. Typically mass-spectrometry measurements take tens of minutes to hours, while counting methods require days to weeks.

The precision of the techniques ultimately depend on the number of nuclear decays (particle counting) or ionized particles detected (mass-spectrometry). Both are related to the number of atoms of the nuclides of interest present in the sample, but with a different functionality. As discussed above, the natural decay rate is the product of the number of atoms present and the activity constant (the reciprocal of half-life multiplied by the natural log of 2, $\lambda = \ln 2/t_{1/2}$). The activities of all nuclides in the U-series chain are equal at secular equilibrium, and generally within a factor of 3, except in cases of extreme disequilibrium (see Section 3.14.3.3). Thus all U-series nuclides can potentially be measured with similar precision using nuclear counting techniques. The number of charged particles produced by mass-spectrometry is the product of the number of atoms of a nuclide multiplied by its ionization efficiency. The number of atoms of a nuclide equals the nuclide's activity divided by its activity constant. Since the activities of all U-series nuclides are generally similar but activity constants vary by orders of magnitude, the atomic abundances of the U-series nuclides also vary by orders of magnitude. Longer-lived nuclides (with smaller activity coefficients) are present in higher atomic abundances than the shorter-lived nuclides, and so are likely more amenable to analysis by mass-spectrometry. The ionization efficiencies of the U-series nuclides vary by orders of magnitude and so this too must be considered (Figure 2).

"Traditional" thermal ionization mass-spectrometry efficiently ionizes Ra (>1%) (Cohen and Onions, 1991; Volpe *et al.*, 1991). Elements with higher first ionization energies, such as Th and U, give significantly lower ion yields, typically <1‰ (see Edwards *et al.*, 1987) although work by Yokoyama *et al.* (2001) reported ionization efficiencies up 5‰ for U. Yet, even given typical thermal ionization efficiencies of 0.5–0.03‰ for 10–100 ng Th loads

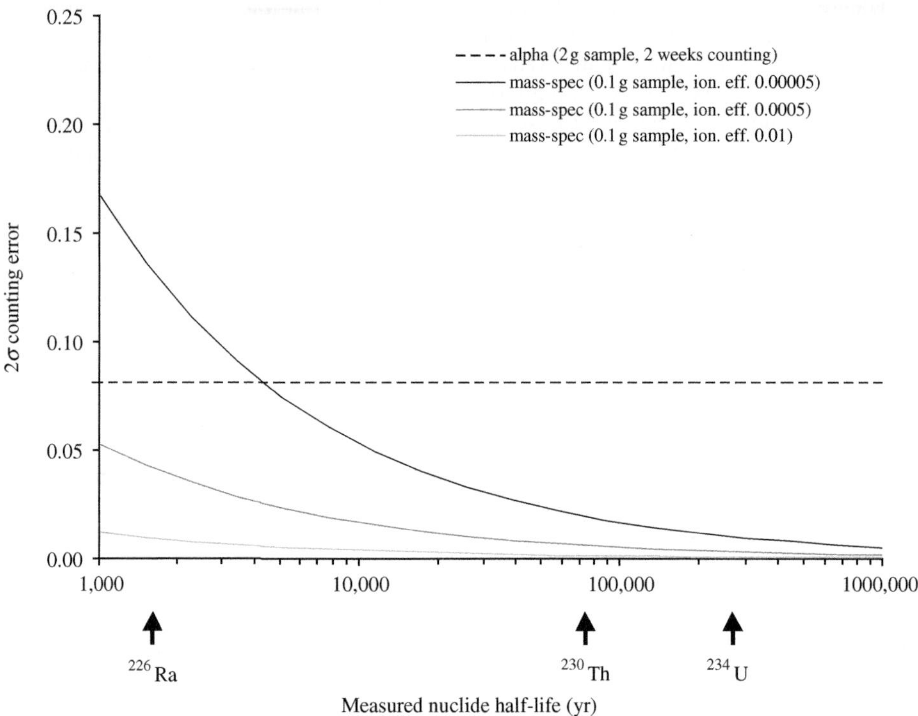

Figure 2 Variations of counting statistical error (2σ or 95% confidence) for nuclides of different half-lives using different measurement techniques. Four different scenarios are shown with details listed in legend, aimed to cover a typical range of conditions. The values of ionization efficiencies span a full range of values appropriate for elements difficult (Th) and easy (Ra) to ionize thermally, see text. All scenarios assume a sample with MORB-like U concentration (50 ng g^{-1}), but with all daughter nuclides in secular equilibrium (for illustrative simplicity). Calculations assume (unrealistic) 100% yields for chemical purification of the nuclide of interest and 40% counting efficiency for alpha detection geometry.

(Asmerom and Edwards, 1995; Edwards *et al.*, 1987; Goldstein *et al.*, 1989; Lundstrom *et al.*, 1999; Rubin, 2001), the total ion yield for the longer-lived nuclides of interest here is significantly greater than the number of nuclear decays over reasonable counting times. The corollary of this is, of course, better analytical precision (Figure 2). There are no alpha counting analyses of MORB for the decay chain of ^{235}U, which is ~138 × less abundant than ^{238}U. Yet, mass-spectrometry has provided a valuable data set of ($^{231}Pa/^{235}U$) disequilibrium in MORBs (Goldstein *et al.*, 1993; Lundstrom *et al.*, 1995; Sims *et al.*, 2002). Furthermore, ion yields of 2–5‰ for Th and U have recently been achieved by plasma-ionization mass-spectrometry (Luo *et al.*, 1997; Pietruszka *et al.*, 2002; Turner *et al.*, 2001) and efficiencies as high as ~2% reported for secondary ion mass-spectrometry (England *et al.*, 1992; Layne and Sims, 2000). Such advances should boost the precision of mass-spectrometry (e.g., Pietruszka *et al.*, 2002).

The counting statistical limit on precision is of importance to studies of MORBs due to the generally low abundances of uranium and its daughter nuclides. Naturally, counting statistics are only a best case limit on the precision and accuracy of analyses. A number of additional chemical preparation (e.g., blanks, sample-spike equilibration) and instrumental problems (e.g., detector intercalibration, interferences) can considerably degrade data quality. As discussed by Rubin (2001), comprehensive assessment of data quality is difficult without associated publication of detailed analytical procedures, reproducibility of solution standards and full replicates of samples and international rocks standards with similar matrices and nuclide abundances. Not all MORB studies, whether by particle counting or mass-spectrometry, provide such complete information. But counting statistics or internal precisions are commonly reported. Whilst these data are not robust assessments of total error, it is important to note that data cannot be more accurate than they are precise, and so they provide minimum constraint on uncertainties.

Given the large variability of actinide abundances in MORBs (e.g., Figure 3) and different laboratory procedures, it is difficult to make definitive statements about the precision obtainable by different techniques. As useful rule of thumb, however, $^{238}U-^{230}Th-^{226}Ra$ and $^{235}U-^{231}Pa$

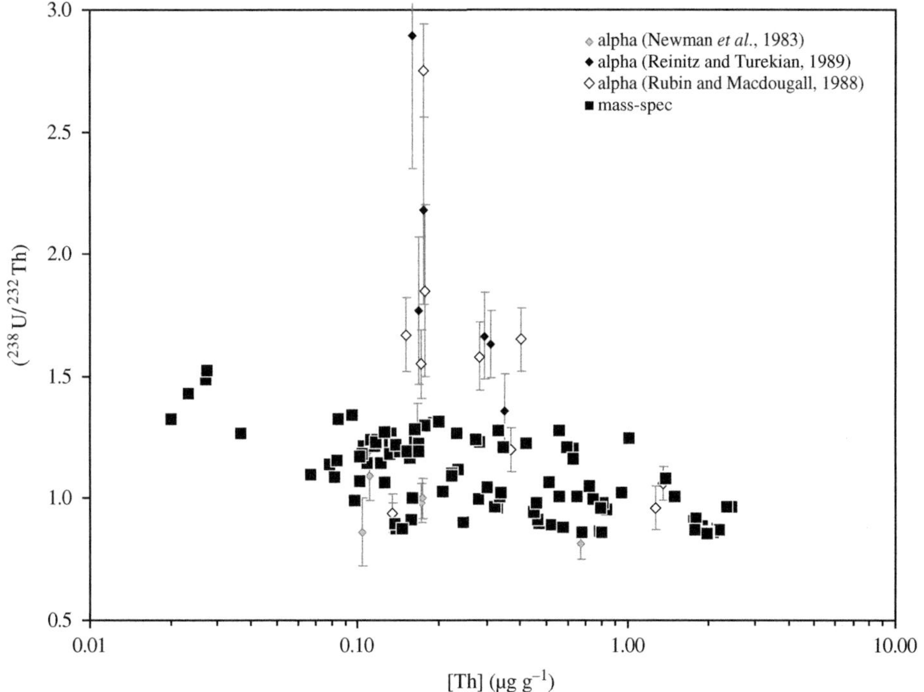

Figure 3 Concentration of Th (μg g^{-1}) versus (^{238}U/^{232}Th) for compiled MORB data set. Note logarithmic scale for [Th]. Mass-spectrometric measurements in filled black symbols, cited errors not shown but typically (2σ) precision comparable or smaller than sample size. Studies with (^{238}U/^{232}Th) measured by alpha counting shown in open symbols and identified individually in legend. Cited counting errors (2σ) shown for (^{238}U/^{232}Th) alpha measurements, but not for [Th] which are comparable to sample size. Note some of the alpha counted samples extend to values of (^{238}U/^{232}Th) much higher than the array defined by mass-spectrometry data.

disequilibria in MORBs are measured with an internal precision (95% confidence or 2σ) of typically ~0.5–2% by mass-spectrometry compared to counting statistical errors of 5–15% (2σ) for the ^{238}U chain disequilibria by alpha counting (Figure 2).

U-series disequilibria are most naturally expressed in terms of activity ratios (e.g., Section 3.14.2.2). Alpha counting measures activity directly, whereas mass-spectrometry yields atomic ratios which need to be converted into activities using activity constants. This introduces an additional component of uncertainty (1–8‰) to the absolute accuracy of mass-spectrometric activity measurements (e.g., Holden, 1989; Jaffey *et al.*, 1971; Meadows *et al.*, 1980). This uncertainty, however, is small compared to the uncertainties in particle counting measurements. Moreover, the high precision of mass-spectrometric measurements has allowed some activity constants to be refined using samples where secular equilibrium can be assumed (Cheng *et al.*, 2000).

Owing to the different measurement techniques, U-series nuclide concentration measurements are reported in a number of ways. Specific activities of each nuclide (e.g., dpm g^{-1}) are reported for alpha counted studies, whereas mass-spectrometric measurements provide weight concentrations of nuclides (e.g., μg g^{-1}) and atomic isotope ratios. As discussed above, disequilibria are expressed in terms of isotope activity ratios, conventionally enclosed in curved brackets to distinguish them from weight ratios (commonly reported in square brackets) and atomic ratios (with no brackets). As well as activity ratios, disequilibria are sometimes expressed as "excesses" or "deficits." An excess is the fraction of unsupported daughter nuclide, e.g., (^{230}Th/^{238}U) = 1.2 can be described as a 20% ^{230}Th-excess. The (^{234}U/^{238}U) ratio is also sometimes reported in a δ-notation or parts per thousand relative deviation from equilibrium.

Concentrations of uranium and thorium are also measured by other techniques (e.g., instrumental neutron activation analysis) and sometimes included in trace element analyses. In geochemical literature the ratio [Th/U] is frequently used. However, ^{232}Th is the only long-lived thorium isotope and ^{238}U is a fixed 99.28% of total uranium, so [Th/U] ratios are closely related (3.3% smaller) to atomic ^{232}Th/^{238}U ratios. The latter ratio has historically been termed κ when used in the Pb isotope literature. To further the confusion, the activity ratio ($^{238}U/^{232}Th$), is used in the common equiline plot. Although all simply interrelated, the various different expressions of Th/U ratios can seem puzzling.

3.14.3 OBSERVATIONS

3.14.3.1 Data Compilation

Since we wish to examine the implications of U-series disequilibria on processes of melting and melt migration, it is necessary to compile a data-set of samples with (near) "primary" mantle signatures. Inevitably, decay of disequilibria after eruption will modify primary values. Indeed, a number of U-series studies of MORBs have specifically targeted off-axis samples and used disequilibrium as a high-resolution chronometer of oceanic plate accretion (e.g., Goldstein *et al.*, 1991; Krishnaswami *et al.*, 1984; Sims *et al.*, in press; Sturm *et al.*, 2000; Zou *et al.*, 2002). For our purposes, however, it is essential to filter out samples from the literature which have likely experienced significant post-eruptive decay. In the following section we describe age constraints on young MORBs, which we have used to guide data compilation for this contribution (Table 1).

A second issue in sample selection is possible crustal or seawater contamination (Section 3.14.3.2). We include in our compilation only analyses from picked, fresh glass. Hand picking of glasses critically allows visual rejection of contaminated material (see Bourdon *et al.*, 2000). We have excluded samples where authors have themselves inferred contamination (e.g., Condomines *et al.*, 1981; Goldstein *et al.*, 1991; Newman *et al.*, 1983), but no further screening has been applied.

Additional sample selection criteria have been applied to try to exclude some samples with anomalous ($^{238}U/^{232}Th$) and ($^{230}Th/^{232}Th$) ratios. Figure 3 illustrates a plot of [Th] against ($^{238}U/^{232}Th$) ratios for MORBs from this compilation. Some of the alpha-counted measurements are clearly elevated relative to the general array defined largely by more precise mass-spectrometric determinations of uranium and thorium concentrations. Reinitz and Turekian (1989) compared the high ($^{238}U/^{232}Th$) obtained by themselves and Rubin and Macdougall (1988) to earlier analyses from a similar location by Newman *et al.* (1983) and concluded that analytical differences most likely explained the differences between the contrasting data sets. Additional samples measured from ~12° N on the EPR (Reinitz and Turekian, 1989; Rubin and Macdougall, 1988) also have much higher ($^{238}U/^{232}Th$) than later isotope dilution mass-spectrometric measurements from the same area (Ben Othman and Allègre, 1990). New mass-spectrometric measurements of Rubin *et al.* (2000) have demonstrated some of his alpha counting measurements on low abundance uranium samples gave inaccurate ($^{238}U/^{232}Th$) and ($^{230}Th/^{232}Th$) due to blank problems (Rubin, personal communication). It is clearly difficult to assess definitively which samples may have been affected. To avoid any of the anomalous samples evident in Figure 3, we exclude from later discussion (Section 3.14.3.5) samples with $[Th] < 0.5$ ppm measured by Rubin and Macdougall (1988) and Reinitz and Turekian (1989). This is an arbitrary cut-off, which may exclude accurate data, but given the known problem, it seems an appropriate approach until further analyses are published. Nevertheless, we include all these samples in the initial general discussion of disequilibria (Section 3.14.3.3), as their ($^{230}Th/^{238}U$) are not obviously anomalous.

We consider exclusively MORB samples, and not OIBs that compromise a smaller portion of the oceanic crust. Both MORBs and OIBs are believed to be derived from decompression melting of the mantle. A comparison of the magma types is interesting but beyond the scope of this contribution; see Bourdon and Sims (2003) for a recent review of the U-series systematics of OIBs. Nevertheless, we do include samples from elevated ridge segments believed to be influenced by OIB-source mantle. In these cases, however, the melting regime should fundamentally resemble that of MORBs. Potentially, a component of active upwelling further complicates interpretation of U-series results from OIBs influenced segments (Bourdon *et al.*, 1996b), as has been discussed for OIBs (e.g., Chabaux and Allègre, 1994; Sims *et al.*, 1999). Yet work on sub-aerial magmatism at the most northerly end of the Icelandic rift suggests that there is little active upwelling this far from the putative plume center (Maclennan *et al.*, 2001). This conclusion also likely applies for the Reykjanes Ridge and Azores platform samples considered here, which lie even further from their respective centers of sub-aerial volcanism.

3.14.3.2 Age Constraints

The effects of post-eruptive decay are illustrated in Figure 4, using disequilibria typical for MORBs (see Section 3.14.3.3) as initial conditions. The problem of recovering samples with "primary" signatures is most acute for ($^{226}Ra/^{230}Th$). Any ^{226}Ra-excesses will be only 1% of their initial values 10 kyr after eruption. Over the same time, the change in typical ($^{230}Th/^{238}U$) is barely resolvable analytically, but

Table 1 Summary of axial MORB samples used in this compilation, grouped by geographical location.

Location	*Lat*	*Depth* (m)	*Half-spreading rate* (cm yr^{-1})	$(^{230}Th/^{238}U)$	$(^{231}Pa/^{235}U)$	$(^{226}Ra/^{230}Th)$	*Equiline gradient*	*err*	*Reference*
Juan de Fuca	44–48° N	2,200	3	13	5	5	0.37 (0.48)	0.05	Goldstein *et al.*, 1991, 1993, Volpe and Goldstein, 1993, Lundstrom *et al.*, 1995
Gorda	41–42° N	3,000	2.75	9	4	4	0.27	0.05	Goldstein *et al.*, 1991, 1993; Volpe and Goldstein, 1993; Cooper *et al.*, 2003
EPR	20–21° N	2,600	3	*6*		*1*	−0.17	0.33	Newman *et al.*, 1983; Rubin and Macdougall, 1988, 1990
EPR	11–13° N	2,600	5.5	*6*		*10*	0.053	0.18	Rubin and Macdougall, 1988; Reinitz and Turekian, 1989, BenOthman and Allègre 1990; Goldstein *et al.*, 1991, 1993
EPR (+Siqueiros transform)	8–10° N	2,600	5.5	40	20	23	0.49 (0.66)	0.08	Volpe and Goldstein, 1993; Lundstrom *et al.*, 1999; Sims *et al.*, 2002
EPR	13–23° S	2,600–2,900	7.7			*10*			Rubin and Macdougall, 1988, 1990
EPR	26–28° and 35° S	2,450	8	*2*		*6*			Rubin and Macdougall, 1988, 1990
MAR (Reykjanes Ridge)	57–63° N	600–1,825	1	7		(7)	0.08	0.08	Peate *et al.*, 2001
MAR (FAZAR)	37–41° N	950–3,000	1	23	1	1	1.31 (1.47)	0.13	Bourdon *et al.*, 1996a,b, 2000
MAR (FAMOUS)	36° 50′ N	2,550	1.1	*3*					Condomines *et al.*, 1981
MAR	29–30° N	3,300–4,000	1.2	2					Bourdon *et al.*, 1996b
MAR	33° S	~3,500 and 2,700	1.8	6	6	(6)	0.44 (0.46)	0.55	Lundstrom *et al.*, 1998a
AAD	50° S	~4,200	3.7	2					Bourdon *et al.*, 1996b

The criteria for selection of samples from full U-series database are described in Section 3.14.3.1. The numbers of suitable samples available for each U-series disequilibrium are listed, and those determined by alpha counting indicated by italics. The equiline gradient is the regressed slope of sample arrays on an equiline diagram (see Figure 7), together with a standard error on the slope. Arrays for locations with less than six samples were not calculated. In some cases two slopes are reported, which include a preferred slope and a regression of all data (in parentheses). Preferred slopes exclude seamount data in JDF and 8–10° N EPR arrays, which potentially have anomalously high melting rates and naturally lower disequilibrium than normal ridge samples. The preferred fit for 8–10° N EPR only includes the 9° 30′–9° 50′ N axial sample set. The preferred fit in the FAZAR samples includes only the samples from the most robust segment (KP4) which are likely best located and have experienced least post-eruptive decay. However, in all cases, the preferred and inclusive regressions are close to error of each other.

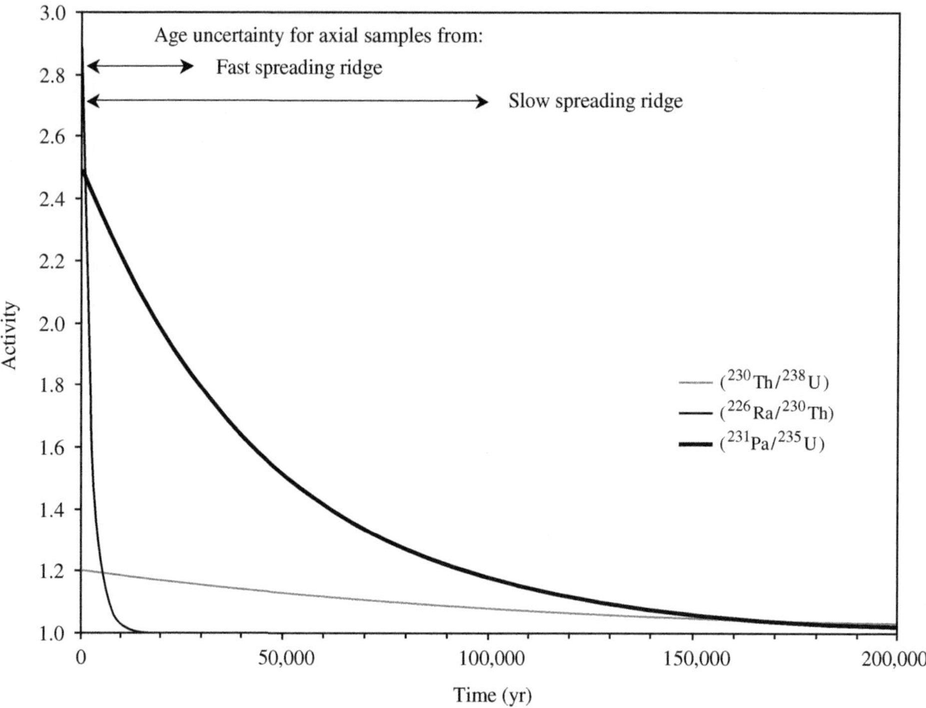

Figure 4 Decay of typical initial U-series excesses in MORB. Indication of age uncertainty for samples located within the axial neovolcanic zone for slow and fast spreading ridges indicated by double headed arrows (see text for details).

is significant for ($^{231}Pa/^{235}U$), largely because of its greater initial excess relative to measurement uncertainty. From Figure 4 it is clear that reliable data on "initial" ($^{226}Ra/^{230}Th$) can only be obtained on samples younger than a few hundred years, whereas samples younger than ~10 kyr provide good estimates on initial ($^{230}Th/^{238}U$) and reasonable values of initial ($^{231}Pa/^{235}U$).

It is, thus, necessary to have good independent controls on the age of samples. In the sub-aerial environment, U-series analyses are generally made on samples whose ages are documented by observations and historical records. In some cases, older samples can be accurately corrected from ^{14}C ages of underlying charcoal, for example. Such constraints are rarely available in studying MORBs, although there are now a few samples from recent eruptions as a result of concerted monitoring programs (Chadwick *et al.*, 1991, 1998) and serendipity (Haymon *et al.*, 1993). There are some qualitative indices of sample age, e.g., state of glass preservation, sediment cover, thickness of Mn crust (Ballard *et al.*, 1981; Haymon *et al.*, 1991; Macdonald *et al.*, 1988). Calibration of such observations with U-series methods is sparse and one study has pointed out inconsistencies between the two approaches (Sturm *et al.*, 2000).

Important age constraints are available, however, from the limited width of the neovolcanic zone on mid-ocean ridges. Assuming lavas are erupted randomly within the axial rift, a measure of the uncertainty in their age can be gauged from the width of the neovolcanic zone divided by the spreading rate (determined over a longer time period from magnetic anomalies). Taking typical neovolcanic zones (see Perfit and Chadwick, 1998) for fast spreading ridges (200 m wide, full spreading rate 80 mm yr^{-1}) and slow spreading ridges (2 km wide, full spreading rate 20 mm yr^{-1}), this approach gives age uncertainties ranging from 2.5 kyr at fast spreading ridges to 100 kyr at slow spreading ridges. Thus for fast-spreading ridges, all axial samples are likely to give a robust measure of eruptive ($^{230}Th/^{238}U$) and ($^{231}Pa/^{235}U$), but not ($^{226}Ra/^{230}Th$). For slow-spreading ridges, even samples well located within the axial rift zone potentially require age correction for all disequilibria. Clearly, this is an oversimplified model that does not consider manifest complexity in crustal accretion (e.g., Perfit and Chadwick, 1998; Sims *et al.*, in press; White *et al.*, 2002), but does highlight the basic issue. It is also worth stressing that good sample location and associated bathymetry is critical to the effectiveness of these basic age constraints. Locations of samples taken by submersible are much more precise than from those from dredging that have a spatial resolution of ~300–500 m at best.

Measured (^{226}Ra/^{230}Th) even on the fastest-spreading ridges can thus only be taken as minimum estimates of primary values. Observations at New Mounds site at ~45° N on the Juan de Fuca (JDF) Ridge (Chadwick *et al.*, 1991), at ~42° 40′ N on the North Gorda Ridge (Chadwick *et al.*, 1998) and BBQ site at ~9° 50′ N on the northern East Pacific Rise (EPR) (Haymon *et al.*, 1993), suggest eruptive activity in the last decade and so samples from these sites should yield reliable initial (^{226}Ra/^{230}Th). ^{210}Po$-^{210}$Pb dating ($t_{1/2}$ ^{210}Po 134 d) has verified the young age of the BBQ and N. Gorda site samples (Rubin *et al.*, 1994, 1998). This technique could be used elsewhere in combination with morphological observations to identify other "zero age" samples. It also might be possible to use ^{210}Pb$-^{226}$Ra disequilibrium ($t_{1/2}$ ^{210}Pb ~ 25 yr) to identify additional samples that are young enough to preserve primary (^{226}Ra/^{230}Th). ^{210}Pb$-^{226}$Ra disequilibrium is observed in some sub-aerial basalts (Gauthier and Condomines, 1999; Sigmarsson, 1996), and like ^{210}Po$-^{210}$Pb disequilibrium can be related to preferential loss of a more volatile nuclide during shallow degassing (Gauthier and Condomines, 1999). A promising initial study of ^{210}Pb$-^{226}$Ra disequilibrium in MORB was reported by Rubin *et al.* (2001).

An additional uncertainty in interpreting the (^{226}Ra/^{230}Th) data is the time a magma batch resided in the oceanic crust, prior to eruption. Cooper *et al.* (2003) have analyzed plagioclase separates from the historic N. Gorda flow to place constraints of <1.5 ka on the crustal residence time of magmas before eruption. Nevertheless, it would be useful to have yet tighter constraints on the residence time to define better initial (^{226}Ra/^{230}Th). It is possible that shorter-lived, mantle-derived disequilibria (e.g., ^{231}Pa$-^{228}$Ac) could provide such information.

For the longer-lived disequilibria, crustal residence time is likely unimportant and age constraints less critical. Crucially, the presence of any (^{226}Ra/^{230}Th) disequilibrium implies samples have near initial (^{230}Th/^{238}U) and (^{231}Pa/^{235}U), Figure 3. (^{226}Ra/^{230}Th) disequilibrium is thus an especially valuable tool for screening (^{230}Th/^{238}U) and (^{231}Pa/^{235}U) on intermediate and slow spreading ridges. In compiling our database, we have included well located samples from within 1 km of the axis of intermediate or fast-spreading ridges, and any samples that show significant (>3%) ^{226}Ra-excesses. The latter constraint has been used to exclude a number of samples from slow-spreading ridges. We have also included two samples sets from slow-spreading ridges for which there are no ^{226}Ra data (Bourdon *et al.*, 1996b). These suites are notable in one case for high (^{230}Th/^{238}U) and in the other for a modest ^{230}Th-deficit, neither of which can be attributed to post-eruptive decay.

3.14.3.3 General Signatures of Disequilibrium

Data and interest in disequilibrium in MORB historically started with (^{230}Th/^{238}U), followed by (^{226}Ra/^{230}Th) and finally (^{231}Pa/^{235}U). We follow a similar path in reviewing the general magnitude of disequilibria in MORB (Figure 5). Subsequently we explore finer scale structure and variability of disequilibria with other geochemical and geophysical parameters (Section 3.14.3.5).

Before considering disequilibrium, however, we examine (^{234}U/^{238}U), that should not be fractionated by high temperature processes (Section 3.14.2.1). Figures 5(a) and (b) illustrate that this is indeed the case for our MORB compilation. Both mass-spectrometric and alpha-counted data sets are dispersed around unity by an amount consistent with their respective typical analytical uncertainty, although some individual studies provide considerably more precise data (e.g., Sims *et al.*, 2002).

It is now well established that MORBs typically show modest ^{230}Th-excesses of 10–25% (Figures 5(c) and (d)). This signature, evident in the first U-series measurements of MORBs from the FAMOUS region of the Mid-Atlantic Ridge (MAR), was initially a surprise (Condomines *et al.*, 1981). Subsequent work verified significant disequilibrium in MORBs (Ben Othman and Allègre, 1990; Newman *et al.*, 1983; Reinitz and Turekian, 1989; Rubin and Macdougall, 1988). As discussed above, alpha-counting techniques were challenged by the low uranium concentrations of MORBs. Typical precisions (5–15%) were often close to the overall degree of disequilibrium and the case for primary disequilibrium was debated (Krishnaswami *et al.*, 1984). Mass-spectrometric (^{230}Th/^{238}U) data typically have a precision of ~1–2%, which provides a finer resolution picture of disequilibrium. For example, there is only a single mass-spectrometrically analyzed sample with a ^{230}Th-deficit, and very few have ^{230}Th-excesses less than 5%. Yet the overall picture provided by both alpha and mass-spectrometric techniques is reassuringly similar.

^{226}Ra$-^{230}$Th disequilibrium in MORBs is much more dramatic than ^{230}Th$-^{238}$U disequilibrium, with (^{226}Ra/^{230}Th) as great as 4.2 (Figures 5(d) and (e)). Alpha-counted and mass-spectrometric data both document (^{226}Ra/^{230}Th) ratios in excess of 2. Alpha-counted errors on (^{226}Ra/^{230}Th) are ~10% (Rubin and Macdougall, 1988) and larger than the ~1% errors by mass-spectrometry (e.g., Volpe and Goldstein, 1993). In contrast to (^{230}Th/^{238}U) disequilibrium, the difference in

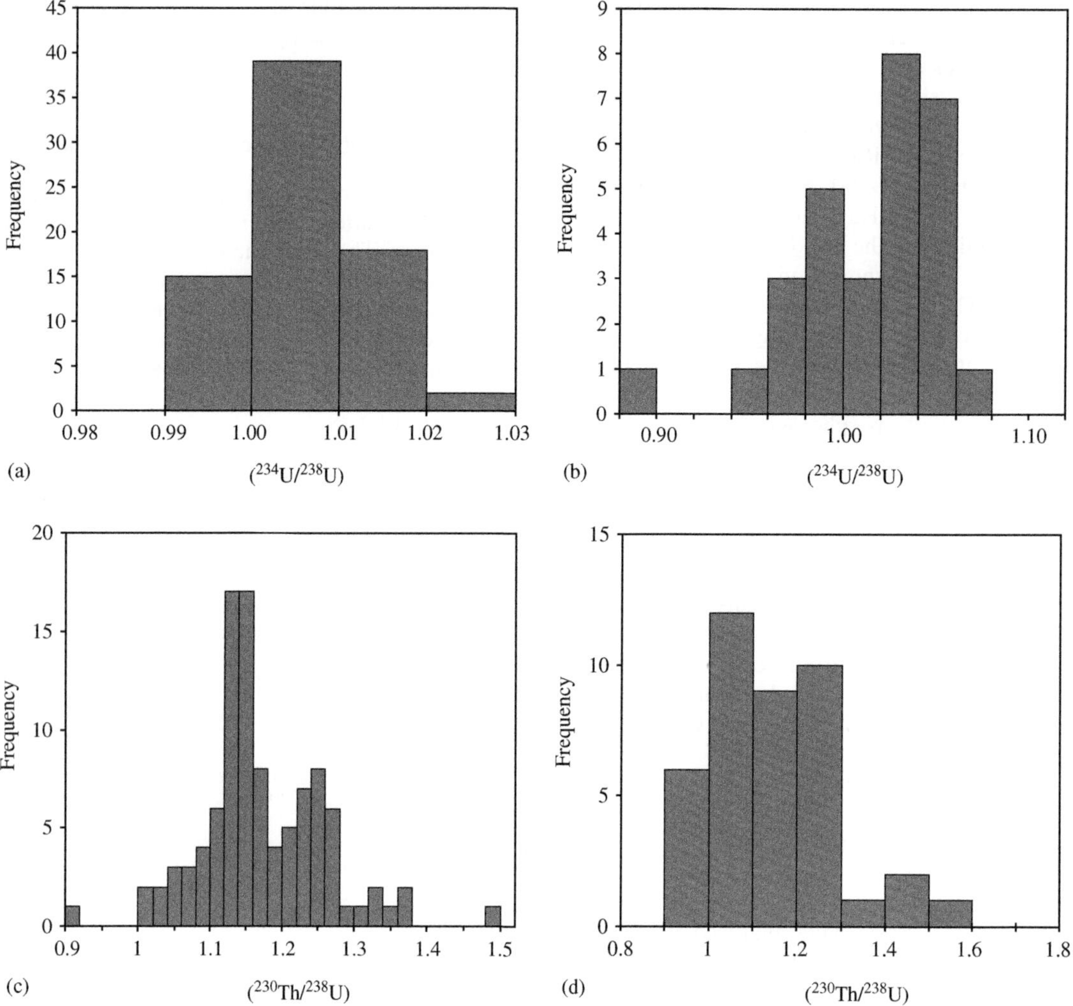

Figure 5 Histograms of activity ratios for key U-series nuclide pairs in compiled MORB data set: (a) $(^{234}U/^{238}U)$ by mass spectrometry, using reported $(^{234}U/^{238}U)$. In the absence of citied λ^{234} for all studies we did not try correcting to a common λ^{234}, (b) $(^{234}U/^{238}U)$ by alpha counting, note expanded scale relative to (a), (c) $(^{230}Th/^{238}U)$ by mass spectrometry, (d) $(^{230}Th/^{238}U)$ by alpha counting, (e) $(^{226}Ra/^{230}Th)$ by mass spectrometry in all samples from intermediate to fast spreading ridges (dark gray) and definitive historic events (light gray), (f) $(^{226}Ra/^{230}Th)$ by alpha counting, (g) $(^{231}Pa/^{235}U)$ in all samples (dark gray) and those from definitively historic eruptions (light gray).

the precision of the techniques is not highly significant given the much larger degrees of $^{226}Ra-^{230}Th$ disequilibrium. Much more important are age uncertainties, and as discussed above (Section 3.14.3.2) most of these ^{226}Ra data are minimum values. Only the historic samples highlighted (Figure 5(e)) are clearly reliable indicators of initial $(^{226}Ra/^{230}Th)$. Interestingly, the historic samples do not show strikingly higher $(^{226}Ra/^{230}Th)$ than a sizable proportion of the mass-spectrometry data set (Figure 5(e)). This suggests that many of the lavas sampled are under a few hundred years old.

^{235}U is 137.88 times less abundant than ^{238}U, and so $(^{231}Pa/^{235}U)$ measurements are analytically very challenging. Only a small number of mass-spectrometric measurements have been reported, but the results are striking. ^{231}Pa-excesses are large (Figure 5(g)), although not quite as extreme as some $(^{226}Ra/^{230}Th)$ ratios (Figure 5(f)). The current data set shows small variations of $(^{231}Pa/^{235}U)$ around a rather well-defined median of $\sim$2.6. Historic samples are also shown and reassuringly the less precisely dated samples do not appear distinct from the historic eruptions (Figure 5(g)).

The presence of significant $^{230}Th-^{238}U$ and large $^{226}Ra-^{230}Th$ and $^{231}Pa-^{235}U$ disequilibria in MORBs is a first-order observation. Accounting for the magnitude of these disequilibria places valuable constraints on melt migration and required development of a new series of melting models (Section 3.14.4.3). However, before ascribing disequilibria to the melting process, it is first necessary to discuss the possible role of crustal contamination.

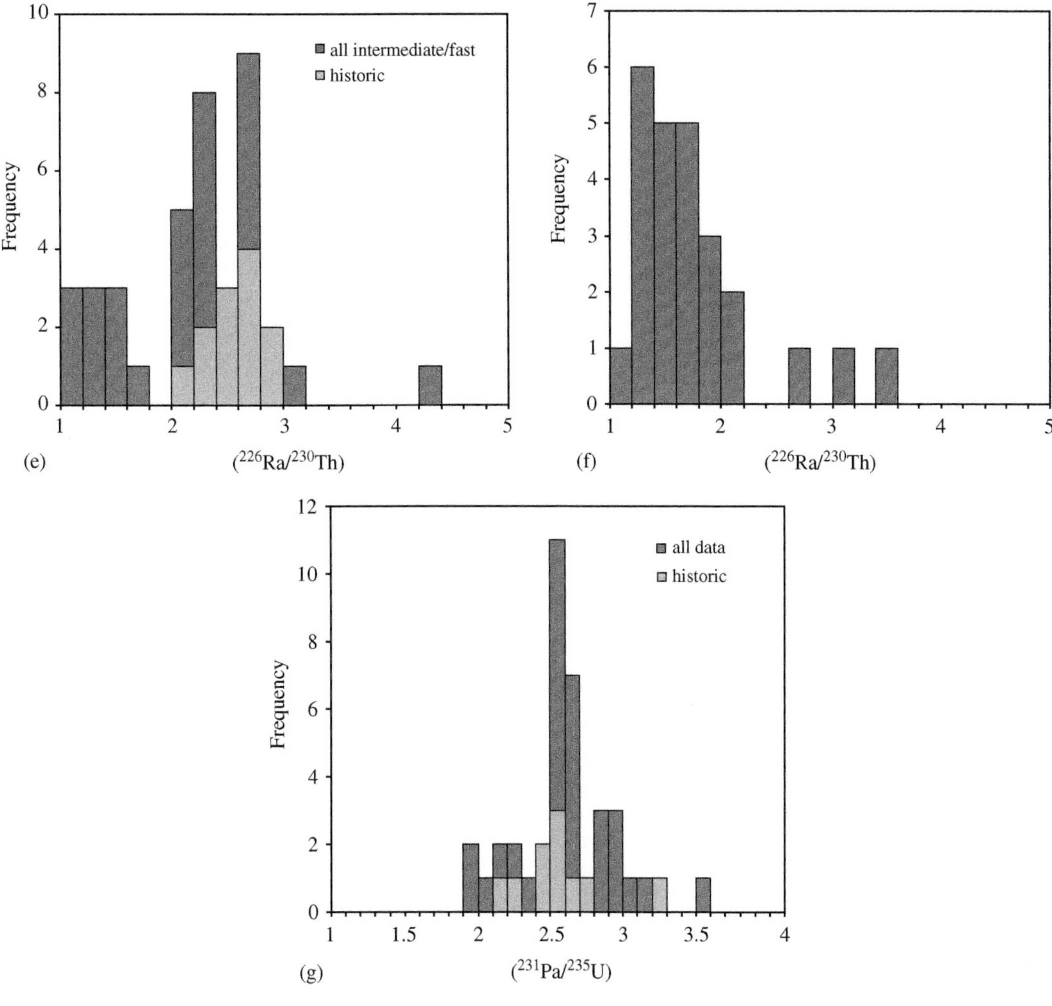

Figure 5 (continued).

3.14.3.4 Assessment of Shallow-level Contamination

Assimilation of near-surface material that has low-temperature-induced disequilibrium is a spectre frequently raised in the discussion of U-series disequilibria in MORBs. For example, ferro-manganese coatings and hydrothermal precipitates on seafloor samples sequester significant amounts of unsupported ^{234}U, ^{230}Th, and ^{231}Pa from seawater (German *et al.*, 1993; Lalou *et al.*, 1993; Shimmield and Price, 1988). Inclusion of even small amounts of such material in the measured lavas can induce significant disequilibrium (Figure 6). To minimize this potential problem, analyses are nearly always made on carefully handpicked glasses. Picking the large amounts of material required for older alpha-counting measurements was an onerous task, and it was therefore difficult to screen rigorously for contaminant on such large samples. Newman *et al.* (1983) noted that some of their samples had distinctly higher (^{230}Th/^{232}Th) and (^{230}Th/^{238}U) than others. They attributed this to contamination and did not further consider these analyses. Similarly, ^{234}U and ^{230}Th contamination of some samples was implicated in a study of mid-Atlantic MORB samples by Condomines *et al.* (1981). In addition to hand picking glass, various leaching procedures have been developed (e.g., Goldstein *et al.*, 1989; Reinitz and Turekian, 1989) in order to remove any ferro-manganese coating not spotted during picking (e.g., in small cracks). It is clearly important that such preparation techniques do not themselves cause fractionation of U-series nuclides in leached glass.

Contamination may also occur before eruption, by magmatic assimilation of altered material. Chemical proxies are required to assess such contamination. Bourdon *et al.* (2000) have recently comprehensively reassessed the issue of contamination using ^{10}Be. The presence of any of this short-lived ($t_{1/2} \sim 1.5$ Myr), cosmogenically produced nuclide in MORBs should be attributable to near-surface contamination alone. Using typical values of ^{10}Be and U-series nuclide

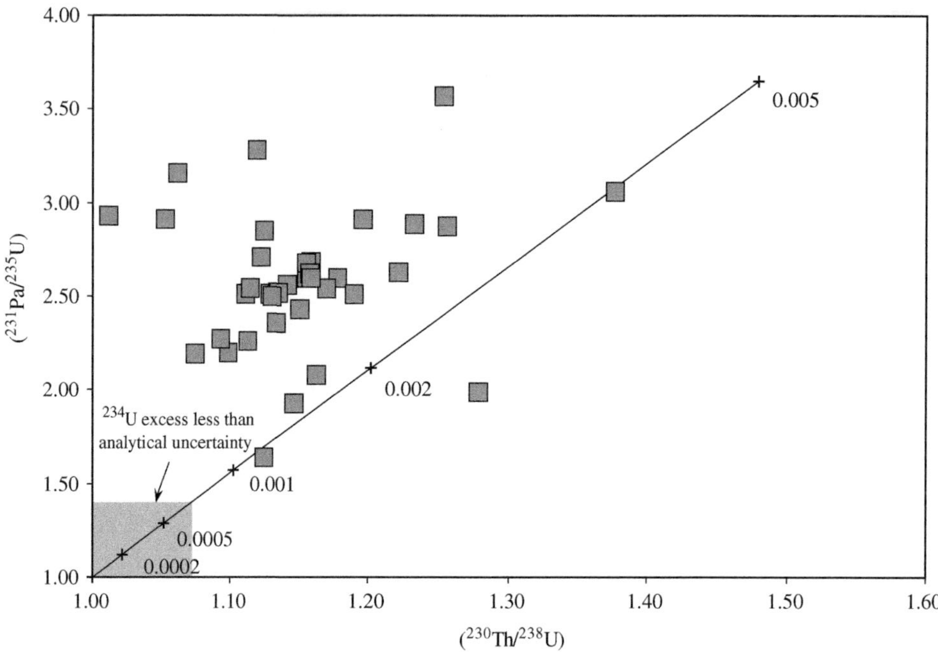

Figure 6 Illustration of the effects of contamination. (^{230}Th/^{238}U) and (^{231}Pa/^{235}U) for compiled MORB data set plotted together with a mixing line for different weight fractions (labeled) for ferro-managanese coating added to a putative MORB with no primary U-series disequilibria. The contaminant calculation is after Bourdon *et al.* (2000) with 50 ng g^{-1} [U] MORB in secular equilibrium and 1 μg g^{-1} [U] ferro-managanese coating with (^{234}U/^{238}U) = 1.15, (^{230}Th/^{238}U) = 6.25, (^{231}Pa/^{235}U) = 30. The shaded box indicates the maximum values that could be ascribed to contamination alone for the ~9° 50′ N East Pacific Rise data set of (Sims *et al.*, 2002), where equilibrium (^{234}U/^{238}U) values have been measured with an average precision of 2‰.

concentrations for ferro-manganese materials, a maximum of between 15% and 1% of the ^{231}Pa and ^{230}Th-excesses, respectively, can be accounted for by contamination (Bourdon *et al.*, 2000). The study also highlighted the effectiveness of hand picking in producing reliable samples.

(^{234}U/^{238}U) ratios in MORBs are also perturbed by addition of ferro-manganese material. Unfortunately, (^{234}U/^{238}U) is not routinely measured with a suitable precision to clearly rule out contamination. The recent work of Sims *et al.* (2002), however, document high precision (^{234}U/^{238}U) measurements of EPR samples that are generally within 3‰ of equilibrium. Figure 6 illustrates the maximum effect this uncertainty implies for possible contamination, and shows that the maximum permissible contamination cannot account for the high measured (^{231}Pa/^{235}U) in particular. Sims *et al.* (2002) also investigated other sensitive geochemical tracers of seawater related contamination, δ^{11}B and C1/K ratios, and found no evidence that these were elevated or correlated with disequilibria.

The issue of shallow-level processes in generating ^{226}Ra$-^{230}$Th disequilibrium is somewhat different to that for ^{238}U$-^{234}$U$-^{230}$Th and ^{235}U$-^{231}$Pa. Ra is not strongly particle reactive and not strongly concentrated in ferro-manganese coatings. Thus to invoke contamination to explain ^{226}Ra-excesses at the same time as ^{230}Th- and ^{231}Pa-excesses requires two unrelated processes.

Hydrothermal processes at the ridge axis result in precipitation of barite particles onto the ocean floor, which initially have extreme ^{226}Ra-excesses (e.g., Moore and Stakes, 1990). Minor addition of recent hydrothermal barite into an MORB flow would dramatically increase any (^{226}Ra/^{230}Th). Although this seems an attractive mechanism to produce what were initially puzzlingly large ^{226}Ra-excesses in MORB, there is a strong chemical argument against this explanation. The effect of barite contamination should also be evident in the barium contents of affected MORB, but not on other highly incompatible elements such as Rb. Unless the amount of barite contamination is constant worldwide, in locations with and without active hydrothermal venting, then it becomes very difficult to account why Ba/Rb remains near constant over a global MORB data set (Hofmann and White, 1983). It becomes even more difficult to account for the near identical ratio of Ba/Rb in MORB and OIB (Hofmann and White, 1983), as the latter are erupted in a setting where barite production is significantly lower. It is also worth noting that some of the highest ^{226}Ra-excesses occur in the most incompatible element depleted samples. One would expect the addition of barite barium to be prominent in these samples.

It has recently been suggested that plagioclase accumulation can account for ^{226}Ra-excesses (Saal *et al.*, 2002a,b; Van Orman *et al.*, 2002b). Such a mechanism seems unlikely to explain the general presence of ^{226}Ra-excesses in recent MORBs. The partition coefficient of radium in plagioclase is greater than thorium but is nevertheless less than 1 (Blundy and Wood, 2003b). Thus, the dominant reservoir of radium will be the melt and a large amount of plagioclase contamination is required to influence a given amount of magma (Cooper *et al.*, 2003). In any individual case, the role of plagioclase contamination can be further explored using other geochemical tracers, e.g., Sr (Saal *et al.*, 2002a,b).

Given the arguments, most observed disequilibria can be ascribed with some confidence to primary mantle processes. They, thus, form first order tests of any model of melting and melt migration.

3.14.3.5 Further Observations

In addition to the basic observations of U-series excesses (Section 3.14.3.3), further insight into processes in the melting regime is gained by investigating how the disequilibria co-vary with each other and additional geochemical and geophysical parameters. The quality of correlations is assessed using a Spearman rank correlation coefficient, r_s, and a probability factor, p. r_s ranges from -1 to 1; -1 is a perfect negative correlation, 0 is no correlation, and 1 is a perfect positive correlation. The rank correlation is more robust than the standard linear correlation as it is insensitive to outliers (see Press *et al.*, 1992). Moreover, it is possible to calculate a probability p that a larger values of r_s could *not* be produced at random (i.e., large values of p imply high confidence that the correlation could not be better for a random distribution).

The systematics discussed below help distinguish between a range of different melting models (see Section 3.14.4) that can explain the basic magnitudes of observed disequilibria (see Section 3.14.3.3), but have rather different implications for melt migration.

3.14.3.5.1 Equiline diagram

^{230}Th–^{238}U disequilibrium data is frequently plotted on the so-called *equiline diagram* (Allègre, 1968). This plot displays ^{230}Th–^{238}U disequilibria in a manner analogous to a traditional isochron diagram. The x-axis (^{238}U/^{232}Th) is affected only by chemical fractionation (on timescales short compared to the half-life of ^{238}U and ^{232}Th), while the y-axis (^{230}Th/^{232}Th) is sensitive to the decay or production of ^{230}Th and will change on the ^{230}Th timescale (Figure 7(a)). Samples in secular equilibrium will all plot on the *equiline* which has a slope of unity, i.e., (^{230}Th) = (^{238}U); therefore, (^{230}Th/^{232}Th) = (^{238}U/^{232}Th). Contours of constant disequilibrium are represented by straight lines that fan out from the origin (Figure 7(a)).

Variable fractionation of uranium and thorium generates a horizontal array of samples (Figure 7(a)). If the melt and residue remain in chemical equilibrium, this flat array can persist indefinitely (Section 3.14.2.2.2). However, if the melts and residues become chemically separated, the array rotates with age, pivoting around a point on the equiline, until it ultimately returns to equilibrium with a slope of 1. As in any isochron plot, a linear array may have age significance, or it may simply reflect mixing between two unrelated samples.

The x-axis of the equiline diagram is an inverted form of the Th/U ratio, which is frequently used as an index of "enrichment." Enriched samples, with higher incompatible element contents tend to have higher Th/U, or lower (^{238}U/^{232}Th). The overall trend can be observed in our MORB data set (Figure 3).

Figure 7(b) shows the global MORB data set plotted on an equiline diagram. The global array cuts across lines of equal–disequilibrium such that "enriched" samples with low (^{238}U/^{232}Th) tend to have high degrees of ^{230}Th–^{238}U disequilibrium and "depleted" samples with high (^{238}U/^{232}Th) generally have lower degrees of disequilibrium. "Local" data sets from limited geographical regions (Figures 7(c)–(f)) show similar arrays: samples with the lowest (^{238}U/^{232}Th) display the greatest degrees of disequilibrium (Goldstein *et al.*, 1991). In some instances, local arrays define a significant portion of the total global array (e.g. Figures 7(c)–(f) and Lundstrom *et al.*, 1999). The higher precision of mass-spectrometric measurements has identified rather linear sample arrays in the equiline plots for some locations (Lundstrom *et al.*, 1998b). The best-fit slopes of these arrays are variable (Lundstrom *et al.*, 1998b) and values for the better defined arrays in our data set are reported in Table 1.

3.14.3.5.2 Source enrichment

In both the local and global arrays on the equiline, there is a general trend from low degrees of disequilibrium at high (^{238}U/^{232}Th), or (^{230}Th/^{232}Th), to higher degrees of disequilibrium at low (^{238}U/^{232}Th), see Figure 7(b). Decreasing (^{238}U/^{232}Th) ratios are associated

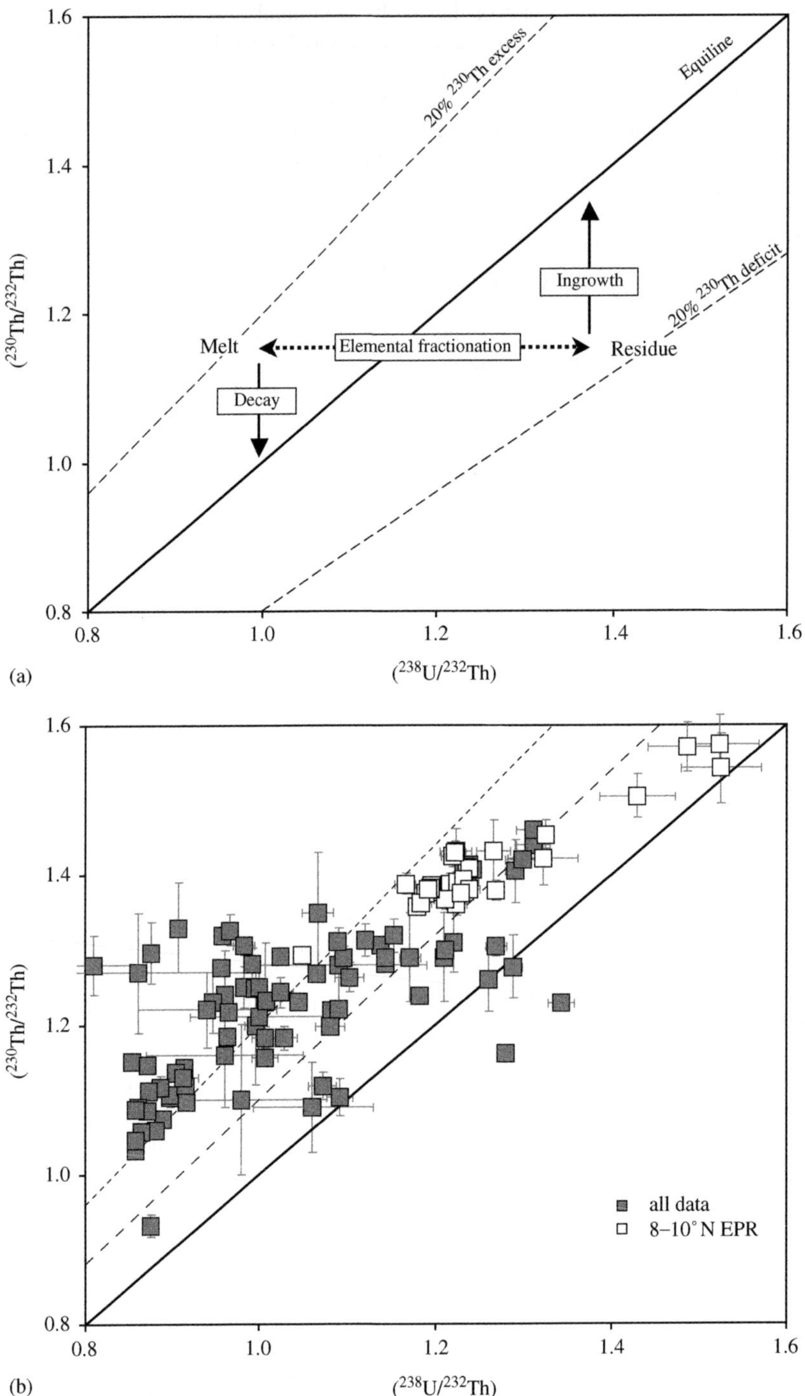

Figure 7 Equiline plots. (a) Cartoon of systematics of equiline diagram, (b) all compiled MORB data, open symbols highlight samples from 8–10° N EPR to indicate substantial fraction of global array evident in a single geographically restricted sample set, (c) samples from Gorda and Juan de Fuca ridges, (d) samples from 12° and 21° N EPR, (e) samples from 8–10° N EPR, including Siqueiros transform, (f) samples from Azores platform (FAZAR), shown by segment and Reykjanes Ridge. Data sources can be found in Table 1. Typical precision for mass-spectrometric measurement comparable to sample size and for clarity errors are only shown when significantly larger. Cited counting errors (2σ) shown on alpha counting measurements. Short dashed and long dashed lines indicate ^{230}Th excesses of 20% and 10% respectively.

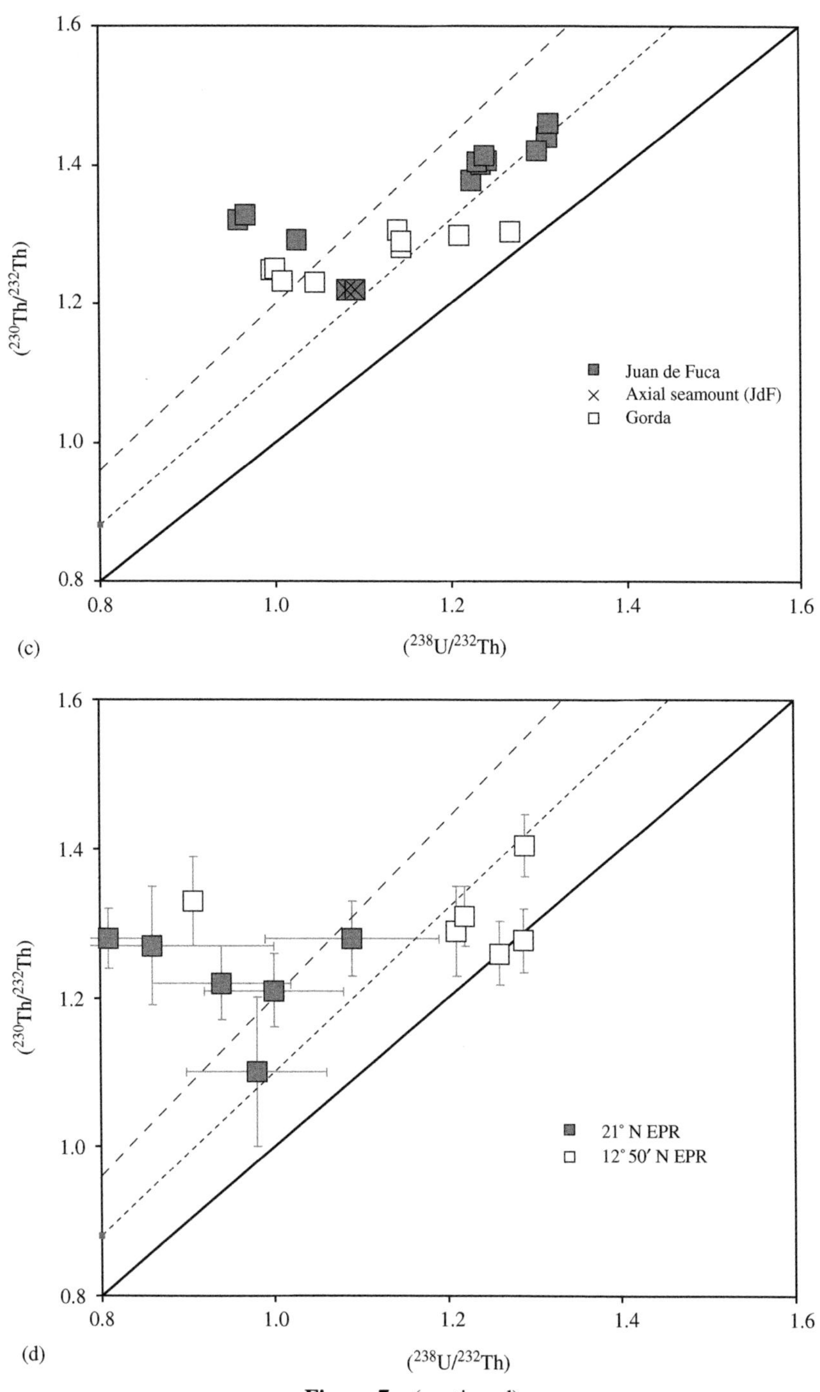

Figure 7 (continued).

with "enrichment" of incompatible element abundances and ratios of more to less incompatible element ratios (e.g., Figure 3). There is, thus, some evidence for the co-variation of the degree of (^{230}Th/^{238}U) disequilibrium with "enrichment." A key question is what caused the enrichment and when did it occur?

(^{238}U/^{232}Th) ought to be a rather robust index of source enrichment. As highly incompatible elements (Section 3.14.4.1), uranium and thorium are likely to be fractionated by the melting process only at very low degrees of melting (Section 3.14.2.2). Average MORB is thought to be the aggregate of ~10% melting (Klein and Langmuir, 1987; McKenzie and Bickle, 1988), in which there should be no net elemental fractionation of uranium from thorium. Whilst average MORB is the product of a large degree of melting, individual MORB samples likely represent an incomplete blend of melts produced at different

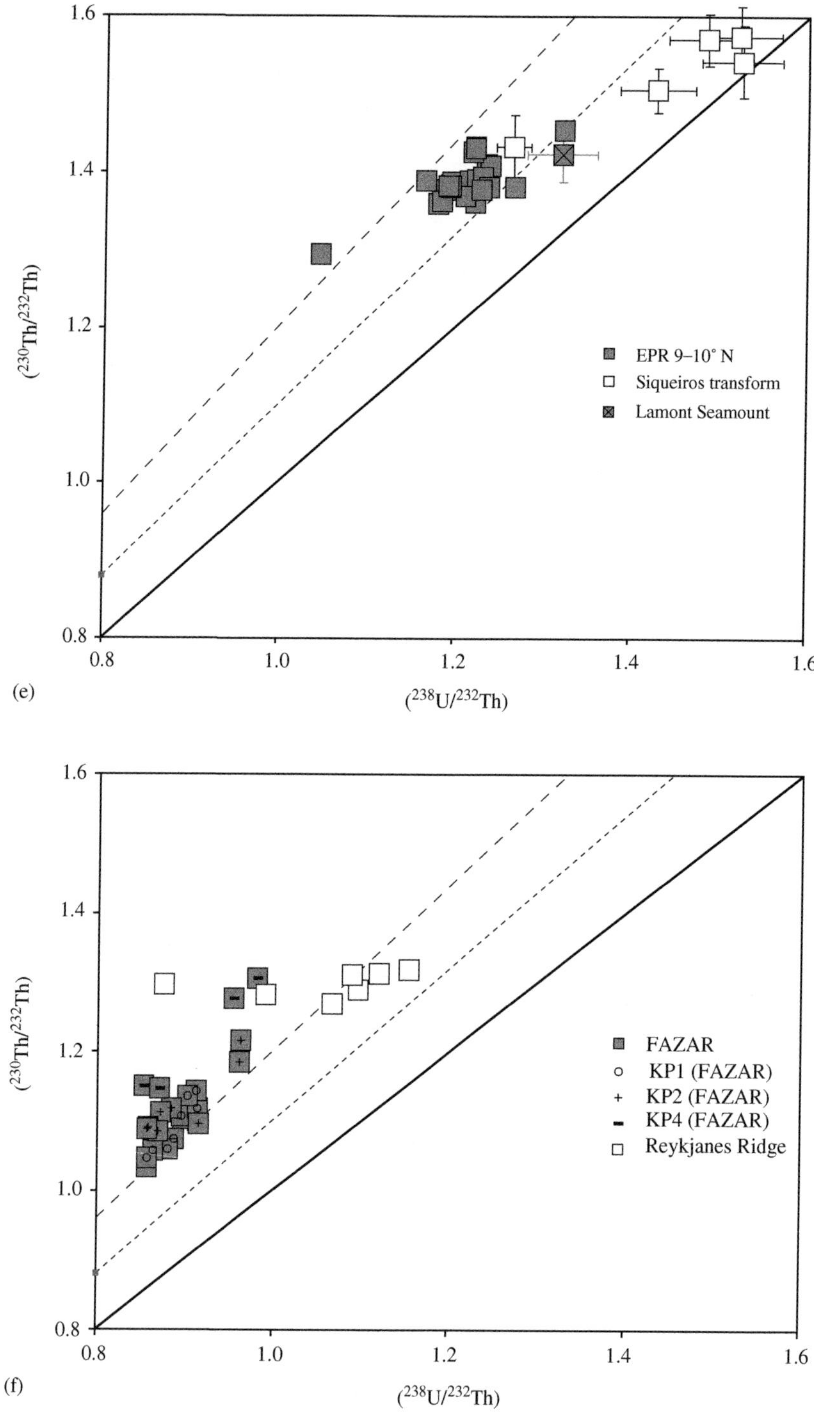

Figure 7 (continued).

locations in the melting regime (Langmuir *et al.*, 1977; Sobolev and Shimizu, 1993). The presence of $^{230}Th-^{238}U$ disequilibrium indeed testifies to the fractionation of thorium from uranium some where during the melting process, even if net elemental fractionation should not be evident in a true mean of all the melts produced (Section 3.14.4).

In order to distinguish between source heterogeneity and recent melt induced fractionation, it would be useful to have tracers that are not fractionated by "melting process." Long-lived radiogenic isotope ratios are such tracers. Sadly, long-lived radiogenic isotope data are rarely available on the same samples analyzed for ($^{230}Th/^{238}U$). Since ($^{238}U/^{232}Th$) is available for every sample with measured ($^{230}Th/^{238}U$), this imperfect trace element ratio is most commonly used as a proxy for source composition.

The few data sets with long-lived radiogenic isotopes as well as $^{230}Th-^{238}U$ data can be used to assess how closely ($^{238}U/^{232}Th$) variations correlate with more robust tracers of heterogeneity (Figure 8). This is hence not a comprehensive plot of all ($^{238}U/^{232}Th$) and $^{87}Sr/^{86}Sr$ data for MORBs, but only for those samples in our data set for which the systematics of disequilibria are also well constrained. A general anticorrelation of $^{87}Sr/^{86}Sr$ with ($^{238}U/^{232}Th$) is evident, which represents the so-called Sr–Th array (Condomines *et al.*, 1981; Rubin and Macdougall, 1992). The samples in Figure 8 comprise subsets from six main locations. Thus, in the same way that variations of slopes on equiline were examined on a "local" scale in Figure 7, it is important to assess correlations on ($^{238}U/^{232}Th$) with $^{87}Sr/^{86}Sr$ on a similar scale.

The FAZAR samples from the Azores platform (Bourdon *et al.*, 1996a) show a rather significant isotopic gradient along the sample section, but variations of ($^{238}U/^{232}Th$) that occur on a segment length are comparable to those across the whole geochemical anomaly. The Reykjanes Ridge samples analyzed for U-series nuclides (Peate *et al.*, 2001) are taken from a region of limited isotopic variability (Murton *et al.*, 2002) on the general trend from "normal" Atlantic MORB to enriched Icelandic compositions (Schilling, 1973). Despite the minor $^{87}Sr/^{86}Sr$ variation in the Reykjanes Ridge samples, there is significant, unrelated ($^{238}U/^{232}Th$) variability.

The suite from 12° 50′ N EPR (Ben Othman and Allègre, 1990) shows the clearest covariation of ($^{238}U/^{232}Th$) with $^{87}Sr/^{86}Sr$ and indeed was used as a case example of the marble-cake mantle (Prinzhoffer *et al.*, 1989). In contrast, samples from ~9° 30′–9° 50′ N EPR (Goldstein *et al.*, 1993; Sims *et al.*, 2002) show much lower ($^{238}U/^{232}Th$) than nearby depleted MORB from the Siqueiros transform (Lundstrom *et al.*, 1999) but all these samples have comparable $^{87}Sr/^{86}Sr$ (Sims *et al.*, 2002). We note, however, that one E-MORB from the Siqueiros transform shows both low ($^{238}U/^{232}Th$) and elevated $^{87}Sr/^{86}Sr$ (Lundstrom *et al.*, 1999; Sims *et al.*, 2002). This sample is not plotted on Figure 8 as it does not meet the age criteria of our compilation, and so interpretation of its disequilibria is uncertain.

Samples from both the JDF and Gorda ridges are striking in the lack of correlation between ($^{238}U/^{232}Th$) and $^{87}Sr/^{86}Sr$ (Goldstein *et al.*, 1991; Sims *et al.*, 1995). Yet, the sample suites from these localities define some of the better arrays on equiline plots (Figure 7(c)). Notably, the most enriched sample from the JDF, with lowest ($^{238}U/^{232}Th$) and highest ($^{230}Th/^{238}U$) has a markedly unradiogenic $^{87}Sr/^{86}Sr$ (Figure 8).

A clear link between ($^{238}U/^{232}Th$) and ancient source enrichment is thus not readily evident on

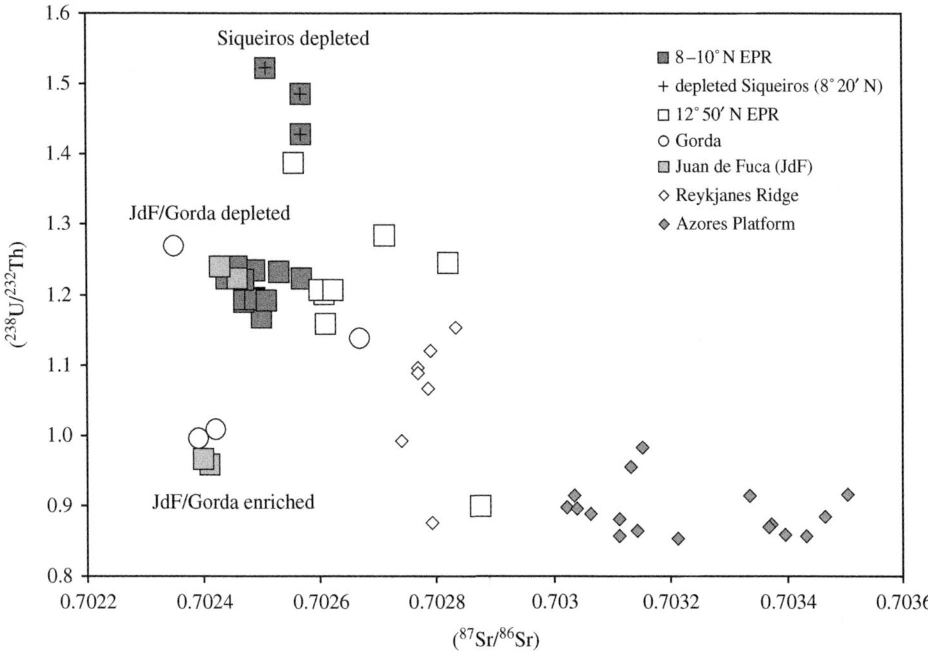

Figure 8 $^{87}Sr/^{86}Sr$ versus ($^{238}U/^{232}Th$) for the MORB compilation of this study. Hence only samples with accompanying U-series data, not effected by post-eruptive decay, are plotted. Despite an overall trend of decreasing ($^{238}U/^{232}Th$) with increasing $^{87}Sr/^{86}Sr$, on a local scale, samples span a wide range of ($^{238}U/^{232}Th$) at similar $^{87}Sr/^{86}Sr$. Additional sources for $^{87}Sr/^{86}Sr$ data: (Davis and Clague, 1987; Dosso *et al.*, 1999; Murton *et al.*, 2002; Sims *et al.*, 1995).

a local scale. Whilst a few enriched samples have elevated $^{87}Sr/^{86}Sr$, more frequently this is not the case.

Moreover, there are large ranges of $(^{238}U/^{232}Th)$ in samples with similar $^{87}Sr/^{86}Sr$. Since Lundstrom *et al.* (1998b) noted that samples with lower $(^{238}U/^{232}Th)$ had higher $(^{230}Th/^{238}U)$ (i.e., plotted further to the left of the equiline) and inferred $(^{238}U/^{232}Th)$ variations to be related to source heterogeneity, they required a mechanism to produce larger degrees of disequilibrium when melting a more enriched source. Lundstrom *et al.* (1995) accounted for this by invoking small-scale, enriched heterogeneities with a significantly different lithology (garnet–pyroxenite) to the surrounding peridotite (Allègre and Turcotte, 1986), and later expanded on this model (Lundstrom *et al.*, 2000). In the light of Figure 8, however, additional processes are likely active. The possible fractionation of $(^{238}U/^{232}Th)$ as a result of MORB generation is discussed in Section 3.14.4.

3.14.3.5.3 Variations of disequilibria with spreading rate

The local arrays of data on equiline plots show variations in slope from region to region (Figure 7), with a much steeper trend for samples from the 8–10° N EPR for example (Figure 7(e)) than the near horizontal array for the Reykjanes Ridge (Figure 7(f)). This was originally noted by Lundstrom *et al.* (1998b), who suggested a relationship between spreading rate and such local equiline slopes. Slow-spreading ridges apparently define flatter arrays than fast-spreading ridge segments.

Using our data compilation, we fitted straight lines to the local arrays illustrated in Figure 7. The results are reported in Table 1 and plotted against spreading rate in Figure 9(a). Correlation of local equiline slopes with spreading rate is very poor ($r_s = -0.25$, $p = 0.4$), unless the data set from the Azores platform (FAZAR samples) and other less well defined arrays are excluded. One key data set used by Lundstrom *et al.* (1998b), from the ultrafast spreading S. EPR (Rubin and Macdougall, 1998), is not plotted here. These data define the steepest normal ridge segment array, lying close to the equiline. However, all bar two of these samples have been excluded from this study due to potential problems of blank contamination (Section 3.14.3.1), that result in very high $(^{230}Th/^{232}Th)$ and $(^{238}U/^{232}Th)$. Given the doubts raised over such extreme values above, we have not included the S. EPR samples in this compilation. The recent repeat analyses of some of these samples by mass-spectrometry (Rubin *et al.*, 2000) will, thus, be of significant interest. Our filtering of some alpha-counted data has also resulted in a different slope for our compiled ~12° N samples compared to the value of (Lundstrom *et al.*, 1998b), 0.05 versus 0.16, respectively.

The Reykjanes Ridge and Azores platform regions represent the slow-spreading end-members of the global data set. Both of these locations are anomalously shallow and inferred to be affected by mantle that feeds nearby ocean islands (Iceland and the Azores, respectively). Despite these similarities in their geophysical setting, the systematics of samples from these two locations are strikingly different (Figure 7(f)). The Reykjanes data pin the Lundstrom *et al.* (1998b) global trend, whereas the FAZAR data greatly degrade any correlation (Figure 9(a)). It is not clear why the FAZAR samples should be different if the Lundstrom *et al.* (1998b) model is globally applicable. Whilst the FAZAR samples do cover a wider range of axial depths than other sample suites shown in Figure 9 (see implications of this in Section 3.14.4), individual ridge segments span a much smaller range of axial depth and display similar slopes to the overall trend (Figures 7(f) and 9(a)).

It is also worth noting, however, that it is the *slope* of arrays on an equiline diagram that Lundstrom *et al.* (1998b) related to spreading rate and not absolute ^{230}Th-excesses. Lundstrom *et al.* (1998b) pointed out that the slope of the equiline represented one way of characterizing a range of compositions from a single location. Interestingly, a simple correlation of $(^{230}Th/^{238}U)$ with spreading rate is more significant ($r_s = -0.57$, $p = 1$) than for the equiline slopes. $(^{231}Pa/^{235}U)$ variations are uncorrelated ($r_s = -0.16, p = 0.68$) with spreading rate (Figures 9(b) and (c)).

The degree of disequilibrium produced during "in-growth" models is inversely dependent on melting rate (see Section 3.14.4), which in the MORB can be linked to upwelling rate. It is, thus, perhaps encouraging to see some apparent inverse spreading rate dependence in the disequilibria data (Figure 9(b)). However, any relationship between spreading rate and erupted disequilibria is not likely to be straightforward (see Section 3.14.4.3.3). It is thus not necessarily surprising that $(^{230}Th/^{238}U)$ shows a weak correlation with spreading rate but $(^{231}Pa/^{235}U)$ does not. It may be significant that the slowest-spreading ridges are also the shallowest (see next section), and that several effects may combine to produce correlations observed.

3.14.3.5.4 Variations of disequilibria with axial depth

Bourdon *et al.* (1996b) presented a data set that indicated a general correlation of segment averaged $(^{230}Th/^{238}U)$ and axial depth. Key data

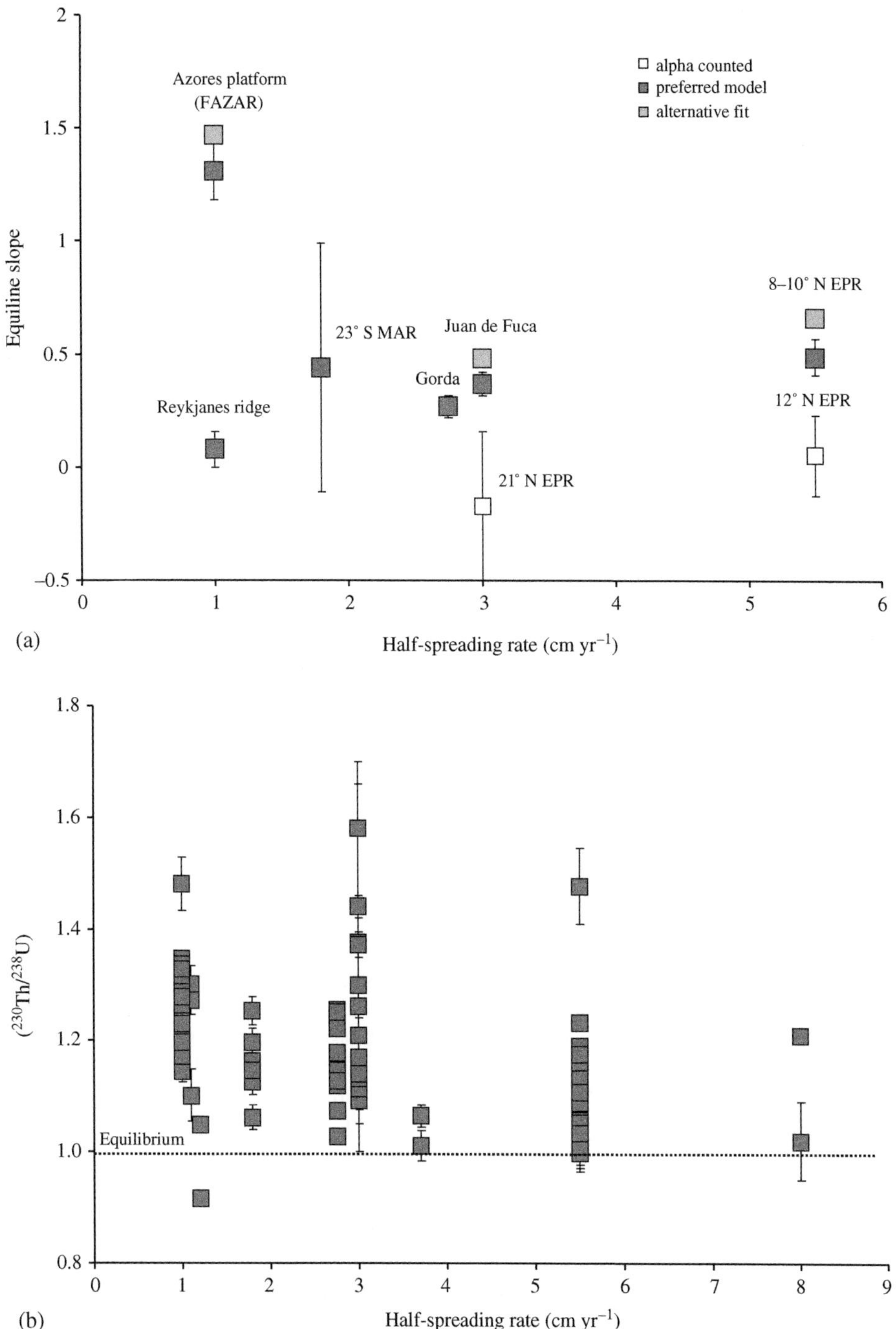

Figure 9 (a) Equiline slope versus half-spreading rate for localities plotted in Figure 7, (b) (^{230}Th/^{238}U) versus half spreading rate for all compiled MORB samples, (c) (^{231}Pa/^{235}U) versus half spreading rate for compiled MORB samples. Data for the regressed equiline slopes and half spreading rates are reported in Table 1. Where reported in Table 1, preferred and inclusive regression are plotted.

on this plot were the shallow FAZAR samples with high degrees of (^{230}Th/^{238}U) and samples from two deep ridge settings, the Australian Antarctic Discordance (AAD) and 30° N MAR, with (^{230}Th/^{238}U) close and even less than unity. Unfortunately none of these samples have measured (^{226}Ra/^{230}Th) excesses to assess post-eruptive decay. Yet, as well located samples from an intermediate spreading center, it is unlikely that uncertainty in age accounts for the very low disequilibrium in the AAD samples. Moreover, the ^{238}U-excess in a single sample

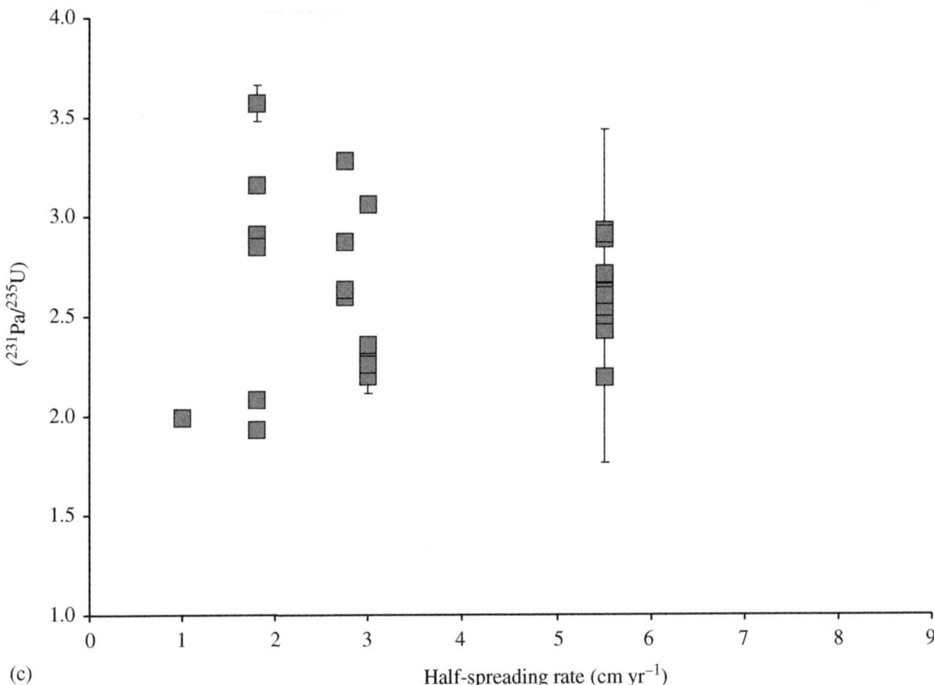

Figure 9 (continued).

from 30° N MAR cannot simply be the result of post-eruptive decay. The FAZAR samples have some of the highest $(^{230}Th/^{238}U)$ observed on the ridge system. Any post-eruptive decay in these samples should only lower their disequilibrium and so not grossly influence the global correlation.

Bourdon *et al.* (1996b) accounted for their correlation as the result of a deeper onset of melting of hotter (or more enriched) mantle. Deeper melting regimes have larger amounts of melting in the garnet stability field, with higher D_U/D_{Th} and this can result in higher $(^{230}Th/^{238}U)$ of the erupted melts (see Section 3.14.4). Melting regimes beneath the deepest ridges potentially do not extend into the garnet stability field and so modest ^{230}Th-excesses or even deficits could be expected (again see Section 3.14.4). Since the publication on Bourdon *et al.* (1996b), the understanding of uranium and thorium partitioning in clinopyroxene has improved considerably. A continuous increase in D_U/D_{Th} in clinopyroxene with increasing pressure provides a further mechanism to account for the global correlation (e.g. Landwehr *et al.*, 2001).

Figure 10 shows an updated plot of $(^{230}Th/^{238}U)$ against axial depth, that includes data from the shallow Reykjanes Ridge (Peate *et al.*, 2001) and deep southern MAR (Lundstrom *et al.*, 1998a) published since Bourdon *et al.* (1996b). A correlation persists ($r_s = -0.58$, $p = 1$), although the new Reykjanes Ridge data do not clearly substantiate the original trend. A single Reykjanes sample shows very high $(^{230}Th/^{238}U)$, but this is not true of the whole data set. Moreover, recent unpublished data (not shown) from the shallow Kolbeinsey Ridge (Sims *et al.*, 2001) is also not generally supportive of the Bourdon *et al.* (1996b) correlation. Kolbeinsey samples define a near horizontal slope on the equiline diagram. While the highest degree of disequilibrium is quite high (^{230}Th-excesses of 24%), the Kolbeinsey ridge also boasts samples with ^{238}U-excesses (Sims *et al.*, 2001). More data from the Kolbeinsey Ridge are needed to see if the ^{238}U-excesses reported by Sims *et al.* (2001) significantly influence the segment average.

As discussed above (Section 3.14.3.5.3) ^{230}Th-excesses show a negative correlation with spreading rate of similar significance to the negative correlation with axial depth. Moreover, the samples that define the shallowest ridges are from the slowest spreading ridges. In order to disentangle these effects it would be useful to have more data from locations such as the southern EPR, which is fast but not anomalously deep, and the (AAD) which is deep but does not have an extreme spreading rate.

A key point about Figure 10 (and Figure 9(b)) is that $(^{230}Th/^{238}U)$ variability at a given depth (or spreading rate) can be a major proportion of the total global range. This further emphasizes the importance of local variations discussed in the previous section. Whilst the global trend is still significant despite this scatter, it is clear that local variations in the melting process have as

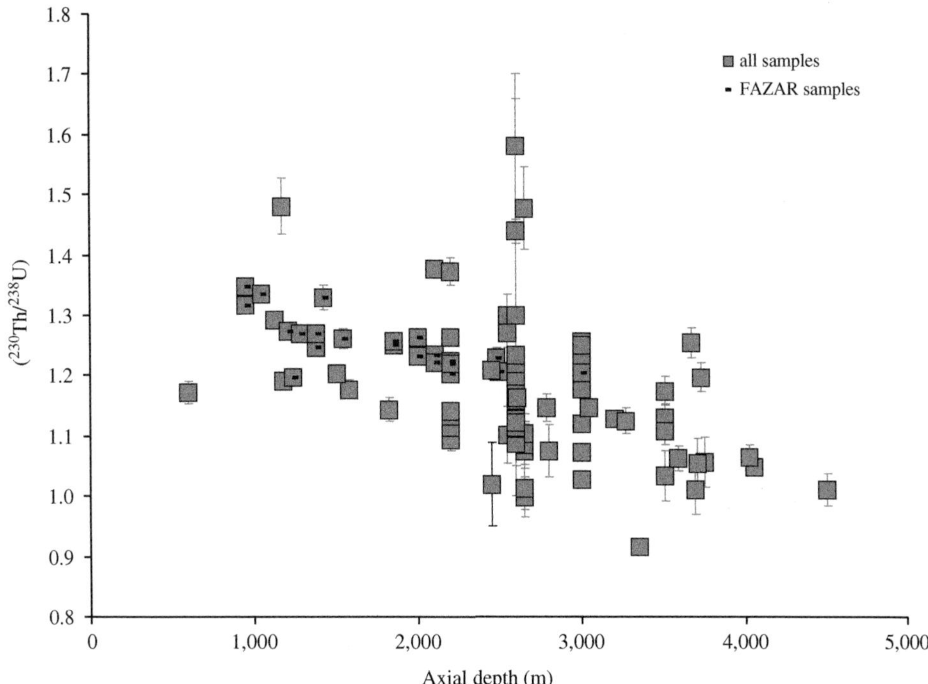

Figure 10 (^{230}Th/^{238}U) against axial depth for all compiled MORB samples. Horizontal dashes further indicate FAZAR samples which cover a significant portion of the array (Bourdon *et al.*, 1996a), but are not the sole source of the correlation.

important an effect on ^{230}Th–^{238}U disequilibrium as the ~200 °C difference in mantle temperature invoked to account for changes in axial depth (Bourdon *et al.*, 1996a; Klein and Langmuir, 1987). This observation has resonances with local variations in the major element systematics of MORB (Klein and Langmuir, 1989; Langmuir *et al.*, 1992).

3.14.3.5.5 Mutual covariations of disequilibria

So far, we have only considered ^{230}Th–^{238}U disequilibrium, for which there is the largest body of data, least effected by post-eruptive decay. It is now worth exploring the mutual variations of the different disequilibria.

As discussed above (Section 3.14.3.2) and illustrated in Figure 4, it is difficult to obtain robust primary estimates of (^{226}Ra/^{230}Th) and even to a lesser extent (^{231}Pa/^{235}U). In the case of ^{226}Ra-excesses, three localities where documented eruptions have occurred (Section 3.14.3.2) provide key constraints, and we will discuss these in turn. A sample from the New Mounds site on the JDF Ridge has (^{226}Ra/^{230}Th) ~ 2.5 (Lundstrom *et al.*, 1995), which is typical of many MORB (Figure 5(e)). Lundstrom *et al.* (1995) pointed out that when combined with other young samples from the area, there was in inverse correlation of (^{226}Ra/^{230}Th) and (^{230}Th/^{238}U). However, the age constraints on the samples with lower (^{226}Ra/^{230}Th) in this data set are not good enough for this to be a firm conclusion. This general inverse correlation ($r_s = -0.68, p = 1$) is also apparent in the larger database of "young" but not precisely dated samples we have compiled (Figure 11(a)), as has been pointed out by others (Kelemen *et al.*, 1997; Volpe and Goldstein, 1993). The significance of the inverse correlation is significantly strengthened by recent data from the tube worm BBQ site ~9° 50′ N EPR (Sims *et al.*, 2002) and the nearby Siquieros transform (Lundstrom *et al.*, 1999). Although the samples from the Siqueiros transform are not well dated, these primitive depleted samples have extreme (^{226}Ra/^{230}Th) as high as 4.2, which are consequently minimum values. The samples from the 1991–1992 New BBQ eruption can be usefully contrasted with the Siqueiros samples, forming a robust inverse correlation of (^{226}Ra/^{230}Th) against (^{230}Th/^{238}U). Significantly, a negative correlation is evident even in the more restricted range of (^{230}Th/^{238}U) of the definitive historic samples (Sims *et al.*, 2002). These high precision analyses, from samples with very similar long-lived radiogenic nuclides, are thus a particularly well constrained data set and place critical constraints on the behavior of U-series nuclides during melting beneath ridges.

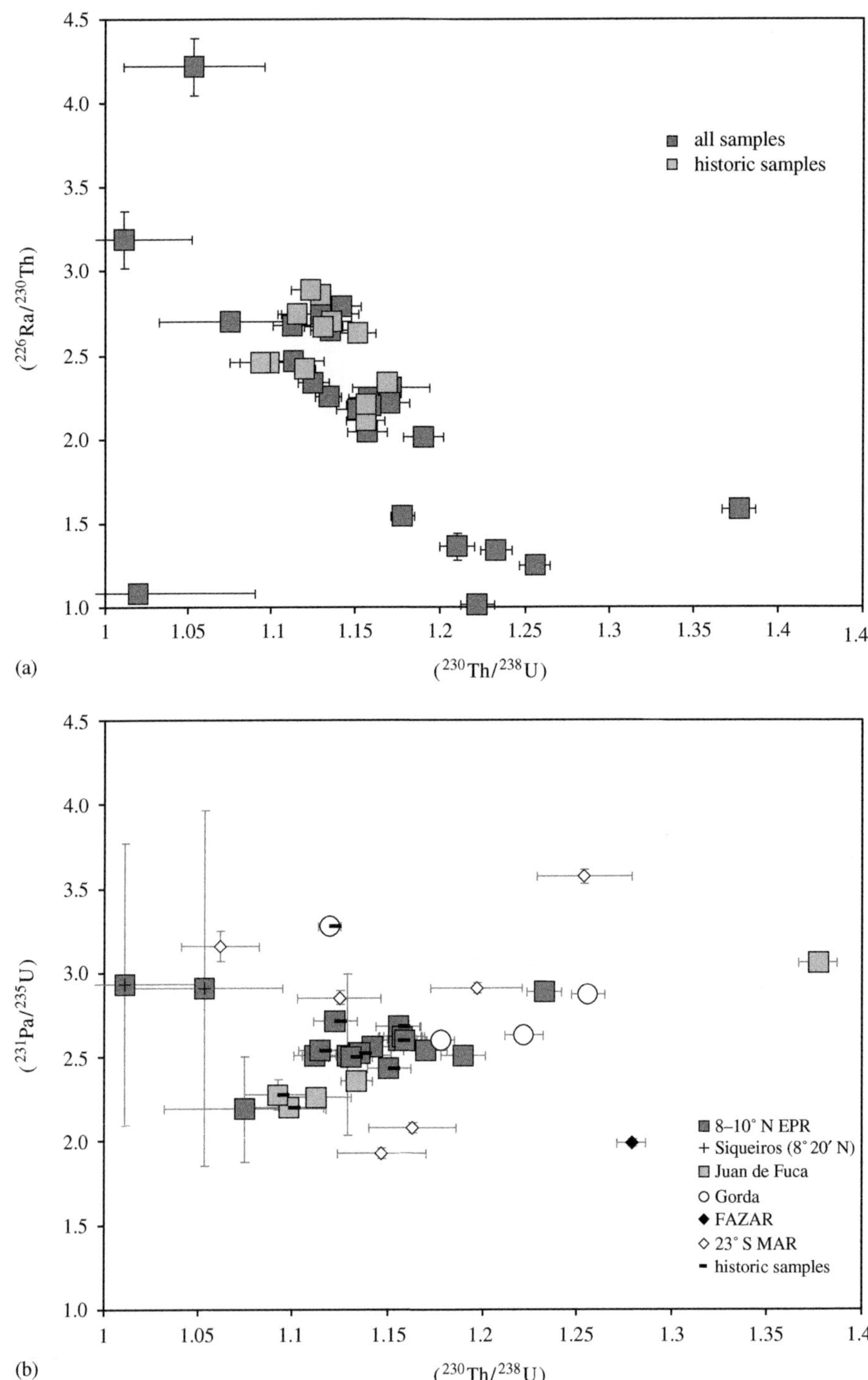

Figure 11 (^{230}Th/^{238}U) versus: (a) (^{226}Ra/^{230}Th) for all fast and intermediate spreading ridge samples, dark gray symbols, and definitively historic eruptions, light gray symbols, (b) (^{231}Pa/^{235}U) for all samples with (^{226}Ra/^{230}Th) excesses (for which the effects of post-eruptive decay are a minimum), dark gray symbols, and definitively historic samples, light gray symbols. Different symbols also used to indicate different geographical regions. Despite a general lack of correlation, in some locations (^{230}Th/^{238}U) and (^{231}Pa/^{235}U) show some covariation.

There is no clear co-variation ($r_s = 0.16$, $p = 0.67$) of (^{230}Th/^{238}U) with (^{231}Pa/^{235}U) for our compilation as a whole (Figure 11(b)). In keeping with earlier discussion, however, there is a sense of some co-variation at a local scale. Sample from 9° 30′–9° 50′ N EPR define a weak positive correlation ($r_s = 0.48$, $p = 0.97$) whilst samples from JDF Ridge show a better but less significant correlation ($r_s = 0.7$, $p = 0.81$).

Additional intriguing correlations are evident within the comprehensive, high-density geochemical data set of the $\sim 9° 50'$ N lavas (Sims *et al.*, 2002). For example, these samples show correlations of Mg#, molar Mg/(Mg + Fe), with $^{226}Ra-^{230}Th$ and $^{230}Th-^{238}U$ disequilibria. Sims *et al.* (2002) strongly argue these trends results from variable mixtures of melts from different depths in the melting regime. Unfortunately, there are no other comparably detailed studies with which to compare these results, but this approach promises to be highly rewarding.

3.14.3.6 Summary of Observations

The previous sections have described a number of key observations that need to be satisfied by successful models of melting and melt migration beneath ridges:

- U-series disequilibria in young MORB are near ubiquitous.
- Most measured disequilibria are analytically significant and unlikely to result from shallow level contamination.
- Modest ^{230}Th-excesses ($\sim$10–20%) and large ($\sim$100%) ^{226}Ra and ^{231}Pa excesses are common in MORB unaffected by post-eruptive decay.
- On a ridge segment scale ($^{230}Th/^{238}U$) disequilibrium is frequently anticorrelated with ($^{238}U/^{232}Th$). This results in "local" linear arrays on equiline diagrams that represent a significant fraction of the global variability in ($^{230}Th/^{238}U$).
- ($^{238}U/^{232}Th$) can show significant variation on a local scale unrelated to $^{87}Sr/^{86}Sr$ and should not be used as a robust proxy of long-lived mantle heterogeneity.
- Historic samples show a negative correlation of ($^{226}Ra/^{230}Th$) with ($^{230}Th/^{238}U$), a trend further defined by less-well-dated samples.
- Apart from the $^{226}Ra-^{230}Th$ excess anticorrelation, global variability of disequilibria with themselves and physical parameters are not well defined, perhaps masked to some extent by large local variability.
  - Local equiline arrays have variable slopes. These gradients do not show a clear variation with spreading rate as has been proposed (Lundstrom *et al.*, 1998b), although ^{230}Th-excesses show a weak, negative correlation with spreading rate. ^{231}Pa-excesses show no spreading rate dependence.
  - ($^{230}Th/^{238}U$) shows a weak negative correlation with axial depth.
  - ($^{231}Pa/^{235}U$) shows a weak correlation with ($^{230}Th/^{238}U$) in two individual sample suites, but a negligible correlation for the global data set.

3.14.4 U-SERIES MELTING MODELS

Geochemical and geophysical models are required that both explain the observations (Section 3.14.3.6) and provide better insight into the properties and processes occurring in partially molten regions beneath mid-ocean ridges. Because of the richness of the observations, this has proved to be reasonably difficult and this section reviews a range of models that have been proposed to explain U-series from the simplest closed-system models to fluid-mechanically consistent models of channelized melt transport. We compare model behavior to the geochemical systematics and discuss the inferences that are driven by the models. All of these models, however, require knowledge of the partitioning behavior of the U-series nuclides in mantle minerals (and some require information about their diffusivities) so we begin by discussing the constraints on these parameters from experimental petrology.

3.14.4.1 Preliminaries: Partition Coefficients and Diffusivity

The initial reports of ($^{230}Th/^{238}U$) disequilibrium in MORB (Condomines *et al.*, 1981) preceded reliable partitioning data for uranium and thorium. However, a plethora of ion-microprobe determined trace element partition coefficients has substantially changed this situation in the 1990s (Blundy and Wood, 2003a) and partition coefficients of uranium and thorium are now some of the best determined (Blundy and Wood, 2003b).

A key feature is that all the nuclides of interest are highly incompatible in common mantle mineral phases (Table 2). Clinopyroxene and garnet (present in normal peridotitic mantle at depths greater than $\sim$80 km) are the principal host minerals for uranium and thorium in the solid phase, although even in these phases partition coefficients do not exceed 0.05. An important consideration is the sense of uranium and thorium fractionation imparted by the presence of different minerals. This can be conveniently expressed in terms of D_U/D_{Th}. Minerals with $D_U/D_{Th} > 1$ retain uranium over thorium and contribute to ^{230}Th-excesses in a coexisting melt, whereas those with $D_U/D_{Th} < 1$ help to create ^{230}Th-deficits. The effects of different minerals are then weighted by their absolute partition coefficients and modal abundance to control the bulk partition coefficient of the mantle, and thus determine the sense of fractionation of thorium from uranium in a coexisting melt.

In spinel lherzolite, which likely comprises the main mantle lithology of the MORB melting regime, clinopyroxene largely controls

Table 2 Illustrative U and Th partition coefficients for common mantle minerals and bulk assemblages.

Pressure, GPa	*Temperature (°C)*		*Clinopyroxene*		*Garnet*		*Orthopyroxene*		*Olivine*		*Bulk*			*References*
			D_U	D_{Th}	D_U	D_{Th}	D_U	D_{Th}	D_U	D_{Th}	D_U	D_{Th}	D_U/D_{Th}	
		Mineral mode	0.2				0.2		0.6					
1.5	1,260		0.029	0.027			0.0017	0.0008	1.80×10^{-5}	1.20×10^{-5}	6.15×10^{-3}	5.57×10^{-3}	1.1	Wood *et al.* (1999)
1.5	1,302		0.0043	0.0036			0.0015	0.0007	3.80×10^{-5}	5.00×10^{-5}	1.18×10^{-3}	8.92×10^{-4}	1.3	Salters *et al.* (2002)
		Mineral mode	0.33		0.02		0.12		0.53					
3	1,495		0.0049	0.0036	0.015	0.002	0.001	0.001	1.80×10^{-5}	1.20×10^{-5}	2.05×10^{-3}	1.35×10^{-3}	1.5	McDade *et al.* (2003)
3	1,566		0.0113	0.0057	0.028	0.009	0.0015	0.0007	3.80×10^{-5}	5.00×10^{-5}	4.49×10^{-3}	2.16×10^{-3}	2.1	Salters *et al.* (2002)

Estimates are given for two assemblages at different pressures (1.5 GPa and 3 GPa). These are "snap shots" which allow direct comparison of approaches from two different laboratories (Blundy and Wood, 2003b; Salters *et al.*, 2002), but should not be taken as representative values for the whole MORB melting regime. Nevertheless these figures provide a good indication of the magnitude of these partition coefficients and current differences of opinion.

the fractionation of uranium and thorium. At "low pressures" (<1 GPa) thorium partitions preferentially into clinopyroxene relative to uranium. Early partition coefficient determinations of uranium and thorium in clinopyroxene documented such $D_U/D_{Th} < 1$ (Benjamin *et al.*, 1980; LaTourrette and Burnett, 1992), and this was initially puzzling in the context of common ^{230}Th-excesses in MORB. Lundstrom *et al.* (1994) illustrated that large variations in the absolute values of partition coefficients of D_{Th}, could be related to compositional variations in the clinopyroxene. This observation rationalized the large variations in D_{Th} of the previous partitioning experiments as a result of different clinopyroxene compositions (e.g., LaTourrette and Burnett, 1992; Beattie, 1993a), but not the sense of fractionation.

A further important effect on clinopyroxene partitioning is pressure. Crystal chemical changes with increasing pressure (notably decreasing CaO and increasing Al_2O_3) result in a decrease in the M2 site size, which hosts thorium and uranium. Thus the clinopyroxene increasingly prefers the smaller U(IV) relative to thorium (Wood *et al.*, 1999). Thus, the sense of U–Th fractionation between clinopyroxene and melt changes with depth, such that between 1–1.5 GPa, D_U becomes greater than D_{Th} (Landwehr *et al.*, 2001; Wood *et al.*, 1999).

Garnet strongly incorporates uranium into its structure relative to thorium (LaTourrette *et al.*, 1993; Beattie, 1993b; Salters and Longhi, 1999). The magnitude of D_U and the ratio D_U/D_{Th} for garnet are larger than for clinopyroxene (Table 2) and so garnet dominates U–Th fractionation when present. Both the changing sense of U–Th fractionation in clinopyroxene and the eventual appearance of garnet result in "deep" melts having higher Th/U than the solid residue. Thus melting at depth is required to generate the correct sense of U–Th fractionation to account for ^{230}Th-excesses in MORB.

A more minor consideration is the role of orthopyroxene. At 1.5 GPa orthopyroxene has D_U an order of magnitude less than clinopyroxene, and the small lattice site strongly favors uranium over thorium (Blundy and Wood, 2003a; Wood *et al.*, 1999). The D_U/D_{Th} for orthopyroxene is consequently high (~2.5, Table 2) and may influence the bulk partition coefficient despite its low absolute partition coefficients, especially in highly depleted peridotites.

The general picture outlined above is that the bulk $D_U/D_{Th} < 1$ for shallow mantle melting (<1 GPa) and so should result in ^{230}Th-deficits (if the melt remains in equilibrium with the solid), whereas in the deeper mantle the converse is true. This overall picture seems secure. Some controversy persists, however, over the details of the depth at which this change in sense of U–Th fractionation occurs and its magnitude. This is largely attributable to differences in the clinopyroxene partitioning data determined from experiments in high-pressure laboratories in Bristol (Landwehr *et al.*, 2001; Wood *et al.*, 1999) and Lamont (Salters and Longhi, 1999; Salters *et al.*, 2002). Partition coefficients from the two different groups are reported in Table 2, using the two identical scenarios to facilitate direct comparison. Despite the ongoing debate, in a recent review of U-series partitioning, Blundy and Wood (2003b) point out some convergence between the parameterization of D_{Th} and D_U by Salters *et al.* (2002) and recent Bristol clinopyroxene partitioning data (McDade *et al.*, 2003, accepted). There remain some differences for the "snap-shot" assemblages presented in Table 2, notably the lower overall clinopyroxene partition coefficient in the spinel stability field for the Salters *et al.* (2002) model. Salters *et al.* (2002) suggest that the aluminum contents of the clinopyroxenes of the Bristol group are unrealistically high, and that this may account for the high absolute values of their D_{Th} and D_U. On the other hand the 1.5 GPa experiments of Salters *et al.* (2002), which form an important part of their parametrization, were conducted at temperatures (e.g., ~1,400 °C at 1.5 GPa) well in excess of most estimates of the mantle solidus (Hirschmann, 2000). Such elevated temperatures could account for anomalously low partition coefficients. It would be useful to resolve these differences, as the models presented below (Section 3.14.4.3) are quite sensitive to the values of partition coefficients used (e.g., see Jull *et al.*, 2002). It should also be stressed that the values cited in Table 2 are only illustrative values for specific fertile mantle compositions at the onset of melting. As melting continues, the compositions and modes of minerals change as the peridotite becomes less fertile. Thus, mapping out the variation in bulk distribution coefficients is a complex task (Salters *et al.*, 2002), and the values cited should not be treated as absolute values for the MORB melting regime.

Fractionation of the other nuclide pairs of interest ($^{226}Ra/^{230}Th$) and ($^{231}Pa/^{235}U$) rely in part on the partition coefficients of thorium and uranium discussed above. The partition coefficients of radium and protactinium have not been directly determined but can be estimated from applying a lattice strain theory to extrapolate data from homovalent cations (Blundy and Wood, 1994, 2003b). This approach suggests that radium and protactinium are indeed highly incompatible in the likely MORB source mineral assemblage, as is inferred empirically from the large degrees of disequilibrium. In both cases, the daughter nuclides are likely more than an order of magnitude more incompatible than their short-lived parent (Cooper *et al.*, 2001; Blundy and Wood, 2003a). In this case, the actual value of the distribution coefficient of the daughter nuclide is not critical and fractionation is then dominantly controlled by the value of the less incompatible parent, which are fortunately much better constrained. For once, Nature smiles on us.

3.14.4.1.1 Diffusion coefficients for U-series nuclides

Use of partition coefficients alone to model element fractionation assumes a melt process controlled by equilibrium partitioning. If melting is rapid relative to the diffusion of elements through mantle minerals, equilibrium may not be maintained (e.g., Qin, 1992; Van Orman *et al.*, 2002a) and this has been invoked to be a potential cause of U-series disequilibrium in MORB (e.g., Rubin and Macdougall, 1988; Volpe and Goldstein, 1993). It is, thus, also important to consider U-series element diffusivities. Seitz (1973) first presented data on uranium and thorium diffusion in clinopyroxene. Recently, the diffusivities of thorium and uranium in clinopyroxene have been redetermined and are so low ($\sim 3 \times 10^{-21}\ m^2\ s^{-1}$ at 1,200 °C) to suggest equilibrium is unlikely during MORB melting (Van Orman *et al.*, 1998). However, differences in diffusivity between thorium and uranium are small and in a sense that should produce minor ^{238}U rather than ^{230}Th excesses during melting (Van Orman *et al.*, 1998). Application of an elastic diffusion model (Van Orman *et al.*, 2001) for divalent cation diffusivities suggests that radium should diffuse several orders of magnitude more rapidly than thorium, and so this can potentially account for ^{226}Ra-excesses in a nonequilibrium model. Predicting the likely opposing effects of smaller ionic radius but larger ionic charge of Pa^{5+} relative to U^{4+} is difficult, but it seems improbable that procactinium diffusivities are sufficiently large to explain 100% ^{231}Pa-excesses by disequilibrium melting. Unfortunately, there are no directly measured data for uranium and thorium diffusivity in garnet, and so full characterization of the state of equilibrium during MORB melting remains to be determined.

3.14.4.2 Implications for Closed-system Models

The very small bulk partition coefficients for the U-series nuclides in the mantle ($D \ll 0.01$) puts a major constraint on the viability of simple closed-system models (Section 3.14.2.2). In these models, element fractionation is only efficient when the degree of melting F is comparable to the bulk partition coefficient. The degree of melting is not well known in many melting environments. In the MORB case, however, the thickness of the oceanic crust coupled with an extensive set of melting experiments constrains melting beneath normal ridges to be ~10% (e.g., Klein and Langmuir, 1987; McKenzie and Bickle, 1988; see Chapter 3.13). Such large amounts of melting would not be expected to produce any U-series disequilibria from closed-system models. Because Osmond (2003) assumed that MORB melting was governed by such a model, he was forced to conclude that there were highly refractory accessory minerals in the mantle that retained uranium and thorium with *D*s comparable to the degree of melting in MORB. This inference is entirely model dependent, however, and is easily relaxed if one considers *open-system* models (which are still based on fundamental statements of conservation of mass for radioactive tracers).

3.14.4.3 Open-system "Ingrowth Models"

The closed-system models are not sufficient to explain U-series excesses in MORB; however, they form a useful precursor to open-system models that allow the melt to move relative to the solid. Melt separation and extraction must occur at some scale for volcanism to be observed. Once melt can move, however, additional processes come into play that can produce significant radioactive disequilibrium, even at large degrees of melting. These models are often described as "ingrowth models" because they produce excess daughter products over that in the unmelted source by ingrowth from the parent. Spiegelman and Elliott (1993) provide a discussion of many of these models and stress that the key to understanding their behavior is that radioactive disequilibrium will arise from any process that produces different *residence times* for parent and daughter nuclides in the melting zone.

As an extreme example, consider an open melting system where the parent is completely compatible in the solid and is never extracted while the daughter partitions completely into the melt. The residence time of the parent in the system would be the time it takes the solid to cross the melting regime. The residence time of the daughter, however, would be the melt extraction time (i.e., once it is produce it leaves the system in the melt). During the extra time the solid spends in the system; however, it is continually decaying and producing *extra* daughter product, which is observed in the melt. If the melt is continually allowed to separate from the system, it will contain only daughter product and have an infinite excess. This process can be kept up indefinitely as long as the radioactive production can keep pace with extraction. In this case, the excesses are controlled by the *rate* of melting and extraction and less by the total amount of melting *F*. At the other extreme, if the unmelted solid is in secular equilibrium at the initiation of melting and all elements spend the same account of time in the system, then they will remain in secular equilibrium because they continue to all decay at the same rate.

The important (and confusing) feature of radioactive decay chains is that these systems do not have to conserve mass for any *individual* nuclide because it can always be produced by the parent or decay depending on the relative concentration of parent and daughter nuclides. For example, there is always *less* of the principal parent (e.g., ^{238}U) leaving the system than entering it, simply because of the radioactive decay of the parent over the timescale of melt extraction. For a long-lived nuclide like ^{238}U the resulting differences in concentration are negligible. However, for short-lived nuclides like ^{226}Ra, the actual ^{226}Ra atoms that leave the system may not even be the same ones that entered. Any initial ^{226}Ra in the melt can decay and be replaced by later production or "ingrowth" from longer-lived ^{230}Th. For further discussion, see Spiegelman and Elliott (1993) or Elliott (1997).

The following sections consider a host of ingrowth models and discuss how each one produces radioactive disequilibria. We first discuss the different mechanisms inherent in the models and then compare their behavior to each other and to the observations.

3.14.4.3.1 "Dynamic melting models"

All the models in this section are "solid-centric" in the sense that they track the behavior of a small parcel of solid undergoing melting. In the equilibrium dynamic melting models, this solid is assumed to be in equilibrium with a small fraction of melt in contact with the solid. The diffusion limited models simply relax the assumption of complete chemical equilibrium between solid and retained melt. The local system of solid plus retained melts acts much the same as the closed-system models described in Section 3.14.2.2.2; however, the excesses are controlled by the retained melt fraction ϕ_0 (usually assumed to be the porosity) relative to the partition coefficients (ϕ_0/D). This differs from the closed-system models where the porosity is assumed to be equal to the degree of melting *F* and the excesses are controlled by *F*/*D*. The important feature of all open-system models is that the porosity must be significantly smaller than the degree of melting if the melt is to separate from the solid (see below).

In the dynamic melting models, all melt fractions produced in excess of ϕ_0 are assumed to be mixed and extracted, although the dynamics of these processes are not included in the models. For nonradiogenic elements, these models are identical to commonly used dynamic melting models for trace elements (e.g., Langmuir *et al.*, 1977; Johnson and Dick, 1992; Sobolev and Shimizu, 1993; Slater *et al.*, 2001; see Chapter 3.13). The principal differences between all of the dynamic ingrowth models are the degree of equilibration maintained between solid and retained melt and how the extracted melts are integrated.

Equilibrium models. Equilibrium dynamic-melting U-series models were first introduced by McKenzie (1985) and Williams and Gill (1989). For stable elements, these models are all identical to "near-fractional" melting models (e.g., Johnson and Dick, 1992). For U-series elements however, they can produce significant radioactive disequilibrium by ingrowth if the *melting rate* is slow with respect to the decay rate of the short-lived daughter. As long as there is sufficient time for new daughter to be produced and concentrated in the equilibrium porosity before it is extracted, then reasonable excesses can be produced throughout the melting process. In the limit of extremely slow melting, the maximum excess in the retained melt (and, therefore, the accumulated melt if there is no time for decay) is

$$\left(\frac{^{230}\mathrm{Th}}{^{238}\mathrm{U}}\right)_{\mathrm{f}} = \frac{(1-\phi_0)D_{\mathrm{U}} + \phi_0}{(1-\phi_0)D_{\mathrm{Th}} + \phi_0} \quad (8)$$

Therefore, these models put constraints on the retained porosity ϕ_0 in the model *relative* to the partition coefficients. If melting is much faster than ingrowth, then these models produce accumulated near-fractional melts that reflect the source concentrations (and therefore will be close to secular equilibrium).

Williams and Gill (1989) actually present several equilibrium models. Their "dynamic melting" model is the same as McKenzie's but they also introduce an "accumulated continuous melting" (ACM) model. The only difference between these two models is how the incremental or "continuous" melts are pooled and extracted. The accumulated continuous melting model is somewhat easier to understand. It follows a piece of solid through the melting process and simply accumulates the extracted melt in a chemically isolated reservoir. Any excesses in this reservoir are unsupported and decay toward secular equilibrium. Thus, there is a trade-off between melting rate and excesses in this model. If the melting rate is too slow, the time required to melt to degree F is long compared to the half-life of the daughter nuclide and the accumulated melt will have negligible excesses. Too rapid melting also leads to secular equilibrium because both parent and daughter are stripped from the solid by chemical fractionation before any ingrowth can occur. It is reasonably difficult to produce significant excesses (particularly radium) using this model.

In contrast, the dynamic melting model assumes a continuous "melting column" appropriate for adiabatic decompression melting. This column is composed of a series of solids all in different stages of melting from $F = 0$ at the bottom to $F = F_{\max}$ at the top, with a melting rate proportional to the solid upwelling rate. In this model, the incremental melts from all depths in the melting column are mixed and assumed to be extracted instantaneously. This model produces significant radioactive disequilibrium because none of the excesses in the accumulated melt have time to decay. As emphasized in Spiegelman and Elliott (1993), most of the excesses in this model are produced near the bottom of the melting column because these melts are most enriched and because phases capable of fractionating thorium from uranium are more prevalent at higher pressures. This model can produce U-series excesses comparable in magnitude to those that are observed in MORB if the retained porosity near the bottom of the melting regime is small compared to the partition coefficients. Because this model assumes instantaneous extraction, all excesses produced by the calculations should be treated as upper bounds.

The dynamic melting model was extended to two-dimensional systems by Richardson and McKenzie (1994). In their model, each solid flow line is treated as a single dynamic melting column and the melts from the entire region are mixed and extracted instantaneously. This model tends to produce larger excesses than the single column model because it is dominated by smaller degree, slower melting regions from off-axis. How these melts actually migrate to the ridge axis and whether they do so without losing their radioactive disequilibrium is unclear because melt transport is not included in these models. An interesting feature of their simplified ridge model, however, is that the excesses produced are nearly *independent* of spreading rate. The argument is best developed in Iwamori (1994), who assumes a passive solid flow field beneath the ridge driven by a thickening wedge-shaped lithosphere that is thicker for slow-spreading ridges than fast-spreading ridges (e.g., see Spiegelman and McKenzie, 1987). Thicker lithosphere drives a faster upwelling which offsets the reduction in spreading rate such that the actual upwelling rate (and, therefore, melting rate) is reasonably constant at all spreading rates. They offer this as an explanation for the relatively weak (or negligible) correlation between ^{230}Th-excesses and spreading rate (Figure 9).

More recently, Zou and Zindler (2000) provide a tractable analytic solution to both McKenzie's and Williams and Gill's equations for the limited problem of constant partition coefficients and porosities. By making all adjustable parameters constant, these models exploit the analytic solutions afforded to linear systems of ordinary differential equations like Equations (1)–(3) (a similar approximation for the transport models described in Section 3.14.4.3.2 can be found in the appendix of Spiegelman and Elliott (1993)). However, because the solutions of these models are sufficiently sensitive to parameters such as D/ϕ_0 it can be challenging to choose specific constants that are actually appropriate to a problem where the properties of the system are expected to change in space (see Section 3.14.4.1).

Disequilibrium/Diffusion limited models. The principal parameters controlling solutions to the equilibrium "dynamic melting models" are the melting rate with respect to the half-life of the daughter product ($\Gamma/[\rho_s t_{1/2}]$) and the retained porosity with respect to the partition coefficient (D/ϕ_0). Given the very small bulk partition coefficients expected for the U-series nuclides ($D \ll 0.01$), these models place significant constraints on the retained porosity ϕ_0. As pointed out by Spiegelman and Elliott (1993), ϕ_0 is not necessarily the physical melt fraction present, rather it is the fraction of melt that remains in equilibrium with the solid. Nevertheless, these models require that this number is very small, i.e., comparable to the partition coefficients of the U-series nuclides. Using a similar model for stable elements to estimate ϕ_0 from the depletion of abyssal peridotites give $\phi_0 \sim 1\%$ (Johnson *et al.*, 1990; Johnson and Dick, 1992) (which would produce no U-series excesses). Similarly, Slater *et al.* (2001) require $\phi_0 \sim 3$–4% to model the variability of melt-inclusions in Iceland. The inconsistencies between inferred porosities given different chemical systems suggests either different models are required or perhaps that strict equilibrium partitioning needs to be relaxed for the U-series.

One possibility that would lead to larger inferred porosities for the U-series was introduced by Qin (1992, 1993), who proposed that the retained melt was only in complete equilibrium with the *surface* of minerals and that solid-state diffusion limited the re-equilibration of the retained melt with the solid. In other respects, this model is identical to the ACM model of Williams and Gill (1989). Qin introduced a specific microscopic melting/diffusion model for spherical grains and coupled it to the larger-scale dynamic melting models. The net affect of this addition was to drive the effective partition coefficient of all elements toward 1 so melting produced less fractionation between elements. Again there is a trade-off between melting rate and diffusion rates. If the melting rate is rapid with respect to diffusion, the problem reduces to disequilibrium melting of the particular phase and the melts will simply pick up the activity ratio of the solid phase (e.g., as in Section 3.14.2.2.2). Slow melting or rapid diffusion returns the equilibrium dynamic melting models. However, beyond changing the effective partition coefficient and potentially allowing larger porosities, the disequilibrium models have the same mixing and transport assumptions as the equilibrium models and, therefore, similar inferences and caveats.

3.14.4.3.2 Single porosity transport models

The physics of melt extraction and mixing are not included explicitly in the dynamic melting models. However, melts must move faster than the upwelling solid, otherwise melt separation would not occur, and the system would reduce to pure batch melting to degree of melting F_{max} (which produces no excesses at large F). By explicitly including the production and transport of melt, *transport models* of U-series excesses add an additional velocity field that can affect the overall residence times of parent and daughter nuclides as well as affect the mixing of melts. These models also produce excesses for large degrees of melting and place constraints on the rates of melting and extraction processes. However, the inferences drawn from these models can be quite different from the dynamic melting models.

The simplest transport model was introduced by Spiegelman and Elliott (1993) and consists of a one-dimensional, steady-state melting column where it is assumed that the melt and solid remain in chemical equilibrium at all times. These models are readily extended to spatially variable melting functions $F(P)$ and partition coefficients $D(P)$ (Lundstrom *et al.*, 1995) and a more user friendly web-based version of these models is available at http://www.ldeo.columbia.edu/~mspieg/UserCalc/. The web-site and U-series calculator is described in detail by Spiegelman (2000). Figure 12 shows some output from UserCalc for a one-dimensional upwelling column and the bulk partition coefficients given in Table 2.

The melt velocity in these models at any height z in the column is of order

$$w(z) \approx W_0 \frac{F(z)}{\phi(z)} \tag{9}$$

where W_0 is the solid upwelling velocity, $F(z)$ is the degree of melting at height z, and $\phi(z)$ is

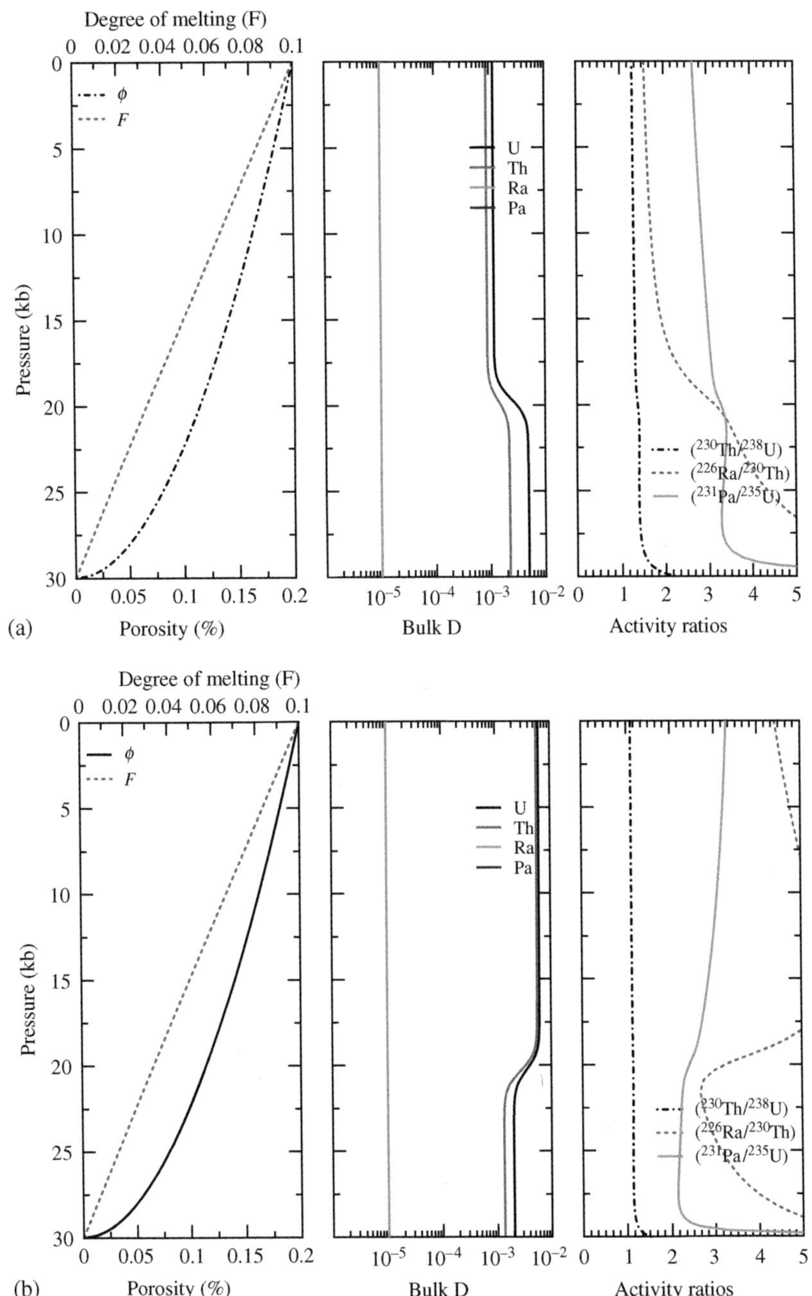

Figure 12 Example output of UserCalc (Spiegelman, 2000) using bulk partition coefficients from Table 2. Physical parameters for these calculations are a melting column of height 90 km, an upwelling rate $W_0 = 7$ cm yr^{-1}, a maximum degree of melting $F_{max} = 0.1$ and a maximum porosity $\phi_{max} = 0.002$. These parameters give a maximum melt velocity that is 50 times faster than the upwelling velocity. This permeability is consistent with estimates from Von Bargen and Waff (1986). Both calculations assume a two-level partition coefficient scheme where the "Garnet stability field" extends from 3–2 GPa and spinel is stable for pressures shallower than 2 GPa. (a) Calculations using Bulk Ds from Table 2 after Salters *et al.* (2002). $D_{Ra} = D_{Pa} = 10^{-5}$. (b) Bulk Ds from Table 2 after McDade *et al.* (2003) (garnet) and Wood *et al.* (1999) (spinel). The first panel is the same in both cases and just shows degree of melting $F(z)$ and porosity $\phi(z)$ as a function of pressure. The second panel shows the bulk-partition coefficient structure with pressure and the final panel shows activity ratios for ($^{230}Th/^{238}U$), ($^{226}Ra/^{230}Th$), and ($^{231}Pa/^{235}U$). The activity ratios in (a) are comparable to those observed for 9–10° N EPR although the ^{226}Ra-excess is too small. The activity ratios in (b) are low for ($^{230}Th/^{238}U$) but quite large for Ra and Pa. This results from the large D_U and D_{Th} in the upper region. Note the large *increase* of Pa due to *ingrowth* in the upper layer of (b). These plots illustrate the effect of partition coefficients on modeling U-series but assume a rather simple D structure. More realistic models should include more variable D with depth that accounts for pressure and changing mineral modes.

the physical porosity at that same point. For example, if the melt velocity is 50 times faster than the upwelling solid velocity at the top of the column ($F_{max} = 0.1$), then it requires that the porosity in this region to be 50 times smaller than the degree of melting or $\phi_{max} = 0.002$. When all the melt extraction is through this single porosity, conservation of mass simply requires that the faster the melt moves, the less melt can be retained. While a porosity of 0.2% seems small, it is consistent with estimates of permeability by Von Bargen and Waff (1986) that give the appropriate velocities at this porosity for a grain size of ~3 mm (see Spiegelman and Kelemen, 2003, for details). If the porosities are also small with respect to the partition coefficients of the U-series nuclides, these models can also produce significant U-series disequilibrium. Figure 12 shows calculated porosity structure and activity ratios for a typical problem.

For stable elements or nuclides with half-lives much longer than the melt extraction times (such as ^{238}U and ^{232}Th), the melts that leave the top of this column have concentrations equivalent to that for a *batch* melt with degree of melting F_{max} independent of porosity (see Ribe, 1985; Spiegelman and Elliott, 1993). Therefore, these models cannot fractionate highly incompatible elements or produce dispersion on an equiline plot. They can, however, produce large U-series excesses if the parent spends longer in the system than the daughter. As discussed at length in Spiegelman and Elliott (1993), the residence times are controlled by the *effective velocity* of parent and daughter elements which is approximately

$$w^i_{eff} = \frac{w + WD'}{1 + D'} \tag{10}$$

where

$$D' = \frac{\rho_s(1-\phi)D_i}{\rho_f \phi} \approx \frac{D_i}{\phi} \tag{11}$$

is the effective partition coefficient and depends (again), principally on D/ϕ (ρ_s and ρ_f are the densities of solid and melt, respectively). Elements with partition coefficients much smaller than the porosity travel near the melt velocity, while those with larger partition coefficients travel more slowly. As discussed in Spiegelman and Elliott (1993), these models have two regimes, one where the excesses depend on the differences in effective velocities, the other which depends on the *ratio* of effective velocities. While the details of these models are somewhat complicated (see Figure 13), their essence is straightforward. Any region where the parent nuclide moves more slowly than the daughter will cause ingrowth and excess activities of the daughter.

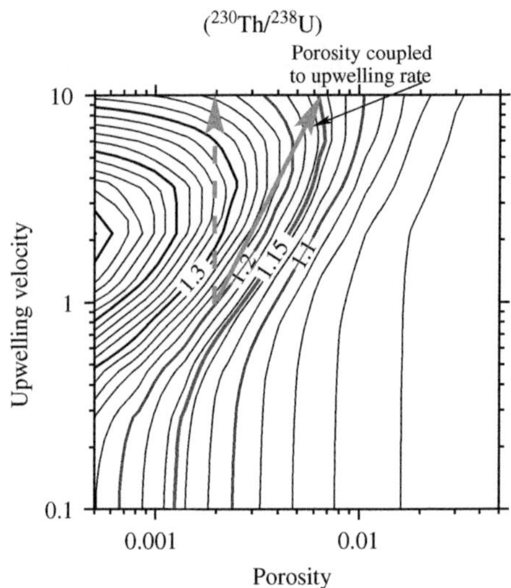

Figure 13 Possible variability of ($^{230}Th/^{238}U$) as a function of spreading rate. This figure shows contours of ($^{230}Th/^{238}U$) as a function of upwelling rate and maximum porosity (i.e., melt velocity) from UserCalc. These contours are the activity of the top of the melting column for the model shown in Figure 12(a) (see Spiegelman and Elliott (1993), for a detailed description of these plots). If the porosity were independent of upwelling velocity (dashed arrow), this specific calculation model would predict increasing then decreasing ($^{230}Th/^{238}U$) with increasing spreading rate. However, for a fixed permeability structure, faster upwelling implies larger porosities. The solid arrow shows a possible set of consistent upwelling rates and porosities for a series of runs for $1 \leq W_0 \leq 10$ cm yr^{-1}. Along this trajectory, variations in ($^{230}Th/^{238}U$) are small.

One of the interesting features of these models is that observable excesses for different parent/daughter pairs can be produced in different parts of the melting region. For example ^{230}Th has a sufficiently long half-life such that excesses produced at the bottom of the column in the garnet stability field can be preserved at the surface. Unless melt transport is extremely fast, however, excess ^{226}Ra produced at the bottom will decay away before eruption. Nevertheless, significant ^{226}Ra can be produced near the top of the melting column if the thorium and radium velocities are sufficiently different (e.g., Figure 12(b)). Thus thorium excesses and radium excesses can be decoupled. If ^{226}Ra is observable at the surface, it is really providing information about the ^{230}Th that produced it within the last ~8 kyr. In the dynamic melting models, that ^{230}Th is at the bottom of the melting zone and requires rapid extraction to preserve it. In the equilibrium transport models, the parent ^{230}Th is near the surface. However, for these models to produce large ^{226}Ra-excesses requires that the porosity near the surface is comparable to D_{Th}.

Disequilibrium single porosity models. In addition to complete equilibrium transport, several other variations to the basic model have been proposed. The first relaxes the assumption that moving melt remains in chemical equilibrium with the solid at all times (Spiegelman and Elliott, 1993), although instantaneous melts are assumed to be in chemical equilibrium with the mantle that produced them. For stable elements, this *disequilibrium transport* produces a residue that reflects perfect fractional melting and the melts have compositions identical to accumulated fractional melts. These models are similar to the dynamic melting models in the limit that $\phi_0 \rightarrow 0$ but include the time it takes the melt to leave the system. These models produce most of the disequilibrium at the bottom of the melting column and produce negligible U-series excesses due to decay during transport.

It is also straightforward to extend the equations to allow for only partial equilibration during transport. Iwamori (1993a) presents a one-dimensional steady-state single-porosity model for stable elements that includes diffusive re-equilibration between melt and solid. He does not extend it to radioactive nuclides in this paper but includes this effect in his "two porosity model" (Iwamori, 1994) (see Section 3.14.4.3.4). The expected effects of chemical disequilibrium should be similar to those in the Qin (1992) dynamic melting model, namely he effective bulk partition coefficients of all elements will be driven towards unity.

Linear kinetic approximations to melt/solid diffusion are also possible, but all of these models should simply produce results that are intermediate between full equilibrium and disequilibrium transport. In general, all of the transport and "dynamic" melting models produce a single melt composition at the top of the melting column for stable elements that has highly incompatible trace element ratios that are indistinguishable from the source.

3.14.4.3.3 Comparison with observations

At this point, it is worth comparing the behavior of both the dynamic melting models and the transport models to each other and to the key observations listed in Section 3.14.3.6 to see how well these models explain the observations and to suggest that more complicated models may still be needed. Both dynamic melting and transport models can produce the magnitude of observed excesses for the large degrees of melting expected at mid-ocean ridges. However, these excesses are produced in different parts of the melting regime and have different implications for the rate of melt transport. The dynamic melting models produce most of their excesses near the bottom of the melting column and require rapid melt extraction (on timescales short with respect to the half-life of ^{226}Ra) to preserve them at the surface. The equilibrium transport models however, produce their excesses throughout the column and in particular can produce the ^{226}Ra-excesses near the surface thus reducing some of the requirement for fast melt transport. Nevertheless, these models still require that the porosity near the top of the column is small with respect to D_{Th} and this indirectly puts constraints on the melt velocity.

To produce significant ^{226}Ra-excesses by transport, requires that the porosity ϕ is comparable to D_{Th}. In the single-porosity model, however, the entire melt flux is required to pass through this porosity which implies a large melt velocity via Equation (9) (which is strictly valid only for one-dimensional, steady-state melting columns with a single $\phi(z)$). For example, if the degree of melting $F_{max} = 10\%$ and $D_{Th} \sim 0.001$ near the top of the melting column, then the melt velocity will be ~ 100 times larger than the upwelling velocity or $1-10$ m yr^{-1} to produce significant ^{226}Ra-excesses. These rates are not excessive, however, and can easily be attained by porous flow using the parametrization of Von Bargen and Waff (1986). Spiegelman and Kelemen (2003) provide an example and details.

The melt velocity estimated from transport models is a bit slower but still comparable to estimates from the dynamic melting models. For example, if we assume that ^{226}Ra-excesses are produced at the bottom of a column 90 km deep and need to move to the surface in ~ 3 half-lives, then $w_0 \sim 20$ m yr^{-1}. It should be stressed that this is a constraint on the *average* melt velocity across the entire melting column rather than a constraint on the maximum velocity near the surface. Moreover, the constraint from Equation (9) assumes that there is only a single porosity near the surface. Two-porosity models (next section) relax this constraint somewhat.

Figure 12 gives a sense of the magnitude of ^{226}Ra-excesses possible for a single column with $F_{max}/\phi_{max} \sim 50$ ($\phi_{max} = 0.002$) but with different partition coefficients. In Figure 12(a), $D_{Th} = 8.9 \times 10^{-4} < \phi_{max}$ and produces small ^{226}Ra-excesses (($^{226}Ra/^{230}Th$) ~ 1.6). Figure 12(b), however, has $D_{Th} = 5.6 \times 10^{-3} > \phi_{max}$, which produces a significantly lager excess (($^{226}Ra/^{230}Th$) ~ 4.3).

For simply producing the magnitude of observed excesses, both dynamic melting models and transport models are viable. The largest uncertainty in either type of model is the appropriate partition coefficients (and D as a function of pressure) as these control the inferences on the porosity through the parameter D/ϕ. The more stringent tests on these models, however, come from the observed correlations between nuclides and excesses. In these tests, both sets of models have difficulty which has driven a need for more complex models.

We begin by discussing the behavior of these models on an equiline diagram and compare it to the observation that individual ridges appear to produce correlated arrays on these diagrams, even for an isotopically similar source. In their usual forms, both dynamic melting models and transport models simply cannot fractionate highly incompatible elements for degrees of melting $F > D_{\mathrm{U}}$, D_{Th} (as is expected at ridges) if the melts that are "erupted" from the model are assumed to be the melts from the top of the column. For dynamic melting models, this is because usually all of the incremental melts are assumed to be efficiently mixed. Thus, for stable elements, the melt composition is similar to an accumulated fractional melt which has ratios of highly incompatible elements such as [U/Th] that are close to source. Similarly, the equilibrium transport models produce melts with compositions identical to batch melts (which are hardly distinguishable from accumulated fractional melts for elements with $D \ll F$). Thus both models produce final erupted melts with (^{238}U/^{232}Th) very close to the source ratios, which implies only vertical motion on an equiline diagram. That is, if the unmelted source lies on the equiline at a given (^{238}U/^{232}Th), then the final melt moves to a higher (^{230}Th/^{232}Th) but at roughly the same (^{238}U/^{232}Th). Simple ingrowth models, with complete mixing, cannot produce an array on an equiline diagram without invoking source variability which was done Lundstrom *et al.* (1998a,b, 2000) to try to explain the local arrays. Nevertheless, one of the key observations for at least 9° N EPR and the Reykjanes Ridge is that individual arrays can show significant variations in (^{238}U/^{232}Th) for uniform ^{87}Sr/^{86}Sr suggesting that this variation cannot be entirely attributed to source.

It is possible for the basic ingrowth models to produce arrays, however, if one allows incomplete mixing over the full range of melt compositions. For example, for dynamic melting models, if the incremental melts are not assumed to mix, they form a very large array of melt compositions from very enriched small-degree melts to extremely depleted fractional melts. For the very small porosities required to produce the U-series excesses, however, the range in concentrations and the stable trace element patterns produced by these models are quite different from what is observed (e.g., see Slater *et al.*, 2001). For transport models, if the full trajectory of melts from the bottom of the melting column to the top are plotted on an equiline, this will produce an array of melt compositions (e.g., see figure 4 in Spiegelman and Elliott, 1993). However, since these melts are automatically mixtures of all melts produced up to that height, they very rapidly attain (^{238}U/^{232}Th) ratios comparable to source for any $F \gg D_{\mathrm{U}}$, D_{Th}. Moreover, the *most* depleted melts are the mixed melts at the top of the column. Thus, this model can never produce a melt more depleted than a batch melt with degree of melting F_{max}. The instantaneous "continuous" melts from the dynamic melting model, however, can become extremely depleted.

Another problem with the transport models is that they assume complete equilibrium between melt and solid throughout the melting regime. This implies that the solid residues at the top of the column (e.g., near the Moho), should be in chemical equilibrium with MORB. However, another key observation of MORBs is that they are out of equilibrium with abyssal peridotities near the Moho for both major and trace elements. A quick fix for the transport models is to assume that melts remain in chemical equilibrium up to some depth and then melt fractionally for the remaining distance (e.g., see Kelemen *et al.*, 1997). However, different attempts to explain both U-series and stable elements in melts and residues in the single-porosity transport models has motivated much of the development of the two-porosity models in next section.

The basic models also have difficulties accounting for the observed correlations (or anticorrelations) among radioactive excesses. In the case of the observed anticorrelation of (^{226}Ra/^{230}Th) versus (^{230}Th/^{238}U) at intermediate to fast-spreading ridges (Figure 11), the dynamic melting models have considerable difficulty because all U-series excesses are produce in the same narrow region at the bottom of the melting regime. This does not necessarily rule the model out because variations in melting rate can cause changes in ^{230}Th without affecting ^{226}Ra if the melting rate is slow with respect to radium decay, but fast with respect to thorium (e.g., Spiegelman and Elliott, 1993; Figure 7(a)). However, it is still difficult to produce an anticorrelation using these models. Moreover, for these variations to be observed in closely spaced samples would suggest a melting regime where the solid velocity varies dramatically over small length scales. So far, no solid flow model has been developed with this property.

As for the transport models, because most ^{230}Th-excesses are produced at the bottom of the column while ^{226}Ra is produced near the top, it is possible to decouple them. Nevertheless, for many systems this model still produces positive correlations. For example, all excesses will tend to increase with decreasing porosity. However, it is possible to produce large ^{226}Ra-excesses with low (^{230}Th/^{238}U) if a long time is spent in the spinel field, while low (^{226}Ra/^{230}Th) and high (^{230}Th/^{238}U) implies more rapid extraction from depth. The problem is how to produce the anticorrelation between the two in closely associated samples in a systematic way with these models (Figure 11).

Similar arguments can be made for the apparent lack of correlation of (^{231}Pa/^{235}U) with (^{230}Th/^{238}U) if the large ^{230}Th-excesses are coming from depth and the ^{231}Pa-excesses are being ingrown from shallower parts of the melting column. Nevertheless, one might expect a positive correlation between these two systems from the models because parameters that increase ^{230}Th ingrowth by retaining ^{238}U should also increase ^{231}Pa ingrowth by retaining ^{235}U. More data probably need to be collected to explore the possible positive correlation between (^{231}Pa/^{235}U) and (^{230}Th/^{238}U) hinted at in some local assemblages.

Perhaps one of the more surprising observations from U-series studies is the weakly negative, to possibly nonexistent correlation between spreading rate and U-series excesses (Figure 9). In the simplest dynamic melting models one might expect a direct correlation because all excesses are produced in a narrow region and are directly influenced by melting rate which is assumed to scale with spreading rate. In these models, independent of source, slower melting produces larger excesses which is only hinted at in the date. For transport models, the problem is slightly more complicated because upwelling rate affects both the melting rate and porosity (melt velocity). Figure 13 shows contours of (^{230}Th/^{238}U) at the top of the model in Figure 12(a) as a function of porosity and upwelling rate as calculated by UserCalc (see Spiegelman and Elliott, 1993, for a detailed discussion of these contour plots). If porosity is independent of upwelling rate, then one might expect to see variations in (^{230}Th/^{238}U) with upwelling rate but they need not be monotonic. The dashed arrow in Figure 13 shows a trajectory through porosity-upwelling velocity space for constant porosity ($\phi_0 = 0.002$) and $1 \leq W_0 \leq 10$ cm yr^{-1}. For this particular run, this trajectory first shows an increase in (^{230}Th/^{238}U) that is a consequence of spending less time in the shallow spinel field followed by a decrease in (^{230}Th/^{238}U) from the melting rate exceeding the ingrowth rate. However, if the functional form of the permeability is independent of upwelling rate, then faster upwelling (faster melting) systems should produce larger porosities. A scaling argument for the maximum porosity as a function of upwelling rate is given in Spiegelman (2000), which for high-permeability systems is approximately

$$\phi_{\max} \propto W_0^{1/n} \tag{12}$$

when n is the exponent in the permeability/porosity relationship which is approximated in this model by $k_\phi = k_0\phi^n$. Such a coupled set of velocities and porosities is shown in Figure 13 by the solid arrow. For this calculation, at least, the contours of activity are subparallel to the arrow suggesting that the overall variation of ^{230}Th-excesses would show little change over an order of magnitude spreading rate variation which is consistent with the data. More specifically, this particular trajectory suggests that ^{230}Th-excesses should be reasonably constant (or slightly increasing) for slow to intermediate upwelling rates and only show a noticeable decrease for faster upwelling. This qualitatively agrees with the variation of ^{230}Th-excesses shown in Figure 9 but it could be entirely coincidental.

The final test of the models is against axial depth (Figure 10). This issue is discussed in detail by Bourdon *et al.* (1996a) using a smaller data set from the Azores platform (FAZAR cruise). Again, in their simplest form, the dynamic melting models would suggest that ^{230}Th-excesses should be nearly independent of axial depth. If we assume that axial depth reflects the total amount of melt production and, therefore, the depth of the melting column, then the simplest dynamic melting model produces all excesses in a narrow region near the bottom. If the partition coefficients do not change significantly within such a small depth range, then all dynamic models with instantaneous transport should produce roughly the same excesses for all columns that start in the garnet field. The only systematic change would be for melts that begin melting in the spinel field. This argument needs to be modified somewhat by the recent results on the pressure dependence of partition coefficients for the U-series (e.g., Wood *et al.*, 1999). If in general, D_U/D_{Th} is an increasing function of pressure, then it is possible that dynamic melting models would show a weak positive correlation with melt production (or negative correlation with axial depth). If the actual transit time is factored in, however, then it is possible that these models would produce *smaller* ^{230}Th-excesses with larger crustal thickness because the deeper produced melts have further to travel and will decay during transport. Equilibrium transport models, however, should show a negative correlation of ^{230}Th-excess with axial depth (assuming that axial depth is a proxy for depth of melting) because deeper melting columns imply a longer time spent in the garnet stability field and higher degrees of melting tend to produce larger porosities and, therefore, faster transport through the spinel field. Bourdon and Langmuir (1996) show this calculation using the equilibrium transport model of Spiegelman and Elliott (1993) for a series of melting columns with constant upwelling rate but increased depth of melting. The calculation can become more complex, however, if upwelling rate is somehow coupled to depth of melting.

In summary, both the dynamic melting model and the transport models can usefully explain some subset of the observations but remain problematic for explaining all of them. To move forward, many authors have developed hybrid models that combine aspects of both dynamic and transport models by invoking some form of channelized melt flow. These "two-porosity models" vary in degree of complexity but have promise for reconciling many of the observations. We now discuss them in detail.

3.14.4.3.4 Two-porosity models

A fundamental observation of melt transport at mid-ocean ridges is that in both major and trace elements, MORBs are out of chemical equilibrium with the shallow upper mantle (e.g., see Kelemen *et al.*, 1997, for review). One mechanism that has been suggested is that melt migration is localized into some form of channel network fed by porous flow at the grain scale. This basic idea has been incorporated into several different "two-porosity models" to try to reconcile U-series observations with those of stable trace elements.

The first model combining porous and "channel flow" was developed by Iwamori (1993a,b) for stable elements and then extended to U-series elements (Iwamori, 1994). This model combines aspects of the full transport models with the instantaneous melt extraction of the dynamic melting models. In some sense, this model is a more complex version of the dynamic melting models where the porous flow regime produces excesses that are extracted into chemically isolated channels, mixed completely and extracted *instantaneously*. That is, like the dynamic melting models, the actual time spent in the channels is not included in the calculations of excesses. However, the porous ingrowth regime is a full transport model with diffusion controlled re-equilibration, rather than a simple constant-porosity regime that moves with the solid. The relative amount of porous and channel flow is controlled by a "suction parameter," S, such that $S = 0$ is essentially the equilibrium transport model of Spiegelman and Elliott (1993) and $S = 1$ is the purely fractional dynamic melting model of McKenzie (1985) in the limit $\phi_0 \rightarrow 0$. This model is a continuous hybrid of all the previous models. Its principal weakness is that it still assumes instantaneous extraction and mixing in the channels and thus the channel melts provide an upper bound on the excesses.

Lundstrom (2000) remedies this problem by assuming the channels are chemically inert dunites undergoing porous flow at a different porosity as suggested by Kelemen *et al.* (1995a) (see also Kelemen *et al.*, 1995b, 1997; Kelemen and Dick, 1995; Aharonov *et al.*, 1995). The dunites are assumed to occupy a fraction χ of the whole system and the flux into the channels is controlled by the same suction parameter as in Iwamori (1994). This model was principally developed to try to reconcile the important observation that abyssal peridotites are extremely depleted, implying near-fractional melting (Johnson *et al.*, 1990; Johnson and Dick, 1992; Hellebrand *et al.*, 2002), while U-series excesses, particularly short-lived excesses of ^{226}Ra and ^{231}Pa, are more easily made with equilibrium transport models. Lundstrom only actually considered the behavior of ^{226}Ra and ^{231}Pa in these models because he assumed that U/Th variations and ^{230}Th variations reflect fundamental source differences (Lundstrom *et al.*, 1995). Moreover, all the melts produced in this model are mixed at the top of the column according to flux and thus this model only produces a single-melt composition. Using this model, he found that ^{231}Pa produced the strongest constraints on the suction parameter and required near equilibrium flow ($S \sim 0.05$). However, the depletion of rare-earth elements (REEs) in abyssal peridotites required near fractional melting with $S \sim 0.5$–0.8 depending on how depleted the solid source is assumed to be. In addition, because this model produces a single-melt composition at the top and was not used to explore ^{230}Th variations, this version of the model was not used to address the local correlations on equiline diagrams.

Nevertheless, it is possible to use this model to explore the behavior on an equiline diagram. Jull *et al.* (2002) developed a very similar model to that of Lundstrom (2000) and used it to try to model the observed correlations of (^{230}Th/^{232}Th) with (^{238}U/^{232}Th) on equiline diagrams and the anticorrelation between (^{226}Ra/^{230}Th) and (^{230}Th/^{238}U) both seen at 9° N on the EPR.

The principal difference between the Jull *et al.* (2002). and the Lundstrom (2002) approach is that Jull *et al.* (2002). use a more realistic partition coefficient structure that allows for garnet melting at the base of the model to produce significant ^{230}Th-excesses. More importantly, they *do not mix* the channel melts and the interchannel melts at the top of the model. Instead they assume that these melts form mixing end-members with the channel melts preserving high ^{230}Th-excesses from the base of the melting zones (but having negligible (^{226}Ra/^{230}Th)), while the interchannel melts are high in ^{226}Ra produced near the surface by porous flow at low porosities but have negligible ^{230}Th-excesses. Jull *et al.* (2002) also allowed the depth where channels begin to be a variable and systematically explored a reasonably large region of parameter space in both physical parameters and partition coefficients. For a narrow parameter regime, Jull *et al.* (2002) were able to produce both (^{230}Th/^{238}U) and ^{226}Ra variations that were comparable to observations for melts at 9° N EPR

(Sims *et al.*, 2002) that had the same long-lived isotopic signature. For this regime, the channels were required to extend into the garnet stability field at depth and the calculation still required ~60% of the melt to flow through the interchannel regions. Thus, the solution is still principally an equilibrium porous flow model with some accelerated channeling. Jull *et al.* (2002) also attempt to explain the very depleted REE signatures of abyssal peridotites but have the same problem as Lundstrom (2002) in that the parameters required to match the U-series excesses in the model do not produce peridotites that are as depleted as observed. (However, as a small source of confusion, Jull *et al.* (2002) use the opposite convention for the suction parameter S. In their paper, $S = 1$ is pure equilibrium flow and $S = 0$ is fractional melting.)

Nevertheless, the hybrid two-porosity models are promising and combine the important feature that transport models produce excesses in different parts of the melting regime, while channel models increase the transport rate of ^{230}Th from depth. Nevertheless, the actual formation and structure of channels in these models is *ad hoc* as is the parametrization of "melt suction" S and the mechanics of mixing. While Jull *et al.* (2002) provide mixing end-members for their model, the model still only produces two melt compositions at any depth and the model does not include any specific mechanism for mixing. To go further requires consistent fluid-mechanical models that explicitly calculate the behavior of the melt channels. Preliminary work on such models (next section) suggests that the results of the simpler hybrid models will hold qualitatively and that U-series and stable element variability may finally be used to address the dynamics and structure of partially molten regimes.

3.14.4.3.5 Full reactive transport models

Motivated by observations of reactive dunite channels in ophiolites, models for reactive flow in deformable permeable media have been developed that show that fluid flow along a solubility gradient will lead to flow localization and the development of a coalescing network of high-porosity melt channels surrounded by extremely low porosity regions (Aharonov *et al.*, 1995; Spiegelman *et al.*, 2001) (Figure 14). The key to this process is the inclusion of compaction. Because the channels are lower-pressure than the surrounding region, they do suck melt into them as was suggested by Iwamori (1993a) (although the suction is not coupled to the melting rate). The consequence of melt suction is to reduce the porosity of the interchannel region by compaction. If there is no resistance to compaction at small porosities, these models will actually compact to

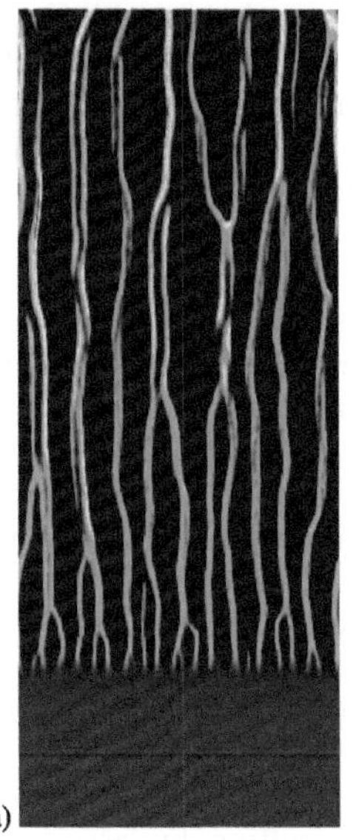

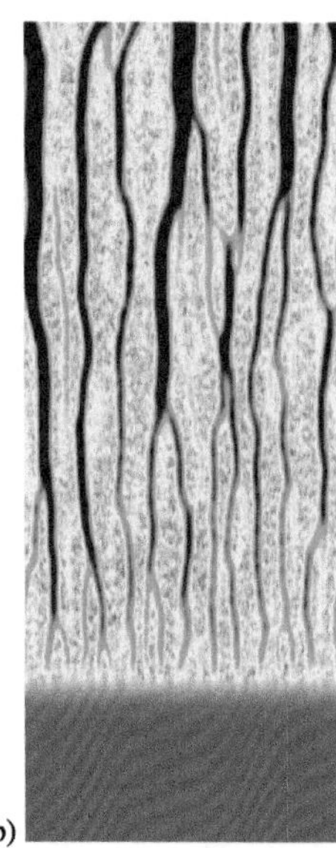

Figure 14 (a) Porosity and (b) residual channel structure for the reactive flow problem described in Spiegelman *et al.* (2001). This solution assumes a *static* solid composed of an insoluble and soluble phase (e.g., ol and opx) where the solubility of the soluble phase increases upwards. This material is fluxed by a corrosive fluid that interacts with the solubility gradient and produces a localized network of high-porosity melt channels surrounded by compacted low porosity regions. In this run the soluble phase becomes exhausted from the channels leaving a residual network (dark regions) that qualitatively resembles residual dunites seen in ophiolites (e.g., Kelemen *et al.*, 1995a, 2000; Braun and Kelemen, 2002).

near impermeability. The very small porosities in the interchannel region can cause significant chemical effects, particularly for U-series nuclides.

Spiegelman and Kelemen (2003) explore the chemical consequences of these channel systems for stable trace elements and show that channel systems can produce significant diversity even from a *homogeneous* source in complete chemical equilibrium. The behavior is similar to that predicted from the two-porosity models. The interchannel regions become extremely depleted because the low permeabilities prevent deeper enriched melts from interacting with the solid. Meanwhile, the solid continues to melt, producing near fractional instantaneous melts in equilibrium with the solid (but extremely depleted). These melts are then sucked into the *edges* of the channel systems but not mixed. The end result is highly variable channel chemistry where the centers of the channels are deep, enriched melts that have mixed from across the entire column, while the edges are extremely depleted shallow melts. Thus, each channel delivers the entire range of chemical variability to the crust on the scale of the channels (10–100 m). The melts produced show variability comparable to that seen in lavas and melt inclusions (Figure 16), while the residues (which are dominated by the interchannel regions) are extremely depleted and have compositions comparable to the abyssal peridotites (Figure 17). Figures 15–17 show results from a high-resolution single-channel calculation showing the porosity structure and synthetic data sets sampled from the melts and residues that show good qualitative agreement with observations. The question remains, however, what these channel systems will do to U-series elements.

Solving for U-series excesses directly in these systems is significantly harder to do accurately than in the one-dimensional column problems. However, we can use the same approach as in Spiegelman and Elliott (1993) to separate the effects of ingrowth from melting by assuming

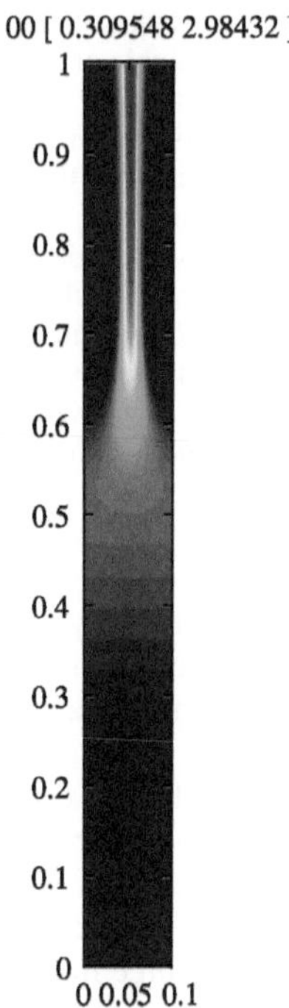

Figure 15 Porosity structure of a high-resolution single-channel calculation for an *upwelling* system undergoing melting by both adiabatic decompression and reactive flow (see Spiegelman and Kelemen, 2003). Colors show the porosity field at late times in the run where the porosity is quasi steady-state. The maximum porosity at the top of the column (red) is ~0.8% while the minimum porosity at the bottom (dark blue) is ~10 times smaller. Axis ticks are height and width relative to the overall height of the box. In the absence of channels this problem is identical to the equilibrium one-porosity transport model of Spiegelman and Elliott (1993). Introduction of channels, however, produces interesting new chemical effects similar to the two porosity models. See Spiegelman and Kelemen (2003) for details.

that the total concentration of a nuclide can be written as

$$c_i^{\rm f} = \alpha_{\rm i} c_{is}^{\rm f} \tag{13}$$

where $c_{is}^{\rm f}$ is the concentration in the melt of a fictive "stable" element with the same partitioning behavior as the radioactive nuclide and $\alpha_{\rm i}$ is the "ingrowth factor." We then solve for the log of the concentrations of U, Th, Ra, Pa as if they were stable elements using

$$\frac{\partial U_i^{\rm f}}{\partial t} + \boldsymbol{v}_{\rm eff}\cdot\boldsymbol{\nabla} U_i^{\rm f} = -\frac{\Gamma(1-D_i)+\rho_{\rm s}(1-\phi)\dot{D}_i}{\rho_{\rm f}\phi+\rho_{\rm s}(1-\phi)D_i} \tag{14}$$

where $U_i^{\rm f} = \ln(c_{is}^{\rm f}/c_0^{\rm f})$ is the log of the concentration of element i relative to the first instantaneous melt. D_i is the bulk partition coefficient of element i, Γ is the total melting rate which includes contribution from both adiabatic and reactive melting, $\rho_{\rm f}$, $\rho_{\rm s}$ are the melt and solid densities, ϕ is the porosity, and $\dot{D}_i = \partial D_i/\partial t + \boldsymbol{v}\cdot\boldsymbol{\nabla} D_i$ is the change in time of the partition coefficient in a frame following the solid:

$$\boldsymbol{v}_{\rm eff} = \frac{\rho_{\rm f}\phi\boldsymbol{v} + \rho_{\rm s}(1-\phi)D_i\boldsymbol{v}}{\rho_{\rm f}\phi+\rho_{\rm s}(1-\phi)D_i} \tag{15}$$

is the *effective* velocity of a tracer in chemical equilibrium with both melt and solid (e.g., see Spiegelman, 1996). Completely incompatible elements ($D \rightarrow 0$) travel at the melt speed $\boldsymbol{v}$, while completely compatible elements ($D \rightarrow \infty$) travel with the solid. The velocity field of intermediate elements depends principally on D/ϕ, as usual.

We then use the concentration ratios of the stable elements to solve for the excesses ($\alpha_{\rm i}$) using

$$\frac{\partial \alpha_{\rm i}}{\partial t} + \boldsymbol{v}_{\rm eff}\cdot\boldsymbol{\nabla}\alpha_{\rm i} = \lambda_i[R_i\alpha_{\rm i-1} - \alpha_{\rm i}] \tag{16}$$

where

$$R_i = \left(\frac{D_{i0}}{D_{(i-1)0}}\right)\frac{\rho_{\rm f}\phi+\rho_{\rm s}(1-\phi)D_{i-1}}{\rho_{\rm f}\phi+\rho_{\rm s}(1-\phi)D_i}{\rm e}^{U_{i-1}^{\rm f}-U_i^{\rm f}} \tag{17}$$

controls the amount of ingrowth by weighting the importance of the parent nuclide. This term depends principally on the relative size of the porosity and partition coefficients as well as the partition coefficients at the first instance of melting and the ratio of concentrations of the "stable" form of the parent and daughter elements. While it looks somewhat complicated, this two-step process is much more accurate than solving for the concentrations directly and gives information that is readily plotted on an equiline diagram. For example, by this technique $[^{230}{\rm Th}/^{232}{\rm Th}]$ is simply $\alpha_{\rm Th}$ because ^{232}Th behave principally as a stable element for these problems.

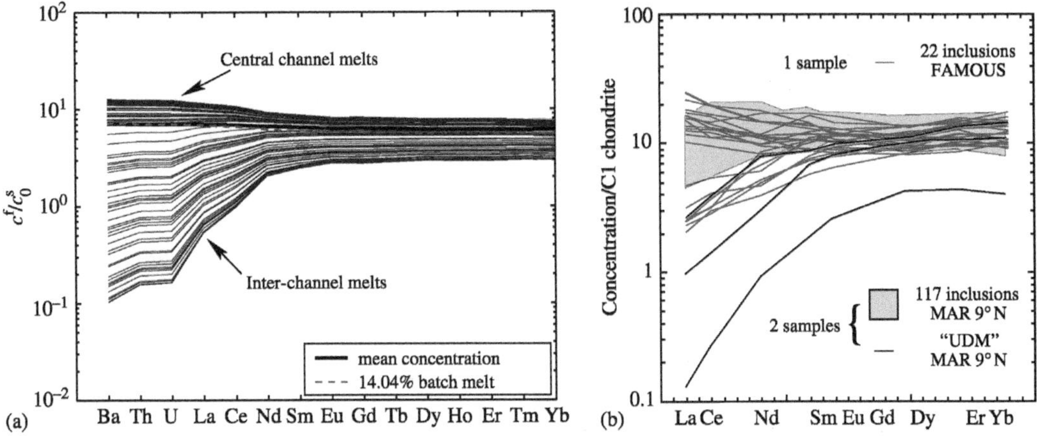

Figure 16 Comparison of model output and melt inclusion data from Spiegelman and Kelemen (2003). (a) Synthetic spidergram for highly incompatible elements plus rare-earths sampled from melts produced at the top of the melting calculation in Figure 15. The mean concentration is comparable to the maximum degree of melting but the distribution is similar to dynamic melting calculations. Unlike the single porosity models, these models produce a *distribution* of melt compositions at the top (rather than a single melt composition) even from a homogeneous source. The spidergrams here are sampled from this distribution in proportion to the melt flux (i.e., enriched melts from the center of the channel are more common than depleted melts). Note, this variability is produced within the channel on a small scale. (b) REE patterns from olivine hosted melt inclusions in three hand samples from the mid-Atlantic ridge (Sobolev and Shimizu, 1993, 1992; Shimizu and Grove, 1998). These samples show a similar range in variability. Most of the melt inclusions are similar in composition to the host glasses but there is a distinct tail of "ultra-depleted melts."

Likewise, $[^{238}U/^{232}Th]$ is calculated from the stable components. Thus, an equiline diagram simply plots the excess due to ingrowth against any chemical fractionation of parent and daughter. The equiline diagram in Figures 18(b) and 19(b) were constructed by this method.

Figure 18 shows a preliminary calculation for U-series excesses in the single-channel problem shown in Figure 15 and is solved using the parameters in Table 3 with self-consistent bulk partition coefficients intended to span two end-member scenarios (Table 2). This problem has not been tuned to match the observations. Significantly, the bulk *D*s are illustrative snap shots at a single pressure and compositions inappropriate for the whole melting column. Moreover, the channel is much larger than observed to improve numerical resolution. Nevertheless, as a first cut, this model can produce U-series distributions that are qualitatively similar to those observed. The important feature of these models that is different from most geochemical models, is that they produce a *distribution* of melt compositions at the top of the system, rather than a single-melt composition. These distributions can then be compared to spatially localized sets of samples. For example, Figure 18 shows the output of this model compared to U-series data from 9° 50′ N EPR (Sims *et al.*, 2002). While the match is not ideal, the patterns are intriguing. This particular run produces a positive correlation on an equiline diagram even for a homogeneous source. The fractionation of $(^{238}U/^{232}Th)$ is solely due to melting with the melts from the center of the channels being enriched mixtures with deep sources and large ^{230}Th-excesses, while the edges of the channels are depleted *shallow* melts with smaller excesses. A key prediction from these models is that there should be some correlation between U-series excesses and general trace element abundances for systems that can be demonstrated to have the same source.

The partition coefficients used here are reasonably good to explain both ^{230}Th-excesses as well as the range in $(^{238}U/^{232}Th)$ although this run produces somewhat greater depletion than is seen in the data. Figure 18(c) shows a negative correlation between $(^{226}Ra/^{230}Th)$ and $(^{230}Th/^{238}U)$ although the model has lower excesses than observed. Finally, Figure 18(d) shows a positive excess between $(^{231}Pa/^{235}U)$ and $(^{230}Th/^{238}U)$ which does not appear to be present in the data. As with the single-porosity transport models, these calculations are sensitive to the parameters, particularly the partition coefficients. Figure 19 shows the same plots but for the bulk partition coefficients of Wood *et al.* (1999) from Table 2. As before, these *D*s produce larger ^{226}Ra- and ^{231}Pa-excesses but negligible ^{230}Th-excesses.

The models have not been explored yet to understand whether we might expect spreading rate dependence or not. These calculations suggest that much of the observed chemical variability of local ridge chemistry reflects the structure and

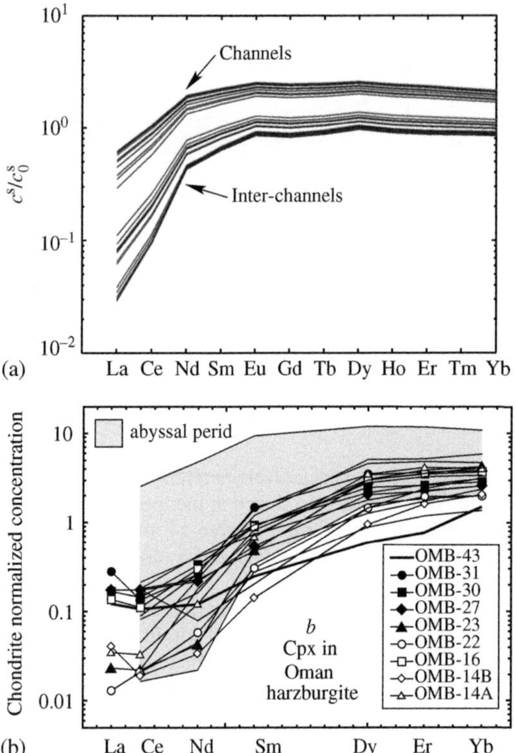

Figure 17 Comparison of model output for distribution of residue compositions with abyssal peridotites and harzburgites from Oman. (a) Distribution of clinopyroxene compositions in equilibrium with melts at the top of the melting column corresponding to Figure 15. This distribution is sampled uniformly by area (as if we were sampling residues from the top of the column) and is dominated by depleted inter-channel samples. (b) Clinopyroxene compositions in harzburgites from the mantle section of the Oman ophiolite (Kelemen *et al.*, 1995a) superimposed on the range of cpx compositions from abyssal peridotites (Johnson *et al.*, 1990; Johnson and Dick, 1992), showing the predominance of highly depleted samples.

plumbing of the partially molten region. How spreading rate affects channeling is still undetermined. To go further clearly requires a systematic exploration of parameter space to understand the full sensitivity of the model and compare its outputs to both U-series and stable trace element data. However, the framework now exists and the data sets are just beginning to come together that will allow for future exploration of both theory and data.

3.14.5 SUMMARY OF MODEL BEHAVIOR

The challenge of explaining observed U-series excess in MORBs has led to the development of a large number of models of differing levels of complexity. The previous section has attempted to clarify their pros and cons. We summarize here:

- Closed-system models cannot account for observed ^{230}Th, ^{226}Ra, and ^{231}Pa-excesses for degrees of melting much larger than the bulk partition coefficients. As $F \sim 10\%$ for MORB while D_U, $D_{Th} \ll 0.01$, these models are not viable for MORB generation.
- Both "dynamic melting" and "transport" models can produce the gross level of observed U-series excesses in MORB for appropriate parameters. However, they lead to different inferences on melt transport rates
  - Dynamic melting models produce all correlated excesses at the bottom of the melting regime and require melt transport rates sufficient to transport radium from the bottom of the melting regime on timescales short compared to the half-life of radium (1,600 yr).
  - Equilibrium transport models produce excesses throughout the melting column. In particular, thorium excesses are produced at depth in the presence of garnet and high-pressure pyroxenes, while potentially observable radium excesses are produced near the top of the melting column. These models still require rapid melt transport near the top so that the porosity in the radium production zone is comparable to D_{Th}.
- Both models assume a single porosity at any height in the melting region and produce well mixed melts at the top. Thus, these models cannot fractionate elemental uranium from thorium and cannot produce (^{230}Th/^{232}Th) versus (^{238}U/^{232}Th) correlations at constant ^{87}Sr/^{86}Sr as is observed.
- Two-porosity models can begin to explain both excesses and correlations by allowing multiple melt compositions to be produced for the channel and interchannel regions. The actual mechanism and structure of channel formation is somewhat *ad hoc* but the underlying behavior seems justified qualitatively.
- Full fluid-mechanically consistent melt transport models with reactive channeling extend the results of the two-porosity models and produce distributions of compositions for both stable and radiogenic tracers in melts and residues. These models suggest that much information on the structure and rates of magmatic process might be contained in the observed variability of mantle melts but they need to be explored more rigorously.
- We have stressed accounting for the striking local variations evident in MORB U-series systematics. There are some less robust variations with geophysical parameters. Equilibrium melting one-porosity models

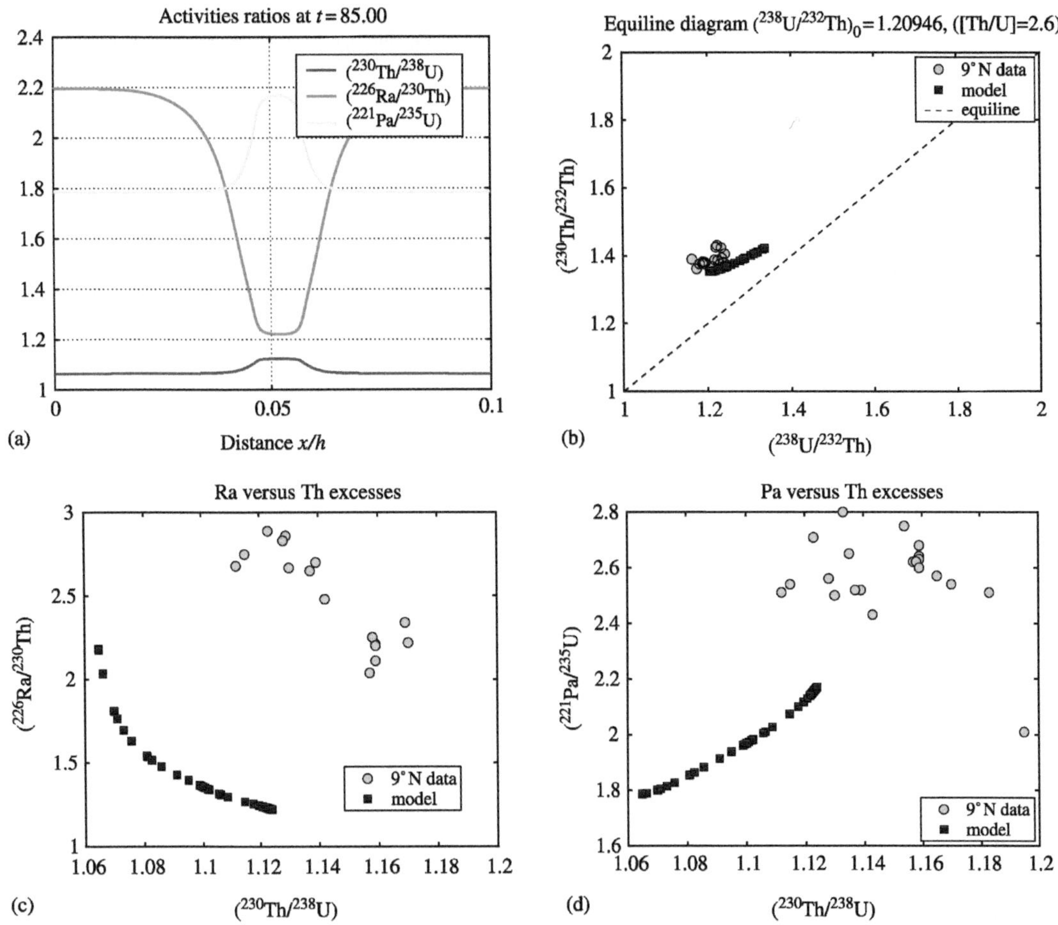

Figure 18 (a) profiles of activity ratios as a function of distance across the top of the calculation shown in Figure 15 for (^{230}Th/^{238}U), (^{226}Ra/^{230}Th), and (^{231}Pa/^{235}U). Partition coefficients are after Salters *et al.* (2002), see Table 2. (b) U, Th behavior of this model on an equiline diagram. Blue squares show model results, gray circles are data from 9° 50′ N, EPR (Sims *et al.*, 2002). These models actually produce correlated distributions on an equiline diagram because the channel melts tend to be more enriched deep melts with significant (^{230}Th/^{238}U) while the interchannel melts are shallow depleted melts with negligible ^{230}Th-excesses. (c) Anticorrelation of (^{226}Ra/^{230}Th) and (^{230}Th/^{238}U) in these systems. Again, the interchannel melts have large Ra excesses while the channel melts have small ones. Ra excesses for this set of parameters are smaller than observed but show the observed anticorrelation. (d) Pa versus Th excesses. Pa is lower and shows a more positive correlation than observed.

appear to be able to reproduce the weak correlation of axial depth with ^{230}Th-excess, and this is likely to extend to two-porosity models. The effect of spreading rate is still unclear observationally and the lack of clear correlation of absolute excesses potentially an interesting test of models.

3.14.6 CONCLUDING REMARKS

U-series data place stringent constraints on melting and melt migration models and successful input parameters. Sophistication of models and analytical data have advanced considerably in the last decade. It is evident that a significant fraction of U-series variability is present on a local scale. Detailed data sets that have a full range of geochemical tracers on the same samples analyzed for U-series disequilibrium will be particularly valuable for further developing the field (e.g. Sims *et al.*, 2002). A limiting factor is finding sample suites clearly young enough that post-eruptive decay is insignificant (a problem most acute for ^{226}Ra–^{230}Th disequilibrium). Progress in developing shorter lived disequilibrium to screen for suitably young samples is thus very welcome (Rubin *et al.*, 2001). Generating high precision U-series data sets is also time consuming. New plasma ionization multi-collector mass-spectrometers, however, should help improve both speed and precision of analysis. The results of recently developed melting models that can calculate both abundances and distributions of

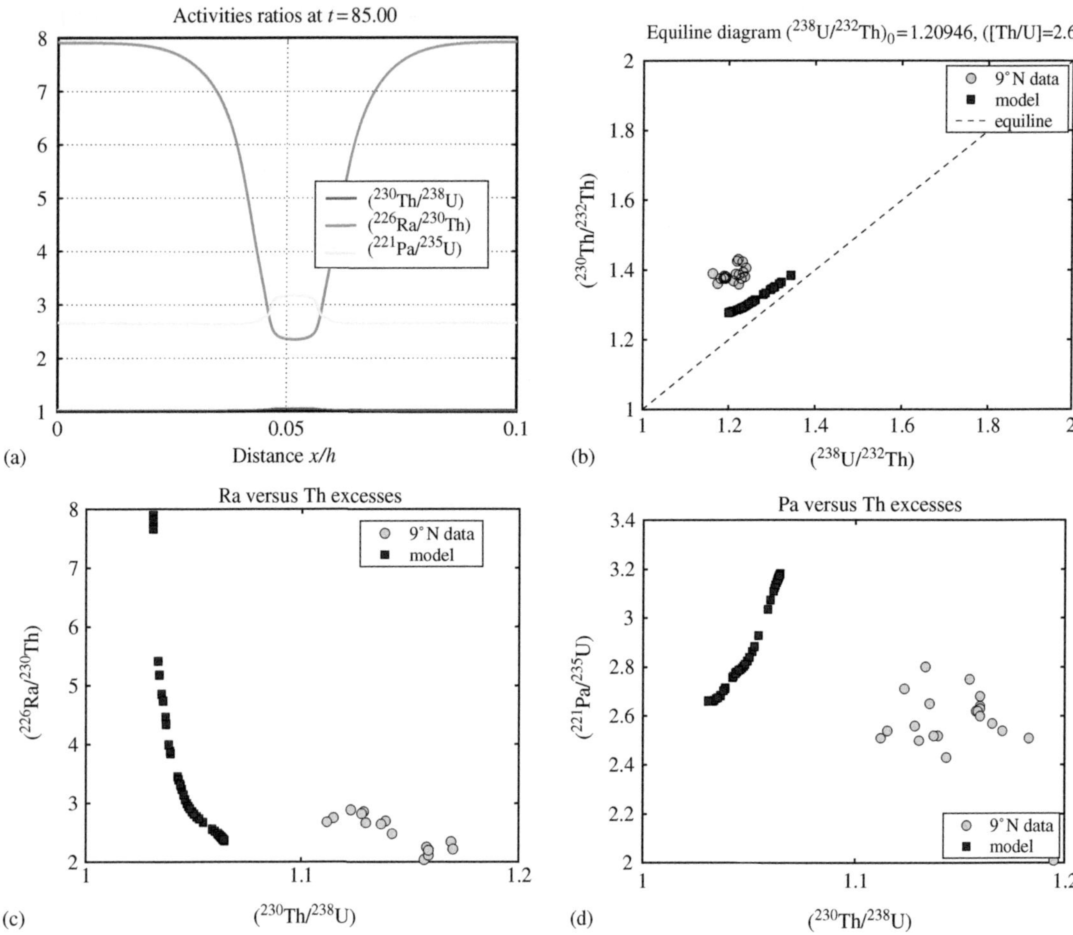

Figure 19 The same figures as in Figure 18 but using the partition coefficients after Blundy and Wood (2003b) see Table 2. (a) horizontal profiles at surface; (b) an equiline diagram showing a small ^{238}U-excess (but most samples are close to the equiline); (c) (^{226}Ra/^{230}Th) versus (^{230}Th/^{238}U); (d) (^{231}Pa/^{235}U) versus (^{230}Th/^{238}U). The large ^{226}Ra- and ^{231}Pa excesses are due to the considerably larger D_U and D_{Th} used in the upper part of this calculation.

Table 3 Parameters for full channel U-series model in Figure 18. Partition coefficients for these runs are taken from Table 2 with Salters *et al.* (2002) used for Figure 18 and McDade *et al.* (2003) (garnet) and Wood *et al.* (1999) (spinel) used in Figure 19. $D_{Ra} = D_{Pa} = 10^{-5}$ used in both figures.

Variable	*Value*
Upwelling rate	7 cm yr^{-1}
Column height	80 km
Max. degree of melting	10%
w_0/W_0	50
% box in Gt field	25%

chemical variability are encouraging. Further exploration of such models in conjunction with more comprehensive data sets holds promise for better understanding the processes of melt migration beneath ridges.

ACKNOWLEDGMENTS

Discussions with many people over the past fifteen years have helped shape ideas in this contribution. More specifically, we are very grateful to Ken Rubin for a valuable review on a less-than-perfect initial version, which helped significantly improve the manuscript. Additional comments by Ken Sims, Chris Hawkesworth, and John Blundy were much appreciated. Discussions with Vincent Salters, Jon Blundy, and Bernie Wood were very useful in trying to synthesize partitioning data. We thank the editor for her patience and faith in our final delivery. The formative role of D. House in this endeavor is recognized. T. E. thanks the Leverhulme Trust for generous funding during the period of writing. NERC and NSF funding of T.E. and M.S., respectively, is also acknowledged.

REFERENCES

Aharonov E., Whitehead J. A., Kelemen P. B., and Spiegelman M. (1995) Channeling instability of upwelling melt in the mantle. *J. Geophys. Res.: Solid Earth* **100**, 20433–20450.

Allègre C. J. (1968) ^{230}Th dating of volcanic rocks: a comment. *Earth Planet. Sci. Lett.* **5**, 209–210.

Allègre C. J. and Condomines M. (1982) Basalt genesis and mantle structure studied through Th-isotopic geochemistry. *Nature* **299**, 21–24.

Allègre C. J. and Turcotte D. L. (1986) Implications of a 2-component marble-cake mantle. *Nature* **323**, 123–127.

Asmerom Y. and Edwards R. L. (1995) U-series isotope evidence for the origin of continental basalts. *Earth Planet. Sci. Lett.* **134**, 1–7.

Ballard R. D., Francheteau J., Juteau T., Rangan C., and Normark W. (1981) East Pacific Rise at 21-degrees-N—the volcanic, tectonic, and hydrothermal processes of the central axis. *Earth Planet. Sci. Lett.* **55**, 1–10.

Bateman H. (1910) Solution of a system of differential equations occurring in the theory of radioactive transformations. *Proc. Cambridge Phil. Soc.* **15**, 423–427.

Beattie P. (1993a) The generation of uranium series disequilibria by partial melting of spinel peridotite-constraints from partitioning studies. *Earth Planet. Sci. Lett.* **117**, 379–391.

Beattie P. (1993b) Uranium thorium disequilibria and partitioning on melting of garnet peridotite. *Nature* **363**, 63–65.

Ben Othman D. and Allègre C. J. (1990) U–Th isotopic systematics at 13-degrees-N East Pacific Ridge segment. *Earth Planet. Sci. Lett.* **98**, 129–137.

Benjamin T. M., Heuser D. S., Burnett D. S., and Seitz M. G. (1980) Actinide crystal-liquid partitioning for clinopyroxene and $Ca_3(PO_4)_2$. *Geochim. Cosmochim. Acta* **44**, 1251–1264.

Blundy J. and Wood B. (1994) Prediction of crystal-melt partition-coefficients from elastic-moduli. *Nature* **372**, 452–454.

Blundy J. and Wood B. J. (2003a) Partitioning of trace elements between crystals and melts. *Earth Planet. Sci. Lett.* **210**, 387–397.

Blundy J. D. and Wood B. J. (2003b) Mineral-melt partitioning of uranium–thorium and their daughters. In *Uranium-series Geochemistry*, Reviews in Mineralogy and Geochemistry, 52 (eds. B. Bourdon, G. M. Henderson, C. C. Lundstrom, and S. P. Turner). Mineralogical Society of America, Washington, DC.

Bourdon B. and Sims K. W. W. (2003) U-series constraints on intraplate basaltic magmatism. In *Uranium-series Geochemistry* (eds. B. Bourdon, G. M. Henderson, C. C. Lundstrom, and S. P. Turner). Mineralogical Society of America, Washington, DC, vol. 52, pp. 215–254.

Bourdon B., Langmuir C. H., and Zindler A. (1996a) Ridge-hotspot interaction along the mid-Atlantic ridge between 37 degrees 30′ and 40 degrees 30′N: the U–Th disequilibrium evidence. *Earth Planet. Sci. Lett.* **142**, 175–189.

Bourdon B., Zindler A., Elliott T., and Langmuir C. H. (1996b) Constraints on mantle melting at mid-ocean ridges from global ^{238}U–^{230}Th disequilibrium data. *Nature* **384**, 231–235.

Bourdon B., Goldstein S. J., Bourlès D., Murrell M. T., and Langmuir C. H. (2000) Evidence from ^{10}Be and U-series disequilibria on the possible contamination of mid-ocean ridge basalt glasses by sedimentary material. *Geochem. Geophys. Geosys.* **1**, article no. 0047.

Bourdon B., Henderson G. M., Lundstrom C. C., and Turner S. P. (2003) Uranium-series geochemistry. In *Reviews in Mineralogy and Geochemistry*. Mineralogical Society of America, Washington, DC, vol. 52. 656pp.

Braun M. G. and Kelemen P. B. (2002) Dunite distribution in the Oman ophiolite: implications for melt flux through porous dunite conduits. *Geochem. Geophys. Geosys.* **3**, article no. 8603.

Chabaux F. and Allègre C. J. (1994) ^{238}U–^{230}Th–^{226}Ra disequilibria in volcanics—a new insight into melting conditions. *Earth Planet. Sci. Lett.* **126**, 61–74.

Chadwick W. W., Embley R. W., and Fox C. G. (1991) Evidence for volcanic-eruption on the Southern Juan-De-Fuca Ridge between 1981 and 1987. *Nature* **350**, 416–418.

Chadwick W. W., Embley R. W., and Shank T. M. (1998) The 1996 Gorda Ridge eruption: geologic mapping, sidescan sonar, and SeaBeam comparison results. *Deep-Sea Res. Part II: Top. Stud. Oceanogr.* **45**, 2547–2569.

Cheng H., Edwards R. L., Hoff J., Gallup C. D., Richards D. A., and Asmerom Y. (2000) The half-lives of uranium-234 and thorium-230. *Chem. Geol.* **169**, 17–33.

Cohen A. S. and Onions R. K. (1991) Precise determination of femtogram quantities of radium by thermal ionization mass-spectrometry. *Anal. Chem.* **63**, 2705–2708.

Condomines M. and Sigmarsson O. (2000) ^{238}U–^{230}Th disequilibria and mantle melting processes: a discussion. *Chem. Geol.* **162**, 95–104.

Condomines M., Morand P., and Allegre C. J. (1981) ^{230}Th–^{238}U radioactive disequilibria in tholeiites from the FAMOUS zone (mid-Atlantic ridge, 36°50′ N)—Th and Sr isotopic geochemistry. *Earth Planet. Sci. Lett.* **55**, 247–256.

Condomines M., Hémond C., and Allègre C. J. (1988) U–Th–Ra radioactive disequilibria and magmatic processes. *Earth Planet. Sci. Lett.* **90**, 243–262.

Cooper K. M., Reid M. R., Murrell M. T., and Clague D. (2001) Crystal and magma residence at Kilauea volcano, Hawaii: ^{230}Th–^{226}Ra dating of the 1955 east rift eruption. *Earth Planet. Sci. Lett.* **184**, 703–718.

Cooper K. M., Goldstein S. J., Sims K. W. W., and Murrell M. T. (2003) Uranium-series chronology of Gorda Ridge volcanism: new evidence from the 1996 eruption. *Earth Planet. Sci. Lett.* **206**, 459–475.

Davis A. S. and Clague D. A. (1987) Geochemistry, mineralogy, and petrogenesis of basalt from the Gorda Ridge. *J. Geophys. Res.* **92**, 10476–10483.

Dosso L., Bougault H., Langmuir C. H., Bollinger C., Bonnier O., and Etoubleau J. (1999) The age and distribution of mantle heterogeneity along the mid-Atlantic ridge (31–41°N). *Earth Planet. Sci. Lett.* **170**, 269–286.

Edwards R. L., Chen J. H., Ku T. L., and Wasserburg G. J. (1987) Precise timing of the last interglacial period from mass-spectrometric determination of ^{230}Th in corals. *Science* **236**, 1547–1553.

Elliott T. (1997) Fractionation of U and Th during mantle melting: a reprise. *Chem. Geol.* **139**, 165–183.

England J. G., Zindler A., Reisberg L. C., Rubenstone J. L., Salters V., Marcantonio F., Bourdon B., Brueckner H., Turner P. J., Weaver S., and Read P. (1992) The Lamont-Doherty-Geological-Observatory Isolab-54 isotope ratio mass-spectrometer. *Int. J. Mass Spectrom. Ion Process.* **121**, 201–240.

Feineman M. D., DePaolo D. J., and Ryerson F. J. (2002) Steady-state ^{226}Ra–^{230}Th disequilibrium in hydrous mantle minerals. *Geochim. Cosmochim. Acta* **66**, A228–A228.

Fleischer R. L., Price P. B., and Walker R. M. (1975) *Nuclear Tracks in Solids*. University of California Press, Berkeley, CA.

Forsyth D. W., Scheirer D. S., Webb S. C., Dorman L. M., Orcutt J. A., Harding A. J., Blackman D. K., Morgan J. P., Detrick R. S., Shen Y., Wolfe C. J., Canales J. P., Toomey D. R., Sheehan A. F., Solomon S. C., and Wilcock W. S. D. (1998) Imaging the deep seismic structure beneath a mid-ocean ridge: the MELT experiment. *Science* **280**, 1215–1218.

Gauthier P. J. and Condomines M. (1999) ^{210}Pb–^{226}Ra radioactive disequilibria in recent lavas and radon degassing: inferences on the magma chamber dynamics at Stromboli and Merapi volcanoes. *Earth Planet. Sci. Lett.* **172**, 111–126.

German C. R., Higgs N. C., Thomson J., Mills R., Elderfield H., Blusztajn J., Fleer A. P., and Bacon M. P. (1993) A geochemical study of metalliferous sediment from the tag hydrothermal mound, 26° 08′ N, mid-Atlantic ridge. *J. Geophys. Res.: Solid Earth* **98**, 9683–9692.

Goldstein S. J. and Stirling C. H. (2003) In *Uranium-series Geochemistry*, Reviews in Mineralogy and Geochemistry, 52 (eds. B. Bourdon, G. M. Henderson, C. C. Lundstrom, and S. P. Turner). Mineralogical Society of America, Washington, DC, pp. 23–57.

Goldstein S. J., Murrell M. T., and Janecky D. R. (1989) Th and U isotopic systematics of basalts from the Juan-De Fuca and Gorda Ridges by mass-spectrometry. *Earth Planet. Sci. Lett.* **96**, 134–146.

Goldstein S. J., Murrell M. T., Janecky D. R., Delaney J. R., and Clague D. A. (1991) Geochronology and petrogenesis of morb from the Juan De Fuca and Gorda Ridges by $^{238}U-^{230}Th$ disequilibrium. *Earth Planet. Sci. Lett.* **107**, 25–41.

Goldstein S. J., Murrell M. T., and Williams R. W. (1993) ^{231}Pa and ^{230}Th chronology of mid-ocean ridge basalts. *Earth Planet. Sci. Lett.* **115**, 151–159.

Haymon R. M., Fornari D. J., Edwards M. H., Carbotte S., Wright D., and Macdonald K. C. (1991) Hydrothermal vent distribution along the East Pacific Rise crest (9°09′–54′ N) and its relationship to magmatic and tectonic processes on fast-spreading mid-ocean ridges. *Earth Planet. Sci. Lett.* **104**, 513–534.

Haymon R. M., Fornari D. J., Vondamm K. L., Lilley M. D., Perfit M. R., Edmond J. M., Shanks W. C., Lutz R. A., Grebmeier J. M., Carbotte S., Wright D., McLaughlin E., Smith M., Beedle N., and Olson E. (1993) Volcanic-eruption of the mid-ocean ridge along the East Pacific Rise crest at 9 °45–52′N—direct submersible observations of sea-floor phenomena associated with an eruption event in April, 1991. *Earth Planet. Sci. Lett.* **119**, 85–101.

Hellebrand E., Snow J. E., Hoppe P., and Hofmann A. W. (2002) Garnet-field melting and late-stage refertilization in 'residual' abyssal peridotites from the Central Indian Ridge. *J. Petrol.* **43**, 2305–2338.

Hirschmann M. M. (2000) Mantle solidus: experimental constraints and the effects of peridotite composition. *Geochem. Geophys. Geosys.* **1**, article no. 0070.

Hofmann A. W. and White W. M. (1983) Ba, Rb, and Cs in the earths mantle. *Z. Naturfors. Sec. A: J. Phys. Sci.* **38**, 256–266.

Holden N. E. (1989) Total and spontaneous fission half-lives for uranium, plutonium, americium, and curium nuclides. *Pure Appl. Chem.* **61**, 1483–1504.

Ivanovich M. and Harmon R. S. (1992) *Uranium-series Disequilibrium: Applications to Earth, Marine, and Environmental Sciences*. Clarendon Press, Oxford, 910pp.

Iwamori H. (1993a) Dynamic disequilibrium melting model with porous flow and diffusion-controlled chemical equilibration. *Earth Planet. Sci. Lett.* **114**, 301–313.

Iwamori H. (1993b) A model for disequilibrium mantle melting incorporating melt transport by porous and channel flows. *Nature* **366**, 734–737.

Iwamori H. (1994) $^{238}U-^{230}Th-^{226}Ra$ and $^{235}U-^{231}Pa$ disequilibria produced by mantle melting with porous and channel flows. *Earth Planet. Sci. Lett.* **125**, 1–16.

Jaffey A. H., Flynn K. F., Glendenin L. E., Bentley W. C., and Essling A. M. (1971) Precision measurement of half-lives and specific activities of ^{235}U and ^{238}U. *Phys. Rev. C* **4**, 1889–1906.

Johnson K. T. M. and Dick H. J. B. (1992) Open system melting and temporal and spatial variation of peridotite and basalt at the Atlantis-II fracture-zone. *J. Geophys. Res.: Solid Earth* **97**, 9219–9241.

Johnson K. T. M., Dick H. J. B., and Shimizu N. (1990) Melting in the oceanic upper mantle—an ion microprobe study of diopsides in abyssal peridotites. *J. Geophys. Res.: Solid Earth* **95**, 2661–2678.

Jull M., Kelemen P. B., and Sims K. (2002) Consequences of diffuse and channelled porous melt migration on uranium series disequilibria. *Geochim. Cosmochim. Acta* **66**, 4133–4148.

Kelemen P. B. and Dick H. J. B. (1995) Focused melt flow and localized deformation in the upper mantle: Juxtaposition of replacive dunite and ductile shear zones in the Josephine peridotite, SW Oregon. *J. Geophys. Res.* **100**, 423–438.

Kelemen P. B., Shimizu N., and Salters V. J. M. (1995a) Extraction of mid-ocean-ridge basalt from the upwelling mantle by focused flow of melt in dunite channels. *Nature* **375**, 747–753.

Kelemen P. B., Whitehead J. A., Aharonov E., and Jordahl K. A. (1995b) Experiments on flow focusing in soluble porous-media, with applications to melt extraction from the Mantle. *J. Geophys. Res.: Solid Earth* **100**, 475–496.

Kelemen P. B., Hirth G., Shimizu N., Spiegelman M., and Dick H. J. B. (1997) A review of melt migration processes in the adiabatically upwelling mantle beneath oceanic spreading ridges. *Phil. Trans. Roy. Soc. London Ser. A: Math. Phys. Eng. Sci.* **355**, 283–318.

Kelemen P. B., Braun M. G., and Hirth G. (2000) Spatial distribution of melt conduits in the mantle beneath oceanic spreading ridges: observations from the Ingalls and Oman ophiolites. *Geochem. Geophys. Geosys.* **1**, article no. 0012.

Klein E. M. and Langmuir C. H. (1987) Global correlations of ocean ridge basalt chemistry with axial depth and crustal thickness. **92**, 8089–8115.

Klein E. M. and Langmuir C. H. (1989) Local versus global variations in ocean ridge basalt composition—a reply. **94**, 4241–4252.

Krishnaswami S., Turekian K. K., and Bennett J. T. (1984) The behavior of ^{232}Th and the ^{238}U decay chain nuclides during magma formation and volcanism. *Geochim. Cosmochim. Acta* **48**, 505–511.

Lalou C., Reyss J. L., and Brichet E. (1993) Actinide-series disequilibrium as a tool to establish the chronology of deep-sea hydrothermal activity. *Geochim. Cosmochim. Acta* **57**, 1221–1231.

Landwehr D., Blundy J., Chamorro-Perez E. M., Hill E., and Wood B. (2001) U-series disequilibria generated by partial melting of spinel lherzolite. *Earth Planet. Sci. Lett.* **188**, 329–348.

Langmuir C. H., Bender J. F., Bence A. E., Hanson G. N., and Taylor S. R. (1977) Petrogenesis of basalts from the famous area, mid-Atlantic ridge. *Earth Planet. Sci. Lett.* **36**, 133–156.

Langmuir C. H., Klein E. M., and Plank T. (1992) Petrological systematics of mid-ocean ridge basalts: constraints on melt generation beneath ocean ridges. In *Mantle Flow and Melt Generation at Mid-ocean Ridges* (eds. J. P. Morgan, D. K. Blackman, and J. Sinton). American Geophysical Union, Washington, DC, vol. 71, pp. 183–280.

LaTourrette T. Z. and Burnett D. S. (1992) Experimental-determination of U-partitioning and Th-partitioning between clinopyroxene and natural and synthetic basaltic liquid. *Earth Planet. Sci. Lett.* **110**, 227–244.

LaTourrette T. Z., Kennedy A. K., and Wasserburg G. J. (1993) Thorium-uranium fractionation by garnet: evidence for a deep source and rapid rise of oceanic basalts. *Science* **261**, 739–742.

Layne G. D. and Sims K. W. (2000) Secondary ion mass spectrometry for the measurement of $^{232}Th/^{230}Th$ in volcanic rocks. *Int. J. Mass Spectrom.* **203**, 187–198.

Lundstrom C. C. (2000) Models of U-series disequilibria generation in MORB: the effects of two scales of melt porosity. *Phys. Earth Planet. Int.* **121**, 189–204.

Lundstrom C. C. (2003) Disequilibria in mid-ocean ridge basalts: observations and models of basalt genesis. In *Uranium-series Goechemistry* (eds. B. Bourdon, G. M. Henderson, C. C. Lundstrom, and S. P. Turner). Mineralogical Society of America, Washington, DC, vol. 52, pp. 175–214.

Lundstrom C. C., Shaw H. F., Ryerson F. J., Phinney D. L., Gill J. B., and Williams Q. (1994) Compositional controls on the partitioning of U, Th, Ba, Pb, Sr, and Zr between clinopyroxene and haplobaslatic melts: implications for uranium series disequilibria in basalts. *Earth Planet. Sci. Lett.* **128**, 407–423.

Lundstrom C. C., Gill J., Williams Q., and Perfit M. R. (1995) Mantle melting and basalt extraction by equilibrium porous flow. *Science* **270**, 1958–1961.

Lundstrom C. C., Gill J., Williams Q., and Hanan B. B. (1998a) Investigating solid mantle upwelling beneath mid-ocean ridges using U-series disequilibria: II. A local study at 33 degrees mid-Atlantic ridge. *Earth Planet. Sci. Lett.* **157**, 167–181.

Lundstrom C. C., Williams Q., and Gill J. B. (1998b) Investigating solid mantle upwelling rates beneath mid-ocean ridges using U-series disequilibria: 1. A global approach. *Earth Planet. Sci. Lett.* **157**, 151–165.

Lundstrom C. C., Sampson D. E., Perfit M. R., Gill J., and Williams Q. (1999) Insights into mid-ocean ridge basalt petrogenesis: U-series disequilibria from the Siqueiros Transform, Lamont Seamounts, and East Pacific Rise. *J. Geophys. Res.: Solid Earth* **104**, 13035–13048.

Lundstrom C. C., Gill J., and Williams Q. (2000) A goechemically consistent hypothesis for MORB generation. *Chem. Geol.* **162**, 105–126.

Luo X. Z., Rehkamper M., Lee D. C., and Halliday A. N. (1997) High precision $^{230}Th/^{232}Th$ and $^{234}U/^{238}U$ measurements using energy-filtered ICP magnetic sector multiple collector mass spectrometry. *Int. J. Mass Spectrom.* **171**, 105–117.

Macdonald K. C., Haymon R. M., Miller S. P., Sempere J. C., and Fox P. J. (1988) Deep-tow and sea beam studies of dueling propagating ridges on the East Pacific Rise near 20° 40′ S. *J. Geophys. Res.: Solid Earth Planets* **93**, 2875–2898.

Maclennan J., McKenzie D., and Gronvold K. (2001) Plume-driven upwelling under central Iceland. *Earth Planet. Sci. Lett.* **194**, 67–82.

McDade P., Wood B. J., Blundy J. D., and Dalton J. A. (2003) Trace element partitioning at 3.0 GPa on the anhydrous garnet peridotite solidus. *J. Petrol.* (accepted).

McKenzie D. (1985) $^{230}Th-^{238}U$ disequilibrium and the melting processes beneath ridge axes. *Earth Planet. Sci. Lett.* **72**, 149–157.

McKenzie D. and Bickle M. J. (1988) The volume and composition of melt generated by extension of the lithosphere. *J. Petrol.* **29**, 625–679.

Meadows J. W., Armani R. J., Calis E. L., and Essling A. M. (1980) Half-life of ^{230}Th. **22**, 750–754.

Moore W. S. and Stakes D. (1990) Ages of barite-sulfide chimneys from the mariana trough. *Earth Planet. Sci. Lett.* **100**, 265–274.

Murton M. J., Taylor R. N., and Thirlwall M. F. (2002) Plume-ridge interaction: a geochemical perspective from the Reykjanes ridge. *J. Petrol.* **43**, 1987–2012.

Newman S., Finkel R. C., and Macdougall J. D. (1983) Th-230–U-238 disequilibrium systematics in oceanic tholeiites from 21° N on the East Pacific Rise. *Earth Planet. Sci. Lett.* **65**, 17–33.

Osmond J. K. (2003) Uranium-series fractionation in mafic extrusives. *Appl. Geochem.* **18**, 127–134.

Peate D. W., Hawkesworth C. J., van Calsteren P. W., Taylor R. N., and Murton B. J. (2001) $^{238}U-^{230}Th$ constraints on mantle upwelling and plume-ridge interaction along the Reykjanes ridge. *Earth Planet. Sci. Lett.* **187**, 259–272.

Perfit M. R. and Chadwick W. W. (1998) Magmatism at mid-ocean ridges: constraints from volcanological and geochemical investigations. In *Faulting and Magmatism at Mid-ocean Ridges* (eds. W. R. Buck, P. T. Delaney, J. A. Karson, and Y. Lagabrielle). American Geophysical Union, Washington, DC, vol. 106, pp. 59–115.

Pietruszka A. J., Carlson R. W., and Hauri E. H. (2002) Precise and accurate measurement of $^{226}Ra-^{230}Th-^{238}U$ disequilibria in volcanic rocks using plasma ionization multicollector mass spectrometry. *Chem. Geol.* **188**, 171–191.

Press W. H., Flannery B. P., Teukolosky S. A., and Vetterling W. T. (1992) *Numerical Recipes*. Cambridge University Press, New York.

Prinzhoffer A., Lewin E., and Allègre C. J. (1989) Stochastic melting of the marble cake mantle: evidence from local study of the East Pacific Rise at 12°50′N. *Earth Planet. Sci. Lett.* **92**, 189–206.

Qin Z. (1992) Disequilibrium partial melting model and its implications for trace element fractionations during mantle melting. *Earth Planet. Sci. Lett.* **112**, 75–90.

Qin Z. W. (1993) Dynamics of melt generation beneath mid-ocean ridge axes—theoretical-analysis based on $^{238}U-^{230}Th-^{226}Ra$ and $^{235}U-^{231}Pa$ disequilibria. *Geochim. Cosmochim. Acta* **57**, 1629–1634.

Reinitz I. and Turekian K. K. (1989) $^{230}Th/^{238}U$ and $^{226}Ra/^{230}Th$ fractionation in young basaltic glasses from the East Pacific Rise. *Earth Planet. Sci. Lett.* **94**, 199–207.

Ribe N. M. (1985) The generation and composition of partial melts in the Earths mantle. *Earth Planet. Sci. Lett.* **73**, 361–376.

Richardson C. and McKenzie D. (1994) Radioactive disequilibria from 2d models of melt generation by plumes and ridges. *Earth Planet. Sci. Lett.* **128**, 425–437.

Rubin K. H. (2001) Analysis of $^{232}Th/^{230}Th$ in volcanic rocks: a comparison of thermal ionization mass spectrometry and other methodologies. *Chem. Geol.* **175**, 723–750.

Rubin K. H. and Macdougall J. D. (1988) ^{226}Ra-excesses in mid-ocean-ridge basalts and mantle melting. *Nature* **335**, 158–161.

Rubin K. H. and Macdougall J. D. (1990) Dating of neovolcanic MORBA using ($^{226}Ra-^{230}Th$) disequilibrium. *Earth Planet. Sci. Lett.* **101**, 313–322.

Rubin K. H. and Macdougall J. D. (1992) Th–Sr isotopic relationships in morb. *Earth Planet. Sci. Lett.* **114**, 149–157.

Rubin K. H., Macdougall J. D., and Perfit M. R. (1994) $^{210}Po-^{210}Pb$ dating of recent volcanic-eruptions on the seafloor. *Nature* **368**, 841–844.

Rubin K. H., Smith M. C., Perfit M. R., Christie D. M., and Sacks L. F. (1998) Geochronology and goechemistry of lavas from the 1996 North Gorda Ridge eruption. *Deep-sea Res. Part II: Top. Stud. Oceanogr.* **45**, 2571–2597.

Rubin K. H., Macdougall J. D., Schilling J.-G., and Sacks L. F. (2000) A role for active MOR processes in mantle heterogeneity during plume-ridge interaction at the Easter microplate. *EOS, Trans., AGU* **81** (abstract V21D-09).

Rubin K. H., Smith C. W., Sacks L. F., Sinton J., and Bergmanis P. (2001) $^{238}U-^{230}Th-^{226}Ra-^{210}Pb$ constraints on eruption timing for discrete lava flows on the S-EPR *EOS, Trans., AGU* **82**, (abstract V12A-0957).

Saal A. E., Van Orman J. A., Hauri E. H., Langmuir C. H., and Perfit M. R. (2002a) An alternative hypothesis for the origin of the high Ra-226 excess in MORBs. *EOS, Trans., AGU* **83**, A659 Fall Meet. Suppl., Abstr. V71C-01.

Saal A. E., Van Orman J. A., Hauri E. H., Langmuir C. H., and Perfit M. R. (2002b) An alternative hypothesis for the origin of the high Ra-226 excess in mid-ocean ridge basalts. *Geochim. Cosmochim. Acta* **66**, A659.

Salters V. J. M. and Hart S. R. (1989) The hafnium paradox and the role of garnet in the source of mid-ocean-ridge basalts. *Nature* **342**, 420–422.

Salters V. J. M. and Longhi J. E. (1999) Trace element partitioning during the initial stages of melting beneath mid-ocean ridges. *Earth Planet. Sci. Lett.* **166**, 15–30.

Salters V. J. M., Longhi J. E., and Bizimis M. (2002) Near mantle solidus trace element partitioning at pressures up to 3.4 GPa. *Geochem. Geophys. Geosys.* **3**, 1038.

Schilling J.-G. (1973) The Icelandic plume, geochemical evidence along the Reykjanes ridge. **242**, 565–571.

Seitz M. G. (1973) Uranium and thorium diffusion in diopside and fluorapatite. *Carnegie Inst. Wash. Yearb.* **72**, 586–588.

Shimizu N. and Grove T. L. (1998) Geochemical studies of olivine-hosted melt inclusions from ridges and arcs. *EOS, Trans., AGU* **79**, F1002.

Shimmield G. B. and Price N. B. (1988) The scavenging of U, ^{230}Th, and ^{231}Pa during pulsed hydrothermal activity at 20-degrees-S, East Pacific Rise. *Geochim. Cosmochim. Acta* **52**, 669–677.

Sigmarsson O. (1996) Short magma chamber residence time at an Icelandic volcano inferred from U-series disequilibria. *Nature* **382**, 440–442.

Sims K. W. W., Depaolo D. J., Murrell M. T., Baldridge W. S., Goldstein S. J., and Clague D. A. (1995) Mechanisms of magma generation beneath Hawaii and mid-ocean ridges-uranium/thorium and samarium/neodymium isotopic evidence. *Science* **267**, 508–512.

Sims K. W. W., DePaolo D. J., Murrell M. T., Baldridge W. S., Goldstein S., Clague D., and Jull M. (1999) Porosity of the melting zone and variations in the solid mantle upwelling rate beneath Hawaii: inferences from ^{238}U–^{230}Th–^{228}Ra and ^{235}U–^{231}Pa disequilibria. *Geochim. Cosmochim. Acta* **63**, 4119–4138.

Sims K. W. W., Mattielli N., Elliott T., Kelemen P., DePaolo D. J., Mertz D. F., Devey C., and Murrell M. T. (2001) ^{238}U and ^{230}Th excesses in Kolbeinsey Ridge basalts. *EOS, Trans., AGU* **82**, (abstract V12A-0925).

Sims K. W. W., Goldstein S. J., Blichert-Toft J., Perfit M. R., Kelemen P., Fornari D. J., Michael P., Murrell M. T., Hart S. R., DePaolo D. J., Layne G., Ball L., Jull M., and Bender J. (2002) Chemical and isotopic constraints on the generation and transport of magma beneath the East Pacific Rise. *Geochim. Cosmochim. Acta* **66**, 3481–3504.

Sims K. W. W., Blichert-Toft J., Fornari D. J., Perfit M. R., Goldstein S. J., Johnson P., DePaolo D. J., Hart S. R., Murrell M. T., Michael P., Layne G., and Ball L. (xxxx) Aberrant youth: chemical and isotopic constraints on the origin of off-axis lavas from the East Pacific Rise, 9°–10°N. *Geochem. Geophys. Geosys.* (in press).

Slater L., McKenzie D., Gronvold K., and Shimizu N. (2001) Melt generation and movement beneath Theistareykir, NE Iceland. *J. Petrol.* **42**, 321–354.

Sobolev A. V. and Shimizu N. (1992) Super depleted melts and ocean mantle permeability. *DokladyRossiyskoy Akademii Nauk* **326**, 354–360.

Sobolev A. V. and Shimizu N. (1993) Ultra-depleted primary melt included in an olivine from the mid-Atlantic ridge. 363, 151–154.

Spiegelman M. (1996) Geochemical consequences of melt transport in 2-D: the sensitivity of trace elements to mantle. *Earth Planet. Sci. Lett.* **139**, 115–132.

Spiegelman M. (2000) UserCalc: a web-based U-series calculator for mantle melting problems. *Geochim. Geophys. Geosys.* **1**, article no. 0030.

Spiegelman M. and Elliott T. (1993) Consequences of melt transport for uranium series disequilibrium in young lavas. *Earth Planet. Sci. Lett.* **118**, 1–20.

Spiegelman M. and Kelemen P. B. (2003) Extreme chemical variability as a consequence of channelized melt transport. *Geochem. Geophys. Geosys.* **4**, No. 72002GC000336.

Spiegelman M. and McKenzie D. (1987) Simple 2-D models for melt extraction at mid-ocean ridges and island arcs. *Earth Planet. Sci. Lett.* **83**, 137–152.

Spiegelman M., Kelemen P. B., and Aharonov E. (2001) Causes and consequences of flow organization during melt transport: the reaction infiltration instability in compactible media. *J. Geophys. Res.: Solid Earth* **106**, 2061–2077.

Strang G. (1998) *Introduction to Linear Algebra.* Cambridge Press, Wellesley.

Sturm M. E., Goldstein S. J., Klein E. M., Karson J. A., and Murrell M. T. (2000) Uranium-series age constraints on lavas from the axial valley of the mid-Atlantic ridge, MARK area. *Earth Planet. Sci. Lett.* **181**, 61–70.

Thompson W. G., Spiegelman M. W., Goldstein S. L., and Speed R. C. (2003) An open-system model for U-series age determinations of fossil corals. *Earth Planet. Sci. Lett.* **210**, 365–381.

Turner S., van Calsteren P., Vigier N., and Thomas L. (2001) Determination of thorium and uranium isotope ratios in low-concentration geological materials using a fixed multi-collector-ICP-MS. *J. Anal. Atom. Spectrom.* **16**, 612–615.

Van Orman J. A., Grove T. L., and Shimizu N. (1998) Uranium and thorium diffusion in diopside. *Earth Planet. Sci. Lett.* **160**, 505–519.

Van Orman J. A., Grove T. L., and Shimizu N. (2001) Rare earth element diffusion in diopside: influence of temperature, pressure, and ionic radius, and an elastic model for diffusion in silicates. *Contrib. Mineral. Petrol.* **141**, 687–703.

Van Orman J. A., Bourdon B., and Hauri E. H. (2002a) A new model for U-series isotope fractionation during igneous processes, with finite diffusion and multiple solid phases. *EOS, Trans., AGU* **83** V71C-02.

Van Orman J. A., Grove T. L., and Shimizu N. (2002b) Diffusive fractionation of trace elements during production and transport of melt in Earth's upper mantle. *Earth Planet. Sci. Lett.* **198**, 93–112.

Volpe A. M. and Goldstein S. J. (1993) ^{226}Ra–^{230}Th disequilibrium in axial and off-axis mid-ocean ridge basalts. *Geochim. Cosmochim. Acta* **57**, 1233–1241.

Volpe A. M., Olivares J. A., and Murrell M. T. (1991) Determination of radium isotope ratios and abundances in geologic samples by thermal ionization mass-spectrometry. *Anal. Chem.* **63**, 913–916.

von Bargen N. and Waff H. S. (1986) Permeabilities, interfacial-areas and curvatures of partially molten systems-results of numerical computation of equilibrium microstructures. *J. Geophys. Res.: Solid Earth* **91**, 9261–9276.

White S. M., Macdonald K. C., and Sinton J. M. (2002) Volcanic mound fields on the East Pacific Rise, 16°–19° S: low effusion rate eruptions at overlapping spreading centers for the past 1 Myr. *J. Geophys. Res.: Solid Earth* **107** article no. 2240.

Williams R. W. and Gill J. B. (1989) Effects of partial melting on the uranium decay series. *Geochim. Cosmochim. Acta* **53**, 1607–1619.

Wood B. J., Blundy J. D., and Robinson J. A. C. (1999) The role of clinopyroxene in generating U-series disequilibrium during mantle melting. *Geochim. Cosmochim. Acta* **63**, 1613–1620.

Yokoyama T., Makishima A., and Nakamura E. (2001) Precise analysis of ^{234}U/^{238}U ratio using UO^{2+} ion with thermal ionization mass spectrometry for natural samples. *Chem. Geol.* **181**, 1–12.

Zou H. B. and Zindler A. (2000) Theoretical studies of ^{238}U–^{230}Th–^{226}Ra and ^{235}U–^{231}Pa disequilibria in young lavas produced by mantle melting. *Geochim. Cosmochim. Acta* **64**, 1809–1817.

Zou H. B., Zinder A., and Niu Y. L. (2002) Constraints on melt movement beneath the East Pacific Rise from ^{230}Th–^{238}U disequilibrium. *Science* **295**, 107–110.

3.15
Hydrothermal Alteration Processes in the Oceanic Crust

H. Staudigel

University of California at San Diego, La Jolla, CA, USA

3.15.1 INTRODUCTION

Hydrothermal alteration processes occurring in oceanic crust impact the physical, chemical, and biological processes of the Earth system. These hydrothermal systems are manifested in vents ranging from 350 °C black smokers, found exclusively in the axial zone of some ridge segments, to 20 °C low-temperature vents at the ridge axis or flanks. Collectively, these systems are responsible for ~20% of Earth's total heat loss (11 TW; C. A. Stein and S. Stein (1994a,b)) and have major impact on ocean and solid earth chemistry. Elderfield and Schultz (1996) estimate black-smoker water fluxes to be $\sim 3.5 \times 10^{12}$ kg yr^{-1} and low-temperature fluxes to be $\sim 6.4 \times 10^{14}$ kg yr^{-1} (at 20 °C). These hydrothermal fluxes also carry substantial elemental flux between seawater and the oceanic crust. Combined with ocean-crust generation and recycling, these processes produce a two-way geochemical pathway between the oceans and the mantle. Recycling of altered oceanic crust into the mantle is likely to produce some of the mantle's chemical heterogeneity (e.g., Hofmann, 1988; see Chapter 2.04) and the delivery of mantle-derived materials to seawater through hydrothermal systems has profound effects on seawater chemistry (e.g., Wheat and Mottl, 2000; Chapters 3.15 and 6.07). Hydrothermal vents in mid-ocean ridges offer a unique habitat for very diverse biological communities that derive much of their energy needs from chemical energy in vent fluids (Jannasch and Mottl, 1985; Jannasch, 1995).

The interior of the oceanic crust is likely to host a deep-ocean biosphere that reaches to at least 500 m depth (Furnes and Staudigel, 1999).

It is important to quantify hydrothermal chemical fluxes because they bear on the chemical and biological evolution of the Earth, the chemical composition of seawater, geochemical mass balance at arcs, and the heterogeneity of the mantle. Hydrothermal fluxes can be independently determined by analyzing the composition of hydrothermal fluids or by analyzing the alteration-related chemical changes in the oceanic crust. Ideally these two methods should yield the same results, but a comparison of data shows that there are major discrepancies between these types of estimates (e.g., Hart and Staudigel, 1982; Chapter 3.15). Reconciling these discrepancies is important for improving our understanding of this central theme in Earth system sciences.

This review focuses on chemical flux estimates derived from studies of the oceanic crust, exploring in detail how such estimates are made, and the underlying assumptions and uncertainties. Three main themes will be covered. The first focuses the role of the original igneous characteristics of the crust in determining the nature of hydrothermal alteration processes. This includes how primary lithology and composition influence alteration, and difficulties encountered in determining an unaltered "fresh-rock" baseline composition for any particular ocean-crust section. The second theme focuses on the methods by which the bulk-altered oceanic composition is determined, and the attendant uncertainties. These include the difficulty of determining an average composition of a very heterogeneous medium by the analyses of rather small samples, and the limitations imposed by an incomplete sampling process on the ocean floor. Finally, hydrothermal fluxes inferred from ocean-crust data are compared to fluxes from hydrothermal vent studies and the reasons behind their differences are explored.

3.15.2 THE UNALTERED OCEANIC-CRUST PROTOLITH

Determining the original igneous characteristics of altered oceanic crust is an important step in understanding its alteration. There are two very different reasons for this.

First, the primary igneous characteristics of the oceanic crust determine important aspects of alteration by controlling heat exchange, fluid flow, and the reactivity of crustal materials with hydrous fluids. These primary characteristics may vary as a function of spreading rate, ridge morphology, or position in a volcano, producing a range of alteration types. Effectively integrating these different crust types is an important challenge in producing a meaningful global integration of alteration fluxes. This problem is addressed here by defining a "standard" section for normal oceanic crust based on results from available sample materials.

Second, the original composition of oceanic crust plays an important role in determining chemical fluxes. Uncertainties in the original composition translate directly into uncertainties of chemical fluxes, and in many cases these differences are the main sources of errors. Both of these features are addressed in the following sections.

3.15.2.1 A "Standard Section" for the Oceanic Crust

The most comprehensive understanding of processes occurring in the oceanic crust comes from the study of subaerially exposed ocean-crust sections in ophiolites, which offer extensive and often continuous exposure over very large areas. The observations from ophiolites complement the significantly more limited data from submerged, *in situ* oceanic crust, which in turn serves as an important test for the validity of these subaerial analogues. For this reason, ophiolites are key to understanding the relationships between alteration behavior and primary crust characteristics. The main ophiolites and ocean-crust drill holes discussed in this paper include the Troodos ophiolite in Cyprus, the Samail ophiolite in Oman, and ocean-crust drilling sites 332B and 417/418 in the Atlantic Ocean and sites 504B and 801C in the eastern and western Pacific, respectively (Figure 1, Table 1).

The classic view of the oceanic crust is based on the "Penrose" ophiolite assemblage (Penrose Conference Participants, 1972) that includes a characteristic sequence of rock types: basal mafic cumulates are overlain by gabbro-norites, gabbros, sheeted dikes, and pillow lavas on the top. Pillow lavas may be overlain by hydrothermal metalliferous sediments, that are commonly overlain by pelagic, typically siliceous fine-grained sediments (cherts). This three-decade-old ophiolite model of the oceanic crust remains effectively unchanged in terms of the main rock types found in oceanic crust and in terms of their stratigraphic succession. The main modifications to this model include changes in the average thicknesses of units. In addition, there is some *in situ* oceanic crust that appears to lack any extrusives, but this type of ocean crust is mostly confined to very slow spreading ridges and represents probably ~5% of the total crust produced (e.g., Bach *et al.*, 2001).

Study of the global impact of ocean-crust generation and alteration ideally requires integration of data from all major crustal types, if possible, including several studies that offer

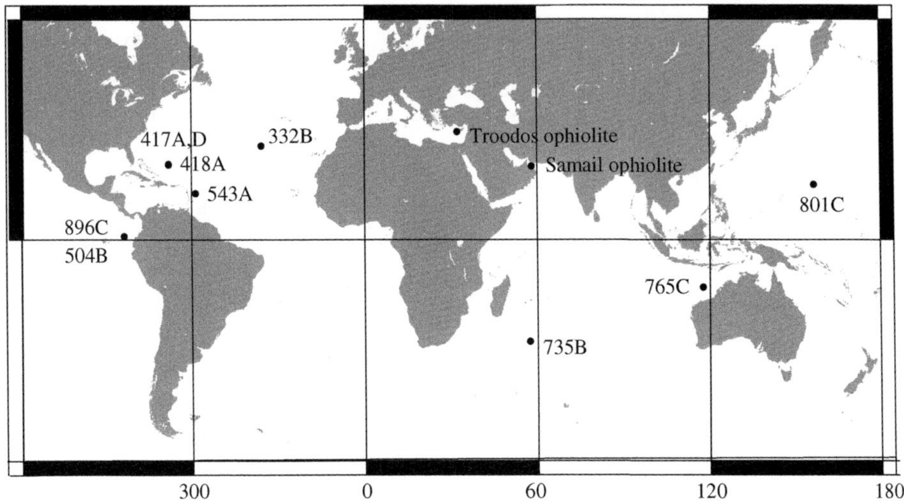

Figure 1 Locations of the drill sites and ophiolites discussed in the text.

Table 1 Ophiolite and ocean-crust drill hole locations.

Site	*Lat.*	*Lon.*	*Age* (Ma)	*Ocean/Region*	*Depth* (m)	*Recovery* (%)	*Comments*
332B	36.945	33.641	3.5	Atlantic	580	21	W MAR
417A	25.106	−68.041	120	Atlantic	206	70	Highly altered 150 m abyssal hill
417D	25.106	−68.041	120	Atlantic	366	70	Normal crust
418A	25.035	−68.057	120	Atlantic	544	72	Normal crust
504B	1.227	−83.734	5.9	E. Pacific	2,000	20	MORB extrusives/dikes
543A	15.712	−58.654	82	Atlantic	44	81	MORB extrusives
735B	−32.722	57.264	9	Indian Ocean	1,000		MORB gabbros/troctolites
765C	−15.976	117.592	155	Indian Ocean	247	31	Normal MORB extrusives
801C	18.642	156.360	157	W. Pacific	135	60	Alkalic basalts and normal MORB extrusives
896C	1.217	−83.723	5.9	E. Pacific	290	28	Normal MORB extrusives
Troodos	34.800	32.000					
CY 1	35.048	33.179	92	Tethys/Cyprus	1,150	95	Suprasubduction zone extrusives
CY 1A							
CY 4			92	Tethys	2,300	95	Dikes gabbros, ultramafics
Samail	24.500	58.000	120	Tehtys/Oman			Normal MORB

All locations given in decimal degrees, negative values give southern latitudes and western longitudes.

independent constraints. Unfortunately, there are too few sections studied for such coverage, and for this reason one has to combine all observations into one "general" type of ocean-crust section. This is clearly an over-simplification, but to a first order a good approximation for the most abundantly produced oceanic crust. Using the stratigraphy of the classic Penrose ophiolite and more modern estimates of unit thicknesses, our standard section (Figure 2) has a total crustal thickness of 7.1 km (White *et al.*, 1992; C. Z. Mutter and J. C. Mutter, 1993), which includes 1,000 m of extrusives, 1,100 m of sheeted dikes, and 5,100 m of gabbros. This "standard" section translates into a volume production rate of 24.14 $km^3\ yr^{-1}$ using Parsons' (1984) estimate of the ocean-crust surface area production rate of 3.4 $km^2\ yr^{-1}$. This rate of oceanic-crust production can be translated into a mass production rate by using *in situ* densities measured in the oceanic crust. These values range from ~2,700 $kg\ m^{-3}$ for extrusives (Salisbury *et al.*, 1979), to ~3,000 $kg\ m^{-3}$ for gabbros (Dick *et al.*, 2000). The sheeted dikes are extrapolated to have an average density of 2,850 $kg\ m^{-3}$. This yields a total ocean-crust production rate of 7.05×10^{16} $g\ yr^{-1}$. However, it must be emphasized that there are still significant uncertainties in these total fluxes and production rates.

The Troodos and Samail ophiolites (Figure 1) contain all the essential components of our standard section in Figure 2. Both of these

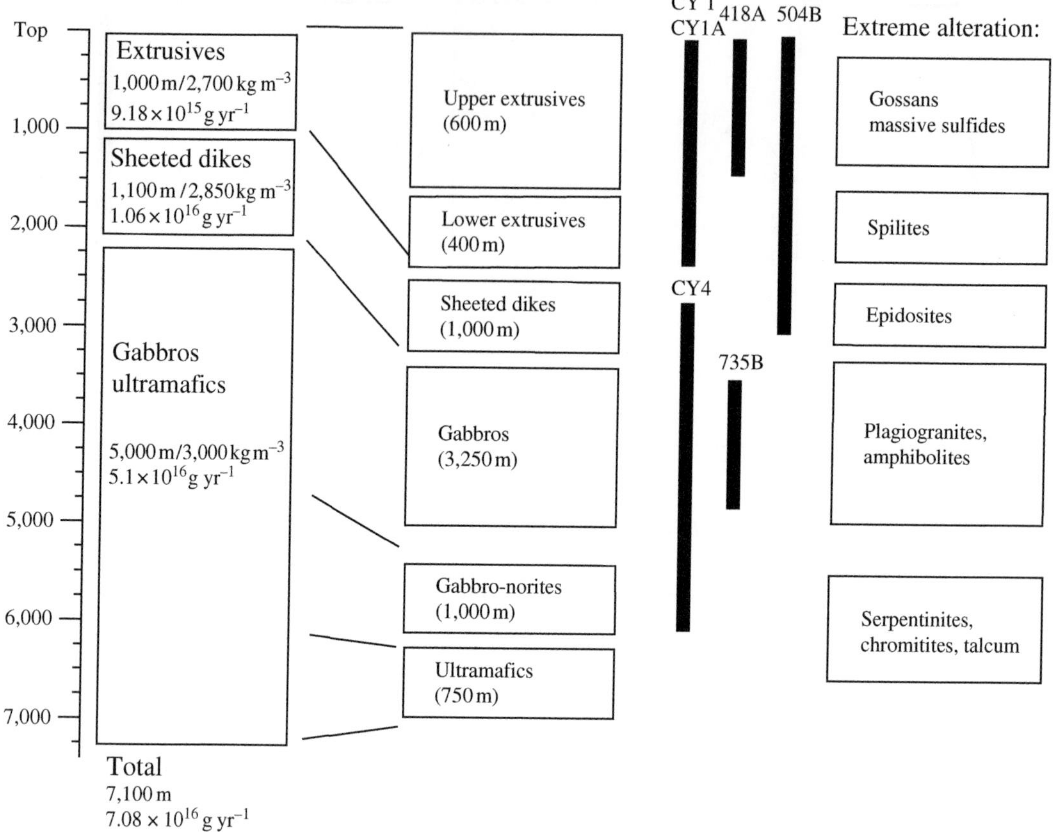

Figure 2 The ocean crust reference section used in this paper, using a standard "Penrose" style ophiolite assemblage (Penrose Conference Participants, 1972) and a crustal thickness after White *et al.* (1992) and C. Z. Mutter and J. C. Mutter (1993). Various modeling parameters used are also indicated, including densities and fluxes, the approximate positions of drill holes discussed in this paper, and some typical extreme alteration environments found at various depths of the oceanic crust.

ophiolites were formed in the ancient Tethys Ocean (92 Ma for the Troodos Ophiolite and 100–130 Ma for the Samail ophiolite). The Samail ophiolite represents crust that is quite similar to "normal" oceanic crust (Kelemen *et al.*, 1997), but it has not been studied for geochemical mass balances. The Troodos was studied extensively for its primary and hydrothermal characteristics but it reflects an oceanic crust that is more typical of supra-subduction zone settings, having geochemical characteristics distinct from normal mid-ocean ridge basalt (MORB) (Robinson *et al.*, 1983; Schmincke *et al.*, 1983). For this reason, the bulk composition of the Troodos ophiolite is not typical of average oceanic crust, even though it is clearly a part of the spectrum that is produced and recycled. The Troodos ophiolite is well studied, partly because of an intense drilling campaign that yielded a total of 8,000 m of core material from all structural levels in the crust. Key drill sites include CY1 and CY1A, with a total penetration of 1,175 m though extrusives (Gibson *et al.*, 1991a), and CY 4, that penetrated 2.3 km of sheeted dikes, gabbros, and pyroxenites (Gibson *et al.*, 1991b). Their stratigraphic position is plotted in the reference section in Figure 2. Recovery from all of these drill cores was found to be ~100%.

There are a large number of drill sites in oceanic crust, but four sites stand out for their impact on studies of hydrothermal alteration processes. These include sites 417 and 418, which were studied to constrain the chemical mass fluxes of the upper crust, site 504B for the mid crust or sheeted dikes and site 735B for the gabbro section (Figures 1 and 2; Table 1).

Three drill holes at sites 417 and 418 probe upper oceanic crust of the western Atlantic. This crust formed at the relatively slowly spreading mid-Atlantic ridge, which may be more fractured than fast spreading crust, and contain more pillows and less massive flows. Site 417A, was drilled into a 150 m high abyssal hill with a total penetration of 208 m, whereas sites 417D and 418A were drilled in the nearby abyssal plain, with penetrations of 360 m and 544 m, respectively. Recovery rate was about 70% at all of the sites. Site 417A contains, on average, 20%

volcaniclastics, whereas sites 417D and 418A contain only ~6% volcaniclastics (Robinson *et al.*, 1979). The higher abundance of volcaniclastics and the more exposed nature of Site 417A resulted in more intense alteration than at sites 417D and 418A (Donnelly *et al.*, 1979a,b,c). Lithologies at Sites 417D and 418A are quite similar, but the crust was formed at different mid-ocean ridge volcanoes, offering a comparison between two separate but quite similar crustal sections. Staudigel *et al.* (1995, 1996) combined these sites into one representative section of almost 600 m length, including 20% materials from 417A and 40% each from 417D and 418A in the upper 150 m, all adjusted to 6% volcaniclastics.

Additional sites that sample upper oceanic crust include 332B, 504B, and 801C (Figure 1; Table 1). Site 332B was also drilled in the Atlantic on very young crust, and 504B and 801C were drilled in the eastern and western Pacific, respectively. Amongst those, site 504B is the deepest drill hole into ocean crust, sampling both extrusives and the underlying sheeted dikes, thus offering a unique opportunity to study alteration processes in the deeper reaches of the crust (Alt *et al.*, 1993a,b). Site 801C is located in the western Pacific, on 170 Myr old oceanic crust and penetrated almost 500 m of basalt with average recovery rate of 47% (Plank *et al.*, 2000). This hole recovered alkalic basalts near the top, overlying a substantial section of "normal" oceanic crust below. The deeper portion of this hole is still under investigation.

Samples of the deepest oceanic crust are accessible in only one ocean drill core: site 735B was drilled on 11 Myr old crust on the SW Indian Ridge (Dick *et al.*, 2000). This hole penetrated a few tens of meters of pillow lavas and ~1,200 m of gabbros, with ~100% recovery. The site never reached cumulates (Dick *et al.*, 2000). The lack of sheeted dikes and the near absence of pillow lavas is clearly quite different from the "normal" ophiolitic crust, but site 735 rock types are not very different from materials found in the upper plutonic section of ophiolites. Thus, site 735B is used to represent the deeper crust for the composite crust section.

3.15.2.2 Estimating Unaltered Oceanic Crust Compositions

Estimating the unaltered composition of the oceanic crust is necessary for flux estimates because these estimates are determined as the difference between original and altered compositions. Table 2 displays an average unaltered mid-ocean ridge basalt composition (MORB, Hofmann, 1988) and estimate of an unfractionated MORB composition (Sun and McDonough, 1989), but it is important to emphasize that the composition of MORB can vary substantially (e.g., Chapter 3.13 or http://petdb.ldeo.columbia.edu; Lehnert *et al.*, 2000). This variation is too large for most elements to allow the use of a single fresh "average MORB" as a starting composition in chemical flux estimates. Thus, fresh compositions have to be constrained for any particular suite of altered basalts studied.

There are two main ways in which the "fresh-rock" composition of an altered rock is determined. One is a petrographic–geochemical approach, which involves re-assembling a rock from the chemical analyses of residual igneous phases still present in the rock using their original modal abundances. Another is a purely geochemical approach that uses the chemical systematics in fresh rocks and placement of an altered rock in such fresh-rock systematics using alteration insensitive parameters. Deviations from the extrapolated fresh-rock composition are then interpreted as being caused by alteration processes. In both methods, there are significant errors. Interpretations of chemical trends have to contend with the natural scatter of sample distributions in correlation diagrams, as well as the cumulative analytical errors introduced from regressions, normalizations, and analytical uncertainty. Petrographic–geochemical reconstructions are affected by small-scale heterogeneities in mineral distribution and uncertainties in modal estimates. These errors are especially large for major elements or trace elements that are compatible in olivine and plagioclase, including magnesium, calcium, silicon, sodium, aluminum, iron, nickel, and barium. These errors are the lowest for incompatible elements that are present in very low abundances in the original igneous rock, such as H_2O, CO_2, uranium, potassium, rubidium, caesium, REEs, and thorium.

A third approach that is commonly used to constrain chemical fluxes compares differently altered materials, such as altered pillow margins and less altered pillow interiors, or samples with or without alteration haloes around veins (e.g., Alt *et al.*, 1986), mineralized and unmineralized zones or differently altered gabbros (e.g., Bach *et al.*, 2001) in order to constrain chemical changes associated with alteration. However, "least" altered samples only rarely reflect the original composition reliably. A second problem in this approach is the relatively small sample sizes typically analyzed from ocean drilling materials. Typical sample sizes are about 15 cm^3, which is small when compared with local variability in modal mineralogy. Indeed, individual phenocryst phases can be several millimeters in size. Local variability in modal mineralogy is particularly common in pillow lavas where phenocryst abundances can vary as a function of radial distance from the center or vertically within the center

Table 2 Hydrothermal fluxes.

	Fresh MORB	*Units*	*Bulk rock gains (+) and losses (−) for extrusives*			*Dikes*	*Gabbros*	*Average crustal gains/losses*	*Units*	*Total flux from crust*	*Units*	*Hydrothermal fluxes from submarine vents* (g yr^{-1})			*River fluxes* (g yr^{-1})
			0–600 m	*Error*	*600–1000*							*Flank*	*Axis low*	*Axis high*	
SiO_2	50.45	wt.%	1.18	1.5	0.5	0.25	0.1	0.23704225	g/100g	−1.67E + 14	g yr^{-1}	1.08E + 12	1.80E + 13	6.61E + 13	3.85E + 14
Al_2O_3	15.26	wt.%	1.94	3						−9.65E + 13	g yr^{-1}				
FeO_{tot}	10.43	wt.%	−0.01	0.03								0.00E + 00	1.65E + 12	1.37E + 13	1.65E + 12
MnO	0.19	wt.%	0	0.01	0.75	−0.035						3.69E + 10	4.97E + 11	3.97E + 12	3.48E + 11
MgO	7.58	wt.%	−0.19	1	0	0	−0.2	−0.15690141	g/100 g	1.11E + 14	g yr^{-1}	−2.18E + 14	−4.03E + 13	−1.05E + 14	2.18E + 14
CaO	11.3	wt.%	2.14	1.5	0.3	0.04	−0.1	0.13352113	g/100 g	−9.43E + 13	g yr^{-1}	2.64E + 14	3.03E + 11	1.23E + 14	6.73E + 14
Na_2O	2.679	wt.%	0.15	0.2	0.15	0.15	0.03	0.06549296	g/100 g	−4.63E + 13	g yr^{-1}	−3.84E + 13			3.53E + 14
K_2O	0.11	wt.%	0.54	0.15	0.05	0	0	0.0484507	g/100 g	−1.71E + 13	g yr^{-1}	−3.11E + 13	6.12E + 12	3.01E + 13	1.22E + 14
Rb	1.26	ppm	10.3	3	1	0	0	0.92676056	mg/kg	−6.55E + 09	g yr^{-1}	−2.22E + 09	1.37E + 10	2.39E + 10	3.16E + 10
Cs	0.0141	ppm	0.183	0.050	0.03	0	0	0.01715493	mg/kg	−1.21E + 08	g yr^{-1}				6.40E + 08
CO_2	0.15	wt.%	3.26	0.5	0.5	0.06	0.06	0.35521127	g/100 g	−2.51E + 14	g yr^{-1}	−7.04E + 12	−1.76E + 12	−7.04E + 12	1.41E + 15
H_2O	0.2	wt.%	2.81	1	2.09	1.5	0.11	0.44873239	g/100 g	−3.17E + 14	g yr^{-1}				
S	960	ppm	−0.06		−0.06	−0.06	0.02	−0.0037	g/100 g	2.59E + 12	g yr^{-1}	−3.53E + 13	−1.60E + 13	−4.49E + 13	2.85E + 13
Li	4.5	ppm	2.8		2.8	−0.5	−2	−2	mg/kg	7.71E + 09	g yr^{-1}	−1.25E + 10	4.8587E + 10	4.37E + 11	9.72E + 10
B	0.5	ppm	25.7	10	4	5.6	2.2	4.81	mg/kg	−3.40E + 11	g yr^{-1}	1.84E + 11	6.49E + 09	7.89E + 10	5.84E + 11
Sr	113	ppm	22	5	3	0.3	−1	1.4	mg/kg	−9.68E + 10	g yr^{-1}	2.19E + 11	−9.00E + 08	4.60E + 09	2.02E + 11
U	0.0711	ppm	0.3	0.05	0.1	0.01	0.007	0.03746479	mg/kg	−2.65E + 09	g yr^{-1}				9.60E + 09
							0.3			−3.60E + 14	g yr^{-1}	1.90E + 14	−1.14E + 13	1.02E + 14	2.23E + 15

of a pillow. Staudigel *et al.* (1996) estimated errors in chemical composition on the basis of variation in phenocryst modal abundances and found relatively large errors for some elements, particularly MgO (e.g., 7.7 ± 0.2 wt.%).

One of the key objectives in estimating fresh rock compositions is to determine the chemical inventory present before alteration in a given reference volume of altered rock. This is different from the "fresh-rock composition," because water–rock interaction occurs in an open system. The pitfalls arising from open-system behavior can be illustrated in two examples: secondary mineral precipitation and rock dissolution.

In rocks with significant pore space, minerals may precipitate in these pore spaces during alteration. For example, precipitation of 5 g per 100 g of calcium carbonate in the pore spaces of a fresh rock dilutes the abundances of all other chemical components by 5%. In this example, carbonate precipitation causes an apparent loss of ~2.5 wt.% SiO_2, when in fact SiO_2 remained immobile during the alteration.

The opposite effect may be caused by wholesale (congruent) dissolution of the rock whereby massive amounts of basalt could be lost from a given rock volume, without any trace of concentration change. However, truly congruent dissolution is quite unlikely; dissolution typically leaves behind some residual immobile material. Titanium is generally assumed to be immobile, and the abundance of titanium in a rock can therefore be used to quantify open-system chemical behavior. The first step in addressing open-system behavior is to estimate the titanium concentration in the fresh rock equivalent, which is taken as the mass of titanium originally present in a rock per 100 g of the material analyzed. The fraction of titanium originally present, together with an independent estimate of the fresh rock composition, allows an estimation of the concentration of all the other elements in the fresh rock. Chemical fluxes, then, are calculated as the differences between elemental concentration in the unaltered rock (per 100 g analyzed) and the concentration of the element in the altered material. The most important sources of uncertainties in this procedure arise from the uncertainties in titanium determination, errors in the estimate of the titanium concentration in the unaltered rock (e.g., Staudigel *et al.*, 1995, 1996) and errors due to small-scale titanium variations when using the titanium from a "fresh"-altered sample pair (e.g., Bach *et al.*, 2001).

It is possible to estimate the validity of the constant titanium assumption through a comparison with other, independent methods, such as working with constant reference volumes and estimating the amount of initial rock present based on density and pore space consideration. Specifically, if the density is known for the fresh and the altered rocks, fresh and altered composition can be expressed on a volume percent basis, and fluxes are calculated relative to a constant volume. This method was evaluated at sites 417 and 418 and shown to produce almost identical results as the constant titanium assumption (Staudigel *et al.*, 1996).

3.15.3 DETERMINING THE ALTERED COMPOSITION OF THE OCEANIC CRUST

Determining the composition of altered oceanic crust is also not very straightforward. Several key steps are required in obtaining a meaningful estimate of altered rock compositions. First, the choice of a study site is important. Is the section old enough to have experienced the bulk of its alteration processes? Are the alteration patterns representative of most oceanic crust, or are they only of local importance? Is core recovery sufficient to allow a representative estimate of the section drilled? Once these three questions are answered positively, a method has to be worked out to determine the bulk composition of the crust and chemical fluxes on scale lengths that are meaningful for global chemical budgets. The following sections evaluate problems related to these questions, in particular the role of recovery, the range of "typical" types of alteration, constraints on the duration of alteration in the crust and techniques for determining representative compositions of a heterogeneous medium. These problems are illustrated with examples from particular ocean drill sites or ophiolites.

3.15.3.1 Recovery Rate

It is extremely rare in ocean drilling to have 100% recovery of basement materials. Most drill cores from basement consist of variously rounded fragments that typically don't match up with the neighboring core fragments. Recovery rates are low, particularly in the upper and young oceanic crust, which has not been effectively sealed through mineral precipitation and fractured on the same scale length as the drill bit (and smaller). Competent rock fragments, resistant to the cutting action of a drill bit, are held together by relatively soft materials like clays, carbonates, or chlorite. The stress imposed on resistant rock fragment is likely to break up the loosely cemented formation, and grind up the soft material filling the void and fracture spaces in between. This material is then ejected from the hole with the drilling fluid on the ocean floor as mud, sand, or chips and not in the core barrel. This problem also affects the inner walls of the drill hole, where competent rocks are more likely to be exposed than soft rocks. Most alteration materials are not found in distinct horizontal layers, but occur as pockets between or within pillows, or as veins that commonly cut the hole at very steep angles, which can be completely eroded, even though the neighboring material is actually recovered. A similar sampling bias occurs in outcrops of ophiolite sections, where clays are much more likely to be eroded than fresh, well-cemented basalt. These problems can be overcome only by complete recovery of drill core.

The recovery rates of various ocean-crust drill holes are given in Table 1. Recovery rates are determined as the ratio of the cumulative length of recovered fragments to the actual length of the section drilled. This estimate is quite meaningful at high recovery rates, but not at low recovery rates, where individual pieces tend to be rounded off and have a smaller diameter, resulting in volume recovery rates that are substantially smaller than the linear rates. Thus, low recovery rates practically eliminate substantial portions of a drill hole from direct study and they can make it very difficult, if not impossible, to estimate bulk compositions from bulk rock data, particularly in pillow sections and breccia zones. Drilling in oceanic crust often has very low recovery, on the average of 25–30%, with some sections having nearly zero recovery (e.g., a substantial fraction of 504B). At an average recovery rate of 30%, proportionately more material is recovered from massive units (50–100%), than breccia zones (0–10%) complicating the estimates of alteration inventories. Ocean drill sites with the highest recovery rates include site 801C, with almost 50% recovery, 417A, 417D, and 418A, with over 70% recovery, and 735, with 86% recovery. The Cyprus drill cores CY 1 and CY 4 have nearly 100% recovery. The other deep drill sites discussed here, 504B and 332B have substantially lower recovery rates, on the order of ~20–30%.

3.15.3.2 Types of Alteration

One of the key goals in quantifying alteration is to determine the respective contributions of different types of alteration. The focus of these efforts have to be on what is considered "normal" or "most representative," but it is also important to explore the complete range of alteration behavior in the oceanic crust. Even unusual alteration environments may have a significant impact on the geochemical behavior of some elements, and they may cause distinct geochemical behavior in subduction zones or in the mantle. For this reason, it is important to focus on the "average" geochemical behavior, but keep an eye on the compositional diversity as well.

Studies of ophiolites have been particularly useful in identifying types of alteration in oceanic crust and for understanding their relative significance. Many of these alteration features have also been found in ocean drill cores, but many aspects of seafloor hydrothermal alteration remain unexplored by ocean drilling.

The upper extrusive oceanic crust (0–600 m, Figure 2) is primarily altered at low temperatures (<100 °C). Alteration is commonly *not* pervasive, whereby igneous phases (glass, phenocrysts) may coexist with alteration phases (clays, zeolites, and carbonates). At high water–rock ratios, oxidative mineral assemblages (e.g., celadonitic clays) form intensely altered zones, but most ocean-crust alteration is accomplished by more reducing fluids at lower fluid–rock ratios (Table 3). However, in most low-temperature altered basalts, highly reducing minerals (like pyrite) may coexist in close proximity with oxidizing minerals like celadonite or hematite. The distinctive differences between mostly oxidized and reduced alteration environments are well illustrated by comparing the oxidative upper (abyssal-hill) portion of site 417A to the more reduced sections at sites 417D and 418A (e.g., Donnelly *et al.*, 1979a,b,c; Alt and Honnorez, 1984). A similar contrast exists between Cyprus drill cores CY 1 and CY 1A, whereby CY 1, from the uppermost Akaki canyon, closely resembles the oxidized alteration in site 417A and the remaining core shows alteration behavior more like site 417D and 418A (Gillis and Robinson, 1988; Bednarz and Schmincke, 1989). Based on the distribution of alteration features in the northern portion of the Troodos ophiolite, Staudigel *et al.* (1995)

Table 3 Secondary minerals in the oceanic crust.

Upper extrusive crust		*Deeper extrusives*
Oxidizing conditions	*Nonoxidizing*	
Aragonite, analcite, calcite, celadonite, chalcedony Fe-hydroxide, hematite, philipiste, K-feldspar, saponite	Anhydrite, analcite, calcite, celadonite, Fe-hydroxide, mixed layer chlorite–smectite, Na zeolite pyrite, saponite	Albite, calcite, chlorite, epidote, pumpellyite, prehnite, quartz, sphene

suggested that the 417A style of oxidative alteration makes up ~20% (±10%) of the upper 150 m of oceanic crust, while the reduced style of alteration seen at sites 417D and 418A are representative of 80%.

The upper oceanic crust can show extreme compositional variation near hydrothermal vents. Based on the abundance of hydrothermal deposits in the Troodos ophiolite, and the frequency of black-smoker-type deposits in mid-ocean ridges, these highly mineralized zones are likely to represent less than one-tenth of a percent of the volume of the total upper oceanic crust. Thus for most elements, these deposits are not very important in the total mass balance. The exception to this is elements that are enriched in vent deposits by several orders of magnitude relative to basalt (e.g., copper, zinc, lead, manganese, iron, nickel, cobalt, platinum, silver, gold; see also Chapter 3.12). This limits the significance of these deposits to trace-element mass balances in the oceanic crust. A convincing case for this was made by Peuker-Ehrenbrink *et al.* (1994), who suggested that global lead cycles in seawater might be substantially influenced by the precipitation of a small amount of hydrothermal metalliferous sediments on top of the oceanic crust.

In the deeper extrusive oceanic crust and in the sheeted dikes, alteration temperatures increase (>100 °C), water–rock ratios decrease, and most primary igneous phases tend to be almost entirely replaced by secondary phases (the exception again is near hydrothermal conduits, where temperature and water–rock ratios are high). In this depth range glass, olivine, and calcic plagioclase are typically replaced by greenschist-facies mineral assemblages (Table 3). Typical phase assemblages in deep extrusives and dikes for *in situ* oceanic crust are best described at site 504B (Alt *et al.*, 1986, 1993, 1996), offering important insights into alteration processes in this depth range.

Several important alteration environments, however, are not observed for *in situ* oceanic crust. These include spilites (Cann, 1969), a rock type that displays almost complete exchange of calcium for sodium, leading to formation of an albite-rich rock, and epidosites, the metal depleted epidote–quartz–chlorite assemblage that is likely to be characteristic of the reaction zones, or at least last equilibration zones of black-smoker fluids (e.g., Schiffman and Smith, 1988; Richardson *et al.*, 1987; Bettison-Varga *et al.*, 1992). In particular, the importance of spilites in chemical mass balances could be large, but they are effectively unknown for *in situ* oceanic crust. Spilites are therefore not considered in any mass balances. Epidosites are likely to be important for the mass balance of some trace metals in the oceans, but are unlikely to influence the ocean crust generation–subduction budget because they are probably just as uncommon as the massive sulfide deposits. Another potentially important alteration assemblage is predicted from experimental studies and observations from black-smoker chemical compositions. Heating seawater to 450 °C results in massive precipitation of sulfate (mostly anhydrite; Bischoff and Seyfried, 1978) and may result in boiling, leaving behind brines (Butterfield *et al.*, 1990). Anhydrite and brines are likely to be stored only temporarily in the oceanic crust and will be dissolved by seawater circulating through the crust at a later time after much of the magmatic heat is exchanged with seawater (e.g., Alt, 1994, 1995).

Deeper levels in the oceanic crust display higher-grade hydrothermal alteration, ranging from amphibolite grade to anatexis (Table 3), where the formation of plagiogranites has been associated with the partial ingestion of hydrothermally altered ocean crust materials. Hydrothermal alteration processes in this regime have been studied in materials recovered from oceanic fracture zones, particularly in site 735B (Dick *et al.*, 2000), as well as in ophiolites (Nehlig *et al.*, 1994). Overall, it appears that the upper 200 m section of site 735B displays unusually high water–rock ratios due to its proximity to seawater, but the bulk of this 1.5 km section are more typical of normal crust (Dick *et al.*, 2000). However, 735B does not display any of the extreme varieties of alteration observed in some ophiolites or in near-surface exposures on the ocean floor, such as serpentinites, or other hydrous equilibrium assemblages such as talcum deposits (e.g., Bonatti, 1976; Baer, 1963). Such compositional domains are likely to be formed at relatively slow spreading centers where brittle deformation may carry water to great depth, deeper than in the faster-spreading oceanic crust. Such deposits could potentially be very important for mass balances, particularly for recycling of water, but are not included in this review due to lack of data.

3.15.3.3 Duration of Alteration

Determining that alteration is complete or near complete is one of the major pre-requisites for determining alteration fluxes at any study site, and for this reason, it is important to determine the duration of alteration in the oceanic crust. There are two main constraints for this: measurements of heat flow, which indicate the duration of convective heat loss, and isotopic dating of secondary minerals that precipitate during alteration of the oceanic crust.

Heat-flow measurements can be used to determine the duration of seafloor alteration by comparing the absolute estimate of heat flow to heat flow calculated from plate-cooling models. The difference between the two is then assigned to convective heat flow. The difference between theoretical and measured heat flow suggests total convective heat loss of approximately 11 TW (Stein *et al.*, 1995). About one-third of this heat loss occurs within the first million years after crust formation, the second one-third occurs between 1 Ma and 8 Ma, and the last one-third occurs between 8 Ma and 65 Ma (Stein *et al.*, 1995; Figure 3). However, the "termination" of convective heat loss at 65 Ma also coincides with the time beyond which the simple relationship between ocean-floor depth and plate-cooling models break down. Thus, a comparison of measured and theoretical heat flow is not very meaningful at this age.

Alternatively, the duration of hydrothermal convection in the oceanic crust can be estimated by mapping the distribution of nonlinear temperature profiles taken during heat-flow measurements as a function of oceanic-crustal age. Purely conductive heat loss (i.e., no hydrothermal circulation) results in linear temperature profiles in sediments, while convective heat loss results in concave or convex profiles, depending on whether the water penetrates into or comes out of the sediments.

Both heat-flow methods deliver similar results, suggesting that convective heat loss, hence hydrothermal alteration, may occur for time periods of up to ~65 Ma (Parsons and Sclater, 1977; Langseth *et al.*, 1988; Anderson *et al.*, 1979; Stein *et al.*, 1995; Pelayo *et al.*, 1994).

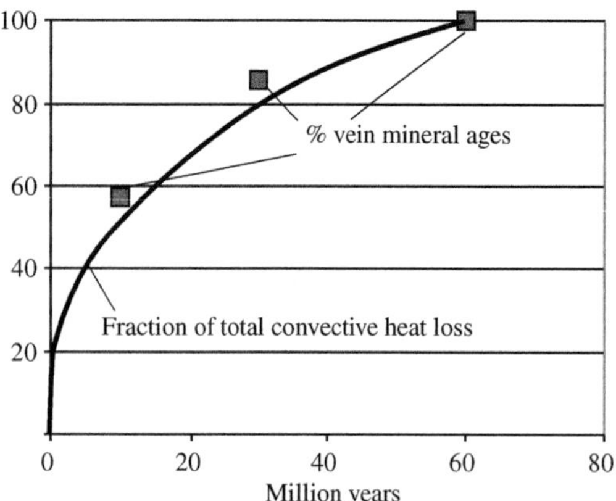

Figure 3 Cumulative convective heat loss (Booij *et al.*, 1995) of the oceanic crust and completion of vein mineral deposition (Stein *et al.*, 1995) in the oceanic crust. Heat flow curve after C. A. Stein and S. Stein (1994b), including vein mineral ages from site 261 after Hart and Staudigel (1986) site 417/418 after Hart and Staudigel (1978, 1986) and Richardson *et al.* (1980), site 462A after Hart and Staudigel (1986) site 516 (the 18 Ma Rio Grande Rise), and 597 (South Pacific) after Hart and Staudigel (1978, 1986) and the Troodos Ophiolite after Staudigel *et al.* (1986), Booij *et al.* (1995) and Gallahan and Duncan (1994). Vein mineral age cumulative curve is binned in 20 Myr age groups.

The duration of chemical exchange between seawater and basalt can also be determined by isotopic studies of hydrothermal minerals. A variety of techniques have been used, ranging from direct dating by K/Ar and Rb/Sr isochron techniques to comparisons of the initial strontium isotopic composition of alteration minerals with the isotopic evolution of seawater (Gallahan and Duncan, 1994; Richardson *et al.*, 1980).

Compiled isochron data for the duration of alteration at a series of DSDP and ODP sites are shown in Figure 3. The ages thus calculated are the differences between crustal age and vein-mineral age. Due to the rather large spread of dates, ages were binned into 20 Ma intervals, with averages at the mid point of each interval shown in Figure 3. Even though rather crude, this distribution shows a remarkable resemblance to the heat-loss curve. While the resolution of sampling does not allow any strong constraints on the actual age distribution in the (most critical) first 20 Ma, they do show very clearly that vein-mineral precipitation can continue for a rather long time in this global distribution of sampling sites. Collectively, these observations suggest that oceanic crust younger than 10 Ma is unlikely to have experienced the complete cycle of seafloor alteration.

3.15.3.4 Determining the Composition of Extremely Heterogeneous Altered Crust

Oceanic-crust alteration involves formation of distinct compositional domains, which range in size from a ridge segment to a submilimeter sized vesicle filling. Obtaining a robust average of such chemically heterogeneous material is critical for understanding the chemical fluxes associated with seafloor alteration. Averaging heterogeneities with length scales smaller than the size of a typical geochemical sample is relatively easy, but becomes more difficult as the size of the heterogeneity increases. For large-scale heterogeneities, multiple samples need to be taken and characterized with respect to the proportion each sample contributes to the average. Such samples may be mixed in the correct proportions to produce composite samples that represent an average for a given section or crust type, or they may all be analyzed individually and then averaged arithmetically.

Large-scale geochemical sampling requires a strategy that bridges the gap between sample size and the scales of key heterogeneities. Heterogeneities on the order of several meters to tens of meters are probably the smallest relevant large-scale heterogeneities that need to be averaged. A compositional domain of this size is relevant geochemically because it may reflect a single chemical system, with its own characteristic phase assemblages that display consistent behavior during dehydration or melting events. This length scale is also a practical size for analyses of compositional domains through local observation of continuous exposure or core sections. Data from such a length scale can be relatively easily extrapolated to larger scales by visual integration of (>kilometer scale) cores, or field outcrop areas.

The selection of individual samples for large-scale compositional study may follow different strategies, depending on the type of crust studied and personal preferences. One of the major differences in strategy is the use of composite samples, versus the analysis of individual samples. Procedurally, these two approaches are identical, as long as the sampling strategy is comprehensive and the modal abundance data are determined in a consistent fashion. The two approaches each have advantages and disadvantages.

Analysis of individual samples has the advantage that it offers the potential for resolving the chemical properties of different alteration types, before the samples are combined arithmetically into one average. The disadvantage of this approach is that it requires a large number of analyses and may be too labor intensive for some geochemical parameters. The advantage of composite samples is the significantly lower number of analyses required, which makes it much more likely to obtain a complete geochemical characterization of a representative set of samples at a particular crustal section.

However, it is also possible to find a compromise between the use of composite or the individual samples. Staudigel *et al.* (1995, 1996) mixed one "Super" composite that represents the average of three drill sites, but they also analyzed two types of "subcomposites" to understand uncertainties and the internal structure of the "Super" composite. One type of subcomposite contrasts the compositions of volcaniclastics with flows that allow to contrast extreme with moderately altered regions. In addition, they made independent "depth" composites for two sites with similar alteration types to obtain an estimate of how variable such estimates can be for different sections. These subcomposites provide some insights into the chemical makeup of a site, without having to analyze all samples individually. Alternatively, it is possible to analyze all individual samples for geochemical parameters that are relatively easy to determine, and using composites only for the more involved analytical steps. The latter was done for all the highly altered volcaniclastic samples in the composites from sites 417 and 418 (Staudigel *et al.*, 1996).

Strategies for selection of individual samples may also vary. One approach is to use relatively large, representative samples that include veins, vugs, and variously altered haloes or other materials (Staudigel *et al.*, 1989, 1995, 1996). Because of the complexity of alteration and the large number of permutations of primary and secondary features, several samples are typically required to approximate the chemical inventory of a particular compositional domain. This approach has the benefit of averaging out local heterogeneities or chemical gradients between veins and host rock, and minimizing uncertainties and contamination problems during sample handling. Another approach is to separate all major alteration types and vein materials, study them individually and recombine them in their respective proportions (e.g., Alt and Teagle, 1999; Bach *et al.*, 2001). The advantage of this method is that it sheds light on the alteration processes and their impact on chemical fluxes more fully than composite samples do. The disadvantage is that it produces larger uncertainties in the overall averages, and requires substantially more sample handling and analyses.

Any attempt to produce a bulk compositional estimate of altered ocean crust is critically dependent on accurate determination of the proportions of different rock types in the volume of interest. Simply averaging all analyses published for a particular drill site does not yield a realistic bulk compositional estimate for two reasons. First, low recovery during drilling biases the average towards the least altered material, whereas high recovery rates include a greater proportion of altered material. A hypothetical case of several holes drilled into the same crust at different recovery rates will show a positive correlation between degree of alteration and recovery rate. Secondly, the choice of samples for a particular study depends strongly on the scientific goals of that study. Sampling for igneous geochemistry studies typically focuses on the least altered samples. Such studies often further bias the sample selection by crushing the rocks and selecting the freshest rock chips out of a sample for analysis, and leaching carbonates out of the rock to obtain a maximally pristine chemical composition of an igneous rock. Such studies are obviously not useful in constraining the amount of alteration at a site. On the other hand, studies of alteration mineralogy of the crust focus their sampling on highly altered sections, especially extreme alteration types. Thus, a simple average of published geochemical data from a particular site is likely to offer insights mostly into the recovery rate and the types of studies performed rather than the actual average composition of oceanic crust at a particular site.

3.15.4 CHEMICAL CHANGES IN ALTERED CRUST COMPOSITION DUE TO HYDROTHERMAL PROCESSES

3.15.4.1 Time Dependence of Crust Hydration and Carbonate Addition

Uptake of water and CO_2 is one of the most sensitive indicators of alteration, and produces some of the most profound chemical changes in altered oceanic crust. These chemical changes are cumulative and offer the opportunity to evaluate the age dependency of alteration in drill sites of various ages. Presently, there are three sites in the upper oceanic crust that can be used for such comparisons: the 3.5 Myr old DSDP site 332B, the 5.5 Myr old site 504B, and the 120 Myr old sites 417A, 417D, and 418A (Figures 4(a) and (b)). Sites 417 and 418 are in oceanic crust that is substantially older than the predicted duration of hydrothermal alteration derived from heat-flow data (Figure 3) and thus these sites represent mature oceanic crust, which has experienced the complete history of hydrothermal alteration. Crust penetrated in site 504B is much younger than that in sites 417/418, and is in the age range of crust that is expected to have active hydrothermal alteration, based on the global heat-flow data set, even though local heat-flow measurements suggest that alteration is completed at this site (Langseth *et al.*, 1988). Crust of site 332B is even closer to the mid-ocean ridge and is expected to be experiencing active hydrothermal circulation. Unfortunately, sites 332B and 504B have very low recovery rates, and for this reason, some effort is needed to determine the true differences in time-variant alteration behavior.

The highest contents of H_2O are found in the upper 300 m of crust from sites 417A, 417D, and 418A (Figure 4(a)). At the same depth, the youngest crust at site 332B shows the least intense hydration, while crust at 504B has intermediate water content. Water contents of crust from 417/418 decrease with depth, while water contents at 332B and 504B increase slightly with depth before decreasing again. All trends converge at depths greater than 300 m. Differences in recovery rates is an unlikely cause for the trends observed in the upper 500 m of this diagram, and it is most likely that 332B and 504B have not reached a mature degree of hydration in their upper portion, while 504B is likely to have completed its hydration at depths >600 m. Thus, the formation of layer silicates in the upper portion of 504B and 332B is probably an ongoing process.

CO_2 shows behavior similar to water. Its concentration is also highest in the upper 300 m of crust at sites 417 and 418 and lowest in 504B throughout the depth intervals covered by all holes. It is interesting to note that crust of site

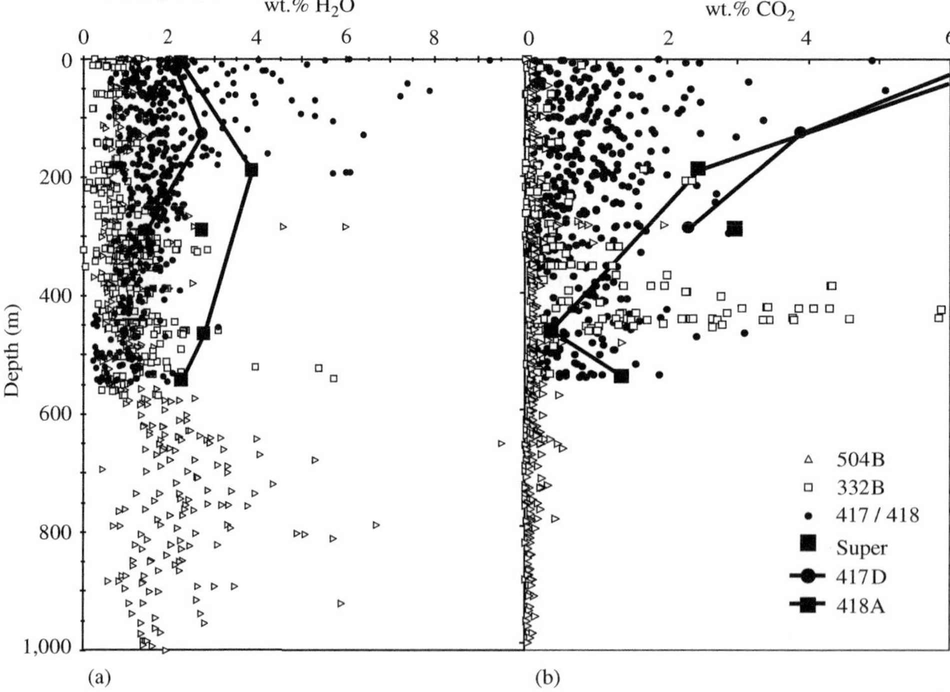

Figure 4 (a) Water versus depth and (b) CO_2 versus depth for major deep drill sites in the Atlantic and Pacific Oceans. Data compilations are taken from the Initial Report Volumes of the Deep Sea Drilling Project or the Ocean Drilling Project (site 332B: Aumento and Melson, 1977; 417/418 Donnelly *et al.*, 1979b; site 504B: Alt *et al.*, 1993; bold lines connect depth composite data from sites 417D and 418A (after Staudigel *et al.*, 1995).

332B has higher CO_2 values than that of 504B, particularly in the depth interval 400–500 m, even though it samples younger crust. With increasing depth, CO_2 values of 417/418 crust show a systematic decrease to values that are substantially above the values from site 504B for some depth. These observations indicate that CO_2 uptake in oceanic crust is highly variable in younger crust, and both young crustal sections are distinct from mature ocean. Site 504B appears to have particularly low CO_2 abundances. This, combined with incomplete hydration, suggests that 504B crust has not experienced all of the carbon uptake it will see throughout its lifetime, and for that reason it is not a reliable reference point for extrapolating to global mass balances. This is acknowledged in the global carbon budget of Alt and Teagle (1999) who ignored the data from the upper portion of 504B.

3.15.4.2 Chemical Fluxes between Oceanic Crust and Seawater: Methods and Uncertainties

Geochemical exchange between seawater and the oceanic crust varies substantially as a function of depth and lithology, and for this reason alteration processes are considered separately for different portions of the crust in the 7.1 km ocean-crust reference section in Figure 2. Chemical fluxes are given as Δ values, the difference in measured and original chemical inventory in g/100 g fluxes (at constant Ti), independently for each depth interval in the oceanic crust. An average altered MORB may be obtained by adding this flux to the fresh MORB listed in Table 2.

The flux estimates in Table 2 primarily derive from the data from sites 417D and 418A for the upper crust, and site 735B for gabbros of the deep crust, because these are the only drill sites with high recovery rates and total depths greater than 300 m that sample crust older than 10 Ma, which is considered a minimum age for "mature" or near-mature ocean crust alteration. Study of sites with a critical depth of penetration (>100 m?) is more likely to help us understand their true behavior than studies of very shallow sites where it is difficult to establish the overall alteration behavior. In particular, sites 417A, 417D, and 418A allow for an in-depth understanding of localized anomalies.

For sites 417/418, the fluxes are calculated from the "super" composite, which is dominantly composed of materials from sites 417 and 418A and offer a grand average for these sites. In its upper 200 m, this composite includes about 20% of the oxidatively weathered materials from site 417A. Sites 417/418 and 735 are the only deep drill holes displaying average recovery rates >70%, allowing a quantitative reconstruction of the alteration inventory. The main sources of data for these sites come from Staudigel *et al.*, 1989, 1995, 1996; Smith *et al.*, 1995; Spivack and Staudigel, 1994; Hart *et al.*, 1999; Bach *et al.*, 2001. However, for the intermediate crustal sections, data from site 504B was also utilized (e.g., Alt *et al.*, 1993a,b; Alt and Teagle, 1999; Chan *et al.*, 2002). This procedure was adopted because there are no other data from *in situ* crust. It is important to note that use of these data leads to increased uncertainties as alterations may not be complete and the recovery rates are very low for 504B. The results from these deep-sea cores are also compared to evidence from the Troodos and Samail ophiolites (Bednarz and Schmincke, 1989, 1994; Bednarz, 1989; Bickle and Teagle, 1992; Spooner *et al.*, 1977). However, in many cases, fluxes were extrapolated between site 735 and Sites 417/418, using ophiolite data or data from 504B only for comparison. This procedure was necessary for a variety of reasons including lack of data, incomplete alteration due to young oceanic-crustal age, and to produce the most conservative estimate.

Obtaining realistic errors is one of the most difficult, yet most crucial problems in all flux estimates. Such errors can be approximated through an independent error analysis for several factors that are involved in estimating fresh and altered rock composition. There are uncertainties arising from petrographic observations, in the choices of representative samples, recovery rate biases, and analytical errors. In most cases analytical errors are a relatively minor source of uncertainty, and they are typically rather well documented. Probably the most crucial analytical uncertainty is in acurately determining the titanium concentration that is used as a normalizing factor to account for open-system behavior. This uncertainty directly relates to an error in the fluxes, and thus fluxes are difficult to constrain to better than ~1% of the whole rock abundance of a particular element.

Uncertainties in phenocryst abundance estimates are a significant source of error, particularly for major and highly compatible elements in the main phenocryst phases. In most cases related to ocean drilling samples, phenocryst abundances in a particular sample are rarely known to better than about 5 vol.%. The most abundant phenocrysts are olivine and plagioclase, which contain the elements magnesium, silicon, calcium, aluminum, and nickel.

Detailed uncertainty estimates for major elements are given in Staudigel *et al.* (1996, table 6). An indication of the overall reproducibility of flux estimates can be obtained by comparing independent data from different sites having similar alteration behavior, like site 417D and 418A. The H_2O and CO_2 contents for composites at sites 417D and 418A as a function of depth are compared to the large scale, bulk composition at these sites in Figure 4. CO_2 abundances at these two sites track each other very well, with an uncertainty between these sites of ~0.5%. In contrast, the two H_2O curves vary more widely (±1%). Close agreement between composites from these sites suggest a high degree of reproducibility, while a large scatter suggests substantial variation between sites or a problem with reproducibility. In the following analysis this scatter within and between sites is used to indicate how robust these flux estimates are for the upper oceanic crust.

3.15.4.3 Chemical Fluxes

H_2O. When discussing the water contents of the oceanic crust it is important to know that it may be found in three forms that are measured differently and that are relevant for different applications. Ocean crust contains (i) formation waters filling pore spaces, (ii) exchangeable water in minerals like clays or zeolites, and (iii) water bound to crystalline structures or dissolved in silicate glass. Formation water can be estimated from the pore space available in the formation, which is typically done by borehole logging. Exchangeable water (H_2O^-) is typically defined as the water expelled at 110 °C, even though the release of this type of water is rather continuous and some fraction may occur at temperatures higher than this. Crystal-bound water is referred to as H_2O^+ and it is determined in rock powders by titration or chromatography following ignition at temperatures >800 °C, or, *in situ* by spectroscopic methods. In most geochemical studies H_2O is understood to be H_2O^+, in particular when H_2O^- is not reported. Non-degassed, fresh MORB has approximately 0.2 wt.% H_2O. The amount of formation water in the upper crust probably scales directly with the porosity of the crust, as it can be determined by borehole logging. The upper crust has about 10 vol.% porosity (e.g., 13% in site 417D; Salisbury *et al.*, 1979), which would yield roughly 4–5 wt.% formation water for the upper crust. There are insufficient numbers of H_2O^- measurements to estimate the loosely bound water in these sections.

Staudigel *et al.* (1995) estimated a H_2O^+ flux for the upper crust of about 2.81 g/100 g. It is interesting that the average H_2O^+ of composites from the very top of cores 417D and 418A have values that are significantly below the values at 100–200 m depth, and that site 418A has values that are substantially above the values of 417D, giving an approximate uncertainty of ~0.5 wt.% for the supercomposite. Taking this average H_2O^+ as typical of the upper crust and including the estimate of formation water given above yields a total of 7–8 wt.% H_2O. Assuming ~2–3 wt.% for H_2O^- suggests that the upper oceanic crust stores close to 10 wt.% water, a water content that is not much different from those of pelagic sediments. This value is likely to decrease substantially with depth, because pore space decreases with depth and because higher-temperature metamorphic assemblages contain less water.

An estimate of water content of the gabbro section can be made using data from site 735B. Bach *et al.* (2001) suggest that gabbros from site 735B took up about 0.11 g H_2O^+/100 g of rock. Formation water can be constrained from porosities in this site. Core materials from site 735B have porosities of 0.7 vol.%, and *in situ* porosity is likely to be twice as high, probably about 1.5 vol.% (Dick *et al.*, 2000), which is substantially less than that of the upper crust. Filling 1.5 vol.% of the rock with water yields about 0.5 wt.% of H_2O, suggesting that the gabbros are likely to store slightly less than 1 wt. % water—an order of magnitude less than that present in the upper crust. The mass balance uses the H_2O^+ flux of Bach *et al.* (2001), which should be close to a total average for the lower crust (even though 735B penetrated only through its upper portion).

Water contents at intermediate depths are extrapolated from the top of the crust downward using the internal variation in sites 417/418 and the data for minimal alteration from site 504B and site 735B. The H_2O^+ fluxes into the lower extrusives and sheeted dikes are estimated at 2.10 g/100 g and 1.5 g H_2O^+/100 g, respectively (see Figure 4(a)). Adding up and averaging these individual estimates, the oceanic crust as a whole takes up ~0.45 g H_2O^+/100 g of rock. The total water inventory of the whole crust (including H_2O^+, H_2O^-, and formation water) is derived simply by scaling (in a similar fashion as in the upper crust), at ~1.5 wt.%.

CO_2. Degassed MORB has ~0.12 wt.% CO_2, whereas undegassed MORB has ~0.45 wt.% CO_2 (Table 2; Gerlach, 1989). The higher CO_2 in altered MORB is due to CO_2 addition to the crust from seafloor alteration subsequent to initial outgassing. This alteration-related addition of CO_2 to the oceanic crust brings its total inventory up to levels exceeding the original non-outgassed carbon contents (Staudigel *et al.*, 1989). Most of this hydrothermal CO_2 inventory in the oceanic crust is added in the form of carbonates, particularly in the upper 600 m (Figure 3(b), Staudigel *et al.*, 1989). In the estimate of carbon fluxes (Table 2), data for the upper 600 m of sites 417/418 yielded a CO_2 uptake of 3.26 g/100 g (Staudigel *et al.*, 1989). Based on the scatter of the depth composites and the discrepancies between sites 417D and 418A, the error on these

estimates is roughly 0.3 g/100 g. This estimate from sites 417 and 418 was confirmed by Alt and Teagle (1999) using data from sites 843 (90 m penetration) and 801(150 m penetration), which yielded 2.4 wt.% and 3.55 wt.% CO_2., respectively.

For the lower crust we use carbon uptake estimated by Bach *et al.* (2001) for the gabbros at site 735B, yielding a CO_2 flux of 0.06 g/100 g into the 1,500 m gabbro section. Lower extrusives and sheeted dikes from site 504B apparently have not seen the complete cycle of alteration, and therefore the very low CO_2 values from these sites are not robust indicators of the total carbon budget of the intermediate crust. To estimate the carbon inventory of the intermediate levels of the oceanic crust, we interpolate between the extrusives and the gabbros, which yields an estimate of 0.5 g CO_2/100 g for the lower extrusive section, slightly above the value of Alt and Teagle's (1999) transition zone of site 504B. For the sheeted dikes we use 0.06 g/100 g, the value of the gabbro section, as a conservative minimum estimate.

The total uptake of the crust is calculated here as 0.355 g/100 g, which is slightly above the estimate made by Staudigel *et al.* (1989), and also similar to the total carbon uptake of the oceanic crust inferred by Alt and Teagle (1999) and for the Troodos ophiolite (Bednarz and Schmincke, 1989). The above extrapolation of data for intermediate depths (lower extrusive crust, sheeted dikes) does not contribute major uncertainties to the total flux estimate, because most of the carbon inventory is located in the upper 600 m of the crust, and, therefore, most of the uncertainties lie in this depth interval.

Oxygen isotopes. $\delta^{18}O$ in ocean-crust studies is typically defined as the per mil deviation in $^{18}O/^{16}O$ ratio of a rock relative to a standard mean ocean water ($\delta^{18}O_{SMOW}$) and it is widely used to understand ocean-crust alteration processes. Fresh MORB has an $\delta^{18}O_{SMOW}$ value of +5.7‰, and water–rock interaction with seawater ($\delta^{18}O_{SMOW}$ = 0‰) at low temperatures increases the value, while high-temperature alteration decreases it. Muehlenbachs and Clayton (1972) drew attention to this relationship and suggested that hydrothermal alteration of the crust may buffer the oxygen isotopic composition of seawater. Oxygen is the major component in the oceanic crust, and therefore, changes in $\delta^{18}O$ are a rather profound indicator of hydrothermal alteration.

The silicate portion of the upper oceanic crust at sites 417/418 has a $\delta^{18}O$ of ~+10‰ (+9.98‰, Staudigel *et al.*, 1995), substantially elevated from the original magmatic value of +5.7‰. The values for the composite crust decrease with depth from ~+11.5‰ in the upper 100 m, down to ~+8.5‰ between 400 m and 600 m depth, roughly defining an uncertainty band of ~±1$\delta^{18}O$ units. This uncertainty is rather small when compared to data from other ocean-crust sections. For example, the same depth interval in the Troodos and Samail ophiolites displays values ranging from ~8‰ to 14‰ (e.g., Spooner *et al.*, 1974; Gregory and Taylor, 1981; Stakes and Taylor, 1992), which is substantially larger than the 417/418 composite range, but these ophiolites also display a general decrease of $\delta^{18}O$ with depth. Site 735B shows a rather interesting variation, even though the data are more scattered. Its uppermost 500 m display a range of 3–7‰, with an average $\delta^{18}O$ of 4.35‰ (Hart *et al.*, 1999), well below the value of fresh MORB, and clearly indicating relatively high reaction temperatures. Deeper sections of 735B display heavier $\delta^{18}O$ values, also with a significant scatter, but with a higher average value, on the order of 5.8‰. This is not very different from the original magmatic value (Bach *et al.*, 2001). This pattern of very low $\delta^{18}O$ values at the top of the gabbros that slightly increase with depth is quite similar to the variation seen in the Troodos and Samail ophiolites (Spooner *et al.*, 1974; Gregory and Taylor, 1981; Stakes and Taylor, 1992).

The $\delta^{18}O$ of the average crustal section in Table 3 derives from the average $\delta^{18}O$ for the top 600 m of sites 417/418 ($\delta^{18}O$ = 10‰), with $\delta^{18}O$ decreasing to 8‰ in the lower extrusives, 4‰ in the sheeted dikes and then increasing again to 5.8‰ in the gabbros from site 735B. This overall variation is solely derived from drill holes in normal oceanic crust, but it is remarkably similar to the patterns observed in ophiolites. This suggests that 735B is actually quite an appropriate analogue for hydrothermal alteration in "true," deeply buried gabbro sections in normal oceanic crust. It appears that most hydrothermal systems penetrate into the top few hundred meters of gabbros, whether or not the section carries any overlying extrusives and dikes.

Strontium. The $^{87}Sr/^{86}Sr$ ratio is also a very powerful indicator of hydrothermal alteration in the oceanic crust, because fresh crust is unradiogenic ($^{87}Sr/^{86}Sr$ = 0.7025–0.703; Chapter 3.13) while seawater is radiogenic (0.7092; Chapter 6.02). Seawater $^{87}Sr/^{86}Sr$ also changes with time. One-hundred-million year old seawater was less radiogenic (0.70735) than today's, and $^{87}Sr/^{86}Sr$ has been continuously increasing since then. The isotopic composition of reaction products quantitatively reflects the strontium contributions from seawater and from basalt. The reference sites 417/418 and 735B both display primary strontium ratios close to 0.7029 (Staudigel *et al.*, 1981; Hart *et al.*, 1999; Bach *et al.*, 2001). The average $^{87}Sr/^{86}Sr$ ratio at sites 417/418 is 0.704575, with a rather systematic down-hole variation that closely mirrors the behavior of H_2O. Alteration-related increases in $^{87}Sr/^{86}Sr$ at the top of the crust are relatively modest, become more pronounced at a depth of ~200 m, and then decrease with depth again, to its lowest values at 550 m (Staudigel *et al.*, 1995). Hart *et al.* (1999) and Bach *et al.* (2001) showed that site 735B has generally very low $^{87}Sr/^{86}Sr$ (0.70295), with some small-scale variation that might resemble several cycles of decreasing $^{87}Sr/^{86}Sr$ with depth. $^{87}Sr/^{86}Sr$ values from sites 417/418 are high relative to those from site 504B (e.g., Barret and Friedrichsen, 1982; Friedrichsen, 1984; Kawahata *et al.*, 1987), but are relatively low when compared with $^{87}Sr/^{86}Sr$ of the Troodos ophiolite (Bickle and Teagle, 1992). The former might be due to incomplete alteration at 504B and the latter due to higher initial isotopic ratios in fresh basalts in the Troodos ophiolite. The trends in $^{87}Sr/^{86}Sr$ at sites 417/418 and in the Troodos ophiolite are used to extrapolate the $^{87}Sr/^{86}Sr$ of the lower extrusives ($^{87}Sr/^{86}Sr = 0.704$) and the sheeted dikes (0.7035). For the gabbros, $^{87}Sr/^{86}Sr$ is assumed to be 0.70295, which is only slightly elevated from the original values (Hart *et al.*, 1999; Bach *et al.*, 2001).

Strontium isotope ratios and abundances in samples from the oceanic crust may be used to determine the complete chemical mass balance of strontium exchange between seawater and basalt, including the loss of basaltic strontium to hydrothermal solutions, and uptake of basaltic or seawater strontium from hydrothermal solutions. A mass balance of this exchange can be made in four steps. (i) The relative amount of basaltic and seawater strontium in altered basalt can be determined from the measured $^{87}Sr/^{86}Sr$ of an altered sample as a (linear) mixture of strontium from the two end members, the (contemporaneous) seawater and basalt. (ii) The inventory of the basaltic and seawater strontium in an altered sample (in mg/kg) may then be determined from the above ratio of seawater and basalt in the sample and the total strontium abundance measured for this sample. (iii) Seawater strontium addition to the basalt is given directly by the seawater strontium inventory calculated in Step (ii). (iv) The determination of flux of basaltic strontium in or out of an altered sample is more complicated because it has to be related to the original inventory of strontium. It is determined as the difference between the original basaltic inventory and the basaltic strontium present in the altered sample.

Using this type of mass balance, it can be shown that on average, basalts from the upper crust at sites 417 and 418 have lost 10 mg/kg of basalt strontium to seawater and gained 32 mg/kg of seawater strontium, resulting in a total flux of 22 mg/kg of strontium into the basalt (Staudigel *et al.*, 1996). Gabbros at 735B show very large changes in strontium abundance between fresh and altered sample pairs, but these changes show both gains and losses of strontium by up to 25% of the total inventory (Bach *et al.*, 2001). Hart *et al.* (1999) estimated from strip composite samples that the upper portion of 735B shows no or very little change in strontium abundances. In Table 2, strontium fluxes are derived from the data of sites 417/418 and 735B for the upper and lower crust, respectively, effectively tapering the rather high strontium fluxes into the upper crust to the near zero strontium fluxes in the gabbros. A slight strontium loss is assumed for the gabbros, because the breakdown of calcic plagioclase is likely to liberate substantial quantities of strontium, some of which may be removed by hydrothermal solutions. Over the total crust, these elemental changes add up to a slight uptake of strontium, of 1.4 mg/kg of rock, which is extremely small when compared to the overall fluxes of strontium into and out of the crust. Thus, seafloor alteration can substantially influence the isotopic composition of seawater, but not significantly change the strontium inventories in either reservoir (see Palmer and Edmond, 1989; Elderfield *et al.*, 1999).

Sulfur. Sulfur is a key element in hydrothermal processes. Seawater introduces sulfate into the hydrothermal system that precipitates anhydrite during heating (e.g., Shanks *et al.*, 1981). Basalt is very reduced and contains sulfur dissolved in melt and immiscible sulfur metal globules (Mathez, 1976, 1980). The reaction of seawater with basalt results in the precipitation of sulfides, mostly pyrite, as one of the most common reaction products. There are only very few systematic sulfur and isotope studies of the oceanic crust, in particular those by Alt (1995) on ODP site 504B, Bach *et al.* (2001) on site 735B, and Alt (1994) on the Troodos ophiolite. The pre-alteration sulfur abundance at site 504B is 960 ppm. Overall, basalts appear to be a source of sulfur to hydrothermal solutions, delivering ~0.1–2 g of sulfur per 100 g of basalt (recalculated from Alt, 1995) and sulfur is precipitated in veins (sulfide and anhydrite), in particular in mineralized zones at the bottom of the extrusives and top of the dikes (Alt, 1995). Alt (1995) determined the volumes of veins in 504B and estimated a total flux of 0.15×10^{12} g yr^{-1} from the oceanic crust, cautioning that this is the only site where such a mass balance has been done and citing relatively large errors. However, the total sulfur flux in altered oceanic crust is relatively small when compared with the sulfur in high-temperature hydrothermal fluxes ((10.3–25.7) $\times 10^{12}$ g yr^{-1}; Von Damm *et al.*, 1985; Sleep, 1991) and river fluxes (28.5×10^{12} g yr^{-1}; Holser *et al.*, 1988). Alt (1995) suggests that the discrepancy in the hydrothermal fluxes is probably best explained by anhydrite precipitation during black-smoker activity, and subsequent dissolution from low-temperature fluids. In Table 2 a sulfur loss of 0.06 g/100 g (Alt, 1995) is estimated for the extrusive section and a sulfur gain of 0.02 is estimated for the gabbros at site 735B (Bach *et al.*, (2001). The total sulfur flux is 0.0037 g/100 g from basalt to seawater.

The δ^{34}S of fresh MORB is +0.1‰ (Sakai *et al.*, 1984). The calculated δ^{34}S of −1.9‰ for the extrusives, +1.3‰ for the dikes (anhydrite veins not included) and +0.13‰ for the gabbros (Table 2) is taken from Alt (1995), who estimates the average δ^{34}S for the total oceanic crust as +0.9‰ per mil.

Lithium. Lithium is a very sensitive indicator of hydrothermal alteration processes, because it is added during low-temperature alteration, mostly in smectites, and is leached during high-temperature alteration (e.g., Seyfried *et al.*, 1984; Von Damm *et al.*, 1985). In addition, lithium isotopes reflect seawater additions to the upper crust, much like strontium and oxygen isotopes do. The most comprehensive study of lithium abundances and isotopic ratios was recently presented by Chan *et al.* (2002) for site 504B. Considering the low recovery rates at this site, this study should be considered to provide a minimum baseline of lithium isotopic alteration, with an emphasis on the early portion of alteration. Unaltered MORB has an average lithium abundance of 4.5 ± 1.5 ppm (Chan *et al.*, 1992; Ryan and Langmuir, 1987). The least altered rock from site 504B has ~3 ppm lithium, the average lithium inventory in the volcanics is ~5.8 ppm, indicating an uptake of 2.8 mg/kg. The average lithium content of the sheeted dike (+ the transition zone) is ~2.5 ppm, indicating a loss of 0.5 mg/kg. Chan *et al.* (2002) suggest that the initial lithium content of site 735 is ~3 ppm and the measured value is 1.0, suggesting a loss of 2 mg Li/kg. In total, the crust loses ~1.09 mg Li/kg of basalt.

δ^7Li, the per mil deviation of the ^{7}Li/^{6}Li ratio from the NBS SRM L-SVEC, of fresh MORB is +3.4–4.7‰ (Chan *et al.*, 1992; Ryan and Langmuir, 1987) and rises steeply with the degree of low-temperature alteration (Chan *et al.*, 1992), reflecting uptake of heavy seawater lithium (+32‰). δ^7Li decreases with depth in site 504B, from an average of ~6‰ in the extrusives to ~2‰ in the sheeted dikes. This reflects addition of seawater lithium to the upper crust and removal and fractionation of lithium in the deeper crust (Chan *et al.*, 2002).

Boron. Boron concentrations in fresh MORB are ~0.5 ppm (Spivack and Edmond, 1987; Ryan and Langmuir, 1987) and 26.2 ppm in average upper crust at sites 417 and 418 (Smith *et al.*, 1995). Boron concentrations in altered crust show a significant decrease with depth, from ~50 ppm in the upper 100 m of 417/418 to 8.5 ppm at 500 m (Smith *et al.*, 1995). Subtracting the initial basaltic boron inventory from the altered extrusive inventory yields a boron uptake of 25.7 mg/kg for the upper 600 m of extrusives. Extrapolating the 417/418 data downward, the lower extrusives have ~4.5 ppm boron, and an uptake of 4 mg/kg (Table 2). An uptake of 5.6 mg/kg for the sheeted dikes and 2.2 mg/kg for the gabbros is estimated using Layer 2B and Layer 3 estimates for boron from Smith *et al.* (1995; 5.9 ppm and 2.7 ppm, respectively). The total uptake is thus 4.81 mg/kg for the whole oceanic crust. δ^{11}B (i.e., the per mil deviation of the ^{11}B/^{10}B ratio from the NBS SRM 951 standard) in the upper crust can increase substantially during seafloor alteration. Hyaloclastites may reach δ^{11}B values up to +5.4‰, but the average crust is only slightly elevated (+0.8‰) relative to the least altered composites of flows (0.5‰, Smith *et al.*, 1995). Gabbros from the upper portion of site 735B have markedly elevated δ^{11}B of 7.35‰ (Hart *et al.*, 1999). δ^{11}B for intermediate depths are extrapolated from these values to be 1‰ for the lower extrusives and 2‰ for the sheeted dikes, resulting in an overall value of δ^{11}B = 5.6‰ for the total crustal section.

Potassium, rubidium, and caesium. They are amongst the most sensitive indicators of water–rock exchange. Like their sister alkali element, lithium, these elements are taken up by basalts during low-temperature alteration reactions and leached from basalts during high-temperature reactions. The upper 600 m of crust at sites 417 and 418 records gains of 0.54 g/100 g K_2O, 10.3 mg/kg rubidium and 0.183 mg/kg caesium (Staudigel *et al.*, 1995, 1996; Hart and Staudigel, 1989). Surprisingly, the gabbros at site 735B all show gains of these alkalis as well: 0.01 g/100 g K_2O, 0.7 mg/kg rubidium and 003 mg/kg caesium (Bach *et al.*, 2001). This may reflect the anomalous nature of the crust at 735B, because it is directly exposed to seawater, or it may be a "typical" late-stage low-temperature alteration overprint that affects all of the oceanic crust, even if it was previously depleted in potassium, rubidium, and caesium from high-temperature reactions at the ridge axis. For this reason it is not clear whether the lower oceanic crust displays slight gains or near-complete losses in these elements. Given the rather low initial inventories of potassium, rubidium, and caesium in oceanic gabbros and the very large gains in the upper crust, the above uncertainties in the gabbro inventories have a relatively small effect on the alteration budget of these elements for the whole crust. For these reasons, the data from sites 417/418 are used for the upper oceanic crust and depth trends at these sites are used to extrapolate the values for the lower 400 m of the extrusives. The sheeted dikes and the gabbros are proposed to have experienced no net change in potassium, rubidium, and caesium. Using average values of 0.05 g/100 g K_2O, 1 g/kg rubidium, and 0.03 g/kg for caesium results in total fluxes of 0.0485 g/100 g of K_2O uptake, 0.927 mg/kg of rubidium uptake, and 0.0172 mg/kg of caesium uptake for the whole crust.

Uranium. Uranium content is also enhanced in the upper oceanic crust during alteration. The uranium content of the altered upper crust is 0.30 ppm (Hart and Staudigel, 1989; Staudigel *et al.*, 1996). Uranium is slightly depleted in the upper part of site 735B (Hart *et al.*, 1999), but slightly enriched overall in 735B (Bach *et al.*, 2001). Data for the top 600 m of site 417/418 are therefore used for the upper crust, and the grand average of 735B is used for the lower crust. As in the previous estimates, the internal variation in 417/418 and 735B are used to extrapolate the intermediate crust, yielding a total uptake of 0.037 mg/kg of uranium in the crust.

Rhenium and osmium. There are no data on behavior of Re/Os during alteration of MORB, but some work has been done on pillow lavas from the submarine seamount sequence in the basement complex of La Palma (Canary Islands; Marcantonio *et al.*, 1995). There, altered pillow rinds have $^{187}Os/^{188}Os$ ratios of 0.40–0.43, while the pillow interiors generally have $^{187}Os/^{188}Os$ values between 0.14–0.19. Schiano *et al.* (1997) found highly variable $^{187}Os/^{188}Os$ ratios in mid-ocean ridge glasses and cannot rule out potential assimilation of manganese crusts with high $^{187}Os/^{188}Os$. Gabbros from site 735B suggest isotopic variations consistent with these changes (Hart *et al.*, 1999). Thus, seafloor alteration is likely to increase the $^{187}Os/^{188}Os$ significantly in some rocks, in particular in the upper crust (Marcantonio *et al.*, 1995), but it is too early to determine whether this effect is of any more than localized importance.

SiO_2 is amongst the most difficult element to track during alteration of the oceanic crust because it shows major variations due to igneous processes and because the majority of alteration reactions involve silicates. The upper extrusive crust at 417/418 gains about 1.2 g/100 g SiO_2. The uncertainties are expectedly very large, including ±0.18 g/100 g in the estimate of the fresh rock composition, and a variance of nearly ±4 g/100 g in fluxes at various depths of sites 417D and 418A (Staudigel *et al.*, 1995, 1996). Bach *et al.* (2001) estimated that gabbros of site 735B gained about 0.1 g SiO_2/100 g gabbro, and the scatter in their data suggest that this value is also rather uncertain. We extrapolate between these values and estimate the fluxes to be 0.5 g/100 g for the lower extrusive crust and 0.25 g/100 g for the dikes (Table 2). However, these values are likely to have errors on the order of ±2 g/100 g, which are extremely large when compared to fluxes at hydrothermal vents and in the ocean in general.

Al_2O_3. Fresh MORB contains, on average, 15.26 wt.% Al_2O_3 (Hofmann, 1988; see also Chapter 3.13). Alteration of the upper crust causes a gain of about 1.94 g/100 g of Al_2O_3 (Staudigel *et al.*, 1996), based on individual fluxes at 417D and 418A that vary from −1 g/100 g to 5.0 g/100 g. This range indicates with significant uncertainty, partly due to the large uncertainty in the fresh estimate (±0.37 wt.%, mostly from uncertainties in plagioclase phenocryst distribution) and to local redistribution of aluminum within the holes. An analytical uncertainty of ~1% in the estimate of the altered rock composition would contribute an error of ±0.15 g/100 g to the fluxes. The lower crust Al_2O_3 content is unlikely to be affected by alteration.

FeO_{tot} in unaltered crust is 10.43 wt.% (Hofmann, 1988; see also Chapter 3.13). The upper crust at 417 and 418 loses about 0.01 g/100 g; individual composites may show losses to 0.85 g/100 g or gains of up to 1.01 g/100 g of up (Staudigel *et al.*, 1996). These variations are most likely to represent uncertainties in the estimates rather than true fluxes. Iron concentrations in MORB are very high and variable while seawater contains very little iron. The analytical uncertainty in the determination of iron alone is on the order of 0.1 wt.%, which is almost an order of magnitude higher than the flux. This shows that iron is essentially immobile and that the fluxes are too low to detect beyond the uncertainties of crustal mass balances. However, due to the extremely low-iron inventories in seawater, even the smallest fluxes would be quite significant for the seawater budget, and, thus, the oceanic crust may deliver substantial fluxes to seawater. Given the concentration gradient, these fluxes are likely to move iron from the crust to the oceans but there is no convincing data from the oceanic crust that can prove this.

MnO shows no change in the upper extrusives, but fluxes for individual composites range from 0.04 g/100 g to 0.05 g/100 g (Staudigel *et al.*, 1996) suggesting very large uncertainties that do not allow us to estimate meaningful fluxes independently. The very low manganese concentration in seawater and the enrichment of manganese in hydrothermal waters suggests that it is likely that manganese leaves the crust, even though local enrichments of manganese are likely near the top of the extrusive oceanic crust due to deposition of manganese crusts and Fe–Mn hydrothermal deposits.

MgO is a key element in the discussion of hydrothermal systems. It is an important element in many igneous phases and in alteration phases. The depletion of magnesium in hydrothermal vent fluids was originally seen as evidence that the oceanic crust is a major sink for magnesium from riverine input to the oceans (Edmond *et al.*, 1979; Van Damm *et al.*, 1985). However, flux estimates from rock analyses do not to support this contention, at least not in a very consistent manner. Data from sites 417/418 and 735B suggest that, on average, rocks lose more magnesium than they take up (0.2 g/100 g in both cases, Staudigel *et al.*, 1996; Bach *et al.*, 2001). Dredged samples of abyssal peridotite suggest that they lose up to 5 wt.% magnesium due to alteration (Snow and Dick, 1995). Magnesium uptake was reported at the Troodos ophiolite (Bednarz, 1989; Bednarz and Schmincke, 1989), and for site 504B (Alt *et al.*, 1996). Part of this discrepancy may be due to the very large uncertainties in these fluxes. At sites 417/418, fluxes of individual depth composites vary from −1.1 g/100 g to +2.01 g/100 g, and the uncertainty in the fresh rock estimate is 0.19 g/100 g. Thus, the data from 417/418 actually permit a small gain in MgO, but the uncertainties are large. The estimate in Table 2 uses the 417/418 data as representative of the upper oceanic crust and makes the most conservative estimate of lower crust/sheeted dikes of a magnesium flux of zero (Table 2). The crustal budget of magnesium is similarly as poorly constrained as the budget for iron, but it is likewise clear that there really are no very large fluxes. A major difference between magnesium and iron is that magnesium has a much larger inventory in seawater, which requires much larger fluxes to make it significant for the seawater–ocean-crust budget. These large fluxes are not observed.

CaO. Calcium is also an element that is very important to global geochemical cycles, as the oceanic crust is considered to be an important source of Ca^{2+} (e.g., Berner *et al.*, 1983). There are several competing processes that control the calcium content in the upper crust. Calcium is lost during glass alteration (Staudigel and Hart, 1983) and calcic feldspar breakdown, and is taken up during calcium carbonate precipitation. Sites 417/418 display an uptake of 2.1 g/100 g ± 1 g/100 g CaO (Staudigel *et al.*, 1996), with an overall decrease with depth. Alt *et al.* (1996) report a small loss of CaO at site 504B, which is consistent with the limited precipitation of carbonate there, while much of the calcic feldspar is replaced. Extrusives in the Troodos ophiolite drill cores consistently show loss of CaO: 8.08 g/100 g in the upper crust and 3.34 g/100 g in the lower crust (Bednarz, 1989). The flux values for CaO in Table 2 use site 417/418 data for the top of the crust, and site 735B data for the gabbros. These sites are considered the most representative of the oceanic crust. Intermediate values in the lower extrusives and sheeted dikes are extrapolated from these data and tapered to zero fluxes, like CO_2, largely because much of the calcium deposition appears to be linked to carbonate precipitation. This yields an uptake of 0.03 g/100 g in the lower extrusives and an uptake of 0.04 g/100 g in the sheeted dikes. Altogether these calcualtions suggest a slight positive flux of calcium into the crust of 0.134 g/100 g. Most evidence suggests that calcium contents do not show much net change, even though it is quite likely that the exchange of calcium between seawater and basalt is very high, probably quite similar to the behavior of strontium.

Na_2O is a mobile element that is involved in important alteration reactions in the shallow crust (loss during glass alteration, gain from formation of zeolite), in the deeper crust (formation of zeolites, albite, and halite) and has a substantial inventory in seawater, which is at times enhanced by boiling of water in high-temperature hydrothermal systems followed by the precipitation of halite. The oceanic crust appears to gain small quantities of sodium, on average, particularly in the upper crust (0.15 g/100 g in 417/418; Staudigel *et al.*, 1996). Bach *et al.* (2001) suggest a gain of 0.03 g/100 g for the gabbros at site 735B. A gain of 0.15 g/100 g for the lower extrusives and the dikes is used in Table 2. This is intermediate between the smaller uptake at site 504B (0.09 g/100 g; Alt *et al.*, 1996) and the higher uptake at the Troodos ophiolite (1.9 g/100 g lower extrusives; 0.421 g/100 g sheeted dikes; Bednarz, 1989). Like other major elements, however, it is obvious that the gains or losses of sodium in basalts are quite variable and uncertainties are high.

A large number of elements are immobile during alteration of the oceanic crust. These are generally elements that are insoluble in seawater and include titanium, which is commonly used as an immobile reference element to constrain overall losses or additions in a chemically open system and REEs, hafnium, niobium, zirconium, and thorium. Similarly, the neodymium and hafnium isotopic composition of oceanic crust is rarely affected by alteration. However, significant mobility of otherwise immobile elements has been identified in extremely altered basalts dredged from some seamounts (e.g., Cheng *et al.*, 1987). Staudigel *et al.* (1996) observed distinct but minor changes in $^{143}Nd/^{144}Nd$ and in Ce/Ce^* in the most altered composite samples from sites 417 and 418, consistent with the addition of REE in very large quantities of seawater or moderate amounts of sediment particulates (Staudigel *et al.*, 1996).

3.15.5 DISCUSSION

3.15.5.1 Hydrothermal Fluxes: Rock Data versus Fluid Data

The estimates of elemental changes due to hydrothermal alteration of oceanic crust in Table 2 can be re-cast into global geochemical fluxes using the ocean-crust production rate discussed above (see Figure 2). These fluxes can then be compared with river fluxes and two types of hydrothermal fluxes: axial hydrothermal fluxes or flank fluxes following Wheat and Mottl (2000).

3.15.5.1.1 Uncertainties

Errors in elemental fluxes derived from crustal estimates are larger than, or similar to, the value of the actual flux estimate for many of the major elements: silicon, aluminum, iron, manganese, magnesium, and sodium. The fluxes for these elements are thus poorly constrained, but these estimates do serve as conservative bounds on the fluxes. Unfortunately, these bounds overlap the fluxes derived from hydrothermal fluid data and river data. For this reason, current ocean-crust flux estimates do not provide independent evidence for the magnitude of hydrothermal fluxes in the geochemical cycle for these elements. Within the bounds of these uncertainties, the data indicate that ocean floor hydrothermal processes may balance (or compound!) missing global fluxes of these elements. For these reasons, these elements are not discussed here in any detail.

The flux estimates of CaO are slightly better known than the forgoing major elements, but the uncertainty of CaO flux in the upper oceanic crust is still almost two-third of its value. Much larger uncertainties are typical of the deeper extrusives and sheeted dikes, where spilitization is characterized by near-complete loss of calcium. All other element fluxes reported here have uncertainties that are less than 30% of their value.

3.15.5.1.2 Bulk fluxes

A comparison can be made between fluxes derived from the rock record versus those obtained from hydrothermal fluids for elements for which well-constrained rock data exist. These elements are calcium, potassium, CO_2, and sulfur (Table 2). This is a somewhat random collection of elements and the comparisons thus derived may not be "typical" of all elemental fluxes, but they do give some first-order comparisons. The crustal fluxes of these elements add up to 3.6×10^{14} g of seawater calcium, potassium, CO_2, and sulfur per year, into the crust, while the sum of the flank and axis hydrothermal fluxes give a net flux out of the oceanic crust ranging from 1.8×10^{14} g yr^{-1} to 2.92×10^{14} g yr^{-1}, depending on whether one uses the high or low estimates for axial fluxes. Overall, the rock record yields fluxes into the crust of the same order of magnitude but opposite sign to the fluid data, illustrating a rather fundamental disconnect between hydrothermal fluid data and ocean crust alteration data. Ideally, fluxes derived from different and complementary data sets should be the same. What are the reasons for this discrepancy? Several contributing factors may be considered:

(i) The two methods sample different types of processes. When Edmond *et al.* (1979) published the first "global" geochemical fluxes based on black-smoker data, Hart and Staudigel (1982) pointed out that there are major discrepancies with the fluxes of rubidium and caesium from ocean-crust alteration data that appear to be largely controlled by low-temperature alteration. Fluxes based on hydrothermal fluids are biased towards high-temperature processes, which may underestimate the total flux.

(ii) The two methods sample processes occurring at very different times in the hydrothermal history of the oceanic crust. The oldest hydrothermal vent samples analyzed to date come from Baby Bare seamount near the Juan de Fuca Ridge, on 3.5 Ma crust (Wheat and Mottl, 2000). At 3.5 Ma, the oceanic crust has lost only half of its total convective heat loss (C. A. Stein and S. Stein, 1994b). Hydrothermal fluxes from the rock record at sites 417 and 418, for example, offer a complete set of chemical changes occurring during the entire hydrothermal history of the oceanic crust.

(iii) Oceanic crust has a substantial amount of pore space produced during its initial emplacement, and new pore space is generated as the crust cools. All these pores are filled with seawater and some additional water is taken up through hydration of ocean-crust materials. The chemical inventory of these pore waters are transferred into the crust, producing a one-way flux of water into the crust that is not accounted for through the measurement of hydrothermal fluids. This flux represents a baseline flux that should be subtracted from hydrothermal fluxes in order to calculate net fluxes out of the crust. It was estimated above that oceanic crust may take up to ~1.5 wt.% water. This translates into addition of $\sim 2 \times 10^{12}$ g MgO yr^{-1}, 0.5×10^{12} g CaO yr^{-1}, 1.5×10^{13} g Na_2O yr^{-1}, 1×10^{8} g Rb yr^{-1}, 0.3×10^{8} g CO_2 yr^{-1}, and 1×10^{13} g S yr^{-1}. Overall, these fluxes are relatively small when compared to the fluxes considered in this mass balance and are thus not likely to explain the discrepancy between hydrothermal- and ocean-crust-derived fluxes.

(iv) Another issue in reconciling fluid versus crust fluxes relates to the nature of fluid circulation in the oceanic crust and the uncertainty in fluid pathways. With the exception of pore waters, fluids sampled from hydrothermal systems reflect relatively high water volumes, and high water–rock ratios. Most of the crust however, is altered at low water–rock ratios, in particular deep within the crust (>100m). Thus, it is possible that the alteration reflected in hydrothermal fluid data only represents hydrothermal activity occurring in high water–rock ratio reaction zones and in high water-throughput aquifers. By far the greatest volume of oceanic crust is altered at much lower water–rock ratios and by processes that may largely be diffusive. In addition, drilling recovery rates are particularly low in regions that are most porous, and thus high water–rock ratio regions, which are typically very porous, may be under-represented in crustal studies

(v) The role of sediments in mass balances of seawater-ocean-crust chemical exchange remains largely unexplored. Pore-water studies in sediments on the ocean floor demonstrate that there is chemical exchange between sediments and the basaltic oceanic crust (Lawrence and Gieskes, 1981). The impact of these processes is not quantified but it is quite possible that this exchange has a significant affect on the bulk composition of the oceanic crust. This is particularly important for sedimented ridges and later stage alteration of the oceanic crust, when the crust is typically covered by sediments. Chemical exchange between sediment pore waters and the ocean crust may have a profound impact on the chemical fluxes into the oceanic crust, but it is not included in the hydrothermal spring data.

At least four out of these five contributing factors may substantially contribute to the discrepancies between fluid- and crust-derived hydrothermal flux estimates.

3.15.5.1.3 Reconciling hydrothermal fluxes from fluid and rock data

The discrepancies discussed above suggest that both methods of flux determination may have intrinsic problems. Flux data from hydrothermal vents are compromised for determining global flux estimates because these fluids are derived from young crust, mostly from deeper crustal levels. They selectively sample high water–rock ratio alteration processes and they ignore low-temperature fluid fluxes (<20 °C). Rock data are particularly unreliable or not available for the deeper crust and for high water–rock ratio reactions. Low recovery rates, and large primary variability in many important elements (magnesium, silicon) produce significant uncertainties in the flux estimates for elements whose fluxes are particularly well determined in hydrothermal fluids. The most important steps in reconciling these differences are to evaluate the weaknesses of both approaches and try to arrive at fluxes that use both methods in a complementary fashion. Furthermore, a critical error analysis will highlight the weaknesses of each of the methods, allowing the design of new experiments or approaches that resolve these issues by reducing a particular source of uncertainty, possibly separately for different element groups.

3.15.5.2 Impact of Ocean-crust Composition on Arc Processes and Mantle Heterogeneity

Using average MORB or the range of compositions of oceanic basalts (e.g., Hofmann, 1988; Chapter 3.13 and http://petdb.ldeo.columbia.edu Lehnert *et al.*, 2000), the fluxes derived here can be applied to determine the average compositions of oceanic crust that is subducted and recycled into the mantle. These compositions thus influence the composition of subduction zone magmas (see Chapter 3.18) and bear on the chemical mass balance of the mantle.

Whereas average fluxes are useful for defining the global mass balance between mantle and the oceans, understanding the compositional diversity of subducted crust is important in constraining its dehydration or partial melting processes during recycling (see Chapter 3.17). Such diversity is reflected in the more extensively altered compositional domains in the oceanic crust such as volcaniclastics, and the moderately altered flow composites of sites 417A, 417D, and 418B (Staudigel *et al.*, 1995, 1996). Other extreme compositional domains include umbers, ophicalcites (calcite-basalt breccia), massive sulfides, epidosites, talcum deposits, or serpentinites, but these are probably best studied in ophiolites. Particular mineralogical assemblages may show distinct phase relationships during prograde metamorphism. For example, ophicalcite or talcum has been shown to display distinct phase relations that is likely to control its dehydration or decarbonation behavior during subdution (Wyllie, 1978; Kerrick and Connelly, 2001). Volatile-rich or alkali-rich compositional domains may contribute preferentially to fluids that are extracted during prograde metamorphism in subduction zones, or they may melt prior to the surrounding less alkalic, and volatile-poor rock. The latter scenario has been considered in mantle melting models as a process that explains the isotopic variation in mantle-derived melts (Phipps-Morgan, 2001). MORB are generated by massive melting events that mix melts derived from depleted mantle and relatively enriched "plums" of recycled materials, while the smaller degrees of melting of ocean island basalts are capable of extracting more of these heterogeneities, displaying the fuller extent of mantle heterogeneity in bulk rock analyses. The geochemical characteristics of extremely altered compositional domains may help decipher the origin of volatiles or melts that may be related to the subduction of oceanic crust.

Recycling of oceanic crust into the Earth's mantle may profoundly influence the uranium budget, and the evolution of U/Th/Pb isotope systematics in the mantle (Hart and Staudigel, 1989; Elliott *et al.*, 1999). Uranium is readily taken up by oceanic crust during hydrothermal alteration and recycled into the mantle while thorium concentrations remain relatively unchanged during ocean-crust alteration. The uptake of uranium in the oceanic crust is restricted to the upper alteration zones, having relatively high water–rock ratios. There, oxidizing seawater enters the crust and loses its dissolved oxygen from reactions with the highly reducing oceanic crust. Most of the uranium dissolved in seawater is in the oxidized form of $UO_2(CO_3)_2^{2-}$ (i.e., U^{6+}), and is reduced to U^{4+}, which is as insoluble in hydrous solutions as thorium (Langmuir, 1978). This fixation of uranium in reducing hydrothermal environments is critically dependent on its mobilization by oxidation under the present-day atmospheric conditions. However, the earth's atmosphere has only been oxidizing for the last 2.2 Gyr (Holland, 1984, 1994), and for this reason, uranium recycling is unlikely to have occurred prior to 2.2 Ga. This process may explain the "kappa-conundrum," whereby MORB and the upper mantle appears to have much lower $^{232}Th/^{238}U$ ratios (="kappa") than required by modeling of lead isotope ratios (Elliott *et al.*, 1999 and references therein). These present-day low kappas may be caused by the recycling of uranium relatively recently in Earth's history.

One of the major fluxes associated with the recycling of oceanic crust involves water and CO_2. Most of the volatile inventory of altered oceanic crust is located in the uppermost 600 m, which is also the section first exposed to the top-down heating of the slab during subduction. For this reason, extraction of volatiles during subduction is particularly efficient in the upper part of the slab and much of this inventory is likely to be extracted (e.g., Kerrick and Connolly, 2001). However, there are several mechanisms that could allow these elements to survive passage through the "subduction zone filter (see also Chapter 3.17)." (i) In particular old and dense crust may be subducted relatively rapidly, which greatly reduces the geotherms in the subducting slab, increasing volatile subduction (Staudigel and King, 1992). (ii) Uneven topography on the surface of subducting upper oceanic crust (due to horst-graben structures and collapsed seamounts) can produce vertical throw of highly altered materials to levels substantially below the average top of the remaining slab, thus isolating these materials from top-down heating. (iii) Some volatiles from dehydration may be retained as fluid inclusions in newly formed minerals that remain stable past the subduction zone, and may make up several percent of a rock (Touret and Olsen, 1985).

3.15.6 CONCLUSIONS

Seafloor hydrothermal alteration processes are important for the global geochemical cycles of many elements, and the record of these processes in the oceanic crust reveals much information about these cycles. Rather robust flux information can be obtained from a variety of elements that have rather low initial abundances in basalt (H_2O, CO_2, K_2O, rubidium, caesium, uranium) or that are rather sensitive to alteration ($^{87}Sr/^{86}Sr$) and $\delta^{18}O$. Fluxes of many other elements are rather poorly constrained because of substantial primary magmatic variation.

This rock record yields results that are inconsistent with fluxes inferred from fluid data from seafloor hydrothermal springs. An in-depth analysis of data methods and uncertainties in fluxes based on rock data and fluid data suggest that these discrepancies are due to profound problems with both types of flux determination.

Fluxes from fluid data are fundamentally limited by their near-exclusive focus on reactions involved in large fluid fluxes, high-temperature reactions and in particular in young crust near the ridge axis. A few studies have recently begun exploring low-temperature vents in crust up to 3.5 Myr old. Hydrothermal fluxes up to this age involves only about half of the total convective heat flow lost in the oceans, and by far the largest volume of fluids passes through oceanic crust older than 3.5 Ma. Furthermore, much, if not most of the oceanic crust is altered at low fluid/rock ratios, and thus, most of the alteration of the oceanic crust is not accounted for by in these fluxes.

Data from the rock record are limited to the rather small number of drill sites that are in crust old enough and have sufficiently high recovery rates to allow reliable estimates to be made. None of the drill holes available so far reaches into the reaction zone of a black smoker, which is where much of the hydrothermal flux data derive from. Most importantly, the uncertainties of some element fluxes (i.e., silicon, aluminum, magnesium, etc.) derived from oceanic crust studies are substantially larger than the fluxes expected from balancing other flux data. This is due to the rather high concentrations of these elements in the oceanic crust, their large magmatic variation, and their complex behavior during alteration. It is thus unlikely that the current approach will yield fluxes for these elements that are sufficiently constrained to be meaningful in the context of global fluxes in the hydrosphere.

There is no simple solution to "fix" the intrinsic problems of fluid or rock-based estimates of hydrothermal fluid fluxes. For this reason, both methods have to be used in a complementary fashion. The first goal in such a complementary analysis is to develop a reference model for ocean-crust hydrothermal alteration, with a clear definition of reservoirs, reaction zones, types of alteration, etc. The goal of such a reference model is to define the portions of oceanic crust that can be constrained by various methods. Fluxes derived from studies of ophiolites need to be used to constrain those of the deeper oceanic crust, until all major fractions of the oceanic crust can be reliably recovered by drilling. Above all, studies should evaluate uncertainties and place all flux determinations in this context.

REFERENCES

Alt J. (1994) A sulfur isotopic profile through the Troodos ophiolite, Cyprus: primary composition and the effects of seawater hydrothermal alteration. *Geochim. Cosmochim. Acta* **58**, 1825–1840.

Alt J. C. (1995) Sulfur isotopic profile through the oceanic crust: sulfur mobility and seawater–crustal sulfur exchange during hydrothermal alteration. *Geology* **23**(7), 585–588.

Alt J. and Honnorez J. (1984) Alteration of the upper oceanic crust: DSDP Site 417: mineralogy and processes. *Contrib. Mineral. Petrol.* **87**, 149–169.

Alt J. C. and Teagle D. A. H. (1999) The uptake of carbon during alteration of ocean crust. *Geochim. Cosmochim. Acta* **63**, 1527–1535.

Alt J. C., Honnorez J., Laverne C., and Emmermann R. (1986) Hydrothermal Alteration of a 1 km section through the upper oceanic crust DSDP hole 504B: the mineralogy, chemistry and evolution of seawater-basalt interactions. *J. Geophys. Res.* **91**, 10309–10335.

Alt J. C., Kinoshita H., and Stokking L. (1993a) *Leg 148 Preliminary Report: Hole 504B*. Texas A and M University, Ocean Drilling Program.

Alt J. C., Kinoshita H., and Stokking L. B., *et. al.* (1993b) *ODP, Proc. Initial Reports*. Ocean Drilling Program.

Alt J. C., Teagle D. A. H., Laverne C., Vanko D. A., Bach W., Honnorez J., Becker K., Ayadi M., and Pezard P. A. (1996) Ridge flank alteration of upper oceanic crust in the Eastern Pacific: synthesis of results fro volcanic rocks of holes 504B and 896A. *Proc. ODP Sci. Results* **148**, 435–450.

Anderson R. N., Hobarth M. A., and Langseth M. G. (1979) Geothermal convection through oceanic crust and sediments in the Indian Ocean. *Science* **204**, 828–832.

Aumento F. and Melson W. (1977) *Initial Reports of the Deep Sea Drilling Project*. US Government Printing Office, Washington, DC, vol. 37, 1008pp.

Bach W., Alt J. C., Niu Y., Humphris S. E., Erzinger J. O., and Dick H. J. B. (2001) The geochemical consequences of late-stage low-grade alteration of lower ocean crust at the SW Indian Ridge: results from ODP Hole 735B (Leg 176). *Geochim. Cosmochim. Acta* **65**, 3267–3287.

Baer L. M. (1963) The mineral resources and mining industry of Cyprus. *Cyprus Geol. Surv. Depart. Bull.* **1**, 208.

Barret T. J. and Friedrichsen H. (1982) Strontium and oxygen isotopic composition of some basalts from Hole 504B, Costa Rica Rift, DSDP Legs 69 and 70. *Earth Planet. Sci. Lett.* **60**, 27–38.

Bednarz U. (1989) Volcanological, geochemical, and petrological evolution and sub-seafloor alteration in the northeastern Troodos ophiolite (Cyprus). PhD, Ruhr Universität.

Bednarz U. and Schmincke H. U. (1989) Mass transfer during sub-seafloor alteration of the upper Troodos crust (Cyprus). *Contrib. Mineral. Petrol.* **102**, 93–101.

Bednarz U. and Schmincke H.-U. (1994) The petrological and chemical evolution of the northeastern Troodos extrusive series, Cyprus, *J. Petrol.* **35**, 489–523.

Berner R. A., Lasaga A. C., and Garrels R. M. (1983) The Carbonate-silicate geochemical cycle and its effect on atmospheric carbon dioxide over the past 100 million years. *Am. J. Sci.* **283**, 611–683.

Bettison-Varga L., Varga R. J., and Schiffman P. (1992) Relation between ore-forming hydrothermal systems and extensional deformation in the Solea graben spreading center, Troodos ophiolite, Cyprus. *Geology* **20**, 987–990.

Bickle M. J. and Teagle D. A. H. (1992) Strontium alteration in the Troodos ophiolite: implications for fluid fluxes and geochemical transport in mid-ocean ridge hydrothermal systems. *Earth Planet. Sci. Lett.* **113**, 219–237.

Bischoff J. L. and Seyfried W. E. (1978) Hydrothermal chemistry of seawater from 25°–350 °C. *Am. J. Sci.* **278**, 838–860.

Bonatti E. (1976) Serpentinite protrusions in the oceanic crust. *Earth Planet. Sci. Lett.* **32**, 107–113.

Booij E., Gallahan W. E., and Staudigel H. (1995) Duration of low temperature alteration in the Troodos Ophiolite. *Chem. Geol.* **126**, 155–167.

Butterfield D. A., Massoth G. J., McDuff R. E., Lupton J. E., and Lilley M. D. (1990) Geochemistry of hydrothermal fluids from axial seamount hydrothermal emissions study vent field, Juan de Fuca Ridge: subseafloor boiling and subsequent fluid-rock interaction. *J. Geophys. Res.* **95**, 12895–12921.

Cann J. R. (1969) Spilites form the Carlsberg Ridge, Indian Ocean. *J. Petrol.* **10**, 1–19.

Chan L.-H., Alt J. C., and Teagle D. A. H. (2002) Lithium and lithium isotope profiles through the upper oceanic crust: a study of seawater-basalt exchange at ODP Sites 504B and 896A. *Earth Planet. Sci. Lett.* **201**, 187–201.

Chan L. H., Edmond J. M., Thompson G., and Gillis K. (1992) Lithium isotopic composition of submarine basalts: implications for the lithium cycle in the oceans. *Earth Planet. Sci. Lett.* **108**, 151–160.

Cheng Q., Park K.-H., MacDougall J. D., Zindler A., Lugmair G. W., Hawkins J., Lonsdale P., and Staudigel H. (1987) Isotopic evidence for a hot spot origin of the Louisville seamount chain. In *American Geophysical Union Monograph* (eds. B. H. Keating, P. Fryer, R. Batiza, and G. W. Boehlert). American Geophysical Union, Washington, DC, vol. 43, pp. 283–296.

Dick H. J. B., Natland J. H., Alt J. C., Bach W., Bideau D., Gee J. S., Haggas S., Hertogen J. G. H., Hirth G., Holm P. M., Ildefonse B., Iturrino G. J., John B. E., Kelley D. S., Kikawa E., Kingdon A., LeRoux P. J., Maeda J., Meyer P. S., Jay Miller D, Naslund H. R., Niu Y.-L., Robinson P. T., Jonathan Snow C., Stephen R. A., Trimby P. W., Worm H.-U., and Yoshinobu A. (2000) A long *in situ* section of the lower ocean crust: results of ODP Leg 176 drilling at the southwest Indian Ridge. *Earth Planet. Sci. Lett.* **179**, 31–51.

Donnelly T., Francheteau A. J., Bryan W. B., Robinson P. T., Flower M. F. J., and Salisbury M. (1979a) *Initial Reports of the Deep Sea Drilling Project*. US Government Printing Office, Washington, DC, vol. 51–53-1, 710pp.

Donnelly T., Francheteau A. J., Bryan W. B., Robinson P. T., Flower M. F. J., Salisbury M., *et al.* (1979b) *Initial Reports of the Deep Sea Drilling Project*. US Government Printing Press, Washington, DC, vol. 51–53-2, 1613pp.

Donnelly T. J., Thompson G., and Salisbury M. (1979c) The chemistry of altered basalts at Site 417, Deep Sea Drilling Project Leg 51. In *Initial Reports of Deep Sea Drilling Project* (eds. T. Donnelly, J. Francheteau, W. Bryan, P. T. Robinson, M. F. J. Flower, and M. Salisbury). US Government Printing Office. Washington, DC, vol. 51–53-2, pp. 1319–1330.

Edmond J. M., Measures A. C., McDuff R. E., Chan L. H., Collier R., Grant B., Gordon L. I., and Corliss J. B. (1979) Ridge Crest hydrothermal activity and the balances of the major and minor elements in the ocean: the Galapagos data. *Earth Planet. Sci. Lett.* **46**, 1–18.

Elderfield H. and Schultz A. (1996) Mid-ocean ridge hydrothermal fluxes and the chemical composition of the ocean. *Ann. Rev. Earth Planet. Sci.* **24**, 191–224.

Elderfield H., Wheat C. G., Mottl M. J., Monnin C., and Spiro B. (1999) Fluid and geochemical transport through oceanic crust: a transect across the eastern flank of the Juan de Fuca Ridge. *Earth Planet. Sci. Lett.* **172**, 151–165.

Elliott T., Zindler A., and Bourdon B. (1999) Exploring the kappa conundrum: the role of recycling in the lead isotope evolution of the mantle. *Earth Planet. Sci. Lett.* **169**, 129–145.

Friedrichsen H. (1984) Strontium, oxygen and hydrogen isotope studies on primary and secondary minerals in basalts from the Costa Rica Rift, Deep Sea drilling project hole 504B, Leg 83. *Initial Rep. Deep Sea Drill. Proj.* **83**, 289–295.

Furnes H. and Staudigel H. (1999) Biological mediation in ocean crust alteration: how deep is the deep biosphere? *Earth Planet. Sci. Lett.* **166**(3–4), 97–103.

Gallahan W. E. and Duncan R. A. (1994) Spatial and temporal variability in crystallization of celadonites within the Troodos ophiolite, Cyprus: implications for low temperature alteration of the oceanic crust. *J. Geophys. Res.* **99**, 3147–3162.

Gerlach T. M. (1989) Degassing of carbon dioxide from basaltic magma at spreading centers: 2. Mid-ocean ridge basalts. *J. Volcanol. Geotherm. Res.* **39**, 221.

Gibson I., Malpas J., Robinson P. T., and Xenophontos C. E. (1991a) *Cyprus Crustal Study Project Initial Report, Holes CY-1 and CY-1A*. Canadian Government Publishing Center.

Gibson I. L., Malpas J., Robinson P. T., and Xenophontos C. E. (1991b) *Cyprus Crustal Study Project Initial Report, Hole CY-4*. Canadian Government Publishing Center.

Gillis K. M. and Robinson P. T. (1988) Distribution of alteration zones in the oceanic crust. *Geology* **16**, 262–266.

Gregory R. T. and Taylor H. P., Jr. (1981) An oxygen isotope profile in a section of Cretaceous oceanic crust, Samail ophiolite, Oman: evidence for $\delta^{18}O$ buffering of the oceans by deep (>5 km) seawater-hydrothermal circulation at mid-ocean ridges. *J. Geophys. Res.* **86**, 2737–2755.

Hart S. R. and Staudigel H. (1978) Oceanic crust: age of hydrothermal alteration. *Geophys. Res. Lett.* **5**, 1009–1012.

Hart S. R. and Staudigel H. (1982) The control of alkalies and uranium in sea water by ocean crust alteration. *Earth Planet. Sci. Lett.* **58**, 202–212.

Hart S. R. and Staudigel H. (1986) Ocean crust vein mineral deposition: Rb/Sr ages, U–Th–Pb geochemistry, and duration of circulation at DSDP Sites 261, 462, and 516. *Geochim. Cosmochim. Acta* **50**, 2751–2761.

Hart S. R. and Staudigel H. (1989) *Isotopic Characterization and Identification of Recycled Components*, vol. 258, pp. 15–24.

Hart S. R., Blusztajn J., Dick H. J. B., Meyer P. S., and Muehlenbachs K. (1999) The fingerprint of seawater circulation in a 500-meter section of ocean crust gabbros. *Geochim. Cosmochim. Acta* **63**, 4059–4080.

Hofmann A. W. (1988) Chemical differentiation of the earth: the relationship between mantle, continental crust, and oceanic crust. *Earth Planet. Sci. Lett.* **90**, 297–314.

Holland H. D. (1984) *The Chemical Evolution of the Atmosphere and Oceans*. Princeton University Press, Princeton, NJ.

Holland H. D. (1994) Early Proterozoic atmospheric change. In *Early Life on Earth* (ed. S. Bengtson). Columbia University Press, New York.

Holser W. T., Schidlowski M., Mackenzie F. T., and Maynard J. B. (1988) Geochemical Cycles of Carbon and Sulfur. In *Chemical Cycles in the Evolution of the Earth* (eds. C. B. Gregor, R. M. Garrels, F. T. Mackenzie, and J. B. Maynard). Wiley, pp. 105–173.

Jannasch H. (1995) Microbial interactions with hydrothermal fluids. *AGU Geophys. Monogr.* **91**, 273–296.

Jannasch H. W. and Mottl M. J. (1985) Geomicrobiology of deep-sea hydrothermal vents. *Science* **229**, 717–725.

Kawahata H., Kusakabe M., and Kikuchi Y. (1987) Strontium, oxygen, and hydrogen isotope geochemistry of hydrothermally altered and weathered rocks in DSDP Hole 504B, Costa Rica Rift. *Earth Planet. Sci. Lett.* **85**, 343–355.

Kelemen P. B., Koga K., and Shimizu N. (1997) Geochemistry of gabbro sills in the crust-mantle transition zone of the Oman Ophiolite; implications for the origin of the oceanic lower crust. *Earth Planet. Sci. Lett.* **146**(3–4), 475–488.

Kerrick D. M. and Connolly J. A. D. (2001) Metamorphic devolatilization of subducted oceanic metabasalts: implications for seismicity, arc magmatism and volatile recycling. *Earth Planet. Sci. Lett.* **189**, 19–29.

Langmuir D. (1978) Uranium solution-mineral equilibria at low temperatures with applications to sedimentary ore deposits. *Geochim. Cosmochim. Acta* **42**, 547–569.

Langseth M. G., Mottl M. J., Hobart M. A., and Fisher A. (1988) The distribution of geothermal and geochemical gradients near site 501/504: implications for hydrothermal circulation in the oceanic crust. *Ocean Drill. Initial Rep.* **111**, 23–32.

Lawrence J. R. and Gieskes J. M. (1981) Constraints on water transport and alteration in the oceanic crust from the isotopic composition of pore water. **86**, 7924–7934.

Lehnert K., Su Y., Langmuir C. H., Sarbas B., and Nohl U. (2000) A global geochemical database structure for rocks. *Geochem. Geophys. Geosyst.* **1**.

Marcantino F., Zindler A., Elliot T., and Staudigel H. (1995) Os isotope systematics of La Palma, Canary Islands: evidence for recycled crust in the mantle source of HIMU ocean islands. *Earth Planet. Sci Lett.* **133**, 397–410.

Mathez E. A. (1976) Sulfur solubility and magmatic sulfides in submarine basalt glass. *J. Geophys. Res.* **81**, 4269–4276.

Mathez E. A. (1980) Sulfide relations in Hole 418A flows and sulfur contents of glasses. In *Initial Reports DSDP* (eds. T. Donnelly, J. Francheteau, W. Bryan, P. Robison, M. Flower, and M. Salisbury). US Government Printing Office, vol. 51–53, pp. 1069–1085.

Muehlenbachs K. and Clayton R. N. (1972) Oxygen isotope studies of fresh and weathered submarine basalts. *Can. J. Earth Sci.* **9**, 172–184.

Mutter C. Z. and Mutter J. C. (1993) Variations in thickness of Layer 3 dominated oceanic crustal structure. *Earth Planet. Sci. Lett.* **117**, 295–317.

Nehlig P., Juteau T., Bendel V., and Cotten J. (1994) The root zones of oceanic hystrothermal systems: constraints from the samail ophiolite (Oman). *J. Geophys. Res.* **99**, 4703–4713.

Palmer H. R. and Edmond J. M. (1989) The strontium isotope budget of the modern ocean. *Earth Planet. Sci. Lett.* **92**, 11–26.

Parsons B. (1984) The rates of plate creation and consumption. *Geophys. J. Roy. Astron. Soc.* **67**, 437–448.

Parsons B. and Sclater J. G. (1977) An analysis of the variation of ocean floor bathymetry and heat flow with age. *J. Geophys. Res.* **82**, 803–829.

Pelayo A. M., Stein S., and Stein C. A. (1994) Estimation of oceanic hydrothermal heat flux from heat flow and depths of midocean ridge seismicity and magma chambers. *Geophys. Res. Lett.* **21**, 713–716.

Penrose Conference Participants (1972) Penrose field conference on ophiolites. *Geotimes* **17**, 24–25.

Peuker-Ehrenbrink B., Hofmann A. W., and Hart S. R. (1994) Hydrothermal lead transfer from mantle to continental crust: the role of metalliferous sediments. *Earth Planet. Sci. Lett.* **125**, 129–142.

Phipps-Morgan J. (2001) Thermodynamics of pressure release melting of a veined plum pudding mantle. *Geochem. Geophys. Geosyst.* **2**, paper number 2000GC000049.

Plank T., Ludden J. N., and Escutia C. (2000) Proc. ODP, Initial Reports, 185 [Online] http://www-odp.tamu.edu/publications/185_IR/185ir.htm

Richardson C. J., Cann J. R., Richards H. G., and Cowan J. G. (1987) Metal depleted root-zones of the Troodos ore-forming hydrothermal systems, Cyprus. *Earth Planet. Sci. Lett.* **84**, 243–254.

Richardson S. H., Hart S. R., and Staudigel H. (1980) Vein mineral ages of old oceanic crust. **85**, 7195–7200.

Robinson P. T., Flower M. F., Staudigel H., and Swanson D. A. (1979) Lithology and eruptive stratigraphy of Cretaceous oceanic crust, western Atlantic. In *Init. Repts. DSDP* (eds. T. In Donnelly and J. Francheteau). US Government Printing Office, Washington, DC, vol. 51–53, pp. 1535–1556.

Robinson P. T., Melson W., and Schmincke H. U. (1983) Volcanic glass compositions of the Troodos ophiolite, Cyprus. *Geology* **11**, 400–404.

Ryan J. G. and Langmuir C. H. (1987) The systematics of lithium abundances in young volcanic rocks. *Geochim. Cosmochim. Acta* **51**, 1727–1741.

Sakai H., Desmarais D. J., Uyeda A., and Moore J. G. (1984) Concentration and isotope ratios of carbon nitrogen and sulfur in ocean floor basalts. *Geochim. Cosmochim. Acta* **48**, 2433–2441.

Salisbury M. H., Donnelly T. W., and Francheteau J. (1979) Geophysical Logging in Deep Sea drilling project hole 417D. *Initial Rep. Deep Sea Drill. Proj.* **51–53**(1), 705–713.

Schiano P., Birck J.-L., and Allegre C. (1997) Osmium–strontium–neodymium–lead isotopic covariations in mid-ocean ridge basalt glasses and the heterogeneity of the upper mantle. *Earth Planet. Sci. Lett.* **150**, 363–379.

Schiffman P. and Smith B. M. (1988) Petrology and oxygen isotope geochemistry of a fossil seawater hydrothermal system within the Solea Graben, northern Troodos ophiolite, Cyprus. *J. Geophys. Res.* **93**, 4612–4624.

Schmincke H. U., Rautenschlein M., Robinson P. T., and Mehegan J. (1983) Volcanic glass compositions of the Troodos ophiolite, Cyprus. *Geology* **11**, 400–404.

Seyfried W. E., Janecky D. R., and Mottl M. J. (1984) Alteration of the oceanic crust: implications for geochemical cycles of lithium and boron. *Geochim. Cosmochim. Acta* **48**, 557–569.

Shanks W. C., Bischoff J. L., and Rosenbauer R. J. (1981) Seawater sulfate reduction and sulfur isotope fractionation in basaltic systems: interaction of seawater with fayalite and magnetite at 200–300°C. *Geochim. Cosmochim. Acta* **45**, 1977–1995.

Sleep N. H. (1991) Hydrothermal circulation, anhydrite precipitation, and thermal structure at ridge axes. *J. Geophys. Res.* **96**, 2375–2387.

Smith H. J., Spivack A. J., Staudigel H., and Hart S. R. (1995) The boron isotopic composition of altered oceanic crust. *Chem. Geol.* **126**, 119–135.

Snow J. E. and Dick H. J. B. (1995) Pervasive magnesium loss by marine weathering of peridotite. *Geochim. Cosmochim. Acta* **59**(20), 4219–4235.

Spivack A. J. and Edmond J. M. (1987) Boron isotope exchange between seawater and the oceanic crust. *Geochim. Cosmochim. Acta* **51**, 1033–1043.

Spivack A. J. and Staudigel H. (1994) Low-temperature alteration of the upper oceanic crust and the alkalinity budget of seawater. *Chem. Geol.* **115**, 239–247.

Spooner E. T. C., Beckinsdale R. D., Fyfe W. S., and Smewing J. (1974) O18 enriched ophiolitic metabasic rocks from E. ligura (Italy) Pindos (Greece) and Troodos (Cyprus). *Contrib. Mineral. Petrol.* **47**, 41–62.

Spooner E. T. C., Chapman H. J., and Smewing J. D. (1977) Strontium isotopic contamination, and oxidation during ocean floor hydrothermal metamorphism of the ophiolitic rocks of the Troodos massif, Cyprus. *Geochim. Cosmochim. Acta* **41**, 873–890.

Stakes D. S. and Taylor H. P., Jr. (1992) The northern Samail ophiolite: an oxygen, microprobe, and field study. *J. Geophys. Res.* **97**, 7043–7080.

Staudigel H. and Hart S. R. (1983) Alteration of basaltic glass: Mechanisms and significance of the oceanic crust–seawater budget. *Geochim. Cosmochim. Acta* **47**, 337–350.

Staudigel H. and King S. D. (1992) Ultrafast subduction: the key to slab recycling efficiency and mantle differentiation? *Earth Planet. Sci. Lett.* **109**, 517–530.

Staudigel H., Muehlenbachs K., Richardson S. H., and Hart S. R. (1981) Agents of low temperature ocean crust alteration. *Contrib. Mineral. Petrol.* **77**, 150–157.

Staudigel H., Gillis K., and Duncan R. (1986) K/Ar and Rb/Sr ages of celadonites from the Troodos ophiolite, Cyprus. *Geology* **14**, 72–75.

Staudigel H., Hart S. R., Schmincke H. U., and Smith B. M. (1989) Cretaceous ocean crust at DSDP sites 417 and 418: Carbon uptake from weathering versus loss by magmatic outgassing. *Geochim. Cosmochim. Acta* **53**, 3091–3094.

Staudigel H., Davies G. R., Hart S. R., Marchant K. M., and Smith B. M. (1995) Large-scale isotopic Sr, Nd, and O isotopic composition of altered oceanic crust at DSDP/ODP Sites 417/418. *Earth Planet. Sci. Lett.* **130**, 169–185.

Staudigel H., Plank T., White W. M., and Schmincke H.-U. (1996) Geochemical fluxes during seafloor alteration of the upper oceanic crust: DSDP sites 417 and 418. In *SUBCON: Subduction from Top to Bottom* (eds. G. E. Bebout and S. H. Kirby). American Geophysical Union, Washington, DC, vol. 96, pp. 19–38.

Stein C. A. and Stein S. (1994a) Comparison of plate and asthenospheric flow models for the thermal evolution of oceanic lithosphere. *Geophys. Res. Lett.* **21**, 709–712.

Stein C. A. and Stein S. (1994b) Constraints on hydrothermal heat flux through the oceanic lithosphere from global heat flow. *J. Geophys. Res.* **99**, 3081–3096.

Stein C. A., Stein S., and Pelayo A. (1995) Heat flow and hydrothermal circulation. In *Seafloor Hydrothermal Processes*, Geophysical Monograph (eds. S. E. Humphris, R. A. Zierenberg, L. S. Mullineaux, and R. E. Thomson). American Geophysical Union, Washington, DC, vol. 91, pp. 425–445.

Sun S.-S., and McDonough W. F. (1989) Chemical and Isotopic systematics of oceanic basalts: implications for mantle composition and processes. In *Magmatism in the Ocean Basins*, Geological Society Special Publication 42 (eds. A. D. Saunders and M. J. Norry). Geological Society of America, Washington, DC, vol. 42, pp. 313–345.

Touret J. and Olsen S. N. (1985) Fluid inclusions in migmatites. In *Migmatites* (ed. J. R. Ashworth). Blackie and Son, Glasgow, Scotland, pp. 265–288.

Von Damm K. L., Edmond J. M., Grant B., and Measures C. I. (1985) Chemistry of submarine hydrothermal solutions at 21°N East Pacific Rise. *Geochim. Cosmochim. Acta* **49**, 2197–2220.

Wheat C. G. and Mottl M. J. (2000) Composition of pore and spring waters from Baby Bare: global implications of geochemical fluxes from a ridge flank hydrothermal system. *Geochim. Cosmochim. Acta* **64**(4), 629–642.

White R. S., McKenzie D., and O'Nions R. K. (1992) Ocean crustal thickness from seismic measurements and rare earth element inversions. *J. Geophys. Res.* **97**, 19683–19715.

Wyllie P. J. (1978) Mantle fluid compositions buffered in peridotite-CO_2-H_2O by carbonates, amphiboles, and phlogopite. *J. Geol.* **86**, 687–713.

3.16
Oceanic Plateaus

A. C. Kerr
Cardiff University, Wales, UK

3.16.1 INTRODUCTION

Although the existence of large continental flood basalt provinces has been known for some considerable time, e.g., Holmes (1918), the recognition that similar flood basalt provinces also exist below the oceans is relatively recent. In the early 1970s increasing amounts of evidence from seismic reflection and refraction studies revealed that the crust in several large portions of the ocean floor is significantly thicker than "normal" oceanic crust, which is 6–7 km thick. One of the first areas of such over-thickened crust to be identified was the Caribbean plate (Edgar *et al.*, 1971) which Donnelly (1973) proposed to be an "oceanic flood basalt province". The term oceanic plateau was coined by Kroenke (1974), and was prompted by the discovery of a large area of thickened crust (>30 km) in the western Pacific known as the Ontong Java plateau (OJP). As our knowledge of the ocean basins has improved over the last 25 years, many more oceanic plateaus have been identified (Figure 1). Coffin and Eldholm (1992) introduced the term "large igneous provinces" (LIPs) as a generic term encompassing oceanic plateaus, continental flood basalt provinces, and those provinces which form at the continent–ocean boundary (volcanic rifted margins).

LIPs are generally believed to be formed by decompression melting of upwelling hotter mantle, known as mantle plumes. Although ideas about hotpots and mantle plumes have been around for almost 40 years (Wilson, 1963), it is only in the past 15 years that LIPs have become the focus of major research. One of the main reasons for the increased research activity into LIPs is the realization that significant proportions of these LIPs erupted over a relatively short time, often less than 2–3 Myr (see review in Coffin, 1994).

This has important implications for mantle processes and source regions (Hart *et al.*, 1992; Stein and Hofmann, 1994), as well as environmental effects on the global biosphere (e.g., Caldeira and Rampino, 1990; Courtillot *et al.*, 1996; Kerr, 1998). Oceanic plateaus can also become accreted to continental margins, and it has been proposed that these plateaus have been significant contributors to the growth of continental crust (e.g., Abbott, 1996; Albarede, 1998).

The most recent major phase of oceanic plateau formation was in the Cretaceous when the Ontong Java, Manihiki, Hess Rise, and the Caribbean–Colombian plateaus formed in the Pacific, while in the Indian Ocean the Kerguelen plateau was developing. The areas, volume maximum thicknesses and ages of the larger of these plateaus are given in Table 1. The Ontong Java is the largest of the Cretaceous plateaus. It covers an area of 1.9×10^6 km^2, and has an estimated total volume of 4.4×10^7 km^3 (Eldholm and Coffin, 2000). Although early seismic refraction data suggested that the OJP was as thick as 43 km (Furomoto *et al.*, 1976), a more recent synthesis based on existing seismic and new gravity data (Gladczenko *et al.*, 1997) has indicated the average thickness to be ~32 km.

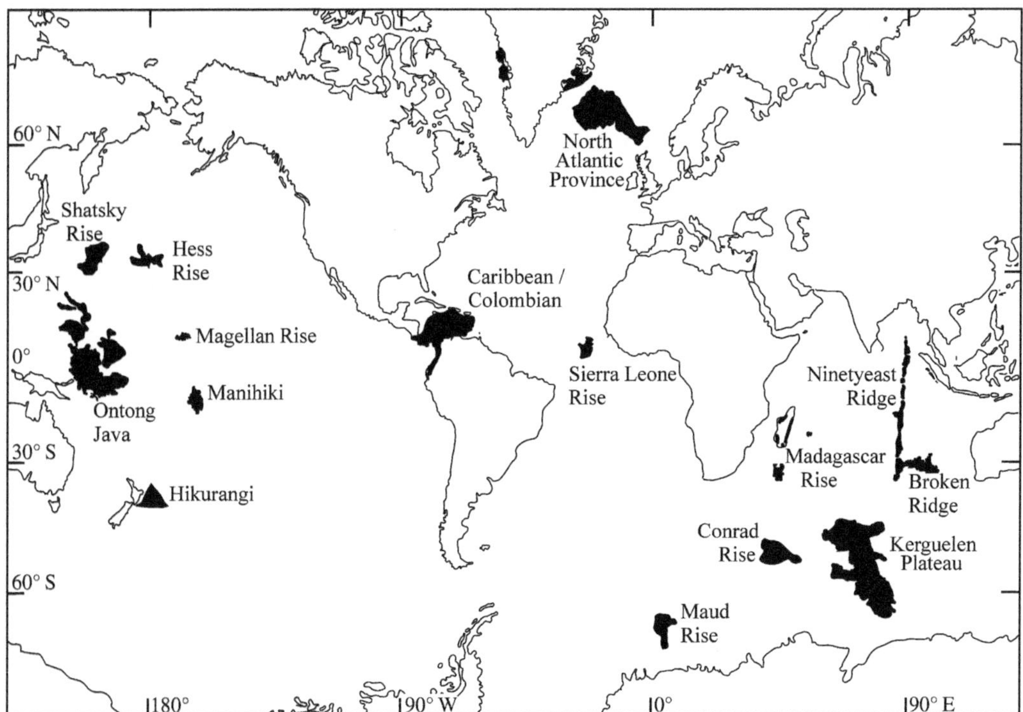

Figure 1 Map showing all major oceanic plateaus, and other large igneous provinces discussed in the text (after Saunders *et al.*, 1992).

Table 1 Ages and dimensions of Jurassic–Cretaceous oceanic plateaus.

Oceanic plateau	*Mean age* (Ma)	*Area* (10^6 km^2)	*Thickness range* (km)	*Volume* (10^6 km^3)
Hikurangi	early-mid Cretaceous	0.7	10–15	2.7
Shatsky Rise	147	0.2	10–28	2.5
Magellan Rise	145	0.5	10	1.8
Manihiki	123	0.8	>20	8.8
Ontong Java	121(90)	1.9	15–32	44.4
Hess Rise	99	0.8	>15	9.1
Caribbean	88	1.1	8–20	4.4
South Kerguelen	110	1.0	~22	6.0
Central Kerguelen/Broken Ridge	86	1.0	19–21	9.1
Sierra Leone Rise	~73	0.9	>10	2.5
Maud Rise	~>73	0.2	>10	1.2

After Eldholm and Coffin (2000).

3.16.2 FORMATION OF OCEANIC PLATEAUS

The production of large volumes (>10^6 km^3) of melt in a period as short as 2–3 Myr implies magma production rates up to 25% higher than those observed at present-day midocean ridges (Eldholm and Coffin, 2000), and is generally believed to necessitate a high flux of hotter-than-ambient asthenospheric mantle below these provinces (e.g., McKenzie and Bickle, 1988). Numerical and physical models show that this hotter mantle commonly takes the form of a mantle plume which ascends by thermal buoyancy through the overlying mantle (Loper, 1983; McKenzie and Bickle, 1988; Campbell *et al.*, 1989; Farnetani and Richards, 1995). Physical constraints demand that mantle plumes must ascend from a boundary layer within the Earth, either the core–mantle boundary (D'') or the 670 km discontinuity. Large ascending mantle plumes are, on average, 200 °C hotter than the ambient upper mantle (McKenzie and Bickle, 1988) and undergo decompression melting as they approach the base of the lithosphere. Physical modeling experiments by Griffiths and Campbell (1990) have shown that mantle plumes are likely to ascend through the mantle from their source boundary layer in the form of a large semi-spherical "head" fed from the source region by a narrower plume tail (Figure 2). Alternatively, numerical modeling by Farnetani and Richards (1995) suggested that plume heads starting in the mantle only rise about three plume head diameters before spreading out. In either case, as the plume approaches the base of the lithosphere, it spreads out over a broadly circular area (which can be as much as 1000 km in diameter) and undergoes adiabatic decompression, producing melt over most of the area covered by the flattened-out plume head (Campbell and Griffiths, 1990). The amount of melt produced is critically dependent on the thickness of the preexisting lithosphere, since the base of the rigid, nonconvecting lithosphere will act as a "lid" on the upwelling plume mantle and on the extent of decompression melting. Thus, a mantle plume ascending below thick continental lithosphere (>50 km) will produce a smaller thickness of melt than a plume which ascends beneath oceanic lithosphere (≤7 km) (Figure 3). Another significant factor in determining the amount of melt generated by a mantle plume is the temperature of the plume: generally the higher the temperature, the more melt will be produced (Figure 3).

The initial $^{40}Ar/^{39}Ar$ step-heating ages for LIPs support models of rapid formation and eruption, often in less than 2–3 Myr (Richards *et al.*, 1989). As more age data have become available, a wider age-range has emerged for some LIPs (e.g., the Caribbean plateau, 95–86 Ma: Kerr *et al.*, 1997a; Sinton *et al.*, 1998; Hauff *et al.*, 2000b). Nevertheless, it still appears that substantial proportions of these provinces were formed over geologically

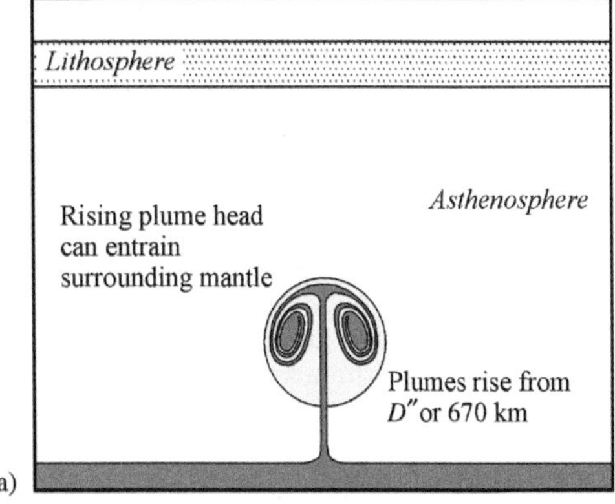

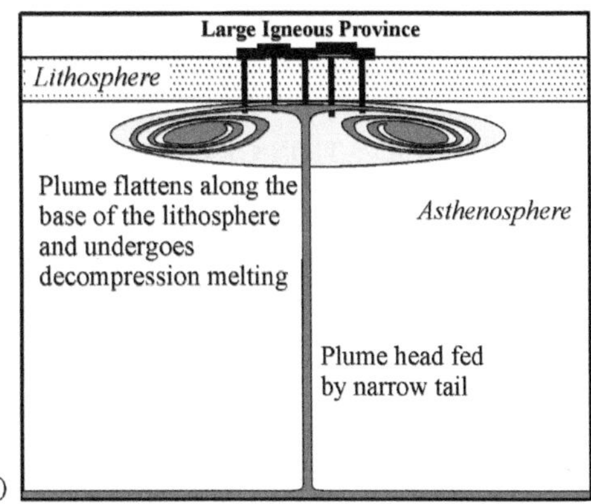

Figure 2 Cartoon to illustrate how mantle plumes are believed to (a) rise through the asthenosphere from either the 670 km discontinuity or D″ and (b) flatten along the base of the lithosphere and undergo decompression melting to produce a LIP (after Saunders *et al.*, 1992).

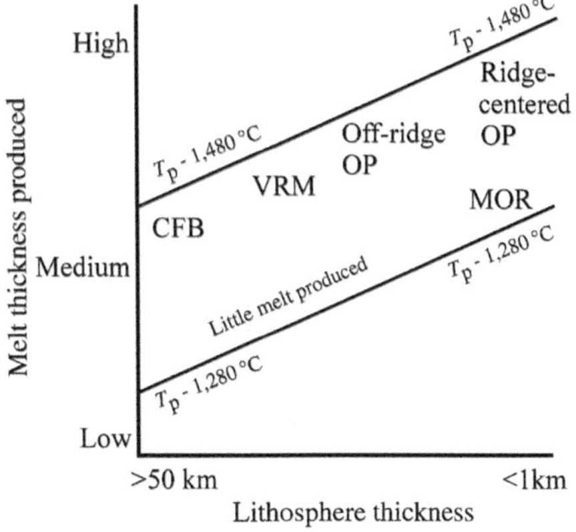

Figure 3 Schematic diagram showing how original lithospheric thickness and mantle potential temperature affect the amount of melt produced (melt thickness) and how these factors relate to continental flood basalts (CFB), volcanic rifted margins (VRM), off–ridge and ridge–centered oceanic plateaus (OP), and midocean ridges (MOR).

short time periods (e.g., the Ontong Java plateau formed on two occasions: 122±3 and 90±4 Ma; Neal *et al.*, 1997).

Theoretically, oceanic plateaus can form anywhere in the ocean basins; however, most oceanic plateaus appear to have formed at or near midocean ridges (e.g., Kerguelen, Manihiki and Ontong Java), i.e., regions that are conducive to voluminous decompression melting (Eldholm and Coffin, 2000). At first glance, it appears somewhat coincidental that plumes of deep mantle origin reach the base of the lithosphere at a midocean ridge. However, as pointed out by Saunders *et al.* (1996), a likely explanation for this observation is that mantle plumes can "capture" oceanic spreading centers (cf. present-day Iceland).

3.16.3 PRESERVATION OF OCEANIC PLATEAUS

The oldest *in situ* oceanic crust is Jurassic in age (Pringle, 1992), because within 200 Myr of its formation at a midocean ridge, the oceanic crust was recycled back into the asthenosphere. Although many of the Cretaceous and late Jurassic oceanic plateaus still form part of the ocean basins, the preservation potential of oceanic plateaus older than Jurassic is low. Fortunately, however, oceanic plateaus are much more buoyant than oceanic crust of "normal" thickness formed at a midocean ridge (e.g., Cloos, 1993). This excess buoyancy is primarily due to the greater crustal thickness of oceanic plateaus if the plateaus are relatively young; residual heat from their formation can also contribute to their buoyancy. Recent measurements of upper mantle shear wave splitting and shear wave velocity structure (Klosko *et al.*, 2001) reveal that the Ontong Java plateau is underlain by a 300 km thick, long-lived, rheologically strong and chemically depleted root. Klosko *et al.* (2001) propose that this root represents the residue from mantle melting and that its consequent lower density contributes significantly to the buoyancy of both the Ontong Java and other oceanic plateaus. The net result of this excess buoyancy is that oceanic plateaus, in contrast to normal oceanic crust, are much less easily subducted (Ben-Avraham *et al.*, 1981; Cloos, 1993; Kimura and Ludden, 1995). Instead of being completely recycled back into the mantle, their upper layers can be "peeled off" (Kimura and Ludden, 1995) and accreted on to the margin of the subduction zone. This plateau accretion can occur either at an Andean-type continental margin, or an island arc.

Although many of the in situ Cretaceous oceanic plateaus have been drilled by the Deep Sea Drilling (DSDP) and Ocean Drilling Programs (ODP), the insight that these drill holes can provide is relatively limited compared to the accreted oceanic plateau sections. For example, the Ontong Java plateau collided with the westward-dipping Solomon Islands subduction zone at 10–20 Ma, resulting in a reversal in the polarity of subduction from west to east, and the uplift and exposure of the deeper sections of the plateau on the Solomon Islands (Neal *et al.*, 1997; Petterson *et al.*, 1999). A second example is the Caribbean–Colombian oceanic plateau, which formed in the Pacific at ~90 Ma (Sinton *et al.*, 1998; Hauff *et al.*, 2000a). Within 10 Myr the eastward-moving Farallon plate had brought the southern portion of the plateau into collision with continental northwestern South America, resulting in the accretion of slices of the plateau on to the continental margin (Kerr *et al.*, 1997b). This accretion was accompanied by back-stepping of the subduction zone west of the accreted plateau slices. Shortly after its formation the northern portion of the plateau began to move into the proto-Caribbean seaway between North and South America (Burke, 1988; Kerr *et al.*, 1999; White *et al.*, 1999). In doing so the plateau encountered the eastward-dipping "Great Arc of the Caribbean". Unable to subduct, the thick plateau clogged the subduction zone, resulting in a reversal in the polarity of subduction from east to west. This reversal in subduction polarity in conjunction with a back-stepping of subduction behind the advancing plateau (Burke, 1988), effectively isolated the Caribbean as a separate plate. Over the past ~80 Ma the northern portion of the Caribbean plateau has moved into the gap between North and South America and has been uplifted and subaerially exposed around its margins (Figure 4), thus making it available for detailed study. It is in ways such as these that remnants of these thick, buoyant oceanic plateaus can be preserved and incorporated into the continental crust. The identification of these older preserved plateaus within the geological record will be explored in a later section.

3.16.4 GEOCHEMISTRY OF CRETACEOUS OCEANIC PLATEAUS

3.16.4.1 General Chemical Characteristics

Table 2 shows representative analyses of *in situ* and accreted portions of Cretaceous oceanic plateaus. Cretaceous oceanic plateaus are predominantly basaltic (<10 wt.% MgO) in composition (Figure 5), although more-MgO-rich lava flows are found in the Caribbean–Colombian oceanic plateau (CCOP). Typically, oceanic plateaus possess generally low levels of compatible elements (i.e., Ni < 300 ppm and Cr < 1,000 ppm; see Figure 6). In terms of incompatible trace elements the majority of oceanic plateau lavas and intrusive sheets possess relatively flat rare earth

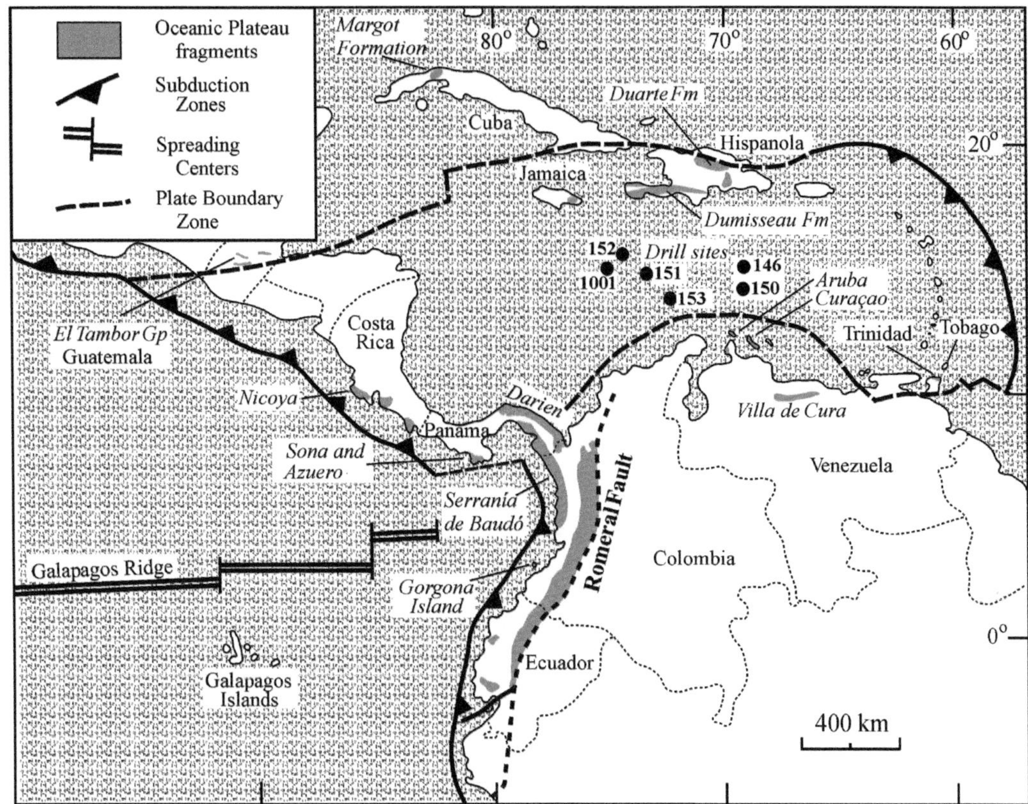

Figure 4 Map to show the main accreted outcrops of the Caribbean–Colombian oceanic plateau along with the locations of DSDP/ODP drill holes which penetrated the thickened crust of the Caribbean plate.

primitive mantle-normalized patterns with abundances varying between 5 and 10 times primitive mantle values (Figure 7).

The radiogenic isotope compositions of oceanic plateaus have been well characterized: initial εNd values for oceanic plateaus generally range from +6.0 to +9.0, whereas initial $^{87}Sr/^{86}Sr$ ratios fall mostly between 0.703 and 0.704 (Figure 8), i.e., the $^{87}Sr/^{86}Sr$ and ε>Nd values are generally less depleted in terms of their radiogenic isotopes than "normal" (N)-MORB as typified by MORB from the East Pacific Rise (Figure 8(a)). Elevated initial $^{87}Sr/^{86}Sr$ ratios (>0.7040) are most likely due to secondary alteration by hydrothermal fluids. It is noteworthy that the high-MgO rocks generally possess more extreme $(\varepsilon Nd)_i$ values than the basalts (Figure 8(a)). Figure 8(b), a plot of $^{207}Pb/^{204}Pb$ against $^{206}Pb/^{204}Pb$, reveals that most oceanic plateau rocks range between 18.5 and 19.5 $^{206}Pb/^{204}Pb$ and between 15.525 and 15.625 $^{207}Pb/^{204}Pb$. An interesting feature of Figure 8(b) is that many of the basalts lie on a trend between the proposed mantle components of HIMU and DMM (Zindler and Hart, 1986) while the high-MgO lavas form a trend between DMM and the enriched mantle component EM2. Zindler and Hart (1986) proposed four mantle components: depleted MORB mantle (DMM), two types of enriched mantle EM1 and EM2, and HIMU, so-called because it has a high $^{238}U/^{204}Pb$ ratio, or μ value.

In recent years our knowledge of the radiogenic isotope systematics of oceanic plateaus has been augmented by the analysis of Hf and Os isotopes. Although data for the Kerguelen and Ontong Java plateaus are still relatively sparse, more data exist for the CCOP. Initial εHf values for the CCOP and OJP range from 10 to 18 (Figure 9(a)). The CCOP samples form a trend between MORB source mantle and the HIMU component, while the OJP samples possess lower initial εNd values at equivalent εHf than the CCOP and appear to form a trend towards EM2. The much lower εHf and εNd for the Kerguelen plateau (Figure 9(a)) will be discussed in Section 3.16.5.2. Initial γOs for high-MgO rocks from the CCOP range from 0 to +18, whereas the basalts range from −7 to +10 (Figure 9(b)). How representative these ranges are for γOs, requires the acquisition of more data from other oceanic plateaus.

3.16.4.2 Mantle Plume Source Regions of Oceanic Plateaus

Since the pioneering study of Hoffman and White, it has become widely accepted that one of

Table 2 Representative analyses of Cretaceous oceanic plateau lavas.

Plateau	*Location*	*Sample*	*Data sources*	*Zr*	*Nb*	*Mo*	*Hf*	*Pb*	*Ta*	*Th*	*U*	$^{87}Sr/^{86}Sri$	εNdi	$^{206}Pb/^{204}Pb$	$^{207}Pb/^{204}Pb$	$^{208}Pb/^{204}Pb$
Kerguelen	ODP site 747	16-5,103-6	1	50.90	1.21	17.23	10.62	0.16	7.47	9.59	1.88	2.00	0.16	101.22	3.63	
Kerguelen	ODP site 748	79-6,90-4	1	49.24	2.74	17.50	8.77	0.09	7.54	6.71	4.72	0.56	1.12	98.99	7.14	
Kerguelen	ODP site 749	15-5,125-7	1	52.73	1.52	15.18	11.95	0.17	7.22	8.31	3.42	0.54	0.18	101.22	1.35	
Kerguelen	ODP site 750	17-3,23-26	1	49.21	1.17	15.77	14.62	0.18	8.36	9.01	1.99	0.19	0.11	100.61	4.31	
Kerguelen	ODP site 738	34-1,88-92	2	52.17	1.73	15.44	10.82	0.18	5.51	9.17	2.98	1.95	0.21	100.16	0.58	
Ontong Java	ODP site 807	75-4,46-48	3	48.75	1.62	14.16	13.43	0.21	6.74	12.13	2.38	0.13	0.14	99.69	0.45	
Ontong Java	ODP site 807	88-3,76-79	3	49.74	1.13	14.27	13.44	0.22	7.43	12.00	2.06	0.28	0.09	100.66	−0.24	
Ontong Java	Santa Isabel	I96	4	49.28	1.31	14.20	12.44	0.16	8.03	12.50	1.89	0.15	0.10	100.06	1.61	
Ontong Java	Maliata	SG1	5	50.38	1.65	13.54	14.01	0.19	7.30	11.64	2.32	0.15	0.14	101.32	0.30	6.0
Ontong Java	Maliata	ML407	5	49.67	1.35	13.64	13.47	0.19	7.45	11.38	3.08	0.26	0.13	100.62	2.14	5.0
CCOP	Gorgona	GOR160	6	45.13	0.64	11.78	12.52	0.18	18.25	10.10	1.30	0.06	0.05	100.01		
CCOP	Gorgona	GOR117	6	50.40	0.77	13.81	12.62	0.23	8.65	12.39	2.43	0.02	0.06	101.38		
CCOP	Gorgona	GOR94-35	7	46.58	0.37	10.57	11.41	0.17	21.96	8.42	0.55	0.19	0.03	100.25	3.26	
CCOP	Colombia	SDB18	8	50.63	1.47	14.19	13.34	0.23	6.77	10.94	2.89	0.15	0.12	100.72	1.57	
CCOP	Colombia	VIJ1	8	51.45	2.02	12.85	15.20	0.19	5.78	9.57	2.29	0.07	0.17	99.59	0.90	
CCOP	Colombia	COL472	9	48.81	0.89	12.86	10.66	0.17	11.55	11.61	2.91	0.08	0.09	99.64	3.03	
CCOP	Curaçao	CUR14	10	46.42	0.57	9.53	11.06	0.17	22.86	8.20	0.90	0.02	0.05	99.79	2.23	
CCOP	Curaçao	CUR20	10	52.13	0.78	12.85	9.80	0.20	8.60	14.30	1.43	0.03	0.06	100.18	1.41	
CCOP	DSDP site 150	11-2, 63-67	11	49.46	1.27	16.60	10.30	0.10	8.58	9.87	2.60	0.13	0.10	99.02		
CCOP	Ecuador	EQ1	12	48.88	1.23	14.39	12.75	0.20	8.96	10.81	1.89	0.72	0.18	100.01	4.45	
				Rb	Cs	Sr	Ba	Sc	V	Cr	Ni	Cu	Zn	Ga	La	Ce
Kerguelen	ODP site 747	16-5,103-6	1	18.5	<0.8	234	229	33.8	201	401	72		66	22.0	12.30	25.50
Kerguelen	ODP site 748	79-6,90-4	1	7.8	2.81	1131	1661	18.7	170	166	182		79	18.7	105.00	224.00
Kerguelen	ODP site 749	15-5,125-7	1	12.3	<0.5	214	114	34.5	271	260	30		115	22.7	6.80	16.10
Kerguelen	ODP site 750	17-3,23-26	1	9.0	<0.9	193	30	38.9	269	193	120		93	19.8	4.00	8.90
Kerguelen	ODP site 738	34-1,88-92	2	37.4		273	336	36.9	267	95	28		102	20.7	17.10	39.30
Ontong Java	ODP site 807	75-4,46-48	3	1.0		174		45.9	313	162	99				6.15	14.59
Ontong Java	ODP site 807	88-3,76-79	3	10.0		107		52.3	349	163	87				2.96	8.07
Ontong Java	Santa Isabel	I96	4	2.0		115	37	50.0	341	238	122				3.40	9.90
Ontong Java	Maliata	SG1	5	1.3	0.01	108	28	41.0	392	56	61	124	93	17.0	4.75	12.60
Ontong Java	Maliata	ML407	5	1.8	0.01	100	23	40.0	295	61	62	229	57	17.0	3.86	10.60
CCOP	Gorgona	GOR160	6	1.0		64	8	28.3	227	1373	723			13.2	0.65	2.17
CCOP	Gorgona	GOR117	6	0.3		107	21	41.1	371	194	112			15.7	1.02	2.99
CCOP	Gorgona	GOR94-35	7	4.2		34	20	27.3	166	82	968	2030	56	10.8	0.22	0.57
CCOP	Colombia	SDB18	8	2.5		156	37	42.2	331	208	97			19.1	3.80	9.80
CCOP	Colombia	VIJ1	8	0.4		89	27	44.4	531	63	58	133	124	20.2	6.05	16.55
CCOP	Colombia	COL472	9	1.9		398	85	30.8	334	1393	264	125	70	14.3	2.42	6.33

CCOP	Curaçao	CUR14	10	1.7		46	6	29.7	186	2017	1032	75	70	9.1	1.20	4.20
CCOP	Curaçao	CUR20	10	0.4		69	11	43.3	258	552	178	102	66	12.2	2.84	7.59
CCOP	DSDP site 150	11-2, 63-67	11	3.2		117	16		335	373	127	150	85	17.0	3.16	8.55
CCOP	Ecuador	EQ1	12	5.2		111	15	61.5	353	285	101				3.50	9.63
				Pr	Nd	Sm	Eu	Gd	Tb	Dy	Ho	Er	Tm	Yb	Lu	Y
Kerguelen	ODP site 747	16-5,103-6	1		13.50	3.50	1.03	0.00	0.65					1.89	0.28	23.2
Kerguelen	ODP site 748	79-6,90-4	1		103.00	14.20	3.31	8.90	1.25					1.80	0.26	27.7
Kerguelen	ODP site 749	15-5,125-7	1		11.90	3.64	1.24		0.69					2.71	0.39	29.3
Kerguelen	ODP site 750	17-3,23-26	1		6.30	2.32	0.84	3.00	0.58					2.57	0.34	24.6
Kerguelen	ODP site 738	34-1,88-92	2		21.20	5.28	1.73		0.94					2.87	0.44	28.8
Ontong Java	ODP site 807	75-4,46-48	3		11.65	3.39	1.31	4.51	0.81					2.71	0.42	30.0
Ontong Java	ODP site 807	88-3,76-79	3		6.50	2.30	0.93	3.39	0.70					2.40	0.38	24.0
Ontong Java	Santa Isabel	I96	4	1.56	7.84	2.77	1.02	3.50	0.63	4.20	0.91	2.64	0.37	2.30	0.35	22.0
Ontong Java	Maliata	SG1	5	1.98	4.61	1.69	1.13	4.14	0.73	4.56	1.00	2.90	0.41	2.83	0.43	25.0
Ontong Java	Maliata	ML407	5	1.70	5.67	2.11	0.94	3.54	0.68	4.22	0.92	2.57	0.36	2.47	0.37	23.0
CCOP	Gorgona	GOR160	6	0.46	2.86	1.25	0.48	1.58	0.29	2.10		1.43	0.22	1.30	0.21	13.8
CCOP	Gorgona	GOR117	6	0.60	3.53	1.74	0.63	2.07	0.41	3.13		2.18	0.37	2.20	0.32	22.8
CCOP	Gorgona	GOR94-35	7	0.13	0.90	0.65	0.34	1.50	0.30	2.18	0.48	1.44	0.23	1.42	0.21	15.3
CCOP	Colombia	SDB18	8		9.30	2.85	1.11	2.94	0.80					2.75	0.43	29.4
CCOP	Colombia	VIJ1	8		12.0	3.97	1.47	4.76	0.93					4.25	0.63	41.7
CCOP	Colombia	COL472	9	0.92	5.58	1.76	0.74	2.67		2.92	0.58	1.44		1.36	0.21	19.0
CCOP	Curaçao	CUR14	10		3.16		0.38							0.50		11.5
CCOP	Curaçao	CUR20	10	0.79	4.80	1.39	0.55	1.87		2.11	0.45	1.38		1.36	0.21	15.8
CCOP	DSDP site 150	11-2, 63-67	11	1.36	6.94	2.32	0.89	2.87	0.52	3.36	0.69	1.89	0.28	1.82	0.26	21.0
CCOP	Ecuador	EQ1	12	1.56	8.11	2.57	0.96	3.46	0.66	4.40	0.90	2.60	0.39	2.53	0.38	22.3
				Zr	Nb	Mo	Hf	Pb	Ta	Th	U	$^{87}Sr/^{86}Sr_i$	εNdi	$^{206}Pb/^{204}Pb$	$^{207}Pb/^{204}Pb$	$^{208}Pb/^{204}Pb$
Kerguelen	ODP site 747	16-5,103-6	1	97	7.00		2.34		0.46	0.20	0.000	0.705783	−4.0	17.65	15.51	38.16
Kerguelen	ODP site 748	79-6,90-4	1	599	121.90		12.40	8.00	9.08	13.20	2.200	0.705319	−3.3	18.19	15.64	38.38
Kerguelen	ODP site 749	15-5,125-7	1	91	5.80		2.16	1.63	0.40	1.80	0.148	0.704260	1.8	18.03	15.55	38.16
Kerguelen	ODP site 750	17-3,23-26	1	47	3.33		1.16	0.82	0.19	1.40	0.048	0.706165	1.4	17.53	15.49	38.01
Kerguelen	ODP site 738	34-1,88-92	2	166	8.90		3.89		0.54	2.27		0.709730	−8.2	17.82	15.75	39.01
Ontong Java	ODP site 807	75-4,46-48	3	98	5.70		2.64		0.34	0.56		0.704330	5.0	18.40	15.53	38.38
Ontong Java	ODP site 807	88-3,76-79	3	64	3.32		1.74		0.20	0.34		0.703560	5.9	18.67	15.55	38.54
Ontong Java	Santa Isabel	I96	4	65	4.30			0.23		0.27	0.120	0.703690	5.4	18.64	15.54	38.60
Ontong Java	Maliata	SG1	5	81	4.90	0.59	2.50	0.17	0.35	0.28	0.097	0.704040	3.9	17.85	15.47	37.91
Ontong Java	Maliata	ML407	5	83	5.00	0.36	2.63	0.15	0.25	0.16	0.047	0.704130	5.7	18.32	15.51	38.23
CCOP	Gorgona	GOR160	6	29	0.48		1.00	1.30		0.02	0.011	0.703041	9.5	18.32	15.50	37.82

(continued)

Table 2 (continued).

Plateau	Location	Sample	Data sources	Zr	Nb	Mo	Hf	Pb	Ta	Th	U	$^{87}Sr/^{86}Sri$	εNdi	$^{206}Pb/^{204}Pb$	$^{207}Pb/^{204}Pb$	$^{208}Pb/^{204}Pb$
CCOP	Gorgona	GOR117	6	39	0.87		1.05	12.10		0.08	0.034	0.703283	8.6	18.86	15.58	38.56
CCOP	Gorgona	GOR94-35	7	13	0.80					1.20		0.704767	9.4	18.68	15.54	38.19
CCOP	Colombia	SDB18	8	70	4.29		2.05		0.30	0.30		0.703380	8.5	18.87	15.54	38.46
CCOP	Colombia	VIJ1	8	119	8.00		3.14		0.48	0.62	0.210	0.703207	8.1	19.22	15.58	38.91
CCOP	Colombia	COL472	9	40	2.16			2.37				0.513197	11.0	19.33	15.58	38.86
CCOP	Curaçao	CUR14	10	30	2.50			1.20		0.10		0.702961	6.6	19.31	15.59	38.90
CCOP	Curaçao	CUR20	10	40	3.40			1.00		0.90		0.703215	7.0	19.08	15.55	38.82
CCOP	DSDP site 150	11-2, 63-67	11	68	11.00		1.92	0.23	0.23	0.26	0.094	0.703546	7.2	19.07	15.60	38.70
CCOP	Ecuador	EQ1	12	60	4.16		1.89	0.14	0.58	0.25	0.080	0.703200	10.1	18.16	15.53	37.84

Sources: 1 Salters *et al.* (1992), 2 Mahoney *et al.* (1995), 3 Mahoney *et al.* (1993a), 4 Tejada *et al.* (1996), 5 Tejada *et al.* (2002), 6 Aitken and Echeverría (1984), Dupré and Echeverría (1984), Jochum *et al.* (1991), 7 AC Kerr unpublished data, 8 Kerr *et al.* (1997), Hauff *et al.* (2000b), 9 Kerr *et al.* (2002), 10 Kerr *et al.* (1996b), 11 Hauff *et al.* (2000b), 12 Reynaud *et al.* (1999).

the principal contributors to the source regions of deep mantle plumes are subducted oceanic slabs and their sediments which descended through the asthenospheric mantle and ponded at either the 670 km discontinuity or the core–mantle boundary (D″). Zindler and Hart (1986) identified three main mantle plume components: HIMU (proposed as being derived from subducted oceanic crust); and two enriched components, EM1 and EM2. In addition, the upper asthenosphere was proposed as consisting of depleted MORB mantle (DMM).

More recently, it has been shown that depleted signatures found in some LIPs represent a component which, rather than being due to entrainment of depleted upper mantle material (DMM), is derived from depth, and so is an integral part the plume itself (Kerr *et al.*, 1995a,b; Fitton *et al.*, 1997; Kempton *et al.*, 2000). Kerr *et al.* (1995a,b) and Walker *et al.* (1999) proposed that this depleted plume component was ultimately derived from subducted oceanic lithosphere, unlike HIMU, which has its source in more-enriched upper oceanic crust.

Evidence for recycled oceanic crust and lithosphere in the mantle plume source regions of oceanic plateaus has been presented by several authors (e.g., Walker *et al.*, 1999; Hauff *et al.*, 2000a), and mixing trends between a depleted component and HIMU are clearly seen on most of the radiogenic isotope plots (Figures 8 and 9).

3.16.4.3 Caribbean–Colombian Oceanic Plateau (~90 Ma)

The Caribbean–Colombian Oceanic Plateau (CCOP) is exposed around the margins of the Caribbean and along the northwestern continental margin of South America (Figure 4). The thickened nature of the bulk of the Caribbean plate (8–20 km; Edgar *et al.*, 1971; Mauffret and Leroy, 1997) testifies to its origin as an oceanic plateau. The plateau has been drilled by DSDP Leg 15 and ODP Leg 165 (Figure 4; Bence *et al.*, 1975; Sinton *et al.*, 2000). The accreted plateau material in Colombia, Ecuador, Costa Rica and Hispaniola consists of fault-bounded slices of basaltic, and occasionally picritic lavas and sills with relatively few intercalated sediments and ash layers (Kerr *et al.*, 1997a). Although they preserve layered and isotropic gabbros and ultramafic rocks, unlike accreted ophiolites generated at spreading centers, these accreted sequences of oceanic plateau do not possess sheeted dyke complexes (Kerr *et al.*, 1998).

Several other exposures are worthy of special mention: firstly, the 5 km thick section on the island of Curaçao, 70 km north of the coast of Venezuela (Figure 4). The sequence consists of

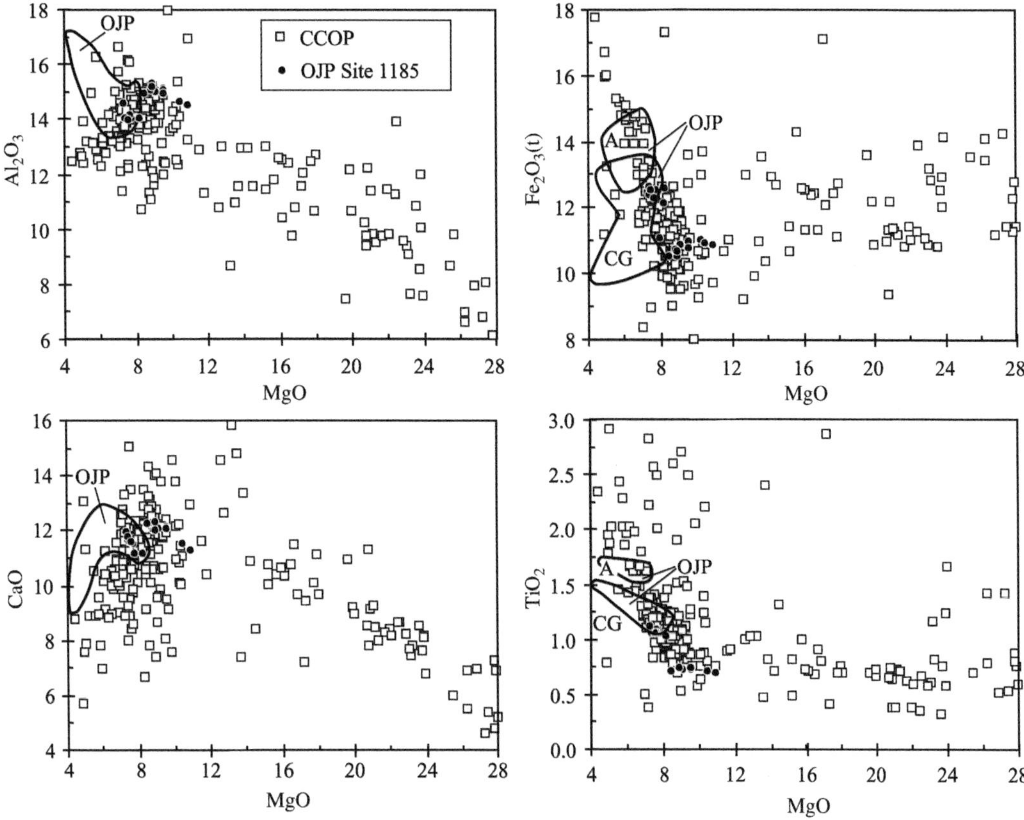

Figure 5 Plots of major elements (wt.%) against wt.% MgO for lavas from throughout the CCOP. All lavas with >18 wt.% MgO contain accumulated olivine. Data sources: Colombia—Kerr *et al.* (1997a), Gorgona—Echeverría (1980), Kerr *et al.* (1996a), Arndt *et al.* (1997), Costa Rica—Hauff *et al.* (2000b), Curaçao—Kerr *et al.* (1996b). Circled fields are from the Ontong Java plateau (Mahoney *et al.*, 1993a,b). Where the A and CG types differ markedly in composition they are plotted as separate fields, otherwise the OJP is plotted as a single field. OJP Site 1185 unpublished data were kindly provided by Godfrey Fitton.

pillowed picrites (MgO >12 wt.%) low in the succession that gradually give way to more basaltic pillow lavas nearer the top. These pillow lavas are intercalated with hyaloclastite horizons and intrusive sheets (Klaver, 1987; Kerr *et al.*, 1996b). The second noteworthy locality is the island of Gorgona, 50 km off the western coast of Colombia (Figure 4). This small island (2.5 × 8 km) is the site of the youngest komatiites (MgO-rich lava flows: >15 wt.%), which possess platy and blade-shaped olivines, giving the rocks a characteristic "spinifex" texture (Echeverría, 1980; Kerr *et al.*, 1996a). Komatiites are relatively common in the pre-Cambrian, however, the only known Phanerozoic komatiites occur as part of the CCOP on Gorgona Island. The formation of these Cretaceous komatiites in the CCOP has led to the suggestion that pre-Cambrian komatiites formed in ancient oceanic plateaus (Storey *et al.*, 1991).

The lavas of the CCOP are classified as tholeiitic. The most magnesian lavas found in the province contain up to 28 wt.% MgO (Figure 5). However, as shown by Kerr *et al.* (1996b), it is likely that these lavas contain substantial accumulated olivine, and so the whole rock compositions of these high-MgO rocks cannot represent those of parental mantle melts. Estimates of the MgO content of the parental melts for various parts of the province vary from 18 wt.% MgO to about 12 wt.% MgO (Kerr *et al.*, 1996a,b; Revillon *et al.*, 1999). Although picritic lavas are more common than in other Cretaceous oceanic plateaus, basalts are by far the most common rock type preserved in the CCOP. The vast majority of samples contain between 6 and 10 wt.% MgO (Figure 5). Al_2O_3 contents broadly increase with decreasing MgO, reflecting the importance of the addition and removal of olivine during the petrogenetic history of the CCOP lavas. CaO increases with decreasing MgO until MgO reaches 8–10 wt.% beyond which the CCOP lavas display a scattered but discernible downward trend. Fe_2O_3(t) and TiO_2 display broadly horizontal trends until about 8–10 wt.% MgO, below which both increase markedly. These trends can be modeled by the initial fractional crystallization or accumulation of

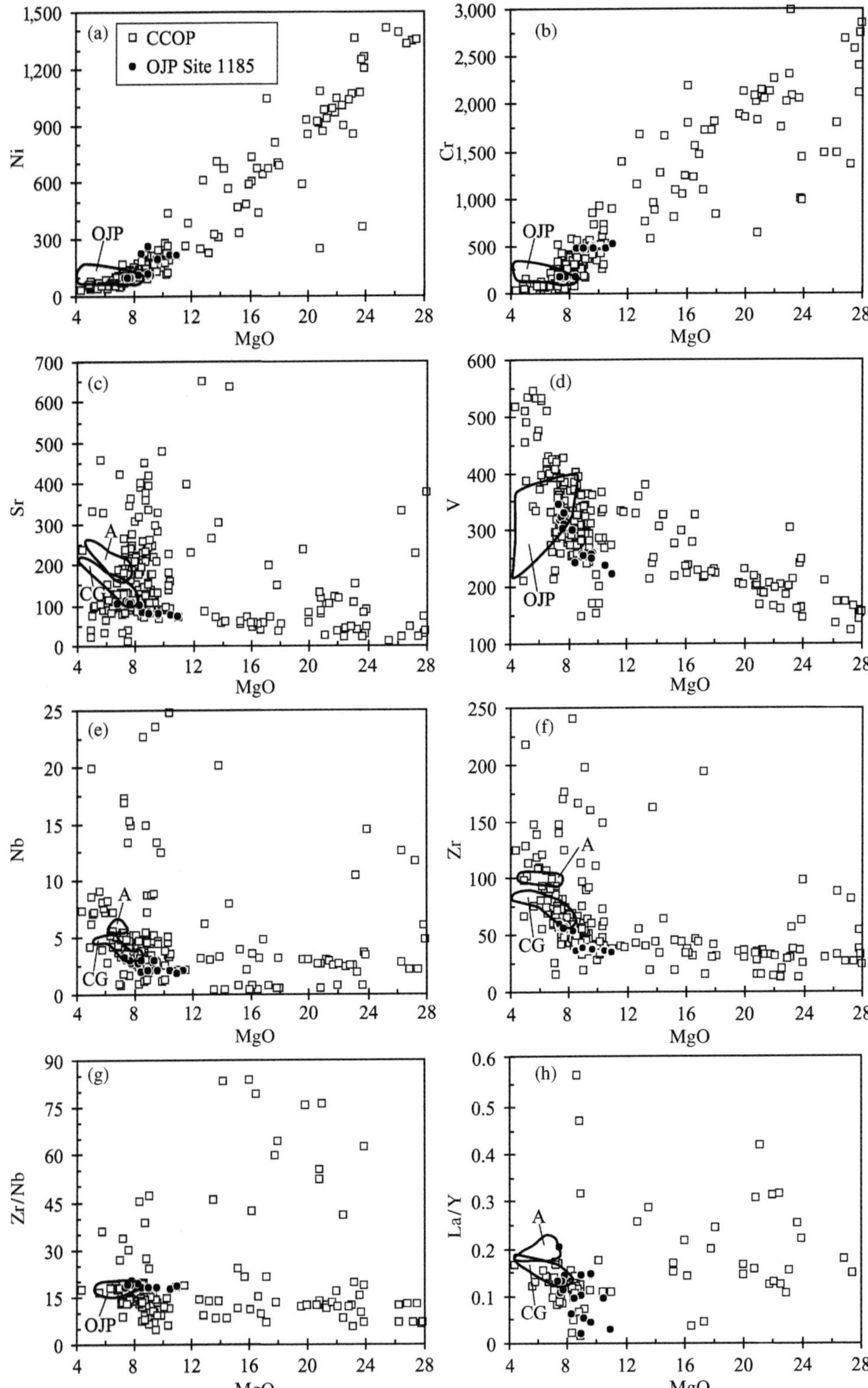

Figure 6 Plots of trace elements (ppm) and ratios of trace elements against wt.% MgO for lavas from the CCOP. The OJP is plotted as a single field except where the A and CG types differ markedly in composition. Data sources are as for Figure 5.

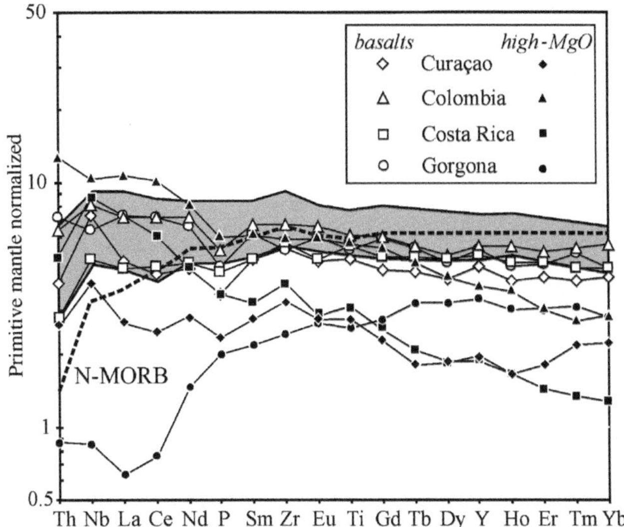

Figure 7 Primitive mantle normalized multi-element plot showing the average composition of high-MgO lavas (komatiites and picrites) and basalts from various parts of the CCOP, plotted along with average N-MORB values (dashed line) and a compositional field for the OJP. Data sources are as for Figure 5. Primitive mantle normalizing values and N-MORB from Sun and McDonough (1989).

olivine (plus minor Cr-spinel) followed by the commencement of crystallization of plagioclase and clinopyroxene between 8 wt.% MgO and 10 wt.% MgO (Kerr *et al.*, 1996b).

Trace element data (Figure 6) also support the proposed fractional crystallization model: Ni and Cr contents fall with decreasing MgO contents, and Sr, although commonly disturbed by subsolidus alteration processes, is generally reduced below 10 wt.% MgO. Despite some scatter, the content of incompatible trace elements, e.g., Nb and Zr, generally increases with decreasing MgO contents (Figure 6). Ratios of highly incompatible trace elements are not normally affected by moderate degrees of fractional crystallization or mantle melting, and they therefore have the potential to reveal heterogeneities in the mantle source region of the plateau. These ratios are plotted against MgO in Figure 6(g)–(h). One of the most interesting aspects of the trace element data for the CCOP is that the basaltic lavas possess a narrower range of incompatible trace element ratios than the picrites. For instance, well over 80% of the basaltic samples (<12 wt.% MgO) from the CCOP possess La/Y ratios between 0.05 and 0.2 and Zr/Nb ratios between 7 and 20 (Figure 6(g)–(h)). In contrast, the picritic and komatiitic lavas possess much more variable ratios of incompatible trace elements, with La/Y ranging from 0.05 to 0.45 and Zr/Nb from 5 to 85. This is also shown on primitive mantle normalized multielement plots, where it can be seen that the CCOP basalts possess broadly flat patterns whereas the high-MgO picrites and komatiites are generally much more variable, with some being more depleted, and some more enriched than the basalts, particularly for the most highly incompatible trace elements such as Th, Nb, La, Ce & Nd (Figure 7).

The heterogeneity of the high-MgO rocks is also reflected in the radiogenic isotope ratios, particularly εNd (Figures 8 and 9). Virtually all the analyzed basalts from the CCOP possess initial εNd ranging from +6 to +9, whereas the high MgO lavas generally fall outside this range (εNd >+9 and <+6; Figure 8). Elevated initial $^{87}Sr/^{86}Sr$ ratios found in several parts of the province have been attributed either to contamination with altered oceanic crust (Curaçao: Kerr *et al.*, 1996b) or to secondary alteration (Gorgona: Revillon *et al.*, 1999).

The wide range of isotopic data for the CCOP reveals that the enriched and depleted lavas are not simply formed by variable melting of a homogeneous source region, but rather reflect melting of long-term depleted and enriched components from a markedly heterogeneous plume source region. (Kerr *et al.*, 1996a, 2002; Arndt *et al.*, 1997; Hauff *et al.*, 2000a; Thompson *et al.*, 2004).

The greater heterogeneity of the high MgO rocks in comparison to the basalts, has been interpreted to reflect the formation of these lower MgO magmas through mixing and fractional crystallisation of the high-MgO magmas in large magma chambers. The heterogeneous high-MgO rocks thus represent magmas that passed relatively quickly through the lithosphere without being trapped in magma chambers (Kerr *et al.*, 1998). The extent of partial melting required to produce

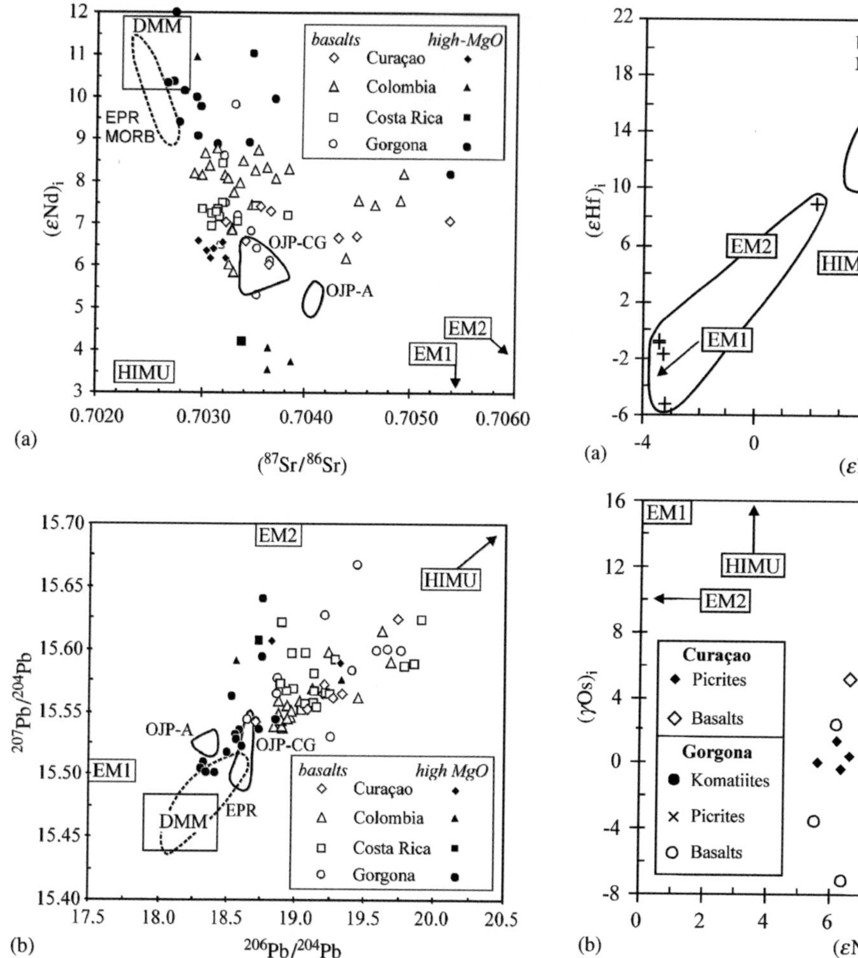

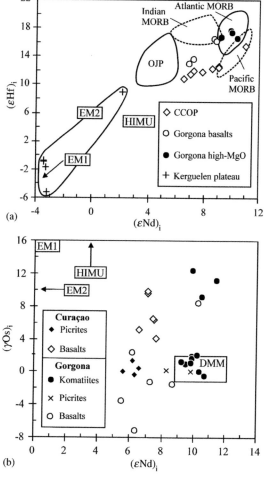

Figure 8 Plot of (a) initial εNd against $^{87}Sr/^{86}Sr$ for high-MgO lavas and basalts from the CCOP and (b) $^{207}Pb/^{204}Pb$ against $^{206}Pb/^{204}Pb$. Shown on both diagrams are fields for the OJP- A and CG types (Mahoney *et al.*, 1993a,b) and East Pacific Rise (EPR) MORB (Mahoney *et al.*, 1995). Other data sources are as for Figure 5. Mantle end-member compositions are from Zindler and Hart (1986).

Figure 9 Plots of (a) initial εHf against initial εNd and (b) initial γOs against initial εNd for Cretaceous oceanic plateaus. Data sources: Atlantic and Indian MORB—Salters (1996), Pacific—Nowell *et al.* (1998), Kerguelen plateau—Salters and Hart (1991), CCOP and Gorgona—Walker *et al.* (1999), Thompson *et al.* (2004), OJP—Babbs (1997).

the parental magmas of the CCOP has been calculated to be of the order of 20% (Kerr *et al.*, 1997a).

3.16.4.4 Ontong Java Plateau (~122 and ~90 Ma)

The Ontong Java plateau (OJP) in the western Pacific (Figure 1) has been tectonically uplifted and exposed along its south eastern margin, at the Solomon Islands arc, mostly on the Islands of Maliata and Santa Isabel. In contrast to the CCOP, which has numerous exposed sections, these are currently the only known subaerial exposures of the OJP. The rest of our knowledge of the OJP comes from a series of drill holes: DSDP Site 289 and ODP Sites 803 and 807 (Mahoney *et al.*, 1993a). Additional data are also now becoming available from ODP Leg 192 which recently penetrated the OJP at four sites (Mahoney *et al.*, 2001).

On Maliata the stratigraphic thickness of the accreted plateau reaches 3–4 km, and the succession is dominated by pillowed and massive basaltic flows (Petterson *et al.*, 1997; Babbs, 1997). Like the CCOP, dykes are volumetrically minor. The DSDP/ODP drill holes have penetrated into the plateau to a depth of 216 m (Site 1185B; Mahoney *et al.*, 2001). The sampled sections consist predominantly of pillowed and massive basalts with occasional thin interlava sediments (Neal *et al.*, 1997; Mahoney *et al.*, 2001)

In general, the OJP lavas are more homogeneous than those of the CCOP (Figures 5, 6 and 8). All the lavas from the province analyzed thus far are basaltic in composition, with most of the samples possessing 6–8 wt.% MgO (Figures 5 and 6). Although the lavas of the OJP possess a restricted compositional range, they nonetheless fall into two compositionally distinct groups. These groups were first noted in the lavas from ODP Site 807, where Mahoney *et al.* (1993b) divided the lavas into different units (A and C-G). Unit A (A-Type) is chemically distinct from Units C-G (C-G-Type) and possesses higher levels of both incompatible elements (e.g., TiO_2; Sr, Zr Nb and the LREE; Figures 5–7) and ratios of highly incompatible to moderately incompatible trace elements (e.g., La/Y). The A-Type basalts also have lower initial εNd values and higher initial $^{87}Sr/^{86}Sr$ ratios than the C-G-Type (Figure 8(a)). However, the total range of εNd (+6.5 – +4.9) and $^{87}Sr/^{86}Sr$ (0.7034–0.7041) is relatively small in comparison with those in the CCOP (Figure 8(a)). The same is true for incompatible trace element ratios, with the OJP basalts only varying between 0.12 and 0.22 for La/Y and between 15.5–19.7 for Zr/Nb (Figure 6). However, recent analyses from ODP Leg 192 (Sites 1185 and 1187; on the eastern edge of the plateau) have revealed the occurrence of more-MgO-rich lavas (up to 11 wt.%) with higher Ni and Cr contents (Mahoney *et al.*, 2001; G. Fitton unpublished data; Figures 5 and 6). In addition to their higher MgO, preliminary geochemical data reveals that these basalts possess lower levels of incompatible elements (e.g., TiO_2: 0.72–0.77 and Zr: 36–43), than the A and C-G Types (Figures 5 and 6).

Neal *et al.* (1997) have concluded on the basis of geochemical modeling that the major and trace element compositions of the A- and C-G-type lavas of the OJP are consistent with 20–30% partial melting of a peridotite source. The more enriched nature of the A-Type lavas implies derivation form a slightly more enriched source region, possibly in conjunction with smaller degrees of melting. Mahoney *et al.* (2001) have proposed that the more MgO-rich, incompatible element-poor lavas discovered during Leg 192 represent more extensive melting of the plume source region. However, an alternative explanation is that these lavas were derived from a more depleted mantle source region, and radiogenic isotope data are required in order to resolve this issue. None of the compositions sampled thus far are magnesian enough to represent possible parental melts, and so are believed to have undergone 30–45% fractional crystallization, involving olivine, plagioclase and clinopyroxene (Neal *et al.*, 1997).

Although the deeper crustal and lithospheric levels of the OJP are not exposed, seismic velocity data has been used to model the crustal structure (Farnetani *et al.*, 1996; Gladczenko *et al.*, 1997). These authors have proposed that the magma chambers that fed the plateau are represented in the midcrust by olivine gabbros. These models also suggest that high compressional P-wave velocities of >7.1 km s^{-1} deep within the OJP are due to the presence of olivine and pyroxene cumulates produced by the fractionation of primary picritic melts. Alternatively, the high P-wave velocities could be due to the presence of garnet granulite deep in the plateau, which Gladczenko *et al.* (1997) suggested may have formed by deformation and hydrothermal alteration of lower crustal cumulates.

3.16.5 THE INFLUENCE OF CONTINENTAL CRUST ON OCEANIC PLATEAUS

Initially it may seem odd that the composition of oceanic plateaus should be influenced by continental crust, and certainly for the CCOP and the OJP, which apparently formed well away from continental margins, there is no evidence of the involvement of continental crust in their petrogenesis. However, LIPs can also form at the continent–ocean boundary as well as erupting onto either oceanic or continental lithosphere, and the formation of a LIP in such a tectonic setting is often related to continental break-up. The role played by mantle plumes in continental break-up (causal or consequential) remains controversial (White and McKenzie, 1989; Hill, 1991; Coffin and Eldholm, 1992; Saunders *et al.*, 1992; Barton and White, 1995). However, whether mantle plumes are the reason for or a result of continental break-up, the associated erupted lavas and intruded sills form thick magmatic sequences on the margins of the rifted continents: the so-called seaward-dipping reflector sequences (SDRS). These LIPs may also erupt on the adjacent continents to form continental flood basalt provinces. Furthermore, continuing plume-related magmatism combined with further separation of the continents ultimately results in the formation of oceanic plateaus. Two examples of provinces such as these are explored below: the North Atlantic Igneous Province (NAIP) and the Kerguelen plateau.

3.16.5.1 The North Atlantic Igneous Province (~60 Ma to Present Day)

The opening of the North Atlantic ~60 Ma is closely associated with magmatism from the "head" phase of the Icelandic plume. (For a comprehensive review of the NAIP see Saunders *et al.*, 1997). Much of the initial volcanism (Phase 1: 62–58 Ma; Saunders *et al.*, 1997) was confined to the continental margins, i.e., the on-land sequences in western Britain, the Faroe

Islands and east and west Greenland, as well as the seaward-dipping reflector sequences of the southeast Greenland margin and the Hatton Bank (Figure 10). Most of these lavas are contaminated with Archean-age continental crust and thus possess low εNd and high Ba/Nb (Figure 11), along with low $^{206}Pb/^{204}Pb$. As the North Atlantic continued to open, a second intense burst of magmatism occurred (beginning at 56 Ma; Phase 2, Saunders *et al.*, 1997). The lavas from this magmatism are preserved in the upper portions of the SDRS, off the coast of southeast Greenland and Western Europe and have been drilled by ODP Legs 104, 152 and 163 (Viereck *et al.*, 1988; Fitton *et al.*, 2000). In contrast to the Phase 1 lavas, these lavas show few signs of contamination by continental crust (low Ba/Nb; $\varepsilon Nd > 6$; (Figure 11) $^{206}Pb/^{204}Pb > 17$), indicating that by this time the NAIP was an entirely oceanic LIP. The Icelandic plume has been producing melt over most of the past 60 Myr, as evidenced by 55–15 Ma volcanism along the Greenland–Iceland ridge and the Faroes–Iceland ridge, and the 15 Ma–present volcanism on Iceland.

3.16.5.2 The Kerguelen Igneous Province (~133 Ma to Present Day)

The initial volcanism of the Kerguelen plume is closely associated with the break-up of Gondwana in the early–mid-Cretaceous, i.e., the separation of India, Australia and Antarctica (Morgan, 1981; Royer and Coffin, 1992). Like the NAIP, much initial volcanism is found on the margins of the rifted continents (Figure 12): the Rajmahal basalts in northeastern India (Kent *et al.*, 1997) and the Bunbury basalts in western Australia (Frey *et al.*, 1996). Not surprisingly, these basalts are extensively contaminated by continental lithosphere and yield an initial $^{87}Sr/^{86}Sr$ ratio of >0.7042 and $\varepsilon Nd < 4.0$ (Figure 13).

The geographical components of the plateau (Figure 12) and the geochronology are briefly outlined below; however, a more detailed review can be found in Frey *et al.* (2000); Coffin *et al.* (2002). The first massive pulse of Kerguelen plume magmatism created the Southern Kerguelen plateau (118–110 Ma; Figure 12). Later melting of the plume was responsible for

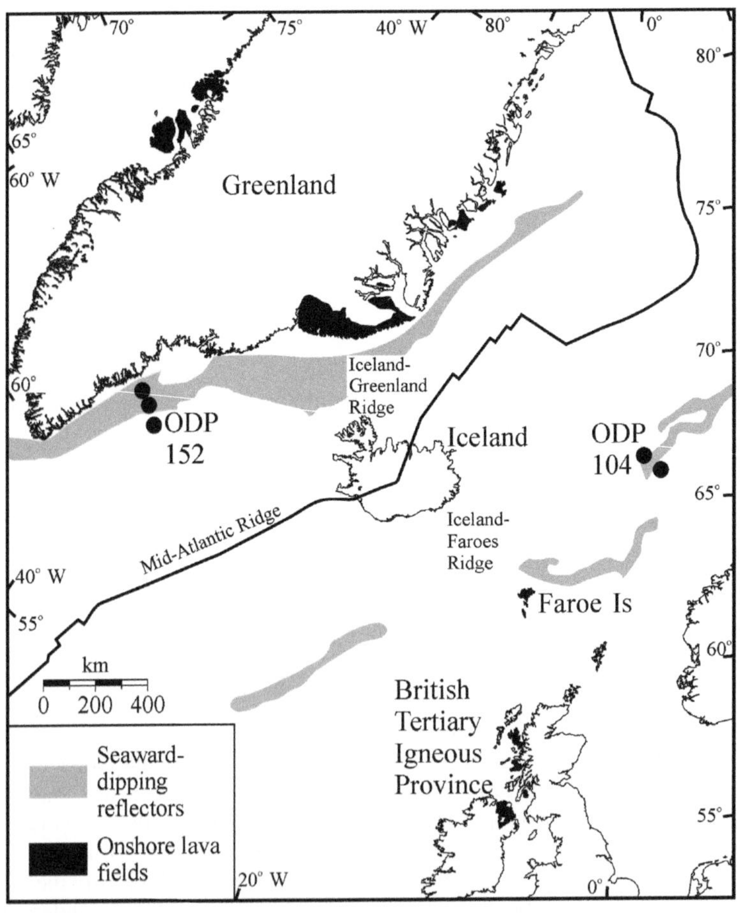

Figure 10 Map showing the locations of the principal on-land exposures of the North Atlantic Igneous Province and the seaward-dipping reflector sequences.

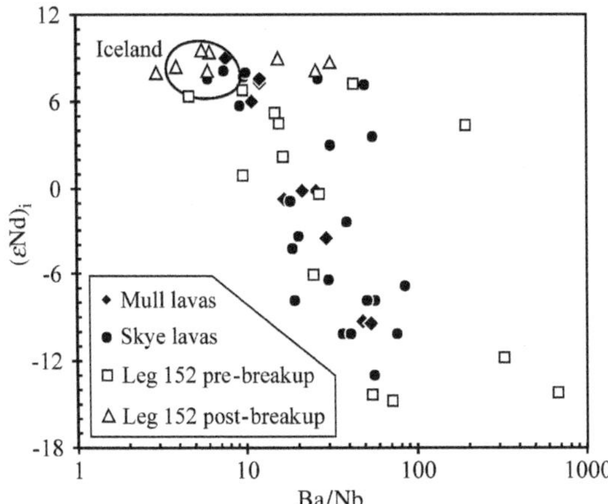

Figure 11 Plot of Ba/Nb vs. initial εNd lavas from the NAIP. Shown on the diagram are lavas from Skye (Thompson *et al.*, 1982; Dickin *et al.*, 1987), Mull (Kerr *et al.*, 1995), Iceland (Hémond *et al.*, 1993) and both pre- and post- continental break-up lavas from ODP Leg 152 (Fitton *et al.*, 1998a).

the formation of the Elan Bank (108–107 Ma), the Central Kerguelen Plateau (101–100 Ma), Broken Ridge (95–94 Ma), the Ninetyeast Ridge (82–37 Ma), and the Northern Kerguelen plateau (35–34 Ma) (Figure 12). Volcanism continues to the present-day and has produced the Kerguelen Archipelago and Heard and MacDonald Islands.

The lavas from the Southern Kerguelen plateau and Broken Ridge have initial $^{87}Sr/^{86}Sr$ ratios and εNd values (Figure 11) which range from 0.7037 to 0.7102 and +4.0 to −9.4, respectively (Salters *et al.*, 1992; Mahoney *et al.*, 1995). Some of this variation can be interpreted as mixing between Southeast Indian Ridge MORB and the Kerguelen plume (Weis and Frey, 1996). However, elevated La/Nb ratios (Figures 13 and 14(a)) and the extreme isotopic compositions of basalts drilled at ODP Site 738 and dredge samples from the eastern Broken Ridge ($^{87}Sr/^{86}Sr$ 0.710; εNd −9.0; Figures 8 and 13) cannot be explained by such mixing processes. It has been proposed that these signatures are due to contamination by continental lithosphere (Storey *et al.*, 1989; Mahoney *et al.*, 1995; Hassler and Shimizu, 1998). Operto and Charvis, 1996 have imaged a seismically reflective transition zone beneath the crust/mantle interface of the Southern Kerguelen plateau, interpreted as fragments of continental crust. This crust appears to have isotopic similarities to Archean crust found on the margins of Gondwana, which raises the possibility that fragments of such crust have become incorporated into the Indian Ocean basin during continental break-up. Recently, during drilling at Site 1137 on the Elan Bank (part of the Kerguelen plateau; Figure 12) clasts of garnet-biotite gneiss have been discovered in a fluvial conglomerate

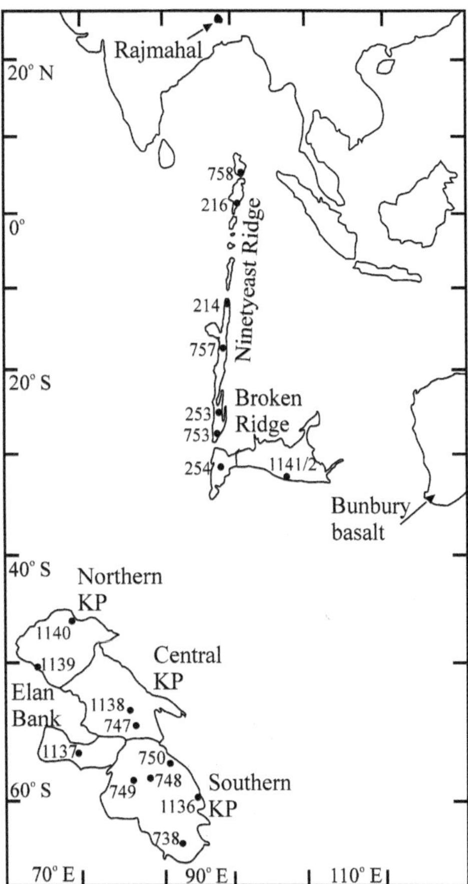

Figure 12 Map showing the main components of the Kerguelen plateau (KP) discussed in the text (after Frey *et al.*, 1996).

intercalated with basalt flows (Frey *et al.*, 2000). This discovery has confirmed the presence of pre-Cambrian crustal rocks within the Kerguelen plateau (Nicolaysen *et al.*, 2001), thus supporting the lithospheric contamination model for the high La/Nb, low εNd basalts of the Kerguelen plateau.

3.16.6 IDENTIFICATION OF OCEANIC PLATEAUS IN THE GEOLOGICAL RECORD

The rationale for this section is summed up by this question: If the CCOP or OJP were accreted on to a continental margin and preserved in the geologic record for 1 billion years, what features could we use to identify them as oceanic plateaus? This section will review diagnostic geochemical and geological characteristics of oceanic plateaus, and then will show, illustrated by examples, how these criteria can be used to identify plateau sequences in the geological record. Table 3 provides a summary of the diagnostic features of Cretaceous oceanic plateaus and mafic sequences

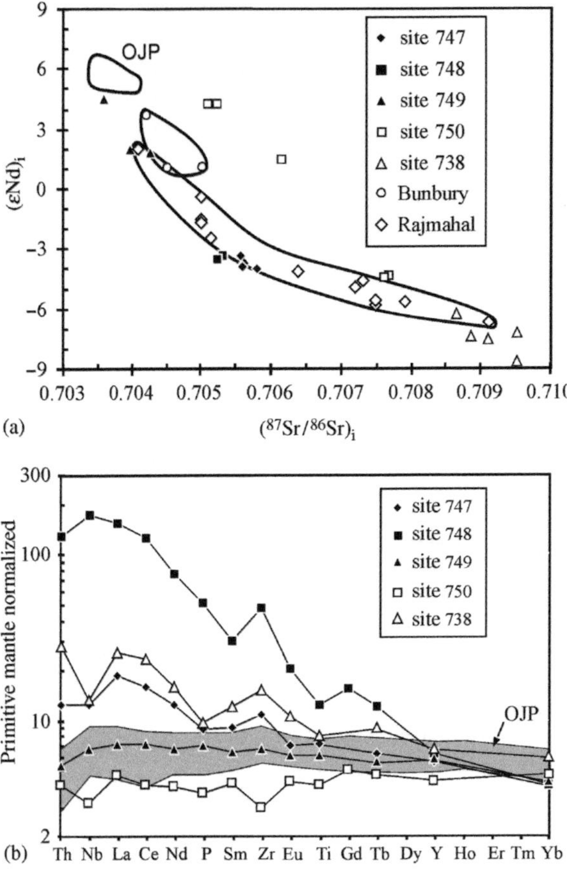

Figure 13 Plots to show the geochemical variation of lavas from the early Cretaceous lavas derived from the Kerguelen plume. (a) Initial $^{87}Sr/^{86}Sr$ vs. initial εNd and (b) primitive mantle normalized multi-element plots showing averaged data for ODP drill sites. A compositional field for the OJP (Mahoney *et al.*, 1993a,b) is shown on both diagrams. Data sources: Rajmahal—Kent *et al.* (1997); Bunbury—Frey *et al.* (1996); ODP sites—Salters *et al.* (1992); Mahoney *et al.* (1995).

within the continental crust, which have been interpreted as oceanic plateaus. Details of accreted oceanic plateaus thus far identified in the geological record are summarized in Table 4.

3.16.6.1 Diagnostic Features of Oceanic Plateaus

Both chemical and geological features can be useful in the identification of oceanic plateaus. Condie (1999) and Kerr *et al.* (2000) have discussed the diagnostic features of oceanic plateaus in detail, and only a brief account will be given here. Table 3 summarizes the characteristics which are useful in distinguishing igneous rocks formed in an oceanic plateau from those which originated in other tectonic settings.

Igneous rocks produced in an island arc, or a continental subduction zone setting, are relatively easily distinguished from oceanic plateau sequences (Table 3), because arcs generally possess more evolved lavas, with ubiquitous high $(La/Nb)_{pmn}$ ratios (Figure 14(b)), and only very rarely contain high-MgO lavas. Additionally, oceanic plateaus do not possess the abundant volcanic ash layers present in volcanic arc sequences. However, as Figure 14 shows, a low $(La/Nb)_{pmn}$ ratio is not an entirely robust signature of an oceanic plateau sequence, since samples of the Kerguelen oceanic plateau often possess high $(La/Nb)_n$ values, due to magma interaction with fragments of continental lithosphere beneath the plateau. This example highlights the importance of not relying solely on chemical discriminants of tectonic environment, without also considering the geological evidence. In the case of Kerguelen, the lack of volcaniclastic horizons helps confirm its oceanic plateau affinity. As discussed by Kerr *et al.* (2000), many of the geological discriminants between oceanic plateaus and midocean ridges may be ambiguous (Table 3). Geochemical characteristics must, therefore, be used to distinguish lavas from these two tectonic settings. Most Cretaceous oceanic plateau lavas possess relatively flat normalized REE patterns (Figure 7), whereas most midocean ridge basalts possess light REE-depleted patterns reflecting a more depleted

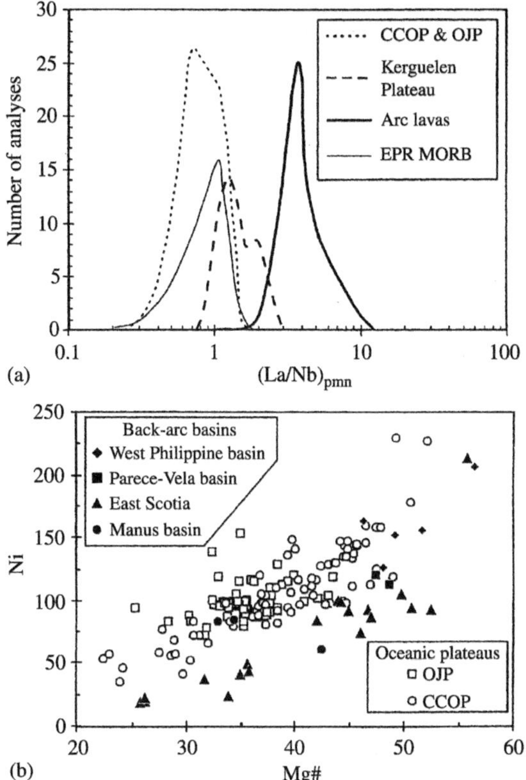

Figure 14 (a) Frequency diagram showing the range in $(La/Nb)_{pmn}$ for lavas from the CCOP–OJP, the Kerguelen plateau, EPR MORB, and arc lavas. (Arc data from Thirlwall *et al.*, 1996.) Other data sources as in Figures 5, 8 and 12). (b) Plot of Ni vs. Mg–number for the CCOP–OJP and lavas from back arc basins. Data from Wood *et al.* (1980), Woodhead *et al.* (1998), Leat *et al.* (2000). Both diagrams modified from Kerr *et al.* (2000).

mantle source region. Furthermore, high-MgO lavas can be found in oceanic plateaus, but are largely absent from oceanic crust generated at midocean ridges.

Incompatible trace elements are only of limited use in distinguishing between volcanic successions formed in back-arc basins and those formed in oceanic plateaus (Table 3). However, the lower mantle temperature below a back-arc basin ($T_p \sim 1{,}280\,°C$) compared to a mantle plume ($T_p > 1{,}400\,°C$) results in the eruption of few high-MgO lavas. An additional consequence of this lower mantle temperature is that back-arc basin lavas generally possess lower Ni and Cr contents for a given Mg number than oceanic plateau lavas (Figure 14(b)). Furthermore, because of their proximity to active subduction sites, back-arc basin sequences are also more likely to contain abundant volcaniclastic horizons than oceanic plateaus.

Continental flood basalts are not easy to preserve for long periods of time in the geological record simply because they are easily eroded away, unless they are buried by sediments. Quite often the only remaining indications of a continental flood basalt province are the dykes and vents through which the lavas erupted. The 200 Ma Central Atlantic Magmatic Province, which formed during the break-up of South America, Africa, and North America has been identified largely on the basis of its remnant dyke swarms (e.g., Marzoli *et al.*, 1999).

3.16.6.2 Mafic Triassic Accreted Terranes in the North American Cordillera

Significant proportions of the North American Cordillera consist of mafic sequences of accreted oceanic terranes (Figure 15). Some of these have been identified as oceanic plateau material ranging in age from Permian to Eocene (see review in Condie, 2001). At least three of these oceanic plateau terranes are predominantly Triassic in age (Wrangellia, Cache Creek and Angayucham; Pallister *et al.*, 1989; Lassiter *et al.*, 1995; Tardy *et al.*, 2001) and obviously represent a major phase of oceanic plateau volcanism at this time. These plateau sequences are characterized by pillow basalts and intrusive sheets, with occasional intercalated tephra and hyaloclastite layers, indicating formation in shallow water, or by subaerial eruption. In the Wrangellia terrane there is considerable evidence for rapid uplift of the sea floor (presumably by the plume head) immediately prior to eruption (Richards *et al.*, 1991).

The basalts of the Cache Creek and Angayucham terranes display a restricted range in MgO with most of the basalts ranging from 5.0 wt.% to 8.5 wt.%. These basalts possess low $(La/Nb)_{pmn}$ ratios (<1.2), essentially flat REE patterns (Pallister *et al.*, 1989; Tardy *et al.*, 2001) and εNd values that range mostly from +9.9 to +4.5 (Figure 16). As Figure 16 shows, all these features are similar to the OJP. However, some of the basalts from the Wrangellia Terrane, despite showing a similar range in MgO content, have $(La/Nb)_{pmn}$ ratios >1 and steeper REE patterns (Figure 16) than those from Cache Creek and Angayucham. Lassiter *et al.* (1995) suggested that this is due to the magmas erupting through, and being contaminated by preexisting island-arc lithosphere. However, it is also possible that, as in the case of the Kerguelen plateau, large fragments of ancient continental lithosphere were incorporated in the proto-Pacific Ocean, and the lavas of the Wrangellia oceanic plateau were contaminated by this lithosphere. The contamination of the most evolved Wrangellia basalts with either arc crust or ancient continental lithosphere is also supported by a broadly negative correlation between εNd and $(La/Nb)_{pmn}$ (Figure 16).

Table 3 Diagnostic geochemical and geological characteristics of volcanic sequences from different tectonic settings.

Tectonic setting	*High-MgO lavas (>14%)*	*Low-MgO lavas (<3%)*	$(La/Nb)^{a}_{pmn}$	*Chondrite normalized REE pattern*	*Pillow lavas*	*Tephra layers*	*Subaerial eruption*	*Intercalated pelagic sediments*
Oceanic plateau	frequent	rare	≤1	Predominantly flat	yes	rare	occasional	yes
Midocean ridge	rare	rare	≤1	LREE-depleted	yes	rare	rare	no
Marginal basin	rare	rare	≤1	Predominantly flat	yes	yes	no	no
Oceanic island basalt	rare	rare	≤1	Predominantly LREE-enriched	yes	rare	frequent	rare
Volcanic rifted margin	frequent	rare	Varies from ≤1 to ≫1	Flat to LREE-enriched	not all lavas are pillowed	occasional	frequent	no
Arc (continental & oceanic)	rare	frequent	≫1	LREE-enriched	not all lavas are pillowed	yes	frequent	Rare
Continental flood basalt	frequent	frequent	mostly ≫1 < 10% of flows ≤1	Flat to LREE-enriched	no	occasional	yes	no

[a] pmn—primitive mantle normalized (after Kerr *et al.* 2000).

3.16.6.3 Carboniferous to Cretaceous Accreted Oceanic Plateaus in Japan

The Japanese islands are essentially composed of a series of terranes that have been accreted to the continental margin of the Eurasian plate during the past 400 Myr. These terranes consist of trench-filling terrigenous sediments with variable quantities of accreted oceanic crust that are intruded and partly overlain by the products of subsequent subduction-related volcanism. Within Japan the ages of the accreted complexes become younger from north to south and from west to east (Kimura *et al.*, 1994). However, relatively little is known about the trace element chemistry of these oceanic accreted terranes.

The Chugoku and Chichibu belts in southwest Japan contain up to 30% basaltic material (greenstones) in thrust contact with limestones, cherts and mudstones (Tatsumi *et al.*, 2000). This lithological association, combined with preliminary major element data and a small range of trace element data, suggests that these basaltic assemblages are remnants of a plume-derived oceanic plateau. Tatsumi *et al.* (2000) proposed that this oceanic plateau formed in the Panthalassan Ocean (proto-Pacific) in the Carboniferous (350–300 Ma).

The Permian Yakuno ophiolite complex in southern Honshu Island is a sequence of submarine basalts, gabbros and ultramafic rocks (Isozaki, 1997). The presence of pelagic sediments, the lack of a sheeted dyke complex and the fact that the sequence is of considerable thickness, all suggest that it is part of an oceanic plateau (Isozaki, 1997). In contrast to the other accretionary belts in Japan, the Sorachi-Anivia terrane in Hokkaido and Sakhalin is dominated by oceanic crust and lithosphere (Kimura *et al.*, 1994; Tatsumi *et al.*, 1998). It comprises pillow lavas and dolerite sills with intercalated pelagic sediments containing Tithonian (150–145 Ma) radiolaria, along with a lower unit in which ultramafic rocks, including serpentinite, harzburgite and dunite, are more common (Kimura *et al.*, 1994). Major element data have shown that some picrites (>12 wt.% MgO) are found within the succession. This, combined with high CaO/Al_2O_3 ratios (indicative of a high degree of mantle melting), led Kimura *et al.* (1994) and Tatsumi *et al.* (1998) to propose an oceanic plateau origin for the Sorachi-Anivia terrane. This Jurassic–early Cretaceous plateau (named the Sorachi plateau; Kimura *et al.*, 1994) is of the same age as the Shatsky Rise (Figure 1, Table 1). In combination with paleomagnetic data, this suggests that the Sorachi plateau and the Shatsky Rise were originally a single plateau which formed near the Kula–Pacific–Farallon triple junction ~150 Ma (Kimura *et al.*, 1994).

Table 4 Proposed accreted oceanic plateaus found within continents.

Name	*Location*	*Age (Ga)*	*References*
Coonterunah and Warrawoona Groups	Pilbara Craton, Australia	~3.5	Green *et al.* (2000)
Southern Barberton Belt	Kaapvaal Craton, S Africa	3.5–3.2	De Wit *et al.* (1987)
Pietersberg Belt	Kaapvaal Craton, S Africa	~3.4	De Wit *et al.* (1987)
Opapimiskan-Markop Unit, North Caribou Belt[a]	Superior Province	~3.0	Hollings and Wyman (1999)
Olondo Belt	Aldan Shield, Siberia	3.0	Puchtel and Zhuravlev (1993); Bruguier (1996)
South Rim Unit, North Caribou Belt	Superior Province	~3.0	Hollings and Wyman (1999)
Sumozero-Kenozero Belt	Baltic Shield	3.0–2.8	Puchtel *et al.* (1999)
Steep Rock & Lumby Lake Belts[a]	Superior Province	3.0–2.9	Tomlinson *et al.* (1999)
Balmer Assemblage, Red Lake Greenstone Belt[a]	Superior Province	2.99–2.96	Tomlinson *et al.* (1998)
Kostomuksha Belt	Baltic Shield	2.8	Puchtel *et al.* (1998b)
Vizien Belt	Superior Province	2.79	Skulski and Percival (1996)
Malartic-Val d'Or Area	Superior Province	2.7	Kimura *et al.* (1993); Desrochers *et al.* (1993)
Tisdale Group, Abitibi Belt	Superior Province	~2.7	Fan and Kerrich (1997)
Schreiber-Hemlo-White River Dayohessarah	Superior Province	2.8–2.7	Polat *et al.* (1998)
Vetreny Belt[a]	Baltic Shield	2.44	Puchtel *et al.* (1997)
Birimian Province	West Africa	2.2	Abouchami *et al.* (1990); Boher *et al.* (1992)
Povungnituk & Chukotat Groups[a]	Cape Smith Fold Belt Northern Québec	2.04	Francis *et al.* (1983); Dunphy *et al.* (1995)
Onega Plateau[a]	Baltic Shield	1.98	Puchtel *et al.* (1998a)
Jormua Ophiolite[a]	NE Finland	1.95	Peltonen *et al.* (1996)
Flin Flon Belt	Central Canada	1.92–1.90	Lucas *et al.* (1996); Stern *et al.* (1995)
Arabian-Nubian Shield	NE Africa-Middle East	0.90-0.87	Stein and Goldstein (1996)
Chichibu & Chugoku Belts	SW Japan	Carboniferous	Tatsumi *et al.* (2000)
Yakuno Ophiolite	SW Japan	0.285	Isozaki (1997)
Mino Terrane	Central Japan	L Permian	Jones *et al.* (1993)
Cache Creek Terrane	Canadian Cordillera	Triassic	Tardy *et al.* (2001)
Angayucham Terrane	Alaska	Triassic	Pallister *et al.* (1989)
Wrangellia Terrane	Western North America	0.227	Lassiter *et al.* (1995).
Sorachi Plateau	Northern Japan	0.152–0.145	Kimura *et al.* (1994); Tatsumi *et al.* (1998)

[a] These sequences display evidence of contamination by continental crust and are interpreted as having formed during continental break-up or, close to a continental margin (see text).

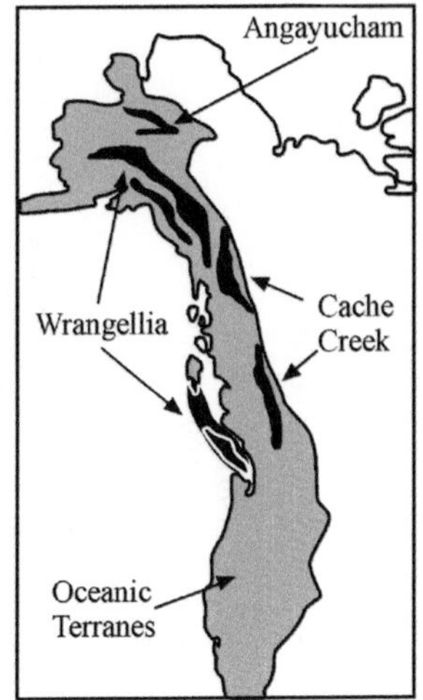

Figure 15 Map showing North American accreted oceanic terranes including the oceanic plateau sequences discussed in the text (after Tardy *et al.*, 2001; Condie, 2001).

The Sorachi part of the plateau was carried northwestwards on the Kula plate and collided with Japan ~110 Ma. Limited trace element data for the Sorachi plateau lavas support a common plume source for these two plateaus. The data cover the same compositional range as that of dredged samples from the Shatsky Rise (Figure 17). Furthermore, the data reveal that the plume source region of the Sorachi plateau was markedly heterogeneous and contained both enriched and depleted components (Kimura *et al.*, 1994; Tatsumi *et al.*, 1998) (Figure 17).

3.16.7 PRECAMBRIAN OCEANIC PLATEAUS

The identification of accreted pre-Cambrian oceanic plateaus, particularly in greenstone belts, has important implications for the generation of continental crust (Abbott, 1996; Albarede, 1998; Condie, 1999). Kerr *et al.* (2000) have presented

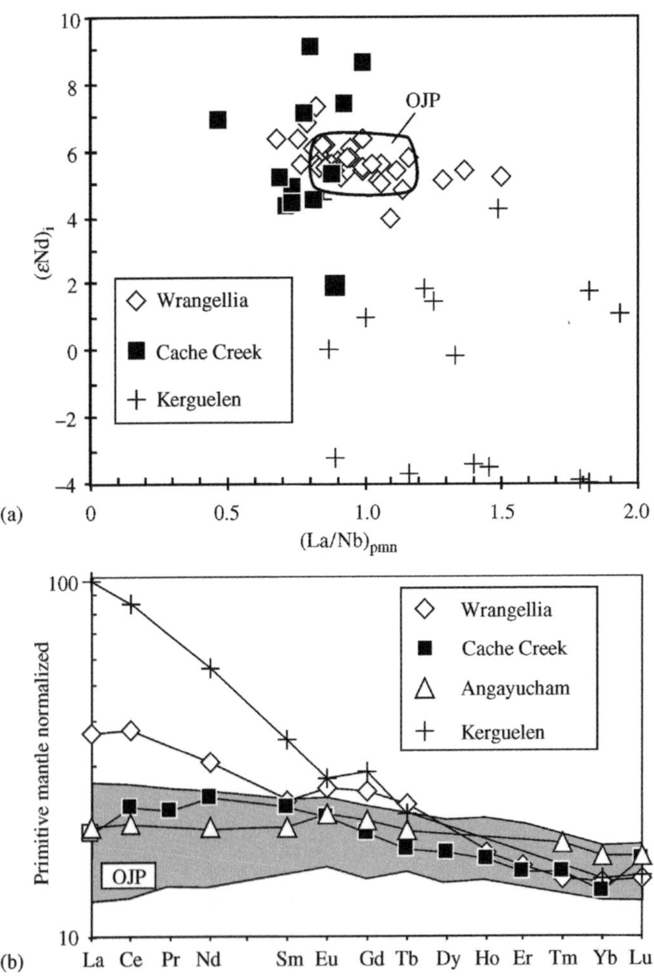

Figure 16 Plots of (a) $(La/Nb)_{pmn}$ against initial εNd for Cache Creek, Wrangellia, OJP and the Kerguelen plateau and (b) chondrite normalized (Sun and McDonough, 1989) REE plot showing averages for Wrangellia, Angayucham, Cache Creek, and the Kerguelen plateau, with the range for the OJP. Data sources are as in Figures 5 and 12; North American data from Pallister *et al.* (1989); Lassiter *et al.* (1995); and Tardy *et al.* (2001).

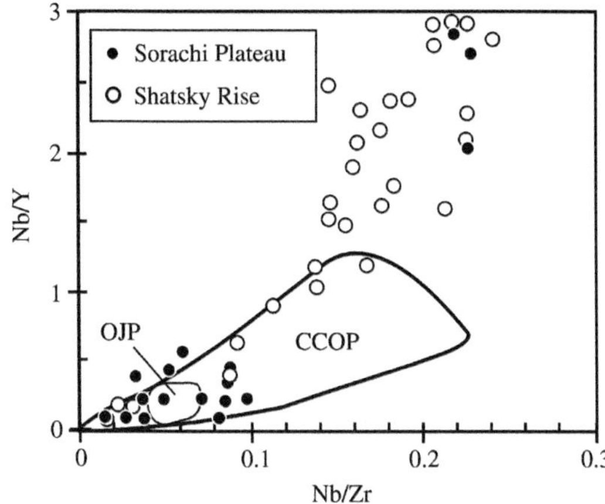

Figure 17 Plot of Nb/Y against Nb/Zr showing data from dredge samples from the Shatsky Rise and the accreted Sorachi plateau (data from Tatsumi *et al.*, 1998). Also shown are fields for the CCOP and OJP. Data sources are as in Figure 5.

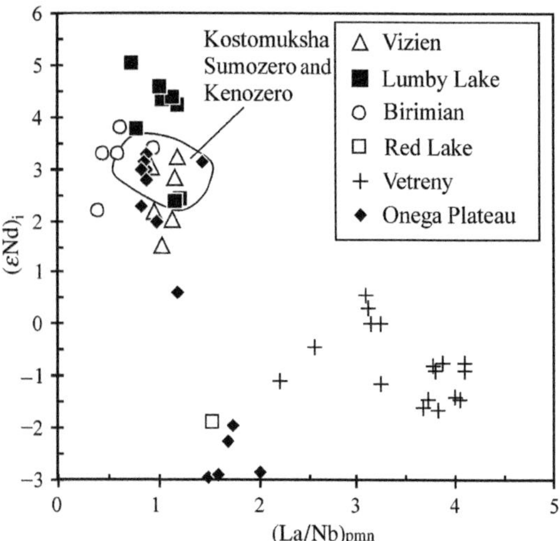

Figure 18 Plot of $(La/Nb)_{pmn}$ vs. initial εNd for data from Archean and Proterozoic accreted oceanic plateaus and provinces proposed to have formed during continental rifting. Data sources: Abouchami *et al.* (1990); Skulski and Percival (1996); Puchtel *et al.* (1998a); Tomlinson *et al.* (1998); Hollings and Wyman (1999).

a summary of accreted pre-Cambrian oceanic plateaus and the reader is referred to their paper for more detailed information.

Some of the oldest preserved oceanic plateau sequences are those found in ~3.5 Ga Barberton and Pietersberg belts of the Kaapvaal Shield of southern Africa (De Wit *et al.*, 1987; Smith and Erlank, 1982). These belts contain pillow basalts and komatiites, with chemical signatures (Lahaye *et al.*, 1995) suggesting a likely origin as part of an oceanic plateau. The Pilbara craton of Australia appears to possess some of the oldest oceanic plateau material so far identified (Green *et al.*, 2000) in the ~3.5 Ga Coonterunah and Warrawoona Groups.

Greenstone belts of the Canadian Superior province, ranging in age from 3.0 to 2.7 Ga also contain lava groups that have been interpreted to be remnants of accreted oceanic plateaus. These belts include the South Rim unit of the North Caribou belt (Hollings and Wyman, 1999), the Vizien belt (Skulski and Percival, 1996), the Malartic-Val d'Or (Desrochers *et al.*, 1993), the Tisdale Group of the Abitibi belt (Fan and Kerrich, 1997), and the Schreiber-Hemlo and the White River-Dayohessarah belts (Polat *et al.*, 1998). The evidence for an oceanic plateau origin is based on the occurrence of pillow basalts and komatiites without terrestrial sedimentary intercalations or sheeted dyke swarms, possessing low $(La/Nb)_{pmn}$ and the low positive εNd (Figure 18) that are characteristic of Cretaceous oceanic plateaus. Several of these sequences within the Superior province are in stratigraphic contact with basalts and komatiites that possess a signature of continental lithosphere contamination, i.e., negative εNd and $(La/Nb)_{pmn} > 1$ (Figure 18). These include the Balmer assemblage, Red Lake belt (Tomlinson *et al.*, 1998), Steep Rock-Lumby Lake belts (Tomlinson *et al.*, 1999), and Opapimiskan-Markop unit, North Caribou belt (Hollings and Wyman, 1999). These sequences have been interpreted as having formed in tectonic settings related to continental break-up, similar to the North Atlantic Tertiary igneous province and parts of the Cretaceous Kerguelen plateau.

Puchtel *et al.* (1998b, 1999) have proposed that the 3.0–2.8 Ga Kostomuksha and Sumozero-Kenozero greenstone belts of the Baltic Shield represent remnants of oceanic plateaus. This interpretation is based on the occurrence of crustally uncontaminated pillow basalts (Figure 18) and komatiites without terrestrial sedimentary intercalations. In contrast the 2.4 Ga Vetreny greenstone belt and the 1.98 Ga Onega plateau, also part of the Baltic Shield, display chemical evidence of crustal contamination (negative εNd; Figure 18). These sequences were interpreted by Puchtel *et al.* (1997, 1998a) as having formed during continental break-up.

Other Proterozoic oceanic plateau terranes have been identified in the Birimian province of western Africa (Figure 18) (Abouchami *et al.* 1990; Boher *et al.*, 1992), the Arabian-Nubian Shield (Stein and Goldstein, 1996), and the Flin Flon belt in Canada (Stern *et al.*, 1995).

3.16.8 ENVIRONMENTAL IMPACT OF OCEANIC PLATEAU FORMATION

Although the potential environmental impact of continental flood basalt provinces has been documented by many authors (e.g., Hallam, 1987a; McLean, 1985; Renne and Basu, 1991; Courtillot *et al.*, 1996), the possible effects of oceanic plateau eruptions on the atmosphere, biosphere and hydrosphere have received comparatively little attention (see, for example, Courtillot, 1994). This omission is surprising since the inclusion of oceanic plateau events actually strengthens the correlation between LIP eruptions and mass extinction events (Kerr, 1997), by providing a feasible terrestrial causal mechanism for several second order extinction events (Sepkoski, 1986).

3.16.8.1 Cenomanian–Turonian Boundary (CTB) Extinction Event

Several of these second order extinction events occurred in the mid-Cretaceous. One of these, the CTB event (~93 Ma) has been linked by several authors to the formation of oceanic plateaus (Sinton and Duncan, 1997; Kerr, 1998). The CTB event is characterized by the world-wide deposition of organic-rich black shales (Jenkyns, 1980; Schlanger *et al.*, 1987). The formation of black shale implies a widespread reducing environment ("anoxia") in the oceans at this time (Figure 19). In addition to this, the CTB was a time of major sea level transgression (Hallam, 1989) and is marked by a positive carbon isotopic anomaly ($\delta^{13}C$ excursion) of up to +4–5‰ (Arthur *et al.*, 1987), indicating an increase in organic carbon burial rate (Figure 19). Sea water $^{87}Sr/^{86}Sr$ (Figure 19) reaches a maximum of 0.70753 in the late-Cenomanian, and drops steadily to a value of 0.70737 in the mid-Turonian, before starting to rise again. Average global surface temperatures (including oceanic temperatures) around the CTB were 6–14 °C higher than present (Kaiho, 1994), and this is most likely due to an increase in global atmospheric CO_2 content which may have been >10 times present-day levels (Figure 19) (Arthur *et al.*, 1987).

These phenomena were accompanied by an extinction event that resulted in the demise of 26% of all known genera (Sepkoski, 1986). Although the overall extinction rate is much lower than that at the Cretaceous–Tertiary boundary, deep water marine invertebrates fared much worse in the CTB event (Kaiho, 1994). This difference supports the view that anomalous oceanic volcanism around the CTB may have played a significant role in the environmental and biotic crisis at this time (Kerr, 1998).

Siderophile and compatible lithophile trace elements such as Sc, Ti, V, Cr, Mn, Co, Ni, Pt, Ir and Au are enriched in CTB black shales (Leary and Rampino, 1990; Orth *et al.*, 1993). Kerr (1998) has shown that trace element abundances and ratios found in CTB black shales are similar to plume-derived volcanic rocks and midocean ridge basalts. For example, in mafic volcanic rocks Ni/Ir × 10^4 values range from 70 to 190, and in CTB sediments this ratio averages 180. In contrast, average sedimentary rocks possess Ni/Ir × 10^4 ratios of ~100 (Orth *et al.*, 1993).

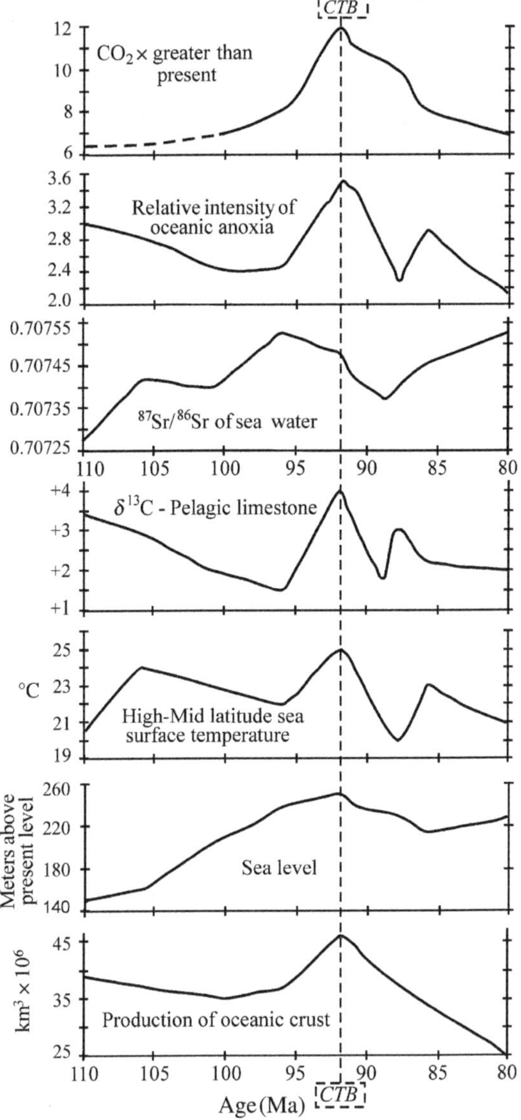

Figure 19 Graphs showing how various parameters discussed in the text vary from 110 Ma to 80 Ma. The dotted vertical line represents the Cenomanian–Turonian boundary (after Kerr, 1998).

3.16.8.2 Links between CTB Oceanic Plateau Volcanism and Environmental Perturbation

The most extensive plume-related volcanism around the CTB occurred in the oceans, with the formation of the CCOP along with portions of the OJP and Kerguelen plateau. In addition to this oceanic volcanism, a continental flood basalt province related to the Marion hotspot also erupted at this time, as Madagascar rifted from India (Storey *et al.*, 1995). The estimated erupted volume of oceanic plateau lavas around this time is $\sim 1.0 \times 10^7$ km^3 and may be much higher (Kerr, 1998). The potential physical and chemical effects of oceanic plateau volcanism on the global environment are summarized in Figure 20 and discussed below.

An obvious physical effect of oceanic mantle plume volcanism is to raise sea level by lava extrusion onto the ocean floor through the buoyant plume head uplifting the oceanic lithosphere and displacing seawater (Courtney and White, 1986) and by the thermal expansion of seawater due to heating. The steady rise in global sea level throughout the Late Albian and Cenomanian (Figure 19) may reflect the arrival of the Caribbean, Ontong Java and Kerguelen plume heads below the oceanic lithosphere, prior to extensive volcanism (Vogt, 1989; Larson, 1991). This plume-related uplift of oceanic lithosphere may also have caused the disruption of important oceanic circulation systems such that cool, polar (oxygenated) waters were not circulated to lower latitudes, resulting in increased oceanic anoxia. Additionally, hydrothermal fluids from oceanic plateau volcanism could have contributed to warmer oceans, and thus to anoxia, since oxygen

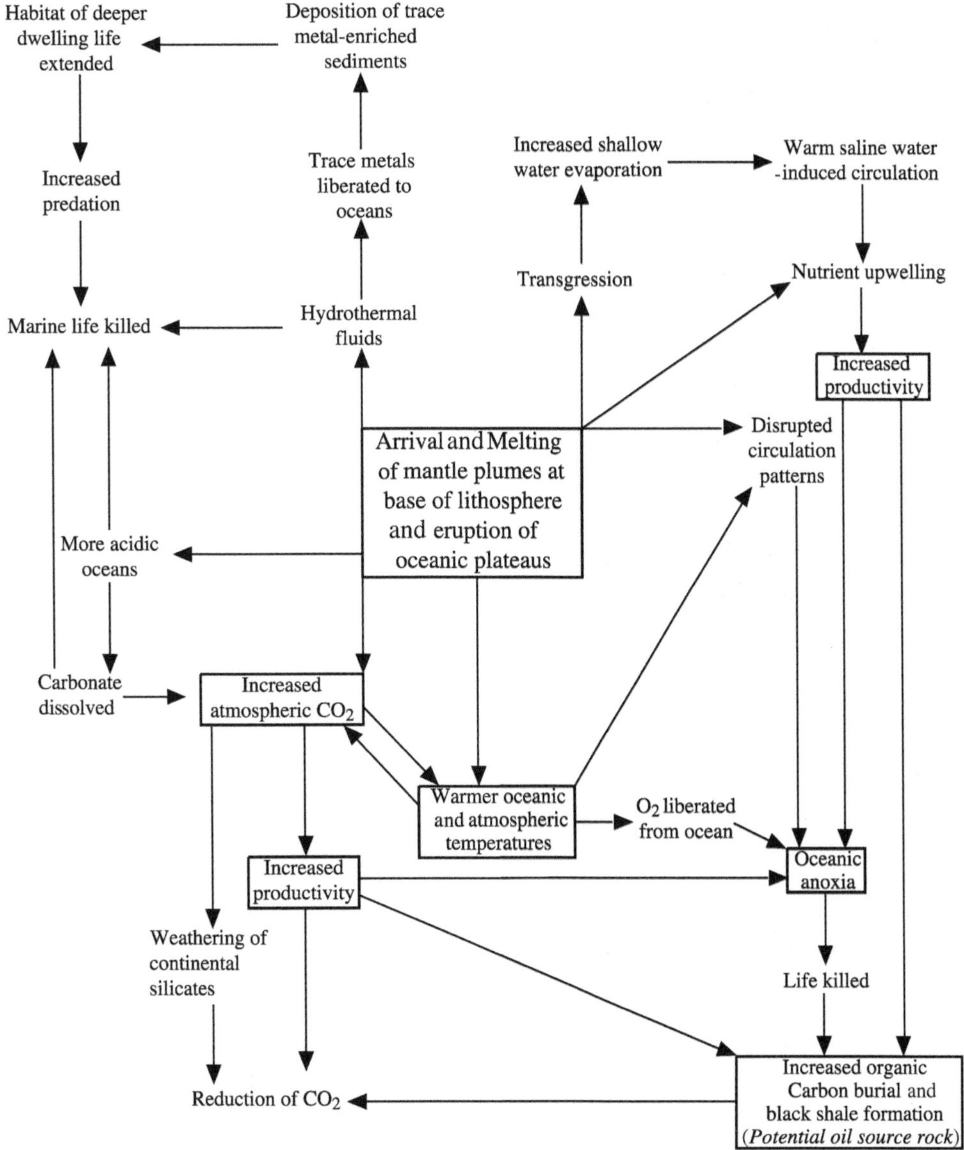

Figure 20 Flow diagram summarizing the possible physical and chemical environmental effects of the formation of large igneous provinces around the Cenomanian–Turonian boundary (after Kerr, 1998).

solubility is considerably reduced in warmer seawater (Sinton and Duncan, 1997).

Positive $\delta^{13}C$ anomalies at the CTB reflect increased rates of organic carbon burial as a result of high productivity and more effective preservation of organic material (Arthur *et al.*, 1987). Such increased productivity means the supply of deep ocean nutrients, such as phosphates, into surface waters must also increase and this process may have been induced by oceanic plateau volcanism (Vogt, 1989; Sinton and Duncan, 1997).

Elevated CO_2 levels at the CTB may also be due to increased volcanic activity. Kerr (1998) has calculated that approximately 10^{17} kg of CO_2 would have been released as a result of oceanic plateau volcanism around the CTB (Arthur *et al.*, 1987). Additionally, LIP volcanism also releases substantial amounts of SO_2, chlorine, fluorine and H_2S which, when released into seawater, would have made the CTB oceans much more acidic (Kerr, 1998). The lack of carbonate at the CTB may be the result of increased dissolution by more acidic seawater, which would also release more CO_2 to the atmosphere (Arthur *et al.*, 1987).

These additions of CO_2 to the atmosphere would have resulted in significant global warming. The solubility of CO_2 in seawater decreases as the temperatures rise; so the warmer the oceans get, the less CO_2 will dissolve in them. Thus, with this positive CO_2 feedback mechanism it is possible that a "runaway greenhouse" climate may have developed quite rapidly (Figure 20) (Kerr, 1998). Increased weathering of continental silicates can reduce CO_2 levels. However, the rate of CO_2 release at the CTB was much greater than its uptake by slow weathering processes. Increased atmospheric CO_2 and the upwelling of nutrients from the deep ocean could have resulted in increased productivity in ocean surface waters (Figure 20), leading to the widespread deposition of black shales and thus a reduction in CO_2 levels.

Increased concentration of toxic trace metals in the oceans, liberated by hydrothermal fluids from oceanic plateau lava piles, may well have been a contributory factor to the demise of some marine organisms around the CTB (Wilde *et al.*, 1990). The upwelling of deep ocean trace metals and nutrients may result in the enlargement of the trace metal-restricted habitat of deeper dwelling organisms (Wilde *et al.*, 1990), leading to increased predation by deeper dwelling creatures on those living in shallower water.

Throughout the past 250 Myr significant black shale deposits occur during several other periods and, like the CTB, these other black shales are associated with, sometimes severe, environmental disruption (Jenkyns, 1980; Hallam, 1987b; Arthur and Sageman, 1994). It is interesting, and probably highly significant, that these black shale events correlate with the formation of oceanic plateaus or plume-related volcanic rifted margins (Table 5). The Aptian–Albian (121–99 Ma) appears to have been a period of persistent environmental disturbance with three distinct oceanic anoxia events (with associated black shales) during this period (Bralower *et al.*, 1993). A causal link between black shale formation, environmental perturbation and oceanic volcanism is given further credence by the fact that a major period of oceanic plateau formation occurred during the Aptian–Albian (121–98 Ma) (see above).

Finally, Condie *et al.* (2001) have presented evidence that the correlation between black shale deposition, paleoclimatic disturbance and mantle superplume events can be extended back to the pre-Cambrian. Particularly good correlations between environmental disturbance and mantle plume activity occur at 1.9 Ga and 2.7 Ga

3.16.9 CONCLUDING STATEMENTS

Oceanic plateaus represent overthickened areas of oceanic crust (10–35 km), which appear to have formed as a result of decompression melting of a large mantle plume head, often (although not always) within 1–2 Ma. Geological and geochemical evidence suggests that oceanic plateaus have formed throughout a considerable period of Earth's history.

The thickness of the crustal sections of oceanic plateaus implies that they are not easily subducted. Thus, when these plateaus encounter a subduction zone, their top-most portions tend to become

Table 5 Correlations between black shale events and oceanic plateau volcanism over the last 250 Ma.

Age	*Black shales*	*Oceanic plateau or volcanic rifted margin*
Aptian–Albian (121–99 Ma)	Extensive world-wide deposits	Ontong Java & other Pacific plateaus, Kerguelen
Tithonian (150–144 Ma)	Extensive deposits in Europe and west Asia	Sorachi plateau & Shatsky Rise
Toarcian (190–180 Ma)	Extensive world-wide deposits	Karoo, Ferrar & Weddell Sea
Carnian (227–220 Ma)	Few deposits	Wrangellia

Data sources: Hallam (1987b); Jenkyns (1980); Riley and Knight (2001); Hergt and Brauns (2001).

accreted to a continent or an island arc. In this way fragments of oceanic plateaus have become incorporated into continental margins and preserved in the geological record.

Two recent examples of such accreted oceanic plateaus are the CCOP and the OJP. All of the lavas sampled from the OJP and most of the lavas from the CCOP are relatively homogeneous basalts with initial εNd values between +5 and +8 and broadly chondritic trace element ratios. In contrast, high-MgO lavas found in the CCOP reveal evidence of a more heterogeneous plume source region, containing both enriched (εNd < +5; $(La/Nd)_{cn} > 1$) and depleted (εNd < +8; $(La/Nd)_{cn} < 1$) components.

The formation of LIPs during continental break-up (e.g., the NAIP and the Kerguelen plateau) results in the formation of seaward-dipping reflector sequences and oceanic plateaus, both of which show chemical evidence of contamination by ancient continental lithosphere (initial εNd < 0; $(La/Nb)_{pmn} > 1$).

The correlation between oceanic plateau formation and marine environmental catastrophes (characterized by mass extinction, oceanic anoxia and black shale deposition) throughout the Mesozoic period suggests a causal link between oceanic plateau formation and environmental crises.

REFERENCES

Abbott D. H. (1996) Plumes and hotspots as sources of greenstone belts. *Lithos* **37**, 113–127.

Abouchami W., Boher M., Michard A., and Albarede F. (1990) A major 2.1 Ga event of mafic magmatism in West Africa—an early stage of crustal accretion. *J. Geophys. Res.* **95**, 17605–17629.

Aitken B. G. and Echeverría L. M. (1984) Petrology and geochemistry of komatiites and tholeiites from Gorgona Island, Colombia. *Contrib. Mineral. Petrol.* **86**, 94–105.

Albarede F. (1998) The growth of continental crust. *Tectonophysics* **296**, 1–14.

Arndt N. T., Kerr A. C., and Tarney J. (1997) Dynamic melting in plume heads: the formation of Gorgona komatiites and basalts. *Earth Planet. Sci. Lett.* **146**, 289–301.

Arthur M. A. and Sageman B. B. (1994) Marine black shales—depositional mechanisms and environments of ancient deposits. *Ann. Rev. Earth Planet. Sci.* **22**, 499–551.

Arthur M. A., Schlanger S. O., and Jenkyns H. C. (1987) The Cenomanian–Turonian Oceanic Anoxic Event: II. In *Paleoceanographic Controls on Organic-matter Production and Preservation,* Geological Society of London Special Publication 26 (eds. J. Brooks and A. J. Fleet), pp. 401–420.

Babbs T. L. (1997) Geochemical and petrological investigations of the deeper portions of the Ontong Java Plateau: Maliata, Solomon Islands. PhD Thesis, University of Leicester, UK, (unpublished).

Barton A. J. and White R. S. (1995) The Edoras Bank margin: continental break-up in the presence of a mantle plume. *J. Geol. Soc. London* **152**, 971–974.

Ben-Avraham Z., Nur A., Jones D., and Cox A. (1981) Continental accretion: from oceanic plateaus to allochthonous terranes. *Science* **213**, 47–54.

Bence A. E., Papike J. J., and Ayuso R. A. (1975) Petrology of submarine basalts from the Central Caribbean: DSDP Leg 15. *J. Geophys. Res.* **80**, 4775–4804.

Boher M., Abouchami W., Michard A., Albarede F., and Arndt N. T. (1992) Crustal Growth in West Africa at 2.1 Ga. *J. Geophys. Res.-Solid Earth* **97**, 345–369.

Bralower T. J., Sliter W. V., Arthur T. J., Lekie R. M., Allard D. J., and Schlanger S. O. (1993) Dysoxic/anoxic events in the Aptian–Albian (Early Cretaceous). In *The Mesozoic Pacific: Geology, Tectonics, and Volcanism,* American Geophysical Union Monograph 77 (eds. M. S. Pringle, W. W. Sager, W. V. Sliter, and S. Stein), pp. 5–37.

Bruguier O. (1996) U–Pb ages on single detrital zircon grains from the Tasmiyele group: implications for the evolution of the Olekma block (Aldan shield, Siberia). *Precamb. Res.* **78**, 197–210.

Burke K. (1988) Tectonic evolution of the Caribbean. *Ann. Rev. Earth Planet. Sci.* **16**, 201–230.

Caldeira K. and Rampino M. R. (1990) Deccan volcanism, greenhouse warming, and the Cretaceous/Tertiary boundary. In *Global Catastrophes in Earth History,* Geological Society of America Special Paper 247 (eds. V. L. Sharpton and P. D. Ward), pp. 117–123.

Campbell I. H. and Griffiths R. W. (1990) Implications of mantle plume structure for the evolution of flood basalts. *Earth Planet. Sci. Lett.* **99**, 79–93.

Campbell I. H., Griffiths R. W., and Hill R. I. (1989) Melting in an Archaean mantle plume: heads it's basalts, tails it's komatiites. *Nature* **339**, 697–699.

Cloos M. (1993) Lithospheric buoyancy and collisional orogenesis: subduction of oceanic plateaus, continental margins, island arcs, spreading ridges, and seamounts. *Geol. Soc. Am. Bull.* **105**, 715–737.

Coffin M. F. (1994) Large igneous provinces: crustal structure, dimensions, and external consequences. *Rev. Geophy.* **32**, 1–36.

Coffin M. F. and Eldholm O. (1992) Volcanism and continental break-up: a global compilation of large igneous provinces. In *Magmatism and the Causes of Continental Breakup,* Geological Society of London Special Publication 68 (eds. B. C. Storey, T. Alabaster, and R. J. Pankhurst), pp. 17–30.

Coffin M. F., Pringle M. S., Duncan R. A., Gladczenko T. P., Storey M., Muller R. D., and Gahagan L. A. (2002) Kerguelen hotspot magma output since 130 Ma. *J. Petrol.* **43**, 1121–1139.

Condie K. C. (1999) Mafic crustal xenoliths and the origin of the lower continental crust. *Lithos* **46**, 95–101.

Condie K. C. (2001) *Mantle Plumes and their Record in Earth History.* Cambridge University Press.

Condie K. C., Marais D. J. D., and Abbott D. (2001) Precambrian superplumes and supercontinents: a record in black shales, carbon isotopes, and paleoclimates? *Precamb. Res.* **106**, 239–260.

Courtillot V. (1994) Mass extinctions in the last 300 million years: one Impact and seven flood basalts? *Isr. J. Earth Sci.* **43**, 255–266.

Courtillot V., Jaeger J. J., Yang Z. Z., Feraud G., and Hofmann C. K. B. (1996) The influence of continental flood basalts on mass extinctions: where do we stand? In *The Cretaceous–Tertiary Event and Other Catastrophes in Earth History,* Geological Society of America Special Paper 307 (eds. G. Ryder, D. Fastovsky, and S. Gartner), pp. 513–525.

Courtney R. and White R. (1986) Anomalous heat flow and geoid across the Cape Verde Rise: evidence for dynamic support from a thermal plume in the mantle. *Geophys, J. Roy. Astr. Soc.* **87**, 815–867.

De Wit M. J., Hart R. A., and Hart R. J. (1987) The Jamestown ophiolite complex Barberton mountain belt: a section through 3.5 Ga oceanic crust. *J. Afr. Earth Sci.* **6**, 681–730.

Desrochers J. P., Hubert C., Ludden J. N., and Pilote P. (1993) Accretion of Archean Oceanic Plateau fragments in the Abitibi Greenstone Belt, Canada. *Geology* **21**, 451–454.

Dickin A. P., Jones N. W., Thirlwall M. F., and Thompson R. N. (1987) A Ce/Nd isotope study of crustal contamination processes affecting Palaeocene magmas in Skye, Northwest Scotland. *Contrib. Mineral. Petrol.* **96**, 455–464.

Donnelly T. W. (1973) Late Cretaceous basalts from the Caribbean, a possible flood basalt province of vast size. *EOS* **54**, 1004.

Dunphy J. M., Ludden J. N., and Francis D. (1995) Geochemistry of mafic magmas from the Ungava Orogen, Quebec, Canada and implications for mantle reservoir compositions at 2.0 Ga. *Chem. Geol.* **120**, 361–380.

Dupré B. and Echeverría L. M. (1984) Pb Isotopes of Gorgona Island (Colombia): isotopic variations correlated with magma type. *Earth Planet. Sci. Lett.* **67**, 186–190.

Echeverría L. M. (1980) Tertiary or Mesozoic komatiites from Gorgona Island, Colombia: field relations and geochemistry. *Contrib. Mineral. Petrol.* **73**, 253–266.

Edgar N. T., Ewing J. I., and Hennion J. (1971) Seismic refraction and reflection in the Caribbean Sea. *Am. Assoc. Petrol. Geol.* **55**, 833–870.

Eldholm O. and Coffin M. F. (2000) Large igneous provinces and plate tectonics. *AGU Monograph* **121**, 309–326.

Fan J. and Kerrich R. (1997) Geochemical characteristics of aluminum depleted and undepleted komatiites and HREE-enriched low-Ti tholeiites, western Abitibi greenstone belt: a heterogeneous mantle plume convergent margin environment. *Geochim. Cosmochim. Acta* **61**, 4723–4744.

Farnetani D. G. and Richards M. A. (1995) Thermal entrainment and melting in mantle plumes. *Earth Planet. Sci. Lett.* **136**, 251–267.

Farnetani C. G., Richards M. A., and Ghiorso M. S. (1996) Petrological models of magma evolution and deep crustal structure beneath hotspots and flood basalt provinces. *Earth Planet. Sci. Lett.* **143**, 81–94.

Fitton J. G., Saunders A. D., Norry M. J., Hardarson B. S., and Taylor R. N. (1997) Thermal and chemical structure of the Iceland plume. *Earth Planet. Sci. Lett.* **153**, 197–208.

Fitton J. G., Hardarson B. S., Ellam R., and Rogers G. (1998a) Sr-, Nd-, and Pb-isotopic composition of volcanic rocks from the Southeast Greenland margin at 63 degrees N: temporal variation in crustal contamination during continental breakup. *Proc. ODP Sci. Results* **152**, 351–357.

Fitton J. G., Saunders A. D., Larsen H. C., Hardarson B. S., and Norry M. J. (1998b) Volcanic rocks from the East Greenland margin at 63°N: composition, petrogenesis and mantle sources. In *Proceedings of the Ocean Drilling Program, Scientific Results* 152 (eds. A. D. Saunders, H. C. Larsen, S. Wise, and J. R. Allen), pp. 503–533.

Fitton J. G., Larsen L. M., Saunders A. D., Hardarson B. S., and Kempton P. D. (2000) Palaeogene continental to oceanic magmatism on the SE Greenland continental margin at 63 degrees N: a review of the results of ocean drilling program legs 152 and 163. *J. Petrol.* **41**, 951–966.

Francis D., Ludden J., and Hynes A. (1983) Magma evolution in a Proterozoic rifting environment. *J. Petrol.* **24**, 556–582.

Frey F. A., McNaughton N. J., Nelson D. R., deLaeter J. R., and Duncan R. A. (1996) Petrogenesis of the Bunbury Basalt, western Australia: interaction between the Kerguelen plume and Gondwana lithosphere? *Earth Planet. Sci. Lett.* **144**, 163–183.

Frey F. A., Coffin M. F., Wallace P. J., Weis D., Zhao X., Wise S. W., Wahnert V., Teagle D. A. H., Saccocia P. J., Reusch D. N., Pringle M. S., Nicolaysen K. E., Neal C. R., Muller R. D., Moore C. L., Mahoney J. J., Keszthelyi L., Inokuchi H., Duncan R. A., Delius H., Damuth J. E., Damasceno D., Coxall H. K., Borre M. K., Boehm F., Barling J., Arndt N. T., and Antretter M. (2000) Origin and evolution of a submarine large igneous province: the Kerguelen Plateau and Broken Ridge, southern Indian Ocean. *Earth Planet. Sci. Lett.* **176**, 73–89.

Furomoto A. S., Webb J. P., Odegaard M. E., and Hussong D. M. (1976) Seismic studies on the Ontong-Java Plateau, 1970. *Tectonophysics* **34**, 71–90.

Gladczenko T. P., Coffin M. F., and Eldholm O. (1997) Crustal structure of the Ontong Java Plateau: modeling of new gravity and existing seismic data. *J. Geophys. Res.-Solid Earth* **102**, 22711–22729.

Green M. G., Sylvester P. J., and Buick R. (2000) Growth and recycling of early Archaean continental crust: geochemical evidence from the Coonterunah and Warrawoona Groups, Pilbara Craton, Australia. *Tectonophysics* **322**, 69–88.

Griffiths R. W. and Campbell I. H. (1990) Stirring and structure in mantle starting plumes. *Earth Planet. Sci. Lett.* **99**, 66–78.

Hallam A. (1987a) End-Cretaceous mass extinction event: argument for terrestrial causation. *Science* **238**, 1237–1242.

Hallam A. (1987b) Mesozoic marine organic-rich shales. In *Marine Petroleum Source Rocks,* Geological Society of London, Special Publication 26 (eds. J. Brooks and A. J. Fleet), pp. 251–261.

Hallam A. (1989) The case for sea-level change as a dominant causal factor in mass extinction of marine invertebrates. *Phil. Trans. R. Soc. Lond. B* **325**, 437–455.

Hart S. R., Hauri E. H., Oschmann L. A., and Whitehead J. A. (1992) Mantle plumes and entrainment: isotopic evidence. *Science*, **256.**

Hassler D. R. and Shimizu N. (1998) Osmium isotopic evidence for ancient subcontinental lithospheric mantle beneath the Kerguelen Islands, southern Indian Ocean. *Science* **280**, 418–421.

Hauff F., Hoernle K., Tilton G., Graham D. W., and Kerr A. C. (2000a) Large volume recycling of oceanic lithosphere over short time scales: geochemical constraints from the Caribbean Large Igneous Province. *Earth Planet. Sci. Lett.* **174**, 247–263.

Hauff F., Hoernle K., van den Bogaard P., Alvarado G. E., and Garbe-Schonberg C. D. (2000b) Age and geochemistry of basaltic complexes in western Costa Rica: contributions to the geotectonic evolution of Central America. *Geochem. Geophys. Geosys.* **1**, Paper number 1999GC000020.

Hergt J. M. and Brauns C. M. (2001) On the origin of Tasmanian dolerites. *Aust. J. Earth Sci.* **48**, 543–549.

Hill R. I. (1991) Starting plumes and continental break-up. *Earth Planet. Sci. Lett.* **104**, 398–416.

Hollings P. and Wyman D. (1999) Trace element and Sm-Nd systematics of volcanic and intrusive rocks from the 3 Ga Lumby Lake Greenstone belt, Superior Province: evidence for Archean plume-arc interaction. *Lithos* **46**, 189–213.

Holmes A. (1918) The basaltic rocks of the Arctic region. *Min. Mag.* **18**, 180–222.

Hémond C., Arndt N. T., Lichtenstein U., Hofmann A. W., Oskarsson N., and Steinthorsson S. (1993) The heterogeneous Iceland plume: Nd–Sr–O isotopes and trace element constraints. *J. Geophys. Res.* **98**, 15833–15850.

Isozaki Y. (1997) Jurassic accretion tectonics of Japan. *Island Arc* **6**, 25–51.

Jenkyns H. C. (1980) Cretaceous anoxic events: from continents to oceans. *J. Geol. Soc. London* **137**, 171–188.

Jochum K. P., Arndt N. T., and Hofmann A. W. (1991) Nb–Th–La in komatiites and basalts: constraints on komatiite petrogenesis and mantle evolution. *Earth Planet. Sci. Lett.* **107**, 272–289.

Jones G., Sano H., and Valsamijones E. (1993) Nature and Tectonic setting of accreted basalts from the Mino terrane central Japan. *J. Geol. Soc. London* **150**, 1167–1181.

Kaiho K. (1994) Planktonic and benthic foraminiferal extinction events during the last 100 m.y. *Palaeogeog. Palaeoclim. Palaeoecol.* **111**, 45–71.

Kempton P. D., Fitton J. G., Saunders A. D., Nowell G. M., Taylor R. N., Hardarson B. S., and Pearson G. (2000) The Iceland plume in space and time: a Sr–Nd–Pb–Hf study of the North Atlantic rifted margin. *Earth Planet. Sci. Lett.* **177**, 255–271.

Kent R. W., Saunders A. D., Kempton P. D., and Ghose N. C. (1997) Rajmahal basalts, eastern India: mantle sources and melt distribution at a volcanic rifted margin. In *Large*

Igneous Provinces: Continental, Oceanic and Planetary Volcanism, American Geophysical Union Monograph 100 (eds. J. J. Mahoney and M. Coffin), pp. 144–182.

Kerr A. C. (1997) Asteroid impact and mass extinction at the K-T boundary: an extinct red herring? *Geol. Today* **13**, 157–159.

Kerr A. C. (1998) Oceanic plateau formation: a cause of mass extinction and black shale deposition around the Cenomanian–Turonian boundary. *J. Geol. Soc. London* **155**, 619–626.

Kerr A. C., Kempton P. D., and Thompson R. N. (1995a) Crustal assimilation during turbulent magma ascent (ATA): new isotopic evidence from the Mull Tertiary lava succession, NW Scotland. *Contrib. Mineral. Petrol.* **119**, 142–154.

Kerr A. C., Saunders A. D., Tarney J., Berry N. H., and Hards V. L. (1995b) Depleted mantle plume geochemical signatures: no paradox for plume theories. *Geology* **23**, 843–846.

Kerr A. C., Marriner G. F., Arndt N. T., Tarney J., Nivia A., Saunders A. D., and Duncan R. A. (1996a) The petrogenesis of komatiites, picrites and basalts from the Isle of Gorgona, Colombia: new field, petrographic and geochemical constraints. *Lithos* **37**, 245–260.

Kerr A. C., Tarney J., Marriner G. F., Klaver G. T., Saunders A. D., and Thirlwall M. F. (1996b) The geochemistry and petrogenesis of the late-Cretaceous picrites and basalts of Curaçao Netherlands Antilles: a remnant of an oceanic plateau. *Contrib. Mineral. Petrol.* **124**, 29–43.

Kerr A. C., Marriner G. F., Tarney J., Nivia A., Saunders A. D., Thirlwall M. F., and Sinton C. W. (1997a) Cretaceous basaltic terranes in western Colombia: elemental, chronological and Sr–Nd constraints on petrogenesis. *J. Petrol.* **38**, 677–702.

Kerr A. C., Tarney J., Marriner G. F., Nivia A., and Saunders A. D. (1997b) The Caribbean–Colombian Cretaceous igneous province: the internal anatomy of an oceanic plateau. In *Large Igneous Provinces; Continental, Oceanic and Planetary Flood Volcanism,* American Geophysical Union Monograph 100 (eds. J. J. Mahoney and M. Coffin), pp. 45–93.

Kerr A. C., Tarney J., Nivia A., Marriner G. F., and Saunders A. D. (1998) The internal structure of oceanic plateaus: Inferences from obducted Cretaceous terranes in western Colombia and the Caribbean. *Tectonophysics* **292**, 173–188.

Kerr A. C., Iturralde-Vinent M. A., Saunders A. D., Babbs T. L., and Tarney J. (1999) A new plate tectonic model of the Caribbean: implications from a geochemical reconnaissance of Cuban Mesozoic volcanic rocks. *Geol. Soc. Am. Bull.* **111**, 1581–1599.

Kerr A. C., White R. V., and Saunders A. D. (2000) LIP reading: recognizing oceanic plateaux in the geological record. *J. Petrol.* **41**, 1041–1056.

Kerr A. C., Tarney J., Kempton P. D., Spadea P., Nivia A., Marriner G. F., and Duncan R. A. (2002) Pervasive mantle plume head heterogeneity: evidence from the late Cretaceous Caribbean–Colombian Oceanic Plateau. *J. Geophys. Res.* **107**(7), DOI. 10.1029, 2001JB000790.

Kimura G. and Ludden J. (1995) Peeling oceanic crust in subduction zones. *Geology* **23**, 217–220.

Kimura G., Ludden J. N., Desrochers J. P., and Hori R. (1993) A model of ocean-crust accretion for the Superior Province, Canada. *Lithos* **30**, 337–355.

Kimura G., Sakakibara M., and Okamura M. (1994) Plumes in central Panthalassa? deductions from accreted oceanic fragments in Japan. *Tectonics* **13**, 905–916.

Klaver, G. T. (1987) The Curaçao lava formation an ophiolitic analogue of the anomalous thick layer 2B of the mid-Cretaceous oceanic plateaus in the western Pacific and central Caribbean. PhD Thesis, University of Amsterdam, The Netherlands.

Klosko E. R., Russo R. M., Okal E. A., and Richardson W. P. (2001) Evidence for a rheologically strong chemical mantle root beneath the Ontong-Java Plateau. *Earth Planet. Sci. Lett.* **186**, 347–361.

Kroenke L. W. (1974) Origin of continents through development and coalescence of oceanic flood basalt plateaus. *EOS* **55**, 443.

Lahaye Y., Arndt N., Byerly G., Chauvel C., Fourcade S., and Gruau G. (1995) The influence of alteration on the trace-element and Nd isotopic compositions of komatiites. *Chem. Geol.* **126**, 43–64.

Larson R. L. (1991) Geological consequences of super plumes. *Geology* **19**, 963–966.

Lassiter J. C., DePaolo D. J., and Mahoney J. J. (1995) Geochemistry of the Wrangellia flood basalt province: implications for the role of continental and oceanic lithosphere in flood basalt genesis. *J. Petrol.* **36**, 983–1009.

Leary P. N. and Rampino M. R. (1990) A multicausal model of mass extinctions: increase in trace metals in the oceans. In *Extinction Events in Earth History, Lecture Notes in Earth Science*, 30 (eds. E. G. Kauffman and O. H. Walliser). Springer, Berlin, pp. 45–55.

Leat P. T., Livermore R. A., Millar I. L., and Pearce J. A. (2000) Magma supply in back-arc spreading centre segment E2 East Scotia Ridge. *J. Petrol.* **41**, 845–866.

Loper D. E. (1983) The dynamical and thermal structure of deep mantle plumes. *Phy. Earth Planet. Int.* **33**, 304–317.

Lucas S. B., Stern R. A., Syme E. C., Reilly B. A., and Thomas D. J. (1996) Intraoceanic tectonics and the development of continental crust: 1.92–1.84 Ga evolution of the Flin Flon belt, Canada. *Geol. Soc. Am. Bull.* **108**, 602–629.

Mahoney J. J., Storey M., Duncan R. A., Spencer K. J., and Pringle M. (1993a) Geochemistry and age of the Ontong Java Plateau. In *The Mesozoic Pacific: Geology, Tectonics, and Volcanism,* American Geophysical Union Monograph 77 (eds. M. S. Pringle, W. W. Sager, W. V. Sliter and S. Stein), pp. 233–261.

Mahoney J. J., Storey M., Duncan R. A., Spencer K. J., and Pringle M. (1993b) Geochemistry and geochronology of Leg 130 basement lavas: nature and origin of the Ontong Java Plateau. In *Proceedings of the Ocean Drilling Program, Scientific Results* 130 (eds. W. H. Berger, L. W. Kroenke, and L. A. Mayer), pp. 3–22.

Mahoney J. J., Jones W. B., Frey F. A., Salters V. J. M., Pyle D. G., and Davies H. L. (1995) Geochemical characteristics of lavas from Broken Ridge, the Naturaliste Plateau and southernmost Kerguelen Plateau: cretaceous plateau volcanism in the Southeast Indian Ocean. *Chem. Geol.* **120**, 315–345.

Mahoney, J. J., Fitton, J. G., Wallace, P. J., *et al.* (2001). *Proceedings of the ODP, Initial Reports.*, 192 [Online]. Available from World Wide Web: <http://www.odp.tamu.edu/publications/192_IR/192ir.htm >. [Cited 2002-02-20].

Marzoli A., Renne P. R., Piccirillo E. M., Ernesto M., Bellieni G., and De Min A. (1999) Extensive 200-million-year-old continental flood basalts of the Central Atlantic Magmatic Province. *Science* **284**, 616–618.

Mauffret A. and Leroy S. (1997) Seismic stratigraphy and structure of the Caribbean igneous province. *Tectonophysics* **283**, 61–104.

McKenzie D. P. and Bickle M. J. (1988) The volume and composition of melt generated by extension of the lithosphere. *J. Petrol.* **29**, 625–679.

McLean D. M. (1985) Deccan traps and mantle degassing in the terminal Cretaceous marine extinctions. *Cret. Res.* **6**, 235–259.

Morgan W. J. (1981) Hotspot tracks and the opening of the Atlantic and Indian Oceans. In *The Oceanic Lithosphere* (ed. C. Emiliani), Wiley-Interscience, pp. 443–487.

Neal C. R., Mahoney J. J., Kroenke L. W., Duncan R. A., and Petterson M. G. (1997) The Ontong Java Plateau. In *Large Igneous Provinces; Continental, Oceanic and Planetary Flood Volcanism,* American Geophysical Union Monograph 100 (eds. J. J. Mahoney and M. Coffin), pp. 183–216.

Nicolaysen K., Bowring S., Frey F., Weis D., Ingle S., Pringle M. S., and Coffin M. F. (2001) Provenance of Proterozoic garnet-biotite gneiss recovered from Elan Bank Kerguelen Plateau, southern Indian Ocean. *Geology* **29**, 235–238.

Nowell G. M., Kempton P. D., and Noble S. R. (1998) High precision Hf isotope measurements of MORB and OIB by thermal ionisation mass spectrometry: insights into the depleted mantle. *Chem. Geol.* **149**, 211–233.

Operto S. and Charvis P. (1996) Deep structure of the southern Kerguelen Plateau (southern Indian Ocean) from ocean bottom seismometer wide-angle seismic data. *J. Geophys. Res.* **101**, 25077–25103.

Orth C. J., Attrep M., Quintana L. R., Elder W. P., Kauffman E. G., Diner R., and Villamil T. (1993) Elemental abundance anomalies in the late Cenomanian extinction interval: a search for the source(s). *Earth Planet. Sci. Lett.* **117**, 189–204.

Pallister J. S., Budahn J. R., and Murchey B. L. (1989) Pillow basalts of the Angayucham terrane—oceanic plateau and island crust accreted to the Brooks Range. *J. Geophys. Res* **94**, 15901–15923.

Peltonen P., Kontinen A., and Huhma H. (1996) Petrology and geochemistry of metabasalts from the 1.95 Ga Jormua Ophiolite, northeastern Finland. *J. Petrol.* **37**, 1359–1383.

Petterson M. G., Neal C. R., Mahoney J. J., Kroenke L. W., Saunders A. D., Babbs T. L., Duncan R. A., Tolia D., and McGrail B. (1997) Structure and deformation of north and central Malaita, Solomon Islands: tectonic implications for the Ontong Java Plateau—Solomon arc collision, and for the fate of oceanic plateaus. *Tectonophysics* **283**, 1–33.

Petterson M. G., Babbs T., Neal C. R., Mahoney J. J., Saunders A. D., Duncan R. A., Tolia D., Magu R., Qopoto C., Mahoa H., and Natogga D. (1999) Geological-tectonic framework of Solomon Islands, SW Pacific: crustal accretion and growth within an intra-oceanic setting. *Tectonophysics* **301**, 35–60.

Polat A., Kerrich R., and Wyman D. A. (1998) The late Archean Schreiber-Hemlo and White River Dayohessarah greenstone belts, Superior Province: collages of oceanic plateaus, oceanic arcs, and subduction-accretion complexes. *Tectonophysics* **289**, 295–326.

Pringle M. S. (1992) Radiometric ages of basaltic basement recovered at Sites 800, 801, and 802, Leg 129, western Pacific Ocean. In *Proceedings of the Ocean Drilling Program, Scientific Results* 129 (eds. R. L. Larson, Y. Lancelot, A. Fisher and E. L. Winterer). Ocean Drilling Program, Texas A&M University, pp. 389–404.

Puchtel I. S. and Zhuravlev D. Z. (1993) Petrology of mafic-ultramafic metavolcanics and related rocks from the Olondo greenstone belt, Aldan Shield. *Petrol.* **1**, 308–348.

Puchtel I. S., Haase K. M., Hofmann A. W., Chauvel C., Kulikov V. S., Garbe Schonberg C. D., and Nemchin A. A. (1997) Petrology and geochemistry of crustally contaminated komatiitic basalts from the Vetreny Belt, southeastern Baltic Shield: evidence for an early Proterozoic mantle plume beneath rifted Archean continental lithosphere. *Geochim. Cosmochim. Acta* **61**, 1205–1222.

Puchtel I. S., Arndt N. T., Hofmann A. W., Haase K. M., Kroner A., Kulikov V. S., Kulikova V. V., Garbe Schonberg C. D., and Nemchin A. A (1998a) Petrology of mafic lavas within the Onega plateau, central Karelia: evidence for 2.0 Ga plume-related continental crustal growth in the Baltic Shield. Contrib. Miner. *Petrol.* **130**, 134–153.

Puchtel I. S., Hofmann A. W., Mezger K., Jochum K. P., Shchipansky A. A., and Samsonov A. V. (1998b) Oceanic plateau model for continental crustal growth in the archaean, a case study from the Kostomuksha greenstone belt, NW Baltic Shield. *Earth Planet. Sci. Lett.* **155**, 57–74.

Puchtel I. S., Hofmann A. W., Amelin Y. V., Garbe-Schonberg C. D., Samsonov A. V., and Schipansky A. A. (1999) Combined mantle plume-island arc model for the formation of the 2.9 Ga Sumozero-Kenozero greenstone belt, SE Baltic Shield: isotope and trace element constraints. *Geochim. Cosmochim. Acta* **63**, 3579–3595.

Renne P. R. and Basu A. R. (1991) Rapid eruption of the Siberian traps flood basalts at the Permo-Triassic boundary. *Science* **253**, 175–178.

Revillon S., Arndt N. T., Hallot E., Kerr A. C., and Tarney J. (1999) Petrogenesis of picrites from the Caribbean Plateau and the North Atlantic magmatic province. *Lithos* **49**, 1–21.

Reynaud C., Jaillard E., Lapierre H., Mamberti M., and Mascle G. H. (1999) Oceanic plateau and island arcs of southwestern Ecuador: their place in the geodynamic evolution of northwestern South America. *Tectonophysics* **307**, 235–254.

Richards M. A., Duncan R. A., and Courtillot V. E. (1989) Flood basalts and hot spot tracks: plume heads and tails. *Science* **246**, 103–107.

Richards M. A., Jones D. L., Duncan R. A., and DePaolo D. J. (1991) A mantle plume initiation model for the formation of Wrangellia and other oceanic flood basalt plateaus. *Science* **254**, 263–267.

Riley T. R. and Knight K. B. (2001) Age of pre-break-up Gondwana magmatism. *Antar. Sci.* **13**, 99–110.

Royer J.-Y. and Coffin M. F. (1992) Jurassic to Eocene plate tectonic reconstructions in the Kerguelen Plateau region. In *Proceedings of the Ocean Drilling Program, Scientific Results*, 120 (eds. J. S. W. Wise, A. P. Julson, R. Schlich, and E. Thomas), Ocean Drilling Program, Texas A&M University, pp. 917–930 .

Salters V. J. M. (1996) The generation of mid-ocean ridge basalts from the Hf and Nd isotope perspective. *Earth Planet. Sci. Lett.* **141**, 109–123.

Salters V. J. M. and Hart S. R. (1991) The mantle sources of ocean ridges, islands and arcs: the Hf-isotope connection. *Earth Planet. Sci. Lett.* **104**, 364–380.

Salters V. J. M., Storey M., Sevigny J. H., and Whitechurch H. (1992) Trace element and isotopic characteristics of Kerguelen-Heard Plateau basalts. In *Proceedings of the Ocean Drilling Program, Scientific Results* (eds. J. S. W. Wise, A. P. Julson, R. Schlich, and E. Thomas). Ocean Drilling Program, Texas A&M University, vol. 120, pp. 55–62. .

Saunders A. D., Storey M., Kent R. W., and Norry M. J. (1992) Consequences of plume-lithosphere interactions. In *Magmatism and the Causes of Continental Breakup* (eds. B. C. Storey, T. Alabaster, and R. J. Pankhurst). Geological Society of London, London, vol. 68, pp. 41–60.

Saunders A. D., Tarney J., Kerr A. C., and Kent R. W. (1996) The formation and fate of large igneous provinces. *Lithos* **37**, 81–95.

Saunders A. D., Fitton J., Kerr A. C., Norry M. J., and Kent R. W. (1997) The North Atlantic Igneous Province. In *Large Igneous Provinces: Continental, Oceanic and Planetary Volcanism,* American Geophysical Union Monograph 100 (eds. J. J. Mahoney and M. Coffin), pp. 45–93.

Schlanger S. O., Arthur M. A., Jenkyns H. C., and Scholle P. A. (1987) The Cenomanian–Turonian oceanic anoxic events: I. Stratigraphy and distribution of organic carbon-rich beds and the marine ^{13}C excursion. In *Marine Petroleum Source Rocks,* Geological Society of London, Special Publication 26 (eds. J. Brooks and A. J. Fleet), pp. 371–399.

Sepkoski J. J. (1986) Phanerozoic overview of mass extinction. In *Pattern and Processes in the History of Life* (eds. D. Raup and D. Japlonski). Springer-Verlag, pp. 277–295.

Sinton C. W. and Duncan R. A. (1997) Potential links beween ocean plateau volcanism and global ocean anoxia at the Cenomanian–Turonian boundary. *Econ. Geol.* **92**, 836–842.

Sinton C. W., Duncan R. A., Storey M., Lewis J., and Estrada J. J. (1998) An oceanic flood basalt province within the Caribbean plate. *Earth Planet. Sci. Lett.* **155**, 221–235.

Sinton C. W., Sigurdsson H., and Duncan R. A. (2000) Geochronology and petrology of the igneous basement at the lower Nicaraguan Rise, Site 1001 In *Proceedings of the Ocean Drilling Program, Scientific Results* 165 (ed. P. Garman). Texas A & M University, Ocean Drilling Program, College Station, TX, United States, pp. 233–236.

Skulski T. and Percival J. A. (1996) Allochthonous 2.78 Ga oceanic plateau slivers in a 2.72 Ga continental arc sequence: vizien greenstone belt, northeastern Superior Province, Canada. *Lithos*, **37**, 163–179.

Smith H. S. and Erlank A. J. (1982) Geochemistry and petrogenesis of komatiites from the Barberton greenstone belt. In *Komatiites* (eds. N. T. Arndt and E. G. Nisbet). Allen and Unwin, pp. 347–398.

Stein M. and Goldstein S. L. (1996) From plume head to continental lithosphere in the Arabian-Nubian shield. *Nature* **382**, 773–778.

Stein M. and Hofmann A. W. (1994) Mantle plumes and episodic crustal growth. *Nature* **372**, 63–68.

Stern R. A., Syme E. C., and Lucas S. B. (1995) Geochemistry of 1.9 Ga MORB- and OIB-like basalts from the Amisk collage, Flin Flon Belt, Canada: evidence for an intra-oceanic origin. *Geochim. Cosmochim. Acta* **59**, 3131–3154.

Storey M., Saunders A. D., Tarney J., Gibson I. L., Norry M. J., Thirlwall M. F., Leat P., Thompson R. N., and Menzies M. A. (1989) Contamination of Indian Ocean asthenosphere by the Kerguelen-Heard mantle plume. *Nature* **338**, 574–576.

Storey M., Mahoney J. J., Kroenke L. W., and Saunders A. D. (1991) Are oceanic plateaus sites of komatiite formation? *Geology* **19**, 376–379.

Storey M., Mahoney J. J., Saunders A. D., Duncan R. A., Kelley S. P., and Coffin M. F. (1995) Timing of hot spot-related volcanism and the breakup of Madagascar and India. *Science* **267**, 852–855.

Sun S.-S. and McDonough W. F. (1989) Chemical and isotope systematics of oceanic basalts: implications for mantle composition and processes. in *Magmatism in the Ocean Basins,* Geological Society of London, Special Publication 42 (eds. A. D. Saunders and M. J. Norry), pp. 313–345.

Tardy M., Lapierre H., Struik L. C., Bosch D., and Brunet P. (2001) The influence of mantle plume in the genesis of the Cache Creek oceanic igneous rocks: implications for the geodynamic evolution of the inner accreted terranes of the Canadian Cordillera. *Can. J. Earth Sci.* **38**, 515–534.

Tatsumi Y., Shinjoe H., Ishizuka H., Sager W. W., and Klaus A. (1998) Geochemical evidence for a mid-Cretaceous superplume. *Geology* **26**, 151–154.

Tatsumi Y., Kani T., Ishizuka H., Maruyama S., and Nishimura Y. (2000) Activation of Pacific mantle plumes during the Carboniferous: evidence from accretionary complexes in southwest Japan. *Geology* **28**, 580–582.

Tejada M. L. G., Mahoney J. J., Duncan R. A., and Hawkins M. P. (1996) Age and geochemistry of basement and alkalic rocks of Maliata and Santa Isabel, Solomon Islands, southern margin of Ontong Java Plateau. *J. Petrol.* **37**, 361–394.

Tejada M. L. G., Mahoney J. J., Neal C. R., Duncan R. A., and Petterson M. G. (2002) Basement geochemistry and geochronology of Central Malaita, Solomon Islands, with implications for the origin and evolution of the Ontong Java Plateau. *J. Petrol.* **43**, 449–484.

Thirlwall M. F., Graham A. M., Arculus R. J., Harmon R. S., and Macpherson C. G. (1996) Resolution of the effects of crustal assimilation, sediment subduction and fluid transport in island arc magmas: Pb–Sr–Nd–O isotope geochemistry of Grenada, Lesser Antilles. *Geochim. Cosmochim. Acta* **60**, 4785–4810.

Thompson R. N., Dickin A. P., Gibson I. L., and Morrison M. A. (1982) Elemental fingerprints of isotopic contamination of Hebridean Palaeocene mantle-derived magmas by Archaean sial. *Contrib. Mineral. Petrol.* **79**, 159–168.

Thompson P. M. E., Kempton P. D., White R. V., Kerr A. C., Tarney J., Saunders A. D., and Fitton J. G. (2004) Hf-Nd isotope constraints on the origin of the Cretaceous Caribbean plateau and its relationship to the Galapagos plume. *Earth Planet. Sci. Lett.* (in press).

Tomlinson K. Y., Stevenson R. K., Hughes D. J., Hall R. P., Thurston P. C., and Henry P. (1998) The Red Lake greenstone belt, Superior Province: evidence of plume-related magmatism at 3 Ga and evidence of an older enriched source. *Precamb. Res.* **89**, 59–76.

Tomlinson K. Y., Hughes D. J., Thurston P. C., and Hall R. P. (1999) Plume magmatism and crustal growth at 2.9 to 3.0 Ga in the Steep Rock and Lumby Lake area, Western Superior Province. *Lithos* **46**, 103–136.

Viereck L. G., Taylor P. N., Parson L. M., Morton A. C., Hertogen J., Gibson I. L., and Party O. S. (1988) Origin of the Palaeogene Vøring Plateau volcanic sequence. In *Early Tertiary Volcanism and the Opening of the NE Atlantic,* Geological Society London Special Publication 39 (eds. A. C. Morton and L. M. Parson), pp. 69–83.

Vogt P. R. (1989) Volcaniogenic upwelling of anoxic, nutrient-rich water: a possible factor in carbonate-bank/reef demise and benthic faunal extinctions. *Bull. Geol. Soc. Amer.* **101**, 1225–1245.

Walker R. J., Storey M. J., Kerr A. C., Tarney J., and Arndt N. T. (1999) Implications of 187Os isotopic heterogeneities in a mantle plume: evidence from Gorgona Island and Curaçao. *Geochim. Cosmochim. Acta* **63**, 713–728.

Weis D. and Frey F. A. (1996) Role of the Kerguelen Plume in generating the eastern Indian Ocean seafloor. *J. Geophy. Res.* **101**, 13831–13849.

White R. S. and McKenzie D. P. (1989) Magmatism at rift zones: the generation of volcanic continental margins. *J. Geophy. Res.* **94**, 7685–7729.

White R. V., Tarney J., Kerr A. C., Saunders A. D., Kempton P. D., Pringle M. S., and Klaver G. T. (1999) Modification of an oceanic plateau, Aruba, Dutch Caribbean: implications for the generation of continental crust. *Lithos* **46**, 43–68.

Wilde P., Quinby-Hunt M. S., and Berry B. N. (1990) Vertical advection from oxic or anoxic water from the pycnocline as a cause of rapid extinction or rapid radiations. In *Extinction Events in Earth History. Lecture Notes in Earth Science* (eds. E. G. Kauffman and O. H. Walliser). Springer, vol. 30, pp. 85–97.

Wilson J. T. (1963) A possible origin of the Hawaiian Islands. *Can. J. Phy.* **41**, 863–870.

Wood D. A., Mattey D. P., Joron J. L., Marsh N. G., Tarney J., and Treuil M. (1980) A geochemical study of 17 selected samples from basement cores recovered at Sites 447, 448, 449, 450, and 453, Deep Sea Drilling Project Leg 59. In *Initial Reports of the Deep Sea Drilling Project* 59 (eds. L. Kroenke, R. Scott, and *et al.*) U.S. Government Printing Office, pp. 743–752.

Woodhead J. D., Eggins S. M., and Johnson R. W. (1998) Magma genesis in the New Britain island arc: further insights into melting and mass transfer processes. *J. Petrol.* **39**, 1641–1668.

Zindler A. and Hart S. R. (1986) Chemical geodynamics. *Ann. Rev. Earth Planet. Sci.* **14**, 493–571.

3.17
Generation of Mobile Components during Subduction of Oceanic Crust

M. W. Schmidt

ETH Zürich, Switzerland

and

S. Poli

Universita di Milano, Italy

3.17.1 INTRODUCTION

Subduction zones are the geotectonic settings where the Earth's mantle is refertilized. Whereas the various magmatic geotectonic settings produce oceanic and continental crust (including the supra-subduction arc magmatism), subduction itself consumes oceanic crust. This recycling process replenishes the mantle with most of the element inventory that otherwise would be, with time, strongly depleted in the mantle. Thus, subduction has a major role in maintaining the Earth's magmatic environments and tectonic style over the geological history. For understanding the recycling process it is necessary to understand the reactions that occur during subduction, and within the subduction setting, in particular those that transfer material from the subducting lithosphere to the mantle wedge.

Prograde metamorphism of subducting oceanic crust causes a series of mineralogical reactions that inevitably result in eclogites that may or may not contain hydrous phases and/or carbonates. An alternation of continuous and discontinuous reactions causes devolatilization, i.e., production of a fluid or, more generally, a mobile phase. The mobile phase is either a low-density fluid, a high-density solute-rich fluid, a silicate melt, or a carbonatite melt. This contribution reviews the reaction mechanisms and conditions resulting in the generation of the various mobile phases and also examines the restite(s) that are subducted to great depths.

In general, four different regimes producing a mobile phase can be recognized in the oceanic crust to depth-equivalents of 10 GPa (Figure 1):

(i) *High dehydration rates* at low-to-medium P, low T (<2.5 GPa, <600 °C) where hydrous phases are abundant and dehydration reactions are often perpendicular (in P–T space) to typical subduction geotherms. All subducted lithosphere goes through this first stage.

(ii) *Medium to low dehydration rates* at medium-to-high P, low T (2.5–10 GPa, 500–850 °C) where hydrous phases are already largely reduced in volume and dehydration reactions are often subparallel to possible P–T paths. In this range fluids become increasingly rich in dissolved matter.

(iii) *Melting* where the amount of melt depends mostly on H_2O-availability and the composition of melts is, in addition, strongly pressure sensitive:

(a) *Flush melting* (1–4 GPa, 650–850 °C) at temperatures between the wet granite solidus and the fluid-absent amphibole, biotite, and phengite melting curves. In this case, additional fluid is provided from underlying dehydrating lithologies.

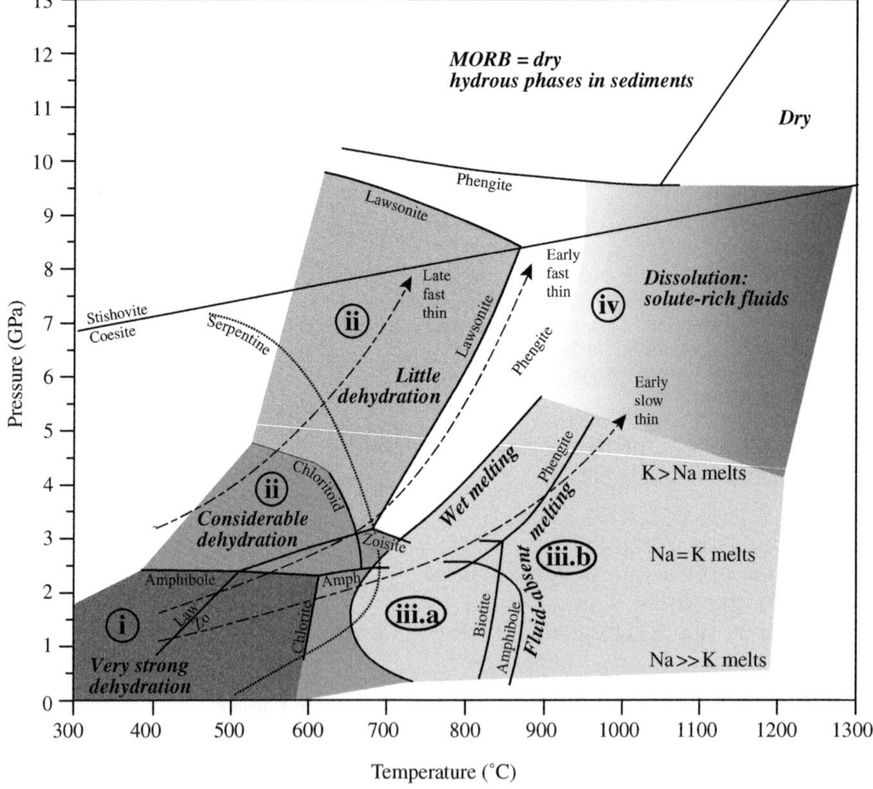

Figure 1 Devolatilization regimes during subduction, mainly based on phase relations in MORB (compare to Figure 3). The P–T region with high dehydration rates is limited by the amphibole stability and the wet solidus, the P–T regions with less dehydration is limited by zoisite and lawsonite stabilities. H_2O-saturated melting takes place in MORB and sediments at the wet solidus, fluid-absent melting to 2.5 GPa is dominated by amphibole (MORB) or biotite (pelite), and above 2.5 GPa by phengite (only in pelites). The classic melting regime is replaced by a continuous dissolution of hydrous phases in a solute-rich fluid at higher pressures (compare to Figure 4). The serpentine stability field in peridotite (dotted line) is given for reference (compare to Figure 5). The stippled arrows represent cold-to-intermediate subduction geotherms (after Kincaid and Sacks, 1997).

(b) *Fluid-absent* melting at high temperatures (>800–900 °C). At relatively low P, amphibole and biotite (1–2.5 GPa and 1–3 GPa, respectively) are the principal hydrous phases to melt. Adakitic magmas, i.e., slab melts (with Na > K), are likely to occur. At medium P (2.5–5 GPa), phengite is the principal hydrous phase to melt, and melts have K > Na.

(iv) *Dissolution* at high *P*, high *T* (>5 GPa, >800 °C) where the solvus between fluid and melt is closed and a solute-rich fluid might dissolve hydrous phases.

In the following discussion some mechanisms for devolatilization reactions are illustrated, including *real-world* aspects such as chemically heterogeneous protoliths and failure to reach equilibrium (compared to a chemically homogeneous protolith and equilibrated ideal situation). We describe and quantify, as far as possible, the four different dehydration/melting regimes, principally investigating three bulk compositions, i.e., pelite, MORB, and harzburgite and H_2O as the major volatile component. Subsequently, carbonates and CO_2 transfer are discussed. Phase petrology is then applied to understand the behavior of trace elements. Finally, we argue for the necessity of integrating fluid/melt-producing processes over the entire oceanic lithosphere, as pressure–temperature conditions of the different lithologies within a given column of subducted lithosphere are strictly related to each other.

The purpose of this chapter is to give an overview of the possible mechanisms responsible for fluid or melt transfer into the mantle wedge, rather than to provide a catalog of minerals (see Table 1) and reactions in subducted lithosphere.

3.17.2 SETTING THE SCENE

3.17.2.1 The Oceanic Lithosphere before Subduction

In the following, we summarize some aspects that are crucial for the production and quantification of mobile phases in the oceanic lithosphere (Figure 2). The unaltered igneous oceanic crust is described in detail in Chapter 3.13, the overlying sediments are described in Chapter 7.01 and alteration processes are described in Chapter 3.15.

A well-stratified oceanic crust is composed of a sedimentary cover layer (including pelites, carbonates, cherts, and volcaniclastics), a basaltic layer built of pillows and sheeted dikes, and a gabbroic layer with an upper layer chemically close to MORB but a lower part mostly composed of differentiated high-magnesium gabbros, Fe–Ti-gabbros, troctolites, norites, and other cumulates (Nicolas, 1989). The crust is underlain by a partly serpentinized, depleted, mostly harzburgitic peridotite (Snow and Dick, 1995).

All sediments have some natural porosity and begin expelling fluids during compaction. They generally remain fluid saturated during prograde metamorphism, and equilibration of major elements can be generally assumed (Carlson, 2002)

Table 1 Major volatile-carrying phases in subduction zones.

Mineral		*Chemical formula*	*H_2O* (wt.%)	*Pelite graywackes*	*Basalts*	*Mg-gabbros*	*Peridotite*
Phengite	phe	$K(Mg,Fe)_{0.5}Al_2Si_{3.5}O_{10}(OH)_2$	4.3	+++	+	−	−
Biotite/phlogopite	bt-phl	$K(Mg,Fe)_{2.8}Al_{1.4}Si_{2.8}O_{10}(OH)_2$	4.1	++	+	−	+
Paragonite	par	$NaAl_3Si_3O_{10}(OH)_2$	4.6	+	++	+	−
K-richterite	K-rich	$KCa(Mg,Fe)_4AlSi_8O_{22}(OH)_2$	2.0	−	−	−	+
Glaucophane–barroisite	amp	$NaCa(Mg,Fe)_3Al_3Si_7O_{22}(OH)_2$–$Na_2(Mg,Fe)_3Al_2Si_8O_{22}(OH)_2$	2.2	+	+++	+++	−
Hornblende–pargasite	amp	$Ca_2(Mg,Fe)_4Al_2Si_7O_{22}(OH)_2$–$NaCa_2(Mg,Fe)_4Al_3Si_6O_{22}(OH)_2$	2.2	−	+++	+++	+
Lawsonite	law	$CaAl_2Si_2O_7(OH)_2{\cdot}H_2O$	11.2	+	++	++	−
Zoisite/epidote	zo/epi	$CaAl_2(Al,Fe^{3+})Si_3O_{12}(OH)$	2.0	+	++	++	−
Chloritoid	cld	$(Mg,Fe)_2Al_4Si_2O_{10}(OH)_4$	7.5	++	+	++	−
Chlorite	chl	$(Fe,Mg)_5Al_2Si_3O_{10}(OH)_8$	12.5	++	+++	+++	+
Talc	tc	$(Mg,Fe)_3Si_4O_{10}(OH)_2$	4.8	++	+	++	+
Talc in Si-rich veins							+++
Serpentine	serp	$(Mg,Fe)_{48}Si_{34}O_{85}(OH)_{62}$	12.3	−	−	+	+++
Phase A	"A"	$(Mg,Fe)_7Si_2O_8(OH)_6$	11.8	−	−	?	+++
Phase E	"E"	$(Mg,Fe)_{2.2}Si_{1.1}O_{2.8}(OH)_{3.2}$	11–18	−	−	?	++
10 Å phase	10A	$(Mg,Fe)_4Si_3O_{10}(OH)_2{\cdot}H_2O$	8–14	Likely	−	Likely	+
Aragonite/calcite	ara/cc	$CaCO_3$		+	+	+	−
Dolomite	dol	$CaMg(CO_3)_2$		+	+	+	+
Magnesite	mgs	$MgCO_3$		+	+	+	+

+ : <5 vol%, ++: 5–20%, +++: >20%.

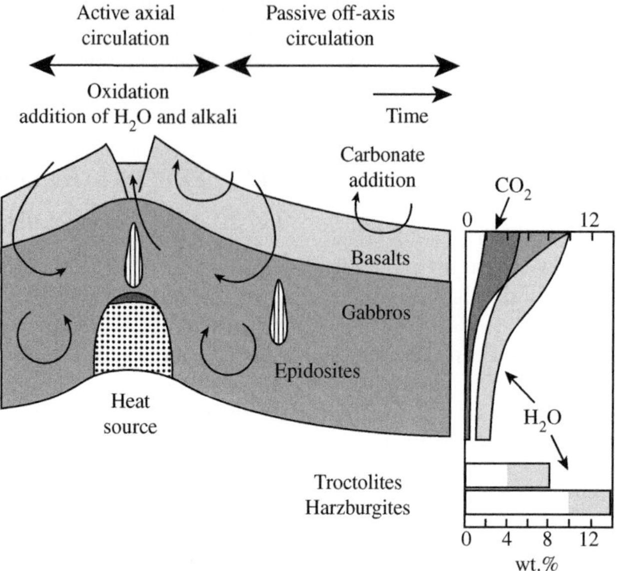

Figure 2 Schematic representation of hydrothermal alteration of the igneous oceanic crust. The inset to the right gives H_2O and CO_2 contents in a depth profile of the oceanic crust, the two horizontal bars indicate typical H_2O contents in troctolites and in intensively serpentinized harzburgite.

as field evidence for prograde disequilibrium is essentially absent. Also, eclogites and blueschists having a basaltic precursor generally appear to be fully hydrated during high-pressure metamorphism, thus assuming fluid saturation and equilibrium during subduction is a reasonable simplification. In contrast, gabbros, which are only partly hydrated (in veins and adjacent alteration zones), may not experience fluid-saturated conditions and may thus fail to reach equilibrium. When pervasive alteration of coarse-grained gabbros takes place, hydration reactions are often limited to grain boundaries. Like the gabbros, serpentinization of peridotite is very heterogeneous. Veins and fractures represent important infiltration pathways for fluids and zones of intensive serpentinization probably alternate with very weakly altered peridotite. Mammerickx (1989) estimated that 20% of the Pacific Ocean floor is affected by fracturing. For both peridotite and gabbros it remains difficult to provide quantitative estimates of the volume affected by hydration. In addition, the amount of hydration/carbonatization also depends on the spreading velocity at the mid-ocean ridges (MORs); oceanic lithosphere produced at slow-spreading ridges appears to undergo higher amounts of hydration than that produced at fast-spreading ridges.

Alteration of the igneous parts of the oceanic crust also forms veins containing carbonates, and although pervasive carbonatization is rare, carbonate contents might be locally high in veins and their immediate surroundings.

The well-stratified oceanic-crust paradigm applies to the circum-Pacific, where 53% of the total length of present Earth oceanic-crust subduction takes place (own compilation). Fast spreading Pacific-type oceanic crust, however, is not found in the Atlantic and Indian oceans where 4% and 16%, respectively, of present subduction occurs (the remainder being subduction of oceanic crust from comparatively small basins). The slow-spreading Atlantic has an oceanic lithosphere that is more complex structurally, and in which lateral heterogeneity is an important feature (Gente *et al.*, 1995; Cannat, 1993). Large transforms offer a preferential site for the emplacement of ultramafics in the shallowest portions of the lithosphere (Constantin, 1999). In the Atlantic it is common to have serpentinites cropping out on the ocean floor; such bodies are rare in the Pacific. The Indian Ocean has a crustal architecture intermediate between the Atlantic and Pacific oceans, it currently subducts in the Macran arc and from Burma to Sunda. Thus, caution is necessary when applying a simple layered oceanic-crust model to infer the geometry of subducted lithologies in the past and in some present subduction zones as, for example, S-Sandwich, Antilles, or Sumatra.

Determining the state of the oceanic lithosphere before subduction is crucial for understanding how the different lithologies evolve. Most of the alteration takes place close to the mid-ocean ridges, however, recent reports show that some hydrothermal activity takes place also in older oceanic crust away from the ridge axis (Kelley *et al.*, 2001). Fracturing at the seafloor and fractures caused by bending of the subducting lithosphere at the trench provide pathways for fluids and hence, further possibilities of hydration. Once the oceanic lithosphere is subducted, continued alteration of the upper layers results from any fluid or melt expelled from deeper levels. Thus, during ongoing subduction the igneous oceanic crust will interact with fluids passing through it from the serpentinized peridotite and the sediments will interact with fluids coming from the basaltic and the serpentinized peridotite layers (Gieskes *et al.*, 1990).

The building blocks of the oceanic crust define the major lithologies involved in devolatilization and melting. To a first approximation, the major players are pelites, basalts, and serpentinized harzburgites. We consider here pelites and carbonates as the main components of the sedimentary layer. From a phase petrological point of view, graywackes and volcaniclastic sediments of broadly andesitic to dacitic composition can be regarded as intermediate between pelites and basalts. In terms of major elements and dehydration, the sediment layer is thin and volumetrically unimportant (with the exception of K_2O). Its major geochemical significance is as a source of incompatible trace elements, which are concentrated in the sediments, and also in contributing to carbonate cycling.

3.17.2.2 Continuous versus Discontinuous Reactions

The production of fluids and melts during subduction (the latter occurring in strictly prograde, relatively low-temperature–high-pressure conditions compared to typical crustal anatepis) is dominated by a succession of continuous and discontinuous reactions that are of comparatively equal importance in terms of fluid productivity in both the sedimentary and mafic layers of the oceanic crust. Discontinuous reactions signify the appearance and/or disappearance of phases, whereas continuous reactions only change composition(s) and proportions of the phases already present.

Extensive solid solutions (amphiboles, micas, pyroxenes, garnet) result in continuous reactions that release fluids over several tens of kilometers depths. This can be exemplified by the disappearance reactions of amphibole (Figure 3(a)) within the amphibole-eclogite facies (Poli, 1993). At the minimum pressure necessary for the formation of omphacite in basaltic bulk compositions (1.5 GPa at 650 °C), more than 50% amphibole remains in the eclogitic assemblage omphacite–garnet–amphibole–epidote–quartz ± paragonite. Within this assemblage, amphibole decomposes progressively, mainly forming omphacite and garnet, until 22% amphibole remains at its upper pressure stability at ~2.3 GPa. Within this pressure interval, amphibole composition changes from calcic and tschermakite-rich to sodic–calcic and barroisite-rich, and the continuous reaction from 1.5 GPa to 2.2 GPa produces more fluid (i.e., 0.7 wt.% H_2O) than the discontinuous terminal amphibole-breakdown at 2.2 GPa (i.e., 0.4 wt.% H_2O), which results in chloritoid as an additional hydrous phase. As in mafic compositions, amphibole decomposition and abundance in peridotites are controlled by continuous reactions that decompose 50–70% of the amphibole present at low pressures before its terminal-pressure breakdown at 2.5–3.0 GPa (e.g., Niida and Green, 1999). These experimental studies demonstrate that continuous reactions and discontinuous terminal breakdown

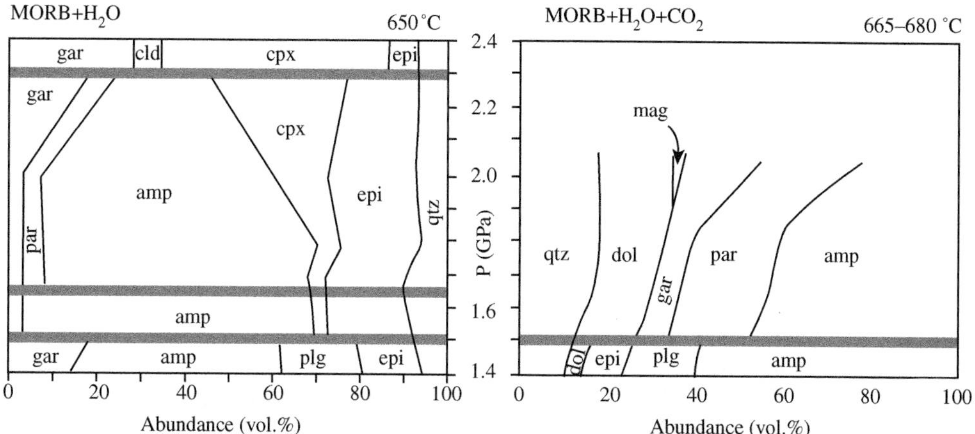

Figure 3 Continuous versus discontinuous reactions contributing to the disappearance of amphibole in (a) H_2O-saturated, CO_2-free MORB (Poli, 1993) and (b) MORB saturated in an H_2O–CO_2-fluid (Molina and Poli, 2000). Modal proportions of minerals in vol.% as a function of pressure, H_2O-contents for each mineral given in Table 1. Note that in both cases continuous reactions produce at least as much fluid (i.e., decompose as much amphibole) as discontinuous reactions (gray bars). In the CO_2-bearing system, amphibole abundances are similar to the CO_2-free system, epidote is replaced by dolomite at 1.5 GPa, and paragonite is more abundant as the appearance of omphacite is retarded to >2.1 GPa. See Table 1 for mineral abbreviations.

reactions are equally important in fluid generation in subduction zones, in contrast to earlier models (e.g., Tatsumi, 1986), which related fluid flow and the position of the volcanic front within a subduction zone directly to amphibole breaking down exclusively at a given depth.

In sediments and the mafic portion of the subducted crust, all of the reactions involving hydrous phases and carbonates involve solid solutions whose compositions depend on the bulk composition, in addition to pressure and temperature. Different bulk compositions cause different phase compositions and thus cause reactions to shift in P–T space, i.e., to start shallower or deeper. In peridotites, amphibole and to some extent chlorite are controlled by continuous and discontinuous reactions; however, the other volumetrically important hydrous phases (e.g., brucite, serpentine, talc, and "phase A") in altered harzburgites display a relatively restricted compositional range, at least compared to those present in mafic eclogites. As a result, breakdown reactions of hydrous phases in harzburgites are dominated (in a first approximation) by discontinuous reactions, and take place over a restricted depth range of only a few kilometers.

3.17.2.3 Fluid Production

There has been a misconception in the literature that hydrous phases break down in the absence of a free fluid phase and thus the stability fields of hydrous phases are not very relevant, as they decompose when the fluid leaves the rock. This is fundamentally wrong: H_2O and CO_2 are chemical species just like any other species (e.g., SiO_2, Al_2O_3, MgO). The only difference is that the phase corresponding to the composition of such chemical species (i.e., H_2O, CO_2, etc.) happens to have a physical state (i.e., fluid) that is different from other phases on composition of chemical species (e.g., quartz, corundum, periclase). It should be remembered that the thermodynamical treatment of all these phases is identical for the entropy and enthalpy terms and that they only differ for the P–V–T (pressure–volume–temperature) relation adopted. Just as SiO_2 saturation is not a prerequisite for the stability of olivine or enstatite, so H_2O saturation is not required for the stability of hydrous phases. In fact, for a given bulk composition it is quite possible (when following a suitable P–T trajectory) to pass from a fluid-absent to a fluid-present regime and back again to a fluid-absent regime.

A fluid is only produced if a given rock volume is already completely hydrated (fluid saturated). If fluid saturation is not realized at the beginning of subduction, a number of fluid-absent reactions will take place. These reactions are of the type $A + V_1 = B + V_2$ (where A, B are volatile free phases and V_1, V_2 are hydrous phases or carbonates), involve hydrates and/or carbonates and change the mineralogy of a rock volume according to the stability fields of the minerals, but do not liberate a fluid. Prograde subduction zone metamorphism (as is true for any type of prograde metamorphism) generally reduces the amount of H_2O that can be stored in hydrous minerals with depth. Thus, almost any part of the oceanic crust sooner or later becomes fluid saturated. In an equilibrium situation, the volatile content bound in hydrous phases and carbonates remains constant until fluid saturation occurs. Either continuous or discontinuous reactions may lead to fluid saturation in a rock. The point at which this occurs depends on initial water content, and pressure and temperature, and somewhat counter-intuitively, initial low water contents do not cause early complete dehydration, but delay the onset of fluid production to high pressures.

Due to heterogeneous alteration (and thus varying initial H_2O and CO_2 contents) there is a wide depth range over which different volumes of the oceanic lithosphere become fluid saturated. A second complication arises from the scale at which equilibrium is effective. A few grains may locally form a fluid-saturated environment, but it is questionable whether the fluid produced on a local grain scale is able to escape. Field evidence argues for equilibration of fluids in eclogites on a centimeter to meter scale until they are able to collect in veins where these fluids might escape (Philippot, 1993; Widmer and Thompson, 2001; Zack *et al.*, 2001; see Chapter 3.06), possibly through fractures in the overlying lithologies without affecting them much, possibly through pervasive infiltration. Thus, considerable uncertainties regarding the quantification of fluid- or melt-producing processes result from the scale of equilibration and the way fluid migrates (Austrheim and Engvik, 1997) through the overlying layers. The ways fluids migrate also influence the phase equilibria and reactions taking place. Most reactions during subduction occur in response to an increase in pressure and temperature, where fluid is produced and expelled. However, the ascending fluid may change the fluid composition in the overlying layers and thus cause reactions.

3.17.2.4 Fluid Availability versus Multicomponent Fluids

In calculations of phase equilibria that appear to be H_2O undersaturated, it is commonly assumed that a fluid phase is present and H_2O activity is lowered by CO_2 in the fluid. The latter is not necessarily true, and this section examines the differences between the limited availability of an aqueous fluid compared to the unlimited availability of a mixed H_2O-CO_2 fluid. Note that both cases are described by the term "H_2O undersaturated."

This can be illustrated by a natural example. In the coarse-grained Allanin magnesium-gabbro, infiltration of fluid caused the formation of reaction rims around olivine (Chinner and Dixon, 1973). The succession is olivine → anthophyllite (2 wt.% H_2O) → talc (4wt.% H_2O + kyanite → chloritoid (8wt.% H_2O+ talc + kyanite. This reaction rim is H_2O undersaturated, and the succession of mineral assemblages corresponds to an increase of H_2O content towards the rim and can only be modeled by an increase in the availability of water towards the rim. The H_2O-undersaturated character of the inner rim zones does not necessitate (or justify) a CO_2 component in the fluid, but rather reflects limited availability of an H_2O fluid.

Thermodynamic calculations based on measured compositions of solid phases (as commonly performed) result in an evaluation of the chemical potential of H_2O in these phases (μ_{H_2O}). Under equilibrium, the chemical potential of H_2O is equal in all phases and only if some additional constraint implies the presence of a fluid phase (e.g., fluid inclusions), the composition of this fluid phase can be calculated from P, T, and μ_{H_2O}. This can be illustrated in the simple system $CaO-Al_2O_3-SiO_2-H_2O-CO_2$ where a given chemical potential of H_2O at a given P and T (4 GPa, 600 °C) may correspond to a fluid-absent situation (with lawsonite + zoisite + aragonite + coesite + kyanite present) or to a situation with a mixed fluid phase (with aragonite + coesite + kyanite + fluid present) (Poli and Schmidt, 1998).

The actual consequence of H_2O undersaturation in the context discussed here is that mixed fluid phases, i.e., solutions, will shift phase equilibria in $P-T-X$ space, whereas the presence or absence of a single component fluid (e.g., a purely aqueous fluid) determines whether or not a reaction takes place, without changing the reaction's position in $P-T$ space.

3.17.2.5 Real World Effects

The compilation of available phase equilibria aims at understanding equilibrium situations in typical (and homogeneous) average bulk compositions. However, different *real-world effects* with amplitudes that may depend on the rock type, are to be expected.

In the real world, the following factors contribute to the continuous character of the devolatilization signal from the downgoing slab: highly variable bulk compositions in the sedimentary and gabbroic layers as well as possibly different degrees of depletion (caused by different amounts of melt extracted) in the hydrated peridotitic layer (Constantin, 1999); heterogeneous distribution of carbonates versus hydrous minerals resulting in an inhomogeneous X_{CO_2} in the fluid phase (Gillis and Robinson, 1990); large temperature gradients within the subducting lithosphere (Kincaid and Sacks, 1997), and finally, possible kinetic effects that inhibit reactions and thus widen reaction zones, which are related both to the effects of fluid availability and deformation history (Austrheim and Engvik, 1997; Molina *et al.*, 2002) and to thermal retardation of sluggish solid–solid transformations (Schmeling *et al.*, 1999).

All of the above effects lead to smearing out the release of a fluid/melt *pulse* over a broader depth range. In contrast, the single focusing mechanism for fluid/melt flow is mechanical: fluids/melts may ascend through channelized flow, hydrofractures or pre-existing fractures, which may focus a broadly distributed fluid into distinct pathways. The relative importance of these mechanisms is largely conjectural, and whether macroscopical fluid/melt focusing is achieved depends on their extent and spacing.

3.17.3 DEVOLATILIZATION REGIMES IN MORB

Based on phase relations in H_2O-bearing MORBs, four distinct $P-T$ regions with characteristic mobile-phase production mechanisms can be identified and are discussed sequentially.

3.17.3.1 High Dehydration Rates and Fluid Production (Typically up to 600 °C and 2.4 GPa)

Once the oceanic crust starts subducting, most of its remnant porosity will be immediately lost by compaction and its pore fluids get expelled. At this stage, zeolites, pumpelleyite, and prehnite are the major H_2O-bearing minerals and H_2O contents stored in hydrous minerals amount to 8–9 wt.% H_2O in the bulk rock (Peacock, 1993). Beyond depths of ~15 km the oceanic crust enters into the blueschist facies in which the major hydrous minerals are chlorite, sodium-rich, calcium-poor amphiboles (glaucophane to barroisite), phengite (white mica), lawsonite or zoisite, and paragonite (e.g., Sorensen, 1986). Water contents of the bulk rock at the beginning of the blueschist facies are ~6 wt.% (Figure 4). Initially abundant chlorite has high H_2O contents (12 wt.% H_2O) and decomposes completely in the depth range to 70 km through various continuous and discontinuous reactions. Lawsonite (11 wt.% H_2O) has a maximum abundance of 25 vol.% at the onset of blueschist-facies metamorphism and decreases to ~10 vol.%

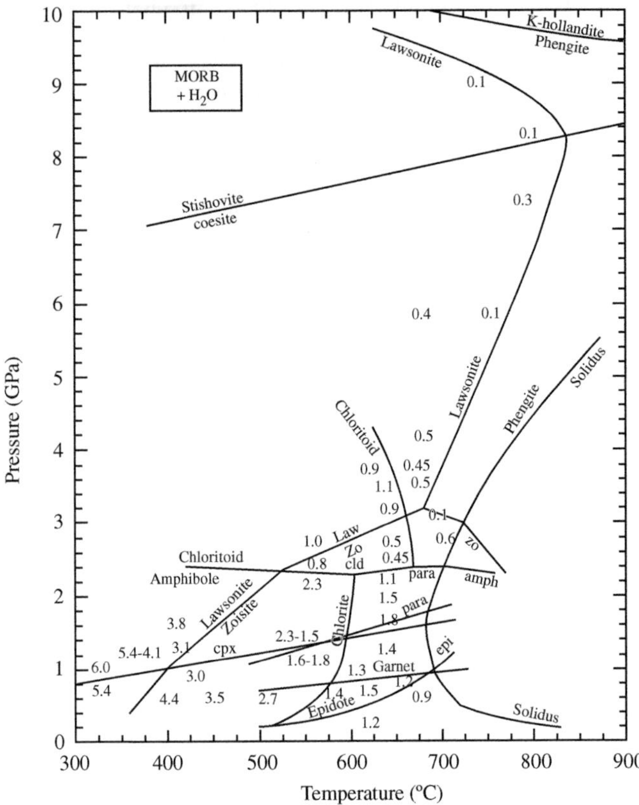

Figure 4 Major phase stability boundaries in H_2O-saturated MORB and H_2O contents (numbers in wt.%) stored in hydrous phases (Schmidt and Poli, 1998). Data below 550 °C are based on natural blueschists and greenschists, all other data are based on experiments (compare with Figure 1).

at the amphibole-out reaction. Anhydrous minerals typically comprise 5–25 vol.% (e.g., Thurston, 1985; Okay, 1980) at 5–10 km depth and some 100 °C, grow in volume to ~50 vol.% just before the amphibole-out reaction, and increase to >70 vol.% at pressures beyond amphibole stability. As a consequence, dehydration rates are considerable within this P–T regime, where a fully hydrated MORB loses ~4–6 wt.% H_2O when passing through the blueschist stage within the fore-arc region.

At pressures of 2.2–2.4 GPa, i.e., the maximum pressure stability of amphibole in MORB (65–70 km), dehydration reactions are numerous and their orientations in P–T-space are mostly oblique to a typical subduction-type P–T path, resulting in high dehydration rates. Although fully hydrated oceanic crust loses about two-thirds of its initial water content in this interval, this fluid will either serpentinize the cold corner of the mantle wedge or eventually pass, through veins, to the ocean floor in the fore-arc region (e.g., Mariana arc, Fryer *et al.*, 1999). The high dehydration rates in conjunction with a—close to the trench—thin mantle wedge should lead to rapid, full serpentinization of the cold corner (i.e., within 0.8–3.0 Ma) if fluids pass pervasively through the mantle wedge (see also Gerya *et al.*, 2002). As a serpentinized cold corner has a much lower viscosity than anhydrous mantle, the serpentinized cold corner may develop a counter-convection cell and stagnate in its corner, whereas the rest of the mantle wedge undergoes large-scale convection (examples of cold corner counter-flow are found in Davies and Stevenson, 1992).

Most information concerning blueschists is obtained from studying natural occurrences; it is only at and beyond the chlorite- and amphibole-out reactions that reaction rates are sufficiently high to study these systems experimentally. Thermodynamic modeling of phase relations in MORB is generally accurate, however, it becomes difficult when incorporating the amphibole solid solution. Both calculated amounts and stabilities of amphiboles are far from what is observed in nature and experiments (e.g., the pressure stability of amphiboles in MORB is calculated to 4–5 GPa instead of the 2.4–2.8 GPa observed in experiments and deduced from natural eclogites), leaving us to rely almost entirely on natural occurrences to deduce relevant dehydration reactions.

The study of blueschists also demonstrates that many of their minerals contain significant amounts of Fe^{3+} (e.g., Maruyama *et al.*, 1986; Brown and Forbes, 1986). The inhomogeneous degrees of oxidation within the altered oceanic crust add another compositional variable, shift reactions in $P-T$-space, and complicate the geochemistry of trace elements with variable oxidation state (e.g., uranium). The fraction of Fe^{3+}/Fe^{tot} is generally highest at the onset of subduction-zone metamorphism and decreases with increasing grade.

3.17.3.2 Low Dehydration Rates and Little Fluid Production (2.4–10 GPa and 500–800 °C)

Until the late 1980s it was believed that oceanic crust is fully dehydrated after the pressure-induced amphibole breakdown. However, experiments (Poli, 1993; Pawley and Holloway, 1993; Poli and Schmidt, 1995; Schmidt and Poli, 1998; Okamoto and Maruyama, 1999; Forneris and Holloway, 2001) and natural occurrences of epidote/zoisite, lawsonite, talc, chloritoid, phengite, staurolite, OH-rich topaz, and many other minor hydrous minerals in eclogites that have already lost their amphibole, clearly demonstrate that dehydration continues above 2.4 GPa. In particular, since the early 1990s, the frenetic chase after coesite- and diamond-bearing hydrated eclogites showed that most of the above hydrous minerals are stable at pressures beyond 2.5 GPa (e.g., the coesite-lawsonite bearing eclogites found in xenoliths of Colorado Plateau, Helmstaedt and Schulze, 1988; Usui *et al.*, 2003). Generally, the natural occurrences of coesite- and sometimes diamond-bearing hydrous eclogites (the most famous being Dora Maira, Norway, Dabie Shan, and the Kokchetav massif, see Liou *et al.* (1998) and Chapter 3.09 for a concise review) are in good agreement with experimental findings, and the discussion below is based on experimental $P-T$ conditions and mineral abundances.

Above 2.4 GPa, a maximum of 1.5 wt.% H_2O remains stored in hydrous phases of the oceanic crust. The major water hosts are lawsonite, zoisite, chloritoid, talc, and phengite. Lawsonite is the most water-rich of these phases (12 wt.% H_2O) and may host >50% of the water present; however it is restricted to relatively low temperatures (Figure 4). Zoisite is stable to ~3.3 GPa and occurs at the wet solidus until 3 GPa. Talc is only a minor phase in MORB but becomes abundant in bulk compositions with high X_{Mg}, i.e., magnesium gabbros. Phengite has an important role, as virtually any K_2O present in the oceanic crust at pressures below 5 GPa will be stored in phengite, irrespective of the bulk composition (Schmidt, 1996). Above this pressure, potassium also enters into clinopyroxene (Okamoto and Maruyama, 1998; Schmidt and Poli, 1998) and the relative amounts of phengite, clinopyroxene, and the coexisting fluid will depend upon the quantity of dissolved potassium and its dissolution rate into the fluid. Above 2.4 GPa continuous reactions dominate over a few discontinuous reactions and most (including the wet solidus) are subparallel to typical subduction $P-T$ paths. As a consequence, a particular rock volume crosses reactions over a wide depth interval. Thus, dehydration rates are rather low, and the (post-amphibole) remnant 1.5 wt.% of H_2O is lost over a wide depth range. The pressure–temperature range that defines this "low dehydration rate" regime corresponds to the depth interval of the subarc region.

The hydrous phases in oceanic crust with the widest pressure stability range are lawsonite and phengite, which both reach into the stishovite field. The breakdown of these two phases corresponds to the *end* of the major devolatilization history of oceanic crust.

Because many reactions are parallel to typical $P-T$ paths during subduction, effective H_2O contents, mineral assemblages, and mineral compositions are fairly sensitive to temperature. A K_2O-free MORB passing 600–650 °C at a depth of 100 km has 1 wt.% H_2O stored in lawsonite and chloritoid and will arrive at 200 km with ~0.4 wt.% H_2O in lawsonite. Passing 100 km at a temperature of 700 °C causes the loss of the last hydrous phase (zoisite) near this depth and our MORB volume becomes a dry eclogite and thus irrelevant for any further production of mobile phases. Now, if the 100 km depth is achieved at a temperature of 750 °C, a MORB is already above the wet solidus and any fluid infiltrating from below should cause partial melting. As is evident from these considerations, the major difficulty in predicting what type and quantity of mobile phase is produced stems from the uncertainty in temperature distribution in the subducting oceanic crust. Unfortunately, a difference in temperature of 100 °C has a significant impact.

3.17.3.3 Melting Regimes (650–950 °C; to 5–6 GPa)

Melting of the oceanic crust is evidenced in so-called adakites which form volcanic suites of andesitic to rhyodacitic composition (Kay, 1978; Drummond *et al.*, 1996). These are interpreted to be partial melts of the oceanic crust more or less modified (mainly magnesium enriched) during their ascent through the mantle wedge. Adakitic suites are distinguished from melts originating from a fluid-metasomatized mantle wedge by distinctive trace-element ratios (i.e., high La/Yb and Sr/Y, review by Martin, 1999), by anomalously high X_{Mg} values and MgO concentrations at comparative SiO_2 contents (thought to be acquired when slab melts absorb olivine on their way through the mantle wedge) and by the absence of volcanic rocks more basic than andesite. Melting of subducted crust was probably predominant on the Archean Earth (Martin, 1999) where mantle temperatures were higher and oceanic plates probably were thinner and hotter. There is consensus that on the modern Earth dehydration dominates in subducting lithosphere. However, when young, hot crust is subducted, or when the thermal field within a subduction zone is disturbed at plate boundaries (i.e., at the lateral end of a subducting plate) or due to flat slabs caused by ridge subduction, melting of the oceanic crust may occur.

Melting of subducted crust can occur under fluid-saturated or fluid-absent conditions. The principal hydrous phases involved are phengite, biotite, epidote/zoisite, and amphibole. If melting of the oceanic crust occurs at pressures below 2.5 GPa, the oceanic crust will go through a greenschist and epidote-amphibolite facies stage at low pressures (both stages not discussed above).

The compositions of slab melts principally depend on the bulk composition, the degree of melting and the pressure of melting. However, it is generally true that at low- to moderate-melt ratios, where cpx is a major residual mineral, Na/K ratios mostly depend on the pressure of melting. At any pressure, potassium strongly partitions into the melt. However, $Na^{cpx/melt}$ partition coefficients are small (<0.5) below 2 GPa, close to unity around 3 GPa and increase to 3–5 at 4 GPa. The latter corresponds to ~50–70% jadeite component in residual clinopyroxene, which retains most of the Na_2O, but leaves all of the K_2O in the melt. As a consequence, melts from amphibole-dominated fluid-absent melting have Na > K, whereas melts resulting from fluid-present or fluid-absent phengite melting at higher pressures are strongly potassic peraluminous granites (with K > Na). The latter, which would migrate into the mantle wedge at >70 km depth are not observed at the surface. This is either because they never form or because it is virtually impossible for them to traverse the thick mantle wedge and retain their peculiar chemistry, which is far out of equilibrium with peridotite. Absorption reactions with mantle minerals would largely modify the major-element composition of these melts, although a sediment-melt contribution is sometimes postulated in arc magmas on the basis of certain trace elements (see Section 3.17.5).

3.17.3.3.1 *Fluid-saturated (flush) melting*

Fluid-saturated melting of basaltic crust begins at temperatures of ~650 °C at 1.5 GPa to ~750 °C at 3 GPa (Figure 5). It should be noted that the wet-solidus temperature is elevated by at least 100–200 °C, if the fluid is in equilibrium with carbonates (an X_{CO_2} of 0.3–0.6 would be expected in the appropriate P–T range). Although a small quantity of free fluid (<0.1 vol.%) is likely to be present in any lithology affected by dehydration, this would not be sufficient to produce a significant melt portion through fluid-saturated melting. However, fluid-saturated melting at relatively low temperatures could be

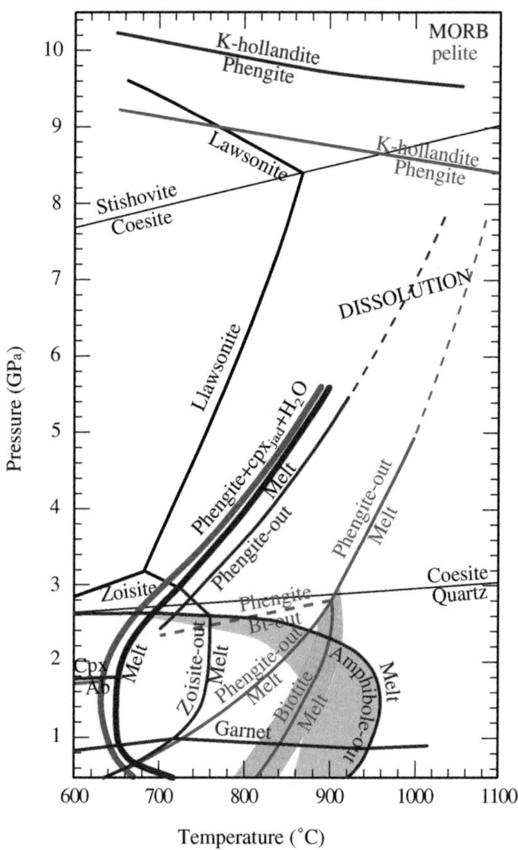

Figure 5 Compilation of melting reactions in MORB and average pelite. Black lines represent reactions with similar P–T locations in both bulk compositions, blue lines: MORB, red lines: pelite. The red and blue areas are P–T fields of biotite-dominated and amphibole-dominated fluid-absent melting respectively. At subduction-zone melting pressures (i.e., >1 GPa), the first fluid-absent melt in pelite appears at the phengite-out reaction and in MORB at the zoisite/epidote-out reaction. The bold lines represent the H_2O-saturated solidus, which involves amphibole or biotite in addition to phengite + cpx + coesite/quartz below 2.5 GPa and plagioclase instead of cpx below 1.5–1.7 GPa. Upto 5–6 GPa classical melting occurs where a solidus separates a crystal + fluid field from a melt-present field (the fluid having distinctively higher H_2O contents than the melt). A continuous increase of solute in the fluid is observed at high pressures (>5.5 GPa) and at least the K-bearing phases (phengite or at higher pressures K-hollandite) dissolve in a fluid of continuously evolving composition.

achieved in subducted crust through flush melting, i.e., by addition of aqueous fluid from below. In this case, the melt productivity depends on the amount of fluid added to the system. The possibilities to obtain volumetrically significant melt fractions in (i) MORB through flushing with fluid originating from dehydration of serpentinized peridotite situated below the MORB or (ii) sediments through flushing with fluids from underlying MORB or serpentinized peridotites are discussed in Section 3.17.5.4.

3.17.3.3.2 Fluid-absent melting

Most adakite suites contain andesites that are consistent with fluid-absent melting of a basaltic source (but not with melting of mica-dominated sediments only). Fluid-absent melting is defined as the production of a silicate melt with low-water contents from an assemblage that contains hydrous phases but no free fluid phase. At pressures below 2.5 GPa, both metapelites and metabasalts contain a pair of hydrous phases with equal water contents (Figure 5): in pelites, phengite (4.3 wt.% H_2O) has a lower modal abundance than biotite (4.1 wt.% H_2O) and also a 100–150 °C lower melting temperature than biotite. In basaltic compositions, epidote/zoisite (2.0 wt.% H_2O, typically 10 vol.%) melts ~100 °C below amphibole (2.2 wt.% H_2O, typically 50–30 vol.%, decreasing with increasing pressure) (Vielzeuf and Schmidt, 2001, and references therein). In both cases, the first fluid-absent melts are produced through the volumetrically less-important phase, i.e., phengite in metapelites and epidote/zoisite in metabasalts. These first melts are thought to be dacitic and probably amount to less than 10 vol.%. In the absence of detailed experimental investigations (at fluid-absent conditions, experimental reaction rates are too sluggish below ~800 °C and equilibration does not take place for the fluid-absent melting of epidote and phengite below 2.5–3 GPa), it is likely that the primary melts of adakites result from 20–35% amphibole-dominated fluid-absent melting of MORB, geochemical modeling is consistent with a garnet + cpx ± amph residue (Martin, 1999). Nevertheless, the temperature distribution in oceanic crust is such that subducted sediments are at higher temperatures than the igneous oceanic crust. Thus, if amphibolites are melted, the overlying sediments must also melt through mica-dominated fluid-absent melting. This complexity needs to be taken into account in geochemical slab-melting models. Fluid-absent melting of amphibolite or metapelite cannot be fully described in a closed system or a system open only to melt extraction. The temperatures necessary for fluid-absent melting of the MORB layer (800–900 °C) are such that the hydrated peridotite below would reach the serpentine stability limit, thus possibly providing a significant amount of fluid. This fluid would strongly increase the melt fractions in amphibolites and metapelites and might be required to produce a sufficient quantity of primary melts that then give rise to adakite suites.

3.17.3.4 Dissolution Regime (>5–6 GPa)

The typical concept of low-density H_2O–CO_2 fluids (with small to moderate amounts of solute) in the sub-solidus and high-density silicate liquids (with typically 1–15 wt.% H_2O dissolved) above the solidus does not apply to the subduction environment at pressures above 5–6 GPa. At higher pressures, a chemical continuum between fluids and melts exists (Boettcher and Wyllie, 1969) and, depending on fluid–rock ratios, a continuous dissolution process leaches hydrophile species out of sediments, basalts, and serpentine.

Considerable attention has focused on the role of high-pressure fluids/melts that exist beyond the *second critical end point* (Ricci, 1951). At crustal pressures and temperatures, a large miscibility gap exists between a low-density aqueous (already supercritical) fluid and a high-density hydrous silicate melt. Consequently, a *fluid* and a *melt* have quite distinct compositions and physical properties (e.g., viscosity, compressibility). This is reflected in the common terminology of dehydration/hydration versus melting reactions. However, at higher pressures, the solubilities of silicates or silicate components in fluid and of H_2O in silicate melts increase. Beyond a certain pressure (depending on the chemical system, between 1.5 GPa and >12 GPa), the miscibility gap between classical fluid and melt shrinks, intersects the solidus (this exact locus is defined as the second critical end point) and a chemical continuum between the two extremes, a dry silicate melt and a pure H_2O-fluid, is possible (see Stalder *et al.*, 2000, figure 1). This implies that a continuum exists between the physical properties of former melt and fluid; it does not imply that a mobile phase with melt chemistry and physical properties characteristic of silicate melts may not exist (or with fluid chemistry and properties). The term *supercritical fluid* is not distinct or adequate enough to describe this phase, as any aqueous/carbonic fluid above the first critical end point (at a few hundred bars and degree centigrade) is already supercritical.

At the conditions described above, the wet solidus vanishes, the concept of melting loses its definition, and solid assemblages continuously dissolve in first, a volatile-rich, and with increasing temperature, silicate-rich *nonsolid* phase.

The exact pressures (at temperatures relevant for subduction) for the disappearance of the wet solidus are somewhat uncertain and strongly depend on the chemical system. In the simple system SiO_2-H_2O, the second critical end point is only at 1 GPa, 1,100 °C (Kennedy *et al.*, 1962), in albite-H_2O it moves to ~1.5 GPa (Stalder *et al.*, 2000), in $CaO-SiO_2-H_2O-CO_2$ it occurs at 3.2 GPa, 500 °C (Boettcher and Wyllie, 1969), whereas in the model ultramafic system $MgO-SiO_2-H_2O$ (MSH) the solidus terminates at ~12 GPa, 1,100 °C (Stalder *et al.*, 2001). In potassium-enriched MORB, and in graywackes and pelites, the miscibility gap closes somewhere between 4 GPa and 6.5 GPa (Schmidt and Vielzeuf, 2001). At 4 GPa, a classical sequence of melting reactions and quenched melts are observed in experiments, whereas at 6.5 GPa, initially abundant phengite becomes less and less abundant until it disappears and a mostly $K_2O-Al_2O_3-SiO_2$-bearing fluid-quench precipitate becomes more and more abundant at grain boundaries. This finding is in agreement with an experimental study on carbonaceous pelites (Domanik and Holloway, 2000), where phengite disappears in a similar fashion at 6.5 GPa and 8 GPa. Thus, there is little doubt that at high pressures the solvus closes and a chemical continuum is realized.

Estimates of the amount of solute in such fluids above 6–7 GPa range from 50 wt.% to 70 wt.% (a molar H_2O : K_2O ratio slightly above unity is in fact sufficient for completely dissolving micas), being close to the solubility of H_2O in water-saturated melts near 3 GPa (solubilities of H_2O in natural silicate melts at higher pressures are unknown). Due to the high solubility of K_2O (and probably other components that are less soluble at low pressures), it is doubtful that much of the subducted potassium and related trace elements could reach great depths. In fact, most of the potassium is likely to be transferred to the overlying mantle wedge, where in a MgO-rich, SiO_2-poor chemical environment it may reprecipitate in phlogopite or potassium-richterite (potassic amphibole) (Sudo and Tatsumi, 1990; Konzett and Ulmer, 1999; Trønnes, 2002) and then be dragged to greater depths in the mantle directly overlying the subducted slab. The little potassium remaining in the oceanic crust after dehydration and leaching is stored in the anhydrous phase potassium-hollandite ($KAlSi_3O_8$), which is stable to >25 GPa.

3.17.4 HOW MUCH H_2O SUBDUCTS INTO THE TRANSITION ZONE?

It is well known that refertilization of the mantle takes place through transfer of fluids or melts from the subducting lithosphere to the overlying mantle wedge, as described above. A portion of the elements transferred into the mantle wedge is partitioned into partial melts, which ultimately form arc volcanism and thus do not refertilize the deep mantle. The residue of oceanic crust remaining from this process is subducted to depths >300 km where it may ultimately be mechanically mixed with mantle material (Allegre and Turcotte, 1986). Dixon *et al.* (2002) deduced from the geochemistry of ocean-island basalts, which are widely believed to contain some recycled oceanic crust in their sources, that dehydration is >92% efficient during subduction.

In order to evaluate how much water is subducted to great depths in the mantle, it is necessary to (i) determine the amount of H_2O stored in peridotite after pressure induced decomposition of serpentine and (ii) understand the state of the oceanic crust at pressures just above the phengite breakdown reaction. Any H_2O stored in oceanic peridotites that pass beyond 220 km depth and any oceanic crust that passes beyond 300 km depth is unlikely to be mobilized within the direct subduction context and will participate in mechanical mixing with deep mantle. Only a few hydrous phases may exist beyond 220 km and 300 km depth in peridotite and oceanic crust, respectively, and temperature stabilities for these phases ("phase A," "phase E," and "phase D" in peridotite: Angel *et al.* (2001), Ulmer and Trommsdorff (1999), Frost (1999), and Ohtani *et al.* (2000); and the hydrous aluminosilicates topaz-OH and "phase egg" in aluminous sediments, Ono (1998)) increase faster than any subduction geotherm, at least to 20 GPa. Furthermore, hydrous phases become much less important at pressures beyond 10 GPa: the volumetrically dominant nominally anhydrous phases (NAMs) such as olivine, wadsleyite, garnet, and cpx (3,000 ppm OH in natural cpx from 6 GPa, 1,000 °C; Katayama and Nakashima (2003); up to 3.3 wt.% H_2O in wadsleyite, Kohlstedt *et al.* (1996)) dissolve considerable amounts of hydrogen at these pressures and become the principal hydrogen reservoirs at greater depths.

Whether a significant amount of water is subducted beyond 200 km in peridotitic compositions depends on the exact $P-T$ path. As can be seen in Figure 6, any serpentine-bearing peridotite descending along geotherms that are cooler than 580 °C at 6 GPa (termed the *choke point* by Kawamoto *et al.* (1995)) will conserve H_2O and form phase A (and subsequently phase E and phase D). In oceanic lithosphere subducting along geotherms that pass between 580 °C, 6 GPa and 720 °C, 7 GPa the 10 Å phase forms upon serpentine breakdown (Fumagalli and Poli, 1999; Fumagalli *et al.*, 2001), and holds 0.6 wt.% H_2O in the peridotite. Subsequent entering into the stability field of phase A at greater depths will not lead to

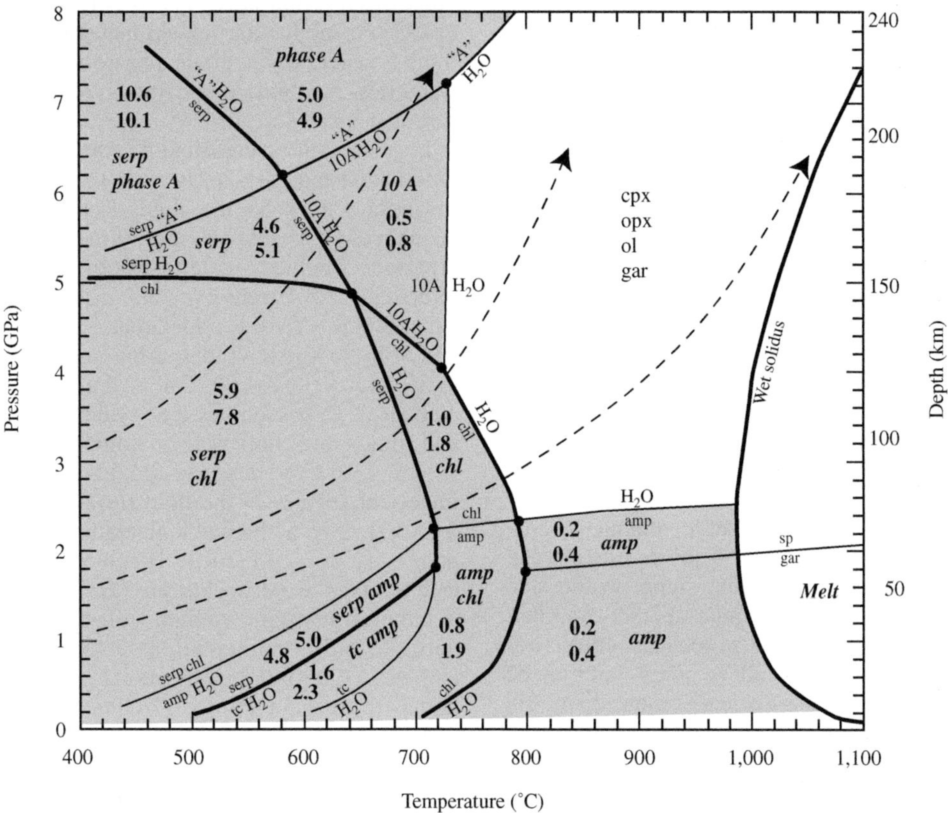

Figure 6 Major phase stability boundaries in H_2O-saturated peridotite and H_2O contents stored in hydrous phases for harzburgite and lherzolite (upper and lower number, respectively; after Schmidt and Poli, 1998 and Fumagalli *et al.*, 2001). The gray field denotes stability fields of hydrous phases. Serpentinized peridotite following the coldest $P-T$-path will not dehydrate up to transition zone depths, as phase boundaries of phase A (and "phase E" at >11 GPa, Frost, 1999) are flatter in $P-T$ space than typical subduction $P-T$ paths.

a significant increase of the bulk H_2O content as most of the free fluid produced from serpentine breakdown would have escaped already.

3.17.5 DEVOLATILIZATION IN SEDIMENTS

3.17.5.1 Pelites

Whereas subduction-zone metamorphism leads to a continuous decrease of H_2O stored in MORB, this is not necessarily true for pelites at pressures up to 2.5–3.0 GPa. By far the most abundant hydrous minerals in metapelites are the potassic micas phengite and biotite. The amount of H_2O stored in micas (containing 9–10 wt.% K_2O and 4–4.5 wt.% H_2O) is easily calculated from the bulk K_2O content. Other important hydrous phases in subduction zone metapelites are talc, chloritoid, and chlorite (see compilation in Poli and Schmidt, 2002). The stabilities of these phases are strongly dependant on the X_{Mg} which varies widely in oceanic pelagic sediments, from 0.2 to 0.8 (e.g., Plank and Langmuir, 1998). Biotite is expected to transform completely to garnet + phengite-bearing assemblages at pressures above 2.5–3.0 GPa (Poli and Schmidt, 2002) but under high X_{Mg} biotite may persist to a maximum of 4 GPa (Hermann, 2002a). In contrast, phengite is stable to considerably higher pressures (up to 8–9 GPa, Domanik and Holloway, 1996; Ono, 1998).

The absolute amount of hydrous phases, and therefore the maximum amount of H_2O in metapelites, is also strongly controlled by the amount of quartz, which is highly variable in mica schists. An average of 9 metapelites (containing between 12 vol.% and 46 vol.% quartz) from the Nome Blueschist Terrane, Alaska (Thurston, 1985) yields 2.7 wt.% H_2O stored in phengite, chlorite, paragonite, epidote, and glaucophane. The same average bulk composition would contain ~2.0 wt.% H_2O stored in the hypothetical assemblage phengite—lawsonite—chloritoid (+gar + cpx + coesite), which is stable near 3 GPa and 600 °C, and 1.1 wt.% H_2O if phengite is the only hydrous phase at higher pressures. Thus, during subduction from 20 km to 80 km depth, 300 m of mica schist would yield only 2–6% of the mass of H_2O released from the underlying 2 km of MORB. These numbers are fairly approximate, as our current knowledge of phase relationships in metapelites metamorphosed at high-pressure, low-temperature conditions is still fragmentary, and further experimental work is required to explore the significance of talc, chlorite, and chloritoid in the critical pressure range between 2 GPa and 4 GPa. Nevertheless, metapelites contain generally less H_2O than MORB in a low-pressure blueschist stage but more H_2O at pressures beyond 2.5 GPa.

Typically, only a few hundred meters of sediments may be subducted, and the quantity of volatiles stored in pelites is small when compared to MORB and serpentinized peridotite. The importance of pelites to the subduction factory lies in their relatively high concentration of K_2O and other highly incompatible minor and trace elements, which may be concentrated in accessory phases (typically rutile, allanite, zircon, phosphates, ellenbergerite). Thus sediments can impart a strong trace-element signal to the slab fluids (or melts), which is quite distinct from that derived from the igneous oceanic crust. For example, if present, allanite (a phase that may be residual during melting) contains more than 90% of the whole-rock LREE and thorium; rutile contains more than 95% of titanium, niobium, and tantalum; zircon contains 95% of the whole-rock zirconium and hafnium, and phengite, at a modal abundance of 20–35% (Figure 8), incorporates more than 95% of the bulk-rock rubidium, barium, and caesium (see also Hermann, 2002b). Finally, as discussed above, pelites also impact melting, as carbonate-free pelites have the lowest melting temperature of all subducted lithologies (see Section 3.17.3.4) and, within the oceanic crust, are situated at the highest temperatures.

3.17.5.2 Carbonates

Most of the discussion above centers on the breakdown of hydrous phases during subduction. The subducted mass of H_2O in most subduction zones is much larger than that of CO_2, and at many trenches, the sediments do not contain any significant carbonates. Near-trench sediment columns have molar $CO_2 : H_2O$ ratios mostly below 1 : 3, and a ratio above 1 : 1 was only found along the middle-American to Peruvian margin (Rea and Ruff, 1996; Plank and Langmuir, 1998; compiled in Poli and Schmidt, 2002). If carbonates are confined to the sediments, they do not alter phase relations in the igneous part of the crust, as fluid flow is generally buoyancy driven away from the igneous crust. However, some carbonate precipitates during hydrothermal alteration (see Chapter 3.15) of the igneous crust. In these instances, carbonate is mainly found in veins and is much less pervasively distributed than H_2O. For a more detailed understanding of fluid-producing processes it is thus necessary to evaluate the effect of CO_2 on subduction-related metamorphism and fluid transfer, on hydrous phase and carbonate stabilities, on fluid compositions, and how deep carbonates can subduct.

In general, the succession of carbonate minerals with increasing pressure is calcite → dolomite → magnesite. X_{CO_2} in the fluid decreases in this sequence. For low temperature P–T paths, aragonite replaces calcite. Both calculations (Kerrick and Connolly, 1998, 2001a,b) and experiments (Yaxley and Green, 1994; Molina and Poli, 2000) show that at low to intermediate temperatures (up to 800 °C at 4 GPa), a very small amount of CO_2 saturates the system in carbonate and fluids are buffered to compositions with ≤10 mol.% CO_2. Carbon dioxide becomes an important species in the fluid (to X_{CO_2} values of 0.4–0.7) only at higher temperatures and comparatively low pressures in coexistence with calcite. Few experiments are available on the maximum pressure stability of carbonates in natural bulk compositions. Domanik and Holloway (2000) found magnesite in a calcareous pelite from 6.5 GPa to 11 GPa. In synthetic systems it is well known that carbonates are very stable at subsolidus conditions: between 5 GPa and 9 GPa dolomite breaks down along a curved reaction line (Luth, 2001) to magnesite and $CaCO_3$ polymorphs, which are stable down to lower mantle pressures (Biellmann *et al.*, 1993).

Carbonates present within the subducting oceanic crust influence dehydration reactions, but contrary to what is intuitively believed, can enlarge the stability field of hydrous phases. This is especially true for melting reactions in mafic and pelitic compositions where a mixed H_2O–CO_2 fluid shifts melting reactions to higher temperatures and thus enlarges the amphibole and mica-stability fields. Carbonate minerals may remain stable in the presence of a siliceous melt and with increasing temperature decompose forming an immiscible carbonatite melt coexisting with a silicate melt.

Molina and Poli (2000) demonstrated the effect of CO_2 on phase relations in mafic compositions (Figure 3). With increasing pressure, the stable carbonates at 665–730 °C are calcite (<1.4 GPa), dolomite, and dolomite + magnesite (>2 GPa). Fluids coexisting with calcite are CO_2 rich ($X_{CO_2} = 0.4$–0.7) but at higher pressures, fluids coexisting with dolomite have much lower CO_2 contents ($X_{CO_2} = 0.02$–0.2). This implies that carbon tends to fractionate into the solid with increasing pressure. The effect of CO_2 on the stability of hydrous phases is surprising: in the carbonated system, plagioclase disappears at the same pressures as in the pure-H_2O system (~1.5 GPa at 650 °C). However, its breakdown does not cause omphacite formation, which is delayed by ~0.5 GPa. Amphibole breaks down at ~2.5–2.6 GPa (Yaxley and Green, 1994; Crottini *et al.*, 2003) and the fluid saturated solidus is located at ~730 °C at 2.2 GPa, as expected by the presence of H_2O-rich fluids at such conditions.

As X_{CO_2} values are low (<0.15) in fluids produced by subduction zone metamorphism, the only efficient decarbonatization processes are either flushing with aqueous fluids from below or melting at relatively high temperatures. A scenario describing decarbonation reactions caused by fluid-infiltration during subduction would be a layer of carbonaceous sediment overlying hydrated oceanic lithosphere. At any time, the metacarbonate would probably contain a very small amount of equilibrated fluid that would have 5–15 mol.% CO_2 at depths beyond 60 km (the exact CO_2-fraction depending on pressure and temperature, Kerrick and Connolly (2001a,b) and Molina and Poli (2000)). Aqueous fluid produced in the serpentinite and basalt below the carbonate layer will migrate upwards, replacing the CO_2-bearing fluid, and then will locally equilibrate, i.e., consume some carbonates in order to increase the CO_2 content in the fluid. If the aqueous fluid passes pervasively through the limestone, the entire carbonate layer will eventually be dissolved and the fluid migrating into the wedge will transport a significant amount of CO_2 over time. For example, 100 m of average siliceous limestone (15.0 wt.% initial CO_2; Plank and Langmuir (1998)) will be completely decarbonated from reaction with aqueous fluids derived from complete dehydration of 150–400 m of serpentinite (ignoring the basaltic crust) or 900–1,600 m of dehydrating MORB (ignoring the peridotite) at depths beyond 60 km, assuming pervasive infiltration and equilibration of the fluid. The entire CO_2 thus derived will flow out of the carbonaceous sediment and into the mantle wedge. However, if the aqueous fluid is efficiently channeled, a very small fraction of the limestone would be affected (depending on channel-spacing) and >90% of the carbonates may survive complete dehydration of the underlying oceanic lithosphere. In this case, these carbonates will be subducted to depths of the transition zone.

3.17.5.3 Graywackes and Volcaniclastics

A significant component of sedimentary columns at trenches are graywackes and volcaniclastic sediments of andesitic to dacitic composition (Plank and Langmuir, 1998). Phase diagrams of graywackes and andesites (Vielzeuf and Montel, 1994; Schmidt, 1993) suggest that, for our present purpose, these systems can be viewed as being intermediate between pelites and MORB. The significant K_2O contents in graywackes and volcaniclastics of intermediate composition lead to abundant micas (phengite or biotite), and thus reactions and phases similar to those in pelites. However, the CaO contents are significant enough for the formation of amphiboles, thus leading to phases and reactions similar to those in intrinsically CaO-rich MORB. Obviously, compared to MORB and pelites, reactions are shifted in $P-T$ space due to different phase compositions, and the amounts of phases are highly variable.

3.17.5.4 Melting of Sediments Compared to Melting of MORB

In hot subduction zones, melting of the downgoing oceanic crust might be achieved. In order to understand which lithologies might melt under what conditions it is necessary to compare melting relations of sediments and MORB under the various fluid-availability conditions. In this context, it should be noted that of all average sediments, a pelagic clay has the highest melt productivity. However, the initial melting temperature of most carbonate free sediments remains virtually identical as long as mica and quartz saturation is maintained. As for MORB, any carbonate addition to sediments results in a mixed H_2O-CO_2 fluid phase and an increase in melting temperatures by at least 100–200 °C. The melting temperatures for MORB and pelitic sediments at water-saturated conditions are similar, the wet MORB solidus being situated typically 20–50 °C above the wet pelite solidus (Figure 5). As pointed out above, fluid-saturated melting only becomes efficient (in terms of melt productivity) if fluid is added from an external source.

In order to achieve flush-melting of the sediments and concomitant dehydration of the MORB layer, a very particular thermal field is necessary: first, the sediments have to pass 650–800 °C at 1.5–3.0 GPa; secondly, the temperature within the lower part of the MORB layer must be 100–200 °C lower than in the sediments in order to provide a significant quantity of fluid through chlorite- and amphibole-decomposition reactions. Within the temperature range where flush melting would occur in the sediments, and considering a *normal* temperature distribution (i.e., temperature decreasing with depth in the oceanic crust), serpentinized peridotite would remain in the chlorite + serpentine stability field and thus not produce any fluid. For the same reason, flush melting of MORB near the wet solidus by fluids derived from underlying serpentinite is highly unlikely in the depth range of interest, as temperatures in the serpentinized peridotite should remain 50–200 °C lower than in the MORB (compare to Figure 1). Thus, conditions necessary for flush melting might be realized in a particular subduction zone; however, flush melting cannot be treated as the general case and is not expected for the fast subduction zones dominant in the Pacific rim. For these, thermal models predict temperatures far too low for any melting in the 1.5–3.0 GPa depth range (Davies and Stevenson, 1992; Kincaid and Sacks, 1997; Gerya *et al.*, 2002).

Fluid-absent melting involving biotite in sediments or amphibole in MORB produces melts with distinct compositions at different $P-T$ conditions (Figure 5), with different pressure dependencies of the melting reactions. The amphibole-melting reaction bends strongly back around 2 GPa and near the solidus, amphibole disappears at 2.4 GPa (Figure 5). In MORB, epidote/zoisite remains stable at the solidus to ~3 GPa, leaving a possibility for minor fluid-absent melting up to these pressures. Any melting at higher pressures depends on the K_2O content of MORB and the related presence of minor phengite. In average metapelites and graywackes, biotite is completely replaced by phengite at 2.5–2.8 GPa (Auzanneau, unpublished experiments, see also Vielzeuf and Schmidt (2001)) but at high Mg/(Mg+Fe) biotite stability extends to ~4 GPa (Hermann, 2002a).

An interesting feature of melting at pressures above 3 GPa arises from the fact that MORB, pelites, and also intermediate andesite or graywacke compositions all contain garnet, cpx, phengite, and coesite (Schmidt, 1996; Okamoto and Maruyama, 1999; Hermann and Green, 2001; Schmidt and Vielzeuf, 2001). Peraluminous graywackes and pelites have kyanite in addition. Thus, all the lithologies of the oceanic crust contain the same assemblage and fluid-saturated melting occurs through the identical reaction: phengite + coesite + cpx + H_2O = melt. If a significant amount of free water is not available, melting occurs through phengite + cpx = garnet + melt ± kyanite. This reaction takes place ~150–200 °C above the wet solidus (at 3 GPa to at least 5 GPa) and leads to 20–30 wt.% melt in the metasediments and to a few percent melt in the mafic rocks (dependent on the bulk K_2O content). In MORB, phengite is immediately consumed upon melting and the temperature must rise by >100 °C to significantly increase melt fractions through the reaction: cpx = garnet + melt (25–30% melt). Nevertheless, temperatures necessary for such fluid-absent, high-pressure melting are high and unrealistic in most subduction zones.

The most likely melting scenarios for subducted oceanic lithosphere strongly depend on the thermal gradient within the subducted crust. In the depth range of interest, most thermal models (e.g., Davies and Stevenson, 1992; Furukawa, 1993; Kincaid and Sacks, 1997; Gerya *et al.*, 2002) predict a temperature difference of ≤200 °C from top of the sediments to the bottom of the crust (which at the same time is the top of the serpentinized peridotite layer). With such a gradient, significant fluid-saturated sediment flush melting cannot take place due to fluids originating from the directly adjacent MORB layer. However, the temperatures necessary to obtain fluids from the serpentinized peridotite cause the MORB layer (intercalated between sediments and serpentinite) to be at temperatures above the fluid-saturated solidus. Thus it appears likely to achieve either no significant fluid-saturated melting at all, or flush melting from both sediments and MORB. For depths of 50–80 km, achieving the temperatures necessary for fluid-absent melting of sediments and MORB results in a temperature that causes dehydration in the serpentinized peridotite layer. This again would enhance melt productivities through flush melting.

3.17.6 SERPENTINIZED PERIDOTITE

The serpentinized peridotite layer situated just below the igneous oceanic crust (or often brought to the surface in slow spreading oceans like the Atlantic) constitutes a major H_2O reservoir in subducted lithosphere, of comparative size to the oceanic crust. It is difficult to estimate the amount of serpentinization within this layer. The only certainty is that the degree of serpentinization is highly variable both on a regional and local scale. Hydrothermal systems near the ridge and transform faults of all dimensions are the primary sites of hydration, which is mostly serpentinization. Recently, hydration has also been suggested to occur in extensional faults that run parallel to the trench and are caused by bending of the subducting plate at the onset of subduction (Peacock, 2001). Our best estimate gives an average of 20% serpentinization to a few kilometer depth (Schmidt and Poli, 1998). Although there might be some localized serpentinization along faults much deeper in the oceanic lithosphere, large-scale serpentinization and the resulting low-density peridotite (2.3–2.5 g cm^{-3}) would cause a buoyancy problem during subduction.

Whereas a number of reactions at intermediate to elevated temperatures are important for hydration of peridotite in the overlying mantle wedge directly adjacent to the top of the oceanic crust (Figures 6 and 7), almost any subduction $P-T$-path will keep the slab in the serpentine stability field to at least >2 GPa (Ulmer and Trommsdorff, 1995). As a consequence, hydrated peridotite in the downgoing lithosphere will remain as serpentine and chlorite (+olivine + clinopyroxene) while a multitude of reactions is taking place in the oceanic crust above. Thus, the serpentinized peridotite layer of the oceanic lithosphere does not produce any significant fluid up to pressures of 3–6 GPa, where, depending on temperature, serpentine breakdown may occur (Figure 6). The H_2O contents of Figure 6 are calculated for average harzburgite. However, oceanic alteration does not only add H_2O, but also removes MgO. Peridotite sitting just above the slab surface experiences alteration by fluids derived from generally quartz-saturated sediments and MORB and thus is likely to become somewhat SiO_2 enriched (Manning, 1994). If the MgO/ (MgO + SiO_2) ratio in the peridotite is shifted from lying between olivine and serpentine to a value between serpentine and talc, then talc becomes stable to higher pressures (talc + serpentine reaction; Ulmer and Trommsdorff (1999)). In monomineralic veins, talc might persist to its maximum-pressure stability limit at 4.5–5.0 GPa (Pawley and Wood, 1995).

Subduction zones can be divided into two types: those where the serpentine-out reaction is crossed below 6 GPa (and major dehydration of serpentinized peridotite occurs), and those where the serpentine stability boundary is crossed at >6 GPa. As discussed above, almost no fluid production is expected when serpentine reacts to phase A (see Figures 6 and 7) and the H_2O stored in the hydrated peridotite is expected to subduct deeply. In contrast, if the stabilities of serpentine and phase A do not overlap along a given slab geotherm, a significant flushing zone (over 20–30 km depth) is expected at the transition between serpentine-chlorite peridotite → chlorite peridotite → anhydrous garnet peridotite. The width of such a zone is controlled by the actual stability of chlorite, which extends ~100 °C higher than the stability of serpentine in complex systems approaching natural rocks (Fumagalli, 2001).

This picture is somewhat modified when the $P-T$-path produces the sequence serpentine → 10 Å phase → phase A. In this case, a moderate amount (max. 0.8 wt.%) of H_2O subducts to great depth, while >75% of the initially subducted H_2O of the serpentinized peridotite is lost via dehydration.

In a number of arcs (NE Japan–Kuriles–Kamchatka, Aleutians, N. Chile), so-called double seismic zones are observed. Whereas the upper seismic zone correlates with the oceanic crust, it has been suggested (Seno and Yamanaka, 1996; Peacock, 2001) that the lower seismic zone corresponds to the limit of serpentine stability in the lower part of the oceanic lithosphere (see Figure 7). It has been argued that the lower seismic zone earthquakes are triggered by reactivation of ancient faults through fluid saturation, where the fluids derive from serpentine dehydration.

3.17.7 IMPLICATIONS FOR TRACE ELEMENTS AND AN INTEGRATED VIEW OF THE OCEANIC LITHOSPHERE

Trace elements and isotopes such as ^{10}Be, B, Li, Ba, Cs, and elements of the U–Th-series provide important information on main element-transfer processes in arcs and are reviewed by Morris (see Chapter 2.11). Such studies allow one to distinguish between a sediment and an altered MORB signal in island arc volcanics (Morris *et al.*, 1990), to deduce across-arc variations as a function of slab depth (Ishikawa and Nakamura, 1994; Ryan *et al.*, 1995; Moriguti and Nakamura, 1998), yield time constraints on element transfer (Sigmarsson *et al.*, 1990; see Turner *et al.*, 2003, for review), and allow one to distinguish between dehydration and melting processes (Sigmarsson *et al.*, 1998; Martin, 1999). However, the correct interpretation of their concentrations and spatial distributions rely on major element phase relations and on the understanding of mobile phase production discussed in this chapter.

3.17.7.1 Mobile Phase Production and Trace-element Transfer

Dehydration, or more generally, devolatilization of the oceanic crust is a process that combines continuous and discontinuous reactions in a variety of heterogeneous bulk compositions. In addition, within a vertical column—the sedimentary, mafic, and serpentinized peridotite layers—each experience a significant thermal gradient. The result is a continuous, but not constant, production of a fluid or melt, with the rate of mobile phase production generally decreasing with depth. Peaks in the volatile flux result from significant discontinuous reactions. However, despite the continuous fluid flux, trace elements may not necessarily be released continuously.

All dehydration reactions in oceanic lithosphere take place at temperatures where diffusion rates in most minerals are insignificant compared to the available time span for fluid production in subduction zones. In terms of trace-element partitioning it is thus necessary to distinguish between the mineral mode in a given composition and the reactive volume, which will be much smaller. For example: a MORB at 4 GPa, 700 °C has 48 vol.% garnet, 39 vol.% cpx, 5 vol.% lawsonite, 2 vol.% phengite, and 6 vol.% coesite. The breakdown of lawsonite can be modeled via the reaction lawsonite + clinopyroxene + $garnet_1$ = $garnet_2$ (grossular enriched) + H_2O, which produces ~8 vol.% garnet in the rock (Schmidt and Poli, 1998). The garnet that grows from the breakdown of lawsonite is in equilibrium with the fluid. However, diffusion rates at 700 °C

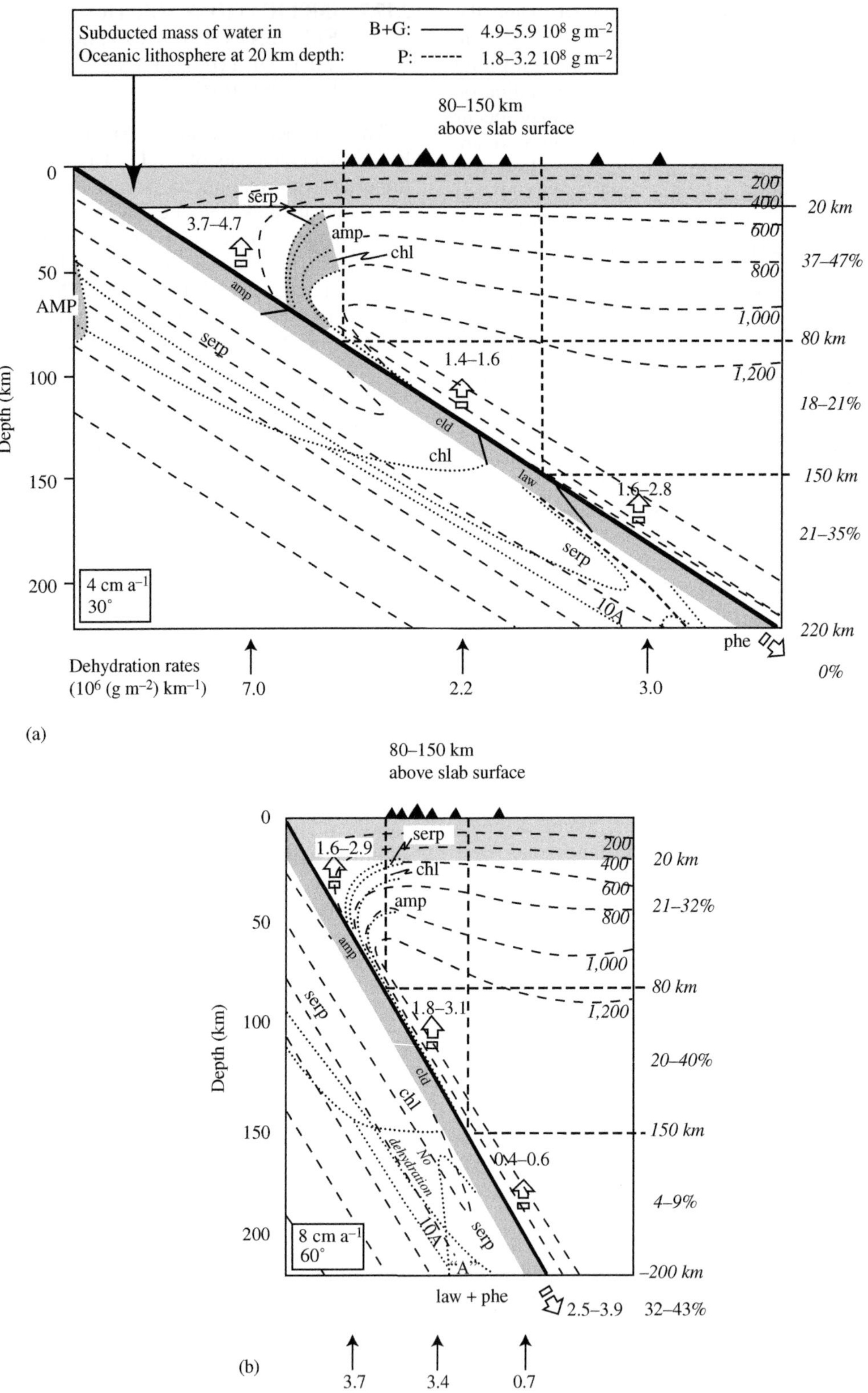

Figure 7 Stabilities of hydrous phases and masses of H_2O in a unit section (in 10^8 g m^{-2}) for two typical temperature fields during intra-oceanic subduction (a) for an intermediate burying velocity (4 cm yr^{-1} convergence rate and 30° dip angle of the slab) and (b) a fast and cold subduction zone (8 cm yr^{-1} convergence rate and 60° dip angle; isotherms from Furukawa, 1993). Dark gray: oceanic crust, light gray: lithosphere of the overriding plate in the Furukawa (1993) model. The dotted lines denote maximum stability fields and not effectively hydrated peridotite. In a subduction zone with a fast subduction rate, serpentine and phase A fields overlap, preventing significant dehydration

for garnet and cpx are so slow that in the available time span (several tens of thousands of years, based on radioactive disequilibrium, see review by Turner *et al.*, 2003) the volumes of the non-reacting garnet and cpx affected by diffusion are negligible (<0.1 vol.%). Thus, the trace elements formerly residing in lawsonite and the 3 vol.% of cpx that decomposed with lawsonite, will be redistributed between the newly formed garnet and fluid. However, >85 vol.% of the rock is not equilibrated and thus should not be included when calculating geochemical residua or the trace-element content of the fluid. As a consequence of slow solid-state diffusion, in most cases the only elements that may be mobilized are those that are hosted in minerals that decompose. It is thus necessary to establish trace-element residence in subduction-zone lithologies (Figure 8).

The following cases may be distinguished:

- A given trace element, for example, boron (Ryan *et al.*, 1995), is extremely soluble in the fluid and any mineral/fluid-partition coefficient is ≪1. Such elements may show a continuously decreasing concentration in the fluid with increasing depth. Concentrations might already be very low in fluids produced at moderate depths due to a shallow effective removal from the subducting crust.
- A given trace element has partition coefficients close to unity and thus its concentration doesn't change in the fluid as subduction progresses.
- A given trace element is strongly partitioned into a particular hydrous phase (e.g., cerium, strontium into lawsonite or epidote; Figure 8) and has only moderate-to-low cpx/fluid and garnet/fluid partition coefficients. These elements will quantitatively enter into the fluid at the breakdown reaction(s) of the given mineral and cause a variation in their concentrations that is not at all proportional to the fluid-flux.
- A given trace element (e.g., barium, and to some extent beryllium in the sediments) partitions into mica and as mica is dissolved away with increasing depth and temperature, enters into the fluid only at greater depth, its flux at low temperature/pressure being small.
- A given trace element strongly partitions into cpx/garnet and will be returned to replenish the deep-mantle trace-element reservoir.

3.17.7.2 Integrating Fluid Flux over the Entire Subducted Oceanic Crust: An Example

The layered structure of oceanic lithosphere originating from fast-spreading ridges, combined with temperature gradients within the subducting lithosphere define the relationship between the amount and depth range of fluids and/or melts generated in the sedimentary and mafic layers and fluids generated from the ultramafic layer. The interdependence between sediment and

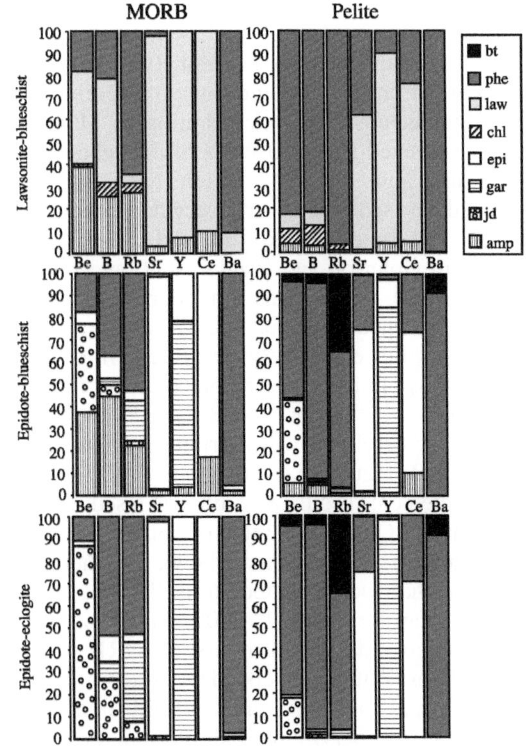

Figure 8 Distribution of Be, B, Rb, Sr, Y, Ce, and Ba between minerals of average MORB and pelite in blueschist and eclogite facies (employing representative mineral modes for natural blueschists and experimental epidote-eclogite, and trace element concentration data mostly from Domanik *et al.*, 1993). At subsolidus temperatures, diffusive equilibration is ineffective (except for micas) and the equilibrating volume that needs to be taken into account for trace-element modeling is defined by the reacting minerals. Thus, a given trace element equilibrates with the fluid only when its host phase(s) break(s) down.

---

of the serpentinized peridotite of the oceanic lithosphere. In such a thermal situation, a thin layer of hydrated peridotite is stable above the oceanic crust and drags some H_2O downwards. Details of the calculations: the input is set to H_2O-amounts at 20 km subduction depth (upper left values), the oceanic lithosphere is composed of 2 km of fully hydrated basalt, 5 km of gabbro hydrated to 20 vol.% and 5 km of serpentinized peridotite with 10–20 vol.% hydrated. Labels on open arrows and percentages at the right side are masses of H_2O (fluxing the mantle wedge) integrated over the depth ranges 20–80 km, 80–150 km, and 150–220 km, and the mass (10^8 g m^{-2} of subducted crust) of H_2O subducting beyond 220 km. The arrows below (a) and (b) give dehydration rates in 10^6 g m^{-2} per (vertical) kilometer of subduction at 50 km, 100 km, and 180 km depth. The highest uncertainty in terms of H_2O masses is in the estimate of initially subducted H_2O in the peridotite of the oceanic lithosphere.

MORB melting and dehydration in serpentinized peridotite was illustrated in Section 3.17.5.4. Here, an example of the effects on trace-element transfer as a function of fluid : rock ratios is illustrated. The layered structure of the oceanic lithosphere may cause the fluid : rock ratio in the sedimentary layer to be greater than one (as fluids from the underlying mafic and peridotitic layers must rise through the sediments). Thus, some of the trace elements (e.g., beryllium, thorium) commonly considered to be only efficiently mobilized by melts could also be quite effectively mobilized by fluids, if the entire subducted lithosphere is considered.

Recently, an apparent contradiction was put forward to argue for melting of sediments contemporaneously with dehydration of MORB. It was estimated that >30–40% of the subducted beryllium and thorium, which are strongly enriched in sediments, are recycled into the mantle and extracted to the surface via arc volcanism (Johnson and Plank, 1999, and references therein). At the same time, boron, which is strongly enriched in altered MORB, and uranium appear to be effectively recycled into arc magmas by fluids. Based on bulk partition coefficients ${}^{xtl/fluid}D$ of 2–4.8 and ${}^{xtl/melt}D$ of 0.7–1.5 for beryllium and thorium, Johnson and Plank (1999) argued that recycling of >30–40% of these elements is only possible when sediments melt while MORB dehydrates (mobilizing boron and uranium). It was then shown by thermal modeling that such a temperature distribution is possible (Van Keken *et al.*, 2002). However, this scenario considers sediments, MORB, and hydrated peridotite as independent systems. Considering that the bulk crystal-fluid partition coefficients for beryllium and thorium are not extremely high, i.e., 2–5, and considering that most of the beryllium partitions into phengite, and that phengite is the only phase where diffusive re-equilibration at sub-solidus temperatures is possible, it follows that an elevated fluid/rock ratio can dissolve >80% of the beryllium (and probably thorium) in the sediments, if pervasive fluid infiltration from the lower layers occurs. Employing the above partition coefficients, the fluid produced from an average thickness MORB layer through dehydration below 3 GPa would be sufficient for leaching 60% of beryllium (and possibly thorium) out of 200 m of pelagic sediments. Thus, sufficient beryllium could be mobilized before typical melting depths are reached. This estimate is conservative because partition coefficients for relatively low-pressure fluids are employed. Solute-rich, high-pressure fluids and continuous dissolution of phengite would greatly facilitate the transfer of beryllium to the mantle wedge. The trace-element transport capacities of such solute-rich fluids are expected to increase significantly compared to low-pressure, low-density fluids; however, the little experimental data that are available are not enough to suggest this. In fact, significant mobilization of beryllium could simply be taken as evidence that micas dissolve away in a dissolution regime.

3.17.8 CONCLUSIONS AND OUTLOOK

The complexity of natural processes—i.e., heterogeneous bulk compositions, heterogeneous volatile distribution, equilibrium on different scales, and kinetic effects—may fairly complicate individual subduction zones. Nevertheless, there is no reason for pessimism. The dehydration behavior of the two volumetrically predominant lithologies, basalt and peridotite, are fairly predictable. The resulting mobile component will then have a sediment signal added. This signal can be reasonably well defined when the subducting column of sediments is well known; there is ample geochemical evidence that the efflux in arc magmas is fairly proportional to the influx in terms of subducted sediment component (e.g., Plank and Langmuir, 1993; Morris, 2003; Turner *et al.*, 2003).

The complexity of natural processes in subducting slabs is reflected by the complex distribution of volcanic emissions in subduction zones. First, it should be emphasized that 69% of modern subduction zones on the Earth (40,900 km, unpublished compilation) show active volcanism in the Quaternary. The rest (18,500 km) are not volcanic because of either unfavorable thermo-mechanical environments (e.g., flat slabs, initiation of subduction, etc.) or possibly the lack of a sufficient amount of volatiles released at depths where melting could take place.

Even though we believe that there is no straightforward relationship between the location of fluid release in the slab and volcanic emissions, the variability of reaction patterns illustrated in this chapter is recorded by variability in the distribution of volcanic arcs on the Earth's surface (Figure 9). The spatial onset of volcanism, the arc, is probably primarily controlled by the thermal structure in the mantle wedge, i.e., when a sufficient thickness and convection intensity in the mantle wedge is reached in order to allow temperatures necessary for the formation of primitive arc magmas (Kushiro, 1987; Schmidt and Poli, 1998). The depth of the slab below the volcanic front is often regarded as a relevant parameter to characterize petrologic processes occurring in subduction zones (Gill, 1981; Tatsumi, 1986). This parameter has often been expressed as single-valued with some sort of standard deviation (128 ± 38 km Gill, 1981; 110 ± 38 km Tatsumi, 1986; 108 ± 14 km for the trench side of the chain, 173 ± 12 km for the back-arc side of the chain, Tatsumi and Eggins, 1995) on the basis of a fairly arbitrary *volcano counting* in selected arcs. On the contrary, our compilation is based on recent geophysical acquisitions (slab surface tomography) and on the spatial extent of volcanic activity and shows that, although some maximum is found at ~100 km depth, a continuum in "depth of the slab surface below the volcanic front" is observed (Figure 9(a)). Moreover, despite a majority of volcanic arcs being fairly well focused on the surface (less than 50 km wide), volcanic activity over more than 100 km arc width is not unusual (Figure 9(b)). Such an arc width corresponds to a comparative depth range of the slab surface. In some continental arcs (e.g., Sumatra, the Southern Volcanic Zone in the Andes), a narrow volcanic arc is the result of a position of virtually all volcanoes on a major fault subparallel to the trench. In such a case, it appears likely that the volcano distribution is not directly related to

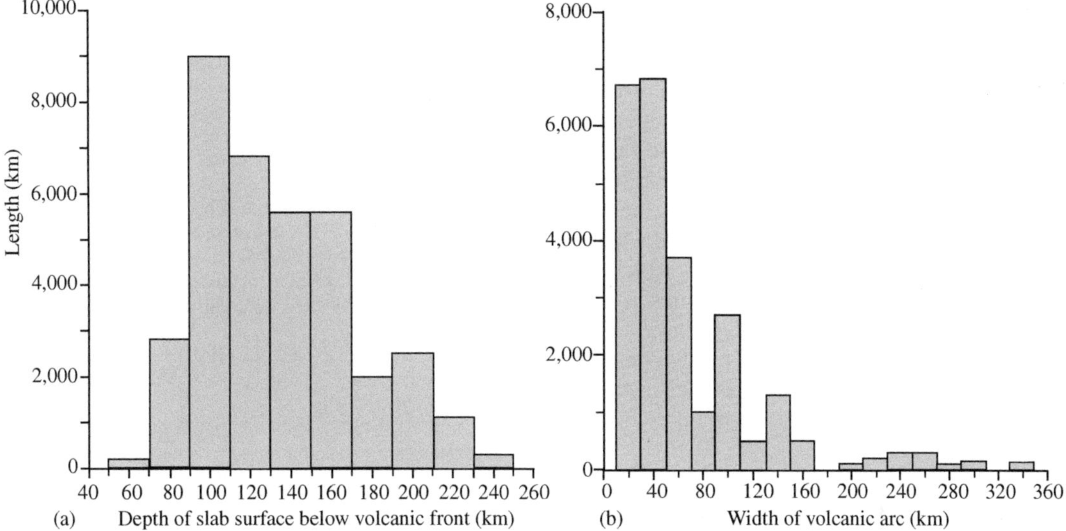

Figure 9 Subduction zone statistics: histograms of depths of (a) the slab surface below the volcanic front and (b) the width of volcanic arcs. The vertical axis denotes arc lengths in km measured at the trench. This is our own compilation (unpublished) based on locations of quaternary volcanoes and slab surfaces from tomography (if available), otherwise earthquake depths.

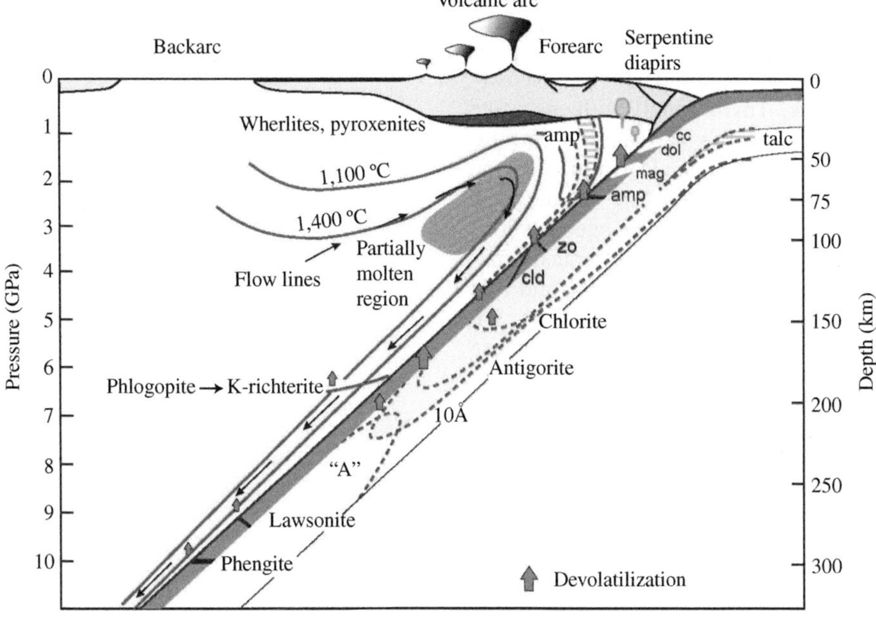

Figure 10 Cartoon of events in a typical cold subduction zone in which melting of the oceanic crust does not occur (after Poli and Schmidt, 2002).

processes taking place below the lithosphere of the overriding plate.

It is beyond the scope of this review to discuss in detail the statistical parameters of subduction zones, and Figure 10 is intended to demonstrate that conditions attained in subducting slabs are highly variable, even for similar convergence parameters, and that interplay between thermo-mechanical properties and reaction paths are responsible for a complex pattern of fluid release and magma genesis. Furthermore, any correlation of kinematic subduction parameters with volcano location tacitly assumes steady state, which is not necessarily the case.

At present, the resolution of thermal models for the oceanic crust with its large temperature gradient is not sufficient. This is because the thermal field strongly depends on the flow field (and on the degree of mechanical coupling between the subducting slab and the down-dragged mantle wedge), which in turn depends on the viscosity, which is a function of P and T and the materials present. The latter then depend on the chemical reactions taking place, which in turn depend on the temperature field. A temperature distribution model is thus a fairly complex problem and needs input from a large variety of disciplines.

An example of feedback between temperature field and phase petrology is serpentinization in the mantle wedge directly overlying the subducted slab (Gerya *et al.*, 2002). First, the calculated temperature distribution from thermomechanical models results in a pressure–temperature region where serpentine would be potentially stable. Second, dehydration rates and fluid transport mechanisms (pervasive versus channeled flow) allow one to model the amount of serpentine formed. However, the serpentinized peridotite has four to six orders of magnitude lower viscosity than dry peridotite. This strongly influences the coupling of the downgoing slab and the convection of the mantle wedge. It also influences the possible amount of shear heating at the slab surface. If convection patterns change due to a modified rheology caused by phase transformations, the temperature field changes and the permissible region for serpentine changes. We thus need more complex models in which these parameters are varied simultaneously.

Upwards directed interaction between the different lithologies largely depends on the temperature gradient within the oceanic lithosphere, which is among the most difficult parameters to model. Thus, at present, a purely forward model for mass transfer in a given subduction zone is not reliable, geochemical information on the subduction output (see Chapters 2.11 and 3.18) is necessary to constrain likely mass-transfer processes.

REFERENCES

Allegre C. J. and Turcotte D. L. (1986) Implications of a 2-component marble-cake mantle. *Nature* **323**, 123–127.

Angel R. J., Frost D. J., Ross N. L., and Hemley R. (2001) Stabilities and equations of state of dense hydrous magnesium silicates. *Phys. Earth Planet. Inter.* **127**, 181–196.

Austrheim H. and Engvik A. (1997) Fluid transport, deformation and metamorphism at depth in a collision zone. In *Fluid Flow and Transport in Rocks* (eds. B. Jamtveit and B. W. D. Yardley). Chapman and Hall, London, pp. 123–135.

Biellmann C., Gillet P., Guyot F., Peyronneau J., and Reynard B. (1993) Experimental evidence for carbonate stability in the Earth's lower mantle. *Earth Planet. Sci. Lett.* **118**, 31–41.

Boettcher A. L. and Wyllie P. J. (1969) The system CaO–SiO_2–CO_2–H_2O: III. Second critical end-point on the melting curve. *Geochim. Cosmochim. Acta* **33**, 611–632.

Brown E. H. and Forbes R. B. (1986) Geology of the shuksan suite, North Cascades, Washington, USA. In *Memoir 164, Blueschists and Eclogites* (eds. B. W. Evans and E. H. Brown). Geological Society of America, pp. 143–167.

Cannat M. (1993) Emplacement of mantle rocks in the seafloor at mid-ocean ridges. *J. Geophys. Res.* **98**, 4163–4172.

Carlson W. D. (2002) Scales of disequilibrium and rates of equilibration during metamorphism. *Am. Mineral.* **87**, 185–204.

Chinner G. A. and Dixon J. E. (1973) Some high-pressure paragenesis of the allalin gabbro, Valais, Switzerland. *J. Petrol.* **14**, 185–202.

Constantin M. (1999) Gabbroic intrusions and magmatic metasomatism in harzburgites from the Garrett transform fault: implications for the nature of the mantle–crust transition at fast spreading ridges. *Contrib. Min. Petrol.* **136**, 111–130.

Crottini A., Poli S., and Molina J. F. (2003) Carbon cycle in subduction zones: experimental constraints in fluid-saturated MORB eclogites. EGS–AGU–EUG joint assembly, Nice, EAE03-A-12638.

Davies H. J. and Stevenson D. J. (1992) Physical model of source region of subduction zone volcanics. *J. Geophy. Res.* **97**, 2037–2070.

Dixon J. E., Leist L., Langmuir C., and Schilling J. G. (2002) Recycled dehydrated lithosphere observed in plume-influenced mid-ocean-ridge basalt. *Nature* **420**, 385–389.

Domanik K. J. and Holloway J. R. (1996) The stability and composition of phengitic muscovite and associated phases from 5.5 to 11 GPa: implications for deeply subducted sediments. *Geochim. Cosmochim. Acta* **60**, 4133–4150.

Domanik K. J. and Holloway J. R. (2000) Experimental synthesis and phase relations of phengitic muscovite from 6.5 to 11 GPa in a calcareous metapelite from the Dabie mountains. *China. Lithos* **52**, 51–77.

Domanik K. J., Hervig R. L., and Peacock S. M. (1993) Beryllium and boron in subduction zone minerals: an ion microprobe study. *Geochim. Cosmochim. Acta* **57**, 4997–5010.

Drummond M. S., Defant M. J., and Kepezhinskas P. K. (1996) Petrogenesis of slab-derived trondhjemite-tonalite-dacite/adakite magmas. *Trans. Roy. Soc. Edinburgh: Earth Sci.* **87**, 205–215.

Forneris J. and Holloway J. R. (2001) Experimental determination of chloritoid stability in subducting oceanic crust. *EOS, Trans., AGU* **82**(47) (Fall Meet. suppl., abstr. T22D-09).

Frost D. J. (1999) The stability of dense hydrous magnesium silicates in earth's transition zone and lower mantle. In *Special Publication No. 6, Mantle Petrology: Field Observations and High-pressure Experimentation* (eds. Y. Fei, C. M. Bertka, and B. O. Mysen). Geochemical Society, pp. 283–296.

Fryer P., Wheat C. G., and Mottl M. J. (1999) Mariana blueschist mud volcanism: implications for conditions within the subduction zone. *Geology* **27**, 103–106.

Fumagalli P. (2001) Processi di trasporto e rilascio di H_2O nelle zone di subduzione: uno studio sperimentale su sistemi ultrafemici ad alta pressione. PhD Thesis, University Milano, Milano, 178pp.

Fumagalli P. and Poli S. (1999) Phase relationships in hydrous peridotites at high pressure: preliminary results of multianvil experiments. *Periodico di Mineralogia* **68**, 275–286.

Fumagalli P., Stixrude L., Poli S., and Snyder D. (2001) The 10Å phase: a high-pressure expandable sheet silicate stable during subduction of hydrated lithosphere. *Earth Planet. Sci. Lett.* **186**, 125–141.

Furukawa Y. (1993) Magmatic processes under arcs and formation of the volcanic front. *J. Geophy. Res.* **98**, 8309–8319.

Gente P., Pockalny R. A., Durand C., Deplus C., Maia M., Ceuleneer G., Mevel C., Cannat M., and Laverne C. (1995) Characteristics and evolution of the segmentation of the Mid-Atlantic ridge between 20° N and 24° N during the last 10 million years. *Earth Planet Sci. Lett.* **129**, 55–71.

Gerya T. V., Stöckhert B., and Perchuk A. L. (2002) Exhumation of high-pressure metamorphic rocks in a subduction channel: a numerical simulation. *Tectonics* **21**(6), 1056, doi:10.1029/2002TC001406.

Gieskes J. M., Vrolijk P., and Blanc G. (1990) Hydrogeochemistry of the northern Barbados accretionary complex transect: ocean drilling project leg 110. *J. Geophys. Res.* **95**, 8809–8818.

Gill J. (1981) *Orogenic Andesites and Plate Tectonics*. Springer, New York, 390pp.

Gillis K. M. and Robinson P. T. (1990) Patterns and processes of alteration in the lavas and dykes of the troodos ophiolite, Cyprus. *J. Geophy. Res.* **95**, 21523–21548.

Helmstaedt H. and Schulze D. J. (1988) Eclogite-facies ultramafic xenoliths from Colorado Plateau diatreme breccias: comparison with eclogites in crustal environments, evalation of the subduction hypothesis, and implications for eclogite xenoliths from diamondiferous kimberlites. In *Eclogite and Eclogite Facies Rocks* (ed. D. C. Smith). Elsevier, New York, pp. 387–450.

Hermann J. (2002a) Experimental constraints on phase relations in subduction continental crust. *Contrib. Min. Petrol.* **143**, 219–235.

Hermann J. (2002b) Allanite: thorium and light rare earth element carrier in subducted crust. *Chem. Geol.* **192**, 289–306.

Hermann J. and Green D. H. (2001) Experimental constraints on high pressure melting in subducted crust. *Earth Planet. Sci. Lett.* **188**, 149–168.

Ishikawa T. and Nakamura E. (1994) Origin of the slab component in arc lavas from across-arc variation of B and Pb isotopes. *Nature* **370**, 205–208.

Johnson M. C. and Plank T. (1999) Dehydration and melting experiments constraints the fate of subducted sediments. *Geochem. Geophys. Geosys. Electron. J.* **1**, paper no. 1999GC000014.

Katayama I. and Nakashima S. (2003) Hydroxyl in clinopyroxene from the deep subducted crust: evidence for H_2O transport into the mantle. *Am. Mineral.* **88**, 229–234.

Kawamoto T., Leinenweber K., Hervig R. L., and Holloway J. R. (1995) Stability of hydrous phases in an H_2O-saturated KLB-1 peridotite up to 15 GPa. In *Volatiles in the Earth and Solar System*, pp. 229–239.

Kay R. W. (1978) Aleutian magnesian andesites: melts from subducted pacific ocean crust. *J. Volcanol. Geotherm. Res.* **4**, 117–132.

Kelley D. S., Karson J. A., Blackman D. K.,Früh-Green G., Butterfield D. A., Lilley M. D., Olson E. J., Schrenk M. O., Roe K. K., Lebon G. T., and Rivizzigno P. (2001) An off-axis hydrothermal vent field near the Mid-Atlantic ridge at 30° N. *Nature* **412**, 145–149.

Kennedy G. C., Wasserburg G. J., Heard H. C., and Newton R. C. (1962) The upper three-phase region in the system SiO_2–H_2O. *Am. J. Sci.* **260**, 501–521.

Kerrick D. M. and Connolly J. A. D. (1998) Subduction of ophicarbonates and recycling of CO_2 and H_2O. *Geology* **26**, 375–378.

Kerrick D. M. and Connolly J. A. D. (2001a) Metamorphic devolatilization of subducted oceanic metabasalts: implications for seismicity, arc magmatism and volatile recycling. *Earth Planet. Sci. Lett.* **189**, 19–29.

Kerrick D. M. and Connolly J. A. D. (2001b) Metamorphic devolatilization of subducted marine sediments and the transport of volatiles into the Earth's mantle. *Nature* **411**, 293–296.

Kincaid C. and Sacks I. S. (1997) Thermal and dynamical evolution of the upper mantle in subduction zones. *J. Geophys. Res.* **102**, 12295–12315.

Kohlstedt D. L., Keppler H., and Rubie D. C. (1996) Solubility of water in the α, β, and γ phases of $(Mg, Fe)_2SiO_4$. *Contrib. Mineral. Petrol.* **123**, 345–357.

Konzett J. and Ulmer P. (1999) The stability of hydrous potassic phases in lherzolite mantle; an experimental study to 9.5 GPa in simplified and natural bulk compositions. *J. Petrol.* **40**, 629–652.

Kushiro I. (1987) A petrological model for the mantle wedge and lower crust in the Japanese island arcs. In *Magmatic Processes: Physicochemical Principle,* Geochemical Society Special Publication 1 (ed. B. O. Mysen), pp. 165–181.

Liou J. G., Zhang R. Y., Ernst W. G., Rumble D., III, and Maruyama S. (1998) High pressure minerals from deeply subducted metamorphic rocks. In *Ultrahigh-pressure Mineralogy: Physics and Chemistry of the Earth's Deep Interior/ Reviews in Mineralogy* (ed. R. J. Hemley). The Mineralogical Society of America, Washington, vol. 37, pp. 33–96.

Luth R. W. (2001) Experimental determination of the reaction aragonite + magnesite = dolomite at 5 to 9 GPa. *Contrib. Mineral. Petrol.* **141**, 222–232.

Mammerickx J. (1989) Large scale undersea features on the north-east pacific. In *The Eastern Pacific Ocean and Hawaii, The Geology of North America* (eds. E. L. Winterer, D. M. Hussong, and R. W. Decker). Geolological Society of America, pp. 5–13.

Manning C. E. (1994) The solubility of quartz in H_2O in the lower crust and upper mantle. *Geochim. Cosmochim Acta* **58**, 4831–4839.

Martin H. (1999) Adakitic magmas: modern analogues of archaean granitoids. *Lithos* **46**, 411–429.

Maruyama S., Cho M., and Liou J. G. (1986) Experimental investigations of blueschist-greenschist transition equilibria: pressure dependence of Al_2O_3 contents in sodic amphiboles—a new geobarometer. In *Memoir 164, Blueschists and Eclogites* (eds. B. W. Evans and E. H. Brown). Geological Society of America, pp. 1–16.

Molina J. F. and Poli St. (2000) Carbonate stability and fluid composition in subducted oceanic crust: an experimental study on H_2O–CO_2-bearing basalts. *Earth Planet. Sci. Lett.* **176**, 295–310.

Molina J. F., Austrheim H., Glodny J., and Rusin A. (2002) The eclogites of the Marun-Keu complex, Polar Urals (Russia): fluid control on reaction kinetics and metasomatism during high P metamorphism. *Lithos* **61**, 55–78.

Moriguti T. and Nakamura E. (1998) Across-arc variation of Li isotopes in lavas and implications for crust/mantle recycling at subduction zones. *Earth Planet. Sci. Lett.* **163**, 167–174.

Morris J. D., Leeman W. P., and Tera F. (1990) The subducted component in island arc lavas: constraints from Be isotopes and B–Be systematics. *Nature* **344**, 31–36.

Nicolas A. (1989) *Structures of Ophiolites and Dynamics of Oceanic Lithosphere*. Kluwer, Dordrecht.

Niida K. and Green D. H. (1999) Stability and chemical composition of paragasitic amphibole in MORB pyrolite under upper mantle conditions. *Contrib. Mineral. Petrol.* **135**, 18–40.

Ohtani E., Mizobata H., and Yurimoto H. (2000) Stability of dense hydrous magnesium silicate phases in the systems Mg_2SiO_4–H_2O and $MgSiO_3$–H_2O at pressures up to 27 GPa. *Phys. Chem. Mineral.* **27**, 533–544.

Okamoto K. and Maruyama S. (1998) Multi-anvil re-equilibration experiments of a Dabie Shan ultrahigh pressure eclogite within the diamond-stability fields. *The Island Arc* **7**, 52–69.

Okamoto K. and Maruyama S. (1999) The high-pressure synthesis of lawsonite in the MORB + H_2O system. *Am. Mineral.* **84**, 362–373.

Okay A. I. (1980) Mineralogy, petrology, and phase relations of glaucophane-lawsonite zone blueschists from the Tavsanh region, northwest Turkey. *Contrib. Mineral. Petrol.* **72**, 243–255.

Ono S. (1998) Stability limits of hydrous minerals in sediment and mid-ocean ridge basalt compositions: implications for water transport in subduction zones. *J. Geophy. Res.* **103**, 18253–18267.

Pawley A. R. and Holloway J. R. (1993) Water sources for subduction zone volcanism: new experimental constraints. *Science* **260**, 664–667.

Pawley A. R. and Wood B. J. (1995) The high-pressure stability of talc and 10 Å phase: potential storage sites for H_2O in subduction zones. *Am. Mineral.* **80**, 998–1003.

Peacock S. M. (1993) The importance of blueschist to eclogite dehydration reactions in subducting oceanic crust. *Geol. Soc. Am. Bull.* **105**, 684–694.

Peacock S. (2001) Are the lower planes of double seismic zones caused by serpentine dehydration in subducting oceanic mantle? *Geology* **29**, 299–302.

Philippot P. (1993) Fluid-melt-rock interaction in mafic eclogites and coesite bearing sediments: constraints on volatile recycling during subduction. *Chem. Geol.* **108**, 93–112.

Plank T. and Langmuir C. H. (1993) Tracing trace elements from sediment input to volcanic output at subduction zones. *Nature* **362**, 739–743.

Plank T. and Langmuir C. H. (1998) The chemical composition of subducting sediment and its consequences for the crust and mantle. *Chem. Geol.* **145**, 325–394.

Poli S. (1993) The amphibolite-eclogite transformation: an experimental study on basalt. *Am. J. Sci.* **293**, 1061–1107.

Poli S. and Schmidt M. W. (1995) H_2O transport and release in subduction zones-experimental constraints on basaltic and andesitic systems. *J. Geophys. Res.* **100**, 22299–22314.

Poli S. and Schmidt M. W. (1998) The high-pressure stability of zoisite and phase relationships of zoisite-bearing assemblages. *Contrib. Mineral. Petrol.* **130**, 162–175.

Poli S. and Schmidt M. W. (2002) Petrology of subducted slabs. *Ann. Rev. Earth Planet. Sci.* **30**, 207–235.

Rea D. K. and Ruff L. J. (1996) Composition and mass flux of sediment entering the world's subduction zones: implications for global sediment budgets, great earthquakes, and volcanism. *Earth Planet. Sci. Lett.* **140**, 1–12.

Ricci J. E. (1951) *The Phase Rule and Heterogeneous Equilibrium.* Dover Publications, New York, 504pp.

Ryan J. G., Morris J., Tera F., Leeman W. P., and Tsvetkov A. (1995) Cross-arc geochemical variations in the kurile arc as a function of slab depth. *Science* **270**, 625–627.

Schmeling H., Monz R., and Rubie D. C. (1999) The influence of olivine metastability on the dynamics of subduction. *Earth Planet. Sci. Lett.* **165**, 55–66.

Schmidt M. W. (1993) Phase relations and compositions in tonalite as a function of pressure: an experimental study at 650 °C. *Am. J. Sci.* **293**, 1011–1060.

Schmidt M. W. (1996) Experimental constraints of recycling of potassium from subducted oceanic crust. *Science* **272**, 1927–1930.

Schmidt M. W. and Poli S. (1998) Experimentally based water budgets for dehydrating slabs and consequences for arc magma generation. *Earth Planet. Sci. Lett.* **163**, 361–379.

Schmidt M. W. and Vielzeuf D. (2001) How to generate a mobile component in subducting crust: melting vs. dissolution processes. In *11th Ann. V. M. Goldschmidt Conf.*, Abstr. 3366. LPI Contrib. No. 1088, Lunar Planet. Inst., Houston (CD-ROM).

Seno T. and Yamanaka Y. (1996) Double seismic zones, compressional deep trench-outer rise events, and superplumes. In *Subduction; Top to Bottom,* AGU Geophysics Monograph 96 (eds. E. Bebout, D. W. Schol, S. H. Kirby, and J. P. Blatt). AGU, pp. 347–355.

Sigmarsson O., Condomines M., Morris J. D., and Harmon R. S. (1990) Uranium and ^{10}Be enrichments by fluids in the Andean arc magmas. *Nature* **346**, 163–165.

Sigmarsson O., Martin H., and Knowles J. (1998) Melting of a subducting oceanic crust from U–Th disequilibria in austral Andean lavas. *Nature* **394**, 566–569.

Snow J. E. and Dick H. J. B. (1995) Pervasive magnesium loss by marine weathering of peridotite. *Geochim. Cosmochim. Acta* **59**, 4219–4235.

Sorensen S. S. (1986) Petrologic and geochemical comparison of the blueschist and greenschist units of the catalina schist terrane, southern California. In *Memoir 164, Blueschists and Eclogites* (eds. B. W. Evans and E. H. Brown). Geological Society of America, pp. 59–75.

Stalder R., Ulmer P., Thompson A. B., and Günther D. (2000) Experimental approach to constrain second critical end points in fluid/silicate systems: near-solidus fluids and melts in the system albite-H_2O. *Am. Mineral.* **85**, 68–77.

Stalder R., Ulmer P., Thompson A. B., and Gunther D. (2001) High pressure fluids in the system MgO–SiO_2–H_2O under upper mantle conditions. *Contrib. Mineral. Petrol.* **140**, 607–618.

Sudo A. and Tatsumi Y. (1990) Phlogopite and K-amphibole in the upper mantle: implication for magma genesis in subduction zones. *Geophy. Res. Lett.* **17**, 29–32.

Tatsumi Y. (1986) Formation of the volcanic front in subduction zones. *Geophys. Res. Lett.* **13**, 717–720.

Tatsumi Y. and Eggins St. (1995) Subduction zone magmatism. In *Frontiers in Earth Sciences.* Blackwell, Cambridge, 211pp.

Thurston S. P. (1985) Structure, petrology, and metamorphic history of the nome group blueschist terrane, Salmon Lake area, Seward Peninsula, Alaska. *Geol. Soc. Am. Bull.* **96**, 600–617.

Trønnes R. G. (2002) Stability range and decomposition of potassic richterite and phlogopite end members at 5–15 GPa. *Mineral. Petrol.* **74**, 129–148.

Turner S., Bourdon B., and Gill J. (2003) Insights into magma genesis at convergent margins from U-series isotopes. *Rev. Mineral. Geochem.* **52**, 255–315.

Ulmer P. and Trommsdorff V. (1995) Serpentine stability to mantle depths and subduction-related magmatism. *Science* **268**, 858–861.

Ulmer P. and Trommsdorff V. (1999) Phase relations of hydrous mantle subducting to 300 km. In *Mantle Petrology: Field Observations and High Pressure Experimentation: A Tribute to Francis R. (Joe) Boyd.* (eds. Y. W. Fei, C. Bertka, and B. O. Mysen), Geochemical Society Special Publication No. 6, pp. 259–281.

Usui T., Nakamura E., Kobayashi K., Maruyama S., and Helmstaedt H. (2003) Fate of the subducted Farallon plate inferred from eclogite xenoliths in the Colorado Plateau. *Geology* **31**(7), 589–592.

Van Keken P. E., Kiefer B., and Peacock S. (2002) High-resolution models of subduction zones: implications for mineral dehydration reactions and the transport of water into the deep mantle. *Geochem. Geophys. Geosys. Electron. J.* **3**(10), 1056, doi: 10.1029/2001GC000256.

Vielzeuf D. and Montel J. M. (1994) Partial melting of Al-metagraywackes: Part I. Fluid-absent experiments and phase relationships. *Contrib. Mineral. Petrol.* **117**, 375–393.

Vielzeuf D. and Schmidt M. W. (2001) Melting relations in hydrous systems revisited: application to metapelites, metagreywackes, and metabasalts. *Contrib. Mineral. Petrol.* **141**, 251–267.

Widmer T. and Thompson A. B. (2001) Local origin of high pressure vein material in eclogite facies rocks of the zermatt-saas zone, Switzerland. *Am. J. Sci.* **301**, 627–656.

Yaxley G. M. and Green D. H. (1994) Experimental demonstration of refractory carbonate-bearing eclogite and siliceous melt in the subduction regime. *Earth Planet. Sci. Lett.* **128**, 313–325.

Zack T., Rivers T., and Foley S. F. (2001) Cs–Rb–Ba systematics in phenite and amphibole: an assesment of fluid mobility at 2.0 GPa in eclogites from Tescolmen (Central Alps). *Contrib. Mineral. Petrol.* **140**, 651–669.

3.18
One View of the Geochemistry of Subduction-related Magmatic Arcs, with an Emphasis on Primitive Andesite and Lower Crust

P. B. Kelemen and K. Hanghøj

Woods Hole Oceanographic Institution, MA, USA

and

A. R. Greene

Western Washington University, Bellingham, WA, USA

3.18.1 INTRODUCTION

This chapter has four main aims.

(i) We wish to provide a comprehensive picture of the composition of volcanic rocks from subduction-related magmatic arcs. There are several recent reviews of the geochemistry of arc basalts. This chapter differs in including andesites as well as basalts, in focusing on major elements as well as trace elements and isotopes, and in using elemental abundance in "primitive" lavas, rather than trace-element ratios, to investigate enrichments of incompatible elements in arc magmas relative to primitive mid-ocean ridge basalts (MORBs).

(ii) We review evidence in favor of the existence of andesitic as well as basaltic primary magmas in arcs. While we have recently reviewed evidence for this in the Aleutian arc, in this chapter we broaden our data set to arcs worldwide, and concentrate on whether mixing of lower crustal melts with primitive basalts offers a viable alternative to the hypothesis that there are "primary" andesites, i.e., andesites in Fe/Mg equilibrium with mantle olivine, passing from the mantle into the crust beneath arcs.

(iii) We present new data on the composition of arc lower crust, based mainly on our ongoing work on the Talkeetna arc section in south central Alaska. To our knowledge, this is the first complete ICP-MS data set on an arc crustal section extending from the residual mantle to the top of the volcanic section.

(iv) We summarize evidence from arc lower crustal sections that a substantial proportion of the dense, lower crustal pyroxenites and garnet granulites produced by crystal fractionation are missing. These lithologies may have been removed by diapirs descending into less dense upper mantle.

In order to achieve these aims, and to limit the length of the chapter, we have not provided a detailed review of theories regarding the origin of primitive arc basalts, the mixing of magmatic components derived from the upper mantle, aqueous fluids, and sediment melts, or open-system processes in the crust including mixing and assimilation. For a more complete view of these theories, we refer the reader to the many excellent review papers and individual studies that are, all too briefly, cited below.

3.18.1.1 Definition of Terms Used in this Chapter

The following terms are used extensively throughout this chapter, and/or in the recent arc literature. We have tried to define each term where it first arises in the text. However, we realize that not all readers will read every section. Thus, we include the following brief definitions here.

Accumulated minerals: These are crystals in lavas that are in excess of the amounts that could crystallize from a melt in a closed system. Like cumulates, some cases of accumulated crystals may be impossible to detect, but others stand out because they result in rock compositions different from all or most terrestrial melts.

Adakite: This term is justly popular, but unfortunately it means many different things to different people, so we try not to use it. It generally is used for andesites and dacites with extreme light rare earth element (REE) enrichment (e.g., $La/Yb > 9$), very high Sr/Y ratios (e.g., $Sr/Y > 50$), and low yttrium and heavy REE concentrations (e.g., $Y < 20$ ppm, $Yb < 2$ ppm).

Andesite: For the purposes of this chapter, andesites are simply lavas (or bulk compositions) with >54 wt.% SiO_2. For brevity, we have not subdivided relatively SiO_2-rich magmas into dacite, rhyodacite, etc.

Basalt: In this chapter, basalt means lavas or bulk compositions with <54 wt.% SiO_2. Note that we have eliminated alkaline lavas (nepheline or kalsilite normative) from the data compilation. In a few places, we use terms picrite and komatiite to refer to basaltic melts with more than 15 wt.% and 18 wt.% MgO, respectively.

Boninite: By boninites, we mean andesites with $TiO_2 < 0.5$ wt.%, plus lavas identified as boninites in the original papers presenting geochemical data (includes some basalts!).

Calc-alkaline: Magmas having both high Na + K at high Mg# (Irvine and Baragar, 1971) and high SiO_2 at high Mg# (i.e., high SiO_2 at low Fe/Mg, Miyashiro (1974)); "tholeiitic magmas" have lower SiO_2 and Na + K at the same Mg#, when compared to calc-alkaline magmas.

Compatible element: An element with equilibrium solid/melt partition coefficient >1. In garnet (and zircon), heavy REE, scandium, vanadium, and yttrium are compatible. These elements are incompatible in all other rock-forming minerals involved in igneous fractionation processes discussed in this chapter.

Cumulate: A rock formed by partial crystallization of a melt, after which the remaining melt was removed. While some cumulates may be difficult to recognize, others are evident because the cumulate mineral assemblage has major and/or trace-element contents distinct from all or most terrestrial melts, easily understood as the result of crystal/melt partitioning.

Eclogite facies: This term represents high-pressure and relatively low temperature, metamorphic parageneses with omphacitic clinopyroxene and pyrope-rich garnet.

EPR: East Pacific Rise.

Evolved: Lavas, melts, liquids with Mg# < 50.

JDF: Juan de Fuca Ridge.

High Mg#: Lavas, melts, liquids with Mg# from 50 to 60.

Incompatible element: Equilibrium solid/melt partition coefficient <1.

LILEs: Large ion lithophile elements, rubidium, radium, barium, potassium (and caesium—but we do not use caesium data in this chapter). Although they are not, strictly speaking, LILE, we sometimes group thorium and uranium with the LILE when referring to elements that are highly incompatible.

Mantle wedge: "Triangular" region underlying arc crust, overlying a subduction zone, extending to perhaps 400 km depth.

Mg#: 100 × molar MgO/(MgO + FeO), where all iron is treated as FeO.

MORB: Mid-ocean ridge basalt.

Primary: Lavas, melts, and liquids derived solely via melting of a specific, homogeneous source. In practice, it is hard to recognize or even conceive of a truly primary melt. Strictly speaking, even mantle-derived MORBs may be mixtures of primary melts derived from a variety of sources, including polybaric melts of variably depleted peridotites and/or "basaltic veins." For brevity we have used the term "primary" in a few cases in this chapter. Where it is used without qualification, we refer to melts that are, or could be, in equilibrium with mantle olivine with Mg# = 90–93.

Primitive: Lavas, melts, and liquids with Mg# > 60. Primitive cumulates have Mg# > 85.

REE groups: Light—lanthanum to samarium; middle—europium to dysprosium; heavy—holmium to lutetium.

Subduction-related magmatic arc (or simply, arc): Chains of volcanoes on the overthrust plate parallel to, and ~100–200 km horizontally away from, the surface expression of a subduction zone, together with coeval, underlying plutonic rocks.

3.18.2 ARC LAVA COMPILATION

Data in this and subsequent sections are from the Georoc database at http://georoc.mpch-mainz.gwdg.de/ (arcs worldwide), our Aleutian arc compilation, including all data available from the database compiled by James Myers and Travis McElfrish and available at http://www.gg.uwyo.edu/aleutians/index.htm, supplemented by additional data cited in Kelemen *et al.* (2003b), a new Costa Rica and Panama compilation (Abratis and Worner, 2001; Carr *et al.*, 1990; Cigolini *et al.*, 1992; de Boer *et al.*, 1991, 1988, 1995; Defant *et al.*, 1991a,b, 1992; Drummond *et al.*, 1995; Hauff *et al.*, 2000; Herrstrom *et al.*, 1995; Patino *et al.*, 2000; Reagan and Gill, 1989; Tomascak *et al.*, 2000), and Central American data compiled by Mike Carr and available at http://www.rci.rutgers.edu/~carr/index.html. We also included a very complete data set on lavas from Mt. Shasta in the southern Cascades (Baker *et al.*, 1994; Grove *et al.*, 2001). We compare compositions of arc lavas to data on MORB glasses downloaded from PetDB, online at http://petdb.ldeo.columbia.edu/petdb/.

Lava data come from intra-oceanic arcs (Tonga, $n = 704$; Kermadec, 189; Bismark/New Britain, 165; New Hebrides, 252; Marianas, 834; Izu-Bonin, 878; oceanic Aleutians, 1082; South Sandwich, 328; Lesser Antilles, 356) and arcs which are, or may be, emplaced within older continental material or thick sequences of continentally derived sediment (Philippines, 221; Indonesia, 380; Papua New Guinea, 78; SW Japan, 92; NW Japan, 2314; Kuriles, 721; Kamchatka, 447; Cascades, 202; Central America, 857; the Andes, 1156; Greater Antilles, 175). Notably missing are data from New Zealand, the Alaska Peninsula, and Mexico. We apologize to authors whose work is not cited here, but whose analyses we compiled using large, online databases. It is simply not practical to cite the sources of all the data compiled for this chapter. We urge readers to contact us, and to visit the online databases, in order to check on the provenance of specific data.

In comparing arc lavas to MORB glasses, it is important to keep in mind that none of our arc data sets discriminate between true liquid compositions, and compositions of lavas potentially including abundant, accumulated phenocrysts. Many of the lavas in our complete compilation had MgO contents >20 wt.% at 100 MgO/(MgO + FeOt), or Mg#, >65. Samples with more than 20 wt.% MgO have been eliminated from all of our plots; we believe these, and possibly many other lavas with 10–20 wt.% MgO, reflect the effect of accumulated olivine. In addition, many high Mg# lavas with lower MgO contents contain abundant phenocrysts of clinopyroxene and/or olivine. The high Mg#s of such samples could be due to accumulated clinopyroxene or olivine, at least in part. High Mg# andesites play a large role in the interpretive sections of this chapter, and thus this problem should be kept in mind. With this said, the high SiO_2 and alkali contents, and the low MgO and CaO contents, of these samples, and the similarity of nearly aphyric primitive andesites from Mt. Shasta

(Baker *et al.*, 1994; Grove *et al.*, 2001) to other, similar compositions worldwide, does not allow for much accumulated olivine and clinopyroxene. Finally, abundant accumulated plagioclase probably accounts for some lava compositions with very high Al_2O_3 contents (e.g., Brophy, 1989; Crawford *et al.*, 1987). However, because this may remain controversial, we did not eliminate lavas on the basis of Al_2O_3 content.

Another issue is that when relying on compiled information from databases, one has to be aware of the possibility that data may have been incorrectly entered, transferred, or normalized. Thus, for visual clarity, we eliminated some outliers, including lavas with more than 80 wt.% SiO_2 and Mg# > 50, lavas with more than 70 wt.% SiO_2 and Mg# > 60, lavas with less than 10 wt.% MgO and Mg# > 80, and lavas with less than 5 wt.% MgO and Mg# > 75. We also eliminated alkali basalts (normative nepheline or kalsilite), especially in data compilations for the Sunda and Honshu arcs, and placed boninitic lavas (>54 wt.% SiO_2, <0.5 wt.% TiO_2, plus samples described in the original data sources as boninites) in a group separate from other lavas.

A third issue in using large numbers of compiled data is quality. Analytical methods have varied over time, and some labs are more reliable than others. Outliers appear on many of our plots, particularly those involving trace elements. In some cases, where outliers are orders of magnitude from the bulk of the data, we have adjusted axis limits in plots so that outliers are no longer visible but the variation in the bulk of the data is easily seen. However, other than this, we have not made any attempt in this chapter to discriminate between "good" and "bad" data. This approach is deliberately different from other recent reviews of arc data (e.g., Elliott, 2003; Plank, 2003). We are not critical of these other reviews, but we think an alternative, more inclusive approach may be useful until a truly large number of high-quality ICP-MS data become available for a fully representative set of arc magma compositions worldwide. In particular, one focus of this chapter, on the origin of primitive andesites and calc-alkaline magma series, would be all but impossible if we restricted attention to data sets including ICP-MS analyses. We urge readers to be cautious in interpreting our data plots, and to check key points for themselves. Also, we believe this chapter indicates several areas in which additional data would be very valuable.

Finally, a substantial limitation of our compilation is that it includes very sparse data on chlorine, fluorine, boron, beryllium, and lithium, and essentially no data on volatile contents (H_2O, CO_2, sulfur, noble gases). While data on H_2O in glass inclusions are beginning to become available, it is not yet clear, for example, to what extent H_2O contents in primitive arc magmas correlate with other compositional characteristics. We look forward to learning more about these topics.

3.18.3 CHARACTERISTICS OF ARC MAGMAS

3.18.3.1 Comparison with MORBs

In this section, we compare arc lava compositions (on an anhydrous basis) with compositions of MORB glasses; also see Chapter 3.13. The contrasts are remarkably distinct (Figure 1). Before we go on to describe these contrasts, it is convenient to define "primitive andesites" (lavas with SiO_2 > 54 wt.% and Mg# > 60, exclusive of boninites) and "high Mg# andesites" (SiO_2 > 54 wt.% and Mg# > 50). These classifications include primitive and high Mg# lavas that are dacites and even rhyolites, as well as true andesites. However, we group them all for brevity.

Implicit in our definition of lavas with Mg# > 60 as "primitive," and lavas with lower Mg# as "evolved" is the assumption that crystallization processes always produce a lower Mg# in derivative liquids, as compared to parental liquids. As far as we know, this assumption is justified on the basis of all available experimental data on crystallization/melting of igneous rocks with Mg# higher than 40 (see compilation in Kelemen (1995)) at oxygen fugacities within 2 log units of Ni-NiO (typical for arcs, Blatter and Carmichael (1998), Brandon and Draper (1996), Gill (1981), and Parkinson and Arculus (1999)), or lower. Oxygen fugacity more than two log units above Ni-NiO may facilitate early and abundant crystallization of FeTi oxides, and thus nearly constant Mg# with decreasing temperature and liquid mass (Kawamoto, 1996). However, this is unlikely beneath most arcs.

3.18.3.1.1 Major elements

We first examine major elements as a function of Mg# (Figure 1). A very small fraction of MORB glasses have Mg# less than 35, whereas lavas with Mg# < 35 are common in arcs. While SiO_2 in primitive (Mg# > 60) MORB is restricted to 48–52 wt.%, primitive arc lavas range from 45 wt.% to more than 60 wt.% SiO_2. The contrast for evolved (Mg# < 60) compositions is even more striking. Overall, Mg# versus SiO_2 for MORB glasses closely approximates a single liquid line of descent, involving olivine + plagioclase + clinopyroxene, with cumulate SiO_2 ~ liquid SiO_2 (see Chapter 3.13). The arc lavas show a much broader trend of Mg# versus SiO_2, consistent with crystallization of SiO_2-poor assemblages (less plagioclase, added hornblende and/or FeTi oxides) from a range of parental melts with SiO_2 contents of 45 wt.% to more than 60 wt%. SiO_2 contents from 45 to greater than 60 wt% are found in arc lavas with a Mg# of 70 (close to Fe/Mg exchange equilibrium with residual mantle peridotite having an olivine Mg# of 90–91). This is not the result of crystal fractionation from basalts with Mg# ~ 70. However, with this said, the manner in which these parental melts acquire their differing SiO_2 contents is uncertain and controversial (see Sections 3.18.3.2.4 and 3.18.3.2.5).

TiO_2 contents of arc magmas are generally lower than in MORB glasses. TiO_2 versus Mg# in the entire arc lava compilation, and in most individual suites, shows a sharp transition from increasing TiO_2 with decreasing Mg#, for Mg# > ~50, to decreasing TiO_2 with decreasing Mg#, for Mg# < ~50, which is due to fractionation of FeTi oxides from evolved melts. It is probably safe to conclude that primitive arc magmas have not undergone FeTi oxide fractionation, and thus their low TiO_2, compared to MORB, is a primary feature or the result of

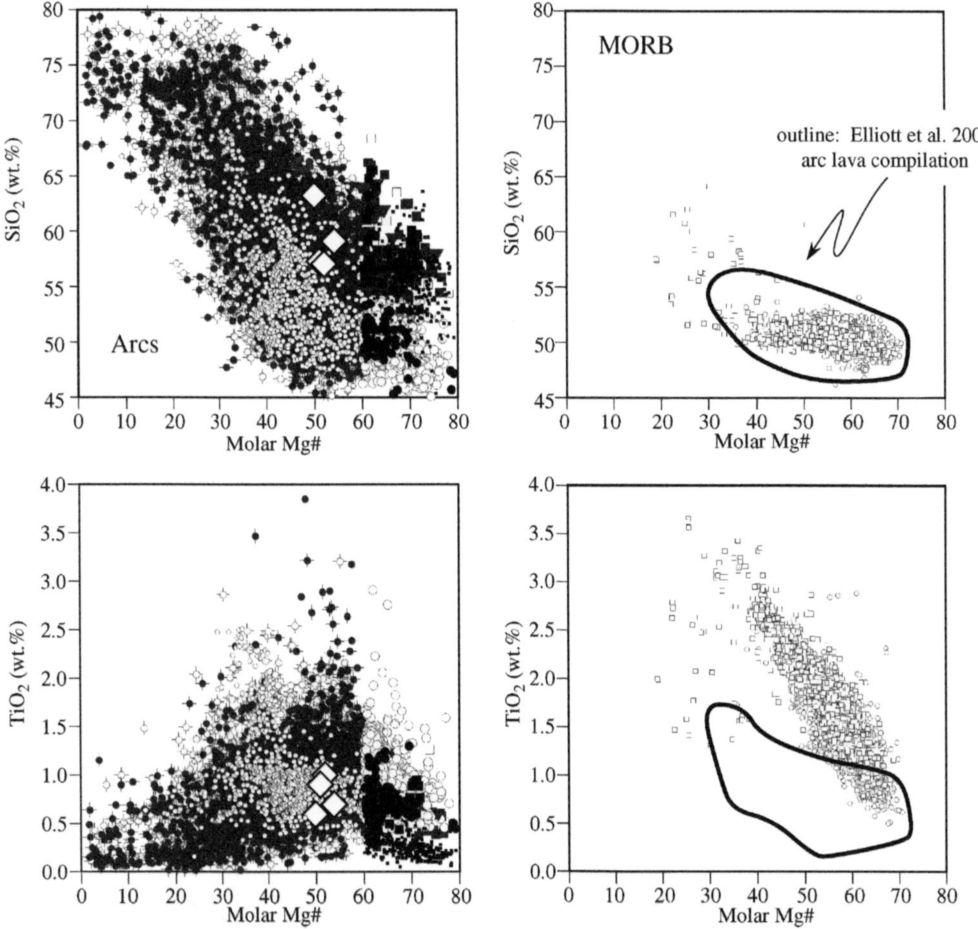

Figure 1 Molar Mg# versus concentrations of major-element oxides, in weight percent, in arc lavas (left) and MORB glasses (right), both on an anhydrous basis. Outlined fields on the right-hand diagrams show the range of variation in arc lava compilation of Elliott (2003). Arc and MORB data sources described in text. For MORB, squares are EPR and circles are Atlantic, JDF, and Indian Ocean Ridges. For primitive arc lavas with Mg# > 60, circles are basalts and squares are andesites, open symbols are for oceanic arcs, gray symbols are for continental arcs, and black symbols are for the oceanic Aleutian arc. Boninites are shown with small black squares. For evolved lavas with Mg# < 60, open circles with barbs are for oceanic arcs, gray circles with barbs are continental arcs, and small open circles are oceanic Aleutian samples. Inverted gray triangles are for primitive andesites from Mt. Shasta, southern Cascades (Grove *et al.*, 2001). Intra-oceanic arcs in our data set are from Tonga, Kermadec, Bismark/New Britain, New Hebrides, Marianas, Izu-Bonin, South Sandwich, and the Lesser Antilles. Samples from arcs in our compilation which are, or may be, emplaced within older continental material or thick sequences of continentally derived sediment are from the Philippines, Indonesia, Papua New Guinea, SW Japan, NW Japan, Kuriles, Kamchatka, Cascades, Central America, the Andes, and the Greater Antilles. Large filled diamonds are estimated compositions of the continental crust from Christensen and Mooney (1995), McLennan and Taylor (1985), Rudnick and Fountain (1995), and Weaver and Tarney (1984), including Archean estimate of Taylor and McLennan.

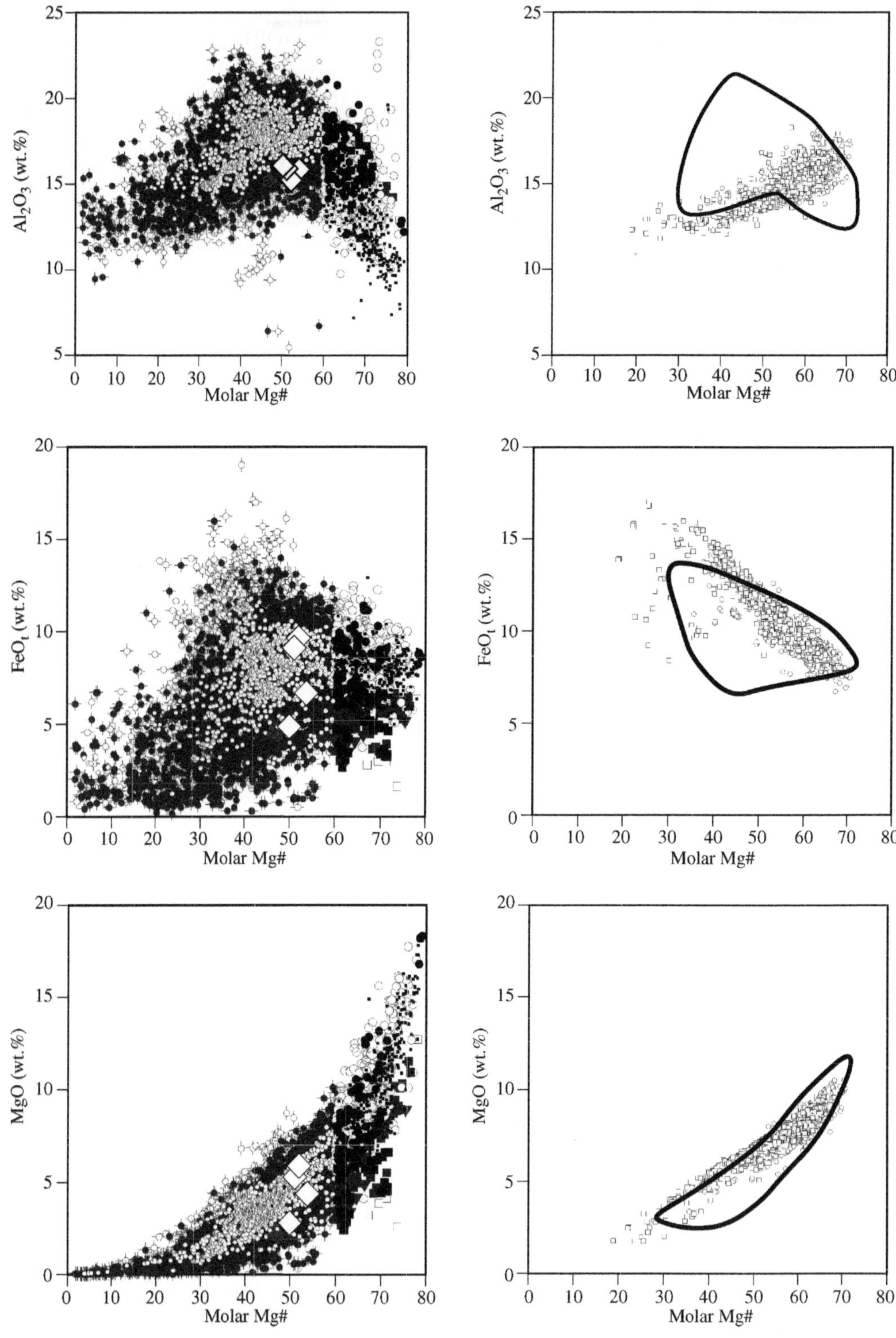

Figure 1 (continued).

magma mixing. Primitive andesites and boninites have the lowest TiO_2 contents in our compilation.

Al_2O_3 contents of primitive arc lavas range from 10 wt.% to 19 wt.%, extending to much lower and higher values than primitive MORB glasses. Al_2O_3 contents of arc lavas increase with decreasing Mg#, for lavas with Mg# > ~50, for the entire data set and for most individual arc suites. For Mg# less than ~55, Al_2O_3 decreases with decreasing Mg#, reflecting plagioclase fractionation. This suggests that plagioclase fractionation may play a minor role in differentiation of most arc melts

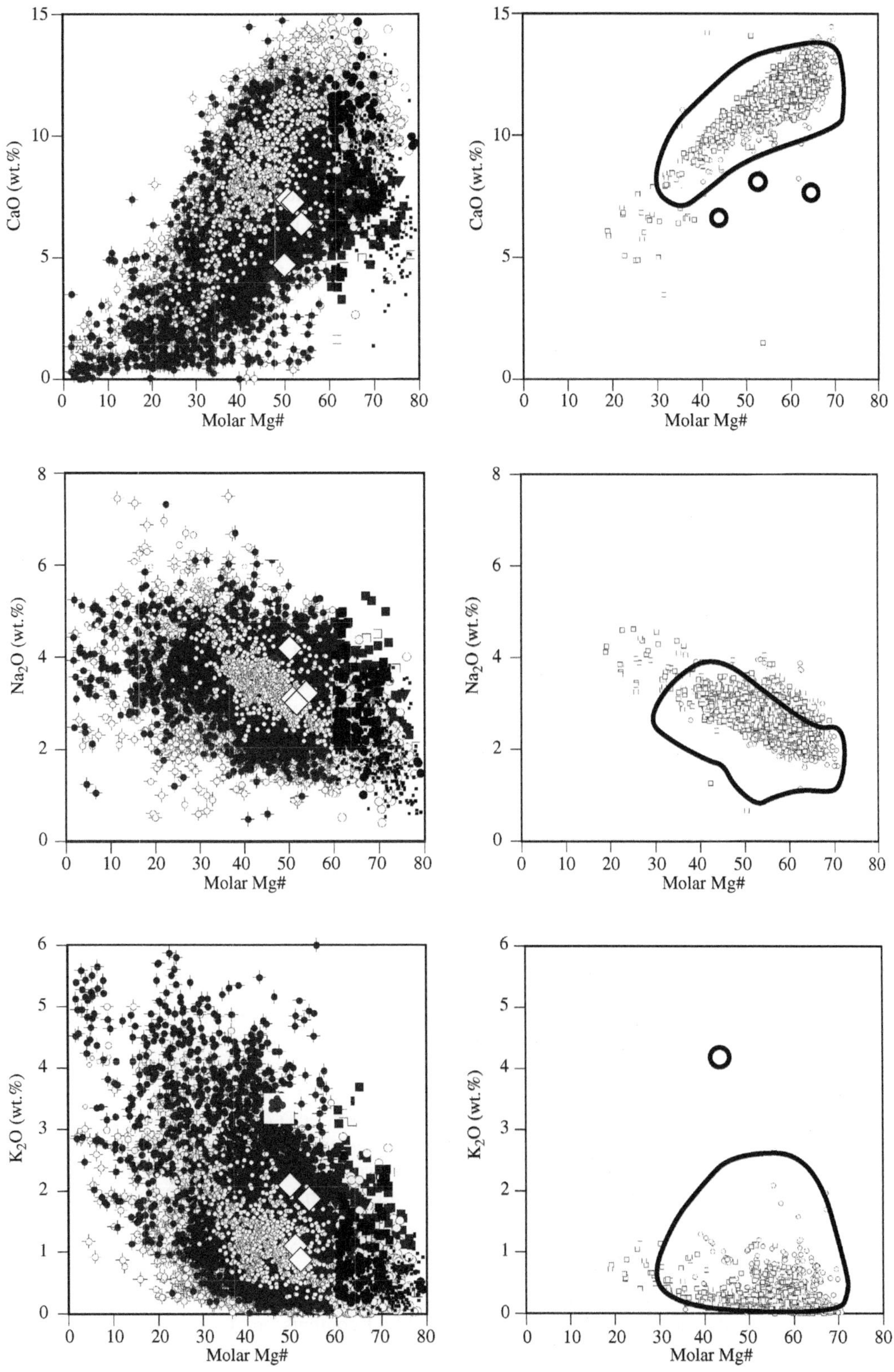

Figure 1 (continued).

from Mg# of 70 or more to Mg# of ~55. Instead, olivine, pyroxene, and/or hornblende fractionation may predominate. Given that primitive arc lavas are generally H_2O-rich compared to anhydrous mantle melts at arc Moho depths (e.g., Anderson, 1974; Falloon and Danyushevsky, 2000; Kamenetsky *et al.*, 1997; Macdonald *et al.*, 2000; Pichavant *et al.*, 2002; Roggensack *et al.*, 1997;

Sisson and Layne, 1993; Sobolev and Chaussidon, 1996), many primary mantle melts are probably in a reaction relationship with olivine (e.g., Müntener *et al.*, 2001), forming by reactions such as orthopyroxene + clinopyroxene + spinel = olivine + melt. Such melts are in equilibrium with olivine, but will not crystallize olivine upon isobaric cooling, and instead will crystallize websterites (two pyroxene pyroxenites). Other primitive, hydrous arc magmas are olivine saturated; these commonly produce olivine clinopyroxenite cumulates (e.g., Conrad and Kay, 1984; Conrad *et al.*, 1983). In addition, primitive arc magmas probably have temperatures >1,100 °C (e.g., Elkins Tanton *et al.*, 2001; Gill, 1981; Kelemen *et al.*, 2003b), above the thermal stability of hornblende, although this might be uncertain for very H_2O-rich compositions. Thus, pyroxenite (clinopyroxenite, websterite) fractionation is most likely responsible for the increase in Al_2O_3 with decreasing Mg#, for liquid Mg# of more than 70 to ~55 (e.g., Conrad and Kay, 1984; Conrad *et al.*, 1983).

Pyroxenites have seismic velocities similar to or only slightly less than residual mantle peridotites. The temperature and melt content of the sub-arc mantle are poorly constrained, and seismologists commonly report sub-Moho P-wave velocities less than 8 km s^{-1} (e.g., Fliedner and Klemperer, 1999; Holbrook *et al.*, 1999; Suyehiro *et al.*, 1996). Thus, the igneous crust may extend well below the seismic Moho in arcs. However, by the same token, seismic data certainly do not require abundant pyroxenite. Pyroxenites, while well represented among arc plutonic xenoliths, comprise a very small proportion of exposed arc crustal sections (e.g., DeBari and Coleman, 1989; Miller and Christensen, 1994). Thus, pyroxenites may be removed by viscous "delamination" during or after arc magmatism (e.g., Arndt and Goldstein, 1989; DeBari and Sleep, 1991, 1996; Herzberg *et al.*, 1983; R. W. Kay and S. M. Kay, 1988, 1991, 1993; S. M. Kay and R. W. Kay, 1985; Turcotte, 1989). We will return to this topic in Sections 3.18.4.2 and 3.18.5.1.

Rare primitive arc lavas, including well-studied compositions that almost certainly represent liquid compositions, have Al_2O_3 higher than primitive MORB. These are generally interpreted as products of equilibration of nearly anhydrous basaltic melts with residual mantle peridotite at 1–1.2 GPa, just beneath the base of arc crust in the Cascades and Indonesia (e.g., Bartels *et al.*, 1991; Elkins Tanton *et al.*, 2001; Sisson and Bronto, 1998), although see also Turner and Foden (2001). Although these compositions are not common (Figure 1), they are nonetheless of great importance because well-known phase equilibria for nearly anhydrous melt/mantle equilibration allows these lavas to be used to place fairly tight constraints on sub-arc temperature at specific depths in the mantle wedge (e.g., Elkins Tanton *et al.*, 2001).

Al_2O_3 contents in evolved arc magmas range up to more than 20 wt.%, much higher than in evolved MORB. Interpretation of Al_2O_3 contents of evolved arc lavas is notoriously difficult, because of the potential for incorporation of accumulated plagioclase in porphyritic lavas (Brophy, 1989; Crawford *et al.*, 1987). However, some carefully studied "high alumina basalts" are probably liquid compositions (Baker and Eggler, 1983, 1987; Sisson and Grove, 1993a,b). In addition, some authors have maintained that high Al_2O_3 arc lavas with Mg# < 0.5 might represent primary melts from diapirs of subducted basalt that rise into the mantle wedge, but do not chemically equilibrate with residual mantle peridotite (e.g., Brophy and Marsh, 1986; Johnston and Wyllie, 1988; Marsh, 1976; Myers *et al.*, 1986a,b, 1985). This topic is visited again in Section 3.18.3.2.3. Alternatively, moderate Mg# lavas may be derived via crystal fractionation from primitive melts, or may be partial melts of arc lower crust. From this perspective, Al_2O_3 increases due to crystallization of plagioclase-free cumulates, where plagioclase saturation is suppressed by abundant H_2O (greater than ~2 wt.% H_2O) in primitive arc melts (e.g., Baker and Eggler, 1983, 1987; Kelemen *et al.*, 1990a; Müntener *et al.*, 2001, Sisson and Grove, 1993a,b).

In this review, using compiled data on arc lava compositions, it is impossible to improve upon previous estimates of oxygen fugacity or Fe^{3+}/Fe^{2+} in arc magmas (generally close to Ni-NiO, e.g., Blatter and Carmichael (1998), Brandon and Draper (1996), Gill (1981), and Parkinson and Arculus (1999)). Thus, here we concentrate on total Fe as FeO (FeOt). Both primitive and evolved arc magmas extend to FeOt contents much lower than in MORB glasses. While some of this difference among evolved compositions results from FeTi oxide fractionation in relatively oxidizing arc magmas, the differences between primitive arc lavas and primitive MORB are clear, and are not attributable to FeTi oxide fractionation because FeTi oxides are not saturated in primitive arc lavas. Instead, this reflects a difference in the FeOt content of primary magmas, and/or the effects of magma mixing.

MgO contents among arc lavas show a much broader range than in MORB glasses. Arc lava MgO contents higher than MORB glasses may be due, in part, to incorporation of compositions with accumulated clinopyroxene and/or olivine into our data set. However, boninite suites clearly include high MgO liquids (e.g., Falloon and Green, 1986; Sobolev and Danyushevsky, 1994), and some picritic and ankaramitic arc lava compositions are also believed to be liquid compositions (e.g., Eggins, 1993; Nye and Reid, 1986; Ramsay *et al.*, 1984). Primitive arc lavas also extend to MgO contents much lower than primitive MORB glasses. These have MgO too low to be derived from high MgO, primitive arc lavas via olivine fractionation. Alternatively, primitive andesites could be derived via olivine fractionation from picritic or komatiitic primary melts. However, as we will show in Sections 3.18.3.2 and 3.18.3.3, primitive andesites generally have trace-element characteristics that are distinct from primitive basalts, so the two cannot be related by crystal fractionation alone. Instead, as for low FeOt, low MgO in primitive arc andesites reflects either low MgO in primary melts equilibrated with residual mantle peridotite, or the effects of magma mixing.

Like FeO and MgO, both primitive and evolved arc lavas extend to CaO contents much lower than in primitive and evolved MORB glasses, and andesites are the low CaO end-member among primitive lavas. It is ironic that low CaO is one of the characteristics of these, end-member "calc-alkaline" lavas; this suggests that the "tholeiitic" versus "calc-alkaline" terminology should be changed.

Na_2O, K_2O and, to a lesser extent, P_2O_5 in both primitive and evolved arc lavas extend to higher and lower concentrations than in primitive and evolved MORB glasses. Again, primitive andesites stand out, having the highest Na_2O, K_2O, and P_2O_5 among primitive compositions. And, as for the other elements discussed above, these characteristics are not due to crystal fractionation, but must be characteristics of primary magmas or must arise via magma mixing.

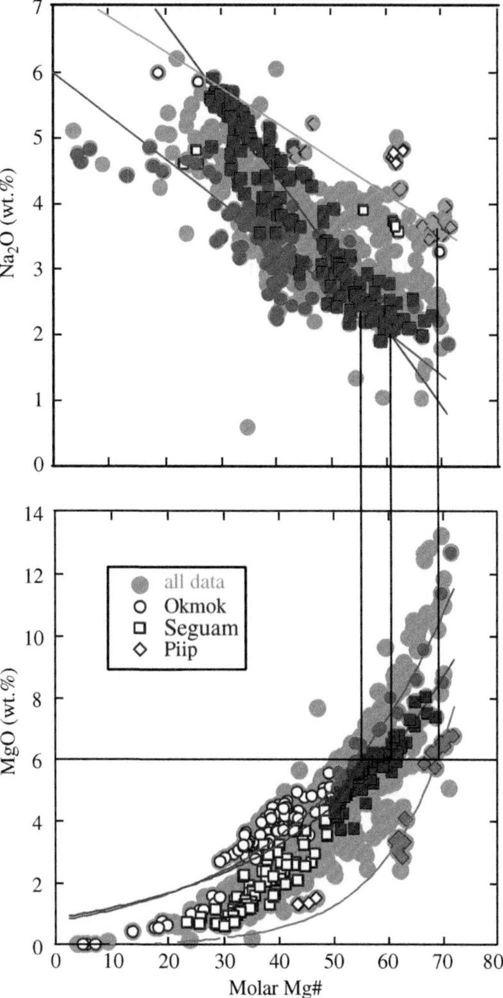

Figure 2 Molar Mg# versus wt.% Na_2O for Aleutian lavas (compiled by Kelemen *et al.*, 2003b), illustrating that variation trends have different slopes for different volcanoes, and that these trends cross. As a result, it would be unwise to infer Na_2O concentrations in primary Aleutian magmas using an "average" fractionation trend and the compositions of evolved lavas. Data for Seguam volcano mainly from Singer *et al.* (1992a,b) and data compiled by James Myers and Travis McElfrish and available at http://www.gg.uwyo.edu/aleutians/index.htm. Data for Okmok volcano mainly from Class *et al.* (2000), Miller *et al.* (1992, 1994), Nye (1983), and Nye and Reid (1986, 1987). Okmok data with Mg# ~ 80 omitted. Data for Piip volcano from Yogodzinski *et al.* (1994). Light grey circles, all Aleutian data (compiled by Kelemen *et al.*, 2003). Other shapes are for individual volcanoes, as shown in legend. Darker symbols used in fits (linear in top panel, exponential in bottom panel); open symbols omitted from fits.

3.18.3.1.2 We are cautious about fractionation correction of major elements

Na_2O concentration in basalts, corrected to 6 wt.% MgO, has been used as an indicator of the relative degree of mantle melting in the arc mantle source (e.g., Plank and Langmuir, 1988, 1993). This method involves two explicit assumptions. First, it is assumed that that primary melts have a common MgO content. While it is beyond the scope of this chapter to investigate this assumption on a volcano-by-volcano basis, in Figure 2, we show that this assumption seems questionable for Aleutian lavas, which show a wide variation in MgO at high Mg#. Lavas from Piip volcano with 6 wt.% MgO are primitive, so that their Na_2O contents may be close to those of the "primary" magma, whereas lavas from Okmok and Seguam volcanoes with 6 wt.% MgO are fractionated, with Na_2O contents higher than in corresponding, primary melts in equilibrium with mantle olivine. Because the trends of Na_2O versus Mg# for Okmok and Seguam volcanoes cross, correction to 6 wt.% MgO yields higher Na_2O in Okmok compared to

Seguam, whereas primary Okmok melts (Mg# ~70) probably have lower Na_2O compared to primary Seguam melts.

Second, it is assumed that Na_2O in arc magmas is derived primarily from the mantle, without a significant contribution from subducting sediment and oceanic crust. We think this also is uncertain, because fluids and melts in equilibrium with subducting, eclogite facies metasediment and metabasalt contain abundant dissolved sodium. This dissolved sodium could be carried into the mantle wedge together with H_2O and other components derived from subducting material (see Chapter 3.17, and references therein). For example, primitive andesites in the Aleutians that have exceptionally high Na_2O also have exceptionally high La/Yb, Sr/Nd, Dy/Yb, and $^{143}Nd/^{144}Nd$, and exceptionally low lead and strontium isotope ratios (Kelemen *et al.*, 2003b), which likely reflects incorporation of an eclogite melt component. We will return to this point in Section 3.18.3.2.

If partial melts of eclogite have ~5 wt.% Na_2O (Rapp *et al.*, 1999; Rapp and Watson, 1995), addition of 5% eclogite melt to some part of the mantle wedge adds 0.25 wt.% Na_2O to that region. This can be compared to, e.g., 0.4 wt.% Na_2O in fertile "pyrolite" (Ringwood, 1966) or ≤0.05 wt.% in depleted oceanic peridotites (Dick, 1989). Thus, it is plausible that a "slab melt" component might add significant and potentially variable amounts of sodium to the mantle "source" of arc magmas, particularly if the arc mantle has been previously depleted by melt extraction beneath a mid-ocean ridge and/or a back-arc basin.

In order to more quantitatively constrain the relative contributions of sodium from melt or fluid derived from subducting material, versus the pre-existing peridotite in the mantle wedge, it would be useful to know the degree of depletion of the pre-existing peridotite, due to previous melt extraction events prior to arc magmatism. One can try to use the concentration of "immobile" elements such as niobium, together with mantle/melt distribution coefficients, to estimate the degree of depletion of the mantle "source" of each arc magma, assuming that "immobile" elements are not added by fluids or melts of subducting material (Langmuir, personal communication, 2003). However, in addition to sodium, eclogite melts also contain ~5 ppm niobium (Rapp and Watson, 1995), while primitive mantle is estimated to contain <1 ppm niobium (Hofmann, 1988; Sun and McDonough, 1989). Therefore, determining the extent of prior depletion in the mantle wedge is also problematic for systems open to eclogite melt.

Third, use of Na_2O as an indicator of mantle melting processes involves the implicit assumption that Na_2O contents in arc magmas have not been affected by open-system processes in the crust, such as magma mixing and assimilation. Again, this assumption is violated in some cases, particularly when considering evolved, rather than primitive, lava compositions (e.g., Grove *et al.*, 1982, 1988; Hildreth and Moorbath, 1988; McBirney *et al.*, 1987).

3.18.3.1.3 Distinctive, primitive andesites

As will be seen throughout this chapter, despite limited data, primitive andesites define an end-member on almost all compositional variation diagrams. The primitive andesite end-member is distinct from primitive MORB glasses in ways that epitomize the overall difference between arc lavas and MORB (Table 1 and Figure 3).

High Mg# andesites, and their plutonic equivalents, are end-member calc-alkaline lavas, as distinct from the tholeiitic magma series. Here we define calc-alkaline magmas as having both high Na + K at high Mg# (Irvine and Baragar, 1971) and high SiO_2 at high Mg# (i.e., high SiO_2 at low Fe/Mg, Miyashiro, 1974); tholeiitic magmas have lower SiO_2 and Na + K at the same Mg#, when compared to calc-alkaline magmas. (Please note that in this chapter we do not consider alkaline lavas, i.e., nepheline or kalsilite normative compositions) While primitive basalts, and evolved, tholeiitic lavas, are found in a variety of plate tectonic settings, it is plain that calc-alkaline andesite lavas are found almost exclusively in arcs (Gill, 1981). Thus, one could argue that the genesis of primitive andesite is the defining process of arc magmatism.

While high Mg# andesites are clearly less voluminous than tholeiitic lavas in most arcs (Figure 4; see also, e.g., White and McBirney (1978)), we consider them very important in other ways. Explaining the difference between calc-alkaline and tholeiitic magma series has been one of the central topics of igneous petrology for almost a century (e.g., Baker *et al.*, 1994; Bowen, 1928; Brophy, 1987; Fenner, 1929, 1937; Green, 1976; Green and Ringwood, 1967, 1966; Grove and Kinzler, 1986; Grove *et al.*, 1982, 2001; Kay, 1978, 1980; S. M. Kay and R. W. Kay, 1994; Kay *et al.*, 1982; Kelemen, 1986, 1990; Kuno, 1950, 1968; Kushiro, 1969, 1974; Kushiro and Yoder, 1972; McBirney *et al.*, 1987; Miller *et al.*, 1992; Nicholls and Ringwood, 1973; Nicholls, 1974; Osborn, 1959; Sisson and Grove, 1993a,b; Tatsumi, 1981, 1982; Tatsumi and Ishizaka, 1981, 1982; Wilcox, 1944), and there is still no community-wide consensus on this.

Similarly, the estimated bulk composition of the continental crust (e.g., Christensen and Mooney, 1995; McLennan and Taylor, 1985;

Table 1(a) Average primitive MORB and arc basalts (Molar Mg# > 60).

	MORB	*n*	*Average oceanic*	*n*	*Average continental*	*n*	*Kermadec*	*n*	*Lesser Antilles*	*n*	*Marianas*	*n*	*New Hebrides*	*n*	*Scotia*	*n*	*Tonga*	*n*	*Aleutian*	*n*	*Andean*	*n*	*Cascades*	*n*	*Central America*	*n*	*Greater Antilles*	*n*	*Honshu*	*n*	*Kamchatka*	*n*	*Luzon*	*n*
n	203		503		497		36		84		168		65		41		70		66		56		60		78		21		137		78		24	
SiO_2	50.51		50.46		51.33		51.12		48.25		51.04		50.26		51.50		50.57		50.50		52.58		51.62		50.27		50.23		51.13		52.22		50.85	
TiO_2	1.22		0.91		0.98		0.81		0.85		1.01		0.69		1.06		0.94		0.79		1.03		0.87		1.02		1.04		1.03		0.92		0.78	
Al_2O3	15.97		15.72		15.70		15.65		14.48		16.64		13.67		16.97		15.59		16.51		16.66		16.83		14.60		15.54		16.14		14.76		14.20	
FeO(T)	8.85		8.52		8.72		8.77		9.17		8.02		8.74		8.05		8.60		8.58		8.11		7.89		9.50		9.20		8.79		8.78		8.92	
MnO	0.16		0.17		0.17		0.16		0.17		0.15		0.19		0.15		0.17		0.16		0.15		0.15		0.17		0.20		0.16		0.17		0.18	
MgO	8.57		9.84		9.48		9.07		13.69		8.57		11.33		7.74		9.39		9.22		8.28		9.12		10.63		10.57		9.29		9.39		10.86	
CaO	11.85		11.44		9.93		11.50		10.79		11.51		11.96		11.34		11.99		11.09		8.80		9.86		10.49		9.57		9.68		9.96		10.75	
Na_2O	2.57		2.35		2.61		2.27		2.01		2.62		2.06		2.72		2.16		2.39		2.97		2.93		2.31		2.50		2.53		2.73		2.24	
K_2O	0.16		0.45		0.88		0.53		0.45		0.31		0.90		0.32		0.43		0.61		1.19		0.54		0.77		0.91		0.99		0.87		0.99	
P_2O_5	0.14		0.15		0.22		0.12		0.15		0.13		0.18		0.15		0.16		0.14		0.21		0.18		0.23		0.23		0.26		0.19		0.22	
Molar Mg#	63.23		66.29		65.24		64.48		70.97		65.05		69.53		63.06		65.81		64.96		64.12		67.17		64.49		67.64		65.00		65.09		66.05	
La	4.13	59	7.01	168	11.85	159	6.69	10	8.72	50	5.00	46	8.38	21	6.49	14	8.95	12	5.98	27	18.79	28	11.29	24	14.75	29	5.60	16			6.92	41	15.99	13
Ce	11.46	62	15.67	181	25.87	157	15.27	10	18.94	56	11.73	51	19.26	20	15.19	16	18.11	12	14.39	27	41.23	26	24.22	24	30.54	29	14.06	16			16.89	41	32.76	13
Pr	1.84	6	2.11	55	2.85	65			0.72	7			2.51	21	2.28	11			2.21	5	7.10	10	1.90	5	1.35	17	1.70	6			2.64	23		
Nd	9.30	60	10.14	168	14.88	152	10.76	10	10.93	55	9.10	40	11.82	21	9.88	16	10.83	12	8.80	27	20.60	26	13.74	24	15.83	29	9.88	16			12.60	41	18.26	12
Sm	2.96	66	2.70	172	3.43	155	2.61	10	2.68	57	2.72	40	2.79	21	2.86	16	2.74	12	2.33	27	4.37	28	3.16	24	3.42	29	2.80	16			3.33	41	2.80	9
Eu	1.13	66	0.95	181	1.07	157	0.88	10	0.91	56	1.01	51	0.90	21	1.03	16	1.00	11	0.80	27	1.24	28	1.02	24	1.08	29	0.93	16			1.09	39	0.94	13
Gd	4.04	27	3.10	134	3.55	107			2.98	45	3.47	36	2.86	20	3.34	15	3.49	5	2.28	12	4.79	10	3.37	11	3.46	29	3.09	6			3.49	41	3.38	8
Tb	0.68	42	0.52	97	0.51	122	0.48	10	0.43	18	0.55	16	0.47	20	0.61	11	0.49	9	0.43	20	0.60	26	0.49	20	0.43	17	0.51	16			0.51	27	0.38	9
Dy	4.70	25	3.31	133	3.32	91			2.93	41	3.86	37	2.77	21	3.87	16	3.41	6	2.81*	25	4.04	10	2.93	5	3.29	29	2.39	6			3.51	31	2.76	8
Ho	0.93	6	0.62	58	0.68	65			0.40	10			0.58	20	0.83	11	0.58	5	0.52	5	0.75	10	0.65	5	0.60	17	0.74	6			0.70	23		
Er	3.00	29	2.00	136	1.95	84			1.69	43	2.38	37	1.65	21	2.33	16	2.20	6	1.36	11	2.14	10	1.79	5	1.88	29	2.15	6			2.02	24	1.62	8
Tm	0.47	4	0.28	33	0.29	63			0.20	11			0.24	6	0.36	11					0.31	9	0.29	5	0.25	17	0.32	6			0.29	22		
Yb	2.72	62	1.86	171	1.82	159	1.88	10	1.59	56	2.29	40	1.57	21	2.27	16	1.62	12	1.40	27	1.83	28	1.88	24	1.68	29	1.95	16			1.89	41	1.38	13
Lu	0.41	40	0.27	97	0.28	138	0.30	10	0.25	42	0.37	7	0.23	6	0.35	11	0.19	8	0.21	19	0.28	26	0.27	24	0.25	17	0.29	16			0.30	40	0.22	7
Sc	36.75	45	36.37	112	32.51	110	35.80	10	34.76	44	33.64	15	38.00	14	34.45	11	44.58	9	38.71	20	26.54	28	34.24	5	33.20	13	33.21	14			35.33	40	36.32	5
V	245.77	39	254.01	119	246.59	107	255.10	10	235.38	39	245.57	21	336.00	16	237.00	11	222.60	10	294.38	8	198.36	22	224.00	5	268.53	29	253.20	15	186.65	20	260.45	31		
Cr	357.10	48	575.68	132	397.96	145	317.10	10	974.55	47	322.50	22	420.11	18	266.33	15	716.67	9	449.93	27	344.42	26	358.75	24	491.51	29	269.13	15	507.53	45	442.26	39	339.57	7
Co	41.76	33	44.17	56	41.20	80			61.22	13	34.79	14	44.67	6	35.94	10	45.07	6	44.09	16	41.06	9	43.08	22					57.13	15	41.72	36	35.63	7
Ni	135.24	46	239.67	135	158.74	146	110.00	10	442.69	52	128.30	20	131.72	18	84.67	15	182.22	9	130.08	26	130.49	28	151.29	24	245.05	29	141.53	15	191.52	57	135.08	40	96.57	7
Cu	69.87	30	84.86	74	91.91	40	82.40	10	78.21	34			126.00	6	65.30	10			78.50	4			46.80	5	118.17	13	108.86	7			75.00	9		
Zn	67.27	30	72.22	97	81.30	68	77.60	10	74.59	40	60.08	12	72.17	6	77.06	16			91.86	7	83.86	22	63.00	5			81.89	9			83.52	27		
Rb	2.93	33	9.89	179	18.63	140	8.30	10	10.22	57	5.61	50	14.83	20	7.24	16	24.96	10	9.88	21	29.23	28	16.42	23	10.63	29	12.48	12	22.10	39	14.45	33	37.07	11
Sr	141.42	55	306.74	181	425.70	153	274.90	10	314.97	57	231.78	51	499.81	21	202.00	16	451.64	11	445.09	23	532.34	28	469.38	24	437.96	29	284.60	16	715.20	38	345.68	41	566.18	11
Y	27.48	46	19.47	145	18.69	141	20.40	10	17.45	54	24.14	21	17.86	21	23.29	16	16.64	11	12.90	18	19.74	28	20.91	22	18.67	29	13.15	16	22.54	21	19.54	32	16.10	10
Zr	92.63	54	62.21	145	92.70	144	65.90	10	59.65	53	65.14	22	51.58	21	86.22	16	50.91	11	54.94	18	118.96	28	105.96	24	79.94	29	63.38	16	122.55	22	87.14	32	96.00	11
Nb	6.04	20	3.99	129	6.23	135	2.30	10	4.79	52	3.81	14	1.83	20	6.71	16	5.88	6	4.05	17	11.76	24	6.70	21	7.03	29	3.51	14	11.80	21	2.90	32	5.05	11
Cs	0.05	6	0.32	54	0.71	87			0.32	12	0.08	10	0.14	6	0.11	11	1.34	5	0.46	21	1.00	19	0.38	22	0.22	10	0.43	2			0.49	27		
Ba	30.70	44	132.96	175	295.04	155	124.00	10	137.67	54	65.47	48	257.24	21	84.23	16	274.91	11	195.54	26	317.81	28	271.33	24	315.53	29	284.88	16	498.13	24	279.70	40	335.80	12
Hf	2.31	41	1.65	87	2.14	133	1.73	10	1.54	13	1.61	14	1.58	18	2.15	11	1.53	9	1.58	22	2.76	26	2.22	22	1.53	25	1.70	15			2.39	32	1.88	6
Ta	0.31	28	0.24	64	0.45	83	0.26	10	0.26	11	0.14	12	0.16	6	0.50	11	0.04	5	0.25	12	0.99	20	0.48	17	0.29	17					0.13	22	0.16	5
Pb	0.38	14	2.48	109	3.36	83	2.80	10	2.40	32	1.08	26	4.12	10	1.63	12	5.99	8	3.95	10	7.21	19	2.26	10	1.40	26	3.20	5			2.77	22		
Th	0.28	34	1.52	111	2.03	139	1.11	10	2.96	38	0.41	12	1.19	20	0.89	12	0.78	8	1.21	23	3.77	26	1.33	22	1.08	26	0.75	15			0.88	32	5.80	13
U	0.23	26	0.59	105	0.53	102	0.28	8	1.07	39	0.17	11	0.41	20	0.23	11	0.51	5	0.58	19	1.35	11	0.39	22	0.37	26					0.43	31	0.62	7
$^{87}Sr/^{86}Sr$	0.70274	104	0.70389	141	0.70401	133	0.70419	19	0.70482	46	0.70303	45	0.70392	4	0.70337	14	0.70406	7	0.70315	19	0.70515	11	0.70382	27	0.70388	25	0.70432	1	0.70437	27	0.70344	28	0.70442	4
$^{143}Nd/^{144}Nd$	0.51310	90	0.51298	124	0.51292	104	0.51293	11	0.51291	42	0.51306	41	0.51300	4	0.51301	5	0.51292	3	0.51303	16	0.51262	6	0.51285	27	0.51301	25	0.51293	1	0.51277	9	0.51307	26	0.51263	5
$^{206}Pb/^{204}Pb$	18.38	86	18.80	75	18.75	76	18.84	2	19.30	26	18.47	30	18.40	4	18.99	2	18.89	2	18.69	14	18.50	5	18.94	26	19.02	21	19.99	1			18.26	21	18.20	1
$^{207}Pb/^{204}Pb$	15.50	86	15.59	75	15.57	76	15.62	2	15.71	26	15.52	30	15.52	4	15.62	2	15.55	2	15.53	14	15.61	5	15.62	26	15.56	21	15.73	1			15.49	21	15.51	1
$^{208}Pb/^{204}Pb$	38.01	86	38.49	75	38.46	75	38.75	2	38.90	26	38.17	30	38.31	4	38.78	2	38.58	2	38.20	14	38.73	5	38.61	26	38.68	21	39.27	1			37.94	21	38.14	1

Major element oxides in wt.%, trace elements in ppm.
All analyses are included in major element average. Analyses with values for at least four of the REEs are included in the trace element averages, and only if there are ten of these analyses, is an individual arc included in the table. Only elements for which at least five analyses are available are included in the table.* For Aleutian samples without Dy analyses, we estimated Dy ~7.12*Tb

Table 1(b) Average primitive arc andesites (molar Mg# > 60).

	Continental		Oceanic		Aleutian		Boninites	
n majors	142		32		47		348	
SiO_2	58.05		57.72		59.03		56.83	
TiO_2	0.79		0.64		0.69		0.25	
Al_2O3	15.96		15.16		16.61		13.22	
FeO(T)	6.14		6.69		5.22		7.93	
MnO	0.12		0.14		0.10		0.15	
MgO	6.56		7.95		5.65		10.64	
CaO	7.20		7.32		7.35		8.35	
Na_2O	3.31		2.95		3.64		2.02	
K_2O	1.67		1.27		1.50		0.56	
P_2O_5	0.22		0.14		0.20		0.06	
Molar Mg#	65.18		66.29		65.62		69.50	
		n		*n*		*n*		*n*
La	18.89	59			16.02	28	1.88	74
Ce	37.44	53			37.27	28	4.44	73
Pr	5.38	13					0.92	23
Nd	20.89	42			20.20	28	2.79	63
Sm	3.92	56			3.95	28	0.77	71
Eu	1.08	59			1.12	28	0.28	76
Gd	3.92	26			3.61	5	0.92	55
Tb	0.51	47			0.43	23	0.17	38
Dy	3.09	23			3.06*	28	1.01	60
Ho	0.55	15					0.31	21
Er	1.63	22			1.77	7	0.70	55
Yb	1.54	57			1.32	28	0.80	69
Lu	0.23	51			0.19	27	0.14	53
Li	6.73	3					7.61	14
Be	0.87	3					0.26	16
Sc	20.98	41	26.14	18	17.55	25	36.45	54
V	158.27	19	196.66	21	170.00	2	188.01	70
Cr	326.83	55	260.75	21	252.76	27	696.03	76
Co	31.36	36	30.69	8	22.07	21	43.10	38
Ni	137.89	52	118.63	22	95.14	28	191.76	75
Cu	91.75	8	75.44	18	63.67	3	65.28	25
Zn	74.43	14	69.02	18	72.33	3	57.13	31
Rb	45.66	47	29.52	25	20.52	21	9.47	76
Sr	586.66	48	358.80	25	1,035.88	27	141.84	77
Y	17.13	37	20.34	25	14.85	22	7.59	68
Zr	137.19	41	91.04	25	115.33	15	39.05	74
Nb	7.94	34	4.08	20	4.95	21	2.17	31
Cs	2.27	20	1.61	4	0.44	23	0.19	14
Ba	501.74	41	273.27	27	309.62	28	54.53	73
Hf	3.56	32	1.83	5	3.06	23	0.70	43
Ta	0.85	22	0.08	4	0.25	20	0.13	22
Pb	8.45	20	6.00	13	4.70	13	1.83	15
Th	4.51	35	3.90	16	1.99	23	0.42	28
U	1.57	29	0.37	6	0.82	24	0.26	19
$^{87}Sr/^{86}Sr$	0.70469	31	0.70493	14	0.70291	13	0.70423	55
$^{143}Nd/^{144}Nd$	0.51277	27	0.51288	6	0.51308	18	0.51294	50
$^{206}Pb/^{204}Pb$	18.53	11	18.81	4	18.30	18	18.68	37
$^{207}Pb/^{204}Pb$	15.56	11	15.58	4	15.47	18	15.53	37
$^{208}Pb/^{204}Pb$	38.36	11	38.54	4	37.76	18	38.31	37

Major element oxides in wt.%, trace elements in ppm. Trace-element averages are calculated as in Table 1(a), except for oceanic andesites, where REE averages are not calculated (too few and variable analyses) and where all analyses are included for other trace elements.

Rudnick and Fountain, 1995; Weaver and Tarney, 1984) is almost identical to some high Mg# andesites in both major and trace-element concentrations, and some authors have proposed that the genesis of continental crust involved processes similar to the generation of high Mg# andesites today (e.g., Defant and Kepezhinskas, 2001; Drummond and Defant, 1990; Ellam and Hawkesworth, 1988b; Kelemen, 1995; Kelemen *et al.*, 1993, 2003b; Martin, 1986, 1999; Rapp and

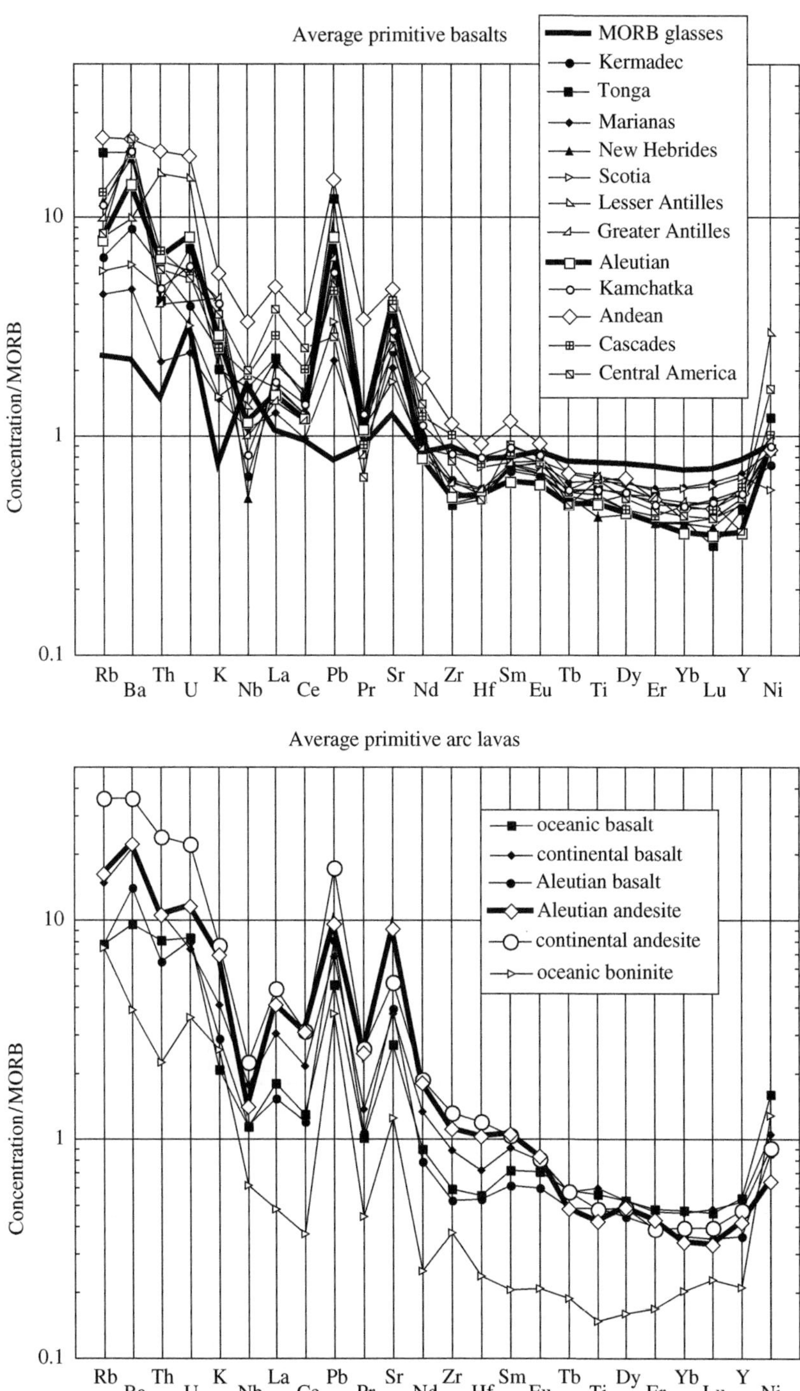

Figure 3 Extended trace-element diagrams for average arc lavas (Table 1). Concentrations are normalized to N-MORB (Hofmann, 1988). Primitive arc basalts are remarkably similar from one arc to another, and consistently distinct from MORB. In the oceanic Aleutian arc, and in continental arcs, primitive andesites are more enriched than primitive basalts. For plotting purposes some REE abundances are extrapolated from neighboring REEs with more analyses (Pr in Lesser Antilles, Dy in Greater Antilles, Er in Aleutian).

Watson, 1995; Taylor, 1977). Again, however, there is no consensus on this.

Finally, plutonic rocks with high Mg# andesite compositions probably form the bulk of the major calc-alkaline plutons in orogenic belts, such as the Mesozoic batholiths along the Pacific margins of North and South America. For example, the average composition of the Peninsular Ranges batholith in southern California is essentially identical to that of continental crust (Gromet and Silver, 1987; Silver and Chappell, 1988). Similarly, the average composition of exposed, Eocene

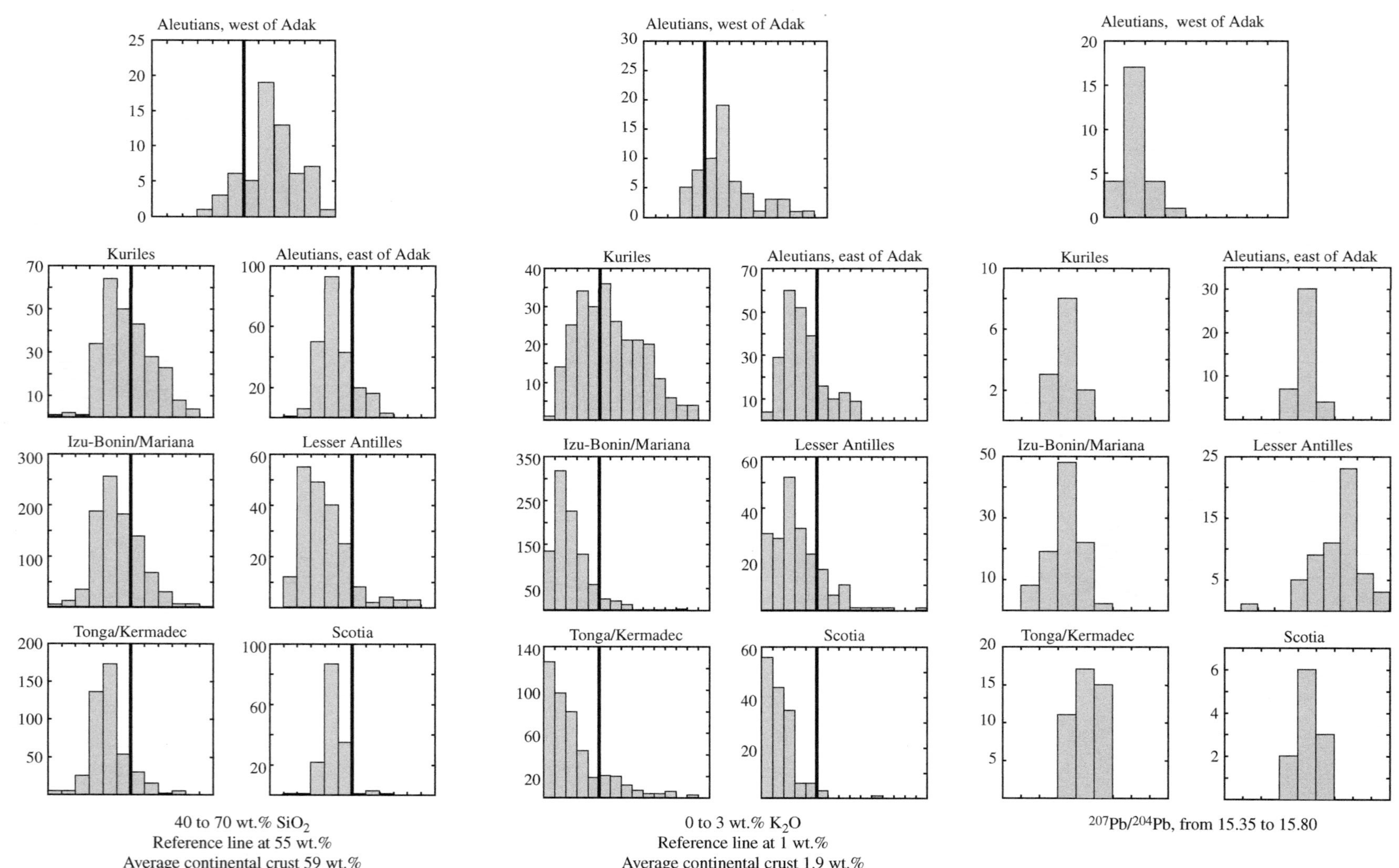
Aleutians, west of Adak
Kuriles
Aleutians, east of Adak
Izu-Bonin/Mariana
Lesser Antilles
Tonga/Kermadec
Scotia
40 to 70 wt.% SiO_2
Reference line at 55 wt.%
Average continental crust 59 wt.%
0 to 3 wt.% K_2O
Reference line at 1 wt.%
Average continental crust 1.9 wt.%
$^{207}Pb/^{204}Pb$, from 15.35 to 15.80

to Miocene plutonic rocks in the Aleutian arc is that of high Mg# andesite (Kelemen *et al.*, 2003b). Thus, high Mg# andesite magmas may be more commonly emplaced as plutonic rocks in the middle and upper crust (Kay *et al.*, 1990; Kelemen, 1995), and may be under-represented among erupted lavas. For these reasons, and because this topic has not received a recent, comprehensive review, one emphasis in this chapter is documentation of the difference between primitive arc basalts and andesites, and evidence bearing on its origin.

3.18.3.1.4 Major elements in calc-alkaline batholiths

> ...there is no reason to suppose that the relative amounts of magmas of different compositions erupted on the surface should be proportional to their amounts...at depth.
>
> Kuno (1968, p. 168)

This section is based on a limited compilation of plutonic rock compositions from "intermediate, calc-alkaline batholiths," such as are common among Mesozoic and early Tertiary exposures associated with circum-Pacific arcs. Section 3.18.4 describes the composition of plutonic rocks from exposed arc sections that extend into the lower crust, including a much larger proportion of mafic gabbros.

Our data include the few data available in GeoRoc, our previous compilation of Aleutian plutonic rock compositions (Kelemen *et al.*, 2003b), plutonic rocks from the Tanzawa complex, interpreted as accreted mid-crustal rocks from the Izu–Bonin arc (Kawate and Arima, 1998), and a compilation of limited data from highly "calc-alkaline" batholiths such as the Mt. Stuart and Chilliwack batholiths in the North Cascades (Erikson, 1977; Kelemen and Ghiorso, 1986; Tepper *et al.*, 1993), the Peninsular Ranges batholith in Baja California (Gromet and Silver, 1987; Larsen, 1948; Silver and Chappell, 1988), and the Ladakh batholith in northwestern India (Honegger *et al.*, 1982). Unfortunately, because this data set is small, it is not clear to what extent the compiled compilations are representative of intermediate arc plutons in general.

A second problem, in interpreting plutonic rock compositions, is determining the extent to which they represent liquid versus "cumulate" compositions, where "cumulate" is taken to mean a component formed by partial crystallization of a melt, after which the remaining melt was extracted from the system of interest. A wide array of possible plutonic compositions can be envisioned, lying between these two extremes. For example, some plutonic rocks may be cumulates plus a small amount of "trapped melt." Others may be cumulates affected by interaction with unrelated, migrating melts.

Cumulates with abundant plagioclase should generally have high Sr/Nd, since strontium is much more compatible than neodymium in plagioclase, and Eu/Sm, since europium is generally much more compatible than samarium in plagioclase (depending on oxygen fugacity). In general, plutonic rocks with more than 55–60 wt.% SiO_2 closely resemble liquid compositions in many ways, often containing abundant incompatible elements and lacking anomalously high Sr/Nd and Eu/Sm. We quantify this for intermediate to felsic plutonic rocks from the Jurassic Talkeetna arc section in Section 3.18.4.1.1.

Our compiled plutonic rock compositions are generally more SiO_2-rich, at a given Mg#, than the compiled arc lavas (Figure 5). This is particularly clear in comparing Aleutian lavas to Aleutian plutons (Kelemen *et al.*, 2003b). TiO_2 is highest in plutonic rocks with Mg# of ~60. Al_2O_3 is low in plagioclase-poor, high Mg# pyroxenites, and then—generally—similar to the lower Al_2O_3 arc lavas at a given Mg#. Arc plutons have lower FeO, MgO, and CaO, at a given Mg#, than the bulk of arc lavas. Although some primitive cumulates in our compilation have very low alkali contents, in general Na_2O contents in arc plutons and lavas are comparable, while K_2O contents are generally higher in arc plutons compared to arc lavas at the same Mg#. To summarize, in all of their major-element characteristics, the samples in our compilation of arc plutons are more strongly calc-alkaline, and include more high Mg# andesite compositions, than typical arc lavas.

It seems that high Mg# andesite liquids may be better represented among intermediate plutonic rocks than among arc lavas. This is certainly the case for exposed Aleutian plutons compared to Aleutian lavas. In order to explain this, following Kay *et al.* (1990), we have suggested that this

Figure 4 Histograms of wt.% SiO_2, wt.% K_2O, and $^{207}Pb/^{204}Pb$ for intra-oceanic arc lavas in our compilation. SiO_2 and K_2O are for lavas with Mg# > 50, while Pb isotopes are for all lavas. This diagram shows that relatively low K basalts predominate over relatively K-rich, high Mg# andesites in all intra-oceanic arcs except the western Aleutians. The western Aleutian arc also has the least radiogenic Pb isotopes of any intra-oceanic arc. Thus, the predominance of primitive andesites in the western Aleutians is probably not due to recycling of components from subducting, continental sediment, nor to crustal contamination involving pre-existing continental material.

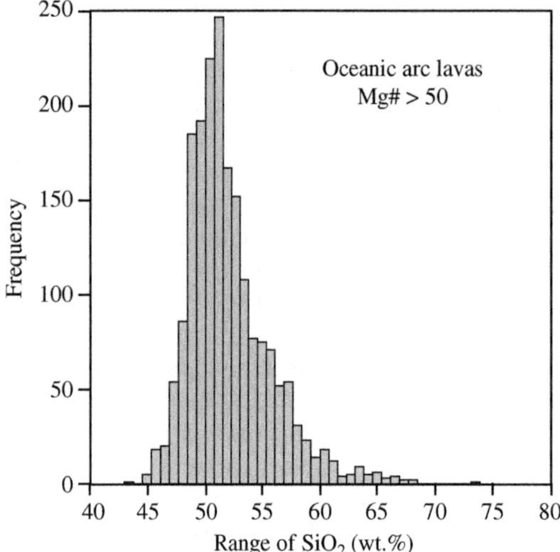

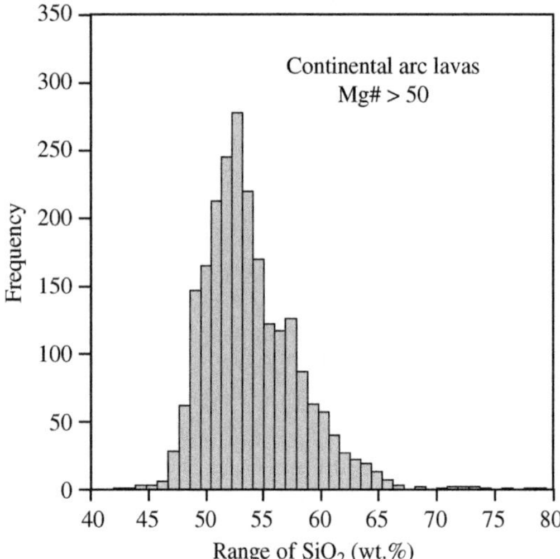

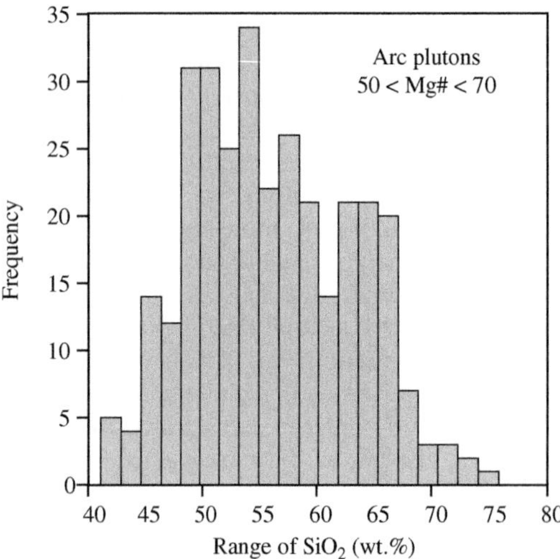

Figure 5 Histograms of wt.% SiO_2 for arc lavas and a limited compilation of samples from arc plutons and batholiths. In general, arc plutonic rocks have higher SiO_2 content at similar Mg#. Data sources as in Figure 1.

difference arises due to the relatively high viscosity of intermediate to felsic magmas as they lose H_2O by degassing in the mid- and upper crust (Kelemen, 1995; Kelemen *et al.*, 2003b). Lower H_2O, low-SiO_2 basaltic melts continue to rise to the surface and erupt, whereas higher SiO_2 magmas, with initially higher H_2O, rise more slowly and crystallize faster.

3.18.3.2 Major and Trace-element Characteristics of Primitive Arc Magmas

> [A few scientists] seem to want to cling to the possibility that andesites are primary mantle-derived melts despite overwhelming evidence from experimental petrology, trace-element geochemistry, mineral chemistry, petrography, textural, and field relations to the contrary. It would be hard to find very many students of arc petrology who would argue...that andesites, even Mg-rich ones, are... primary mantle melts.
>
> Anonymous (2003, personal communication)

> ...primitive magnesian andesites and basaltic andesites from the Mt. Shasta region, N. California...form by hydrous mantle melting
>
> Grove *et al.* (2003)

We turn now to the chemical characteristics of primitive arc lavas (Mg# $>$ 60). First, we examine major-element variation as a function of Na_2O and TiO_2 contents. Although Na_2O contents of primitive basalts and andesites overlap, plots of Na_2O versus TiO_2, FeO, MgO, and CaO clearly discriminate between boninites (very low TiO_2 and Na_2O) primitive basalts (high TiO_2, FeO, MgO, and CaO at a given Na_2O), and primitive andesites (low TiO_2, FeO, MgO, and CaO at a given Na_2O) (Figure 6). It is evident from the trends of these elements versus Mg# (Figure 1) that these variations do not arise from crystal fractionation. For example, arc lavas with Mg# of ~70 or more, in Fe/Mg equilibrium with mantle olivine having Mg# of 90–91 or more, have SiO_2 contents ranging from 45 wt.% to 63 wt.%. These cannot be related by fractional crystallization. The antithetical behavior of sodium and titanium (high sodium and low titanium in primitive andesites, high titanium and low sodium in primitive basalts; ~2 wt.% Na_2O in both primitive MORB and primitive arc basalts, but 0.3–1 wt.% TiO_2 in primitive arc basalts compared to 1.2 wt.% in primitive MORB, Table 1) suggests that sodium contents of arc magmas may not be a good indicator of the degree of partial melting in the sub-arc mantle. (Also see Figure 2, and the last three paragraphs of Section 3.18.3.1.1.)

Although there is some overlap, plots of rubidium, barium, thorium, strontium, lead, zirconium, hafnium, and light REE versus Na_2O also

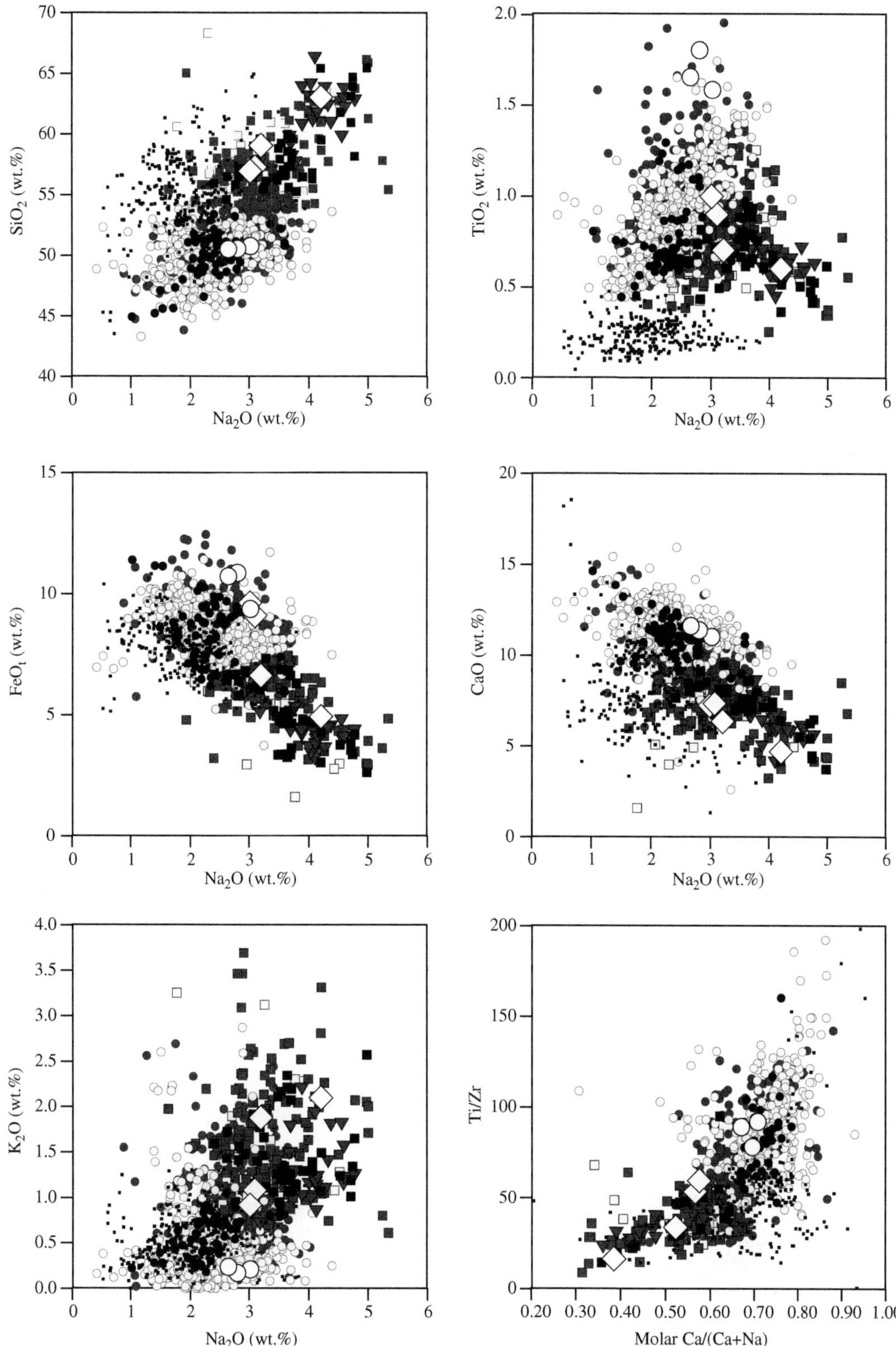

Figure 6 Wt.% Na_2O versus other major-element oxides, in wt.%, plus molar Ca/(Ca+Na) versus ppm Ti/Zr, for primitive arc lavas (Mg# > 60). Many of these plots clearly show distinct compositional fields for primitive basalts, primitive andesites, and boninites. While most of the primitive andesites are from "continental" arcs, they plot together with western Aleutian primitive andesites, which are from an intra-oceanic arc and have MORB-like Sr, Pb, and Nd isotope ratios. Thus, assimilation of older, continental material is not essential to producing the distinctive composition of primitive andesites. Large filled circles show values for average MORB glasses from the East Pacific Rise, Juan de Fuca Ridge, and Indian Ocean. Other symbols and data as for Figure 1.

discriminate between these groups (for example, lanthanum versus Na_2O, Figure 7), with high LILEs, light REEs, and other highly incompatible elements in primitive andesites as compared to primitive basalts at a given Na_2O concentration. Middle REE concentrations are similar in both andesites and basalts, while heavy REE and related elements (holmium, erbium, thulium, ytterbium, lutetium, yttrium, scandium, vanadium) are lower in andesites than in basalts at a given Na_2O concentration (e.g., scandium and ytterbium versus Na_2O, Figure 7). Low heavy REE and titanium contents in arc magmas have sometimes been considered to be indicative of a highly refractory mantle source and/or high degrees of melting beneath arcs, and scandium may be more compatible than sodium during mantle melting. However, it seems to us that either heavy REE, yttrium, titanium, and vanadium are poor indicators of the extent of mantle melting in the source of primitive lavas, or Na_2O is a poor indicator, or neither is a good proxy for the extent of mantle melting.

Positive correlation between FeO, scandium, vanadium, and titanium, and negative correlations of each of these with SiO_2, for all primitive arc lavas raises, once again, the question of whether the concentrations of all these elements are related to FeTi oxide fractionation. However, the plot of TiO_2 versus Mg# shows that

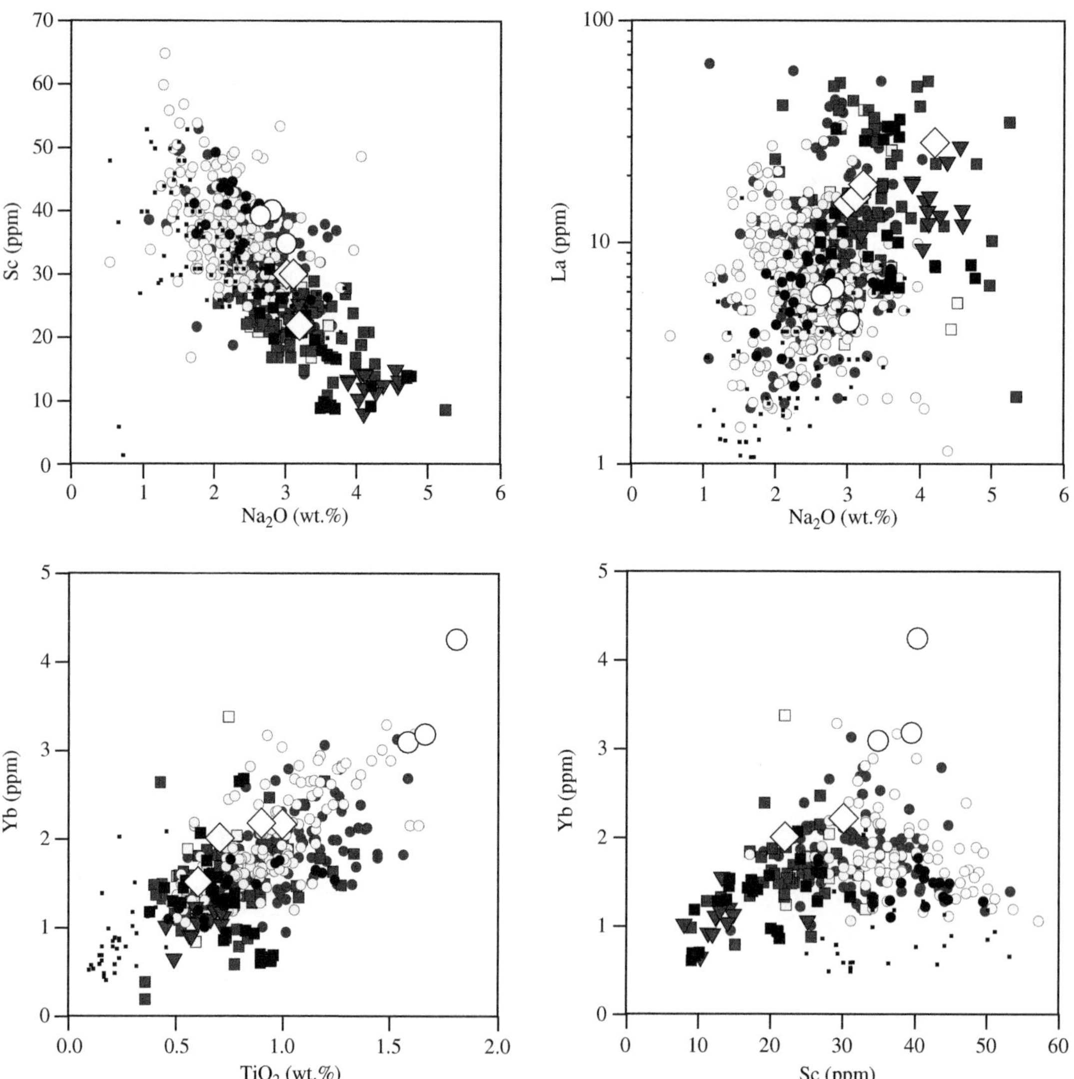

Figure 7 Wt.% Na_2O, wt.% TiO_2, and ppm Sc versus other trace-element concentrations in ppm, for primitive arc lavas. Again, these plots clearly separate andesites from basalts. Primitive andesites from the oceanic, western Aleutians plot with other primitive andesites from continental arcs. Combined heavy REE, Y, Sc, Ti, and V depletion are indicative of either an important role for garnet fractionation, or a more depleted mantle source, in the genesis of primitive andesites as compared to primitive basalts. Data and symbols as for Figure 6.

concentration of titanium *increases* with decreasing Mg# in primitive lavas (Figure 1), in accord with experimental studies showing that primitive lavas at oxygen fugacities typical for arcs are not saturated in FeTi oxides. In addition, heavy REE concentrations are correlated with TiO_2 in primitive arc lavas. Heavy REE are incompatible in FeTi oxides and silicates other than garnet (e.g., EarthRef database at http://earthref.org/ and references therein). These two observations almost certainly rule out an important role for FeTi oxide fractionation in controlling the major-element compositions of primitive lavas. Instead, the similar behavior of vanadium, titanium and heavy REE suggests either high degrees of mantle melting, or an important role for garnet fractionation, in the genesis of primitive andesites (see Section 3.18.3.2.5).

3.18.3.2.1 Primitive basalts predominate

Primitive arc basalts have trace-element characteristics that are very distinct from primitive MORB (Figure 3). Figure 4 shows that primitive basalts are more commonly sampled than any other primitive magma type in most oceanic arcs. Biased sampling of picturesque and dangerous strato-volcanoes composed of calc-alkaline andesite, rather than low lying basaltic shields, may have complicated this picture for continental arcs, but basalts may predominate among primitive lavas in continental arcs as well (e.g., White and McBirney, 1978). The geochemical characteristics of primitive basalts have been the subject of numerous recent reviews (Elliott, 2003; Davidson, 1996; Elliott *et al.*, 1997; Hawkesworth *et al.*, 1993a,b, 1997; Pearce and Peate, 1995; Plank, 2003; Plank and Langmuir, 1988, 1993, 1998; Tatsumi and Eggins, 1995; Turner *et al.*, 2003, 2001).

Considerable uncertainty remains regarding the relative importance of various processes in producing primitive arc basalts. For example, in some arcs the presence of nearly anhydrous, primitive basalts suggests a large role for decompression melting (Bartels *et al.*, 1991; Draper and Johnston, 1992; Elkins Tanton *et al.*, 2001; Sisson and Bronto, 1998), although see also Turner and Foden (2001), with major and trace-element systematics that might be similar to those beneath mid-ocean ridges (e.g., Plank and Langmuir, 1988, 1993). Possible mechanisms for decompression melting include near vertical, diapiric upwelling of low density mixtures of melt + mantle peridotite (e.g., Davies and Stevenson, 1992; Iwamori, 1997) or diagonal upwelling of peridotite in return flow due to viscous entrainment of the mantle wedge with the subducting plate (Conder *et al.*, 2002; Furukawa, 1993a,b; Kelemen *et al.*, 2003a; Kincaid and Sacks, 1997; van Keken *et al.*, 2002).

Other evidence implies that decompression may be minor or absent, and "fluxed melting" of the mantle is controlled mainly by addition of H_2O and other fluxes from subducting material into the mantle wedge. This has generally been modeled as "fluid-fluxed melting," resulting from addition of an aqueous fluid to initially solid, but hot peridotite (Abe *et al.*, 1998; Eiler *et al.*, 2000; Grove *et al.*, 2001, 2003; Ozawa, 2001; Ozawa and Shimizu, 1995; Stolper and Newman, 1992). Thus, static or even descending mantle peridotite in the wedge could partially melt if sufficient aqueous fluid were added, provided aqueous fluid reached parts of the mantle where temperature exceeded the fluid-saturated peridotite solidus. Substantial H_2O contents in primitive arc basalts, commonly ~3 wt.% (Anderson, 1974; Baker and Eggler, 1987; Falloon and Danyushevsky, 2000; Kamenetsky *et al.*, 1997; Kelemen *et al.*, 1990b; Macdonald *et al.*, 2000; Müntener *et al.*, 2001; Pichavant *et al.*, 2002; Roggensack *et al.*, 1997; Sisson and Grove, 1993b; Sisson and Layne, 1993; Sobolev and Chaussidon, 1996), but more than 4.5 wt.% in primitive andesites (Grove *et al.*, 2003) are often taken as evidence for addition of H_2O-rich fluid to the mantle wedge, and for fluxed melting.

A variant on the flux melting hypothesis is the idea of "melt-fluxed melting," in which reaction between hydrous partial melts of subducting sediment and/or basalt and overlying mantle peridotite leads to increasing melt mass, producing a hybrid "primary melt" in which more than 90% of the compatible elements, such as magnesium, iron, and nickel, are derived from the mantle, while most of the alkalis and other incompatible trace elements come from the initial, melt of subducted material (e.g., Kelemen, 1986, 1990, 1995; Kelemen *et al.*, 1993, 2003b; Myers *et al.*, 1985; Yogodzinski *et al.*, 1995; Yogodzinski and Kelemen, 1998). If this process occurs, it would be facilitated by the fact that melts migrating upwards in the mantle wedge must heat as they decompress; under such circumstances, even anhydrous melts will be able to dissolve solid mantle minerals comprising tens of percent of the initial liquid mass (Kelemen, 1986, 1990, 1995; Kelemen *et al.*, 1993). Similarly, Grove and co-authors propose that fluid-saturated partial melts of peridotite form at ~950 °C, very close to the subducting plate, and then these fluid-saturated peridotite melts cause melt-fluxed melting higher in the wedge (Grove *et al.*, 2003). Aqueous-fluid-saturated melts of eclogite facies sediment or basalt, and fluid-saturated melts of mantle peridotite, would have 25–50 wt.% H_2O at 3–5 GPa (Dixon and Stolper, 1995; Dixon *et al.*, 1995; Kawamoto and Holloway, 1997;

Mysen and Wheeler, 2000). In this regard, they would be efficient fluxing agents, causing additional melt to form via melt/rock reaction in the mantle wedge (Eiler *et al.*, 2000). In principle, addition of such H_2O-rich melts to the mantle wedge could explain the substantial water contents in primitive arc magmas, without additional H_2O from a fluid.

It is worth noting that melt-fluxed melting is distinct from most open-system processes proposed to explain melting beneath mid-ocean ridges, including batch melting (e.g., Gast, 1968; Presnall and Hoover, 1984; Shaw, 1970), fractional melting (e.g., Gast, 1968; Johnson *et al.*, 1990; Langmuir *et al.*, 1977; Richardson and McKenzie, 1994; Shaw, 1970), incremental melting (e.g., Johnson *et al.*, 1990; Kinzler and Grove, 1992; Kinzler and Grove, 1993; Klein and Langmuir, 1987; Langmuir *et al.*, 1977), and continuous melting (Iwamori, 1994; Johnson and Dick, 1992; Langmuir *et al.*, 1977; Sobolev and Shimizu, 1992), because in the latter processes melt forms due to decompression, not as a result of reaction between solid phases and migrating melt. In zone refining (e.g., Harris, 1957; Langmuir *et al.*, 1977), melt mass is constant, so this too is different from melt-fluxed melting. Models of melt generation beneath mid-ocean ridges that include increasing melt mass due to reactive porous flow (e.g., Asimow and Stolper, 1999; Iwamori, 1994; Jull *et al.*, 2002; Lundstrom *et al.*, 1995, 2000; Spiegelman and Elliot, 1992; Spiegelman *et al.*, 2001) are similar to proposed processes of melt-fluxed melting beneath arcs. (Note that Langmuir *et al.*, 1977 mentioned but did not model this process for MORB genesis.) However, beneath arcs—unlike ridges—melt-fluxed melting may be extensive, even in regions that are not simultaneously undergoing decompression melting.

Understanding the different mantle melting processes, and determining their relative importance in the generation of primary arc basalts, is an active area of research, with much sponsorship from the US National Science Foundation's MARGINS Initiative. We anticipate rapid developments in increasingly refined theories on this topic.

Another very active area of recent research is the identification of several different source components in primary arc basalts, including (i) fluids derived by dehydration of subducting metabasalt, (ii) fluids derived by dehydration of subducting metasediment, (iii) partial melts of subducting basalt, (iv) partial melts of subducting sediment, (v) fertile mantle peridotite similar to the MORB source, (vi) mantle peridotite depleted by melt extraction beneath a mid-ocean ridge and/or a back-arc basin, and (vii) enriched mantle similar to the source of ocean island basalt. We return to this topic in Section 3.18.3.3.

Nonetheless, as we show in previous and subsequent sections of this chapter, focusing exclusively on arc basalts risks missing end-members whose characteristics epitomize the difference between arc versus mid-ocean ridge magmas. Thus, in the following sections we focus on other types of primitive arc lavas. However, some detailed characteristics of primitive arc basalts, together with other primitive arc magmas, are described in Section 3.18.3.3.

3.18.3.2.2 Are some low Mg# basalts primary melts? perhaps not

Hypothetical derivation of primary, low Mg# basalts and andesites from partial melting of subducted basalt—without major-element equilibration with the overlying mantle—remains controversial for arcs (e.g., Brophy and Marsh, 1986; Johnston and Wyllie, 1988; Marsh, 1976; Myers *et al.*, 1986a,b, 1985) as well as hotspots (e.g., Chauvel and Hemond, 2000; Hauri, 1995; Korenaga and Kelemen, 2000; Lassiter and Hauri, 1998; Sobolev *et al.*, 2000) and even mid-ocean ridges (e.g., Schiano *et al.*, 1997). Based on the criteria outlined by Gill (1981, 1974, 1978), we see no evidence for direct partial melts of subducted, eclogite facies sediment or basalt in our data compilation. It may be that diapirs of melting basalt always rise to depths at which garnet is no longer stable, prior to separation of melt from residue (Brophy and Marsh, 1986), but there seems to be no direct evidence for this. Instead, arc lavas with a trace-element signature consistent with derivation via partial melting of eclogite (e.g., high middle/heavy REE ratios) are primitive, with Mg# > 60 (e.g., Grove *et al.*, 2001; Kay, 1978; Kelemen *et al.*, 2003b; Yogodzinski *et al.*, 1995; Yogodzinski and Kelemen, 1998). Thus, in this chapter we make the simplifying assumptions that lavas with Mg# < 60 are derived from primitive melts via crustal differentiation, and that all melts passing from the mantle wedge into arc crust have Mg# > 65 (depending on Fe^{3+}/Fe^{2+}), and are close to Fe/Mg exchange equilibrium with mantle peridotite (Mg# ~70).

Related to this topic are questions about the genesis of tonalites, trondhjemites, and granodiorites (TTGs), that are common in Archean cratons (see Chapter 3.11). Although it is difficult to be certain, we believe that TTGs are probably not *primary* melts of subducting eclogite as has been proposed (Defant and Kepezhinskas, 2001; Martin, 1986, 1999; Rapp and Watson, 1995; Rapp *et al.*, 1991), simply because it does not seem likely that H_2O-rich, low temperature melts could traverse the high temperature mantle wedge without substantial reaction with peridotite. Instead, we infer that TTGs are probably the products of intracrustal differentiation, with felsic melts rising to the upper crust, and mafic residues remaining in the lower crust. (Note that, while seismic and petrologic data on continental crust clearly establish that it is differentiated, the intracrustal differentiation process could have modified an initially andesitic *or* basaltic bulk composition.) They may have evolved by crystal fractionation from a parental, primitive andesite melt. However, because the process of intracrustal differentiation may have involved residual garnet, it is difficult to discern which TTGs with heavy REE depletion inherited their trace-element characteristics from primitive andesites, and which reflect crustal garnet fractionation.

*3.18.3.2.3 **Boninites, briefly***

In determining the characteristics of "primary" arc magmas—melts that pass from residual mantle into the overlying, igneous crust—most recent reviews of arc geochemistry have concentrated on the characteristics of primitive basaltic magmas. In doing so, these reviews have implicitly incorporated the assumption that primary arc magmas are invariably basaltic. In our view, there are two types of "andesitic" primitive magmas in arcs, boninites and primitive andesites. These two types of magmas extend to end-members having Mg# > 0.7, and carry olivine phenocrysts with Mg# > 90 (typical mantle values). While these are less common than primitive basalts, we think they are important for the reasons enumerated in Section 3.18.3.1.3.

While there are many far more detailed definitions and subdivisions of boninite lava compositions (e.g., Crawford, 1989), we found it convenient to simply define boninites as lavas with >54 wt.% SiO_2, <0.5 wt.% TiO_2 (*plus samples described in the original data sources as boninites, including some basalts*). As can be seen in Figures 1, 6, and 7, lavas defined in this way share many other distinctive characteristics, including high MgO and low alkali contents at a given SiO_2 and Mg#. Boninites are largely restricted to western Pacific island arcs, and in those arcs they are apparently more abundant in the early stages of magmatism (e.g., Bloomer and Hawkins, 1987; Falloon *et al.*, 1989; Stern and Bloomer, 1992). Their high MgO contents (some >10 wt.%) and the presence of clinoenstatite phenocrysts, probably reflect both high temperatures and high water contents in the mantle wedge, with a highly depleted, harzburgite residue, consistent with generally low REE concentrations and flat to light REE depleted patterns (e.g., Falloon and Green, 1986; Falloon *et al.*, 1989; Pearce *et al.*, 1992; Sobolev and Danyushevsky, 1994). Most authors accept that most boninites are derived by crystal fractionation from primary andesite melts derived by high degrees of relatively low pressure melting, with a harzburgite residue. However, some lavas termed boninites could conceivably be derived via substantial olivine ± low calcium pyroxene fractionation from very high Mg#, incompatible-element-depleted, primary picrites or komatiites.

The abundance of boninites in early stages of western Pacific arc magmatism, combined with the high magmatic fluxes inferred for the early stages of magmatism in those arcs, may have led to bulk crustal compositions that remain dominantly boninitic. Relatively low bulk crustal seismic velocities in the Izu–Bonin arc (Suyehiro *et al.*, 1996), compared to the central Aleutian arc (e.g., Holbrook *et al.*, 1999), might reflect higher SiO_2 in the Izu–Bonin crust. Nonetheless, boninitic crust with >54 wt.% SiO_2 would be depleted in alkalis and light REEs, and thus very different from continental crust, and from calc-alkaline magma series.

*3.18.3.2.4 **Primitive andesites: a select group***

We turn now to primitive andesites (Mg# > 0.6) and high Mg# andesites (Mg# > 0.5). High Mg# andesites have been the subject of much attention in recent years because of their unique major and trace-element characteristics (e.g., Baker *et al.*, 1994; Defant *et al.*, 1991a; Defant and Drummond, 1990; Defant *et al.*, 1992, 1989, 1991b; Defant and Kepezhinskas, 2001; Grove *et al.*, 2001, 2003; Kay, 1978; Kelemen, 1995; Kelemen *et al.*, 2003b; Rogers *et al.*, 1985; Shimoda *et al.*, 1998; Stern and Kilian, 1996; Tatsumi, 1981, 1982, 2001a; Tatsumi and Ishizaka, 1981, 1982; Yogodzinski and Kelemen, 1998, 2000; Yogodzinski *et al.*, 1995, 2001, 1994).

Some high Mg# andesites—particularly some lavas on Adak Island in the Aleutians (Kay, 1978)—have been called "adakites" (e.g., Defant and Drummond, 1990) as well as "sanukitoids," "high Mg andesites," and "bajaites." The term "adakite" is used in a variety of contexts by different investigators, but generally refers to andesites and dacites with extreme light REE enrichment (e.g., La/Yb > 9), very high Sr/Y ratios (e.g., Sr/Y > 50), and low yttrium and heavy REE concentrations (e.g., Y < 20 ppm, Yb < 2 ppm). In the Aleutians, all lavas with these characteristics are high Mg# andesites and dacites. However, the *de facto* definition of "adakite" does not specify a range of Mg#. Worldwide, many evolved lavas have been termed adakites. Thus, not all adakites are high Mg#

andesites. Similarly, most high Mg# andesites, in the Aleutians and worldwide, have La/Yb < 9 and Sr/Y < 50, and so not all high Mg# andesites are adakites. Finally, for some authors adakite has a genetic connotation. Some investigators infer that all andesites and dacites with extreme light REE enrichment, very high Sr/Y ratios, and low yttrium and heavy REE concentrations formed via partial melting of subducted basalt in eclogite facies, and use the term adakite to refer to both composition and genesis interchangeably. While we believe that many high Mg# andesites do indeed include a component derived from partial melting of eclogite, we feel it is important to separate rock names, based on composition, from genetic interpretations. For this reason, we do not use the term adakite in this chapter.

3.18.3.2.5 Three recipes for primitive andesite

Loosely speaking, the difference between primitive basalts and primitive andesites might arise in several ways.

(i) Both may arise from melting of different sources, with primitive andesites incorporating a relatively large proportion of melts of subducted basalt and/or sediment, compared to primitive basalts.

(ii) They might arise from the same mantle source, with different degrees of melting, related to different extents of enrichment via fluids derived from subducting sediment and/or oceanic crust.

(iii) High Mg# andesites might arise via mixing of primitive basalts with evolved, high SiO_2 melts, or assimilation of "granitic" rocks in primitive basalts.

We briefly expand on each of these in the next few paragraphs.

(i) Primary andesite magma with an eclogite melt component. High Mg# andesites may incorporate a component formed by partial melting of subducted basalt or sediment in eclogite facies, which subsequently reacted with the overlying mantle peridotite to form a hybrid melt (Carroll and Wyllie, 1989; Kay, 1978; Kelemen, 1986, 1995; Kelemen *et al.*, 1993, 2003b; Myers *et al.*, 1985; Yogodzinski *et al.*, 1995, 1994). In the hybrid melt, high incompatible-element contents reflect eclogite melting, and major-element concentrations reflect melt/mantle equilibration. In this view, high H_2O, K_2O, and Na_2O contents stabilize high SiO_2 melt in equilibrium with mantle olivine at ~1 GPa, as demonstrated experimentally for simple systems (Hirschmann *et al.*, 1998; Kushiro, 1975; Ryerson, 1985), peridotite melting experiments (Hirose, 1997; Kushiro, 1990; Ulmer, 2001), and phase equilibrium experiments on natural primitive andesite compositions (e.g., Baker *et al.*, 1994; Grove *et al.*, 2003; Tatsumi, 1981, 1982).

In some cases the entire incompatible trace-element budget of these hybrid melts might be derived from eclogite melting, with only major elements and compatible trace elements (nickel, chromium) affected by interaction with peridotite. However, there are few primitive lavas in our compilation with a clear eclogite melting signature. Since heavy REEs and yttrium are compatible in garnet, heavy REE and yttrium concentrations in eclogite melts should be low, and middle to heavy REE ratios in eclogite melts should be high (e.g., chondrite normalized Dy/Yb > 1.5). The few primitive lavas in our compilation that do have chondrite normalized Dy/Yb > 1.5 are mainly primitive andesites from the western Aleutian arc, at and west of Adak Island. Thus (Figure 8), not all light REE enriched, high Sr/Nd arc lavas have high Dy/Yb.

Lack of a clear eclogite melting signature in heavy REE and yttrium contents does not rule out a role for eclogite melt in producing high Mg# andesites. In hypothesis (i), the concentrations of heavy REE and yttrium in primitive arc andesites are interpreted as having been raised by reaction of eclogite melt with mantle peridotite at moderate melt/rock ratios (Kelemen, 1995; Kelemen *et al.*, 1993, 2003b). Modeling shows that this process can produce a very close match to most high Mg# andesite compositions (e.g., Kelemen *et al.*, 2003b; figures 21D and 21E). In this interpretation, primitive arc basalts incorporate a finite but smaller amount of an eclogite melt component (e.g., Kelemen *et al.*, 2003b; figure 21F).

As far as we know, there are few if any petrological or geochemical arguments that can be used to rule out this hypothesis. In fact, excluding heavy REEs and yttrium, the incompatible trace-element abundances in the "subduction component" (McCullouch and Gamble, 1991) inferred from inversion of major and trace elements in Marianas back-arc lavas (Stolper and Newman, 1992), an array of western Pacific arc lavas focused on the Vanuatu arc (Eiler *et al.*, 2000), and primitive lavas from Shasta volcano in the southern Cascades (Grove *et al.*, 2001) closely resemble trace-element concentrations experimental and predicted partial melts of eclogite (Kelemen, 1995; Kelemen *et al.*, 1993, 2003b; Rapp *et al.*, 1999), and of erupted high Mg# andesites. One inverse model that is consistent with all available major-element, trace-element and isotopic constraints (Eiler *et al.*, 2000, p. 247) involves reaction between mantle peridotite and a silicate melt derived from subducting eclogite, with 30 wt.% H_2O (appropriate for fluid-saturation at 3–5 GPa, Dixon and Stolper (1995), Dixon *et al.* (1995), Kawamoto and Holloway (1997), Mysen and Wheeler (2000)).

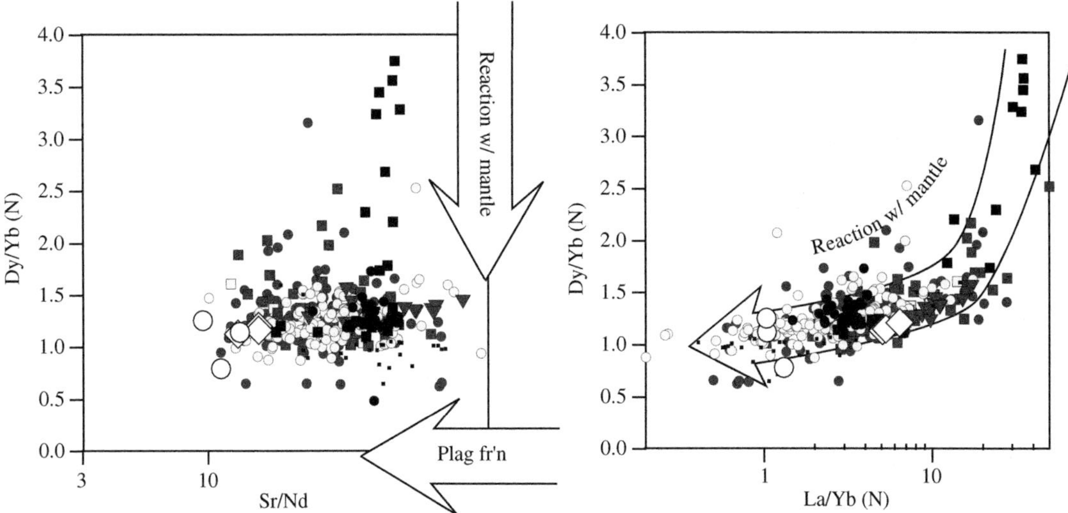

Figure 8 Relationship between Sr/Nd, La/Yb, and Dy/Yb for primitive arc lavas. High Dy/Yb is probably indicative of an important role for residual garnet in the genesis of some lavas. Not all lavas with very high La/Yb and Sr/Nd have high Dy/Yb. Thus, it may be unwise to use La/Yb and high Sr as indications that a given igneous rock is derived from a source with abundant, residual garnet. Arrows labeled "reaction with mantle" show results for trace-element models of reaction of a partial melt of MORB in eclogite facies with upper mantle peridotite (Kelemen *et al.*, 2003b). Arrow marked "plag fr′n" emphasizes that, because Sr is much more compatible than Nd in plagioclase, crystal fractionation of plagioclase, or crystal assemblages with cotectic proportions of plagioclase, leads to decreasing Sr/Nd. Data and symbols as for Figure 6. Sr/Nd on x-axis in left panel is not normalized, and is on a logarithmic scale.

Nevertheless, this hypothesis has been unpopular since the 1980s, because geodynamic models, incorporating either constant mantle viscosity or a rigid upper plate of prescribed thickness, predicted that solidus temperatures could not be reached in basalt or sediment at the top of the subducting plate, except under unusual circumstances (see reviews in Kelemen *et al.*, 2003a; Peacock, 1996, 2003; Peacock *et al.*, 1994). For example (Eiler *et al.*, 2000, p. 247), discounted their successful model involving H_2O-rich silicate melt because "[the] successful melt-fluxed [model]...require[s]...temperatures...that are...not obviously compatible with...thermal models... (Peacock, 1996)."

Recently, as discussed in Section 3.18.3.3.4, subduction zone thermal models that incorporate thermally dependent viscosity and/or non-Newtonian viscosity in the mantle wedge predict temperatures higher than the fluid-saturated solidus near the top of the subducting plate beneath arcs at normal subduction rates and subducting plate ages (Kelemen *et al.*, 2003a; van Keken *et al.*, 2002). While this is an area of active research, it is no longer the case that thermal models "rule out" partial melting of subducted material in eclogite facies.

(ii) Primary andesite magma from fluxed melting. A more popular model for arc magma genesis is "fluid-fluxed melting" (Section 3.18.3.2.1). An aqueous fluid derived from subducted basalt and sediment enriches the mantle source of arc magmas in "mobile elements," while simultaneously causing partial melting of that source (e.g., Abe *et al.*, 1998; Eiler *et al.*, 2000; Grove *et al.*, 2001, 2003; Ozawa, 2001; Ozawa and Shimizu, 1995; Stolper and Newman, 1992; Vernieres *et al.*, 1997). In this interpretation, high LILEs and light REEs are seen as the result of fluid enrichment, while low heavy REEs, titanium, and scandium are seen as the results of high degrees of melting. Again, high H_2O, K_2O, and Na_2O contents derived from the fluxing fluid may lead to relatively high SiO_2 melt in equilibrium with peridotite. Following this reasoning, primitive andesites could be the extreme products of fluxed melting (Grove *et al.*, 2003).

A series of recent papers describes how this process can produce a very close match to high Mg# andesites from Mt. Shasta volcano in the southern Cascades (Grove *et al.*, 2001, 2003). However, as in previous inversions based on a variety of arc lavas (Eiler *et al.*, 2000; Stolper and Newman, 1992), the required fluids have dissolved light REEs and thorium contents larger than predicted from experimental fluid/rock partitioning studies (Ayers *et al.*, 1997; Brenan *et al.*, 1996, 1995a,b; Kogiso *et al.*, 1997; Stalder *et al.*, 1998; Tatsumi and Kogiso, 1997). If the fluxing agent in flux melting were a silicate melt rather than an aqueous fluid, predicted REE and LILE contents would be much higher, potentially resolving this discrepancy. We return to this point in Section 3.18.3.3.1. From this perspective, "melt-fluxed melting" can be considered more or less identical to (i).

(iii) Mixing of primary basalt and granitic lower crustal melts. Although there are significant fluid mechanical barriers to such a process (e.g., Campbell and Turner, 1985), many chemical features of high Mg# andesites could be explained as the result of mixing of primitive arc basalt with evolved, silica-rich melt with high LILEs and light REEs, and low heavy REEs, titanium, and scandium.

In the western Aleutians, high Mg# andesites have abundant, zoned phenocrysts which probably do reflect magma mixing processes. However, the most light REE enriched, heavy REE depleted magmas have the highest Mg# (Kelemen *et al.*, 2003b; Yogodzinski and Kelemen, 1998), which is inconsistent with the hypothesis of mixing primitive basalt with enriched granitic melt outlined in the previous paragraph (Kelemen *et al.*, 2003b; tables 18 and 19). Instead, mixing apparently combined primitive, light REE enriched andesites with more evolved, less enriched andesites.

With the exception of the western Aleutian arc, primitive and high Mg# andesites (excluding boninites) are rare in intra-oceanic arcs (Figure 4). High Mg# andesites are most common in continental arcs, where interaction between basalt and pre-existing crust might be important. Primitive andesites have been reported from the Cascades (Baker *et al.*, 1994; Grove *et al.*, 2001, 2003; Hughes and Taylor, 1986), Baja California (Rogers *et al.*, 1985), southeast Costa Rica and western Panama (de Boer *et al.*, 1988, 1995; Defant *et al.*, 1991a,b, 1992), Ecuador (Beate *et al.*, 2001; Bourdon *et al.*, 2002; Monzier *et al.*, 1997), Argentina (R. W. Kay and S. M. Kay, 1991, 1993), southern Chile (Sigmarsson *et al.*, 2002, 1998; Stern and Kilian, 1996), the Philippines (Defant *et al.*, 1989; Maury *et al.*, 1992; Schiano *et al.*, 1995), Papua New Guinea ((Arculus *et al.*, 1983), SW Japan (Shimoda *et al.*, 1998; Tatsumi, 1982, 2001a,b; Tatsumi and Ishizaka, 1981, 1982), and Kamchatka (Kepezhinskas *et al.*, 1997).

In all of these localities, other than the Aleutians, most high Mg# andesites have elevated $^{208}Pb/^{204}Pb$, compared to MORB (Figure 9). Thus, lead isotope data suggest the presence of a component derived either from recycling of lead from subducting sediment, or from crustal interaction of primitive basalts with older, continental crust and continentally derived sediment. Given the fact that so many primitive andesites are in "continental" arcs, crustal interaction processes must be considered. That said, primitive andesites and basalts have overlapping $^{87}Sr/^{86}Sr$ and $^{143}Nd/^{144}Nd$, which restricts the range of crustal sources that could be involved in mixing or assimilation to create primitive andesites from basalts.

Lower crustal anatexis in arcs like the Andes (e.g., Babeyko *et al.*, 2002), that have thick crust and probably garnet at the base of the crust, might be expected to yield appropriate mixing end-members. Indeed, our compilation includes dacitic to rhyolitic lavas from the Andes with high Dy/Yb (Bourdon *et al.*, 2000; S. M. Kay and R. W. Kay, 1994; Matteini *et al.*, 2002). Mixtures of these compositions with primitive basalt have most of the major and trace-element characteristics of high Mg# andesites (e.g., Figure 10).

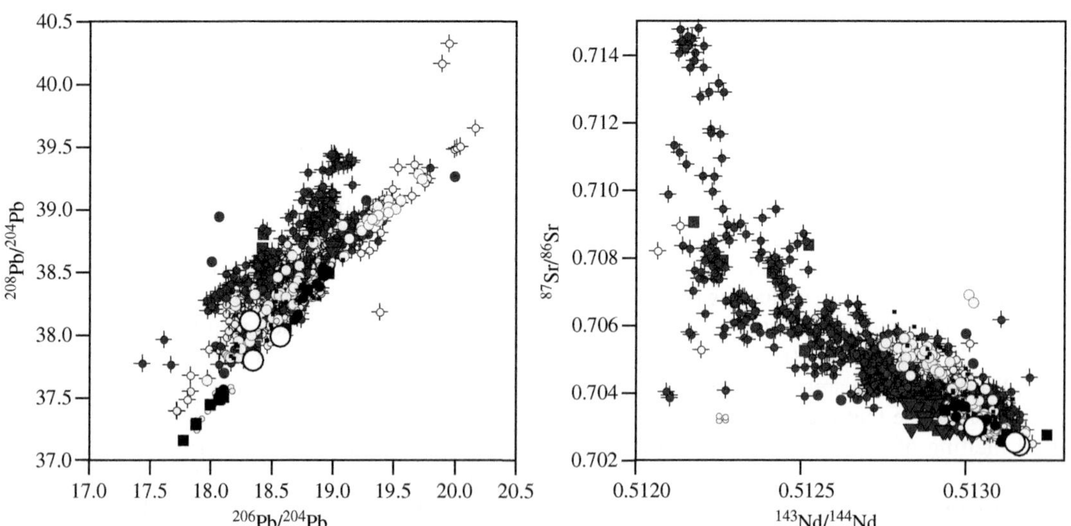

Figure 9 Pb, Sr, and Nd isotopes in primitive lavas from our arc compilation. In general, primitive andesites do not have distinctive isotopic characteristics, compared to primitive basalts. Primitive andesites from the western Aleutians have the most depleted values in our data compilation. Data and symbols as for Figures 1 and 6.

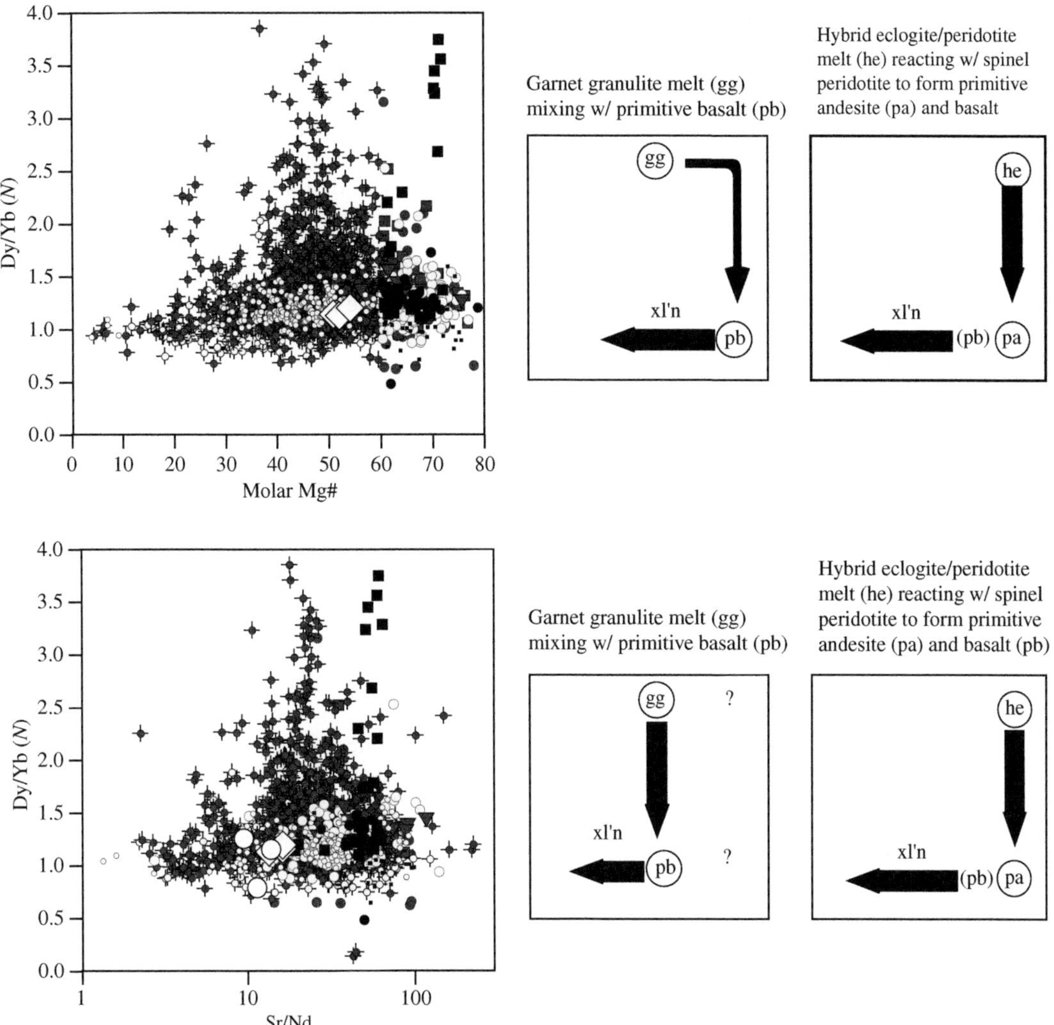

Figure 10 Relationship between Mg#, Dy/Yb, and Sr/Nd in arc lavas from our compilation. Partial melts of subducted, eclogite facies MORB have high Dy/Yb and Sr/Nd (e.g., Gill, 1974, 1978; Kay, 1978; Kelemen *et al.*, 1993, 2003b; Rapp *et al.*, 1999; Yogodzinski and Kelemen, 1998). When these melts react with overlying mantle, the hybrid liquid acquires high Mg# (Kay, 1978; Kelemen *et al.*, 1993, 2003b; Rapp *et al.*, 1999; Yogodzinski and Kelemen, 1998). Western Aleutian primitive andesites may form in this way. Continued reaction increases heavy REE and Y contents to values in equilibrium with depleted mantle peridotite as melt/rock ratios decrease to ~0.1 or less (Kelemen *et al.*, 1993, 2003b). High Sr/Nd, primitive andesites and even primitive basalts Dy/Yb ~1.5 may form in this way. Alternatively, primitive andesites could be mixtures of lower crustal melts and primitive basalt (e.g., S. M. Kay and R. W. Kay, 1994). Partial melts of lower crustal garnet granulites, as exemplified by some Andean andesite and dacites (Bourdon *et al.*, 2000; S. M. Kay and R. W. Kay, 1994; Matteini *et al.*, 2002), have high Dy/Yb at lower Mg# and Sr/Nd than Aleutian primitive andesites. Mixing with primitive basalt (e.g., average Marianas primitive basalt, Table 1) yields high Dy/Yb, high Mg# melts similar to Aleutian primitive andesites. However, such mixtures have low Sr/Nd, unlike Aleutian and other primitive andesites. Because Sr is more compatible than Nd in plagioclase, this result is likely to be general. Thus, while magma mixing may have played some role in the genesis of many or most arc lavas, mixing is an unlikely explanation for the genesis of primitive andesite compositions.

Thus, in this chapter we re-examine the hypothesis that primitive andesites are produced by crustal mixing or assimilation using global data, rather than simply data from the Aleutians. This re-examination could be extended to the western Aleutians as well; perhaps, small proportions of high SiO_2, small-degree melts of garnet granulite in arc lower crust are commonly produced but rarely erupted. Could such mixing explain the composition of primitive andesites in general?

Key data to address this question come from ratios of trace elements that are fractionated differently by partial melting of eclogite compared to garnet granulite. For example, because strontium is compatible in plagioclase, strontium is more incompatible than neodymium in melting of plagioclase-free eclogite, whereas strontium is more compatible than neodymium in partial melting of lower crustal granulite with substantial plagioclase. Because temperatures at the base of the crust in active arcs are high (>800 °C) and crustal thickness rarely exceeds 60 km, eclogite facies assemblages will not form in arc lower crust. However, formation of garnet granulites is possible where crustal thickness exceeds 25–30 km (e.g., Jull and Kelemen, 2001). Although garnet formation might be inhibited by slow kinetics in H_2O-poor lower crustal cumulates, observations from the Talkeetna and Kohistan arc crustal sections show that garnet did form at pressures where it was thermodynamically stable (Section 3.18.4). While some garnet in the Talkeetna lower crustal section is entirely metamorphic in origin, other samples include residual or cumulate, igneous garnet (Section 3.18.4.3). As noted above, our arc lava compilation includes high Dy/Yb, evolved lavas from the Andes, which are probably partial melts of garnet granulite. However, while these lavas are appropriate in other ways as mixing end-members to produce high Mg# andesites, they have low Sr/Nd (Figure 10). Therefore, these evolved Andean lavas could not mix with primitive arc basalt to produce high Mg# andesites, which have higher Sr/Nd (Figure 10), and it is unlikely that any other lower crustal melts would be appropriate.

In many ways, the process of "assimilation," or melt/rock reaction, is comparable to magma mixing. Reaction between hot, primitive basalt and granitic wall rock, in particular, shares many characteristics with mixing of primitive basalt and a granitic partial melt of wall rock. For example, both processes tend to produce high, nearly constant compatible-element concentrations and ratios (magnesium, iron, nickel, Mg#) together with substantial enrichment in incompatible trace elements, over much of the range of mixing or reaction progress (compare, e.g., DePaolo, 1981; and Kelemen, 1986 with O'Hara and Mathews, 1981).

In specific cases the outcome of melt/rock reaction may be quite distinct from magma mixing. Selective dissolution of plagioclase in hydrous, plagioclase-undersaturated melt could enrich resulting liquids in Sr/Nd, potentially producing trends distinct from the mixing trends in Figure 10. However, note that selective dissolution of plagioclase, alone, might create a telltale anomaly with high Eu/Sm, and would not explain other characteristics of high Mg# andesites, including heavy REE, titanium, scandium, and vanadium depletion. Instead, selective dissolution of plagioclase would have to be coupled with crystallization of garnet in order to explain high Mg# andesite genesis via crustal melt/rock reaction. In view of the fact that garnet is not saturated in primitive arc melts at pressures of ~1–1.5 GPa (e.g., Müntener *et al.*, 2001), we view this as unlikely. Furthermore, on a global basis, high Sr/Nd is negatively correlated with $^{87}Sr/^{86}Sr$, ruling out assimilation of continental granitoids, or sediments derived from continental crust.

(iv) Primitive andesite summary. In summary, primitive andesites are probably derived from primary andesite magmas, produced by processes below the base of igneous arc crust, which are different from primary magmas of primitive arc basalts. Mixing of primitive basalt and evolved partial melts of lower crustal garnet granulite probably cannot produce end-member high Mg# andesite lava compositions. Similarly, lower crustal assimilation probably cannot produce typical high Mg# andesites. Instead, we conclude that most primitive andesites are probably produced via process (i), described above, by reaction of small degree partial melts of subducted, eclogite-facies sediment and/or basalt with the overlying mantle wedge. In our view, process (ii), melting of the mantle fluxed by an enriched "fluid" component derived from subducted sediment and/or basalt, is only viable if the fluxing "fluid" is, in fact, a melt or a supercritical fluid with partitioning behavior similar to melt/rock partitioning. Thus, viable versions of process (ii) are the same as process (i).

Primary magmas parental to primitive andesites may also be parental to calc-alkaline, evolved arc magmas. It is evident that fractionation of olivine, pyroxene, and/or plagioclase from a primitive andesite melt leads to higher SiO_2 and alkali contents, at a given degree of crystallization, compared to fractionation of the same phases from primitive basalt (Figure 11). The key here is that olivine + pyroxene cumulates, and olivine + pyroxene + plagioclase cumulates, have ~50% SiO_2, as determined by mineral stoichiometry and cotectic proportions. Crystallizing these solid assemblages from a basalt with ~50% SiO_2 does not change the SiO_2 content of the resulting liquid, whereas removing the same cumulates from an andesite with 55 wt.% or even 60 wt.% SiO_2 leads to an increase in SiO_2 in the derivative liquid. (Note that the variation in Figure 11, taken out of context, might also be attributed to fractionation from a common parental magma with Mg# ~0.8; however, trace-element and isotope variation precludes this possibility.) The high SiO_2 and—probably—H_2O contents of primitive andesites may make them difficult to erupt. As they reach the mid-crust, become saturated in H_2O, and degas, their viscosity must rise abruptly, leading to slower melt transport and enhanced rates of crystallization. For these reasons,

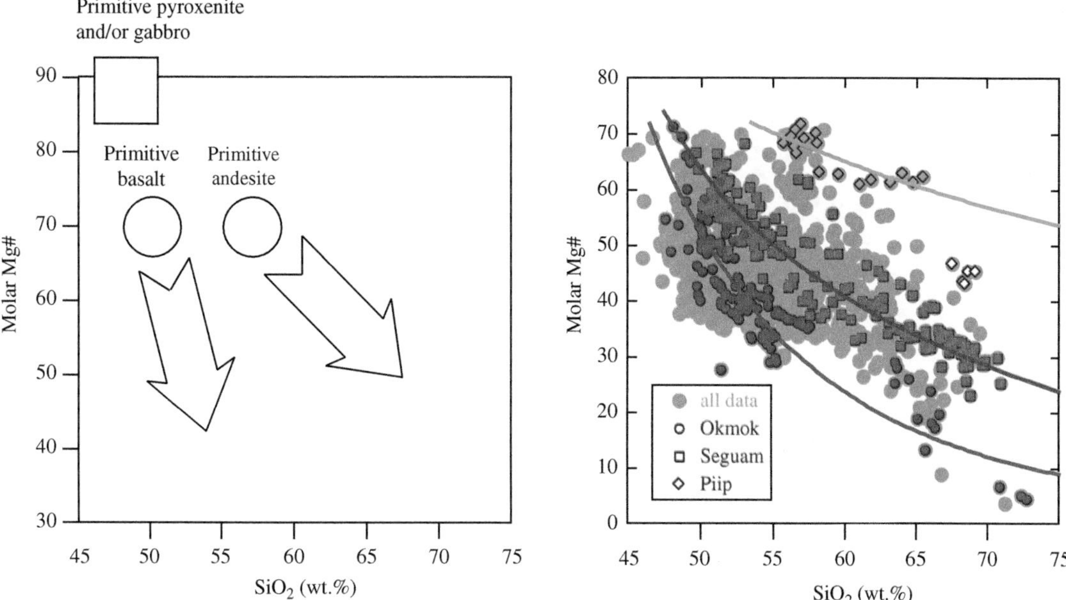

Figure 11 On the left, a schematic illustration showing how fractionation of primitive pyroxenite or gabbro from primitive basalt leads to decreasing Mg# at nearly constant SiO_2, while fractionation of the same crystal assemblages from primitive andesite leads to increasing SiO_2. On the right, data from the oceanic Aleutian arc (compiled by Kelemen *et al.*, 2003b) show how variation between lava series from different volcanoes might arise as a result of this effect. Sources of data for Seguam, Okmok and Piip volcanoes are given in the caption for Figure 2. Filled symbols for each volcano were used in power law curve fits. Open symbols were omitted.

primitive andesites may be more common among melts entering the base of the igneous crust than they are among erupted lavas.

It is not clear to what extent this analysis can be extended to high Mg# andesite compositions typical of plutonic rocks in calc-alkaline batholiths. Some Aleutian and Cascades plutonic rocks are high Sr/Nd, high Mg# andesites which cannot be produced via crustal mixing, but high Sr/Nd in these rocks could arise via incorporation of cumulate plagioclase. Many calc-alkaline plutonic rocks have relatively low Sr/Nd, and could be mixtures of lower crustal melts and primitive basalt. Because plagioclase crystallization leads to decreasing Sr/Nd with decreasing Mg#, however, most calc-alkaline plutonic rocks could also be derived via crystal fractionation from primitive andesite. Therefore, the Sr/Nd discriminant between lower crustal melts (low Sr/Nd) and eclogite melts (high Sr/Nd) is only useful for lavas that retain high Dy/Yb. Unfortunately, this means that Sr/Nd cannot be used to determine the extent to which the high Mg# andesite composition of continental crust is due to crystal fractionation from a primitive andesite parent, versus the extent to which it is due to mixing of primitive basalt and lower crustal melts.

(v) Why are primitive andesites rare? In Sections 3.18.3.2.5 and 3.18.3.3, we argue that a component derived from partial melting of subducting sediment and/or basalt is included in most arc magmas. This is consistent with a substantial body of work calling upon partial melts of subducting sediment to explain trace-element enrichments in arc basalts, but inconsistent with the theory that primitive andesites with "adakite" trace-element signatures (Section 3.18.3.2.4), which apparently include tens of percent eclogite melt (Kelemen *et al.*, 2003b), are only found in arcs with unusually hot subduction zones—due to subduction of young oceanic crust, very slow convergence rates allowing substantial time for conductive heating, and/or discontinuous "tears" in the subducting plate which enhance mantle convection and allow conductive heating from the side as well as the top and bottom (e.g., de Boer *et al.*, 1991, 1988; Defant and Drummond, 1990; Yogodzinski *et al.*, 1995, 2001, 1994).

If partial melts of subducted material are ubiquitous in arcs, why do they form large proportions of arc magma in a few places, and very small proportions (a few percent, e.g., Class *et al.* (2000)) in most arcs? Following (Kelemen *et al.*, 2003b), we offer the following tentative explanation. Arcs that have primitive andesites are similar in having slow convergence rates, and it may be true that many are situated above "tears" in the subducting plate. Most (except the Aleutians) are in regions of young plate subduction. Thus, the subduction

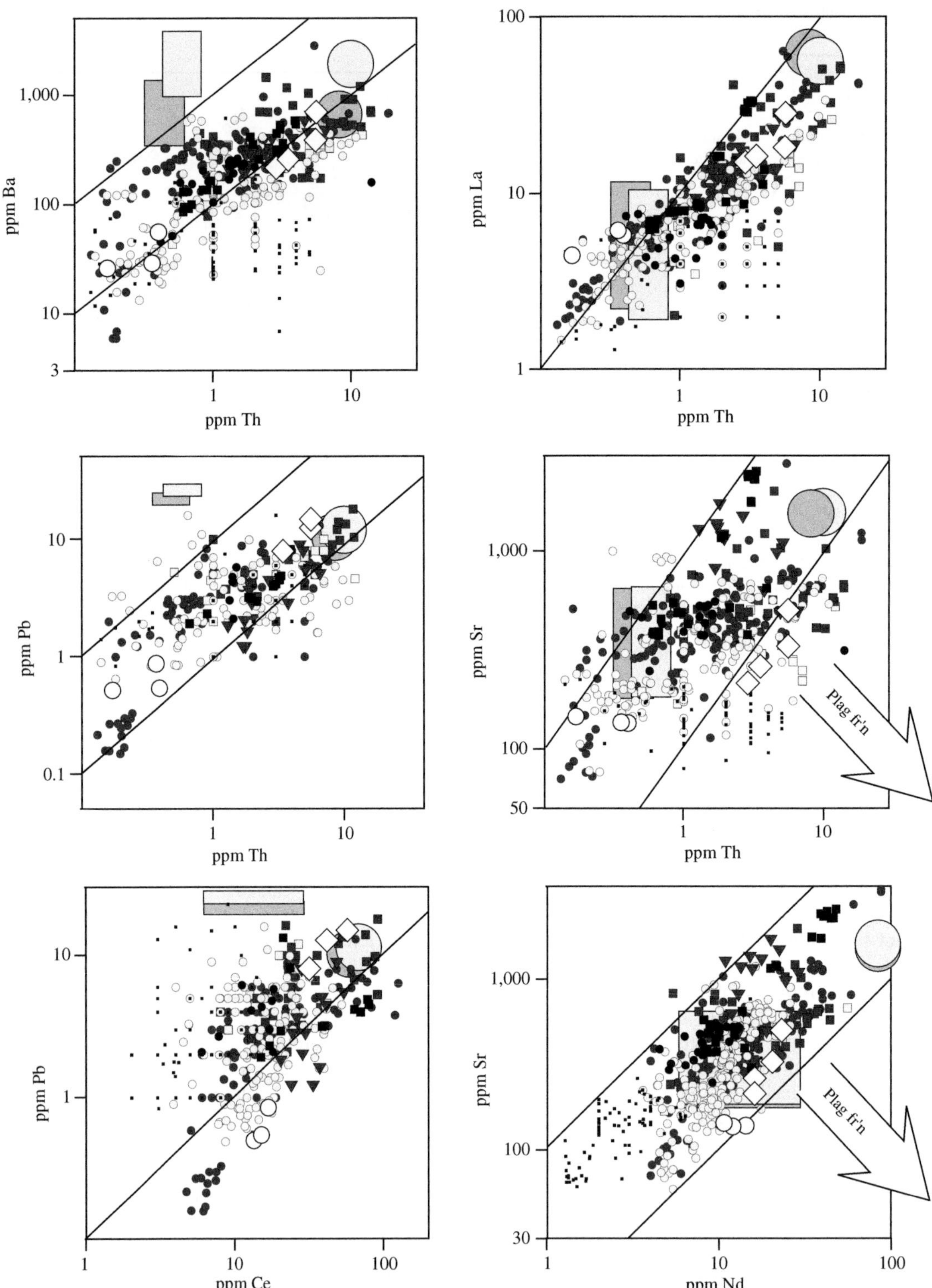

Figure 12 Trace-element concentrations, in ppm, for primitive lavas in our compilation. In general, primitive arc lavas are enriched in Th, Ba, La, Pb, Ce, Sr, and Nd, compared to average MORB. (Concentrations of these elements in primitive MORB are even lower than the average values). This is somewhat at odds with theories invoking separate enrichment processes involving aqueous fluids for Ba, Pb, and Sr, and partial melts of subducting sediment for Th, La, Ce, and Nd. Diagonal lines illustrate constant trace-element ratios. The overall trends of the compiled data do not diverge dramatically from constant ratios of these elements (diagonal lines). If anything, "fluid-mobile" Ba, Pb, and Sr show less variation and less enrichment relative to MORB than "immobile" Th and La. Since all of these elements are highly incompatible during melting of eclogite facies sediment and basalt, and in peridotite/melt equilibria,

zone may be unusually hot, and may yield a larger proportion of partial melt. However, it is hard to imagine that this can account for a factor-of-ten difference in the proportion of subduction zone melt in arc lavas. Instead, other factors may be involved. Slow convergence leads to slow convection in the mantle wedge, enhancing the amount of conductive cooling. This could be very important, because of positive feedback: increased viscosity due to cooling could slow wedge convection still further. Thus, we suggest that primitive andesites are found in areas in which the subducting plate is relatively hot, producing more partial melt, and the overlying wedge is relatively cold, producing less partial melt. In contrast, in normal arcs, abundant melts derived from the mantle wedge overwhelm the subduction zone melt signature, except for distinctive isotope ratios and thorium contents carried in partial melts of sediment.

In addition, as outlined in Sections 3.18.3.1.3 and 3.18.3.1.4, primitive andesites may contain more H_2O than primitive basalts (e.g., Grove *et al.*, 2003). If so, they will become H_2O saturated and degas at mid-crustal depths. As a result of degassing, these magmas will undergo rapid crystal fractionation, so that only evolved andesites and dacites reach the surface. Also, as a result of degassing together with their relatively high SiO_2 contents, primitive andesites will become very viscous in the mid-crust, and this may inhibit their eruption.

3.18.3.3 Trace Elements, Isotopes, and Source Components in Primitive Magmas

3.18.3.3.1 Incompatible trace-element enrichment

Lava compositions from arcs worldwide share many characteristics that are, in turn, very distinct from those of MORBs. These distinctions have been known for decades, and we cannot hope to provide a comprehensive review of the entire literature on this topic. However, we do wish to use our data compilation to quantify the differences between arc and MORB lavas. Thus, caesium, rubidium, radium, barium, thorium, uranium, and potassium are enriched in most arc lavas relative to MORB, together with light REEs, potassium, lead, and strontium (Figure 12). Moderately incompatible elements (middle- to heavy-REE and titanium) are generally depleted relative to MORB (Figure 6). These characteristics are present in both oceanic and continental arcs. Globally, these characteristics are observed even in the most primitive lavas, and some of the most enriched lavas in our data set are western Aleutian primitive andesites with low $^{87}Sr/^{86}Sr$ and lead isotope ratios (Kelemen *et al.*, 2003b; Yogodzinski *et al.*, 1995, 1994). Thus, it seems likely that the enrichments do not arise mainly as a result of crustal processes, and instead are present in melts entering the base of arc crust, and have a sub-Moho origin, at least in part.

Based on similar reasoning, a host of studies in the 1980s and 1990s were designed to decipher the subcrustal source of enriched incompatible trace-element contents in arc magmas. Possible sources of enrichment, relative to the MORB mantle source, include (i) aqueous fluids derived by dehydration of subducting metabasalt, (ii) aqueous fluids derived by dehydration of subducting metasediment, (iii) aqueous fluids derived by dehydration of partially serpentinized mantle peridotite, (iv) hydrous partial melts of subducting basalt, (v) hydrous partial melts of subducting sediment, and (vi) the presence of "enriched mantle" similar to the various mantle source components inferred for ocean island basalt. These could act upon, or mix with (vii) fertile mantle peridotite similar to the MORB source, or (viii) mantle peridotite depleted by melt extraction beneath a mid-ocean ridge and/or a back arc basin. Given the relatively poorly known (and probably variable) compositions of these various components, their poorly known (and variable) proportions in a given primary arc magma, and the poorly known processes through which different components might interact with additional trace-element fractionation, there may be numerous combinations of these components which could account for the composition of a given primary magma. As a result, most authors have tried to simplify the geochemical interpretation of arc petrogenesis by concentrating on just a few of these components, whose interaction might account for much of the trace-element variability in arc basalts.

similar enrichment of all of them is a natural consequence of melting and melt transport. Conversely, because Th, La, Ce, and Nd are less soluble in aqueous fluids than Ba, Pb, and Sr, dehydration and aqueous fluid transport should fractionate these elements. Large gray symbols show estimated compositions of fluid (rectangles) and melt (circles) in equilibrium with eclogite for Marianas (dark gray) and Aleutians (light gray) at 2 wt.% fluid or melt extracted (Table 2(b)). Size of rectangle reflects results obtained using the range of different partition coefficients. Nd concentration is not calculated and is assumed to be the same as Ce concentration (not shown in Table 2). Arrow outline labeled "plag fr′n" in diagrams involving Sr reminds readers that even small amounts of plagioclase crystallization will lead to decreasing Sr together with increasing concentrations of Th and Nd. Symbols as in Figures 1 and 6.

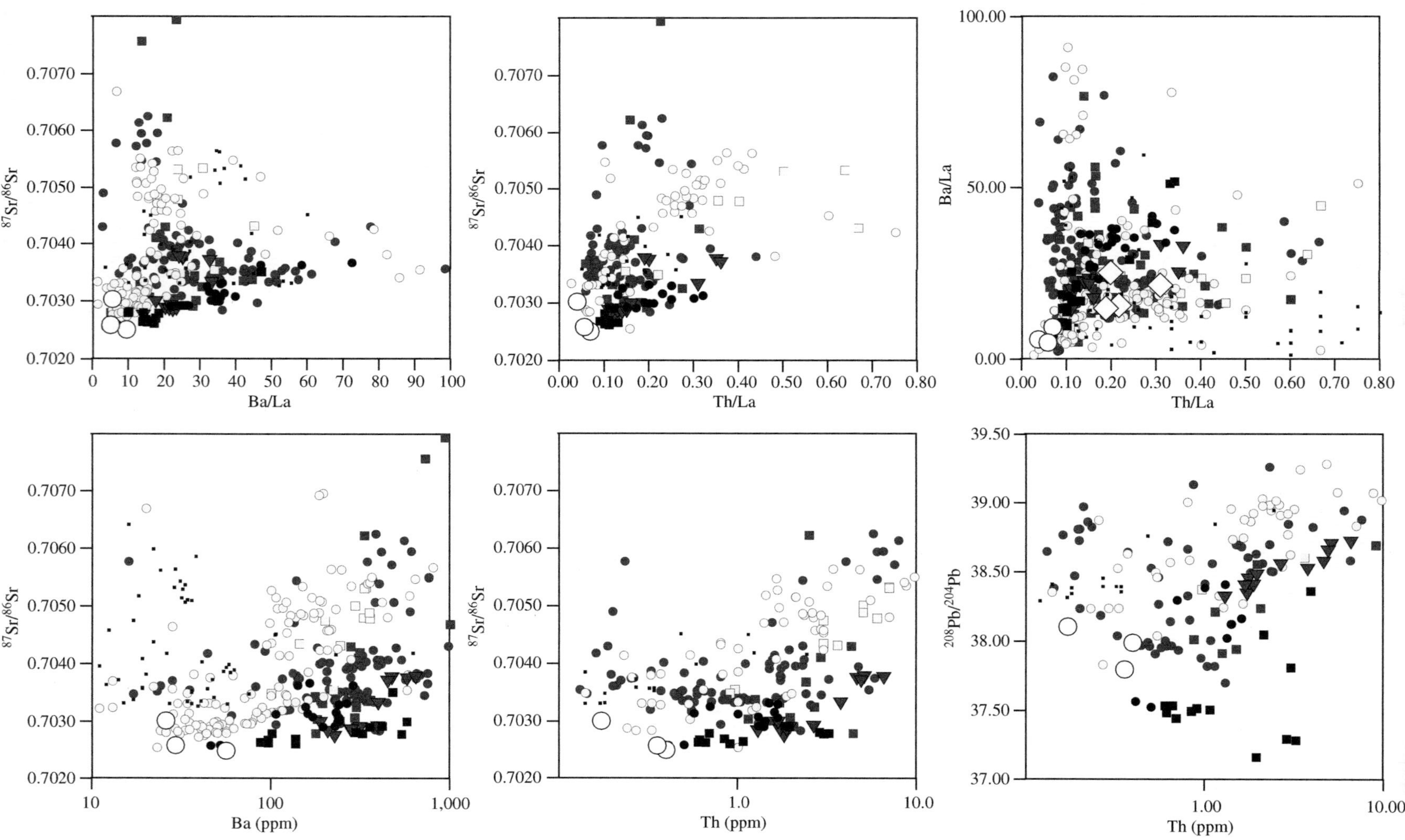

$^{87}Sr/^{86}Sr$
Ba/La
$^{87}Sr/^{86}Sr$
Th/La
Ba/La
Th/La
$^{87}Sr/^{86}Sr$
Ba (ppm)
$^{87}Sr/^{86}Sr$
Th (ppm)
$^{208}Pb/^{204}Pb$
Th (ppm)

(i) Three main components? Using trace-element ratios. Particularly influential have been studies which call upon three principal components in arc basalt petrogenesis: (a) aqueous fluids derived from metabasalt, (b) partial melts of subducting sediment, and (c) MORB source mantle. (e.g., Class *et al.*, 2000; Ellam and Hawkesworth, 1988a; Elliott, 2003; Elliott *et al.*, 1997; Hawkesworth *et al.*, 1997; Johnson and Plank, 1999; Miller *et al.*, 1994; Plank and Langmuir, 1993, 1998; Turner *et al.*, 2003, 2000a,c). These studies generally rely on analyses of both primitive and evolved basalts, and so they use incompatible trace-element ratios in an attempt to remove the effects of crystal fractionation. In these studies, one component, with relatively low Ba/La and Th/La, and MORB-like isotope ratios is interpreted as the pre-existing mantle source, prior to fluid and sediment melt enrichment.

Also in these studies, enrichments in "fluid mobile" elements relative to light REE (e.g., high U/Th, Ra/Th, Ba/La, Pb/Ce, and Sr/Nd), are attributed to an aqueous fluid component. In some arcs and in Elliott's worldwide compilation of ICP-MS data on arc basalts, this component has isotopic characteristics similar to hydrothermally altered MORB (e.g., $^{87}Sr/^{86}Sr \sim 0.7035$, $^{143}Nd/^{144}Nd \sim 0.5132$; $^{208}Pb/^{204}Pb$ down of 38).

High boron concentrations, and high boron and oxygen isotope ratios in the aqueous fluid component, relative to the MORB source, are attributed to hydrothermal alteration in the fluid source (Bebout *et al.*, 1993, 1999; Domanik *et al.*, 1993; Eiler *et al.*, 2000; Ishikawa and Nakamura, 1994; Leeman, 1987, 1996; Ryan and Langmuir, 1992; Ryan *et al.*, 1989; You *et al.*, 1993, 1995). Our data compilation includes few data on boron and concentration, or on Boron and oxygen isotopes, and so we have not attempted to re-visit these topics.

In contrast, enrichments in "fluid immobile" Th relative to light REE (e.g., Th/La) and enrichments of light/middle REE (e.g., La/Sm) are attributed to a partial melt of subducted material. This component is thought to be a sediment melt because in some arcs it has isotopic characteristics similar to subducting sediment (e.g., $^{87}Sr/^{86}Sr$ up to 0.706; $^{143}Nd/^{144}Nd$ down to 0.5127; $^{208}Pb/^{204}Pb \sim 39.0$) and because thorium enrichments at a given $^{143}Nd/^{144}Nd$ are larger than can be accounted for by simple mixing of mantle peridotite and sediment. The fact that basalt mainly contributes aqueous fluid whereas the sediment rich component is mainly transported in melt might be attributed to the steep thermal gradient at the top of a subducting plate, in which only the topmost layers are heated above their fluid-saturated solidus, giving rise to the aphorism "sediments melt, basalts dehydrate."

The role of subducted sediment is particularly well documented for selected, high quality data on arc basalts in which the high Th/La component has Th/La identical to that in the subducting sediment column (Plank, 2003). Very efficient recycling of subducted thorium, together with subducted ^{10}Be (present only in surficial sediments), has also been taken as evidence for transport of sediment-derived thorium and beryllium in a partial melt, rather than an aqueous fluid (e.g., Johnson and Plank (1999), Kelemen *et al.* (1995a); but apparently in disagreement with Morris (see Chapter 2.11) and Schmidt and Poli (see Chapter 3.17).

(ii) Two main components? Using trace-element abundance. In our data compilation, there are many primitive lavas whose trace-element abundance is minimally affected by crystal fractionation, so we decided to dispense with the use of ratios such as Ba/La and Th/La. We did this because many arc lavas are enriched in both light REEs (the denominator) and barium, thorium, lead, and strontium (the numerator) in commonly used trace-element ratios. Thus, for example, a lava with 1 ppm lanthanum, Ba/La of 100, and Th/La of 0.1 has less lanthanum, barium, and thorium than a lava with 50 ppm lanthanum, Ba/La of 20, and Th/La of 0.3. Thus, in Figures 12 and 13 we plot incompatible trace-element abundances in primitive lavas. These data present a somewhat different picture from the three component hypothesis outlined in Section 3.18.3.3.1.

In Figure 13, one can see that oceanic lavas with high Ba/La generally have lower $^{87}Sr/^{86}Sr$ than lavas with high Th/La, and that in most oceanic arcs, lavas with high Ba/La have low Th/La and lavas with high Th/La have low Ba/La, as outlined in Section 3.18.3.3.1. However, almost all data for the oceanic Aleutian arc have low $^{87}Sr/^{86}Sr$ (<0.7035) despite a wide range of Ba/La and

Figure 13 Trace-element concentrations in ppm, trace-element ratios, and isotope ratios for primitive arc lavas in our compilation. A high Ba/La, low Th/La component has $^{87}Sr/^{86}Sr \sim 0.704$, while a distinct low Ba/La, high Th/La component in most oceanic arcs has $^{87}Sr/^{86}Sr \sim 0.706$. However, continental arcs, the oceanic Aleutian arc, and some other primitive arc lavas have relatively high Ba/La *and* Th/La. For most oceanic arcs, Ba and Th concentration in primitive lavas are correlated (Figure 12) and both are correlated with $^{87}Sr/^{86}Sr$. With very few exceptions, primitive andesites from the oceanic Aleutian arc and from continental arcs show lower $^{87}Sr/^{86}Sr$ and $^{208}Pb/^{204}Pb$, at a given Ba or Th concentration, compared to lavas from other oceanic arcs at the same Ba or Th concentration. Symbols as in Figures 1 and 6.

Th/La, and the Aleutians show strong positive correlation between Ba/La and Th/La. Figure 13 also shows that most oceanic arc lavas show positively correlated barium, thorium, lanthanum, and $^{87}Sr/^{86}Sr$, while oceanic Aleutian arc lavas, and continental arc lavas, show wide variation in barium, thorium, and lanthanum that is not correlated with $^{87}Sr/^{86}Sr$ or, e.g., $^{208}Pb/^{204}Pb$. Thus, for example, enrichments in thorium can occur in primitive arc lavas without incorporation of radiogenic strontium and lead, (and unradiogenic neodymium) from subducted sediments (Kelemen *et al.*, 2003b).

High thorium in primitive arc magmas with MORB-like lead, neodymium, and strontium isotope ratios, such as in primitive andesites from the western Aleutians, is probably *not* attributable to incorporation of a sedimentary thorium component, yet it has very similar thorium enrichment (though lower Th/La) compared to magmas, which *are* thought to incorporate thorium derived from subducted sediment. However, experimental studies of eclogite/aqueous fluid and peridotite/aqueous fluid partitioning show that thorium is not very soluble (Ayers *et al.*, 1997; Brenan *et al.*, 1996, 1995a,b; Johnson and Plank, 1999; Kogiso *et al.*, 1997; Stalder *et al.*, 1998; Tatsumi and Kogiso, 1997). In contrast, thorium is strongly enriched in small degree melts of basalt or sediment (e.g., Johnson and Plank, 1999; Rapp *et al.*, 1999; Ryerson and Watson, 1987). For this reason, we attribute thorium enrichment in arc magmas generally to incorporation of thorium in a partial melt of eclogite facies, subducted basalt *and/or* sediment (e.g., Kelemen *et al.*, 1993, 2003b). We return to this topic in Section 3.18.3.3.1.

Somewhat to our surprise, over the past few years several investigators have found that small degree partial melts of eclogite may also have some characteristics normally attributed to an aqueous fluid component in arc magmas, such as high Ba/La, Pb/Ce, and Sr/Nd (e.g., Kelemen *et al.*, 2003b; Rapp *et al.*, 1999; Tatsumi, 2000). Thus, it may be difficult to distinguish between partial melts of eclogite and an aqueous fluid component using these trace-element ratios. Although we do not doubt that—for instance—barium, lead, and strontium *can* be fluid mobile, fluid enrichment of the arc magma source may not be the dominant control on barium, lead, and strontium enrichment in arc magmas in our compilation. Correlation of barium, lead, and strontium with thorium suggests that either "fluid" and "melt" components generally combine in similar proportions world wide, or that some process can produce both barium and thorium enrichment simultaneously. This observation and its interpretation are very similar to observations and interpretation of ^{10}Be versus B/Be systematics (e.g., Morris *et al.*, 1990).

Most primitive arc lavas have elevated barium, thorium, lanthanum, lead, caesium, strontium, *and* neodymium, compared to MORBs. Because all of these components are incompatible during melting of eclogite or peridotite, while they are fractionated from each other by fluid/rock partitioning, it may be that the difference between arc lavas and MORBs is primarily due to incorporation of a partial melt of subducted, eclogites facies sediment and/or basalt into the arc magma source. Where sediments are present, high sediment thorium and lead concentrations, together with distinctive sediment isotope ratios, are likely to impart a sedimentary signature to melts of subducted material. Where sediments are absent, Th/La and Pb/Ce are likely to be somewhat lower, and isotope ratios in subduction zone melts are likely to be closer to those in the MORB source.

We do not mean to imply that dehydration reactions in subducting material do not evolve aqueous fluid. However, much of that fluid may escape into the fore arc region. And, certainly, scatter in Figure 12—with logarithmic axes—is indicative of a wide range of Ba/Th, Pb/Ce, and Sr/Nd. These variations may result from additional enrichment of barium, lead, and strontium (and boron) relative to thorium and light REE in some arcs due to an aqueous fluid component, as described in Section 3.18.3.3.1.

(iii) Melt and fluid compositions and arc mass balance. As already mentioned, it is particularly important to consider enrichments in lanthanum and thorium, since these are relatively insoluble in aqueous fluids under subduction zone conditions (Ayers *et al.*, 1997; Brenan *et al.*, 1996, 1995a,b; Johnson and Plank, 1999; Kogiso *et al.*, 1997; Stalder *et al.*, 1998; Tatsumi and Kogiso, 1997). Given likely values for arc magma flux, average arc lava concentrations of these elements, assumptions about arc lower crustal composition, experimentally measured solubilities for these elements in aqueous fluids, estimates of the subducting flux of H_2O, and assumptions about the mantle source of arc magmas, it has often been concluded that aqueous fluids derived from subducting material cannot explain the magmatic flux of lanthanum and thorium in arcs (e.g., Elliott, 2003; Elliott *et al.*, 1997; Hawkesworth *et al.*, 1997; Johnson and Plank, 1999; Kelemen *et al.*, 1993, 2003b; Plank, 2003; Plank and Langmuir, 1988, 1993, 1998). If true, this is crucial because it requires that *almost all arcs require enrichment of the mantle source in lanthanum and thorium via some mechanism other than aqueous fluid transport from subducting material.*

Because this is so crucial, we offer our own flux calculations in Table 2. These calculations

Table 2(a) Arc inputs and outputs assuming 5 wt.% H_2O or 5 wt.% melt extracted.

		Aleutians	*Marianas*	*Izu-Bonin*	*Kermadec*	*Tonga*	*ref. D*
Age	my	55	45	45	30	24	
Thickness arc crust	km	20	20	20	18	12	
Material subducted	10^6 kg km^{-1} yr^{-1}	1,321	592	896	1,154	1,564	
Magmatic flux	10^6 kg km^{-1} yr^{-1}	115	140	140	150	113	
Magmatic flux	km^3 km^{-1} yr^{-1}	38	47	47	50	38	
Excess La	kg km^{-1} yr^{-1}	292	155	95	503	569	
Excess Th	kg km^{-1} yr^{-1}	117	31	209	166	67	
Excess Ba	kg km^{-1} yr^{-1}	20,809	7,224	5,023	19,823	29,367	
Excess Sr	kg km^{-1} yr^{-1}	38,016	16,601	12,361	29,106	38,075	
Excess Pb	kg km^{-1} yr^{-1}	396	83	542	416	619	
La flux in fluid	kg km^{-1} yr^{-1}	133	67	106	131	171	1
La flux in fluid	kg km^{-1} yr^{-1}	679	345	541	673	876	2
Th flux in fluid	kg km^{-1} yr^{-1}	27	10	16	22	23	1
Th flux in fluid	kg km^{-1} yr^{-1}	52	20	31	42	44	3
Ba flux in fluid	kg km^{-1} yr^{-1}	99,692	15,605	22,725	37,121	37,576	3
Ba flux in fluid	kg km^{-1} yr^{-1}	47,136	7,378	10,745	17,551	17,766	2
Sr flux in fluid	kg km^{-1} yr^{-1}	11,827	5,215	7,738	10,152	13,616	1
Sr flux in fluid	kg km^{-1} yr^{-1}	33,566	14,800	21,963	28,813	38,646	3
Sr flux in fluid	kg km^{-1} yr^{-1}	37,606	16,581	24,606	32,281	43,297	2
Pb flux in fluid	kg km^{-1} yr^{-1}	861	323	547	1,175	1,399	1
Pb flux in fluid	kg km^{-1} yr^{-1}	957	360	609	1,307	1,556	3
La flux in melt	kg km^{-1} yr^{-1}	2,689	1,364	2,143	2,662	3,465	4
Th flux in melt	kg km^{-1} yr^{-1}	356	134	209	287	300	4
Ba flux in melt	kg km^{-1} yr^{-1}	72,557	11,358	16,540	27,017	27,348	4
Sr flux in melt	kg km^{-1} yr^{-1}	79,353	34,988	51,922	68,117	91,363	4
Pb flux in melt	kg km^{-1} yr^{-1}	573	215	364	782	931	4

Arc ages from Jarrard (1986), except Aleutians which is from Scholl *et al.* (1987). Material subducted is calculated assuming 7 km thick subducting oceanic crust with desities of 3.0 g/cm^3; sediment thicknesses and densities from Plank and Langmuir (1998); subduction rates from England *et al.*, 2003. Magmatic flux is calculated using arc crust thickness from Holbrook *et al.* (1999) (Aleutians), Suyehiro *et al.* (1996) (Izu-Bonin) and Plank and Langmuir (1988) (Marianas, Kermadec and Tonga), subtracting 6 km preexcisting oceanic crust, and assuming a 150 km arc width. Excesses of selected trace elements are calculated subtracting abundances in N-MORB (Hofmann, 1988) from average abundances in primitive arc basalt (Table 1). Trace element fluxes in fluid and melt are calculated assuming modal batch melting (5% in Table 2(a), 2% in Table 2(b)). Because estimates of partition coefficients (Ds) between eclogite and fluid are highly variable, more than one value was used for each element. References for the values used are listed in the column labelled Ref. D. Partition coefficients are from 1: Ayers (1998), 2: Stalder *et al.* (1998), 3: Brenan *et al.* (1995b), 4: Kelemen *et al.* (2003b). Average composition for Izu Bonin primitive basalt is not included in Table 1 (too few analyses) and is (in ppm): La = 4.57, Th = 1.68, Ba = 49.75, Sr = 201.49, Pb = 4.36).

Table 2(b) Arc inputs and outputs assuming 2 wt.% H_2O or 2 wt.% melt extracted.

		Aleutians	*Marianas*	*Izu-Bonin*	*Kermadec*	*Tonga*	*Ref. D*
La flux in fluid	kg km^{-1} yr^{-1}	52	26	42	52	67	1
La flux in fluid	kg km^{-1} yr^{-1}	284	144	226	281	366	2
Th flux in fluid	kg km^{-1} yr^{-1}	11	4	6	9	9	1
Th flux in fluid	kg km^{-1} yr^{-1}	22	8	13	17	18	3
Ba flux in fluid	kg km^{-1} yr^{-1}	99,012	15,499	22,570	36,868	37,320	3
Ba flux in fluid	kg km^{-1} yr^{-1}	25.678	4,019	5,853	9,561	9,679	2
Sr flux in fluid	kg km^{-1} yr^{-1}	4,811	2,121	3,148	4,130	5,539	1
Sr flux in fluid	kg km^{-1} yr^{-1}	15,009	6,618	9,820	12,883	17,280	3
Sr flux in fluid	kg km^{-1} yr^{-1}	17,131	7,553	11,209	14,705	19,723	2
Pb flux in fluid	kg km^{-1} yr^{-1}	621	233	395	848	1,010	1
Pb flux in fluid	kg km^{-1} yr^{-1}	764	287	486	1,044	1,243	3
La flux in melt	kg km^{-1} yr^{-1}	1,466	744	1,168	1,451	1,889	4
Th flux in melt	kg km^{-1} yr^{-1}	264	99	155	212	223	4
Ba flux in melt	kg km^{-1} yr^{-1}	50,570	7,916	11,528	18,830	19,061	4
Sr flux in melt	kg km^{-1} yr^{-1}	44,852	19,776	29,347	38,501	51,640	4
Pb flux in melt	kg km^{-1} yr^{-1}	321	121	204	438	522	4

Data sources and calculations as in Table 2(a). Concentration ranges in fluids and concentrations in melts shown in Figure 12, are calculated as flux divided by water/melt mass.

support previous work on the subject. Using experimental eclogite/fluid distribution coefficients, 2–5% fluid equilibrating with the entire section of subducted oceanic crust + sediments beneath an arc (e.g., see Chapter 3.17) could carry the entire excess magmatic flux (primitive arc basalt—primitive MORB) of barium, lead, and strontium. Some experimental data have lanthanum solubilities in subduction fluids just high enough to account for the excess magmatic lanthanum flux in arcs. However, our results are consistent with previous calculations showing that aqueous fluid transport cannot account for the excess magmatic thorium flux. Further, our simple calculations support the idea that aqueous fluid transport should result in large fractionations of barium from thorium, lead from caesium, and strontium from neodymium. Thus, calculated fluid compositions lie at high Ba/Th, Pb/Ce, and Sr/Nd compared to barium, thorium, lead, strontium, and light REE enriched primitive arc lavas (Figure 12). In contrast, transport in 2–5% melt of basalt + sediment in eclogite facies can account for excess magmatic flux of barium, thorium, lanthanum, lead, and strontium, and this mechanism will produce relatively small fractionations between these different elements. As a result, calculated melt compositions plot at the enriched end of the trend from MORB to barium, thorium, lead strontium, and light REE enriched primitive arc lavas (Figure 12).

(iv) Fluids, melts, or goo above the solvus? The suggestion that aqueous fluids might play a minor role in arc magma genesis is apparently odds with interpretations of data on oxygen isotopes versus trace-element enrichment in several arcs (Eiler *et al.*, 2000). These data were inferred to indicate that enrichment in arcs is via an aqueous fluid, with a relatively high O/Ti ratio, rather than a silicate liquid with a much lower O/Ti ratio. However, the Eiler *et al.* result applies mainly to their relatively large data set for the Vanuatu arc, and depends on the composition of depleted boninite magmas, and questionable assumptions about mantle source composition (e.g., initial Cr/(Cr + Al) = 0.1). Further, although they did make a successful model involving melt transport rather than fluid transport, Eiler *et al.* discounted the result because it was apparently at odds with thermal models that rule out melting of subducting material. In their successful melt transport model, the melt has 30 wt.% H_2O, within the range of 25–50 wt.% H_2O inferred for aqueous fluid saturated melts at 3–5 GPa (Dixon and Stolper, 1995; Dixon *et al.*, 1995; Kawamoto and Holloway, 1997; Mysen and Wheeler, 2000).

The suggestion that aqueous fluids may not play a key role in subduction zone petrogenesis may seem at first to be at odds with the decades-old inference that addition of H_2O to the mantle wedge is one of the key causes of mantle melting and arc magmatism. However, this is certainly not what we wish to propose. Instead, as elegantly shown in calculations by Eiler *et al.* (2000) the effect of adding H_2O to peridotite is very similar whether the added H_2O is in an aqueous fluid or dissolved in silicate melt. Thus, assuming that the effect of other possible fluxes such as K_2O is second order, if 10% fluid-fluxed melting requires addition of ~1 wt.% fluid with ~90 wt.% H_2O, then 10% melt-fluxed melting might require addition of ~3 wt.% melt with 30 wt.% H_2O, or ~2 wt.% melt with 45 wt.% H_2O.

About 30–50 wt.% H_2O in a silicate melt corresponds to molar H/Si ~3–6, which raises the question, is a fluid-saturated melt at 3–5 GPa more like an anhydrous melt, or more like an aqueous fluid? It is possible that the H_2O-rich phase generated via dehydration reactions in subducting plates at 3–5 GPa might form at conditions where there is no longer a solvus separating distinct melt and fluid phases (Bureau and Keppler, 1999; Keppler, 1996). In this interpretation, the differences in experimentally constrained partitioning behavior at subduction zone pressures, for example between fluid/eclogite (Ayers *et al.*, 1997; Brenan *et al.*, 1996, 1995a,b; Johnson and Plank, 1999; Kogiso *et al.*, 1997; Stalder *et al.*, 1998; Tatsumi and Kogiso, 1997) and melt/eclogite (Rapp *et al.*, 1999), might arise as a result of the H_2O/silicate ratio in a given experimental bulk composition, rather the existence of distinct melt and fluid phases. Alternatively, the traditional interpretation, in which distinct fluid and melt phases can be present in H_2O-eclogite at, e.g., 3–5 GPa and 700–900 °C, may well be correct.

The presence or absence of a solvus between melt and fluid in equilibrium with eclogite at high pressure and moderate temperature is likely to be controversial for several years to come. Meanwhile, the message from our flux calculations (Table 2) and calculated fluid versus melt compositions (Figure 12) remains clear: It is easiest to understand the range of trace-element enrichment in arc lavas, relative to MORBs, if transport of barium, thorium, lead, strontium, and light REEs from subducted sediment and basalt is mainly in a phase whose partitioning characteristics are similar to those measured for relatively H_2O-poor "melt"/rock, and different from those measured for relatively H_2O-rich "fluid"/rock.

3.18.3.3.2 Tantalum and niobium depletion

Tantalum and niobium in arc magmas are depleted relative to REEs, so that Nb/La and Ta/La are lower than in the primitive mantle and in MORBs. Depletion of primitive arc magmas in tantalum and niobium relative to lanthanum and thorium (Figure 14) is ubiquitous (a few niobium-enriched lavas—e.g., Kepezhinskas *et al.* (1997)—form a distinct anomaly which will not be discussed in this chapter). For this reason, and because thorium, tantalum, and niobium are relatively immobile in low temperature alteration of basalts, low Nb/Th and Ta/Th have been used as discriminants between arc magmas and both

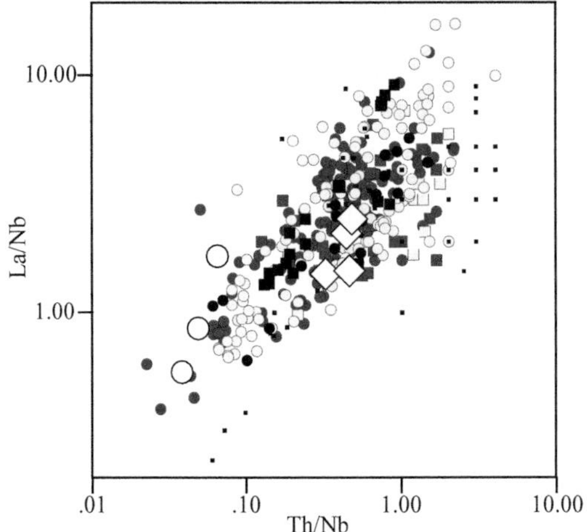

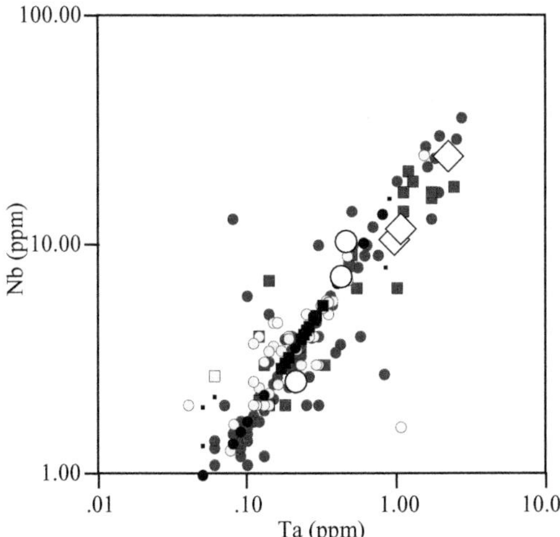

Figure 14 Relative depletion of Nb relative to Th and La in primitive arc lavas compared to MORB. For samples without Nb analyses (mostly, Aleutian data), we estimated Nb ~17*Ta. Primitive andesites have among the highest Th/Nb and La/Nb in the compiled data. Nb, Ta, and La concentrations are all higher than MORB in some continental arc lavas; in most oceanic arc lavas, including Aleutian primitive andesites, Nb and Ta concentrations are similar to or less than in MORB, while La is enriched compared to MORB. Symbols as in Figures 1 and 6.

ocean island basalts and MORBs, to constrain the provenance of lavas where tectonic accretion has obscured their original setting (e.g., Pearce, 1982; Pearce and Peate, 1995).

While tantalum and niobium are relatively depleted, compared to other incompatible elements, many primitive arc magmas have higher tantalum and niobium concentrations than MORB (Figure 14). This is particularly true of primitive, continental arc lavas, both basalts and andesites. Oceanic arc lavas tend to have tantalum and niobium concentrations as low as, or lower than MORBs, and elevated lanthanum concentrations compared to MORBs.

Depletion of tantalum and niobium relative to other incompatible elements in arc lavas has been ascribed to many processes (review in Kelemen *et al.* (1993)) including (i) crystal fractionation of Fe–Ti oxides in the crust, (ii) fractionation of titanium-rich, hydrous silicates such as phlogopite or hornblende in the mantle or crust, (iii) extensive, chromatographic interaction between migrating melt and depleted peridotite, (iv) the presence of phases such as rutile or sphene in the mantle wedge (Bodinier *et al.*, 1996), (v) relative immobility of tantalum and niobium relative to REE and other elements in aqueous fluids derived from subducting material, (vi) inherited, low Ta/Th and Nb/Th from subducted sediment (Plank, 2003), and (vii) the presence of residual rutile during partial melting of subducted material.

Primitive basalts have not been affected by extensive FeTi oxide fractionation in the crust, and they have magmatic temperatures too high for amphibole or biotite crystallization, ruling out (i) and (ii) in the previous paragraph. Chromatographic fractionation of niobium and tantalum from thorium and lanthanum requires melt/rock ratios $\sim 10^{-3}$, and thus—given estimated arc fluxes—requires that parental arc magmas react with the entire mass of the mantle wedge during ascent (Kelemen *et al.*, 1993), so (iii) seems unlikely. Consistently low Nb/La and Nb/Th is observed in arc magmas, even in primitive basalts which are hot, and far from rutile saturation at mantle pressures (Kelemen *et al.*, 2003b), eliminating (v) for most arc basalts. Finally, as noted above, lanthanum and thorium enrichment in arc magmas probably occurs via addition of a melt from subducting sediment or basalt in eclogite facies, and not via aqueous fluid metasomatism, so high La/Nb and Th/Ta does not arise as a result of fluid/rock fractionation (vi). In some arc lavas, it appears that niobium, tantalum, and lanthanum concentrations are equally enriched, compared to MORBs, and thus low Nb/La and Ta/La in these may be due to transport of all these elements in a melt of subducted, continental sediment without residual rutile (Johnson and Plank, 1999). If so, however, this raises the question of how low Nb/La and Ta/La originally formed in the continents. Suites without a strong signature of recycled sediment (many oceanic arc lavas, and particularly primitive Aleutian andesites) generally show lanthanum enrichment without niobium and tantalum enrichment, relative to MORBs.

For these reasons, if a single explanation for niobium and tantalum depletion in primitive arc lavas is to be sought, we prefer hypothesis (vii), fractionation of niobium and tantalum from other highly incompatible elements via partial melting of subducting, eclogite facies basalt or sediment with residual rutile (Elliott *et al.*, 1997; Kelemen *et al.*, 1993; Ryerson and Watson, 1987; Turner *et al.*, 1997). If this inference is correct, it follows that nearly all arc magmas include a component derived from partial melting subducted material in eclogite facies, with residual rutile.

3.18.3.3.3 U-series isotopes

There have been numerous recent papers and reviews on U/Th, U/Pa, and Ra/Th isotopic disequilibrium in arc lavas (Bourdon *et al.*, 1999, 2000; Clark *et al.*, 1998; George *et al.*, 2003; Gill and Condomines, 1992; Gill and Williams, 1990; Newman *et al.*, 1984, 1986; Reagan and Gill, 1989; Reagan *et al.*, 1994; Regelous *et al.*, 1997; Sigmarsson *et al.*, 2002, 1990, 1998; Thomas *et al.*, 2002; Turner *et al.*, 2003, 2000a,b,c, 2001, 1997; Turner and Foden, 2001). This is a complicated topic, and we cannot provide sufficient background information to make it accessible to a nonspecialist. However, because these data have bearing on other topics covered in this chapter, we summarize our understanding of recent work in this section. An explanation of the basic principles governing U-series fractionation and isotopic evolution is given by Spiegelman and Elliott (see Chapter 3.14).

Several recent papers have emphasized the presence of substantial ^{226}Ra excess (over parent ^{230}Th) in arc lavas. In the Marianas and Tonga arcs, ^{226}Ra excess correlates with Ba/La, Ba/Th, and Sr/Th (George *et al.*, 2003; Sigmarsson *et al.*, 2002; Turner *et al.*, 2003, 2000a,b,c, 2001; Turner and Foden, 2001). As a result, ^{226}Ra excess is linked in these papers to transport of a fluid component from subducted material to arc volcanoes in less than a few thousand years.

The argument that ^{226}Ra excess is generated by deep, subduction zone processes is particularly compelling for lavas from southern Chile, in which ^{10}Be/^{9}Be correlates with ^{226}Ra excess, suggesting that a component derived from young, subducted sediment may reach arc volcanoes in one or two thousand years (Sigmarsson *et al.*, 2002). In contrast, a data set on beryllium- and uranium-series isotopes from the Aleutian arc does not show correlation of ^{226}Ra excess and ^{10}Be/^{9}Be (George *et al.*, 2003). Unfortunately Sigmarsson *et al.* (2002), do not report sufficient geochemical data to evaluate whether Ba/La is high in lavas with high ^{226}Ra excess, as in other data sets, and/or whether Th/La and Th/Ba are high, as in the proposed sediment melt component (Section 3.18.3.3.1) which might transport ^{10}Be. If the high ^{226}Ra component in the southern Andes has high barium *and* thorium, we would propose that radium, barium, thorium, and beryllium are all carried in a partial melt of subducted sediment or basalt, not in an aqueous fluid.

Transport of fluid and/or melt from the subduction zone directly beneath an arc to the surface in approximately one half-life of ^{226}Ra requires transport rates of order ~100 m yr^{-1}, and this cannot be sustained during diffuse porous flow of melt through peridotite at porosities less than ~0.03. Instead, melt flow must be focused into high porosity conduits or cracks (Sigmarsson *et al.*, 2002; Turner *et al.*, 2001). It has been claimed that velocities ~100 m yr^{-1} require flow of melt in fractures rather than via porous flow (e.g., Sigmarsson *et al.*, 2002). However, simple calculations show that if melt/fluid viscosity is ~2 Pa s, density contrast between mantle and fluid/melt is ~500 kg m^{-3}, and mantle grain size is between 4 mm and 10 mm, the velocity of buoyancy driven porous flow of melt through mantle peridotite will exceed 100 m/yr^{-1} at porosities of 0.09–0.035 (Kelemen *et al.*, 1997a). Estimates of porous flow velocity depend on uncertain parameterizations of mantle permeability, poorly constrained grain size in the mantle wedge, and so on. Nonetheless, it is apparent that—even if all ^{226}Ra excess arises from dehydration and/or partial melting in subduction zones at a depth of ~100 km, and ^{226}Ra excess data require transport in less than one half life of ^{226}Ra (1,600 yr)—this result cannot be used to discriminate between transport in fractures versus focused flow in high porosity conduits.

In addition, while the currently accepted interpretation may well be correct, the present understanding of U-series data in arcs is reminiscent of early work on ^{226}Ra excess in MORB, in which it was suggested that ^{226}Ra excess forms during the initial stages of decompression melting, ~100 km below the seafloor, and is transported to ridge lavas in less than a few thousand years (e.g., McKenzie, 1985; Richardson and McKenzie, 1994). Currently available data on young MORB show a negative correlation between ^{226}Ra excess and ^{230}Th excess (e.g., Sims *et al.*, 2002). Because ^{230}Th excess is probably formed by melting or melt/rock reaction involving garnet peridotite (e.g., McKenzie, 1985; Spiegelman and Elliot, 1992), garnet pyroxenite (e.g., Hirschmann and Stolper, 1996; Lundstrom *et al.*, 1995), or fertile, high pressure clinopyroxene (e.g., Turner *et al.*, 2000b; Wood *et al.*, 1999), the negative correlation between ^{230}Th and ^{226}Ra excesses may be indicative of a role for shallow level processes in the generation of ^{226}Ra excess (e.g., melt/rock reaction in the shallow mantle, interaction with lower crustal plagioclase (Jull *et al.*, 2002; Lundstrom, 2000; Lundstrom *et al.*, 2000, 1995, 1999; Saal *et al.*, 2002; Spiegelman and Elliot, 1992; Van Orman *et al.*, 2002). As a result, ^{226}Ra excess may not be a reliable indicator of melt transport velocities in the mantle beneath mid-ocean ridges (see Chapter 3.14).

Arc lavas, like MORB, also show a negative correlation between ^{226}Ra excess and ^{230}Th excess (or 1/^{238}U excess), allowing for some ^{226}Ra decay (George *et al.*, 2003; Reagan *et al.*, 1994; Sigmarsson *et al.*, 2002; Turner *et al.*, 2003, 2000a,c, 2001) (Figure 15). Since the generation of ^{230}Th excess probably involves garnet and/or fertile clinopyroxene at pressures of 2 GPa or more, we anticipate the evolution of theories in which ^{230}Th excess forms deep, while shallow processes play a role in generating ^{238}U and ^{226}Ra excess in arc lavas as well as in MORB. However, because radium, thorium, and uranium concentrations are higher in most arc lavas than in MORB, shallow level processes capable of generating ^{238}U and ^{226}Ra excess in arcs may differ from those beneath mid-ocean ridges. Thus, reaction between ascending melt and anhydrous mantle peridotite may not be capable of generating large ^{226}Ra and ^{238}U excesses in primitive arc lavas (Thomas *et al.*, 2002), and assimilation of young plagioclase may also be an unlikely explanation for ^{226}Ra excess in arc lavas (e.g., George *et al.*, 2003). However, steady state diffusive gradients involving radium- and barium-rich minerals such as phlogopite or biotite could produce ^{226}Ra excess correlated with high Ba/La in melts interacting with the upper mantle or lower arc crust (Feineman and DePaolo, 2002). In this interpretation, relatively high Sr/Th in Marianas and Tonga lavas with high ^{226}Ra excess (Turner *et al.*, 2003) might be due to interaction with both biotite and plagioclase. Because radium and barium are geochemically similar to each other, and very different from thorium and lanthanum, there may be other processes that result in enrichment of radium and barium relative to thorium and lanthanum. From this perspective, not all lavas with high Ba/La and Ba/Th necessarily record selective enrichment of barium via aqueous fluid metasomatism in the mantle source.

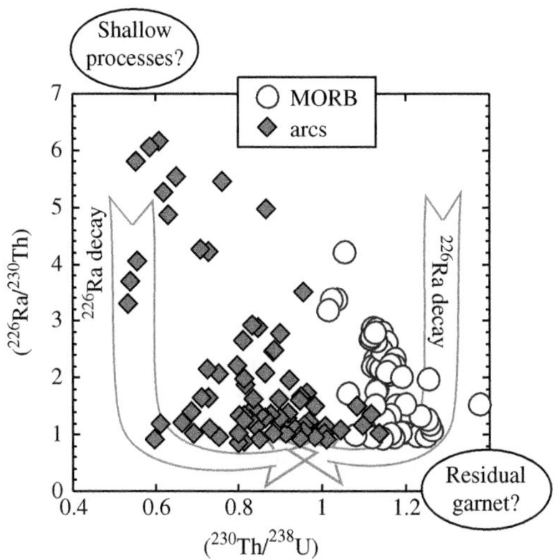

Figure 15 ^{226}Ra excess versus ^{230}Th excess in MORB and arc magmas. MORB data compiled by Sims *et al.* (2002). Arc data compiled by Turner *et al.* (2003). ^{226}Ra decay may explain very low ^{226}Ra excess in some lavas. However, the highest ^{226}Ra excess at any given (^{230}Th/^{238}U) shows a negative correlation with ^{230}Th excess. In the MORB data, ^{230}Th excess may arise at depths of 2–3 GPa, via U/Th exchange between melt and garnet, or melt and high pressure, fertile clinopyroxene compositions. Lavas with high ^{226}Ra excess do not show this high-pressure signature, and so many hypotheses call upon shallow melt/mantle interaction, or even lower crustal processes, to explain the genesis of ^{226}Ra excess. Similar theories have been advanced to explain the arc data. See text for discussion and references.

If ^{226}Ra excess were *always* the result of relatively shallow processes, then correlation of high ^{10}Be/^{9}Be with high ^{226}Ra excess in southern Chilean lavas would imply a shallow source for ^{10}Be enrichment as well. Some of the Chilean samples analyzed for ^{226}Ra (Sigmarsson *et al.*, 2002) are evolved (Stern and Kilian, 1996), and not enough data are presented to determine whether any of the other Chilean samples are primitive, or not. While excess ^{226}Ra and ^{10}Be may be transported from subduction zone depths in some arcs, it may also be worthwhile to re-evaluate the extent to which some evolved magmas interact with meteoric water, or assimilate alteration products, that have high ^{226}Ra excess and high ^{10}Be/^{9}Be.

It is not yet clear whether high thorium, primitive lavas worldwide have high ^{230}Th excess, or not. ^{230}Th excess has been observed in a few arc lavas, notably primitive andesites from Mt. Shasta in the southern Cascades (Newman *et al.*, 1986), and high Mg# andesites from the Austral Andes in southernmost Chile (Sigmarsson *et al.*, 2002). This is consistent with the hypothesis that primitive andesites contain a substantial proportion of partial melt from a source rich in residual garnet, such as subducting, eclogite facies sediment and/or basalt (Section 3.18.3.2.5).

Substantial ^{230}Th excess is also observed in some primitive basalts from Central America (Thomas *et al.*, 2002). If primitive basalts do not include a substantial eclogite melt component, then the ^{230}Th excess in Central American basalts might reflect melting or melt/rock reaction in the presence of garnet in the mantle wedge. (Another alternative is that ^{230}Th excess arises as a result of melting, or melt/rock reaction, in the presence of fertile, high-pressure clinopyroxene (Turner *et al.*, 2000b; Wood *et al.*, 1999).) Two factors favor a potentially large role for garnet in the mantle wedge, compared to the melting region beneath mid-ocean ridges. First, relatively high H_2O fugacity in the mantle beneath most arcs lowers mantle solidus temperatures at a given pressure. Because of the positive pressure/temperature slope of reactions such as pyroxene + spinel = olivine + garnet, lower solidus temperatures lowers the minimum depth at which garnet can be stable on the arc mantle solidus (e.g., Gaetani and Grove, 1998). Second, theoretical considerations suggest both porous and/or fracture transport of melt in the mantle wedge to the base of arc crust might be diagonal, beginning at depths of 150 km or more beneath the back arc region (Davies, 1999; Spiegelman and McKenzie, 1987). Thus the maximum pressure of mantle melting beneath an arc may not be constrained by the depth to the subduction zone directly beneath the arc. Because the maximum pressure of melting might be as large or larger beneath arcs, and the minimum pressure of garnet stability on the mantle solidus is probably lower beneath arcs, the interval in which garnet could react with arc melts is probably larger beneath arcs compared to ridges. We will return to this point in Section 3.18.3.3.4.

More generally, arc lavas show ^{238}U excess, usually interpreted to be indicative of high U/Th solubility in aqueous fluids derived from subducting material. ^{226}Ra excess is correlated with ^{238}U excess in most data sets on historically erupted lavas. Thus, if ^{226}Ra excess is formed at great depth, it is likely (though not required; see Thomas *et al.* (2002)) that ^{238}U excess is also formed at depth. Conversely, shallow processes that fractionate thorium from radium could also fractionate thorium from uranium.

3.18.3.3.4 Geodynamic considerations

> Theory and observation were in conflict, and theory seemed to get the better of it.
>
> Fenner (1937, p. 166)

Since the 1980s, most geochemists and petrologists working on arcs have considered partial melting of subducted material to be relatively unusual, and absent beneath most modern arcs. This is based on three main lines of reasoning.

First, very few arc lavas show strong heavy REE/middle REE depletion, but such depletion is predicted and observed in melts of eclogite or garnet amphibolite (e.g., Gill, 1974, 1978; Kelemen *et al.*, 2003b; Rapp *et al.*, 1999). In addition, partial melts of subducted material are likely to be granitic (e.g., Johnson and Plank, 1999; Nichols *et al.*, 1994; Rapp *et al.*, 1999; Rapp and Watson, 1995), and close to H_2O saturation at 3 GPa or more, with ~25–50 wt.% H_2O (Dixon and Stolper, 1995; Dixon *et al.*, 1995; Kawamoto and Holloway, 1997; Mysen and Wheeler, 2000). In these ways, no arc magma resembles a fluid-saturated melt of sediment or basalt in eclogite facies.

Second, because many dehydration reactions in the subducting plate may be complete by 2 GPa, it is not evident that free fluid will be available to facilitate fluid-saturated melting (e.g., Davies and Stevenson, 1992; Peacock *et al.*, 1994; Rapp and Watson, 1995).

Third, thermal models for arcs published between 1980 and 2002 uniformly indicate that the top of the subducting plate in "normal" subduction zones (convergence rate >0.03 m yr^{-1}, subducting oceanic crust older than 20 Ma) does not reach temperatures above the fluid-saturated solidus for metabasalt or metasediment (see reviews in Kelemen *et al.* (2003a), Peacock (1996, 2003), and Peacock *et al.* (1994)).

Recently, all three of these lines of reasoning have been challenged.

First, modeling of reaction between heavy REE depleted melts of eclogite facies basalt or sediment and mantle peridotite shows that heavy REE abundances rise to levels in equilibrium with spinel peridotite at melt/rock ratios less than ~0.1, while the light REEs and other highly incompatible elements are almost unaffected (Kelemen *et al.*, 1993, 2003b). Thus, primitive arc lavas with flat, middle to heavy REE patterns at low abundance, and light REE enrichment, could be formed by reaction between a partial melt of subducted material and the overlying mantle wedge. In addition, anatectic, H_2O-rich melts of subducting sediment or basalt will both decompress and heat up as they rise into the overlying mantle. This "super-adiabatic" ascent may enhance melt/rock reaction, leading to a net increase in the liquid mass (Grove *et al.*, 2001, 2003; Kelemen, 1986, 1990, 1995; Kelemen *et al.*, 1993, 2003b). Thus, major elements in hybrid melts may be primarily derived from the mantle wedge, while highly incompatible trace elements may reflect the original, eclogite facies residue.

(Note that the conclusion of the previous paragraph seems to be somewhat at odds with Section 3.18.3.3.3, in which it was proposed that ^{226}Ra and ^{238}U excesses might arise via some shallow process. If this were so, then presumably other incompatible element concentrations would also be affected by this shallow process. However, based on data from (Newman *et al.*, 1984; Sigmarsson *et al.*, 2002), we anticipate that primitive andesites with a substantial eclogite melt component will have ^{230}Th excess and little or no ^{226}Ra excess).

Second, it is now apparent that continuous dehydration reactions involving hydrous phases in metasediment and upper oceanic crust with higher pressure stability than glaucophane plus extensive solid solution, such as lawsonite, chloritoid, phengite, and zoisite, provide a small but nearly continuous source of fluid from shallow depths to depths exceeding ~250 km (review in Chapter 3. 17). In addition, hydrous phases such as serpentine and talc in the uppermost mantle of the subducting plate have an extensive stability field extending to high pressure, and will continue to dehydrate due to conductive heating of the cold interior of the subducting plate to depths up to ~200 km (review in Chapter 3.17). Finally, aqueous fluid may not be wetting in eclogite facies assemblages with abundant clinopyroxene. H_2O-rich fluid in clinopyroxenite does not become interconnected until fluid fractions exceeding 7 vol.% (Watson and Lupulescu, 1993). Thus, some of the H_2O evolved by dehydration reactions may remain within the metamorphic protolith until melting increases the permeability and permits H_2O dissolved in the melt to escape the subducting slab by porous flow (Kelemen *et al.*, 2003b), or until enough low density, interconnected aqueous fluid is present to fracture the overlying rock due to fluid overpressure (Davies, 1999). Thus, although some of the processes described in this paragraph are highly speculative, it is likely that through some combination of these processes, aqueous fluid is present to flux melting of subducting material at depths shallower than ~200–300 km.

Third, computational and theoretical advances have made it possible for thermal models to incorporate temperature-dependent viscosity, and/or non-Newtonian viscosity, in the mantle wedge (Conder *et al.*, 2002; Furukawa, 1993b; Furukawa and Tatsumi, 1999; Kelemen *et al.*, 2003a; Kincaid and Sacks, 1997; Rowland and Davies, 1999; van Keken *et al.*, 2002). This has several important effects, among them eliminating the necessity for prescribing the thickness of a rigid upper plate (Kelemen *et al.*, 2003a). When the lithosphere in the upper plate is allowed to "find its own thickness," this results in upwelling of the mantle to shallow depths near the wedge corner, so that asthenospheric potential temperatures extend to depths as shallow as 40 km beneath the arc. We think these models are preferable to previous, isoviscous models because they provide a much closer match to the high metamorphic temperatures recorded in exposures of arc Moho and lower crust (Kelemen *et al.*, 2003a) and to calculated arc melt/mantle equilibration conditions (~1,300 °C at 1.2–1.5 GPa, Bartels *et al.* (1991), Draper and Johnston (1992), Elkins Tanton *et al.* (2001), Sisson and Bronto (1998), Tatsumi *et al.* (1983)). They also provide an explanation for the observed, anomalously slow seismic structure observed in the uppermost mantle beneath arcs (e.g., Zhao *et al.*, 1992, 1997), for the high heat flow in arcs (Blackwell *et al.*, 1982; Furukawa, 1993b), and for the sharp gradient in the transition to very low heat flow in fore arcs (Kelemen *et al.*, 2003a).

The most recent of these models predict temperatures in the wedge and the top of the slab that are significantly higher compared to isoviscous models. Predicted temperatures are higher than the fluid-saturated solidus for both basalt and sediment (Johnson and Plank, 1999; Lambert and Wyllie, 1972; Nichols *et al.*, 1994; Schmidt and Poli, 1998; Stern and Wyllie, 1973) near the top of the subducting plate directly beneath arcs at normal subduction rates and subducting plate ages (Kelemen *et al.*, 2003a; van Keken *et al.*, 2002). It would be premature to conclude that the tops of most subducting plates cross the fluid-saturated solidus directly beneath arcs, because this is an area of ongoing research. Also, it is not clear that subducting metasediment and/or metabasalt are fluid-saturated at these depths. And, even if fluids are present, natural rocks might evolve fluids with H_2O activities lower, and melting temperatures higher, than in most melting experiments (Becker *et al.*, 1999, 2000; Johnson and Plank, 1999). Nonetheless, it is clear that thermal models should no longer be invoked to "rule out" partial melting of subducted material in eclogite facies.

Our community has been focused for a long time on very simple pictures of subduction. While this is expedient for maintaining sanity, it is intriguing to speculate briefly on the possibility of solid material transfer across the Benioff zone (Figure 16). Several decades ago, geochemists suggested that physical mixing of a few percent sediment with peridotite in the mantle wedge beneath arcs could account for many geochemical features of arc lavas (Armstrong, 1981; Kay, 1980). Indeed, by analogy with predicted gravitational instability of dense lower crust (e.g., Arndt and Goldstein, 1989; DeBari and Sleep, 1991; Ducea and Saleeby, 1996; Herzberg *et al.*, 1983; R. W. Kay and S. M. Kay, 1988, 1991, 1993; S. M. Kay and R. W. Kay, 1985; Turcotte, 1989), subducting sediment may "delaminate" and rise into the overlying mantle wedge. Subducting sediment in eclogite facies is likely to be substantially less dense than the overlying mantle. Thus, viscous density instabilities will arise provided subducting sediment layers have thicknesses of 100 m to 1 km, once the overlying mantle viscosity becomes less than some critical value. For example, our calculations (Jull and Kelemen, 2001) suggest that, for density contrasts of 50–150 kg m^{-3} and background strain rates $\sim 10^{-14}$, a 1 km thick layer of subducting sediment would form unstable diapirs and rise into the overlying mantle at ~750 °C. This process will lead to mechanical mixing of sediment and mantle peridotite. In addition, rising, heating diapirs of sediment would certainly undergo partial melting in the mantle wedge.

In another mechanism of solid transfer across the Benioff zone, imbrication of the subduction thrust at shallow depth, or downward migration of the subduction shear zone at greater depth, may transfer material from the top of the downgoing

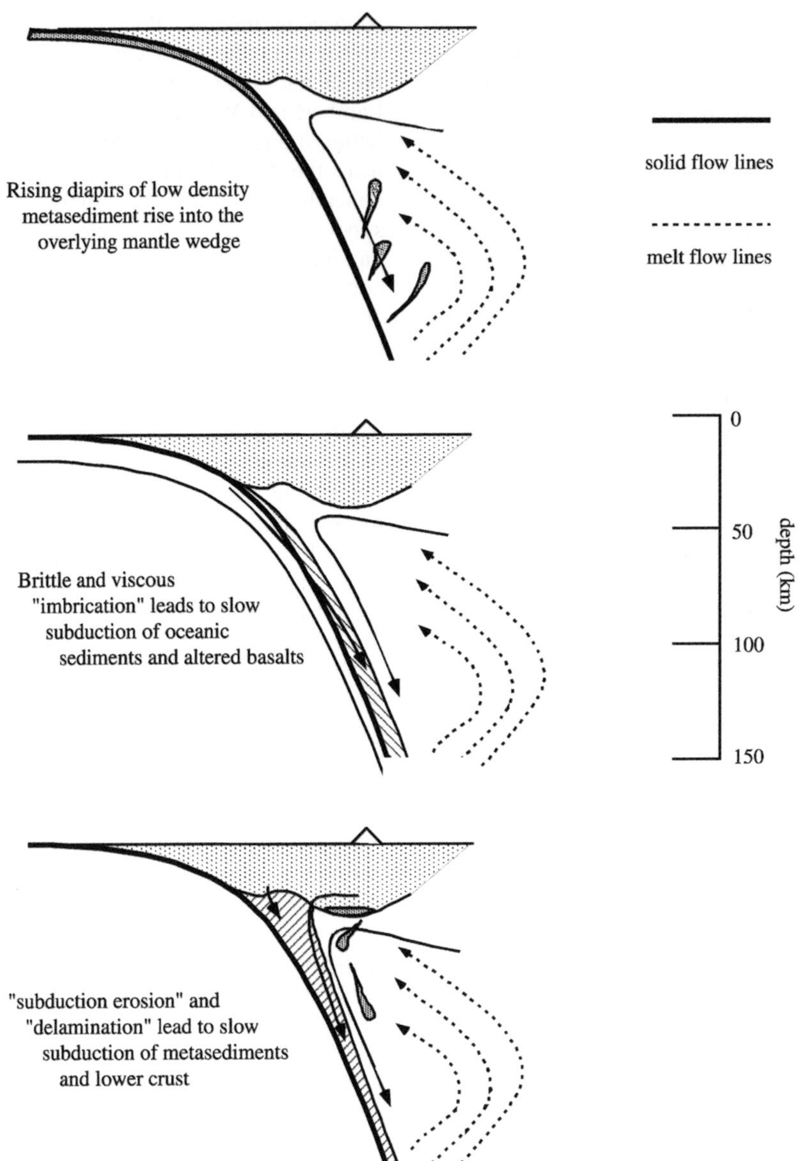

Figure 16 Schematic illustration of mechanisms for transfer of sediments, volcanics, and/or lower crustal gabbros into the mantle wedge from the subducting plate and the base of arc crust. Dark black line indicates position of the "subduction zone"; below this line, material subducts at the convergence velocity. Above this line, material is carried downward more slowly. Any process leading to slow transport of low-melting point metasediment, metabasalt, or metagabbro into the mantle wedge would lead to partial melting of this material beneath an arc.

plate into the hanging wall, reducing its convergence velocity and allowing more time for conductive heating (Figure 16).

"Subduction erosion" may transfer relatively hot, middle to lower crust from the fore-arc (metasediment, etc.) or the arc (garnet granulite, pyroxenite) to the top of the subducting plate (Clift and MacLeod, 1999; Ranero and von Huene, 2000; Vannucchi *et al.*, 2001; von Huene and Scholl, 1991, 1993) (Figure 16). Because this material has a low melting point, compared to the mantle, but is already hot compared to the top of the subducting oceanic crustal section, conductive heating will take it above its fluid-saturated solidus in a relatively short time, while fluid may supplied from the underlying, subducting plate. Similarly, although it is not "subduction," delamination of dense crustal cumulates could lead to transformation of gabbroic rocks into eclogites, followed by heating, and partial melting (Gromet and Silver, 1987; R. W. Kay and S. M. Kay, 1993), as descending diapirs become entrained in ductile flow of the mantle wedge (Figure 16).

Another important consideration is that partial melts of subducted material do not necessarily rise vertically through the mantle wedge from the subduction zone to the base of arc crust. One might expect initial trajectories to trend diagonally upward, away from the trench, due to the sum of buoyancy-induced upwelling and diagonal, downward solid flow in the mantle wedge. Higher in the wedge, theoretical considerations suggest that both porous flow and fracture transport of melt would be diagonally upward, toward the trench (Davies, 1999; Spiegelman and McKenzie, 1987). This diagonal, trenchward upwelling would be enhanced if solid flow in the upper part of the mantle wedge is also diagonally upward toward the trench, as is predicted from models incorporating a temperature dependent mantle viscosity (Conder *et al.*, 2002; Furukawa, 1993b; Furukawa and Tatsumi, 1999; Kelemen *et al.*, 2003a; Kincaid and Sacks, 1997; Rowland and Davies, 1999; van Keken *et al.*, 2002). As a result, it may be a mistake to concentrate solely on whether subducted material exceeds its fluid-saturated solidus directly beneath an arc. Instead, if subducted material crosses the solidus at 150 km, or even 250 km, the resulting melt might be transported diagonally to the base of arc crust.

In summary, for a variety of reasons, it is unwarranted to use geodynamical models to rule out partial melting of eclogite facies sediment, basalt, or lower crustal gabbro beneath arcs. Meanwhile, we believe that geochemical data on primitive arc lavas are best understood if a partial melt of subducted sediment and/or basalt (and/or gabbro) is the primary medium for transport of in compatible trace elements into the mantle wedge.

3.18.4 ARC LOWER CRUST

> Ever clearer became the danger of restricting attention to the 'observed facts'. Direct observations are usually restricted to…a two-dimensional field. …Three dimensional it must be, in any case, even at the cost of one's peace of mind—even at the cost of risking the quagmire of speculation about the invisible and intangible. There is, indeed, no other way. By declining Nature's own invitation to think intensively about her third dimension, petrologists have 'lost motion' and have held back the healthy progress of their science. What petrology needs is controlled speculation about the depths of the Earth.
>
> Reginald A. Daly,
> *Igneous Rocks and the Depths of the Earth*
> (Daly, 1933)

> To reason without data is nothing but delusion.
>
> Arthur Holmes,
> *The Age of the Earth*
> (Holmes, 1937, p. 152)

3.18.4.1 Talkeetna Arc Section

Although there are outcrops of middle to lower crustal plutonic rocks with arc provenance in many places (e.g., DeBari, 1994; DeBari *et al.*, 1999; Pickett and Saleeby, 1993, 1994), in this part of the chapter we will concentrate on data from the Talkeetna (south-central Alaska) and Kohistan (Pakistan Himalaya) arc crustal sections (e.g., Bard, 1983; Barker and Grantz, 1982; Burns, 1985; Coward *et al.*, 1982; DeBari and Coleman, 1989; DeBari and Sleep, 1991; Jan, 1977; Miller and Christensen, 1994; Plafker *et al.*, 1989; Tahirkheli, 1979). Both regions preserve tectonically dissected but relatively complete sections, from volcanics and sediments at the top to residual mantle peridotites near the base. Metamorphic equilibria at the Moho in both sections record conditions of ~1,000 °C and 1–1.2 GPa. In addition, high P/T metamorphism, with pressure perhaps as high as 1.8 GPa, may be recorded by the Jijal complex, along the Main Mantle thrust at the base of the Kohistan section (Anczkiewicz and Vance, 2000; Ringuette *et al.*, 1999; Yamamoto, 1993) and by late veins in the Kamila amphibolites (Jan and Karim, 1995); this is probably related to continental collision and exhumation of high P rocks along the Indian–Asian suture zone (Gough *et al.*, 2001; Treloar, 1995; Treloar *et al.*, 2001). Because we are most familiar with the Talkeetna section, we will emphasize data from our recent studies there, with supporting data from the Kohistan section.

The Talkeetna section represents an arc fragment, ranging in age from ~200 Ma to ~175 Ma that was accreted along the North American margin and is now exposed in south central Alaska and along the Alaska Peninsula (Barker and Grantz, 1982; Detterman and Hartsock, 1966; Grantz *et al.*, 1963; Martin *et al.*, 1915; Millholland *et al.*, 1987; Newberry *et al.*, 1986; Nokleberg *et al.*, 1994; Palfy *et al.*, 1999; Plafker *et al.*, 1989; Rioux *et al.*, 2002b, 2001b; Roeske *et al.*, 1989). The general geology and petrology of the Talkeetna section has been summarized by (Burns, 1983, 1985; Burns *et al.*, 1991; DeBari, 1990; DeBari and Coleman, 1989; DeBari and Sleep, 1991; Newberry *et al.*, 1986; Nokleberg *et al.*, 1994; Pavlis, 1983; Plafker *et al.*, 1989; Winkler *et al.*, 1981). It is bounded to the north, along a contact of uncertain nature, by the accreted Wrangellia terrane. To the south, the Talkeetna arc section is juxtaposed along the Border Ranges Fault with accretionary wedge mélanges, the Liberty Creek, McHugh and Valdez complexes. This major fault has been a thrust and a right lateral strike-slip fault. Although the Border Ranges Fault is near vertical at present, a flat-lying klippe of gabbroic rocks, almost certainly derived from the Talkeetna section, overlies the McHugh complex in the Chugach Mountains north of Valdez. This is called the Klanelneechina klippe.

Preliminary data suggest no inheritance in 200–180 Ma zircons from Talkeetna plutonic rocks (Rioux *et al.*, 2002a, 2001a). The small contrast between neodymium isotopes in Talkeetna gabbros with neodymium isotopes for Jurassic MORB resembles the small neodymium isotope difference between Marianas arc lavas and present-day MORB (Greene *et al.*, 2003). Small bodies of metaquartzite and marble intruded by Talkeetna plutonic rocks contain little or no zircon, and are interpreted as pelagic sediments (J. Amato, personal communication, 2003), while amphibolite rafts have andesitic compositions and the trace-element signatures of arc magmas (e.g., high La/Nb and Th/Nb; our unpublished data). Thus, recent work is consistent with the hypothesis that the entire Talkeetna arc section is composed of late Triassic to middle Jurassic rocks that formed in an intra-oceanic arc.

After reconnaissance mapping and sampling (Barker *et al.*, 1994; Barker and Grantz, 1982; Newberry *et al.*, 1986), the volcanic section of the Talkeetna arc has received relatively little study until now. Our preliminary data agree with earlier estimates that the volcanics are 5–7 km thick. High Mg# basalts are relatively common, though 11 of our 87 samples are high Mg# andesites. More evolved lavas range from mainly tholeiitic andesites through tholeiitic and calc-alkaline dacites, to calc-alkaline rhyodacites and rhyolites. The volcanics are underlain and intruded by felsic to gabbroic plutons.

Excellent descriptions of the petrology and major-element composition of the Talkeetna lower crust (e.g., Burns, 1985; DeBari and Coleman, 1989; DeBari and Sleep, 1991) and our recent work (Greene *et al.*, 2003) show that much of the section is composed of compositionally monotonous gabbronorites. Some, but not all, include abundant magnetite. Prismatic hornblende of obvious igneous origin is rare, though most samples have hornblende rims—probably of deuteric origin—around prismatic pyroxene crystals. Olivine is extremely rare, even at the base of the crustal section. Both Talkeetna and Kohistan lower crustal gabbroic rocks have very high aluminum contents, compared to gabbroic rocks from ophiolites and to the average composition of continental lower crustal xenoliths (Rudnick and Presper, 1990). Because they have such high aluminum, these rocks can form more than 30% garnet at garnet granulite facies conditions, and are thus denser than mantle peridotite at the same pressure and temperature (Jull and Kelemen, 2001).

DeBari and Coleman (1989), building on previous mapping by Burns, Newberry, Plafker, Coleman and others, concentrated much of their work in the Tonsina area, where several small mountains preserve a laterally continuous Moho section, with relatively mafic gabbronorites overlying a thin but regionally continuous horizon of mafic garnet granulite (orthogneiss), overlying ~500 m of pyroxenite (mostly clinopyroxene-rich websterite), in turn overlying residual mantle harzburgite with ~10% dunite (Figure 17). The harzburgite, in turn, is bounded by the Border Ranges fault to the south. It is important to emphasize that gabbronorites, garnet granulites, pyroxenites, and harzburgites show "conformable," high temperature contacts which extend across intermittent outcrop for several kilometers. In this area, garnet granulites record conditions of ~1,000 °C and 1 GPa (DeBari and Coleman, 1989; Kelemen *et al.*, 2003a), indicating that at the time of garnet growth, the crustal thickness was probably ~30 km. Previous studies (Burns, 1985; DeBari and Coleman, 1989; DeBari and Sleep, 1991) and our recent work (Greene *et al.*, 2003) have found that Talkeetna gabbronorites and lavas form a cogenetic, igneous differentiation sequence.

3.18.4.1.1 Geochemical data from the Talkeetna section

We turn now to new geochemical data on the Talkeetna section (averages in Table 3). As in pilot data from Barker *et al.* (1994), REE patterns in all but two lavas are relatively flat, just slightly more light REE enriched than MORB (Figure 18). Extended trace-element "spidergrams" show the distinctive characteristics of arc magmas, such as high Th/Nb, La/Nb, Ba/La, Pb/Ce, and Sr/Nd. Many lavas, particularly the ones with the highest abundance of incompatible trace elements, show low Eu/Sm, consistent with plagioclase fractionation, and low Ti/Dy, which is probably due, at least in part, to magnetite fractionation. Many of these lavas are pervasively altered. Nonetheless, the parallelism of most trace-element patterns suggests that few of the geochemical characteristics summarized here have been substantially modified by alteration. Felsic plutonic rocks, primarily calc-alkaline tonalites, have spidergram patterns similar to the lavas.

The gabbroic rocks, mostly gabbronorites, have spidergram patterns that are remarkably similar to the lavas in many ways, with high Th/Nb, La/Nb, Ba/La, Pb/Ce, and Sr/Nd (Figure 19). While DeBari and Coleman (1989) emphasized the presence of a few gabbroic samples with REE patterns reminiscent of MORB, suggesting that they might represent older oceanic crust into which the Talkeetna arc was emplaced, the extended trace-element patterns of all our gabbroic samples show arc-like signatures. For example, Talkeetna

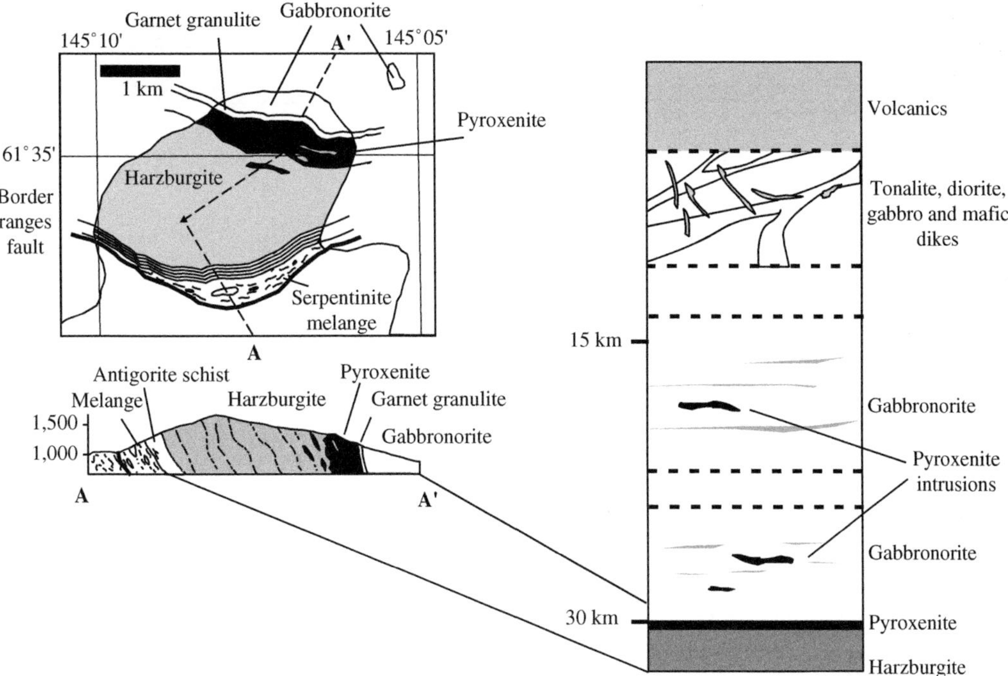

Figure 17 Schematic cross-section of the accreted, Jurassic Talkeetna arc in south-central Alaska based on new data and compilations by Greene *et al.* (2003), together with a geologic map and cross-section of the Moho exposure at Bernard Mountain, in the Tonsina region of the Talkeetna arc section, simplified from DeBari and Coleman (1989).

gabbronorites have average Th/Nb of 0.4, substantially higher than in gabbroic rocks in the Oman ophiolite (averaging 0.1 in our unpublished data) and lower oceanic crust (averaging 0.05; Hart *et al.* (1999)).

High Ti/Dy in many gabbronorite samples is indicative of the presence of cumulate magnetite, which fractionates titanium from dysprosium during igneous crystallization, and high Eu/Sm reflects plagioclase accumulation. Similarly, high Pb/Ce and Sr/Nd in the gabbroic rocks may arise in part from lead and strontium enrichments in parental melts, and in part from the presence of cumulate plagioclase. In order to constrain which additional characteristics of the Talkeetna gabbronorites are inherited from trace-element features in parental, arc magmas, and which characteristics reflect igneous fractionation, we divide the average Talkeetna gabbronorite composition by the average lava composition to yield a highly approximate bulk crystal/liquid distribution coefficient pattern (Figure 20). This shows that, despite complementary lead and strontium enrichments in the lavas, high Pb/Ce and Sr/Nd in the gabbroic rocks also reflect the presence of cumulate plagioclase. In addition, note that thorium and uranium are more incompatible than rubidium, barium and potassium in plagioclase.

Spidergrams for mafic garnet granulites from the Tonsina area show no evidence for the presence of cumulate igneous garnet, which would be reflected in high heavy REE/middle REE ratios (Figure 21). Instead, they are very similar to typical Talkeetna gabbronorites. Thus, following DeBari and Coleman (1989), we believe that the garnets in this area are entirely of metamorphic origin. In contrast, gabbronorite, two-pyroxene quartz diorite and tonalite in garnet granulite facies from the Klanelneechina klippe, which record ~700 °C and 0.7 GPa, do show evidence for the presence of cumulate, igneous garnet (Figure 22). These rocks all have evolved compositions (whole-rock Mg# less than 50). They could be cumulates from evolved melts crystallizing in the lower crust, and/or restites produced via partial melting of high temperature lower crust.

Trace-element contents of pyroxenites from the Tonsina area are shown in Figure 23. Although data on the most depleted of these may be very imprecise, they show high Pb/Ce and Sr/Nd, and generally low niobium and tantalum relative to potassium, lanthanum, and thorium. All of these characteristics are probably inherited from equilibrium with primitive arc magmas. Müntener *et al.* (2001) emphasized the strikingly low Al_2O_3 contents (1–6.5 wt.%, averaging 3.4 wt.% in our data) in many Talkeetna and Kohistan pyroxenites, despite their high pressures of crystallization. These low Al_2O_3 contents were not produced in experiments on hydrous,

Table 3 New geochemical data on the Talkeetna section.

	Tonsina pyroxenite	*N*	*Std. error*	*Talkeetna gabbronorite*	*N*	*Std. error*	*Tonsina garnet granulite*	*N*	*Std. error*	*Klanelneechina garnet diorite and tonalite*	*N*	*Std. error*	*Intermediate to felsic plutons*	*N*	*Std. error*	*Lavas, tuffs and volcano-clastic*	*N*	*Std. error*
Normalized oxides (wt.%)																		
SiO_2	49.95	17	0.296	47.86	95	0.047	46.55	6	1.234	51.60	7	0.880	68.54	28	0.210	59.64	114	0.069
TiO_2	0.07	17	0.003	0.66	95	0.004	0.79	6	0.122	0.89	7	0.086	0.50	28	0.011	0.86	114	0.003
Al_2O_3	3.42	17	0.126	19.00	95	0.030	18.66	6	0.153	18.92	7	0.352	15.19	28	0.057	16.53	114	0.017
FeO*	8.31	17	0.229	9.94	95	0.035	10.04	6	0.722	10.30	7	0.248	4.06	28	0.092	7.78	114	0.026
MnO	0.17	17	0.002	0.18	95	0.001	0.19	6	0.014	0.21	7	0.005	0.11	28	0.003	0.19	114	0.001
MgO	27.95	17	0.343	7.78	95	0.027	7.86	6	0.486	4.89	7	0.187	1.67	28	0.035	3.57	114	0.017
CaO	9.88	17	0.355	12.53	95	0.025	14.58	6	0.903	10.41	7	0.275	4.66	28	0.063	6.30	114	0.033
Na_2O	0.22	17	0.006	1.82	95	0.009	1.08	6	0.081	2.51	7	0.111	4.38	28	0.025	3.82	114	0.011
K_2O	0.02	17	0.001	0.16	95	0.002	0.18	6	0.067	0.13	7	0.033	0.76	28	0.019	1.03	114	0.007
P_2O_5	0.01	17	0.000	0.08	95	0.001	0.07	6	0.019	0.14	7	0.013	0.13	28	0.003	0.18	114	0.001
Mg#	86	17	0.2	58	95	0.1	58	6	2.8	45	7	0.6	43	28	0.4	43	114	0.1
XRF data (ppm)																		
Ni	516	17	15	42	31	1	41	6	3	11	7	3	4.8	13	0.2	17.2	86	0.3
Cr	3239	17	105	167	31	4	188	6	21	55	7	10	6.3	13	0.4	45.2	86	0.7
Sc	31.5	17	1.0	39.5	31	0.5	52	6	2	38.7	7	1.4	21.4	13	0.9	29.3	86	0.1
V	106	17	4	250	31	5	295	6	30	242	7	11	94	13	7	168.0	86	1.2
Ba	15.5	17	0.5	75	31	3	55	6	13	36	7	3	519	13	17	409	86	5
Rb	1.1	17	0.1	2.0	31	0.1	3.2	6	1.4	0.7	7	0.2	17.8	13	0.7	15.2	86	0.2
Sr	18.4	17	1.5	303	31	2	225	6	9	277	7	4	235	13	5	249	86	2
Zr	6.2	17	0.2	17.4	31	0.5	18	6	3	45	7	11	112	13	4	92.0	86	0.6
Y	2.8	17	0.1	9.8	31	0.2	16	6	2	17	7	2	31	13	1	28.8	86	0.1
Nb	1.8	17	0.0	1.9	31	0.0	1.9	6	0.2	3.6	7	0.4	4.2	13	0.2	4.2	86	0.0
Ga	4.2	17	0.1	15.7	31	0.1	16.2	6	0.6	17.9	7	0.4	15.3	13	0.2	16.5	86	0.0
Cu	43.9	17	2.9	88	31	2	152	6	27	34	7	2	25	13	4	45.1	86	0.5
Zn	43.4	17	1.8	55.8	31	0.7	60	6	8	85	7	2	48	13	2	96.6	86	0.9
Pb	0.6	17	0.1	1.6	31	0.1	2.3	6	0.5	0.4	7	0.2	3.0	13	0.1	3.3	86	0.0
La	6.6	17	0.4	5.6	31	0.2	9.7	6	2.2	5.3	7	0.8	10.1	13	0.9	9.8	86	0.1
Ce	2.9	17	0.4	8.3	31	0.2	11.3	6	2.5	14	7	2	24.4	13	0.9	23.3	86	0.1
Th	1.2	17	0.1	1.3	31	0.0	1.2	6	0.2	1.0	7	0.2	2.0	13	0.1	1.6	86	0.0

ICP-MS data (ppm)															
La	0.192	16	0.019	1.153	31	0.040	1.565	6	0.377	2.985	7	0.611	7.815	13	0.331	7.490	42	0.103
Ce	0.417	16	0.040	2.901	31	0.095	3.734	6	0.769	7.616	7	1.475	17.358	13	0.714	16.825	42	0.219
Pr	0.059	16	0.005	0.466	31	0.014	0.605	6	0.103	1.161	7	0.197	2.387	13	0.096	2.360	42	0.028
Nd	0.328	16	0.025	2.625	31	0.076	3.594	6	0.552	6.133	7	0.911	11.506	13	0.450	11.573	42	0.127
Sm	0.149	16	0.009	1.020	31	0.027	1.519	6	0.209	2.140	7	0.258	3.650	13	0.148	3.702	42	0.037
Eu	0.063	16	0.003	0.474	31	0.008	0.654	6	0.070	0.941	7	0.077	0.966	13	0.024	1.095	42	0.007
Gd	0.244	16	0.013	1.350	31	0.033	2.160	6	0.282	2.667	7	0.274	4.264	13	0.174	4.260	42	0.041
Tb	0.049	16	0.002	0.253	31	0.006	0.410	6	0.052	0.484	7	0.046	0.777	13	0.033	0.776	42	0.008
Dy	0.362	16	0.017	1.701	31	0.041	2.793	6	0.356	3.167	7	0.281	5.092	13	0.220	5.083	42	0.050
Ho	0.081	16	0.004	0.367	31	0.009	0.616	6	0.080	0.673	7	0.055	1.110	13	0.048	1.097	42	0.011
Er	0.236	16	0.011	1.020	31	0.025	1.716	6	0.232	1.861	7	0.146	3.174	13	0.138	3.094	42	0.032
Tm	0.034	16	0.001	0.150	31	0.004	0.243	6	0.033	0.273	7	0.021	0.481	13	0.021	0.462	42	0.005
Yb	0.219	16	0.009	0.941	31	0.024	1.500	6	0.212	1.714	7	0.124	3.104	13	0.134	2.951	42	0.032
Lu	0.035	16	0.001	0.151	31	0.004	0.232	6	0.033	0.271	7	0.019	0.510	13	0.021	0.473	42	0.005
Ba	4.254	16	0.357	65	31	3	43	6	13	27	7	4	516	13	18	434	42	10
Th	0.024	16	0.002	0.071	31	0.004	0.152	6	0.043	0.097	7	0.037	1.716	13	0.101	1.268	42	0.023
Nb	0.096	16	0.011	0.399	31	0.019	0.749	6	0.190	1.882	7	0.468	2.407	13	0.131	2.407	42	0.040
Y	2.030	16	0.095	9.472	31	0.240	15.728	6	2.143	17.415	7	1.450	30.467	13	1.356	29.107	42	0.303
Hf	0.052	16	0.004	0.385	31	0.014	0.463	6	0.100	0.978	7	0.247	3.381	13	0.143	2.610	42	0.038
Ta	0.012	16	0.001	0.028	31	0.001	0.050	6	0.012	0.116	7	0.036	0.174	13	0.009	0.147	42	0.003
U	0.008	16	0.000	0.043	31	0.003	0.064	6	0.022	0.048	7	0.016	0.877	13	0.044	0.669	42	0.012
Pb	0.456	16	0.038	1.212	31	0.031	1.525	6	0.373	0.649	7	0.038	3.331	13	0.085	4.260	42	0.083
Rb	0.271	16	0.010	1.744	31	0.131	3.617	6	1.472	0.633	7	0.199	16.850	13	0.702	13.732	42	0.313
Cs	0.256	16	0.022	0.469	31	0.017	0.607	6	0.166	0.202	7	0.025	1.301	13	0.064	0.932	42	0.046
Sr	13.0	16	1.4	300	31	2	229	6	10	285	7	4	228	13	4	248	42	4
Sc	35.4	16	1.1	43.7	31	0.5	57	6	2	45	7	1	20.4	13	1.0	26.9	42	0.3
Zr	1.49	16	0.13	10.7	31	0.5	13	6	4	37	7	11	108	13	4	84.3	42	1.3

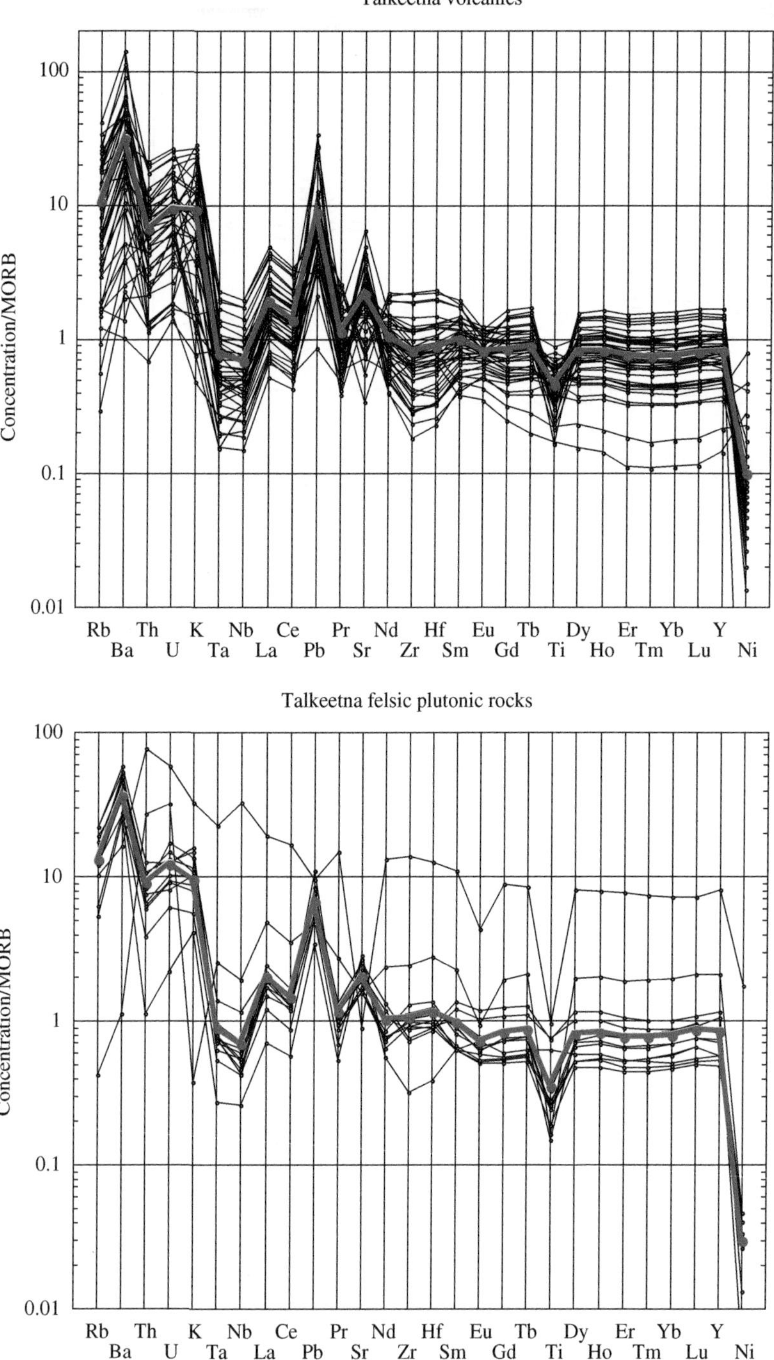

Figure 18 Extended trace-element diagrams (hereafter, spidergrams) for volcanics and felsic plutonic rocks from the Talkeetna arc section, south central Alaska. Concentrations are normalized to N-MORB (Hofmann, 1988). Bold red lines are average values from Table 3. Talkeetna lavas, and plutonic rocks interpreted as liquid compositions, are only slightly enriched in light REE compared to MORB, but show depletion of Nb and Ta, and enrichment of Pb and Sr, typical for arc lavas worldwide. Their trace-element contents are similar to, for example, lavas from the modern Tonga arc. Data from Greene *et al.* (2003) and our unpublished research.

fluid-undersaturated arc basalt at 1.2 GPa, but were reproduced in experiments on primitive arc andesite at the same conditions. However, preliminary trace-element and neodymium isotope data on Talkeetna pyroxenites suggest that they equilibrated with parental melts like the tholeiitic basalts that formed the overlying gabbroic and volcanic sections of the arc.

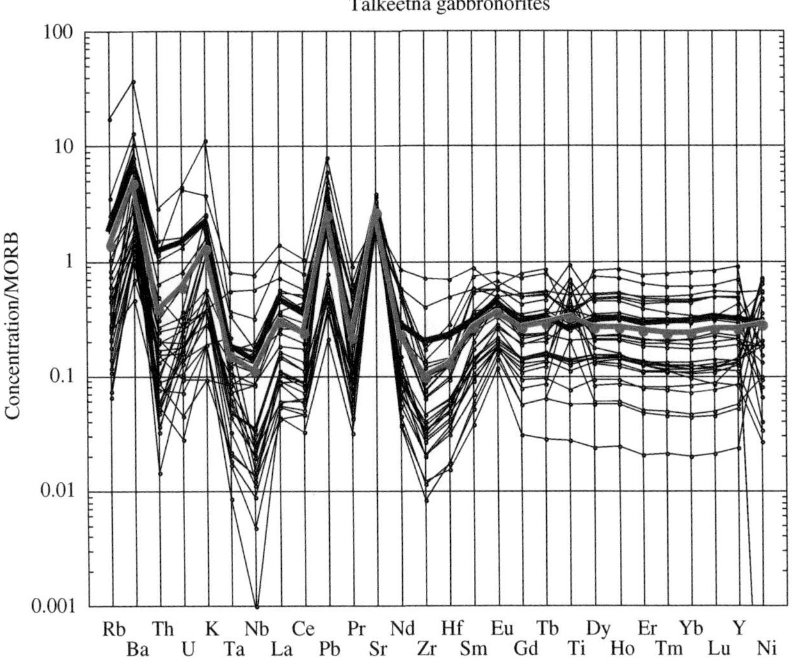

Figure 19 MORB-normalized spidergrams for lower crustal gabbronorites in the Talkeetna arc section, south-central Alaska. Bold red line is average from Table 3. In these cumulate gabbros, some of the Nb and Ta depletion, and Pb and Sr enrichment, is inherited from parental, arc magmas, but the pattern is modified by high Ti (and probably Nb and Ta) in cumulate magnetite and high Pb and Sr in cumulate plagioclase. Data from Greene *et al.* (2003) and our unpublished research.

3.18.4.1.2 Composition, fractionation, and primary melts in the Talkeetna section

DeBari and Sleep (1991) estimated the bulk composition of the Talkeetna arc crust in the Tonsina area by adding compositions of different samples in proportions determined by their abundance in outcrop. They then added olivine, to calculate a primary magma composition for the Talkeetna arc in equilibrium with mantle olivine with Mg# of 90. This approach requires several assumptions. For example, the structural thickness of the Tonsina section cannot be more than ~15 km, whereas the garnet granulites at the base of the section record an original thickness of ~30 km. Thus, DeBari and Sleep assumed that tectonic thinning was homogeneously distributed over the entire section. Also, DeBari and Sleep assumed that outcrop exposures provided a representative estimate of the proportions of different rock types. This could be questioned. Up-section from the Moho level exposures, outcrop exposures in the Tonsina area are poor due to subdued topography and the presence of a major, Tertiary sedimentary basin to the NE. However, our new data on the composition of the most primitive Talkeetna lavas, high alumina basalts with Mg# ~ 60, are very similar to the bulk composition of the crust derived by DeBari and Sleep on the basis of observed rock compositions and proportions.

In addition, DeBari and Sleep assumed that the entire crust could be derived from a single, parental magma composition. Many arcs have isotopically and compositionally heterogeneous primitive magmas. However, our preliminary analyses show remarkable homogeneity in initial $^{143}Nd/^{144}Nd$ ratios (Greene *et al.*, 2003). Also, with two exceptions, the lavas show nearly parallel spidergram patterns (Figure 18). In all these respects, our new data are consistent with the first-order assumption that most of the Talkeetna arc section was derived from a single type of primitive magma. Thus, the bold first-order approach of DeBari and Sleep (1991) has been largely vindicated by more extensive data.

In order to better constrain the possible proportions of different igneous rock types prior to tectonic thinning, we recently completed least-squares modeling of crystal fractionation, using observed compositions and proportions of minerals in Talkeetna gabbronorites to reproduce the liquid line of descent from the most primitive Talkeetna high alumina basalts to average Talkeetna andesites and basaltic andesites (Greene *et al.*, 2003). A striking result is that this modeling requires extensive crystallization of high Mg# clinopyroxene which is, in fact, very rare in the Talkeetna gabbronorites. Modeling predicts that ~20–30%

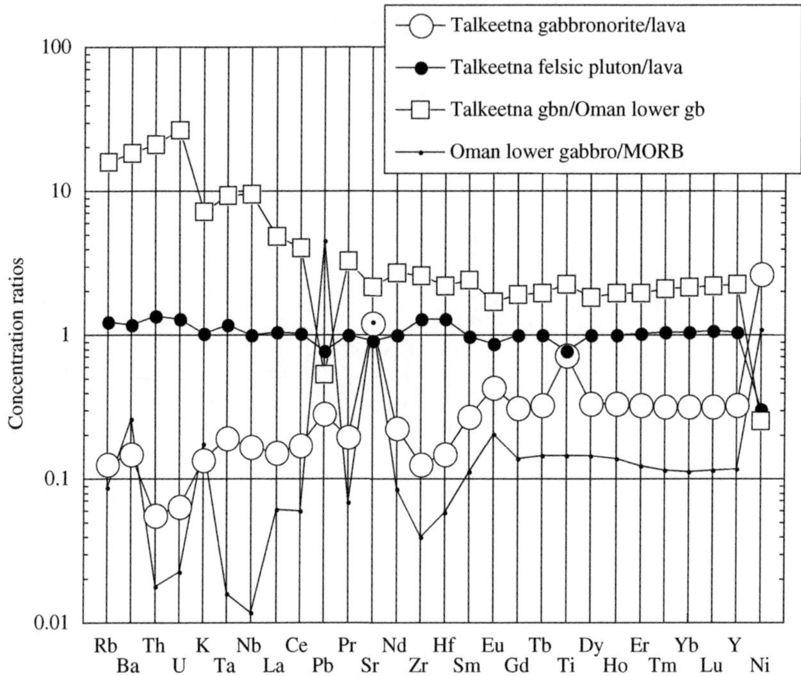

Figure 20 Spidergrams showing ratios of trace-element concentration in Talkeetna arc gabbronorites/lavas (approximating bulk crystal/liquid distribution coefficients), Talkeetna felsic plutons/lavas, Talkeetna gabbronorite (gbn)/lower crustal gabbros from the Oman ophiolite, and Oman lower crustal gabbros/MORB. The gabbronorite/lava "distribution coefficients" show the effects of cumulate plagioclase (high Pb/Ce, Sr/Nd, Eu/Gd, and high Rb, Ba and K relative to Th and U) and magnetite (high Ti/Dy and very slightly high Ta and Nb/La). Talkeetna intermediate to felsic plutons have trace-element contents virtually identical to those in lavas, supporting the idea that the plutons represent melt compositions. Talkeetna gabbronorites are richer in REE and other incompatible elements, except Pb, than Oman lower crustal gabbros. Higher Pb in Oman versus Talkeetna lower crust probably reflects the influence of sulfide/sulfate equilibria during igneous crystallization. Oman cumulate gabbros show very depleted, smooth REE patterns extending to low Nb, Ta, U, and Th. This depleted pattern is interrupted by marked enrichments in Eu, Sr, Pb, K, Ba, and Rb, all of which reflect the presence of cumulate plagioclase. Talkeetna data from Greene *et al.* (2003) and our unpublished research. Oman data are samples from the Khafifah crustal section, Wadi Tayin massif, analyzed by ICP-MS at the Université de Montpellier by Marguerite Godard, from Garrido *et al.* (2001) and our unpublished research.

of the gabbroic lower crust should have clinopyroxene Mg#> 85. In contrast, Figure 24 illustrates that none of our gabbroic samples have such high clinopyroxene Mg#s.

In addition, most primitive Talkeetna lavas have Mg#'s that are too low for Fe/Mg exchange equilibrium with mantle olivine and pyroxene (incorporating reasonable assumptions about oxygen fugacity and Fe^{2+}/Fe^{3+} in arcs). Following Müntener *et al.* (2001), we infer that many hydrous, primary arc magmas are in a reaction relationship with olivine, forming by reactions such as orthopyroxene + clinopyroxene + spinel = olivine + melt. Such melts are in equilibrium with olivine, but will not crystallize olivine upon isobaric cooling, and instead will crystallize websterites (two pyroxene pyroxenites). This is consistent with the observation that cumulate dunites (olivine Mg# < 90) are absent and olivine pyroxenites are rare in the Talkeetna section. Thus, to constrain the composition of a primary magma in equilibrium with mantle peridotite, we performed pyroxenite addition calculations, using the observed phase proportions in Talkeetna pyroxenites. Approximately 20–30% crystallization of pyroxenites from a primary melt in Fe/Mg equilibrium with mantle peridotite was required to produce the most primitive Talkeetna basalts (Greene *et al.*, 2003). In other words, this modeling predicts that ~20–30% of the arc section should be composed of pyroxenite.

3.18.4.2 Missing Primitive Cumulates: Due to Delamination

> The great volume of andesite... in the orogenic belts is often taken up as a serious objection against the idea of its derivation from basalt magma by fractionation.
>
> Kuno (1968, p. 165).

The large proportions of pyroxenite and primitive gabbronorite predicted by fractionation modeling of primitive arc magmas contrast

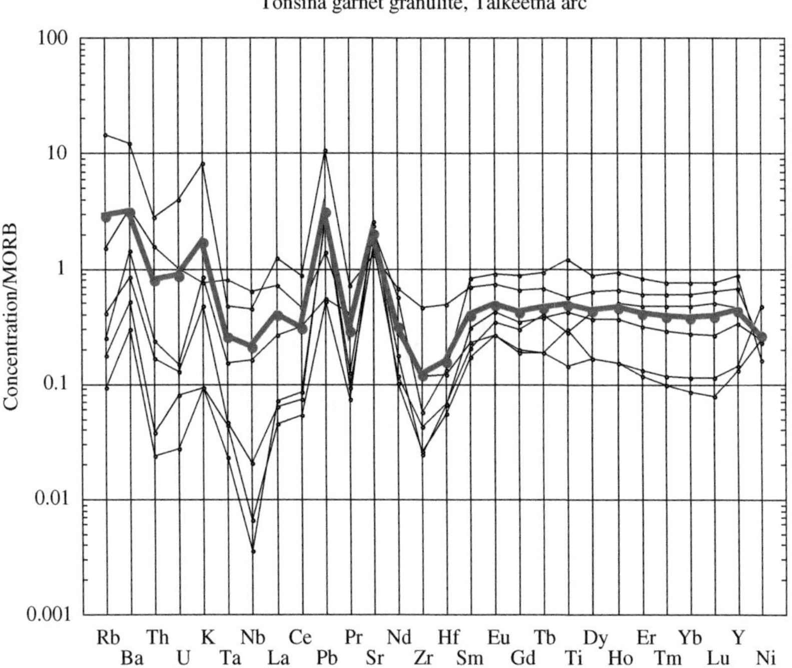

Figure 21 MORB normalized spidergrams for mafic garnet granulites from the Tonsina area, at the base of the Talkeetna arc section at ~1 GPa, 1,000 °C (DeBari and Coleman, 1989; Kelemen *et al.*, 2003a). Bold red line is average from Table 3. These orthogneisses formed via metamorphic recrystallization of protoliths with major and trace-element contents identical to gabbronorites from higher in the Talkeetna section. Garnet in these rocks is metamorphic, as previously proposed (DeBari and Coleman, 1989). Data from our unpublished research.

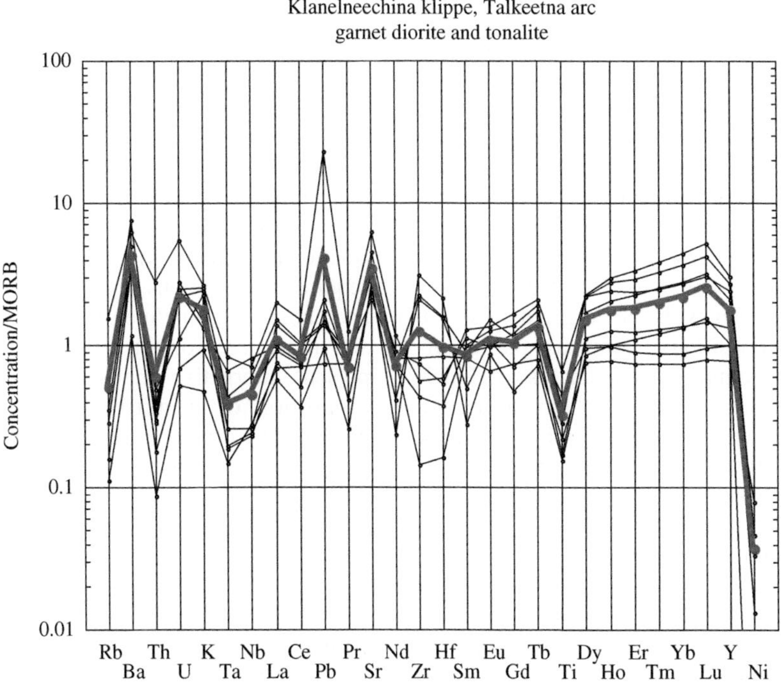

Figure 22 MORB normalized spidergrams for garnet diorites and tonalites from the Klanelneechina klippe, recording lower crustal depths (~0.7 GPa, 700 °C; Kelemen *et al.*, 2003a) from the Talkeetna arc section. Bold red line is average from Table 3. These evolved rocks (Mg# < 50) all include cumulate, igneous garnet, as indicated by their high heavy REE contents. They probably record partial melting of older arc lithologies under lower crustal conditions. Data from our unpublished research.

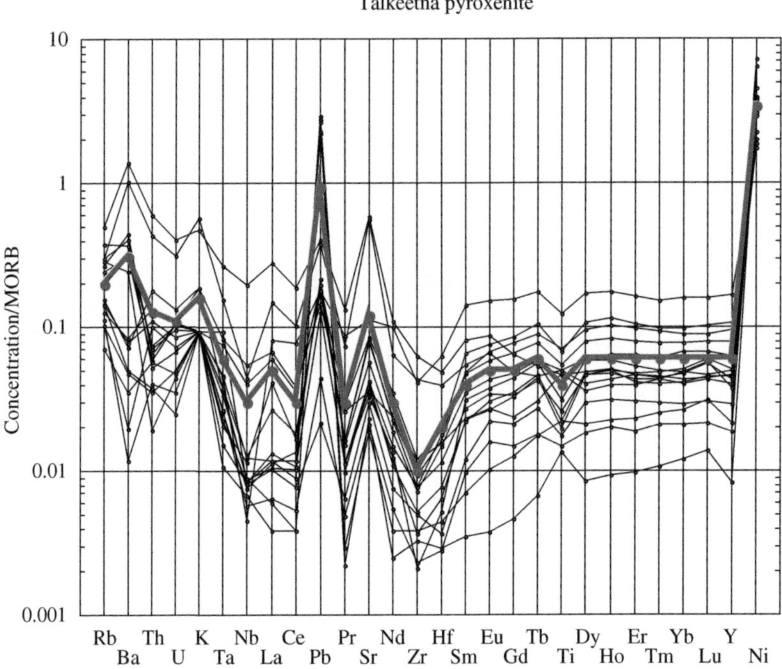

Figure 23 MORB normalized spidergrams for pyroxenites from the Tonsina area, at the base of the Talkeetna arc section. Bold red line is average from Table 3. Trace-element concentrations in these rocks are low, and data for, for example, U and Th may be very imprecise. Nonetheless, it is apparent that the pyroxenites inherited high Pb/Ce and Sr/Nd, and low Nb and Ta relative to K, Th, and U, from parental arc magmas. Data from our unpublished research.

dramatically with the observed proportion of primitive gabbronorites and pyroxenites in the Tonsina area. We have found no gabbronorite or garnet granulite (and only a very small outcrop of plagioclase pyroxenite) with clinopyroxene Mg#s >85. Moreover, the thickness of the pyroxenite layer, between overlying gabbroic rocks and underlying residual mantle peridotites, is less than ~500 m.

The relationship between clinopyroxene Mg# and whole-rock Mg# in Talkeetna gabbronorites (Figure 24), together with our larger data set on whole-rock compositions, confirms the observations based on clinopyroxene analyses. Modeling predicts that there should be a large proportion of gabbronorites and pyroxenites with whole-rock Mg# > 80, but in fact there are very few. This observation is strikingly similar to compiled data on whole-rock Mg# in the Kohistan arc section, in which there are no gabbroic rocks with Mg# >80, and pyroxenites with Mg# > 80 are mainly found in a narrow band, less than 3 km thick, immediately above the Moho. There are small, ultramafic intrusions within the gabbroic lower crust in both the Talkeetna and Kohistan arc sections. However, they comprise less than 5% of the outcrop area. Also, at least in the Talkeetna crustal section, these bodies generally have clinopyroxene Mg# < 85.

Thus, the proportions of igneous rocks calculated from modeling of the liquid line of descent in the lavas are strikingly different from those observed in the Talkeetna and Kohistan section. The bulk of the predicted, primitive cumulates, with clinopyroxene Mg# between 92 and 85, are apparently missing. This result is both uncertain and important, and so we provide additional constraints on the modeling here. The crystal fractionation modeling of Greene *et al.* (2003) requires assumptions about the Fe^{2+}/Fe^{3+} ratio, and about pyroxene/melt Fe/Mg equilibria, which are imprecise. For this reason, in Figure 25 we present alternative methods for estimating the proportion of primitive cumulates that were produced by crystal fractionation in the Talkeetna arc section. The left-hand panel of Figure 25 shows clinopyroxene Mg# in equilibrium with Talkeetna lavas versus ytterbium concentration in the same lavas. If ytterbium were a completely incompatible element, quantitatively retained in melts during crystal fractionation, then a doubling of the ytterbium concentration would indicate 50% crystallization. In fact, ytterbium is only moderately incompatible, so doubling of ytterbium indicates more than 50% crystallization. It can be seen from these data that a decrease in clinopyroxene Mg# from 85 to 75 is accompanied by more than 50% crystallization. If this trend can be

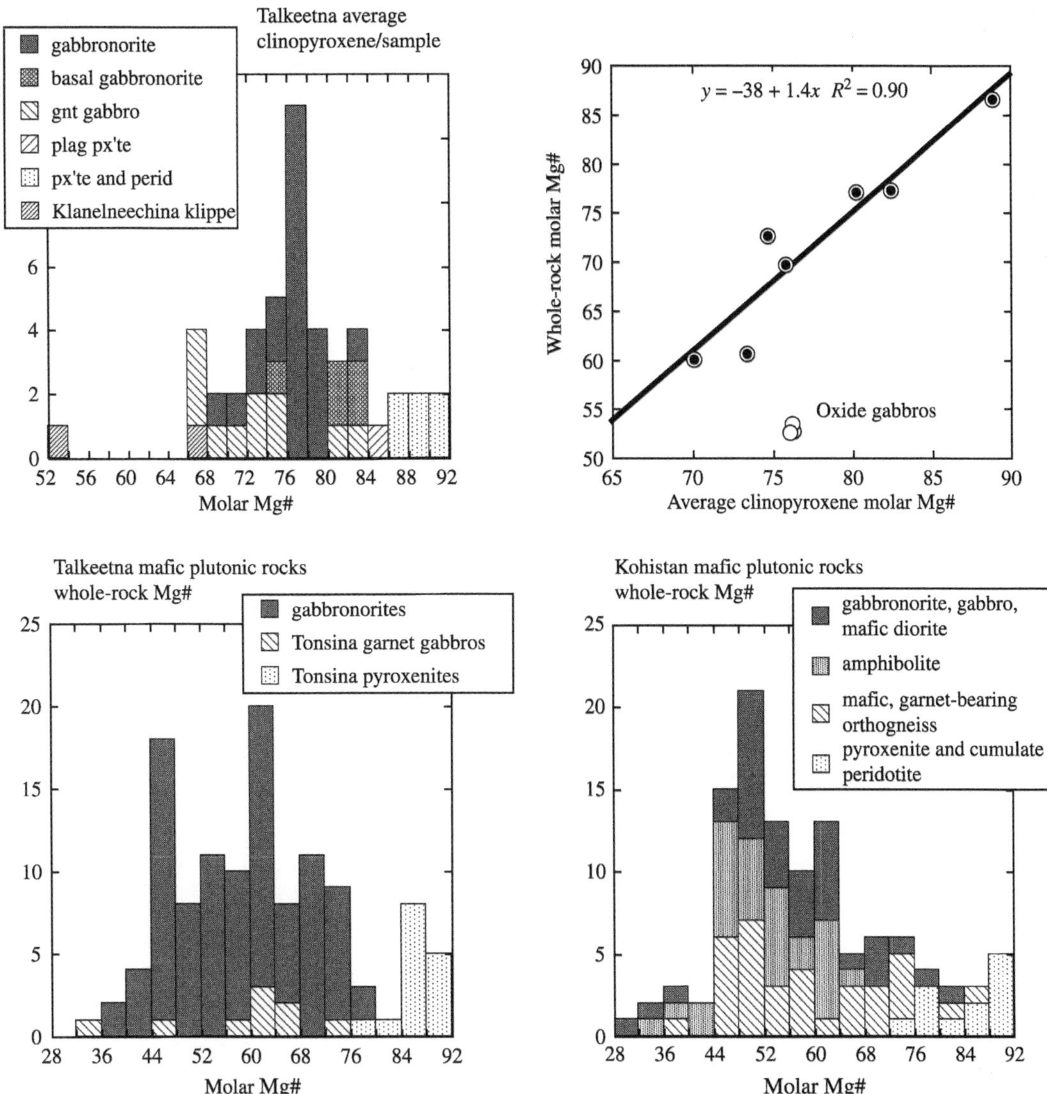

Figure 24 Histograms of clinopyroxene Mg# in mafic plutonic rocks from the Talkeetna arc section, whole-rock Mg# in mafic plutonic rocks from the Talkeetna arc section, and whole-rock Mg# in mafic plutonic rocks from the Kohistan arc section, NW Pakistan. Also shown is average clinopyroxene Mg# versus whole-rock Mg# for samples from the Talkeetna section. Assuming that the number of samples analyzed is representative of the proportion of different compositions present in the Talkeetna and Kohistan sections, these data show that no gabbroic rocks in the Talkeetna section have clinopyroxene Mg# > 86 or whole-rock Mg# > 80, and only 3 of 100 gabbroic rocks in the Kohistan section have whole-rock Mg# > 80. Pyroxenites in the Tonsina area in the Talkeetna section, with clinopyroxene Mg# from 92 to 86 and whole-rock Mg# from 92 to 76, form a layer ~500 m thick between underlying residual peridotite and overlying garnet granulite and gabbronorite. Pyroxenites in the Kohistan section, with whole-rock Mg# from 92 to 72, mainly occur in a band a few km thick between underlying residual mantle peridotites and overlying garnet granulite and gabbroic rocks in the Kohistan section. More evolved pyroxenites form volumetrically insignificant intrusions into gabbroic rocks in both sections. Crustal thickness in both sections is estimated to have been ~30 km, based on metamorphic equilibria (see text). Thus, pyroxenites and primitive gabbroic rocks with Mg# from 92 to 85 comprise a very small proportion of both the Talkeetna and Kohistan arc sections. These observed proportions are in striking contrast to estimates based on fractionation models and experimental data requiring ~30% crystallization of primitive cumulates with clinopyroxene Mg# between 92 and 85. Talkeetna data from Greene *et al.* (2003) and our unpublished research. Kohistan whole-rock compositions from George *et al.* (1993), Hanson (1989), Jan (1988), Jan and Howie (1981), Jan *et al.* (1982), Khan *et al.* (1993, 1989), Miller and Christensen (1994), Petterson *et al.* (1991, 1993), Petterson and Windley (1985, 1992), Shah and Shervais (1999), Sullivan *et al.* (1993), Treloar *et al.* (1996), Yamamoto and Yoshino (1998), and Yoshino and Satish-Kumar (2001).

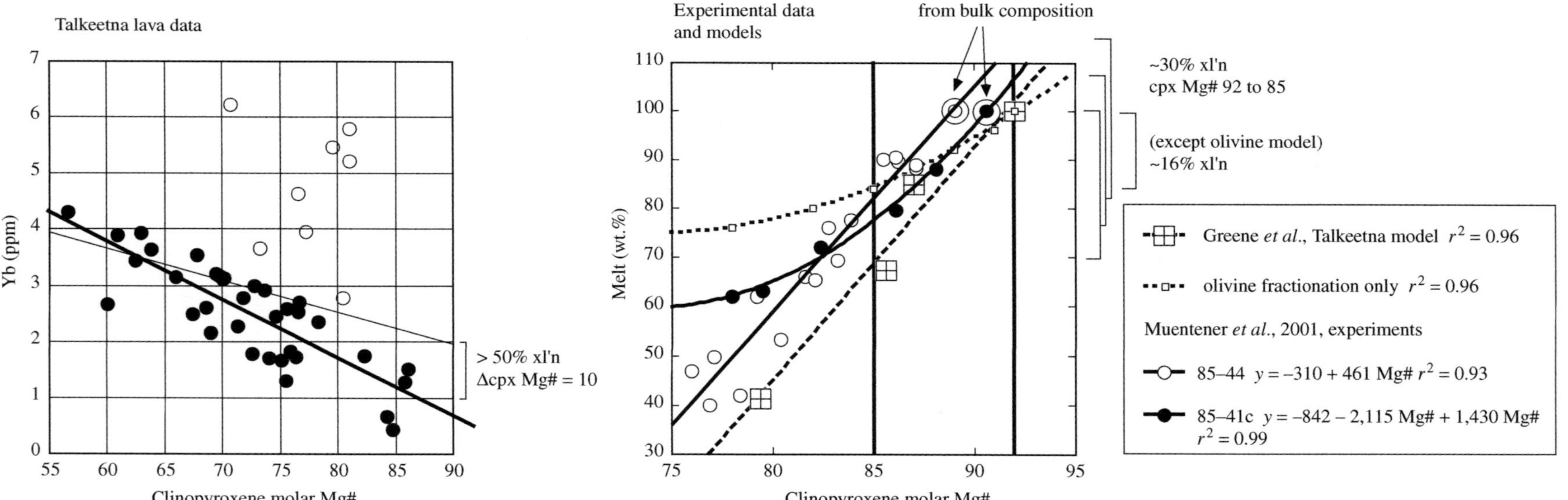

Figure 25 Estimates of melt fraction versus clinopyroxene Mg# from a variety of different methods. On the left, Talkeetna arc lava data are used to establish an empirical relationship between measured Yb concentration, and calculated clinopyroxene Mg# based on a clinopyroxene/melt Fe/Mg Kd of 0.23 (Sisson and Grove, 1993a). If Yb were a perfectly incompatible element, entirely retained in the melt, doubling of the Yb concentration would reflect 50% crystallization. Since Yb is not perfectly incompatible, doubling of Yb indicates more than 50% crystallization. Thus, a change in clinopyroxene Mg# from ~85 to ~75 is associated with more than 50% crystallization (Filled symbols used in linear fit; open symbols omitted from fit). On the right, experimental and model results are used to constrain similar relationships. Data on experimental crystallization of hydrous primitive arc basalt (85-44) and primitive arc andesite (85–41c) at 1.2 GPa (Müntener *et al.*, 2001) yield two empirical relationships between melt fraction and clinopyroxene Mg#, in which a change in clinopyroxene Mg# from 92 to 85 required ~30% crystallization. Least squares modeling of the Talkeetna liquid line of descent, based on Talkeetna lava, pyroxenite, and gabbronorite compositions, provides a similar result (Greene *et al.*, 2003). Olivine is rare in Talkeetna pyroxenite and gabbroic rocks. Olivine in dunites and residual mantle harzburgites underlying the pyroxenites in the Tonsina area has Mg#s ~ 90 (in equilibrium with clinopyroxene Mg# ~ 92), so it does not record substantial crystal fractionation. Nonetheless, the right-hand panel in this figure also shows an olivine-only fractionation calculation, using an estimated Talkeetna primary melt (Greene *et al.*, 2003), in terms of equivalent clinopyroxene Mg# versus melt fraction. This model provides a minimum bound of ~15% crystallization to shift clinopyroxene Mg# from 92 to 85.

extrapolated to crystallization of more primitive melts, it suggests that the decrease in clinopyroxene Mg# from mantle values (~92) to typical Talkeetna gabbronorite values (<85) was produced by ~35% crystallization.

The right-hand panel in Figure 25 illustrates results of experimental crystallization of pyroxenite and primitive gabbronorite from hydrous arc basalt and andesite at 1.2 GPa (Müntener *et al.*, 2001). Again, these data indicate that ~30% crystallization is required for clinopyroxene Mg# to decrease from ~92 to ~85, consistent with the modeling of (Greene *et al.*, 2003). There is no evidence for extensive fractionation of olivine fractionation from primitive Talkeetna magmas. Dunites are present in the mantle section, but these are probably replacive melt conduits (see Kelemen, 1990; Kelemen *et al.*, 1997a,b; and references therein). In any case, mantle dunites have olivine Mg# > 90, and thus it is apparent that they do not record extensive crystal fractionation. Nonetheless, in Figure 25 we also illustrate a model for olivine fractionation alone. This yields a lower bound of ~15 wt.% olivine crystallization required to decrease melt Mg# so that equilibrium clinopyroxene Mg# is decreased from 92 to 85. Finally, our results are similar to those of earlier least-squares fractionation models, which require 21% (Conrad and Kay, 1984) and 16–26% (Gust and Perfit, 1987) crystallization of ultramafic cumulates to produce high-aluminum basalt in island arcs.

There are at least four possible explanations for the discrepancy between the proportion of high Mg# cumulates inferred from crystal fractionation modeling and the proportion of high Mg# cumulates observed in the Talkeetna and Kohistan sections.

(i) The exposed section is not representative of the original arc crustal section. In the Talkeetna area, outcrop is discontinuous, due to numerous faults and subdued topography. It is evident that the present-day structural thickness of the section (~20 km from Moho to volcanics) cannot be as great as the thickness inferred on the basis of thermobarometry, ~30 km. However, the missing section is unlikely to be high Mg# pyroxenites near the Moho, because the Tonsina area gabbronorites and garnet granulites with clinopyroxene Mg# < 85 are in high temperature, conformable contact with high Mg# pyroxenites, which in turn are interfingered along high temperature contacts with residual mantle peridotites. Thus, in this section, the original thickness of cumulates with clinopyroxene Mg# between 85 and 92 as exposed in continuous outcrop, is only ~500 m.

(ii) The missing Talkeetna arc high Mg# plutonic rocks crystallized beneath the "Moho" exposed in the Tonsina area. While arc mantle is rarely exposed, worldwide, it is apparent from dredging at mid-ocean ridges and from ophiolite studies that gabbroic and ultramafic intrusions can form lenses within the residual mantle beneath oceanic spreading ridges (e.g., Cannat, 1996). In the Tonsina area of the Talkeetna section, only a 1–2 km of residual mantle is exposed. Thus, we cannot rule out the presence of high Mg# pyroxenites emplaced as plutons or sills within mantle peridotites at greater depth.

(iii) Equilibrium crystallization of primitive magmas (as distinct from fractional crystallization) occurred until ~70–50% of the initial liquid remained. The remaining liquid was then efficiently extracted, leaving relatively pure gabbronorite cumulates with clinopyroxene Mg#s less than 85. This might happen if, for example, dense, rising melts ponded near the Moho, underwent partial crystallization, and then less dense, evolved melts were extracted to form the overlying crust. Calculated densities for Talkeetna average gabbronorite at pressures from 0.1 GPa to 0.8 GPa are ~3,000 kg m^{-3} (Jull and Kelemen, 2001). Calculated densities for the primary and primitive melts for the Talkeetna section (Greene *et al.*, 2003) are ~2,800 kg m^{-3} on an anhydrous basis, and would be less if—as we surmise—they contained several weight percent H_2O. Thus, primitive Talkeetna melts would have been buoyant with respect to the igneous crust and the Moho would not have been a zone of neutral buoyancy.

Alternatively, melts may have ponded at depth beneath a permeability barrier, in a transition from porous flow to flow in melt induced fractures, as proposed for the base of the crust at mid-ocean ridges based on observations in the Oman ophiolite (Kelemen and Aharonov, 1998; Kelemen *et al.*, 1997b; Korenaga and Kelemen, 1997) and geophysical data from the East Pacific Rise (Crawford and Webb, 2002; Crawford *et al.*, 1999; Dunn and Toomey, 1997). However, the idea that primary Talkeetna magmas underwent 30–50% equilibrium crystallization to form gabbronorites with clinopyroxene Mg#s < 85 is not consistent with the presence of plagioclase-free pyroxenites with pyroxene Mg#s from 92 to 85, nor with experimentally determined phase equilibria for hydrous, primitive arc magmas that indicate a substantial interval for plagioclase-free, pyroxenite crystallization.

(iv) Gravitational instability of dense cumulates overlying less dense upper mantle peridotites may have induced viscous "delamination" at the base of the crust. This is consistent with the observation that the Mg# "gap" occurs near the Moho, and the few rocks with clinopyroxene Mg#s from 90 to 85 are pyroxenites. As pointed out by Arndt and Goldstein (1989), and quantified by Jull and Kelemen (2001) and Müntener *et al.* (2001), pyroxenites are generally denser than upper mantle peridotites, because olivine and pyroxene densities are very similar at similar Mg#, but igneous pyroxenites have lower Mg# than residual peridotites. Furthermore, the density contrast between ultramafic cumulates and mantle peridotites is sufficient to drive viscous "delamination" of a pyroxenite layer 1–2 km thick in ~10 Ma at ~750 °C and ~10^5 yr at ~1,000 °C (stress dependent, non-Newtonian olivine rheology at a background strain rate of $10^{-14}\ s^{-1}$; see Jull and Kelemen (2001) for details and other estimates).

Although all four hypotheses outlined above remain possible, for the reasons outlined in the previous paragraphs we favor the fourth hypothesis, in which pyroxenites foundered into the underlying mantle as a result of density instabilities. Even if some of pyroxenites were emplaced into residual mantle beneath the Moho, the high temperature of sub-arc mantle, and the high strain inferred from Talkeetna peridotite ductile deformation fabrics, combined with the short instability times calculated by Jull and Kelemen (2001), suggest that any significant thickness of pyroxenite near the base of the crust or within the mantle would have been viscously removed during the duration of arc magmatism.

Garnet granulites with the composition of Talkeetna gabbronorites would also be denser than the underlying mantle (Jull and Kelemen, 2001). We suggest that it is no coincidence that the "Mg# gap" coincides with the garnet-in isograd at the base of the Talkeetna gabbroic section. Thus, along with the missing pyroxenites, we infer that garnet granulites may also have been removed from the base of the Talkeetna arc section via density instabilities. In this view, the narrow horizons of pyroxenite and garnet granulite along the Moho in the Talkeetna section are a small remnant of a much larger mass of primitive cumulates, most of which were removed by delamination.

3.18.4.3 Garnet Diorites and Tonalites: Igneous Garnet in the Lower Crust

In the Klanelneechina klippe, which was thrust south of the Talkeetna arc section over younger, accretionary complexes, we found that most rocks are pyroxene quartz diorites to tonalites (bulk Mg# <50, plagioclase anorthite content <50 mol.%), and include some garnet-bearing assemblages. Metamorphic equilibria in garnet + pyroxene bearing rocks record ~7 GPa, 700 °C (Kelemen *et al.*, 2003a). While these rocks will be discussed elsewhere in detail, they are pertinent to the question of continental genesis, and so we illustrate their trace-element contents in Figure 22. While their LILE, HFSE, and light REE patterns vary widely, these samples consistently show heavy REE enrichment, and low Ti/Dy. These characteristics are probably indicative of the presence of cumulate or residual, igneous garnet. Felsic melts extracted from these rocks were garnet-saturated at a depth of ~20 km in the Talkeetna arc crust. They were light REE enriched, heavy REE depleted, high SiO_2, low Mg# melts that could have mixed with primitive melts to produce high Mg# andesite compositions.

3.18.5 IMPLICATIONS FOR CONTINENTAL GENESIS

3.18.5.1 Role of Lower Crustal Delamination in Continental Genesis

The similarity of trace-element patterns in arc magmas and continental crust has led to the inference that most continental crust is derived from igneous, arc crust. We know of three potentially viable explanations for the generation of andesitic, continental crust with an Mg# of ~0.5 via processes involving arc crust:

(i) Crystal fractionation from, or partial melting of, a primitive arc basalt composition at high f_{O_2} and high f_{H_2O}, forming high Mg# andesite melt and corresponding, low SiO_2 cumulates, followed by delamination of the resulting solid residue.

(ii) Magma mixing or simply juxtaposition of primitive arc basalt and evolved, granitic rocks, together with delamination of the solid residue left after granite generation.

(iii) Crystal fractionation from, or partial melting of, a primitive arc andesite composition, followed by delamination of the corresponding cumulates.

In (i)–(iii), delamination would be unnecessary if primitive cumulates or residues of lower crustal melting were ultramafic, and remained present below the seismic Moho. However, the absence of pyroxenite layers ~10 km in thickness in the Talkeetna and Kohistan arc sections suggests that large proportions of ultramafic, igneous rocks are not present at the base of arc crust.

Thus, all three explanations for the genesis of continental crust require delamination of garnet granulite, eclogite and/or pyroxenite. Moreover, recent, dynamical calculations support the hypothesis that delamination is possible—even likely—where Moho temperatures exceed ~750 °C, where crustal thicknesses reach 30 km or more, and where ultramafic cumulates are present (Jull and Kelemen, 2001). The base of arc crust fulfills all of these criteria. Finally, our data on the Talkeetna arc section, and more limited data on the Kohistan arc section, support the hypothesis that substantial proportions of pyroxenite, and perhaps also garnet granulite, were removed by viscous delamination from the base of the arc crust.

If delamination of dense lower crustal rocks has been essential to continental genesis, and the delaminated rocks constitute ~20–40% of the mass of the continents, one might expect to see evidence for this component in magmas derived from the convecting mantle. However, continental crust comprises only ~0.5% of the silicate Earth, so that—even if it represents 40% of the original crustal mass—recycled lower crust might comprise a very small fraction of the convecting mantle.

Tatsumi (2000) argued that recycled pyroxenite or cumulate gabbro, added to the convecting mantle via delamination from the base of arc crust, has produced the EM I isotopic end-member observed in some ocean island basalts, because he inferred on the basis of modeling that this recycled material would have high Rb/Sr, U/Pb, and Th/Pb, and low Sm/Nd compared to bulk Earth and the primitive mantle. However, although the Tonsina pyroxenites have variable trace-element patterns, the average and median compositions have U/Pb and Th/Pb ratios less than in MORBs and primitive mantle, and average Sm/Nd and Lu/Hf greater than in MORBs and primitive mantle. In general, one might expect delaminated Talkeetna pyroxenite to evolve isotope ratios similar to, or more depleted than, the MORB source. Such compositions could, in fact, be present in the source of the more depleted end-members of the MORB isotope spectrum.

If a delaminated lower crustal component included substantial amounts of garnet granulite with the composition of garnet granulites or gabbronorites from the Tonsina region in the Talkeetna arc section, the trace-element ratios and abundances in this reservoir would be different from those in delaminated pyroxenite. However, it remains true that Rb/Sr, U/Pb, and Th/Pb in most of our garnet granulite and gabbronorite samples are lower than in the MORB source, while Sm/Nd and Lu/Hf are higher than in the MORB source. Again, as a consequence, long term isotopic evolution followed by melting of this component during upwelling in the convecting mantle, would yield a melt with isotope ratios similar to, or more depleted than, the MORB source. While incompatible trace-element concentrations in pyroxenite are really very low, trace-element concentrations in the garnet granulites and gabbronorites are comparable to those in MORBs. Thus, given the large strontium and lead anomalies in the garnet granulites, removal of substantial amounts of garnet granulite from the base of continental crust would tend to decrease Pb/Ce and Sr/Nd in the remaining crust. This provides one possible explanation for the fact that Sr/Nd in continental crust is lower than in otherwise geochemically similar arc magmas, as also noted by Kemp and Hawkesworth (see Chapter 3.11) and Rudnick (1995).

3.18.5.2 Additional Processes are Required

The process of delamination—if it occurred—apparently did not produce an andesitic bulk composition in the remaining Talkeetna arc crust. Instead, the Talkeetna arc section probably has a basaltic bulk composition (DeBari and Sleep, 1991; Greene *et al.*, 2003) even after removal of dense, primitive cumulates. This inference is consistent with seismic data, and the composition of primitive arc basalts from the central Aleutian arc and the Izu–Bonin–Marianas arc system, all of which appear to have bulk crust compositions that are basaltic rather than andesitic (Fliedner and Klemperer, 1999; Holbrook *et al.*, 1999; Kerr and Klemperer, 2002; Suyehiro *et al.*, 1996). As noted above, some additional processes must be required to produce continental crust.

3.18.5.2.1 Andesitic arc crust at some times and places

Our favored hypothesis is that continental crust was mainly produced by fractionation of olivine and clinopyroxene from primitive andesite. Following many others (Defant and Kepezhinskas, 2001; Drummond and Defant, 1990; Martin, 1986, 1999; Rapp and Watson, 1995; Rapp *et al.*, 1991), we believe that higher mantle temperatures in the Archean led to more common, larger degrees of partial melting of subducting, eclogite facies basalt. Alternatively, or in addition, dense lower crustal rocks foundering into the hot upper mantle may have commonly undergone partial melting (Zegers and van Keken, 2001). Also, due to higher degrees of melting at hot spots and beneath spreading ridges, more depleted peridotite was present in the upper mantle, including the mantle wedge above Archean subduction zones. Thus, interaction between eclogite melts and highly depleted mantle peridotite yielded primitive andesite (e.g., Kelemen *et al.*, 1998; Rapp *et al.*, 1999; Ringwood, 1974; Rudnick *et al.*, 1994), but produced little additional, basaltic melt. Delamination removed mostly ultramafic cumulates from the base of the crust. Intracrustal differentiation of an andesitic bulk composition formed a felsic upper crust and a residual, mafic lower crust.

3.18.5.2.2 Arc thickening, intracrustal differentiation, and mixing

However, fractionation of primitive andesite is clearly just one of many possibilities for the genesis of continental crust with the composition of high Mg# andesite. An alternative view is that intracrustal differentiation, perhaps involving residual garnet and/or magnetite, has converted

Figure 26 Schematic illustration of the process of progressive burial of early formed plutonic and volcanic rocks within a growing arc edifice (also see Kuno, 1968, figure 9). Pyroxenites near the base of the crust are always denser than the underlying mantle, and temperatures are high near the Moho, so that these ultramafic cumulates may delaminate repeatedly whenever their thickness exceeds some critical value (e.g., Jull and Kelemen, 2001). Increasing pressure forms abundant garnet in Al-rich, mafic, gabbroic rocks near the base of the section, and these too may delaminate. Intermediate to felsic plutonic rocks, and even volcanics, may be buried to lower crustal depths, where they undergo partial melting. Mixing of lower crustal melts with primitive basalt could produce high Mg# andesite magmas. This last process, together with delamination, provides a possible explanation for the formation of andesitic continental crust from initially basaltic arc crust.

basaltic arc crust to andesitic continental crust. One mechanism for this is suggested by intermediate to felsic, garnet "cumulates" in the Talkeetna lower crust. These may have been derived from intermediate plutonic rocks and/or evolved volcanics that were gradually buried in the growing arc edifice (Figure 26, also see figure 9 in Kuno (1968)). Later, heating of the lower crust formed light REE enriched, heavy REE depleted melts that were extracted from these rocks. Mixing between primitive basalt and crustal melts with garnet-bearing residues could form high Mg# andesite (although we see little, if any, evidence for such mixing among Talkeetna arc volcanics). Then, delamination of the residues of lower crustal melting could yield an andesitic bulk composition for the entire crust.

3.18.6 CONCLUSIONS

Arc magmas are dramatically different, on average, from MORBs. Together with boninites, primitive and high Mg# andesites exemplify these differences. Unfortunately, primitive andesites are rare in intra-oceanic arcs, and so there is some possibility that they are produced by special intracrustal differentiation processes specific to continental arcs. However, there are primitive andesites in the oceanic Aleutian arc, and these lack any evidence for crustal assimilation, or even for recycling of components from subducted, continental sediments. Aleutian primitive andesites, and other primitive andesites with high Sr/Nd, cannot be produced by the mixing of primitive arc basalts and lower crustal melts. In most respects, the Aleutian primitive andesites are similar to all other primitive andesites, indicating that these magmas can be derived from primary, andesite melts.

Primary andesites are probably produced by interaction between mantle peridotite and a partial melt of eclogite facies, subducting sediment and/or basalt. Thermal modeling cannot be used to rule out partial melting of subducting material beneath arcs, as has previously been supposed. Some characteristics of all primitive arc magmas, such as enrichment in thorium and lanthanum, and large depletions of niobium and tantalum relative to thorium and lanthanum, are best explained by a similar process, in which thorium and lanthanum are carried into the mantle wedge in a partial melt of subducting material in eclogite facies. Thus, the difference between primitive basalt and andesite may be largely determined by the relative proportions of mantle versus subduction zone melt components, together with a fluid component derived from relatively shallow dehydration of subducting materials.

Although we favor the notion that fractionation of olivine and/or pyroxene from primitive andesites has played a key role in the formation of continental crust, which has a high Mg# andesite composition, this process alone is not sufficient to change the Mg# of the crust from a value of ~70, typical for primary melts in equilibrium with mantle olivine, to the value of ~50 estimated for bulk continental crust. Instead, removal of the crystalline products of fractionation from the crust into the mantle is also required. In fact, all "recipes" for continental crust probably require delamination. We show that the Talkeetna and Kohistan arc sections provide evidence for such a process, in the sense that they only contain a small proportion of the primitive, lower crustal cumulates that must have originally formed. The remaining primitive cumulates, which are pyroxenites and garnet granulites, were denser than the underlying mantle during arc crustal formation, while overlying gabbronorites were less dense. Thus, the Mg# gap in both sections is observed only where rocks become denser than residual peridotite. Removal of pyroxenite and garnet granulite from the base of arc crust leaves remaining crust with lower Sr/Nd and Pb/Ce than in primary arc magmas. Delaminated, recycled primitive arc cumulates would evolve to isotope signatures as depleted as, or more depleted than, the MORB source.

ACKNOWLEDGMENTS

We thank Sue DeBari for advice and support for Andrew Greene, Bärbel Sarbas for information from the GeoRoc database, Nate Hart for help with data entry, John Eiler for discussions, Tim Grove, Simon Turner, and Terry Plank for sharing preprints and sane advice, and Roberta Rudnick for forbearance during our long struggle to complete this chapter. Formal and informal reviews of a draft manuscript by Terry Plank, Steve Eggins, Simon Turner, Chris Hawkesworth, Bernard Bourdon, and Charlie Langmuir were both helpful and tolerant. We hope they find the finished product substantially improved. Research by Kelemen and Hanghøj for this chapter was supported in part by National Science Foundation Research Grants EAR-0125919, EAR-0087706, EAR-9910899, and OCE-9819666, plus the Charles Francis Adams Senior Scientist Chair at Woods Hole Oceanographic Institution.

REFERENCES

Abe N., Arai S., and Yurimoto H. (1998) Geochemical characteristics of the uppermost mantle beneath the Japan island arcs: implications for upper mantle evolution. *Phys. Earth Planet. Int.* **107**, 233–248.

Abratis M. and Worner G. (2001) Ridge collision, slab-window formation, and the flux of Pacific asthenosphere into the Caribbean realm. *Geology* **29**, 127–130.

Anczkiewicz R. and Vance D. (2000) Isotopic constraints on the evolution of metamorphic conditions in the Jijal-Patan Complex and Kamila Belt of the Kohistan Arc, Pakistan Himalaya. In *Tectonics of the Nanga Parbat syntaxis and the Western Himalaya*, Geological Society Special Publications, vol. 170, pp. 321–331.

Anderson A. T. (1974) Evidence for a picritic, volatile-rich magma beneath Mt. Shasta, California. *J. Petrol.* **15**, 243–267.

Arculus R. J., Johnson R. W., Chappell B. W., McKee C. O., and Sakai H. (1983) Ophiolite-contaminated andesites, trachybasalts, and cognate inclusions of Mount Lamington, Papua New Guinea: anhydrite-amphibole-bearing lavas and the 1951 cumulodome. *J. Volcanol. Geotherm. Res.* **18**, 215–247.

Armstrong R. L. (1981) Radiogenic isotopes: the case for crustal recycling on a near-steady-state no-continental-growth Earth. *Phil. Trans. Roy. Soc. London* **301**, 443–472.

Arndt N. T. and Goldstein S. L. (1989) An open boundary between lower continental crust and mantle: its role in crust formation and crustal recycling. *Tectonophysics* **161**, 201–212.

Asimow P. and Stolper E. (1999) Steady-state mantle-melt interactions in one dimension: I. Equilibrium transport and melt focusing. *J. Petrol.* **40**, 475–494.

Ayers J. (1998) Trace element modeling of aqueous fluid: peridotite interaction in the mantle wedge of subduction zones. *Contrib. Min. Petrol.* **132**, 390–404.

Ayers J. C., Dittmer S. K., and Layne G. D. (1997) Partitioning of elements between peridotite and H_2O at 2.0–3.0 GPa and 900–1100°C, and application to models of subduction zone processes. *Earth Planet. Sci. Lett.* **150**, 381–398.

Babeyko A. Y., Sobolev S. V., Trumbull R. B., Oncken O., and Lavier L. L. (2002) Numerical models of crustal scale convection and partial melting beneath the Altiplano-Puna plateau. *Earth Planet. Sci. Lett.* **199**, 373–388.

Baker D. R. and Eggler D. H. (1983) Fractionation paths of Atka (Aleutians) high-alumina basalts: constraints from phase relations. *J. Volcanol. Geotherm. Res.* **18**, 387–404.

Baker D. R. and Eggler D. H. (1987) Compositons of anhydrous and hydrous melts coexisting with plagioclase, augite and olivine or low-Ca pyroxene form 1 atm to 8 kbar: application to the Aleutian volcanic center of Atka. *Am. Mineral.* **72**, 12–28.

Baker M. B., Grove T. L., and Price R. (1994) Primitive basalts and andesites from the Mt. Shasta region N. California: products of varying melt fraction and water content. *Contrib. Mineral. Petrol.* **118**, 111–129.

Bard J. P. (1983) Metamorphism of an abducted island arc: example of the Kohistan Sequence (Pakistan) in the Himalayan collided range. *Earth Planet. Sci. Lett.* **65**(1), 133–144.

Barker F. and Grantz A. (1982) Talkeetna formation in the southeastern Talkeetna Mountains, southern Alaska: an early Jurassic andesitic intraoceanic island arc. *Geol. Soc. Am. Abstr.* **14**, 147.

Barker F., Aleinikoff J. N., Box S. E., Evans B. W., Gehrels G. E., Hill M. D., Irving A. J., Kelley J. S., Leeman W. P., Lull J. S., Nokleberg W. J., Pallister J. S., Patrick B. E., Plafker G., and Rubin C. M. (1994) Some accreted volcanic rocks of Alaska and their elemental abundances. In *The Geology of North America*, The Geology of Alaska (ed. G. Plafker and H. C. Berg). Geological Society of America, vol. G-1.

Bartels K. S., Kinzler R. J., and Grove T. L. (1991) High pressure phase relations of primitive high-alumina basalts from Medicine Lake volcano, northern California. *Contrib. Mineral. Petrol.* **108**, 253–270.

Beate B., Monzier M., Spikings R., Cotten J., Silva J., Bourdon E., and Eissen J.-P. (2001) Mio-Pliocene adakite generation related to flat subduction in southern Ecuador: the Quimsacocha volcanic centre. *Earth Planet. Sci. Lett.* **192**, 561–570.

Bebout G. E., Ryan J. G., and Leeman W. P. (1993) B–Be systematics in subduction-related metamorphic rocks: characterization of the subducted component. *Geochim. Cosmochim. Acta* **57**, 2227–2237.

Bebout G. E., Ryan J. G., Leeman W. P., and Bebout A. E. (1999) Fractionation of trace elements by subduction-zone metamorphism—effect of convergent-margin thermal evolution. *Earth Planet. Sci. Lett.* **171**, 63–81.

Becker H., Jochum K. P., and Carlson R. W. (1999) Constraints from high-pressure veins in eclogites on the composition of hydrous fluids in subduction zones. *Chem. Geol.* **160**, 291–308.

Becker H., Jochum K. P., and Carlson R. W. (2000) Trace element fractionation during dehydration of eclogites from high-pressure terranes and the implications for element fluxes in subduction zones. *Chem. Geol.* **163**, 65–99.

Blackwell D. D., Bowen R. G., Hull D. A., Riccio J., and Steele J. L. (1982) Heat flow, arc volcanism, and subduction in northern Oregon. *J. Geophys. Res.* **87**, 8735–8754.

Blatter D. L. and Carmichael I. S. E. (1998) Hornblende peridotite xenoliths from central Mexico reveal the highly oxidized nature of subarc upper mantle. *Geology* **26**, 1035–1038.

Bloomer S. H. and Hawkins J. W. (1987) Petrology and geochemistry of boninite series volcanic rocks from the Mariana trench. *Contrib. Mineral. Petrol.* **97**, 361–377.

Bodinier J.-L., Merlet C., Bedini R. M., Simien F., Remaidi M., and Garrido C. J. (1996) Distribution of niobium, tantalum, and other highly incompatible trace elements in the lithospheric mantle: the spinel paradox. *Geochim. Cosmochim. Acta* **60**(3), 545–550.

Bourdon B., Turner S., and Allegre C. (1999) Melting dynamics beneath the Tonga-Kermadec island arc inferred from ^{231}Pa-^{235}U systematics. *Science* **286**, 2491–2493.

Bourdon B., Woerner G., and Zindler A. (2000) U-series evidence for crustal involvement and magma residence times in the petrogenesis of Parinacota Volcano, Chile. *Contrib. Mineral. Petrol.* **139**, 458–469.

Bourdon E., Eissen J.-P., Monzier M., Robin C., Martin H., Cotten J., and Hall M. L. (2002) Adakite-like lavas from Antisana Volcano (Ecuador): evidence from slab melt metasomatism beneath the Andean northern volcanic zone. *J. Petrol.* **192**, 561–570.

Bowen N. L. (1928) *The Evolution of the Igneous Rocks*. Princeton University Press, Princeton, NJ.

Brandon A. D. and Draper D. S. (1996) Constraints on the origin of the oxidation state of mantle overlying subduction zones: an example from Simcoe, Washington, USA. *Geochim. Cosmochim. Acta* **60**, 1739–1749.

Brenan J. M., Shaw H. F., and Ryerson R. J. (1995a) Experimental evidence for the origin of lead enrichment in convergent margin magmas. *Nature* **378**, 54–56.

Brenan J. M., Shaw H. F., Ryerson R. J., and Phinney D. L. (1995b) Mineral—aqueous fluid partitioning of trace elements at 900°C and 2.0 GPa: constraints on trace element chemistry of mantle and deep crustal fluids. *Geochim. Cosmochim. Acta* **59**, 3331–3350.

Brenan J. M., Shaw H. F., Ryerson F. J., and Phinney D. L. (1996) Erratum to "Experimental determination of trace-element partitioning between pargasite and a synthetic hydrous andesitic melt" [*Earth Planet Sci. Lett.* **135** (1995) 1–11]. *Earth Planet. Sci. Lett.* **140**, 287–288.

Brophy J. G. (1987) The Cold Bay volcanic center, Aleutian volcanic arc: II. Implications for fractionation and mixing mechanisms in calc-alkaline andesite genesis. *Contrib. Mineral. Petrol.* **97**, 378–388.

Brophy J. G. (1989) Basalt convection and plagioclase retention: a model for the generation of high-alumina arc basalt. *J. Geol.* **97**, 319–329.

Brophy J. G. and Marsh B. D. (1986) On the origin of high alumina arc basalt and the mechanics of melt extraction. *J. Petrol.* **27**, 763–789.

Bureau H. and Keppler H. (1999) Complete miscibility between silicate melts and hydrous fluids in the upper mantle: experimental evidence and geochemical implications. *Earth Planet. Sci. Lett.* **165**, 187–196.

Burns L. E. (1983) The border ranges ultramafic and mafic complex: plutonic core of an intraoceanic island arc. PhD Thesis, Stanford University.

Burns L. E. (1985) The border ranges ultramafic and mafic complex, south-central Alaska: cumulate fractionates of island-arc volcanics. *Can. J. Earth Sci.* **22**, 1020–1038.

Burns L. E., Pessel G. H., Little T. A., Pavlis T. L., Newberry R. J., Winkler G. R., and Decker J. (1991) Geology of the northern Chugach Mountains, southcentral Alaska. *State of Alaska Division of Geological and Geophysical Surveys, Professional Report* **94**, 1–63.

Campbell I. H. and Turner J. S. (1985) Turbulent mixing between fluids with different viscosities. *Nature* **313**, 39–42.

Cannat M. (1996) How thick is the magmatic crust at slow spreading oceanic ridges? *J. Geophys. Res.* **101**, 2847–2857.

Carr M. J., Feigenson M. D., and Bennett E. A. (1990) Incompatible element and isotopic evidence for tectonic control of source mixing and melt extraction along the Central American Arc. *Contrib. Mineral. Petrol.* **105**, 369–380.

Carroll M. R. and Wyllie P. J. (1989) Experimental phase relations in the system tonalite-peridotite-H_2O at 15 kb: implications for assimilation and differentiation processes near the crust-mantle boundary. *J. Petrol.* **30**, 1351–1382.

Chauvel C. and Hemond C. (2000) Melting of a complete section of recycled oceanic crust: trace element and Pb isotopic evidence from Iceland. *Geochem. Geophys. Geosys. (G-cubed)*, **1**, February 14, 2000, Paper number 1999GC000002.

Christensen N. I. and Mooney W. D. (1995) Seismic velocity structure and composition of the continental crust: a global view. *J. Geophys. Res.* **100**, 9761–9788.

Cigolini C., Kudo A. M., Brookins D. G., and Ward D. (1992) The petrology of Poas Volcano lavas: basalt-andesite relationship and their petrogenesis within the magmatic arc of Costa Rica. *J. Volcanol. Geotherm. Res.* **48**, 367–384.

Clark S. K., Reagan M. K., and Plank T. (1998) Trace element and U-series systematics for 1963–1965 tephras from Irazu Volcano, Costa Rica: implications for magma generation processes and transit times. *Geochim. Cosmochim. Acta* **62**, 2689–2699.

Class C., Miller D. L., Goldstein S. L., and Langmuir C. H. (2000) Distinguishing melt and fluid components in Umnak Volcanics, Aleutian Arc. *Geochem. Geophys. Geosys. (G-cubed)*, **1**, June 1, 2000, Paper number 1999GC000010.

Clift P. D. and MacLeod C. J. (1999) Slow rates of subduction erosion estimated from subsidence and tilting of the Tonga forearc. *Geology* **27**, 411–414.

Conder J. A., Weins D. A., and Morris J. (2002) On the decompression melting structure at volcanic arcs and back-arc spreading centers. *Geophys. Res. Lett.* **29**(15), 4pp.

Conrad W. K. and Kay R. W. (1984) Ultramafic and mafic inclusions from Adak Island: crystallization history, and implications for the nature of primary magmas and crustal evolution in the Aleutian Arc. *J. Petrol.* **25**, 88–125.

Conrad W. K., Kay S. M., and Kay R. W. (1983) Magma mixing in the Aleutian arc: evidence from cognate inclusions and composite xenoliths. *J. Volcanol. Geotherm. Res.* **18**, 279–295.

Coward M. P., Jan M. Q., Rex D., Tarney J., Thirlwall M. F., and Windley B. F. (1982) Structural evolution of a crustal section in the western Himalaya. *Nature* **295**(5844), 22–24.

Crawford A. J. (1989) *Boninites*. Unwin Hyman, London, UK, 446pp.

Crawford A. J., Falloon T. J., and Eggins S. (1987) The origin of island arc high-alumina basalts. *Contrib. Mineral. Petrol.* **97**, 417–430.

Crawford W. C. and Webb S. C. (2002) Variations in the distribution of magma in the lower crust and at the Moho beneath the East Pacific Rise at 9 degrees–10 degrees N. *Earth Planet. Sci. Lett.* **203**, 117–130.

Crawford W. C., Webb S. C., and Hildebrand J. A. (1999) Constraints on melt in the lower crustal and Moho at the East Pacific Rise, 9 degrees 48′N, using seafloor compliance measurements. *J. Geophys. Res.* **104**, 2923–2939.

Daly R. A. (1933) *Igneous Rocks and the Depths of the Earth.* McGraw-Hill, New York, NY, 508pp.

Davidson J. P. (1996) Deciphering mantle and crustal signatures in subduction zone magmatism. In *Subduction, Top to Bottom* (eds. G. Bebout *et al.*). *AGU Geophys. Monogr. Ser.*, vol. 96, pp. 251–262.

Davies J. H. (1999) The role of hydraulic fractures and intermediate-depth earthquakes in generating subduction-zone magmatism. *Nature* **398**, 142–145.

Davies J. H. and Stevenson D. J. (1992) Physical model of source region of subduction zone volcanics. *J. Geophys. Res.* **97**, 2037–2070.

de Boer J. Z., Defant M. J., Stewart R. H., Restrepo J. F., Clark L. F., and Ramirez A. H. (1988) Quaternary calc-alkaline volcanism in western Panama: regional variation and implication for the plate tectonic framework. *J. South Am. Earth Sci.* **1**, 275–293.

de Boer J. Z., Defant M. J., Stewart R. H., and Bellon H. (1991) Evidence for active subduction below western Panama. *Geology* **19**, 649–652.

de Boer J. Z., Drummond M. S., Bordelon M. J., Defant M. J., Bellon H., and Maury R. C. (1995) Cenozoic magmatic phases of the Costa Rican island arc (Cordillera de Talamanca). In *Geologic and Tectonic Development of the Caribbean Plate Boundary in Southern Central America*, GSA Spec. Pap. 295 (ed. P. Mann). Geological Society of America, Boulder, CO, pp. 35–55.

DeBari S. M. (1990) Comparative field and petrogenetic study of arc magmatism in the lower crust: exposed examples from a continental margin and an intraoceanic setting. PhD, Stanford University.

DeBari S. M. (1994) Petrogenesis of the Fiambalá gabbroic intrusion, northwestern Argentina, a deep crustal syntectonic pluton in a continental magmatic arc. *J. Petrol.* **35**, 679–713.

DeBari S. M. and Coleman R. G. (1989) Examination of the deep levels of an island arc: evidence from the Tonsina ultramafic-mafic assemblage, Tonsina, Alaska. *J. Geophys. Res.* **94**(B4), 4373–4391.

DeBari S. M. and Sleep N. H. (1991) High-Mg, low-Al bulk composition of the Talkeetna island arc, Alaska: implications for primary magmas and the nature of arc crust. *Geol. Soc. Am. Bull.* **103**, 37–47.

DeBari S. M., Anderson R. G., and Mortensen J. K. (1999) Correlation among lower to upper crustal components in an island arc: the Jurassic Bonanza Arc, Vancouver Island, Canada. *Can. J. Earth Sci.* **36**, 1371–1413.

Defant M. J. and Drummond M. S. (1990) Derivation of some modern arc magmas by melting of young subducted lithosphere. *Nature* **347**, 662–665.

Defant M. J. and Kepezhinskas P. (2001) Evidence suggests slab melting in arc magmas. *EOS* **82**, 65–69.

Defant M. J., Jacques D., Maury R. C., de Boer J. Z., and Joron J.-L. (1989) Geochemistry and tectonic setting of the Luzon arc, Philippines. *Geol. Soc. Am. Bull.* **101**, 663–672.

Defant M. J., Clark L. F., Stewart R. H., Drummond M. S., de Boer J. Z., Maury R. C., Bellon H., Jackson T. E., and Restrepo J. F. (1991a) Andesite and dacite genesis via contrasting processes: the geology and geochemistry of El Valle Volcano, Panama. *Contrib. Mineral. Petrol.* **106**, 309–324.

Defant M. J., Richerson P. M., de Boer J. Z., Stewart R. H., Maury R. C., Bellon H., Drummond M. S., Feigenson M. D., and Jackson T. E. (1991b) Dacite genesis via both slab melting and differentiation: petrogenesis of La Yeguada Volcanic Complex, Panama. *J. Petrol.* **32**, 1101–1142.

Defant M. J., Jackson T. E., Drummond M. S., de Boer J. Z., Bellon H., Feigenson M. D., Maury R. C., and Stewart R. H. (1992) The geochemistry of young volcanism throughout western Panama and southeastern Costa Rica: an overview. *J. Geol. Soc. London* **149**, 569–579.

DePaolo D. J. (1981) Trace element and isotopic effects of combined wallrock assimilation and fractional crystallization. *Earth Planet. Sci. Lett.* **53**, 189–202.

Detterman R. and Hartsock J. (1966) Geology of the Iniskin-Tuxedni region, Alaska. *USGS Prof. Pap.* **512**, 1–78.

Dick H. J. B. (1989) Abyssal peridotites, very slow spreading ridges and ocean ridge magmatism. *Geol. Soc. Spec. Publ. (Magmatism in the Ocean Basins)* **42**, 71–105.

Dixon J. and Stolper E. (1995) An experimental study of water and carbon dioxide solubilities in mid-ocean ridge basaltic liquids: II. Applications to degassing. *J. Petrol.* **36**, 1633–1646.

Dixon J., Stolper E., and Holloway J. (1995) An experimental study of water and carbon dioxide solubilities in mid ocean ridge basaltic liquids: I. Calibration and solubility models. *J. Petrol.* **36**, 1607–1631.

Domanik K. J., Hervig R. L., and Peacock S. M. (1993) Beryllium and boron in subduction zone minerals: an ion microprobe study. *Geochim. Cosmochim. Acta* **57**, 4997–5010.

Draper D. S. and Johnston A. D. (1992) Anhydrous P-T phase relations of an Aleutian high-MgO basalt: an investigation of the role of olivine-liquid reaction in the generation of arc high-alumina basalts. *Contrib. Mineral. Petrol.* **112**, 501–519.

Drummond M. S. and Defant M. J. (1990) A model for trondjhemite-tonalite-dacite genesis and crustal growth via slab melting. *J. Geophys. Res.* **95**, 21503–21521.

Drummond M. S., Bordelon M., Boer J. Z. d., Defant M. J., Bellon H., and Feigenson M. D. (1995) Igneous petrogenesis and tectonic setting of plutonic and volcanic rocks of the Cordillera de Talamanca, Costa Rica—Panama, Central American Arc. *Am. J. Sci.* **295**, 875–919.

Ducea M. N. and Saleeby J. B. (1996) Buoyancy sources for a large, unrooted mountain range, the Sierra Nevada, California: evidence from xenolith thermobarometry. *J. Geophys. Res.* **101**, 8229–8244.

Dunn R. A. and Toomey D. R. (1997) Seismological evidence for three-dimensional melt migration beneath the East Pacific Rise. *Nature* **388**, 259–262.

Eggins S. (1993) Origin and differentiation of picritic arc magmas, Ambae (Aoba), Vanuatu. *Contrib. Mineral. Petrol.* **114**, 79–100.

Eiler J., Crawford A., Elliott T., Farley K. A., Valley J. V., and Stolper E. M. (2000) Oxygen isotope geochemistry of oceanic arc lavas. *J. Petrol.* **41**, 229–256.

Elkins Tanton L. T., Grove T. L., and Donnelly-Nolan J. (2001) Hot shallow mantle melting under the Cascades volcanic arc. *Geology* **29**, 631–634.

Ellam R. M. and Hawkesworth C. (1988a) Elemental and isotopic variations in subduction related basalts: evidence for a three component model. *Contrib. Mineral. Petrol.* **98**, 72–80.

Ellam R. M. and Hawkesworth C. J. (1988b) Is average continental crust generated at subduction zones? *Geology* **16**, 314–317.

Elliott T. (2003) Tracers of the slab. In *AGU Monograph* (ed. J. Eiler). American Geophysical Union (in press).

Elliott T., Plank T., Zindler A., White W., and Bourdon B. (1997) Element transport from slab to volcanic front at the Mariana Arc. *J. Geophys. Res.* **102**, 14991–15019.

England P., Engdahl R., and Thatcher W. (2003) Systematic variation in the depths of slabs beneath arc volcanoes. *Geophys. J. Int.* (in press).

Erikson E. H., Jr. (1977) Petrology and petrogenesis of the Mount Stuart Batholith—Plutonic equivalent of the high-alumina basalt association? *Contrib. Mineral. Petrol.* **60**, 183–207.

Falloon T. J. and Danyushevsky L. V. (2000) Melting of refractory mantle at 1.5, 2.0 and 2.5 GPa under H_2O-undersaturated conditions: implications for the petrogenesis of high-Ca boninites and the influence of subduction components on mantle melting. *J. Petrol.* **41**, 257–283.

Falloon T. J. and Green D. H. (1986) Glass inclusions in magnesian olivine phenocrysts from Tonga: evidence for highly refractory parental magmas in the Tongan arc. *Earth Planet. Sci. Lett.* **81**, 95–103.

Falloon T. J., Green D. H., and McCulloch M. T. (1989) 14: petrogenesis of high-Mg and associated lavas from the north Tonga Trench. In *Boninites* (ed. A. J. Crawford). Unwin Hyman, London, UK, pp. 357–395.

Feineman M. D. and DePaolo D. J. (2002) A diffusion-decay model for steady-state U-series disequilibrium in the mantle with implications for island arc lavas. *EOS, Trans., AGU* **83**(47) *Fall Meet. Suppl.*

Fenner C. N. (1929) The crystallization of basalts. *Am. J. Sci.* **XVIII**, 223–253.

Fenner C. N. (1937) A view of magmatic differentiation. *J. Geol.* **45**, 158–168.

Fliedner M. and Klemperer S. L. (1999) Structure of an island arc: wide-angle seismic studies in the eastern Aleutian Islands, Alaska. *J. Geophys. Res.* **104**, 10667–10694.

Furukawa Y. (1993a) Depth of the decoupling plate interface and thermal structure under arcs. *J. Geophys. Res.* **98**, 20005–20013.

Furukawa Y. (1993b) Magmatic processes under arcs and formation of the volcanic front. *J. Geophys. Res.* **98**, 8309–8319.

Furukawa Y. and Tatsumi Y. (1999) Melting of a subducting slab and production of high-Mg andesite magmas: unusual magmatism in SW Japan at 13 approximately 15 Ma. *Geophys. Res. Lett.* **26**(15), 2271–2274.

Gaetani G. A. and Grove T. L. (1998) The influence of water on melting of mantle peridotite. *Contrib. Mineral. Petrol.* **131**, 323–346.

Garrido C. J., Kelemen P. B., and Hirth G. (2001) Variation of cooling rate with depth in lower crust formed at an oceanic spreading ridge: plagioclase crystal size distributions in gabbros from the Oman ophiolite. *Geochem. Geophys. Geosys. (G-cubed)*, **2**, Paper number 2000GC000136.

Gast P. W. (1968) Trace element fractionation and the origin of tholeiitic and alkaline magma types. *Geochim. Cosmochim. Acta* **32**, 1057–1089.

George M. T., Harris N. B. W., Butler R. W. H., Treloar P. J. e., and Searle M. P. e. (1993) The tectonic implications of contrasting granite magmatism between the Kohistan island arc and the Nanga Parbat-Haramosh Massif, Pakistan Himalaya. *Himalayan tectonics, Seventh Himalaya–Karakoram–Tibet workshop* **74**, 173–191.

George R., Turner S., Hawkesworth C., Morris J., Nye C., Ryan J., and Zheng S.-H. (2003) Melting processes and fluid and sediment transport rates along the Alaska–Aleutian arc from an integrated U–Th–Ra–Be isotope study. *J. Geophys. Res.* **108**(B5), 2252, doi: 10.1029/2002JB001916.

Gill J. (1981) *Orogenic Andesites and Plate Tectonics*. Springer.

Gill J. B. (1974) Role of underthrust oceanic crust in the Genesis of a Fijian Calc-alkaline suite. *Contrib. Mineral. Petrol.* **43**, 29–45.

Gill J. B. (1978) Role of trace element partition coefficients in models of andesite genesis. *Geochim. Cosmochim. Acta* **42**, 709–724.

Gill J. and Condomines M. (1992) Short-lived radioactivity and magma genesis. *Science* **257**, 1368–1376.

Gill J. B. and Williams R. W. (1990) Th isotope and U-series studies of subduction-related volcanic rocks. *Geochim. Cosmochim. Acta* **54**, 1427–1442.

Gough S. J., Searle M. P., Waters D. J., and Khan M. A. (2001) Igneous crystallization, high pressure metamorphism, and subsequent tectonic exhumation of the Jijal and Kamila complexes, Kohistan. *Abstracts: 16th Himalaya–Karakorum–Tibet Workshop, Austria, in J. Asian Earth Sciences* **19**, 23–24.

Grantz A., Thomas H., Stern T., and Sheffey N. (1963) Potassium-argon and lead-alpha ages for stratigraphically bracketed plutonic rocks in the Talkeetna Mountains, Alaska. *USGS Prof. Pap.* **475-B**, B56–B59.

Green D. H. (1976) Experimental testing of "equilibrium" partial melting of peridotite under water-saturated, high-pressure conditions. *Can. Min.* **14**, 255–268.

Green D. H. and Ringwood A. E. (1967) The genesis of basaltic magmas. *Contrib. Mineral. Petrol.* **15**, 103–190.

Green T. H. and Ringwood A. E. (1966) Origin of the calc-alkaline igneous rock suite. *Earth Planet. Sci. Lett.* **1**, 307–316.

Greene A. R., Kelemen P. B., DeBari S. M., Blusztajn J., and Clift P. (2003) A detailed geochemical study of island arc crust: the Talkeetna arc section, south-central Alaska. *J. Petrol.* (submitted).

Gromet L. P. and Silver L. T. (1987) REE variations across the Peninsular Ranges batholith: implications for batholithic petrogenesis and crustal growth in magmatic arcs. *J. Petrol.* **28**, 75–125.

Grove T. L. and Kinzler R. J. (1986) Petrogenesis of andesites. *Ann. Rev. Earth Planet. Sci.* **14**, 417–454.

Grove T. L., Gerlach D. C., and Sando T. W. (1982) Origin of calc-alkaline series lavas at Medicine Lake Volcano by fractionation, assimilation and mixing. *Contrib. Mineral. Petrol.* **80**, 160–182.

Grove T. L., Kinzler R. J., Baker M. B., Donnelly-Nolan J. M., and Lesher C. E. (1988) Assimilation of granite by basaltic magma at Burnt Lava flow, Medicine Lake volcano, California: decoupling of heat and mass transfer. *Contrib. Mineral. Petrol.* **99**, 320–343.

Grove T. L., Parman S. W., Bowring S. A., Price R. C., and Baker M. B. (2001) The role of H_2O-rich fluids in the generation of primitive basaltic andesites and andesites from the Mt. Shasta region N. California. *Contrib. Mineral. Petrol.* **142**, 375–396.

Grove T. L., Elkins Tanton L. T., Parman S. W., Chatterjee N., Müntener O., and Gaetani G. A. (2003) Fractional crystallization and mantle melting controls on calc-alkaline differentiation trends. *Contrib. Mineral. Petrol.* (in press).

Gust D. A. and Perfit M. R. (1987) Phase relations of a high-Mg basalt from the Aleutian island arc: implications for primary island arc basalts and high-Al basalts. *Contrib. Mineral. Petrol.* **97**, 7–18.

Hanson C. R. (1989) The northern suture in the Shigar Valley, Baltistan, northern Pakistan. *Tect. West. Himalayas, Spec. Pap.: GSA* **232**, 203–215.

Harris P. G. (1957) Zone refining and the origin of potassic basalts. *Geochim. Cosmochim. Acta* **12**, 195–208.

Hart S. R., Blusztajn J., Dick H. J. B., Meyer P. S., and Muehlenbachs K. (1999) The fingerprint of seawater circulation in a 500-meter section of ocean crust gabbros. *Geochim. Cosmochim. Acta* **63**, 4059–4080.

Hauff F., Hoernle K., Bogaard P. v. d., Alvarado G., and Garbe-Schonberg D. (2000) Age and geochemistry of basaltic complexes in western Costa Rica: contributions to the geotectonic evolution of Central America. *Geochem. Geophys. Geosys. (G-cubed)*, **1**, 41pp, Paper number 1999GC000020.

Hauri E. H. (1995) Major element variability in the Hawaiian mantle plume. *Nature* **382**, 415–419.

Hawkesworth C. J., Gallagher K., Hergt J. M., and McDermott F. (1993a) Mantle and slab contributions in arc magmas. *Ann. Rev. Earth Planet. Sci.* **21**, 175–204.

Hawkesworth C. J., Gallagher K., Hergt J. M., and McDermott F. (1993b) Trace element fractionation processes in the generation of island arc basalts. *Phil. Trans. Roy. Soc. London A* **342**, 179–191.

Hawkesworth C. J., Turner S. P., McDermott F., Peate D. W., and van Calsteren P. (1997) U–Th isotopes in arc magmas: implications for element transfer from the subducted crust. *Science* **276**, 551–555.

Herrstrom E. A., Reagan M. K., and Morris J. D. (1995) Variations in lava composition associated with flow of asthenosphere beneath southern Central America. *Geology* **23**, 617–620.

Herzberg C. T., Fyfe W. S., and Carr M. J. (1983) Density constraints on the formation of the continental Mohoand crust. *Contrib. Mineral. Petrol.* **84**, 1–5.

Hildreth W. and Moorbath S. (1988) Crustal contributions to arc magmatism in the Andes of central Chile. *Contrib. Mineral. Petrol.* **98**, 455–489.

Hirose K. (1997) Melting experiments on lherzolite KLB-1 under hydrous conditions and generation of high-magnesian andesitic melts. *Geology* **25**, 42–44.

Hirschmann M. M. and Stolper E. M. (1996) A possible role for garnet pyroxenite in the origin of the "garnet signature" in MORB. *Contrib. Mineral. Petrol.* **124**, 185–208.

Hirschmann M. M., Baker M. B., and Stolper E. M. (1998) The effect of alkalis on the silica content of mantle-derived melts. *Geochim. Cosmochim. Acta* **62**, 883–902.

Hofmann A. W. (1988) Chemical differentiation of the Earth: the relationship between mantle, continental crust, and oceanic crust. *Earth Planet. Sci. Lett.* **90**, 297–314.

Holbrook W. S., Lizarralde D., McGeary S., Bangs N., and Diebold J. (1999) Structure and composition of the Aleutian island arc and implications for continental crustal growth. *Geology* **27**, 31–34.

Holmes A. (1937) *The Age of the Earth*. Holmes, Arthur, Thomas Nelson & Sons Ltd., London, UK, 263pp.

Honegger K., Dietrich V., Frank W., Gansser A., Thoni M., and Trommsdorff V. (1982) Magmatism and metamorphism in the Ladakh Himalayas (the Indus-Tsangpo suture zone). *Earth Planet. Sci. Lett.* **60**, 253–292.

Hughes S. S. and Taylor E. M. (1986) Geochemistry, petrogenesis, and tectonic implications of central High Cascade mafic platform lavas. *Geol. Soc. Am. Bull.* **97**, 1024–1036.

Irvine T. N. and Baragar W. R. (1971) A guide to the chemical classification of the common volcanic rocks. *Can. J. Earth Sci.* **8**, 523–548.

Ishikawa T. and Nakamura E. (1994) Origin of the slab component in arc lavas from across-arc variation of B and Pb isotopes. *Nature* **370**, 205–208.

Iwamori H. (1994) $^{238}U-^{230}Th-^{226}Ra$ and $^{235}U-^{231}Pa$ disequilibria produced by mantle melting with porous and channel flows. *Earth Planet. Sci. Lett.* **125**, 1–16.

Iwamori H. (1997) Heat sources and melting in subduction zones. *J. Geophys. Res.* **102**, 14803–14820.

Jan M. Q. (1977) The Kohistan basic complex: a summary based on recent petrological research. *Bulletin of the Centre of Excellence in Geology. University of Peshawar* **9–10**(1), 36–42.

Jan M. Q. (1988) Relative abundances of minor and trace elements in mafic phases from the southern part of the Kohistan Arc. *Geological Bulletin, University of Peshawar* **21**, 15–25.

Jan M. Q. and Howie R. A. (1981) The mineralogy and geochemistry of the metamorphosed basic and ultrabasic rocks of the Jijal Complex, Kohistan, NW Pakistan. *J. Petrol.* **22**(1), 85–126.

Jan M. Q. and Karim A. (1995) Coronas and high-P veins in metagabbros of the Kohistan island arc, northern Pakistan: evidence for crustal thickening during cooling. *J. Metamorph. Geol.* **13**(3), 357–366.

Jan M. Q., Wilson R. N., and Windley B. F. (1982) Paragonite paragenesis from the garnet granulites of the Jijal Complex, Kohistan N. Pakistan. *Min. Mag.* **45**(337), 73–77.

Jarrard R. D. (1986) Relations among subduction parameters. *Rev. Geophys.* **24**, 217–284.

Johnson K. T. M. and Dick H. J. B. (1992) Open system melting and temporal and spatial variation of peridotite and basalt at the Atlantis II Fracture Zone. *J. Geophys. Res.* **97**, 9219–9241.

Johnson K. T. M., Dick H. J. B., and Shimizu N. (1990) Melting in the oceanic upper mantle: an ion microprobe study of diopsides in abyssal peridotites. *J. Geophys. Res.* **95**, 2661–2678.

Johnson M. C. and Plank T. (1999) Dehydration and melting experiments constrain the fate of subducted sediments. *Geochem. Geophys. Geosys. (G-cubed)*, **1**, Paper number 1999GC000014.

Johnston A. D. and Wyllie P. J. (1988) Constraints on the origin of Archean trondhjemites based on phase relationships of Nuk gneiss with H_2O at 15 kbar. *Contrib. Mineral. Petrol.* **100**, 35–46.

Jull M. and Kelemen P. B. (2001) On the conditions for lower crustal convective instability. *J. Geophys. Res.* **106**, 6423–6446.

Jull M., Kelemen P. B., and Sims K. (2002) Consequences of diffuse and channelled porous melt migration on U-series disequilibria. *Geochim. Cosmochim. Acta* **66**, 4133–4148.

Kamenetsky V. S., Crawford A. J., Eggins S., and Muhe R. (1997) Phenocryst and melt inclusion chemistry of near-axis seamounts, Valu Fa Ridge, Lau Basin: insight into mantle wedge melting and addition of subduction components. *Earth Planet. Sci. Lett.* **151**, 205–223.

Kawamoto T. (1996) Experimental constraints on differentiation and H_2O abundance of calc-alkaline magmas. *Earth Planet. Sci. Lett.* **144**, 577–589.

Kawamoto T. and Holloway J. R. (1997) Melting temperature and partial melt chemistry of H_2O saturated mantle peridotite to 11 Gigapascals. *Science* **276**, 240–243.

Kawate S. and Arima M. (1998) Petrogenesis of the Tanzawa plutonic complex, central Japan: exposed felsic middle crust of the Izu—Bonin–Mariana arc. *The Island Arc* **7**, 342–358.

Kay R. W. (1978) Aleutian magnesian andesites: melts from subducted Pacific ocean crust. *J. Volcanol. Geotherm. Res.* **4**, 117–132.

Kay R. W. (1980) Volcanic arc magmas: implications of a melting-mixing model for element recycling in the crust-upper mantle system. *J. Geol.* **88**, 497–522.

Kay R. W. and Kay S. M. (1988) Crustal recycling and the Aleutian arc. *Geochim. Cosmochim. Acta* **52**, 1351–1359.

Kay R. W. and Kay S. M. (1991) Creation and destruction of lower continental crust. *Geol. Rundsch.* **80**, 259–278.

Kay R. W. and Kay S. M. (1993) Delamination and delamination magmatism. *Tectonophysics* **219**, 177–189.

Kay S. M. and Kay R. W. (1985) Role of crystal cumulates and the oceanic crust in the formation of the lower crust of the Aleutian Arc. *Geology* **13**, 461–464.

Kay S. M. and Kay R. W. (1994) Aleutian magmas in space and time. In *The Geology of Alaska: The Geology of North America* (eds. G. Plafker and H. C. Berg). Geological Society of America, Boulder, CO, vol. G-1, pp. 687–722.

Kay S. M., Kay R. W., and Citron G. P. (1982) Tectonic controls on tholeiitic and calc-alkaline magmatism in the Aleutian Arc. *J. Geophys. Res.* **87**, 4051–4072.

Kay S. M., Kay R. W., and Perfit M. R. (1990) Calc-alkaline plutonism in the intra-oceanic Aleutian arc, Alaska. In *Plutonism from Antarctica to Alaska*, Geological Society of America Special Paper 241 (eds. S. M. Kay and C. W. Rapela). Geological Society of America, Boulder, CO, pp. 233–255.

Kelemen P. B. (1986) Assimilation of ultramafic rock in subduction-related magmatic arcs. *J. Geol.* **94**, 829–843.

Kelemen P. B. (1990) Reaction between ultramafic rock and fractionating basaltic magma: I. Phase relations, the origin of calc-alkaline magma series, and the formation of discordant dunite. *J. Petrol.* **31**, 51–98.

Kelemen P. B. (1995) Genesis of high Mg# andesites and the continental crust. *Contrib. Mineral. Petrol.* **120**, 1–19.

Kelemen P. B. and Aharonov E. (1998) Periodic formation of magma fractures and generation of layered gabbros in the lower crust beneath oceanic spreading ridges. In *Faulting and Magmatism at Mid-Ocean Ridges*, Geophysical Monograph 106 (eds. W. R. Buck, P. T. Delaney, J. A. Karson, and Y. Lagabrielle). Am. Geophys. Union, Washington, DC, pp. 267–289.

Kelemen P. B. and Ghiorso M. S. (1986) Assimilation of peridotite in calc-alkaline plutonic complexes: evidence from the Big Jim Complex, Washington Cascades. *Contrib. Mineral. Petrol.* **94**, 12–28.

Kelemen P. B., Joyce D. B., Webster J. D., and Holloway J. R. (1990a) Reaction between ultramafic rock and fractionating basaltic magma: II. Experimental investigation of reaction between olivine tholeiite and harzburgite at 1150–1050C and 5kb. *J. Petrol.* **31**, 99–134.

Kelemen P. B., Joyce D. B., Webster J. D., and Holloway J. R. (1990b) Reaction between ultramafic rock and fractionating basaltic magma: II. Experimental investigation of reaction between olivine tholeiite and harzburgite at 1150–1050°C and 5 kb. *J. Petrol.* **31**, 99–134.

Kelemen P. B., Shimizu N., and Dunn T. (1993) Relative depletion of niobium in some arc magmas and the continental crust: partitioning of K, Nb, La, and Ce during melt/rock reaction in the upper mantle. *Earth Planet. Sci. Lett.* **120**, 111–134.

Kelemen P. B., Hirth G., and Shimizu N. (1995a) Be and B partitioning in high pressure pelites, metabasalts, and peridotites: potential sources for Be and B in arc magmas. In *V. M. Goldschmidt Conference Program and Abstracts vol. 60, May 24–26, 1995* (ed. H. L. Barnes). Geochemical Society, State College, PA.

Kelemen P. B., Shimizu N., and Salters V. J. M. (1995b) Extraction of mid-ocean-ridge basalt from the upwelling mantle by focused flow of melt in dunite channels. *Nature* **375**, 747–753.

Kelemen P. B., Hirth G., Shimizu N., Spiegelman M., and Dick H. J. B. (1997a) A review of melt migration processes in the asthenospheric mantle beneath oceanic spreading centers. *Phil. Trans. Roy. Soc. London* **A355**, 283–318.

Kelemen P. B., Koga K., and Shimizu N. (1997b) Geochemistry of gabbro sills in the crust-mantle transition zone of the Oman ophiolite: implications for the origin of the oceanic lower crust. *Earth Planet. Sci. Lett.* **146**, 475–488.

Kelemen P. B., Hart S. R., and Bernstein S. (1998) Silica enrichment in the continental upper mantle lithosphere via melt/rock reaction. *Earth Planet. Sci. Lett.* **164**, 387–406.

Kelemen P. B., Rilling J. L., Parmentier E. M., Mehl L., and Hacker B. R. (2003a) Thermal structure due to solid-state flow in the mantle wedge beneath arcs. In *AGU Monograph* (ed. J. Eiler). American Geophysical Union, (in press).

Kelemen P. B., Yogodzinski G. M., and Scholl D. W. (2003b) Along strike variation in lavas of the Aleutian island arc: implications for the genesis of high Mg# andesite and the continental crust. In *AGU Monograph* (ed. J. Eiler). American Geophysical Union, (in press).

Kepezhinskas P., McDermott F., Defant M. J., Hochstaedter A., Drummond M. S., Hawkesworth C. J., Koloskov A., Maury R. C., and Bellon H. (1997) Trace element and Sr–Nd–Pb isotopic constraints on a three-component model of Kamchatka Arc petrogenesis. *Geochim. Cosmochim. Acta* **61**, 577–600.

Keppler H. (1996) Constraints from partitioning experiments on the composition of subduction-zone fluids. *Nature* **380**, 237–240.

Kerr B. C. and Klemperer S. (2002) Wide-angle imaging of the Mariana Subduction Factory. *EOS, Trans., AGU* **83**(47), *Fall Meet. Suppl.*

Khan M. A., Jan M. Q., Windley B. F., Tarney J., and Thirlwall M. F. (1989) The Chilas mafic-ultramafic igneous complex: the root of the Kohistan island arc in the Himalaya of northern Pakistan. *Tectonics of the western Himalayas, Special Paper—Geological Society of America* **232**, 75–94.

Khan M. A., Jan M. Q., Weaver B. L., Treloar P. J. e., and Searle M. P. e. (1993) Evolution of the lower arc crust in Kohistan N, Pakistan: temporal arc magmatism through early, mature and intra-arc rift stages. *Himalayan tectonics, Seventh Himalaya–Karakoram–Tibet workshop* **74**, 123–138.

Kincaid C. and Sacks I. S. (1997) Thermal and dynamical evolution of the upper mantle in subduction zones. *J. Geophys. Res.* **102**, 12295–12315.

Kinzler R. J. and Grove T. L. (1992) Primary magmas of mid-ocean ridge basalts: 2. Applications. *J. Geophys. Res.* **97**, 6907–6926.

Kinzler R. J. and Grove T. L. (1993) Corrections and further discussion of the primary magmas of mid-ocean ridge basalts, 1 and 2. *J. Geophys. Res.* **98**, 22339–22347.

Klein E. and Langmuir C. H. (1987) Global correlations of ocean ridge basalt chemistry with axial depth and crustal thickness. *J. Geophys. Res.* **92**, 8089–8115.

Kogiso T., Tatsumi Y., and Nakano S. (1997) Trace element transport during dehydration processes in the subducted oceanic crust: 1. Experiments and implications for the origin of ocean island basalts. *Earth Planet. Sci. Lett.* **148**, 193–205.

Korenaga J. and Kelemen P. B. (1997) The origin of gabbro sills in the Moho transition zone of the Oman ophiolite: implications for magma transport in the oceanic lower crust. *J. Geophys. Res.* **102**, 27729–27749.

Korenaga J. and Kelemen P. B. (2000) Major element heterogeneity in the mantle source of the North Atlantic igneous province. *Earth Planet. Sci. Lett.* **184**, 251–268.

Kuno H. (1950) Petrology of Hakone volcano and the adjacent areas, Japan. *Geol. Soc. Am. Bull.* **61**, 957–1020.

Kuno H. (1968) Origin of andesite and its bearing on the island arc structure, Bull. *Volc.* **32**(1), 141–176.

Kushiro I. (1969) The system forsterite-diopside-silica with and without water at high pressures. *Am. J. Sci.* **267-A**, 269–294.

Kushiro I. (1974) Melting of hydrous upper mantle and possible generation of andesitic magma: an approach from synthetic systems. *Earth Planet. Sci. Lett.* **22**, 294–299.

Kushiro I. (1975) On the nature of silicate melt and its significance in magma genesis: regularities in the shift of the liquidus boundaries involving olivine, pyroxene, and silica minerals. *Am. J. Sci.* **275**, 411–431.

Kushiro I. (1990) Partial melting of mantle wedge and evolution of island arc crust. *J. Geophys. Res.* **95**, 15929–15939.

Kushiro I. and Yoder H. S., Jr. (1972) Origin of calc-alkalic peraluminous andesite and dacites. *Carnegie Inst. Geophys. Yearb.* **71**, 411–413.

Lambert I. B. and Wyllie P. J. (1972) Melting of gabbro (quartz eclogite) with excess water to 35 kilobars, with geological applications. *J. Geol.* **80**, 693–708.

Langmuir C. H., Bender J. F., Bence A. E., Hanson G. N., and Taylor S. R. (1977) Petrogenesis of basalts from the Famous area, Mid-Atlantic Ridge. *Earth Planet. Sci. Lett.* **36**, 133–156.

Larsen E. S., Jr. (1948) *Batholith and Associated Rocks of Corona, Elsinore, and San Luis Rey Quadrangles Southern California.* Harvard University, Cambridge, MA.

Lassiter J. C. and Hauri E. H. (1998) Osmium-isotope variations in Hawaiian lavas: evidence for recycled oceanic lithosphere in the Hawaiian plume. *Earth Planet. Sci. Lett.* **164**, 483–496.

Leeman W. P. (1987) Boron geochemistry of volcanic arc magmas: evidence for recycling of subducted oceanic lithosphere. *EOS* **68**, 462.

Leeman W. P. (1996) Boron and other fluid-mobile elements in volcanic arc lavas: implications for subduction processes.

In *Subduction Top to Bottom, Geophysical Monograph*, (ed. J. P. Platt). American Geophysical Union, Washington, DC, vol. 96, pp. 269–276.

Lundstrom C. (2000) Models of U-series disequilibria generation in MORB: the effects of two scales of melt porosity. *Phys. Earth Planet. Int.* **121**, 189–204.

Lundstrom C. C., Gill J., Williams Q., and Perfit M. R. (1995) Mantle melting and basalt extraction by equilibrium porous flow. *Science* **270**, 1958–1961.

Lundstrom C. C., Sampson D. E., Perfit M. R., Gill J., and Williams Q. (1999) Insights into mid-ocean ridge basalt petrogenesis: U-series disequilibria from the Siqueiros Transform, Lamont Seamounts, and East Pacific Rise. *J. Geophys. Res.* **104**, 13035–13048.

Lundstrom C. C., Gill J., and Williams Q. (2000) A geochemically consistent hypothesis for MORB generation. *Chem. Geol.* **162**, 105–126.

Macdonald R., Hawkesworth C. J., and Heath E. (2000) The Lesser Antilles volcanic chain: a study in arc magmatism. *Earth Sci. Rev.* **49**, 1–76.

Marsh B. D. (1976) Some Aleutian andesites: their nature and source. *J. Geol.* **84**, 27–45.

Martin G., Johnson B., and Grant U. (1915) Geology, and mineral resources of Kenai Peninsula, Alaska. *USGS Bull. Report B* **0587**, 1–243.

Martin H. (1986) Effect of steeper Archean geothermal gradient on geochemistry of subduction-zone magmas. *Geology* **14**, 753–756.

Martin H. (1999) Adakitic magmas: modern analogues of Archaean granitoids. *Lithos* **46**, 411–429.

Matteini M., Mazzuoli R., Omarini R., Cas R., and Maas R. (2002) The geochemical variations of the upper cenozoic volcanism along the Calama–Olacapato–El Toro transversal fault system in central Andes (~24°S): petrogenetic and geodynamic implications. *Tectonophysics* **345**, 211–227.

Maury R. C., Defant M. J., and Joron J.-L. (1992) Metasomatism of the sub-arc mantle inferred from trace elements in Philippine xenoliths. *Nature* **360**, 661–663.

McBirney A. R., Taylor H. P., and Armstrong R. L. (1987) Paricutin re-examined: a classic example of crustal assimilation in calc-alkaline magma. *Contrib. Mineral. Petrol.* **95**, 4–20.

McCullouch M. T. and Gamble J. A. (1991) Geochemical and geodynamical constraints on subduction zone magmatism. *Earth Planet. Sci. Lett.* **102**, 358–374.

McKenzie D. (1985) $^{230}Th-^{238}U$ disequilibrium and the melting processes beneath ridge axes. *Earth. Planet. Sci. Lett.* **72**, 149–157.

McLennan S. M. and Taylor S. R. (1985) *The Continental Crust: Its Composition and Evolution: An Examination of the Geochemical Record Preserved in Sedimentary Rocks*. Blackwell Scientific, Oxford.

Miller D. J. and Christensen N. I. (1994) Seismic signature and geochemistry of an island arc: a multidisciplinary study of the Kohistan accreted terrane, northern Pakistan. *J. Geophys. Res.* **99**, 11623–11642.

Miller D. M., Langmuir C. H., Goldstein S. L., and Franks A. L. (1992) The importance of parental magma composition to calc-alkaline and tholeiitic evolution: evidence from Umnak Island in the Aleutians. *J. Geophys. Res.* **97**(B1), 321–343.

Miller D. M., Goldstein S. L., and Langmuir C. H. (1994) Cerium/lead and lead isotope ratios in arc magmas and the enrichment of lead in the continents. *Nature* **368**, 514–520.

Millholland M., Graubard C., Mattinson J., and McClelland W. (1987) U–Pb age of zircons from the Talkeetna Formation, Johnson River area, Alaska. *Isochron/West* **50**, 9–11.

Miyashiro A. (1974) Volcanic rock series in island arcs and active continental margins. *Am. J. Sci.* **274**, 321–355.

Monzier M., Robin C., Hall M. L., Cotten J., Mothes P., Eissen J. P., and Samaniego P. (1997) Les Adakites d'Equateur: modele preliminaire. *Comptes Rendus de l'Academie des Sciences, Serie II. Sciences de la Terre et des Planetes* **324**, 545–552.

Morris J. D., Leeman W. P., and Tera F. (1990) The subducted component in island arc lavas: constraints from Be isotopes and B–Be systematics. *Nature* **344**, 31–35.

Müntener O., Kelemen P. B., and Grove T. L. (2001) The role of H_2O and composition on the genesis of igneous pyroxenites: an experimental study. *Contrib. Mineral. Petrol.* **141**, 643–658.

Myers J. D., Marsh B. D., and Sinha A. K. (1985) Strontium isotopic and selected trace element variations between two Aleutian volcano centers (Adak and Atka): implications for the development of arc volcanic plumbing systems. *Contrib. Mineral Petrol.* **91**, 221–234.

Myers J. D., Frost C. D., and Angevine C. L. (1986a) A test of a quartz eclogite source for parental Aleutian magmas: a mass balance approach. *J. Geol.* **94**, 811–828.

Myers J. D., Marsh B. D., and Sinha A. K. (1986b) Geochemical and strontium isotopic characteristics of parental Aleutian Arc magmas: evidence from the basaltic lavas of Atka. *Contrib. Mineral. Petrol.* **94**, 1–11.

Mysen B. and Wheeler K. (2000) Solubility behavior of water in haploandesitic melts at high pressure and high temperature. *Am. Mineral.* **85**, 1128–1142.

Newberry R., Burns L., and Pessel P. (1986) Volcanogenic massive sulfide deposits and the "missing complement" to the calc-alkaline trend: evidence from the Jurassic Talkeetna island arc of southern Alaksa. *Econ. Geol.* **81**, 951–960.

Newman S., Macdougall J. D., and Finkel R. C. (1984) 230th-238U disequilibrium in island arcs: evidence from the Aleutians and the Marianas. *Nature* **308**, 268–270.

Newman S., MacDougall J. D., and Finkel R. C. (1986) Petrogenesis and $^{230}Th-^{238}U$ disequilibrium at Mt. Shasta, California and in the Cascades. *Contrib. Mineral. Petrol.* **93**, 195–206.

Nicholls I. A. (1974) Liquids in equilibrium with peridotite mineral assemblages at high water pressures. *Contrib. Mineral. Petrol.* **45**, 289–316.

Nicholls I. A. and Ringwood A. E. (1973) Effect of water on olivine stability in tholeiites and the production of silica-saturated magmas in the island-arc environment. *J. Geol.* **81**, 285–300.

Nichols G. T., Wyllie P. J., and Stern C. R. (1994) Subduction zone melting of pelagic sediments constrained by melting experiments. *Nature* **371**, 785–788.

Nokleberg W., Plafker G., and Wilson F. (1994) Geology of south-central Alaska. In *The Geology of Alaska, The Geology of North America vol. G-1*. Geological Society of America, Boulder, CO, pp. 311–366.

Nye C. J. (1983) Petrology and geochemistry of Okmok and Wrangell Volcanoes, Alaska. PhD thesis. University of California at Santa Cruz, CA, 215pp.

Nye C. J. and Reid M. R. (1986) Geochemistry of primary and least fractionated lavas from Okmok Volcano, Central Aleutians: implications for arc magmagenesis. *J. Geophys. Res.* **91**, 10271–10287.

Nye C. J. and Reid M. R. (1987) Corrections to "Geochemistry of primary and least fractionated lavas from Okmok volcano, central Aleutians: implications for magmagenesis". *J. Geophys. Res.* **92**, 8182.

O'Hara M. J. and Mathews R. E. (1981) Geochemical evolution in an advancing, periodically replenished, periodically tapped, continuously fractionated magma chamber. *J. Geol. Soc. London* **138**, 237–277.

Osborn E. F. (1959) Role of oxygen pressure in the crystallization and differentiation of basaltic magma. *Am. J. Sci.* **257**, 609–647.

Ozawa K. (2001) Mass balance equations for open magmatic systems: trace element behavior and its application to open system melting in the upper mantle. *J. Geophys. Res.* **106**, 13407–14434.

Ozawa K. and Shimizu N. (1995) Open-system melting in the upper mantle: constraints from the Hayachine-Miyamori ophiolite, northeastern Japan. *J. Geophys. Res.* **100**(No. B11), 22315–22335.

Palfy J., Smith P., Mortensen J., and Friedman R. (1999) Integrated ammonite biochronology and U–Pb geochronometry from a basal Jurassic section in Alaska. *GSA Bull.* **111**, 1537–1549.

Parkinson I. J. and Arculus R. J. (1999) The redox state of subduction zones: insights from arc peridotites. *Chem. Geol.* **160**, 409–423.

Patino L. C., Carr M. J., and Feigenson M. D. (2000) Local and regional variations in Central American arc lavas controlled by variations in subducted sediment input. *Contrib. Mineral. Petrol.* **138**, 265–283.

Pavlis T. (1983) Pre-cretaceous crystalline rocks of the western Chugach Mountains, Alaska: nature of the basement of the Jurassic Peninsular terrane. *Geol. Soc. Am. Bull.* **94**, 1329–1344.

Peacock S. M. (1996) Thermal and petrologic structure of subduction zones. In *Subduction Zones, Top to Bottom*, Geophys. Monogr. 96 (eds. G. E. Bebout, D. W. Scholl, S. H. Kirby, and J. P. Platt). American Geophysical Union, Washington, DC, pp. 119–133.

Peacock S. M. (2003) Thermal structure and metamorphic evolution of subducting slabs. In *AGU Monograph* (ed. J. Eiler). American Geophysical Union (in press).

Peacock S. M., Rushmer T., and Thompson A. B. (1994) Partial melting of subducting oceanic crust. *Earth Planet. Sci. Lett.* **121**, 227–244.

Pearce J. A. (1982) Trace element characteristics of lavas from destructive plate boundaries. In *Andesites: Orogenic Andesites and Related Rocks* (ed. R. S. Thorpe). John Wiley & Sons, Chichester, UK, pp. 526–547.

Pearce J. A. and Peate D. W. (1995) Tectonic implications of the composition of volcanic arc magmas. *Ann. Rev. Earth Planet. Sci.* **23**, 251–285.

Pearce J. A., van der Laan S. R., Arculus R. J., Murton B. J., Ishii T., Peate D. W., and Parkinson I. J. (1992) 38: Boninite and Harzburgite from Leg 125 (Bonin-Mariana Forearc): a case study of magmagenesis during the initial stages of subduction. *Proc. ODP, Sci. Res.* **125**, 623–659.

Petterson M. G. and Windley B. F. (1985) Rb–Sr dating of the Kohistan arc-batholith in the Trans-Himalaya of North Pakistan, and tectonic implications. *Earth Planet. Sci. Lett.* **74**(1), 45–57.

Petterson M. G. and Windley B. F. (1992) Field relations, geochemistry and petrogenesis of the Cretaceous basaltic Jutal dykes, Kohistan, northern Pakistan. *J. Geol. Soc. London* **149**(part 1), 107–114.

Petterson M. G., Windley B. F., and Sullivan M. (1991) A petrological, chronological, structural and geochemical review of Kohistan Batholith and its relationship to regional tectonics. *Phys. Chem. Earth* **17**(part II), 47–70.

Petterson M. G., Crawford M. B., and Windley B. F. (1993) Petrogenetic implications of neodymium isotope data from the Kohistan Batholith, North Pakistan. *J. Geol. Soc. London* **150**(part 1), 125–129.

Pichavant M., Mysen B. O., and Macdonald R. (2002) Source and H_2O content of high-MgO magmas in island arc settings: an experimental study of a primitive calc-alkaline baslt from St. Vincent, Lesser Antilles arc. *Geochim. Cosmochim. Acta* **66**, 2193–2209.

Pickett D. A. and Saleeby J. B. (1993) Thermobarometric constraints on the depth of exposure and conditions of plutonism and metamorphism at deep levels of the Sierra Nevada Batholith, Tehachapi Mountains, California. *J. Geophys. Res.* **98**, 609–629.

Pickett D. A. and Saleeby J. B. (1994) Nd, Sr, and Pb isotopic characteristics of Cretaceous intrusive rocks from deep levels of the Sierra Nevada Batholith, Tehachapi Mountains, California. *Contrib. Mineral. Petrol.* **118**, 198–215.

Plafker G., Nokleberg W. J., and Lull J. S. (1989) Bedrock geology and tectonic evolution of the Wrangellia, Peninsular, and Chugach terranes along the trans-Alaska crustal transect in the Chugach Mountains and Southern Copper River Basin, Alaska. *J. Geophys. Res.* **94**(B4), 4255–4295.

Plank T. (2003) Constraints from Th/La on the evolution of the continents. *J. Petrol.* (submitted).

Plank T. and Langmuir C. H. (1988) An evaluation of the global variations in the major element chemistry of arc basalts. *Earth Planet. Sci. Lett.* **90**, 349–370.

Plank T. and Langmuir C. H. (1993) Tracing trace elements from sediment input to volcanic output at subduction zones. *Nature* **362**, 739–743.

Plank T. and Langmuir C. H. (1998) The chemical composition of subducting sediment and its consequences for the crust and mantle. *Chem. Geol.* **145**, 325–394.

Presnall D. C. and Hoover J. D. (1984) Composition and depth of origin of primary mid-ocean ridge basalts. *Contrib. Mineral. Petrol.* **87**, 170–178.

Ramsay W. R. H., Crawford A. J., and Foden J. D. (1984) Field setting, mineralogy, chemistry and genesis of arc picrites, New Georgia, Solomon Islands. *Contrib. Mineral. Petrol.* **88**, 386–402.

Ranero C. R. and von Huene R. (2000) Subduction erosion along the Middle America convergent margin. *Nature* **404**, 748–755.

Rapp R. P. and Watson E. B. (1995) Dehydration melting of metabasalt at 8–32 kbar: implications for continental growth and crust–mantle recycling. *J. Petrol.* **36**(4), 891–931.

Rapp R. P., Watson E. B., and Miller C. F. (1991) Partial melting of amphibolite/eclogite and the origin of Archean trondhjemites and tonalites. *Precamb. Res.* **51**, 1–25.

Rapp R. P., Shimizu N., Norman M. D., and Applegate G. S. (1999) Reaction between slab-derived melts and peridotite in the mantle wedge: experimental constraints at 3.8 GPa. *Chem. Geol.* **160**, 335–356.

Reagan M. K. and Gill J. B. (1989) Coexisting calcalkaline and high-niobium basalts from Turrialba Volcano, Costa Rica: implications for residual titanates in arc magma sources. *J. Geophys. Res.* **94**, 4619–4633.

Reagan M. K., Morris J. D., Herrstrom E. A., and Murrell M. T. (1994) Uranium series and beryllium isotope evidence for an extended history of subduction modification of the mantle below Nicaragua. *Geochim. Cosmochim. Acta* **58**, 4199–4212.

Regelous M., Collerson K. D., Ewart A., and Wendt J. I. (1997) Trace element transport rates in subduction zones: evidence from Th, Sr and Pb isotope data for Tonga-Kermadec arc lavas. *Earth Planet. Sci. Lett.* **150**, 291–302.

Richardson C. and McKenzie D. (1994) Radioactive disequilibria from 2D models of melt generation by plumes and ridges. *Earth Planet. Sci. Lett.* **128**, 425–437.

Ringuette L., Martignole J., and Windley B. F. (1999) Magmatic crystallization, isobaric cooling, and decompression of the garnet-bearing assemblages of the Jijal Sequence (Kohistan Terrane, western Himalayas). *Nature* **27**(2), 139–142.

Ringwood A. E. (1966) The chemical composition and origin of the Earth. In *Advances in Earth Science* (ed. P. M. Hurley). MIT Press, Cambridge, MA, pp. 287–356.

Ringwood A. E. (1974) The petrological evolution of island arc systems. *J. Geol. Soc. London* **130**, 183–204.

Rioux M., Mehl L., Hacker B., Mattinson J., Gans P., and Wooden J. (2001a) Understanding island arc evolution through U–Pb and ^{40}Ar–^{39}Ar geochronology of the Talkeetna arc, south-central Alaska. *EOS, Fall Meet. Abstr.* **82**, 1200.

Rioux M., Mehl L., Hacker B., Mattinson J., and Wooden J. (2001b) Understanding island arc thermal structure through U–Pb and ^{40}Ar–^{39}Ar geochronology of the Talkeetna arc section, south central Alaska. *Geol. Soc. Am. Abstr.* **33**, 256–257.

Rioux M., Mattinson J., Hacker B., and Grove M. (2002a) Growth and evolution of the accreted Talkeetna arc, south-central Alaska: solutions to the "arc paradox". *Geol. Soc. Am. Abstr.* **34**, 269–270.

Rioux M., Mattinson J., Hacker B., and Grove M. (2002b) Growth and evolution of the accreted Talkeetna arc, south-central Alaska: solutions to the "arc paradox". *EOS, Fall Meet. Abstr.* **83**, 1482.

Roeske S., Mattinson J., and Armstrong R. (1989) Isotopic ages of glaucophane schists on the Kodiak Islands, southern Alaska, and their implications for the Mesozoic tectonic history of the Border Ranges fault system. *Geol. Soc. Am. Bull.* **101**, 1021–1037.

Rogers G., Saunders A. D., Terrell D. J., Verma S. P., and Marriner G. F. (1985) Geochemistry of Holocene volcanic rocks associated with ridge subduction in Baja California, Mexico. *Nature* **315**, 389–392.

Roggensack K., Hervig R. L., McKnight S. B., and Williams S. N. (1997) Explosive basaltic volcanism from Cerro Negro volcano: influence of volatiles on eruptive style. *Science* **277**, 1639–1642.

Rowland A. D. and Davies H. J. (1999) Buoyancy rather than rheology controls the thickness of the overriding mechanical lithosphere at subduction zones. *Geophys. Res. Lett.* **26**, 3037–3040.

Rudnick R. L. (1995) Making continental crust. *Nature* **378**, 571–577.

Rudnick R. L. and Fountain D. M. (1995) Nature and composition of the continental crust: a lower crustal perspective. *Rev. Geophys.* **33**(3), 267–309.

Rudnick R. L. and Presper T. (1990) Geochemistry of intermediate- to high-pressure granulites. In *Granulites and Crustal Evolution.* NATO ASI Series C: Mathematical and Physical Sciences (eds. D. Vielzeuf and P. Vidal). D. Reidel Publishing Company, Dordrecht-Boston, vol. 311, pp. 523–550.

Rudnick R. L., McDonough W. F., and Orpin A. (1994) Northern Tanzania peridotite xenoliths: a comparison with Kapvaal peridotites and inferences on metasomatic reactions. In *Kimberlites, Related Rocks and Mantle Xenoliths: Vol. 1. Proc. 5th Int. Kimberlite Conf.* (eds. Henry O. A. Meyer, Leonardosand Othon H). CRPM-Special Publication, Araxa, Brazil, vol. 1A. pp. 336–353, Companhia de Pesquisa de Recursos Minerais, Rio de Janeiro, Brazil.

Ryan J. G. and Langmuir C. H. (1992) The systematics of boron abundances in young volcanic rocks. *Geochim. Cosmochim. Acta* **57**, 1489–1498.

Ryan J. G., Leeman W. P., Morris J. D., and Langmuir C. H. (1989) B/Be and Li/Be systematics and the nature of subducted components in the Mantle. *EOS* **70**, 1388.

Ryerson F. J. (1985) Oxide solution mechanisms in silicate melts: systematic variations in the activity coefficient of SiO_2. *Geochim. Cosmochim. Acta* **49**, 637–649.

Ryerson F. J. and Watson E. B. (1987) Rutile saturation in magmas: implications for Ti–Nb–Ta depletion in island-arc basalts. *Earth Planet. Sci. Lett.* **86**, 225–239.

Saal A. E., Van Orman J. A., Hauri E. H., Langmuir C. H., and Perfit M. R. (2002) An alternative hypothesis for the origin of the high 226Ra excess in MORBs. *EOS, Trans., AGU* **83**(47) *Fall Meet. Suppl.*

Schiano P., Clocchiatti R., Shimizu N., Maury R. C., Jochum K. P., and Hofmann A. W. (1995) Hydrous, silica-rich melts in the sub-arc mantle and their relationship with erupted arc lavas. *Nature* **377**, 595–600.

Schiano P., Birck J.-L., and Allègre C. J. (1997) Osmium-strontium-neodymium-lead isotopic covariations in mid-ocean ridge basalt glasses and the heterogeneity of the upper mantle. *Earth Planet. Sci. Lett.* **150**, 363–379.

Schmidt M. W. and Poli S. (1998) Experimentally based water budgets for dehydrating slabs and consequences for arc magma generation. *Earth Planet. Sci. Lett.* **163**(1–4), 361–379.

Scholl D. W., Vallier T. L., and Stevenson A. J. (1987) Geologic evolution and petroleum geology of the Aleutian Ridge. In *Geology and Resource Potential of the Continental Margin of Western North America and Adjacent Ocean Basins*, Beaufort Sea to Baja California, Circum-Pacific Council for Energy and Mineral Resources. Earth Sci. Series (eds. D. W. Scholl, A. Grantz, and J. G. Vedder). Circum-Pacific Council for Energy and Mineral Resources, Houston, TX, vol. 6, pp. 123–155.

Shah M. T. and Shervais J. W. (1999) The Dir-Utror metavolcanic sequence, Kohistan arc terrane, northern Pakistan. *J. Asian Earth Sci.* **17**(4), 459–475.

Shaw D. M. (1970) Trace element fractionation during anatexis. *Geochim. Cosmochim. Acta* **34**, 237–243.

Shimoda G., Tatsumi Y., Nohda S., Ishizaka K., and Jahn B. M. (1998) Setouchi high-Mg andesites revisited: geochemical evidence for melting of subducting sediments. *Earth Planet. Sci. Lett.* **160**, 479–492.

Sigmarsson O., Condomines M., Morris J. D., and Harmon R. S. (1990) Uranium and ^{10}Be enrichments by fluids in Andean arc magmas. *Nature* **346**, 163–165.

Sigmarsson O., Martin H., and Knowles J. (1998) Melting of a subducting oceanic crust from U–Th disequilibria in austral Andean lavas. *Nature* **394**, 566–569.

Sigmarsson O., Chmele J., Morris J., and Lopez-Escobar L. (2002) Origin of 226Ra/230Th disequilibria in arc lavas from southern Chile and implications for magma transfer time. *Earth Planet. Sci. Lett.* **196**, 189–196.

Silver L. T. and Chappell B. W. (1988) The Peninsular Ranges batholith: an insight into the evolution of the Cordilleran batholiths of southwestern North America. *Trans. Roy. Soc. Edinburgh: Earth Sci.* **79**, 105–121.

Sims K. W., Goldstein S. J., Blichert-Toft J., Perfit M. R., Kelemen P., Fornari D. J., Michael P., Murrell M. T., Hart S. R., DePaolo D. J., Layne G., Ball L., Jull M., and Bender J. (2002) Chemical and isotopic constraints on the generation and transport of magma beneath the East Pacific Rise. *Geochim. Cosmochim. Acta* **66**, 3481–3504.

Singer B. S., Myers J. D., and Frost C. D. (1992a) Mid-Pleistocene basalts from the Seguam Volcanic Center, Central Aleutian arc, Alaska: local lithospheric structures and source variability in the Aleutian arc. *J. Geophys. Res.* **97**, 4561–4578.

Singer B. S., Myers J. D., and Frost C. D. (1992b) Mid-Pleistocene lavas from the Seguam Island volcanic center, central Aleutian arc: closed-system fractional crystallization of a basalt to rhyodacite eruptive suite. *Contrib. Mineral. Petrol.* **110**, 87–112.

Sisson T. W. and Bronto S. (1998) Evidence for pressure-release melting beneath magmatic arcs from basalt at Galunggung, Indonesia. *Nature* **391**, 883–886.

Sisson T. W. and Grove T. L. (1993a) Experimental investigations of the role of H_2O in calc-alkaline differentiation and subduction zone magmatism. *Contrib. Mineral. Petrol.* **113**, 143–166.

Sisson T. W. and Grove T. L. (1993b) Temperatures and H_2O contents of low MgO high-alumina basalts. *Contrib. Mineral. Petrol.* **113**, 167–184.

Sisson T. W. and Layne G. D. (1993) H_2O in basalt and basaltic andesite glass inclusions from four subduction-related volcanoes. *Earth Planet. Sci. Lett.* **117**, 619–635.

Sobolev A. V. and Chaussidon M. (1996) H_2O concentrations in primary melts from supra-subduction zones and mid-ocean ridges. *Earth Planet. Sci. Lett.* **137**, 45–55.

Sobolev A. V. and Danyushevsky L. V. (1994) Petrology and geochemistry of boninites from the north termination of the Tonga Trench: constraints on the generation conditions of primary high-Ca boninite magmas. *J. Petrol.* **35**, 1183–1211.

Sobolev A. V. and Shimizu N. (1992) Superdepleted melts and ocean mantle permeability. *Doklady Rossiyskoy Akademii Nauk* **326**(2), 354–360.

Sobolev A. V., Hofmann A. W., and Nikogosian I. K. (2000) Recycled oceanic crust observed in 'ghost plagioclase' within the source of Mauna Loa lavas. *Nature* **404**, 986–990.

Spiegelman M. and Elliot T. (1992) Consequences of melt transport for uranium series disequilibrium in young lavas. *Earth Planet. Sci. Lett.* **118**, 1–20.

Spiegelman M. and McKenzie D. (1987) Simple 2-D models for melt extraction at mid-ocean ridges and island arcs. *Earth Planet. Sci. Lett.* **83**, 137–152.

Spiegelman M., Kelemen P. B., and Aharonov E. (2001) Causes and consequences of flow organization during melt transport: the reaction infiltration instability. *J. Geophys. Res.* **106**, 2061–2078.

Stalder R., Foley S. F., Brey G. P., and Horn I. (1998) Mineral-aqueous fluid partitioning of trace elements at 900–1200°C and 3.0–5.7 GPa: new experimental data for garnet, clinopyroxene and rutile and implications for mantle metasomatism. *Geochim. Cosmochim. Acta* **62**, 1781–1801.

Stern C. R. and Kilian R. (1996) Role of the subducted slab, mantle wedge and continental crust in the generation of adakites from the Andean Austral Volcanic Zone. *Contrib. Mineral. Petrol.* **123**, 263–281.

Stern C. R. and Wyllie P. J. (1973) Melting relations of basalt-andesite-rhyolite H_2O and a pelagic red clay at 30 kb. *Contrib. Mineral. Petrol.* **42**, 313–323.

Stern R. J. and Bloomer S. H. (1992) Subduction zone infancy: examples from the Eocene Izu-Bonin-Mariana and Jurassic California arcs. *Geol. Soc. Am. Bull.* **104**, 1621–1636.

Stolper E. and Newman S. (1992) The role of water in the petrogenesis of Mariana Trough magmas. *Earth Planet. Sci. Lett.* **121**, 293–325.

Sullivan M. A., Windley B. F., Saunders A. D., Haynes J. R., and Rex D. C. (1993) A palaeogeographic reconstruction of the Dir Group: evidence for magmatic arc migration within Kohistan N. Pakistan. *Himalayan tectonics, Seventh Himalaya–Karakoram–Tibet workshop* **74**, 139–160.

Sun S.-S. and McDonough W. F. (1989) Chemical and isotopic systematics of oceanic basalts: implications for mantle composition and processes. *Geol. Soc. Spec. Publ.* **42**, 313–345.

Suyehiro K., Takahaski N., Ariie Y., Yokoi Y., Hino R., et al. (1996) Continental crust, crustal underplating, and low-Q upper mantle beneath an oceanic island arc. *Science* **272**, 390–392.

Tahirkheli R. A. K. (1979) Geotectonic evolution of Kohistan. In *Geology of Kohistan, Karakoram Himalaya, Northern Pakistan*. Geol. Bull. (eds. R. A. K. e. Tahirkheli and M. Q. e. Jan). University of Peshawar, Peshawar, Pakistan, vol. 11, pp. 113–130.

Tatsumi Y. (1981) Melting experiments on a high-magnesian andesite. *Earth Planet. Sci. Lett.* **54**, 357–365.

Tatsumi Y. (1982) Origin of high-magnesian andesites in the Setouchi volcanic belt, southwest Japan: II. Melting phase relations at high pressures. *Earth Planet. Sci. Lett.* **60**, 305–317.

Tatsumi Y. (2000) Continental crust formation by delamination in subduction zones and complementary accumulation of the enriched mantle I component in the mantle. *Geochem. Geophys. Geosys. (G-cubed)* **1**, 1–17.

Tatsumi Y. (2001) Geochemical modelling of partial melting of subducting sediments and subsequent melt-mantle inteaction: generation of high-Mg andesites in the Setouchi volcanic belt, Southwest Japan. *Geology* **29**(4), 323–326.

Tatsumi Y. and Ishizaka K. (1981) Existence of andesitic primary magma: an example from southwest Japan. *Earth Planet. Sci. Lett.* **53**, 124–130.

Tatsumi Y. and Eggins S. (1995) *Subduction Zone Magmatism*. Blackwell, Oxford, 211pp.

Tatsumi Y. and Ishizaka K. (1982) Origin of high-magnesian andesites in the Setouchi volcanic belt, southwest Japan: I. Petrographical and chemical characteristics. *Earth Planet. Sci. Lett.* **60**, 293–304.

Tatsumi Y. and Kogiso T. (1997) Trace element transport during dehydration processes in the subducted oceanic crust: 2. Origin of chemical and physical characteristics in arc magmatism. *Earth Planet. Sci. Lett.* **148**, 207–221.

Tatsumi Y., Sakuyama M., Fukuyama H., and Kushiro I. (1983) Generation of arc basalt magmas and thermal structure of the mantle wedge in subduction zones. *J. Geophys. Res.* **88**, 5815–5825.

Taylor S. R. (1977) Island arc models and the composition of the continental crust. In *Island Arcs, Deep Sea Trenches, and Back-Arc Basins*, Geophys. Monogr. 1 (eds. M. Talwani and W. C. Pitman). American Gephysical Union, Washington, DC, pp. 325–335.

Tepper J. H., Nelson B. K., Bergantz G. W., and Irving A. J. (1993) Petrology of the Chilliwack batholith, North Cascades, Washington: generation of calc-alkaline granitoids by melting of mafic lower crust with variable water fugacity. *Contrib. Mineral. Petrol.* **113**, 333–351.

Thomas R. B., Hirschmann M. M., Cheng H., Reagan M. K., and Edwards R. L. (2002) (231Pa/235U)–(230Th/238U) of young mafic volcanic rocks from Nicaragua and Costa Rica and the influence of flux melting on U-series systematics of arc lavas. *Geochim. Cosmochim. Acta* **66**, 4287–4309.

Tomascak P. B., Ryan J. G., and Defant M. J. (2000) Lithium isotope evidence for light element decoupling in the Panama subarc mantle. *Geology* **28**, 507–510.

Treloar P. J. (1995) Pressure-temperature-time paths and the relationship between collision, deformation and metamorphism in the north-west Himalaya. *Geol. J.* **30**, 333–348.

Treloar P. J., Petterson M. G., Jan M. Q., and Sullivan M. A. (1996) A re-evaluation of the stratigraphy and evolution of the Kohistan Arc sequence, Pakistan Himalaya: implications for magmatic and tectonic arc-building processes. *J. Geol. Soc. London* **153**, 681–693.

Treloar P. J., O'Brien P. J., and Khan M. A. (2001) Exhumation of early Tertiary, coesite-bearing eclogites from the Kaghan valley, Pakistan Himalaya. *Abstracts: 16th Himalaya–Karakorum–Tibet Workshop, Austria, in J. Asian Earth Sciences* **19**, 68–69.

Turcotte D. L. (1989) Geophysical processes influencing the lower continental crust. In *Properties and Processes of Earth's Lower Crust*, Geophys. Monogr. Ser. 51 (eds. R. F. Mereu, S. Mueller, and D. M. Fountain). American Geophysical Union, Washington, DC, pp. 321–329.

Turner S. and Foden J. (2001) U, Th and Ra disequilibria, Sr, Nd and Pb isotope and trace element variations in Sunda arc lavas: predominance of a subducted sediment component. *Contrib. Mineral. Petrol.* **142**, 43–57.

Turner S., Hawkesworth C., Rogers N., Bartlett J., Worthington T., Hergt J., Pearce J., and Smith I. (1997) 238U–230Th disequilibria, magma petrogenesis, and flux rates beneath the depleted Tonga-Kermadec island arc. *Geochim. Cosmochim. Acta* **61**, 4855–4884.

Turner S., Bourdon B., Hawkesworth C., and Evans P. (2000a) $^{226}Ra-^{230}Th$ evidence for multiple dehydration events, rapid melt ascent and the timescales of differentiation beneath the Tonga-Kermadec island arc. *Earth Planet. Sci. Lett.* **179**, 581–593.

Turner S., Blundy J., Wood B., and Hole M. (2000b) Large ^{230}Th excesses in basalts produced by partial melting of spinel lherzolite. *Chem. Geol.* **162**, 127–136.

Turner S., Evans P., and Hawkesworth C. J. (2001) Ultrafast source-to-surface movement of melt at island arcs from ^{226}Ra-^{230}Th systematics. *Science* **292**, 1363–1366.

Turner S., Bourdon B., and Gill J. (2003) Insights into magma genesis at convergent margins from U-series isotopes,

Chapter 7. In *Uranium Series Geochemistry*, Rev. Mineral. Geochem. 52 (eds. B. Bourdon, G. Henderson, C. Lundstrom, and S. Turner). Mineralogical Society of America and Geochemical Society, Washington, DC, pp. 255–312.

Turner S. P., George R. M. M., Evans P. J., Hawkesworth C. J., and Zellmer G. F. (2000c) Timescales of magma formation, ascent and storage beneath subduction-zone volcanoes. *Phil. Trans. Roy. Soc. London* **A358**, 1443–1464.

Ulmer P. (2001) Partial melting in the mantle wedge: The role of H_2O in the genesis of mantle derived, arc-related, magmas. *Phys. Earth Planet. Int.* **127**, 215–232.

van Keken P. E., Kiefer B., and Peacock S. M. (2002) High resolution models of subduction zones: implications for mineral dehydration reactions and the transport of water into the deep mantle. *Geochem. Geophys. Geosys (G-cubed)* **3**, Page number 2001GC000256.

Van Orman J., Saal A., Bourdon B., and Hauri E. (2002) A new model for U-series isotope fractionation during igneous processes, with finite diffusion and multiple solid phases. *EOS, Trans., AGU* **83**(47) *Fall Meet. Suppl.*

Vannucchi P., Scholl D. W., Meschede M., and McDougall-Reid K. (2001) Tectonic erosion and consequent collapse of the Pacific margin of Costa Rica: combined implications from ODP Leg 170, seismic offshore data, and regional geology of the Nicoya Peninsula. *Tectonics* **20**, 649–668.

Vernieres J. G., Marguerite Bodinier, and Jean-Louis. (1997) A plate model for the simulation of trace element fractionation during partial melting and magma transport in the Earth's upper mantle. *J. Geophys. Res.* **102**, 24771–24784.

von Huene R. and Scholl D. W. (1991) Observations at convergent margins concerning sediment subduction, subduction erosion, and the growth of continental crust. *Rev. Geophys.* **29**, 279–316.

von Huene R. and Scholl D. W. (1993) The return of sialic material to the mantle indicated by terrigeneous material subducted at convergent margins. *Tectonophysics* **219**, 163–175.

Watson E. B. and Lupulescu A. (1993) Aqueous fluid connectivity and chemical transport in clinopyroxene-rich rocks. *Earth Planet. Sci. Lett.* **117**, 279–294.

Weaver B. L. and Tarney J. (1984) Empirical approach to estimating the composition of the continental crust. *Nature* **310**, 575–577.

White C. and McBirney A. R. (1978) Some quantitative aspects of orogenic volcanism in the Oregon Cascades. *Geol. Soc. Am. Mem.* **153**, 369–388.

Wilcox R. E. (1944) Rhyolite-basalt complex on Gardiner River, Yellowstone Park, Wyoming. *Geol. Soc. Am. Bull.* **55**, 1047–1080.

Winkler G. R., Silberman M. L., Grantz A., Miller R. J., and E. M. MacKevett J. (1981) *Geologic Map and Summary Geochronology of the Valdez Quadrangle, Southern Alaska.* US Geol. Surv. Open File Report, 80–892-A.

Wood B. J., Blundy J. D., and Robinson J. A. C. (1999) The role of clinopyroxene in generating U-series disequilibrium during mantle melting. *Geochim. Cosmochim. Acta* **63**, 1613–1620.

Yamamoto H. (1993) Contrasting metamorphic P-T-time paths of the Kohistan granulites and tectonics of the western Himalayas. *J. Geol. Soc. London* **150**(part 5), 843–856.

Yamamoto H. and Yoshino T. (1998) Superposition of replacements in the mafic granulites of the Jijal Complex of the Kohistan Arc, northern Pakistan: dehydration and rehydration within deep arc crust. *Lithos* **43**(4), 219–234.

Yogodzinski G. M. and Kelemen P. B. (1998) Slab melting in the Aleutians: implications of an ion probe study of clinopyroxene in primitive adakite and basalt. *Earth Planet. Sci. Lett.* **158**, 53–65.

Yogodzinski G. M. and Kelemen P. B. (2000) Geochemical diversity in primitive Aleutian magmas: evidence from an ion probe study of clinopyroxene in mafic and ultramafic xenoliths. *EOS* **81**, F1281.

Yogodzinski G. M., Volynets O. N., Koloskov A. V., Seliverstov N. I., and Matvenkov V. V. (1994) Magnesian andesites and the subduction component in a strongly calc-alkaline series at Piip Volcano, Far Western Aleutians. *J. Petrol.* **35**(1), 163–204.

Yogodzinski G. M., Kay R. W., Volynets O. N., Koloskov A. V., and Kay S. M. (1995) Magnesian andesite in the western Aleutian Komandorsky region: implications for slab melting and processes in the mantle wedge. *Geol. Soc. Am. Bull.* **107**(5), 505–519.

Yogodzinski G. M., Lees J. M., Churikova T. G., Dorendorf F., Woeerner G., and Volynets O. N. (2001) Geochemical evidence for the melting of subducting oceanic lithosphere at plate edges. *Nature* **409**, 500–504.

Yoshino T. and Satish-Kumar M. (2001) Origin of scapolite in deep-seated metagabbros of the Kohistan Arc, NW Himalayas. *Contrib. Mineral. Petrol.* **140**(5), 511–531.

You C.-F., Spivack A. J., Smith J. H., and Gieskes J. M. (1993) Mobilization of boron in convergent margins: implications for the boron geochemical cycle. *Geology* **21**, 207–210.

You C. F., Spivack A. J., Gieskes J. M., Rosenbauer R., and Bischoff J. L. (1995) Experimental study of boron geochemistry: implications for fluid processes in subduction zones. *Geochim. Cosmochim. Acta* **59**, 2435–2442.

Zegers T. E. and van Keken P. E. (2001) Middle Archean continent formation by crustal delamination. *Geology* **29**, 1083–1086.

Zhao D., Horiuchi S., and Hasegawa A. (1992) Seismic velocity structure of the crust beneath the Japan Islands. *Tectonophysics* **212**(3–4), 289–301.

Zhao D., Yingbiao X., Weins D. A., Dorman L., Hildebrand J., and Webb S. (1997) Depth extent of the Lau back-arc spreading center and its relation to subduction processes. *Science* **278**, 254–257.

Volume Subject Index

The index is in letter-by-letter order, whereby hyphens and spaces within index headings are ignored in the alphabetization (e.g. Arabian–Nubian Shield precedes Arabian Sea). Terms in parentheses are excluded from the initial alphabetization. In line with normal materials science practice, compound names are not inverted but are filed under substituent prefixes.

The index is arranged in set-out style, with a maximum of three levels of heading. Location references refer to the page number. Major discussion of a subject is indicated by bold page numbers. Page numbers suffixed by *f* or *t* refer to figures or tables.

Printed and bound by CPI Group (UK) Ltd, Croydon, CR0 4YY
08/05/2025
01864784-0002